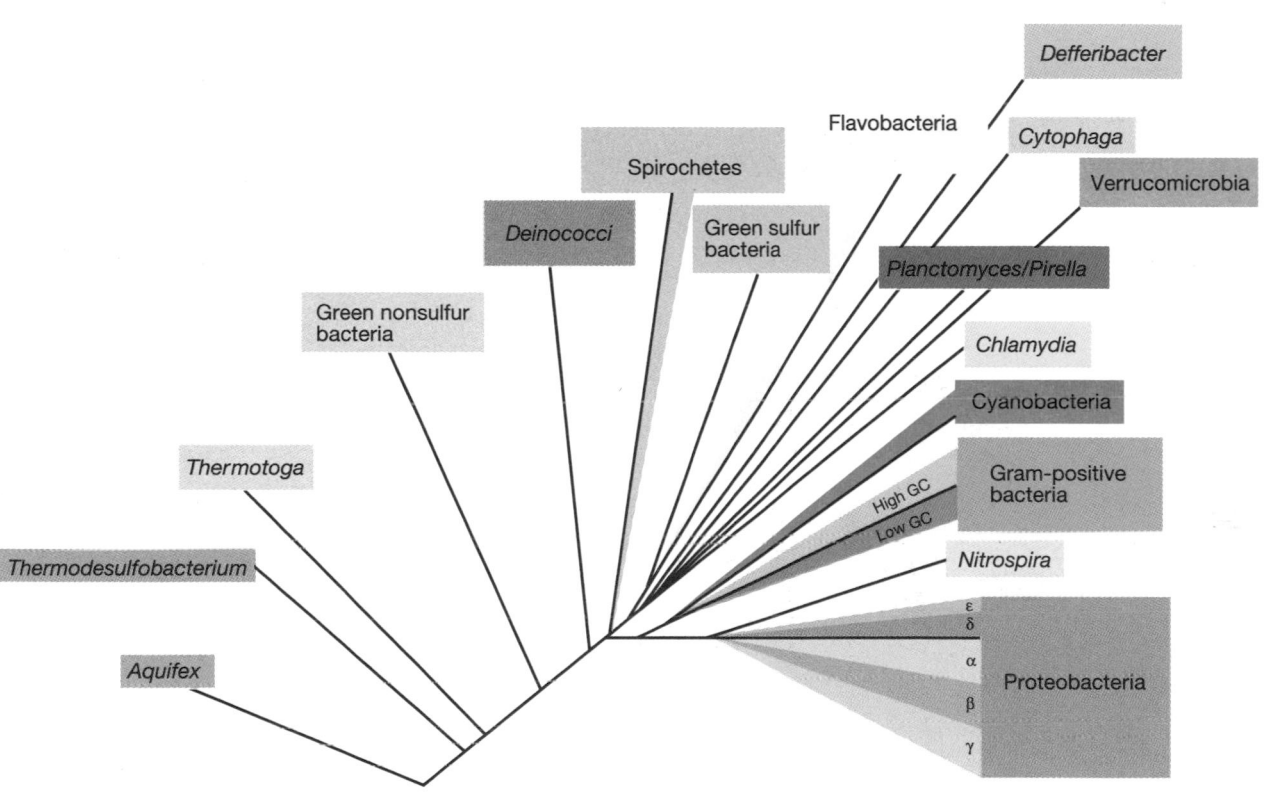

PHYLOGENETIC TREE OF *BACTERIA*. This tree is derived from 16S ribosomal RNA sequences. At least 17 major groups of *Bacteria* can be defined as indicated. See Sections 11.4–11.8 for further information on ribosomal RNA-based phylogenies. *Data for the tree obtained from the Ribosomal Database project* **http://rdp.cme.msu.edu**

17

7

2

PHYLOGENY OF THE LIVING WORLD–OVERVIEW

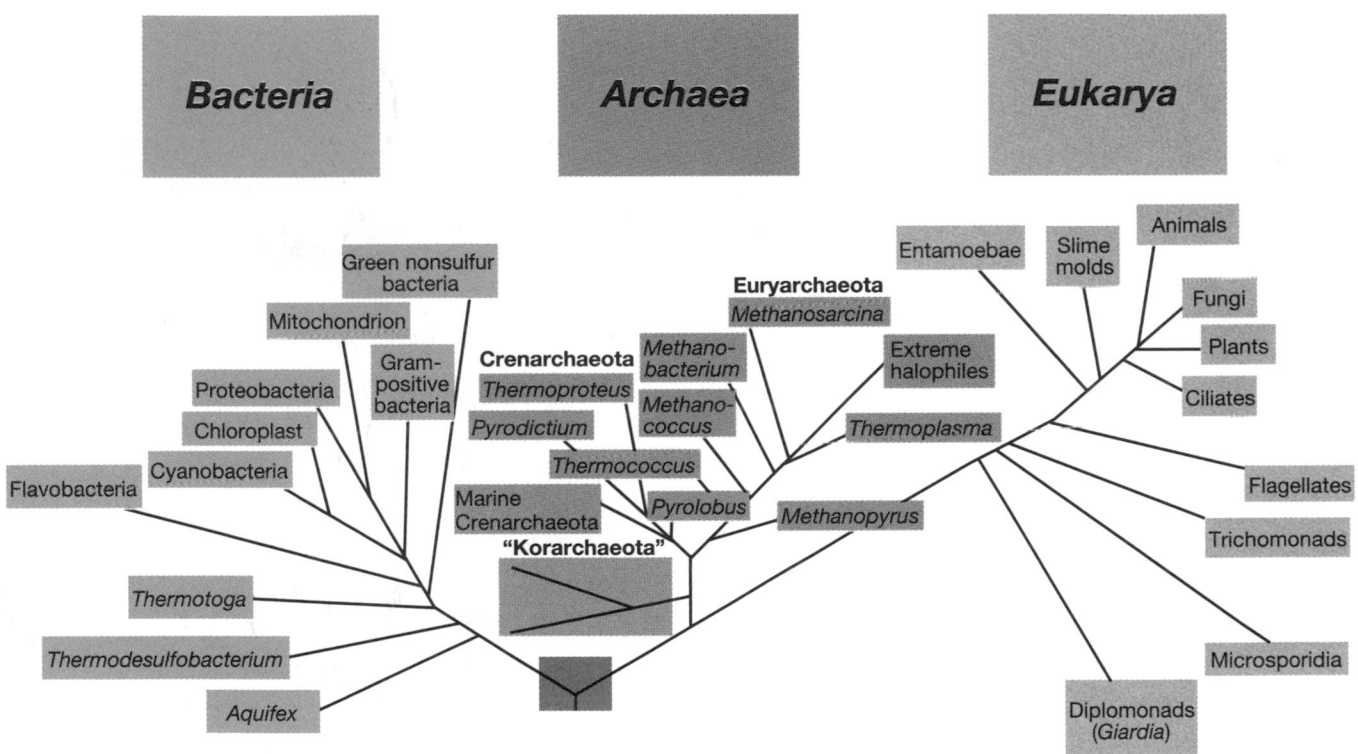

UNIVERSAL PHYLOGENETIC TREE. This tree is derived from comparative sequencing of 16S or 18S RNA. Note the three major domains of living organisms: the *Bacteria*, the *Archaea*, and the *Eukarya*. The evolutionary distance between two groups of organisms is proportional to the cumulative distance between the end of the branch and the node that joins the two groups. See Sections 11.4–11.8 for further information on ribosomal RNA-based phylogenies. *Data for the tree obtained from the Ribosomal Database project* **http://rdp.cme.msu.edu**

BROCK

BIOLOGY OF MICROORGANISMS

Michael T. Madigan dedicates this book to four individuals who departed this life in 2001. First, my mother Myrtle (February 28, 2001); she was a loving mother who raised a good family and encouraged me to go to college at a rather critical stage in my upbringing. Second, Charles Maas (the Colonel) (March 8, 2001), my best friend for over 35 years; oh how I wish you could have enjoyed those retirement years that we dreamed about over a beer on more than one occasion. Third, my father-in-law Bart Spear (June 29, 2001); a man of so many talents who left behind so many good memories. And finally, King, the wonder horse (November 9, 2001); his spirit and zest for life were inspirational to everyone who knew him. I miss all of you guys a lot.

> *"Wishing you were somehow here again; wishing you were somehow near.*
> *Sometimes it seemed, if I just dreamed, somehow you would be here."*
> Christine Daaé, *The Phantom of the Opera*, Act II, Scene 5.

John M. Martinko dedicates this book to his mother, Lottie Martinko. Lottie has shared with me, and all of her children, the most important lessons of a successful life. Her unflagging optimism, good humor, critical insights, common sense, and especially her persistence, continue to be an inspiration and a guide. Thank you for your guiding example and gentle encouragement for over a half of a century.

Jack Parker dedicates this book to his mother, Ruth M. Parker, and to the memory of his father, Hayden John Parker.

Michael T. Madigan received a bachelor's degree in biology and education from Wisconsin State University at Stevens Point in 1971 and M.S. and Ph.D. degrees in 1974 and 1976, respectively, from the University of Wisconsin, Madison, Department of Bacteriology. His graduate work involved study of hot spring phototrophic bacteria under the direction of Thomas D. Brock. Following three years of postdoctoral training in the Department of Microbiology, Indiana University, where he worked on phototrophic bacteria with Howard Gest, he moved to Southern Illinois University Carbondale, where he is now Professor of Microbiology. He has been a coauthor of *Biology of Microorganisms* since the fourth edition (1984) and teaches courses in introductory microbiology and bacterial diversity. In 1988 he was selected as the outstanding teacher in the College of Science, and in 1993 its outstanding researcher. In 2001 he was selected as the university's outstanding scholar. His research has dealt almost exclusively with anoxygenic phototrophic bacteria, especially those species that inhabit extreme environments. He has published 95 research papers, has coedited a major treatise on phototrophic bacteria, and is Chief Editor for North America of the journal *Archives of Microbiology*. His nonscientific interests include reading, hiking, tree planting, and caring for his dogs and horses. He lives beside a quiet lake about five miles from the SIU campus with his wife, Nancy, two dogs, Willie and Plum, and Springer and Feivel (horses).

John M. Martinko attended The Cleveland State University and majored in biology. As an undergraduate student he participated in a cooperative education program, gaining experience in several microbiology and immunology laboratories. He then worked for two years at Case Western Reserve University as a laboratory manager, conducting research on the structure, serology, and epidemiology of *Streptococcus pyogenes*. He did his graduate work at the State University of New York at Buffalo, investigating antibody specificity and antibody idiotypes for his M.A. and Ph.D. (1978) in microbiology. As a postdoctoral fellow, he worked at Albert Einstein College of Medicine in New York on the structure of major histocompatibility complex proteins. Since 1981, he has been in the Department of Microbiology at Southern Illinois University Carbondale where he is currently the Chair and Associate Professor. His research interests include the effects of growth hormone in the immune response and the development of immunodiagnostic tests for soybean brown stem rot disease. His teaching interests include undergraduate and graduate courses in immunology. He also teaches a portion of a general microbiology course, with responsibility for immunology, host defense, and infectious diseases. He lives in Carbondale with his wife, Judy, a high school science teacher, and their daughters, Martha and Helen.

Jack Parker received his bachelor's degree in biology and also received his doctoral degree in a biology program (Ph.D., Purdue University, 1973). His research project dealt with bacterial physiology and he completed his Ph.D. research while in the microbiology department at the University of Michigan. Following this, he spent four years studying bacterial genetics at York University in Toronto, Ontario. He has taught courses in bacterial genetics, general genetics, human genetics, molecular biology, and molecular genetics, and has participated in courses in introductory microbiology, medical microbiology, and virology primarily at Southern Illinois University Carbondale, where he is now a Professor in the Department of Microbiology and Dean of the College of Science. His research has been in the broad area of molecular genetics and gene expression and has been focused most specifically on studies of how cells control the accuracy of protein synthesis. He is the author of approximately 50 research papers. His home is on the edge of the Shawnee National Forest in deep southern Illinois.

Tenth Edition

BROCK

BIOLOGY OF
MICROORGANISMS

Michael T. Madigan

John M. Martinko

Jack Parker

Southern Illinois University Carbondale

Pearson Education International

Editions of Biology of Microorganisms
First Edition, 1970, Thomas D. Brock
Second Edition, 1974, Thomas D. Brock
Third Edition, 1979, Thomas D. Brock
Fourth Edition, 1984, Thomas D. Brock, David W. Smith,
    and Michael T. Madigan
Fifth Edition, 1988, Thomas D. Brock and
    Michael T. Madigan
Sixth Edition, 1991, Thomas D. Brock and
    Michael T. Madigan

Seventh Edition, 1994, Thomas D. Brock,
    Michael T. Madigan, John M. Martinko, and Jack Parker
Eighth Edition, 1997, Michael T. Madigan, John M.
    Martinko, and Jack Parker
Ninth Edition, 2000, Michael T. Madigan, John M. Martinko,
    and Jack Parker
Tenth Edition, 2003, Michael T. Madigan, John M. Martinko,
    and Jack Parker

Executive Editor: Gary Carlson
Editor-in-Chief, Life and Geosciences: Sheri L. Snavely
Production Editor: Debra A. Wechsler
Vice President of Production and Manufacturing:
    David W. Riccardi
Executive Managing Editor: Kathleen Schiaparelli
Senior Marketing Manager for Biology: Martha McDonald
Executive Marketing Manager for Biology:
    Jennifer Welchans
Assistant Manufacturing Manager: Michael Bell
Manufacturing Manager: Trudy Pisciotti
Director of Creative Services: Paul Belfanti
Creative Director: Carole Anson
Art Director: Kenny Beck
Interior Designer: Susan Anderson
Assistant Managing Editor, Science Media: Nicole Bush

Development Editor: Carol Pritchard-Martinez
Editor-in-Chief, Development: Carol Truehart
Assistant Managing Editor, Supplements: Dinah Thong
Media Editor: Andrew Stull
Managing Editor, Audio/Video Assets: Patricia Burns
Art Editor: Adam Velthaus
Copy Editor: Jane Loftus
Editorial Assistants: Susan Zeigler, Lisa Tarabokjia,
    Nancy Bauer
Illustrator: Imagineering
Cover Designer: Maureen Eide
Cover Photos: Reproduced by permission of the American
    Society for Microbiology from A. T. Nielsen et al. 1999.
    Applied Environmental Microbiology 65:1251–58, Fig. 5B.
    Image courtesy of Alex T. Nielsen, Technical University
    of Denmark, Lyngby, Denmark.

© 2003 Pearson Education, Inc.
Pearson Education, Inc.
Upper Saddle River, NJ 07458

© 1984, 1979 by Thomas D. Brock

First through seventh editions titled *Biology of Microorganisms*

Printed in the United States of America

10 9 8 7 6 5 4 3 2 1

ISBN 0-13-049147-0

Pearson Education LTD., *London*
Pearson Education Australia PTY, Limited, *Sydney*
Pearson Education Singapore, Pte. Ltd.
Pearson Education North Asia Ltd., *Hong Kong*
Pearson Education Canada Ltd., *Toronto*
Pearson Education de Mexico, S.A. de C.V.
Pearson Education—Japan, *Tokyo*
Pearson Education Malaysia, Pte. Ltd.
Pearson Education, *Upper Saddle River, New Jersey*

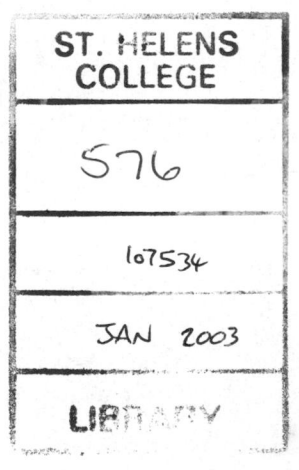

Brief Contents

BROCK BIOLOGY OF MICROORGANISMS

New Organization

The Tenth Edition is completely reorganized into six units that group chapters together within a major theme, allowing you to easily relate thematic coverage to core materials and to the goals of your course. This edition also features seven new chapters

NEW! Unit I forms the heart of the general microbiology course as envisioned by the Education Division of the ASM.

NEW! Chapter 2 provides an **early overview of microbial diversity.**

Chapter 9 emphasizes the **core concepts of virology.** (Chapter 16 explores viral diversity in more detail.)

Chapter 10 was rewritten to better reflect **bacterial genetics as it is actually practiced today**— a blend of in vivo and in vitro science.

NEW! Chapter 15 covers the essentials of the **microbial genomics projects**, from bacteria to yeast.

NEW! Chapter 16 allows instructors to easily choose those **viral examples** they wish to emphasize as a supplement to the essentials material in Chapter 9.

NEW! Chapter 18 emphasizes **methods**.

I. PRINCIPLES OF MICROBIOLOGY

1. Microorganisms and Microbiology
2. An Overview of Microbial Life
3. Macromolecules
4. Cell Structure/Function
5. Nutrition, Laboratory Culture, and Metabolism of Microorganisms
6. Microbial Growth
7. Principles of Microbial Molecular Biology
8. Regulation of Gene Expression
9. Essentials of Virology
10. Bacterial Genetics

II. EVOLUTIONARY MICROBIOLOGY AND MICROBIAL DIVERSITY

11. Microbial Evolution and Systematics
12. Prokaryotic Diversity: *Bacteria*
13. Prokaryotic Diversity: The *Archaea*
14. Eukaryotic Cell Biology and Eukaryotic Microorganisms
15. Microbial Genomics
16. Bacterial, Plant, and Animal Viruses

III. METABOLIC DIVERSITY AND MICROBIAL ECOLOGY

17. Metabolic Diversity
18. Methods in Microbial Ecology

19. Microbial Habitats, Nutrient Cycles, and Interactions with Plants and Animals

IV. IMMUNOLOGY, PATHOGENICITY, AND HOST RESPONSES

20. Microbial Growth Control
21. Human–Microbe Interactions
22. Essentials of Immunology
23. Molecular Immunology
24. Clinical Microbiology and Immunology

V. MICROBIAL DISEASES

25. Epidemiology
26. Person-to-Person Microbial Diseases
27. Animal-Transmitted, Arthropod-Transmitted, and Soilborne Microbial Diseases
28. Wastewater Treatment, Water Purification, and Waterborne Microbial Diseases
29. Food Preservation and Foodborne Microbial Diseases

VI. MICROORGANISMS AS TOOLS FOR INDUSTRY AND RESEARCH

30. Industrial Microbiology/Biocatalysis
31. Genetic Engineering and Biotechnology

Chapter 19 covers **habitats and microbial ecology** without including material on methods.

NEW! **Two new chapters on immunology**: Chapter 22 covers the essentials and Chapter 23 covers the molecular details.

Chapter 24 includes expanded coverage of **immunoassays**.

Coverage of **diseases** now spans five chapters instead of two.

NEW! Three new chapters on these important topics in microbiology.

Chapters 30 and 31 have been grouped into their own unit to better reflect their common goals and contrast their industrial production methods.

Section Numbers keyed to page numbers provides easy reference points.

New "Superheads" more effectively organize chapter contents.

The working glossary is the students' guide to the language of microbiology. Understanding these terms is key to mastery of the concepts in the chapter.

To study microorganisms in the laboratory, it is usually necessary to culture (grow) them. This can be done in a number of ways, including the use of solid culture media and the Petri plate, as shown here. Individual cells that were spread out on the plate grow and divide to form *colonies*. Although some common nutritional principles apply to all organisms, many microorganisms have special nutritional and cultural requirements. In addition, in order to sustain life processes, all microorganisms must carry out one or more series of reactions that conserve energy. Thus, an understanding of nutrition and bioenergetics is necessary to understand how microbial cells "make a living" in their natural habitats.

NUTRITION, LABORATORY CULTURE, AND METABOLISM OF MICROORGANISMS

WORKING GLOSSARY

Activation energy the energy required to bring substrates to the reactive state

Aerobe a microorganism able to use O_2 in respiration

Anabolism the sum total of all biosynthetic reactions in the cell

Anaerobe a microorganism that either can or must grow in the absence of O_2

Aseptic technique the series of manipulations used to prevent contamination during the handling of sterile objects or microbial cultures

ATP synthase (ATPase) A multiprotein enzyme complex embedded in the membrane that catalyzes the synthesis of ATP coupled to dissipation of the proton motive force

Autotroph an organism capable of biosynthesizing all cell material from CO_2 as the

Coenzyme a small nonprotein molecule that participates in a catalytic reaction as part of an enzyme

Complex medium a culture medium composed of digests of chemically undefined substances such as yeast and meat extracts

Culture medium an aqueous solution of various nutrients suitable for the growth of microorganisms

Defined medium a culture medium whose precise chemical composition is known

Electron acceptor a substance that can accept electrons from some other substance, thereby becoming reduced in the process

Electron donor a substance that can donate electrons to some electron acceptor, thereby becoming oxidized in the process

glucose is fermented yielding energy (ATP) and various fermentation products. Also called the Embden-Meyerhof pathway

Oxidative phosphorylation the production of ATP at the expense of a proton motive force formed by electron transport

Photophosphorylation the production of ATP from a proton motive force formed from photosynthetic reactions

Proton motive force an energized state of the membrane resulting from the separation of charge and the elements of water (H^+ versus OH^-) across the membrane

Pure culture a culture that contains a single kind of microorganism

Reduction potentiale the inherent tendency (measured in volts) of a compound to donate electrons; E_0' is reduction p

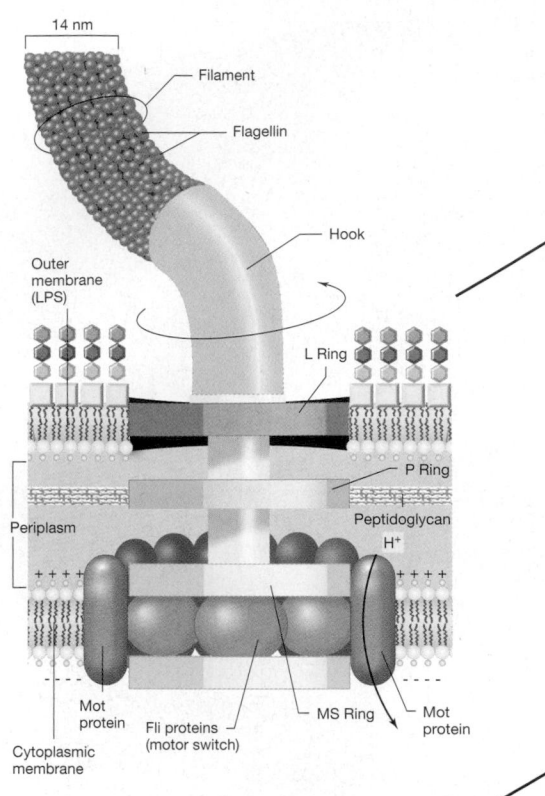

14 nm

Filament

Flagellin

Hook

Outer membrane (LPS)

L Ring

P Ring

Periplasm

Peptidoglycan

H⁺

Mot protein

Fli proteins (motor switch)

MS Ring

Mot protein

Cytoplasmic membrane

The exceptional art program has been thoroughly reviewed and adjusted to maximize its impact and clarity.

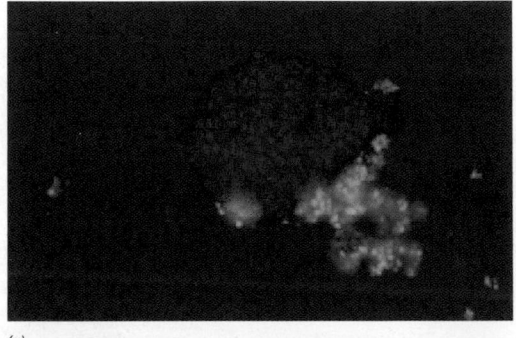

David A. Stahl

(a)

Outstanding micrographs, most obtained from researchers in the field.

Jiri Snaidr

(b)

608 ■ Chapter 17 ■ METABOLIC DIVERSITY

A Focus On ... The Power of Metabolic Diversity: A Novel Nitrogenase

Nitrogenases have been characterized from a wide variety of prokaryotes, including some *Archaea*, and all of them show significant sequence homology at both the gene and polypeptide level: All of them, that is, until the nitrogenase from the streptomycete *Streptomyces thermoautotrophicus* was characterized.[1]

S. thermoautotrophicus is a thermophilic (optimum temperature 65°C), gram-positive, filamentous prokaryote that occurs naturally in burning compost and charcoal piles (Figure 1). The organism is an aerobic chemolithotrophic H₂ bacterium that can also use carbon monoxide (CO) as an electron donor. Although *S. thermoautotrophicus* has been known to be a nitrogen fixer for some time, some unusual properties of its nitrogen fixation system (like the fact that ammonia did not repress nitrogenase synthesis and that the enzyme did not reduce acetylene) prompted a more detailed examination of its nitrogenase. What was found represents a totally new paradigm for N₂ fixation.

The *S. thermoautotrophicus* nitrogenase contains Mo, but unlike classic Mo nitrogenase, it is completely *insensitive* to O₂. The dinitrogenase component of the *S. thermoautotrophicus* nitrogenase, called *St1*, contains three different polypeptides that show some structural similarity to dinitrogenase polypeptides from other nitrogen-fixers, but the dinitrogenase reductase component, called *St2*, shows no similarity to other dinitrogenase reductases. However, St2 shows *very high* sequence similarity to manganese-containing superoxide dismutases. In fact St2 *is a superoxide dismutase!* Recall from Chapter 6 (Section 6.13) that superoxide dismutases function in the cell to consume superoxide ¹O₂⁻, forming O₂ in the process and thus preventing oxidative damage to cell components. But what does superoxide have to do with nitrogen fixation?

It has been shown that St2 supplies electrons to St1. The source of the electrons is O₂⁻, and the O₂⁻ is formed from the reduction of O₂ by a CO dehydrogenase (Figure 2). Thus, in analogy to the pyruvate : flavodoxin : dinitrogenase reductase : dinitrogenase sequence in classical nitrogen-fixation (see Figure 17.70), in *S. thermoautotrophicus* the sequence is CO : O₂⁻ : St2 : St1. And, astonishingly, instead of O₂ *inhibiting* nitrogenase (as it does in every nitrogenase that has ever been examined), in *S. thermoautotrophicus* O₂ is actually *required* in the reaction mechanism of the enzyme!

Clearly, *S. thermoautotrophicus* nitrogenase is a structurally and functionally unique nitrogen-fixing system. How widespread such a system is and whether its primary function in the cell is actually to fix N₂, is as yet unknown. However, the *S. thermoautotrophicus* nitrogenase system is a good example of the power of metabolic diversity in prokaryotes. Even well-studied systems in which conformity prevails can occasionally yield big surprises and totally new concepts. Discovery of the *S. thermoautotrophicus* nitrogenase system has also renewed hope for the eventual genetic engineering of nitrogenase into agricultural crops such as corn. The fact that this nitrogenase is not oxygen labile and its energy requirements are much lower than those of classical nitrogenases (measurements show the *S. thermoautotrophicus* system to use only 25–50% of the ATP of classic Mo nitrogenases; see Figure 17.70) could make the dream of nitrogen-fixing row crops a reality someday. ■

Figure 1 Two burning charcoal piles in the Bavarian forest, Germany, containing cells of the N₂-fixing bacterium, *Streptomyces thermoautotrophicus*. The scientist shown is doing temperature measurements at various points in the piles. The piles has gases of CO₂, CO, CH₄, and C₂H₂, and vary in temperature with depth. The surface to about 15 cm into the piles have temperatures of less than 100°C but deeper into the piles, temperatures can be over 300°C.

[1]Ribbe, M., D. Gadkari, and O. Meyer. 1997. *J. Biol. Chem.* 272: 26627–26633.

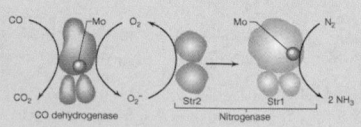

CO

Mo

O₂

Mo

N₂

CO₂

O₂⁻

Str2

Str1

2 NH₃

CO dehydrogenase

Nitrogenase

Figure 2 Reactions of nitrogen fixation in *S. thermoautotrophicus.*

Interesting scientific asides in boxed inserts—Techniques and Application boxes describe methods in microbiology and their application in the real world; Learning from the Past boxes describe historical developments in microbiology and their implications today; A Focus On boxes address text topics in greater detail.

Completely redesigned graphs facilitate student comprehension.

---

152 ■ Chapter 6 ■ MICROBIAL GROWTH

The maximum growth temperature of a given organism most likely reflects the inactivation of one or more key proteins in the cell. However, the factors controlling an organism's *minimum* growth temperature are not as clear. As mentioned earlier (Section 4.5), the cytoplasmic membrane must be in a fluid state for proper functioning. Perhaps the minimum temperature of an organism results from "freezing" of the cytoplasmic membrane so it no longer functions properly in nutrient transport or proton gradient formation. This explanation is supported by experiments in which the minimum temperature for an organism is altered to some extent by adjustments in membrane lipid composition (see Section 6.9). It is also observed that the cardinal temperatures of different microorganisms differ widely; some organisms have temperature optima as low as 4°C and some higher than 100°C. The temperature range throughout which growth occurs is even wider than this, from below freezing to greater than boiling. (The archaeon *Pyrolobus fumarii* has a temperature maximum of 113°C!) However, no single organism can grow over this whole temperature range, and the typical range for a given organism is about 30°, although some have a broader temperature range than others.

Temperature Classes of Organisms

Although there is a continuum of organisms, from those with very low temperature optima to those with high temperature optima, it is possible to broadly distinguish *four groups* of microorganisms in relation to their temperature optima: **psychrophiles**, with low temperature optima, **mesophiles**, with midrange temperature optima, **therm-**

ophiles, with high temperature optima, and **hyperthermophiles**, with very high temperature optima (Figure 6.17). Mesophiles are found in warm-blooded animals and in terrestrial and aquatic environments in temperate and tropical latitudes. Psychrophiles and thermophiles are found in unusually cold and unusually hot environments, respectively. Hyperthermophiles are found in extremely hot habitats such as hot springs, geysers, and deep-sea hydrothermal vents (see Sections 6.10 and 19.8).

In *Escherichia coli*, a typical mesophile, a detailed study of growth as a function of temperature has precisely defined its cardinal temperatures. The optimum temperature of *E. coli* in a complex medium is 39°C, the maximum is 48°C, and the minimum is 8°C. These values are subject to slight strain differences, and in general, the maximum and minimum temperatures supporting growth of an organism are higher and lower, respectively, when tested in complex rather than defined media.

✓ **6.8 Concept Check**

Temperature is a major environmental factor controlling microbial growth. The cardinal temperatures describe the minimum, optimum, and maximum temperatures at which each organism grows. Microorganisms can be grouped by the temperature ranges they require.

✓ What are the approximate cardinal temperatures for *Escherichia coli*? To what temperature class does it belong?

✓ How does a *hyperthermophile* differ from a *psychrophile*?

✓ *Escherichia coli* can grow at a higher temperature in complex than in defined medium. Why?

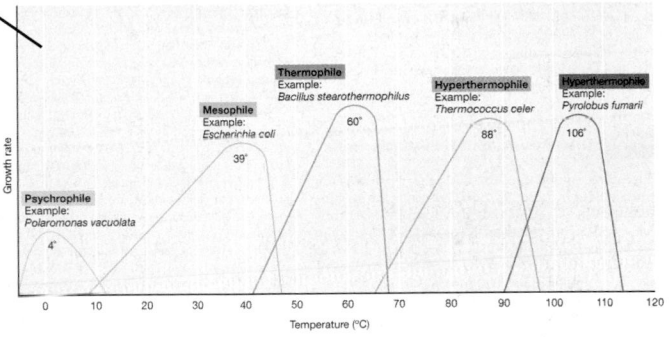

Figure 6.17 Relation of temperature to growth rates of a typical psychrophile, a typical mesophile, a typical thermophile, and two different hyperthermophiles. The temperature optima of the example organisms are shown on the graph.

Concept Checks summarize each section and provide quiz questions so students can evaluate their understanding as they progress through the chapter.

✓ **6.8 Concept Check**

Temperature is a major environmental factor controlling microbial growth. The cardinal temperatures describe the minimum, optimum, and maximum temperatures at which each organism grows. Microorganisms can be grouped by the temperature ranges they require.

✓ What are the approximate cardinal temperatures for *Escherichia coli*? To what temperature class does it belong?

✓ How does a *hyperthermophile* differ from a *psychrophile*?

✓ *Escherichia coli* can grow at a higher temperature in complex than in defined medium. Why?

called **exons**, and the intervening noncoding regions, **introns**. Both intron and exon regions are transcribed into the **primary transcript**, or **pre-mRNA**, and the functional mRNA is subsequently formed by removal of noncoding regions. A summary contrasting genetic phenomena in prokaryotes and eukaryotes is given in Figure 7.2.

✓ 7.1 Concept Check

The three key processes of macromolecular synthesis are DNA replication; transcription, the synthesis of RNA from a DNA template; and translation, the synthesis of proteins using messenger RNA as template. Although the basic processes are the same in both prokaryotes and eukaryotes, the organization of genetic information is more complex in eukaryotes. Many eukaryotic genes have both coding regions (exons) and noncoding regions (introns).

✓ What three informational macromolecules are involved in genetic information flow?

✓ In all cells there are three processes involved in genetic information flow. What are they?

II DNA STRUCTURE

We dealt with the general structure of nucleic acids in Chapter 3. In the next few sections of this chapter we shall discuss the details of DNA structure necessary for an understanding of molecular genetics and the types of genetic elements containing DNA that are found in cells. With this information as a basis, we can then discuss how DNA is replicated, transcribed into RNA, and translated into protein.

7.2 DNA Structure: The Double Helix

The genetic information for all cellular processes is stored in DNA in the *sequence* of bases along the polynucleotide chain. As we have noted, only four different nucleic acid bases are found in DNA: adenine (A), guanine (G), cytosine (C), and thymine (T). As already shown in Figure 3.11, the backbone of the DNA chain consists of alternating units of phosphate and the sugar *deoxyribose*; connected to each sugar is one of the nucleic acid *bases*. Recall especially the numbering system for the positions of sugar and base; the phosphate connecting two sugars spans from the 3′-carbon of one sugar to the 5′-carbon of the adjacent sugar (see Figure 7.14). At one end of the DNA molecule the sugar has a phosphate on the 5′-hydroxyl, whereas at the other end the sugar has a free hydroxyl at the 3′-position.

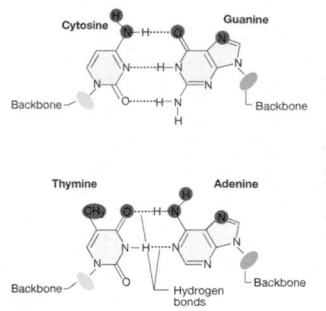

Figure 7.3 Specific pairing between adenine (A) and thymine (T) and between guanine (G) and cytosine (C) via hydrogen bonds. These two base pairs are the base pairs typically found in double-stranded DNA. Atoms that are found in the major groove of the double helix and that interact with proteins are highlighted in red. The deoxyribose phosphate backbones of the two strands of DNA are also indicated.

DNA as a Double Helix

As we will discuss in Chapter 9, the chromosomes of some viruses are single-stranded. In all **cellular organism chromosomes**, however, DNA exists as two polynucleotide strands whose base sequences are **complementary**. The complementarity of DNA arises because of the specific pairing of the purine and pyrimidine bases: Adenine always pairs with thymine, and guanine always pairs with cytosine (Figure 7.3). The two strands in the resulting **double-stranded** molecule are arranged in an *antiparallel* fashion (see Figure 7.4). This means the two strands are in a "head-to-toe" arrangement. In Figure 7.4 the strand on the left is arranged 5′ to 3′ top to bottom, whereas the other strand is 5′ to 3′ bottom to top. The two strands are wrapped around each other forming a **double helix** (Figure 7.5). In this double helix, DNA forms two distinct grooves, the *major groove* and the *minor groove*. Of the many proteins that interact specifically with DNA (as we shall see in Chapter 8), most engage predominantly with the major groove, where there is a considerable amount of space. Because of the regularity of the double helix, some atoms of the bases are always exposed in the major groove (and some in the minor groove). Atoms in the major groove that are known to be important in interactions with proteins are shown in Figure 7.3.

The size of a DNA molecule can be expressed in terms of its *molecular weight*, but because a single nucleotide has a molecular weight of about 330, and because DNA molecules are many nucleotides long, the molecular weight mounts up rapidly. (The nucleic acid in

Media Tutorials available on the Companion Web Site are indicated by icons next to key figures.

Media Tutorials consist of animations, interactive exercises, review quizzes, links to important web sites, and much more.

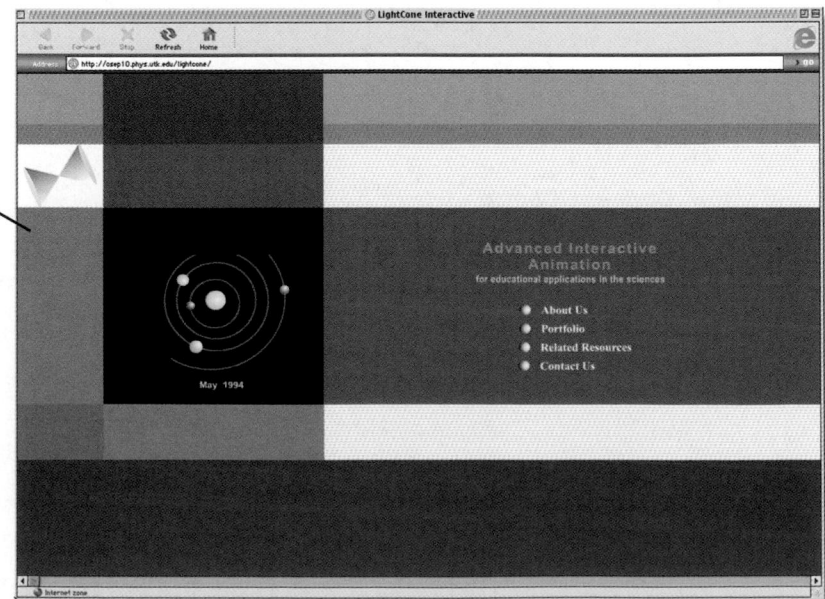

Complete Instructor Package

The Instructor Resource CD will contain PowerPoint presentations as well as an image bank of all illustrations and most of the photographs in the text along with the Instructor's Manual & Test Item File, and reproduction of the student's Companion Web Site.

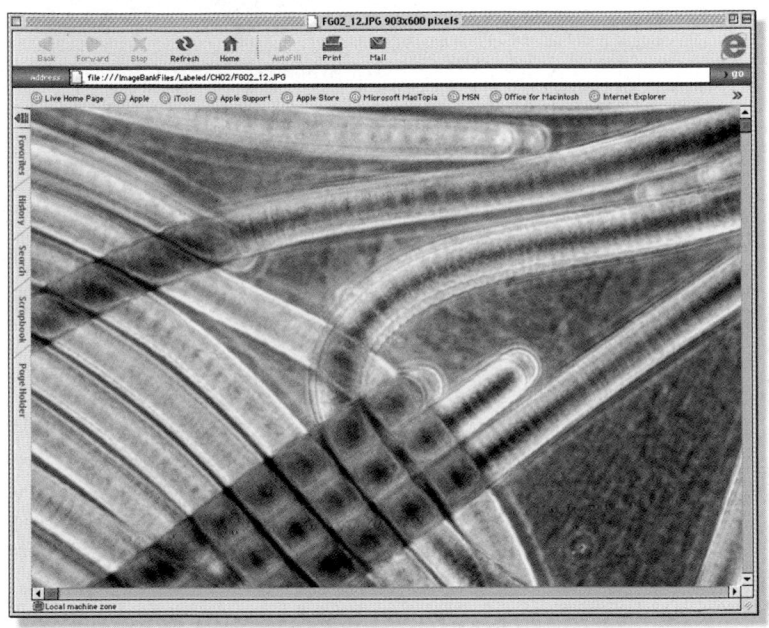

TRANSPARENCIES

BROCK
BIOLOGY
OF
MICROORGANISMS

TENTH EDITION

MICHAEL T. MADIGAN
JOHN M. MARTINKO
JACK PARKER

INSTRUCTOR'S MANUAL

BROCK
BIOLOGY
OF
MICROORGANISMS

TENTH EDITION

MICHAEL T. MADIGAN
JOHN M. MARTINKO
JACK PARKER

Contents

Microbiology is a biological science that has effectively wedded the old and the new. Some of the basic techniques of microbiology discovered over 100 years ago—the isolation of pure cultures, for example—are still practiced with regularity in the laboratory today. But today's microbiologists are also armed with sophisticated tools that facilitate detailed molecular analyses of microbial cells. These tools have fueled new discoveries that are thrusting microbiology into the limelight of disciplines as diverse as medicine, agriculture, and ecology. It is within this exciting period in microbiology that we present the tenth edition of *Brock Biology of Microorganisms (BBOM)*, a textbook of microbiology that blends fundamental principles (the old) with state-of-the-art science (the new) in a format that will appeal to both students and instructors.

What's New?
Organization

This edition of *BBOM* contains many new organizational features that will help students better master the material and help instructors prepare stimulating presentations for the classroom. First, the book has been extensively reorganized into six major units: (1) Principles of Microbiology; (2) Evolutionary Microbiology and Microbial Diversity; (3) Metabolic Diversity and Microbial Ecology; (4) Immunology, Pathogenicity, and Host Responses; (5) Microbial Diseases; and (6) Microorganisms as Tools for Industry and Research. Each unit consists of several chapters whose content define the themes. Unit 1, The Principles, forms the heart of the general microbiology course as currently envisioned by the Education Division of the American Society for Microbiology (ASM); in other words, Unit 1 is the "core" material that every student should know. Included for the first time in this unit is an overview chapter (Chapter 2) on microbial diversity that introduces the major groups of microorganisms and their evolutionary relationships. This chapter will allow instructors who emphasize the medical or molecular aspects of microbiology the opportunity to give their students a taste of microbial diversity without the details. This chapter also covers some of the basic aspects of cell structure/function and is written in such a way that only a minimal background in chemistry and biology is necessary to follow the story.

Other chapters that have been revamped in Unit 1 include Chapters 9 and 10. Compared with previous editions of *BBOM*, Chapter 9 (Essentials of Virology) has been restructured and downsized to place emphasis on the *essential concepts* of virology, instead of viral diversity. The latter material still exists, however, in the new Chapter 16 (Bacterial, Plant and Animal Viruses), for those instructors who wish to explore viral diversity in more detail. Another substantially reorganized chapter is Chapter 10, Bacterial Genetics. This chapter has been rewritten from two major standpoints, microbial genetics as it occurs in the intact organism (*in vivo*) and microbial genetics as it is practiced *in vitro*. To accomplish the latter, some material from the biotechnology chapter has been reworked and moved to this new chapter on genetics. Thus, Chapter 10 better reflects bacterial genetics as it is actually practiced today—a blend of *in vivo* and *in vitro* science.

As has been a tradition with *BBOM*, the material in each chapter is broken into several *numbered heads* to assist instructors in assigning reading material. But in addition, in parallel to the unit concept that pervades organization at the level of the entire book, the numbered heads within a chapter are themselves grouped into major themes. The latter are signaled by red headings set in all caps, and were introduced in this edition to better group related material within a chapter into logical pieces.

In summary then, the tenth edition of *BBOM* is organized to capture and distill the basics while deploying the full story of the science at those points where it will have maximum impact. The authors and publishers are confident that this new format will make *BBOM 10/e* an even stronger resource for students and instructors alike.

Content

Every three years the authors of this book face one major question: how do we add new material and still keep the book within bounds? Longtime users of *BBOM* will immediately recognize that the tenth edition is essentially no longer than the ninth. This feat was accomplished by balancing the needs of the new material with a careful reevaluation of the old. Nothing essential to a fundamental understanding of microbiology has been deleted from *BBOM*; the tenth edition is still a book built on basic principles and strong science. But streamlining of some chapters along with a first-class art program has given the authors the space necessary to paint an up-to-the-minute picture of the science of microbiology in a volume that does not require weight training to lift off the table.

Several *totally new* chapters will be found in *BBOM 10/e*. The new overview of microbial diversity chapter (Chapter 2) and the viral diversity chapter (Chapter 16) have already been mentioned in this regard. Also new to this edition are Chapter 15

(Microbial Genomics); Chapter 18 (Methods in Microbial Ecology); Chapter 22 (Essentials of Immunology); Chapter 23 (Molecular Immunology); Chapter 27 (Animal-Transmitted, Arthropod-Transmitted, and Soilborne Microbial Diseases); Chapter 28 (Wastewater Treatment, Water Purification, and Waterborne Microbial Diseases); and Chapter 29 (Food Preservation and Foodborne Microbial Diseases). All of these areas are "hot topics" in microbiology today and needed increased visibility and expanded coverage. These new chapters should accomplish just that.

The genomic revolution has transformed microbiology into a new science almost overnight. For the first time, scientists can inspect, almost in a routine fashion now, the entire genetic blueprint of a microorganism, and then compare the blueprint with those of other organisms, from viruses to humans. Genomics has revealed the great genetic unity and diversity of living organisms and has opened the door to new advancements in every discipline of biology. And combined with proteomic analyses, scientists can now ask sophisticated questions about *gene expression* in ways never before possible. Chapter 15 in *BBOM 10/e* tells the genomic story, but goes well beyond just listing organisms whose genomes have been sequenced. The chapter explains what genomics is, how the reams of DNA sequence data that are being generated can be used, and what the genomic revolution has revealed thus far in terms of both the genomic and proteomic capacities of key microorganisms.

The new chapters in immunology were written to provide both the basics and the details of this important science. Chapter 22 (Essentials of Immunology) presents the basic principles of immunology without delving into too much detail. This chapter should therefore be a very student-friendly and readily teachable overview of immunology. We reserve the molecular details of immunology for the rather short Chapter 23. This chapter places the essentials material (Chapter 22) within a molecular context for those students and instructors whose background and interests support the study of immunology at this level. In Chapter 24 (Clinical Microbiology and Immunology) we have expanded our coverage of immunoassays to include more information about the basic mechanisms behind precipitation, agglutination, and antibody production. For those instructors who teach immunology only as a diagnostic or investigative tool, we have also included a very short summary of immune principles here. Thus, with the material on immunology organized as it is, the science of immunology can be integrated into introductory microbiology classes at all levels.

The new chapters in medical microbiology are expansions of this material originally covered in only two chapters in previous editions. This has given the authors the opportunity to develop this important material in a more thorough way. And in this day and age where foodborne and waterborne illnesses are major public health problems (even in developed countries), and new threats to health and security, such as bioterrorism, are a fact of life, the unit on medical microbiology and immunology will be both a source of basic principles and a reference for keeping up with events in the news.

In summary, long-time users of *BBOM* will find the tenth edition to be the reliable friend they've always known. New users will find it to be the most current, accurate, and complete coverage of microbiology available in a textbook today. Coupled with an excellent set of teaching aids (see below) *BBOM 10/e* should set the standard in the field for years to come.

Pedagogical Aids

Art and photos are the mainstay of any textbook in the biological sciences. And frankly speaking, we think *BBOM 10/e* has the best in the business in both regards. The art program has once again been delivered by Imagineering of Toronto, Canada. Virtually every piece of art has seen some modification in order to maximize its impact and clarity. Some stylistic improvements have also been introduced into the art program, including a beautiful new rendering of all graphs in the book. High quality photos and photomicrographs have been a mainstay in *Biology of Microorganisms* since the first edition appeared in 1970, and *BBOM 10/e* proudly carries on this tradition with the inclusion of nearly 50 new B&W and color photos. And, as usual, these photos have been supplied by top researchers in the field.

BBOM 10/e once again employs a variety of student study aids to weave together the concepts and strengthen the learning experience. Instead of placing summary and quiz material only at the *end* of a chapter (as many textbooks like to do), *BBOM 10/e* contains two review tools—concept checks and concept links—*built right into* each chapter. Each **Concept Check** consists of a short summary of the material in the previous numbered head along with a short series of questions that together, reviews the major points in that section. **Concept Links** (signaled by the blue link icon, ⚬⚬) are the ties between the current text and related material found elsewhere in the book. In addition, readers will find the popular **Working Glossary**—a dictionary of essential terms—at the opening of each chapter. The Working Glossary is the student's lifeline to the *language* of microbiology, an understanding of which is a key to mastering the concepts. Finally, and as in previous editions, the end of each chapter contains a number of review and application questions, many new to this edition; the questions are designed to probe a student's retention of important concepts and ability to solve problems.

Supplements

A number of supplements accompany this book. Totally new to the tenth edition of *BBOM* are a series of online media tutorials that are found in the **Companion Website** (*www.prenhall.com/brock*). These cover a number of conceptually challenging topics in microbiology, including basic processes in molecular biology, genetics, medical microbiology and immunology, and microbial metabolism. These unique instructional resources include animations, interactive exercises, and self-quizzes, and will be a major supporting feature for the material in Unit I of *BBOM 10/e*, the heart of the introductory course in microbiology.

These tutorials are designed to guide students' understanding of various fundamental concepts through animations, interactive exercises, and self-assessment. Each concept that is the subject of an Online Media Tutorial is identified by an icon similar to the one in the margin next to this paragraph, placed alongside the relevant figure in the text. Also on the website are additional materials for each chapter along with the popular **Virtual Exam**, first introduced with *BBOM 9/e*. The Virtual Exam is a large pool of questions (written in an objective format, multiple choice, true-false, matching, and the like) keyed to individual chapters that students can use as a resource to help prepare for their real exams in the classroom. Virtual Exam questions have been assembled from actual examinations given in introductory microbiology courses in the United States that assign *BBOM* as a textbook. With the Virtual Exam, students can take an exam online and receive instant feedback on their readiness for exam day.

A variety of supplements are available for instructors. First, a set of over **350 Full Color Transparencies**, far more than is available with any other textbook of microbiology, accompany every adoption of *BBOM 10/e*. Although computer lectures are becoming the norm in many classrooms, the transparency is still the visual aid workhorse for many instructors. To help instructors in this regard, all of the most teachable figures in the book are covered in the transparency set. Second, a first-rate **Instructor's Resource CD-ROM** is available that contains virtually all of the art and photos from *BBOM 10/e* in Microsoft PowerPoint® to assist instructors in tailoring computer presentations to the goals and objectives of their particular course. In addition, all of the animations and exercises that appear on the student website will be on the Instructor's Resource CD-ROM. Whether one uses transparencies or CD-ROMs, the *BBOM 10/e* instructor package offers all of the necessary tools for developing clear, compelling, and stimulating presentations for the classroom.

Acknowledgments

Although the authors have put a strong effort into assembling this classic tenth edition of *BBOM*, the book would never have come together in the way it did without the help of many other individuals. These include all the people at Prentice Hall/Pearson, including in particular our editor, Gary Carlson and his editorial assistant, Susan Zeigler, and our production editor, Debra Wechsler. Gary offered key editorial input into this project, including the commissioning of several very useful reviews. Susan kept the project moving along in the critical reviewing stages, and was extremely helpful in dealing with author requests. Debra, in her usual thorough and professional way, is largely responsible for the beautiful appearance of the final product. Through everyone's gallant efforts in the publishing side of this book, the authors had more freedom than usual to contemplate issues such as organization, content, and pedagogy.

We also wish to acknowledge the excellent input of the developmental editor Carol Pritchard-Martinez (Orinda, CA). Carol saw this project through experienced eyes and gave the authors a critical outside perspective on several editorial and production issues throughout the book. We also thank Jane Loftus (Clackamas, OR) for excellent copy editing input, Steele/Katigbak (Amherst, MA) for composing the index, Toni Huppert, Southern Illinois University, for expert word processing assistance, and Deborah Jung, Southern Illinois University, for her excellent Photoshop® skills and input into the cover design. Finally, we wish to acknowledge those individuals who gave valuable reviewer input on the first draft of *BBOM 10/e* or who supplied color photographs, directly from their research. We are extremely grateful for their efforts and list them below. Special thanks go to Alex T. Nielsen, Technical University of Denmark, whose beautiful photos grace the cover and back cover of this book.

Charles Abella, *University of Girona, Spain*
Robert E. Andrews, Jr., *Iowa State University*
Gernot Arp, *Universität Göttingen, Germany*
Hans-Dietrich Babenzien, *Institut für Gewässerökologie und Binnenfischerrei, Berlin, Germany*
Tim Baker, *Purdue University*
Mary Bateson, *Montana State University*
Stephen Bentley, *The Sanger Center*
T. den Blaauwen, *University of Amsterdam*
Nicholas Blackburn, *Marine Biological Laboratory, Helsingør, Denmark*

Christian Boeker, *Universität Göttingen, Germany*
Antje Boetius, *Max Planck Institute for Marine Microbiology, Bremen, Germany*
David R. Boone, *Portland State University*
Derrick Brazill, *Hunter College*
John Breznak, *Michigan State University*
Cheryl Broadie, *Southern Illinois University*
Michael Cassidy, *University of Waterloo*
Richard W. Castenholz, *University of Oregon*
Randall Cohrs, *University of Colorado Health Sciences Center*
David Crowley, *University of California, Riverside*
Michael J. Daly, *Uniformed Services University, Maryland*
Frank B. Dazzo, *Michigan State University*
Eileen Dimalanta, *University of Wisconsin, Madison*
Katrina J. Edwards, *Woods Hole Oceanographic Institution*
Richard Ellis, *Bucknell University*
Jen Fagg, *Montana State University*
Michael Ferris, *Montana State University*
Douglas Fix, *Southern Illinois University*
Bernard L. Frye, *University of Texas, Arlington*
John Fuerst, *University of Queensland, Australia*
Tiffany Full, *Southern Illinois University*
George M. Garrity, *Michigan State University and Bergey's Manual Trust*
Eric Grafman, *Centers for Disease Control and Prevention, the Public Health Image Library (PHIL)*
William D. Grant, *University of Leicester, England*
Amy Grunden, *North Carolina State University*
Constance Holden, *Science magazine*
Johannes F. Imhoff, *Universität Kiel, Germany*
Michael Jetten, *University of Nijmegen, The Netherlands*
Deborah Jung, *Southern Illinois University*
Irena Kaczmarska, *Mount Allison University, Canada*
Sam Kaplan, *University of Texas, Houston*
Steve Keating, *Penn State University*
Joseph E. Kleinman, *North Carolina State University Center for Applied Aquatic Ecology*
Michael E. Konkel, *Washington State University*
Robert G. Kranz, *Washington University, St. Louis*
Alex Lim, *University of Wisconsin, Madison*
Juergen Marquardt, *Penn State University*
Mark J. McBride, *University of Wisconsin, Milwaukee*

William McCleary, *Brigham Young University*
Joan McCune, *Idaho State University*
Maura J. Meade, *Allegheny College*
Susan Merkel, *Cornell University*
Nanne Ninninga, *University of Amsterdam*
Alex T. Nielsen, *Technical University of Denmark*
Bo Normander, *National Environmental Research Institute, Roskilde, Denmark*
Norm Olson, *Purdue University*
Aharon Oren, *Hebrew University, Jerusalem*
Norman R. Pace, *University of Colorado*
Brian Palenik, *Scripps Institute of Oceanography*
Ron Parejko, *Northern Michigan University*
Julian Parkhill, *The Sanger Centre*
Norbert Pfennig, *Überlingen, Germany*
Neil Quigley, *University of Tennessee*
Reinhard Rachel, *Universität Regensburg, Germany*
Jackie Reynolds, *Richland College*
Sue Rhee, *Carnegie Institution of Washington, Stanford, CA*
Timberley Roane, *University of Colorado, Denver*
Francisco Rodriguez-Valera, *Universidad Miguel Hernandez, Spain*
Susan Rogers, *Southern Illinois University*
Wael Sabra, *GBF, Braunschweig, Germany*
Heinz Schlesner, *Universität Kiel, Germany*
Steven J. Schmitt, *Southern Illinois University*
David C. Schwartz, *University of Wisconsin, Madison*
James P. Shapleigh, *Cornell University*
Linda Sherwood, *Montana State University*
Jolynn F. Smith, *Southern Illinois University*
Kristine M. Snow, *Fox Valley Technical College*
Gary A. Sojka, *Bucknell University*
Catherine L. Squires, *Tufts University School of Medicine*
James T. Staley, *University of Washington*
Karl O. Stetter, *Universität Regensburg, Germany*
Richard Stouthamer, *University of California, Riverside*
Marc Strous, *University of Nijmegen, The Netherlands*
Michael A. Sulzinski, *University of Scranton*
James L. Van Etten, *University of Nebraska*
Lori Van Waasbergen, *University of Texas, Arlington*
Thomas M. Wahlund, *California State University, San Marcos*
David M. Ward, *Montana State University*

Frances Westall, *Lunar and Planetary Institute,*
  Houston
Carl R. Woese, *University of Illinois*
Alexandra Z. Worden, *Scripps Institute of*
  Oceanography
Xiaodong Yan, *Purdue University*
John M. Zamora, *Middle Tennessee State University*
Mark Zelman, *Harper College*

Any errors in this book, either those of commission or omission, are, of course, solely the responsibility of the authors. In this regard, we list author contact information and would greatly appreciate receiving comments, suggestions, and corrections from users. The entire author team has been with the book through a number of editions now and have through the years developed rather "thick skins" to criticism. Thus, we would sincerely enjoy receiving the frank input of users concerning how the book could be made even better in the future.

Michael T. Madigan (madigan@micro.siu.edu)

John M. Martinko (martinko@micro.siu.edu)

Jack Parker (parker@cos.siu.edu)

The field of microbiology grew out of the pioneering studies of a handful of early scientists, including the great Russian microbiologist, Sergei Winogradsky. While other giants of the early era focused on the importance of microorganisms as agents of disease, Winogradsky studied bacteria that link key nutrient cycles in nature. His remarkably accurate artistic depictions of microorganisms, such as the phototrophic purple sulfur bacteria shown here, helped other scientists of the day begin to grasp the vast metabolic diversity of microorganisms that we now know exists on Earth.

MICROORGANISMS AND MICROBIOLOGY

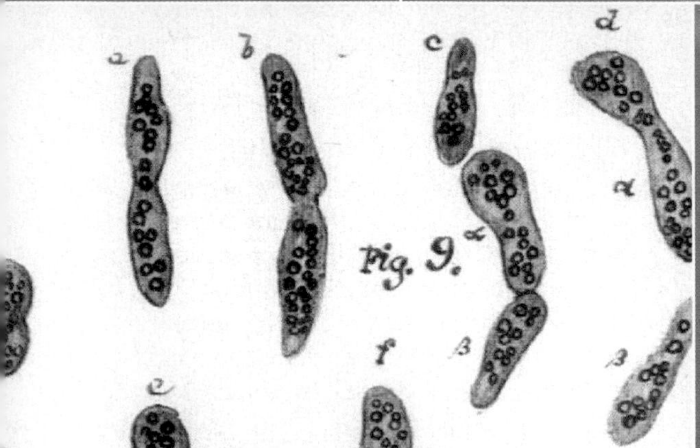

Working Glossary

Cell the fundamental unit of living matter

Cytoplasm the fluid portion of a cell, bounded by the cell membrane but excluding the nucleus (if present)

DNA deoxyribonucleic acid, the hereditary material of cells and some viruses

Ecology the study of organisms in their natural environments

Ecosystem organisms plus their nonliving environment

Enrichment culture a method for isolating microorganisms from nature using specific culture media and incubation conditions

Enzymes protein catalysts that function to speed up chemical reactions

Evolution change in a line of descent over time leading to the production of new species or varieties within a species

Habitat the location in an environment where a microbial population resides

Metabolism all biochemical reactions in a cell

Microorganism a microscopic organism consisting of a single cell or cell cluster, including the viruses

Pathogen a disease-causing microorganism

Prokaryote a cell lacking a nucleus and other organelles

Pure culture a culture containing a single kind of microorganism

RNA ribonucleic acid, involved in protein synthesis as messenger RNA, transfer RNA, and ribosomal RNA

Spontaneous generation the hypothesis that living organisms can originate from nonliving matter

Sterile absence of all living organisms and viruses

With this introductory chapter we begin our journey through the field of microbiology. Here we will discover what microorganisms are and why we should study them. We will also take a little time to place the science of microbiology in historical perspective, highlighting some of the landmark contributions of early as well as more recent microbiologists. Microbiology today is a dynamic science with ramifications in virtually all aspects of human life, including medicine, agriculture, and the environment. Welcome to the study of microbiology!

I INTRODUCTION TO MICROBIOLOGY

In the first four sections of this chapter we introduce the subject of microbiology, look at microorganism as cells, examine where and how microorganisms live in nature, and review the impact that microorganisms have had and continue to have on human affairs.

1.1 Microbiology

Microbiology is the study of microorganisms, a large and diverse group of microscopic organisms that exist as single cells or cell clusters; it also includes viruses, which are microscopic but not cellular. Microbial cells are thus distinct from the cells of animals and plants, which are unable to live alone in nature and can exist only as parts of multicellular organisms (Figure 1.1). In contrast to macroorganisms, microorganisms are generally able to carry out their life processes of growth, energy genera-

tion, and reproduction independently of other cells, either of the same kind or of a different kind.

The Science of Microbiology

Microbiology is about living cells and how they work. It is about microorganisms, especially about bacteria, a large group of cells of enormous basic and practical significance. It is about microbial diversity and evolution, about how different kinds of microorganisms arose and why. It is about what microorganisms do in the world at large, in human society, in the human body, and in the bodies of animals and plants. One way or the other, microorganisms affect all other life forms on Earth, and thus the science of microbiology is of enormous importance.

Microbiology revolves around two basic themes; basic and applied:

1. As a *basic* biological science, microbiology provides some of the most accessible research tools for probing the nature of life processes. Our most sophisticated understanding of the chemical and physical basis of life has arisen from studies of microorganisms. This is in large part because microbial cells share many biochemical properties with cells of multicellular organisms; indeed, *all* cells have much in common. And this, coupled with the fact that microbial cells can grow to extremely high densities in laboratory culture and are readily amenable to biochemical and genetic study, makes them excellent models for understanding cell function in higher organisms.

2. As an *applied* biological science, microbiology deals with many important practical problems in medicine, agriculture, and industry. Some of the most important diseases of humans, other animals, and plants are caused by microorganisms. Microorganisms also play major roles in soil fertility and

Figure 1.1 Living organisms are composed of cells. (a) Plants and (b) animals are composed of many cells arranged to form tissues, and from these various organs; they are thus *multicellular*. A single plant or animal cell cannot have an independent existence; each of its cells is dependent on the other. (c, d) Microorganisms are free-living cells. A single microbial cell can have an independent existence. Shown are photomicrographs of photosynthetic microorganisms called (c) *purple bacteria*, and (d) *cyanobacteria*. Cyanobacteria were the first O_2-evolving organisms on Earth and were responsible for oxygenating the atmosphere.

domestic animal production. In addition, many large-scale industrial and biotechnology processes, such as the production of antibiotics or human proteins, are microbially based.

The Importance of Microorganisms

As this book unfolds, the central role that microorganisms play in both human activities and the whole web of life on Earth will become readily apparent. We will see how, in the absence of microorganisms, higher life forms would never have arisen and could not now be sustained; consider, for instance, that the very oxygen we breathe is the result of microbial activity (Figure 1.1*d*). Moreover, we will see how humans, plants, and animals are intimately tied to microbial activities for the recycling of key nutrients and for degrading organic matter. Indeed, no other life forms approach the importance of microorganisms in supporting and maintaining all life on

Earth. We will also learn how microorganisms existed on Earth for billions of years before plants and animals appeared, how their collective physiological capacities rank them as Earth's greatest chemists, and how certain microorganisms have established relationships with higher organisms that can be either very beneficial or extremely harmful.

We begin our journey with a consideration of microorganisms as cellular entities.

1.2 Microorganisms as Cells

The **cell** is the fundamental unit of life. A single cell is an entity, isolated from other cells by a cell membrane (and perhaps a cell wall) and containing within it a variety of chemicals and subcellular structures (Figure 1.2). The **cell membrane** is the barrier that separates the inside of the cell from the outside. Inside the cell membrane are the various structures and chemicals that make it possible for the cell to function. Key structures are the **nucleus** or **nucleoid**, where the *genetic information*, deoxyribonucleic acid (DNA), needed to make more cells is stored, and the cytoplasm, where the *machinery* for cell growth and function is present.

All cells are made up of four chemical components: **proteins, nucleic acids, lipids,** and **polysaccharides**. Collectively, these are called *macromolecules*. It is the chemical nature and arrangement of macromolecules in a cell of one organism that makes it distinct from those of another. Although each kind of cell has a definite structure and size, a cell is a dynamic unit, constantly undergoing change and replacing its parts. Even when it is not growing, a cell may be acquiring materials from its environment and incorporating them into its own fabric. At the same time, it discards waste products into its environment. A cell is thus an *open system*, forever changing yet generally remaining the same.

Where did the first cells come from? In some way the first cell must have come from a noncell, something before the cell, a procellular structure. Although evolution of the first cell over 3.8 billion years ago was an improbable event that may have taken several hundred million years to occur, once the first cell arose, a series of highly probable events followed, including growth and division to form populations of cells from which evolution could select for improvements and diversification. Then, through billions of years of evolutionary change, the tremendous diversity of extant cell types that exist today arose. And, because all cells are constructed from the four basic classes of macromolecules mentioned earlier and share many other traits in common, it is hypothesized that all cells have descended from a common ancestor, the *universal ancestor* of all life (Chapter 11).

Characteristics of Living Systems

What are the essential characteristics of life? What differentiates cells from inanimate objects? Our concept of what is "alive" is constrained by what we can observe on Earth today or can deduce from the fossil record. But from our knowledge of biology thus far, we can identify several characteristics shared by most living systems, and these are summarized in Figure 1.3.

All cellular organisms are highly organized structures that show some form of **metabolism**. That is, cells take up chemical substances from their environment and transform them, conserving some of the energy of these substances in a form the cell can use, and then eliminate waste products. All cells show **reproduction**, that is, they are capable of directing the series of biochemical events that results in their own synthesis. As a result of metabolic processes a cell grows and divides to form two cells. Many cells undergo **differentiation**, a process by which new substances or structures are formed. Cell differentiation is often part of a cellular life cycle in which cells form special structures involved in reproduction, dispersal, or survival.

Cells respond to chemical signals in their environment, including those produced by other cells. Cells can thus undergo **communication** and even assess their own numbers in the surrounding environment by way of small diffusible molecules passed between neighboring cells. Living organisms are often capable of **movement** by self-propulsion, and in the microbial world we will see several different mechanisms that confer motility on cells. Finally, unlike nonliving structures, cells can *evolve*. Through the process of **evolution**, cells can permanently change their characteristics and transmit these new properties to their offspring.

Cells as Machines and as Coding Devices

Cells can be viewed in two ways. On one hand, cells can be considered *chemical machines* that carry out chemical transformations within the confines of a cellular structure. The catalysts of this chemical machine are **enzymes**, proteins capable of greatly accelerating the rate of specific chemical reactions. On the other hand, cells can be considered *coding devices*, analogous to computers, which store and process genetic information (DNA) that is eventually passed on to offspring during reproduction (Figure 1.4). Replication of the stored genetic information and its processing will be considered in Chapter 7 where we cover the core cellular functions of *DNA replication, transcription,* and *translation* in detail.

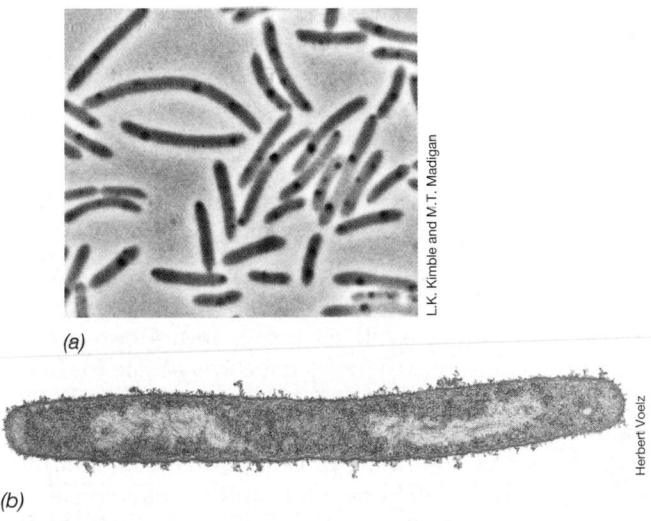

(a)

(b)

Figure 1.2 Cells. (a) Photomicrograph of rod-shaped bacterial cells as seen in the light microscope; a single cell is about 1 μm in diameter. (b) Longitudinal section through a bacterial cell as viewed with an electron microscope. The two lighter areas are the nucleoid, regions in the cell containing aggregated DNA.

1. Metabolism
Uptake of chemicals from the environment, their transformation within the cell, and elimination of wastes into the environment. The cell is thus an *open* system.

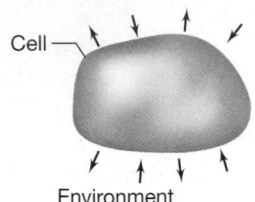

Cell

Environment

2. Reproduction (growth)
Chemicals from the environment are turned into new cells under the direction of preexisting cells.

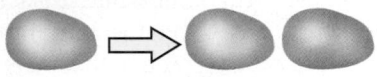

3. Differentiation
Formation of a new cell structure such as a spore, usually as part of a cellular *life cycle*.

Spore

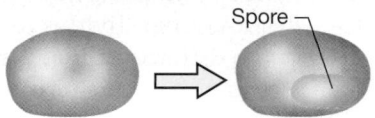

4. Communication
Cells *communicate* or *interact* primarily by means of chemicals that are released or taken up.

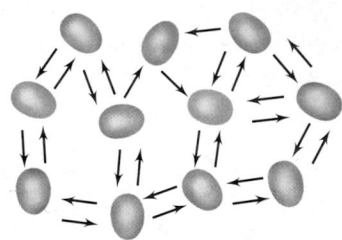

5. Movement
Living organisms are often capable of self-propulsion.

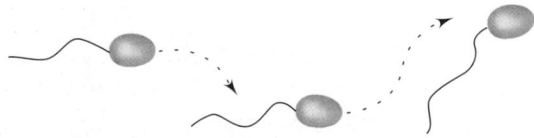

6. Evolution
Cells *evolve* to display new biological properties. Phylogenetic trees show the evolutionary relationships between cells.

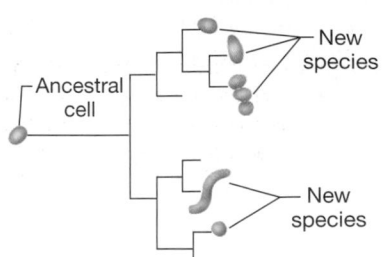

Ancestral cell

New species

New species

In reality, cells are both chemical machines *and* coding devices, and the link between these two attributes of a cell is *growth*. Under proper conditions, a viable cell will grow larger and then divide to form two cells (Figure 1.4). In the orderly process that results in cell division, the amount of all the constituents of the cell must *double*. This requires the chemical machinery of the cell function to supply energy and precursors for biosynthesis of macromolecules. But when a cell divides, each of the two resulting cells must contain all of the genetic information necessary for the formation of more cells, and so during the growth process there must also be a replication of DNA (Figure 1.4). Thus, the machine and coding functions of the cell must be highly coordinated in order for the cell to reproduce itself faithfully. We will see later that this is indeed true, and that in addition to *coordination*, the various machine and coding functions of the cell are subject to *regulation*, such that the cell stays optimally attuned to its environment.

✓ 1.2 Concept Check

The cell has a barrier, the cytoplasmic membrane, that separates the cytoplasm from the environment. Other cell features include the nucleus or nucleoid, cell wall, and cytoplasm. Certain key features such as metabolism and reproduction are associated with the living state, and cells can be thought of conceptually as both chemical machines and as biological coding devices.

✓ List six features associated with living organisms. Why is each feature important to the survival of a cell?

✓ Compare the machine and coding functions of a microbial cell. Why is neither of value to a cell without the other?

✓ List the four classes of cellular macromolecules.

1.3 Microorganisms and Their Natural Environments

Cells live in nature in association with other cells in assemblages called **populations**. Such populations are composed of groups of related cells, generally derived by successive cell divisions from a single parent cell. The location in an environment where a microbial population lives is called the **habitat**. We will see microbial habitats in familiar as well as quite unfamiliar places, including some so extreme as to be unsuitable for higher life forms. In nature, populations of cells rarely live alone. Rather, they live and interact with other populations of cells in assemblages called **microbial communities** (Figure 1.5). These communities may consist of free-swimming cells in aquatic environments, but very often form on living or nonliving surfaces where they are called *biofilms* (∞ Section 19.3).

Figure 1.3 The hallmarks of cellular life. Differentiation and motility are not properties of all microbial cells.

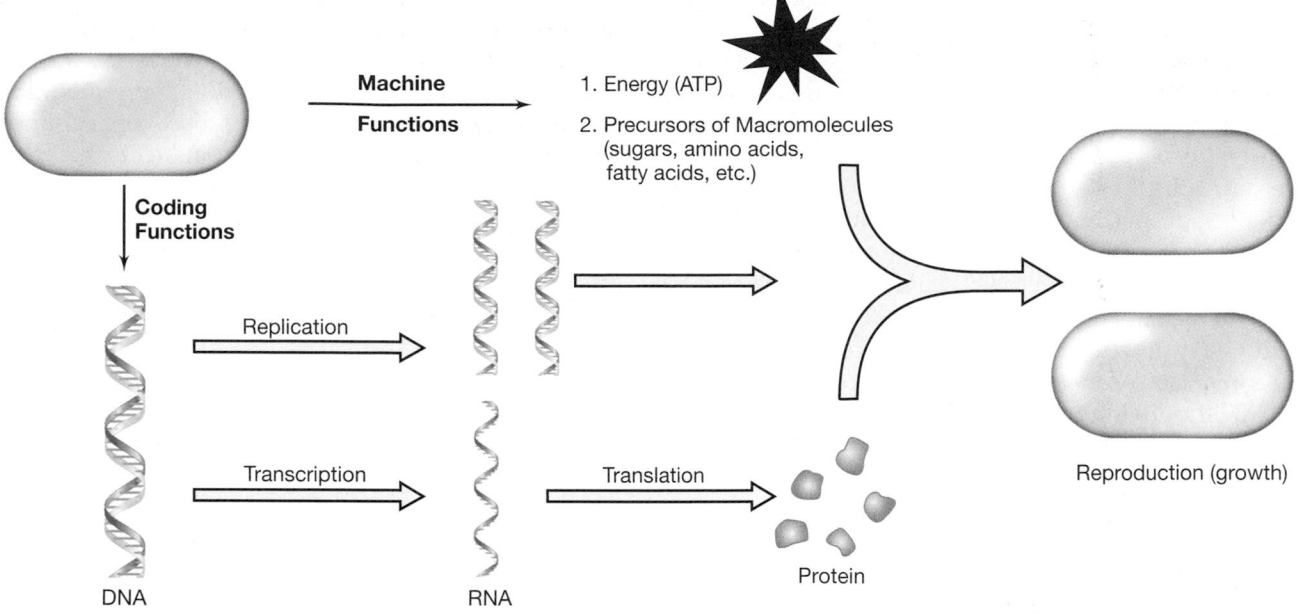

Figure 1.4 The machine and coding functions of the cell. In order for a cell to reproduce itself there must be an adequate supply of energy and precursors for the synthesis of new macromolecules, the genetic instructions must be replicated such that upon division each cell receives a copy, and genes must be expressed (the processes of transcription and translation) to form the proper amounts of necessary proteins and other macromolecules that will make up the new cell.

The Effect of Organisms on Each Other and on Their Habitats

Populations in microbial communities interact in various ways, and these interactions may be either harmful or beneficial. In many cases the populations interact and cooperate in their feeding efforts, with the waste products of the metabolic activities of some cells serving as nutrients for others. Organisms in a habitat also interact with their physical and chemical environment. Habitats differ markedly in their characteristics, and a habitat that is favorable for the growth of one organism may actually be harmful for another organism. Thus, the makeup of the

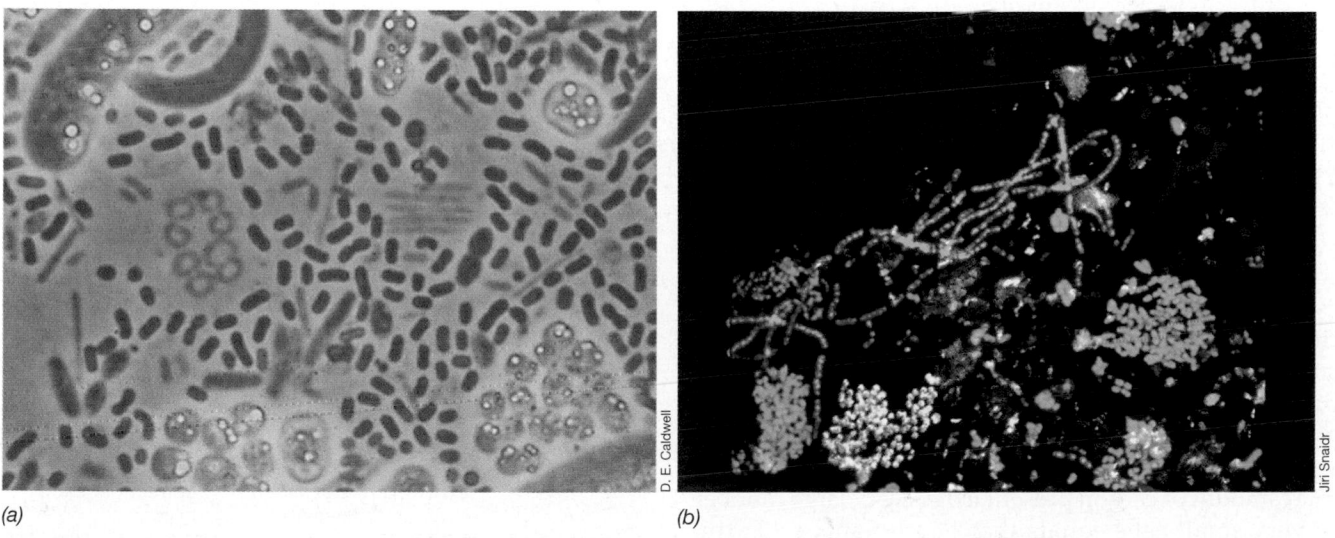

(a) *(b)*

Figure 1.5 Examples of microbial communities. (a) Photomicrograph of a bacterial community that developed in the depths of a small lake (Wintergreen Lake, Michigan), showing cells of various sizes. (b) A bacterial community in a sewage sludge sample. The sample was stained with a series of dyes, each of which stained a different bacterial group (∞ Section 18.4 and Figure 18.11*b* for details of how the staining was done). From R. Amann, J. Snaidr, M. Wagner, W. Ludwig, and K.-H. Schleifer, 1996. Journal of Bacteriology 178: 3496–3500, Fig. 2b. ©1996 American Society for Microbiology.

microbial community in a given habitat is determined to a great extent by the physical and chemical characteristics of that environment. Collectively, we speak of the living organisms together with the physical and chemical constituents of their environment as an **ecosystem**. Major microbial ecosystems include aquatic (oceans, ponds, lakes, streams, hot springs), terrestrial (soil, rock), and higher organisms, both plant and animal.

The properties of an ecosystem are often controlled to a significant extent by microbial activities. Organisms carrying out metabolic processes remove nutrients from the environment and use them to build new cells. At the same time, organisms excrete waste products of their metabolism into the environment. Therefore, over time, a microbial ecosystem can gradually change, both chemically and physically, through living processes. Gaseous oxygen is a good example of this. We will see later that molecular oxygen, O_2, is a vital nutrient for some microorganisms while being a poison to others. However, the oxygen-consuming activities of one group of organisms can make an oxic habitat anoxic (O_2-free) and suitable for growth of organisms that were previously kept in check.

Since single microbial cells are too small to be seen with the unaided eye, our knowledge of microorganisms in nature begins with studies using the microscope. Examination of natural material such as soil or water always reveals microbial cells. Although such tiny cells may seem inconsequential, single cells are capable of multiplying rapidly and producing huge populations that may have a major effect on the habitat. Thus, although microorganisms may seem to be minor components in nature, they are extremely important parts of virtually every ecosystem. In later chapters, after we have learned some of the details of microbial structure and function, genetics, evolution, and diversity, we will return to a discussion of the ways in which microorganisms affect animals, plants, and the whole global ecosystem.

The Extent of Microbial Life

It is easy to assume that since microorganisms are so small, their biomass on Earth must also be small when compared with the biomass of higher organisms. Interestingly, however, this is probably not true. Careful estimates of total microbial cell numbers on Earth, specifically, the numbers of **prokaryotes** (also called **bacteria**, rather small cells that lack a nucleus and about which we will have much more to say in later chapters), show this number to be on the order of 5×10^{30} cells. The total amount of carbon present in this very large number of very small cells equals that of all plants on Earth, while the collective contents of nitrogen and phosphorous in prokaryotic cells is over 10 times that of all plant biomass. Thus prokaryotic cells, small as they are, *constitute the major portion of biomass on Earth*, and are key reservoirs of essential nutrients for life. An equally star-

tling revelation is the discovery that most prokaryotic cells do not reside on the Earth's surface but instead lie underground in the oceanic and terrestrial subsurfaces. Because these habitats are relatively unexplored, there is much left for microbiologists to discover and understand about the life forms that dominate Earth.

✓ 1.3 Concept Check

Microorganisms exist in nature in populations that interact with other populations in microbial communities. The activities of microbial communities can greatly affect the chemical

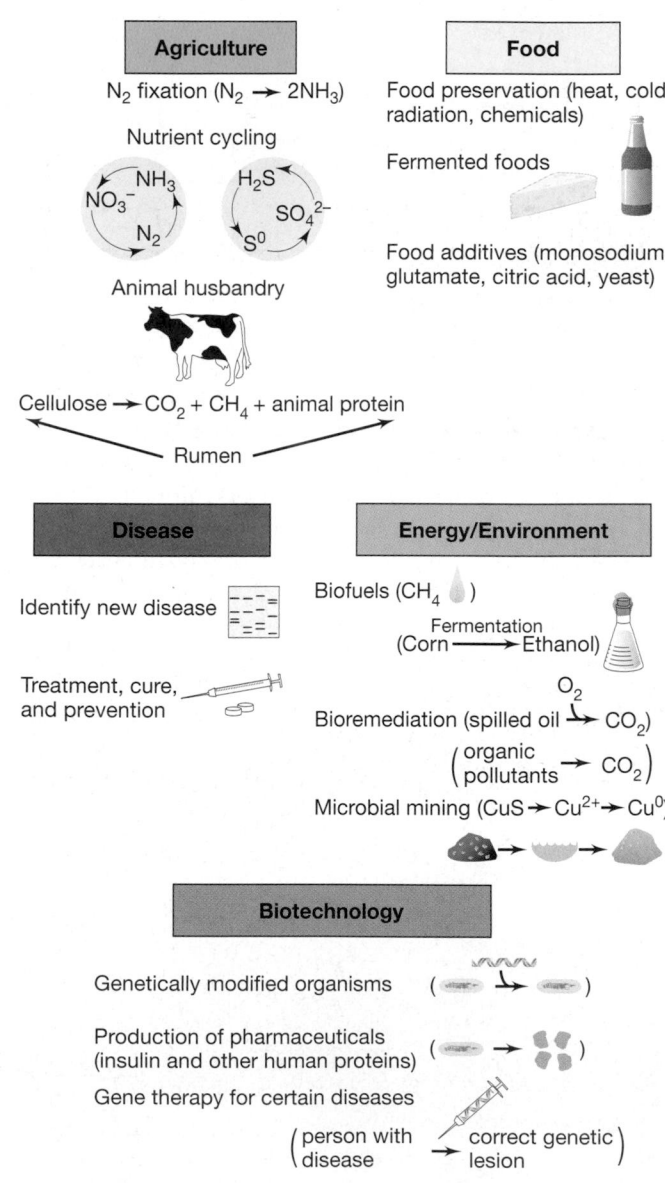

Figure 1.6 The impact of microorganisms on human affairs. Although many people think of microorganisms in the context of infectious diseases, few microorganisms actually cause disease. Microorganisms affect many aspects of our lives in addition to playing a role as disease agents.

and physical properties of their habitats. Most of the biomass on Earth is microbial.

✓ What is a *microbial habitat?*
✓ How do microorganisms change the chemical and physical properties of their habitats?
✓ Where are most prokaryotic cells located on Earth?

1.4 The Impact of Microorganisms on Humans

One goal of the microbiologist is to understand how microorganisms work and, through this understanding, to devise ways in which benefits of microorganisms may be increased and their harmful effects curtailed. Microbiologists have been highly successful in achieving these goals, and microbiology has played a major role in the advancement of human health and welfare. An overview of the impact of microorganisms on human affairs is shown in Figure 1.6.

Microorganisms as Disease Agents

One measure of the microbiologist's success in controlling microorganisms is shown by the statistics in Figure 1.7, which compare the present causes of death in the United States to those of over 100 years ago. At the beginning of the twentieth century, the major causes of death were infectious diseases; currently, such diseases are of much less importance. Control of infectious disease has come as a result of our comprehensive understanding of disease processes as well as improved sanitary practices and the

discovery and use of antimicrobial agents. As we will see later in this chapter, microbiology as a science had its beginnings in these studies of disease.

However, although we now live in a world where many pathogenic microorganisms are under control, for the individual dying slowly of a microbial infection as a consequence of acquired immune deficiency syndrome (AIDS), the cancer patient whose immune system has been devastated as a result of treatment with an anticancer drug, or the individual infected with a multiple-drug-resistant pathogen, microorganisms can still be a major threat to survival. Further, microbial diseases still constitute the major causes of death in many of the developing countries of the world. While eradication of smallpox from the world has been a stunning triumph for medical science, millions still die yearly from such pervasive microbial diseases as malaria, tuberculosis, cholera, African sleeping sickness, and severe diarrheal syndromes.

Clearly, microorganisms are still serious threats to human existence. However, we wish to emphasize that most microorganisms are *not* harmful to humans. In fact, most microorganisms cause no harm at all and instead are actually *beneficial*, carrying out processes that are of immense value to human society. We consider some of these now.

Microorganisms and Agriculture

Our whole system of *agriculture* depends in many important ways on microbial activities. A number of major crops are members of a plant group called the **legumes**, which live in close association with special bacteria that

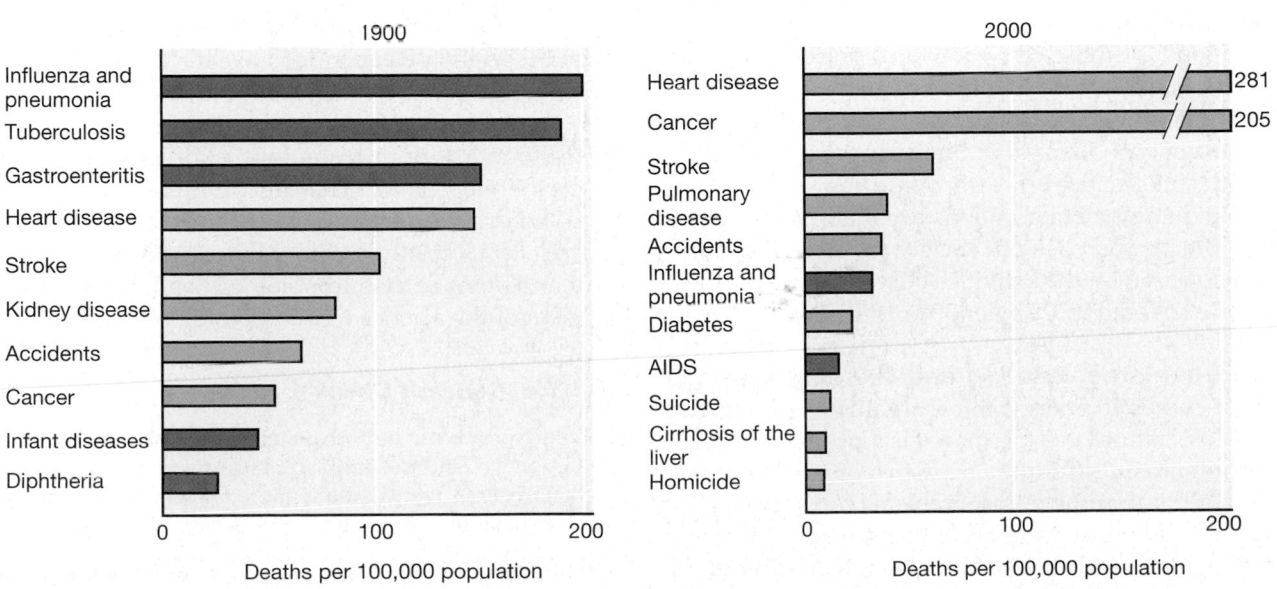

Figure 1.7 Death rates for the 10 leading causes of death in the United States: 1900 and 2000. Infectious diseases were the leading causes of death in 1900, whereas today they are much less important. Microbial diseases are shown in red, nonmicrobial diseases in green. Data from the United States National Center for Health Statistics.

form structures called **nodules** on their roots. In these root nodules, atmospheric nitrogen (N_2) is converted to fixed nitrogen compounds that the plants can use for growth. In this way, the activities of the root nodule bacteria reduce the need for costly plant fertilizer. Also of major agricultural importance are the microorganisms that are essential for the digestive process in ruminant animals such as cattle and sheep. These important farm animals have a special digestive organ called the **rumen** in which microorganisms carry out the digestive process. Without these microorganisms, cattle and sheep could not digest their food and thus could not thrive on nutrient-poor substances like grass and hay. Microorganisms also play key roles in the cycling of important nutrients in plant nutrition, particularly carbon, nitrogen, and sulfur. Microbial activities in soil and water convert these elements to forms that are readily accessible to plants. In addition to benefits to agriculture, microorganisms also have harmful effects. Animal and plant diseases due to microorganisms have major economic impact.

Microorganisms and Food

Once food crops and animals are produced, they must be delivered in wholesome form to consumers. Microorganisms play important roles in the *food industry*. We note first that food spoilage results in immense economic loss every year. The canning, frozen-food, and dried-food industries exist to prepare foods in such ways that they will not undergo microbial spoilage. Foodborne *disease* is also a consideration. Because food fit for human consumption can support the robust growth of many microorganisms, foods must be properly prepared and monitored to avoid transmission of disease.

However, not all microorganisms have harmful effects on foods or those who eat them. For example, dairy products manufactured, at least in part, via microbial activity include cheese, yogurt, and buttermilk, all products of major economic value. Similarly, sauerkraut, pickles, and some sausages also owe their existence to microbial activity. Baked goods are made using yeast. Even more pervasive in our society are alcoholic beverages, also based on the activities of yeast. Many of these topics are covered in Chapter 30 of this book.

Microorganisms, Energy, and the Environment

When it comes to energy, microorganisms play major roles. Most natural gas (methane) is a product of bacterial action, arising from the activities of methanogenic bacteria. Phototrophic microorganisms can harvest light energy for the production of **biomass**, energy stored in living organisms. Microbial biomass and existing waste materials such as domestic refuse, surplus grain, and animal wastes, can be converted to "biofuels," such as methane and ethanol, by the degradative activities of microorganisms.

Microorganisms can also be used to help clean up pollution created by human activities, a process called *bioremediation*. Various organisms have now been isolated from nature that consume spilled oil, solvents, pesticides, and other environmentally toxic pollutants, either directly at the site of the spill or later on after the toxic materials have pervaded soils or entered the groundwater. The great diversity of microorganisms on Earth contains vast genetic resources for solutions to cleaning up the environment, and much research in this area is taking place at present.

Microorganisms and the Future

Biotechnology entails the use of microorganisms in large-scale industrial processes, usually using genetically modified microorganisms capable of synthesizing specific products of high commercial value (∞Chapter 31).

Biotechnology is highly dependent on **genetic engineering**, the discipline that concerns the artificial manipulation of genes and their products. Genes from any source can be broken into pieces and modified in various ways, using microorganisms and their enzymes as molecular tools. It is even possible to make completely artificial genes using genetic engineering techniques. Once the desired gene has been selected or created, it can be inserted into a microorganism where it can be expressed to make the desired gene products. For instance, human insulin, a hormone found in abnormally low amounts in people with the disease diabetes, can be produced microbiologically from a human insulin gene engineered into a microorganism. We discuss genetic engineering and biotechnology in detail in Chapter 31.

The overwhelming influence of microorganisms in human society is clear. Indeed, we have many reasons to be aware of microorganisms and their activities (Figure 1.6). As the eminent French scientist Louis Pasteur, one of the founders of microbiology, expressed it: "The role of the infinitely small in nature is infinitely large." Therefore, before we begin our study of microbiology in earnest, let us briefly consider the contributions that Pasteur and other early microbiologists made to the development of the science of microbiology.

✓ 1.4 Concept Check

Microorganisms can be both beneficial and harmful to humans. Although we tend to emphasize harmful microorganisms (infectious disease agents), many more microorganisms in nature are beneficial than harmful.

- ✓ In what ways are microorganisms important in the food and agricultural industries?
- ✓ What fuels can be made by microorganisms?
- ✓ What is biotechnology and how might it improve the lives of humans?

II PATHWAYS OF DISCOVERY IN MICROBIOLOGY

Like any science, modern microbiology owes much to its past. Although claiming early roots, the science of microbiology didn't really develop until the nineteenth century. Since that time, the field has exploded and spawned several new but related fields. We retrace these pathways of discovery now.

1.5 The Historical Roots of Microbiology

Although the existence of creatures too small to be seen with the naked eye had long been suspected, their discovery was linked to the invention of the microscope. Robert Hooke described the fruiting structures of molds in 1664 (Figure 1.8), but the first person to see microorganisms in any detail was the Dutch amateur microscope builder Antoni van Leeuwenhoek, who in 1684, used extremely simple microscopes of his own construction (Figure 1.9a). Leeuwenhoek's microscopes were crude by today's standards, but by careful manipulation and focusing he was able to see organisms as small as bacteria. He reported his observations in a series of letters to the Royal Society of London, which published them in 1684 in English translation. Drawings of some of Leeuwenhoek's "wee animalcules," as he referred to them, are shown in Figure 1.9b. His observations were confirmed by other workers, but progress in understanding the nature and importance of these tiny organisms came only slowly. Only in the nineteenth century did improved microscopes become available and widely distributed, and about this time the extent and nature of microbial life forms became more apparent.

Ferdinand Cohn and the Science of Bacteriology

Microbiology did not develop as a science until improvements in microscopy allowed closer observation of bacterial cells and basic laboratory techniques for the study of microorganisms had been devised. Two perplexing biological questions led to the development of these essential laboratory techniques in the nineteenth century. One of these questions involved the issue of *spontaneous generation*. For centuries the question of whether nonliving matter can give rise to living organisms had perplexed serious thinkers. The second question concerned the nature of infectious disease. Disease had been observed to spread from one individual to another but the mechanism of this transfer was unknown. Although answers to these questions are associated with Louis Pasteur and Robert Koch,

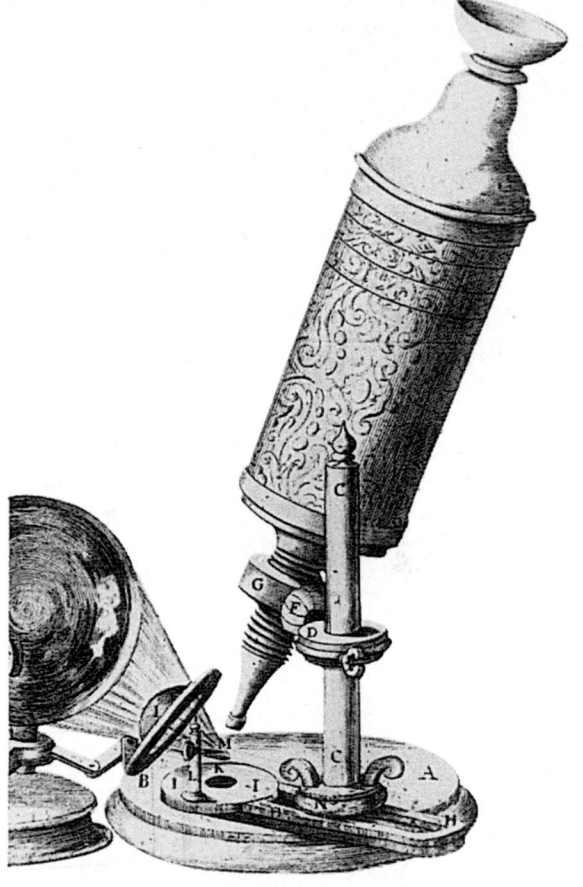

(a)

(b)

Figure 1.8 (a) The microscope used by Robert Hooke. The objective lens was fitted at the end of an adjustable bellows (G), with illumination focused on the specimen by a single lens (1). (b) A drawing by Robert Hooke, which represents one of the first microscopic descriptions of microorganisms: a blue mold growing on the surface of leather; the round structures contain spores of the mold.

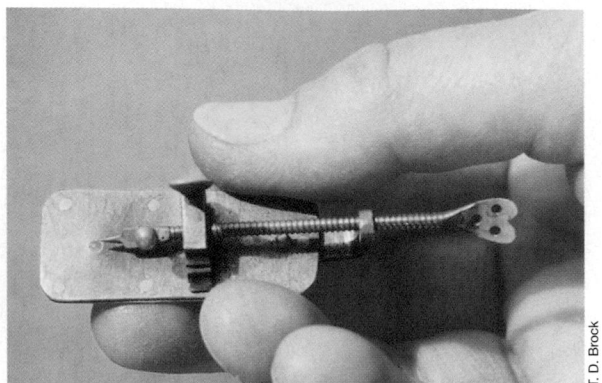

(a)

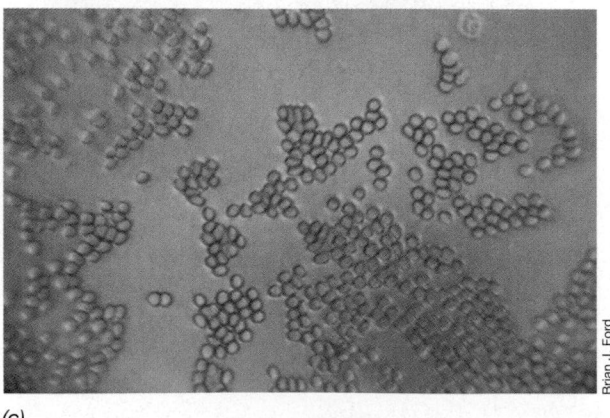

(b)

respectively, it was a German botanist, Ferdinand Cohn (1828–1898), a contemporary of Pasteur and Koch, who is credited with the founding of bacteriology (the study of bacteria) and with getting the fledgling field of microbiology off to a good start.

Cohn was born in what is now Wroclaw, Poland, in 1828. He was trained as a botanist and by 1850 had become very interested in microscopy, being fortunate to have available the best microscopes of his era for the study of growth and division of plant cells. His interests in microscopy naturally led him toward the study of unicellular plants, the algae, and later to photosynthetic life forms that are actually bacteria, the cyanobacteria. Cohn believed that all bacteria, even those lacking photosynthetic pigments, were members of the plant kingdom, and his microscopic studies gradually drifted away from plants and algae to a variety of different bacteria, including the large sulfur bacterium, *Beggiatoa* (Figure 1.10).

Cohn became particularly interested in heat-resistant forms of bacteria, which led him to discover the genus *Bacillus*, and the process of spore formation. We now know that bacterial endospores are extremely heat-resistant structures, in fact more so than all known microbial cells, except for a few that grow best at extremely high temperatures. Cohn described the entire life cycle of *Bacillus* (vegetative cell → endospore → vegetative cell; ∞ Section 4.15) and discovered that vegetative cells but not endospores are killed by boiling. Indeed, Cohn's findings helped explain why earlier scientists, such as John Tyndall, had found boiling to be usually but not always an effective technique for sterilization.

Cohn continued to work with bacteria until his retirement, and during this time contributed to the development of bacteriology in many ways, including laying the groundwork for a scheme of bacterial classification and founding a major scientific journal. During this time Cohn was a strong advocate of the techniques and research of the pioneer medical microbiologist, Robert Koch. Cohn also is credited with helping devise simple but very effective methods for preventing the contamination of sterile culture media, such as the use of cotton for closing flasks and tubes. These methods were later used by Koch and allowed him to make rapid progress in the isolation and characterization of several disease-causing bacteria (see later in this section).

Pasteur and the Downfall of Spontaneous Generation

In the nineteenth century, a major controversy erupted concerning the theory of **spontaneous generation**. The basic idea of spontaneous generation can easily be understood. If food is allowed to stand for some time, it putrefies. When the putrefied material is examined microscopically, it is found to be teeming with bacteria. Where do these bacteria come from, since they are not seen in fresh food? Some people said they developed

Figure 1.9 (a) Photograph of a replica of Leeuwenhoek's microscope. The lens is mounted in the brass plate adjacent to the tip of the adjustable focusing screw. (b) Leeuwenhoek's drawings of bacteria, published in 1684. Even from these crude drawings we can recognize several morphological types of common bacteria. A, C, F, and G, rod-shaped; E, spherical or coccus-shaped; H, cocci packets (∞ Figure 4.11). (c) Photomicrograph of a human blood smear taken through a van Leeuwenhoek microscope. Red blood cells are clearly apparent.

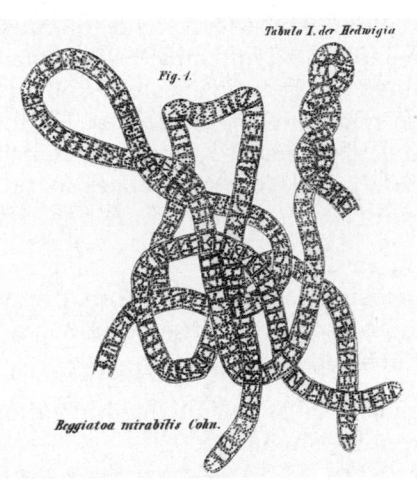

Figure 1.10 Drawing by Ferdinand Cohn made in 1866 of the filamentous sulfur-oxidizing bacterium *Beggiatoa mirabilis*. The small granules inside the cell consist of elemental sulfur, produced from the oxidation of hydrogen sulfide (H_2S). Cohn was the first to identify the granules as sulfur.

from seeds or germs that had entered the food from the air, whereas others said that they arose spontaneously from nonliving materials.

The most powerful opponent of spontaneous generation was the French chemist Louis Pasteur (1822–1895), whose work on this problem was the most exacting and convincing. Pasteur first showed that structures were present in air that closely resembled the microorganisms seen in putrefying materials. He found that in ordinary air there constantly exists a variety of microbial cells and that they could not be distinguished from the organisms found in much larger numbers in putrefying materials. Pasteur concluded that the organisms found in putrefying materials originated from microorganisms present in the air. He postulated that these cells are constantly being deposited on all objects. If this conclusion was correct, Pasteur deduced that food treated in such a way as to destroy all living organisms contaminating it, should not putrefy.

Since it had already been established that heat effectively kills living organisms, Pasteur used heat to eliminate contaminants. In fact, other workers had shown that when a nutrient solution was sealed in a glass flask and heated to boiling, it did not support microbial growth. Proponents of spontaneous generation criticized such experiments by declaring that fresh air was necessary for spontaneous generation and that the air itself inside the sealed flask was affected in some way by heating so that it could no longer support spontaneous generation. Pasteur skirted this objection simply and brilliantly by constructing a swan-necked flask, now called a *Pasteur flask* (Figure 1.11). In such a flask nutrient solutions could be heated to boiling; after the flask was cooled, air could reenter but the bends in the neck prevented particulate matter containing bacteria

or other microorganisms from getting into the main body of the flask. Broth sterilized in such a flask did not putrefy, and no microorganisms ever appeared in the flask as long as the neck of the flask did not contact the sterile liquid. If, however, the flask was tipped to allow the sterile liquid to contact the neck of the flask (Figure 1.11c), putrefaction occurred and the liquid soon teemed with microorganisms. This simple experiment effectively settled the controversy surrounding the theory of spontaneous generation.

Killing all the bacteria or other microorganisms in or on objects is a process we now call **sterilization**, and the procedures that Pasteur, Cohn, and others used were eventually refined and carried over into microbiologi-

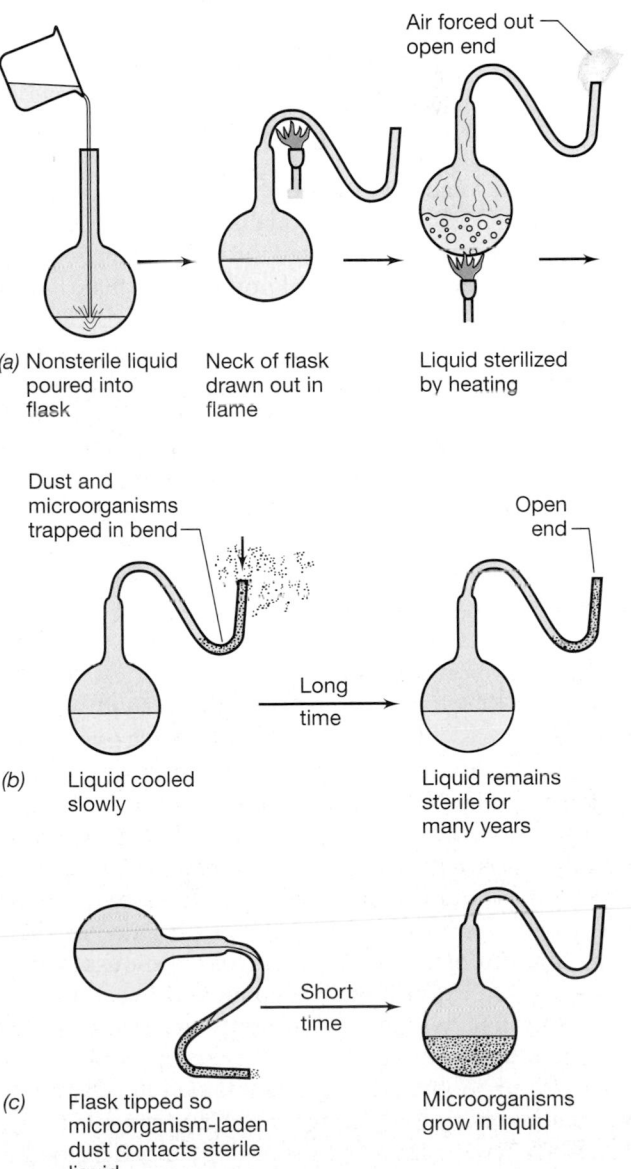

(a) Nonsterile liquid poured into flask — Neck of flask drawn out in flame — Liquid sterilized by heating — Air forced out open end

(b) Liquid cooled slowly — Dust and microorganisms trapped in bend — Long time — Open end — Liquid remains sterile for many years

(c) Flask tipped so microorganism-laden dust contacts sterile liquid — Short time — Microorganisms grow in liquid

Figure 1.11 Pasteur's experiment with the swan-necked flask. (a) Sterilizing the contents of the flask. (b) If the flask remained upright, no microbial growth occurred. (c) If microorganisms trapped in the neck reached the sterile liquid, microbial growth ensued.

cal research. Disproving the theory of spontaneous generation thus led to the development of effective sterilization procedures, without which microbiology as a science could not have developed. Food science also owes a debt to Pasteur, as his principles are applied in the canning and preservation of many foods.

Pasteur went on to many other triumphs in microbiology and medicine. Chief among these was his development of vaccines for the diseases anthrax, fowl cholera, and rabies during a very productive period from 1880 to 1890. These medical and veterinary breakthroughs were not only highly significant in their own right, but helped solidify the concept of the *germ theory of disease* whose principles were being developed at this time by a contemporary of Pasteur, Robert Koch.

Koch and the Germ Theory of Disease

Proof that microorganisms could cause disease provided the greatest impetus for the development of the science of microbiology. Indeed, even in the sixteenth century it was thought that something could be transmitted from a diseased person to a healthy person that induced the disease. Many diseases seemed to spread through populations and were called *contagious*; the unknown agent that did the spreading was called the *contagion*. After the discovery of microorganisms, it was widely held that these organisms were responsible for contagious diseases, but proof was lacking. Discoveries in sanitation by Ignaz Semmelweis and Joseph Lister provided indirect evidence for the importance of microorganisms in causing human diseases, but it was not until the work of Robert Koch (1843–1910), a physician by training, that the *germ theory of disease* was clearly conceptualized and given experimental support.

In his early work Koch studied *anthrax*, a disease of cattle that occasionally occurs in humans. Anthrax is caused by a spore-forming bacterium called *Bacillus anthracis*, and the blood of an animal infected with anthrax teems with cells of this large bacterium. Koch established by careful microscopy that the bacteria were always present in the blood of an animal that was succumbing to the disease. However, mere association of the bacterium with the disease did not prove that it actually *caused* the disease; it might instead be a *result* of the disease. Therefore, Koch demonstrated that it was possible to take a small amount of blood from a diseased mouse, which he used as an experimental animal, and inject it into a second mouse, which subsequently became diseased and died. He took blood from this second animal, injected it into another, and again obtained the characteristic disease symptoms. In repeated experiments of this sort, Koch demonstrated by microscopy that the blood of the diseased animal contained large numbers of the spore-forming bacterium.

Koch carried this experiment even further. He found that the bacteria could also be cultivated in nutrient fluids outside the animal body and that even after many transfers in culture the bacteria could still cause the disease when reinoculated into an animal. Bacteria from a diseased animal and bacteria in culture both induced the same disease symptoms upon injection. On the basis of these and other experiments Koch formulated the following criteria, now called **Koch's postulates**, for proving that a specific type of microorganism causes a specific disease:

Koch's Postulates

1. The organism should be constantly present in animals suffering from the disease and should not be present in healthy individuals.

2. The organism must be cultivated in a pure culture away from the animal body.

3. Such a culture, when inoculated into susceptible animals, should initiate the characteristic disease symptoms.

4. The organism should be reisolated from these experimental animals and cultured again in the laboratory, after which it should still be the same as the original organism.

Koch's postulates are summarized in Figure 1.12. Koch's postulates not only supplied a means of demonstrating that specific organisms cause specific diseases, but also catalyzed the development of the science of microbiology by stressing the importance of laboratory culture. With Koch's postulates as a guide, subsequent investigators discovered the causes of many important diseases of humans and other animals. These discoveries led to the development of successful treatments for the prevention and cure of many infectious diseases, thereby greatly improving the scientific basis of clinical medicine.

Koch and Pure Cultures

To link a specific microorganism to a specific process, such as a disease, the organism must first be isolated in a culture; that is, the culture must be *pure*. This concept was not lost on Robert Koch in formulating his famous postulates (Figure 1.12), and he developed several ingenious methods of obtaining bacteria in pure culture (see the box, Solid Media, the Petri Plate, and Pure Cultures).

Koch started in a crude way by using solid nutrients such as a potato slice to culture bacteria, but quickly developed more reliable methods, many of which are still in use today. Koch observed that when a solid surface such as a potato slice was incubated in air, bacterial colonies developed, each having a characteristic shape and color. He inferred that each colony had arisen from a single bacterial cell that had fallen on the surface, found suitable nutrients, and multiplied; that is, each colony represented a *pure culture*. Koch realized that this discovery provided a simple way of obtaining pure cultures. Since not all organisms grow on potato slices, however, Koch devised more uniform and reproducible nutrient solutions solidi-

KOCH'S POSTULATES:

1. **The suspected pathogenic organism should be present in *all* cases of the disease and absent from healthy animals.**

2. **The suspected organism should be grown in pure culture.**

3. **Cells from a pure culture of the suspected organism should cause disease in a healthy animal.**

4. **The organism should be reisolated and shown to be the same as the original.**

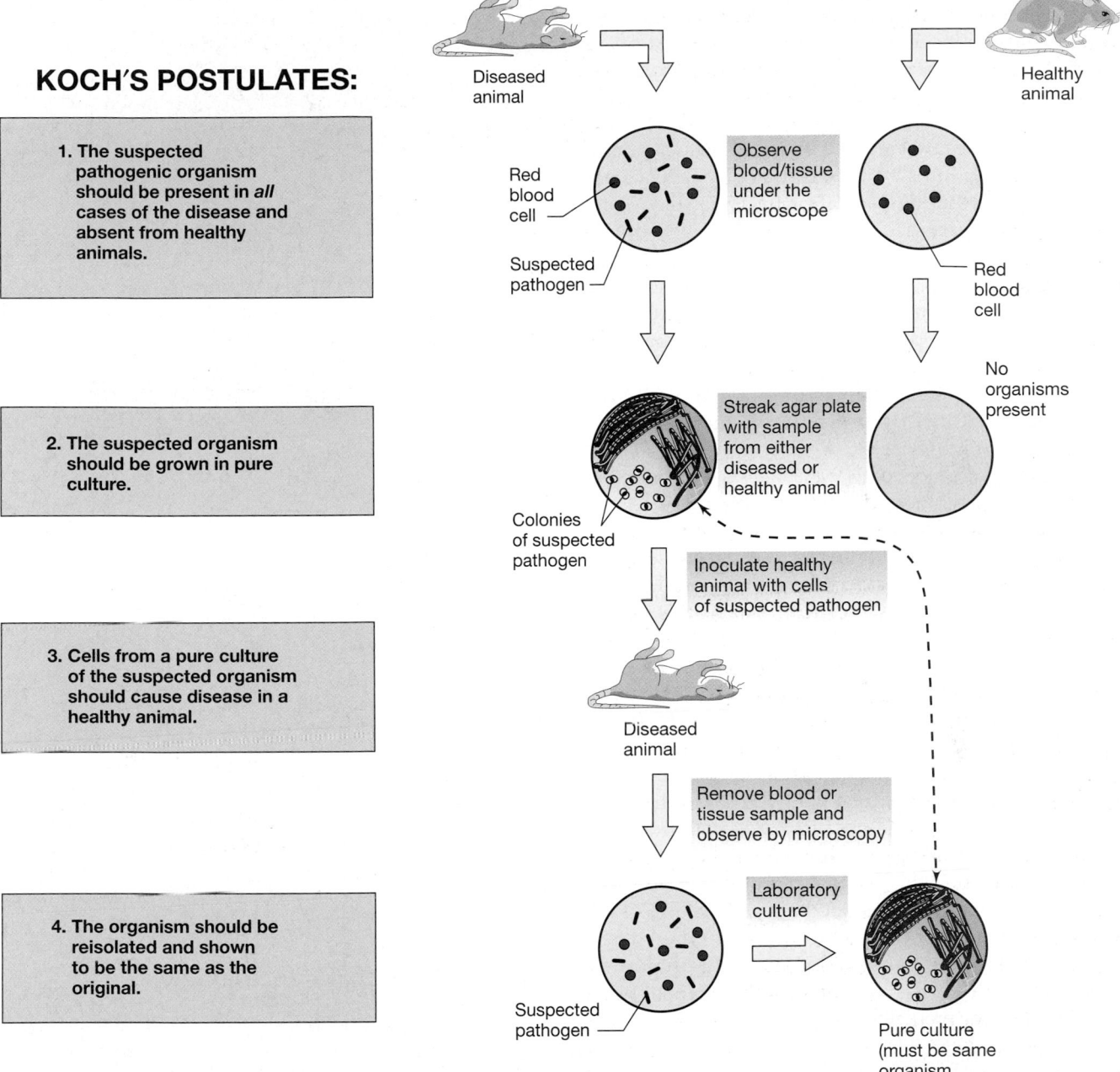

Diseased animal

Healthy animal

Red blood cell

Observe blood/tissue under the microscope

Suspected pathogen

Red blood cell

No organisms present

Streak agar plate with sample from either diseased or healthy animal

Colonies of suspected pathogen

Inoculate healthy animal with cells of suspected pathogen

Diseased animal

Remove blood or tissue sample and observe by microscopy

Suspected pathogen

Laboratory culture

Pure culture (must be same organism as before)

Figure 1.12 Koch's postulates for proving that a specific microorganism causes a specific disease. Note how it is essential that following isolation of a pure culture of the suspected pathogen, a laboratory culture of the organism should both initiate the disease *and* be recovered from the diseased animal. Establishing the correct conditions for growing the pathogen is essential, otherwise it will be missed.

fied with gelatin and later agar (see the box). Agar as a solidifying agent is widely used in the microbiology laboratory today for obtaining and maintaining pure cultures of various microorganisms, especially bacteria.

Koch and Tuberculosis

Koch's greatest accomplishment in medical bacteriology was with tuberculosis. At the time Koch began this work (1881), one-seventh of all reported human deaths were

caused by tuberculosis. Even at this time there was strong evidence that tuberculosis was a contagious disease, but the suspected causal organism had never been seen, either in diseased tissues or in culture. Koch's aim from the beginning of his work on tuberculosis was to demonstrate the causal agent of tuberculosis, and to this end he employed all the methods he had so carefully developed previously: microscopy, staining of tissues, pure culture isolation, and animal inoculation.

As is now well known, *Mycobacterium tuberculosis*, the "tubercle bacillus," is very difficult to stain because of large amounts of waxy lipid present on its outer surface. But Koch devised a staining procedure for *M. tuberculosis* in tissue samples using alkaline methylene blue in conjunction with a second stain (bismarck brown) that stained only the tissue (Koch's method was the forerunner of the Ziehl-Nielsen stain used today for staining acid-fast bacteria like *M. tuberculosis*; ∞Section 12.23). Using his newly developed staining method, Koch observed bright-blue, rod-shaped cells of *M. tuberculosis* in tuberculous tissues, the latter staining a light brown (Figure 1.13). However, from his previous work on anthrax, Koch realized that simply *identifying* an organism associated with tuberculosis was not enough; he must *culture* the organism in order to prove that it was the specific cause of tuberculosis.

Producing cultures of *M. tuberculosis* was not easy, but eventually Koch was successful in obtaining colonies of this organism on coagulated blood serum. Later he used agar, which had just been introduced as a solidifying agent (see the box). Under the best of conditions, *M. tuberculosis* grows slowly in culture, but Koch's persistence and patience eventually led to pure cultures of this organism from a variety of human and animal sources. From here it was relatively easy to obtain definitive proof that the organism he had isolated was the true cause of the disease tuberculosis. Guinea pigs can be readily infected with *M. tuberculosis* and eventually succumb to systemic tuberculosis. Koch showed that diseased guinea pigs contained masses of *M. tuberculosis* cells in their tissues and that pure cultures obtained from such animals transmitted the disease to uninfected animals. Thus, Koch successfully satisfied all four criteria of his postulates (Figure 1.12), and the cause of tuberculosis was understood. For his contributions in this important arena, Robert Koch was awarded the 1905 Nobel Prize for physiology or medicine.

✓ 1.5 Concept Check

Ferdinand Cohn founded the field of bacteriology and discovered bacterial endospores. Louis Pasteur's work on spontaneous generation led to the development of methods for control of the growth of microorganisms. Robert Koch developed criteria for the study of infectious microorganisms and developed the first methods for the growth of pure cultures and microorganisms.

✓ How did Pasteur's famous experiment defeat the theory of spontaneous generation?

✓ How can Koch's postulates prove cause and effect in a disease?

✓ What advantages do solid media offer for the culture of microorganisms?

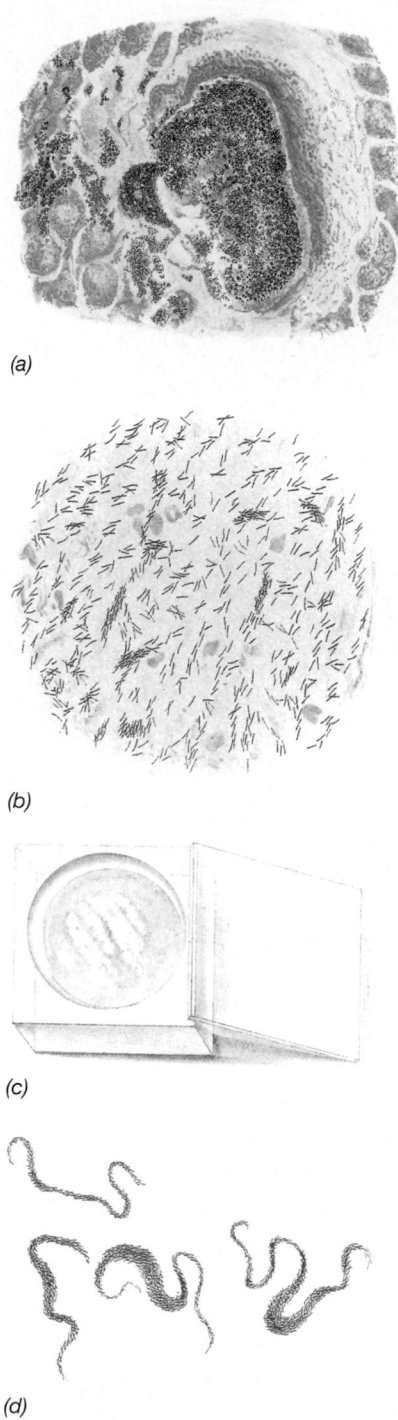

(a)

(b)

(c)

(d)

Figure 1.13 Robert Koch's drawings of cells of *Mycobacterium tuberculosis* in tissues and in laboratory culture. (a) Section through a tubercle from lung tissue. Cells of *M. tuberculosis* stain blue, whereas the lung tissue stains brown. (b) Cells of *M. tuberculosis* in a sputum sample of a tuberculous patient. (c,d) Growth of *M. tuberculosis* in pure culture. (c) Growth on a glass plate of coagulated blood serum inside a glass box (with lid open). (d) A colony of *M. tuberculosis* cells taken from the plate in (c) and observed microscopically at 700×; cells appear as long "cordlike" forms (compare with Figure 12.70b). Original drawings appeared in Koch, R. 1884. "Die Aetiologie der Tuberkulose." *Mittheilungen aus dem Kaiserlichen Gesundheitsamte* 2:1–88.

Learning from the Past | Solid Media, the Petri Plate and Pure Cultures

Robert Koch was the first to grow bacteria on solid culture media. Koch initially employed gelatin as a solidifying agent for the various nutrient fluids he used to culture pathogenic bacteria and developed a method for preparing horizontal slabs of solid media that were kept free of contamination by covering them with a bell jar or glass box (see Figure 1.13c).

Nutrient gelatin was a marvelous culture medium for the isolation and study of various bacteria, but it had several drawbacks, the most important being that gelatin does not remain solid at body temperature (37°C), the optimum temperature for growth of most human pathogens. Thus, a more versatile solidifying agent was needed, and this turned out to be agar.

Agar is a polysaccharide derived from red algae. It was used widely in the nineteenth century, especially in tropical countries, as a gelling agent. Walter Hesse first used agar as a solidifying agent for bacteriological culture media. The actual suggestion that agar be used instead of gelatin was made by Hesse's wife, Fannie. Fannie Hesse had used agar in the preparation of fruit jellies, and when it was tried as a solidifying agent in nutrient media, its superior qualities were immediately evident. Hesse wrote to Koch about this discovery, and Koch quickly adapted agar to his own studies, including his classic studies on the isolation

of the bacterium *Mycobacterium tuberculosis,* the cause of the disease tuberculosis (see text and Figure 1.13).

In 1887 Richard Petri published a brief paper describing a modification of Koch's flat plate technique. Petri's enhancement, which turned out to be amazingly useful, was the development of the double-sided dishes that bear his name. The advantage of Petri dishes was apparent; they could be easily stacked and sterilized separately from the medium, and, following the addition of molten medium to the smaller of the two double dishes, the larger dish could be used as a cover to prevent contamination. Colonies that formed on the surface of the agar in the Petri dish remained fully exposed to air and could easily be manipulated for further study. The original idea of Petri has not been improved on to this day, as the Petri dish, made either of reusable glass and sterilized by dry heat or of disposable plastic and sterilized by ethylene oxide (a gaseous sterilant), is a mainstay of the microbiology laboratory.

Finally, it should also be noted that Koch was keenly aware of the implications his pure culture methods had for the study of microbial systematics. Koch observed that different colonial forms (differing in color, colony morphology, size, and the like, see Figure 1) developed on solid media exposed to a contaminated object, and that these colonial forms bred true and could be distinguished from one another by their colony characteristics. Cells from different colonies also differed microscopically and often in their temperature or nutrient requirements as well. Koch realized that these differences among microorganisms met all the requirements that taxonomists had established for the classification of larger organisms, such as plant and animal species. In Koch's own words: "*All bacteria which maintain the characteristics which differentiate one from another when they are cultured on the same medium and under the same conditions, should be designated as species, varieties, forms, or other suitable designation.*" Koch also realized from the study of pure cultures that one could show that specific organisms have specific effects. Such insightful thinking was significant in the relatively rapid acceptance of microbiology as an independent biological science.

Koch's discovery of solid culture media and his emphasis on pure culture microbiology reached far beyond the realm of medical bacteriology; his discoveries supplied critically needed tools for development of the field of bacterial taxonomy, genetics, and several related disciplines. Indeed, the entire field of microbiology owes a huge debt of gratitude to Koch and his associates for the intuition they displayed in grasping the significance of pure cultures and developing some of the most basic methods in microbiology. ■

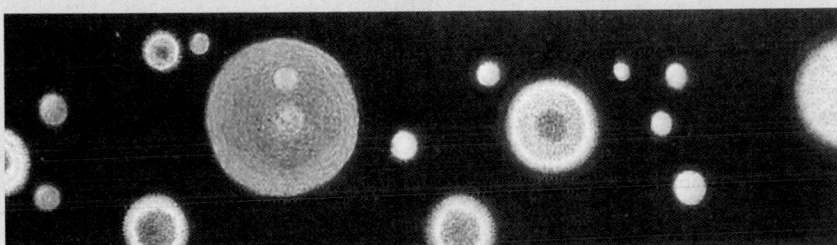

Figure 1 A hand-colored photograph of colonies formed on agar taken by Walter Hesse, an associate of Robert Koch. The colonies include those of fungi (molds) and bacteria and were obtained during studies Hesse initiated on the microbiological content of air in Berlin, Germany, in 1882. From Hesse, W. 1884. "Ueber quantitative Bestimmung der in der Luft enthaltenen Mikroorganismen," in Struck (ed.), *Mittheilungen aus dem Kaiserlichen Gesundheitsamte.* August Hirschwald.

1.6 Microbial Diversity and the Advent of Molecular Microbiology

As microbiology advanced from the nineteenth to the twentieth century, our understanding of microbial diversity improved significantly and several subdisciplines arose in the field of microbiology, leading to the present era of "molecular microbiology." Two giants in the field led this transition, Martinus Beijerinck, who was Dutch, and Sergei Winogradsky, who was Russian. Both of these early microbiologists were interested in bacteria that inhabit soil and water, and both are remembered primarily for their contributions to bacterial diversity.

Beijerinck and Winogradsky

Martinus Beijerinck (1851–1931) was a professor at the Delft Polytechnic School in his later years (van Leeuwenhoek was also from Delft), but was originally trained in botany and began his career in microbiology studying the microbiology of plants. Perhaps Beijerinck's greatest contribution to the field of microbiology was his clear formulation of the concept of the **enrichment culture**. Instead of isolating microorganisms from nature in a nonselective fashion, Beijerinck proposed *selecting* specific microorganisms from a natural sample through the use of specific culture media and incubation conditions that favored growth of only one type or a physiologically related group of organisms. Using his enrichment culture (or "selective culture," as he called it) technique, Beijerinck isolated the first pure cultures of many soil and aquatic microorganisms, including aerobic nitrogen-fixing bacteria (Figure 1.14), sulfate-reducing and sulfur-oxidizing bacteria, nitrogen-fixing root nodule bacteria, *Lactobacillus* species, green algae, and many other microorganisms. And from his studies of tobacco mosaic disease, Beijerinck showed, using selective filtration techniques, that the infectious agent (a virus) was not bacterial but somehow became incorporated into the cells of the host plant and required the living plant to reproduce; in essence, Beijerinck described the basic tenets of virology.

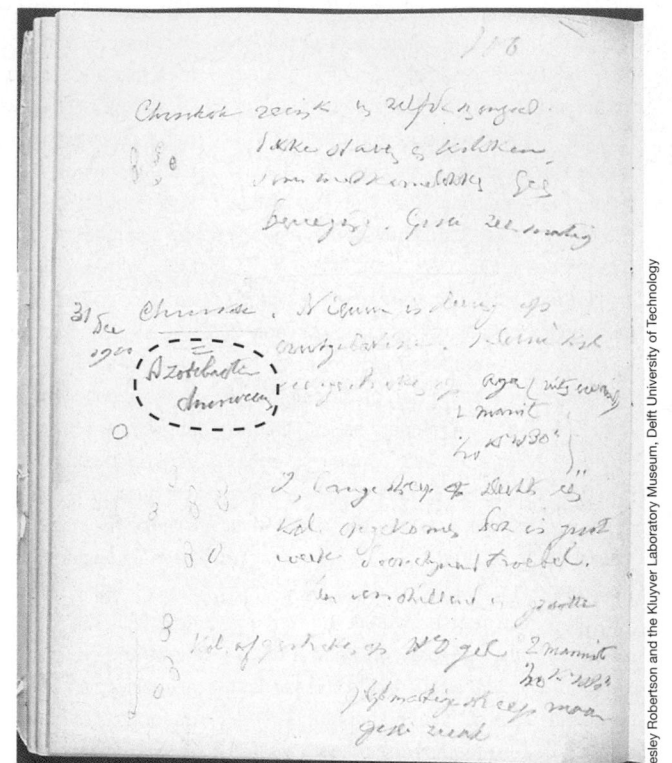

(a)

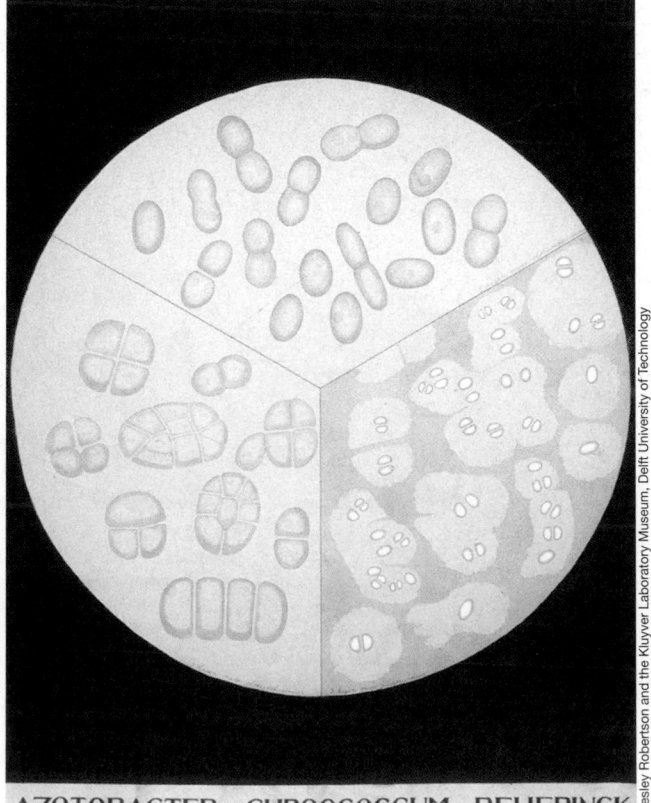

AZOTOBACTER CHROOCOCCUM BEIJERINCK

(b)

Figure 1.14 Martinus Beijerinck and *Azotobacter*. (a) A page from the laboratory notebook of M. Beijerinck dated December 31, 1900, that describes his observations on the aerobic nitrogen-fixing bacterium *Azotobacter chroococcum* (name circled in red). It is on this page that Beijerinck uses this name for the first time. Compare Beijerinck's drawings of pairs of *A. chroococcum* cells with a photomicrograph of cells of *Azotobacter* shown in Figure 12.19a. (b) A painting by M. Beijerinck's sister, Henriëtte Beijerinck, showing cells of *Azotobacter chroococcum*. Beijerinck used such paintings to illustrate his lectures, because this was long before the days of the overhead, slide, and computer projectors used in lectures and seminars today.

Sergei Winogradsky (1856–1953) had interests similar to Beijerinck's and was also successful in isolating several key bacteria for the first time. Winogradsky was interested in soil bacteria, particularly those involved in the cycling of nitrogen and sulfur compounds (Figure 1.15). In this connection, Winogradsky isolated pure cultures of nitrifying bacteria, clearly showing that the process of nitrification (the oxidation of ammonia to nitrate) was the result of bacterial action, and studied the oxidation of hydrogen sulfide by sulfur-oxidizing bacteria directly in their natural habitats.

Besides demonstrating that bacteria can be biogeochemical agents, Winogradsky is also remembered for his keen insight into the *metabolic significance* of these processes. For example, from his studies of the sulfur-oxidizing bacteria, Winogradsky proposed the concept of *chemolithotrophy*, the oxidation of *inorganic* compounds coupled to the release of energy (∞ Sections 2.4, 5.14, and 17.8). And from studies of the nitrifying bacteria Winogradsky concluded that these organisms obtained their carbon from CO_2 in air, that is, that they were - *autotrophs*. Although neither of these concepts was readily accepted at the time, we now know that bacterial chemolithotrophy and autotrophy are extremely important processes on Earth and can even support the growth of higher organisms (∞ Section 19.8). Using an enrichment method, Winogradsky also isolated the first nitrogen-fixing bacterium (the anaerobe *Clostridium pasteurianum*) and by so doing developed the concept of bacterial N_2 fixation. Winogradsky lived to be almost 100, publishing many scientific papers, along with a major monograph, *Microbiologie du Sol (Soil Microbiology)*; the latter work, a true milestone in microbiology, contained his original drawings of many of the organisms he had isolated or otherwise studied in enrichment culture or natural material during his career (Figure 1.15).

Table 1.1 summarizes some of the important discoveries in the field of microbiology from van Leeuwenhoek to the present.

Development of the Major Subdisciplines of Microbiology

In the twentieth century the field of microbiology developed rapidly in two separate directions—applied microbiology and basic microbiology. On the applied side, the practical advances of Koch led to extensive developments in *medical microbiology* and *immunology* in the early part of the twentieth century, with the discovery of many new bacterial pathogens (∞ the box, Discoverers of the Main Bacterial Pathogens, Chapter 26) and determination of the mechanisms by which these pathogens infect the body and are in turn resisted by the body's defenses. Other early practical advances, bolstered by the discoveries of Beijerinck and Winogradsky, were in the field of *agricultural microbi-*

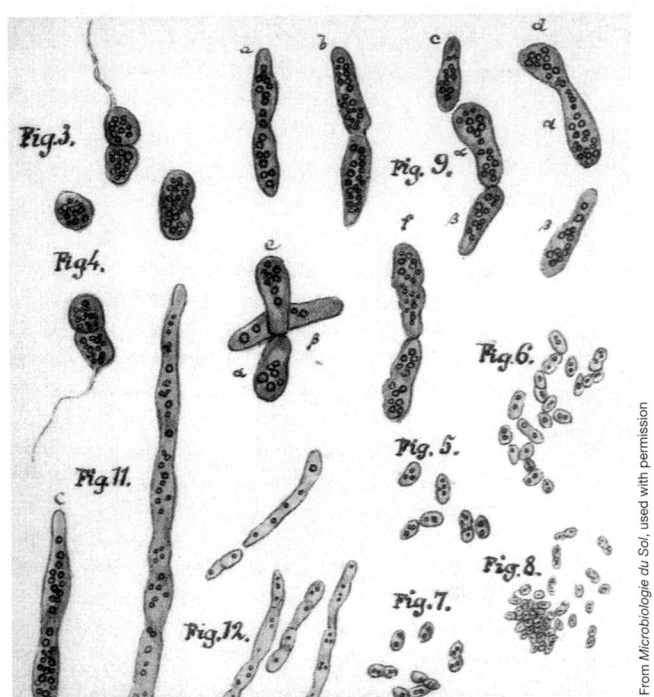

From *Microbiologie du Sol*, used with permission

Figure 1.15 Hand-colored drawings of cells of purple sulfur phototrophic bacteria included in the monograph *Microbiologie du Sol*, by Sergei Winogradsky. The original drawings were made by S. Winogradsky about 1887 and then copied and hand-colored by his wife Hélène for publication in the monograph. These drawings included cells of the genus *Chromatium*, such as *C. okenii* (Figs. 3 and 4) and *C. vinosum* (Figs. 5–8). These species are still recognized today. Note the prominent flagella on cells of *C. okenii*. Compare Figs. 3 and 4 with a photomicrograph of living cells of *C. okenii* shown in Figure 12.4a of this book. From Sergei Winogradsky, *Microbiologie du Sol*, portion of Plate IV. Paris, France: Masson et Cie Editeurs, 1949. Reproduced by permission of Dunod Editeur, Paris, France.

ology, which led to an understanding of microbial processes in the soil that are beneficial or harmful to plant growth. Later in the twentieth century, such studies on soil microbiology led to the discovery of important uses of microorganisms, such as in the synthesis of antibiotics and commodity chemicals. This led, especially after World War II, to the field of *industrial microbiology*.

Advances in soil microbiology also provided an important foundation for studies on microbial processes in water bodies such as lakes, rivers, and oceans, studies classified under the field of *aquatic microbiology*. One branch of aquatic microbiology deals with processes for treating sewage and providing safe water for humans. To provide safe drinking water, procedures for eliminating harmful bacteria from water supplies were developed and are now in widespread practice worldwide. As interest in the biodiversity and activities of microorganisms in their

Table 1.1 Three hundred years of microbiology: Some key papers in microbiology, 1684–2000[a]

Year	Investigator(s)	Discovery
1684	Antoni van Leeuwenhoek	Discovery of bacteria
1798	Edward Jenner	Smallpox vaccination
1857	Louis Pasteur	Microbiology of the lactic acid fermentation
1860	Louis Pasteur	Role of yeast in alcoholic fermentation
1864	Louis Pasteur	Settled spontaneous generation controversy
1867	Robert Lister	Antiseptic principles in surgery
1876	Ferdinand Cohn	Discovery of endospores
1881	Robert Koch	Methods for study of bacteria in pure culture
1882	Robert Koch	Discovery of cause of tuberculosis
1882	Élie Metchnikoff	Phagocytosis
1884	Robert Koch	Koch's postulates
1884	Christian Gram	Gram-staining method
1885	Louis Pasteur	Rabies vaccine
1889	Sergei Winogradsky	Concept of chemolithotrophy
1889	Martinus Beijerinck	Concept of a virus
1890	Emil von Behring and Shibasaburo Kitasato	Diphtheria antitoxin
1890	Sergei Winogradsky	Autotrophic growth of chemolithotrophs
1901	Martinus Beijerinck	Enrichment culture method
1901	Karl Landsteiner	Human blood groups
1908	Paul Ehrlich	Chemotherapeutic agents
1911	Francis Rous	First cancer virus
1928	Frederick Griffith	Discovery of pneumococcus transformation
1929	Alexander Fleming	Discovery of penicillin
1931	Cornelius van Niel	H_2S (sulfide) as electron donor in anoxygenic photosynthesis
1935	Gerhard Domagk	Sulfa drugs
1935	Wendall Stanley	Crystallization of tobacco mosaic virus
1941	George Beadle and Edward Tatum	One gene–one enzyme hypothesis
1943	Max Delbruck and Salvador Luria	Inheritance of genetic characters in bacteria
1944	Oswald Avery, Colin Macleod, Maclyn McCarty	Explanation of Griffith's work—DNA is genetic material
1944	Selman Waksman and Albert Schatz	Discovery of streptomycin
1946	Edward Tatum and Joshua Lederberg	Bacterial conjugation
1951	Barbara McClintock	Discovery of transposable elements
1952	Joshua Lederberg and Norton Zinder	Bacterial transduction
1953	James Watson, Francis Crick, Rosalind Franklin	Structure of DNA
1959	Arthur Pardee, François Jacob, Jacques Monod	Gene regulation by a repressor protein
1959	Rodney Porter	Immunoglobulin structure
1959	F. Macfarlane Burnet	Clonal selection theory
1960	François Jacob, David Perrin, Carmon Sanchez, Jacques Monod	Concept of an operon
1960	Rosalyn Yalow and Solomon Bernson	Development of radioimmunoassay (RIA)
1961	Sydney Brenner, François Jacob, and Matthew Meselson	Messenger RNA and ribosomes as the site of protein synthesis
1966	Marshall Nirenberg and H. Gobind Khorana	Discovery of the genetic code
1967	Thomas Brock	Discovery of bacteria growing in boiling hot springs
1969	Howard Temin, David Baltimore, Renato Dulbecco	Discovery of retroviruses/reverse transcriptase
1969	Thomas Brock and Hudson Freeze	Isolation of *Thermus aquaticus*, source of *Taq* DNA polymerase
1970	Hamilton Smith	Specificity of action of restriction enzymes
1973	Stanley Cohen, Annie Chang, Robert Helling, and Herbert Boyer	Recombinant DNA
1975	Georges Kohler, Cesar Milstein	Monoclonal antibodies
1976	Susumu Tonegawa	Rearrangement of immunoglobulin genes
1977	Carl Woese and George Fox	Discovery of the *Archaea*
1977	Fred Sanger, Steven Niklen, Alan Coulson	Methods for sequencing DNA
1981	Stanley Prusiner	Characterization of prions
1982	Karl Stetter	Isolation of first prokaryote with temperature optimum > 100°C
1983	Luc Montagnier	Discovery of HIV, the cause of AIDS
1988	Kary Mullis	Discovery of the polymerase chain reaction (PCR)
1995	Craig Venter and Hamilton Smith	Complete sequence of a bacterial genome
1999	The Institute for Genomic Research (TIGR), and others	Over 100 microbial genomes sequenced or in progress
2000	Edward Delong	Discovery of marine *Archaea*, proteorhodopsin, and other aspects of prokaryotic marine life

[a]Major reference sources here include Brock, T. D. (1961), *Milestones in Microbiology*, Prentice Hall, Englewood Cliffs, NJ; Brock, T. D. (1990). *The Emergence of Bacterial Genetics*, Cold Spring Harbor Press, Cold Spring Harbor, NY. *Year* refers to the year in which the discovery was published.

natural environments grew, the field of *microbial ecology* emerged as a major discipline in microbiology, and is currently enjoying a second "golden era" at present (∞Chapters 18 and 19).

In addition to advances in *applied* areas of microbiology that have provided such important advances for human society, the twentieth century witnessed extensive developments in our understanding of the *basic* principles of microbial function. For example, many new kinds of microorganisms have been discovered and classified, resulting in considerable refinement of *microbial systematics*. Study of the nutrients that microorganisms require and the products that they make has advanced the field of *microbial physiology*. Enhanced understanding of the physical and chemical structure of microorganisms (*cytology*) and the discovery of the complement of microbial enzymes and the chemical reactions they carry out (*microbial biochemistry*), have also influenced how microbiology is approached today.

A key area of basic research that moved forward rapidly in the mid-twentieth century was the study of heredity and variation in bacteria, the discipline of *bacterial genetics*. Although some aspects of bacterial variation were known early in the twentieth century, it was not until the discovery of genetic exchange in bacteria around 1950 that bacterial genetics really became a major field of study. Bacterial genetics, biochemistry, and physiology developed mainly during the 1950s, leading by the early 1960s to an advanced understanding of DNA, RNA, and protein synthesis. The field of *molecular biology* arose to a great extent from these bacterial studies.

The study of viruses also blossomed in the twentieth century. Although Beijerinck discovered the first virus over 100 years ago, it was not until the middle of the twentieth century that the true nature of viruses was understood. Much of this work involved the study of viruses that infect bacteria, called *bacteriophages*. An important development was the realization that virus infection was analogous to genetic transfer, and the relationships between viruses and other genetic elements was worked out primarily from research on bacteriophages.

By the 1970s, our knowledge of the basic processes of bacterial physiology, biochemistry, and genetics had advanced to such an extent that it was possible to manipulate the genetic material of cells experimentally, using bacteria as tools. It also became possible to introduce genetic material (DNA) from foreign sources into bacteria and control its replication and characteristics. This led to development of the field of *biotechnology*. Although biotechnology originally arose from basic studies, its use in promoting human welfare required application of the principles of physiology and industrial microbiology, a good example of how basic and applied research advance together. Also at about this same time, nucleic acid sequencing was developed and used as a tool to discern phylogenetic (evolutionary) relationships among prokaryotes, which led to revolutionary new concepts in the field of biological classification and to the first true understanding of the evolutionary history of microorganisms. Now, in the new millenium, entire genomes can be rapidly sequenced and the age of genomic analysis is clearly upon us. The huge amounts of genomic information now in hand is fueling startling advances in medicine, microbial ecology, industrial microbiology, and many related areas. It is easy to see that the science of microbiology has come a long way in 300 years. But the best is yet to come!

✓ 1.6 Concept Check

Unlike Koch and Pasteur, Beijerinck and Winogradsky studied bacteria in soil and water and developed the enrichment culture technique for the specific isolation of representatives of various physiological groups. In the middle to latter part of the twentieth century, basic and applied microbiology worked hand in hand to usher in the current era of molecular microbiology.

✓ What is the enrichment culture technique and why was it a useful new method in microbiology?

✓ List the subdisciplines of microbiology whose focus is the following: metabolism; enzymology; nucleic acid and protein synthesis; microorganisms and their natural environments.

Review Questions

1. List six key properties associated with the living state. Which of these are properties of *all* cells? Which are properties of only *some types* of cells?

2. Cells can be thought of as both chemical machines and coding devices. Explain how these two attributes of a cell differ.

3. What is needed for translation to occur in a cell? What is the product of the translational process?

4. What is an ecosystem? Do microorganisms live in pure cultures in an ecosystem? What effects can microorganisms have on their ecosystems?

5. How would you convince a friend that microorganisms are much more than just agents of disease?

6. What is a pure culture and how can one be obtained? Why was knowledge of how to obtain a pure culture important for development of the science of microbiology?

7. Explain the principle behind the use of the Pasteur flask in studies on spontaneous generation.

8. Explain why the invention of solid culture media was of great importance to the development of microbiology as a science.

9. How did Ferdinand Cohn contribute to bacteriology?

10. What are Koch's postulates and how did they influence the development of microbiology?

11. Describe a major contribution to microbiology of the early microbiologist Martinus Beijerinck.

12. Using Table 1.1 as a guide, contrast the focus of microbiological research *before* and *after* World War II.

Application Questions

1. Observe the organisms shown in Figure 1.1. Describe how the cells in the organisms shown in parts (a) and (b) differ from the organisms in parts (c) and (d). List as many differences as you can.

2. Pasteur's experiments on spontaneous generation were of enormous importance for the advance of microbiology, having an impact on the methodology of microbiology, ideas on the origin of life, and the preservation of food, to name just a few. Explain briefly how the impact of his experiments was felt on each of the topics listed.

3. Describe the various lines of proof Robert Koch used to definitively associate the bacterium *Mycobacterium tuberculosis* with the disease tuberculosis. How would his proof have been flawed if any of the tools he developed for studying bacterial diseases had not been available for his study of tuberculosis?

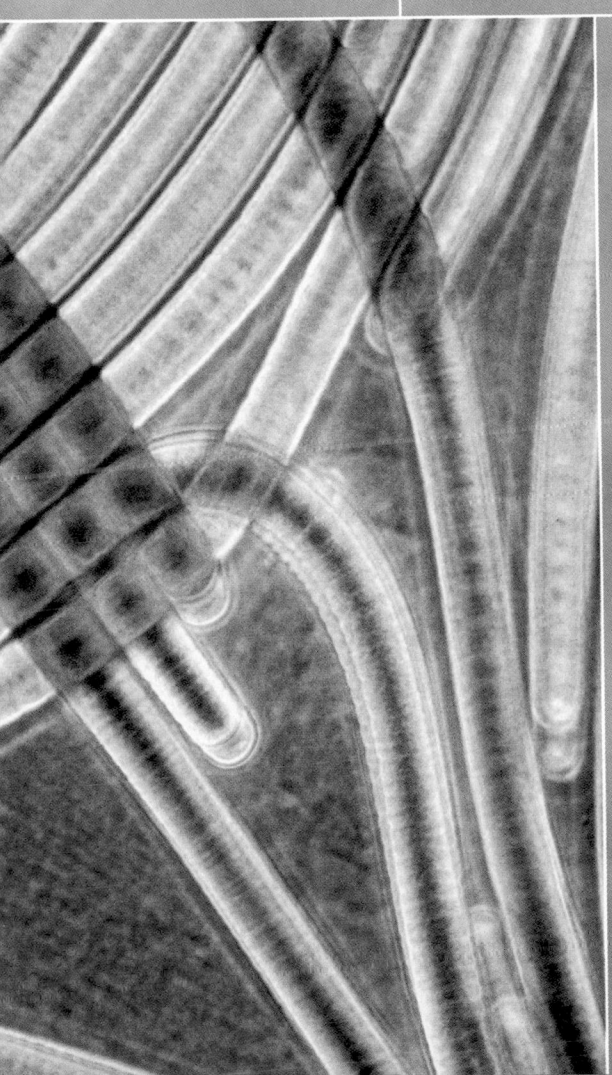

yanobacteria, such as the large filamentous genus *Oscillatoria*, shown here, are phototrophic bacteria whose oxygen-evolving metabolism paved the way for higher life forms to appear on Earth. Interestingly, cyanobacteria form a major evolutionary lineage that eventually spawned the chloroplast of phototrophic eukaryotes, such as algae and oak trees. Molecular methods have opened the door to microbial diversity, allowing microbiologists to construct a universal tree of life and to reveal the many hidden relationships that exist between these highly diverse organisms.

AN OVERVIEW OF MICROBIAL LIFE

2

Working Glossary

Archaea phylogenetically related prokaryotes distinct from *Bacteria*

Bacteria phylogenetically related prokaryotes distinct from *Archaea*

Chemolithotroph an organism obtaining its energy from the oxidation of inorganic compounds

Chemoorganotroph an organism obtaining its energy from the oxidation of organic compounds

Chromosome a genetic element carrying genes essential to cell function

Cytoplasm cellular contents inside the cytoplasmic membrane, excluding the nucleus (if present)

Domain the highest level of biological classification

Endosymbiosis the process by which mitochondria and chloroplasts originated from descendants of *Bacteria*

Eukaryote a cell having a membrane-bound nucleus and usually other organelles

Extremophile an organism that grows optimally under one or more environmental extremes

Genome the complement of genes in an organism

Morphology cell shape

Nucleoid the aggregated mass of DNA that constitutes the chromosome of prokaryotic cells (*Bacteria* and *Archaea*)

Nucleus a membrane-enclosed structure that contains the chromosomes in eukaryotic cells

Organelle a unit membrane-enclosed structure present in the cytoplasm of eukaryotic cells

Phototroph an organism that obtains its energy from light

Phylogeny the evolutionary relationships between organisms

Plasmid an extrachromosomal genetic element nonessential for growth

Prokaryote a cell that lacks a membrane-enclosed nucleus and other organelles

Ribosome a cytoplasmic particle that carries out the process of protein synthesis

CELL STRUCTURE AND EVOLUTIONARY HISTORY

This chapter introduces concepts of cell structure and function and microbial diversity that will have bearing throughout the remainder of this book. Here we will compare the general architecture of microbial cells, differentiate cells from viruses, explore the essential features of the evolutionary tree of life, and consider some of the major groups of microorganisms that affect our lives and our planet.

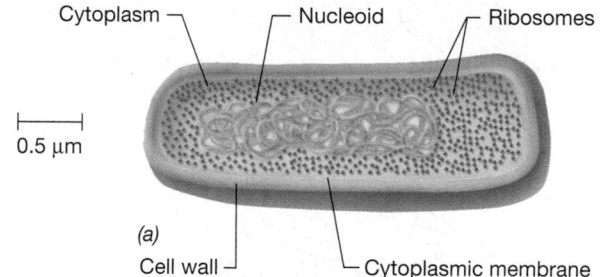

(a)

2.1 Elements of Cell and Viral Structure

What is the structure of a cell? All cells have a barrier that separates the inside from the outside called the *cytoplasmic (cell) membrane* (Figure 2.1). It is through the cell membrane that nutrients and other substances needed by the cell enter, and it is through this same membrane that waste materials and other cell products exit. Within a cell, and bounded by the cytoplasmic membrane, is a complex mixture of substances and structures called the **cytoplasm**. These materials and structures, either dissolved or suspended in water, carry out the functions of the cell.

The major components of the cytoplasm, other than water, include *macromolecules* (two especially important classes of which are the *proteins* and *nucleic acids*), ribosomes, small organic molecules (mainly precursors of

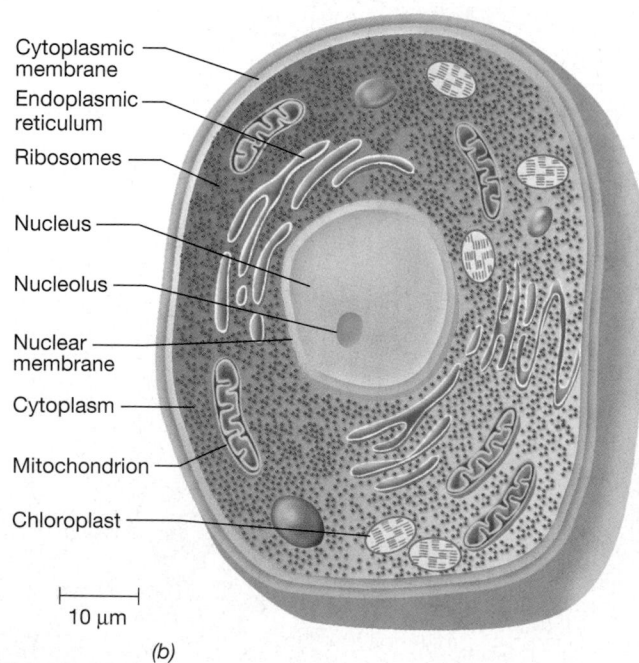

(b)

Figure 2.1 Internal structure of microbial cells. (a) Diagram of a prokaryote. (b) Diagram of a eukaryote.

macromolecules), and various inorganic ions. **Ribosomes** the cell's protein-synthesizing factories, are particle-like structures composed of ribonucleic acid (RNA) and various proteins that interact with several soluble proteins and messenger RNA in the central process of *protein synthesis*.

A **cell wall** gives structural strength to a cell. The cell wall is relatively permeable and located outside the membrane (Figure 2.1*a*); it is a much stronger layer than the membrane itself. Plant cells and most microorganisms have cell walls, while animal cells for the most part do not. (Animal cells are instead reinforced by scaffolding within the cytoplasm called the *cytoskeleton*.)

Prokaryotic and Eukaryotic Cells

Careful examination of internal structure will differentiate two structural types of cells: the *prokaryote* and the *eukaryote* (Figure 2.1). Eukaryotic cells are generally larger and structurally more complex than prokaryotic cells, and a major feature of eukaryotic cells, absent from prokaryotic cells, is the presence of membrane-enclosed structures called *organelles*. These include the *nucleus, mitochondria,* and *chloroplasts* (in photosynthetic cells only) (Figure 2.1*b*). Mitochondria and chloroplasts play specific roles in energy generation by carrying out respiration and photosynthesis, respectively. Eukaryotic microorganisms include **algae, fungi,** and **protozoa** (see Figures 2.23 and 2.24). All metazoans (plants and animals) are constructed of eukaryotic cells. We consider eukaryotic cells in more detail in Chapter 14.

In contrast to eukaryotic cells, prokaryotic cells have a simpler internal structure, lacking membrane-enclosed organelles (Figures 2.1*a* and 2.2). **Prokaryotes** consist of the **Bacteria** and the **Archaea** (Figure 2.2*a, b*). Although species of *Bacteria* and *Archaea* share a prokaryotic cell structure, they differ dramatically in their evolutionary history. In this book, the term *bacteria*, written with a lower case "b," is synonymous with the term *prokaryote*. The term *Bacteria*, written with a capital "B," refers to species of this large evolutionarily related group of prokaryotes distinct in this regard from the *Archaea*.

In general, microbial cells are very small. A rod-shaped prokaryote is typically about 1–5 micrometers (μm) long and about 1 μm wide (a micrometer is 10^{-6} of a meter) and thus is invisible to the naked eye. To conceive of how small a bacterium is, consider that 500 bacteria each 1 μm long could be placed end-to-end across the period at the end of this sentence. Eukaryotic cells are typically much larger than prokaryotic cells, but the range can vary dramatically (Figure 2.3*c*). We revisit the subject of cell size in more detail in Chapter 4.

Viruses

Viruses, a major class of microorganisms, are not cells (Figure 2.3). Viruses lack many of the attributes of cells, the most important of which is that they are not dynamic open systems, taking in nutrients and expelling wastes. Instead, a virus particle is a static structure, quite stable and unable to change or replace its parts. Only when it infects a cell does a virus acquire the key attribute of a

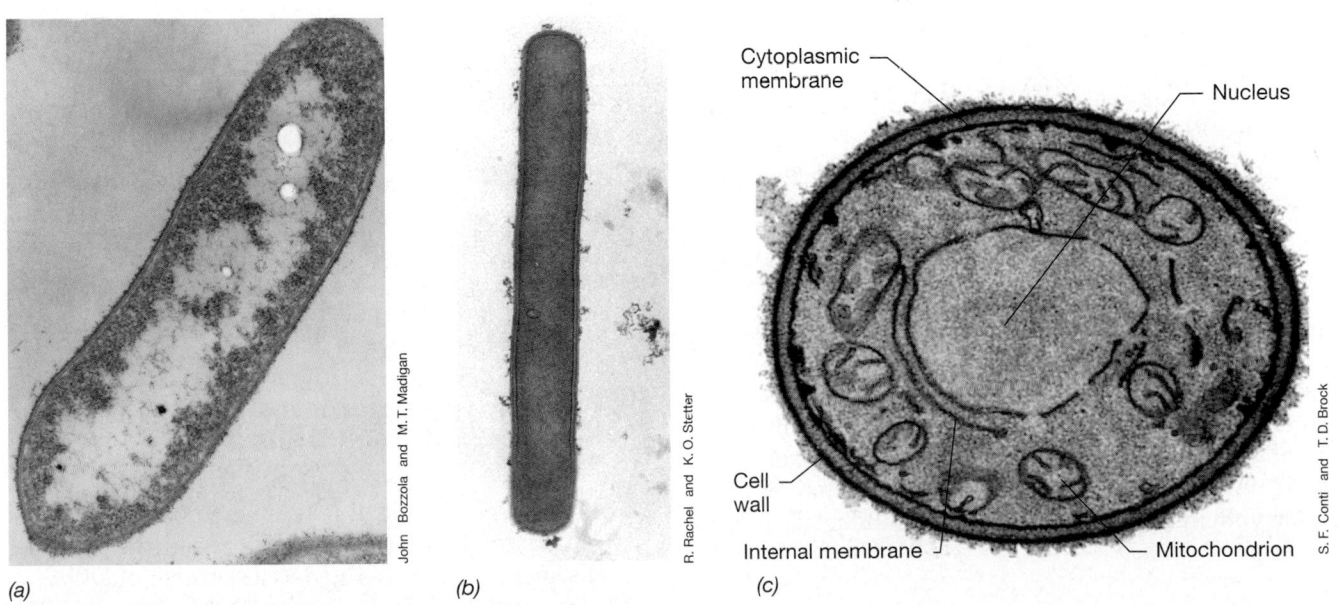

(a)   John Bozzola and M.T. Madigan   *(b)*   R. Rachel and K.O. Stetter   *(c)*

Cytoplasmic membrane   Nucleus   Cell wall   Internal membrane   Mitochondrion   S.F. Conti and T.D. Brock

Figure 2.2   Electron micrographs of sectioned cells from each of the domains of living organisms. (a) *Heliobacterium modesticaldum* (*Bacteria*); the cell measures 1 × 3 μm. (b) *Methanopyrus kandleri* (*Archaea*); the cell measures 0.5 × 4 μm. [Reinhold Rachel and Karl O. Stetter, 1981. *Archives of Microbiology* 128:288–293. ©1981 by Springer-Verlag GmbH & Co. KG.] (c) *Saccharomyces cerevisiae* (*Eukarya*); the cell measures 8 μm in diameter.

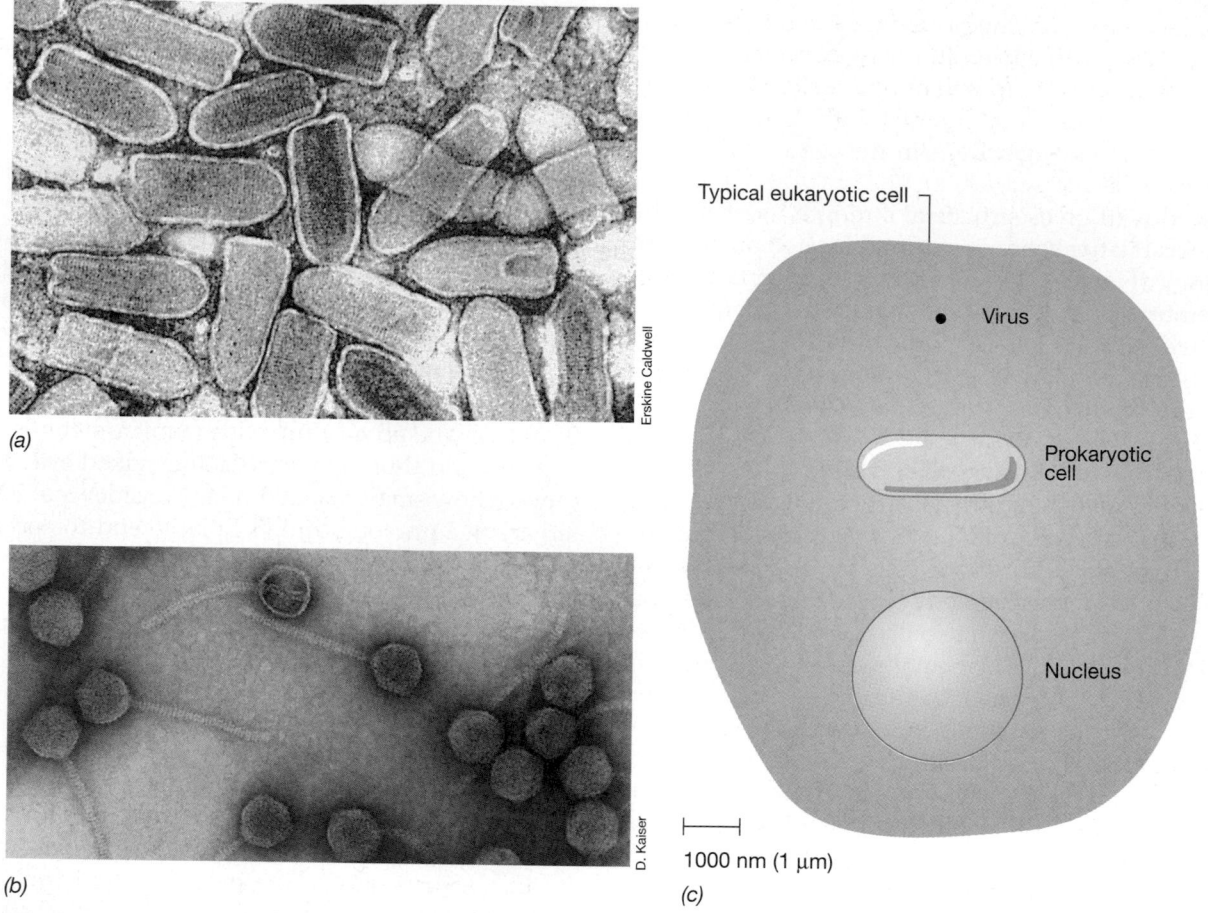

Erskine Caldwell

D. Kaiser

Figure 2.3 Virus structure, and size comparisons of viruses and cells. (a) Rhabdovirus (a virus of eukaryotes) particles. A single virus particle is about 65 nm (0.065 μm) in diameter. (b) Bacterial virus (bacteriophage) lambda. The head of each particle is about 65 nm in diameter. (c) The size of the viruses shown in (a) and (b) in comparison to a bacterial and eukaryotic cell.

living system, reproduction (Figure 1.3). Unlike cells, viruses have no metabolic abilities of their own. And, although they contain their own genes, viruses lack ribosomes and therefore depend on the cell's biosynthetic machinery for protein synthesis.

Viruses are known to infect all cells, including microbial cells. Many viruses cause disease in the organisms they infect, but virus infection does not always lead to disease. We will see in Chapters 9 and 16 that besides causing disease, viruses can have other profound effects on cells, including causing genetic alterations that can actually improve the capabilities of the cell. Viruses are much smaller than cells, even much smaller than prokaryotic cells. Figure 2.3 illustrates the relative sizes of cells and viruses.

✓ 2.1 Concept Check

All microbial cells have certain basic structures in common such as a cytoplasmic membrane, ribosomes, and (usually) a cell wall. Two structural categories of cells are recognized: the

prokaryote and the eukaryote. Viruses are not cells but depend on cells for their replicative functions.

- ✓ By looking inside a cell how could you tell if it were *prokaryotic* or *eukaryotic*?
- ✓ What important function do *ribosomes* play in cells?
- ✓ How long is a typical rod-shaped bacterial cell? How much larger are you than this single cell?

2.2 Arrangement of DNA in Microbial Cells

The living processes of all cells are governed by their genetic makeup, their complement of genes (the **genome**). In cells, a gene can be defined as a segment of DNA that encodes a protein (via messenger RNA) or another RNA molecule, such as ribosomal RNA. In Chapter 15 we will consider the rapid advances that have been made in sequencing and analyzing the genomes of living organisms, from bacteria to humans, which have yielded detailed genetic blueprints of hundreds of different organisms. Here we just consider how genomes are organized in prokaryotic and eukaryotic cells.

Nucleus vs. Nucleoid

The genomes of prokaryotic and eukaryotic cells are organized differently. In prokaryotic cells, DNA is present in a large double-stranded molecule called the *bacterial chromosome* that aggregates to form a visible mass called the **nucleoid** (Figure 2.4). We will see in Chapter 7 that DNA is *circular* in most prokaryotes and that most prokaryotes have only a *single* chromosome. Because of this, most prokaryotes thus contain only a *single copy* of each gene and are therefore genetically *haploid*. Most prokaryotes also contain small amounts of extrachromosomal DNA, again usually arranged in a circular fashion, called **plasmids**. Plasmids typically contain genes that confer special properties (such as unique metabolic properties) on a cell, as opposed to carrying essential ("housekeeping") genes, which are needed for basic survival and are located on the chromosome.

In eukaryotes, DNA is present in linear molecules within the nucleus, packaged in a very organized state called **chromosomes**. Chromosome number varies with the organism. For example, the baker's yeast *Saccharomyces cerevisiae* contains 16 chromosomes arranged in 8 pairs, while human cells contain 46 (23 pairs). Chromosomes in eukaryotes contain more than just DNA; they include proteins that assist in folding and packing the DNA and other proteins that are required for gene expression. A key genetic difference between prokaryotes and eukaryotes is that eukaryotes typically contain *two copies* of each gene and are thus genetically *diploid*. During cell division in eukaryotic cells the nucleus divides (following a doubling of chromosome number) in the well-known process of **mitosis** (Figure 2.5). Two identical daughter cells result from a mitotic division, and each daughter cell receives a nucleus with a full complement of genes.

The diploid complement of genetic material of eukaryotic cells is halved in the process of **meiosis** to form

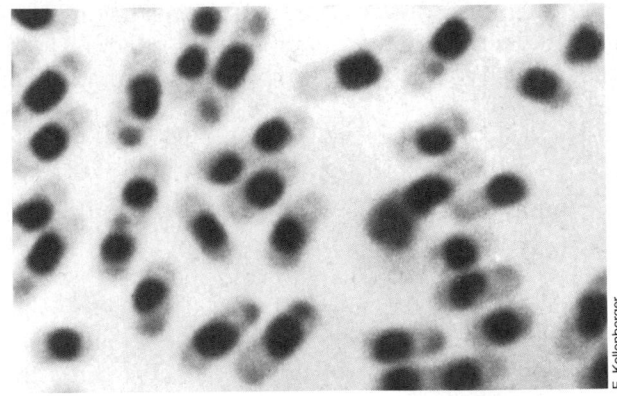

(a)

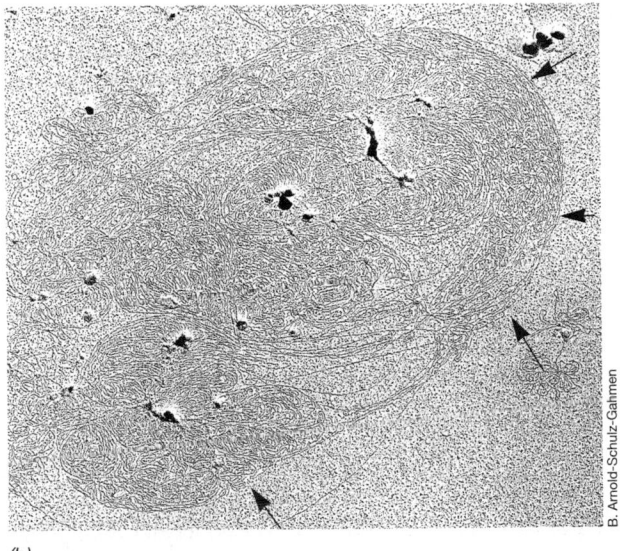

(b)

Figure 2.4 The nucleoid. (a) Light photomicrograph of cells of *Escherichia coli* treated in such a way as to make the nucleoid visible. (b) Electron micrograph of an isolated nucleoid released from a cell of *E. coli*. The cell was gently lysed to allow the highly compacted nucleoid to emerge intact. Arrows point to the edge of DNA strands. Most bacterial nucleoids consist of a single circular molecule (the bacterial chromosome), although linear genomes are present in certain species (∞ Section 7.4).

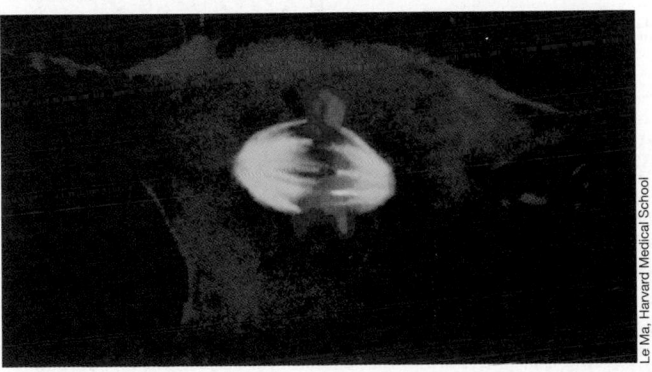

Figure 2.5 Mitosis in stained kangaroo rat cells. The cell was photographed while in the *metaphase* stage of mitotic division. The green color stains a protein called *tubulin*, important in pulling chromosomes apart (∞ Section 14.5). The blue color is from a DNA-binding dye and shows the chromosomes. Although an integral part of the cell cycle of eukaryotic cells, mitosis does not occur in prokaryotic cells.

haploid gametes for sexual reproduction. Fusion of two gametes during zygote formation restores the cell to the diploid state. We discuss these processes in more detail in Chapter 14.

Genes

How many genes and proteins does a cell have? *Escherichia coli*, a typical bacterium, contains a single chromosome of about 4.6 million base pairs of DNA. Because the *E. coli* chromosome has been completely sequenced, we also know that it contains about 4,300 genes. Some bacterial cells have over three times this number of genes while some have as few as one-eighth this number. Eukaryotic cells generally have many more genes than prokaryotes. A human cell, for example, contains over 1,000 times the DNA of *E. coli* and about 7 times as many genes (we will see later that much of the DNA in eukaryotic cells is noncoding DNA). A single cell of *E. coli* contains about 1,900 *different kinds* of proteins and about 2.4 million *total* protein molecules. Some proteins, however, are very abundant, others only moderately abundant, and some are present in only one or a few copies. Thus, *E. coli* has mechanisms for *controlling the expression* of its genes so that not all genes are expressed to the same extent or at the same time. This is a common observation in all cells, prokaryotic and eukaryotic; we focus on mechanisms of gene expression in Chapter 8.

✓ 2.2 Concept Check

Genes govern the properties of cells, and a cell's complement of genes is referred to as its genome. DNA is arranged in cells to form chromosomes. In prokaryotes there is usually a single circular chromosome, while in eukaryotes, several linear chromosomes exist.

✓ Differentiate between the *nucleus* and the *nucleoid*.

✓ How do *plasmids* differ from *chromosomes*?

✓ Why does it make sense that a human cell would have more genes than a bacterial cell?

2.3 The Tree of Life

Is cell structure a function of evolutionary relationship? The answer to this question is both "yes" and "no." The evolutionary ties between life forms are the subject of the science of **phylogeny**. On the one hand, we can say that all known prokaryotic cells are phylogenetically distinct from eukaryotic cells. But on the other hand, it is clear that not all prokaryotic cells are closely related in an evolutionary sense. This conclusion has emerged from molecular evolution studies of prokaryotes—specifically, from phylogenetic relationships deduced from comparative sequencing of key macromolecules. For reasons to be detailed in Chapter 11, macromolecules in the ribosome, in particular *ribosomal RNAs*, are excellent barometers of evolutionary relationship. And because all organisms contain ribosomes and thus ribosomal RNA, this molecule can and has been used to construct a phylogenetic tree of all life forms, prokaryotic and eukaryotic. Recognition of ribosomal RNA as a tool for constructing phylogenetic relationships was first made by Carl Woese, an American microbiologist. The technology employed in these measurements has now become routine and is outlined in Figure 2.6.

From comparative ribosomal RNA sequences three phylogenetically distinct lineages of cells have been identified; two of these lineages contain only prokaryotes, while the third is composed of eukaryotes. The lineages, referred to as evolutionary **domains**, are called

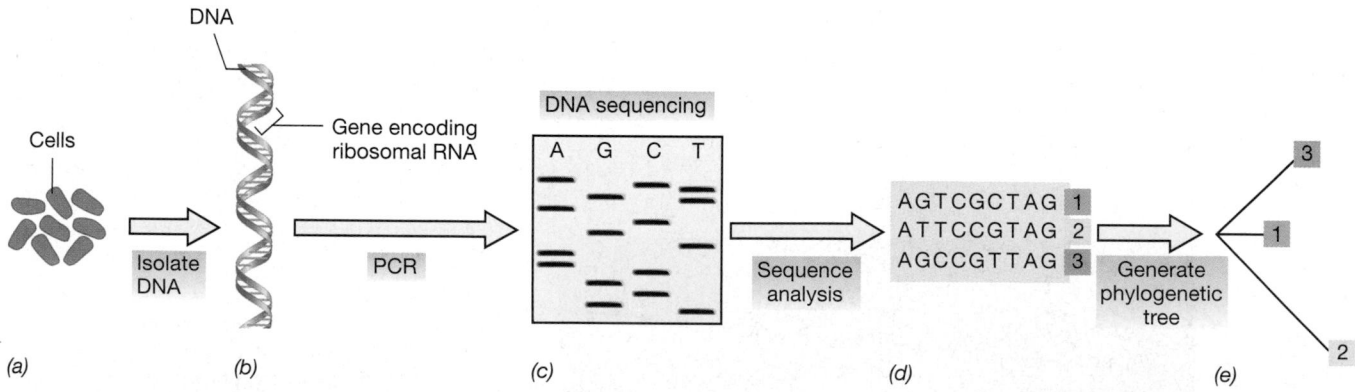

Figure 2.6 Ribosomal RNA gene sequencing and phylogeny. (a) Cells, either from a pure culture or from an environmental sample, are broken open; (b) the gene-encoding ribosomal RNA is isolated, and many identical copies made by the technique called the polymerase chain reaction (PCR); ∞ Section 10.17. (c) The gene is sequenced (∞ Section 10.13), and (d) the sequences obtained are aligned in a computer. An algorithm makes pair-wise comparisons and generates a tree (e) that depicts the differences in ribosomal RNA sequence between the organisms analyzed. If an environmental sample is used, the isolated ribosomal RNA genes from the different microorganisms in the sample must first be sorted out (cloned) before copies are made and sequencing done. For further discussion of these methods, see Sections 11.5 and 18.5.

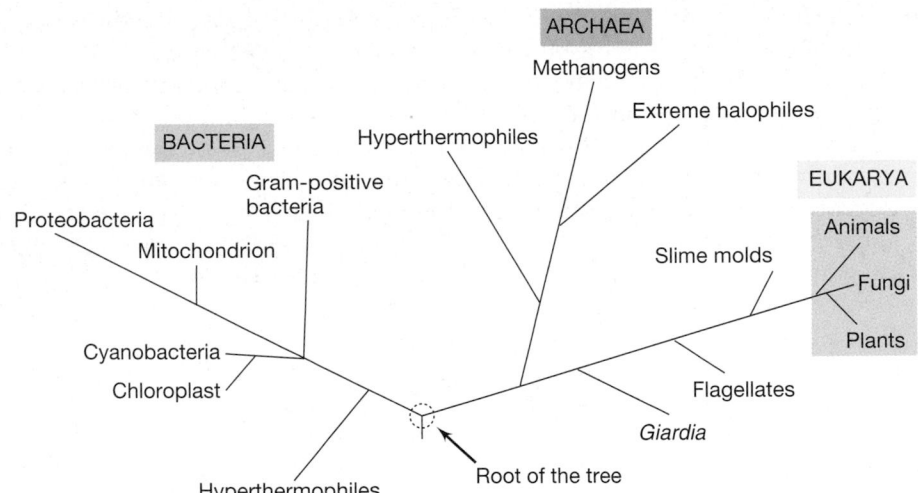

Figure 2.7 The phylogenetic tree of life as defined by comparative ribosomal RNA sequencing. The tree consists of three domains of organisms: the *Bacteria* and the *Archaea*, cells of which are prokaryotic, and the *Eukarya* (eukaryotes). Only a few of the groups of organisms within each domain are shown. See more detailed trees of each domain in Figures 2.9, 2.18, and 2.22, and the phylogenetic trees in Chapters 11–14. Hyperthermophiles are prokaryotes that grow best at temperatures of 80°C or higher. The group shaded in red are *macro*organisms. All other organisms on the tree of life are *micro*organisms.

the *Bacteria*, the *Archaea*, and the *Eukarya* (Figure 2.7). These domains are thought to have diverged from a common ancestral organism, the "universal ancestor," early in the history of life on Earth.

Besides clearly showing that all prokaryotes are not phylogenetically closely related, the tree of life reveals another important evolutionary fact: Species of *Archaea* are more closely related to eukaryotes than are species of *Bacteria* (Figure 2.7). This surprising conclusion has gained considerable support from the comparative study of several other macromolecules from species of each of the three domains. Thus, evolutionary diversification from the universal ancestor originally went in two directions, *Bacteria* versus "other," with the latter eventually diverging to yield the separate domains of the *Archaea* and the *Eukarya*.

Because the cells of higher animals and plants are all eukaryotic, it follows that eukaryotic microorganisms were the ancestors of multicellular organisms. The tree of life clearly bears this out—as you would expect, microbial eukaryotes branch off early on the tree while plants and animals reside near the apex (Figure 2.7). Moreover, we have learned that eukaryotic cells contain genomes from cells of two domains of organisms. Besides containing a genome of their own packed away in the chromosomes of the cell nucleus, some organelles (specifically the mitochondria and chloroplasts) of eukaryotes contain their own DNA (usually arranged in circular fashion, as in *Bacteria*) and ribosomes. Using the technology described in Figure 2.6, these organelles can be shown to be the highly derived ancestors of specific lineages of *Bacteria* (Figure 2.7). Presumably, organelles were once free-living cells that, perhaps for protection or cooperative metabolic reasons, established a stable residence in cells of *Eukarya* eons ago. The process by which this arrangement developed has come to be known as *endosymbiosis* (∞ Sections 11.3 and 14.4).

Ribosomal RNA-based phylogenies have revealed the evolutionary connections between all cells. But most importantly, the technology has created an evolutionary framework for the prokaryotes, a crucial breakthrough that the science of microbiology had done without since its very beginnings. The universal tree (Figure 2.7) can be made increasingly more detailed by adding additional ribosomal RNA sequences to the comparisons (see Figure 2.9). Also, as we will see later, the technologies developed to solve the problem of prokaryotic phylogeny have had significant application in microbial ecology and clinical microbiology. We will consider these developments in later chapters (∞ Chapters 18 and 24).

✓ 2.3 Concept Check

Ribosomal RNA sequencing has revolutionized microbiology and has yielded an evolutionary framework for the prokaryotes. The three domains of life are the *Bacteria*, the *Archaea*, and the *Eukarya*. Ribosomal RNA sequencing has shown that the major organelles of *Eukarya* have evolutionary roots in the *Bacteria* and has yielded new tools for microbial ecology and diagnostics.

✓ What evidence supports the idea that *Bacteria* and *Archaea* are different? In what ways are they similar?

✓ What molecular evidence supports the theory of endosymbiosis?

II MICROBIAL DIVERSITY

Microbial diversity is a function of microbial evolution. Because evolution has molded all life on Earth, the structural/functional diversity we see in microbial cells today is the result of billions of years of evolutionary experimentation. Microbial diversity can be seen in many ways; for example, in variations in cell size and shape (morphology), metabolic strategies, motility, mechanisms of cell division, developmental biology, adaptation to environmental extremes, and many other aspects of cell biology. In the following sections we paint a picture of microbial diversity with a broad brush. We return to the theme of microbial diversity in greater detail in Chapters 12–14, but for now, give just a taste of what is to come. We preface our discussion of microbial diversity with a brief consideration of metabolic diversity, since microbial diversity has been significantly driven by microbial exploitation of every conceivable means of "making a living" consistent with the laws of chemistry and physics. Metabolic diversity will be covered in more detail in Chapters 5, 6, and 17.

2.4 Physiological Diversity of Microorganisms

Energy and Carbon

All cells require energy. This energy can be obtained in three ways, and Figure 2.8 summarizes the options: *organic* chemicals, *inorganic* chemicals, or *light*.

Many of the thousands of different organic chemicals present on Earth can be used by one microorganism or another to obtain energy. All natural and even most syn-

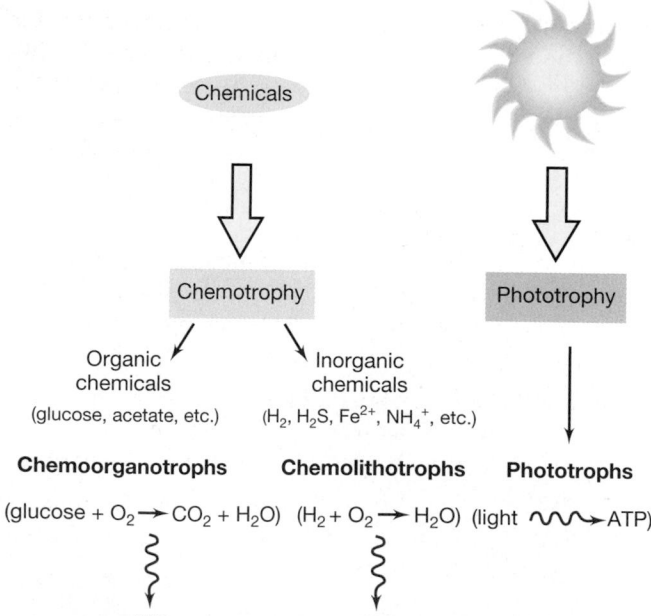

Figure 2.8 Metabolic options for obtaining energy. The organic and inorganic chemicals listed here are just a few of the many different chemicals used by various chemotrophic organisms. Oxidation of the organic or inorganic chemicals yields ATP in chemotrophic organisms while conversion of solar energy to chemical energy (again, in the form of ATP) occurs in phototrophic organisms.

thetic organic compounds can be broken down by one or more microorganisms. Energy is obtained by *oxidizing* (removing electrons from) the compound and is conserved in the cell as the high-energy compound, **adenosine triphosphate (ATP)** (Figure 2.8). Some microorganisms can extract energy from the compound only in the presence of oxygen; these organisms are called **aerobes**. Others can extract energy only in the absence of oxygen (**anaerobes**). Still others can break down organic compounds in either the presence or absence of oxygen. Organisms that obtain energy from *organic* compounds are called **chemoorganotrophs** (Figure 2.8). Most microorganisms that have been cultured are chemoorganotrophs.

A number of prokaryotes can tap the energy available in *inorganic compounds*. This is a form of metabolism called *chemolithotrophy* and is carried out by organisms called **chemolithotrophs** (Figure 2.8). This form of energy-yielding metabolism is found only in prokaryotes and is widely distributed among species of both *Bacteria* and *Archaea*. The spectrum of different inorganic compounds used is quite broad, but as a rule, a particular prokaryote specializes in the use of one or a related group of inorganic compounds. It should be obvious why the capacity to extract energy from inorganic chemicals might be advantageous: Competition with chemoorganotrophs is not a consideration. In addition, many of the inorganic compounds oxidized, for example H_2 and H_2S, are actually the *waste products* of chemoorganotrophs. Thus, chemolithotrophs have evolved strategies for exploiting resources that many other organisms find unusable.

Phototrophic microorganisms contain pigments that allow them to use light as an energy source and thus their cells are highly colored (see Figure 2.10a). Unlike chemotrophic organisms, phototrophs do not require chemicals as a source of energy; ATP is made at the expense of sunlight. This is obviously a significant advantage, since competition for energy with chemotrophs is not a problem, and light is available in a wide variety of microbial habitats.

All cells require *carbon* as a major nutrient. Microbial cells are either **heterotrophic**, requiring one or more organic compounds as their carbon source, or **autotrophic**, where CO_2 is the source of cell carbon. Chemoorganotrophs are also clearly heterotrophs. By contrast, many chemolithotrophs and virtually all phototrophs are autotrophs. Autotrophs are also called *primary producers* because they synthesize organic matter from CO_2 for both their own benefit and that of chemoorganotrophs. The latter either feed directly on the primary producers or live off products they excrete.

Tolerance of Environmental Extremes

Another aspect of the physiological diversity of microorganisms, particularly that of the prokaryotes, is the remarkable ability of some species to thrive in habitats characterized by one or more environmental extremes. For example, not only are there prokaryotes that grow at pH 7 and 25°C—ideal conditions for humans—but prokaryotes also abound in boiling hot springs, in ice, in extremely salty bodies of water, and in soils and waters having a pH lower than 0 or as high as 12. Indeed, species of prokaryotes that inhabit such environments define the biological limits to physiochemical extremes. Moreover, these prokaryotes are not just *tolerant* of these extremes, but actually *require* the environmental extreme in order to grow. Because of this, these prokaryotes have been referred to as *extremophiles* (the combining form "phile" means "loving") to emphasize their requirement for one or more extreme conditions. Rarely in the eukaryotic world do we find such requirements for extreme growth conditions, and when we do, the extremes are generally more modest. Thus, extreme habitats are those in which the prokaryote is preeminent. Table 2.1 summarizes the current "record holders" for extremophilic prokaryotes and lists the types of habitats in which they reside.

✓ 2.4 Concept Check

Sources of energy and carbon are needed by all cells. The terms chemoorganotroph, chemolithotroph, and phototroph refer to cells that use organic chemicals, inorganic chemicals, or light, respectively, as their source of energy. Autotrophic organisms use CO_2 as their carbon source. Many prokaryotes thrive under environmental conditions that humans would call extreme.

✓ How might you distinguish a *phototrophic* microorganism from a *chemotrophic* one by simply looking at it under a microscope?

✓ What are *extremophiles*?

2.5   Prokaryotic Diversity

As we have seen, prokaryotes form two evolutionary domains, the *Archaea* and the *Bacteria* (Figure 2.7). In this section we will climb around the phylogenetic tree and briefly consider some major organisms found there. Most of the prokaryotes that become familiar to the beginning student of microbiology are found in the domain *Bacteria*, and we begin here.

Bacteria

The domain *Bacteria* contains an enormous variety of prokaryotes. All known disease-causing (pathogenic) prokaryotes are *Bacteria*, as well as thousands of nonpathogenic species, and a large variety of morphologies and physiologies occur in this domain. The **Proteobacteria** is the largest division (phylum) of *Bacteria* (Figure 2.9). Within the Proteobacteria are found many chemoorganotrophic bacteria, like *Escherichia coli*, the model organism of microbial physiology, biochemistry, and molecular biology, as well as several phototrophic and chemolithotrophic species. Many of the latter groups use hydrogen sulfide (H_2S, the smell from rotten eggs) in their metabolism, producing elemental sulfur that is stored within or outside the cell (Figure 2.10). The sulfur is an oxidation product of H_2S and can be further oxidized to sulfate (SO_4^{2-}). The sulfide and sulfur are oxidized to fuel important metabolic functions such as CO_2 fixation (autotrophy) or energy generation (Figure 2.8).

Several other common prokaryotes of soil and water, or species that live in or on plants and animals in both casual and disease-causing ways, are members of the Proteobacteria. These include species of *Pseudomonas*, many of which can degrade complex and otherwise toxic

TABLE 2.1	Classes and examples of extremophiles[a]						
Extreme	Descriptive term	Genus/species	Domain	Habitat	Minimum	Optimum	Maximum
Temperature							
High	Hyperthermophile	*Pyrolobus fumarii*	*Archaea*	Hot, undersea hydrothermal vents	90°C	**106°C**	113°C
Low	Psychrophile	*Polaromonas vacuolata*	*Bacteria*	Sea-ice	0°C	**4°C**	12°C
pH							
Low	Acidophile	*Picrophilus oshimae*	*Archaea*	Acidic hot springs	−0.06	**0.7**[b]	4
High	Alkaliphile	*Natronobacterium gregoryi*	*Archaea*	Soda lakes	8.5	**10**[c]	12
Pressure	Barophile	MT41 (Mariana Trench-41)[d]	*Bacteria*	Deep ocean sediments	500 atm	**700 atm**	> 1000 atm
Salt (NaCl)	Halophile	*Halobacterium salinarum*	*Archaea*	Salterns	15%	**25%**	32% (saturation)

[a]In each category the organism listed is the current "record holder" for requiring a particular extreme condition for growth.

[b]*P. oshimae* is also a thermophile, growing optimally at 60°C.

[c]*N. gregoryi* is also an extreme halophile, growing optimally at 20% NaCl.

[d]Strain MT41 does not yet have a formal genus and species name and is also a psychrophile, growing best below 10°C.

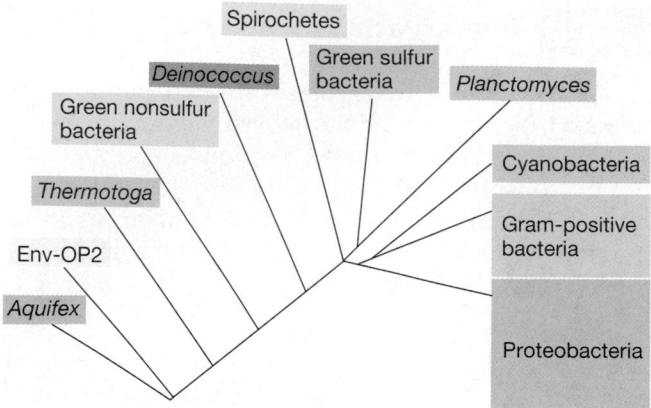

Figure 2.9 Detailed phylogenetic tree of *Bacteria*. Not all known groups of *Bacteria* are depicted on this tree. The relative sizes of the colored boxes are an indication of the number of genera and species in each of the groups. The Proteobacteria are currently the largest group of *Bacteria* known. The lineage on the tree labeled "Env" (for Environmental) does not represent a cultured organism but instead is a sequence of ribosomal RNA genes isolated from an organism in a natural sample (see text). Although not shown, there are many such other Env's known, located all over the tree.

natural and synthetic organic compounds, and *Azotobacter*, a nitrogen-fixing bacterium.

Staining properties of bacteria will be discussed in Chapter 4. Here it will suffice to know that some bacteria can be distinguished by use of the *Gram-staining* procedure. The gram-positive lineage of *Bacteria* contains a number of species united by a common phylogeny and cell wall structure. Here we find the endospore-forming *Bacillus* (discovered by Ferdinand Cohn, ∞ Section 1.5) (Figure 2.11*a*) and *Clostridium*, and related spore-forming prokaryotes such as the antibiotic-producing *Streptomyces*. Also included here are the lactic acid bacteria, common inhabitants of decaying plant material and dairy products, that include such organisms as *Streptococcus* (Figure 2.11*b*) and *Lactobacillus*. Other interesting gram-positive relatives are the mycoplasmas. These prokaryotes lack a cell wall, have very small genomes, and are often pathogenic; *Mycoplasma* is a major genus of organisms in this medically important group (∞ Section 12.21).

The **Cyanobacteria** (Figure 2.12) are phylogenetic relatives of gram-positive bacteria (Figure 2.9) and are oxygenic phototrophs, meaning that in their metabolism, molecular oxygen (O_2) is evolved, just as it is in plants. Cyanobacteria were critical in the evolution of life, as they were the first oxygenic phototrophs to appear on Earth (∞ Figure 1.1), and their production of O_2 paved the way for the evolution of prokaryotes that could respire O_2. The development of "higher organisms," such as the plants and animals, of course, followed from this.

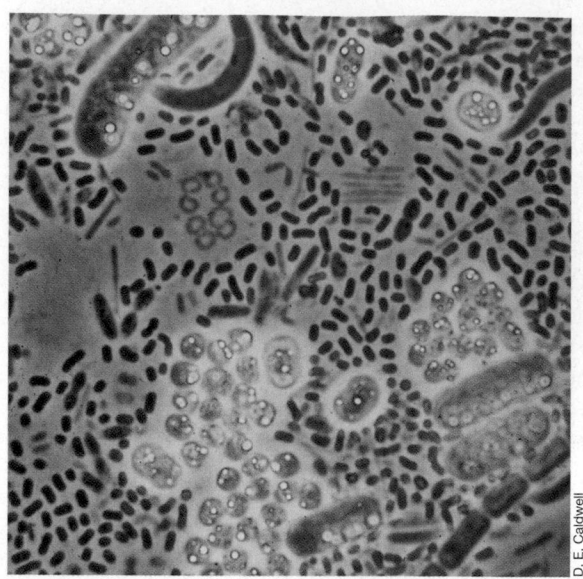

(a)

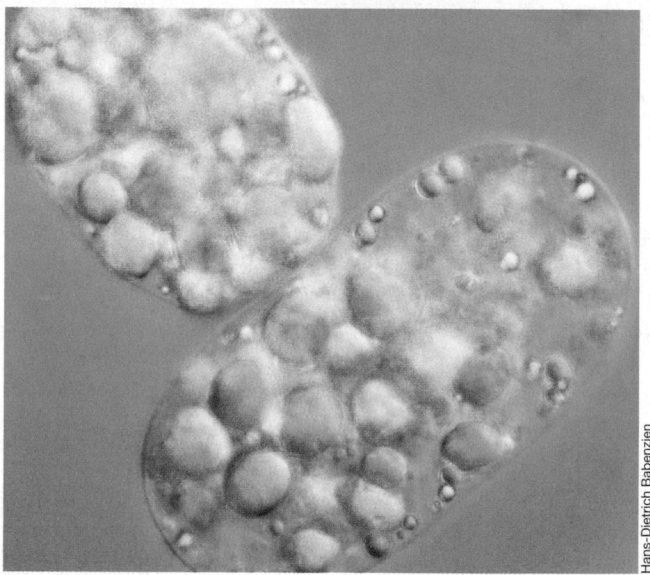

(b)

Figure 2.10 Phototrophic and chemolithotrophic Proteobacteria. (a) The phototrophic purple sulfur bacterium, *Chromatium* (large, red, rod-shaped cells in this photomicrograph of a natural microbial community). A cell is about 10 μm in diameter. (b) The large chemolithotrophic sulfur-oxidizing bacterium, *Achromatium*. A cell is about 20 μm in diameter. Globules of elemental sulfur can be seen in both cells. Both of these organisms oxidize hydrogen sulfide (H_2S) produced by sulfate-reducing bacteria. The latter are chemoorganotrophs that oxidize organic compounds or H_2 and couple this to the reduction of sulfate (SO_4^{2-}) to H_2S, thus completing the sulfur cycle (∞ Section 19.13).

Several lineages of *Bacteria* contain species with unique morphologies. These include the aquatic **Planctomyces** group, characterized by cells with a distinct stalk that allows the organisms to attach to a solid substratum (Figure 2.13), and the helically shaped **Spirochetes**

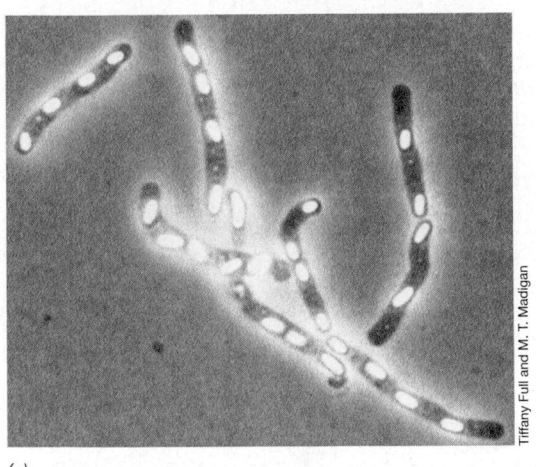

(a)

Tiffany Full and M. T. Madigan

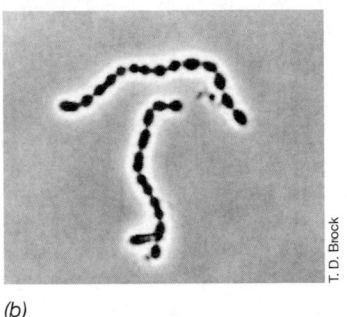

(b)

T. D. Brock

Figure 2.11 Gram-positive bacteria. (a) The rod-shaped endospore-forming bacterium *Bacillus*, here shown as cells in a chain. Note the presence of endospores (bright refractile structures) inside the cells. Endospores are extremely resistant to heat, chemicals, and radiation. (b) *Streptococcus*, a spherical cell that exists in chains. Streptococci are widespread in dairy products and some are potent pathogens.

(Figure 2.14). Several diseases, most notably syphilis and Lyme disease (∞ Sections 26.12 and 27.4), are caused by spirochetes.

Two other major lineages of *Bacteria* are phototrophic: the **Green Sulfur Bacteria** and the **Green Nonsulfur Bacteria** (*Chloroflexus* group) (Figure 2.15). Species in both lineages contain similar photosynthetic pigments and can grow as autotrophs. *Chloroflexus* is a filamentous prokaryote that inhabits hot springs and shallow marine bays, forming stratified microbial mats containing a community of microorganisms. *Chloroflexus* is also noteworthy because it is believed to be an important link in the evolution of photosynthesis (∞ Sections 12.35 and 17.7).

Other major lineages of *Bacteria* include the **Chlamydia** and **Deinococcus** groups (Figure 2.9). The genus *Chlamydia*, most species of which are pathogens, cause a variety of respiratory and venereal diseases in humans (∞ Sections 12.27 and 26.13). Chlamydia are *obligate intracellular parasites*, meaning that they live inside the cells of higher organisms, in this case, human cells. Several other pathogenic prokaryotes (for example,

(a)

R. W. Castenholz

(b)

R. W. Castenholz

Figure 2.12 Filamentous cyanobacteria. (a) *Oscillatoria*, (b) *Spirulina*. Eons ago, cyanobacteria produced the oxygen now present on Earth. Many other morphological forms of cyanobacteria are known, including unicellular, colonial, and heterocystous. The latter contain special structures called *heterocysts* that carry out nitrogen fixation (∞ Sections 12.25 and 17.28).

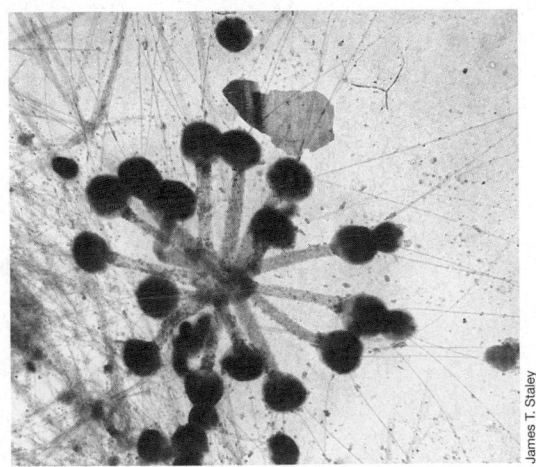

James T. Staley

Figure 2.13 The morphologically unusual stalked bacterium *Planctomyces*. Shown are several cells attached by their stalks to form a rosette.

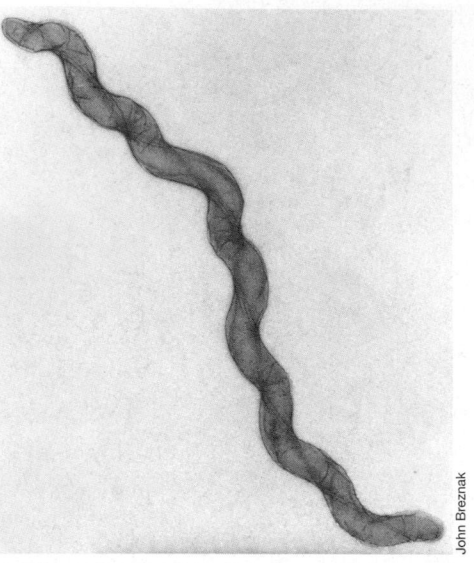

John Breznak

Figure 2.14 Spirochetes. A cell of *Spirochaeta zuelzerae*. These morphologically distinct prokaryotes are also phylogenetically distinct (see Figure 2.9). Spirochetes are widespread in nature and some cause diseases such as syphilis and Lyme disease. Reproduced by permission from J. A. Breznak, 1973. *CRC Critical Reviews of Microbiology* 2:457–489. Original micrographs from R. Joseph and E. Canale-Parola, 1972. *Archives of Microbiology* 81:146–168.

species of *Rickettsia*, a member of the Proteobacteria that cause diseases like typhus and Rocky Mountain spotted fever, or *Mycobacterium tuberculosis*, a gram-positive bacterium that causes tuberculosis) have also evolved the strategy of living and reproducing *inside* eukaryotic cells.

The intracellular location of these pathogens provides a means by which they can avoid destruction by the host immune response. The *Deinococcus* lineage contains species with unusual cell walls and an innate resistance to high levels of radiation; *Deinococcus radiodurans* (Figure 2.16) is a major species in this group.

Finally, several lineages of *Bacteria* branch off very early on the phylogenetic tree, very near the root (Figure 2.9). Although phylogenetically distinct from one another, these groups are unified by the common property of growth at high temperature (*thermophily*). Organisms like *Aquifex* (Figure 2.17) and *Thermotoga* grow in environments that are near the boiling point of water; as you might imagine, their habitats are hot springs. The early branching nature of these lineages on the tree (Figures 2.7 and 2.9) thus makes good sense, since it is strongly suspected that the early Earth was very hot (⌘ Section 11.1) and thus that life first evolved on a very warm planet. Organisms such as *Aquifex* and its close relatives are therefore modern descendants of very ancient cell lineages.

Archaea

Inspection of the domain *Archaea* (Figure 2.18) shows that two major subdivisions of these prokaryotes exist. Many *Archaea* are extremophiles, with species capable of growth at the highest temperatures and extremes of pH of all known organisms (Table 2.1). All *Archaea* are chemotrophic, although *Halobacterium* (to be discussed) can use light to make ATP, but not in the typical way of phototrophic organisms. Some *Archaea* use organic compounds in their energy metabolism while many are

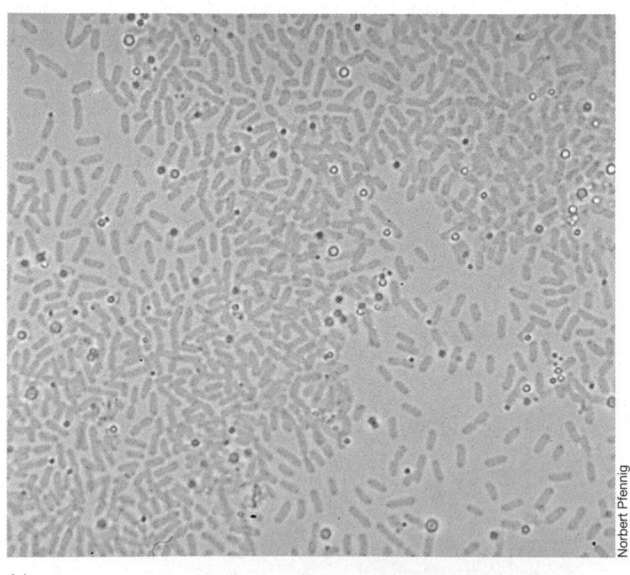

Norbert Pfennig

(a)

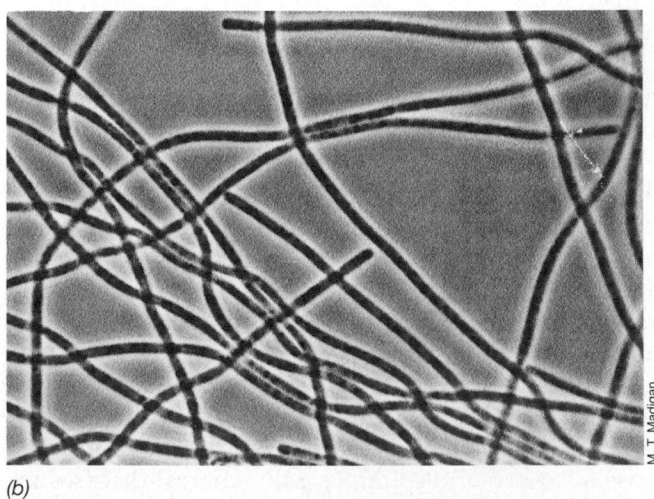

M. T. Madigan

(b)

Figure 2.15 Phototrophic green bacteria. (a) *Chlorobium* (green sulfur bacteria); (b) *Chloroflexus* (green nonsulfur bacteria). Despite sharing many features like pigments and membrane structures in common (⌘ Section 17.2), these organisms are phylogenetically quite distinct (Figure 2.9).

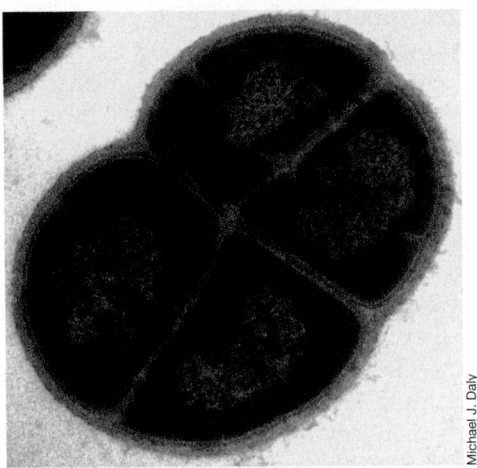

Michael J. Daly

Figure 2.16 The highly radiation-resistant bacterium *Deinococcus radiodurans*. This organism can withstand radiation levels far above that sufficient to kill a human.

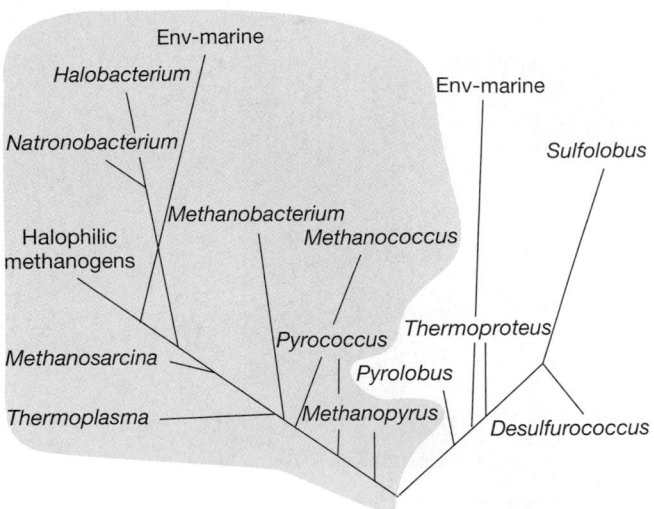

Figure 2.18 Detailed phylogenetic tree of *Archaea*. Not all known groups of *Archaea* are depicted on this tree. There are two major subgroups of *Archaea*. The organisms shown in yellow are primarily hyperthermophiles, growing at very high temperatures. In red are shown methanogens and the extreme halophiles and acidophiles. Each major group has its own Env lineages (see legend to Figure 2.9 and text) as well, most of which are marine species. There are roughly the same number of species in both subgroups of *Archaea*, but the total number of cultured *Archaea* is far lower than for *Bacteria*.

chemolithotrophs, with hydrogen gas (H_2) being a favorite energy source (Figure 2.8). Many *Archaea* grow at high temperatures. And just as we saw for the heat-loving *Aquifex* (Figure 2.17), such species tend to branch near the root of the domain (Figures 2.9 and 2.17). The archaeon *Pyrolobus* (Figures 2.18 and 2.19), for example, is the most heat-loving of all known prokaryotes (Table 2.1).

The other major branch on the tree of *Archaea* (Figure 2.18) contains three groups of organisms with dramatically different physiologies. Some species require O_2 while others loathe it, and some grow at the upper and lower extremes of pH (Table 2.1). *Methanogens* like *Methanobacterium* are strict anaerobes. Their metabolism is unique in biology in that energy is obtained by the production of natural gas (methane)! Methanogens are ecologically important prokaryotes in the biodegradation of organic matter in nature (∞ Sections 13.4, 17.17, and 19.10), and most if not all of the natural gas found on Earth has accumulated from their metabolism.

The extreme halophiles are relatives of the methanogens (Figure 2.18), but are physiologically distinct from them. Unlike methanogens, which are killed by

oxygen, extreme halophiles *require* oxygen, and are unified by their requirement for very large amounts of salt (NaCl) for metabolism and reproduction. These organisms are therefore called *halophiles* (salt lovers). In fact, organisms like *Halobacterium* are so salt loving that they can actually grow on and within salt crystals

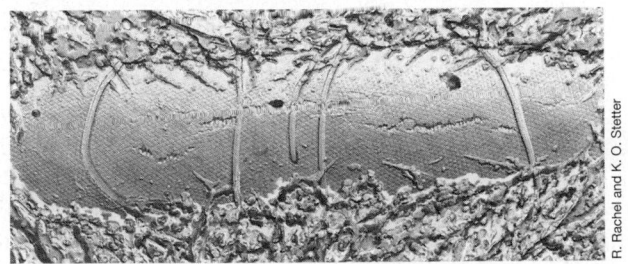

R. Rachel and K. O. Stetter

Figure 2.17 *Aquifex*. This "early-branching" species of *Bacteria* (see Figure 2.9) is a hyperthermophile, growing optimally above 80°C.

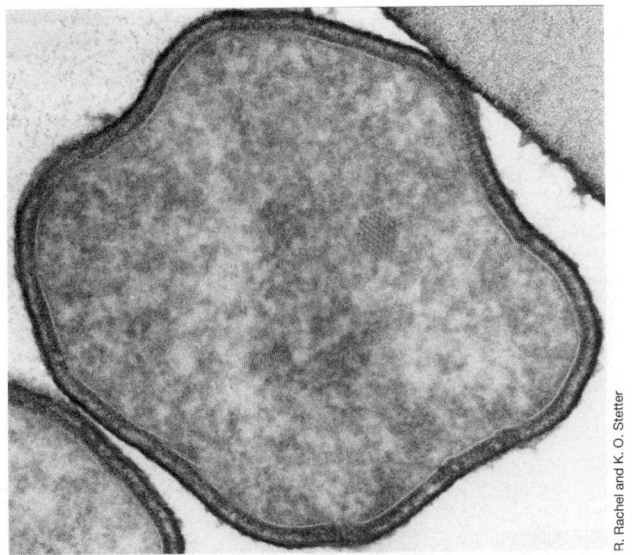

R. Rachel and K. O. Stetter

Figure 2.19 *Pyrolobus*. A hyperthermophilic archaeon that grows best above the boiling point of water!

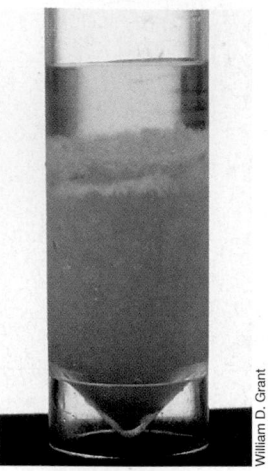

William D. Grant

Figure 2.20 Extremely halophilic *Archaea*. A vial of brine at the point of NaCl precipitation and containing cells of the extreme halophile, *Halobacterium*. The organism contains pigments that absorb light and lead to ATP production. Cells of *Halobacterium* can also live *within* salt crystals themselves (👓 box, Chapter 4, How Long Can an Endospore Survive?).

(Figure 2.20). As previously mentioned (see Section 2.4), many prokaryotes can generate ATP from light. Although they do not produce chlorophyll like true phototrophs, *Halobacterium* species contain light-sensitive pigments that can absorb light and trigger ATP synthesis (👓 Section 13.3). Extremely halophilic *Archaea* inhabit salt lakes, salterns, and other very salty environments. Some extreme halophiles, like *Natronobacterium*, inhabit soda lakes, environments characterized by both high levels of salt and pH. Such organisms are therefore *alkaliphilic* and grow at the highest pH of all known organisms (Table 2.1).

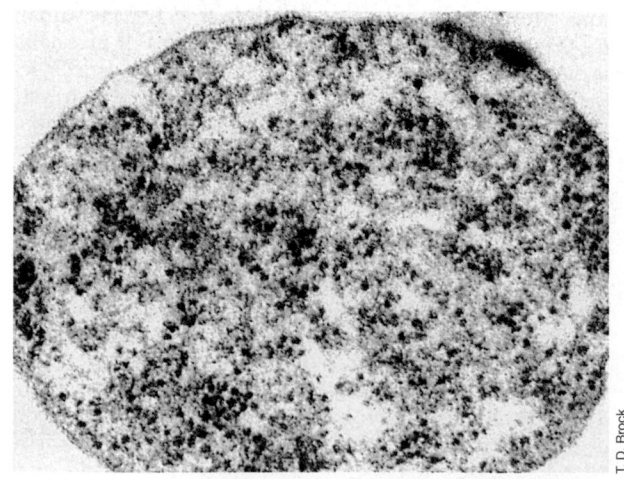

T. D. Brock

Figure 2.21 The cell wall-less archaeon, *Thermoplasma*, shown here, and its close relative *Picrophilus* (see Table 2.1) grow at moderately high temperature and extremely low pH. Although members of the *Bacteria*, the genus *Mycoplasma* also contains species lacking a cell wall. Cell wall-less prokaryotes are discussed in Sections 12.21 and 13.5.

The final groups of *Archaea* we will consider are the thermoacidophiles, such as *Thermoplasma* (Figure 2.21). These are cell wall-less prokaryotes (resembling *Mycoplasma* in this respect) that grow best at moderately high temperatures and extremely low pH. This group includes *Picrophilus*, the most acidophilic of all known prokaryotes (Table 2.1).

Lest one get the incorrect notion that all *Archaea* are extremophiles, many are not and can be found in lakes, soils, and oceans. Unfortunately, however, most of these *Archaea* have thus far defied laboratory culture. So, how do we know they are there? We know this because we can easily isolate ribosomal RNA genes from cells existing in a natural sample, for example, in soil. Just as the old saying "where there's smoke, there's fire," is a truism, if a given sample of soil or water contains ribosomal RNA, the organisms that made this ribosomal RNA were obviously present in the sample. Thus, we can process such a sample for ribosomal RNA genes, sequence them, and then place a branch on the phylogenetic tree that represents the organism from which the genes were obtained, even if that organism was never cultured (see Figures 2.9 and 2.18). Using these methods of molecular microbial ecology initially devised by Norman Pace, an American microbiologist, we now know that the extent of prokaryotic diversity is far greater than originally thought and that various nonextreme habitats, such as marine waters (see Figure 2.18), contain a wealth of *Archaea*. The challenge now is to understand the biology of these *Archaea* sufficiently so that we can obtain them in laboratory culture; this is currently a major challenge for microbiologists.

✓ 2.5 Concept Check

Several lineages are present in the domains *Bacteria* and *Archaea*, and an enormous diversity of cell morphologies and physiologies are represented there. Retrieval and analysis of ribosomal RNA genes from cells in natural samples have shown that many phylogenetically distinct but as yet uncultured prokaryotes exist in nature.

✓ What important bacterial species that resides in your gut is a member of the Proteobacteria?

✓ Why can it be said that the cyanobacteria prepared the Earth for the evolution of higher life forms?

✓ What is unusual about species of the genus *Halobacterium*?

✓ How do we know a particular lineage of prokaryote exists in nature without growing a culture of it in the laboratory?

2.6 Eukaryotic Microorganisms

Eukaryotic microorganisms, or phylogenetically speaking, the *Eukarya*, are unified by their distinct cell structure (Figure 2.1) and phylogenetic history. Inspection of the domain *Eukarya* (Figure 2.22) shows a long branch culminating in the most recent of eukaryotes, the plants and animals. Interestingly, and consistent with their phy-

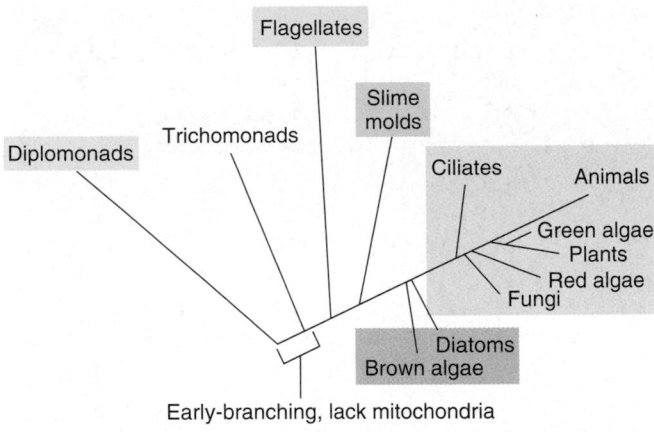

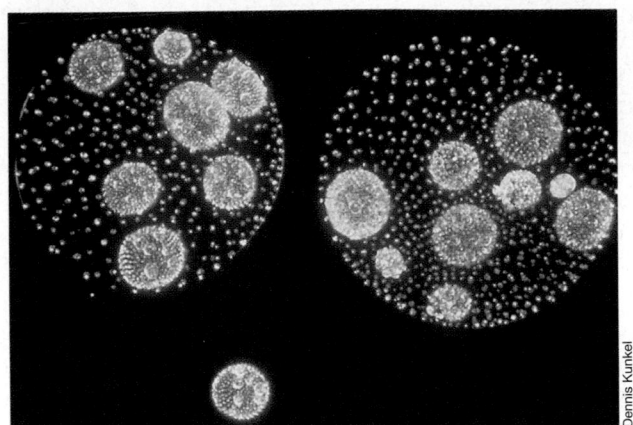

(a)

Figure 2.22 Detailed tree of *Eukarya*. Not all known lineages of *Eukarya* are depicted. Some early-branching species of *Eukarya* lack organelles other than the nucleus. Note how "higher organisms" (plants and animals) branch near the apex of the tree.

logenetic location on the tree, some of the "early-branching" *Eukarya* are structurally simple eukaryotes, lacking mitochondria and some other key organelles. These cells, such as diplomonads like *Giardia* (Figure 2.22), appear to be modern descendents of primitive eukaryotic cells that did not engage in endosymbiosis (⊂⊃ Sections 2.3 and 11.3) or for some reason did not permanently merge with a partner organism. Most of these early eukaryotes are metabolically deficient cells and are pathogenic parasites of humans and other animals.

Like prokaryotes, there exists a diverse array of eukaryotic microorganisms. Some, such as algae (Figure 2.23*a*), are phototrophic; they contain chlorophyll-rich organelles called *chloroplasts* and can live in environments containing only a few minerals, CO_2, and light. Algae inhabit both soil and aquatic habitats and are major primary producers in nature. Fungi (Figure 2.23*b*) lack photosynthetic pigments and are either unicellular (yeasts) or filamentous (molds). Fungi are major agents of biodegradation in nature and are major recyclers of organic matter in soils and other ecosystems.

Cells of algae and fungi have cell walls whereas the protozoa (Figure 2.23*c*) do not. Most protozoans are motile, and different species are widespread in nature in aquatic habitats and as pathogens of humans and other animals. Different protozoa are also spread about the phylogenetic tree of *Eukarya*. Some, like the flagellates, are fairly early branching species, while others, like the ciliated species *Paramecium* (Figure 2.23), are phylogenetically more derived (Figure 2.22). The *slime molds* resemble protozoa in that they are motile and lack cell walls. Slime molds differ from protozoa, however, in phylogeny and by the fact that their cells undergo a life cycle where motile cells aggregate to form a multicellu-

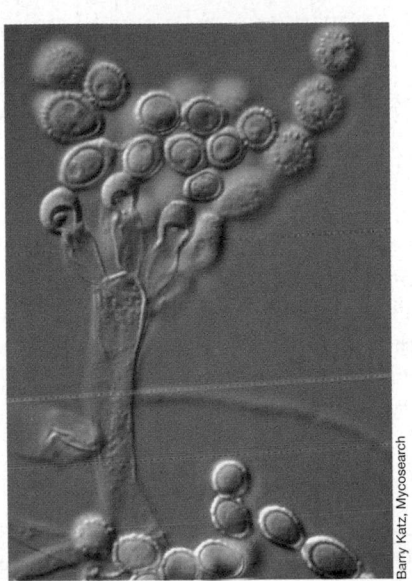

(b)

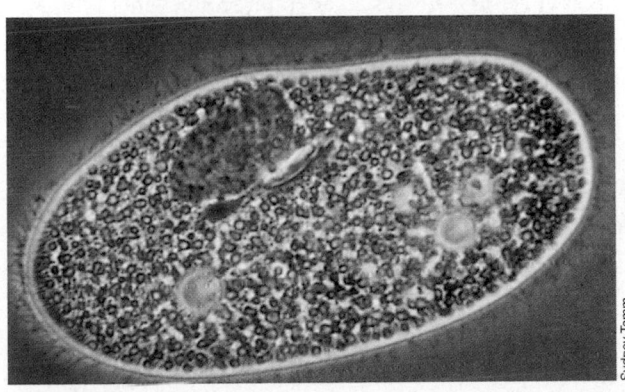

(c)

Figure 2.23 Photomicrographs of microbial *Eukarya*. (a) Algae; the colonial green alga, *Volvox* (⊂⊃ Section 14.11). Each spherical-shaped cell contains several chloroplasts, the photosynthetic organelle of phototrophic eukaryotes. (b) Fungi; the spore-bearing structures of a typical mold. Each spore can give rise to a new filamentous fungus (⊂⊃ Section 14.9). (c) Protozoa; the ciliated protozoan *Paramecium* (⊂⊃ Section 14.8). Cilia function like oars in a boat, conferring motility on the cell.

lar structure called a *fruiting body* from which spores are produced that yield new motile cells (Section 14.11).

Lichens are leaf-like structures often found growing on rocks, trees, and other surfaces (Figure 2.24). Lichens are an example of microbial *mutualism*, a situation where two organisms live together for both their benefits. Lichens consist of a fungus and a phototrophic partner, either an alga (eukaryote) or a cyanobacterium (a prokaryote). The phototrophic component is the primary producer while the fungus provides the phototroph with an anchor and protection from the elements. The lichen is thus a dynamic organism that has evolved a successful strategy of mutualistic interaction between two quite different microorganisms.

This tour of microbial diversity has by necessity given only a brief overview of microbial life. The story is actually much larger and will be continued in Chapters 12–14. Note also that the viruses were intentionally left out here. Recall that viruses are not cells (see Section 2.1), yet require cells for their replicative processes. Cells in all domains of life contain viral parasites, and we will deal with viral diversity in later chapters (Chapters 9 and 16). First, however, we must become familiar with the molecular features of cells, especially prokaryotic cells. Along the way we will learn of the great chemical diversity of living organisms, a direct result of the evolutionary breadth of cells summarized in this chapter.

✓ 2.6 Concept Check

Microbial eukaryotes are a diverse group that includes algae, protozoa, fungi, and slime molds. Some algae and fungi have developed mutualistic associations called *lichens*.

✓ List at least two ways algae differ from cyanobacteria.
✓ List at least two ways algae differ from protozoa.
✓ How do the components of a lichen benefit each other?

(a)

(b)

Figure 2.24 Lichens. (a) An orange-pigmented lichen growing on a rock, and (b) a yellow-pigmented lichen growing on a dead tree stump, Yellowstone National Park, USA. The color of the lichen comes from the pigmented (algal) component of the lichen structure.

Review Questions

1. Why does a cell need a cytoplasmic membrane? What properties do you suppose this membrane should have?
2. Which domains of life have a prokaryotic cell structure? Does prokaryotic cell structure correlate with evolutionary relationships?
3. How do viruses resemble cells? How do they differ from cells?
4. What is meant by the word *genome*? How does the arrangement of the genome of prokaryotes differ from that of eukaryotes?
5. For what reason do the processes of mitosis and meiosis occur in eukaryotic cells?
6. What is the theory of endosymbiosis?
7. What is meant by the word *phylogeny*? Why do you think that the phylogeny of prokaryotes had to await development of molecular tools for this purpose?
8. How many genes does an organism like *Escherichia coli* have? How does this compare with the number of genes in one of your cells?

9. What terms are used to describe the three domains of life? Members of which two domains are most similar from a *structural* standpoint? From a *phylogenetic* standpoint?

10. Molecular studies have shown many macromolecules in species of *Archaea* resemble their counterparts in various eukaryotes more closely than in species of *Bacteria*. Explain.

11. How do *chemoorganotrophs* differ from *chemolithotrophs* from the standpoint of energy metabolism? What carbon sources do members of each group use? Are they therefore *heterotrophs* or *autotrophs*?

12. What is unusual about the organism *Pyrolobus*?

13. What similarities and differences exist between the following three organisms: *Pyrolobus*, *Halobacterium*, *Thermoplasma*?

14. Examine Figure 2.18. What does the lineage "Env-marine" mean?

Application Questions

1. Prokaryotic cells containing plasmids can often be "cured" of their plasmids (that is, the plasmids can be permanently removed) with no ill effects, while removal of the chromosome would be lethal. Explain.

2. It has been said that knowledge of the evolution of *macroorganisms* greatly preceded that of *microorganisms*. Why do you think that reconstruction of the evolutionary lineage of horses, for example, might have been an easier task than doing the same for bacteria?

3. Examine the phylogenetic tree shown in Figure 2.6. Using the sequence data shown, describe why the tree would be incorrect if it remained the same shape but the positions of organisms 2 and 3 on the tree were switched?

4. Microbiologists have cultured a great diversity of microorganisms but know that an even greater diversity exists, despite the fact that they have never seen these organisms or grown them in the laboratory. Explain.

5. What data from this chapter could you use to convince your friend that extremophiles are not just organisms that were "hanging on" in their respective habitats?

6. Defend this statement: If cyanobacteria had never evolved, life on Earth would have remained strictly microbial.

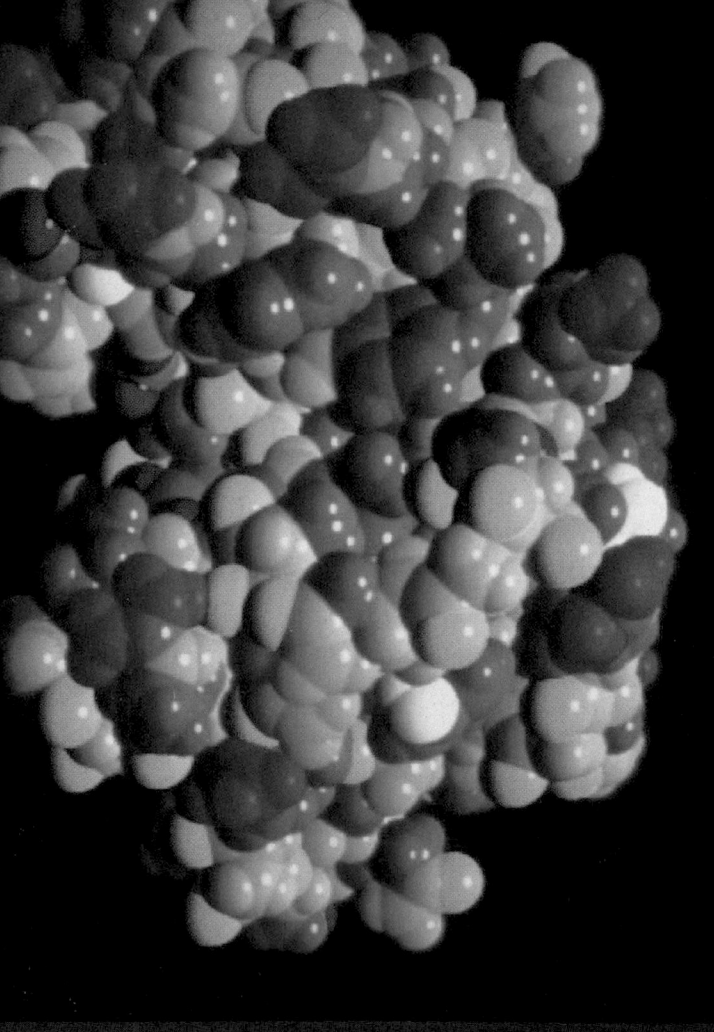

Cells consist of highly specific arrangements of macromolecules. These chemically diverse molecules include the polysaccharides, lipids, nucleic acids, and proteins. Besides playing a structural role, proteins, as shown here in a molecular model, often function in the cell as catalysts called enzymes. The structure of macromolecules to a large degree defines their function. Thus, knowledge of macromolecular structure is important for understanding key cell functions to be described in later chapters.

MACROMOLECULES 3

Covalent bond a chemical bond in which electrons are shared between two atoms

Denaturation destruction of the folding properties of a protein leading (usually) to loss of biological activity

Enantiomer one form of a molecule that is the mirror image of another form of the same molecule

Glycosidic bond a type of covalent bond that links sugar units together in a polysaccharide

Hydrogen bond a weak chemical bond between a hydrogen atom and a second, more electronegative element, usually an oxygen or nitrogen atom

Lipid glycerol bonded to fatty acids or other hydrophobic molecules by ester or ether linkage. Often contain other groups as well, such as phosphate

Macromolecule polymer of covalently linked monomeric units

Molecule two or more atoms chemically bonded to one another

Nonpolar possessing hydrophobic (water-repelling) characteristics and not easily dissolved in water

Nucleoside a nucleotide without its phosphate group

Nucleotide a monomer of a nucleic acid containing a nitrogen base (adenine, guanine, cytosine, thymine, or uracil), a molecule of phosphate, and a sugar, either ribose (in RNA) or deoxyribose (in DNA)

Peptide bond a type of covalent bond joining amino acids in a polypeptide

Phosphodiester bond a type of covalent bond linking nucleotides together in a polynucleotide

Polar possessing hydrophilic characteristics and generally water-soluble

Polymer a chemical compound formed by polymerization and consisting of repeating units called monomers

Polynucleotide a polymer of nucleotides bonded to one another by phosphodiester bonds

Polypeptide a polymer of amino acids bonded to one another by peptide bonds

Polysaccharide a polymer of sugar units bonded to one another by glycosidic bonds

Primary structure in an informational macromolecule, such as a polypeptide, the precise sequence of monomeric units

Protein a polypeptide or group of polypeptides that form a molecule of specific biological function

Quaternary structure in proteins, the number and arrangement of individual polypeptides in the final protein molecule

Secondary structure the initial pattern of folding of a polypeptide or a polynucleotide, usually dictated by opportunities for hydrogen bonding

Tertiary structure the final folded structure of a polypeptide that has previously attained secondary structure

---

I CHEMICAL BONDING AND WATER IN LIVING SYSTEMS

To understand microbiology and how cells work, one must have some understanding of the molecules present and chemical processes that take place within cells. Molecules, especially *macromolecules,* are the "guts" of the cell and are the subject of this chapter. It is assumed that the reader has some background in elementary chemistry, especially regarding the chemical nature of atoms and atomic bonding. In this chapter we will expand on this background with a primer on relevant biochemical bonds, followed by a detailed discussion of the structure and function of the four classes of macromolecules: **polysaccharides, lipids, nucleic acids,** and **proteins.**

3.1 Strong and Weak Chemical Bonds

The major chemical elements of life include hydrogen, oxygen, carbon, nitrogen, phosphorus, and sulfur; these can bond in various ways to form the molecules of life. What is a molecule? A **molecule** is two or more atoms chemically bonded to one another. Thus, two oxygen (O) atoms can combine to form a molecule of oxygen (O_2). The chemical elements of life are able to form strong bonds in which electrons are shared more or less equally between atoms; these are called **covalent bonds.** To envision a covalent bond, consider the formation of a molecule of water from its constituent elements, O and H:

$$\ddot{O} + 2H\cdot \longrightarrow H\!:\!\ddot{O}\!:\!H$$

Oxygen contains six electrons in its outermost shell, while hydrogen has but a single electron. When they combine to form H_2O they do so by way of covalent bonds that maintain the three atoms in tight association. In similar fashion, depending on the elements, of course, double and triple covalent bonds can form, and the strength of these bonds increases dramatically with their number (Figure 3.1).

Hydrogen Bonds and Other Weak Associations

Besides covalent bonds, a variety of much weaker chemical bonds also plays an important role in biological molecules. Foremost among these are **hydrogen bonds.** Hydrogen bonds (Figure 3.2) form between hydrogen atoms and more electronegative elements, like oxygen or nitrogen. An individual hydrogen bond is very weak. However, when many hydrogen bonds are formed within and between molecules, the stability of the molecules may be greatly increased.

Water molecules readily undergo hydrogen bonding (Figure 3.2*a*), which contributes to the well-known

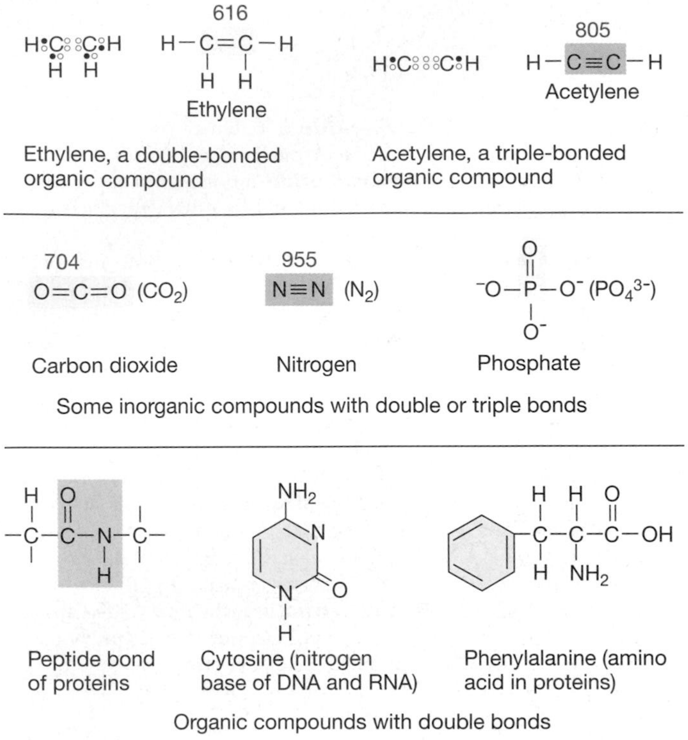

Ethylene
Ethylene, a double-bonded organic compound

Acetylene
Acetylene, a triple-bonded organic compound

O=C=O (CO₂)
Carbon dioxide

N≡N (N₂)
Nitrogen

⁻O—P—O⁻ (PO₄³⁻)
Phosphate

Some inorganic compounds with double or triple bonds

Peptide bond of proteins

Cytosine (nitrogen base of DNA and RNA)

Phenylalanine (amino acid in proteins)

Organic compounds with double bonds

Figure 3.1 Covalent bonding of some biologically important molecules containing double or triple bonds. For acetylene and ethylene, the electronic configuration of the molecules is shown. Bond strength (the numbers in red) is measured in units of kilojoules (kJ), which is the amount of heat needed to break a bond.

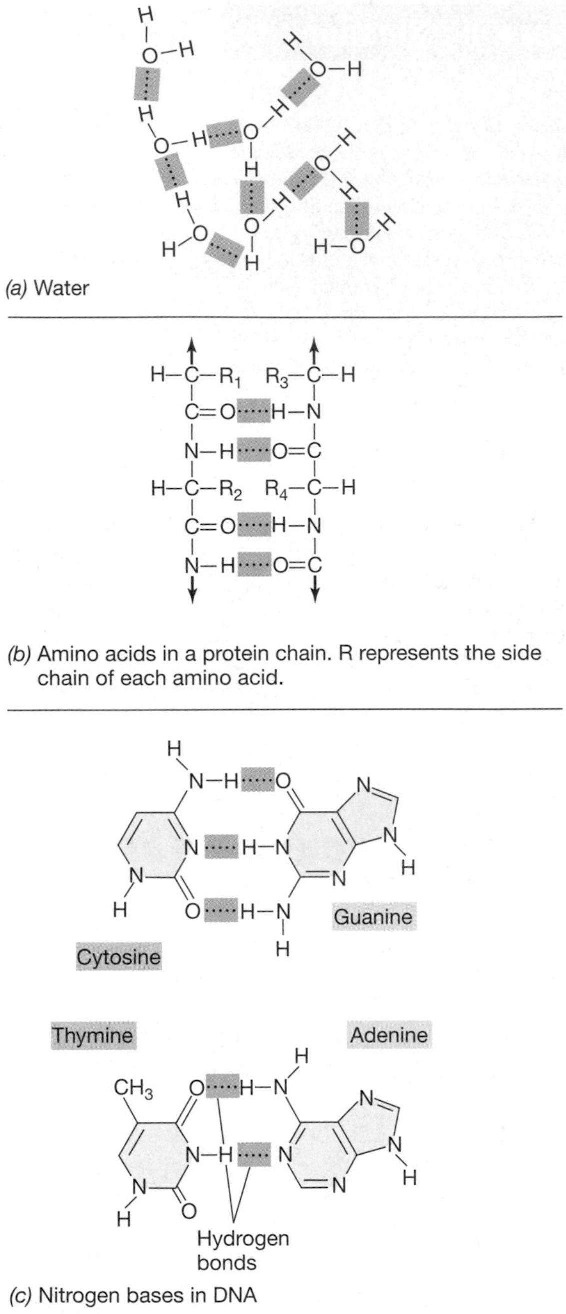

(a) Water

(b) Amino acids in a protein chain. R represents the side chain of each amino acid.

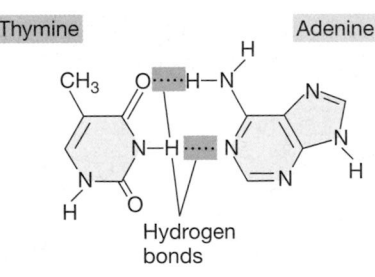

Cytosine — Guanine

Thymine — Adenine

Hydrogen bonds

(c) Nitrogen bases in DNA

Figure 3.2 Hydrogen bonding. In nucleic acids, hydrogen bonds are often depicted as lines rather than dots, with two lines between adenine/thymine pairs and three lines between guanine/cytosine pairs (see Figure 3.11).

polarity of water. Since an oxygen atom is rather electronegative (electron withdrawing), while a hydrogen atom is not, the covalent bond between oxygen and hydrogen is one in which the shared electrons in the outer shells actually orbit slightly nearer the oxygen nucleus than the hydrogen nucleus. Because electrons carry a negative charge, this creates a slight charge separation, oxygen negative and hydrogen positive (Figure 3.2*a*). As water molecules orient themselves in solution, the slight positive charge on the hydrogen atom can bridge negative charges on two oxygen atoms; this bridge is the *hydrogen bond*. Hydrogen bonds also form between atoms in macromolecules (Figure 3.2*b* and *c*). When these weak electrical forces accumulate in a large molecule like a protein, they improve the stability of the molecule and can also affect its overall structure. We will see later that hydrogen bonds play major roles in the biological properties of macromolecules, especially in proteins (Figure 3.2*b*) and nucleic acids (Figure 3.2*c*) (see Sections 3.5, 3.7, and 3.8).

Biomolecules form other types of weak interactions. For instance, *van der Waals forces* are weak attractive forces that occur between atoms when they become closer than about 3–4 angstroms (Å). The action of van der

Waals forces can play a significant role in the binding of substrates to enzymes (∞ Section 5.5) and in protein-nucleic acid interactions.

Hydrophobic interactions are also important in biomolecules. Hydrophobic interactions occur because nonpolar (water-repelling) molecules tend to cluster in an aqueous environment. Because of this, most nonpolar portions of a macromolecule will associate. Hydrophobic interactions can play a major role in the folding patterns of proteins, and, along with van der Waals interactions, play important roles in the binding of substrates to enzymes. In addition, hydrophobic interactions often control how different subunits in a multisubunit protein associate with one another to form the biologically active molecule.

Bonding Patterns in Biomolecules

The element **carbon** is a major component of all macromolecules. Carbon is able to combine not only with itself, but with many other elements as well, to yield large structures of considerable diversity and complexity. In different organic (carbon containing) compounds, a variety of chemical combinations reappear with high frequency, and an awareness of these combinations will make our later discussion of macromolecular structure, cell physiology, and biosynthesis easier to follow. Table 3.1 lists several of these bonding patterns and the molecules or macromolecules they appear in most often. Each of these *functional groups*, as they are called, has unique chemical properties that may be important in their biological role within the cell.

TABLE 3.1   Functional groups of biochemical importance

Chemical species	Structure[a]	Biological relevance
Carboxylic acid	—C(=O)—OH	Organic, amino, and fatty acids; lipids; proteins
Aldehyde	—C(=O)—H	Functional group of reducing sugars such as glucose; polysaccharides
Alcohol	—C(H)(H)—OH	Lipids; carbohydrates
Keto	—C(=O)—	Pyruvate, citric acid cycle intermediates
Ester	—C(H)—O—C(=O)—	Lipids of *Bacteria* and *Eukarya*; amino acid attachment to tRNAs
Phosphate ester	$^-$O—P(O$^-$)(=O)—O—C—	Nucleic acids, DNA and RNA
Thioester	—C(=O)~S—	Energy metabolism; biosynthesis of fatty acids
Ether	—C(H)(H)—O—C(H)(H)—	Lipids of *Archaea*; sphingolipids
Acid anhydride	—C(=O)~O—P(=O)(O$^-$)(O)	Energy metabolism, for example, acetyl phosphate
Phosphoanhydride	$^-$O—P(=O)(O$^-$)~O—P(=O)(O$^-$)—O$^-$	Energy metabolism, for example, ATP

[a]A squiggle-type bond depiction (~) indicates a "high energy" bond (⬭ Section 5.8).

Covalent bonds are strong bonds that hold together elements in macromolecules. Weak bonds, such as hydrogen bonds, van der Waals forces, and hydrophobic interactions, also affect macromolecular structure, but through more subtle atomic interactions. A variety of functional groups containing carbon atoms are common in important biomolecules.

✓ Why are *covalent* bonds stronger than *hydrogen* bonds?

✓ How can a hydrogen bond play a role in macromolecular structure?

3.2 An Overview of Macromolecules and Water as the Solvent of Life

If you were to biochemically dissect a prokaryotic cell like the common intestinal bacterium *Escherichia coli*, what would you find? You would find water as the major constituent, but after removing the water you would find large amounts of *macromolecules*, smaller amounts of *monomers* (the precursors of macromolecules), and a variety of inorganic ions (Table 3.2). In fact, 96% of the dry weight of a cell consists of macromolecules, and of these, proteins are by far the most abundant class (Table 3.2). **Proteins** are polymers of monomers called *amino acids.* Proteins are found throughout the cell, in both structural as well as catalytic (enzymatic) roles (Figure 3.3a).

TABLE 3.2   Chemical composition of a prokaryotic cell[a]

Molecule	Percent of dry weight[b]	Molecules per cell (different kinds)
Total macromolecules	96	24,610,000 (~2500)
Protein	55	2,350,000 (~1850)
Polysaccharide	5	4,300 (2)[c]
Lipid	9.1	22,000,000 (4)[d]
Lipopolysaccharide	3.4	1,430,000 (1)
DNA	3.1	2.1 (1)
RNA	20.5	255,500 (~660)
Total monomers	3.0	—[e] (~350)
Amino acids and precursors	0.5	— (~100)
Sugars and precursors	2	— (~50)
Nucleotides and precursors	0.5	— (~200)
Inorganic ions	1	— (18)
Total	100%	—

[a]Data from Neidhardt, F.C., et al. (eds.), 1996. *Escherichia coli* and *Salmonella typhimurium—Cellular and Molecular Biology,* 2nd edition. American Society for Microbiology, Washington, DC.

[b]Dry weight of an actively growing cell of *E. coli* ≅ 2.8×10^{-13} g; total weight (70% water) = 9.5×10^{-13} g.

[c]Assuming peptidoglycan and glycogen to be the major polysaccharides present.

[d]There are several classes of phospholipids, each of which exists in many kinds because of variability in fatty acid composition between species and because of different growth conditions.

[e]Reliable estimates of monomer and inorganic ion composition are lacking.

Nucleic acids are polymers of *nucleotides* and are found in the cell in two forms, **RNA** and **DNA**. Next to protein, ribonucleic acid (RNA) is the next most abundant macromolecule in an actively growing prokaryotic cell (Table 3.2 and Figure 3.3b). This is because there are thousands of ribosomes (the "machines" that make new proteins) in each cell, and ribosomes are composed of RNA and protein. In addition, smaller amounts of RNA are present in the form of messenger and transfer RNAs, other key players in protein synthesis. In contrast to RNA, DNA makes up a relatively insignificant (by weight) fraction of the bacterial cell (Table 3.2). However, although quantitatively minor, DNA is of course critical to cell function as the repository of genetic information needed to make a new cell.

Lipids have hydrophobic (water-repelling) properties and play crucial roles in membrane structure and as storage depots for excess carbon (Figure 3.3d). **Polysaccharides** are polymers of *sugars* and are present primarily in the cell wall; however, as for lipids, polysaccharides like glycogen (see later in this chapter) can be major forms of carbon and energy storage in the cell (Figure 3.3c).

Water as a Biological Solvent

Macromolecules and all other molecules in cells are bathed in water. Water has several important chemical features that make it an ideal biological solvent; indeed, water is a necessary prerequisite for life as we know it. The key chemical properties of water that make it such a good solvent are its *polarity* and *cohesiveness*.

The polar properties of water are important because many biologically important molecules are themselves polar and thus readily dissolve in water. As we will see in Chapter 4, dissolved substances are continually passing into and out of the cell through transport activities of the cytoplasmic membrane (∞ Sections 4.6 and 4.7). The polar properties of water also promote hydrogen bonding. Water forms three-dimensional networks, both with itself (Figure 3.2a) and within macromolecules, and by so doing, helps to position atoms within biomolecules for potential interactions. The high polarity of water is also beneficial to the cell because it tends to force *nonpolar* substances to aggregate and remain together. Membranes, for example, contain large amounts of nonpolar macromolecules such as lipids, and these tend to aggregate in such a way as to prevent the free flow of polar molecules into and out of the cell.

In addition to hydrogen bonding, the polar nature of water makes it highly cohesive, meaning that water molecules tend to have a high affinity for one another and form chemically ordered arrangements in which hydrogen bonds (Figure 3.2) are constantly forming, breaking, and re-forming. The cohesive na-

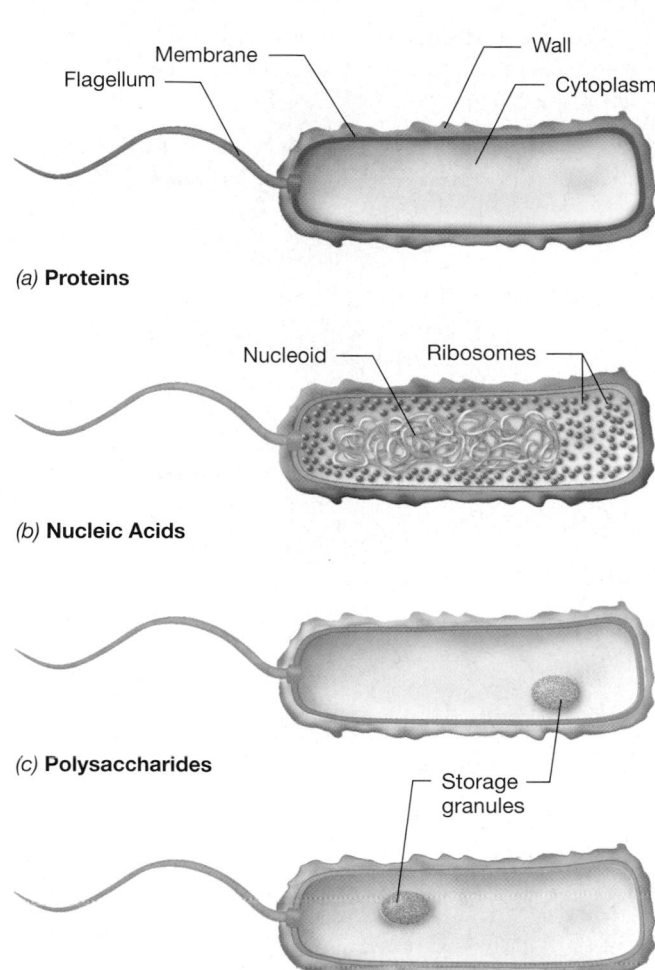

Membrane — Wall
Flagellum — — Cytoplasm

(a) **Proteins**

Nucleoid — Ribosomes

(b) **Nucleic Acids**

(c) **Polysaccharides**

Storage granules

(d) **Lipids**

Figure 3.3 The locations of macromolecules in the cell. (a) *Proteins* (brown) are found throughout the cell both as parts of cell structures and as enzymes. The flagellum is a structure involved in swimming motility. (b) *Nucleic acids.* DNA (green) is found in the nucleoid of prokaryotic cells and in the nucleus of eukaryotic cells. RNA (orange) is found in the cytoplasm (mRNA, tRNA) and in ribosomes (rRNA). (c) *Polysaccharides* (yellow) are located in the cell wall and occasionally in internal storage granules. (d) *Lipids* (blue) are found in the cytoplasmic membrane, the cell wall, and in storage granules. Note the color-coding scheme used here; these same colors will be used to depict these macromolecules throughout this book. For DNA, see also the legend to Figure 3.11.

ture of water is responsible for many of its biologically important properties, such as *high surface tension* and *high specific heat* (heat required to raise the temperature 1°C). Also, the fact that water expands on freezing to yield a less dense solid form (ice) has a profound effect on life in temperate and polar aquatic environments. In a lake, for example, ice on the surface insulates the water beneath the ice and prevents its freezing, thus allowing aquatic organisms to survive despite the overlying ice.

Life originated in water, and anywhere on Earth where liquid water exists, microorganisms are likely to be found (and we will see some spectacular examples of this in Section 6.10 and Chapter 13). We now consider the structure of the major macromolecules of life (Table 3.2).

✓ **3.2 Concept Check**

Proteins are the most abundant class of macromolecule in the cell. Other macromolecules include the nucleic acids (DNA and RNA), lipids, polysaccharides, and lipopolysaccharides. Water is a particularly good solvent for living organisms because of its polarity and cohesiveness.

✓ Why does RNA make up such a large proportion of an actively growing cell?

✓ Why does the high polarity of water make it useful as a biological solvent?

II MACROMOLECULES

We now dissect the macromolecules listed in Table 3.2 and examine the components of each.

3.3 Polysaccharides

Carbohydrates (sugars) are organic compounds containing carbon, hydrogen, and oxygen in a ratio of 1:2:1. The structural formula for glucose, the most abundant of all sugars, is $C_6H_{12}O_6$ (Figure 3.4). The most biologically relevant carbohydrates are those containing 4, 5, 6, and 7 carbon atoms (designated as C_4, C_5, C_6, and C_7). C_5 sugars (pentoses) are of special significance because of their role as structural backbones of nucleic acids. Likewise, C_6 sugars (hexoses) are the monomeric constituents of cell wall polymers and energy reserves. Figure 3.4 shows the structural formulas of a few common sugars.

Derivatives of simple carbohydrates can be formed by replacing one or more of the hydroxyl groups by other chemical species. For example, the important bacterial cell wall polymer **peptidoglycan** (∞ Section 4.8) contains the glucose derivative *N*-acetylglucosamine (Figure 3.5). Besides sugar derivatives, sugars having the same *structural* formula can still differ in their *stereoisomeric properties* (see Section 3.6). Hence, a large number of different sugars are potentially available to the cell for the construction of polysaccharides.

The Glycosidic Bond

Polysaccharides are carbohydrates containing many (sometimes hundreds or even thousands) monomeric units connected by covalent bonds referred to as

MEDIA TUTORIAL Macromolecules

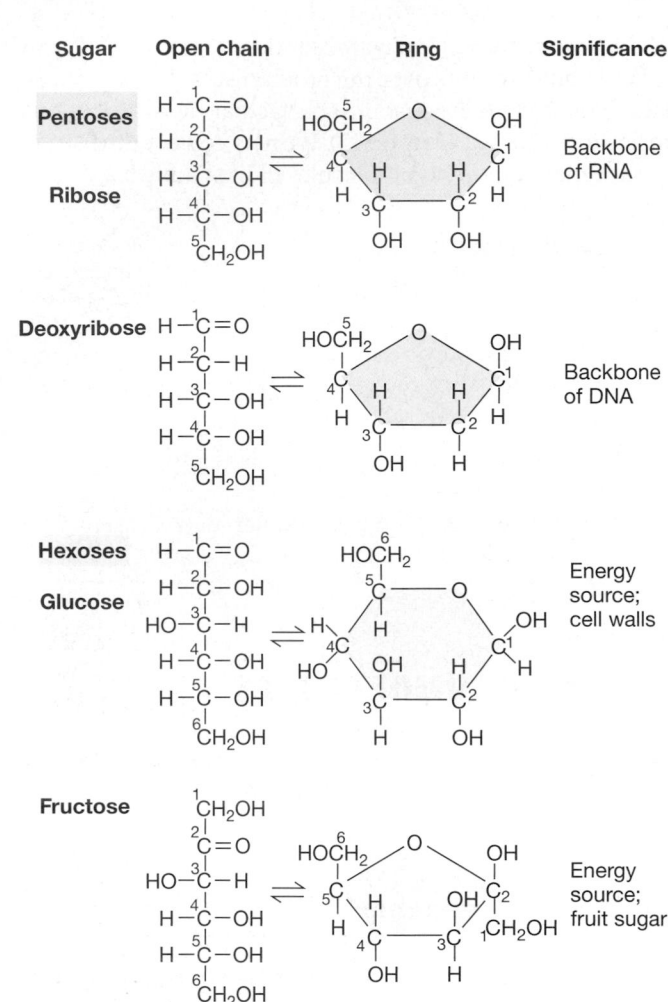

Figure 3.4 Structural formulas of a few common sugars. The formulas can be represented in two alternate ways, open chain and ring. The open chain is easier to visualize, but the ring form is the commonly used structure. Note the numbering system on the ring.

glycosidic bonds (Figure 3.6). If two sugars (monosaccharides) are joined by a glycosidic linkage, the resulting molecule is called a *disaccharide*. The addition of one more monosaccharide yields a *trisaccharide*, and several more an *oligosaccharide*; an extremely long chain is called a **polysaccharide**.

Figure 3.5 *N*-acetylglucosamine, a sugar derivative.

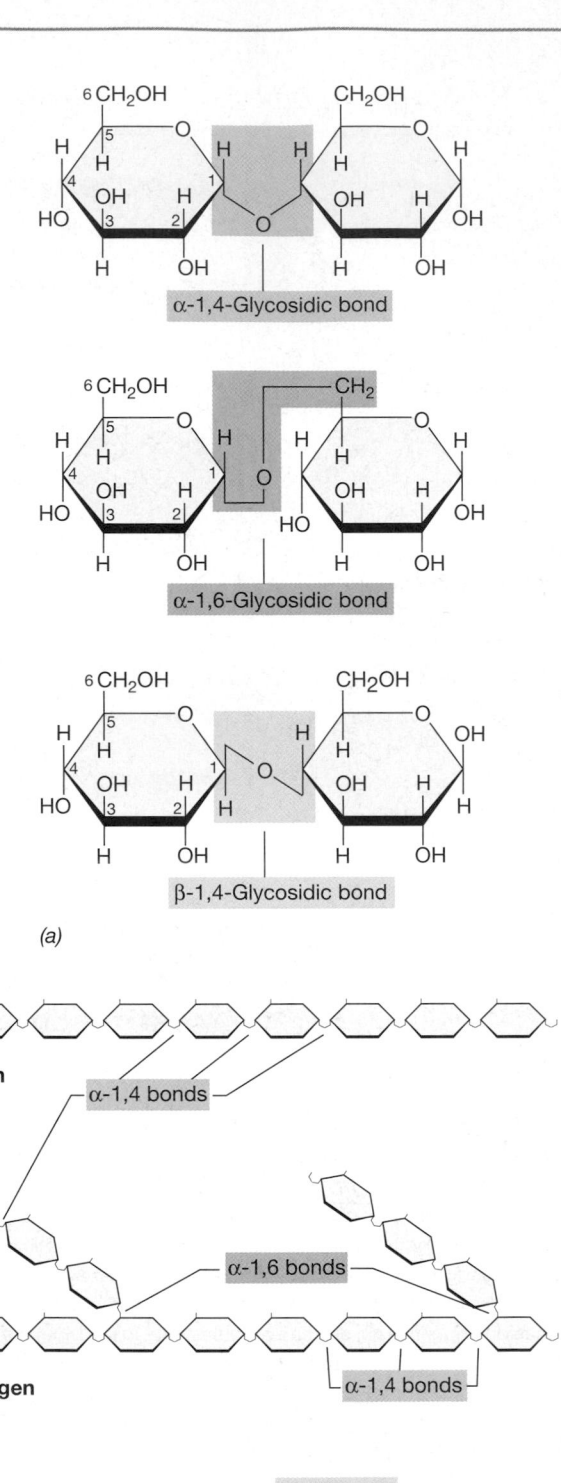

Figure 3.6 Polysaccharides (a) Structure of different glycosidic bonds. Note that both the *linkage* and the *geometry* (α or β) of the linkage can vary about the glycosidic bond. (b) Structures of some common polysaccharides. Compare color coding to (a).

The glycosidic bond can exist in two different geometric orientations, referred to as alpha (α) and beta (β) (Figure 3.6a). Polysaccharides with a repeating structure composed of glucose units linked between carbons 1 and 4 in the *alpha* orientation (for example, glycogen and starch, Figure 3.6b) function as important carbon and energy reserves in bacteria, plants, and animals. Alternatively, glucose units joined by *beta*-1,4 linkages are present in cellulose (Figure 3.6b), a stiff plant and algal cell wall component. Thus, even though starch and cellulose are both composed solely of glucose units, their functional properties are entirely different because of the different configurations, α or β, of their glycosidic bonds. Polysaccharides can also combine with other classes of macromolecules, such as protein and lipid, to form complex polysaccharides such as **glycoproteins** and **glycolipids**. These compounds play important roles in cell membranes as cell surface receptor molecules. The compounds reside on the external surfaces of the membrane where they are in contact with the environment. Glycolipids also constitute a major portion of the cell wall of gram-negative bacteria, and as such, impart a number of unique surface properties to these organisms (∞ Section 4.9).

✓ 3.3 Concept Check

Sugars combine into long polymers called polysaccharides. The two different orientations of the glycosidic bonds that link sugar residues impart different properties to the resultant molecules. Polysaccharides can also contain other molecules such as protein or lipid, forming complex polysaccharides.

✓ How can glycogen and cellulose differ so much in their physical properties when they both consist of 100% glucose?

3.4 Lipids

Fatty acids are the main constituents of the **lipids** of *Bacteria* and *Eukarya*. The lipids of *Archaea*, by contrast, are constructed of the hydrophobic molecule phytane (∞ Section 4.5). Fatty acids have interesting chemical properties because they contain both highly hydrophobic (water-repelling) and highly hydrophilic (water-soluble) regions. Palmitate,* for example (Figure 3.7), is a 16-carbon fatty acid composed of a chain of 15 saturated (fully hydrogenated) carbon atoms and a single carboxylic acid group. Other common fatty acids found in lipids include saturated or monounsaturated forms from C_{14} to C_{20} (Figure 3.7).

*Fatty acids can exist in both protonated (RCOOH) and unprotonated (RCOO⁻) forms, depending on pH. Because at pH 7 fatty acids are generally unprotonated, this is indicated by adding the suffix *-ate* to the root term for the fatty acid. Thus, palmitic acid is $C_{15}H_{31}COOH$, and palmitate is $C_{15}H_{31}COO^-$.

Triglycerides and Complex Lipids

Simple lipids (fats) consist of fatty acids bonded to the C_3 alcohol *glycerol* (Figure 3.7). Simple lipids are also referred to as **triglycerides** because three fatty acids are linked to the glycerol molecule.

Complex lipids are simple lipids that contain additional elements such as phosphorus, nitrogen, or sulfur, or small hydrophilic carbon compounds such as sugars,

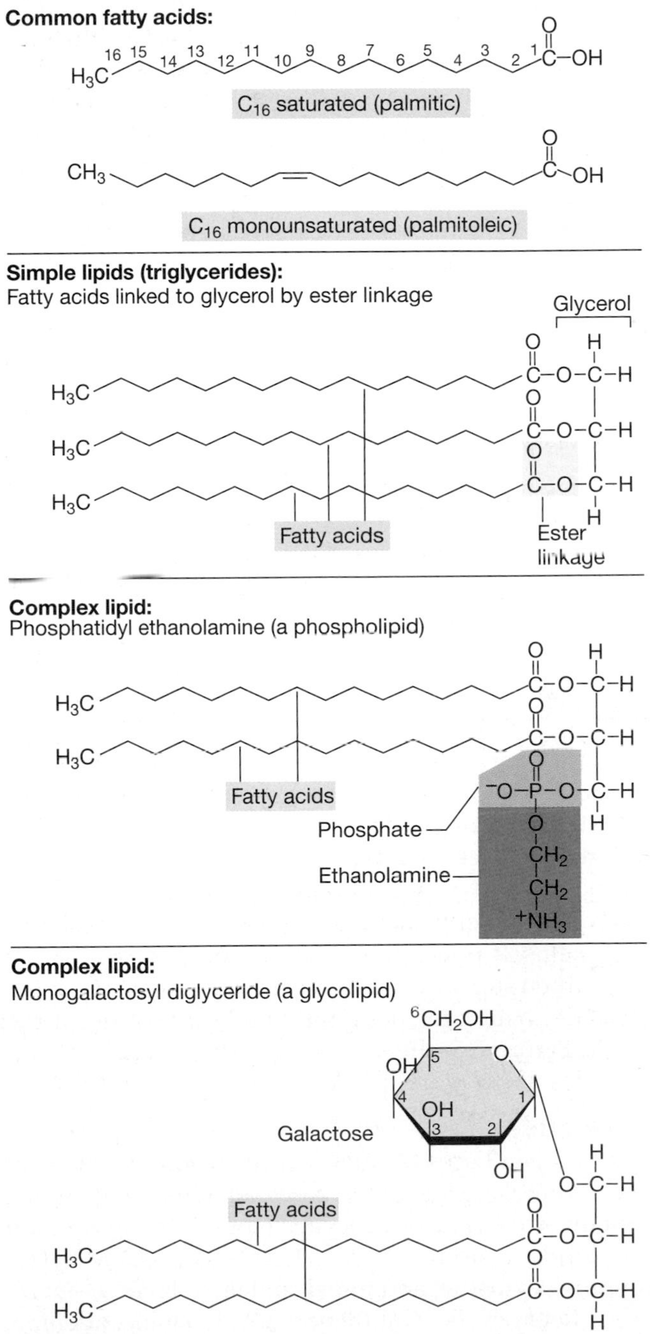

Figure 3.7 Fatty acids, simple lipids (fats), and complex lipids. Simple lipids are formed by a dehydration reaction between fatty acids and glycerol to yield the ester linkage.

ethanolamine (Figure 3.7), serine, or choline. Lipids containing a phosphate group, called **phospholipids**, are an important class of complex lipids because they play a major structural role in the cytoplasmic membrane (∞ Section 4.5).

The chemical properties of lipids make them ideal structural components of membranes. Because they are *amphipathic*, that is, show properties of hydrophobicity and hydrophilicity, lipids aggregate in membranes with the hydrophilic portions toward the external or internal (cytoplasmic) environment, while maintaining their hydrophobic portions away from the aqueous milieu (∞ Section 4.5). Such structures are ideal permeability barriers because of the inability of water-soluble substances to flow through the hydrophobic fatty acid portion of the lipids. Indeed, the major function of the cytoplasmic membrane is to serve as a barrier to the diffusion of substances into or out of the cell.

✓ **3.4** *Concept Check*

Lipids contain both hydrophobic and hydrophilic components; their chemical properties make them ideal structural components for cell membranes.

✓ What part of a fatty acid molecule is hydrophobic? Hydrophilic?
✓ How does a *phospholipid* differ from a *triglyceride*?
✓ Draw the chemical structure of butyrate, a C_4 fully saturated fatty acid.

3.5 Nucleic Acids

The nucleic acids, deoxyribonucleic acid (DNA), and ribonucleic acid (RNA), are macromolecules of monomers called **nucleotides**. DNA and RNA are thus both **polynucleotides**. DNA carries the genetic blueprint for the cell, and RNA acts as an intermediary molecule to convert the blueprint into defined amino acid sequences in proteins. A nucleotide is composed of three units: a five-carbon sugar, either ribose (in RNA) or deoxyribose (in DNA), a nitrogen base, and a molecule of phosphate, PO_4^{3-}. Figure 3.8 shows the general structure of nucleotides of DNA and RNA.

Nucleotides

The nitrogen bases of nucleic acids belong to either of two chemical classes. *Purine* bases, **adenine** and **guanine**, contain two fused heterocyclic (containing more than one kind of atom) rings, whereas *pyrimidine* bases, **thymine**, **cytosine**, and **uracil** contain a single six-membered heterocyclic ring (Figure 3.9). Guanine, adenine, and cytosine are found in both DNA and RNA; thymine is present (with minor exceptions) only in DNA, and uracil is present only in RNA. In a nucleotide, a base is

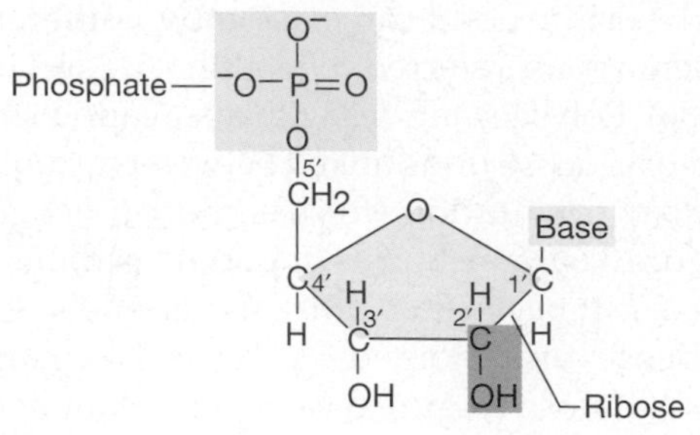

Figure 3.8 Nucleotides. Shown is a *ribo*nucleotide, found in RNA. *Deoxyribo*nucleotides, found in DNA, have an H atom instead of an OH group on the 2′ carbon. The numbers on the sugar contain a prime (′) after them because the ring structure in the nitrogen base is also numbered (1, 2, 3, etc.) (see Figure 3.9). Both deoxyribonucleotides and ribonucleotides contain a 5′-phosphate.

attached to a pentose sugar by a glycosidic linkage between carbon atom 1 of the sugar and a nitrogen atom of the base, either the nitrogen atom labeled 1 (pyrimidine base) or 9 (purine base). Without a phosphate, a base bonded to its sugar is referred to as a **nucleoside**. Nucleotides are thus nucleosides containing one or more phosphates (Figure 3.10).

Nucleotides play other roles in the cell besides their major role as constituents of nucleic acids. Nucleotides, especially adenosine triphosphate (ATP) (Figure 3.10), function as sources of chemical energy, releasing sufficient energy during the cleavage of a

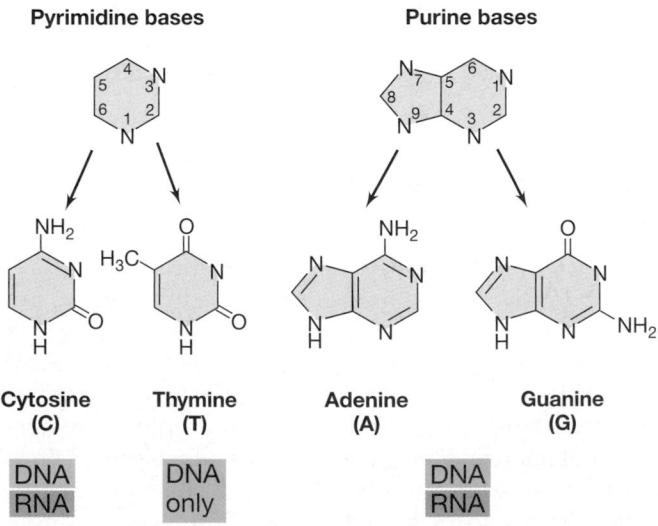

Figure 3.9 Structure of the bases of DNA and RNA. Uracil (U) is not shown, but its structure is the same as thymine, except that it lacks the methyl group on C-5. Note the numbering system of the rings. In attaching the base to the 1′ carbon of the sugar phosphate shown in Figure 3.8, pyrimidine bases are bonded through N-1 of the ring and purine bases through N-9 of the ring.

phosphate bond to drive energy-requiring reactions in the cell (∞ Section 5.8). Other nucleotides or nucleotide derivatives function in oxidation-reduction reactions in the cell (∞ Section 5.7) as carriers of sugars in the biosynthesis of polysaccharides (∞ Section 5.15) and as regulatory molecules inhibiting or stimulating the activities of certain enzymes or metabolic events. However, here we are discussing the role of a nucleotide as a building block of nucleic acid, the major *informational* function of nucleotides.

Nucleic Acids

The backbone of a nucleic acid is a polymer in which sugar and phosphate molecules alternate (Figure 3.11).

Polynucleotides contain nucleotides covalently bonded via phosphate from carbon 3 [referred to as the 3′ (3 prime) carbon] of one sugar to carbon 5 (5′) of the adjacent sugar (Figure 3.11*a*). The phosphate linkage is chemically a **phosphodiester** since a single phosphate is connected by ester linkage to two separate sugars.

The sequence of nucleotides in a DNA or RNA molecule is referred to as its *primary structure*. As we have discussed, the sequence of bases in a DNA or RNA molecule is *informational*, representing the genetic information necessary to reproduce an identical copy of the organism. We will see later that the replication of DNA and the production of RNA are highly complex processes (∞ Chapter 7) and that a virtually error-free mechanism is necessary to ensure the faithful transfer of genetic traits from one generation to another.

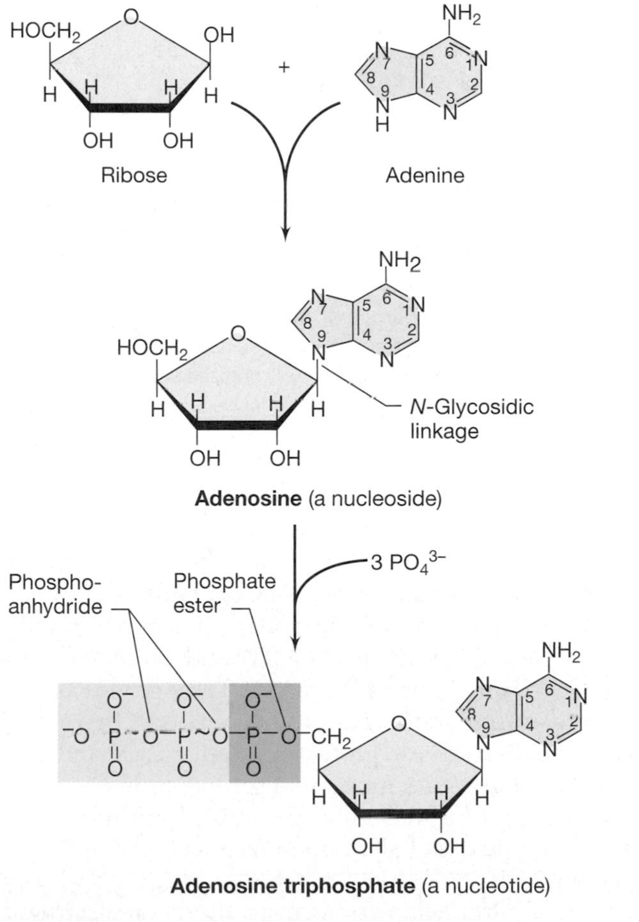

DNA

In cells, DNA is present in double-stranded form. Each chromosome contains two strands of DNA, each strand containing several million nucleotides linked by phosphodiester bonds. The strands themselves associate with one another by hydrogen bonds that form between the nucleotides of one strand and the nucleotides of the other. When positioned adjacent to one another, purine and pyrimidine bases can undergo hydrogen bonding (see Figure 3.2*c*). The most stable hydrogen bonding occurs when guanine (G) forms hydrogen bonds with cytosine (C), and adenine (A) forms hydrogen bonds with thymine (T) (see Figure 3.2*c*). Specific base pairing, A with T and C with C, means that the two strands of DNA are *complementary* in base sequence; wherever a G is found in one strand, a C is found in the other, and wherever a T is present in one strand, its complementary strand has an A (Figure 3.11*b*).

RNA

With a few exceptions, all ribonucleic acids are *single-stranded* molecules. However, RNA molecules usually fold back upon themselves in regions where complementary base pairing can occur to form folded structures. This pattern of folding observed in RNA is referred to as its *secondary structure* (Figure 3.11*c*).

RNA plays three crucial roles in the cell. **Messenger RNA** (mRNA) contains the genetic information of DNA in a single-stranded molecule *complementary* in base sequence to a portion of the base sequence of DNA. **Transfer RNA** (tRNA) molecules are the "adaptor" molecules in protein synthesis. Transfer RNAs adapt the genetic information from the language of nucleotides to the language of amino acids, the building blocks of proteins. **Ribosomal RNAs** (rRNAs), of which several distinct types are known, are important structural and catalytic components of the ribosome, the protein-synthesizing system of the cell. These various RNA molecules are discussed in detail in Chapters 7 and 11.

Figure 3.10 Components of the important nucleotide, adenosine triphosphate. The energy of hydrolysis of a phosphoanhydride bond (shown as squiggles) is greater than that of a phosphate ester and will have significance in Chapter 5 (∞ Section 5.8).

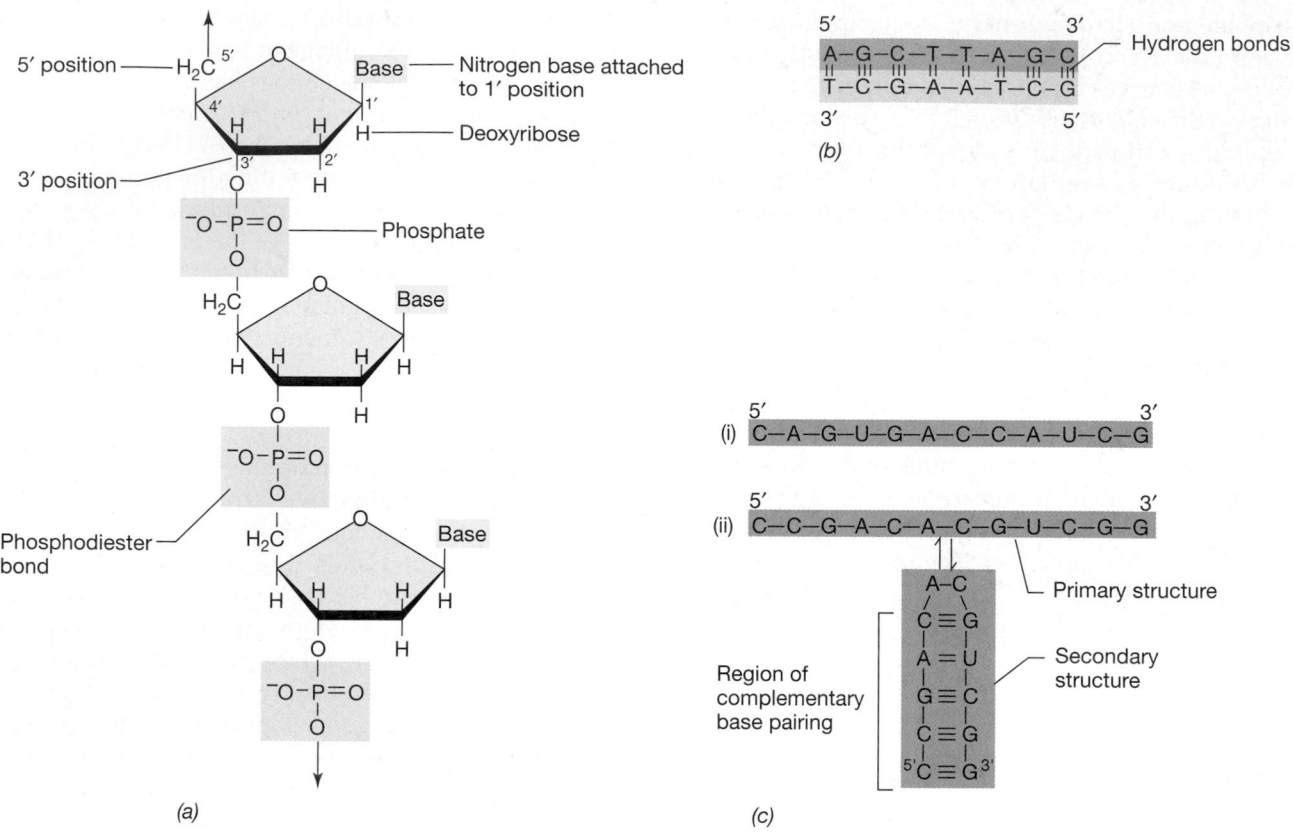

Figure 3.11 DNA and RNA. (a) Structure of part of a DNA chain. The nitrogen bases can be adenine, guanine, cytosine, or thymine. In RNA, an OH group is present on the 2′ carbon of the pentose sugar (see Figure 3.8), and uracil replaces thymine. (b) Simplified structure of DNA in which only the nitrogen bases are shown. Note how the two strands are complementary in base sequence (A=T; G≡C) and bonded by hydrogen bonds. Note also how the two strands of DNA are shown in two different shades of green; this convention is used throughout this book. (c) RNA: (i) A sequence showing only primary structure; (ii) A sequence that allows for secondary structure. In RNA, secondary structures form when opportunities for *intra*strand base pairing arise, as shown here. In certain very large RNA molecules, such as ribosomal RNA (⟨⟩ Sections 7.15 and 11.4), some parts of the molecule contain only primary structure while others contain both primary and secondary structure. This can lead to highly twisted or contorted molecules (⟨⟩ Figure 11.8c) whose biological function is critically dependent on their final three-dimensional shape. RNA is shown in orange throughout this book.

✓ 3.5 Concept Check

The informational content of a nucleic acid is determined by the sequence of nitrogen bases along the polynucleotide chain. Both RNA and DNA are informational macromolecules. RNA can fold into various configurations to obtain secondary structure.

✓ What is a *nucleotide*?

✓ How does a nucleo*side* differ from a nucleo*tide*?

✓ Distinguish between the *primary* and *secondary* structure of RNA.

3.6 Amino Acids and the Peptide Bond

Amino acids are the monomeric units of proteins. Most amino acids consist of only carbon, hydrogen, oxygen, and nitrogen, but two of the 21 common amino acids found in cells also contain sulfur and one contains selenium. All amino acids contain two important functional groups, a *carboxylic acid* group (—COOH) and an *amino* group (—NH$_2$) (Table 3.1 and Figure 3.12). These groups are functionally important because covalent bonds, between the carbon of the carboxyl group of one amino acid and the nitrogen of the amino group of a second amino acid (with elimination of a molecule of water), form the **peptide bond**, a type of covalent bond characteristic of proteins (Figure 3.13).

All amino acids conform to the general structure shown in Figure 3.12. Amino acids differ in the nature of the side group (abbreviated R in Figure 3.12) attached to the α-carbon. The α-carbon is the carbon atom *immediately adjacent* to the carboxylic acid group. The side chains vary considerably, from as simple as a hydrogen atom in the

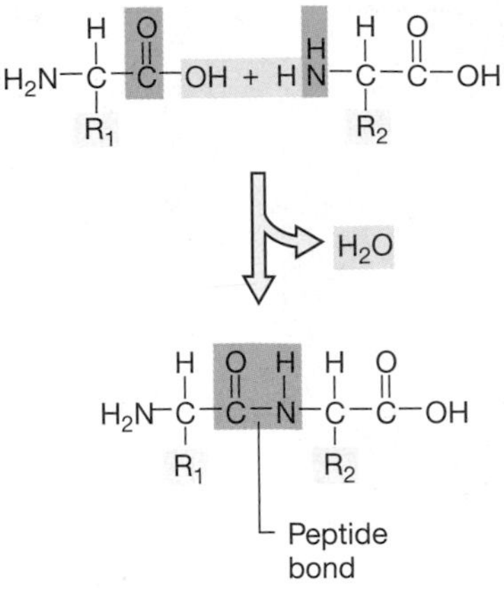
H₂N–C–C–OH + H N–C–C–OH
R₁
R₂
H₂O
Peptide bond

Figure 3.13 Peptide bond formation. R_1 and R_2 refer to the variable portion (side chain) of the amino acid (see Figure 3.12).

amino acid glycine, to aromatic ringed structures in amino acids such as phenylalanine (Figure 3.12). The chemical properties of an amino acid are to a major degree governed by the nature of the side chain, and thus amino acids that show similar chemical properties can be grouped into amino acid "families" as shown in Figure 3.12. For example, the side chain may itself contain a carboxylic acid group, such as in aspartic acid or glutamic acid, rendering the amino acid acidic. Alternatively, several amino acids contain hydrophobic side chains and are grouped together as nonpolar amino acids. The amino acid cysteine contains a sulfhydryl group (—SH), which can connect one chain of amino acids to another by *disulfide linkage* (R—S—S—R). In a nutshell, the diversity of chemically distinct amino acids makes it possible for cells to produce an enormous number of chemically distinct proteins with widely different biochemical properties.

Isomers

Two molecules may have the same molecular formula but exist in different structural forms. These related but not identical molecules are referred to as **isomers**. Louis Pasteur, the famous early microbiologist who crushed the theory of spontaneous generation (Section 1.5), began his scientific career as a chemist studying a class of isomers called *optical isomers*. The asymmetry that Pasteur first discovered in tartaric acid crystals (Figure 3.14) was the foundation for his later work that showed that living organisms could produce optically active molecules (such as amino acids and sugars) as well.

Many isomers of common sugars are found as constituents of the cell walls of *Bacteria* and *Archaea* (Section 4.8). Isomers that contain the same molecular and structural formulas, except that one is a "mirror

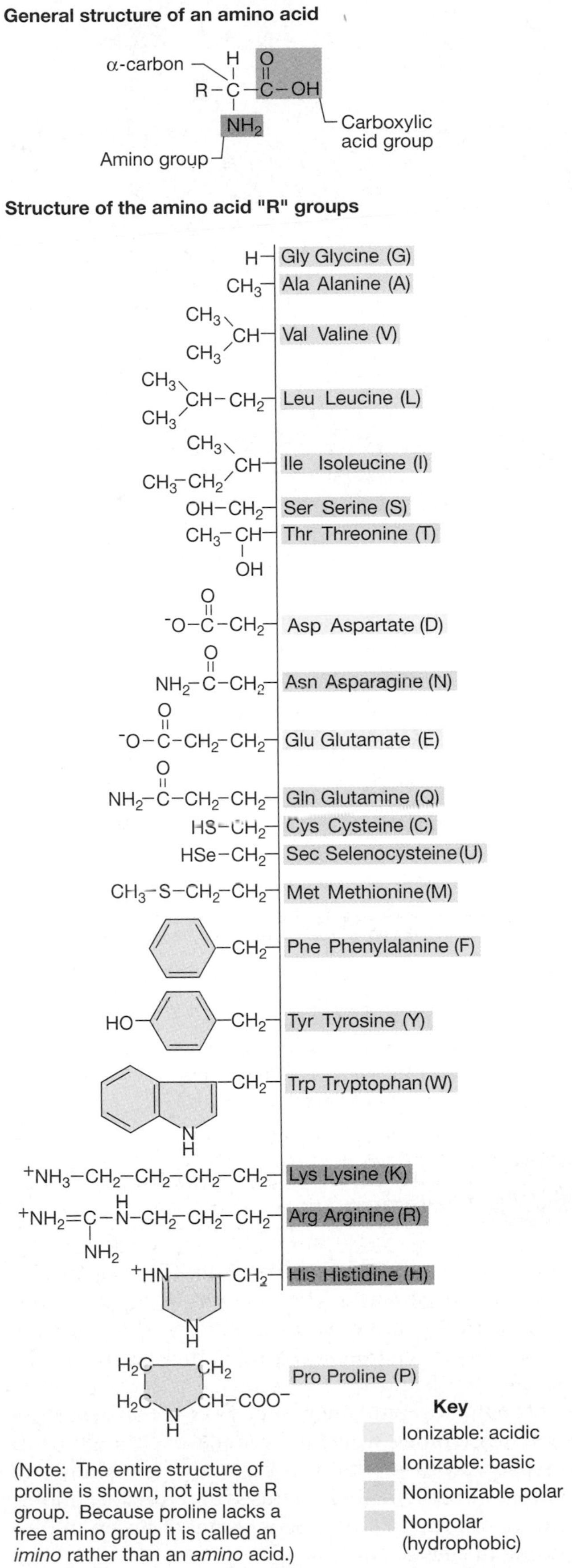
General structure of an amino acid
α-carbon
Carboxylic acid group
Amino group
Structure of the amino acid "R" groups
Gly Glycine (G)
Ala Alanine (A)
Val Valine (V)
Leu Leucine (L)
Ile Isoleucine (I)
Ser Serine (S)
Thr Threonine (T)
Asp Aspartate (D)
Asn Asparagine (N)
Glu Glutamate (E)
Gln Glutamine (Q)
Cys Cysteine (C)
Sec Selenocysteine (U)
Met Methionine (M)
Phe Phenylalanine (F)
Tyr Tyrosine (Y)
Trp Tryptophan (W)
Lys Lysine (K)
Arg Arginine (R)
His Histidine (H)
Pro Proline (P)
Key
Ionizable: acidic
Ionizable: basic
Nonionizable polar
Nonpolar (hydrophobic)
(Note: The entire structure of proline is shown, not just the R group. Because proline lacks a free amino group it is called an *imino* rather than an *amino* acid.)

Figure 3.12 Structure of the 21 common amino acids. The three-letter codes for the amino acids are to the left of the names, and the one-letter codes are in parentheses to the right of the names.

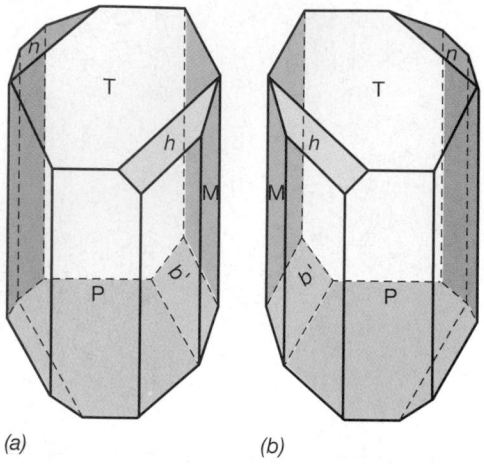

(a)   (b)

Figure 3.14 Louis Pasteur's drawings of tartaric acid ($C_4H_6O_6$) crystals, used to illustrate his famous paper on optical activity. (a) Left-handed crystal (L form). (b) Right-handed crystal (D form). Note that the two crystals are mirror images (that is, they are enantiomers). The letters on the faces of the crystals were Pasteur's way of labeling the mirror image faces of the two crystals. Color has been added here to show the mirror image faces more clearly.

image" of the other, just as the left hand is a mirror image of the right, are called **enantiomers** and have been given the designations D and L (Figure 3.15*b*). D-Sugars predominate in biological systems.

Like sugars, amino acids can exist as D or L enantiomers. However, in the case of protein, life has evolved to use the L-amino acid rather than the D (Figure 3.15*c*). Nevertheless, D-amino acids are found occasionally in nature, most commonly in the cell wall polymer peptidoglycan (👓 Section 4.8) and in certain peptide antibiotics (👓 Section 20.9). Prokaryotes are equipped to handle the conversions of D-amino acids to the L form or L-sugars to the D form by way of enzymes that specifically catalyze this transformation. Cells contain enzymes called **racemases** whose function is to convert the unusual form (L-sugar or D-amino acid) to the readily metabolizable form (D-sugar or L-amino acid).

✓ 3.6 Concept Check

Twenty-one common amino acids are found in cells and can bond to each other via the *peptide bond*. Mirror image (enantiomeric) forms of sugars and amino acids exist, but only one optical isomer of each is found in cell polysaccharides and proteins, respectively.

- ✓ Why can it be said that all amino acids are structurally similar yet different at the same time?
- ✓ Draw the structure of a dipeptide containing the amino acids alanine and tyrosine. Outline the peptide bond.
- ✓ What enantiomeric form of sugars and amino acids are commonly found in living organisms? Why doesn't the amino acid *glycine* have different enantiomers?

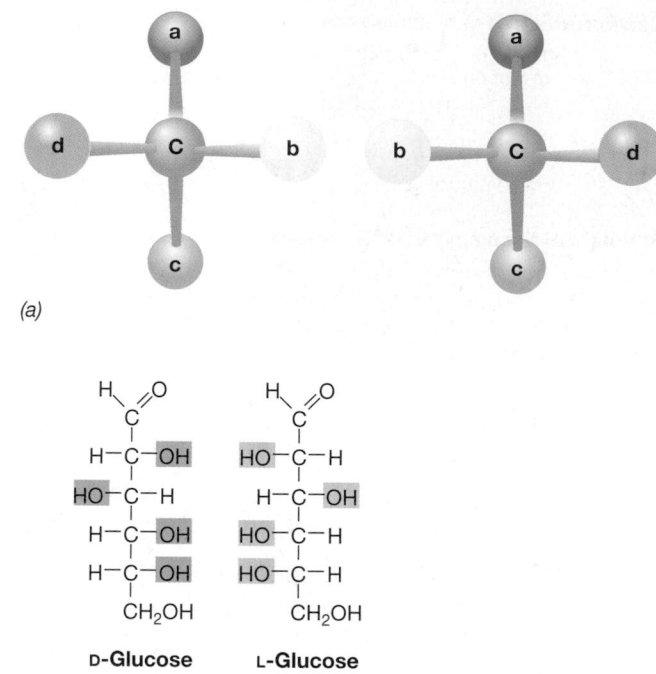

(a)

(b)

D-Glucose   **L-Glucose**

(c)

Planar
projection

Three-dimensional
projection*

L-Alanine   **D-Alanine**

* In the three-dimensional projection the arrow should be understood as coming toward the viewer whereas the dashed line indicates a plane away from the viewer.

Figure 3.15 Isomers. (a) Ball-and-stick model showing mirror images. (b) Enantiomers of glucose. (c) Enantiomers of the amino acid alanine. Note that no matter how the three-dimensional views are rotated, the L and D forms can never be superimposed.

3.7 Proteins: Primary and Secondary Structure

Proteins play key roles in cell function. Two major classes of proteins are *catalytic* proteins (enzymes) and *structural* proteins. Enzymes are the catalysts for the wide variety of chemical reactions that occur in cells (👓 Chapters 5 and 17). By contrast, structural proteins are those that become integral parts of the structures of cells in membranes, walls, and cytoplasmic components. In essence, a cell is what it is because of the kinds and amounts of proteins it contains. Therefore, an understanding of protein structure is essential for an understanding of cell function.

Primary Structure

Proteins are polymers of amino acids covalently bonded by peptide bonds (Figure 3.13). Two amino acids bonded together constitute a *dipeptide,* three amino acids a *tripeptide,* and so on. When *many* amino acids are covalently linked via peptide bonds, they form a **polypeptide,** and proteins consist of one or more polypeptides. The number of amino acids in a protein varies from one protein to another. Proteins with as few as 15 and as many as 10,000 amino acids are known. Since proteins may vary in their composition, sequence, and number of amino acids, it is easy to see that enormous variation in protein structure (and thus function) is possible.

The linear array of amino acids in a polypeptide is referred to as its **primary structure**. The primary structure of a polypeptide is very important, because a given primary structure is consistent with only certain types of folding. It is the final, folded polypeptide that takes on biological activity. Thus, one can say that the primary structure of a polypeptide dictates to a major degree it's final biological properties.

Secondary Structure

The juxtaposition of the R groups on individual amino acids in a polypeptide forces the molecule to twist and fold in a specific way. This leads to formation of the **secondary structure** of the protein (Figure 3.16). Hydrogen bonds, the weak noncovalent linkages discussed earlier (see Section 3.1), play important roles in the type of secondary structure that a protein attains. One type of structure for polypeptides is the *α-helix*. To envision a protein helix, imagine a linear polypeptide wound around a cylinder (Figure 3.16a). Under these conditions, oxygen and nitrogen atoms from different amino acids become positioned close enough in the twisted structure to allow for hydrogen bonding to occur. This opportunity for hydrogen bonding (and the inherent stability associated with it) helps direct many polypeptides to take on an α-helix secondary structure (Figure 3.16a).

Other polypeptides conform to a different type of secondary structure called a *β-sheet*. In the β-sheet, the chain of amino acids in the polypeptide folds back and forth upon itself instead of forming a helix; this type of folding exposes hydrogen atoms that can undergo hydrogen bonding (Figure 3.16b).

Many polypeptides contain both regions of α-helix and regions of β-sheet secondary structure, the type of folding being determined by the available opportunities for hydrogen bonding and hydrophobic interactions. Since β-sheet secondary structure generally yields a rather rigid structure, whereas α-helical secondary structures are usually more flexible, the secondary structure of a given polypeptide to a large degree dictates the functional role for the protein in the cell. Many polypeptides fold into two or more segments, each displaying α-helix

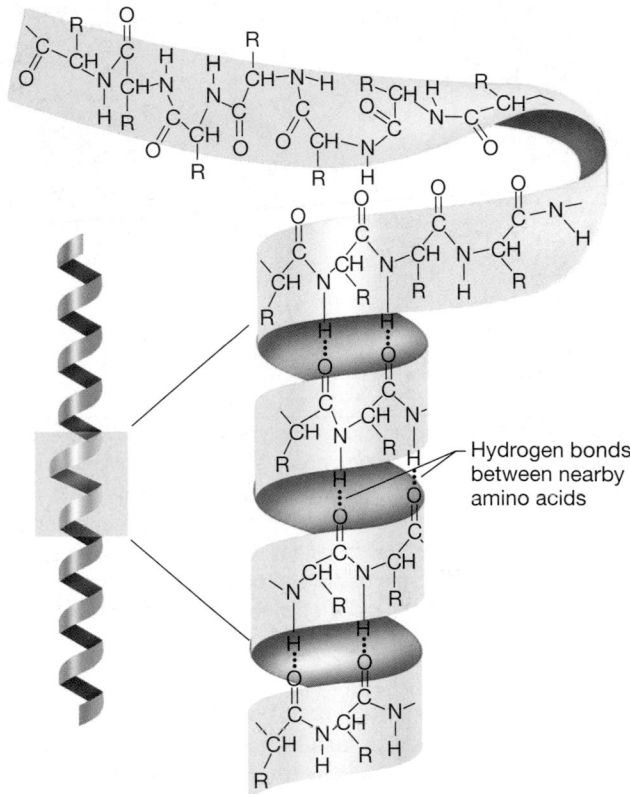

(a)

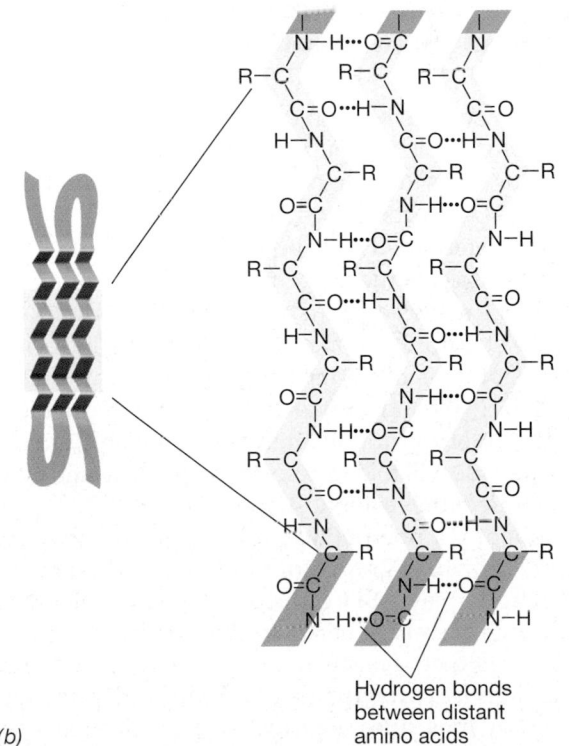

(b)

Hydrogen bonds between nearby amino acids

Hydrogen bonds between distant amino acids

Figure 3.16 Secondary structure of polypeptides. (a) α–Helix secondary structure. Note that hydrogen bonding does not involve the R groups but instead occurs between atoms in the peptide bonds. (b) β–Sheet secondary structure.

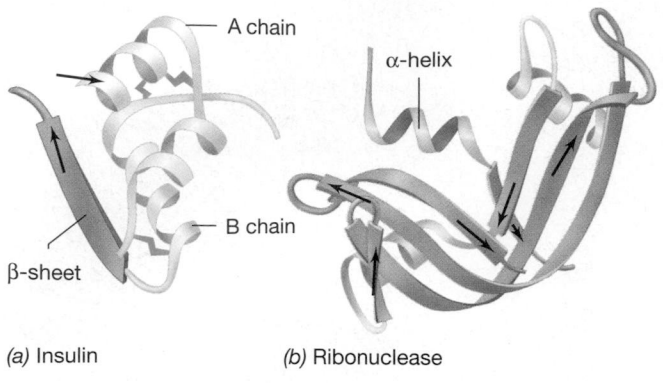

(a) Insulin   *(b)* Ribonuclease

Figure 3.17 Tertiary structure of polypeptides showing where regions of α-helix or β-sheet secondary structure might be located. (a) Insulin, a protein containing two polypeptide chains (Section 31.9); note how the B chain contains both α-helix and β-sheet secondary structure and how disulfide linkages (shown in blue) help in dictating folding patterns (tertiary structure). (b) Ribonuclease, a large protein with several regions of α-helix and β-sheet secondary structure.

or β-sheet secondary structure (see Figure 3.17). These segments, called *domains*, are regions of the polypeptide that have specific functions in the final protein molecule.

3.8 Proteins: Higher Order Structure and Denaturation

Once a polypeptide has achieved secondary structure it continues to fold to form an even more stable molecule. This folding leads to formation of the **tertiary structure** of the protein. Like secondary structure, tertiary structure is ultimately determined by primary structure, but tertiary structure is also governed to some extent by the secondary structure of the molecule. As a result of the formation of secondary structure, the side chain of each amino acid in the polypeptide is positioned in a specific way. If additional hydrogen bonds, covalent bonds, hydrophobic interactions, or other atomic interactions are able to form, the polypeptide will fold to accommodate them and attain a unique three-dimensional shape (Figure 3.17).

Frequently a polypeptide folds in such a way that adjacent sulfhydryl groups of cysteine residues are exposed. These free — SH groups can join covalently to form a disulfide bridge between the two amino acids. If the two cysteine residues are located in different polypeptides in a protein, the disulfide bond physically links the two molecules (Figure 3.17a). In addition, a single polypeptide can spontaneously fold and bond to itself covalently if two cysteine residues form a disulfide linkage within the molecule. The tertiary folding of the polypeptide ultimately forms exposed regions or grooves in the molecule (Figures 3.17 and 3.18), which

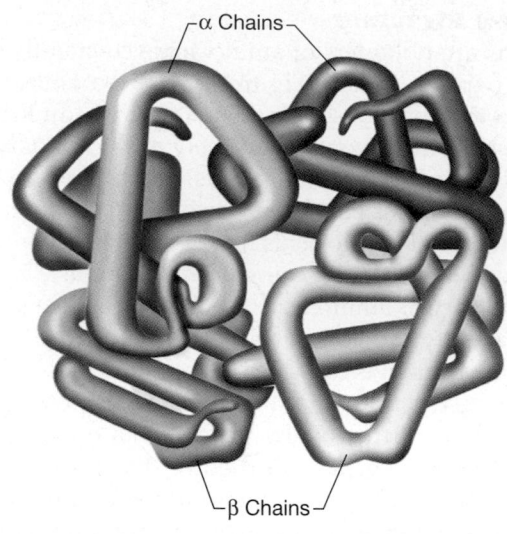

Figure 3.18 Quaternary structure of hemoglobin. There are two *kinds* of polypeptide in hemoglobin, α chains (shown in blue and red) and β chains (shown in orange and yellow), but a total of four polypeptides in the final protein molecule (α and β refer here to chain names, not to polypeptide structures). Separate colors are used to distinguish the four distinct chains.

are often important for binding other molecules (for example, in the binding of a substrate to an enzyme) (Section 5.5).

If a protein consists of two or more polypeptides, and many proteins do, the number and type of polypeptides that form the final protein molecule is referred to as its **quaternary structure** (Figure 3.18). In proteins showing quaternary structure, each subunit contains primary, secondary, and tertiary structure. Some proteins contain multiple copies of a single subunit (a protein containing two identical subunits would be referred to as a *homodimer*), while others contain nonidentical subunits, each present in one or more copies. The subunits are held together either by noncovalent interactions (hydrogen bonding, van der Waals forces, or hydrophobic interactions) or by covalent linkages, generally intersubunit disulfide bonds.

Denaturation

When proteins are exposed to extremes of heat or pH, or to certain chemicals or metals that affect their folding, they are said to undergo **denaturation** (Figure 3.19). In general, the biological properties of a protein are lost when it is denatured; however, peptide bonds (Figure 3.13) are unaffected. Denaturation causes the polypeptide chain to unfold, destroying the higher order structure of the molecule. The denatured polypeptide retains its primary structure because it is held together by covalent (peptide) bonds. Depending on the severity of the denaturing conditions, refolding of the polypeptide may occur after removal of the denaturant

(Figure 3.19). However, the fact that denaturation of a protein is usually associated with loss of its biological activity shows that biological activity is not inherent in the primary structure of a protein but instead is a result of the unique folding of the molecule (as ultimately directed by primary structure). Thus, folding of a polypeptide confers on it a unique shape that is compatible with a *specific* biological function.

Denaturation of proteins is of more than just academic interest, since this process is a major means by which microorganisms can be destroyed. We will see in Chapter 20 how several chemicals kill microorganisms because of their ability to denature proteins. For example, alcohols such as phenol and ethanol are effective disinfectants because they readily penetrate cells and irreversibly denature their proteins. Such chemical agents are extremely useful for disinfecting inanimate objects, such as surfaces, and have enormous practical value in household, hospital, and industrial disinfectant applications.

The Road Ahead

Now that we have a firmer grasp of the chemistry of life, we proceed to Chapter 4, which deals with the structural relationships of cells. In this chapter we will see how the different macromolecules reviewed here come together to form major structures of the cell, including the membrane, wall, flagellum, and so on. From there we will consider the basic metabolic properties of cells in Chapter 5. Although perhaps not obvious at this point, the material in Chapters 4 and 5 is actually quite related, because metabolism, the machine function of a cell (∞ Section 1.2), drives the synthesis and assembly of macromolecules into cell structures. The culmination of these processes is cell growth (Chapter 6).

As we travel through this book, it may be wise to return to Chapter 3 from time to time to reinforce the principles surrounding the chemistry of life. Although microorganisms have evolved a nearly limitless chemical diversity of macromolecules, this diversity almost always boils down to variations on the structural themes of the four classes of macromolecules discussed in this chapter. And as any successful microbiologist knows, a conceptual understanding of the chemistry and properties of proteins, lipids, nucleic acids, and polysaccharides pays big dividends in grasping the basic principles of microbiology.

✓ 3.7 and 3.8 Concept Check

The primary structure of a protein is determined by its amino acid sequence, but it is the folding (higher order structure) of the polypeptide that determines how the protein functions in the cell.

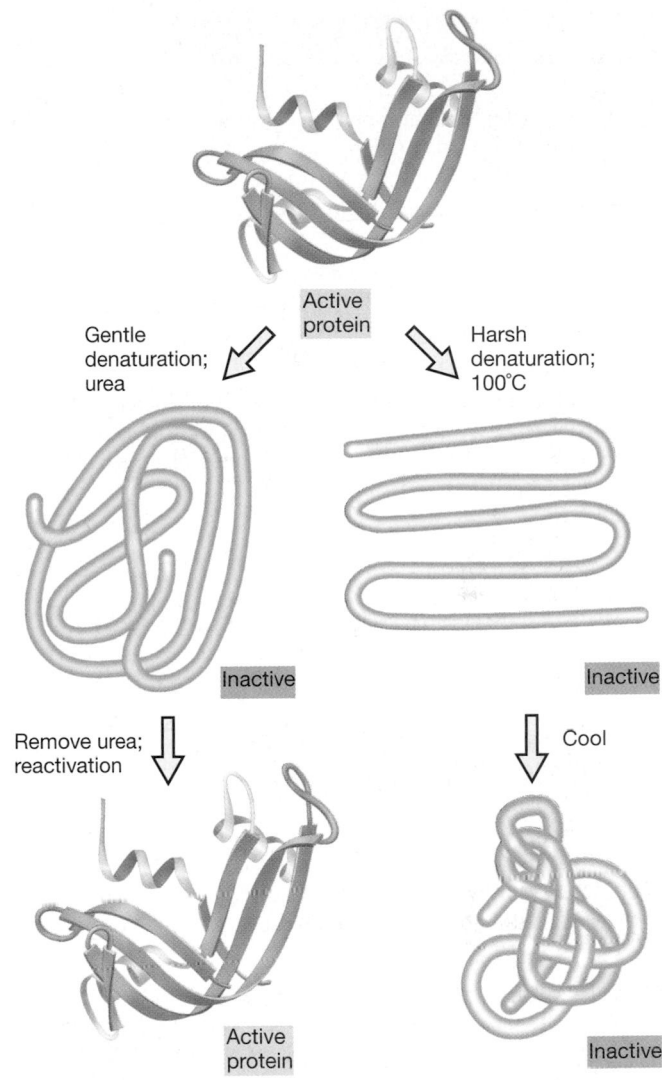

Figure 3.19 Denaturation of the protein ribonuclease (whose structure was discussed in Figure 3.17*b*). Note how harsh denaturation yields a permanently destroyed molecule (from the standpoint of biological function) because of improper folding.

✓ Define the terms *primary, secondary,* and *tertiary* with respect to protein structure.

✓ How does a *polypeptide* differ from a *protein*?

✓ What secondary structural features tend to make β-sheet proteins more rigid than α-helices?

✓ What would be the quaternary structure of a *homodimeric* protein? A *heterodimeric* protein?

✓ Describe the structural and biological effects of the denaturation of a protein. Of what *practical* value is knowledge of protein denaturation?

Review Questions

1. Which are the major elements found in living organisms? Why are oxygen and hydrogen particularly abundant in living organisms?

2. Define the word *molecule*. How many atoms are in a molecule of hydrogen gas? In a molecule of glucose?

3. Refer to the structure of the nitrogen base *cytosine* shown in Figure 3.1. Draw this structure and then label the positions of all single bonds and double bonds in the cytosine molecule.

4. Compare and contrast the words *monomer* and *polymer*. Give three examples of biologically important polymers and list the monomers of which they are composed.

5. List the components that would make up a simple lipid. How does a triglyceride differ from a complex lipid?

6. Examine the structures of the triglyceride and of phosphatidyl ethanolamine shown in Figure 3.7. How might the substitution of phosphate and etha- nolamine for a fatty acid alter the chemical properties of the lipid?

7. RNA and DNA are similar types of macromolecules but show distinct differences as well. List three ways in which RNA differs chemically or physically from DNA. What is the cellular function of DNA and RNA?

8. Why are *amino acids* so named? Write a general structure for an amino acid. What is the importance of the R group to final protein structure? Why does the amino acid *cysteine* have special significance for protein structure?

9. Chemically, what type of reaction between two amino acids leads to formation of the peptide bond? (You may wish to review Figure 3.13 before answering.)

10. Draw the peptide bond. Now redraw this structure in atomic form (see Figure 3.1) showing the arrangement of electrons within the atoms of the peptide bond.

Application Questions

1. Observe the following nucleotide sequences of RNA: (a) GUCAAAGAC, (b) ACGAUAACC. Can either of these RNA molecules have secondary structure? If so, draw the potential secondary structure(s).

2. A few soluble (cytoplasmic) proteins contain a high content of hydrophobic amino acids. How would you predict these proteins would fold as to their tertiary structure and why?

3. Cells of the genus *Halobacterium,* an archaeon that lives in very salty environments, contain over 5 molar (M) potassium (K^+). Because of this high K^+ content, many cytoplasmic proteins of *Halobacterium* cells are enriched in two specific amino acids that are present in much higher proportions in *Halobacterium* proteins than in functionally similar proteins from *Escherichia coli* (which has only very low levels of K^+ in its cytoplasm). Which amino acids are enriched in *Halobacterium* proteins and why? *Hint:* Which amino acids could best neutralize the *positive* charges due to K^+?

4. It is often the case that proteins that show α-helix secondary structure are more flexible than proteins showing β-sheet secondary structure. Discuss why this could be the case.

5. When an egg is placed in a beaker of boiling water, changes in the egg occur almost immediately. Describe what happens and why the contents of a boiled egg look so different from the contents of a fresh egg.

6. In light of your answer to the preceding question, explain how it can be that certain prokaryotes, called *hyperthermophiles,* thrive (and indeed grow optimally) in boiling hot springs. How must the proteins of hyperthermophiles differ from proteins in the egg?

M icroorganisms contain a number of structures that allow them to coexist with other microorganisms and with higher organisms. One of these structures is the flagellum, a propeller-like device that confers motility on a cell. Shown here is a colony of the highly motile phototrophic bacterium *Rhodospirillum centenum*. This organism can sense light (in this case, coming in from the right), and the actively swimming cells drag the entire colony towards the light, a response called *phototaxis*. Phototaxis and related taxes are examples of behavioral responses and clearly show that even organisms as "simple" as bacteria can respond to stimuli in their environment.

CELL STRUCTURE/FUNCTION 4

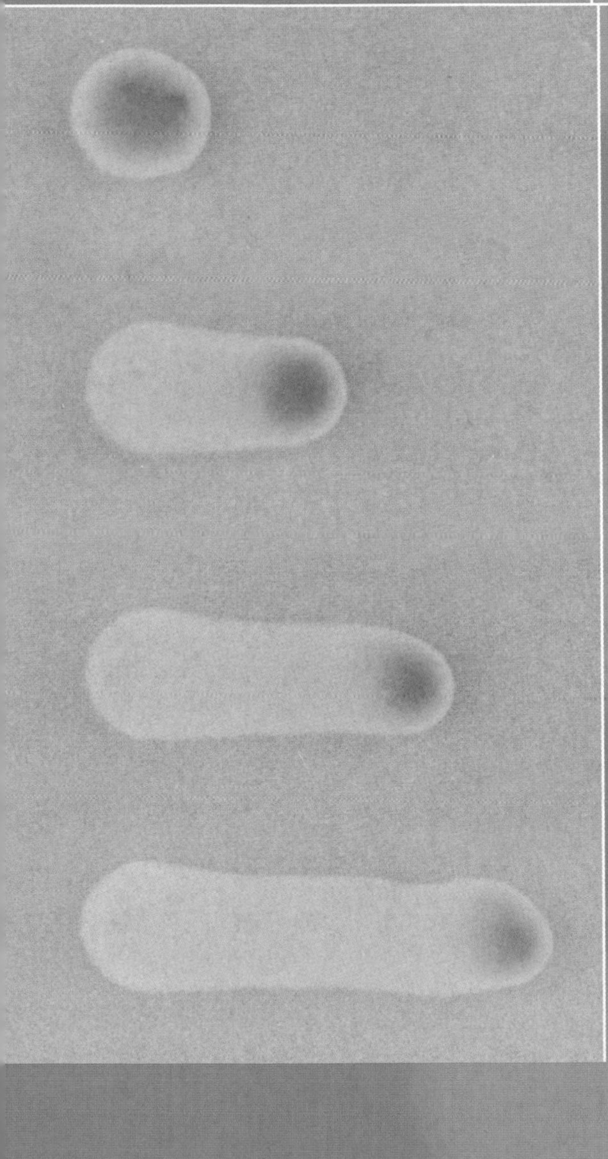

Working Glossary

ABC (ATP-Binding Cassette) transporter a membrane transport system consisting of three proteins, one of which hydrolyzes ATP as an energy source to drive the transport event, one of which binds the substrate on the outside of the cell, and one of which functions as the transport channel through the membrane

Chemotaxis movement of an organism toward *(positive)* or away from *(negative)* a chemical gradient

Cytoplasmic membrane the permeability barrier of the cell, separating the cytoplasm from the environment; consists of lipid and protein

Endospore a highly heat-resistant, thick-walled, differentiated cell produced by certain gram-positive *Bacteria*

Flagellum a long, thin cellular appendage capable of rotation in prokaryotic cells and responsible for swimming motility

Gas vesicles gas-filled cytoplasmic structures bounded by protein and conferring buoyancy on cells

Gram-negative a prokaryotic cell whose cell wall contains relatively little peptidoglycan but contains an outer membrane composed of lipopolysaccharide, lipoprotein, and other complex macromolecules

Gram-positive a prokaryotic cell whose cell wall consists chiefly of peptidoglycan and lacks the outer membrane of gram-negative cells

Group translocation an energy-dependent transport process in which the substance transported is chemically modified during the transport process

Lipopolysaccharide (LPS) lipid in combination with polysaccharide and protein forming the major portion of the cell wall in gram-negative *Bacteria*

Magnetosomes particles of magnetite (Fe_3O_4) organized into nonunit membrane-enclosed structures in the cytoplasm of magnetotactic *Bacteria*

Peptidoglycan a polysaccharide composed of alternating repeats of acetylglucosamine and acetylmuramic acid with the latter in adjacent layers cross-linked by short peptides

Periplasm a gel-like region between the outer surface of the cytoplasmic membrane and the inner surface of the lipopolysaccharide layer of gram-negative *Bacteria*

Peritrichous in reference to flagellation pattern; flagella located in many places around the surface of the cell

Phototaxis movement of an organism toward light

Polar in reference to flagellation, having flagella emanating from one or both poles of the cell

Poly-β-hydroxybutyrate (PHB) a common storage material of prokaryotic cells consisting of a polymer of β-hydroxybutyrate or another β-alkanoic acid

Protoplast an osmotically protected cell whose cell wall has been removed

Sterols hydrophobic multiringed structures that strengthen the cytoplasmic membrane of eukaryotic cells and a few prokaryotes

I MICROSCOPY AND CELL MORPHOLOGY

In this chapter we describe the basic structure and function of the components that make up microbial cells, in particular, prokaryotic cells (*Bacteria* and *Archaea*). We reserve Chapter 14 for a detailed discussion of the structure and diversity of eukaryotic cells. This chapter is organized into four sections dealing with microscopy, cell membranes and walls, locomotion, and cell surface structures and inclusions, respectively.

Like houses, cells are built by connecting simple building blocks in various ways to create more complex structures. Monomers (such as amino acids and nucleotides) are used to construct macromolecules (such as proteins and nucleic acids), and macromolecules form into complexes to yield structures with defined functions, such as ribosomes, membranes, cell walls, and the like. Interestingly, however, despite great diversity in the *chemistry* of the macromolecules that make up different cells, we will see that all cells face similar structural problems and from an architectural standpoint, solve them in common ways.

We begin with a discussion of the microscope. Historically it was the microscope that first revealed the secrets of cell structure, and today it remains a powerful tool in cell biology.

4.1 Light Microscopy

Microscopic examination of microorganisms makes use of either the **light microscope** or the **electron microscope**. In general, light microscopes are used to look at intact cells, whereas electron microscopes are used to look at internal structure or details of cell surfaces.

All microscopes employ the principle that specific lenses magnify the image of a cell such that details of its structure are more apparent. In addition to magnification, however, is *resolution*, the ability to distinguish two adjacent points as separate. Although magnification can be increased virtually without limit, resolution cannot; resolution is dictated by the physical properties of light. It is thus resolution and not magnification that ultimately defines the limits of what we are able to see with a microscope. We begin our discussion with the light microscope, for which the limits of resolution are about 0.2 μm [0.2 micrometer or 200 nanometers (nm)], and then proceed to describe the electron microscope, for which resolution is improved over that of the light microscope by about 1000-fold.

The Compound Light Microscope

The **light microscope** has been of crucial importance for the development of microbiology as a science and remains a basic tool of routine microbiological research. Several types of light microscopes are commonly used in microbiology: *bright-field, phase contrast, dark-field,* and *fluorescence.* The **bright-field microscope** is most commonly used in biology and microbiology courses and consists of two series of lenses (objective lens and ocular lens), which function together to resolve the image (Figure 4.1). With this microscope, specimens are made visible because of the differences in contrast that exist between them and the surrounding medium. Contrast differences arise because cells absorb or scatter light in varying degrees. Many bacterial cells are difficult to see well with the bright-field microscope because of their lack of contrast with the surrounding medium. Pigmented organisms are an exception, because the color of the organism adds contrast, thus improving visualization of the cells (Figure 4.2).

Magnification and Resolution

The total magnification of a compound microscope is the *product* of the magnification of its objective and ocular lenses (Figure 4.1*b*). Magnifications of about 1500× are near the upper limit obtainable with a compound light microscope. This limit is set because of a property of a lens called **resolution**. Resolving power is a function of the wavelength of light used and an innate property of the objective lens known as its *numerical aperture* (a measure of light-gathering ability). In general, there is a correspondence between the magnification of a lens and its numerical aperture: Lenses with higher magnification usually have higher numerical apertures (the numerical aperture of a lens is stamped alongside the magnification). The diameter of the smallest resolvable object is equal to 0.5λ/numerical aperture, where λ is the wavelength of light used. Based on this formula, resolution is greatest when blue light is used to illuminate a specimen and the objective that is used has a very high numerical aperture.

As mentioned, the highest resolution possible in a compound light microscope is about 0.2 μm. This means that two objects closer together than 0.2 μm are not resolvable as distinct and separate. Most microscopes used in microbiology have oculars that magnify 10–15× and

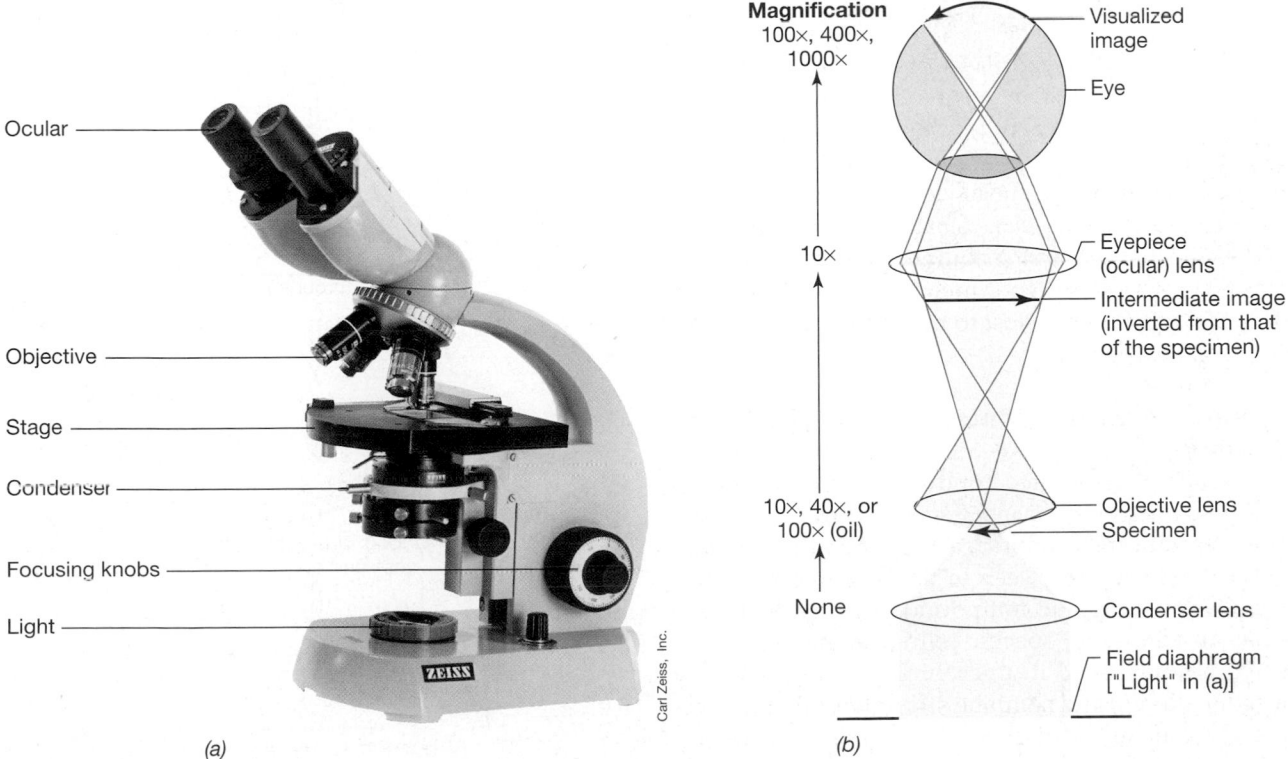

(a)

(b)

Figure 4.1 (a) A compound light microscope. Various key parts of the microscope are labeled. (b) Path of light through a compound light microscope. Besides 10×, eyepieces (oculars) are available in 15–30×.

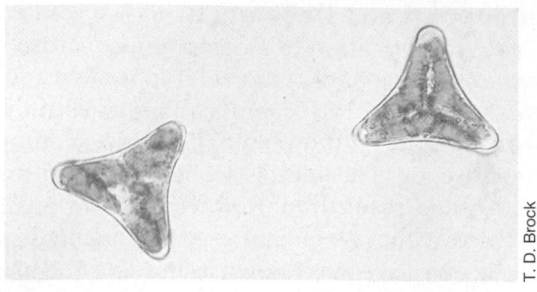

(a)

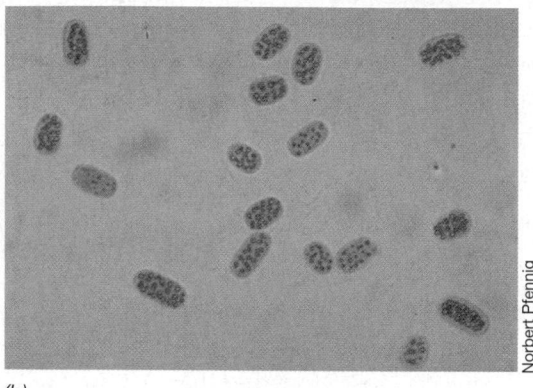

(b)

T. D. Brock

Norbert Pfennig

Figure 4.2 Photomicrographs of pigmented microorganisms by bright-field microscopy. (a) A green alga (eukaryote). (b) A purple phototrophic bacterium (prokaryote). The algal cells are about 15 μm in diameter, and the bacterial cells are about 5 μm in diameter.

objectives of 10–100× (Figure 4.1*b*); at 1000× , objects 0.2 μm in diameter can just be resolved. With the 100× objective, and with certain other objectives of very high numerical aperture, a high-grade optical oil is used between the specimen and the objective. Lenses on which oil is used are called *oil-immersion lenses.* Immersion oil increases the light-gathering ability of a lens by allowing rays emerging from the specimen at higher angles (and that would otherwise be lost to the objective lens) to be collected and viewed.

Staining: Increasing Contrast for Bright-Field Microscopy

As previously mentioned, one of the limitations to bright-field microscopy is insufficient contrast. Dyes can be used to stain cells and increase their contrast so that they can be more easily seen in the bright-field microscope. Dyes are organic compounds, and each class of dye has an affinity for specific cellular materials. Many dyes commonly used in microbiology are positively charged (cationic) and combine strongly with negatively charged cellular constituents such as nucleic acids and acidic polysaccharides. Examples of cationic dyes include *methylene blue, crystal violet,* and *safranin.* Because cell surfaces are also generally negatively charged, these

dyes combine with structures on the surfaces of cells and hence are excellent general-purpose stains.

The simplest staining procedures are done with dried preparations (Figure 4.3). A slide containing a dried suspension of microorganisms is flooded for a minute or two with a dilute solution of a dye, rinsed several times in water, and blotted dry. It is usual to observe dried stained preparations of bacteria with a high-power (oil-immersion) lens (Figure 4.3).

Differential stains are so named because they are used in procedures that do not stain all kinds of cells equally. An important differential staining procedure widely used in bacteriology is the *Gram stain* (Figure 4.4*a*). On the basis of their reaction to the Gram stain, bacteria can be divided into two major groups: **gram-positive** and **gram-negative**. After Gram staining, gram-positive bacteria appear purple and gram-negative bacteria appear red (Figure 4.4*b*). This difference in reaction to the Gram

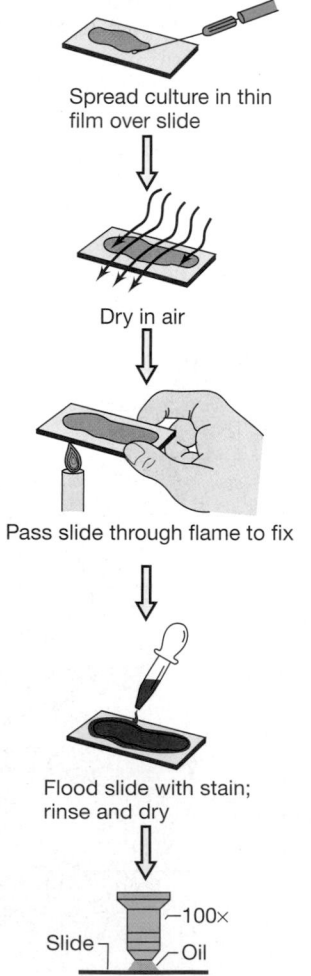

Spread culture in thin film over slide

Dry in air

Pass slide through flame to fix

Flood slide with stain; rinse and dry

Slide — ⌐100×
— Oil

Place drop of oil on slide; examine with 100× objective

Figure 4.3 Staining cells for microscopic observation.

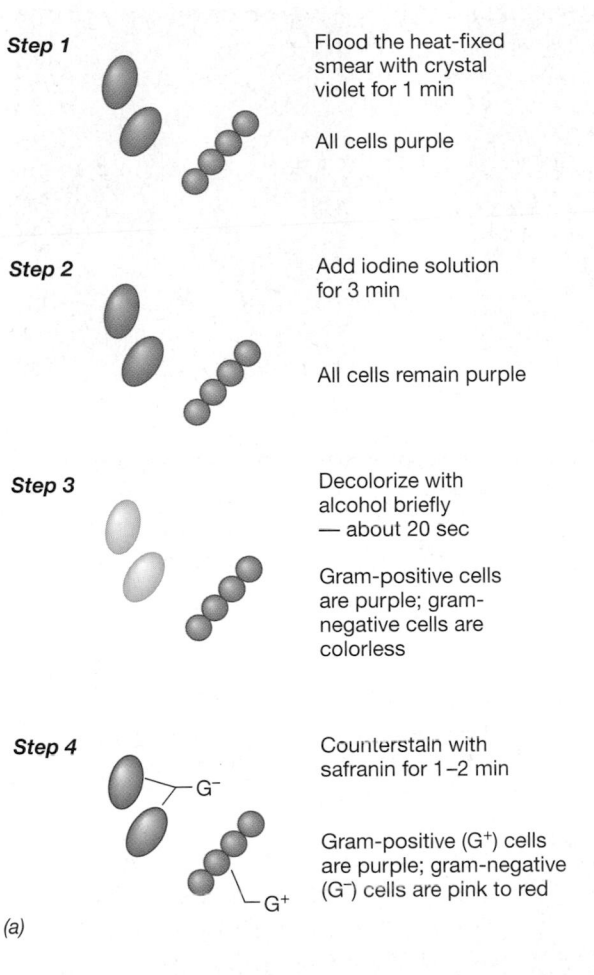

Step 1 — Flood the heat-fixed smear with crystal violet for 1 min

All cells purple

Step 2 — Add iodine solution for 3 min

All cells remain purple

Step 3 — Decolorize with alcohol briefly — about 20 sec

Gram-positive cells are purple; gram-negative cells are colorless

Step 4 — Counterstain with safranin for 1–2 min

Gram-positive (G⁺) cells are purple; gram-negative (G⁻) cells are pink to red

(a)

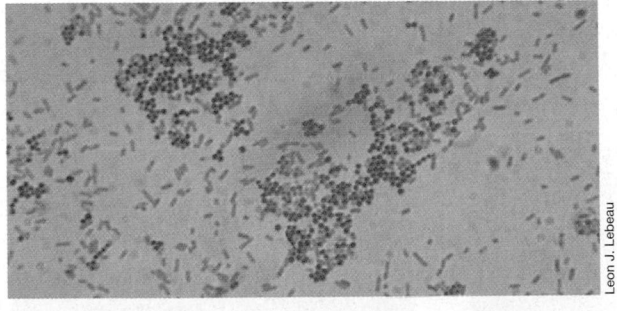

(b)

Leon J. Lebeau

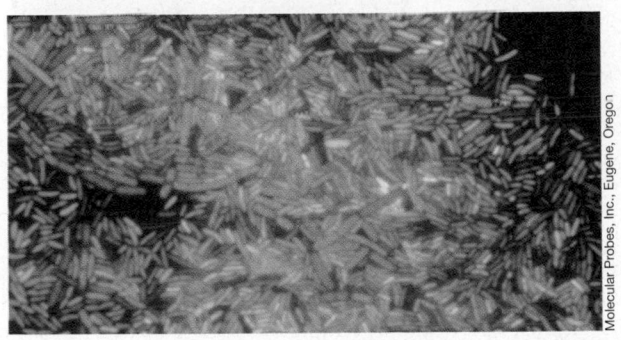

(c)

Molecular Probes, Inc., Eugene, Oregon

Figure 4.4 The Gram stain. (a) Steps in the Gram stain procedure. (b) Photomicrograph of Gram-stained *Bacteria* that are gram-positive (blue-purple) and gram-negative (pink-red). The species are *Staphylococcus aureus* and *Escherichia coli*, respectively. (c) Photomicrograph of cells of *Pseudomonas aeruginosa* (gram-negative, green) and *Bacillus cereus* (gram-positive, orange) stained with the one step fluorescent staining method **LIVE *Bac* Light**™. This method allows for differentiating gram-positive from gram-negative cells in a single staining step.

stain arises because of differences in the cell wall structure of gram-positive and gram-negative cells (as discussed later in this chapter), which leads to ethanol decolorizing gram-negative, but not gram-positive cells (Figure 4.4). The Gram stain is one of the most useful staining procedures in the bacteriological laboratory; it is almost essential in identifying an unknown bacterium to determine first whether it is gram-positive or gram-negative. If a fluorescent microscope is available (see later discussion of fluorescence microscopy), the Gram stain can be reduced to a one-step procedure where gram-positive and gram-negative cells fluoresce different colors (Figure 4.4*c*).

Phase Contrast, Dark-Field, and Fluorescence Microscopy

The **phase contrast microscope** was developed to improve contrast differences between cells and the surrounding medium, making it possible to see cells without staining them (Figure 4.5). Phase contrast microscopy is based on the principle that cells differ in refractive index from their surroundings and hence bend some of the light rays that pass through them. Light passing through a specimen of refractive index different from that of the surrounding medium is retarded. This effect is amplified by a special ring in the objective lens of a phase contrast microscope, leading to the formation of a dark image on a light background (Figure 4.5*b*). The phase contrast microscope is widely employed in research applications because it can be used to observe wet-mount (living) preparations. Staining, on the other hand, although a widely used procedure in light microscopy, kills cells and can distort their features.

The **dark-field microscope** is a light microscope in which the lighting system has been modified to reach the specimen from the sides only. The only light reaching the lens is light scattered by the specimen, and thus the specimen appears light on a dark background (Figure 4.5*c*). Resolution by dark-field microscopy is quite high, and objects can frequently be resolved by dark-field that are not resolvable in bright-field or phase contrast microscopes. Dark-field microscopy is also an excellent way to observe the motility of microorganisms, as bundles of flagella are often resolvable with this technique (see Figure 4.40*a*).

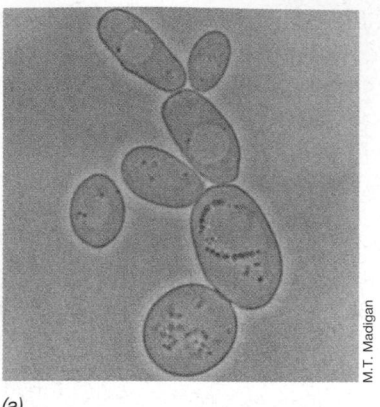
(a)

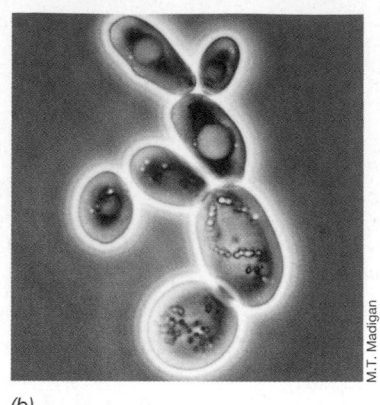
(b)

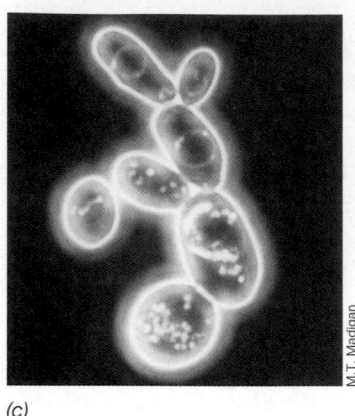
(c)

Figure 4.5 Photomicrographs of the same field of cells of the baker's yeast *Saccharomyces cerevisiae*, taken by different types of light microscopy. (a) Bright-field. (b) Phase contrast. (c) Dark-field. Cells average 8–10 μm in diameter.

The **fluorescence microscope** is used to visualize specimens that *fluoresce,* that is, emit light of one color when light of another color shines upon them (Figure 4.6). Fluorescence occurs either because of the presence within cells of naturally fluorescent substances such as chlorophyll or other fluorescing components (*autofluorescence*) (see Figure 4.6*a*, *b*) or because the cells have been treated with a fluorescent dye (Figures 4.4*c* and 4.6*c*). Fluorescence microscopy is widely used in clinical diagnostic microbiology and also in microbial ecology (∞ Chapters 19, 20, and 24).

✓ 4.1 Concept Check

Microscopes are essential for microbiological studies. Various types of light microscopes exist, including bright-field, phase contrast, and fluorescence microscopes. In bright-field microscopy, stains are necessary to increase contrast.

✓ Define the term *resolution.*

✓ What is the upper limit of magnification for a light microscope?

✓ What light microscopic techniques can sometimes improve resolution?

✓ What color would a gram-negative bacterium be after Gram staining by the conventional method?

4.2 Three-Dimensional Imaging: Interference Contrast, Atomic Force, and Confocal Scanning Laser Microscopy

One of the drawbacks to the forms of light microscopy just considered is that the images obtained are essentially two dimensional. How can this limitation be over-

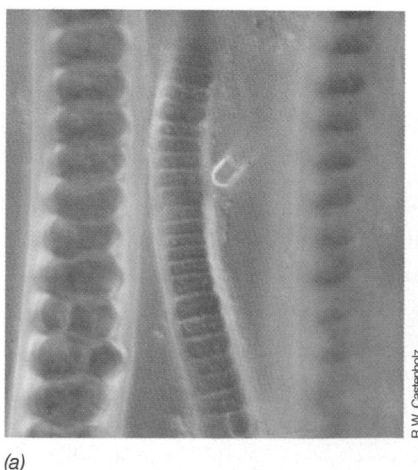
(a)

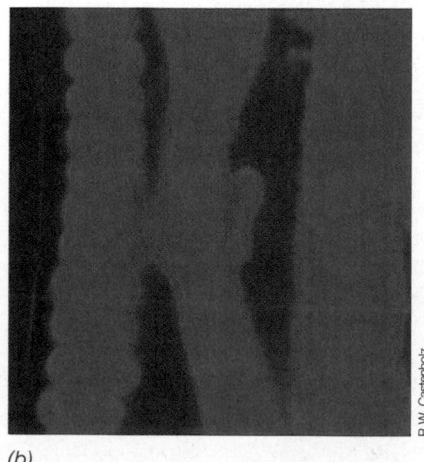
(b)

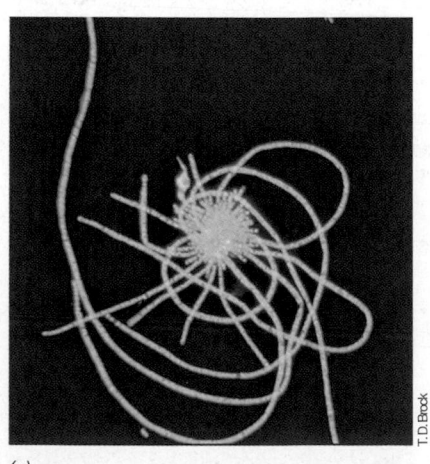
(c)

Figure 4.6 Photomicrographs of various microorganisms as visualized by fluorescence microscopy. (a, b) Cyanobacteria. (a) Cells observed by bright-field microscopy. (b) Same cells observed by fluorescence after shining light of 546 nm on them. The red color is due to autofluorescence of chlorophyll and other pigments. (c) Cells of the filamentous bacterium *Leucothrix mucor* stained with the fluorescent dye, acridine orange, which fluoresces green. Cells are 3 μm in diameter and may reach lengths of greater than 100 μm.

come? We will see in the next section that the scanning electron microscope offers one solution to this problem, but certain forms of light microscopy do as well, and we discuss these options here.

Differential Interference Contrast Microscopy

Differential interference contrast (DIC) is a form of light microscopy that employs a polarizer to produce polarized light. The polarized light then passes through a prism that generates two distinct beams, and these are what traverse the specimen and enter the objective lens. Here the two beams are recombined into one, and because of slight differences in refractive index of the substances each beam passed through, the combined beams are not totally in phase but instead create an interference effect. This effect intensifies subtle differences in cell structure, and thus, by DIC microscopy, things like the nucleus of eukaryotic cells (Figure 4.7a), spores, vacuoles, granules, and the like attain a three-dimensional appearance. DIC microscopy is particularly useful for observing *unstained* cells because of its ability to generate images that reveal internal cell structures that are less apparent (or even invisible) by bright-field techniques (compare Figures 4.5a and b with Figure 4.7a).

Atomic Force Microscopy

Another form of microscopy useful for three-dimensional imaging of biological structures is the **atomic force microscope (AFM)**. In atomic force microscopy, a tiny stylus is positioned extremely close to the specimen such that weak repulsive atomic forces are established between the probe and the specimen. As the specimen is scanned in both the horizontal and vertical directions, the stylus rides up and down the hills and valleys, constantly recording its interactions with the surface. This pattern is monitored by a series of detectors that feed the digital information into a computer that generates an image (Figure 4.7b). Although the images obtained from an atomic force microscope appear similar to those from the scanning electron microscope (compare Figure 4.7b with Figure 4.10b), the AFM has the big advantage that specimen preparation is similar to that for light microscopy (that is, no fixatives or coatings are required). The AFM also allows living and hydrated specimens to be viewed, something that is generally not possible with electron microscopes.

Confocal Scanning Laser Microscopy

Confocal scanning laser microscopy (CSLM) is a computerized microscope that couples a laser light source to a light microscope; this technique allows for the generation of three-dimensional digital images of microorganisms and other biological specimens (Figure 4.8). In CSLM, a laser beam is bounced off a mirror that directs the beam through a scanning device. Then the laser

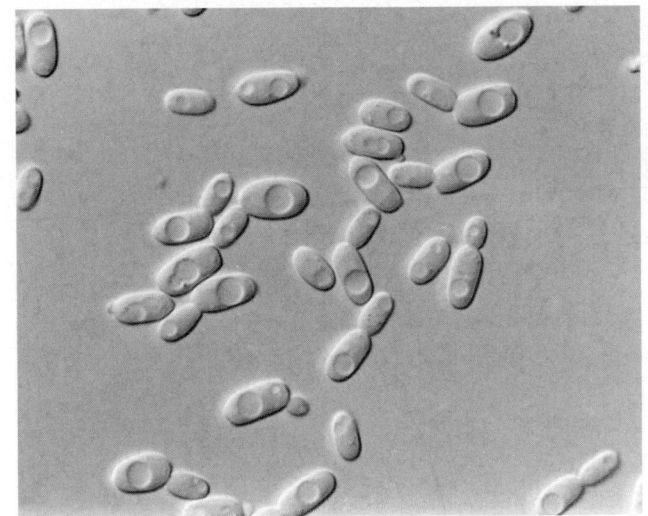

(a)

Linda Barnett and James Barnett

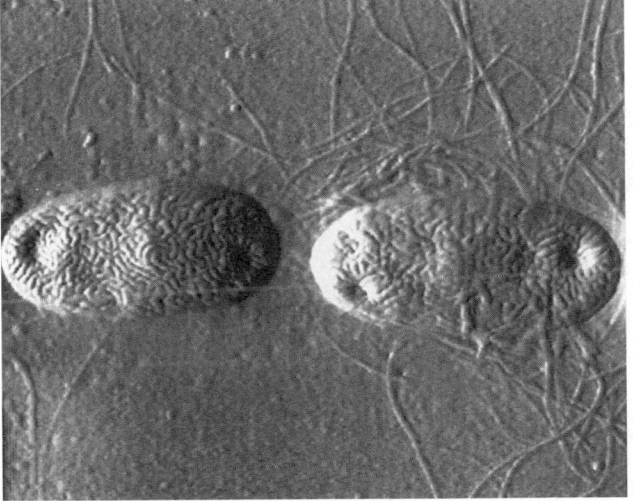

(b)

Suzanne Kelly

Figure 4.7 Three-dimensional imaging of cells using (a) interference contrast microscopy and (b) atomic force microscopy. The yeast cells in (a) are about 8 μm in diameter. Note how the nucleus is clearly visible here (compare Figure 4.7a with Figure 4.5a). The bacterial cells in (b) are about 2.2 μm in length, and the micrograph was taken from a natural biofilm that developed on the surface of a glass slide immersed for 24 h in a dog's water bowl. The slide was air dried before viewing with an atomic force microscope.

beam is directed through a pinhole that precisely adjusts the plane of focus of the beam to a given vertical layer within a specimen. By precisely illuminating only a single plane of the specimen, illumination intensity drops off rapidly above and below the plane of focus, and because of this, stray light from other planes of focus are minimized. Thus, in a relatively thick specimen such as a microbial biofilm, for example (Figure 4.8a), not only are cells on the surface of the biofilm apparent, as would be the case with conventional light microscopy, but cells

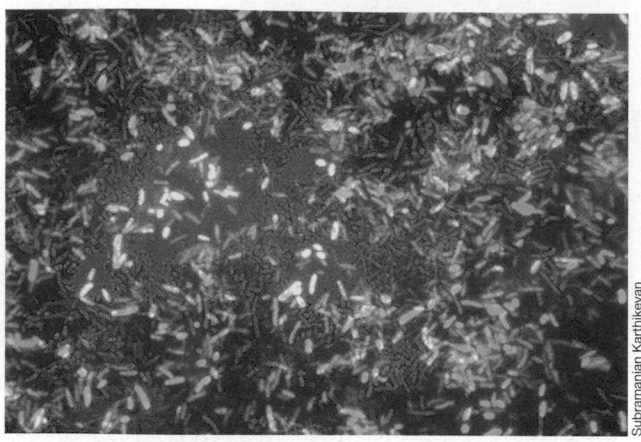

(a)

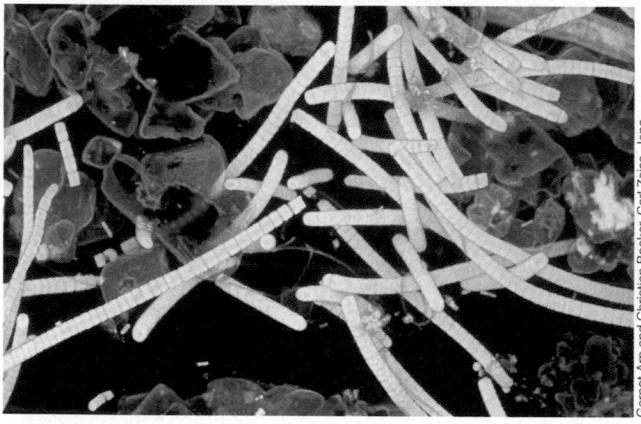

(b)

Figure 4.8 Confocal scanning laser microscopy. (a) Confocal image of a mixed microbial biofilm community cultivated in the laboratory. The green, rod-shaped cells are *Pseudomonas aeruginosa* experimentally introduced into the biofilm. Other cells that are different colors are present at different depths in the biofilm. (b) Confocal micrograph of a filamentous cyanobacterium growing in a soda lake.

in the various layers can also be observed by adjusting the plane of focus of the laser beam.

Cells in CSLM preparations are frequently stained with fluorescent dyes to make them more visible (Figure 4.8a). Alternatively, false color images can be generated by adjusting the microscope in such a way as to make different layers take on different colors. The laser confocal microscope is equipped with computer software to assemble digital images for subsequent image processing. Thus, images obtained from different layers can be stored and then digitally overlaid to reconstruct a three-dimensional image of the entire specimen (Figure 4.8a). CSLM has found widespread use in microbial ecology, especially for identifying phylogenetically distinct populations of cells present in a microbial habitat (see for example, Figure 18.11b), but is useful anywhere thick specimens need to be examined for their microbial content with depth.

✓ **4.2 Concept Check**

Interference contrast (DIC) and confocal scanning (CSLM) are forms of light microscopy that allow for greater three-dimensional imaging than other forms of light microscopy, and confocal microscopy allows imaging through thick specimens. The atomic force microscope yields a detailed three-dimensional image of live preparations.

✓ What structure in eukaryotic cells is more easily imaged using DIC microscopy?

✓ How is CSLM able to view different layers in a thick preparation?

4.3 Electron Microscopy

Electron microscopes are widely used for studying the detailed structure of cells. To study the internal structure of cells, a **transmission electron microscope (TEM)** is essential. In the TEM, electrons are used instead of light rays and electromagnets function as lenses, the whole system operating in a high vacuum (Figure 4.9). The resolving power of the electron microscope is much greater than that of the light microscope, enabling one to view structures as small as proteins and nucleic acids (∞ Figure 2.4b). Electron beams do not penetrate very well, however; even a single cell is too thick to be viewed directly. Consequently, special techniques of *thin sectioning* are needed to prepare specimens for the electron

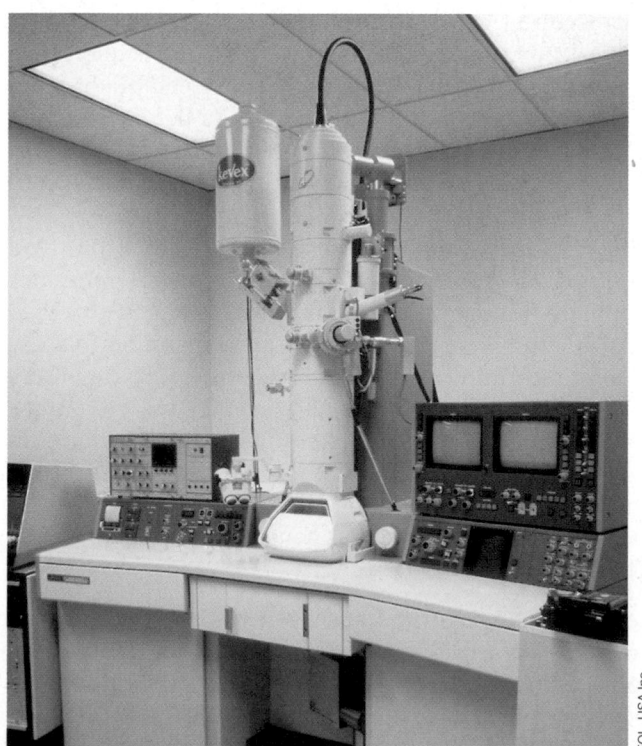

Figure 4.9 An electron microscope. This instrument encompasses both transmission and scanning electron microscope functions.

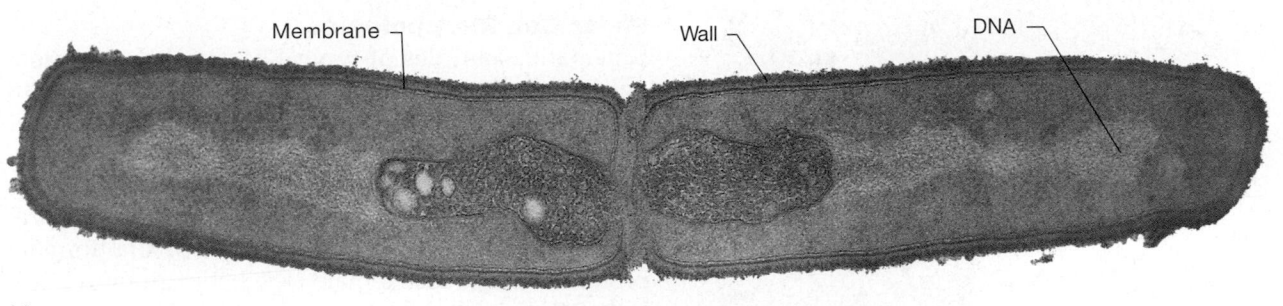

Membrane   Wall   DNA

Stanley C. Holt

(a)

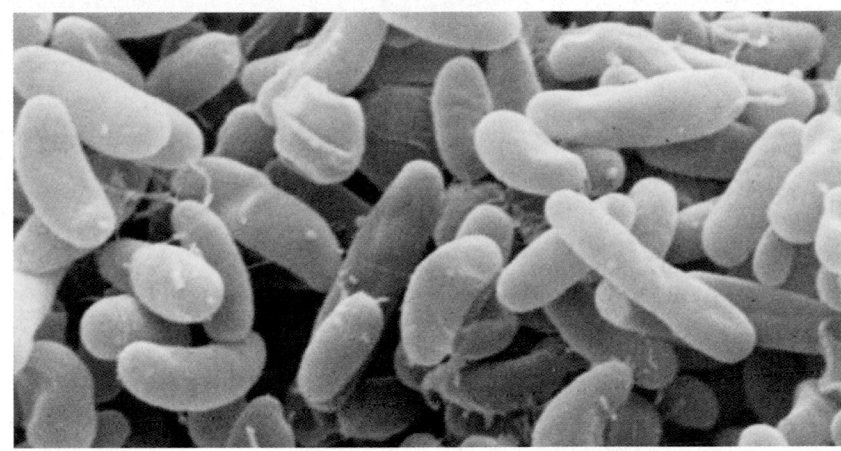

(b)

Figure 4.10 Electron micrographs of bacterial cells taken with (a) transmission and (b) scanning electron microscopes. (a) Thin section of a typical gram-positive bacterium, *Bacillus subtilis.* The cell has just divided, and two membrane-containing structures are attached to the cross-wall. Note the light region in the middle (DNA or the *nucleoid*). The cell is about 0.8 μm in diameter. (b) Cells of the phototrophic bacterium *Rhodovibrio sodomensis.* A single cell is about 0.75 μm wide. Note how SEM allows for great depth of field and thus excellent three-dimensional imaging.

F. R. Turner

microscope. A single bacterial cell, for instance, is cut into many very thin (20–60 nm) slices with a special knife, which are then examined individually with the electron microscope (Figure 4.10*a*). To obtain sufficient contrast, the preparations are treated with a special electron microscope stain such as osmic acid, permanganate, uranium or lanthanum salts, or lead. Because these substances are composed of atoms of high atomic weight, they scatter electrons well and thus improve contrast (Figure 4.10*a*).

Scanning Electron Microscopy

If only the *external* features of an organism need be observed, thin sections are not necessary, and intact cells or cell components can be observed directly by TEM with a technique called *negative staining* (see, for example, Figure 4.39). Alternatively, one can use the **scanning electron microscope (SEM)** (Figures 4.9 and 4.10*b*). With this tool, the specimen is coated with a thin film of a heavy metal such as gold. An electron beam from the SEM is then directed down on the specimen and scans back and forth across it. Electrons scattered by the metal are collected, and they activate a viewing screen to produce an image (Figure 4.10*b*). In the SEM, even fairly large specimens can be observed, and the depth of field is extremely good. A wide range of magnifications can be obtained with the SEM, from as low as 15× up to about 100,000× , but only the *surface* of an object can be visualized. All electron micro-

scopes are fitted with cameras to allow a photograph, called an *electron micrograph,* to be taken.

✓ 4.3 Concept Check

Electron microscopes have far greater resolving power than do light microscopes, the limits of resolution being about 0.2 nm. Two major types of electron microscopy are performed: transmission electron microscopy, for observing internal cell structure down to the molecular level, and scanning electron microscopy, useful for three-dimensional imaging and for examing surfaces.

✓ What is an *electron micrograph?*

✓ Keeping in mind that chemical fixatives are necessary and that electron microscopes must operate at high vacuum, what major *disadvantage* do electron microscopes have compared with light microscopes?

✓ What type of electron microscope would you use to observe the bacterial nucleoid?

4.4 Cell Morphology and the Significance of Being Small

In biology, the term *morphology* refers to the *shape* of an organism. Several morphologies are known among prokaryotes and most have well-accepted terms to describe them. We explore these terms and shapes here and then look at some of the biological benefits of small cells.

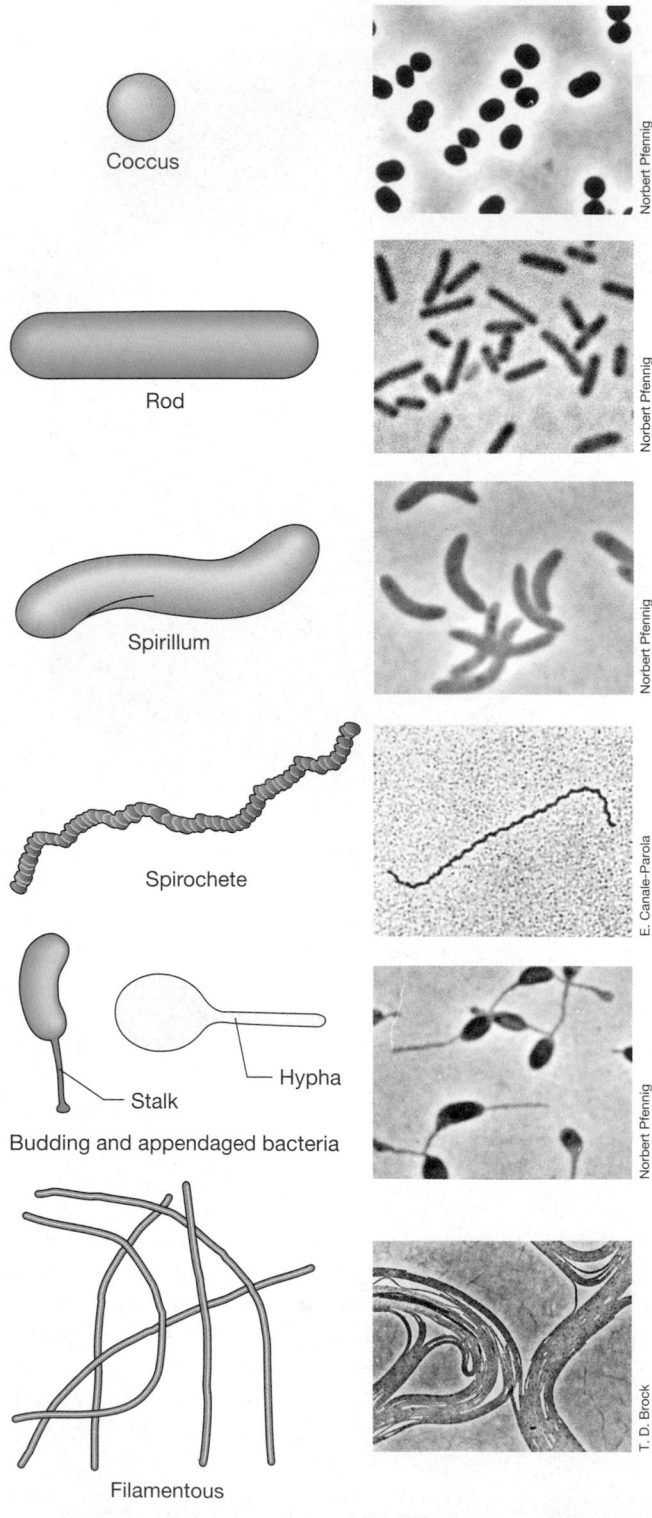

Figure 4.11 Representative cell shapes (morphology) in prokaryotes. Next to each drawing is a phase photomicrograph showing an example of that morphology. Organisms are coccus, *Thiocapsa roseopersicina* (diameter of a single cell = 1.5 μm); rod, *Desulfuromonas acetoxidans* (diameter = 1 μm); spirillum, *Rhodospirillum rubrum* (diameter = 1 μm); spirochete, *Spirochaeta stenostrepta* (diameter = 0.25 μm); budding and appendaged, *Rhodomicrobium vannielii* (diameter = 1.2 μm); filamentous, *Chloroflexus aurantiacus* (diameter = 0.8 μm).

Major Cell Morphologies

Schematic examples of typical bacterial shapes along with phase photomicrographs of actual example organisms are shown in Figure 4.11. A bacterium that is spherical or ovoid in morphology is called a **coccus** (plural, **cocci**). A bacterium with a cylindrical shape is called a **rod**. Some rods are curved, frequently forming spiral-shaped patterns, and are then called **spirilla**. In many prokaryotes, the cells remain together in groups or clusters after division, and the arrangements in these groups are often characteristic of different organisms. For instance, cocci or rods may occur in long chains. Some cocci form thin sheets of cells, whereas others occur in three-dimensional cubes or irregular cube-like clusters.

Several groups of bacteria are immediately recognizable by their unusual shapes. Examples include **spirochetes**, which are tightly coiled bacteria, **appendaged bacteria**, which possess extensions of their cells as long tubes or stalks, and **filamentous bacteria**, which form long, thin cells or chains of cells (Figure 4.11). Figure 4.11 displays *representative* morphologies; many variations of these basic morphological types are possible.

The Size of Microbial Cells and the Significance of Being Small

Prokaryotes vary in size from cells as small as 0.1–0.2 μm in diameter to those more than 50 μm in diameter; a few very large prokaryotes, such as the surgeonfish symbiont *Epulopiscium fishelsoni* (Figure 4.12), are up to 50 μm in diameter and can be more than 0.5 millimeters (mm) in

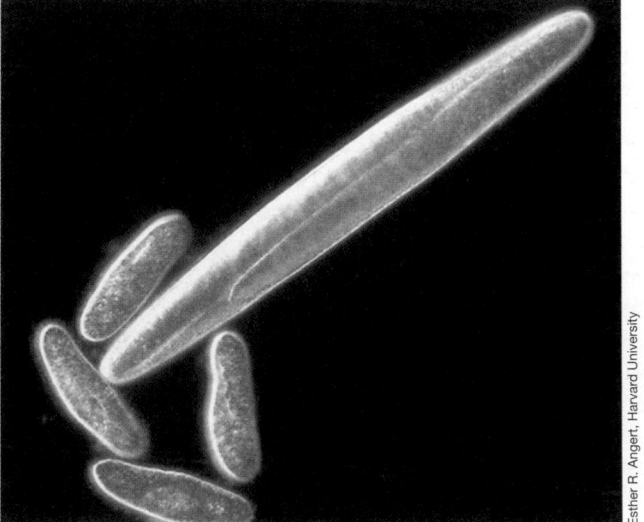

Figure 4.12 Darkfield photomicrograph of a giant prokaryote, the surgeonfish symbiont *Epulopiscium fishelsoni*. The rod-shaped *E. fishelsoni* cell in this field is about 600 μm (0.6 mm) long and is shown with four cells of the protozoan (eukaryote) *Paramecium*, each of which measures about 150 μm in length. *E. fishelsoni* is a member of the *Bacteria* and phylogenetically related to *Clostridium* species.

length. However, the dimensions of an average rod-shaped prokaryote, the bacterium *Escherichia coli*, for example, are about 1×3 µm (Figure 4.13). For comparison, typical eukaryotic cells may be 2 µm to more than 200 µm in diameter. Prokaryotes are thus very small cells compared with eukaryotes, and the small size of prokaryotes affects a number of their biological properties. For example, the rate at which nutrients and waste products pass into and out of a cell, a factor that can greatly affect cellular metabolic rates and growth rates, is in general *inversely* proportional to cell size. This is because transport rates are to some degree a function of the *membrane surface area* available, and relative to cell volume, small cells have more surface available than do large cells. This point can be seen most readily in the case of a sphere, in which the *volume* is a function of the cube of the radius ($V = \frac{4}{3}\pi r^3$), whereas the *surface area* is a function of the square of the radius ($SA = 4\pi r^2$). The surface-to-volume (S/V) ratio of a sphere can thus be expressed as $3/r$ (Figure 4.14). A cell with a smaller *r* value therefore has a *higher* S/V ratio than a larger cell and thus can have a more efficient exchange of nutrients with its surroundings than can a large cell.

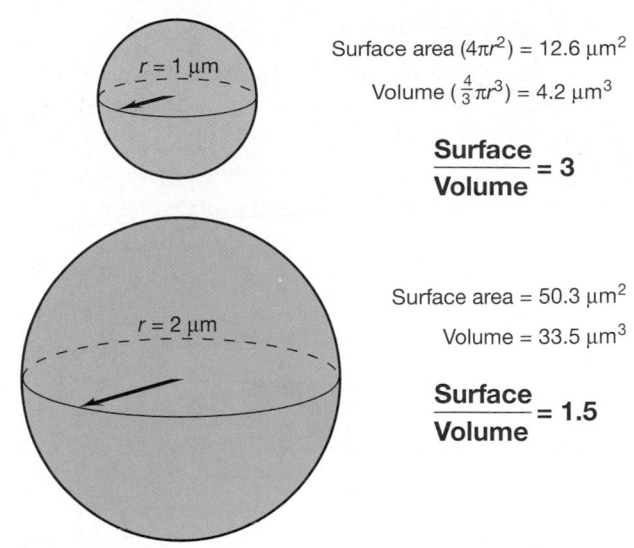

$$\text{Surface area } (4\pi r^2) = 12.6 \text{ µm}^2$$
$$\text{Volume } (\tfrac{4}{3}\pi r^3) = 4.2 \text{ µm}^3$$

$$\frac{\textbf{Surface}}{\textbf{Volume}} = 3$$

$$\text{Surface area} = 50.3 \text{ µm}^2$$
$$\text{Volume} = 33.5 \text{ µm}^3$$

$$\frac{\textbf{Surface}}{\textbf{Volume}} = 1.5$$

Figure 4.14 Surface area and volume relationships in cells. As a cell *increases* in size, its surface area-to-volume ratio *decreases*.

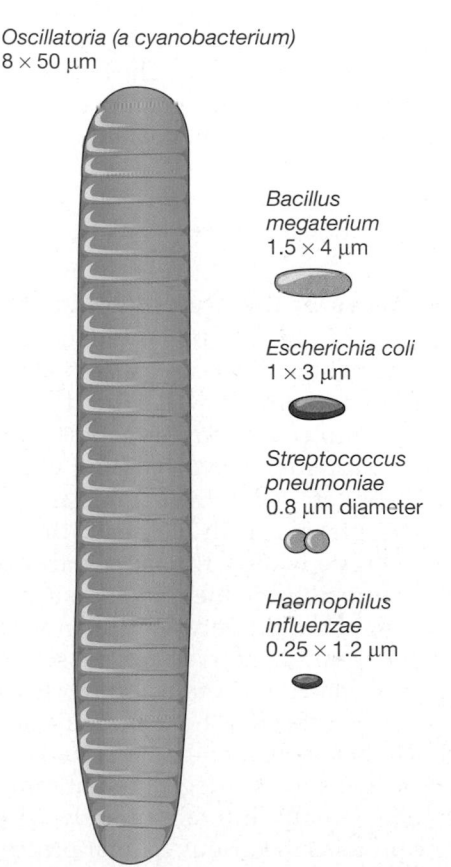

Oscillatoria (a cyanobacterium)
8×50 µm

Bacillus megaterium
1.5×4 µm

Escherichia coli
1×3 µm

Streptococcus pneumoniae
0.8 µm diameter

Haemophilus influenzae
0.25×1.2 µm

Figure 4.13 Comparison of sizes of a variety of prokaryotes. Most known prokaryotes have cell diameters in the range of 0.5–2 µm.

This S/V differential typically manifests itself in more rapid growth rates and the formation of larger cell populations (per unit of resources) for smaller cells than for larger cells. These parameters, rapid growth and high cell numbers, in turn greatly affect microbial ecology. This is so because high numbers of rapidly metabolizing cells can cause major physiochemical changes in an ecosystem even over very short periods of time. We will return to the theme of microorganisms in their natural habitats in Chapters 18, 19, 21, and 26–29.

Some microbiologists have proposed that *very small bacteria* exist in nature, cells referred to as *nanobacteria.* Most reports of nanobacteria have been associated with their supposed formation of precipitates and biofilms (∞ Section 19.3) in environments as diverse as mineral surfaces and human tissues. The size of putative nanobacteria are on the order of 0.1 µm diameter for coccus-shaped structures, extremely small, even by prokaryotic standards. Opponents of the nanobacteria theory claim that nanobacteria are simply artifacts of chemical or geochemical reactions of nonliving materials, and that even the smallest known bacterial cells are significantly larger than reported nanobacteria. Opponents also claim that if one considers the space needed to store all of the essential biomolecules of life, it is highly unlikely that these could exist within the volume available to a structure of 0.1 µm or less. Thus, the issue of nanobacteria is an unsettled one. If such very small cells actually exist, they would be the smallest known living structures.

✓ **4.4** **Concept Check**

Prokaryotes are typically smaller in size than eukaryotes, and prokaryotic cells can have a wide variety of morphologies. The small size of prokaryotic cells affects their physiology, growth rate, and ecology.

✓ List three morphological types of prokaryotes.

✓ What physical property of cells *increases* as cells become smaller?

II CELL MEMBRANES AND CELL WALLS

We now consider two extremely important cell structures: the cell membrane and the cell wall. Each carries out well-defined and critical functions for the cell, including the transport of nutrients (membrane) and prevention of osmotic lysis (wall).

4.5 Cytoplasmic Membrane: Structure

The **cytoplasmic membrane** is a thin structure that completely surrounds the cell. Only about 8 nm thick, this vital structure is the barrier separating the inside of the cell (the cytoplasm) from its environment. If the membrane is broken, the integrity of the cell is destroyed, the internal contents leak into the environment, and the cell dies. The cytoplasmic membrane is also a *highly selective barrier*, enabling a cell to concentrate specific metabolites and excrete waste materials.

Chemical Composition of Membranes

The general structure of biological membranes is a **phospholipid bilayer** (Figure 4.15). As discussed in Section 3.4, phospholipids contain both highly hydrophobic (fatty acid) and relatively hydrophilic (glycerol) moieties and can exist in many different chemical forms as a result of variation in the nature of the fatty acids or phosphate-containing groups attached to the glycerol backbone. As phospholipids aggregate in an aqueous solution, they tend to form bilayer structures spontaneously—the fatty acids point inward toward each other in a hydrophobic environment, and the hydrophilic portions remain exposed to the aqueous external environment (Figure 4.15). The bilayer character of membranes probably represents the most stable arrangement of lipid molecules in an aqueous environment.

Thin sections of the cytoplasmic membrane can be seen with the electron microscope; a representative example is seen in Figure 4.16*a*. By careful high resolution electron microscopy, the cytoplasmic membrane

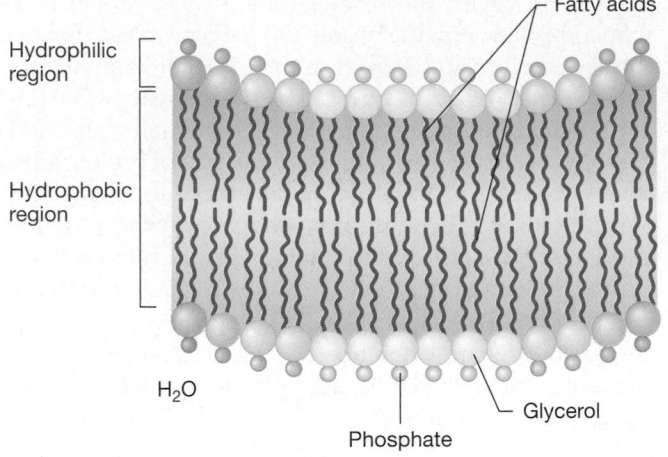

Figure 4.15 Structure of a phospholipid bilayer. The cytoplasmic membrane is about 8 nm (80 Å) wide.

appears as two light-colored lines separated by a darker area (Figure 4.16*a*). This **unit membrane**, as it is called, consists of a phospholipid (∞ Figure 3.7) bilayer and proteins embedded within it (Figure 4.17). The major proteins of the cell membrane generally have very hydrophobic external surfaces in the regions of the protein that span the membrane and have hydrophilic surfaces exposed on both the inside and the outside of the cell (Figure 4.17). The overall structure of the cytoplasmic membrane is stabilized by hydrogen bonds and hydrophobic interactions. In addition, cations such as Mg^{2+} and Ca^{2+} also help stabilize the membrane by combining ionically with negative charges of the phospholipids.

Other Features of the Cytoplasmic Membrane

The outer surface of the cytoplasmic membrane faces the environment and in certain bacteria makes contact with a variety of proteins that bind substrates or process large molecules for transport into the cell (periplasmic proteins) (see discussion in Sections 4.7 and 4.9). The inner side of the cytoplasmic membrane faces the cytoplasm and interacts with proteins involved in energy-yielding reactions and other important cellular functions. Some proteins, such as those in the periplasm (a region between the inner and outer membranes of gram-negative bacteria, see Section 4.9) and some cytoplasmic proteins, may associate quite firmly with the surface of the membrane and actually function as if they were membrane-bound proteins. Although not themselves integral membrane proteins, such proteins usually interact directly with integral membrane proteins in various cellular processes. Some of these *peripheral membrane proteins,* as they are called, are lipoproteins and contain a lipid tail on the amino terminus of the protein, which anchors the protein into the membrane.

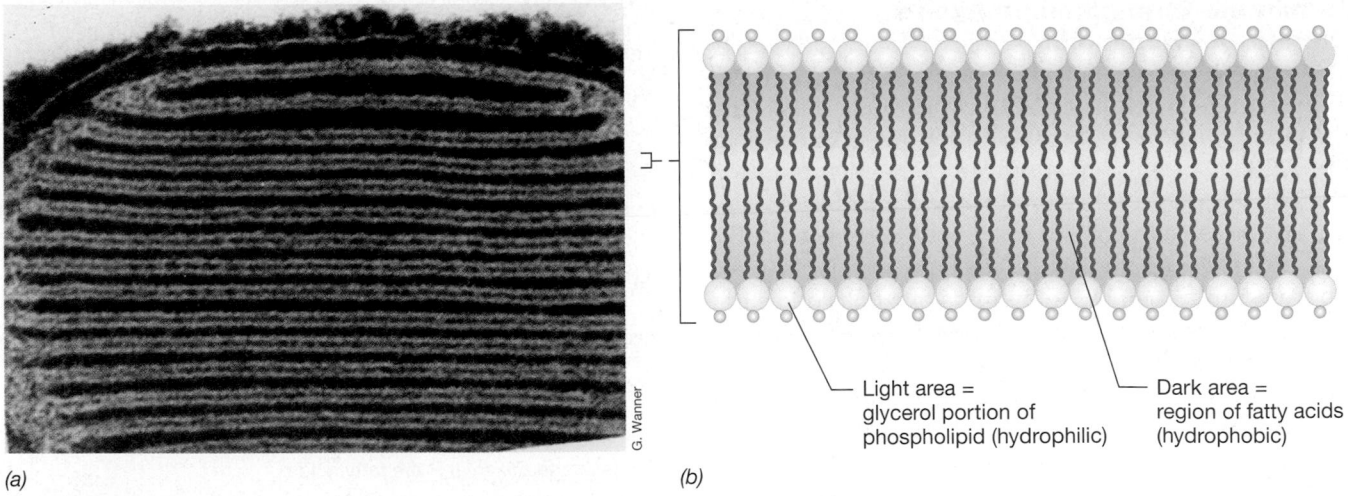

(a)   *(b)*

Light area =
glycerol portion of
phospholipid (hydrophilic)

Dark area =
region of fatty acids
(hydrophobic)

Figure 4.16 The cytoplasmic membrane. (a) Electron micrograph of photosynthetic membrane stacks derived from the cytoplasmic membrane in the phototrophic bacterium *Halorhodospira halochloris.* Note the distinct lipid bilayers (unit membranes). Each bilayer is about 8 nm thick. (b) Enlarged schematic view of a single unit membrane shown in (a).

Although appearing somewhat rigid when viewed diagrammatically (Figures 4.16 and 4.17), the cytoplasmic membrane is actually quite fluid; phospholipid and protein molecules have significant freedom to move about the membrane surface. Measurements of membrane viscosities indicate that membranes have a viscosity approximating that of a light-grade oil. Thus, membranes can be thought of as *fluid mosaics* in which globular proteins oriented in a specific manner span a highly mobile, yet ordered, phospholipid bilayer. This arrangement confers a number of important functional properties on membranes, and we discuss these properties in the next section.

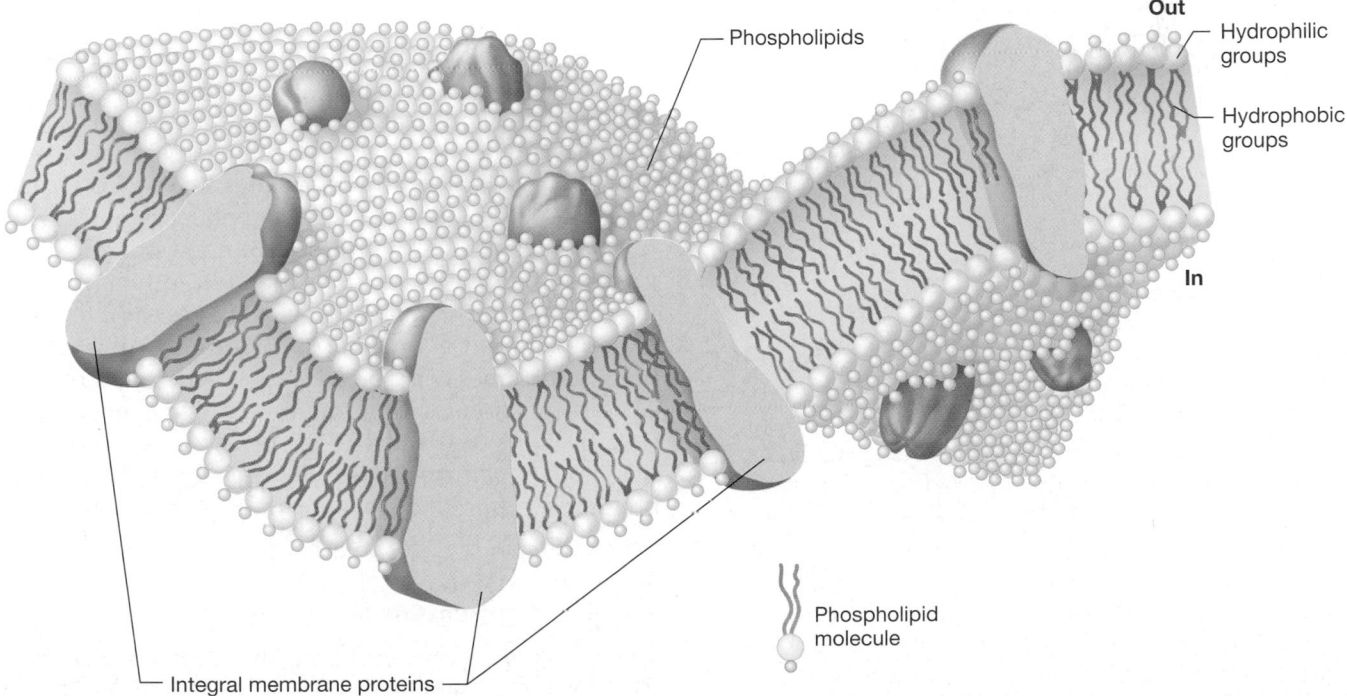

Phospholipids

Out

Hydrophilic
groups

Hydrophobic
groups

In

Phospholipid
molecule

Integral membrane proteins

Figure 4.17 Diagram of the structure of the cytoplasmic membrane; the inner surface (In) faces the cytoplasm and the outer surface (Out) faces the environment. The matrix of the unit membrane is composed of phospholipids, with the hydrophobic groups directed inward and the hydrophilic groups toward the outside, where they associate with water. Embedded in the matrix are proteins that have considerable hydrophobic character in the region that traverses the fatty acid bilayer. Hydrophilic proteins and other charged substances, such as metal ions, may be attached to the hydrophilic surfaces. Although there are some chemical differences, the overall structure of the cytoplasmic membrane shown is similar in both prokaryotes and eukaryotes (but see an exception to the bilayer design in Figure 4.20*d*).

Membrane Strengthening Agents: Sterols and Hopanoids

One major difference in chemical composition of membranes between eukaryotic and prokaryotic cells is that the eukaryotes have **sterols** in their membranes (Figure 4.18a, b). Sterols are absent from the membranes of virtually all prokaryotes (methanotrophic bacteria and the mycoplasmas are major exceptions, ⌢ Sections 12.6 and 12.21). Depending on the cell type, sterols can make up from 5 to 25% of the total lipids of eukaryotic membranes. Sterols are rigid, planar molecules, whereas fatty acids are flexible. The association of sterols with the membrane serves to stabilize its structure and make it less flexible. Molecules similar to sterols, called *hopanoids*, are present in the membranes of many *Bacteria* and may play a role similar to that of sterols in eukaryotic cells. One widely distributed hopanoid is the C_{30} hopanoid *diploptene* (Figure 4.18c). As far as is known, hopanoids are not found in species of *Archaea*.

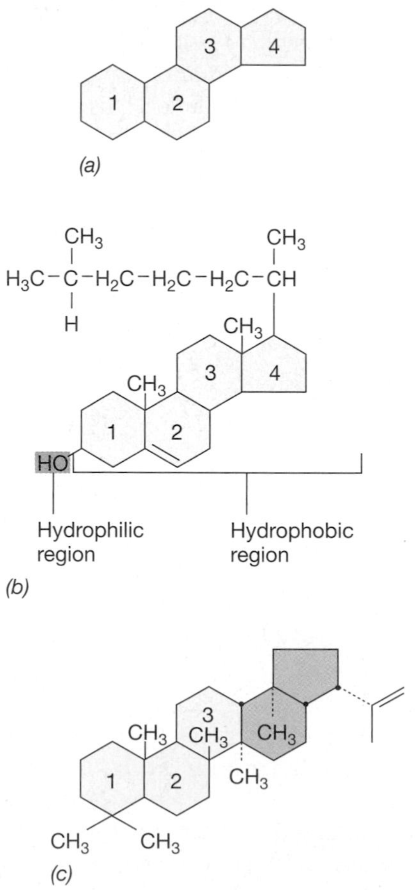

(a)

(b)

(c)

Figure 4.18 Sterols and hopanoids. (a) The general structure of a sterol. All sterols contain the same four rings, labeled 1, 2, 3, and 4. (b) The structure of cholesterol. (c) The structure of the hopanoid diploptene. Note the structural resemblance to cholesterol in rings 1 through 3. Sterols are found in the membranes of eukaryotes and hopanoids in the membranes of some prokaryotes.

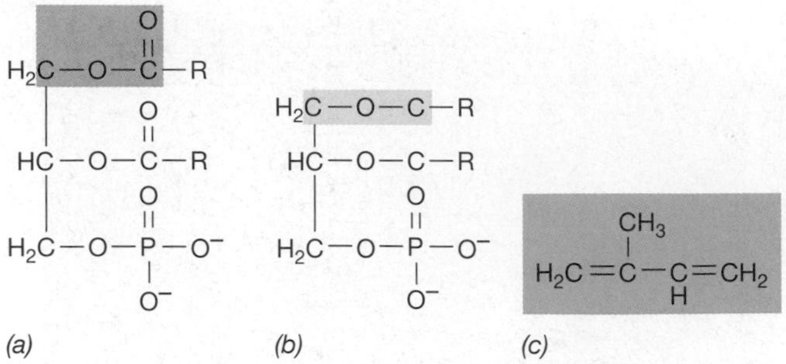

Figure 4.19 Chemical bonds in lipids. (a) The *ester* linkage as found in the lipids of *Bacteria* and *Eukarya*. (b) The *ether* linkage of lipids from *Archaea*. (c) Isoprene, the parent structure of the hydrophobic side chains (R) of archaeal lipids. By contrast, in lipids of *Bacteria* and *Eukarya*, R are fatty acids.

Archaeal Membranes

The lipids in *Archaea* are chemically unique. In contrast to the lipids in *Bacteria* and *Eukarya* in which *ester* linkages bond the fatty acids to the glycerol molecule (Figure 4.19a; ⌢ Section 3.4), lipids from *Archaea* have *ether* linkages between glycerol and their hydrophobic side chains. In addition, archaeal lipids lack fatty acids and instead have side chains composed of repeating units of the five-carbon hydrocarbon *isoprene* (Figure 4.19c). However, the overall structure of archaeal lipid membranes, forming inner and outer hydrophilic surfaces with a hydrophobic interior, is maintained.

Glycerol *diethers* and glycerol *tetraethers* (Figure 4.20a, b) are the major classes of lipids present in *Archaea*. Note that in the tetraether molecule the phytanyl (composed of four linked isoprenes) side chains from each glycerol molecule are covalently bonded together (Figure 4.20b). Employed within a membrane structure, this yields a lipid *monolayer* instead of a lipid bilayer (Figure 4.20d). Lipid monolayers are quite resistant to peeling apart, and it is thus not surprising that this membrane structure is widespread among hyperthermophilic *Archaea*, prokaryotes that grow at very high temperatures (⌢ Sections 6.10 and 13.5–13.10). We will encounter many other features that set *Bacteria* and *Archaea* apart, but the chemistry of membrane lipids is a major defining feature of each phylogenetic group.

✓ 4.5 Concept Check

The cytoplasmic membrane is a highly selective permeability barrier constructed of lipid and protein that forms a bilayer with hydrophilic exteriors and a hydrophobic interior. Other molecules, such as sterols and hopanoids, may strengthen the membrane, and integral proteins involved in transport and other functions traverse it. Unlike *Bacteria* and *Eukarya*, *Archaea* contain ether-linked lipids, and some species have membranes of monolayer instead of bilayer construction.

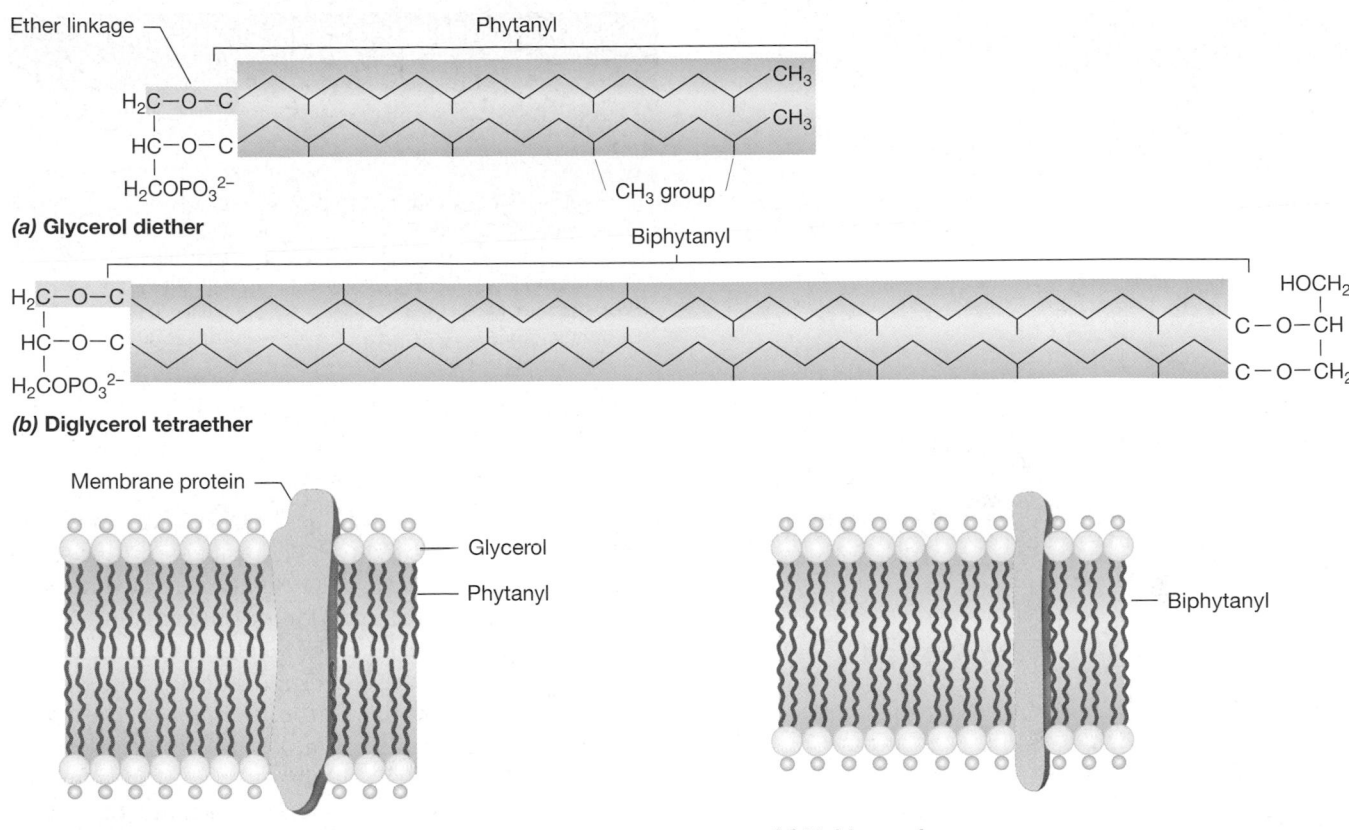

(a) **Glycerol diether**

(b) **Diglycerol tetraether**

(c) **Lipid bilayer**

(d) **Lipid monolayer**

Figure 4.20 Major lipids of *Archaea* and the structure of archaeal membranes. (a) Glycerol diethers. (b) Diglycerol tetraethers. Note that in both cases, the hydrocarbon is attached to the glycerol by *ether* linkages. Hydrocarbon in (a) phytanyl (C_{20}) and (b) dibiphytanyl (C_{40}). (c, d) Membrane structure in *Archaea*. (c) Lipid bilayer. (d) Lipid monolayer.

✓ Draw the basic structure of a lipid bilayer.

✓ Why are compounds like sterols and hopanoids good at stabilizing the cytoplasmic membrane?

✓ Contrast the linkage between glycerol and the hydrophobic portion of lipids in *Bacteria* and *Archaea*.

4.6 Cytoplasmic Membrane: Function

The cytoplasmic membrane is more than just a barrier separating the inside from the outside of the cell. The membrane plays several critical roles in cell function. First and foremost, the membrane functions as a *permeability barrier*, preventing the passive leakage of cytoplasmic constituents into or out of the cell (Figure 4.21). In addition, the membrane is the site of many proteins, some of which are enzymes and many of which are involved in one way or another in the transport of substances into and out of the cell.

We will learn in Chapter 5 that the cytoplasmic membrane is also a site of energy conservation in the cell. The membrane can exist in an energetically "charged" form in which a separation of protons (H^+) from hydroxyl ions (OH^-) occurs across its surface (Figure 4.21). This charge separation is a form of energy, analogous to the potential energy present in a charged battery. This energized state of the membrane, referred to as a *proton motive force* (PMF), is responsible for driving many energy-requiring functions in the cell, including some forms of transport, motility, and the biosynthesis of the cell's energy currency, ATP (Figure 4.21).

The Cytoplasmic Membrane as a Permeability Barrier

The interior of the cell (the cytoplasm) consists of an aqueous solution of salts, sugars, amino acids, vitamins, coenzymes, and a wide variety of other soluble materials. The hydrophobic nature of the cytoplasmic membrane makes it a tight barrier; although some small hydrophobic molecules may pass through the membrane by diffusion, hydrophilic and charged molecules do not readily pass but instead must be specifically transported. Even a substance as small as a hydrogen ion (H^+) does not diffuse across the cytoplasmic membrane. One molecule that does penetrate the membrane is water itself, which is sufficiently small to pass between phospholipid molecules. Water transport through the membrane can be greatly accelerated, however, by water

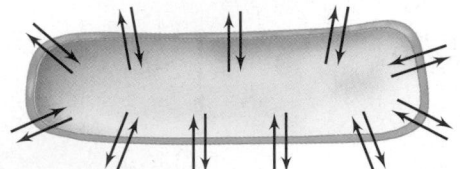

Permeability Barrier — Prevents leakage and functions as a gateway for transport of nutrients into and out of the cell

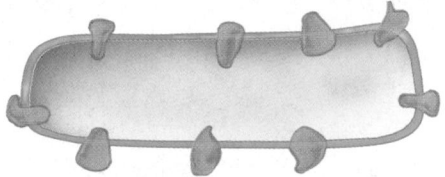

Protein Anchor — Site of many proteins involved in transport, bioenergetics, and chemotaxis

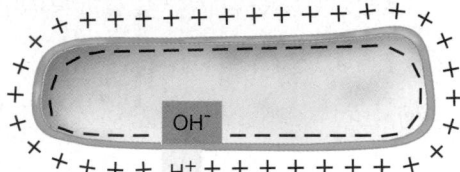

Energy Conservation — Site of generation and use of the proton motive force

Figure 4.21 The major functions of the cytoplasmic membrane.

transport proteins called *aquaporins*. These proteins, present in organisms of all domains of life, consist of membrane-spanning channels that specifically transport water into or out of the cytoplasm. Aquaporin AqpZ of *Escherichia coli*, for example, is a nonessential protein whose synthesis is greatly increased under low osmotic conditions. It is thus likely that AqpZ functions more as a water *exporter* than a water *importer*, existing primarily to prevent the cell from experiencing hypo-osmotic shock.

The relative permeability of a few biologically important substances is shown in Table 4.1. As can be seen, most substances do not passively enter the cell, and thus *transport* processes are critical to cellular function. The data of Table 4.1 should also be viewed with the understanding that water flow in prokaryotic cells may be assisted by aquaporins and is not entirely due to diffusion through the membrane.

The Necessity for Transport Proteins

Transport proteins do more than just ferry things across the membrane; they are able to accumulate solutes inside the cell *against* the concentration gradient. The necessity for carrier-mediated transport in microorganisms is easy to understand. If diffusion were the only way that solutes could enter the cell, cells would never achieve the intracellular concentrations necessary to

Table 4.1 Comparative permeability of membranes to various molecules

Substance	Rate of permeability[a]
Water	100
Glycerol	0.1
Tryptophan	0.001
Glucose	0.001
Chloride ion (Cl^-)	0.000001
Potassium ion (K^+)	0.0000001
Sodium ion (Na^+)	0.00000001

[a] Relative scale—permeability with respect to permeability of water, given as 100. Permeability of the membrane to water may be affected by aquaporins (see text).

carry out biochemical reactions. This is because both the rate of uptake and the intracellular level of diffusable solutes are proportional to their external concentration (Figure 4.22). The concentration of nutrients in nature, however, is often very low. Thus, cells must have mechanisms for accumulating nutrients to levels higher than those in nature, and this is the function of transport systems. Moreover, unlike simple diffusion, carrier-mediated transport shows a *saturation effect*; if the concentration of substrate is high enough to saturate the carrier, which is usually the case even at very low substrate concentration, the rate of uptake becomes maximal and the addition of more substrate cannot increase the rate (Figure 4.22).

One characteristic of carrier-mediated transport processes is the *highly specific* nature of the transport event. Many carrier proteins react only with a single molecule while others show affinities for a chemical class of molecules. For instance, there are carriers that transport a variety of related sugars or amino acids. This economy

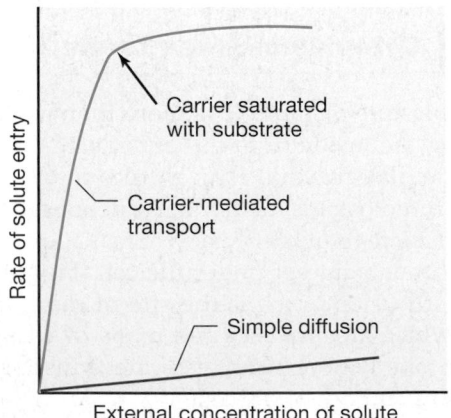

Figure 4.22 Relationship between uptake rate and external concentration in diffusion and transport. Note that in the carrier-mediated process, the uptake rate shows saturation at relatively low external concentrations.

in uptake reduces the need for separate transport proteins for every single amino acid or every single sugar the cell needs to transport. In addition, the synthesis of transport proteins is *regulated* by the cell such that the specific complement of transporters present in the membrane is a function of both the nutrients present and their concentration. The latter is a factor because oftentimes transport of a particular nutrient occurs via one type of transporter when the nutrient is present at high concentration and by a different, higher affinity transporter when present at low concentration.

✓ 4.6 Concept Check

The major function of the cytoplasmic membrane is as a permeability barrier, preventing leakage of cytoplasmic metabolites into the environment. Selective permeability also prevents the diffusion of most solutes. To accumulate nutrients against the concentration gradient, specific transport mechanisms are employed.

- ✓ Besides permeability, what other functions does the cytoplasmic membrane have?
- ✓ List two reasons why a cell cannot depend on diffusion as a means of getting nutrients into the cell.
- ✓ Why is physical damage to the cytoplasmic membrane a more critical problem for the cell than damage to some other cell component?

4.7 Membrane Transport Systems

As just discussed, transport of nutrients and expulsion of wastes are key events in the life of any cell. Different transport systems have evolved in prokaryotes, each with its own unique features. We explore the major classes of transport here.

Structure and Function of Membrane Transport Proteins

There are at least three classes of membrane-transporting systems: those involving only a membrane-spanning component, those involving a periplasmic-binding component plus a membrane-spanning component, and those, like the phosphotransferase system, that involve a series of proteins that cooperate to mediate the transport event (Figure 4.23). All of these transport systems require energy, either in the form of the proton motive force, ATP, or some other high-energy compound.

The membrane-spanning proteins of virtually all bacterial transport systems show significant homologies in both their primary and secondary structure, a testament to their common evolutionary roots. Structurally, these transporters often form 12 alpha helices (↝ Section 3.7) that wind back and forth through the membrane to form a channel through which the transported substance is carried into the cell (Figure 4.24). The actual

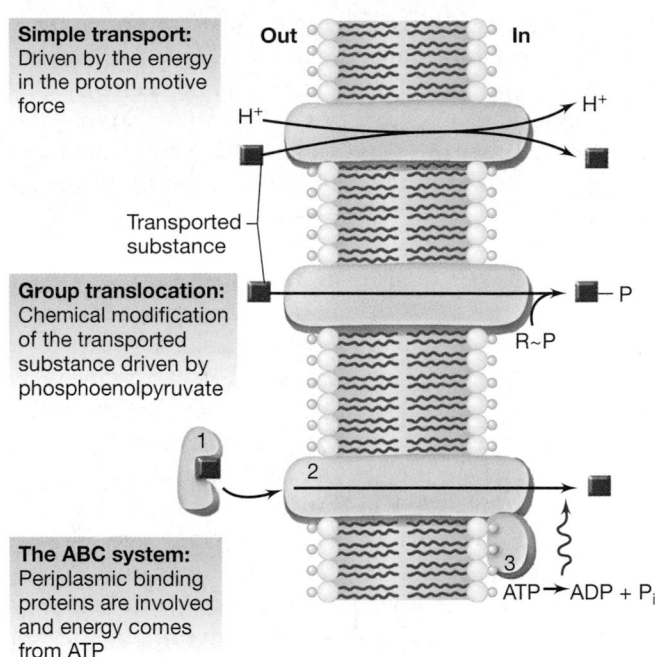

Figure 4.23 The three classes of membrane-transporting systems. Note how simple transporters and the ABC system transport substances *without* chemically modifying them, while group translocation results in the chemical modification (phosphorylation) of the transported substance.

transport event involves a conformational change in the protein following binding of its specific substrate and this event shuttles the compound across the membrane.

The various *types* of transport events that can occur are also summarized in Figure 4.24. *Uniporters* are proteins that simply transport a molecule in a unidirection-

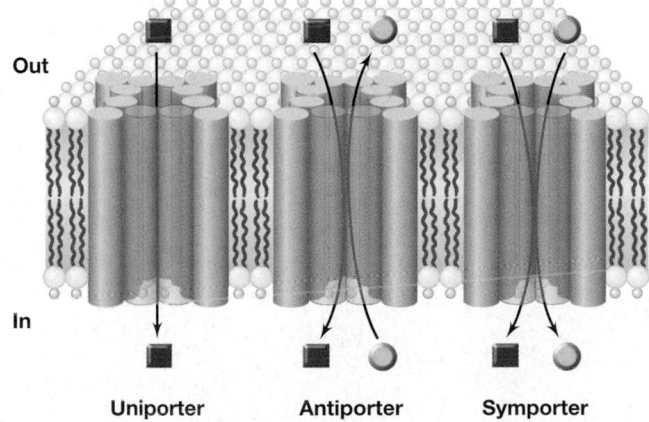

Figure 4.24 Structure of membrane-spanning transporters and types of transport events. In prokaryotes, membrane-spanning transporters typically contain 12 alpha helices that align with each other in a circle to form a channel through the membrane. Shown here are three individual transporters, each showing a different type of transport event. For antiporters and symporters, the cotransported molecule is shown in yellow.

al fashion across the membrane. *Symporters* are proteins that transport a substance *along with* another substance, frequently a proton (H⁺). *Antiporters* are proteins that, as their name implies, transport a substance across the membrane in one direction while at the same time transporting a second substance in the *opposite* direction (Figure 4.24).

Lac Permease: A Simple Transporter

The bacterium *Escherichia coli* can grow on the disaccharide lactose. Lactose is taken up by cells of *E. coli* by a symporter called the Lac permease. The Lac permease is an example of a typical symporter, taking up one molecule of lactose along with one proton. This is shown in Figure 4.25, where the activity of the permease is compared with other simple transporters, including uniporters and antiporters. Note that as each lactose molecule is transported by the Lac permease, the energy in the proton motive force is slowly diminished by the influx of protons into the cell. However, the proton motive force is being constantly reestablished in the cell through energy-yielding reactions that we will describe in later chapters (∞ Chapters 5 and 17). The final result

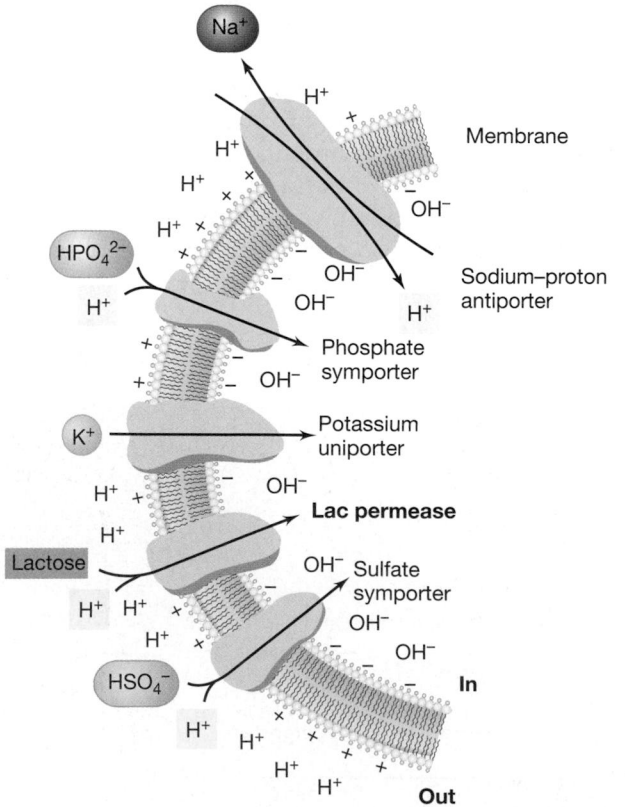

Figure 4.25 Function of the Lac permease (a symporter) of *Escherichia coli*, and several other well-characterized simple transporters. Although for simplicity the membrane-spanning proteins are drawn here in globular form, note that their structure is actually as depicted in Figure 4.24.

of a simple transporter such as the Lac permease is the accumulation of a solute, in this case lactose, to high enough concentrations where its metabolism yields energy for the cell.

Group Translocation

Group translocation is a transport process in which the transported substance is *chemically altered* during passage across the membrane. The best-studied cases of group translocation involve transport of the sugars glucose, mannose, and fructose, which are *phosphorylated* during transport by the **phosphotransferase system**.

The phosphotransferase system in the bacterium *Escherichia coli* is composed of a family of proteins, at least four of which are necessary to transport a given sugar. Before the sugar is transported into the cell, the proteins in the phosphotransferase system are themselves alternately phosphorylated and dephosphorylated in a cascading fashion until the membrane-spanning protein, called Enzyme II$_c$, receives the phosphate group and phosphorylates the sugar in the actual transport event (Figure 4.26). A small protein called HPr, the enzyme that phosphorylates it (Enzyme I), and Enzyme II$_a$ are cytoplasmic proteins, while Enzyme II$_b$ lies on the inner membrane surface, and Enzyme II$_c$ is an integral membrane protein (Figure 4.26). HPr and Enzyme I are nonspecific components of the phosphotransferase system and participate in the uptake of various sugars, while specific Enzymes II exist for each individual sugar.

The high-energy phosphate bond that supplies the energy for the phosphotransferase system comes from a key metabolic intermediate called *phosphoenolpyruvate*. However, it should be noted that although energy in the form of one high energy phosphate bond is consumed in the process of transporting the glucose molecule (Figure 4.26), the phosphorylation of glucose to glucose-6-P is the first step in its intracellular metabolism anyway (∞ Section 5.10). Thus, the phosphotransferase system prepares glucose for immediate entry into a central metabolic pathway.

Periplasmic Binding Protein-Dependent Transport: The "ABC" System

We will learn a bit later in this chapter (see Section 4.9) that gram-negative bacteria contain a space between the cytoplasmic membrane and a lipid-rich outer membrane layer, called the *periplasm* (see Figure 4.36). The periplasm contains various proteins, many of which function in transport; the latter are called *periplasmic-binding proteins*. Transport systems that employ periplasmic-binding proteins also have membrane-spanning components that actually mediate the transport event and a third component that supplies the necessary energy, obtained by hydrolysis of ATP. These types of transporters

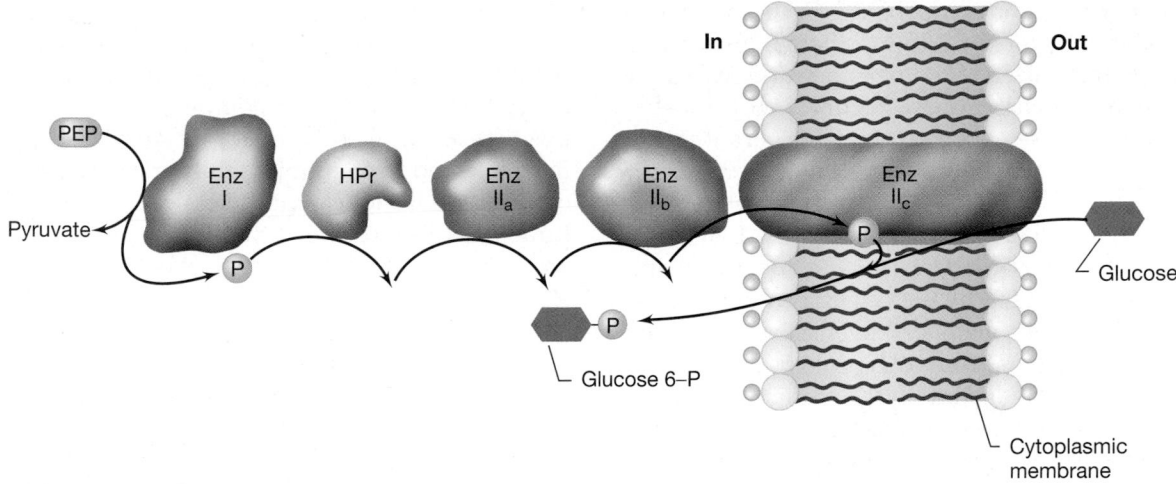

Figure 4.26 Mechanism of the phosphotransferase system of *Escherichia coli*. For glucose uptake, the system consists of five proteins: Enzyme (Enz) I; Enzymes II$_a$, II$_b$, and II$_c$, and HPr. Sequential phosphate transfer occurs from phosphoenolpyruvate (PEP) through the proteins shown to Enzyme II$_c$. The latter actually transports (and phosphorylates) the sugar.

have been called **ABC transport systems**, the *ABC* standing for *ATP-binding cassette*. The mechanism of action of an ABC transporter is shown in Figure 4.27.

More than 200 different ABC transport systems have been identified in prokaryotes, and analyses of the structures of functionally related components from the different systems show that they are clearly a *family* of related proteins. One of the interesting properties of

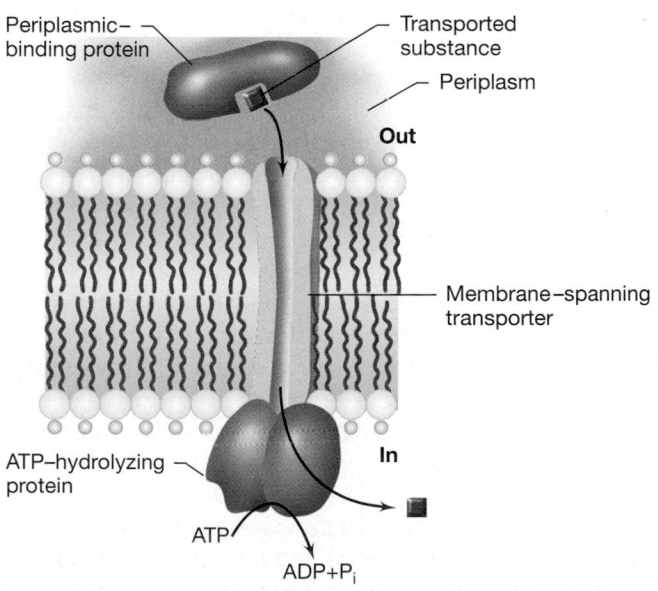

Figure 4.27 Mechanism of an ATP-Binding Cassette (ABC-type) transporter. The periplasmic binding protein has high affinity for substrate, the membrane-spanning protein is the transport channel, and the cytoplasmic ATP-hydrolyzing protein supplies the energy for the transport event. In *Escherichia coli*, the maltose (a disaccharide sugar) transport system is an example of an ABC system.

ABC-type transport systems is the extremely high substrate affinity of the periplasmic-binding proteins. These proteins are mobile within the periplasm and are able to bind their substrates even when substrates are present at extremely low concentration; for example, substrate concentrations of 1 micromolar (10^{-6} M) or less can be bound and transported. Once trapped by the binding protein, the complex interacts with its respective membrane-spanning component, and the actual transport event occurs driven by the energy of ATP hydrolysis (Figure 4.27). Interestingly, even though gram-positive bacteria lack a periplasm, binding protein-dependent transport systems are also present in many of these organisms as well. In gram-positive bacteria, specific binding proteins are not mobile but instead are anchored to the cytoplasmic membrane. However, as in gram-negative bacteria, once these binding proteins bind their substrate, they interact with a membrane-spanning component where, at the expense of ATP, transport across the membrane occurs.

Protein Export

Thus far our discussion of transport has focused on small molecules. What about the transport of large molecules, such as proteins? To function properly, many proteins need to be transported *outside* the unit membrane or inserted into the membrane in specific ways. Protein translocation occurs in prokaryotic cells through the activities of proteins called *translocases,* a key one being the Sec (for secretory) system. SecYEG is a membrane-associated translocase that exports certain proteins while inserting others into the membrane in a specific orientation consistent with their function. Some translocases are very specific in their action, but SecYEG is widely distributed among prokaryotes and is capable of translocating a variety of different proteins. How proteins destined for transport are recognized is another story, and we discuss this issue in a later chapter (∞ Section 7.16).

Protein export is important to bacteria because many bacterial enzymes function outside the cell. For example, hydrolytic enzymes such as amylase or cellulase are excreted directly into the environment where they cleave starch or cellulose (∞ Figure 3.6b), respectively, into glucose that can be used by the cell as a carbon and energy source. Moreover, many pathogenic bacteria excrete protein toxins or other deleterious proteins into the host during infection. All of these large molecules need to move through the cytoplasmic membrane, and translocases like SecYEG assist in these transport events.

✓ 4.7 Concept Check

At least three types of transporters are known: simple transporters, phosphotransferase-type transporters, and ABC systems, which contain three interacting components. Transport requires energy from either the proton motive force or from ATP.

✓ Contrast the energy requirements of simple transporters, the phosphotransferase system, and the ABC transport system.

✓ Contrast the three types of transport systems in terms of any chemical alterations that occur in the transported molecule.

✓ Which transport system is best suited for the transport of nutrients present in the environment in extremely low amounts, and why?

✓ How are proteins exported from the cell?

4.8 The Cell Wall of Prokaryotes: Peptidoglycan and Related Molecules

Because of the concentration of dissolved solutes inside the bacterial cell, a considerable turgor pressure develops, estimated at 2 atmospheres (atm) in a bacterium like *Escherichia coli*; this is roughly the same as the pressure in an automobile tire. To withstand these pressures, bacteria contain **cell walls**, which also function to give shape and rigidity to the cell. The prokaryotic cell wall is difficult to see well with the light microscope but can be readily seen in thin sections of cells with the electron microscope.

Bacteria can be divided into two major groups, called **gram-positive** and **gram-negative**. The original distinction between gram-positive and gram-negative was based on a special staining procedure, the *Gram stain* (see Section 4.1), but differences in cell wall structure are at the base of these differences in the Gram-staining reac-

tion. The appearance of the cell walls of gram-positive and gram-negative cells differ markedly, as is shown in Figure 4.28. The gram-negative cell wall is a multilayered structure and quite complex, whereas the gram-positive cell wall consists of primarily a single type of molecule and is often much thicker. Close examination of Figure 4.28 shows that there is also a significant textural difference between the surfaces of gram-positive and gram-negative *Bacteria,* as revealed by the scanning electron microscope.

The focus of this section is on the polysaccharide component of the cell walls of prokaryotes, both *Bacteria* and *Archaea*. These include, in particular, peptidoglycan, but also a variety of related and unrelated polysaccharides found in *Archaea*. In Section 4.9 we describe the special wall components found in gram-negative *Bacteria*.

Peptidoglycan

The cell walls of *Bacteria* have one rigid layer that is primarily responsible for the strength of the wall. In gram-negative *Bacteria,* additional layers are present outside this rigid layer. The rigid layer of both gram-negative and gram-positive *Bacteria* is very similar in chemical composition. Called **peptidoglycan** (or **murein**), each layer that makes up peptidoglycan is a thin sheet composed of two sugar derivatives, *N-acetylglucosamine* and *N-acetylmuramic acid*, and a small group of amino acids consisting of L-alanine, D-alanine, D-glutamic acid, and either lysine or diaminopimelic acid (DAP) (Figure 4.29). These constituents are connected to form a repeating structure, the *glycan tetrapeptide* (Figure 4.30).

The basic structure of peptidoglycan is a sheet in which the glycan chains formed by the sugars are connected by *peptide cross-links* formed by the amino acids. The glycosidic bonds connecting the sugars in the glycan chains are very strong, but these chains alone cannot provide rigidity in all directions. The full strength of the peptidoglycan structure is realized only when these chains are cross-linked by amino acids. This cross-linking occurs to characteristically different extents in different *Bacteria,* with greater rigidity coming from more complete cross-linking. In gram-negative Bacteria, cross-linkage usually occurs by direct peptide linkage of the amino group of diaminopimelic acid to the carboxyl group of the terminal D-alanine (Figure 4.31a). In gram-positive *Bacteria,* cross-linkage is usually by a peptide interbridge, the kinds and numbers of cross-linking amino acids varying from organism to organism. In *Staphylococcus aureus,* a well-studied gram-positive organism, each interbridge peptide consists of five molecules of the amino acid glycine connected by peptide bonds (Figure 4.31b). The overall structure of a peptidoglycan molecule is shown in Figure 4.31c.

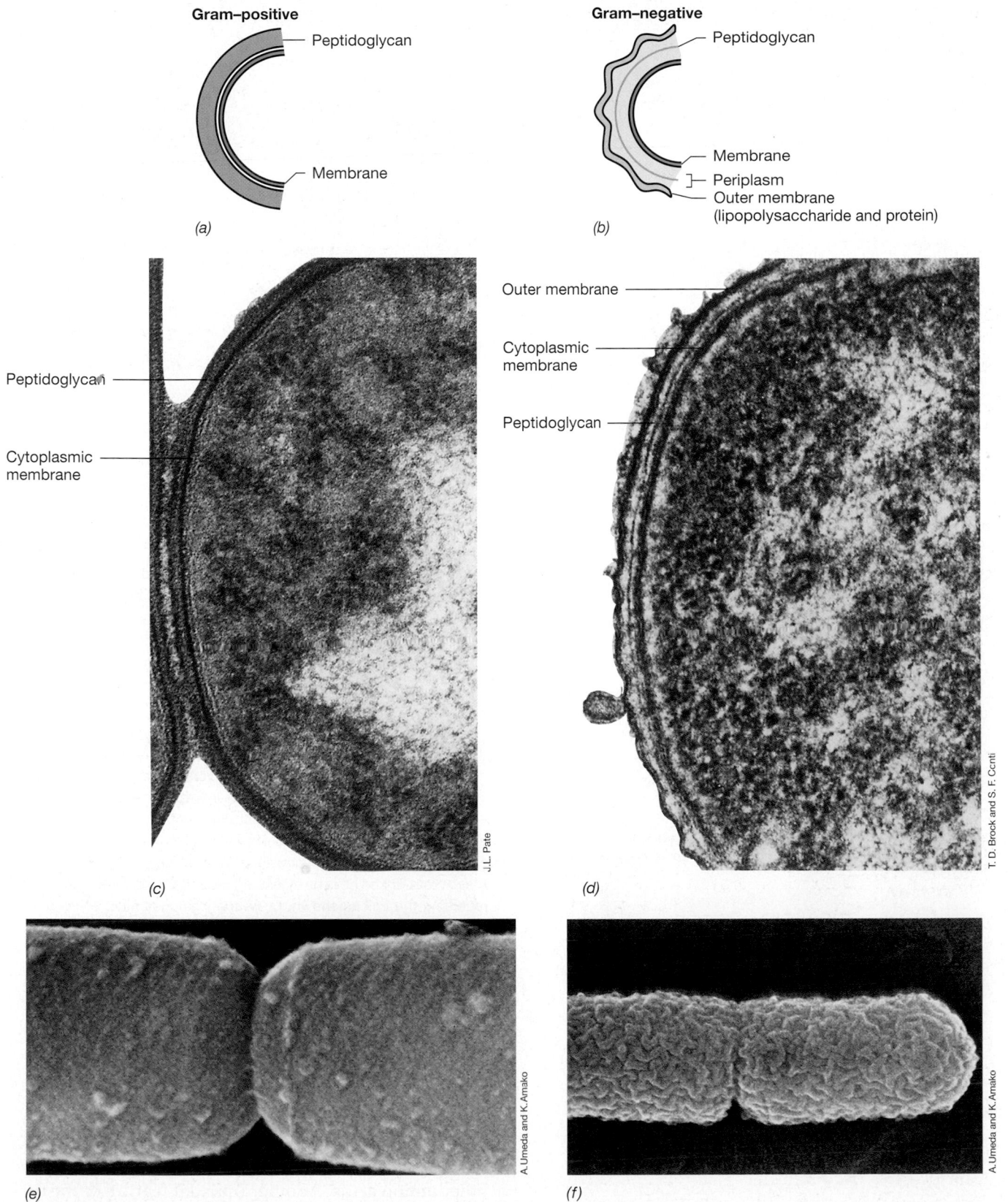

Figure 4.28 Cell walls of *Bacteria*. (a,b) Schematic diagrams of gram-positive and gram-negative cell walls. (c) Electron micrograph showing the cell wall of a gram-positive bacterium, *Arthrobacter crystallopoietes*. (d) Gram-negative bacterium, *Leucothrix mucor*. (e,f) Scanning electron micrographs of gram-positive (*Bacillus subtilis*) and gram-negative (*Escherichia coli*) *Bacteria*. Note the surface texture in the cells shown in (e) and (f). A single cell of *B. subtilis* or *E. coli* is about 1 μm in diameter.

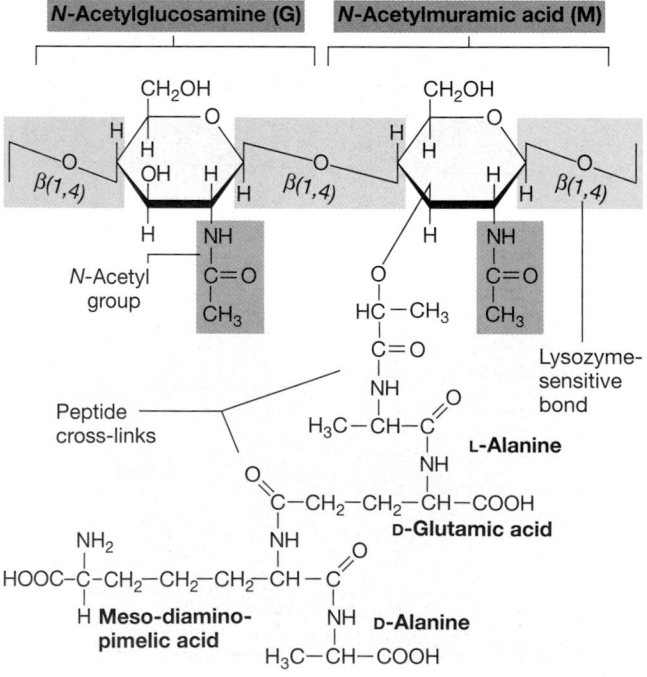

Figure 4.29 (a) Diaminopimelic acid. (b) Lysine. The only difference in the two molecules is highlighted in green.

In gram-positive *Bacteria*, as much as 90% of the cell wall consists of peptidoglycan, although another kind of constituent, teichoic acid (see discussion later in this section), is usually present in small amounts. And, although some bacteria have only a single layer of peptidoglycan surrounding the cell, many *Bacteria*, especially gram-positive *Bacteria*, have several (up to about 25) layers of peptidoglycan. In gram-negative *Bacteria* only about 10% of the wall is peptidoglycan, the majority of the wall consisting of the outer membrane as discussed in Section 4.9. However, the shape of both gram-positive and gram-negative cells is thought to be determined by the lengths of the peptidoglycan chains and by the manner and extent of cross-linking of the chains.

Figure 4.30 Structure of one of the repeating units of the peptidoglycan cell wall structure, the glycan tetrapeptide. The structure given is that found in *Escherichia coli* and most other gram-negative *Bacteria*. In some *Bacteria*, other amino acids are found.

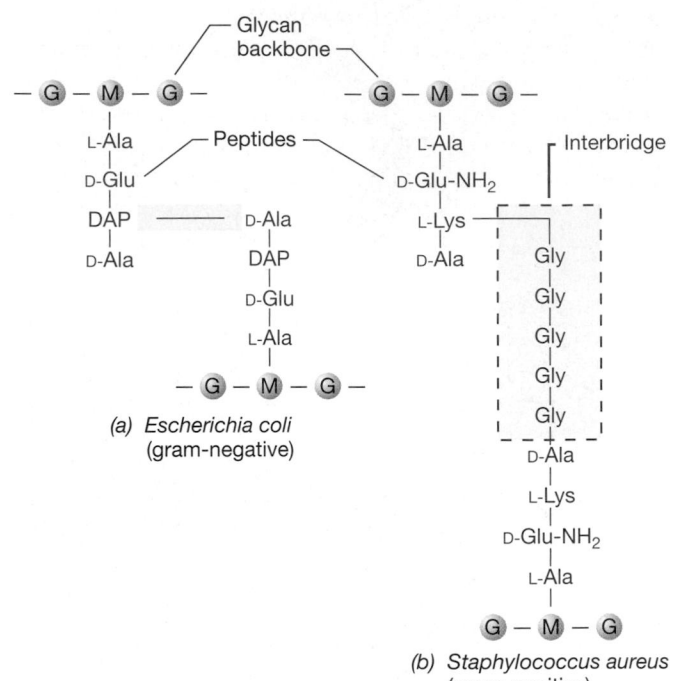

(a) *Escherichia coli*
(gram-negative)

(b) *Staphylococcus aureus*
(gram-positive)

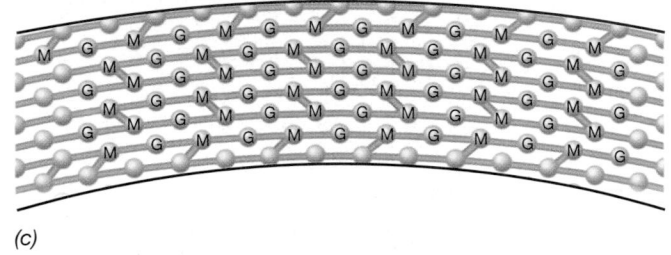

(c)

Figure 4.31 Manner in which the peptide and glycan units are connected in formation of the peptidoglycan sheet. (a) No interbridge in gram-negative *Bacteria*. (b) Glycine interbridge in *Staphylococcus aureus* (gram-positive). (c) Overall structure of peptidoglycan. The diagram depicts several ribbons of peptidoglycan cross-linked to one another. To visualize an entire single layer of peptidoglycan, imagine these cross-linked ribbons extending around a cylinder or sphere representing the cell as shown. G, *N*-acetylglucosamine; M, *N*-acetylmuramic acid.

Diversity in Peptidoglycan

Peptidoglycan is present only in species of *Bacteria*; the sugar *N*-acetylmuramic acid and the amino acid diaminopimelic acid are never found in the cell walls of *Archaea* or *Eukarya*. However, not all *Bacteria* have DAP in their peptidoglycan. This amino acid is present in all gram-negative *Bacteria* and in some gram-positive species, but most gram-positive cocci have lysine instead of DAP, and a few other gram-positive *Bacteria* have other amino acids. Another unusual feature of the bacterial cell wall is the presence of two amino acids that have the D configuration, D-alanine and D-glutamic acid. As we saw in Chapter 3, in proteins amino acids are always of the L isomeric form (⌖ Section 3.6).

Several generalizations regarding peptidoglycan structure can be made. The glycan portion is uniform, with only the sugars N-acetylglucosamine and N-acetylmuramic acid being present, and these sugars are always connected in β-1,4 linkage. The tetrapeptide of the repeating unit shows major variation in only one amino acid, the lysine–diaminopimelic acid alternation. However, the D-glutamic acid at position 2 can be hydroxylated in some organisms, whereas substitutions occur in amino acids at positions 1 and 3 in a few others.

More than 100 different peptidoglycan types are known, and the greatest variation among them occurs in the interbridge. Any of the amino acids present in the tetrapeptide can also occur in the interbridge, but in addition, a number of other amino acids can be found there, such as glycine, threonine, serine, and aspartic acid. However, certain amino acids are never found in the interbridge: branched-chain amino acids, aromatic amino acids, sulfur-containing amino acids, and histidine, arginine, and proline. Thus, it can be stated that although the precise chemistry of peptidoglycan can vary, the structural makeup of peptidoglycan is the same in all forms of the molecule: Glucosamine and muramic acid form the backbone, and the muramic acid molecules are cross-linked with amino acids.

Teichoic Acids and a Summary of the Gram-Positive Wall

Many gram-positive *Bacteria* have acidic polysaccharides called **teichoic acids** embedded in their cell wall.

The term *teichoic acids* encompasses all wall, membrane, or capsular polymers containing glycerophosphate or ribitol phosphate residues. These polyalcohols are connected by phosphate esters and usually have other sugars and D-alanine attached (Figure 4.32*a*). Because they are negatively charged, teichoic acids are partially responsible for the negative charge of the cell surface as a whole and may function to effect passage of ions through the cell wall. Certain glycerol-containing acids are bound to membrane lipids of gram-positive *Bacteria*; because these teichoic acids are intimately associated with lipids, they have been called *lipoteichoic acids*.

Figure 4.32*b* summarizes the structure of the cell wall of gram-positive *Bacteria* and shows how teichoic acids and lipoteichoic acids are arranged in the overall wall structure.

Cells with No Walls

Peptidoglycan, the signature molecule of species of *Bacteria*, can be destroyed by certain agents. One such agent is the enzyme **lysozyme**, a protein that breaks the β-1,4-glycosidic bonds between N-acetylglucosamine and N-acetylmuramic acid in peptidoglycan (Figure 4.30), thereby weakening the wall. Water then enters the cell, and the cell swells and eventually bursts, a process called **lysis** (Figure 4.33*a*). Lysozyme is found in animal secretions including tears, saliva, and other body fluids and presumably functions as a major line of defense against infection by *Bacteria*.

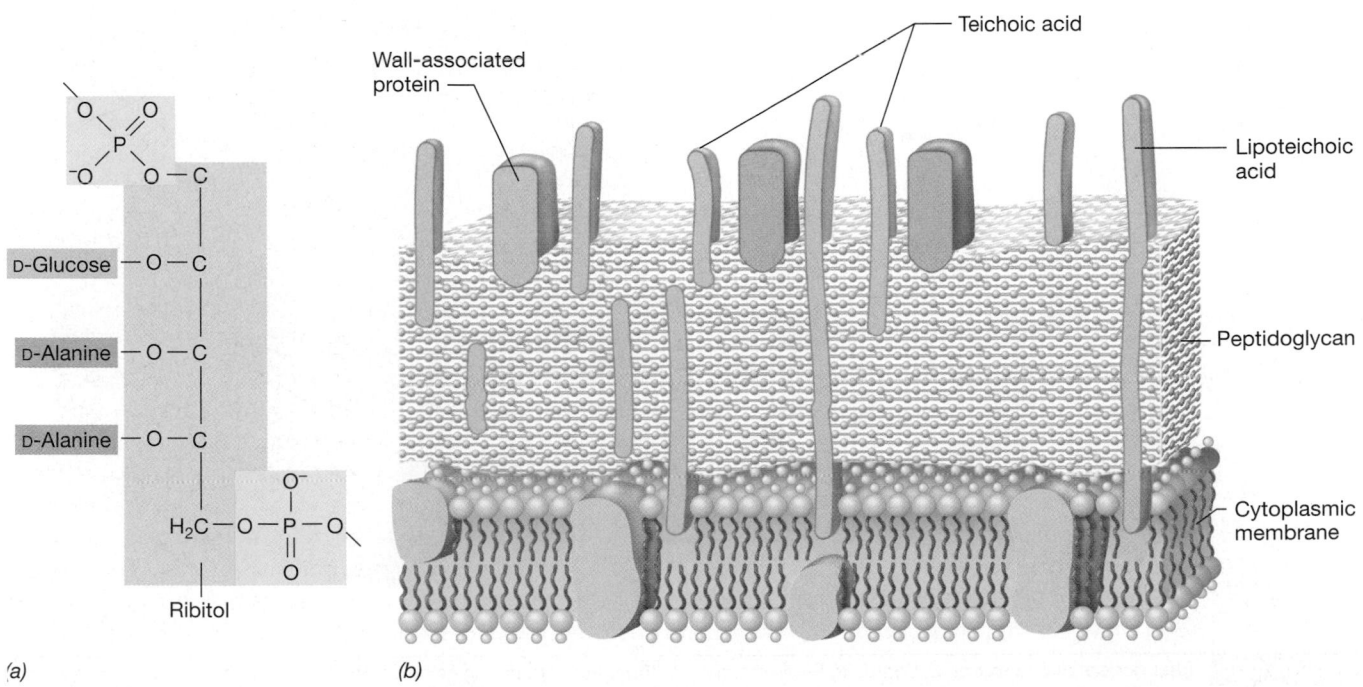

(a) (b)

Figure 4.32 Teichoic acids and the overall structure of the gram-positive cell wall. (a) Structure of the ribitol teichoic acid of *Bacillus subtilis*. The teichoic acid is a polymer of the repeating ribitol units shown here. (b) Summary diagram of the gram-positive cell wall.

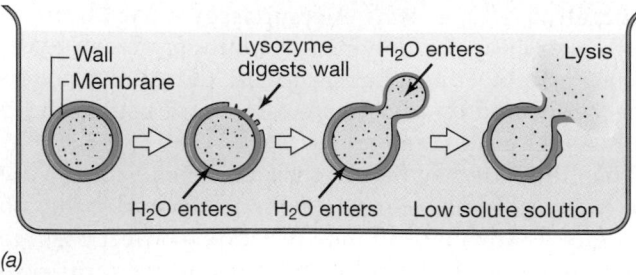

(a)

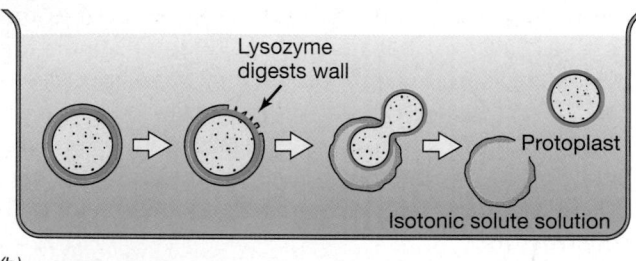

(b)

Figure 4.33 Protoplasts. (a) In dilute solution breakdown of the cell wall releases the protoplast, but it immediately lyses because the cytoplasmic membrane is very weak. (b) In a solution containing an isotonic concentration of a solute such as sucrose, water does not enter the protoplast and it remains stable. Lysozyme breaks the β-1,4 glycosidic bonds in peptidoglycan (see Figure 4.30).

If the proper concentration of a solute that does not penetrate the cell, such as sucrose, is added to the medium, the solute concentration outside the cell balances that inside (conditions are isotonic). Under these conditions, lysozyme still digests peptidoglycan, but water does not enter the cell and lysis does not occur, and instead a **protoplast** (a bacterium that has lost its wall) is formed (Figure 4.33b). If such sucrose-stabilized protoplasts are placed in water, lysis occurs immediately. The word *spheroplast* is often used as a synonym for protoplast, although the two words have slightly different meanings: Protoplasts are generally free of residual cell wall material, whereas spheroplasts usually contain pieces of wall material attached to the otherwise membrane-enclosed structure.

Although most prokaryotes cannot survive without their cell walls, several are able to do so. These include the mycoplasmas, a group that cause certain infectious diseases (∞ Section 12.21), and the *Thermoplasma* group, *Archaea* that naturally lack cell walls (∞ Section 13.5). These prokaryotes are free-living protoplasts and are able to survive without cell walls either because they have unusually tough membranes or because they live in osmotically protected habitats, such as the animal body. Certain mycoplasmas have sterols (see Section 4.5) in their cell membranes, which lends strength and rigidity to this structure.

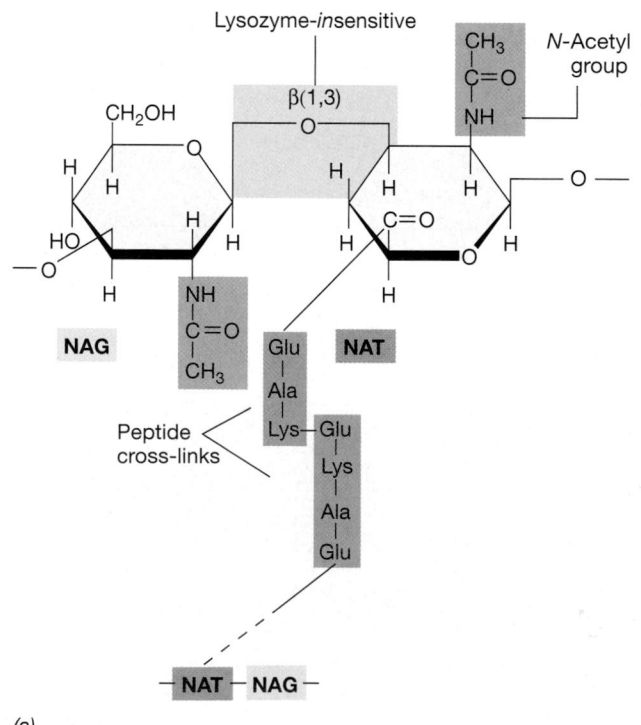

(a)

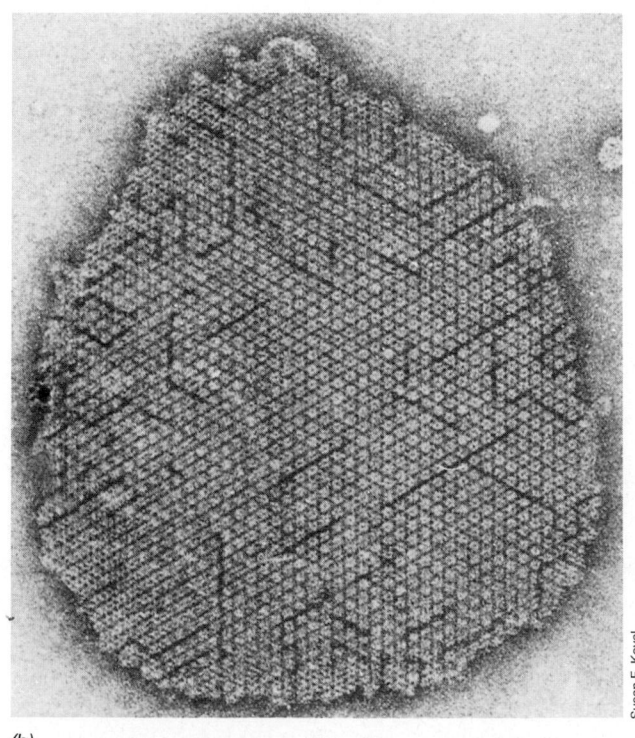

(b)

Susan F. Koval

Figure 4.34 Pseudopeptidoglycan and S-layers. (a) Structure of pseudopeptidoglycan, the cell wall polymer of *Methanobacterium* species. Note the resemblance to the structure of peptidoglycan shown in Figure 4.30, especially the peptide cross-links, in this case between *N*-acetyltalosaminuronic acid (NAT) residues instead of muramic acid residues. NAG, *N*-Acetylglucosamine. (b) Transmission electron micrograph of a portion of an S-layer showing the paracrystalline nature of this cell wall layer. Shown is the S-layer from the prokaryote *Aquaspirillum serpens* (a member of the *Bacteria*); this S-layer displays hexagonal symmetry as do many of the S-layers found in *Archaea*.

Pseudopeptidoglycan and Other Cell Walls of *Archaea*

Some species of *Archaea* contain cell walls constructed of a polysaccharide very similar to that of peptidoglycan. This material is called *pseudopeptidoglycan* (Figure 4.34*a*). The backbone of pseudopeptidoglycan is composed of alternating repeats of *N*-acetylglucosamine and *N*-acetyltalosaminuronic acid (the latter replaces the *N*-acetylmuramic acid of peptidoglycan) (compare Figures 4.30 and 4.34*a*). The backbone of pseudopeptidoglycan also varies from peptidoglycan in that the glycosidic bonds are β-1,3 instead of β-1,4 found in true peptidoglycan (compare Figures 4.30 and 4.34*a*).

Cell walls of other *Archaea* lack both peptidoglycan and pseudopeptidoglycan and consist of polysaccharide, glycoprotein, or protein. For example, *Methanosarcina* species contain thick polysaccharide walls composed of glucose, glucuronic acid, galactosamine, and acetate. Extremely halophilic (salt-loving) *Archaea* such as *Halococcus* contain cell walls similar to that of *Methanosarcina* but that contain, in addition, an abundance of sulfate (SO_4^{2-}) residues. However, the most common wall type among *Archaea* is the paracrystalline surface layer (S-layer) (see Section 4.13 for more discussion) consisting of protein or glycoprotein, generally of hexagonal symmetry. S-layers have been found among species of all groups of *Archaea*, the extreme halophiles, the methanogens, and the hyperthermophiles. Several species of *Bacteria* also have S-layers on their outer surfaces (Figure 4.34*b*).

We thus see in species of *Archaea* a great variety of cell wall types, varying from molecules that closely resemble peptidoglycan to cell walls totally lacking a polysaccharide component. But with rare exception, all *Archaea* contain a cell wall of some sort, and as in *Bacteria*, the archaeal cell wall functions to prevent osmotic lysis and to define cell shape. In addition, because they lack peptidoglycan in their cell walls, all *Archaea* are naturally resistant to the action of lysozyme (see earlier) and penicillin, agents that destroy this molecule or prevent its proper synthesis (∞ Section 6.2), respectively.

✓ 4.8 Concept Check

The cell walls of *Bacteria* contain a polysaccharide called peptidoglycan. This material consists of strands of alternating repeats of *N*-acetylglucosamine and *N*-acetylmuramic acid, with the latter cross-linked between strands by short peptides. *Archaea* lack peptidoglycan but contain walls made of other polysaccharides or of protein. The enzyme lysozyme destroys peptidoglycan, leading to cell lysis.

- ✓ List the monomeric components of peptidoglycan.
- ✓ Why is peptidoglycan such a strong macromolecule?
- ✓ How is a protoplast generated?
- ✓ How do some cells live without cell walls?
- ✓ How does pseudopeptidoglycan resemble peptidoglycan? How do the two molecules differ?

4.9 The Outer Membrane of Gram-Negative *Bacteria*

Besides peptidoglycan, gram-negative *Bacteria* contain an additional wall layer made of **lipopolysaccharide**. This layer is effectively a second lipid bilayer, but it is not constructed solely of phospholipid, as is the cytoplasmic membrane; instead it contains polysaccharide and protein. The lipid and polysaccharide are intimately linked in the outer layer of the outer membrane to form specific lipopolysaccharide structures. Because of the presence of lipopolysaccharide, this outer layer is frequently called the **lipopolysaccharide layer**, or simply **LPS**. Another term in widespread use for this structure is the **outer membrane**.

Chemistry of LPS

Although complex, the chemistry of the LPS of several bacteria is now understood. As seen in Figure 4.35, the polysaccharide consists of two portions, the core polysaccharide and the O-polysaccharide. In *Salmonella*, where it has been best studied, the **core polysaccharide** consists of

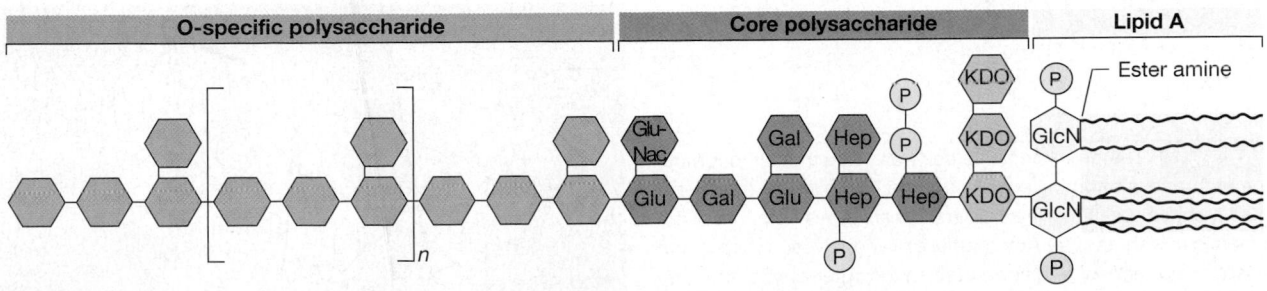

Figure 4.35 Structure of the lipopolysaccharide of gram-negative *Bacteria*. The precise chemistry of lipid A and the polysaccharide components varies among species of gram-negative *Bacteria*, but the sequence of major components (lipid A–KDO–core–O-specific) is generally uniform. The O-specific polysaccharide varies among species. KDO, ketodeoxyoctonate; Hep, heptose; Glu, glucose; Gal, galactose; GluNac, *N*-acetylglucosamine; GlcN, glucosamine; P, phosphate. Glucosamine and the lipid A fatty acids are linked by an ester amine bond. The lipid A portion of LPS can be toxic to animals and comprises the *endotoxin complex* (∞ Section 21.12). Compare this figure with Figures 4.36 and 4.37, and note the color coding of different portions of the LPS in Figures 4.35 and 4.36.

ketodeoxyoctonate (KDO), seven-carbon sugars (heptoses), glucose, galactose, and *N*-acetylglucosamine. Connected to the core is the *O-polysaccharide*, which usually contains galactose, glucose, rhamnose, and mannose (all six-carbon sugars), as well as one or more unusual dideoxy sugars such as abequose, colitose, paratose, or tyvelose. These sugars are connected in four- or five-membered sequences, which often are branched. When the sequences are repeated, the long *O*-polysaccharide is formed.

The relationship of the *O*-polysaccharide to the rest of the LPS is shown in Figure 4.36. The lipid portion of the lipopolysaccharide, referred to as **lipid A** (Figure 4.35) is

not a glycerol lipid, but instead the fatty acids are connected by ester amine linkage to a disaccharide composed of *N*-acetylglucosamine phosphate (Figure 4.35). The disaccharide is attached to the core *O*-polysaccharides through KDO (Figure 4.35). Fatty acids commonly found in lipid A include caproic, lauric, myristic, palmitic, and stearic acids. In the outer membrane, the LPS associates with various proteins to form the *outer* half of the unit membrane structure. A **lipoprotein** complex is found on the *inner* side of the outer membrane of a number of gram-negative *Bacteria* (Figure 4.36a). This lipoprotein is a small protein that functions as an anchor between the outer

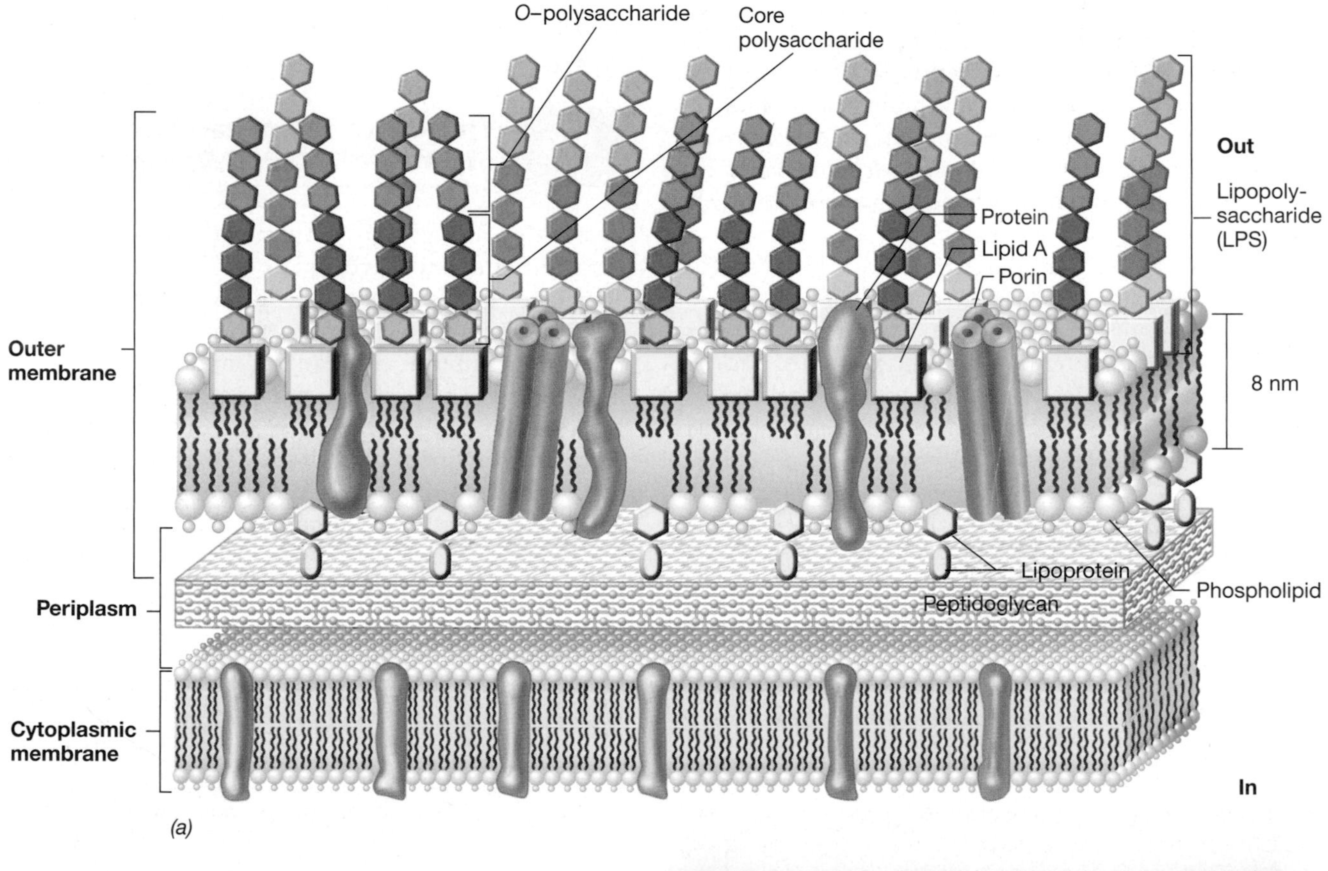

Figure 4.36 The gram-negative cell wall. Note that although the outer membrane is often called the "second lipid bilayer," the chemistry and architecture of this layer differs in many ways from that of the cytoplasmic membrane. (a) Arrangement of lipopolysaccharide, lipid A, phospholipid, porins, and lipoprotein in the outer membrane. See Figure 4.35 for details of the structure of LPS. Lipid A can be toxic in humans, and if so, is referred to as endotoxin (⊙⊙ Section 21.12). (b) Molecular model of porin proteins. Note the three pores present, one formed from each of the proteins forming a porin molecule. The view is perpendicular to the plane of the membrane. Model based on X-ray diffraction studies of *Rhodobacter blasticus* porin.

Georg E. Schulz

membrane and peptidoglycan. In the *outer* leaf of the outer membrane, LPS replaces phospholipids; the latter are found predominantly in the inner leaf (Figure 4.36*a*).

Endotoxin

Although the major function of the outer membrane is structural, one of its important biological properties is that it is frequently *toxic* to animals. Gram-negative *Bacteria* that are pathogenic for humans and other mammals include members of the genera *Salmonella, Shigella,* and *Escherichia,* among others.

Some of the symptoms these pathogens elicit in their hosts are due to their toxic outer membrane. The toxic properties are associated with part of the lipopolysaccharide layer, in particular, lipid A, of these organisms. The term **endotoxin** is used to refer to this toxic component of LPS, as we will discuss in Section 21.11. Interestingly, LPS from several nonpathogenic bacteria have been shown to have endotoxin activity. Thus, the organism itself need not be pathogenic to contain toxic cell wall components.

Porins and the Periplasm

Unlike the cytoplasmic membrane, the outer membrane of gram-negative *Bacteria* is relatively permeable to small molecules even though it is basically a lipid bilayer. This is because proteins called **porins** are present in the outer membrane of gram-negative *Bacteria* and function as channels for the entrance and exit of hydrophilic low-molecular-weight substances (Figure 4.36). Several porins exist, and both specific and nonspecific classes are known. *Nonspecific porins* form water-filled channels through which small substances of any type can pass. By contrast, some porins are highly specific because they contain a specific binding site for one or a group of structurally related substances.

Structural studies have shown that most porins are proteins containing *three* identical subunits. Porins are transmembrane proteins (Figure 4.36*a*) and associate to form small membrane holes about 1 nm in diameter (Figure 4.36*b*). Through the action of porins, the outer membrane is relatively permeable to small molecules. However, the outer membrane is *not* permeable to enzymes or other large molecules. In fact, one of the major functions of the outer membrane is to keep certain enzymes, which are present outside the cytoplasmic membrane, from diffusing away from the cell. These enzymes are present in a region called the **periplasm** (see Figures 4.36 and 4.37). In *Escherichia coli* this space between the outer surface of the cytoplasmic membrane and the inner surface of the LPS-containing outer membrane occupies a distance of about 12–15 nm and is gel-like in consistency because of the abundance of periplasmic proteins found there (Figure 4.37). The periplasm contains several proteins including *hydrolytic enzymes,* which function in the initial degradation of food molecules, *binding proteins,* which begin the process of transporting sub-

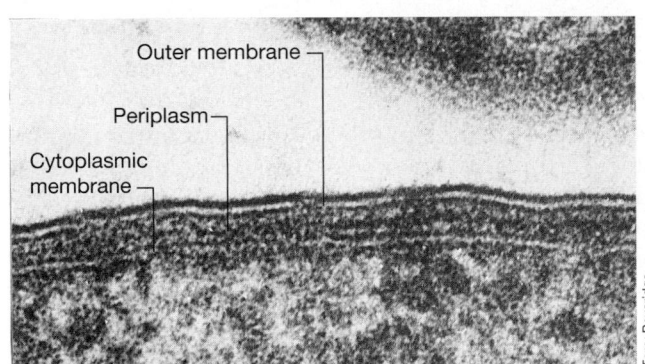

Figure 4.37 High magnification thin section of the cell envelope of *Escherichia coli* showing the periplasmic gel bounded by the outer membrane and the cytoplasmic membrane. The large, dark particles in the cytoplasm are ribosomes.

strates (see Section 4.7), and *chemoreceptors,* which are proteins involved in the chemotaxis response (see Sections 4.12 and 8.11). As previously discussed, most of these proteins reach the periplasm via transport by the SecYEG system (see Section 4.7).

Relationship of Cell Wall Structure to the Gram Stain

Are the structural differences between the cell walls of gram-positive and gram-negative *Bacteria* responsible in any way for the Gram stain reaction? In the Gram stain (see Section 4.1), an insoluble crystal violet-iodine complex is formed inside the cell, and this complex is extracted by alcohol from gram-*negative* but not from gram-*positive Bacteria*. Gram-positive *Bacteria*, which have very thick cell walls consisting of several layers of peptidoglycan, become dehydrated by the alcohol. This causes the pores in the walls to close, preventing the insoluble crystal violet-iodine complex from escaping. In gram-negative *Bacteria*, alcohol readily penetrates the lipid-rich outer layer, and the thin peptidoglycan layer also does not prevent solvent passage, thus, the crystal violet-iodine complex is easily removed.

✓ 4.9 Concept Check

In addition to peptidoglycan, gram-negative *Bacteria* contain an outer membrane consisting of lipopolysaccharide, protein, and lipoprotein. Proteins called porins allow for permeability across the outer membrane. The space between the membranes is the periplasm, which contains various proteins involved in important cellular functions.

- ✓ What components constitute the LPS layer of gram-negative *Bacteria*?
- ✓ What is the function of porins and where are they located in a gram-negative cell wall?
- ✓ What component of the cell has endotoxin properties?
- ✓ Why does alcohol readily decolorize gram-negative *Bacteria*?

III MICROBIAL LOCOMOTION

Many cells are able to move. Although movement requires an energy expenditure, the ability to move about its habitat may have profound ecological consequences for a cell and may spell the difference between life and death. In the next few sections we examine the different types of cell movement including swimming and gliding, and then consider how the ability to move toward or away from particular stimuli can be of benefit to the cell.

4.10 Flagella and Motility

Many prokaryotes are motile, and this function is usually due to a special structure, the **flagellum** (plural, **flagella**) (Figure 4.38). Certain bacterial cells can move along solid surfaces by *gliding* (see Section 4.11), and certain aquatic microorganisms can regulate their position in a water column by gas-filled structures called gas vesicles (see Sections 4.14 and 12.25). The majority of motile prokaryotes, however, move by means of flagella. Motility allows the cell to reach different regions of its environment. In the struggle for survival, movement to a new location may offer a cell new resources and opportunities. But, as in any physical process, cell movement is closely tied to an energy expenditure, and the movement of flagella is no exception. We begin now with a detailed consideration of flagellar motility in prokaryotes.

Bacterial Flagella

Bacterial flagella are long, thin appendages free at one end and attached to the cell at the other end. Flagella are so thin (about 20 nm) that a single flagellum can never

be seen directly with the light microscope, but only after staining with special flagella stains that increase their diameter (Figure 4.38). Flagella are readily seen with the electron microscope (Figure 4.39).

Flagella are arranged differently on different bacteria. In **polar flagellation**, the flagella are attached at one or both ends of the cell (Figures 4.38b and 4.39a). Occasionally a tuft (group) of flagella may arise at one end of the cell, an arrangement called **lophotrichous** (*lopho* means "tuft"; *trichous* means "hair") (Figure 4.38c). Tufts of flagella of this type can be seen in living cells by dark-field microscopy (see Section 4.1 and Figure 4.40a), where the flagella appear light and are attached to light-colored cells against a dark background. In extremely large prokaryotes, tufts of flagella can also be observed by phase contrast microscopy (Figure 4.40b). In **peritrichous flagellation** (Figures 4.38a and 4.39b), the flagella are inserted at many places around the cell surface (*peri* means "around"). Besides their usefulness in motil-

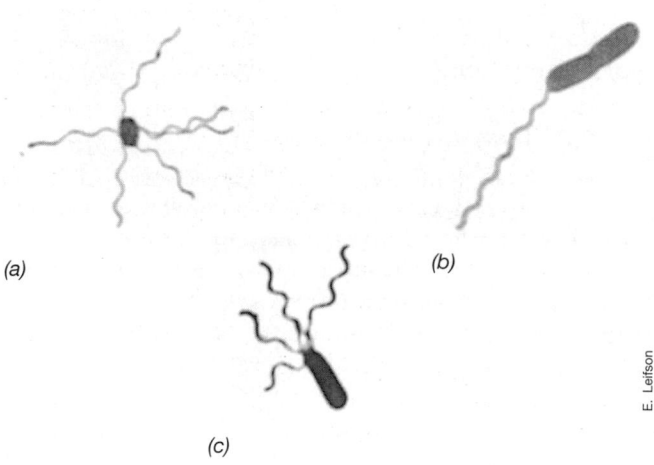

(a)

(b)

(c)

E. Leifson

Figure 4.38 Light photomicrographs of prokaryotes containing different flagellar arrangements. Cells are stained with Leifson flagella stain. (a) Peritrichous. (b) Polar. (c) Lophotrichous.

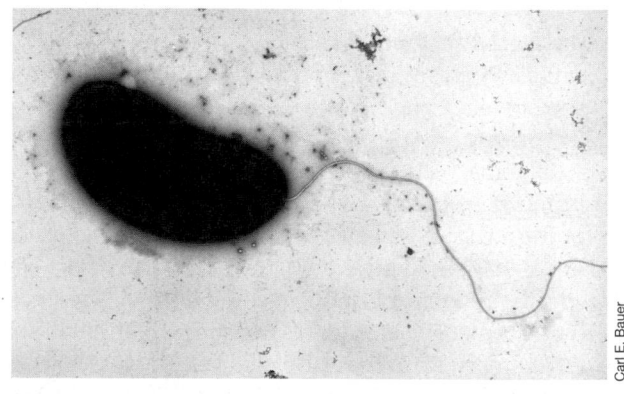

(a)

Carl E. Bauer

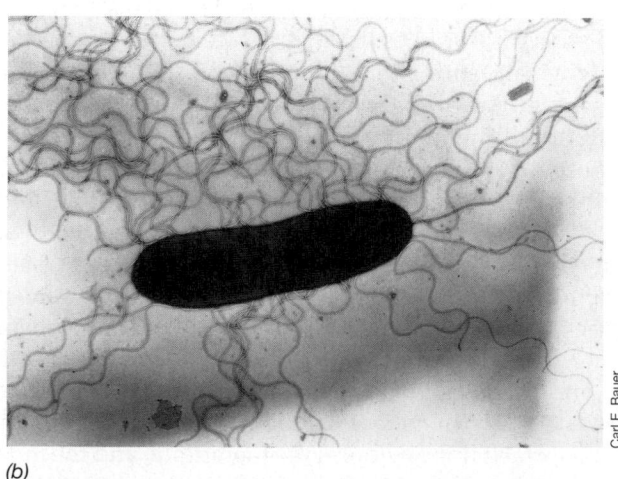

(b)

Carl E. Bauer

Figure 4.39 Bacterial flagella as observed by negative staining in the transmission electron microscope. (a) Polar flagella. (b) Peritrichous flagella. Both micrographs are of cells of the phototrophic bacterium *Rhodospirillum centenum*. Cells of *R. centenum* are normally polarly flagellated but under certain growth conditions form peritrichously flagellated "swarmer" cells. See also Figure 4.48b.

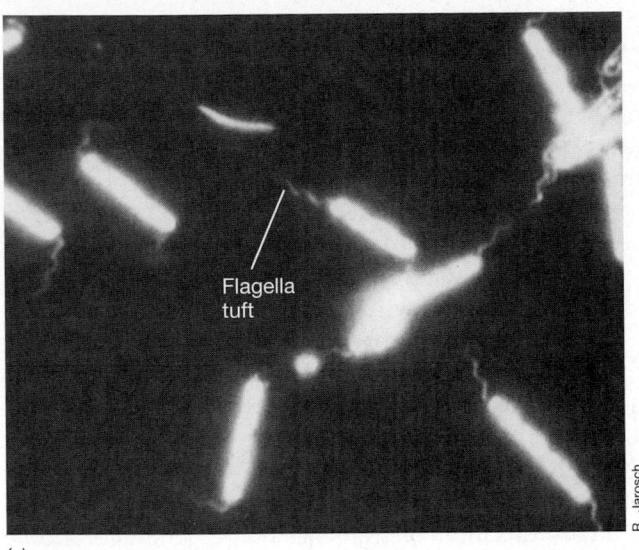

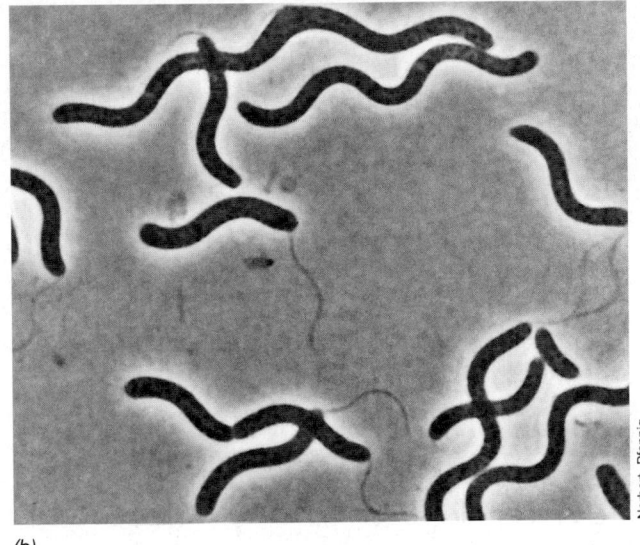

(a)

(b)

R. Jarosch

Norbert Pfennig

Figure 4.40 Bacterial flagella as observed in living cells. (a) Dark-field photomicrograph of a group of large rod-shaped bacteria with flagellar tufts at each pole. A single cell is about 2 μm wide. Dark-field microscopy uses horizontal illumination to yield reflected light (see Section 4.1 and Figure 4.5c). (b) Phase contrast photomicrograph of the large phototrophic purple bacterium *Rhodospirillum photometricum*. A single cell measures about 3 × 30 μm. Note the lophotrichous flagella that emanate from one of the poles.

ity, the type of flagellation, polar or peritrichous, that a cell shows, is used as a characteristic in the classification of bacteria.

Flagellar Structure

Flagella are not straight but helically shaped; when flattened, they show a constant distance between two adjacent curves, called the *wavelength,* and this wavelength is constant for a given organism (Figures 4.38–4.40). The filament of bacterial flagella is composed of subunits of a protein called **flagellin**. The shape and wavelength of the flagellum are in part determined by the structure of the flagellin protein and also to some extent by the direction of rotation of the filament. The basic flagellar structure to be described here varies little among species of *Bacteria,* however, in *Archaea* several different flagellins are known, and the flagellar structure is probably quite different from that of *Bacteria*. In species of the latter, flagellin is highly conserved, suggesting that flagellar motility has deep evolutionary roots within this prokaryotic group.

The base of the flagellum is different in structure from that of the filament (Figure 4.41). There is a wider region at the base of the flagellum called the *hook*. The hook consists of a single type of protein and functions to connect the filament to the motor portion of the flagellum. The *motor* is anchored in the cytoplasmic membrane and cell wall. The motor consists of a small central rod that passes through a system of rings. In gram-negative *Bacteria,* an outer ring is anchored in the lipopolysaccharide layer and another in the peptidoglycan layer of

the cell wall, and an inner ring is located within the cytoplasmic membrane (Figure 4.41). In gram-positive *Bacteria,* which lack the outer lipopolysaccharide layer, only the inner pair of rings is present. Surrounding the inner ring and anchored in the cytoplasmic membrane are a pair of proteins called Mot (Figure 4.41). These proteins actually drive the motor causing a torque that rotates the filament. A final set of proteins, called the Fli proteins (Figure 4.41) function as the motor switch, reversing rotation of the flagella in response to intracellular signals.

Flagellar Synthesis

Several genes are required for flagellar synthesis and subsequent motility. In *Escherichia coli* and *Salmonella typhimurium,* where studies have been most extensive, over 40 genes are necessary for motility. These genes have several functions, including encoding structural proteins of the flagellar apparatus, export of flagellar components through the membrane to the outside of the cell, and regulation of the many biochemical events surrounding the synthesis of new flagella.

An individual flagellum grows not from its base, as does an animal hair, but from the tip. The MS ring is synthesized first and inserted into the membrane. Then other anchoring proteins are synthesized along with the hook before filament formation occurs (Figure 4.42). Flagellin molecules synthesized in the cytoplasm pass up through a 3-nm channel inside of the filament and add on at the terminus to form the final flagellum. At the end of the growing flagellum a protein "cap" exists, and cap proteins assist flagellin molecules that have

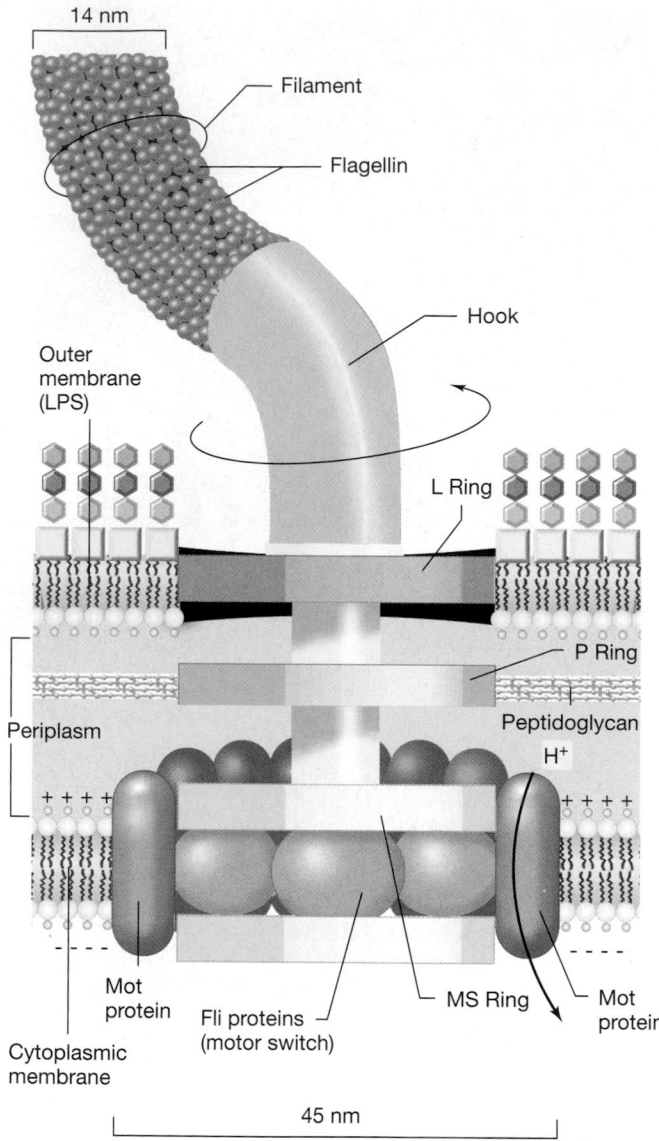

Figure 4.41 Structure of the prokaryotic flagellum and attachment to the cell wall and membrane in gram-negative *Bacteria.* The L ring is embedded in the LPS, and the P ring in peptidoglycan. The MS ring is embedded in the cytoplasmic membrane. A narrow channel exists in the filament through which flagellin molecules diffuse to reach the site of flagellar synthesis. The Mot proteins function as the flagellar motor, whereas the Fli proteins function as the motor switch. The flagellar motor rotates the filament to propel the cell through the medium.

diffused through the channel organize at the flagellum termini to form a new portion of filament (Figure 4.42). Growth of the flagellum occurs more-or-less continuously until the flagellum reaches its final length. Broken flagella still rotate and can be repaired with new flagellin units passed through the filament channel to replace the lost ones.

Flagellar Movement

How is motion imparted to the flagellum? Each individual flagellum is actually a semirigid structure that does not flex but, as mentioned previously, moves by rotation, in the manner of a propeller. Visual evidence of this can be obtained by observing the behavior of motile cells tethered by their flagella to microscope slides. Such cells rotate around the point of attachment at rates of revolution consistent with those inferred for flagellar movement in free-swimming cells.

The rotary motion of the flagellum is imparted from the motor. The energy required for rotation of the flagellum comes from the proton motive force (see Sections 4.6 and 5.12). Proton movement across the membrane through the Mot complex (Figure 4.41) drives rotation of the flagellum, and calculations have shown that about 1000 protons must be translocated per single rotation of the flagellum.

Flagella do not rotate at a constant speed but instead can increase or decrease their rotational speed in relation to the strength of the proton motive force. Flagella rotation can move bacteria through liquid media at speeds of up to 60 cell lengths/second (sec). Although this is only about 0.00017 kilometer/hour (km/h), when comparing this speed with that of higher organisms in terms of the number of lengths moved per second, it is extremely fast. The fastest animal, the cheetah, moves at a maximum rate of about 110 km/h, but this represents only about 25 body lengths/sec. Thus, when size is accounted for, prokaryotic cells swimming at 50–60 lengths/sec are actually moving much faster than larger organisms.

The motions of polarly and lophotrichously flagellated organisms are different from those of peritrichously flagellated organisms. Peritrichously flagellated organisms generally move in a straight line in a slow, stately fashion. Polarly flagellated organisms, on the other hand, move more rapidly, spinning around and dashing from place to place. The different behavior of flagella on polar and peritrichous organisms, including differences in reversibility of the flagellum, is illustrated in Figure 4.43.

✓ 4.10 Concept Check

Motility in microorganisms is usually associated with flagella. In prokaryotes the flagellum is a complex structure made of several proteins, most of which are anchored in the cell wall and membrane. The flagellum filament, which is made of a single kind of protein, rotates at the expense of the proton motive force, which drives the flagellar motor.

✓ What is *flagellin* and where is it found?
✓ How does a bacterial flagellum move a cell forward?
✓ How does *polar flagellation* differ from *peritrichous flagellation?*

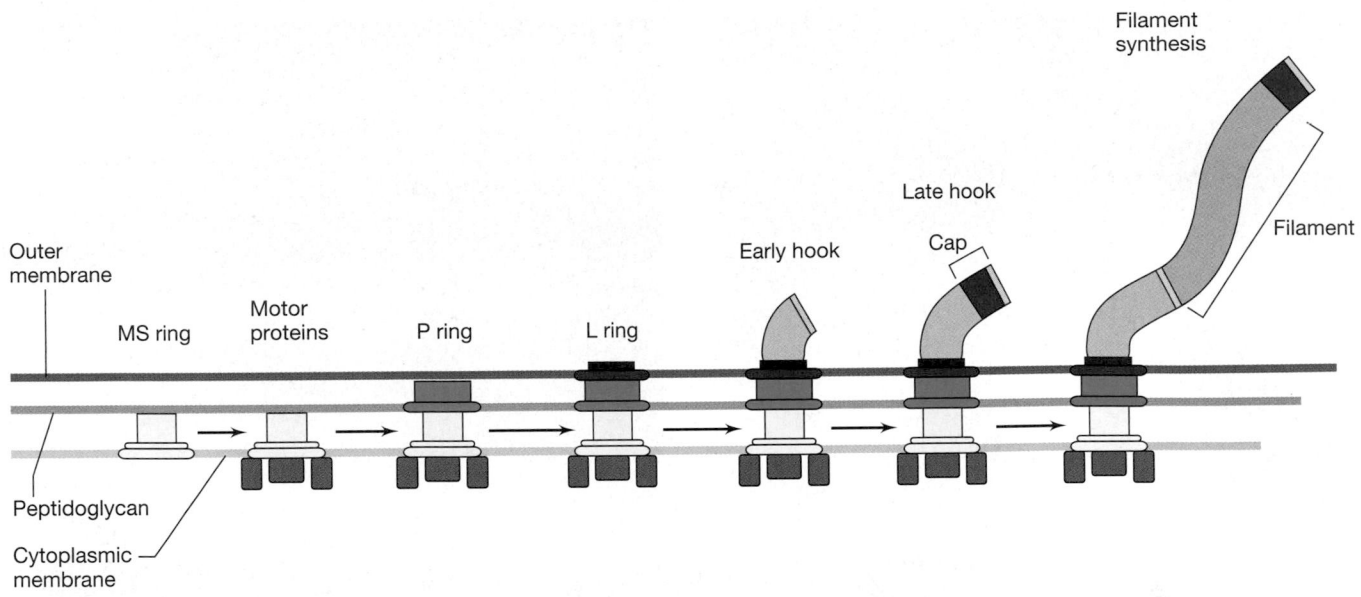

Figure 4.42 Summary of steps in flagella biosynthesis. Synthesis begins with MS ring assembly in the membrane. This is followed by formation of the other rings, the hook and the cap. At this point, flagellin protein (approximately 20,000 copies are needed to make one filament) flows through the hook to form the filament. Flagellin molecules are guided into position by cap proteins to ensure that the growing filament develops evenly.

4.11 Gliding Motility

Many prokaryotes are motile but lack flagella. These nonswimming bacteria move across solid surfaces in a process called **gliding**. Gliding motility is widely distributed among *Bacteria* but has been well studied in only a few groups. The gliding movement itself—up to 10 μm/sec in some gliding bacteria—is considerably slower than propulsion by flagella, but still offers the cell a means of moving about its habitat.

Gliding prokaryotes are filamentous or rod-shaped cells (Figure 4.44), and the gliding process requires that the cells be in contact with a solid surface. The morphology of colonies of a typical gliding bacterium (colonies are masses of bacterial cells that form from successive cell divisions of a single cell, ◀▶ Section 5.3), are distinctive, since cells glide out and move away from the center of the colony (Figure 4.44c). Perhaps the most well-known gliding bacteria are the filamentous cyanobacteria (Figure 4.44a, b and ◀▶ Section 12.25), certain

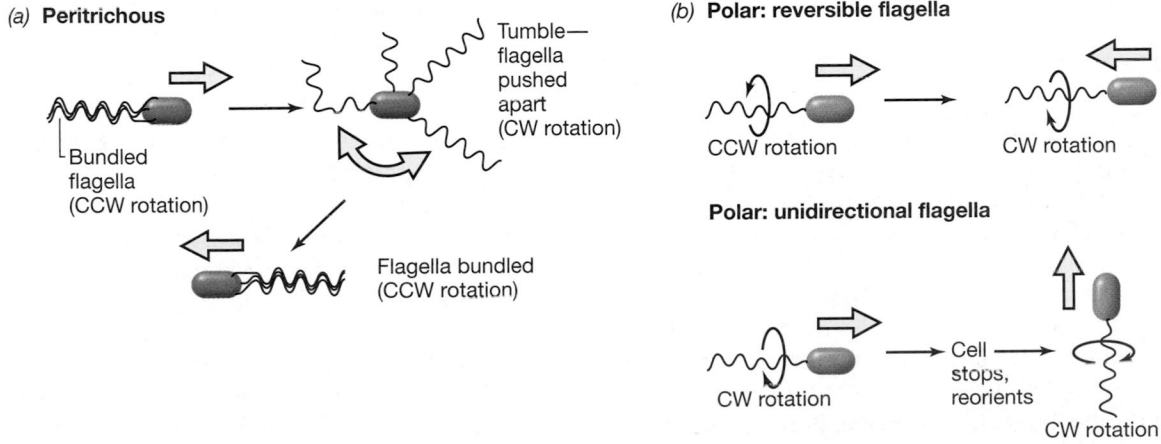

Figure 4.43 Manner of movement in polarly and peritrichously flagellated prokaryotes. (a) Peritrichous: Forward motion is imparted by all flagella rotating counterclockwise (CCW) in a bundle. Clockwise (CW) rotation causes the cell to tumble, and then a return to counterclockwise rotation leads the cell off in a new direction. (b) Polar: Cells change direction by reversing flagellar rotation (thus pulling instead of pushing the cell), or in unidirectional flagella, by stopping periodically to reorient, and then moving forward by clockwise rotation of its flagella. The yellow arrows show the direction the cell is traveling.

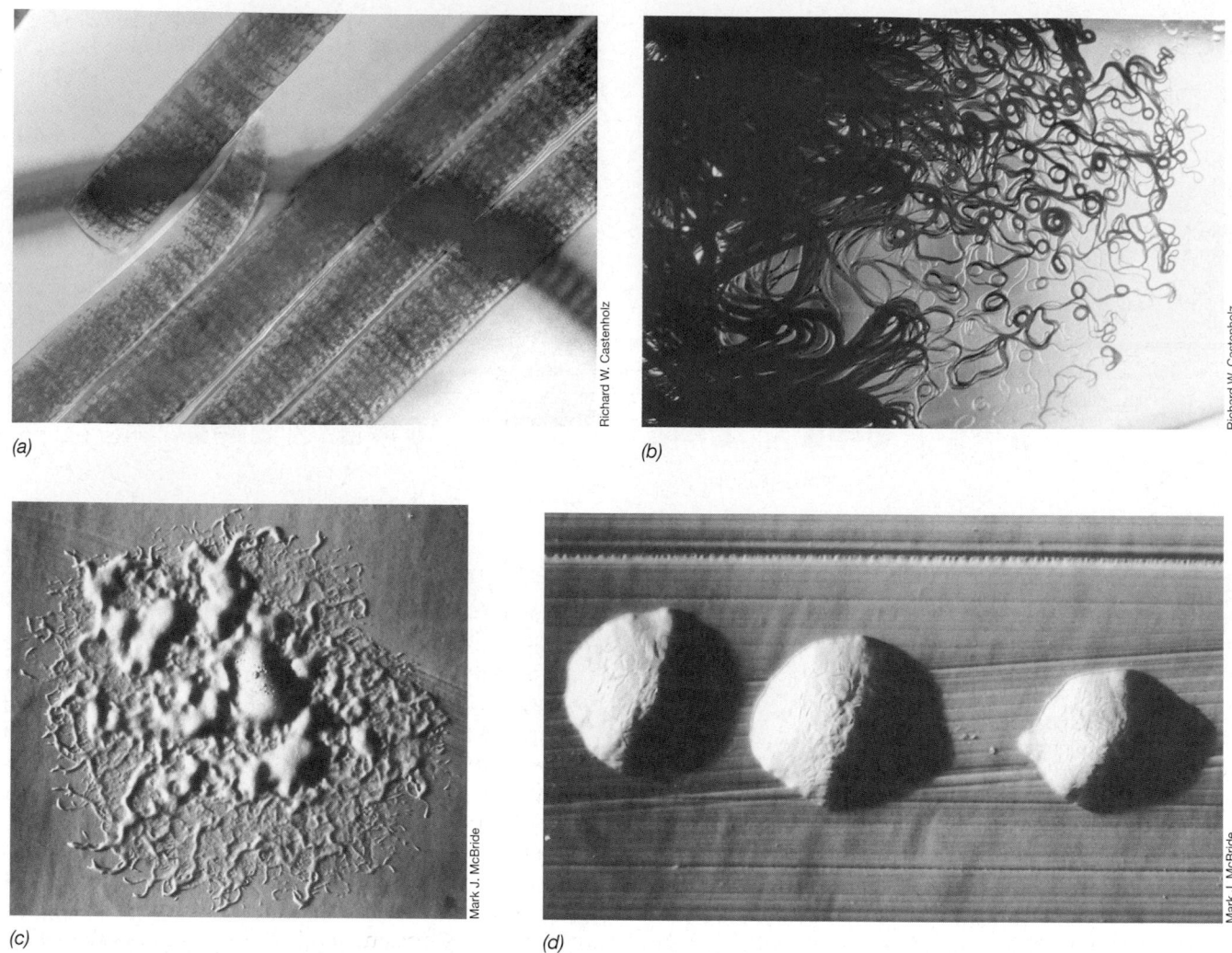

(a)

(b)

(c)

(d)

Richard W. Castenholz (a, b)

Mark J. McBride (c, d)

Figure 4.44 Gliding bacteria. (a, b) The large filamentous cyanobacterium *Oscillatoria princeps*. (a) Photomicrograph. A cell is about 35 μm wide. (b) Photograph of filaments gliding on an agar surface. Cells can move by gliding against a solid surface or one filament can glide using a second filament as the solid surface. (c, d) The gram-negative gliding bacterium *Flavobacterium johnsoniae*. (c) Masses of cells gliding away from the center of the colony (the colony is about 2.7 mm wide). (d) Nongliding mutant strain showing typical colony morphology of nongliding bacteria (the colonies are 0.7–1 mm in diameter). For the proposed mechanism of gliding in *F. johnsoniae* cells, see Figure 4.45.

gram-negative *Bacteria* such as *Myxococcus xanthus* and other myxobacteria (∞ Section 12.17), and species of *Cytophaga* and *Flavobacterium* (Figure 4.44*c, d* and ∞ Section 12.31).

Mechanisms of Gliding Motility

Although no gliding mechanism has been positively identified, there are models for bacterial gliding and evidence that more than one mechanism is responsible. In cyanobacteria (Figure 4.44*a, b*) it is known that a polysaccharide slime is secreted on the outer surface of the cell as it glides. The slime appears to contact both the cell surface and the solid surface against which the gliding cell moves; as the excreted slime adheres to the surface, the cell is pulled along. This hypothesis is supported by the observation of slime-

excreting pores on the cell surface of several filamentous cyanobacteria.

Slime extrusion is clearly not the mechanism of gliding in nonphototrophic gliding bacteria. In *Flavobacterium johnsoniae* (Figure 4.45), for example, the movement of proteins in the cell surface is the likely mechanism of gliding. In the *F. johnsoniae* model, specific motility proteins anchored in the cytoplasmic and outer membranes are thought to propel the cell forward in a type of continuous ratcheting mechanism (Figure 4.45). The movement of the cytoplasmic membrane proteins are likely driven by energy released from the proton motive force (∞ Sections 4.6 and 5.12), and they somehow transmit this energy to outer membrane proteins located along a "race track" on the cell surface (Figure 4.45). It is thought that movement of the race track

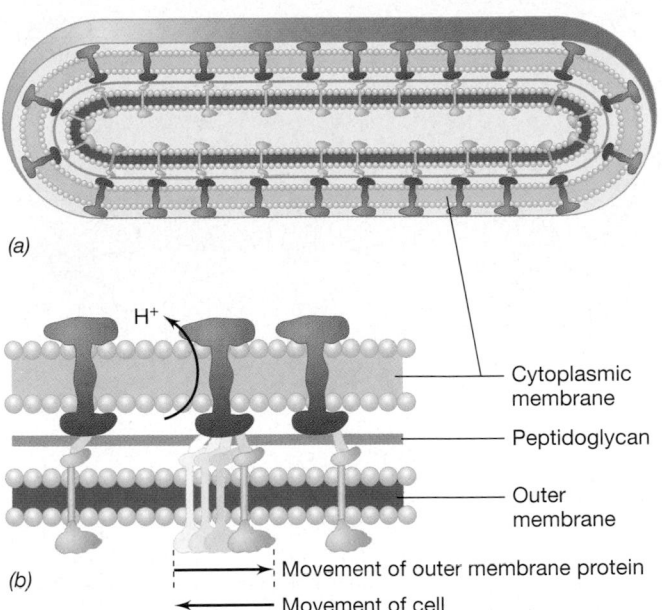

(a)

H⁺

Cytoplasmic membrane

Peptidoglycan

Outer membrane

(b)

Movement of outer membrane protein

Movement of cell

Figure 4.45 Proposed model for gliding motility in *Flavobacterium johnsoniae* and some other gliding bacteria (model courtesy of Dr. Mark J. McBride). (a) Cut away view of a gram-negative cell showing specific gliding proteins in the cytoplasmic membrane and outer membrane. (b) Detailed view. Tracks (yellow) are thought to exist in the peptidoglycan that connect cytoplasmic proteins (brown) to outer membrane proteins (orange) and propel them along the solid surface. Note the difference in direction of movement of outer membrane proteins and of the cell proper.

proteins against the solid surface literally pulls the cell forward.

Like other forms of motility, gliding motility has significant ecological relevance. Gliding allows a cell to exploit new resources or interact in some beneficial way with other cells. In the latter connection, it is of interest that the myxobacteria, classic examples of gliding bacteria, have a very social and cooperative lifestyle where gliding motility may play important roles in cell-to-cell interactions (ᴼᴼᴼ Section 12.17).

4.12 Behavioral Responses: Chemotaxis, Phototaxis, and Other Taxes

Although not all prokaryotes are motile, many are, and it is reasonable to assume that motility confers a selective advantage on cells under certain environmental conditions. Prokaryotes often encounter *gradients* of physical and chemical agents in nature, and the motility machinery in the cell has evolved to respond in a positive or negative way to these gradients by directing movement of the cell either toward or away from the signal molecule, respectively. Such directed movements are called *taxes,* and a variety of such responses occur in microorganisms. **Chemotaxis,** a response to chemicals, and **phototaxis,** a response to light, are two well-known taxes on which we focus here.

Chemotaxis as a phenomenon has been well studied in swimming bacteria, and much is known at the genetic level concerning how the chemical status of the external environment is communicated to the flagellar assembly. Our discussion here will thus deal solely with swimming bacteria. However, some gliding bacteria have also been shown to be chemotactic, and phototactic movements in filamentous cyanobacteria have been known for many years. Thus, although the mechanisms are unknown, like swimming bacteria, gliding bacteria must also be able to communicate with their motility apparatus.

Chemotaxis

To understand chemotaxis we can focus on the behavior of a single bacterial cell faced with a chemical gradient of an attractant (Figure 4.46). Unlike larger organisms, prokaryotes are too small to sense a gradient along their body length. They must instead, while moving, compare the chemical or physical state of their environment with that sensed a few seconds before. In other words, bacteria respond to the *temporal* (rather than *spatial*) *gradient* of signal molecules as they swim along.

Much research on chemotaxis has been done with the peritrichously flagellated bacterium, *Escherichia coli.* In the absence of a gradient, cells of *E. coli* move in a random fashion that includes **runs**, where the cell is swimming forward in a smooth fashion, and **tumbles** when the cell stops and jiggles about (Figure 4.46a). Following a tumble, the direction of the next run is random (Figure 4.46a). Thus, by means of runs and tumbles the cell moves about randomly in its environment but does not really go anywhere. However, if a gradient of a chemical attractant is present, these random movements become biased. As the organism senses higher concentrations of the attractant (through periodic sampling of the concentration of the chemical in its environment), runs become longer and tumbles less frequent. The net result of this behavior is that the organism moves up the concentration gradient of the attractant (Figure 4.46b). If the organism is sensing a repellent, the same general mechanism applies, although in this case it is the *decrease* in concentration of the repellent (rather than the *increase* in concentration of an attractant) that promotes runs. Forward movement in a run occurs when the flagellar motor is rotating counterclockwise. When the flagella rotate clockwise, the bundle pushes apart, forward motion ceases, and the cells tumble (Figure 4.43).

The situation with polarly flagellated cells is somewhat different. Many polarly flagellated bacteria, such as *Pseudomonas* species, can reverse the direction of

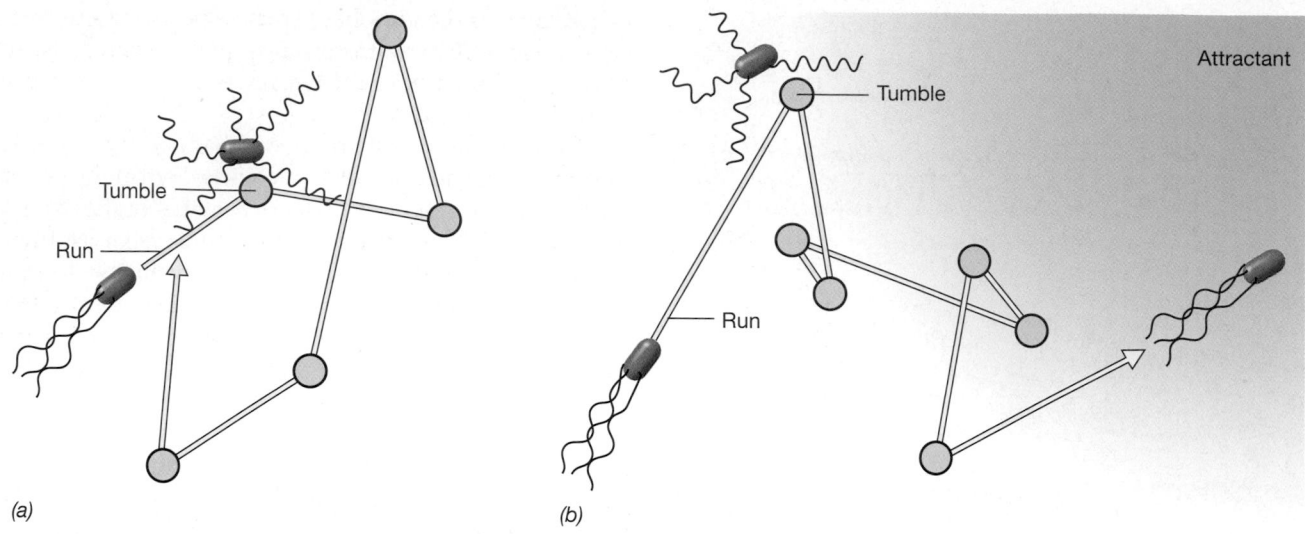

(a)

(b)

Figure 4.46 Chemotaxis in a peritrichously flagellated bacterium like *Escherichia coli*. (a) In the absence of a chemical attractant the cell swims randomly in runs, changing direction during tumbles. (b) In the presence of an attractant runs become biased, and the cell moves up the gradient of the attractant.

rotation of their flagella like peritrichously flagellated cells and reverse direction in this fashion (Figure 4.43*b*). However, some polarly flagellated bacteria, such as the phototrophic bacterium *Rhodobacter sphaeroides*, have unidirectional flagella that rotate only in a clockwise direction. How do such cells change direction and are they chemotactic? In *R. sphaeroides*, which has only a single flagellum, rotation of the flagellum stops periodically. During this time the cell becomes randomly reoriented by Brownian motion. As the flagellum begins to rotate once again, the cell moves off in a new direction (Figure 4.43*b*). Cells of *R. sphaeroides* are strongly chemotactic to a variety of carbon compounds and also show tactic responses to oxygen and light (see later discussion of these taxes). And although it cannot reverse its flagellar motor, chemotaxis in *R. sphaeroides* works in a similar way to that of *E. coli:* An increased gradient of attractants tend to keep the *R. sphaeroides* flagellum rotating while a decreased gradient of an attractant or an increased gradient of a repellant tends to make the cells stop and reorient.

Measuring Chemotaxis

How do bacteria use temporal changes in chemical concentrations to control flagellar rotation? This is a complex story and involves several regulatory events at the genetic and biochemical levels. We therefore reserve detailed discussion of the mechanism of chemotaxis for Chapter 8 (Section 8.11) where this topic can be better understood in a biochemical-genetic context. Suffice it for now to say that the molecular mechanism of chemotaxis involves sensory proteins in the membrane called *chemoreceptors* that sense the chemical gradient with time and interact with cytoplasmic

proteins to affect flagellar motor direction. Thus, because the direction of flagellar rotation governs whether the cell runs or tumbles, chemotaxis can be thought of as a chemically driven *sensory response system* affecting flagellar function.

Bacterial chemotaxis can be demonstrated by immersing a small glass capillary containing an attractant in a suspension of motile bacteria that does not contain the attractant. From the tip of the capillary, a gradient is set up into the surrounding medium, with the concentration of chemical gradually decreasing with distance from the tip (Figure 4.47*a*). If the capillary contains an attractant, the bacteria will move toward it, forming a swarm around the open tip (Figure 4.47*b*); subsequently, many of the motile bacteria will move into the capillary. Of course some bacteria will move into the capillary even if it contains a solution of the same composition as the medium because of random movements (Figure 4.47*c*). But if an attractant is present, the concentration of bacteria within the capillary can be many times higher than the external concentration. On the other hand, if the capillary contains a repellent, the concentration of bacteria within the capillary will be considerably less than the concentration outside (Figure 4.47*d*). Using this simple method, it is possible to rapidly screen chemicals for their ability to act as attractants or repellents for a given bacterium.

Chemotaxis can also be observed microscopically. Using a video camera that captures the position of bacterial cells with time and shows the tracks of each cell, it is possible to see the chemotactic movements of cells (Figure 4.47*f*). This method has been adapted to studies of chemotaxis of bacteria in natural environments. In nature it is thought that the major chemotactic agents for bacteria are nutrients excreted from larger microbial cells, such as protozoans or algae (Figure 4.47*f*) or from live or dead macroorganisms.

Phototaxis

Many phototrophic microorganisms move toward light, a process called *phototaxis*. The advantage of phototaxis is that it allows a phototrophic organism to orient itself most efficiently for photosynthesis. This can be shown if a light spectrum is spread across a microscope slide on which there are motile phototrophic bacteria; the bacteria accumulate at the wavelengths at which their photosynthetic pigments absorb (Figure 4.48*a*; ∞ Sections 17.2 and 17.3 for a discussion of photosynthetic pigments).

Two different taxes are observed in phototrophic prokaryotes. One, called *scotophobotaxis*, can be observed only microscopically and occurs when a phototrophic bacterium by chance happens to swim outside the illuminated field of view of the microscope into darkness. Entering darkness negatively affects the energy state of the cell and signals the cell to tumble, reverse direction, and once again swim in a run, thus reentering the light.

In addition to scotophobotaxis, phototrophic microorganisms can carry out true phototaxis, a directed movement up a light gradient toward an increasing intensity of light. This is analogous to positive chemotaxis except that the attractant is not a chemical but instead is light. In some species, such as the highly motile phototrophic organism *Rhodospirillum centenum* (see Figure 4.39), *entire colonies* of cells show phototaxis and move in unison toward the light (Figure 4.48*b*). Although the molecular details of phototaxis are not yet known, there is good evidence that several parts of the regulatory system that govern chemotaxis are also involved in phototaxis. These include in particular cytoplasmic proteins (Che proteins) that control the direction of rotation of the flagella (∞ Section 8.11). This evidence has emerged from the study of mutants of phototrophic bacteria defective in phototaxis; such mutants frequently have defective chemotaxis systems as well. A *photoreceptor*,

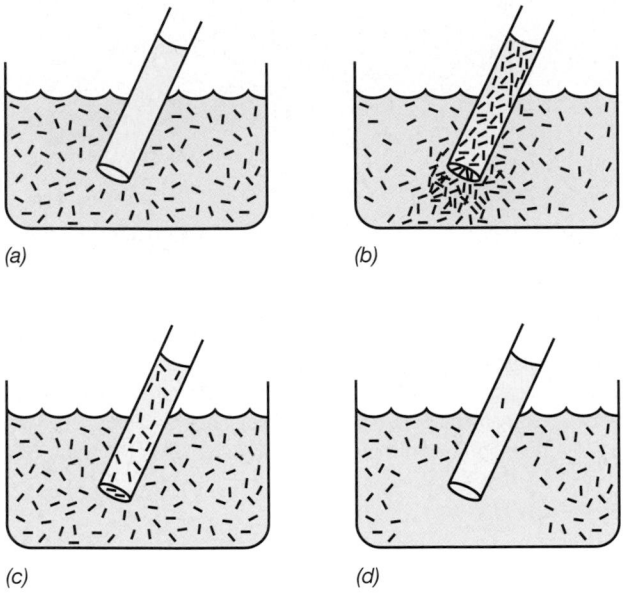

(a) (b)

(c) (d)

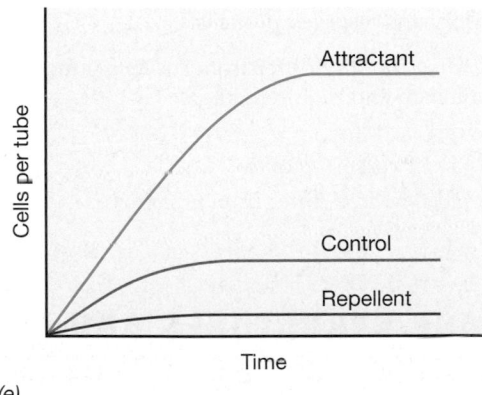

(e)

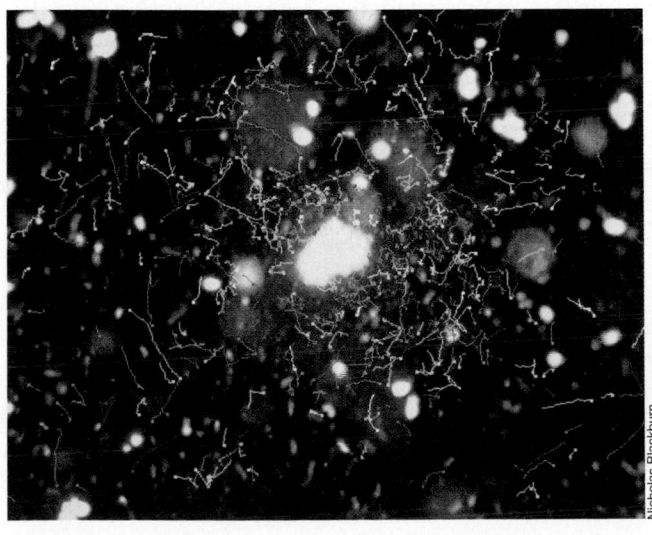

(f)

Figure 4.47 Chemotaxis. (a–e) Techniques for measuring chemotaxis in bacteria. (a) Insertion of capillary into a bacterial suspension. As the capillary is inserted, gradient formation begins. (b) Accumulation of bacteria in a capillary containing an attractant. (c) Control capillary contains a salt solution that is neither an attractant nor a repellent. Cell concentration inside the capillary becomes the same as that outside. (d) Repulsion of bacteria by a repellent. (e) Time course showing cell numbers in capillaries containing various chemicals. (f) Tracks of motile bacteria in seawater swarming around an algal cell (large white spot, center) taken using a tracking video camera system attached to a microscope. Note how the bacterial cells are responding positively to oxygen (that is, showing *aerotaxis*) produced through photosynthesis by the alga. The average velocity of the cells was about 25 μm/sec. The alga is about 60 μm in diameter.

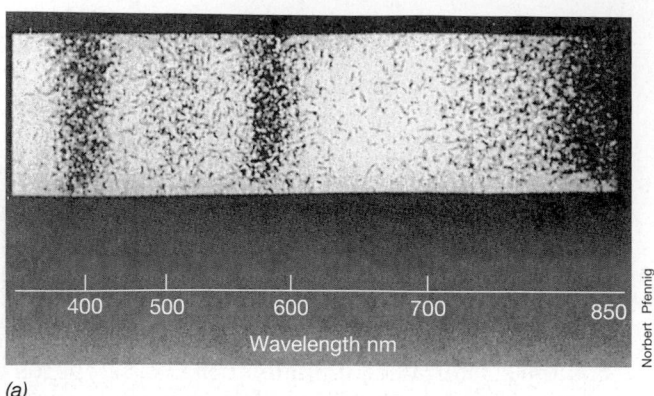

(a)

Norbert Pfennig

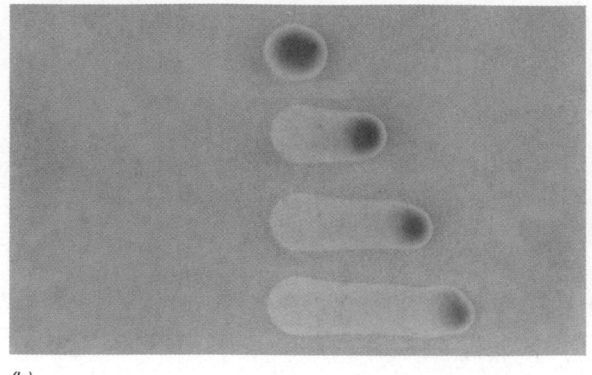

(b)

Carl E. Bauer

Figure 4.48 Phototaxis. (a) Scotophobic accumulation of the phototrophic bacterium *Thiospirillum jenense* at light wavelengths at which its pigments absorb. A light spectrum was displayed on a microscope slide containing a dense suspension of the bacteria; after a period of time, the bacteria had accumulated selectively and the photomicrograph was taken. The wavelengths at which accumulations occur are those at which bacteriochlorophyll *a* absorbs (compare with Figure 17.3*b*). (b) Phototaxis of an entire colony of the purple phototrophic bacterium *Rhodospirillum centenum* over a 2-h time course (time 0 at top). These strongly phototactic cells move in unison toward the light source on the right. See Figure 4.39 for electron micrographs of *R. centenum* cells.

analogous to a chemoreceptor but able to sense a gradient of *light* instead of chemicals, is responsible for orchestrating the phototaxis response. It is hypothesized that the photoreceptor interacts with the proteins that affect flagella rotation to maintain the cell in a run if it is swimming toward an increasing intensity of light.

Other Taxes

Other bacterial taxes, such as movement toward or away from oxygen (*aerotaxis*, see Figure 4.47*f*) or toward or away from conditions of high ionic strength (*osmotaxis*), are also beginning to be understood in molecular terms now that some of the general principles of sensory response systems have been elucidated, primarily from work on chemotaxis. In most of these cases a common mechanism applies: Cells periodically sample their environment and process this information through a signal transduction pathway (∞ Section 8.11) that leads to control of the direction of flagellar rotation. These can be considered simple behavioral responses, and a rationale for elucidating the molecular mechanisms of bacterial taxes is to gain a better understanding of similar responses, such as nerve transmission, in higher organisms. Thus, from a behavioral point of view, motile prokaryotes are well attuned to the chemical and physical state of their environment, and as such, can move toward or away from various stimuli presumably as a means of remaining competitively successful.

✓ 4.11–4.12 Concept Check

Some bacteria are motile by gliding means while others are motile by swimming. Motile bacteria can respond to chemical and physical gradients in their environment. In the processes of chemotaxis and phototaxis, random movement of a prokaryotic cell can

be biased either toward or away from a chemical (or toward light in a phototrophic microorganism) by controlling the degree to which runs or tumbles occur. The latter are controlled by the direction of rotation of the flagellum, which in turn is controlled by a network of sensory and response proteins.

✓ How does *gliding* motility differ from *swimming* motility in both mechanism and requirements?

✓ Define the word *chemotaxis*.

✓ What causes a *run* versus a *tumble*?

✓ How does *scotophobotaxis* differ from *phototaxis*?

IV SURFACE STRUCTURES AND INCLUSIONS OF PROKARYOTES

In addition to the cell wall and related surface layers discussed earlier in this chapter, some prokaryotic cells have other outer layers that comprise the surfaces that contact the environment. Moreover, most cells can form cellular inclusions of one sort of the other. Usually these inclusions are composed of some nutrient present in the environment in abundance that can be broken down to survive periods of nutrient starvation. We examine some of these structures here.

4.13 Bacterial Cell Surface Structures and Cell Inclusions

Prokaryotes can produce a variety of structures that are attached to or in some way protrude from the cell surface. Not all bacteria contain such structures, and thus they are *optional,* produced by some kinds of prokaryotes but not others.

Fimbriae and Pili

Fimbriae and *pili* are structurally similar to flagella but are not involved in motility. **Fimbriae** are considerably shorter than flagella and are more numerous (Figure 4.49) but, like flagella, consist of protein. The functions of fimbriae are not known for certain, but there is good evidence that they enable organisms to stick to surfaces, including animal tissues in the case of some pathogenic bacteria, or to form pellicles or biofilms (⟁ Section 19.3) on surfaces.

Pili are similar structurally to fimbriae but are generally longer, and only one or a few pili are present on the surface. Pili can be seen under the electron microscope because they serve as receptors for certain types of virus particles, and when coated with virus they can be easily seen (Figure 4.50). There is strong evidence that pili are involved in the process of conjugation in prokaryotes, as will be discussed in Section 10.9. Pili are also involved in attachment to human tissues by some pathogenic bacteria.

Paracrystalline Surface Layers (S-Layers)

Many prokaryotes contain a cell surface layer composed of a two-dimensional array of protein. These layers are called *S-layers.* S-layers have been detected in representatives of virtually every phylogenetic grouping of *Bacteria* and are widespread among *Archaea.* In some species of *Archaea* the S-layer is also the cell wall (see Section 4.8). S-layers have a crystalline appearance and show various symmetries, such as hexagonal, tetragonal, or trimeric, depending upon the number and structure of the protein or glycoprotein subunits of which they are composed (see Figure 4.34*b* to view an electron micrograph of an S-layer).

The major function of S-layers is unknown. As the interface between the cell and its environment it is likely that the S-layer at least functions as an external permeability barrier, allowing the passage of low-molecular-weight substances while excluding large molecules. Evidence also exists that in pathogenic (disease-causing) bacteria that contain S-layers, this structure may confer protection on the bacterium against certain host defense mechanisms.

Capsules and Slime Layers: The Glycocalyx

Many prokaryotic organisms secrete on their surfaces slimy or gummy materials (Figure 4.51). A variety of these structures consist of polysaccharide, and a few consist of protein. The terms **capsule** and **slime layer** are frequently used to describe polysaccharide layers; the more inclusive term **glycocalyx** is also used. Glycocalyx is defined as the polysaccharide-containing material

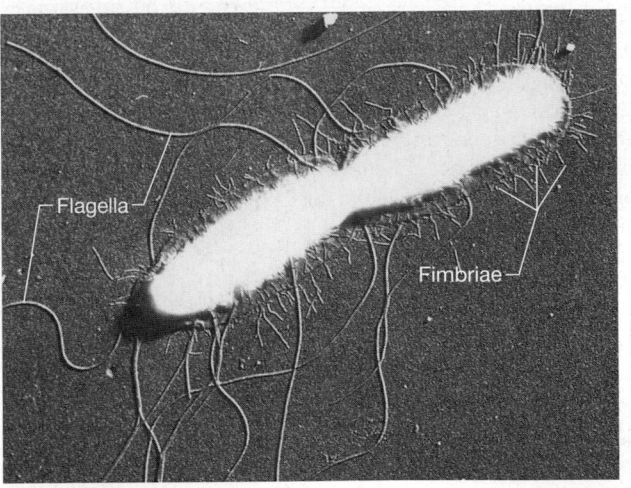

Figure 4.49 Electron micrograph of a dividing cell of *Salmonella typhi,* showing flagella and fimbriae. A single cell is about 0.9 μm in diameter.

J. P. Duguid and J. F. Wilkinson

lying outside the cell. The composition of these layers varies in different organisms but may be thick or thin, rigid or flexible, depending on their chemical nature in a specific organism. The rigid layers are organized in a tight matrix that excludes particles, such as india ink; this form is referred to as a *capsule* (Figure 4.51*a*). If the glycocalyx is more easily deformed, it will not exclude particles and is more difficult to see; this arrangement is referred to as a *slime layer.*

Glycocalyx layers have several functions in bacteria. Outer polysaccharide layers play an important role in the *attachment* of certain pathogenic microorganisms to their hosts. As we will see (⟁ Section 21.6), pathogenic microorganisms that enter the animal body by specific routes usually do so because of binding reactions that occur between outer cell surface components (such as the glycocalyx) and specific host tissues. The glycocalyx plays other roles as well. There is some evidence that encapsulated bacteria are more difficult for phagocytic cells of the immune system (⟁ Section 22.2) to

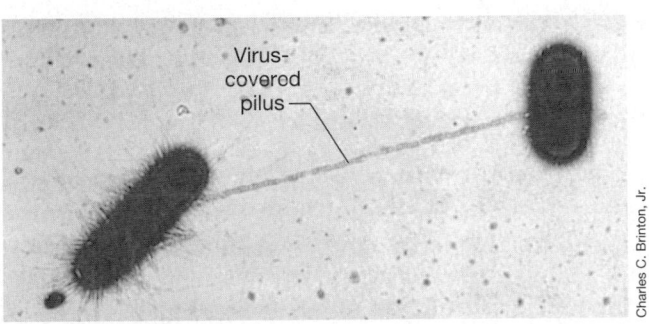

Figure 4.50 The presence of pili on an *Escherichia coli* cell is revealed by the use of viruses that specifically adhere to the pilus. The cell is about 0.8 μm in diameter.

Charles C. Brinton, Jr.

Figure 4.51 Bacterial capsules. (a) Demonstration of the presence of a capsule in an *Acinetobacter* species by negative staining with india ink observed by phase contrast microscopy. The india ink does not penetrate the capsule, and so it is revealed in outline as a light structure on a dark background. (b) Electron micrograph of a thin section of a *Rhizobium trifolii* cell stained with ruthenium red to reveal the capsule. The diameter of the cell proper (not including the capsule) is about 0.7 μm. Although most capsules consist of polysaccharide, some bacteria contain protein capsules. Cells of *Bacillus anthracis*, for example, an organism associated with both animal disease and bioterrorism (⊂⊃ Section 25.11), contain a poly-ᴅ-glutamic acid capsule that is effective in preventing cell destruction by host defenses.

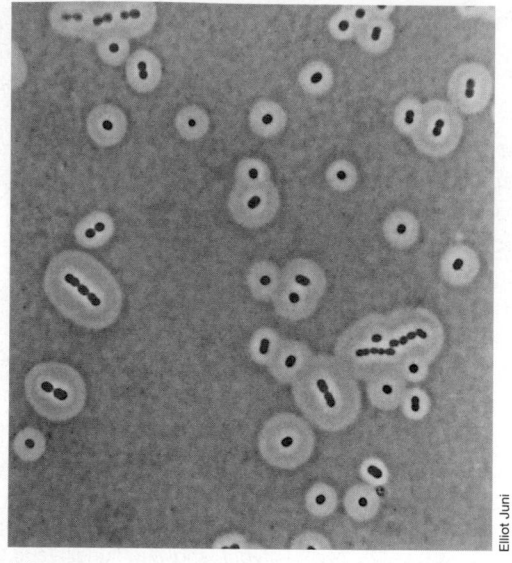
(a)

Elliot Juni

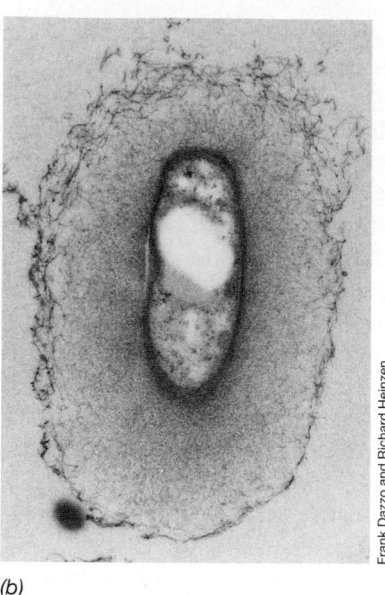
(b)

Frank Dazzo and Richard Heinzen

recognize and subsequently destroy. In addition, because outer polysaccharide layers probably bind a significant amount of water, there is reason to believe a glycocalyx layer plays some role in resistance to desiccation.

Carbon Storage Polymers

Granules or other inclusions are often seen within cells. Their nature differs in different organisms, but they almost always function in the storage of energy or as a reservoir of structural building blocks. Inclusions can often be seen directly with the phase contrast microscope, but their contrast can usually be increased by using dyes. Inclusions often show up very well with the electron microscope (Figure 4.52). Most cellular inclusions are bounded by a thin *nonunit* membrane consisting of lipid separating the inclusion from the cytoplasm proper.

In prokaryotic organisms, one of the most common inclusion bodies consists of **poly-β-hydroxybutyric acid (PHB)**, a lipidlike compound that is formed from β-hydroxybutyric acid units (Figure 4.52*a*). The monomers of this acid are connected by ester linkages, forming the long PHB polymer, and these polymers aggregate into granules (Figure 4.52*b*). The length of the monomer in the polymer can vary considerably, from as short as C_4 to as long as C_{18} in certain organisms. Thus, the collective term *poly-β-hydroxyalkanoate* (PHA) is used to describe this whole class of carbon/energy storage polymers. A wide variety of prokaryotes, including representatives of both the *Bacteria* and the *Archaea*, produce PHAs. Another storage product formed by prokaryotes is **glycogen**, which is a polymer of glucose (Section 3.3 and Figure 3.6). Like PHAs, glycogen is a storage depot

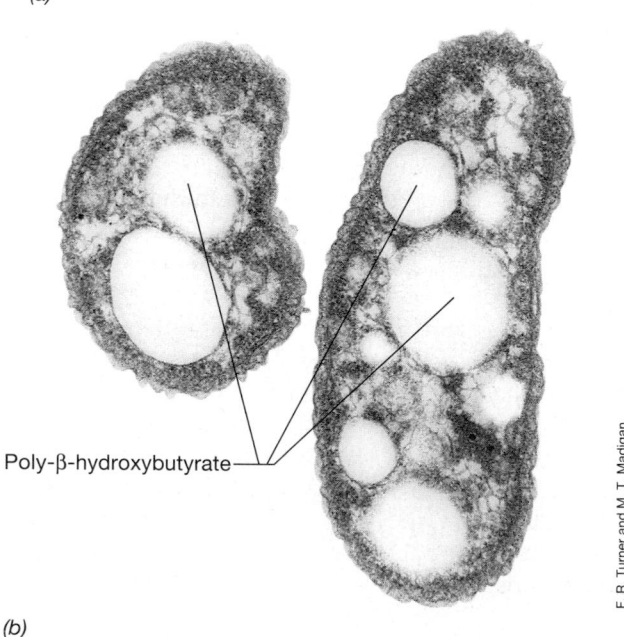

(a)

Poly-β-hydroxybutyrate

(b)

F. R. Turner and M. T. Madigan

Figure 4.52 Poly-β-hydroxybutyrate (PHB). (a) Chemical structure of PHB, a common poly-β-hydroxyalkanoate. A monomeric unit is shaded. Other alkanoate polymers are made by substituting longer-chain hydrocarbons for the –CH₃ group on the β carbon. (b) Electron micrograph of a thin section of cells of the phototrophic bacterium *Rhodovibrio sodomensis* containing granules of PHB.

for carbon and energy and is produced when carbon is in excess in the environment.

Other Storage Materials and Inclusions

Many microorganisms accumulate large reserves of inorganic phosphate in the form of granules of **polyphosphate**. These granules can be degraded and used as sources of phosphate for nucleic acid and phospholipid synthesis to support growth. In addition, a variety of prokaryotes are capable of oxidizing reduced sulfur compounds such as hydrogen sulfide and thiosulfate. These oxidations are linked to either reactions of energy metabolism (∞ Sections 17.8 and 17.10) or biosynthesis (∞ Section 17.6), but in both instances **elemental sulfur** may accumulate inside the cell in large, readily visible granules (Figure 4.53). The granules of elemental sulfur remain as long as the source of reduced sulfur is still present. However, as the reduced sulfur source becomes limiting, the sulfur in the granules is oxidized, usually to sulfate, and the granules slowly disappear as this reaction proceeds.

Magnetosomes are intracellular crystal particles of the iron mineral magnetite, Fe_3O_4 (Figure 4.54). Magnetosomes impart a permanent magnetic dipole to a cell, allowing it to respond to a magnetic field. Bacteria that produce magnetosomes (Figure 4.54a) exhibit *magnetotaxis*, the process of orienting and migrating along geomagnetic field lines (∞ Section 12.14) Although the combining form -*taxis* is used in the word *magnetotaxis*, there is no evidence that magnetotactic bacteria employ the sensory systems of chemotactic or phototactic bacteria (see Sections 4.12 and 8.11). Instead, the alignment of magnetosomes in the cell simply

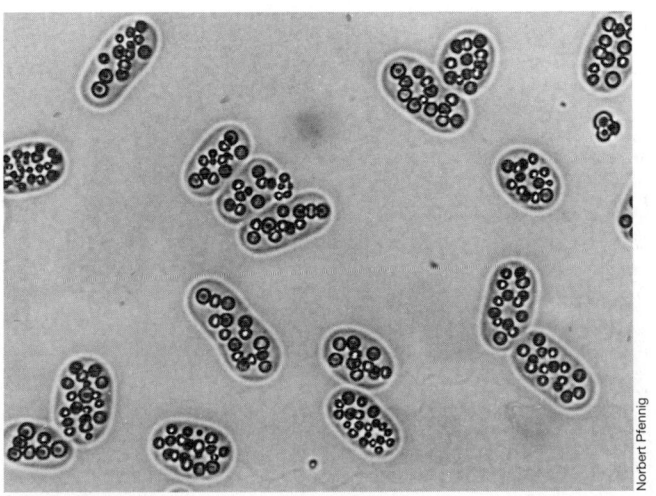

Figure 4.53 Bright-field photomicrograph of cells of the purple sulfur bacterium *Isochromatium buderi*. Note the sulfur globules inside the cell. A single cell measures about 4×7 μm.

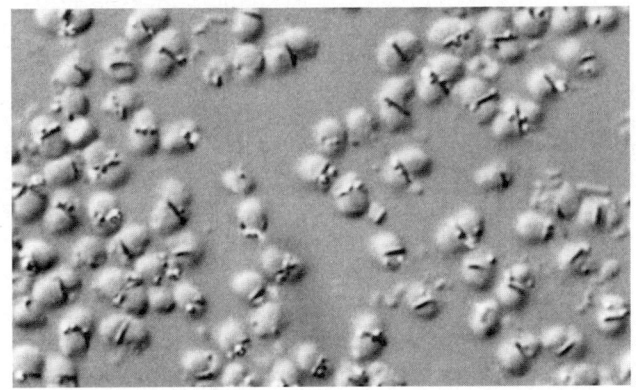

(a)

Stefan Spring

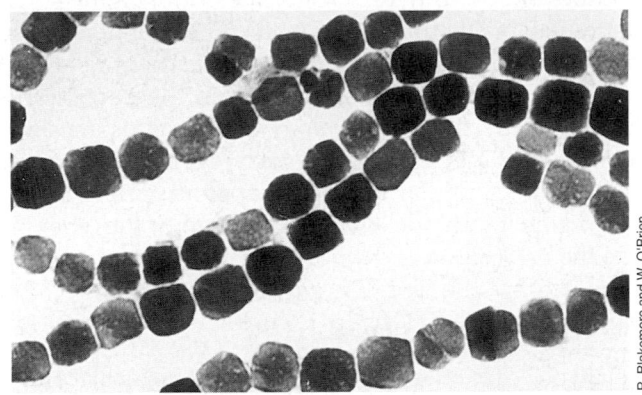

(b)

R. Blakemore and W. O'Brien

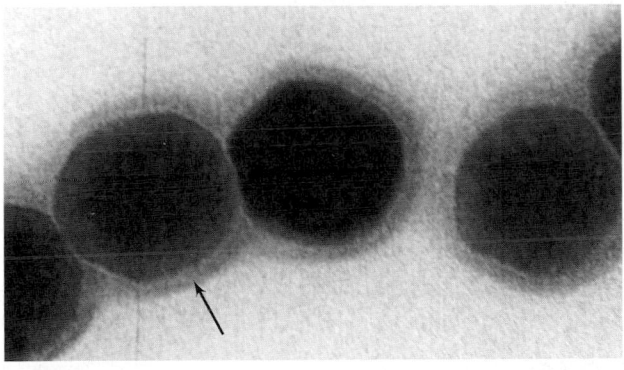

(c)

Dennis Bazylinski

Figure 4.54 Magnetotactic bacteria and magnetosomes. (a) Interference contrast micrograph of coccoid magnetotactic bacteria. Note magnetosomes. A single cell is about 2.2 μm in diameter. (b) Magnetosomes isolated from the magnetotactic bacterium *Magnetospirillum magnetotacticum*. Each particle is about 50 nm in length (∞ Figure 12.32). (c) High magnification electron micrograph of magnetosomes from a magnetic coccus-shaped bacterium. The arrow points to the membrane surrounding each magnetosome. A single magnetosome is about 90 nm wide. Although containing lipids and proteins, the magnetosome membrane, like the PHB membrane (see Figure 4.52), is a monolayer and not a true "unit" membrane.

imparts magnetic properties to it, which then orient the cell in a particular direction in its environment. Magnetosomes have been described from a variety of primarily aquatic *Bacteria* (Figure 4.54a). Many of these grow best at very low O_2 concentrations, and it seems likely that the major function of magnetosomes is to guide aquatic cells toward the sediments where O_2 levels are lower.

Magnetosomes are surrounded by a membrane containing phospholipids, proteins, and glycoproteins (Figure 4.54b, c). This membrane is not a unit membrane, as is the cytoplasmic membrane (Figure 4.17), but instead, is a monolayer membrane, like that surrounding granules of poly-β-hydroxybutyrate (PHB, Figure 4.52). Magnetosome membrane proteins probably play a role in precipitating Fe^{3+} (brought into the cell in soluble form by chelating agents) as Fe_3O_4 in the developing magnetosome. The morphology of magnetosomes appears to be species-specific, varying in shape from square to rectangular to spike-shaped in certain bacteria, and most form into chains of one sort or the other inside the cell (Figure 4.54).

✓ 4.13 Concept Check

Prokaryotic cells often contain various structures that either emerge from or are present inside the cell. These include fimbriae and pili, S-layers, capsules or slime layers, carbon and other storage polymers, and magnetosomes. However, unlike the organelles of eukaryotic cells, these structures are not bounded by a unit membrane.

- ✓ How do fimbriae differ from pili, both structurally and functionally?
- ✓ What is *poly-β-hydroxybutyrate* and what is its function in the cell? Under what conditions is this material made?
- ✓ What are *magnetosomes*, what are they made of, and what property do they confer on cells that contain them?

4.14 Gas Vesicles

A number of prokaryotic organisms that live a floating existence in lakes and the sea produce **gas vesicles**, which confer buoyancy on the cells. Gas vesicles are a means of motility, allowing cells to float up and down in a water column in response to environmental factors. The most dramatic instances of flotation due to gas vesicles are seen in cyanobacteria that form massive accumulations (blooms) in lakes (Figure 4.55). Gas-vesiculate cells rise to the surface of the lake and are blown by winds into dense masses. Gas vesicles are also present in certain purple and green phototrophic bacteria (∞ Sections 12.2 and 12.32) and in some non-phototrophic bacteria that live in lakes and ponds. Some species of *Archaea* also contain gas vesicles.

Figure 4.55 Flotation of gas vesiculate cyanobacteria from a bloom on a nutrient-rich lake, Lake Mendota, Madison, Wisconsin.

Structure of Gas Vesicles

Gas vesicles are spindle-shaped structures made of protein, hollow but rigid, that are of variable lengths and diameters (Figure 4.56). Gas vesicles in different organisms vary in length from about 300 to over 1000 nm and in width from 45 to 120 nm, but the vesicles of any given organism are more or less of constant size. They are present in the cytoplasm and may number from few to hundreds per cell. The gas vesicle membrane is composed only of protein, is about 2 nm thick, and is impermeable to water and solutes but permeable to gases; thus, gas vesicles exist as gas-filled structures surrounded by the constituents of the cytoplasm (Figure 4.56). The presence of gas vesicles in cells can be determined either by light microscopy (where groups of vesicles show up as irregular bright inclusions) or by electron microscopy (Figure 4.57).

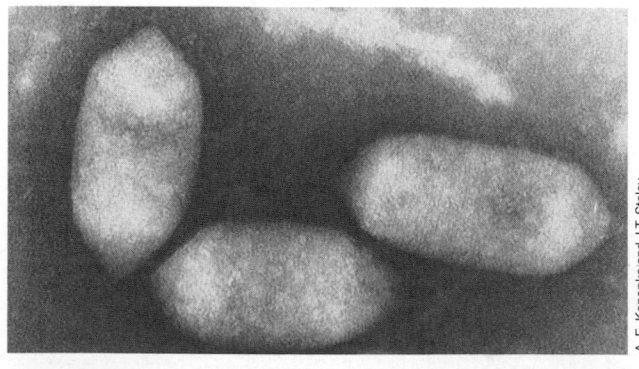

Figure 4.56 Transmission electron micrographs of gas vesicles purified from the bacterium *Ancyclobacter aquaticus* and examined in negatively stained preparations. A single gas vesicle is about 100 nm in diameter. Reproduced with permission from Allan E. Konopka, J. C. Lara, and James T. Staley, 1977. *Archives of Microbiology* 112:133–140. ©1977 by Springer-Verlag GmbH & Co. KG.

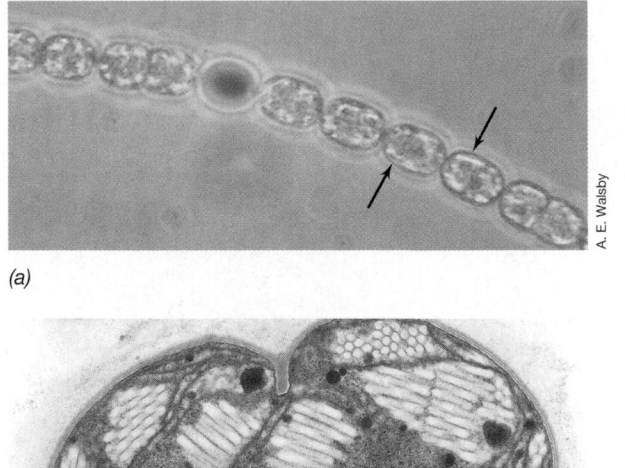

(a)

A. E. Walsby

(b)

S. Pellegrini and M. Grilli Caiola

Figure 4.57 Gas vesicles of the cyanobacteria *Anabaena* and *Microcystis*. (a) *Anabaena flos-aquae.* The dark cell in the center (a heterocyst) lacks gas vesicles. In the other cells, the vesicles group together as phase-bright objects that scatter light (arrows). (b) Transmission electron micrograph of the cyanobacterium *Microcystis*. Gas vesicles are arranged in bundles, here observable in both longitudinal and cross section.

Molecular Structure of Gas Vesicles

Gas vesicles contain only two types of protein (Figure 4.58). The major gas vesicle protein, called GvpA, is a small, highly hydrophobic and very rigid protein. The rigidity of the gas vesicle membrane is essential for the

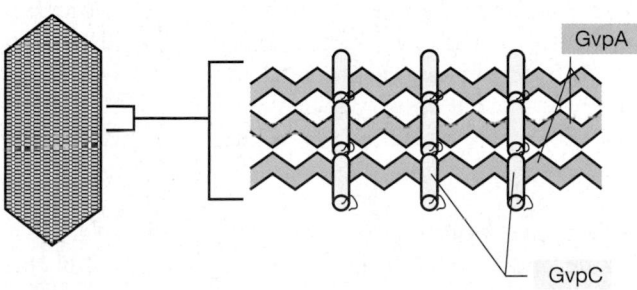

GvpA

GvpC

Figure 4.58 Model of how the two proteins that make up the gas vesicle, GvpA and GvpC, interact to form a watertight but gas-permeable structure. GvpA makes up the rib and is a rigid β-sheet. GvpC is the cross-linker and is of an α-helix structure (⚬⚬ Section 3.7 and Figure 3.16).

structure to resist the pressures exerted on it from outside. GvpA is the shell protein and makes up 97% of the total protein of the gas vesicle. The second protein, called GvpC, is present in much smaller amounts; the function of GvpC protein is to strengthen the shell of the gas vesicle (Figure 4.58).

Gas vesicles are constructed of several copies of the GvpA protein aligned as parallel "ribs" forming a watertight surface. The GvpA ribs are strengthened by GvpC, which acts as a cross-linker, binding several GvpA ribs together like a clamp (Figure 4.58). The final shape of the gas vesicle, which can vary in different organisms from long and thin to short and fat (compare Figures 4.56 and 4.57*b*), is a function of how the GvpA and GvpC proteins are arranged to form the intact vesicle.

Because the gas vesicle membrane is freely permeable to gases, the composition and pressure of the gas inside a gas vesicle is the same as that of the gas in which the organism is suspended. And, because the gas vesicle attains a density of about 5–20% of that of the cell proper, intact gas vesicles decrease the density of the cell, thereby increasing its buoyancy. Aquatic phototrophic organisms in particular benefit from this because it allows them to adjust their position vertically in a water column to regions where the light intensity for photosynthesis is optimal.

✓ 4.14 Concept Check

Gas vesicles are small gas-filled structures made of protein that function to confer buoyancy on cells. Gas vesicles contain two different proteins arranged to form a gas permeable, but watertight structure.

- ✓ What might be the benefit of gas vesicles to phototrophic cells?
- ✓ How are the two proteins that make up the gas vesicle, GvpA and GvpC, arranged to form such a water impermeable structure?

4.15 Endospores

Certain species of *Bacteria* produce special structures called **endospores** within their cells (Figure 4.59) during a process called *sporulation* (see Figure 4.63). Endospores are differentiated cells that are very resistant to heat and cannot be destroyed easily, even by harsh chemicals. Endospore-forming bacteria are found most commonly in the soil, and virtually any sample of soil has some endospores present. The genera *Bacillus* and *Clostridium* are the best studied of endospore-forming bacteria.

The discovery of bacterial endospores was of immense importance to microbiology because the knowledge of such remarkably heat-resistant forms was

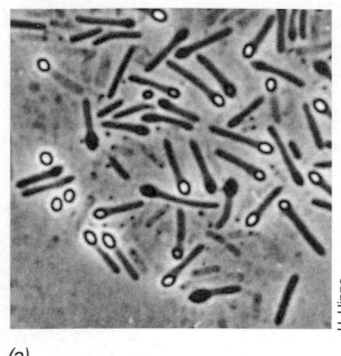
(a)

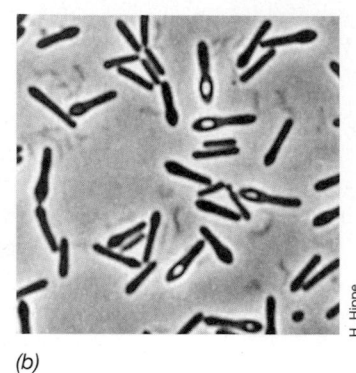
(b)

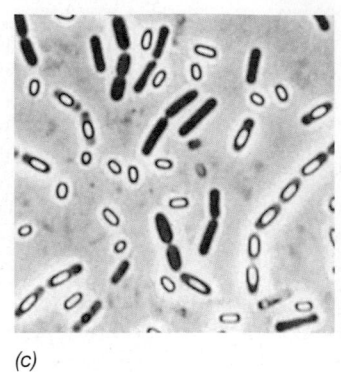
(c)

Figure 4.59 The bacterial endospore. Phase contrast photomicrographs illustrating several types of endospore morphologies and intracellular locations. (a) Terminal. (b) Subterminal. (c) Central.

essential for the development of adequate methods of sterilization, not only of culture media but also of foods and other perishable products. Although many organisms other than bacteria form spores, the bacterial endospore is unique in its degree of heat resistance. Endospores are also resistant to other harmful agents such as drying, radiation, acids, and chemical disinfectants, and can remain dormant for extremely long periods of time (see the box, How Long Can an Endospore Survive?).

Endospore Structure

Endospores (so called because the spore is formed *within* the cell) are readily seen under the light microscope as strongly refractile bodies (see Figure 4.59). Spores are very impermeable to dyes, so occasionally they are seen as unstained regions within cells that have been stained with basic dyes such as methylene blue. To stain spores specifically, special spore-staining procedures must be used. The structure of the spore as seen with the electron microscope is vastly different from that of the vegetative cell (Figure 4.60); the spore is structurally much more complex in that it has many layers that are absent from the vegetative cell. The outermost layer is the **exosporium**, a thin, delicate covering made of protein. Within this are the **spore coats**, composed of layers of spore-specific proteins (Figure 4.60*b*). Below the spore coat is the **cortex**, which consists of loosely cross-linked peptidoglycan, and inside the cortex is the **core** or **spore protoplast**, which contains the usual cell wall (core wall), cytoplasmic membrane, cytoplasm, nucleoid, and so on (⌀ Section 2.1 and Figure 2.1*a*). Thus the spore differs structurally from the vegetative cell primarily in the kinds of structures found *outside* the core wall.

One chemical substance that is characteristic of endospores but not present in vegetative cells is **dipicolinic acid** (Figure 4.61). This substance has been found in all endospores examined and is located in the core. Spores are also high in calcium ions, most of which are com-

bined with dipicolinic acid. The calcium-dipicolinic acid complex of the core represents about 10% of the dry weight of the endospore.

Properties of the Endospore Core

The core of a mature endospore differs greatly from the vegetative cell from which it was formed. Besides having an abundant calcium dipicolinate (Figure 4.61) content, the core is in a partially dehydrated state. The core of a mature endospore contains only 10–30% of the water content of the vegetative cell, and thus the consistency of the core cytoplasm is that of a thick gel. Dehydration of the core greatly increases the heat resistance of the endospore but has also been shown to confer resistance to chemicals, such as hydrogen peroxide (H_2O_2), and causes enzymes remaining in the core to become inactive.

In addition to the low water content of the spore, the pH of the core cytoplasm is about one unit lower than that of the vegetative cell and contains high levels of core-specific proteins called *small acid-soluble spore proteins* (SASPs). These are made during the sporulation process and have at least two functions. SASPs bind tightly to DNA in the core and protect it from potential damage from ultraviolet radiation, dessication, and dry heat. However, in addition, SASPs function as a carbon and energy source for the outgrowth of a new vegetative cell from the endospore, a process called *germination* (discussed later in this section).

Endospore Formation

During endospore formation, a vegetative cell is converted to a nongrowing, heat-resistant structure (Figure 4.62). As previously described and as summarized in Table 4.2, the differences between the endospore and the vegetative cell are profound. Sporulation involves a very complex series of events in *cellular differentiation*. Bacterial sporulation does not occur when cells are

How Long Can an Endospore Survive?

In this chapter we have discussed the dormancy and resistance properties of bacterial endospores and have pointed out that endospores can survive for long periods in a dormant state. But how long is long?

Published evidence for endospore longevity has shown that endospores can remain viable (that is, capable of germination into vegetative cells) for at least several decades and probably for much longer than that. A suspension of spores of the bacterium *Clostridium aceticum* (Fig. 1*a*) prepared in 1947 was placed in growth medium in 1981, 34 years later, and in less than 12 h growth commenced, leading to a robust culture. *Clostridium aceticum* was originally isolated by the Dutchman K. T. Wieringa in 1940 but was thought to have been lost until this vial of *C. aceticum* spores was found in a storage room at the University of California at Berkeley and revived.[1]

Other more extreme examples of endospore longevity have been documented. Bacteria of the genus *Thermoactinomyces* are thermophilic endospore formers that are widespread in nature in soil, plant litter, and fermenting plant material. Microbiological examination of a Roman archaeological site in the United Kingdom that was dated to over *2000* years ago yielded significant numbers of viable *Thermoactinomyces* spores in various pieces of debris. Additionally, *Thermoactinomyces* spores were recovered in fractions of sediment cores from a Minnesota lake known to be over *7000* years old. Although contamination is always a possibility in such studies, samples in both of these cases were processed in such a way as to virtually rule out contamination with "recent" spores.[2]

What factors could limit the age of an endospore? Cosmic radiation has been considered a major factor because it can introduce mutations in DNA.[2] It has been hypothesized that over periods of thousands of years, the cumulative effects of cosmic radiation could introduce so many mutations into the genome of an organism that even

(a)   *(b)*

Gerhard Gottschalk   William D. Grant

Figure 1 Longevity of endospores. **(a) Photograph of a tube containing spores of the bacterium *Clostridium aceticum* prepared on May 7, 1947. After remaining dormant for over 30 years, the spores were suspended in a culture medium after which growth occurred within 12 h.[1] (b) Halophilic bacteria trapped within salt crystals. These crystals (about 1 cm in diameter) were grown in the laboratory in the presence of *Halobacterium* cells (orange) that remain viable in the crystals. Crystals similar to these but of Permian age (≈250 million years old) were reported to contain viable halophilic endosporulating bacteria.[4]**

highly radiation-resistant structures such as endospores would succumb to the genetic damage. However, extrapolations from actual experimental assessments of the effect of natural radiation on endospores suggest that if suspensions of endospores were partially shielded from cosmic radiation, for example, by being embedded in layers of organic matter, they could retain viability over periods as great as *several hundred thousand years* and perhaps even longer. Amazing, but is this the upper limit?

In 1995 a group of scientists reported the revival of bacterial spores they claimed were 25–40 million years old.[3] The spores were allegedly preserved in the gut of an extinct bee

trapped in amber of known geological age. The presence of endospore-forming bacteria in these bees was previously suspected from electron microscopic studies of the insect gut which showed endospore-like structures and because *Bacillus*-like DNA was recovered from the insect. Incredibly, samples of bee tissue incubated in a sterile culture medium quickly yielded endospore-forming bacteria. Rigorous precautions were taken to demonstrate that the endospore-forming bacterium revived from the amber-encased bee was not a modern-day contaminant. An even more spectacular claim was made that halophilic (salt-loving) endospore-forming bacteria had been isolated from fluid inclusions in salt crystals of Permian age, over 250 million years old.[4] Presumably these cells were trapped in brines within the crystal (Fig. 1*b*) as it formed eons ago, and remained viable for over a quarter billion years.

If these claims of almost unbelievable endospore longevity are supported by repetition of the results in independent laboratories (and such confirmation is crucial for verifying such a highly controversial finding), then endospores stored under the proper conditions can remain viable indefinitely. This is a remarkable testimony to the endospore, a structure that undoubtedly evolved to help cells remain viable for relatively short periods but that turned out to be such a well-designed structure that dormancy for hundreds of thousands, if not millions, of years may be possible. ■

[1]Braun, M., F. Mayer, and G. Gottschalk. 1981. *Clostridium aceticum* (Wieringa), a microorganism producing acetic acid from molecular hydrogen and carbon dioxide. *Arch. Microbiol. 128:* 288–293.

[2]Gest, H., and J. Mandelstam. 1987. Longevity of microorganisms in natural environments. *Microbiol. Sci. 4:* 69–71.

[3]Cano, R. J., and M. K. Borucki. 1995. Revival and identification of bacterial spores in 25- to 40-million-year-old Dominican amber. *Science 268:* 1060–1064.

[4]Vreeland, R.H., W.D. Rosenzweig, and D.W. Powers. 2000. Isolation of a 250 million-year-old halotolerant bacterium from a primary salt crystal. *Nature 407:* 897–900.

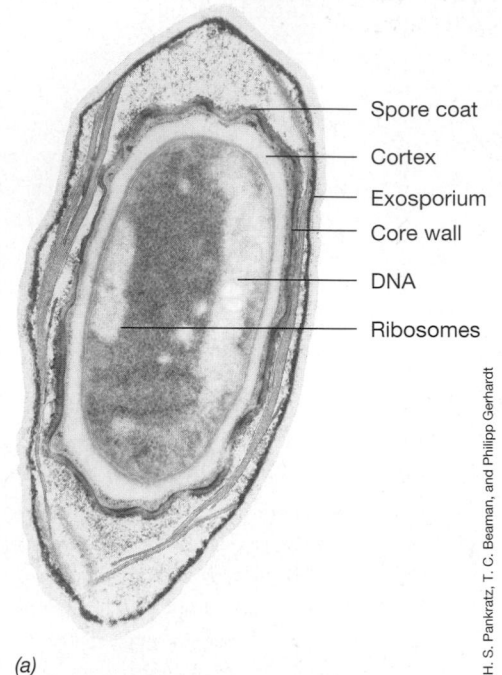

Spore coat
Cortex
Exosporium
Core wall
DNA
Ribosomes

(a)

H. S. Pankratz, T. C. Beaman, and Philipp Gerhardt

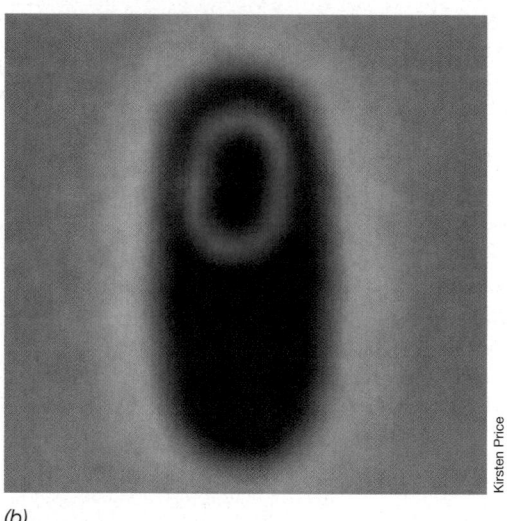

(b)

Kirsten Price

Figure 4.60   The bacterial endospore. (a) Transmission electron micrograph of a mature endospore from *Bacillus megaterium*. (b) Fluorescent photomicrograph of a cell of *Bacillus subtilis* undergoing sporulation. The green area is due to a dye that specifically stains a sporulation protein in the spore coat.

dividing exponentially, but only when growth ceases owing to the exhaustion of an essential nutrient. Thus, cells of *Bacillus*, a typical endospore-forming bacterium, cease vegetative growth and begin sporulation when a key nutrient such as the carbon or nitrogen source becomes limiting.

Many genetically directed changes in the cell underlie the conversion from vegetative growth to sporulation. The structural changes occurring in sporulating cells of *Bacillus* are shown in Figure 4.63, and the process can be

(a)

(b)   Carboxylic acid groups

Figure 4.61   Dipicolinic acid (DPA). (a) Structure of DPA. (b) How Ca^{2+} cross-links DPA molecules to form a complex.

divided into several stages. In *Bacillus subtilis*, where detailed studies of sporulation have been done, the entire sporulation process takes about 8 hours. Genetic studies of mutants of *Bacillus*, each blocked at one of the various stages of sporulation shown in Figure 4.63, have shown that as many as 200 genes are involved in the sporulation process. Sporulation requires that the synthesis of some proteins involved in vegetative cell functions cease and that specific spore proteins be made (see Figure 4.60*b*). This is accomplished by activation of a variety of spore-specific genes including *spo, ssp* (which encodes SASPs), and many other genes in response to an environmental trigger to sporulate. The proteins encoded by these genes catalyze the series of events leading from a moist, metabolizing vegetative cell to a dry, metabolically inert but extremely resistant endospore (Table 4.2 and Figure 4.63).

Germination

An endospore can remain dormant for many years (see the box), but it can convert back to a vegetative cell relatively rapidly. This process involves three steps: activation, germination, and outgrowth (Figure 4.64).

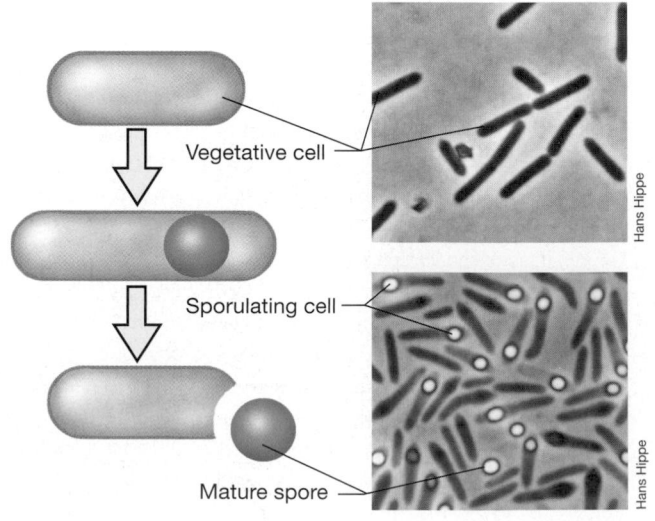

Vegetative cell
Sporulating cell
Mature spore

Hans Hippe
Hans Hippe

Figure 4.62   Formation of the endospore. Phase contrast photomicrographs are of cells of *Clostridium pascui*.

Table 4.2 Differences between endospores and vegetative cells

Characteristic	Vegetative cell	Endospore
Structure	Typical gram-positive cell; a few gram-negative cells	Thick spore cortex Spore coat Exosporium
Microscopic appearance	Nonrefractile	Refractile
Calcium content	Low	High
Dipicolinic acid	Absent	Present
Enzymatic activity	High	Low
Metabolism (O_2 uptake)	High	Low or absent
Macromolecular synthesis	Present	Absent
mRNA	Present	Low or absent
Heat resistance	Low	High
Radiation resistance	Low	High
Resistance to chemicals (for example, H_2O_2) and acids	Low	High
Stainability by dyes	Stainable	Stainable only with special methods
Action of lysozyme	Sensitive	Resistant
Water content	High, 80–90%	Low, 10–25% in core
Small acid-soluble proteins (product of *ssp* genes)	Absent	Present
Cytoplasmic pH	About pH 7	About pH 5.5–6.0 (in core)

Activation is most easily accomplished by heating freshly formed endospores for several minutes at a sublethal but elevated temperature. Activated spores are then conditioned to germinate when placed in the presence of specific nutrients. *Germination*, usually a rapid process (on the order of several minutes), involves loss of microscopic refractility of the spore, increased ability to be stained by dyes, and loss of resistance to heat and chemicals. Loss from the spores of calcium dipicolinate and cortex components occurs during this stage, and the

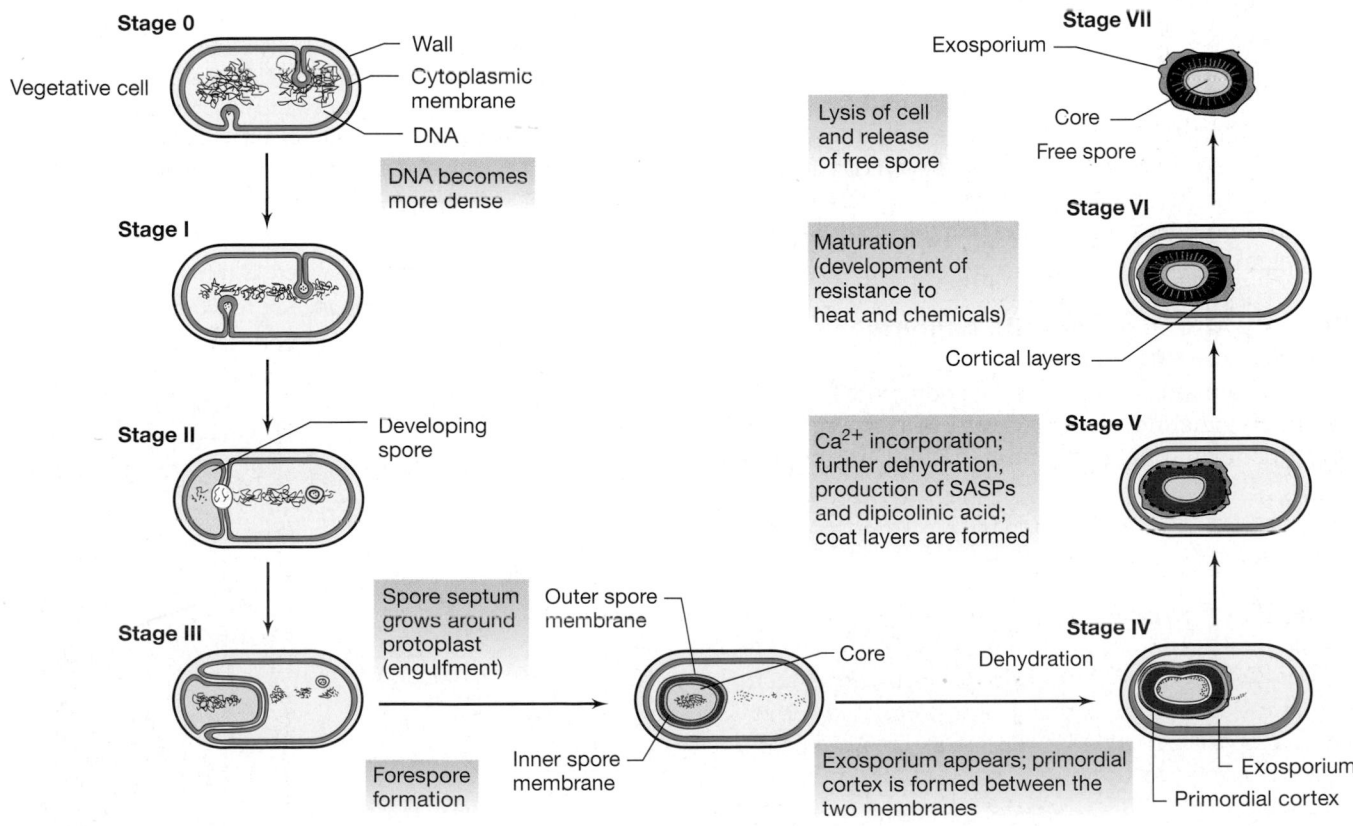

Figure 4.63 Stages in endospore formation. The stages listed (0 through VII) are defined from both genetic studies and microscopic analyses.

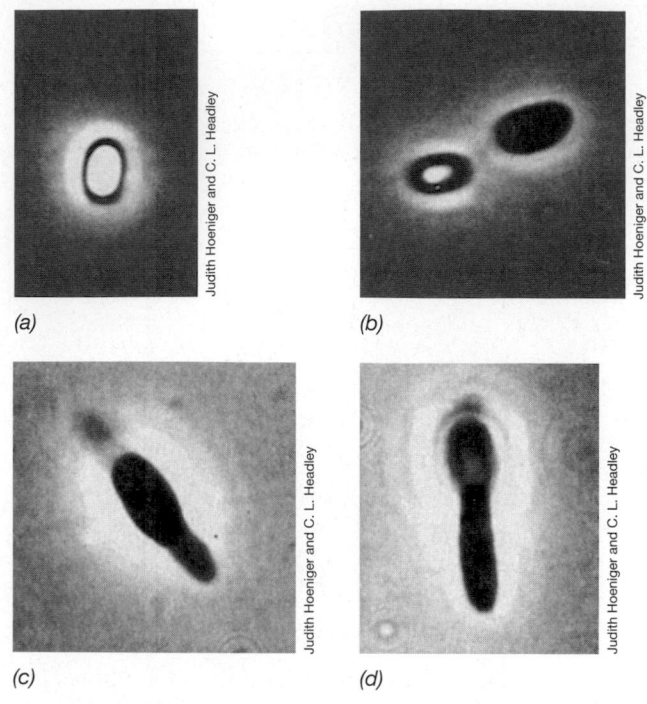

(a)   *(b)*

(c)   *(d)*

Figure 4.64 Endospore germination in *Bacillus*. Conversion of the mature endospore (a) to a vegetative cell (d); photomicrographs showing the sequence of events starting from a highly refractile mature spore. In (b) (activation) refractility is being lost, while in (c) and (d) the new vegetative cell is emerging (outgrowth).

SASPs are degraded. The next stage, *outgrowth,* involves visible swelling due to water uptake and synthesis of new RNA, proteins, and DNA. The cell emerges from the broken spore coat and eventually begins to divide (Figure 4.64). The cell then remains in vegetative growth until environmental signals that trigger sporulation are once again sensed.

Diversity and Phylogenetic Aspects of Endospore Formation

How widespread is the process of endospore formation? Nearly 20 genera of *Bacteria* have been shown capable of endosporulation (Table 12.25), although the process has only been studied in detail in a few species of *Bacillus* and *Clostridium*. Nevertheless, many of the secrets to endospore survival, like the formation of calcium dipicolinate complexes (Figure 4.61) and the possession of spore-specific genes, seem to be universal. It is thus likely that the general principles that guide endospore construction are the same in all organisms capable of this process.

From a phylogenetic perspective, the capacity to produce endospores is uniquely tied to a particular sublineage of the gram-positive bacteria (Sections 2.5 and 12.21). Despite this, the physiologies of endospore-forming bacteria are highly diverse, and include anaerobes, aerobes, phototrophs, and chemolithotrophs. Considering this physiological diversity, the actual triggers for endospore formation may vary with different species, and could include signals other than simple nutrient starvation, the major trigger for endospore formation in *Bacillus* and *Clostridium* species. Interestingly, no species of *Archaea* have been shown capable of endosporulation, and if true, suggests that the capacity to produce endospores evolved *after* the major prokaryotic lineages diverged billions of years ago (Section 2.3 and Figure 2.7).

✓ 4.15 Concept Check

The endospore is a highly resistant differentiated bacterial cell produced by certain types of primarily gram-positive *Bacteria*. Spore formation leads to a nearly dehydrated spore coat that contains essential macromolecules and a variety of substances such as calcium dipicolinate and small acid-soluble proteins, absent from vegetative cells. Spores can remain dormant indefinitely but germinate quickly when the appropriate trigger is applied.

✓ What is *dipicolinic acid* and where is it found?
✓ What are *SASPs* and what is their function?
✓ What happens when an endospore germinates?

Review Questions

1. What is the function of staining in light microscopy? Why are cationic dyes used for general staining purposes?

2. What is the advantage of a *differential interference contrast* microscope over a bright-field microscope? A *phase-contrast* microscope over a bright-field microscope?

3. What is the major advantage of electron microscopes over light microscopes? What type of electron microscope would be used to view the three-dimensional features of a cell?

4. What are the major morphologies of prokaryotes? Draw cells for each morphology you list.

5. Describe in a single sentence the manner in which a unit membrane is formed from phospholipid molecules.

6. Explain in a single sentence why ionized molecules do not readily pass through the membrane barrier of a cell. How do such molecules get through the cytoplasmic membrane?

7. Describe a major chemical difference between membranes of *Bacteria* and of *Archaea*.

8. Cells of *Escherichia coli* take up lactose via the Lac permease system, glucose via the phosphotransferase system, and maltose via an ABC-type transporter. For each of these sugars describe: (1) the components of their transport system, and (2) the source of energy that drives the transport event.

9. Why is the bacterial cell wall rigid layer called *peptidoglycan*? What are the chemical reasons for the rigidity that is conferred on the cell wall by the peptidoglycan structure?

10. Since a single peptidoglycan molecule is very thin, explain in chemical terms how the very *thick* peptidoglycan-containing cell wall of gram-positive *Bacteria* is formed.

11. List several functions for the outer membrane in gram-negative *Bacteria*. What is the chemical composition of the outer membrane?

12. Write a clear explanation (two or three sentences) of why sucrose is able to stabilize bacterial cells from lysis by lysozyme.

13. Describe the structure and function of a bacterial flagellum. What is the energy source for the flagellum?

14. How does the mechanism for gliding motility in *Flavobacterium* differ from the motility in *Escherichia coli*?

15. In a few sentences, write an explanation for how a motile bacterium is able to sense the direction of an attractant and move toward it.

16. What types of cytoplasmic inclusions are formed by prokaryotes? How does an inclusion of poly-β-hydroxybutyric acid (PHB) differ from a magnetosome in composition and metabolic role?

17. What is the function of gas vesicles? How are these structures made such that they can remain gas tight?

18. In a few sentences, indicate how the bacterial endospore differs from the vegetative cell in structure, chemical composition, and ability to resist extreme environmental conditions.

19. The discovery of the bacterial endospore was of great practical importance. Why?

20. Define the following terms: mature endospore, vegetative cell, and germination.

21. How long may endospores remain viable in a state of "suspended animation," so to speak, and what is the evidence for this?

Application Questions

1. Calculate the size of the smallest resolvable object if 600-nm light is used to observe a specimen with a 100× oil-immersion lens having a numerical aperture of 1.32. How could resolution be improved using this same lens?

2. Calculate the surface-to-volume ratio of a spherical cell 15 μm in diameter and a cell 2 μm in diameter. What are the consequences of these differences in surface-to-volume ratio for cell function?

3. Imagine a planet where life evolved in a totally nonaqueous environment. Cells on this planet contain highly hydrophobic cytoplasm and live in water-free environments. Predict and draw the structure of the type of cytoplasmic membranes organisms on this planet would have and discuss why such a membrane would be best suited to these organisms.

4. From what you know about the nature of the bacterial cell wall and membrane, explain why a rod-shaped bacterial cell becomes a spherical structure when its wall is removed under conditions such that cell lysis cannot occur.

5. Assume you are given two cultures, one of a species of gram-negative *Bacteria* and one of a species of *Archaea*. Other than by sequencing ribosomal RNA (∞ Section 2.3), discuss at least four different ways you could tell which culture was which.

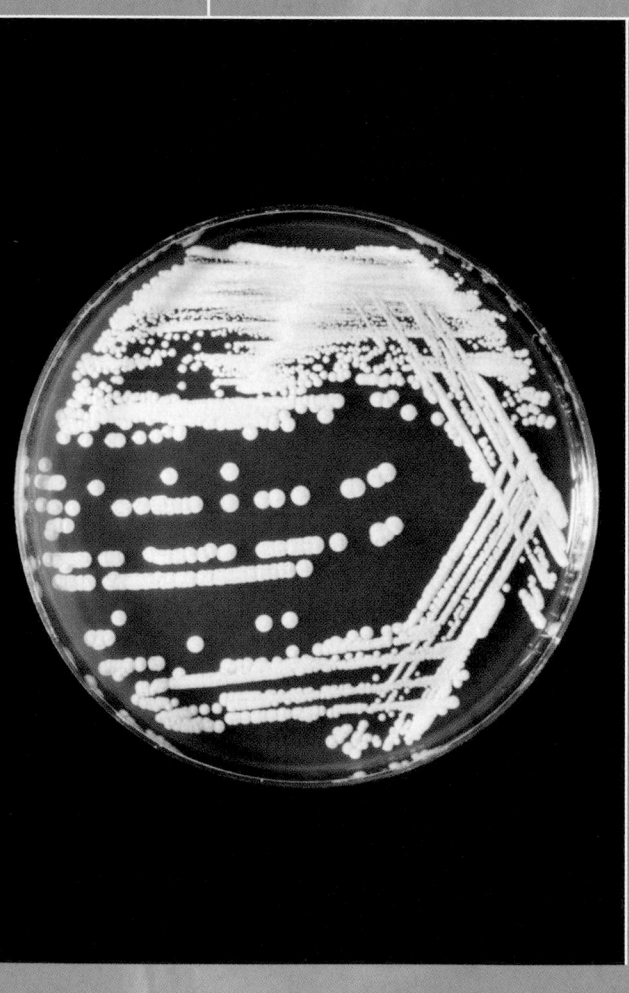

To study microorganisms in the laboratory, it is usually necessary to culture (grow) them. This can be done in a number of ways, including the use of solid culture media and the Petri plate, as shown here. Individual cells that were spread out on the plate grow and divide to form *colonies*. Although some common nutritional principles apply to all organisms, many microorganisms have special nutritional and cultural requirements. In addition, in order to sustain life processes, all microorganisms must carry out one or more series of reactions that conserve energy. Thus, an understanding of nutrition and bioenergetics is necessary to understand how microbial cells "make a living" in their natural habitats.

NUTRITION, LABORATORY CULTURE, AND METABOLISM OF MICROORGANISMS

WORKING GLOSSARY

Activation energy the energy required to bring substrates to the reactive state

Aerobe a microorganism able to use O_2 in respiration

Anabolism the sum total of all biosynthetic reactions in the cell

Anaerobe a microorganism that either can or must grow in the absence of O_2

Aseptic technique the series of manipulations used to prevent contamination during the handling of sterile objects or microbial cultures

ATP synthase (ATPase) A multiprotein enzyme complex embedded in the membrane that catalyzes the synthesis of ATP coupled to dissipation of the proton motive force

Autotroph an organism capable of biosynthesizing all cell material from CO_2 as the sole carbon source

Catabolism biochemical reactions leading to the production of usable energy (usually ATP) by the cell

Catalyst a substance that accelerates a chemical reaction but that is not consumed in the reaction

Chemiosmosis a theory describing linkage of ATP synthesis to dissipation of a proton motive force

Citric acid cycle a cyclical series of reactions resulting in the conversion of acetate to two CO_2

Coenzyme a small nonprotein molecule that participates in a catalytic reaction as part of an enzyme

Complex medium a culture medium composed of digests of chemically undefined substances such as yeast and meat extracts

Culture medium an aqueous solution of various nutrients suitable for the growth of microorganisms

Defined medium a culture medium whose precise chemical composition is known

Electron acceptor a substance that can accept electrons from some other substance, thereby becoming reduced in the process

Electron donor a substance that can donate electrons to some electron acceptor, thereby becoming oxidized in the process

Endergonic energy requiring

Enzyme a protein that has the ability to speed up (catalyze) a specific chemical reaction

Exergonic energy releasing

Fermentation anaerobic catabolism in which an organic compound serves as both an electron donor and an electron acceptor and in which ATP is produced by substrate-level phosphorylation

Free energy (G) energy available to do work; $G^{0\prime}$ is free energy under standard conditions

Glycolysis a biochemical pathway in which glucose is fermented yielding energy (ATP) and various fermentation products. Also called the Embden-Meyerhof pathway

Oxidative phosphorylation the production of ATP at the expense of a proton motive force formed by electron transport

Photophosphorylation the production of ATP from a proton motive force formed from photosynthetic reactions

Proton motive force an energized state of the membrane resulting from the separation of charge and the elements of water (H^+ versus OH^-) across the membrane

Pure culture a culture that contains a single kind of microorganism

Reduction potential ($E_0\prime$) the inherent tendency (measured in volts) of a compound to donate electrons; $E_0\prime$ is reduction potential under standard conditions

Respiration the process in which a compound is oxidized with O_2 (or an O_2 substitute) functioning as the terminal electron acceptor, usually accompanied by ATP production by oxidative phosphorylation

Siderophores iron chelators that can bind iron present at very low concentrations

Substrate-level phosphorylation production of ATP by the direct transfer of a high-energy phosphate molecule from a phosphorylated organic compound to ADP

I NUTRITION AND CULTURE OF MICROORGANISMS

A key feature of cells is their ability to direct chemical reactions and organize molecules into specific structures. The ultimate expression of this organization is growth (replication). Before cell replication can take place, a variety of chemical reactions must occur in the cell; collectively, these reactions are referred to as **metabolism**. Metabolic reactions are either *energy releasing*, called **catabolic reactions**, or *energy requiring*, called **anabolic reactions**. Several classes of catabolic and anabolic reactions occur in cells, and we will examine some of the key ones in this and future chapters.

Before we consider some of the major reactions of metabolism, we will discuss how microorganisms are grown in the laboratory. Indeed, most of what we know about the metabolic activities of microorganisms has emerged from the study of laboratory cultures. Our focus here will be on *chemoorganotrophs*, those organisms that use organic compounds of one sort or another as their carbon and energy sources (∞ Section 2.4). Later in this chapter we will consider mechanisms other than chemoorganotrophy that generate energy. Organisms that employ these metabolic alternatives can also be grown in the laboratory, and many of the basic principles of nutrition and cell culture discussed now will apply to them as well.

5.1 Microbial Nutrition

Recall from Chapter 3 that cells consist mainly of macromolecules and water and that macromolecules consist of smaller units called monomers (∞ Section 3.2). Thus, microbial nutrition is really all about supplying cells with the chemical tools they need to make monomers. These chemical tools are called **nutrients**. Different organisms need different sets of nutrients and often need these nutrients in one or another specific form. And not all nutrients are required in the same amounts; some nutrients, called *macronutrients* (Table 5.1), are required in large amounts, while others, called *micronutrients*, are

TABLE 5.1 Macronutrients in nature and in culture media

Element	Usual form of nutrient found in the environment	Chemical form supplied in culture media
Carbon (C)	CO_2, organic compounds	Glucose, malate, acetate, pyruvate, amino acids, hundreds of other compounds, or complex mixtures (yeast extract, peptone, and so on)
Hydrogen (H)	H_2O, organic compounds	H_2O, organic compounds
Oxygen (O)	H_2O, O_2, organic compounds	H_2O, O_2, organic compounds
Nitrogen (N)	NH_3, NO_3^-, N_2, organic nitrogen compounds	*Inorganic:* NH_4Cl, $(NH_4)_2SO_4$, KNO_3, N_2 *Organic:* Amino acids, nitrogen bases of nucleotides, many other N-containing organic compounds
Phosphorus (P)	PO_4^{3-}	KH_2PO_4, Na_2HPO_4
Sulfur (S)	H_2S, SO_4^{2-}, organic S compounds, metal sulfides (FeS, CuS, ZnS, NiS, and so on)	Na_2SO_4, $Na_2S_2O_3$, Na_2S, cysteine, or other organic sulfur compounds
Potassium (K)	K^+ in solution or as various K salts	KCl, KH_2PO_4
Magnesium (Mg)	Mg^{2+} in solution or as various Mg salts	$MgCl_2$, $MgSO_4$
Sodium (Na)	Na^+ in solution or as NaCl or other Na salts	NaCl
Calcium (Ca)	Ca^{2+} in solution or as $CaSO_4$ or other Ca salts	$CaCl_2$
Iron (Fe)	Fe^{2+} or Fe^{3+} in solution or as FeS, $Fe(OH)_3$, or many other Fe salts	$FeCl_3$, $FeSO_4$, various chelated iron solutions (Fe^{3+} EDTA, Fe^{3+} citrate, and so on)

required in lesser, sometimes even trace, amounts. We begin with a consideration of the major macronutrients *carbon* and *nitrogen*.

Carbon and Nitrogen

Many prokaryotes require an organic compound of some sort as their source of **carbon**. Nutritional studies have shown that bacteria can assimilate various organic carbon compounds and use them to make new cell material. Amino acids, fatty acids, organic acids, sugars, nitrogen bases, aromatic compounds, and countless other organic compounds have been shown to be used by one bacterium or another. Some prokaryotes are *autotrophs,* able to build all of their organic structures from carbon dioxide (CO_2) with energy obtained from either light or inorganic chemicals. On a dry weight basis, a typical cell is about 50% carbon and carbon is the major element in all classes of macromolecules.

After carbon, the next most abundant element in the cell is **nitrogen**. A typical bacterial cell is about 12% nitrogen (by dry weight), and nitrogen is an important element in proteins, nucleic acids, and several other constituents in the cell. Nitrogen is present in nature in both organic and inorganic forms (Table 5.1). However, the bulk of available nitrogen in nature is in *inorganic* form, either as ammonia (NH_3), nitrate (NO_3^-), or N_2. Most bacteria are capable of using ammonia as the sole nitrogen source, and many can also use nitrate. However, nitrogen gas (N_2) can only be a nitrogen source for certain bacteria, the *nitrogen-fixing bacteria*, and we discuss the properties of these organisms in detail later (Sections 12.9, 17.28, and 19.12).

Other Macronutrients: P, S, K, Mg, Ca, Na

Phosphorus occurs in nature in the form of organic and inorganic phosphates and is required by the cell primarily for synthesis of nucleic acids and phospholipids. **Sulfur** is required because of its structural role in the amino acids cysteine and methionine (Section 3.6) and because it is present in a number of vitamins, such as thiamine, biotin, and lipoic acid, as well as in coenzyme A. Sulfur undergoes a number of chemical transformations in nature, many of which are carried out exclusively by microorganisms (Section 19.13) and is available to organisms in a variety of forms. Most cell sulfur originates from inorganic sources, either sulfate (SO_4^{2-}) or sulfide (HS^-) (Table 5.1).

Potassium is required by all organisms. A variety of enzymes, including some of those involved in protein synthesis, specifically require potassium. **Magnesium** functions to stabilize ribosomes, cell membranes, and nucleic acids and is also required for the activity of many enzymes. **Calcium** (which is not an essential nutrient for the growth of many microorganisms) helps stabilize the bacterial cell wall and plays a key role in the heat stability of endospores (Section 4.15). **Sodium** is required by some but not all organisms, and its need often reflects the habitat of the organism. For example, seawater has a high sodium content and marine microorganisms generally require sodium for growth, whereas closely related freshwater species are usually able to grow in the absence of sodium.

Iron

Iron plays a major role in cellular respiration, being a key component of the cytochromes and iron-sulfur proteins involved in electron transport (see Section 5.11 and

TABLE 5.2 Micronutrients (trace elements) needed by living organisms[a]

Element	Cellular function
Chromium (Cr)	Required by mammals for glucose metabolism; no known microbial requirement
Cobalt (Co)	Vitamin B_{12}; transcarboxylase (propionic acid bacteria)
Copper (Cu)	Respiration, cytochrome c oxidase; photosynthesis, plastocyanin, some superoxide dismutases
Manganese (Mn)	Activator of many enzymes; present in certain superoxide dismutases and in the water-splitting enzyme in oxygenic phototrophs (Photosystem II)
Molybdenum (Mo)	Certain flavin-containing enzymes; nitrogenase, nitrate reductase, sulfite oxidase, DMSO-TMAO reductases, some formate dehydrogenases
Nickel (Ni)	Most hydrogenases; coenzyme F_{430} of methanogens; carbon monoxide dehydrogenase; urease
Selenium (Se)	Formate dehydrogenase; some hydrogenases; the amino acid selenocysteine
Tungsten (W)	Some formate dehydrogenases; oxotransferases of hyperthermophiles
Vanadium (V)	Vanadium nitrogenase; bromoperoxidase
Zinc (Zn)	Carbonic anhydrase; alcohol dehydrogenase; RNA and DNA polymerases; and many DNA-binding proteins
Iron (Fe)[b]	Cytochromes; catalases; peroxidases; iron-sulfur proteins; oxygenases; all nitrogenases

[a]Not every micronutrient listed is required by all cells; some metals listed are found in enzymes present in only specific microorganisms.

[b]Needed in greater amounts than other trace metals.

Table 5.2). Under anoxic conditions, iron is generally in the +2 oxidation state and soluble. However, under oxic conditions, iron is generally in the +3 oxidation state and forms various insoluble minerals. In order to obtain iron from such minerals, cells produce iron-binding agents called **siderophores** that solubilize iron and transport it into the cell. One major group of siderophores consists of derivatives of hydroxamic acid, which chelate ferric (Fe^{3+}) iron very strongly (Figure 5.1a). Once the iron-hydroxamate complex has passed into the cell, the iron is released and the hydroxamate can exit the cell and be used again for iron transport. Bacteria such as *Escherichia coli* and *Salmonella typhimurium* produce structurally complex phenolic siderophores called **enterobactins** (Figure 5.1b). These siderophores are derivatives of the aromatic compound catechol and have an extremely high binding affinity for iron. Without such tenacious iron-binding agents, many disease-causing bacteria (pathogens) would be unable to initiate an infection due to iron limitation (↝ Sections 21.7 and 21.8).

In marine waters, iron can be virtually undetectable (surface ocean waters typically contain only a few *picograms* of iron per milliliter), and certain marine bacteria have been found to produce structurally complex siderophores that can sequester iron present in these vanishingly small amounts. These siderophores contain a peptide head group that complexes Fe^{3+} and a lipid tail that can associate with the cell membrane. Compounds such as *aquachelin* (Figure 5.1c) bind iron, aggregate into lipidlike micelles, and then transport the bound iron into the cell. Some prokaryotes can grow in the total absence of iron. For example, the bacteria *Lactobacillus plantarum* and *Borrelia burgdorferii* (the latter is the causative agent of Lyme disease, ↝ Section 27.4) do not contain iron. In these bacteria, Mn^{2+} substitutes for iron as the metal component of enzymes that normally contain Fe^{2+}.

Micronutrients (Trace Elements)

Although required in just tiny amounts, micronutrients are nevertheless just as critical to cell function as are macronutrients. Micronutrients are metals, many of which play a structural role in various enzymes, the cells' catalysts. Table 5.2 summarizes the major micronutrients of living systems and gives examples of enzymes in which each plays a role.

Because the requirement for trace elements is so small, for the laboratory culture of microorganisms it is frequently unnecessary to add trace elements to the culture medium. However, if a culture medium contains highly purified chemicals dissolved in high purity distilled water, a trace element deficiency can occur. In such cases a small amount of a solution of trace metals (Table 5.2) is added to the medium to make available the necessary metals.

Growth Factors

Growth factors are *organic* compounds that, like micronutrients, are required in very small amounts and only by some cells. Growth factors include vitamins,

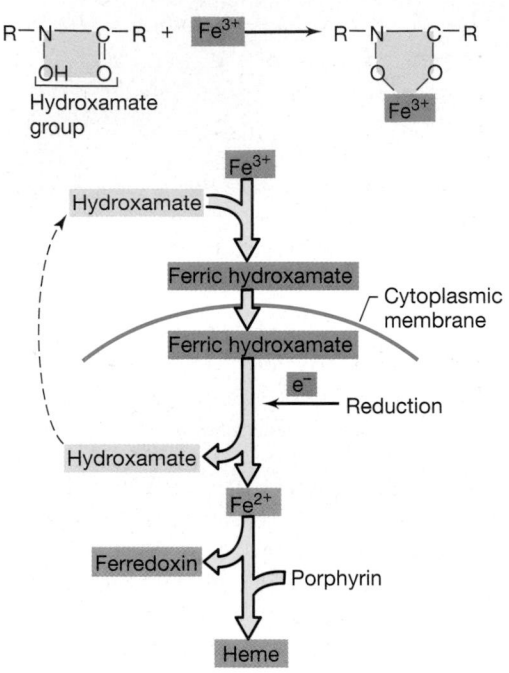

(a)

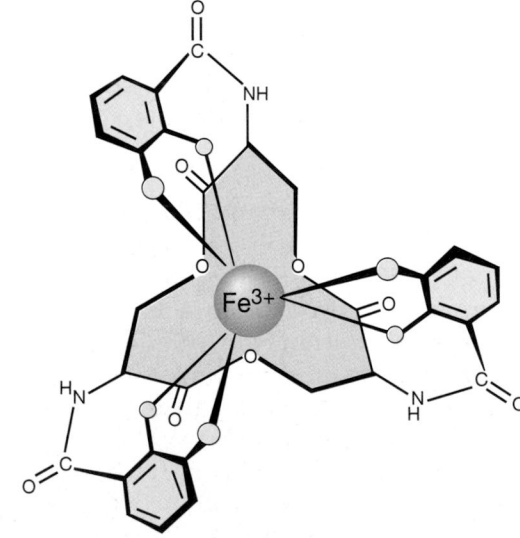

(b)

(c)

TABLE 5.3 Growth factors: Vitamins and their functions

Vitamin	Function
p-Aminobenzoic acid	Precursor of folic acid
Folic acid	One-carbon metabolism; methyl group transfer
Biotin	Fatty acid biosynthesis; β-decarboxylations; some CO_2 fixation reactions
Cobalamin (B_{12})	Reduction of and transfer of single carbon fragments; synthesis of deoxyribose
Lipoic acid	Transfer of acyl groups in decarboxylation of pyruvate and α-ketoglutarate
Nicotinic acid (niacin)	Precursor of NAD^+ (see Figure 5.10); electron transfer in oxidation-reduction reactions
Pantothenic acid	Precursor of coenzyme A; activation of acetyl and other acyl derivatives
Riboflavin	Precursor of FMN (see Figure 5.15), FAD in flavoproteins involved in electron transport
Thiamine (B_1)	α-Decarboxylations; transketolase
Vitamins B_6 (pyridoxal-pyridoxamine group)	Amino acid and keto acid transformations
Vitamin K group; quinones	Electron transport; synthesis of sphingolipids
Hydroxamates	Iron-binding compounds; solubilization of iron and transport into cell

amino acids, purines, and pyrimidines. Although most microorganisms are able to synthesize all of these compounds, some microorganisms require one or more of them preformed from the environment.

Vitamins are the most commonly needed growth factors. Most vitamins function as parts of coenzymes (see, for instance, Figures 5.10, 5.12, and 5.15), and these are summarized in Table 5.3. Many microorganisms are able to synthesize all the components of their coenzymes, but some are unable to do so and must be provided with certain parts of these coenzymes in the form of vitamins. Lactic acid bacteria, which include the genera *Streptococcus*, *Lactobacillus*, *Leuconostoc*, and others (⬤ Section 12.19), are renowned for their complex vitamin requirements, which are even more extensive than those of humans (see Table 5.4)! The vitamins most commonly required by microorganisms are thiamine (vitamin B_1), biotin, pyridoxine (vitamin B_6), and cobalamin (vitamin B_{12}).

Figure 5.1 Iron-chelating agents produced by microorganisms. (a) Hydroxamate. Iron is bound as Fe^{3+} and released inside the cell as Fe^{2+}. The hydroxamate then exits the cell and repeats the cycle. (b) Ferric enterobactin of *Escherichia coli*. The oxygen atoms of each catechol molecule are shown in yellow. (c) The peptidic and tailed siderophore aquachelin.

✓ 5.1 Concept Check

The hundreds of chemical compounds present inside a living cell are formed from starting materials called nutrients. Elements required in fairly large amounts are called macronutrients, while metals and organic compounds needed in very small amounts are called micronutrients and growth factors, respectively.

✓ What two classes of macromolecules contain the bulk of the *nitrogen* in a cell?

✓ Why is an element like Co^{2+} considered a *micro*nutrient whereas an element like C is considered a *macro*nutrient?

✓ What roles do iron play in cellular metabolism?

✓ What mechanism(s) are responsible for sequestering iron?

5.2 Culture Media

We summarize our discussion of microbial nutrition by examining the chemical composition of several culture media (Table 5.4). **Culture media** are the nutrient solutions used to grow microorganisms in the laboratory. Two broad classes of culture media are used in microbiology: **chemically defined** and **undefined (complex)**. Chemically defined media are prepared by adding precise amounts of highly purified inorganic or organic chemicals to distilled water. Therefore, the *exact chemical composition* of a defined medium is known. In many cases, however, knowledge of the exact composition of a medium is not critical. In these instances complex media may suffice or for various reasons may even be advantageous. Complex media often employ digests of casein (milk protein), beef, soybeans, yeast cells, or any of a number of other highly nutritious (yet chemically undefined) substances. Such digests are available commercially in powdered form and can be weighed out rapidly and dissolved in distilled water to give a medium. However, a major concession in using a complex medium is loss of control of its precise nutrient composition.

Nutritional Requirements and Biosynthetic Capacity

Table 5.4 shows three recipes for culture media, two defined and one complex. The complex medium is easiest to prepare and supports good growth of either organism shown in the table, the enteric bacterium *Escherichia coli* or the lactic acid bacterium *Leuconostoc mesenteroides*, an extremely fastidious (nutritionally demanding) bacterium. The simple defined medium supports excellent growth of *E. coli* but not of *L. mesenteroides*; growth of the latter organism in defined medium requires the addition of several organic nutrients and growth factors not needed by *E. coli* (Table 5.4). With this in mind, which

TABLE 5.4 Examples of culture media for microorganisms with simple and demanding nutritional requirements[a]

Defined culture medium for *Escherichia coli*	Defined culture medium for *Leuconostoc mesenteroides*	Complex culture medium for either *E. coli* or *L. mesenteroides*
K_2HPO_4 7 g KH_2PO_4 2 g $(NH_4)_2SO_4$ 1 g $MgSO_4$ 0.1 g $CaCl_2$ 0.02 g Glucose 4–10 g Trace elements (Fe, Co, Mn, Zn, Cu, Ni, Mo) 2–10 µg each Distilled water 1000 ml pH 7	K_2HPO_4 0.6 g KH_2PO_4 0.6 g NH_4Cl 3 g $MgSO_4$ 0.1 g Glucose 25 g Sodium acetate 20 g Amino acids (alanine, arginine, asparagine, aspartate, cysteine, glutamate, glutamine, glycine, histidine, isoleucine, leucine, lysine, methionine, phenylalanine, proline, serine, threonine, tryptophan, tyrosine, valine) 100–200 µg of each Purines and pyrimidines (adenine, guanine, uracil, xanthine) 10 mg of each Vitamins (biotin, folate, nicotinic acid, pyridoxal, pyridoxamine, pyridoxine, riboflavin, thiamine, pantothenate, *p*-aminobenoic acid) 0.01–1 mg of each Trace elements (see first column) 2–10 µg each Distilled water 1000 ml pH 7	Glucose 15 g Yeast extract 5 g Peptone 5 g KH_2PO_4 2 g Distilled water 1000 ml pH 7

(a)

(b)

[a]The photos are tubes of (a) the defined medium described, and (b) the complex medium described. Note how the complex medium is colored from the various organic extracts and digests that it contains. *Photos courtesy of Cheryl L. Broadie and John Vercillo, Southern Illinois University at Carbondale.*

organism, *E. coli* or *L. mesenteroides*, has a greater *biosynthetic* capacity? Obviously, *E. coli,* since its ability to grow on a simple defined culture medium means that it has the ability to synthesize *all* its organic cellular constituents from a single carbon compound, in this case glucose (Table 5.4). By contrast, *L. mesenteroides* has multiple growth factor requirements indicative of limited biosynthetic capacity. The complex nutritional needs of *L. mesenteroides* can be satisfied by either preparing a defined medium as shown in Table 5.4 (although in this case it could take quite a bit of time to do so) or by using a complex medium (Table 5.4), which can usually be prepared quickly.

It is important to understand when examining recipes for culture media such as those shown in Table 5.4 that *different microorganisms can have vastly different nutritional requirements.* Thus, for successful culture of a given microorganism it is necessary to understand its nutritional requirements and then supply it with its essential nutrients in the proper form and proportions in a culture medium. If care is taken in preparing culture media, it is usually fairly easy to culture many different types of microorganisms in the laboratory. We discuss the procedures involved in culturing microorganisms now.

✓ 5.2 Concept Check

Culture media supply the nutritional needs of microorganisms and can be either chemically defined or undefined (complex).

✓ Why is the routine culture of *Leuconostoc mesenteroides* easier in a complex medium than in a chemically defined medium?

✓ In which medium, simple, defined, or complex (shown in Table 5.4), do you think *Escherichia coli* would grow more quickly? Why?

5.3 Laboratory Culture of Microorganisms

Once a culture medium has been prepared, it can be inoculated (that is, organisms added) and incubated under conditions that will support microbial growth. In most cases growth will be of a **pure culture**, a culture con-

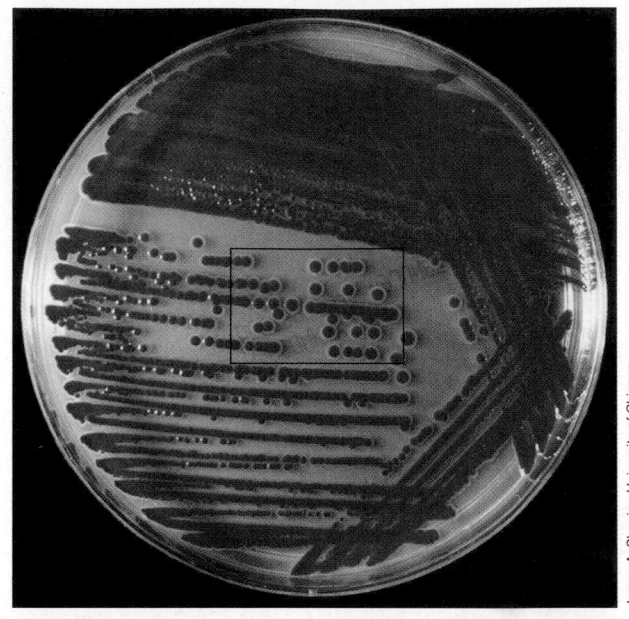

(a)

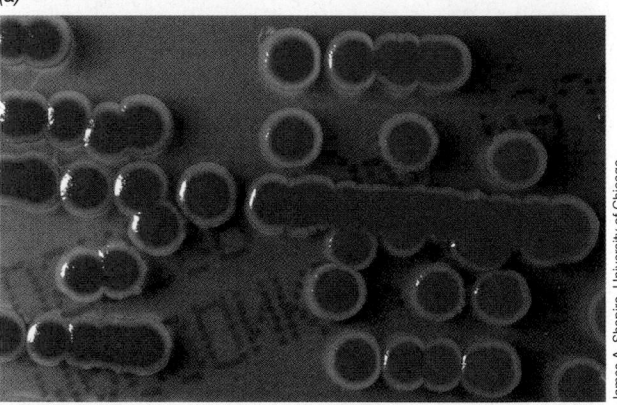

(b)

(c)

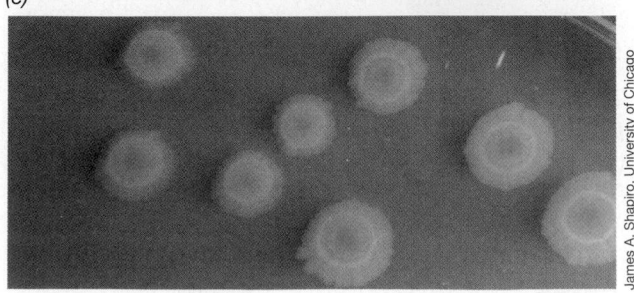

(d)

Figure 5.2 Examples of bacterial colonies. Colonies are visible masses of cells formed from the subsequent division of one or a few cells. The size, shape, texture, and color of a bacterial colony is a function of the organism that made them. Depending on its size and the arrangement of cells within the colony, a colony can have highly variable cell numbers; colonies containing over one *billion* individual cells are not uncommon. (a) *Serratia marcesens*, grown on MacConkey agar. (b) Close-up of colonies outlined in (a). (c) *Pseudomonas aeruginosa*, grown on Trypticase-Soy agar. (d) *Shigella flexneri*, grown on MacConkey agar.

James A. Shapiro, University of Chicago

taining only a *single kind* of microorganism. To obtain or maintain a pure culture it is essential that other organisms are prevented from entering it. Such unwanted organisms, called *contaminants,* are ubiquitous (as Pasteur discovered over 125 years ago, ⚭ Section 1.5), and microbiological techniques are designed to avoid contaminants. A major method for obtaining pure cultures and for assessing the purity of a culture is the use of solid media, specifically, solid media in the Petri plate.

Solid versus Liquid Culture Media

Thus far we have only considered the preparation of liquid culture media, such as the defined and complex media shown in tubes in Table 5.4. However, culture media are frequently prepared in a semisolid form by the addition of a gelling agent to liquid media. Such solid culture media immobilize cells, allowing them to grow and form visible, isolated masses called *colonies* (Figure 5.2). Bacterial colonies can be of various shapes and sizes depending on the organism, the culture conditions, the nutrient supply (including the amount of oxygen present), and several other physiological parameters. Some bacteria produce pigments that cause the colony to be colored (Figure 5.2). Regardless of whether they are pigmented, colonies permit the microbiologist to visualize the purity of the culture; plates that contain more than one colony type were not streaked from a pure culture. In this way, the Petri plate has been used as a criterion of culture purity for over 100 years (⚭ Section 1.5).

Solid media are prepared the same way as for liquid media except that before sterilizing the medium, *agar* is added as a gelling agent, usually at about 1.5%. The agar melts during the sterilization process, and the molten medium is then poured into sterile glass or plastic plates and allowed to harden before use (Figure 5.2).

Aseptic Technique

Because microorganisms are everywhere, culture media must be sterilized before use. For most culture media this is done by heating, typically by moist heat in a large pressure cooker called an *autoclave.* We discuss the operation and principles of the autoclave later, along with other methods of sterilization (⚭ Section 20.1).

Once a sterile culture medium has been prepared, it can receive an *inoculum* from a previously grown pure culture to start the growth process once again. This requires the practice of **aseptic technique**. Aseptic technique is the series of procedures used to prevent contamination during manipulations of cultures and sterile culture media (Figures 5.3 and 5.4). Its mastery is required for success in the microbiology laboratory, and it is one of the first methods learned by the novice microbiologist. Airborne contaminants are the most common problem because the air always contains dust particles that generally have a community of microorganisms on them. When containers are opened, they must be han-

dled in such a way that contaminant-laden air does not enter (Figures 5.3 and 5.4). Aseptic transfer of a culture from one tube of medium to another is usually accomplished with an inoculating loop or needle that has been sterilized by incineration in a flame (Figure 5.3). Cultures in which growth has taken place can also be transferred to the surface of agar plates (Figure 5.4), where colonies

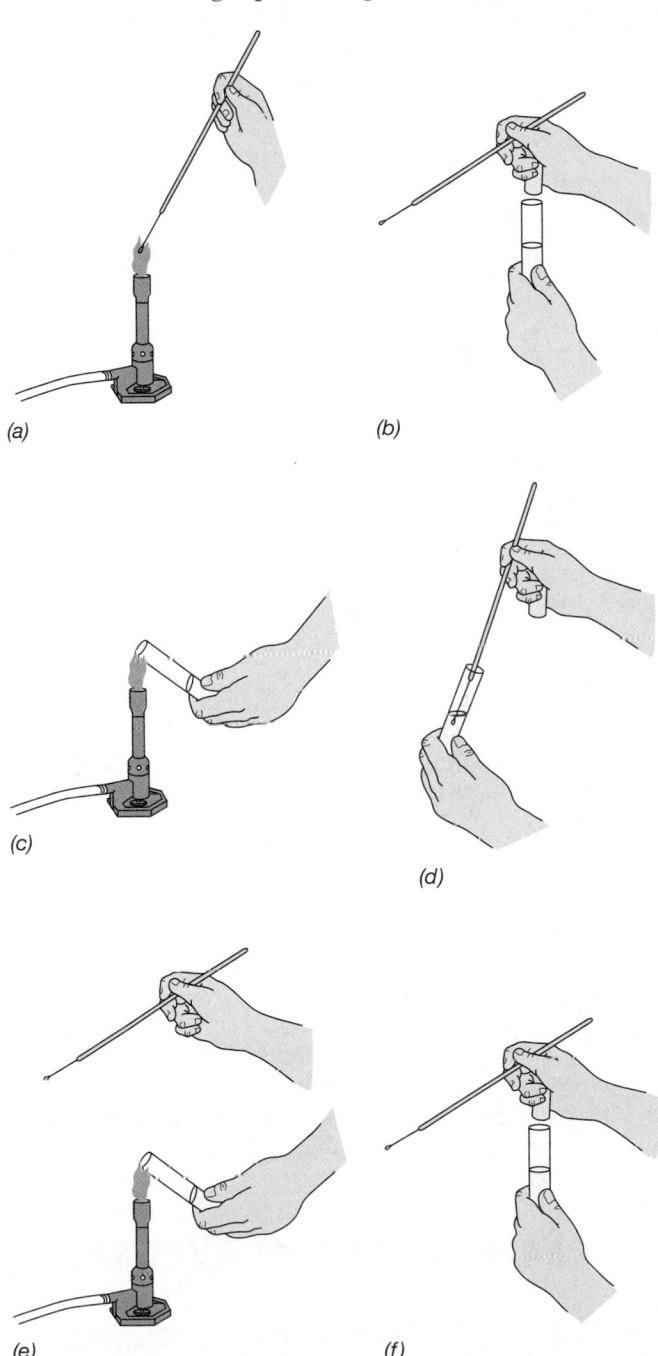

(a)

(b)

(c)

(d)

(e)

(f)

Figure 5.3 Aseptic transfer. (a) Loop is heated until red-hot and cooled in air briefly. (b) Tube is uncapped. (c) Tip of tube is run through the flame. (d) Sample is removed on sterile loop. (e) After removing sample on loop, the tube is reflamed and the sample transferred to a sterile medium. (f) The tube is recapped. Loop is reheated before being taken out of service.

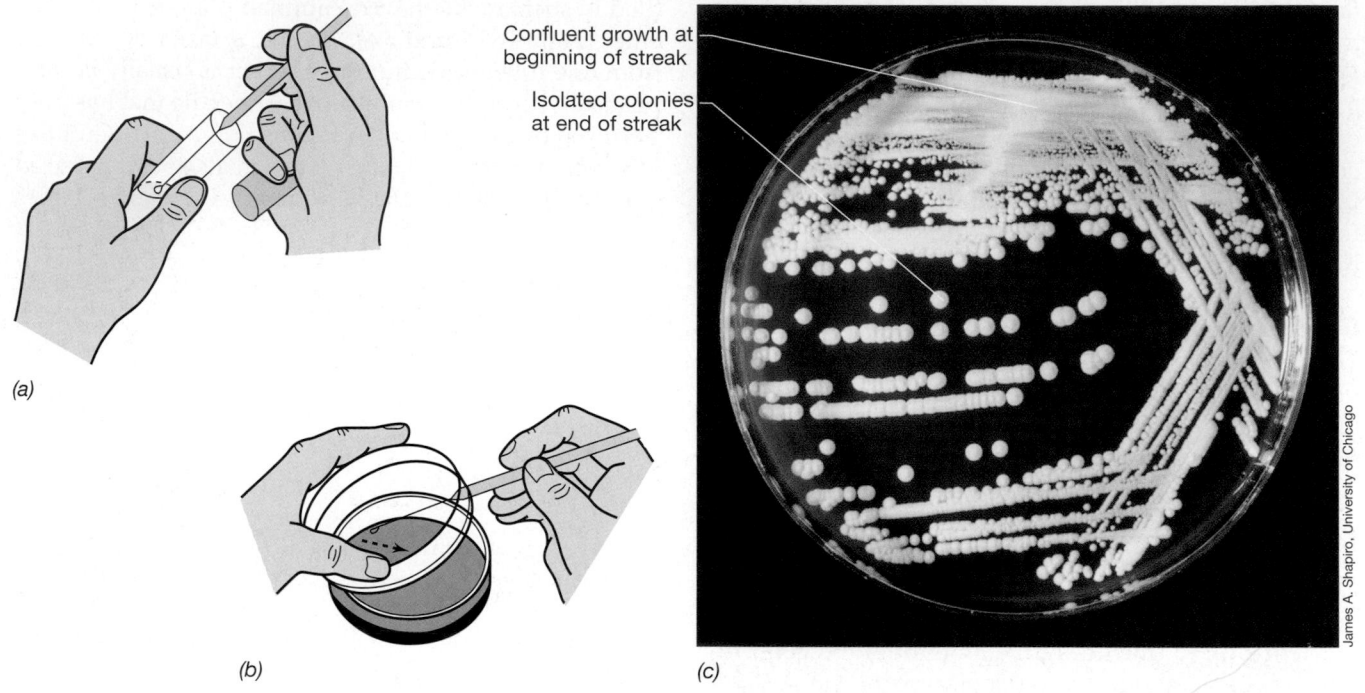

Confluent growth at beginning of streak

Isolated colonies at end of streak

(a)

(b)

(c)

James A. Shapiro, University of Chicago

Figure 5.4 Method of making a streak plate to obtain pure cultures. (a) Loop is sterilized, and then a loopful of inoculum is removed from tube. (b) Streak is made over a sterile agar plate, spreading out the organisms. Following the initial streak, subsequent streaks are made at angles to it, the loop being resterilized between streaks. (c) Appearance of the streaked plate after incubation. Colonies of the bacterium *Micrococcus luteus* grown on blood agar plates. It is from such well-isolated colonies that pure cultures can usually be obtained.

develop from the growth and division of single cells. Picking and restreaking from an isolated colony is a major method of obtaining pure cultures from microbial communities containing many different organisms.

✓ 5.3 Concept Check

Microorganisms can be grown in the laboratory in culture media containing the nutrients they require. Successful cultivation of pure cultures of microorganisms can be done only if aseptic technique is practiced.

✓ What is meant by the word *sterile?* What would happen if freshly prepared culture media were not sterilized?

✓ Why is aseptic technique necessary for successful cultivation of pure cultures in the laboratory?

II   ENERGETICS AND ENZYMES

We learned in Chapter 2 of the different energy classes of microorganisms, *chemoorganotrophs, chemolithotrophs,* and *phototrophs* (⌦ Section 2.4). However, regardless of how an organism makes a living, it must be able to obtain energy from either chemical compounds or from light, and then conserve the energy as ATP. Here we discuss the principles of energy conservation, using some

simple laws of chemistry and physics to guide our understanding, and then consider the action of enzymes, the cell's biocatalysts.

5.4   Bioenergetics

Energy is defined as the ability to do work. In microbiology, energy is measured in units of *kilojoules* (kJ), a measure of heat energy. Chemical reactions are accompanied by changes in energy. Although in any chemical reaction some energy is lost as heat, in microbiology we are interested in **free energy** (abbreviated G), which is defined as the energy released *that is available to do useful work.* The change in free energy during a reaction is expressed as $\Delta G^{0\prime}$, where the symbol Δ should be read "change in." The "0" and "prime" mean that the free-energy value was obtained under "standard" conditions: pH 7, 25°C, all reactants and products initially at 1 M concentration.*

---

*Free energy calculations using standard conditions are estimations of the free energy changes that actually occur when a reaction takes place in nature, despite the fact that nutrient concentrations of 1 M rarely occur in nature. Although for now, calculations of $\Delta G^{0\prime}$ are reasonable estimates, we will see later that the actual concentration of products and reactants can occasionally alter the bioenergetics of reactions in microbiologically important ways (⌦ Sections 17.21 and 19.10). These issues are also discussed in more detail in Appendix 1.

If in the reaction:

$$A + B \rightarrow C + D$$

the $\Delta G^{0\prime}$ is *negative*, the reaction will proceed with the *release* of free energy, energy that the cell may be able to conserve in the form of ATP. Such energy-yielding reactions are called **exergonic**. However, if $\Delta G^{0\prime}$ is *positive*, the reaction *requires* energy in order to proceed; such reactions are called **endergonic**. Thus, from the standpoint of the microbial cell, exergonic reactions *yield* energy while endergonic reactions *require* energy.

Free Energy of Formation and Calculating $\Delta G^{0\prime}$

In order to calculate the free energy yield of a reaction, one needs to know the free energy of its reactants and products. This is the *free energy of formation*, the energy yielded or energy required for the *formation* of a given molecule from its constituent elements. By convention, the free energy of formation ($G^0{}_f$) of the elements (for instance, C, H_2, N_2) is zero. If the formation of a *compound* from elements proceeds exergonically, then the free energy of formation of the compound is negative (energy is released), whereas if the reaction is endergonic (energy is required), then the free energy of formation of the compound is positive.

A few examples of free energies of formation are given in Table 5.5. For most compounds $G^0{}_f$ is *negative*, reflecting the fact that compounds tend to form spontaneously from elements. However, the positive $G^0{}_f$ for nitrous oxide (+104.2 kJ/mol) tells us that this molecule does not form spontaneously, but rather decomposes to nitrogen and oxygen. The free energies of formation of a variety of compounds of microbiological interest are given in Appendix 1.

Using free energies of formation, it is possible to calculate the *change* in free energy occurring in a given reaction. For a simple reaction such as $A + B \rightarrow C + D$, $\Delta G^{0\prime}$ is calculated by subtracting the *sum* of the free energies

of formation of the reactants (in this case A and B) from that of the products (C and D). Thus,

$\Delta G^{0\prime}$ of $A + B \rightarrow C + D$

$$= G^0{}_f[C + D] - G^0{}_f[A + B]$$

The phrase "products minus reactants" encompasses the necessary steps for calculating changes in free energy during chemical reactions. However, it is first necessary to balance the reaction chemically before free-energy calculations can be made. Appendix 1 details the steps involved in calculating free energies for any hypothetical reaction.

✓ 5.4 Concept Check

The chemical reactions of the cell are accompanied by changes in energy, expressed in kJ. A chemical reaction can occur with the release of free energy (exergonic) or with the consumption of free energy (endergonic).

- ✓ What is free energy?
- ✓ In general, are *catabolic* reactions exergonic or endergonic?
- ✓ Using the data in Table 5.5, calculate $\Delta G^{0\prime}$ for the reaction $CH_4 + \frac{1}{2}O_2 \rightarrow CH_3OH$.

5.5 Catalysis and Enzymes

A free-energy calculation tells us only whether energy is released or required in a given reaction; it tells us nothing about the *rate* of the reaction. Consider the formation of water from gaseous oxygen and hydrogen. The energetics of this reaction is quite favorable: $H_2 + \frac{1}{2}O_2 \rightarrow H_2O$, $\Delta G^{0\prime} = -237$ kJ. However, if we were to simply mix O_2 and H_2 together, no measurable formation of water would occur for many years. This is because the rearrangement of oxygen and hydrogen atoms to form water requires that the chemical bonds of the reactants be broken first. The breaking of bonds requires energy, and this energy is referred to as **activation energy**. Activation energy is the amount of energy required to bring all molecules in a chemical reaction to the reactive state. For a reaction that proceeds with a net release of free energy (that is, an exergonic reaction), the situation is as diagrammed in Figure 5.5.

Enzymes

The idea of activation energy leads us to the concept of catalysis. A **catalyst** is a substance that *lowers* the activation energy of a reaction, thereby *increasing* the rate of reaction. Catalysts facilitate reactions but are themselves not consumed or transformed by these reactions. It is important to note that catalysts do not affect the energetics or the equilibrium of a reaction; catalysts affect only the *speed* at which reactions proceed.

TABLE 5.5	Free energy of formation for a few compounds of biological interest
Compound	**Free energy of formation**[a]
Water (H_2O)	−237.2
Carbon dioxide (CO_2)	−394.4
Hydrogen gas (H_2)	0
Oxygen gas (O_2)	0
Ammonium ($NH_4{}^+$)	−79.4
Nitrous oxide (N_2O)	+104.2
Acetate ($C_2H_3O_2{}^-$)	−369.4
Glucose ($C_6H_{12}O_6$)	−917.3
Methane (CH_4)	−50.8
Methanol (CH_3OH)	−175.4

[a]The free energy of formation values ($G_f{}^0$) are in *kJ/mol*.

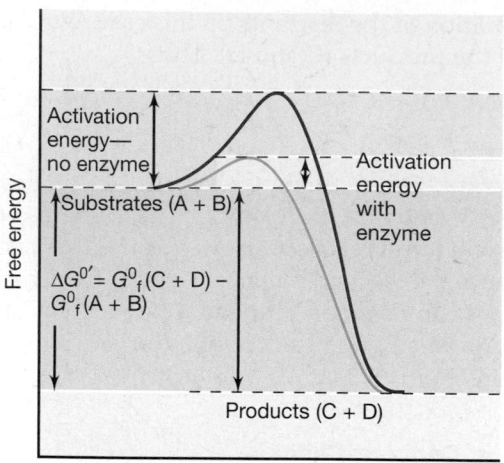

Figure 5.5 Progress of a hypothetical exergonic reaction: A + B → C + D and the concept of activation energy. Chemical reactions may not proceed spontaneously even though energy would be released, because the reactants must first be activated. Once activation has occurred, the reaction then proceeds spontaneously. Catalysts such as enzymes lower the required activation energy.

Most reactions in living organisms would not occur at appreciable rates without catalysis. The catalysts of biological reactions are proteins called **enzymes**. Enzymes are highly specific in the reactions that they catalyze. That is, each enzyme catalyzes only a *single type* of chemical reaction, or in the case of certain enzymes, a *class* of closely related reactions. This specificity is related to the precise three-dimensional structure of the enzyme molecule. In an enzyme-catalyzed reaction, the enzyme temporarily combines with the reactant, which is termed a **substrate** (S) of the enzyme, forming an **enzyme-substrate complex**. Then, as the reaction proceeds, the **product** (P) is released and the enzyme (E) is returned to its original state:

$$E + P \rightleftharpoons E\!-\!S \rightleftharpoons E + P$$

The enzyme is generally much larger than the substrate(s), and the combination of enzyme and substrate(s) usually depends on weak bonds, such as hydrogen bonds, van der Waals forces, and hydrophobic interactions (∞ Section 3.1) to join the enzyme to the substrate. The small portion of the enzyme to which substrates bind is referred to as the **active site** of the enzyme.

Enzyme Catalysis

The catalytic power of enzymes is impressive. Enzymes typically increase the rate of chemical reactions from 10^8 to 10^{20} times the rate that would occur spontaneously. To catalyze a specific reaction, an enzyme must do two things: (1) bind the correct substrate, and (2) position the substrate relative to the catalytically active groups at the enzyme's active site. Binding of substrate to enzyme produces the enzyme-substrate complex (Figure 5.6). This serves to align reactive groups and places strain on specific bonds in the substrate(s). The result of en-

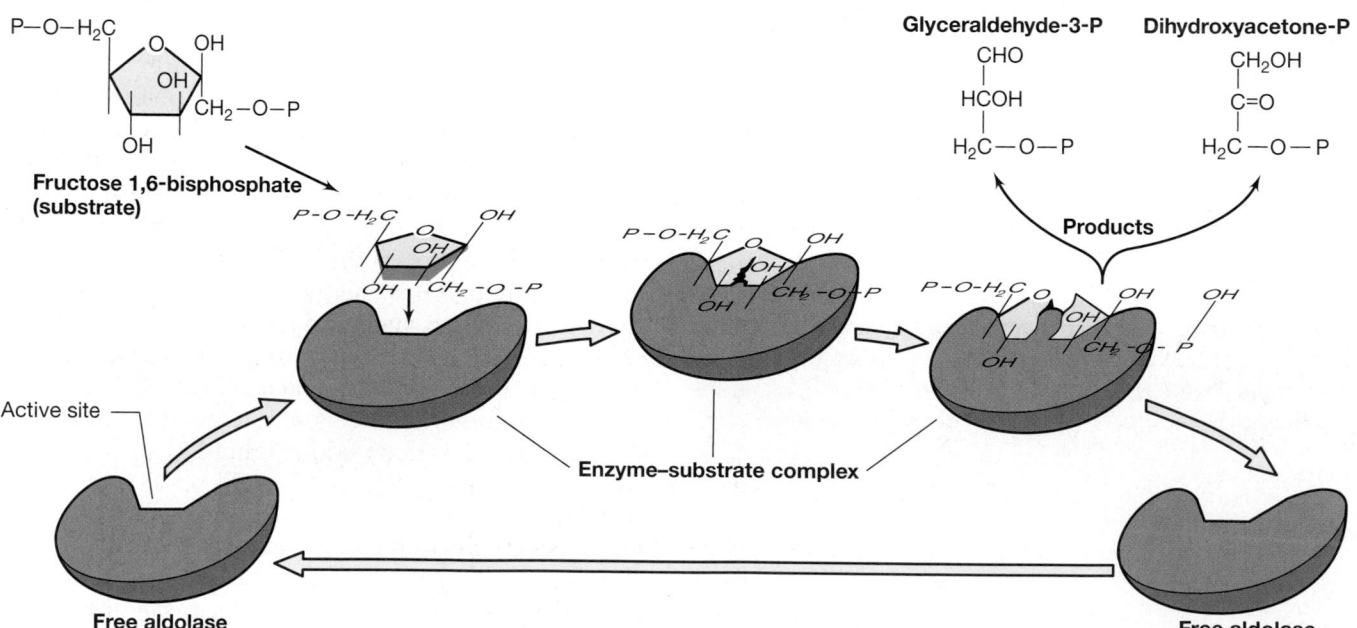

Figure 5.6 The catalytic cycle of an enzyme as depicted for the enzyme fructose bisphosphate aldolase. This enzyme catalyzes the reaction: fructose 1,6-bisphosphate → glyceraldehyde 3-phosphate + dihydroxyacetone phosphate in glycolysis (see Figure 5.14). Following binding of fructose 1,6-bisphosphate in the formation of the enzyme–substrate complex, the conformation of the enzyme is altered, placing strain on certain bonds of the substrate, which break and yield the two products.

zyme-substrate complex formation is a reduction in the activation energy required to make the reaction proceed (Figure 5.5) with the conversion of substrate(s) to product(s). These steps are summarized diagrammatically in Figure 5.6 for the glycolytic enzyme *fructose bisphosphate aldolase* (see Section 5.10).

Note that the reaction depicted in Figure 5.5 is exergonic because the free energy of formation of the substrate is *greater* than that of the product; that is, product formation proceeds with the release of energy. Enzymes can also catalyze energy-requiring reactions, converting energy-poor substrates to energy-rich products. In this case, not only must an activation energy barrier be overcome, but sufficient free energy must also be put *into* the system to raise the energy level of the substrates to that of the products. Although theoretically all enzymes are reversible in their action, in practice, enzymes catalyzing highly exergonic or highly endergonic reactions are essentially unidirectional. If a particularly exergonic reaction needs to be reversed during cellular metabolism, a distinctly different enzyme is frequently involved in the reaction.

Structure of Enzymes

As we have discussed, enzymes are proteins, polymers of amino acids (∞ Sections 3.7–3.8). Each enzyme has a specific three-dimensional shape. The linear array of amino acids (primary structure) folds and twists into a specific configuration to achieve secondary and tertiary structure. A specifically folded protein thus assumes specific binding and physical properties. The precise conformation of an enzyme may be seen more easily in a computer-generated space-filling model (Figure 5.7). In this example of the peptidoglycan-cleaving enzyme *lysozyme* (∞ Section 4.8), the large cleft is the site where the substrate binds (the active site).

Many enzymes contain small nonprotein molecules that participate in the catalytic function but are not considered substrates in the usual sense. These small enzyme-associated molecules are divided into two categories on the basis of the nature of their association with the enzyme: *prosthetic groups* and *coenzymes*. **Prosthetic groups** are bound very tightly to their enzymes, usually permanently. The heme group present in cytochromes is an example of a prosthetic group, and cytochromes will be described in detail later in this chapter. **Coenzymes** are bound rather loosely to enzymes, and a single coenzyme molecule may associate with a number of different enzymes at different times during growth. Coenzymes serve as intermediate carriers of small molecules from one enzyme to another (see Figure 5.11). Most coenzymes are derivatives of vitamins (see Table 5.3).

Enzymes are named either for the substrate they bind or for the chemical reaction they catalyze, by addition of the combining form *-ase*. Thus, cellul*ase* is an

Figure 5.7 Computer-generated space-filling model of the enzyme lysozyme. The substrate (peptidoglycan) binding site (active site) is in the large cleft on the left side of the model (∞ Section 4.8).

Richard Feldmann

enzyme that attacks cellulose, glucose oxid*ase* is an enzyme that catalyzes the oxidation of glucose, and ribonucle*ase* is an enzyme that decomposes ribonucleic acid. A more formal nomenclature system employing a specific numbering system is used to classify enzymes more precisely.

✓ 5.5 *Concept Check*

The reactants in a chemical reaction must first be activated before the reaction can take place, and this often requires a catalyst. Enzymes are catalytic proteins that speed up the rate of biochemical reactions. Enzymes are highly specific in the reactions they catalyze, and this specificity resides in the folding pattern of the polypeptide(s) in the protein.

✓ What is the function of a *catalyst*?

✓ What *class* of macromolecules are enzymes?

✓ Where on an enzyme does its substrate bind?

✓ What is *activation energy*?

III OXIDATION–REDUCTION AND HIGH-ENERGY COMPOUNDS

In biological systems, energy conservation involves oxidation–reduction reactions. The result of energy released in these reactions is the production of *high-energy* compounds such as ATP. We now consider oxidation–reduction reactions and the major electron carriers present in both the cytoplasm and the cytoplasmic membrane. Finally, we examine the nature of the high-energy compounds that actually conserve the energy released in oxidation–reduction reactions.

5.6 Oxidation–Reduction

The conservation of energy from chemical reactions in living organisms involves **oxidation–reduction (redox)** reactions. Chemically, an oxidation is defined as the *removal* of an electron or electrons from a substance. A reduction is defined as the *addition* of an electron (or electrons) to a substance. In biochemistry—the chemistry of cells—oxidations and reductions frequently involve the transfer of not just electrons, but whole hydrogen atoms. A hydrogen atom (H) consists of an electron plus a proton. When its electron is removed, the hydrogen atom becomes a *proton* (or hydrogen ion, H^+). We will, on occasion, need to distinguish between oxidation–reduction reactions involving electrons only or hydrogen atoms only, but reserve this distinction for the appropriate time (see Sections 5.11 and 5.12).

Electron Donors and Acceptors

Oxidation–reduction reactions involve electrons being donated by an electron donor and being accepted by an electron acceptor. For example, hydrogen gas, H_2, can release electrons and hydrogen ions (protons) and become oxidized:

$$H_2 \rightarrow 2\,e^- + 2\,H^+$$

However, electrons cannot exist alone in solution; they must be part of atoms or molecules. The equation as drawn provides chemical information but does not itself represent a real reaction. The above reaction is only a *half reaction,* a term that implies the need for a second half reaction. This is because for any *oxidation* to occur, a subsequent *reduction* must also occur. For example, the oxidation of H_2 could be coupled to the reduction of many different substances including O_2 in a second reaction:

$$\tfrac{1}{2}O_2 + 2\,e^- + 2\,H^+ \rightarrow H_2O$$

This half reaction, which is a reduction, when coupled to the oxidation of H_2 above, yields the following overall balanced reaction:

$$H_2 + \tfrac{1}{2}O_2 \rightarrow H_2O$$

In reactions of this type, we will refer to the substance *oxidized,* in this case H_2, as the **electron donor**, and the substance *reduced,* in this case O_2, as the **electron acceptor** (Figure 5.8). The key to understanding biological oxidations and reductions is to keep straight the proper

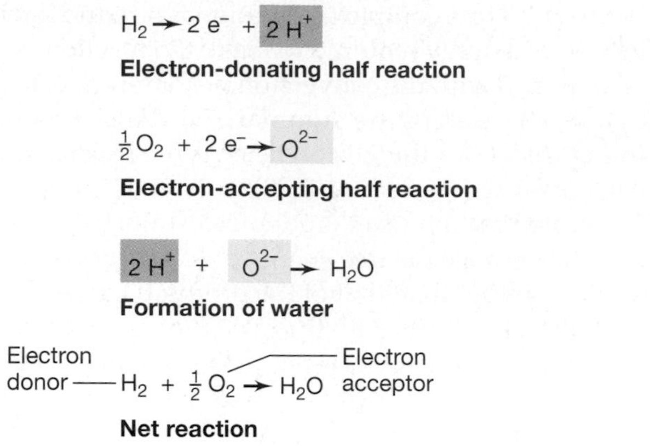

$$H_2 \rightarrow 2\,e^- + 2\,H^+$$
Electron-donating half reaction

$$\tfrac{1}{2}O_2 + 2\,e^- \rightarrow O^{2-}$$
Electron-accepting half reaction

$$2\,H^+ + O^{2-} \rightarrow H_2O$$
Formation of water

Electron donor —— $H_2 + \tfrac{1}{2}O_2 \rightarrow H_2O$ —— Electron acceptor
Net reaction

Figure 5.8 Example of an oxidation–reduction reaction: The formation of H_2O from the electron donor H_2 and the electron acceptor O_2.

half reactions: There must always be one reaction involving an electron *donor* and another reaction involving an electron *acceptor.*

Reduction Potentials

Substances vary in their tendency to become oxidized or to become reduced. This tendency is expressed as the **reduction potential** (E_0', standard conditions) of the substance. This potential is measured electrically in units of volts in reference to a standard substance, H_2. By convention, reduction potentials are expressed for half reactions written as *reductions.* Thus, oxidized form + $e^- \rightarrow$ reduced form. If protons are involved in the reaction, as is often the case, then the reduction potential is to some extent influenced by the hydrogen ion concentration (pH). By convention in biology, reduction potentials are given for neutrality (pH 7) because the cytoplasm of most cells is neutral or nearly so. Using these conventions, at pH 7 the reduction potential (E_0') of

$$\tfrac{1}{2}O_2 + 2\,H^+ + 2\,e^- \rightarrow H_2O$$

is +0.816 volts (V), and that of

$$2\,H^+ + 2\,e^- \rightarrow H_2$$

is –0.421 V.

Oxidation–Reduction Couples and Complete Redox Reactions

Most molecules can be either electron donors or electron acceptors under different circumstances, depending on the substances with which they react. The same atom on each side of the arrow in the half reactions can be thought of as representing a *redox couple,* such as $2\,H^+/H_2$ or $\tfrac{1}{2}O_2/H_2O$. When writing a redox couple, the *oxidized* form is always placed on the left.

In constructing complete oxidation–reduction reactions from their constituent half reactions, it is simplest to remember that the reduced substance of a redox couple whose reduction potential is more negative *donates* electrons to the oxidized substance of a redox couple whose potential is more positive. Thus, in the couple 2 H^+/H_2, which has a potential of -0.42 V, H_2 has a great tendency to *donate* electrons. On the other hand, in the couple $\frac{1}{2}O_2/H_2O$, which has a potential of $+0.82$ V, H_2O has a very slight tendency to donate electrons, but O_2 has a great tendency to *accept* electrons. It follows then that in a reaction of H_2 and O_2, H_2 will be the electron *donor* and become oxidized, and O_2 will be the electron *acceptor* and become reduced (Figure 5.8). Even though by chemical convention both half reactions are written as reductions, in an actual redox reaction one of the two half reactions must be written as an oxidation and therefore proceeds in the reverse direction. Thus, note that in the reaction shown in Figure 5.8, the oxidation of H_2 to $2 H^+ + 2 e^-$ is reversed from the formal half reaction, written as a reduction.

The Electron Tower

A convenient way of viewing electron transfer in biological systems is to imagine a vertical tower (Figure 5.9). The tower represents the range of reduction potentials for redox couples from the most negative at the top to the most positive at the bottom. The reduced substance in the redox pair at the top of the tower has the greatest tendency to donate electrons, whereas the oxidized substance in the couple at the bottom of the tower has the greatest tendency to accept electrons.

As electrons from the electron donor at the top of the tower fall, they can be "caught" by acceptors at various levels. The difference in potential between two substances is expressed as $\Delta E_0'$. The farther the electrons drop from a donor before they are caught by an acceptor, the greater the amount of energy released; that is, $\Delta E_0'$ *is proportional to* $\Delta G^{0'}$ (Figure 5.9). O_2, at the bottom

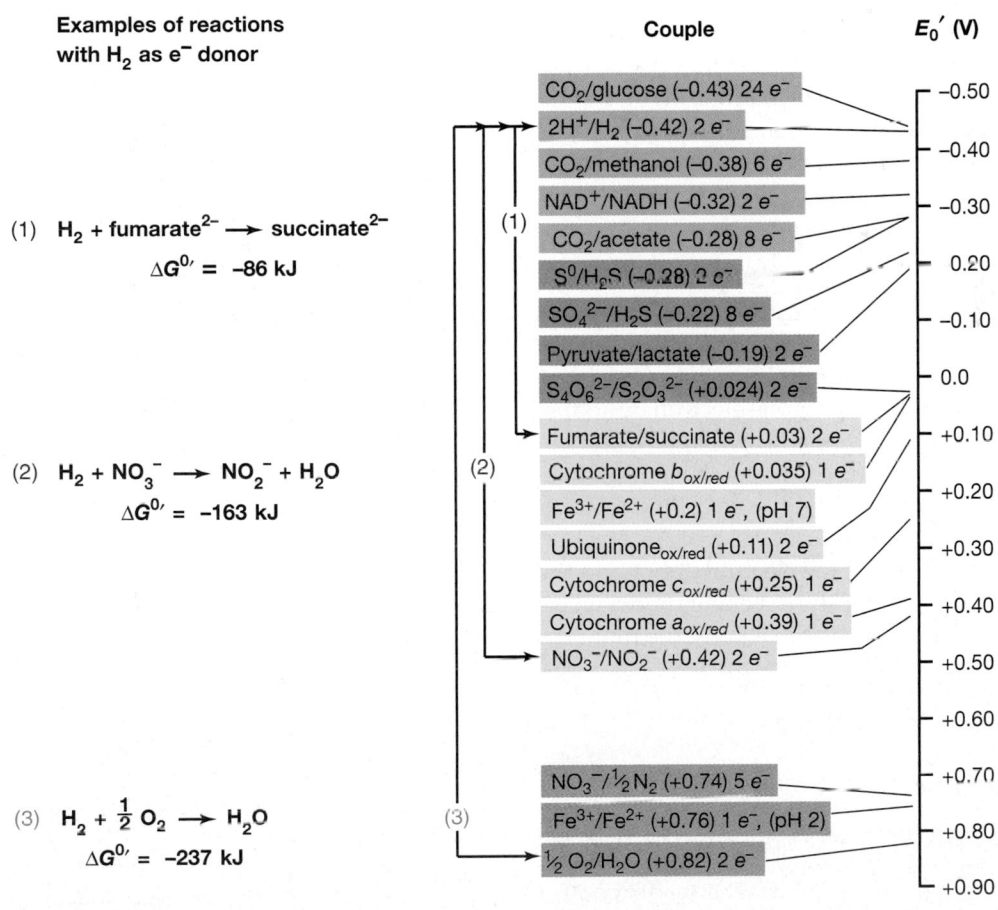

Figure 5.9 The electron tower. Redox couples are arranged from the strongest reductants (negative reduction potential) at the top to the strongest oxidants (positive reduction potentials) at the bottom. As electrons are donated from the top of the tower, they can be "caught" by acceptors at various levels. The farther the electrons fall before they are caught, the greater the difference in reduction potential between electron donor and electron acceptor and the more energy that is released. As an example of this, on the left is shown the differences in energy released when a single electron donor, H_2, reacts with any of three different electron acceptors, fumarate, nitrate, and oxygen.

of the tower, is the most favorable electron acceptor used by organisms. In the middle of the tower, redox couples can act as either electron donors or acceptors. For instance, the $2 H^+/H_2$ couple has a reduction potential of -0.42 V. The fumarate–succinate couple has a potential of $+0.02$ V. Hence, the oxidation of hydrogen (the electron donor) can be coupled to the reduction of fumarate (the electron acceptor):

$$H_2 + fumarate^{2-} \rightarrow succinate^{2-}$$

On the other hand, the oxidation of succinate to fumarate can be coupled to the reduction of NO_3^- or $\frac{1}{2}O_2$

$$Succinate^{2-} + NO_3^- \rightarrow fumarate^{2-} + NO_2^- + H_2O$$
$$Succinate^{2-} + \tfrac{1}{2}O_2 \rightarrow fumarate^{2-} + H_2O$$

Hence, under conditions where oxygen is absent (called *anoxic*) in the presence of H_2, fumarate can be an electron acceptor (producing succinate), and under other conditions (for example, anoxic in the presence of NO_3^-, or aerobic) succinate can be an electron donor (producing fumarate). Indeed, all the transformations involving fumarate and succinate described here are actually carried out by various microorganisms under certain nutritional and environmental conditions.

Electron Donor ⇌ Energy Source

In catabolism the electron donor is often referred to as an **energy source**. Many potential electron donors exist in nature (Chapters 17 and 19), but for now it is essential to understand that it is not the electron donor *per se*, that contains energy, but it is the *chemical reaction* in which the electron donor gets oxidized, that actually releases energy. As discussed in the context of the electron tower, the amount of energy released in a redox reaction depends on the nature of *both* the electron donor and the electron acceptor: The greater the difference between reduction potentials of the two half reactions, the more energy there will be released when they react (Figure 5.9) (see also Appendix 1).

✓ 5.6 Concept Check

Oxidation–reduction reactions, which are involved in the energy-yielding reactions of cells, involve the transfer of electrons from one substance to another. The tendency of a compound to accept or release electrons is expressed quantitatively by its reduction potential.

✓ In the reaction $H_2 + \frac{1}{2}O_2 \rightarrow H_2O$, what is the electron *donor* and what is the electron *acceptor*?

✓ What is the E_0' of the $2 H^+/H_2$ couple?

✓ Why is NO_3^- a better electron acceptor than fumarate?

5.7 NAD as a Redox Electron Carrier

In the cell, the transfer of electrons in an oxidation–reduction reaction from donor to acceptor usually involves one or more intermediates referred to as **carriers**. When such carriers are used, we refer to the initial donor as the **primary electron donor** and to the final acceptor as the **terminal electron acceptor**. The net energy change of the complete reaction sequence is determined by the *difference* in reduction potentials between the primary donor and the terminal acceptor.

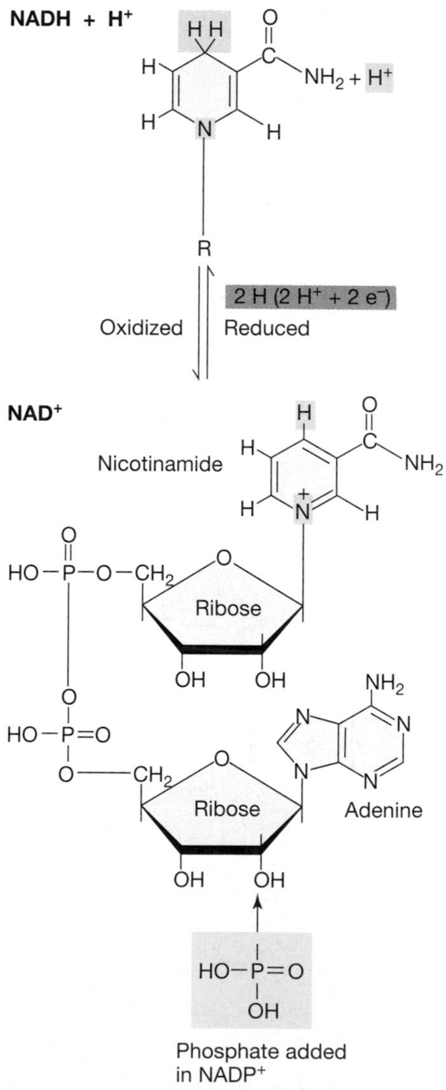

Figure 5.10 Structure of the oxidation–reduction coenzyme nicotinamide adenine dinucleotide (NAD^+). In $NADP^+$, a phosphate group is present, as indicated. Both NAD^+ and $NADP^+$ undergo oxidation–reduction as shown, are freely diffusable, and are hydrogen atom carriers.

Electron carriers can be divided into two general classes: those freely diffusible and those firmly attached to enzymes in the cytoplasmic membrane. The fixed carriers function in membrane-associated electron transport reactions and are discussed in Section 5.11. Freely diffusible carriers include the coenzymes nicotinamide-adenine dinucleotide (NAD$^+$) and NAD-phosphate (NADP$^+$) (Figure 5.10). NAD$^+$ and NADP$^+$ are *hydrogen atom* carriers and transfer two hydrogen atoms to the next carrier in the chain. Such hydrogen atom transfer is referred to as a dehydrogenation.[†]

The reduction potential of the NAD$^+$/NADH (or NADP$^+$/NADPH) couple is –0.32 V, which places it fairly high on the electron tower; that is, NADH (or NADPH) is a good electron *donor*. However, although the NAD$^+$ and NADP$^+$ couples have the same reduction potentials, they generally function in different capacities in the cell. NAD$^+$/NADH is directly involved in energy-generating (catabolic) reactions, whereas NADP$^+$/NADPH is involved primarily in biosynthetic (anabolic) reactions.

[†]Strictly speaking NAD$^+$ or NADP$^+$ carries two electrons and one proton, the second H$^+$ being released to solution. Therefore, NAD$^+$ + 2 e$^-$ + 2 H$^+$ actually yields NADH + H$^+$. However, for simplicity, we write NADH + H$^+$ as NADH.

NAD/NADH Cycling

Coenzymes increase the diversity of redox reactions by making it possible for chemically dissimilar molecules to interact as primary electron donor and terminal electron acceptor, with the coenzyme acting as intermediary. As we have discussed, most biological reactions are catalyzed by specific enzymes that can react with only one or a very limited range of substrates. Oxidation–reduction reactions may be considered to proceed in three stages: removal of electrons from the primary donor, transfer of electrons through one or a series of electron carriers, and addition of electrons to the terminal acceptor. Each step in the reaction is catalyzed by a different enzyme, each of which binds to its substrate and to its specific coenzyme. Figure 5.11 is a schematic diagram showing the functioning of the coenzyme NAD$^+$ in a two-part reaction. Note that after a coenzyme has performed its chemical function in one reaction, it can diffuse through the cytoplasm until it collides with another enzyme that requires the coenzyme in that form. Following conversion of the coenzyme back to its original form, the whole process can be repeated (Figure 5.11).

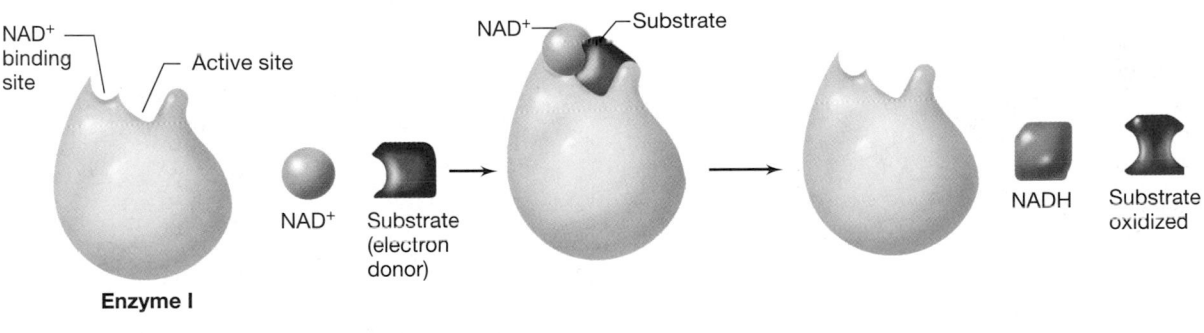

Reaction 1. Enzyme I reacts with substrate (electron donor) and oxidized form of coenzyme, NAD$^+$.

Reaction 2. Enzyme II reacts with substrate (electron acceptor) and reduced form of coenzyme, NADH.

Figure 5.11 Schematic example of an oxidation–reduction reaction involving the oxidized and reduced forms of the coenzyme nicotinamide-adenine dinucleotide, NAD$^+$ and NADH.

✓ 5.7 Concept Check

The transfer of electrons from donor to acceptor in a cell often involves the participation of one or more electron carriers. Some electron carriers are membrane-bound, whereas others, such as $NAD^+/NADH$, are freely diffusible, transferring electrons from one place to another in the cell.

✓ What is the difference between an *electron* and a *hydrogen* atom?

✓ Is NADH a better electron donor than H_2? Why or why not?

5.8 High-Energy Compounds and Energy Storage

Energy released as a result of oxidation–reduction reactions must be conserved for cell functions. In living organisms, chemical energy released in redox reactions is usually conserved in the form of **high-energy phosphate bonds**; these compounds then function as the energy source to drive energy-requiring reactions in the cell.

In phosphorylated compounds, phosphate groups are attached via oxygen atoms by *ester* or *anhydride* bonds, as illustrated in Figure 5.12. Not all phosphate bonds, however, are high-energy bonds. The energy of phosphate bonds is expressed in terms of the free energy released when the phosphate is hydrolyzed. As seen in

Figure 5.12, the $\Delta G^{0\prime}$ of hydrolysis of the phosphate bond in glucose 6-phosphate is only -13.8 kJ/mol, whereas the $\Delta G^{0\prime}$ of hydrolysis of the phosphate bond in phosphoenolpyruvate is -51.6 kJ/mol, almost four times that of glucose 6-phosphate. Thus, phosphoenolpyruvate, a phosphoanhydride (∞ Table 3.1), is considered a *high-energy compound* and glucose 6-phosphate, a phosphate ester, is not.

Adenosine Triphosphate (ATP)

The most important high-energy phosphate compound in living organisms is adenosine triphosphate (ATP). ATP consists of the ribonucleoside adenosine to which three phosphate molecules are bonded in series (Figure 5.12). ATP serves as the prime energy carrier in living organisms, being generated during exergonic reactions and being used to drive endergonic reactions. From the structure of ATP (Figure 5.12) it can be seen that two of the phosphate bonds of ATP are phosphoanhydrides and thus, have high free energies of hydrolysis.

It should be emphasized that although we express the energy of high-energy phosphate bonds in terms of the free energy of hydrolysis, in actuality it is undesirable for these bonds to hydrolyze in cells in the absence of a second reaction that can use the energy released because the free energy of hydrolysis would then be lost to the cell as heat. The free energy of high-energy phosphate

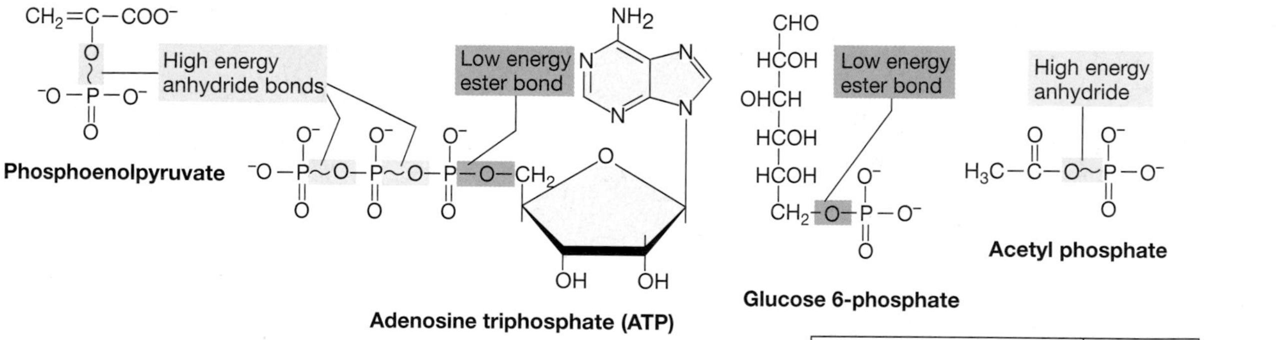

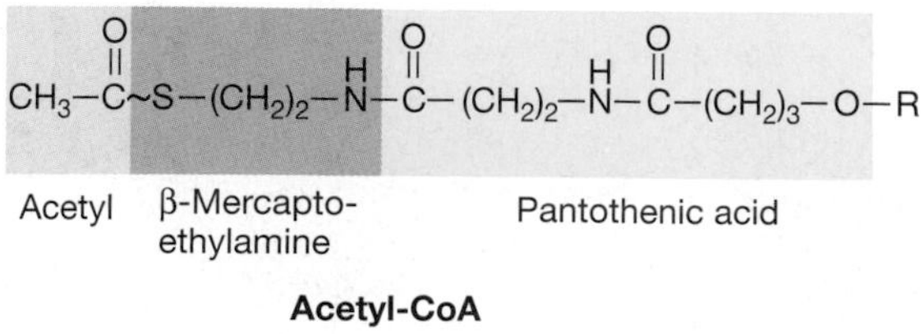

Compound	$G^{0\prime}$ kJ/mol
High energy	
Phosphoenolpyruvate	-51.6
1,3-Bisphosphoglycerate	-52.0
Acetyl phosphate	-44.8
ATP	-31.8
ADP	-31.8
Low energy	
AMP	-14.2
Glucose 6-phosphate	-13.8

Figure 5.12 High-energy bonds. The table shows the free energy of hydrolysis of some of the key phosphate esters and anhydrides, indicating that some of the phosphate ester bonds are of higher energy than others. Structures of four of the compounds are given to indicate the position of low-energy and high-energy bonds. ATP contains three phosphates, but only two of them are high energy (shown in blue). ADP contains two phosphates of which only one is high energy. AMP does not contain a high-energy phosphate bond. Also shown is the structure of the coenzyme acetyl-CoA. The C–S bond between the acetyl portion and the β-mercaptoethylamine portion is a high-energy thioester bond (∞ Table 3.1). The "R" group of acetyl-CoA is a 3′ phospho ADP group.

bonds is generally used to drive biosynthetic reactions and other aspects of cell function through carefully regulated processes in which the energy released from ATP hydrolysis is coupled to energy-requiring reactions.

Coenzyme A

In addition to high-energy *phosphate* compounds, certain other high-energy compounds are produced in the cell and can conserve the energy released in exergonic reactions. These include derivatives of coenzyme A (for example, acetyl-CoA; see structure in Figure 5.12). Coenzyme A derivatives contain *sulfo*anhydride (thioester) instead of *phospho*anhydride (Figure 5.12) bonds and yield sufficient free energy on hydrolysis to drive the synthesis of a high energy phosphate bond (∞ Table 3.1). For example, in the reaction: acetyl-S-CoA + H_2O + ADP + P → acetate$^-$ + HS-CoA + ATP + H$^+$, the energy released from the hydrolysis of coenzyme A is conserved in the synthesis of ATP. Coenzyme A derivatives (acetyl-CoA is just one of many) are especially important to the energetics of anaerobic microorganisms, in particular those whose energy metabolism involves fermentation (∞ Sections 5.10, 17.19, and 17.20); we will thus return to the importance of these compounds in Chapter 17.

Energy Storage

ATP, the cell's major energy currency, is present in relatively low concentration in the cell, about 2 millimolar (mM) in an actively growing cell. ATP is continuously being broken down to drive biosynthetic reactions and resynthesized at the expense of catabolic reactions. For energy *storage,* microorganisms produce insoluble polymers that can later be oxidized for the production of ATP. Examples include the polyglucose polymers starch and glycogen (∞ Figure 3.6), the lipid polymer poly-β-hydroxybutyrate and other polyhydroxyalkanoates (∞ Figure 4.52), and elemental sulfur, stored by many sulfur chemolithotrophs. These polymers are deposited within the cell as large granules that can often be seen with the light or electron microscope (∞ Section 4.13). In the absence of an external energy source, the cell may oxidize these polymers and thus be able to make new cell material or simply supply itself with maintenance energy, even when energy-yielding substances are temporarily unavailable in the environment.

✓ 5.8 Concept Check

The energy released in oxidation–reduction reactions is conserved in the formation of certain compounds that contain high-energy phosphate or sulfur bonds. The most common high-energy phosphate compound is ATP, which serves as a prime energy carrier in the cell. Long-term storage of energy is linked to polymers, which can be consumed to yield ATP.

✓ Why are ATP and ADP considered high-energy phosphate compounds, whereas AMP is not?

✓ Following periods of nutrient abundance, how can cells prepare for periods of nutrient starvation?

IV MAJOR CATABOLIC PATHWAYS, ELECTRON TRANSPORT, AND THE PROTON MOTIVE FORCE

In the following sections we look at well-studied and widely distributed examples of catabolic reactions in microorganisms. We will compare the processes of fermentation and respiration, following up with a more detailed look at each process and a discussion of the proton motive force. The latter is the key not only to classical (that is, *aerobic*) respiration, but to many metabolic processes, including the energy-yielding reactions of anaerobic respiration, photosynthesis, and chemolithotrophy. The proton motive force also drives a number of other functions in the cell, including many different transport reactions and flagellar and gliding motility (∞ Sections 4.7 and 4.10–4.12).

5.9 Energy Conservation: Options

For chemotrophs, those organisms that use chemicals as electron donors in energy metabolism, two mechanisms for energy conservation are known, **fermentation** and **respiration**. The final result in each case is the same, the synthesis of ATP, an endergonic reaction. This energy-requiring reaction is coupled to one or another energy-releasing (exergonic) reactions that occur during catabolism of the electron donor. In terms of redox reactions, fermentation and respiration differ as follows: (1) in *fermentation* the redox process occurs in the *absence* of usable terminal electron acceptors, while (2) in *respiration,* molecular oxygen or some other electron acceptor is present as a terminal electron acceptor. The oxidation in a fermentation is coupled to the reduction of a compound generated from the initial substrate; thus, no externally supplied electron acceptor is involved.

In addition to differences involving redox, fermentation and respiration differ in the mechanism by which ATP is synthesized. In fermentation, ATP is produced by a process called **substrate-level phosphorylation**. In this process, ATP is synthesized during steps in the catabolism of an organic compound (Figure 5.13*a*). This is in contrast to **oxidative phosphorylation** (to be discussed later) in which ATP is produced at the expense of the proton motive force (Figure 5.13*b*). A third form of

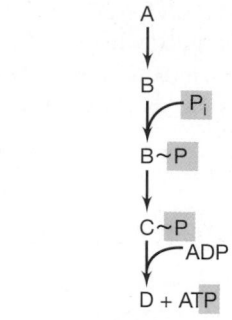

(a) **Substrate-level phosphorylation**

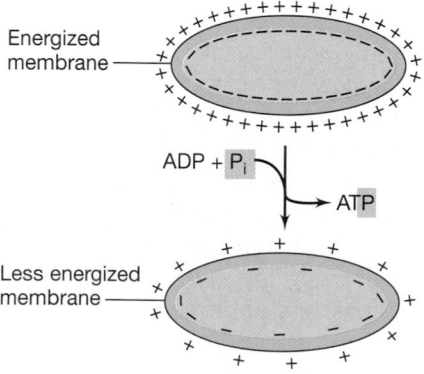

Energized membrane

ADP + P$_i$

ATP

Less energized membrane

(b) **Oxidative phosphorylation**

Figure 5.13 Energy conservation in fermentation and respiration. (a) In fermentation, ATP synthesis occurs as a result of *substrate-level phosphorylation*; a phosphate group gets added to some intermediate in the biochemical pathway where it becomes a "high-energy" phosphate group and eventually gets transferred to ADP to form ATP. (b) In respiration, the cytoplasmic membrane, energized by the proton motive force, dissipates some of that energy in the formation of ATP from ADP and inorganic phosphate (P$_i$) in the process called *oxidative phosphorylation*. The coupling of the proton motive force to ATP synthesis occurs by way of a membrane protein enzyme complex called ATP synthase (ATPase) (see Section 5.12 and Figure 5.21).

ATP synthesis, **photophosphorylation**, occurs in photosynthetic organisms, but its basic mechanism is similar to that of oxidative phosphorylation except that light rather than a chemical compound is involved in driving oxidation–reduction reactions (⌖ Section 17.2).

✓ 5.9 Concept Check

Fermentation and respiration are the two means by which chemoorganotrophs can conserve energy from the oxidation of organic compounds. During these catabolic reactions, ATP synthesis occurs by way of either substrate-level phosphorylation (fermentation) or oxidative phosphorylation (respiration).

✓ Which form of ATP synthesis directly employs the cytoplasmic membrane?

✓ How does substrate-level phosphorylation differ from oxidative phosphorylation?

5.10 Glycolysis as an Example of Fermentation

A fermentation is an internally balanced oxidation–reduction reaction in which some atoms of the energy source (electron donor) become more reduced whereas others become more oxidized, and energy is produced by substrate-level phosphorylation. A common biochemical pathway for the fermentation of glucose is **glycolysis**, also named the **Embden-Meyerhof pathway** for its major discoverers. Several other fermentations are known, and many of these are covered in Chapter 17.

Glycolysis can be divided into three major stages, each involving a series of individually catalyzed enzymatic reactions (Figure 5.14). Stage I of glycolysis is a series of preparatory rearrangements, reactions that do not involve oxidation–reduction and do not release energy but that lead to the production from glucose of two molecules of the key intermediate, *glyceraldehyde 3-phosphate*. In Stage II, oxidation–reduction occurs, energy is conserved in the form of ATP, and two molecules of pyruvate are formed. In Stage III, a second oxidation–reduction reaction occurs and *fermentation products* (for example, ethanol and CO$_2$, or lactic acid) are formed (Figure 5.14).

Stages I and II: Preparatory and Redox Reactions

In Stage I, glucose is phosphorylated by ATP yielding glucose 6-phosphate; this is then converted to an isomeric form, fructose 6-phosphate, and a second phosphorylation leads to the production of *fructose 1,6-bisphosphate*, which is a key intermediate product of glycolysis. If hexoses *other* than glucose are fermented, they are converted to fructose 1,6-bisphosphate if the Embden-Meyerhof pathway is used. The enzyme **aldolase** splits fructose 1,6-bisphosphate into two three-carbon molecules, glyceraldehyde 3-phosphate and its isomer, dihydroxyacetone phosphate (see also Figure 5.6).[‡] Note that thus far, there have been no oxidation–reduction reactions and that all the reactions, including the consumption of ATP, proceed without electron transfers.

The first redox reaction of glycolysis occurs in Stage II during the conversion of glyceraldehyde 3-phosphate to 1,3-bisphosphoglyceric acid. In this reaction (which occurs twice, once for each molecule of glyceraldehyde 3-phosphate), an enzyme whose coenzyme is NAD$^+$ accepts two hydrogen atoms and NAD$^+$ is converted to NADH; the enzyme catalyzing this reaction is called

---

[‡]There is an enzyme that catalyzes the interconversion of dihydroxyacetone phosphate and glyceraldehyde 3-phosphate. For simplicity, we consider here only glyceraldehyde 3-phosphate since it is the compound that is further metabolized.

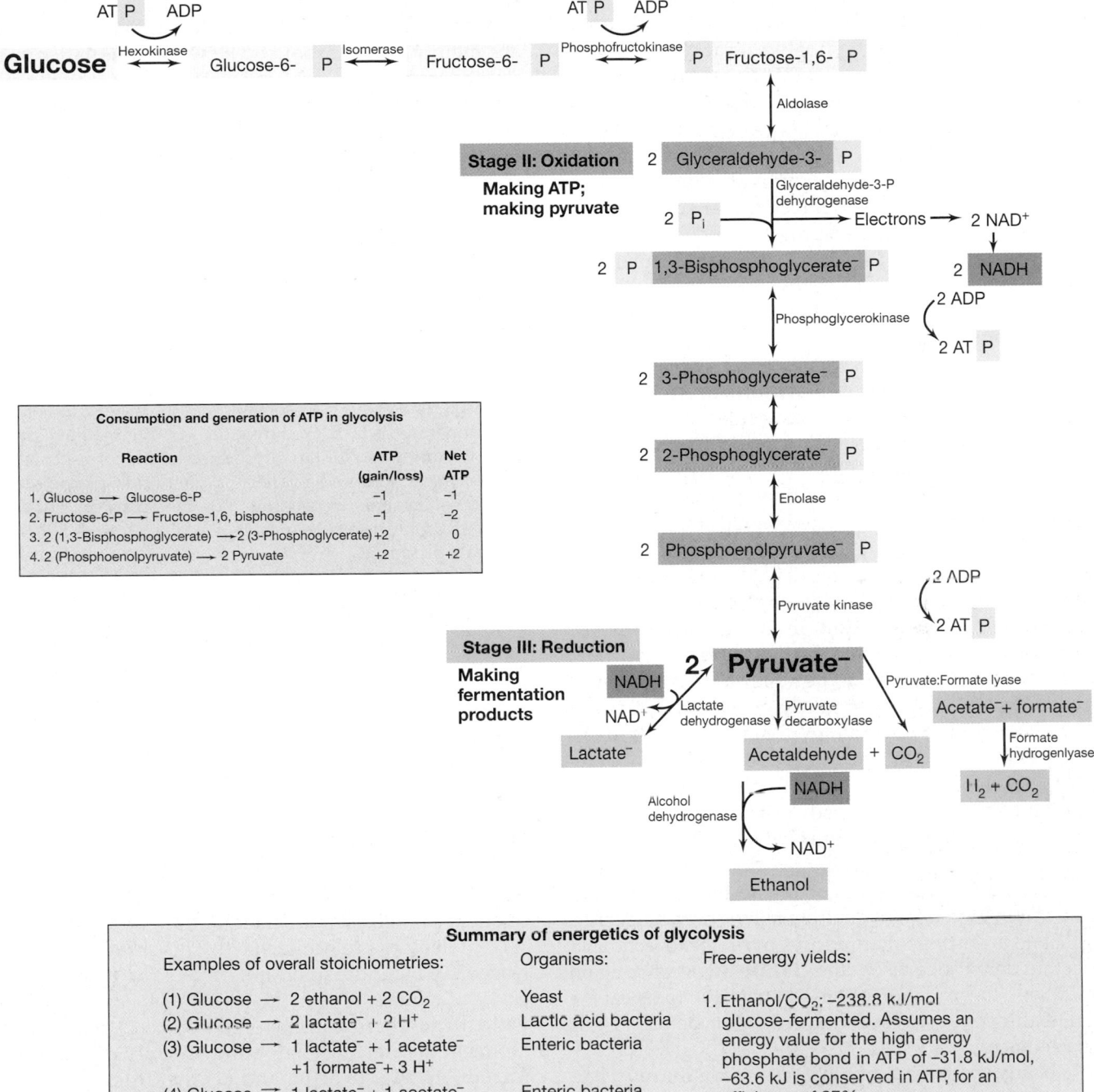

Figure 5.14 Embden-Meyerhof pathway (glycolysis), the sequence of enzymatic reactions in the conversion of glucose to pyruvate and then to fermentation products (enzymes are shown in small type). The product of aldolase is actually glyceraldehyde 3-P and dihydroxyacetone P, but the latter is converted to glyceraldehyde 3-P. Note how pyruvate is the central "hub" of glycolysis—all fermentation products are made from pyruvate, and just a few common examples are given.

glyceraldehyde-3-phosphate dehydrogenase. Simultaneously, each glyceraldehyde-3-P molecule is phosphorylated by the addition of a molecule of inorganic phosphate. This reaction, in which inorganic phosphate is converted to organic form, sets the stage for energy conservation by substrate-level phosphorylation; ATP formation is possible because each of the phosphates on a molecule of 1,3-bisphosphoglyceric acid is a high-energy phosphate (see Figure 5.12). The synthesis of ATP occurs when each molecule of 1,3-bisphosphoglyceric acid is converted to 3-phosphoglyceric acid and later on in the pathway, when each molecule of phosphoenolpyruvate is converted to pyruvate (Figure 5.14).

In glycolysis, *two* ATP molecules are consumed in the two phosphorylations of glucose, and *four* ATP molecules are synthesized (two from each 1,3-bisphosphoglyceric acid converted to pyruvate). Thus, the *net gain* to the organism is two molecules of ATP per molecule of glucose fermented.

Stage III: Production of Fermentation Products

During the formation of two molecules of 1,3-bisphosphoglyceric acid, two molecules of NAD^+ are reduced to NADH (see Figure 5.14). However, a cell contains only a small amount of NAD^+, and if all of it were converted to NADH, the oxidation of glucose would stop; the continued oxidation of glyceraldehyde 3-phosphate can proceed only if there is a molecule of NAD^+ present to accept released electrons. This "roadblock" is overcome in fermentation by the oxidation of NADH back to NAD^+ through reactions involving the reduction of pyruvate to any of a variety of **fermentation products**. In the case of yeast, pyruvate is reduced to ethanol with the release of CO_2. In lactic acid bacteria, pyruvate is reduced to lactate (see lower part of Figure 5.14). Many routes of pyruvate reduction in various fermentative prokaryotes are known (∞ Chapters 12 and 17), but the net result is the same; NADH must be returned to the oxidized form, NAD^+, for the energy-yielding reactions of fermentation to continue. As a diffusible coenzyme, NADH can move away from glyceraldehyde-3-phosphate dehydrogenase, attach to an enzyme that reduces pyruvate to lactic acid (lactate dehydrogenase), and diffuse away, once again following the oxidation of NADH to NAD^+ to repeat the cycle all over again (see Figure 5.11 for details of this mechanism).

In any energy-yielding process, oxidation must balance reduction, and there must be an electron acceptor for each electron removed. In this case, the *reduction* of NAD^+ at one enzymatic step in glycolysis is balanced with its *oxidation* at another. The final product(s) must also be in oxidation–reduction balance with the starting substrate, glucose. Hence, the products discussed here, ethanol plus CO_2 or lactate plus protons, are in electrical and atomic balance with the starting substrate, glucose.

Glucose Fermentation: Net and Practical Results

The ultimate result of glycolysis is the consumption of glucose, the net synthesis of two ATPs, and the production of fermentation products. For the organism the crucial product is ATP, which is used in a wide variety of energy-requiring reactions, and fermentation products are merely waste products. However, the latter substances are hardly considered waste products by the distiller, the brewer, the cheese-maker, or the baker (see the box, The Products of Yeast Fermentation). Thus, fermentation is more than just an energy-yielding process. It is a means of producing natural products useful to humans. We discuss industrial production of fermentation products in more detail in Chapter 30.

✓ 5.10 Concept Check

Glycolysis is a major pathway for fermentation and is widespread in living organisms. The end result of glycolysis is the release of a small amount of energy that is conserved as ATP and used for various cell functions and the production of fermentation products derived from the key intermediate, pyruvate. Common fermentation products of glycolysis include ethanol, lactic acid, and a variety of other acids, alcohols, and gaseous substances, depending upon the organism. For each molecule of glucose consumed in glycolysis, two ATP molecules are produced.

✓ Which reaction(s) in glycolysis involve oxidations and reductions?

✓ What is the role of NAD^+ in glycolysis?

✓ Why are fermentation products made during glycolysis?

5.11 Respiration and Membrane-Associated Electron Carriers

We have just discussed the fermentation of glucose in glycolysis, a process that occurs in the absence of external electron acceptors. A relatively small amount of energy is released in fermentation, and only a few ATP molecules are synthesized. This small energy release may be understood in terms of the formal principles of oxidation–reduction reactions. Fermentations yield little energy for two reasons: (1) the carbon atoms in the starting compound are only partially oxidized (see Figure 5.14), and (2) the difference in reduction potentials between the primary electron donor and terminal electron acceptor is small. However, if O_2 or some other terminal acceptor is present, all the substrate molecules can be oxidized completely to CO_2 and a far higher yield of ATP is theoretically possible. The process by which a compound is oxidized using O_2 as the terminal electron acceptor is called **aerobic respiration.**

Techniques & Applications...   The Products of Yeast Fermentation

The aerobic and anaerobic processes of energy generation may seem dull and prosaic, but they are at the basis of some of the most striking discoveries of the human race: fermented foods and beverages (see photo).

In the production of breads and most alcoholic beverages, it is the yeast *Saccharomyces cerevisiae* that is exploited to produce ethanol plus CO_2. Found in various sugar-rich environments such as fruit juices and nectar, yeasts can carry out the two opposing modes of chemoorganotrophic metabolism discussed in this chapter, fermentation and respiration. When oxygen is present, yeasts grow efficiently on the sugar substrate, making yeast cells and CO_2 (the latter from the citric acid cycle). However, when O_2 is absent, yeasts switch to an anaerobic metabolism, resulting in a reduced cell yield but significant amounts of alcohol plus CO_2.

Every home wine maker or brewer is an amateur microbiologist, perhaps without even realizing it (Home Brew, Chapter 30). When grapes are squeezed to make juice, small numbers of yeast cells present on the grapes in the vineyard are transferred to the must. During the first several days of the wine-making process, these yeast cells grow by respiration and consume O_2, making the juice anoxic. As soon as the oxygen is depleted, fermentation can begin, and the process of alcohol formation takes over. This switch from aerobic to anaerobic metabolism is crucial, and special care must be taken to make sure air is kept out of the fermenting vessel.

Figure 1 **Major products in which fermentation by the yeast** *Saccharomyces cerevisiae* **is critical.**

Wine is only one of many products made with yeast. Others include beer and distilled spirits such as brandy, whisky, vodka, and gin (see Figure 1). In distilled spirits, the ethanol, produced in relatively low amounts (10–15% by volume) by the yeast, is concentrated by distilling to make a beverage containing 40–60% alcohol. Even alcohol for motor fuel is made with yeast in parts of the world where sugar is plentiful but petroleum is in short supply (such as Brazil). In the United States, major production of ethyl alcohol for use as an industrial solvent occurs using corn starch as the fermentable substrate. Yeast also

serves as the leavening agent in bread, although here it is not the *alcohol* that is important but CO_2, the other product of the alcohol fermentation (see Figure 5.14). We discuss yeast and yeast products in some detail in Chapter 30.

We can thus appreciate how the yeast cell, forced to carry out a fermentative lifestyle because the oxygen it needs for respiration is absent, has impacted the lives of humans. Besides being "waste products" of the glycolytic pathway in yeast, ethanol and CO_2 are the key ingredients in the products of the alcoholic beverage and baking industries, respectively. ■

Our discussion of aerobic respiration deals with both the carbon and electron transformations: (1) the biochemical pathways involved in the transformation of organic carbon to CO_2 and (2) the way electrons are transferred from the organic compound to the terminal electron acceptor, driving ATP synthesis at the expense of the proton motive force (Figure 5.13*b*). We begin with a discussion of electron flow.

Electron Transport Carriers

Electron transport systems are composed of *membrane-associated* electron carriers. These systems have two basic functions: (1) to accept electrons from an electron donor

and transfer them to an electron acceptor, and (2) to conserve some of the energy released during electron transfer for synthesis of ATP.

Several types of oxidation–reduction enzymes are involved in electron transport: (1) NADH dehydrogenases, which transfer hydrogen atoms from NADH; (2) riboflavin-containing electron carriers, generally called flavoproteins [which contain flavin mononucleotide (FMN) or flavin-adenine dinucleotide (FAD)]; (3) iron-sulfur proteins; and (4) cytochromes, which are proteins containing an iron-porphyrin ring called *heme*. In addition, one class of *nonprotein* electron carriers is known, the lipid-soluble *quinones*. Quinones can diffuse freely through the membrane, generally transferring electrons from iron-sulfur proteins to cytochromes. We now consider each of these classes of electron transport components in slightly more detail.

NADH dehydrogenases are proteins bound to the inside surface of the cell membrane. They accept hydrogen atoms from NADH (Figure 5.10) generated in various cellular reactions and pass the hydrogen atoms to flavoproteins.

Flavoproteins are proteins containing a derivative of riboflavin (Figure 5.15); the flavin portion, which is bound to a protein, is a prosthetic group that is alternately reduced as it accepts hydrogen atoms and oxidized when electrons are passed on. Note that flavoproteins *accept* hydrogen atoms and *donate* electrons; we will consider what happens to the two protons later. Two flavins are commonly observed in cells, flavin mononucleotide and flavin-adenine dinucleotide, in which FMN is bonded to ribose and adenine through a second phosphate. Riboflavin, also called vitamin B_2, is a required growth factor for some organisms (see Section 5.2 and Table 5.3).

The **cytochromes** are proteins with iron-containing porphyrin ring (heme) prosthetic groups attached to them (Figure 5.16). They undergo oxidation and reduction through loss or gain of a *single electron* by the iron atom at the center of the cytochrome:

$$\text{Cytochrome} - Fe^{2+} \rightleftharpoons \text{Cytochrome} - Fe^{3+} + e^-$$

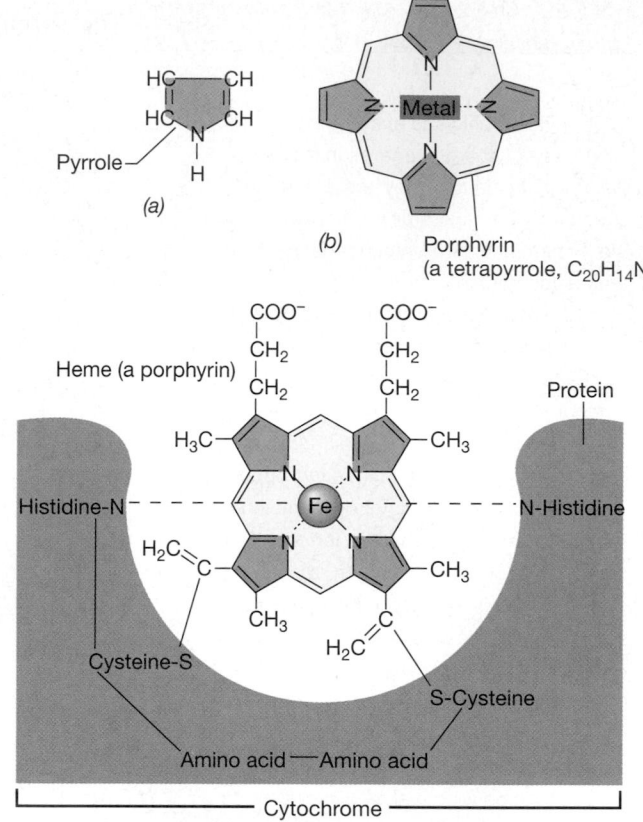

(a)

(b) Porphyrin (a tetrapyrrole, $C_{20}H_{14}N_4$)

(c)

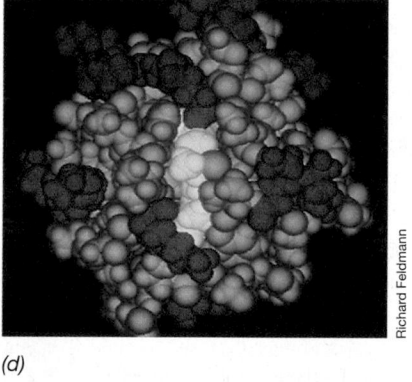

(d)

Figure 5.16 Cytochrome and its structure. (a) Structure of the pyrrole ring. (b) Four pyrrole rings are condensed, leading to formation of the porphyrin ring. Various metals can be incorporated into the porphyrin ring system. For example, in chlorophyll pigments, the metal is Mg^{2+} (∞ Section 17.2 and Figure 17.3); in vitamin B_{12}, it is Co^{2+} (∞ Section 30.7 and Figure 30.12); and in certain unique porphyrin coenzymes, Ni^{2+} may be the central metal atom (∞ Section 17.17 and Figure 17.42). (c) In some cytochromes, like cytochrome *c*, the porphyrin ring is covalently linked via disulfide bridges to cysteine molecules in the protein. Note the presence of iron in the center of the ring. (d) Computer-generated model of cytochrome *c*. The protein completely surrounds the porphyrin ring (light color) in the center. Cytochromes carry electrons only, not entire hydrogen atoms; the redox site is the iron atom, which can alternate between the Fe^{2+} and Fe^{3+} oxidation state.

Figure 5.15 Flavin mononucleotide (FMN) (riboflavin phosphate, a hydrogen atom carrier). The site of oxidation–reduction is the same in FMN and flavin-adenine dinucleotide (FAD).

Several classes of cytochromes are known, differing in their reduction potentials. One cytochrome can transfer electrons to an acceptor (heme, quinone, Fe/S) that has a more positive reduction potential and can itself accept electrons from a donor with a less positive reduction potential. The different cytochromes are designated by letters, such as cytochrome a, cytochrome b, cytochrome c, depending upon the type of heme they contain. The cytochromes of one organism may differ slightly from those of another, and so there are designations such as cytochrome a_1, cytochrome a_2, cytochrome aa_3, and so on among cytochromes of the same class. Occasionally, cytochromes form tight complexes with other cytochromes or with iron-sulfur proteins. An example of such a complex is the cytochrome bc_1 complex, which contains two different b-type cytochromes and one c-type cytochrome. It plays an important role in energy metabolism (see Section 5.12 and Figure 5.20).

In addition to the cytochromes, where iron is bound to heme, several **nonheme iron-sulfur (Fe/S) proteins** are associated with electron transport chains. Various arrangements of sulfur and iron have been found in different nonheme iron-sulfur proteins, but Fe_2S_2 and Fe_4S_4 clusters are the most common (Figure 5.17). The iron atoms are bonded to free sulfur and to the protein via sulfur atoms from cysteine residues (Figure 5.17). *Ferredoxin*, a common iron-sulfur protein in biological systems, has an Fe_2S_2 configuration. The reduction potentials of iron-sulfur proteins vary over a wide range depending on the number of iron and sulfur atoms present and how the iron centers are attached to protein.

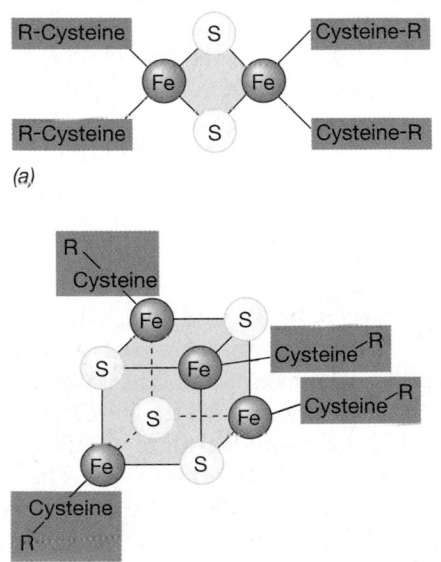

(a)

(b)

Figure 5.17 Arrangement of the iron-sulfur centers of nonheme iron-sulfur proteins. (a) Fe_2S_2 center. (b) Fe_4S_4 center. The cysteine linkages are from the protein portion of the molecule. Iron-sulfur proteins typically carry electrons only.

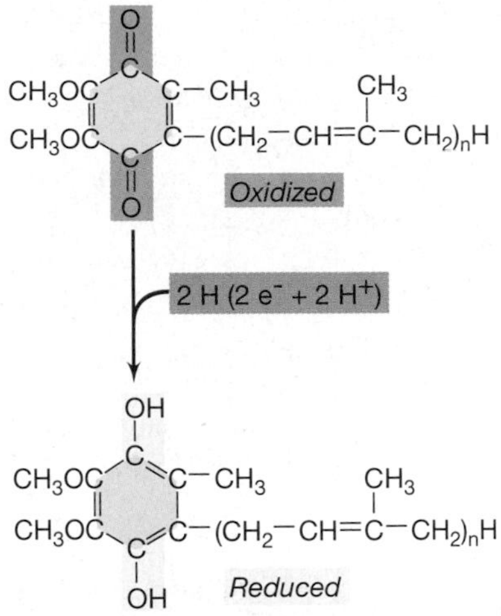

Figure 5.18 Structure of oxidized and reduced forms of coenzyme Q, a quinone. The five-carbon unit in the side chain (an isoprenoid) occurs in a number of multiples. In prokaryotes, the most common number is $n=6$; in eukaryotes, $n=10$. Note that oxidized quinone requires two hydrogen atoms (2 H) to become fully reduced. An intermediate form, the semiquinone (one H more reduced than oxidized quinone), is formed during the reduction of a quinone.

Thus, different iron-sulfur proteins can function at different points in the electron transport process. Like cytochromes, iron-sulfur proteins carry *electrons only*, not hydrogen atoms.

The **quinones** (Figure 5.18) are highly hydrophobic nonprotein-containing molecules involved in electron transport. Some quinones found in bacteria are related to vitamin K, a growth factor for higher animals. Like flavoproteins, quinones serve as *hydrogen atom* acceptors and *electron* donors.

✓ 5.11 Concept Check

Aerobic electron transport systems consist of a series of membrane-associated electron carriers that function in an integrated fashion to carry electrons from the electron donor to oxygen as final (terminal) electron acceptor.

- ✓ How is iron present in cytochromes compared with a protein like ferredoxin?
- ✓ In what major way do quinones differ from other electron carriers in the membrane?
- ✓ Which electron carriers accept hydrogen atoms only? Electrons only?

5.12 Energy Conservation from the Proton Motive Force

The major components of electron transport chains are shown in Figure 5.19. During electron transport, ATP is produced by the process of oxidative phosphorylation. The

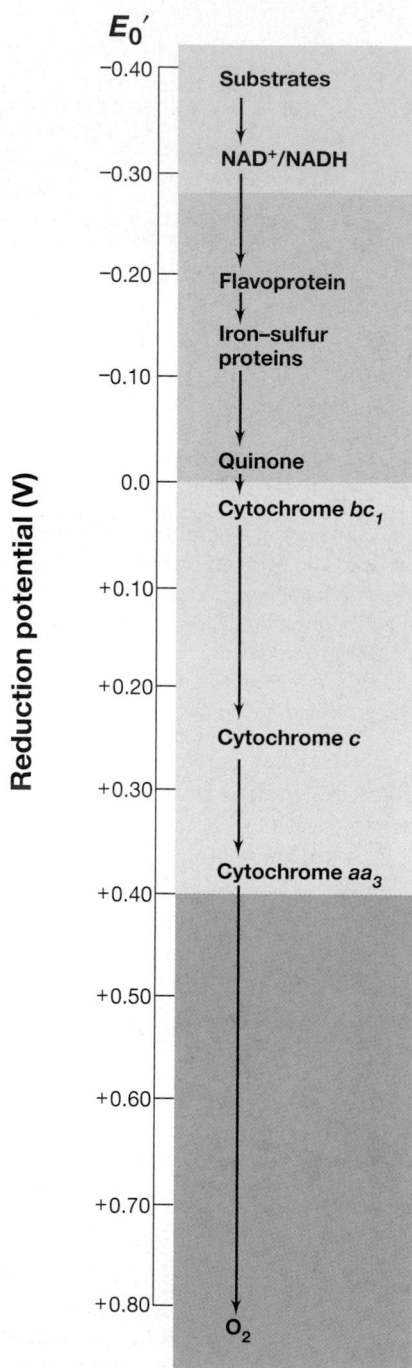

Figure 5.19 Electron transport chains and their relation to E_0'. Shown here is one example of an electron transport system, leading to the transfer of electrons from substrate to O_2. This particular sequence is typical of the electron transport chain of the mitochondrion of eukaryotic cells and that of some *Bacteria* (for example, *Paracoccus denitrificans*). The electron transport chain in *Escherichia coli* lacks cytochromes c and aa_3, and instead electrons go directly from cytochrome b to cytochrome o or d (the latter has a similar E_0' to cytochrome aa_3), which is the terminal oxidase (⇌ Figure 17.37). By breaking up the complete oxidation into a series of discrete steps, energy conservation is possible through proton motive force formation leading to ATP synthesis. Compare color-coding here with those in Figure 5.9.

production of ATP is linked directly to the establishment of a **proton motive force** across the membrane, electron transport reactions serving to establish this energized state of the membrane. We now consider the details of this process.

The Proton Motive Force: Chemiosmosis

To understand the manner in which electron transport is linked to ATP synthesis, we must first discuss the manner in which the electron transport system is oriented in the cell membrane. The overall structure of the membrane was outlined in Section 4.5 (⇌ Figure 4.17). It was shown there that proteins are embedded in the lipid bilayer of membranes and that the orientation of proteins in the membrane is such that most have access to both the outside and the inside of the cell (that is, they are transmembrane proteins).

The electron transport carriers discussed earlier are oriented in the membrane in such a way that a *separation* of protons from electrons occurs across the membrane during the transport process. Hydrogen atoms, removed from hydrogen atom carriers such as NADH, are separated into electrons and protons, the electrons being transported through the chain by specific carriers and the protons being pumped outside the cell into the environment (in gram-negative prokaryotes protons are pumped to the periplasm); this results in a slight acidification of the external surface of the membrane. At the end of the electron transport chain, the electrons are passed to the final electron acceptor (in the case of aerobic respiration, this is O_2) and reduce it.

When O_2 is reduced to H_2O, it requires H^+ from the cytoplasm to complete the reaction, and these protons originate from the dissociation of water into H^+ and OH^-. The use of H^+ in the reduction of O_2 to H_2O and the extrusion of H^+ cause a net accumulation of OH^- on the *inside* of the membrane. Despite their small size, because they are charged, neither H^+ nor OH^- freely passes through the membrane, and so equilibrium cannot be spontaneously restored. Thus, although electron transport to O_2 can be thought of as producing water, what is actually produced are the *components* of water, H^+ and OH^-, which accumulate on opposite sides of the membrane. The net result is the generation of a *pH gradient* and an *electrochemical potential* across the membrane, with the *inside* of the cytoplasm electrically negative and alkaline, and the *outside* of the membrane electrically positive and acidic. This pH gradient and electrochemical potential cause the membrane to be energized (much like a battery), and some of this electrical energy can be conserved by the cell.

In the same way that the energized state of a battery is expressed as its electromotive force (in volts), the energized state of a membrane is expressed as the *proton motive force* (also in volts). The energized state of the membrane induced as a result of electron transport processes can be used directly to do useful work such as ion transport (⊂⊃ Section 4.7) or flagellar rotation (⊂⊃ Section 4.10), or it can be used to drive the formation of high-energy phosphate bonds in ATP, as will be described later. The idea of a proton gradient driving ATP synthesis was first proposed as the *chemiosmotic theory* in 1961 by the English scientist Peter Mitchell; Mitchell later received the Nobel Prize for this important contribution.

Generation of the Proton Motive Force

The key steps in proton motive force formation involve the activities of the flavin enzymes, quinones, and the cytochrome bc_1 complex (Figure 5.20). The series of oxidation–reduction reactions occurring during electron transport may be analyzed by examining each pair of carriers sequentially (Figure 5.20). Following the donation of two hydrogen atoms from NADH to FAD, two H^+ are extruded when FADH donates two electrons (only) to an iron-sulfur protein, part of Complex I, shown in Figure 5.20. Two protons are taken up from the dissociation of water in the cytoplasm when the nonheme iron protein reduces coenzyme Q. Coenzyme Q passes electrons one at a time to the cytochrome bc_1 complex, shown as Complex III in Figure 5.20. The cytochrome bc_1 complex is formed by several proteins containing several hemes or metal centers, such as two b type hemes (b_L and b_H), one c type heme (c_1), and one Fe/S protein (called the *Rieske protein*). The bc_1 complex is present in the electron transport chain of most organisms able to respire. It also plays a role in photosynthetic electron flow (⊂⊃ Sections 17.4 and 17.5).

The major function of the cytochrome bc_1 complex is to transfer electrons from quinones to cytochrome c linked to the translocation of protons across the membrane, resulting in the accumulation of OH^- in the cytoplasm and protons on the outer surface of the membrane (Figure 5.20). Electrons travel from the bc_1 complex to a molecule of cytochrome c that is in the periplasm or attached to the external side of the membrane, and from there to the high potential cytochrome aa_3 complex (Complex IV, Figure 5.20). The latter is the *terminal oxidase* in the system and reduces O_2 to H_2O in the final step of electron transport (Figure 5.20).

The electron transport scheme shown in Figure 5.20 is just one of many different carrier sequences known from different organisms. Several features are characteristic of *all* electron transport chains, however, and can be summarized as follows: (1) the presence of a series of membrane-associated electron carriers arranged in order of increasingly more positive E_0'; (2) an alternation in the chain of electron-only and hydrogen-atom only carriers; and (3) the generation of a proton motive force as a result of charge separation across the membrane, acidic outside and alkaline inside. As we will see now, it is the proton motive force that actually yields ATP.

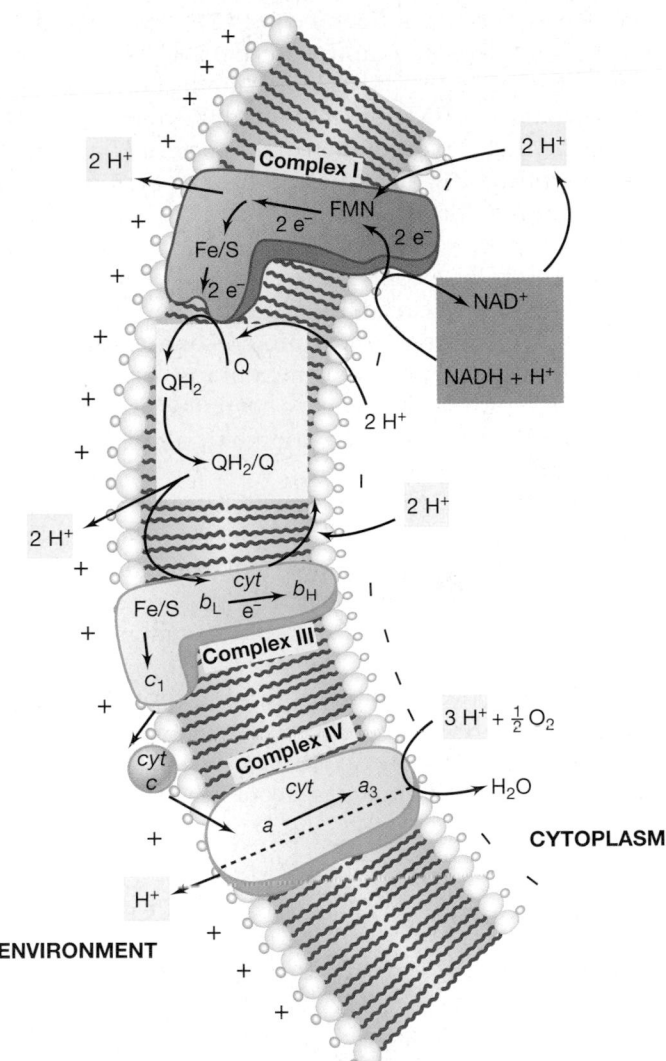

Figure 5.20 Generation of the proton motive force during aerobic respiration. The figure shows the orientation of key electron carriers in the membrane of an organism like *Paracoccus denitrificans*, a model prokaryote for studies of respiration. The + and – charges across the membrane represent H^+ and OH^-, respectively. Abbreviations are as follows: FMN, flavoprotein; Q, quinone; Fe/S, iron sulfur protein; cyt a, b, c, cytochromes (b_L and b_H, low and high potential b-type cytochromes, respectively). At the quinone site a recycling of electrons occurs during the "Q cycle." This is because electrons from QH_2 can be split in the bc_1 complex (Complex III) between the Fe/S protein and the b-type cytochromes. Electrons that travel through the latter reduce Q (in two, one-electron steps) back to QH_2, thus increasing the number of protons pumped at the Q-bc_1 site. Electrons that travel to Fe/S proceed to reduce cytochrome c_1 and then cytochrome c, and then a-type cytochromes in Complex IV, eventually reducing O_2 to H_2O (two hydrogen atoms are required to reduce $\frac{1}{2} O_2$ to H_2O). For simplicity, protein Complex II, the succinate dehydrogenase complex, is not shown in this scheme. The numbers of the complexes shown are those used by convention by scientists in the field of membrane bioenergetics. Compare this electron transport chain with that of *Escherichia coli*, shown in Figure 17.37.

The Proton Motive Force and ATP Formation

How does the proton motive force drive ATP synthesis? Interestingly, there is a strong parallel here between the mechanism of ATP synthesis and the mechanism of the motor that drives rotation of the bacterial flagellum (⚭ Section 4.10). The catalyst for conversion of the proton motive force into ATP is a large membrane enzyme complex called **ATP synthase**, or **ATPase** for short. The ATPase contains two major parts, a multisubunit headpiece called F_1, located on the cytoplasmic side of the membrane, and a proton-conducting channel, called F_0, that spans the membrane (Figure 5.21). The F_1/F_0 complex catalyzes a *reversible* reaction between ATP and ADP + P_i (inorganic phosphate) as shown in Figure 5.21.

The F_1/F_0 ATPase is the smallest known biological motor. According to the current model of how the *Escherichia coli* ATPase functions, proton movement through the a subunit of F_0 drives rotation of the c proteins generating a torque that is transmitted to F_1 by the $\gamma\epsilon$ subunits (Figure 5.21). In essence, energy is transferred to F_1 through the coupled rotation of $\gamma\epsilon$ subunits causing conformational changes in the β subunits; this represents work done on the system (potential energy) that can be tapped to make ATP. This is possible because the conformational changes

in the β subunits allow for binding of ADP + P_i and these are converted to ATP when the β subunits return to their original conformation. Thus, in analogy to the flagellar motor (⚭ Figure 4.41), the primary function of the $b_2\delta$ subunits of F_1 is to serve as a fixture (stator), preventing the α and β subunits from rotating with $\gamma\epsilon$ so that conformational changes in β can be affected. But unlike the flagellum, rotation of the ATPase is not used for cell propulsion, but instead to synthesize ATP (Figure 5.21). ATPase-catalyzed ATP synthesis is referred to as **oxidative phosphorylation** in respiratory systems or **photophosphorylation** in phototrophic organisms. Measurements of the stoichiometry between the number of protons consumed per ATP produced by either type of phosphorylation yield a number of 3 to 4.

The tiny F_1/F_0 molecular motor can be reversed. The hydrolysis of ATP supplies torque for $\gamma\epsilon$ to rotate in the opposite direction, causing protons to be pumped from inside to outside the cell through the a subunit, thereby *creating*, instead of *dissipating*, a proton motive force. This reversibility explains why fermentative organisms unable to carry out oxidative phosphorylation still contain an ATP synthase. Since many reactions in the cell such as motility and transport require the energy of a proton motive force (⚭ Section 4.6), the ATPase in organisms such as the lactic acid bacteria, that do not respire, functions unidirectionally to generate a proton motive force to drive these necessary cell functions.

Inhibitors and Uncouplers

The events of electron transport can be studied with certain chemicals that affect oxidative phosphorylation. Two such classes of chemicals are known: *inhibitors* and *uncouplers*. Inhibitors block electron flow and thus, establishment of the proton motive force. Examples include carbon monoxide (CO) and cyanide (CN^-), both of which bind tightly to certain cytochromes and prevent their functioning. In contrast, uncouplers prevent ATP synthesis *without* affecting electron transport. These lipid soluble substances, such as dinitrophenol and dicumarol, make membranes leaky, thereby destroying the proton motive force and its ability to drive ATP synthesis.

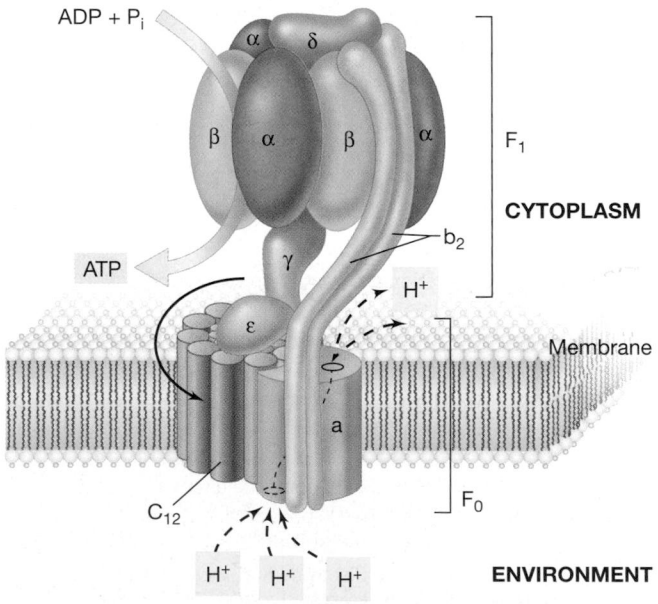

ADP + P_i

α δ

β α β α F_1

CYTOPLASM

b_2

ATP

γ

H^+

ϵ

Membrane

a

F_0

C_{12}

H^+ H^+ H^+

ENVIRONMENT

Figure 5.21 Structure and function of ATP synthase (ATPase). F_1 consists of five different polypeptides present as an $\alpha_3\beta_3\gamma\epsilon\delta$ complex. F_1 is the catalytic complex responsible for the interconversion of ADP + P_i and ATP. F_0 is integrated in the membrane and consists of three polypeptides in an ab_2c_{12} complex. Subunit a is responsible for channeling protons across the membrane while subunit b protrudes outside the membrane and forms, along with the b_2 and δ subunits, the stator. As protons enter, the dissipation of the proton motive force drives ATP synthesis. The ATPase is reversible in its action; that is, ATP hydrolysis can drive formation of a proton motive force.

✓ 5.12 Concept Check

When electrons are transported through a membrane-integrated electron transport system, protons are extruded to the outside of the membrane forming the proton motive force. Key electron carriers include flavins, quinones, the cytochrome bc_1 complex, and other cytochromes, depending on the organism. The cell uses the proton motive force via rotating ATPases to make ATP.

✓ How do electron transport reactions generate the proton motive force?

✓ What structure in the cell converts the proton motive force to ATP? How does it operate?

5.13 Carbon Flow in Respiration: The Citric Acid Cycle

We now consider the metabolic aspects of carbon flow in respiration. The early steps in the respiration of glucose involve the same biochemical steps as those of glycolysis (see Figure 5.14). As we noted, a key intermediate in glycolysis is pyruvate. Whereas in fermentation pyruvate is converted to fermentation products, in respiration pyruvate is oxidized fully to CO_2. One major pathway by which pyruvate is completely oxidized to CO_2 is called the **citric acid cycle** (CAC) as outlined in Figure 5.22.

Pyruvate is first decarboxylated, leading to the production of one molecule of NADH and an acetyl molecule coupled to coenzyme A (acetyl-CoA) (see Figure 5.12). The acetyl group of acetyl-CoA combines with the four-carbon compound oxalacetate, leading to the formation of citric acid, a six-carbon organic acid, the energy of the high-energy acetyl-CoA bond (Figure 5.12) being used to drive this synthesis. Hydration, decarboxylation, and oxidation reactions follow, and two additional CO_2 molecules are released. Ultimately, oxalacetate is regenerated and can function again as an acetyl acceptor, thus completing the cycle.

CO₂ Release and Fuel for Electron Transport

For each pyruvate molecule oxidized through the cycle, three CO_2 molecules are released, one during the formation of acetyl-CoA, one by the decarboxylation of isocitrate, and one by the decarboxylation of α-ketoglutarate (Figure 5.22). As in fermentation, the electrons released during the oxidation of intermediates in the CAC are transferred to enzymes containing the coenzyme NAD^+ or FAD. However, respiration differs from fermentation in the manner in which NADH and FADH are oxidized. In respiration, the electrons from NADH, instead of being used to reduce an intermediate such as pyruvate, are transferred to oxygen or other terminal electron acceptors through the action of the *electron transport system* described in Section 5.12. Thus, unlike the situation in fermentation, the presence of an electron acceptor in respiration allows for the complete oxidation of glucose to CO_2 with a much greater yield of energy.

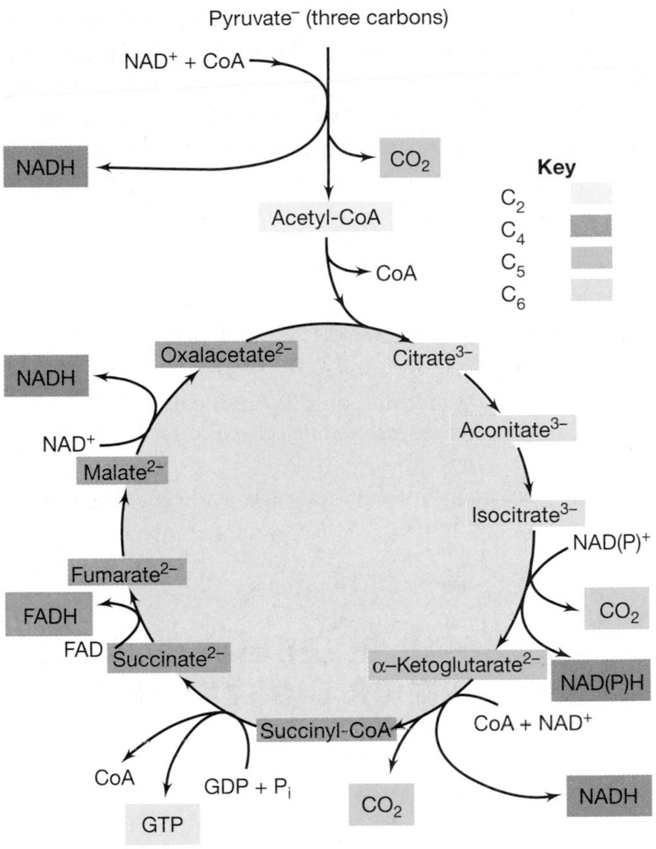

Biosynthesis and the Citric Acid Cycle

Besides playing a key role in catabolic reactions, the citric acid cycle is important to the cell for biosynthetic reasons as well. This is because the cycle is composed of a number of key intermediates that can be drawn off for biosynthetic purposes when needed. Particularly important in this regard are the intermediates α-ketoglutarate and oxalacetate, which are the precursors of a number of amino acids (see Section 5.15), and succinyl-CoA, needed to form the porphyrin ring of the cy-

(a)

Overall reaction: Pyruvate⁻ + 4 NAD⁺ + FAD → 3 CO₂ + 4 NADH + FADH

(1) Substrate-level phosphorylation
GDP + P$_i$ → GTP
GTP + ADP → GDP + ATP
} 15 ATP

(2) Electron transport phosphorylation
4 NADH ≡ 12 ATP
FADH ≡ 2 ATP

(3) Sum: CAC plus glycolysis → 38 ATP per glucose

(b)

Figure 5.22 The citric acid cycle (CAC). (a) The CAC begins when the two-carbon compound acetyl-CoA (formed from pyruvate) condenses with the four-carbon compound oxalacetate to form the six-carbon compound citrate. Through a series of oxidations and transformations, this six-carbon compound is ultimately converted back to the four-carbon compound oxalacetate, which then begins another cycle with addition of the next molecule of acetyl-CoA. (b) The overall balance sheet of fuel (NADH/FADH) for the electron transport chain and CO_2 generated in the CAC. Compare with Figure 5.14 and note the major energy difference between fermenting and respiring glucose.

tochromes, chlorophyll, and several other tetrapyrrole compounds (see Figure 5.16). Oxalacetate is also important because it can be converted to phosphoenolpyruvate, a precursor of glucose (see Figure 5.25). In addition to these, acetyl-CoA provides the starting material for fatty acid biosynthesis (see Section 5.15). Thus, the citric acid cycle plays two major roles in the cell: *bioenergetic* and *biosynthetic*. Much the same can be said about the glycolytic pathway, as intermediates from this pathway are drawn off for various biosynthetic needs as well.

✓ 5.13 Concept Check

Respiration involves the complete oxidation of an organic compound with much greater energy release than during fermentation. The citric acid cycle plays a major role in the respiration of organic compounds.

- ✓ How many molecules of CO_2 and pairs of hydrogen atoms are released *per acetate consumed* in the citric acid cycle?
- ✓ What two major roles do the citric acid cycle and glycolysis have in common?

V CATABOLIC DIVERSITY AND AN OVERVIEW OF BIOSYNTHESIS

As emphasized in Chapter 2, at the heart of microbial diversity is *metabolic diversity*, in particular, the diverse strategies microorganisms have evolved for making ATP. Thus far in this chapter, we have dealt only with the reactions of chemoorganotrophs. We now briefly consider the alternatives to organic compounds as energy sources and then close the chapter with an overview of biosynthetic reactions that yield the monomers required for the assembly of macromolecules.

5.14 Catabolic Alternatives

Figure 5.23 summarizes the methods by which cells may generate energy other than via fermentation and aerobic respiration. These include anaerobic respiration, chemolithotrophy, and phototrophy.

Anaerobic Respiration

One alternate mode of energy generation is a variation on respiration in which electron acceptors *other than* oxygen are used. In contrast to aerobic respiration, these processes are called **anaerobic respiration**. Electron acceptors used in anaerobic respiration include nitrate (NO_3^-), ferric iron (Fe^{3+}), sulfate (SO_4^{2-}), carbonate (CO_3^{2-}), and even certain organic compounds. Because of their positions on the electron tower (none of these acceptors have as electropositive an E_0' as the O_2/H_2O couple) (see Figure 5.9), less energy is released when these alternate electron acceptors are used instead of oxygen. However, the use of these alternate electron acceptors permits microorgan-

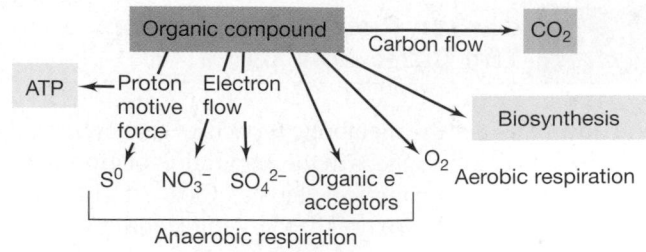

(a) **Chemoorganotrophic metabolism**

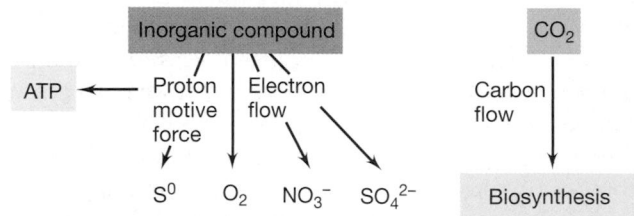

(b) **Chemolithotrophic metabolism**

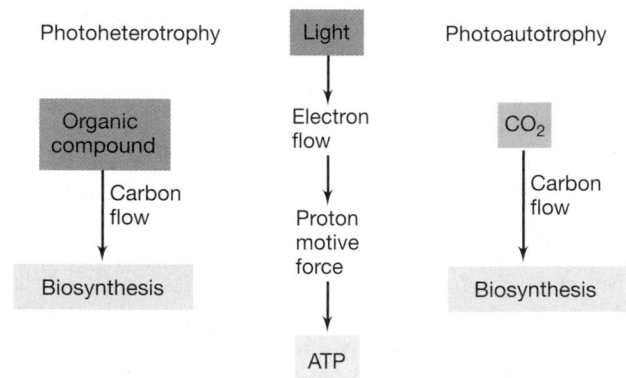

(c) **Phototrophic metabolism**

Figure 5.23 Energetics and carbon flow in (a) chemoorganotrophic respiratory metabolism, (b) chemolithotrophic metabolism, and (c) phototrophic metabolism. Note how in phototrophic metabolism carbon for biosynthesis can come from CO_2 (photoautotrophy) or organic compounds (photoheterotrophy). Note also the importance of electron transport leading to proton motive force formation in each case.

isms to respire in environments where oxygen is absent. Because the solubility of O_2 in water is rather low, and because O_2 is in such high demand as an electron acceptor, anaerobic respirations are thus ecologically extremely important. We discuss anaerobic respiration in more detail in Section 17.13.

Chemolithotrophy

A second mode of energy generation involves the use of *inorganic* rather than organic chemicals. Organisms able to use inorganic chemicals as electron donors are a type of chemotroph called **chemolithotrophs**. Examples of inorganic electron donors include hydrogen sulfide

(H_2S), hydrogen gas (H_2), ferrous iron (Fe^{2+}), and ammonia (NH_3). Chemolithotrophic metabolism usually involves aerobic respiratory processes such as those described in this chapter but using an inorganic energy source rather than an organic one (Figure 5.23). Chemolithotrophs have electron transport components similar to those of chemoorganotrophs and form a proton motive force, which drives ATP synthesis. One important distinction between chemolithotrophs and chemoorganotrophs is in their sources of *carbon* for biosynthesis. Chemoorganotrophs can generally use compounds such as glucose as carbon sources as well as energy sources, but chemolithotrophs obviously cannot use their inorganic electron donors as sources of carbon. Most chemolithotrophs use *carbon dioxide* (CO_2) as a carbon source and are, hence, **autotrophs**. We discuss many forms of chemolithotrophy in Chapter 17.

Phototrophy

A large number of microorganisms, as well as the green plants, are *phototrophic*, using light as an energy source in the process of photosynthesis. The mechanisms by which light is used as an energy source are unique and complex, but the underlying result is the generation of a proton motive force that can be used in the synthesis of ATP. Most phototrophs use energy conserved in ATP for the assimilation of carbon dioxide as the carbon source for biosynthesis; they are called **photoautotrophs**. However, some phototrophs use organic compounds as carbon sources with light as the energy source; these are the **photoheterotrophs** (Figure 5.23). As we will see in Chapter 17, photosynthesis in microorganisms has some special features and complications. There are, for instance, two types of photosynthesis in microorganisms, one form similar to that of higher plants in which O_2 is evolved and a unique type of photosynthesis found only in certain prokaryotes in which O_2 evolution does not occur.

Importance of the Proton Motive Force to Alternate Bioenergetic Strategies

It should now be clear that in terms of energy metabolism, microorganisms show an amazing diversity of bioenergetic strategies. Thousands of organic compounds, many inorganic compounds, and light can be used by one or another microorganism as an energy source. However, with the exception of fermentations, where substrate-level phosphorylation occurs, metabolic diversity in respiration and photosynthesis revolves around a common theme: *generation of a proton motive force*. Whether electrons come from the oxidation of organic or inorganic chemicals or from phototrophic processes, they all traverse a membrane-associated electron transport chain and in so doing, generate a proton motive force (Figure 5.20); energy conservation in all cases occurs through function of the ATPase (Figure 5.21). In Chapter 17 we will examine some of the details of the different bioenergetic strategies that result in generation of the proton motive force.

✓ 5.14 Concept Check

Electron acceptors other than oxygen can function as terminal electron acceptors for energy generation. Because oxygen is absent, this is called anaerobic respiration. Chemolithotrophs use inorganic compounds as electron donors, while phototrophs use light to form a proton motive force. The proton motive force is involved in all forms of respiration and photosynthesis.

✓ In terms of their electron donor(s), how do chemoorganotrophs differ from chemolithotrophs?

✓ What is the carbon source for autotrophic organisms?

✓ How do *photoautotrophs* differ from *photoheterotrophs*?

5.15 Biosynthesis of Key Monomers

We have just discussed **catabolism**, the processes by which microorganisms obtain energy from organic compounds. We now briefly consider the other major reaction series that occur in the cell, **anabolism**, the processes by which microorganisms build up the vast array of chemical substances of which they are composed.

Energy for anabolism is provided by ATP or the proton motive force, which are different forms of chemical energy (Figure 5.24). Energy conserved during catabolic reactions as ATP or the proton motive force is consumed in both the formation of monomers and during the polymerization of monomers to form macromolecules. Our

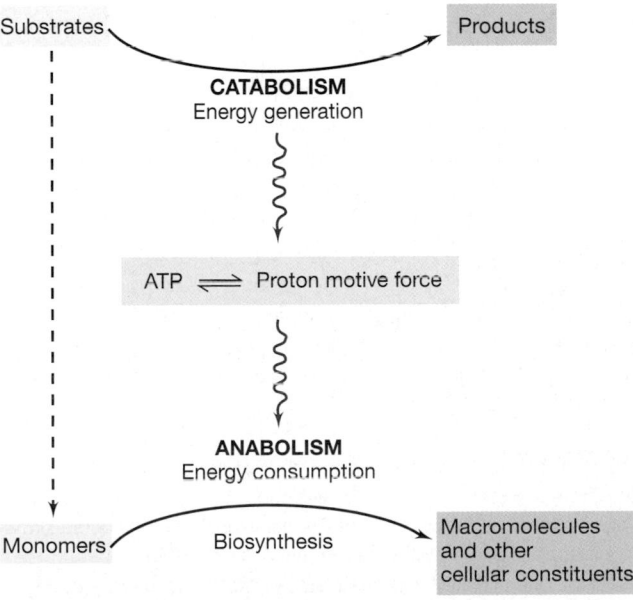

Figure 5.24 Scheme of anabolism and catabolism showing the key role of ATP and the proton motive force in integrating the processes. Monomers can come preformed as nutrients from the environment or from catabolic pathways like glycolysis and the citric acid cycle.

coverage of biosynthetic pathways here is intended only to be an overview of monomer synthesis; the biosynthesis of macromolecules is considered in Chapter 7.

Monomers of Polysaccharides: Sugars

Polysaccharides are key constituents of the cell walls of many organisms, and in *Bacteria*, the peptidoglycan cell wall (⬯ Section 4.8) has a polysaccharide backbone. In addition, cells often store carbon and energy in the form of the polysaccharides *glycogen* and *starch* (⬯ Sections 3.3 and 4.13). The monomeric units of these polysaccharides are six-carbon sugars called *hexoses*, in particular, glucose or glucose derivatives. In addition to hexoses, five-carbon sugars called *pentoses* are common in the cell. Most notably, these include ribose and deoxyribose, present in the backbone of RNA and DNA, respectively.

In prokaryotes, polysaccharides are synthesized from either *uridine diphosphoglucose* (UDPG, Figure 5.25*a*) or *adenosine diphosphoglucose* (ADPG), both of which are *activated* forms of glucose. ADPG is the precursor for the biosynthesis of glycogen while UDPG is the precursor of various glucose derivatives needed for the biosynthesis of other polysaccharides in the cell, such as peptidoglycan or the lipopolysaccharide of the gram-negative outer membrane (⬯ Sections 4.8 and 4.9).

When a cell is growing on a hexose such as glucose, obtaining glucose for polysaccharide synthesis is obviously not a problem. But when the cell is growing on other carbon compounds, glucose must be synthesized. This process, called *gluconeogenesis,* uses as a starting material the compound *phosphoenolpyruvate,* one of the key intermediates in glycolysis (see Figure 5.14). Phosphoenolpyruvate can be synthesized from oxalacetate, a citric acid cycle intermediate. An overview of hexose metabolism is shown in Figure 5.25.

Pentoses are formed by the removal of one carbon atom from a hexose, commonly as CO_2 (Figure 5.25*c*). The pentoses needed for nucleic acid synthesis, ribose and deoxyribose, are formed as shown in Figure 5.25*c*. The important enzyme *ribonucleotide reductase* converts ribose into deoxyribose by reduction of the 2′ carbon on the ring. Interestingly, this reaction occurs *after*, not before, synthesis of nucleotides, such that ribonucleotides are biosynthesized first, and later some of each are reduced to deoxyribonucleotides for use as precursors of DNA.

Figure 5.25 Sugar metabolism. (a) Polysaccharides are synthesized from activated forms of hexoses such as UDPG, whose structure is shown here. (b) Glycogen is biosynthesized from adenosine-phosphoglucose (ADPG) by the sequential addition of glucose. (c) Pentoses for nucleic acid synthesis are formed by decarboxylation of hexoses like glucose-6-phosphate. Note how the precursors of DNA are produced from the precursors of RNA by the enzyme ribonucleotide reductase. (d) Gluconeogenesis. When glucose is needed it can be biosynthesized from other carbon compounds, generally by the reversal of steps in glycolysis.

Monomers of Proteins: Amino Acids

Organisms that cannot obtain some or all of their amino acids preformed from the environment must synthesize them from other sources. Amino acids can be grouped into structurally related *families* that share biosynthetic features (Figure 5.26). The carbon skeletons for amino

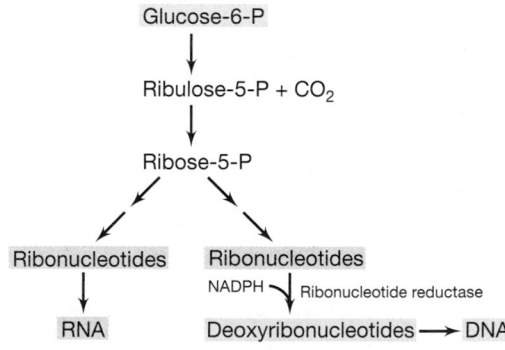

Uridine diphosphoglucose (UDPG)

(a)

ADPG + Glycogen ⟶ ADP + Glycogen-Glucose

(b)

Glucose-6-P
↓
Ribulose-5-P + CO_2
↓
Ribose-5-P
↓   ↓
Ribonucleotides   Ribonucleotides
                    NADPH ⟍ Ribonucleotide reductase
↓   ↓
RNA   Deoxyribonucleotides ⟶ DNA

(c)

Other Pathways
↓ ↓ ↓ ↓
Citric acid cycle
↓
Oxalacetate
↓
Phosphoenolpyruvate + CO_2
↓
Reversal of glycolytic steps
↓
Glucose-6-P

(d)

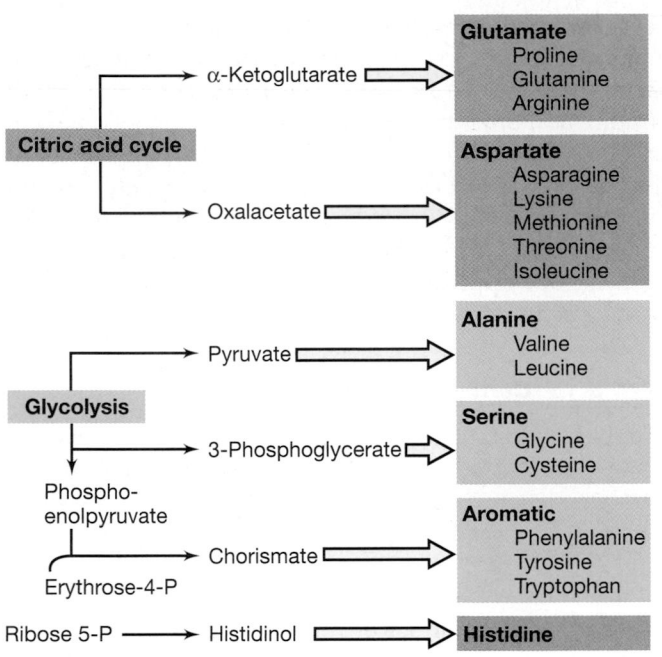

Figure 5.26 Amino acid families. Note how the carbon skeletons for most amino acids are derived from either the citric acid cycle or from glycolysis. Synthesis of the various amino acids in a family frequently requires many separate enzymatically catalyzed steps starting from the parent amino acid (shown in bold).

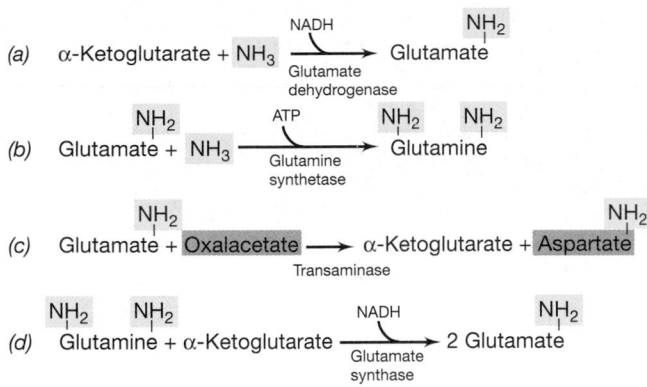

Figure 5.27 Ammonia incorporation in bacteria. Free ammonia as well as the amino groups of all amino acids are shown in blue. Two major pathways for NH_3 assimilation in bacteria are those catalyzed by the enzymes (a) glutamate dehydrogenase and (b) glutamine synthetase. (c) Transaminase reactions simply transfer an amino group from an amino acid to an organic acid. (d) In the reaction catalyzed by the enzyme glutamate synthase, two glutamates are formed from one glutamine and one α-ketoglutarate.

acids come almost exclusively from intermediates of glycolysis or the citric acid cycle (Figure 5.26).

The *amino group* of amino acids is often derived from some inorganic nitrogen source in the environment, such as ammonia (NH_3). Ammonia is typically incorporated in the formation of the amino acids *glutamate* or *glutamine* by the enzymes *glutamate dehydrogenase* or *glutamine synthetase,* respectively (Figure 5.27a, b). Once ammonia is incorporated, it can be transferred to form other needed nitrogenous compounds. For example, glutamate can donate its amino group to oxalacetate in a *transaminase* reaction to yield α-ketoglutarate and aspartate (Figure 5.27c). Alternatively, glutamine can react with α-ketoglutarate to form two molecules of glutamate in an *aminotransferase* reaction (Figure 5.27d). The end result of these types of reactions is the shuttling of ammonia into various key carbon skeletons from which further biosynthetic reactions can occur to form all the 20 amino acids needed to make proteins.

Monomers of Nucleic Acids: Nucleotides

The biosynthesis of the purines and pyrimidines is quite complex. Purines are constructed, literally atom by atom, from several carbon and nitrogen sources, including CO_2 (Figure 5.28a). The first key purine, *inosinic acid* (Figure 5.28b), is the precursor of all of the other purine nucleotides; once these are synthesized (in their triphosphate forms), they are ready to be incorporated into DNA or RNA (∞ Sections 7.5–7.11). Like the purine ring, the pyrimidine ring is also constructed from several sources (Figure 5.28c), and the first key pyrimidine is the compound *uridylate,* from which all other pyrimidines are derived (Figure 5.28d).

Monomers of Lipids: Fatty Acids

Fatty acids are required for the biosynthesis of lipids in *Bacteria* and *Eukarya* (∞ Sections 3.4 and 4.5). Fatty acids are biosynthesized two carbon atoms at a time with the help of a small protein called the *acyl carrier protein* (ACP). ACP holds the growing fatty acid as it is being synthesized and releases it once it has reached the required length, such as the very common C_{16} fatty acid palmitate (Figure 5.29; ∞ Figure 3.7). Interestingly, although the fatty acid chain is constructed *two* carbons at a time, the two carbons are donated from a *three*-carbon compound called *malonate* attached to ACP to form malonyl-ACP; as each malonyl residue is donated, one molecule of CO_2 is released (Figure 5.29).

The fatty acid composition of cells vary from species to species and can also vary somewhat within a species due to temperature (growth at low temperatures favors shorter-length fatty acids), but C_{14}–C_{18} are the most common fatty acids found in *Bacteria*. In addition to saturated, even-carbon-number fatty acids, fatty acids can also be unsaturated, contain branches, or have an odd number of carbon atoms. *Unsaturated* fatty acids contain one or more double bonds in the long hydrophobic portion of the molecule. The number and position of these double bonds is often species or group specific. *Branched chain* and *odd-carbon-number* fatty acids are made using an initiating molecule that

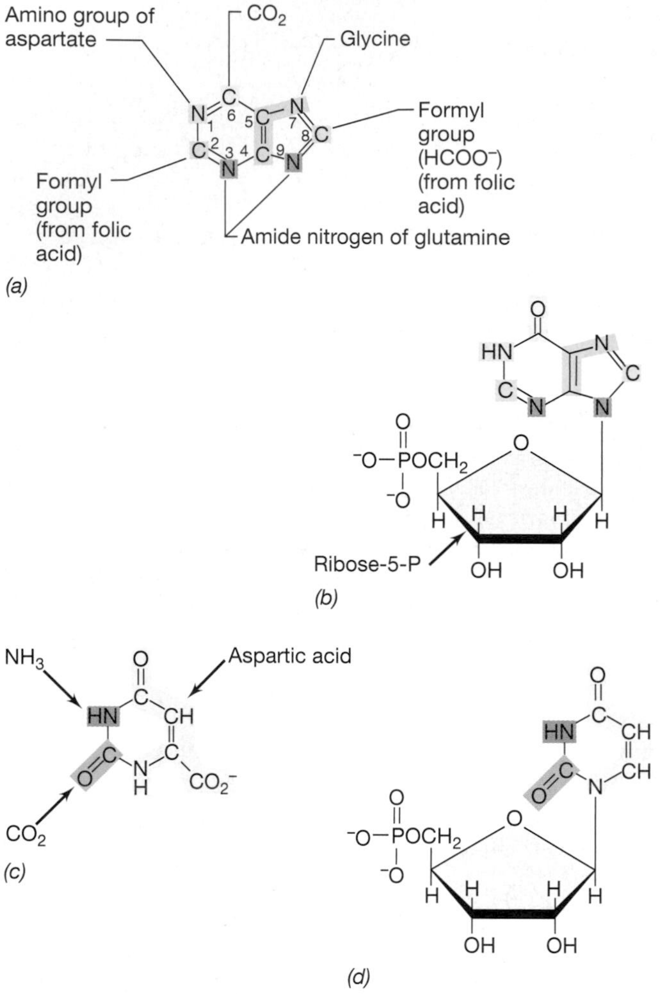

(a)

(b)

(c)

(d)

Figure 5.28 Biosynthesis of purines and pyrimidines. (a) The precursors of the purine skeleton. (b) Inosinic acid, the precursor of all purine nucleotides. (c) The precursors of the pyrimidine skeleton, orotic acid. (d) Uridylate, the precursor of all pyrimidine nucleotides. Uridylate is formed from orotate following a decarboxylation and the addition of ribose-5-phosphate.

contains a branched chain fatty acid or a propionyl (C_3) group, respectively.

The final assembly of lipids in *Bacteria* and *Eukarya* involves the addition of fatty acids to a glycerol molecule. For simple triglycerides, all three glycerol carbons are esterified with fatty acids, but for complex lipids, one of the carbons of glycerol contains a molecule such as phosphate, ethanolamine, a sugar, or some other polar substance (Figure 3.7). In *Archaea,* lipids contain phytanyl side chains instead of fatty acids (Section 4.5) but, as in *Bacteria* or *Eukarya,* the third carbon of the glycerol backbone usually contains a polar group of some sort.

Biosynthesis and Growth

We have now discussed the basic principles of biosynthesis of the monomers needed for the synthesis of macromolecules, the bulk of a cell's substance (Table 3.2). How-

ever, before we consider how the cell directs synthesis of its macromolecules, in particular its informational macromolecules, we examine the phenomenon of cell growth. Growth (multiplication) is the final result of all the catabolic and anabolic reactions we have considered in this chapter. Once we have mastered the principles of microbial growth and understand the environmental factors that control it, we will be ready to study the details of "information flow" and genetics and appreciate how the syntheses of nucleic acids and proteins are coordinated in the prokaryotic cell.

✓ **5.15 Concept Check**

The biosynthesis of monomers from either nutrients in the environment or from catabolic intermediates prepares the cell for the final step in biosynthesis, production of macromolecules. A variety of enzyme systems and biosynthetic pathways exist for the biosynthesis of sugars, amino acids, nucleotides, and fatty acids.

✓ What activated forms of glucose are involved in polysaccharide synthesis or in the synthesis of other hexoses?

✓ Where do most of the carbon skeletons for amino acid biosynthesis come from?

✓ Explain why, in fatty acid synthesis, fatty acids get built *two* carbon atoms at a time while the immediate donor for these carbons contains *three* carbon atoms.

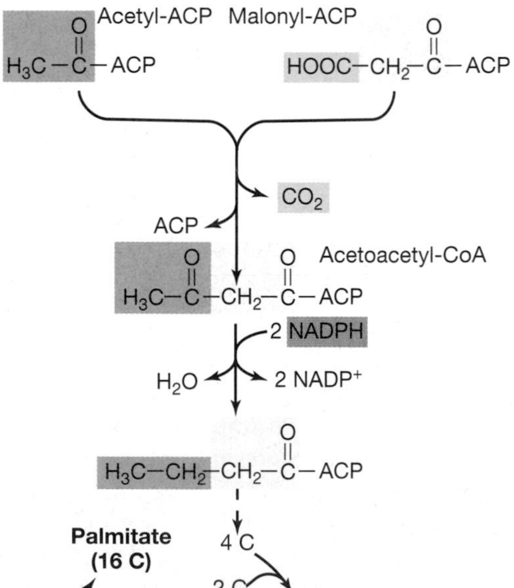

Figure 5.29 The biosynthesis of fatty acids; shown is the biosynthesis of the C_{16} fatty acid, *palmitate* (Figure 3.7). The condensation of acetyl-ACP and malonyl-ACP forms acetoacetyl-CoA. Each successive addition of an acetyl unit comes from malonyl-CoA.

Review Questions

1. Define the terms *chemoorganotroph, chemolithotroph, photoautotroph,* and *photoheterotroph.*

2. Why are *carbon* and *nitrogen* macronutrients while cobalt is a micronutrient?

3. What are siderophores and why are they necessary?

4. Why would the following medium *not* be considered a chemically defined medium: glucose, 5 grams (g); NH_4Cl, 1 g; KH_2PO_4, 1 g; $MgSO_4$, 0.3 g; yeast extract, 5 g; distilled water, 1 liter.

5. Describe how you would calculate $\Delta G^{0\prime}$ for the reaction: glucose + 6 $O_2 \rightarrow$ 6 CO_2 + 6 H_2O. If you were told that this reaction is highly *exergonic*, what would be the sign (negative or positive) of the $\Delta G^{0\prime}$ you would expect for this reaction?

6. Distinguish between $\Delta G^{0\prime}$ and G^0_f.

7. Why are enzymes needed by the cell?

8. Describe a rationale for why an enzyme from the bacterium *Escherichia coli* loses its catalytic ability after being boiled.

9. Describe the difference between a *coenzyme* and a *prosthetic group.*

10. The following is a series of coupled electron donors and electron acceptors. Using just the data given in Figure 5.9, order this series from most energy-yielding to least energy-yielding. H_2/Fe^{3+}, H_2S/O_2, methanol/NO_3^- (producing NO_2^-), H_2/O_2, Fe^{2+}/O_2, NO_2^-/Fe^{3+}, H_2S/NO_3^-.

11. What is an electron carrier? Give three examples of electron carriers, and indicate their oxidized and reduced forms.

12. Is the reaction glucose 6-phosphate + ADP → glucose + ATP exergonic or endergonic. (Refer to Figure 5.12 to help answer this question.)

13. Where in glycolysis is NADH *produced?* Where is NADH *consumed?*

14. Iron plays an important role within the cell in energy-generating processes. Give three examples in which iron plays a role as an electron carrier. How is iron provided as a nutrient in culture media?

15. What is meant by the term *proton motive force* and why is this concept so important in biology?

16. How is rotational energy in the ATPase used to produce ATP?

17. The chemicals dinitrophenol and cyanide both act as cellular poisons but in quite different ways. Compare and contrast the modes of action of these two chemicals.

18. Give, in simplified form, the equation for the oxidation–reduction reaction that takes place in the cytochrome molecule. Can you suggest a role for the portions of the cytochrome molecule that are *not* involved in oxidation–reduction?

19. Knowing the function of the electron transport chain, can you imagine an organism that could live if it completely lacked the components (for example, cytochromes) needed for an electron transport chain? (*Hint:* Focus your answer on the mechanism of ATPase.)

20. Work through the energy balance sheets for fermentation and respiration, and account for all sites of ATP synthesis. Organisms can obtain nearly 20 times more ATP when growing aerobically on glucose than anaerobically. Write one sentence that accounts for this difference.

21. Why can it be said that the citric acid cycle plays *two* major roles in the cell?

22. What are the similarities and differences in aerobic respiration in *Escherichia coli*, a chemoorganotroph, and *Thiobacillus thioparus*, a sulfur chemolithotroph?

23. Figure 5.26 indicates that there are only a few intermediate compounds that serve as the starting points for amino acid biosynthesis. For each family of amino acids, identify the starting compound and from what major pathway it originates.

24. Describe the process by which a fatty acid such as palmitate (C_{16} saturated) is synthesized in a cell.

25. What is meant by the term "aseptic technique?" Describe the steps involved in successfully transferring a pure culture from its culture tube to a tube of sterile medium.

26. If you had a liquid culture containing three different species of bacteria, how could you use a Petri plate to confirm this? How could you use the Petri plate to generate pure cultures of each organism? Is there any other way you might be able to detect contamination in a culture? (*Hint:* Quickly review material in the beginning of Chapter 4 before you answer.)

27. Observe the model of the enzyme lysozyme in Figure 5.7. What is the substrate for this enzyme? (*Hint:* A quick review of the material in Section 4.8 will help you here.) Where does the substrate bind? Why might lysozyme be found in natural body secretions such as the fluids that bathe the eye?

28. Why can it be said that NADH is an "electron shuttle?" Which figure in this chapter shows this shuttling action?

29. For a reaction to yield ATP, what is the minimal amount of energy that must be released to drive the synthesis of ATP from ADP? In Figure 5.14, which enzymatic reactions release energy that is conserved as ATP?

30. What is the carbon source for most chemolithotrophic bacteria? Why does this make sense considering the nature of their electron donors?

Application Questions

1. Design a defined culture medium for an organism that can grow aerobically on acetate as a carbon and energy source. Make sure all the nutrient needs of the organism are accounted for and in the correct relative proportions.

2. *Desulfovibrio* can grow anaerobically with H_2 as electron donor and SO_4^{2-} as electron acceptor (which is reduced to H_2S). Based on this information and the data in Table A1.2 (Appendix 1), indicate which of the following components *could not* exist in the electron transport chain of this organism and why: cytochrome c, ubiquinone, cytochrome c_3, cytochrome aa_3, ferredoxin.

3. Again, using the data in Table A1.2, predict the sequence of electron carriers in an organism growing aerobically and producing the following electron carriers: ubiquinone, cytochrome aa_3, cytochrome b, NADH, cytochrome c, FAD.

4. Explain the following observation: cells of *Escherichia coli* fermenting glucose grow faster when NO_3^- is supplied to the culture (NO_2^- is produced), and then grow even faster (and stop producing NO_2^-) when the culture is highly aerated.

5. Applying what you know about glycolysis, the citric acid cycle, and gluconeogenesis, explain why a mutant of *Escherichia coli* containing a defective malate dehydrogenase (that is, cannot carry out the reaction malate^{2-} + NAD$^+$ → oxaloacetate^{2-} + NADH + H$^+$) (see Figure 5.22) is able to grow on glucose but not on acetate, whereas nonmutated (wild-type) *E. coli* can be grown on either substrate.

C ell division is the process that generates two cells from one. In rod-shaped Bacteria this process typically occurs by *binary fission*, the gradual enlargement of a cell followed by its equal partitioning into two cells. Cell division is assisted by specific proteins, such as the division protein *FtsZ* stained in the cells shown here. To initiate the division process, copies of FtsZ are laid down in a ring along the mid-cell ridge, and in so doing, identify the midpoint in the cell. Once this has occurred, other cell division proteins are activated that trigger a series of events that culminate in the actual division process.

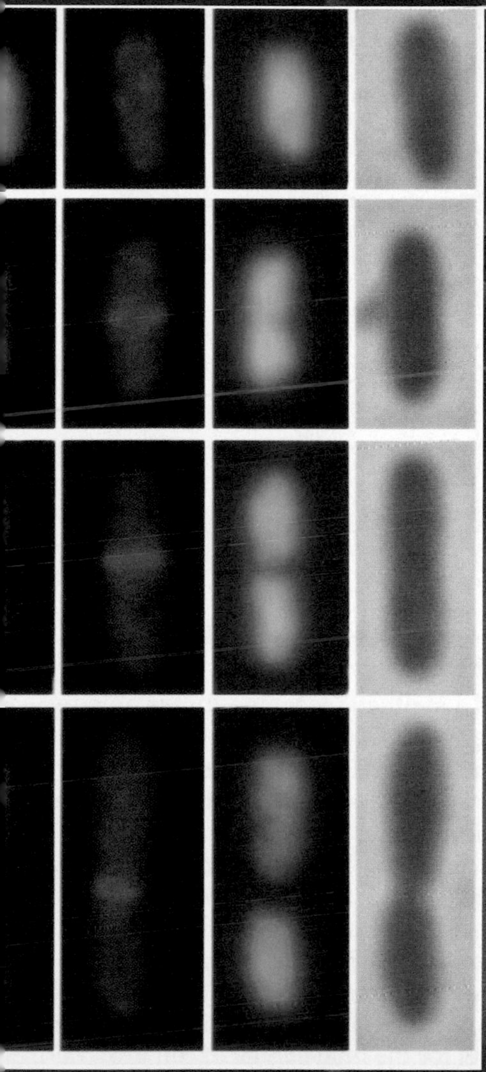

MICROBIAL GROWTH 6

Working Glossary

Acidophile an organism that grows best at low pH

Aerobe an organism that can use O_2 in respiration; some require O_2 for growth

Aerotolerant anaerobe a microorganism unable to respire molecular oxygen (O_2) but whose growth is unaffected by the presence of O_2

Alkaliphile an organism that grows best at high pH

Anaerobe an organism that cannot use O_2 in respiration and whose growth may be inhibited by O_2

Autolysis spontaneous cell lysis, usually due to the activity of lytic proteins called autolysins

Batch culture a closed-system microbial culture of fixed volume

Binary fission cell division following enlargement of a cell to twice its minimum size

Cardinal temperatures the minimum, maximum, and optimum growth temperatures for a given organism

Chemostat a device that allows for the continuous culture of microorganisms in which both growth rate and cell number can be controlled independently

Compatible solute a molecule that is accumulated in the cytoplasm for adjustment of water activity but that does not inhibit biochemical processes

Divisome a complex of proteins involved in cell division processes in prokaryotes

Exponential growth growth of a microorganism where the cell number doubles within a fixed time period

Extreme halophile a microorganism that requires very large amounts of salt (NaCl), usually greater than 10% and in some cases near to saturation, for growth

Extremophile an organism that grows optimally under one or more chemical or physical extremes, such as high or low temperature or pH

Facultative with respect to O_2, an organism that can grow in either its presence or absence

FtsZ a key cell division protein that forms a ring along the division plane to initiate cell elongation

Generation time the time required for a population of microbial cells to double

Growth an increase in cell number

Halophile a microorganism that requires NaCl for growth

Halotolerant an organism that does not require salt (NaCl) for growth but which can grow in the presence of salt, in some cases, substantial levels of salt

Hyperthermophile a microorganism that has a growth temperature optimum of 80°C or greater

Lag phase a period preceding the exponential growth phase when cells may be metabolizing but are not yet growing

Mesophile an organism that grows best at temperatures between 20 and 45°C

Microaerophile an aerobic organism that can grow only when oxygen tensions are reduced from that in air

pH the negative logarithm of the hydrogen ion (H^+) concentration of a solution

Psychrophile an organism with a growth temperature optimum of 15°C or lower and a maximum growth temperature below 20°C

Psychrotolerant an organism capable of growth at low temperatures but whose growth temperature optimum is above 20°C

Stationary phase the period immediately following exponential growth when the growth rate of the population falls to zero

Thermophile an organism whose growth temperature optimum lies between 45 and 80°C

Transpeptidation formation of peptide cross-links between muramic acid residues in peptidoglycan synthesis

Viable capable of reproducing

Xerophile an organism that is able to live, or that lives best, in very dry environments

■ THEORY AND PRACTICE OF MICROBIAL GROWTH

We have thus far discussed the structure of cellular macromolecules (∞ Chapter 3), cell structure/function (∞ Chapter 4), and the general principles of microbial nutrition and metabolism (∞ Chapter 5). Before we begin our study of the biosynthesis of macromolecules and the molecular genetics of microorganisms (∞ Chapter 7), we consider here some aspects of microbial growth.

In microbiology, the word **growth** is defined as *an increase in the number of cells*. Growth is an essential component of microbial function, as any given cell has a finite life span in nature and the species is maintained only as a result of continued growth of the population. In addition to understanding the basic science of microbial growth, many practical situations call for the *control* of microbial growth. Knowledge of how microbial populations can rapidly expand is quite useful in designing methods to control microbial growth; we will study these methods in Chapter 20.

6.1 Cell Growth and Binary Fission

The bacterial cell is a synthetic machine that is able to duplicate itself. The synthetic processes of bacterial cell growth involve as many as 2000 chemical reactions of a wide variety of types. Some of these reactions involve energy transformations. Other reactions involve biosynthesis of small molecules—the building blocks of macromolecules—as well as the various cofactors and coenzymes needed for enzymatic reactions. However, the main reactions of cell synthesis are *polymerization reactions*, the processes by which polymers (macromolecules) are made from monomers. Once polymers are made, the stage is set for the final events of cell growth: assembly of macromolecules and formation of cellular structures such as the cell wall, cytoplasmic membrane, flagella, ribosomes, inclusion bodies, enzyme complexes, and so on.

Binary Fission

In most prokaryotes, growth of an individual cell continues until the cell divides into two new cells, a process called **binary fission** (*binary* to express the fact that *two* cells have arisen from one cell). In a growing culture of a rod-shaped bacterium such as *Escherichia coli*, for example, cells are observed to elongate to approximately twice the length of the smallest cell and then form a partition that eventually separates the cell into two daughter cells (Figure 6.1). This partition is referred to as a *septum* and is a result of the inward growth of the cytoplasmic membrane and cell wall from opposing directions until the two daughter cells are pinched off (Figure 6.1).

During the growth cycle all cellular constituents increase in number such that each daughter cell receives a complete chromosome and sufficient copies of all other macromolecules, monomers, and inorganic ions to exist as an independent cell. Partitioning of the replicated DNA molecule between the two daughter cells depends on the DNA remaining attached to membranes during division, with septum formation leading to separation of chromosome copies, one going to each daughter cell (Figure 6.1).

The time required for a complete growth cycle in bacteria is highly variable and is dependent on a number of factors, both nutritional and genetic. Under the best nutritional conditions the bacterium *Escherichia coli* can complete the cycle in about 20 min; a few bacteria

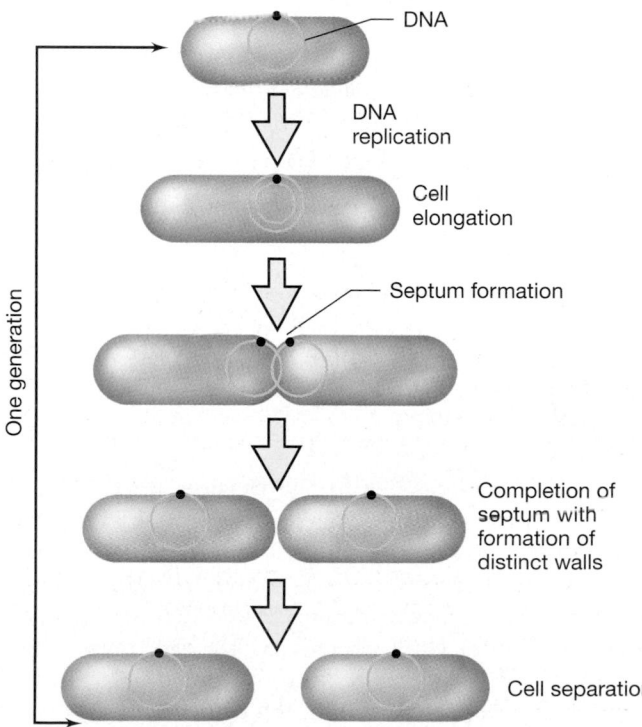

DNA

DNA replication

Cell elongation

Septum formation

Completion of septum with formation of distinct walls

Cell separation

One generation

Figure 6.1 The general process of binary fission in a rod-shaped prokaryote. For simplicity, the nucleoid is depicted as a single circle in green.

can grow even faster than this, but many grow much slower. As we will see, the control of cell division is a complex process and appears to be intimately tied to chromosomal replication events.

Fts Proteins and the Cell Division Plane

Several proteins have been identified as being important in the cell division process of prokaryotes. These proteins are essential for normal cell division and have been called **Fts proteins** (Fts stands for "*f*ilamentous *t*emperature *s*ensitive, which describes the properties of cells harboring mutations in the genes that encode Fts proteins). FtsZ, a key protein in the group, has been well studied in *Escherichia coli* and several other bacteria. Fts proteins are universally distributed among prokaryotes, including the *Archaea,* and FtsZ-type proteins have even been found in mitochondria and chloroplasts, further emphasizing their evolutionary ties to the *Bacteria* (∞ Sections 2.3 and 14.5). Interestingly, FtsZ shows structural similarities to tubulin, the important cell division protein in eukaryotes (∞ Section 14.1).

Fts proteins interact to form a division apparatus in the cell called the **divisome**. Formation of the divisome begins with the attachment of molecules of FtsZ in a ring around the cell cylinder in the center of the cell (Figure 6.2); this spot defines the cell division plane. FtsZ molecules polymerize to form an intact ring, and then the ring attracts other Fts proteins (Figure 6.2). The divisome is also thought to contain proteins involved in peptidoglycan synthesis (see Section 6.2). How actual cell division occurs from activities of the divisome is an active area of research. However, the divisome is thought to orchestrate synthesis of new membrane and wall material in both directions until a cell becomes approximately twice its original length; following this, constriction occurs to form two daughter cells as shown in Figure 6.1.

DNA replication occurs prior to the laying down of the FtsZ ring. Cessation of DNA synthesis appears to be the signal for FtsZ ring formation, and the ring itself forms in the space between the duplicated nucleoids. Location of the actual cell midpoint by FtsZ may be assisted by a series of proteins called Min, especially MinE, which interacts in some way with the duplicated nucleoids. At any rate, as cell elongation occurs, the two copies of the chromosome are pulled apart, each to its own daughter cell (Figure 6.1). As constriction occurs, the FtsZ ring begins to depolymerize, triggering the inward growth of wall materials to eventually seal off one daughter cell from the other. FtsZ has enzymatic activity and can hydrolyze guanosine triphosphate (GTP) to yield energy. It is thought that this energy fuels the polymerization and depolymerization of FtsZ and subsequent formation and destruction of the FtsZ ring (Figure 6.2).

Fts proteins are crucial cogs in the bacterial cell division process. There is great interest in understanding

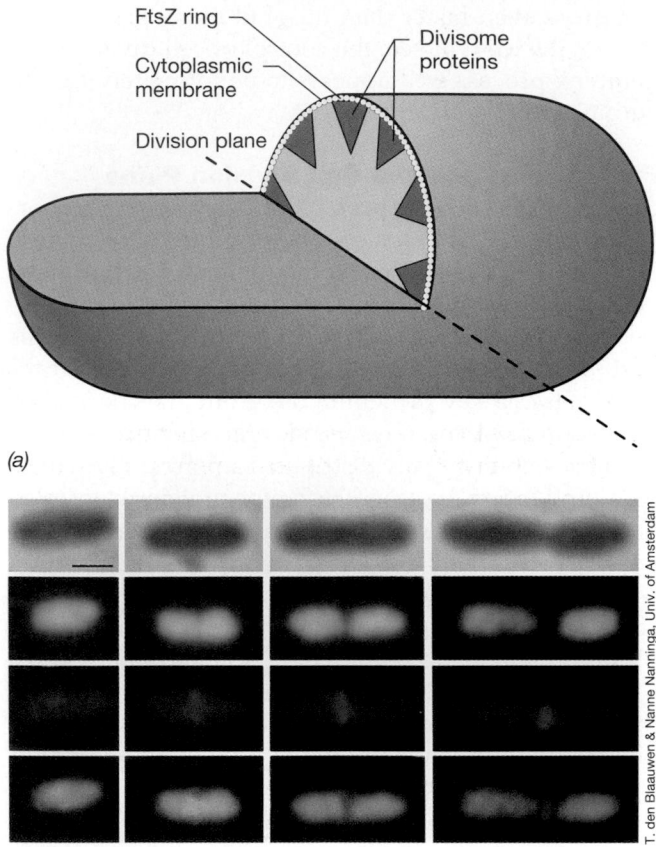

(a)

(b)

T. den Blaauwen & Nanne Nanninga, Univ. of Amsterdam

Figure 6.2 The FtsZ ring and cell division. (a) Cut-away view of a rod-shaped cell showing the ring of FtsZ molecules around the division plane. (b) Appearance and breakdown of the FtsZ ring during the cell cycle of *Escherichia coli*. Microscopy: upper row, phase contrast; second row, nucleoid stain; third row, cells stained with a specific reagent against FtsZ; fourth row, combination nucleoid and FtsZ staining. Cell division events: first column, FtsZ ring not yet formed; second column, FtsZ ring appears as nucleoids start to segregate; third column, full FtsZ ring forms as cell elongates; fourth column, breakdown of the FtsZ ring and cell division. Marker bar in upper left photo, 1 μm.

bacterial cell division at the molecular level, not only for reasons of basic research but also because such understanding could lead to the development of new drugs targeting specific steps in the division process. Like penicillin, a drug that targets bacterial cell wall synthesis (see Section 6.2), drugs that interfere with bacterial cell division could have specific effects on pathogenic bacteria without harming humans or other animals whose cells do not produce Fts proteins.

Cell Shape and Actin in Prokaryotes

Although FtsZ plays a major role in the cell *division* process, what factor(s) affect the *shape* of a prokaryotic cell? For many years it was thought that peptidoglycan was synthesized in such a way as to define or at least help define the morphology of a cell. Now it is clear that specific proteins exist in prokaryotes that de-

fine cell shape. Interestingly, these proteins show significant homology to the eukaryotic cell protein *actin*, a key component of the cytoskeleton of eukaryotic cells (∞ Section 14.4).

The key shape-determining protein in prokaryotes is called MreB. This protein forms an actin-like cytoskeleton in species of *Bacteria* and probably also in *Archaea*. MreB forms into filamentous spiral-shaped bands around the inside of the cell, just underneath the cytoplasmic membrane. Presumably, the MreB cytoskeleton defines cell shape by in some way generating a force against the cytoplasmic membrane.

Interestingly, coccus-shaped bacteria appear to lack MreB and the gene that encodes it. This suggests that the "default" shape for a bacterium is a sphere, and that variations in the arrangement of MreB filaments in cells of nonspherical species of bacteria yield the rod and other cell morphologies typical of various prokaryotes (∞ Figure 4.11). Thus, between FtsZ and MreB, prokaryotic cells possess proteins structurally similar to tubulin and actin, respectively, the key eukaryotic proteins involved in cell division and internal cell scaffolding. It is of interest that biological solutions to these key processes in eukaryotic cells appear to have their evolutionary roots in prokaryotic cells.

✓ 6.1 Concept Check

Microbial growth involves an increase in the *number* of cells rather than in the size of individual cells. Growth of most microorganisms occurs by binary fission. Cell division and chromosome replication are usually coordinately regulated, and the Fts proteins are the keys to these processes.

✓ Why is it essential that the bacterial chromosome be replicated before binary fission occurs?

✓ What is the function of the protein FtsZ in cell division in prokaryotes?

✓ What is the role of the MreB protein in prokaryotes?

6.2 Peptidoglycan Synthesis and Cell Division

When a cell enlarges before cell division, new cell wall synthesis must take place. This new wall material must be added to the preexisting wall without loss of structural integrity. This process occurs as shown in Figure 6.3. Beginning at the FtsZ ring (Figure 6.2*a* and Figure 6.3), small openings in the wall are created by enzymes called *autolysins*, similar in function to lysozyme (∞ Section 4.8), which are present within the divisome. New wall material is then added across the openings (Figure 6.3*a*). The junction between new and old peptidoglycan forms a ridge on the cell surface of gram-positive *Bacteria* (Figure 6.3*b*), analogous to a scar. It is essential that new pep-

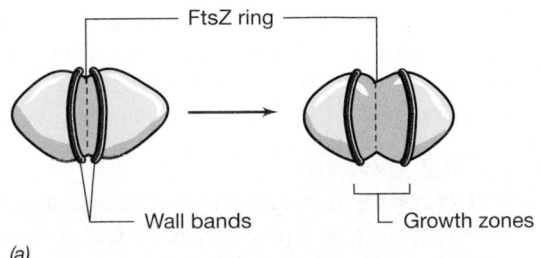

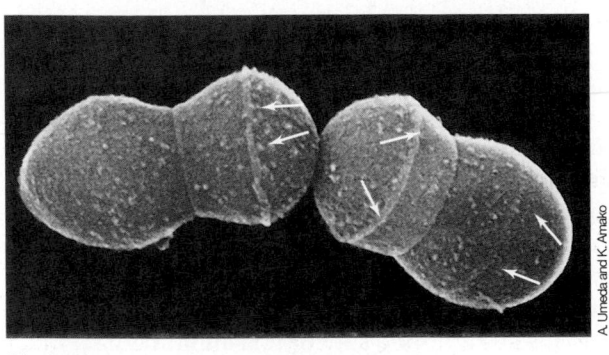

Figure 6.3 Cell wall synthesis in gram-positive *Bacteria*. (a) Localization of new cell wall synthesis during cell division. In cocci, new cell wall synthesis (shown in green) is localized at only one point. The FtsZ ring (see Figure 6.2) defines the cell division plane. (b) Scanning electron micrograph of cells of *Streptococcus hemolyticus* showing wall bands (arrows). A single cell is about 1 μm in diameter.

tidoglycan be spliced onto preexisting peptidoglycan *before* severing bonds within the latter to ensure that cell turgor pressure does not burst the cell wall at a splice point. If this does not take place, a process of spontaneous cell lysis called **autolysis** can occur.

Biosynthesis of Peptidoglycan

We discussed the general structure of peptidoglycan in Chapter 4 (⌦ Section 4.8). The peptidoglycan layer can be thought of as a stress-bearing fabric, much like a sheet of rubber. Synthesis of new peptidoglycan during cell growth involves controlled cutting by autolysins of preexisting peptidoglycan with the simultaneous insertion of peptidoglycan precursors. A carrier molecule, a lipid called **bactoprenol** (Figure 6.4), plays a major role in this process. Bactoprenol is a very hydrophobic C_{55} alcohol that bonds to the *N*-acetyl glucosamine/*N*-acetylmuramic acid/pentapeptide peptidoglycan precursor (Figure 6.5a). Bactoprenol transports peptidoglycan building blocks across the membrane by rendering the precursors sufficiently hydrophobic to pass through the interior of the cytoplasmic membrane. Once in the periplasm, bactoprenol interacts with enzymes that insert cell wall precursors into the growing point of the cell wall and catalyze glycosidic bond formation (Figure 6.5b). ■

Figure 6.4 Bactoprenol (undecaprenolphosphate), the lipid carrier of the cell wall peptidoglycan building blocks.

Transpeptidation: The Penicillin Target

The final step in cell wall synthesis, a process known as **transpeptidation**, is formation of the peptide cross-links between muramic acid residues in adjacent glycan chains. Transpeptidation is medically noteworthy because it is the reaction inhibited by the antibiotic *penicillin*. Several

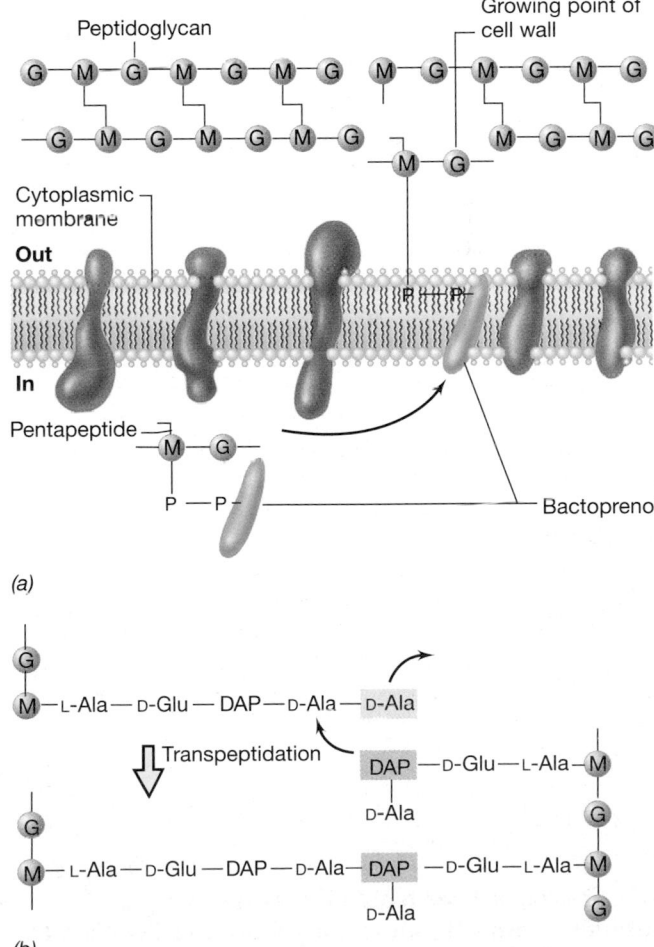

Figure 6.5 Peptidoglycan synthesis. (a) Transport of peptidoglycan precursors across the cytoplasmic membrane to the growing point of the cell wall. (b) The transpeptidation reaction that leads to the final cross-linking of two peptidoglycan chains. Penicillin inhibits this reaction.

penicillin-binding proteins have been identified in the periplasm of gram-negative *Bacteria*, at least one of which is an Fts protein (see Section 6.1); when penicillin binds to these proteins, they are no longer catalytically active. In the absence of new wall synthesis by these proteins, the continued action of autolysins eventually weakens the cell wall and leads to cell lysis.

Transpeptidation involves peptide bond formation with one of several different amino acids, depending on the cell wall structure of the organism involved. In gram-negative *Bacteria*, cross-linking usually occurs between diaminopimelic acid on one peptide and D-alanine on the adjacent peptide (Figure 6.5b). Initially, there are *two* D-alanine residues at the end of the peptidoglycan precursor, but one D-alanine group is removed during the transpeptidation reaction (Figure 6.5b) supplying the energy necessary to drive the reaction forward (recall that transpeptidation occurs outside the cell membrane where ATP is unavailable). In gram-positive *Bacteria*, where a glycine interbridge is commonly present (∞ Section 4.8 and Figure 4.31), cross-links occur across the interbridge, usually from an L-lysine of one peptide to a D-alanine on the other.

✓ 6.2 Concept Check

New cell wall is synthesized during bacterial growth by inserting new glycan units into preexisting wall material. A long-chain alcohol called bactoprenol facilitates transport of new glycan units through the cytoplasmic membrane to become part of the growing cell wall. Transpeptidation seals the precursors into the peptidoglycan fabric.

- ✓ What are *autolysins* and why are they necessary?
- ✓ What is the function of bactoprenol?
- ✓ What is *transpeptidation* and why is it important?

6.3 Population Growth

As we have mentioned, *growth* is defined as an increase in the *number* of microbial cells in a population, which can also be measured as an increase in microbial *mass*. **Growth rate** is the change in cell number or cell mass *per unit time*. During this cell division cycle, all the structural components of the cell double. The interval for the formation of two cells from one is called a **generation**, and the time required for this to occur is called the **generation time** (see Figure 6.1). The generation time is thus the time required for the cell population to double. Because of this, the generation time is also called the *doubling time*. Note that during a single generation, both the cell number and cell mass double. Generation times vary widely among organisms. Many bacteria have generation times of 1–3 h, but a few very rapidly growing organisms are known that divide in

as little as 10 min or as long as several days. Also, the generation time of any given organism is dependent to some extent on the growth medium used and the incubation condition employed.

Exponential Growth

A growth experiment beginning with a single cell having a doubling time of 30 min is presented in Figure 6.6. This pattern of population increase, where the number of cells *doubles* during each unit time period, is referred to as **exponential growth**. When the cell number from such an experiment is graphed on arithmetic coordinates as a function of elapsed time, one obtains a curve with a constantly increasing slope (Figure 6.6b). However, deriving growth rate information from such curves is difficult. The number of cells on a logarithmic ($\log_{10}$) scale is presented in Figure 6.6b in a graph in which cell number is plotted logarithmically and time is plotted arithmetically (a *semilogarithmic* graph), resulting in a straight line. This straight-line function is an immediate indicator that the cells are growing exponentially. Semilogarithmic graphs are also convenient and simple to use for estimating generation times from a set of results. The doubling time may be read directly from the graph (Figure 6.7; see also Figure 6.12b).

Time (h)	Total number of cells	Time (h)	Total number of cells
0	1	4	256
0.5	2	4.5	512
1	4	5	1,024
1.5	8	5.5	2,048
2	16	6	4,096
2.5	32	. .	.
3	64	.	.
3.5	128	10	1,048,576

(a)

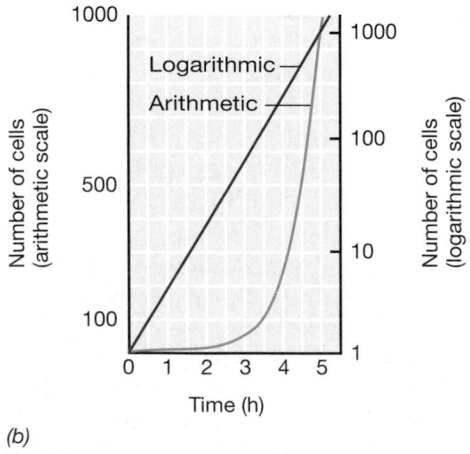

(b)

Figure 6.6 The rate of growth of a microbial culture. (a) Data for a population that doubles every 30 min. (b) Data plotted on an arithmetic (left ordinate) and a logarithmic (right ordinate) scale.

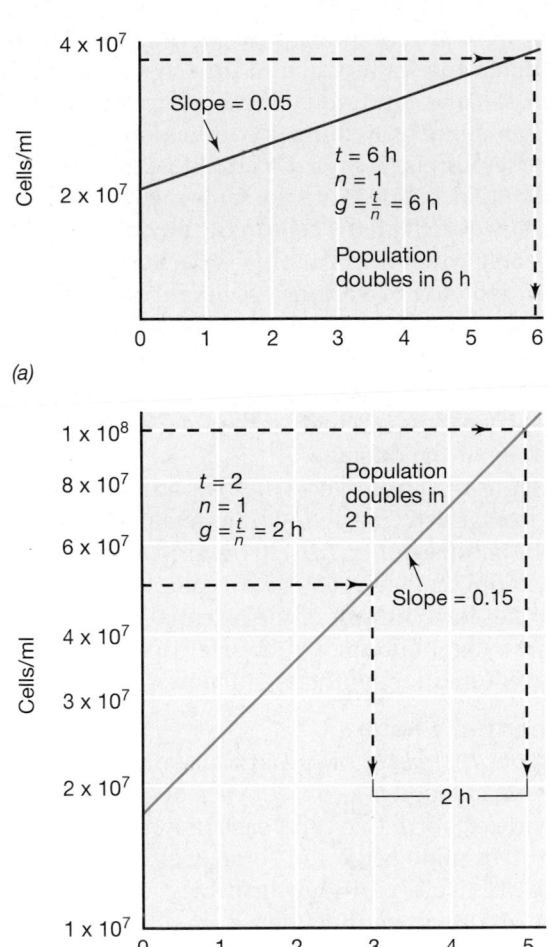

(a)

(b)

Figure 6.7 Method of estimating the generation times (*g*) of exponentially growing populations with generation times of (a) 6 h and (b) 2 h from data plotted on semilogarithmic graphs. The slope of each line is equal to 0.301/*g* and *n* equals the number of generations that have occurred in the time, *t*. All numbers are expressed in scientific notation; that is, 10,000,000 is 1×10^7, 60,000,000 is 6×10^7, and so on.

One of the characteristics of exponential growth is that the *rate* of increase in cell number is slow initially but increases at an ever faster rate. This results, in the later stages, in an explosive increase in cell numbers. For example, in the experiment in Figure 6.6, the *rate* of cell production in the first 30 min of growth is one cell per 30 min. However, between 4 and 4.5 h of growth, the rate of cell production is considerably faster at 256 cells per 30 min (Figure 6.6). A practical implication of exponential growth is that when a nonsterile product such as milk is allowed to stand under conditions such that microbial growth can occur, a few hours during the early stages of exponential growth are not detrimental, whereas standing *for the same length of time* during the later stages is disastrous.

Growth Parameters

The increase in cell number that occurs in an exponentially growing bacterial culture is a geometric progression of the number 2. As two cells double (to become four cells), we can express this as $2^1 \rightarrow 2^2$. As four cells become eight, we express this as $2^2 \rightarrow 2^3$, and so on (Figure 6.6). Because of this geometric progression, there is a direct relationship between the number of cells present in a culture initially and the number present after a period of exponential growth:

$$N = N_0 2^n$$

where N = final cell number, N_0 = initial cell number, and n = *number of generations* that have occurred during the period of exponential growth. The generation time g of the cell population is calculated as t/n, where t is simply the hours or minutes of exponential growth. Thus, from a knowledge of the initial and final cell numbers in an exponentially growing cell population, it is possible to calculate n, and from n and knowledge of t, the generation time g.

To express the equation $N = N_0 2^n$ in terms of n, the following transformations are necessary:

$$N = N_0 2^n$$
$$\log N = \log N_0 + n \log 2$$
$$\log N - \log N_0 = n \log 2$$
$$n = \frac{\log N - \log N_0}{\log 2} = \frac{\log N - \log N_0}{0.301}$$
$$= 3.3 (\log N - \log N_0)$$

With n now expressed in terms of readily measurable quantities, N and N_0, generation times can be calculated. As an example of how to perform a calculation, we use actual data from the lower graph in Figure 6.7. The generation time of 2 h, which in this case was determined directly from the graph, can also be derived from the facts that $N = 10^8$, $N_0 = 5 \times 10^7$, and $t = 2$. Thus,

$$n = 3.3 [\log 10^8 - \log (5 \times 10^7)] = 3.3(8 - 7.69)$$
$$= 3.3(0.301) = 1$$

Thus, the generation time $t/n = 2/1 = 2$ h. The generation time g can also be calculated from the slope of the line obtained in the semilogarithmic plot of exponential growth, as slope = 0.301/g.

Another index of growth rate is the *growth rate constant*, abbreviated k. The growth rate constant can be calculated as $k = \frac{\ln 2}{g} = \frac{0.693}{g}$ and has units of reciprocal hours (h^{-1}). While the term g is a measure of the *time* it takes for a population to double in cell number, k is a measure of the *number of generations* that occur per unit time in an exponentially growing culture.

Armed with knowledge of n and t, one can calculate g and k for different microorganisms growing under different culture conditions. This is often useful for optimizing culture conditions for a particular organism and also for testing the positive or negative effect of some treatment on the bacterial culture.

✓ 6.3 Concept Check

Microbial populations show a characteristic type of growth pattern called exponential growth, which is best seen by plotting the number of cells over time on a semilogarithmic graph. From knowledge of the initial and final cell numbers and the time of exponential growth, the generation time of the cell population can be calculated directly.

- ✓ Distinguish between the terms *growth rate* and *generation time*.
- ✓ Why does exponential growth lead to large cell populations in so short a period of time?
- ✓ What is a *semilogarithmic* plot?

6.4 The Growth Cycle

The data presented in Figure 6.6 reflect only part of the growth cycle of a microbial population, the part called *exponential growth*. In an enclosed vessel, referred to as a *batch culture*, a typical *growth curve* for a population of cells is obtained as illustrated in Figure 6.8. This growth curve can be divided into several distinct phases called the **lag phase**, **exponential phase**, **stationary phase**, and **death phase**.

Lag Phase

When a microbial population is inoculated into a fresh medium, growth usually does not begin immediately but only after a period of time called the *lag phase*, which may be brief or extended depending on the history of

the culture and growth conditions. If an exponentially growing culture is inoculated into the same medium under the same conditions of growth, a lag is not seen and exponential growth begins immediately. However, if the inoculum is taken from an old (stationary phase) culture and inoculated into the same medium, a lag usually occurs even if all the cells in the inoculum are *viable*, that is, able to reproduce. This is because the cells are usually depleted of various essential constituents and time is required for their resynthesis. A lag also ensues when the inoculum consists of cells that have been damaged (but not killed) by treatment with heat, radiation, or toxic chemicals because of the time required for the cells to repair the damage.

A lag is also observed when a population is transferred from a rich culture medium to a poorer one. This happens because for growth to occur in a particular culture medium the cells must have a complete complement of enzymes for synthesis of the essential metabolites not present in that medium. On transfer to a new medium, time is required for synthesis of the new enzymes.

Exponential Phase

The *exponential phase* of growth has already been discussed. As noted, it is a consequence of the fact that each cell divides to form two cells, each of which also divides to form two more cells, and so on. Cells in exponential growth are usually in their healthiest state, and thus, cells in "mid-exponential" phase are often desirable for studies of enzymes or other cell components.

Most unicellular microorganisms grow exponentially, but rates of exponential growth vary greatly. The rate of exponential growth is influenced by environmental conditions (temperature, composition of the culture medium), as well as by genetic characteristics of the organism itself. In

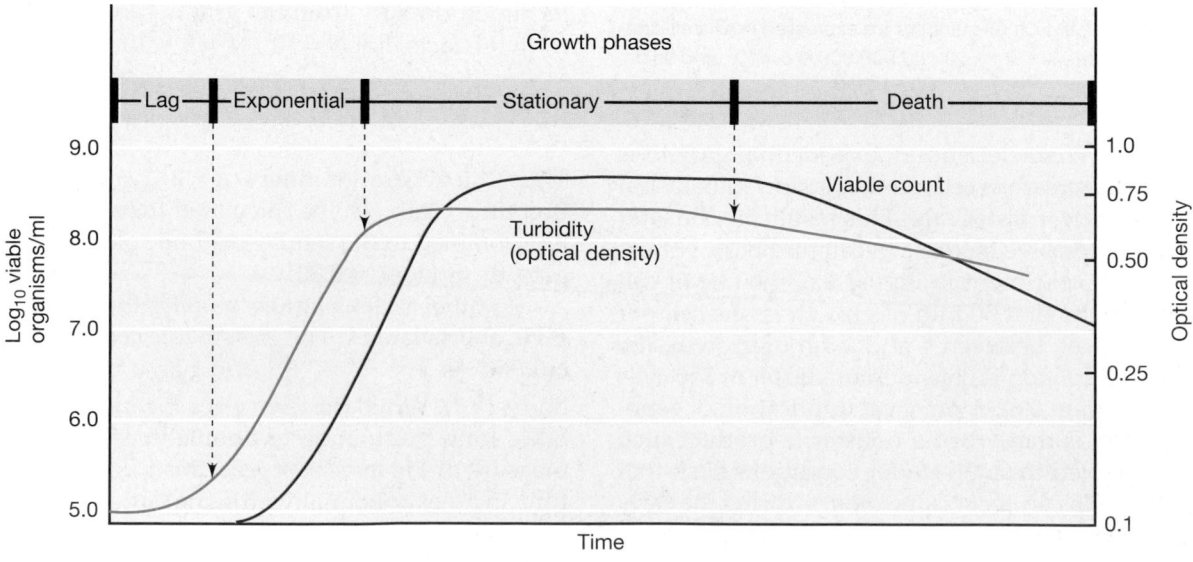

Figure 6.8 Typical growth curve for a bacterial population. See Sections 6.5 and 6.6 for a description of the counting methods employed.

general, prokaryotes grow faster than eukaryotic microorganisms, and small eukaryotes grow faster than large ones.

Stationary Phase

In a batch culture, such as in a tube or a flask, exponential growth cannot occur indefinitely. One can calculate that a single bacterium with a generation time of 20 min would, if it continued to grow exponentially for 48 h, produce a population that weighed about 4000 times the weight of Earth! This is particularly impressive because a single bacterial cell weighs only about one-trillionth (10^{-12}) of a gram (∞ Table 3.2). Obviously, something must happen to limit growth of the population long before this time. What generally happens is that either (1) an essential nutrient of the culture medium is used up or (2) some waste product of the organism builds up in the medium to an inhibitory level and exponential growth ceases, or both. At this point the population has reached the **stationary phase**.

In the stationary phase, there is no net increase or decrease in cell number. Although growth usually does not occur in the stationary phase, many cell functions may continue, including energy metabolism and some biosynthetic processes. In some organisms, slow growth may occur during the stationary phase; some cells in the population grow, whereas others die, the two processes balancing out so that no net increase or decrease in cell number occurs (this is a phenomenon called *cryptic growth*).

Death Phase

If incubation continues after a population reaches the stationary phase, the cells may remain alive and continue to metabolize, but they may also die. If the latter occurs, the population is said to be in the *death phase*. In some cases death is accompanied by actual cell **lysis**. Figure 6.8 indicates that the death phase of the growth cycle is also exponential; in most cases, however, the rate of cell death is much slower than that of exponential growth.

In closing, we reiterate that the phases of the bacterial growth curve shown in Figure 6.8 are reflections of the events in a *population* of cells, not in individual cells.

The terms *lag phase, exponential phase, stationary phase,* and *death phase* do not apply to individual cells but only to *populations* of cells.

Now we turn to methods for determining cell numbers in microbial cultures.

✓ 6.4 Concept Check

Microorganisms typically show a characteristic growth pattern when inoculated into a fresh culture medium. There is usually a lag phase, and then growth commences in an exponential fashion. As essential nutrients are depleted, or toxic products build up, growth ceases and the population enters the stationary phase. If incubation continues, cells may begin to die.

✓ In what phase of the growth curve are cells dividing in a regular and orderly process?

✓ When does a lag phase usually *not* occur?

✓ Why do cells enter the stationary phase?

6.5 Direct Measurements of Microbial Growth: Total and Viable Counts

Population growth is measured by following changes in the number of cells or in the weight of some component of cell mass (for example, protein) or total dry weight of cells themselves. There are several methods for counting cell numbers or estimating cell mass that are suited to different organisms or different problems.

Total Cell Count

The number of cells in a population can be measured by counting a sample under the microscope, a method called the **direct microscopic count**. Two kinds of direct microscopic counts are done, either on samples dried on slides or on samples in liquid. With liquid samples, special *counting chambers* must be used. In such a counting chamber, a grid is marked on the surface of the glass slide, with squares of known small area (Figure 6.9).

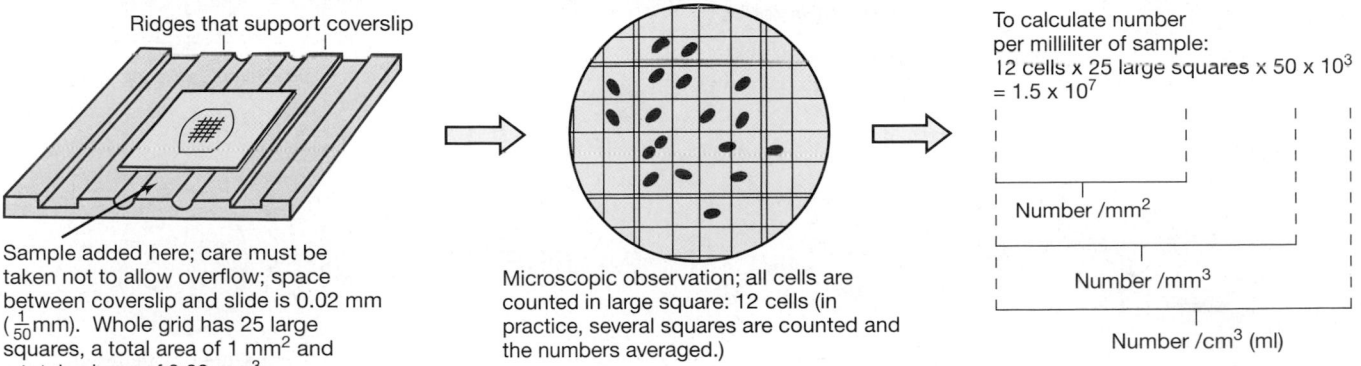

Ridges that support coverslip

Sample added here; care must be taken not to allow overflow; space between coverslip and slide is 0.02 mm ($\frac{1}{50}$mm). Whole grid has 25 large squares, a total area of 1 mm² and a total volume of 0.02 mm³.

Microscopic observation; all cells are counted in large square: 12 cells (in practice, several squares are counted and the numbers averaged.)

To calculate number per milliliter of sample: 12 cells x 25 large squares x 50 x 10³ = 1.5 x 10⁷

Number /mm²

Number /mm³

Number /cm³ (ml)

Figure 6.9 Direct microscopic counting procedure using the Petroff-Hausser counting chamber.

Over each square on the grid is a volume of known size, very small but precisely measured. The number of cells per unit area of grid can be counted under the microscope, giving a measure of the number of cells per small chamber volume. Converting this value to the number of cells per milliliter of suspension is easily done by multiplying by a conversion factor based on the volume of the chamber sample (Figure 6.9).

Direct microscopic counting is a quick way of estimating microbial cell number. However, it has certain limitations: (1) Dead cells are not distinguished from living cells. (2) Small cells are difficult to see under the microscope, and some cells are probably missed. (3) Precision is difficult to achieve. (4) A phase contrast microscope is required when the sample is not stained. (5) The method is not usually suitable for cell suspensions of low density. With bacteria, if a cell suspension has less than 10^6 cells/milliliter (ml), few if any bacteria will be seen in the microscope field. However, dilute suspensions may be counted if a sample is first concentrated and then resuspended in a small volume. (6) Motile cells must be immobilized before counting.

Viable Count

In direct microscopic counting, both living and dead cells are counted. In many cases we are interested in counting only live cells, and for this purpose **viable** cell counting methods have been developed. A viable cell is defined as one that is able to divide and form offspring, and the usual way to perform a viable count is to determine the number of cells in the sample capable of forming *colonies* on a suitable agar medium. For this reason, the viable count is often called the **plate count**, or **colony count**. The assumption made in this type of counting procedure is that *each viable cell can yield one colony.*

There are two ways of performing a plate count: the spread plate method and the pour plate method (Figure 6.10). With the **spread plate method**, a volume of an appropriately diluted culture usually no greater than 0.1 ml is spread over the surface of an agar plate using a sterile glass spreader. The plate is then incubated until the colonies appear, and the number of colonies is counted. It is important that the surface of the plate be dry so that the liquid that is spread soaks in. Volumes greater than 0.1 ml are rarely used because the excess liquid does not soak in and may cause the colonies to coalesce as they form, making them difficult to count. In the **pour plate method** (Figure 6.10), a known volume (usually 0.1–1.0 ml) of culture is pipetted into a sterile Petri plate; melted agar medium is then added and mixed well by gently swirling the plate on the table top. Because the sample is mixed with the molten agar medium, a larger volume can be used than with the spread plate; however, with the pour plate method the organism to be counted must be able to briefly withstand the temperature of melted agar, 45°C.

Diluting Cell Suspensions Before Plating

With both the spread plate and pour plate methods, it is important that the number of colonies developing on the plates not be too large because on crowded plates some cells may not form colonies, and some colonies may fuse, leading to erroneous measurements. It is also essential that the number of colonies not be too small, or the statistical significance of the calculated count will be low. The usual practice, which is the most valid statistically, is to count colonies only on plates that have between 30 and 300 colonies.

To obtain the appropriate colony number, the sample to be counted must almost always be *diluted*. Since one rarely knows the approximate viable count ahead of

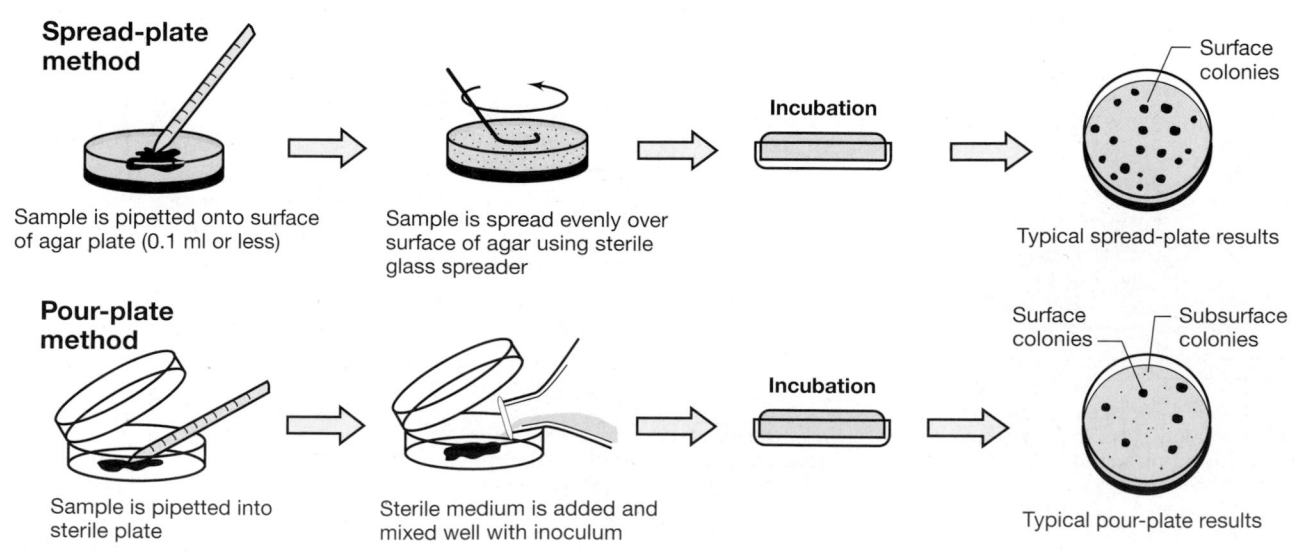

Spread-plate method

Sample is pipetted onto surface of agar plate (0.1 ml or less)

Sample is spread evenly over surface of agar using sterile glass spreader

Incubation

Surface colonies

Typical spread-plate results

Pour-plate method

Sample is pipetted into sterile plate

Sterile medium is added and mixed well with inoculum

Incubation

Surface colonies — Subsurface colonies

Typical pour-plate results

Figure 6.10 Two methods of performing a viable count (plate count). In either case the sample must usually be diluted before plating.

time, it is usually necessary to make more than one dilution. Several 10-fold dilutions of the sample are commonly used (Figure 6.11). To make a 10-fold (10^{-1}) dilution, one can mix 0.5 ml of sample with 4.5 ml of diluent, or 1.0 ml sample with 9.0 ml diluent. If a 100-fold (10^{-2}) dilution is needed, 0.05 ml can be mixed with 4.95 ml diluent, or 0.1 ml with 9.9 ml diluent. Alternatively, a 10^{-2} dilution can be made by making two successive 10-fold dilutions. In most cases, such *serial dilutions* are needed to reach the final dilution desired. Thus, if a 10^{-6} ($1/10^6$) dilution is needed, it can be achieved by making three successive 10^{-2} ($1/10^2$) dilutions or six successive 10^{-1} dilutions (Figure 6.11).

Sources of Error in Plate Counting

The number of colonies obtained in a viable count depends not only on the inoculum size but also on the suitability of the culture medium and the incubation

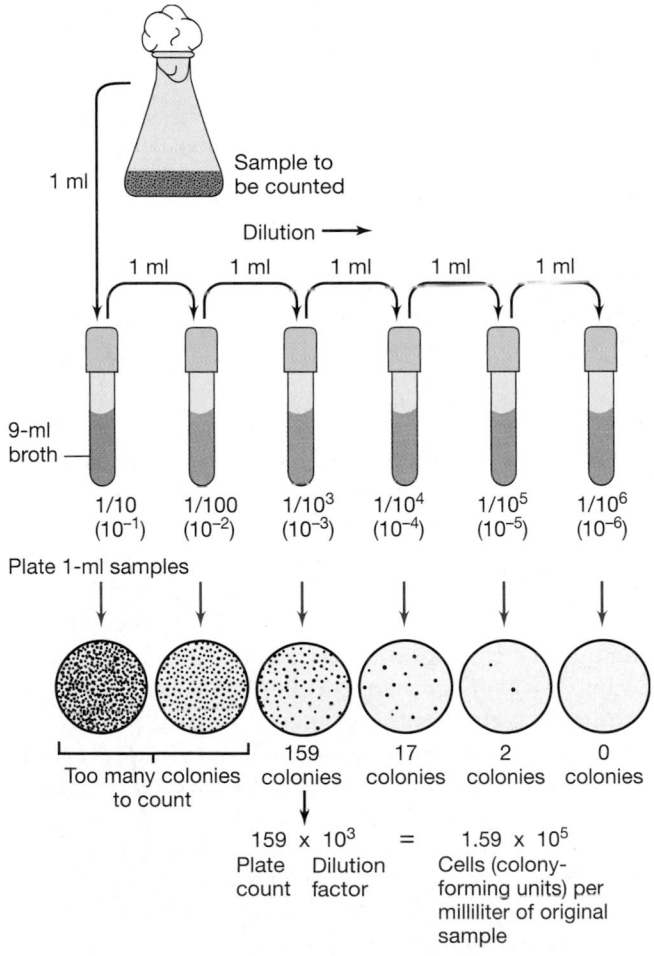

Figure 6.11 Procedure for viable counting using serial dilutions of the sample and the pour plate method. The sterile liquid used for making dilutions can simply be water, but a balanced salt solution or growth medium may yield a higher recovery. The dilution factor is the reciprocal of the dilution. For spread plating (Figure 6.10), further dilutions may be necessary in order to spread samples of 0.1 ml.

conditions; it also depends on the length of incubation. The cells deposited on the plate will not all develop into colonies at the same rate, and if a short incubation time is used, less than the maximum number of colonies will be obtained. Furthermore, the size of colonies often varies. If some tiny colonies develop, they may be missed during the counting. It is usual to determine the incubation conditions (medium, temperature, time) that will give the maximum number of colonies of a given organism and then use these conditions throughout. Viable counts can be subject to large error, and if accurate counts are desired, great care must be taken and replicate plates of key dilutions must be prepared. Note that two or more cells in a clump form only a single colony, and so a viable count may be erroneously low. To more clearly state the result, viable counts are often expressed as the number of *colony-forming units* obtained rather than as the number of *viable cells* (since a colony-forming unit may contain one or more cells).

Despite the difficulties associated with viable counting, the procedure gives the best information on the number of viable cells and so is widely used. In food, dairy, medical, and aquatic microbiology, viable counts are employed routinely. The method has the virtue of high sensitivity: Samples containing very few cells can be counted, thus permitting sensitive detection of microbial contamination of products or materials. Moreover, the use of highly selective culture media and growth conditions (∞ Section 24.2) in viable counting procedures allows for the counting of only particular cell types in a mixed population of microorganisms.

The Great Plate Count Anomaly

Although a very sensitive technique, plate counts can be highly unreliable when used to assess total cell numbers of natural samples, such as soil and water. Indeed, direct microscopic counts of natural samples typically reveal far more organisms than are recoverable on plates of *any* given culture medium (∞ Sections 18.3 and 18.4). Some microbiologists have referred to this as "the great plate count anomaly." Why do plate counts reveal lower numbers of cells than direct microscopic counts? This is likely due to a combination of factors, including the realization that microscopic methods count dead cells, whereas viable methods do not, and to the fact that different organisms in even a very small sample may have vastly different requirements for resources and culture conditions (∞ Chapters 5, 17, 18, and 19). Thus, although targeted plate counts using highly selective media, as in, for example, the microbial analysis of sewage or food, can yield rather reliable data (∞ Sections 28.1 and 29.1), "total cell counts" of the same habitats may be, and usually are, underestimates by several orders of magnitude.

✓ 6.5 Concept Check

Growth is measured by the change in number of cells with time. Cell counts done microscopically measure the total number of cells in a population, whereas viable cell counts (plate counts) measure only the living population.

✓ Why is a *viable count* more sensitive than a *microscopic count*?

✓ What is the major assumption made in relating plate count results to cell number?

✓ Describe how you would dilute a bacterial culture by 10^{-7}.

✓ What is the "great plate count anomaly"?.

6.6 Indirect Measurements of Microbial Growth: Turbidity

A rapid and quite useful method of obtaining an estimate of cell number is by use of *turbidity measurements.* A cell suspension looks cloudy (turbid) to the eye because cells scatter light passing through the suspension. The more cells present, the more light scattered and hence the more turbid the suspension. Turbidity can be measured with a *photometer* or a *spectrophotometer,* devices that pass light through a cell suspension and detect the amount of unscattered light that emerges (Figure 6.12). The major difference between these two instruments is that a photometer employs a simple filter to generate incident light of relatively narrow wavelength, whereas a spectrophotometer employs a prism or diffraction grating to generate incident light in a very narrow band of wavelengths to impinge on the sample (Figure 6.12*a*). Commonly used wavelengths for bacterial turbidity measurements include 540 nm (green), 600 nm (orange), or 660 nm (red). However, both spectrophotometers and photometers measure only *unscattered* light, and readings are recorded in photometer units (for example, "Klett units" for the Klett-Summerson photometer) (see Figure 6.12*b*) or optical density (OD) units for a spectrophotometer.

Generating a Standard Curve

For unicellular organisms, photometer units or OD are proportional (within certain limits) to cell number; therefore, turbidity readings can be used as a substitute for direct counting methods. However, before using turbidity as an estimate of cell number, a *standard curve* must first be prepared for each organism to be studied, relating some direct measurement of cell number (microscopic or viable count) or mass (dry weight) to the *indirect* measurement obtained from turbidity (Figure 6.12*c*). Such a curve can contain data for both cell number and cell mass, allowing for an estimate of both parameters from a single turbidity reading (Figure 6.12*c*).

At high concentrations of cells, light scattered away from the photocell by one cell can be rescattered back by another (and thus appear to the photocell as if it had

never been scattered), and when this occurs, the one-to-one correspondence between cell number and turbidity loses linearity (Figure 6.12*c*). Nevertheless, within limits turbidity measurements can be reasonably accurate, and they have the virtue of being quick and easy to perform. In addition, turbidity measurements can usually be made without destroying or significantly disturbing the sample. For these reasons, turbidity measurements are widely employed to follow the growth rate of microbial cultures; the same sample can be checked repeatedly, and the measurements plotted on a semilogarithmic plot versus time (Figure 6.12*b*) and used to calculate the generation time of the growing culture.

✓ 6.6 Concept Check

Turbidity measurements are an indirect but very rapid and useful method of measuring microbial growth. However, in order to relate a direct cell count to a turbidity value, a standard curve must first be established.

✓ List two advantages of using turbidity as a measure of cell growth.

✓ Describe how you could use a turbidity measurement to tell how many colonies you would get from plating a culture of a given OD.

6.7 Continuous Culture: The Chemostat

Our discussion of population growth thus far has been confined to **batch cultures**, growth occurring in a fixed volume of a culture medium that is continually being altered by the actions of the growing organisms until it is no longer suitable for growth. In the early stages of exponential growth in batch cultures, conditions may remain relatively constant, but in later stages when cell numbers become quite large, drastic changes in the chemical composition of the culture medium usually occur. However, for many studies, it is desirable to keep cultures in constant environments for long periods, and this is done by employing *continuous cultures.* A continuous culture is a flow system of constant volume to which fresh medium is added continuously and spent culture medium removed continuously at a constant rate. Once such a system is in equilibrium, cell number and nutrient status remain *constant,* and the system is said to be in **steady state**.

The Chemostat

The most common type of continuous culture device used is a **chemostat** (Figure 6.13), which controls both the population density and the growth rate of the culture. Two factors are important in the control of a

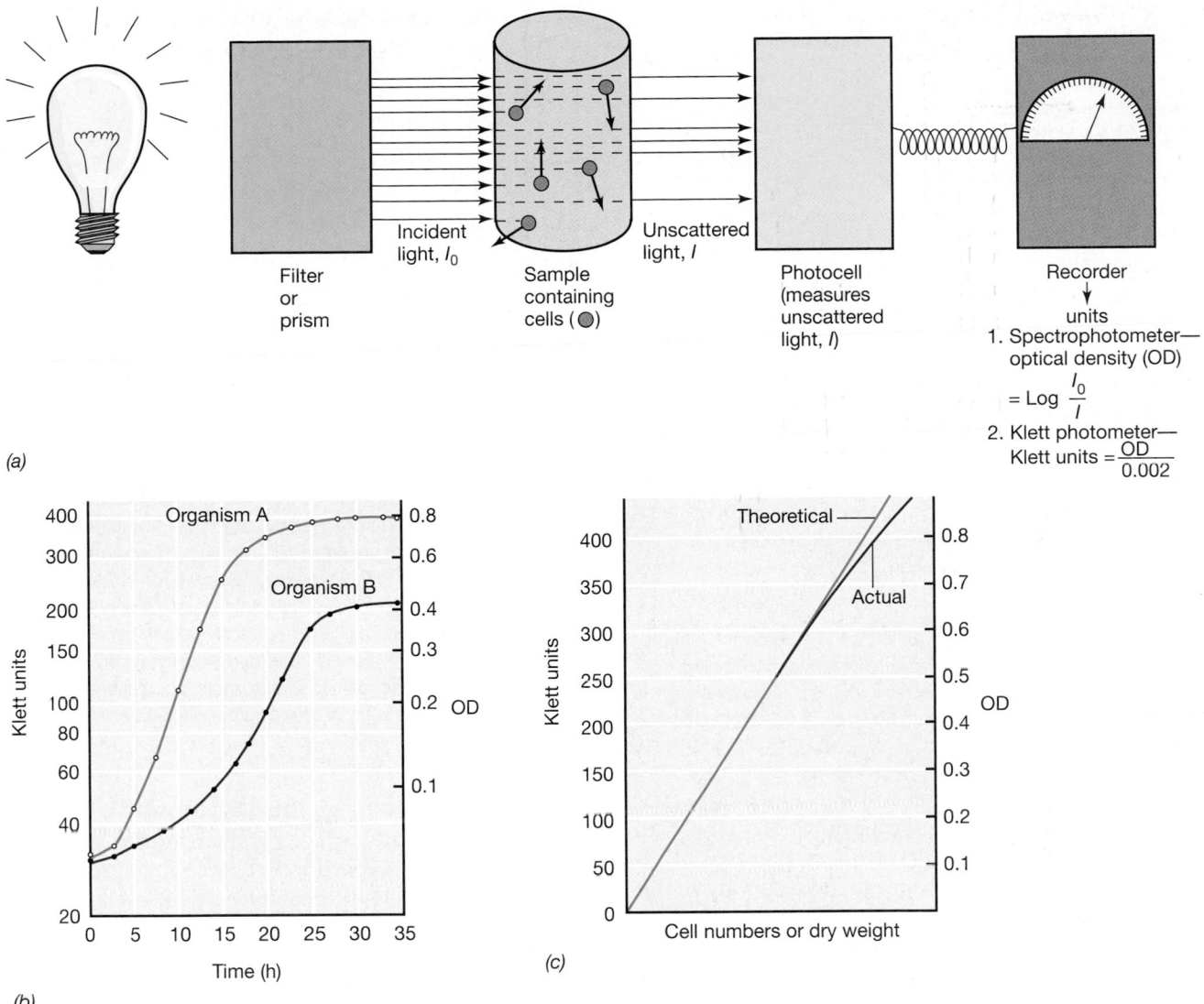

Figure 6.12 Turbidity measurements of microbial growth. (a) Measurements of turbidity are made in a spectrophotometer or photometer. The photocell measures incident light unscattered by cells in suspension and gives readings in optical density or photometer units. (b) Typical growth curve data obtained in Klett units or optical density (OD) for two organisms growing at different growth rates. For practice, calculate the generation time (*g*) of the two cultures using the formula $n = 3.3$ (log N – log N_0) where N and N_0 are two different Klett values taken between a time interval *t*. Which organism is growing faster, A or B? (c) Relationship between cell number or dry weight and turbidity readings. Note that the one-to-one correspondence between these relationships breaks down at high turbidities.

chemostat: the *dilution rate* and the *concentration of a limiting nutrient*, such as a carbon or nitrogen source. In a batch culture nutrient concentration can affect both the growth rate and the growth yield of a microorganism (Figure 6.14). At very low concentrations of a given nutrient, the growth rate is reduced, probably because the nutrient cannot be transported into the cell fast enough to satisfy metabolic demand, whereas at moderate or higher nutrient levels, growth *rates* may not be affected while cell *yield* continues to increase (Figure 6.14). In contrast to a batch culture, in a chemostat, growth rate and growth yield can be controlled *independently of each other*, the former by adjusting the dilution rate and the latter by varying the concentration of a nutrient present in a limiting amount.

Effects of varying dilution rate and concentration of the growth-limiting nutrient are given in Figure 6.15. As seen, there are rather wide limits over which the dilution rate controls growth rate, although at both very low and very high dilution rates, the steady state breaks down. At high dilution rates, the organism cannot grow fast enough to keep up with its dilution, and the culture is washed out of the chemostat. At the other extreme, at very low dilution rates, a large fraction of the cells may die from starvation because the limiting nutrient is not being added fast enough to permit maintenance of cell metabolism.

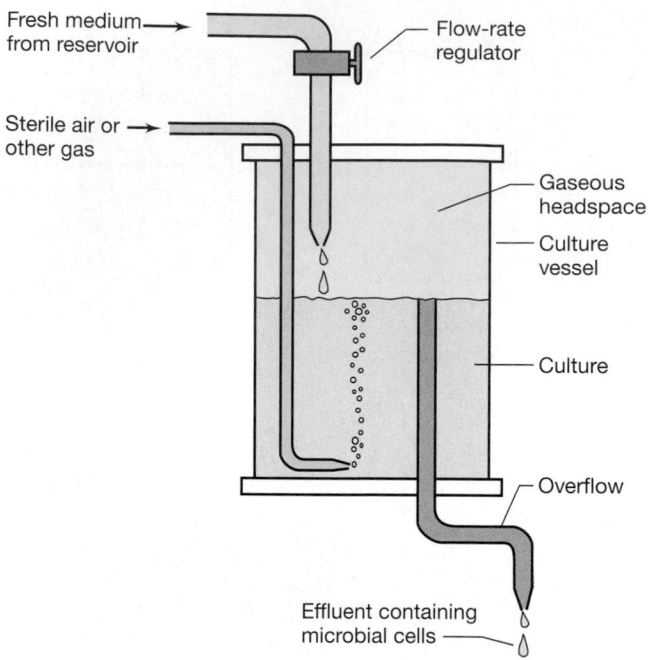

Figure 6.13 Schematic for a continuous culture device (chemostat). In such a device, the population density is controlled by the concentration of limiting nutrient in the reservoir, and the growth rate is controlled by the flow rate (see Figure 6.15). Both parameters can be set by the experimenter.

The *cell density* (cells/ml) in the chemostat is controlled by the level of the limiting nutrient, just as cell yield was controlled in a batch culture (Figure 6.14). If the concentration of this nutrient in the incoming medium is raised, with the dilution rate remaining constant, the cell density will increase. Thus, by adjusting dilution rate and nutrient level, the experimenter can obtain at will a variety of population densities growing at a variety of growth rates.

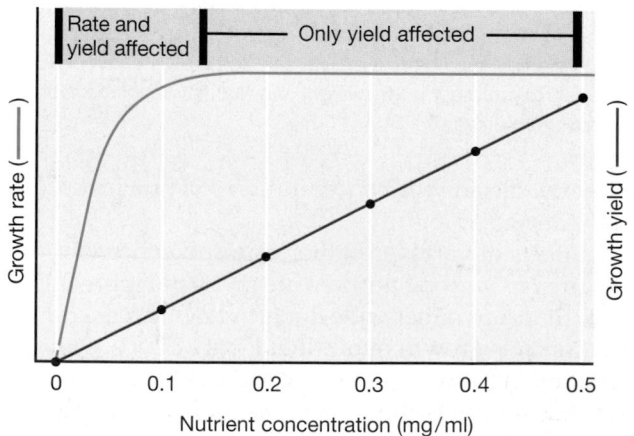

Figure 6.14 Relationship between nutrient concentration, growth rate (green curve), and growth yield (red curve) in a batch culture (closed system). At low nutrient concentrations both growth rate and growth yield are affected.

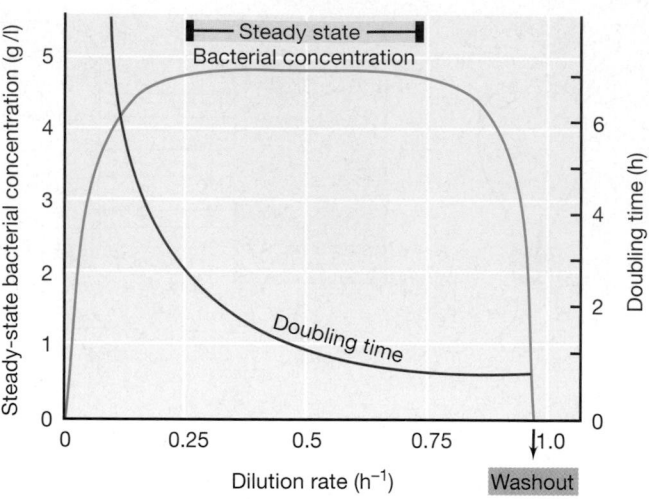

Figure 6.15 Steady-state relationships in the chemostat. The dilution rate is determined from the flow rate and the volume of the culture vessel. Thus, with a vessel of 1000 ml and a flow rate through the vessel of 500 ml/h, the dilution rate would be 0.5 h⁻¹. Note that at high dilution rates, growth cannot balance dilution, and the population washes out. Note also that although the population density remains constant during steady state, the growth rate (doubling time) can vary over a wide range. Thus, the experimenter can obtain populations with widely varying growth rates without affecting population density.

Experimental Uses of the Chemostat

One of the major advantages of a chemostat is that this device allows the experimenter to control growth rate and population density *independently* of each other. As was shown in Figure 6.15, even over rather wide ranges, any desired growth rate can be obtained in the chemostat by simply varying the dilution rate. Similarly, the population density may be set by varying the concentration of a single nutrient (the growth-limiting nutrient) in the medium reservoir. Independent control of these two crucial growth parameters is impossible with batch cultures because the batch culture is a *closed system* from the standpoint of nutrient resources and waste product removal, and thus growth conditions are constantly changing with time.

A practical advantage to the chemostat is that a population may be maintained in the exponential growth phase for long periods, for days and even weeks. Because exponential phase cells are usually most desirable for physiological experiments, the experimenter using the chemostat can have such cells available at any time. Thus, experiments can be planned in detail and then performed whenever most convenient. Moreover, repetition of experiments can be done with the knowledge that the cell population will be as close to being the same each time as possible.

The chemostat has also been very useful in microbial ecology studies. For example, because the chemostat can mimic the low substrate concentrations that often prevail in nature, it is possible to study mixed bacterial populations in a chemostat and ask questions about the competitiveness of one organism or another at particular nutrient concentrations. Using cultural methods as well as the powerful tools of molecular ecology such as phylogenetic stains and gene tracking (∞ Chapter 18), changes in the microbial community can be monitored as a function of chemostat conditions. Such experiments often reveal interactions among individual components of the population that are not obvious from growth studies in batch culture. Chemostats have also been used for enrichment and isolation of bacteria (∞ Sections 1.5 and 18.1 for discussion of enrichment and isolation of bacteria). From a mixed inoculum, one can select a stable population under the nutrient and dilution rate conditions chosen, and then slowly increase the dilution rate until a single organism remains. In this way, microbiologists recently isolated a bacterium with a 6-minute doubling time; the fastest growing bacterium known.

✓ 6.7 Concept Check

Continuous culture devices (chemostats) are a means of maintaining cell populations in exponential growth for long periods. In a chemostat, the rate at which the culture is diluted governs the growth rate, and the population size is governed by the concentration of the growth-limiting nutrient entering the vessel.

✓ How do microorganisms in a *chemostat* differ from microorganisms in a *batch culture*?

✓ What happens in a chemostat if the dilution rate exceeds the maximal growth rate of the organism?

✓ Do pure cultures have to be used in a chemostat?

II ENVIRONMENTAL EFFECTS ON MICROBIAL GROWTH

Up to now we have described growth of microorganisms under essentially ideal laboratory conditions. However, the activities of microorganisms are greatly affected by the chemical and physical conditions of their environments. Understanding environmental influences helps us to explain the distribution of microorganisms in nature and makes it possible for us to devise methods for controlling or enhancing microbial activities. Many environmental factors could be considered in this connection. However, *four* main factors have been identified that clearly play major roles in controlling microbial growth: temperature, pH, water availability, and oxygen. We consider each of these factors in detail here.

6.8 Effect of Temperature on Growth

Temperature is one of the most, if not *the* most, important environmental factors affecting growth and survival of microorganisms. At either too cold or too hot a temperature, microorganisms will not be able to grow. But what absolute values these minimum and maximum temperatures take vary greatly among different microorganisms and are usually a reflection of the temperature range and average temperature of their habitats. We examine the effect of temperature on microbial growth here.

Cardinal Temperatures

Temperature can affect living organisms in either of two opposing ways. As the temperature rises, chemical and enzymatic reactions in the cell proceed at more rapid rates, and growth becomes faster. However, *above* a certain temperature, particular proteins may be irreversibly damaged. Thus, as the temperature is increased within a given range, growth and metabolic function increase up to a point where inactivation reactions set in. Above this point, cell functions fall sharply to zero. Thus, we find that for every organism there is a *minimum* temperature below which growth no longer occurs, an *optimum* temperature at which growth is most rapid, and a *maximum* temperature above which growth is not possible (Figure 6.16). The optimum temperature is always nearer the maximum than the minimum. These three temperatures, called the **cardinal temperatures**, are generally characteristic of each type of organism but are not completely fixed, as they can be modified slightly by other factors of the environment; in particular, by the composition of the growth medium.

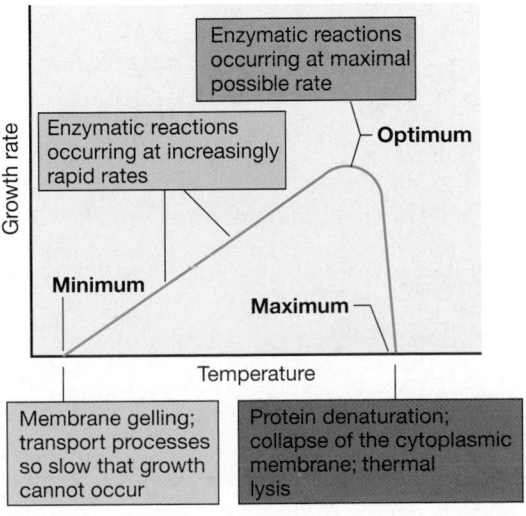

Figure 6.16 Effect of temperature on growth rate and the molecular consequences for the cell. The three cardinal temperatures vary by organism.

The maximum growth temperature of a given organism most likely reflects the inactivation of one or more key proteins in the cell. However, the factors controlling an organism's *minimum* growth temperature are not as clear. As mentioned earlier (∞ Section 4.5), the cytoplasmic membrane must be in a fluid state for proper functioning. Perhaps the minimum temperature of an organism results from "freezing" of the cytoplasmic membrane so it no longer functions properly in nutrient transport or proton gradient formation. This explanation is supported by experiments in which the minimum temperature for an organism is altered to some extent by adjustments in membrane lipid composition (see Section 6.9). It is also observed that the cardinal temperatures of different microorganisms differ widely; some organisms have temperature optima as low as 4°C and some higher than 100°C. The temperature range throughout which growth occurs is even wider than this, from below freezing to greater than boiling. (The archaeon *Pyrolobus fumarii* has a temperature maximum of 113°C!) However, no single organism can grow over this whole temperature range, and the typical range for any given organism is about 30°, although some have a broader temperature range than others.

Temperature Classes of Organisms

Although there is a continuum of organisms, from those with very low temperature optima to those with high temperature optima, it is possible to broadly distinguish *four groups* of microorganisms in relation to their temperature optima: **psychrophiles**, with low temperature optima, **mesophiles**, with midrange temperature optima, **thermophiles**, with high temperature optima, and **hyperthermophiles**, with very high temperature optima (Figure 6.17). Mesophiles are found in warm-blooded animals and in terrestrial and aquatic environments in temperate and tropical latitudes. Psychrophiles and thermophiles are found in unusually cold and unusually hot environments, respectively. Hyperthermophiles are found in extremely hot habitats such as hot springs, geysers, and deep-sea hydrothermal vents (see Sections 6.10 and 19.8).

In *Escherichia coli*, a typical mesophile, a detailed study of growth as a function of temperature has precisely defined its cardinal temperatures. The optimum temperature of *E. coli* in a complex medium is 39°C, the maximum is 48°C, and the minimum is 8°C. These values are subject to slight strain differences, and in general, the maximum and minimum temperatures supporting growth of an organism are higher and lower, respectively, when tested in complex rather than defined media.

✓ 6.8 Concept Check

Temperature is a major environmental factor controlling microbial growth. The cardinal temperatures describe the minimum, optimum, and maximum temperatures at which each organism grows. Microorganisms can be grouped by the temperature ranges they require.

✓ What are the approximate cardinal temperatures for *Escherichia coli*? To what temperature class does it belong?

✓ How does a *hyperthermophile* differ from a *psychrophile*?

✓ *Escherichia coli* can grow at a higher temperature in a complex medium than in a defined medium. Why?

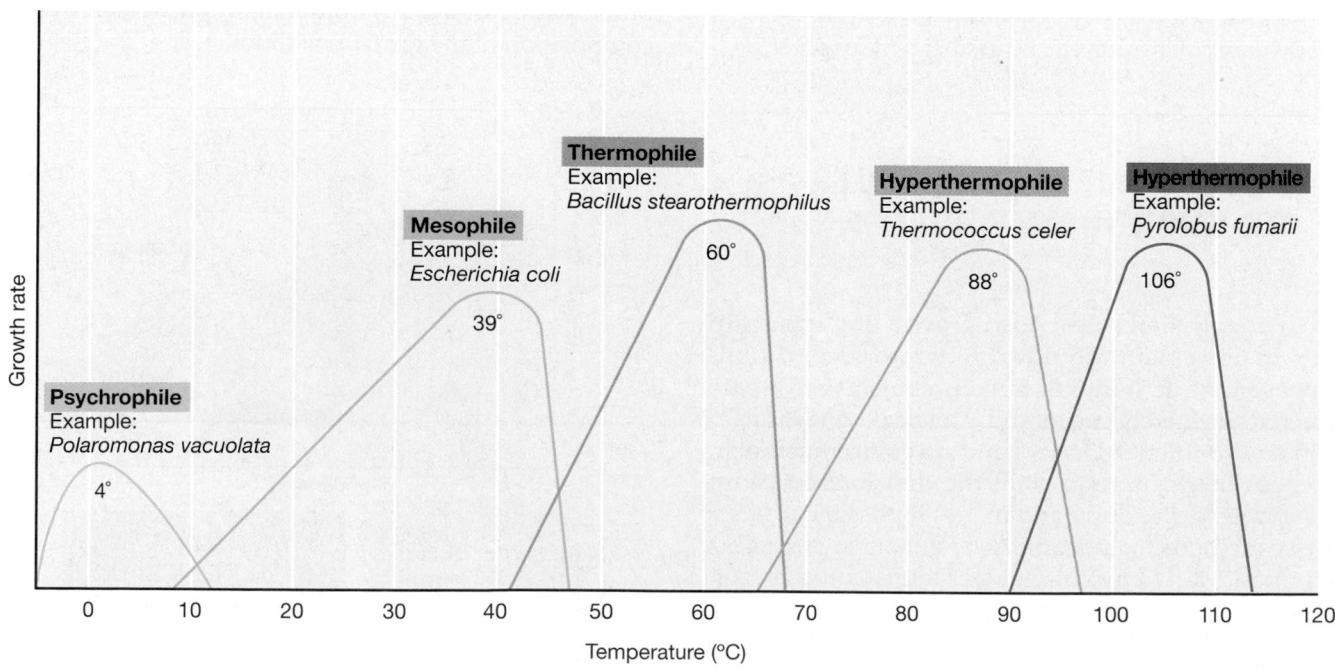

Figure 6.17 Relation of temperature to growth rates of a typical psychrophile, a typical mesophile, a typical thermophile, and two different hyperthermophiles. The temperature optima of the example organisms are shown on the graph.

6.9 Microbial Growth at Cold Temperatures

Because humans live and work on the surface of Earth where temperatures are generally moderate, it is natural to consider very hot and very cold environments as being "extreme." And they are extreme for human habitation because humans would die quickly if immersed in boiling or freezing water. However, the natural habitats of many microorganisms can be either extremely hot or extremely cold, and the organisms that live there, referred to as **extremophiles** (∽ Section 2.4 and Table 2.1), have evolved to grow *optimally* under these conditions. We consider organisms here that grow at cold temperatures.

Cold Environments

Much of Earth's surface experiences fairly low temperatures. The oceans, which make up over half of Earth's surface, have an average temperature of 5°C, and the depths of the open oceans have constant temperatures of about 1–3°C. Vast land areas of the Arctic and Antarctic are permanently frozen or are unfrozen for only a few weeks in summer (Figure 6.18a). These cold environments are not sterile, and some microorganisms can be found alive and growing at any low temperature at which liquid water still exists. Even in many frozen materials there are usually microscopic pockets of liquid water present where microorganisms can metabolize and grow.

It is important to distinguish between environments that are cold *throughout* the year and those that are cold *only* in winter. The latter, characteristic of continental temperate climates, may have summer temperatures as high as 40°C and winter temperatures of –20°C or colder. Such highly variable environments are much less favorable for cold-adapted organisms than are the constantly cold environments found in polar regions, at high altitudes, and in the depths of the oceans.

Psychrophilic and Psychrotolerant Microorganisms

As noted earlier, organisms with low temperature optima are called **psychrophiles**. A psychrophile can be defined as an organism with an optimal temperature for growth of 15°C or lower, a maximum growth temperature below 20°C, and a minimal temperature for growth at 0°C or lower. Organisms that grow at 0°C but have optima of 20–40°C are called **psychrotolerant**.

Psychrophiles are found in environments that are constantly cold, such as in polar regions or in marine sediments from polar regions, and they may be rapidly killed by even brief warming to room temperature. For this reason, their laboratory study requires that great care be taken to ensure that they never warm up during sampling, transport to the laboratory, isolation, or other manipulations.

(a)

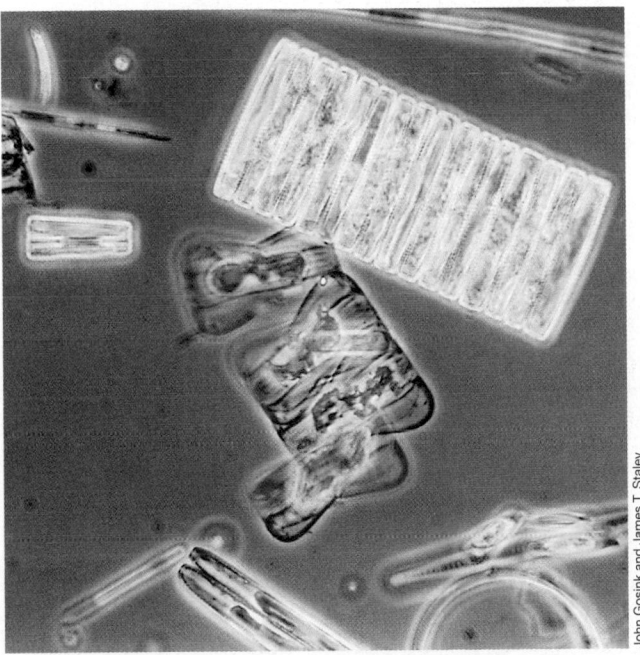

(b)

Figure 6.18 Microorganisms from Antarctic sea ice. (a) A core of permanently frozen seawater from McMurdo Sound, Antarctica. Note the dense coloration due to pigmented microorganisms (the white boot shows scale for the ice core). (b) Phase contrast micrograph of phototrophic microorganisms from the core shown in (a). Most organisms are either diatoms or green algae (both eukaryotic microorganisms).

Some of the best-studied psychrophiles have been algae that grow in dense masses within and under the ice in polar regions (Figure 6.18b). Psychrophilic algae are also often seen on the surfaces of snowfields and glaciers in such large numbers that they impart a distinctive red or green coloration to the surface (Figure 6.19a). The most common snow alga is *Chlamydomonas nivalis;* its brilliant red spores are responsible for the red color (Figure 6.19b). This green alga grows within the snow as a green-pigmented vegetative cell and then sporulates; as the snow dissipates by melting, erosion, and vaporization, the spores become concentrated on the surface. Snow algae are most commonly seen on melting permanent snowfields in midsummer to late summer and are especially common in sunny, dry areas, probably because in more rainy areas they are washed away from the snowfields. In addition to snow algae, several psychrophilic chemoorganotrophic bacteria are known, many from Antarctic sea ice (Figure 6.18a). Some of these show the lowest maximum growth temperature of any known microorganisms.

Psychrotolerant microorganisms are much more widely distributed than psychrophiles and can be isolated from soils and water in temperate climates as well as from meat, milk and other dairy products, cider, vegetables, and fruit stored under refrigeration (4°C). As noted, psychrotolerant microorganisms grow best at a temperature between 20 and 40°C. Because temperate environments warm up in summer, it is understandable that they cannot support the heat-sensitive psychrophiles, the warming essentially providing a selective force favoring psychrotolerant species and excluding psychrophilic forms. It should be emphasized that although psychrotolerant microorganisms do grow at 0°C, they do not grow very well, and one must often wait several weeks before visible growth is seen in culture media. Various genera of *Bacteria*, fungi, algae, and protozoa have members that are psychrotolerant.

Molecular Adaptations to Psychrophily

Psychrophiles produce enzymes that function optimally in the cold and that are often denatured or otherwise inactivated at even very moderate temperatures. The molecular basis for this is not entirely understood, but it has been observed that, on average, cold-active enzymes have greater amounts of α-helix and lesser amounts of β-sheet secondary structure (⚭ Section 3.7 and Figure 3.16) than enzymes that are inactive in the cold. Because the β-sheet tends to form a more rigid structure, the greater α-helix content of cold-active enzymes allows these proteins greater flexibility in the cold. Cold-active enzymes also tend to have greater polar and lesser hydrophobic amino acid contents than their counterparts from mesophiles and thermophiles, and this may also assist in keeping the protein flexible (and thus enzymatically active) at cold temperatures.

(a)

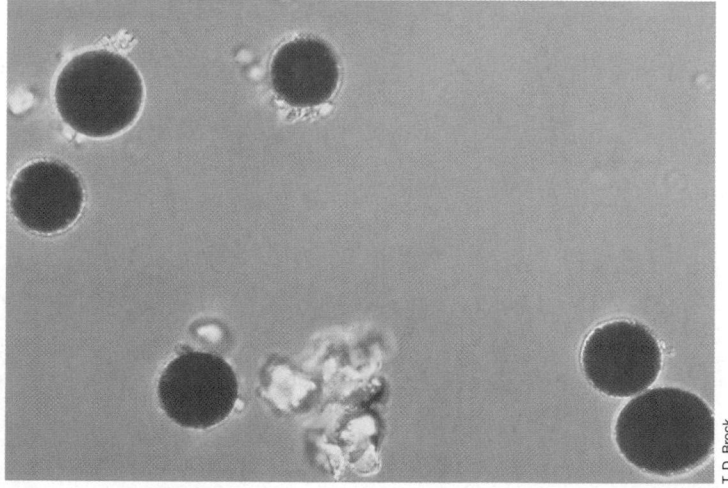

(b)

Figure 6.19 Snow algae. (a) Snow bank in the Sierra Nevada, California, with red coloration caused by the presence of snow algae. Pink snow such as this is common on summer snow banks at high altitudes throughout the world. (b) Photomicrograph of red-pigmented spores of the snow alga *Chlamydomonas nivalis*. The spores germinate to yield motile green algal cells. Related species of snow algae contain different carotenoid pigments (⚭ Section 17.3), and thus, fields of snow algae can also be green, orange, brown, or purple in color.

Another feature of psychrophiles is that compared to mesophiles, active transport (∞ Section 4.7) occurs optimally at low temperature, an indication that the cytoplasmic membranes of psychrophiles are constructed in such a way that low temperatures do not inhibit membrane phenomena. Studies on the composition of cytoplasmic membranes from psychrophiles have shown them to contain a higher content of *unsaturated* fatty acids (∞ Section 5.15), which help to maintain a semifluid state of the membrane at low temperatures (membranes composed of predominantly saturated fatty acids would become waxy and nonfunctional at low temperatures). The lipids of some psychrophilic bacteria also contain *polyunsaturated* fatty acids and long-chain hydrocarbons with multiple double bonds. In the latter connection, a hydrocarbon with nine double bonds ($C_{31:9}$) has been identified from the lipids of some Antarctic bacteria, and the bacterium *Psychroflexus* has been shown to contain fatty acids with 4 and 5 double bonds.

Freezing

Despite the ability of some organisms to grow at low temperatures, there is a lower limit below which reproduction is impossible. Pure water freezes at 0°C and seawater at –2.5°C, but freezing is not continuous and microscopic pockets of water continue to exist at these and even much lower temperatures. Although freezing prevents microbial growth, it does not necessarily cause microbial death. In addition, the medium in which the cells are suspended considerably affects sensitivity to freezing. Water-miscible liquids such as glycerol and dimethylsulfoxide (DMSO), when added at about 10% (final concentration) to the suspending medium, penetrate the cells and protect them by reducing the severity of dehydration effects and preventing ice crystal formation. In fact, the addition of such agents, called *cryoprotectants*, is a common way of *preserving* microbial cultures at very low temperatures (usually at –70 to –196°C). Properly prepared frozen cells can remain viable for long periods (at least several decades).

✓ 6.9 Concept Check

Organisms with cold temperature optima are called *psychrophiles*, and the most extreme examples inhabit permanently cold environments. Psychrophiles have evolved biomolecules that function best at cold temperatures but that can be unusually sensitive to warm temperatures.

✓ How do *psychrotolerant* organisms differ from *psychrophilic* organisms?

✓ What molecular adaptations to the cytoplasmic membrane are seen in psychrophiles, and why are they necessary?

6.10 Microbial Growth at High Temperatures

Microbial life flourishes in high-temperature environments, up to and including boiling water. And above about 65°C, only *prokaryotic* life forms survive, but even here, a huge diversity of both *Bacteria* and *Archaea* exist. We consider some of these hot environments and their microbial life here.

Thermal Environments

Recall that organisms whose growth temperature optimum is above 45°C are called **thermophiles,** and those whose optimum is above 80°C are called **hyperthermophiles** (Figure 6.17). Temperatures as high as these are found in nature only in certain restricted areas. For example, soils subject to full sunlight are often heated to temperatures above 50°C at midday, and some soils may become warmed to even 70°C, although a few centimeters under the surface the temperature is much lower. Fermenting materials such as compost piles and silage usually reach temperatures of 60–65°C. However, the most extensive and extreme high-temperature environments found in nature are in association with volcanic phenomena.

Many hot springs have temperatures near or at boiling, and steam vents (fumaroles) may reach 150–500°C. Hydrothermal vents in the bottom of the ocean have temperatures of 350°C or greater (∞Section 19.8). Hot springs occur throughout the world, but are especially concentrated in the western United States, New Zealand, Iceland, Japan, Italy, Indonesia, Central America, and central Africa. However, the area with the largest single concentration of hot springs in the world is Yellowstone National Park, Wyoming (USA). Although some hot springs vary in temperature, others are very constant, not varying more than 1–2°C over many years. In addition, these springs have variable chemical compositions and pH values, generally containing sufficient levels of nutrients to support populations of both chemoorganotrophs and chemolithotrophs.

Hyperthermophiles in Hot Springs

Many hot springs are at the boiling point for the altitude (92–93°C at Yellowstone, 99–100°C at locations where the springs are close to sea level). In boiling hot springs (Figure 6.20), a variety of hyperthermophiles are typically present. The growth of such organisms can be studied by immersing microscope slides into the spring and retrieving them after a few days. Microscopic examination of the slides reveals colonies of prokaryotes (Figure 6.20*b*) that have developed from single bacterial cells that attached to and grew on the glass surface.

Ecological studies of organisms living in boiling springs have shown that growth rates are fairly rapid,

(a)

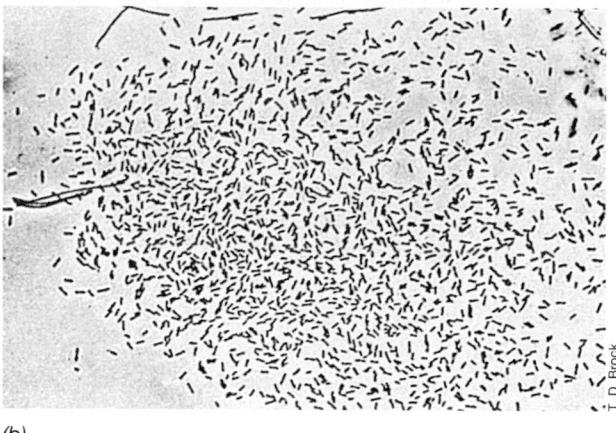

(b)

Figure 6.20 Growth of hyperthermophiles in boiling water. (a) Boulder Spring, a small boiling spring in Yellowstone National Park. This spring is superheated, having a temperature 1–2°C above the boiling point. The mineral deposits around the spring consists mainly of silica and sulfur. (b) Photomicrograph of a microcolony of prokaryotes that developed on a microscope slide immersed in a boiling spring such as that shown in (a).

and doubling times of as short as 1 h have been recorded. Cultures of many of these prokaryotes have been obtained and a variety of morphological and physiological types exist (⚭ Chapter 13). Phylogenetic studies using ribosomal RNA sequencing (⚭ Sections 2.3 and 11.5) have shown great evolutionary diversity among these hyperthermophiles including various *Bacteria* and, especially, *Archaea*. Some of these hyperthermophiles show growth-temperature optima *greater than 100°C* and thus are grown in the laboratory in pressurized vessels to reach temperatures above the boiling point.

Thermophiles

Many thermophiles (optima 45–80°C) are present in hot springs as well as other thermal environments. In hot springs, as boiling water overflows the edges of the spring and flows away from the source, it gradually cools, setting up a *thermal gradient*. Along this gradient,

various microorganisms grow (Figure 6.21), with different species growing in the different temperature ranges. By studying the species distribution along such thermal gradients and by examining hot springs and other thermal habitats at different temperatures around the world, it is possible to determine the upper temperature limits for each kind of organism (Table 6.1). From this information we conclude that (1) prokaryotic organisms in general are able to grow at temperatures higher than those at which eukaryotes can grow; (2) the most thermophilic of all prokaryotes are certain species of *Archaea;* and (3) nonphototrophic organisms are able to grow at higher temperatures than can phototrophic forms.

Thermophilic prokaryotes have also been found in artificial thermal environments. The hot water heater, domestic or industrial, has a temperature of 55–80°C and is therefore a favorable habitat for the growth of thermophilic prokaryotes. Organisms resembling *Thermus aquaticus*, a common hot spring thermophile, have been isolated from hot water heaters. Electric power plants, hot industrial process water, and other artificial thermal sources also provide sites where thermophiles can grow and such organisms can be easily isolated using the appropriate enrichment techniques.

Molecular Adaptations to Thermophily

How can thermophiles and hyperthermophiles thrive at high temperatures? First, their enzymes and other proteins are much more stable to heat than are those of

Figure 6.21 Growth of thermophilic cyanobacteria in a hot spring in Yellowstone National Park. Characteristic pattern formed by cyanobacteria at the upper temperature for phototrophic life, 70–74°C, in the thermal gradient formed from a boiling hot spring. The pattern develops because the water cools more rapidly at the edges than in the center of the channel. The spring flows from the back of the picture toward the foreground. Just above where the scientist is kneeling it is too hot for cyanobacteria. The light-green color is from high temperature forms of the organism *Synechococcus*. As water flows down the gradient, the standing crop of cells increases, and the color becomes more intensely green.

TABLE 6.1 Presently known upper temperature limits for growth of living organisms

Group	Upper temperature limits (°C)
Animals	
Fish and other aquatic vertebrates	38
Insects	45–50
Ostracods (crustaceans)	49–50
Plants	
Vascular plants	45
Mosses	50
Eukaryotic microorganisms	
Protozoa	56
Algae	55–60
Fungi	60–62
Prokaryotes	
Bacteria	
Cyanobacteria	70–74
Anoxygenic phototrophs	70–73
Chemoorganotrophic/chemolithotrophic *Bacteria*	95
Archaea	
Chemoorganotrophic/chemolithotrophic *Archaea*	113

mesophiles, and their macromolecules actually function *optimally* at high temperatures. How is heat stability achieved? Amazingly, studies of several thermostable enzymes have shown that they often differ very little in amino acid sequence from an enzyme that catalyzes the same reaction in a mesophile. It appears that a critical amino acid substitution in one or a few locations in the enzyme allows it to fold in a way that is consistent with heat stability. Heat stability of proteins in hyperthermophiles is also improved as a result of an increased number of ion pairs (ionic bonds between the positive and negative charges of various amino acids) present and the densely packed highly hydrophobic interiors of the proteins, which naturally resist unfolding in the aqueous cytoplasm. Finally, from studies of hyperthermophiles it appears that certain solutes such as di-inositol phosphate, diglycerol phosphate, and manosylglycerate are produced in significant quantities and help to stabilize proteins against thermal degradation.

In addition to enzymes and other components of the cell, the cytoplasmic membranes of thermophiles and hyperthermophiles need to be heat-stable. We mentioned earlier that psychrophiles have membrane lipids rich in *unsaturated* fatty acids, thus making the membranes fluid and functional at low temperatures. Conversely, thermophiles typically have lipids rich in *saturated* fatty acids, thus allowing the membranes to remain stable and functional at high temperatures. Saturated fatty acids form a stronger hydrophobic environment than do unsaturated fatty acids, which helps account for the membrane stability. Hyperthermophiles, most of which are *Archaea*, do not contain fatty acids at all in their membranes, but instead have C_{40} hydrocarbons composed of repeating units of the five-carbon compound isoprene (∞ Figure 4.19c) bonded by ether linkage to glycerol phosphate. In addition, the overall structure of these membranes forms a *lipid monolayer* (∞ Figure 4.20d), and this structure is undoubtedly much more heat resistant than the lipid bilayer of species of *Bacteria* and *Eukarya*. We discussed the details of the unique membrane architecture of hyperthermophilic *Archaea* in Section 4.5 and will consider other aspects of heat stability, including that of DNA, in hyperthermophiles in Section 13.11.

Biotechnological Aspects of Thermophily

Thermophilic and hyperthermophilic microorganisms are interesting for more than just basic biological reasons. These organisms offer some major advantages for industrial and biotechnological processes, many of which run more rapidly and efficiently at high temperatures. Enzymes from thermophiles and hyperthermophiles are capable of catalyzing biochemical reactions at high temperatures (∞ Section 30.9 and Figure 30.15b) and are generally more stable than enzymes from mesophiles, thus prolonging the shelf life of enzyme preparations. A practical example of a heat-stable enzyme of great applied importance is the DNA polymerase isolated from the thermophile *Thermus aquaticus*. *Taq polymerase*, as this enzyme is known, has been used to automate the repetitive steps involved in the *polymerase chain reaction (PCR)* technique, an extremely important method for amplifying specific DNA sequences (∞ Section 10.17). Several other uses of heat-stable enzymes and other thermostable cell products are known or are being developed for industrial applications.

✓ 6.10 Concept Check

Organisms with growth temperature optima between 45 and 80°C are called *thermophiles*, while those with optima >80°C are called *hyperthermophiles*. These organisms inhabit hot environments up to and including boiling hot springs, as well as undersea hydrothermal vents that can have temperatures in excess of 100°C. Thermophiles and hyperthermophiles produce heat stable macromolecules.

✓ What is the current upper temperature limit for growth of a prokaryote? Is the organism that grows at this temperature a member of the *Bacteria* or the *Archaea*?

✓ What is the structure of membranes of hyperthermophilic *Archaea*, and why might this structure be useful for growth at high temperature?

✓ What is *Taq polymerase* and why has it been of use to biotechnology?

6.11 Microbial Growth at Low or High pH

Acidity or alkalinity of a solution is expressed by its **pH** on a scale on which neutrality is pH 7 (Figure 6.22). Those pH values that are less than 7 are said to be *acidic,* and those greater than 7 are *alkaline* (or *basic*). It is important to remember that pH is a *logarithmic function;* a change of 1 pH unit represents a *10-fold* change in hydrogen ion concentration. Thus, vinegar (pH near 2) and household ammonia (pH near 11) differ in hydrogen ion concentration by a billionfold.

pH and Microbial Growth

Each organism has a pH range within which growth is possible and usually has a well-defined pH optimum. Most organisms show a growth pH range of 2–3 units. Most natural environments have pH values between 5 and 9, and organisms with optima in this range are most common. Only a few species can grow at pH values of less than 2 or greater than 10. Organisms that grow best at low pH are types of extremophiles called **acidophiles**. Fungi as a group tend to be more acid-tolerant than bacteria. Many fungi grow optimally at pH 5 or below, and a few grow well at pH values as low as 2. Several bacteria are also acidophilic. In fact, some of

these bacteria are *obligate* acidophiles, unable to grow at all at neutral pH. Obligately acidophilic bacteria include several species of *Thiobacillus* (◌◌ Section 12.4) and several genera of *Archaea*, including *Sulfolobus, Thermoplasma*, and *Ferroplasma* (◌◌ Sections 13.5 and 13.9).

The most critical factor for obligate acidophily is the stability of the cytoplasmic membrane. When the pH is raised to neutrality, the cytoplasmic membrane of strongly acidophilic bacteria actually dissolves and the cells lyse, suggesting that high concentrations of hydrogen ions are required for membrane stability. For example, the most acidophilic prokaryote known, *Picrophilus oshimae,* is also a thermophile (growth temperature optimum, 60°C) and grows optimally at pH 0.7; above pH 4 cells of *P. oshimae* lyse! *P. oshimae* inhabits extremely acidic thermal soils associated with volcanic activity.

A few extremophiles have very high pH optima for growth, sometimes as high as pH 10; these are known as **alkaliphiles**. Alkaliphilic microorganisms are usually found in highly basic habitats such as soda lakes and high carbonate soils. However, the most well-studied alkaliphilic prokaryotes have been aerobic nonmarine bacteria, and many are *Bacillus* species. Some extremely alkaliphilic bacteria are also halophilic (salt-loving), and most of these are *Archaea* (◌◌ Section 13.3). Some alkaliphiles have found industrial uses because they produce hydrolytic enzymes, such as proteases and lipases, which function well at alkaline pH and are used as supplements for household detergents (◌◌ Section 30.9).

Alkaliphiles are also of interest because of the bioenergetic problems they face living at such high pH. For example, can a proton motive force (◌◌ Section 5.12) be established when the external surface of the membrane is so alkaline? From studies of an alkaliphilic *Bacillus* species it has been shown that a Na^+ gradient (instead of the usual proton motive force) supplies the energy for transport and motility but that a proton motive force can indeed be established and is responsible for driving respiratory ATP synthesis. Exactly how this occurs is an interesting problem in alkaliphile research today.

Internal Cell pH

Despite the requirements of a particular organism for a specific pH for growth, the optimal growth pH represents the pH of the *extracellular* environment only; the *intracellular* pH must remain near neutrality in order to prevent destruction of acid- or alkali-labile macromolecules in the cell. In extreme acidophiles or extreme alkaliphiles, the intracellular pH may vary by several pH units from neutrality, but for the majority of microorganisms, whose pH optimum for growth is between pH 6 and 8 (referred to as **neutrophiles**), the cytoplasm remains neutral or very nearly so (Figure 6.22).

In the previously mentioned acidophile *P. oshimae,* the internal pH has been measured at pH 4.6, and in extreme

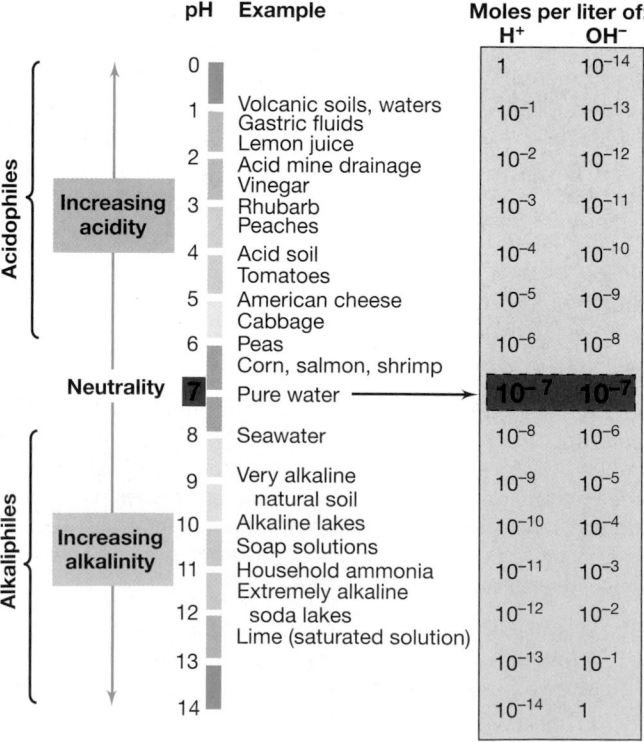

pH	Example	Moles per liter of:	
		H^+	OH^-
0		1	10^{-14}
1	Volcanic soils, waters / Gastric fluids	10^{-1}	10^{-13}
2	Lemon juice / Acid mine drainage / Vinegar	10^{-2}	10^{-12}
3	Rhubarb / Peaches	10^{-3}	10^{-11}
4	Acid soil / Tomatoes	10^{-4}	10^{-10}
5	American cheese / Cabbage	10^{-5}	10^{-9}
6	Peas / Corn, salmon, shrimp	10^{-6}	10^{-8}
7	Pure water →	10^{-7}	10^{-7}
8	Seawater	10^{-8}	10^{-6}
9	Very alkaline natural soil	10^{-9}	10^{-5}
10	Alkaline lakes / Soap solutions	10^{-10}	10^{-4}
11	Household ammonia	10^{-11}	10^{-3}
12	Extremely alkaline soda lakes / Lime (saturated solution)	10^{-12}	10^{-2}
13		10^{-13}	10^{-1}
14		10^{-14}	1

Acidophiles — Increasing acidity
Neutrality — 7
Alkaliphiles — Increasing alkalinity

Figure 6.22 The pH scale. Note that although some microorganisms can live at very low or very high pH, the cell's internal pH remains near neutrality.

alkaliphiles an intracellular pH of as high as 9.5 has been measured. If these are not the lower and upper limits of cytoplasmic pH, respectively, they must be very close to the limits, since macromolecular stability would almost certainly be compromised above or below these pH values.

Buffers

In a batch culture, the pH can change during growth as the result of metabolic reactions that consume or produce acidic or basic substances. Thus, chemicals called *buffers* are frequently added to microbial culture media to keep the pH relatively constant. Such pH buffers generally work over only a narrow pH range; hence, different buffers must be used to buffer at different pH values. For near-neutral pH ranges (pH 6–7.5), phosphate, usually supplied as KH_2PO_4, is an excellent buffer. Many other buffers for use in microbial growth media or for the assay of enzymes extracted from microbial cells are available, and the best buffering system for one organism or enzyme may be considerably different from that of another. Thus, the optimal buffer for use in a particular situation must usually be determined empirically, although for assaying enzymes *in vitro*, a certain buffer that works well in an assay for the enzyme from one organism will usually work well for assaying this same enzyme from other organisms.

✓ 6.11 Concept Check

The acidity or alkalinity of an environment can greatly affect microbial growth. Some organisms have evolved to grow best at low or high pH, but most organisms grow best between pH 6 and 8. The internal pH of a cell must stay relatively close to neutrality even though the external pH is highly acidic or basic.

✓ What is the increase in concentration of protons (H^+) when going from pH 7 to pH 4?

✓ What are *buffers* and why are they needed?

6.12 Osmotic Effects on Microbial Growth

Water is the solvent of life. All organisms require water, and water availability is an important factor affecting the growth of microorganisms in nature. Water availability not only depends on the water content of an environment, that is, how moist or dry a solid microbial habitat may be, but is also a function of the concentration of solutes such as salts, sugars, or other substances that are dissolved in water. This is because dissolved substances have an affinity for water, which makes the water associated with solutes unavailable to organisms.

Water Activity, Osmosis, and Halophiles

Water availability is generally expressed in physical terms such as **water activity**. Water activity, abbreviated a_w, is a

ratio of the vapor pressure of the air in equilibrium with a substance or solution to the vapor pressure of pure water. Thus, values of a_w vary between 0 and 1, and some representative values are given in Table 6.2. Water activities in agricultural soils generally range between 0.90 and 1.

Water diffuses from a region of high water concentration (low solute concentration) to a region of lower water concentration (higher solute concentration) in the process of *osmosis*. In most cases, the cytoplasm of a cell has a higher solute concentration than the environment, so water tends to diffuse into the cell, and the cell is said to be in *positive water balance*. However, when a cell is in an environment of low water activity, there is a tendency for water to flow out of the cell.

In nature, osmotic effects are of interest mainly in habitats with high concentrations of salts. Seawater contains about 3% sodium chloride (NaCl) plus small amounts of many other minerals and elements. Microorganisms found in the sea usually have a specific requirement for the sodium ion in addition to growing optimally at the water activity of seawater (Figure 6.23). Such organisms are called **halophiles**. The growth of halophiles requires at least some NaCl, but the optimum varies with the organism; thus the terms *mild halophile* and *moderate halophile* are used to describe halophiles with low (1–6%) and moderate (6–15%) NaCl requirements, respectively (Figure 6.23).

Most microorganisms are unable to cope with environments of very low water activity and either die or become dehydrated and dormant under such conditions. **Halotolerant** organisms can tolerate some reduction in the a_w of their environment but generally grow best in the absence of the added solute (Figure 6.23). By contrast, some organisms thrive at very low water activity, and these organisms are of interest not only from the

TABLE 6.2 Water activity of several substances

Water activity (a_w)	Material	Example organisms[a]
1.000	Pure water	*Caulobacter, Spirillum*
0.995	Human blood	*Streptococcus, Escherichia*
0.980	Seawater	*Pseudomonas, Vibrio*
0.950	Bread	Most gram-positive rods
0.900	Maple syrup, ham	Gram-positive cocci such as *Staphylococcus*
0.850	Salami	*Saccharomyces rouxii* (yeast)
0.800	Fruit cake, jams	*Saccharomyces bailii, Penicillium* (fungus)
0.750	Salt lakes, salted fish	*Halobacterium, Halococcus*
0.700	Cereals, candy, dried fruit	*Xeromyces bisporus* and other xerophilic fungi

[a] Examples, of which in most cases many others are known, of prokaryotes or fungi capable of growth in culture media adjusted to the stated water activity.

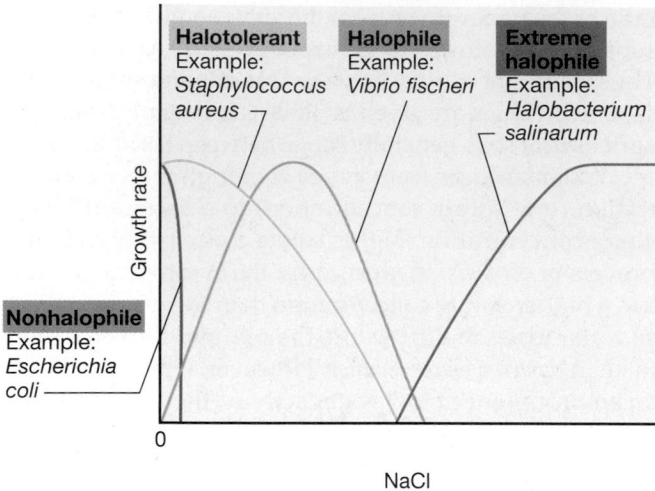

Figure 6.23 Effect of sodium ion concentration on growth of microorganisms of different salt tolerances or requirements. The optimum NaCl concentration for marine microorganisms such as *V. fischeri* is about 3%; for extreme halophiles, it is between 15 and 30%, depending on the organism.

standpoint of their adaptation to life under these conditions, but also from an applied standpoint, such as that of the food industry, where solutes such as salt and sucrose are commonly used as preservatives to inhibit microbial growth. Organisms capable of growth in very salty environments are called **extreme halophiles** (Figure 6.23); extreme halophiles require 15–30% NaCl, depending on the species, for optimum growth (Section 13.3). Organisms able to live in environments high in sugar are called **osmophiles**, and those able to grow in very dry environments (made dry by lack of water) are called **xerophiles**. Examples of these various organisms are given in Table 6.2.

Compatible Solutes

How do organisms grow under conditions of low water activity? When an organism grows in a medium with a low water activity, it can obtain water from its environment only by increasing its *internal* solute concentration. An increase in internal solute concentration can be accomplished by either pumping inorganic ions into the cell from the environment or synthesizing or concentrating an organic solute. Organisms are known that employ either of these mechanisms, and several examples are given in Table 6.3.

The solute used inside the cell for adjustment of cytoplasmic water activity must be noninhibitory to biochemical processes within the cell; such compounds are called **compatible solutes**. Several different compatible solutes are known in microorganisms (Table 6.3 and Figure 6.24). These substances are all highly water-soluble sugars or sugar alcohols, other alcohols, amino acids or their derivatives, or in the case of extremely halophilic *Archaea* and a very few extremely halophilic *Bacteria*, potassium ions (from KCl) (Table 6.3). Compatible solutes are either synthesized by the microorganisms directly or in some cases (such as glycine betaine or KCl) accumulated from the environment. The concentration of compatible solutes in the cell is a function of the level of external solutes, and in each organism the maximal amount of compatible solute(s) made or that can be accumulated is a genetically directed characteristic; this results in different organisms tolerating different ranges of water potential (Tables 6.2 and 6.3). Thus, nonhalotolerant, halotolerant, halophilic, and extremely halophilic microorganisms are to a major extent defined by their genetic capacity to produce or accumulate compatible solutes.

Gram-positive cocci of the genus *Staphylococcus* are notoriously halotolerant (in fact, a common isolation procedure for them is to use media containing 7.5% NaCl),

TABLE 6.3 Compatible solutes of microorganisms

Organism	Major solute(s) accumulated	Minimum a_w for growth
Bacteria, nonphototrophic	Glycine betaine, proline (mainly gram-positive), glutamate (mainly gram-negative)	0.97–0.90
Freshwater cyanobacteria	Sucrose, trehalose	0.98
Marine cyanobacteria	α-Glucosylglycerol	0.92
Marine algae	Mannitol, various glycosides, proline, dimethylsulfoniopropionate	0.92
Salt lake cyanobacteria	Glycine betaine	0.90–0.75
Halophilic anoxygenic phototrophic *Bacteria* (*Ectothiorhodospira*/*Halorhodospira* and *Rhodospirillum* species)	Glycine betaine, ectoine, trehalose	0.90–0.75
Extremely halophilic *Archaea* (for example, *Halobacterium*) and some *Bacteria* (for example, *Haloanaerobium*)	KCl	0.75
Dunaliella (halophilic green alga)	Glycerol	0.75
Xerophilic yeasts	Glycerol	0.83–0.62
Xerophilic filamentous fungi	Glycerol	0.72–0.61

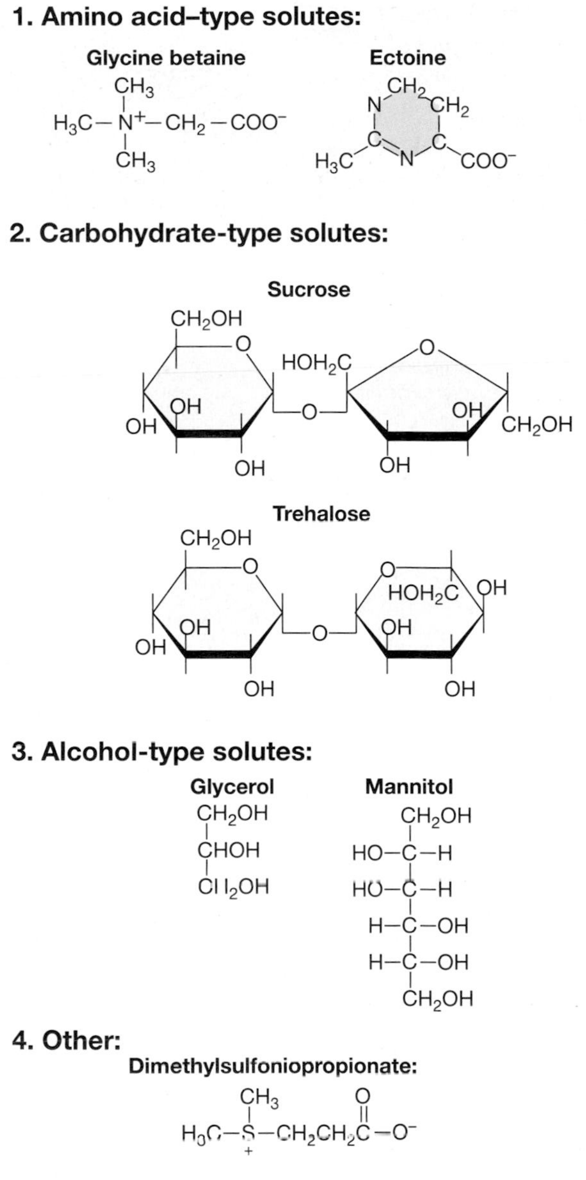

1. Amino acid–type solutes:

Glycine betaine

$$H_3C-\overset{\overset{\displaystyle CH_3}{|}}{\underset{\underset{\displaystyle CH_3}{|}}{N^+}}-CH_2-COO^-$$

Ectoine

2. Carbohydrate-type solutes:

Sucrose

Trehalose

3. Alcohol-type solutes:

Glycerol

$$\begin{array}{c} CH_2OH \\ | \\ CHOH \\ | \\ CH_2OH \end{array}$$

Mannitol

$$\begin{array}{c} CH_2OH \\ | \\ HO-C-H \\ | \\ HO-C-H \\ | \\ H-C-OH \\ | \\ H-C-OH \\ | \\ CH_2OH \end{array}$$

4. Other:

Dimethylsulfoniopropionate:

$$H_3C-\overset{\overset{\displaystyle CH_3}{|}}{\underset{+}{S}}-CH_2CH_2\overset{\overset{\displaystyle O}{||}}{C}-O^-$$

Figure 6.24 Structures of some common compatible solutes in microorganisms. The structures of glutamate and proline, other common solutes, were shown in Figure 3.12. The formal name of ectoine is 1,4,5,6-tetrahydro-2-methyl-4-pyrlmidine carboxylate. Note that all the compounds shown here are very polar (water soluble) molecules.

and these organisms use the amino acid *proline* as a compatible solute. *Glycine betaine* is a derivative of the amino acid glycine in which the protons on the amino group are replaced by three methyl groups; this leaves a permanent positive charge on the N atom (Figure 6.24), which increases its solubility. Glycine betaine is widely distributed as a compatible solute, especially among halophilic *Bacteria* and cyanobacteria (Table 6.3). Some extremely halophilic bacteria produce a novel compatible solute called *ectoine* (Figure 6.24), which is a cyclic derivative of the amino acid aspartate. A variety of glycosides are pro-

duced by marine algae, but with rare exception they accumulate only in low amounts because the cells are not very halophilic. Xerophilic yeasts and halophilic green algae produce mainly *glycerol* as a compatible solute. Other examples of compatible solutes are listed in Table 6.3, and structures are shown in Figure 6.24.

✓ **6.12 Concept Check**

Water activity becomes limiting to an organism when the dissolved solute concentration in its environment increases. To counteract this situation organisms produce or accumulate intracellular compatible solutes that function to maintain the cell in positive water balance. Some microorganisms have evolved to grow best at reduced water potential, and some even require high levels of salts in order to grow.

✓ What is the a_w of pure water?

✓ What is a *compatible solute* and why is it needed?

✓ What is the compatible solute for *Halobacterium* species?

6.13 Oxygen and Microbial Growth

Since humans absolutely require molecular oxygen (O_2) for life, it is easy for us to assume that all life forms require O_2. However, this is not true, because many microorganisms can (and some must) live in the total absence of O_2. Oxygen is weakly soluble in water, and because of the respiratory activities of microorganisms in aquatic or other moist habitats (especially those containing an abundance of organic materials), O_2 can quickly become exhausted. Thus, anoxic (O_2-free) microbial habitats abound on Earth, including muds and other sediments, bogs and marshes, water-logged soils, intestinal tracts of animals, sewage sludge, the deep subsurface of the Earth, and many other environments. In these anoxic habitats, microorganisms, particularly various prokaryotes, thrive.

Oxygen Classes of Microorganisms

Microorganisms vary in their need for, or tolerance of, oxygen. In fact, microorganisms can be divided into several groups depending on the effect of oxygen, as outlined in Table 6.4. **Aerobes** are species capable of growth at full oxygen tensions (air is 21% O_2), and many can even tolerate elevated concentrations of oxygen (hyperbaric oxygen). **Microaerophiles**, by contrast, are aerobes that can use O_2 only when it is present at levels reduced from that in air, usually because of their limited capacity to respire or because they contain some oxygen-sensitive molecule such as an oxygen-labile enzyme. Many aerobes are **facultative**, meaning that, under the appropriate nutrient and culture conditions, they can grow under *either* aerobic or anaerobic conditions.

TABLE 6.4 Oxygen relationships of microorganisms

Group	Relationship to O_2	Type of metabolism	Example[a]	Habitat[b]
Aerobes				
Obligate	Required	Aerobic respiration	*Micrococcus luteus* (B)	Skin, dust
Facultative	Not required, but growth better with O_2	Aerobic, anaerobic respiration, fermentation	*Escherichia coli* (B)	Mammalian large intestine
Microaerophilic	Required but at levels lower than atmospheric	Aerobic respiration	*Spirillum volutans* (B)	Lake water
Anaerobes				
Aerotolerant	Not required, and growth no better when O_2 present	Fermentation	*Streptococcus pyogenes* (B)	Upper respiratory tract
Obligate	Harmful or lethal	Fermentation or anaerobic respiration	*Methanobacterium formicicum* (A)	Sewage sludge digestors, anoxic lake sediments

[a]Letters in parentheses indicate phylogenetic status (B, *Bacteria*; A, *Archaea*). Representatives of either domain of prokaryotes are known in each category. Most eukaryotes are obligate aerobes, but facultative aerobes (for example, yeast) and obligate anaerobes (for example, certain protozoa and fungi) are known.
[b]Listed are typical habitats of the example organism.

Some organisms cannot respire oxygen (O_2); such organisms are called **anaerobes**. There are two kinds of anaerobes: **aerotolerant anaerobes**, which can tolerate oxygen and grow in its presence even though they cannot use it, and **obligate** (or **strict**) **anaerobes**, which are inhibited or even killed by oxygen (Table 6.4). The reason obligate anaerobes are killed by oxygen is unknown, but it may be because they are unable to detoxify some of the products of oxygen metabolism (see later discussion).

So far as is known, obligate anaerobiosis occurs in three groups of microorganisms: a wide variety of prokaryotes, a few fungi, and a few protozoa. One of the best-known groups of obligately anaerobic *Bacteria* belongs to the genus *Clostridium*, a group of gram-positive spore-forming rods. Clostridia are widespread in soil, lake sediments, and intestinal tracts and are often responsible for spoilage of canned foods (∞ Sections 12.21 and 29.2). Other obligately anaerobic bacteria are found among the methanogens and many other species of *Archaea* (∞ Chapter 13), the sulfate-reducing and homoacetogenic bacteria (∞ Chapters 12 and 13), and many of the bacteria that inhabit the animal gut (∞ Section 21.4). Among obligate anaerobes, however, the sensitivity to oxygen varies greatly; some organisms are able to tolerate traces of oxygen, whereas others are not.

Microbial Culture and the Effects of Oxygen

For the growth of many aerobes, it is necessary to provide extensive aeration. This is because O_2 is only poorly soluble in water and the O_2 used up by the organisms during growth is not replaced fast enough by diffusion from the air. Forced aeration of cultures is therefore frequently desirable and can be achieved either by vigorously shaking the flask or tube on a shaker or by bubbling sterilized air into the medium through a fine glass tube or porous glass disc. Usually aerobes grow

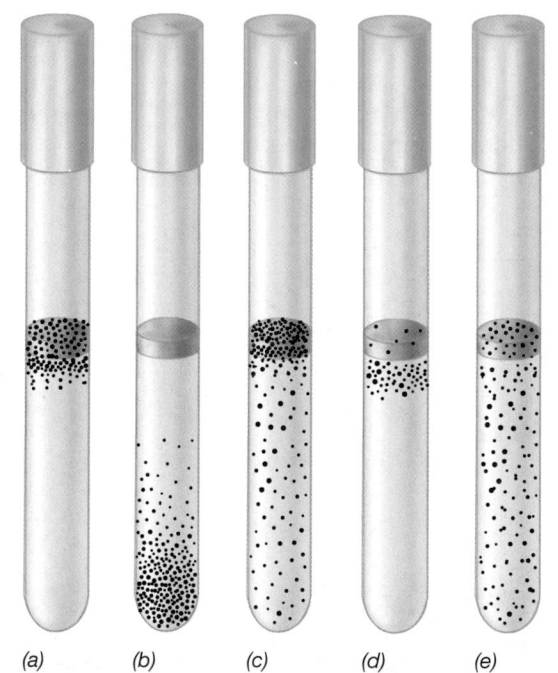

(a) (b) (c) (d) (e)

Figure 6.25 Aerobic, anaerobic, facultative, microaerophilic, and aerotolerant anaerobe growth, as revealed by the position of microbial colonies (depicted here as black dots) within tubes of thioglycolate broth culture medium. A small amount of agar has been added to keep the liquid from becoming disturbed and the redox dye, resazurin, which is pink when oxidized and colorless when reduced, is added as a redox indicator. (a) Oxygen penetrates only a short distance into the tube, so obligate aerobes grow only at the surface. (b) Anaerobes, being sensitive to oxygen, grow only away from the surface. (c) Facultative aerobes are able to grow in either the presence or the absence of oxygen and thus grow throughout the tube. However, better growth occurs near the surface because these organisms can respire. (d) Microaerophiles grow away from the most oxic zone. (e) Aerotolerant anaerobes grow throughout the tube. However, growth is no better near the surface because these organisms can only ferment.

much better with forced aeration than when O_2 is provided by simple diffusion.

For anaerobic culture, the problem is to *exclude*, not provide, oxygen. And because oxygen is present in the air, special methods are needed to culture anaerobic microorganisms. Obligate anaerobes vary in their sensitivity to oxygen, and a number of procedures are available for reducing the O_2 content of cultures; some are simple and suitable mainly for less sensitive organisms, while others are more complex but necessary for growth of strict anaerobes.

Bottles or tubes filled completely to the top with culture medium and provided with tightly fitting stoppers provide anoxic conditions for organisms not too sensitive to small amounts of oxygen. It is also possible to add a chemical called a *reducing agent* that reacts with oxygen and reduces it to H_2O. A good example is *thioglycolate*, which is added to a medium called *thioglycolate broth*, commonly used to test an organism's requirements for O_2 (Figure 6.25). After thioglycolate reacts with oxygen throughout the tube, oxygen can penetrate only near the top of the tube where the medium contacts air. Obligate aerobes grow only at the top of such tubes. Facultative organisms grow throughout the tube but best near the top. Microaerophiles grow near the top but not right at the top. Anaerobes grow only near the bottom of the tube, where oxygen cannot penetrate (Figure 6.25). A redox indicator dye called *resazurin* is added to the medium because the dye changes color in the presence of oxygen and thereby indicates the degree of penetration of oxygen into the medium (Figure 6.25).

To remove all traces of O_2 for the culture of anaerobes, it is possible to place an O_2-consuming gas in a jar holding the tubes or plates. One of the simplest devices for this is an *anoxic jar*, a heavy-walled jar with a gastight seal within which tubes, plates, or other containers to be incubated are placed (Figure 6.26a). The air in the jar is replaced with a mixture of H_2 and CO_2, and in the presence of a chemical catalyst the traces of O_2 left in the vessel or culture medium are consumed, thus leading to anoxic conditions.

For strict anaerobes, such as the methanogens (∞ Section 13.4), it is necessary not only to carefully remove all traces of O_2 but also to carry out all manipulations of cultures in an anoxic atmosphere, as these organisms can be killed by even a brief exposure to O_2. In these cases, a culture medium is first boiled to render it oxygen-free, and then a reducing agent such as H_2S is added, and the mixture is sealed under an oxygen-free gas. All manipulations are carried out under a jet of oxygen-free hydrogen or nitrogen gas that is directed into the culture vessel when it is open, thus driving out any O_2 that might enter. For extensive research on anaerobes, special boxes fitted with gloves, called *anoxic glove boxes* (*bags*), permit work with open cultures in completely anoxic atmospheres (Figure 6.26b).

Toxic Forms of Oxygen

Oxygen is a powerful oxidant and an excellent electron acceptor for respiration (∞ Section 5.11). Oxygen in its normal ground state is referred to as **triplet oxygen ($^3O_2$)**.

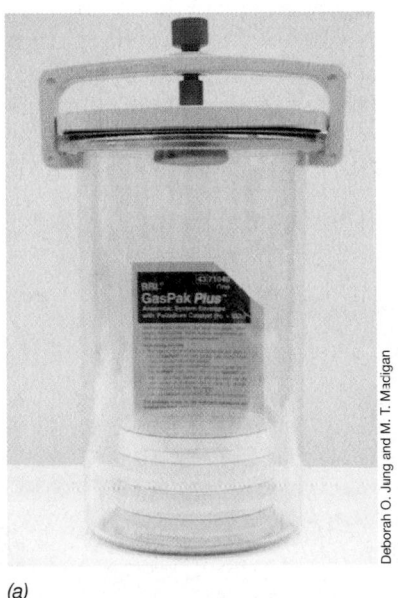

(a)

Deborah O. Jung and M. T. Madigan

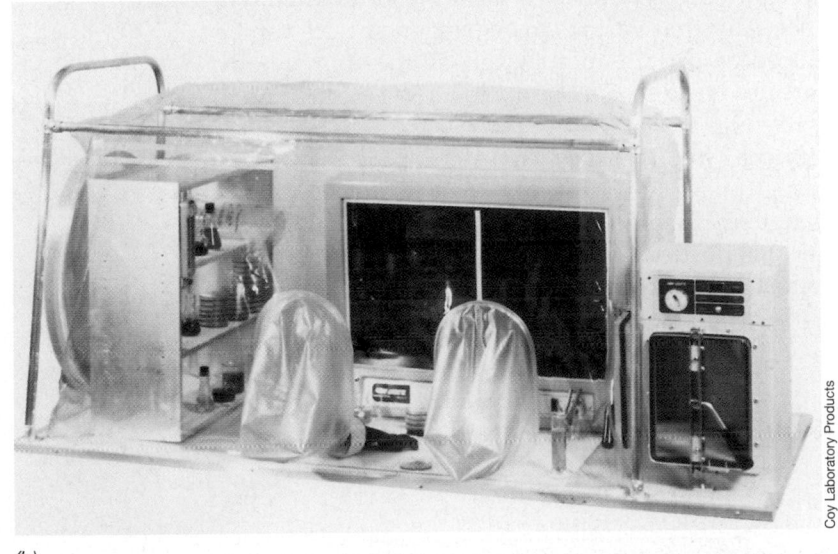

(b)

Coy Laboratory Products

Figure 6.26 Incubation under anoxic conditions. (a) Anoxic jar. A chemical reaction in the envelope in the jar generates $H_2 + CO_2$. The H_2 reacts with O_2 in the jar on the surface of a palladium catalyst to yield H_2O; the final atmosphere contains N_2, H_2, and CO_2. (b) Anoxic glove bag for manipulating and incubating cultures under anoxic conditions. The airlock on the right, which can be evacuated and filled with oxygen-free gas, serves as a port for adding and removing materials to and from the glove bag.

However, one major form of toxic oxygen is called **singlet oxygen ($^1O_2$)**, a higher energy form of oxygen in which outer shell electrons surrounding the nucleus become highly reactive and are able to carry out a variety of spontaneous and undesirable oxidations within the cell. Singlet oxygen is produced both photochemically and biochemically, the latter through the action of various peroxidase enzymes. Organisms that frequently encounter singlet oxygen, such as airborne bacteria and phototrophic microorganisms, often contain pigments called **carotenoids**, which function to convert singlet oxygen to nontoxic forms (Section 17.3).

Other highly toxic forms of oxygen include **superoxide anion (O_2^-)**, **hydrogen peroxide (H_2O_2)**, and **hydroxyl radical (OH·)**, all of which are produced as inadvertent by-products during the reduction of O_2 to H_2O in respiration (Figure 6.27). Flavoproteins, quinones, thiols, and iron-sulfur proteins (Section 5.11) can also carry out the reduction of O_2 to O_2^-. Superoxide is highly reactive and can oxidize virtually any organic compound in the cell, including macromolecules. Peroxides such as H_2O_2 can damage cell components but are generally not as toxic to the cell as superoxide or hydroxyl radical. The latter is the most reactive of all toxic oxygen species and can, like superoxide, instantly oxidize any organic substance in the cell. However, the hydroxyl radical is only a transient species in most cells because the major source of OH· is ionizing radiation, to which most cells are not commonly exposed. Small amounts of OH· can also be produced from H_2O_2 (Figure 6.27), but when peroxides do not accumulate in the cell (because of the action of catalase, which is discussed later), this source of hydroxyl radical is virtually eliminated. We will see later that various toxic oxygen species can be produced by certain immune cells in the animal body and are used to kill microbial invaders (Section 22.2).

Enzymes That Destroy Toxic Oxygen

With such an array of toxic oxygen derivatives, it is perhaps not surprising that organisms have evolved enzymes that destroy toxic oxygen products (Figure 6.28). The most common enzyme in this category is **catalase**, which attacks hydrogen peroxide; the activity of cata-

(a) **Catalase:**
$$H_2O_2 + H_2O_2 \rightarrow 2\,H_2O + O_2$$

(b) **Peroxidase:**
$$H_2O_2 + NADH + H^+ \rightarrow 2\,H_2O + NAD^+$$

(c) **Superoxide dismutase:**
$$O_2^- + O_2^- + 2\,H^+ \rightarrow H_2O_2 + O_2$$

(d) **Superoxide dismutase/catalase in combination:**
$$4\,O_2^- + 4\,H^+ \rightarrow 2\,H_2O + 3\,O_2$$

(e) **Superoxide reductase:**
$$O_2^- + 2\,H^+ + cyt\,c_{reduced} \rightarrow H_2O_2 + cyt\,c_{oxidized}$$

Figure 6.28 Enzymes that destroy toxic oxygen species. (a) Catalases and (b) peroxidases are porphyrin-containing proteins, although some flavoproteins may consume toxic oxygen species as well. (c) Superoxide dismutases are metal-containing proteins, the metals being copper and zinc, manganese, or iron. (d) Combined reaction of superoxide dismutase and catalase. (e) Superoxide reductase catalyzes the one electron reduction of O_2^- to H_2O_2 using reduced cytochrome c as the electron donor.

lase is illustrated in Figures 6.28*a* and 6.29. Another enzyme that destroys hydrogen peroxide is **peroxidase** (Figure 6.28*b*), which differs from catalase in requiring a reductant, usually NADH, producing H_2O as a product. Superoxide is destroyed by the enzyme **superoxide dismutase** (Figure 6.28*c*), which combines two molecules of superoxide to form one molecule of hydrogen peroxide and one molecule of oxygen. Superoxide dismutase and catalase working together can thus bring about the conversion of superoxide back to oxygen (Figure 6.28*d*).

Aerobes and facultative aerobes generally contain both superoxide dismutase and catalase, although a few obligate aerobes lack catalase. Superoxide dismutase is

$$O_2 + e^- \rightarrow O_2^- \quad \textbf{Superoxide}$$
$$O_2^- + e^- + 2\,H^+ \rightarrow H_2O_2 \quad \textbf{Hydrogen peroxide}$$
$$H_2O_2 + e^- + H^+ \rightarrow H_2O + OH· \quad \textbf{Hydroxyl radical}$$
$$OH· + e^- + H^+ \rightarrow H_2O \quad \textbf{Water}$$

Overall: $O_2 + 4\,e^- + 4\,H^+ \rightarrow 2\,H_2O$

Figure 6.27 Four-electron reduction of O_2 to water by stepwise addition of electrons. All the intermediates formed are reactive and toxic to cells except for water, of course.

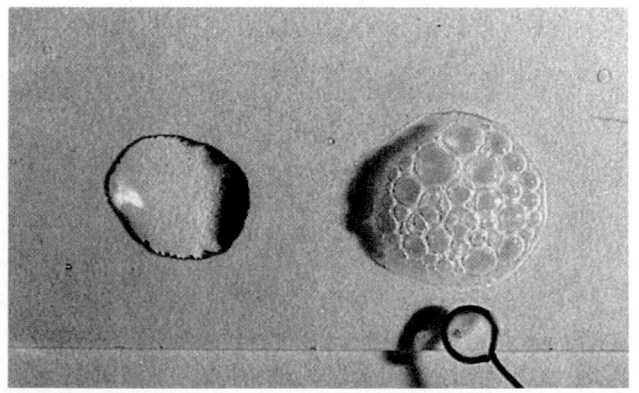

T. D. Brock

Figure 6.29 Method for testing a microbial culture for the presence of catalase. A heavy loopful of cells from an agar culture was mixed on a slide with a drop of 30% hydrogen peroxide. The immediate appearance of bubbles is indicative of the presence of catalase. The bubbles are O_2 produced by the reaction $H_2O_2 + H_2O_2 \rightarrow 2\,H_2O + O_2$.

indispensable to aerobic cells, and the absence of this enzyme in obligate anaerobes has been suggested as the reason why oxygen is toxic to them (but see next paragraph). Some aerotolerant anaerobes, such as lactic acid bacteria, also lack superoxide dismutase, but they use protein-free Mn^{2+} complexes to carry out the dismutation of O_2^- to H_2O_2 and O_2. Such a reaction may have functioned as a primitive form of superoxide dismutase in ancient organisms. This is supported by the fact that all known superoxide dismutases contain a metal cofactor, usually Mn^{2+}, but also Fe^{2+} or Cu^{2+} plus Zn^{2+}, at the enzyme's active site.

An interesting means of superoxide disposal has been discovered in certain obligately anaerobic prokaryotes. In *Pyrococcus furiosus* (a member of the *Archaea*), for example, superoxide dismutase is absent but a unique enzyme, *superoxide reductase*, is present and is responsible for superoxide removal. Unlike superoxide dismutase, superoxide reductase reduces superoxide to H_2O_2 *without* the production of O_2 (Figure 6.28*e*), thus avoiding exposure of the organism to O_2. *P. furiosus* also lacks catalase, an enzyme that like superoxide dismutase, also generates O_2 (Figure 6.28*a*). In *P. furiosus*, the

H_2O_2 produced by superoxide reductase is removed by the activity of peroxidase-like enzymes that yield H_2O as a final product (Figure 6.28*b*). Superoxide reductase may be widely distributed among obligate anaerobes, since genomic sequencing studies have revealed superoxide reductase-like genes in several other obligate anaerobes. This means that these organisms, previously thought to be strict anaerobes because of a lack of superoxide dismutase, indeed do have a mechanism to deal with the highly toxic superoxide anion.

✓ 6.13 Concept Check

Aerobes require oxygen to live, whereas anaerobes do not and may even be killed by O_2. Several toxic forms of oxygen can be formed in the cell, but enzymes are present that can neutralize most of them. Special methods may be necessary to grow strictly aerobic or anaerobic bacteria.

✓ What is a *facultative aerobe*?

✓ How does a reducing agent work?

✓ How does superoxide dismutase protect a cell?

✓ How does the activity of superoxide *dismutase* differ from that of superoxide *reductase*?

Review Questions

1. Describe the key molecular processes that occur when a cell grows and divides into two. What proteins assist in this cell division process?

2. In what way do derivatives of the rod-shaped bacterium *Escherichia coli* that carry mutations in the gene that encodes the protein MreB look different microscopically from wild-type (unmutated) cells? What is the reason for this?

3. How does the antibiotic penicillin kill bacterial cells?

4. What is the difference between the growth rate constant (k) of an organism and its generation time (g)?

5. Why is it useful to plot growth data from a growing microbial culture?

6. Describe the growth cycle of a population of bacterial cells from the time this population is first inoculated into fresh medium. How can the growth pattern differ when it is measured by total count or by viable count?

7. Describe one direct and one indirect method by which microbial growth can be measured. Make sure that the methods you choose agree with your definition.

8. Briefly describe the process by which a single cell develops into a visible colony on an agar plate. With this explanation as a background, describe the principle behind the viable count method.

9. How can a chemostat regulate growth rate and cell numbers independently?

10. Examine the graph describing the relationship between growth rate and temperature (Figure 6.17). Give an explanation, in biochemical terms, of why the optimum temperature for an organism is usually closer to its maximum than its minimum.

11. Concerning the pH of the environment and of the cell, in what ways are acidophiles and alkaliphiles different? In what ways are they similar?

12. Write an explanation in molecular terms for how a cell of a halophile is able to make water molecules flow *into* itself.

13. List three *chemical classes* of compatible solutes produced by various microorganisms. List at least two things they all have in common.

14. Contrast an aerotolerant and an obligate anaerobe in terms of sensitivity and ability to grow in the presence of oxygen (O_2). How does an aerotolerant anaerobe differ from a microaerophile?

15. Compare and contrast the enzymes *catalase, superoxide dismutase,* and *superoxide reductase* from the following points of view: substrates, oxygen products, organisms containing them, role in oxygen tolerance of the cell.

Application Questions

1. Starting with four bacterial cells per milliliter in a rich nutrient medium, with a 1-h lag phase and a 20-min generation time, how many cells will there be in 1 l of this culture after 1 h? After 2 h? After 2 h if one of the initial four cells was dead?

2. Calculate g and k in a growth experiment in which a medium was inoculated with 5×10^6 cells/ml of *Escherichia coli* cells and, following a 1-h lag, grew exponentially for 5 h, after which the population was 5.4×10^9 cells/ml.

3. Return to Chapter 3 and locate a figure that best describes what happens to enzymes from a cell of a mesophile like *Escherichia coli* when placed in a culture medium at 80°C. Contrast this with a figure from Chapter 6 that best describes what would happen if cells of *Pyrolobus fumarii* were placed under the same conditions. Describe why neither organism would grow.

4. Would you expect to find a psychrophilic microorganism alive in a hot spring? Why? It is frequently possible to isolate thermophilic microorganisms from cold-water environments. Give an explanation of how this can be.

The genetic material of all cells is double-stranded deoxyribonucleic acid (DNA), the molecular structure of which is shown here. In order for an organism to grow and reproduce, the genetic information stored in the double helix must be precisely replicated and also used to make specific ribonucleic acid (RNA) and protein molecules. The processes used to carry out these functions were first elucidated in microorganisms, but the basic mechanisms are very similar in all cellular organisms, from bacteria to humans.

PRINCIPLES OF MICROBIAL MOLECULAR BIOLOGY

7

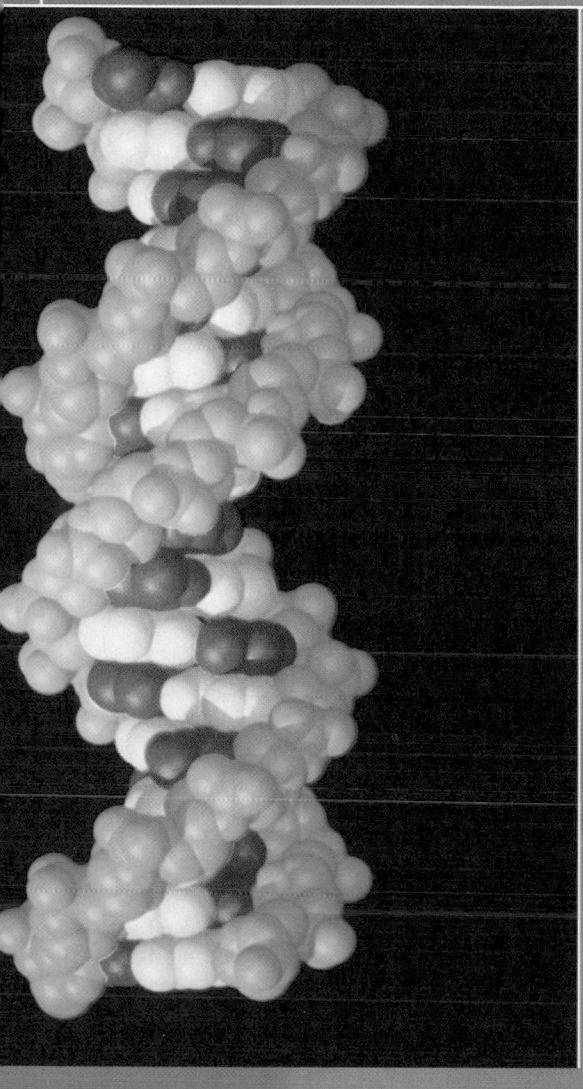

Working Glossary

Anticodon a sequence of three bases in a tRNA molecule that base-pairs with a codon during protein synthesis

Aminoacyl-tRNA synthetases a group of enzymes each one of which catalyzes the attachment of the correct amino acid to a tRNA

Antiparallel in reference to double-stranded DNA, one strand runs 5′ → 3′ and the other 3′ → 5′

Chromosome a genetic element, usually circular in prokaryotes and linear in eukaryotes, carrying genes essential to cellular function

Codon a sequence of three bases in mRNA that encodes an amino acid

Complementary nucleic acid sequences that can base-pair with each other

DNA gyrase an enzyme found in most prokaryotes that introduces negative supercoils in DNA

DNA polymerase an enzyme that synthesizes a new strand of DNA in the 5′ → 3′ direction using an antiparallel DNA strand as a template

Exon the coding DNA sequences in a split gene (contrast with intron)

Gene a segment of DNA specifying a protein (via mRNA), a tRNA, or an rRNA

Genome the total complement of genes contained in a cell or virus

Intron the intervening noncoding DNA sequences in a split gene (contrast with exon)

Messenger RNA (mRNA) an RNA molecule that contains the genetic information necessary to encode a particular protein

Molecular chaperones a group of proteins that help other proteins fold or refold from a partially denatured state

Operon a cluster of genes whose expression is controlled by a single operator

Primary transcript an unprocessed RNA molecule that is the direct product of transcription

Primer a molecule (usually a polynucleotide) to which DNA polymerase can attach the first deoxyribonucleotide during DNA replication

Promoter a site on DNA to which RNA polymerase can bind and begin transcription

Replication synthesis of DNA using DNA as a template

Ribosomal RNA (rRNA) types of RNA found in the ribosome; some participate actively in the process of protein synthesis

Ribosome a cytoplasmic particle composed of ribosomal RNA and protein, which is a central part of the protein-synthesizing machinery of the cell

Ribozyme an RNA molecule that can catalyze chemical reactions

RNA polymerase an enzyme that synthesizes RNA in the 5′ → 3′ direction using an antiparallel DNA strand as a template

RNA processing the conversion of a precursor RNA to its mature form

Semiconservative replication DNA synthesis yielding new double helices, each consisting of one parental and one progeny strand

Splicesome a complex of several proteins and small RNAs that functions to remove introns and join adjacent exons to form mature mRNA in eukaryotic cells

Splicing the RNA processing step by which introns are removed and exons joined

Telomerase an enzyme that contains a short RNA template and that functions to replicate the ends (telomeres) of eukaryotic chromosomes

Transcription the synthesis of RNA using a DNA template

Transfer RNA (tRNA) an adaptor molecule used in translation that has specificity for both a particular amino acid and for one or more codons

Translation the synthesis of protein using the genetic information in messenger RNA as a template

---

The term *molecular biology* typically applies to the study of the structure and function of the diverse array of **macromolecules** found in living cells. This chapter examines only those macromolecules that are directly involved in the flow of information in microorganisms.

In Chapter 1 we characterized cells as "chemical machines" and "coding devices" (⌾ Figure 1.4). In Chapter 3 we examined the basic structure of cellular macromolecules. In this chapter we will look at how cells code genetic information and the mechanisms they use to process and use this information. In the next chapter we shall describe how the expression of genetic information is regulated. Subsequent chapters concern the genetic information of some noncellular forms, the transfer of genetic information, and how genetic information can be manipulated.

I OVERVIEW OF GENES AND GENE EXPRESSION

Genetics is the discipline that deals with the mechanisms by which traits are passed from one organism to another and how they are expressed. Since biological information flow is the basis of cellular function, genetics is a major research tool. Coupled with molecular biology, genetics allows us to understand the molecular mechanisms by which cells function. The study of genetics at the molecular level is also central to an understanding of the variability of organisms and the evolution of species.

Understanding how information is transmitted through biological systems also has tremendous practical applications. For instance, genetics provides us with techniques to specifically modify organisms for our own purposes. Many important recent advances in agriculture, industry and medicine are dependent on molecular biology and genetics. An understanding of biological information flow in a cell not only allows us to understand how normal organisms function, but also how these functions are altered by disease. Armed with such knowledge, scientists are developing not only methods for controlling important infectious diseases, but also genetic-based cures for metabolic disorders once thought incurable. We will have much to say about the application of genetics (including the genetics of microorganisms) to human affairs in subsequent chapters.

7.1 Macromolecules and Genetic Information

Many of the molecules involved in genetic information flow are **macromolecules**. Long before we knew what these molecules were, or even what steps in the flow might be, it was clear that there was a functional unit of genetic information. This unit has come to be called the **gene**.

What Is a Gene and What Is Its Function?

Most genes are entities that specify the structure of a single polypeptide or *protein* chain. We discussed the chemistry of proteins in Chapter 3 and noted that they consist of one or more polypeptide chains. A polypeptide is composed of a series of amino acids connected in peptide linkage. There are usually 20 different amino acids present in proteins (∞ Figure 3.12), and a single protein molecule typically has several hundred amino acid residues (∞ Sections 3.6–3.8). The gene is the element of information that specifies the *sequence* of amino acids of the protein. Genes are stored information, whereas proteins are the cell's functional entities.

In all cells genes are composed of *deoxyribonucleic acid* (DNA). The information in the gene is present as the sequence of bases in the DNA. Interestingly, the information stored in the DNA specifies the sequence of a protein only through an intermediary macromolecule, *ribonucleic acid* (RNA). RNA can serve either as a true informational intermediate (a messenger), or, in some cases, as a more active part of the cell's machinery. Because all three of these molecules, DNA, RNA, and protein, contain biological information in their sequences, they are often called **informational macromolecules** (∞ Section 3.2).

In DNA and RNA the information is encoded in the *base sequence* of the purine and pyrimidine bases of the polynucleotide chain. When we discuss the information content of a nucleic acid, we thus speak of the *coding* properties of this material. The amino acid sequence of the polypeptide is *coded* by the sequence of purine and

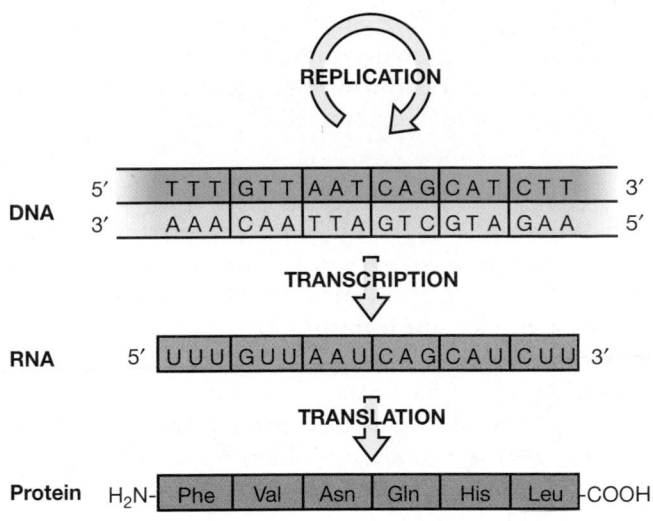

Figure 7.1 Synthesis of the three types of informational macromolecules. Note that in any particular region, only one of the two strands of the DNA double helix is transcribed.

pyrimidine bases within the nucleic acid. We will discuss this coding function in detail in this chapter.

The Steps in Information Flow

The molecular processes underlying genetic information flow can be divided into three stages, which we summarize briefly here (Figure 7.1).

1. *Replication.* The DNA molecule, which serves as the cell's genetic material, is a **double helix** of two long chains (∞ Section 3.5). During replication, DNA is duplicated, producing two double helices.

2. *Transcription.* DNA does not participate directly in protein synthesis but through an RNA intermediate. The transfer of the information to RNA is called **transcription**, and the RNA molecule carrying the information to encode a protein is called **messenger RNA** (mRNA). In most cases, at any particular location on the chromosome, only one strand of the DNA is transcribed (Figure 7.1). Some regions of DNA that are transcribed do not encode proteins but rather contain information for other types of RNA, such as **transfer RNA** (tRNA) and **ribosomal RNA** (rRNA). Therefore, we must expand our definition of a gene to include regions of DNA that encode one of these types of RNA.

3. *Translation.* The specific sequence of amino acids in each protein is determined by a specific sequence of bases in the mRNA (which was transcribed from the DNA). There is a linear correspondence between the base sequence of a gene and the amino acid sequence of a polypeptide (Figure 7.1). It takes *three* bases on the mRNA to encode a single amino acid, and each such triplet of bases is called a **codon**. This

genetic code is actually translated into protein by means of the protein-synthesizing system. This system consists of **ribosomes** (which are themselves made up of proteins and rRNA), tRNA, and a number of enzymes.

In the processes of replication, transcription, and translation, the information in the sequence of the nucleic acid may specify either the sequence of another nucleic acid or of a protein. However, the transfer of sequence information from nucleic acid to protein is unidirectional; the sequence of a protein **does not** specify the sequence of a nucleic acid. This one-way transfer of genetic information from nucleic acid to protein is sometimes called the *central dogma of molecular biology* because it holds true for all life forms on our planet.

The three transfer steps shown in Figure 7.1 are those used in all cells. In Chapter 9 we will learn that information transfer in the reproduction of some viruses can involve two other types of transfer. One is *RNA replication,* where RNA is used as a template for RNA synthesis. The other is *reverse transcription,* where the use of sequence information in RNA specifies a sequence in DNA. Note that in both cases the information transfer is from nucleic acid to nucleic acid, and therefore neither violates the central dogma.

Prokaryotic and Eukaryotic Genetics

Replication, transcription, and translation occur in all cellular organisms. However, there are some differences in the mechanisms of these processes in prokaryotes and in eukaryotes. In part this is due to differences in the organization of the genetic information and to the fact that eukaryotes have a nucleus.

We emphasized in Chapter 2 the basic differences in the organization of DNA in prokaryotes and eukaryotes. In summary, the typical prokaryotic genome consists of a single, covalently closed *circular* molecule of DNA found in the cytoplasm of the cell, and the eukaryotic genome consists of several *linear* pieces of DNA present in individual chromosomes in the cell nucleus.

In prokaryotes there is no membrane separating the chromosome from the cytoplasm (Section 2.2). In eukaryotes, however, the chromosomes are located inside the nucleus and the ribosomes in the cytoplasm, so transcription and translation are spatially separated.

In all types of cells the definition of a gene is the same: a segment of DNA specifying a protein (via mRNA), a tRNA, or an rRNA. However, in eukaryotes protein-encoding genes are frequently split into two or more coding regions, with noncoding regions separating coding regions. The coding sequences are

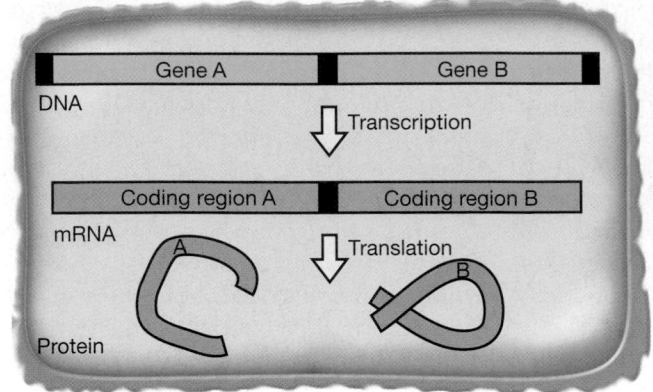

(a) **Prokaryote**

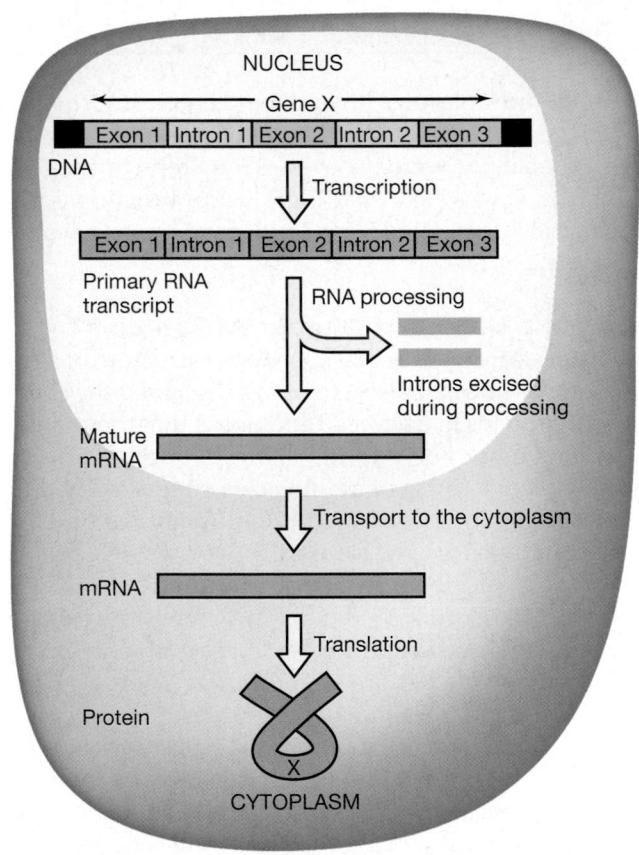

(b) **Eukaryote**

Figure 7.2 Contrast of information transfer in prokaryotes and eukaryotes. (a) Prokaryote. A single mRNA often contains more than one coding region (such mRNAs are called *polycistronic*). (b) Eukaryote. Noncoding regions (*introns*) are removed from the primary RNA transcript before translation. The mRNAs of eukaryotes are almost always *monocistronic*. Please note the two types of cells are not drawn to scale. A typical prokaryotic cell would be 1 to 2 μm in diameter, and a typical eukaryotic (animal) cell about 25 μm in diameter (Figure 2.3).

called **exons**, and the intervening noncoding regions, **introns**. Both intron and exon regions are transcribed into the **primary transcript**, or **pre-mRNA**, and the functional mRNA is subsequently formed by removal of noncoding regions. A summary contrasting genetic phenomena in prokaryotes and eukaryotes is given in Figure 7.2.

✓ 7.1 Concept Check

The three key processes of macromolecular synthesis are DNA replication; transcription, the synthesis of RNA from a DNA template; and translation, the synthesis of proteins using messenger RNA as template. Although the basic processes are the same in both prokaryotes and eukaryotes, the organization of genetic information is more complex in eukaryotes. Many eukaryotic genes have both coding regions (exons) and noncoding regions (introns).

✓ What three informational macromolecules are involved in genetic information flow?

✓ In all cells there are three processes involved in genetic information flow. What are they?

II DNA STRUCTURE

We dealt with the general structure of nucleic acids in Chapter 3. In the next few sections of this chapter we shall discuss the details of DNA structure necessary for an understanding of molecular genetics and the types of genetic elements containing DNA that are found in cells. With this information as a basis, we can then discuss how DNA is replicated, transcribed into RNA, and translated into protein.

7.2 DNA Structure: The Double Helix

The genetic information for all cellular processes is stored in DNA in the *sequence* of bases along the polynucleotide chain. As we have noted, only four different nucleic acid bases are found in DNA: adenine (A), guanine (G), cytosine (C), and thymine (T). As already shown in Figure 3.11, the backbone of the DNA chain consists of alternating units of phosphate and the sugar *deoxyribose*; connected to each sugar is one of the nucleic acid *bases*. Recall especially the numbering system for the positions of sugar and base; the phosphate connecting two sugars spans from the 3′-carbon of one sugar to the 5′-carbon of the adjacent sugar (see Figure 7.14). At one end of the DNA molecule the sugar has a phosphate on the 5′-hydroxyl, whereas at the other end the sugar has a free hydroxyl at the 3′-position.

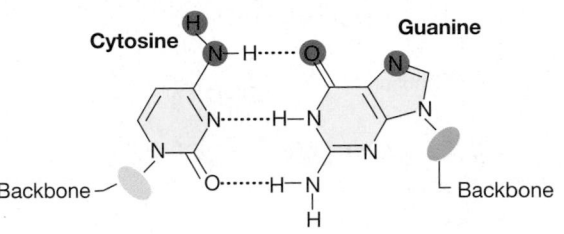

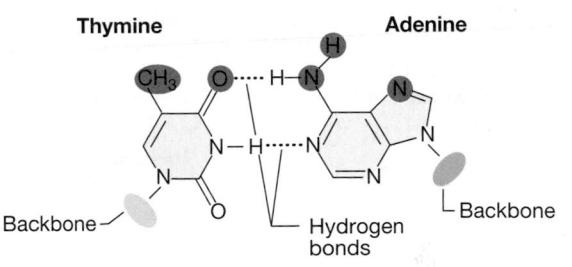

Figure 7.3 Specific pairing between adenine (A) and thymine (T) and between guanine (G) and cytosine (C) via hydrogen bonds. These two base pairs are the base pairs typically found in double-stranded DNA. Atoms that are found in the major groove of the double helix and that interact with proteins are highlighted in red. The deoxyribose phosphate backbones of the two strands of DNA are also indicated.

DNA as a Double Helix

As we will discuss in Chapter 9, the chromosomes of some viruses are single-stranded. In all **cellular organism chromosomes**, however, DNA exists as two polynucleotide strands whose base sequences are **complementary**. The complementarity of DNA arises because of the specific pairing of the purine and pyrimidine bases: Adenine always pairs with thymine, and guanine always pairs with cytosine (Figure 7.3). The two strands in the resulting **double-stranded** molecule are arranged in an *antiparallel* fashion (see Figure 7.4). This means the two strands are in a "head-to-toe" arrangement. In Figure 7.4 the strand on the left is arranged 5′ to 3′ top to bottom, whereas the other strand is 5′ to 3′ bottom to top. The two strands are wrapped around each other forming a **double helix** (Figure 7.5). In this double helix, DNA forms two distinct grooves, the *major groove* and the *minor groove*. Of the many proteins that interact specifically with DNA (as we shall see in Chapter 8), most engage predominantly with the major groove, where there is a considerable amount of space. Because of the regularity of the double helix, some atoms of the bases are always exposed in the major groove (and some in the minor groove). Atoms in the major groove that are known to be important in interactions with proteins are shown in Figure 7.3.

The size of a DNA molecule can be expressed in terms of its *molecular weight*, but because a single nucleotide has a molecular weight of about 330, and because DNA molecules are many nucleotides long, the molecular weight mounts up rapidly. (The nucleic acid in

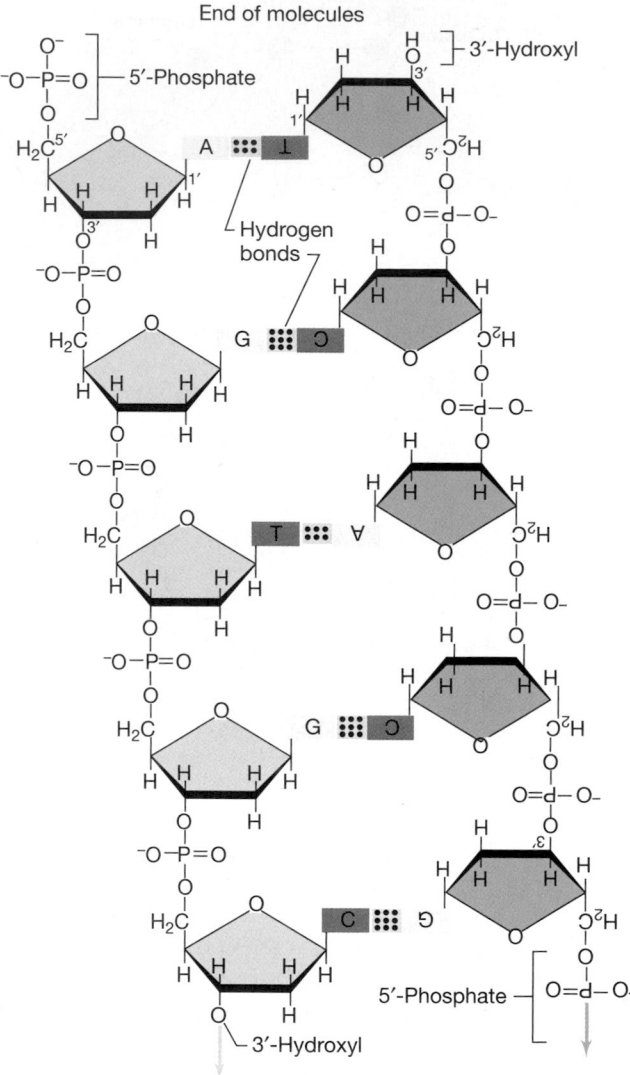

Figure 7.4 DNA structure. Complementary and antiparallel nature of DNA. Note that one chain ends in a 5'-phosphate group, whereas the other ends in a 3'-hydroxyl. The red bases represent the pyrimidines cytosine (C) and thymine (T), and the yellow bases represent the purines adenine (A) and guanine (G).

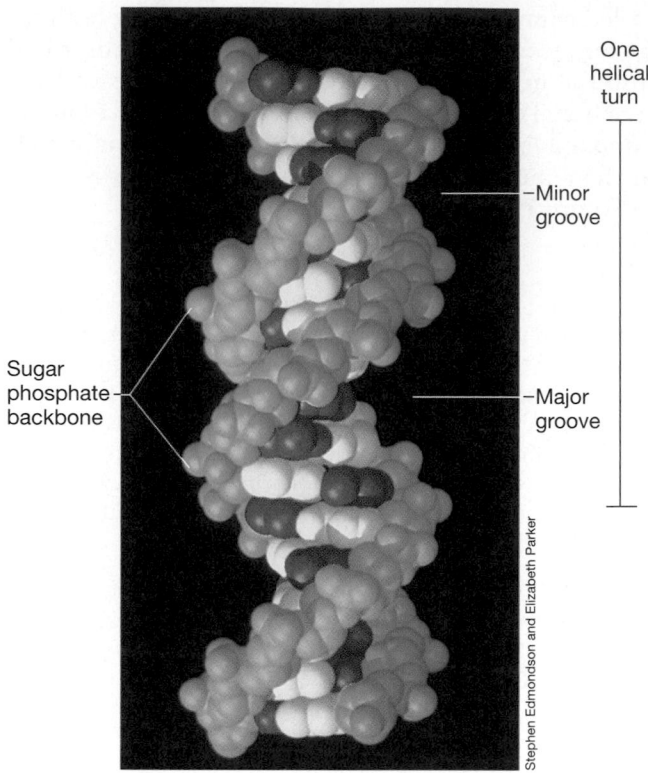

Figure 7.5 A computer model of a short segment of DNA showing the overall arrangement of the double helix. One of the sugar-phosphate backbones is shown in blue and the other in green. The pyrimidine bases are shown in red and the purines in yellow. Note the locations of the major and minor grooves. The model was produced using software from the Computer Graphics Laboratory, University of California at San Francisco.

even small viruses, for instance, may have a molecular weight in the millions, the DNA in cells in the billions.) A more convenient way of expressing the sizes of DNA molecules is in terms of the *number of thousands* of nucleotide bases, or base pairs, per molecule. Thus, a DNA molecule with 1000 bases contains 1 *kilobase* of DNA. If the DNA is a double helix, then one speaks of *kilobase pairs*. Thus, a double helix 5000 base pairs in length would have a length of 5 kilobase pairs. The bacterium *Escherichia coli* has about 4600 kilobase pairs of DNA in its chromosome. Given the huge amount of genomic sequence information now becoming available, it is often useful to speak of *millions* of base pairs, or *megabase pairs*. The genome of *E. coli* is 4.6 megabase pairs.

Each base pair takes up 0.34 nanometers (nm) in length along the helix, and each turn of the helix contains 10 base pairs. Therefore, 1 kilobase of DNA has 100 turns of the helix and is 0.34 μm long. Calculations such as this can be very interesting, as we shall soon see.

Note that unlike molecular weight, which relates to mass, the measurement of nucleic acids by the number of bases or base pairs is a measurement of *length*. While a single-stranded nucleic acid molecule may have less structure than a double-stranded one, either type is best described by the number of bases in a single chain. Therefore, molecular biologists use the abbreviation *kb*, for *kilobase*, in describing both double-stranded and single-stranded molecules.

Sequence-Specific Features of DNA Structure

In regions of the chromosome that encode proteins, the sequence of the DNA is constrained in large measure by the amino acid sequence of the encoded proteins and the nature of the genetic code (see Section 7.13). However, there are frequently base sequences in DNA that are present not because of their coding properties but because they influence the *secondary structure* of DNA, or the way in which DNA interacts with proteins.

Long DNA molecules are quite flexible, but stretches of DNA less than 100 base pairs are much more rigid. Some short segments of DNA can be bent by proteins that interact with them. However, certain sequences themselves result in bends in the DNA. The sequences of this **bent DNA** often involve several runs of five or six adenines (in the same strand), each separated by four or five bases. In addition, some sequences contribute to the ability of DNA to bend when certain proteins interact with the DNA. DNA bending seems to be commonly involved in the regulation of gene expression, as we shall discuss in Chapter 8.

Short, repeated sequences are often found in DNA molecules. Many proteins interact with regions of DNA containing repeated sequences (⊸ Chapter 8) but that are repeated in inverse orientation. This type of repeat is called an **inverted repeat**. Inverted repeats give the DNA sequence a twofold symmetry. As shown in Figure 7.6, nearby inverted repeats could theoretically lead to the formation of **stem-loop** (cruciform) structures in DNA.

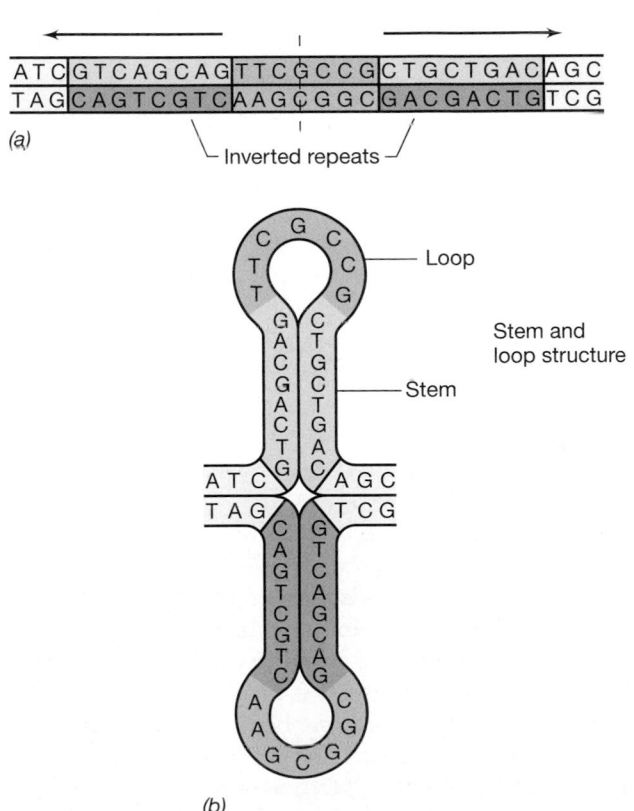

(a)

Inverted repeats

(b)

Loop

Stem and loop structure

Stem

Figure 7.6 Inverted repeats and the formation of a stemloop structure. (a) Nearby inverted repeats in DNA. The arrows indicate the symmetry around the imaginary axis (dashed line). (b) Formation of stem-loop structures (cruciform structures) by pairing of complementary bases on the *same* strand.

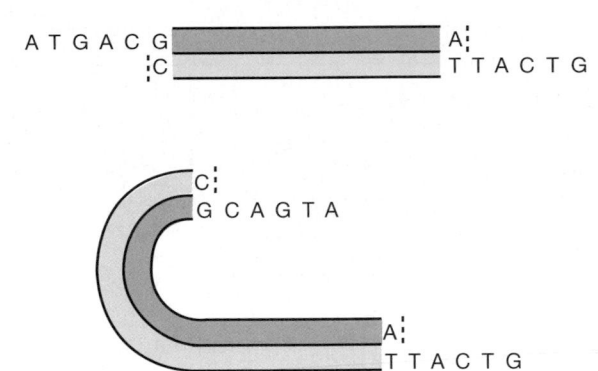

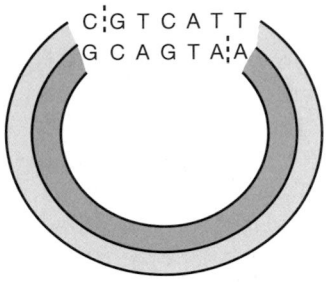

Figure 7.7 Linear DNA with complementary single-stranded ends ("sticky ends") can cyclize by base pairing of the complementary ends.

(Note that the stem of a stem-loop is a short double helix with normal base pairing and antiparallel strands.)

By examining Figure 7.6 it can be seen that either strand from this region could form a stem-loop. Therefore, some inverted repeats found in DNA that is transcribed may be important because of the stem-loops found in the RNA. Such *secondary structure* (⊸ Section 3.5) formed by base pairing within a single strand of nucleic acid is very important in both transfer RNA (see Section 7.14) and ribosomal RNA (⊸ Section 11.4 and Figure 11.8).

The *ends* of linear DNA molecules can also have interesting sequences that lead to changes in structure. Some DNA molecules have single-stranded regions at each end that are complementary. This leads to the possibility that the two ends can find each other and associate by complementary base pairing, as illustrated in Figure 7.7 for the formation of a circle. The genome of the bacterial virus lambda circularizes in this fashion (⊸ Section 9.10). DNA with single-strand complementary sequences at the ends is said to have "sticky ends." Some linear DNA molecules have **hairpin** structures at each end. A hairpin is like a stem-loop but with almost no loop. Hairpins can be formed from a single-stranded region at the end of a molecule that contains an inverted repeat, as illustrated in Figure 7.8. We shall discuss other important sequences found at the ends of the linear DNA in eukaryotic chromosomes in Section 7.7.

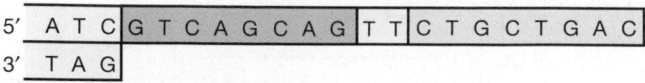

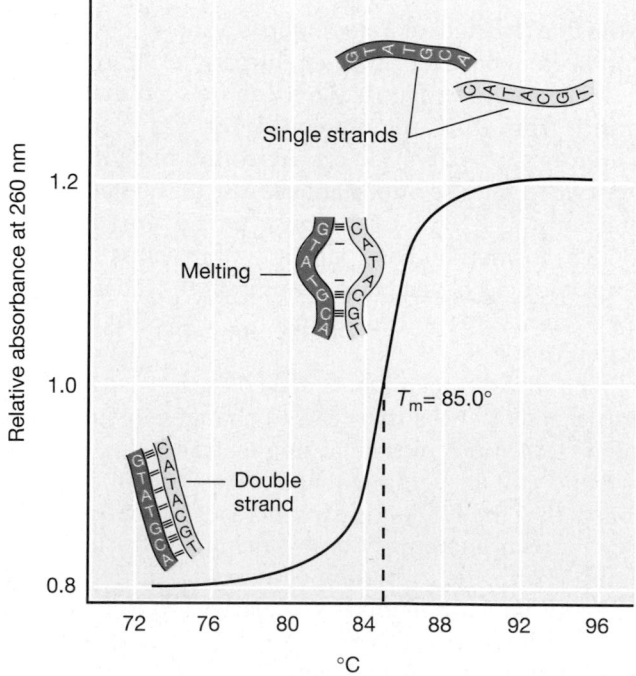

Figure 7.8 A hairpin structure at one end of a linear DNA molecule. If the linear DNA had been completely double-stranded, the sequences shown in green would have been inverted repeats.

The Effect of Temperature on DNA Structure

Although the hydrogen bonds between the base pairs are individually very weak (∞ Section 3.1), many such bonds hold together the strands of a typical double-stranded DNA molecule. There may be millions or even hundreds of millions of hydrogen bonds, depending on the number of base pairs in the molecule. Recall from Figure 7.3 that each adenine-thymine base pair has two

Figure 7.9 Thermal denaturation of DNA. The DNA absorbs more ultraviolet light as the double helix is denatured. The transition is quite abrupt, and the temperature of the midpoint, T_m, is directly related to the GC content of the DNA. Although the denatured DNA can be renatured by a slow cooling, the process does not follow a similar curve. Renaturation becomes progressively more complete at temperatures well below the T_m, and then only after a considerable incubation time.

such bonds, while each guanine-cytosine base pair has three. This makes GC pairs stronger than AT pairs.

When isolated from cells and kept near room temperature and at physiological salt concentrations, DNA remains in a double-stranded form. However, if the temperature is raised, the hydrogen bonds will break, but the covalent bonds holding a chain together will not, and so the DNA strands will separate. This process is generally called *melting* and can be measured experimentally because single-stranded and double-stranded nucleic acid differ in their ability to absorb certain wavelengths of ultraviolet light, even when they have the same sequence (Figure 7.9). DNA with high numbers of GC pairs melts at a higher temperature than a similar-sized molecule with more AT pairs. If the heated DNA is allowed to cool slowly, the double-stranded native DNA can reform. Interestingly, such a process can be used not only to reform native DNA but also to form *hybrid* molecules whose two strands come from separate sources. **Hybridization**, the artificial construction of a double-stranded nucleic by complementary base pairing of two single-stranded nucleic acids, can be a powerful and useful technique (∞ Section 10.12).

✓ 7.2 Concept Check

DNA is generally arranged as a double-stranded molecule that assumes a helical configuration. It is measured in terms of base pairs. The two strands in the double helix are antiparallel. The strands of a double-helical DNA molecule can be separated experimentally by heat in a process called melting.

- ✓ Explain what antiparallel means in regard to the structure of double-stranded DNA.
- ✓ Define "complementary" as it is used to refer to two strands of DNA.

7.3 DNA Structure: Supercoiling

A *relaxed* DNA molecule is one with exactly the number of turns one would predict by knowing the number of base pairs. An example is shown in Figure 7.10. If you consider the length of a simple, relaxed double helix and the size of microbial cells and viruses that must accommodate this DNA, however, it is clear that there must be some higher order structure, because such a molecule could not otherwise be packed into a cell. For instance, if we calculate the length of DNA in the *Escherichia coli* chromosome, we will find it to be more than 1 mm, about 500 times longer than the *E. coli* cell itself! Such packaging problems also occur in viral DNA and the DNA of eukaryotic cells. How is it possible to pack so much DNA into such a little space? The solution: *supercoiling.*

(a) Relaxed, covalently closed
circular DNA

Break one strand ⇅ Seal

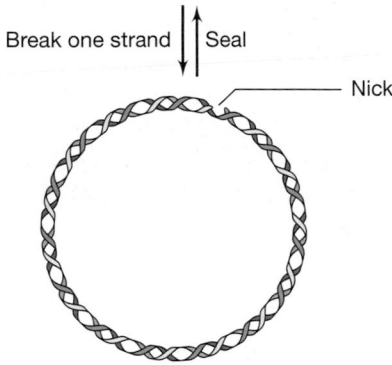

Nick

(b) Relaxed, nicked circular DNA

Break one strand ↕ Rotate one end of
broken strand around
helix and seal

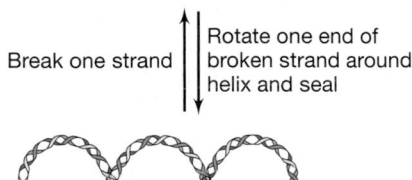

(c) Supercoiled circular DNA

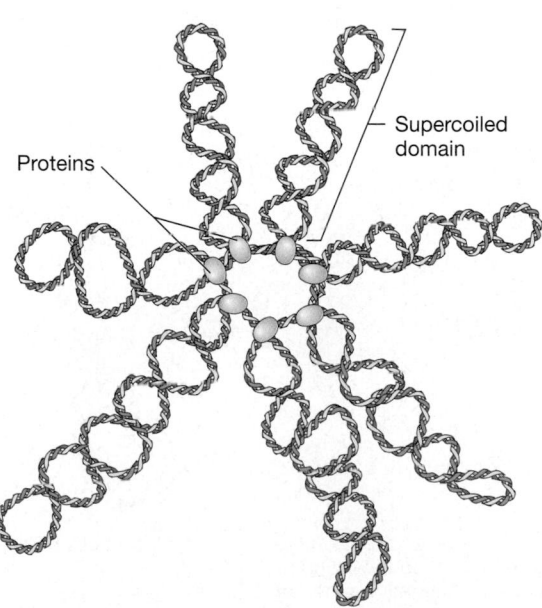

Proteins

Supercoiled
domain

(d) Chromosomal DNA with supercoiled
domains

Supercoiled DNA

Supercoiling is a state in which double-stranded DNA molecules are further twisted. Figure 7.10 diagrams how this could happen in a circular DNA duplex. Supercoiling puts the DNA molecule under torsion. DNA can be supercoiled in either a *positive* or *negative* direction. **Negative supercoiling** occurs when the DNA is twisted about its axis in the *opposite* direction from that of the right-handed double helix. It is in this form that supercoiled DNA is predominantly found in nature.

Although there are proteins associated with the DNA of prokaryotes, one can almost consider the chromosomes of prokaryotes to be "naked" DNA. This is not the case for the chromosomes of eukaryotes in which large amounts of protein are bound to the DNA in a very regular fashion. As has been mentioned (see Section 7.1) each eukaryotic chromosome contains a linear double-stranded DNA molecule. This linear DNA molecule is wound around proteins called **histones** in a very regular way to form structures called **nucleosomes** (Figure 7.11). In eukaryotic chromosomes, the formation of the nucleosome introduces negative supercoils. Histones are positively charged proteins that can neutralize the negative charge of DNA (resulting from the phosphate groups). The nucleosomes are spaced along the double helix at very regular intervals, but can aggregate to form a fibrous material called *chromatin.* Chromatin itself can be further compacted by folding and looping to eventually form a very dense structure. It is these compact structures that are most easily visible during cell division (see Figure 7.23).

Topoisomerases

Supercoiling of DNA in prokaryotes is typically brought about in a much different manner than it is in eukaryotes. In *Bacteria* and most *Archaea,* there is an enzyme called **DNA gyrase**, which introduces negative supercoils. The process can be thought to occur in several stages. First, the circular DNA molecule is twisted, then a break occurs where the two chains come together, and then the broken double helix is resealed on the opposite side of the intact strand (Figure 7.12). DNA gyrase is a *topoisomerase,* specifically a topoisomerase II. Note the derivation of the name *topoisomerase. Topology* is the branch of mathematics that deals with the properties of geometric figures that are unaltered when the figures

Figure 7.10 Supercoiled DNA. Parts a, b and c show supercoiled circular DNA and relaxed, nicked circular DNA interconversions. A nick is a break in a phosphodiester bond of one strand. (d) In actuality, the double-stranded DNA in the bacterial chromosome is arranged not in one supercoil but in several *supercoiled domains,* as shown here. In *Escherichia coli* over 50 supercoiled domains are thought to exist, each of which is stabilized by binding to specific proteins.

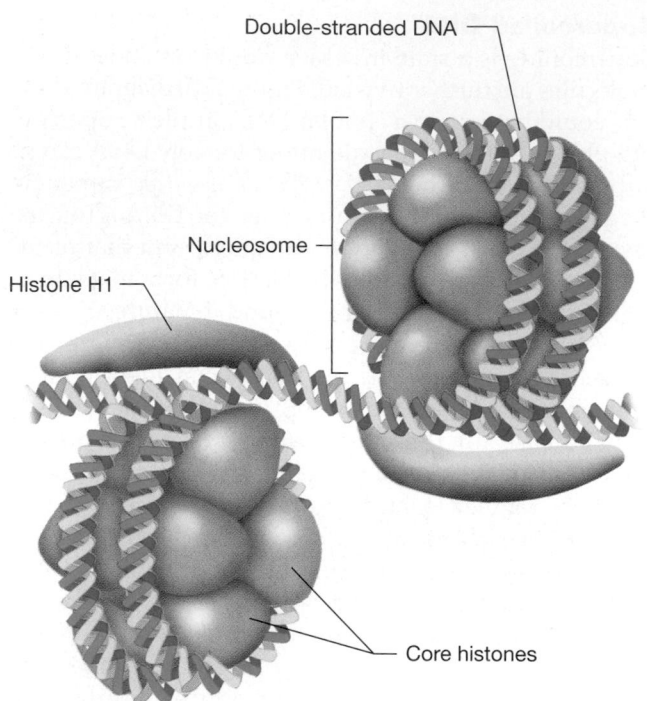

Figure 7.11 Packaging of DNA around a core of histone proteins to form a nucleosome. Nucleosomes are arranged along the DNA strand somewhat like beads on a string. This arrangement is typical of DNA in eukaryotic cells.

are twisted or contorted. We are dealing here with the topology of DNA, and a topoisomerase is an enzyme that affects this topology. It is interesting that some of the antibiotics that act on *Bacteria,* such as the quinolones (e.g., *nalidixic acid*), fluoroquinolones (e.g., *ciprofloxacin*) and *novobiocin,* inhibit the action of DNA gyrase. Novobiocin is also effective against several species of *Archaea,* where it also seems to inhibit DNA gyrase.

Another enzyme is able to *remove* supercoiling in DNA. This enzyme, called *topoisomerase I,* introduces a single-strand break in the DNA and causes the rotation of one single strand of the double helix around the other. As was shown in Figure 7.10, a break in the backbone (a **nick**) of either strand allows the DNA to return to the relaxed state. This is true whether the supercoiling is positive or negative. Such enzymes are found in both prokaryotes and eukaryotes. Linear DNA, as in eukaryotic chromosomes, is prevented from returning to the relaxed state by the proteins bound to it. To prevent the entire bacterial chromosome from becoming relaxed every time a nick is made, the chromosome contains *supercoiled domains* as shown in Figure 7.10. A nick in the DNA in one of these domains does not relax the DNA in the others. It is unclear what holds the DNA in these domains, but it is likely to involve proteins.

Through the action of these topoisomerases, the DNA molecule can be alternately supercoiled and relaxed. Because supercoiling is necessary for packing the DNA into the confines of a cell and relaxing is necessary so DNA can be replicated, these two complementary processes clearly play an important role in the behavior of DNA in the cell. In most prokaryotes, the actual level of negative supercoiling is the result of a balance between the activity of DNA gyrase and topoisomerase I. In addition, however, supercoiling is known to affect gene expression. Certain genes are more actively transcribed when DNA is supercoiled, whereas transcription of other genes is inhibited by excessive supercoiling.

A few prokaryotes contain an enzyme called *reverse gyrase.* This topoisomerase introduces *positive* supercoils in DNA. The organisms that contain this enzyme are mostly species of *Archaea* among those that grow at extremely high temperatures (hyperthermophiles, ∞ Sections 6.10

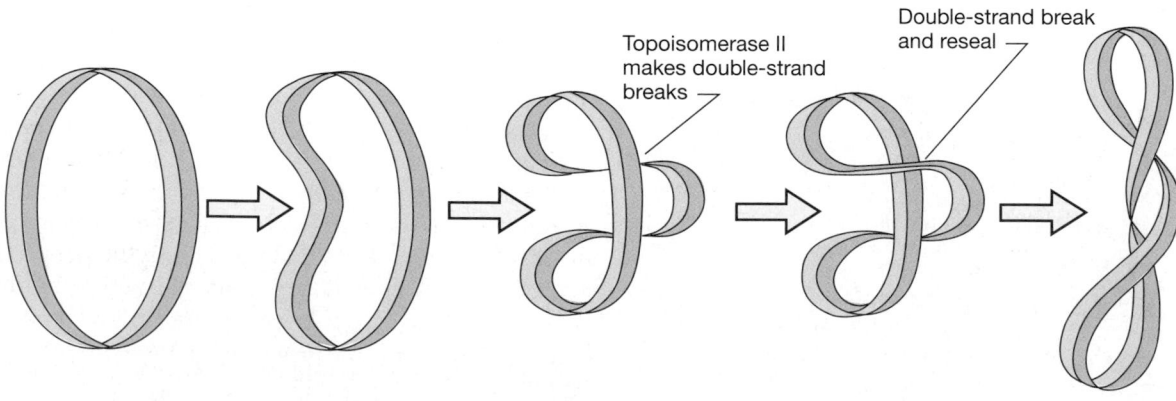

1. Relaxed circle

2. One part of circle is laid over the other

3. Result is contact between the helix in two places. Note that no twisting has as yet been introduced.

4. After topoisomerase II action, twisting (a negative supercoil) has been introduced.

5. Supercoiled DNA

Figure 7.12 Introduction of supercoiling into a circular DNA by activity of topoisomerase II, which makes double-strand breaks.

and 13.11). Some DNA in these organisms seems to be "relaxed" that is, with neither positive nor negative supercoils. This may be a matter of a balance between reverse gyrase and the activity of histonelike proteins. Interestingly, one organism, the hyperthermophilic archaeon *Methanothermus fervidus*, has nucleosome-like structures in which the DNA is positively supercoiled (∞ Section 13.11).

Positive supercoiling, brought about by reverse gyrase or other means, is a kind of "overwinding" and might play an important role in protecting the DNA from being denatured. However, in order to be of use, the information in DNA must be accessible to the cell's machinery. Therefore, in all cells, the structure of DNA is likely to be quite dynamic.

✓ 7.3 Concept Check

The very long DNA molecule can be packaged into the cell because it is supercoiled. In most prokaryotes this supercoiling is brought about by enzymes called topoisomerases. In eukaryotic chromosomes, DNA is wound around proteins called histones, forming structures called nucleosomes.

- ✓ Why is supercoiling important?
- ✓ What function do topoisomerases serve inside cells?

7.4 Genetic Elements

Before proceeding to describe how cells replicate their DNA, we consider the kinds of structures that must be replicated. Structures containing genetic material (DNA in most organisms but RNA in some viruses) can be called *genetic elements*. The **genome** is the total complement of genes in a cell or virus. Although the main genetic element is the *chromosome*, other genetic elements are known and play important roles in gene function in both prokaryotes and eukaryotes (Table 7.1).

The Chromosome

In Section 2.2 we discussed the fact that a typical prokaryote has a single chromosome containing all (or most) of the genes found inside the cell. Eukaryotes have multiple chromosomes as a part of their genome. Also, the typical prokaryotic chromosome is a circular DNA molecule, whereas the DNA in all known eukaryotic chromosomes is linear. In Table 7.2 the number, size, and configuration of chromosomes in a few known microorganisms, both prokaryotic and eukaryotic, are given. Note that the chromosome of the bacterium *Borrelia burgdorferi*, the causative agent of Lyme disease (∞ Section 27.4), is linear. The ends of this chromosome may have hairpin repeats like those shown in Figure 7.8. Although uncommon, linear chromosomes are now known to exist in several other *Bacteria*. Members of the genus *Streptomyces* also have linear chromosomes, but these

TABLE 7.1	Kinds of genetic elements
Element	**Description**
Prokaryote	
Chromosome	Extremely long, usually circular, double-stranded DNA molecule
Plasmid	Typically a relatively short, usually circular, double-stranded DNA molecule which is extrachromosomal
Viral genome	Single- or double-stranded DNA or RNA molecule
Transposable element	Double-stranded DNA molecule always found within another DNA molecule
Eukaryote	
Chromosome	Extremely long, linear, double-stranded DNA molecule
Plasmid[a]	Typically a relatively short circular or linear double-stranded DNA molecule which is extrachromosomal
Mitochondrion or chloroplast	Intermediate-length DNA molecules, usually circular
Viral genome	Single- or double-stranded DNA or RNA molecules
Transposable element	Double-stranded DNA molecule always found within another DNA molecule

[a]Plasmids are very uncommon in eukaryotes.

have proteins covalently bound to their ends. We will discuss the importance of such proteins in Section 7.7.

The few examples of prokaryotes given in Table 7.2 are not random choices. They include species of *Archaea* as well as *Bacteria*, and examples of among the smallest and largest known prokaryotic chromosomes. We shall discuss the fact later, that some prokaryotes, such as *Rhodobacter sphaeroides*, have more than one chromosome.

Only a few examples of eukaryotic microorganisms are also given in Table 7.2, but all have linear DNA and multiple chromosomes even in the haploid state. Both these conditions are the rule in all eukaryotic organisms, whatever their size. In yeast (and probably many other eukaryotic microorganisms), the length of the DNA in a single chromosome is actually shorter than that of a linearized prokaryotic chromosome. For instance, the total amount of DNA per yeast cell is only three times that in *Escherichia coli*, but yeast has 16 chromosomes. In higher organisms, however, the length of the DNA molecule in a single chromosome is many times greater than that in the prokaryotic chromosome if it were opened and linearized. The haploid human genome has only a few more chromosomes than yeast but has over 200 times more DNA.

The linear DNA molecule in a eukaryotic chromosome has a special DNA sequence called a *telomere* at each end and a *centromere* somewhere between the telomeres. Centromeres are important for partitioning the chromosomes during cell division. As we shall see, telomeres play an important role in the replication of

TABLE 7.2 Sizes, shapes, and numbers of chromosomes in selected microorganisms from each domain of life

Organism	Comments	Chromosome Size (Mb)[a]	Number	Geometry
Bacteria				
Mycoplasma genitalium	Smallest known cellular genome	0.58	1	
Borrelia burgdorferi	Causes Lyme disease (Chapter 27)	0.91[b]	1	
Haemophilus influenzae	Gram-negative, can cause disease (Chapter 26)	1.83[c]	1	
Rhodobacter sphaeroides	Gram-negative, phototrophic	4.00[d]	2	
Bacillus subtilis	Gram-positive, genetic model	4.21	1	
Escherichia coli K-12	Gram-negative, genetic model	4.64[e]	1	
Streptomyces coelicolor	Actinomycete, produces antibiotics (Chapter 12)	8.66	1	
Archaea				
Methanococcus jannaschii	Methanogen, which grows at high temperature (Chapters 6 and 13)	1.66	1	
Pyrococcus abyssi	Grows at high temperature (Chapters 6 and 13)	1.77	1	
Halobacterium sp. NRC1	Grows in high salt (Chapters 6 and 13)	2.57[f]	3	
Sulfolobus solfatarius	Grows at high temperature and high acidity (Chapters 6 and 13)	2.99	1	
Eukarya[g]				
Giardia lamblia	Flagellated protozoan that causes acute gastroenteritis (Chapter 14)	12.00	4	
Saccharomyces cerevisiae	Yeast, widely used in science and industry (Chapters 14 and 30)	12.06[h]	16	
Dictyostelium discoideum	Cellular slime mold, developmental model (Chapter 14)	34.0	6	
Tetrahymena thermophila	Ciliated protozoan (Chapter 14)	210.0	5	

[a] Mb is megabase pairs. In the case of the *Eukarya* the genome sizes and chromosome number are for the haploid form. (All the prokaryotic genomes listed have been completely sequenced.)

[b] This is for the linear chromosome. The genome of this organism also contains at least 17 circular and linear plasmids, which themselves have a combined size of more than 0.5 Mb.

[c] *Haemophilus influenzae* Rd was the first cellular organism to have its genome entirely sequenced.

[d] Chromosome I is 3.1 Mb and Chromosome II is 0.90 Mb. The strain sequenced also contains five plasmids.

[e] The reported sequence does not contain that of the F-plasmid (Section 10.8) nor that of bacteriophage lambda (Section 9.10), both of which would be present in a typical K-12 strain.

[f] There is one large chromosome of 2.01 Mb and two minichromosomes of 0.19 Mb and 0.37 Mb.

[g] All the organisms listed are single-celled.

[h] *Saccharomyces cerevisiae* was the first eukaryote to have its genome completely sequenced. The number given here does not include the mitochondrial genome and all the copies of some repeated sequences.

these molecules (see Section 7.7). The number of chromosomes is constant within a species but varies widely among species.

We mentioned in Section 7.3 an important structural difference between eukaryotic and prokaryotic chromosomes, namely that the DNA in eukaryotic chromosomes is packaged in nucleosomes. There are also interesting differences between eukaryotes and prokaryotes with regard to DNA sequence organization.

One difference is that many eukaryotic protein-encoding genes are interrupted or split by noncoding DNA sequences present between the sequences that actually code for a single polypeptide (see Figure 7.2). These noncoding intervening sequences are called **introns**, and the coding sequences are called **exons**. The number of introns per gene is variable and ranges from none to more than 50. During transcription, both introns and exons are copied, and the intron sequences are sub-

sequently cut out and removed when the messenger RNA is processed into its final form.

Essentially all the protein-encoding genes of the higher eukaryotes, such as mammals, have introns, and some of these are very large, often far larger than the exons. The lower eukaryotes have fewer and smaller introns. About 40% of the protein-encoding genes of the fission yeast *Schizosaccharomyces pombe* have introns, while only 4% of protein-encoding genes in the yeast *Saccharomyces cerevisiae* have introns. Introns are also known in the protein-encoding genes of prokaryotes but they are quite rare and are removed from the transcript by a different process.

The other major difference related to sequence organization is that eukaryotes generally contain much more DNA per genome than is needed to encode all the proteins required for cell function. For instance, in the human genome only about 3% of the total DNA sequence is found in recognizable genes, whereas in *Bacteria* this fraction can exceed 90%. The "extra" DNA in eukaryotes is most often found in repetitive sequences, some repeated hundreds of thousands of times. The function of these sequences is not clear. Lower eukaryotic microorganisms seem to have a higher coding density than the higher eukaryotes. An extreme example seems to be the yeast *Saccharomyces cerevisiae*; about 70% of its DNA encodes for protein. Although this gene density may seem quite high, it is lower than that of any known prokaryote.

Eukaryotes also often have many copies of certain genes, such as those that encode transfer RNAs and ribosomal RNAs. These latter may also be found repeated in prokaryotes but usually only a few copies are present at most. However, there are even a few short but more highly repetitive sequences known in prokaryotes. For instance, there is a 38-base-pair sequence found in *Escherichia coli* K-12 that is repeated 581 times, making up over 0.5% of the genome. However, the two generalizations are still correct: Eukaryotic genomes are characterized by having interrupted genes and repetitive sequences.

Nonchromosomal Genetic Elements

A number of other genetic elements are known. Although some of these have been called "chromosomes" they are quite distinct from cellular chromosomes. Therefore, we will refer to them collectively as *nonchromosomal genetic elements*. Most of these elements are found only in cells, but this is not true of viral genomes.

Viruses contain genomes, either DNA or RNA, that control their own replication and transfer from cell to cell. The viral genome is also referred to as a chromosome. However, it contains genes essential to the virus, not the host cell, and is therefore clearly functionally distinct from the cellular chromosomes. (Interestingly, some viral genomes can become integrated into cellular chromosomes, ⌀ Sections 9.10, 9.12, and 16.10.) Both linear and circular viral chromosomes are known. Viruses are of special interest because they are often (but not always) responsible for disease states. We discuss viruses in Chapters 9 and 16, and virus diseases in Chapters 26 and 27.

Plasmids are typically small genetic elements that exist and replicate separately from the chromosome. The great majority of plasmids are double-stranded DNA, and although most plasmids are circular, some are linear. Plasmids differ from viruses in two ways: (1) they do not cause cellular damage (generally they are beneficial), and (2) plasmids do not have extracellular forms, whereas viruses do. Although plasmids have been recognized in only a few eukaryotes, they have been found in most prokaryotic species. We discuss plasmids in Chapter 10.

Many prokaryotes seem to contain one or more plasmids in addition to their chromosome. Some plasmids contain genes whose protein products can confer important properties on the host cell, such as resistance to antibiotics. The definition of what constitutes a chromosome in *Bacteria*, however, is a genetic element that contains genes whose products are involved in essential metabolic steps *under all growth conditions*. Such genes are sometimes referred to as *housekeeping genes*. For instance, a gene encoding DNA gyrase is always required by a cell, whereas a gene that enables a bacterium to be resistant to an antibiotic is required only under certain conditions (the presence of the antibiotic). Therefore, proof that a prokaryote has more than one chromosome requires evidence that each "chromosome" contains single-copy genes that are essential. However, even using such a stringent definition, there are several prokaryotes that clearly seem to have more than one chromosome, including *Rhodobacter sphaeroides* and *Halobacterium* sp. NRC1 (see Table 7.2) and members of the genus *Brucella*. This may also be true of the spirochete *Borrelia burgdorferi* (see Table 7.2), which has a complex genome containing a large, linear chromosome and several circular and linear plasmids.

Mitochondria and **chloroplasts** contain nonchromosomal genetic elements and are found in eukaryotes. The mitochondrion is the site of respiratory enzymes and plays a major role in energy generation in most eukaryotes (⌀ Section 14.2). The chloroplast is a green, chlorophyll-containing structure that is the site of phototrophic ATP formation (⌀ Section 14.3). The chromosomes of mitochondria and chloroplasts can be viewed as independently replicating genetic elements. However, these organelles are much more complex than plasmids and viruses because they contain not only DNA but also a complete machinery for protein synthesis, including 70S ribosomes, transfer RNA, and all the other components necessary for translation and formation of functional proteins. Although the chromosomes found in mitochondria and chloroplasts may contain many genes and a complete translation system, their existence is not independent of the cellular chromosomes because most proteins in them are coded not by organelle DNA but by the cell's chromosomal DNA.

Transposable elements are pieces of DNA that can move from one site on a chromosome to another. Transposable elements are found in prokaryotes and eukaryotes and play important roles in genetic variation. In prokaryotes there are three types of transposable elements: insertion sequences, transposons, and some special viruses. *Insertion sequences* are the simplest type and carry no genetic information other than that required for them to move into new locations. *Transposons* are larger and contain other genes. We discuss both of these types in more detail in Chapter 10. In Chapter 16 we discuss a bacterial virus, Mu, that is also a transposable element. The unique feature of transposable elements is that *they all replicate as part of some other molecule of DNA*.

✓ 7.4 Concept Check

In addition to the chromosomes, a number of other genetic elements exist in cells. Plasmids are DNA molecules that exist separately from the chromosome of the cell. Mitochondria and chloroplasts contain their own DNA chromosomes. Viruses contain a genome, either DNA or RNA, that controls their own replication. Transposable elements exist as a part of other genetic elements.

✓ What is a *genome*?

✓ What genetic material is found in all cellular chromosomes?

✓ What defines a chromosome in prokaryotes?

III DNA REPLICATION

In the remainder of this chapter we shall examine in more detail the three steps in biological information flow. The first of these is DNA replication. In all organisms DNA replication is necessary for cells to divide, whether to reproduce new organisms, as in unicellular microorganisms, or to produce new cells as part of a multicellular organism. The process must be accurate so that these new cells or organisms will have essentially the same genetic information as the parent cell. Therefore, the nucleotide base sequence residing in each long molecule of the DNA double helix must be precisely duplicated to form a copy of the original molecule.

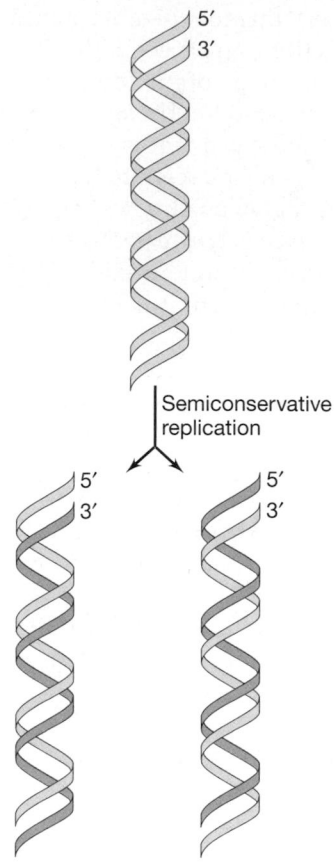

Figure 7.13 DNA replication is a semiconservative process in both prokaryotes and eukaryotes. Note that the new double helices each contain one new and one old strand.

7.5	**DNA Replication: Templates and Primers**

Just how DNA could be replicated became apparent when it was shown that DNA was a double helix with *complementary base pairing*. As we have discussed (see Figure 7.3), adenine pairs specifically with thymine and guanine pairs with cytosine. If the DNA double helix is opened up, a new strand can be synthesized as the complement of each of the parental strands. As shown in Figure 7.13, replication is **semiconservative**, with the two resulting double helices consisting of one progeny and one parental strand.

The DNA molecule that is copied to form a complement is called a *template*. A template is a preformed pattern that is copied. The *new* DNA molecule is not covalently connected to the *old* DNA molecule.

The chemistry of DNA, the nature of its precursors, and the activities of the enzymes involved in replication place some important restrictions on the manner in which this new strand is synthesized. The precursor of

each new nucleotide in the chain is a nucleoside 5′-*tri*phosphate, from which the two terminal phosphates are removed and the internal phosphate is attached covalently to deoxyribose of the growing chain (Figure 7.14). The addition of the nucleotide to the growing chain requires the presence of a free hydroxyl group, and such a free hydroxyl group is available only at the 3′-end of the molecule. This chemical restriction leads to an important "law" that is at the basis of many facets of DNA replication: *DNA replication always proceeds from the 5′ end to the 3′-hydroxyl end, the 5′-phosphate of the incoming nucleotide being attached to the 3′-hydroxyl of the previously added nucleotide.*

The enzymes that catalyze addition of the nucleotides are called **DNA polymerases**. *All DNA polymerases synthesize new DNA in the 5′ → 3′ direction. However, no known DNA polymerase can begin a new chain. All these enzymes can only add a nucleotide onto a preexisting 3′-OH group.* Therefore, for a *new* chain to be started, there must be a **primer**, a site at which the DNA polymerase can attach the first nucleotide. In most cases this primer is a short stretch of *RNA*.

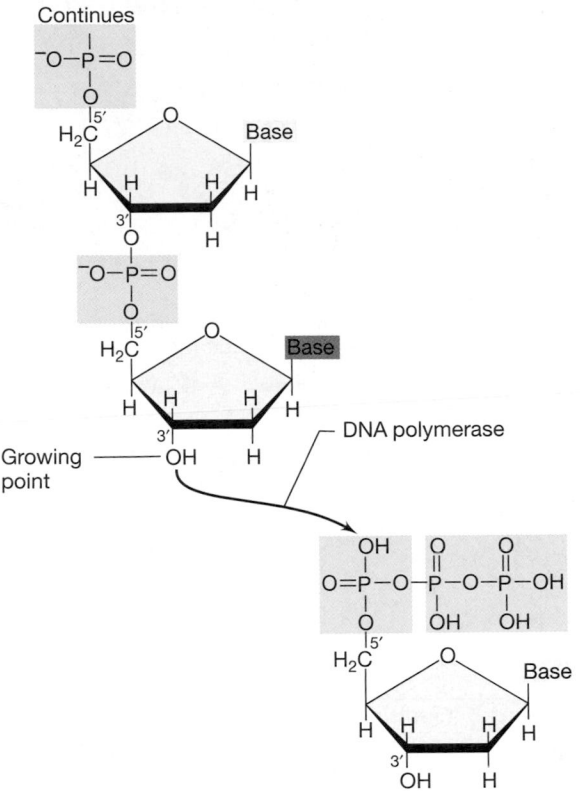

Figure 7.14 Structure of the DNA chain and mechanism of growth by addition from a deoxyribonucleoside triphosphate at the 3′-end of the chain. Growth always proceeds from the 5′-phosphate to the 3′-hydroxyl ond. The enzyme DNA polymerase catalyzes the addition reaction. The four deoxyribonucleotides that serve as precursors are deoxythymidine triphosphate (dTTP), deoxyadenosine triphosphate (dATP), deoxyguanosine triphosphate (dGTP), and deoxycytidine triphosphate (dCTP). The two terminal phosphates of the triphosphate are split off as pyrophosphate (PP$_i$). Thus, two high energy phosphate bonds are consumed on the addition of a single nucleotide.

When the double helix is opened up at the beginning of replication, an RNA-polymerizing enzyme acts first, resulting in formation of this RNA primer. A specific RNA-polymerizing enzyme, called *primase,* participates in primer synthesis by synthesizing a short stretch of RNA. At the growing end of this RNA primer is a 3′-OH group to which DNA polymerase can add the first deoxyribonucleotide. Therefore, continued extension of the molecule occurs as *DNA* rather than RNA. Thus, the newly synthesized molecule has a structure like that shown in Figure 7.15. The primer must eventually be removed, as we shall see.

✓ **7.5 Concept Check**

Both strands of the DNA helix serve as templates for the synthesis of two new strands. The two progeny double helices each contain one parental strand and one new strand. The new strands are elongated by always adding on to the 3′-end. DNA polymerases cannot start new strands. Therefore, new strands must start with a primer, which is usually RNA.

✓ To which end of a newly synthesized strand of DNA does polymerase add a base?

✓ Why is a primer required for DNA replication?

7.6 **DNA Replication: The Replication Fork**

To understand the complete replication of a double-stranded DNA molecule, it is easiest to choose an actual example and see how it is replicated. Much of the information on the mechanism of DNA replication was originally obtained from the bacterium *Escherichia coli*, and the following discussion deals primarily with this organism.

Initiation of DNA Synthesis

As is the case for most prokaryotes, the chromosome of *Escherichia coli* is a circular DNA molecule. Also like most *Bacteria*, there is a single location on this chromosome where DNA synthesis is initiated, the so-called **origin of replication**. The origin of replication consists of a specific sequence of about 300 bases that is recognized by specific initiation proteins. At the origin of replication, the DNA double helix is opened up and the initiation of DNA replication occurs on the two single strands. As replication proceeds, the site of replication, called the **replication fork**, appears to move down the DNA.

Replication is frequently bidirectional from the origin of replication, as shown in Figure 7.16, and therefore there are *two* replication forks replicating in opposite directions. In circular DNA, bidirectional replication leads to the formation of characteristic structures called **theta structures**. Most large DNA molecules, whether from prokaryotes or eukaryotes, have bidirectional replication from fixed origins. A single eukaryotic chromosome has many origins. This is not simply because the DNA is longer because, as we have seen, this is not always the case (see Section 7.4). It may reflect the fact that DNA polymerases from eukaryotes do not replicate as fast as the prokaryotic enzymes. DNA replication is carefully regulated, and the site where this regulation takes place is the origin.

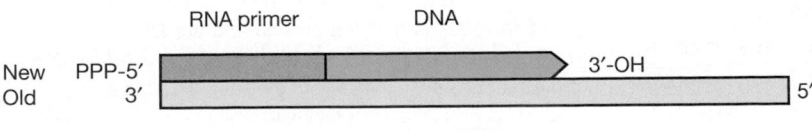

Figure 7.15 Structure of the RNA–DNA combination that results at the initiation of DNA synthesis.

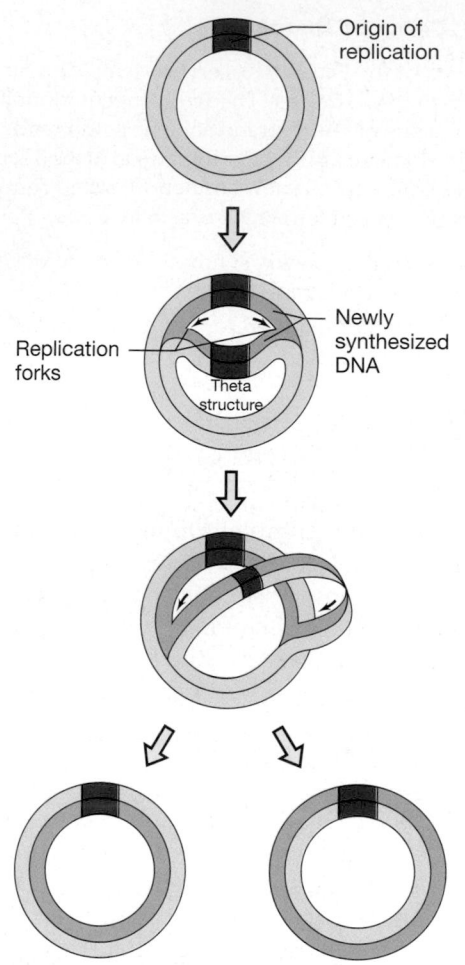

Figure 7.16 In circular DNA, bidirectional replication from an origin leads to the formation of replication intermediates resembling the Greek letter theta (Θ).

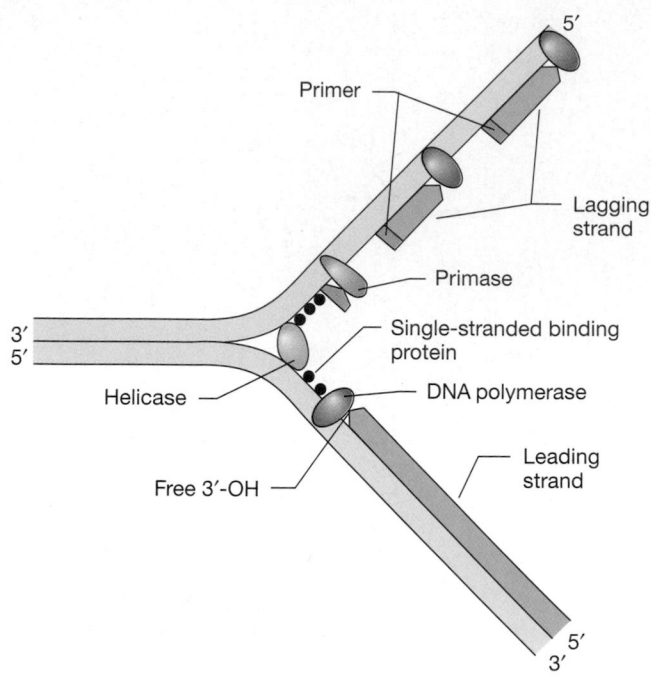

Figure 7.17 Events at the DNA replication fork. Note the polarity and antiparallel nature of the DNA strands. The substrates for primase are ribonucleotide triphosphates, while for DNA polymerase, they are *deoxy*ribonucleotide triphosphates.

Leading and Lagging Strands

There are five different DNA polymerases in *Escherichia coli*, called DNA polymerases I through V. It is DNA polymerase III that is the primary enzyme of replication at the replicating forks. However, several other enzymes are also involved. The details of events at the replication fork are illustrated in Figure 7.17. At the replication fork, the DNA double helix is unwound and a small single-stranded region is formed by the action of specific proteins called *helicases*. Helicases are ATP-dependent enzymes that hydrolyze ATP as they move down the helix in advance of the replicating fork (Figure 7.18). The single-stranded region generated is complexed with a special protein, the *single-strand binding protein*, which stabilizes the single-stranded DNA, preventing the formation of intrastrand hydrogen bonds.

Note in Figure 7.17 an important difference between replication of the two strands, which arises from the fact that DNA replication always proceeds from 5′-phosphate to 3′-hydroxyl (always adding a *new* nu-

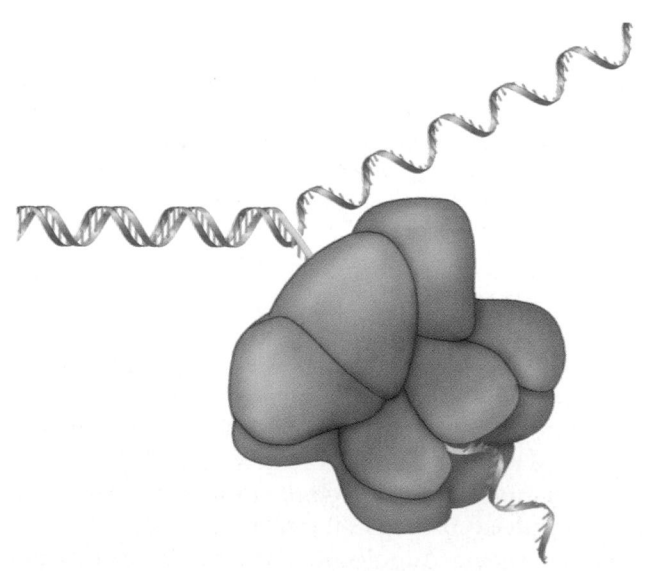

Figure 7.18 A DNA helicase unwinding a double helix. This figure is adapted from the work of Smita S. Patel and shows a model of a multisubunit DNA helicase moving along DNA. What makes this figure so strikingly different than the diagrams shown in Figures 7.17, 7.19, and 7.21 is that the protein and the DNA molecules here are drawn to scale. Simple diagrams are often used to give an idea of molecular events in the cell but they may give the incorrect impression that most proteins are relatively small compared to DNA. This scale model shows otherwise.

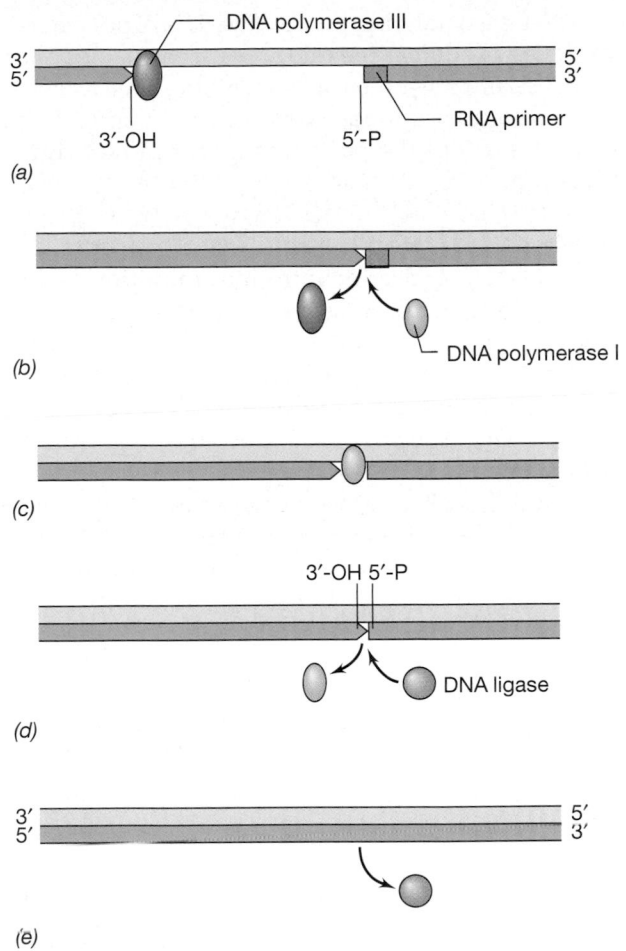

(a)

(b)

(c)

(d)

(e)

Figure 7.19 Sealing two fragments on the lagging strand. (a) DNA polymerase III is synthesizing DNA in the 5′ → 3′ direction toward the RNA primer of a previously synthesized fragment on the lagging strand. (b) On reaching the fragment, DNA polymerase I replaces III. (c) DNA polymerase I continues synthesizing DNA while removing the RNA primer from the previous fragment. (d) DNA ligase replaces DNA polymerase I after the primer has been removed. (e) DNA ligase seals the two fragments together.

cleotide to the 3′-OH of the growing chain). On the strand growing from the 5′-phosphate to the 3′-hydroxyl, called the **leading strand**, DNA synthesis can occur *continuously* because there is always a free 3′-OH at the replication fork to which a new nucleotide can be added. But on the opposite strand, called the

lagging strand, DNA synthesis must occur *discontinuously* (because there is no 3′-OH at the replication fork to which a new nucleotide can attach). Where is the 3′-OH on this strand? At the *opposite* end, *away* from the replication fork. Therefore, on the lagging strand, a small (11-base) RNA primer must be synthesized by primase to provide free 3′-OH groups. After synthesizing the primer, primase is replaced by the enzyme DNA polymerase III. Then deoxyribonucleotides are added until DNA polymerase III reaches the previously synthesized DNA.

At this point, DNA polymerase III stops. The next enzyme that is involved, *DNA polymerase I,* has more than one activity. It can clearly synthesize DNA. However, at the same time it is adding nucleotides on to the 3′-OH, it has a 5′ → 3′ *exonuclease* activity that removes the RNA primer from in front of it (Figure 7.19). When the primer has been removed and replaced with DNA, DNA polymerase I is then released. The last phosphodiester bond is made by an enzyme called *DNA ligase.* (This enzyme can seal any nicks made in DNAs that have a 5′-phosphate and 3′-OH and along with DNA polymerase I is also involved in DNA repair.)

Each short stretch of DNA made by DNA polymerase III on the lagging strand is called an *Okazaki fragment* and is about 1000 bases long. Each of these must be primed individually. By contrast, the leading strand is primed only once, at the origin. Because of bidirectionality there are two replicating forks going in opposite directions (see Figure 7.16), which means that at the origin there are two leading strands and two lagging strands started (Figure 7.20).

While DNA synthesis is continuing at the replication fork, changes in the coiling of the DNA are occurring, modified by unwinding enzymes and topoisomerases (see Section 7.3). Unwinding is obviously an essential feature of DNA replication, and because supercoiled DNA is under strain, it unwinds more easily than DNA that is not supercoiled. Thus, by regulating the degree of supercoiling, topoisomerases regulate the process of replication (and also transcription, as discussed later).

Figure 7.17 shows the differences in replication of the leading and the lagging strands and the various

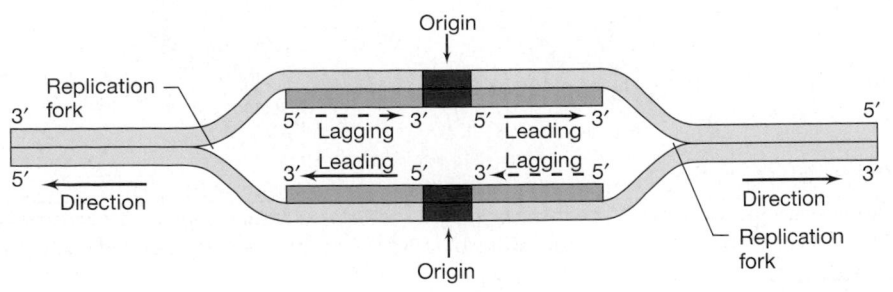

Figure 7.20 At an origin of replication that directs bidirectional replication, two replication forks must start. Therefore, two leading strands must be primed, one in each direction.

Replisome

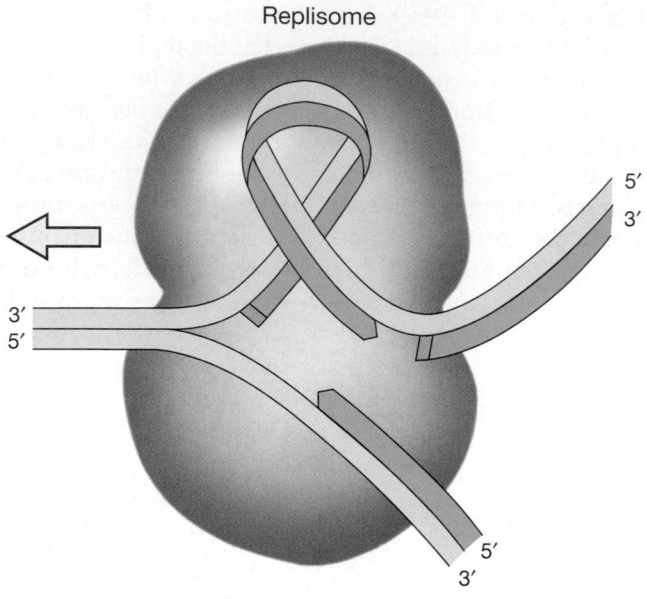

Figure 7.21 Apparent movement of the replisome, a complex of proteins and enzymes involved in DNA replication, on the double helix. All the reactions shown in Figure 7.17 are taking place, but the looping out of the lagging strand allows the complex to move forward smoothly at the replication fork.

enzymes involved. It would appear from such a simplified drawing that each replication fork must consist of DNA polymerase moving smoothly along, synthesizing the leading strand, and one or more polymerases jumping about synthesizing the lagging strand. Actually the two strands are being synthesized by a duplex of DNA polymerases. This is made possible by a "looping" of the lagging strand as shown in Figure 7.21. The *replisome* is a complex containing helicases, primase, two DNA polymerase III molecules,

and other associated proteins. The DNA polymerase complex is fixed near the midpoint of the bacterial cell and serves as a *replication factory*, pulling the DNA template through it as replication occurs. Therefore, it is the DNA and not the polymerase that moves during replication.

Fidelity of DNA Replication: Proofreading

Errors in DNA replication introduce mutations. Mutation rates in living organisms are remarkably low, between 10^{-8} and 10^{-11} errors per base pair inserted. This accuracy is possible partly because DNA polymerase actually gets *two* chances to incorporate the correct base at a given site. The first chance occurs when complementary bases are inserted by base-pairing rules, A with T and G with C, using the template strand as the pattern. The second chance occurs because of a second enzymatic activity, referred to as **proofreading**, associated with both DNA polymerase I and DNA polymerase III (Figure 7.22). In addition to inserting nucleotides in the replicating strand, these DNA polymerases also contain a $3' \rightarrow 5'$ *exonuclease* activity that can remove a misinserted nucleotide and allow it to be replaced with the correct nucleotide. Proofreading activity is summoned if an incorrect base has been inserted because misinsertion creates unstable base pairing. This proofreading activity gives the polymerase activity a second chance to insert the correct base (Figure 7.22). (Note that the proofreading exonuclease activity is the *opposite* of the $5' \rightarrow 3'$ exonuclease activity of DNA polymerase I used to remove the primer from "in front" of the polymerase. Only DNA polymerase I has this latter activity.)

Exonuclease proofreading occurs in prokaryotes, eukaryotes, and viral DNA replication systems. However,

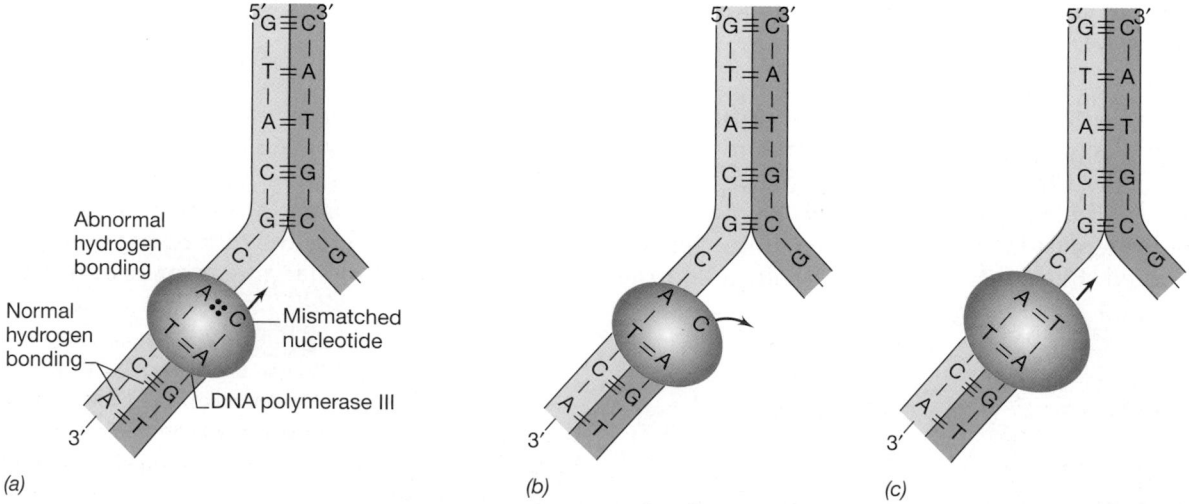

(a) (b) (c)

Figure 7.22 Proofreading by the $3' \rightarrow 5'$ exonuclease activity of DNA polymerase III. (a) A mismatch in base pairing at the terminal base pair causes the polymerase to pause briefly. This is a signal for the proofreading activity (b) to excise the mismatched nucleotide, after which the correct base is incorporated (c) by polymerase activity.

many organisms have other mechanisms for reducing errors made during DNA replication. We will discuss some of these in Chapter 10.

Termination of Replication

The details about the process of termination of the replicating forks are not completely known, but it is quite clear that the two replicating forks do not just smash into each other like runaway trains. Certain DNA sequences and specific proteins are involved in slowing down the replication forks and allowing replication to be completed. When the replication of the circular molecule is complete, the two circular molecules are linked together, much like the links of a chain. These can be unlinked by a topoisomerase.

Another problem remains that we will not discuss in any detail: the partitioning of the replicated double helices into the daughter cells. In Chapter 6 we outlined how cell wall synthesis is coupled with cell division (☜ Section 6.2). Obviously it is critical that, after DNA replication, the DNA is partitioned so that each daughter cell has a copy of the chromosome.

This process has been investigated for a much longer time in higher eukaryotes because much of the cellular apparatus involved is visible in the light microscope. In eukaryotic cells, the nucleus divides following a doubling of chromosome number in the process called **mitosis**, which yields two cells, each with a full complement of chromosomes. It is during mitosis that the eukaryotic chromosomes are most compacted (see Section 7.3) and most easily visible (Figure 7.23). Small protein tubes called *microtubules* play important roles in the mitotic process. Microtubules and other protein assemblies attach to specific sequences on the chromosomes and function to form the spindle apparatus, which is the actual structure that moves the chromosomes to the two poles of the dividing cell (see Figure 7.23). There is now some evidence that in prokaryotic cells there may be proteins that attach to specific sequences of the DNA and help pull the chromosomes into the daughter cells.

✓ 7.6 Concept Check

DNA synthesis begins at a unique location called the origin of replication. The double helix is unwound by helicase and is stabilized by single-stranded binding protein. Extension of the DNA occurs continuously on the leading strand but discontinuously on the lagging strand. Most errors in base pairing are corrected by proofreading functions associated with the action of DNA polymerase.

✓ Why are there *leading* and *lagging* strands?

✓ Why is proofreading so important to the cell?

7.7 DNA Replication: Linear Genetic Elements

Circular DNA molecules are common; most prokaryotic chromosomes are circular, as are most plasmids and some viruses. Almost all the steps in replication are identical whether the chromosome is linear or circular. However, there is one problem with replication of linear genetic elements that does not occur with circular ones, and that problem is at the extreme 5′-end of each strand. To understand the problem, refer back to Figure 7.15. Imagine that the left end of the DNA in this diagram is actually one end of a linear chromosome. Even if the RNA primer is very short and there is a special enzyme to remove it, no DNA polymerase can replace it with DNA since *all* DNA polymerases require a primer. Therefore, if nothing is done, the DNA molecule will become shorter each time it is replicated. Linear genetic elements have clearly solved this problem!

In fact, there are many solutions to this problem. Some viruses having linear chromosomes actually circularize themselves by their sticky ends, as shown in Figure 7.7. Some other viruses have direct repeats at each end of their chromosomes. A recombination process (a joining together of different DNA molecules) uses the repeats to join several partially replicated DNA molecules together into a very large molecule

(a)

(b)

Carolina Biological Supply Co.

Figure 7.23 Mitosis, as seen in the light microscope. These are onion root tip cells that have been stained to reveal nucleic acid and chromosomes. (a) Metaphase. Chromosomes are paired in the center of the cell. (b) Anaphase. Chromosomes are separating.

from which perfect copies are cut by endonucleases (∞ Section 9.9). Several types of viruses and many linear plasmids solve the problem of replicating linear DNA by using not an *RNA* primer but rather a *protein* primer. Although all DNA polymerases must add each nucleotide to a free —OH group, some DNA polymerases can add the first base onto an —OH group found on specific proteins that bind to the ends of these linear chromosomes (Figure 7.24). These proteins are encoded by the plasmid or virus, and they function to recognize the ends of the chromosomes. These protein primers are not removed, so these particular types of plasmids and viruses have proteins covalently attached to the 5′-ends of their DNA. This may also be the means by which some linear chromosomes of *Bacteria,* such as those of *Streptomyces lividans,* are replicated.

None of these methods of replicating linear DNA are used to complete the ends of eukaryotic chromosomes (telomeres). Telomeres of eukaryotic chromosomes contain repetitive DNA: a short sequence (often six base pairs) tandemly repeated from 20 to several hundred times (Figure 7.25*a*). The sequences from different eukaryotes are closely related, and one strand always has several guanines. This guanine-rich sequence can be added onto the 3′-end of a DNA molecule by an interesting enzyme called **telomerase** (see Figure 7.25). Telomerases add onto the 3′-ends of linear DNA. *They do not need a DNA template because they contain a small RNA template as a cofactor.* These enzymes can work repetitively to make a long extension. Once this extension is

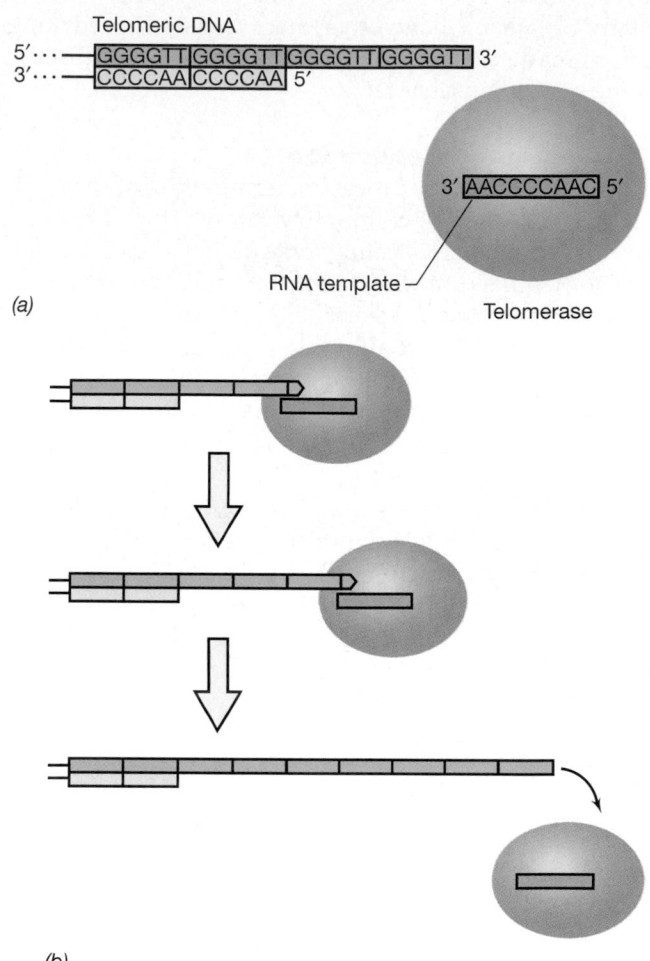

(a)

(b)

Figure 7.25 Model for the action of the telomerase at one end of a eukaryotic chromosome. (a) A diagram of the sequence of the end of the DNA in a telomere, with four of the guanine-rich repeats, and the enzyme telomerase, which contains a short RNA template. (b) Steps in elongation of the guanine-rich strand catalyzed by telomerase. After telomerase finishes, the lagging strand can be primed with an RNA primer by primase (not shown) followed by completion of the lagging strand by DNA polymerase and ligase.

long enough, the other strand can be primed with an RNA primer in the normal fashion. The telomeres do not need to be a precise number of repeats long, just long enough to ensure that no genetic information becomes lost during DNA replication.

✓ 7.7 Concept Check

The ends of linear genetic elements present a problem to the replication machinery that circular genetic elements do not. Some prokaryotic linear elements solve this problem using a protein primer. Eukaryotes solve the problem using a special enzyme called telomerase to extend one strand of the DNA.

✓ What is a *protein primer*?

✓ What is *telomerase*?

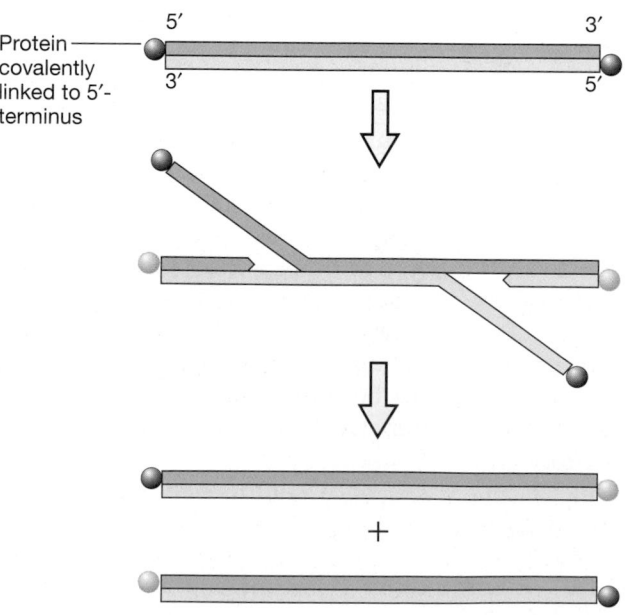

Figure 7.24 Replication of linear DNA using protein primers. The new strands of DNA are primed by proteins that stay covalently attached to the 5′-ends.

IV RNA SYNTHESIS AND PROCESSING

Ribonucleic acid (RNA) plays a number of important roles in the expression of genetic information in the cell. Three major types of RNA have been recognized: **messenger RNA (mRNA), transfer RNA (tRNA),** and **ribosomal RNA (rRNA).** These are all products of *transcription* of the information in an organism's DNA. There are three key differences between the chemistry of RNA and that of DNA: (1) RNA has the sugar *ribose* instead of *deoxyribose;* (2) RNA has the base *uracil* instead of the base *thymine;* and (3) except in certain viruses, RNA is not double-stranded. A change from *deoxyribose* to *ribose* affects some of the chemical properties of a nucleic acid, and enzymes that affect DNA in general have no effect on RNA, and vice versa. The change from *thymine* to *uracil* does not affect base pairing, as the two nucleotide bases pair with adenine equally well.

It should be emphasized that RNA acts at two levels, genetic and functional. At the *genetic* level, RNA can carry the genetic information from DNA (mRNA) (or in the case of RNA viruses, play a direct genetic function). At the *functional* level, RNA acts as a macromolecule in its own right, serving a functional and structural role in ribosomes (rRNA) or an amino acid transfer role in protein synthesis (tRNA). Some RNA even has catalytic (enzymatic) activity. In this section we focus our discussion on how RNA is synthesized.

7.8 Overview of Transcription

The transcription of genetic information from DNA to RNA is carried out through the action of the enzyme **RNA polymerase,** which catalyzes the formation of phosphodiester bonds between ribonucleotides. RNA polymerase requires DNA as a template. The precursors of RNA are the ribonucleoside triphosphates ATP, GTP, UTP, and CTP. The chemistry of RNA synthesis is much like the chemistry of DNA synthesis (see Figure 7.14). During elongation of an RNA chain, the nucleotides are added to the 3′-OH of the ribose of the preceding nucleotide, which are polymerized with the release of the two high-energy phosphate bonds. Thus, in RNA synthesis (as in DNA synthesis), the overall direction of chain growth is from the 5′-end to the 3′-end, and the *template* strand is antiparallel. Unlike DNA polymerase, however, *RNA polymerase can start chains* (the initial nucleotide in an RNA chain then retains all three phosphates). The first base in the RNA is almost always a purine, either adenine or guanine.

In most cases, the DNA template for RNA polymerase is a double-stranded DNA molecule, but only *one* of the two strands is transcribed for any given gene.

While these principles are true for RNA polymerases from all organisms, RNA polymerase differs markedly among *Bacteria, Archaea,* and *Eukarya. Bacteria* and *Archaea* each have a single RNA polymerase while the eukaryotic nucleus contains three such enzymes: RNA polymerase I, RNA polymerase II, and RNA polymerase III. All are multisubunit enzymes and all contain some subunits that are evolutionarily conserved (⚬ Figure 11.16). The following discussion deals only with RNA polymerase from *Bacteria,* which has the simplest structure (and about which the most is known).

All RNA polymerases that have been studied from *Bacteria* are complex enzymes with closely related subunit structures. The enzyme from *Escherichia coli* has four different types of protein subunits, designated β, β′, α, and σ (sigma), with α present in two copies. The subunits interact to form the active enzyme, called the *holoenzyme,* but the sigma factor is not as tightly bound as the others and easily dissociates, leading to the formation of what is called the *core enzyme* ($\alpha_2\beta\beta'$). (Interestingly, it is the subunits in the core that are also conserved in RNA polymerases of *Archaea* and *Eukarya.*) The core enzyme alone can catalyze the formation of RNA, and the role of sigma is in *recognition* of the appropriate site on the DNA for the initiation of RNA synthesis. The process of RNA synthesis involving RNA polymerase and sigma is illustrated in Figure 7.26.

RNA polymerase is a large protein and forms contacts with the DNA over many bases simultaneously. As noted (see Section 7.2), proteins can interact specifically with DNA because parts of the base pairs are exposed in the major groove. In order to *start* an RNA chain correctly, RNA polymerase must first recognize the proper region on the DNA. These particular sites on the DNA where RNA polymerase binds are called **promoters.** Note that only *one* strand of the DNA double helix is transcribed at a time. Which strand is transcribed is determined by the orientation of the promoter sequence. RNA polymerase travels away from the promoter region, synthesizing RNA as it moves.

Once the RNA polymerase has bound, the process of transcription can proceed. In this process, the DNA double helix at the promoter is *opened up* by the RNA polymerase (Figure 7.26). As the polymerase moves, it causes the DNA to unwind in short segments, transcription of these segments occurs, and the DNA double helix closes up again. As a result of this transient unwinding, the bases of the template strand are *exposed* and can then be copied into the RNA complement. Thus, the promoter *points* the RNA polymerase in one or the other direction. When a region of DNA has two nearby promoters pointing in opposite directions, then transcription from one of the promoters occurs in one direction (on one of the strands) and transcription from the other occurs in the opposite direction (on the other strand) (⚬ Figure 15.4).

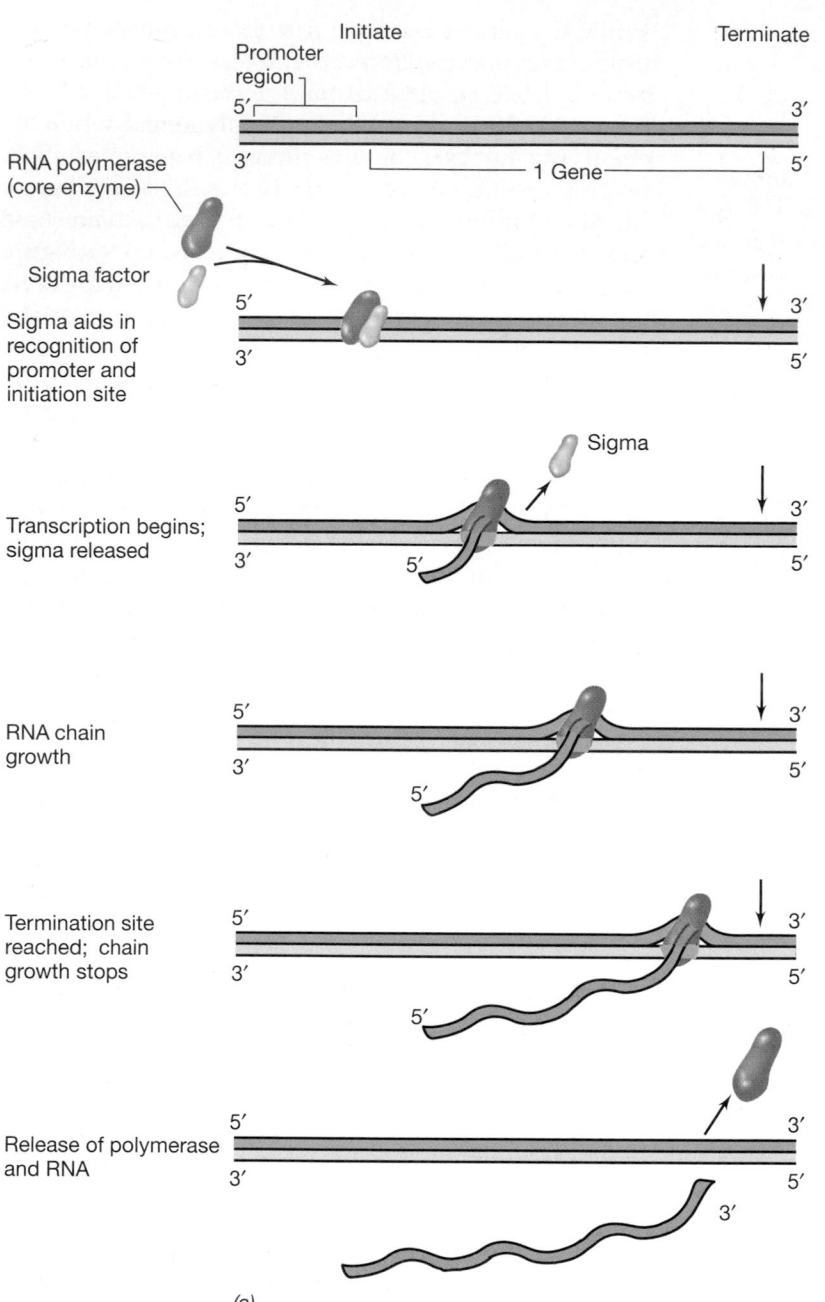

RNA polymerase
(core enzyme)

Sigma factor

Sigma aids in
recognition of
promoter and
initiation site

Transcription begins;
sigma released

RNA chain
growth

Termination site
reached; chain
growth stops

Release of polymerase
and RNA

Promoter
region
Initiate
Terminate
1 Gene
Sigma

(a)

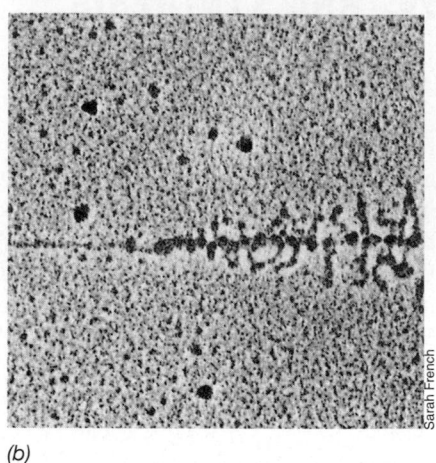

(b)

Figure 7.26 Transcription. (a) Steps in RNA synthesis. The initiation and termination sites are specific nucleotide sequences on the DNA. The sigma factor allows RNA polymerase to recognize the initiation site (the promoter). The sigma factor is released during elongation. RNA polymerase moves down the DNA chain, causing temporary opening of the double helix and transcription of one of the DNA strands. When a termination site is reached, chain growth stops, and the mRNA and polymerase are released. (b) Electron micrograph of transcription occurring along a gene on the *Escherichia coli* chromosome. The region of active transcription represents about two kilobase pairs of DNA. Transcription proceeds from left to right.

Once a small portion of RNA has been formed, the sigma factor dissociates; most of the elongation is therefore carried out by the core enzyme alone (Figure 7.26). Thus, sigma is involved in the formation of only the initial RNA polymerase-DNA complex. As the newly synthesized RNA dissociates from the DNA, the opened DNA closes into the original double helix. Transcription stops at specific regions called **transcription terminators**.

Therefore, unlike replication, which involves copying an entire genome, transcription usually involves much smaller units of DNA, often a single gene. This allows the cell to transcribe different genes at very different frequencies. As we shall see in Chapter 8, regulation of the transcription of specific genes can be a very efficient mechanism of controlling gene expression.

✓ 7.8 Concept Check

The three major types of RNA are messenger RNA (mRNA), transfer RNA (tRNA), and ribosomal RNA (rRNA). The transcription of RNA from the DNA involves the enzyme RNA polymerase, which adds bases onto 3′-ends of growing chains. Unlike DNA polymerase, RNA polymerase can start a chain. RNA polymerase recognizes a specific start site on the DNA called the promoter. RNA synthesis stops at a transcription terminator.

✓ What is a *promoter*?

✓ What is a *transcription terminator*?

7.9 Promoters

As we have noted, the promoter plays a key role in the initiation of RNA synthesis. Promoters are specific DNA sequences where RNA polymerase enzymes attach. The sequences of a large number of promoters from a variety of organisms have been determined. Figure 7.27 shows the sequence of a few promoters from *Escherichia coli*. It is the sigma factor, as part of the RNA polymerase, that is primarily involved in recognition of these promoters.

A single organism can have several different sigma factors; *E. coli* encodes seven, *Bacillus subtilis* encodes 17. These alternative sigma factors allow the RNA polymerase to recognize several different types (sequences) of promoters. Even so, in a given bacterium the majority of genes typically require only a single species of sigma factor.

All the sequences in Figure 7.27 are recognized by the same sigma factor, the major sigma factor in *E. coli*. If you examine the sequences, you will see that they are not identical. However, two sequences within the promoter region are *highly conserved* between promoters, and it is these that are recognized by sigma. Both sequences precede (are *upstream* of) the site where transcription starts. One is a region 10 bases before the start of transcription, the −10 region (called the *Pribnow box*). Notice that although each promoter is slightly different, many bases are the same. When comparing the −10 regions of all the promoters rec-

ognized by this sigma to determine which base occurs most often at each position, one arrives at the *consensus sequence* TATAAT. In our example, each promoter has from three to five matches for these bases. The second region of conserved sequence is about 35 bases from the start of transcription. The consensus sequence in the −35 region is TTGACA. Once again, most of the sequences are not *exactly* the same as the consensus sequence.

Although *E. coli* has seven different sigma factors, each sigma factor recognizes a different consensus sequence. Therefore, these sigma factors will direct RNA polymerase to promoters with sequences different from those shown in Figure 7.27. For the most part these will be associated with different genes (although some genes have more than one promoter). Note that in Figure 7.27 the sequence of only one strand is given. By convention among geneticists the strand shown is the one oriented with its 5′-end upstream (therefore, it is *not* the strand used as the template by RNA polymerase). Showing only the sequence of one strand is simply "shorthand" to save the space of writing the other strand. It is essential, though, to remember that promoters are double-stranded; that is, RNA polymerase recognizes and binds to double-stranded DNA. However, only one strand will serve as a template for transcription.

Other sigma factors in other organisms are sometimes much more specific; very little leeway is allowed in the critical bases that are recognized. In *E. coli*, promoters that are most like the consensus are usually more effective in

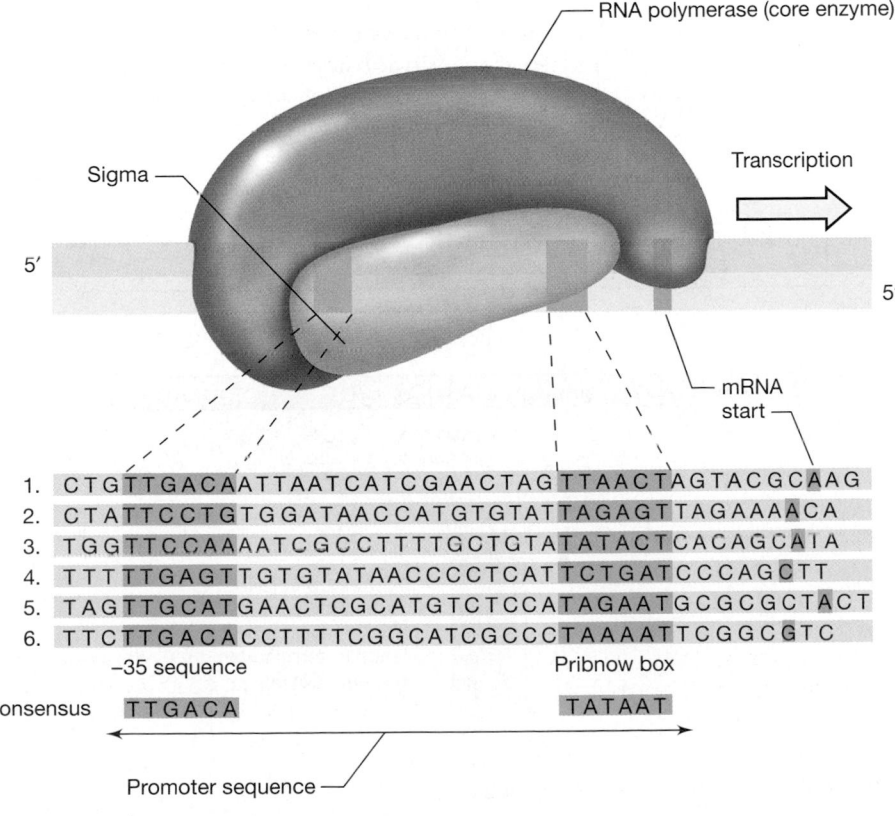

Figure 7.27 The interaction of RNA polymerase with the promoter. Shown below the diagram are six different promoter sequences identified in *Escherichia coli*, a species of *Bacteria*. The contacts of the RNA polymerase with the −35 sequence and the Pribnow box (−10 sequence) are shown. Transcription begins at a unique base just downstream from the Pribnow box. Below the actual sequences at the −35 and Pribnow box regions are consensus sequences derived from comparing many promoters.

binding RNA polymerase. The more effective promoters are called *strong promoters* and are of considerable value in genetic engineering, as will be discussed in Chapter 31.

Recall that the eukaryotic nucleus contains three different RNA polymerases. Each of these enzymes recognizes a promoter that is associated with a particular class of gene. **RNA polymerase I** *synthesizes most types of rRNA;* **RNA polymerase II** *synthesizes all the mRNA,* and **RNA polymerase III** *synthesizes tRNA* (and one type of rRNA). The reason for this specificity is that each type of RNA polymerase recognizes only those promoters that occur with the particular class of gene. In *Bacteria* the promoter for a gene encoding a protein could well be identical to a promoter for a gene encoding a tRNA. That does not happen in eukaryotes. As is the case in *Bacteria*, the great majority of genes in eukaryotes encode proteins, and therefore the great majority of genes are transcribed by RNA polymerase II. Eukaryotic RNA polymerases also require special accessory factors to recognize specific promoters, but, unlike the case in *Bacteria*, the eukaryotic initiation factors (and those of the *Archaea*) recognize the promoter elements independently, not as part of a polymerase holoenzyme. Although the *Archaea* have only a single RNA polymerase, like the *Bacteria*, it is more closely related to the eukaryotic enzymes.

✓ 7.9 Concept Check

In *Bacteria*, promoters are recognized by the sigma subunit of RNA polymerase. Promoters recognized by a specific sigma factor have very similar sequences. In the *Eukarya* the major classes of RNA are transcribed by different RNA polymerases.

✓ What is a *consensus sequence*?

✓ In a eukaryote, which type of RNA polymerase transcribes genes that encode proteins?

7.10 Transcription Terminators

As important as initiation of transcription is *termination* of transcription. **Termination** of RNA synthesis occurs at specific base sequences on the DNA. In *Bacteria* a common termination sequence on the DNA is one containing an inverted repeat with a central nonrepeating segment (see Section 7.2 and Figure 7.6 for an explanation of inverted repeats). When such a DNA sequence is transcribed, the RNA can form a stem-loop structure by intrastrand base pairing (Figure 7.28). When such stem-loop structures *in the RNA* are followed by runs of uridines, they are effective transcription terminators. Other termination sites are regions where a GC-rich sequence is followed by an AT-rich sequence. Such kinds of sequences lead to termination without addition of any extra factors and are sometimes termed *intrinsic terminators*.

Other types of terminator sequences have been discovered that require specific protein factors in addition to RNA polymerase in order to function. In *Escherichia coli* one type of transcription terminator requires a protein called *Rho*. Rho does not bind to RNA polymerase or to DNA but binds tightly to RNA and moves down the chain toward the RNA polymerase–DNA complex. Once RNA polymerase has paused at a *Rho-dependent termination site*, Rho can then cause the RNA and polymerase to leave the DNA, thus terminating transcription. Other proteins involved in transcription termination are, like Rho, RNA-binding proteins. In all cases the sequences involved in termination operate at the level of RNA. However, remember that RNA is transcribed from DNA, and so transcription termination is ultimately determined by *specific nucleotide sequences on the DNA*.

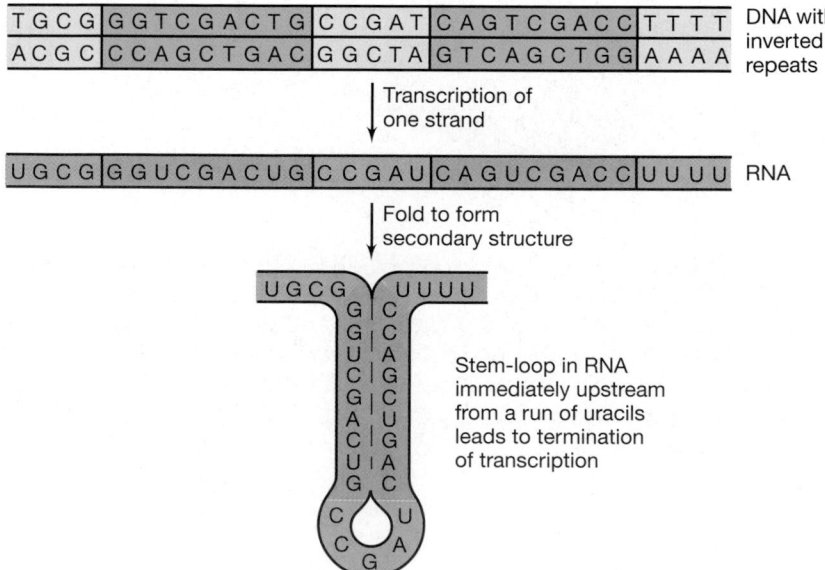

Figure 7.28 Inverted repeats in transcribed DNA lead to formation of a stem-loop structure in the RNA, which can result in termination of transcription.

Transcription termination has not been as well studied in *Archaea* and *Eukarya*. In genes transcribed by the eukaryotic RNA polymerase II there seems to be a connection between transcription termination and a processing event that takes place at the 3′-end of the mRNA (see Section 7.12).

Less is known about the transcription termination signals in the *Archaea*, but in some genes it seems clear that inverted repeats followed by an AT-rich sequence are involved, sequences very similar to those found in many bacterial transcription terminators. However, such sequences are not found in other archaeal genes. One other type of possible transcription terminator contains no inverted repeats, but rather the nucleotide sequence contains repeated stretches with runs of T's.

✓ 7.10 Concept Check

RNA polymerase stops transcription at specific sites called transcription terminators. Although encoded by DNA these signals function at the level of RNA. Some are intrinsic terminators and require no accessory proteins beyond the polymerase. In *Bacteria* these sequences are often stem-loops followed by a run of U's.

✓ What is an *intrinsic terminator*?

✓ What is a *stem-loop structure*?

7.11 The Unit of Transcription

Chromosomes are organized into units that are bound by sites where transcription of DNA into RNA is initiated and terminated: units of transcription. One might assume that each unit of transcription includes only a single gene. While this is common, it is not always the case.

Some transcription units contain two or more genes. These genes are then *cotranscribed*, giving a single RNA molecule.

As we saw in Section 7.1, most genes encode proteins, but others encode RNAs that are not translated, such as ribosomal RNA (rRNA) and transfer RNA (tRNA). There are several different types of rRNA in an organism (with a ribosome having one copy of each type; see Section 7.15). Prokaryotes have three types: 16 S rRNA, 23 S rRNA, and 5 S rRNA. As shown in Figure 7.29, there are clusters containing one gene for each of these rRNAs, and the genes in such a cluster are cotranscribed. A similar situation occurs in eukaryotes (although there is one type of eukaryotic rRNA not found in such clusters). Therefore, in all organisms the unit of transcription for most rRNA is longer than a single gene. In prokaryotes tRNA genes are often cotranscribed with each other or even, as shown in Figure 7.29, with genes for rRNA. All such transcripts must be processed before mature rRNA or tRNA is produced (see Section 7.12).

Recall that genes that encode proteins do so via an intermediary called messenger RNA (mRNA). Most mRNA, in both prokaryotes and eukaryotes, is unstable and is degraded by cellular nucleases. This is in contrast to rRNA and tRNA, which are sometimes referred to as stable RNA. In prokaryotes, a single mRNA molecule often encodes more than one protein (see Figure 7.2). In *prokaryotic* genetic elements, genes coding for related enzymes are often clustered together (⌧ Figure 10.48). In these situations the RNA polymerase proceeds down the chain and transcribes the whole series of genes into a single long mRNA molecule. An mRNA coding for such a group of cotranscribed genes is called a **polycistronic mRNA** (⌧ Section 10.10). Subsequently, when this polycistronic mRNA participates in protein

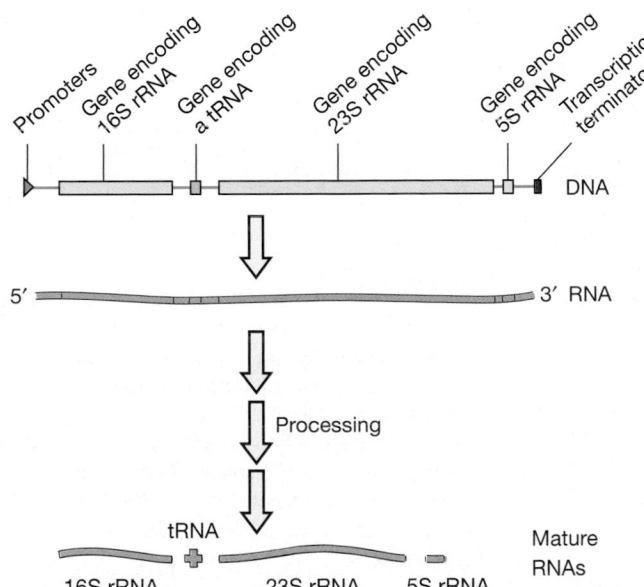

Figure 7.29 A ribosomal rRNA transcription unit from *Bacteria*. This type of transcription unit is called an "rRNA operon." In *Bacteria* all such operons have the genes for the rRNAs in the order 16S rRNA, 23S rRNA, and 5S rRNA (and they are shown approximately to scale). Note that in this particular operon the "spacer" between 16S and 23S rRNA genes contains a tRNA gene. In other operons this region may contain more than one tRNA gene and often one or more tRNA genes follow the 5S rRNA gene, which are cotranscribed. *Escherichia coli* contains seven such operons. Nonribosomal RNAs are not found in such units in eukaryotes.

synthesis (see Section 7.15), several polypeptides coded by a single mRNA can be synthesized at one time.

We will discuss regulation of mRNA synthesis in Chapter 8, but introduce here the concept of the operon. An **operon** is a complete unit of gene expression, often involving genes encoding several polypeptides on a polycistronic mRNA or genes encoding rRNA. In some cases, the transcription of the mRNA for an operon is under the control of a specific region of the DNA, the **operator**, which is adjacent to the coding region of the first gene in the operon. As we shall see in Chapter 8, the operator functions by being able to bind certain regulatory proteins.

For the most part, polycistronic mRNA does not exist in eukaryotes. However, this is because of differences in *translation,* not in transcription. We will discuss these differences later in this chapter after we deal with additional steps often required to convert a transcript from a eukaryotic protein-encoding gene into usable mRNA.

✓ 7.11 Concept Check

The unit of transcription often contains more than a single gene. Transcription of several genes into a single mRNA molecule may occur in prokaryotes, and so the mRNA may contain the information for more than one polypeptide. Such genes that are transcribed together from a single promoter constitute an operon. Genes encoding rRNA are cotranscribed in both prokaryotes and eukaryotes.

✓ What is *messenger RNA?*

✓ What is a *polycistronic mRNA?*

7.12 RNA Processing and Ribozymes

As we discussed (see Section 7.8) transcription can produce several types of RNA: messenger RNA, transfer RNA, and ribosomal RNA. Remember that each of these types of RNA has an important function in the cell. However, to be functional, many of these RNAs must first be processed to a mature form after transcription. Indeed, the only functional RNA that is the direct product of transcription is mRNA in prokaryotes. All the other RNAs require some type of processing. **RNA processing** is the conversion of a *precursor RNA* to a *mature RNA.* There are many different types of processing.

In prokaryotes and eukaryotes, tRNAs and rRNAs are made initially as long precursor molecules, which are then cut to make the final mature RNAs (see Figure 7.29). In addition, many of the bases in tRNA are modified after transcription (see Section 7.14).

In eukaryotes, and much less commonly in prokaryotes, mRNA is also the result of processing a pre-mRNA.

As discussed in Section 7.4, the genes of eukaryotes are often interrupted, with noncoding intervening sequences, *introns,* separating the coding regions, *exons.* The *primary transcript* from such a gene must be extensively processed to remove the noncoding regions before the translation process can be initiated. Only a few introns have been discovered in protein-encoding genes in prokaryotes and in certain bacteriophage. The processing step by which introns are removed and exons are joined is called **splicing**.

Although introns are found in different types of genes in both prokaryotes and eukaryotes, the splicing machinery that removes introns from eukaryotic pre-

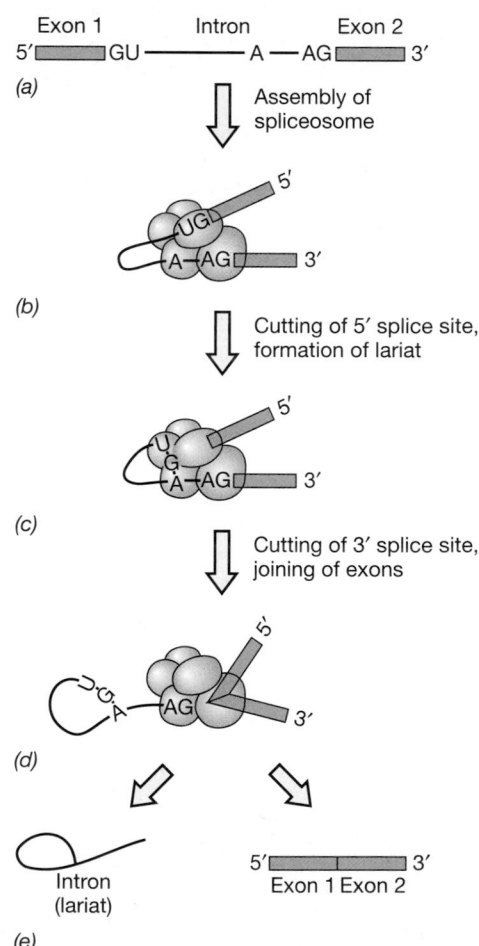

Figure 7.30 Removal of an intron from the transcript of a eukaryotic protein-encoding gene. (a) The pre-mRNA with a single intron. The sequence GU is conserved at the 5′ splice site and AG at the 3′ splice site. There is also an interior A which serves as a branch point. (b) Several small ribonucleoprotein particles (shown in brown) assemble on the RNA to form a spliceosome. Each of these particles contains distinct small RNA molecules that are involved in the splicing mechanism. (c) The 5′ splice site has been cut with the simultaneous formation of a branch point. (d) The 3′ splice site has been cut, while the two exons were joined. Note that overall two phosphodiester bonds were broken but two others were formed. (e) The final products are the joined exons, the mRNA, and the released intron.

mRNA is unique. The process involves a complex containing several different ribonucleoproteins (each contains both a small RNA and several proteins) called a **spliceosome**. The spliceosome is a highly complex structure capable of removing introns and joining adjacent exons to form a mature mRNA. Figure 7.30 diagrams the two-step reaction by which an intron found in eukaryotic pre-mRNA is removed while it is still in the nucleus. Note that there are some conserved bases at the splice junctions and that the intron is removed as a *lariat structure*. These removed introns are degraded by the cell. In higher eukaryotes there are often many introns in a single gene, and so it is clearly important not only that they be removed but they be removed in the correct order. Some introns (particularly those found in tRNA genes and in genes of the mitochondria or chloroplasts) are removed by a different process involving just proteins. Several introns, including all of those that are found in *Bacteria* and bacteriophage, are *self-splicing* (ribozymes; see next subsection).

Two other unique steps occur in the processing of eukaryotic mRNA. Both steps also take place in the nucleus before transport of the mature mRNA into the cytoplasm. The first step is called **capping** and actually occurs before transcription is complete. Capping consists of adding a methylated guanine nucleotide at the 5′-phosphate end. Occasionally other nucleotides near the 5′ end of the eukaryotic mRNA are also modified.

The remaining processing step consists of trimming the 3′-end of the pre-mRNA and adding a *poly-A tail*. This step is called **polyadenylation** or **tailing** and, as we discussed (see Section 7.10), may occur in conjunction with termination. All three steps leading to the formation of eukaryotic mRNA are shown in Figure 7.31.

We thus see that the synthesis of a mature, functional RNA is a complex and dynamic process that involves considerably more than the simple transcription of a DNA template.

Ribozymes

We emphasized in Chapter 3 the role of proteins as biochemical catalysts. Many important cellular processes involve *ribonucleoproteins*, complexes that include both RNA and protein. The role of RNA in such complexes was once *assumed* to be structural (a place for the proteins to bind) or involved in base pairing with other nucleic acids. As we have seen, there is a short RNA molecule in the enzyme *telomerase* that functions as a template. However, it has now been shown that certain types of RNA can function as *enzymes* as well. Catalytic RNAs, referred to as **ribozymes**, are involved in a number of important cellular reactions. RNA enzymes work like protein enzymes in that they have an "active site" that binds the substrate and catalyzes formation of a product (∞ Section 5.5). Ribozymes have been dis-

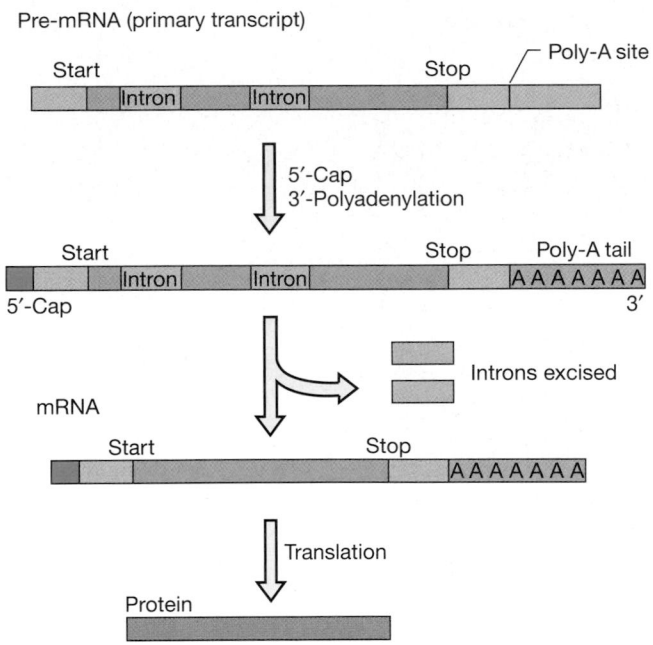

Pre-mRNA (primary transcript)

Figure 7.31 An overview of the processing of the pre-mRNA into mature mRNA in eukaryotes. The processing steps include adding a cap at the 5′-end, removing the introns, and clipping of the 3′-end of the transcript while adding a poly-A tail. All these steps are carried out in the nucleus. The location of the start and stop codons to be used during translation are also indicated.

covered in both prokaryotes and eukaryotes, and in organelles, and others have now been synthesized in laboratories. Studies show that some very short RNAs, with as few as 19 bases, can function as ribozymes.

Most ribozymes are **self-splicing introns**. They are *RNA-splicing enzymes* that remove themselves from an RNA molecule while joining adjacent exons together. In one well-studied case of splicing in a ribosomal RNA in *Tetrahymena* (a eukaryote), a 413-nucleotide *intron* acts as a ribozyme and splices itself out of a longer precursor rRNA, joining two adjacent exons to form the final rRNA (Figure 7.32). The intron ribozyme acts as a sequence-specific endoribonuclease and, once removed from the precursor RNA, circularizes with the further removal of a short oligonucleotide fragment (Figure 7.32). Self-splicing introns are widespread in nature and are the only type known in *Bacteria* and bacteriophage. Absolute proof that these ribozymal transformations occur in the absence of specific protein has come from experiments in which the gene for the entire precursor rRNA from *Tetrahymena* has been transferred to *Escherichia coli* where the segment can be transcribed. This transcribed segment carries out the splicing reaction in the complete absence of *Tetrahymena* proteins. However, there is evidence that proteins play some role in the splicing reaction of certain types of self-splicing introns in the cell.

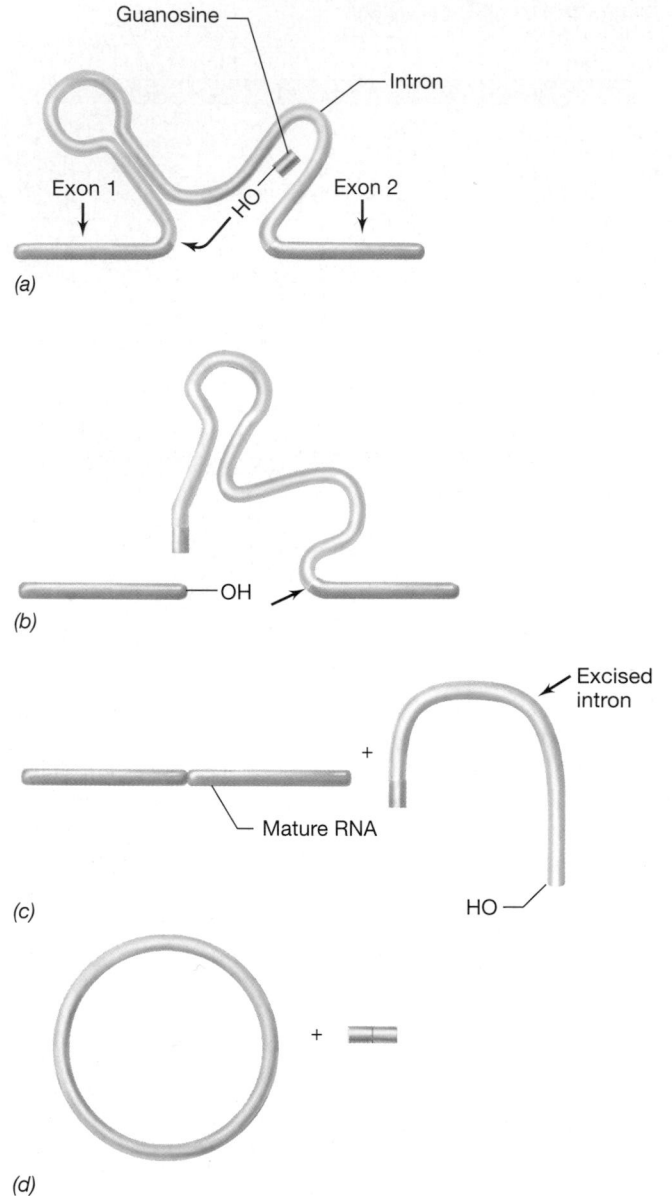

Figure 7.32 Self-splicing ribozymal intron of the protozoan *Tetrahymena*. There is considerable secondary structure in such molecules, which is critical for the splicing reaction. (a) A ribosomal RNA precursor contains a 413-nucleotide intron. (b) Following the addition of the nucleoside guanosine, the intron splices itself out and joins the two exons. (c) The intron is spliced out. (d) The intron circularizes with the loss of a 15-nucleotide fragment.

Self-splicing introns differ from most protein enzymes in a major way in that they normally can catalyze their reaction only once. However, there is another ribozyme, RNase P, that can act repeatedly on many different substrate molecules because it does not digest itself in the reaction. RNase P is a ribonucleoprotein, but the small RNA (377 nucleotides in *E. coli*) is the catalytic component, not the protein. As is the case for proteins with enzymatic activity, all ribozymes must be folded

into the proper structure for activity. In some cases, this structure might be supplied by the secondary structure of the RNA itself. In others, like RNase P, specific proteins may help keep the RNA in the active conformation. RNase P functions in the cell to modify primary transcripts coding for transfer RNAs (see Section 7.14).

The discovery of ribozymes has caused a reevaluation of other cellular processes that involve RNA. For instance, it is now clear that ribosomal RNA plays an active role in protein synthesis, even catalyzing the formation of the peptide bonds that link amino acids together in a protein (see Section 7.15). But clearly most enzymes are proteins. Why do ribozymes exist? It has been proposed that they are the vestigial remains of a simpler form of life, "RNA life," which may have predated the era of proteins as the cell's major catalysts. We discuss this concept in more detail in Chapter 11.

✓ 7.12 Concept Check

RNA molecules are often modified after transcription, an operation called RNA processing. All tRNAs and rRNAs are the result of processing of a longer precursor. The processing of eukaryotic pre-mRNAs is unique and can involve three distinct processing steps: splicing, capping, and tailing. Introns found in some other transcripts are self-splicing, and the RNA itself catalyzes the reaction. RNA molecules with catalytic activity are called ribozymes and play an important role in the cell.

✓ What is *splicing*?

✓ What is a *ribozyme*?

V PROTEIN SYNTHESIS

The first two steps in biological information transfer, *replication* and *transcription*, involve synthesis of nucleic acids using nucleic acid templates. The last step, *translation*, involves a nucleic acid template but in this case the final product is a protein. Proteins are made up of amino acid residues and not bases, so one can imagine that information transfer in translation is considerably more complicated than base pairing. In the next sections we shall deal with translation.

7.13 The Genetic Code

Before we discuss the mechanism of translation, we will discuss the correspondence between the nucleic acid template and the amino acid sequence of the protein product: the **genetic code**.

As we mentioned in Section 7.1, a triplet of three bases called a **codon** encodes a specific amino acid. It is

TABLE 7.3 The genetic code as expressed by triplet base sequences of mRNA[a]

Codon	Amino acid	Codon	Amino acid	Codon	Amino acid	Codon	Amino acid
UUU	Phenylalanine	UCU	Serine	UAU	Tyrosine	UGU	Cysteine
UUC	Phenylalanine	UCC	Serine	UAC	Tyrosine	UGC	Cysteine
UUA	Leucine	UCA	Serine	UAA	None (stop signal)	UGA	None (stop signal)
UUG	Leucine	UCG	Serine	UAG	None (stop signal)	UGG	Tryptophan
CUU	Leucine	CCU	Proline	CAU	Histidine	CGU	Arginine
CUC	Leucine	CCC	Proline	CAC	Histidine	CGC	Arginine
CUA	Leucine	CCA	Proline	CAA	Glutamine	CGA	Arginine
CUG	Leucine	CCG	Proline	CAG	Glutamine	CGG	Arginine
AUU	Isoleucine	ACU	Threonine	AAU	Asparagine	AGU	Serine
AUC	Isoleucine	ACC	Threonine	AAC	Asparagine	AGC	Serine
AUA	Isoleucine	ACA	Threonine	AAA	Lysine	AGA	Arginine
AUG (start)[b]	Methionine	ACG	Threonine	AAG	Lysine	AGG	Arginine
GUU	Valine	GCU	Alanine	GAU	Aspartic acid	GGU	Glycine
GUC	Valine	GCC	Alanine	GAC	Aspartic acid	GGC	Glycine
GUA	Valine	GCA	Alanine	GAA	Glutamic acid	GGA	Glycine
GUG	Valine	GCG	Alanine	GAG	Glutamic acid	GGG	Glycine

[a]The boxes of codons are colored according to the scheme: ▪ ionizable: acidic, ■ ionizable: basic, ▪ nonionizable polar, and ▪ nonpolar (⮂ Figure 3.12). The nucleotide on the left is at the 5′-end of the triplet.
[b]AUG encodes N-formylmethionine at the beginning of mRNAs of *Bacteria*.

conventional to present the genetic code as mRNA rather than as DNA because it is with mRNA that the translation process occurs. The 64 possible codons of mRNA are presented in Table 7.3. Note that in addition to the codons specifying the various amino acids, there are also special codons for starting (AUG) and for stopping (UAA, UAG, UGA) translation.

Perhaps the most interesting feature of the genetic code is that most amino acids are encoded by several different but related base triplets. This means that in most cases there is no one-to-one correspondence between the amino acid and the codon—knowing the amino acid at a given location does not mean that the codon at that location is automatically known.* The property of a code in which there is no one-to-one correspondence between word and code is called **degeneracy**.

As we shall see (Section 7.15), in the cell a codon is *read* by base-pairing with a tRNA at a sequence of three bases called an **anticodon**. If the base-pairing involved was always the standard pairing of A with U and G with C, then one would expect that there must be at least one specific tRNA for each codon and, therefore, for some amino acids there must be several tRNAs. For instance, there are six different tRNAs in *Escherichia coli* that carry the amino acid leucine. However, it is also true that some tRNAs can read more than one codon. For instance, there is only one tRNA in *E. coli* that carries the amino acid ly-

sine and it can read either AAA or AAG (see Table 7.3). This is possible because in some cases, tRNA molecules form *standard* base pairs at only the first two positions of the codon, tolerating unusual base pairs at the third position. This apparent mismatch phenomenon, called **wobble** is illustrated in Figure 7.33. The pairing between G and U is allowed at the wobble position.

Start and Stop Codons

As seen in Table 7.3, a few codons do not correspond to an amino acid. These triplets (UAA, UAG, UGA) are the **nonsense**, or **stop, codons**, and they signal the termination of translation of the gene encoding a specific protein (see Section 7.15).

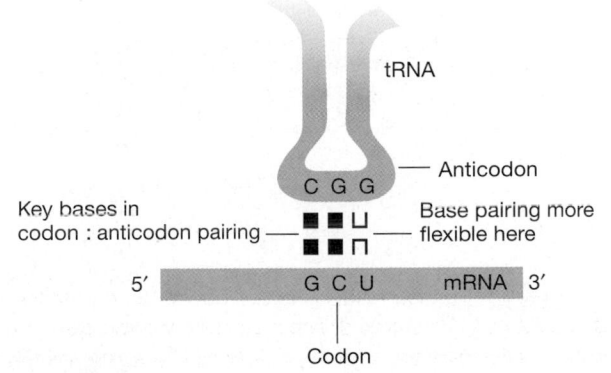

Figure 7.33 The wobble concept: base pairing is more flexible for the third base of the codon. Only a portion of the tRNA is shown (see Figure 7.35).

---

*The reverse is true, however. Knowing the DNA codon, one can specify the amino acid in the protein (assuming the proper reading frame is known). This permits the determination of amino acid sequences from DNA base sequences.

What is the mechanism by which the proper *starting point* for translation is found? The message is read by reading the **start codon**, **AUG**, which, at the beginning of the message, codes for the amino acid *N*-formylmethionine (or methionine in *Eukarya* and *Archaea*). The importance of having a well-defined starting point is readily understood if we consider that with a triplet code it is absolutely essential that translation begin at the correct location because if it does not, the whole **reading frame** will be shifted and an entirely different protein (or no protein at all) will be formed. By convention the "correct" reading frame, that is the one that can be translated to give the protein encoded by the gene, is called the 0 frame. As can be seen in Figure 7.34, the other two possible reading frames (–1 and +1) do not encode the same amino acid sequence. Therefore it is imperative that the ribosome find the *correct start codon* and that once it has, move down the mRNA exactly three bases at a time.

As we will discuss in Section 7.15, ribosomes from *Bacteria* recognize a specific AUG as a start codon with the aid of an upstream sequence on the mRNA called the Shine–Dalgarno sequence. This extra help at initiation explains why a few messages from *Bacteria* actually use other codons, such as GUG, for a start codon. However, even these unusual start codons specify *N*-formylmethionine.

Open Reading Frames

Today the genomes of many organisms (including the human genome) have been sequenced and many more are being sequenced. This sequencing information would be useless unless the scientists can relatively easily determine the location of protein-encoding genes. How can this be determined?

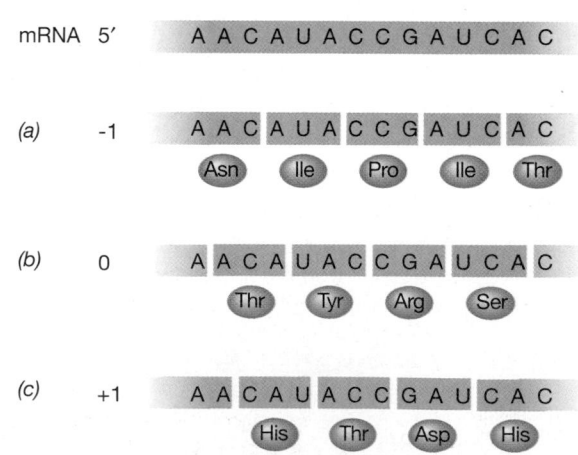

Figure 7.34 Possible reading frames in an mRNA. An interior sequence of an mRNA is shown. The "correct" (or 0) reading frame is determined by the start codon of the mRNA. (a) The amino acids that would be encoded by this region of the mRNA if the ribosome were in the –1 reading frame. (b) The amino acids that would be encoded if the ribosome were in the correct reading frame. (c) The amino acids that would be encoded if the ribosome were in the +1 reading frame.

One common method is to examine each strand of the DNA sequence for *open reading frames*. Remember that mRNA is transcribed from DNA so that if one knows the sequence of DNA, one also knows the sequence of RNA that could be transcribed from it. If this RNA can be translated, it must contain an **open reading frame (ORF)**: a start codon (typically AUG) followed by some number of codons and then a stop codon in the same frame as the start codon. A computer can be programmed to scan long base sequences in DNA databases to look for open reading frames. The search for ORFs is very useful in genomics (🔗 Chapter 15) and genetic engineering (🔗 Chapter 31) when one has isolated and sequenced an unknown piece of DNA and is not certain whether or not it encodes protein. The computer search for ORFs permits the researcher to identify putative genes that were previously unsuspected.

Other Genetic Codes

When the genetic code was cracked during the 1960s, all prokaryotes and eukaryotes examined were found to use the same code. When the mRNA for mammalian hemoglobin was given to the *Escherichia coli* protein-synthesizing machinery (such as ribosomes and tRNA), mammalian hemoglobin was synthesized. Therefore, the genetic code appeared to be a **universal code** in that the same code was used by all living systems. This view was changed when the sequences of genes and proteins were obtained from a large number of sources. It was discovered that some organelles and some cells use genetic codes that are slight variations of the "universal" genetic code (see the box, Selenocysteine: The Twenty-first Amino Acid).

These alternative codes were originally discovered in the genomes of mitochondria. So far as is known, only the mitochondria of plants use the universal code without change. The other organelles in plants, the chloroplasts, also use this standard code. The mitochondria of all other eukaryotes use codes with one or a few slight differences. Note that there is *not* simply a mitochondrial code, although there are a few common themes, such as the general use of UGA as a tryptophan codon instead of as a stop codon. It is also clear that these alternate codes are very closely related to the universal code and are almost certainly derived from it evolutionarily. Several cells are now known whose chromosomes also use slightly different codes such as *Bacteria* in the genus *Mycoplasma* and *Eukarya* in the genus *Paramecium*. All these alternative chromosomal genetic codes have different assignments for what are normally stop codons. These organisms simply have fewer nonsense codons because one or two are read as sense codons.

If every codon has an assignment, you might imagine that it is very difficult to change the genetic code in an organism. For instance, the change of AUA from an isoleucine codon to a methionine codon means that every protein that

A Focus On ...   Selenocysteine: The Twenty-first Amino Acid

The genetic code has codons for 20 amino acids that are assembled into proteins during translation. However, many proteins contain other amino acids. In fact, there are well over 100 different amino acids found in at least a few proteins. Until recently, it was thought that these "extra" amino acids were made by modifying one of the standard amino acids *after* it was incorporated into protein, a process called *posttranslational modification.* However, it is now clear that one of these extra amino acids is put into protein by the translational machinery itself. This one exception is *selenocysteine.*

Selenocysteine has the same structure as cysteine, but it has a selenium atom rather than a sulfur atom. It was known for some time that a few proteins contained this unusual amino acid. For example, *Escherichia coli* makes two different formate dehydrogenase enzymes and both contain a single selenocysteine residue. When the gene encoding one of these enzymes was sequenced, it was found that the codon corresponding to the selenocysteine

was a UGA. UGA is normally an efficient stop codon in *E. coli*, but it has now been demonstrated that it can be translated directly as selenocysteine in certain mRNA molecules, not only in *E. coli*, but also in other prokaryotes and in eukaryotes, including humans. Therefore, selenocysteine is the twenty-first amino acid known to be encoded by the genetic code.

How can a codon sometimes be a stop codon and sometimes a sense codon *in the same chromosome?* The answer lies in the *context* of the codon, the sequence of the bases surrounding the UGA codon and in their secondary structure. In certain contexts, the translational machinery interprets UGA as "selenocysteine." In all other contexts, UGA means "stop translation." Selenocysteine has its own tRNA (as do all the standard amino acids) and also has a special protein factor that brings only this tRNA to the ribosome.

Selenocysteine is even more readily oxidized than cysteine. Therefore, enzymes that contain this amino acid must be protected from oxygen. It has been proposed that UGA might once have been a normal sense codon, calling only for selenocysteine, but that the increase in oxygen in our environment following the evolution of photosynthesis (Chapter 11) selected for proteins that contain cysteine (whose codons are UGU and UGC). This allowed the coding assignment of UGA to be altered except in a few special cases. ■

$$^-OOC - \underset{\underset{H}{|}}{\overset{\overset{NH_3^+}{|}}{C}} - CH_2 - SH$$

Cysteine

$$^-OOC - \underset{\underset{H}{|}}{\overset{\overset{NH_3^+}{|}}{C}} - CH_2 - SeH$$

Selenocysteine

once had an isoleucine encoded by AUA now has a methionine at this position. Such a protein may not function normally. Changing a codon assignment may not be as severe a problem if *codon usage* is not random. After the genetic code had been worked out by biochemists and before any genes had actually been sequenced, it was assumed that the degenerate codons for an amino acid would be used at an equal frequency. This is another assumption that DNA sequencing has shown to be incorrect! Codon usage is highly biased, and this bias changes from organism to organism. In *Escherichia coli*, for instance, only about 1 out of 20 isoleucine residues is encoded by an AUA, the other 19 being encoded by AUU and AUC. It is thought that one of the steps that could have led to codon reassignment was that the codon became rarely used in a genome. This would have been easier to achieve in mitochondria because they have very small genomes (Section 15.7).

✓ 7.13 Concept Check

The genetic code is expressed in terms of RNA, and a single amino acid may be encoded by several different but related codons. In addition to the stop codons, there is also a specific start codon that signals where the translation process should begin.

✓ Why is it important for the ribosome to read "in frame"?
✓ Describe an *open reading frame.*

7.14 Transfer RNA

Recall from Section 7.13, that it is the anticodon on the tRNA that "reads" (base pairs with) the codon. However, a tRNA is much more than simply an anticodon (Figure 7.35). The tRNA not only has a specificity for the codon on the mRNA, it also has specificity for the appropriate amino acid. The transfer RNA and its specific amino acid are brought together by means of specific enzymes that ensure that a particular tRNA receives its correct amino acid. These enzymes, called **amino acid activating enzymes** or **aminoacyl-tRNA synthetases**, have the important function of recognizing *both* the amino acid *and* the specific tRNA for that amino acid.

Structure of tRNA

There are about 60 different specific tRNAs in bacterial cells and 100–110 in mammalian cells. Transfer RNA molecules are short, single-stranded molecules with lengths (among different tRNAs) of 73–93 nucleotides. When compared, it has been found that certain bases and secondary structures are constant for all tRNAs, and there are other parts that are variable. Transfer RNA molecules also contain some purine and pyrimidine bases differing slightly from the normal bases found in RNA in that they are chemically modified, often methylated. These

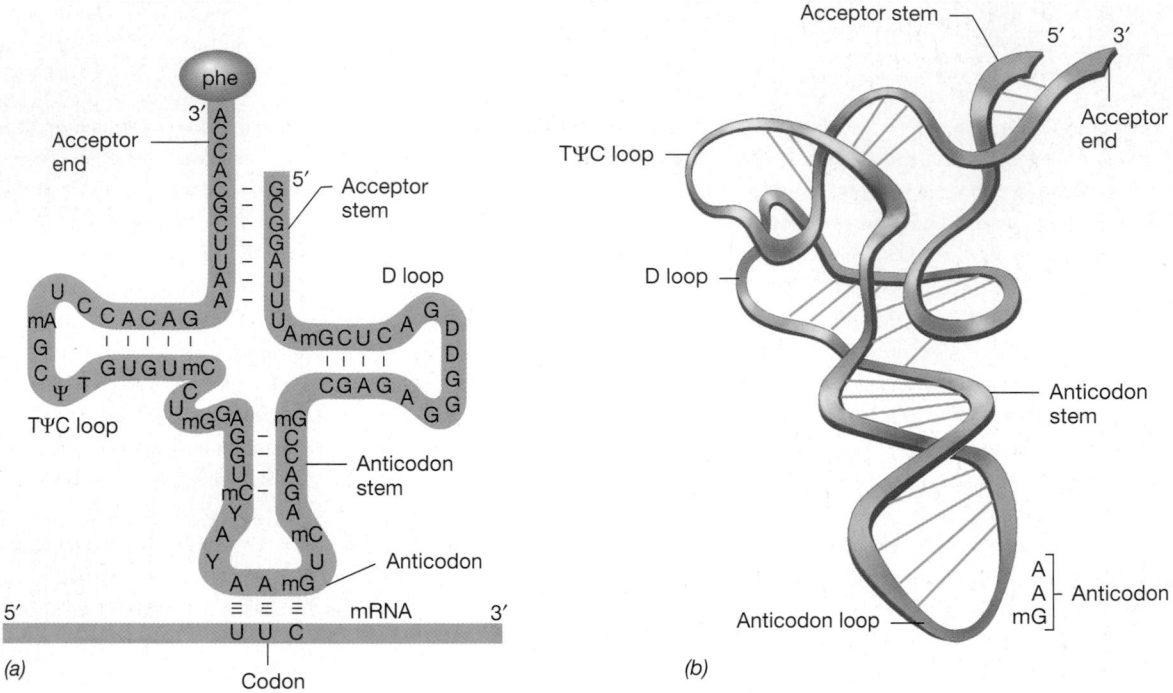

Figure 7.35 Structure of a transfer RNA, yeast phenylalanine tRNA. (a) The conventional cloverleaf structure. The amino acid is attached to the ribose of the terminal A at the acceptor end. A, adenine; C, cytosine; U, uracil; G, guanine; ψ, pseudouracil; D, dihydrouracil; m, methyl; Y, a modified purine. (b) In actuality the molecule folds so that the D loop and TψC loops are close together and associate by hydrophobic interactions.

modifications are made to the bases after transcription. Some of these unusual bases are pseudouridine, inosine, dihydrouridine, ribothymidine, methyl guanosine, dimethyl guanosine, and methyl inosine. Base modifications as well as other types of processing (see Section 7.12) are necessary to make a functional tRNA from the transcript of a tRNA-encoding gene. Although tRNA is a single-strand, there are extensive double-stranded regions within the molecule as a result of internal base pairing when the molecule folds back on itself.

The structure of tRNA can be drawn in a cloverleaf fashion, as in Figure 7.35a. Some regions of secondary structures are given names having to do either with the bases most often found there (the TψC loop and the D loop) or with specific functions (anticodon loop and acceptor end). The three-dimensional structure of a tRNA is more clearly shown in Figure 7.35b. Note that bases that appear widely separated in the cloverleaf model are actually close together when viewed in three dimensions. This means that some of the bases in the "loops" are actually paired with bases in other loops.

One of the variable parts of the tRNA molecule contains the **anticodon**, the site that recognizes the codon on the mRNA. The anticodon is found in the *anticodon loop*, shown in Figure 7.35. There are just *three* nucleotides in the anticodon loop that are specifically involved in the recognition process and that base-pair with the codon (see Section 7.13). Other portions of the tRNA interact with the ribosome (both rRNA and protein), other protein factors, and the activating enzyme. At the 3′-end, or **acceptor end**, of all tRNAs, are three unpaired nucleotides. The sequence of these three nucleotides is always cytosine-cytosine-adenine (CCA), and it is to the ribose sugar of the terminal A that the amino acid is covalently attached via an ester linkage. From this acceptor portion of the tRNA, the amino acid is transferred to the growing polypeptide chain on the ribosome by a mechanism that will be described in the next section.

Recognition, Activation, and Charging of tRNA
Recognition of the correct tRNA by an aminoacyl-tRNA synthetase involves specific contacts between key regions of the nucleic acid and particular amino acids of its respective synthetase (Figure 7.36). As might be expected because of the unique sequence in this region, the *anticodon* of the tRNA is important in recognition by the synthetase. However, other contact sites between the tRNA and the synthetase are also important. Studies of tRNA binding to aminoacyl-tRNA synthetases in which specific bases in the tRNA have been changed by genetic mutation have shown that only a small number of key nucleotides in a tRNA besides the anticodon region are involved in recognition; these other key recognition nucleotides are often part of the acceptor stem of the tRNA

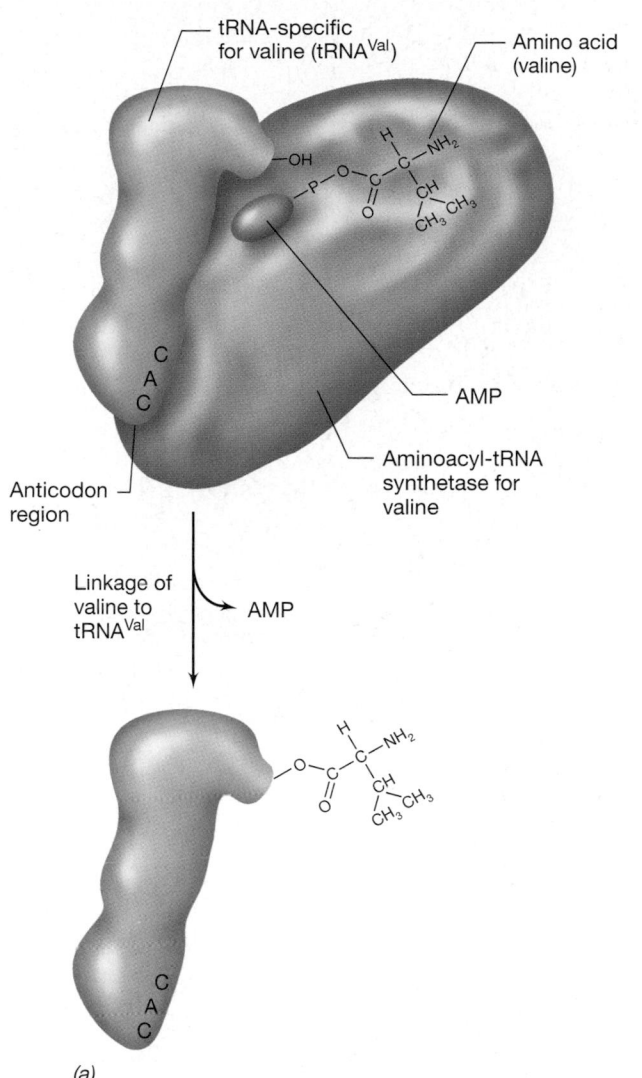

(a)

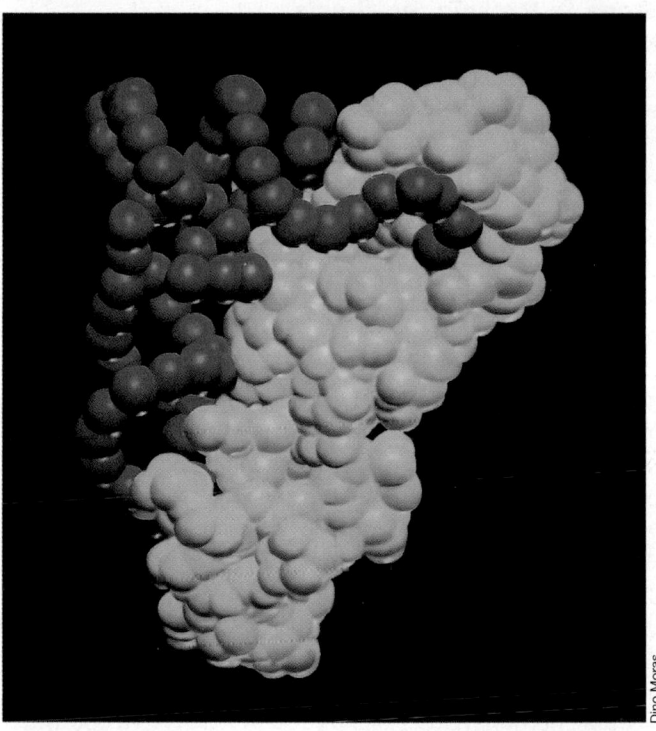

(b)

Dino Moras

Figure 7.36 Aminoacyl-tRNA synthetases. (a) Action of an aminoacyl-tRNA synthetase. Recognition of the correct tRNA by a particular synthetase involves contacts between specific nucleic acid sequences and specific amino acids of the synthetase. In this diagram, valyl-tRNA synthetase is shown catalyzing the final step of the reaction, where the valine in valyl-AMP is transferred to tRNA. (b) A computer model showing the interaction of glutaminyl-tRNA synthetase (blue) with its tRNA (red). Reprinted with permission from M. Ruff et al. 1991. *Science* 252: 1682–1689. © 1991, AAAS.

molecule (see Figure 7.35). In a few cases, recognition of a tRNA by its cognate synthetase is totally independent of the anticodon region. It should be emphasized at this point that the fidelity of this recognition process is crucial, for if the wrong amino acid is attached to the tRNA, it may be inserted in the improper place in the polypeptide, leading to the synthesis of a faulty protein.

The specific chemical reaction between amino acid and tRNA catalyzed by the aminoacyl-tRNA synthetase first involves *activation* of the amino acid by reaction with ATP:

$$\text{Amino acid} + \text{ATP} \rightleftharpoons \text{aminoacyl–AMP} + \text{P–P}$$

The aminoacyl-AMP intermediate formed normally remains bound to the enzyme until collision with the appropriate tRNA molecule, and, as shown in Figure 7.36a, the activated amino acid is then transferred to the tRNA to form a *charged* tRNA:

$$\text{Aminoacyl–AMP} + \text{tRNA} \rightleftharpoons$$
$$\text{aminoacyl–tRNA} + \text{AMP}$$

The pyrophosphate (P–P) formed in the first reaction is split by a pyrophosphatase, forming two molecules of inorganic phosphate. Since ATP is used and AMP is formed, a total of *two* high-energy phosphate bonds are required for the activation of an amino acid and charging a tRNA. Once activation and charging have occurred, the aminoacyl-tRNA (AA-tRNA) leaves the synthetase and is brought to the ribosome by a protein factor. The mechanism of protein synthesis is discussed in the next section.

✓ 7.14 Concept Check

One or more transfer RNAs exist for each amino acid found in protein. Enzymes called aminoacyl-tRNA synthetases function to attach an amino acid to a tRNA. Once the correct amino acid is attached to its tRNA, further specificity resides only in the codon-anticodon interaction.

✓ What is the function of the *anticodon* of a tRNA?

✓ What is the function of the *acceptor end* of a tRNA? ■

7.15 Translation: The Process of Protein Synthesis

It is the amino acid *sequence* that determines the structure (and ultimately the function) of the final active protein. Thus, it is critical that the proper amino acid be inserted at the proper place in the polypeptide chain. This is the role of the protein-synthesizing machinery of the cell.

Steps in Protein Synthesis

Ribosomes are the site of protein synthesis. Each ribosome is constructed of two subunits. In prokaryotes, the ribosome subunits are of 30S (Svedberg units) and 50S, yielding intact 70S ribosomes.[†] Each subunit is itself a ribonucleoprotein complex made up of specific ribosomal RNAs and ribosomal proteins. The 30S subunit contains 16S rRNA and about 21 proteins, while the 50S subunit contains 5S and 23S rRNA and about 34 proteins (Table 7.4 and Figure 7.37a). In *Escherichia coli*, there are at least 53 different ribosomal proteins, most present at one copy per ribosome.

Since the genome of *E. coli* has been completely sequenced and the ribosome has been studied for some time, one might think that we should know exactly how many ribosomal proteins there are overall and in each subunit. However, some proteins are tightly associated with the ribosome, some less so, some are associated only with one subunit, some with both. In addition, the ribosome is a dynamic structure that interacts with many other proteins in the cell. There are protein factors absolutely essential to the ribosomes function and that interact with the ribosome but are not considered "ribosomal proteins."

The actual synthesis of a protein involves a complex cycle in which the various ribosomal components play specific roles. Although a continuous process, protein synthesis can be broken into a number of discrete steps: **initiation**, **elongation**, and **termination-release**. The first two steps are outlined in Figure 7.37b. In addition to mRNA, tRNA, and ribosomes, the process involves a number of proteins designated initiation, elongation,

[†] The numbers 30S, 50S, and 70S refer to *Svedberg units*, which are units of sedimentation coefficients of ribosome subunits or intact ribosomes when subjected to centrifugal force in an ultracentrifuge.

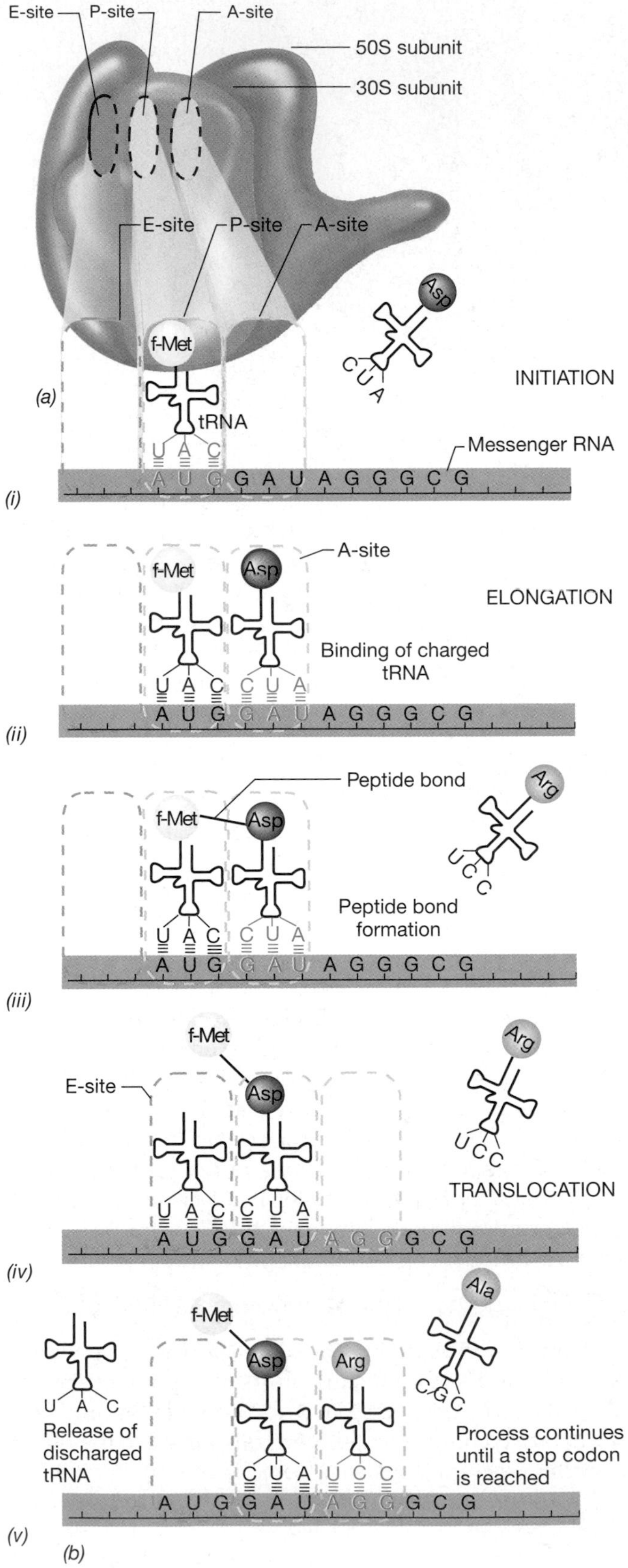

Figure 7.37 (a) Structure of the ribosome, showing the position of the A- or acceptor site (right), the P- or peptide site, and the E- or exit site. (b) Translation of the information from the messenger RNA into the amino acid sequence of a protein. Only the sites on the ribosome are shown. (i, ii) Interaction between the codon and anticodon brings into the position the correct charged tRNA—in this case the initiator tRNA and then the tRNA for the second amino acid. (iii) The formation of a peptide bond between amino acids on adjacent tRNA molecules. (iv) The translocation of the ribosome from one codon to the next with the release of the tRNA, now free of an amino acid, from the E-site. (v) The next charged tRNA binds to the A-site.

TABLE 7.4   Ribosome structure[a]

Property	Prokaryote	Eukaryote
Overall size	70S	80S
Small subunit	30S	40S
Number of proteins	~21	~30
RNA size	16S (1500)	18S (2300)
(number of bases)		
Large subunit	50S	60S
Number of proteins	~34	~50
RNA size	23S (2900)	28S (4200)
(number of bases)	5S (120)	5.8S (160)
		5S (120)

[a]Ribosomes of mitochondria and chloroplasts of eukaryotes are similar to prokaryotic ribosomes (⚭ Section 14.4).

and termination factors; guanosine triphosphate provides energy for the process.

Initiation of Protein Synthesis

In prokaryotes initiation always begins with a free 30S ribosome subunit, and an **initiation complex** forms consisting of a 30S ribosome subunit, mRNA, formylmethionine tRNA, and initiation factors. Guanosine triphosphate is required for this step. To this initiation complex a 50S ribosome subunit is added to make the active 70S ribosome. At the end of the translational process, the released ribosome separates again into 30S and 50S subunits. Just preceding the initiation codon on the mRNA is a sequence of from three to nine nucleotides (the so-called **Shine–Dalgarno sequence**) that is involved in the binding of the mRNA to the ribosome. This *ribosome binding site* at the 5′-end of the mRNA is complementary to the 3′-end of the 16S RNA of the ribosome, and it is this base pairing that ensures effective formation of the ribosome–mRNA complex.

The presence of the Shine–Dalgarno site on the mRNA and its specific interaction with 16S rRNA allow prokaryotic ribosomes to use *polycistronic* mRNA because the ribosome can find each initiation site within a message (see Section 7.11). Eukaryotic ribosomes typically recognize an mRNA by its 5′-cap and initiate only at the first possible initiation codon. Therefore, they normally cannot translate polycistronic mRNA.

Initiation always begins with a special initiator aminoacyl-tRNA binding to the **start codon**, AUG. In *Bacteria* this is **formylmethionine** tRNA. Subsequently, the formyl group at the N-terminal end of the polypeptide is removed; the terminal amino acid of the completed protein is hence methionine. Because the Shine–Dalgarno sites (and other possible interactions between the rRNA and the mRNA) are also involved in directing the ribosome to a start site, prokaryotic messages sometimes use a start codon other than AUG. The most common alternative start codon is GUG. When used in

this unusual way, however, GUG calls for formylmethionine initiator tRNA (and not valine). In *Eukarya* and *Archaea*, initiation begins with methionine instead of formylmethionine. (Although all proteins are *initiated* with a methionine, this amino acid is usually removed by a specific protease after translation.)

Elongation and Termination

The mRNA is threaded through the ribosome primarily bound to the 30S subunit. The ribosome contains other sites where the tRNAs interact. Two of these sites are located primarily on the 50S subunit, and they are termed the P-site and the A-site (see Figure 7.37b). The A-site, the **acceptor** site, is the site where the new AA-tRNA first attaches. The P-site, the **peptide** site, is the site where the growing peptide is held by a tRNA. During peptide bond formation, the peptide moves to the tRNA at the A-site as a new peptide bond is formed. Several soluble (nonribosomal) elongation factors are required for **elongation**, as well as additional molecules of GTP (to simplify Figure 7.37b, the elongation factors are omitted and only a portion of the ribosome is shown). The tRNA that holds the peptide must now be moved (translocated) from the A-site to the P-site, thus opening up the A-site for another AA-tRNA.

Translocation requires a specific elongation factor and one molecule of GTP per each tRNA translocated. At each translocation step the message is advanced three nucleotides, exposing a new codon at the ribosome A-site. Translocation pushes the now empty tRNA to a third site, called the E-site. It is from this **exit** site that the tRNA is actually released from the ribosome. The precision of the translocation step is critical to the accuracy of protein synthesis. The ribosome must move exactly one codon at each step. The ribosome is a dynamic structure: Both mRNA and tRNA move through the ribosome as it carries out each of the stages of protein synthesis. Thus, the three sites we have discussed are not simply static locations on the surface but are *moving parts* of a complex biomolecular machine.

When several ribosomes are simultaneously translating a single message, the complex is called a **polysome** (Figure 7.38). Polysomes increase the speed and efficiency of mRNA translation, and because each ribosome acts independently of the others, each ribosome in a polysome complex makes a complete polypeptide. Note in Figure 7.38 how ribosomes closest to the 5′-end (the beginning) of the mRNA molecule have short polypeptides attached to them because only a few codons have been read, while ribosomes closest to the 3′-end of the message have nearly finished polypeptides.

The **termination** of protein synthesis occurs when a codon is reached that does not specify an AA-tRNA: a **stop codon** (see Section 7.13). No tRNA binds to a stop codon; instead, proteins called *release factors* read the chain-terminating signal and cleave the attached

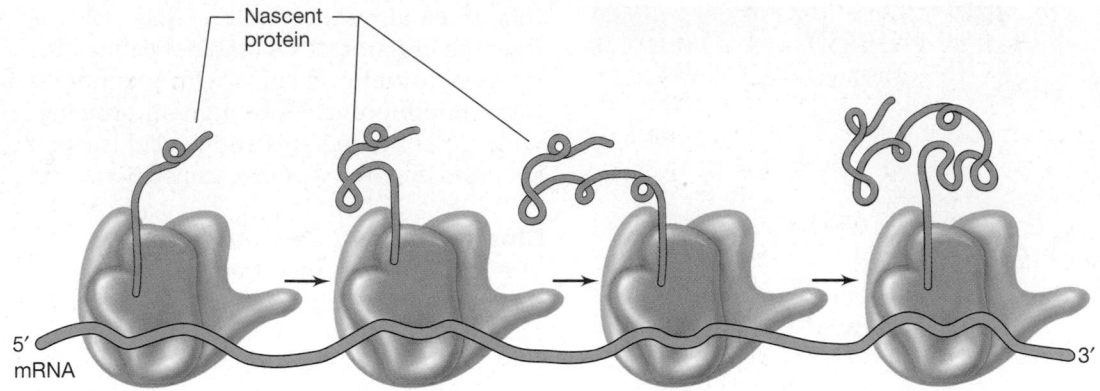

Figure 7.38 Translation by several ribosomes on a single messenger RNA (polysome). Note how the ribosomes nearest the 5′-end of the message are at an earlier stage in the translation process.

polypeptide from the terminal tRNA, releasing the completed protein. Following this, the ribosome dissociates, and the subunits are then free to form new initiation complexes.

Role of Ribosomal RNA in Protein Synthesis

Ribosomes are composed of a series of proteins and ribosomal RNAs (see Table 7.4). Two decades ago it was assumed that ribosomal proteins were probably the functional components of the ribosome, whereas the role of the ribosomal RNAs was largely structural, that is, serving as a support for ribosomal proteins. It is now clear that rRNA plays a critical *functional* role in all stages of protein synthesis, from initiation to termination. The role of the many proteins present in the ribosome, although less clear, may be as facilitators of RNA function by stabilizing or positioning the key functional sequences in the various ribosomal RNAs.

In prokaryotes, it is clear that 16S rRNA is involved in initiation through base pairing between the ribosome binding sequence (the Shine–Dalgarno sequence). There is strong evidence that mRNA and rRNA interactions also occur during elongation.

Strong evidence also exists for a role for rRNA in ribosome subunit association, as well as for tRNA positioning in the decoding (A- or P-, see Figure 7.37b) sites on the ribosome and even in catalyzing peptide bond formation. Charged tRNAs that enter the ribosome recognize the correct codon by codon:anticodon base pairing, but they are also physically attached to the ribosome by interactions of the anticodon stem-loop of the tRNA with specific locations within 16S rRNA. The acceptor end of the tRNA interacts with the 23S rRNA.

The peptidyl transferase reaction (the actual formation of peptide bonds that occurs on the 50S subunit of the ribosome) is associated with 23S rRNA. For a long time it seemed possible that this reaction might be catalyzed by some combination of proteins. However, it is now clear that the reaction itself is catalyzed by ribozymal activity of 23S

rRNA itself (see Section 7.12). The 23S rRNA also seems to play a role in translocation, and the elongation factors are known to interact with 23S rRNA. Finally, the 16S rRNA is also involved in termination, possibly interacting with the mRNA or through interactions with the release factors.

Ribosomal RNA thus plays a major role in translation. Ribosomal function is clearly dependent on the major RNA species present.

Effect of Antibiotics on Protein Synthesis

A large number of antibiotics inhibit protein synthesis by interacting with the ribosome. These interactions are quite specific, and many have been shown to involve rRNA. Several of these antibiotics are medically useful, and several are also effective research tools because they are specific for different steps in protein synthesis. For instance, *streptomycin* inhibits initiation, whereas *puromycin, chloramphenicol, cycloheximide,* and *tetracycline* inhibit elongation.

Adding to their clinical usefulness is the fact that many antibiotics specifically inhibit ribosomes of organisms from only one or two of the phylogenetic domains. Of the antibiotics just listed, chloramphenicol and streptomycin are specific for the ribosomes of *Bacteria* and cycloheximide for ribosomes of *Eukarya*. Many of these antibiotics will be discussed in Chapter 20.

✓ 7.15 Concept Check

The ribosome plays a key role in the translation process, bringing together mRNA and amino acid-charged tRNAs. There are three sites on the ribosome: the acceptor site, where the charged tRNA first combines; the peptide site, where the growing polypeptide chain is held; and an exit site. During each step of amino acid addition, the message advances three nucleotides (one codon) and the tRNA moves from the acceptor to the peptide site. Termination of protein synthesis occurs when a stop codon, which does not code for any amino acid, is reached.

✓ What are the components of a ribosome?

✓ What functional roles does rRNA play in protein synthesis?

7.16 Folding and Secreting Proteins

We have now discussed how the information present in the sequence of bases in DNA is replicated, transcribed into a sequence of bases in RNA, and translated into a sequence of amino acids in protein. Our discussion of the flow of biological information is not quite finished, though. In Section 7.1 we mentioned that proteins were the cell's "functional entities"; for example, as enzymes, proteins catalyze most of the thousands of reactions that occur in a typical cell. However, for a protein to function it must be *folded* correctly and it must also be in the correct location. Here we briefly deal with these two processes.

Protein Folding

It was long thought that all proteins folded spontaneously into their active form while they were being synthesized (Figure 7.38). However, we now know this is not the case. Many proteins require the assistance of other proteins called **molecular chaperones** for proper folding or for assembly into larger complexes. The chaperones themselves do not become part of the assembled proteins. One important activity of the chaperones is to prevent improper aggregation of proteins. There are many different kinds of these chaperones, some are even associated with ribosomes. Some are also extremely abundant in the cell. These proteins seem to be both extremely widespread, and their sequences highly conserved.

One type of molecular chaperone, called *chaperonins*, is probably present in all living organisms. The activity of a chaperonin is shown in Figure 7.39. The unfolded or improperly folded protein enters the molecular chaperone where it is folded and then released. Energy for the folding comes from ATP. Other cellular chaperones are involved in carrying the unfolded protein to the chaperonin. In addition to folding newly synthesized proteins, chaperones also can refold proteins that have partially denatured in the cell. Such protein denaturation can occur because the organism has temporarily experienced high temperatures in its environment. Refolding is not always successful, and cells also contain proteases that specifically target and destroy misfolded proteins.

Protein Secretion

Many proteins function in the cell's membrane or even *outside* the cell and must somehow get from the site of synthesis on ribosomes into and sometimes through the cytoplasmic membrane. In prokaryotes, periplasmic enzymes and extracellular enzymes are secreted or *secretory* proteins (⌾ Section 4.7). In eukaryotes there are a large number of membrane-enclosed organelles, for example, mitochondria, into which proteins must be transported, as well as proteins that must be secreted outside the cell. It has been estimated that almost

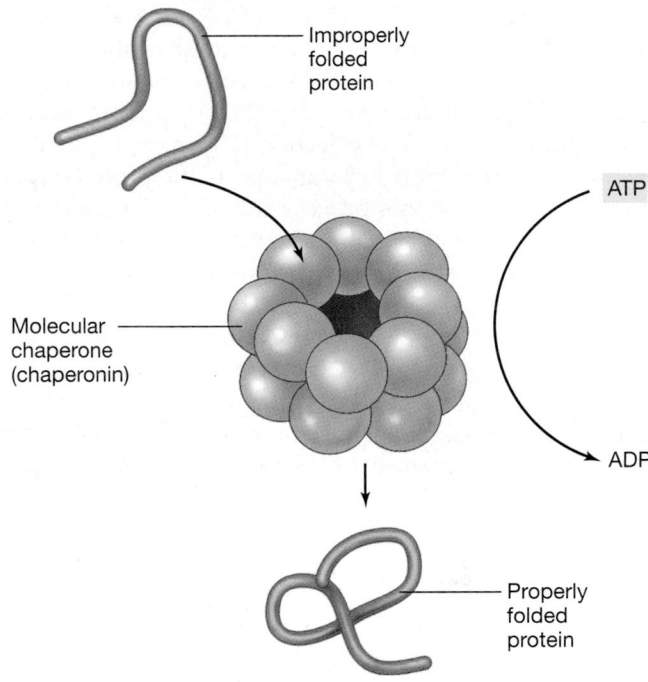

Figure 7.39 The action of a molecular chaperone. An improperly folded protein is taken into the barrel-like structure of the chaperonin. ATP is used to supply the energy required to fold the protein properly.

half the proteins of a typical eukaryotic cell are transported into or across a membrane.

How is it possible for a cell to selectively transfer some proteins across a membrane while leaving most proteins in place in the cytoplasm? Most proteins that must be transported through membranes are synthesized with an extra N-terminal peptide sequence, about 15–20 amino acids long, which is called the **signal sequence**. Signal sequences are quite variable but typically each has a few positively charged residues at the beginning, a central region of hydrophobic residues, and then a more polar region. The signal sequence serves to help other proteins in the cell recognize that this particular protein is to be exported and also can help prevent the protein of which it is a part from completely folding. Since the signal sequence is the first part of the protein to be synthesized, the early steps in export may begin before the protein is completely synthesized. In eukaryotes, transport is tightly coupled to translation itself and the newly synthesized protein is not released from the ribosome free into the cytoplasm. The ribosomes themselves are associated with membranes. Even in prokaryotes there are ribosome-associated chaperones involved in this process.

The principal role in the selection of most proteins to be secreted is played by the **signal recognition particle** (**SRP**). SRPs are found in all cells. In *Bacteria*, they contain a single protein and a small RNA molecule called 4.5S RNA. (This small RNA is not a transfer RNA, a ribosomal

RNA, or a messenger RNA. Most cells contain some small RNA molecules that do not belong to these major classes.) The SRP recognizes the signal sequence-containing protein and delivers it to a special membrane protein complex (for example, SecYEG, ∞ Section 4.7) where, in processes involving transport through a pore, the protein is secreted. As part of this process the signal sequence is usually removed by a peptidase enzyme, an example of the process of **posttranslational modification**.

The study of protein secretion has important practical implications for genetic engineering (∞ Chapter 31). If bacteria are genetically engineered to serve as agents for the production of foreign proteins, it is desirable to manipulate the signal sequence in order to arrange for the desired protein to be excreted so it can be readily isolated and purified.

✓ **7.16 Concept Check**

Proteins must be properly folded in order to function correctly. Folding may occur spontaneously but may also involve other proteins called molecular chaperones. Many proteins also need to be transported into or through cell membranes.

Such proteins are synthesized with a signal sequence that is recognized by the cellular export apparatus and is removed after export.

✓ What is a *molecular chaperone*?

✓ Why do some proteins have a *signal sequence*?

✓ What is a *signal recognition particle*?

In this chapter we have seen that biological information flows from DNA to RNA, and finally to protein. The mechanisms of the three main steps in information transfer, replication, transcription, and translation are quite similar in prokaryotes and eukaryotes. The information in a gene can be accurately replicated so that when a cell divides the progeny has the same genetic information. The information in the gene can also be transcribed into RNA, and for genes that encode proteins, translated into proteins. Although some RNA has catalytic activity, it is the proteins that are the components of the cell that carry out the thousands of reactions making up the cell's metabolism. We next turn our attention to how cells regulate the expression of a particular gene or set of genes.

Review Questions

1. Describe the *central dogma* of molecular biology.

2. Genes were discovered before their chemical nature was known. Define a gene without mentioning its chemical nature. Of what is a gene composed?

3. Inverted repeats can give rise to stem-loops. Show this by giving the sequence of double-stranded DNA containing an inverted repeat and show how the transcript from this region can form a stem-loop.

4. Is the sequence 5′-GCACGGCACG-3′ referred to as an inverted repeat? Explain your answer.

5. DNA molecules that are AT-rich separate into two strands more easily when the temperature is raised than do DNA molecules that are GC-rich. Write an explanation for this observation based on the properties of AT and GC base pairing.

6. A structure commonly seen in circular DNA during replication is called a *theta structure*. Draw a diagram of the replication process and show how a theta structure could arise.

7. Why are errors in DNA replication so rare? What additional enzyme activity (other than polymerization) is associated with DNA polymerase III and how does it serve to reduce errors?

8. What are ribozymes and what types of biochemical reactions are they generally associated with?

9. There are three processing steps in producing most eukaryotic mRNA but not prokaryotic mRNA. Write a short description of each of these three steps.

10. What are aminoacyl-tRNA synthetases and what types of reactions do they carry out? Approximately how many different types of these enzymes are present in the cell? How does a synthetase recognize its correct substrates?

11. Do genes for tRNAs have promoters? Do they have start codons? Explain.

12. The start and stop sites for mRNA synthesis (on the DNA) are different from the start and stop sites for protein synthesis (on the mRNA). Explain.

13. The activity that forms peptide bonds on the ribosome is called *peptidyl transferase*. What molecule in the cell catalyzes this reaction?

14. Sometimes misfolded proteins can be correctly refolded but sometimes they cannot and are destroyed. What kinds of proteins are involved in refolding misfolded proteins? What kind of enzymes are involved in destroying misfolded proteins?

Application Questions

1. The genome of the bacterium *Neisseria gonorrhoeae* consists of a single double-stranded DNA molecule that contains 2220 kilobase pairs. Calculate the length of this DNA molecule in centimeters. If 85% of this DNA molecule is made up of the open reading frames of genes encoding proteins and the average protein is 300 amino acids long, how many protein-encoding genes does *Neisseria* have? What kind of information do you think might be present in the other 15% of the DNA?

2. Circular DNA molecules, such as those of most bacterial chromosomes, circumvent one problem encountered in replication of linear DNA. What is this problem? Also, having a circular chromosome results in a new problem: The two daughter chromosomes are interlocked after replication is completed. Is there any type of enzyme discussed in this chapter that might help separate these molecules so they can be partitioned?

3. Many bacterial mRNA molecules are polycistronic, each mRNA coding for more than one protein. Imagine an mRNA that codes for two proteins with an intervening noncoding region between the two coding regions. From your understanding of how the translation works, explain why the end result would be two separate proteins rather than one mixed (hybrid) protein.

4. What would be the result (in terms of protein synthesis) if RNA polymerase initiated *transcription* one base upstream of its normal starting point? Why? What would be the result (in terms of protein synthesis) if *translation* began one base downstream of its normal starting point? Why?

5. In the bacterium *Salmonella enterica*, glutamyl-phosphate reductase and glutamate kinase, two of the enzymes involved in the synthesis of proline, are apparently translated from a single polycistronic message. Draw a diagram of the region of the chromosome containing the two genes encoding these enzymes. Show the correct relative position(s) of that portion of the DNA that contains or encodes all of the following: promoter(s), Shine–Dalgarno sequence(s), start and stop codons, and transcription terminator(s). Do you believe these genes probably contain introns? Why or why not?

6. If the genes you diagrammed in answering Question 5 were actually two genes encoding two different tRNAs, how would your diagram be different?

M̲ost genes contain the information to encode a single protein. However, in order for this protein to be made, the gene must first be expressed. Organisms regulate the expression of their genes so that proteins, and other molecules, are made in the correct amounts and at the correct times in the cell cycle. Regulation of gene expression can be influenced by the cell's environment. This regulation often involves very precise interactions between particular genes in the DNA and specific regulatory proteins, as shown here for the bacteriophage lambda repressor protein (a transcriptional regulatory protein) bound to its DNA. Binding of proteins to DNA can be a mechanism for both turning "on" and turning "off" gene expression.

REGULATION OF GENE EXPRESSION

8

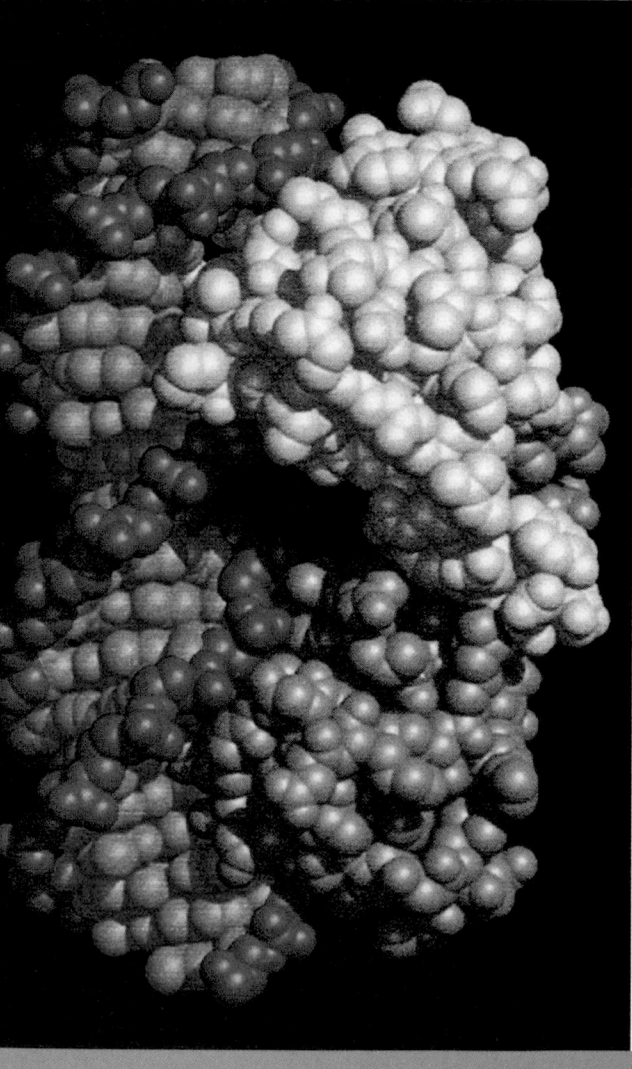

Working Glossary

Activator protein a regulatory protein that binds to specific sites on DNA and stimulates transcription; involved in positive control

Allosteric enzyme an enzyme that contains two binding sites, the active site (where the substrate binds) and the allosteric site (where an effector molecule binds)

Attenuation a mechanism for controlling gene expression; typically transcription is terminated after initiation but before a full-length mRNA is produced

Feedback inhibition a decrease in the activity of the first enzyme of a pathway caused by the final product of the pathway

Kinase an enzyme that adds a phosphoryl group to a compound

Negative control a mechanism for regulating gene expression in which a *repressor protein* functions to prevent transcription of a gene or genes

Operon one or more genes transcribed into a single RNA and under the control of a single regulatory site. Typically, however, the term is used to describe a transcription unit containing more than one gene

Positive control a mechanism for regulating gene expression in which an *activator protein* functions to promote transcription of a gene or genes

Quorum sensing regulatory pathways in *Bacteria* that respond to population density

Repressor protein a regulatory protein that binds to specific sites on DNA and blocks transcription; involved in negative control

Response regulator protein one of the members of a two-component system; a regulatory protein that is phosphorylated by a sensor kinase protein (see sensor kinase protein)

Sensor kinase protein one of the members of a two-component system; a kinase that is found in the cell membrane and that phosphorylates itself in response to an external signal and then passes the phosphoryl group to a response regulator protein (see response regulator protein)

Two-component system a regulatory system containing a sensor kinase protein and a response regulator protein (see sensor kinase protein and response regulator protein)

In the previous chapter we saw how the information stored as a sequence of bases in a gene can be transcribed into RNA and then translated to yield a specific protein. Most proteins are enzymes (∞ Secion 5.5) and carry out the reactions responsible for the cell's metabolism. Hundreds of different enzymatic reactions occur simultaneously during a single cycle of cell growth. *Micro*organisms also respond rapidly to changes in their environment, and many organisms have complicated developmental pathways. To efficiently orchestrate the numerous chemical reactions in a cell, make maximal use of available resources, and carry out developmental processes, cells need to *control* the expression of their genetic information.

▌ OVERVIEW OF REGULATION

Some enzymes may be needed in about the same amounts under all growth conditions and are said to be *constitutive*. Constitutive enzymes are generally key cellular enzymes required for growth under all nutritional conditions and are thus synthesized continuously in the growing cell. However, it is far more common for a particular reaction to be needed under some conditions but not under others. Indeed, under many growth conditions a cell might not need to carry out a particular reaction at all. For instance, enzymes required for the breakdown of the sugar lactose are useful to the cell only if lactose is present in its environment. Most microorganisms have the genes to encode many more different kinds of proteins than are actually present in the cell under any particular condition (∞ Section 2.2). Thus, the need to regulate gene expression in response to changing growth conditions, or as part of a developmental process, is clear. How does this type of regulation occur?

8.1 Major Modes of Regulation

There are two major modes of regulation in the cell. One controls the *activity* of preexisting enzyme and one controls the *amount* (or even the complete presence or absence) of an enzyme (Figure 8.1). Regulation of the activity of an enzyme obviously happens *after* the protein has been synthesized (that is, posttranslationally). By contrast, regulation of the amount of enzyme synthesized can occur at the level of transcription (how much messenger RNA [mRNA] is made) or at the level of translation (whether or not the mRNA is translated to make the protein). Regulation of the synthesis of an enzyme is a coarser level of control than regulating activity. The regulation of activity is typically very rapid (seconds or less), while the regulation of enzyme synthesis is a relatively slow process (a few minutes). If a new enzyme needs to be synthesized, it will take some time before that enzyme is present in the cell in sufficient amounts to affect metabolism. Alternatively, if synthesis of an enzyme is stopped, a considerable amount of time may elapse before the existing enzyme is diluted out sufficiently to no longer affect metabolism. Working together, these mechanisms can result in an efficient regulation of cell metabolism.

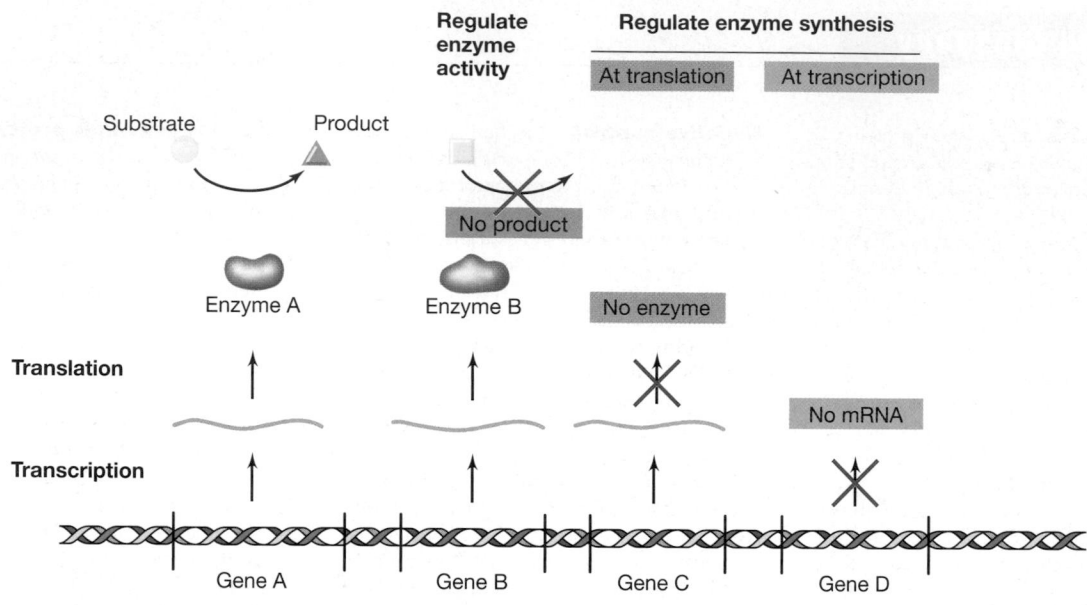

Figure 8.1 An overview of the mechanisms that can be used in regulation. The product of gene A is enzyme A, which is synthesized constitutively and carries out its reaction. Enzyme B is also synthesized constitutively but its *activity* can be inhibited. The *synthesis* of the product of gene C can be prevented by control at the level of translation. The *synthesis* of the product of gene D can be prevented by control at the level of transcription.

Control systems that vary the level of *expression* of particular genes are the main subject of this chapter. However, the actual number of different regulatory mechanisms is vast, and most genes seem to be regulated by more than one. We begin by briefly discussing the processes involved in regulating the *activity* of preformed enzymes before considering how the synthesis of enzymes is controlled.

✓ 8.1 Concept Check

Most genes encode proteins and most proteins are enzymes. The expression of such a gene can be regulated by controlling the activity of the enzyme or controlling the amount of enzyme produced.

✓ What steps in the synthesis of protein might be subject to regulation?

✓ Which is likely to be more rapid, the regulation of activity or the regulation of synthesis?

II REGULATION OF ENZYME ACTIVITY

There are many mechanisms of posttranslational regulation. In some cases an enzyme is synthesized as part of a larger inactive precursor protein, and the enzyme must be activated by removing a portion of the precursor protein (see the box, Protein Processing). Another

mechanism is to reduce the level of activity by actually degrading the enzyme molecules. In the next two sections we discuss reversible and temporary forms of regulation involving less drastic changes to the enzyme molecule.

8.2 Inhibiting Enzyme Activity

As mentioned above, some proteins have no enzymatic activity until they are processed. It is more common, however, for a protein to be synthesized with full enzymatic activity, and for this activity to subsequently be reduced, or *inhibited*, by certain specific compounds in the cell. These compounds are usually related to the metabolic pathway in which the enzyme functions.

A major mechanism for the control of enzymatic activity involves the phenomenon of **feedback inhibition**. Feedback inhibition is seen primarily in the regulation of entire biosynthetic pathways, such as the pathway involved in the synthesis of an amino acid or purine. As we have seen, such pathways involve many enzymatic steps, and the final product, the amino acid or nucleotide, is many steps removed from the starting substrate (◯◯◯ Section 5.15). Yet, this final product is able to feed back to the first step in the pathway and regulate its own biosynthesis. How?

In feedback inhibition the amino acid or other end product of the biosynthetic pathway inhibits the activity of the *first* enzyme in this pathway. Thus, as the end product builds up in the cell, its further synthesis is inhibited. If the end product is used up, however, synthesis can resume (Figure 8.2).

How is it possible for the end product to inhibit the activity of an enzyme that acts on a substrate quite unrelated to it? This occurs because of a property of the inhibited enzyme known as **allostery**. An allosteric enzyme has two important binding sites, the *active* site, where the substrate binds, and the *allosteric* site, where the inhibitor (sometimes called an "effector") binds reversibly. When an inhibitor binds, generally noncovalently, at the allosteric site, the conformation of the enzyme molecule changes so that the substrate no longer binds efficiently at the active site (Figure 8.3). When the concentration of the inhibitor falls, equilibrium favors dissociation of the inhibitor from the allosteric site, returning the active site to its catalytic shape. Allosteric enzymes are very common in both anabolic and catabolic pathways and are especially important in branched pathways. For example, the amino acids proline and arginine are both synthesized from glutamic acid. Figure 8.4 shows that these two amino acids can control the first enzyme unique to their own synthesis without affecting the other so that a surplus of proline, for example, does not cause the organism to be starved for arginine.

In addition, some biosynthetic pathways are regulated by the use of **isozymes** (short for isofunctional enzymes: *iso* means "same" or "constant"). These enzymes catalyze the same reaction but are subject to different regulatory control. An example is synthesis of the aromatic amino acids (Figure 8.5; ∞ Figure 5.26). Three different isozymes catalyze the first reaction in this pathway, and each enzyme is regulated independently by

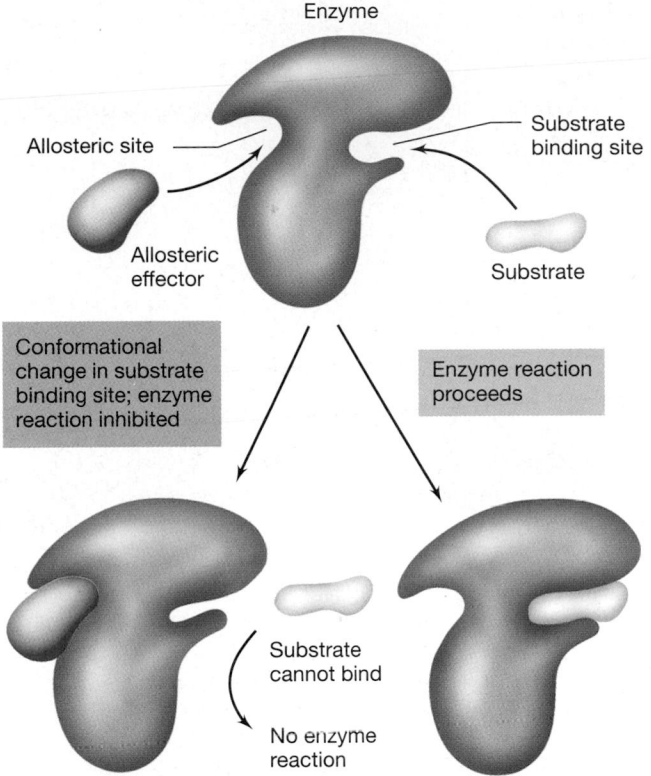

Figure 8.3 Mechanism of enzyme inhibition by an allosteric effector. When the effector combines with the allosteric site, the conformation of the enzyme is altered so that the substrate can no longer bind.

each of the three different end product amino acids. Unlike the earlier examples of feedback inhibition where inhibitors completely stopped an enzyme activity, in this case the total amount of the initial enzyme activity is

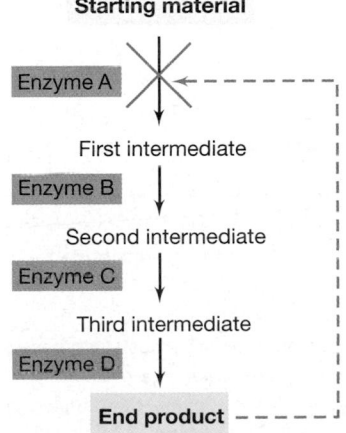

Figure 8.2 Feedback inhibition of enzyme activity. The activity of the first enzyme of the pathway is inhibited by the end product, thus controlling production of end product.

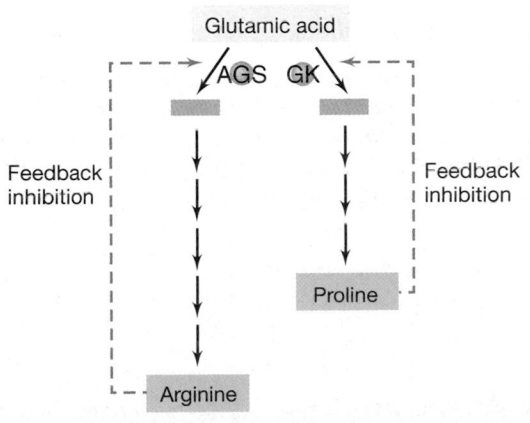

Figure 8.4 Feedback inhibition (dashed red arrows) in a branched biosynthetic pathway. A key intermediate in each pathway is shown in pink. The enzymes being inhibited are *N*-acetyl glutamate synthase (AGS) and γ-glutamyl kinase (GK).

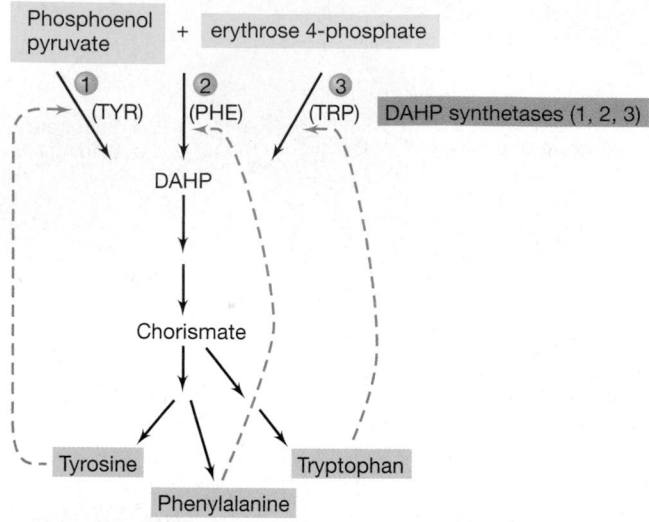

Figure 8.5 The common pathway leading to the synthesis of the aromatic amino acids contains three isozymes of DAHP synthetase (DAHP is 3-deoxy-D-*arabino*-heptulosonate 7-phosphate). Each of these enzymes is specifically feedback-inhibited by one of the aromatic amino acids. Note how an excess of all three amino acids is required to completely shut off the synthesis of DAHP.

diminished in a stepwise fashion and falls to zero only when all three products are present in excess.

The mechanism of feedback inhibition is of more than just academic interest. We will see in Chapter 30 how an understanding of the biochemistry of feedback inhibition has allowed industrial microbiologists to isolate mutants that have lost the ability to feedback inhibit the production of specific amino acids. These mutants are then used for the large scale commercial production of amino acids as a food supplement (∞ Section 30.7 and Figure 30.13).

✓ 8.2 Concept Check

Metabolic reactions can be regulated through control of the activities of the enzymes that catalyze these reactions. An important type of regulation of enzyme activity is feedback inhibition, in which the final product of a biosynthetic pathway feeds back and inhibits the first enzyme unique to that pathway.

✓ What is *feedback inhibition*?
✓ What is an *allosteric enzyme*?

8.3 Modification of Enzymes

Several examples are known in bacteria in which an enzyme is regulated by being *covalently modified*, usually by addition or deletion of some small organic molecule. As in the case of allosteric proteins, the covalent

binding of the modifying group changes the conformation of the protein and can dramatically affect its catalytic activity. Removal of the modifying group returns the enzyme to an active state. Many examples of covalent modification regulatory systems are known, but the best characterized are enzymes whose activity is affected by attachment of the nucleotide adenosine monophosphate (AMP) or adenosine diphosphate (ADP), by attachment of inorganic phosphate, or by methylation.

One of the most extensively studied examples of an allosteric enzyme regulated by adenylylation (addition of AMP) is glutamine synthetase (GS), an enzyme that plays a key role in nitrogen metabolism in the cell (∞ Section 5.15). GS is feedback inhibited by at least eight different compounds, including several amino acids and compounds involved in nucleotide metabolism. The susceptibility of GS to feedback inhibition is greatly increased, and its overall activity is lowered when cells are grown in media with abundant

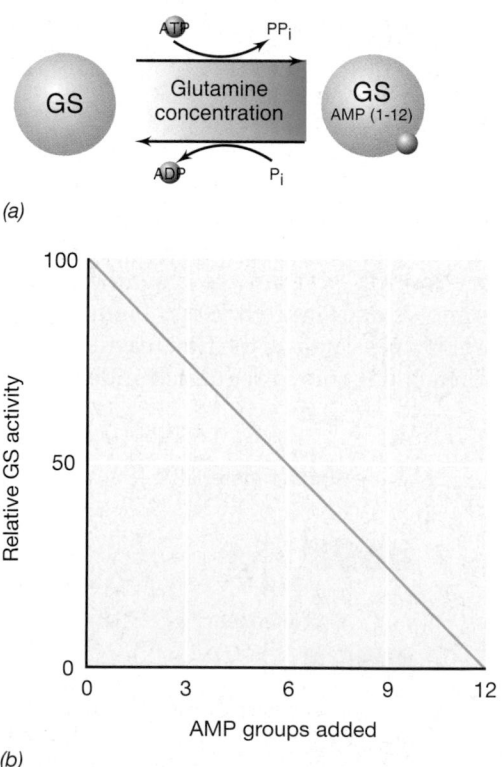

(a)

(b)

Figure 8.6 Regulation of glutamine synthetase by covalent modification. (a) When cells are grown in a medium rich in fixed nitrogen, glutamine synthetase (GS) is covalently modified by becoming progressively adenylylated. As many as 12 adenyl (AMP) groups can be added. When the medium becomes nitrogen poor, the groups are removed forming ADP. (b) Adenylylated GS subunits are inactive, so the overall GS activity *decreases* as more subunits are adenylylated. Any remaining unadenylylated subunits remain active but they have increased sensitivity to feedback inhibition.

nitrogen (glutamine). This is caused by adenylylation of GS (Figure 8.6). Each GS molecule contains 12 identical subunits and each can be adenylylated at a particular site. When the enzyme is completely adenylylated, it is essentially inactive; when it is partially adenylylated, it is partially active. Interestingly, the enzyme that adds and removes AMP is itself regulated by covalent modification.

Covalent modification of enzymes for regulatory purposes, such as the adenylylation of GS, is readily reversible. However, not all posttranslational modification of proteins is reversible. In some cases, the newly translated polypeptide must be extensively processed to become an active enzyme (see the box, Protein Processing).

✓ 8.3 Concept Check

Covalent modification is a regulatory mechanism for temporarily changing the activity of an enzyme. Enzymes regulated in this way can be modified and the modification subsequently removed. One type of modification is adenylylation (the addition of AMP).

✓ What does covalent modification do to enzyme activity?

✓ How can covalent modification be a mechanism that "temporarily" affects an enzyme?

III REGULATION OF TRANSCRIPTION: NEGATIVE AND POSITIVE CONTROL

Control of enzyme activity by feedback inhibition or covalent modification may seem sufficient for regulation of gene expression, but it is not. Even though glutamine synthetase (GS) is subject to both types of regulation, the level of the GS protein itself is also increased or decreased in relation to the level of nitrogen in the medium.

A Focus On ...   Protein Processing

In Chapter 7 we discussed the processing of RNA (Section 7.12). We mentioned that *all* tRNAs and rRNAs are transcribed as part of a larger RNA molecule and are then cut to size. Processing also occurs in proteins. As we also mentioned in Chapter 7, all proteins are synthesized beginning with a methionine, a formylated methionine in *Bacteria*. Therefore, one might expect that any protein isolated from the cell would have a methionine at its amino terminal. However, this is not the case, most proteins have the initiating methionine removed by a specific enzyme. (In *Bacteria* the formyl group of the methionine is removed first, and it is removed from all proteins.)

We also discussed the fact that proteins that will be secreted are synthesized with a leader sequence that is removed as part of transport (Section 7.16). In some cases the final active product is the result of extensive processing of the original product of translation. This is often the case for the peptide hormones made by higher organisms. A reasonably simple example is the hormone insulin. Insulin is important in regulating carbohydrate metabolism in mammals and is used to treat diabetes. Insulin is derived from

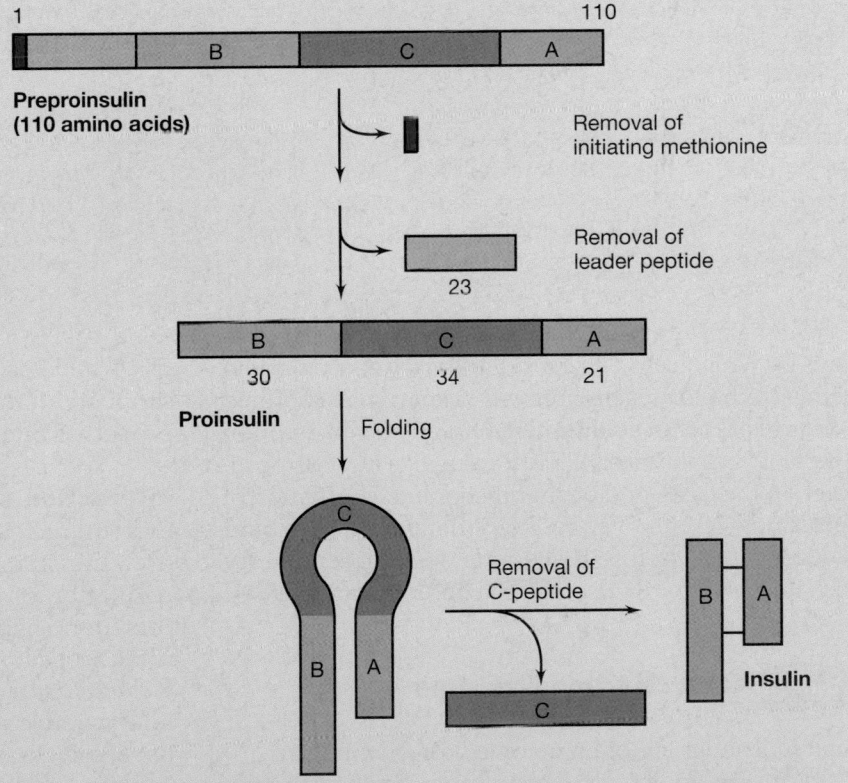

Figure 1 Processing of preproinsulin. The initiating methionine is actually cleaved from the chain before the entire preproinsulin molecule is synthesized. After the leader peptide is removed, the proinsulin molecule folds, and the C-peptide is then removed leaving the A and B chains on insulin. These chains are held together by disulfide bridges (Section 3.8).

continues

a 110 amino acid residue protein called pre-proinsulin, as is shown in Figure 1. The "pre" leader peptide is removed during secretion leaving proinsulin, which is further cleaved to yield the final active insulin molecule which has an A-chain and a B-chain.

Some genes are translated to yield a "polyprotein," that is, a long polypeptide chain that is processed to give several different active products. We will see examples of this in our discussion of poliovirus (⟲ Section 16.7) and retroviruses (⟲ Section 9.12).

When introns are removed during the processing of RNA, the flanking sequences, the exons, are ligated together to form a single molecule (⟲ Section 7.12). Recently a number of instances have been uncovered where the "extra" information in the gene is removed at the level of the *protein* and not at the level of RNA. The process has been termed "protein splicing" and the interior peptide removed is called an **intein**. This process has been shown to occur in specific proteins from *Archaea*, *Bacteria*, and *Eukarya*. Figure 2 shows the process in the product of the *gyrA* gene of the bacterium *Mycobacterium leprae*. Note that the flanking sequences, referred to as *exteins*, are ligated together to form a single protein, in this case a subunit of DNA gyrase. Interestingly,

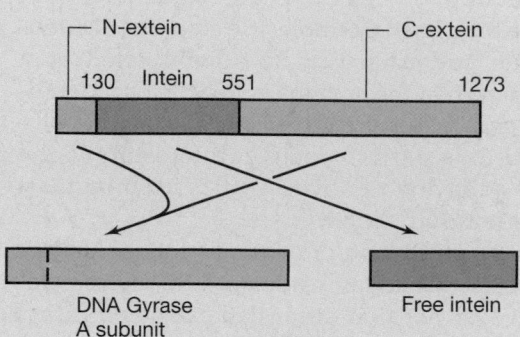

Figure 2 Protein splicing. The protein synthesized from the *gyrA* mRNA in *Mycobacterium leprae* is 1273 amino acid residues in length (10 times the length of preproinsulin, shown in Fig. 1). The N-extein is the amino terminal extein and the C-extein is the carboxyl terminal extein. Residues 131 to 550 make up an intein, which removes itself in a self-splicing reaction that generates the free intein and the DNA Gyrase A subunit.

inteins are self-splicing and the intein protein is itself a specific protease (an enzyme that cleaves proteins). Remember that many introns are also self-splicing (⟲ Section 7.12).

Many other "processing" steps are possible with proteins just as they are with RNA. We mentioned that tRNA contains many modified bases and that these bases are modified after transcription (⟲ Section 7.14). Similarly there are many more than 20 amino acids (or 21, ⟲ box in Chapter 7, Selenocysteine: The Twenty-First Amino Acid) known in proteins that are the result of post-translational covalent modifications. Covalent modifications are sometimes reversible and can be important means by which cells regulate enzyme activity (see Section 8.3). ■

In the remainder of this chapter we will discuss mechanisms by which cells can control the *amount* of a protein synthesized. Our discussion will primarily be confined to control at the level of transcription. For transcription to occur, RNA polymerase must recognize a specific promoter on the DNA and begin functioning. Regulation of transcription commonly involves proteins that can bind to DNA. Thus, before discussing specific regulatory mechanisms, we will discuss proteins that bind to DNA.

8.4 DNA Binding Proteins

Small molecules are often involved directly in the regulation of protein activity. For instance, in the example given in Figure 8.4, the amino acids proline and arginine bind directly to the enzyme and inhibit it. The situation with regard to regulating enzyme *synthesis* is quite different. Although small molecules are often involved in regulating transcription, they rarely do so directly. Instead they typically influence the binding of certain pro-

teins, called *regulatory proteins,* to specific sites on the DNA, and it is these proteins that actually regulate transcription. In this section we shall discuss a few general properties of proteins that bind to DNA.

Interaction of Proteins with Nucleic Acids

Protein-nucleic acid interactions are central to replication, transcription, and translation, as well as to the regulation of these processes. Two general kinds of protein-nucleic acid interactions are noted: nonspecific and specific, depending on whether the protein attaches *anywhere* along the nucleic acid or whether the interaction is sequence-specific. As an example of proteins that do *not* interact in a sequence-specific fashion, we mention the **histones,** proteins that are extremely important in the structure of the eukaryotic chromosome (⟲ Section 7.3), although less significant in prokaryotes. Histones are relatively small proteins that have a high proportion of positively charged amino acids (arginine, lysine, histidine). DNA has a large number of negatively charged phosphate groups on the outside of the DNA double helix. His-tones, because of their positive charge, combine strong-ly and relatively nonspecifically with the negatively charged DNA. Association of histones with DNA leads to the formation of nucleosomes, the unit particles of the eukaryotic chromosome (⌾ Section 7.3). These nucleosomes are further compacted into higher order structures that make the DNA inaccessible. If the DNA is covered with histones, other proteins such as RNA polymerase will not be able to bind and transcription cannot take place. Certain proteins can disrupt the con-densed structure of DNA. However, loss of histones need not automatically lead to transcription but sim-ply leaves the gene capable of being activated by other factors. Even after binding and beginning transcrip-tion, eukaryotic RNA polymerases need specific pro-tein factors to help them elongate through the protein-coated DNA.

There are also a number of proteins that interact with DNA in a *sequence-specific* manner. These interac-tions occur by association of the amino acid side chains of the proteins with the bases as well as with the phos-phate and sugar molecules of the DNA. The major groove in DNA, because of its size, is an important site of protein binding. Figure 7.3 identifies several of the atoms of the base pairs found in the major groove that are known to interact with proteins. To achieve *specificity* in such interactions, the protein must interact simulta-neously with more than one nucleic acid base, frequently several. This means that a protein with a specific amino acid sequence will interact with DNA with a specific base sequence.

We have already described a structure in DNA called an *inverted repeat* (⌾ Figure 7.6). Such inverted repeats are frequently the locations at which protein molecules combine specifically with DNA (Figure 8.7). Note that this interaction does not involve the formation of cruci-form structures in the DNA. Proteins that interact specif-ically with DNA are frequently *dimers*, composed of two identical polypeptide chains. On each polypeptide chain is a region, called a *domain*, that interacts specifically with a region of DNA in the major groove. When protein dimers interact with inverted repeats, *each* of the polypeptides of the protein dimer combines with each of the DNA strands (Figure 8.7). Because the protein rec-ognizes *contact points* associated with specific base pairs, its binding is sequence-specific.

Structure of DNA Binding Proteins

Studies of the structure of several DNA binding pro-teins from both prokaryotes and eukaryotes have re-vealed a few types of common protein substructures that are apparently critical for proper binding of many of these proteins to DNA. One of these is termed the *helix-turn-helix motif* (Figure 8.8). The helix-turn-helix

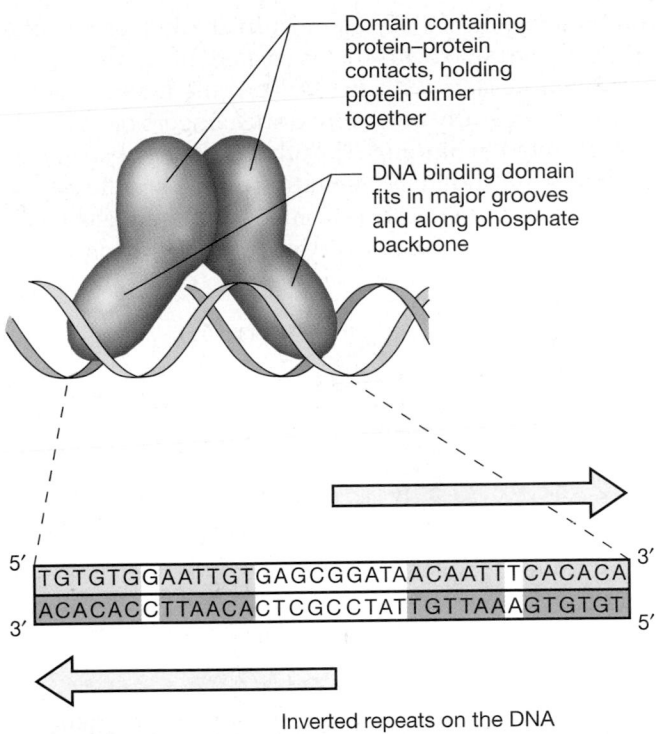

Domain containing protein–protein contacts, holding protein dimer together

DNA binding domain fits in major grooves and along phosphate backbone

5' TGTGTGGAATTGTGAGCGGATAACAATTTCACACA 3'
3' ACACACCTTAACACTCGCCTATTGTTAAAGTGTGT 5'

Inverted repeats on the DNA

Figure 8.7 DNA binding proteins. Many such proteins are dimers that combine specifically with *two sites* on the DNA. The specific DNA sequences that interact with the protein are *inverted repeats*. The nu-cleotide sequence of the operator gene of the lactose operon is shown and the inverted repeats, which are sites at which the *lac* repressor makes contact with the DNA, are shown in shaded boxes.

consists of a stretch of amino acids that form an α-helix secondary structure (the so-called recognition helix that interacts with DNA), which is joined to a short stretch of three amino acids, the first of which is usually a glycine that functions to "turn" the protein (Figure 8.8a). The other end of the "turn" is connected to a second helix, which stabilizes the first by interacting hy-drophobically with it and participates in dimerization (see above). Recognition of specific DNA sequences oc-curs by a combination of noncovalent interactions in-cluding hydrogen bonds and van der Waals contacts (⌾ Section 3.1) between the protein and base pairs on the DNA. Many different DNA binding proteins from *Bacteria* show the helix-turn-helix structure, including many repressor proteins such as the bacteriophage lambda repressor (Figure 8.8b) and the *lac* and *trp* re-pressors of *Escherichia coli* (see Section 8.5). Indeed, there are over 250 different proteins known with this motif that bind to DNA to regulate transcription.

Two other types of protein substructures are com-monly found in DNA binding proteins. One of these, the *zinc finger*, is frequently found in eukaryotic regulatory proteins that bind to DNA. The zinc finger is a sub-structure of protein that, as its name implies, binds a zinc ion (Figure 8.9a). It seems most likely that part of the "finger" of amino acids that is created forms an α-helix, and this interacts with the DNA in the major groove. There are typically at least two such fingers on the protein involved in binding. The other protein substructure commonly found in DNA binding proteins is the *leucine zipper.* This substructure is formed by the side chains on leucine residues spaced every seven amino acids, and it

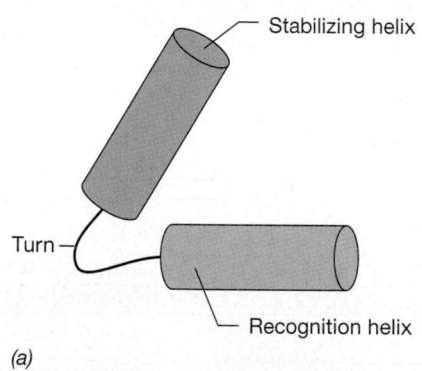

(a)

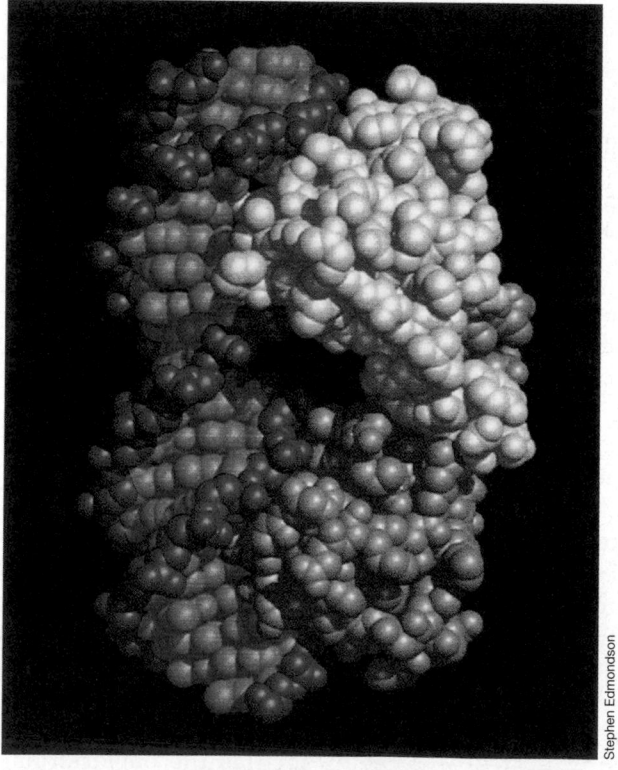

(b)

Figure 8.8 The helix-turn-helix structure of some DNA binding proteins. (a) A simple model of the helix-turn-helix elements. (b) A computer model of the bacteriophage lambda repressor, a typical helix-turn-helix protein, bound to its operator gene. One subunit of the dimeric repressor is shown in dark brown and the other in dark yellow. Each subunit contains a helix-turn-helix structure. The coordinates used to generate this image were downloaded from the Protein Data Base, Brookhaven, NY.

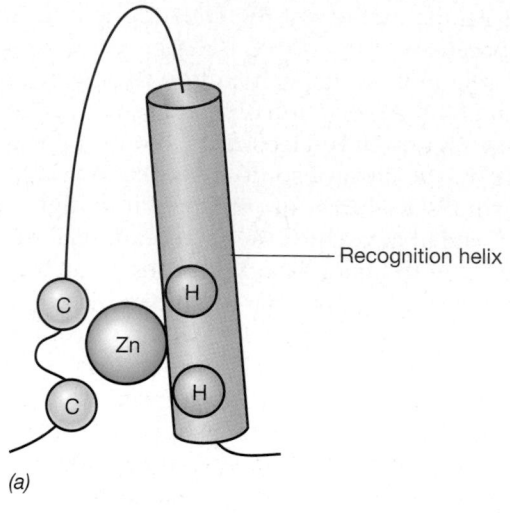

(a)

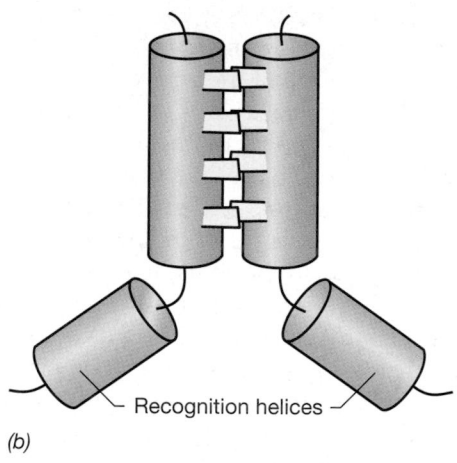

(b)

Figure 8.9 Simple models of protein substructures found in eukaryotic DNA binding proteins. α-Helices are represented by cylinders. Recognition helices are the domains involved in DNA binding. (a) The zinc finger structure. The amino acids holding the Zn^{2+} ion always include at least two cysteine residues (C) with the other residues being histidine (H). (b) The leucine zipper structure. The leucine residues (shown in yellow) are always spaced exactly every seven amino acids. The interaction of the leucine side chains helps hold the two helices together.

somewhat resembles a zipper. Unlike the helix-turn-helix and the zinc finger, the leucine zipper does not seem to interact with DNA itself but serves to hold two other α-helices in the correct position to bind DNA (Figure 8.9b).

Once a protein combines at a specific site on the DNA, a number of outcomes can occur. In some cases, the protein is an enzyme that carries out some specific action on the DNA, such as RNA polymerase, which makes RNA using DNA as the template. In other cases the protein that binds can *block* transcription (negative regulation, see Section 8.5) or can *activate* it (positive regulation, see Section 8.6).

8.4 Concept Check

Certain proteins can bind to DNA because of specific interactions between certain regions of the proteins and specific regions of the DNA molecule. In some cases the interactions are not sequence-specific, but in other cases they are. Proteins that bind to nucleic acid may be enzymes that use nucleic acid as substrates, or they may be regulatory proteins that affect how genes function.

✓ What is a *protein domain*?

✓ Why are some interactions specific to certain DNA sequences?

8.5 Negative Control of Transcription: Repression and Induction

Transcription is the first step in biological information flow where it is relatively simple to increase the expression of one gene relative to another. If one gene is transcribed more frequently than another, there will be a greater abundance of its mRNA in the cell and therefore a greater amount of its protein product. Even constitutive proteins can be required in different amounts. High levels of transcription can result because a gene has a promoter very close to the *consensus sequence* (∞ Section 7.9) and high levels of translation can result if the gene's Shine–Dalgarno has extensive complementarity with 16S rRNA (∞ Section 7.15). These sequences are fixed, however, and serve to set expression at a fixed level.

The remainder of this chapter will not, however, deal with the relative strength of different promoters or the efficiency of translation initiation of a particular mRNA but with how the level of transcription of an individual gene can be changed in a regulated manner.

In Section 8.2 we considered how cells regulate enzyme *activity*. We now begin a discussion on how cells regulate enzyme *synthesis*.

Several different mechanisms for controlling enzyme synthesis are known in bacteria, and all of them are greatly influenced by the *environment* in which the organism is growing, in particular by the presence or absence of specific small molecules. These molecules can interact with specific proteins like the DNA binding proteins just described to control transcription or, more rarely, translation.

We begin our discussion by describing repression and induction, simple forms of regulation that govern gene expression at the level of *transcription*. In this section we will deal only with **negative control** of transcription; a kind of control whose regulatory mechanism involves stopping transcription.

Enzyme Repression

Often the enzymes catalyzing the synthesis of a specific product are not synthesized if this product is present in the medium. For example, the enzymes involved in formation of the amino acid arginine are synthesized only when arginine is *absent* from the culture medium; external arginine *represses* the synthesis of these enzymes. As can be seen in Figure 8.10, if arginine is added to a culture growing exponentially in a medium devoid of arginine, growth continues at the previous rate, but the formation of the enzymes involved in arginine synthesis stops. Note that this is a *specific* effect, as the syntheses of all other enzymes in the cell continue at the same rates as previously.

Enzyme repression is a very widespread phenomenon in bacteria—it occurs as a means of controlling the synthesis of a wide variety of enzymes involved in the biosynthesis of amino acids, purines, and pyrimidines. In almost all cases it is the final product of a particular biosynthetic pathway that represses the enzymes of this pathway. In these cases repression is quite specific, and the process usually has no effect on the synthesis of enzymes other than those involved in the specific biosynthetic pathway. The value to the organism of enzyme repression is obvious because it effectively ensures that the organism does not waste energy synthesizing unneeded enzymes.

Enzyme Induction

A phenomenon complementary to repression is *enzyme induction,* the synthesis of an enzyme only when its substrate is present. Figure 8.11 shows this process in the case of the enzyme β-galactosidase, which is involved in the use of the sugar lactose. If lactose is absent from the medium the enzyme is not synthesized, but synthesis begins almost immediately after lactose is added.

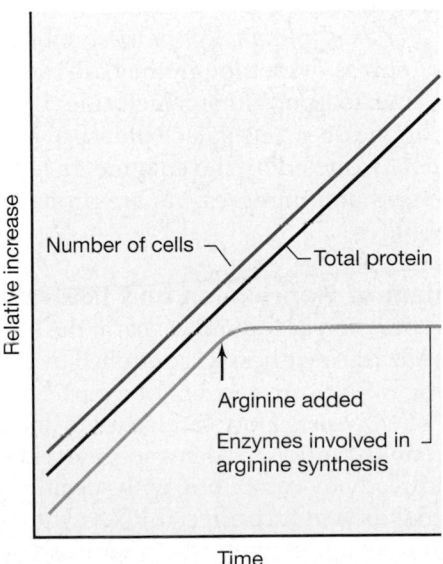

Figure 8.10 Repression of enzymes involved in arginine synthesis by addition of arginine to the medium. Note that the rate of total protein synthesis remains unchanged.

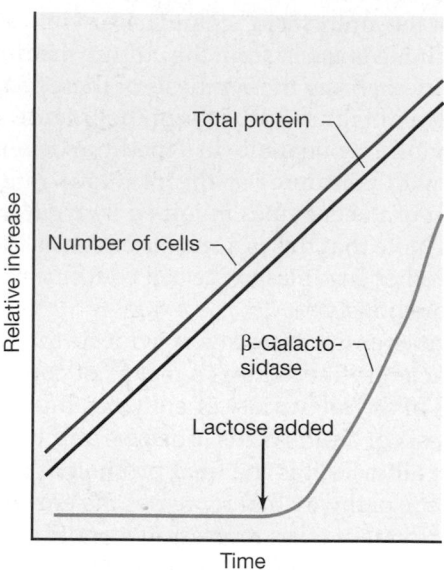

Figure 8.11 Induction of the enzyme β-galactosidase on the addition of lactose to the medium. Note that the rate of total protein synthesis remains unchanged.

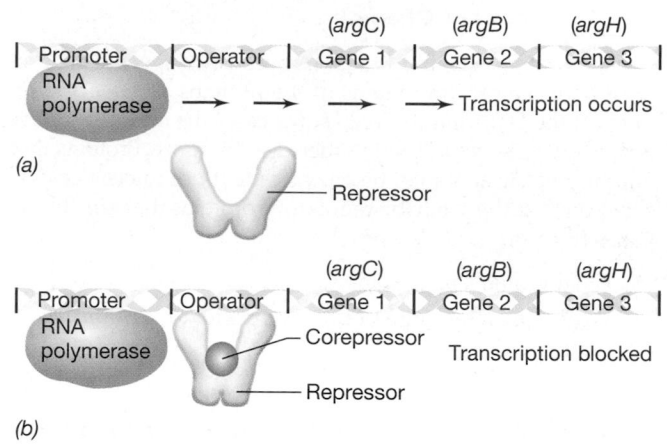

Figure 8.12 The process of enzyme repression. (a) Transcription of the operon occurs because the repressor is unable to bind to the operator. (b) After a corepressor (small molecule) binds to the repressor, the repressor binds to the operator and blocks transcription; mRNA and the proteins it encodes are not made. In the case of the *argCBH* operon the repressor would be the arginine repressor and the corepressor would be the amino acid arginine.

Enzymes involved in the catabolism of carbon and energy sources are often inducible. Again, one can see the value to the organism of such a mechanism, as it provides a means whereby the organism does not synthesize an enzyme until it is needed.

The substance that initiates enzyme induction is called an **inducer**, and a substance that represses production is called a **corepressor**; these substances, which are always small molecules, are often collectively called **effectors**. Not all inducers and corepressors are substrates or end products of the enzymes involved. For example, *analogs* of these substances may induce or repress even though they are not substrates of the enzyme. Isopropylthiogalactoside (IPTG), for instance, is an inducer of β-galactosidase even though it cannot be hydrolyzed by the enzyme. In nature, however, inducers and corepressors are probably normal cell metabolites.

Mechanism of Repression and Induction

Enzyme repression or induction acts at the level of transcription; enzyme synthesis is controlled by initiating or terminating mRNA production for a particular enzyme or group of enzymes. How can inducers and corepressors affect transcription in such a specific manner? They do this indirectly by combining with specific regulatory proteins that then in turn affect mRNA synthesis. In the case of a repressible enzyme, the corepressor (for example, arginine) combines with a specific **repressor protein**, the arginine repressor, that is present in the cell (Figure 8.12). The repressor protein is an allosteric protein (see Sections 8.2 and 8.4), its conformation being altered when the corepressor combines with it. This altered repressor protein can then combine with a specific region of the DNA near the promoter of the gene, the **operator region**. This region gave its name to the **operon** which, as we have discussed (∞ Section 7.11), is a cluster of genes whose expression is under the control of a single operator. All the genes in an operon are transcribed as a single unit yielding a single mRNA. The operator is adjacent to the promoter where synthesis of mRNA is initiated. If the repressor binds to the operator, the synthesis of mRNA is blocked because the RNA polymerase cannot either bind or proceed (depending on the exact location of the operator). Thus the protein or proteins specified by this mRNA cannot be synthesized. If the mRNA is polycistronic, *all* the proteins encoded by this mRNA will be repressed.

Enzyme induction can also be controlled by a repressor. In this case, the situation previously described is reversed. The specific repressor protein is active in the *absence* of the inducer, completely blocking the synthesis of mRNA, but when the inducer is added, it combines with the repressor protein and inactivates it. Inhibition of mRNA synthesis being overcome, the enzyme or enzymes can be made (Figure 8.13). All systems involving repressors have the same underlying mechanism, *inhibition* of the synthesis of mRNA by the action of specific repressor proteins that are themselves under the control of specific small-molecule inducers and repressors. Because the repressor's role is inhibitory, regulation involving repressors is often referred to as *negative control*.

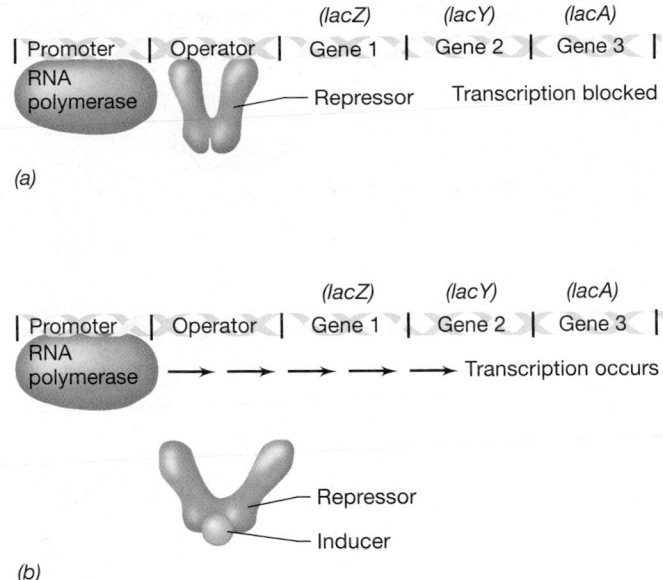

(a)

(b)

Figure 8.13 The process of enzyme induction using a repressor. (a) A repressor protein binds to the operator region and blocks the action of RNA polymerase. (b) An inducer molecule binds to the repressor and inactivates it so that it no longer can bind to the operator. Transcription by RNA polymerase occurs and an mRNA for that operon is formed. In the case of the *lac* operon the repressor would be the *lac* repressor, and the inducer would be allolactose.

✓ 8.5 Concept Check

The amount of an enzyme in the cell can be controlled by increasing (induction) or decreasing (repression) the amount of mRNA that encodes the enzyme. This transcriptional regulation involves regulatory proteins that bind to DNA and to small molecules called effectors. For negative control of transcription the regulatory protein is called a repressor and it functions by inhibiting mRNA synthesis.

✓ What is "negative control" of transcription?

✓ How does a repressor inhibit the synthesis of a specific mRNA?

8.6 Positive Control of Transcription

Repression constitutes a kind of regulation called **negative control**. The controlling element—the repressor protein—brings about the *repression* of mRNA synthesis. Even though the repressor has a negative role, a system using a repressor can control enzyme induction, as we saw with β-galactosidase. However, another type of control, called *positive control*, has also been recognized. In **positive control** of transcription a regulator protein *promotes* the binding of RNA polymerase, thus acting to *increase* mRNA synthesis. We will now consider a system that involves positive regulation, the regulation of maltose catabolism in *Escherichia coli*.

The Maltose Regulon

The enzymes for the utilization of the sugar maltose in *Escherichia coli* are synthesized only after the addition of maltose to the medium. The pattern of induction of these enzymes follows that shown for β-galactosidase in Figure 8.11, but in this case it is maltose, not lactose, that is the inducer. The synthesis of the enzymes for maltose utilization is controlled at the level of transcription, but by an **activator protein**, not by a repressor. The *maltose activator protein* cannot bind to the DNA unless the protein first binds maltose, the effector. When the activator protein does bind to DNA, it allows RNA polymerase to begin transcription (Figure 8.14). Activators, like repressors, recognize specific sequences on the DNA. The sequence that serves as the binding site of the activator is not called an operator but an *activator binding site*. Nonetheless, the genes controlled by this activator binding site *are* called an operon.

In negative control, the repressor binds to the operator and blocks transcription. How does an activator protein work? Positively controlled promoters have nucleotide sequences that are not close matches to the consensus sequence (∞ Figure 7.27). Even with the correct sigma factor, the RNA polymerase has difficulty recognizing these promoters. The activator protein, when bound to DNA, helps the RNA polymerase either recognize the promoter or begin transcription. The activator

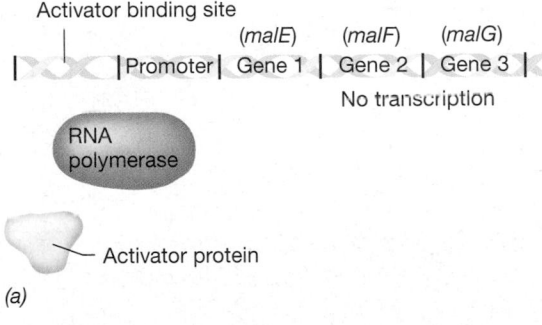

(a)

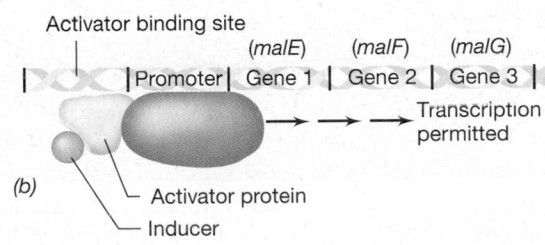

(b)

Figure 8.14 Positive control of enzyme induction. (a) In the absence of an inducer, neither the activator protein nor the RNA polymerase can bind to the DNA. (b) An inducer molecule binds to the activator protein, which in turn binds to the activator binding site. This allows RNA polymerase to bind to the promoter and begin transcription. In the case of the *malEFG* operon, the activator protein would be the maltose activator protein and the inducer would be the sugar maltose.

protein may cause a change in the structure of the DNA, perhaps by bending it (Figure 8.15), allowing the RNA polymerase to make the correct contacts with the DNA. The activator protein may also interact directly with the RNA polymerase. This can happen either when the activator binding site is close to the promoter (Figure 8.16*a*) or when it is several hundred base pairs away from the promoter (Figure 8.16*b*).

The genes needed for maltose utilization are spread out in several operons, each of which has an activator binding site to which the maltose activator protein can bind. Therefore, the maltose activator protein actually controls more than one operon. When more than one operon is under the primary control of the same regulatory protein, these operons are collectively known as a **regulon**. Therefore, the enzymes for maltose utilization are encoded by the *maltose regulon*. Regulons are also known for operons under negative control. The arginine biosynthetic enzymes (see Section 8.5) are encoded by the *arginine regulon* whose operons are all under the control of the arginine repressor protein. One of these operons was shown in Figure 8.12.

Many genes in *Escherichia coli* have promoters under positive control, and many have promoters under negative control. However, there are other types of regulation known. In addition, many genes (perhaps most genes) either have a promoter with multiple types of control or have more than one promoter, each with its

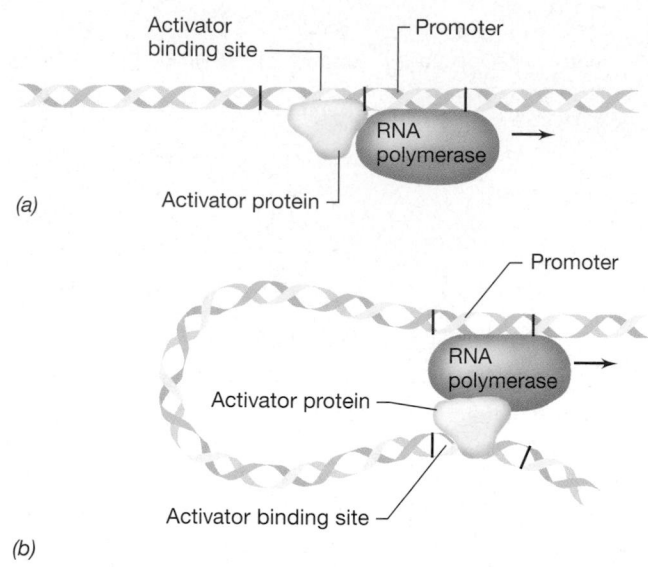

(a)

(b)

Figure 8.16 Some activator proteins interact with RNA polymerase. (a) The activator binding site is near the promoter. (b) The activator binding site is several hundred base pairs from the promoter. In this case, the DNA must be looped to allow the activator and the RNA polymerase to contact.

own control system! In many cases there are layers of control with gene expression regulated both by specific effectors related to a particular gene's function (such as the maltose regulon responding to maltose) and also by more general effectors responding to some more global aspect of metabolism. Indeed, both the lactose operon and the maltose regulon respond to such a global regulatory system. We discuss this system in the next section.

✓ 8.6 Concept Check

Positive regulators of transcription are called activator proteins. They bind to activator binding sites on the DNA and stimulate transcription by RNA polymerase. Activator protein activity, like repressor protein activity, is modified by effectors. For positive control of enzyme induction, the effector promotes the binding of the activator protein and thus stimulates mRNA synthesis.

- ✓ Compare and contrast the activities of an activator protein and a repressor protein.
- ✓ Distinguish between an *operon* and a *regulon*.

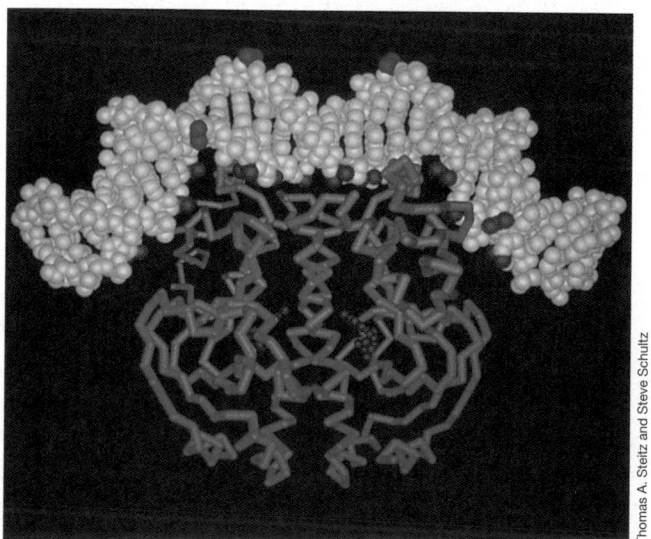

Figure 8.15 Computer model of the interaction of a positive regulatory protein with DNA. This figure shows the cyclic AMP binding protein, a regulatory protein involved in the control of several operons. The α-carbon backbone of this protein is shown in blue and purple. The protein is shown binding to a DNA double helix, which is shown in yellow and light blue. Note that binding of this protein to DNA has caused the DNA to be bent by almost 90°. Reprinted with permission from S. Schultz, G. Shields, and T. Steitz. 1991. *Science 253*: 1001–1007. © 1991, AAAS.

8.7 Global Control and the *lac* Operon

Often an organism needs to regulate many different genes simultaneously in response to a change in its environment. Regulatory mechanisms that respond to environmental signals by regulating expression of many different genes are called *global control systems*.

Our discussions above (Sections 8.5 and 8.6) did not consider the possibility that the environment in which the cells are growing might contain several different carbon sources that the bacteria could use. Many bacteria, such as *Escherichia coli*, can use a very wide array of carbon sources.

It would be wasteful to induce enzymes for lactose or maltose metabolism if the cells were already growing on a carbon source that they could use more efficiently. In fact, one of the global regulatory networks, **catabolite repression,** prevents this problem.

Catabolite Repression

In catabolite repression the syntheses of a variety of unrelated enzymes, primarily catabolic, are inhibited when cells are grown in a medium that contains a preferred energy source such as glucose. Catabolite repression has been called the **glucose effect** because glucose was the first substance shown to initiate it, although in some organisms carbon sources other than glucose cause this form of enzyme repression. Catabolite repression ensures that the organism uses the more readily catabolizable carbon and energy source, such as glucose, first.

One consequence of catabolite repression is that it can lead to *diauxic growth* if the two energy sources are present in the medium at the same time and if the enzyme needed for utilization of one of the energy sources is subject to catabolite repression. In **diauxic growth,** the organism grows first on one energy source, and there is then a lag period before growth resumes on the other energy source. This phenomenon is illustrated in Figure 8.17 for growth on a mixture of glucose and lactose. The enzyme β-galactosidase, which is re-

sponsible for utilization of lactose, is inducible, but its synthesis is also subject to catabolite repression. Thus, as long as glucose is present in the medium, β-galactosidase is not synthesized; the organism grows only on the glucose and leaves the lactose untouched. When the glucose is exhausted, catabolite repression is abolished. After a lag, β-galactosidase is synthesized and growth on lactose occurs. Notice that Figure 8.17 shows that the cells grow more rapidly on glucose. Thus, catabolite repression ensures that the cells use the *best* carbon source first.

How does catabolite repression work? Catabolite repression involves control of transcription by an activator protein (see Section 8.6) and is therefore a type of positive control. In the case of catabolite-repressible enzymes, binding of RNA polymerase to DNA occurs only if another protein, called **catabolite activator protein (CAP)**, has bound first. An allosteric protein, CAP binds to DNA only if it has first bound a small molecule called *cyclic adenosine monophosphate* or **cyclic AMP** (see Figure 8.15). For this reason some scientists refer to this protein as the *cAMP-receptor protein (CRP)*. Cyclic AMP (Figure 8.18) has been shown to be a key element in a variety of control systems, not only in bacteria but in eukaryotes also. Cyclic AMP is synthesized from ATP by an enzyme called *adenylate cyclase*. Glucose entry into the cell (∞ Figure 4.26) inhibits the synthesis of cyclic AMP and stimulates cyclic AMP transport out of the cell. When glucose is transported into the cell, the cyclic AMP level in the cell is lowered, and binding of RNA polymerase to the promoter does not occur. Thus, catabolite repression is really a result of a deficiency of cyclic AMP and can be overcome by adding this compound to the medium.

Although this may sound like a simple positive regulatory system (as in Figure 8.14), each of the operons that CAP controls is *also* under control of a specific regulatory protein. Therefore, catabolite repression modulates several unrelated regulatory systems and thus is a prime example of global control. As long as glucose is present, catabolite repression prevents expression of all other catabolic operons under this global controlling element. The complete regulatory region of the lactose operon (*lac* operon) is shown in Figure 8.19. For transcription to occur, two requirements must be met: (1) the

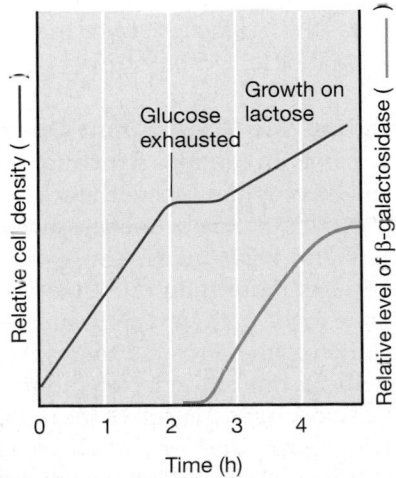

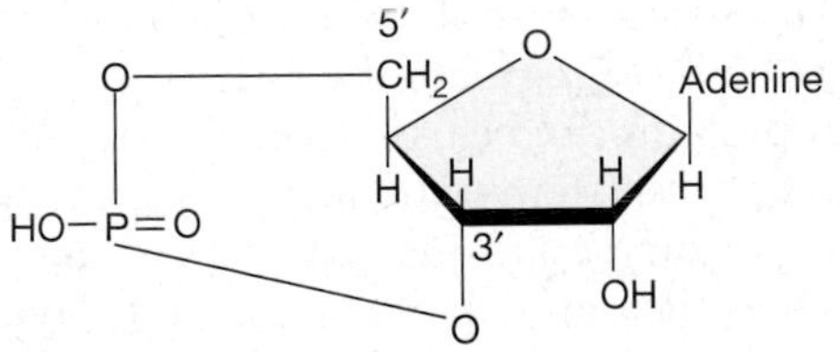

Figure 8.17 Diauxic growth on a mixture of glucose and lactose. Glucose represses the synthesis of β-galactosidase. After glucose is exhausted, a lag occurs until β-galactosidase is synthesized, and then growth can resume on lactose but at a slower rate.

Figure 8.18 Cyclic adenosine monophosphate (cyclic AMP, cAMP) is produced from ATP by the enzyme adenylate cyclase.

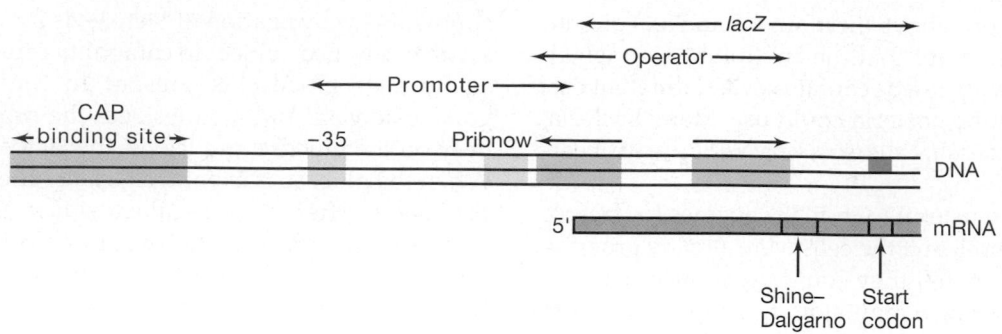

Figure 8.19 The genetic elements involved in regulation of the lactose operon. The first gene in this operon, *lacZ*, encodes the enzyme β-galactosidase, which breaks down lactose (see Figure 8.13). The operon contains two other genes that are also involved in lactose metabolism. This figure shows the control region of this operon. The two halves of the operator (where the repressor would bind) are almost perfect inverted repeats. There are also inverted repeats in the CAP binding site although these are less perfect. The transcriptional start site would be located on the DNA exactly at the 5′-end of the mRNA. The location of the −35 sequence and the Pribnow box, which are part of the promoter (∞ Figure 7.27) are also shown. In addition, the location of the base pairs encoding the Shine–Dalgarno sequence and the start codon are also given. These two sequences would function on the mRNA (∞ Section 7.13).

level of cyclic AMP must be high enough so that the CAP protein binds to the CAP binding site (positive control), and (2) there must be an inducer such as lactose present so that the lactose repressor does not block transcription by binding to the operator (negative control).

✓ 8.7 Concept Check

Global control systems respond to signals in the environment and regulate the expression of many genes simultaneously. Catabolite repression is a global control system, and it serves to help cells make the most efficient use of carbon sources. The *lac* operon is under the control of catabolite repression as well as its own specific regulatory system.

- ✓ Explain how catabolite repression can involve an activator protein.
- ✓ Explain how the *lac* operon is both positively and negatively controlled.

IV REGULATION OF TRANSCRIPTION: OTHER MECHANISMS

We have discussed a few of the most fundamental mechanisms cells use to regulate the transcription of genes. Repressor proteins can be used to inhibit transcription in negative control systems, and activator proteins can activate transcription in positive control systems. We have also seen that there may be more than one layer of control in the regulation of the expression of specific genes. In the next sections we will discuss a few other mechanisms cells use to regulate transcription.

8.8 Attenuation

Some control systems do not involve regulatory proteins that bind to the DNA. The regulatory process called **attenuation** is such a system. The word *attenuation* means "to lessen in amount." Previously, we described regulating transcription at *initiation;* that is, repressors block the synthesis of RNA whereas activator proteins encourage synthesis. In transcription attenuation the control occurs *after* initiation of RNA synthesis but before its completion. That is, the number of *completed* transcripts from a gene or an operon is reduced, even though the number of initiated transcripts is not. Most of the first known examples of attenuation involved regulating genes controlling the biosynthesis of certain amino acids in gram-negative *Bacteria*. The first such system to be described was the *tryptophan operon* in *Escherichia coli*, and we focus on it here.

Attenuation and the Tryptophan Operon

The tryptophan operon contains structural genes for five proteins of the tryptophan biosynthetic pathway, plus the promoter and regulatory sequences at the beginning of the operon (Figure 8.20). Like many operons, the tryptophan operon has more than one type of regulation. The first enzyme in the pathway, anthranilate synthetase (a multisubunit enzyme encoded by *trpD* and *trpE*) is subject to feedback inhibition by tryptophan (see Section 8.2). In addition, transcription of the operon is under control of a repressor, and one of the regulatory sequences is an operator to which the tryptophan repressor can bind.

In addition to promoter (P) and operator (O) regions, there is a sequence called the **leader sequence**, which encodes a polypeptide that contains tandem tryptophan

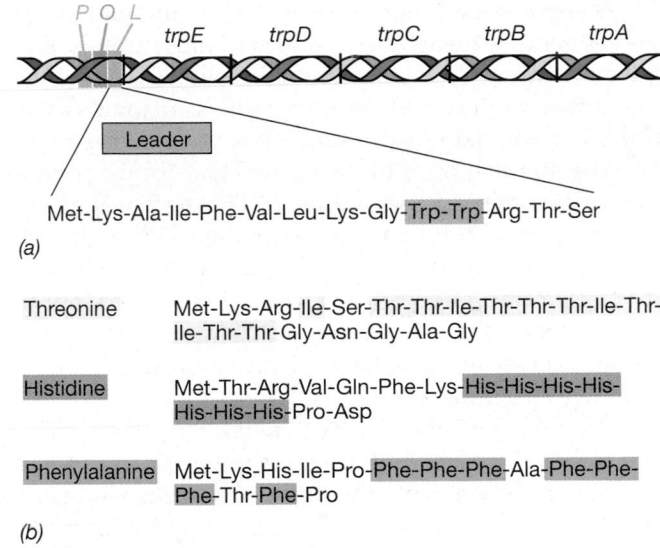

(a)

Met-Lys-Ala-Ile-Phe-Val-Leu-Lys-Gly-Trp-Trp-Arg-Thr-Ser

Threonine	Met-Lys-Arg-Ile-Ser-Thr-Thr-Ile-Thr-Thr-Thr-Ile-Thr-Ile-Thr-Thr-Gly-Asn-Gly-Ala-Gly
Histidine	Met-Thr-Arg-Val-Gln-Phe-Lys-His-His-His-His-His-His-His-Pro-Asp
Phenylalanine	Met-Lys-His-Ile-Pro-Phe-Phe-Phe-Ala-Phe-Phe-Phe-Thr-Phe-Pro

(b)

Figure 8.20 Structure of the tryptophan operon and of tryptophan and other leader peptides in *Escherichia coli*. (a) Arrangement of the tryptophan operon. Note that the leader (L) encodes a short peptide containing two tryptophan residues near its terminus (there is a stop codon following the Ser codon). The promoter is labeled P and the operator is labeled O. The genes labeled *trpE* through *trpA* encode the enzymes involved in tryptophan biosynthesis. (b) Amino acid sequence of leader peptides synthesized in some other amino acid biosynthetic operons. Because isoleucine is made from threonine, it is an important constituent of the threonine leader peptide.

codons near its terminus and functions as an **attenuator** (Figure 8.20). If tryptophan is plentiful in the cell, the leader peptide will be synthesized. On the other hand, if tryptophan is in short supply, the tryptophan-rich leader peptide will *not* be synthesized. The striking fact is that synthesis of the leader peptide results in *termination* of transcription of the remainder of the *trp* operon, which includes the structural genes for the biosynthetic enzymes. If synthesis of the leader peptide is blocked by tryptophan deficiency, transcription of the rest of the operon can occur.

How does *translation* of the leader peptide regulate *transcription* of the tryptophan genes downstream? This can be explained by considering that these two processes in prokaryotic cells are occurring virtually simultaneously (Figure 8.21). Thus, while *transcription* of downstream DNA sequences is still proceeding, *translation* of sequences already transcribed has begun. This is because, as the mRNA is released from the DNA, the ribosome binds to it and translation begins. Attenuation occurs (RNA polymerase stops transcription) because a portion of the newly formed mRNA folds into a double-stranded loop followed by a run of uracils that signals cessation of RNA polymerase action (∞ Figure 7.28). The stem-loop structures formed by mRNA are brought about because two stretches of nucleotide bases near each other are complementary

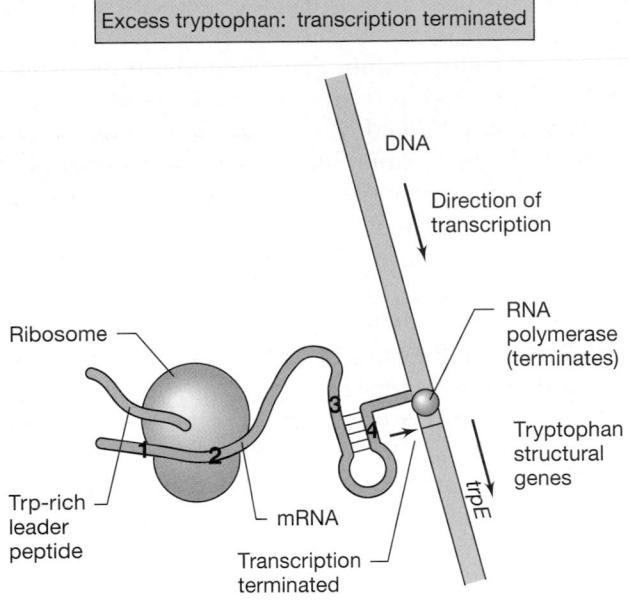

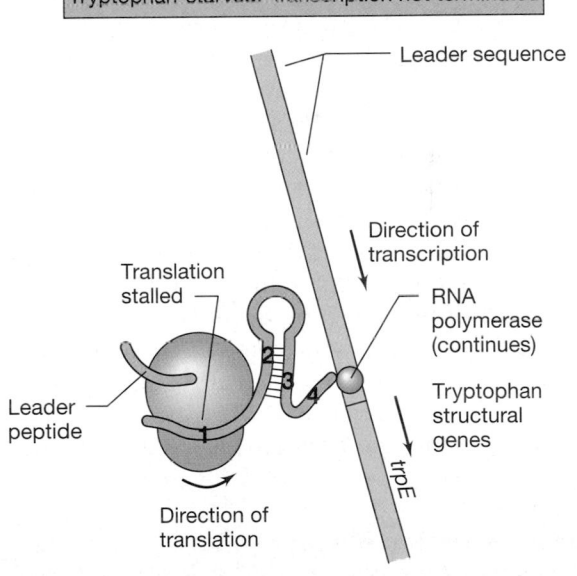

Figure 8.21 Control of transcription of tryptophan operon structural genes by attenuation in *Escherichia coli*. The leader peptide is coded by regions 1 and 2 of the mRNA. Two regions of the growing mRNA chain are able to form double-stranded loops, shown as 2:3 and 3:4. Under conditions of excess tryptophan, the ribosome translates the complete leader peptide, and so region 2 cannot pair with region 3. Regions 3 and 4 then pair to form a loop that terminates RNA polymerase. If translation is stalled because of tryptophan starvation, loop formation via 2:3 pairing occurs, loop 3:4 does not form, and transcription proceeds past the leader sequence. If the ribosome cannot begin translation of the leader because of some translational block *other* than tryptophan starvation, then loop 3:4 will be free to form, and transcription is also terminated.

and can thus base-pair. If tryptophan is plentiful, the ribosome will translate the leader sequence until it comes to the stop codon. The remainder of the leader RNA can then assume a stem-loop, a *transcription pause site,* which is followed by a uracil-rich sequence that actually causes termination. However, if tryptophan is in short supply, the ribosome pauses at a tryptophan codon; the presence of the stalled ribosome *at this position* allows an alternative stem-loop to form (sites 2 and 3 in Figure 8.21). This stem-loop is *not* a termination signal, and it effectively prevents the terminator (sites 3 and 4 in Figure 8.21) from forming. RNA polymerase then moves past the nonfolded termination site and begins transcription of the tryptophan structural genes. Thus, we see that in attenuation there is a highly integrated system in which transcription and translation interact, with the rate of transcription being influenced by the rate of translation.

In the tryptophan operon, two distinct mechanisms for the regulation of transcription exist, repression and attenuation. Repression is a mechanism that has large effects on the rate of enzyme synthesis, whereas attenuation brings about a finer control. Working together, these two mechanisms precisely regulate the synthesis of tryptophan biosynthetic enzymes, and hence the biosynthesis of tryptophan. Attenuation has also been shown to occur in *Escherichia coli* in the biosynthetic pathways for histidine, threonine-isoleucine, phenylalanine, and several other amino acids and essential metabolites as well. As shown in Figure 8.20*b*, the leader peptide for each amino acid biosynthetic operon is rich in that particular amino acid. The *his* operon is dramatic in this regard because its leader contains *seven* histidines in a row near the end of the peptide (Figure 8.20*b*). This longer stretch of regulatory codons gives attenuation a larger effect, which may compensate for the fact that the *his* operon is not under the control of a repressor.

Other Attenuation Mechanisms

Gram-positive *Bacteria,* such as *Bacillus,* also use attenuation to regulate certain amino acid biosynthetic operons. As in gram-negative *Bacteria,* the mechanisms also involve attenuation of transcription and alternative secondary structures, which in one configuration lead to termination. However, the mechanisms are translation-independent, and rather than a translating ribosome, an RNA binding protein is involved. This protein binds as a result of interacting with an *effector,* in some cases a tRNA, but in other cases the amino acid itself. In the *Bacillus subtilis* tryptophan operon the protein is called the *trp attenuation protein,* and in the presence of the amino acid tryptophan binds to the leader and favors transcription termination. If tryptophan is limiting, the protein does not bind and transcription proceeds.

Many cases are now known where attenuation involves genes unrelated to amino acid biosynthesis, and the mechanisms obviously do not involve measuring the amount of amino acid. In *Escherichia coli* some of the operons involved in pyrimidine biosynthesis are regulated by attenuation, and the same is true for the pyrimidine biosynthetic genes of *Bacillus.* The mechanisms are, however, quite different from each other, although each monitors the level of pyrimidine nucleotides in the cell. In *E. coli* a translated mRNA leader is involved, but in *Bacillus* there is no coupling of transcription and translation. Instead, an RNA binding protein controls the alternative structures of the mRNA.

Finally, a type of regulation called *translational attenuation* is also known. In these cases the translation of the leader peptide prevents the *translation* of the next gene on the polycistronic mRNA. The mechanism apparently involves accessibility of the Shine-Dalgarno sequence of the regulated gene (Section 7.15). Translational attenuation is known to regulate expression of several antibiotic resistance genes in gram-positive *Bacteria.*

✓ 8.8 Concept Check

Attenuation is a mechanism whereby gene expression (typically at the level of transcription) is controlled *after* initiation of RNA synthesis. Most attenuation mechanisms involve a coupling of transcription and translation and can therefore occur only in prokaryotes.

- ✓ Why can control systems involving coupled transcription and translation occur only in *prokaryotes?*
- ✓ Explain how the formation of one stem-loop in the RNA can block the formation of another.

8.9 Other Global Control Networks

In Section 8.7 we discussed the fact that organisms have *global control systems* that regulate many genes simultaneously in response to a specific environmental signal. In that section we discussed catabolite repression in *Escherichia coli.* However, there are several different global control systems; a few of those in *E. coli* are shown in Table 8.1.

Global control systems often include more than one regulon (see Section 8.7). Sometimes the term *modulon* is used to describe a group of genes that can respond to a common regulatory protein even though they may also be members of different regulons (and therefore have at least one other type of control). The term *stimulon* is sometimes used to describe a group of genes that all respond to the same environmental signal. Such a group of genes can be very large and the regulatory pathways can be very complex. Indeed, many genes belong to more than one global control system.

TABLE 8.1 A few of the global control systems known in *Escherichia coli*[a]

System	Signal	Primary activity of regulatory protein	Number of genes regulated
Aerobic respiration	Presence of O_2	Repressor (ArcA)	50+
Anaerobic respiration	Lack of O_2	Activator (FNR)	70+
Catabolite repression	Cyclic AMP concentration	Activator (CAP)	300+
Heat shock	Temperature	Alternative sigma (σ^{32})	36
Nitrogen utilization	NH_3 limitation	Activator (NR_1)/alternative sigma (σ^{54})	12+
Oxidative stress	Oxidizing agent	Activator (OxyR)	30+
SOS response	Damaged DNA	Repressor (LexA)	20+

[a]For many of the global control systems, regulation is complex. A single regulatory protein can play more than one role. For instance, the regulatory protein for aerobic respiration is a repressor for many promoters but an activator for others, whereas the regulatory protein for anaerobic respiration is an activator protein for many promoters but a repressor for others. Regulation can also be indirect or require more than one regulatory protein. Some of the regulatory proteins involved are members of two-component systems (see Section 8.10). Many genes are regulated by more than one global system. (For a discussion of the SOS response, ∞ Section 10.3.)

Alternative Sigma Factors

Genes belonging to global control systems do not all use a simple combination of repressors or activators to achieve regulation. Several involve *alternative sigma factors,* including some shown in Table 8.1, and in these cases, regulation is brought about by changing either the amount or the activity of these factors.

Recall that sigma factor is the subunit of RNA polymerase responsible for promoter recognition (∞ Section 7.8). Most genes in *Escherichia coli* require the sigma factor referred to as σ^{70} (the superscript 70 indicates the size of this protein, 70 kilodaltons) for transcription and have promoters like those shown in Figure 7.27. The genes that are induced by an increase in temperature (heat shock) have promoters with a quite different sequence, and RNA polymerase requires a different sigma factor (σ^{32}) to recognize them. It is the *amount* of this alternative sigma factor in the cell that regulates the *heat shock response,* and the amount of σ^{32} itself is controlled not by transcription but by the stability of the factor, its rate of translation, and its activity. In total there are seven different sigma factors known in *E. coli* (Table 8.2). Most

of these have counterparts in other *Bacteria,* such as *Bacillus subtilis.* However, *B. subtilis* also has many alternative sigma factors specific for endospore formation (∞ Section 4.15).

Heat Shock Response

Most proteins are very stable; once made, they continue to perform their functions and are passed along at cell division. However, some proteins are unstable. They are recognized by enzymes in the cell called **proteases** and are rapidly degraded. In *E. coli,* σ^{32} is normally degraded within a minute or two after it is synthesized. However, when cells experience a heat shock, this degradation process is inhibited. This means there will be more σ^{32}, which can therefore direct more RNA polymerase to more heat shock promoters. There is also *translational control* of σ^{32} involving sequences on its mRNA, but the mechanism is unclear.

Global regulatory systems must have a common way to transmit a signal from the environment to the gene(s). We have seen how this is accomplished with catabolite repression, but in the case of the heat shock response, how does a bacterium know what the temperature is? This mechanism seems to involve the *heat shock proteins,* including a protein called DnaK.

DnaK is essential for the normal growth of *E. coli* at any temperature, but the amount that is synthesized is increased by heat shock (recall that genes under the control of a global control system are also usually regulated in other ways). The protein DnaK is a **chaperonin,** one of a group of proteins called *molecular chaperones* (∞ Section 7.16). DnaK helps other proteins fold properly and is also involved in the degradation of σ^{32}. Possibly when the temperature is increased, the activity of DnaK is directed more toward folding proteins and is less available for the pathway leading to degradation of σ^{32}. (DnaK is also involved in inhibiting the *activity* of σ^{32}.)

TABLE 8.2 Sigma factors in *Escherichia coli*

Name[a]	Function
σ^{70}	For most genes, major factor during normal growth
σ^{54}	Nitrogen assimilation
σ^{38}	Major factor during stationary phase, also for genes involved in oxidative and osmotic responses
σ^{32}	Heat shock response
σ^{28}	For genes involved in flagellum synthesis
σ^{24}	Response to misfolded proteins in periplasm (∞ Section 4.8)
σ^{19}	For certain genes in iron transport

[a] Superscript number in name indicates size of protein in kilodaltons. Most factors also have other names, i.e., σ^{70} is also sometimes called σ^D.

Certainly an increase in temperature could influence formation of the correct secondary and tertiary structures of proteins or even cause them to denature slightly (∞ Section 3.8). This would result in an increased level of σ^{32}, and the genes encoding the heat shock proteins would be transcribed. However, since the amount of DnaK increases as part of the heat shock response, it eventually builds up and σ^{32} is degraded again, bringing the cell back to its normal state.

Quorum Sensing

Global control systems allow an organism to respond effectively to signals in its environment. One interesting "signal" is the presence of other organisms of the same species. It has been discovered that certain *Bacteria* have regulatory pathways that are controlled in response to the density of cells within their own population. This type of control is called **quorum sensing.**

Each gram-negative bacterium that has this type of regulation has an enzyme that synthesizes a specific acylated homoserine lactone (AHL). This molecule is diffusible to the outside of the cell. Therefore, it can only reach high concentrations in the cell if there are a number of cells nearby each making the same AHL. These AHL molecules are the inducer that combines with an activator protein. Quorum sensing was first discovered as the form of regulation of bioluminescence in a number of bacteria (∞ Section 12.12 and Figure 12.28). Figure 8.22 shows colonies of the bacterium *Vibrio fischeri* which are glowing because of the

production of bacterial luciferase. The *lux* operons, which encode the proteins involved in bioluminescence, are under control of the activator protein LuxR and are induced when the concentration of *N*-3-oxohexanoyl homoserine lactone becomes high enough. This AHL is synthesized by the protein encoded by the *luxI* gene. Other genes are also controlled by this system. In some bacteria, for instance, *Pseudomonas aeruginosa*, it is clear that quorum sensing is a global response leading to the expression of a large number of different genes when the population density becomes sufficiently high. This assists in the formation of biofilms by *P. aeruginosa* (∞ Section 19.3), an important aspect of its pathogenesis.

Pathogenesis of *Staphylococcus aureus* (∞ Sections 25.7 and 26.9) involves the production and secretion of numerous cell surface and extracellular proteins that damage the host cells or tissues or that interfere with the immune system. The genes encoding these virulence factors are under the control of a quorum system that responds to a peptide produced by the organism itself. The regulation of these genes is quite complex and, in part, involves a *regulatory RNA molecule*. Although we have discussed only regulatory proteins in this chapter, regulatory RNA molecules also exist (see box, Antisense Nucleic Acid). The *S. aureus* quorum sensing system also involves regulatory proteins that are part of a two-component, signal-transducing regulatory system. We discuss this type of regulation in the next section.

✓ **8.9 Concept Check**

Cells have many different global control systems that allow the cell to regulate the expression of many genes in response to changes in the environment. These changes can include nutritional factors, temperature, and even the number of bacteria present. Some of these regulatory systems involve alternative sigma factors.

✓ What is an *alternative sigma factor?*
✓ What is meant by *quorum sensing?*

8.10 Signal Transduction and Two-Component Regulatory Systems

Bacteria regulate cell metabolism in response to a wide variety of environmental fluctuations, including temperature changes, changes in pH and oxygen availability, changes in the availability of nutrients, and even changes in the number of cells present. Therefore, there must be mechanisms by which bacteria receive signals from the environment and transmit them to the specific target to be regulated. We have seen in preceding sec-

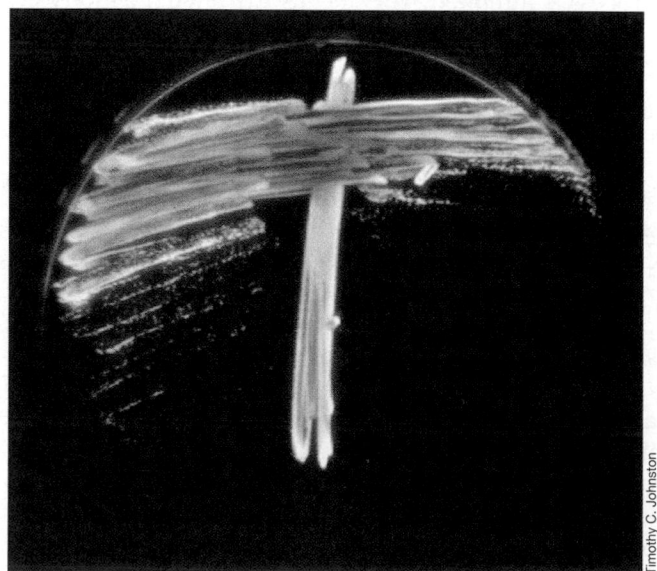

Figure 8.22 Bioluminescent bacteria, which are producing the enzyme luciferase. Cells of the bacterium *Vibrio fischeri* were spread on nutrient agar in a Petri dish and allowed to grow overnight. The photograph was taken using only the light generated by the bacteria.

Timothy C. Johnston

A Focus On ... Antisense Nucleic Acid

Regulation of the synthesis of proteins often involves transcriptional control. Less often, genes are controlled at the level of translation. Most control networks, whether they are transcriptional or translational, use regulatory proteins. It is known, however, that in some cases it is a regulatory *RNA*, not a regulatory *protein,* that is involved.

One type of regulatory RNA, called **antisense RNA**, is known to be used in the regulation of several different bacterial genes. Antisense RNA acts by forming base pairs with a complementary, or sense, strand of RNA (Fig. 1). When the sense RNA is mRNA, the resulting double-stranded structure can prevent translation. In some regulatory systems this antisense RNA is synthesized from a very small gene whose sequence is essentially the same as that of the target gene but in which transcription is in the opposite direction (see Fig. 1). If this antisense RNA is synthesized it can base-pair with the mRNA and prevent its translation. Note that the syn-

thesis or the stability of the antisense RNA must also be regulated.

Antisense RNA does not have to be used only to regulate the synthesis of a protein. In some plasmids, it controls the initiation of DNA synthesis. In eukaryotes there is also a phenomenon called "RNA silencing" in which small double-stranded RNA molecules apparently target certain mRNAs for rapid destruction. (Although even in eukaryotes mRNA is less stable than rRNA or tRNA, it is typically much more stable than mRNA in prokaryotes.)

Antisense nucleic acids can be specifically designed and synthesized by scientists in the laboratory and delivered directly to cells. These short (15–25 nucleotides) synthetic chains (oligonucleotides) are usually made of DNA rather than RNA. Their sequence can be made to allow them to bind to a specific mRNA and prevent translation (or allow the molecule to be recognized by nucleases). Antisense nucleic acid can also bind to the DNA in the nucleus and prevent

transcription. The latter is possible because some DNA can form a *triple* helix! The "extra" strand (the oligonucleotide) forms specific interactions with those parts of the bases that are in the major groove of a normal double helix to give **triple DNA**. (Not all DNA sequences can form triple helices, at least not without the aid of special enzymes.)

Synthetic antisense oligonucleotides can be designed to be extremely specific, whether they bind to a message or to the regulatory region of a gene. The specificity arises because a sequence of only 20 bases should occur no more often than once in 10^{12} bases of "random" DNA. Therefore, it is unlikely that the antisense RNA would bind to anything other than its known target in any cell. This specificity might allow antisense nucleic acids to become an important new class of antibiotic. Antisense nucleic acids could be designed to be used against specific viruses or to regulate particular genes in either disease-causing (pathogenic) organisms or human tumor cells. ■

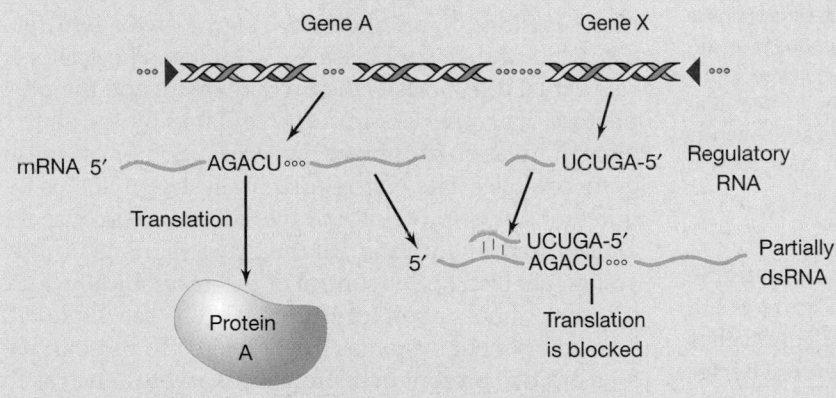

Figure 1 Gene A is transcribed from its promoter (⟳ Section 7.9) to give an mRNA that can be translated to yield protein A. Gene X is a small gene whose sequence is identical to that of Gene A but has its promoter at the opposite end. Therefore, if it is transcribed, the resulting RNA will be complementary to the mRNA of gene A. If these two RNAs base-pair, translation will be blocked.

tions that some signals can be small molecules that enter the cell (often by specific uptake mechanisms) and act as *effectors.* For instance, in the case of the maltose regulon (see Section 8.6) the sugar maltose binds to the maltose activator protein, causing the protein to bind to specific DNA sequences and activate transcription. However, in many cases the external signal is not transmitted directly to the regulatory protein. Instead, a signal is first detected by a sensor and then transmitted in a changed form to the rest of the regulatory machinery, a process called **signal transduction**.

Sensor Kinases and Response Regulators

Many of the regulatory systems by which cells sense and then respond to environmental signals are called **two-component systems**. Characteristically, such systems include two different proteins: (1) a specific **sensor protein** located in the cell membrane, and (2) a partner **response regulator protein**. The *sensor protein* has *kinase* activity and is usually referred to as a **sensor kinase**. A **kinase** is an enzyme that phosphorylates compounds. Sensor kinases detect a signal from the environment on their outer surface and in response phosphorylate

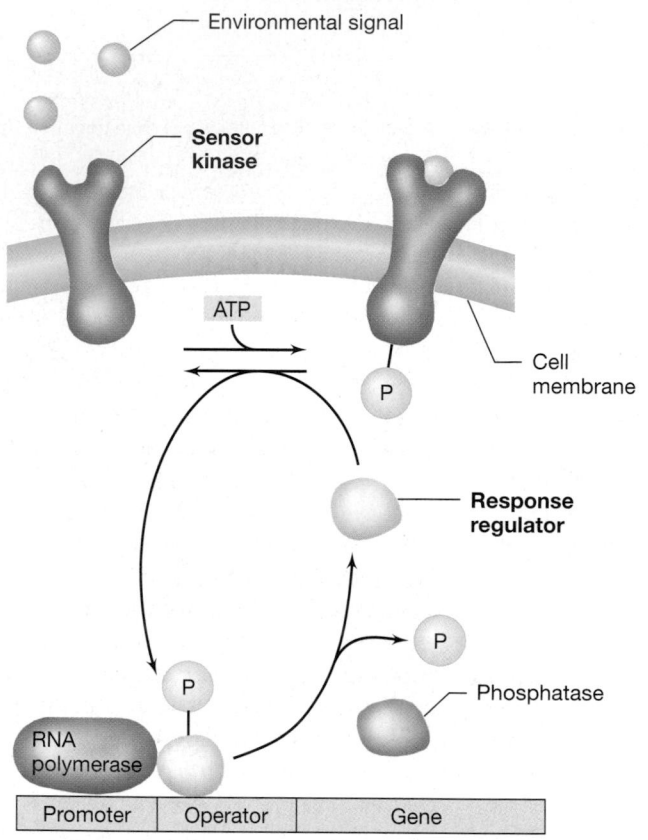

Figure 8.23 The control of gene expression by a two-component system. The main components of the system include a *sensor kinase* in the cell membrane that phosphorylates itself in response to an environmental signal. The phosphoryl group is then transferred to the other main component, a *response regulator.* In the system diagrammed in this figure the phosphorylated response regulator serves as a repressor. There must also be a phosphatase in the system to cycle the response regulator.

themselves (autophosphorylation) at a specific histidine residue on their cytoplasmic surface (see Figure 8.23). This phosphoryl group is then transmitted to another protein inside the cell, the *response regulator*. The response regulator is typically a DNA binding protein that regulates transcription.

In Figure 8.23 the phosphorylated response regulator is acting as a repressor protein, while the unphosphorylated response regulator does not bind to DNA. Often, however, the phosphorylated response regulator acts as an activator protein. The mechanism used by the response regulator to control transcription depends on the system being described.

In order to complete a regulatory circuit, there must be a way to terminate the signal. Typically, this involves a *phosphatase*, an enzyme that can remove the phosphoryl group from the response regulator protein. In some cases this reaction is carried out by the sensor kinase itself, while in other systems there is a third protein that carries out this

reaction. Therefore, there are "two-component" systems with three components! Actually, some systems have even more components, as the signal may be processed through several steps. However, in all cases two-component systems have a sensor kinase and a response regulator.

Examples of Two-Component Regulatory Systems

Two-component systems are now known to regulate a large number of genes in many different bacteria. A few key examples include phosphate assimilation in *Escherichia coli*, nitrogen fixation in *Klebsiella* and *Rhizobium*, and sporulation in *Bacillus* (which has a very complex regulatory system). In *E. coli* alone it is estimated that almost 50 different two-component systems operate, and a few of these are listed in Table 8.3. Many of these systems involve global regulation (see also Table 8.1). In *E. coli* the osmolarity of the environment controls which of two proteins, OmpC or OmpF, is synthesized as part of the outer membrane. The response regulator of this system is OmpR. When OmpR is phosphorylated, it acts as an *activator* of transcription of the *ompC* gene and a *repressor* of transcription of the *ompF* gene.

In some cases more regulatory elements are involved. As an example, in the *Ntr system,* which regulates nitrogen assimilation, the response regulator is an activator protein, NR_I (Nitrogen Regulator I), which works by allowing transcription from promoters recognized by RNA polymerase using σ^{54}, an alternative sigma factor (see Table 8.2). While the sensor kinase, NR_{II} (Nitrogen Regulator II), fills a dual role, both as the protein kinase and the phosphatase, its activity is in turn regulated by the state of phosphorylation of another protein, P_{II}. Some systems are quite complex. The Nar regulatory system involves two different sensor proteins and two different response regulators, and in addition, all the genes regulated by this system are also under control of the anaerobically active transcriptional regulatory protein FNR (see Table 8.1). Two-component systems closely related to those in bacteria are also present in microbial eukaryotes, such as the yeast *Saccharomyces cerevisiae*. Higher eukaryotes also use phosphorylation as a mechanism of signal transduction in order to respond to environmental changes.

✓ 8.10 Concept Check

Signal transduction systems transmit environmental signals to the cell. In prokaryotes signal transduction typically involves two-component systems, which include a sensor protein located in the membrane and a cytoplasmic response regulator protein. The sensor protein is a kinase, and the activity of the response regulator depends on its state of phosphorylation.

✓ What are *kinases* and what is their role in two-component regulatory systems?

✓ Could a response regulator be an activator or a repressor?

TABLE 8.3 Some two-component regulatory systems from *Escherichia coli* that regulate transcription

System	Environmental signal	Sensor kinase	Response regulator	Activity of response regulator[a]
Arc system	O_2	ArcB	ArcA	Repressor/Activator
Nitrate and nitrite anaerobic regulation (Nar)	Nitrate and nitrite	NarX and NarQ	NarL	Activator/Repressor
			NarP	Activator/Repressor
Nitrogen utilization (Ntr)	NH_4^+	NR_{II}, the product of *glnL*	NR_I, the product of *glnG*	Activates RNA polymerase at promoters requiring σ^{54}
Pho regulon	Inorganic phosphate	PhoR	PhoB	Activator
Porin regulation	Osmotic pressure	EnvZ	OmpR	Activator/Repressor

[a]Note that several of the response regulator proteins act as both activators and repressors depending on the genes being regulated. Although ArcA can function as either an activator or a repressor, it functions as a repressor on most operons that it regulates.

8.11 Regulation of Chemotaxis

Not all response regulators regulate transcription. We have previously discussed the fact that bacteria can move toward or away from particular chemicals, a process called *chemotaxis* (∞ Section 4.12). We noted that bacteria are too small to actually sense *spatial* gradients of a chemical, but rather they respond to *temporal* gradients. That is, they can sense the change in concentration of a chemical outside the cell *over time*. Bacteria use a two-component system to sense the temporal changes in chemical concentration and regulate flagellar motion.

Mechanism of Chemotaxis: Response to Signal

The mechanism of chemotaxis is quite complex and involves a variety of different proteins. A number of *sensory proteins* are in the cell membrane, and these sense the presence of attractants and repellants. These proteins allow the cell to sense whether, over time, the concentration of the substance increases or decreases as the cell moves. The cell thus responds to the *change* in concentration rather than the *absolute* concentration of the chemical stimulus. The sensory proteins are called *methyl-accepting chemotaxis proteins (MCPs),* or *receptor-transducer proteins,* or simply **transducers**. In *Escherichia coli*, five different MCPs have been identified, and each is a transmembrane protein (Figure 8.24). Each MCP can sense a variety of compounds. For example, the *Tar* transducer of *E. coli* can sense the attractants aspartate and maltose as well as repellents such as the heavy metals cobalt and nickel.

MCPs bind attractants or repellents directly, or in some cases indirectly, through interactions with periplasmic binding proteins. Binding of an attractant or repellent sets in play a series of interactions with cytoplasmic proteins that eventually affects flagellar ro-

tation. If rotation of the flagellum is *counterclockwise,* the cell will continue to move in a run. If the flagellum rotates *clockwise,* however, the cell will tumble (∞ Section 4.11).

The current model for flagellar control shows that the transducers are in contact with the cytoplasmic proteins CheW and CheA (Figure 8.24). CheA is a *sensor kinase*. When a transducer has bound a chemical, it changes conformation and (with CheW) causes a change in the autophosphorylation of CheA (forming CheA-P). *Attractants decrease* the rate of autophosphorylation, whereas *repellents increase* this rate. Phosphorylated CheA (CheA-P) then phosphorylates CheY (forming CheY-P), a *response regulator*. CheA-P can also phosphorylate CheB, another response regulator, but this is a much slower reaction than the phosphorylation of CheY. We will discuss the activity of CheB-P later.

CheY is the central protein in the system because it serves as the *response regulator* for chemotaxis, governing the direction of rotation of the flagellum. CheY-P interacts with the flagellar motor to induce clockwise flagellar rotation and tumbling (the motor switch itself consists of proteins encoded by *fla* genes; ∞ Section 4.10). If unphosphorylated, CheY cannot bind, the flagellar motor continues counterclockwise rotation, and the cell undergoes a run. Another protein, CheZ, dephosphorylates CheY, returning it to a form that allows runs instead of tumbles. Because repellents increase the level of CheY-P, they lead to tumbling, whereas attractants lead to a lower level of CheY-P and smooth swimming.

Mechanism of Chemotaxis: Adaptation

Note that the system described can *signal* that a chemical has been bound and regulate flagellar rotation but seems to be unable to note a change with the passage of time. There is a second component to chemotaxis, and this is **adaptation**.

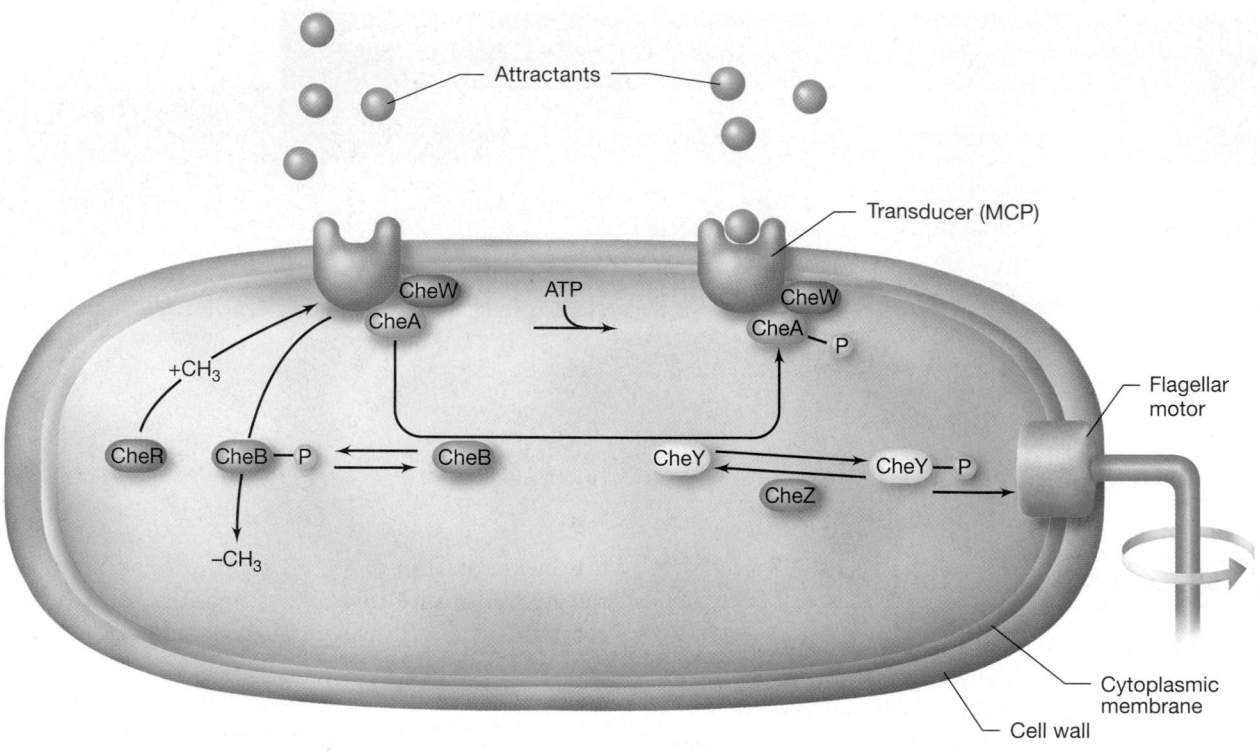

Figure 8.24 Interactions of transducers, chemotaxis (Che) proteins, and the flagellar motor in bacterial chemotaxis. The transducer (MCP) forms a complex with the *sensor kinase* CheA and the coupling protein CheW. This combination results in a signal-regulated autophosphorylation of CheA to CheA-P. CheA-P can then phosphorylate the *response regulators* CheB and CheY. Phosphorylated CheY (CheY-P) interacts directly with the flagellar motor switch. CheZ dephosphorylates CheY-P. CheR continually adds methyl groups to the transducer. CheB-P (but not CheB) removes them. The degree of methylation of the transducers controls their ability to respond to attractants and repellants and leads to adaptation. The structure of the flagellar motor was shown in Figure 4.41.

As their name implies, MCPs can be methylated. There is a cytoplasmic protein, CheR (Figure 8.24), that continually adds methyl groups to the MCPs at a slow rate using S-adenosylmethionine as a methyl donor. The phosphorylated form of the response regulator CheB is a demethylase that removes methyl groups from the MCPs. The level of methylation of the MCPs affects their conformation and controls adaptation to a sensory signal. It allows resetting of the signaling state of the receptor even though the concentration of the chemical remains unchanged.

If the level of an attractant remains high, the level of phosphorylation of CheA (and, therefore, of CheY and CheB) will remain low, the cell will swim smoothly, and the level of methylation of the MCPs will increase (because CheB-P is not present to demethylate). However, the MCPs no longer respond to the attractant when they are fully methylated. Therefore, even though the level of attractant might remain high, the level of CheA-P (and CheB-P) increases and the cell begins to tumble. However, now the MCPs can be demethylated by CheB-P, and when this happens, the receptors can once again respond to attractants. The situation is the opposite with regard to repellants (fully methylated MCPs respond best to repellants).

The control of chemotaxis is obviously quite complicated and involves a number of regulatory switches.

Unlike the case in many other two-component systems, in chemotaxis the signal transduction system regulates the activity of the gene products, not their synthesis.

✓ 8.11 Concept Check

Chemotaxis is under complex regulation involving signal transduction and two-component systems. However, these systems regulate the activity of the proteins involved rather than their synthesis. Adaption allows the system to reset itself to the continued presence of a signal.

✓ What is the primary *response regulator* and the primary *sensor kinase* involved in regulating chemotaxis?

✓ Why is adaptation important?

V CONCLUDING REMARKS ON REGULATION

We have discussed only a few of the mechanisms by which cells can control the activity of a protein and also only a few major mechanisms that can regulate the synthesis of a protein. Most of the mechanisms we considered for regulating synthesis operate at the level of transcription, and all involve regulatory proteins. However, as we mentioned, regulatory RNA also exists (see the box, Antisense Nucleic Acid).

8.12 Contrasts in Gene Expression between Prokaryotes and Eukaryotes

Although many of the major regulatory patterns are shared between prokaryotes and eukaryotes, there are many differences. Because of the lack of compartmentation in prokaryotes, the processes of transcription and translation are *coupled*. Also, the messenger RNA of prokaryotes is frequently polycistronic, with more than one protein being translated from the same message.

In eukaryotes, on the other hand, transcription and translation take place in separate compartments in the cell, and the integration of these processes seen in prokaryotes is lacking. How about induction and repression in eukaryotes? Although many eukaryotes do exhibit repression, there is little evidence for the kind of negative control so commonly found in prokaryotes. However, positive control mechanisms are common in eukaryotes. If operons exist in eukaryotes, they involve the control of only single enzymes, rather than the multienzyme control systems so commonly seen in prokaryotes, and there is no evidence for polycistronic RNA molecules in eukaryotes except in a few viruses. These are translated inefficiently. Posttranslational protein modification is quite common in eukaryotes. Eukaryotes also have regulation involving splicing of mRNA, regulation that does not exist in prokaryotes.

It is the regulation of genes that is the basis for *development* of multicellular eukaryotic organisms, which begin life as single cells and develop into complex organisms with many very different and highly specialized cell types. This process of differentiation requires that specific sets of genes become active at precisely the correct time during the development of the organism. The complex regulatory pathways involved in development are currently under intense investigation. Understanding and controlling these pathways would allow for tremendous advances in medicine.

✓ 8.12 Concept Check

Prokaryotes and eukaryotes share many similar regulatory mechanisms. Regulation in both involves proteins that recognize DNA sequences, although in eukaryotes positive control is more common. In prokaryotes transcription and translation can be coupled, but in eukaryotes these processes take place in separate compartments. Prokaryotes also have polycistronic messages and eukaryotes do not.

✓ What regulatory control mechanism was discussed that involves coupling of transcription and translation?

✓ What is a polycistronic mRNA?

Review Questions

1. If an enzyme can be effectively inhibited by feedback inhibition, why would cells also have mechanisms to regulate its synthesis?

2. Compare the processing of proinsulin and the product of the *gyrA* gene of *Mycobacterium leprae* (see box, Protein Processing). In both cases an "interior" peptide is removed. However, only the latter case is considered protein splicing. Explain.

3. Compare and contrast *introns* and *inteins*.

4. Describe why a protein that binds to a specific sequence of double-stranded DNA is unlikely to bind to the same sequence if the DNA is single-stranded.

5. Some operons are under the control of an operator, but others are not. Explain.

6. Describe the regulation of two different operons, one having an effector that is an *inducer* and the other having an effector that is a *corepressor*.

7. The maltose regulon is inducible and is regulated by an activator protein. The lactose operon is inducible but is regulated by a repressor protein. Explain how induction can be brought about by either positive control (activator protein) or negative control (repressor protein).

8. In most cases operators are very close to the promoters they control, while activator binding sites can be some distance away. Explain why this should be so.

9. Describe how transcriptional attenuation works. What is actually being "attenuated"? Why hasn't the type of attenuation that controls several different amino acid biosynthetic pathways in *Escherichia coli* also been found in eukaryotes?

10. Describe the mechanism by which catabolic activator protein (CAP), the regulatory protein for catabolite repression, functions using the lactose operon as an example. For this operon the CAP protein is not a repressor. Describe the regulatory region of a gene for which the CAP protein *is* a repressor. (*Hint:* Think about your answer to Question 8.)

11. What are the two components that give the name to signal transduction regulation in prokaryotes? What is the function of each of the components?

12. One of the members of a two-component system is typically located in the cell membrane. Why might this be so?

13. Many genes are under multiple control systems. In the lactose operon, there is lactose-specific regulation and regulation by a global control system. Describe how each of the controls on the lactose operon actually functions. Why do you think both systems are necessary?

14. Adaptation allows the regulatory mechanism controlling flagellar rotation to be reset. How is this accomplished?

Application Questions

1. The amino acids isoleucine and valine share a common pathway for most steps in their biosynthesis. In *Escherichia coli* the first common step can be subject to feedback inhibition by valine but not by isoleucine. In most strains, though, the addition of valine does not cause isoleucine deprivation. However, in other strains it does (that is, adding valine causes isoleucine starvation and the cells stop growing). What explanation can you give for the difference between the normal "valine-resistant" strains and those whose growth is sensitive to valine?

2. What would happen to regulation from a promoter under negative control if the region where the regulatory protein binds were deleted? What if the promoter were under positive control?

3. Promoters from *Escherichia coli* under positive control are not close matches to the DNA consensus sequence for *E. coli* (∞ Section 7.9). Why?

4. Interestingly, the attenuation control of some of the pyrimidine biosynthetic pathway genes in *Escherichia coli* actually involves coupled transcription and translation. Can you describe a mechanism whereby the cell could somehow make use of translation to help it measure the level of pyrimidine nucleotides?

5. Most of the regulatory systems described in this chapter involve regulatory proteins. However, regulatory RNA is known to exist. Describe how one could achieve negative control of the *lac* operon using a regulatory RNA molecule.

V iruses are infectious agents that require cells for their reproduction. Like cells, viruses have a genome, but the viral genome contains insufficient genetic information for independent replication. Although the genetic material of many viruses is double-stranded deoxyribonucleic acid (DNA), for other viruses the genetic material is single-stranded DNA and for still others, the genome is ribonucleic acid (RNA). The infectious form of a virus is a particle called a *virion* (shown here for human papilloma virus), which contains the genetic material surrounded by a protein coat.

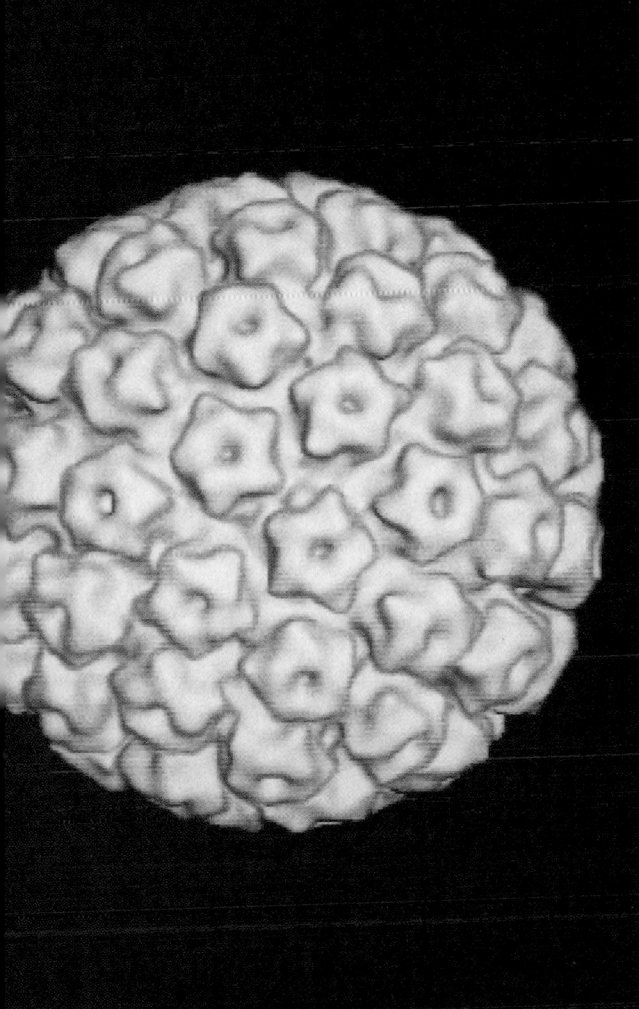

ESSENTIALS OF VIROLOGY

9

Working Glossary

Bacteriophage a virus that infects prokaryotic cells

Early proteins proteins synthesized soon after virus infection

Late proteins proteins synthesized toward the end of virus infection

Lysogen a bacterium containing a prophage

Lysogenic pathway a series of steps that, after virus infection, lead to a state (lysogeny) where the viral genome is replicated as a prophage along with that of the host

Lytic pathway a series of steps after virus infection that lead to virus replication and the destruction (lysis) of the host cell

Minus (negative)-strand nucleic acid an RNA or DNA strand that has the opposite sense of (is complementary to) the mRNA of a virus

Oncogene a gene whose expression causes formation of a tumor

Plaque a zone of lysis or cell inhibition caused by virus infection of a lawn of sensitive cells

Plus (positive)-strand nucleic acid an RNA or DNA strand that has the same sense as the mRNA of a virus

Prion an infectious agent whose extracellular form may contain no nucleic acid

Provirus (prophage) the genome of a temperate virus when it is replicating with, and usually integrated into, the host chromosome

Retrovirus a virus whose RNA genome has a DNA intermediate as part of its replication cycle

Reverse transcription the process of copying information found in RNA into DNA

Temperate virus a virus whose genome is able to replicate along with that of its host and not cause cell death in a state called lysogeny

Transformation a process by which a normal cell becomes a cancer cell (but see alternative usage in Chapter 10)

Virion the complete virus particle; the nucleic acid surrounded by a protein coat and in some cases other material

Virulent virus a virus that lyses or kills the host cell after infection; a nontemperate virus

Virus a genetic element containing either RNA or DNA that replicates in cells but is characterized by having an extracellular state

Viroid small, circular, single-stranded RNA

V**iruses** are genetic elements that can replicate independently of a cell's chromosomes but not independently of cells themselves (∞ Section 7.4). In order to multiply, viruses must enlist a cell in which they can replicate. Such a cell is called a *host*. Viruses are characterized by also having a mature infectious form that is typically extracellular.

Viruses, like plasmids and some other genetic elements (∞ Section 7.4), take advantage of the metabolic machinery encoded by the host cell's own chromosomes. Like these other elements, viruses can confer important new properties on their host cell. These properties will be inherited when the host cell divides if each new cell also inherits the viral genome. These changes are often not harmful and may even be beneficial. However, viruses, unlike genetic elements such as plasmids (∞ Sections 7.4 and 10.8), have an extracellular form that enables them to be easily transmitted from one host to another. This extracellular form has enabled some viruses to replicate themselves in a host in a way that is destructive to the host cell. This destructive replication accounts for the fact that some viruses are agents of disease. In many cases, whether a virus causes disease or hereditary change depends on the host cell and on the environmental conditions. Because viruses have a form independent of cells some people refer to them as "living organisms" or "life forms." As we will discuss below, however, some properties that serve as characteristics of living systems (∞ Section 1.2) do not occur in the extracellular form of the virus. Without host cells in which to replicate, there could be no viruses.

In this chapter we shall discuss some of the ways in which viruses can redirect the metabolism of the host cell in order to replicate. This chapter is divided into four parts. The first part introduces basic concepts of virus structure. The second part deals with the host cell and how viruses can be quantitated. The third part deals with some of the basic molecular biology of virus multiplication. The fourth part gives an overview of a few of the viruses that infect bacteria and animals (further material on specific viruses can be found in Chapter 16). These discussions expand on the concepts of macromolecular synthesis and gene regulation we covered in Chapters 7 and 8.

Viruses are among the most numerous "microorganisms" on our planet and infect all types of cellular organisms. Therefore, they are interesting to study in their own right. However, scientists have also studied and continue to study viruses for what they can tell us about the genetics and biochemistry of cellular metabolism and, in the case of some viruses, the development of disease. Furthermore, as we shall see in Chapters 10, 15 and 31, viruses are also important tools for the microbial geneticist and the genetic engineer.

I VIRUS AND VIRION

9.1 General Properties of Viruses

Here we describe both the extracellular and intracellular states of viruses. In the **extracellular** state, a virus is a minute particle containing nucleic acid surrounded by protein and occasionally, depending on the specific virus, containing other macromolecular components. In this typically extracellular form, the **virus particle**, also called the **virion**, is metabolically inert and does not carry out respiratory or biosynthetic functions. The virion is the structure by which the **virus genome** is carried from the cell in which it has been produced to another cell where the viral nucleic acid can be introduced. Once in the new cell, the **intracellular state** is initiated. In the intracellular state, **virus replication** occurs: New copies of the virus genome are produced, and the components that make up the virus coat are synthesized. When a virus genome is introduced into a host cell and reproduces, the process is called **infection**. A cell that a virus can infect and in which it can replicate is called a **host**. Viral genomes are very limited in size, and they encode primarily those functions that they cannot adapt from their hosts. Therefore, during replication inside a cell, viruses depend heavily on host cell structural and metabolic components. The virus redirects preexisting host machinery and metabolic functions to support virus replication and the assembly of new virions. Therefore, for most viruses, virions can also be found inside the cell late in infection.

Viral Genomes

As we have seen (∞ Section 7.1), all cells have double-stranded deoxyribonucleic acid (DNA) as their genetic material. In contrast, viruses can have either DNA or ribonucleic acid (RNA) as their genetic material, and it can be either single-stranded or double-stranded. Viruses are sometimes divided into two types based on whether they have DNA or RNA as their genetic material, and *all* viruses contain either one or the other in the virion. However, there is a third group of viruses that use *both* DNA and RNA as their genetic material but at different stages of their reproductive cycle (Figure 9.1). This last group includes the retroviruses, which contain an RNA genome in the virion but replicate through a DNA intermediate, and the human hepatitis B virus, which contains DNA in the virion but has an RNA intermediate in replication. These classes can be further subdivided according to whether the nucleic acid in the

virion is single- or double-stranded (Figure 9.1). The classification of viruses based on the type of nucleic acid in the virions and the associated replication strategies have been formalized as the Baltimore Classification System, which we will discuss in Section 9.7.

Despite the diversity of genome structure, viruses obey the *central dogma* of molecular biology (∞ Section 7.1): All genetic information flows from nucleic acid to protein. In addition, all viruses use the cell's translational machinery, and so no matter what the genome structure of the virus, messenger RNA (mRNA) must be generated that can be translated on the host's ribosomes.

Viral Hosts and Taxonomy

Viruses can also be classified on the basis of the hosts they infect. Thus, we have animal viruses, plant viruses, and bacterial viruses. Bacterial viruses, sometimes called **bacteriophages** (or *phage* for short, from the Greek *phagein* meaning "to eat"), have been studied primarily as convenient model systems for research on the molecular biology and genetics of virus reproduction.

Many of the basic concepts of virology were first worked out with bacterial viruses and subsequently applied to viruses of higher organisms. Because of their frequent medical importance, *animal viruses* also have been extensively studied. The two groups of animal viruses most studied are those infecting insects and those infecting warm-blooded animals. *Plant viruses* are important in agriculture but have been less studied than animal viruses.

Finally, there is also a formal system of viral taxonomy that organizes viruses within the hierarchical taxon levels: order, family (and subfamily), genus, and species. Although at times such a formal taxonomic system can seem quite arbitrary (and common names are still in wide use), the system is increasingly useful because of the increased phylogenetic data being accumulated. The family taxon seems particularly useful. Members of a familiy of viruses have distinct virion morphology, genome structure, and/or strategies of replication. Virus families have names including the suffix *-viridae* (as in *Poxviridae,*) and we will discuss a few of these in Chapter 16.

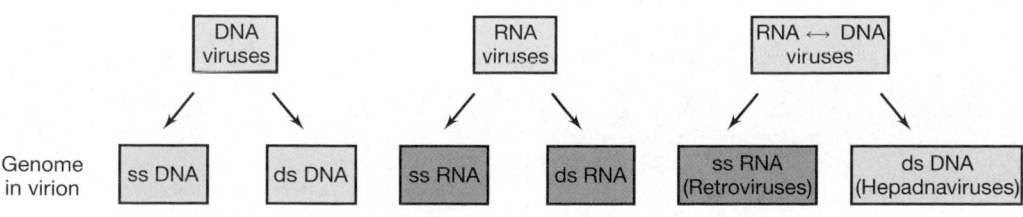

Figure 9.1 Viral genomes. The genomes of viruses can be composed of either DNA or RNA, and some use both as their genomic material at different stages in their life cycle. However, only one type of nucleic acid is found in the virion of any particular type of virus. This can be single-stranded (ss), double-stranded (ds), or in the case of the hepadnaviruses, partially double-stranded.

✓ 9.1 Concept Check

A virion is the extracellular form of a virus and contains either an RNA or a DNA genome. The virus genome is introduced into a new host cell by infection. The virus redirects the host metabolism in order to support virus replication. Viruses are roughly classified by replication strategy as well as by type of host.

✓ How does a *virus* differ from a *plasmid*?
✓ How does a *virion* differ from a *cell*?

9.2 Nature of the Virion

Virus particles (virions) vary widely in size and shape. Viruses are smaller than cells, ranging in size from 0.02 to 0.3 μm. A common unit of measure for viruses is the *nanometer*, which is 1000 times smaller than 1 μm and 1 million times smaller than 1 mm. Smallpox virus, one of the largest viruses, is about 200 nm in diameter (a bit smaller than the size of the smallest *Bacteria*); poliovirus, one of the smallest, is only 28 nm in diameter (about the size of a ribosome).

As we have stated, some viruses contain RNA, others DNA, and the nucleic acid can be either double- or single-stranded, depending on the virus. Viral *genomes* are also smaller than those of most cells. Most bacterial genomes are between 1000 and 5000 kilobase pairs of DNA, with the smallest known being 580 kilobase pairs. (Interestingly, *Bacteria* with the smallest genomes are, like viruses, parasites that replicate in other cells; ∞ Sections 12.13 and 12.27.) However, the largest known viral genome, that of bacteriophage G, is 670 kilobase pairs. This virus, which infects *Bacillus megaterium*, is, however, one of only a few viruses currently known whose genome is larger than a cellular genome. More typical virus genome sizes are listed in Table 9.1. Some viruses have genomes so small they contain less than five genes. As can be seen in the table, the genome of some viruses, such as reovirus, is *segmented* into more than one molecule.

The structures of virions are quite diverse, varying widely in size, shape, and chemical composition. The nucleic acid of the virion is always located within the particle, surrounded by a protein coat called the *capsid*. The terms *coat, shell,* and *capsid* are often used interchangeably to refer to this outer layer. The protein coat is always formed of a number of individual protein molecules, called *structural subunits,* which are arranged in a precise and highly repetitive pattern around the nucleic acid (Figure 9.2). The small genome size of most viruses restricts the number of different viral proteins. A few viruses have only a single kind of protein in their capsid, but most viruses have several chemically distinct structural subunits that are themselves associated in specific ways to form larger assemblies called *capsomers*. The capsomer is the morphological unit that can be seen with the electron microscope.

The information for proper aggregation of the structural subunits into capsomers is contained within the structure of the proteins themselves, and the overall process of assembly is thus called **self-assembly**. For many viruses, this self-assembly process is assisted by *molecular chaperones*, proteins that assist in folding and assembly but that themselves are not a part of the final structure (∞ Section 7.16). A single virion generally has a large number of morphological units. The complete complex of nucleic acid and protein, packaged in the virus particle, is called the virus **nucleocapsid**. Inside the viri-

TABLE 9.1 Some types of viral genomes[a]					
		Viral genome			
Virus	Host	Type of nucleic acid in virion	Structure	Number of molecules	Size
H-1 parvovirus	Animals	Single-stranded DNA	Linear	1	5,176 bases
φX174	*Bacteria*	Single-stranded DNA	Circular	1	5,386 bases
Simian virus 40 (SV40)	Animals	Double-stranded DNA	Circular	1	5,243 base pairs
Poliovirus	Animals	Single-stranded RNA	Linear	1	7,433 bases
Cauliflower mosaic virus	Plants	Double-stranded DNA	Circular	1	8,025 base pairs
Cowpea mosaic virus	Plants	Single-stranded RNA	Linear	2 different	9,370 bases (total)
Reovirus type 3	Animals	Double-stranded RNA	Linear	10 different	23,549 base pairs (total)
Bacteriophage lambda	*Bacteria*	Double-stranded DNA	Linear	1	48,514 base pairs[b]
Herpes simplex virus type I	Animals	Double-stranded DNA	Linear	1	152,260 base pairs
Bacteriophage T4	*Bacteria*	Double-stranded DNA	Linear	1	168,903 base pairs
Human cytomegalovirus	Animals	Double-stranded DNA	Linear	1	229,351 base pairs

[a]The sizes of the viral genomes chosen for this table are known accurately because they have been sequenced. However, this accuracy can be misleading because only a particular strain or isolate of a virus was sequenced. Therefore, the sequence and exact number of bases for other isolates may be slightly different. No attempt has been made to choose the largest and smallest viruses known, but rather to give a fairly representative sampling of the sizes and structures of the genomes of viruses containing both single- and double-stranded RNA and DNA.

[b]Thus total includes single-stranded extensions of 12 nucleotides at either end of the linear form of the DNA (see Section 9.10).

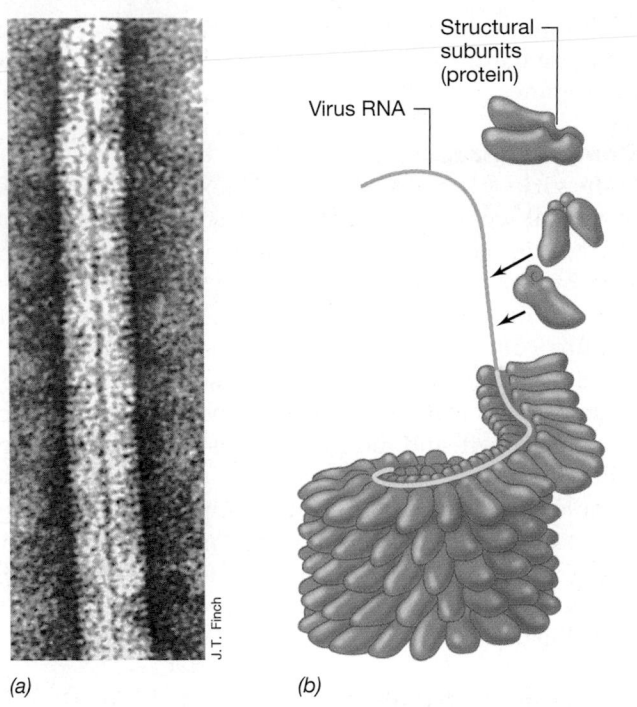

(a)　　　　　*(b)*

J.T. Finch

Figure 9.2 An example of the arrangement of virus nucleic acid and protein coat in a simple virus, tobacco mosaic virus. (a) Electron micrograph at high resolution of a portion of the virus particle. (b) Assembly of the tobacco mosaic virion. The RNA assumes a helical configuration surrounded by the protein capsid. The center of the particle is hollow.

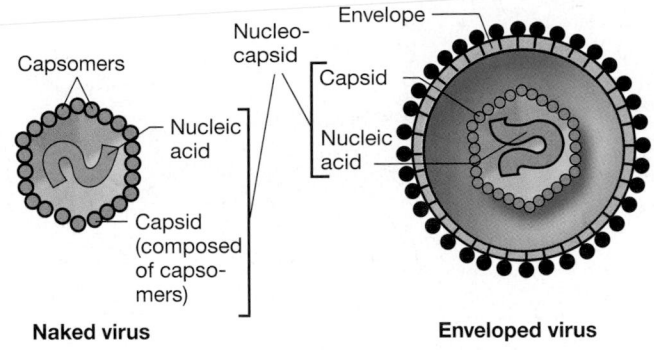

Naked virus　　　　　**Enveloped virus**

Figure 9.3 Comparison of naked and enveloped virus, two basic types of virus particles.

on are often one or more virus-specific *enzymes.* Such enzymes usually play a role during the infection and replication process, as we will discuss later in this chapter.

Although the virus structure just described is frequently the total structure of a virus particle, a number of viruses have more complex structures. In *enveloped* viruses the nucleocapsid is enclosed in a membrane (Figure 9.3). (Viruses without membranes are sometimes called *naked* viruses.) *Virus membranes* are generally lipid bilayer membranes (⊂⊃ Section 4.5), but associated with these membranes are often *virus-specific* proteins.

Virus Symmetry

The nucleocapsids of viruses are constructed in highly symmetric ways. Symmetry refers to the way in which the protein morphological units are arranged in the virus shell. When a symmetric structure is rotated around an axis, the same form is seen again after a certain number of degrees of rotation. Two kinds of symmetry are recognized in viruses, which correspond to the two primary shapes, rod and spherical. Rod-shaped viruses have helical symmetry, and spherical viruses have icosahedral symmetry. In all cases, the characteristic structure of the virus is determined by the structure of the protein subunits of which it is constructed.

A typical virus with **helical symmetry** is the tobacco mosaic virus (TMV) illustrated in Figure 9.2. It is an

RNA virus in which the 2130 identical protein subunits are arranged in a helix. The overall dimensions of the TMV virion are 18×300 nm. The lengths of helical viruses are determined by the length of the nucleic acid, but the width of the helical virus particle is determined by the size and packaging of the protein subunits.

An **icosahedron** is a symmetric structure roughly spherical in shape that has 20 faces. Icosahedral symmetry is the most efficient arrangement for subunits in a closed shell because it uses the smallest number of units to build a shell. The simplest arrangement of morphological units is 3 per face, for a total of 60 units per virus particle. The 3 units at each face can be either identical or different. Most viruses have more nucleic acid than can be packed into a shell made of just 60 morphological units. The next possible structure that permits close packing contains 180 units, and many viruses have shells with this configuration. Other known configurations involve 240 units and 420 units. Figure 9.4*a* shows a model of an icosahedron. Figure 9.4*b* shows an electron micrograph of a typical icosahedral virus (human papilloma virus; HPV), and Figure 9.4*c* shows a computer model of the same virus.

Enveloped Viruses

Many viruses have complex membranous structures surrounding the nucleocapsid (Figure 9.5*a*). Enveloped viruses are common in the animal world (for example, influenza virus), but enveloped bacterial and plant viruses are also known. The virus envelope consists of a lipid bilayer with proteins, usually glycoproteins, embedded in it. The lipids of the membrane are derived from the membranes of the host cell but the proteins are encoded by the virus. The symmetry of enveloped viruses is expressed not in terms of the virion as a whole but in terms of the nucleocapsid present inside the virus membrane.

What is the function of the membrane in a virus particle? We will discuss this in detail later but note that the membrane is the structural component of the virus particle that interacts first with the cell. The specificity of virus infection, and some aspects of virus penetration, are controlled in part by characteristics of virus membranes.

Complex Viruses

Some virions are even more complex, being composed of several separate parts, each with separate shapes and symmetries. The most complicated viruses in terms of structure are some of the bacterial viruses, which possess not only icosahedral heads but also helical tails (Figure 9.5*b*). In some bacterial viruses, such as the T4 virus of *Escherichia coli,* the tail itself has a complex structure. For instance, T4 has almost 20 different proteins in the tail, and the T4 head has several more proteins. In such complex viruses, assembly is also complex. For instance, in T4 the complete tail is formed as a subassembly, and then the tail is added to the DNA-containing head. Finally, tail fibers formed from another protein are added to make the mature, infectious virus particle.

Enzymes in Virions

We have stated that virions do not carry out metabolic processes. Outside a host cell, a virion is metabolically inert. However, some virions do contain enzymes that play roles in the infection process. Some of these enzymes are required for very early events in the infection process. For example, virions infecting some bacteria possess an enzyme, *lysozyme* (☞ Section 4.8), that makes a small hole in the bacterial cell wall that allows the viral nucleic acid to enter. (Lysozyme is produced in large amounts in the later stages of infection, causing lysis of the host cell and release of the virions.) Also, some viruses contain their own nucleic acid polymerases that transcribe the viral nucleic acid into messenger RNA (without using cellular enzymes) or because they require an enzyme the cell does not have. For example, retroviruses are RNA viruses that replicate inside the cell as DNA intermediates. These viruses possess an enzyme, an RNA-dependent DNA polymerase called *reverse transcriptase,* that transcribes the information in the incoming RNA into a DNA intermediate. Other viruses contain enzymes that aid in release of the virus from the host cells in which they were produced. One group of such enzymes, called *neuraminadases,* breaks down glycosidic bonds of glycoproteins and glycolipids of the connective tissue of animal cells, thus aiding in the liberation of the virus.

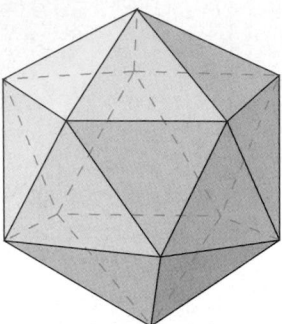

(a)

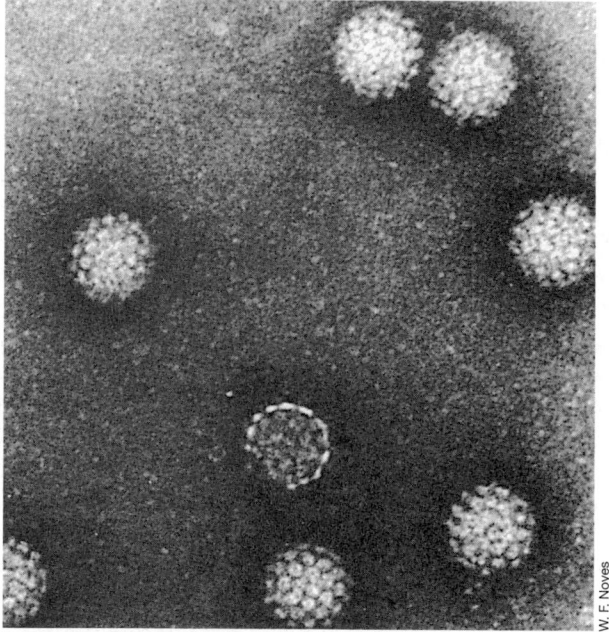

(b)

W. F. Noyes

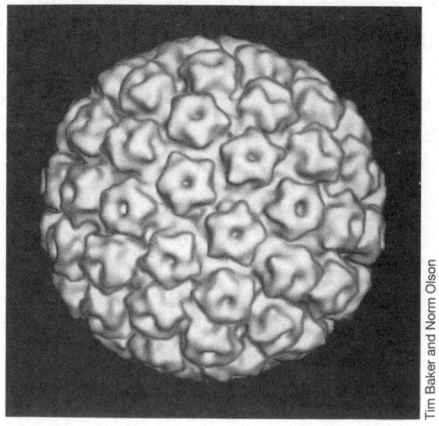

(c)

Tim Baker and Norm Olson

Figure 9.4 Icosahedral symmetry. (a) A model of an icosahedron. (b) Electron micrograph of human papilloma virus, a virus with icosahedral symmetry. The individual particles are about 55 nm in diameter. (c) Three-dimensional reconstruction of human papilloma virus calculated from images of frozen hydrated virions.

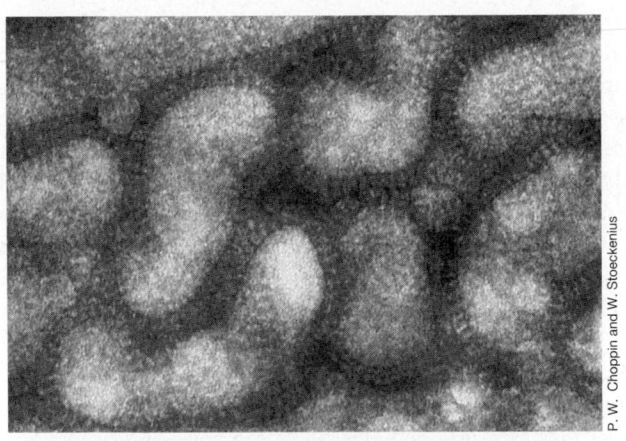

(a)

P. W. Choppin and W. Stoeckenius

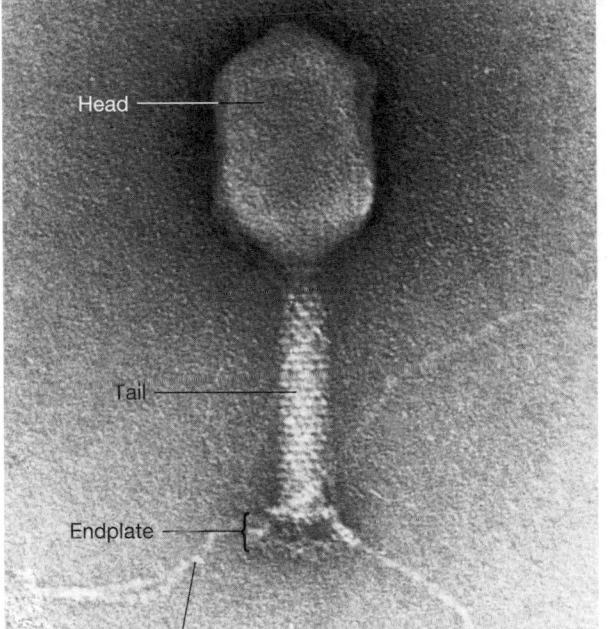

Head

Tail

Endplate

Tail fibers

M. Wurtz

(b)

Figure 9.5 Electron micrographs of animal and bacterial viruses. (a) Influenza virus, an enveloped virus. The individual particles are about 80 nm in diameter, but have no defined shape (∞ Figure 16.15). Influenza virus is discussed in Section 16.8. (b) Bacterial virus (bacteriophage) T4 of *Escherichia coli*. Note the complex structure. The tail components are involved in attachment of the virion to the host and injection of the nucleic acid (see Figure 9.10). The head is about 85 nm in diameter. Bacteriophage T4 is discussed in more detail in Section 9.9

✓ 9.2 Concept Check

In the virion of the naked virus, only nucleic acid (DNA or RNA) and protein are present, with the nucleic acid on the inside; the whole unit is called the nucleocapsid. Enveloped viruses have one or more lipoprotein layers surrounding the nucleocapsid. The nucleocapsid is arranged in a symmetric fashion, with a precise number and arrangement of structural subunits surrounding the virus nucleic acid. Although viruses are metabolically inert, in some viruses, one or more enzymes are present within the virion. Such enzymes play a role in the initial stages of the infection process.

✓ What is the difference between a *naked* virus and an *enveloped* virus?

✓ What kinds of enzymes can be found within the virions of specific viruses?

II GROWTH AND QUANTIFICATION

9.3 The Virus Host

Because viruses replicate only inside living cells, research on viruses requires use of appropriate hosts. Of the three types of hosts, bacteria, plants, and animals, viruses infecting bacteria are typically the easiest to grow in the laboratory.

For the study of bacterial viruses, pure cultures are used either in liquid or on semisolid (agar) media. Because many bacteria are so easy to culture, it is quite easy to study bacterial viruses, and this is why such detailed knowledge of bacterial virus multiplication is available.

Plant viruses can be more difficult to work with, since their study sometimes requires use of the whole plant. Plant viruses also often require a break in the thick plant cell wall in order to infect. Sometimes these viruses are assayed by mechanically damaging a plant leaf and applying virus to the damaged area. However, some plant viruses can be studied using protoplasts (∞ Section 4.8) or other cell culture techniques.

Although initially an animal virus may be grown in a whole animal that is susceptible to the virus, for research purposes a more manageable host is desirable. Fortunately many animal viruses can be cultivated in *tissue* or *cell cultures,* and the use of such cultures has enormously facilitated research on animal viruses.

Cell Cultures

A cell culture is obtained by promoting growth of cells taken from an organ of the experimental animal. Cell cultures are generally obtained by aseptically removing pieces of the tissue in question, dissociating the cells by treatment with an enzyme that breaks apart the intercellular cement, and spreading the resulting suspension out on the bottom of a flat surface, such as a bottle or a Petri dish. The cells generally produce glycoprotein-like materials that permit them to adhere to glass surfaces. The thin layer of cells adhering to the glass or plastic dish, called a *monolayer,* is overlaid with a suitable culture medium and the culture incubated. The culture media used for cell cultures are generally quite complex, employing a number of amino acids and vitamins, salts, glucose, and a bicarbonate buffer system. To obtain best growth, addition of a small amount of blood serum is usually necessary, and several antibiotics are generally added to prevent bacterial contamination. A colored pH indicator is added to help detect this contamination.

Some cell cultures prepared in this way grow indefinitely and can be established as *permanent cell lines.* Such cell cultures are most convenient for virus research because cell material is continuously available for research purposes. In other cases, indefinite growth does not occur, but the culture may remain alive for a number of days. Such cultures, called *primary cell cultures,* may still be useful for virus research, although new cultures will have to be prepared from fresh sources from time to time.

In some cases, cell culture monolayers cannot be obtained, but whole organs, or pieces of organs can be cultured. Such **organ cultures** may still be useful in virus research because they permit growth of viruses under more-or-less controlled laboratory conditions.

There is greater specificity between the virus and the host than simply determining to which kingdom or phylum the host belongs. In some cases the specificity is extreme; the virus may only be able to infect and/or multiply in a single species or subspecies and sometimes only some tissues within the host. In other cases, however, the specificity is broader. We will discuss some of the reasons for this specificity in the following sections and in Chapter 16.

✓ 9.3 Concept Check

Viruses can only replicate in certain types of cells or organisms. Bacterial viruses have proved useful model systems because the host cells are easy to grow and manipulate in culture. Many animal viruses can be grown in cultured cells.

✓ What is a host organism?

✓ Why is it helpful to use cell culture for viral research?

9.4 Quantification of Viruses

In order to obtain any significant understanding of the nature of viruses and virus replication, it is necessary to be able to *quantify* the number of virus particles. Virions are almost always too small to be seen under the light microscope. Although they can be observed under the electron microscope, the preparation of samples for ob-

servation can be too cumbersome to use simply for quantification. In general, viruses are quantified by measuring their effects on the host cells that they infect. In this context a *virus infectious unit* is the smallest unit that causes a detectable effect when placed with a susceptible host. By determining the number of infectious units per volume of fluid, a measure of virus quantity can be obtained. We discuss here several approaches to assessment of the virus infectious unit, limiting our discussion to bacterial and animal viruses.

Plaque Assay

When a virion initiates an infection on a layer or lawn of host cells growing on a flat surface, a zone of *lysis* or a zone of *growth inhibition* may occur that results in a clear area in the lawn of growing host cells. This clearing is called a **plaque**, and it is assumed that each plaque has originated from replication events that began with one virion.

Plaques are essentially "windows" in the lawn of confluent cell growth. With bacterial viruses, plaques may be obtained when virus particles are mixed into a thin layer of host bacteria that is spread out as an agar overlay on the surface of an agar medium (Figure 9.6*a*). During incubation of the culture, the bacteria grow and form a turbid layer that is visible to the naked eye. However, wherever a successful viral infection has been initiated, lysis of the cells occurs, resulting in the formation of a *plaque* (Figure 9.6*b*).

The plaque procedure also permits the isolation of pure virus strains because if a plaque has arisen from a single virion, all the virions in this plaque are probably genetically identical. Some of the virions from this plaque can be picked and inoculated into a fresh bacterial culture to establish a pure virus line. The development of the plaque assay technique was as important for the advance of virology as Koch's development of solid media (⟳ Section 1.5) was for bacteriology.

Plaques may be obtained for animal viruses by using animal cell culture systems as hosts. A monolayer of cultured animal cells is prepared on a plate or flat bottle, and the virus suspension overlaid. Plaques are revealed by zones of destruction of the animal cells (Figure 9.7).

In some cases, the virus may not actually destroy the cells but may cause changes in morphology or growth rate that can be recognized. For instance, tumor viruses may not destroy cells but may cause the cells to grow faster than uninfected cells, a phenomenon called **transformation**. As we have noted, the general arrangement of cells in a tissue culture is a monolayer. This is because growth generally ceases when the cells, as a result of growth, come in contact with each other (a phenomenon known as *contact inhibition*). Transformed cells have altered growth requirements and continue to grow, piling up to form a small *focus of growth* (called a *focus of infection* when the transformation has been brought about

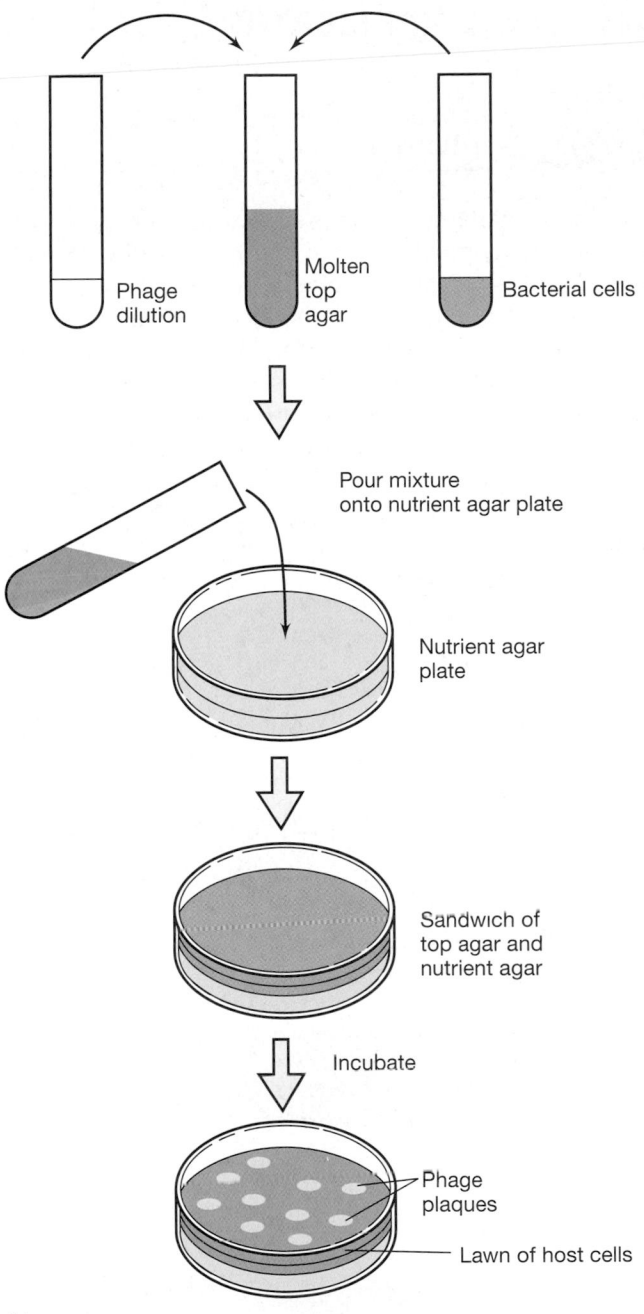

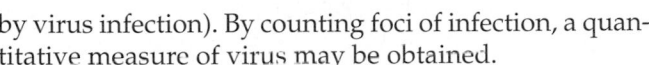

(a)

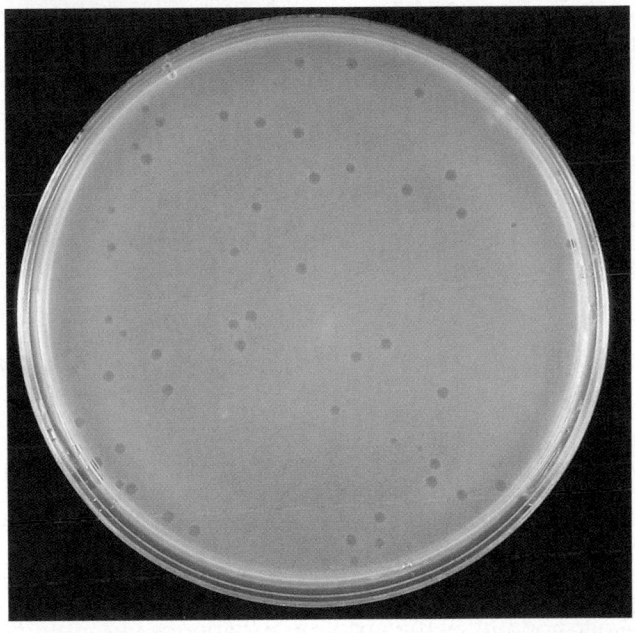

(b)

Figure 9.6 Quantification of bacterial virus by plaque assay using the agar overlay technique. (a) A dilution of a suspension containing the virus material is mixed in a small amount of melted agar with the sensitive host bacteria, and the mixture poured on the surface of a nutrient agar plate. The host bacteria, which have been spread uniformly throughout the top agar layer, begin to grow, and after overnight incubation form a *lawn* of confluent growth. Each virus particle that attaches to a cell and reproduces may cause cell lysis, and the virus particles released can spread to adjacent cells in the agar, infect them, be reproduced, and again lead to lysis and release. The size of the plaque formed depends on the virus, the host, and conditions of culture. (b) Photograph of a plate showing plaques formed by bacteriophage on a lawn of sensitive bacteria. The plaques shown are about 1–2 mm in diameter.

by virus infection). By counting foci of infection, a quantitative measure of virus may be obtained.

Efficiency of Plating

One important concept in quantitative virology involves the idea of *efficiency of plating*. Counts made by plaque assay are always lower than counts made with the electron microscope. The efficiency with which virions infect host cells is rarely 100% and may often be considerably less. This does not mean that virions that have not caused infection are inactive, although this is sometimes the case. It may merely mean that under the conditions used, suc-

cessful infection with these particles has not occurred. Although with bacterial viruses, efficiency of plating is often higher than 50%, with many animal viruses it may be very low, 0.1 or 1%. Why virus particles vary in infectivity is not well understood. In some cases it is possible that the conditions used for quantification are not optimal. Because the electron microscope is not routinely used to count virions, it is sometimes difficult to assess the actual efficiency of plating, but the concept is important in both research and medical practice. Because the efficiency of plating is rarely close to 100%, when the plaque method is used to quantify virus, it is accurate to

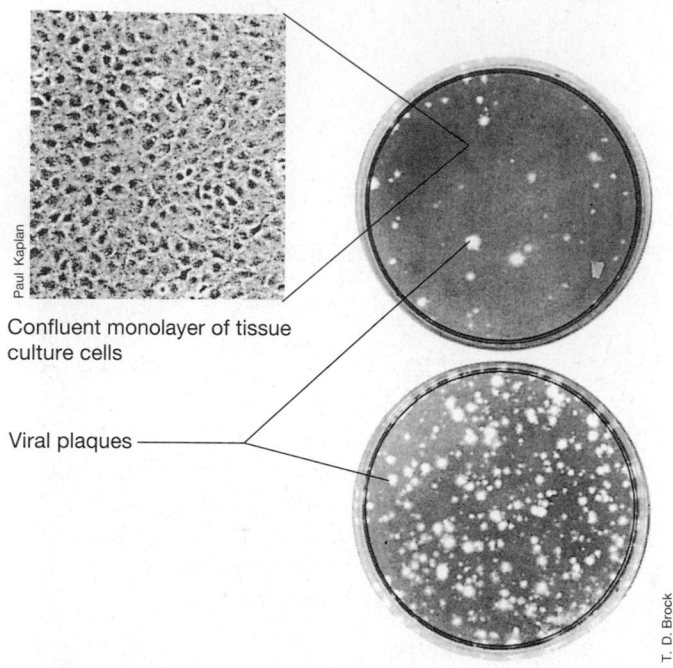

Confluent monolayer of tissue culture cells

Viral plaques

Paul Kaplan

T. D. Brock

Figure 9.7 Cell cultures in monolayers within Petri plates. Note the presence of plaques where virus-induced cell lysis has occurred. Also shown is a photomicrograph of a cell culture.

express the concentration (called the *titer*) of the virus suspension not as the absolute number of virion units but as the number of *plaque-forming units*.

Animal Infectivity Methods

Some viruses do not cause recognizable effects in cell cultures but cause death in the whole animal. In such cases, quantification can be done only by some sort of titration in infected animals. The general procedure is to carry out a serial dilution of the unknown sample (Section 6.5), generally at 10-fold dilutions, and to inject samples of each dilution into numbers of sensitive animals. After a suitable incubation period, the fraction of dead and live animals at each dilution is tabulated and an *end point dilution* is calculated. This is the dilution at which, for example, *half* of the injected animals die. Although such serial dilution methods are much more cumbersome and much less accurate than cell culture methods, they may be essential for the study of certain types of viruses.

✓ 9.4 Concept Check

Although it requires only a single virion to initiate an infectious cycle, not all virus particles are equally infectious. One of the most accurate ways of measuring virus infectivity is by the plaque assay. Plaques are clear zones that develop on layers or lawns of host cells, each plaque due to infection by a single virus particle. The virus plaque is analogous to the bacterial colony.

✓ Give a definition of *efficiency of plating*.

✓ What is a *plaque-forming unit*?

III VIRAL REPLICATION

9.5 General Features of Virus Replication

The basic challenge in virus replication can be simply put: The virus must induce a living host cell to synthesize all the essential components needed to make more virus particles. These components must then be assembled into the proper structure, and the new virions must escape from the cell in order to infect other cells. The various phases of this replication process in a bacteriophage can be categorized in five steps (Figure 9.8).

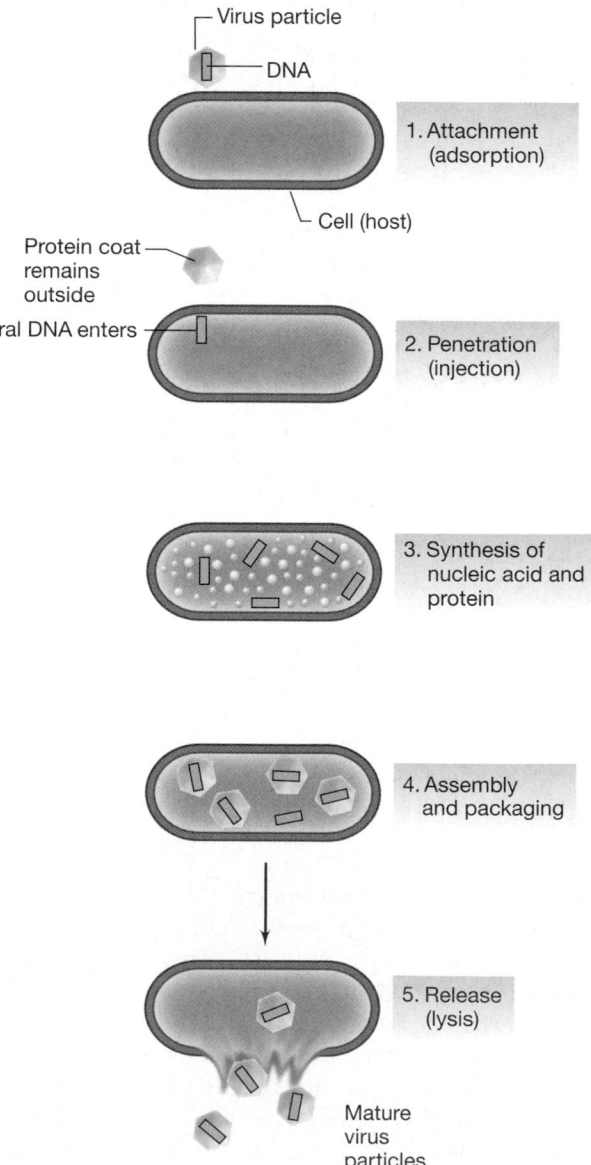

Virus particle

DNA

1. Attachment (adsorption)

Cell (host)

Protein coat remains outside

Viral DNA enters

2. Penetration (injection)

3. Synthesis of nucleic acid and protein

4. Assembly and packaging

5. Release (lysis)

Mature virus particles

Figure 9.8 The replication cycle of a bacterial virus. The general stages of virus replication are indicated.

1. *Attachment* (adsorption) of the virion to a susceptible host cell.

2. *Penetration* (injection) of the virion or its nucleic acid into the cell.

3. *Synthesis of nucleic acid and protein* takes place from early to late times during infection. Early in infection the virus redirects cell metabolism to synthesize new viral nucleic acid and proteins. Late in infection, structural proteins that are subunits of the virus coat are synthesized.

4. *Assembly* of structural subunits (and membrane components in enveloped viruses) and packaging of nucleic acid into new virus particles.

5. *Release* of mature virions from the cell.

These stages in virus replication are recognized when virus particles infect cells in culture and are illustrated in Figure 9.9, which exhibits what is called a **one-step growth curve**. In the first few minutes after infection the virus is said to undergo an *eclipse*. The virus nucleic acid has become separated from its protein coat, and so even if the infected cell had broken open, the virion no longer exists as an infectious entity. Although virus nucleic acid may be infectious, the infectivity of virus nucleic acid is many times lower than that of whole virions because the machinery for bringing the virus genome into the cell is lacking. Also, outside the virion the nucleic acid is no longer protected from deleterious activities of the environment as it was when it was inside the protein coat.

The eclipse occurs during the early stages of virus replication. Maturation begins as the newly synthesized nucleic acid molecules become packaged inside protein coats. During the *maturation* phase, the titer of active virions inside the cell rises dramatically. The period of time when no infectious virions are present extracellularly is called the *latent period*. At the end of maturation, *release* of mature virions occurs, either as a result of cell *lysis* or because of some budding or excretion process. The number of virions released, called the *burst size*, varies with the particular virus and the particular host cell and can range from a few to a few thousand. The timing of this overall virus replication cycle varies from 20–60 min in many bacterial viruses to 8–40 h in most animal viruses. In the next two sections we will consider a few of the steps of the virus multiplication cycle in more detail.

✓ 9.5 Concept Check

The virus life cycle can be divided into five stages: attachment (adsorption), penetration (injection), protein and nucleic acid synthesis, assembly and packaging, and virus release.

✓ What is *packaged* into the virions?

✓ To what does *eclipse* refer?

9.6 Virus Multiplication: Attachment and Penetration

In this section we discuss in more detail virus attachment and penetration, the first steps in virus multiplication. In addition, we consider the mechanism by which some bacteria react to penetration to bacteriophage DNA.

Attachment

High specificity characterizes the interaction between virus and host. The most common basis for host specificity involves the attachment process. The virus particle itself (whether it is naked or enveloped) has one or more proteins on the outside that interact with specific cell surface components called *receptors*. The receptors on the cell surface are normal surface components of the host, such as proteins, carbohydrates, glycoproteins, lipids, lipoproteins, or complexes of these, to which the virion attaches. The receptors carry out normal functions for the cell. For example, the receptor for the bacteriophage T1 is an iron-uptake protein and that for the bacteriophage lambda is involved in maltose uptake. Animal virus receptors may include macromolecules involved in cell–cell contact or the immune system.

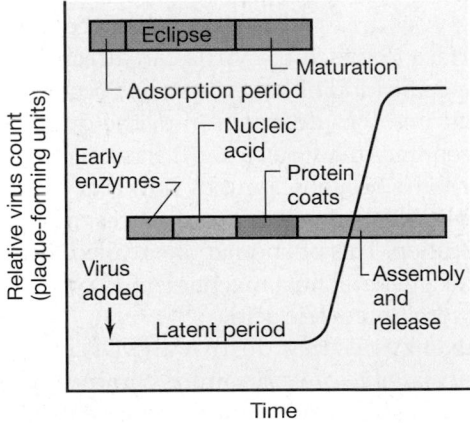

Figure 9.9 The one-step growth curve of virus replication. This graph displays the results of a single round of viral multiplication in a population of cells. Following adsorption, the infectivity of the virus particles disappears, a phenomenon called *eclipse*. This is due to the uncoating of the virus particles. During the *latent period*, replication of viral nucleic acid and protein occurs. The *maturation period* follows, when virus nucleic acid and protein are assembled into mature virus particles. At this time, if the cells are broken open, active virus can be detected. Finally, *release* occurs, either with or without cell lysis. The timing of the one-step growth cycle varies with the virus and host. Compare this general picture and color scheme with specific replication events shown for bacteriophage T4 in Figure 9.15.

Closely related viruses do not necessarily share closely related receptors. For example, the receptor for bacteriophage T2 is a membrane protein while that for the very closely related bacteriophage T4 is a polysaccharide component of the outer cell membrane (see below). Some receptors are recognized by a number of different viruses.

Receptors determine which cells will be **susceptible** to infection. In multicellular organisms receptors may be expressed only in certain cell types. The receptor for influenza virus is a specific carbohydrate, sialic acid, found as the terminal part of an oligosaccharide covalently attached to a membrane protein. This glycoprotein is found on many cell types, including red blood cells, so influenza virus can attach to many different cells. Other receptors are found on only a few types of cells, such as the CD4 protein found only on certain cells of the immune system and used as a receptor by human immunodeficiency virus (HIV), which restricts the types of cells that HIV can infect.

In the absence of the receptor site, the virus cannot adsorb, and hence cannot infect. If the receptor site is altered, the host may become resistant to virus infection. However, mutants of the virus can also arise that are able to adsorb to resistant hosts. In addition, some animal viruses may be able to use more than one receptor, so the loss of one may not prevent attachment.

Penetration

The attachment of a virus to the cell results in changes to the virus and/or the cell that in turn result in *penetration*. Viruses must replicate within cells, therefore, at a minimum the viral genome must enter the cell. However, we mentioned (see Section 9.2) that for some viruses to replicate, certain proteins must also enter the host cell. It is important to note that attachment to and even penetration of a susceptible cell will not lead to virus multiplication if the information in the viral genome cannot be read. A cell that allows multiplication of a virus to take place is said to be **permissive** for that virus.

Different viruses have different strategies for penetration. For some animal viruses the enveloped virus is uncoated at the cell membrane. In the case of many animal viruses the entire virion enters the cell by endocytosis. In such cases the virus must subsequently be *uncoated* (partially or completely) so that the genome has access to the cell and replication can proceed. In some this uncoating takes place in the cytoplasm, while in others uncoating occurs at the nuclear membrane.

Cells with cell walls, such as bacteria, are infected in a different manner from animal cells, which lack a cell wall. The most complicated penetration mechanisms have been found in viruses that infect bacteria. The bacteriophage T4, which infects *Escherichia coli,* can be used as an example.

The structure of the bacterial virus T4 was shown in Figure 9.5*b*. The virion has a **head**, within which the viral DNA is folded, and a long, fairly complex **tail**, at the end of which is a series of tail fibers. During the attachment process, the virions first attach to cells by means of the tail fibers (Figure 9.10). The ends of the fibers interact specifically with core polysaccharides that are part of the outer layer of the gram-negative cell wall (⚭ Section 4.8). These tail fibers then retract, and the core of the tail makes contact with the cell wall of the bacterium. The action of a lysozyme-like enzyme results in the formation of a small hole. The tail sheath contracts, and the DNA of the virus passes into the cell through a hole in the tip of the tail, the majority of the coat protein remaining outside.

Virus Restriction and Modification by the Host

Multicellular animals can often deal with and eliminate invading viruses by a variety of immune defense mechanisms before the infection becomes widespread or sometimes before the virus has penetrated target cells. We will discuss such mechanisms in Chapters 21 and 22. Prokaryotes lack these defenses. While the extremely thick cell walls of fungi and plants also offer some protection, the relatively thinner cell walls of prokaryotes offer much less protection. Prokaryotes do, however, also develop resistance to viruses.

We have already seen that one form of host resistance to virus arises when there is no receptor site on the cell surface to which the virus can attach. Another and more specific kind of host resistance occurs in prokaryotes and involves destruction of the double-stranded DNA genome of a virus after it has been injected. This destruction is brought about by host enzymes that cleave the viral DNA at one or several places, thus preventing its replication. This phenomenon is called **restriction** and is part of a general host mechanism to prevent the invasion of foreign nucleic acid.

The enzymes that destroy the DNA are called *restriction endonucleases* or, more commonly, *restriction enzymes*. Restriction enzymes are highly specific, attacking only certain sequences (generally four or six base pairs). For such a system to be effective, the host must have a mechanism for protecting its own DNA. This is accomplished by specific **modification** of its DNA at the sites where the restriction enzymes act. Modification of host DNA is brought about by methylation of purine or pyrimidine bases (in such a way that their base-pairing properties are not altered). DNA is usually modified at specific bases on each strand. During semiconservative replication (⚭ Section 7.5) there will be a period in which only the

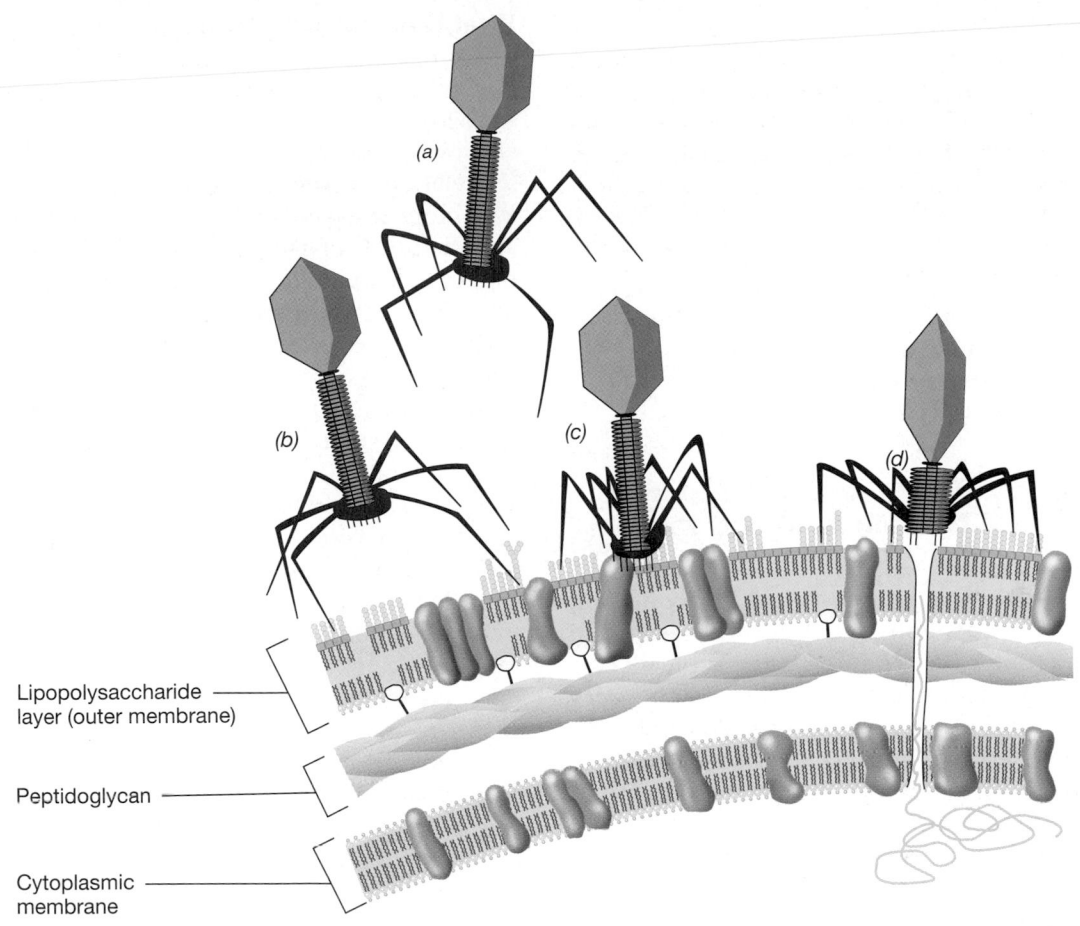

Lipopolysaccharide
layer (outer membrane)

Peptidoglycan

Cytoplasmic
membrane

Figure 9.10 Attachment of T4 bacteriophage virion to the cell wall of *Escherichia coli* and injection of DNA: (a) Unattached virion. (b) Attachment to the wall by the long tail fibers interacting with core polysaccharide. (c) Contact of cell wall by the tail pins. (d) Contraction of the tail sheath and injection of the DNA. For a detailed description of the gram-negative cell wall, see Section 4.9.

template strand will be modified. A typical restriction/modified enzyme system recognizes this DNA as needing further modification, not as DNA that should be restricted.

Restriction enzymes are specific for double-stranded DNA, and bacteriophages containing double-stranded DNA are fairly widespread. Consequently, restriction enzymes are also widely distributed in prokaryotes. We will discuss the great practical uses of restriction enzymes in genetics and genetic engineering later (Section 10.12 and Chapter 31).

Some viruses can overcome host restriction mechanisms by modifications of their nucleic acids so that they are no longer subject to enzymatic attack. Two kinds of chemical modifications of viral DNA have been recognized: *glucosylation* and *methylation*. For instance, the T-even bacteriophages (T2, T4, and T6) have their DNA glucosylated to varying degrees, and the glucosylation prevents or greatly reduces endonuclease attack. Many other viral nucleic acids have been found to be modified by methylation, but glucosylation has been found only

in the T-even bacteriophages. It should be emphasized that modification of viral nucleic acid occurs after replication has occurred. Other viruses, such as the bacteriophages T3 and T7, avoid restriction by encoding proteins that inhibit the host restriction systems. Some hosts have multiple restriction and methylation systems that help in preventing infection by viruses that can circumvent only one of them.

However, not all restriction systems recognize unmodified DNA. Host restriction systems are also known that restrict only *modified* DNA! Clearly the host containing this enzyme does *not* contain the modification enzyme. However, this host is protected from infection by a virus that was modified during reproduction in its previous host strain.

Hosts also contain other DNA methylases in addition to those involved in protecting the host from its own restriction enzymes. Some of these methylases may be involved in DNA repair or in gene regulation, but others may offer protection to *host* DNA. This is because *some viruses themselves encode restriction systems.*

✓ **9.6** *Concept Check*

The attachment of a virion to a host cell is a highly specific process involving the interaction of proteins on the surface of the virus particle to receptors on the surface of a susceptible host cell. Only after attachment has occurred can the virion or its genome penetrate the host cell. Resistance of the host to infection by the virus can occur in a number of ways. Prokaryotic hosts may contain restriction-modification systems that recognize and destroy foreign DNA.

✓ How does the attachment process contribute to virus–host specificity?

✓ Why do some viruses need to be uncoated after penetration while others do not?

9.7 Virus Multiplication: Production of Viral Nucleic Acid and Protein

New copies of the viral genome must be replicated, and virus-specific proteins must be synthesized in order for virus multiplication to occur. Typically the production of at least some viral proteins begins very early after the viral genome has been taken up by the cell (and uncoated if required). The production of these proteins requires virus-specific messenger RNA. For certain types of RNA viruses the genome itself can serve as mRNA. In the case of most viruses, however, the mRNA must first be synthesized. Many viruses have double-stranded DNA genomes (like those of the host cell) and we have discussed the essential features of producing mRNA from double-stranded DNA (∞ Sections 7.8 to 7.11).

However, a great many viruses have other types of genomes, and these include not only single-stranded DNA but both single- and double-stranded RNA. Furthermore, we have mentioned that some viruses have one type of nucleic acid in the virion but use another as a replicative intermediate. All these "unusual" genomes present problems in understanding virus multiplication because they involve information transfers, such as RNA to RNA and RNA to DNA, that host enzymes do not perform.

Viral Replication Schemes

The virologist David Baltimore, one of the winners of the Nobel Prize in Medicine in 1975 for the discovery of retroviruses and reverse transcriptase, developed a classification scheme for viruses that is based primarily on the relationship of the viral genome to its mRNA. In the Baltimore Classification scheme (Table 9.2), double-stranded (ds) DNA viruses are in Class I. The production of mRNA from such viruses can occur as it would from the host genome, although different viruses use different strategies to ensure that viral mRNA is made in preference to host mRNA. Class II viruses are single-stranded (ss) DNA viruses. Before mRNA can be produced from such viruses, a complementary DNA strand must be synthesized since RNA polymerase uses double-stranded DNA (∞ Sections 7.8 and 7.9). These viruses have a double-stranded DNA intermediate during replication, and it is this intermediate that is used for transcription (Figure 9.11). The synthesis of the ds DNA intermediate and subsequent transcription can be carried out by cellular enzymes (although viral proteins may also be involved). The situation is much different with viruses with RNA genomes (Classes III–VI).

Before discussing these latter in more detail, some nomenclature related to nucleic acid strand orientation is needed. Remember that mRNA is complementary to the strand of DNA that was used as its template and that the mRNA can be translated to yield protein (∞ Chapter 7). By convention, mRNA is said to be in the *plus* (+) configuration. Its complement is said to be in the *minus* (–) configuration. This nomenclature can be used to discuss the orientation of any single-stranded virus (RNA or DNA) but is primarily used for RNA viruses. A virus that has a single-stranded RNA genome with the same orientation as its mRNA is said to be a *positive-strand RNA* virus. A virus whose single-stranded RNA genome is complementary to its mRNA is said to be a *negative-strand RNA* virus. (Note positive and negative are not absolute terms, but are used only relative to the mRNA of the virus.)

In addition, consider that cellular RNA polymerases do not generally catalyze the formation of RNA from

---

TABLE 9.2 The Baltimore Classification system of viruses

Class	Description of genome and replication strategy	Examples	
		Bacterial viruses	**Animal viruses**
I	Double-stranded DNA genome	Lambda, T4	Herpesvirus, pox virus
II	Single-stranded DNA genome	φX174	Chicken anemia virus
III	Double-stranded RNA genome	φ6	Reoviruses (∞ Section 16.9)
IV	Single-stranded RNA genome of plus sense	MS2	Poliovirus
V	Single-stranded RNA genome of minus sense		Influenza virus, rabies virus
VI	Single-stranded RNA genome that replicates with DNA intermediate		Retroviruses
VII	Double-stranded DNA genome that replicates with RNA intermediate		Hepatitis B virus

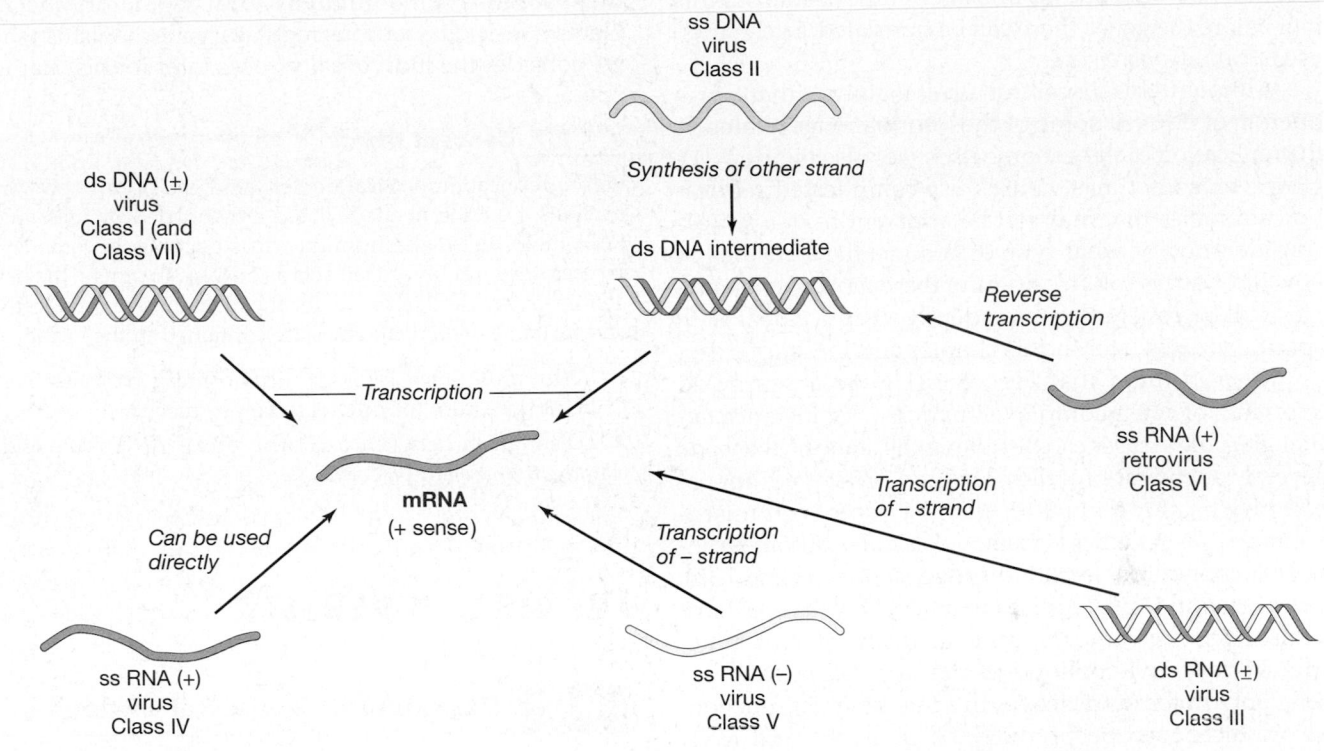

Figure 9.11 Formation of mRNA after infection of cells by viruses of different types. The chemical sense of the mRNA is considered as plus (+). The senses of the various virus nucleic acids are indicated as + if the same as mRNA, as − if opposite, or as ± if double-stranded. Almost all single-stranded DNA viruses are of the + sense, although in a few cases apparently either the + or − strand can be packaged. It is not completely clear if a virion with a − strand is infectious. The different classes of viruses in the Baltimore Classification are shown (see Table 9.2).

an RNA template, but instead require a DNA template. Therefore, RNA viruses require a specific RNA-dependent RNA polymerase. Let us consider this requirement in the light of the strandedness of RNA virus genomes. The simplest case is the positive-strand RNA viruses (Class IV) in which the single incoming viral RNA strand is the plus strand and hence serves directly as mRNA. In addition to the other required proteins, this mRNA encodes a virus-specific RNA polymerase. This polymerase first makes complementary minus strands and then uses them as templates to make more plus strands. For negative-strand RNA viruses (Class V) (whose virion contains only the minus strand) or double-stranded RNA viruses (Class III), the situation is more complicated. In neither case can the incoming RNA serve as mRNA, and therefore mRNA must be synthesized first. However, as mentioned earlier, cells do not typically have an RNA polymerase capable of this. To circumvent this problem, these viruses contain some of this enzyme in their virions, and it is injected into the cell along with the genomic RNA. Therefore, in these cases, the complementary plus strand is synthesized by this RNA-dependent RNA polymerase and used as message.

Retroviruses (causal agents of certain kinds of cancers and acquired immunodeficiency syndrome, AIDS) are RNA viruses that replicate through a DNA intermediate (Class VI). The process of copying the information found in RNA into DNA is called **reverse transcription**, and thus these viruses require an enzyme called **reverse transcriptase**. (Telomerase is a type of reverse transcriptase; ⟳ Section 7.7.) Although the incoming RNA of retroviruses is the plus strand, it is not used as message, and therefore these viruses must carry reverse transcriptase in their virions. After infection, the virion ss RNA is copied to a double-stranded DNA (through an ss DNA intermediate), and the ds DNA then serves as the template for mRNA synthesis (thus, ss RNA → ss DNA → ds DNA). Another class of viruses, Class VII, are viruses that have double-stranded DNA in their virions but replicate through an RNA intermediate and, therefore, also use reverse transcriptase (⟳ Section 16.14). The strategy these viruses use to produce mRNA is, however, the same as that of Class I viruses (Figure 9.11).

While this may seem to cover all possibilities, there are viruses known that have a single-stranded RNA genome, half of which is in the plus orientation (and can be used as mRNA) and half in the minus configuration.

A complementary strand must be synthesized from this half before the genes there can be translated. Such viruses are called *ambiviruses.*

Although this discussion dealt mainly with the production of mRNA, some of the information is related to the replication of the genome itself (see also Figure 9.11). However, sometimes viruses use complicated replication strategies that may not be apparent from a simple consideration of what type of genome the virus has or how it produces mRNA. Some of these replication strategies will be covered in more depth when we deal with specific viruses later in this chapter and in Chapter 16.

Keep in mind that Figure 9.11 gives a simplified overview of often complex situations. For instance, although some viruses can be quite small, almost all encode several functional proteins. This can present additional complexities for animal (or plant) viruses with unsegmented RNA genomes because eukaryotic ribosomes cannot read polycistronic mRNA (∞ Sections 7.11 and 7.15). Positive-strand RNA viruses that infect eukaryotes have viruses that overcome this difficulty with different strategies. For example, poliovirus translates the mRNA as a long polyprotein and cleaves this into different functional products (∞ Section 16.7). Negative-strand RNA viruses usually produce subgenomic mRNAs, that is, short plus strands are synthesized that encode single proteins. Of course, a full-length plus strand must be synthesized as part of genome replication.

Viral Proteins

Once viral mRNA is made, viral proteins (for example, enzymes and structural subunits) can be synthesized. The proteins synthesized as a result of virus infection can be grouped into two broad categories:

1. Proteins (usually enzymes) synthesized soon after infection, called the **early proteins**, which are necessary for the replication of virus nucleic acid

2. Proteins synthesized later, called the **late proteins**, which include the proteins of the virus coat

Generally, both the time of appearance and the amount of virus proteins are regulated. The early proteins are enzymes that, because they act catalytically, are synthesized in smaller amounts, and the late proteins, often structural, are made in much larger amounts.

Virus infection upsets the regulatory mechanisms of the host because there is a marked overproduction of viral nucleic acid and protein in the infected cell. In some cases, virus infection causes a complete shutdown of host macromolecular synthesis, whereas in other cases host synthesis proceeds concurrently with virus synthesis. In either case, the regulation of virus synthesis is under the control of the virus rather than the host. There are several elements of this control that are similar to the host regulatory mechanisms discussed in Chapter 8, but there are also some uniquely viral regulatory mechanisms. We discuss various regulatory mechanisms when we consider the individual viruses later in this chapter.

✓ 9.7 Concept Check

Before replication of viral nucleic acid can occur, new virus proteins are often needed. These are encoded by messenger RNA molecules made from the virus genome. In some RNA viruses, the viral RNA itself acts as mRNA. In others, the virus genome serves as a template for the formation of viral mRNA and certain essential enzymes are contained in the virion.

✓ Why must some types of virus contain enzymes in the virion in order for mRNA to be produced?

✓ Distinguish between a *positive-strand RNA virus* and a *negative-strand RNA virus.*

IV VIRAL DIVERSITY

9.8 Overview of Bacterial Viruses

Bacterial viruses, also called bacteriophages or phages, are quite diverse. Various kinds of bacterial viruses are illustrated in Figure 9.12. Most of the bacterial viruses that have been studied in detail infect *Bacteria* of the enteric group, such as *Escherichia coli* and *Salmonella typhimurium.* However, viruses are known that infect a variety of prokaryotes, both *Bacteria* and *Archaea,* and as we mentioned, the virus with the largest known genome, bacteriophage G (for giant), infects *Bacillus megaterium* (see Section 9.2). Most well-studied bacteriophages contain double-stranded DNA genomes, and this type of bacteriophage is thought to be the most common type in the environment. However, there are certainly many other kinds known, including those with single-stranded RNA genomes, segmented double-stranded RNA genomes, and single-stranded DNA genomes.

A few bacterial viruses have lipid envelopes but most do not. However, many bacterial viruses are structurally complex. All examples of bacteriophages with double-stranded DNA genomes shown in Figure 9.12 have tails. The tails of bacteriophages T2, T4, and Mu are contractile and are involved in nucleic acid penetration of the host (Figure 9.10). The tail of bacteriophage lambda is flexible.

Although these bacterial viruses were first studied as *model systems* for understanding general features of virus multiplication, some of them now serve as convenient tools for *genetic engineering* (discussed in (∞ Chapter 31). Thus, the information on bacterial viruses is not only valuable as background for the discussion of animal viruses but also is essential for

The Phage Group

Historically, the T-even phages provided the study material for early research of the "Phage Group," a group of research workers from various universities and research institutions who spent their summers working together at the Cold Spring Harbor Laboratory on Long Island. The key members of the Phage Group, Max Delbrück, Salvador Luria, and A. D. Hershey, subsequently shared the Nobel Prize for their pioneering work. Among important concepts first uncovered from research on the T-even phages: Only the nucleic acid of the virus entered the cell during infection (a discovery that provided key support for the hypothesis that DNA is the genetic material); the existence of genetic recombination in viruses; the phenomenon of restriction and modification (which led to the discovery of the restriction enzymes so important for genetic engineering); the presence in viruses of unique virus-encoded gene functions; the distinction between early and late viral functions. The first ideas of how viruses cause killing of host cells were also developed from research on the T-even phages. Selecting the T-even phages as a model system and concentrating work on them were greatly responsible for the remarkable success of the Phage Group. ■

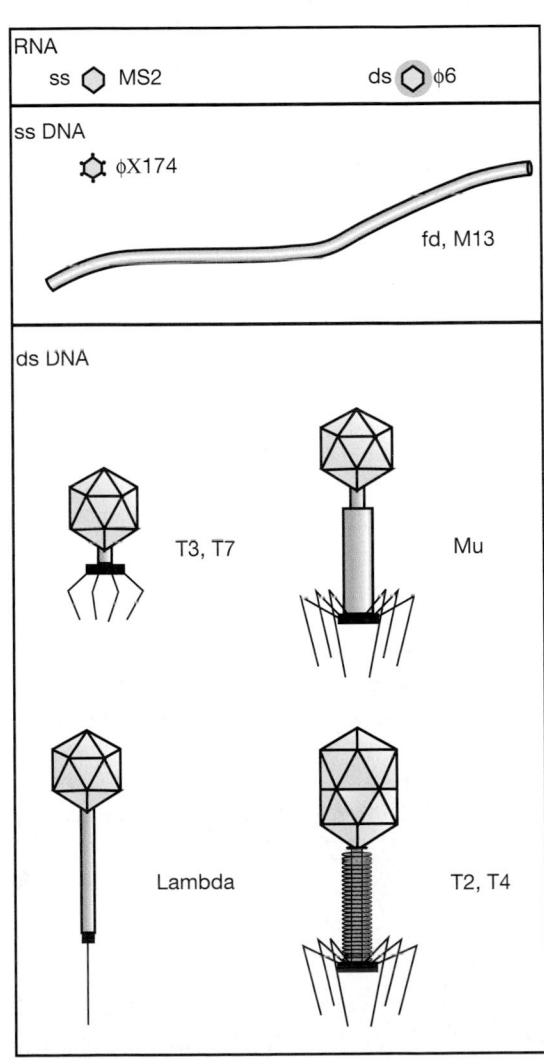

Figure 9.12 Schematic representations of the main types of bacterial viruses. Those discussed in detail are M13, φX174, MS2, T4, lambda, T7, and Mu. Sizes are to approximate scale. The nucleocapsed of φ6 is surrounded by a membrane.

the material presented in the chapters on microbial genetics (∞ Chapter 10) and genetic engineering (∞ Chapter 31).

In the next two sections of this chapter we will briefly examine two different types of viral life cycles: **virulent** and **temperate**. Virulent viruses lyse or kill their hosts after infection, while temperate viruses are able to achieve a state where their genome replicates along with the host genome without killing their hosts.

✓ 9.8 Concept Check

Bacterial viruses, or bacteriophages, are very diverse. The best-studied bacteriophages infect enteric *Bacteria* such as *Escherichia coli* and are structurally quite complex.

✓ What is thought to be the most common type of bacteriophage genome?

9.9 Virulent Bacteriophage: T4

As we have mentioned, a virulent virus is one whose replication cycle results in the destruction of the cell, typically by cell lysis. The first viruses to be studied in any detail were a number of bacteriophages with linear, double-stranded DNA genomes that infect *Escherichia coli* and a number of related *Bacteria*. A group of scientists began studying these viruses as model systems and used them to establish many of the fundamental principles of molecular biology and genetics (see the box, The Phage Group). These phages were given designations of T1, T2, and so on, up to T7. Already in this chapter we have briefly mentioned how one of these viruses, T4, attaches to host and how its DNA penetrates the host (see Section 9.6, Figure 9.10). In this section we discuss this virus in more detail to illustrate the replication cycle of virulent viruses.

MEDIA TUTORIAL Phage Genetics

Bacteriophages T2, T4, and T6 are very closely related, but T4 is the most extensively studied. The virion of phage T4 is structurally complex (see Figure 9.5b). It consists of an elongated icosahedral head whose overall dimensions are 85 × 110 nm. To this head is attached a complex tail consisting of a helical tube (25 × 110 nm) to which are connected a sheath, a connecting "neck" with "collar," and a complex endplate, to which are attached long, jointed tail fibers (see Figure 9.5b). All together, the virus particle has over 25 distinct types of proteins.

The genome of T4 is a double-stranded, linear DNA molecule of 168,903 base pairs. The genome encodes over 250 different proteins, and although no known virus encodes its own translational apparatus, T4 does encode several different tRNAs.

While the T4 genome has been determined to have a unique linear sequence, the genome in a given individual virion actually differs from the sequence one would find in another virion. First, comparison of the DNA from various T4 virions shows that the DNA is *circularly permuted*. Molecules that are related by circular permutation all appear to have been cut from a circle, but in each case the cut was at a different place. In addition, the DNA in each virion has repeated sequences at each end (*terminal repeats*) of about 3000 to 6000 base pairs. The packaging mechanism of T4 DNA involves cutting DNA from a long DNA molecule containing several genome equivalents linked end-to-end, a molecule known as a *concatemer* (Figure 9.13). This concatemer is cut after the head is full and not at a specific sequence, a headful mechanism. Since the T4 head holds slightly more than a genome length, this headful mechanism leads to circular permutation and the terminal redundancy. Circular permutation leads to the interesting result that although the T4 genome is linear, the genetic map of this organism is circular. T4 DNA contains the modified base 5-hydroxymethylcytosine instead of cytosine (Figure 9.14). It is these residues that are glucosylated (see Section 9.6), and DNA with this modification is resistant to virtually all known restriction enzymes; therefore the incoming DNA is well protected from host defenses.

Events During T4 Infection

Early in infection T4 must direct the synthesis of RNA and then protein from its own genome and also replicate DNA that contains a base unlike that found in the host. One minute after attachment and penetration by T4, the synthesis of host DNA, RNA, and protein has ceased, but transcription of certain phage genes has begun. Translation of these proteins begins soon after, and by 4 min phage DNA replication has begun. The time course of events during T4 infection is shown in Figure 9.15.

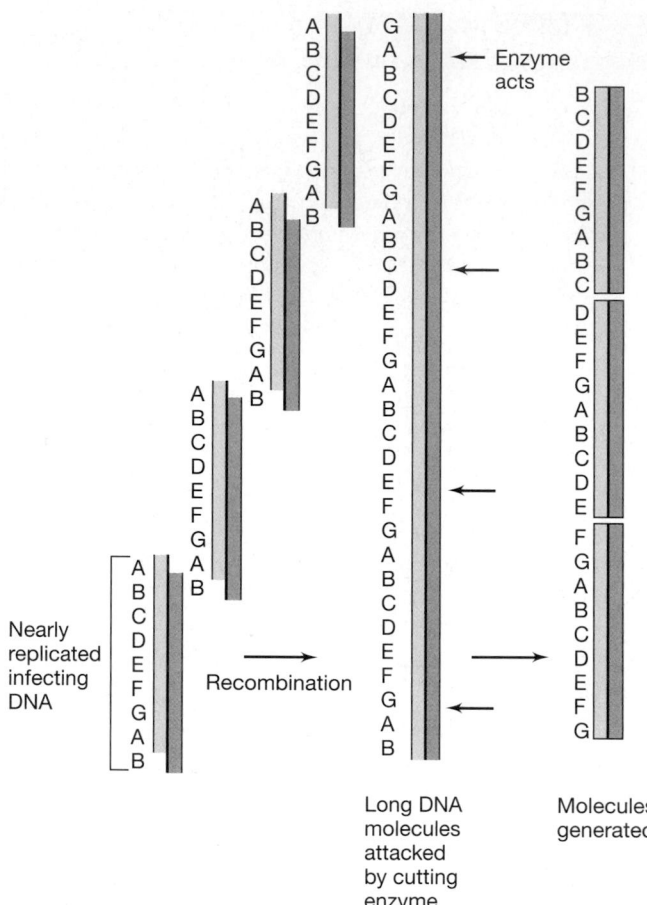

Figure 9.13 Generation in T4 of virus length DNA molecules with permuted sequences by a cutting enzyme, which cuts off constant lengths of DNA irrespective of the sequence. Left: nearly replicated copies of infecting T4 genome, arrow, recombination (◯◯ Section 10.5); middle: arrows, sites of enzyme attack; right: molecules generated.

Overall, the T4 genes can be divided into three groups, one encoding early proteins, one middle proteins, and the other, late proteins (Figure 9.15). The **early** and **middle proteins** are the enzymes involved in DNA replication and transcription. The **late proteins** are the head and tail proteins and the enzymes involved in liberating the mature phage particles from the cell.

Although T4 has a very large genome, it does not encode its own RNA polymerase. The control of T4 mRNA synthesis involves the production of proteins that

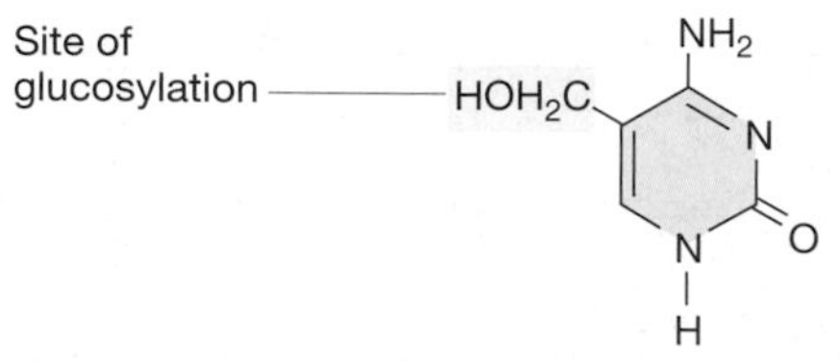

Figure 9.14 The unique base in the DNA of the T-even bacteriophages, 5-hydroxymethylcytosine. The site of glucosylation is shown.

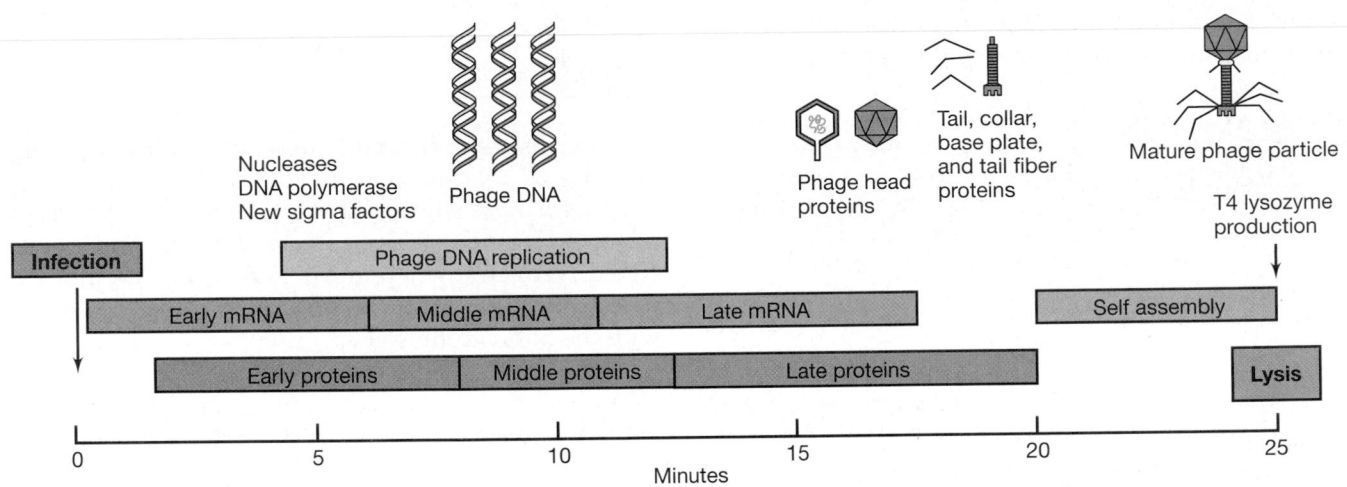

Figure 9.15 Time course of events in phage T4 infection. Following injection of DNA, early and middle mRNA is produced that codes for nucleases, DNA polymerase, new phage-specific sigma factors, and various other proteins involved in DNA replication. Late mRNA codes for structural proteins of the phage virion and for T4 lysozyme, needed to lyse the cell and release new phage particles.

sequentially modify the specificity of the host RNA polymerase so that it recognizes different phage promoters. The early promoters are read directly by the host RNA polymerase and involve the function of host *sigma* factor. (The first T4 genes transcribed, the early genes, are sometimes called *immediate early genes* to distinguish them from *delayed early genes,* whose transcription requires the synthesis of some phage proteins, possibly involved in blocking normal transcription termination.) Phage-specific proteins synthesized from the early genes carry out covalent modifications on the host RNA polymerase α subunits (⚭ Section 7.8), and a few phage-encoded proteins also bind to the polymerase. These modifications change the specificity of the polymerase so it now recognizes T4 middle promoters. One of the T4 early proteins, called MotA, apparently recognizes a particular DNA sequence in these promoters. Transcription from the late promoters requires a new T4-encoded sigma factor. Interestingly, it also requires T4 DNA synthesis. Sequential modification of host cell RNA polymerase is used to regulate gene expression by many bacteriophages.

T4 encodes over 20 new proteins that are synthesized early after infection. These include enzymes for the synthesis of the unusual base S-hydroxymethylcytosine, and for its glucosylation, as well as an enzyme that breaks down the normal DNA precursor deoxycytidine triphosphate. In addition, T4 encodes a number of enzymes that have functions similar to those of host enzymes in DNA replication but are formed in larger amounts, thus permitting faster synthesis of T4-specific DNA.

As we have mentioned, T4 encodes over 250 proteins, and almost one-quarter of these have some role in the synthesis or processing of newly replicated phage DNA. (Several are also nucleases that destroy the host DNA to obtain building blocks for viral DNA.) Recall that for packaging the DNA for each virion is cut from a much longer molecule and cut in such a way that the DNA packaged is linear, circularly permuted, terminally redundant, and slightly longer than the minimal genome length. Such an arrangement is not merely a complication but the result of the mechanism that T4 uses to fully replicate its linear genome.

Remember that when we discussed DNA replication in Chapter 7, we mentioned that there was a difficulty involved in replicating linear DNA completely and that linear genetic elements had different solutions to this problem (⚭ Section 7.7). The terminally redundant T4 DNA infecting a single host cell is first replicated as a unit, and then these units are recombined (recombination is discussed in Chapter 10) into a large concatemer (see Figure 9.13). It is from this concatemer that new DNA molecules are cut to be packaged in new virions. Because the recombination takes place in the repetitive sequences at the end, no information is lost. We shall see other mechanisms viruses with linear DNA genomes use to overcome this difficulty (see Sections 9.10, 16.4, and 16.5).

Many of the late genes encode structural proteins for the virion, including those for the head and tail. Assembly of heads and tails occurs independently, DNA is packaged into the assembled head, and the tail and tail fibers are added later. Exit of the virus from the cell occurs as a result of cell lysis. The phage codes for a lytic enzyme, the *T4 lysozyme*, which attacks the peptidoglycan of the host cell.

After a lytic cycle, which takes only about 25 min (Figure 9.15), over 100 new virus particles will be released from the host cell, which itself has been almost completely destroyed. T4 is a good example, but not an extreme example, of a virulent virus. We will now examine a virus that has another option in its life cycle.

✓ 9.9 *Concept Check*

After a virulent virus attaches to a host cell and penetrates it, the expression of the viral genes are regulated so as to redirect the host synthetic machinery to the reproduction of viral nucleic acid and protein. New virions are then assembled and released from the cell, typically by cell lysis. T4 is an example of a virulent bacteriophage with a double-stranded DNA genome.

✓ What is a virulent virus?

✓ Give an example of a mechanism used by T4 to ensure that its genes rather than those of the host are transcribed.

9.10 Temperate Bacteriophage: Lambda

Many viruses are **virulent**, but other viruses, although also able to kill cells, have the option of a different life cycle with more subtle effects on the host. Such viruses are called **temperate**. These viruses can enter into a state called **lysogeny**, where most virus genes are not expressed, and the virus genome is replicated in synchrony with the host chromosome.

The temperate phage genome can be duplicated along with that of the host and during cell division be passed from one generation of bacteria to the next.

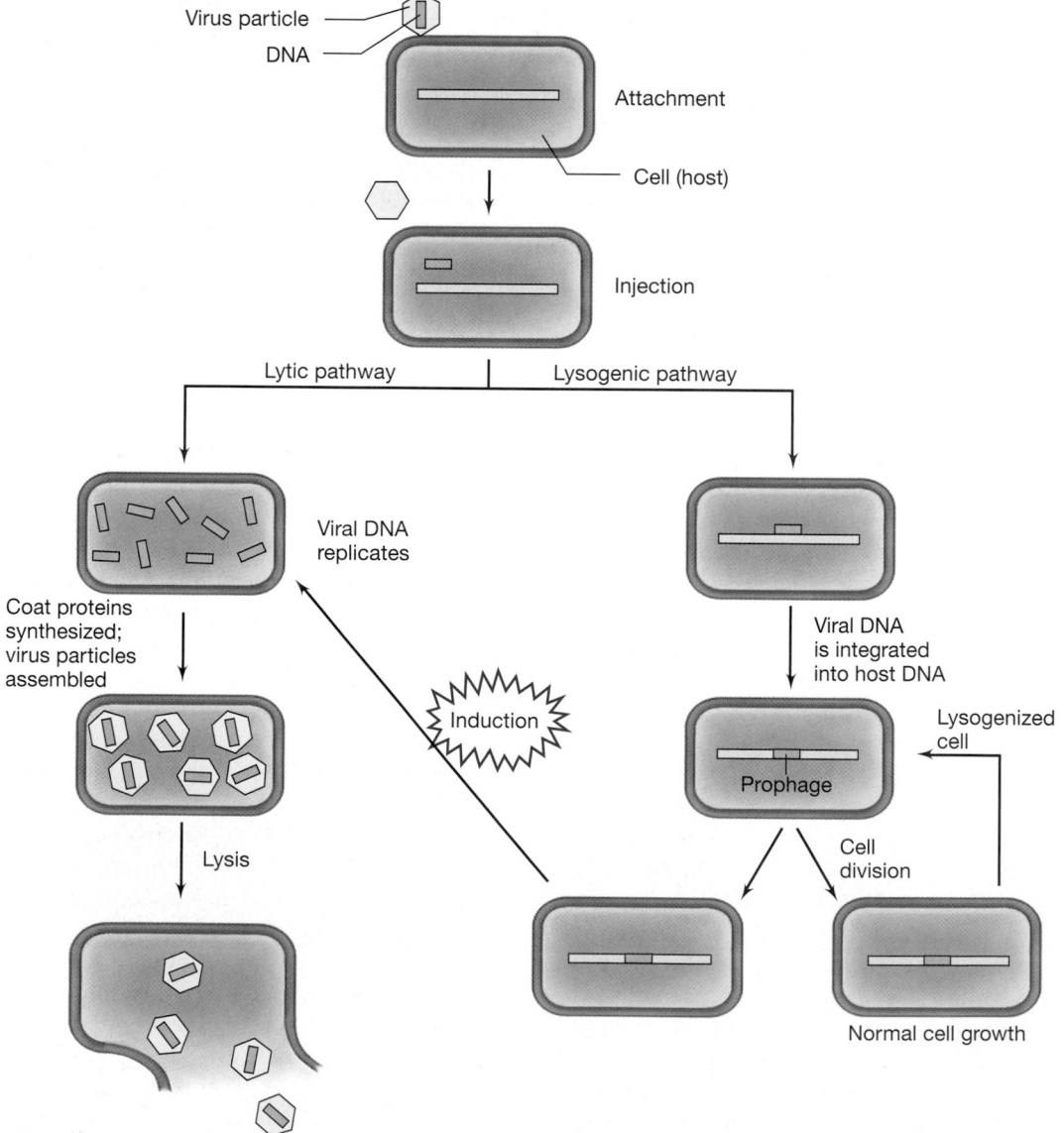

Figure 9.16 The consequences of infection by a temperate bacteriophage. The alternatives on infection are replication and release of mature virus (lysis) or integration of the virus DNA into the host DNA (lysogenization). The lysogenic cell can also be induced to produce mature virus and lyse.

Under certain conditions these bacteria, called **lysogens,** can spontaneously produce virions of the temperate virus. Lysogeny is probably of ecological importance because most bacteria isolated from nature are lysogens for one or more bacteriophages. Lysogeny is not limited to bacteriophages. Many animal viruses set up similar relationships with their hosts.

Overview of the Life Cycle of a Temperate Phage

It is not the presence, or even the replication, of viral DNA that leads to the production of new virions and host cell death. Rather it is *expression* of the viral genome that is deleterious. One could imagine that host cells can harbor virus genomes without harm if the expression of the viral genes can be controlled. This is the situation found in lysogens. However, once this control has been lost, the virus enters the lytic pathway, it produces new virions, and then the host cell lyses. In a culture of lysogens at any one time, only a small fraction of the cells, 0.1–0.0001%, produce virus and lyse, while the majority of the cells neither produce virus nor lyse. Although only rarely do cells of a lysogenic strain actually produce virus, every cell has the potential for virus production. Lysogeny can thus be considered a genetic trait of a bacterial strain.

An overall view of the life cycle of a temperate bacteriophage is shown in Figure 9.16. The temperate virus does not exist in its mature, infectious state inside the cell but rather in a latent form called the **provirus** or **prophage** state. In the example shown in Figure 9.16, the prophage is integrated into the bacterial chromosome. The prophage replicates along with the host cell as long as the genes controlling its lytic pathway are not expressed. Typically this control is maintained by a phage-encoded repressor protein (indicating that at least this gene is being expressed). The virus repressor protein not only controls the lytic genes on the prophage but also prevents the expression of any incoming genomes of the same virus. This results in the lysogens having **immunity** to infection by the same type of virus.

However, if this repressor is inactivated, or if its synthesis is prevented, the prophage is induced (center, Figure 9.16). This induction results in the production of new virions and the lysis of the host cell. In some cases (as we shall see later), induction can be brought about by environmental conditions. If the virus loses the ability to leave the host genome (because of mutation), it becomes a cryptic virus. Sequence studies have shown that many bacterial chromosomes contain stretches of DNA that were clearly once part of a viral genome.

Note that Figure 9.16 shows that infection of a normal cell by a temperate virus can lead to either the lytic pathway or the lysogenic pathway. We next discuss the factors that favor one or the other pathways during infections by the bacteriophage lambda.

The Bacteriophage Lambda

One of the best-studied temperate phages is lambda, which infects *Escherichia coli,* and our knowledge of the molecular mechanisms involved in lysogenization and lytic processes in this phage is very advanced. Morphologically, lambda particles look like those of many other bacteriophages (Figure 9.17; see also Figure 9.12).

The genome of lambda consists of a linear double-stranded DNA molecule, but at the 5′-terminus of each of the strands is a single-stranded tail 12 nucleotides long. These single-stranded ends are complementary (the ends of the DNA are said to be *cohesive*). Thus, when the two ends of the DNA are free in the host cell, they associate and the genome forms a double-stranded circle. Therefore, with lambda, the problem of replicating linear DNA is solved much different than with T4. In the circular form the DNA contains 48,502 base pairs. Figure 9.18 is a representation of the genetic map of lambda after circularization.

Before considering the organization and expression of these genes, note that for lambda, as for other temperate viruses, both the **lytic** and the **lysogenic pathways** are available. Lambda typically takes the lytic route (that is to say, most of the time it behaves very much like a virulent virus). However, the regulation of gene expression in lambda is such that sometimes the lytic pathway is not chosen. We shall now discuss infection of a cell by lambda to see how the options are presented.

Lambda Infection and the Lytic Pathway

The lambda virion attaches to a specific receptor protein in the cell wall of *Escherichia coli* (see Section 9.6) and injects its DNA. The DNA circularizes almost immediately, and, if the cell is not a lambda lysogen (and therefore not immune), expression of the phage genome begins. The first steps in gene expression are the same whether the final result is lysis or lysogeny.

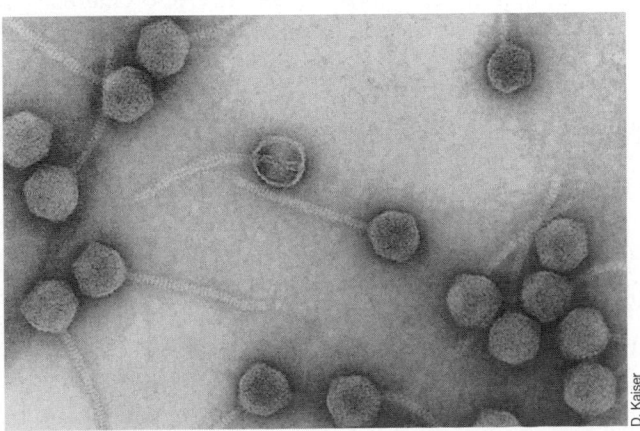

Figure 9.17 Electron micrograph by negative staining of bacteriophage lambda particles. The head of each particle is about 65 nm in diameter.

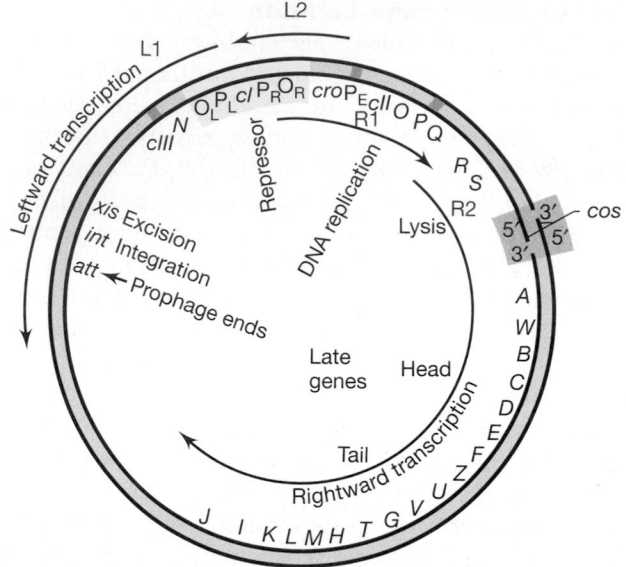

Figure 9.18 Genetic and molecular map of lambda. The genes are designated by letters; *att*, attachment site for phage to host chromosome. Genes of special interest: *cl*, repressor protein; O_R, operator right; P_R, promoter right; O_L, operator left; P_L, promoter left; *cro*, gene for second repressor; N, positive regulator counteracting rho-dependent termination. J through U are genes that encode tail proteins. Genes Z through A encode head proteins. The regulatory region of lambda (shown in yellow) is positioned at the top of this circular map. It is also known as the immunity region and contains the *cl* gene. The site created when the cohesive ends of the lambda genome join is called *cos* (shown in blue). Early transcription in lambda is primarily leftward (counterclockwise) from P_L and rightward (clockwise) from P_R. The main leftward transcript is labeled L1, and the main early rightward transcript is labeled R1. The three transcription terminators affected by the N-protein are shown as gray boxes on the DNA. The late rightward transcript, which encodes head and tail proteins and proteins for lytic function, is labeled R2 and begins at a promoter near the Q gene. The transcript labeled L2 is the positively regulated transcript from P_E that encodes the repressor protein.

Production of RNA using host RNA polymerase begins at a few promoters, two of which, called P_L (promoter left) and P_R (promoter right), are on either side of the regulatory region shown in Figure 9.18. These yield short transcripts that are translated to give the products of the *N* and the *cro* genes. Both the proteins encoded by these genes are involved in regulatory events. The Cro protein (the product of the *cro* gene) participates in the selection between the lytic and lysogenic pathways, and we shall discuss its function later. The N protein is an *antiterminator* protein that allows the RNA polymerase to transcribe past specific terminators (marked on Figure 9.18), making the transcripts from P_L and P_R longer. These longer transcripts can be translated to yield more proteins, including the products of the *cII* and *cIII* genes. The antiterminator is not completely effective at the terminator before the *Q* gene, and so only a small amount of the Q protein is made.

Early DNA synthesis is bidirectional from a single origin and gives rise to typical theta-like intermediates (∞ Section 7.5). At this stage it is still possible for lambda to switch to a lysogenic cycle. However, let us consider the situation that would result if this switch were *not* made.

The Q protein is also an antiterminator protein. If its concentration becomes high enough it will allow the transcript from a nearby promoter to synthesize the transcript labeled R2 in Figure 9.18. This transcript is translated to yield the late proteins, all the necessary structural proteins to construct a virion, and the proteins necessary for cell lysis. At the same time the Q protein has built up to this level, the Cro protein (see earlier discussion) has also reached levels where it can block transcription from both P_L and P_R by binding to both O_L (operator left) and O_R (operator right). Therefore, Cro operates as a repressor protein (∞ Section 8.5).

The mechanism of Cro protein action at O_R is diagrammed in more detail in Figure 9.19. Note that there are three similar but nonidentical sites at this operator where the Cro protein can bind. It does so first at site 3, and then site 2, and only when those two sites are filled, at site 1. Note that only when bound to site 1 does it block P_R. Note also that once P_L and P_R are blocked, no more cII or cIII proteins can be synthesized. These proteins are needed to enter the lysogenic pathway (see later), and so when Cro is made in high amounts, lambda is irrevocably committed to the lytic pathway.

The shutoff of these promoters also results in a change of lambda DNA replication. At this stage when the late proteins are being made, long, linear concatamers of DNA are synthesized by **rolling circle replication**. In this mechanism, replication proceeds in only one direction and can result in very long chains of repli-

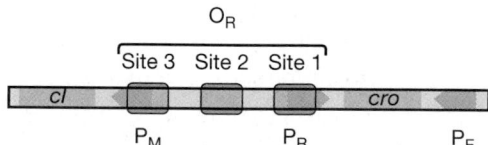

Figure 9.19 Both regulatory proteins Cro and the lambda repressor bind to operator right (O_R) on the lambda genome to carry out regulatory functions. The Cro protein (product of the *cro* gene) binds to the three sites in the order site 3, then site 2, and then site 1. The lambda repressor binds to these sites in the opposite order. The promoter P_R is transcribed immediately on phage entry into the cell. Rightward transcription from this promoter is necessary to produce Cro protein and other downstream genes (see Figure 9.18 for a complete map of the lambda genome). Leftward transcription from either of the promoters P_E or P_M is necessary to synthesize the lambda repressor (product of the *cl* gene). Both these promoters require activation in order to function.

cated DNA (Figure 9.20). This mechanism is efficient in permitting extensive, rapid, relatively uncontrolled DNA replication; thus, it is of value in the later stages of the phage replication cycle when large amounts of DNA are needed to form mature virions. The long concatamers formed are then cut into virus-sized lengths by a DNA-cutting enzyme. In the case of lambda (unlike that in T4; see Section 9.9), the cutting enzyme makes staggered breaks at *specific* sites on the two strands, 12 nucleotides apart, which provide the cohesive ends involved in the cyclization process. These DNA molecules are packaged in phage heads, and then the tails and other proteins are added. The cell is then lysed by the action of phage-encoded proteins.

Although there are many differences between the lytic pathway of lambda and that of T4, the final results is the production of new virions and lysis of the cell. However, in the case of lambda infection, the host cell's metabolism is not irreversibly subverted early in the process, ensuring that lysogenization can take place if events favor it.

Lysis or Lysogenization?

We have seen that phage genes are controlled in such a way that viral proteins and nucleic acids are made in appropriate amounts and at appropriate times. For many viruses the patterns of expression always proceed in the same programmed manner. However, lambda and other temperate viruses have a *genetic switch* that controls whether the lytic pathway or the lysogenic pathway is followed. So far the steps we have outlined for lambda are those for the lytic pathway. We now consider how the genetic switch can be thrown to lead to lysogeny.

In order to establish lysogeny, two events must happen: The production of all late proteins must be prevented, and a copy of the lambda genome must be integrated into the host chromosome. In order to prevent synthesis of the late proteins, the product of the

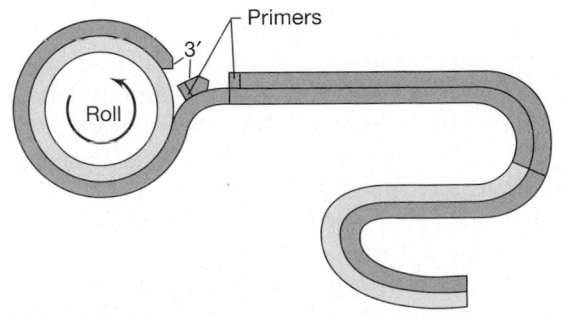

Figure 9.20 A late stage in the rolling circle replication of lambda. Both strands of DNA are being copied at the replicating fork, and two copies of the genome have already been synthesized. Note that this synthesis is *asymmetric* because one of the parental strands continues to serve as a template and the other is used only once.

cI gene must be produced. This protein is the **lambda repressor**. If it is synthesized *it will repress the synthesis of all other lambda-encoded proteins*. It is needed to establish lysogeny and to maintain the lysogenic state. The *cI* gene is located between P_L and P_R (see Figure 9.18), but these promoters are oriented in such a way that neither transcribes the *cI* gene. The promoter that can produce mRNA from the *cI* gene during infection is called P_E (promoter establishment) and is located on the map slightly to the right of the *cro* gene but facing the direction opposite that of P_R. Therefore, transcription is in the direction opposite that promoted by P_R (Figures 9.18 and 9.19). Unlike the other promoters we have previously mentioned, P_E must be *activated*. Once it is, lambda repressor protein is synthesized and the lysogenic pathway is followed.

The product of the *cII* gene is an *activator protein* (∞ Section 8.6) that activates promoter P_E (and another promoter required for the production of integrase) (see later). Although the cII protein is made early after infection, it is typically unstable in *Escherichia coli* because it is degraded by a host protease (an enzyme that degrades proteins). If the cII protein is degraded, then there is no possibility that the lysogenic pathway can be chosen. However, this protein can be stabilized by the phage-encoded cIII protein if there is no excess of host protease (or if there is an excess of cIII protein). If the cII protein is stabilized, then it will activate P_E and lambda repressor protein will be made. This rather complicated process monitors conditions in the host.

Lambda repressor binds to O_L and O_R, as does the Cro protein, but it binds to the sites within these operators in the order opposite that of Cro (see Figure 9.19). That is, it first binds to site 1, turning off P_R (and P_L by a similar mechanism). When this happens, the synthesis of all other lambda proteins is stopped, and lambda cannot enter the lytic pathway.

However, without the cII protein P_E no longer functions. Therefore, if the lysogenic state is to be maintained, there must be another way to transcribe the *cI* gene. Note that Figure 9.19 shows yet another promoter, P_M (promoter maintenance). This promoter is facing toward the *cI* gene (in the same direction as P_E). It is *activated* when lambda repressor binds to site 1 and is repressed only when lambda repressor is bound to all three sites. Therefore, the lambda repressor is both a *repressor* and an *activator* when it binds to site 1, repressing P_R and activating P_M. This type of regulation continues to occur even after lysogenization. Only the lambda repressor is made after lambda is integrated as a prophage.

Integration

Integration of lambda DNA occurs at a unique site on the *Escherichia coli* chromosome and is required for lysogeny. Integration occurs by insertion of the virus

DNA into the host genome (thus effectively lengthening the host genome by the length of the virus DNA). As illustrated in Figure 9.21, on injection, the cohesive ends of the linear lambda molecule find each other and form a circle, and it is this circular DNA that becomes integrated into the host genome (the site created when these ends join is called *cos*). To establish lysogeny, genes *cI* and *int* (encoding *integrase*) must be expressed as we discussed. The integration process requires integrase, which is a site-specific topoisomerase catalyzing recombination of the phage and bacterial attachment sites (labeled *att* in Figures 9.18 and 9.21). The *int* gene has a promoter that, like P_E, is activated by the cII protein.

During cell growth, the lambda repression system prevents expression of the integrated lambda gene ex-

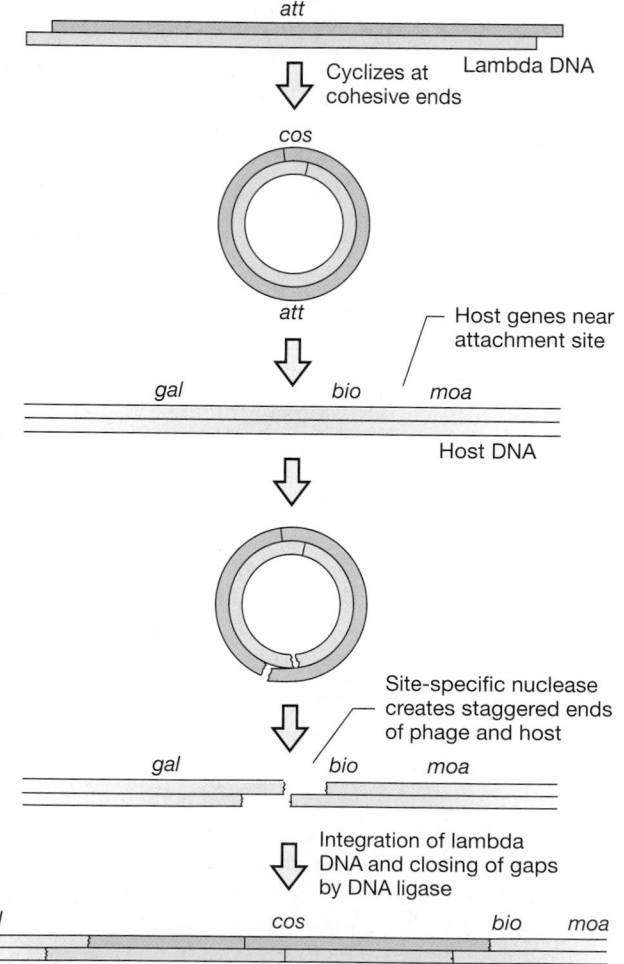

Figure 9.21 Integration of lambda DNA into the host. See the genetic map, Figure 9.18 for details of the gene order. Integration always occurs at a specific site on the host DNA, involving a specific attachment site (*att*) on the phage. Some of the host genes near the attachment site are given: *gal* operon, galactose utilization; *bio* operon, biotin biosynthesis; *moa* operon, molybdenum cofactor biosyntheses. A site-specific enzyme (integrase) is involved, and specific pairing of the complementary ends results in integration of phage DNA.

cept for the gene *cI*, which codes for the lambda repressor. During host DNA replication, the integrated lambda DNA is replicated along with the rest of the host genome and transmitted to progeny cells. When release from repression occurs (see later), the lambda lytic cycle occurs. In order to be excised from the chromosome, *excisionase* (the product of the *xis* gene) and the *int* gene product are required.

Lytic Growth of Lambda After Induction

Agents that induce lambda lysogens (cells containing lambda as a prophage) to produce phage are agents that damage DNA. These include ultraviolet irradiation, X-rays, and DNA-damaging chemicals such as the nitrogen mustards. When DNA damage occurs a host defense mechanism called the SOS response (⚬⚬ Section 10.3) is brought into play. An array of bacterial genes is turned on, some of which help the bacterium survive radiation. One result of DNA damage is that a bacterial protein called RecA (normally involved in genetic recombination) is turned into a special kind of protease that participates in the destruction of the lambda repressor. With the lambda repressor destroyed, the inhibition of expression of lambda lytic genes is abolished. We should note that the protease activity of RecA, brought about by DNA damage, normally plays an important role in the cell's response to DNA-damaging agents (⚬⚬ Section 10.3). Induction of bacteriophage lambda is thus an indirect consequence of the SOS response.

Once the lambda repressor has been inactivated, the control exerted by this repressor is abolished and new transcriptional events can be initiated. These almost inevitably lead to lysis because even if the lambda repressor is made, it is inactivated. Rarely, such treatment may "cure" a cell. That is, the prophage was induced but did not replicate and was lost during subsequent growth.

Other Strategies Used by Temperate Viruses

Lambda provides one of the best-studied examples of how a "decision" is made at the molecular level. It is also an important genetic tool and is widely used in genetic engineering as a vector for carrying recombinant DNA. We describe these uses of lambda in Chapter 10. Other types of temperate viruses are known in bacteria, and some have also been widely studied. The virus P1 (which we will also mention in Chapter 10) is a temperate virus that maintains itself in the lysogenic state not as an integrated prophage but replicates as a circular DNA molecule in the cytoplasm, resembling a plasmid. Many animal viruses can also exist in a provirus state. We will briefly discuss animal viruses in the next section, before discussing the retroviruses in some detail. Retroviruses insert a DNA copy into the host genome as part of their replication cycle.

✓ 9.10 Concept Check

Temperate viruses, such as lambda, do not always cause the death of the cells they infect. The infected cell sometimes survives because the virus genome becomes a prophage (and replicates with the host chromosome), and the lytic genes of the prophage are kept under the control of a virus-encoded repressor. Sometimes this regulatory system is circumvented and prophage induction occurs, resulting in virus multiplication and lysis of the host cell. Host cells carrying temperate bacteriophages are called lysogens.

- ✓ What are the two pathways available to a temperate virus?
- ✓ Describe how a single protein like the lambda repressor can act both as an activator and a repressor.

9.11 Overview of Animal Viruses

The first few sections of this chapter were devoted to general properties of all viruses, and little was said about animal viruses. Here we will discuss some of the attributes of animal viruses, setting the stage for a discussion of the retroviruses, an interesting group of animal viruses, one of whose members causes *acquired immunodeficiency syndrome* (AIDS). In Chapter 16 we discuss several types of animal viruses in more detail.

In our discussions of viral reproduction in Sections 9.5–9.7 we also dealt with the virus "host" in a very general way. However, it is important to remember that the host of a bacteriophage is a prokaryotic cell and the host of an animal virus is a eukaryotic cell. The differences between these cell types cause some of the differences in the replication strategies of the viruses that infect them.

Differences between Prokaryotes and Eukaryotes That Affect Virus Multiplication

We have already discussed the fact that the presence of cell walls in *Bacteria* and plants and their absence in animal cells lead to differences in viral attachment and penetration (see Section 9.6). For many bacteriophages only the genome and perhaps one or two proteins penetrate into the cytoplasm itself. For animal viruses, however, the entire virion or at least the nucleocapsid typically enters the cytoplasm by endocytosis and then must be uncoated. There are a few bacteriophages, such as the enveloped phage φ6 (see Figure 9.12) whose entire nucleocapsid also enters the cell.

However, differences between virus replication in prokaryotic and eukaryotic cells go far beyond the mechanics of viral penetration. Prokaryotes do not show compartmentation of the biosynthetic processes. The genome of a bacterium relates directly to the cytoplasm of the cell. Transcription into mRNA can lead directly to translation since the processes of transcription and translation are not carried out in separate compartments (⚌ Figure 7.2a). On the other hand, animal cells, being eukaryotic, show compartmentation. DNA replication and transcription of the genome into mRNA occur in the nucleus, whereas translation occurs in the cytoplasm (⚌ Figure 7.2b). This compartmentation has an impact on where animal viruses replicate. For instance, a DNA virus that uses host polymerases must replicate in the nucleus. Therefore, we can expect differences in replication strategies between viruses that multiply in the nucleus (for example, Herpesviruses, ⚌ Section 16.11) and those that multiply in the cytoplasm (for example, Poxviruses, ⚌ Section 16.12.) In addition, viruses that replicate in the nucleus must be transported there.

Furthermore, the transcripts from eukaryotic genes must be processed and transported to the cytoplasm before they can be used as mRNA (⚌ Section 7.12). This processing usually involves **splicing** out *introns* as well as adding a **poly-A tail** (polyadenylation) to the 3′-end and a methylated guanosine triphosphate, called the **cap**, to the 5′-end. The cap is required for binding of the mRNA to the ribosome. The difference between the way ribosomes recognize mRNA allows prokaryotes to use polycistronic mRNA, whereas eukaryotes use monocistronic mRNA (⚌ Section 7.11 and 7.15).

Note, therefore, that the *genomes* of positive-strand RNA viruses of eukaryotes must typically be in a processed state within the virion if this RNA is to serve directly as mRNA. In most cases, however, the poly-A tail in a positive-strand RNA virus genome is not added by a processing event but is actually encoded by the genome. Capping may be achieved in a number of ways. Of course, many viral RNAs are capped normally in the nucleus during synthesis from a DNA template, just as the host mRNA is capped (⚌ Section 7.12). When viral mRNAs are synthesized in the cytoplasm, however, the capping is carried out by proteins encoded by the virus itself. It is fascinating that influenza virus, whose RNA genome is replicated in the nucleus (⚌ Section 16.8), actually has a mechanism of "snatching" caps from newly synthesized host mRNA. However, splicing of introns from eukaryotic pre-mRNA can only take place in the nucleus using host machinery (⚌ Section 7.12). Also, all the protein-synthesizing machinery of the eukaryotic cell—the ribosomes, tRNA molecules, and accessory components—is in the cytoplasm.

Classification of Animal Viruses

Various types of animal viruses are illustrated in Figure 9.22. We have discussed the principles by which viruses are classified in Sections 9.1, 9.2, and 9.7. Note that there are many more kinds of enveloped animal viruses than enveloped bacterial viruses (see

Section 9.7). This doubtless relates to the differences in host cell exteriors. Most of the animal viruses that have been studied in any detail are those that have been amenable to cultivation in cell cultures. Animal viruses are classified according to the same schemes as bacterial viruses, including the Baltimore Classification System (see Table 9.2), which classifies viruses on the basis of genome types and reproductive strategy. Animal viruses are known in all categories and many of these will be discussed in Chapter 16.

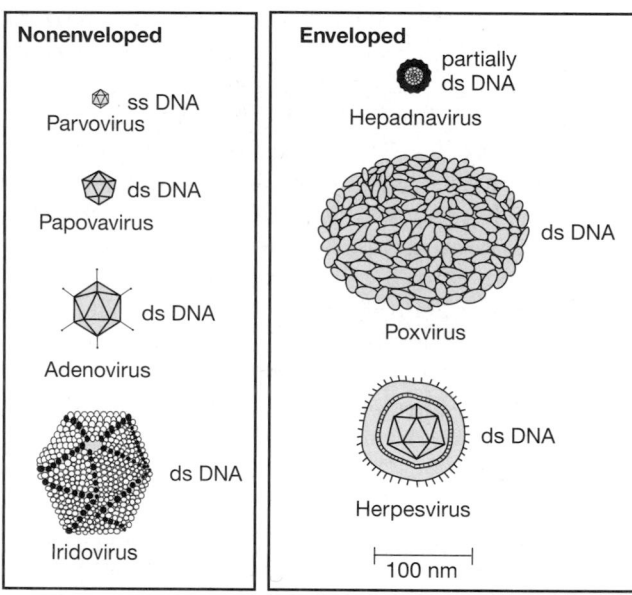

(a) DNA viruses

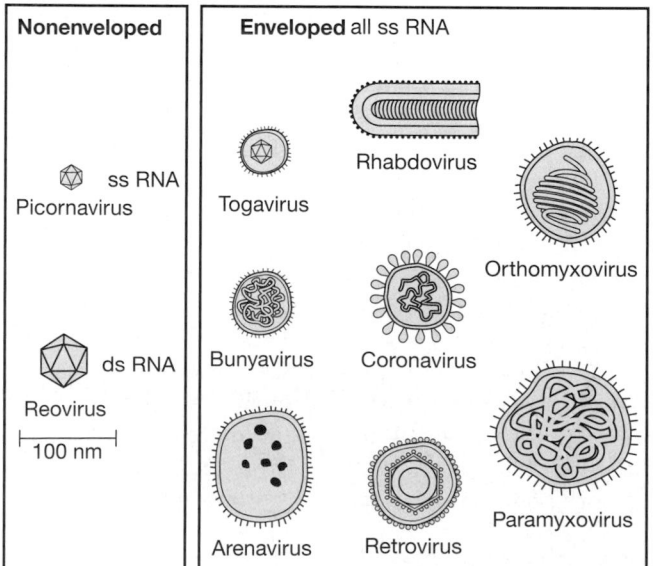

(b) RNA viruses

Figure 9.22 The shapes and relative sizes of vertebrate viruses of the major taxonomic groups. The hepadnavirus genome has one complete DNA strand and part of its complement.

Consequences of Virus Infection in Animal Cells

Viruses can have varied effects on cells. **Lytic infection** results in the destruction of the host cell (Figure 9.23). Several other possible effects may follow viral infection of animal cells. In the case of enveloped viruses, release of virions, which occurs by a kind of budding process, may be slow, and the host cell may not be lysed. The cell may remain alive and continue to produce virus over a long period of time. Such infections are referred to as **persistent infections** (Figure 9.23). Viruses may also cause **latent infection** of a host. In a latent infection, there is a delay between infection by the virus and the appearance of symptoms. Fever blisters (cold sores), caused by the herpes simplex virus (∞ Section 16.11), result from a latent viral infection; the symptoms reappear sporadically as the virus emerges from latency. The latent stage in viral infection of an animal cell is generally not due to integration of the viral genome into the genome of the animal cell, as is the case with latent infections by temperate bacteriophages.

Viruses and Cancer

A number of animal viruses participate in the events that change a cell from a normal one to a cancer or tumor cell (Figure 9.23 and Table 9.3). **Cancer** is a cellular phenomenon of uncontrolled growth. Many cells in a mature animal, although alive, do not divide extensively, apparently because of the presence of growth-inhibiting factors that prevent them from initiating cell division. These factors are under genetic control. As noted in Section 9.4, infection by certain types of animal viruses leads to a process called **transformation** during which growth becomes uncontrolled. Rapidly growing cells pile up into accumulations that are visible in culture as **foci of infection**. Cancerous cells in the animal body grow profusely, leading to the formation of large masses of cells, called *tumors*. The term **neoplasm** is often used in the medical literature to describe malignant tumors.

Not all tumors are seriously harmful. The body is able to wall off some tumors so that they do not spread; such noninvasive tumors are said to be **benign**. Other tumors, called **malignant**, invade the body and destroy normal body tissues and organs. In advanced stages of cancer, malignant tumors may develop the ability to spread to other parts of the body and initiate new tumors, a process called **metastasis**.

How does a normal cell become cancerous? The development of cancer is clearly a multistep process. There seem to be many different causes of cancer, including mutations arising either spontaneously or as the result of exposure to certain chemicals, called *carcinogens* (∞ Section 10.4), or by physical stimuli, such as ultraviolet radiation or X-rays. Certain viruses also

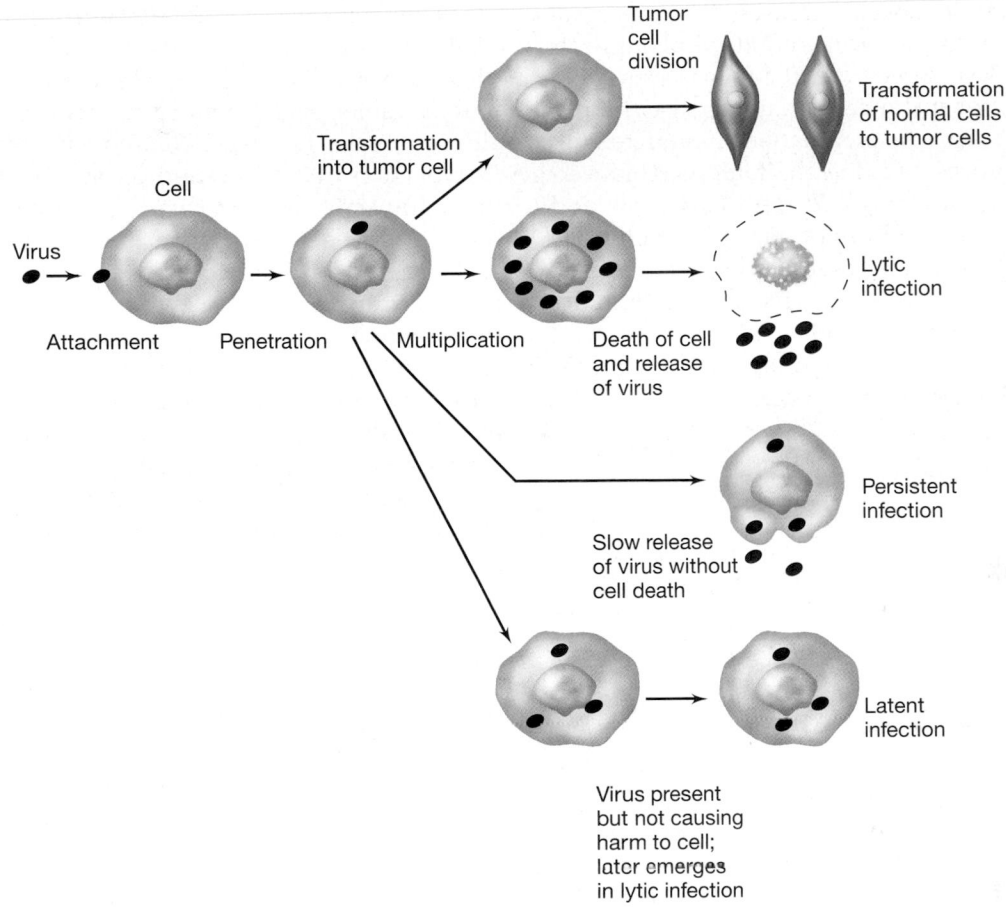

Figure 9.23 Possible effects that animal viruses may have on cells they infect. Note that unlike bacteriophages, with animal viruses the *entire virion* is taken up into the cell.

bring about the genetic change that results in initiation of tumor formation.

Events that cause cancer lead to a loss of the cell's normal control of its growth. The growth and division of normal cells is regulated by at least two types of genes. The first type, called *proto-oncogenes*, promote growth, but they are controlled by the second type, the growth-restraining *tumor suppressor genes*. Changes in either or both types can lead to uncontrolled cell growth and therefore to cancer.

The *initiation* event can be the activation of a proto-oncogene into an **oncogene** (a gene that causes a tumor), or the inactivation of a tumor suppressor gene. Once initiation has occurred, the potentially cancerous cell may

remain dormant, but under certain conditions, generally involving some environmental alteration, it may become converted to a tumor cell, a process called **promotion**. Once a cell has been promoted to the cancerous condition, continued cell division can result in the formation of a tumor.

Although the ability of viruses to cause tumors in animals was proven many years ago, the relationship of viruses to cancer in humans has, in most cases, been uncertain. It is difficult to prove the viral origin of a human cancer because of the difficulties of carrying out the necessary experimentation. However, it is well established that certain kinds of human tumors are strongly associ-

TABLE 9.3 Some human cancers where viruses play a role

Cancer	Virus	Family	Genome in virion
Adult T-cell leukemia	Human T-cell leukemia virus (type I)	*Retroviridae*	RNA
Burkitt's lymphoma	Epstein-Barr virus	*Herpesviridae*	DNA
Nasopharyngeal carcinoma	Epstein-Barr virus	*Herpesviridae*	DNA
Hepatocellular carcinoma (liver cancer)	Hepatitis B virus	*Hepadnaviridae*	DNA
Skin and cervical cancers	Papilloma virus	*Papillomaviridae*	DNA

ated with infection by specific viruses. A summary of some of the human cancers with definite viral origins is given in Table 9.3. In some cases the role a virus plays may be indirect; for instance, it may increase the frequency of a particular type of mutation or interfere with normal cellular processes in the cell it infects. This is almost certainly the case of the Epstein-Barr virus and Burkitt's lymphoma. In addition, some viral infections can lead indirectly to an increased risk of cancer, apparently by weakening the immune system's ability to detect and destroy transformed cells. This might be why infection with the retrovirus that causes AIDS increases the risk for developing certain cancers.

✓ 9.11 Concept Check

Multiplication of animal viruses differs in significant ways from the multiplication of bacterial viruses. Many of these differences arise because eukaryotes exhibit compartmentation of macromolecule synthesis whereas prokaryotes do not. Not all infections of animal host cells result in cell lysis or death. In some cases, latent infection occurs, the virus remaining infectious but dormant inside the host and appearing spontaneously at a later time. Some animal viruses cause transformation of host cells.

✓ Which macromolecules are normally synthesized in the nucleus and which are synthesized in the cytoplasm of eukaryotic cells?

✓ Contrast the mechanisms by which animal viruses enter cells with those used by bacterial viruses.

9.12 Retroviruses

As we have mentioned, **retroviruses** contain an RNA genome in the virion but replicate through a DNA intermediate (see Section 9.1). The term *retro* means "backward," and the name of this class of virus is derived from the fact that these viruses appear to transfer information backward from RNA to DNA (∞ Section 7.1). These viruses use the enzyme **reverse transcriptase** to carry out this interesting information transfer.

The retroviruses are interesting for several other reasons. For example, they were the first viruses shown to cause *cancer* and have been studied most extensively for their carcinogenic characteristics. Also, one retrovirus, *human immunodeficiency virus* (HIV) causes *acquired immunodeficiency syndrome (AIDS)* and although it has been known only since the early 1980s, it has become a major public health problem worldwide. In addition, the retrovirus genome can become integrated into the host genome by way of the DNA intermediate, and this integration process is being studied as a means of introducing *foreign* genes into a host, a process called *gene therapy*.

Some properties of the retroviruses are like those of RNA viruses and others are like those of DNA viruses. Retroviruses resemble to a considerable extent movable genetic elements and are sometimes considered to be *escaped cellular transposable elements* (∞ Section 10.11). We should note that the use of reverse transcriptase by viruses is not restricted to the retroviruses because hepatitis B virus (a human virus) and cauliflower mosaic virus (a plant virus) also use reverse transcription in their replication processes (∞ Section 16.14). But in contrast to the retroviruses, these latter viruses encapsidate the DNA genome rather than the RNA genome as retroviruses do. Some transposable elements of eukaryotes, called *retrotransposons*, also encode and use reverse transcriptase as part of their replication cycle. In addition, reverse transcriptases capable of producing small multicopy DNA (ms DNA) with an RNA template have been discovered in *Bacteria*.

The retroviruses are enveloped viruses (Figure 9.24a). There are a number of proteins in the virus coat and typically seven internal proteins, four of which are structural and three enzymatic. The enzymatic activities found in the virus particle are *reverse transcriptase, DNA endonuclease (integrase),* and a *protease*. The virion also contains specific cellular tRNA molecules used in *replication* (see later in this section).

Features of Retroviral Genomes and Replication

The genome of the retrovirus is unique. It consists of *two* identical single-stranded RNA molecules of plus complementarity, each 8.5–9.5 kilobases in length. The 5′-terminus of the RNA is capped and the 3′-terminus is polyadenylated, so the RNA is capable of acting directly as mRNA but is *not* used as such. A genetic map of a typical retrovirus is shown in Figure 9.24b. Although there are differences between the genetic maps of different types of retroviruses, all contain the following regions and in the same order: *gag,* encoding internal structural proteins; *pol,* encoding reverse transcriptase and integrase; and *env,* encoding envelope proteins. Some, such as Rous sarcoma virus, carry a fourth gene downstream from *env* that is involved in cellular transformation and cancer (Figure 9.24c). The terminal repeats shown on the map play an essential role in the replication process (see later).

The overall process of replication of a retrovirus can be summarized in the following steps (Figure 9.25):

1. **Entrance** into the cell via fusion with cell membrane at sites of specific receptors.

2. **Virion uncoated** at membrane but genome and enzymes remain in a core particle.

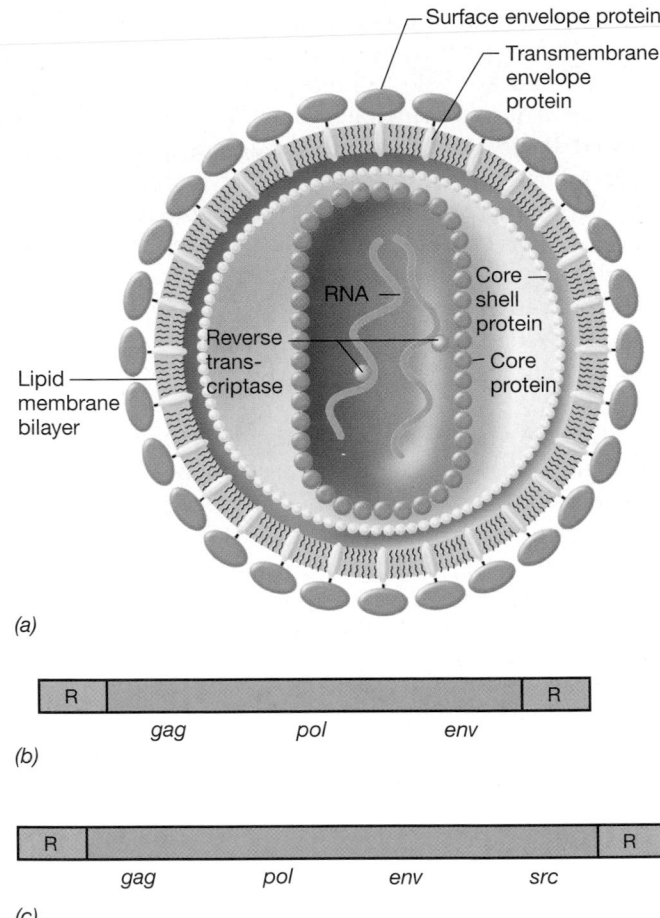

(a)

(b)

R				R
	gag	pol	env	

(c)

R					R
	gag	pol	env	src	

Figure 9.24 Retrovirus structure and function. (a) Structure of a retrovirus. (b) Genetic map of a typical retrovirus genome. (c) Genetic map of Rous sarcoma virus, a retrovirus that causes malignant tumors in birds. Each end of the genomic RNA contains direct repeats (R), and this RNA also has a 5′-cap and a 3′-poly-A tail. See text for more details.

3. **Reverse transcription** of *one* of the two RNA genomes into a single-stranded DNA that is subsequently converted to a linear double-stranded DNA by reverse transcriptase, which then enters the nucleus.

4. **Integration** of the DNA copy into the host genome.

5. **Transcription** of the viral DNA, leading to the formation of viral mRNAs and progeny viral RNA.

6. **Encapsidation** of the viral RNA into nucleocapsids in the cytoplasm.

7. **Budding** of enveloped virions at the cytoplasmic membrane and release from the cell.

A very early step after the entry of the RNA genome into the cell is reverse transcription: conversion of RNA into a DNA copy using the enzyme reverse transcrip-

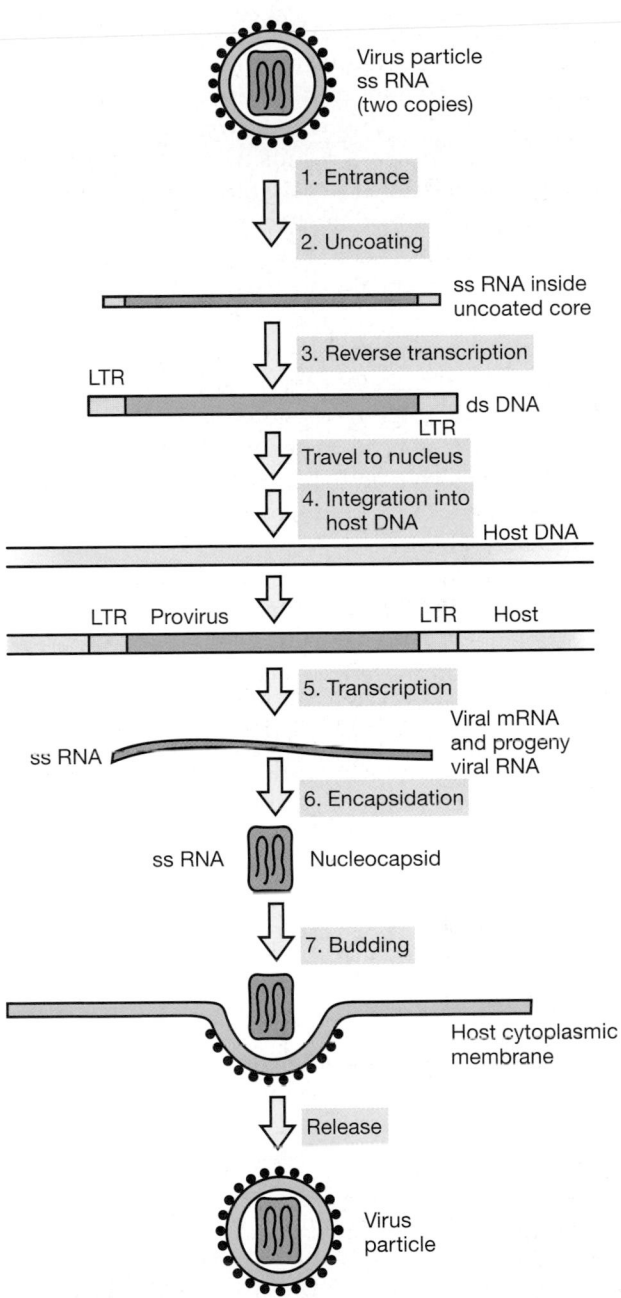

Figure 9.25 Replication process of a retrovirus. For more detail on conversion of RNA to DNA (step 3) refer to Figure 16.23.

tase present in the virion. The DNA formed is a linear double-stranded molecule and is synthesized in the cytoplasm within an uncoated viral core particle.

A detailed figure showing this complicated process can be found in Chapter 16 (∞ Figure 16.23). The process depends on the several different activities of reverse transcriptase (∞ Section 16.14). However, reverse transcriptase is a DNA polymerase and like all DNA polymerases must have a primer (∞ Section 7.5). The primer for retrovirus reverse transcription is a specific *cellular transfer RNA (tRNA).* The type of tRNA used as primer depends on the virus and is brought into the viron from the previous host cell. In the case of Rous sarcoma virus, the tRNA used is the *tryptophan* tRNA.

The overall process results in a final product that has long terminal repeats (LTRs), which are longer than the terminal repeats on the genome itself (Figure 9.24). This entire double-stranded DNA molecule, with the integrase protein, enters the nucleus where it is integrated into the host DNA. The LTRs contain strong promoters of transcription and are also involved in the integration process. The *integration* of the viral DNA into the host genome is analogous to the integration of a bacterial transposon into a bacterial genome (Section 10.11). Integration can occur anywhere in the cellular DNA, and once integrated, the element, now called a *provirus*, is a stable genetic element.

If the promoters in the right LTR are activated, the integrated proviral DNA is transcribed by a cellular RNA polymerase into transcripts that are capped and polyadenylated. These RNA transcripts either may be encapsidated into virus particles or may be processed and translated into virus proteins. Some virus proteins are made initially as a large primary *gag* protein, which is split by proteolytic action into the capsid proteins. Occasionally, read-through past the *gag* region (involving either inserting an amino acid in response to a stop codon or a shift in reading frame by the ribosome) leads to the translation of *pol,* the reverse transcriptase gene. Other proteins are synthesized from spliced transcripts.

When the virus proteins have accumulated in sufficient amounts, assembly of the nucleocapsid can occur. This encapsidation process leads to the formation of nucleocapsids, which move to the cytoplasmic membrane for final assembly into the enveloped virus particles.

Not all retroviruses cause cancer, but tumorigenic retroviruses are quite common. Infection with one of these viruses can cause cellular transformation, leading to the formation of a tumor. Why are these viruses tumorigenic? It appears that they possess a transforming gene, or viral *oncogene,* that encodes a protein that brings about cellular transformation (see Section 9.11). This gene, known in Rous sarcoma virus as the *src* gene (*src* for *sarcoma*) (see Figure 9.24*c*), encodes a phosphoprotein that possesses protein kinase activity. Protein kinases bring about the phosphorylation of proteins, and protein phosphorylation is one mechanism for regulating the activity of proteins (Section 8.10).

Transforming genes analogous to *src* have also been detected in human cancer cells. Interestingly, similar genes have also been detected in *normal* (that is, noncancerous) cells. These cell sequences are the proto-oncogenes (see Section 9.11) and have been found not only in mammalian cells but also in the cells of insects

and yeast, suggesting that these sequences are of fundamental importance in the regulation of cell growth. Retroviruses are able to incorporate such normal sequences, which become altered and are abnormally expressed. Retroviruses are thus the agents by which such genes are transferred from cell to cell.

As previously mentioned, one notable retrovirus is HIV, the virus causing AIDS. This virus infects a specific cell type in the human, a kind of T lymphocyte (T-helper cell) that is vital for proper functioning of the immune system. In later chapters we discuss the medical and immunological aspects of AIDS (Sections 25.6 and 26.14).

Because viruses are not cells but depend on cells for their replication, viral diseases pose serious medical problems; it is frequently difficult to prevent antiviral drugs from doing some damage to host cells. Despite this, certain chemotherapeutic strategies have been devised for use against viral pathogens, including retroviruses. We discuss these in some detail later along with the chemotherapy of other viral diseases (Section 20.10).

✓ *9.12 Concept Check*

Retroviruses are enveloped viruses that have complex life cycles since they are RNA viruses that replicate by means of a DNA intermediate. The retrovirus called human immunodeficiency virus causes acquired immunodeficiency syndrome. The retrovirus virion contains an enzyme, reverse transcriptase, that copies the information from its RNA genome into DNA. The DNA becomes integrated into the host chromosome in the manner of a temperate virus. The retrovirus DNA can be transcribed to yield mRNA (and new genomic RNA) or may remain in a latent state.

✓ Why are these viruses called retroviruses?

✓ How does the life cycle of a temperate bacteriophage differ from that of a retrovirus?

9.13 Viroids and Prions

So far in this chapter, we have discussed the principles of viral reproduction emphasizing viruses that infect bacteria and those that infect animals. The bacteriophages represent model genetic systems and they infect prokaryotes, organisms that are the major focus of this text. Animal viruses were discussed because several of them cause important human diseases. Some plant viruses also cause plant diseases, which have considerable impact on human affairs. However, it is not possible to cover all viruses here. In Chapter 16 we will discuss some other bacterial and animal viruses and a few plant viruses.

We also have not discussed fungal viruses. These "viruses" do not have an extracellular stage in their life cycle. Although some do become packaged into virion-like structures (sometimes referred to as "virus-like particles") these particles are not infectious if released from a cell. Therefore, these elements are very difficult to fit into our working definition of a virus. Recall that our definition of a virus is a genetic element that subverts the normal cellular process for its own replication and that has a mature infectious form that is typically extracellular. Although these "viruses" are not infectious by mechanisms similar to those of other viruses, they are transmitted by cell-to-cell fusion. This method of transmission may be because fungi have very thick cell walls or because fungal cell fusion is a common event in nature. Most of these genetic elements replicate benignly within the cells that carry them. The yeast *Saccharomyces cerevisiae* is known to have both double-stranded RNA elements and retroviral-like elements.

There are a few known entities whose properties are at variance with this definition and that most scientists would not actually consider to be viruses. However, they seem closely related to viruses and are not considered plasmids. Two of the most important of these are *viroids* and *prions*.

Viroids

Viroids are small, circular, single-stranded RNA molecules that are the smallest known pathogens (ranging from the coconut cadang-cadang viroid, which is 246 nucleotides in size, to citrus exocortis viroid, which has 375 nucleotides). Viroids cause a number of very important crop diseases. The extracellular form of the viroid is naked RNA—there is no capsid of any kind. Even more interestingly, *the RNA molecule contains no protein-encoding genes*, and therefore the viroid is totally dependent on host function for its replication. Although the viroid RNA is a single-stranded circle, there is such considerable secondary structure possible that it resembles a short double-stranded molecule with closed ends (Figure 9.26). The viroid molecule seems to be replicated in the host cell nucleus, and its structure, which somewhat mimics DNA, apparently allows it to be replicated by the host RNA polymerase.

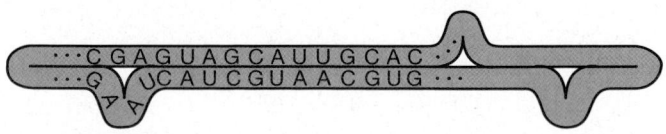

Figure 9.26 Structure of viroids, showing how single-stranded circular RNA can form a seemingly double-stranded structure by intrastrand base-pairing.

Viroids are sometimes considered "escaped" introns and, like self-splicing introns (∞ Section 7.12), appear to be remnants of an RNA world (see the discussion of the "RNA World" in Section 11.2).

Prions

Prions represent the other extreme from viroids. They have a distinct extracellular form, but the extracellular form seems to be *entirely protein*. It apparently does not contain any nucleic acid, or if it does, the molecule is not long enough to encode the single kind of protein of which the prion is composed. However, the prion protein particle is infectious, and various prions are known to cause a variety of diseases in animals, such as scrapie in sheep, bovine spongiform encephalopathy in cattle (BSE or "mad cow disease"), and kuru and Creutzfeldt-Jakob disease (CJD) in humans. In 1996, information became available from England that indicated that the prion that causes BSE in cattle might infect humans, resulting in a variant of CJD, called variant CJD or vCJD. Subsequent evidence seems to confirm that the BSE prion did "jump" the species barrier; however, the transmission seems to be very inefficient. Despite its inefficient transmission and the fact that relatively few cases of human infection have been identified, in just a few years vCJD has become a worldwide health concern with a major impact on the animal industry in several countries.

In addition to serious disease, prion infection results in production of more copies of the prion protein. Unless prions violate the central pattern of genetic information flow discussed in Chapter 7, this protein *must* be encoded by nucleic acid. Indeed, it has been discovered that the host cell contains a gene on one of its chromosomes that encodes a protein very similar to the prion protein. The host protein is normally produced and is found mostly in neurons. Apparently, the incoming prion modifies this host protein either during or after synthesis. The modification involves an alternative pattern of folding and causes the protein to lose its normal function, to become partially resistant to proteases, and to become insoluble. Therefore, prions do not simply subvert host enzymes but somehow cause a normal host gene to produce more copies of the pathogenic protein itself.

In 1997 the American scientist Stanley B. Pruisner won the Nobel Prize for his pioneering work with these diseases and with the prion proteins. Interestingly, there is now a model system for prion diseases that is much easier to study than following the disease in mammals. This model system is the yeast *Saccharomyces cerevisiae*.

It has now been shown that two characteristics of yeast seem to be transmitted by an "infectious" prion like protein. One of these involves the PSI$^+$ phenotype. PSI$^-$ (normal) strains of yeast synthesize a protein called sup35, which is involved with accurate protein synthesis. In PSI⁺ strains this protein does not function correctly, even though the gene is normal. In PSI⁺ strains the protein is misfolded. This misfolding is caused by previously misfolded protein (hence the infectious nature of the process) and a molecular chaperone. This situation seems to be the same as that found in the prion diseases of mammals.

Viroids and prions do more than stretch our definition of a virus. They also demonstrate both the unexpected ways that genetic elements can replicate and the unexpected ways they can subvert the host cells. Of course, they are also of interest to us because they cause disease.

✓ **9.13 Concept Check**

Viroids are circular single-stranded RNA molecules that encode no proteins and are completely dependent on host-encoded enzymes. They are the smallest known pathogens. Unlike viruses, their extracellular form is the same as their intracellular form, and they have no protein coat. Prions have an extracellular form that does contain protein, but it does not contain the nucleic acid that encodes the protein. The gene that encodes the prion protein is found in the host cell, and the prion somehow modifies this protein product.

Review Questions

1. Define the term *host* as it relates to viruses.

2. Define *virus*. What are the minimal features needed to fit your definition?

3. Under some conditions, it is possible to obtain nucleic acid-free protein coats (*capsids*) of certain viruses. Under the electron microscope these capsids look very similar to complete virions. What does this fact tell you about the role of the virus nucleic acid in the virus assembly process? Would you expect such virions to be infectious? Why?

4. Write a paragraph describing the events that occur on an agar plate containing a bacterial lawn when a single bacteriophage particle causes the formation of a *bacteriophage plaque*.

5. Why does one usually speak of plaque-forming units (pfu)/ml rather than viruses/ml?

6. Describe how a *restriction endonuclease* might play a role in resistance to bacteriophage infection. Why could a restriction endonuclease play such a role whereas a generalized DNase could not?

7. One can divide the replication process of a virus into five steps. Describe the events associated with each of these steps.

8. While some cells may be *susceptible* to infection by a specific virus, they may not be *permissive* for that virus. Explain.

9. Specifically, why are both the life cycle and the virion of a positive-strand RNA virus likely to be simpler than those of a negative-strand RNA virus?

10. Many animal viruses with ds DNA genomes replicate their genomes in the host nucleus. Why might they use this replication strategy?

11. A bacterium that is missing the outer membrane protein responsible for maltose uptake is *resistant* to lambda infection. A lambda lysogen is *immune* to lambda infection. Describe the functional difference between resistance and immunity. Explain how these conditions are brought about in the examples given.

12. Many of the viruses we discussed have *early* genes and *late* genes. What is meant by these two classifications? What types of proteins tend to be encoded by early genes? What type of proteins by late genes? For any three viruses we discussed, describe how expression of the late genes is controlled.

13. The genome of the bacteriophage T4 is both *circularly permuted* and *terminally redundant*. Describe what is meant by each of these two terms.

14. Typically, transfer RNA is used in translation. However, it also plays a role in the replication of retroviral nucleic acid. Explain this role.

15. *Provirus* and *prophage* mean almost the same thing. Give a definition of these terms. When does one typically use the word *provirus*, and when *prophage*?

16. What are the similarities and differences between prions, viruses, and viroids?

Application Questions

1. Can you imagine any advantage for a virus having a metabolically inert extracellular stage rather than having one that is metabolically active?

2. What causes the viral plaques that appear on a bacterial lawn to stop growing larger?

3. One characteristic of *temperate bacteriophages* is that they cause turbid rather than clear plaques on bacterial lawns. Can you think why this might be? (Remember the process by which a bacterial plaque develops.)

4. There are three lambda genes that, when rendered nonfunctional, turn lambda from a temperate to a virulent virus. What are these three genes and how do they normally function?

5. Figure 9.19 shows two promoters, P_R and P_E, on either side of the *cro* gene. Both transcribe through the *cro* gene, but only the transcript from P_R can be translated to yield the Cro protein. Explain.

6. The lambda cIII protein will be produced inside an infected cell in relatively high amounts if the cell is simultaneously infected by more than one lambda virion (because there will be multiple copies of the lambda chromosome being simultaneously expressed). High amounts of the cIII protein lead to lysogeny (and you should be able to explain how this is accomplished). From the point of view of the virus the "decision" to switch to the lysogenic pathway under these conditions is a good one. Explain.

7. Knowledge of the type of RNA carried in the genome of retroviruses would lead to a prediction that the virion would not carry any enzymes. Explain why one could make this prediction and why in this case it is wrong.

8. The promoters for mRNA encoding early proteins in viruses sometimes have a much different sequence than the promoters for mRNA encoding late proteins in the same virus. Explain why this might be true. (*Hint:* What type of RNA polymerase must recognize the "early" promoters?)

9. *Chemotherapeutic agents* are lacking for most virus diseases. From what you know about the stages of virus multiplication, give an explanation of why you think that may be so.

10. There are several genetic elements in yeast that have most of the characteristics of viruses, including the formation of virus-like particles containing the genome of the element. However, these genetic elements do not have an extracellular stage in their life cycle. Rather, they are transferred by cell fusion. See if you can write a simple definition of the word "virus" that would include these elements but still not include elements like plasmids. (Please note that plasmids can also be transferred from cell-to-cell, ⟲ Section 10.9.)

To understand an organism's growth, development, and reproduction, it is necessary to understand its complement of genes and how they function. In order to gain this kind of knowledge, it is necessary to be able to manipulate the genetic material of an organism. A large array of techniques, both those that operate within the organism (*in vivo*) and those that can be performed in the test tube (*in vitro*), are available to study the genetics of the prokaryotes. Mutations that occur in the genes of an organism can lead to a variety of changes. Some mutations lead to alterations in the characteristics of an organism that are easy to recognize, like those that affect the color of colonies of *Halobacterium*, a member of the *Archaea*, as shown here.

BACTERIAL GENETICS 10

Working Glossary

Auxotroph an organism that has developed a nutritional requirement through mutation

Cloning vector genetic element into which genes can be recombined and replicated

Conjugation transfer of genes from one prokaryotic cell to another by a mechanism involving cell-to-cell contact and a plasmid

Electroporation the use of an electric pulse to induce cells to take up free DNA

Gene disruption use of genetic techniques to inactivate a gene by inserting within it a DNA fragment containing an easily selectable marker. The inserted fragment is called a *cassette*, and the process of insertion, *cassette mutagenesis*

Genetic map the arrangement of genes on a chromosome

Genotype the precise genetic makeup of an organism

Hybridization formation of a duplex nucleic acid molecule with strands derived from different sources by complementary base pairing

Molecular cloning isolation and incorporation of a fragment of DNA into a vector where it can be replicated

Mutagens agents that cause mutation

Mutant an organism whose genome carries a mutation

Mutation an inheritable change in the base sequence of the genome of an organism

Nucleic acid probe a strand of nucleic acid that can be labeled and used to hybridize to a complementary molecule from a mixture of other nucleic acids

Phenotype the observable characteristics of an organism

Plasmid an extrachromosomal genetic element that has no extracellular form

Point mutation a mutation that involves one or only a very few base pairs

Polymerase chain reaction (PCR) a method used to amplify a specific DNA sequence *in vitro* by repeated cycles of synthesis using specific primers and DNA polymerase

Probe see *Nucleic Acid Probe*

Recombination the process by which parts or all of the DNA molecules from two separate sources are exchanged or brought together into a single unit.

Restriction enzyme an enzyme that recognizes and makes double-stranded breaks at specific DNA sequences

Screening any of a number of procedures that permit the sorting of organisms by phenotype or genotype, but allow the growth of those possible

Selection placing organisms under conditions where the growth of those with a particular genotype will be favored

Shotgun cloning making a gene library by closing random DNA fragments

Site-directed mutagenesis a technique whereby a gene with a specific mutation can be constructed *in vitro*

Synthetic DNA a DNA molecule made by a chemical process in a laboratory

Transduction transfer of host genes from one cell to another by a virus

Transformation transfer of bacterial genes involving free DNA (but see alternative usage in Chapter 9)

Transposable element a genetic element that has the ability to move (transpose) from one site on a chromosome to another

Transposon a type of transposable element that carries genes in addition to those involved in transposition

---

Now that we have introduced the main features of the molecular biology of cells and viruses, we can turn to a discussion of specific aspects of microbial genetics. In this chapter, we present the basic principles and techniques of bacterial genetics

First we will discuss mutation and genetic recombination, and then we will explain how genetic material can be transferred from one organism to another. The transfer of genetic material from one organism to another can occur in a number of different ways in a microorganism, and if it is accompanied by genetic recombination, it can lead to profound changes in the organism. Finally we will discuss some of the basic techniques of *in vitro* DNA manipulation that have revolutionized the study of genetics of all organisms, from bacteria to humans.

I MUTATION AND RECOMBINATION

Mutation is an inherited change in the base sequence of the nucleic acid comprising the genome of an organism. **Genetic recombination** is the process by which genetic

elements contained in two separate genomes are brought together in one unit. Through this mechanism, new combinations of genes can arise even in the absence of mutation. Since the genetic elements brought together may enable the organism to carry out some new function, genetic recombination can result in adaptation to changing environments. Whereas mutation usually brings about only a very small amount of genetic change in a cell, genetic recombination usually involves much larger changes. Entire genes, sets of genes, or even whole chromosomes, are transferred between organisms.

Prokaryotes, unlike many eukaryotes, do not reproduce sexually (∞ Section 14.6). However, there are mechanisms of genetic exchange in prokaryotes that, although considerably different from those involved in eukaryotic sexual reproduction, allow for both gene transfer and recombination.

To detect genetic exchange between two organisms, it is necessary to employ *genetic markers* whose transfer can be detected. Genetically altered strains are used for this purpose, the alteration(s) being due to one or more mutations in the DNA of the organism. These mutations may involve changes in only one or a few base pairs or even the insertion or deletion of entire genes. We begin this chapter on microbial genetics with a consideration of the molecular mechanism of mutation and the properties of mutant microorganisms as a prelude to our discussion of genetic exchange.

·10.1 Mutations and Mutants

As previously mentioned, a *mutation* is a heritable change in the base sequence of the nucleic acid genome of an organism. In all cells this nucleic acid is double-stranded DNA (∞ Section 7.1). A strain carrying such a change is called a **mutant**. A mutant by definition differs from its parental strain in **genotype**, a precise description of the genes an organism has. But in addition, the observable properties of the mutant, its **phenotype**, may also be altered relative to the parental strain. One would refer to this altered phenotype as a *mutant phenotype*. It is common to refer to a strain isolated from nature as a *wild-type* strain. Mutant derivatives can be obtained either from wild-type strains or from a strain derived from the wild type, for example, another mutant.

Designating Genotype and Phenotype

Depending on the mutation, a mutant may or may not have a phenotype different from that of its parent. By convention in bacterial genetics, the *genotype* of an organism is designated by three lowercase letters followed by a capital letter (all in italics) indicating the particular gene involved. For example, the *hisC* gene of *Escherichia coli* codes for a protein that could be called the HisC protein. In this case, this protein (an enzyme in the biosynthetic pathway of histidine) is usually referred to by the name histidinol-phosphate aminotransferase, which describes its enzymatic activity. However, some proteins, such as the RecA protein (∞ Sections 9.10, 10.3, and 10.5), do not have other names, because enzyme functions can be difficult to describe in a few words. Mutations in the *hisC* gene would be designated as *hisC1*, *hisC2*, and so on, the numbers referring to the order of isolation of the mutant strains.

The *phenotype* of an organism is usually designated by a capital letter followed by two lowercase letters, with either a plus or minus superscript to indicate the presence or absence of that property. For example, a His$^+$ strain of *E. coli* is capable of making its own histidine whereas a His$^-$ strain is not. Mutations in the *hisC* gene may lead to a His$^-$ phenotype if they eliminate the function of the gene product.

Isolation of Mutants

Virtually any characteristic of an organism can be changed by mutation. However, some mutations are

selectable, conferring some type of advantage on organisms possessing them, while others are *nonselectable*, even though they may lead to a very clear change in the phenotype of an organism. An example of such a nonselectable mutation is that of loss of color in a pigmented organism (Figure 10.1*b*). Such colonies usually have neither an advantage nor a disadvantage over the pigmented parent colonies when grown on agar plates (there may be a selective advantage for pigmented organisms in nature, however). We can detect such mutations only by examining large numbers of colonies and looking for the "different" ones, a process called **screening**. Note that the apparent color of a colony need not be because of pigment production. Mutant colonies of *Halobacterium* lacking gas vesicles (∞ Section 4.14) transmit light differently than wild-type colonies and appear strikingly different on plates (Figure 10.1*c*).

Nonselectable mutants must be found by screening a large population of organisms, and the mutant phenotype may not be as easy to recognize as the difference between pigmented and nonpigmented colonies. A *selectable* mutation, on the other hand, confers on the mutant an advantage under certain environmental conditions, so the progeny of the mutant cell are able to outgrow and replace the parent. An example of a selectable mutation is drug resistance: An antibiotic-resistant mutant can grow in the presence of antibiotic concentrations that inhibit or kill the parent (Figure 10.1*a*). However, the antibiotic-sensitive phenotype cannot be directly selected for by eliminating the antibiotic from the medium. It is relatively easy to detect and isolate selectable mutants by choosing the appropriate environmental conditions. Therefore, **selection** is an extremely powerful genetic tool, allowing the isolation of a single mutant from a population containing millions or even billions of parental organisms.

Although screening is always more tedious than selection, for certain types of mutations methods are available for screening large numbers of colonies. For instance, nutritional mutants can be detected by the technique of **replica plating** (Figure 10.2*a*). With the use of sterile velveteen cloth or filter paper, an imprint of colonies from a master plate is made onto an agar plate lacking the nutrient. The colonies of the parental type will grow normally, whereas those of the mutant will not. Thus, the inability of a colony to grow on the replica plate (Figure 10.2*b*) is a signal that it is a mutant. The colony on the master plate corresponding to the vacant spot on the replica plate (Figure 10.2*b*) can then be picked, purified, and characterized. A nutritional mutant that has a requirement for a growth factor is often called an **auxotroph**, and the wild-type

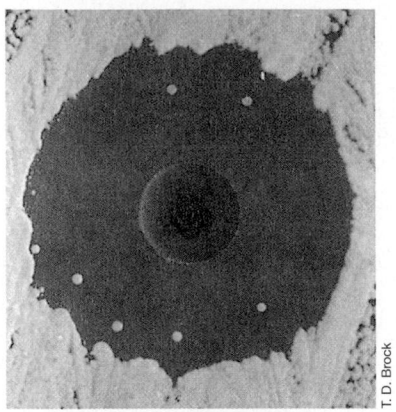

(a)

T. D. Brock

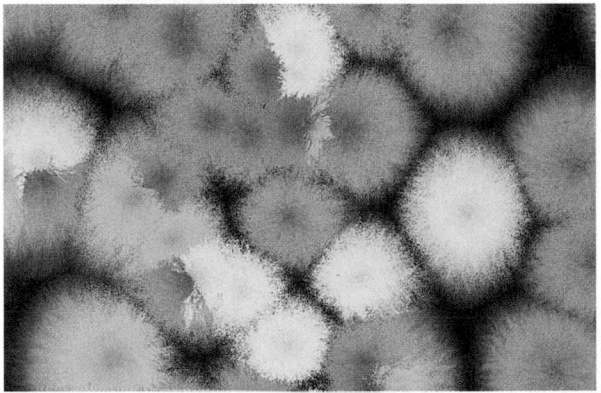

(b)

Peter T. Borgia

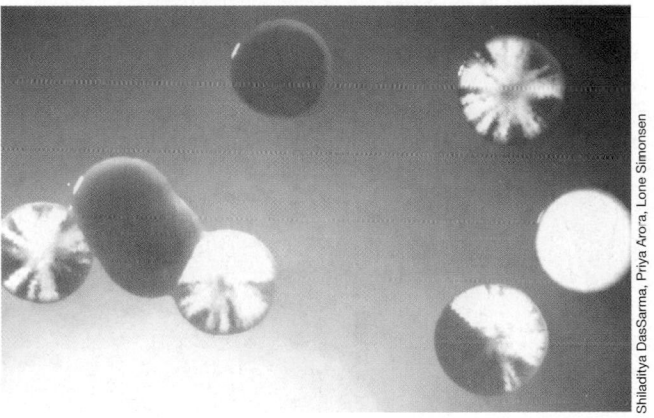

(c)

Shiladitya DasSarma, Priya Arora, Lone Simonsen

Figure 10.1 Observation of several kinds of mutants. (a) Development of antibiotic-resistant mutants within the inhibition zone of an antibiotic assay disc. (b) Pigmented mutants and nonpigmented mutants of the fungus *Aspergillus nidulans*. The wild type has a green pigment. The white or colorless mutants make no pigment, whereas the yellow mutants cannot convert the pigment they do make to the normal color. (c) Colonies of mutants of a species of *Halobacterium*, a member of the *Archaea*. The colonies that appear white are the wild type. The orange/brown colonies are mutants that are missing gas vesicles (∞ Section 4.14). Sectored colonies are the result of the mutagenic activities of transposable elements (see Section 10.11).

parent from which the auxotroph was derived is called a **prototroph**. For instance, mutants of *Escherichia coli* with a His⁻ phenotype are said to be *histidine auxotrophs*. Although of great utility, replica plating is a screening process, and it can be laborious to isolate mutants by screening.

Penicillin Selection

An ingenious method widely used to isolate auxotrophs is the **penicillin-selection method**. Ordinarily, mutants that require growth factors are at a disadvantage in competition with the parent cells, and so there is no direct way of isolating them. However, penicillin kills only *growing* cells, and if penicillin is added to a population growing in a medium lacking the growth factor required by the desired mutant, the parent cells will be killed, whereas the nongrowing mutant cells will be unaffected. Thus, after preliminary incubation in the absence of growth factor in a penicillin-containing medium, the population is washed free of penicillin and transferred to plates containing the growth factor. Among the colonies that grow up (including some wild-type cells that have escaped penicillin killing) should be some growth factor mutants. Penicillin selection is a kind of *negative selection*; the selection is not for the mutant but against the parental type.

Some of the most common kinds of mutants and the means by which they are detected are listed in Table 10.1.

✓ 10.1 Concept Check

Mutation, a heritable change in DNA, can lead to a change in phenotype. Selectable mutations are those that give the mutant a growth advantage under certain environmental conditions and are especially useful in genetics research.

✓ Distinguish between the words *mutation* and *mutant*.

✓ Distinguish between the words *screening* and *selection*.

10.2 Molecular Basis of Mutation

As previously mentioned, mutations arise in cells because of changes in the *base sequence* of an organism's genetic material. In many cases, mutations lead to phenotypic changes in the organism; these changes are mostly harmful or neutral, although beneficial changes do occur occasionally.

Mutation can be either spontaneous or induced. **Spontaneous mutations** can occur as a result of the action of natural radiation (cosmic rays, and so on) that alters the structure of bases in the DNA. Spontaneous

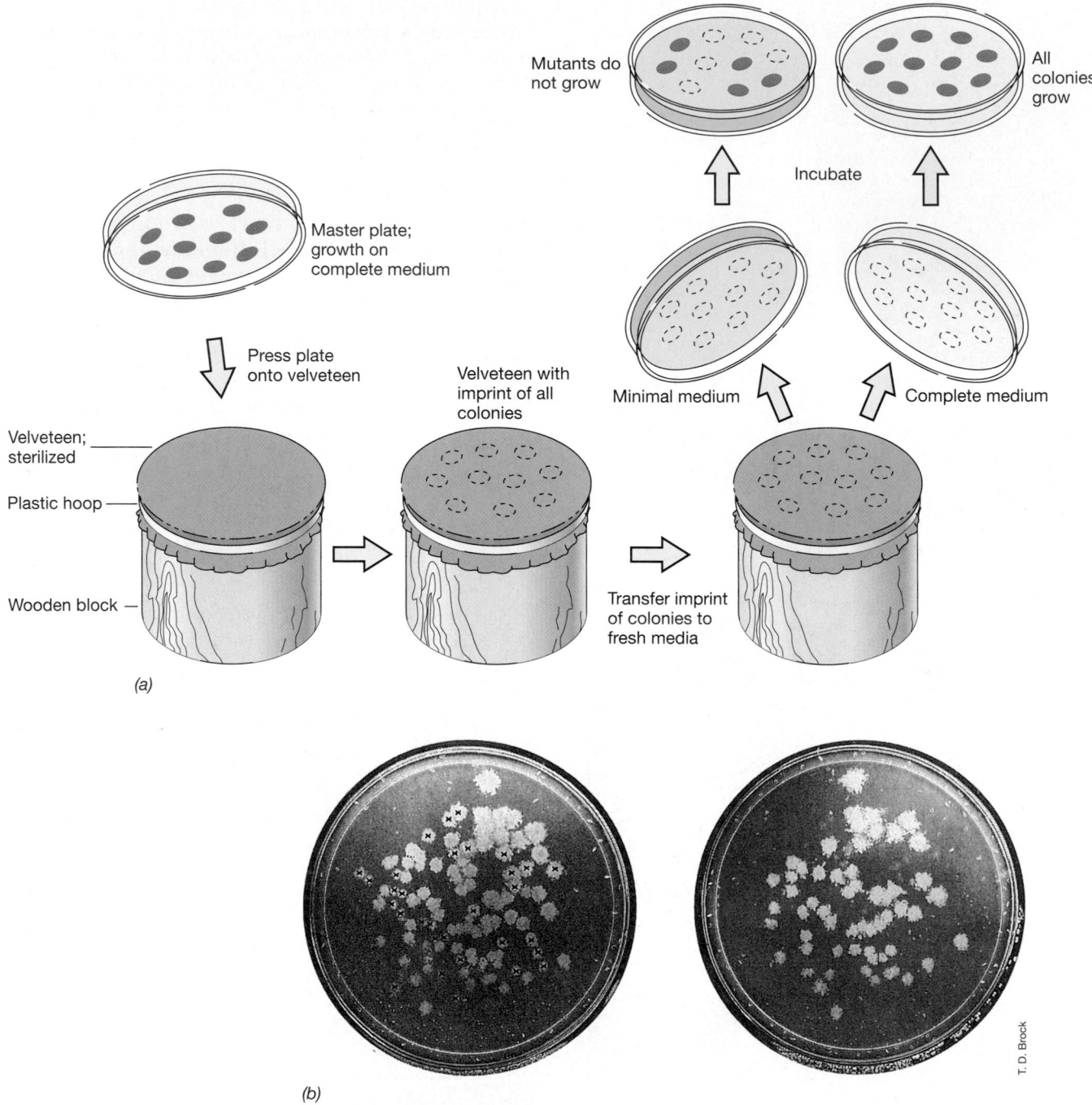

Figure 10.2 (a) Replica plating method for detection of nutritional mutants. (b) Nutritional mutants, as revealed by the replica plating method. The photograph on the left shows the master plate. The colonies not appearing on the replica plate are marked with an X. The replica plate lacked one nutrient (leucine) present in the master plate. Therefore, the colonies marked with an X are leucine auxotrophs.

mutations can also occur during replication, as a result of errors in the pairing of bases, leading to changes in the replicated DNA.

Mutations involving one (or a very few) base pairs are sometimes referred to as **point mutations**. Point mutations can result in *base-pair substitutions* in the DNA or in the insertion or deletion of a base pair (called *micro-insertions* and *microdeletions*). As is the case with all mutations, the phenotypic change that comes about because of a point mutation depends on exactly where the mutation took place in the gene, what the nucleotide change was, and what product the gene normally encodes.

TABLE 10.1	Kinds of mutants	
Phenotype	Nature of change	Detection of mutant
Auxotroph	Loss of enzyme in biosynthetic pathway	Inability to grow on medium lacking the nutrient
Cold-sensitive	Alteration of an essential protein so it is inactivated at low temperature	Inability to grow at a low temperature (for example, 20°C) that normally supports growth
Drug-resistant	Alteration of permeability to drug or drug target or detoxification of drug	Growth on medium containing a growth-inhibitory concentration of the drug
Noncapsulated	Loss or modification of surface capsule	Small, rough colonies instead of larger, smooth colonies
Nonmotile	Loss of flagella; nonfunctional flagella	Compact colonies instead of flat, spreading colonies
Pigmentless	Loss of enzyme in biosynthetic pathway leading to loss of one or more pigments	Presence of different color or lack of color
Rough colony	Loss or change in lipopolysaccharide outer layer	Granular, irregular colonies instead of smooth, glistening colonies
Sugar fermentation	Loss of enzyme in degradative pathway	Lack of color change on agar containing sugar and a pH indicator
Temperature-sensitive	Alteration of an essential protein so it is more heat-sensitive	Inability to grow at a temperature normally supporting growth (for example, 40°C) but still growing at a lower temperature (for example, 30°C)
Virus-resistant	Loss of virus receptor	Growth in presence of large amounts of virus

Base-Pair Substitutions

If a point mutation occurs within the coding region of a gene that encodes a protein, any change in the phenotype of the cell is almost certainly the result of a change in the amino acid *sequence* of the protein being produced. Figure 10.3 shows a number of base-pair substitutions that can occur in a short region of DNA within a gene that encodes a protein. The error in the DNA is transcribed into mRNA, and this erroneous mRNA in turn is used as a template and translated into protein. (Because only one strand of the DNA is used as template for the mRNA, an AT base pair does not have the same meaning as a TA base pair.) The triplet code that directs the insertion of an amino acid via a transfer RNA will thus be incorrect. What are the consequences of base-pair substitutions?

In interpreting the results of mutation, we must first recall that the genetic code is degenerate (∞ Section 7.13). Because of this degeneracy, not all mutations in protein-encoding genes result in changes in protein. This is illustrated in Figure 10.3, which shows several possible results when the DNA that encodes a single *tyrosine codon* undergoes mutation. As seen, a change in the RNA from UAC to UAU would have no apparent effect because UAU is also a tyrosine codon. Mutations that give rise to such changes are called **silent mutations**. Note that silent mutations in coding regions almost always occur in the *third base* of the codon (arginine and leucine can also have silent mutations in the first position). As seen in Chapter 7 (∞ Table 7.3), changes in the first or second base of the triplet much more often lead to significant changes in the protein. For instance, a single base change from UAC to AAC (Figure 10.3) would result in a change in the protein from tyrosine to asparagine. This is referred to as a **missense mutation** because the chemical "sense" (sequence of amino acids) in the ensuing polypeptide has changed. If the change occurred at a critical point in the polypeptide chain, the protein could be inactive, or have reduced activity. However, not all mutations that cause amino acid substitution necessarily lead to nonfunctional proteins. The outcome depends on where in the polypeptide chain the substitution has occurred and on how it affects the folding and the catalytic activity of the protein. A missense mutation can lead to an enzyme that is temperature-sensitive. **Temperature-sensitive mutants** of

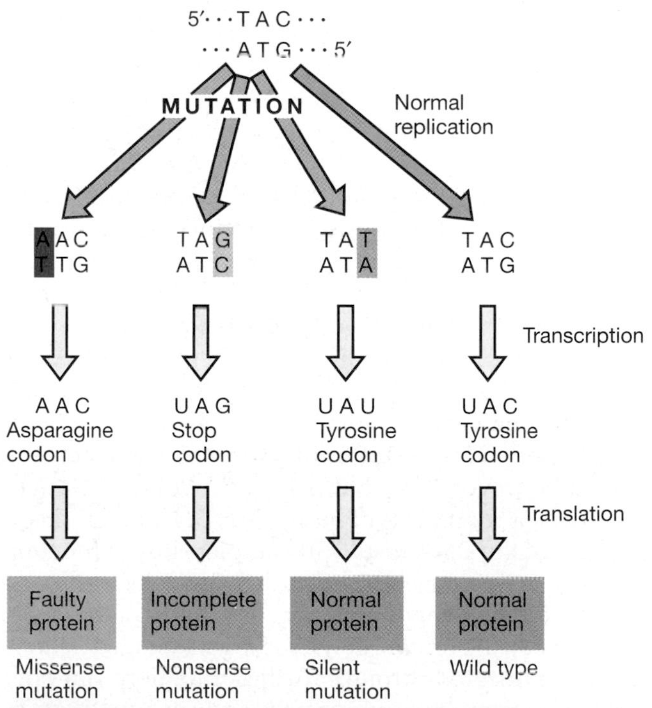

Figure 10.3 Possible effects of base-pair substitution in a gene encoding a protein: three different protein products from changes in the DNA for a single codon.

bacteria are known, for example, that function normally at 30°C but cannot grow at 40°C , although the wild type grows well at both temperatures. Such mutations are also called **conditionally lethal** because the bacteria cannot grow under one condition but can under another. Temperature-sensitive phenotypes often occur because the mutant protein can maintain its correct conformation at the low temperature but becomes partially unfolded (denatured) at the high temperature.

Another possible result of a base pair substitution is the formation of a *stop codon*, which would result in premature termination of translation, leading to an incomplete protein that would almost certainly not be functional (Figure 10.3). Mutations of this type are called **nonsense mutations** because the change is from a codon for an amino acid (sense codon) to a stop codon (nonsense codon). Unless the nonsense mutation occurs very near the end of the gene, the incomplete protein would be completely inactive.

Frameshift Mutations

Because the genetic code is read from one end in consecutive blocks of three bases, any deletion or insertion of a base pair results in a **reading frame shift**, and the translation of the gene is completely upset (Figure 10.4). Partial restoration of gene function can often be accomplished by insertion of another base pair near the one deleted (one kind of suppressor mutation; see later in this section). After correction, depending on the exact amino acids coded by the still faulty region and the region of the pro-

tein involved, the protein formed may have some biological activity or even be completely normal.

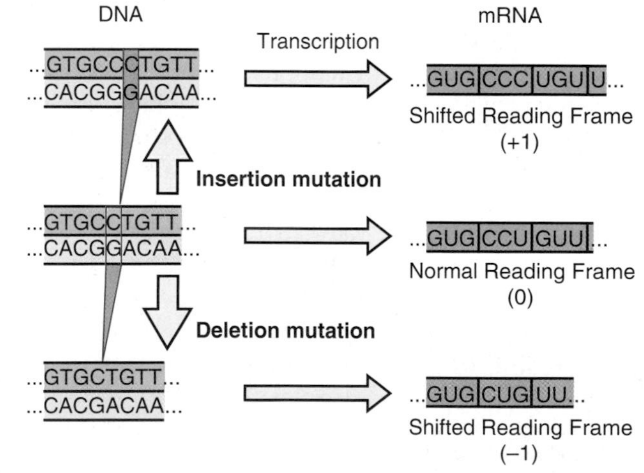

It is important to remember that microinsertions or microdeletions are frameshift mutations only if they occur in the part of a protein-encoding gene that includes the reading frame. A single base-pair insertion in the promoter of a gene could lead to a dramatic change in the ability of the gene to function (because of increased or decreased expression or a change in regulation), but it would not be a frameshift mutation. (Similarly, base-pair substitutions that are not within the reading frame are not missense or nonsense mutations.)

Back Mutations or Reversions

Point mutations are reversible. A *revertant* is operationally defined as a strain in which the wild-type phenotype that was lost in the mutant is restored. Revertants can be of two types. In *same-site revertants*, the mutation that restores activity occurs at the same site at which the original mutation occurred. (If the back mutation is not only at the same site but also leads to the wild-type sequence, it is called a *true revertant*.) In *second-site revertants*, the mutation occurs at a different site in the DNA.

Second-site mutations may cause restoration of a wild-type phenotype because of several types of **suppressor mutations** that restore the original phenotype. Suppressor mutations are new mutations that compensate for the effect of the original mutation. Several types of suppressor mutations are known: (1) a mutation somewhere else in the same gene can restore enzyme function, such as in a frameshift mutation; (2) a mutation in another gene may restore the function of the original mutated gene and the wild-type phenotype; and (3) a mutation in another gene may result in the production of another enzyme that can replace the mutant one by introducing a metabolic pathway different from that used by the mutant enzyme. In this last type, no production of the original enzyme occurs although it does in the other types.

Mutations Involving Many Base Pairs

Deletions are mutations in which a region of the DNA has been eliminated. As we have discussed, microdeletions, the removal of one or a very few base pairs (Figure 10.4), are often frameshift mutations and can inactivate a gene. However, deletions can also involve the loss of hundreds or thousands of base pairs. Deletion of a large segment of the DNA results in complete loss of function of any gene that may be involved. Some deletions are so large that they involve several genes (if any of the genes are essential, the mutation will be lethal). Such deletions cannot be restored through further mutations but only through genetic recombination. Indeed, one way in which large deletions are distinguished from point mutations is that the latter are reversible through further mutations, whereas the former are not.

Figure 10.4 Shifts in the reading frame of messenger RNA caused by insertion or deletion mutations in DNA. The reading frame in mRNA is established by the ribosome that begins at the 5′ end (toward the left in the figure) and precedes by units of three bases (codons, ↝ Sections 7.13 and 7.15). The normal reading frame is referred to as the 0 frame, that missing a base the −1 frame, and that with an extra base the +1 frame. To determine the effects of a frameshift, translate the codons using Table 7.3.

Insertions occur when new bases are added to the DNA. As we have discussed, insertions, like deletions, can involve only a single base or many bases. Microinsertions result from replication errors as do deletions, but larger insertions arise as a result of mistakes that occur during genetic recombination. Insertions inactivate the gene in which they occur. Many insertion mutations are due to the insertion of specific identifiable DNA sequences 700–1400 base pairs in length called **insertion sequences**, a type of transposable element (∞ Section 7.4). The behavior of such insertion sequences is discussed in detail in Section 10.11. Even large insertion mutations can revert by further mutation, that is, by a deletion that removes the insertion.

Other types of large-scale mutations seem to involve rearrangements brought about by mistakes in recombination. These include **translocations**, in which a large section of chromosomal DNA is moved to a new location (and in eukaryotes often to a different chromosome), and **inversions**, in which the orientation of a particular segment of DNA is reversed with respect to the surrounding DNA.

Rates of Mutation

The rates at which various kinds of mutations occur vary widely. Some types of mutations occur so rarely that they are almost impossible to detect, whereas others occur so frequently that they often present difficulties for an experimenter trying to maintain a genetically stable stock culture.

Errors in DNA replication occur at a frequency of about 10^{-7}–10^{-11} per base pair during a single round of replication. A typical gene has about 1000 base pairs; therefore, the frequency of these errors in a gene would be 10^{-4} to 10^{-8} per generation. Thus, in a normal, fully grown culture of organisms having approximately 10^8 cells/ml, there are probably a number of different mutants in each milliliter of culture.

However, not all mutations happen at the same frequency. Transposition events may occur more frequently, at about 10^{-4}. On the other hand, the occurrence of a nonsense mutation is less frequent, 10^{-6}–10^{-8}, because only a few codons can mutate to nonsense codons. There are also DNA sequences, usually involving short repeats, which are *hot spots* for mutations because the error frequency of DNA polymerase can be quite high at such sequences.

Unless the mutant is selectable, the experimental detection of events of such rarity is difficult, and much of the skill of the microbial geneticist is applied to increasing the efficiency of mutation detection. As we will see in the next section, it is possible to significantly increase the rate of mutation by the use of mutagenic treatments. In addition, we will also see that the mutation rate of a gene may change in certain situations.

Mutations in RNA Genomes

Whereas all cells have *DNA* as their genetic material, some viruses have *RNA* genomes (for examples, ∞ Sections 9.12, 16.1, 16.7–16.9, and 16.14. These genomes can also mutate. Importantly, the mutation rate in RNA genomes is about *1000-fold higher* than in DNA genomes. At least some RNA polymerases have *proof-reading* activities like those of DNA polymerases (∞ Section 7.6). However, while there are several repair systems for DNA that can correct many changes before they become fixed as mutations (see Section 10.3), there seem to be no comparable RNA repair mechanisms. This very high rate of mutation in RNA viruses is not merely of academic interest. The RNA genomes of viruses that cause disease can mutate very rapidly, presenting a constantly changing and evolving population of viruses.

✓ 10.2 Concept Check

Mutations, which can be either spontaneous or induced, arise because of changes in the base sequence of the nucleic acid of an organism's genome. A point mutation, which is due to a change in a single base pair, can lead to a single amino acid change in a protein or to no change at all, depending on the particular codon involved. In a nonsense mutation, the codon becomes a stop codon and an incomplete protein is made. Deletions and insertions cause more dramatic changes in the DNA, including frameshift mutations, and often result in complete loss of gene function.

✓ What does it mean to say that point mutations can spontaneously revert?

✓ Do missense mutations occur in genes encoding transfer RNAs (tRNAs)?

10.3 Mutagenesis

While the spontaneous rate of mutation is very low, there are a variety of chemical, physical, or biological agents that can increase the mutation rate, and are therefore said to *induce* mutations. These agents are referred to as **mutagens**. We discuss some of the major categories of mutagens and their actions here.

Chemical Mutagens

An overview of some of the major chemical mutagens and their modes of action is given in Table 10.2. Several classes of chemical mutagens exist. One variety of chemical mutagens are the **base analogs**, which resemble DNA purine and pyrimidine bases in structure yet show faulty pairing properties (Figure 10.5). When one of these base analogs is incorporated into DNA, replication may occur normally most of the time, but occasional copying errors occur, resulting in incorporation of the wrong base into the copied strand. During subsequent segregation of this strand, the mutation is revealed.

TABLE 10.2 Chemical and physical mutagens and their modes of action

Agent	Action	Result
Base analogs		
5-Bromouracil	Incorporated like T; occasional faulty pairing with G	AT pair → GC pair Occasionally GC → AT
2-Aminopurine	Incorporated like A; faulty pairing with C	AT → GC Occasionally GC → AT
Chemicals reacting with DNA		
Nitrous acid (HNO_2)	Deaminates A and C	AT → GC and GC → AT
Hydroxylamine (NH_2OH)	Reacts with C	GC → AT
Alkylating agents		
Monofunctional (for example, ethyl methane sulfonate)	Put methyl on G; faulty pairing with T	GC → AT
Bifunctional (for example, nitrogen mustards, mitomycin, nitrosoguanidine)	Cross-link DNA strands; faulty region excised by DNase	Both point mutations and deletions
Intercalative dyes		
Acridines, ethidium bromide	Insert between two base pairs	Microinsertions and microdeletions
Radiation		
Ultraviolet	Pyrimidine dimer formation	Repair may lead to error or deletion
Ionizing radiation (for example, X-rays)	Free-radical attack on DNA, breaking chain	Repair may lead to error or deletion

Some other chemical mutagens react directly on DNA, causing chemical changes in one base or another, which results in faulty pairing or other changes (Table 10.2). *Alkylating agents* such as nitrosoguanidine, for example, are powerful mutagens and generally induce mutations at higher frequency than base analogs. Such chemicals differ in their action from the base analogs in that the chemicals reacting on DNA are able to introduce direct changes even in nonreplicating DNA, whereas the base analogs act only after incorporation during replication. Both base analogs and alkylating agents tend to induce base-pair substitutions (see Section 10.2).

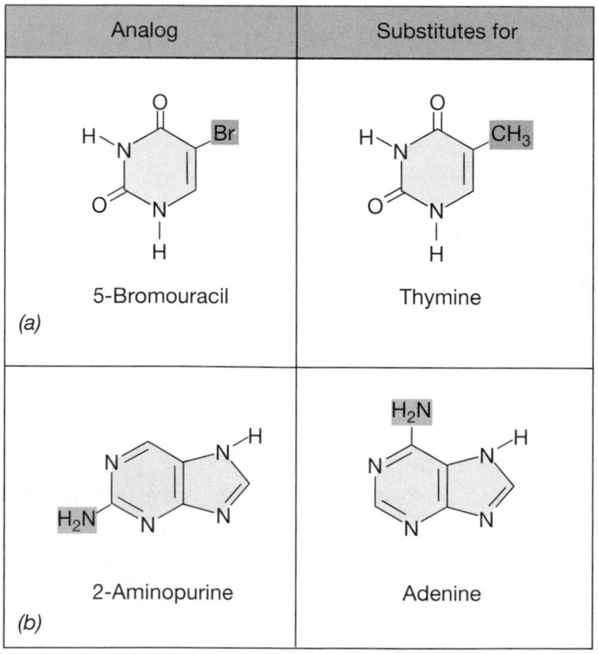

One interesting group of chemicals, the acridines, are planar molecules that act as *intercalating agents*. These mutagens become inserted between two DNA base pairs, thereby pushing them apart. During replication, this abnormal conformation can lead to microinsertions or microdeletions in acridine-treated DNA. Thus, acridines can induce frameshift mutations (see Section 10.2). Ethidium bromide, which is often used to detect DNA in electrophoresis (see Section 10.12), is also an intercalating agent that acts as a mutagen.

Analog	Substitutes for
5-Bromouracil	Thymine
2-Aminopurine	Adenine

(a) *(b)*

Figure 10.5 Structure of two common nucleotide base analogs used to induce mutations and the normal nucleic acid bases they substitute for. (a) 5-Bromouracil can base pair with guanine causing AT to GC substitutions. (b) 2-Aminopurine can base pair with cytosine, causing AT to GC substitutions.

Radiation

Several forms of radiation are highly mutagenic. We can divide mutagenic radiation into two main categories, ionizing and nonionizing (electromagnetic) (Figure 10.6). Although both kinds of radiation are used in microbial genetics, *nonionizing* radiations find the widest use and will be discussed first.

The purine and pyrimidine bases of the nucleic acids absorb ultraviolet (UV) radiation strongly, and the absorption maximum for DNA and RNA is at 260 nm (∞ Figure 7.9). Proteins also absorb UV but have a peak at 280 nm due to absorption of the aromatic amino acids (tryptophan, phenylalanine, tyrosine). It is now well established that killing of cells by UV radiation is due primarily to its action on DNA, and so UV radiation at 260 nm is most effective as a lethal agent. Although several

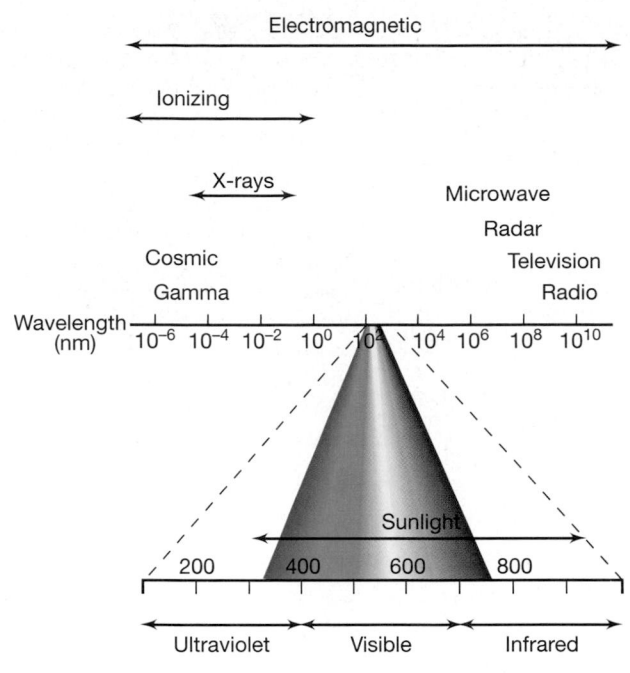

Figure 10.6 Wavelengths of radiation.

effects are known, one well-established effect is the induction in DNA of **pyrimidine dimers**, a state in which two adjacent pyrimidine bases (cytosine or thymine) become covalently joined so that during replication of the DNA, the probability of DNA polymerase inserting an incorrect nucleotide at this position is greatly increased.

The type of UV radiation source most frequently used for mutagenesis is the germicidal lamp, which emits large amounts of UV radiation in the 260-nm region. A dose of UV radiation is used that brings about 90–95% killing of the cell population (Section 20.2), and mutants are then looked for among the survivors. If much higher doses of radiation are used, the number of viable cells will be too low, whereas if lower doses are used, insufficient damage to the DNA will be induced. UV radiation is a very useful tool in isolating mutants of microbial cultures.

Ionizing Radiation

Ionizing radiation is a more powerful form of radiation and includes short-wavelength rays such as X-rays, cosmic rays, and gamma rays (Figure 10.6). These radiations cause water and other substances to ionize, and mutagenic effects are brought about indirectly through this ionization. Among the potent chemical species formed by ionizing radiation are chemical free radicals, of which the most important is the hydroxyl radical, $OH \cdot$. Free radicals react with and inactivate macromolecules in the cell, of which the most important is DNA. DNA is probably no more sensitive to ionizing radiation than other macromolecules, but because DNA is

the genetic material, changes induced in it can have a permanent effect. At low doses of ionizing radiation, only a few hits on DNA occur, but at higher doses multiple hits occur, leading to the death of the cell. In contrast to UV radiation, ionizing radiation penetrates readily through glass and other materials. Because of this, ionizing radiation is used frequently to induce mutations in animals and plants (where its penetrating power makes it possible to reach the gamete-producing cells of these organisms readily), but because ionizing radiation is more dangerous to use and is less readily available, it finds less use with microorganisms (where penetration with UV is not a problem).

Mutations That Arise from DNA Repair

Recall that a mutation is an *inheritable* change in the genetic material. Therefore, if an error in DNA synthesis can be corrected before the cell divides, there will be no mutation. Furthermore, some DNA damage clearly cannot be replicated and, therefore, cannot itself be a mutation. For instance, if a DNA molecule contains pyrimidine dimers, it will *not* be replicated to give two DNA molecules, each containing pyrimidine dimers. If such DNA damage cannot be repaired, cells often die. Most cells have a variety of different DNA repair processes to correct mistakes or repair damage. Many of these DNA repair systems do not make mistakes. However, some processes seem to be *error-prone*, and it is the repair process itself that introduces the mutation. Many kinds of mutations arise as a result of faulty repair of damage induced in DNA by some of the various agents just discussed.

Often DNA damage itself can *induce* DNA repair systems. A complex cellular mechanism, called the **SOS regulatory system**, is activated as a result of some types of DNA damage, initiating a number of DNA repair processes. However, in the SOS system, some DNA repair occurs in the absence of template instruction, that is, without base-pairing, which results in the creation of many errors, hence many mutations.

In the SOS regulatory system, DNA damage serves as a distress signal to the cell, resulting in the coordinate derepression (induction) of a number of cellular functions involved in DNA repair. The SOS system is normally repressed by a protein called LexA. However, this repressor protein is inactivated by RecA, a protease that is activated as a result of DNA damage (Figure 10.7). Since one of the DNA repair mechanisms of the SOS system is inherently error-prone, many mutations arise. Thus, through the SOS regulatory system, DNA damage by various agents such as chemicals and radiation leads to mutagenesis. Much of this is brought about by DNA polymerase V, which is encoded by the *umuC* and *umuD* genes (see Figure 10.7). The error-prone DNA polymerase IV is also induced as part of the SOS

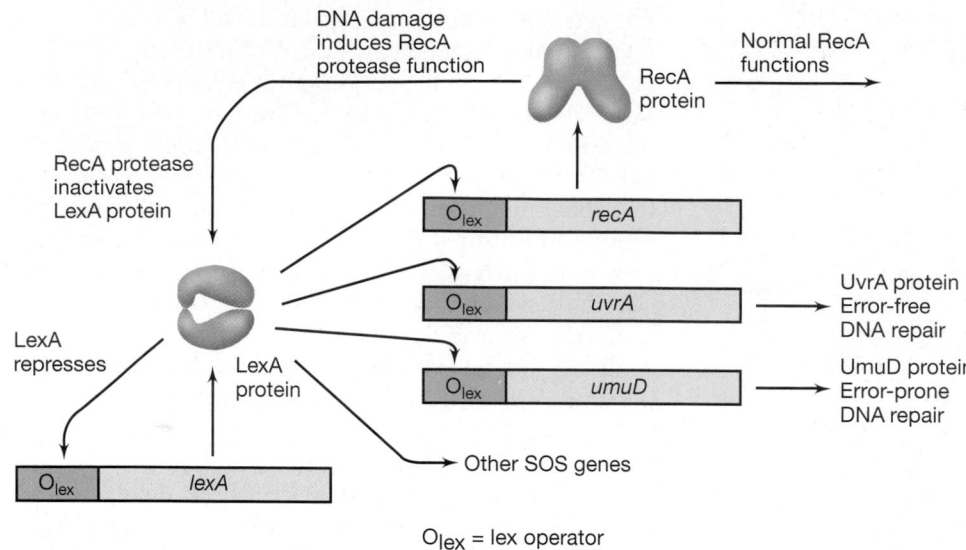

Figure 10.7 Mechanism of the SOS response. DNA damage results in conversion of RecA protein into a protease that cleaves LexA protein. LexA protein normally represses the activities of the *recA* gene and the DNA repair genes *uvrA* and *umuD*. (The UmuD protein is part of DNA polymerase V.) Note, however, that repression is not complete. Some RecA protein is produced even in the presence of LexA protein. With LexA inactivated, these genes become active. As a protease, the RecA protein also cleaves the lambda repressor protein (∞ Section 9.10).

response. These enzymes are sometimes referred to as **DNA mutases**.

Once the DNA damage has been repaired, the SOS system is switched off and further mutagenesis ceases. In addition to its effect on the cellular mutation rate, the SOS regulatory system plays a central role in the regulation of temperate virus replication (∞ Section 9.10).

It should be emphasized that not all DNA repair occurs in the absence of template instruction. Cells generally have many DNA repair systems that require template instruction and lead to proper DNA repair. These systems apparently work most of the time but are not sufficient to repair the large amounts of damage done by some of the agents previously mentioned.

Changes in Mutation Rate

The mutation rate of an organism is a phenotype under selection pressure like any other. The fact that hyperthermophilic *Archaea* (∞ Chapter 13) have about the same mutation rate as laboratory strains of *Escherichia coli* might lead one to imagine that this pressure has led to the selection of organisms with the lowest possible mutation rates. However, this is not so. Mutants of some organisms are known that are hyperaccurate. In these organisms, though, the proofreading or repair mechanisms may have such a high metabolic cost that the organism is typically at a disadvantage. Of course, certain organisms may benefit from such mechanisms to occupy certain environmental niches. A good example is the bacterium *Deinococcus radiodurans* (∞ Section 12.34), which is 20 times more resistant to ultraviolet light and 200 times more resistant to ionizing radiation than *E. coli*. This resistance, dependent in part upon redundant DNA repair systems and on a mechanism for exporting damaged nucleotides, allows the organism to survive in environments where other organisms cannot. Perhaps even more interesting, the opposite seems also to be true; that is, some organisms may benefit from an *increased* mutation rate.

DNA repair systems are themselves encoded by genes and thus subject to mutation. For example, the subunit of DNA polymerase III involved in proofreading (∞ Section 7.6) is encoded by the gene *dnaQ*. Certain mutations in this gene lead to cells that are still viable but have an increased rate of mutation. A strain with an increased mutation rate is said to be hypermutable or a **mutator**. Mutations leading to a mutator phenotype are known in a number of DNA repair systems. The mutator phenotype is apparently selected for in complex and changing environments, since strains of bacteria with mutator phenotypes are more abundant under these conditions. Presumably, in such an environment whatever disadvantage an increased mutation rate may have is offset by the ability to generate new and useful mutations. (Note that when the environmental conditions stabilize or return to "normal," the advantage the mutator strains have will be lost.)

As we have indicated earlier, a mutator phenotype may be induced in wild-type strains by stress situations, not because of mutations in the genes involved in DNA repair, but because of the normal regulation or function of these genes. For instance, the SOS response induces error-prone repair. Therefore, when the SOS response is activated, the mutation rate increases. While in some cases this may be a necessary by-product of DNA repair, in other cases the increased mutation rate may itself be of value. That is, the organism may generate *adaptive mutations*. **Adaptive mutations** lead to a phenotype in the mutant that allows it to survive a particular stress.

Biological and Site-Directed Mutagenesis

Mutations can be introduced without the use of chemical or physical agents through the process of *transposon mutagenesis*. We discuss the details of transposon mutagenesis later in this chapter (see Section 10.11) and note here only that if insertion of a transposable element occurs *within* a gene, loss of gene function generally results (see Figure 10.1c). Because transposable elements can enter the chromosome at various locations, transposons are widely used by microbial geneticists as mutagenic agents.

So far, the mutations that we have been discussing have been randomly directed at the genome of the microbial cell. Recombinant DNA technology and the use of synthetic DNA make it possible to induce *specific* mutations in *specific* genes. The procedures for carrying out mutagenesis of specific sites in the genome are called *site-directed mutagenesis* and will be discussed in detail at the end of this chapter (see Section 10.18).

✓ 10.3 Concept Check

Mutagens are chemical, physical, or biological agents that increase the mutation rate. Mutagens can alter DNA in many different ways. However, alterations in DNA are not mutations unless they can be inherited. Some DNA damage can lead to cell death if not repaired.

✓ How do mutagens work?
✓ Differentiate between a mutation and DNA damage.

10.4 Mutagenesis and Carcinogenesis: The Ames Test

A practical use of mutant bacterial strains has been developed to identify potentially hazardous chemicals in the environment. Because the sensitivity with which selectable mutants can be detected in large populations of bacteria is very high, bacteria can be used as screening agents for the potential mutagenicity of chemicals. This is relevant because it has been found that many mutagenic chemicals are also carcinogenic, capable of causing cancer in animals or humans.

The variety of chemicals, both natural and artificial, that the human population comes into contact with through agricultural and industrial exposure is enormous. There is considerable need for simple tests to ascertain the safety of such compounds. There is good evidence that a large proportion of human cancers have environmental causes, most likely through the agency of various chemicals, making the detection of chemical carcinogens urgent. It does not necessarily follow that because a compound is mutagenic it is also carcinogenic. The correlation, however, is very high, and the knowledge that a compound is mutagenic in a bacterial sys-

tem serves as a warning of possible danger. Similarly, the fact that a compound is not mutagenic in a bacterial system does not mean that it is not carcinogenic, because the bacterial system cannot detect all compounds active in higher animals. The development of bacterial tests for carcinogenic screening was carried out primarily by a group at the University of California in Berkeley under the direction of Bruce Ames, and the mutagenicity test for carcinogens is sometimes called the **Ames test** (Figure 10.8).

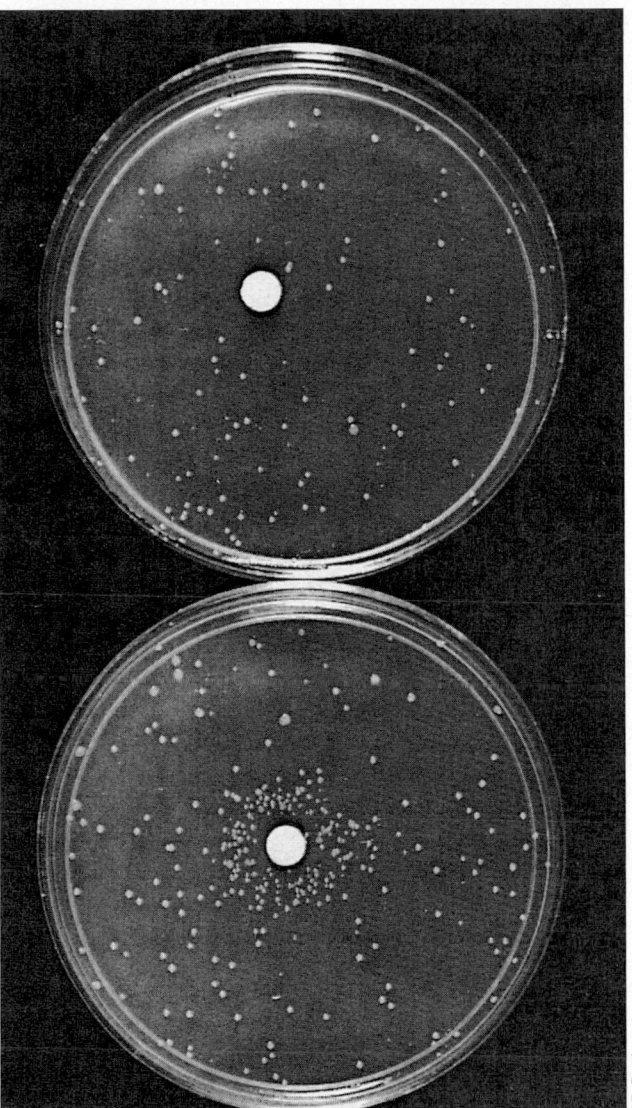

Figure 10.8 The Ames test is used to evaluate the mutagenicity of a chemical. Both plates were inoculated with a culture of a histidine-requiring mutant of *Salmonella enterica*. The medium does not contain histidine, so only cells that revert back to wild type can grow. Spontaneous revertants appear on both plates, but the chemical on the filter paper disc in the test plate (bottom) has caused an increase in the mutation rate, as shown by the large number of colonies surrounding the disc. Revertants are not seen very close to the disc because the concentration of the mutagen is so high there that it is lethal.

Protocol for an Ames Test

The standard way to test chemicals for mutagenesis has been to determine if the rate of *back* mutation (reversion) in strains of bacteria that are auxotrophic for some nutrient is increased by the suspected mutagen. It is important that the original mutation be a point mutation so reversion can occur. When cells of such an auxotrophic strain are spread on a medium lacking the required nutrient (for example, an amino acid), no growth occurs, and even very large populations of cells can be spread on the plate without formation of visible colonies. However, if back mutants (revertants) are present, those cells will be able to form colonies. Thus, if 10^8 cells are spread on the surface of a single plate, even as few as 10–20 revertants can be detected by the 10–20 colonies they form. If the reversion rate has been increased by a chemical mutagen, the number of revertant colonies will also increase. Histidine auxotrophs of *Salmonella enterica* (Figure 10.8) and tryptophan auxotrophs of *Escherichia coli* have been the major tools of the Ames test.

Although the simple testing of chemicals for mutagenesis in bacteria has been carried out for a long time, two elements have been introduced in the Ames test to make it much more powerful. The first of these is the use of strains of bacteria that almost exclusively use errorprone pathways to repair DNA damage (see Section 10.3). The second important element in the Ames test is the use of liver enzyme preparations to convert the chemicals into their active mutagenic (and carcinogenic) forms. It has been well established that many potent carcinogens are not directly carcinogenic or mutagenic but undergo chemical changes in the human body that convert them into active substances. These changes take place primarily in the liver, where enzymes (mixed-function oxygenases) normally involved in detoxification cause formation of epoxides or other activated forms of the compounds, which are then highly reactive with DNA.

In the Ames test, a preparation of enzymes from rat liver is first used to activate the compound. Then the activated complex is taken up on a filter-paper disk, which is placed in the center of a plate on which the proper bacterial strain has been overlaid. After overnight incubation, the mutagenicity of the compound can be detected by looking for a halo of back mutations in the area around the paper disk (Figure 10.8). It is always necessary to carry out this test with several different concentrations of the compound and with appropriate positive and negative controls because compounds vary in their mutagenic activity and are lethal at higher levels. A wide variety of chemicals have been subjected to the Ames test, and it has become one of the most useful prescreens for determining the potential carcinogenicity of a compound.

✓ **10.4** *Concept Check*

The Ames test employs a sensitive bacterial assay system for detecting chemical mutagens in the environment.

✓ Why does the Ames test measure the rate of *back* mutation rather than the rate of *forward* mutation?

✓ Of what significance is the detection of mutagens to the prevention of cancer?

10.5 Genetic Recombination

Genetic recombination involves the physical exchange of genetic material between genetic elements. In this section, we focus on **general** or **homologous recombination**, which results in genetic exchange between *homologous* DNA sequences from two different sources (although such recombination can occur between homologous sequences on the same chromosome). Homologous DNA sequences have the same or nearly the same sequence; therefore, base pairing can occur over an extended length of the two DNA molecules. This type of recombination is involved in the process referred to as "crossing over" in classical genetics.

Homologous recombination is extremely important to all organisms. However, it is also very complex. Even in the bacterium *Escherichia coli* there are at least 25 genes involved. In addition, homologous recombination seems to be of such importance that there are several redundant pathways. Therefore, if one pathway is inhibited or nonfunctional, another may be able to supply necessary functions.

Molecular Events in Homologous Recombination

At the molecular level, recombination has been studied mostly in prokaryotes and viruses. In *Bacteria*, general recombination involves the participation of a specific protein called the RecA protein, which is specified by the *recA* gene. The RecA protein has been shown to be essential in nearly every homologous recombination pathway. RecA-like proteins have been identified in all prokaryotes examined, including the *Archaea*, as well as in yeast and in the higher *Eukarya*.

An overall molecular mechanism of general recombination is shown in Figure 10.9. The process begins with a *nick* (usually generated by a nuclease) in one of the DNA molecules. This nicked strand must be displaced from the other strand by proteins having helicase activity (∞ Section 7.6). In some pathways specialized enzymes, such as the RecBCD enzyme of *E. coli*, have both nuclease and helicase activities. Single-stranded binding protein (∞ Section 7.6) then binds to the resulting single-stranded segment. Next, the RecA protein binds to the single-stranded fragment, forming

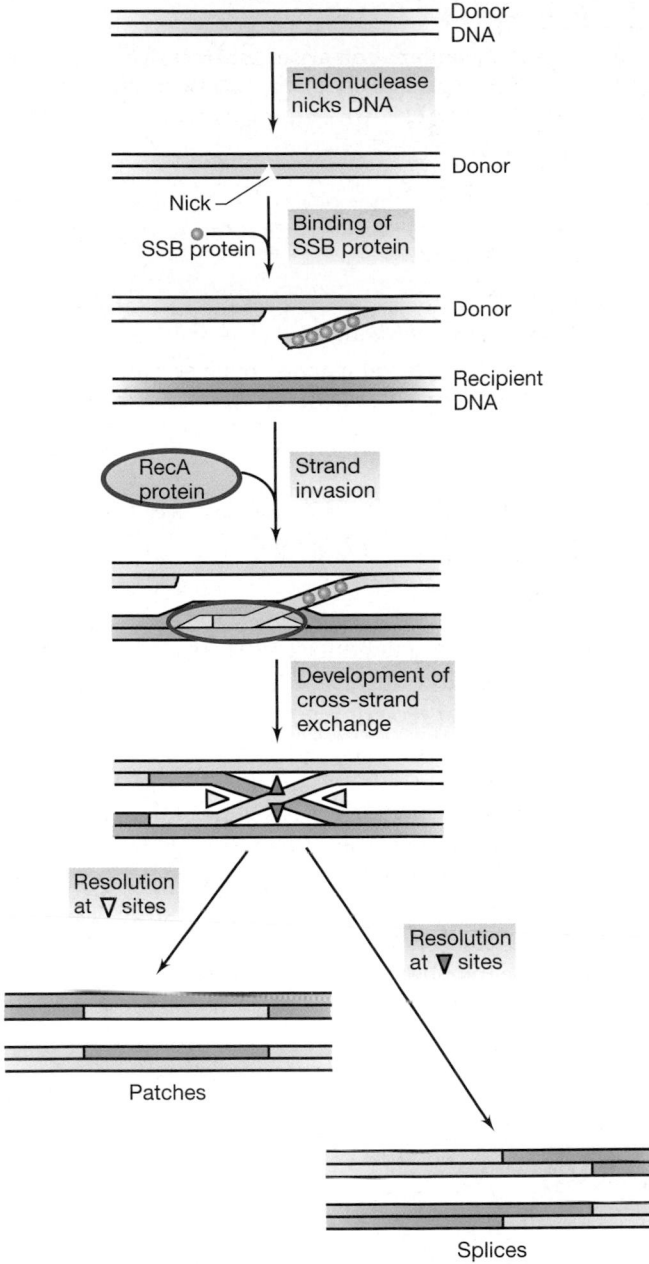

Figure 10.9 A simplified version of one molecular mechanism of genetic recombination. Homologous DNA molecules pair and exchange DNA segments. The mechanism involves breakage and reunion of paired segments. Two of the proteins involved, a single-stranded binding (SSB) protein and the RecA protein, are shown. The other proteins involved are not shown. The diagram is not to scale: Pairing can occur over hundreds or thousands of bases. Resolution occurs by cutting and ligating the cross-linked DNA molecules. Note that there are two possible outcomes, depending on which strands are cut during the resolution process. In one outcome the recombinant molecules have patches, whereas in the other the two parental molecules appear to have been cut and then spliced together.

a complex that facilitates annealing with a complementary sequence in the adjacent duplex, simultaneously displacing the resident strand (Figure 10.9). This process is often referred to as *strand invasion*. As noted, recombination involves the *pairing* of DNA molecules over long stretches. Following pairing, *exchange* of homologous DNA molecules can occur, leading to the formation of recombination intermediates containing extensive **heteroduplex** regions, where each strand has originated from a different chromosome. This process also involves DNA polymerase and ligase. These regions may be extended by the RecA protein. Finally, the linked molecules are *resolved* by nucleases and DNA ligase to form two recombinant molecules.

This mechanism for the formation of recombinant DNA structures is completely natural and occurs extensively within the cell. Whether it leads to the formation of new genotypes depends on whether the two molecules undergoing recombination differ genetically in regions outside the region of recombination. Within limits, general recombination can be thought of as occurring at random sites throughout the genome. Thus, the probability of recombination occurring between two genes is proportional to their distance. This fact is useful for genetic mapping. By recombinational analysis it is possible to map the position of genes on chromosomes because the *farther* two genes are apart, the more likely they are to show recombination.

For new genotypes to arise as a result of general recombination, it is essential that the two homologous sequences be genetically distinct. This is the case in a diploid eukaryotic cell (∞ Section 14.6), which has two sets of chromosomes, one from each parent. The two distinct molecules are brought together as the result of sexual reproduction, a process that occurs as part of the regular life cycles of most eukaryotic organisms. In prokaryotes, genetically distinct but homologous DNA molecules are brought together in different ways, but the process of genetic recombination is no less important. Recombination can also be critical in the life cycle of some viruses. In Chapters 9 and 16, we discuss how certain bacteriophages, such as T4 and T7 (∞ Sections 9.9 and 16.4), require homologous recombination as a step during DNA replication.

In prokaryotes genetic recombination is observed because *fragments* of homologous DNA from a donor chromosome are transferred to a recipient cell by one of three processes: *transformation* (see Section 10.6), *transduction* (see Section 10.7) or *conjugation* (see Section 10.9). It is *after* the transfer, when the DNA fragment from the host is in the recipient cell, that homologous recombination may occur. Because in prokaryotes only a chromosomal fragment is transferred, if recombination does not occur, the fragment will be lost because it cannot replicate independently. Therefore, it is important to remember that in prokaryotes transfer is just the first step in obtaining recombinant organisms.

Detection of Recombination

In order to detect physical exchange of DNA segments, the cells resulting from recombination must be phenotypically different from the parents. In crosses involving microorganisms, one must usually use as recipients strains that lack some selectable characteristic that the recombinants will possess. For instance, the recipient may not be able to grow on a particular medium, and genetic recombinants are selected that can. Various kinds of selectable and nonselectable markers (such as drug resistance, nutritional requirements, and so on) were discussed in Section 10.1. The exceedingly great sensitivity of the selection process is shown by the fact that 10^8 or more bacterial cells can be spread on a single plate and, if proper selective conditions are used, no parental colonies will appear, whereas even a few recombinants can form colonies (Figure 10.10). The only requirement is that the *reverse* mutation rate for the selected characteristic be low, because revertants will also form colonies. This problem can often be overcome by using double mutants because it is very unlikely that two back mutations will occur in the same cell. Much of the skill of the bacterial geneticist is exhibited in the choice of proper mutants and selective media for efficient detection of genetic recombination. Because selection is so powerful and because crosses can be made using billions of individual cells, recombinational analysis is a very important tool to the microbial geneticist.

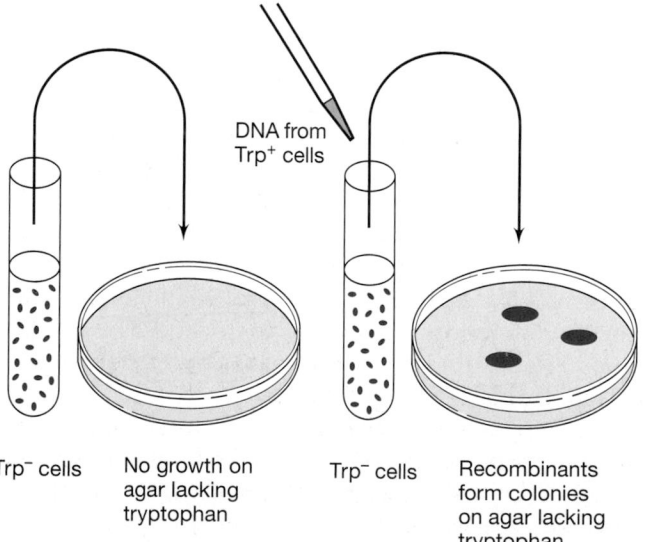

Trp⁻ cells   No growth on agar lacking tryptophan   **Trp⁻ cells**   Recombinants form colonies on agar lacking tryptophan

Figure 10.10 How a selective medium can be used to detect rare genetic recombinants among a large population of nonrecombinants. On the selective medium only the rare recombinants form colonies. Procedures such as this, which offer high resolution for genetic analyses, can ordinarily be used only with microorganisms. The type of genetic exchange being illustrated is transformation, which is discussed in Section 10.6.

✓ 10.5 Concept Check

Homologous recombination arises when closely related sequences from two genetically distinct elements are combined together in the same element. Recombination is an important evolutionary process, and cells have specific mechanisms for ensuring that recombination takes place. Mechanisms of recombination that occur in prokaryotes involve DNA transfer during the processes of transformation, transduction, and conjugation.

✓ What protein, found in all prokaryotes, facilitates the pairing required for homologous recombination?

✓ In eukaryotes, recombination involves entire chromosomes, but this is not true in prokaryotes. Explain.

II TECHNIQUES OF BACTERIAL GENETICS: *IN VIVO*

To "do genetics" the geneticist must cross strains of an organism that have different genotypes (and phenotypes) and look for recombinants. For much of the last 50 years, the bacterial geneticist wishing to carry out these crosses was dependent on the natural processes by which bacteria exchange genetic material. These three *in vivo* processes are (1) **transformation**, which involves donor DNA free in the environment (see Section 10.6), (2) **transduction**, in which the donor DNA transfer is mediated by a virus (see Section 10.7), and (3) **conjugation**, in which the transfer involves cell-to-cell contact and a *conjugative plasmid* in the donor cell (see Section 10.9). These processes are contrasted in Figure 10.11 and will be discussed in detail in the following sections.

10.6 Genetic Transformation

As we have noted, genetic transformation is a process by which free DNA is incorporated into a recipient cell and brings about genetic change. The discovery of genetic transformation in bacteria was one of the outstanding events in biology, as it led to experiments demonstrating that DNA is the genetic material (see the box, Origins of Bacterial Genetics). This discovery became the keystone of molecular biology and modern genetics.

A number of prokaryotes have been found to be naturally transformable, including certain species of both gram-negative and gram-positive *Bacteria* and some species of *Archaea*. However, even within transformable genera, only certain strains or species are transformable. Since the DNA of prokaryotes is present in the cell as a large single molecule, when the cell is gently lysed, the DNA pours out (Figure 10.12, page 282). Because of its

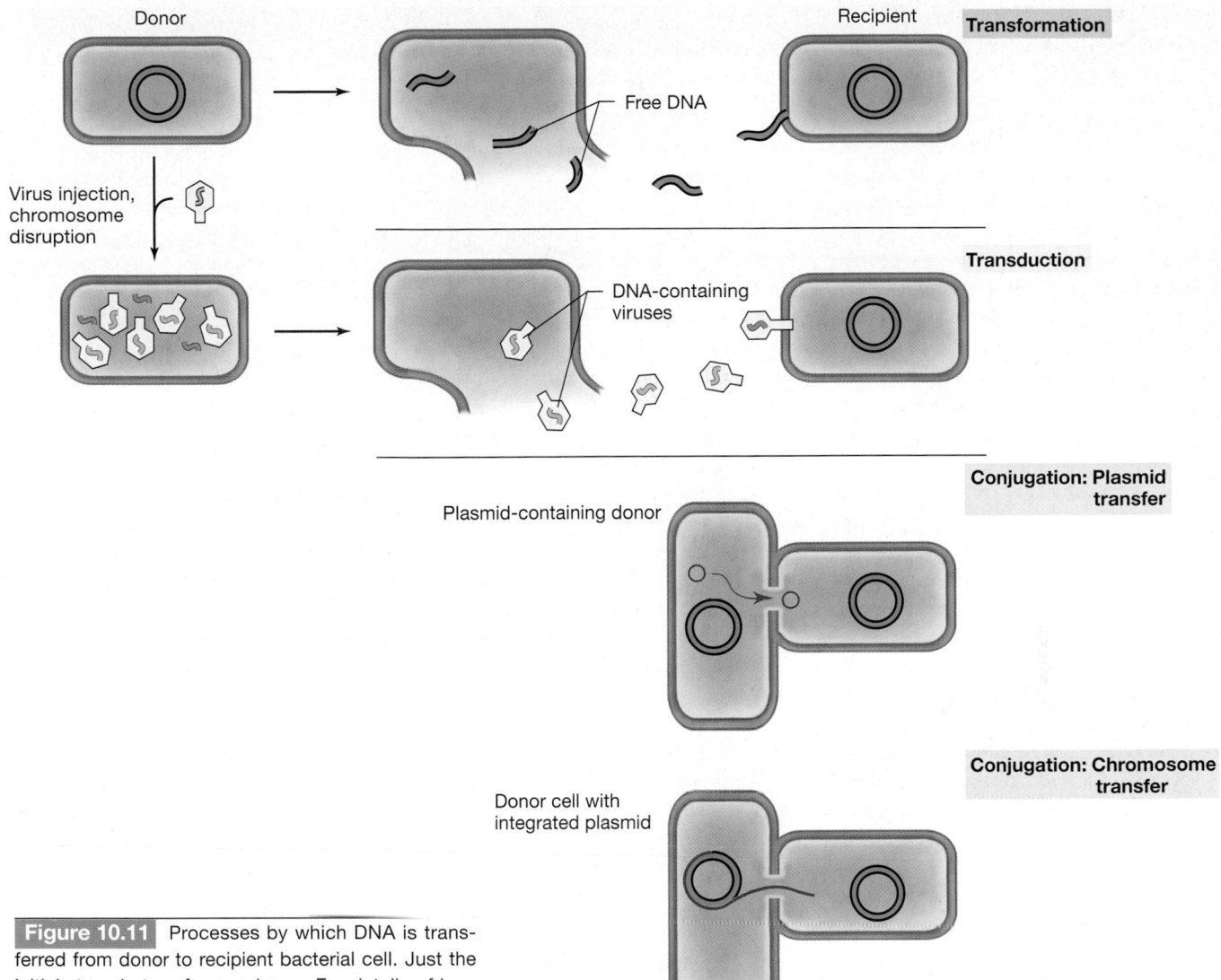

Figure 10.11 Processes by which DNA is transferred from donor to recipient bacterial cell. Just the initial steps in transfer are shown. For details of how the DNA is integrated into the recipient, see text.

extreme length (1700 μm in *Bacillus subtilis*), the DNA molecule breaks easily; even after gentle extraction the *B. subtilis* chromosome of 4.2 megabase pairs is converted to fragments of about 15 kilobase pairs. Because the DNA that corresponds to an average gene is about 1000 nucleotides, each of the fragments of purified DNA has about 15 genes. A single cell usually incorporates only one or a few DNA fragments so only a small proportion of the genes of one cell can be transferred to another by a single transformation event.

Competence

A cell that is able to take up a molecule of DNA and be transformed is said to be **competent**. Only certain strains are competent; the ability seems to be genetically determined. Competence in most naturally transformable bacteria is regulated, and special proteins play a role in the

uptake and processing of DNA. These competence-specific proteins may include a membrane-associated DNA binding protein, a cell wall autolysin, and various nucleases. One pathway to natural competence in *Bacillus subtilis* is part of a quorum-sensing system (a regulatory system that responds to cell number, ∞ Section 8.9) regulated by a two-component regulatory system (∞ Section 8.10). Cells produce and excrete a small peptide during growth and become competent in high concentrations of this peptide, through the action of ComP, a sensor kinase, and ComA, a response regulator protein. ComA is an activator protein (∞ Section 8.6) that regulates a number of genes affecting transformation. In *Bacillus*, about 20% of the cells become competent and stay that way for several hours. However, in *Streptococcus*, 100% of the cells can become competent, but only for a few minutes during the growth cycle.

Learning From the Past ... | Origins of Bacterial Genetics

Although genetic recombination in eukaryotes had been known for a long time, the discovery of genetic recombination in bacteria following transformation, transduction, and conjugation has been a more recent event. Of the three processes, the discovery of transformation was the most significant as it provided the first direct evidence that DNA is the genetic material. The first evidence of bacterial transformation was obtained by the British scientist Fred Griffith in the late 1920s. Griffith was working with *Streptococcus pneumoniae* (pneumococcus), a bacterium that owes its ability to invade the body in part to the presence of a polysaccharide capsule. Mutants can be isolated that lack this capsule and are thus unable to cause infection; such mutants are called R strains because their colonies appear rough on agar, in contrast to the smooth appearance of capsulated strains. A mouse infected with only a few cells of a smooth (S) strain succumbs in a day or two to pneumococcus infection, whereas even large numbers of R cells do not cause death when injected. Griffith showed that if heat-killed S cells were injected along with living R cells, a fatal infection ensued and the bacteria isolated from the dead mouse were S types. A number of different polysaccharide capsules were known in different pneumococcus S strains, and it was possible to do this experiment with heat-killed S cells from a type different from that from which the R strain was derived. Since the isolated living

S cells always had the capsule type of the heat-killed S cells, the R cells had been transformed into a new type, and the process had all the properties of a genetic event. The molecular explanation for the transformation of pneumococcus types was provided by Oswald T. Avery and his associates at Rocke-feller Institute in New York in a series of studies carried out during the 1930s, culminating in the now classic paper by Avery, C. M. MacLeod, and M. McCarty in 1944. Avery and his coworkers showed that under certain conditions the transformation process could be carried out in the test tube rather than the mouse and that a cell-free extract of heat-killed cells could induce transformation. In a long series of painstaking biochemical experiments, the active fraction of cell-free extracts was purified and was shown to consist of DNA. The transforming activity of purified DNA preparations was very high, and only very small amounts of material were necessary. Subsequently, others at Rockefeller showed that transformation could occur in pneumococcus not only for capsular characteristics but for other genetic characteristics of the organism as well, such as antibiotic resistance and sugar fermentation.

In 1953, James Watson and Francis Crick announced their model of the structure of DNA, providing a theoretical framework for how DNA could serve as the genetic material. Thus, two types of studies, the bacteriological and biochemical ones of

Avery and the physical-chemical ones of Watson and Crick, solidified the concept of DNA as the genetic material. In the subsequent years, this work has led to the whole field of molecular genetics.

Although bacterial transformation resulted from an essentially accidental discovery, bacterial conjugation was initially shown to occur by Joshua Lederberg and E. L. Tatum in 1946 in experiments carefully designed to determine if a sexual process might occur in bacteria. Because it appeared that the process, if present, would be quite rare (no microscopic evidence for bacterial mating had ever been seen, although such evidence can easily be obtained in eukaryotes), Lederberg developed a method involving the use of nutritional mutants of *Escherichia coli*. Fortunately, these mutants had been isolated in strain K-12, one of the few wild-type strains now known to contain the F plasmid. The principle was to mix two strains, one requiring biotin and methionine, the other requiring threonine and leucine, and plate the mixture on a minimal medium lacking all four growth factors. Neither parental type could grow on this medium, but any recombinants could, and when about 10^8 cells were plated, a small but significant number of colonies was obtained. Strains with two separate nutritional requirements were employed because it would be unlikely that spontaneous back mutation of both genes would occur in a single cell. Thus, the only explanation for the

Uptake of DNA

Bacteria differ in the form in which DNA is taken up, although in both cases only single-stranded DNA enters the cytoplasm. In *Haemophilus*, which is gram-negative, for example, only double-stranded DNA is taken up into the cell despite the fact that only single-stranded segments actually become incorporated into the genome by recombination. In gram-negative *Bacteria*, it seems that the double-stranded DNA is degraded to a single-stranded form in the periplasmic space (∞ Section 4.9) and only a single strand is transported to the cytoplasm. In the gram-positive *Bacteria Streptococcus* and *Bacillus*, by contrast, only a single DNA strand is taken up, while the complementary strand is simultaneously degraded.

However, in all cases, double-stranded DNA binds more effectively to the cells.

During transformation, competent bacteria first bind DNA reversibly; soon, however, the binding becomes irreversible. Competent cells bind much more DNA than do noncompetent cells—as much as 1000 times more. As we noted earlier, the sizes of the transforming fragments are much smaller than that of the whole genome, and this DNA is further degraded during the uptake process. In *Streptococcus pneumoniae* each cell can bind only about 10 molecules of double-stranded DNA of 15–20 kilobase pairs each. However, as they are taken up, they are converted to single-stranded pieces of about 8 kilobases. The DNA fragments in the mixture compete with each other for uptake,

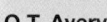

O.T. Avery

E.L. Tatum (left) and J. Lederberg in 1947

phenomenon was some sort of genetic recombination. To show that the process required cell-to-cell contact, and hence, could not be a type of transformation, it was shown that when culture filtrates or extracts were separated by a sintered glass disc, permeable to macromolecules but not to cells, recombination did not occur. Although initially conjugation appeared to be a very rare event, by the early 1950s a strain of *E. coli* had been isolated by the Italian scientist L. L. Cavalli-Sforza, while he was working in Lederberg's laboratory, that showed a high frequency of recombination. The British physician William Hayes, who independently isolated an Hfr strain, then showed that ge-

netic transfer during mating between Hfr and F⁻ was a one-way event, with the Hfr serving as donor. The interrupted mating experiment and the demonstration of the circular genetic map of *E. coli* were then carried out by Elie Wollman and Francois Jacob, working with Jacques Monod at the Pasteur Institute in Paris. The distinction between Hfr and F⁺ was made by Lederberg, who also showed that F⁺ behaved in an infectious manner. Lederberg coined the term *plasmid* in the 1950s to describe such apparently extrachromosomal genetic elements, although it did not find wide usage until the 1970s when infectious drug resistance became a major medical problem.

Bacterial transduction was discovered by the American scientist Norton Zinder when he was a graduate student working with Lederberg at the University of Wisconsin on genetic recombination in *Salmonella typhimurium.* The original motivation for this work was to show that conjugation occurred in an organism other than *E. coli,* and the techniques involved isolation of mutants and quantification of recombination by observing colony growth on minimal medium. However, although evidence of recombination was obtained, it could be shown that cell-to-cell contact was *not* required. Although this suggested a type of transformation, the process was not affected by DNase, and the gene transfer agent behaved like a bacteriophage. The gene transfer agent could be purified by the same procedures used to purify virus particles, and transduction occurred only with recipient cells that had receptor sites for the virus in question. Further, transducing activity could be eliminated by treatment of a lysate with substances able to adsorb the virus, such as sensitive cells or antibodies. Thus, in all cases, transducing activity and virus activity behaved in similar ways. Zinder and Lederberg coined the word *transduction* to refer to any genetic recombinational process that was only fragmentary and did not involve cell-to-cell contact, intending in this way to encompass processes involving either free DNA (transformation) or phages, but subsequently the word *transduction* has been applied only to virus-mediated genetic transfer. ■

and if excess DNA that does not contain the genetic marker is added, a decrease in the number of transformants occurs. In preparations of transforming DNA, only about 1 out of 100–200 DNA fragments contains the marker being studied. Thus, at high concentrations of DNA, the competition between DNA molecules results in saturation of the system so even under the best conditions it is impossible to transform all the cells in a population for a given genetic marker. The maximum frequency of transformation that has so far been obtained is about 20% of the population; actually the values usually obtained are between 0.1 and 1.0%. The minimum concentration of DNA yielding detectable transformants is about 0.00001 μg/ml ($1 \times 10^{-5} \mu$g/ml), which is so low that it is undetectable chemically.

Interestingly, in *Haemophilus influenzae* there is a requirement that the DNA fragment have a particular 11-base pair sequence for irreversible binding and uptake to occur. This sequence is found at an unexpectedly high frequency in the *Haemophilus* chromosome. Evidence such as this, and the fact that at least certain bacteria become competent in their natural environment, suggest that transformation is not a laboratory artifact but plays an important role in gene transfer in nature.

Integration of Transforming DNA

Transforming DNA is bound at the cell surface by a DNA binding protein, after which either the entire double-stranded fragment is taken up, or a nuclease degrades one

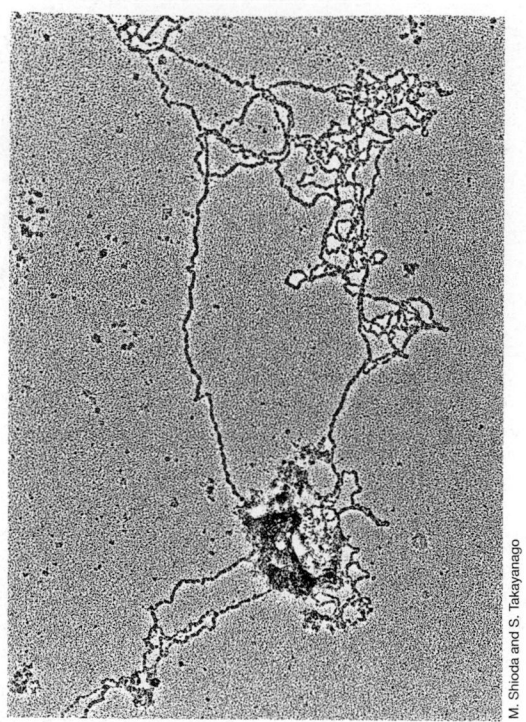

Figure 10.12 The prokaryotic chromosome, as shown in the electron microscope. The circular chromosome is from the hyperthermophile *Sulfolobus*, a member of the *Archaea* (⌀⌀ Section 13.9). See also Figure 10.16.

strand and the other is taken up (Figure 10.13). After uptake, the DNA associates with a competence-specific protein that remains attached to the DNA, presumably preventing it from nuclease attack, until it reaches the chromosome where RecA protein takes over. The DNA is then integrated into the genome of the recipient by recombinational processes (Figure 10.13; see also Figure 10.9). During replication of this heteroduplex DNA, one parental and one recombinant DNA molecule are formed. On segregation at cell division, the latter is present in the transformed cell, which is now genetically altered as compared to the parental type. The preceding discussion pertains to only small pieces of *linear* DNA. Many naturally transformable *Bacteria* are transformed only poorly by plasmid DNA because the plasmid must remain double-stranded and circular in order to replicate.

Transfection

Bacteria can be transformed with DNA extracted from a *bacterial virus* rather than from another bacterium, a process known as **transfection**. If the DNA is from a lytic bacteriophage, transfection leads to normal virus production and can be measured by the standard phage plaque assay (⌀⌀ Section 9.4). Transfection has become a useful tool in studying the mechanism of transformation and recombination because the small size of phage genomes allows for the isolation of a nearly homoge-

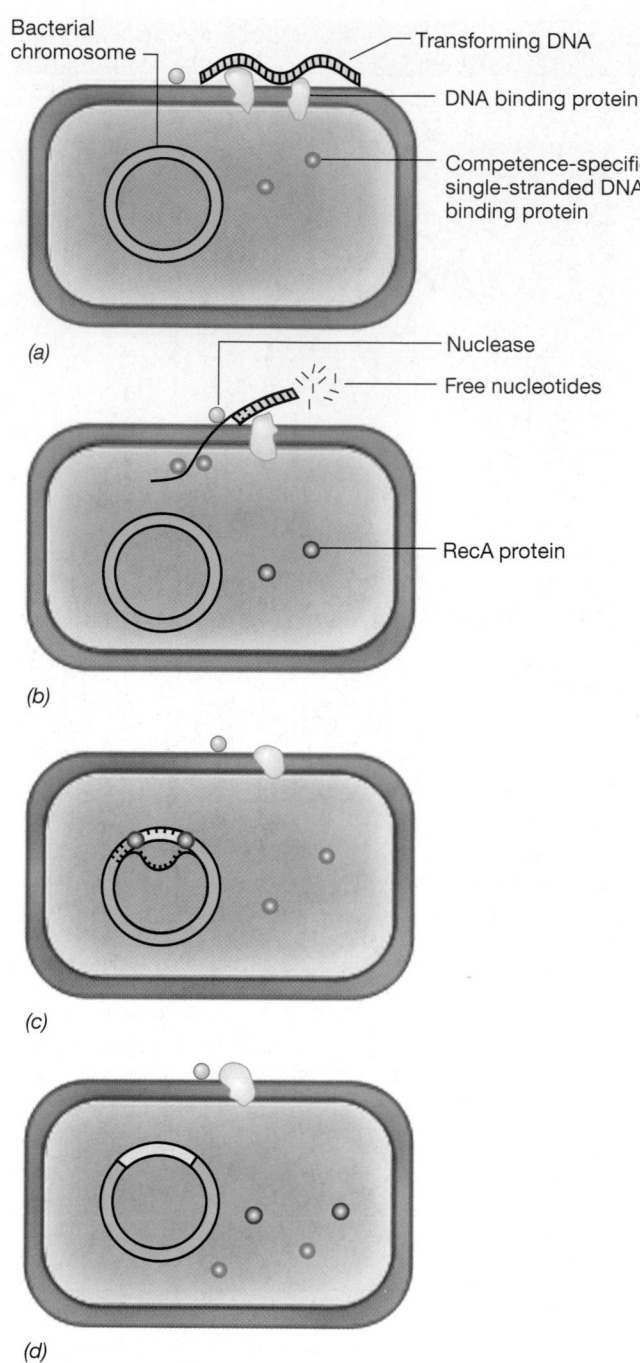

Figure 10.13 Mechanism of DNA transfer by transformation in a gram-positive bacterium. (a) Binding of free doubled-stranded DNA by a membrane-bound DNA binding protein. (b) Passage of one of the two strands into the cell while nuclease activity degrades the other strand. (c) The single strand in the cell is bound by specific proteins, and recombination with homologous regions of the bacterial chromosome mediated by RecA protein occurs. (d) Transformed cell. Note that if recombination does not occur, the incoming DNA cannot replicate and will be lost.

neous population of DNA molecules. By contrast, in conventional transformation, the transforming DNA is generally a random assortment of chromosomal DNA of various lengths, and this tends to complicate experiments designed to study the mechanism of transformation.

Artificially Induced Competence

High efficiency natural transformation is found only in a few bacteria; *Acinetobacter, Azotobacter, Bacillus, Streptococcus, Haemophilus, Neisseria*, and *Thermus*, for example, are easily transformed. Many prokaryotes are transformed only poorly or not at all under natural conditions. Determination of how to induce competence in such bacteria may involve considerable empirical study, with variation in culture medium, temperature, and other factors. However, to transfer DNA into cells for genetic engineering (see later in this chapter and also ∞ Chapter 31), it was necessary to find a way to make *Escherichia coli*, a gram-negative organism, competent. It has been found that when *E. coli* is treated with high concentrations of calcium ions and then stored in the cold, it becomes transformable at low efficiency. *Escherichia coli* treated in this manner takes up double-stranded DNA, and therefore transformation by plasmid DNA is relatively efficient. Why the calcium treatment works is not known, but this procedure also works with some other gram-negative *Bacteria*. However, methods such as this for artificial induction of competence are being supplanted by a new method termed *electroporation*.

DNA Transfer by Electroporation

Electroporation is a technique in which cells are exposed to pulsed electrical fields to open small pores in their membranes. When DNA molecules are present outside the cells during the electric pulse, they can then enter the cells through these pores. Electroporation requires a sophisticated power supply because the pulses must be carefully controlled and last for only milliseconds. This technique has been used to transport DNA into a large number of different species of prokaryotes, both *Archaea* and *Bacteria*, and also into many eukaryotic cells. Additionally, electroporation allows an experimenter to transfer a plasmid directly from one cell to another if both are present during electroporation. Therefore, electroporation allows small molecules of DNA to come out of cells as well as to go in! This type of "transformation" eliminates the steps required to isolate the plasmid from the first strain before introducing it into the second.

✓ *10.6 Concept Check*

Certain prokaryotes exhibit competence, a state in which cells are able to take up free DNA released by other bacteria. This process is called transformation. Relatively few species of prokaryotes can be naturally transformed. However, certain laboratory procedures have been developed that make it possible to introduce DNA into completely unrelated organisms. Electroporation involves modification of the cytoplasmic membrane by treatment with an electric field to facilitate DNA uptake.

✓ The donor bacterial cell in a transformation is probably dead. Explain.

✓ Even in naturally transformable cells competency is usually inducible. What does this mean?

10.7 Transduction

In transduction, DNA is transferred from cell to cell through the agency of viruses. Genetic transfer of host genes by viruses can occur in two ways. In the first, called **generalized transduction**, host DNA derived from virtually any portion of the host genome becomes a part of the DNA of the mature virus particle in place of the virus genome. The second, called **specialized transduction**, occurs only in some temperate viruses; DNA from a specific region of the host chromosome is integrated directly into the virus genome—usually replacing some of the virus genes. The transducing virus particle in both specialized and generalized transduction is usually *defective* as a virus because bacterial genes have replaced some necessary viral genes.

In generalized transduction, if the donor genes do not undergo homologous recombination with the recipient bacterial chromosome, they will be lost. They cannot replicate independently and are not part of a viral genome. In specialized transduction homologous recombination may also occur. However, since the donor bacterial DNA is now actually a part of a temperate phage genome, there are two other possibilities: (1) the DNA may be integrated into the host chromosome during lysogenization (∞ Section 9.10), and (2) the DNA may be replicated in the recipient as part of a lytic infection.

Transduction occurs in a variety of prokaryotes, including certain species of the *Bacteria: Desulfovibrio, Escherichia, Pseudomonas, Rhodococcus, Rhodobacter, Salmonella, Staphylococcus*, and *Xanthobacter*, as well as the archaeon *Methanothermobacter thermoautotrophicus*. Not all phages can transduce, and not all bacteria are transducible; but the phenomenon is sufficiently widespread that it likely plays an important role in genetic transfer in nature.

Generalized Transduction

In generalized transduction, virtually any genetic marker can be transferred from donor to recipient. Generalized transduction was first discovered and extensively studied in the bacterium *Salmonella enterica* with phage P22 and has also been studied with phage P1 in *Escherichia coli*. An example of how *transducing particles* may be formed is given in Figure 10.14. When the population of sensitive bacteria is infected with a phage, the events of the phage lytic cycle may be initiated. During a lytic infection, the enzymes responsible for packaging viral DNA into the bacteriophage sometimes accidentally package host DNA. The resulting particle is called a *transducing particle*. On lysis of the cell, these particles are released along with normal virions, and so the lysate contains a mixture of normal virions and transducing particles. Because transducing particles cannot lead to a normal viral infection (they contain no viral DNA), they are said to be *defective*. When this lysate is used to infect a population of recipient cells, most of the cells become infected with normal virus. However, a small proportion of the population receives transducing particles that inject the DNA they received from the previous host bacterium. While this DNA cannot replicate, it can undergo genetic recombination with the DNA of the new host. Because only a small proportion of the particles in the lysate are of the defective transducing type and each of these contains only a small fragment of donor DNA, the probability of a transducing particle containing a particular gene is quite low, and usually only about 1 cell in 10^6–10^8 is transduced for a given marker.

Phages that form transducing particles can be either temperate or virulent, the main requirements being that they have a DNA packaging mechanism that permits accidental recognition of host DNA and that packaging occurs before the host genome is completely degraded. The detection of transduction is most certain when the multiplicity of phage to host is low, so a host cell is infected with only a single phage particle; with multiple infection, the cell will likely be killed by the normal virions in the lysate.

Specialized Transduction

Generalized transduction allows the transfer of DNA from one bacterium to another at a low frequency. However, specialized transduction can allow extremely efficient transfer while also allowing a small region of a bacterial chromosome to be replicated independently of the rest. The example we shall use to discuss specialized transduction was the first to be discovered and involves transduction of the galactose genes by the temperate phage lambda of *Escherichia coli*.

As we discussed (∞ Section 9.10), when a cell is lysogenized by lambda, the phage genome becomes integrated into the host DNA at a specific site. The region in which

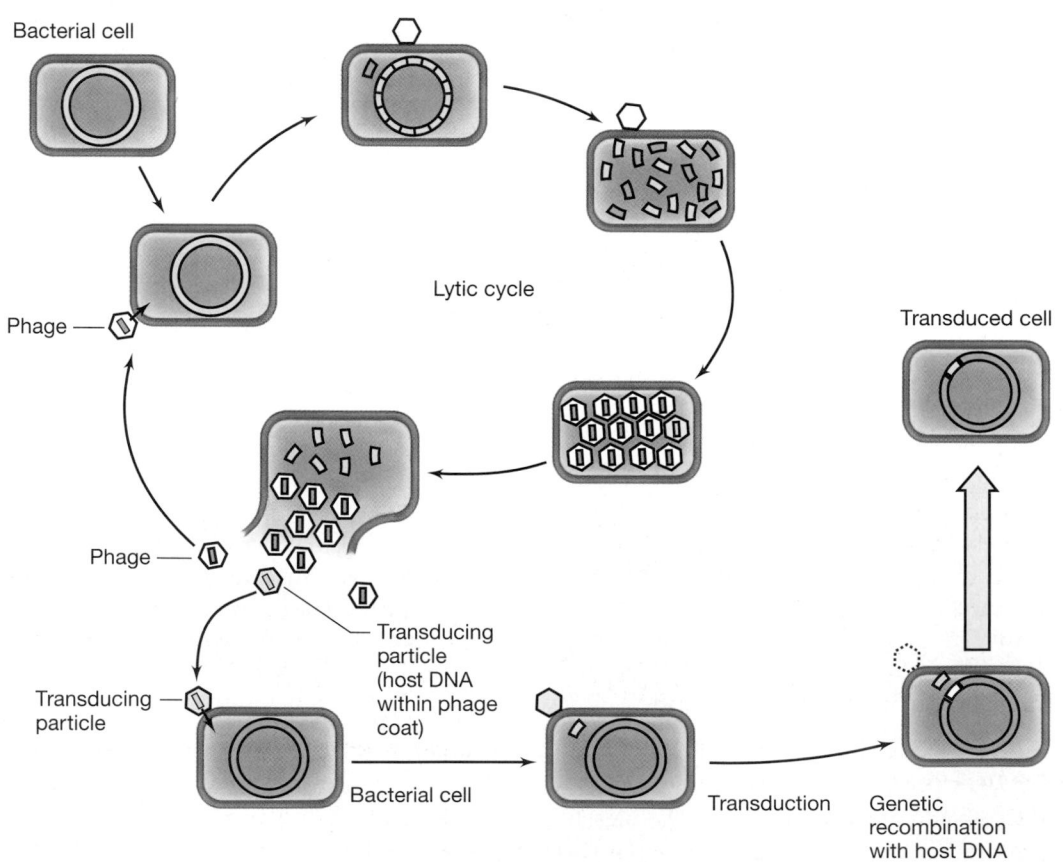

Figure 10.14 Generalized transduction: one possible mechanism by which virus (phage) particles containing host DNA can be formed.

lambda integrates is immediately adjacent to the cluster of host genes that control the enzymes involved in galactose utilization (Figure 9.21), and the DNA of lambda is inserted into the host DNA at that site. From then on, viral DNA replication is under host control. On induction (for example, by ultraviolet radiation), the viral DNA separates from the host DNA by a process that is the reverse of integration (Figure 10.15). Ordinarily when the lysogenic cell is induced, the lambda DNA is excised as a unit. Under rare conditions, however, the phage genome is excised incorrectly. Some of the adjacent bacterial genes (for example, the galactose operon) are excised along with phage DNA. At the same time, some phage genes are left behind. One type of altered phage particle, called **lambda dgal** or λ*dgal* (*dgal* means "defective, galactose"), is defective because of the phage genes lost and does not make mature phage. However, a **helper phage** can provide those functions missing in the defective particles. This "helper" is identical to the original lambda. Thus, the culture lysate obtained contains a few λ*dgal* particles mixed in with a large number of normal lambda virions.

When a galactose-negative bacterial culture is infected at high multiplicity with such a lysate and *gal*$^+$ transductants are selected, many are double lysogens, carrying both lambda and λ*dgal*. (Note then that the bacterium has become a diploid for the *gal* region. This will have significance in performing complementation tests in bacteria; see Section 10.10.) When such a double lysogen is induced, a lysate is produced containing about equal numbers of lambda and λ*dgal*. Such a lysate can transduce at high efficiency, although only for a restricted group of *gal* genes.

If the phage is to be viable, there is a maximum limit to the amount of phage DNA that can be replaced with host DNA, because sufficient phage DNA must be retained in order to provide the information for production of the phage protein coat and for other phage proteins needed for lysis and lysogenization. However, if a helper phage is used together with the defective phage in a mixed infection, then even less information is needed in the defective phage for transduction. Only the *att* (attachment) region, the *cos* site (cohesive ends, for packaging), and the replication origin of the lambda genome are needed for production of a transducing particle, provided a helper is used (see the genetic map of lambda, (Figure 9.18 and Section 9.10).

One important distinction between specialized and generalized transduction is in how the transducing lysate can be formed. In specialized transduction this *must* occur by induction of a lysogen, whereas in generalized transduction it can occur either in this way or by infection of a nonlysogen by the phage, with subsequent phage replication and cell lysis.

Although we have discussed specialized transduction only in the lambda-*gal* system, phage lambda and its

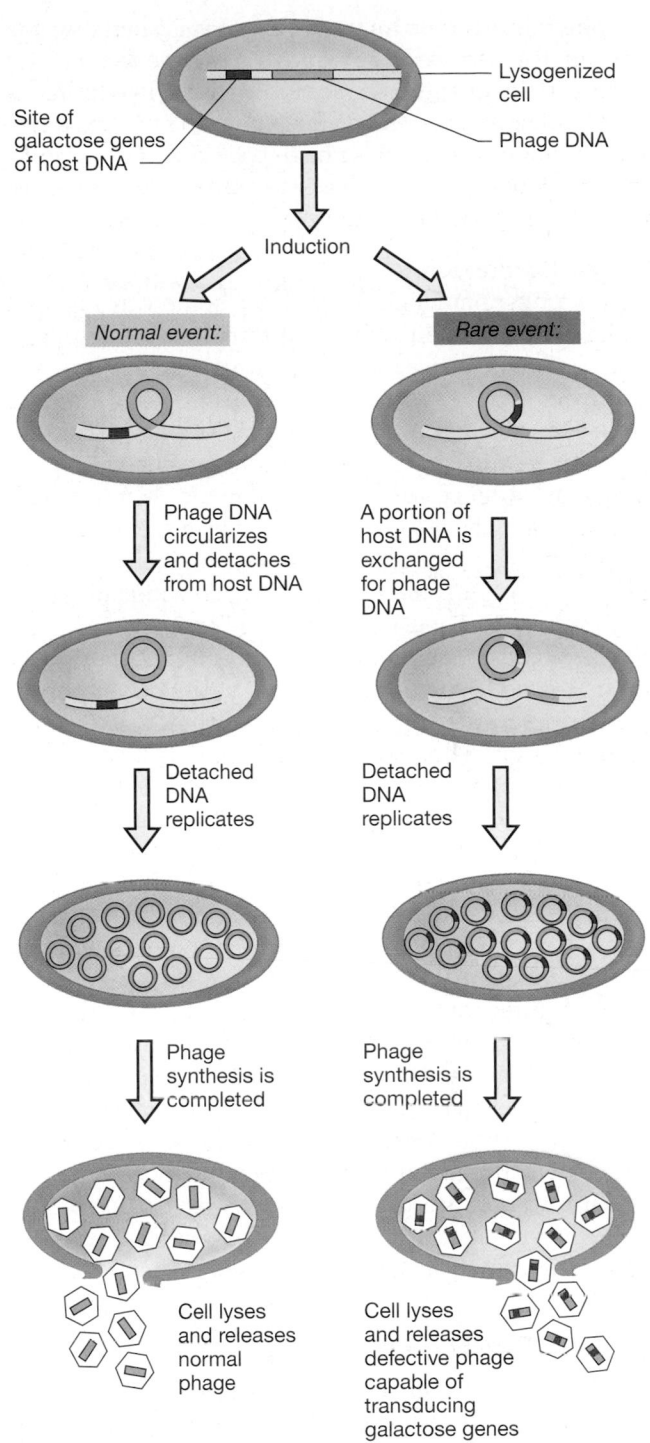

Figure 10.15 Normal lytic events and the production of particles transducing the galactose genes in an *Escherichia coli* cell containing a lambda prophage.

relative, φ80, have been widely used to form specialized transducing phages covering many specific regions of the *E. coli* genome. In addition, lambda specialized transducing phage can be constructed by the techniques of genetic engineering to contain genes from any organism (see Section 10.16).

Specialized transduction in *Bacteria* is similar to the situation we discussed with retroviruses (∞ Section 9.12) where certain host genes can be incorporated into the retroviral genome and be passed along (often in modified form) to subsequent hosts. These have been detected in retroviruses because the genes involved are *oncogenes*, involved in the development of cancer.

Phage Conversion

This is a phenomenon analogous in some ways to specialized transduction. When a normal temperate phage (that is, a nondefective one) lysogenizes a cell and its DNA is converted to the prophage state, the lysogen is immune to further infection by the same type of phage. This acquisition of immunity can be considered a change in phenotype. In certain cases, other phenotypic alterations can be detected in the lysogenized cell, which seem to be unrelated to the phage immunity system. Such a change, which is brought about through lysogenization by a normal temperate phage, is called **phage conversion**.

Two cases of conversion have been especially well studied. One involves a change in structure of a polysaccharide on the cell surface of *Salmonella anatum* on lysogenization with phage ϵ^{15}. The second involves the conversion of nontoxin-producing strains of *Corynebacterium diphtheriae* (the bacterium that causes the disease diphtheria) to toxin-producing (pathogenic) strains on lysogenization with phage β (∞ Section 26.3). In these situations, the information for production of these new materials is apparently an integral part of the phage genome and hence is automatically and exclusively transferred on infection by the phage and lysogenization.

Lysogeny probably carries a strong selective value for the host cell because it confers resistance to infection by viruses of the same type. Phage conversion may also be of considerable evolutionary significance because it results in efficient genetic alteration of host cells. Many bacteria isolated from nature are lysogens. It seems reasonable to conclude, therefore, that lysogeny is the normal state of affairs and may often be essential for survival of the host in nature.

✓ 10.7 Concept Check

Transduction involves transfer of host genes from one bacterium to another by viruses. In generalized transduction, defective virus particles randomly incorporate fragments of the cell's chromosomal DNA; virtually any gene of the donor can be transferred, but the efficiency is low. In specialized transduction, the DNA of a temperate virus excises incorrectly and brings adjacent host genes along with it; only genes close to the integration point of the virus are transduced, but the efficiency may be high.

✓ What is the important difference between generalized transduction and transformation?

✓ In specialized transduction, the donor DNA can replicate inside the recipient cell without homologous recombination taking place, but this is not true in generalized transduction. Explain.

10.8 Plasmids

Before we discuss the third method of genetic transfer, **conjugation**, we must discuss *plasmids*. **Plasmids** are genetic elements that replicate independently of the host chromosome (∞ Section 7.4).

Unlike viruses, plasmids do not have an extracellular form and exist inside cells simply as nucleic acid. However, distinguishing between viruses and plasmids can sometimes be difficult. As we have discussed, the prophage form of some temperate viruses, such as bacteriophage P1, replicates independently of the host chromosome in a fashion analogous to plasmid replication (∞ Section 9.10).

We noted earlier (∞ Section 7.4) that it is possible to differentiate between a plasmid and a host chromosome in prokaryotes because plasmids do not carry genes required by the host under all conditions. This can be difficult to prove, and therefore it may sometimes be difficult to distinguish between chromosomes and very large plasmids in prokaryotes.

Despite these few difficulties, literally thousands of different types of plasmids are known. Indeed, over 300 different naturally occurring plasmids have been isolated from strains of *Escherichia coli* alone. In this section we shall discuss the properties of a few of them.

Physical Nature of Plasmids

Almost all of the known plasmids are double-stranded DNA. Most plasmids are circular, but many linear plasmids are also known. Naturally occurring plasmids vary in size from approximately 1 to more than 1000 kilobase pairs. The typical plasmid is a circular double-stranded DNA molecule less than $\frac{1}{20}$ the size of the chromosome (Figure 10.16). Most of the plasmid DNA isolated from cells is in the supercoiled configuration, which is the most compact form within the cell (∞ Figure 7.10).

Isolation of plasmid DNA can generally be readily accomplished by making use of the physical properties of supercoiled DNA molecules. Although chromosomes are also supercoiled inside the cell, isolation of chromosomal DNA almost always leads to breakage of the strands and consequent loss of supercoiling. Separation can then proceed by a variety of techniques, including ultracentrifugation and electrophoresis on agarose gels (see Section 10.12).

Replication of Plasmids

The enzymes involved in plasmid replication are normal cell enzymes, so the genes carried by the plasmid itself that control its replication are concerned primarily with control of the timing of the initiation process and with apportionment of the replicated plasmids between daughter cells. Also, different plasmids are present in cells in a particular number of plasmid molecules per cell; this is called the *copy number*. Some plasmids are

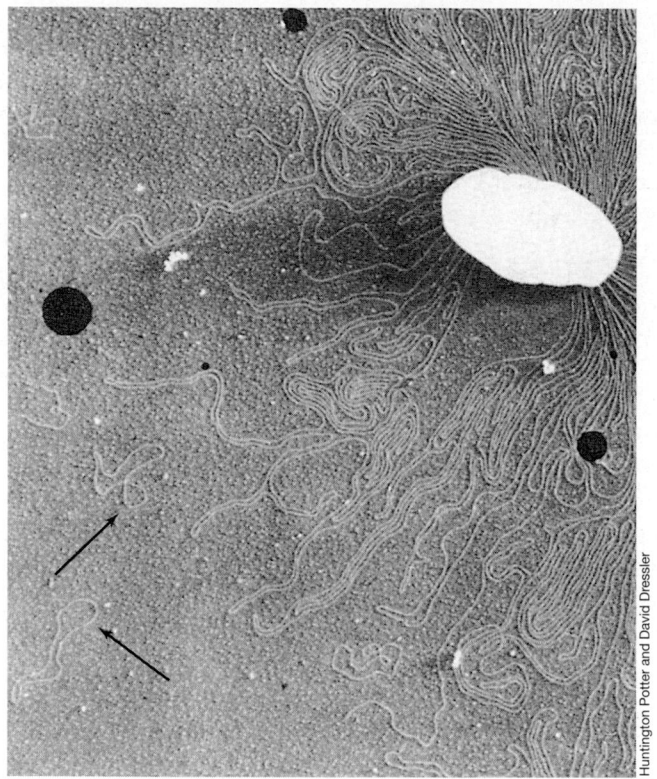

Figure 10.16 The bacterial chromosome and bacterial plasmids, as shown in the electron microscope. The plasmids (arrows) are the circular structures, much smaller than the main chromosomal DNA. The cell (large, white structure) was broken gently so the DNA would remain intact.

present in the cell in only 1–3 copies, whereas others may be present in over 100 copies. Copy number is controlled by genes on the plasmid and by interactions between the host and the plasmid.

Most plasmids in gram-negative *Bacteria* replicate in a manner similar to that already described for the chromosome (◯◯ Section 7.6). This involves initiation of replication at an origin and bidirectional replication around the circle, giving a *theta* intermediate. However, some plasmids have *unidirectional* replication. Because of the small size of plasmid DNA relative to the chromosome, the whole replication process occurs very quickly, perhaps in $\frac{1}{10}$ or less of the total time of the cell division cycle.

Most plasmids of gram-positive *Bacteria* replicate by a rolling circle mechanism similar to that used by the phage ϕX174 (◯◯ Section 16.2 and Figure 16.4). This mechanism gives rise to a single-stranded intermediate, and thus these plasmids are sometimes referred to as *single-stranded DNA plasmids*. Most of the linear plasmids now known replicate using a mechanism involving a protein bound to the 5'-end of each strand that is used in priming DNA synthesis (◯◯ Figures 7.24 and 16.22).

Some individual bacterial cells may also contain several different types of plasmids; for example, *Borrelia*

burgdorferi (the Lyme pathogen, ◯◯ Section 27.4) contains 17 different circular and linear plasmids. The ability of two different plasmids to both replicate in the same cell is controlled by plasmid genes involved in regulating DNA replication. When a plasmid is transferred into a cell that already carries another plasmid, a common observation is that the second plasmid may not be maintained and is lost during subsequent cell replication. The two plasmids are then said to be **incompatible**. A number of incompatibility (Inc) groups have been recognized, the plasmids of one incompatibility group excluding each other but being able to coexist with plasmids from other groups. Plasmids of one incompatibility group share a common mechanism of regulating their replication and are thus *related* to one another. Therefore, although a bacterial cell may contain different kinds of plasmids, they are not closely related because they must be compatible.

Some plasmids, called *episomes*, have the ability to become integrated into the chromosome, and under such conditions their replication comes under control of the chromosome. This situation is remarkably like that of several viruses whose genomes can become incorporated into the host genome (for example, ◯◯ Sections 9.10, 16.5, 16.10, and 16.14). Plasmids can sometimes be eliminated from host cells by various treatments. This process, termed **curing**, apparently results from inhibition of plasmid replication without parallel inhibition of chromosome replication, and as a result of cell division the plasmid is diluted out. Curing may occur spontaneously, but it is greatly increased by use of acridine dyes, which become inserted into DNA, or other treatments that seem to interfere more with plasmid replication than with chromosome replication. Electroporation may also be used to cure a cell of plasmids (see Section 10.6).

Many of these characteristics are exemplified in a very well-characterized plasmid called the *F plasmid*. The F plasmid is a circular DNA molecule of 99,159 base pairs. Cells containing it can be easily cured with acridine orange. Figure 10.17 shows a genetic map of the F plasmid. One region of the plasmid contains genes involved in regulating DNA replication. It also contains a number of transposable elements (see Section 10.11) involved in its ability to function as an episome. Last, it has a large region of DNA, the *tra* region, containing genes that permit it to be transferred from one cell to another.

Cell-to-Cell Transfer of Plasmids

Because one of the defining characteristics of a plasmid is the lack of a distinct extracellular form, one might imagine that plasmids are confined almost exclusively to transferring only to daughter cells during cell division. Since some prokaryotic cells can take up free DNA from the environment (see Section 10.6), it is possible that lysis of the host, however it may happen, may bring the plasmid in contact with a new host. However, this process occurs naturally in only a few bacterial species and is unlikely

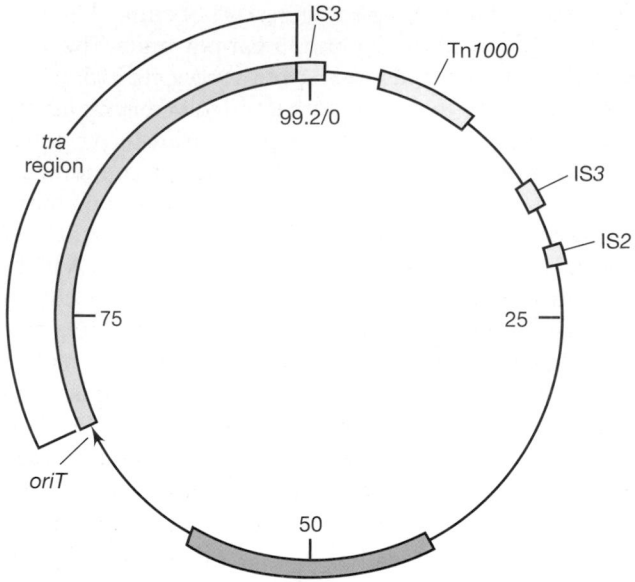

Figure 10.17 Genetic map of the F (fertility) plasmid of *Escherichia coli*. The numbers on the interior show the size of the plasmid in kilobase pairs (the exact size is 99,159 bp). The region shown in dark green at the bottom of the map contains genes primarily responsible for the replication and segregation of the F plasmid in normally growing cells. The light green region, the *tra* region, contains the genes involved in conjugative transfer. The *oriT* sequence is the origin of transfer during conjugation. The arrow indicates the direction of transfer (the *tra* region would be transferred last). The regions shown in yellow on F are transposable elements where integration into identical elements on the bacterial chromosome can occur and lead to the formation of different Hfr strains (see Section 10.9).

to account for much cell-to-cell plasmid transfer. The main mechanism of cell-to-cell transfer is **conjugation**, and it is a function encoded by some plasmids themselves. Conjugation is a replicative process, and both cells end up with copies of the plasmid (Figure 10.18).

Plasmids that govern their own transfer by cell-to-cell contact are called **conjugative**, but not all plasmids are conjugative. Transmissability by conjugation is controlled by a set of genes within the plasmid called the *tra region*. The *tra* region contains genes encoding proteins that function in DNA transfer and replication and others that function in mating pair formation. The presence of a *tra* region in a plasmid can have another important con-

sequence if the plasmid becomes integrated into the chromosome. In that case, the plasmid can *mobilize* the transfer of chromosomal DNA from one cell to another. Strains of bacteria that transfer large amounts of chromosomal DNA during conjugation are called *Hfr* (high frequency of recombination). The use of conjugation to transfer host genes is discussed further in the next section.

Some conjugative plasmids from *Pseudomonas* have a broad host range, that is, they are transferrable to a wide variety of other gram-negative *Bacteria*. Some conjugative plasmids can transfer genetic information between distantly related organisms. Conjugative plasmids have been shown to transfer between gram-negative and gram-positive *Bacteria*, between *Bacteria* and plant cells, and between *Bacteria* and fungi. Even if the plasmid cannot replicate in the new host, the transfer of the DNA itself could have important evolutionary consequences, as well as being involved in pathogenic processes, if it can recombine into the genome of the new host.

Types of Plasmids and their Biological Significance

Clearly all plasmids must carry genes that ensure their own replication. However, for many plasmids we know very little about what other genes they carry. These are called *cryptic plasmids* and were discovered by physical means (such as examining a cell extract using gel electrophoresis.) As we have seen, some plasmids also carry genes necessary for conjugation, and they can sometimes be detected biologically, either by the transfer functions themselves or by sensitivities to certain viruses (∞ Section 16.1). Although plasmids do not carry genes that are essential to the host under all conditions, the presence of plasmids in a cell can have a profound influence on the cell's phenotype. In some cases, plasmids encode properties that we think of as fundamental to the bacterium in question, for example, the ability of *Rhizobium* to interact with plants (∞ Section 19.22). Some plasmids in *Pseudomonas* have been shown to transfer the genetic information for biochemical pathways that degrade unusual organic compounds such as camphor, octane, and naphthalene.

Plasmids can carry a very wide variety of genes. Indeed, the only limitation is that the genes they carry do

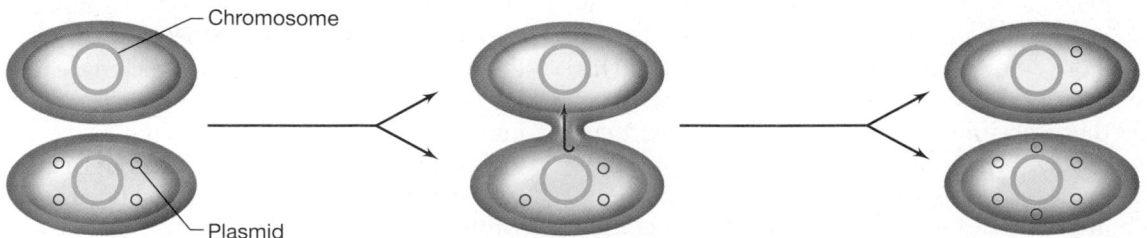

Figure 10.18 Plasmid transfer from cell to cell during conjugation.

not interfere either with their own replication or with the survival of the host. Because plasmids can be large and may carry many different genes, it is not always a simple matter to classify a plasmid into a simple phenotypic category. As we shall see, a single plasmid may confer many different phenotypes on its host cell. A few of the phenotypes conferred on prokaryotes by plasmids are given in Table 10.3. It is likely that virtually all prokaryotic groups possess plasmids. In the remainder of this section we will discuss a few of the many phenotypes plasmids may confer on cells.

Resistance Plasmids

Among the most widespread and well-studied groups of plasmids are the *resistance plasmids* (*R plasmids*), which confer resistance to antibiotics and various other inhibitors of growth. R plasmids were first discovered in Japan in strains of enteric bacteria that had acquired resistance to a number of antibiotics (multiple resistance) and have since been found throughout the world. The emergence of bacteria resistant to several antibiotics is of considerable medical significance and was correlated with the increasing use of antibiotics for the treatment of infectious diseases. Soon after these resistant strains were isolated it was shown that they could transfer resistance to sensitive strains via cell-to-cell contact. The infectious nature of the conjugative R plasmids permits rapid spread of the characteristic through populations.

A variety of antibiotic resistance genes can be carried by an R plasmid. In general, these genes encode proteins that either inactivate the antibiotic or affect its uptake into the cell. Plasmid R100, for example, is a 94.3-kilobase-pair plasmid (Figure 10.19) that carries resistance genes for sulfonamides, streptomycin and spectinomycin, fusidic acid, chloramphenicol, and tetracycline. R100 also carries several genes conferring resistance to mercury (Section 19.17). R100 can transfer itself between enteric bacteria of the genera *Escherichia*, *Klebsiella*, *Proteus*, *Salmonella*, and *Shigella* but does not transfer to the nonenteric bacterium *Pseudomonas*. R plasmids with genes for resistance to most antibiotics are known. Many drug-resistant elements on R plasmids, such as those on R100, are transposable elements (see Section 10.11) and this, plus the fact many of these plasmids are conjugative, make these plasmids a threat to traditional antibiotic therapies.

Toxins and Other Virulence Characteristics

We will discuss in Chapter 21 the physiological and genetic characteristics of microorganisms that enable them to colonize hosts and set up infections, which can lead to harm. In the present context, we merely note the two major characteristics involved in virulence: (1) the ability of microorganisms to attach to and colonize specific sites in the host; and (2) the formation of substances (toxins, enzymes, and other molecules) that cause damage to the

Table 10.3 Some phenotypes conferred by plasmids in prokaryotes

Phenotype class[a]	Organisms[b]
Antibiotic production	*Streptomyces*
Conjugation	*Escherichia, Pseudomonas, Rhizobium, Staphylococcus, Streptococcus, Sulfolobus, Vibrio*
Physiological functions	
Degradation of octane, camphor, naphthalene	*Pseudomonas*
Degradation of herbicides	*Alcaligenes*
Formation of acetone and butanol (Section 30.11)	*Clostridium*
Lactose, sucrose or urea utilization and nitrogen fixation	Enteric bacteria
Nodulation and symbiotic nitrogen fixation (Section 19.22)	*Rhizobium*
Pigment production	*Erwinia, Staphylococcus*
Resistance	
Antibiotic resistance (Section 20.12)	*Campylobacter*, Enteric bacteria, *Neisseria, Staphylococcus*
Resistance to cadmium, cobalt, mercury, nickel, and/or zinc (Section 19.16)	*Acidocella, Alcaligenes, Listeria, Pseudomonas, Staphylococcus*
Bacteriocin resistance (and production)	*Bacillus*, Enteric bacteria, *Lactococcus, Propionibacterium*
Virulence	
Host cell invasion	*Salmonella, Shigella, Yersinia*
Coagulase, hemolysin, enterotoxin (Sections 21.9 and 21.11)	*Staphylococcus*
Enterotoxin, K antigen (Sections 12.11 and 21.11)	*Escherichia*
Tumorigenicity in plants (Section 19.21)	*Agrobacterium*

[a] Only a few of the many phenotypes known to be associated with plasmids are given.

[b] Only a few well-characterized examples are given. All of the organisms given in the list are *Bacteria* except for *Sulfolobus*, which is a member of the *Archaea*.

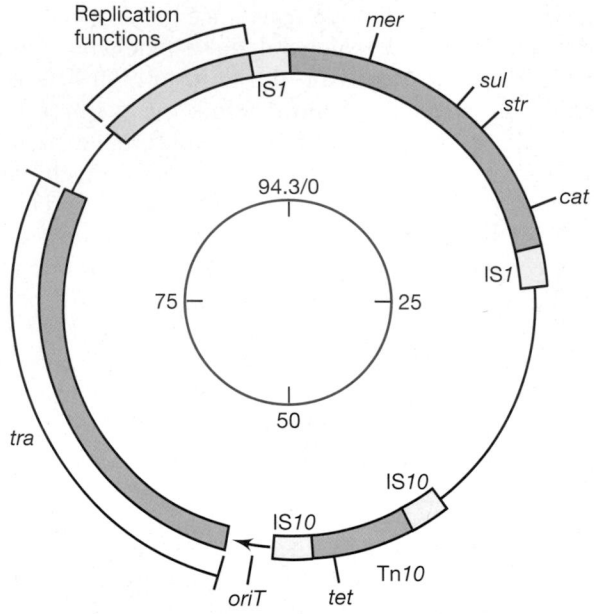

Figure 10.19 Genetic map of the resistance plasmid R100. The inner circle shows the size of the plasmid in kilobase pairs. The outer circle shows the location of major antibiotic resistance genes and other key functions: *cat*, chloramphenicol resistance; *oriT*, origin of conjugative transfer; *mer*, mercuric ion resistance; *sul*, sulfonamide resistance; *str*, streptomycin resistance; *tet*, tetracycline resistance; *tra*, transfer functions. The locations of insertion sequences (IS) and the transposor Tn*10* are also shown. Several genes related to plasmid replication are found in the region from 88–92 kilobase pairs.

host. It has now been well established that in several pathogenic bacteria each of these virulence characteristics is carried on plasmids. For example, enteropathogenic strains of *Escherichia coli* are characterized by an ability to colonize the small intestine and to produce a toxin that causes symptoms of diarrhea. Colonization requires the presence of a cell surface protein called the colonization factor antigen (CFA), encoded by a plasmid, which confers on the cells the ability to attach to epithelial cells of the intestine. At least two toxins in enteropathogenic *E. coli* are known to be coded for by a plasmid: the *hemolysin*, which lyses red blood cells, and the *enterotoxin*, which induces extensive secretion of water and salts into the bowel. It is the enterotoxin that is responsible for the induction of diarrhea, as will be discussed in Chapter 21. Pathogenesis in the genus *Yersinia*, the causative agent of plague (∞ Section 27.6), requires the expression and secretion of *Yersinia* outer proteins (Yops) and the V antigen. These proteins have functions that allow the bacterium to subvert host defense functions. The genes that encode and regulate these proteins are carried by a virulence plasmid.

Some virulence factors are known to be encoded on plasmids, while others are known to be encoded by other types of *mobile genetic elements*: transposons and bacteriophages. Several examples are now known where the genes producing a particular type of infection are present on different genetic elements within the same cell. For instance, the genes encoding the virulence determinants of the shigatoxin-producing *E. coli* are located on the chromosome, a bacteriophage, as well as on a virulence plasmid.

Bacteriocins

Many bacteria produce agents that inhibit or kill closely related species or even different strains of the same species; these agents are called **bacteriocins** to distinguish them from the antibiotics, which have a wider spectrum of activity. Bacteriocins are ribosomally synthesized peptides (although several require extensive posttranslational modification for activity). The structural gene for the bacteriocin and the genes encoding proteins involved with processing and transporting the bacteriocin (and for conferring immunity to its action) are often carried by a plasmid or a transposon. Bacteriocins are named in accordance with the species of organism that produces them. Thus, in *Escherichia coli* we have *colicins*, coded by Col plasmids, *Bacillus subtilis* produces *subtilisin*, and so on.

The Col plasmids of *Escherichia coli* encode various colicins. Colicins released from a producing cell bind to specific receptors on the surface of susceptible cells. The receptors for colicins are generally entities whose normal function is to transport some substance, frequently a growth factor or micronutrient, through the outer membrane (the lipopolysaccharide layer) of the cell. Colicins kill cells by disrupting some critical cell function. Many colicins form channels in the cell membrane that allow potassium ions and protons to leak out, leading to a loss of the cell's energy-forming ability. However, colicin E2 is a DNA endonuclease that can cleave cellular DNA, and colicin E3 is a nuclease that cuts at a specific site in 16S rRNA and inactivates ribosomes. Col plasmids can be either conjugative or nonconjugative.

The bacteriocins or bacteriocin-like agents of gram-positive bacteria are quite different from the colicins but are also often encoded by plasmids, and some even have commercial value. For instance, the lactic acid bacteria produce the bacteriocin Nisin A, which strongly inhibits the growth of a wide range of gram-positive bacteria and is used as a preservative in the food industry.

Engineered Plasmids

The techniques of genetic engineering, discussed later in this chapter, have made possible the construction in the laboratory of a limitless number of new, artificial plasmids. Incorporation into such plasmids of genes from a wide variety of sources has made possible the transfer of genetic material across any species barrier. It is even possible to synthesize completely new genes and introduce them into plasmids.

We now turn our attention to the details of conjugation and show how certain plasmids can act to mobilize the bacterial chromosome, allowing transfer of chromosomal genes from donor to recipient.

✓ 10.8 Concept Check

The genetic information that plasmids carry is not essential for cell function under all conditions but may confer selective growth advantage under certain conditions. Examples include antibiotic resistance, enzymes for degradation of unusual organic compounds, and special metabolic pathways. Closely related plasmids cannot both replicate in the same cell. Some plasmids can transfer from cell to cell by a mechanism called conjugation.

✓ Are two incompatible plasmids likely to be very similar or very different?

✓ Conjugative plasmids tend to be large. Why?

10.9 Conjugation and Chromosome Mobilization

Bacterial conjugation (mating) is a process of genetic transfer that involves cell-to-cell contact. As we discussed previously (see Section 10.8), conjugation is a plasmid-encoded mechanism. A conjugative plasmid uses this mechanism to transfer a copy of itself to a new host. However, sometimes other genetic elements can be *mobilized* during conjugation. These other genetic elements can be other plasmids, or the host chromosome. Indeed, conjugation was discovered because the F plasmid of *Escherichia coli* (see Figure 10.17) can mobilize the host chromosome. Mechanisms of conjugative transfer may differ depending on the plasmid involved, but most plasmids in gram-negative *Bacteria* seem to employ a mechanism similar to that used by the F plasmid.

Conjugation involves a *donor* cell, which contains a particular type of conjugative plasmid, and a *recipient* cell, which does not. Because conjugation was discovered by performing genetic crosses, the donors were called *males* and the recipients were called *females*. The genes that control conjugation are contained in the *tra* region of the plasmid (see Section 10.8). Many genes in the *tra* region are involved in mating pair formation and most of these have to do with the synthesis of a surface structure, the **sex pilus** (Figure 10.20). Only donor cells have these pili, explaining why RNA bacteriophages that attach to them are called "*male-specific*" (◯◯◯ Section 16.1). Different conjugative plasmids may have slightly different *tra* regions, and the pili may also be different. The F plasmid and its relatives encode *F pili*.

Pili allow specific pairing to take place between the donor cell and the recipient cell. All conjugation in gram-negative *Bacteria* is thought to depend on cell pairing brought about by pili. The pili make specific contact with a receptor on the recipient and then retract, pulling the two cells together. The contacts between the donor and recipient cells then become stabilized, probably from fusion of the outer membranes, and the DNA is then transferred from one cell to another.

Mechanism of DNA Transfer During Conjugation

DNA synthesis is necessary for DNA transfer to occur, and the evidence suggests that one of the DNA strands is derived from the donor cell and the other is newly synthesized in the recipient during the transfer process. Some genes in the *tra* region have to do with DNA transfer and replication. A mechanism of DNA synthesis in certain bacteriophages, called **rolling circle replication**, is presented in Figures 9.20 and 16.4. This model best explains DNA transfer during conjugation, and a possible

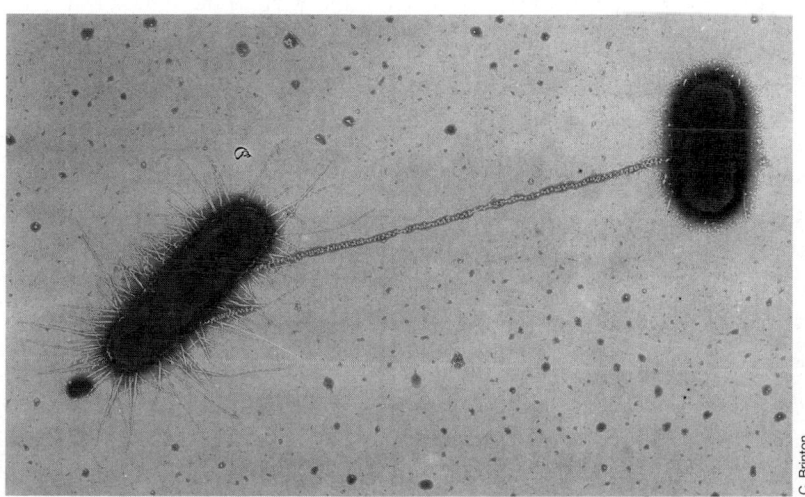

Figure 10.20 Direct contact between two conjugating bacteria is first made via a pilus. The cells are then drawn together to form a mating pair for the actual transfer of DNA. This occurs by retraction (depolymerization) of the pilus within the donor cell. Note the F-specific bacteriophages on the pilus (◯◯◯ Section 16.1).

C. Brinton

mechanism for this process is outlined in Figure 10.21. The whole series of events is triggered by cell-to-cell contact, at which time one strand of the plasmid DNA circle is nicked and one parental strand is transferred. The nicking enzyme required to initiate the process, TraI, is encoded by the *tra* operon of the F plasmid. This protein also has helicase activity and is thus also involved in unwinding the strand to be transferred. As this transfer occurs, DNA synthesis by the rolling circle mechanism replaces the transferred strand in the donor. A complementary DNA strand is also made in the recipient. Therefore, at the end of the process, both donor and recipient possess completely formed plasmids.

The plasmid DNA transfer process is highly efficient; under appropriate conditions virtually every recipient cell that pairs acquires a plasmid. When the plasmid genes can be expressed in the recipient, the recipient itself becomes a donor and can transfer the plasmid to other recipients.

In this fashion, conjugative plasmids can spread rapidly between populations, behaving like infectious agents. The infectious nature of this phenomenon is of major ecological significance because a few plasmid-positive cells introduced into an appropriate population of recipients can, if they contain genes that confer a selective advantage, convert the whole recipient population into a plasmid-bearing population in a short period of time. The widespread occurrence of drug resistance carried by conjugative plasmids (see Section 10.8) has led to some serious problems for the chemotherapy of infectious disease (⌖ Section 20.12). However, as mentioned above, plasmids can also be lost from a cell by a process called *curing*. This could happen spontaneously in natural populations when there is no selection pressure to maintain the plasmid. For example, plasmids conferring antibiotic resistance can be lost without affecting the cell's viability if there are no antibiotics in the cell's environment.

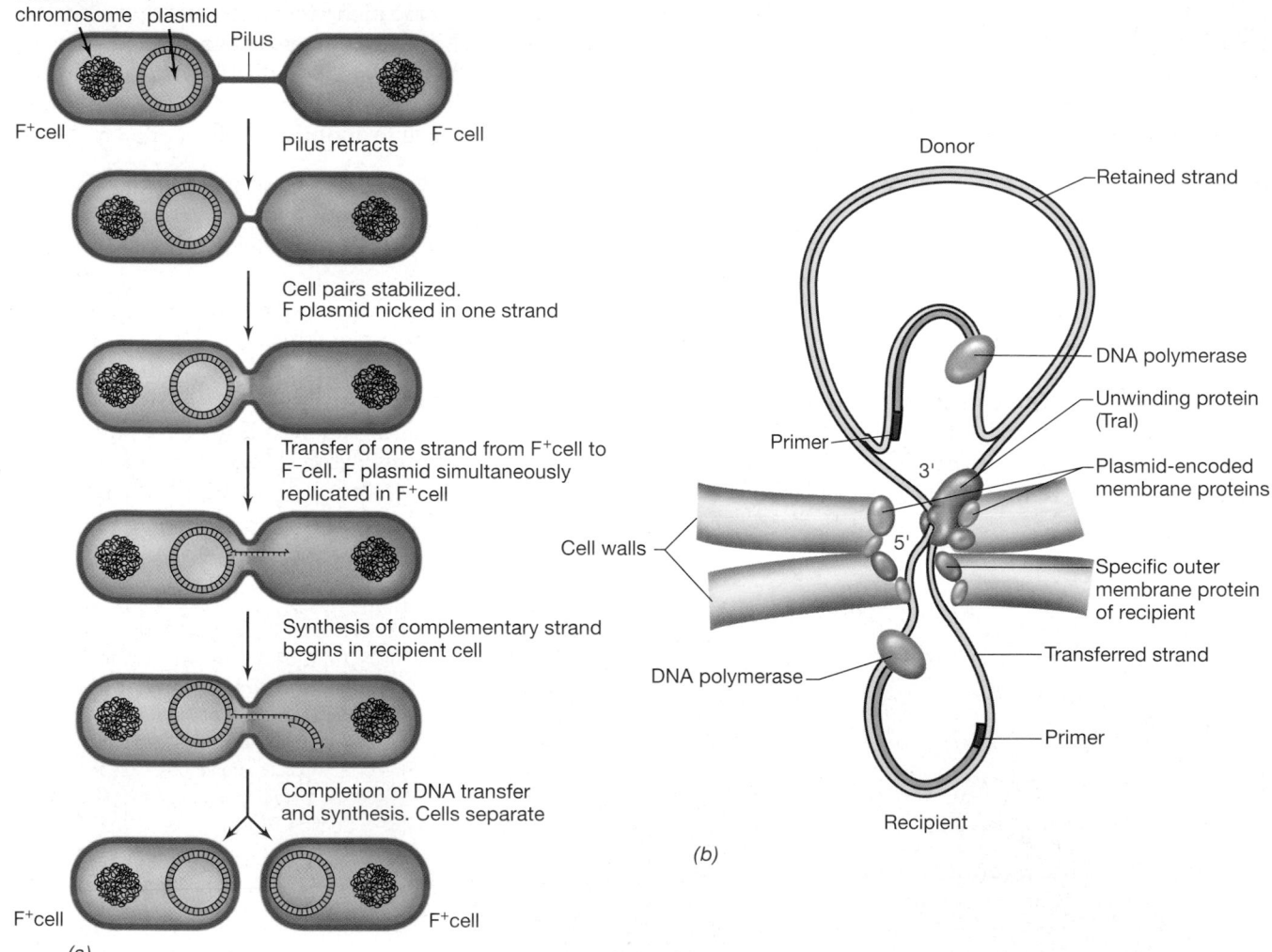

Figure 10.21 Transfer of plasmid DNA by conjugation. (a) In this example, the F plasmid of an F⁺ cell is being transferred to an F⁻ recipient cell. Note the mechanism of rolling circle replication (⌖ Figure 9.20 and Figure 16.4). (b) Details of the replication and transfer process.

The Formation of Hfr Strains and Chromosome Mobilization

The F plasmid of *Escherichia coli* (see Section 10.8) is not only conjugative but also can mobilize the chromosome for transfer during cell-to-cell contact. The F plasmid is an episome, a plasmid that can integrate into the host chromosome (see Section 10.8). When the F plasmid is integrated into the chromosome, conjugation can lead to transfer of large regions of the host chromosome and genetic recombination between donor and recipient can then be very extensive.

Cells possessing an unintegrated F plasmid are called **F⁺**, and those that have a chromosome-integrated F plasmid are called **Hfr** (for high frequency of recombination). Both F⁺ and Hfr cells act as donors, but conjugation using an Hfr donor leads to transfer of the host chromosome because the plasmid is part of the chromosome. Plasmid integration is a very simple mechanism of mobilizing other genetic elements. The term "high frequency of recombination" refers to genetic recombination between the donor chromosome and that of the recipient. Cells without an F plasmid are called F⁻ and act as recipients. In general, cells that contain a conjugative plasmid are very poor recipients for the same or closely related plasmids. This is because the plasmid is blocked from entering, not because of replication incompatibility (see Section 10.8).

We thus see that the presence of the F plasmid results in three distinct alterations in the properties of a cell: (1) ability to synthesize the F pilus, (2) mobilization of DNA for transfer to another cell, and (3) alteration of surface receptors so the cell is no longer able to behave as a recipient in conjugation.

The integration of the F plasmid into the host chromosome can occur at several specific sites, called IS (for *insertion sequences*). These sites are regions of homology between chromosome and F plasmid DNA (see Section 10.11 for a discussion of insertion sequences). Figure 10.22 shows the integration of an F plasmid, which involves insertion in the chromosome at such a site. In the particular Hfr shown, the integration site is between the chromosomal genes *pro* and *lac*. Once integrated, the plasmid no longer controls its own replication, but the *tra* operon still functions normally and the strain synthesizes pili. When a recipient is encountered, conjugation is triggered just as in an F⁺ cell, and DNA transfer is initiated at *oriT*. However, since the plasmid is now part of the chromosome, after part of the plasmid DNA is transferred, chromosomal genes begin to be transferred (Figure 10.23). (As is the case of conjugation with just the F plasmid itself, DNA transfer involves replication, so after transfer, the Hfr strain

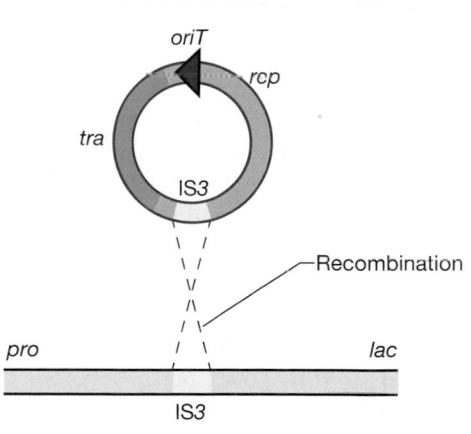

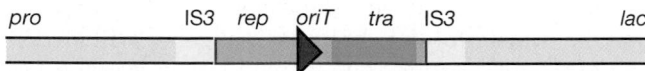

Figure 10.22 Integration of an F plasmid into the chromosome with the formation of an Hfr. The insertion of the F plasmid occurs at a variety of specific sites where IS elements are located, the one here being an IS3 located between the chromosomal genes *pro* and *lac*. Some of the genes on the F plasmid are shown. The arrow indicates the origin of transfer, *oriT*, with the arrow as the leading end. Thus, in this Hfr *pro* would be the first chromosomal gene to be transferred and *lac* would be among the last.

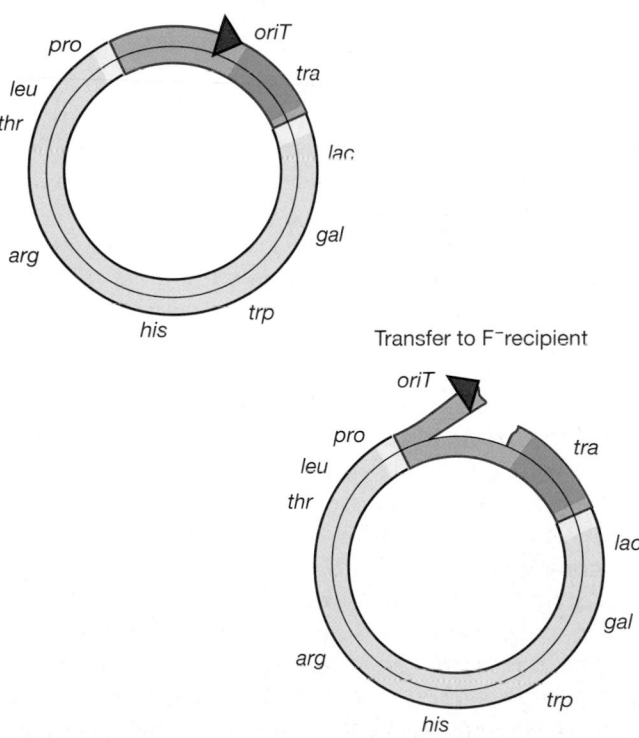

Figure 10.23 Breakage of the Hfr chromosome at the origin of transfer and the beginning of DNA transfer to the recipient. Replication occurs during transfer (see Figure 10.21). Please note that the figure is not drawn to scale. The inserted F plasmid is less than 3% of the size of the *Escherichia coli* chromosome.

still remains Hfr because it has retained a copy of the transferred genetic material.)

Thus, the chromosome is said to be *mobilized*. As we mentioned, integration of a plasmid is a simple mechanism of mobilization. Other mechanisms of mobilization are known, involving not only the host chromosome but other plasmids. One such mechanism is for the mobilizable plasmid to contain an *oriT* sequence like that found on the F plasmid. Such a plasmid will be transferred by conjugation if it is in a cell with a conjugative plasmid that can supply all the other necessary genes.

Since a number of distinct insertion sites are present, a number of distinct Hfr strains are possible. A given Hfr strain always donates genes in the same order, beginning with the same position, but Hfr strains of independent origin transfer genes in different sequences.

Usually, because of breakage of the DNA strand during transfer, only a *part* of the donor chromosome is transferred. Since only part of the chromosome is transferred, it cannot replicate in the recipient cell. Therefore, donor genes normally cannot be detected unless recombination between the incoming fragment and the recipient chromosome takes place.

Although Hfr strains transmit chromosomal genes at high frequency, they usually do not convert F⁻ cells to F⁺ or Hfr because the entire F plasmid is only rarely transferred. On the other hand, F⁺ cells efficiently convert F⁻ to F⁺ because the entire F plasmid is transferred.

At some insertion sites, the F plasmid is integrated with the origin in one direction, whereas at other sites the origin is in the opposite direction. The direction in which the F plasmid is inserted determines which of the chromosomal genes will be transferred into the recipient first. The manner in which a variety of Hfr strains can arise is illustrated in Figure 10.24. By use of various Hfr strains, it has been possible in *Escherichia coli* to determine the arrangement and orientation of a large number of chromosomal genes, as will be described in Section 10.19.

Use of Hfr Strains and the Phenomenon of Interrupted Mating

As is the case for any system of bacterial gene transfer, one **selects** recombinants from conjugation. However, unlike the situation in transformation and transduction, during conjugation both the donor and recipient cells are viable, so it is necessary to chose selection media where the desired recombinants can grow but where neither of the parental strains can form colonies. Typically a recipient is used that is resistant to an antibiotic but is auxotrophic for some substance, and a donor that is sensitive to the antibiotic but is prototrophic for the same substance.

For instance, in the experiment shown in Figure 10.25, an Hfr donor that is sensitive to streptomycin (Str^s) and contains wild-type genes encoding enzymes needed for synthesis of the amino acids threonine and leucine (Thr⁺

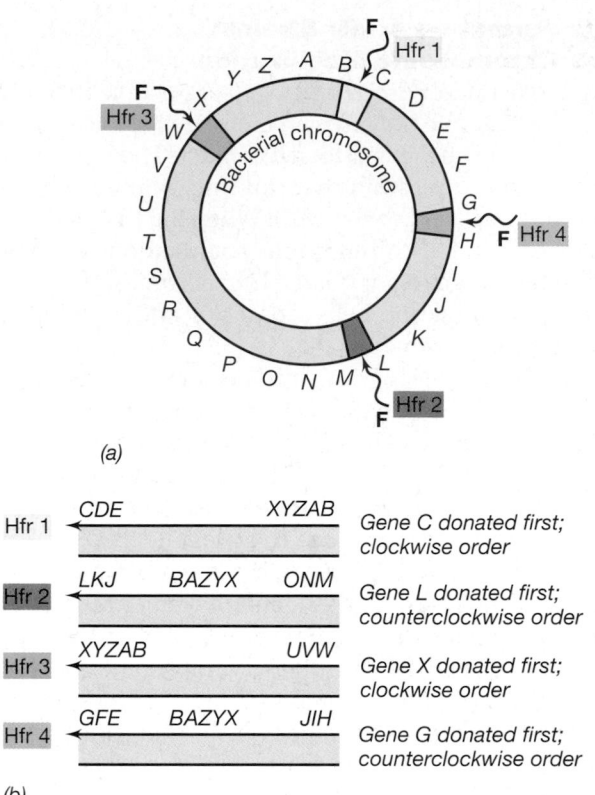

(a)

(b)

Figure 10.24 Manner of formation of different Hfr strains, which donate genes in different orders and from different origins. The bacterial chromosome is a circle (a) that can open at various insertion sequences, at which F plasmids become inserted. The gene orders are shown in part (b).

and Leu⁺) and for utilization of the energy source lactose (Lac⁺) is mated with a recipient cell that is mutant for these genes but is resistant to streptomycin (Str^r) . The selective medium is a minimal medium containing streptomycin so that only recombinant cells can grow. The composition of each selective medium is varied depending on which genotypic characteristics are desired in the recombinant, as shown in Figure 10.25. The frequency of the process is measured by counting the colonies grown on the selective medium.

Understanding the mechanism of conjugation and knowing that an Hfr strain is a cell with an F plasmid integrated into its chromosome, should lead you to see that the transfer of the chromosomal genes will always be in the same order from a fixed site in a given Hfr strain. The nature of the oriented, fragmentary transfer of host genes during conjugation was first shown by a procedure called **interrupted mating**. The mating pairs are rather weakly joined and can be separated by agitation in a mixer or blender. If mixtures of Hfr and F⁻ cells are agitated at various times after mixing and the genetic recombinants scored, it is found that the longer the time between pairing and agitation, the greater the num-

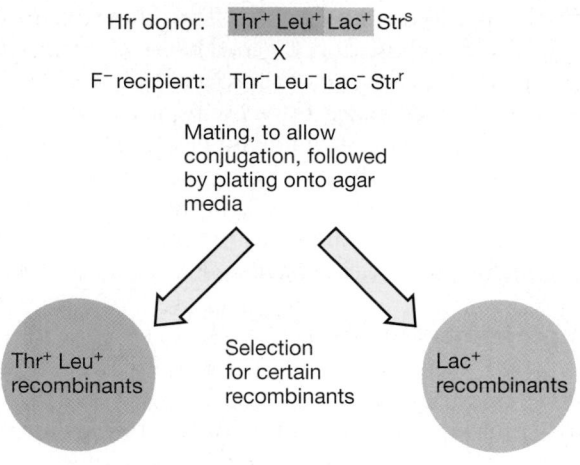

Hfr donor: Thr⁺ Leu⁺ Lac⁺ Strˢ

X

F⁻ recipient: Thr⁻ Leu⁻ Lac⁻ Strʳ

Mating, to allow conjugation, followed by plating onto agar media

Thr⁺ Leu⁺ recombinants

Selection for certain recombinants

Lac⁺ recombinants

Agar mimimal medium with streptomycin and glucose; selects for markers Thr⁺ Leu⁺; does not select for Lac

Agar mimimal medium with streptomycin, lactose, threonine, leucine; selects for marker Lac⁺; does not select for Thr or Leu

Figure 10.25 Laboratory procedure for the detection of genetic conjugation. Thr, threonine; Leu, leucine; Lac, lactose; Str, streptomycin. Note that each medium selects for specific classes of recombinants. The controls for the experiment are to plate samples of the donor and the recipient before they are mixed. Neither should be able to grow on the selective media used.

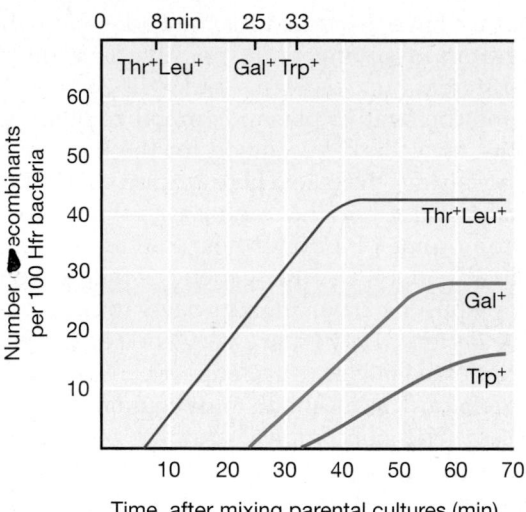

Figure 10.26 Rate of formation of recombinants containing different genes after mixing Hfr and F⁻ bacteria by the process known as interrupted mating. The location of the genes along the Hfr chromosome is shown at the upper left. Note that the genes closest to the origin (0 min) are the first ones detected in the recombinants. The experiment is done by mixing Hfr and F⁻ cells under conditions in which essentially all Hfr cells find mates. The F⁻ recipient was streptomycin-resistant but auxotrophic for the markers being scored. The Hfr donor was streptomycin-sensitive. At various times, samples of the mixture are shaken violently to separate the mating pairs and plated on a selective medium in which only the recombinants can grow and form colonies.

ber of genes of the Hfr that will appear in the F⁻ recombinant. As shown in Figure 10.26, genes present closer to the origin enter the F⁻ first and are always present in a higher percentage of the recombinants than genes that enter late. In addition to showing that gene transfer from donor to recipient is a sequential process, experiments of this kind provide a method of determining the order of the genes on the bacterial DNA (genetic mapping). The arrangement of gene loci on the chromosome is called a **genetic map** (see Section 10.19).

As in transformation and transduction, genetic recombination between Hfr genes and F⁻ genes involves homologous recombination in the recipient cell. This has been shown by the isolation of mutants of F⁻ strains that are unable to form recombinants when mated with Hfr. These mutants are Rec⁻ (recombination minus) and are deficient in the RecA protein because of a mutation in the *recA* gene (see Section 10.5). It is important to remember that recombination is not the same as DNA transfer. DNA transfer is unaffected by the fact that the recipient is a Rec⁻ cell, even though recombination does not take place after transfer.

Transfer of Chromosomal Genes to the F Plasmid

Occasionally integrated F plasmids may be excised from the chromosome, and the possibility exists for the incorporation at that time of *chromosomal* genes into the liberated F plasmid. This can happen because both the

integrated F plasmid and the chromosome contain a number of identical IS at which recombination can occur. Such F plasmids containing chromosomal genes are called *F' (F prime) plasmids*. These F' plasmids differ from normal F plasmids in that they contain identifiable chromosomal genes, and they transfer these genes at high frequency to recipients. F'-mediated transfer resembles specialized transduction in that only a restricted group of chromosomal genes can be transferred Transferring a known F' into a recipient allows one to establish diploids for a limited region of the chromosome (such partial diploids are called *merodiploids*). This is important for doing complementation tests (see Section 10.10).

Other Conjugation Systems

Although we have discussed conjugation almost exclusively as it occurs in *Escherichia coli*, conjugative plasmids have been found in many other gram-negative *Bacteria*. Indeed, conjugative plasmids of the incompatibility group (see Section 10.8) IncP can be maintained in practically all gram-negative species, and DNA transfer between species and genera can occur. Conjugative plasmids are also known in gram-positive *Bacteria* (for example, in *Streptococcus*, *Enterococcus*, and *Staphylococcus*). As a rule, the mechanism of conjugation is very similar in the other gram-negative *Bacteria* to what we have discussed for the F plasmid, whereas conjugation involving gram-positive *Bacteria* can be quite different.

Some conjugative plasmids mobilize other genetic elements, as we have discussed for the F plasmid, but this is not always the case. There are also elements called *conjugative transposons* that can transfer themselves from the chromosome of a donor host to that of a recipient and can also mobilize other genetic elements. Conjugative transposons have mostly been found in gram-positive *Bacteria*. They have a very wide host range and can be involved in gene transfer between bacteria of different genera. The mechanism of conjugation used by these elements is not completely known, but it does seem to involve circular, plasmid-like intermediates. Several conjugative plasmids have also been found in *Sulfolobus*, a genus of *Archaea*. Very little is known about conjugation in *Sulfolobus*, although it is known that cell-pairing occurs before plasmid transfer and that transfer is unidirectional. However, with one exception, the genes involved seem to have little similarity to those in gram-negative *Bacteria*. The exception is a gene similar to *traG*, whose protein product in F plasmid-mediated conjugation seems to be involved in stabilizing mating pairs. It thus seems likely that the mechanism of conjugation in *Archaea* is quite different from that in *Bacteria*.

✓ 10.9 Concept Check

Conjugation is a mechanism of DNA transfer in prokaryotes that requires cell-to-cell contact. Conjugation is controlled by genes carried by certain plasmids (such as the F plasmid) and typically involves transfer of the plasmid from a donor cell to a recipient cell. However, other genetic elements, including the donor cell chromosome, can sometimes be mobilized and also transferred. Transfer of the host chromosome is rarely complete but can be used to map the order of the genes on the chromosome.

✓ How are donor and recipient cells brought into contact with each other?

✓ In conjugation involving the F plasmid of *Escherichia coli*, how is the host chromosome mobilized?

10.10 Complementation

The methods of bacterial gene transfer we have discussed in the preceding sections involve the transfer of only a *portion* of the donor chromosome. Therefore, unless recombination takes place with the recipient chromosome this incoming donor DNA will be lost because it cannot replicate independently. In only two instances have we seen that a state of partial diploidy can be stably maintained. One was in the case of a specialized transducing phage where the donor genes are maintained as part of

the viral genome (see Section 10.7). The other was the use of F′ plasmids where donor genes have become a part of the F plasmid genome (see Section 10.9). Since it is possible to create specialized transducing phage or recombinant plasmids using recombinant DNA techniques (see Sections 10.15 and 10.16), it is now possible to put essentially any portion of the bacterial chromosome on a phage or plasmid. This can be quite useful for a number of purposes, one being the *complementation test*.

Complementation Test

When two mutant strains are genetically crossed (mated), homologous recombination can yield a wild-type recombinant unless both of the mutations include changes in exactly the same base pairs. Therefore, if two different Trp⁻ *Escherichia coli* (strains that require the amino acid tryptophan in the medium) are crossed and Trp⁺ recombinants are obtained, it is clear that the mutations in the two strains did not include the same base pairs. However, this experiment cannot detect whether the mutations were in the same gene. This can be determined by a type of experiment called a **complementation test**.

Complementation was first used in *diploid* eukaryotic organisms. In diploid organisms the cell has *pairs* of chromosomes, one member from each parent (⟳ Section 14.6). When the two mutations are present on separate members of a pair (that is, one mutation from each parent), they are said to be in **trans** configuration. On the other hand, if the two mutations are on the *same* chromosome (both from the same parent), then they are said to be in **cis** configuration. As we have discussed, true diploidy does not exist in prokaryotes, but a diploid state can be a achieved for a region of the chromosome by transfer of a chromosomal fragment from a donor. However, the principle of the test and the nomenclature (*cis* and *trans*) is the same.

If one of the tryptophan genes has been inactivated in a particular strain by a mutation, leading to the Trp⁻ phenotype, then putting a copy of the wild-type gene from another strain into the same cell should restore the wild-type phenotype of the cell; that is, the resulting partially diploid cell should be Trp⁺. (This will be true unless the mutation is *dominant*.) One could then say that the wild-type gene *complements* the mutation.

Even though in prokaryotes only a chromosomal fragment is transferred, this fragment will typically be large enough to contain many genes. Remember that the genes for the biosynthesis of tryptophan form an operon (⟳ Figure 8.20), and therefore one can easily transfer the entire operon from the donor cell. Note then that even if the operon in the donor also has a mutation, it will complement the mutated gene in the donor *if the two mutations are in different genes*. The two mutations are then said to *complement* each other. This is shown diagrammatically in Figure 10.27. Mutations can complement because the genes encode proteins that can diffuse through the cytoplasm of the cell. (One cannot complement mutations in regulatory sites such as promoters because these function at the DNA level.) If each homologous DNA molecule contributes a different required gene, then the cell will have all the enzymes it requires to synthesize tryptophan. Notice that complementation *does not* involve recombination. To do the test, the mutations must be in *trans*. (If one molecule has both mutations, the other is wild type and should be sufficient itself to confer the wild-type phenotype unless one of the mutations is *dominant*. Therefore, having the mutations in *cis* serves as a control.)

The Cistron

This type of complementation test, called a *cis-trans test*, is used to define whether two mutations are in the same genetic (functional) unit. The genetic unit defined by the cis-trans test is sometimes called a **cistron** (a term essentially equivalent to a gene). As noted, two mutations in the *same* cistron *cannot* complement each other, and so when complementation is found to exist, this implies that the two mutations lie in *different* cistrons (that is, different genes). The term *cistron* is now rarely used except when describing whether an mRNA has the genetic information from one gene (monocistronic mRNA) or from more than one gene (polycistronic mRNA) (👁 Section 7.11).

✓ **10.10 Concept Check**

If a cell can be made to contain two copies for a region of its genome that is under study, one can do complementation tests to determine if two mutations are in the same or different genes. This is often necessary because mutations in different genes in the same pathway may give the same phenotype. Complementation tests do not involve recombination.

✓ What is a *partial diploid*?

✓ Complementation tests have been referred to as *cis-trans tests*. Explain.

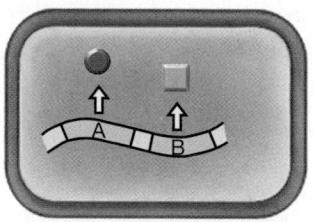
Wild-type cell: both genes are functional and cell is Trp⁺

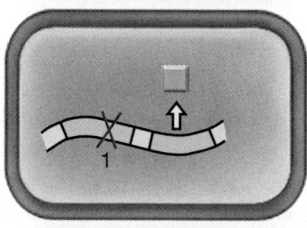
Mutant X: cell contains mutation 1 and is Trp⁻ (requires tryptophan for growth)

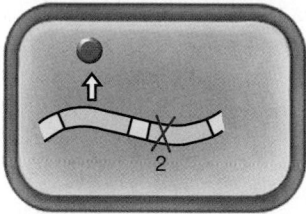
Mutant Y: cell contains mutation 2 and is Trp⁻

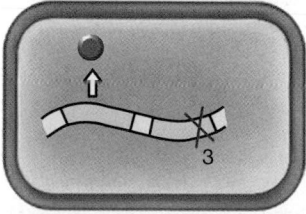
Mutant Z: cell contains mutation 3 and is Trp⁻

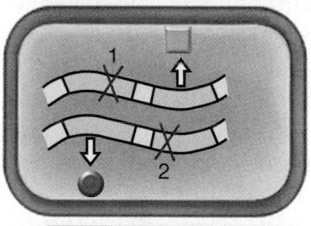
Trans test of mutations 1 and 2: complementation occurs (cell is Trp⁺), therefore mutations are in separate genes

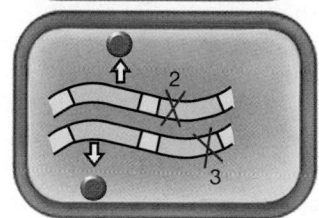
Trans test of mutations 2 and 3: no complementation occurs (cell is Trp⁻), therefore mutations are in the same gene

Figure 10.27 Complementation analysis. The protein products of both genes (A and B) are required to synthesize tryptophan. Mutations 1, 2, and 3 each lead to the same phenotype, a requirement for tryptophan. Complementation analysis indicates that mutations 2 and 3 are in one gene and that mutation 1 is in another.

10.11 Transposons and Insertion Sequences

The order of genes on a bacterial chromosome can be determined by the methods of gene transfer we have considered, just as genes in eukaryotic chromosomes can be mapped by mating experiments. However, the exact arrangement of the genes along a chromosome is not necessarily permanently fixed; some genes are capable of moving under certain conditions. The process by which a gene moves from one place to another in the genome is called **transposition** and is an important process in evolution and in genetic analysis.

We should emphasize that transposition typically is a *rare* event, occurring at frequencies of 10^{-5}–10^{-7} per generation. Thus, the genes of living organisms are relatively stable.

In addition, not all genes are capable of transposition. Rather, transposition of genes is linked to the presence of special genetic elements called **transposable elements**.

Transposition was originally discovered in corn (maize) and then later in *Bacteria* owing to the extremely sensitive types of genetic analysis available in these organisms. It has now been shown that DNA sequences with the properties of transposable elements are widespread in nature. However, in organisms with poorly characterized genetic systems, it can be difficult to detect transposition (because it is a rare event) and actually prove that a sequence is a transposable element.

Transposable Elements

As we discussed earlier (⟢ Section 7.4), there are three types of transposable elements in *Bacteria: insertion sequences, transposons*, and some special viruses (such as Mu) (⟢ Section 16.5). In this section we shall confine our discussion primarily to insertion sequences and transposons. Both these elements have two important features in common: They both carry genes encoding a **transposase**, the enzyme necessary for transposition, and both have short *inverted terminal repeats* at the ends of their DNA (remember that these "ends" are continuous with whatever DNA the element is inserted into). These repeats range in length from fewer than 20 base pairs in simple IS elements to more than 1000 base pairs in some transposons; each IS has a specific number of base pairs in its terminal repeats. Such inverted terminal repeats are involved in the transposition process (see later). Figure 10.28 shows a genetic map of an insertion element named IS2 and of a transposon named Tn5.

Insertion sequences are the simplest type of transposable element and carry no genetic information other than that required for them to move to new locations. Insertion sequences are short segments of DNA, about 1000 nucleotides long, that can become integrated at specific sites on the genome. Insertion sequences, abbreviated IS, are found in both chromosomal and plasmid DNA, as well as in certain bacteriophages. Several hundred distinct IS elements have been characterized, and most are designated by a number identifying its type: IS1, IS2, IS3, and so on. Because so many IS elements have been discovered, some receive designations related to the organism in which they were first identified. For example, IS Mt1 is from *Mycobacterium tuberculosis*. IS elements are scattered about the chromosome, and strains vary in the number and frequency of these elements. For instance, one strain of *Escherichia coli* has five copies of IS2 and five copies of IS3. Many plasmids also carry these insertion sequences (see Figures 10.17 and 10.19), and it is homologous recombination between identical sequences on the F plasmid and the chromosome (*not* transposition) that allows the F plasmid to integrate into the bacterial chromosome and mobilize it (see Section 10.9). Some of the *Archaea* have large numbers of IS elements in their chromosomes.

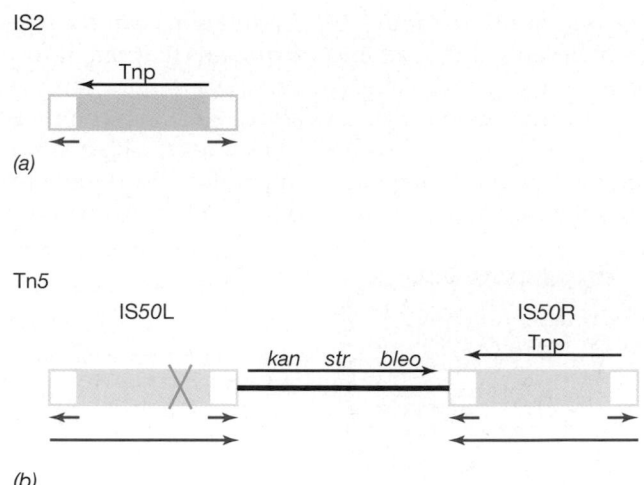

Figure 10.28 Maps of the transposable elements IS2 and Tn5. The red arrows underneath each map indicate the inverted repeats. The arrows above the maps show the direction of transcription of any genes on the elements. *Tnp* is the gene encoding the transposase. The transposase genes of these two elements are not closely related. (a) IS2 is an insertion sequence of 1327 base pairs with inverted repeats of 41 base pairs at its ends. (b) Tn5 is a composite transposon of 5.7 kilobase pairs with the insertion sequences IS50L and IS50R at its left and right ends, respectively. IS50L is not capable of independent transposition because there is a *nonsense mutation* (see Section 10.2) marked by a blue cross in its transposase gene. Otherwise, the two IS50 elements are very nearly identical. Note that these two IS50 elements are inverted with respect to each other. The genes *kan*, *str*, and *bleo*, confer resistance to the antibiotics kanamycin (and neomycin), streptomycin, and bleomycin. Interestingly, streptomycin resistance is not expressed in *Escherichia coli*.

Transposons are larger than insertion sequences and carry other genes, some of them conferring important properties on the organism carrying them. These often include drug resistance markers and other easily selectable genes. In addition, as we have mentioned (see Section 10.9), there are *conjugative transposons*. These transposons have genes allowing them not only to move from one location on a bacterial genome to another but also to transfer themselves from one bacterium to another.

Some transposons are actually composite structures containing a gene or group of genes lying between two identical insertion sequences. The existence of such *composite transposons* indicates that new transposons probably continue to arise in cells that have insertion sequences.

The Mechanism of Transposition

As mentioned previously, the inverted repeats found at the ends of transposable elements are essential for transposition. The other essential component is an enzyme called *transposase*, which recognizes these repeats. This enzyme is usually encoded by the transposable element, although some very simple IS elements use an enzyme encoded by another genetic element. The transposase apparently recognizes, cuts, and eventually ligates the DNA during transposition (Figure 10.29).

When a transposable element becomes inserted into another DNA (the target DNA), a short sequence in the target DNA at the site of integration is duplicated. This target DNA sequence was not present in the element, but the transposable element has brought about a duplication of this DNA by the insertion process. The duplication of the target sequence arises because single-stranded breaks are generated by the transposase. The transposon is then attached to the single-stranded ends that have been generated, and repair of the single-strand portions results in the duplication (see Figure 10.30).

Certain transposable elements prefer certain sequences as target sites, but others, including the bacteriophage Mu, can insert themselves almost randomly (for a representation of Mu insertion, see Chapter 16, ∞ Figure 16.9).

Two mechanisms of transposition are known, called *conservative* and *replicative* (Figure 10.30). In conservative transposition, such as can occur in the transposon Tn5, the transposable element is excised from one location in the chromosome and becomes reinserted at a second location. The copy number of a conservative transposon therefore remains at one. By contrast, replicative transposons, such as bacteriophage Mu (∞ Section 16.5), are duplicated, and a new copy is inserted at another location. Thus, after the transposition event is completed, *one* copy of the transposing element *remains* at the original site and *another copy* is found at the new site. During this whole transposition process, the source transposon *remains* at its original site; at no time does the source transposon become free in the cell.

Although the molecular details of transposition remain uncertain, and different transposable elements appear to have somewhat different mechanisms, models

for both conservative and replicative transposition are illustrated in Figure 10.30. As seen, single-strand cuts are made at the ends of the transposon (at the sites of the inverted repeats), and staggered single-strand cuts are made at the target site. The transposon is now joined to the target site via the single-stranded ends. In conservative transposition, the donor site is now cut and replication repair then fills in the single-strand gaps in

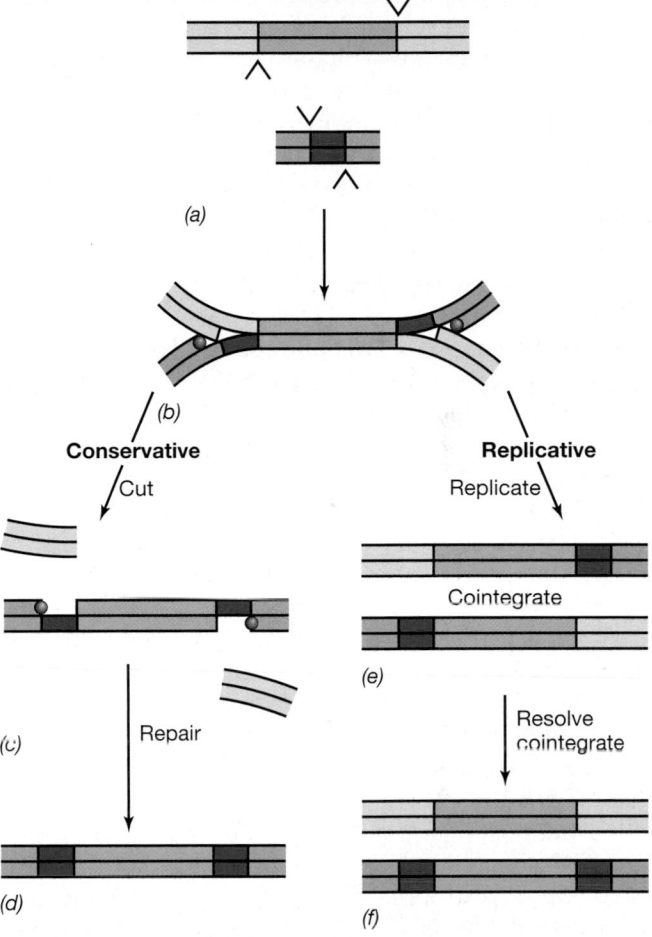

(a)

(b)

Conservative   **Replicative**

Cut   Replicate

(c)   Cointegrate

(e)

Repair   Resolve cointegrate

(d)   *(f)*

Figure 10.30 Mechanisms of transposition. (a) In both conservative and replicative transposition the transpose makes cuts (marked with arrows) in the DNA strands at the end of the transposable element (orange) and at the target site (red). The number and location of the cuts may vary depending on the mechanism. (b) The target site becomes ligated to the transposable element. The black dots indicate free 3′ ends of DNA strands at which replication can occur (∞ Section 7.5). (c) In conservative transposition further cuts are made before DNA replication/repair occurs, and the transposable element is lost from the donor DNA. (d) Repair leads to duplication of the target site and completion of transposition to the new site. (e) In replicative transposition, replication occurs without the cutting of the transposable element from the donor site leading to two copies of the transposable element as part of a cointegrate. Note, however, this has led to the joining of the donor (light green) and target (dark green) DNA molecules together. (f) These molecules are separated (resolved) in a further reaction. Resolution of cointegrates is shown in more detail in Figure 10.31.

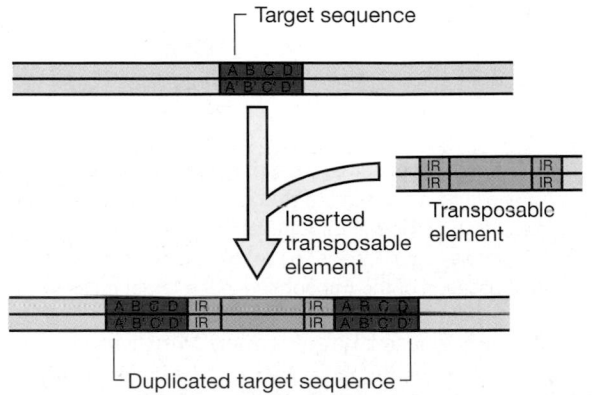

Target sequence

Inserted transposable element

Transposable element

Duplicated target sequence

Figure 10.29 Transposition. Insertion of a transposable element generates a duplication of the target sequence. Note the presence of inverted repeats (IRs) at the ends of the transposable element. Figure 10.30 shows more detailed models of the mechanism of transposition.

the target site. This process results in the formation of *direct repeats* in the target site at the ends of the transposon. In replicative transposition, the replication repair takes place while the transposable element is still attached to both the original and the target sites. This leads to the formation of a composite structure called a *cointegrate*. The final event in this pathway is *resolution* of the cointegrate structure, leading to release of the original transposon and the presence of a new copy of the transposon at the target site (Figure 10.31).

Transposition is essentially a *recombination event*, but one that does not occur between homologous sequences or use the general recombination system of the cell. It involves *transposase* rather than the RecA protein that is involved in general recombination. Because this recombination involves a *specific* base sequence, it is called *site-specific recombination* (in contrast to homologous recombination discussed earlier in this chapter).

Mutagenesis with Transposable Elements

If the insertion site for a transposable element is *within* a gene, insertion of the transposon will result in loss of linear continuity of the gene, leading to mutation (Figure 10.32). Transposons thus provide a facile means of creating mutants throughout the chromosome. The most convenient element for **transposon mutagenesis** is one containing an antibiotic resistance gene. Clones containing the transposon can then be selected by the isolation of antibiotic-resistant colonies. If the antibiotic-resistant clones are selected on rich medium on which all auxotrophs can grow, they can be subsequently screened on minimal medium supplemented with various growth factors to determine if a growth factor is required.

Transposons are also useful for incorporating an auxotrophic gene marker into a wild-type organism. Normally, auxotrophic recombinants cannot be isolated by positive selection, but if the auxotrophic marker to be introduced contains a transposable element with an antibiotic resistance marker, then one can select for antibiotic-resistant clones, a positive selection procedure, and automatically obtain clones that have incorporated the auxotrophic marker.

Two transposons widely used for mutagenesis are Tn*5* (see Figure 10.29), which confers neomycin and kanamycin resistance, and Tn*10*, which contains a marker for tetracycline resistance (see Figure 10.19).

The bacteriophage Mu (⚭ Section 16.5) is also widely used as a *biological mutagen*. Because Mu integrates at a wide variety of host sites, it can be used to induce mutations at many locations. Also, Mu can be used to carry into the cell genes that have been derived from other host cells, a form of *in vivo* genetic engineering. In addition, modified Mu phages have been made artificially in which some of the lytic functions of Mu have

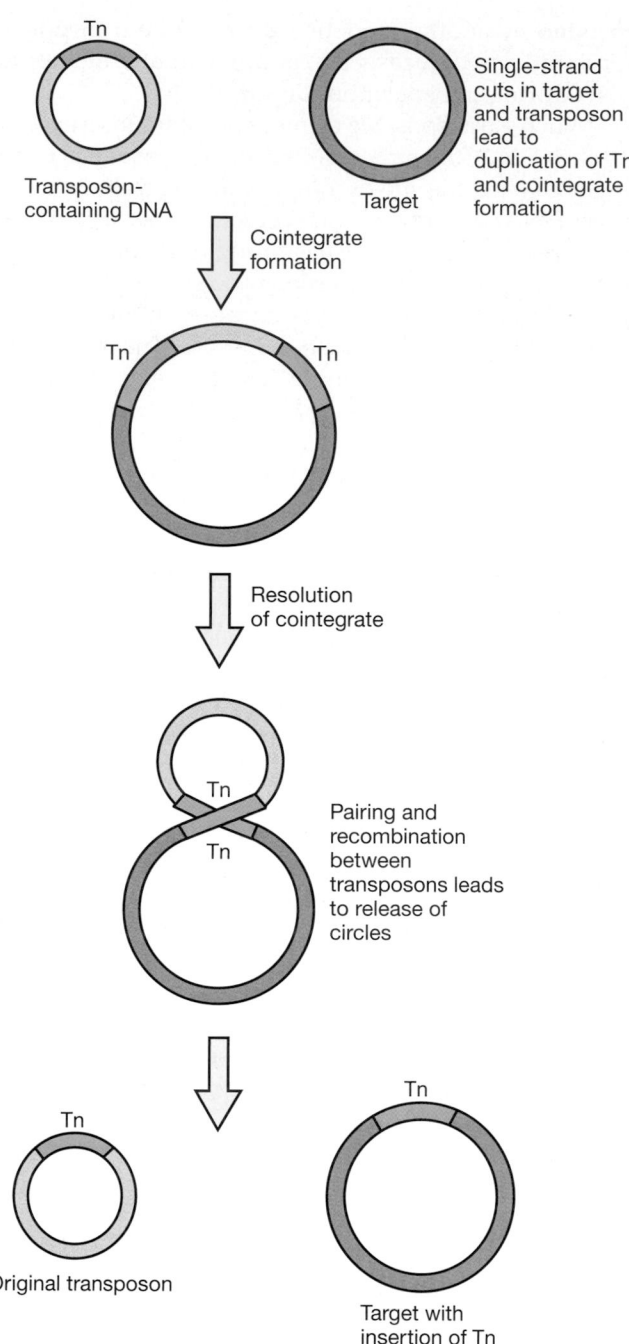

Figure 10.31 Replicative transposition. After the formation of single-strand cuts, a cointegrate structure arises by association of the two molecules (see Figure 10.30). After recombination, resolution of the cointegrate structure leads to the release of the original transposon and duplication of the transposon in the target molecule.

been deleted. These phages, called Mini-Mu, are deleted for significant portions of Mu but have the ends of the phage in normal orientation. Mini-Mu phages are usually defective, unable to form plaques, and their presence must be ascertained by the presence of other genes they carry. One set of Mini-Mu phages containing the β-galactosidase gene of the host (called Mu*d-lac, d* for

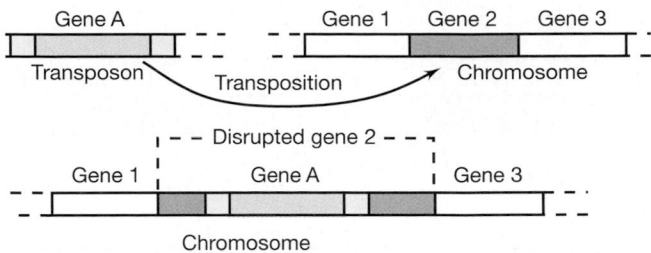

Figure 10.32 Transposon mutagenesis. The transposon moves into the middle of gene 2. Gene 2 is now disrupted by the transposon and is inactivated. Gene A of the transposon will be expressed in both locations.

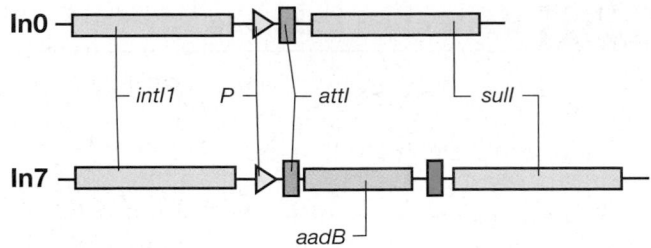

Figure 10.33 Structure of two naturally occurring integrons from *Pseudomonas*. The integron In0 has the basic set of genes: *intl1*, encodes integrase; *attl*, the site where site-specific integration can occur; P, a promoter; and *sull*, a gene conferring sulfonamide resistance that contains its own promoter. The integron In7 contains all of these genes, but in addition, a gene cassette has been integrated. All cassettes contain a site (blue square) for site-specific recombination. This cassette contains *aadB*, which confers resistance to certain aminoglycoside antibiotics.

"defective") can be detected in the integrated state if the *lac* gene is oriented properly in relation to a host promoter. Under these conditions, the host cell forms the enzyme β-galactosidase, which can be detected in colonies by using a special color indicator. β-Galactosidase-positive colonies from a β-galactosidase-negative host are thus an indication that Mud-*lac* infection has occurred.

Integrons

Because many transposons contain genes conferring both antibiotic resistance and the ability to transpose to conjugative plasmids traditional antibiotic therapies in the treatment of bacterial infections can be seriously challenged. Some transposons also contain other elements that make them even more formidable, *integrons*. **Integrons** are genetic elements that can capture and express genes from other sources. Integrons contain a gene that encodes an integrase. The integrase is an enzyme that catalyzes another type of *site-specific recombination*. Recall that the lambda genome becomes integrated into the *Escherichia coli* genome at a specific site by the action of an integrase encoded by lambda (∞ Section 9.10). The integron also contains a specific DNA sequence where the integrase can function to integrate *gene cassettes* (which are located near similar sequences on other elements) and a promoter that can then express the newly integrated gene cassette.

Integrons have been found in a number of *Bacteria*, for example *Acinetobacter*, *Citrobacter*, *Escherichia*, *Klebsiella*, *Pseudomonas*, and *Vibrio*, often in clinical isolates. They can occur in transposons, for example Tn7, on plasmids, or on the bacterial chromosome. Some isolates have as many as five different gene cassettes. Over 40 different antibiotic resistance genes have been identified on such cassettes, as have some genes associated with virulence. Figure 10.33 shows the structure of two integrons from *Pseudomonas aeruginosa*. The clinical significance of integrons and the gene cassettes they capture seems obvious. What is much less obvious is the origin of the gene cassettes themselves. These are not simply random genes that can be captured, but genes bounded by specific DNA sequences that are recognized by the integrase and genes that are apparently not expressed until they become part of an integron and can be transcribed from the promoter on the integron.

✓ 10.11 Concept Check

Transposons and insertion sequences are genetic elements that can move from one location on a chromosome to another by a process called transposition, a type of site-specific recombination. Transposition can be either replicative or conservative. Transposons often carry genes encoding antibiotic resistance. Transposons can be used as biological mutagens.

✓ What features do insertion sequences and transposons have in common?

✓ What are integrons?

III TECHNIQUES OF BACTERIAL GENETICS: *IN VITRO*

The three mechanisms of genetic exchange we have discussed, transformation (see Section 10.6), transduction (see Section 10.7), and conjugation (see Section 10.9), can be used to do bacterial crosses and select recombinants with a goal of making a genetic map of an organism, creating new strains of an organism, or specifically studying the genetics and biochemistry of an organism's metabolism. Some severe limitations are, however, associated with doing all the crosses *in vivo* using these techniques. Many of these limitations can be overcome by manipulating DNA *in vitro*, in a test tube. The next several sections of this chapter discuss some of these *in vitro* techniques.

10.12 Restriction Enzymes

In Chapter 9, we discussed the fact that prokaryotes have restriction and modification systems, the function of which seems to be to protect the cells from foreign DNA, such as viral DNA (∞ Section 9.6). Although such systems are very widespread among prokaryotes (both *Bacteria* and *Archaea*), they are very rare in eukaryotes. Indeed, the only known "eukaryotic" restriction and modification systems are found in viruses that infect certain unicellular plants (∞ Section 16.6). (Some prokaryotic viruses also have their own restriction and modification systems.) Recall that the function of the **restriction enzymes** (restriction endonucleases), which are part of these systems, is to recognize certain sequences of DNA and cut the DNA. One large class of these enzymes, called type II restriction enzymes, not only recognize specific sequences but also make double-strand breaks within these sequences. Such enzymes have proven enormously valuable in many of the techniques of *in vitro* DNA manipulation. Indeed the discovery of these enzymes was the beginning of what came to be called *genetic engineering* (∞ Chapter 31).

Many of the sequences recognized by type II restriction enzymes exhibit twofold symmetry around a given point. Figure 10.34 shows the sequence that is recognized and cleaved by one restriction endonuclease from *Escherichia coli*, called *Eco*RI.

The cleavage sites are indicated by arrows, and the axis of symmetry by a dashed line. Note that the two strands have the same sequence if one is read from the left and the other from the right (or, in terms of polynucleotide strands, if both are read 5′ → 3′ or both read 3′ → 5′). Such a structure is called a **palindrome**. (A palindrome is a sequence of characters that reads the same when read from either right or left—for instance, *Sex at noon taxes*. The term *palindrome* is derived from the Greek meaning "to run back again.")

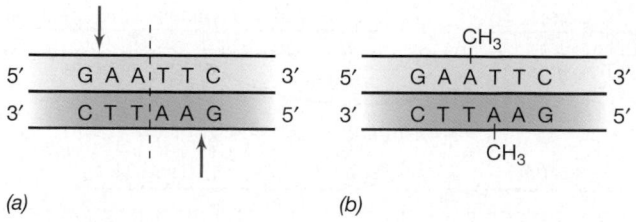

Figure 10.34 Restriction and modification of DNA. (a) The sequence of DNA recognized by the restriction endonuclease *Eco*RI. The red arrows indicate the bonds cleaved by the enzyme. The dashed line indicates the axis of symmetry of the sequence. (b) The same sequence after modification by the *Eco*RI methylase. The methyl groups added by this enzyme are shown.

Many restriction enzymes are composed of two identical polypeptide subunits, and each subunit recognizes and cuts the sequence on one of the strands. Since the sequences recognized by restriction enzymes are relatively short, and frequently palindromic, such enzymes always make *double-stranded* breaks. Interestingly, restriction enzymes that recognize *non*-palindromic sequences cut near the sequence, but not within it.

The recognition sequences and cutting sites for a few restriction enzymes are given in Table 10.4. Note that the recognition sites for the enzymes listed are 4, 6, and 8 base pairs. In a "random" DNA molecule, one would expect any 4-base pair sequence to occur approximately once every 256 base pairs based on the probability of $\frac{1}{4} \times \frac{1}{4} \times \frac{1}{4} \times \frac{1}{4}$ (assuming each base pair is equally probable in the DNA). Therefore, an enzyme that recognizes a 4-base-pair restriction site would cut a large DNA molecule into many specific fragments. A specific 6-base-pair sequence should appear every 4096 base pairs in random DNA, and an 8-base-pair sequence should appear only about once every megabase pair. The *Escherichia coli* chromosome, which is about 4.6 megabase pairs, is cut 21 times by the enzyme *Not*I, which recognizes an 8-base

TABLE 10.4	Recognition sequences of a few restriction endonucleases	
Organism	**Enzyme designation**	**Recognition sequence**[a]
Bacillus subtilis	*Bsu*RI	GG↓C̊C
Brevibacterium albidum	*Bal*I	TGG↓C̊CA
Escherichia coli	*Eco*RI	G↓AÅTTC
Haemophilus haemolyticus	*Hha*I	GC̊G↓C
Haemophilus influenzae	*Hind*II	GTPy↓PuAC̊
Haemophilus influenzae	*Hind*III	A↓AGCTT
Nocardia otitidis-caviarum	*Not*I	GC↓GGC̊CGC
Thermus aquaticus	*Taq*I	T↓CGÅ

[a] Arrows indicate the sites of enzymatic attack. Asterisks indicate the site of methylation (modification). G, guanine; C, cytosine; A, adenine; T, thymine; Pu, any purine; Py, any pyrimidine. Only the 5′ → 3′ sequence is shown.

pair sequence indicating that in this chromosome the *Not*I recognition sequence is present somewhat more often than one might have predicted.

Well over 2000 restriction enzymes with over 200 different specificities are now known. Others are being sought because these enzymes are of great importance in genetic research. Consider that the enzyme *Eco*RI can cut *any* double-stranded DNA that has its recognition sequence and cuts only at that sequence. This enables scientists to cut large DNA molecules into smaller fragments. Such fragments with defined termini, created as a result of the action of specific restriction enzymes have many uses. Restriction enzymes are such important tools in modern molecular genetic research that they have become widely available commercially. A number of companies purify and market restriction enzymes with a variety of specificities.

Modification: Protection from Restriction

An integral part of the cell's restriction-modification system is the modifying enzyme, which chemically **modifies** the specific sequences on its *own* DNA so these sequences are not attacked by the cell's own restriction enzymes. Such modification generally involves *methylation* of specific bases within the recognition sequence so the restriction nuclease can no longer act. Thus, for each restriction enzyme there must also be a modification enzyme, the two enzymes being closely associated. For example, the sequence recognized by the *Eco*RI restriction enzyme (see Figure 10.34*a*) can be modified by methylation of the two most interior adenines (Figure 10.34*b*). The enzyme that performs this modification is called *Eco*RI methylase. If even a single strand is modified, the sequence is no longer a substrate for the restriction enzyme *Eco*RI.

Restriction Enzyme Analysis of DNA

As noted, a DNA molecule will be cut at a specific sequence by a given restriction enzyme. Because the base sequences recognized by many restriction enzymes are four to six nucleotides long, there will generally be only a limited number of such sequences in piece of DNA. After cleaving the DNA, the fragments generated can be separated from each other by **gel electrophoresis**.

Electrophoresis is the procedure by which charged molecules migrate in an electrical field, the rate of migration being determined by the charge on the molecule and by its size. In gel electrophoresis the molecules are separated in a gel, usually composed of agarose or polyacrylamide. The gel is a complex network of fibrils and the pore size of the gel, which controls separation, can be set by the experimenter. Nucleic acid (which is negatively charged) will migrate through the gel, once an electrical current is applied, at rates dependent on the shape and size of the molecule. Small or compact molecules migrate more rapidly than large molecules. After

a defined period of migration time the gel can be "stained" with a compound that binds to DNA. If ethidium bromide (see Section 10.3) is used as a stain, the DNA bands will fluoresce under ultraviolet light.

A typical agarose gel is shown in Figure 10.35. In lane A a standard digestion has been applied, and the size of each band is known. The other lanes contain a purified source of DNA (in this case a plasmid) cut by one or more restriction enzymes. Because a given restriction enzyme

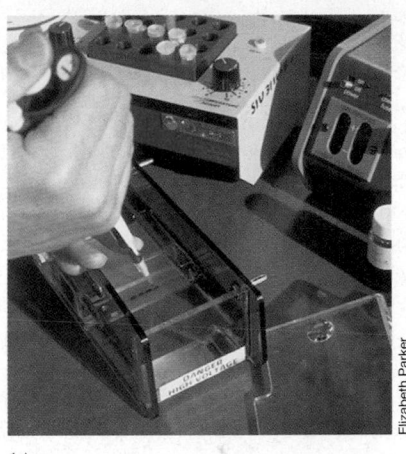

(a)

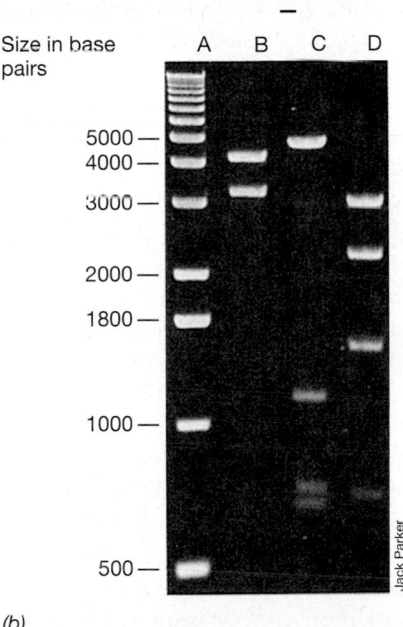

(b)

Figure 10.35 Agarose gel electrophoresis of DNA. (a) DNA samples are loaded into wells in a partially submerged agarose gel. (b) A photograph of a stained agarose gel. The DNA has been loaded into wells toward the top of the gel as shown, and the positive pole of the electrical field is at the bottom. The sample in lane A is used as a standard where the size of the fragments was known. Using the standards, one can determine the sizes of the fragments in the other lanes. Although each band in a lane will contain the same number of fragmented molecules, the bands stain less intensely at the bottom of the gel because the fragments are smaller and chemically there is less DNA to stain.

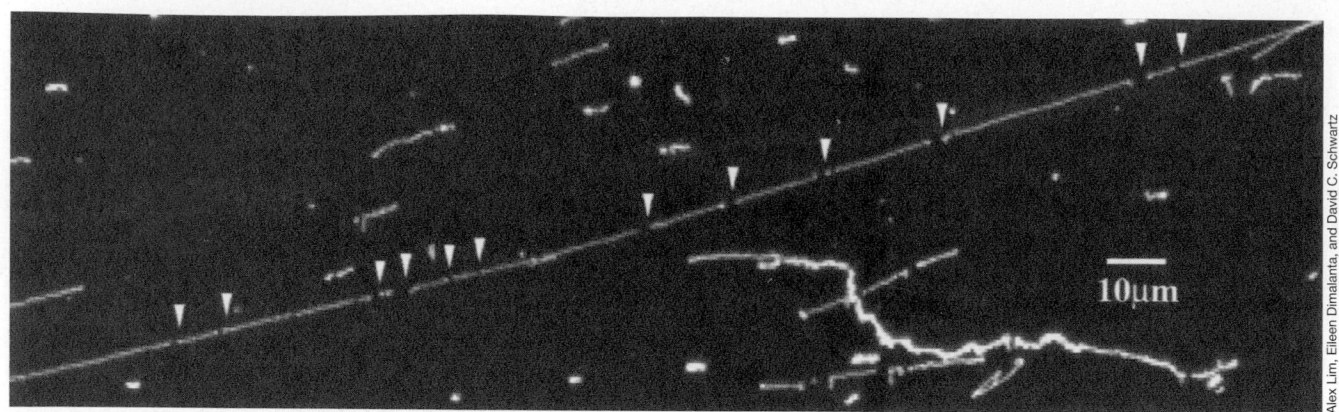

Alex Lim, Eileen Dimalanta, and David C. Schwartz

Figure 10.36 Optical mapping of restriction fragments. The figure shows a digital fluorescence micrograph of a portion of the *Escherichia coli* chromosome digested with the restriction enzyme *Xho*I (originally isolated from *Xanthomonas holica*). Sites of cutting by the restriction enzyme are indicated by arrows. This image was constructed from a series of images. DNA from bacteriophage lambda (∞ Section 9.10) was used as a sizing standard. This system has been used to create restriction maps of the complete chromosomes of *E. coli* and other *Bacteria* such as *Deinococcus radiodurans*.

always cuts at the same site, the banding pattern is reproducible and, using a standard, the size of the fragments can be determined. Such a technique can be used to generate a **physical map** of the DNA where the sites being mapped are not genes or mutations but the location of the restriction enzyme recognition sites. Creating physical maps in this way involves doing several digestions with several restriction enzymes, singly and in combinations, and then determining how the pieces should be ordered, much like solving a puzzle. Each lane in the gel shown in Figure 10.35 contains many millions of identical fragments of DNA. Coupling restriction enzyme digestion of a single molecule of DNA with microscopy allows a restriction map to be made directly from observing the cuts in a single molecule. This technique, called *optical mapping*, is shown in Figure 10.36.

DNA fragments can also be purified from gels, which can be useful in some cloning experiments (see Section 10.15). However, there are other uses for these gels, one of them involves *nucleic acid hybridization* or **hybridization**.

Recall from Section 7.2 that when DNA is denatured (i.e., made single-stranded), it can then be used to form hybrid molecules with other single-stranded DNA (or RNA) molecules having complementary or almost complementary sequences. These latter molecules are called **nucleic acid probes** or, more simply, **probes**. Hybridization can be very useful for finding similar sequences from different genetic elements, or to find the location of a specific gene. A technique was developed by E.M. Southern for hybridizing probes to DNA fragments that have been separated by gel electrophoresis. The DNA fragments in the gel are transferred to a membrane and denatured. This membrane is then exposed to a labeled probe to allow hybrids to form if the probe is complementary to any of the fragments (the fragments themselves are fixed on the membrane and cannot reform the original double helix). Sites of hybridization can be detected by assaying for label that has become

bound to the membrane. Probes can be labeled with radioactivity or a variety of agents that yield colored products or even emit light. The hybridization procedure when DNA is in the gel and RNA or DNA is the probe is called a *Southern blot*, named for the scientist who developed it. When RNA is in the gel and DNA or RNA is the probe, the procedure is called a *Northern blot*. (A *Western blot*, or *immunoblot*, involves protein in the gel and an antibody probe, ∞ Section 24.12.) Figure 10.37 shows how a Southern blot can be used to identify fragments of DNA containing a region (or gene) that hybridizes to the probe.

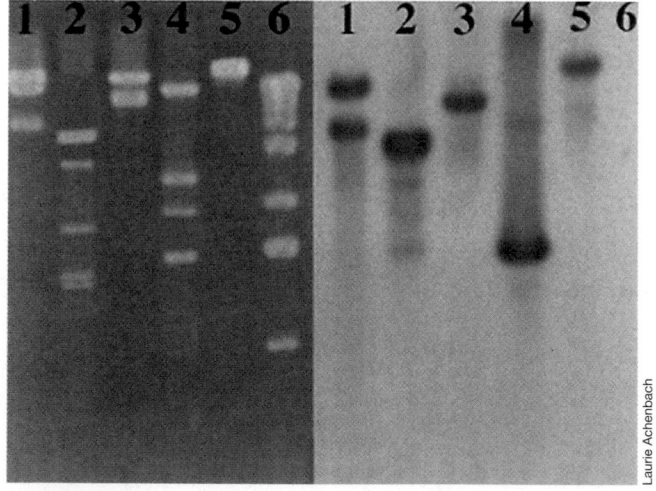

Laurie Achenbach

Figure 10.37 Southern blotting. (Left panel) Agarose gel electrophoresis of DNA molecules. Purified molecules of DNA from several different plasmids were treated with restriction enzymes and then subjected to electrophoresis. (Right panel) Southern blot of the DNA gel shown to the left. After blotting, hybridization with a radioactively labeled probe was carried out. The positions of the bands have been detected by X-ray autoradiography. Note that only some of the DNA fragments have sequences complementary to the labeled probe. Lane 6 contained DNA used as a size marker and none of the bands hybridized to the probe.

✓ 10.12 Concept Check

Restriction enzymes are cellular enzymes that recognize specific short sequences in DNA and make two single-stranded breaks so close together as to result in a double-stranded break in the DNA. The products of restriction enzyme digestion can be separated using gel electrophoresis.

✓ Why are restriction enzymes useful to the geneticist?

✓ What is a Southern blot?

(a)

10.13 Sequencing and Synthesizing DNA

Another technical advance that changed the study of genetics was the development of methods for sequencing DNA. In addition, procedures are available to synthesize DNA molecules with defined sequences for use as probes (see Section 10.12), to aid in DNA sequencing, or even to construct entire genes (∞ Section 31.5).

DNA Sequencing

Two different procedures for DNA sequencing were developed in the mid-1970s: the *Maxim and Gilbert* method and the *Sanger dideoxy* method. As originally developed, both methods generate DNA fragments that end at each of the four bases and that are radioactive. These fragments are then subjected to gel electrophoresis so that molecules with one-nucleotide difference in length are separated on the gel. Both procedures involved four separate lanes for each sequence determination, one for fragments ending at each of the four bases in DNA. adenine, guanine, cytosine, and thymine. The positions of the fragments are located by autoradiography, and the sequences can be read off the gel.

The differences between the two procedures stems from the method used to generate random fragments of the DNA with known nucleotides at the ends. The Maxam and Gilbert procedure uses chemicals that break the DNA preferentially at each of the four nucleotides. In the Sanger dideoxy procedure, the sequence is actually determined by making a *copy* of the single-stranded DNA, using the enzyme *DNA polymerase*. This enzyme uses deoxyribonucleoside triphosphates as substrates and adds them to a *primer* (∞ Section 7.5). In the incubation mixtures (four separate test tubes) are small amounts of each of the dideoxy analogs of the deoxyribonucleoside triphosphates (see Figure 10.38). Because the dideoxy sugar lacks the 3′-hydroxyl, continued lengthening of the chain cannot occur. The dideoxy analog thus acts as a *specific chain-termination reagent*. Fragments of variable length are obtained, depending on the incubation conditions. The nucleic acid fragments formed are radioactive from using either a radioactive primer or a radioactive deoxyribonucleoside triphosphate

(b)

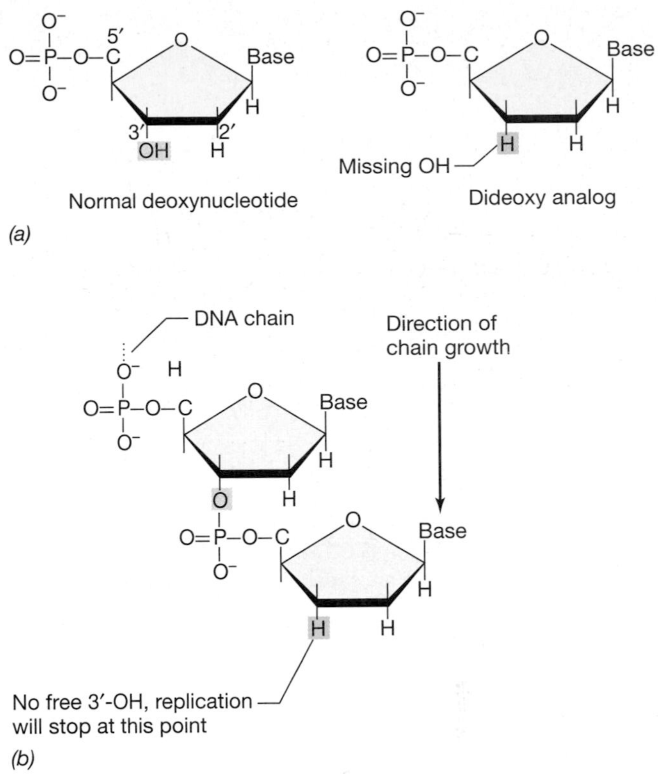

Figure 10.38 Dideoxynucleotides and Sanger sequencing. (a) A normal deoxynucleotide has a hydroxyl group on the 3′ carbon and a dideoxynucleotide does not. (b) Chain termination will occur after incorporation of a dideoxynucleotide.

in the reactions. Electrophoresis of these fragments is then carried out, and the positions of the radioactive bands are determined by autoradiography. By aligning the four dideoxynucleotide lanes and noting the vertical position of each fragment relative to its neighbor, the sequence of the DNA copy can be read directly from the gel (Figure 10.39).

A major advantage of the Sanger method is that it can be used to sequence RNA as well as DNA. To sequence RNA, a single-stranded DNA copy is made (using the RNA as the template) by the enzyme reverse transcriptase. By making the single-stranded DNA in the presence of dideoxynucleotides, various-sized DNA fragments are generated suitable for Sanger-type sequencing. From the sequence of the DNA, the RNA sequence is deduced by base-pairing rules. For determining the DNA sequence of a long molecule, such as a whole gene, it is necessary to proceed in stages. First, the DNA is broken into small overlapping fragments and the sequence of each fragment is determined. Using the overlaps as a guide, the sequence of the whole molecule can be deduced.

The demands of projects that involve sequencing entire genomes (∞ Chapter 15) have led to the development of automated DNA sequencing systems. With such

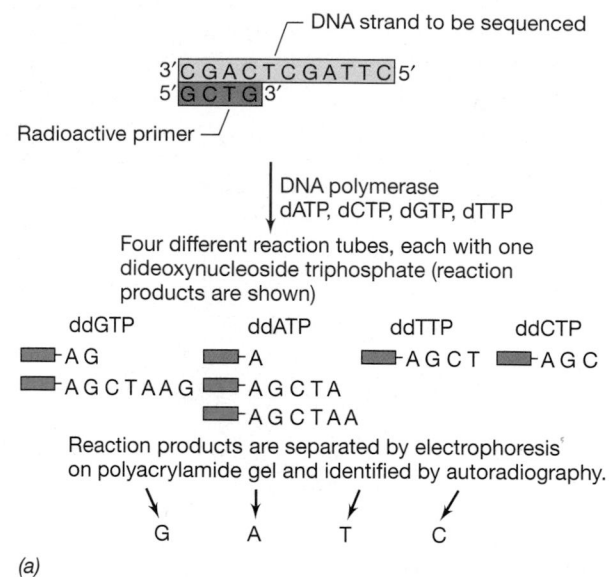

DNA strand to be sequenced

3′ C G A C T C G A T T C 5′
5′ G C T G 3′

Radioactive primer

DNA polymerase
dATP, dCTP, dGTP, dTTP

Four different reaction tubes, each with one
dideoxynucleoside triphosphate (reaction
products are shown)

ddGTP	ddATP	ddTTP	ddCTP
A G	A	A G C T	A G C
A G C T A A G	A G C T A		
	A G C T A A		

Reaction products are separated by electrophoresis
on polyacrylamide gel and identified by autoradiography.

G   A   T   C

(a)

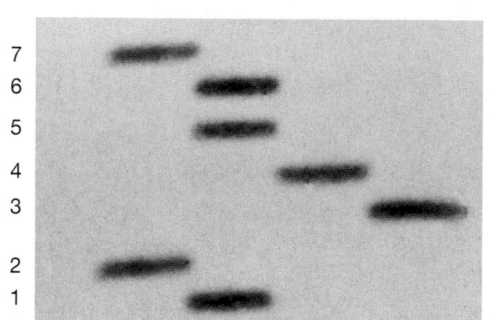

7
6
5
4
3
2
1

Sequence reads from bottom of gel as
A G C T A A G. Sequence of unknown
is 3′ T C G A T T C.

(b)

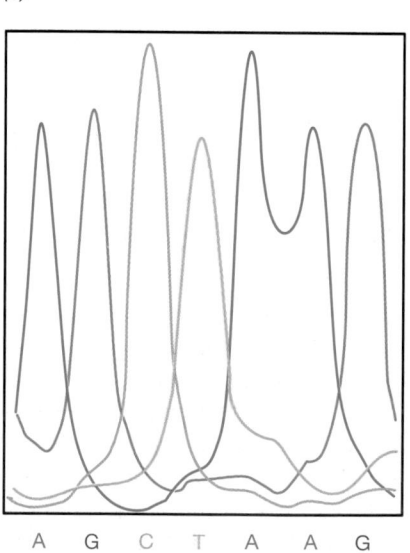

A   G   C   T   A   A   G

(c)

systems, the sequencing reactions are still based on the dideoxy methodology, but fluorescent dye-labeled primers (or bases) are used so that bands can be easily detected. The products are separated by automated electrophoresis and the bands detected by fluorescence spectroscopy. In one procedure each of the four different reactions use a different fluorescent label so that all four reactions can be run on a single lane. The results are analyzed by computer and a sequence printed out with each of the four bases being color coded (Figure 10.39*c*).

Note that Sanger-type sequencing requires a primer of known sequence. That is, the primer must be complementary to a portion of the DNA being sequenced. This obstacle is simply overcome because the DNA to be sequenced has been **cloned** (see Section 10.14); therefore, the known sequence to be used will be from the **cloning vector** (see Sections 10.15 and 10.16). The primer itself can be chemically synthesized.

Synthetic DNA

The procedures for making **synthetic DNA** can be completely automated so an oligonucleotide of 30–35 bases can be easily made in a few hours and oligonucleotides of well over 100 bases in length can be made if necessary. For the synthesis of longer polynucleotides, the oligonucleotide fragments can be joined enzymatically using DNA ligase.

DNA is synthesized in a *solid-phase procedure* in which the first nucleotide in the chain is fastened to an insoluble porous support (such as silica gel with particles about 50 μm in size). The overall procedure, the chemical details of which need not concern us here, is shown in Figure 10.40. Several steps are needed for the addition of each nucleotide. After each step is completed, the reaction mixtures are flushed out of the solid support and the series of reactions repeated for the addition of the next nucleotide. Once the desired length is achieved, the oligonucleotide is removed from the solid-phase support and purified to eliminate by-products and contaminants.

Synthetic DNA molecules are widely used for various purposes. In addition to their use as primers for DNA sequencing, they are used as primers for the polymerase chain reaction (see Section 10.17). In addition, they are used as probes to detect, via nucleic acid hybridization, specific DNA sequences (Figure 10.37). We will describe later (see Section 10.18) how synthetic DNA is also used in a procedure called **site-directed mutagenesis** to create mutations at specific locations on the genome.

Figure 10.39 DNA sequencing using the Sanger method. (a) The template DNA strand is the DNA whose sequence is to be determined. Note that four different reactions must be run, one with each dideoxynucleotide. (b) A portion of a gel containing the reaction products from (a). (c) Results of sequencing the same DNA as shown in (a) and (b), but using an automated sequencer and fluorescent labels.

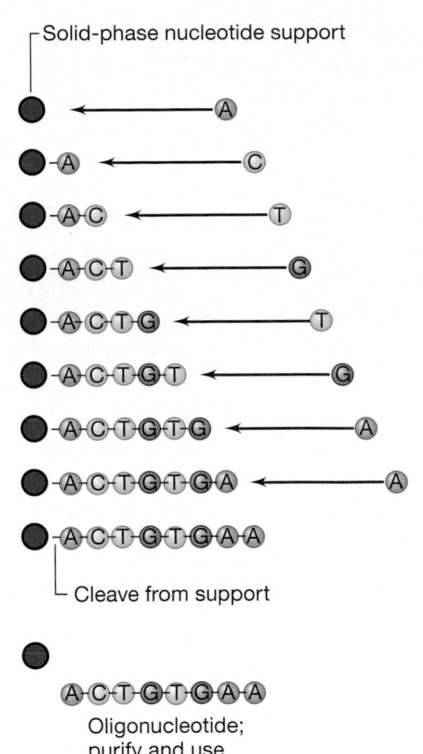

Solid-phase nucleotide support

Cleave from support

Oligonucleotide; purify and use

Figure 10.40 Solid-phase procedure for synthesis of a DNA fragment of defined sequence. Chemical synthesis proceeds by adding one nucleotide at a time to the growing chain.

✓ **10.13 Concept Check**

DNA can be sequenced by the Sanger method, which involves copying the DNA to be sequenced but using chain-terminating dideoxynucleotides. The final products are separated by electrophoresis and the sequence read. The short DNA primers required in this method can be synthesized chemically.

✓ Why do the methods involved in DNA sequencing involve the use of *four* different reactions?

10.14 Molecular Cloning

Molecular cloning is at the base of most genetic-engineering procedures and has greatly facilitated the analysis of any genome. The purpose of molecular cloning (also called *gene cloning*) is to isolate large quantities of specific genes or chromosomal fragments in pure form. While it might be theoretically possible to physically isolate pure DNA fragments with single genes from a restriction enzyme digest of chromosomal DNA (see Section 10.12), a little reflection will demonstrate the impracticality of such an approach. Consider that even for a genetically simple organism like *Escherichia coli*, a specific gene represents 1–2 kilobases (kb) out of a genome of over 4600 kb. An average *E. coli* gene is thus less than 0.05% of the total DNA in the cell. In hu-

mans the problem is even worse because the coding regions of average genes are not much bigger than in *E. coli*, but the genome is almost 1000 times larger! In contrast, the DNA of bacteriophage lambda is only 50 kb, and the DNA of some plasmids is less than 5 kb. In these genetic elements, a single average gene constitutes 2–40% of the DNA.

Thus, the basic strategy of molecular cloning is to move the desired gene or region from a large, complex genome to a small, simple one. Fortunately, our knowledge of DNA chemistry and enzymology allows us to break and join DNA molecules *in vitro*. This process is known as *in vitro recombination*. Restriction enzymes, DNA ligase (see Sections 7.6 and 10.12), and synthetic DNA (see Section 10.13) are important tools used for *in vitro* recombination.

Molecular cloning can be divided into several steps:

1. Isolation and fragmentation of the source DNA. This can be total genomic DNA from an organism of interest, DNA synthesized from an RNA template by reverse transcriptase (⊙⊙ Section 9.12), DNA synthesized by the polymerase chain reaction (see Section 10.17), or even totally synthetic DNA made *in vitro*. If genomic DNA is the source, it is generally cut with restriction enzymes to give a mixture of fragments.

2. Joining the DNA fragments to a *cloning vector* with DNA ligase. The small, independently replicating genetic elements used to replicate genes are known as **cloning vectors**, and most are derived from plasmids or viruses. Cloning vectors are generally designed to allow recombination of foreign DNA at a restriction site that cuts the vector in a way that does not affect its replication. If the source DNA and the vector are cut with the same restriction enzyme, joining can be mediated by annealing of the single-stranded regions called "sticky ends" (see Section 10.12 and Figure 7.7). "Blunt ends" generated by different restriction enzymes can also be joined, and different sticky ends or blunt ends can be joined by the use of synthetic DNA **linkers** or **adapters**. The properties of cloning vectors are discussed in Sections 10.15 and 10.16, as well as in Chapter 31.

3. Introduction and maintenance in a **host** organism. The recombinant DNA molecule made in a test tube is introduced into a host organism, for example, by DNA transformation (see Section 10.6) where it can replicate. Transfer of the DNA into the host usually yields a mixture of clones. Some cells contain the desired cloned gene, whereas other cells contain other clones generated by joining the source DNA to the vector. Such a mixture is known

as a **DNA library** or a **gene library** because many different clones can be purified from the mixture, each containing different cloned DNA segments from the source organism. Making a gene library by cloning random fragments of a genome is called **shotgun cloning**.

✓ 10.14 Concept Check

The isolation of a specific gene or region of a chromosome by molecular cloning is usually done using a plasmid or virus as the cloning vector. Restriction enzymes and DNA ligase are used in an *in vitro* recombination procedure to produce the hybrid DNA molecule. Once introduced into a suitable host, the target DNA can be produced in large amounts under the control of the cloning vector.

✓ What is the purpose of molecular cloning?

✓ What are the roles of a cloning vector, restriction enzymes, and DNA ligase in molecular cloning?

10.15 Plasmids as Cloning Vectors

Plasmids replicate independently of the host chromosome (see Section 10.8). In addition to carrying genes required for their own replication, most plasmids are natural vectors because they often carry other genes that confer important properties on their hosts. Such genes can be acquired by recombination within the host. In genetic engineering, geneticists add genes to a plasmid in a test tube.

Plasmids have very useful properties as cloning vectors. These include (1) *small size*, which makes the DNA easy to isolate and manipulate; (2) *independent origin of replication*, so plasmid replication in the cell proceeds independently from direct chromosomal control; (3) *multiple copy number*, so they can be present in the cell in several or numerous copies, making amplification of the DNA possible; (4) *selectable markers* such as antibiotic resistance genes, making detection and selection of plasmid-containing clones easier.

Although in the natural environment conjugative plasmids are generally transferred by cell-to-cell contact, plasmid cloning vectors generally have been modified to prevent their transfer conjugatively in order to achieve biological containment. However, transfer in the laboratory can be brought about by transformation or electroporation (see Section 10.6). Depending on the host-plasmid system, replication of the plasmid may be under tight cellular control, in which case only a few copies are made, or under relaxed cellular control, in which case a large number of copies are made. Achievement of high copy number is often important in gene cloning, and by proper selection of the host-plasmid system and manipulation of cellular macromolecule synthesis, plasmid copy numbers of several thousand per cell can be obtained.

The Plasmid pBR322

In Section 10.9, we discussed how the F plasmid could obtain genes from the *Escherichia coli* chromosome *in vivo*, becoming F′ plasmids. Therefore, one might imagine using F *in vitro*. However, the F plasmid is far too large (almost 100 kb, see Figure 10.17) to be useful, and it has no easily selectable markers. Even so, the very first plasmid cloning vectors used were natural isolates. These were soon replaced, however, by plasmids that were themselves the result of *in vitro* manipulations. An example of one of these modified plasmid cloning vectors is pBR322, which replicates in *Escherichia coli* (Figure 10.41). Plasmid pBR322 has a number of characteristics that make it suitable as a cloning vehicle:

1. It is relatively small, only 4361 base pairs.

2. It is stably maintained in its host (*Escherichia coli*) in relatively high copy number, 20–30 copies per cell.

3. It can be amplified to a very high number (1000–3000 copies per cell, about 40% of the genome!) by inhibition of protein synthesis by the addition of chloramphenicol.

4. It is easy to isolate in the supercoiled form using a variety of simple techniques.

5. A reasonable amount of foreign DNA can be inserted, although inserts of more than 10 kilobases lead to plasmid instability.

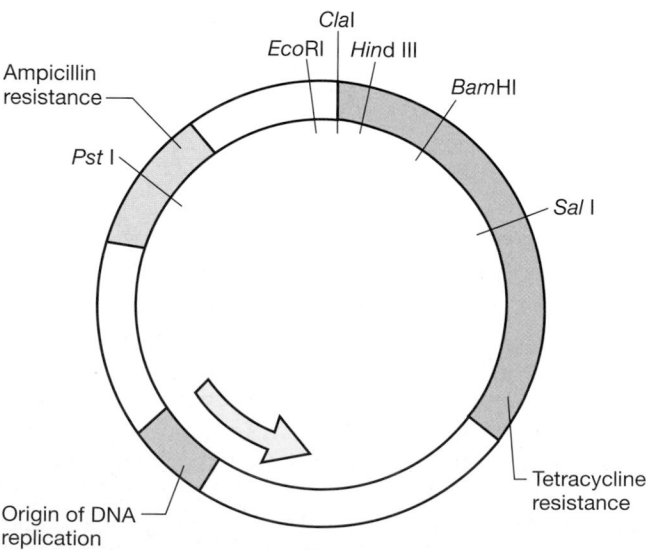

Figure 10.41 The structure of plasmid pBR322, an early and widely used cloning vector, showing the essential features. The arrow indicates the direction of DNA replication from the origin.

6. The complete base sequence of this plasmid is known, making it possible to locate sites where restriction enzymes can act.

7. There are *single* cleavage sites for various restriction enzymes such as *Pst*I, *Sal*I, *Eco*RI, *Hind*III, and *Bam*HI. It is important that only a single recognition site for at least one restriction enzyme is available so treatment with that enzyme opens the plasmid to a full-length linear molecule but does not cut it into pieces.

8. It has a gene conferring ampicillin resistance on the host and another conferring tetracycline resistance. These permit ready selection of hosts containing the plasmid. The sites recognized by some of the restriction enzymes are within one or the other of these resistance genes, facilitating the identification of plasmids carrying cloned DNA (see later).

9. It can be placed into cells easily by transformation.

The use of plasmid pBR322 in gene cloning is shown in Figure 10.42. As seen, the *Bam*HI site is with-

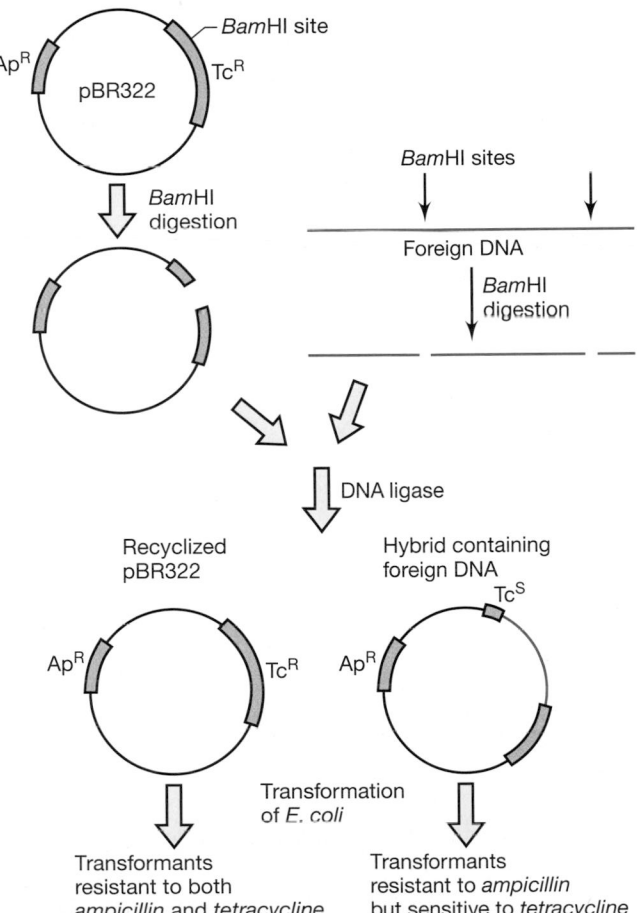

Figure 10.42 The use of plasmid pBR322 as a cloning vector, showing how insertion of foreign DNA causes inactivation of the tetracycline resistance gene, permitting easy identification of transformants containing the cloned DNA fragment.

in the gene for tetracycline resistance and the *Pst*I site is within the gene for ampicillin resistance. If foreign DNA is inserted into one of these sites, the antibiotic resistance conferred by the gene containing this site is lost, a phenomenon called **insertional inactivation**. Insertional inactivation is used to detect the presence of foreign DNA within the plasmid. Thus, when pBR322 is digested with *Bam*HI and linked with foreign DNA, and transformed bacterial clones then isolated, those clones that are both ampicillin-resistant and tetracycline-resistant *lack* the foreign DNA (the plasmid incorporated into these cells represents vector DNA that had recyclized without picking up foreign DNA). However, those cells still *resistant* to ampicillin but *sensitive* to tetracycline *contain* the plasmid with inserted foreign DNA. Since ampicillin resistance and tetracycline resistance can be determined independently on agar plates, isolation of bacteria containing the desired clones and elimination of cells not containing the plasmid can readily be accomplished.

The plasmid pBR322 represents an early generation of cloning vectors which were themselves constructed *in vitro*. There are now newer generations of plasmid vectors that have been engineered to have even more useful features and are even simpler to use. These new features almost always include a **polylinker** or **multiple cloning site**, a short segment of DNA with many different restriction sites, each unique to the vector. This polylinker is usually contained in the coding region of a gene where insertional inactivation is very easy to monitor. Such features are also found in bacteriophage vectors, and a specific example is discussed in Section 15.1. Sometimes insertional inactivation can be used as a positive selection. For example, there is a protocol that allows selection of bacteria that have lost tetracycline resistance. In some of the newly developed vectors, the gene carrying the polylinker normally produces a protein that is lethal to the host cell. Therefore, only cells containing a plasmid in which this gene has been *inactivated* can grow.

Cloning in plasmids is a versatile and fairly general procedure of wide use in genetic engineering, particularly when the fragment to be cloned is fairly small. Plasmids are often used as cloning vectors if *expression* of the cloned gene is desired (∞ Section 31.4).

✓ 10.15 Concept Check

Plasmids are useful cloning vectors because they are easy to isolate and purify and are able to multiply to high copy numbers in bacterial cells. Antibiotic resistance genes of the plasmid are used to identify bacterial cells containing the plasmid.

✓ Explain why it is best to use a restriction enzyme that cuts the plasmid vector only once.

✓ What is *insertional inactivation?*

10.16 Bacteriophage Lambda as a Cloning Vector

Recall that during specialized transduction (see Section 10.7) some host genes become incorporated into a bacteriophage genome. One phage that is used as a specialized transducing phage is bacteriophage lambda (∞ Section 9.10). During specialized transduction, lambda acts as a vector but the recombination occurs in the cell, not in a test tube. Lambda, however, can also be used as a cloning vector for *in vitro* recombination. It is a particularly useful cloning vector because its molecular genetics is well known, it can hold larger amounts of DNA than most plasmids, and DNA can be efficiently packaged into phage particles *in vitro*. These can be used to infect suitable host cells, and infection is much more efficient than transformation (transfection). Lambda has a complex genetic map (∞ Figure 9.18) and a large number of genes. Of particular significance for use as a cloning vector, the central third of the lambda genome, between genes *J* and *N*, is not essential and can be replaced with foreign DNA.

Modified Lambda Phages

Wild-type lambda is not suitable as a cloning vector because it has too many restriction enzyme sites. To avoid this difficulty, modified lambda phages have been constructed that can be used in cloning. In one set of modified lambda phages, the so-called Charon phages, unwanted restriction enzyme sites have been removed by point mutation, deletion, or substitution. In variants that have only a single restriction site, a foreign piece of DNA can be *inserted*, whereas in variants with two sites, the foreign DNA can *replace* a specific segment of the lambda DNA. The latter variants, called **replacement vectors**, are especially useful in cloning large DNA fragments.

Figure 10.43 shows some of the essential features of a wild-type lambda and two of the Charon vectors. Whereas wild-type lambda contains five *Eco*RI sites, Charon 4A contains three and Charon 16 only one. Charon 4A is used as a replacement vector; the two small interior fragments are cut out and discarded during cloning. With Charon 16, the DNA to be cloned is inserted at the single *Eco*RI site. Both Charon 4A and 16 contain deletions (not shown in the figure) that not only remove some sites found in the wild-type lambda but also make the genome smaller. This allows the cloning of larger DNA fragments.

Both vectors also contain substitution mutations, which are shown in Figure 10.43. One of the substitutions is the gene for β-galactosidase. When the vectors replicate on a lactose-negative (Lac⁻) strain of *Escherichia coli*, β-galactosidase is synthesized from the phage gene and the presence of lactose-positive (Lac⁺) plaques can be detected by using a color indicator agar (∞ Section 15.1). If a foreign gene is inserted *into* the β-galactosidase gene, the Lac⁺ character is lost. Such Lac⁻ plaques can be readily detected as colorless plaques among a background of colored plaques.

Steps in Cloning with Lambda

Cloning with lambda replacement vectors involves the following steps (Figure 10.44):

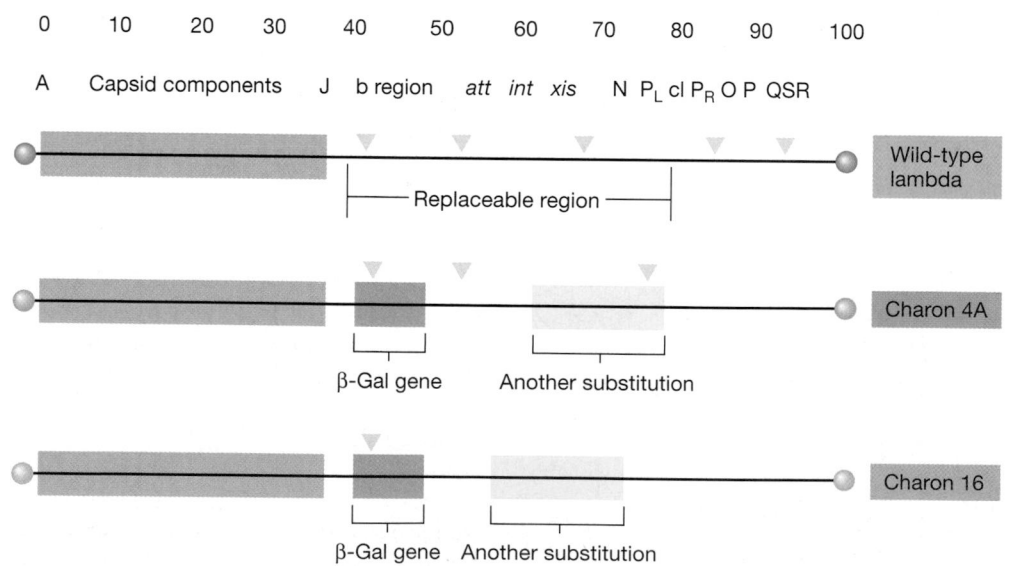

Figure 10.43 Molecular cloning with lambda. Abbreviated genetic map of bacteriophage lambda showing the cohesive ends as circles (∞ Figure 9.18). Charon 4A and 16 are both derivatives of lambda, which have various substitutions and deletions in the nonessential region. One of the substitutions in each case is a gene (β-Gal) that codes for the enzyme β-galactosidase, which permits detection of clones containing this phage. Whereas the wild-type lambda genome is 48.5 kilobase pairs, that for Charon 4A is 45.4 and that for Charon 16 is 41.7 kilobase pairs. The arrows (▼) shown above the maps of each phage indicate the sites recognized by the restriction enzyme *Eco*RI.

1. Isolation of the vector DNA from phage particles and digestion with the appropriate restriction enzyme.

2. Connection of the two lambda fragments to fragments of foreign DNA using DNA ligase. Conditions are chosen so molecules are formed of a length suitable for packaging into phage particles.

3. Packaging of the DNA by adding cell extracts containing the head and tail proteins and allowing the formation of viable phage particles.

4. Infection of *E. coli* and isolation of phage clones by picking plaques on a host strain.

5. Checking recombinant phage for the presence of the desired foreign DNA sequence using nucleic acid hybridization procedures or observation of genetic properties.

Selection of recombinants is less of a problem with lambda replacement vectors (such as Charon 4A) than with plasmids because (1) the efficiency of transfer of recombinant DNA into the cell by lambda is very high, and (2) lambda fragments that have not received new DNA are too small to be incorporated into phage particles.

Although lambda is a useful cloning vector, there are limits on how much DNA can be inserted. Viability of phage particles is low if the DNA is longer than 105% of normal lambda DNA, and some lambda genes cannot be discarded and still maintain the vector's ability to replicate. Therefore, really large DNA fragments (greater than 20 kb) cannot be efficiently cloned.

Cosmids

A related type of vector that employs specific lambda genes is called a **cosmid**. Cosmids are plasmid vectors containing foreign DNA plus only the *cos* (cohesive end) site from the lambda genome. These *cos* sites are required for packaging DNA into lambda virions. Cosmids are constructed from plasmids containing cloned DNA by ligating the lambda *cos* region to the plasmid DNA. The modified plasmid can then be packaged into lambda virions *in vitro* as described previously, and the phage particles used to transduce *Escherichia coli*. Cosmid construction avoids the necessity of having to transform *E. coli*, which at best is an inefficient process (see Section 10.6).

One major advantage of cosmids is that they can be used to clone large fragments of DNA. Therefore, fewer clones are needed to obtain representation of the whole genetic element. Cosmids also permit storage of the DNA in phage particles instead of in plasmids. Phage particles are much more stable than plasmids, so the recombinant DNA can be kept for long periods of time.

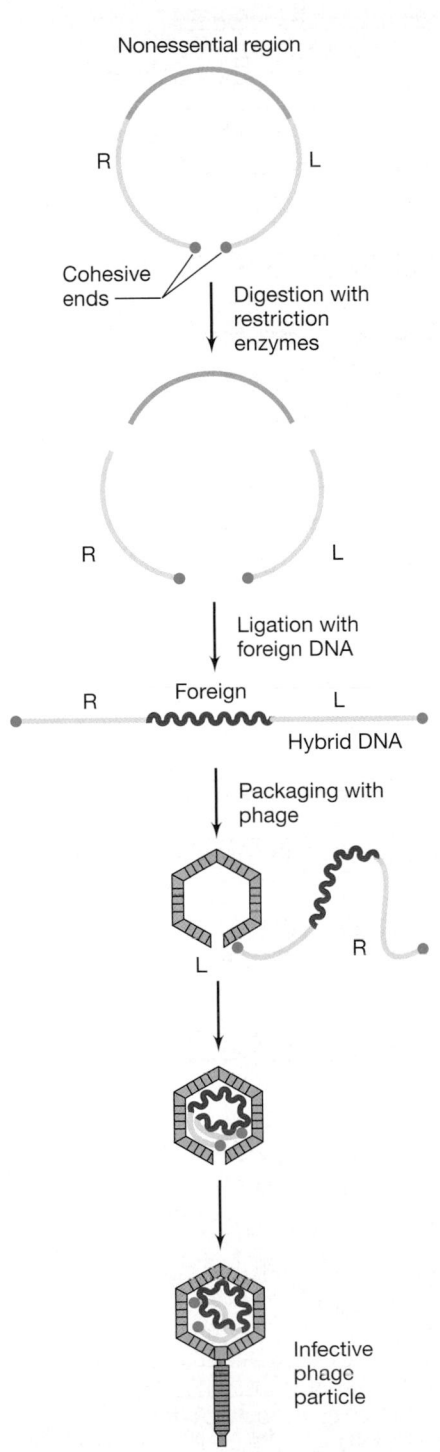

Figure 10.44 The use of bacteriophage lambda as a cloning vector. (See text for details.)

✓ **10.16 Concept Check**

Bacteriophages such as lambda have been modified to make useful cloning vectors. Larger amounts of foreign DNA can be cloned with lambda than with many plasmids. In addition, the recombinant DNA can be packaged *in vitro* for efficient transfer to a host cell. Plasmid vectors containing the lambda *cos* sites are called cosmids, and they can carry a large fragment of foreign DNA.

✓ Why is the ability to package recombinant DNA in a test tube useful?

✓ What is a *replacement vector*?

10.17 Amplifying DNA: The Polymerase Chain Reaction

Conventional molecular cloning methods can be considered *in vivo* DNA-amplifying tools. However, the development of synthetic DNA has spawned a new method for the rapid amplification of DNA *in vitro*, the **polymerase chain reaction (PCR)**. The polymerase chain reaction can multiply DNA molecules by up to a billionfold in the test tube, yielding large amounts of specific genes for cloning, sequencing, or mutagenesis purposes. PCR makes use of the enzyme *DNA polymerase*, which copies DNA molecules (∞ Section 7.5).

The PCR technique requires that the nucleotide sequence of a portion of the desired gene be known. This is necessary because short oligonucleotide *primers* complementary to sequences in the gene or genes of interest must be available for PCR to work. The steps in PCR amplification of DNA are as follows:

1. Two oligonucleotide primers flanking the target DNA (Figure 10.45*a*) are made on an oligonucleotide synthesizer and added in great excess to heat-denatured target DNA.

2. As the mixture cools, the excess of primers relative to the target DNA ensures that most target strands anneal to a primer and not to each other (Figure 10.45*a*).

3. DNA polymerase then extends the primers using the target strands as template (Figure 10.45*b*).

4. After an appropriate incubation period, the mixture is heated again to separate the strands. The mixture is then cooled to allow the primers to hybridize with complementary regions of newly synthesized DNA, and the whole process is repeated (Figure 10.45*c–e*).

Thus, each PCR "cycle" involves the following: (1) heat denaturation of double-stranded target DNA, (2) cooling to allow annealing of specific primers to target DNA, and (3) primer extension by the action of DNA polymerase (Figure 10.45). Note in Figure 10.45 how the

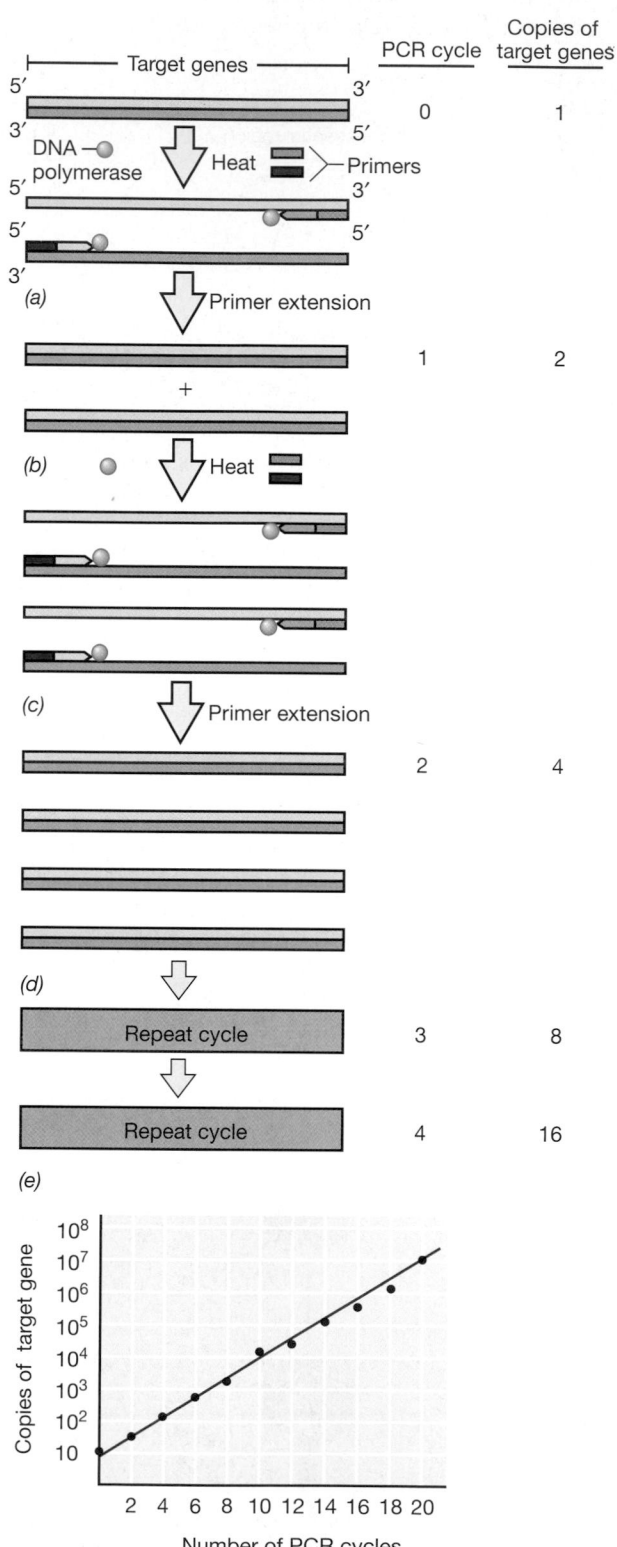

Figure 10.45 The polymerase chain reaction (PCR) for amplifying specific DNA sequences. (a) Target DNA is heated to separate the strands, and a large excess of two oligonucleotide primers, one complementary to the target strand and one to the complementary strand, is added along with DNA polymerase. (b) Following primer annealing, primer extension yields a copy of the original double-stranded DNA. (c) Further heating, primer annealing, and primer extension yields a second double-stranded DNA. (d) The second double-stranded DNA. (e) Two additional PCR cycles yield 8 and 16 copies, respectively, of the original DNA sequence. (f) Effect of running 20 PCR cycles on a DNA preparation originally containing 10 copies of a target gene. Note that the graph is semilogarithmic.

extension products of one primer can serve as a template for the other primer in the next cycle. The beauty of the PCR technique is that each cycle literally *doubles* the content of the original target DNA. In practice, 20–30 cycles are usually run, yielding a 10^6-to-10^9-fold increase in the target sequence (Figure 10.45*f*).

PCR at High Temperature

The original PCR technique employed *Escherichia coli* DNA polymerase, but because of the high temperatures needed to denature the double-stranded copies of DNA being made, the polymerase itself also was denatured and had to be replenished every cycle. This tended to limit the number of cycles that could be run and was very expensive. This problem was solved by employing a thermostable DNA polymerase isolated from the thermophilic bacterium *Thermus aquaticus*. DNA polymerase from *T. aquaticus*, known as *Taq polymerase*, is stable to 95°C and thus is unaffected by the denaturation step employed in the PCR reaction. The use of *Taq* DNA polymerase also increased the *specificity* of the PCR reaction because the DNA is copied at 72°C rather than 37°C. At high temperatures, nonspecific hybridization of primers to nontarget DNA rarely occurs, thus making the product of *Taq* PCR more homogeneous than that obtained using the *E. coli* enzyme.

One problem with the *Taq* polymerase is that it has no proofreading function (Section 7.6) and consequently makes more mistakes than the *Escherichia coli* enzyme. DNA polymerase from the hyperthermophilic archaean *Pyrococcus furiosus* (growth temperature optimum, 100°C) (Sections 6.10 and 13.6), called *Pfu* polymerase (or "vent polymerase"), is also widely used and is even more thermally stable than *Taq* polymerase. *Pfu* polymerase has proofreading activity, making it a particularly good enzyme when accuracy is crucial.

Because a number of highly repetitive steps are involved in the PCR technique, machines have been developed that can be programmed to run through heating and cooling cycles automatically. Because each cycle requires only about 5 min, the automated procedure allows for large amplifications in only a few hours (by contrast, such amplification by *in vivo* cloning methods would take several days). To supply the demand for thermostable DNA polymerase in the growing PCR and DNA sequencing markets, the genes for these enzymes have been cloned into *E. coli* and produced in large quantities; the cost of the PCR method is now just a fraction of what it was when the technique was first introduced.

Applications of PCR

PCR is a powerful tool because it is very simple to perform, extremely specific, and extremely efficient. Remember that during each round of amplification the amount of product *doubles*, leading to an exponential increase. This means not only that one can get large amounts of amplified DNA in a few hours, but that only a few molecules of target DNA need to be present in the sample to start the reaction. The reaction is so specific that, with primers of 15 nucleotides and high annealing temperatures, there will be extremely low levels of "false priming" and therefore the product formed should be relatively homogeneous.

Although automated thermocycling machines are still rather expensive, the technique has found a wide range of applications because what can be done with PCR (and analysis of the amplified product) can sometimes accomplish in a few hours what would have previously taken an entire laboratory several days or even weeks to do. Therefore, PCR can be extremely cost efficient and is now used in a very wide number of practical applications.

PCR is extremely valuable for obtaining DNA for cloning and sequencing because the gene or genes of interest can easily be amplified if flanking sequences are known. Because the primers used do not have to be perfectly complementary, PCR is routinely used in comparative or evolutionary studies to amplify genes from a variety of sources where the DNA has already been cloned (and sequenced) from one organism. In these cases the primers are made to regions of the gene thought to be conserved throughout a wide variety of organisms. Because of the sensitivity of PCR, it has been used to amplify and clone DNA from sources such as mummified human remains and even samples of extinct plants and animals.

PCR can also be used to amplify very small quantities of DNA present in a sample. Using appropriate primers, one can find and identify a single bacterial cell in a sample even if large numbers of other species are present. Please note that when amplifying DNA from a bacterium it is not necessary that the organism be grown in the laboratory. Therefore, PCR is one of the tools most responsible for revealing to scientists the enormous diversity of the microbial world and making it clear that microorganisms that have been cultured make up but a tiny fraction of all microorganisms in the environment (Section 18.5). Because of this ability to amplify and analyze DNA without having to grow the microorganism, PCR has also taken over a very important role in diagnostic microbiology (Section 24.13). PCR has also been used in DNA *fingerprinting*, a powerful forensic tool that permits identification of individuals, or relationships between individuals, from very small samples of their DNA (box, DNA Fingerprinting, Chapter 31). PCR can also be coupled with reverse transcription in a process called RT-PCR. This can be very useful when one wants to make large amounts of DNA using RNA as a template.

✓ 10.17 Concept Check

The polymerase chain reaction, a procedure for amplifying DNA *in vitro*, makes use of a heat-stable DNA polymerase from thermophilic prokaryotes. Heat is used to denature the DNA into two single-stranded molecules, each of which is copied by the polymerase. Beginning with a small oligonucleotide that serves as a primer for the target DNA to be amplified, the polymerase copies the complete DNA to which the primer associates. After a single copy cycle, the newly formed double strands are separated by heat again, and a new round of copying permitted. At each thermal cycle, the amount of target DNA doubles.

✓ Why does PCR require primers?

✓ Why is a primer needed at each "end" of the DNA to be amplified?

10.18 In Vitro and Site-Directed Mutagenesis

Methods of *in vitro* DNA manipulation have opened up a whole new field of mutagenesis. Whereas conventional mutagens (see Section 10.3) act at random, by use of synthetic DNA and recombinant DNA techniques it is possible to introduce mutations at *precisely determined sites* on genes. This process is called **site-directed mutagenesis**. Proteins made from strains carrying such mutations can be expected to have properties different from those of the wild-type proteins, properties that may, in some cases, be predicted from a knowledge of protein structure.

Site-Directed Mutagenesis

The first approaches to site-directed mutagenesis included treating transducing bacteriophage with chemical mutagens and then infecting bacteria and screening for mutants. Such a strategy had the effect of increasing the mutation rate in a limited region of a genome. New techniques are now available that allow the geneticist much greater specificity; a specific base pair in a gene can be changed to another base pair. Some of the techniques are simple and very powerful.

The basic procedure is to synthesize a short oligodeoxyribonucleotide containing the desired base change and to allow this to pair with a single-stranded DNA containing the gene of interest. Pairing is complete except for the short region of mismatch. Then the short single-stranded fragment of the synthetic oligonucleotide is extended using DNA polymerase, thus copying the rest of the gene. The double-stranded molecule obtained is inserted into a cloning host by transformation, and mutants selected by a procedure already described (see Section 10.6). The mutant obtained is then used in production of the modified (mutant) protein.

One procedure for *site-directed mutagenesis* is illustrated in Figure 10.46. It can be convenient to begin with the gene of interest cloned into a single-stranded DNA

vector. A widely used vector for site-directed mutagenesis is bacteriophage M13 (∞ Section 16.3). We shall discuss M13 as a vector in Chapter 15.

Cassette Mutagenesis and Gene Disruption

Because of the large number of restriction enzymes commercially available and therefore the large number of different DNA sequences that can be cut, it is usually possible to find several different restriction sites in the gene of interest. If sites for the appropriate enzyme are not found in the gene, or at the precise location required, they can also be added by site-directed mutagenesis (see Figure 10.46). If restriction sites are close together, the intervening DNA fragment can be cut out and replaced

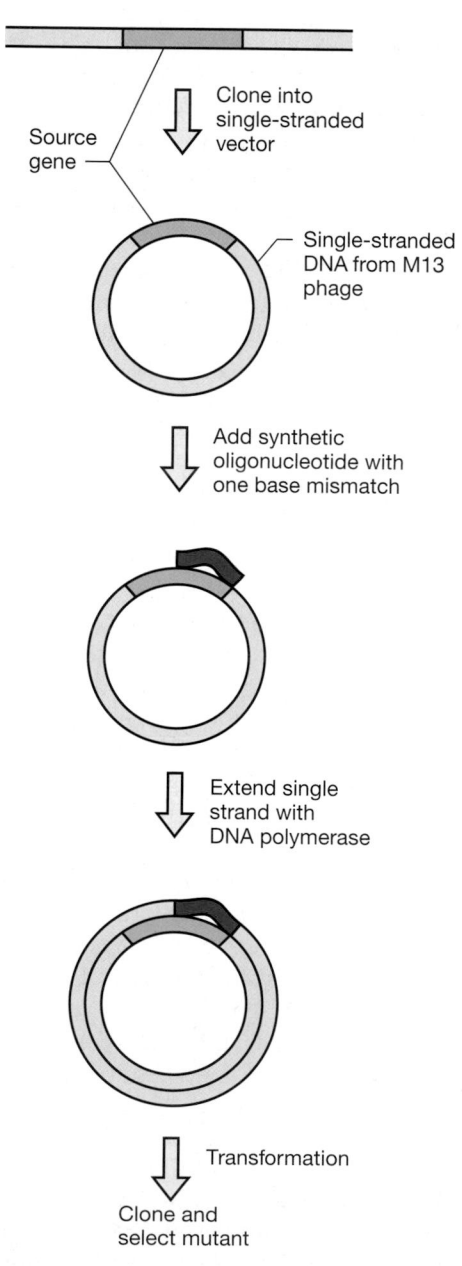

Figure 10.46 Site-directed mutagenesis using short synthetic oligodeoxyribonucleotide fragments.

by a synthetic DNA fragment in which one or more of the bases have been changed. These synthetic fragments are called *cassettes* (or cartridges), and the process is known as **cassette mutagenesis**.

Insertion mutations can also be generated by simply inserting a cassette at a single site. When using cassettes to replace sections of genes, the cassettes are typically the same size as wild-type DNA fragments. However, the cassette used for making insertion mutations can be almost any size and can even be an entire gene. In fact, cassettes that encode proteins that confer a particular antibiotic resistance on the host are commonly used. This type of cassette mutagenesis is used in a process called **gene disruption**. The process of gene disruption is illustrated in Figure 10.47. In this case, a fragment carrying a gene conferring kanamycin resistance, the *kan* cassette, is inserted at a restriction site in a cloned gene. The vector carrying this mutant gene is then linearized by being cut with a different restriction enzyme, and the linear DNA is transformed into the host with kanamycin resistance selected. The linearized plasmid cannot replicate, and so resistant cells likely arise by homologous recombination (see Section 10.5) between the mutated gene on the plasmid and the wild-type gene on the chromosome.

The cells have not only gained kanamycin resistance but have also lost the function of the gene in which the *kan* cassette was inserted. Therefore, these mutations are also called "knockout" mutations. This process is similar to searching for insertion mutations made by transposons (see Section 10.11), but in this case the geneticist chooses exactly which gene will receive the mutation. Note that gene disruption in haploid organisms yields viable cells only if the disrupted gene is unessential. Gene disruption experiments are now used as one mechanism to find out whether a gene is essential. Methods of obtaining gene disruption have been developed for higher organisms, including mice.

✓ 10.18 Concept Check

Synthetic DNA molecules of desired sequence can be made *in vitro* and used to construct a mutated gene directly or to change specific base pairs within a gene via site-directed mutagenesis. Genes can also be disrupted by inserting DNA fragments, called cassettes, into them. The inserted cassette eliminates the function of the wild-type gene while conferring a new, and usually selectable, phenotype on the cell.

✓ Why might site-directed mutagenesis be useful?

✓ What are *knockout mutations*?

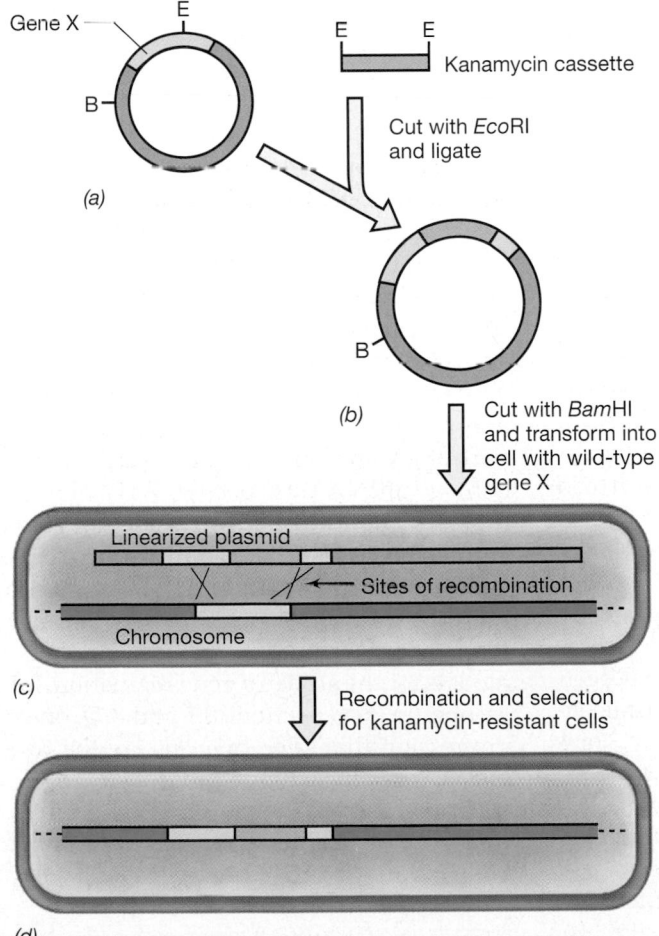

(a)

(b)

(c)

Linearized plasmid

Sites of recombination

Chromosome

(d)

Gene X

E

B

E   E

Kanamycin cassette

Cut with *Eco*RI and ligate

B

Cut with *Bam*HI and transform into cell with wild-type gene X

Recombination and selection for kanamycin-resistant cells

Figure 10.47 Gene disruption using cassette mutagenesis. (a) A plasmid containing a cloned wild-type copy of gene X is cut with the restriction enzyme *Eco*RI and mixed with a DNA fragment (the kanamycin cassette) that contains a gene capable of conferring kanamycin resistance on a cell and that has been obtained using the same restriction enzyme. The cut plasmid and the cassette are ligated. (b) The product of the ligation is a plasmid that now contains the kanamycin cassette as an insertion mutation within gene X. This new plasmid is now cut with a further restriction enzyme, *Bam*HI, and transformed into a cell containing a wild-type gene X. (c) The transformed cell contains the linearized plasmid with a disrupted gene X and its own chromosome with a wild-type copy of the gene. In some cells, homologous recombination occurs between the wild-type and mutant forms of gene X. (d) Cells that can grow in the presence of kanamycin must have the kanamycin cassette recombined into their chromosome because the linearized plasmid cannot replicate. These cells now have only a single disrupted copy of gene X. This disruption typically abolishes all gene X function.

IV THE BACTERIAL CHROMOSOME

The three mechanisms of genetic exchange described in this chapter, transformation, transduction, and conjugation, can be used to map the locations of various genes (actually mutations in genes) on the chromosome. In *Escherichia coli* genes were mapped to a particular region of the chromosome using conjugation. By using Hfr strains with origins at different sites, it is possible to map the whole bacterial gene complement.

Conjugation does not permit ordering genes that are closely linked. Therefore, generalized transduction was used for more fine structure mapping of the *E. coli* chromosome. Bacteriophage P1 can carry fragments of DNA equivalent to about 2 min on the map and has proved very useful for mapping genes. The genetic techniques that were used were dictated by the efficiency with which genetic transfer occurs in the organism to be studied. Transformation is a very inefficient process in *E. coli*, but is more efficient in other organisms and has proved an effective tool for mapping genes in *Bacillus*.

Although all of these "classic" techniques are still useful for the characterization and construction of strains, they have largely been supplanted in constructing genetic maps by molecular cloning (see Section 10.14), restriction mapping (see Section 10.12), and DNA sequencing (see Section 10.13), which have revolutionized the study of the genome organization of prokaryotes (and of eukaryotes).

10.19 Genetic Map of the *Escherichia coli* Chromosome

A circular reference map for *Escherichia coli* strain K-12 is shown in Figure 10.48. The map distances are given in minutes of transfer, with 100 min for the whole chromosome and with "zero time" arbitrarily set as that at which the first genetic transfer (the *threonine* operon) can be detected using the original Hfr strain.

The map shown in Figure 10.48 shows the location of a few restriction enzyme recognition sites, and the size is given both in minutes (the original units determined by conjugation studies) and in kilobase pairs. The *E. coli* chromosome, like that of many other prokaryotes, has been completely sequenced. Indeed, sequencing of the relatively "small" bacterial chromosomes that contain "only" 3 or 4 megabases is now done by automated sequencing of random, or "shotgun," clones (see Section 10.14). Because of the enormous amount of genetic information sequencing makes available, this information can be most effectively accessed through computer databases (∞ Chapter 15).

Escherichia coli as a Model Prokaryote

Many factors have favored the use of *Escherichia coli* as the workhorse for studies of biochemistry, genetics, and bacterial physiology. As we have seen in Chapters 9 and 16, even *E. coli* viruses have served as model systems of study. Therefore, although the chromosome of *E. coli* was not among the first prokaryotic chromosomes sequenced, this organism remains the best-known microorganism. In addition to its important role as a model organism, *E. coli* continues to be the organism of choice for both research and applications of genetic engineering (∞ Chapter 31).

The strain of *E. coli* whose chromosome was originally sequenced, MG1655, is a derivative of *E. coli* K-12, the traditional strain used for genetic studies. The "wild-type" *E. coli* K-12 is a lysogen of bacteriophage lambda (∞ Section 9.10) and also contains the F plasmid (see Section 10.8). However, the strain sequenced had been "cured" of lambda (by radiation) and of the F plasmid (by acridine treatment, see Section 10.8). The chromosome of this strain contains 4,639,221 base pairs. Analysis of the genomic sequence showed there to be 4288 possibly functional *open reading frames* (∞ Sections 7.13 and 15.2), which account for about 88% of the genome. Approximately 1% of the genome is comprised of genes encoding tRNAs and rRNAs, and about 0.5% is comprised of noncoding, repetitive sequences (∞ Section 7.4). The remaining 10% contain all the regulatory sequences: promoters, operators, origin and terminus of DNA replication, and so forth.

Arrangement and Expression of Genes on the *E. coli* Chromosome

Early mapping experiments and studies on the regulation of the genes that control the enzymes of a single biochemical pathway had shown that these genes were often clustered. On the genetic map in Figure 10.48 a few such clusters are shown. Notice, for instance, the *gal* gene cluster at 18 min., the *trp* gene cluster at about 28 min., and the *his* cluster at 44 min. Each of these clusters is part of an operon and is transcribed as a single *polycistronic* mRNA (∞ Section 7.11). However, genes for other biochemical pathways are not clustered, for example, the genes involved in arginine biosynthesis (*arg* genes) are scattered around the chromosome, as part of the *arg* regulon (∞ Section 8.5). Because of the early discovery of multigene operons, and their usefulness as models for the study of gene regulation, for example the *lac* operon (∞ Sections 8.5 and 8.7), one often gets the impression that such operons are the rule in prokaryotes. However, sequence analysis of the *E. coli* chromosome has shown that of the 2584 predicted or known transcriptional units, over 70% have only a *single* gene. Only about 6% of the operons have polycistronic mRNAs encoding four or more genes.

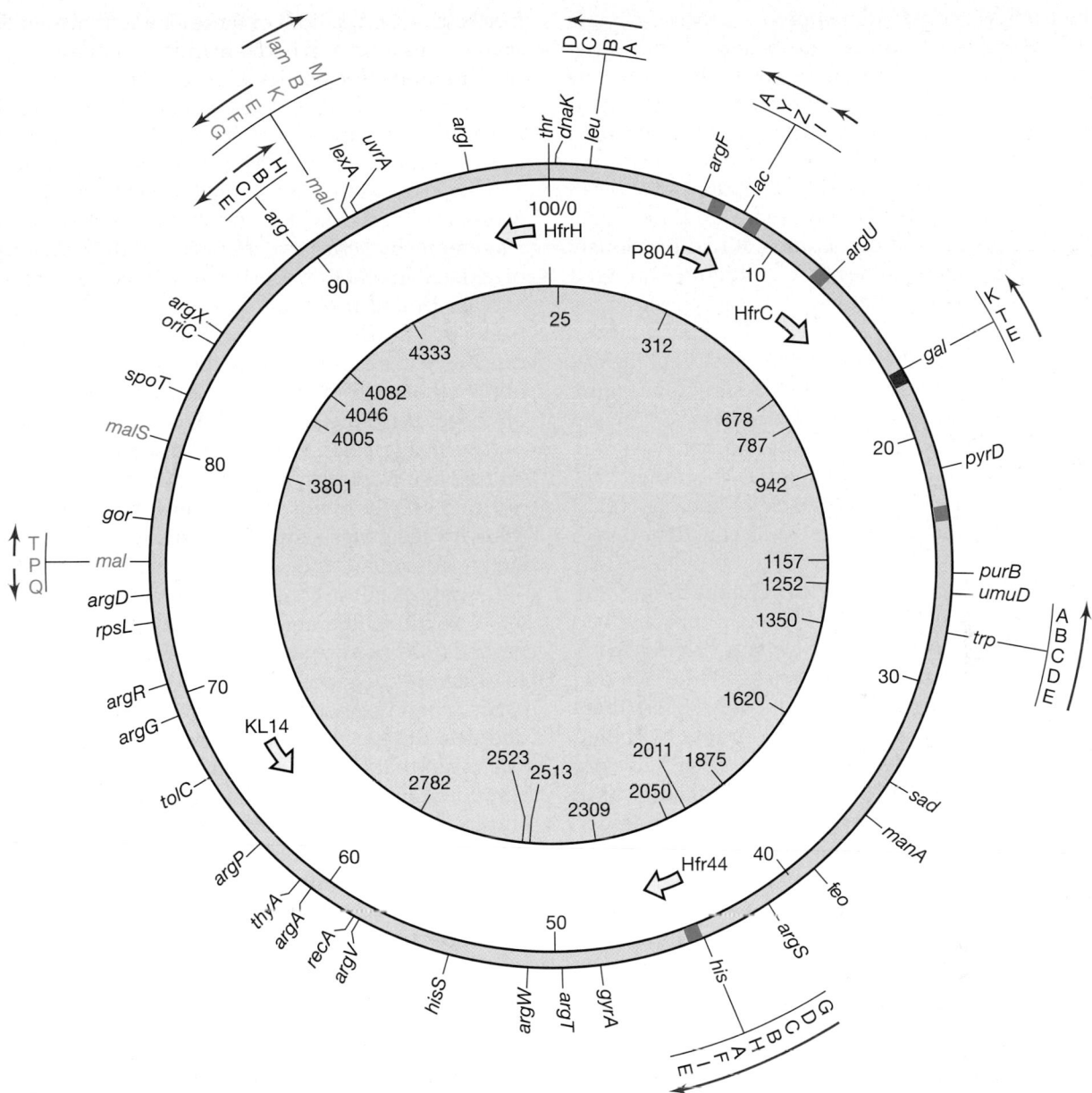

Figure 10.48 Circular linkage map of the chromosome of *Escherichia coli* strain K-12. On the outer edge of the map, the locations of a few of the mapped genes are indicated. A few operons are also shown, along with the direction in which they are transcribed. Along the inner edge of the map, the numbers from 0 to 100 refer to map position in minutes. The origin of DNA replication is marked *oriC* (84.3 min), and replication proceeds bidirectionally from this point (Section 7.6 and Figures 7.16–7.20). The inner circle shows the locations, in kilobase pairs, of the sequences recognized by the restriction enzyme *Not*I. Note that 0 min and 0 kilobase pairs are both, by convention, at the *thr* locus. The origins and directions of transfer of a few Hfr strains are also shown (arrows). The positions where five copies of the transposable element IS3 have been located in a particular strain are shown in blue. This element is also found in two copies on the F plasmid and is involved in Hfr formation (see Section 10.9). The position of the site where the bacteriophage lambda prophage integrates is shown in red (Section 9.10). If the prophage were present, it would add an extra 48.5 kilobase pairs (slightly over 1 min) to the map. The genes of the maltose regulon (Section 8.6), which includes several operons, are shown in green. Although most genes in this regulon have an abbreviation beginning with *mal*, note that one of the genes is *lamB*. This gene encodes a membrane protein involved in maltose uptake by the cell, but the protein is also the receptor for bacteriophage lambda. The gene *rpsL* (73 min) encodes a ribosomal protein. The gene was once called *str* because mutations in this gene lead to streptomycin resistance.

The transcription of some operons proceeds in one direction along the chromosome, whereas with other operons transcription is in the opposite direction. The direction of transcription of a few multigene operons is shown by arrows in Figure 10.48. Because transcription always occurs in a 5′ → 3′ direction (∞ Section 7.8), this implies that transcription off either DNA strand can occur, and evidence indicates that there are about equal numbers of operons on both strands. It had previously been known that many genes that are highly expressed in *E. coli* are oriented so that they are transcribed in the same direction that the DNA replication fork moves through them. (The two replication forks start at the origin, *oriC* at about 84 min. on the map shown in Figure 10.48, and move bidirectionally [∞ Section 7.6] toward the terminus, which is located at approximately 34 min. on the map.) Genomic sequencing has confirmed this, demonstrating that all seven of the rRNA operons and 53 of the 86 tRNA genes are transcribed in the direction of replication.

Almost 2000 *E. coli* proteins, or genes encoding proteins, had been identified before the chromosome had been completely sequenced. We now know that as many as 4288 different proteins can theoretically be encoded by this organism, although approximately 38% of them are of unknown function and/or are simply hypothetical. As was expected, the "average" *E. coli* protein contains slightly more than 300 amino acid residues, but many proteins are smaller and many are much larger. The largest gene should encode a protein of 2383 amino acid residues whose function is unknown, but the gene is similar to those involved in pathogenesis in related organisms. Although *E. coli* has very few duplicated genes, many of the genes that encode proteins seem clearly to have arisen by gene duplication during evolutionary history. There are some large *gene families*, groups of genes with related sequence and with products that have related functions. For example, there is a family of 70 genes whose products are all membrane-transport proteins.

Other Features of the *E. coli* Chromosome

Even though the strain of *E. coli* sequenced had been cured of lambda and the F plasmid, the chromosome contained many other genetic elements that replicate as part of it. There are copies of 10 different IS elements, including seven copies of IS2 and five copies of IS3

(see Section 10.11). Both of these elements are also found on the F plasmid and both are involved in the formation of Hfr strains (see Section 10.9 and Figure 10.22). There are also several different cryptic, defective, prophages (three of which are related to lambda) in the *E. coli* chromosome and several genes that are clearly parts of other prophages now almost completely lost through deletion. Moreover, *E. coli* has likely obtained a portion amount of its genome by *horizontal gene transfer*; that is, genes that originated in other organisms. It has been estimated that at least 18% of the *E. coli* K-12 genome has originated from such transfers. Large scale changes in the genome can still occur by such mechanisms. Strains of *E. coli* are known that contain genes involved in virulence located on large, unstable regions of the chromosome called *pathogenicity islands*, which can be acquired by horizontal transfer. Interestingly, horizontal gene transfer does not necessarily result in an ever-larger genome size. Many of the genes acquired in this way provide no selective advantage and so are lost by deletion.

Analysis of the complete sequence of a genome can yield an incredible amount of information: seemingly limited only by the questions that the investigator wishes to ask (∞ Chapter 15). Since the molecular genetics of *E. coli* have been seriously explored for several decades, and the genome has now been sequenced, does this mean that further analysis will be done mostly by computer and that traditional genetic and biochemical studies of *E. coli* are over? Far from it! Although computer analysis of a sequence can yield much information, in order to understand the *function* of genes, and particularly of regulatory sequences, it is often necessary to isolate mutants, map their mutations, and use biochemical and physiological analyses to determine their effects on the organism.

✓ 10.19 Concept Check

The *Escherichia coli* chromosome has been mapped using conjugation, transduction, molecular cloning, and sequencing. *E. coli* has been a useful model organism, and a considerable amount of information has been obtained from it, not only about gene structure but also gene function and regulation.

✓ Genetic maps of prokaryotic chromosomes are now typically made using only molecular cloning and DNA sequencing. Why were other methods also used in the case of *E. coli*?

Review Questions

1. Write a one-sentence definition of the term *genotype*. Do the same for the term *phenotype*. Does the phenotype of an organism automatically change when a change in genotype occurs? Why or why not? Can phenotype change without a change in genotype? In both cases, give some examples to support your answer.

2. Explain why an *Escherichia coli* strain that is His⁻ is an auxotroph and one that is Lac⁻ is not. (*Hint:* Think about what *E. coli* does with histidine and lactose.)

3. What are silent mutations and why do they occur? From your knowledge of the genetic code, why do you think most silent mutations afect the *third* position of the codon?

4. Microinsertions occur in promoters but are not frameshift mutations. Define the terms *microinsertion, frameshift, mutation,* and *promoter* (⌒⌒ Section 7.9). Explain how the statement can be true.

5. Explain how it is possible for a frameshift mutation early in a gene to be corrected by another frameshift mutation farther along the gene.

6. Give an example of one biological, one chemical, and one physical mutagen and describe the mechanism by which each causes a mutation.

7. What is site-specific mutagenesis? How can this procedure target specific genes for mutagenesis?

8. How does homologous recombination differ from site-specific recombination?

9. Why is it difficult in a single experiment using transformation to transfer a large number of genes to a cell?

10. Explain why in generalized transduction one always refers to a transducing *particle* but in specialized transduction one refers to a transducing *virus* (or transducing phage).

11. Some conjugative plasmids in gram-negative *Bacteria* have the genes involved in conjugation arranged in two regions, not just one. One region is called *Dtr* and the other *Mpf*. What do these acronyms stand for?

13. Explain why the insertion of a transposon leads to mutation.

14. The most useful transposons for isolating a variety of bacterial mutants are transposons containing antibiotic resistance genes. Why are such transposons so useful for this purpose?

15. What are restriction enzymes? What is the probable function of a restriction enzyme in the cell that produces it? How is it that the restriction enzyme in a cell does not cause the degradation of that cell's DNA?

16. What are the essential characteristics of a cloning vector? What characteristics of plasmids make them especially useful as vectors for molecular cloning? What characteristic(s) of the F plasmid makes it less useful for use *in vitro*?

17. If insertional inactivation of an antibiotic resistance gene is used to detect the presence of plasmid containing foreign DNA, why is it desirable to have two antibiotic resistance markers on the plasmid vector?

18. What advantages can there be to using a viral cloning vector rather than a plasmid vector?

19. Describe the basic principles of gene amplification using the polymerase chain reaction (PCR). How have thermophilic bacteria simplified the use of PCR?

20. What are the similarities and differences between *cassette mutagenesis* and *transposon mutagenesis*?

21. Differentiate clearly between the words *genome* and *chromosome* as they might be used referring to a prokaryote. In what cases would the two words have the same meaning?

Application Questions

1. Mutations within a gene encoding a protein can sometimes be suppressed by a mutation in another gene. One type of *suppressor mutation* involves a change in tRNA. Draw a diagram with coding sequences and amino acid sequences indicating how this occurs.

2. A constitutive mutant is a strain that continuously makes a protein that in the wild type is inducible. Describe two ways in which a change in a DNA molecule could lead to the development of a constitutive mutant. How could these two types of constitutive mutants be distinguished genetically?

3. In Chapter 7, we saw that it was critical for the ribosome to translocate with great accuracy in order to maintain the proper reading frame. However, sometimes ribosomes make frameshift errors. Compare the impact on the cell of a ribosome periodically making a frameshift error in the mRNA from a particular gene with the impact of a frameshift mutation in the same gene.

4. Although a large number of mutagenic chemicals are known, no chemical is known to induce mutations in a single gene (gene-specific mutagenesis). From what you know about mutagens, explain why it is unlikely that a gene-specific chemical mutagen will be found. How then is site-specific mutagenesis accomplished?

5. Defective specialized transducing phage can be replicated in a cell where a helper phage is replicating. However, a generalized transducing particle cannot replicate even if the cell is also infected with a wild-type phage. Explain how a helper phage "helps" the transducing phage and why this is not possible for the generalized transducing particle.

6. Transposable elements cause mutations when the element is inserted within the gene. These elements disrupt the continuity of a gene. One could also say that introns (⚭ Section 7.12) disrupt the continuity of a gene, yet the gene is still functional. Explain why the presence of an intron in a gene does not inactivate that gene but why insertion of a transposable element does.

7. Insertion sequences transpose (a type of site-specific recombination) in cells that have a defective *recA* gene. However, the formation of an Hfr strain from an F^+ strain cannot take place in a cell with a defective *recA* gene even though this is an event involving insertion sequences and recombination. Explain how Hfr formation takes place and why the RecA protein is essential.

8. Nucleic acids can be separated in gels by shape and size. They are not differentiated strictly by charge because essentially all DNA fragments will have the same charge per mass ratio. Explain.

9. When employing the plasmid vector pUC18, DNA fragments made using the restriction enzyme *Bam*HI inactivates the gene encoding β-galactosidase. When employing pBR322, cloning the same fragments inactivates the gene conferring tetracycline resistance. Both vectors contain a gene for ampicillin resistance that is used to select for bacterial transformants after making the recombinant molecules. Explain why it is much more efficient to use pUC18 rather than pBR322 as a cloning vector. (*Hint*: The increased efficiency has to do with finding the cells that contain vectors with cloned DNA.)

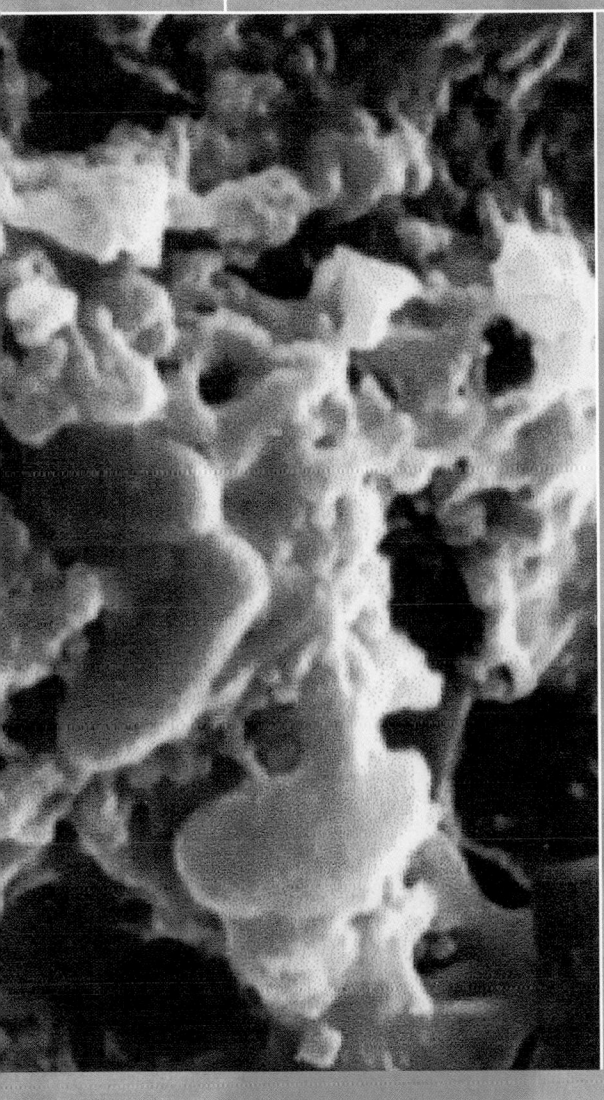

Although we know that prokaryotes are ancient, as evidenced by these microfossil cells found in rocks nearly 3.5 billion years old, until the last 20 years or so, microbiologists have not had effective tools for determining their evolutionary relationships. This can now be done in an almost routine fashion by molecular methods involving comparative nucleic acid sequencing. The evolutionary history of prokaryotes mirrors the evolutionary history of Earth itself, and now, as has been done with multicellular organisms for years, microorganisms, and in particular the prokaryotes, can be placed reliably on the phylogenetic tree of life.

MICROBIAL EVOLUTION AND SYSTEMATICS

Working Glossary

Archaea a group of phylogenetically related prokaryotes distinct from *Bacteria*

Bacteria a group of phylogenetically related prokaryotes distinct from *Archaea*

Binomial system the system devised by Linneaus for naming living organisms by which an organism is given a genus name and a species epithet

Domain in a taxonomic sense, the highest level of biological classification

Ecotype a population of genetically identical cells sharing a particular resource within an ecological niche

Endosymbiosis a theory stating that the mitochondrion and chloroplast were originally free-living *Bacteria* that established stable residence in primitive eukaryotic cells, eventually yielding the modern eukaryotic cell

Eukarya all eukaryotic cells: algae, protozoa, fungi, slime molds, plant and animal cells

Evolutionary chronometer a molecule, such as ribosomal RNA, whose molecular sequence can be used as a comparative measure of evolutionary divergence

Evolutionary distance in phylogenetic trees, the sum of the physical distance on a tree separating organisms; this distance is inversely proportional to evolutionary relatedness

FAME fatty acid methyl ester

Genomic hybridization the experimental determination of genome similarities by measuring the extent of hybridization of DNA from the genome of one organism with that of another

Family in biological classification, an intermediate level of taxonomic hierarchy. Contains one or more genera, each of which consists of one or more species

FISH fluorescent *in-situ* hybridization

GC base ratio in DNA (or RNA) from any organism, the percentage of the total nucleic acid that consists of guanine and cytosine bases

Genus a collection of different species, each sharing one or more (usually several) major properties

Lateral (horizontal) gene transfer the exchange of genes between and among cells in a microbial community

Metazoa multicellular organisms

Organelle unit membrane-enclosed structures of bacterial size and specialized in metabolic function found inside eukaryotic cells

Phylogenetic probe an oligonucleotide, sometimes made fluorescent by attachment of a dye, complementary in sequence to some ribosomal RNA signature sequence

Phylogeny the evolutionary history of organisms

Phylum a major lineage of cells in one of the three domains of life

Proteobacteria a large group of phylogenetically related gram-negative *Bacteria*

Ribosomal Database Project (RDP) a large database of small subunit ribosomal RNA sequences that can be retrieved electronically and used in comparative ribosomal RNA sequencing studies

Ribotyping a means of identifying microorganisms from analysis of DNA fragments generated from restriction enzyme digestion of genes encoding their 16S rRNA

RNA life a life form lacking DNA and protein that may have existed on early Earth and in which RNA served both a genetic coding and a catalytic function

Signature sequence short oligonucleotides of defined sequence in 16S or 18S rRNA characteristic of specific organisms or a group of phylogenetically related organisms and useful for constructing probes

16S rRNA a large polynucleotide (~1500 bases) that functions as part of the small subunit of the ribosome of prokaryotes and from whose sequence evolutionary information can be obtained; eukaryotic counterpart, 18S rRNA

Species in microbiology, a collection of strains that all share the same major properties but differ in one or more significant properties from other collections of strains

Stromatolites laminated microbial mats, typically built from layers of filamentous and other microorganisms, which can become fossilized

Taxonomy the science of identification, classification, and nomenclature

Universal tree a phylogenetic tree that shows the position of representatives of all domains of living organisms

---

I EARLY EARTH, THE ORIGIN OF LIFE, AND MICROBIAL DIVERSIFICATION

An integrating theme in all of biology is the enormous diversity of living organisms on Earth. This is especially true of microorganisms. In this chapter, we travel from a time before cells evolved in through to the modern eukaryotic cell. We will learn that prebiotic chemistry could well have supplied the necessary prerequisites for life, that early self-replicating "life forms" may not have been cellular, that early energy sources may have been inorganic, and that the eukaryotic cell is a genetic chimera consisting of at least two organisms.

11.1 Evolution of Earth and Earliest Life Forms

Origin of Earth

Earth is about 4.6 billion years old as determined by geochemical dating measurements. Our solar system was formed when a large, very hot star exploded, generating a new star (our sun) and the other components of our galaxy. Although no rocks dating to this period have yet been discovered on Earth, presumably because they have been weathered, rocks dating back to nearly 4 *billion* years ago have been found in several locations on Earth.

The oldest rocks discovered thus far are those of the Itsaq gneiss complex in southwestern Greenland, which date to about 3.86 billion years ago. These rocks are of three types: sedimentary, volcanic, and carbonate. The sedimentary rocks are of particular evolutionary interest because from our understanding of how modern sedimentary rocks are formed, the presence of sedimentary rocks 3.86 billion years old strongly suggests that *liquid water* in the form of oceans or large lakes was present at that time. The presence of liquid water in turn implies that conditions on Earth at that time were likely to be compatible with life as we know it. Other rocks of ancient origin include the Warrawoona series, Towers Formation, and Pilbara supergroups in Western Australia and the Swaziland supergroup in southern Africa; all of these rock formations are about 3.5 billion years old (see Figure 11.1).

Evidence for Microbial Life on Early Earth

Evidence for microbial life in the oldest known rocks is scant and rests on the fossilized remains of carbonaceous materials and the isotopically "light" carbon abundant in these rocks (we discuss the use of isotopic analyses of carbon and sulfur as an indication of living processes in Section 18.11). Some ancient rocks contain microfossils that appear bacterial shaped, usually as simple rods or cocci (Figure 11.1). Also of about this same age (~3.5 billion years old) are microbial formations called *stromatolites*. **Stromatolites** are fossilized microbial mats consisting of layers of filamentous prokaryotes and trapped sediment (Figure 11.2*a* and *b*); we discuss some characteristics of microbial mats in Section 18.11. What kind of organisms were these ancient stromatolitic bacteria? By comparing ancient stromatolites with modern stromatolites growing in shallow marine basins (Figure 11.2*c–e*) or in hot springs (Figure 11.2 *f*; ⚭ Figure 18.18*a*), scientists have concluded that ancient stromatolites were formed by filamentous phototrophic bacteria, perhaps

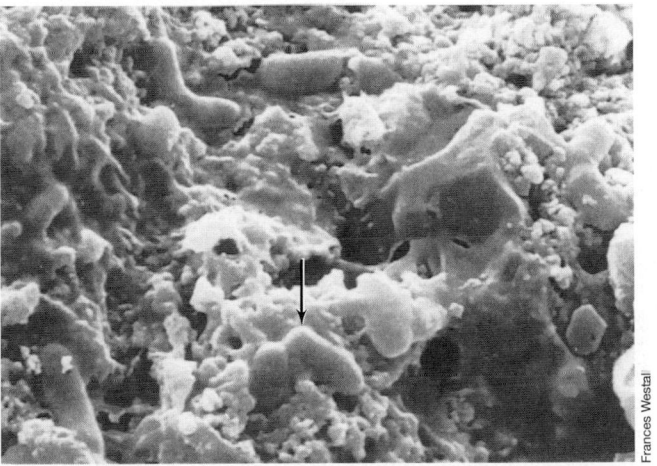

Figure 11.1 Scanning electron micrograph of microfossil prokaryotes from 3.45 billion year old rocks of the Barberton Greenstone Belt, South Africa. Note the rod-shaped bacteria (arrow) attached to particles of mineral matter. The cells are about 0.7μm in diameter.

relatives of the green nonsulfur bacterium *Chloroflexus* (⚭ Section 12.36).

Figure 11.3 shows photomicrographs of thin sections of more recent rocks containing cell-like structures remarkably similar to modern filamentous bacteria and green algae (⚭ Section 14.12). In the oldest stromatolites these organisms were likely anoxygenic (nonoxygen-evolving) phototrophic bacteria (⚭ Sections 12.2, 12.21, 12.33, and 12.36) rather than the O_2-evolving cyanobacteria (⚭ Sections 12.26 and 12.27) that dominate modern stromatolites. Nevertheless, the conclusion is inescapable that prokaryotic microorganisms had evolved an impressive morphological diversity very early in the history of life on Earth.

Conditions on Early Earth

The atmosphere of early Earth was devoid of significant amounts of O_2 and hence constituted a *reducing* environment. Besides H_2O, a variety of gases were present, the most abundant being methane (CH_4), carbon dioxide (CO_2), nitrogen (N_2), and ammonia (NH_3). In addition, trace amounts of carbon monoxide (CO) and hydrogen (H_2) existed, as well as considerable amounts of sulfide, as a mixture of H_2S and FeS. It is also likely that a considerable amount of hydrogen cyanide, HCN, was produced on early Earth when NH_3 and CH_4 reacted chemically to yield HCN. Geochemical estimates of the surface temperature of early Earth also suggest that it was a much hotter planet than it is today. For the first half billion or so years of its existence it is likely that the surface of Earth exceeded 100°C and was bombarded by meteorites; thus free water probably did not exist on early Earth but accumulated only later as Earth cooled. How fast Earth cooled is unknown, but it has been hypothesized that self-replicating entities first appeared at a time when Earth was much hotter than it is now. Thus, early cellular life forms were likely heat-tolerant and may have resembled in this respect the hyperthermophilic prokaryotes that inhabit thermal environments today (⚭ Section 6.10 and Chapter 13). We will consider evolutionary evidence that supports this hypothesis in Section 11.8.

Origin of Life

It is now well established that the synthesis of biologically important molecules can occur if reducing atmospheres containing the aforementioned gases are subjected to intense energy sources. Of the energy sources available on primitive Earth, the most important was probably ultraviolet (UV) radiation from the sun, but lightning discharges, radioactivity, heat from meteoritic impacts, and thermal energy from volcanic activity were also available. If gaseous mixtures resembling those thought to be present on primitive Earth are irradiated with UV or subjected to electric discharges in the laboratory, a wide variety of biochemically important molecules can be

(a)   *(b)*   *(c)*

(d)   *(e)*

Figure 11.2 Ancient and modern stromatolites. (a) The oldest known stromatolite, found in a rock about 3.5 billion years old, from the Warrawoona Group in Western Australia. Shown is a vertical section through a laminated, hemispheroidal structure, which has been preserved in the rock. Scale, 10 cm. (b) Stromatolites of conical shape from 1.6-billion-year-old dolomite rock of the McArthur basin of the Northern Territory of Australia. (c) Modern stromatolites in a warm marine bay, Shark Bay, Western Australia. (d) Another view of large modern stromatolites from Shark Bay. Note the resemblance to the ancient stromatolites shown in (b). (e) Underwater photograph of modern stromatolites growing in Shark bay. The diver indicates the scale. Shown are large columns formed by a complex community of diatoms, cyanobacteria, and green algae, to which are attached various macroscopic algae. (f) Modern stromatolites composed of thermophilic cyanobacteria growing in a thermal pool in Yellowstone National Park. Each structure is about 2 cm high.

(f)

made, such as sugars, amino acids, purines, pyrimidines, various nucleotides, thioesters, and fatty acids. Critical biochemical molecules like pyruvate have also been produced at high pressures and temperatures characteristic of undersea hydrothermal vents (Section 19.8); for this and other reasons, many scientists believe that such vents may have spawned the first life forms.

Under prebiological conditions some of these biochemical building blocks have been shown to *polymerize*, leading to the formation of polypeptides, polynucleotides, and other important macromolecules. We can therefore imagine that by whatever means, a mixture of organic compounds eventually accumulated on the primitive Earth, thus setting the stage for the origin of life.

But how could *macromolecules* have originated from monomeric constituents spontaneously? A possibility favored today is that relatively anhydrous *exposed surfaces* such as clays, pyrite, or basaltic glasses functioned as supports for prebiotic polymerization reactions. Such surfaces would have provided a stable, relatively dry

environment for the synthesis and accumulation of macromolecules into organic films from which primitive, self-replicating structures could have emerged. Pyrite (FeS_2) and Montmorillonite clay have been suggested as possibilities because of the crucial role the former may have played in early energy-generating systems (see next section) and the known ability of the latter to selectively absorb ribonucleic acid monomers and form RNA oligomers from them.

✓ 11.1 Concept Check

Earth is thought to be 4.6 billion years old; the first evidence of microbial life emerges in rocks about 3.86 billion years old. Early Earth was anoxic and much hotter than at present. The first biochemical compounds were made by abiotic syntheses that set the stage for the origin of life.

✓ How did the atmosphere of early Earth compare with that of Earth today?

✓ How were early biochemicals formed?

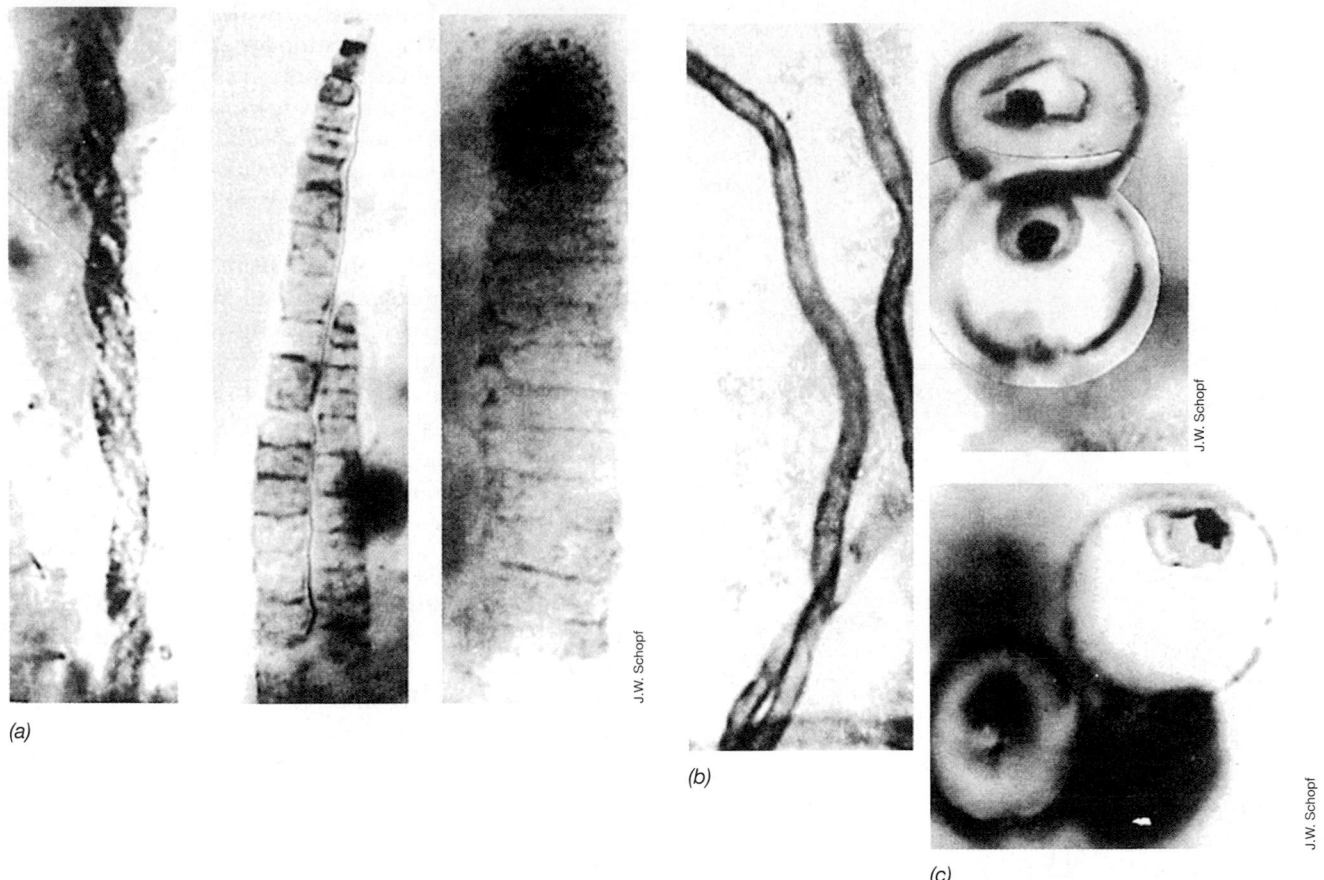

(a)

(b)

(c)

J.W. Schopf

Figure 11.3 Fossil prokaryotes and eukaryotes from more recent rocks than those shown in Figure 11.1. The four photographs in (a) (magnification, 2000×) and (b) (magnification, 920×) show fossil prokaryotic microorganisms found in the Bitter Springs Formation, a rock formation in central Australia about 1 billion years old. These forms bear a striking resemblance to modern filamentous cyanobacteria, anoxygenic phototrophs, or filamentous sulfur chemolithotrophs (⌾ Chapter 12). The two photographs in (c) (magnification, 2000×) show microfossils possibly of a eukaryotic alga. The cellular structure is remarkably similar to that of certain modern green algae, such as *Chlorella* sp. These are from the same rock formation as are the prokaryotic organisms.

11.2 Primitive Life: The RNA World, Molecular Coding, and Energy Generation

What was the first self-replicating *organism* like? This question is impossible to answer at present, but based on what we know about microbial life forms today, we can predict that even the simplest self-replicating entities needed a means of obtaining energy and some form of heredity in order to make copies of itself. But would these processes have required a *cellular* structure? It is tempting to extrapolate backward from the present and postulate that early organisms were much like modern cells, with only a very few genes perhaps containing limited transcriptional and translational abilities. Even a structure like this would have been relatively complex— probably much more complex than the first self-replicating entities. What would these creatures have been like? Following the discovery that certain types of

ribonucleic acid (RNA) are catalytic (⌾ Section 7.12), many scientists now believe that the earliest life forms lacked DNA altogether, contained just a very few, if any, proteins, and consisted primarily of RNA (Figure 11.4). Simply put, this was the age of **RNA life**, where RNA was the agent of both catalysis and genetic coding.

RNA Life

In an RNA world, RNA molecules would have functioned simply to replicate themselves and would likely have carried out only the minimal number of catalytic reactions necessary for this purpose (Figure 11.4). Studies of various catalytic RNAs (ribozymes, ⌾ Section 7.12) have shown that several different reactions can be catalyzed, including the synthesis of nucleotides from a sugar and a nitrogenous base. Thus, an era of RNA life may well have predated cellular life. These RNA life forms may have evolved into the first *cellular* life forms when self-replicating RNAs became enclosed

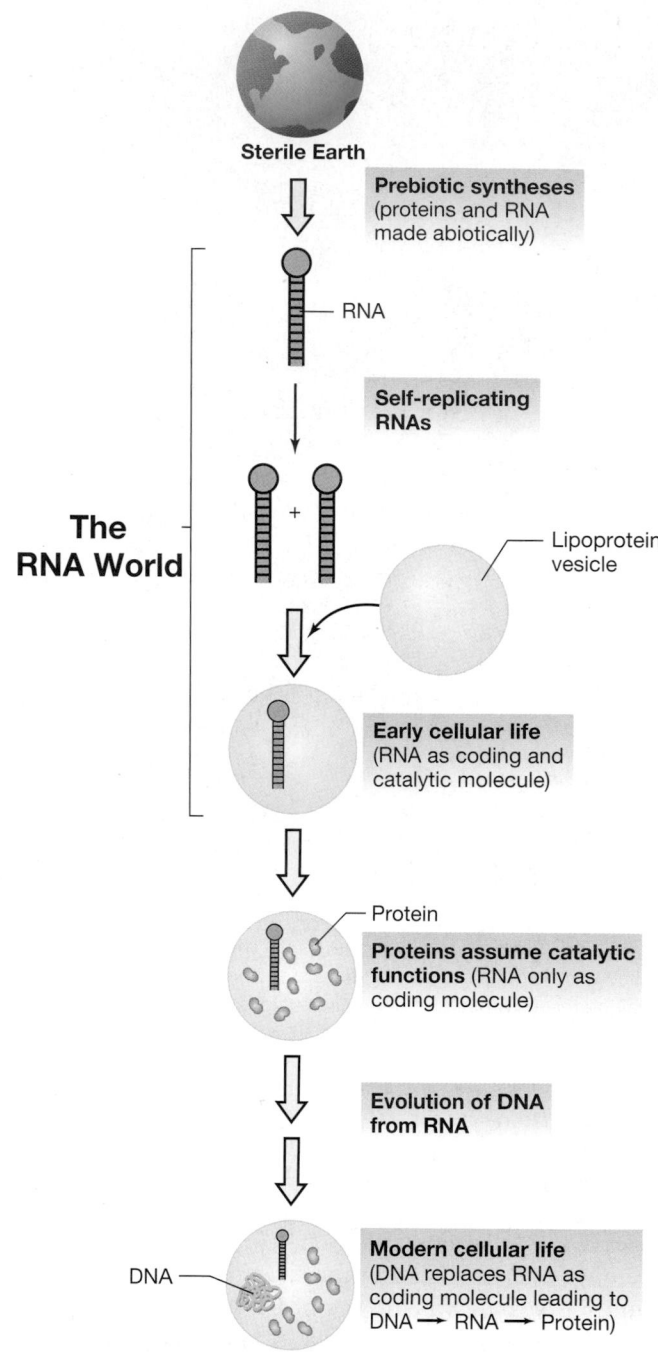

Earth to no avail, but eventually the proper constituents and set of circumstances came together and a primitive cellular organism arose. Although still lacking DNA and possessing only abiotically produced proteins, this cellular life form would otherwise have resembled a modern cell; as such life forms became more widespread, natural selection led to their further evolutionary development.

Biochemists tell us that proteins show a much higher degree of catalytic specificity than do catalytic RNAs. Thus, as primitive organisms became biochemically more complex, an evolutionary push to *proteins* as the major biocatalysts occurred. It is likely that proteins appeared gradually in cells, perhaps at first complexed with RNA, and as evolution selected for more and more precise biochemical catalysts, RNA was eventually replaced by protein as the major cellular enzymes (Figure 11.4).

The Modern Cell: DNA → RNA → Protein

The establishment of DNA as the genome of the cell may have resulted from evolutionary pressure toward greater efficiency and fidelity in replicating genetic information (DNA polymerases are more accurate than RNA polymerases, ⚮ Section 7.8). Also, by storing all the genetic information in one place in the cell and processing only what was needed under a specific set of conditions (that is, *regulating gene expression*), cells would have saved energy, which would have increased their competitive fitness. Somewhere in the early stages of microbial evolution, the three-part system—DNA, RNA, and protein—became fixed in cellular life forms as the best solution to biological information processing. That this system was an evolutionary success is evident in today's cells, all of which, as far as is known, contain all three types of these macromolecules.

Metabolism in Primitive Organisms

Life is a highly ordered process. To render inherently disordered molecules into a complex biological machine (the cell), energy was necessary. How were the energy demands of primitive self-replicating entities met? Until the evolution of cyanobacteria, molecular oxygen was unavailable in any significant quantities on Earth (see Figure 11.6). Thus, only energy-generating mechanisms that could occur under *anoxic* conditions could be exploited to fuel the energy needs of early organisms. However, as we will see in Chapter 17, anoxic conditions place few restrictions on metabolic diversity; a variety of chemoorganotrophic and chemolithotrophic energy-generating mechanisms, as well as several variations of photosynthesis occur anoxically. However, a simple chemical reaction involving ferrous iron (which was known to have been abundant on early Earth), may have been an early method by which primitive organisms derived energy.

Figure 11.4 Possible scenario for the evolution of cellular life forms from RNA life forms. Self-replicating RNAs could have become cellular entities by becoming stably integrated into lipoprotein vesicles. With time, proteins replaced the catalytic functions of RNA and DNA replaced the coding functions of RNA.

within lipoprotein vesicles (Figure 11.4). These cell-like structures could have arisen through the spontaneous aggregation of lipid and protein molecules to form membranous structures within which were trapped RNAs and other precursors of key biomolecules. This step may have occurred countless times on the early

The reaction

$$FeS + H_2S \rightarrow FeS_2 + H_2 \quad \Delta G^{0'} = -42\,kJ/reaction$$

proceeds exergonically with the release of energy. This reaction also yields H_2, and it has been hypothesized that this H_2 could have been used by primitive cells to form a proton-motive force across a membrane from which a primitive ATPase could have recovered chemically useful energy as ATP (Figure 11.5). With H_2 as the electron donor an electron acceptor would also have been required, and this could have been elemental sulfur, S^0. As shown in Figure 11.5, this simple coupled reaction would have required few enzymes and would have been a limitless means of energy conservation as long as FeS was available. Interestingly, many hyperthermophilic *Archaea* (which we will see later in this chapter are the closest extant relatives of Earth's earliest organisms), can carry out this very reaction. Many other forms of anoxic metabolism could also have occurred on early Earth, including fermentations and various forms of anaerobic respiration, but these would likely have required more catalysts and biochemical complexity than that outlined in Figure 11.5.

Primitive organisms may have derived carbon from various sources including organic carbon from abiotic syntheses or even from CO_2, a gas that was abundant on early Earth. Use of carbon dioxide would have had to follow the evolution of autotrophy, the process where CO_2 is converted into all organic components in the cell (∞ Sections 17.6 and 17.7). An "early autotrophy" hypothesis has been supported by microbial genome-sequencing projects (∞ Chapter 15), where it has been shown that autotrophy occurs in a number of small genome-containing hyperthermophiles that branch near to the root of the evolutionary tree of life (see Figure 11.13).

Oxygenation of the Atmosphere

Regardless of the mechanism(s) employed for assimilating carbon, a milestone in Earth's history occurred with the evolution of *oxygenic photosynthesis* in the cyanobacteria. These organisms probably first appeared on Earth some 3 billion years ago, but the O_2 that they produced did not accumulate in the atmosphere because of all of the reducing substances (like FeS) still present that reacted spontaneously with O_2 to produce H_2O. It is highly likely that cyanobacteria evolved from anoxygenic phototrophs through the development of a photosystem that could use H_2O as an electron donor for photosynthetic CO_2 reduction, thus releasing O_2 as a by-product ($CO_2 + H_2O \rightarrow CH_2O + O_2$). We discuss the biochemistry of photosynthesis in Chapter 17.

The evolution of oxygenic photosynthesis had enormous consequences for the environment of Earth because as O_2 accumulated, the atmosphere gradually changed from an anoxic to an oxic one (Figure 11.6). With O_2 now available as an electron acceptor, aerobic organisms could evolve. These organisms were able to obtain more energy from the oxidation of organic compounds than anaerobes could (∞ Chapter 17), allowing higher population densities to develop and increasing the chances for the evolution of new types of organisms and metabolic schemes. There is good evidence from the fossil record that at about the time that Earth's atmosphere became oxidizing there was an enormous burst in the rate of evolution, leading to the appearance of eukaryotic microorganisms with organelles and from them to the rapid diversification of metazoa (multicellular organisms) and eventually to higher animals and plants (see Figure 11.13 and Section 11.8).

Another major consequence of the appearance of O_2 was the formation of ozone (O_3), a substance that provides a barrier preventing the intense ultraviolet radiation of the sun from reaching Earth. When O_2 is subject to short wavelength ultraviolet radiation, it is converted to O_3, which strongly absorbs wavelengths up to 300 nm. Until an ozone shield developed, evolution could have continued only in environments protected from direct radiation from the sun, such as under rocks or in the oceans, because the intense UV radiation would have caused lethal DNA damage. However, after the photosynthetic production of O_2 and subsequent development of an ozone shield, organisms could have ranged generally over the surface of Earth, permitting evolution of a greater diversity of living organisms. A summary of the steps that could

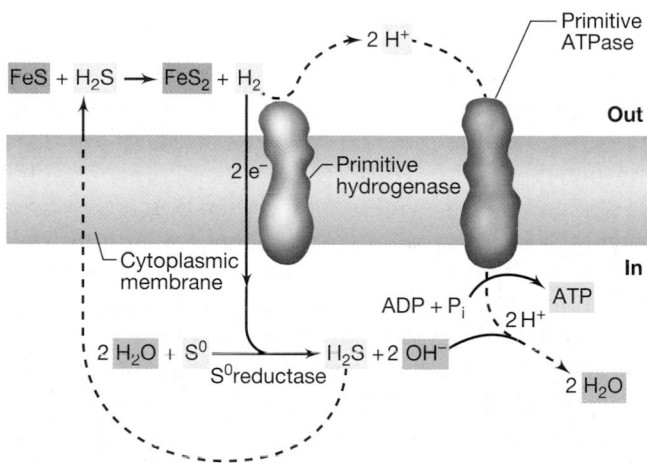

Figure 11.5 A hypothetical energy-generating scheme for primitive cells. Formation of pyrite leads to H_2 production and S^0 reduction, which fuels a primitive ATPase. Note how H_2S plays only a catalytic role; the net substrates would be FeS and S^0. Also note how few different proteins would be required. The $\Delta G^{0'}$ of the reaction $FeS + H_2S \rightarrow FeS_2 + H_2 = -42\,kJ$.

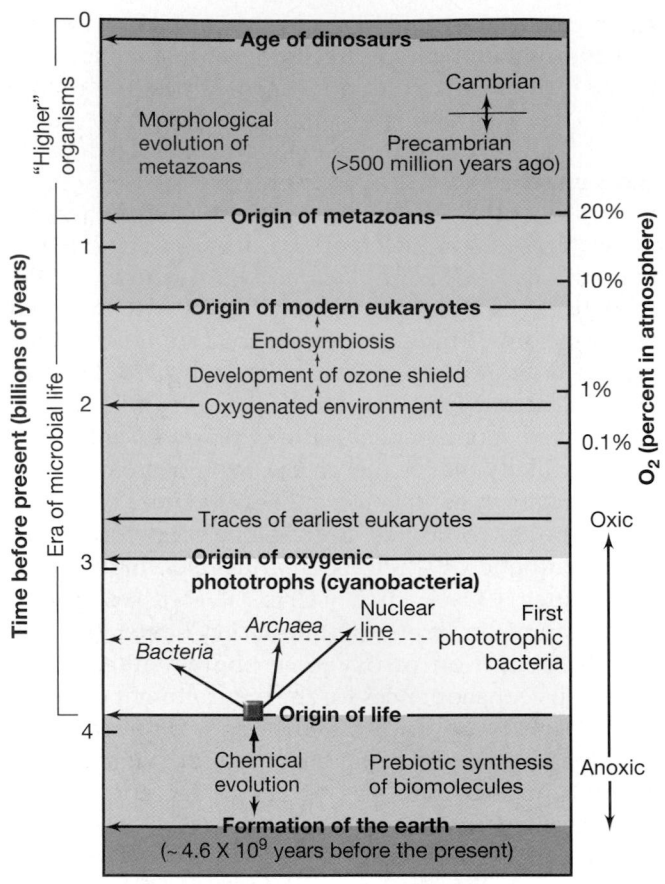

Figure 11.6 Major landmarks in biological evolution. The positions of the stages on the time scale are approximate. Note how the oxygenation of the atmosphere due to cyanobacterial metabolism was a gradual process, occurring over a period of about 2 billion years. Also note that for the bulk of Earth's history, only microbial life forms existed. Although microfossil evidence is lacking, microchemical evidence for eukaryotic cells goes back as far as 2.7 billion years ago.

have occurred in biological evolution and Earth's oxygenation is shown in Figure 11.6.

✓ 11.2 Concept Check

The first "life forms" may have been self-replicating RNAs. The first cellular organisms probably employed a simple strategy to obtain energy. Primitive metabolism was anaerobic and likely chemolithotrophic, exploiting the abundant sources of FeS and H_2S present. Oxygenic photosynthesis led to development of an oxic environment and to great bursts of biological evolution.

✓ What evidence supports the concept of a period of "RNA life," and why would RNA life forms not have survived to the present?

✓ How could energy have been obtained by early cells from FeS + H_2S?

✓ Why was the evolution of cyanobacteria a critical step in evolution?

11.3 Eukaryotes and Organelles: Endosymbiosis

Chapter 2 offered a brief tour of eukaryotic microbial diversity (∞ Section 2.6). Chapter 14 will examine eukaryotic microorganisms in more detail. Here, we consider how the modern eukaryotic cell came by its characteristic internal structure: the membrane-enclosed nucleus and organelles.

Origin of the Nucleus

Primitive eukaryotic cells likely consisted of structurally simple cells, resembling modern-day prokaryotic cells in lacking mitochondria, chloroplasts, and a membrane-enclosed nucleus. Indeed, we will see in Chapter 14 that examples of such cells are still extant. As cells in the eukaryotic line of descent became larger, the nucleus and mitotic apparatus evolved, along with the partitioning of DNA into distinct units (chromosomes). Chromosomes may have arisen to ensure the replication and orderly separation of DNA as primitive genomes enlarged to the point at which replication as a single molecule (as in prokaryotes) was no longer feasible. Development of the eukaryotic nucleus also facilitated the huge genomes needed by large microbial eukaryotes (and much later, multicellular organisms), and also made possible the recombination of genomes through sexual reproduction (∞ Sections 2.2 and 14.6). There is no obvious reason why primitive eukaryotes would have needed other organelles, and these likely arose later. And as previously mentioned, even today eukaryotic microorganisms are known that lack mitochondria and chloroplasts (∞ Section 14.8), indicating that these organelles are not essential for eukaryotic cell function.

Endosymbiosis

Strong evidence supports the theory that the modern (organelle-containing) eukaryotic cell evolved in steps through the stable incorporation of chemoorganotrophic and phototrophic symbionts from the domain *Bacteria* (Figure 11.7). This theory, called the **endosymbiotic** (*endo* means "inside") **theory** of eukaryotic evolution, postulates that an aerobic bacterium established stable residency within the cytoplasm of a primitive eukaryote and supplied the latter with energy in exchange for a protected environment and a ready supply of nutrients (Figure 11.7). This symbiont was the forerunner of the modern mitochondrion. Similarly, the endosymbiotic uptake of an oxygenic phototroph would have conferred photosynthetic properties on a primitive eukaryote, so it need rely no longer on organic compounds for energy production. This phototrophic endosymbiont was the forerunner of the modern day chloroplast (Figure 11.7). Some eukaryotic cells either never incorporated endosymbionts

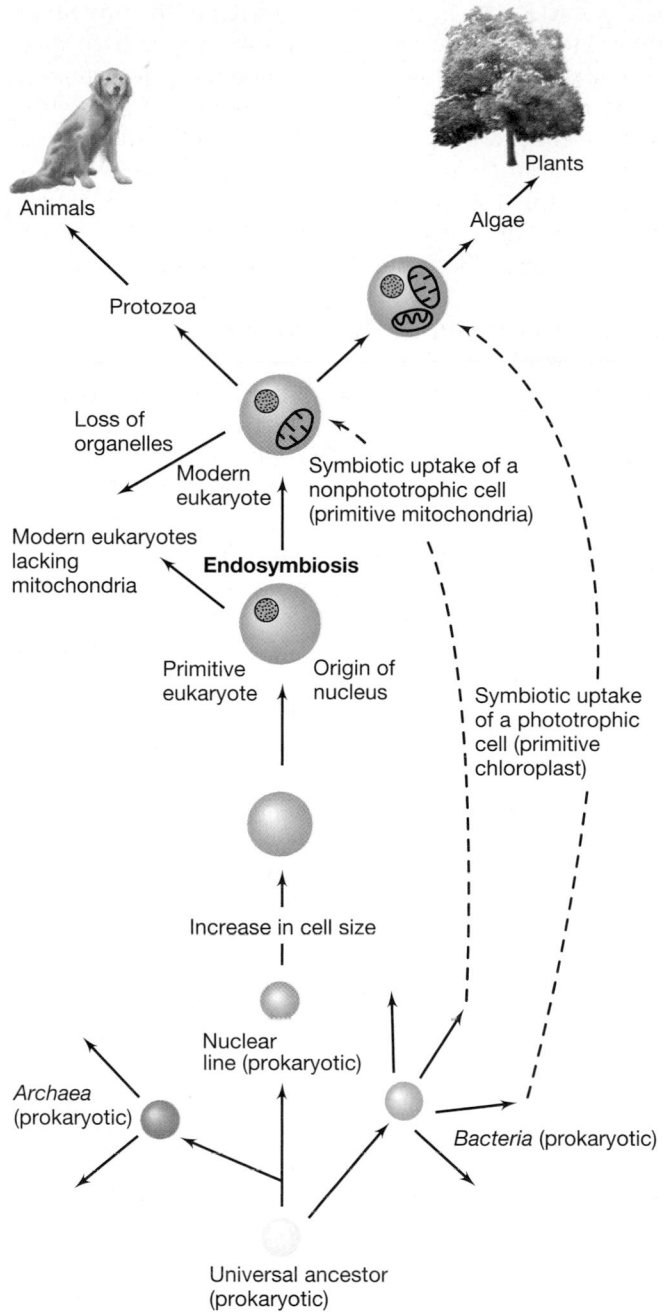

Figure 11.7 Origin of modern eukaryotes by endosymbiotic events. Note how organelles originated from *Bacteria* rather than *Archaea*. Endosymbiosis was unlikely to have been a one-time event and probably occurred in various types of cells of the nuclear line of descent. Note, however, how some primitive eukaryotes either never underwent endosymbiotic events or permanently lost their symbionts, but otherwise maintained the basic properties of eukaryotic cells. Extant examples of such eukaryotes, all microbial, are known today (∞ Chapter 14). The increase in cell size in the nuclear line of descent allowed for larger genomes to evolve, but also likely led to the evolution of the nucleus to affect the orderly replication and partitioning of such genomes. Compare this figure with Figure 11.13 that shows from which lineages of *Bacteria* organelles originated (∞ Section 14.4).

or did and then for some reason disposed of them; either way, these lineages remained competitive and spawned modern relatives that contain a membrane-enclosed nucleus but lack organelles (Figure 11.7; ∞ Section 14.8).

What hard evidence supports the notion that mitochondria and chloroplasts were once *Bacteria*, as postulated in the theory of endosymbiosis? Plenty. First, both chloroplasts and mitochondria contain ribosomes and they are clearly of the prokaryotic type (∞ Table 7.4). Moreover, ribosomes from these organelles show ribosomal RNA sequences (see Section 11.5) characteristic of specific *Bacteria*, and their function is inhibited by the same antibiotics that affect ribosome function in free-living *Bacteria* (∞ Section 14.5). Second, mitochondria and chloroplasts also contain small amounts of DNA arranged in a covalently closed, circular form, typical of prokaryotes (∞ Section 2.2). Although other organelle-*Bacteria* connections are known (∞ Section 14.5), the telltale signs of a previous free-living lifestyle are clearly present in mitochondria and chloroplasts. These energy factories conferred such important new properties on eukaryotic cells that the stage was set for an explosion in biological diversity. Indeed, the period from about 1.5 billion years ago to the present saw the rise and diversification of unicellular eukaryotic microorganisms and the metazoa, culminating in the structurally complex higher plants and animals (see Figures 11.6, 11.7, and 11.13).

Biological Evolution and Geological Time Scales

The period from the origin of the first metazoan to the present day represents about one-sixth of the total time that life has existed on Earth. Put another way, five-sixths of Earth's biological history was restricted to microbial life, the bulk of this period solely to prokaryotes (see Figure 11.6). However, because metazoa have left a considerable and highly diverse fossil record, our understanding of biological evolution from the time of the metazoa to the present vastly exceeds our knowledge of evolutionary relationships among prokaryotes. However, this has changed dramatically within the past 20 years with the advent of molecular methods for inferring microbial phylogenies, and we consider these methods in the next few sections.

✓ 11.3 Concept Check

The eukaryotic nucleus and mitotic apparatus probably arose as a necessity for ensuring the orderly partitioning of DNA in large-genome organisms. Mitochondria and chloroplasts, the principal energy-producing organelles of eukaryotes, arose from symbiotic association of prokaryotes of the domain *Bacteria* within eukaryotic cells, a process called endosymbiosis.

✓ When in geological time did eukaryotes first appear? Metazoa?

✓ What evidence supports the idea that organelles were once free-living species of the domain *Bacteria*?

II EVOLUTIONARY RELATIONSHIPS AMONG MICROORGANISMS

We now consider the **phylogeny** of microorganisms as deduced from molecular studies of nucleic acids. The major topic here will be comparative sequencing of structural RNAs from ribosomes. The results of these studies have generated the first true picture of microbial phylogeny and have yielded new research tools for microbial ecology and clinical diagnostics.

11.4 Evolutionary Chronometers

It is now clear that certain cellular macromolecules are evolutionary chronometers—measures of evolutionary change. It has been convincingly shown that the **evolutionary distance** between two organisms can be measured by differences in the nucleotide or amino acid sequence of monomers in *homologous* macromolecules isolated from them. This is so because the number of sequence differences in a molecule is proportional to the number of stable mutational changes fixed in the DNA encoding that molecule in the two organisms. As mutations become fixed in different populations, evolution occurs, and biodiversity is the end result.

Choosing the Right Chronometer

In order to determine true evolutionary relationships between organisms, it is essential that the correct molecules be chosen for sequencing studies. This is important for several reasons. First, the molecule should be *universally distributed* across the group chosen for study. Second, it must be *functionally homologous* in each organism; phylogenetic comparisons must start with molecules of *identical* function. Third, it is crucial to be able to properly *align* the two molecules in order to identify regions of sequence homology and sequence variance. Finally, the sequence of the molecule chosen should change at a rate commensurate with the evolutionary distance measured. And in fact, the greater the phylogenetic distance being measured, the *slower* must be the rate at which the sequence changes; too much change tends to scramble the evolutionary record.

Many molecules have been tested as molecular chronometers and comparative sequencing studies done on them to generate phylogenetic trees. These molecules include various cytochromes, iron–sulfur proteins like the ferredoxins, and genes for several other proteins and the ribosomal RNAs. However, the genes encoding ribosomal RNAs, key components in the translational system (∞ Sections 7.13–7.16), ATPase, the membrane-bound enzyme complex that can both synthesize as well as hy-drolyze ATP (∞ Section 5.12), and RecA, the protein required for genetic recombination (∞ Section 10.5), have provided the most insightful phylogenetic information about microorganisms. All of these molecules were probably essential for even rather primitive cells, and thus sequence variation in the genes encoding them allows us to probe deeply into the evolutionary record. We focus our discussion here on the most widely used of these chronometers, ribosomal RNA (Figure 11.8).

Ribosomal RNAs as Evolutionary Chronometers

Because of the likely antiquity of the protein-synthesizing machinery and for several other reasons, ribosomal RNAs are excellent molecules for discerning evolutionary relationships among living organisms. Ribosomal RNAs are functionally constant, universally distributed, and moderately well conserved in sequence across broad phylogenetic distances. Also, because the *number* of different possible sequences of large molecules such as ribosomal RNAs is so large, similarity in two sequences always indicates *some* phylogenetic relationship. However, it is the *degree* of similarity in ribosomal RNA sequences between two organisms that indicates their relative evolutionary relatedness. From comparative sequence analyses, molecular genealogies can be constructed leading to phylogenetic trees that show the most probable evolutionary position of organisms relative to one another (see Figure 11.13).

Recall the structure of the ribosome (Figure 11.8). There are three ribosomal RNA molecules, which in prokaryotes have sizes of 5S, 16S, and 23S (the "S" stands for "Svedberg," a unit of mass). The large bacterial rRNAs, 16S (Figure 11.8*c*) and 23S rRNA (approximately 1500 and 2900 nucleotides, respectively) contain several regions of highly conserved sequence useful for obtaining proper sequence alignments, yet contain sufficient sequence variability in other regions of the molecule to serve as excellent phylogenetic chronometers. Because 16S RNA is more experimentally manageable than 23S RNA, it has been used extensively to develop the phylogeny of both prokaryotes and eukaryotes. (In eukaryotes, 18S rRNA, the counterpart of 16S rRNA in the 80S ribosomes of these organisms, is the molecule sequenced.) Because 16S and 18S rRNAs originate from the *small* (30S or 40S) *subunit* of the ribosome (Figure 11.8*b*), the acronym *SSU* (for *s*mall *s*ubunit) sequencing is synonymous with 16S or 18S sequencing. The database of rRNA sequences in the Ribosomal Database Project (RDP) now numbers over 24,000 (16,000 aligned 16S sequences and 8000 aligned 18S sequences) and can be accessed on the Internet (http://rdp.cme.msu.edu/html/). Use of SSU rRNA as a phylogenetic tool was pioneered in the early 1970s by Carl Woese at the University of Illinois, and the method is now widely used.

Figure 11.8 Ribosomal RNA. (a) Electron micrograph of 70S ribosomes from *Escherichia coli*. (b) Parts of the ribosome; 5S, 16S, and 23S refer to different forms of RNA in the small subunit of the ribosome. (c) Primary and secondary structure of 16S ribosomal RNA (rRNA). This is the 16S rRNA from *Escherichia coli* (*Bacteria*); 16S rRNA from *Archaea* has general similarities in secondary structure (folding) but numerous differences in primary structure (sequence). The counterpart to 16S rRNA in eukaryotes is 18S rRNA present in cytoplasmic ribosomes. Using oligonucleotide primers specific for conserved rRNA sequences in organisms of one domain or the other (see Table 11.1), it is possible using PCR techniques to specifically amplify small subunit RNAs of cells of a particular domain present in a mixture with cells of other domains (see Figure 11.9). This has allowed for estimates of the diversity of particular domains of life directly in the habitat by "community sampling" techniques (co Section 18.5 and Figure 18.14). In addition, attaching specific oligonucleotides to fluorescent dyes has generated a new tool for use in microbial ecology (see Figures 11.11, 11.12, and 11.14).

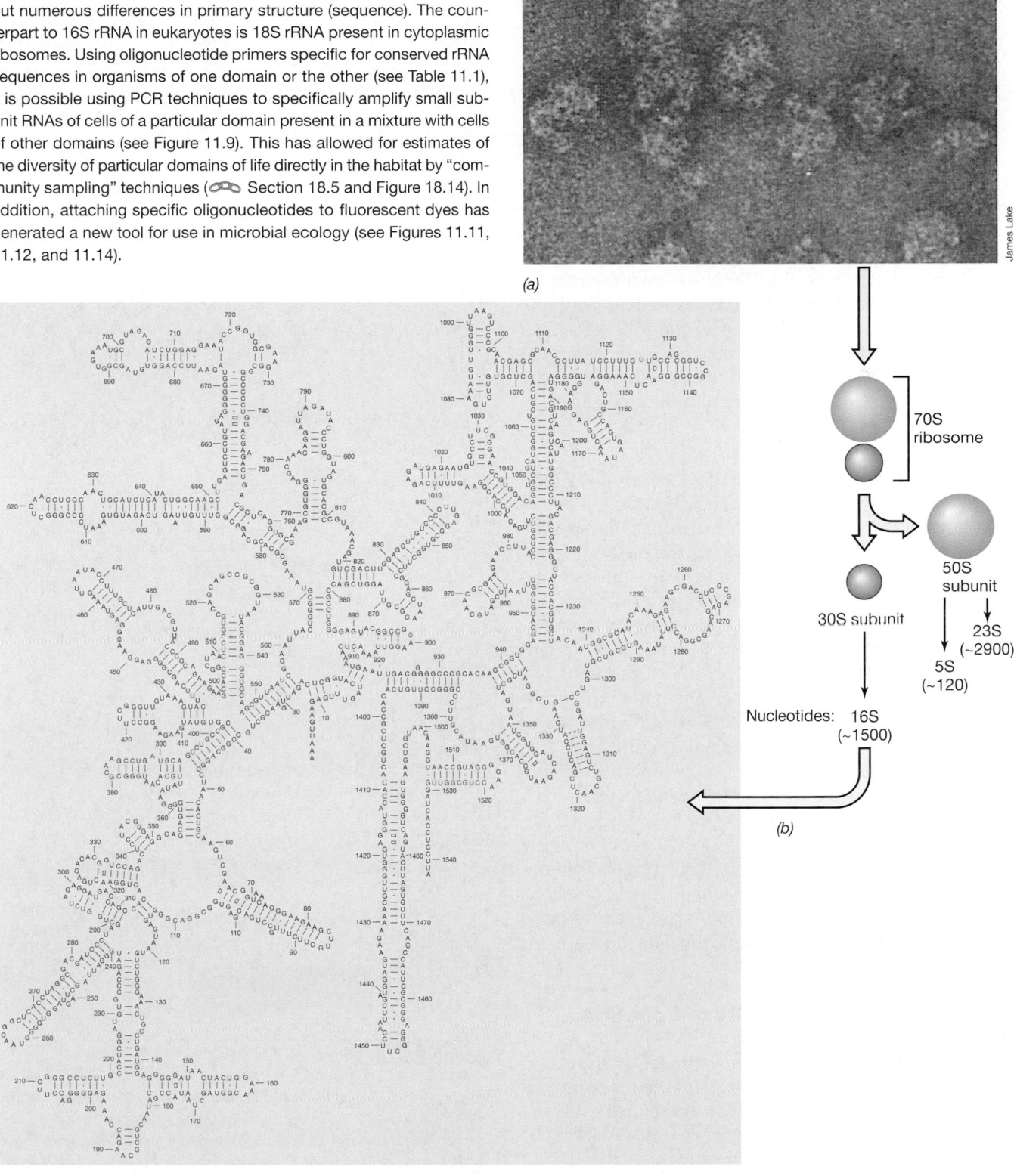

James Lake

(a)

(b)

(c)

✓ **11.4 Concept Check**

Comparisons of sequences of ribosomal RNA can be used to determine the evolutionary relationships between organisms. Phylogenetic trees based on ribosomal RNA have now been prepared for all the major prokaryotic and eukaryotic groups.

✓ Why is ribosomal RNA a good evolutionary chronometer?
✓ What is the RDP?

11.5 Ribosomal RNA Sequences and Cellular Evolution

The methods for obtaining ribosomal RNA sequences and generating phylogenetic trees are now quite routine and involve a combination of molecular biology and computer analyses. Newly generated sequences are compared with sequences in the RDP and/or with sequences obtained from other genetic databases such as GenBank (USA), DDBS (Japan), or EMBL (Germany). Then, using a treeing program, several possible trees are generated and the "best fit" tree selected that best summarizes the evolutionary information inherent in the sequences.

Sequence Methodology

We begin with the assumption that we are working with a pure culture of a microorganism although, as we will see, pure cultures are not necessary for comparative SSU rRNA sequencing. Like life itself, the methods for ribosomal RNA sequencing have evolved through the years, and although there are several ways to obtain the actual sequence, most scientists today employ the polymerase chain reaction (PCR, ∞ Section 10.17) to amplify directly the genes encoding 16S ribosomal RNA from genomic DNA and then sequence the PCR product by standard dideoxy DNA sequencing (that is, *Sanger sequencing*, ∞ Section 10.13) methods (Figure 11.9). This procedure is rapid and specific, and using synthetically produced oligonucleotide primers (available commercially at low cost) complementary to conserved sequences in SSU rRNAs as PCR primers, a tiny amount of DNA from a microbial culture can yield a huge amount of PCR product for sequencing reactions (Figure 11.9). Once the sequencing is done, either manually or by automated sequencers, the data are ready for computer analyses.

Generating Phylogenetic Trees from RNA Sequences

Several different algorithms for sequence analysis and phylogenetic tree formation are available for comparative ribosomal RNA sequencing. However, regardless of which program is used, raw sequence data must first be *aligned* with previously aligned sequences using a sequence editor. The aligned sequences are then imported into the actual treeing program and the comparative analyses done. Two widely used treeing algorithms are *distance* and *parsimony*. Using distance methods, sequences are aligned and an **evolutionary distance** (E_D) is calculated by having the computer count every position in the data set in which there is a *difference* (Figure 11.10). From these data, a distance matrix can be constructed that shows the E_D between any two sequences in the data set. Once this is done, a correction is factored into the E_D that takes into account the possibility that multiple changes have occurred at any given site (Figure 11.10). For example, there is a low, but statistically significant, probability that the base that exists at any given site in two sequences could be the result of two mutational events, one that originally changed the sequence and one that returned it to the same base as before. These possibilities can be estimated and the correction factor does this. Finally, a phylogenetic tree is generated in which the length of the lines in the tree are proportional to evolutionary distances (Figure 11.10).

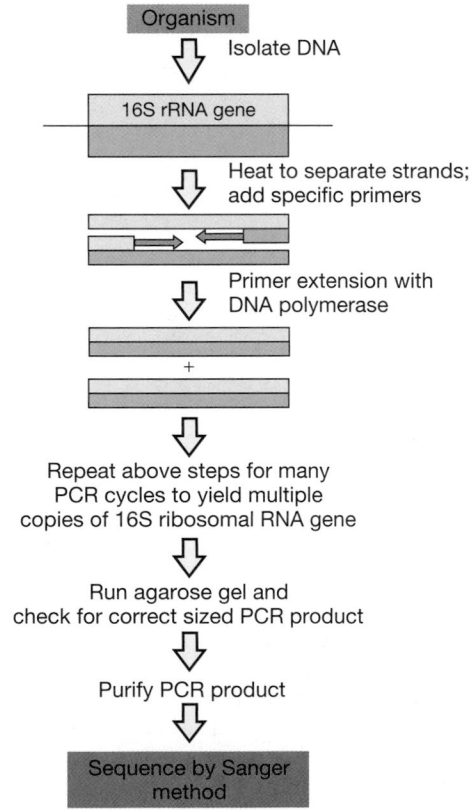

Figure 11.9 Ribosomal RNA sequencing of a pure culture of a microorganism using the polymerase chain reaction (PCR). The 16S rRNA gene is amplified and then sequenced by the Sanger method (∞ Section 10.13). The primers added are complementary to conserved sequences in one of the domains of 16S rRNA (see Figure 11.8c). A cloning step may also be used in this procedure to clone the DNA encoding the 16S rRNA following its PCR amplification.

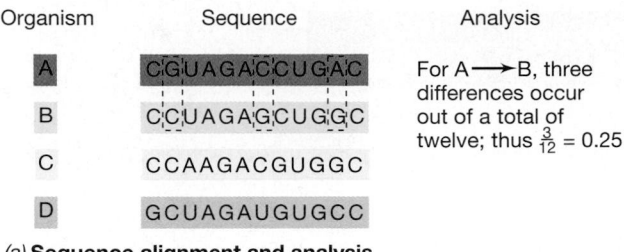

Organism	Sequence	Analysis
A	CGUAGACCUGAC	For A⟶B, three differences occur out of a total of twelve; thus $\frac{3}{12}$ = 0.25
B	CCUAGAGCUGGC	
C	CCAAGACGUGGC	
D	GCUAGAUGUGCC	

(a) **Sequence alignment and analysis**

Evolutionary distance				Corrected evolutionary distance
E_D	A ⟶ B	0.25		0.30
E_D	A ⟶ C	0.33		0.44
E_D	A ⟶ D	0.42		0.61
E_D	B ⟶ C	0.25		0.30
E_D	B ⟶ D	0.33		0.44
E_D	C ⟶ D	0.33		0.44

(b) **Calculation of evolutionary distance**

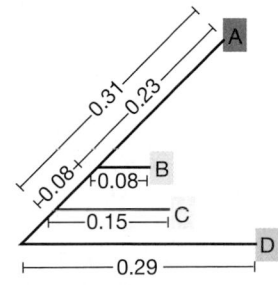

(c) **Phylogenetic tree**

Figure 11.10 Preparing a phylogenetic distance tree from 16S ribosomal RNA sequences. For illustrative purposes, only short sequences are shown. The evolutionary distance (F_D) in (b) is calculated as the percentage of *nonidentical* sequences between the RNAs of any two organisms. The corrected F_D is a statistical correction necessary to account for either back mutations to the original genotype or additional forward mutations at the same site that could have occurred. The tree (c) is ultimately generated by computer analysis of the data to give the best fit. The total length of the branches separating any two organisms is proportional to the calculated evolutionary distance between them. In actual analyses a statistical process called *bootstrapping* is typically used whereby the computer generates hundreds of versions of the tree to confirm that the final tree is indeed the best fit to the data set. In addition, insertions of several nucleotides may separate regions of sequence homology in two organisms' rRNA, and these insertions are "masked" (not considered) in the actual analyses.

Parsimony, another popular method of phylogenetic analysis, creates evolutionary trees based on the assumption that only the *minimal amount* of sequence change necessary to diverge two lineages from a common ancestor actually occurred during their evolutionary divergence. Like distance programs, this method requires summing the number of sequence differences in a particular data set, but the algorithm handles the analysis somewhat differently. Nevertheless, phylogenetic trees based on parsimony appear similar to distance trees, although the branching order within a parsimonious tree can and often does differ from that in a distance tree generated from the exact same data set. Thus, no single phylogenetic tree can provide the "final word" on the evolutionary relationships of the organisms in question. What phylogenetic trees accomplish, however, is as close an approximation as possible to the true phylogeny of the group.

✓ 11.5 Concept Check

Comparative ribosomal RNA sequencing is now a routine procedure involving the amplification of the gene encoding 16S rRNA, sequencing it, and analyzing the sequence in reference to other sequences. Two widely used treeing programs involve distance and parsimony methods.

✓ What is an *evolutionary distance* (E_D)?
✓ How is the *polymerase chain reaction (PCR)* of use in comparative ribosomal RNA sequencing?

11.6 Signature Sequences, Phylogenetic Probes, and Microbial Community Analyses

The next section will introduce the "big picture" of microbial phylogeny, the universal tree of life . However, before we see how SSU ribosomal RNA sequencing has revolutionized our picture of cellular evolution with the three domains of life, we discuss here some methods and applications of the technology itself. These include signature sequences and the design and use of ribosomal RNA probes in microbial ecology and diagnostic medicine.

Signature Sequences

Computer analyses of ribosomal RNA sequences have revealed so-called **signature sequences**, short oligonucleotides unique to certain groups of organisms. For example, signature sequences specific for each of the domains of cellular life are known (Table 11.1). Moreover, signatures defining a specific group within a domain or, in some cases, a particular genus or even a single species, are also known. Because of their exclusivity, signature sequences serve several useful purposes. For example, signatures help us place newly isolated or previously misclassified organisms into their correct phylogenetic group. But the most extensive use of signature sequences is in the design of specific nucleic acid probes called *phylogenetic probes*.

TABLE 11.1 Signature sequences from 16S or 18S rRNA defining the three domains of life

Oligonucleotide signatures[a]	Approximate position[b]	Occurrence among[c]		
		Archaea	Bacteria	Eukarya
CACYYG	315	0	> 95	0
AAACUCAAA	910	3	100	0
AAACUUAAAG	910	100	0	100
YUYAAUUG	960	100	< 1	100
CAACCYYCR	1110	0	> 95	0
UCCCUG	1380	> 95	0	100
UACACACCG	1400	0	> 99	100
CACACACCG	1400	100	0	0

[a] Y, any pyrimidine; R, any purine:

[b] Refer to Figure 11.8c for numbering scheme of 16S rRNA.

[c] Occurrence refers to percentage of organisms examined in any domain that contain that sequence.

Phylogenetic Probes and FISH

Recall that a probe is a strand of nucleic acid that can be labeled and used to hybridize to a complementary nucleic acid from a mixture (∞ Section 10.12). Probes can be general or specific. For example, universal ribosomal RNA probes can be synthesized that will bind to conserved sequences in the ribosomal RNA of all organisms, regardless of domain. Getting more specific, probes can be designed that will react only with cells of the domain *Bacteria* because of unique signatures in their RNA (Table 11.1). Likewise, archaeal-specific and eukaryl-specific probes react only with species of *Archaea* or *Eukarya*, respectively. Even major groups within a domain such as genera or families (see Section 11.11) can be targeted with specific probes.

The binding of probes to cellular ribosomes can be seen microscopically when a fluorescent dye is attached to the probe (Figure 11.11). By treating cells with the appropriate reagents, membranes become permeable and allow penetration of the probe/dye mixture. After hybridization of the probe directly to ribosomal RNA in ribosomes, the cells become uniformly fluorescent and can be observed under a fluorescent microscope (Figure 11.11, see also Figures 11.12 and 11.14). This technique has been nicknamed *FISH*, for *f*luorescent *in-situ* *h*ybridization, because it can be applied directly to cells in culture or in a natural environment.

FISH technology is widely used in microbial ecology and clinical diagnostics. In ecology, FISH can be used for the microscopic identification and tracking of organisms directly in the environment. These methods thus offer insights into the composition of microbial communities and the role a specific organism or related group of organisms plays in ecological processes in na-

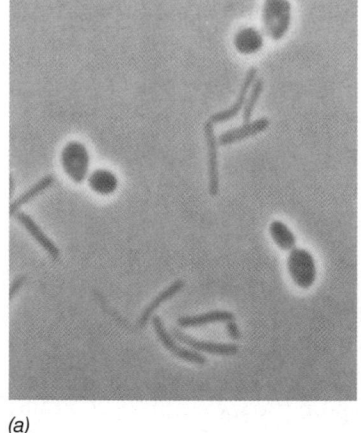
(a)

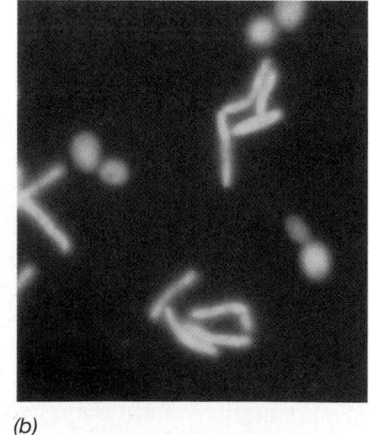
(b)

(c)

Norman Pace

Figure 11.11 Fluorescently labeled ribosomal RNA probes. (a) Phase contrast photomicrograph (no probes present) of cells of *Bacillus megaterium* (member of the *Bacteria*) and the yeast *Saccharomyces cerevisiae* (*Eukarya*). (b) Same field, cells stained with a yellow-green universal rRNA probe (this probe reacts with species from any domain). (c) Same field, cells stained with a eukaryal probe (only cells of *S. cerevisiae* react). Cells of *B. megaterium* are about 1.5 μm in diameter and cells of *S. cerevisiae* are about 6 μm in diameter.

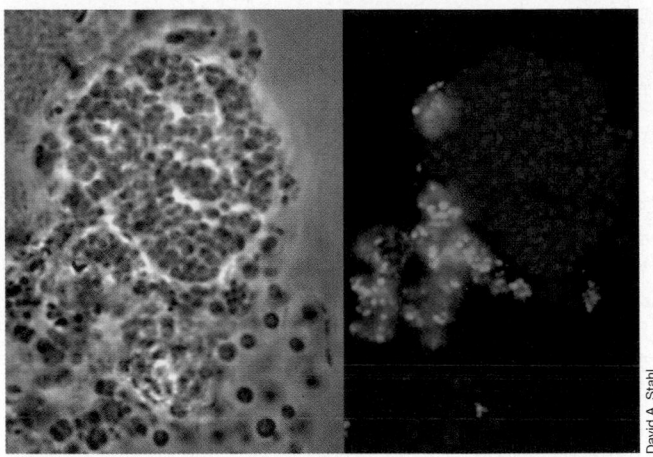

David A. Stahl

Figure 11.12 Use of phylogenetic stains to make nitrifying bacteria visible in a granule of activated sewage sludge. (Left) Phase photomicrograph. (Right) Color photomicrograph of the same field after applying phylogenetic stains. The probe that fluoresces red is specific for a signature sequence (see Table 11.1) in the 16S rRNA of ammonia-oxidizing bacteria, while the probe that fluoresces green is specific for a sequence present only in nitrite-oxidizing bacteria. Both ammonia-oxidizing and nitrite-oxidizing bacteria are phylogenetically closely related members of the *Bacteria* (∞ Sections 12.2 and 17.12) and carry out an interdependent series of chemolithotrophic reactions (∞ Section 19.12).

ture (Figure 11.12 and ∞ Section 18.4). In clinical diagnostics, FISH has been used for the rapid identification of specific pathogens from patient specimens. FISH circumvents the need to grow an organism in culture, making it possible to begin treatment hours to days earlier than would be possible using classic isolation and identification techniques.

Microbial Community Analysis

PCR-amplified ribosomal RNA genes do not need to originate from a pure culture grown in the laboratory. Using methods to be described in detail in Chapter 18, a phylogenetic survey of a natural microbial community can be done using PCR to amplify the genes encoding SSU ribosomal RNA from all members of that community. Such genes can be sorted out, sequenced, aligned, and a phylogenetic tree generated from these "environmental" sequences. The tree reveals the different ribosomal RNAs present in the community and from this the organisms present can be inferred, even though none of them were actually cultivated or otherwise identified. Such *microbial community analyses*, as they have come to be called, are a major thrust of microbial ecology research today and have revealed many key features of microbial community structure and microbial interactions (∞ Chapter 18 for detailed discussion of this topic).

✓ **11.6 Concept Check**

Signature sequences, short oligonucleotides found within a ribosomal RNA molecule, can be highly diagnostic of a particular organism or group of related organisms. Signature sequences can be used to generate specific phylogenetic probes, useful for FISH or microbial community analyses.

✓ Using the sequence data in Table 11.1, choose a specific phylogenetic probe that would allow you to distinguish a cell of *Archaea* from a cell of *Bacteria*.

✓ How can oligonucleotide probes be made visible under the microscope? What is this technology called?

11.7 Microbial Phylogeny Derived from Ribosomal RNA Sequences

We now consider what SSU ribosomal RNA sequencing has revealed about microbial evolution. Biologists previously grouped the living world into five kingdoms, only one of which was prokaryotic, based on structural similarities between organisms. Molecular phylogeny, on the other hand, has revealed that the five kingdoms do not represent five major evolutionary lines. Instead, as previously outlined in Chapter 2, cellular life on Earth has evolved along only *three* major lineages, two of which are exclusively microbial and are composed only of prokaryotic cells. The third line constitutes the eukaryotic lineage (Figure 11.13). The two prokaryotic lines are the **Bacteria** and the **Archaea**. The eukaryotic line is called the **Eukarya** (Figure 11.13). These terms define the three **domains** of life, the domain being the highest of biological taxons. Thus plants, animals, fungi, and protists are all kingdoms within the domain *Eukarya*.

The Universal Tree of Life

The universal phylogenetic tree (Figure 11.13) is the road map of life. It depicts the evolutionary history of all organisms relative to all other organisms and clearly shows the three primary groupings of organisms, the phylogenetic domains. The *root* of the universal tree represents a point in evolutionary history when all extant life on Earth shared a common ancestor, the so called *Universal Ancestor*.

Microbial genome sequencing projects (∞ Chapter 15) have yielded clues about the nature of the Universal Ancestor. Complete genomic sequences have confirmed the concept of the *Archaea*—the latter contain hundreds of genes that have no counterparts in *Bacteria* or *Eukarya*. But equally important has been the finding from the genome projects that many genes are *shared* among species of *Bacteria*, *Archaea*, and *Eukarya*. How can these seemingly conflicting findings be reconciled?

One hypothesis is that early in the history of life, before the primary domains had diverged, *lateral (hori-*

zontal) *gene transfer* was extensive; genes encoding proteins that conferred exceptionalœ fitness (for example, genes for core cellular functions like transcription and translation) were promiscuously transferred among a population of primitive organisms derived from a common ancestral cell. If this hypothesis is true, it would explain why, as genome analyses have shown, *all cells*, regardless of domain, share many key genes in common, as would be expected if all cells shared a common, ancestor (Figure 11.13). But what about the genetic *differences* observed from complete genome sequences? It is further hypothesized that with time, barriers to unrestricted gene flow evolved, perhaps from the selective colonization of habitats (thereby generating reproductive isolation) or as the result of structural or enzymatic (for example, restriction endonuclease) barriers that in some way prevented free genetic exchange. As a result, the previously genetically promiscuous population slowly began to sort out into the primary lines of evolutionary descent (Figure 11.13). As each lineage continued to evolve, certain unique biological traits became genetically fixed within each group such that what we see today, after some 3.5 billion years of microbial evolution, are three domains of cellular life that, on the one hand, share many common features, while on the other, display a distinctive evolutionary history of their own.

Primitive Versus Modern Organisms

A point that deserves emphasis here is that *none* of the organisms living today and depicted on the universal tree (Figure 11.13) are *primitive*. All extant life forms are *modern* organisms, well adapted to, and successful in, their ecological niches. Certain of these organisms may indeed be phenotypically similar to hypothetical primitive organisms, and hyperthermophilic prokaryotes (growth temperature optima > 80°C) are a good case in point. For example, the organisms *Aquifex* and *Methanopyrus* branch relatively close to the root of the tree (Figure 11.13). Both of these organisms are capable of growth at very high temperatures (Sections 12.38 and 13.6), and this is likely the condition that faced the primitive relatives of these organisms, growing on a much hotter, early Earth. From the phylogenetic tree it appears that *Aquifex* and *Methanopyrus* are more closely related to primitive organisms than, let's say, species of Proteobacteria (*Bacteria*) or extreme halophiles (*Archaea*) (Figure 11.13), but it should be understood that these early branching organisms are not themselves primitive.

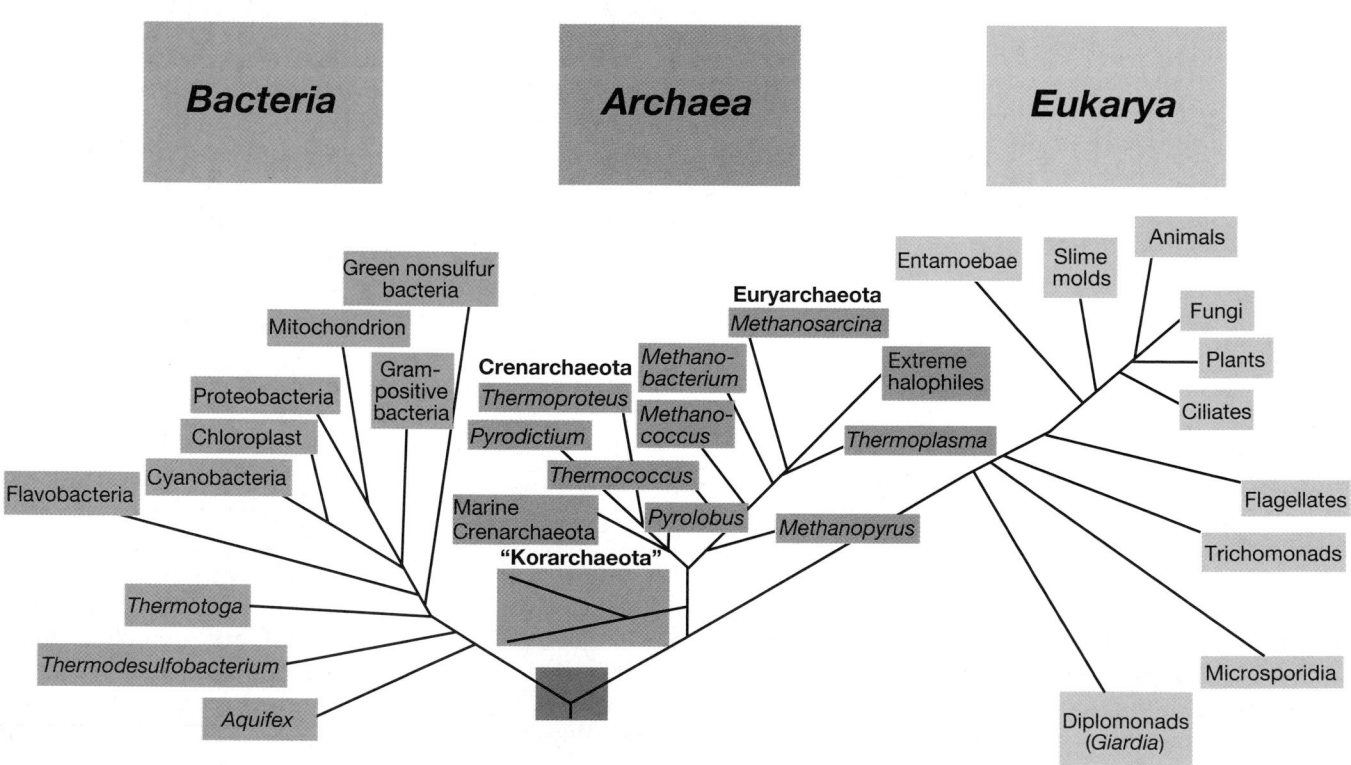

Figure 11.13 Universal phylogenetic tree as determined from comparative ribosomal RNA sequencing. Only a few key organisms or lineages are shown in each domain. For detailed domain trees refer to Figures 12.1, 13.1, and 14.11. Of the three domains, two (*Bacteria* and *Archaea*) contain only prokaryotic representatives. The location highlighted in red is the hypothetical root of the tree, which represents the position of the universal ancestor of all cells.

They are modern organisms that are simply less derived than organisms that branch further up the tree of life.

Bacteria

Among *Bacteria*, at least 40 divisions (called *phyla*) have been discovered and several key ones are shown in the universal tree in Figure 11.13. Many phyla are defined from environmental sequences only (see Section 11.6 and ⌾ Figure 12.1). Some of the lineages in the domain *Bacteria* are ones previously distinguished by some phenotypic property like morphology or physiology; the spirochetes and the cyanobacteria, respectively, are good examples of this. But most of the phyla consist of species that, although specifically related from a phylogenetic standpoint, lack strong phenotypic cohesiveness. The Proteobacteria are a good example here, because the mixture of physiologies present within the group nearly runs the gamut of known microbial physiology (⌾ Chapter 17). This clearly indicates that physiology and phylogeny are not necessarily linked.

Eukaryotic organelles clearly originated from within the domain *Bacteria*. As shown on the universal tree, mitochondria arose from a major group of *Bacteria* called the *Proteobacteria* (Figure 11.13), specifically from relatives like *Agrobacterium*, *Rhizobium*, and the rickettsias. Interestingly, like mitochondria, the latter organisms are also capable of an intracellular existence, either within plants (⌾ Sections 19.21 and 19.22) or animals (⌾ Sections 12.13/27.3). The chloroplast arose from within the cyanobacterial phylum (Figure 11.13), as would be expected since both carry out oxygenic photosynthesis (⌾ Section 17.5).

Archaea

From a phylogenetic perspective, the domain *Archaea* consists of three large phyla, the Crenarchaeota, Euryarchaeota, and a third possible lineage called the "Korarchaeota" (Figure 11.13). Branching very close to the root of the universal tree are hyperthermophilic Crenarchaeota like *Thermoproteus*, *Pyrolobus*, and *Pyrodictium* (Figure 11.13). They are followed by the Euryarchaeota, the methane-producing (methanogenic) prokaryotes and the extreme halophiles; *Thermoplasma*, an acidophilic, thermophilic cell wall-less prokaryote is loosely related to this latter group (Figure 11.13).

There are some branches in the Crenarchaeota lineage (Figure 11.13 and ⌾ Figure 13.1) that are known from community sampling of ribosomal genes from the environment (see Section 11.6) but have not been obtained from actual cultured isolates. Interestingly, however, these sequences come from organisms that inhabit the open oceans, including Antarctic marine waters where temperatures are far colder than in the hot-spring or deep sea hydrothermal-vent habitats of known Crenarchaeota, and also from soil and lake water. We discuss cold adapted Crenarchaeota in more detail in Section 13.8.

Like cold-adapted Crenarchaeota, an entire phylum of *Archaea*, the "Korarchaeota," have been identified from microbial community analysis (see Section 11.6), in this case of a specific hot spring (Obsidian Pool) in Yellowstone National Park (Wyoming, USA). Stable mixed cultures containing "Korarchaeota" (identifiable through phylogenetic staining, see Figure 11.14) can be grown in the laboratory. From the media and incubation conditions supporting these cultures we can deduce that "Korarchaeota" are hyperthermophiles and may have metabolic properties similar to those of hyperthermophilic Crenarchaeota (⌾ Sections 13.8–13.10). We consider all phyla of *Archaea* in more detail in Chapter 13.

Eukarya

Phylogenetic trees of species in the domain *Eukarya* are generated from comparative sequencing of 18S rRNA, the functional equivalent of 16S rRNA in eukaryotic cytoplasmic ribosomes. Early eukaryotes were probably similar to present-day microsporidia and diplomonads. These organisms are all obligate parasites that live in association with representatives of various groups of eukaryotes, from microorganisms to humans [for example, the pathogen *Giardia* (⌾ Section 28.7) is a member of the diplomonad group]. Interestingly, although these organisms contain a membrane-enclosed nucleus, microsporidia and diplomonads lack mitochondria. In this connection they resemble the type of cell that might have first accepted stable endosymbionts (see Figure 11.7 and ⌾ Sections 14.5 and 14.8).

More phylogenetically derived *Eukarya* include the multicellular organisms (the metazoa), culminating with the largest and structurally most complex of the *Eukarya*, the plants and animals (Figure 11.13). When the fossil

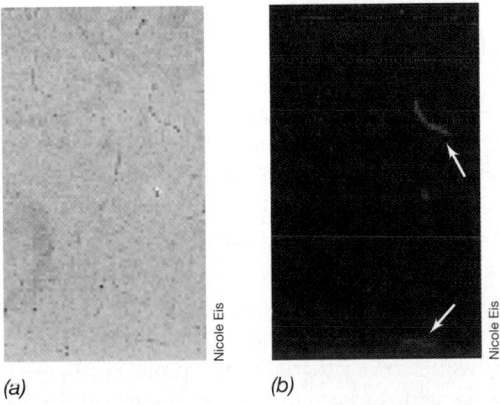

(a)   (b)

Figure 11.14 Identifying cells of "Korarchaeota" with phylogenetic stains. (a) Phase-contrast micrograph of an enrichment culture containing korarchaeotan cells. (b) Fluorescence photomicrograph of the same field as in (a) but treated with a phylogenetic stain designed from a signature sequence (see Section 11.6) from the 16S rRNA of the "Korarchaeota." The two red cells in this field are thus korarchaeotans.

record is compared with the phylogenetic tree of *Eukarya*, rapid evolutionary radiation can be dated to about 1.5 billion years ago. Geochemical evidence suggests that this is the period in Earth's history in which significant oxygen levels had accumulated in the atmosphere (Figure 11.6). It is thus likely that the onset of oxic conditions and subsequent development of an ozone shield (which would have greatly expanded the number of surface habitats available for colonization) also triggered the rapid diversification of *Eukarya*. We consider the major groups of microbial *Eukarya* and their basic biology in Chapter 14.

✓ 11.7 Concept Check

Life on Earth evolved along three major lines, called domains, all derived from a common ancestor. Each domain contains several phyla. Two of the domains, *Bacteria* and *Archaea*, remained prokaryotic, whereas the third line, *Eukarya*, evolved into the modern eukaryotic cell.

✓ How does the universal tree support the idea that early Earth was very hot?

✓ How does the universal tree support the hypothesis of endosymbiosis (see Figure 11.7)?

11.8 Characteristics of the Primary Domains of Life

Although the primary domains—*Bacteria, Archaea*, and *Eukarya*—are defined on the basis of comparative ribosomal RNA sequencing, each domain is also characterized by a number of *phenotypic* properties. Some of these characteristics are unique to one domain, whereas others are found in two of three domains. We present here an overview of major phenotypic traits of phylogenetic value.

Cell Walls

Virtually all *Bacteria* have cell walls containing *peptidoglycan* (∞ Section 4.8). The only known exceptions are members of the *Planctomyces–Pirella* group (∞ Section 12.29), whose cell walls are composed of protein, and the *Mycoplasma–Chlamydia* groups, which lack cell walls altogether (∞ Sections 12.22 and 12.28). Peptidoglycan can thus be considered a "signature" molecule for species of *Bacteria* (when assaying for peptidoglycan, it is *muramic acid* that is actually detected because this is what is unique to peptidoglycan, Table 11.3).

Eukarya and *Archaea* lack peptidoglycan. In eukaryotes, if cell walls are present, they are usually made of cellulose or chitin (∞ Chapter 14). In *Archaea*, as previously discussed (∞ Section 4.8), various cell wall types exist, from the peptidoglycan analog pseudopeptidoglycan to walls made of polysaccharide, protein, or glycoprotein. Thus, great diversity is observed in the chemistry of microbial cell walls, but it is the presence or absence of peptidoglycan that is most useful in distinguishing *Bacteria* from *Archaea*.

Lipids

The chemical nature of membrane lipids is perhaps the most useful of all nongenetic criteria for differentiating *Archaea* from *Bacteria*. *Bacteria* and *Eukarya* synthesize membrane lipids with a backbone consisting of fatty acids hooked in *ester* linkage to a molecule of glycerol (see Figure 11.15). Although the nature of the fatty acid can be highly variable, the *ester link* to glycerol is the key defining feature. By contrast, archaeal lipids consist of *ether-linked* molecules (Figure 11.15). In ester-linked lipids, the fatty acids are usually straight-chain (linear) molecules, whereas in *Archaea*, long-chain, branched hydrocarbons, either of the phytanyl or biphytanyl type, are present in place of fatty acids and are bonded by ether linkage to glycerol.

In terms of overall membrane structure, the membranes of some *Archaea*, especially hyperthermophilic species, form into a lipid *monolayer* instead of a lipid *bilayer* (∞ Section 4.5 and Figure 4.20). A lipid monolayer is less likely than a lipid bilayer to denature at the high temperatures at which these organisms grow. The ether linkage between glycerol and the hydrophobic side chains (Figure 11.15) combined with monolayer membrane construction protect the membranes of these extremophilic *Archaea* from the harsh conditions of their habitats.

RNA Polymerase

Transcription is carried out by DNA-dependent RNA polymerases in all organisms; DNA is the template, and RNA is the product (∞ Section 6.7). Cells of *Bacteria* contain a single type of RNA polymerase of rather simple quaternary structure. This is the classic RNA polymerase containing *four* polypeptides, α, β, β', and σ, combined in a ratio of 2:1:1:1 respectively, in the active polymerase (Figure 11.16) (∞ also Section 7.8).

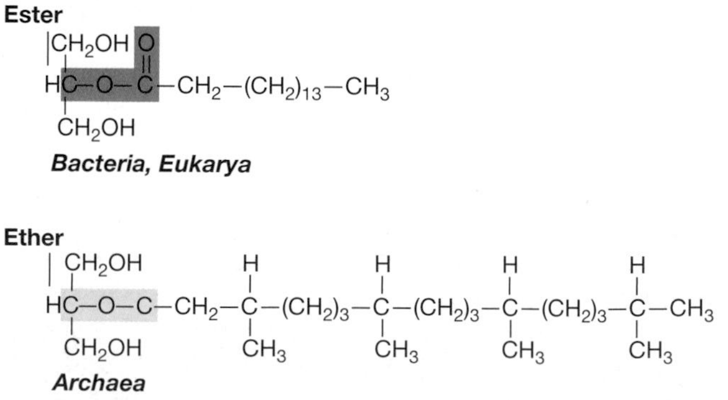

Ester

$$\begin{array}{l} |CH_2OH \quad O \\ HC-O-C-CH_2-(CH_2)_{13}-CH_3 \\ |CH_2OH \end{array}$$

Bacteria, Eukarya

Ether

$$\begin{array}{l} |CH_2OH \\ HC-O-C-CH_2-C-(CH_2)_3-C-(CH_2)_3-C-(CH_2)_3-C-CH_3 \\ |CH_2OH \qquad CH_3 \qquad CH_3 \qquad CH_3 \qquad CH_3 \end{array}$$

Archaea

Figure 11.15 Lipids in *Bacteria, Eukarya*, and *Archaea*. In *Bacteria* and *Eukarya*, lipids contain fatty acids (palmitic acid is shown) bonded by *ester* linkages to glycerol. In *Archaea*, the side chains are branched hydrocarbons (phytanyl, C_{20}, is shown) bonded by *ether* linkages to glycerol. Phytanyl is synthesized from isoprene (∞ Figure 4.19c).

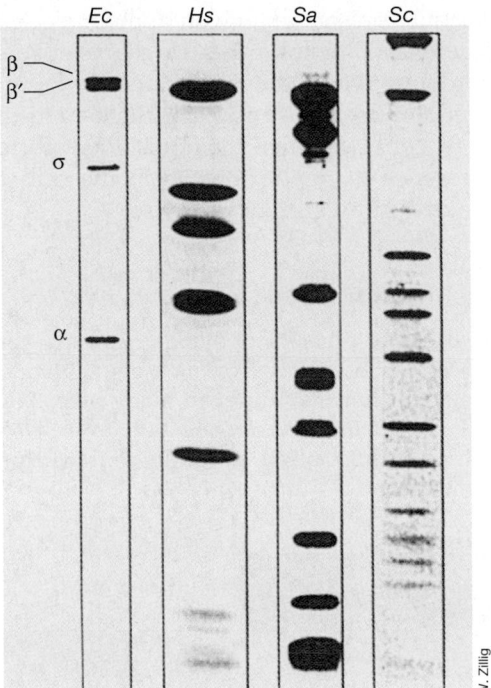

Figure 11.16 RNA polymerases from representatives of the three domains: *Ec, Escherichia coli* (*Bacteria*), *Hs, Halobacterium salinarum* (Euryarchaeota, *Archaea*), *Sa, Sulfolobus acidocaldarius* (Crenarchaeota, *Archaea*), and *Sc, Saccharomyces cerevisiae* (*Eukarya*). The purified components of the RNA polymerases have been separated by electrophoresis on a polyacrylamide gel. The largest polypeptide subunits are on the top, and the smallest subunits are on the bottom. Only members of the *Bacteria* contain the simple (four-polypeptide) RNA polymerase.

Archaeal RNA polymerases are structurally more complex than those of *Bacteria*. The RNA polymerases of species of *Archaea* contain *eight* or more polypeptides, more closely resembling the pattern in *Eukarya* (Figure 11.16). The major RNA polymerase of eukaryotes (there

are three) contains 10–12 polypeptides, and the relative sizes of the peptides coincide most closely with those from species of hyperthermophilic *Archaea* (Figure 11.16). Thus, in terms of phylogenetic signatures, the $(\alpha_2\beta\beta'\sigma)$ polymerase is highly diagnostic of *Bacteria*, whereas the remaining polymerases are too complex to be phylogenetically definitive.

Features of Protein Synthesis

Because of differences in ribosomal RNA sequences and several protein synthesis factors, it is not surprising that certain aspects of the protein-synthesizing machinery differ in representatives of the three domains. Although ribosomes of *Archaea* and *Bacteria* are the same size (70S, as compared with the 80S ribosomes in the cytoplasm of eukaryotes), several steps in archaeal protein synthesis more strongly resemble those in eukaryotes than in *Bacteria*. Recall that translation always begins at a unique codon, the so-called *start codon*. In *Bacteria* this start codon (AUG) calls for the incorporation of an initiator tRNA containing a modified methionine residue, *formyl*methionine (◌ Section 7.15). By contrast, in eukaryotes and in *Archaea*, the initiator tRNA carries an *unmodified* methionine.

The exotoxin produced by *Corynebacterium diphtheriae* is a potent inhibitor of eukaryotic protein synthesis because it ADP-ribosylates (adds ADP to) an elongation factor required to translocate the ribosome along the mRNA; thus, the modified elongation factor is inactive (◌ Section 21.10). Diphtheria toxin also inhibits protein synthesis in *Archaea* but not in *Bacteria*.

Most antibiotics that specifically affect protein synthesis in *Bacteria* do not affect archaeal or eukaryal protein synthesis. The sensitivity of representatives of the three domains to various protein synthesis inhibitors is shown in Table 11.2 (◌ also Section 7.15), where

TABLE 11.2 Sensitivity of representatives of the three domains to various protein synthesis inhibitors[a]

		Archaea		Bacteria	Eukarya
		Euryarchaeota *Methanobacterium thermoautotrophicum*	Crenarchaeota *Sulfolobus acidocaldarius*	*Escherichia coli*	*Saccharomyces cerevisiae*
Antibiotics	Mode of action				
Fusidic acid, sparsomycin	Inhibits elongation steps	+	−	+	+
Anisomycin, narciclasine	Inhibits peptidyl transfer	+	−	−	+
Cycloheximide	Blocks initiation	−	−	−	+
Erythromycin, streptomycin, chloramphenicol	Increases error frequencies and other effects	−	−	+	−
Virginiamycin, pulvomycin	Inhibits elongation steps	+	−	+	+
Neomycin, puromycin	Causes premature termination	+	+	+	+
Rifamycin	Inhibits β subunit of RNA polymerase	−	−	+	−

[a] A + indicates that protein synthesis (and growth) is inhibited.

various antibiotics are grouped according to their modes of action in blocking protein synthesis in various domains of organisms.

Collectively, these results suggest that ribosomal proteins from cells of *Archaea* and *Eukarya* are functionally more similar to each other than they are to ribosomal proteins from *Bacteria*, and further support the position of the domains relative to one another in the universal tree (Figure 11.13).

Other Features Defining the Domains

From microbiological research a number of other features, physiological and otherwise, have been discovered that can be used to differentiate organisms at the domain level; a summary of these is listed in Table 11.3. When examining this table it should be understood that not all features

are universally present in a given domain. For example, chlorophyll-based photosynthesis is characteristic of only some representatives of the *Bacteria* and the *Eukarya* (and in the latter only because of photosynthetic endosymbiotic *Bacteria*). By contrast, other distinguishing features, such as the presence of peptidoglycan in the cell walls of *Bacteria*, may be universal or nearly so.

✓ 11.8 Concept Check

Although the three domains of living organisms were originally defined by ribosomal RNA sequencing, subsequent studies have shown that they differ in many other ways, too. Particularly, the *Bacteria* and *Archaea* differ extensively in cell wall and lipid chemistry and in features of transcription and protein synthesis.

TABLE 11.3 Summary of major differential features among *Bacteria*, *Archaea*, and *Eukarya*[a]

Characteristic	Bacteria	Archaea	Eukarya
Morphological and Genetic			
Prokaryotic cell structure	Yes	Yes	No
DNA present in covalently closed and circular form	Yes	Yes	No
Histone proteins present	No	Yes	Yes
Membrane-enclosed nucleus	Absent	Absent	Present
Cell wall	Muramic acid present	Muramic acid absent	Muramic acid absent
Membrane lipids	Ester-linked	Ether-linked	Ester-linked
Ribosomes (mass)	70S	70S	80S
Initiator tRNA	Formylmethionine	Methionine	Methionine
Introns in most genes	No	No	Yes
Operons	Yes	Yes	No
Capping and poly-A tailing of mRNA	No	No	Yes
Plasmids	Yes	Yes	Rare
Ribosome sensitivity to diphtheria toxin	No	Yes	Yes
RNA polymerases (see Figure 11.16)	One (4 subunits)	Several (8–12 subunits each)	Three (12–14 subunits each)
Transcription factors required (⚙ Section 7.10)	No	Yes	Yes
Promoter structure (⚙ Section 7.9)	−10 and −35 sequences (Pribnow box)	TATA box	TATA box
Sensitivity to chloramphenicol, streptomycin, and kanamycin	Yes	No	No
Physiological			
Methanogenesis	No	Yes	No
Dissimilative reduction of S^0 or SO_4^{2-} to H_2S, or Fe^{3+} to Fe^{2+}	Yes	Yes	No
Nitrification	Yes	No	No
Denitrification	Yes	Yes	No
Nitrogen fixation	Yes	Yes	No
Chlorophyll-based photosynthesis	Yes	No	Yes (in chloroplasts)
Rhodopsin-based energy metabolism	Yes	Yes	No
Chemolithotrophy (Fe, S, H_2)	Yes	Yes	No
Gas vesicles	Yes	Yes	No
Synthesis of carbon storage granules composed of poly-β-hydroxyalkanoates	Yes	Yes	No
Growth above 80°C	Yes	Yes	No

[a] Note that for many features only particular representatives within a domain show the property

✓ How do the lipids from *Bacteria* and *Eukarya* differ from those of the *Archaea*?

✓ Organisms from which domain of prokaryotes synthesize RNA polymerases that most closely resemble those of the *Eukarya*? What phylogenetic evidence supports this observation?

III MICROBIAL TAXONOMY AND ITS RELATIONSHIP TO PHYLOGENY

A major discipline in microbiology is microbial classification (taxonomy). Classification allows microbiologists to see relationships among different microorganisms and develop the necessary framework for synthesis of a natural (evolutionary) taxonomy of these organisms. We consider basic issues of bacterial taxonomy in the next four sections.

11.9 Classical Taxonomy

Taxonomy is the science of classification and consists of two major subdisciplines, *identification* and *nomenclature*. It is important to distinguish between bacterial *taxonomy* and the main topic of this chapter up to this point, bacterial *phylogeny*, for the terms really mean different things. Bacterial taxonomy has traditionally relied on *phenotypic* analyses—taking into account what an organism looks like, its energy metabolism, its enzymes, etc., —as the basis of classification. However, because bacteria are so small and contain relatively few structural clues to their evolutionary roots, the phylogeny of prokaryotes has only emerged from *genotypic* analyses as discussed in the previous sections. Nevertheless, phenotypic analyses have traditionally played an important role in bacterial identification and classification, especially in applied situations where identification may be an end in itself, for example, in clinical diagnostic microbiology. In this section, we discuss classical bacterial taxonomy, and in the next section, summarize some molecular methods that have been found useful in the classification of bacteria.

In classical bacterial taxonomy, several characteristics are measured, and these traits are then used to group the organisms up the taxonomic ladder from species to domain (see Tables 11.5 and 11.6). Characteristics of taxonomic value that are widely used include various aspects of morphology, nutrition and physiology, and habitat. Table 11.4 breaks down these

TABLE 11.4 Some phenotypic characteristics of taxonomic value

Major category	Components
I. Morphology	Shape; size; Gram reaction
II. Motility	Motile by flagella; motile by gliding; motile by gas vessels; nonmotile
III. Nutrition and physiology	Mechanism of energy conservation (phototroph, chemoorganotroph, chemolithotroph); relationship to oxygen; temperature, pH, and salt requirements/tolerances; ability to use various carbon, nitrogen, and sulfur sources
IV. Other factors	Pigments; cell inclusions, or surface layers; pathogenicity; antibiotic sensitivity

major categories into specific topics of taxonomic value.

To identify an organism, classical taxonomists apply a series of criteria from general to specific. An example of this is shown using morphological and physiological criteria in Figure 11.17 (see also Table 11.5). Using a dichotomous key method and testing an isolate for several individual characteristics, the list of possible organisms is narrowed until a positive identification can be made (Figure 11.17).

The determination of the guanine plus cytosine content of an organism's genomic DNA, usually referred to as the *GC ratio*, can be a part of classical bacterial taxonomy. DNA base ratios are expressed as:

$$\frac{G + C}{A + T + G + C} \times 100\%$$

GC ratios can be determined in several ways, including by measuring the melting temperature (∞ Section 7.2), by Southern blotting (∞ Section 10.12), or by chromatographic methods.

GC ratios vary over a wide range, with values as low as 20% and high as nearly 80% known among prokaryotes, a somewhat broader range than for eukaryotes (Figure 11.18). Base compositions of DNA have been determined for a wide variety of organisms. Knowledge of an organism's GC ratio is occasionally useful for identification purposes but is rarely definitive. This is because two organisms can have identical GC ratios and yet be quite unrelated (both taxonomically and phylogenetically) because a variety of base *sequences* is possible with DNA of a given base composition. Thus, although occasionally helpful in taxonomic studies, GC ratios are usually of minimal value in the overall taxonomic characterization of an organism.

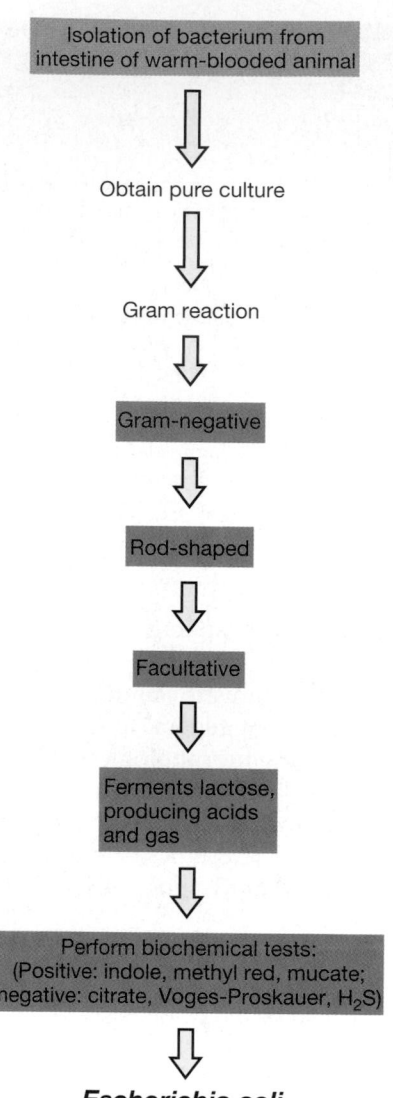

Figure 11.17 Example of methods that would be used for identification of a newly isolated enteric bacterium, using classic microbiological methods (the example given shows the procedures that would be used for identifying *Escherichia coli*). Note that most of the analyses here require that the organisms be grown in pure culture and that solely phenotypic criteria be used in the identification. A description of biochemical tests is presented in Chapter 24 (⟳ Section 24.2, Table 24.3, and Figure 24.7).

✓ 11.9 Concept Check

Conventional bacterial taxonomy places heavy emphasis on descriptive analyses of phenotypic properties of the organism. Determining the guanine plus cytosine (G + C) base ratio of the DNA of the organism can be part of this process.

✓ What phenotypic properties can be used to distinguish bacteria?

✓ How is it possible for two organisms to have very similar DNA GC ratios yet share few genes in common?

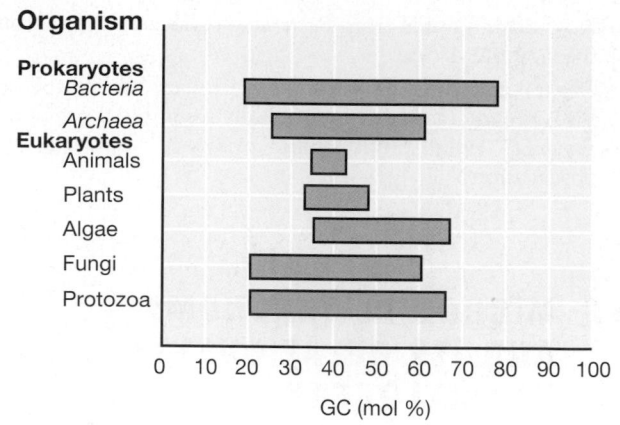

Figure 11.18 Ranges of DNA base composition of various organisms. Note that the greatest range exists with *Bacteria*.

11.10 Molecular Taxonomy

Molecular taxonomy, or *chemotaxonomy* as it is also called, employs molecular analyses of any of a number of biomolecules in the cell. Chief among chemotaxonomic methods have been *DNA:DNA hybridization, ribotyping*, and *lipid analyses*.

DNA:DNA Hybridization

A GC base ratio describes the percentage of each nucleotide present in genomic DNA of a given species but gives absolutely no information on the *sequence* of those nucleotides. Sequences are critical, of course, because if two organisms have many of the same nucleotide sequences in their DNAs, they likely contain many highly similar (if not identical) genes. Two DNAs would thus be expected to *hybridize* to one another in proportion to the similarities in their gene sequences. Genomic hybridization measures the degree of sequence similarity and is thus useful for differentiating very closely related organisms.

We discussed the theory and methodology of nucleic acid hybridization in Section 10.12. In an actual hybridization experiment, DNA isolated from one organism is made radioactive with ^{32}P or ^{3}H, sheared to a relatively small size, heated to denature, and mixed with an excess of unlabeled DNA prepared in the same way from a second organism (Figure 11.19). The DNA mixture is then cooled to allow it to reanneal, and double-stranded duplex DNA is separated from any remaining unhybridized DNA. Following this, the amount of radioactivity in the hybridized DNA is determined and compared with the control, which is taken as 100% (Figure 11.19). In addition to radioactivity, a variety of nonradioactive DNA labels have been introduced in recent years, and these have the advantage that the hybridization experiment generates no radioactive wastes.

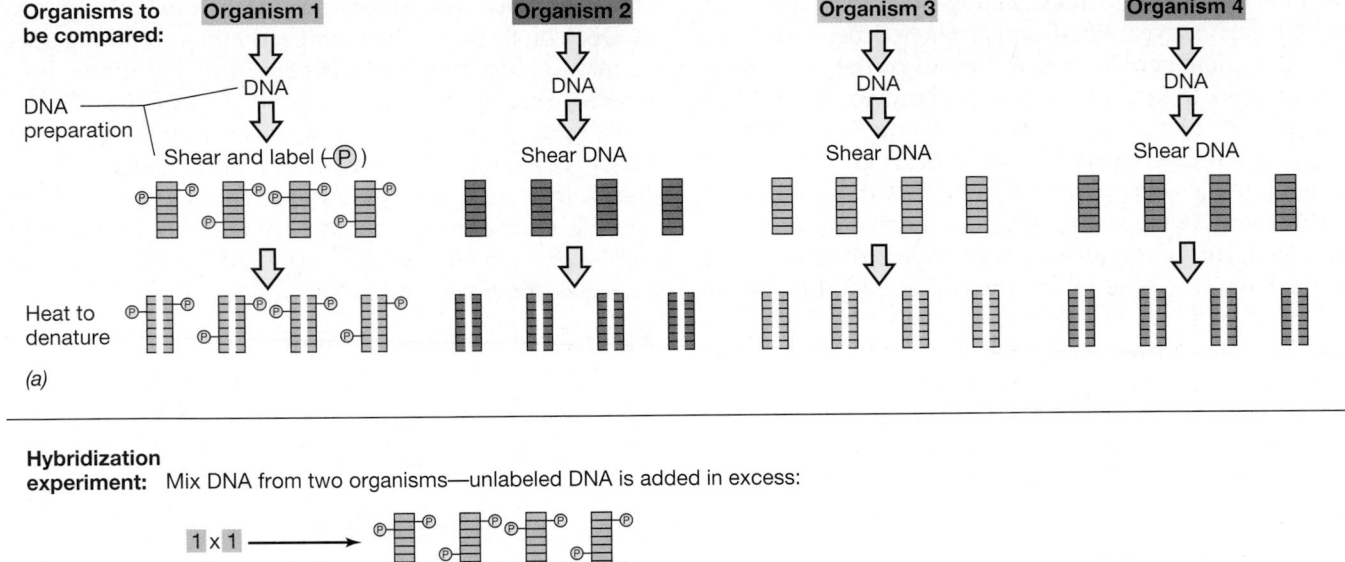

(a)

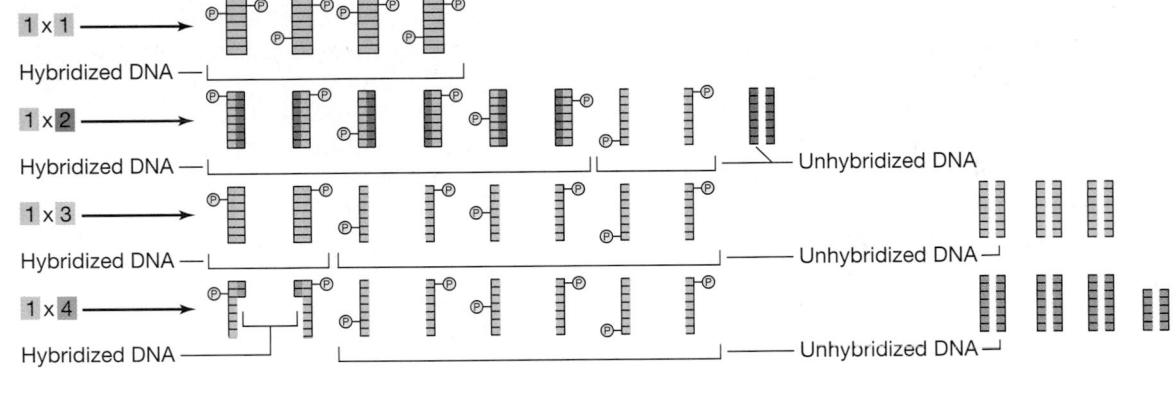

(b)

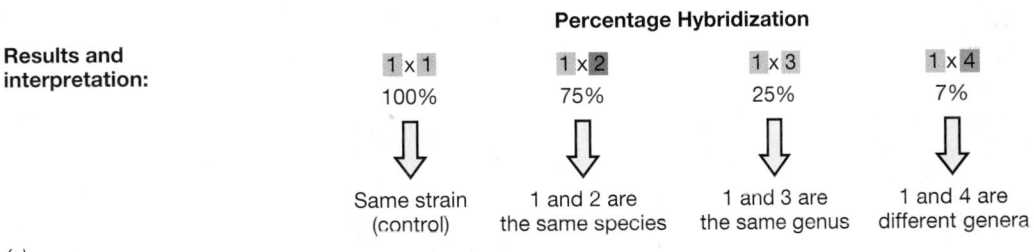

(c)

Figure 11.19 Genomic hybridization as a taxonomic tool. (a) DNA is isolated from test organisms. One of the DNAs is labeled (shown here as radioactive phosphate in the DNA of Organism 1). (b) Actual hybridization experiment. All combinations are tried and excess unlabeled DNA is added in each experiment to prevent labeled DNA from reannealing with itself. Following hybridization, hybridized DNA is separated from unhybridized DNA before measuring radioactivity in the hybridized DNA only. (c) Results. Radioactivity in the control is taken as the 100% hybridization value.

There is no fixed convention as to how much hybridization between two DNAs is necessary to assign two organisms to the same taxonomic rank. However, hybridization values of 70% or above are recommended for considering two isolates to be of the same *species* while values of at least 20–30% are required to argue that two organisms should reside in same *genus* (see Section 11.11 for working definitions of a bacterial genus and species); hybridization of DNAs from to-tally unrelated organisms occurs at less than 10% (Figure 11.19).

DNA:DNA hybridization is a sensitive method for revealing subtle differences in the genetic makeup of two organisms and is therefore a useful technique for differentiating very closely related organisms. Indeed, a common application of genomic hybridization in taxonomic studies is to test two organisms that are suspected to be different species even though SSU ribosomal RNA sequencing (see Figure 11.22) and phenotypic analyses have failed to reveal significant differences between them.

Complete genome sequences, of course, will eventually make hybridization analyses unnecessary. When the genome of a given bacterium is sequenced, its complete assortment of genes can be compared with those of any other bacterium (∞ Chapter 15). At present, with only about 100 prokaryotic genomes completely sequenced, such comparisons are not possible on a grand scale. However, when this time comes, and it will be in the not too distant future, microbiologists will have the ultimate tool for molecular taxonomy. Until taxonomy is based on complete genome sequence comparisons, hybridization will remain useful as a rapid estimate of genetic similarity and thus taxonomic relationship.

Ribotyping

Ribotyping is a technique for bacterial identification that employs some of the methods previously discussed for ribosomal RNA-based phylogenetic characterizations (see Section 11.5). However, unlike comparative *sequencing* methods, ribotyping does not involve sequencing, but instead measures the unique pattern that is generated when DNA from a particular organism is subject to restriction enzyme digestion and the fragments separated and probed with a ribosomal RNA probe (Figure 11.20). Because differences in ribosomal RNA sequences between two organisms translate into the presence or absence of specific restriction enzyme cut sites, the restriction pattern of a particular bacterial species is unique (Figure 11.20); in fact, the method is so specific that it has been given the nickname "molecular fingerprinting."

In practice, ribotyping begins when either bulk DNA or DNA encoding 16S rRNA and related genes within the ribosomal RNA operon is PCR-amplified, treated with one or more restriction enzymes, separated by electrophoresis and then probed (∞ Chapters 7 and 10 for discussion of

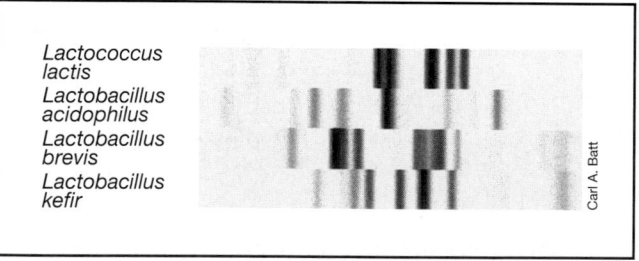

Lactococcus lactis
Lactobacillus acidophilus
Lactobacillus brevis
Lactobacillus kefir

Carl A. Batt

Figure 11.20 Ribotyping. Ribotype results from four different lactic acid bacteria. The pattern of DNA fragments generated from restriction enzyme digestion of DNA taken from a colony of each bacterium and then probed with 16S rRNA genes is unique to a species or even to strains within a species. The patterns generated on a gel with known organisms are digitized and stored in a database for comparisons in identifying environmental or clinical isolates. Variations in both *position* and *intensity* of the bands are important in identification.

these methods). The pattern generated from the fragments of DNA on the gel is then digitized and a computer used to make comparisons of this pattern with patterns from reference organisms in a database (Figure 11.20). Ribotyping is both a *rapid* (since it bypasses the actual sequencing, sequence alignment, and analysis requirements of ribosomal RNA sequencing methods) and *specific* method of bacterial identification and for these reasons has found many applications in clinical diagnostics and for the microbial analyses of food, water, and beverages.

Fatty Acid Analyses: FAME

Another popular method of bacterial identification is through characterization of the types and proportions of fatty acids present in membrane and outer membrane (gram-negative bacteria, ∞ Section 4.9) lipids of cells. Because the fatty acid composition of prokaryotes can be so variable, including differences in the fatty acids in terms of chain length, the presence or absence of unsaturated groups, rings, branched chains, or hydroxy groups (Figure 11.21*a*), the fatty acid profile of a particular bacterium can often be diagnostically useful.

For actual analyses, fatty acids, extracted from cell hydrolysates of a culture grown under standardized conditions, are chemically derivatized to form their corresponding methyl esters. These now volatile derivatives are then identified by gas chromatography. A chromatogram showing the types and amounts of fatty acids from the unknown bacterium are then compared with a database containing the fatty acid profiles of thousands of reference bacteria grown under the same conditions, and the best matches to that of the unknown selected (Figure 11.21*b*).

This technique has been nicknamed *FAME*, for *fatty acid methyl ester*, and is in widespread use in clinical, public health, and food and water inspection laboratories where the identification of pathogens or other bacterial hazards needs to be done on a routine basis. FAME analyses require rigid standardization, since fatty acid profiles in a single organism can vary as a function of temperature, growth phase (exponential versus stationary), and to a lesser extent, growth medium. Thus, for best results, it is necessary to grow the unknown organism on a specific medium and at a specific temperature in order to compare its fatty acid profile with those of organisms from the database that have been grown in the same way.

✓ 11.10 Concept Check

Molecular taxonomy includes molecular analyses of specific cell components. These include, among others, DNA:DNA hybridization, ribotyping, and fatty acid analyses.

✓ Hybridization of less than 10% between two organisms' DNA indicates that they are _____.

✓ How does ribotyping differ from 16S ribosomal RNA gene sequencing?

✓ What is FAME analysis?

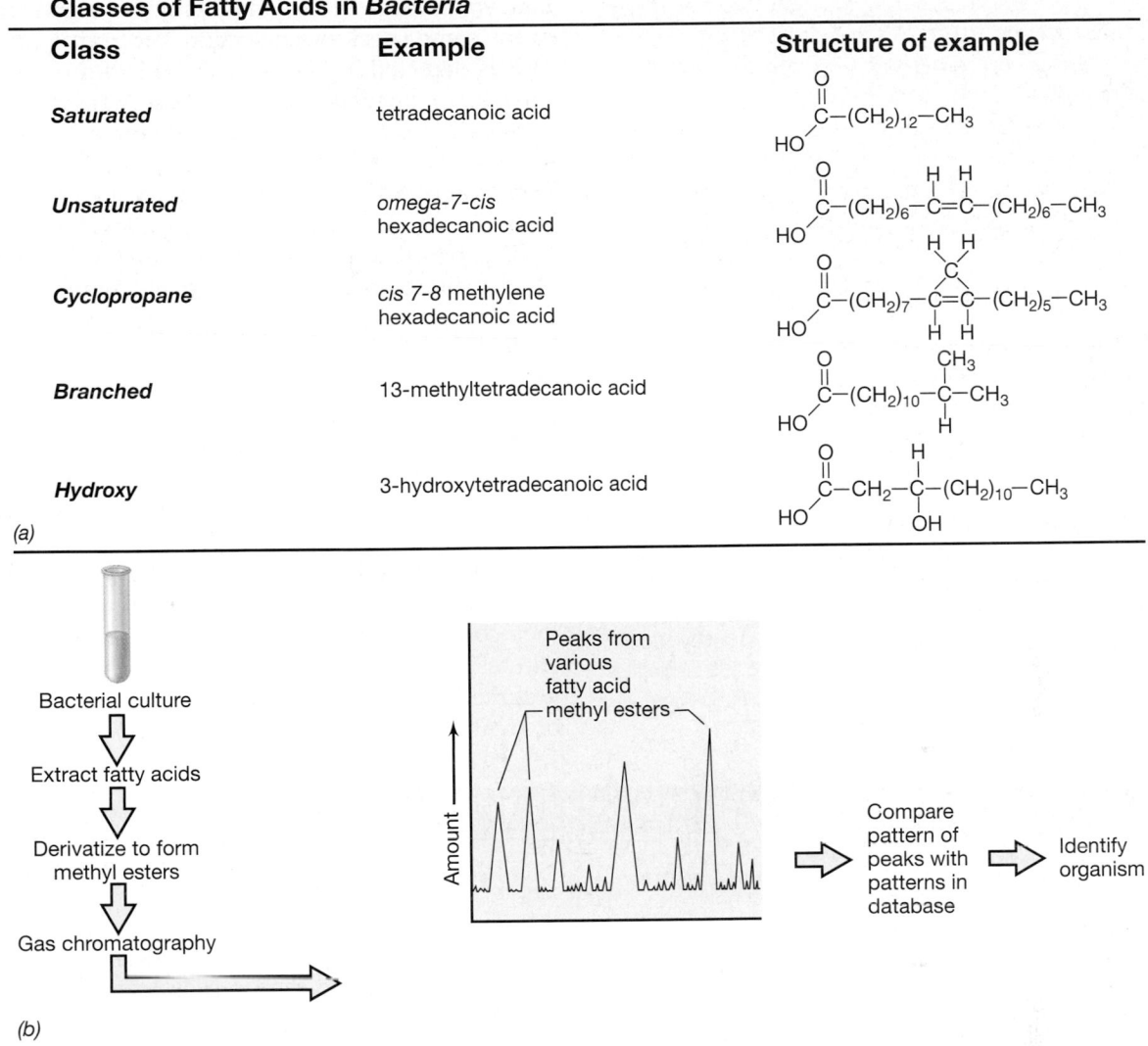

Figure 11.21 Fatty acid methyl ester (FAME) analysis in bacterial identification. (a) Classes of fatty acids in *Bacteria*. Only a single example is given of each class, but in actuality, more than 200 different fatty acids have been discovered from bacterial sources. A methyl ester contains a methyl group (CH_3) in place of the proton on the carboxylic acid group (COOH) of the fatty acid. (b) Procedure. Each peak from the gas chromatograph is due to one particular fatty acid methyl ester and the peak height is proportional to the amount.

11.11 The Species Concept in Microbiology

In the world of plants and animals, a species is considered a population of individuals that can interbreed under natural conditions, produce fertile offspring, and that is reproductively isolated from other populations. However, this definition does not hold for prokaryotes. Prokaryotes are haploid and reproduce asexually, so concepts like "the production of fertile offspring" have no meaning. Nevertheless, microbiologists traditionally refer to "species" of bacteria by formal name and regularly give new isolates of bacteria genus and species names. So what *is* a bacterial species? Although the concept of a prokaryotic species is evolving, microbiologists today have turned to a combination of methods—so-called *polyphasic taxonomy*—to differeniate prokaryotic species on both genetic and phenotypic grounds. In terms of genetic criteria, polyphasic taxonomy includes, in particular, SSU ribosomal RNA sequencing and genomic hybridization.

Bacterial Species and Higher Taxa

It has been proposed that a prokaryote whose 16S ribosomal RNA sequence differs by more than 3% from that of all other organisms (that is, the sequence is less than 97% identical to any other sequence in the databases), should be considered a new species. Support for this proposal emerges from the observation that DNA from two bacteria whose 16S rRNA sequences are less than 97% identical typically hybridize to less than 70%, a min-

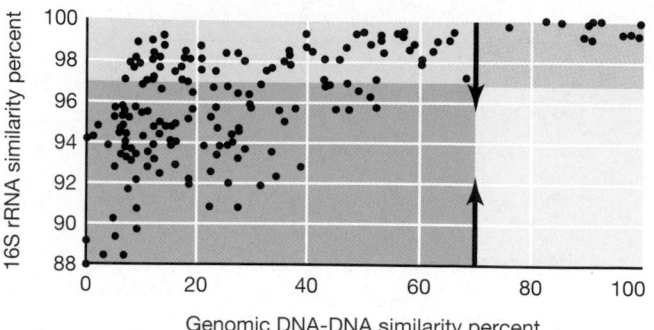

Figure 11.22 Relationship between 16S ribosomal RNA sequence similarity and genomic DNA hybridization between different pairs of organisms. These data are the results from several independent experiments with various species of the domain *Bacteria*. Points in the orange box represent combinations where 16S sequence similarity and genomic hybridization were both very high; thus, in each case, the two organisms tested were clearly the same species. By contrast, points in the green box represent combinations that indicate the two organisms tested were different species, and both methods show this. Points in the blue box indicate that the two test organisms were different species as measured by genomic DNA hybridization but not by 16S sequencing. Note how above 70% DNA hybridization, no 16S similarities were found of less than 97%. Data redrawn from Rosselló-Mora, R., and R. Amann. 2001. *FEMS Microbiol. Revs.* 25:39–67.

imal value considered evidence for two organisms being of the same species (see Section 11.8 and Figure 11.19). This is depicted in Figure 11.22. This figure also shows how 16S sequencing lacks the resolving power of genomic DNA hybridization for differentiating organisms at the species level. Although a 16S sequence that differs from that of all known organisms by greater than 3% is likely a new species as defined by the less than 70% DNA hybridization criterion, some organisms that have very similar ribosomal RNA sequences have genomes that are quite unrelated (Figure 11.22). Thus, new species could easily be missed if 16S sequencing alone were used as a taxonomic marker; therefore, at the species level, a polyphasic taxonomic approach is often needed. *Above the taxonomic rank of species, however, 16S sequencing is a powerful tool for classifying prokaryotes.*

A new species is usually defined from the characterization of several strains. The species concept is important in microbiology because it gives the collected strains formal taxonomic identity. Groups of species are then collected into genera (singular, **genus**). What constitutes a genus by molecular criteria is more a matter of judgment than for species, but 16S sequence differences of greater than 5–7% (93–95% identity) have been taken

TABLE 11.5 Taxonomic hierarchy for the purple sulfur bacterium *Allochromatium warmingii*

Taxonomic division	Name	Properties	Confirmed by
Domain	*Bacteria*	Prokaryotic cells; ribosomal RNA sequences typical of *Bacteria*	Microscopy; 16S ribosomal RNA sequencing; presence of unique biomarkers, for example, peptidoglycan
Phylum	Gamma Proteobacteria	Ribosomal RNA sequence typical of Proteobacteria	16S rRNA sequencing
Class	Zymobacteria	Gram-negative bacteria	Gram-staining, microscopy
Order	Chromatiales	Phototrophic purple bacteria	Characterizing pigments (👓 Figure 17.3)
Family	Chromatiaceae	Purple sulfur bacteria	Ability to oxidize H_2S and store S^0 within cells; observe culture microscopically for presence of S^0 (see photo)
Genus	*Allochromatium*	Rod-shaped purple sulfur bacteria	Microscopy (see photo)
Species	*warmingii*	Cells 3.5–4.0 μm × 5–11 μm; store sulfur mainly in poles of cell (see photo)	Measure cells in microscope using a micrometer; look for position of S^0 globules in cells (see photo)
Photograph of cells of *Allochromatium warmingii*:			

Sulfur (S^0) globules

Norbert Pfennig

as support for a new genus. Groups of genera are collected into **families**, families into **orders**, orders into **classes**, and so on up to the highest level taxon, the **domain** (Tables 11.4 and 11.5). Over 5200 species of *Bacteria* and *Archaea* are currently recognized (Table 11.5).

In the identification of an unknown organism, it is essential that an organism satisfy all the taxonomic criteria of ranks *above* its species designation. Thus, in the example given in Table 11.4 where we show the taxonomic hierarchy for the bacterium *Allochromatium warmingii*, all species of the genus *Allochromatium* must be rod-shaped purple sulfur gram-negative *Bacteria*; if one of these criteria is not met, the organism is not considered a species of *Allochromatium*. In other words, as one descends the taxonomic hierarchy from the level of domain to that of species, the criteria used to distinguish two organisms become less general and more specific (Table 11.4).

Bacterial Speciation

How do new bacterial species arise? To consider this, imagine a population of cells originating from the growth of a single cell and occupying a particular ecological niche. If cells in this population share a particular resource (for example, a key nutrient), this population of cells can be called an *ecotype*. Different ecotypes can coexist in a habitat, but each is only successful within its niche in the habitat. Within a given ecotype, occurrence of an adaptive mutation allows for periodic selection, with the old ecotype being purged by a population of the new ecotype (Figure 11.23). In this way, populations of cells eventually move away from each other, and repeated rounds of mutation and selection eventually leads to an ecotype that is sufficiently genetically distinct from the original ecotype to be recognized as a new species (Figure 11.23). This series of events within an ecotype has no effect on other ecotypes, since they are not competing for the same resources (Figure 11.23).

Bacterial speciation is also affected by lateral (horizontal) gene flow—the transfer of genes between species by conjugation, transduction, and transformation (∞ Chapter 10). As we know, prokaryotes are sexually promiscuous and can exchange genes across broad phylogenetic lines. Thus, a new genetic capability in an ecotype may arise from genes obtained from cells of another ecotype rather than from mutation and selection. However, actual speciation is thought to be more a result of mutation and periodic selection on genes within an organism's genome than the obtaining of new genes by lateral transfer. Unlike the main complement of genes in an organism, genes obtained by lateral transfer are usually few in number, commonly confer only a temporary benefit in the organism's niche, and can often be lost if the selective pressure to retain them decreases.

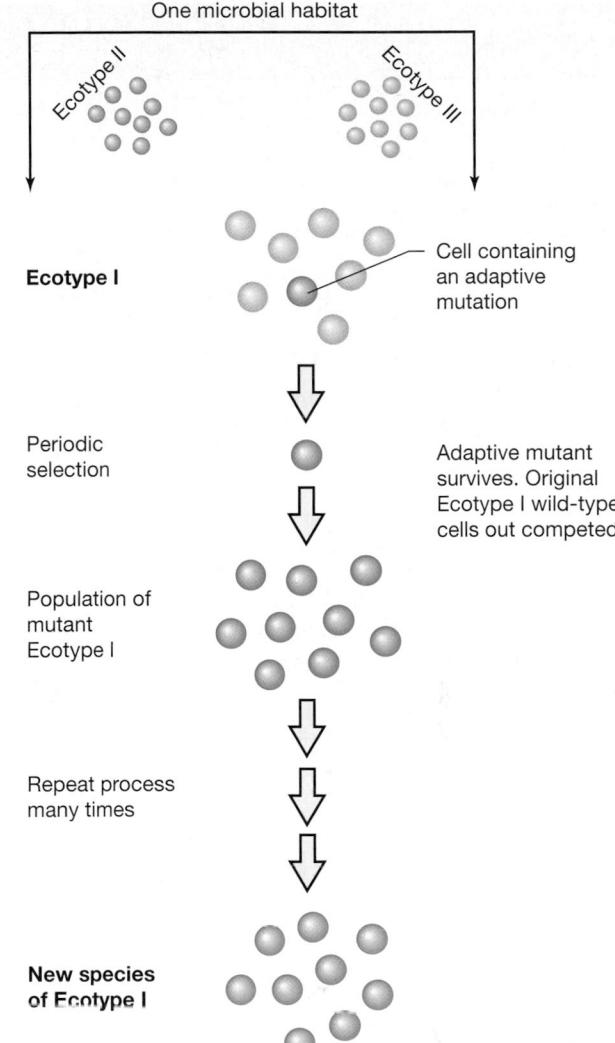

Figure 11.23 Possible mechanism of bacterial speciation. Several ecotypes can coexist in a single microbial habitat, each occupying their own prime ecological niche. When a beneficial mutation occurs within an ecotype, the cell containing the adaptive mutation will eventually form a population that will replace the original ecotype. As this occurs repeatedly within a given ecotype, a genetically distinct population of cells arises that represents a new species. Because other ecotypes do not compete for the same resources, they are unaffected by genetic and selection events that occur outside of their ecotype.

The net result of nearly 4 billion years of bacterial evolution (see Figure 11.6) are the prokaryotic species we see today. Microbiologists pretty much agree that they have no firm idea of how many prokaryotic species exist. Several thousands are already known (Table 11.6), and several thousands more, perhaps as many as 100,000–200,000 in total (or even more by some estimates) are suspected to exist. Under any circumstances, the scope of prokaryotic diversity is indeed enormous.

TABLE 11.6 Taxonomic ranks and numbers of known prokaryotic species[a]

Rank	Bacteria	Archaea	Total
Domains	1	1	2
Phyla	23	3[a]	26
Classes	32	8	40
Orders	77	12	89
Families	182	21	203
Genera	871	69	941
Species	5007	217	5224

[a]Numbers represent validly named genera and species of *Bacteria* and *Archaea* as of 2001. "Korarchaeota" is a provisional phylum.

Source: Garrity, G.M., Boone, D.R., and R.W. Castenholz (eds.). 2001. *Bergey's Manual of Systematic Bacteriology*, 2d ed., Vol. 1. Springer, New York.

✓ 11.11 Concept Check

The species concept applies to prokaryotes as well as eukaryotes, and a similar taxonomic hierarchy exists, with the domain as the highest level taxon. Bacterial speciation may occur due to repeated periodic selection for a favorable trait within an ecotype.

✓ If a given genus includes several _____, a _____ includes several genera.

✓ What is an ecotype?

11.12 Nomenclature and *Bergey's Manual*

Following the **binomial system** of nomenclature used throughout biology, prokaryotes are given genus and species epithets. The names are usually Latin or Latinized Greek derivations of some descriptive property appropriate for the species and are set in print in *italics*. For example, over 100 species of the genus *Bacillus* have been described, including *Bacillus (B.) subtilis, B. cereus, B. stearothermophilus*, and *B. acidocaldarius*. These species epithets mean "slender," "waxen," "heat-loving," and "acid-thermal," respectively, and in each case refer to key morphological, physiological, or ecological traits characteristic of each organism.

Culture Collections and Publication of New Taxa

When a new organism is isolated and thought to be unique, a decision must be made as to whether it is sufficiently different from other species to be described as novel, or perhaps even sufficiently different from all described genera to warrant description as a new genus (in which case a species is automatically created). To achieve formal taxonomic standing as a new genus or species, a detailed description of the isolate and the proposed name is published, and a viable culture of the organism is deposited in two culture collections, such as the American Type Culture Collection (ATCC, Manassas, Virginia, USA) or the Deutsche Sammlung von Mikroorganismen und Zellkulturen (DSMZ, German Collection for Microorganisms, Braunschweig, Germany) (∞ Section 30.1). The deposited strain becomes the *type* strain of the new species and the standard by which other strains thought to be the same can be compared.

Culture collections preserve the deposited culture, generally by freezing at very low temperatures (−80 to −196 °C) or freeze-drying. This practice differs from the botanical or zoological approach. These disciplines employ preserved (dead) specimens (either dried herbarium material or chemically fixed animal specimens) as the basis for comparison with proposed new species. Microbiologists rely on a *living type strain* that can be distributed to the scientific community, grown, and studied, and this approach allows for more detailed and reproducible comparisons, especially at the molecular level.

If the description of a new organism is published in a journal other than the *International Journal of Systematic and Evolutionary Microbiology (IJSEM)*, the official publication of record for the taxonomy and classification of prokaryotes and yeasts, a copy of the published paper must be submitted to this journal and the name validated before it is formally accepted as a new microbiological taxon. In each issue the *IJSEM* publishes an approved list of names, which formalizes newly proposed names and paves the way for their inclusion in *Bergey's Manual of Systematic Bacteriology* (Figure 11.24), a major taxonomic treatment of prokaryotes.

Figure 11.24 *Bergey's Manual of Systematic Bacteriology*, second edition. In five volumes this source describes the major properties of all known prokaryotes, both *Bacteria* and *Archaea*.

Bergey's Manual and The Prokaryotes

Although no source is recognized as *official* in the field of microbial taxonomy, *Bergey's Manual of Systematic Bacteriology* is the reference of choice. Widely used, *Bergey's Manual* (Figure 11.24) is a compendium of standard and molecular information on all recognized species of prokaryotes at the time of publication and contains a number of tables, figures, and other systematic information useful for identification purposes. The second edition of *Bergey's Manual*, published in five volumes to appear over several years beginning in 2001, has incorporated many of the concepts that have emerged from ribosomal RNA sequencing studies and blends this with a wealth of classical taxonomic information. A second major reference is a multivolume treatise called *The Prokaryotes*. This work, with more than 4100 pages in its second edition (1992), is now available in an online edition (http://www.prokaryotes.com) that will be revised periodically to reflect the rapid pace at which new information is being generated on prokaryotic taxonomy and phylogeny. Collectively, *Bergey's Manual* and *The Prokaryotes* offer microbiologists the foundations as well as the details of microbial taxonomy and phylogeny as we know it today, and are typically the first sources the practicing taxonomist seeks out when characterizing a newly isolated prokaryote.

✓ 11.12 Concept Check

Prokaryotes are given descriptive genus names and species epithets. Formal recognition of a new prokaryotic species requires depositing a sample of the organism in a culture collection center and official publication of the new species name and description. *Bergey's Manual of Systematic Bacteriology* is a major taxonomic compilation of *Bacteria* and *Archaea*.

✓ What is the *IJSEM* and what taxonomic function does it fulfill?

✓ Why might living cell material be of more use in taxonomy than preserved specimens?

Review Questions

1. What is the age of planet Earth? What is the age of the earliest known microfossils?

2. What major features would primitive organisms have had to have in order to replicate themselves, and why?

3. Discuss the role that iron–sulfur compounds and H_2 may have played in early life processes.

4. Why was the evolution of cyanobacteria of such importance to the further evolution of life on Earth?

5. What properties of RNA could have made possible an era of RNA life? If RNA life forms ever existed, what remnants of them may exist today?

6. What might have been the advantages of abandoning the era of RNA life for cellular life based on DNA, RNA, and protein?

7. Why are ribosomal RNAs better molecules for phylogenetic studies than proteins like ferredoxin, cytochromes, or specific enzymes?

8. What are signature sequences and of what phylogenetic value are they? How are signature sequences discerned?

9. What is FISH technology? Give an example of how it would be used.

10. Describe the methods involved in obtaining 16S rRNA sequences. What role does the polymerase chain reaction (PCR) play in molecular phylogeny?

11. What is *microbial community sampling* and what can this method tell us about the extent of microbial diversity in a given habitat?

12. What major evolutionary finding has emerged from the study of ribosomal RNA sequences? How did this modify the classic view of evolution? How has this discovery supported previous beliefs on the origin of eukaryotic organisms?

13. What major physiological and biochemical properties do *Archaea* share with *Eukarya*? With *Bacteria*?

14. What major phenotypic properties are used to group organisms in classical bacterial taxonomy? Which, if any, of these properties have phylogenetic predictive value?

15. Why aren't GC base ratios useful for making phylogenetic determinations? In what situations are GC base ratios of use in taxonomic studies?

16. How does ribotyping differ from 16S sequencing as an identification tool?

17. What is measured in FAME analyses?

18. Examine the following bacterial name: *Pseudomonas aeruginosa*. What part of this is the *species* epithet? What is the other name? In reference to taxonomic hierarchy, which of the two names might have several other names listed *under* it?

19. How is it thought that new bacterial species arise?

Application Questions

1. Compare and contrast the physical and chemical conditions on Earth at the time life first arose with conditions today. From a physiological standpoint, discuss at least two reasons why *animals* could not have existed on early Earth.

2. Why is it highly unlikely that life could originate today as it did billions of years ago?

3. Imagine that you are debating someone who is arguing against the theory of endosymbiosis. List five forms of evidence you would use to convince your opponent that endosymbiosis did occur. (You may wish to review Section 14.00 before writing your answer.)

4. On the basis of the following sequences, calculate an evolutionary distance between these three organisms and predict which two of the three are most closely related.

   Organism 1: AGGUACGUUA

   Organism 2: UGCCACGGUU

   Organism 3: AGGUACGGUA

   Sketch a phylogenetic tree that shows the approximate evolutionary relationships of these three organisms.

5. Imagine that you are doing lipid analyses of two prokaryotes along the lines shown in Figure 11.21. Your results on culture A show an abundance of short-chain unsaturated fatty acids. Analyses of cells in culture B show ether-linked phytanyl lipids to be present. Based on this information, to which phylogenetic domain do organisms A and B belong? Also, if you were told that these organisms were both extremophiles (∞ Section 2.4), and that one originated from a boiling hot spring and one from polar sea ice, which would be which? Finally, based on your lipid analyses and knowledge of where most lipid is located in a cell (∞ Section 3.4 and Figure 3.3), describe how the substances you detected might benefit each organism to thrive in its extreme environment. (You will likely need to review material in Sections 6.8–6.10 before answering.)

6. Determine the GC ratio of the following stretch of DNA:

   TAAGCCTGCAAGCTTAGCTA

   ATTCGGACGTTCGAATCGAT

7. What reference resource in your library would you check for information on the taxonomy and phylogeny of prokaryotes? If your library has both sources, compare their tables of contents. Which has the greater emphasis on classification and nomenclature? On enrichment, isolation, and culture?

T he phylogenetic tree of *Bacteria* contains about 20 phyla of cultured representatives and another 20–30 phyla known to exist but not yet cultured. Species of *Bacteria* show enormous diversity in terms of morphology, physiology, and evolutionary history. The giant sulfur bacterium *Achromatium*, shown here, is but one genus of chemolithotrophs, those prokaryotes that use *inorganic* compounds as their electron donors for energy metabolism. Several other physiological types of chemolithotrophs, as well as countless chemoorganotrophs and a major group of phototrophs (the purple bacteria) are scattered among subgroups of Proteobacteria, the largest phylum of *Bacteria*.

PROKARYOTIC DIVERSITY: *BACTERIA*

12

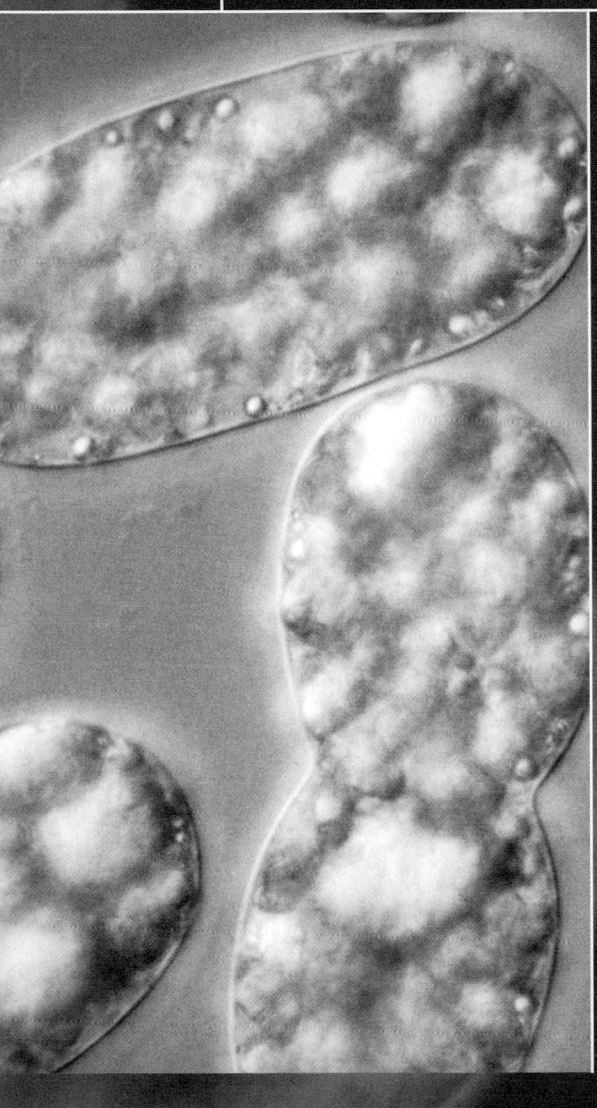

Working Glossary

Acid fastness a property of *Mycobacterium* species in which cells stained with the dye basic fuchsin resist decolorization with acidic alcohol

Carboxysomes polyhedral cellular inclusions of crystalline ribulose bisphosphate carboxylase (RubisCO), the key enzyme of the Calvin cycle

Chemolithotrophs organisms able to oxidize inorganic compounds (such as H_2, Fe^{2+}, S^0, or NH_4^+) as energy sources (electron donors)

Chlorosomes cigar-shaped structures bounded by a nonunit membrane and containing the light-harvesting bacteriochlorophyll (*c, d,* or *e*) in green bacteria and *Chloroflexus*

Consortia two or more-membered association of prokaryotes, usually living in an intimate symbiotic fashion

Cyanobacteria prokaryotic oxygenic phototrophs that contain chlorophyll *a* and phycobilins but not chlorophyll *b*

Enteric bacteria a large group of gram-negative rod-shaped *Bacteria* characterized by a facultatively aerobic metabolism

Green bacteria anoxygenic phototrophs containing chlorosomes and bacteriochlorophyll *c, c_s, d,* or *e* as light-harvesting chlorophyll

Heliobacteria anoxygenic phototrophs containing bacteriochlorophyll *g*

Hyperthermophile an organism with a growth temperature optimum of greater than 80°C

Heterocyst a differentiated cyanobacterial cell that carries out nitrogen fixation but not oxygenic photosynthesis

Heterofermentative in reference to lactic acid bacteria, capable of making more than one fermentation product

Homofermentative in reference to lactic acid bacteria, producing only lactic acid as a fermentation product

Methanotroph an organism capable of oxidizing methane (CH_4) as an electron donor in energy metabolism

Methylotroph an organism capable of oxidizing organic compounds that do not contain carbon-carbon bonds; if able to oxidize CH_4, also a methanotroph

Nitrifying bacteria chemolithotrophs capable of carrying out the transformation $NH_3 \rightarrow NO_2^-$, or $NO_2^- \rightarrow NO_3^-$

Purple nonsulfur bacteria a group of phototrophic prokaryotes containing bacteriochlorophylls *a* or *b* that grow best as photoheterotrophs and have a relatively low tolerance for H_2S

Prochlorophyte a prokaryotic oxygenic phototroph that contains chlorophylls *a* and *b* but lacks phycobilins

Prostheca an extrusion of cytoplasm often forming a distinct appendage, bounded by the cell wall

Proteobacteria a major lineage of *Bacteria* that contains a large number of gram-negative rods and cocci

Pseudomonad member of the genus *Pseudomonas,* a large group of gram-negative, obligately respiratory (never fermentative) *Bacteria*

Purple sulfur bacteria a group of phototrophic prokaryotes containing bacteriochlorophylls *a* or *b* and characterized by the ability to oxidize H_2S and store elemental sulfur inside the cells (or in the genus *Ectothiorhodospira,* outside the cell)

Spirochete a slender, tightly coiled gram-negative prokaryote characterized by possession of endoflagella used for motility

Stickland reaction fermentation of an amino acid pair in which one amino acid serves as an electron donor and a second serves as an electron acceptor

Sulfate-reducing bacteria a large group of anaerobic *Bacteria* that respire anaerobically with SO_4^{2-} as electron acceptor, producing H_2S

▌ THE PHYLOGENY OF BACTERIA

In the preceding chapter we stressed the evolutionary relationships among microorganisms. In this and the next two chapters we expand on these concepts with a discussion of properties of the major microbial groups themselves. With the thousands of different species of microorganisms known, obviously we will not be able to consider them all. So, using a phylogenetic tree as the focus of our discussion, we will examine particularly well-known cultured species and ones in which much phenotypic information is available. For more detailed information on prokaryotic diversity the student should refer to *Bergey's Manual of Systematic Bacteriology* and *The Prokaryotes* (👓 Section 11.10). In this chapter we focus on species of *Bacteria*, while in the next two chapters we focus on *Archaea* and microbial *Eukarya*, respectively.

12.1 Phylogenetic Overview of *Bacteria*

At least 17 major lineages (phyla) of *Bacteria* are known from the study of laboratory cultures, and many others have been identified from retrieval and sequencing of ribosomal RNA genes from *Bacteria* in natural habitats. Figure 12.1 gives a phylogenetic overview of *Bacteria*. The most phylogenetically ancient (least derived) phylum contains the genus *Aquifex* and relatives, all of which are hyperthermophilic H$_2$-oxidizing chemolithotrophs. Other nearby phyla such as *Thermodesulfobacterium*, *Thermotoga*, and the green nonsulfur bacteria (the *Chloroflexus* group), also contain thermophilic species.

Continuing past the green nonsulfur bacteria, we see the deinococci and relatives, the morphologically unique spirochetes, the phototrophic green sulfur bacteria, the chemoorganotrophic *Flavobacteria* and *Cytophaga* groups, the budding *Planctomyces-Pirella* and the *Verrucomicrobium* groups, the *Chlamydia*, and the genera *Nitrospira* and *Defferibacter* (Figure 12.1). Each of these groups is discussed in detail in this chapter.

The remaining phyla of cultured *Bacteria* constitute the major thrust of this chapter. They include the gram-positive bacteria, the cyanobacteria, and the Proteobacteria. Each of these is a large group containing many genera and are *Bacteria* about which much phenotypic information is known. The gram-positive bacteria can be separated into two subgroups, called *low* GC and *high* GC, the terms referring to the fact that the species tend to have DNA GC base ratios (see Section 11.9) either well below or well above 50%, respectively. The gram-positive bacteria are a large group of primarily chemoorganotrophic

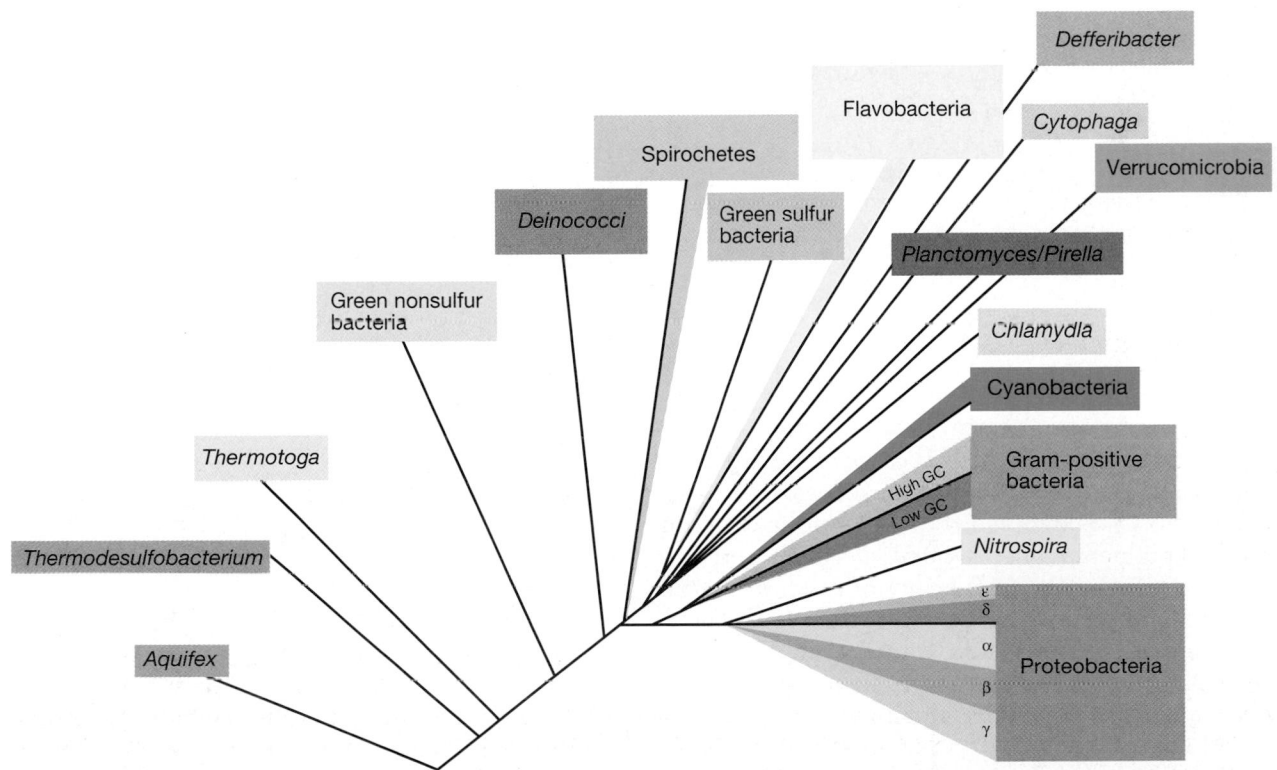

Figure 12.1 Detailed phylogenetic tree of the major lineages (phyla) of *Bacteria* based on 16S ribosomal RNA sequence comparisons.

Bacteria and are discussed in detail in Sections 12.19–12.24. The cyanobacteria are oxygenic phototrophic prokaryotes with evolutionary roots near those of the gram-positive *Bacteria*; these organisms are covered in Sections 12.25 and 12.26.

The final phylum on the tree of *Bacteria* is the *Proteobacteria* (Figure 12.1). This group is the largest and most physiologically diverse of all *Bacteria* (see Sections 12.2–12.18). The Proteobacteria contains five clusters containing several genera each, designated by the Greek letters *alpha*, *beta*, *gamma*, *delta*, and *epsilon* (see Table 12.1). Physiologically, Proteobacteria can be either phototrophic, chemolithotrophic, or chemoorganotrophic. Indeed, we will see in Chapter 17 the great diversity of energy-generating mechanisms characteristic of representatives of this group.

With this introduction to the phylogeny of the domain *Bacteria*, we proceed to a description of each phylum. We begin with the largest group of known *Bacteria*: the Proteobacteria.

II PHYLUM 1: PROTEOBACTERIA

Table 12.1 lists some of the key genera of Proteobacteria. As a group these organisms are all gram-negative, show extreme metabolic diversity, and represent the majority of known gram-negative bacteria of medical, industrial, and agricultural significance. We begin our discussion with *phototrophic* Proteobacteria—the purple bacteria.

12.2 Purple Phototrophic *Bacteria*

Key Genera

Chromatium
Ectothiorhodospira
Rhodobacter
Rhodospirillum

The purple phototrophic bacteria carry out *anoxygenic* photosynthesis; unlike the cyanobacteria (see Section 12.25) no O_2 is evolved. Purple bacteria contain chlorophyll pigments called *bacteriochlorophylls* and additionally contain any of a variety of *carotenoid* pigments. Together, these pigments give purple bacteria their spectacular colors, usually purple, red, or brown (Figure 12.2). We will examine the structure of these pigments and learn how they actually function in light-mediated energy generation (a process called *photophosphorylation*) in Chapter 17. The purple bacteria are a morphologically diverse group, and the taxonomy

TABLE 12.1 Major genera of Proteobacteria[a]

Subdivision	Genera	
Alpha	Acetobacter	Nitrobacter
	Agrobacterium	Paracoccus
	Alcaligenes	Rhodospirillum
	Azospirillum	Rhodopseudomonas
	Beijerinckia	Rhodobacter
	Bradyrhizobium	Rhodomicrobium
	Brucella	Rhodovulum
	Caulobacter	Rhodopila
	Ehrlichia	Rhizobium
	Gluconobacter	Rickettsia
	Hyphomicrobium	Sphingomonas
	Methylocystis	Zymomonas
Beta	Aquaspirillum	Oxalobacter
	Bordetella	Ralstonia
	Burkholderia	Rhodocyclus
	Chromobacterium	Rhodoferax
	Gallionella	Sphaerotilus
	Leptothrix	Spirillum
	Methylophilis	Thiobacillus
	Neisseria	Zoogloea
	Nitrosomonas	
Gamma	Acetobacter	Photobacterium
	Acinetobacter	Pseudomonas
	Azotobacter	Methylococcus
	Chromatium	Methylobacter
	Escherichia	Nitrosococcus
	Ectothiorhodospira	Thiobacillus
	Erwinia	Thiomicrospira
	Francisella	Thiospirillum and
	Halomonas	other purple
	Halorhodospira	sulfur bacteria
	Legionella	Salmonella and other
	Leucothrix	enteric bacteria
	Methylomonas	Vibrio
	Oceanospirillum	Xanthomonas
Delta	Acinetobacter	Geobacter
	Aeromonas	Halomonas
	Bdellovibrio	Moraxella
	Desulfuromonas	Myxococcus and other
	Desulfovibrio and most	myxobacteria
	other sulfate-	Pelobacter
	reducing bacteria	Syntrophobacter
	Francisella	
Epsilon	Campylobacter	Thiovulum
	Helicobacter	Wolinella
	Thiomicrospira	

[a] This table is not meant to be inclusive but only lists well-described genera of Proteobacteria. For a complete list of Proteobacteria and genera of other lineages of *Bacteria*, see Appendix 2.

of these organisms has been established along phylogenetic, morphological, and physiological lines.

Purple bacteria synthesize intracytoplasmic photosynthetic membrane systems into which their pigments are inserted. These membranes can be of various morphologies (Figure 12.3) but in all cases originate from invaginations of the cytoplasmic membrane. These

Norbert Pfennig

Figure 12.2 Photograph of liquid cultures of phototrophic purple bacteria showing the color of species with various carotenoid pigments. The blue culture is a carotenoid-less mutant derivative of *Rhodospirillum rubrum* showing how bacteriochlorophyll *a* is actually *blue* in color. The bottle on the far right (*Rhodobacter sphaeroides* strain G) lacks one of the carotenoids of the wild type and thus is more green in color.

internal membranes allow purple bacteria to increase their specific pigment content and to thus better utilize the available light; when cells are grown at high light intensities, internal membranes are few and pigment contents low, while at low light intensities, the cells are packed with membranes and photopigments.

Purple Sulfur *Bacteria*

Purple bacteria that utilize hydrogen sulfide (H_2S) as an electron donor for CO_2 reduction in photosynthesis are known as *purple sulfur* bacteria (Table 12.2). The sulfide is oxidized to elemental sulfur (S^0) that is stored in globules inside the cells (Figure 12.4); the sulfur later disappears as it is oxidized to sulfate. Many purple sulfur bacteria can also use other reduced sulfur compounds as photosynthetic electron donors, thiosulfate ($S_2O_3^{2-}$) being a key one commonly used to grow laboratory cultures of these organisms. All purple sulfur bacteria discovered thus far group with the gamma Proteobacteria.

Purple sulfur bacteria are generally found in illuminated anoxic zones of lakes and other aquatic habitats where H_2S accumulates, and also in "sulfur springs," where geochemically or biologically produced H_2S can trigger the formation of blooms of purple sulfur bacteria (Figure 12.5). The most favorable lakes for development of purple sulfur bacteria are *meromictic* (permanently stratified) lakes. Meromictic lakes stratify because they have denser (usually saline) water in the bottom and less dense (usually freshwater) nearer the surface. If sufficient sulfate is present to support sulfate reduction, the sulfide, produced in the sediments, diffuses upward into the anoxic bottom waters, and here purple sulfur bacteria can form massive blooms, usually in association with green sulfur phototrophic bacteria (Figure 12.5c).

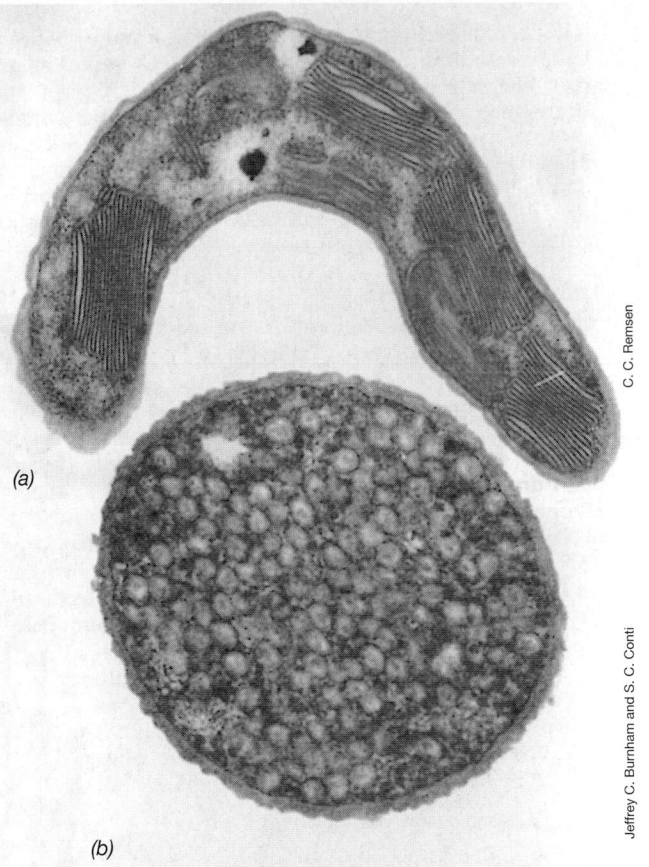

C. C. Remsen

(a)

(b)

Jeffrey C. Burnham and S. C. Conti

Figure 12.3 Membrane systems of phototrophic purple bacteria as revealed by the electron microscope. (a) Purple phototrophic bacterium, *Ectothiorhodospira mobilis*, showing the photosynthetic membranes in flat sheets (lamellae). (b) *Allochromatium vinosum*, another purple phototrophic bacterium, showing the membranes as individual, spherical-shaped vesicles.

The genera *Ectothiorhodospira* and *Halorhodospira* are of special interest because, unlike other purple sulfur bacteria, these organisms oxidize H_2S and produce S^0 *outside* of the cell (Figure 12.4f) but also because some species are extremely halophilic (salt-loving) and are among the most halophilic of all known prokaryotes. These organisms are typically found in marine environments, saline lakes, soda lakes, and salterns.

Purple Nonsulfur *Bacteria*

These bacteria have been called "nonsulfur" because it was originally thought that they were unable to use sulfide as an electron donor for the reduction of CO_2 to cell material. However, sulfide *can* be used by most species, although the levels of sulfide utilized well by purple *sulfur* bacteria are toxic to most purple *nonsulfur* bacteria. Some purple nonsulfur bacteria can also grow anaerobically in the dark using fermentative or anaerobic respiratory metabolism, and most can grow aerobically in darkness by respiration. Under the latter conditions, the electron donor can be an organic com-

TABLE 12.2 Genera and characteristics of purple sulfur bacteria[a]

Characteristics	Genus	Number of species	DNA (mol % GC)
Sulfur deposited externally:			
Spirilla, polar flagella	Ectothiorhodospira	9	62–67
Sprilla, extreme alkaliphiles	Thiorhodospira	1	57
Spirilla, extreme halophiles	Halorhodospira	3	50–69
Sulfur deposited internally:			
Do not contain gas vesicles			
Ovals or rods, polar flagella	Chromatium; Allochromatium; Halochromatium; Rhabdochromatium; Thermochromatium; Isochromatium; Marichromatium	23	48–70
Spheres, alkaliphilic	Thioalkalicoccus	1	64
Spheres, contain bateriochlorophyll *b*	Thioflavicoccus	1	66
Spheres	Thiorhodococcus	1	67
Spheres, diplococci, tetrads, nonmotile; cells 1.2–3 μm in diameter	Thiocapsa	5	63–70
Spheres or ovals, polar flagella; cells 2.5–3 μm in diameter	Thiocystis	4	61–68
Spheres, 1.5–2.5 μm in diameter	Thiohalocapsa	1	66
Spheres, 1–2 μm in diameter	Thiorhodococcus	1	67
Spheres, 1.2–1.5 μm in diameter	Thiococcus	1	69
Large spirilla, polar flagella	Thiospirillum	1	45
Small spirilla	Thiorhodovibrio	1	61–62
Contain gas vesicles			
Irregular spheres, ovals, nonmotile	Amoebobacter	4	63–65
Irregular spheres forming platelets of 4–16 cells	Thiolamprovulum	1	
Rods	Lamprobacter	1	64
Spheres, ovals, polar flagella	Lamprocystis	1	64
Rods, nonmotile; forming irregular network	Thiodictyon	2	65–66
Spheres, nonmotile; forming flat sheets of tetrads	Thiopedia	1	62–64

[a] From a phylogenetic standpoint, all are members of the gamma subdivision of the Proteobacteria (Figure 12.1).

pound or in some species even an inorganic compound, such as H_2. However, it is the great ability of this group to practice *photoheterotrophy* (where light is the energy source and an organic compound is the carbon source, ∞ Figure 17.1), that likely accounts for their competitive success in nature. Purple nonsulfur bacteria are typically nutritionally diverse in this regard, using fatty, organic, or amino acids; sugars; alcohols; and even aromatic compounds like benzoate as carbon sources. Most can also grow photoautotrophically with (CO_2 + H_2) or (CO_2 + low levels of H_2S).

Enrichment and isolation of purple nonsulfur bacteria is easy using a mineral salts medium supplemented with an organic acid as carbon source. Such media, inoculated with a mud, lake water, or sewage sample and incubated anoxically in the light, invariably select for purple nonsulfur bacteria. Enrichment cultures can be made even more selective by omitting fixed nitrogen sources (for example, NH_4^+) or organic nitrogen sources (for example, yeast extract or peptone) from the medium and supplying a gaseous headspace of N_2; virtually all purple nonsulfur bacteria can fix N_2 (∞ Section 17.28)

and will thrive under such conditions, usually outcompeting other organisms.

The morphological diversity of purple nonsulfur bacteria is typical of that of purple sulfur bacteria (Table 12.3 and Figure 12.6), and it is clearly a heterogeneous group in this regard. All purple nonsulfur bacteria isolated thus far are either alpha or beta Proteobacteria (Figure 12.1).

✓ 12.2 Concept Check

Purple bacteria are anoxygenic phototrophs that grow phototrophically, obtaining carbon from CO_2 + H_2S (purple sulfur bacteria) or organic sources (purple nonsulfur bacteria). Some purple nonsulfur bacteria are highly physiologically diverse and, collectively, the photoautotrophic activities of purple bacteria can be of great ecological significance. The purple bacteria reside in the alpha, beta, and gamma subdivisions of the Proteobacteria.

✓ What is meant by the term *anoxygenic*?

✓ Give a major reason why photosynthesis in purple nonsulfur bacteria does not occur under aerobic conditions.

✓ Can purple bacteria grow in the absence of light?

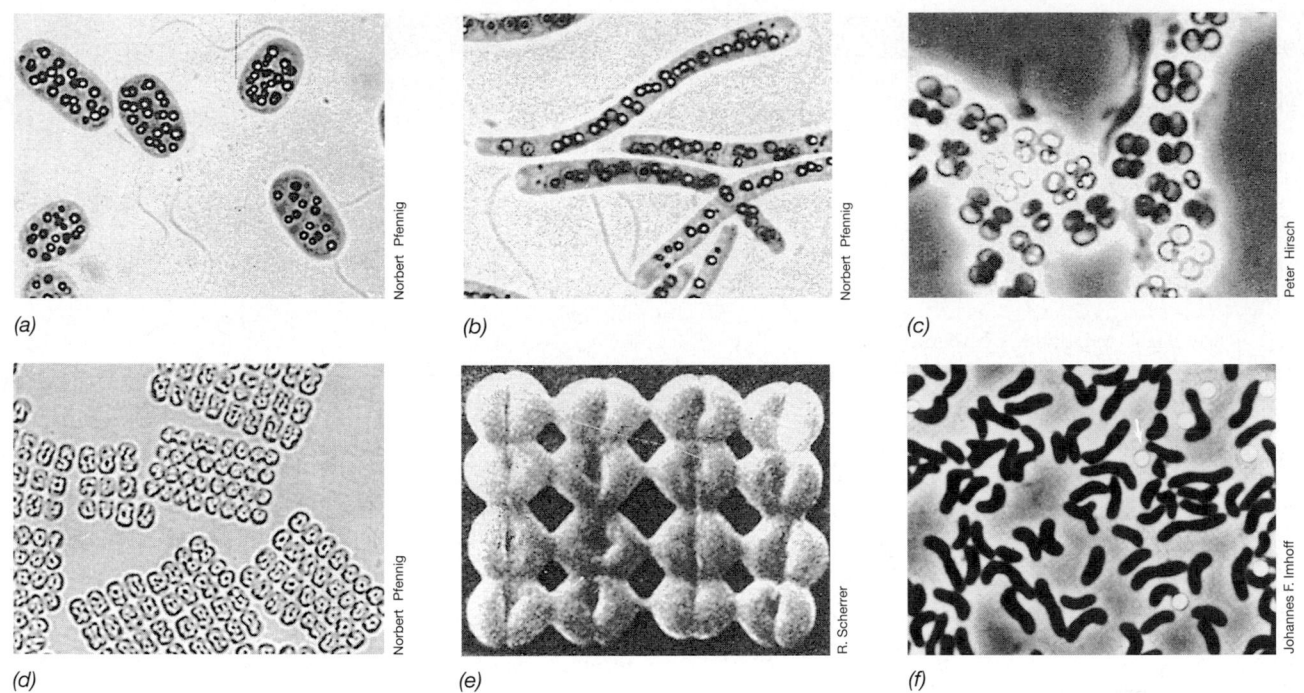

Figure 12.4 Bright-field photomicrographs of purple sulfur bacteria (see also Table 12.2). (a) *Chromatium okenii*; cells are about 5 μm wide. Note the globules of elemental sulfur inside the cells. (b) *Thiospirillum jenense*, a very large, polarly flagellated spiral; cells are about 30 μm long. Note the sulfur globules. (c) *Thiocapsa*; cells are about 2 μm wide. (d) *Thiopedia rosea*; cells are about 1.5 μm wide. (e) Scanning electron micrograph of a sheet of 16 cells of *Thiopedia rosea* showing the major division planes. (f) Phase micrograph of cells of *Ectothiorhodospira mobilis*. Cells are about 0.8 μm wide. Note external sulfur globules (arrow). Compare the photo of *Chromatium okenii* with the drawings of purple sulfur bacteria made by the great Russian microbiologist, Sergei Winogradsky, over 115 years ago (◷◷ Section 1.6 and Figure 1.15). Although Winogradsky never obtained pure cultures of these organisms, he studied their natural history and transformation of sulfur compounds.

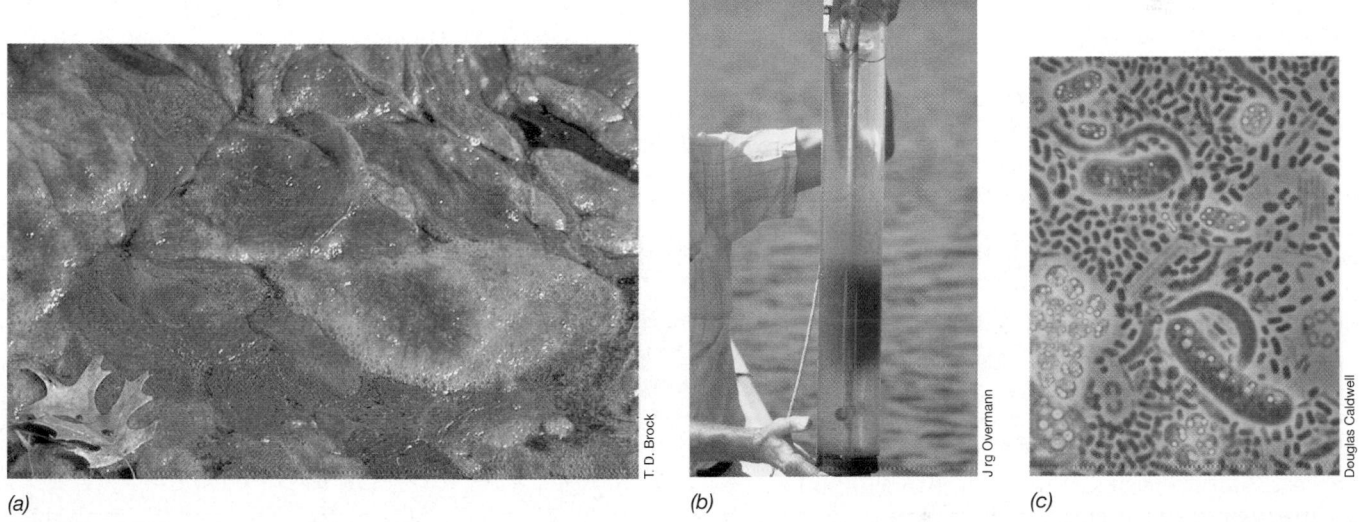

Figure 12.5 Blooms of purple sulfur bacteria. (a) *Thiopedia roseopersicinia*, in a sulfide spring in Madison, Wisconsin. The bacteria grow near the bottom of the spring pool and float to the top (by virtue of their gas vesicles) when disturbed. The green color is from cells of the eukaryotic alga *Spirogyra*. (b) Sample of water from 7 m in Lake Mahoney, British Columbia. The major organism is *Amoebobacter purpureus*. (c) Phase-contrast photomicrograph of layers of purple sulfur bacteria from a small stratified lake in Michigan. The purple sulfur bacteria include *Chromatium* species (large rods) and *Thiocystis* (small cocci).

TABLE 12.3   Genera and characteristics of purple nonsulfur bacteria

Characteristics	Genus	Number of species	16S rRNA group[a]	DNA (mol % GC)
Spirilla, polarly flagellated	*Rhodospirillum;* *Phaeospirillum;* *Rhodovibrio;* *Rhodothalassium;* *Roseospira;*	15	Alpha	62–68
	Rhodospira;	1	Alpha	66
	Roseospirillum	1	Alpha	71
Rods, polarly flagellated; divide by budding	*Rhodopseudomonas;*	15	Alpha	64–72
	Rhodoplanes;	2	Alpha	66–69
	Rhodobium	2	Alpha	61–65
Rods; divide by binary fission	*Rhodobacter*	8	Alpha	62–71
Ovoid to rod-shaped cells	*Rhodovulum*	4	Alpha	64–68
Ovals, peritrichously flagellated; growth by budding and hypha formation	*Rhodomicrobium*	1	Alpha	61–63
Large spheres, acidophilic (pH 5 optimum)	*Rhodopila*	1	Alpha	66
Small spheres, akaliphilic (pH 9 optimum)	*Rhodobaca*	1	Alpha	59
Ring-shaped or spirilla	*Rhodocyclus*	3	Beta	64–66
Curved rods	*Rubrivivax*	1	Beta	70–72
Curved rods	*Rhodoferax*	2	Beta	59–60

[a] All are members of the Proteobacteria (see Figure 12.1 and Table 12.1).

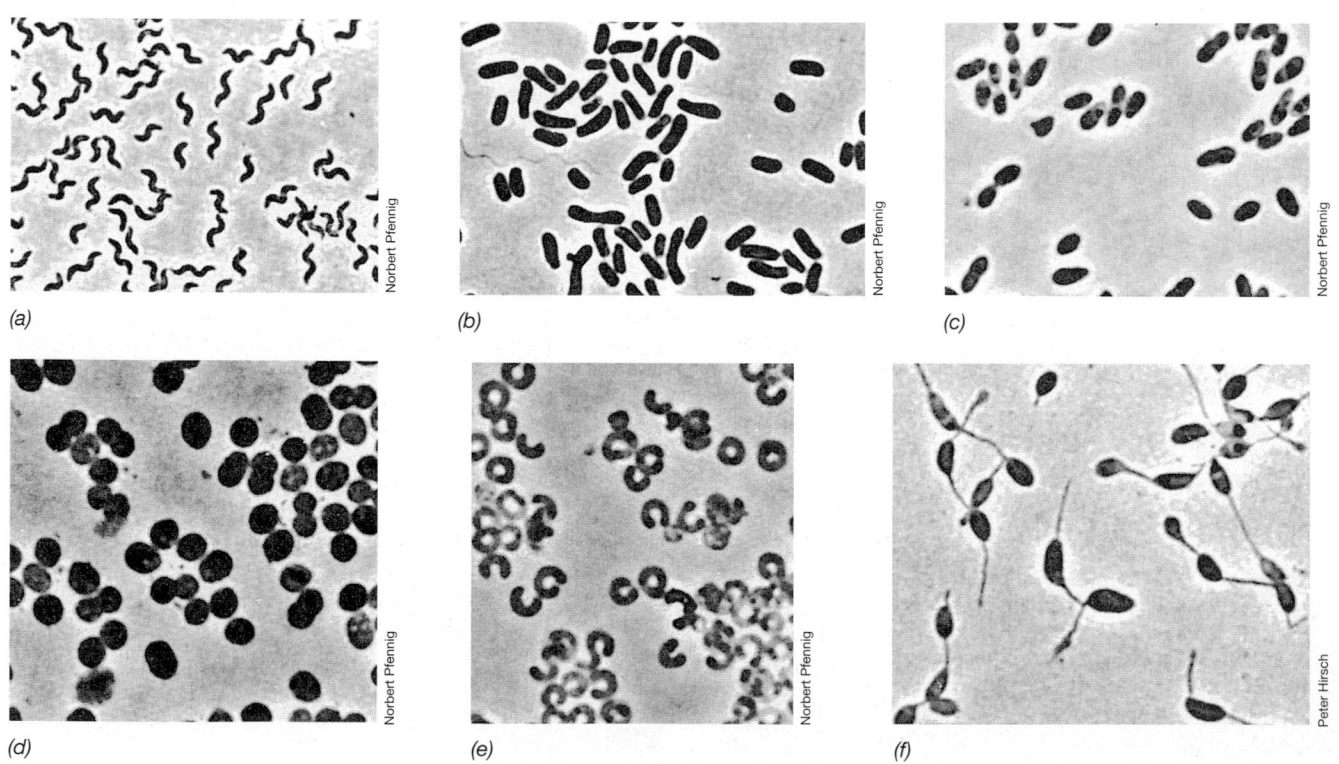

(a)   (b)   (c)

(d)   (e)   (f)

Figure 12.6   Representatives of several genera of purple nonsulfur bacteria (see also Table 12.3). (a) *Phaeospirillum fulvum;* cells are about 3 μm long. (b) *Rhodopseudomonas acidophila;* cells are about 4 μm long. (c) *Rhodobacter sphaeroides;* cells are about 1.5 μm wide. (d) *Rhodopila globiformis;* cells are about 1.6 μm wide. (e) *Rhodocyclus purpureus;* cells are about 0.7 μm in diameter. (f) *Rhodomicrobium vannielii;* cells are about 1.2 μm wide. All purple nonsulfur bacteria have been shown to be capable of aerobic dark as well as phototrophic (anoxic, light) growth, but some species, such as *P. fulvum*, can only respire at reduced oxygen tensions (that is, microaerophilic conditions, ∞ Section 6.13).

TABLE 12.4 Characteristics of the nitrifying bacteria

Characteristics	Genus	Phylogenetic group[a]	DNA (mol% GC)	Habitats
Oxidize ammonia:				
Gram-negative short to long rods, motile (polar flagella) or nonmotile; peripheral membrane systems	*Nitrosomonas*	Beta	45–53	Soil, sewage, freshwater, marine
Large cocci, motile; vesicular or peripheral membranes	*Nitrosococcus*	Gamma	49–50	Freshwater, marine
Spirals, motile (peritrichous flagella); no obvious membrane system	*Nitrosospira*	Beta	54	Soil
Pleomorphic, lobular, compartmented cells; motile (peritrichous flagella)	*Nitrosolobus*	Beta	54	Soil
Slender, curved rods	*Nitrosovibrio*	—	54	Soil
Oxidize nitrite:				
Short rods, reproduce by budding, occasionally motile (single subterminal flagellum); membrane system arranged as a polar cap	*Nitrobacter*	Alpha	59–62	Soil, freshwater, marine
Long, slender rods, nonmotile; no obvious membrane system	*Nitrospina*	Delta	58	Marine
Large cocci, motile (one or two subterminal flagella); membrane system randomly arranged in tubes	*Nitrococcus*	Gamma	61	Marine
Helical to vibrioid-shaped cells, nonmotile; no internal membranes	*Nitrospira*	Nitrospira group	50	Marine, soil

[a] Phylogenetically, all nitrifying bacteria thus far examined are Proteobacteria, except for *Nitrospira*, which constitutes its own phylogenetic lineage (Figure 12.1).

12.3 The Nitrifying *Bacteria*

Key Genera

Nitrosomonas
Nitrobacter

We will discuss in Chapter 17 the conceptual basis of chemolithotrophy. Chemolithotrophic bacteria are physiologically united by their ability to utilize *inorganic* electron donors as energy sources. Most chemolithotrophs are also capable of autotrophic growth and in this way share a major physiological trait with phototrophic bacteria and cyanobacteria. The best-studied chemolithotrophs are those capable of oxidizing reduced sulfur and nitrogen compounds, and the hydrogen-oxidizing bacteria; we focus on these groups here and in the next two sections.

Nitrosifyers and Nitrifyers

Bacteria able to grow chemolithotrophically at the expense of reduced inorganic nitrogen compounds are called **nitrifying bacteria**. Several genera are recognized on the basis of morphology and the particular steps in the oxidation sequences that they carry out (Table 12.4). Although morphologically heterogeneous, they are fairly tightly related phylogenetically, with the exception of the genus *Nitrospira*, which forms its own phylum of *Bacteria* (Figure 12.1 and see Section 12.38).

No chemolithotroph is known that will carry out the complete oxidation of ammonia to nitrate; thus, **nitrification** in nature results from the sequential action of two separate groups of organisms, the **ammonia-oxidizing bacteria**, the **nitrosifyers** (Figure 12.7), and the **nitrite-oxidizing bacteria**, the true **nitrifying** (nitrate-producing) bacteria (Figure 12.8). Nitrosifying bacteria typically have genus names beginning in "Nitroso," while true nitrifyers usually begin with "Nitro";

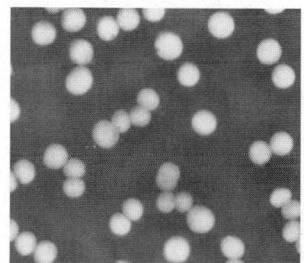
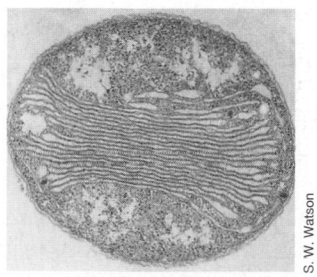

Figure 12.7 Phase contrast photomicrograph (left) and electron micrograph (right) of the nitrosifying bacterium *Nitrosococcus oceani*. A single cell is about 2 μm in diameter.

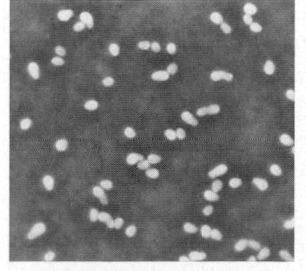
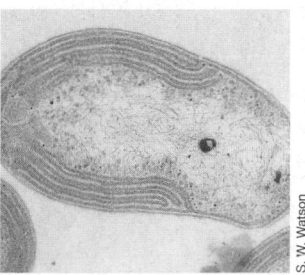

Figure 12.8 Phase contrast photomicrograph (left) and electron micrograph (right) of the nitrifying bacterium *Nitrobacter winogradskyi*. A cell is about 0.7 μm in diameter.

Nitrosomonas and *Nitrobacter* are major genera of nitrifying bacteria (Table 12.4). Historically, the nitrifying bacteria were the first organisms to be shown to grow chemolithotrophically; Winogradsky showed that they were able to produce organic matter and cell mass when provided with CO_2 as sole carbon source (∞ box, Winogradsky's Legacy, Chapter 17).

Many species of nitrifying bacteria have remarkably complex internal membrane systems, in many respects similar to the internal membranes found in their close relatives, the purple phototrophs (see Section 12.2) and the methane-oxidizing (methanotrophic) bacteria (see Section 12.6). The membranes are the location of a key enzyme in NH_3 oxidation, *ammonia monooxygenase*, which oxidizes NH_3 to hydroxylamine; the latter is further oxidized to NO_2^- by the nitrosifying bacteria (Figure 12.9) and we discuss the biochemistry of this process in more detail in Section 17.24. The NO_2^- generated in this reaction is oxidized to NO_3^- by the nitrifying bacteria (Table 12.4 and Figure 12.9).

Ecology, Isolation, and Culture

The nitrifying bacteria are widespread in soil and water. They are present in highest numbers in habitats where considerable amounts of ammonia are present, such as sites where extensive protein decomposition occurs (ammonification) and in sewage treatment facilities (∞ Section 28.2). Nitrifying bacteria develop especially well in lakes and streams that receive inputs of sewage or other wastewaters because these are frequently high in ammonia.

Enrichment cultures of nitrifying bacteria are readily established by using mineral salts media containing ammonia or nitrite as electron donor and bicarbonate (HCO_3^-) as sole carbon source. Because of the inefficiency of growth of these organisms (∞ Section 17.12), visible turbidity may not develop even after extensive nitrification has occurred, and so an easy means of monitoring growth is to assay for the production of nitrite (with ammonia as electron donor) or the disappearance of nitrite or production of nitrate (with nitrite as elec-

---

Nitrosifying bacteria
1. $NH_3 + O_2 + 2\,e^- + 2\,H^+ \longrightarrow NH_2OH + H_2O$
2. $NH_2OH + H_2O + \frac{1}{2}O_2 \longrightarrow NO_2^- + 2\,H_2O + H^+$
Sum: $NH_3 + 1\frac{1}{2}O_2 \longrightarrow NO_2^- + H_2O$
$\Delta G^{0\prime} = -275$ kJ/reaction

Nitrifying bacteria
$NO_2^- + \frac{1}{2}O_2 \longrightarrow NO_3^-$
$\Delta G^{0\prime} = -74.1$ kJ/reaction

Figure 12.9 Reactions involved in the oxidation of inorganic nitrogen compounds by chemolithotrophic nitrifying bacteria (∞ also Figures 17.32 and 17.33).

---

tron donor). Most of the nitrifying bacteria are obligate chemolithotrophs. Species of *Nitrobacter* are an exception and are able to grow chemoorganotrophically on acetate or pyruvate as sole carbon and energy source.

12.4 Sulfur- and Iron-Oxidizing *Bacteria*

Key Genera

Thiobacillus
Achromatium
Beggiatoa

The ability to grow chemolithotrophically on reduced sulfur compounds is a property of a diverse group of Proteobacteria (Table 12.5). Two broad ecological classes of sulfur-oxidizing bacteria can be discerned, those living at neutral pH and those living at acid pH. Some of the acidophiles also have the ability to grow chemolithotrophically using ferrous iron (Fe^{2+}) as electron donor. We discuss the biogeochemistry of acidophilic sulfur- and iron-oxidizing bacteria in Sections 19.13–19.16 and the biochemistry of these processes in Sections 17.10 and 17.11.

Thiobacillus and *Achromatium*

The genus *Thiobacillus* contains several gram-negative, rod-shaped bacteria, indistinguishable morphologically from most other gram-negative rods (Figure 12.10a), and are the best studied of the sulfur chemolithotrophs. Phylogenetically, species of *Thiobacillus* scatter among the Proteobacteria, with different species residing in the alpha, beta, and gamma subdivisions. The sulfur compounds most commonly used as electron donors in chemolithotrophic metabolism of *Thiobacillus* species are H_2S, S^0, and $S_2O_3^{2-}$, and the energy-yielding reactions involved are as follows:

$$H_2S + 2\,O_2 \rightarrow SO_4^{2-} + 2\,H^+$$
$$\Delta G^{0\prime} = -798\,\text{kJ/reaction}$$

$$S^0 + H_2O + 1\tfrac{1}{2}O_2 \rightarrow SO_4^{2-} + 2\,H^+$$
$$\Delta G^{0\prime} = -587\,\text{kJ/reaction}$$

$$S_2O_3^{2-} + H_2O + 2\,O_2 \rightarrow 2\,SO_4^{2-} + 2\,H^+$$
$$\Delta G^{0\prime} = -818\,\text{kJ/reaction}$$

It is obvious that large amounts of energy are released in these reactions, and we will see in Section 17.10 that some of this energy can be trapped as ATP from electron transport reactions leading to a proton motive force. Moreover, these reactions generate large amounts of sulfuric acid and thus several *Thiobacillus* species are acidophilic. One acidophilic species, *T. ferrooxidans*, can also grow chemolithotrophically by the oxidation of ferrous

TABLE 12.5 Physiological characteristics of sulfur-oxidizing chemolithotrophic prokaryotes

Genus and/or species	Inorganic electron donor	Range of pH for growth	Phylogenetic group	DNA (mol % GC)
Thiobacillus species growing poorly if at all in organic media:				
T. thioparus	H_2S, sulfides, S^0, $S_2O_3^{2-}$	6–8	Beta	61–66
T. denitrificans[b]	H_2S, S^0, $S_2O_3^{2-}$	6–8		63–68
T. neapolitanus	S^0, $S_2O_3^{2-}$	6–8		52–56
T. thiooxidans	S^0	2–4		51–53
T. ferrooxidans	S^0, metal sulfides, Fe^{2+}	2–4		55–65
Thiobacillus species growing well in organic media:				
T. novellus	$S_2O_3^{2-}$	6–8	Beta	66–68
T. intermedius	$S_2O_3^{2-}$	3–7		64
Filamentous sulfur chemolithotrophs:				
Beggiatoa	H_2S, $S_2O_3^{2-}$	6–8	Gamma	37–51
Thiothrix	H_2S	6–8	Gamma	52
Thioploca[c]	H_2S, S^0	—	Gamma	—
Other genera:				
Achromatium[c]	H_2S	—	Gamma	—
Thiomicrospira[a]	$S_2O_3^{2-}$, H_2S	6–8	Gamma	36–44
Thiosphaera[b]	H_2S, $S_2O_3^{2-}$, H_2	6–8	Alpha	66
Thermothrix[a]	H_2S, $S_2O_3^{2-}$, SO_3^-	6.5–7.5	Beta	—
Thiovulum	H_2S, S^0	6–8	Epsilon	—

[a] One of its species is capable of using NO_3^- anaerobically.

[b] Facultative aerobes; use NO_3^- as electron acceptor anaerobically.

[c] Pure cultures not yet available.

[d] *Thiosphaera pantotropha* has the exact same 16S rRNA sequence as *Paracoccus denitrificans*.

iron and is a major biological agent for the oxidation of this metal (∞ Section 19.14). Iron pyrite (FeS_2) is a major source of Fe^{2+} (as well as sulfide) and the oxidation of FeS_2, especially in mining operations, can be both beneficial, because leaching of the ore releases the Fe from the sulfide mineral, and ecologically disastrous, from acidification of the environment and the release of other heavy metals associated with the pyrite (∞ Sections 19.14 and 19.16).

Achromatium is a spherical sulfur-oxidizing chemolithotroph that is common in freshwater sediments containing sulfide. Cells of *Achromatium* are large cocci that can have diameters of 10–100 μm (Figure 12.10b). Phylogenetic analyses of natural populations (∞ Sections 11.6 and 18.5) of *Achromatium* have shown that several species likely exist (probably each of distinct size), although pure cultures of this organism have not yet been achieved. Phylogenetically, *Achromatrium* belongs to the gamma Proteobacteria and is specifically related to phototrophic purple bacteria, such as its phototrophic counterpart *Chromatium* (see Section 12.2). Like *Chromatium*, cells of *Achromatium* store elemental sulfur internally (Figure 12.10b); the granules later disappear as sulfur is oxidized to sulfate. Cells of *Achromatrium* also store large granules of calcite ($CaCO_3$) inside the cells (Figure 12.10b), possibly as a carbon source for growth.

Culture

Some sulfur chemolithotrophs are obligate chemolithotrophs, locked into a lifestyle of using inorganic instead of organic compounds as electron donors. When growing in this fashion they are also autotrophs, converting CO_2 into cell material by reactions of the Calvin cycle (∞ Section 17.6). Other sulfur chemolithotrophs are *facultative chemolithotrophs*, facultative in the sense that they can grow chemolithotrophically (and thus, also as autotrophs) or chemoorganotrophically (Table 12.5). In addition, however, there are organisms like *Beggiatoa*, most species of which can obtain energy from the oxidation of inorganic sulfur compounds but lack enzymes of the Calvin cycle and thus require organic compounds as carbon source. Such a nutritional lifestyle is called *mixotrophy*.

Beggiatoa

Organisms of this genus are filamentous, gliding sulfur-oxidizing bacteria, usually quite large in diameter and long, consisting of many short cells attached end to end (Figure 12.11); filaments then flex and twist so that many filaments may become intertwined to form a complex tuft. *Beggiatoa* is found in nature primarily in habitats rich in H_2S, such as sulfur springs (Figure 12.11b), decaying seaweed beds, mud layers of

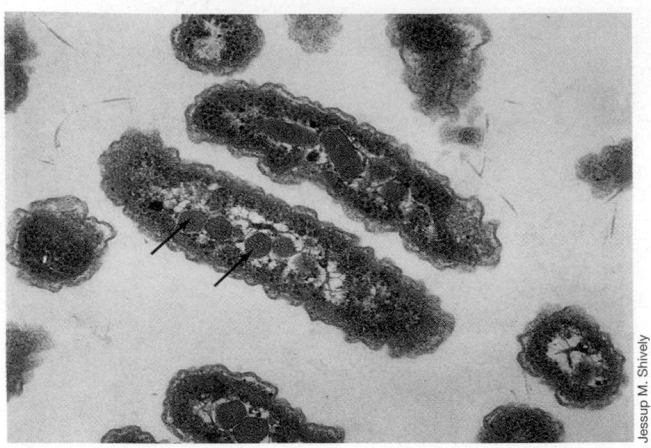

(a)

Jessup M. Shively

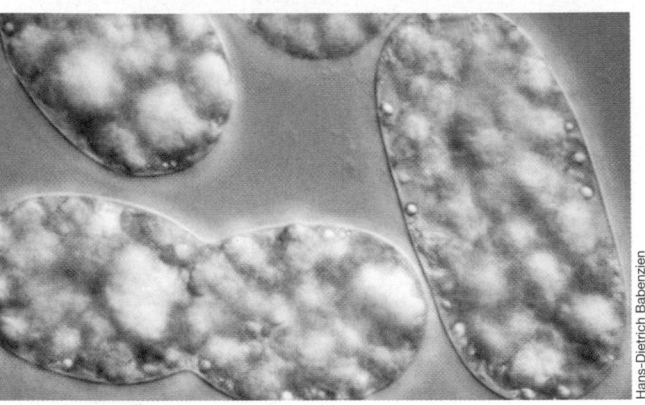

(b)

Hans-Dietrich Babenzien

Figure 12.10 (a) Nonfilamentous sulfur chemolithotrophs. Transmission electron micrograph of cells of the chemolithotrophic sulfur oxidizer *Thiobacillus neapolitanus*. A single cell is about 0.5 μm in diameter. Note the polyhedral bodies (carboxysomes) distributed throughout the cell (arrows). (b) *Achromatium*. Cells isolated from a small lake in Germany and photographed by Nomarski light microscropy. The small globular structures near the periphery of the cells (arrow) are elemental sulfur, while the large granules consist of calcium carbonate. A single *Achromatium* cell is about 25 μm in diameter.

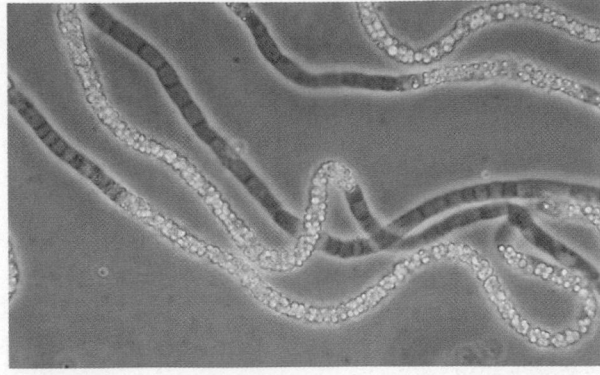

(a)

Michael Richard

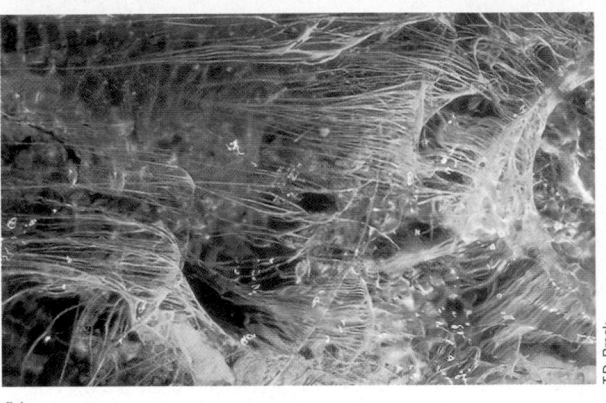

(b)

T.D. Brock

Figure 12.11 Filamentous sulfur-oxidizing bacteria. (a) Phase contrast photomicrograph of a *Beggiatoa* species isolated from a sewage treatment plant. Note the abundant elemental sulfur granules in some of the cells. (b) Sulfur-oxidizing bacteria in a small stream. The filamentous cells twist together to form thick streamers, and the white color is due to the abundant elemental sulfur content of the cells.

lakes, and waters polluted with sewage, and in these habitats the filaments of *Beggiatoa* are usually filled with sulfur granules (Figure 12.11*a*). *Beggiatoa* are also common inhabitants of hydrothermal vents (👓 Section 19.8). It was with *Beggiatoa* that Winogradsky first demonstrated that a living organism could oxidize H_2S to S^0 and then to SO_4^{2-}, leading him to formulate the concept of chemolithotrophy (👓 box, Winogradsky's Legacy, Chapter 17). Although a few strains of *Beggiatoa* are truly chemolithotrophic autotrophs, most grow best mixotrophically with reduced sulfur compounds as electron donors and organic compounds as carbon sources.

An interesting habitat of *Beggiatoa* is the rhizosphere of plants (rice, cattails, and other swamp plants) living in flooded, and hence anoxic, soils. Such plants pump oxygen down into their roots so a sharply defined oxic/anoxic boundary develops between the root and the soil. *Beggiatoa* (and probably other sulfur bacteria) develops at this boundary, and plays a beneficial role for the plant by oxidizing (and thus detoxifying) the H_2S.

Beggiatoa and other filamentous bacteria like *Sphaerotilus* (see Section 12.15) can cause major settling problems in sewage treatment facilities and in industrial waste lagoons such as from canning, paper pulping, brewing, and milling. These problems are generally referred to as *bulking* and occur when filamentous bacteria overgrow the normal flora of the waste system, producing a loose detrital floc instead of the normal and more easily settling tight floc containing organisms like *Zoogloea* (👓 Section 28.2). If bulking occurs, the wastewater remains improperly treated because the effluent discharged is still high in organic matter; in sewage treatment, for example, bulking occurs when *Beggiatoa* or other filamentous bacteria replace *Zoogloea* in the activated sludge process (👓 Section 28.2).

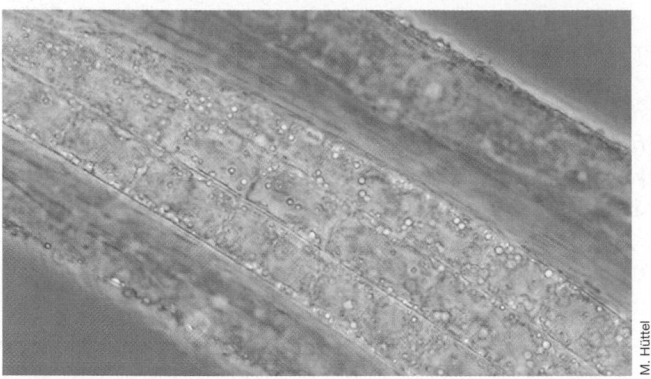

Figure 12.12 Cells of a large marine *Thioploca* species. Cells contain sulfur granules (yellow) and are about 40–50 μm wide.

(a)

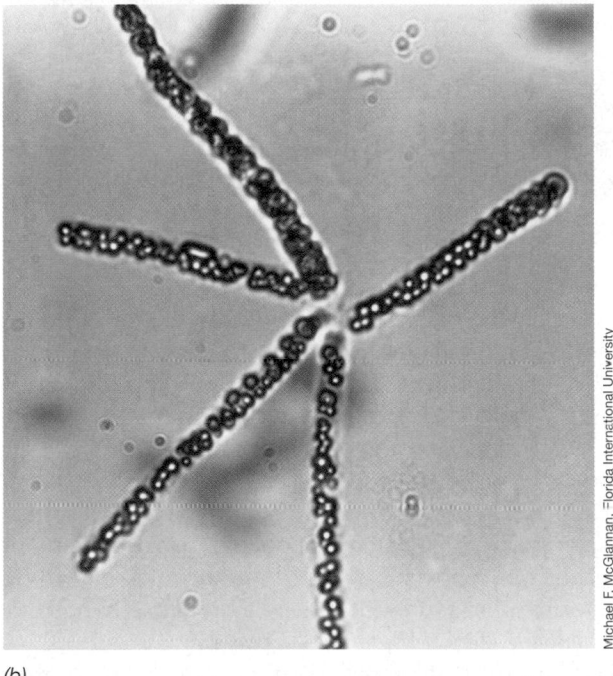

(b)

Figure 12.13 *Thiothrix.* (a) A sulfide-containing artesian spring in Florida (USA). The outside of the spring is coated with a mat of *Thiothrix.* (b) Phase contrast photomicrograph of a rosette of cells of *Thiothrix* isolated from the spring and grown in pure culture. Note the internal sulfur globules produced from the oxidation of sulfide.

Thioploca and *Thiothrix*

Other filamentous sulfur-oxidizing bacteria include *Thioploca* and *Thiothrix*. *Thioploca* is a very large, filamentous sulfur-oxidizing chemolithotroph that forms cell bundles surrounded by a common sheath (Figure 12.12). Thick mats of a marine *Thioploca* species have been found on the ocean floor off the coast of Chile and Peru. Studies on the ecology of these organisms have shown that they carry out the anoxic oxidation of H_2S coupled to the reduction of nitrate (NO_3^-) presumably to N_2 (denitrification) (∞ Sections 17.14 and 19.12). Interestingly, it has been shown that cells of *Thioploca* can accumulate huge amounts of nitrate intracellularly and that this nitrate can then support extended periods of anaerobic respiration with H_2S as electron donor. It is postulated that these marine *Thioploca* mats fix substantial amounts of CO_2 and also play a major role in sulfur and nitrogen cycling. Similar mats consisting primarily of *Beggiatoa* are found near hydrothermal vents (∞ Section 19.8), but the connection with nitrate respiration in these cases is not as well established.

Thiothrix is a filamentous sulfur-oxidizing organism in which the filaments group together at their ends by way of a holdfast to form rosettes (Figure 12.13). Physiologically, *Thiothrix* is an obligately aerobic mixotroph, and in this and most other respects resembles *Beggiatoa*.

12.5 Hydrogen-Oxidizing *Bacteria*

Key Genera

Ralstonia
Alcaligenes

A wide variety of bacteria are capable of growing with H_2 as sole electron donor and O_2 as electron acceptor using the "knallgas" reaction, the reduction of O_2 with H_2, as their energy metabolism:

$$H_2 + \tfrac{1}{2}O_2 \rightarrow H_2O \qquad \Delta G^{0\prime} = -237 \text{ kJ}$$

Many, but not all, of these organisms can also grow autotrophically (using reactions of the Calvin cycle to incorporate CO_2) and are grouped together here as the chemolithotrophic *hydrogen-oxidizing bacteria*. Both gram-positive and gram-negative hydrogen bacteria are known, with the best-studied representatives classified in the genera *Ralstonia* (Figure 12.14), *Pseudo-*

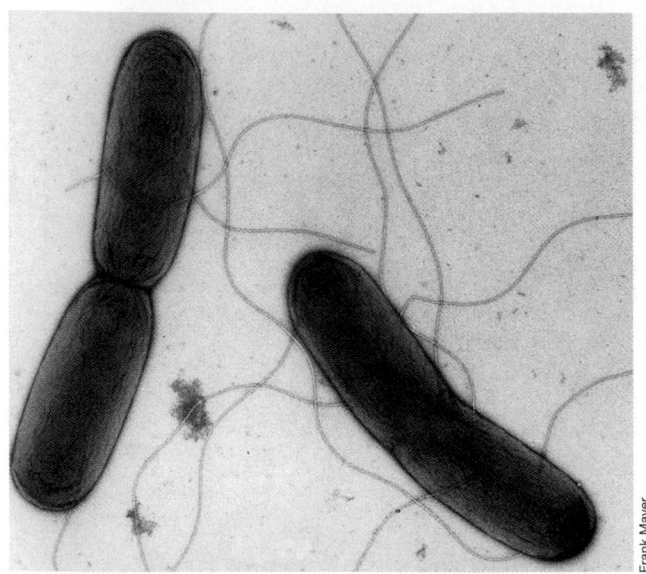

Figure 12.14 Transmission electron micrograph of negatively stained cells of the hydrogen-oxidizing chemolithotroph *Ralstonia eutropha*. A cell is about 0.6 μm in diameter and contains several flagella.

monas, *Paracoccus*, and *Alcaligenes* (Table 12.6). All hydrogen-oxidizing bacteria contain one or more *hydrogenase* enzymes that function to bind H_2 and use it either to produce ATP or for reducing power for autotrophic growth.

Almost all hydrogen bacteria are *facultative chemolithotrophs*, meaning that they can also grow chemoorganotrophically with organic compounds as energy sources. This is a major distinction between hydrogen chemolithotrophs and many sulfur chemolithotrophs or nitrifying bacteria; most representatives of the latter two groups are *obligate chemolithotrophs*—growth does not occur in the absence of the inorganic energy source. By contrast, hydrogen chemolithotrophs can switch between chemolithotrophic and chemoorganotrophic modes of metabolism and presumably do so in nature as nutritional conditions warrant.

Physiology and Ecology of Hydrogen *Bacteria*
Most hydrogen bacteria grow best under microaerobic conditions when growing chemolithotrophically on H_2 because hydrogenases are oxygen-sensitive enzymes. Typically, oxygen levels of about 5–10% support best growth. *Nickel* must be present in the medium for chemolithotrophic growth of hydrogen bacteria because virtually all hydrogenases contain Ni^{2+} as a metal cofactor. A few hydrogen bacteria also fix molecular N_2, and when growing on N_2, the organisms are quite oxygen-sensitive because the enzyme nitrogenase needed for the reduction of molecular nitrogen (∞ Section 17.28) is an oxygen-sensitive enzyme.

Hydrogen-oxidizing bacteria can be enriched if a small amount of mineral salts medium containing trace metals (especially Ni^{2+} and Fe^{2+}) is inoculated with soil or water and incubated in a large, sealed flask containing a head space of 5% O_2, 10% CO_2, and 85% H_2. When the liquid becomes turbid, plates of the same medium are streaked and incubated in a glass jar containing the same gas mixture (one must exercise care here, as mixtures of O_2 and H_2 are potentially explosive).

CO Oxidation
Some hydrogen bacteria can grow on carbon monoxide, CO, as energy source, with electrons from the oxidation of CO to CO_2 entering the electron transport chain to drive ATP synthesis. CO-oxidizing bacteria, which are called *carboxydotrophic* bacteria, grow autotrophically using Calvin cycle reactions (∞ Section 17.6) to fix the CO_2 generated from the oxidation of CO. CO is oxidized to CO_2 by the enzyme *carbon monoxide dehydrogenase*, which is a molybdenum-containing enzyme. The molybdenum in CO dehydrogenase is bound to a small cofactor consisting of a multiringed structure called a *pterin*, similar to the situation in the enzyme nitrate reductase (∞ Section 17.14).

CO consumption by carboxydotrophic bacteria is a significant ecological process. Although much CO is generated from various human and other sources, CO levels in air have not risen significantly over many years. Microbial CO consumption is probably the reason why. Because the most significant releases of CO (primarily from automobile exhaust, incomplete combustion of fossil fuels, and the catabolism of lignin) occur in oxic environments, carboxydotrophic bacteria in the upper layers of soil probably represent the most significant sink for CO in nature. Some of the best-studied carboxydobacteria include *Pseudomonas carboxydovorans*, *Bacillus schlegelii*, and *Alcaligenes carboxydus* (Table 12.6). At least one carboxydobacterium can grow on CO anaerobically with nitrate as electron acceptor, but this does not seem to be a widespread property of the group. Like the hydrogen bacteria, virtually all isolates of carboxydotrophic bacteria also grow chemoorganotrophically on organic substrates as well as on CO.

✓ 12.3–12.5 Concept Check
Chemolithotrophs are prokaryotes that can oxidize inorganic electron donors and in many cases use CO_2 as sole carbon source.

✓ Compare and contrast the nitrifying bacteria with the sulfur, iron, and hydrogen bacteria in terms of inorganic electron donors used, carbon sources, E_0' of electron donors (∞ Chapter 17), and habitats.

✓ What major pathway is present for assimilation of CO_2 in many chemolithotrophs?

TABLE 12.6 Differential characteristics of species of hydrogen-oxidizing bacteria

Genus and/or species	Denitri-fication	Growth on fructose	Motility	Phylogenetic group[a]	DNA (mol % GC)	Other characteristics
Gram-negative						
Acidovorax facilis	−	+	+	Beta	64	Membrane-bound hydrogenase
Ralstonia eutropha[a]	+	+	+	Beta	66	Membrane-bound and cytoplasmic hydrogenases
Alcaligenes xylosoxidans	−	+	+	Beta	—	Membrane-bound and cytoplasmic hydrogenases
Aquaspirillum autotrophicum	−	−	+	Beta	61	Only membrane-bound hydrogenase present
Pseudomonas carboxydovorans	−	−	+	Gamma	60	Only membrane-bound hydrogenase present; also oxidizes CO
Hydrogenophaga flava	−	+	+	Beta	67	Colonies are bright yellow
Seliberia carboxydohydrogena	−	?	+	Alpha	58	Also oxidizes CO
Paracoccus denitrificans	+	+	−	Alpha	66	Only membrane-bound hydrogenase present; strong denitrifier
Aquifex pyrophilus	+	−	+	Aquifex group[b]	65	Hyperthermophile, grows microaerophilically or anaerobically (with NO_3^-), obligate chemolithotroph; also uses S^0 or $S_2O_3^{2-}$ as electron donor
Hydrogenobacter thermophilus	−	−	−	Aquifex group[b]	37–46	As for *Aquifex*, but obligate aerobe (microaerophile)
Gram-positive						
Bacillus schlegelii	−	−	+	Low GC gram-positive[c]	66	Produces endospores; thermophile; also uses CO or $S_2O_3^{2-}$ as electron donor
Arthrobacter sp.	−	+	−	High GC gram-positive[d]	70	Only membrane-bound hydrogenase present
Mycobacterium gordonae	−	?	−	High GC gram-positive[e]	—	Acid-fast; colonies yellow to orange

[a] Aerobic hydrogen bacteria are Proteobacteria except as indicated.
[b] See Section 12.37.
[c] See Section 12.20.
[d] See Section 12.22.
[e] See Section 12.23.

12.6 Methanotrophs and Methylotrophs

Key Genera

Methylomonas
Methylobacter

Methane, CH_4, is found extensively in nature. It is produced in anoxic environments by methanogenic *Archaea* (∞ Sections 13.4 and 17.17) and is a major gas of anoxic muds, marshes (∞ Figure 13.5), anoxic zones of lakes, the rumen, and the mammalian intestinal tract. Methane is the major constituent of natural gas and is also present in many coal formations. It is a relatively stable molecule; but a variety of bacteria, the **methanotrophs**, oxidize it readily, utilizing methane and a few other one-carbon compounds as electron donors for energy generation and as sole sources of carbon. These bacteria are all aerobes and are widespread in nature in soil and water. They exhibit diverse morphologies and are related both phylogenetically and in their ability to oxidize methane.

C₁ Metabolism

In addition to methane, a number of other one-carbon compounds can be utilized by microorganisms. A list of these compounds is given in Table 12.7. From a biochemical viewpoint, these compounds share a key

TABLE 12.7 Substrates used by methylotrophic bacteria[a]

Substrates used for growth	Substrates oxidized but not used for growth
Methane, CH_4[b]	Ammonium, NH_4^+
Methanol, CH_3OH	Ethylene, $H_2C{=}CH_2$
Methylamine, CH_3NH_2	Chloromethane, CH_3Cl
Dimethylamine, $(CH_3)_2NH$	Bromomethane, CH_3Br
Trimethylamine, $(CH_3)_3N$	Higher hydrocarbons (ethane, propane)
Tetramethylammonium, $(CH_3)_4N^+$	
Trimethylamine N-oxide, $(CH_3)_3NO$	
Trimethylsulfonium, $(CH_3)_3S^+$	
Formate, $HCOO^-$	
Formamide, $HCONH_2$	
Carbon monoxide, CO	
Dimethyl ether, $(CH_3)_2O$	
Dimethyl carbonate, $CH_3OCOOCH_3$	
Dimethyl sulfoxide, $(CH_3)_2SO$	
Dimethylsulfide, $(CH_3)_2S$	

[a] A single isolate does not use all of the above, but at least one methylotrophic bacterium has been reported to oxidize each of the listed compounds.

[b] Methylotrophs able to oxidize methane are called *methanotrophs*.

characteristic: *they contain no carbon-carbon bonds*. Thus, all carbon-carbon bonds of the cell must be synthesized de novo. Organisms that can grow using only one-carbon organic compounds are called **methylotrophs**. Many, but not all, methylotrophs are also methanotrophs. But methanotrophs are unique in that they can grow not only on some of the more oxidized one-carbon compounds but also on methane. Methanotrophs pos-

sess a specific enzyme, *methane monooxygenase*, for the introduction of an oxygen atom into the methane molecule, leading to the formation of methanol (Section 17.24). The requirement for O_2 as a reactant in the initial oxygenation of methane thus explains why all methanotrophs are obligate aerobes. All methanotrophs also appear to be *obligate* C_1 utilizers, unable to utilize compounds containing carbon-carbon bonds. By contrast, many nonmethanotrophic methylotrophs are able to utilize organic acids, ethanol, and sugars.

Methane-oxidizing bacteria are unique among prokaryotes in possessing relatively large amounts of **sterols**. As we noted in Section 4.5, sterols are found in eukaryotes as a functional part of the membrane system but are absent from most prokaryotes. In methanotrophs, sterols may be an essential part of the complex internal membrane system (see later) involved in methane oxidation.

Classification of Methanotrophs

An overview of the classification of methanotrophs is given in Table 12.8. These bacteria were initially distinguished on the basis of morphology and formation of resting stages, but it was then found that they could be divided into two major groups based on their internal cell structure and carbon assimilation pathway. *Type I* organisms assimilate one-carbon compounds via a unique pathway, the **ribulose monophosphate cycle**, whereas *Type II* organisms assimilate C_1 intermediates via the **serine pathway**. We discuss the biochemical details of these pathways in Section 17.24.

Both groups of methanotrophs contain extensive internal membrane systems, which appear to be related to

TABLE 12.8 Some characteristics of methanotrophic bacteria

Organism	Morphology	16S rRNA group[a]	Resting stage	Internal membranes[b]	Citric acid cycle[c]	Carbon assimilation pathway[d]	N_2 fixation	DNA (mol % GC)
Methylomonas	Rod	Gamma	Cystlike body	I	Incomplete	Ribulose monophosphate	No	50–54
Methylomicrobium	Rod	Gamma	None	I	Incomplete	Ribulose monophosphate	No	49–60
Methylobacter	Coccus to ellipsoid	Gamma	Cystlike body	I	Incomplete	Ribulose monophosphate	No	50–54
Methylococcus	Coccus	Gamma	Cystlike body	I	Incomplete	Ribulose monophosphate	Yes	62–64
Methylosinus	Rod or vibrioid	Alpha	Exospore	II	Complete	Serine	Yes	63
Methylocystis	Rod	Alpha	Exospore	II	Complete	Serine	Yes	63
Methylocella[e]	Rod	Alpha	Exospore	II	—	Serine	Yes	61

[a] All are Proteobacteria.

[b] Internal membranes: Type I, bundles of disc-shaped vesicles distributed throughout the organism; Type II, paired membranes running along the periphery of the cell. See Figure 12.15.

[c] Organisms with an incomplete citric acid cycle lack the enzyme α-ketoglutarate dehydrogenase and thus cannot oxidize acetate to CO_2.

[d] See Figures 17.59 and 17.60. Unlike other methylotrophs, *Methylococcus* species contain Calvin cycle enzymes.

[e] Acidophilic, growth optimal at pH 5.

their methane-oxidizing ability. Type I methanotrophs are characterized by internal membranes arranged as bundles of disc-shaped vesicles distributed throughout the cell (Figure 12.15b), whereas Type II species possess paired membranes running along the periphery of the cell (Figure 12.15a). Type I methanotrophs are also characterized by a lack of a complete citric acid cycle (the enzyme α-ketoglutarate dehydrogenase is absent), whereas Type II organisms possess a complete cycle. Absence of a complete citric acid cycle (co Figure 5.22) greatly diminishes the ability of an organism to grow chemoorganotrophically. If these reactions cannot be run as a cycle, NADH cannot be generated from reactions of the cycle, thus preventing growth at the expense of organic compounds metabolized through the citric acid cycle.

Ecology and Isolation

Methanotrophs are widespread in aquatic and terrestrial environments, being found wherever stable sources of methane are present. Methane produced in the anoxic regions of lakes rises through the water column, and methanotrophs are often concentrated in a narrow band at the thermocline, where methane from the anoxic zone meets oxygen from the oxic zone. Methane-oxidizing bacteria therefore play an important role in the carbon cycle, converting methane derived from anoxic decomposition back into cell material (and CO_2).

For the enrichment of methanotrophs all that is needed is a mineral salts medium over which an atmosphere of 80% methane and 20% air is maintained. Once good growth is obtained, purification is carried out by repeated streaking on mineral salts agar plates, which are incubated in a jar with the methane-air mixture. Colonies appearing on the plates are of two types, common chemoorganotrophs growing on traces of or-

ganic matter in the medium, which appear in 1–2 days, and methanotrophs, which appear after about a week. The colonies of many methanotrophs are pink in color from the production of various carotenoid pigments, and this feature can help in their isolation.

Methanotrophs and Nitrosofying *Bacteria*

Methanotrophs are able to oxidize ammonia, although they cannot grow chemolithotrophically using ammonia as sole electron donor. In addition to methane oxidation, methane monooxygenase also functions to oxidize ammonia, and a competitive interaction between the two substrates exists. For this reason, ammonia is generally toxic to methanotrophs, and the preferred nitrogen source is nitrate. It has been speculated that methanotrophic bacteria could have evolved from the nitrosifying bacteria via mutations converting the ammonia monooxygenase to a methane monooxygenase. The fact that both groups of bacteria have elaborate internal membrane systems (see Section 12.3) and are phylogenetically closely related supports such a theory. In addition, however, methanotrophic bacteria contain some of the same genes and make some of the same proteins as methanogenic (methane-producing) prokaryotes, which phylogenetically are *Archaea* (co Section 13.4). We will see how the contrasting processes of methanogenesis and methanotrophy are related in this regard in Sections 17.17 and 17.24.

Methanotrophic Symbionts of Animals

A symbiotic association between methanotrophic bacteria and marine mussels and certain types of marine sponges is known to occur. Mussels live in the vicinity of hydrocarbon seeps where methane is released in substantial amounts. Intact mussels, as well as isolated mussel gill tissue, consume methane at high rates in the presence of O_2. In the gill tissue of the mussel, coccoid-shaped bacteria are

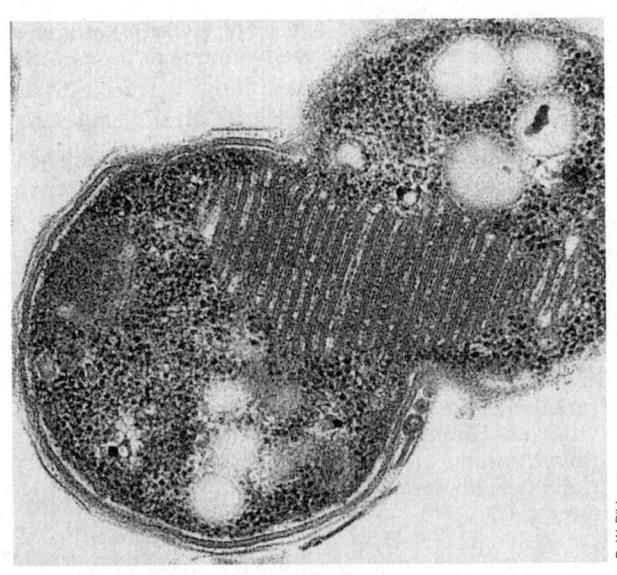

(a)

(b)

Figure 12.15 Electron micrographs of methanotrophs. (a) A *Methylosinus* species, illustrating a Type II membrane system. Cells are about 0.6 μm in diameter. (b) *Methylococcus capsulatus*, illustrating a Type I membrane system. Cells are about 1 μm in diameter.

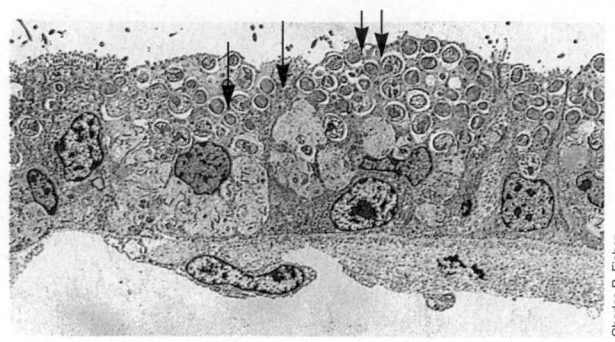

(a)

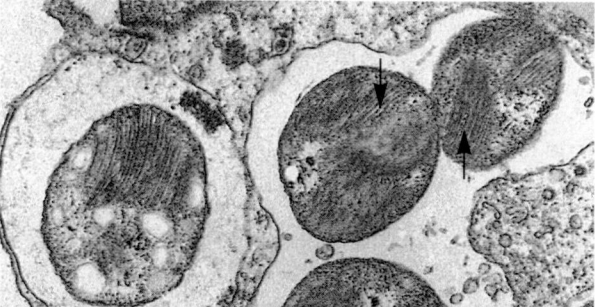

(b)

Figure 12.16 Methanotrophic symbionts of marine mussels. (a) Electron micrograph of a thin section at low magnification of gill tissue of a marine mussel living near hydrocarbon seeps in the Gulf of Mexico. Note the symbiotic methanotrophs (arrows) in the tissues. (b) High magnification view of gill tissue showing Type I methanotrophs. Note membrane bundles (arrows). The methanotrophs are about 1 μm in diameter.

present in high numbers (Figure 12.16*a*). The bacterial symbionts contain stacks of intracytoplasmic membranes (Figure 12.16*b*) typical of Type I methanotrophs. The symbionts are found in vacuoles within animal cells near the gill surface, which probably ensures an effective gaseous exchange with seawater. Presumably methane assimilated by the methanotrophs is distributed throughout the animals by the excretion of carbon compounds by the methanotrophs. The methanotrophic symbiosis is therefore conceptually similar to the symbiosis established between sulfide-oxidizing chemolithotrophs and hydrothermal vent tube worms and giant clams discussed in Section 19.8. Animal-bacteria symbioses, such as the methanotrophic mussel/sponge symbiosis and the sulfide-oxidizing vent animal symbioses, thus show that prokaryotic cells can occasionally constitute the basis of a one-step food chain.

✓ 12.6 Concept Check

Methylotrophs are prokaryotes able to grow on carbon compounds that lack carbon-carbon bonds. Some methylotrophs are also methanotrophs, able to grow on CH_4. Two classes of

methanotrophs are known, each having a number of structural and biochemical properties in common. Methanotrophs reside in water and soil and can also exist as symbionts of marine shellfish.

✓ What is the difference between a *methanotroph* and a *methylotroph*?

✓ What features differentiate Type I from Type II methanotrophs?

✓ What types of animals harbor methanotrophic symbionts and where does the methane the symbionts need come from?

12.7 *Pseudomonas* and the Pseudomonads

Key Genera

Pseudomonas
Burkholderia
Zymomonas
Xanthomonas

All the genera in this group are straight or slightly curved chemoorganotrophic and aerobic rods with *polar* flagella (Figure 12.17*b*). The important genera are *Pseudomonas, Commamonas, Ralstonia,* and *Burkholderia,* discussed in some detail here. Other genera include *Xanthomonas,* primarily a plant pathogen that is responsible for a number of necrotic plant lesions and that is characterized by its yellow-colored pigments; *Zoogloea,* characterized by its formation of an extracellular fibrillar polymer, which causes the cells to aggregate into distinctive flocs (this organism is a dominant component of activated sewage sludge) (∞ Section 28.2), and *Gluconobacter,* characterized by its incomplete oxidation of sugars or alcohols to acids, such as the oxidation of glucose to gluconic acid or ethanol to acetic acid (this organism is discussed briefly with the other acetic acid bacteria in Section 12.8). Phylogenetically, the various genera of pseudomonads scatter within the Proteobacteria (see Table 12.1).

Characteristics of Pseudomonads

The distinguishing characteristics of the pseudomonad group are given in Table 12.9. Also listed in this table are the minimal characteristics needed to identify an organism as a pseudomonad. Key identifying characteristics are the absence of gas formation from glucose, and the positive oxidase test, both of which help to distinguish pseudomonads from enteric bacteria (see Section 12.11).

The species of the genus *Pseudomonas* are defined on the basis of phylogeny and various physiological char-

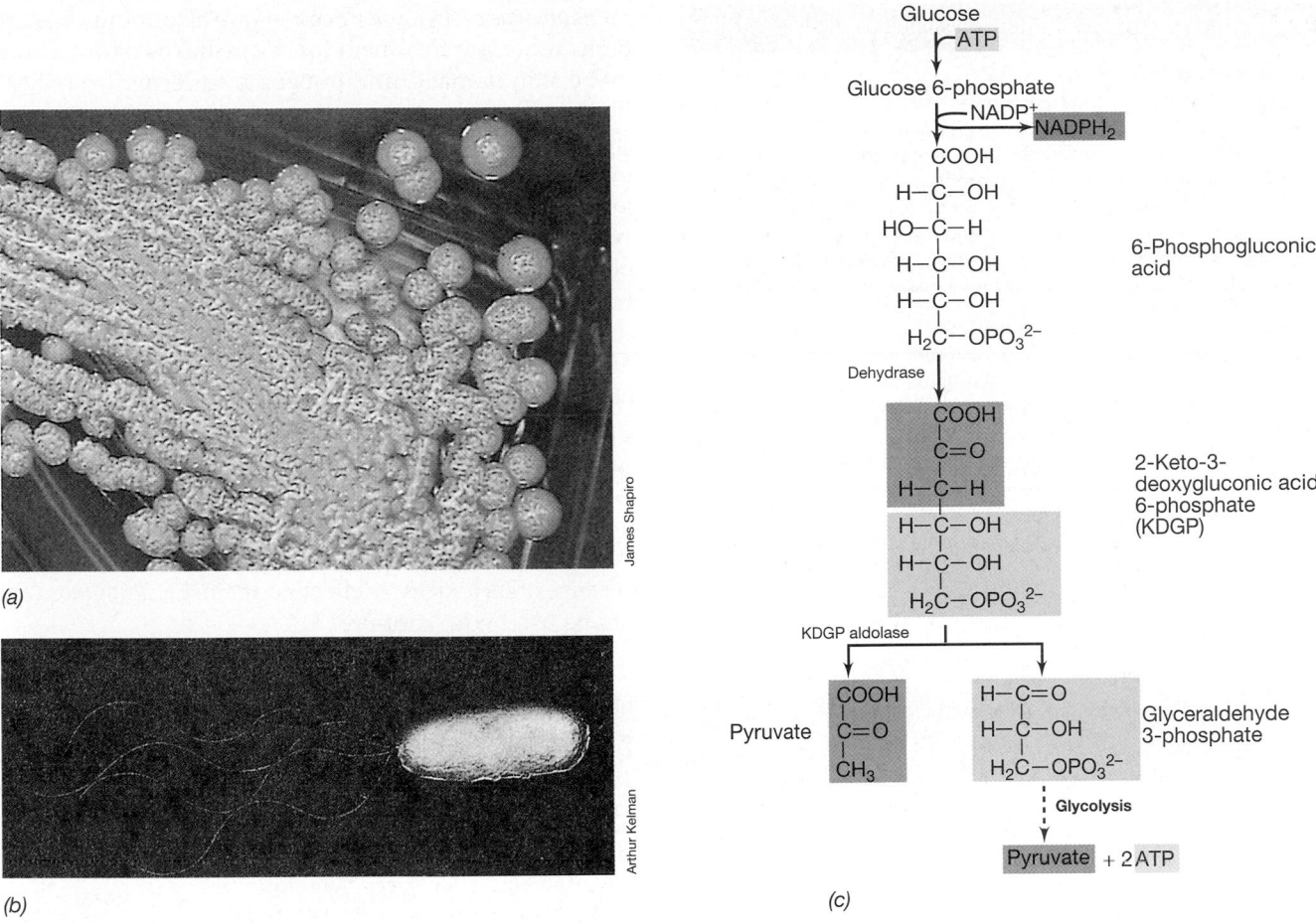

Figure 12.17 Typical pseudomonad colony and cell morphology and a biochemical pathway common in pseudomonads. (a) Photograph of colonies of *Burkholderia cepacia* on an agar plate. (b) Shadow-cast TEM preparation of a *Pseudomonas* species. The cell measures about 1 μm in diameter. (c) The Entner–Doudoroff pathway, the major means of glucose catabolism in pseudomonads.

acteristics, as outlined in Tables 12.10 and 12.11. Pseudomonads have very simple nutritional requirements and grow chemoorganotrophically at neutral pH and at temperatures in the mesophilic range. One of the striking properties of many species of pseudomonads is the wide variety of organic compounds used as carbon and energy sources. Some species utilize over *100* different compounds, and only a few species utilize fewer than 20. As an example of this versatility, a single strain of *Burkholderia cepacia* can use many different sugars, fatty acids, dicarboxylic acids, tricarboxylic acids, alcohols, polyalcohols, glycols, aromatic compounds, amino acids, and amines, plus miscellaneous organic compounds not fitting into any of the preceding categories. On the other hand, pseudomonads generally lack the hydrolytic enzymes necessary to break down polymers into their component monomers. Nutritionally versatile pseudomonads typically contain numerous inducible operons (∞ Section 8.5) because the catabolism of unusual organic substrates often requires the activity of several different enzymes. The pseudomonads are eco-

logically important organisms in soil and water and are probably responsible for the degradation of many soluble compounds derived from the breakdown of plant and animal materials in oxic habitats.

Many pseudomonads, as well as a variety of other gram-negative *Bacteria*, metabolize glucose via the Entner–Doudoroff pathway (Figure 12.17c). Two key enzymes of the Entner–Doudoroff pathway are *6-phosphogluconate dehydrase* and *2-keto-3-deoxyglucosephosphate aldolase* (Figure 12.17c). A survey for the presence of these enzymes in a variety of bacteria has shown that they are absent from gram-positive *Bacteria* (except for a few *Nocardia* isolates) but are generally present in bacteria of the genera *Pseudomonas, Rhizobium, Agrobacterium, Zymomonas*, and several other gram-negative *Bacteria*.

Pathogenic Pseudomonads

A number of pseudomonads are pathogenic (Table 12.11). Among the fluorescent pseudomonads, the species *Pseudomonas aeruginosa* is frequently associated with infections of the urinary and respiratory tracts in humans.

TABLE 12.9 Characteristics of pseudomonads

General characteristics:
Straight or curved rods but not vibrioid; size 0.5–1.0 μm by 1.5–4.0 μm; no spores; gram-negative; polar flagella: single or multiple; no sheaths, appendages, or buds; respiratory metabolism, never fermentative, although may produce small amounts of acid from glucose aerobically; use low-molecular-weight organic compounds, not polymers; some are chemolithotrophic, using H_2 or CO as sole electron donor; some can use nitrate as electron acceptor anaerobically; some can use arginine as energy source anaerobically

Minimal characteristics for identification:
Gram-negative, straight or slightly curved; no spores; motile (always); polar flagella (flagellar stain); oxidative-fermentative medium with glucose: tube open, acid produced; tube sealed, acid not produced; gas not produced from glucose (distinguishes them easily from enteric bacteria and *Aeromonas*); oxidase, almost always positive (enterics are oxidase-negative); catalase always positive; photosynthetic pigments absent (distinguishes them from purple nonsulfur bacteria); indole-negative; methyl red-negative; Voges-Proskauer-negative (for discussion of many of these biochemical tests, see Section 24.2)

Pseudomonas aeruginosa infections are also common in patients receiving treatment for severe burns or other traumatic skin damage, and in people suffering from cystic fibrosis. *Pseudomonas aeruginosa* is not an obligate parasite, however, but appears to be primarily an opportunist, initiating infections in individuals whose resistance is low. In addition to urinary infections, it can also cause systemic infections, usually in individuals who have experienced extensive skin damage. The organism is naturally resistant to many of the widely used antibiotics, so chemotherapy is often difficult. Resistance is frequently due to a *resistance transfer plasmid (R plasmid)* (∞ Sections 10.8 and 20.12), which is a plasmid carrying genes coding for detoxification of various antibiotics. *Pseudomonas aeruginosa* is commonly found in the hospital environment and can easily infect patients receiving treatment for other illnesses (∞ Section 25.7). Polymyxin, an antibiotic not ordinarily used in human therapy because of its toxicity, is effective against *P. aeruginosa* and can be used with caution.

TABLE 12.10 Characteristics of subgroups and species of the genera *Pseudomonas*, *Commamonas*, *Ralstonia*, and *Burkholderia*

Group	Phylogenetic group[a]	Characteristics	DNA (mol % GC)
Fluorescent subgroup	Gamma	**Most produce water-soluble, yellow-green fluorescent pigments; do not form poly-β-hydroxybutyrate; single DNA homology group**	
Pseudomonas aeruginosa		Pyocyanin production, growth at up to 43°C, single polar flagellum, capable of denitrification	67
Pseudomonas fluorescens		Does not produce pyocyanin or grow at 43°C; tuft of polar flagella	59–61
Pseudomonas putida		Similar to *P. fluorescens* but does not liquefy gelatin and does grow on benzylamine	60–63
Pseudomonas syringae		Lacks arginine dihydrolase, oxidase-negative, pathogenic to plants	58–60
Pseudomonas stutzeri		Soil saprophyte; strong denitrifyer and nonfluorescent	62
Acidovorans subgroup	Beta	**Nonpigmented, form poly-β-hydroxybutyrate, tuft of polar flagella, do not use carbohydrates; single DNA homology group**	
Commamonas acidovorans		Uses muconic acid as sole carbon source and electron donor	67
Commamonas testosteroni		Uses testosterone as sole carbon source	62
Pseudomallei-cepacia subgroup	Beta	**No fluorescent pigments, tuft of polar flagella, forms poly-β-hydroxybutyrate; single DNA homology group**	62
Burkholderia cepacia		Extreme nutritional versatility; some strains pathogenic to plants	67
Burkholderia pseudomallei		Causes melioidosis in animals; nutritionally versatile	69
Burkholderia mallei		Causes glanders in animals; nonmotile; nutritionally restricted	69
Diminuta-vesicularis subgroup	Alpha	**Single flagellum of very short wavelength, require vitamins (pantothenate, biotin, B_{12})**	
Pseudomonas diminuta		Nonpigmented, does not use sugars	66–67
Pseudomonas vesicularis		Carotenoid pigment, uses sugars	66
Ralstonia subgroup	Beta		
Ralstonia solanacearum		Plant pathogen	66–68
Ralstonia saccharophila		Grows chemolithotrophically with H_2, digests starch	69
Pseudomonas maltophilia		Requires methionine, does not use NO_3^- as N source, oxidase-negative	67

[a] All pseudomonads are members of the Proteobacteria (see Table 12.1).

TABLE 12.11 Pathogenic species of *Pseudomonas, Burkholderia, Ralstonia,* and *Xanthomonas*

Species	Relationship to disease
Animal pathogens	
P. aeruginosa	Opportunistic pathogen, especially in hospitals; in patients with metabolic, hematologic, and malignant diseases; hospital-acquired (nosocomial) infections from catheterizations, tracheostomies, lumbar punctures, and intravenous infusions; in patients given prolonged treatment with immunosuppressive agents, corticosteroids, antibiotics and radiation; may contaminate surgical wounds, abscesses, burns, ear infections, lungs of patients treated with antibiotics; cystic fibrosis; primarily a soil organism
P. fluorescens	Rarely pathogenic, as does not grow well at 37°C; may grow in and contaminate blood and blood products under refrigeration
P. maltophilia	A ubiquitous, free-living organism that is a common nosocomial pathogen
B. cepacia	Causes onion bulb rot; has also been isolated from humans and from environmental sources of medical importance
B. pseudomallei	Causes melioidosis, a disease endemic in animals and humans in Southeast Asia
B. mallei	Causes glanders, a disease of horses that is occasionally transmitted to humans
P. stutzeri	Often isolated from humans and environmental sources; may live saprophytically in the body
Plant pathogens	
R. solanacearum	Causes wilts of many cultivated plants (for example, potato, tomato, tobacco, peanut)
P. syringae	Attacks foliage, causing chlorosis and necrotic lesions on leaves; rarely found free in soil
P. marginalis	Causes soft rot of various plants; active pectinolytic species
X. campestris	Causes necrotic lesions on foliage, stems, fruits; also causes wilts and tissue rots; rarely found free in soil

Certain species of *Pseudomonas, Ralstonia,* and *Burkholderia* and the genus *Xanthomonas* are well-known plant pathogens (phytopathogens) (see Table 12.11). In many cases these organisms are so highly adapted to the plant environment that they can rarely be isolated from other habitats, including soil. Phytopathogens frequently inhabit nonhost plants (where disease symptoms are not apparent) and from there become transmitted to host plants and initiate infection. Disease symptoms vary considerably depending on the particular phytopathogen and host plant and are generally due to the release by the bacterium of plant toxins, lytic enzymes, plant growth factors, and other substances that destroy or distort plant tissue. In many cases the disease symptoms are highly diagnostic of the type of pseudomonad phytopathogen; thus, *Pseudomonas syringae* is frequently isolated from leaves showing chlorotic (yellowing) lesions, whereas *P. marginalis* is a typical "soft-rot" pathogen, infecting stems and shoots but rarely leaves.

Zymomonas

The genus *Zymomonas* consists of large, gram-negative rods that carry out a vigorous fermentation of sugars to ethanol. Although strictly fermentative, *Zymomonas* shows phylogenetic affiliation with the pseudomonads and contains Entner–Doudoroff pathway (Figure 12.17c) enzymes. *Zymomonas* is a common organism involved in alcoholic fermentation of various plant saps, and in many tropical areas of South and Central America, Africa, and Asia, it occupies a position in the fermented beverage industry similar to that of *Saccharomyces cerevisiae* (yeast) in North America and Europe. *Zymomonas* is involved in the alcoholic fermentation of agave in Mexico, to form the drink, *pulque,* and palm sap in many

tropical areas. It also carries out an alcoholic fermentation of sugarcane juice and honey. Although *Zymomonas* is rarely the sole organism involved in these alcoholic fermentations, it is often the dominant organism and is probably responsible for the production of most of the ethanol in these beverages. *Zymomonas* is also responsible for spoilage of fruit juices such as apple cider and perry and is also a constituent of the bacterial flora of spoiled beer.

Zymomonas is distinguished from *Pseudomonas* by its fermentative metabolism, microaerophilic to anaerobic nature, oxidase negativity, and other molecular taxonomic characteristics. It also resembles the acetic acid bacteria (see Section 12.8) and it is often found in nature associated with these organisms. This is of interest because, like yeast, *Zymomonas* ferments glucose to ethanol, whereas the acetic acid bacteria oxidize ethanol to acetic acid. Thus, the acetic acid bacteria probably depend on the activity of yeast and *Zymomonas* for the production of their growth substrate, ethanol. Unlike yeast, however, which ferments glucose to ethanol via the glycolytic pathway, *Zymomonas* employs the Entner–Doudoroff pathway.

12.8 Acetic Acid *Bacteria*

Key Genera

Acetobacter
Gluconobacter

As originally defined, the *acetic acid bacteria* comprised a group of gram-negative, aerobic, motile rods that carried out *incomplete* oxidation of alcohols and sugars, leading to the accumulation of organic acids as end products. With *ethanol* as a substrate, *acetic acid* is produced; hence, the derivation of the common name for these bacteria. Another property is the relatively high tolerance to acidic conditions, most strains being able to grow well at pH values lower than 5. This acid tolerance should of course be essential for an organism producing large amounts of acid. The acetic acid bacteria are a heterogeneous assemblage, comprising both peritrichously and polarly flagellated organisms. The *polarly* flagellated organisms are classified in the genus *Gluconobacter*, while the peritrichously flagellated species are grouped into the genus *Acetobacter*. All known acetic acid bacteria group phylogenetically with the alpha Proteobacteria (see Table 12.1).

In addition to flagellation, *Acetobacter* differs from *Gluconobacter* in being able to further oxidize the acetic acid it forms to CO_2. This difference in ability to oxidize acetic acid is related to the presence of a complete citric acid cycle. *Gluconobacter*, which *lacks* a complete citric acid cycle, is unable to oxidize acetic acid, whereas *Acetobacter*, which has all enzymes of the cycle, can oxidize it.

Ecology and Industrial Uses

The acetic acid bacteria are frequently found in association with alcoholic juices. Acetic acid bacteria can often be isolated from an alcoholic fruit juice such as hard cider or wine, or from beer. Colonies of acetic acid bacteria can be recognized on $CaCO_3$-agar plates containing ethanol, since the acetic acid produced causes a dissolution and clearing of the otherwise insoluble $CaCO_3$ (Figure 12.18). Cultures of acetic acid bacteria are used in the commercial production of vinegar (⟳ Section 30.10).

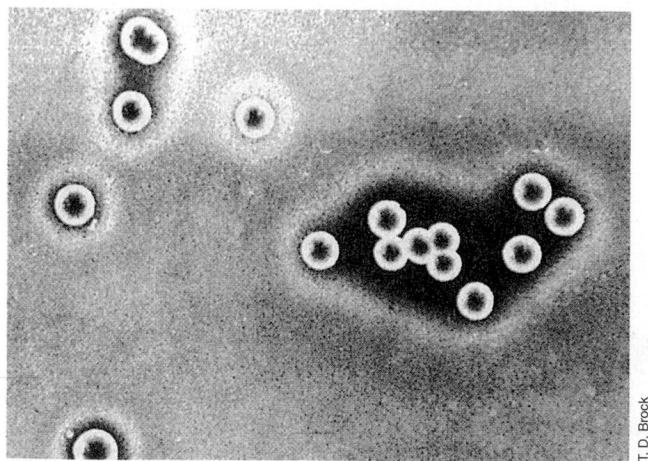

T. D. Brock

Figure 12.18 Photograph of colonies of *Acetobacter aceti* on calcium carbonate agar containing ethanol as energy source. Note the clearing around the colonies due to the dissolution of calcium carbonate by the acetic acid produced by the bacteria.

In addition to ethanol, these organisms carry out an incomplete oxidation of such organic compounds as higher alcohols and sugars. For instance, glucose is oxidized only to gluconic acid, galactose to galactonic acid, arabinose to arabonic acid, and so on. This property of "underoxidation" is exploited in the manufacture of ascorbic acid (vitamin C). Ascorbic acid can be formed from sorbose, but sorbose is difficult to synthesize chemically. It is, however, conveniently obtainable from acetic acid bacteria, which oxidize sorbitol (a readily available sugar alcohol) only to sorbose, a process called *bioconversion* (⟳ Section 30.8).

Another interesting property of some acetic acid bacteria is their ability to synthesize *cellulose*. The cellulose formed does not differ significantly from that of plant cellulose, with the exception that it is pure and not mixed in with other polymers like hemicelluloses, pectin, or lignins, and is formed as a matrix outside the wall where the bacteria become embedded in the tangled mass of cellulose microfibrils. When these species of acetic acid bacteria grow in an unshaken vessel, they form a surface pellicle of cellulose in which the bacteria develop. Since these bacteria are obligate aerobes, the ability to form such a pellicle may be a means by which the organisms remain at the surface of the liquid where oxygen is readily available.

12.9 Free-Living Aerobic Nitrogen-Fixing *Bacteria*

Key Genera

Azotobacter
Azomonas

A variety of organisms that inhabit primarily the soil are capable of fixing N_2 *aerobically* (Table 12.12). The genus *Azotobacter* comprises large, gram-negative, obligately aerobic rods capable of fixing N_2 nonsymbiotically (Figure 12.19). The first species of this genus was discovered by the Dutch microbiologist M. W. Beijerinck early in the twentieth century, using an aerobic enrichment culture technique with a medium containing N_2 (air) but devoid of a combined nitrogen source (⟳ box, Rise of General Microbiology, Chapter 19). Most free-living nitrogen-fixing bacteria are alpha or gamma Proteobacteria.

Taxonomy

The major free-living nitrogen-fixing bacteria that have been studied include *Azotobacter*, *Azospirillum*, and *Beijerinckia*. *Azotobacter* cells are large, many isolates being almost the size of yeasts, with diameters of 2–4 μm or more. Pleomorphism is common, and a variety of cell

TABLE 12.12 Genera of free-living aerobic nitrogen-fixing bacteria

Genus	Number of species	Phylogenetic group[a]	Characteristics	DNA (mol % GC)
Azotobacter	9	Gamma	Large rod; produces cysts; primarily found in neutral to alkaline soils	63–67
Azomonas	3	Gamma	Large rod; no cysts; primarily aquatic	52–59
Azospirillum	4	Alpha	Microaerophilic rod; associates with plants	69–71
Beijerinckia	4	Alpha	Pear-shaped rod with large lipid bodies at each end; produces extensive slime; inhabits acidic soils	54–59
Derxia	1	Alpha	Rods; form coarse, wrinkled colonies	69–73

[a] All species examined are members of the Proteobacteria (Figure 12.1).

shapes and sizes have been described. Some strains are motile by having peritrichous flagella. On carbohydrate-containing media, extensive capsules or slime layers are produced by free-living N_2-fixing bacteria (Figure 12.20; ∞ Figure 17.71). Despite the fact that *Azotobacter* is an obligate aerobe, its nitrogenase, the enzyme that catalyzes biological N_2 fixation (∞ Section 17.28), is O_2-sensitive. It is thought that the high respiratory rate characteristic of *Azotobacter* and the abundant capsular slime helps protect nitrogenase from O_2. *Azotobacter* is able to grow on a wide variety of carbohydrates, alcohols, and organic acids. The metabolism of carbon compounds is strictly oxidative, and acids or other fermentation products are rarely produced. All members fix nitrogen, but growth also occurs on simple forms of combined nitrogen: ammonia, urea, and nitrate.

Azotobacter can form resting structures called *cysts*. Like bacterial endospores, *Azotobacter* cysts (Figure 12.19*b*) show negligible endogenous respiration and are resistant to desiccation, mechanical disintegration, and ultraviolet and ionizing radiation. In contrast to endospores, however, they are *not* especially heat-resistant,

and they are not completely dormant because they rapidly oxidize exogenous energy sources.

The remaining genera of free-living N_2 fixers include *Azomonas*, a genus of large, rod-shaped bacteria that resemble *Azotobacter* except that they do not produce cysts and are primarily aquatic, *Beijerinckia* and *Derxia* (Figure 12.21), two genera that grow well in acidic soils, and *Azospirillum*, a spirillum-shaped nitrogen-fixing bacterium that forms nonspecific symbiotic associations with various plants, in particular, corn.

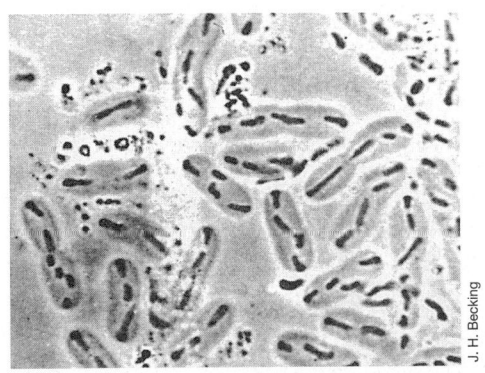

(a)

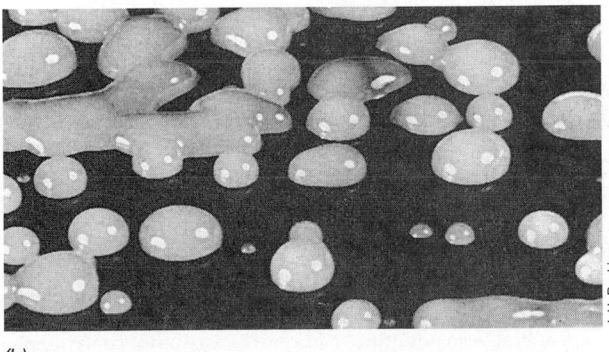

(b)

Figure 12.20 Examples of slime production by free-living N_2-fixing bacteria. (a) Cells of *Derxia gummosa* encased in slime. Cells are about 1–1.2 µm wide. (b) Colonies of *Beijerinckia* species growing on a carbohydrate-containing medium. Note the raised, glistening appearance of the colonies due to abundant capsular slime.

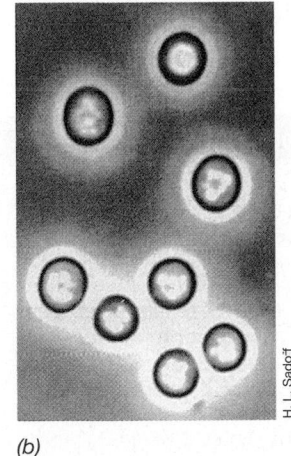

(a)   (b)

Figure 12.19 *Azotobacter vinelandii:* (a) vegetative cells and (b) cysts visualized by phase contrast microscopy. A cell measures about 2 µm in diameter, and a cyst about 3 µm. Compare with Figure 1.14*b*.

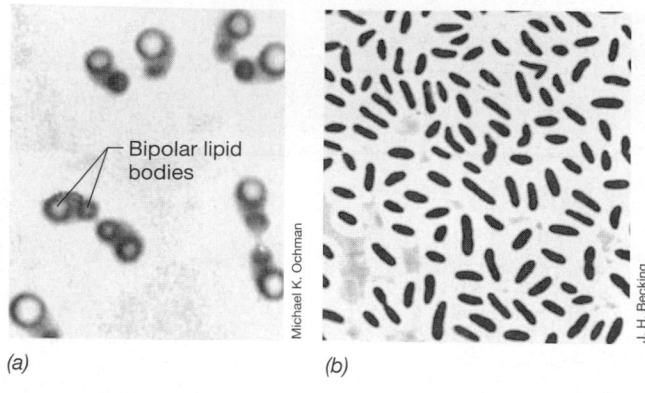

(a)   *(b)*

Figure 12.21 Phase-contrast photomicrographs of two genera of acid-tolerant, free-living N_2-fixing bacteria. (a) *Beijerinckia indica*. The cells are roughly pear-shaped, about 0.8 μm in diameter, and contain a large globule of poly-β-hydroxybutyrate (⬡⬡ Section 4.13) at each end. (b) *Derxia gummosa*. Cells are about 1 μm in diameter.

Azotobacter and Alternative Nitrogenases

We study the important process of biological N_2 fixation in Section 17.28. There we will learn of the central importance of the metals molybdenum (Mo) and iron (Fe) to the enzyme nitrogenase. The species *Azotobacter chroococcum* was the first nitrogen-fixing bacterium shown capable of growth on N_2 in the *absence* of molybdenum. In *A. chroococcum*, it was first shown that in place of Mo nitrogenase, two "alternative nitrogenases" containing either vanadium (V) plus Fe or Fe only can be made under certain growth conditions; subsequent investigations of other nitrogen-fixing bacteria have shown that these "backup" nitrogenase systems, which function only when Mo is absent, are widely distributed among nitrogen-fixing bacteria including *Archaea*, in which a few species fix N_2.

✓ 12.7–12.9 Concept Check

Pseudomonads include many gram-negative chemoorganotrophic aerobic rods; many N_2-fixing species are phylogenetically closely related and can reduce N_2 to NH_3 in the process of nitrogen fixation. The acetic acid bacteria are also phylogenetically related to pseudomonads and are characterized by an ability to oxidize ethanol to acetate aerobically.

✓ Compare and contrast the pseudomonads, *Azotobacter*, and the acetic acid bacteria in terms of O_2 and nitrogen requirements, electron donors, pathogenicity, and habitats.

✓ Compare and contrast the organism *Acetobacter* with the organism *Acetobacterium* (see Section 17.16) in as many ways as you can think of.

12.10 Neisseria, Chromobacterium, and Relatives

Key Genera

Neisseria
Chromobacterium

This group of beta and gamma Proteobacteria comprises a diverse collection of organisms, related phylogenetically as well as by Gram stain, morphology, lack of motility, and aerobic metabolism. The genera *Neisseria*, *Moraxella*, *Branhamella*, *Kingella*, and *Acinetobacter* are distinguished as outlined in Table 12.13.

In the genus *Neisseria*, the cells are always cocci (⬡⬡ Figure 26.28), whereas cells of the other genera are rod-shaped, becoming coccoid only in the stationary phase of growth. This has led to designation of these organisms as **coccobacilli**. Organisms of the genera *Neisseria*, *Kingella*, and *Moraxella* are commonly isolated from animals, and some of them are pathogenic, whereas organisms of the genus *Acinetobacter* are common soil and water organ-

---

TABLE 12.13 Characteristics of the genera of gram-negative cocci[a]

Characteristics	Genus	Number of species	Phylogenetic group[a]	DNA (mol % GC)
I. Oxidase-positive, penicillin-sensitive:				
Cocci; complex nutrition, utilize carbohydrates, obligate aerobes	*Neisseria*	24	Beta	49–55
	Moraxella	8	Gamma	—
Rods or cocci; generally no growth-factor requirements, generally do not utilize carbohydrates; do not contain flagella, but some species exhibit "twitching" motility; many are commensals or pathogens of animals	*Branhamella*	10	Beta	40–47
	Kingella	2	Beta	47–55
II. Oxidase-negative, penicillin-resistant:				
some strains can utilize a restricted range of sugars, and some exhibit "twitching" motility; saprophytes in soil, water, and sewage	*Acinetobacter*	7	Gamma	38–47

[a] All are Proteobacteria.

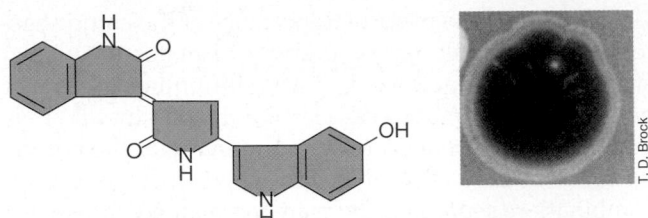

Figure 12.22 *Chromobacterium*. A large colony of *Chromobacterium violaceum* is shown beside the structure of the pigment violacein, produced by cells of *C. violaceum*.

isms, although they are occasionally found as parasites of animals and have been implicated in some nosocomial infections. Some strains of *Moraxella* and *Acinetobacter* possess the interesting property of *twitching motility*, exhibited as brief translocative movements or "jumps" covering distances of about 1–5 μm. We discuss the clinical microbiology of *Neisseria gonorrhoeae* in Section 24.1 and the pathogenesis of gonorrhea itself in Section 26.12.

Chromobacterium is a close phylogenetic relative of *Neisseria* but is rod-shaped in morphology, resembling the pseudomonads or enteric bacteria. The best-known *Chromobacterium* species is *C. violaceum*, a purple-pigmented organism (Figure 12.22) found in soil and water and occasionally in pus-forming infections of humans and other animals. *Chromobacterium violaceum* and a few other chromobacteria produce the purple pigment *violacein* (Figure 12.22), a water-insoluble pigment that has antibiotic-like properties and is produced only in media containing the amino acid tryptophan. *Chromobacterium* is a facultative aerobe, growing fermentatively on sugars and aerobically on a variety of carbon sources.

12.11 Enteric *Bacteria*

Key Genera

Escherichia
Salmonella
Proteus
Enterobacter

The **enteric bacteria** comprise a relatively homogeneous phylogenetic group within the gamma Proteobacteria and are characterized phenotypically as follows: gram-negative, nonsporulating rods, nonmotile or motile by *peritrichous* flagella (Figure 12.23*a*), facultative aerobes, oxidase-*negative* with relatively simple nutritional requirements, fermenting sugars to a variety of end products. The phenotypic characteristics used to separate the enteric bacteria from other bacteria of similar morphology and physiology are given in Table 12.14.

Among the enteric bacteria are many strains pathogenic to humans, animals, or plants as well as other

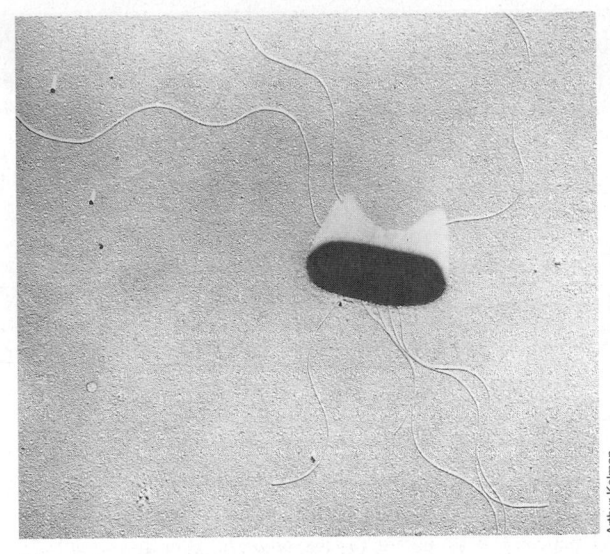

(a)

(b)

Figure 12.23 (a) Electron micrograph of a shadow-cast preparation of cells of the butanediol-producing enteric bacterium *Erwinia carotovora*. Cells are about 0.8 μm wide. Note the peritrichously arranged flagella (⚬⚬ Section 4.10). (b) Biochemical pathway for formation of butanediol from two molecules of pyruvate by butanediol fermenters.

TABLE 12.14 Defining characteristics of the enteric bacteria

General characteristics:

Gram-negative straight rods; motile by peritrichous flagella, or nonmotile; nonsporulating; facultative aerobes, producing acid from glucose; sodium neither required nor stimulatory; catalase-positive; oxidase-negative; usually reduce nitrate to nitrite (not to N_2); 16S rRNA of gamma Proteobacteria (see Table 12.1)

Key tests to distinguish enteric bacteria from other bacteria of similar morphology[a]:

Oxidase test, enterics always negative—separates enterics from oxidase-positive bacteria of genera *Pseudomonas, Aeromonas, Vibrio, Alcaligenes, Achromobacter, Flavobacterium, Cardiobacterium,* which may have similar morphology; nitrate reduced only to nitrite, (assay for nitrite after growth)—distinguishes enteric bacteria from bacteria that reduce nitrate to N_2 (gas formation detected), such as *Pseudomonas* and many other oxidase-positive bacteria; ability to ferment glucose—distinguishes enterics from obligately aerobic bacteria

[a]See Section 24.2 and Figure 24.7.

strains of industrial importance. Undoubtedly more is known about *Escherichia coli* than about any other bacterial species (∞ Section 10.12).

Because of the medical importance of the enteric bacteria, an extremely large number of isolates have been studied and characterized, and a fair number of distinct genera have been defined. Despite the fact that there is marked genetic relatedness among many of the enteric bacteria, as shown by DNA homologies and genetic recombination, separate genera are maintained, largely for practical reasons. Since these organisms are frequently cultured from diseased states, some means of identifying them is necessary to facilitate rapid treatment, and thus phenotypic characteristics have traditionally been very important in distinguishing between genera of enteric *Bacteria*.

Fermentation Patterns in Enteric *Bacteria*

One of the key taxonomic characteristics separating the various genera of enteric bacteria is the type and proportion of *fermentation products* produced by anaerobic fermentation of glucose. Two broad patterns are recognized, the **mixed-acid fermentation** and the **2,3-butanediol fermentation** (Figure 12.24). In mixed-acid fermentation, *three* acids are formed in significant amounts—acetic, lactic, and succinic; ethanol, CO_2, and H_2 are also formed, but *not* butanediol. In butanediol fermentation, *smaller* amounts of acids are formed, and butanediol, ethanol, CO_2, and H_2 are the main products. As a result of a mixed-acid fermentation *equal* amounts

Figure 12.24 Distinction between (a) mixed acid and (b) butanediol fermentation in enteric bacteria. The bold arrows indicate reactions leading to major products. Dashed arrows indicate minor products. The upper photo shows the production of acid (yellow color) and gas (in inverted tube) in a culture of *E. coli* (purple tube was uninoculated). The bottom photo shows the pink-red color in the Voges-Proskauer (VP) test, which indicates butanediol production, following growth of *Enterobacter aerogenes*. The left (yellow) tube was uninoculated. Note the major difference in CO_2 production in the two pathways, butanediol production leading to substantially greater CO_2 yields.

(a) Mixed acid fermentation (for example, *Escherichia coli*)

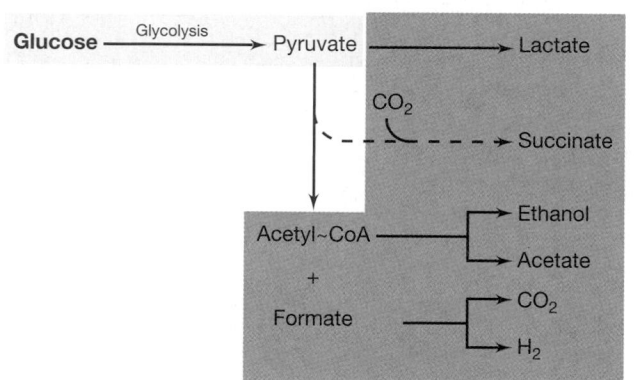

Typical products (molar amounts)

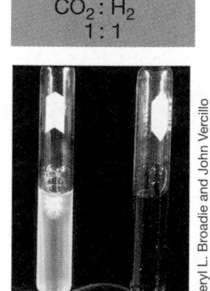

Acidic : neutral
4 : 1
CO_2 : H_2
1 : 1

(b) Butanediol fermentation (for example, *Enterobacter*)

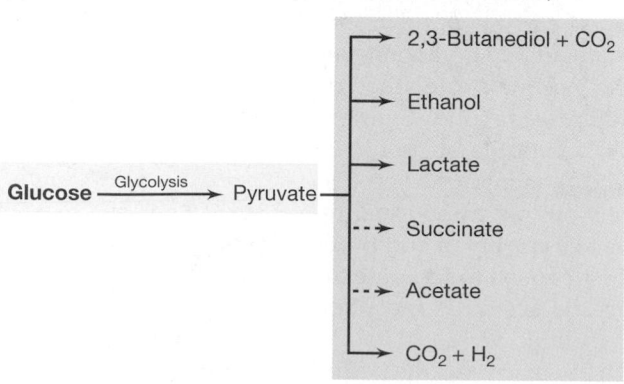

Typical products (molar amounts)

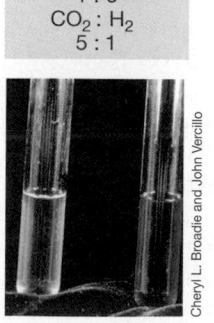

Acidic : neutral
1 : 6
CO_2 : H_2
5 : 1

of CO_2 and H_2 are produced, whereas with a butanediol fermentation considerably *more* CO_2 than H_2 is produced. This is because mixed-acid fermenters produce CO_2 only from formic acid by means of the enzyme system *formate hydrogen lyase*:

$$HCOOH \rightarrow H_2 + CO_2$$

and this reaction results in equal amounts of CO_2 and H_2. The butanediol fermenters also produce CO_2 and H_2 from formic acid, but they produce two additional molecules of CO_2 during the formation of each molecule of butanediol (Figure 12.23*b*).

A variety of diagnostic tests and differential media are used to separate the various genera in the two broad groups of enteric bacteria, and these are described in Tables 24.2 and 24.3. On the basis of these and other tests, genera can be defined as outlined in Tables 12.15 and 12.16.

Because enteric bacteria are genetically very closely related, their positive identification often presents considerable difficulty. In clinical laboratories, identification is frequently based on computer analysis of a large num-

ber of diagnostic tests carried out using miniaturized rapid diagnostic media kits and immunological and nucleic acid probes (∞ Figure 24.1), with consideration being given for variable reactions of exceptional strains. Thus, the separation of genera given in Tables 12.15 and 12.16 must only be considered as approximate, for it is always possible to isolate a strain that does not possess one or another characteristic normally considered positive for the genus as a whole. With these limitations in mind, an even more simplified separation of the key genera is found in Figure 12.25. This key permits a quick decision on the likely *genus* in which to place a new isolate. We now consider some of the properties of key genera.

Escherichia, Salmonella, and Shigella

Members of the genus *Escherichia* are almost universal inhabitants of the intestinal tract of humans and warm-blooded animals, although they are by no means the dominant organisms in these habitats. *Escherichia* may play a nutritional role in the intestinal tract by synthesizing vitamins, particularly vitamin K. As a facultative aerobe, this organism probably also helps consume oxy-

TABLE 12.15 Key diagnostic reactions used to separate the various genera of enteric bacteria[a]

Genus	H₂S(TSI)	Urease	VP[b]	Indole	Motility	Gas from glucose[b]	β-Galactosidase
Escherichia	−	−	−	+	+ or −	+	+
Enterobacter	−	−	+	−	+	+	+
Shigella	−	−	−	+ or −	−	−	+ or −
Edwardsiella	+	−	−	+	+	+	−
Salmonella	+	−	−	−	+	+	+ or −
Klebsiella	−	+	+ or −	−	−	+	+
Arizona	+	−	−	−	+	+	+
Citrobacter	+ or −	−	−	−	+	+	+
Proteus	+ or −	+	−	+ or −	+	+ or −	−
Providencia	−	−	−	+	+	−	−
Yersinia	−	+	−	−	+[c]	−	+
Hafnia	−	−	+	−	+	+	+ or −

Genus	KCN	Citrate	Mucate utilization	Phenyl-methyl red	Tartrate utilization	Alanine deaminase	DNA (mol % GC)
Escherichia	−	−	+	+	+	−	48–52
Enterobacter	+	+	+	−	−	−	52–60
Shigella	−	−	−	+	−	−	50
Edwardsiella	−	−	−	+ or −	−	−	53–59
Salmonella	−	+ or −	+ or −	+	+ or −	−	50–53
Klebsiella	+	+	+	−	+ or −	−	53–58
Arizona	−	+	+ or −	+	−	−	50
Citrobacter	+ or −	+	+	+	+	−	50–52
Proteus	+	+ or −	−	+	+	+	38–41
Providencia	+	+	−	+	+	+	39–42
Yersinia	−	−	−	+	−	−	46–50
Hafnia	+	+	−	+	−	−	48–49

[a] See Table 24.1 for the procedures for these diagnostic reactions.

[b] See Figure 12.24 for a photo of this reaction.

[c] Motile when grown at room temperature; nonmotile at 37°C.

TABLE 12.16 Key diagnostic reactions used to separate the various genera of 2,3-butanediol producers[a]

Genus	Ornithine decarboxyl-ase	Gelatin hydrolysis	Temperature optimum (°C)	Pigmentation	Motility	Lactose	DNase	Sorbitol	DNA (mol % GC)
Klebsiella	−	−	37–40	None	−	+	−	+	53–58
Enterobacter	+	Slow	37–40	Yellow (or none)	+	+	−	+	52–60
Serratia	+	+	37–40	Red (or none)	+	−	+	−	52–60
Erwinia[b]	−	+ or −	27–30	Yellow (or none)	+	+ or −	−	+	50–58
Hafnia	+	−	35	None	+	−	−	−	48–49

[a]See Table 24.3 for a description of these diagnostic tests.
[b]See Figure 12.23*a*.

gen, thus rendering the large intestine anoxic. Wild-type *Escherichia* strains rarely show any growth-factor requirements and are able to grow on a wide variety of carbon and energy sources such as sugars, amino acids, organic acids, and so on. Some strains of *Escherichia* are pathogenic. The latter have been implicated in diarrhea in infants, occasionally occurring in epidemic proportions in children's nurseries or obstetric wards, and *Escherichia* may also cause urinary tract infections in older persons or in those whose resistance has been lowered by surgical treatment or by exposure to ionizing radiation. Enteropathogenic strains of *E. coli* are becoming more frequently implicated in dysentery-like infections and generalized fevers (⊂⊃ Sections 21.6, 21.9, and 29.7). As noted there, these strains form *K antigen*, permitting attachment and colonization of the small intestine, and *enterotoxin*, responsible for the symptoms of diarrhea.

Salmonella and *Escherichia* are quite closely related; the two genera showing about 50% homology by DNA:DNA hybridization (⊂⊃ Section 11.9). However, in contrast to most *Escherichia*, members of the genus *Salmonella* are usually pathogenic, either to humans or to other warm-blooded animals. In humans the most common diseases caused by salmonellas are *typhoid fever* and *gastroenteritis* (⊂⊃ Sections 28.8 and 29.6). The salmonellas are characterized immunologically on the basis of three cell surface antigens, the O, or cell wall (somatic) antigen; the H, or flagellar, antigen; and the Vi (outer polysaccharide layer) antigen, found primarily in strains of *Salmonella* causing typhoid fever. The O antigens are part of the lipopolysaccharides that comprise the outermost layer of the outer membrane of these organisms (⊂⊃ Sections 4.9 and 21.12). We discussed the chemical structure of lipopolysaccharides in Section 4.9 (⊂⊃ Figures 4.35 and 4.36). The genus *Salmonella* contains over 1000 distinct serotypes having different antigenic specificities in their O antigens. Additional antigenic subdivisions are based on the antigenic specificities of the flagellar (H) antigens. There is little or no correlation between the antigenic type of a *Salmonella* and the disease symptoms elicited, but immunological typing permits tracing a single strain involved in an epidemic.

The shigellas are also genetically very closely related to *Escherichia*. Tests for DNA homology show strains of *Shigella* having 70 to nearly 100% homology with *Escherichia coli*. In contrast to *Escherichia*, however, *Shigella* is commonly pathogenic to humans, causing a rather severe gastroenteritis usually called *bacillary dysentery*. *Shigella dysenteriae* is transmitted by food- and waterborne routes and is capable of invading intestinal epithelial cells (⊂⊃ Section 29.10). Once established, it produces both an endotoxin and a neurotoxin that exhibits enterotoxic effects.

Diagnostic test	Go to number
1 MR+; VP − (mixed-acid fermenters)	2
MR −; VP + (butanediol producers)	7
2 Urease +	*Proteus*
Urease −	3
3 H₂S (TSI) +	4
H₂S (TSI) −	6
4 KCN +	*Citrobacter*
KCN −	5
5 Indole +; citrate −	*Edwardsiella*
Indole −; citrate +	*Salmonella*
6 Gas from glucose	*Escherichia*
No gas from glucose	*Shigella*
7 Nonmotile; ornithine −	*Klebsiella*
Motile; ornithine +	8
8 Gelatin+; DNAse +	*Serratia* (red pigment)
Gelatin slow; DNAse −	*Enterobacter*

Key
  Mixed-acid fermenters
  Butanediol producers

Figure 12.25 A simplified key to the main genera of enteric bacteria. Only the most common genera are given. See text for precautions in the use of this key. Diagnostic tests for use with this figure are given in Table 24.3. Other characteristics of the genera are given in Tables 12.14–12.16. Color coding is as in Figure 12.24.

Proteus

The genus *Proteus* is characterized by rapid motility (Figure 12.26) and by production of the enzyme *urease*. By DNA homology it shows only a distant relationship to *Escherichia coli*. *Proteus* is a frequent cause of urinary tract infections in humans, and probably benefits in this regard from its ready ability to degrade urea. Because of the rapid motility of *Proteus* cells, colonies growing on agar plates often exhibit a characteristic **swarming** phenomenon (Figure 12.26b). Cells at the edge of the growing colony are more rapidly motile than those in the center of the colony; the former move a short distance away from the colony in a mass and then undergo a reduction in motility, settle down, and divide, forming a new crop of motile cells that again swarm. As a result, the mature colony appears as a series of concentric rings, with higher concentrations of cells alternating with lower concentrations (Figure 12.26b).

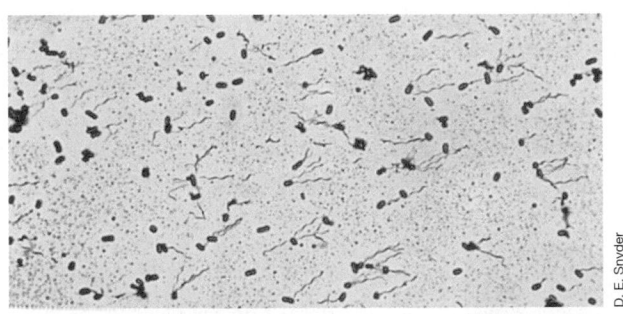

(a)

(b)

Figure 12.26 Swarming in *Proteus*. (a) Cells of *Proteus mirabilis* stained with a flagella stain: the peritrichous flagella of each cell group into a bundle. (b) Photo of a swarming colony of *Proteus vulgaris*. Note the concentric rings.

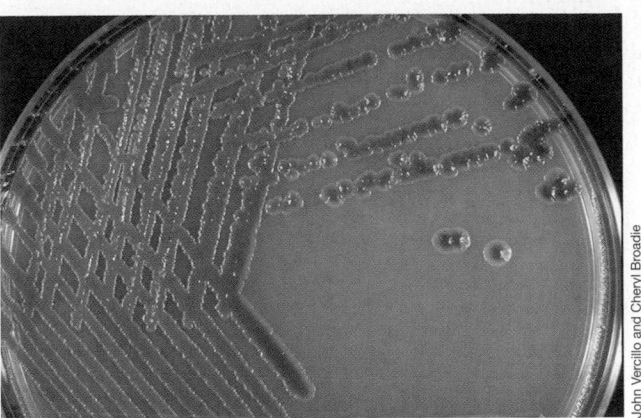

Figure 12.27 Colonies of *Serratia marcesens*. The orange-red pigmentation is due to the pyrrole-containing pigment *prodigiosin*.

Butanediol Fermenters: *Enterobacter, Klebsiella*, and *Serratia*

The butanediol fermenters are genetically more closely related to each other than to the mixed-acid fermenters, a finding that is in agreement with the observed physiological differences. Their DNA base composition is higher, 53–58% GC, and a current classification of this group is outlined in Table 12.16.

Enterobacter aerogenes is a common species in water and sewage as well as the intestinal tract of warm-blooded animals and is an occasional pathogen in urinary tract infections. One species of *Klebsiella, K. pneumoniae*, occasionally causes pneumonia in humans, but klebsiellas are most commonly found in soil and water. Most *Klebsiella* strains also fix N_2 (∞ Section 17.28), a property not found among other enteric bacteria. The genus *Serratia* forms a series of red pyrrole-containing pigments called **prodigiosins** (Figure 12.27). Prodigiosin is produced in stationary phase as a secondary metabolite (∞ Section 30.2), and is of interest because it contains the pyrrole ring also found in the pigments involved in energy transfer: porphyrins, chlorophylls, and phycobilins (∞ Sections 17.2 and 17.3). However, there is no evidence that prodigiosin plays any role in energy transfer, and its exact function is unknown. Species of *Serratia* can be isolated from water and soil as well as from the gut of various insects and vertebrates and occasionally from the intestines of humans.

12.12 *Vibrio* and *Photobacterium*

Key Genera

Vibrio
Photobacterium

The *Vibrio* group contains gram-negative, facultatively aerobic rods and curved rods that possess a fermentative metabolism. Most of the members of the *Vibrio* group are polarly flagellated, although some are peritrichously flagellated. One key difference between the *Vibrio* group and enteric bacteria is that members of the former are oxidase-*positive* (∞ Table 24.3), whereas members of the latter are oxidase-*negative*. Although *Pseudomonas* is also polarly flagellated and oxidase-positive, it is *not* fermentative and hence, can be separated from the vibrios by simple sugar fermentation tests. The best known genera of the *Vibrio* group include *Vibrio* and *Photobacterium*.

Most vibrios and related bacteria are aquatic, found either in freshwater or marine habitats, although one important organism, *Vibrio cholerae*, is pathogenic for humans. *Vibrio cholerae* is the specific cause of the disease *cholera* in humans (∞ Sections 21.11 and 28.6); the organism does not normally infect other hosts. Cholera is one of the most common infectious human diseases in underdeveloped countries and one that has had a long history. The organism is transmitted almost exclusively via water, and studies on its distribution in the nineteenth century played a major role in demonstrating the importance of water purification in urban areas (∞ box, Snow on Cholera, Chapter 28). We discuss the pathogenesis of *V. cholerae* in Section 21.11.

Vibrio parahemolyticus is a marine organism. It is a major cause of gastroenteritis in Japan (where raw fish is widely consumed) and has also been implicated in outbreaks of gastroenteritis in other parts of the world, including the United States. The organism can be frequently isolated from seawater or from shellfish and crustaceans, and its primary habitat is probably marine animals, with human infection being a secondary development (∞ Section 28.1).

Photobacterium and Bioluminescence

A number of gram-negative, polarly flagellated rods possess the interesting property of emitting light (luminescence). Most of these bacteria have been classified in the genus *Photobacterium*, but a few *Vibrio* isolates are also luminescent (Figure 12.28). Most **bioluminescent bacteria** are marine, usually found associated with fish. Some fish possess a special organ in which bioluminescent bacteria grow (Figure 12.28*c*–*f*). Other bioluminescent marine bacteria live saprophytically on dead fish and occasionally form visible colonies on the fish surface. (To see bioluminescence readily, one should observe the material in a completely dark room after the eyes have become adapted to the dark, Figure 12.28*a,b*.)

Although *Photobacterium* isolates are facultative aerobes, they are bioluminescent only when O_2 is present. Several components are needed for bioluminescence: the enzyme **luciferase** and a long-chain aliphatic aldehyde (for example, *dodecanal*); reduced flavin mononucleotide ($FMNH_2$) and O_2 are also involved. The primary elec-

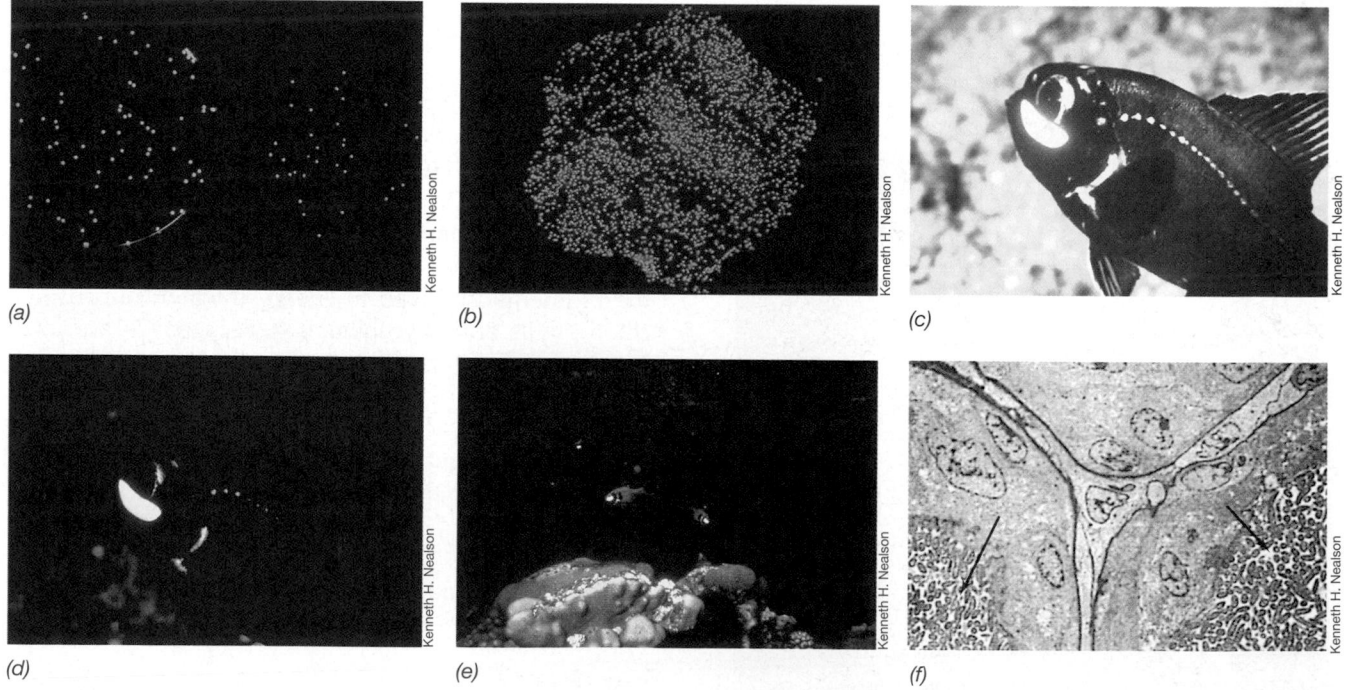

(a)   (b)   (c)

(d)   (e)   (f)

Figure 12.28 Bioluminescent bacteria and their role as light organs in the flashlight fish. (a) Two plates of luminous bacteria photographed by their own light. Note the different colors. Left, *Vibrio fischeri* strain MJ-1, blue light, and right, *V. fischeri* strain Y-1, green light. (b) Colonies of *Photobacterium phosphoreum* photographed by their own light. (c) The flashlight fish *Photoblepharon palpebratus*; the bright area is the light organ containing bioluminescent bacteria. (d) Same fish photographed by its own light. (e) Underwater photograph taken at night of *P. palpebratus* in coral reefs in the Gulf of Eilat. (f) Electron micrograph of a thin section through the light-emitting organ of *P. palpebratus*, showing the dense array of bioluminescent bacteria (arrows).

tron donor is NADH, and the electrons pass through luciferase. The reaction can be expressed as

$$\text{FMNH}_2 + \text{O}_2 + \text{RCHO} \xrightarrow{\text{Luciferase}}$$
$$\text{FMN} + \text{RCOOH} + \text{H}_2\text{O} + \text{Light}$$

The light-generating system constitutes a bypass route for shunting electrons from FMNH_2 to O_2, without involving other electron carriers such as quinones and cytochromes.

Regulation of Bioluminescence

The enzyme luciferase shows a unique kind of regulatory synthesis called **autoinduction**. The luminous bacteria produce a specific substance, the *autoinducer*, which accumulates in the culture medium during growth, and when the amount of this substance reaches a critical level, induction of luciferase occurs. The autoinducer in *Vibrio fischeri* has been identified as *N-β-ketocaproylhomoserine lactone*. Thus, cultures of luminous bacteria at low cell density are not luminous but only become luminous when growth reaches a sufficiently high density that the autoinducer can accumulate and function, a mechanism referred to as **quorum sensing** for the density-dependent nature of the phenomenon (Section 8.9).

Because of autoinduction, it is obvious that free-living luminescent bacteria in seawater are not luminous because the autoinducer cannot accumulate, and luminescence develops only when conditions are favorable for the development of high population densities (Figure 12.28). Although it is not clear why luminescence is density-dependent in free-living bacteria, in symbiotic strains of luminescent bacteria (see Figure 12.28), the rationale for density-dependent luminescence is clear: luminescence develops only when sufficiently high population densities are reached in the light organ of the fish to allow a visible flash of light. The genetics of quorum sensing using bioluminescence as a model experimental system has been actively explored, and it now appears that this type of sensing system is not limited to bioluminescent bacteria but instead is a general regulatory feature of a number of different bacteria for phenomena in which a minimum cell density is required (Section 8.9).

✓ 12.11–12.12 Concept Check

The enteric bacteria are a large group of facultative aerobic rods of great medical and molecular biological significance. *Vibrio* and *Photobacterium* species are marine and many species are bioluminescent.

- ✓ How is *Escherichia coli* distinguished from *Enterobacter aerogenes* based on physiology?
- ✓ Describe two major properties of *Proteus* species that distinguish them from other enteric bacteria.
- ✓ What is necessary for an organism like *Photobacterium* to give off visible light?

12.13 Rickettsias

Key Genera

Rickettsia
Wolbachia

The rickettsias are small, gram-negative, coccoid or rod-shaped Proteobacteria in the size range of 0.3–0.7 μm wide and 1–2 μm long. They are, with one exception, *obligate intracellular parasites* (Figure 12.29a) and have not yet been cultivated in the absence of host cells. Rickettsias are the causative agents of such human diseases as typhus fever, Rocky Mountain spotted fever, and Q fever (Section 27.3). Electron micrographs of thin sections of rickettsias show cells with a normal bacterial morphology (Figure 12.29b); both cell wall and cell membrane are visible. The cell wall contains muramic acid and diaminopimelic acid. Both RNA and DNA are present, and the rickettsias divide by normal binary

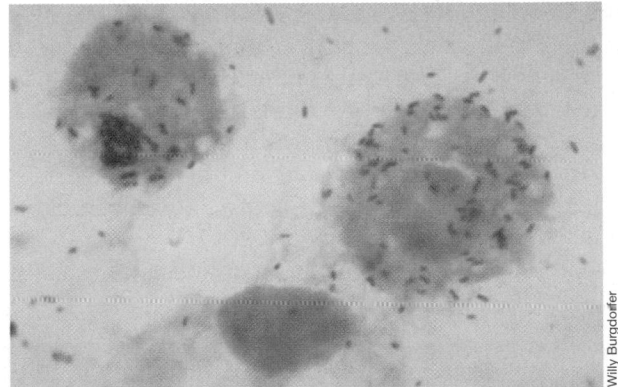

(a)

Willy Burgdorfer

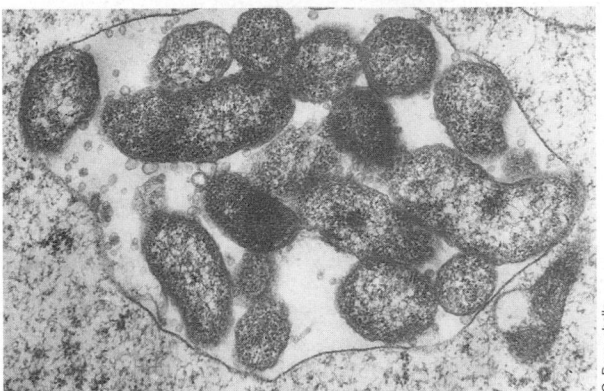

(b)

G. Devauchelle

Figure 12.29 Rickettsias growing within host cells. (a) *Rickettsia rickettsii* in tunica vaginalis cells of the vole, *Microtus pennsylvanicus*. Cells are about 0.3 µm in diameter. (b) Electron micrograph of cells of *Rickettsiella popilliae* within a blood cell of its host, the beetle *Melolontha melolontha*. Notice that the bacteria are growing within a vacuole within the host cell.

fission, with doubling times of about 8 h. The penetration of a host cell by a rickettsial cell is an active process, requiring both host and parasite to be alive and metabolically active. Once inside the host phagocytic cell, the bacteria multiply primarily in the cytoplasm and continue replicating until the host cell is loaded with parasites (see Figures 12.29 and Figure 27.3), at which time the host cell bursts and liberates the bacteria into the surrounding fluid. Several genera of rickettsias are known, and the properties of four key genera are shown in Table 12.17.

Metabolism and Pathogenesis

Much attention has been directed to the metabolic activities and biochemical pathways of rickettsias in an attempt to explain why they are obligate intracellular parasites. Many rickettsias possess a highly distinctive energy metabolism, being able to oxidize only glutamate or glutamine and being unable to oxidize glucose or organic acids. However, *Coxiella burnetii* is able to utilize both glucose and pyruvate as electron donors. Rickettsias possess a respiratory chain complete with cytochromes and are able to carry out electron transport phosphorylation, using NADH as electron donor. They are also able to synthesize at least some of the small molecules needed for macromolecular synthesis and growth, and they obtain the rest of their nutrients from the host cell.

Rickettsias do not survive long outside their hosts, and this may explain why they must be transmitted from animal to animal by arthropod vectors. When the arthropod obtains a blood meal from an infected vertebrate, rickettsias present in the blood are inoculated directly into the arthropod, where they penetrate to the epithelial cells of the gastrointestinal tract, multiply, and appear later in the feces. When the arthropod feeds on an uninfected individual, it then transmits the rickettsias either directly with its mouthparts or by contaminating the bite with its feces. However, the causal agent of Q fever, *Coxiella burnetii* (∞ Section 27.3), can also be transmitted to the respiratory system by aerosols. *Coxiella burnetii* is the most resistant of the rickettsias to physical damage, probably because it produces a resistant, sporelike form, and this explains its ability to survive in air. *Rochalimaea* is an atypical rickettsia because it can be grown in culture and is thus not an obligate intracellular parasite. In addition, when growing in tissue culture, cells of *Rochalimaea* grow on the *outside surface* of the eukaryotic host cells rather than within the cytoplasm or the nucleus. *Rochalimaea quintana* is the causative agent of *trench fever*, a disease that decimated troops in World War I. Species of the genus *Ehrlichia* cause disease in humans and other animals, two of which, *ehrlichiosis* in humans and *Potomac fever* in horses, can be quite severe and debilitating (∞ Section 27.3).

Wolbachia

The genus *Wolbachia* contains species of rod-shaped Proteobacteria that are intracellular parasites of arthropod insects (Figure 12.30). *Wolbachia* are phylogenetically related to the rickettsias and can have any of several effects on their insect hosts, including inducing parthenogenesis (development of unfertilized eggs), the killing of

Table 12.17	Characteristics of rickettsias					
Species	Rickettsial group	Alternate host	Cellular location	DNA (mol % GC)	Phylogenetic group[b]	DNA hybridization to *R. rickettsii* DNA (%)[a]
Rickettsia						
R. rickettsii	Spotted fever	Tick	Cytoplasm and nucleus	32–33	Alpha	100
R. prowazekii[c]	Typhus	Louse	Cytoplasm	29–30		53
R. typhi	Typhus	Flea	Cytoplasm	29–30		36
Rochalimaea						
R. quintana	Trench fever	Louse	Epicellular	39	Alpha	30
R. vinsonii	—	Vole	Epicellular	39		30
Coxiella						
C. burnetii	Q fever	Tick	Vacuoles	43	Gamma	—
Ehrlichia						
	Ehrlichiosis (humans); Potomac fever (horses)	Tick or domestic animals	Mononuclear leukocytes	—	Alpha	—
Wolbachia[d]	—	Arthropods	Cytoplasm	30	Alpha	—

[a] For discussion of DNA:DNA hybridization, see Section 11.9.

[b] All are Proteobacteria.

[c] The genome of this organism has been sequenced and shows several similarities to the mitochondrial genome.

[d] Not a pathogen of humans or other animals.

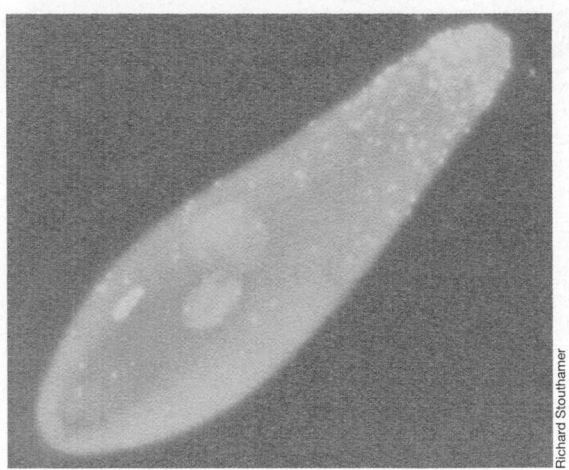

Figure 12.30 Photomicrograph of a 4′,6-diamidine-2′ phenylin-dole dihydrochloride (DAPI; ∽ Section 18.3) stained egg of the par-asitoid wasp, *Trichogramma kaykai* infected with *Wolbachia pipietis* that induces parthenogenesis. The *W. pipietis* cells are primarily lo-cated in the narrow end of the egg.

males, and feminization (the conversion of male insects into females).

Wolbachia pipientis is the best studied species in the genus. Cells colonize the insect egg (Figure 12.30), where they multiply in vacuoles of host cells, surrounded by a membrane of host origin. *W. pipientis* cells are passed from infected females to her offspring through this egg infection. *Wolbachia*-induced parthenogenesis occurs in a number of species of wasps. In these insects, males nor-mally arise from unfertilized eggs (which contain only one set of chromosomes), while females arise from fer-tilized eggs (which contain two sets of chromosomes). However, in unfertilized eggs infected with *Wolbachia*, the organism somehow triggers a doubling of the chro-mosome number, thus yielding only females. That *Wolb-achia* cells are necessary for this phenomenon has been shown by feeding female insects antibiotics that kill *Wolbachia*; in such females, parthenogenesis ceases. In certain species of woodlice (pillbugs), a species of *Wolbachia* is present that causes males to develop as fe-males. This occurs by *Wolbachia* infection and damage to the male hormone-producing glands. In other insects, such as certain lady beetles and butterflies, *Wolbachia* in-fection somehow results in the death of male but not fe-male offspring. In addition to these reproductive anomalies, in some insects *Wolbachia* infection may ac-tually be *essential* for survival. For example, in the ne-matode worms that cause the diseases elephantiasis and river blindness, antibiotic treatment kills the worms, ap-parently by killing their *Wolbachia* symbionts.

Like other rickettsias, *W. pipientis* has a small genome (about 1.5 mB), which has now been sequenced. Although this apparent parasite does not cause disease in either vertebrates or its invertebrate hosts, its various reproductive effects raises the interesting question of whether it has affected speciation in a major class of in-sects (arthropods make up 70% of all known insects). However, because the association between *Wolbachia* and insects is so common, it could turn out that the repro-ductive effects are secondary to a true *Wolbachia*-insect symbiosis, with the *Wolbachia* cells gaining a ready source of nutrients and a protected environment, and the insects benefiting in some as yet unknown ways.

✓ 12.13 Concept Check

The Rickettsias are obligate intracellular parasites, many of which cause disease. Rickettsias are deficient in many meta-bolic functions and obtain key metabolites from their hosts.

✓ Name a disease caused by a *Rickettsia* species.

✓ What is meant by the phrase "obligate intracellular parasite?"

✓ What effects can *Wolbachia* have on its insect hosts?

12.14 Spirilla

Key Genera

Spirillum
Bdellovibrio
Campylobacter

The **spirilla** are gram-negative, motile, spiral-shaped Proteobacteria that show a wide variety of physiologi-cal attributes. Some of the key taxonomic criteria used are cell shape, size, kind of polar flagellation (single or multiple), relation to oxygen (obligately aerobic, micro-aerophilic, facultative), relationship to plants (as sym-bionts or plant pathogens) or animals (as pathogens), fermentative ability, and certain other physiological char-acteristics (nitrogen-fixing ability, halophilic nature, thermophilic nature). The genera to be covered here are given in Table 12.18.

Spirillum, Aquaspirillum, Oceanospirillum, and Azospirillum

These are helically curved rods, which are motile by means of polar flagella (usually tufts at both poles) (see Figure 12.31). The number of turns in the helix may vary from less than one complete turn (in which case the or-ganism looks like a vibrio) to many turns. Spirilla with many turns can superficially resemble spirochetes (see Section 12.33) but differ both phylogenetically and in that they do not have an outer sheath and endoflagella but instead contain typical bacterial flagella.

Some spirilla are very large bacteria and were seen by early microscopists. It is likely that Antoni van

TABLE 12.18 Characteristics of the genera of spiral-shaped bacteria[a]

Genus	Phylogenetic group[b]	Characteristics	DNA (mol % GC)
Spirillum	Beta	Cell diameter 1.7 μm; microaerophilic; freshwater	36–38
Aquaspirillum	Alpha or beta	Cell diameter 0.2–1.5 μm; aerobic; freshwater	49–66
Magnetospirillum	Alpha	Vibrio to spirillum-shaped; cell diameter about 0.3 μm; contains magnetosomes; microaerophilic	65
Oceanospirillum	Gamma	Cell diameter 0.3–1.2 μm; aerobic; marine (require 3% NaCl)	42–51
Azospirillum	Alpha	Cell diameter 1 μm; microaerophilic; soil and rhizosphere; fixes N_2	68–70
Herbaspirillum	Beta	Cell diameter 0.6–0.7 μm; microaerophilic; soil and rhizosphere; fixes N_2	66–67
Campylobacter	Epsilon	Cell diameter 0.2–0.8 μm; microaerophilic to anaerobic; pathogenic or commensal in humans and animals; single polar flagellum	30–38
Helicobacter	Epsilon	Cell diameter 0.5–1 μm; tuft of polar flagella; associated with pyloric ulcers in humans	36–38
Bdellovibrio	Delta	Cell diameter 0.25–0.4 μm; aerobic; predatory on other bacteria; single polar sheathed flagellum	33–52
Ancyclobacter	Alpha	Cell diameter 0.5 μm; curved rods forming rings; nonmotile, aerobic; sometimes gas-vesiculate	66–69

[a] All are gram-negative and respiratory but never fermentative.
[b] All are Proteobacteria.

Leeuwenhoek first observed a *Spirillum* species in the 1670s (◌ Section 1.5), and the genus was first created by the protozoologist Ehrenberg in 1832. The organism seen by these workers is now called *Spirillum volutans* and is a rather large bacterium (Figure 12.31*a*). A phototrophic organism resembling *S. volutans* is *Thiospirillum* (see Figure 12.4*b*). *Spirillum volutans* is microaerophilic, requiring O_2, but is inhibited by O_2 at normal levels. Another prominent characteristic of *S. volutans* is the formation of granules (volutin granules) consisting of polyphosphate (see Figure 12.31*a* and Section 4.13).

Azospirillum lipoferum is a nitrogen-fixing organism, which was originally described and named *Spirillum lipoferum* by Beijerinck in 1922. It has become of considerable interest in recent years because this bacterium has been found to enter into a symbiotic relationship with tropical grasses and grain crops (◌ Section 19.28). The small-diameter spirilla (which are not microaerobic) have been separated into two genera, *Aquaspirillum* and *Oceanospirillum*, the former for freshwater forms and the latter for those living in seawater and requiring NaCl for growth (Table 12.18). Numerous species of *Aquaspirillum* and *Oceanospirillum* have been described, the various species being separated on physiological grounds. These organisms undoubtedly play an important role in the recycling of organic matter in aquatic environments.

Magnetotactic Spirilla

Highly motile microaerophilic magnetic spirilla have been isolated from freshwater habitats. These organisms demonstrate a dramatic directed movement in a mag-

netic field referred to as **magnetotaxis**. In an artificial magnetic field magnetic spirilla quickly orient their long axis along the north-south magnetic moment of the field. Within the cells, chains of 5–40 magnetic particles consisting of magnetite (Fe_3O_4) and greigite (Fe_3S_4) called **magnetosomes** (◌ Section 4.13 and Figure 4.54) are present (Figure 12.32), and these function as internal magnets that orient the cells along a specific magnetic field. Magnetic bacteria can have one of two magnetic polarities depending on the orientation of magnetosomes within the cell. Cells in the Northern Hemisphere have the north-seeking pole of their magnetosomes forward with respect to their flagella and thus move in a northward direction. Cells in the Southern Hemisphere have the opposite polarity and move southward. Although the ecological role of bacterial magnets is unclear, it has been suggested that the ability to orient in a magnetic field is of selective advantage in maintaining these microaerophilic organisms in zones of low O_2 concentration near the oxic/anoxic interface. The spirillum *Magnetospirillum magnetotacticum* (Figure 12.32 and Table 12.18) is a major organism in this group.

Bdellovibrio

These small vibroid organisms have the unusual property of preying on other bacteria, using as nutrients the cytoplasmic constituents of their hosts. These bacterial predators are small, highly motile cells that stick to the surfaces of their prey cells. Because of the latter property, they have been given the name *Bdellovibrio* (*bdello-*is a combining form meaning "leech"). Other predatory bac-

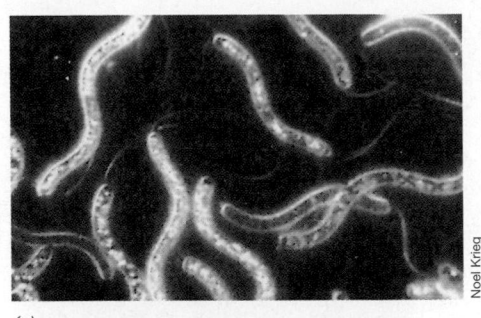

(a)

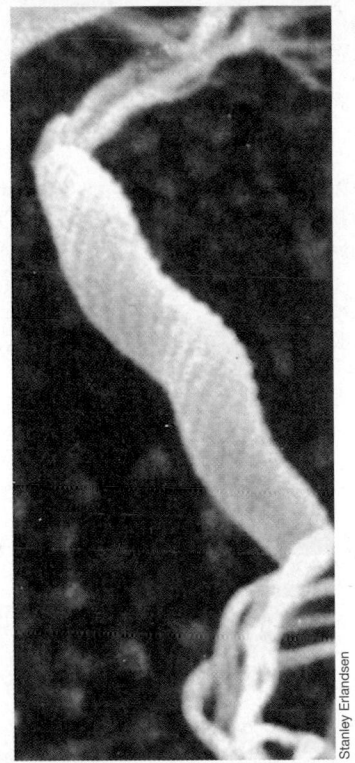

(b)

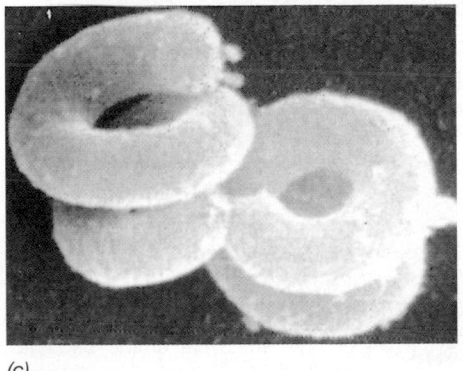

(c)

Figure 12.31 (a) *Spirillum volutans*, visualized by dark-field microscopy, showing flagellar bundles and volutin (polyphosphate) granules. Cells are about 1.5 × 2.5 μm. (b) Scanning electron micrograph of an intestinal spirillum. Note the polar flagellar tufts and the spiral structure of the cell surface. (c) Scanning electron micrograph of cells of *Ancyclobacter linguale*. Cells are about 0.5 μm in diameter.

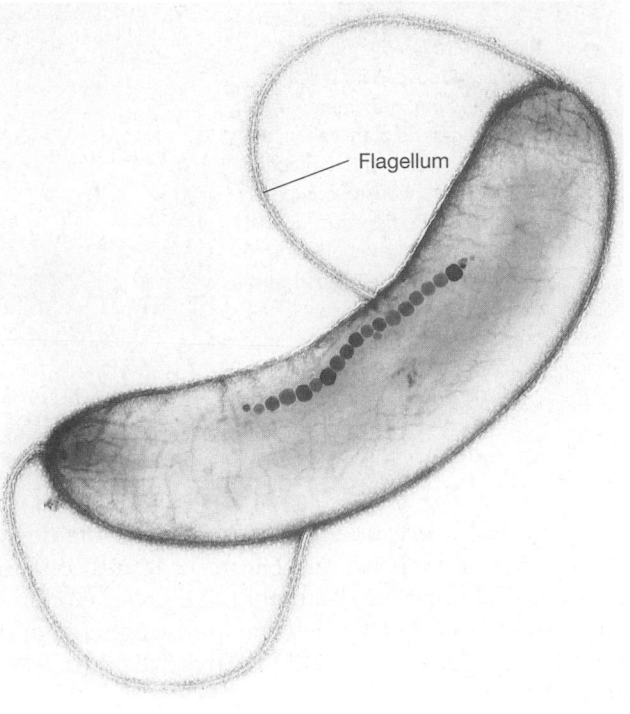

Flagellum

Figure 12.32 Negatively stained electron micrograph of a magnetotactic spirillum, *Magnetospirillum magnetotacticum*. A cell measures 0.3 × 2 μm. This bacterium contains particles of Fe_3O_4 (magnetite) called magnetosomes (Figure 4.54) arranged in a chain; the particles align the cell along geomagnetic lines. The organism was isolated from a water treatment plant in Durham, New Hampshire.

teria have been isolated and given such suggestive names as *Vampirococcus*. However, *Bdellovibrio* has a unique mode of attack and develops intraperiplasmically. After attachment of a *Bdellovibrio* cell to its prey, the predator penetrates through the prey wall and replicates in the space between the prey wall and membrane (the periplasmic space), eventually forming a spherical structure called a **bdelloplast**. The stages of attachment and penetration are shown in electron micrographs in Figure 12.33 and diagrammatically in Figure 12.34. A wide variety of gram-negative *Bacteria* can be attacked by a single *Bdellovibrio* species; gram-positive cells are not attacked.

Bdellovibrio is an obligate aerobe, obtaining its energy from the oxidation of amino acids and acetate. In addition, *Bdellovibrio* assimilates nucleotides, fatty acids, and even whole proteins in some cases directly from its host without first breaking them down. Thus it is clear that the predatory mode of existence has involved the development in *Bdellovibrio* of interesting and unusual biochemical processes. In addition to a predatory lifestyle, however, derivatives of predatory strains of *Bdellovibrio* that are prey-independent can be isolated, showing that predation is not obligatory. Prey-independent strains can be grown on complex media containing yeast extract and peptone.

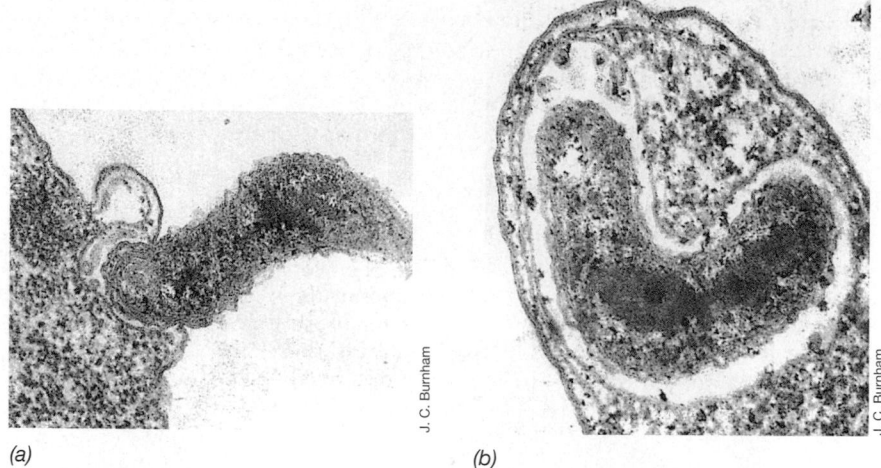

Figure 12.33 Stages of attachment and penetration of a prey cell by *Bdellovibrio*. A *Bdellovibrio* cell measures about 0.3 µm in diameter. (a,b) Electron micrographs of thin sections of *Bdellovibrio* attacking *Escherichia coli*; (a) early penetration; (b) complete penetration. The *Bdellovibrio* cell is enclosed in a membranous infolding of the prey cell (the bdelloplast) and replicates in the periplasmic space between the wall and the membrane (see Figure 12.34).

(a)   *(b)*

Phylogenetically, bdellovibrios fall into the delta subdivision of Proteobacteria (Figure 12.1) and contain a genome about half the size of that of *E. coli*. Taxonomically, three species of *Bdellovibrio* are recognized and supported by genomic DNA:DNA hybridization (∞ Section 11.9).

Members of the genus *Bdellovibrio* are widespread in soil and water, including the marine environment. Their detection and isolation require methods reminiscent of those used in the study of bacterial viruses (∞ Section 9.4). Prey bacteria are spread on the surface of an agar plate to form a lawn and the surface is inoculated with a small amount of soil suspension that has been filtered through a membrane filter; the latter retains most bacteria but allows the small *Bdellovibrio* cells to pass. On incubation of the agar plate, plaques analogous to those produced by bacteriophages are formed at locations where *Bdellovibrio* cells are growing. However, unlike phage plaques, which continue to enlarge only as long as the bacterial host is growing, *Bdellovibrio* plaques continue to enlarge even after the prey has stopped growing, resulting in large plaques on the agar surface. Pure cultures of *Bdellovibrio* can then be isolated from these plaques. *Bdellovibrio* cultures have been obtained from a wide variety of soils and are thus common members of the soil population.

Ancyclobacter

Members of the genus *Ancyclobacter* are ring-shaped, nonmotile, extremely nutritionally diverse chemoorganotrophic bacteria (Figure 12.31*c*). They resemble very tightly coiled vibrios and are widely distributed in aquatic environments. A phototrophic counterpart to *Ancyclobacter* exists in the purple nonsulfur bacterium-*Rhodocyclus purpureus* (see Figure 12.6*e*).

Campylobacter and *Helicobacter*

These two genera are key representatives of the *epsilon* subdivision of the Proteobacteria. They are gram-negative, motile spirilla, most species of which are pathogenic to

humans or other animals. *Campylobacter* and *Helicobacter* are both microaerophilic and are cultured from clinical specimens in media incubated at low (3–15%) O_2 and high (3–10%) CO_2. *Campylobacter* species cause acute enteritis leading to (usually) bloody diarrhea, and pathogenesis is due to several factors including an enterotoxin that is re-

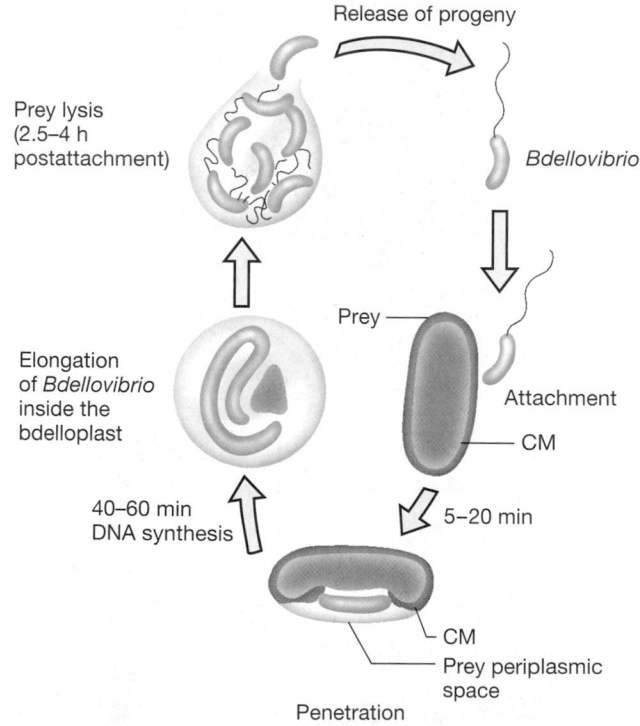

Figure 12.34 Developmental cycle of the bacterial predator *Bdellovibrio bacteriovorus*. Following primary contact between a highly motile *Bdellovibrio* cell and a gram-negative bacterium, attachment and penetration into the prey periplasmic space occurs. Once inside, *Bdellovibrio* elongates and within 4 h progeny cells are released. The number of progeny cells released varies with the size of the prey bacterium. For example, 5–6 bdellovibrios are released from each infected *Escherichia coli* cell, and 20–30 for a larger cell, such as a *Spirillum* sp. CM, Prey cytoplasmic membrane.

lated to cholera toxin (∞ Section 21.11). *Helicobacter py-lori* is closely related to *Campylobacter* species and causes both acute and chronic gastritis, leading to the formation of *peptic ulcers*. We discuss the diseases caused by *Campylobacter* and *Helicobacter*, including modes of transmission of the organisms and clinical symptoms, in more detail in Section 29.8.

✓ 12.14 Concept Check

Spirilla are spiral-shaped, chemoorganotrophic prokaryotes widespread in the aquatic environment. The genera *Helicobacter* and *Campylobacter* are pathogenic spirilla. Spirilla are distributed among the alpha, beta, gamma, delta, and epsilon subdivisions of Proteobacteria.

✓ What is a *volutin granule*?

✓ What is unique about the spirilla *Bdellovibrio* and *Magnetospirillum*?

12.15 Sheathed Proteobacteria: *Sphaerotilus* and *Leptothrix*

Key Genera

Sphaerotilus
Leptothrix

Sheathed bacteria are filamentous beta Proteobacteria (Table 12.1) with a unique life cycle involving formation of flagellated swarmer cells within a long tube or sheath. Under certain (generally unfavorable) conditions, the swarmer cells move out and become dispersed to new environments, leaving behind the empty sheath. Under favorable conditions, vegetative growth occurs within the filament, leading to the formation of long, cell-packed sheaths. Sheathed bacteria are common in freshwater habitats that are rich in organic matter, such as polluted streams, and trickling filters and activated sludge digestors in sewage treatment plants (∞ Section 28.2), being found primarily in flowing waters. In habitats where reduced iron or manganese compounds are present, the sheaths may become coated with ferric hydroxide or manganese oxide (∞ Figure 17.29). Iron precipitation is probably due to chemical reactions, but some sheathed bacteria have the biochemical ability to oxidize manganous ions to manganese oxide. Two genera are the major organisms here: *Sphaerotilus*, in which manganese oxidation does not occur, and *Leptothrix*, whose members do oxidize Mn^{2+}.

Sphaerotilus

The *Sphaerotilus* filament is composed of a chain of rod-shaped cells with rounded ends enclosed in a closely fitting sheath. This thin, transparent sheath is difficult to see when it is filled with cells, but when it is partially empty, the sheath can easily be seen by phase contrast microscopy (Figure 12.35*a*) or by staining. The cells within the sheath divide by binary fission (Figure 12.35*b*), and the new cells pushed out at the end synthesize new sheath material. Thus, the sheath is always formed at the *tips* of the filaments. Individual cells are 1–2 μm wide and 3–8 μm long and stain gram-negatively. Eventually cells are liberated from the sheaths, probably when the nutrient supply is low. These free cells are actively motile, the flagella being arranged lophotrichously (in a bundle at one pole) (Figure 12.35*c*). Probably the flagella are synthesized before the cells leave the sheath and, if so, may even aid in their liberation. It is

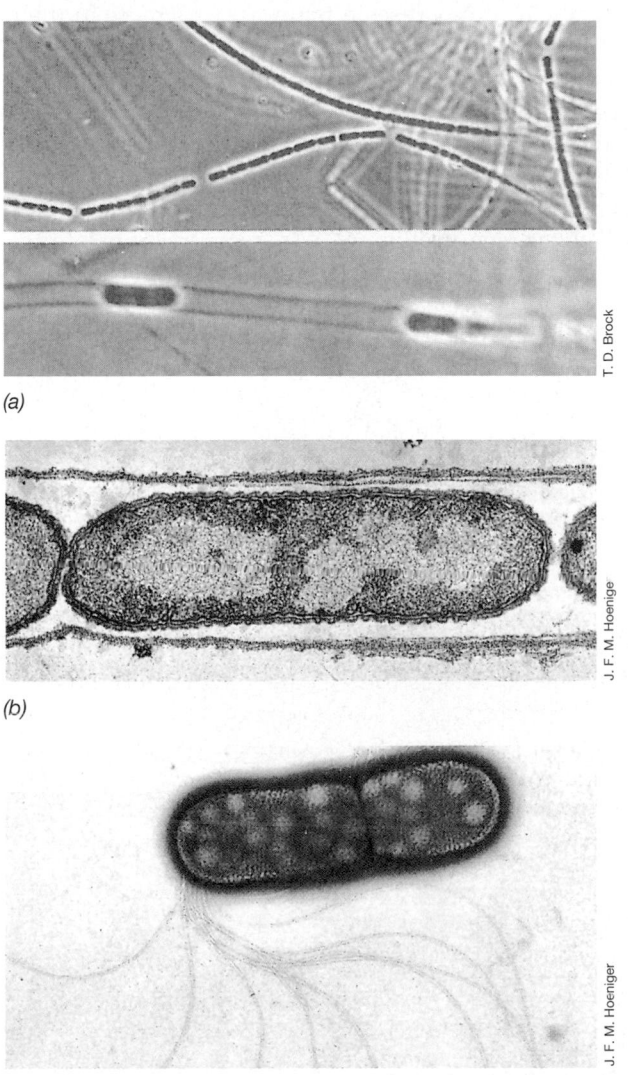

(a)

(b)

(c)

Figure 12.35 *Sphaerotilus natans*. A single cell is about 2 μm wide. (a) Phase contrast photomicrographs of material collected from a polluted stream. Active growth stage (above) and swarmer cells leaving the sheath. (b) Electron micrograph of a thin section through a filament. (c) Electron micrograph of a negatively stained swarmer cell. Notice the polar flagellar tuft.

thought that the swarmer cells then migrate, attach to some solid surface, and begin to grow, each swarmer being the forerunner of a new filament. The sheath, which is devoid of muramic acid or other components of the peptidoglycan cell wall, is a protein-polysaccharide-lipid complex, possibly analogous to the capsules formed by many gram-negative *Bacteria* but differing in that it forms a linear structure.

Sphaerotilus cultures are nutritionally versatile, able to use a wide variety of simple organic compounds as carbon and energy sources, with inorganic nitrogen sources. Befitting its habitat in flowing waters, *Sphaerotilus* is an obligate aerobe. *Sphaerotilus* blooms often occur in the fall of the year in streams and brooks when leaf litter causes a temporary increase in the organic content of the water. Its filaments are the main component of a microbial complex that sanitary engineers call "sewage fungus," which is a filamentous slime found on the rocks in streams receiving sewage pollution. In activated sludge works of sewage treatment plants (⊂⊃ Section 28.2), *Sphaerotilus* growth, like that of *Beggiatoa* (see Section 12.4), is often responsible for a detrimental condition called "bulking." The tangled masses of *Sphaerotilus* filaments so increase the bulk of the sludge that it does not settle properly, thus presenting difficulties in sludge clarification.

Leptothrix

The ability of *Sphaerotilus* and *Leptothrix* to precipitate iron oxides on their sheaths is well established and such iron-encrusted sheaths are frequently seen in iron-rich waters (Figure 12.36). Iron precipitation usually occurs

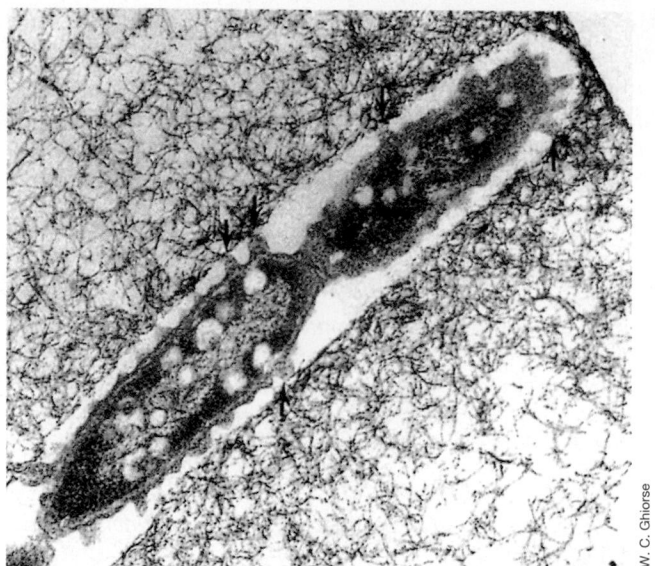

Figure 12.36 Transmission electron micrograph of a thin section of *Leptothrix* sp. in a sample from a ferromanganese film in a swamp in Ithaca, New York. A single cell measures about 0.9 μm in diameter. Note the protuberances of the cell envelope that contact the sheath (arrows).

when iron, chelated to organic materials such as humic or tannic acids, is metabolized; the iron gets precipitated on the sheath while the organic constituents may get taken up and used as a carbon or energy source.

Besides Fe^{2+} oxidation, *Leptothrix* is capable of oxidizing Mn^{2+} to Mn^{4+}. The reaction that occurs:

$$Mn^{2+} + \tfrac{1}{2}O_2 + H_2O \rightarrow MnO_2 + 2\,H^+ \qquad \Delta G^{0\prime} = -68\ kJ$$

TABLE 12.19 Characteristics of stalked, appendaged (prosthecate), and budding bacteria			
Characteristics	**Genus**	**Phylogenetic group**[a]	**DNA (mol% GC)**
Stalked bacteria:			
Stalk an extension of the cytoplasm and involved in cell division	*Caulobacter*	Alpha	62–67
Stalked, fusiform-shaped cells	*Prosthecobacter*	Alpha	54–60
Stalked, but stalk an excretory product not containing cytoplasm:			
Stalk depositing iron, cell vibrioid	*Gallionella*	Beta	55
Laterally excreted gelatinous stalk not depositing iron	*Nevskia*	Gamma	60
Appendaged (prosthecate) bacteria:			
Single or double prosthecae	*Asticcacaulis*	Alpha	55–61
Multiple prosthecae			
Short prosthecae, multiply by fission, some with gas vesicles	*Prosthecomicrobium*	Alpha	64–70
Flat, star-shaped cells, some with gas vesicles	*Stella*	Alpha	69–74
Long prosthecae, multiply by budding, some with gas vesicles	*Ancalomicrobium*	Alpha	70–71
Budding bacteria:			
Phototrophic, produce hyphae	*Rhodomicrobium*	Alpha	61–63
Phototrophic, budding without hyphae	*Rhodopseudomonas*	Alpha	64–72
Chemoorganotrophic, rod-shaped cells	*Blastobacter*	Alpha	59–66
Chemoorganotrophic, buds on tips of slender hyphae			
Single hyphae from parent cell	*Hyphomicrobium*	Alpha	59–65
Multiple hyphae from parent cell	*Pedomicrobium*	Alpha	62–67

[a] All are Proteobacteria.

is exergonic, and there is evidence that Mn^{2+} oxidation is coupled to energy-yielding reactions. The gene encoding the manganese-oxidizing protein of *Leptothrix* has been isolated and from biochemical studies it has been shown that this protein resides in the sheath, not inside the cells. Exactly why *Leptothrix* oxidizes Mn^{2+} is not entirely clear, but it has been hypothesized that this occurs either for energetic benefits or possibly because the Mn^{4+} produced reacts with humic and fulvic acids, thereby releasing organic nutrients for growth.

12.16 Budding and Prosthecate/Stalked *Bacteria*

Key Genera

Hyphomicrobium
Caulobacter

This large and rather heterogeneous group of Proteobacteria contains organisms that form various kinds of cytoplasmic extrusions: *stalks, hyphae,* or *appendages* (Table 12.19). Extrusions of these kinds, which are smaller in diameter than the mature cell, contain cytoplasm, and are bounded by the cell wall, are called **prosthecae** (singular, **prostheca**) (Figure 12.37). Of considerable interest in this group of bacteria is that cell division occurs as a result of *unequal cell growth*. In contrast to cell division in the typical bacterium, which occurs by *binary fission* and results in the formation of two equivalent cells (Figure 12.38), cell division in the stalked and budding bacteria involves the formation of a new daughter cell with the mother cell retaining its identity after the cell division process is completed (Figure 12.38). The genera from Table 12.19 that show this unequal cell division process are indicated in Figure 12.38.

A fundamental difference between these bacteria and conventional bacteria is not the formation of buds or stalks but the formation of a new cell wall from a *single point* (polar growth) rather than throughout the whole cell (intercalary growth). Several genera not normally considered to be budding bacteria show polar growth

without differentiation of cell size (Figure 12.38). An important consequence of **polar growth** is that internal structures, such as membrane complexes, are not involved in the cell division process, thus permitting the formation of more complex internal structures than in cells undergoing intercalary growth. And many budding

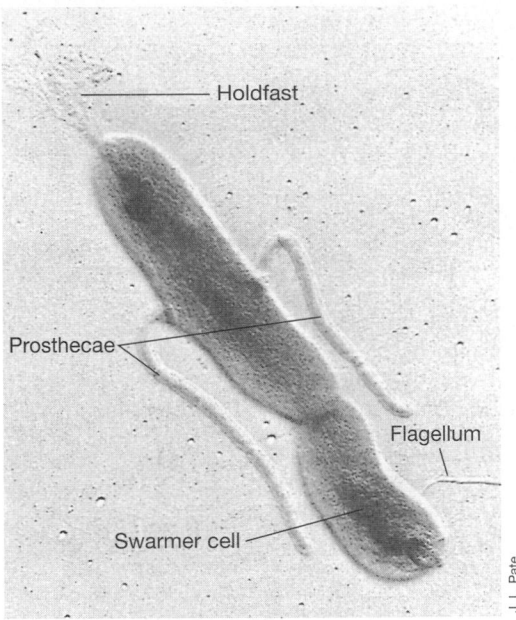

(a)

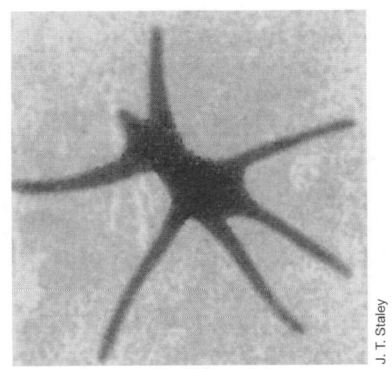

(b)

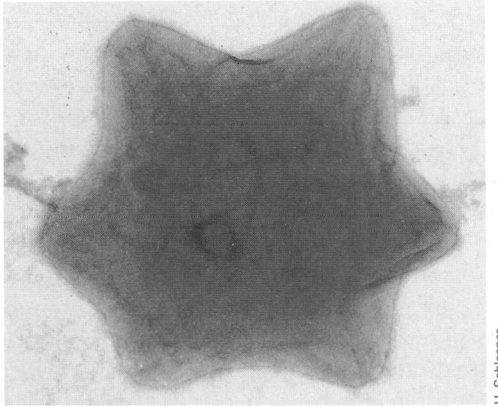

(c)

Figure 12.37 Prosthecate bacteria. (a) Electron micrograph of a shadow-cast preparation of *Asticcacaulis biprosthecum*, illustrating the location and arrangement of the prosthecae. Cells are about 0.6 μm wide. Note also the holdfast material and the swarmer cell in the process of differentiation. (b) Electron micrograph of a negatively stained preparation of a cell of the prosthecate bacterium *Ancalomicrobium adetum*. The appendages are cellular (prosthecae) because they are bounded by the cell wall and contain cytoplasm and are about 0.2 μm in diameter. (c) Electron micrograph of the star-shaped bacterium *Stella*. Cells are about 0.8 μm in diameter.

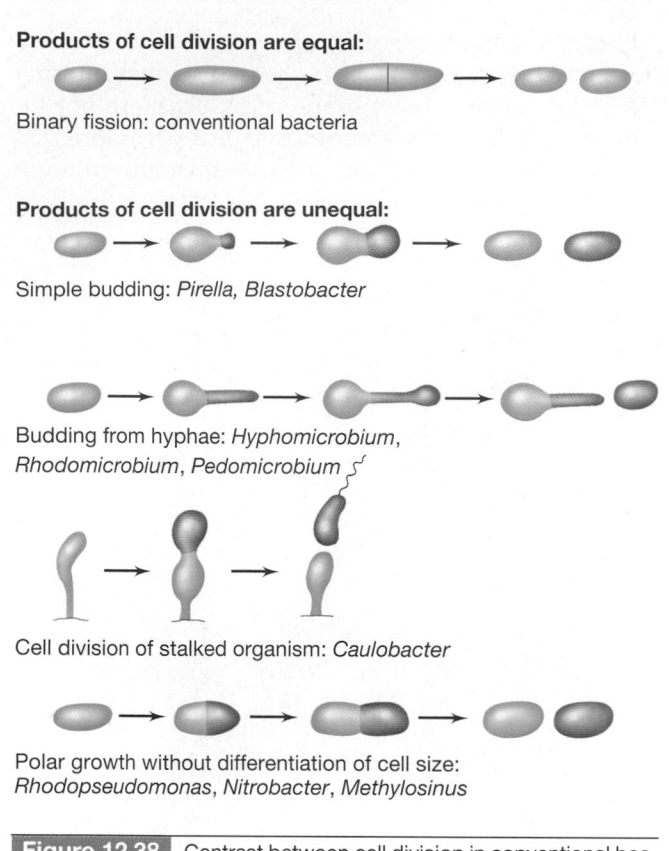

Products of cell division are equal:

Binary fission: conventional bacteria

Products of cell division are unequal:

Simple budding: *Pirella, Blastobacter*

Budding from hyphae: *Hyphomicrobium, Rhodomicrobium, Pedomicrobium*

Cell division of stalked organism: *Caulobacter*

Polar growth without differentiation of cell size: *Rhodopseudomonas, Nitrobacter, Methylosinus*

Figure 12.38 Contrast between cell division in conventional bacteria and in budding and stalked bacteria.

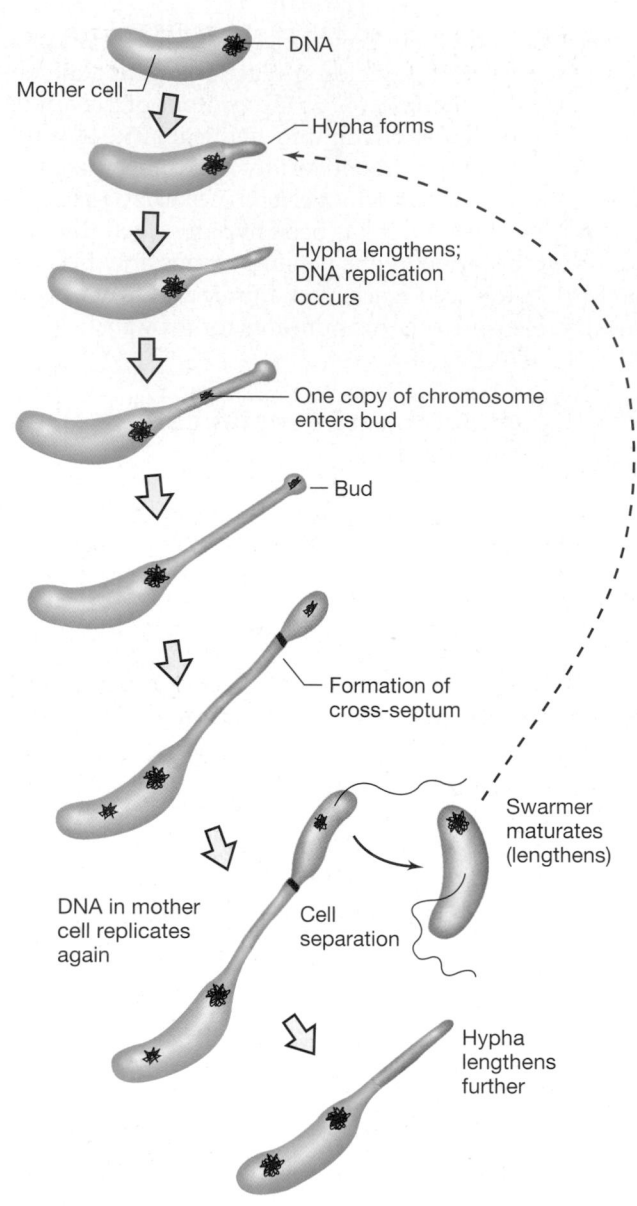

- DNA
- Mother cell
- Hypha forms
- Hypha lengthens; DNA replication occurs
- One copy of chromosome enters bud
- Bud
- Formation of cross-septum
- Swarmer maturates (lengthens)
- DNA in mother cell replicates again
- Cell separation
- Hypha lengthens further

Figure 12.39 Stages in the *Hyphomicrobium* cell cycle. The single chromosome of *Hyphomicrobium* is circular.

bacteria, particularly phototrophic species, contain extensive internal membrane systems.

Budding Bacteria: *Hyphomicrobium*

The two best-studied budding bacteria are *Hyphomicrobium*, which is chemoorganotrophic, and *Rhodomicrobium*, which is phototrophic; both organisms are phylogenetically closely related and release buds from the ends of long, thin hyphae. The hypha is a direct cellular extension of the mother cell (Figures 12.39 and 12.40*b*), containing cell wall, cytoplasmic membrane, ribosomes, and occasionally DNA. The process of reproduction in *Hyphomicrobium* is illustrated in Figure 12.39. The mother cell, which is often attached by its base to a solid substrate, forms a thin outgrowth that lengthens to become a hypha, and at the end of the hypha a bud forms. This bud enlarges, forms a flagellum, breaks loose from the mother cell, and swims away. Later, the daughter cell loses its flagellum and after a period of maturation forms a hypha and buds. Further buds can also form at the hyphal tip of the mother cell leading to arrays of cells connected by hyphae (Figure 12.40*a*). In some cases, a bud begins to form directly from the mother cell without the intervening formation of a hypha, whereas in other cases a single cell forms hyphae from each end (Figure 12.40*a*).

Nucleoid replication events during the budding cycle are of interest (Figure 12.39). The DNA located in the mother cell replicates, and then once the bud has formed, a copy of the circular chromosome is moved down the length of the hypha and into the bud. A cross-septum then forms, separating the still developing bud from the hypha and mother cell.

Hyphomicrobium is a methylotrophic bacterium (see Section 12.6). Preferred carbon sources are *one-carbon* compounds such as methanol, methylamine, formaldehyde, and formate. Growth on acetate, ethanol, or higher aliphatic compounds is usually slow, and growth is poor on sugars or most amino acids. Urea, amides, ammonia, nitrite, and nitrate can be utilized as

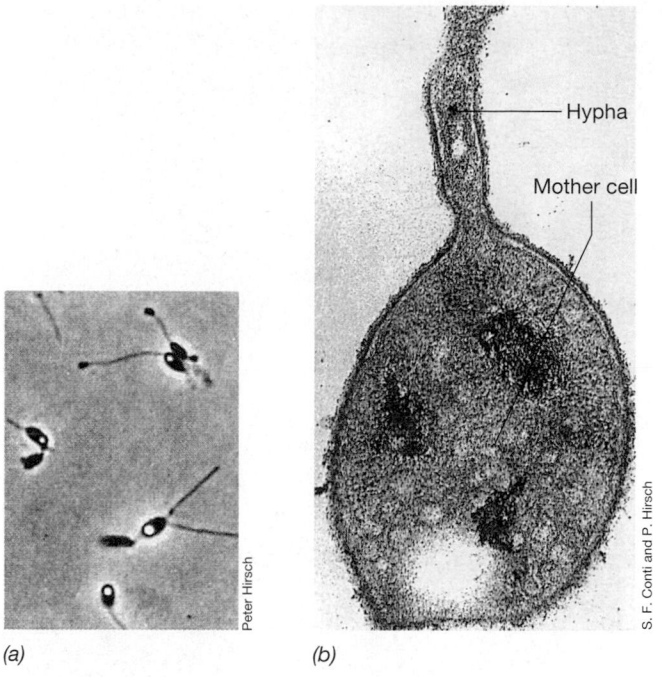

Figure 12.40 *Hyphomicrobium*. (a) Phase micrograph of cells of *Hyphomicrobium*. Cells are about 0.7 μm wide. (b) Electron micrograph of a thin section of a single *Hyphomicrobium* cell. The hypha is about 0.2 μm wide.

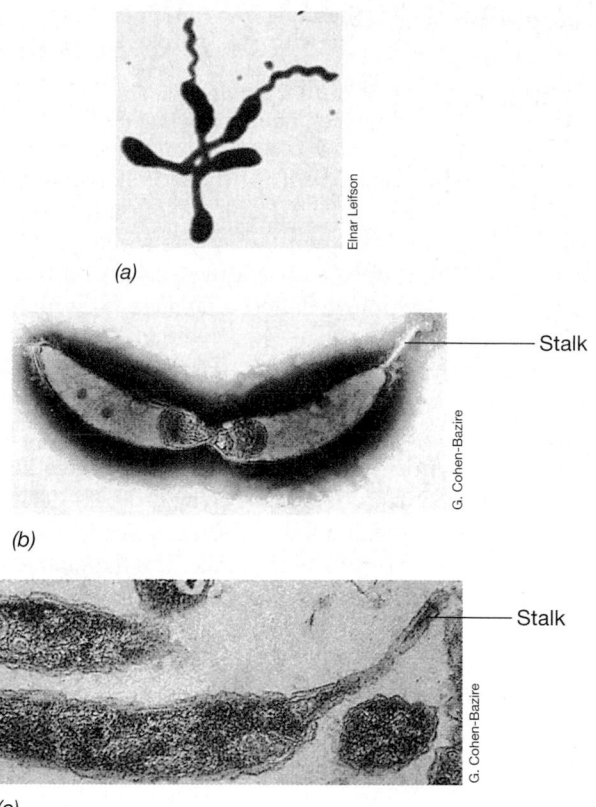

Figure 12.41 Stalked bacteria. (a) A *Caulobacter* rosette. A single cell is about 0.5 μm wide. The five cells are attached by their stalks (prosthecae). Two of the cells have divided, and the daughter cells have formed flagella. (b, c) Electron micrographs of *Caulobacter* cells. (b) Negatively stained preparation of a cell in division. (c) A thin section. Notice that cytoplasmic constituents are present in the stalk region.

nitrogen sources; vitamins are not required. *Hyphomicrobium* is widespread in freshwater, marine, and terrestrial habitats. Initial enrichment cultures can be prepared using a mineral-salts medium lacking organic carbon and nitrogen, to which a sample of natural material is added. After several weeks of incubation, the surface film that develops is streaked out on agar medium containing methylamine or methanol as a sole carbon source. Colonies are then checked microscopically for the characteristic *Hyphomicrobium* cellular morphology. A fairly specific enrichment procedure for *Hyphomicrobium* uses methanol as electron donor with nitrate as electron acceptor under anoxic conditions. Virtually the only denitrifying organism using methanol is *Hyphomicrobium*, and so this procedure selects this organism out of a wide variety of environments.

Prosthecate and Stalked *Bacteria*

Prosthecate and stalked (Figures 12.37 and 12.41) bacteria are appendaged chemoorganotrophic aerobes that attach to particulate matter, plant material, or other microorganisms in aquatic habitats. Although a major function of these appendages is undoubtedly attachment, they also increase significantly the surface-to-volume ratio of the cells. Recall from Chapter 4 that the high surface-to-volume ratio of small cells like bacteria confers an increased ability on them to take up nutrients

and expel wastes. The unusual morphology of appendaged bacteria (Figure 12.37) carries this theme to an extreme and may be an evolutionary adaptation to life in the oligotrophic (nutrient-poor) waters where these organisms are most commonly found. Selective isolation of prosthecate/stalked bacteria can be achieved by mixing an inoculum with a very dilute nutrient solution, for example 0.01% peptone, and leaving it to sit undisturbed. Within a few days a surface film often develops containing prosthecae and stalked bacteria and isolation can be achieved by streaking on agar plates.

Another function of prosthecae may be to decrease sedimentation rates in aquatic environments. It has been shown that centrifugation of prosthecate cells in the laboratory requires greater centrifugal force than for nonprosthecate cells. This inherent tendency to resist sinking is likely due to the prosthecae and might be an advantage for prosthecate cells in their natural habitats. Because these organisms are in general strict aerobes, prosthecae may keep cells from sinking into sediments or other anoxic zones where they would be unable to respire.

Caulobacter and *Gallionella*

Two commonly observed stalked bacteria are *Caulobacter* (Figure 12.41) and *Gallionella* (see Figure 12.43). The former is a chemoorganotroph that produces a cytoplasm-filled stalk, that is, a prostheca, while the latter is a chemolithotrophic iron-oxidizing bacterium whose stalk is composed of ferric hydroxide. *Caulobacter* cells are often seen on surfaces in aquatic environments with the stalks of several cells attached to form *rosettes* (Figure 12.41*a*). At the end of the stalk is a structure called a *holdfast* by which the stalk is attached to a surface.

The *Caulobacter* cell division cycle (Figure 12.42) is of special interest because it involves a process of *unequal binary fission*, and much molecular research has been done on this organism to understand cell-division and developmental events. Cell division of a stalked cell occurs by elongation of the cell followed by fission, a single flagellum forming at the pole opposite the stalk. The flagellated cell so formed, called a *swarmer*, separates from the nonflagellated mother cell, swims around, and attaches to a new surface, forming a new stalk at the flagellated pole; the flagellum is then lost (Figure 12.42). Stalk formation is a necessary precursor of cell division and is coordinated with DNA synthesis (Figure 12.42). The cell division cycle in *Caulobacter* is thus more complex than simple binary fission because the stalked and swarmer cells are structurally different and the growth cycle must include both forms.

Another interesting stalked organism is *Gallionella*, which forms a twisted stalk containing ferric hydroxide (Figure 12.43). However, the stalk of *Gallionella* is not an integral part of the cell but is *excreted* from the cell surface. It contains an organic matrix on which the ferric hydroxide accumulates. *Gallionella* is frequently found in the waters draining bogs, iron springs, and other habitats where ferrous iron (Fe^{2+}) is present, usually in association with sheathed bacteria such as *Sphaerotilus*. *Gallionella* is an au-

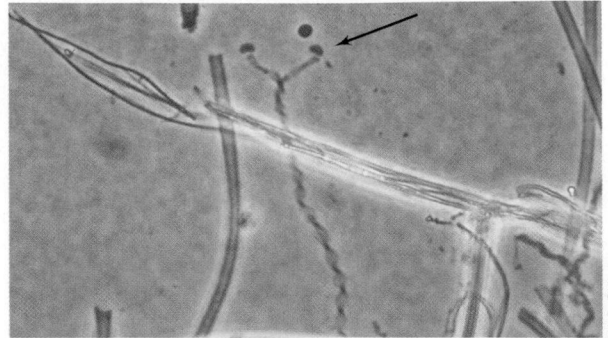

(a)

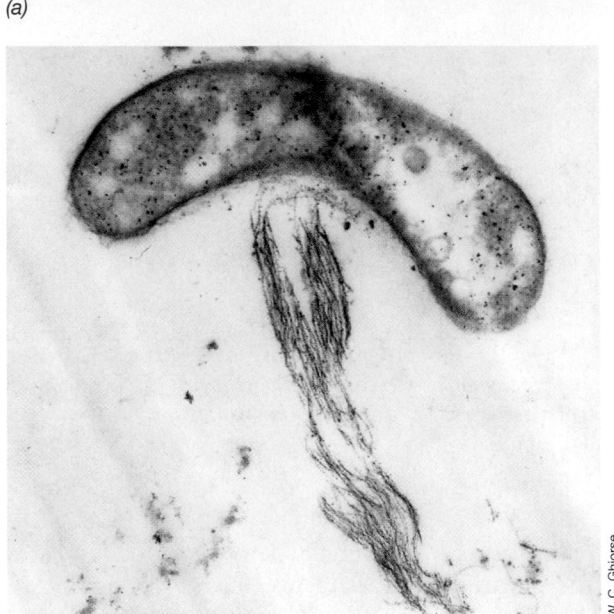

(b)

Figure 12.43 *Gallionella ferruginea.* (a) Photomicrograph of cells from an iron seep near Ithaca, New York. Notice the twisted stalk leading to the two bean-shaped cells (arrow). (b) Transmission electron micrograph of a thin section of a cell. Cells are about 0.6 μm wide. Note the twisted stalk of ferric hydroxide emanating from the center of the cell in both photos.

totrophic chemolithotroph containing enzymes of the Calvin cycle (◌◌ Section 17.6) by which CO_2 is incorporated into cell material with Fe^{2+} as electron donor.

✓ *12.15–12.16 Concept Check*

Sheathed bacteria are filamentous Proteobacteria in which individual cells form chains within an outer layer called the sheath. Budding and prosthecate bacteria are appendaged cells that form stalks or prosthecae used for attachment or nutrient absorption and are primarily aquatic.

- ✓ Physiologically, what is unique about the sheathed bacterium *Leptothrix*?
- ✓ How does *budding* division differ from *binary* fission?
- ✓ What advantage might a prosthecate organism have in a very nutrient-poor environment?

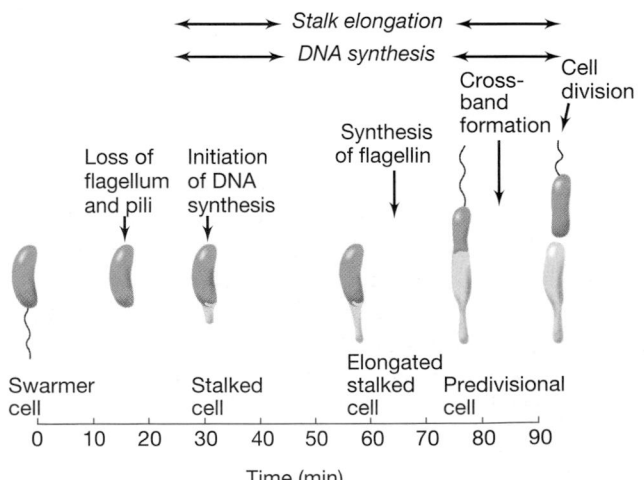

Figure 12.42 Stages in the *Caulobacter* cell cycle beginning with a swarmer cell.

TABLE 12.20 Classification of the fruiting myxobacteria[a]

Characteristics	Genus	DNA (mol% GC)
Vegetative cells tapered		
Spherical or oval myxospores, fruiting bodies usually soft and slimy without well-defined sporangia or stalks	*Myxococcus*	68–71
Rod-shaped myxospores:		
Myxospores not contained in sporangia, fruiting bodies without stalks	*Archangium*	67–68
Myxospores embedded in slime envelope:		
Fruiting bodies without stalks	*Cystobacter*	68
Stalked fruiting bodies, single sporangia	*Melittangium*	—
Stalked fruiting bodies, multiple sporangia	*Stigmatella*	68–69
Fruiting bodies are dark-brown clusters consisting of tiny spherical or disclike sporangia with an outer wall	*Angiococcus*	—
Vegetative cells not tapered (blunt, rounded ends); myxospores resemble vegetative cells; sporangia always produced:		
Fruiting bodies without stalks; myxospores rod-shaped	*Polyangium*	69
Fruiting bodies without stalks; myxospores oval; highly cellulolytic	*Sorangium*	—
Fruiting bodies without stalks; myxospores coccoid	*Nannocystis*	70–72
Stalked fruiting bodies	*Chondromyces*	69–70

[a] Phylogenetically, those species examined fall into the delta subdivision of the Proteobacteria (see Table 12.1).

12.17 Gliding Myxobacteria

Key Genera

Myxococcus
Stigmatella

A variety of prokaryotes exhibit a form of motility called *gliding*. We discussed the process of gliding motility in Section 4.11. Gliding microorganisms, usually long rods or filaments, lack flagella but are able to move when in contact with surfaces. One group of gliding bacteria, the fruiting myxobacteria, possess the interesting property of forming multicellular structures called **fruiting bodies** and show complex developmental lifecycles involving intercellular communication (see Figure 12.47). Phylo-genetically, the gliding myxobacteria are members of the *delta* subdivision of Proteobacteria (Table 12.20).

The fruiting myxobacteria exhibit the most complex behavioral patterns and life cycles of all known prokaryotic organisms. In keeping with this complexity, the chromosome of some myxobacteria is very large. *Myxococcus xanthus*, for example, has a single circular chromosome of 9500 kilobase pairs, *twice* as large as that of *Escherichia coli* (∞ Section 10.12). Indeed, this is two-thirds the size of the entire yeast genome, which is contained on 16 chromosomes (∞ Section 14.6). The vegetative cells of the fruiting myxobacteria are simple, nonflagellated, gram-negative rods (Figure 12.44*a*) that glide across surfaces and obtain their nutrients primarily by causing the lysis of other bacteria. Under appropriate conditions, a swarm of vegetative cells aggregate and construct *fruiting bodies*, within which some of the cells become

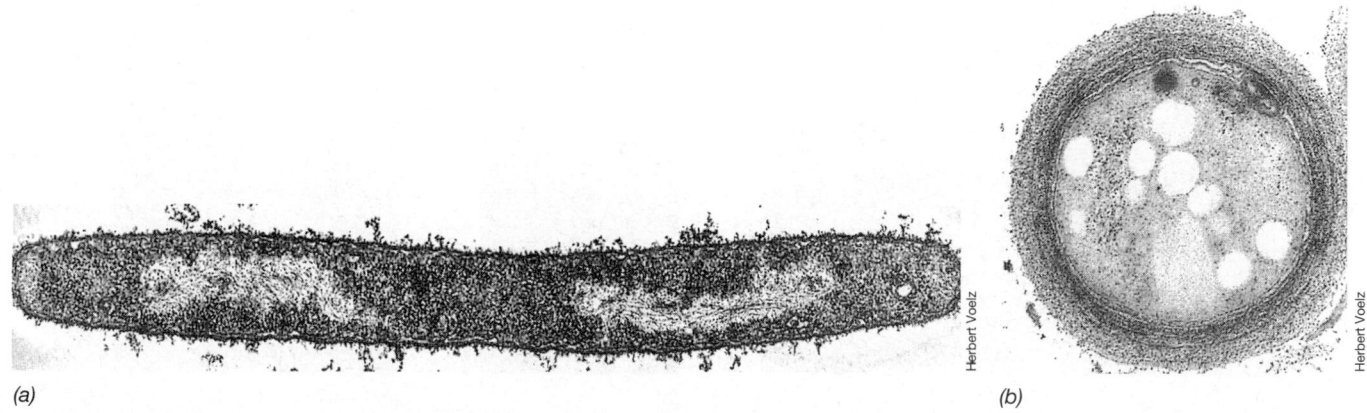

(a) *(b)*

Herbert Voelz   Herbert Voelz

Figure 12.44 (a) Electron micrograph of a thin section of a vegetative cell of *Myxococcus xanthus*. A cell measures about 0.75 μm wide. (b) Myxospore of *M. xanthus*, showing the multilayered outer wall. Myxospores measure about 2 μm in diameter.

converted to resting structures called **myxospores** (Figure 12.44*b*).

The fruiting bodies of the myxobacteria vary from simple globular masses of myxospores in loose slime to complex forms with a fruiting-body wall and a stalk (Figure 12.45). The fruiting bodies are often strikingly colored (Figures 12.45 and 12.46). They can usually be seen with a hand lens or dissecting microscope on moist pieces of decaying wood or plant material. Fruiting bodies of myxobacteria often develop on dung pellets (for example, rabbit pellets) after they have been incubated for a few days in a moist chamber. Another way of isolating fruiting myxobacteria is to prepare Petri plates of water agar (1.5% agar in distilled water with no added nutrients) on to which is spread a heavy suspension of virtually any bacterium. In the center of the plate a small amount of soil, decaying bark, or other natural material is placed. Myxobacteria in the inoculum lyse the bacterial cells and use their liberated products as nutrients; as they grow, they swarm out across the plate from the inoculum site. After several days to a week, the plates are examined under a dissecting microscope for myxobacterial swarms or fruiting bodies, and pure cultures are obtained by transfer of cells from the fruiting bodies or from the edge of the swarm to organic media. Many myxobacteria can be grown in the laboratory on media containing peptone or casein hydrolysate, which provides organic nutrients in the form of amino acids or small peptides. The organisms are typical aerobes with a complete citric acid cycle and cytochrome system.

Life Cycle of a Fruiting Myxobacterium

The life cycle of a typical fruiting myxobacterium is shown in Figure 12.47. A vegetative cell usually excretes slime, and as it moves across a solid surface it leaves a slime trail behind (Figure 12.48*a*). This trail is preferen-

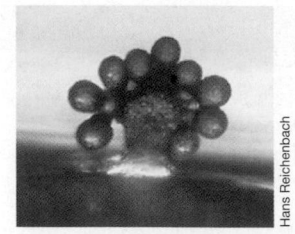

(a)

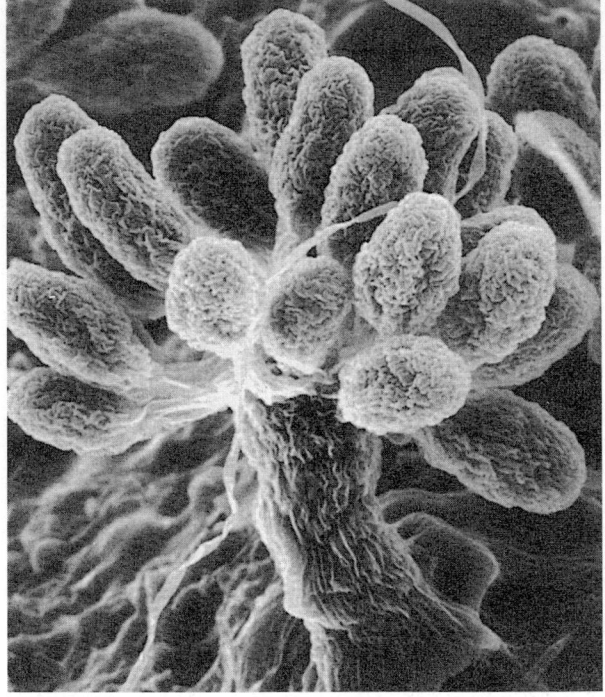

(b)

Figure 12.45 *Stigmatella aurantiaca.* (a) Color photo of a single fruiting body. The structure is about 150 μm high. (b) Scanning electron micrograph of a fruiting body growing on a piece of wood. Note the individual cells visible in each fruiting structure.

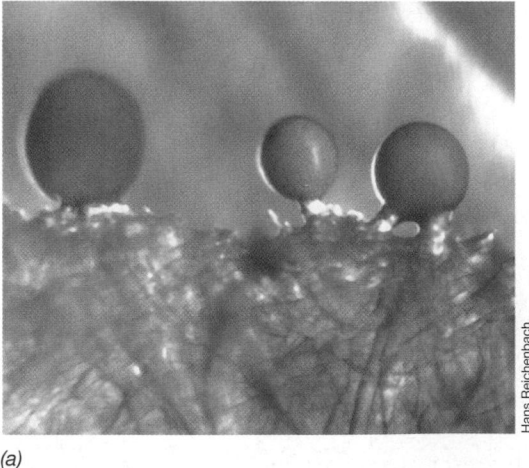

(a)

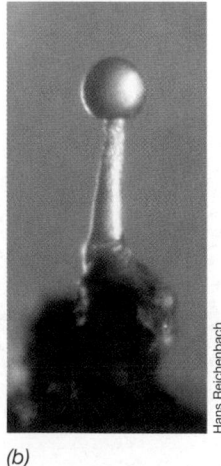

(b)

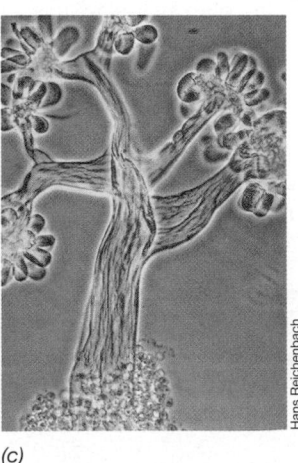

(c)

Figure 12.46 Fruiting bodies of four species of fruiting myxobacteria. (a) *Myxococcus fulvus* (125 μm high). (b) *Myxococcus stipitatas* (170 μm high). (c) *Mellitangium erectum* (50 μm high). (d) *Chondromyces crocatus* (560 μm high).

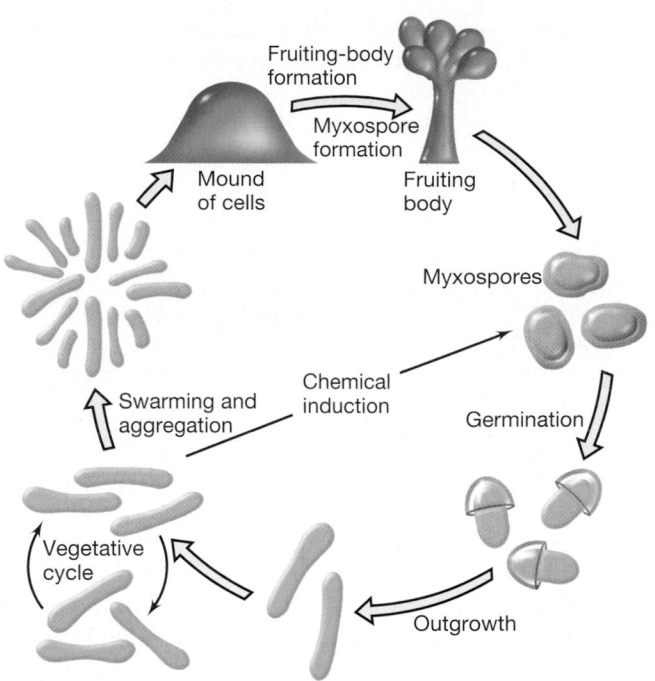

Figure 12.47 Life cycle of *Myxococcus xanthus*. Aggregation serves to assemble vegetative cells for fruiting-body formation. Vegetative cells undergo morphogenesis to resting cells called myxospores. The latter germinate under favorable nutritional and physical conditions to yield vegetative cells. Vegetative cells can be converted directly to myxospores without fruiting-body formation by certain chemical inducers, notably high concentrations of glycerol. See the photograph of *Myxococcus* fruiting bodies in Figure 12.45 and 12.46.

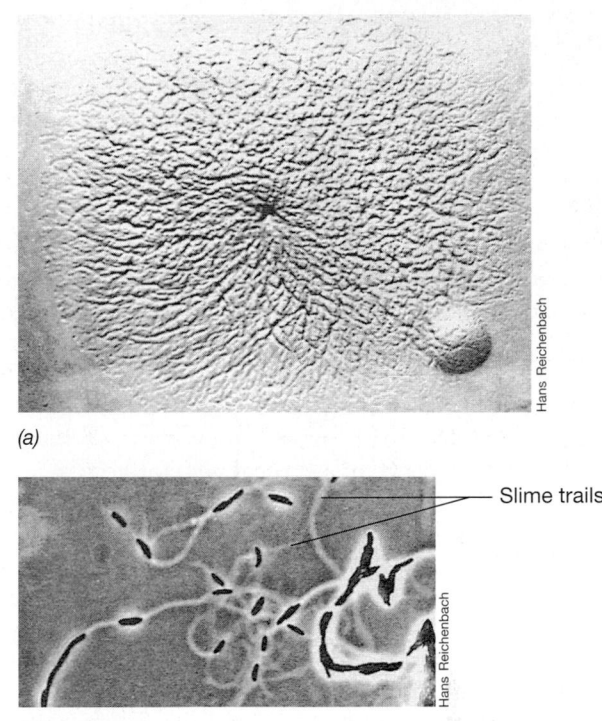

Figure 12.48 (a) Photomicrograph of a swarming colony (9-mm diameter) of *Myxococcus xanthus* on agar. (b) Single cells of *Myxococcus fulvus* from an actively gliding culture, showing the characteristic slime trails on the agar. A cell of *M. fulvus* is about 0.8 μm in diameter.

tially used by other cells in the swarm so that a characteristic radiating pattern is soon created, with cells migrating along slime trails (Figure 12.48*b*). The fruiting body ultimately formed (Figures 12.45 and 12.46) is a complex structure produced by the differentiation of cells in the stalk region and in the myxospore-bearing head.

Fruiting-body formation does not occur so long as adequate nutrients for vegetative growth are present, but upon nutrient exhaustion, the vegetative swarms begin to fruit. Cells aggregate, likely through a chemotactic response, with the cells migrating toward each other and forming mounds or heaps (Figure 12.49); a single fruiting body may have 10^9 or more cells. As the cell mounds become higher, the differentiation of the fruiting body into stalk and head begins (Figure 12.49*b* and *c*). Figure 12.49*d* clearly illustrates the differentiation of the fruiting body into stalk and head. The stalk is composed of slime, within which a few cells may be

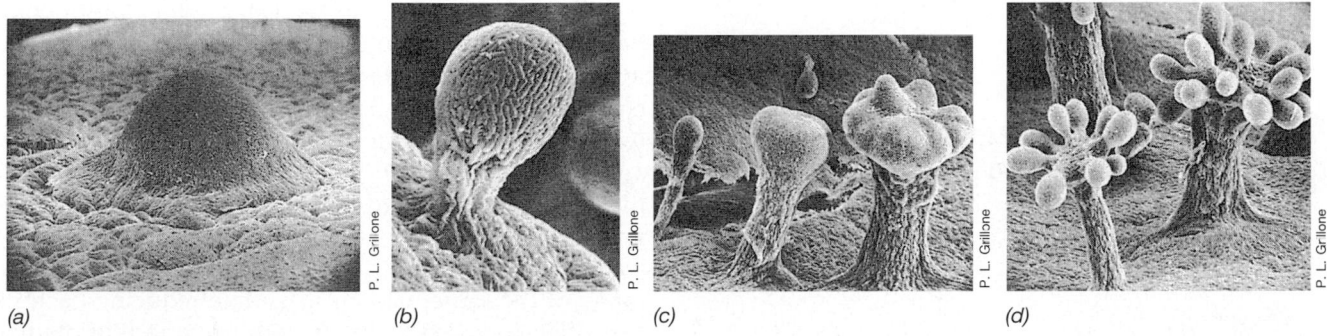

Figure 12.49 Scanning electron micrographs of fruiting-body formation in *Chondromyces crocatus*. (a) Early stage, showing aggregation and mound formation. (b) Initial stage of stalk formation. Slime formation in the head has not yet begun, and so the cells of which the head is composed are still visible. (c) Three stages in head formation. Note that the diameter of the stalk also increases. (d) Mature fruiting bodies. The entire fruiting structure is about 600 μm in height (see Figure 12.46*c*).

trapped. The majority of the cells accumulate in the fruiting-body head and undergo differentiation into *myxospores* (Figures 12.44–12.47). And, in some genera, the myxospores are enclosed in large walled structures called **cysts**. Compared to the vegetative cell, the myxospore is more resistant to drying, sonic vibration, UV radiation, and heat, but the degree of heat resistance is much less than that of the bacterial endospore. It seems likely that the main function of encysted myxospores is to enable the organism to survive desiccation during dispersal or during drying of the habitat. Upon dissemination to a suitable habitat or restoration of adequate growth conditions, the myxospore eventually germinates by a localized rupture of the capsule, with the growth and emergence of a typical vegetative rod.

Myxobacteria are usually colored by carotenoid pigments (see Figures 12.45a and 12.46), and the main pigments are carotenoid glycosides. Pigment formation is promoted by light, and at least one function of the pigment is photoprotection. Since in nature the myxobacteria usually form fruiting bodies in the light, the presence of these photoprotective pigments is understandable. In the genus *Stigmatella* (Figure 12.45), light greatly stimulates fruiting-body formation, and catalyzes production of a lipid pheromone called 2,5,8-trimethyl-8-hydroxy-nonane-4-one that initiates the aggregation step. The fruiting myxobacteria are currently classified primarily on morphological grounds using characteristics of the vegetative cells, the myxospores, and fruiting-body structure (Table 12.20).

✓ 12.17 Concept Check

The fruiting myxobacteria are rod-shaped, gliding bacteria that aggregate to form complex masses of cells called *fruiting bodies*. Myxobacteria are chemoorganotrophic soil bacteria that live by consuming dead organic matter or other bacterial cells.

- ✓ What triggers fruiting body formation in myxobacteria?
- ✓ What is a *myxospore* and how does it compare with an *endospore*?
- ✓ To what specific phylogenetic group do the myxobacteria belong?

12.18 Sulfate- and Sulfur-Reducing Proteobacteria

Key Genera

Desulfovibrio
Desulfobacter
Desulfuromonas

Sulfate (SO_4^{2-}) and sulfur (S^0) function as electron acceptors under anoxic conditions for a large group of delta Proteobacteria that utilize organic compounds or H_2 as electron donors; hydrogen sulfide (H_2S) is the product of both SO_4^{2-} and S^0 reduction. Over 20 genera of these organisms, collectively known as the *dissimilative sulfate-* or *sulfur-reducing bacteria*, are known, and some of the key ones are shown in Table 12.21. The genera in group I, such as *Desulfovibrio* (Figure 12.50a), *Desulfomonas, Desulfotomaculum,* and *Desulfobulbus* (Figure 12.50c), utilize lactate, pyruvate, ethanol, or certain fatty acids as electron donors, reducing sulfate to hydrogen sulfide. The genera in group II, such as *Desulfobacter* (Figure 12.50d), *Desulfococcus, Desulfosarcina* (Figure 12.50e), and *Desulfonema* (Figure 12.50b), specialize in the oxidation of fatty acids, particularly *acetate*, reducing sulfate to sulfide. The sulfate-reducing bacteria are for the most part obligate anaerobes, and strict anoxic techniques must be used in their cultivation.

Sulfate-reducing bacteria are widespread in aquatic and terrestrial environments that become anoxic as a result of microbial decomposition processes. The best-studied genus is *Desulfovibrio* (Figure 12.50a), which is common in aquatic habitats or waterlogged soils containing abundant organic material and sufficient levels of sulfate. *Desulfotomaculum*, phylogenetically a member of the gram-positive *Bacteria*, consists of endospore-forming rods found primarily in soil. Growth and reduction of sulfate by *Desulfotomaculum* in certain canned foods leads to a type of spoilage called *sulfide stinker*. The remaining genera of sulfate reducers are indigenous to anoxic freshwater or marine environments, and can occasionally be isolated from the mammalian intestine.

Sulfur Reduction

The dissimilative sulfur-reducers can reduce elemental sulfur to sulfide but are unable to reduce sulfate to sulfide. Members of the genus *Desulfuromonas* (Figure 12.50f) can grow anaerobically by coupling the *oxidation* of substrates such as acetate or ethanol to the *reduction* of elemental sulfur to hydrogen sulfide. However, the ability to reduce elemental sulfur, as well as other sulfur compounds such as thiosulfate, sulfite, or dimethyl sulfoxide (DMSO), is a widespread property of a variety of chemoorganotrophic, generally facultatively aerobic bacteria (for example, *Proteus, Campylobacter, Pseudomonas,* and *Salmonella*). *Desulfuromonas* differs from the latter in that it is an obligate anaerobe and utilizes *only* sulfur as an electron acceptor (see Table 12.21). Dissimilative sulfur-reducing bacteria like *Desulfuromonas* reside in many of the same habitats as sulfate-reducing bacteria and often form associations with bacteria that oxidize H_2S to S^0, like green sulfur bacteria (see Section 12.33). The sulfur produced is then reduced back to H_2S during metabolism of the sulfur-reducer to complete a very abbreviated anaerobic sulfur cycle (⟳ Section 19.13).

TABLE 12.21 Characteristics of sulfate- and sulfur-reducing bacteria[a]

Genus	Characteristics	DNA (mol % GC)
Group I sulfate reducers: Nonacetate oxidizers		
Desulfovibrio	Polarly flagellated, curved rods, no spores; gram-negative; contain desulfoviridin; twelve species, one thermophilic	46–61
Desulfomicrobium	Motile rods, no spores; gram-negative; desulfoviridin absent; three species	52–57
Desulfobotulus	Vibrios; gram-negative; motile; desulfoviridin absent; one species	53
Desulfofustis	Motile rods, specializes in the degradation of glycolate and glyoxalate	56
Desulfotomaculum	Straight or curved rods; motile by peritrichous or polar flagellation; gram-negative; desulfoviridin absent; produce endospores; four species, one thermophilic; one species capable of utilizing acetate as energy source	37–46
Desulfomonile	Rod; capable of reductive dechlorination of 3-chlorobenzoate to benzoate (∞ Section 17.18)	49
Desulfobacula	Oval to coccoid cells, marine; can oxidize various aromatic compounds including the aromatic hydrocarbon toluene, to CO_2; one species	42
Archaeoglobus	Archaeon; hyperthermophile, temperature optimum, 83°C; contains some unique coenzymes of methanogenic bacteria, makes small amount of methane during growth; H_2, formate, glucose, lactate, and pyruvate are electron donors, SO_4^{2-}, $S_2O_3^{2-}$, or SO_3^{2-}, electron acceptors; two species (∞ Section 13.7)	41–46
Desulfobulbus	Ovoid or lemon-shaped cells; no spores; gram-negative; desulfoviridin absent; if motile, by single polar flagellum; utilizes propionate as electron donor with acetate $+ CO_2$ as products; three species	59–60
Desulforhopalus	Curved rods, gas vacuolate, psychrophile; uses propionate, lactate, or alcohols as electron donor	48
Thermodesulfobacterium	Small, gram-negative rods; desulfoviridin present; thermophilic, optimum growth at 70°C; a member of the *Bacteria* but contains ether-linked lipids (see Section 12.36)	34
Group II sulfate reducers: Acetate oxidizers		
Desulfobacter	Rods; no spores, gram-negative; desulfoviridin absent; if motile, by single polar flagellum; utilizes only acetate as electron donor and oxidizes it to CO_2 via the citric acid cycle; four species	45–46
Desulfobacterium	Rods, some with gas vesicles, marine; capable of autotrophic growth via the acetyl-CoA pathway; three species	41–59
Desulfococcus	Spherical cells; nonmotile; gram-negative; desulfoviridin present, no spores; utilizes C_1 to C_{14} fatty acids as electron donor with complete oxidation to CO_2; capable of autotrophic growth via the acetyl-CoA pathway; two species	57
Desulfonema	Large, filamentous gliding bacteria; gram-positive, no spores; desulfoviridin present or absent; utilizes C_2 to C_{12} fatty acids as electron donor with complete oxidation to CO_2; capable of autotrophic growth via the acetyl-CoA pathway (H_2 as electron donor); two species	35–42
Desulfosarcina	Cells in packets (sarcina arrangement); gram-negative; no spores; desulfoviridin absent; utilizes C_2 to C_{14} fatty acids as electron donor with complete oxidation to CO_2; capable of autotrophic growth via the acetyl-CoA pathway (H_2 as electron donor); one species	51
Desulfoarculus	Vibrios; gram-negative; motile; desulfoviridin absent; utilizes only C_1 to C_{18} fatty acids as electron donor	66
Desulfacinum	Cocci to oval-shaped cells; gram-negative; utilizes C_1 to C_{18} fatty acids, very nutritionally diverse, capable of autotrophic growth; thermophile	64
Desulforhabdus	Rods; no spores; gram-negative; nonmotile; utilizes fatty acids with complete oxidation to CO_2	52
Thermodesulforhabdus	Gram-negative motile rods; thermophilic; uses fatty acids up to C_{18}	51
Dissimilatory sulfur reducers		
Desulfuromonas	Straight rods, single lateral flagellum; no spores; gram negative; does not reduce sulfate; acetate, succinate, ethanol, or propanol used as electron donor; obligate anaerobe; four species, at least one of which is capable of the reductive dechlorination of trichloroethylene (∞ Section 17.18)	50–63
Desulfurella	Motile short rods; gram-negative; requires acetate; thermophilic	31
Sulfurospirillum	Small vibrios, reduces S^0 with H_2 or formate as electron donors	—
Campylobacter	Curved, vibrio-shaped rods; polar flagella; gram-negative; no spores; unable to reduce sulfate but can reduce sulfur, sulfite, thiosulfate, nitrate, or fumarate anaerobically with acetate or a variety of other carbon or electron donor sources; facultative aerobe	40–42

[a] Phylogenetically, most sulfate- and sulfur-reducing bacteria are delta Proteobacteria.

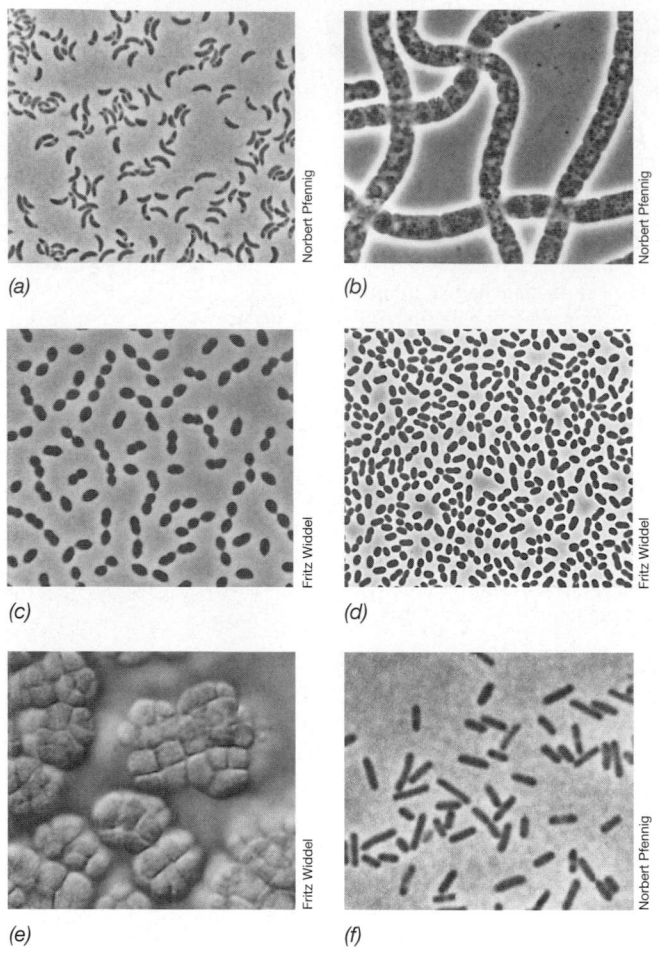

(a)   (b)

(c)   (d)

(e)   (f)

Figure 12.50 Phase contrast photomicrographs of (a–e) representative sulfate-reducing and (f) sulfur-reducing bacteria. (a) *Desulfovibrio desulfuricans*; cell diameter about 0.7 μm. (b) *Desulfonema limicola*; cell diameter 3 μm. (c) *Desulfobulbus propionicus*; cell diameter about 1.2 μm. (d) *Desulfobacter postgatei*; cell diameter about 1.5 μm. (e) *Desulfosarcina variabilis* (interference contrast microscopy); cell diameter about 1.25 μm. (f) *Desulfuromonas acetoxidans*; cell diameter about 0.6 μm.

Physiology of Sulfate-Reducing Bacteria

The range of electron donors used by sulfate-reducing bacteria is broad. H_2, lactate, and pyruvate are almost universally used, and many group I species utilize malate, sulfonates, and certain primary alcohols (for example, methanol, ethanol, propanol, and butanol). Some strains of *Desulfotomaculum* utilize glucose, but this is rather rare among sulfate reducers in general. Group I sulfate reducers oxidize their energy source to the level of acetate and excrete this fatty acid as an end product. Group II organisms differ from those in group I by their ability to oxidize fatty acids, lactate, succinate, and even benzoate in some cases, all the way to CO_2. *Desulfosarcina, Desulfonema, Desulfococcus, Desulfobacterium, Desulfotomaculum*, and certain species of *Desulfovibrio*, are unique among sulfate-reducers in their ability to grow

chemolithotrophically and autotrophically with H_2 as electron donor, sulfate as electron acceptor, and CO_2 as sole carbon source.

In addition to using sulfate as an electron acceptor, many sulfate-reducing bacteria can grow using *nitrate* (NO_3^-) as an electron acceptor, reducing NO_3^- to NH_3, or sulfonates, such as isethionate ($HO-CH_2-CH_2-SO_3^-$), or elemental sulfur (S^0), both of which are reduced to H_2S, or can use certain organic compounds for energy generation by fermentative pathways in the complete absence of terminal electron acceptors. The most common fermentable compound is *pyruvate*, which is converted via the phosphoroclastic reaction to acetate, CO_2, and H_2 (⬤ Figure 17.51). Moreover, although thought for a long time to be *obligate* anaerobes, certain isolates of sulfate-reducing bacteria, primarily ones isolated from microbial mats where they coexist with O_2-producing cyanobacteria, are quite oxygen tolerant and actually respire with O_2 as electron acceptor. However, aerobic respiration does *not* support growth and is probably a means of removing O_2 when it is present in an environment otherwise suitable for growth of sulfate-reducing bacteria.

Isolation

The enrichment of *Desulfovibrio* is relatively easy on an anoxic lactate-sulfate medium to which ferrous iron is added. A reducing agent, such as thioglycolate or ascorbate, is also added to achieve a lower E_0'. When sulfate-reducing bacteria grow, the sulfide formed from sulfate reduction combines with the ferrous iron to form black, insoluble ferrous sulfide. This blackening not only indicates sulfate reduction, but the iron also ties up and detoxifies the sulfide, making possible growth to higher cell yields. After some growth has occurred as evidenced by blackening of the medium, purification is accomplished by streaking onto a tube coated on the inside surface with a thin layer of agar (called *roll tubes*) or on Petri plates in an anoxic glove box. Alternatively, agar shake tubes can be used for purification purposes. In the *shake tube method* a small amount of liquid from the original enrichment is added to a tube of molten agar growth medium, mixed thoroughly, and sequentially diluted through a series of molten agar tubes (⬤ Section 18.1 and Figure 18.3b). On solidification, individual cells distributed throughout the agar form black colonies that can be removed aseptically, and the whole process is repeated until pure cultures are obtained.

✓ *12.18 Concept Check*

Sulfate- and sulfur-reducing bacteria are a large group of delta Proteobacteria unified by their physiological process of reducing either SO_4^{2-} or S^0 to H_2S under anaerobic conditions. Two physiological subgroups of sulfate-reducing bacteria are known: group I, which is incapable of oxidizing acetate to CO_2, and group II, which is capable of doing so.

✓ What organic substrate would you use to enrich and isolate a group I sulfate reducer from nature?

✓ For sulfate-reducing bacteria capable of chemolithotrophic and autotrophic growth: (1) What is the electron donor, (2) what is the electron acceptor, and (3) what is the source of cell carbon?

✓ Physiologically, how does *Desulfuromonas* differ from *Desulfovibrio*?

III PHYLUM 2: GRAM-POSITIVE BACTERIA

12.19 Nonsporulating, Low GC, Gram-Positive *Bacteria*: Lactic Acid Bacteria and Relatives

Key Genera

Staphylococcus
Micrococcus
Streptococcus
Lactobacillus

As mentioned in the introduction to this chapter, gram-positive *Bacteria* fall into two major phylogenetic subdivisions, "low GC" and "high GC." We consider in this section the genera *Staphylococcus*, *Sarcina*, and *Micrococcus* because they are morphologically quite similar even though *Micrococcus* is actually a member of the "High GC" group, along with the lactic acid *Bacteria*, classical nonsporulating gram-positive rods, and cocci. An overview of the properties of these organisms is given in Table 12.22.

Staphylococcus and Micrococcus

Staphylococcus (Figure 12.51) and *Micrococcus* are both aerobic organisms with a typical respiratory metabolism. They are catalase-positive, and this test permits their distinction from *Streptococcus* and some other genera of gram-positive cocci. Gram-positive cocci are relatively resistant to reduced water potential and tolerate drying and high salt fairly well. Their ability to grow in media with high salt provides a selective means for isolation. For example, if an appropriate inoculum is spread on an agar plate with a rich medium containing 7.5% NaCl and the plate incubated aerobically, gram-positive cocci often form the predominant colonies. Often, these organisms are pigmented, and this provides an additional aid in selecting gram-positive cocci. The two genera *Micrococcus* and *Staphylococcus* can easily be separated based on the oxidation-fermentation (O/F) (∞ Table 24.3) test. *Micrococcus* is an obligate aerobe and produces acid from glucose only aerobically, whereas *Staphylococcus* is a facultative aerobe and produces acid from glucose both aerobically and anaerobically.

TABLE 12.22 Distinguishing features of major gram-positive cocci

Genus	Motility	Arrangement of cells	Growth by fermentation	DNA (mol % GC)	Phylogenetic group[a]	Other characteristics
Micrococcus	−	Clusters, tetrads	−	66–73	High GC	Strict aerobe
Staphylococcus	−	Clusters, pairs	+	30–39	Low GC	Only genus of this group to contain teichoic acid in cell wall
Stomatococcus	−	Clusters, pairs	+	56–60	Actinobacterial group	Only genus of this group containing a capsule
Planococcus	+	Pairs, tetrads	−	39–52	Low GC	Primarily marine
Sarcina	−	Cuboidal packets of eight or more cells	+	28–31	Low GC	Extremely acid-tolerant; cellulose in cell wall
Ruminococcus	+	Pairs, chains	+	39–46	Low GC	Obligate anaerobe; inhabits rumen, cecum, and large intestine of many animals
Peptococcus	−	Clusters, pairs	+	50–51	Low GC	Obligate anaerobe; ferments peptone but not sugars
Peptostreptococcus	−	Clumps, short chains	+	28–37	Low GC	Obligate anaerobe; ferments peptone; common member of human normal flora, skin, intestine, vagina; also isolated from vaginal and purulent discharges

[a]All are members of the gram-positive *Bacteria* (see Figure 12.1), except for *Stomatococcus*.

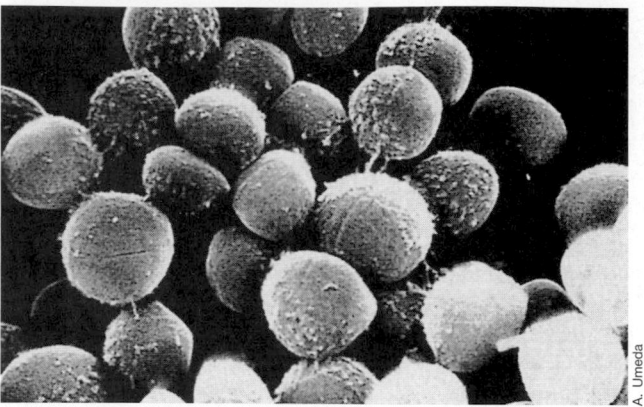

Figure 12.51 Scanning electron micrograph of typical *Staphylococcus*, showing the irregular arrangement of the cell clusters. Individual cells are about 0.8 μm in diameter.

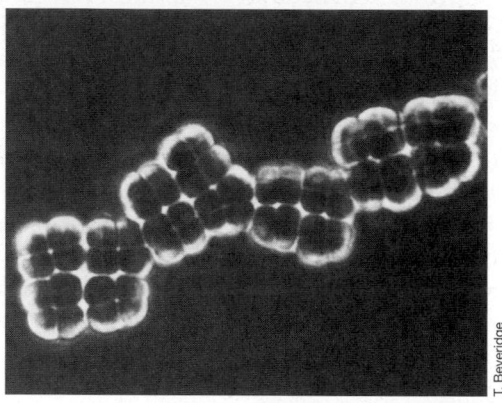

(a)

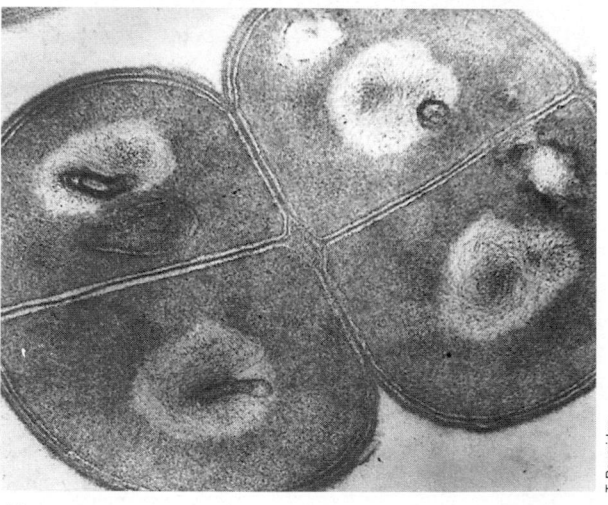

(b)

Figure 12.52 (a) Phase contrast photomicrograph of cells of a typical gram-positive coccus *Sarcina* sp. A single cell is about 2 μm in diameter. (b) Electron micrograph of a thin section. The outermost layer of the cell consists of cellulose.

Staphylococci are common parasites of humans and animals and occasionally cause serious infections. In humans, two major species are recognized, *Staphylococcus epidermidis*, a nonpigmented, nonpathogenic organism usually found on the skin or mucous membranes, and *Staphylococcus aureus*, a yellow pigmented species that is most commonly associated with pathological conditions, including boils, pimples, pneumonia, osteomyelitis, meningitis, and arthritis. We discuss the pathogenesis of *S. aureus* in Chapter 21 and staphylococcal diseases in Chapters 26 and 29. *Micrococcus* species can also be isolated from skin but are much more common on inanimate objects, dust particles, and in soil.

Sarcina

The genus *Sarcina* contains species of bacteria that divide in three perpendicular planes to yield packets of eight cells or more (Figure 12.52*a*). *Sarcina* are obligate anaerobes and are extremely acid-tolerant, being able to ferment sugars and grow down to pH 2. Cells of one species, *Sarcina ventriculi*, contain a thick fibrous layer of cellulose surrounding the cell wall (Figure 12.52*b*). The cellulose layers of adjacent cells become attached, and this functions as a cementing material to hold together packets of *S. ventriculi* cells. *Sarcina* species can be isolated from soil, mud, feces, and stomach contents. Because of its extreme acid tolerance, *S. ventriculi* is one of the only bacteria that can actually grow in the stomach of humans and other monogastric animals. Rapid growth of *S. ventriculi* is observed in the stomach of humans suffering from certain gastrointestinal pathological conditions (such as pyloric ulcerations) that retard the flow of food to the intestine.

Lactic Acid Bacteria and Lactic Acid Fermentations

The lactic acid bacteria are gram-positive rods and cocci that produce lactic acid as a major or sole fermentation product. Members of this group lack porphyrins and cytochromes, do not carry out electron transport phosphorylation, and hence obtain energy only by *substrate-level phosphorylation*. All lactic acid bacteria grow anaerobically. Unlike many anaerobes, however, most lactic acid bacteria are not sensitive to O_2 and can grow in its presence as well as in its absence; thus they are **aerotolerant anaerobes**. Most lactic acid bacteria obtain energy only from the metabolism of sugars and hence are usually restricted to habitats in which sugars are present. They usually have only limited biosynthetic ability, and their complex nutritional requirements include needs for amino acids, vitamins, purines, and pyrimidines (⬅ Table 5.4).

One important difference between subgroups of the lactic acid bacteria lies in the nature of the products formed from the fermentation of sugars. One group, called **homofermentative**, produces a single fermentation product, *lactic acid*, whereas the other group, called **heterofermentative**, produces other products, mainly *ethanol* and CO_2, as well as lactate (Table 12.23). Abbreviated pathways for the fermentation of glucose by a homo- and a heterofermentative organism are shown in Figure 12.53. The differences observed in the fermentation patterns are determined by the presence or absence of the enzyme **aldolase**, the key enzyme in *glycolysis* (Figure 5.14). The heterofermenters, lacking aldolase, cannot break down fructose bisphosphate to triose phosphate. Instead, they oxidize glucose 6-phosphate to 6-phosphogluconate, and then decarboxylate this to pentose phosphate, which is broken down to triose phosphate and acetylphosphate by means of the enzyme **phosphoketolase** (Figure 12.53).

In heterofermenters, triose phosphate is converted ultimately to lactic acid with the production of 1 mol of ATP, while the acetylphosphate accepts electrons from the NADH generated during the production of pentose phosphate and is thereby converted to ethanol *without* yielding ATP. Because of this, heterofermenters produce only *1 mol* of ATP from glucose instead of the 2 mol produced by homofermenters. Because the heterofermenters decarboxylate 6-phosphogluconate, they produce CO_2 as a fermentation product, whereas the homofermenters produce little or no CO_2; therefore, one simple way of detecting a heterofermenter is to observe for production of CO_2 in laboratory cultures.

The various genera of lactic acid bacteria have been defined on the basis of cell morphology, DNA base composition, and type of fermentative metabolism, as is shown in Table 12.23. Members of the genera *Streptococcus, Enterococcus, Lactococcus, Leuconostoc,* and *Pediococcus* have fairly similar DNA base ratio compositions; in addition, there is very little variation from strain to strain. The genus *Lactobacillus*, on the other hand, has members with widely diverse DNA compositions and hence does not constitute a homogeneous group.

Streptococcus and Other Cocci

The genus *Streptococcus* (Figure 12.54) contains a wide variety of homofermentative species with quite distinct habitats, whose activities are of considerable practical importance to humans. Some members are pathogenic to humans and animals (Section 26.2). As producers of lactic acid, certain streptococci play important roles in the production of buttermilk, silage, and other fermented products (Section 29.2), and certain species play a major role in dental caries formation (Section 21.3). To distinguish generally nonpathogenic streptococci from human pathogenic species, three genera are recognized. The genus *Lactococcus* contains those streptococci of dairy significance, whereas the genus *Enterococcus* includes streptococci that are primarily of fecal origin.

Organisms in the genus *Streptococcus* have been divided into two groups of related species on the basis of characteristics enumerated in Table 12.24. Hemolysis on blood agar is of considerable importance in the subdivision of the genus. Colonies of those strains producing streptolysin O or S are surrounded by a large zone of complete red blood cell hemolysis, a condition called **β hemolysis** (Figure 21.17*a*). On the other hand, many streptococci and the lactococci and enterococci do not produce hemolysins, but instead cause the formation of a greenish or brownish zone around colonies on blood agar, which is due not to true hemolysis, but to discoloration and loss of potassium from the red cells. This type of reaction has classically been referred to as **α hemolysis**. Streptococci and related cocci are also divided into *immunological* groups based on the presence of specific carbohydrate antigens. These antigenic groups (or **Lancefield groups** as they are commonly known, named for Rebecca Lancefield, a pioneer in *Streptococcus* taxonomy), are designated by letters; A through O are currently recognized. Those β-hemolytic streptococci found in human beings usually contain the group A antigen, while enterococci contain the group D antigen. Group B streptococci, usually found in association with animals, are a cause of mastitis in cows, and have also been implicated in human infections. Lactococci are of antigen group N and are not pathogenic.

TABLE 12.23 Differentiation of the principal genera of lactic acid bacteria[a]

Genus	Cell form and arrangement	Fermentation	DNA (mol % GC)
Streptococcus	Cocci in chains	Homofermentative	34–46
Leuconostoc	Cocci in chains	Heterofermentative	38–41
Pediococcus	Cocci in tetrads	Homofermentative	34–42
Lactobacillus	(1) Rods, usually in chains	Homofermentative	32–53
	(2) Rods, usually in chains	Heterofermentative	34–53
Enterococcus	Cocci in chains	Homofermentative	38–40
Lactococcus	Cocci in chains	Homofermentative	38–41

[a] Phylogenetically, all organisms are members of the low GC subdivision of the gram-positive *Bacteria*.

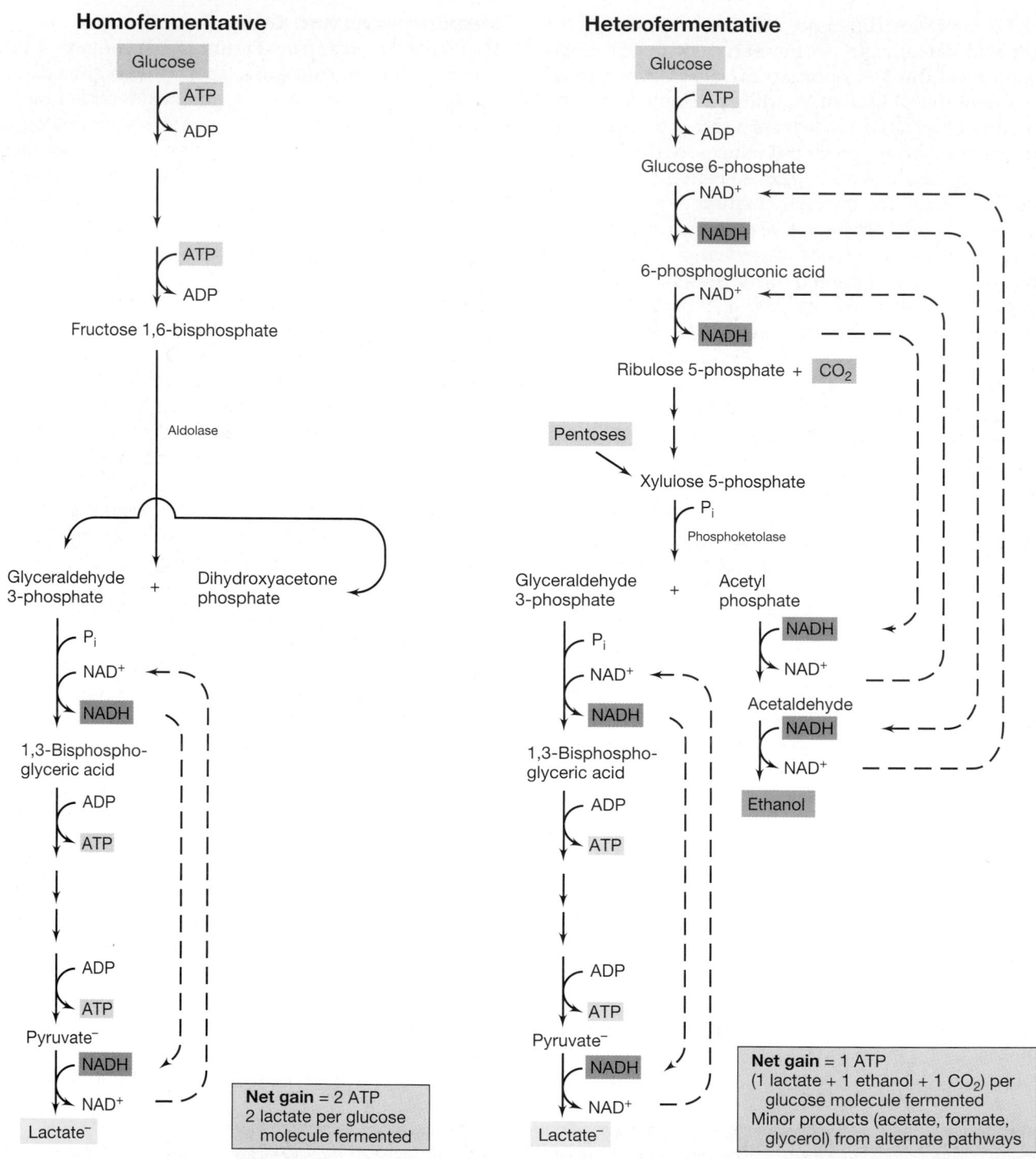

Figure 12.53 The fermentation of glucose in homofermentative and heterofermentative lactic acid bacteria. Note that no ATP is made in reactions leading to ethanol formation. If oxygen is present, many heterofermentative lactic acid bacteria can reduce oxygen with NADH (through flavin enzymes as intermediates), forming water; acetate is then made instead of ethanol, and this allows for the production of one additional ATP.

Placed in the genus *Leuconostoc* are heterofermentative cocci. Strains of *Leuconostoc* also produce the flavoring ingredients diacetyl and acetoin by breakdown of citrate and have been used as starter cultures in dairy fermentations. Some strains of *Leuconostoc* produce large amounts of dextran polysaccharides (α-1,6-glucan) when cultured on sucrose (Figure 17.64), and some of these have found medical use as plasma extenders in blood transfusions. Other strains of *Leuconostoc* produce other polysaccharide polymers such as fructose polymers called *levans*.

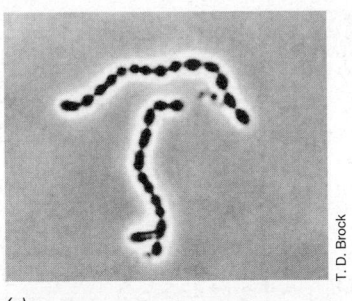

(a)

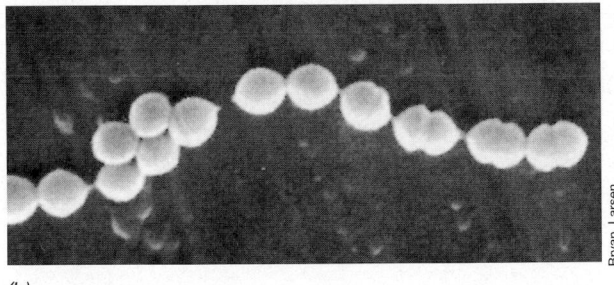

(b)

Figure 12.54 Phase contrast (a) and scanning electron (b) micrographs of *Streptococcus* species. (a) *Streptococcus lactis*. (b) *Streptococcus* sp. Cells in both cases are 0.5–1 µm in diameter.

Lactobacillus

Lactobacilli are typically rod-shaped, varying from long and slender to short, bent rods (Figure 12.55). Most species are homofermentative, but some are heterofermentative. Lactobacilli are common in dairy products, and some strains are used in the preparation of fermented milk products. For instance, *Lactobacillus delbrueckii* (Figure 12.55c) is used in the preparation of yogurt, *L. acidophilus* (Figure 12.55a) in the production of acidophilus milk, and other species in the production of sauerkraut, silage, and pickles (∞ Section 29.2). The lactobacilli are usually more resistant to acidic condi-

(a)

(b)

(c)

Figure 12.55 Phase contrast and electron micrographs of *Lactobacillus* species. (a) *Lactobacillus acidophilus*. Cells are about 0.75 µm wide. (b) *Lactobacillus brevis*, transmission electron micrograph. Cells measure about 0.8 × 2 µm. (c) *Lactobacillus delbrueckii*, scanning electron micrograph. Cells are about 0.7 µm in diameter. Both heterofermentative and homofermentative species of *Lactobacillus* are known (see Figure 12.53).

TABLE 12.24 Differential characteristics of streptococci, lactococci, and enterococci

Group	Antigenic (Lancefield) groups	Representative species	Type of hemolysis on blood agar	Good growth at 10°C	Good growth at 45°C	Survive 60°C for 30 min	Growth in Milk with 0.1% methylene blue	Growth in Broth with 40% bile	Habitat
Streptococci									
Pyogenes subgroup	A,B,C,F,G	*Streptococcus pyogenes*	Lysis (β)	−	−	−	−	−	Respiratory tract, systemic
Viridans subgroup	—	*Streptococcus mutans*	Greening (α)	−	+	−	−	−	Mouth, intestine
Enterococci	D	*Enterococcus faecalis*	Lysis (β), greening (α), or none	+	+	+	+	+	Intestine, vagina, plants
Lactococci	N	*Lactococcus lactis*	None	+	−	+	+	+	Plants, dairy products

TABLE 12.25 Major genera of endospore-forming bacteria[a]

Characteristics	Genus	DNA (mol % GC)
Rods		
Aerobic or facultative, catalase produced	*Bacillus*	32–69
	Paenibacillus	40–54
Microaerophilic, no catalase; homofermentative lactic acid producer	*Sporolactobacillus*	46–47
Anaerobic:		
Sulfate-reducing	*Desulfotomaculum*	38–50
Does not reduce sulfate, fermentative	*Clostridium* (see Figure 12.56)	21–54
Thermophilic, temperature optimum 65–70°C, fermentative	*Thermoanaerobacter*	31–39
Gram-negative; can grow as homoacetogen on $H_2 + CO_2$	*Sporomusa*	41–49
Halophile, isolated from the Dead Sea	*Sporohalobacter*	31
Produces up to five spores per cell; fixes N_2	*Anaerobacter*	29
Acidophile, pH optimum 3	*Alicyclobacillus*	52–60
Alkaliphile, pH optimum 9	*Amphibacillus*	36–38
Phototrophic	*Heliobacterium, Heliophilum*	50–58
Syntrophic, degrades fatty acids but only in coculture with a H_2-utilizing bacterium ( Section 17.21)	*Syntrophospora*	37
Reductively dechlorinates chlorophenols ( Section 17.18)	*Desulfitobacterium*	46
Cocci (usually arranged in tetrads or packets), aerobic	*Sporosarcina* (see Figure 12.60)	40–41

[a]Phylogenetically, all organisms are members of the low GC subdivision of the gram-positive *Bacteria*.

tions than are the other lactic acid bacteria, being able to grow well at pH values as low as 4. Because of this, they can be selectively isolated from natural materials by use of an acidic rich carbohydrate-containing medium such as tomato juice-peptone agar. The acid resistance of the lactobacilli enables them to continue growing during natural lactic fermentations when the pH value has dropped too low for other lactic acid bacteria to grow, and the lactobacilli are therefore responsible for the final stages of most lactic acid fermentations. The lactobacilli are rarely, if ever, pathogenic.

Listeria

Listeria are gram-positive coccobacilli that tend to form chains of three to five cells. *Listeria* is phylogenetically related to *Lactobacillus* species, and like lactic acid bacteria, produce acid but not gas from glucose. But unlike true lactic acid bacteria, organisms that are capable of growth under strictly anoxic conditions and that lack the enzyme catalase, *Listeria* requires microaerobic to aerobic growth conditions and produces catalase. Although several species of *Listeria* are known, the species *L. monocytogenes* is most noteworthy because it causes a major foodborne illness, *listeriosis* ( Section 29.9). The organism is transmitted in contaminated, usually ready-to-eat, foods (cheese is a common vehicle) and can cause anything from a mild illness to a fatal form of meningitis.

12.20 Endospore-Forming, Low GC, Gram-Positive *Bacteria*: *Bacillus*, *Clostridium*, and Relatives

Key Genera

Bacillus
Clostridium
Sporosarcina
Heliobacterium

Several genera of endospore-forming bacteria have been recognized (Table 12.25), distinguished on the basis of cell morphology, shape, and cellular position of the endospore (Figure 12.56), relationship to O_2, and energy metabolism; all endospore-formers show phylogenetic

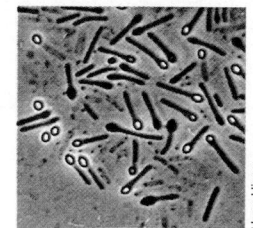
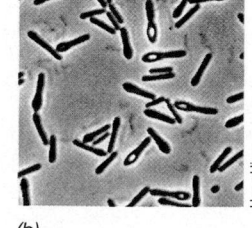
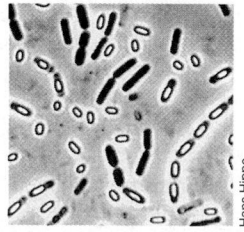

Figure 12.56 Phase contrast photomicrographs of various *Clostridium* species, showing the different locations of the endospore. (a) *Clostridium cadaveris*, terminal spores. Cells are about 0.9 μm wide. (b) *Clostridium sporogenes*, subterminal spores. Cells are about 1 μm wide. (c) *Clostridium bifermentans*, central spores. Cells are about 1.2 μm wide. (a) (b) (c)

affiliation to the "low GC" gram-positive *Bacteria*. The two genera most frequently studied are *Bacillus*, species of which are aerobic or facultatively aerobic, and *Clostridium*, which contains the strictly anaerobic, fermentative species. One group of endospore-formers, the *heliobacteria*, are actually phototrophic (the word *helio* comes from the Greek word for *sun*). The structure and heat resistance of the bacterial endospore along with the process of endospore formation itself was discussed in Section 4.15.

Although considerable genetic heterogeneity exists among endospore-formers—for example, the GC ratios of *Bacillus* species alone vary over a range of nearly 40%—all endospore-forming bacteria are *ecologically* related because they are found in nature primarily in soil. Even those species that are pathogenic to humans or other animals are primarily saprophytic soil organisms, and infect hosts only incidentally. Indeed, the ability to produce endospores should be advantageous for a soil microorganism because soil can be a highly variable environment in terms of nutrient levels, temperature, and water activity. Thus, a heat- and desiccation-resistant structure capable of remaining dormant for long periods (perhaps even millions of years, ⟨⟩ Section 4.15) should offer considerable survival value in nature.

Spore-formers can be selectively isolated from soil, food, dust, and other materials by exposing the sample to 80°C for 10 min (pasteurization), a treatment that effectively kills vegetative cells while the spores present remain viable. Streaking heat-treated samples on plates of the appropriate media and incubating either aerobically or anaerobically readily yield species of *Bacillus* or *Clostridium*, respectively.

Bacillus

Species of *Bacillus* usually grow well on defined media containing any of a number of carbon sources. Many bacilli produce extracellular hydrolytic enzymes that break down complex polymers such as polysaccharides, nucleic acids, and lipids, permitting the organisms to use these products as carbon sources and electron donors. Many bacilli produce antibiotics, of which bacitracin, polymyxin, tyrocidin, gramicidin, and circulin are examples. In most cases, antibiotic production is related to the sporulation process, the antibiotic being released when the culture enters the stationary phase of growth and after it is committed to sporulation. An outline of the subdivision of the genus *Bacillus* is given in Table 12.26.

Several *Bacillus* species, most notably *B. popilliae* and *B. thuringiensis*, produce insect larvicides. *Bacillus popilliae* causes a fatal condition called *milky disease* in Japanese beetle larvae and larvae of closely related beetles of the family Scarabaeidae. *Bacillus thuringiensis* causes a fatal disease of larvae of many different groups of insects, although individual strains are specific as to host affected. Strains exist that are specific for lepidopterans, such as the silkworm, the cabbage worm, the tent caterpillar, and the gypsy moth. Some strains kill dipterans such as mosquitoes and black flies. Others kill coleopterans such as Colorado potato beetles. Strains of *B. thuringiensis* have also been discovered that are toxic

TABLE 12.26 Characteristics of representative species of the genus *Bacillus*

Characteristics	Species	Spore position	DNA (mol % GC)
I. Spores oval or cylindrical, facultative aerobes, casein and starch hydrolyzed; sporangia not swollen, spore wall thin			
Thermophiles and acidophiles	*B. coagulans*	Central or terminal	47
	B. acidocaldarius	Terminal	60
Mesophiles	*B. licheniformis*	Central	46
	B. cereus	Central	35
	B. anthracis	Central	33
	B. megaterium	Central	37
	B. subtilis	Central	43
Insect pathogen	*B. thuringiensis*	Central	34
Sporangia distinctly swollen, spore wall thick			
Thermophile	*B. stearothermophilus*	Terminal	52
Mesophiles	*B. polymyxa*	Terminal	44
	B. macerans	Terminal	52
	B. circulans	Central or terminal	35
Insect pathogens	*B. larvae*	Central or terminal	—
	B. popilliae	Central	41
II. Spores spherical, obligate aerobes, casein and starch not hydrolyzed			
Sporangia swollen	*B. sphaericus*	Terminal	37
Sporangia not swollen	*B. pasteurii*	Terminal	38

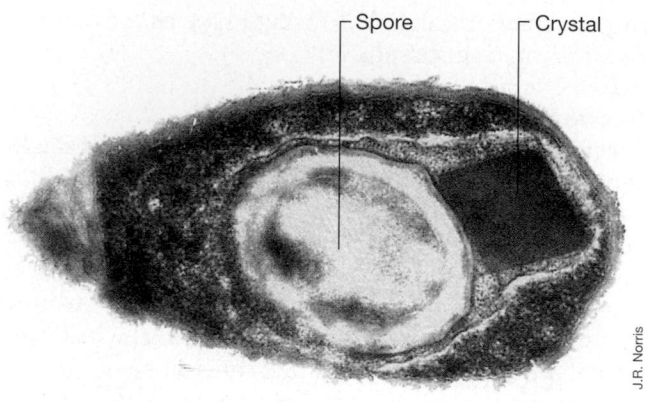

┌─ Spore   ┌─ Crystal

J.R. Norris

Figure 12.57 Formation of the toxic parasporal crystal in the insect pathogen *Bacillus thuringiensis*. Electron micrograph of a thin section of a sporulating cell. The crystalline protein (Bt-toxin) is toxic to certain insects by causing lysis of intestinal cells.

to Japanese beetles. Spore preparations derived from *B. thuringiensis* and *B. popilliae* are commercially available as biological insecticides.

The disease caused by *Bacillus popilliae* is a septicemia, whereas the disease caused by *B. thuringiensis* is essentially an intoxication. Both of these insect pathogens form a crystalline protein during sporulation called the *parasporal body*, which is deposited within the sporangium but outside the spore proper (Figure 12.57). In the case of *B. thuringiensis*, the crystal (parasporal body) protein is a protoxin that is converted to a toxin by proteolytic cleavage in the larval gut. The toxin binds to intestinal epithelial cells and induces pore formation that causes leakage of the host cells followed by lysis.

Genes encoding crystal proteins from several *B. thuringiensis* strains have been isolated. The genes for the *B. thuringiensis* crystal protein (known commercially as "Bt-toxin") have been introduced into plants to render the plants "naturally" resistant to insects. This strategy has been shown to be effective in controlled sit-

TABLE 12.27 Characteristics of some groups of the genus *Clostridium*

Key characteristics	Other characteristics	Species	DNA (mol % GC)
I. Ferment carbohydrates			
Ferment cellulose	Fermentation products: acetate, lactate, succinate, ethanol, CO_2, H_2	*C. cellobioparum* *C. thermocellum*	28 38–39
Ferment sugars, starch, and pectin	Fermentation products: acetone, butanol, ethanol, isopropanol, butyrate, acetate, propionate, succinate, CO_2, H_2; some fix N_2	*C. butyricum* *C. acetobutylicum* *C. pasteurianum* *C. perfringens* *C. thermosulfurogenes*	27–28 28–29 26–28 24–27 33
Ferment sugars primarily to acetic acid	Total synthesis of acetate from CO_2; cytochromes present in some species	*C. aceticum* *C. thermoaceticum* *C. formicoaceticum*	33 54 34
Ferments only pentoses or methylpentoses	Ring-shaped cells form left-handed, helical chains; fermentation products: acetate, propionate, *n*-propanol, CO_2, H_2	*C. methylpentosum*	46
II. Ferment amino acids	Fermentation products: acetate, other fatty acids, NH_3, CO_2, sometimes H_2; some also ferment sugars to butyrate and acetate; may produce exotoxins (⟳ Sections 21.10, 27.8, and 29.5)	*C. sporogenes* *C. tetani* *C. botulinum* *C. tetanomorphum*	26 25–26 26–28 25–28
	Ferments three-carbon amino acids (for example, alanine) to propionate, acetate, and CO_2	*C. propionicum*	35
III. Ferments carbohydrates or amino acids	Fermentation products from glucose: acetate, formate, small amounts of isobutyrate and isovalerate	*C. bifermentans*	27
IV. Purine fermenters	Ferments uric acid and other purines, forming acetate, CO_2, NH_3	*C. acidurici*	27–30
V. Ethanol fermentation to fatty acids	Produces butyrate, caproate, and H_2; requires acetate as electron acceptor; does not use sugars, amino acids, or purines	*C. kluyveri*	30

uations, and a variety of genetically altered Bt-toxins are being developed by genetic engineering in attempts to increase toxicity and reduce resistance (⌥ Section 31.7).

Clostridium

The clostridia lack a cytochrome system and a mechanism for electron transport phosphorylation; hence, unlike *Bacillus* species, they obtain ATP *only* by substrate-level phosphorylation. A wide variety of anaerobic energy-yielding mechanisms are known in the clostridia (fermentative diversity will be discussed in Section 17.20); indeed, the separation of the genus into subgroups is based primarily on these properties and on the nature of the fermentable substrate used (Table 12.27).

A number of clostridia ferment sugars, producing as a major end product *butyric acid*. Some of these also produce *acetone* and *butanol*, and at one time acetone-butanol fermentation by clostridia was of great industrial importance as it was the main commercial source of these products. Some clostridia of the acetone-butanol type fix N_2; the most vigorous N_2 fixer is *Clostridium pasteurianum*, which probably is responsible for most anaerobic nitrogen fixation in the soil. One group of clostridia ferments cellulose with the formation of acids and alcohols, and these are likely the major organisms decomposing cellulose anaerobically in soil.

The biochemical steps in the formation of butyric acid and butanol from sugars are well understood (Figure 12.58). Glucose is converted to pyruvate via the Embden–Meyerhof pathway, and pyruvate is split to acetyl-CoA, CO_2, and hydrogen (reduced ferredoxin) by the phosphoroclastic reaction (⌥ Section 17.19 and Figure 17.51). Acetyl-CoA is then reduced to fermentation products using the NADH derived from glycolytic reactions. The proportions of the various products are influenced by the duration and the conditions of the fermentation. During the early stages, butyric and acetic acids are the predominant products, but as the pH of the medium drops, synthesis of acids ceases and the neutral products acetone and butanol begin to accumulate. However, if the pH of the medium is kept neutral with $CaCO_3$, very little of the neutral products are formed and the fermentation products consist of about three parts butyric and one part acetic acid. This makes good physiological sense because, unlike neutral product formation, acid production allows for additional ATP synthesis (Figure 12.58).

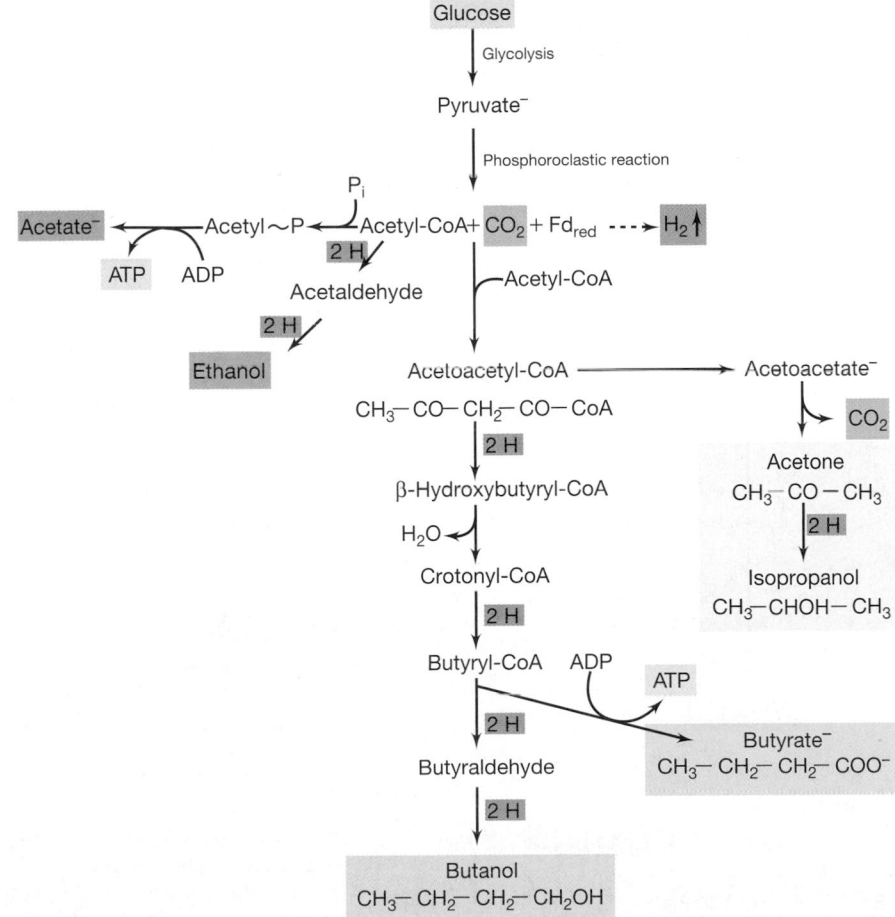

Figure 12.58 Pathway of formation of fermentation products from the butyric acid group of clostridia. The designation "2 H" represents two electrons from one molecule of NADH. Note how the production of acetate and butyrate lead to additional ATP by substrate-level phosphorylation (⌥ Section 17.19). By contrast, formation of butanol and acetone (these products are typically found in a ratio of 1:1 because butanol formation requires two NADH, while acetone formation requires none) reduces the ATP yield because the butyryl-CoA to butyrate step is bypassed.

Another group of clostridia obtain their energy by fermenting *amino acids*. Some species ferment individual amino acids while others ferment only amino acid *pairs*. In this situation one functions as the electron *donor* and is *oxidized*, whereas the other acts as the electron *acceptor* and is *reduced*. This type of coupled decomposition is known as the **Stickland reaction**. For instance, *Clostridium sporogenes* catabolizes a mixture of glycine and alanine, as outlined in Figure 12.59. Various amino acids that can function as either electron donors or acceptors in Stickland reactions are also listed in Figure 12.59. The products of Stickland oxidation are always NH_3, CO_2, and a carboxylic acid with one *less* carbon atom than the amino acid that is oxidized (Figure 12.59).

The amino acids that can be fermented singly are alanine, cysteine, glutamate, glycine, histidine, serine, or threonine. The products are generally acetate, butyrate, CO_2, and H_2. Many of the products of amino acid fermentation by clostridia are foul-smelling substances, and the odor that results from putrefaction is a result mainly of clostridial action. In addition to butyric acid, other odoriferous compounds produced are isobutyric acid, isovaleric acid, caproic acid, hydrogen sulfide, methylmercaptan (from sulfur amino acids), cadaverine (from lysine), putrescine (from ornithine), and ammonia.

The main habitat of clostridia is the soil, where they live primarily in anoxic "pockets," made anoxic by facultative organisms metabolizing various organic compounds present. In addition, a number of clostridia have adapted to the anoxic environment of the mammalian intestinal tract. Also, as is discussed in Section 21.8, several clostridia are capable of causing disease in humans under specialized conditions. Botulism is caused by *Clostridium botulinum*, tetanus by *C. tetani*, and gas gangrene by *C. perfringens* and a number of other clostridia, both sugar and amino acid fermenters. These pathogenic clostridia seem in no way unusual metabolically but are distinct in that they produce specific toxins or, in the case of those causing gas gangrene, a group of toxins (∞ Section 21.10 and Table 21.4). *C. perfringens* and related species can also cause gastroenteritis in humans and domestic animals (∞ Section 29.5), and botulism occurs in sheep and ducks and a variety of other animals. An unsolved ecological problem is what role these toxins play in the natural habitat of the organism.

Sporosarcina

The genus *Sporosarcina* is unique among endospore formers because cells are *cocci* instead of rods. *Sporosarcina* consists of strictly aerobic spherical to oval cells that divide in two or three perpendicular planes to form tetrads or packets of eight or more cells (Figure 12.60). Two species of *Sporosarcina* are known, *S. ureae* and *S. halophila*. *Sporosarcina ureae* (Figure 12.60) can easily be enriched from soil by plating dilutions of a pasteurized soil sample on nutrient agar supplemented with 8% urea and incubating in air. Most soil bacteria are strongly inhibited by as little as 2% urea. However, *S. ureae* actively decomposes urea to CO_2 and NH_3 and in so doing can dramatically raise the pH of unbuffered media (*S. ureae* is remarkably alkaline-tolerant and grows in media up to pH 10–11). *Sporosarcina ureae* is common in soils, and studies on its distribution suggest that numbers of *S. ureae* are greatest in soils that receive

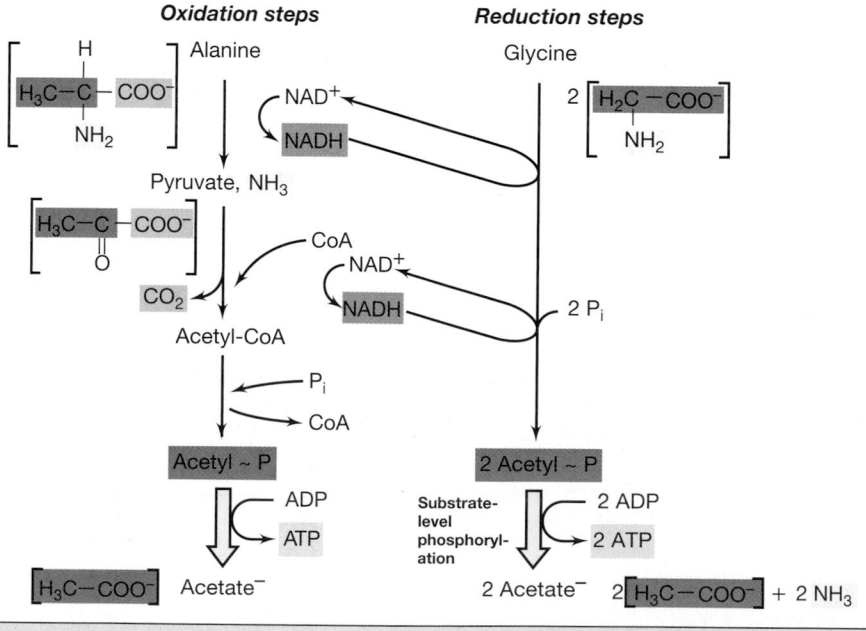

Amino acids participating in coupled fermentations (Stickland reaction)	
Amino acids oxidized:	**Amino acids reduced:**
Alanine	Glycine
Leucine	Proline
Isoleucine	Hydroxyproline
Valine	Tryptophan
Histidine	Arginine

Overall: Alanine + 2 Glycine + 2 H_2O + 3 ADP + 3 P_i ⟶ 3 Acetate⁻ + CO_2 + 3 NH_4^+ + 3 ATP

Figure 12.59 Coupled oxidation-reduction reaction (Stickland reaction) between alanine and glycine in *Clostridium sporogenes*. The structures of the key substrates, intermediates, and products are shown (in brackets) to allow the chemistry of the reaction to be followed. Note how in the reaction shown, alanine is the electron *donor*, while glycine is the electron *acceptor*.

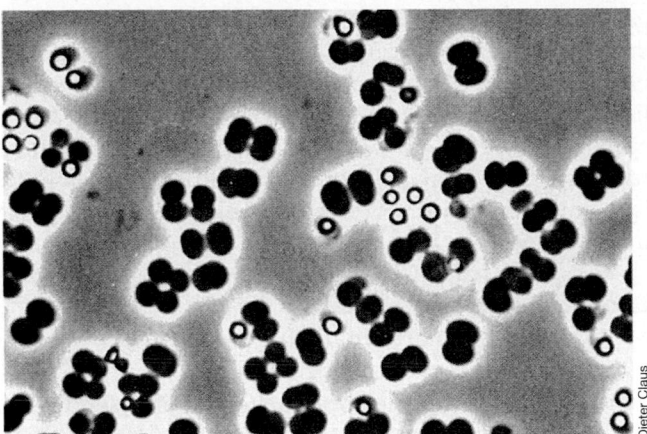

Figure 12.60 Phase contrast photomicrograph of cells of *Sporosarcina ureae*. A single cell is about 2 μm wide. Note bright refractile endospores. Most cell packets contain 8 cells.

inputs of urine (a source of urea), such as soils in which animals periodically urinate. Since many soil organisms are quite urea-sensitive, these results suggest that *S. ureae* is ecologically important as a major urea degrader in nature.

Heliobacteria

Heliobacteria are *phototrophic*, low GC, gram-positive bacteria. They are anoxygenic phototrophs and produce a unique structural form of bacteriochlorophyll ( Section 17.2). The group contains four genera: *Heliobacterium*, *Heliophilum*, *Heliorestis*, and *Heliobacillus*. All known heliobacteria are rod-shaped, either short or long rods, frequently with pointed ends (Figure 12.61). *Heliophilum* is especially interesting because its rod-shaped cells form into bundles (Figure 12.61*b*) that are motile as a unit. Heliobacteria are strict anaerobes, but in addition to anaerobic phototrophic growth, can grow chemotrophically in darkness by pyruvate fermentation (as can many clostridia, close relatives of the heliobacteria). Like the endospores of *Bacillus* or *Clostridium* species, the endospores of heliobacteria (Figure 12.61*c*)

contain elevated Ca^{2+} levels and the signature molecule of the endospore, *dipicolinic acid* ( Section 4.15). Heliobacteria reside in soil, especially paddy field soils, where their strong N_2-fixation activities may benefit rice productivity.

✓ **12.19–12.20 Concept Check**

The "Low GC," gram-positive *Bacteria* are a large phylogenetic group that contains rods and cocci, sporulating and nonsporulating species. Production of endospores is a hallmark of the key genera *Bacillus* and *Clostridium*. Gram-positive bacteria are major agents for the degradation of organic matter in soil and a few species are pathogenic.

✓ What are the major features that differentiate *Staphylococcus* from *Bacillus*?

✓ How could you distinguish between a *heterofermentative* and a *homofermentative* lactic acid bacterium?

✓ What is the major physiological distinction between *Bacillus* and *Clostridium* species?

✓ Among endospore-producing genera, what is unique about the heliobacteria?

12.21 Cell Wall-Less, Low GC, Gram-Positive *Bacteria*: The Mycoplasmas

Key Genera

Mycoplasma
Spiroplasma

The mycoplasmas are organisms without cell walls that do not revert to walled organisms. They are probably the smallest organisms capable of autonomous growth and are of special evolutionary interest because of their extremely simple cell structure and small genomes. And, although not staining gram-positively since they lack cell walls, the mycoplasmas are clearly phylogenetically related to low GC, gram-positive bacteria.

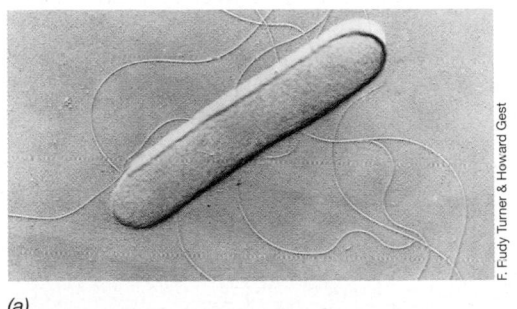

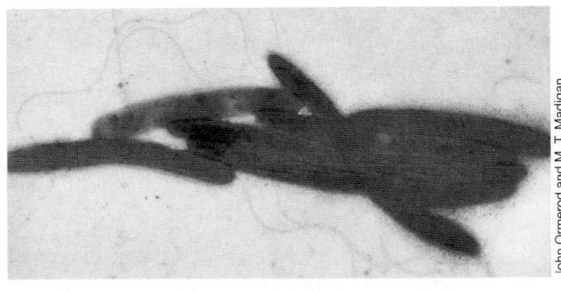

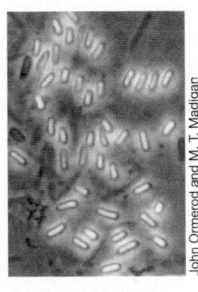

(a) *(b)* *(c)*

Figure 12.61 Cells and endospores of heliobacteria. (a) Electron micrograph of *Heliobacillus mobilis*, a peritrichously flagellated species. (b) *Heliophilum fasciatum*, bundles as observed by electron microscopy. (c) Phase micrograph of spores from *Heliobacterium gestii*.

Properties of Mycoplasmas

The absence of cell walls in the mycoplasmas observed by electron microscopy has been confirmed by chemical analysis, the latter showing that the key components of peptidoglycan, muramic acid, and diaminopimelic acid, are missing. In Chapter 4 we discussed protoplasts and showed how these structures can be formed when cell wall-digesting enzymes act on cells that are in an osmotically protected medium, and that when the osmotic stabilizer is removed, protoplasts take up water, swell, and burst (∞ Figure 4.33). Mycoplasmas resemble protoplasts in their lack of a cell wall, but they are more resistant to osmotic lysis and are able to survive conditions under which protoplasts lyse. This ability to resist osmotic lysis is at least partially determined by the fact that sterols are present in the mycoplasma cytoplasmic membrane, making it more stable than that of other prokaryotes. Indeed, some mycoplasmas require sterols in their growth media and this sterol requirement is a basis for separating the mycoplasmas into two groups (Table 12.28).

In addition to sterols, certain mycoplasmas contain compounds called **lipoglycans** (see Table 12.28). Lipoglycans are long-chain heteropolysaccharides covalently linked to membrane lipids and embedded in the cytoplasmic membrane of many mycoplasmas. Lipoglycans resemble the lipopolysaccharides (LPSs) of gram-negative *Bacteria* (∞ Section 4.9) except that they lack the lipid A backbone and the phosphate typical of bacterial LPSs. Lipoglycans also function to help stabilize the membrane and have also been identified as facilitating attachment of mycoplasmas to cell surface receptors of animal cells. Like LPSs, lipoglycans stimulate antibody production when injected into experimental animals.

Growth of Mycoplasmas

Mycoplasma cells are small and they are highly pleomorphic, a consequence of their lack of a cell wall and hence rigidity. A single culture may exhibit small coccoid elements, larger, swollen forms, and filamentous forms of variable lengths, often highly branched (Figure 12.62).

The small coccoid elements ($0.2-0.3$ μm in size) are the smallest mycoplasma units capable of independent growth. Cellular elements of diameters close to 0.1 μm are occasionally seen in mycoplasma cultures, but these are not capable of growth. Even so, the minimum reproductive unit of $0.2-0.3$ μm probably represents the smallest *free-living* cell. Additionally, the genome size of mycoplasmas is also smaller than that of most prokaryotes, between 500 and 1100 kilobase pairs of DNA in many cases (Table 12.28), which is comparable to that of the obligately parasitic chlamydia and rickettsia (see Sections 12.27 and 12.13)

			DNA	Genome size	Presence of
Genus	**Number of species**	**Properties**	**(mol % GC)**	**(kilobase pairs)**	**lipoglycans**
Require sterols					
Mycoplasma	110	Many pathogenic; require sterols; facultative aerobes (see Figure 12.62)	23–41	600–1350	+
Anaeroplasma	4	May or may not require sterols; obligate anaerobes; degrade starch, producing acetic, lactic, and formic acids plus ethanol and CO_2; inhibited by thallium acetate; found in the bovine and ovine rumen	29–33	1500–1600	+
Spiroplasma	33	Spiral to corkscrew-shaped cells; associated with various phytopathogenic (plant disease) conditions (see Figure 12.64)	25–30	940–2200	–
Ureaplasma	6	Coccoid cells; occasional clusters and short chains; growth optimal at pH 6; strong urease reaction; associated with certain urinary tract infections in humans; inhibited by thallium acetate	27–30	750	–
Entomoplasma	5	Facultative aerobe; associated with insects and plants	27–29	790–1140	?
Do not require sterols					
Acholeplasma	16	Facultative aerobes	27–36	1500	+
Asteroleplasma	1	Obligate anaerobe; isolated from the bovine or ovine rumen	40	1500	+
Mesoplasma	12	Phylogenetically and ecologically related to *Entomoplasma*	27–30	870–1100	?

TABLE 12.28 Major characteristics of mycoplasmas[a]

[a] Phylogenetically, all known mycoplasmas are members of the "Low GC" gram-positive *Bacteria*.

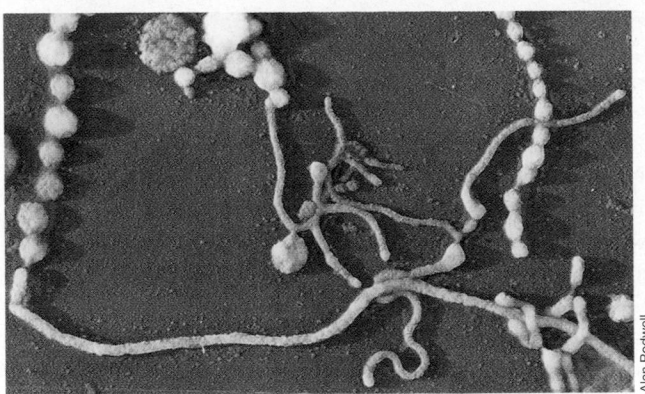

Figure 12.62 Electron micrograph of a metal-shadowed preparation of *Mycoplasma mycoides*. Note the coccoid and hyphalike elements. The average diameter of cells in chains is about 0.5 μm.

and about one-fifth to one-fourth that of *Escherichia coli*. The genome of at least one *Mycoplasma* species, *M. genitalium* contains 580 kilobase pairs and has been completely sequenced (∞ Section 15.3). This very small genome is thought to be near to the lower limit for the amount of DNA any cell must have to carry out life processes.

The mode of growth of mycoplasmas differs in liquid and agar cultures. On agar, there is a tendency for the organisms to grow so that they become embedded in the medium, and colonies of mycoplasmas on agar thus show a characteristic "fried-egg" appearance from a dense central core, which penetrates downward into the agar, surrounded by a circular spreading area that is lighter in color (Figure 12.63). As would be expected, growth of mycoplasmas is not inhibited by penicillin, cycloserine, or other antibiotics that inhibit cell wall synthesis, but the organisms are as sensitive as other *Bacteria* to antibiotics that act on targets other than the cell wall.

Culture media for the growth of mycoplasmas have usually been quite complex. For many species growth is poor or absent even in complex yeast extract-peptone-beef heart infusion media unless fresh serum or ascitic fluid is added; the latter provide unsaturated fatty acids and sterols. Some mycoplasmas can be cultivated on relatively simple media, however, and defined media have been de-

veloped for some strains. Most mycoplasmas use carbohydrates as energy sources and require a range of vitamins, amino acids, purines, and pyrimidines as growth factors. The energy metabolism of mycoplasmas is variable but not unique, as some species are strictly respiratory while others are facultative or even obligate anaerobes (Table 12.28).

Spiroplasma

The genus *Spiroplasma* consists of helical or spiral-shaped cells (Figure 12.64). Although they lack a cell wall and flagella, they are motile by means of a rotary (screw) motion or a slow undulation. Intracellular fibrils that are thought to play a role in motility have been demonstrated. The organism has been isolated from ticks, the hemolymph (Figure 12.64) and gut of insects, vascular plant fluids and insects that feed on fluids, and the surfaces of flowers and other plant parts. *Spiroplasma citri* has been isolated from the leaves of citrus plants, where it causes a disease called *citrus stubborn disease* and from corn plants suffering from *corn stunt disease*. A number of other mycoplasma-like bodies have been detected in diseased plants by electron microscopy, which indicates that a large group of plant-associated mycoplasmas may exist. Four species of *Spiroplasma* are recognized that cause a variety of animal diseases such as *honeybee spiroplasmosis, suckling mouse cataract disease*, and *lethargy disease* of the beetle *Melolontha*.

✓ 12.21 Concept Check

The mycoplasma group are organisms that lack cell walls and contain a very small genome. Many species require sterols to strengthen their membranes, and several are pathogenic for humans and other animals.

- ✓ Why do mycoplasmas need to have stronger cell membranes than other prokaryotes?
- ✓ Where do the mycoplasmas group phylogenetically?
- ✓ Compare and contrast the genus *Mycoplasma* with the genus *Acholeplasma* in terms of growth requirements, genome size, and metabolism.

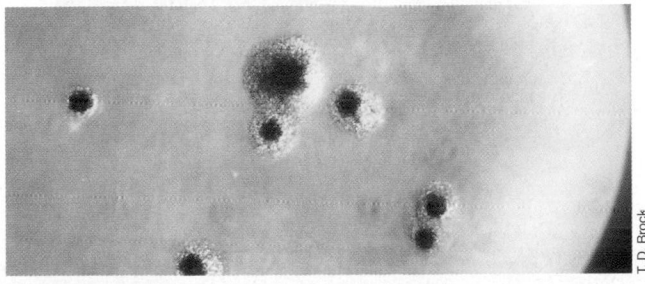

Figure 12.63 Typical "fried egg" appearance of mycoplasma colonies on agar. The colonies are about 0.5 mm in diameter.

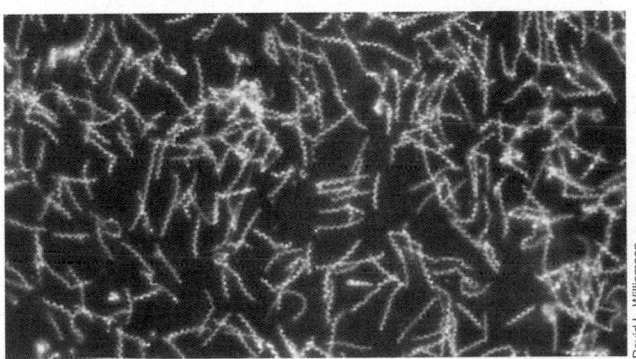

Figure 12.64 Dark-field micrograph of the "sex ratio" spiroplasma removed from the hemolymph of the fly *Drosophila pseudoobscura*. Female flies infected with the sex ratio spiroplasma bear only female progeny. Individual spiroplasma cells are about 0.15 μm in diameter.

High GC, Gram-Positive *Bacteria*: Coryneform and Propionic Acid Bacteria

Key Genera

Corynebacterium
Arthrobacter
Propionibacterium

High GC, gram-positive *Bacteria* are typically rod-shaped to filamentous primarily aerobic prokaryotes that are common inhabitants of the soil and of plant materials. For the most part they are harmless commensals, species of *Mycobacterium* (for example *Mycobacterium tuberculosis*) being notable exceptions, and some are of great economic value in either the production of antibiotics or certain fermented dairy products. We begin with the rod-shaped representatives.

Corynebacteria

The coryneform bacteria are gram-positive, aerobic, non-motile, rod-shaped organisms with the characteristic of forming irregular-shaped, club-shaped, or V-shaped cell arrangements during normal growth. V-shaped cell groups arise as a result of a snapping movement that occurs just after cell division (called postfission snapping movement or, simply, *snapping division*) (Figure 12.65). **Snapping division** has been shown to occur in one species because the cell wall consists of two layers; only the inner layer participates in cross-wall formation, and so after the cross-wall is formed, the two daughter cells remain attached by the outer layer of the cell wall. Localized rupture of this outer layer on one side results in a bending of the two cells away from the ruptured side (Figure 12.66) and thus development of V-shaped forms.

The main genera of coryneform bacteria are *Corynebacterium* and *Arthrobacter*. The genus *Corynebacterium* consists of an extremely diverse group of bacteria, including animal and plant pathogens as well as saprophytes. Some species, such as *C. diphtheriae*, are pathogenic (diphtheria, ∞ Section 26.3). The genus *Arthrobacter*, consisting primarily of soil organisms, is distinguished from *Corynebacterium* on the basis of a developmental cycle in *Arthrobacter* involving conversion from rod to sphere and back to rod again (Figure 12.67). However, some corynebacteria are pleomorphic and form coccoid elements during growth, and so the distinction between the two genera on the basis of life cycle is not absolute. The *Corynebacterium* cell frequently has a swollen end, so it has a club-shaped appearance (hence the name of the genus: *koryne* is the Greek word for "club"), whereas *Arthrobacter* is less commonly club-shaped.

Organisms of the genus *Arthrobacter* are among the most common of all soil bacteria. They are remarkably resistant to desiccation and starvation, despite the fact that they do not form spores or other resting cells. Arthrobacters are a heterogeneous group that have considerable nutritional versatility, and strains have been isolated that decompose herbicides, caffeine, nicotine, phenols, and other unusual organic compounds.

Propionic Acid Bacteria

The propionic acid bacteria (genus *Propionibacterium*) were first discovered as inhabitants of Swiss (Emmentaler) cheese, where their fermentative production of

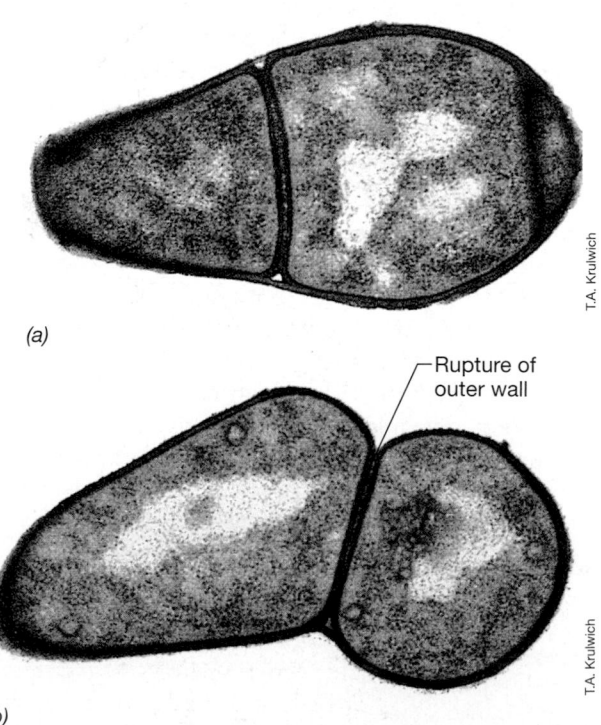

(a)

Rupture of outer wall

(b)

Figure 12.66 Electron micrograph of cell division in *Arthrobacter crystallopoietes*, illustrating how snapping division and V-shaped cell groups arise. (a) Before rupture of the outer cell wall layer. (b) After rupture of the outer layer on one side. Cells are 0.9–1 μm in diameter.

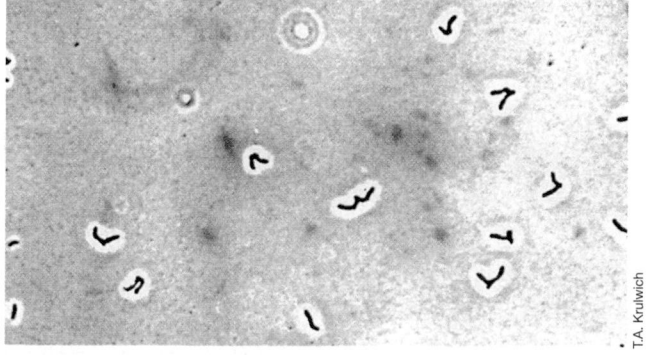

Figure 12.65 Photomicrograph of characteristic V-shaped cell groups in *Arthrobacter crystallopoietes*, resulting from snapping division. Cells are about 0.9 μm in diameter.

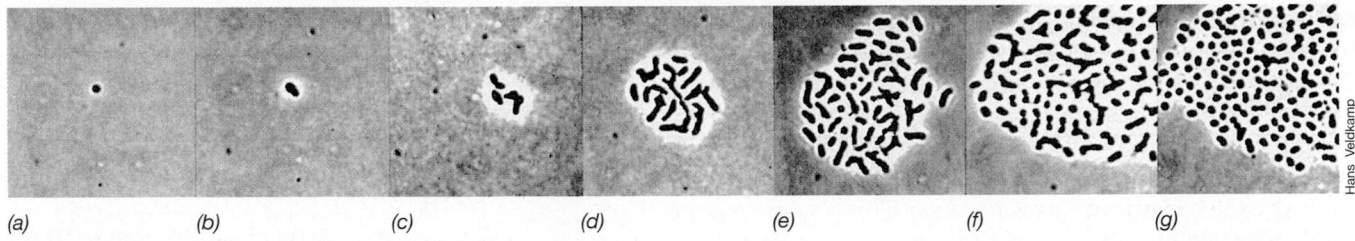

(a)   (b)   (c)   (d)   (e)   (f)   (g)

Hans Veldkamp

Figure 12.67 Stages in the life cycle of *Arthrobacter globiformis* as observed in slide culture: (a) Single coccoid element; (b–e) conversion to rod and growth of a microcolony consisting predominantly of rods; (f–g) conversion of rods to coccoid forms. Cells are about 0.9 µm in diameter.

CO_2 produces the characteristic holes; the presence of propionic acid is at least partly responsible for the unique flavor of the cheese. Although this acid is produced by some other bacteria, its production in large amounts by the propionic acid bacteria is a distinguishing characteristic of the genus. The bacteria in this group are gram-positive anaerobes that ferment lactic acid, carbohydrates, and polyhydroxy alcohols, producing primarily propionic acid, acetic acid, and CO_2. Their nutritional requirements are complex, and they usually grow rather slowly.

The enzymatic reactions leading from glucose to propionic acid are shown in Figure 12.68. The initial catabolism of glucose to pyruvate follows the Embden–Meyerhof pathway as in the lactic acid bacteria, but the NADH formed is oxidized as one part of a cycle in which *propionic acid* is formed. Pyruvate accepts a carboxyl group from methylmalonyl-CoA by a transcarboxylase reaction, leading to the formation of oxalacetate and propionyl-CoA. The latter substance reacts with succinate in a step catalyzed by a CoA transferase, producing succinyl-CoA and propionate. The succinyl-CoA is then isomerized to methylmalonyl-CoA, and the cycle is complete (Figure 12.68). Oxidation of NADH occurs in the steps between oxalacetate and succinate, and the oxidation-reduction balance is restored.

Propionibacterium also ferments lactate with the production of propionate, acetate, and CO_2. The anaerobic fermentation of lactic acid to propionate is of interest because lactic acid itself is an end product of fermentation for many bacteria (see Section 12.19). The propionic acid bacteria are thus able to obtain energy anaerobically from a fermentation product that other bacteria have produced; this metabolic strategy has been called a *secondary fermentation.*

It is the fermentation of lactate to propionate that is important in Swiss cheese manufacture. The starter culture consists of a mixture of homofermentative streptococci and lactobacilli, plus propionic acid bacteria. The initial fermentation of lactose to lactic acid during formation of the curd is carried out by the homofermentative organisms. After the curd (protein and fat) has been drained, the propionic acid bacteria develop rapidly and

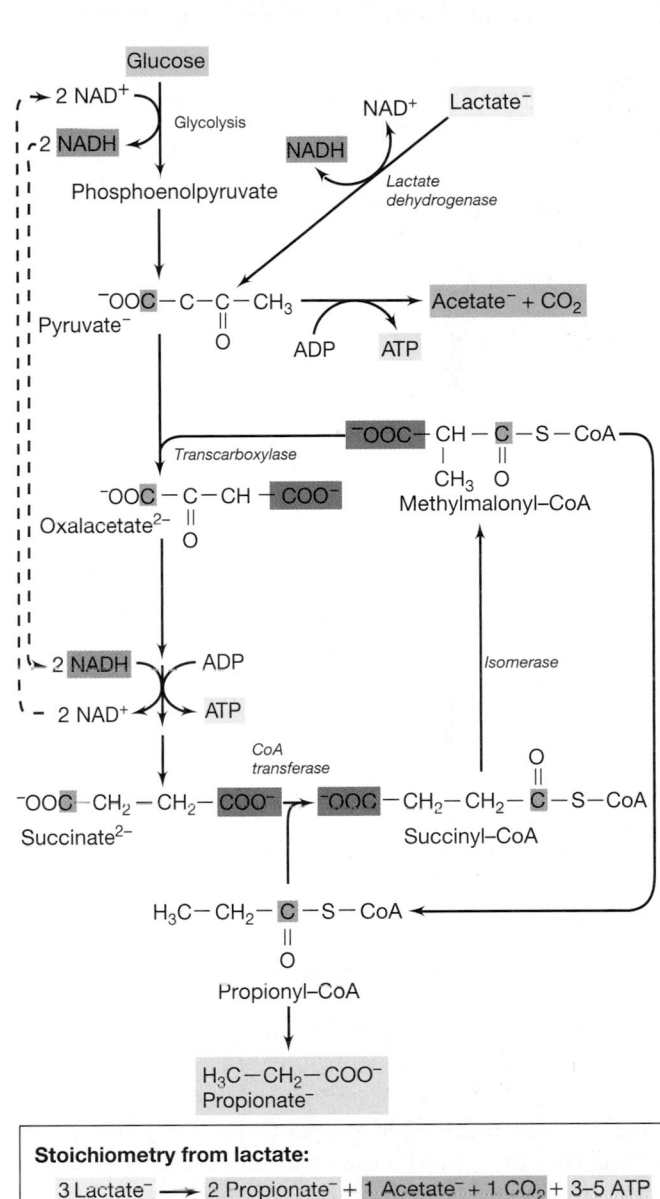

Stoichiometry from lactate:

3 Lactate⁻ ⟶ 2 Propionate⁻ + 1 Acetate⁻ + 1 CO_2 + 3–5 ATP

Figure 12.68 The formation of propionic acid by *Propionibacterium.* Either lactate, produced by the fermentative activities of other bacteria, or glucose can be fermented in the propionate fermentation. ATP synthesis is associated with electron transport reactions occurring during the formation of succinate and by substrate level phosphorylation in the production of acetate.

the eyes (holes) characteristic of Swiss cheese are formed by the accumulation of CO_2, the gas diffusing through the curd and gathering at weak points. In the fermentation, lactate is oxidized to pyruvate from which it is converted to propionate as shown in Figure 12.68.

Propionate is also formed in the fermentation of succinate by the bacterium *Propionigenium*. This organism is phylogenetically and ecologically unrelated to *Propionibacterium* but energetic aspects of its fermentation are of considerable interest. We discuss the mechanism of the *Propionigenium* fermentation in Section 17.20.

12.23 High GC, Gram-Positive *Bacteria*: *Mycobacterium*

Key Genus

Mycobacterium

The genus *Mycobacterium* consists of rod-shaped organisms, which at some stage of their growth cycle possess the distinctive staining property called **acid-fastness**. This property is due to the presence on the surface of the mycobacterial cell of unique lipid components called **mycolic acids** and is found only in the genus *Mycobacterium*. First discovered by Robert Koch during his pioneering investigations on tuberculosis (Section 1.5), this unique staining property permitted the identification of the organism in tuberculous lesions; it has subsequently proved to be of great taxonomic use in defining the genus *Mycobacterium*.

Acid-Fastness (Ziehl–Neelsen Stain)

A mixture of the dye basic fuchsin and phenol is used in this staining procedure, the stain being driven into the cells by slow heating of the smear on the microscope slide to the steaming point for 2–3 min. The role of the phenol is to enhance penetration of the fuchsin into the lipids. After washing in distilled water, the preparation is decolorized with acid-alcohol; after another wash, a final counterstain of methylene blue is used. Acid-fast organisms in the final preparation appear *red*, whereas the background and non-acid-fast organisms appear *blue*.

As noted, the key component necessary for acid-fastness is a unique lipid fraction of mycobacterial cells called *mycolic acid*. Mycolic acid is actually a group of complex branched-chain hydroxy lipids with the general structure shown in Figure 12.69a; in the acid-fast stain the carboxylic acid group of the mycolic acid reacts with the fuchsin dye (Figure 12.69b). The mycolic acid is covalently bound to the peptidoglycan of the mycobacterial wall, and this complex leaves the cell surface with a waxy, hydrophobic consistency. Mycobacteria are not readily stained by the Gram method because of the high

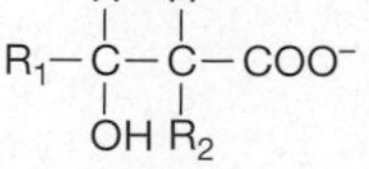

(a) Mycolic acid; R_1 and R_2 are long-chain aliphatic hydrocarbons

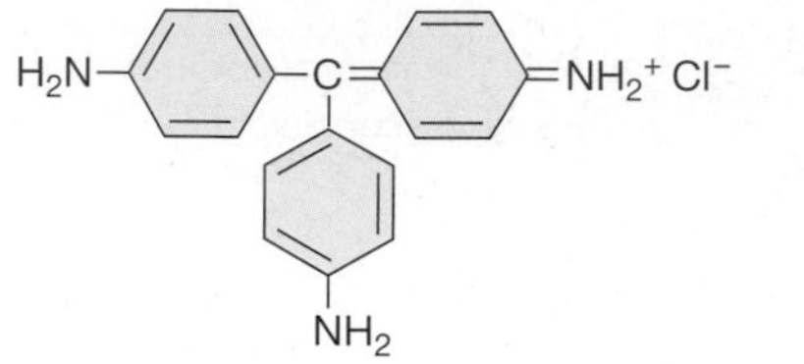

(b) Basic fuchsin

Figure 12.69 Structure of (a) mycolic acid and (b) basic fuchsin, the dye used in the acid-fast stain. The fuchsin dye combines with the mycolic acid via ionic bonds between COO^- and NH_2^+.

surface lipid content, but if the lipoidal portion of the cell is removed with alkaline ethanol, the intact cell remaining is non-acid-fast but instead is gram-positive.

Characteristics of Mycobacteria

Mycobacteria are rather pleomorphic and may undergo branching or filamentous growth. However, in contrast to those of the actinomycetes (see Section 12.24), filaments of the mycobacteria become fragmented into rods or coccoid elements on slight disturbance; a true mycelium is not formed. In general, mycobacteria can be separated into two major groups, *slow growers* and *fast growers* (Table 12.29). *Mycobacterium tuberculosis* is a typical slow grower, and visible colonies are produced from dilute inoculum only after days to weeks of incubation. (The reason Koch was successful in first isolating *M. tuberculosis* was that he waited long enough after inoculating media; Section 1.5.) When growing on solid media, mycobacteria generally form tight, compact, often wrinkled colonies, the organisms piling up in a mass rather than spreading out over the surface of the agar (Figure 12.70a). This formation is probably due to the high lipid content and hydrophobic nature of the cell surface.

For the most part, mycobacteria have relatively simple nutritional requirements. Growth often occurs in simple mineral salts medium with ammonium as nitrogen source and glycerol or acetate as sole carbon source and electron donor incubated in air. Growth of *Mycobacterium tuberculosis* is stimulated by lipids and fatty acids, and egg yolk (a good source of lipids) is often added to culture media to achieve more luxuriant growth. A glycerol-whole egg medium (Lowenstein–Jensen medium) is often used in primary isolation of *M. tuberculosis* from pathological materials. Perhaps because of the high

TABLE 12.29 Some characteristics of representative mycobacteria

Species	Growth in 5% NaCl	Nitrate reduction	Growth at 45°C	Human pathogen	Pigmentation
Slow-growing species					
Mycobacterium tuberculosis	−	+	−	+	None
Mycobacterium avium	−	−	−	+	Old colonies pigmented (see Figure 12.70c)
Mycobacterium bovis	−	−	+	+	None
Mycobacterium kansasii	−	+	−	+	Photochromogenic
Fast-growing species					
Mycobacterium smegmatis	+	+	+	−	None
Mycobacterium phlei	+	+	+	−	Pigmented
Mycobacterium chelonae	+	−	−	+	None
Mycobacterium parafortuitum	+	+	−	−	Photochromogenic

lipid content of its cell walls, *M. tuberculosis* is able to resist such chemical agents as alkali and phenol for considerable periods of time, and this property is used in the selective isolation of the organism from patient sputum and other materials that are grossly contaminated. The sputum is first treated with 1 N NaOH for 30 min and then neutralized and streaked onto an isolation medium.

A characteristic of many mycobacteria is their ability to form yellow carotenoid pigments (Figure 12.70c). Based on pigmentation, the mycobacteria can be classified into three groups: nonpigmented (including *Mycobacterium tuberculosis, M. bovis*); forming pigment only when cultured in the light, a property called *photochromogenesis* (including *M. kansasii, M. marinum*); and forming pigment even when cultured in the dark, a property called *scotochromogenesis* (including *M. gordonae, M. paraffinicum*). Photoinduction of carotenoid formation involves short-wavelength (blue) light and occurs only in the presence of O_2. The evidence indicates that the critical event in photoinduction is a light-catalyzed oxidation event, and it appears that one of the early enzymes in carotenoid biosynthesis is photoinduced. As with other carotenoid-containing bacteria, it has been suggested that carotenoids protect mycobacteria against oxidative damage involving singlet oxygen (⚬⚬ Section 6.13).

The virulence of *Mycobacterium tuberculosis* cultures has been correlated with the formation of long, cord-like structures (Figure 12.70b) on agar or in liquid medium, due to side-to-side aggregation and intertwining of long chains of bacteria. Growth in cords reflects the presence on the cell surface of a characteristic lipid, the **cord factor**, which is a glycolipid (Figure 12.71). The pathogenesis of the disease tuberculosis is discussed in detail in Section 26.5.

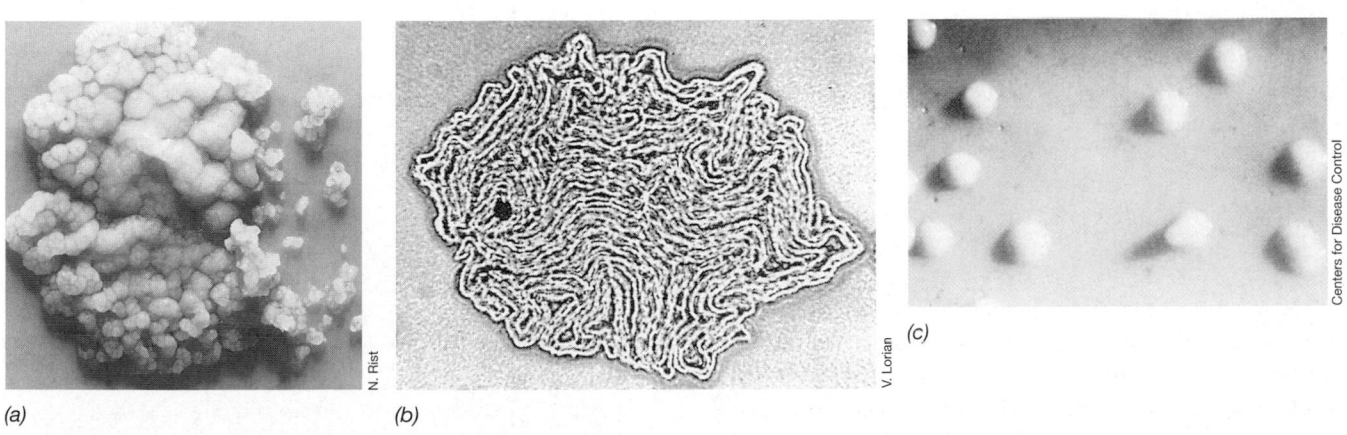

(a) (b) (c)

N. Rist V. Lorian Centers for Disease Control

Figure 12.70 Characteristic colony morphology of mycobacteria. (a) *Mycobacterium tuberculosis*, showing the compact, wrinkled appearance of the colony. The colony is about 7 mm in diameter. (b) A colony of virulent *M. tuberculosis* at an early stage, showing the characteristic cordlike growth. Individual cells are about 0.5 μm in diameter. (See also the historic drawings of *M. tuberculosis* cells made by Robert Koch, ⚬⚬ Figure 1.13). (c) Colonies of *Mycobacterium avium* from a strain of this organism isolated as an opportunistic pathogen from an AIDS patient.

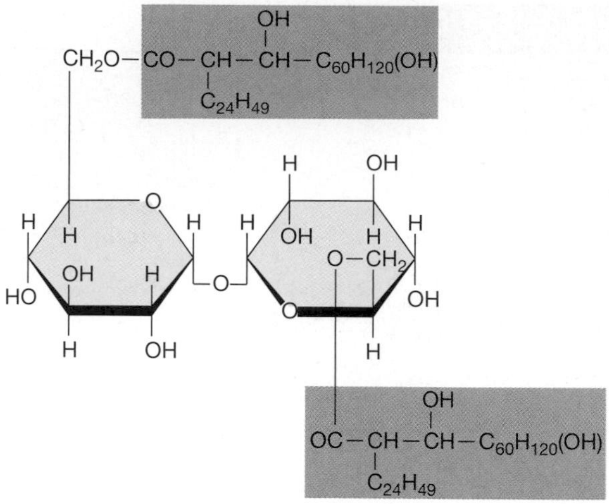

$$CH_2O-CO-CH-CH-C_{60}H_{120}(OH)$$
$$\qquad\qquad\quad C_{24}H_{49}$$

$$OC-CH-CH-C_{60}H_{120}(OH)$$
$$\qquad\quad C_{24}H_{49}$$

Figure 12.71 Structure of cord factor, a mycobacterial glycolipid: 6,6'-dimycolyltrehalose. The two identical long-chain dialcohol groups are shown in purple.

✓ 12.22–12.23 *Concept Check*

"High GC," gram-positive *Bacteria* include such organisms as *Corynebacterium*, *Arthrobacter*, *Propionibacterium*, and *Mycobacterium*. They are mainly harmless soil saprophytes but *M. tuberculosis* is the causative agent of the disease tuberculosis. *M. tuberculosis* cells have a lipid-rich, waxy outer surface layer that requires special staining procedures (the acid-fast stain) in order to observe the cells microscopically.

✓ What is snapping division and what organism practices it?

✓ What organism is involved in the ripening of Swiss cheese and what chemical compounds does it make that helps to flavor the cheese and make the holes?

✓ What is mycolic acid, what organism produces it, and what properties does this substance confer on cells that make it?

12.24 Filamentous, High GC, Gram-Positive *Bacteria: Streptomyces* and Other Actinomycetes

Key Genera

Streptomyces
Actinomyces

The actinomycetes are a large group of filamentous, gram-positive *Bacteria* that form branching filaments. As a result of successful growth and branching, a ramifying network of filaments is formed, called a *mycelium* (Figure 12.72). Although it is of bacterial dimensions, the mycelium is in some ways analogous to the mycelium formed by the filamentous fungi (⊂⊃ Figure 14.19). Most actinomycetes form spores; the manner of spore formation varies and is used in separating subgroups, as out-

lined in Table 12.30. The DNA base compositions of most members of the actinomycetes fall within the range of 63–78% GC, and organisms at the upper end of this range have the highest GC percentage of any bacteria known. Phylogenetically, the filamentous actinomycetes form a coherent group; thus, the mycelial spore-forming habit is of both phylogenetic as well as taxonomic importance. In the present discussion we concentrate on the genus *Streptomyces*.

Streptomyces

Streptomyces is a genus represented by a large number of species and varieties. Over 500 species of *Streptomyces* are recognized, although GC base ratios cluster tightly between 69 and 73 mol %. *Streptomyces* filaments are usually 0.5–1.0 μm in diameter, are of indefinite length, and often lack cross-walls in the vegetative phase. Growth occurs at the tips of the filaments and is often accompanied by branching so that the vegetative phase consists of a complex, tightly woven matrix, resulting in a compact, convoluted mycelium and subsequent colony. As the colony ages, characteristic aerial filaments called *sporophores* are formed, which project above the surface of the colony and give rise to spores (Figure 12.73). *Streptomyces* spores, called **conidia**, are not related in any way to the endospores of *Bacillus* and *Clostridium* because the streptomycete spores are produced simply by the formation of cross-walls in the multinucleate sporophores followed by separation of the individual cells directly into spores (Figure 12.74). Differences in the shape and arrangement of aerial filaments and spore-bearing structures of various species are among the fundamental features used in classifying the *Streptomyces* groups (Figure 12.75). The conidia and sporophores are often pigmented and contribute a characteristic color to the mature colony (Figure 12.76*a*, page 419). The dusty appearance of the mature colony, its compact nature, and its color make detection of *Streptomyces* colonies on agar plates relatively easy (Figure 12.76*b*).

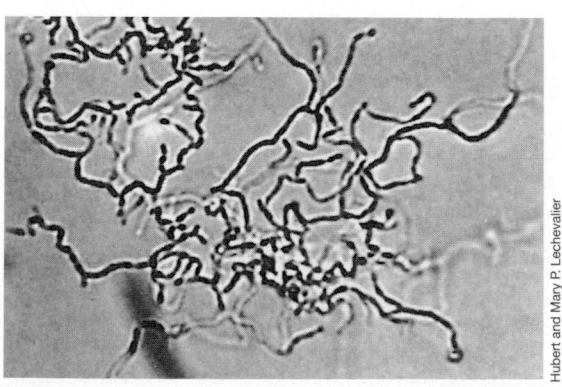

Figure 12.72 A young colony of an actinomycete of the genus *Nocardia*, showing typical filamentous cellular structure (mycelium). Each filament is about 0.8–1 μm in diameter.

TABLE 12.30 Actinomycetes and related genera (all gram-positive)[a]

Major groups	DNA (mol % GC)
Coryneform group of bacteria: rods, often club-shaped, morphologically variable; not acid-fast or filamentous; snapping cell division	
Corynebacterium: irregularly staining segments, sometimes granules; club-shaped swelling frequent; animal and plant pathogens, also soil saprophytes	51–65
Arthrobacter: coccus-rod morphogenesis; soil organisms	59–70
Cellulomonas: coryneform morphology; cellulose digested; facultative aerobe	71–73
Kurthia: rods with rounded ends occurring in chains; coccoid later	36–38
Brevibacterium: coccus-rod morphogenesis; cheese, skin	60–67
Propionic acid bacteria: anaerobic to aerotolerant; rods or filaments, branching	
Propionibacterium: nonmotile; anaerobic to aerotolerant; produce propionic acid and acetic acid; dairy products (Swiss cheese); skin, may be pathogenic	53–68
Eubacterium: obligate anaerobes; produce mixture of organic acids, including butyric, acetic, formic, and lactic; intestine, infections of soft tissue, soil; may be pathogenic; probably the predominant member of the intestinal flora	26–48
Obligate anaerobes	
Bifidobacterium: smooth microcolony, no filaments; coryneform cells common; found in intestinal tract of breast-fed infants	55–67
Acetobacterium: homoacetogen; sediments and sewage	39–43
Butyrivibrio: curved rods; rumen	36–42
Thermoanaerobacter: rods, thermophilic, found in hot springs	37–39
Actinomycetes: filamentous, often branching; highly diverse	
Group I. Actinomycetes: not acid-alcohol-fast; facultatively aerobic; mycelium not formed; branching filaments may be produced; rod, coccoid, or coryneform cells	
Actinomyces: anaerobic to facultatively aerobic; filamentous microcolony, but filaments transitory and fragment into coryneform cells; may be pathogenic for humans or animals; found in oral cavity	57–69
Other genera: *Arachnia, Bacterionema, Rothia, Agromyces*	
Group II. Mycobacteria: acid-fast, filaments transitory	
Mycobacterium: pathogens, saprophytes; obligate aerobes; lipid content of cells and cell walls high; waxes, mycolic acids; simple nutrition; growth slow; tuberculosis, leprosy, granulomas, avian tuberculosis; also soil organisms; hydrocarbon oxidizers	62–70
Group III. Nitrogen-fixing actinomycetes: nitrogen-fixing symbionts of plants; true mycelium produced	
Frankia: forms nodules of two types on various plant roots; probably microaerophilic; grows slowly; fixes N_2	67–72
Group IV. Actinoplanes: true mycelium produced; spores formed, borne inside sporangia	
Actinoplanes, Streptosporangium	69–71
Group V. Dermatophilus group: mycelial filaments divide transversely, and in at least two longitudinal planes, to form masses of motile, coccoid elements; aerial mycelium absent; occasionally responsible for epidermal infections	
Dermatophilus, Geodermatophilus	56–75
Group VI. Nocardias: mycelial filaments commonly fragment to form coccoid or elongate elements; aerial spores occasionally produced; sometimes acid-fast; lipid content of cells and cell wall very high	
Nocardia: common soil organisms; obligate aerobes; many hydrocarbon utilizers	61–72
Rhodococcus: soil saprophytes, also common in gut of various insects; utilize hydrocarbons	59–69
Group VII. Streptomycetes: mycelium remains intact, abundant aerial mycelium and long spore chains	
Streptomyces: Nearly 500 recognized species, many produce antibiotics	69–75
Other genera (differentiated morphologically): *Streptoverticillium, Sporichthya, Kitasatoa, Chainia*	67–73
Group VIII. Micromonosporas group: mycelium remains intact; spores formed singly, in pairs, or short chains; several thermophilic; saprophytes found in soil, rotting plant debris; one species produces endospores	
Micromonospora, Microbispora, Themobispora, Thermoactinomyces, Thermomonospora	54–79

[a] Phylogenetically, all species (except for *Acetobacterium, Butyrivibrio,* and *Thermoanaerobacter*) fall into the high GC subdivision of the gram-positive *Bacteria*.

Ecology and Isolation of *Streptomyces*

Although a few streptomycetes can be found in aquatic habitats, they are primarily *soil* organisms. In fact, the characteristic earthy odor of soil is caused by the production of a series of streptomycete metabolites called *geosmins*. These substances are sesquiterpenoid compounds, unsaturated ring compounds of carbon, oxygen, and hydrogen. A common geosmin is trans-1, 10-dimethyl-trans-9-decalol. Geosmins are also produced by some cyanobacteria (see Section 12.26).

Alkaline and neutral soils are more favorable for the development of *Streptomyces* than are acid soils. Higher numbers of *Streptomyces* are usually found in well-drained soils (such as sandy loams, or soils covering limestone), and there is some evidence to suggest that *Streptomyces* require a lower water potential for growth

(a)

(b)

Peter Hirsch

Hubert and Mary P. Lechevalier

Figure 12.73 Photomicrographs of several spore-bearing structures of actinomycetes. (a) *Streptomyces*, a monoverticillate type. (b) *Streptomyces*, a spiral type. Filaments are about 0.8 μm wide in both cases.

than many other soil bacteria. Isolation of *Streptomyces* from soil is relatively easy: A suspension of soil in sterile water is diluted and spread on selective agar medium, and the plates are incubated at 25°C (∞ Figure 30.7a). Media often selective for *Streptomyces* contain the usual

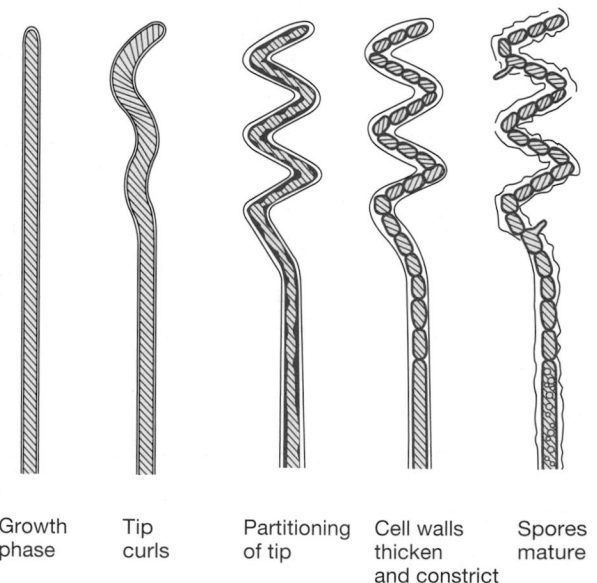

| Growth phase | Tip curls | Partitioning of tip | Cell walls thicken and constrict | Spores mature |

Figure 12.74 Diagram of stages in the conversion of a streptomycete's aerial hypha (sporophores) into spores (conidia).

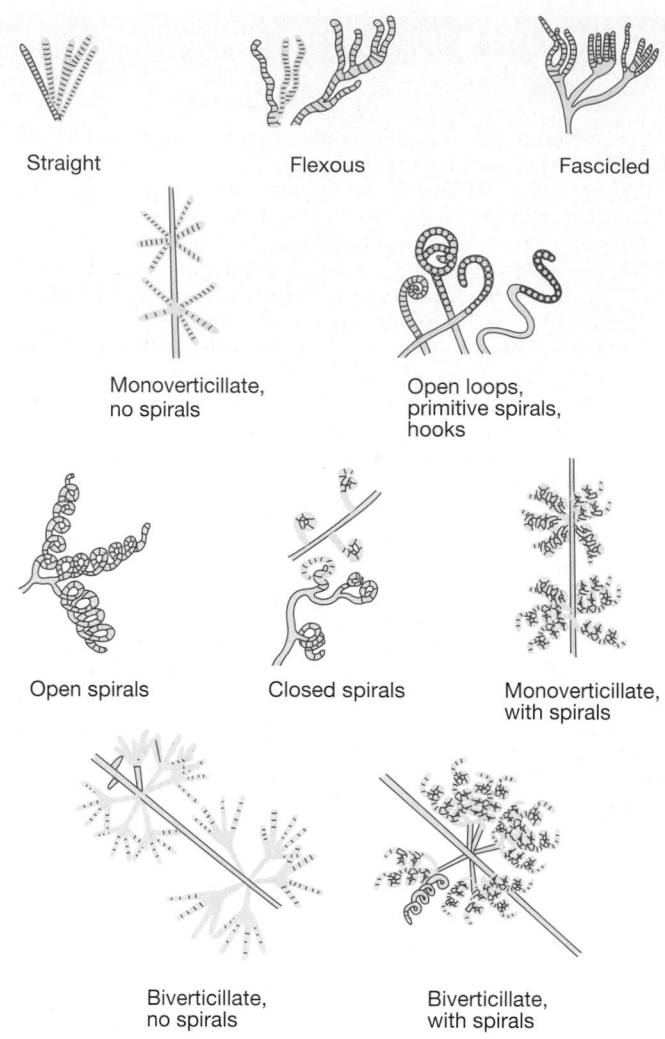

Straight   Flexous   Fascicled

Monoverticillate, no spirals   Open loops, primitive spirals, hooks

Open spirals   Closed spirals   Monoverticillate, with spirals

Biverticillate, no spirals   Biverticillate, with spirals

Figure 12.75 Various types of sporebearing structures in the streptomycetes.

assortment of inorganic salts to which starch, asparagine, or calcium malate is added as a carbon source and undigested casein or potassium nitrate as a nitrogen source. After incubation for 5–7 days in air, the plates are examined for the presence of the characteristic *Streptomyces* colonies (Figures 12.76 and 12.77), and spores of interesting colonies can be streaked to isolate pure cultures.

Nutritionally, the streptomycetes are quite versatile. Growth-factor requirements are rare, and a wide variety of carbon sources, such as sugars, alcohols, organic acids, amino acids, and some aromatic compounds, can be utilized. Most isolates produce extracellular hydrolytic enzymes that permit utilization of polysaccharides (starch, cellulose, hemicellulose), proteins, and fats, and some strains can use hydrocarbons, lignin, tannin, or even rubber. *Streptomyces* can often be obtained by spreading a soil dilution on an alkaline agar medium containing polymers such as casein and starch (Figure 12.76a). A single isolate may be able to break down over 50 distinct carbon sources. Streptomycetes are strict aerobes whose growth

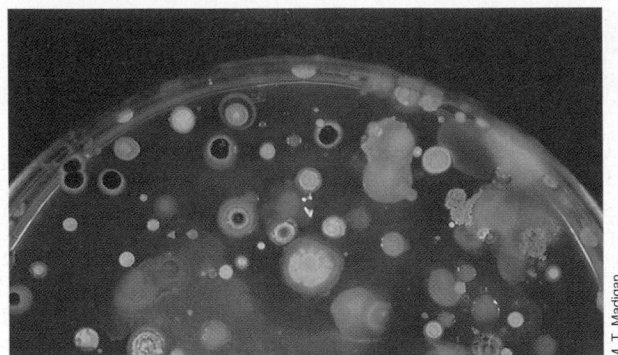

(a)

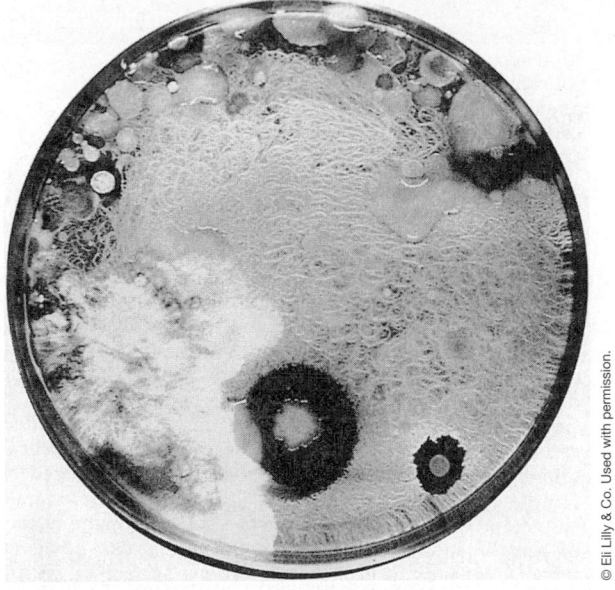

(a)

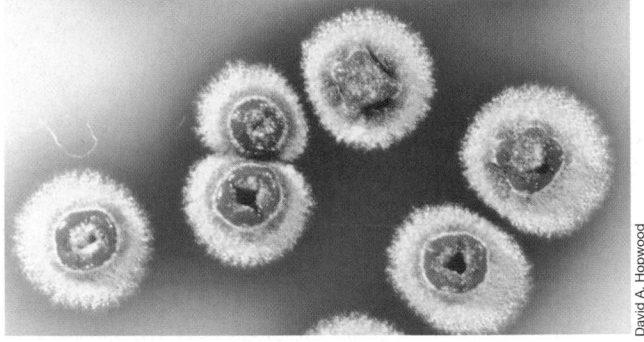

(b)

Figure 12.76 *Streptomyces*. (a) Colonies of *Streptomyces* and other soil bacteria derived from spreading a soil dilution on a casein-starch agar plate. The *Streptomyces* colonies are of various colors (several black *Streptomyces* colonies are in the foreground) but can easily be identified by their opaque, rough, nonspreading morphology. (b) Close-up photo of colonies of *Streptomyces coelicolor*.

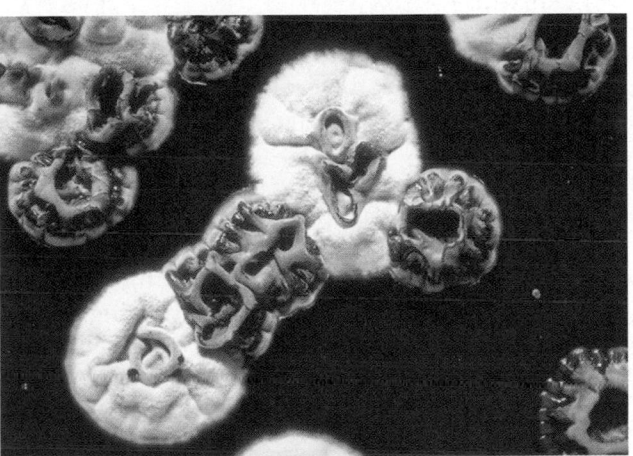

(b)

Figure 12.77 Antibiotics from *Streptomyces*. (a) Antibiotic action of soil microorganisms on a crowded plate. The smaller colonies surrounded by inhibition zones are streptomycetes; the larger, spreading colonies are *Bacillus* species. (b) The red-colored antibiotic undecylprodigiosin is being excreted by colonies of *Streptomyces coelicolor*.

in liquid culture is usually markedly stimulated by forced aeration. Sporulation usually does not take place in liquid culture but only when the organism is growing on the surface of agar or another solid substrate; it can occur, however, when organisms form a pellicle on the surface of an unshaken liquid culture.

Antibiotics of *Streptomyces*

Perhaps the most striking property of the streptomycetes is the extent to which they produce **antibiotics** (Table 12.31). Evidence for antibiotic production is often seen on the agar plates used in the initial isolation of *Streptomyces*: Adjacent colonies of other bacteria show zones of inhibition (Figures 12.76a and 12.77a; ⇨ Figure 30.7a). In some studies, close to 50% of all *Streptomyces* isolated have proved to be antibiotic producers. Because of the great economic and medical importance of many streptomycete antibiotics, an enormous amount of work has been done on these producers. Over 500 distinct antibiotic substances have been shown to be produced by streptomycetes, and many more are suspected (⇨ Section 30.6), and a large number of these have been identified chemically (Figure 12.77b). Some organisms produce more than one antibiotic, and often the several kinds

produced by one organism are not even chemically related. The same antibiotic may be formed by different species found in widely scattered parts of the world. And, although an antibiotic-producing organism is resistant to its own antibiotics, it usually remains sensitive to antibiotics produced by other streptomycetes.

More than 60 streptomycete antibiotics have found practical application in human and veterinary medicine, agriculture, and industry. Some of the more common antibiotics of *Streptomyces* origin are listed in Table 12.31. They are grouped into classes based on the chemical structure of the parent molecule. The search for new streptomycete antibiotics continues because many infectious diseases are still not adequately controlled

TABLE 12.31 Some common antibiotics synthesized by species of *Streptomyces*

Chemical class	Common name	Produced by	Active against[a]
Aminoglycosides	Streptomycin	*S. griseus*	Most gram-negative *Bacteria*
	Spectinomycin	*Streptomyces* spp.	*M. tuberculosis*, penicillinase-producing *N. gonorrhoeae*
	Neomycin	*S. fradiae*	Broad spectrum, usually used in topical applications because of toxicity
Tetracyclines	Tetracycline	*S. aureofaciens*	Broad spectrum, gram-positive and gram-negative *Bacteria*, rickettsias and chlamydias, *Mycoplasma*
	Chlortetracycline	*S. aureofaciens*	As for tetracycline
Macrolides	Erythromycin	*S. erythreus*	Most gram-positive *Bacteria*, frequently used in place of penicillin, *Legionella*
	Clindamycin	*S. lincolnensis*	Effective against obligate anaerobes, especially *Bacteroides fragilis*
Polyenes	Nystatin	*S. noursei*	Fungi, especially *Candida* infections
	Amphocetin B	*S. nodosus*	Fungi
None	Chloramphenicol	*S. venezuelae*	Broad spectrum; drug of choice for typhoid fever

[a] Most antibiotics are effective against several different *Bacteria*. The entries in this column refer to the common clinical application of a given antibiotic. The structures and mode of action of many of these antibiotics are discussed in Sections 20.7–20.9.

by existing antibiotics. Also, the development of antibiotic-resistant strains requires the continual discovery of new agents. Ironically, however, despite the extensive work on antibiotic-producing streptomycetes and the fact that the antibiotic industry is a multibillion dollar enterprise, the ecology of *Streptomyces* remains poorly understood. The ecological rationale for why antibiotics are produced is not clear. However, one hypothesis for why *Streptomyces* species produce antibiotics is that antibiotic production, which is linked to sporulation (a process itself triggered by nutrient depletion), might be a mechanism to inhibit the growth of other organisms competing with differentiating *Streptomyces* cells for limiting nutrients. This would allow the *Streptomyces* to complete the sporulation process, there-by forming a dormant structure with increased chances of survival.

✓ *12.24 Concept Check*

The Streptomycetes are a large group of filamentous, gram-positive *Bacteria* that form spores at the end of aerial filaments. Many clinically useful antibiotics like tetracycline and neomycin have come from *Streptomyces* species.

✓ How do spores and the process of sporulation in a *Streptomyces* species differ from that in a *Bacillus* species?

✓ What energy class of organism is a *Streptomyces* and from what types of compounds do these organisms obtain their energy?

✓ Why might antibiotic production be of advantage to Streptomycetes?

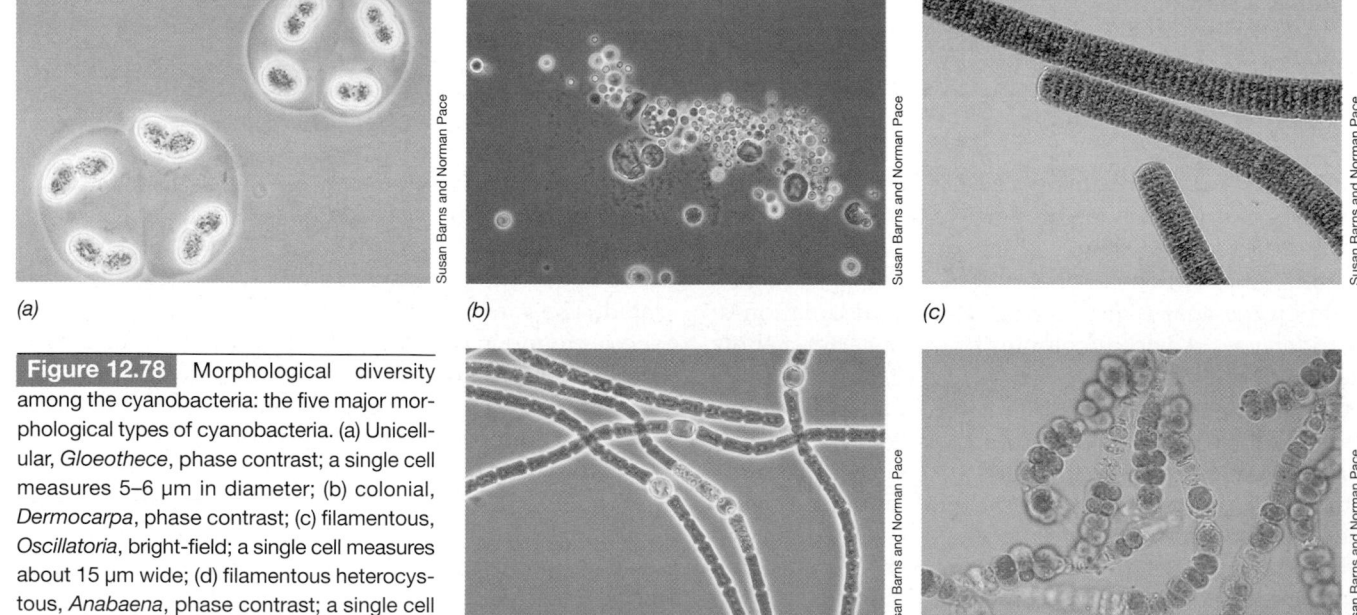

(a)   (b)   (c)

(d)   (e)

Figure 12.78 Morphological diversity among the cyanobacteria: the five major morphological types of cyanobacteria. (a) Unicellular, *Gloeothece*, phase contrast; a single cell measures 5–6 µm in diameter; (b) colonial, *Dermocarpa*, phase contrast; (c) filamentous, *Oscillatoria*, bright-field; a single cell measures about 15 µm wide; (d) filamentous heterocystous, *Anabaena*, phase contrast; a single cell measures about 5 µm wide; (e) filamentous branching, *Fischerella*, bright-field.

Susan Barns and Norman Pace

IV PHYLUM 3: CYANOBACTERIA AND PROCHLOROPHYTES

12.25 Cyanobacteria

Key Genera

Synechococcus
Oscillatoria
Nostoc

Cyanobacteria comprise a large and morphologically heterogeneous group of phototrophic *Bacteria*. Cyanobacteria differ in fundamental ways from purple and green anoxyphototrophs, most notably in the fact that they are *oxygenic* phototrophs. Cyanobacteria represent one of the major kingdoms of *Bacteria* and show a distant relationship to gram-positive *Bacteria* (see Figure 12.1). The evolutionary significance of cyanobacteria was discussed in Section 11.1, and it is likely that these organisms were the first oxygen-evolving phototrophic organisms on Earth and were responsible for the conversion of the atmosphere of the Earth from anoxic to oxic.

Structure and Classification of Cyanobacteria

The morphological diversity of the cyanobacteria is impressive (Figure 12.78). Both unicellular and filamentous forms are known, and considerable variation within these morphological types occurs. Cyanobacteria can be divided into five morphological groups: unicellular dividing by binary fission (see Figure 12.78a); unicellular dividing by multiple fission (colonial) (see Figure 12.78b); filamentous containing differentiated cells called heterocysts that function in nitrogen fixation (see Figures 12.78d and 12.80); filamentous nonheterocystous forms (see Figure 12.78c); and branching filamentous types (see

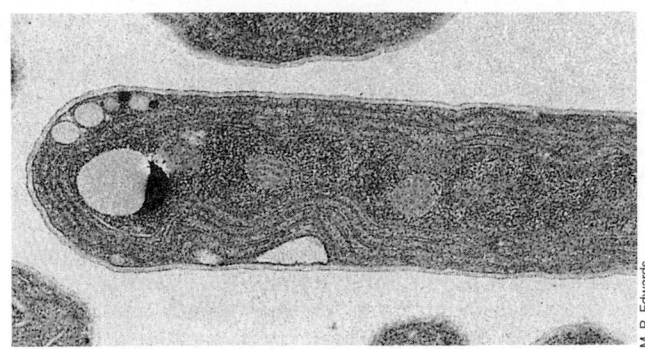

Figure 12.79 Electron micrograph of a thin section of the cyanobacterium *Synechococcus lividus*. A cell is about 5 μm in diameter. Note thylakoid membranes running parallel to the cell wall.

Figure 12.78e). Table 12.32 lists the major genera currently recognized in each group. Cyanobacterial cells range in size from those of typical bacteria (0.5–1 μm in diameter) to cells as large as 40 μm in diameter (in the species *Oscillatoria princeps*, ∞ Figure 4.44a).

The structure of the cell wall of some cyanobacteria is similar to that of gram-negative *Bacteria*, and peptidoglycan is present in the walls (Figure 12.79). Many cyanobacteria produce extensive mucilaginous envelopes, or sheaths, that bind groups of cells or filaments together (see, for example, Figure 12.78a). The photosynthetic lamellar membrane system is often complex and multilayered (∞ Figure 17.10b), although in some of the simpler cyanobacteria the lamellae are regularly arranged in concentric circles around the periphery of the cytoplasm (Figure 12.79). Cyanobacteria have only one form of chlorophyll, chlorophyll *a*, and all of them also have characteristic biliprotein pigments, **phycobilins** (∞ Figure 17.10a), which function as accessory pigments in photosynthesis. One class of phycobilins, *phycocyanins*, are blue, and together with the green chlorophyll *a* are responsible for the blue-green color of the bacteria. However, some cyanobacteria produce

TABLE 12.32	Genera and grouping of cyanobacteria	
Group	**Genera**	**DNA (mol% GC)**
Group I. Unicellular: single cells or cell aggregates	*Gloeothece* (Figure 12.78a), *Gloeobacter, Synechococcus, Cyanothece, Gloeocapsa, Synechocystis, Chamaesiphon, Merismopedia*	35–71
Group II. Pleurocapsalean: reproduce by formation of small spherical cells called baeocytes produced through multiple fission	*Dermocarpa* (Figure 12.78b), *Xenococcus, Dermocarpella, Pleurocapsa, Myxosarcina, Chroococcidiopsis*	40–46
Group III. Oscillatorian: filamentous cells that divide by binary fission in a single plane	*Oscillatoria* (Figure 12.78c), *Spirulina, Arthrospira, Lyngbya, Microcoleus, Pseudanabaena*	40–67
Group IV. Nostocalean: filamentous cells that produce heterocysts	*Anabaena* (Figure 12.78d), *Nostoc, Calothrix, Nodularia, Cylinodrosperum, Scytonema*	38–46
Group V. Branching: cells divide to form branches	*Fischerella* (Figure 12.78e), *Stigonema, Chlorogloeopsis, Hapalosiphon*	42–46

phycoerythrin, a red phycobilin, and species possessing this pigment are red or brown in color.

Structural Variations: Gas Vesicles and Heterocysts

Among the cytoplasmic structures seen in many cyanobacteria are **gas vesicles** (⚬ Section 4.14), which are especially common in species that live in open waters (planktonic species). Their function is to regulate cell buoyancy such that it can remain in a position in the water column where light intensity is optimal for photosynthesis. Some filamentous cyanobacteria form **heterocysts**, which are rounded, seemingly empty cells, usually distributed regularly along a filament or at one end of a filament (Figure 12.80*a*). Heterocysts arise from differentiation of vegetative cells and are the sole sites of *nitrogen fixation* (the reduction of N_2 to NH_3, ⚬ Section 17.28) in heterocystous cyanobacteria. In *Anabaena*, a well-studied heterocystous cyanobacterium, complex gene rearrangements occur within the heterocyst to yield a contiguous cluster of *nif* genes that can be expressed as a unit (*nif* genes encode nitrogenase, ⚬ Section 17.28).

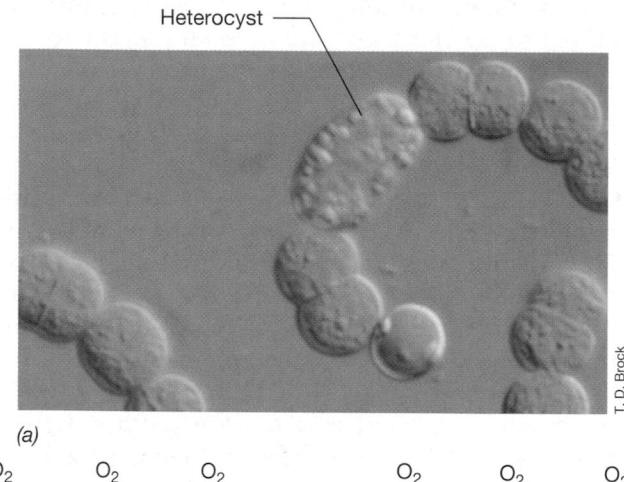

Heterocyst

(a)

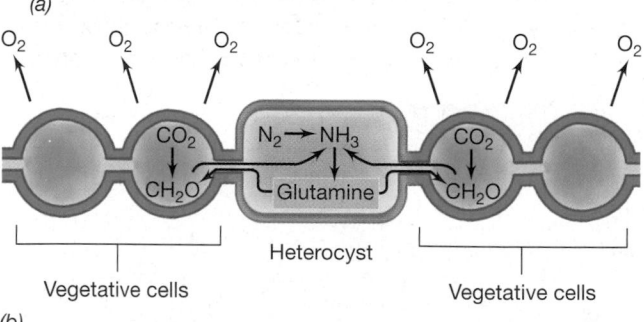

(b)

Figure 12.80 Heterocysts. (a) Heterocysts in the cyanobacterium *Anabaena*. Heterocysts are the sole site of nitrogen fixation in heterocystous cyanobacteria. (b) Model for the operation of a heterocyst. The heterocyst lacks oxygen-producing ability (Photosystem II, ⚬ Section 17.5) and obtains the needed reductant for nitrogen fixation from organic matter produced by adjacent vegetative cells. Glutamine is the form of fixed nitrogen transported from heterocysts to vegetative cells.

Heterocysts have intercellular connections with adjacent vegetative cells, and there is mutual exchange of materials between these cells, with products of photosynthesis moving from vegetative cells to heterocysts and products of nitrogen fixation moving from heterocysts to vegetative cells (Figure 12.80*b*). Heterocysts are low in phycobilin pigments and *lack* photosystem II, the oxygen-evolving photosystem that generates reducing power from H_2O (⚬ Section 17.5). Without photosystem II they are unable to fix CO_2 and thus lack the necessary electron donor to reduce N_2 to NH_3; fixed carbon imported to the heterocyst from an adjacent vegetative cell solves this problem (Figure 12.80*b*).

Heterocysts are surrounded by a thickened cell wall containing large amounts of glycolipid, which serves to slow the diffusion of O_2 into the cell. Because of the oxygen lability of the enzyme nitrogenase (⚬ Section 17.29), it seems likely that the heterocyst, by maintaining an anoxic environment, stabilizes the nitrogen-fixing system in organisms that are not only aerobic but also oxygen-producing. Indeed, some nonheterocystous filamentous cyanobacteria produce nitrogenase and fix nitrogen in normal vegetative cells if they are grown anaerobically by vigorous bubbling with N_2 to remove O_2.

Cyanophycin and Other Structures

A structure called **cyanophycin** can be seen in electron micrographs of many cyanobacteria. This structure is a copolymer of aspartic acid and arginine:

$$Asp—Asp—Asp—Asp—Asp—$$
$$|\quad\quad|\quad\quad|\quad\quad|\quad\quad|$$
$$Arg\quad Arg\quad Arg\quad Arg\quad Arg$$

and can constitute up to 10% of the cell mass. Cyanophycin is a nitrogen storage product in many cyanobacteria, and when nitrogen in the environment becomes deficient, this polymer is broken down and used. Cyanophycin is also an energy reserve in cyanobacteria. Arginine, derived from cyanophycin, can be hydrolyzed to yield ornithine, with the production of ATP through the action of the enzyme *arginine dihydrolase* with carbamyl phosphate (⚬ Section 17.19) occurring as an intermediate:

Arginine $+$ ADP $+$ P$_i$ $+$ H_2O →

Ornithine $+$ 2 NH_3 $+$ CO_2 $+$ ATP

Arginine dihydrolase is present in many cyanobacteria and may function as a source of ATP for maintenance purposes during dark periods.

Many cyanobacteria exhibit gliding motility; true rotating flagella have never been found. Gliding occurs only when the cell or filament is in contact with a solid surface or with another cell or filament. In some cyanobacteria gliding is not a simple translational movement, but is accompanied by rotations, reversals, and flexings of filaments. Most gliding species exhibit directional movement toward light (phototaxis), and chemotaxis (⚭ Section 4.12) may occur as well.

Among the filamentous cyanobacteria, fragmentation of the filaments often occurs by formation of **hormogonia** (Figure 12.81*a* and *b*), which break away from the filaments and glide off. In some species, resting spores or **akinetes** (Figure 12.81*c*) are formed, which protect the organism during periods of darkness, drying, or freezing. These are cells with thickened outer walls; they germinate through the breakdown of the outer wall and outgrowth of a new vegetative filament. However, even the vegetative cells of many cyanobacteria are relatively resistant to drying or low temperatures.

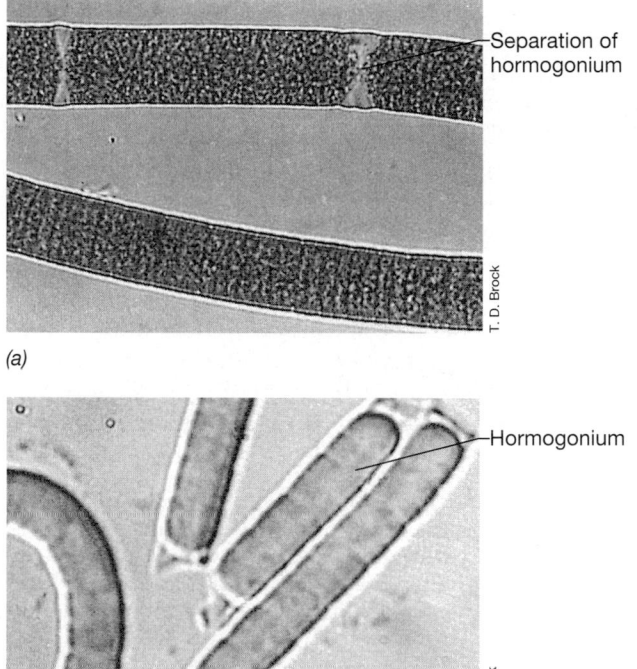

(a)

Separation of hormogonium

Hormogonium

(b)

Akinete

(c)

Figure 12.81 Structural differentiation in filamentous cyanobacteria. (a) Initial stage of hormogonium formation in *Oscillatoria*. Notice the empty spaces where the hormogonium is separating from the filament. (b) Hormogonium of a smaller *Oscillatoria* species. Notice that the cells at both ends are rounded. Nomarski interference contrast microscopy. (c) Akinete (resting spore) of *Anabaena* by phase contrast.

Physiology of Cyanobacteria

The nutrition of cyanobacteria is simple. Vitamins are not required, and nitrate or ammonia is used as nitrogen source. Nitrogen-fixing species are common. Most species tested are obligate phototrophs, being unable to grow in the dark on organic compounds. However, some cyanobacteria can assimilate simple organic compounds such as glucose and acetate if light is present (photoassimilation). Some cyanobacteria, mainly filamentous species, can actually grow in the dark on glucose or other sugars, using the organic material as both carbon and energy source.

Several metabolic products of cyanobacteria are of considerable practical importance. Many cyanobacteria produce potent neurotoxins, and during water blooms when massive accumulations of cyanobacteria may develop, animals ingesting such water may be killed. Many cyanobacteria are also responsible for the production of earthy odors and flavors in fresh waters, and if such waters are used as drinking water sources, aesthetic problems may arise. The major compound produced is *geosmin* (trans-1,10-dimethyl-trans-9-decalol). This substance is also produced by many actinomycetes (see the discussion in Section 12.24) and is responsible for the distinctive "earthy" odor of moist, freshly turned soil.

Ecology and Phylogeny of Cyanobacteria

Cyanobacteria are widely distributed in nature in terrestrial, freshwater, and marine habitats. In general, they are more tolerant to environmental extremes than are algae and are often the dominant or sole oxygenic phototrophic organisms in hot springs (⚭ Table 6.1), saline lakes, and other extreme environments. Many members are found on the surfaces of rocks or soil and occasionally even within rocks themselves (⚭ Figure 14.29). In desert soils subject to intense sunlight, cyanobacteria often form extensive crusts over the surface, remaining dormant during most of the year and growing during the brief winter and spring rains. In shallow marine bays, where relatively warm seawater temperatures exist, cyanobacterial mats of considerable thickness may form. Freshwater lakes, especially those that are fairly rich in nutrients, may develop blooms of cyanobacteria (⚭ Figure 19.10*b*). A few cyanobacteria are symbionts of liverworts, ferns, and cycads; a number are found as the phototrophic component of lichens. In the case of the water fern *Azolla* (⚭ Sections 17.28 and 19.22), it has been shown that the cyanobacterial endophyte (a species of *Anabaena*) fixes nitrogen that becomes available to the plant.

Base compositions of DNA of a variety of cyanobacteria have been determined. Those of the unicellular forms vary from 35 to 71% GC, a range so wide as to suggest that this group contains many members with little genetic relationship to each other. On the other hand, the values for the heterocyst formers vary much less, from 38 to 46% GC. Phylogenetically, cyanobacteria group along morphological lines in most cases. Filamentous heterocystous and nonheterocystous species form distinct groups, as do the branching forms. However, unicellular cyanobacteria are phylogenetically highly diverse, with different representatives showing phylogenetic relationships to different morphological groups.

12.26 Prochlorophytes and Chloroplasts

Key Genera

Prochlorococcus
Prochloron
Prochlorothrix

Prochlorophytes are oxygenic phototrophs that contain chlorophyll *a* and *b* but do *not* contain phycobilins. Prochlorophytes therefore resemble both cyanobacteria (because they are prokaryotic and produce chlorophyll *a*) and the green plant/alga chloroplast (because they contain chlorophyll *b* instead of phycobilins). Phylogenetically, however, different prochlorophytes show specific relationships to cyanobacteria, so the two groups clearly shared a common ancestry.

Prochloron

Prochloron was the first prochlorophyte discovered. It is found in nature as a symbiont of marine invertebrates (didemnid ascidians), and has not been cultured in the laboratory (all studies of the organism to date have relied on material collected from natural samples). Cells of *Prochloron* expressed from the cavities of didemnid tissue are roughly spherical in morphology (Figure 12.82), and

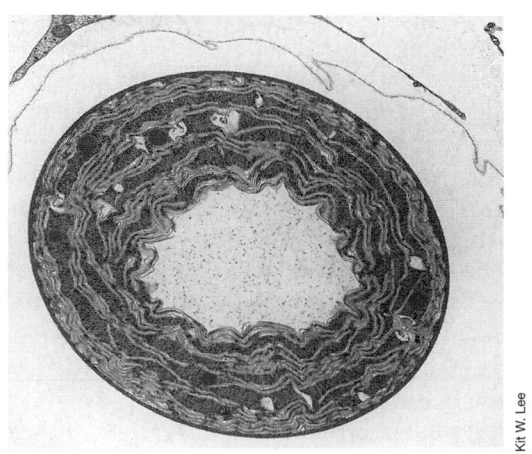

Figure 12.82 Electron micrograph of the prochlorophyte *Prochloron*. Note the extensive intracytoplasmic membranes (thylakoids). Cells are about 10 µm in diameter.

8–10 µm in diameter. Electron micrographs of thin sections (Figure 12.82) show that *Prochloron* has an extensive thylakoid membrane system similar to that observed in the chloroplast (Figure 14.6). Further evidence that *Prochloron* is phylogenetically a member of the *Bacteria* is the presence of muramic acid in the cell walls, indicating that peptidoglycan is present (Section 4.8). The carotenoids of *Prochloron* are similar to those of cyanobacteria, predominantly β-carotene and zeaxanthin. The G + C ratio of different samples of *Prochloron* isolated from different ascidians varies from 31 to 41%, indicating a fair bit of genetic heterogeneity. Thus, different species of *Prochloron* probably exist, but confirmation of this must await laboratory culture and study of pure strains.

Prochlorothrix and *Prochlorococcus*

Prochlorothrix is a filamentous prochlorophyte (Figure 12.83) that can be grown in pure culture. Like *Prochloron*, *Prochlorothrix* contains chlorophylls *a* and *b* and lacks phycobilins, although the thylakoid membranes are less well developed than in *Prochloron* (compare Figures 12.82 and 12.83*b*).

A novel prochlorophyte, *Prochlorococcus*, inhabits the euphotic zone of the open oceans. Cells of these phototrophs are small cocci, measuring less than 1 µm in diameter (Figure 19.11*a*). Like other prochlorophytes, cells of *Prochlorococcus* contain chlorophyll *b*. However, *Prochlorococcus* lacks true chlorophyll *a* and produces instead a modified form of chlorophyll *a* called *divinyl chlorophyll a*. Cells of *Prochlorococcus* also contain α-(instead of β-) carotene, a pigment previously unknown in prokaryotes. Because their numbers in the oceans are relatively large (10^4–10^5 cells/ml), prochlorophytes like *Prochlorococcus* probably have considerable ecological significance as primary producers in open ocean waters. A variety of other prochlorophytes have been isolated including *Acaryochloris* (Figure 12.83*c*), which contains chlorophyll *d* as its major pigment; the latter is present in a variety of algae (eukaryotic cells; Section 14.12).

Prochlorophytes, Chloroplasts, and Evolution

Based on our discussion of endosymbiosis (Sections 2.6, 11.3, and 14.5), the evolutionary significance of prochlorophytes should be apparent. Until the discovery of prochlorophytes, it was always assumed that the chloroplast originated from endosymbiotic association of *cyanobacteria* with a primitive eukaryotic cell. However, this hypothesis has never been scientifically satisfying for at least one major reason: How did the green plant chloroplast evolve the pigment complement it has today if it originated from a *cyanobacterial* endosymbiont that contained *phycobilins* instead of chlorophyll *b*? The hypothesis that *prochlorophytes* instead of cyanobacteria

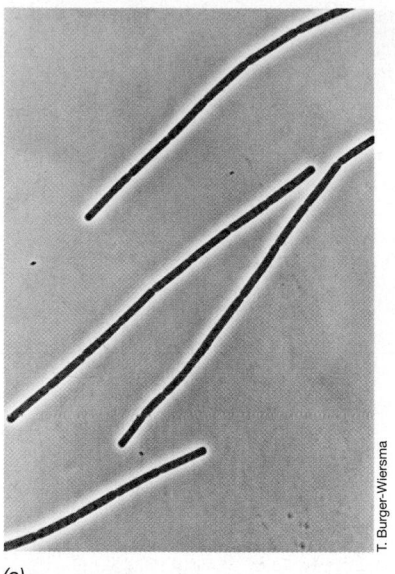 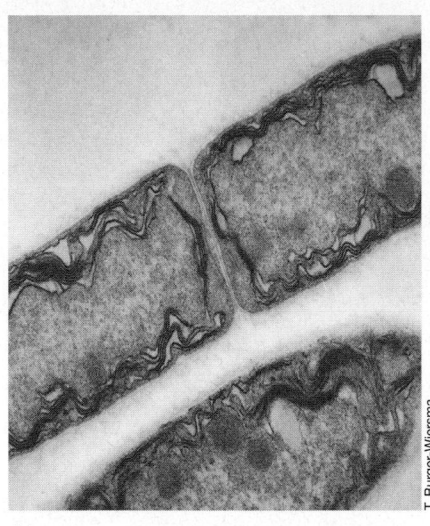 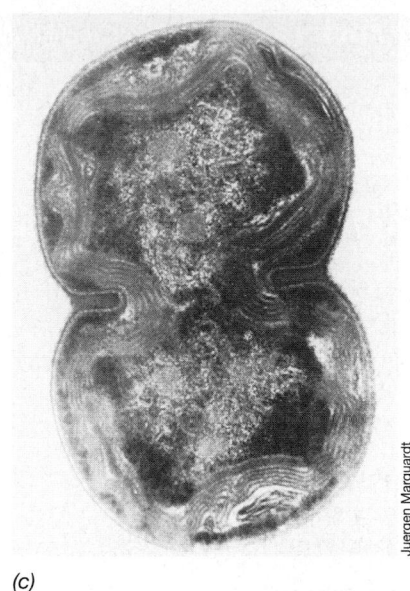

(a)   (b)   (c)

Figure 12.83 Phase and electron micrographs of the filamentous prochlorophyte *Prochlorothrix*. (a) Phase contrast. (b) Electron micrograph of thin section showing arrangement of membranes. The diameter of cells is about 2 μm. (c) *Acaryochloris*. This prochlorophyte contains chlorophyll *d* as its main chlorophyll pigment. A cell is about 1.5 μm in diameter.

were the ancestors of the green plant chloroplast eliminates this major point of contention. However, phylogenetic analyses do not show *Prochloron*, *Prochlorococcus*, or *Prochlorothrix* to be the immediate ancestor of the green plant chloroplast. Instead, prochlorophytes, cyanobacteria, and the plant chloroplast all *shared* a common ancestor.

A hypothesis that accounts for all of these observations is to propose that cyanobacteria and prochlorophytes evolved from ancestors that contained phycobilins and chlorophylls other than just chlorophyll *a*, either chlorophyll *b* or *d*. From here, cyanobacterial lineages evolved to lose accessory chlorophylls while prochlorophyte lineages dispensed with phycobilins. The complement of pigments that we find in each group today, then, likely represents the best combination of photosynthetic pigments for survival in their particular habitat.

✓ 12.25–12.26 Concept Check

Cyanobacteria and prochlorophytes are oxygenic phototrophic prokaryotes. Prochlorophytes differ most clearly from cyanobacteria in that prochlorophytes contain chlorophyll *b* or *d* and lack phycobilins. Oxygen in Earth's atmosphere is thought to have originated from cyanobacterial photosynthesis.

- ✓ Describe at least three ways in which cyanobacteria differ from purple bacteria.
- ✓ What is a *heterocyst* and what is its function?
- ✓ How are cyanobacteria, prochlorophytes, and the chloroplasts of corn plants similar and how do they differ?
- ✓ Of what ecological significance is *Prochlorococcus*?

V PHYLUM 4: CHLAMYDIA

12.27 The Chlamydia

Key Genus

Chlamydia

Organisms of the genus *Chlamydia* are obligately parasitic bacteria with little metabolic capacity that form a distinct phylum of bacteria (see Figure 12.1). Three species of *Chlamydia* are recognized (Table 12.33): *C. psittaci*, the causative agent of the disease *psittacosis*; *C. trachomatis*, the causative agent of *trachoma* and a variety of other human diseases; and *C. pneumoniae*, the cause of a variety of respiratory syndromes (Table 12.33). **Psittacosis** is an epidemic disease of birds that is occasionally transmitted to humans and causes pneumonia-like symptoms. **Trachoma** is a debilitating disease of the eye characterized by vascularization and scarring of the cornea. Trachoma is the leading cause of blindness in humans. Other strains of *C. trachomatis* infect the genitourinary tract, and chlamydial infections are one of the leading sexually transmitted diseases today (⌾ Section 26.13). A comparison of the properties of *C. psittaci*, *C. trachomatis*, and *C. pneumoniae* and the diseases they cause is shown in Table 12.33.

Molecular and Metabolic Properties

Besides being disease entities, the chlamydias are intriguing because of the biological, evolutionary, and metabolic problems they pose. Biochemical studies show

TABLE 12.33 Differential characteristics of species of the genus *Chlamydia*

Characteristic	C. trachomatis	C. psittaci	C. pneumoniae
Hosts	Humans	Birds, mammals, occasionally humans	Humans
Usual site of infection	Mucous membrane	Multiple sites	Respiratory mucosa
Human-to-human transmission	Common	Rare	Probable
Mol % GC of DNA	42–45	39–43	40
Percent homology to *C. trachomatis* DNA by DNA:DNA hybridization[a]	100	10	10
DNA, kilobase pairs/genome (*Escherichia coli* = 4600)	1000	550	~ 1000
Human diseases	Trachoma, otitis media, nongonococcal urethritis (males), urethral inflammation (females), lymphogranuloma venereum, cervicitis	Psittacosis	Respiratory syndromes
Domesticated animal diseases	—	Avian chlamydiosis (parrots, parakeets), pneumonia, synovial tissue arthritis, or conjunctivitis (kittens, lambs, calves, piglets, foals)	—

[a] For discussion of DNA:DNA hybridization, see Section 11.9.

that the chlamydias have gram-negative-type cell walls and they have both DNA and RNA, that is, they are cells, not viruses. Electron microscopy of thin sections of infected cells shows forms that clearly are undergoing binary fission (Figure 12.84). The biosynthetic capacities of the chlamydias are much more limited than even the rickettsias, the other group of obligate intracellular parasites known among the *Bacteria* (see Section 12.13). Indeed, for some time it was thought that chlamydias were "energy parasites," obtaining not only biosynthetic intermediates from their hosts, as do the rickettsias, but also ATP. However, this hypothesis has been questioned

following the sequencing of the genome of *C. trachomatis* (Section 15.3). The approximately 1 Mb chromosome of *C. trachomatis* contains easily recognizable genes for ATP synthesis, and even contains a complement of genes encoding peptidoglycan biosynthetic functions, indicating that this organism may well contain peptidoglycan even though analyses for this cell-wall polymer have been negative. Nevertheless, the chlamydias still probably have the simplest biochemical capacities of all known cellular organisms, and a summary of this is shown in Table 12.34.

Other interesting features of the *C. trachomatis* chromosome include the fact that it lacks a gene encoding the protein FtsZ, a key protein involved in septum formation during cell division (Section 6.1) and previously thought to be indispensable for growth of all prokaryotes, both *Archaea* and *Bacteria*. Moreover, genes are present that have a distinct "eukaryotic look" to them, suggesting the *C. trachomatis* has picked up some host genes that may encode functions that assist it in its pathogenic lifestyle (Table 12.33; Sections 21.7 and 26.13).

Life Cycle of *Chlamydia*

The life cycle of a typical member of the genus *Chlamydia* is shown in Figure 12.85. Two cellular types are seen in a typical life cycle; a small, dense cell, called an **elementary body**, which is relatively resistant to drying and is the means of *dispersal* of the agent, and a larger, less dense cell, called a **reticulate body**, which divides by binary fission and is the *vegetative* form. El-

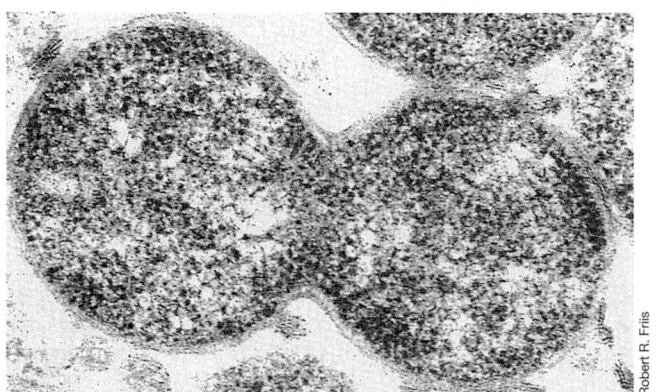

Figure 12.84 | Electron micrograph of a thin section of a dividing cell (reticulate body, see Figure 12.85) of *Chlamydia psittaci*, a member of the psittacosis group, within a mouse tissue culture cell. A single chlamydial cell is about 1 μm in diameter.

TABLE 12.34 Comparison of obligate intracellular parasites: rickettsias, chlamydias, and viruses

Property	Rickettsias	Chlamydias	Viruses
Structural			
Nucleic acid	RNA and DNA	RNA and DNA	Either RNA or DNA (single- or double-stranded), never both
Ribosomes	Present	Present	Absent
Cell wall	Peptidoglycan present	Peptidoglycan present[a]	No wall
Structural integrity during multiplication	Maintained	Maintained	Lost
Metabolic capacities			
Macromolecular synthesis	Carried out	Carried out	Only with use of host machinery
ATP-generating system	Present	Present (?)[a]	Absent
Capable of oxidizing glutamate	Yes	No	No
Sensitivity to antibacterial antibiotics	Sensitive	Sensitive (except for penicillin)	Resistant
Phylogeny	Alpha Proteobacteria	Chlamydial phylum	Not cells

[a] The genome of one chlamydial species, *C. trachomatis*, has been entirely sequenced (⚬⚬ Section 15.3), and genes for peptidoglycan synthesis and ATP synthesis are present. However, the lack of penicillin sensitivity of the chlamydia raises doubt whether peptidoglycan is synthesized.

ementary bodies are nonmultiplying cells specialized for transmission, whereas reticulate bodies are noninfectious forms that specialize in intracellular multiplication. Unlike the rickettsias (see Section 12.13), the chlamydias are not transmitted by arthropods but are primarily *airborne* invaders of the respiratory system—hence the significance of resistance to drying of elementary bodies. When a virus infects a cell, it loses its structural integrity and liberates nucleic acid. When an elementary body enters a cell, however, although it changes form, it remains a structural unit and enlarges and begins to undergo binary fission. A reticulate body is seen in Figure 12.84. After a number of divisions, the vegetative cells are converted to elementary bodies that are released when the host cell disintegrates and can then infect other cells. Generation times of 2–3 h have been measured for reticulate bodies, which are considerably faster than those found for the rickettsias.

In sum, then, the chlamydias appear to have evolved an efficient and effective survival strategy including parasitizing the resources of the host (Table 12.34) and the production of resistant cell forms for transmission. It is thus not surprising that chlamydias have been associated with so many different disease syndromes (Table 12.33; ⚬⚬ Section 26.13).

✓ 12.27 Concept Check

Chlamydia are extremely small parasitic bacteria that cause a variety of human diseases. *Chlamydia* contain a very small genome and are deficient in many metabolic functions.

- ✓ Using the data of Table 12.34 as a guide, how can Chlamydias be differentiated from rickettsias? From viruses?

- ✓ What is the difference between an *elementary* body and a *reticulate* body?

- ✓ What surprises have emerged from sequencing of the chlamydial genome?

VI PHYLUM 5: PLANCTOMYCES/PIRELLULA

12.28 *Planctomyces*: A Phylogenetically Unique Stalked Bacterium

Key Genera

Planctomyces
Pirellula
Gemmata

This phylum contains a number of morphologically unique bacteria including the genera *Planctomyces*, *Pirellula*, *Gemmata*, and *Isosphaera*. The best studied of these has been *Planctomyces* (Figure 12.86). In Section 12.16, we considered stalked bacteria such as *Caulobacter*. *Planctomyces* is also a stalked bacterium; however, unlike *Caulobacter*, the stalk of *Planctomyces* is made of protein and does not contain a cell wall or cytoplasm (compare Figure 12.86 with Figure 12.41). The *Planctomyces* stalk presumably functions in attachment but it is a much narrower and finer structure than the prosthecal stalk of *Caulobacter*.

Other Features of the Group

Planctomyces and relatives are also of interest because they lack peptidoglycan and their cell walls are of an

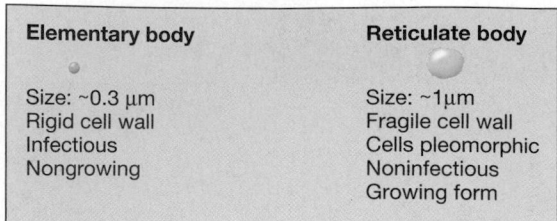

Elementary body	Reticulate body
Size: ~0.3 μm	Size: ~1μm
Rigid cell wall	Fragile cell wall
Infectious	Cells pleomorphic
Nongrowing	Noninfectious
	Growing form

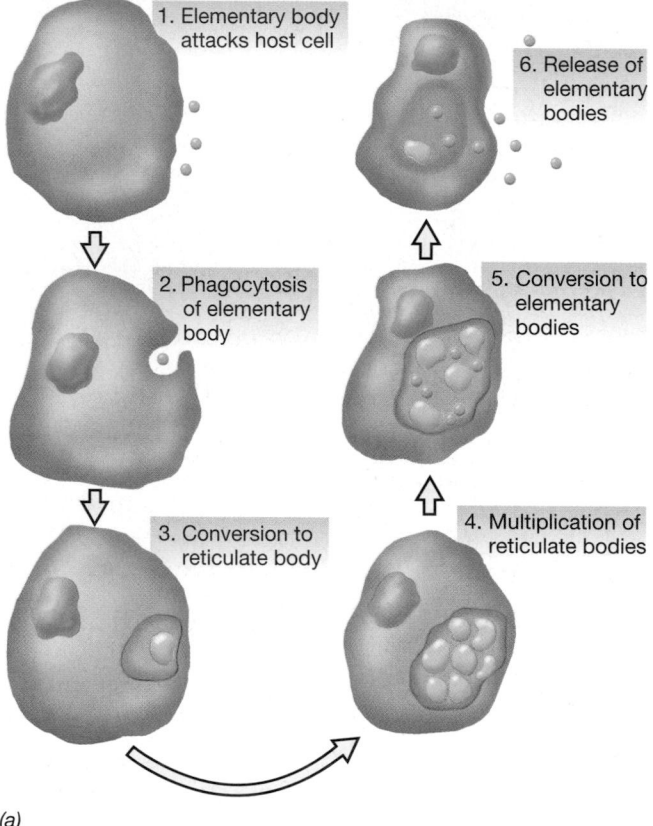

1. Elementary body attacks host cell

2. Phagocytosis of elementary body

3. Conversion to reticulate body

4. Multiplication of reticulate bodies

5. Conversion to elementary bodies

6. Release of elementary bodies

(a)

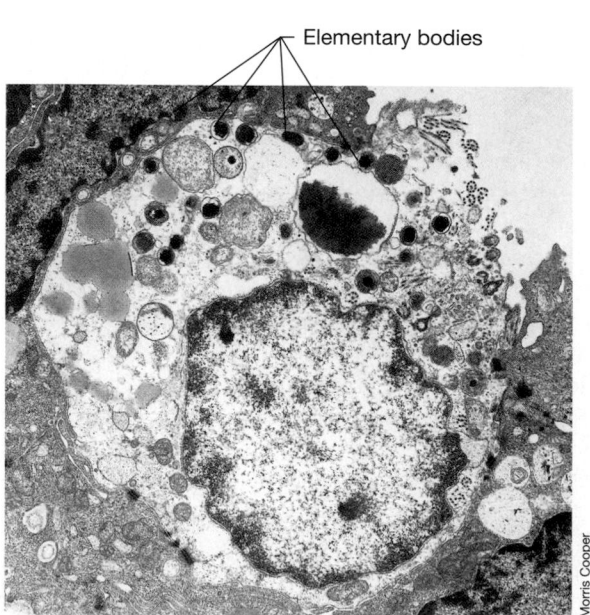

Elementary bodies

Morris Cooper

(b)

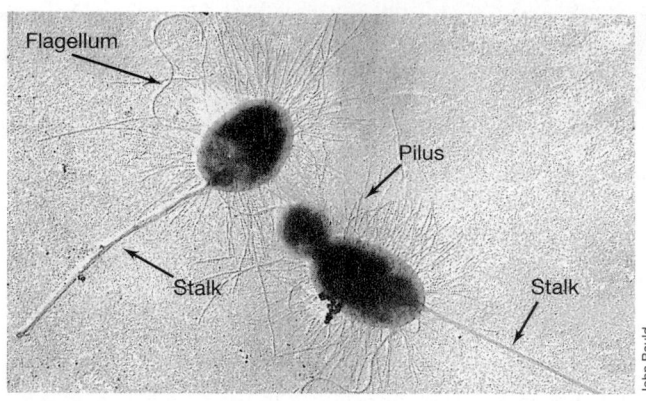

Flagellum

Pilus

Stalk

Stalk

John Bauld

Figure 12.86 An electron micrograph of a metal-shadowed preparation of *Planctomyces maris*. A single cell is about 1–1.5 μm long. Note the fibrillar nature of the stalk. Pili are also abundant. Note also the flagella (curly appendages) on each cell and the bud that is developing from the nonstalked pole of one cell.

S-layer type (Section 4.8) made of protein containing large amounts of cysteine (as cystine) and proline. As would be expected of organisms lacking peptidoglycan, these organisms are resistant to antibiotics, such as penicillin and cephalosporin, drugs that disrupt peptidoglycan synthesis. Like *Caulobacter* (Figure 12.41), *Planctomyces* is also a budding bacterium and shows a type of "life cycle," wherein motile swarmer cells attach to a surface, grow a stalk from the attachment point, and generate a new cell from the opposite pole by budding. This daughter cell grows a flagellum, breaks away from the attached mother cell, and begins the cycle anew. Physiologically, *Planctomyces* species are typical facultatively aerobic chemoorganotrophs, growing either by fermentation or respiration of sugars. The habitat of members of the kingdom *Planctomyces* is primarily aquatic, both freshwater and marine, and the genus *Isosphaera* is a filamentous, gliding hot spring bacterium. Like for *Caulobacter* (see Section 12.16), isolation of *Planctomyces* and relatives requires dilute media and, since all known members of this group lack peptidoglycan, enrichments can be made even more selective by the addition of penicillin.

One of the most unusual features of cells of the planctomycetes phylum is *cell compartmentalization*. We learned in Chapter 2 of the major structural differences between prokaryotic and eukaryotic cells, especially concerning the membrane-bound nucleus of eukaryotes

Figure 12.85 The infectious cycle of chlamydia. (a) Schematic diagram of the cycle: the whole cycle takes about 48 h. (b) Human chlamydial infection. An infected fallopian tube cell is bursting, releasing mature elementary bodies.

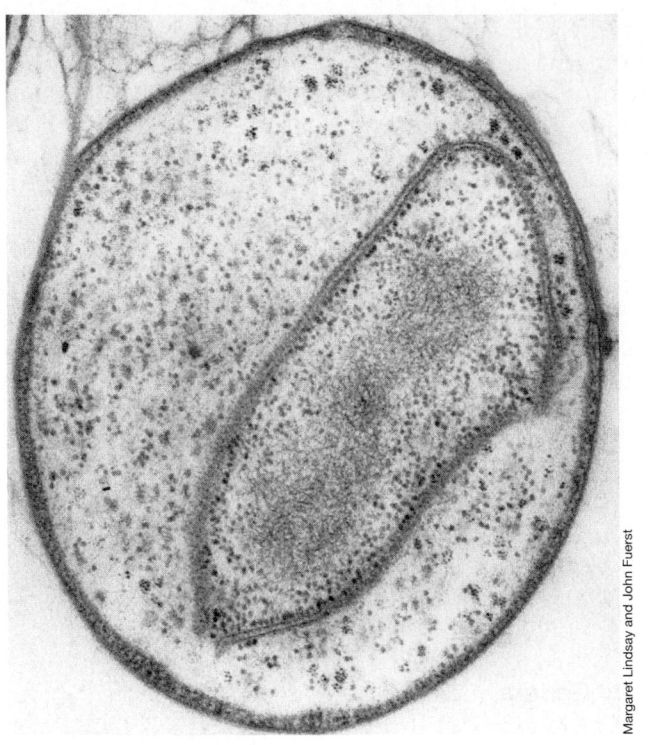

Figure 12.87 Thin section transmission electron micrograph of a cell of *Gemmata obscuriglobus* showing the nucleoid surrounded by a nuclear envelope. The cell is about 1.5 μm in diameter.

(∞ Section 2.2). However, the Planctomycetes are unique among all known prokaryotes in that they show extensive internal cell structure, including what can be considered a membrane-bound nuclear structure. For example, in the bacterium *Gemmata* (Figure 12.87), the nucleoid is surrounded by a nuclear envelope. DNA in *Gemmata* remains in a covalently closed, circular, and supercoiled form, typical of prokaryotes (∞ Section 7.3), but is highly condensed and remains partitioned from the remaining cytoplasm by a true unit membrane (Figure 12.87).

All species of Planctomycetes that have been examined in this regard have been found to contain internal cell compartments of one sort or the other. Some compartments lack DNA and thus have other functions (for example, metabolic functions, ∞ Section 17.12). In no other group of known prokaryotes do internal compartments exist that so closely resemble those of the eukaryotic cell. Indeed, the unique compartmentalization of cells of Planctomycetes tends to blur the distinction between prokaryotes and eukaryotes. Nevertheless, Planctomycetes are distinct in many ways other than cell structure, including in particular, the specific phylogenetic position they occupy within the heart of the domain *Bacteria* (Figure 12.1).

VII PHYLUM 6: THE VERRUCOMICROBIA

12.29 *Verrucomicrobium* and *Prosthecobacter*

This phylum of bacteria shares with prosthecate Proteobacteria (see Section 12.16) the formation of cytoplasmic appendages called *prostheca*. The genera *Verrucomicrobium* and *Prosthecobacter* produce two to several *prosthecae* per cell (Figure 12.88). Also, unlike cells of *Caulobacter* (Figure 12.41), that contain a single prostheca and produce flagellated and nonprosthecate swarmer cells (see Section 12.16), *Verrucomicrobium* and *Prosthecobacter* divide symmetrically, and both mother and daughter cells contain prostheca at the time of cell division. The genus name *Verrucomicrobium* derives from Greek roots meaning "warty," and the appearance of cells of *V. spinosa* with its multiple projecting prosthecae (Figure 12.88) is very much in line with this name.

Members of the Verrucomicrobia share with other prosthecate bacteria the presence of peptidoglycan in their cell walls and are aerobic to facultatively aerobic bacteria, capable of fermenting various sugars. Verrucomicrobia are widespread in nature, inhabiting freshwater and marine environments as well as forest and agricultural soils. From a phylogenetic standpoint, of course, Verrucomicrobia are distinct from all known *Bacteria* (Figure 12.1). The group shows some phylogenetic affiliation with the *Planctomyces* and *Chlamydia* phyla (Figure 12.1), but are clearly sufficiently distinct from both of these groups to form their own separate lineage.

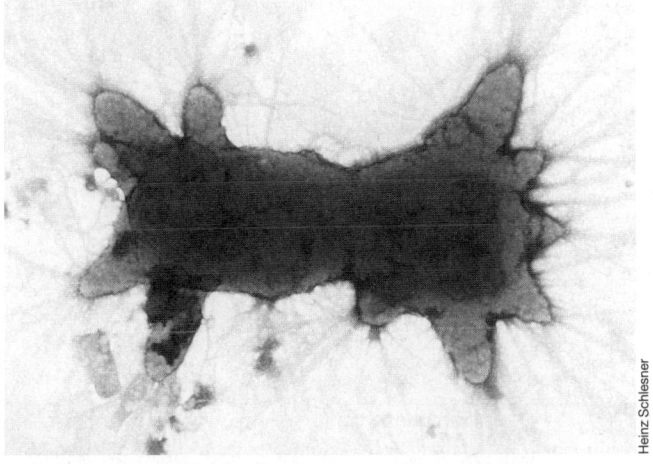

Figure 12.88 Negatively stained transmission electron micrograph of a dividing cell of *Verrucomicrobium spinosum*. Note the warty like prosthecae. A single cell is about 1 μm in diameter.

VIII PHYLUM 7: THE FLAVOBACTERIA

This phylum of *Bacteria* contains a mixture of physiological types from obligate aerobes to obligate anaerobes, unified by a common phylogenetic thread. The organisms inhabit many different types of environments, and we focus here on the two main genera in the group.

12.30 *Bacteroides* and *Flavobacterium*

Key Genera

Bacteroides
Flavobacterium

The genus *Bacteroides* contains obligately anaerobic, nonsporing species that are saccharolytic, fermenting sugars to primarily acetate and succinate as fermentation products. *Bacteroides* are normally commensals, found in the intestinal tract of humans and other animals (∞ Sections 19.11 and 21.4). In fact, *Bacteroides* species are the numerically dominant bacteria in the human large intestine, where measurements have shown 10^{10}–10^{11} cells to be present per gram of human feces. However, species of *Bacteroides* can also be pathogens and are the most important anaerobic bacteria associated with human infections. Species of *Bacteroides* are also unusual among *Bacteria* in that they are one of the few groups of organisms to synthesize *sphingolipids*, a heterogeneous collection of lipids characterized by the long-chain amino alcohol *sphingosine* in place of glycerol (Figure 12.89). Sphingolipids such as sphingomyelin, cerebrosides, and gangliosides are common in mammalian tissues, especially in the brain and other nervous tissues.

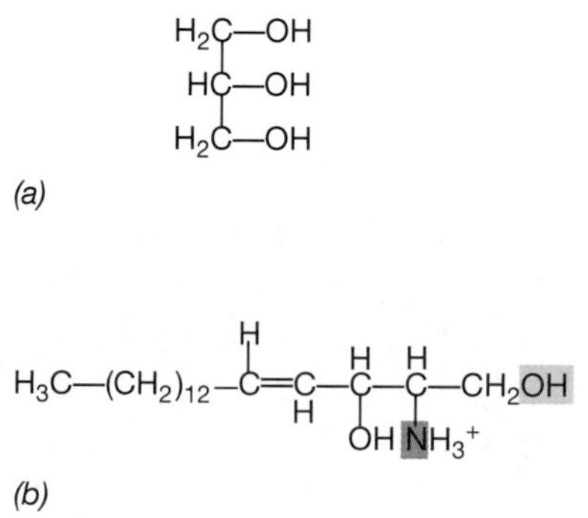

(a)

(b)

Figure 12.89 Comparison of (a) glycerol with (b) sphingosine. In sphingolipids, characteristic of *Bacteroides* species, sphingosine is the esterifying alcohol; a fatty acid is bonded by peptide linkage through the N atom (shown in red), and the terminal —OH group (shown in green) can contain any of a number of compounds including phosphatidyl choline (sphingomyelin) or various sugars (cerebrosides and gangliosides).

In contrast to *Bacteroides*, *Flavobacterium* species are primarily found in aquatic habitats, both freshwater and marine, as well as in foods and food-processing plants. Colonies of *Flavobacterium* are frequently yellow-pigmented and physiologically these organisms are rather restricted, using glucose as carbon and energy source but very few other carbon compounds. Flavobacteria are rarely pathogenic; however, one species, *F. meningosepticum*, has been associated with cases of infant meningitis.

Other important genera in this group are psychrophilic or psychrotolerant (∞ Section 6.9). These include in particular, the genera *Polaribacter* and *Psychroflexus*. Many other genera in the group are also capable of good growth below 20°C.

IX PHYLUM 8: THE CYTOPHAGA GROUP

Key Genera

Cytophaga
Sporocytophaga
Flexibacter

12.31 *Cytophaga* and Relatives

Organisms of the *Cytophaga* group are long, slender rods, often with pointed ends, that move by gliding (Figure 12.90*a,b*). The related genus, *Sporocytophaga*, is similar to *Cytophaga* in morphology and physiology but the cells form resting spherical structures called *microcysts* (Figure 12.90*c,d*) similar to those produced by some fruiting myxobacteria (see Section 12.17). They are widespread in the soil and water, often being present in great abundance. Many cytophagas digest polysaccharides like cellulose (Figure 12.90*c*), agar (Figure 12.90*a*), or chitin. The cellulose decomposers can be easily isolated by placing small crumbs of soil on pieces of cellulose filter paper laid on the surface of mineral salts agar. The bacteria attach to and digest the cellulose fibers, forming spreading colonies (Figure 12.90*c*). The cytophagas do not produce soluble, extracellular, cellulose-digesting enzymes (cellulases); their cellulases remain attached to the cell envelope, which probably accounts for the fact that the cells must adhere to cellulose fibrils in order to digest them. In pure culture, *Cytophaga* can be cultured on agar containing embedded cellulose fibers, the presence of the organism being indicated by the clearing that occurs as the cellulose is digested (Figure 12.90*c*; see also Figure 17.62).

Species of *Cytophaga* and *Sporocytophaga* are obligately aerobic and probably account for much of the cel-

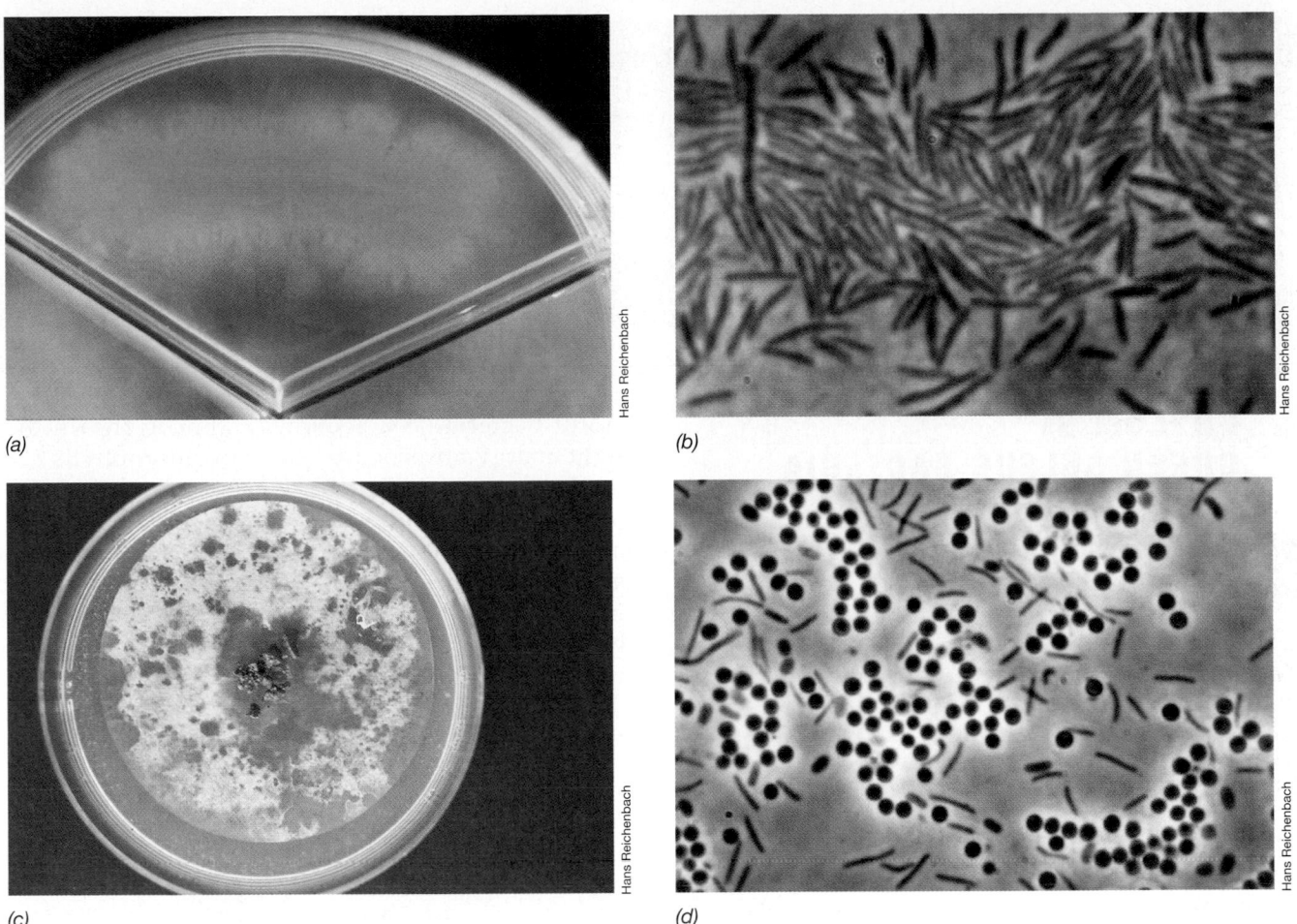

(a)

(b)

(c)

(d)

Hans Reichenbach

Figure 12.90 *Cytophaga* and *Sporocytophaga*. (a) Streak of an agarolytic marine *Cytophaga* species hydrolyzing agar in the Petri dish. (b) Phase contrast photomicrograph of cells of *C. hutchinsonii* grown on cellulose filter paper (cells are about 1.5 µm in diameter). (c) Colonies of *Sporocytophaga* growing on cellulose. Note the clearing zones where the cellulose has been degraded. (d) Phase contrast photomicrograph of the rod-shaped cells and spherical microcysts of *Sporocytophaga myxococcoides* (cells are about 0.5 µm and microcysts about 1.5 µm in diameter).

lulose digestion that occurs by prokaryotes in oxic environments in nature. A number of *Cytophaga* species are also fish pathogens and can cause serious problems in the cultivated fish business. Two of the most important diseases are *columnaris disease*, caused by *C. columnaris*, and *cold-water disease*, caused by *C. psychrophila*. Both diseases preferentially affect stressed fish, such as those living in waters receiving pollutant discharges or living in high density confinement situations such as fish hatcheries and aquaculture operations. Infected fish show tissue destruction, frequently around the gills, and this may stem from the fact that *Cytophaga* species isolated from infected fish are frequently strongly proteolytic.

The genus *Flexibacter* differs from the cytophagas in that the species usually require complex media for good growth and are not cellulolytic. Cells of some *Flexibacter* species also undergo changes in cell morphology from long, gliding, threadlike filaments lacking cross walls, to short, nonmotile rods. Many species are pigmented due to carotenoids located in the cytoplasmic membrane,

or related pigments called *flexirubins*, located in the gram-negative outer membrane. *Flexibacter* species are common soil and freshwater saprophytes, and none have been identified as pathogens.

✓ **12.28–12.31 Concept Check**

The *Planctomyces* group contains stalked, budding bacteria while the Flavobacteria contain a variety of gram-negative *Bacteria* motile by either flagella or by gliding associated with animals, food, and the soil. Members of the Verrucomicrobia are distinguished by their multiple prosthecate cells.

✓ What is unique about the cell wall and arrangement of DNA in *Planctomyces*?

✓ How does the stalk of *Planctomyces* differ from the stalk of *Caulobacter*?

✓ Where might you find large numbers of *Bacteroides* cells in nature?

✓ Describe a method for isolating *Cytophaga* species from nature.

TABLE 12.35   Genera and characteristics of phototrophic green sulfur bacteria

Characteristics	Genus	Color of cultures	Number of species	DNA (mol % GC)
No gas vesicles:				
Straight or curved rods nonmotile (see Figure 12.91*a*)	*Chlorobium*	Green or brown	8	49–58
Spheres and ovals, nonmotile, forming prosthecae (appendages)	*Prosthecochloris*	Green or brown	2	50–56
Contain gas vesicles:				
Branching nonmotile rods, in loose irregular network (see Figure 12.91*b*)	*Pelodictyon*	Green or brown	4	48–58
Spheres with prosthecae	*Ancalochloris*	Green	1	—
Rods, gliding	*Chloroherpeton*	Green	1	47

X PHYLUM 9: GREEN SULFUR BACTERIA

12.32 *Chlorobium* and Other Green Sulfur Bacteria

Key Genera

Chlorobium
Prosthechochloris
"*Chlorochromatium*"

Green sulfur bacteria are a phylogenetically distinct group of nonmotile anoxygenic phototrophic bacteria that contain only obligately anaerobic and phototrophic species among cultured isolates. The group is morphologically diverse and includes short to long rods (Table 12.35 and Figure 12.91). Like purple sulfur bacteria they utilize H_2S as an electron donor, oxidizing it first to S^0 and then to SO_4^{2-}. But unlike purple sulfur bacteria, the sulfur produced by green sulfur bacteria resides *outside* the cell (⚭ Figure 17.17*b*). Most species can also assimilate a few organic compounds in the light (that is, *photoheterotrophy*, ⚭ Section 17.4). Strict autotrophy, however, is supported not by the reactions of the Calvin cycle as in purple bacteria, but instead by a reversal of steps in the citric acid cycle (⚭ Section 17.7, reverse citric acid cycle), a unique means of autotrophy among phototrophic organisms.

Pigments and Ecology

The bacteriochlorophylls found in green sulfur bacteria include bacteriochlorophyll *a* and either bacteriochlorophylls *c*, *d*, or *e*. The latter pigments function only in light-harvesting reactions (⚭ Section 17.2) and are located in a unique structure called the **chlorosome** (Figure 12.92). Chlorosomes are oblong bacteriochlorophyll-rich bodies bounded by a thin, nonunit membrane and lie attached to the cytoplasmic membrane in the periphery of the cell (Figure 12.92; see also Figure 17.7). Studies of energy transfer in green

sulfur bacteria (⚭ Section 17.2) have shown that light energy absorbed by bacteriochlorophylls *c*, *d*, or *e* in the chlorosome gets funneled to bacteriochlorophyll *a*, which resides in the cytoplasmic membrane, and it is here where photosynthetic energy conversion and ATP synthesis actually occur (⚭ Figure 17.7). Both green- and brown-colored species of green sulfur bacteria are known, the brown-colored species containing bacteriochlorophyll *e* and carotenoids that render the cells brown in color (Figure 12.93; ⚭ Figure 17.9).

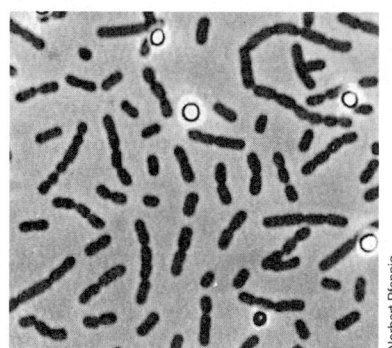

(a)

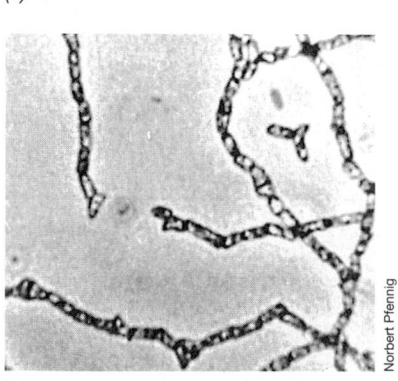

(b)

Figure 12.91   Phototrophic green sulfur bacteria. (a) *Chlorobium limicola*; cells are about 0.8 µm wide. Note the sulfur granules deposited *extra*cellularly. (b) *Pelodictyon clathratiforme*, a bacterium forming a three-dimensional network; cells are about 0.8 µm wide.

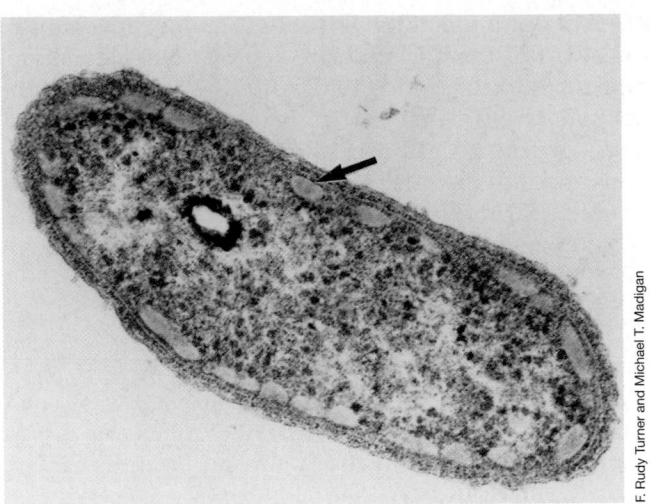

Figure 12.92 Thin section electron micrograph of a cell of the green sulfur bacterium *Chlorobium tepidum*. Note chlorosomes (arrow) in the cell periphery. A cell is about 0.7 μm wide.

Like purple sulfur bacteria (see Section 12.2), green sulfur bacteria live in anoxic aquatic environments, especially where H_2S is abundant (in general, green sulfur bacteria are more tolerant of sulfide than are purple bacteria). Because the chlorosome is such an efficient light-harvesting structure, little light is required to support the photosynthetic activities of green sulfur bacteria and they are thus typically found at the greatest depths in lakes of any phototrophic organisms. One species of the genus *Chlorobium, C. tepidum* (Figure 12.92), is thermophilic and forms dense microbial mats in high sulfide hot springs. *C. tepidum* is also notable because its genome (2.1 megabases) has been completely sequenced, the first genome from an anoxygenic phototroph to have been done so, and for the fact that it grows rapidly and is amenable to genetic manipulation by both conjugation and transformation.

Green Sulfur Bacteria Consortia

Certain green sulfur bacteria can form a tight two-membered association with a chemoorganotrophic bacterium in which each organism benefits the other; such associations are called **consortia**. The phototrophic component, called an *epibiont*, appears physically attached to the nonphototrophic component (Figure 12.94), although the mechanism of attachment is not clear. The name "*Chlorochromatium aggregatum*" has been used to describe one such consortium, although the name has no basis in formal prokaryotic taxonomy because it refers to two organisms rather than one.

The "*C. aggregatum*" consortium is green in color because the epibionts are green sulfur bacteria that contain bacteriochlorophyll *c* or *d* and green-colored carotenoids and surround a central nonphototrophic cell. A structurally similar form referred to as "*Pelochromatium roseum*" is brown in color. Yet other consortia are known in which the epibiont cells are half-moon shaped (Figure 12.94*b,c*); thus, a variety of such consortia probably exist in nature.

Some green bacterial consortia have been grown in laboratory culture. On average, the "*C. aggregatum*" consortium (Figure 12.94) contains about 12 epibionts per central cell while the "*P. roseum*" consortium contains about 20. Solid evidence that the epibionts are indeed green sulfur bacteria comes from the fact that chlorosomes are visible in thin sections of the epibionts (Figure 12.94*d*) and from molecular evidence; treatment of the consortium with a fluorescent oligonucleotide probe specific for green sulfur bacterial 16S rRNA (FISH technology, ∞ Sections 11.6 and 18.4) causes the epibionts, but not the central cell, to fluoresce (Figure 12.94*e*). Studies of laboratory cultures have also shown that the central cell and the epibiont divide in synchrony, suggesting that the two components have some means of intercommunicating.

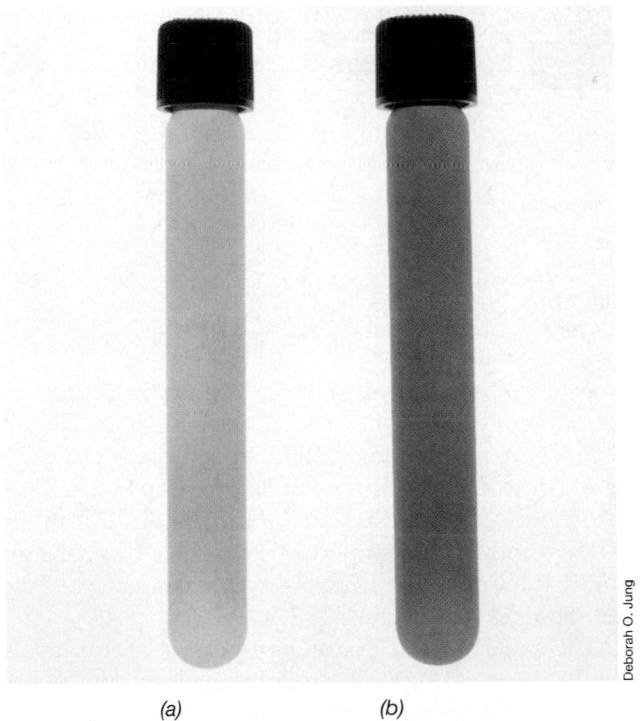

(a)   (b)

Figure 12.93 Green and brown chlorobia. Tube cultures of (a) *Chlorobium tepidum* and (b) *Chlorobium phaeobacteroides*. Cells of *C. tepidum* contain bacteriochlorophyll *c* and a series of green-colored carotenoids, while cells of *C. phaeobacteroides* contain bacteriochlorophyll *e* and isorenieratene, a brown-colored carotenoid. For structures of the specific green and brown carotenoids, see Figure 17.9.

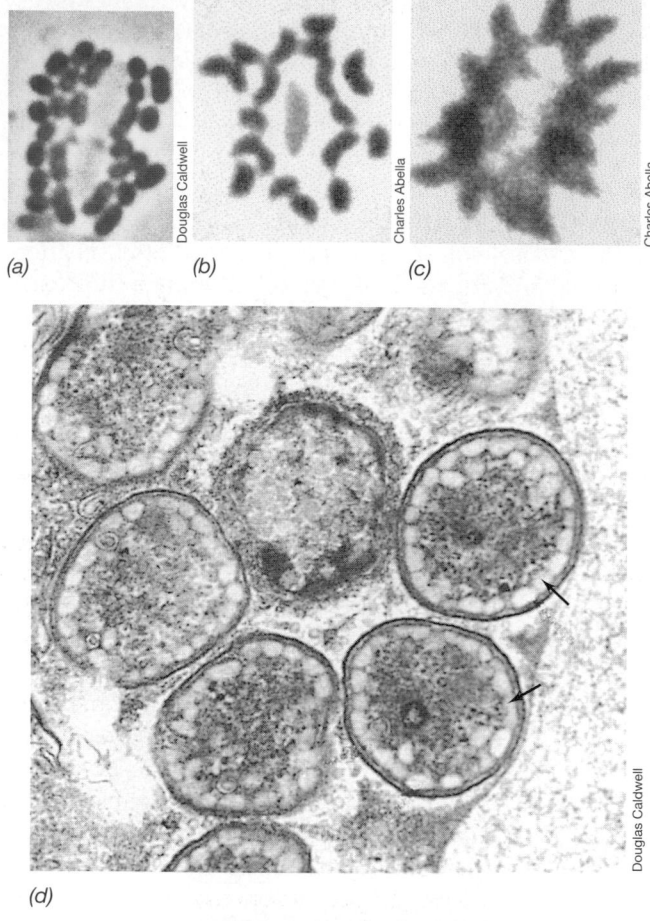

(a)   (b)   (c)

Douglas Caldwell (a)
Charles Abella (b)
Charles Abella (c)

(d)

Douglas Caldwell

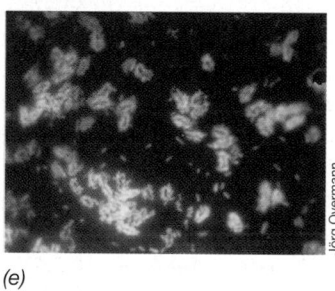

Jörg Overmann

(e)

Figure 12.94 Green sulfur bacteria consortia. (a–c) Phase contrast micrographs and (d) transmission electron micrograph of the green or brown bacterial consortia "*Chlorochromatium aggregatum*," or "*Pelochromatium*." In (a–c) the nonphototrophic central organism is much lighter in color than the pigmented phototrophic bacteria. Note the chlorosomes (arrows) in (d). The entire consortium is about 3×6 μm. (b, c). Half-moon-shaped epibionts in a *Pelochromatium* consortium from a stratified Wisconsin lake. The colorless central cell in both photos is about 2 μm long. (e) Phylogenetic staining (⊙ Section 18.4) of "*Chlorochromatium aggregatum*." The yellow stain contained a nucleic acid probe specific for green sulfur bacteria. Note that only the epibionts are stained. The epibionts number from about 10 to 20 per central (nonphototrophic) cell. Note the intimate contact between the epibionts and the central cell in (d). Such contact may facilitate intercellular communication in the consortium necessary for controlling possible chemotaxis or phototaxis responses.

Although the rationale for why and the mechanism for how these associations form is not yet clear, combined laboratory and field studies suggest that the epibionts in these consortia are adapted to a very narrow regimen of light intensities and sulfide concentrations and that the role of the large central cell may be as a vehicle for these otherwise nonmotile phototrophs to position themselves in a water column where conditions for photosynthesis are optimal.

✓   **12.32 Concept Check**

Green sulfur bacteria are obligately anaerobic anoxygenic phototrophs that produce unique structures called *chlorosomes*. These organisms can grow at very low light intensities and oxidize H_2S to S^0 and SO_4^{2-}.

✓   What pigments are found in the chlorosome?

✓   What is unique about autotrophy in *Chlorobium* (⊙ Section 17.7)?

✓   What evidence supports the idea that the epibionts of green bacterial consortia are truly green sulfur bacteria?

XI   PHYLUM 10: THE SPIROCHETES

12.33   Spirochetes

Key Genera

Spirochaeta
Treponema
Cristispira
Leptospira
Borrelia

Spirochetes are gram-negative, motile, tightly coiled *Bacteria*, typically slender and flexous in shape (Figure 12.95). These morphologically unique prokaryotes form a major phylogenetic lineage of *Bacteria* (see Figure 12.1). Spirochetes are widespread in aquatic environments and in animals, and some of them cause diseases, including the important human sexually transmitted disease syphilis (⊙ Section 26.12).

The spirochete cell is made up of a "protoplasmic cylinder," consisting of the regions enclosed by the cell wall and membrane (Figure 12.96). Motility is conferred by a single to many flagella that emerge from each pole (Figure 12.96). However, unlike typical bacteria flagella (⊙ Section 4.10), spirochete flagella fold back from each pole upon the protoplasmic cylinder and remain located in the periplasm of the cell; thus they have been called *endoflagella*. Both the endoflagella and the protoplasmic

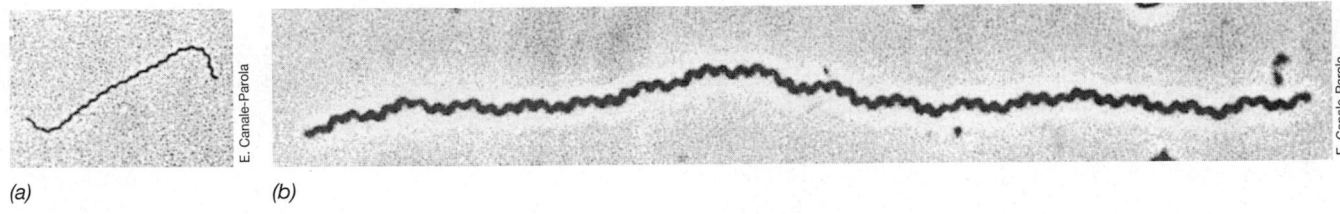

(a)   (b)

Figure 12.95 Two spirochetes at the same magnification, showing the wide size range in the group. (a) *Spirochaeta stenostrepta*, by phase contrast microscopy. A single cell is 0.25 μm in diameter. (b) *Spirochaeta plicatilis*. A single cell is 0.75 μm in diameter and can be up to 250 μm (0.25 mm) in length.

cylinder are surrounded by a multilayered but flexible membrane called the *outer sheath* (Figure 12.96).

Motility of Spirochetes

Each spirochete endoflagellum is anchored at one end and extends about two-thirds of the length of the cell. Endoflagella rotate rigidly, as do typical bacterial flagella (∞ Section 4.10). However, because the protoplasmic cylinder is also rigid whereas the outer sheath is flexible, when both endoflagella rotate in the same direction, the protoplasmic cylinder rotates in the opposite direction, placing torsion on the cell as illustrated in Figure 12.96b. In liquid this causes the spirochete cell to move by flexing or lashing motions due to torque exerted at the ends of the protoplasmic cylinder by the rotating endoflagella (Figure 12.96b). Thus, despite the fact that the endoflagella of spirochetes do not extend away from the cell, but instead reside inside the cell's outer membrane, the rotating action of these structures provide motility, albeit of more an irregular and jerky form than for smooth swimming, just as do the flagella of other prokaryotes.

Classification

Spirochetes are classified into eight genera primarily on the basis of habitat, pathogenicity, ribosomal RNA sequences, and morphological and physiological characteristics. Table 12.36 lists the major genera and their characteristics.

Spirochaeta and *Cristispira*

The genus *Spirochaeta* includes free-living, anaerobic, and facultatively aerobic spirochetes. These organisms are common in aquatic environments, such as the water and mud of rivers, ponds, lakes, and oceans. One species of the genus *Spirochaeta* is *S. plicatilis* (Figure 12.95b), a fairly large spirochete found in freshwater and marine H_2S-containing habitats and is probably anaerobic. The endoflagella of *S. plicatilis* are arranged in a bundle that winds around the coiled protoplasmic cylinder. From 18 to 20 endoflagella are inserted at each pole of this spirochete. Another species, *Spirochaeta stenostrepta*, has been cultured and is shown in Figure 12.95a. It is an obligate anaerobe commonly found in H_2S-rich, black muds. It ferments sugars via the glycolytic pathway to ethanol, acetate, lactate, CO_2, and H_2. The species *Spirochaeta aurantia* is an orange-pigmented facultative aerobe, fermenting sugars via the glycolytic pathway under anaerobic conditions and oxidizing sugars aerobically mainly to CO_2 and acetate.

A metabolically unusual spirochete has been isolated from the hindgut of the termite. This environment is highly cellulolytic, as termites live mostly on wood and

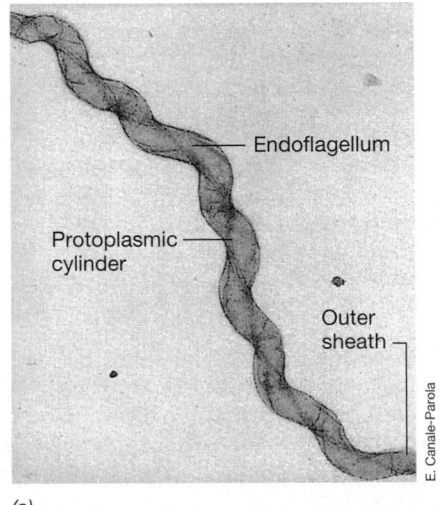

(a)

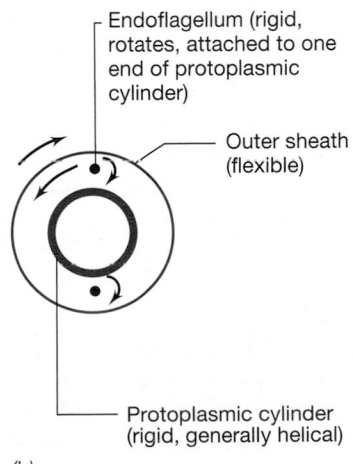

(b)

Figure 12.96 (a) Electron micrograph of a negatively stained preparation of *Spirochaeta zuelzerae*, showing the position of the endoflagellum. A single cell is about 0.3 μm in diameter. (b) Cross section of a spirochete cell, showing the arrangement of the protoplasmic cylinder, endoflagella, and external sheath and the manner in which the rotation of the rigid endoflagellum can generate rotation of the protoplasmic cylinder and (in the opposite direction) rotation of the external sheath. If the sheath is free, the cell will rotate about its longitudinal axis and move along it. If the sheath is in contact with a solid surface, the cell will creep forward.

TABLE 12.36 Genera of spirochetes and their characteristics

Genus	Dimensions (μm)	Number of species recognized	General characteristics	Number of endoflagella	DNA (mol % GC)	Habitat	Diseases
Cristispira	30–150 × 0.5–3.0	1	3–10 complete coils; bundle of endoflagella visible by phase contrast microscopy	> 100	—	Digestive tract of molluscs; has not been cultured	None known
Spirochaeta	5–250 × 0.2–0.75	14	Anaerobic or facultatively aerobic; tightly or loosely coiled	2–40	50–65	Aquatic, free-living, freshwater and marine	None known
Treponema	5–15 × 0.1–0.4	20	Microaerophilic or anaerobic; helical or flattened coil amplitude up to 0.5 μm	2–32	25–53	Commensal or parasitic in humans, other animals	Syphilis, yaws, swine dysentery, pinta
Borrelia	8–30 × 0.2–0.5	31	Microaerophilic; 5–7 coils of approx. 1 μm amplitude	7–20	46	Humans and other mammals, arthropods	Relapsing fever, Lyme disease, ovine and bovine borreliosis
Leptospira	6–20 × 0.1	13	Aerobic, tightly coiled, with bent or hooked ends; requires long-chain fatty acids	2	33–43	Free-living or parasitic in humans, other mammals	Leptospirosis
Leptonema	6–20 × 0.1	1	Aerobic; does not require long-chain fatty acids	2	54	Free-living	None known
Brachyspira	7–10 × 0.35–0.45	8	Anaerobe	8–28	25–27	Intestine of warm-blooded animals	Causes diarrhea in chickens and swine
Brevinema	4–5 × 0.2–0.3	1	Microaerophile, by 16S rRNA sequencing, forms deep branch in spirochete lineage (see Figure 12.1)	2	34–36	Blood and tissue of mice and shrews	Infectious for laboratory mice

wood products, and much H_2 and CO_2 is produced from the fermentation of glucose released from the cellulose. A spirochete isolated from the termite hindgut can convert this H_2 plus CO_2 to acetate, that is, it is a *homoacetogen* (∞ Section 17.16 for a discussion of homoacetogenesis), and this is the first instance of this form of energy metabolism being found in a phylogenetic group of bacteria outside of the clostridia and relatives (see Section 12.20). This spirochete also fixes molecular nitrogen (N_2 fixation, ∞ Section 17.16), a property not previously found in spirochetes.

The genus *Cristispira* (Figure 12.97) contains organisms with a unique distribution, being found in nature primarily in the *crystalline style* of certain molluscs, such as clams and oysters. The crystalline style is a flexible, semisolid rod seated in a sac and rotated against a hard surface of the digestive tract, thereby mixing with and grinding the small particles of food. Being large spirochetes, the cristispiras can readily be seen microscopically within the mollusc style as they rapidly rotate forward and backward in corkscrew fashion. *Cristispira* may occur in both freshwater and marine molluscs, but not all species of molluscs possess them. Unfortunately, *Cristispira* has not been cultured, and so the physiological reason for its restriction to this unique habitat is not known.

Treponema

Anaerobic, host-associated spirochetes that are commensals or parasites of humans and animals are placed in the genus *Treponema*. *Treponema pallidum*, the causal agent of syphilis (∞ Section 26.12), is the best-known species of *Treponema*. It differs in morphology from other spirochetes; the cell is not helical but has a flat wave form. The *T. pallidum* cell is remarkably thin, measuring ap-

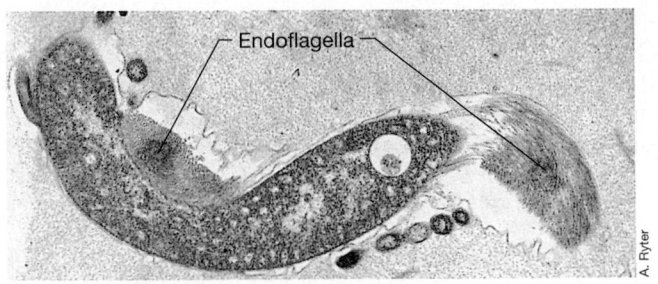

Figure 12.97 Electron micrograph of a thin section of *Cristispira*, a large spirochete. A cell measures about 2 μm in diameter. Notice the numerous endoflagella.

proximately 0.2 μm in diameter. Living cells are clearly visible in the dark-field microscope or after staining with fluorescent antibody; dark-field microscopy has long been used to examine exudates from suspected syphilitic lesions (⟳ Figure 26.28). In nature, *T. pallidum* is restricted to humans, although artificial infections have been established in rabbits and monkeys. Although never grown in laboratory culture, it has been established from animal studies that virulent *T. pallidum* cells (purified from infected rabbits) contain a cytochrome system and are in fact microaerophiles. Such cells have also yielded sufficient DNA for the complete genome of *T. pallidum* (1.14 megabases) to be sequenced (⟳ Section 15.3).

Other species of the genus *Treponema* are common commensal organisms in the oral cavity of humans and can generally be seen in material scraped from between the teeth and from the narrow space between the gums and the teeth. Three species, *Treponema denticola*, *T. macrodentium*, and *T. oralis*, have been described, differing in morphology and physiological characteristics. *Treponema denticola* ferments amino acids such as cysteine and serine, forming acetate as the major fermentation acid as well as CO_2, NH_3, and H_2S. Spirochetes are also found in the rumen, the digestive organ of ruminant animals

(⟳ Section 19.11). *Treponema saccharophilum* (Figure 12.98) is a large, *pectinolytic* spirochete found in the bovine rumen. *Treponema saccharophilum* is an obligate anaerobe that ferments pectin, starch, inulin, and other plant polysaccharides. This and other spirochetes may play an important role in the conversion of plant polysaccharides to volatile fatty acids, usable as energy sources by the ruminant (⟳ Section 19.11). Although the genus *Treponema* is phylogenetically a unit, the true relationship between *T. pallidum* and other *Treponema* species may be a distant one because the GC base ratio of *T. pallidum* is about 53%, whereas other species of this genus cluster between 38 and 40% or 25 and 26%.

Leptospira and *Leptonema*

The genera *Leptospira* and *Leptonema* contain strictly *aerobic* spirochetes that use long-chain fatty acids (for example, oleic acid) as electron donor and carbon sources. With few exceptions, these are the only substrates utilized by leptospiras for growth. The leptospira cell is thin, finely coiled, and usually bent at each end into a semicircular hook. At present, several species are recognized in this group, some free-living and many parasitic. Two major species are *Leptospira interrogans* (parasitic) and *L. biflexa* (free-living). Strains of *L. interrogans* are parasitic for humans and animals. Rodents are the natural hosts of most leptospiras, although dogs and pigs are also important carriers of certain strains. In humans the most common leptospiral syndrome is *leptospirosis*; in this disorder the organism localizes in the kidney and can cause renal failure and death.

Leptospiras ordinarily enter the body through the mucous membranes or through breaks in the skin. After a transient multiplication in various parts of the body, the organism localizes in the kidney and liver, causing nephritis and jaundice. The organism then passes out of the body in the urine, and infection of another individual is most commonly by contact with infected urine. Therapy with penicillin, streptomycin, or the tetracyclines is possible but may require extended courses to eliminate the organism from the kidney. Domestic animals such as dogs are vaccinated against **leptospirosis** with a killed virulent strain in the combined distemper-leptospira-hepatitis vaccine. In humans, prevention is effected primarily by elimination of the disease from animals.

Borrelia

The majority of species in the genus *Borrelia* are animal or human pathogens. *Borrelia recurrentis* is the causative agent of **relapsing fever** in humans and is transmitted via an insect vector, usually by the human body louse. Relapsing fever is characterized by a high fever and generalized muscular pain, which lasts for 3–7 days followed by a recovery period of 7–9 days. Left untreated, the fever returns in two to three more cycles (hence the

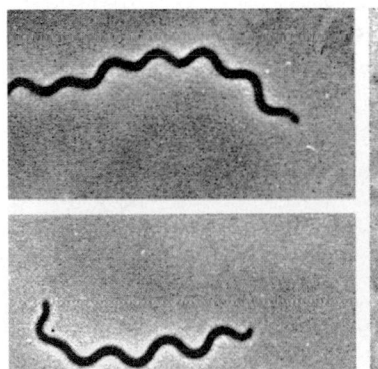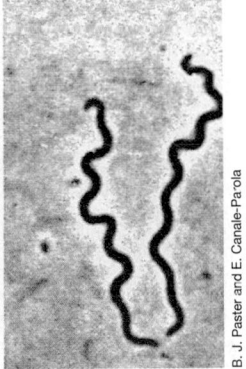

Figure 12.98 Phase contrast photomicrographs of *Treponema saccharophilum*, a large pectinolytic spirochete from the bovine rumen. A cell measures about 0.4 μm in diameter. Left, regularly coiled cells; right, irregularly coiled cells.

name, relapsing fever) and causes death in up to 40% of those infected. Fortunately, the organism is quite sensitive to tetracycline, and if the disease is correctly diagnosed, treatment is straightforward. Other borrelia are of veterinary importance, causing diseases in cattle, sheep, horses, and birds. In most of these diseases the organism is transmitted by ticks. *Borrelia burgdorferi* is the causative agent of the tick-borne disease called *Lyme disease*, which infects humans and animals. Lyme disease is discussed in Section 27.4. *Borrelia burgdorferi* is also of interest because it is as yet one of the only known prokaryotes with a *linear* (as opposed to a *circular*) chromosome. This rather small genome of *B. burgdorferi* (1.44 megabases) has been completely sequenced (⚭ Section 15.3).

✓ 12.33 Concept Check

Spirochetes are tightly coiled, motile, helical prokaryotes that contain both free-living as well as pathogenic species.

✓ How do the endoflagella of spirochetes compare with the flagella of *Escherichia coli*?

✓ Name two diseases of humans caused by spirochetes.

✓ What is the habitat of the spirochete *Cristispira*?

XII PHYLUM 11: DEINOCOCCI

12.34 *Deinococcus/Thermus*

Key Genera

Deinococcus
Thermus

This phylum of *Bacteria* contains only three genera, the best studied being *Deinococcus* and *Thermus*. The latter genus contains thermophilic chemoorganotrophic bacteria including *Thermus aquaticus*, the organism from which *Taq* DNA polymerase is obtained. Because it is so heat stable, this enzyme is the major one used in the polymerase chain reaction (PCR) technique for amplifying DNA, as was discussed in Section 10.17. *Thermus* species stain gram-negatively and contain a rare form of peptidoglycan in which ornithine is present in place of diaminopimelic acid in the muramic acid cross bridges (⚭ Section 4.8). Interestingly, *Deinococcus* also contains ornithine. A number of species of *Thermus* have been described and all grow aerobically by catabolism of sugars, amino and organic acids, or various complex mixtures. We focus the rest of the discussion here on *Deinococcus*.

The genus *Deinococcus* contains four species of gram-positive cocci; *D. radiodurans* is the best-studied species. The *D. radiodurans* cell wall is structurally complex and

consists of several layers including an outer membrane (Figure 12.99), normally present only in gram-negative bacteria (⚭ Section 4.9). However, unlike the outer membrane of gram-negative bacteria like *E. coli*, the outer membrane of *D. radiodurans* lacks lipid A. Physiologically, *D. radiodurans* is an aerobic chemoorganotroph, usually grown on complex media.

Radiation Resistance

Most deinococci are red or pink in color because of the variety of carotenoids found in these organisms, and many strains are highly resistant to ultraviolet radiation and to desiccation. Resistance to radiation can be used to advantage in isolating deinococci. These remarkable organisms can be isolated from soil, ground meat, dust, and filtered air following exposure of the sample to intense ultraviolet (or even gamma) radiation and plating on a rich medium containing tryptone and yeast extract. Because many

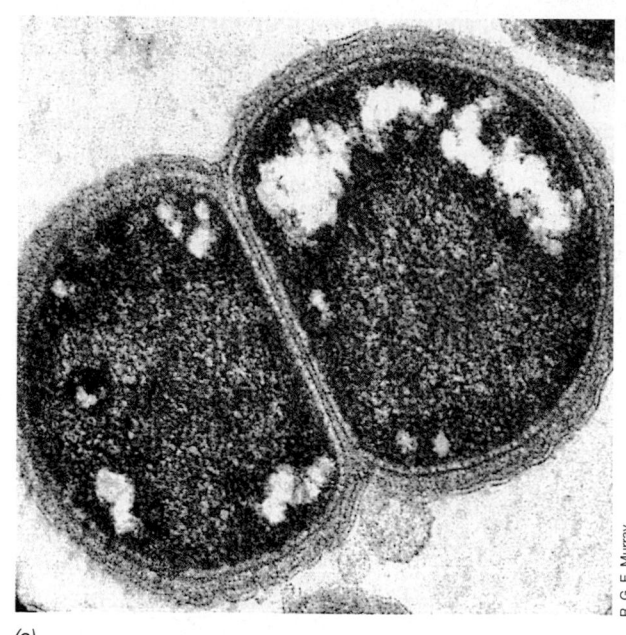

(a)

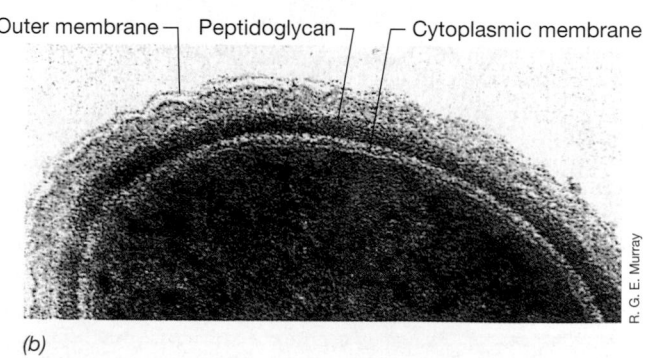

Outer membrane ⌐ Peptidoglycan ⌐ ⌐ Cytoplasmic membrane

(b)

Figure 12.99 The radiation-resistant coccus *Deinococcus radiodurans*. An individual cell is about 2.5 μm in diameter. (a) Transmission electron micrograph of *D. radiodurans*. Note the outer membrane layer. (b) High magnification micrograph of wall layer.

strains of *Deinococcus radiodurans* are even more resistant to radiation than bacterial endospores, treatment of a sample with strong doses of radiation effectively sterilizes the sample of organisms other than *D. radiodurans*, making isolation of deinococci relatively straightforward. For example, *D. radiodurans* cells can survive exposure to up to 30,000 Gy of ionizing radiation (1 Gy = 100 rad), sufficient to shatter the organism's chromosome into hundreds of fragments (by contrast, a human can be killed by exposure to less than 5 Gy). A powerful DNA repair machinery exists in *Deinococcus* cells that is able to repair the organism's chromosome, even from a fragmented state. Consistent with the fact that *D. radiodurans* is an extremely radiation-resistant bacterium, strains of this organism have been isolated from near atomic reactors and other potentially lethal radiation sources.

In addition to impressive radiation resistance, *Deinococcus radiodurans* is resistant to the mutagenic effects of many other mutagenic agents. Studies of the mutability of *D. radiodurans* have shown it to be highly efficient in repairing damaged DNA. Several different DNA repair enzymes exist in *D. radiodurans* for repairing breaks in single- or double-stranded DNA and for excising and repairing thymine dimers formed by the action of ultraviolet light. The only chemical mutagens that seem to work on *D. radiodurans* are agents like nitrosoguanidine, which tend to induce *deletions* in DNA; deletions are apparently not repaired as efficiently as point mutations in this organism and mutants of *D. radiodurans* can be isolated in this way.

XIII PHYLUM 12: THE GREEN NONSULFUR BACTERIA

Key Genera

Chloroflexus
Thermomicrobium

This phylum of *Bacteria* is phylogenetically distinct and contains just a few genera, the best known being the anoxygenic phototroph *Chloroflexus*; all cultured representatives are thermophilic. *Thermomicrobium* is a chemotrophic member of this group and is a strictly aerobic, gram-negative rod, growing optimally in complex media at 75°C. Besides its phylogenetic novelty, *Thermomicrobium* is also of interest because of its membrane lipids. Recall that the lipids of *Bacteria* and *Eukarya* contain fatty acids esterified to glycerol (∞ Sections 3.4 and 4.5). By contrast, the lipids of *Thermomicrobium* contain 1,2-dialcohols instead of glycerol, and have neither ester *nor* ether linkages (Figure 12.100). In addition, cells of *Thermomicrobium* are unusual for species of *Bacteria* in that they lack peptidoglycan, the signature cell-wall polymer of this domain of prokaryotes (∞ Section 11.8).

12.35 *Chloroflexus* and *Heliothrix*

Chloroflexus and most other green nonsulfur bacteria are filamentous prokaryotes that form thick microbial mats in neutral to alkaline hot springs (Figure 12.101; see also Figure 18.18a). *Chloroflexus*-like organisms have also been found in nonthermal marine microbial mats. Although an anoxygenic phototroph, *Chloroflexus* is a "hybrid" phototroph in the sense that its photosynthetic mechanism shows features characteristic of both purple bacteria and green sulfur bacteria. Like the latter, *Chloroflexus* contains bacteriochlorophyll *c* and chlorosomes (see Figure 12.92 for an electron micrograph of chlorosomes). However, the bacteriochlorophyll *a* located in the cytoplasmic membrane of cells of *Chloroflexus* is arranged to form a photosynthetic reaction center structurally similar to those of purple bacteria (by contrast, the reaction center of green sulfur bacteria is structurally quite different, ∞ Figure 17.18). It has thus been proposed that modern *Chloroflexus* may be a vestige of a very early form of phototroph (see also later discussion concerning autotrophy and phylogeny of this organism) that perhaps first evolved a photosynthetic reaction center and then received chlorosome-specific genes by lateral transfer.

Physiologically, *Chloroflexus* resembles purple nonsulfur bacteria in that photoautotrophy can be supported by ($H_2S + CO_2$) or ($H_2 + CO_2$). However, in *Chloroflexus*, phototrophic growth is best with organic compounds as carbon sources (photoheterotrophy). *Chloroflexus* also grows well in the dark as a chemoorganotroph by

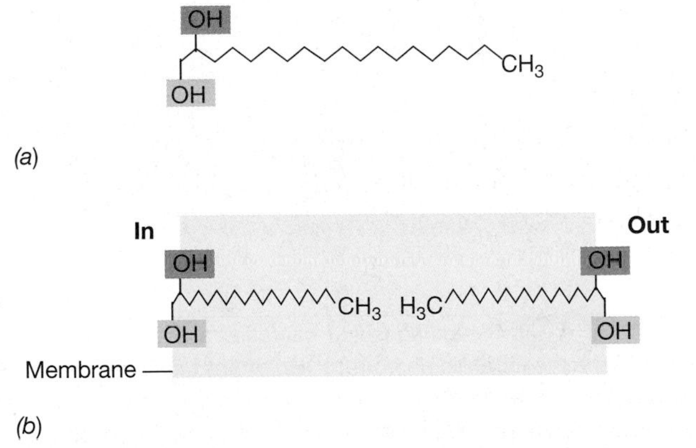

(a)

(b)

Figure 12.100 The unusual lipids of *Thermomicrobium*. (a) Membrane lipids from *T. roseum* contain long chain diols like the one shown here (13-methyl-1,2 nonadecanediol). Note that unlike the lipids of other *Bacteria* or of *Archaea* (∞ Section 4.5), neither ester- nor ether-linked side chains are present. (b) To form a bilayer membrane, dialcohol molecules presumably oppose each other at the methyl groups with the OH groups being the inner and outer hydrophilic surfaces. Small amounts of the diols have fatty acids esterified to the secondary —OH group (shown in red) while the primary —OH group (shown in green) can bond a hydrophilic molecule like phosphate.

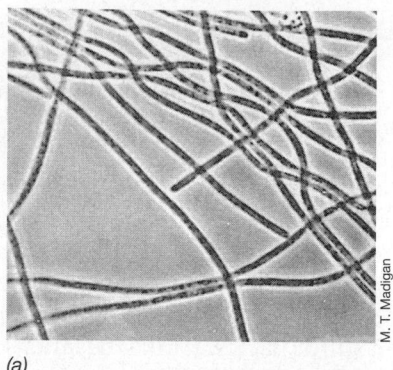

(a)

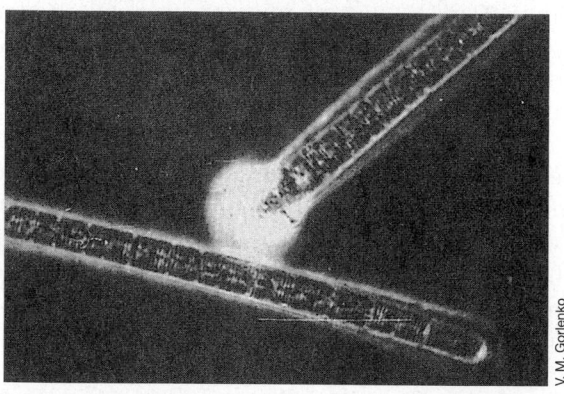

(b)

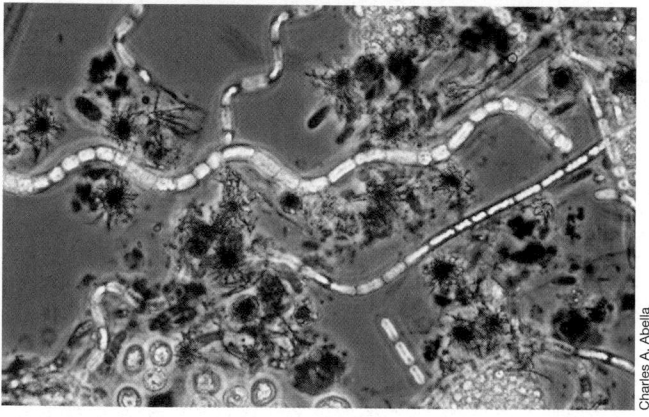

(c)

Figure 12.101 Green nonsulfur bacteria. (a) Phase photomicrograph of the thermophilic phototroph, *Chloroflexus aurantiacus*. Cells are about 1 μm in diameter. (b) Phase micrograph of the large phototroph *Oscillochloris*. Cells are about 5 μm wide. The brightly contrasting material is a holdfast, used for attachment. (c) Color photomicrograph of filaments of *Chloronema species* growing in a stratified Michigan lake. These cells of *Chloronema* are wavy filaments and about 2.5 μm in diameter. Despite being classified as "nonsulfur" bacteria, all of these species are capable of oxidizing H_2S as a photosynthetic electron donor. In addition, *Chloroflexus* can grow chemoorganotrophically in darkness by aerobic respiration.

aerobic respiration. Interestingly, and this should be considered in light of the evolutionary position of *Chloroflexus* as the most phylogenetically ancient of anoxygenic phototrophs (see Figure 12.1), autotrophy in *Chloroflexus* is based on a CO_2 incorporation pathway, the hydroxypropionate cycle, unique to this organism (∞ Figure 17.24*b*). We consider the biochemistry of this novel autotrophic pathway in Section 17.7.

Other Green Nonsulfur Bacteria

In addition to *Chloroflexus*, other phototrophic green nonsulfur bacteria include the thermophile *Heliothrix* and the large-celled mesophiles *Oscillochloris* (Figure 12.101*b*) and *Chloronema* (Figure 12.101*c*). *Heliothrix* is of interest because it is phylogenetically and phenotypically quite similar to *Chloroflexus* except that it lacks the bacteriochlorophyll *c* and chlorosomes of this organism. *Oscillochloris* and *Chloronema* are unusual because they are rather large cells, 2–5 μm wide, and can be up to several hundred micrometers long (Figure 12.101*c*). Both organisms develop in freshwater lakes containing low levels of sulfide, occurring together with other species of green sulfur or purple sulfur bacteria.

✓ *12.34–12.35* Concept Check

Deinococcus and *Chloroflexus* are each key genera in separate major lineages of *Bacteria*. *Deinococcus radiodurans* is the most radiation resistant of all known organisms and *Chloroflexus* is an anoxygenic phototroph that shows photosynthetic properties characteristic of both purple bacteria and green bacteria.

✓ How does *Deinococcus radiodurans* prevent being killed by high levels of radiation?

✓ In what ways does *Chloroflexus* resemble an organism like *Chlorobium*? An organism like *Rhodobacter*?

✓ What is unique about the organism *Thermomicrobium*?

XIV PHYLUM 13 AND 14: DEEPLY BRANCHING HYPERTHERMOPHILIC BACTERIA

Key Genera

Thermotoga
Thermodesulfobacterium
Aquifex
Thermocrinus

These three phyla of *Bacteria* cluster deep in the phylogenetic tree of *Bacteria*, near to the hypothetical "root" (∞ Figures 11.13 and 12.1). Each phylum consists of one or two major genera, and a key physiological feature of most of them is *hyperthermophily*, that is, optimal growth at temperatures *above* 80°C (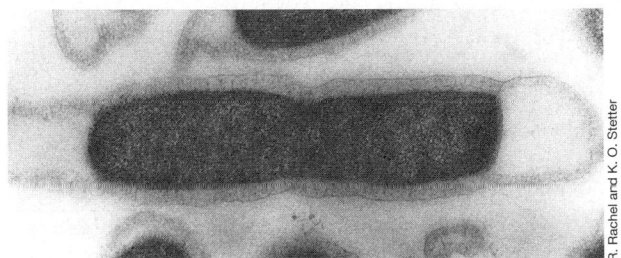 Section 6.10).

12.36 *Thermotoga* and *Thermodesulfobacterium*

Thermotoga is a rod-shaped hyperthermophile capable of growth to 90°C (optimum, 80°C). Cells of *Thermotoga* contain a sheath-like envelope (the "toga," see Figure 12.102*a*), stain gram-negatively, and are nonsporing. *Thermotoga* is an anaerobic, fermentative chemoorganotroph, catabolizing sugars and polymers such as starch and producing lactate, acetate, CO_2, and H_2 as fermentation products. Species of *Thermotoga* have been isolated from terrestrial hot springs as well as marine hydrothermal vents.

Thermodesulfobacterium

Thermodesulfobacterium (Figure 12.103*a*) is a thermophilic, sulfate-reducing bacterium, positioned on the phylogenetic tree between *Thermotoga* and *Aquifex* (Figure 12.1). Although not a true hyperthermophile because its growth temperature optimum is only 70°C,

(a)

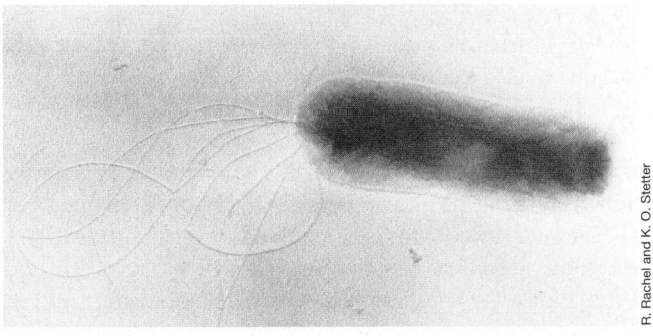

(b)

R. Rachel and K. O. Stetter

Figure 12.102 Hyperthermophilic *Bacteria*. (a) *Thermotoga maritima*—temperature optimum, 80°C. Note the outer covering on the cell (the "toga"). (b) *Aquifex pyrophilus*—temperature optimum, 85°C. Cells of *Thermotoga* (thin section) measure 0.6 × 3.5 μm; cells of *Aquifex* (freeze-fracture micrograph) measure 0.5 × 2.5 μm. Both *Thermotoga* and *Aquifex* form their own phylogenetic lineages on the tree of *Bacteria* (see Figure 12.1).

Thermodesulfobacterium is the most thermophilic of all known sulfate-reducing *Bacteria* (the archaeon sulfate-reducer *Archaeoglobus* is a true hyperthermophile, Section 13.7). Like other "group I" sulfate reducers (see Section 12.18), *Thermodesulfobacterium*, a strict anaerobe, cannot utilize acetate as an electron donor in its energy metabolism and instead uses compounds like lactate, pyruvate, and ethanol, reducing SO_4^{2-} to H_2S.

An unusual biochemical feature of species of *Thermodesulfobacterium* is the presence of *ether-linked lipids*. Recall that the latter are a hallmark of the *Archaea* and that a poly-isoprenoid C_{20} hydrocarbon (phytanyl) replaces fatty acids as the side chains in archaeal lipids (Sections 4.5 and 11.8). A very interesting situation exists in *Thermodesulfobacterium* ether-linked lipids because the glycerol side chains are not phytanyl groups but instead a unique C_{17} hydrocarbon along with some fatty acids (Figure 12.103*b*). Thus we see in *Thermodesulfobacterium* both a deep phylogenetic lineage (Figure 12.1) and a lipid profile that combines features of both the *Archaea* and the *Bacteria* (Figure 12.103*b*). However, the bacterium *Ammonifex*, a thermophilic member of the low GC, gram-positive *Bacteria* that grows anaerobically by H_2 oxidation coupled to the reduction of NO_3^- to NH_3, also contains lipids like those of *Thermodesulfobacterium*; thus, ether-linked lipids may be more common among *Bacteria* than previously thought.

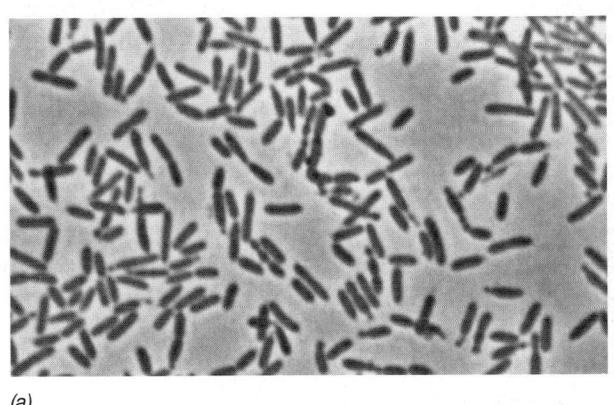

(a)

Fritz Widdel

(b)

Figure 12.103 *Thermodesulfobacterium*. (a) Phase photomicrograph of cells of *T. mobile*. (b) Structure of one of the lipids of *T. mobile*. Note that although these are ether-linked, the two hydrophobic side chains are *not* phytanyl units as in *Archaea* (Section 4.5). The designation "R" is for a hydrophilic residue, such as a phosphate group.

12.37 Aquifex, Thermocrinus, and Relatives

The genus *Aquifex* (Figure 12.102*b*) is an obligately chemolithotrophic and autotrophic hyperthermophile and is the most thermophilic of all known *Bacteria*. Various *Aquifex* species utilize H_2, S^0, or $S_2O_3^{2-}$ as electron donors and O_2 or NO_3^- as electron acceptors and grow up to 95°C (optimum, 85°C). Only very low O_2 concentrations are tolerated by *Aquifex*, but it remains (along with a few archaeans, ⚭ Sections 13.5 and 13.9) one of the few *aerobic* hyperthermophiles known. Nutritional studies of *Aquifex* species have shown them totally unable to grow chemoorganotrophically on organic compounds including complex mixtures like yeast or meat extract. *Hydrogenobacter*, a relative of *Aquifex*, shows most of the same properties as *Aquifex* but is an obligate aerobe.

Autotrophy in *Aquifex* is supported by enzymes of the reverse citric acid cycle, a series of reactions previously found only in green sulfur bacteria (see Sections 12.32 and 17.7) within the domain *Bacteria*. The complete genome sequence of *Aquifex aeolicus* has been determined (⚭ Section 15.3) and its entirely chemolithotrophic/autotrophic lifestyle is supported by an amazingly small genome of only 1.55 megabases (one-third the size of the *E. coli* genome). This fact was previously discussed in Section 11.2 as regards the physiological properties of early life forms. The discovery that so many hyperthermophiles, both species of *Archaea* and *Bacteria* like *Aquifex*, are H_2 chemolithotrophs, coupled with the finding that they branch as very early lineages on their respective phylogenetic trees (⚭ Figures 11.13, 12.1, and 13.1), suggests that H_2 was a key electron donor for living organisms under early Earth conditions. A simple model for the utilization of H_2 to generate a proton motive force and hence, usable energy as ATP was shown in Figure 11.5.

Thermocrinus

An interesting relative of *Aquifex* and *Hydrogenobacter* is the organism *Thermocrinus* (Figure 12.104). This organism is a hyperthermophilic (temperature optimum, 80°C) chemolithotroph capable of oxidizing either H_2, thiosulfate, or sulfur as electron donors, with O_2 as electron acceptor. *Thermocrinus ruber*, the only known species, grows in the outflow of certain hot springs in Yellowstone National Park, where it forms pink "streamers" consisting of a filamentous form of the cells attached to siliceous sinter (Figure 12.104*a*). In static culture, cells of *T. ruber* grow as individual rod-shaped cells (Figure 12.104*b*). However, when cultured in a flowing system in which growth medium is trickled over a solid glass surface to which cells can attach, *Thermocrinus*

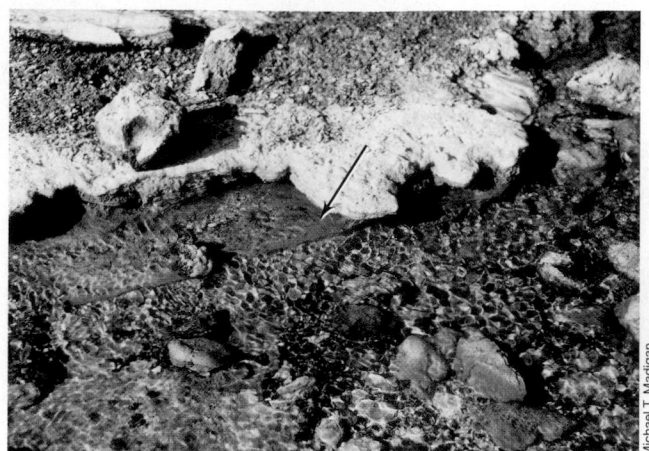

(a)

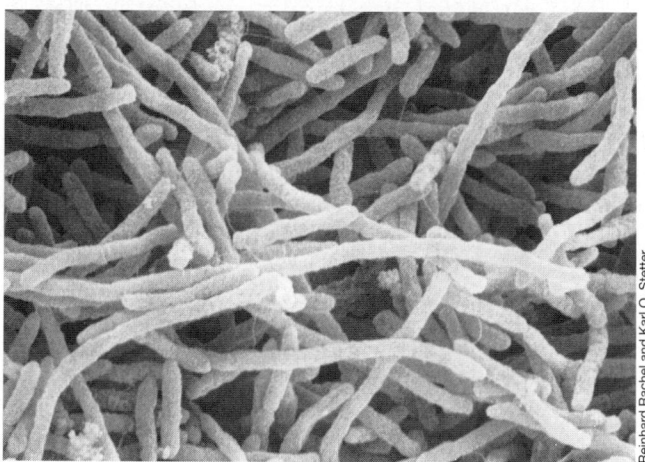

(b)

Figure 12.104 *Thermocrinus*. (a) Cells of *Thermocrinus ruber* growing as filamentous streamers (arrow) attached to siliceous sinter in the outflow (85°C) of Octopus Spring, Yellowstone National Park. The pink color is due to a carotenoid pigment. (b) Scanning electron micrograph of rod-shaped cells of *T. ruber* grown on a silicon-coated cover glass. The hairlike structures are silicon. A single cell of *T. ruber* is about 0.4 μm in diameter and from 1 to 3 μm long.

takes on the streamer form, just as it exists in its constantly flowing habitat in nature (Figure 12.104*a*).

Thermocrinus ruber is of historical significance in microbiology because it was one of the organisms studied in the 1960s by Thomas Brock, a pioneer in the field of thermal biology. The discovery by Brock that the pink streamers (Figure 12.104*a*) contained protein and nucleic acids clearly indicated that they were living organisms and not just mineral debris. Moreover, the presence of streamers in the outflow of hot springs at 80–90°C but not at lower temperatures supported Brock's hypothesis that hot spring microbial life forms actually *required* high temperatures for growth, and were likely to be present in even boiling water. Both of these conclusions were subsequently supported by the discovery of literally dozens of genera of hyperthermophilic prokaryotes inhabiting hot springs, hydrothermal vents, and other thermal environments (∞ Sections 6.10, 12.37 and 12.38, 13.4–13.10, and 19.8).

XV PHYLUM 15 AND 16: NITROSPIRA AND DEFFERIBACTER

12.38 Nitrospira, Defferibacter, and Relatives

Although we have covered a number of phyla of *Bacteria* in this chapter, several other phyla have been identified by ribosomal RNA sequencing about which little else is known. Two such phyla are defined by the organisms *Nitrospira* and *Deferribacter*, respectively (Figure 12.1). Physiologically, these organisms are either chemolithotrophs or chemoorganotrophs, and are mesophiles to thermophiles.

Like nitrifying Proteobacteria (see Section 12.3), *Nitrospira* oxidizes NO_2^- to NO_3^- and grows autotrophically. Despite this close physiological relationship to the classical nitrifying bacteria (for example, members of the genus *Nitrobacter* and its relatives, see Section 12.3, Table 12.4, and Figure 12.1), *Nitrospira* is phylogenetically quite distinct from them. Also, *Nitrospira* lacks the extensive internal membranes found in species of nitrifying Proteobacteria (see Figures 12.7 and 12.8). Nevertheless, *Nitrospira* inhabits many of the same environments as nitrifying Proteobacteria, so it has been suggested that its physiological capacities may have been transferred to it by lateral gene flow from nitrifying Proteobacteria; this mechanism for acquiring physiological traits has likely been widely exploited in the prokaryotic world (∞ Section 11.7).

Other genera in the *Nitrospira* group include *Leptospirillum*, an iron-oxidizing chemolithotroph respon-

sible for much of the acid mine drainage associated with the mining of coal and iron (∞ Section 19.16), and *Thermodesulfovibrio*, a thermophilic sulfate-reducing bacterium that inhabits hot spring microbial mats (∞ Section 18.10).

Defferibacter

The genus *Deferribacter* also forms its own distinct lineage (Figure 12.1) and is composed of species that specialize in anaerobic energy metabolism. Other genera in this group include *Geovibrio* and *Flexistipes*; the latter genus is an obligately anaerobic and fermentative bacterium. *Deferribacter* and *Geovibrio* are extremely versatile at anaerobic respiration, using a variety of electron acceptors, including the metals Fe^{3+} and Mn^{2+}. We discuss anaerobic respiration in Chapter 17 and show that the process can be linked to many different terminal electron acceptors. Members of the *Defferibacter* group appear to be unusual in the vast number of alternative electron acceptors they can use and in the fact that they are obligate anaerobes. Most organisms capable of growing by anaerobic respiration with nitrate or metals as electron acceptors are facultative aerobes, able to grow fully aerobically as well as by anaerobic respiration (∞ Section 17.13).

✓ 12.36–12.38 Concept Check

Thermotoga, Thermodesulfobacterium, and *Aquifex* grow at high temperature and each spearhead a major lineage of *Bacteria*. *Aquifex* is an H_2-oxidizing chemolithotroph, while *Thermotoga* and *Thermodesulfobacterium* are both anaerobic chemoorganotrophs. *Nitrospira* and *Deferribacter* each form their own phylum.

- ✓ Compare the catabolic metabolism of *Thermotoga* and *Thermodesulfobacterium*.
- ✓ What is unusual about the lipids of *Thermodesulfobacterium*?
- ✓ From a genomic perspective, why is the fact that *Aquifex* is able to grow on a diet of $H_2 + CO_2 + O_2$, seem surprising?
- ✓ Contrast the metabolic features of *Nitrospira* and *Defferibacter*.

Review Questions

1. Of all of the phyla of *Bacteria* studied in this chapter, which has groups with the most diverse physiologies? Give examples of this in terms of O_2 requirements, carbon sources used, and energy metabolism.

2. List examples of how the Gram reaction has phylogenetic predictive value.

3. What do cyanobacteria have in common with prochlorophytes? With chloroplasts? How are all of these thought to be phylogenetically related?

4. What do species in the *Planctomyces* phylum share in common with *Archaea*? With eukaryotic cells?

5. In what ways are *Chlorobium* and *Chloroflexus* similar? In what ways do they differ?

6. Describe a key *physiological* feature of the following *Bacteria* that would differentiate each from the others: *Acetobacter, Methanococcus, Azotobacter, Desulfovibrio, Lactobacillus, Nitrobacter, Oscillatoria*.

7. Describe a key *morphological* feature that would differentiate each of the following *Bacteria: Streptococcus, Spirillum, Streptomyces, Verrucomicrobium*, and *Spirochaeta*.

8. What key features could be used to differentiate the following genera of gram-positive *Bacteria: Bacillus, Mycoplasma*, and *Mycobacterium?*

9. What traits do Chlamydia and Rickettsias share in common? In what ways do they differ?

10. What major physiological property unites species of *Thermotoga, Aquifex* and *Thermocrinus?*

11. Compare and contrast the metabolism, morphology, and phylogeny of purple nonsulfur and green nonsulfur bacteria.

12. List an electron donor for energy metabolism for each of the following *Bacteria* and state whether the organism is an aerobe or an anaerobe: *Thiobacillus, Nitrosomonas, Ralstonia eutrophus, Methylomonas, Acetobacter, Gallionella*, and *Propionibacterium*.

Application Questions

1. Defend the following statement using phylogenetic, structural, and physiological arguments: *"Escherichia coli* is a much more evolved bacterium than *Thermodesulfobacterium."*

2. Defend or refute the following statement: "Cell morphology has absolutely no phylogenetic predictive value."

Although some species of *Archaea* inhabit rather innocuous environments, such as the intestine of warm blooded animals, it is in extreme environments that *Archaea* are preeminent. Shown here is a bloom of haloalkaliphilic *Archaea* that developed in an Egyptian soda lake. The red color is due to carotenoid pigments that these organisms produce to protect them from the damaging effects of intense sunlight. Haloalkaliphiles are extremophilic prokaryotes that have evolved to grow optimally under conditions of both high salt and high pH.

PROKARYOTIC DIVERSITY: THE *ARCHAEA*

13

Working Glossary

Acetotrophic acetate consuming. When used to describe a methanogen, an organism capable of splitting acetate into CH_4 and CO_2

Acetyl-CoA (Ljungdahl-Wood) pathway a pathway of autotrophic CO_2 fixation widespread in obligate anaerobes including methanogens, homoacetogens, and sulfate-reducing bacteria

Bacteriorhodopsin a membrane protein containing retinal produced by certain extreme halophiles and capable of light-mediated proton motive force formation

Compatible solutes organic or inorganic substances accumulated in the cytoplasm of halophilic organisms in order to maintain ionic pressure

Crenarchaeota a phylum of *Archaea* that contains both hyperthermophilic and cold-dwelling prokaryotes

Euryarchaeota a phylum of *Archaea* that contains primarily methanogens, the extreme halophiles, and *Thermoplasma*

Extreme halophile an organism whose growth is dependent on large concentrations (generally > 10%) of NaCl

Halorhodopsin a light-driven chloride pump that accumulates Cl^- within the cytoplasm

Hyperthermophile a prokaryote with a growth temperature optimum of 80°C or greater

"Korarchaeota" a phylum of hyperthermophilic *Archaea* that branches close to the archaeal root

Methanogen a methane-producing prokaryote; CH_4 is produced by either reduction of CO_2 with H_2 or from certain organic compounds

Phytanyl a branched-chain hydrocarbon containing 20 carbon atoms and commonly found in the lipids of *Archaea*

Reverse DNA gyrase a protein universally present in hyperthermophiles that introduces positive supercoils into circular DNA

Solfatara a hot, sulfur-rich, generally acidic environment commonly inhabited by hyperthermophilic *Archaea*

Thermosome a heat-shock chaperonin that refolds partially heat-denatured proteins in hyperthermophiles

I PHYLOGENY AND GENERAL METABOLISM

We now consider the domain *Archaea*. In Chapter 11 we emphasized the profound phylogenetic as well as phenotypic differences between *Bacteria* and *Archaea*. Here we examine the organisms themselves. Some major characteristics (described in Table 11.3) of the *Archaea* include absence of peptidoglycan in cell walls, and the presence of ether-linked lipids and complex RNA polymerases. But as we will see, despite these unifying properties, *Archaea* show enormous phenotypic diversity, which we will discuss here. As in Chapter 12 on *Bacteria*, we begin with a phylogenetic overview to show the evolutionary relationships within the domain *Archaea*.

13.1 Phylogenetic Overview of the *Archaea*

A phylogenetic tree of *Archaea* is shown in Figure 13.1. The tree bifurcates into two major phyla called the **Crenarchaeota** and the **Euryarchaeota**. A third phylum, the **"Korarchaeota,"** branches off close to the root (Figure 13.1). Among cultured representatives, the Crenarchaeota contain mostly hyperthermophilic species including those able to grow at the highest temperatures of all known organisms. Many hyperthermophiles are chemolithotrophic autotrophs; and because their habitats are devoid of photosynthetic life, these organisms are thus the only primary producers in these harsh environments.

Hyperthermophilic crenarchaeotes tend to cluster closely together and occupy short branches on the 16S ribosomal RNA-based tree of life (∞ Figures 11.13 and 13.1). This suggests that these organisms have "slow evolutionary clocks" and have evolved the least away from the hypothetical universal ancestor of all life. Such organisms should thus be good models for study of early life on Earth, and we return to this theme at the end of this chapter (see Section 13.12). By contrast, cold-dwelling relatives of hyperthermophilic crenarchaeotes have been identified by community sampling (∞ Sections 18.5 and 19.6) of ocean waters, and from a phylogenetic perspective these are more rapidly evolving and thus occupy longer branches of the tree (Figure 13.1). We consider the Crenarchaeota in more detail in Sections 13.8–13.10.

Euryarchaeota comprise a physiologically diverse group of *Archaea*, many of which, like crenarchaeotes, inhabit extreme environments of one sort or the other. Here we see methanogenic *Archaea*—prokaryotes whose metabolism is linked to the production of methane (CH_4)—related to several genera of extremely halophilic prokaryotes, the "halobacteria" (Figure 13.1). As a study in physiological contrasts, methanogens are obligate anaerobes, arguably *the* most strict of all anaerobes, while extreme halophiles are for the most part obligate aerobes. Other groups of euryarchaeotes include the hyperthermophiles *Thermococcus* and *Pyrococcus* and the methanogen *Methanopyrus*, that branch near to the root (Figure 13.1), and the cell wall-less prokaryote *Thermoplasma*, an organism phenotypically similar in this respect to the mycoplasmas (∞ Section 12.22). Finally, there exists a large group of as yet uncultured marine euryarchaeotes that are positioned at the ends of long branches up near the top of the phylo-

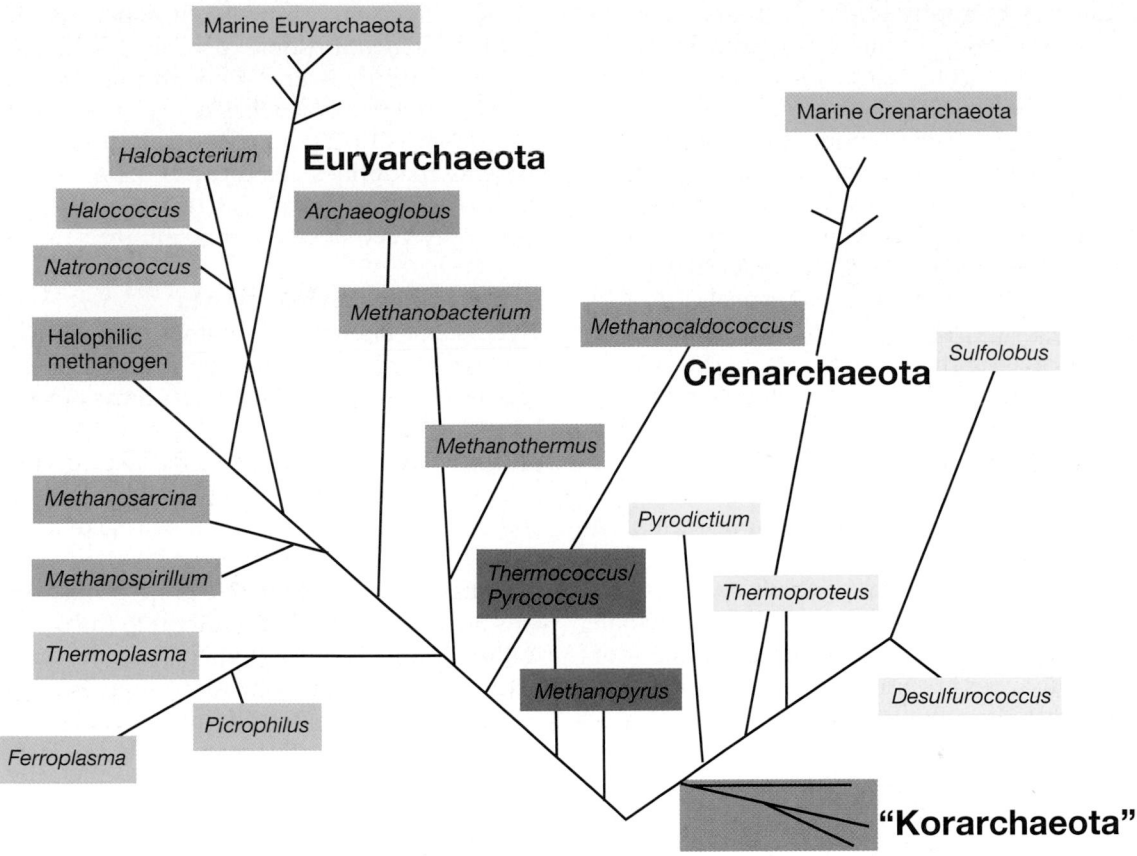

Figure 13.1 Detailed phylogenetic tree of the *Archaea* based on 16S ribosomal RNA sequence comparisons. The marine Euryarchaeota and marine Crenarchaeota are thus far known only from community sampling (∞ Section 11.6, 18.4, and 18.5).

genetic tree (Figure 13.1). We consider Euryarchaeota in more detail in Sections 13.3–13.7.

The "Korarchaeota" (Figure 13.1) were originally discovered by community sampling (∞ Sections 11.6 and 18.5) of an unusual Yellowstone hot spring, but now they exist in laboratory culture (see Section 13.12 and Figure 13.24). The "Korarchaeota," a group that is not yet officially recognized in taxonomy, branch on the archaeal tree close to the root, and for this reason their biological properties may reveal interesting features of ancient organisms.

With this overview to the phylogeny of the *Archaea*, let us proceed to a brief description of the metabolic features of *Archaea* and then to a description of the major properties of cultured representatives.

13.2 Energy Conservation and Autotrophy in *Archaea*

Energy metabolism in methanogens, a major group of *Archaea*, is unlike that of any other group prokaryotes, *Bacteria* or *Archaea*. We reserve our discussion of the many novel features of methanogenesis for Chapter 17 and focus here on the metabolism of nonmethanogenic *Archaea*.

Several *Archaea* are chemoorganotrophic and thus use organic compounds as energy sources for growth. Catabolism of glucose in *Archaea* proceeds via slight modifications of the Entner–Doudoroff (E–D) pathway (∞ Section 12.7 for a description of this pathway in *Bacteria*) or glycolytic pathway (∞ Section 5.10). Oxidation of acetate to CO_2 in *Archaea* proceeds through the citric acid cycle (∞ Section 5.13) or some slight variation of this reaction series, or by the acetyl-CoA pathway (∞ Section 17.16). Little is known concerning biosynthesis of amino acids and other macromolecular precursors in *Archaea*, but presumably key monomers are produced from the central biosynthetic intermediates discussed previously for *Bacteria* (∞ Section 5.15).

Electron transport chains including cytochromes of the *a*, *b*, and *c* types exist in some *Archaea*. Employing these and other electron carriers, chemoorganotrophic metabolism in most *Archaea* probably proceeds by introduction of electrons from organic electron donors into an electron transport chain, leading to the reduction of O_2, S^0, or some other electron acceptor, with concurrent establishment of a proton motive force that drives adenosine triphosphate (ATP) synthesis through membrane-bound ATPases (∞ Section 5.12 for a description of ATPase function). Chemolithotrophy is also well established in the *Archaea*, with H_2 being a common electron donor. We examine chemolithotrophic metabolism of hyperthermophilic *Archaea* later in this chapter (see Section 13.8).

Autotrophy is widespread among *Archaea* and occurs by several different means. In methanogens, and presumably in most chemolithotrophic hyperthermophiles, CO_2 is incorporated via the acetyl-CoA pathway or some modification thereof (∞ Section 17.16). In other hyperthermophiles CO_2 fixation occurs via the reverse citric acid cycle, a reaction series that functions as the autotrophic pathway in the green sulfur bacteria (∞ Sections 12.33 and 17.7), or via the Calvin cycle, the most widespread autotrophic pathway in *Bacteria* and eukaryotes (∞ Section 17.6). In this connection, genes encoding functional and very thermostable RubisCO enzymes (RubisCO catalyzes the first step in the Calvin cycle) have been characterized from the methanogen *Methanocaldococcus jannaschii* and from a *Pyrococcus* species, both hyperthermophiles.

We thus see that many of the catabolic and anabolic sequences in the *Archaea* are familiar ones from our study of these processes in various *Bacteria*, emphasizing the fact that metabolism has a long evolutionary history. With this as background we now begin our study of the diversity of this interesting domain of life.

✓ 13.1–13.2 Concept Check

Archaea form three major phylogenetic phyla, the Euryarchaeota, the Crenarchaeota, and the "Korarchaeota". With the exception of methanogenesis, bioenergetics and intermediary metabolism in species of *Archaea* are much the same as that of *Bacteria*.

✓ Which organism shown in Figure 13.1, *Halobacterium* or *Methanopyrus*, likely contains the slower "evolutionary clock," and how do you know this?

✓ What autotrophic pathways are found in *Archaea*?

II PHYLUM EURYARCHAEOTA

13.3 Extremely Halophilic *Archaea*

Key Genera

Halobacterium
Haloferax
Natronobacterium

Extremely halophilic *Archaea* are a diverse group of prokaryotes that inhabit highly saline environments, such as solar salt evaporation ponds and natural salt lakes, or artificial saline habitats, such as the surfaces of heavily salted foods like certain fish and meats. Such habitats are often called *hypersaline*. The term *extreme* halophile is used to indicate not only that these organ-

isms are halophilic, but also that their requirement for salt is *very high*, in some cases near saturation. An extreme halophile is an organism that requires at least 1.5 *M* (about 9%) NaCl for growth; most species require 2–4 *M* NaCl (12–23%) for optimal growth. Virtually all extreme halophiles can grow at 5.5 *M* NaCl (32%, the limit of saturation for NaCl), although some species grow only very slowly at this salinity.

Hypersaline Environments

Hypersaline habitats are common throughout the world, but *extremely* hypersaline habitats are rather rare. Most such environments are in hot, dry areas of the world. Salt lakes can vary considerably in ionic composition. The predominant ions in a hypersaline lake depend to a major extent on the surrounding topography, geology, and general climatic conditions. Great Salt Lake in Utah (USA) (Figure 13.2*a*), for example, is essentially concentrated seawater because the relative proportions of the various ions are those of seawater, although the overall concentration of ions is much higher. Sodium is the predominant cation in Great Salt Lake, whereas chloride is the predominant anion; significant levels of sulfate are also present at a slightly alkaline pH (Table 13.1). By contrast, another hypersaline basin, the Dead Sea, is relatively low in sodium but contains high levels of magnesium (Table 13.1). The water chemistry of soda lakes resembles that of hypersaline lakes such as Great Salt Lake, but because high levels of *carbonate* minerals are present in the surrounding rocks, the pH of soda lakes is quite high; pH values of 10–12 are not uncommon in these environments (Table 13.1 and Figure 13.2*c*). In addition, Ca^{2+} and Mg^{2+} are virtually absent because they precipitate out at high pH and carbonate concentrations (Table 13.1).

Despite what may seem like rather harsh conditions, salt lakes can be highly productive ecosystems. *Archaea* are not the only microorganisms found. The eukaryotic alga *Dunaliella* is the major, if not sole, oxygenic pho-

TABLE 13.1	Ionic composition of some highly saline environments[a]		
Ion	**Concentration (g/l)**		
	Great Salt Lake[b]	**Dead Sea**	**Lake Zugm**[c]
Na^+	105	40.1	142
K^+	6.7	7.7	2.3
Mg^{2+}	11	44	< 0.1
Ca^{2+}	0.3	17.2	< 0.1
Cl^-	181	225	155
Br^-	0.2	5.3	—
SO_4^{2-}	27	0.5	23
HCO_3^-	0.7	0.2	67
pH	7.7	6.1	11

[a] For comparison, seawater contains (grams per liter): Na^+, 10.6; K^+, 0.38; Mg^{2+}, 1.27; Ca^{2+}, 0.4; Cl^-, 19; Br^-, 0.065; SO_4^{2-}, 2.65; HCO_3^-, 0.14; pH 8.
[b] See Figure 13.2*a*
[c] Wadi El Natroun, Egypt (see Figure 13.2*c*).

(a)

(b)

(c)

(d)

Figure 13.2 Hypersaline habitats for halophilic *Archaea*. (a) Great Salt Lake, Utah, a hypersaline lake in which the ratio of ions is similar to that in seawater but in which absolute concentrations of ions are about 10 times that of seawater. The green color is primarily from cells of the halophilic green alga, *Dunaliella salina*. (b) Aerial view near San Francisco Bay, California, of a series of seawater evaporating ponds where solar salt is prepared. The red-purple color is predominantly due to bacterioruberins and bacteriorhodopsin in cells of *Halobacterium*. (c) Lake Hamara, Wadi El Natroun, Egypt. A bloom of pigmented haloalkaliphiles is growing in this pH 10 soda lake. Note the deposits of trona (Na_2CO_3) around the edge of the lake. (d) Scanning electron micrograph of halophilic prokaryotes including square bacteria present in a Spanish saltern.

totroph in most salt lakes. In highly alkaline soda lakes where *Dunaliella* is absent, anoxygenic phototrophic purple bacteria of the genus *Ectothiorhodospira* and *Halorhodospira* (∞ Section 12.2) predominate. Organic matter originating from primary production by oxygenic or anoxygenic phototrophs then sets the stage for development of the extremely halophilic *Archaea*, all of which are chemoorganotrophic. In addition, a few extremely halophilic anaerobic chemoorganotrophic *Bacteria*, such as *Haloanaerobium* and *Halobacteroides*, thrive in such environments.

Marine salterns are also habitats for extremely halophilic prokaryotes. Marine salterns are small enclosed basins filled with seawater that are left to evaporate yielding solar sea salt (Figure 13.2*b, d*). As salterns approach the minimum salinity limits for extreme halophiles, the waters turn a reddish purple color due to the massive growth (called a *bloom*) of halophilic *Archaea* (the red coloration apparent in Figures 13.2*b* and *c* comes from carotenoids and other pigments discussed later). Morphologically unusual prokaryotes are often present in salterns, including even square bacteria (Figure 13.2*d*)! Extreme halophiles have also been found in high salt foods such as certain sausages, marine fish, and salt pork.

Taxonomy and Physiology of Extremely Halophilic *Archaea*

Table 13.2 lists the currently recognized species of extremely halophilic *Archaea*. 16S ribosomal ribonucleic acid (rRNA) sequencing and other criteria have defined ten genera of

TABLE 13.2   Genera of extremely halophilic *Archaea*

Genus	Morphology	Number of species	DNA (mol % GC)	Habitat
Extreme Halophiles				
Halobacterium	Rods	1	66–71	Salted fish; hides; hypersaline lakes; salterns
Halorubrum	Rods	7	62–71	Dead Sea; salterns
Halobaculum	Rods	1	70	Dead Sea
Haloferax	Flattened discs	4	60–66	Dead Sea; salterns
Haloarcula	Irregular discs	7	63–65	Salt pools, Death Valley, CA; marine salterns
Halococcus	Cocci	3	59–66	Salted fish; salterns
Halogeometricum	Rods	1	59–60	Solar salterns
Haloterrigena	Rods, ovals	1	59–60	Saline soil
Haloalkaliphiles				
Natronobacterium	Rods	1	65	Highly saline soda lakes
Natrinema	Rods	2	70	Salted fish; hides
Natrialba	Rods	2	60–63	Soda lakes; beach sand
Natronomonas	Rods	1	61–64	Soda lakes
Natronococcus	Cocci	2	63–64	Soda lakes
Natronorubrum	Flattened cells	2	59–60	Soda lakes

extreme halophiles (Table 13.2). The extremely halophilic *Archaea* are frequently referred to collectively as "halobacteria," because the genus *Halobacterium* (Figure 13.3) was the first in this group to be described and is still the best-studied representative of the group. *Natronobacterium, Natronomonas*, and their relatives differ from other extreme halophiles in being extremely *alkaliphilic* as well as halophilic. As befits their soda lake habitat (see Table 13.1 and Figure 13.2*c*), growth of natronobacteria is optimal at very low Mg^{2+} concentrations and high pH (9–11).

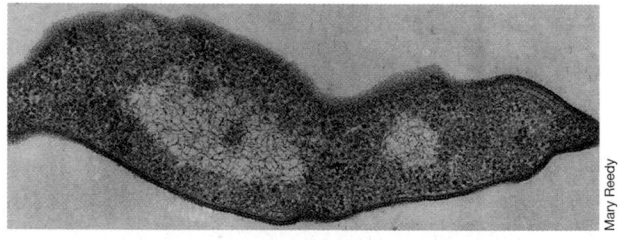

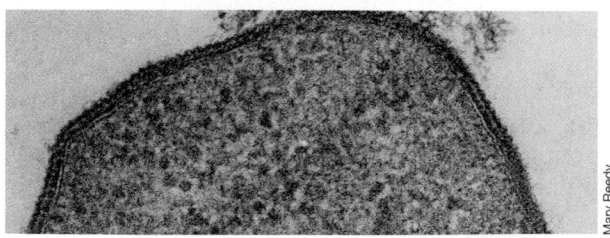

Figure 13.3 Electron micrographs of thin sections of the extreme halophile *Halobacterium salinarum*. A cell is about 0.8 μm in diameter. (a) Longitudinal section. (b) High magnification electron micrograph showing the glycoprotein subunit structure of the cell wall.

Extremely halophilic *Archaea* stain gram *negatively*, reproduce by binary fission, and do not form resting stages or spores. Most halobacteria are nonmotile, but a few strains are weakly motile by flagellar means. The genomic organization of *Halobacterium* and *Halococcus* is unusual in that large plasmids containing up to 25–30% of the total cellular DNA are present and the GC base ratio of these plasmids (57–60% GC) is significantly different from that of chromosomal DNA (66–68% GC). Plasmids from extreme halophiles are among the largest naturally occurring plasmids known (∞ Section 10.8).

Extremely halophilic *Archaea* are chemoorganotrophs, and most species are obligate *aerobes*. Most halobacteria use amino acids or organic acids as energy sources and require a number of growth factors (mainly vitamins) for optimal growth. A few *Halobacterium* species oxidize carbohydrates, but this ability is relatively rare. Electron transport chains containing cytochromes *a*, *b*, and *c* are present in *Halobacterium*, and energy is conserved during aerobic growth via a proton motive force arising from membrane-mediated chemiosmotic events. Some halophilic *Archaea* have been shown to grow anaerobically. Anoxic growth at the expense of sugar fermentation and by anaerobic respiration (∞ Section 17.13) linked to the reduction of nitrate or fumarate has been demonstrated in certain strains.

Water Balance in Extreme Halophiles

Extremely halophilic *Archaea* require large amounts of sodium for growth. In the case of *Halobacterium*, where detailed salinity studies have been performed, the requirement for Na^+ cannot be satisfied by replacement with another ion, even with the chemically related ion K^+. We learned in Section 6.12 that certain microorganisms can withstand the osmotic forces that accompany life in a high solute environment by accumulating or synthesizing organic compounds *intracellularly*; the latter compounds are referred to as **compatible solutes**. These compounds counteract the tendency of the cell to become dehydrated under conditions of high osmotic strength by placing the cell in positive water balance with its surroundings. As a compatible solute, cells of *Halobacterium* pump large amounts of K^+ from the environment into the cytoplasm such that the concentration of K^+ *inside* the cell is considerably greater than the concentration of Na^+ *outside* the cell (Table 13.3).

The *Halobacterium* cell wall (Figure 13.3*b*) is composed of glycoprotein and stabilized by sodium ions. Sodium ions bind to the outer surface of the *Halobacterium* wall and are absolutely essential for maintaining cellular integrity; when insufficient Na^+ is present, the cell wall breaks apart and the cell lyses. This is a consequence of the exceptionally high content of the *acidic* (negatively charged) amino acids aspartate and glutamate in the glycoprotein of the *Halobacterium* cell wall. The negative charges contributed by the carboxyl groups of these amino acids are shielded by Na^+; when Na^+ is diluted away, the negatively charged parts of the proteins actively repel each other, leading to cell lysis.

Halophilic Cytoplasmic Components

Like cell wall proteins, cytoplasmic proteins of *Halobacterium* are also highly acidic, but it is K^+, not Na^+, that is required for activity. This, of course, is not surprising since K^+ is the predominate internal cation in cells of *Halobacterium* (Table 13.3). Besides a high acidic amino acid composition, halobacterial cytoplasmic proteins typically contain very low levels of *hydrophobic* amino acids and lysine (a *basic* amino acid) compared with those of nonhalophiles. This also makes sense since in a highly ionic cytoplasm, highly polar proteins tend to remain in solution, whereas nonpolar molecules would tend to cluster and perhaps lose activity. The ribosomes of *Halobacterium* also require high K^+ levels for stability (ribosomes of nonhalophiles have no K^+ requirement).

It thus appears that the extremely halophilic *Archaea* are highly adapted, both internally and externally, to life in a highly ionic environment. Cellular components exposed to the external environment require high Na^+ for stability, whereas internal components require high K^+. With the exception of a few extremely halophilic members of the *Bacteria* that also use K^+ as a compatible solute, in no other group of prokaryotes do we find this unique requirement for such high amounts of specific cations.

Bacteriorhodopsin and Light-Mediated ATP Synthesis

Certain species of extremely halophilic *Archaea* are capable of a *light-mediated* synthesis of ATP that does not involve chlorophyll (thus it is *not* photosynthesis). The highly pigmented nature of extremely halophilic *Archaea* is visible in Figure 13.2. Pigmentation is due to red- and orange-colored carotenoids, primarily C_{50} carotenoids called *bacterioruberins*, and also to inducible pigments involved in energy metabolism. Under conditions of low aeration, *Halobacterium salinarum* and certain other extreme halophiles synthesize and insert a protein called **bacteriorhodopsin** into their membranes. Bacteriorhodopsin was named because of its structural and functional similarity to the visual pigment of the eye, *rhodopsin*. Conjugated to bacteriorhodopsin is a molecule of *retinal*, a carotenoid-like molecule that can absorb light and catalyze formation of a proton motive force (Figure 13.4). The retinal gives bacteriorhodopsin a purple hue, and cells of *Halobacterium* switched from growth under conditions of high aeration to oxygen-limiting conditions (a trigger of bacteriorhodopsin synthesis) gradually change from an orange or red color to a more

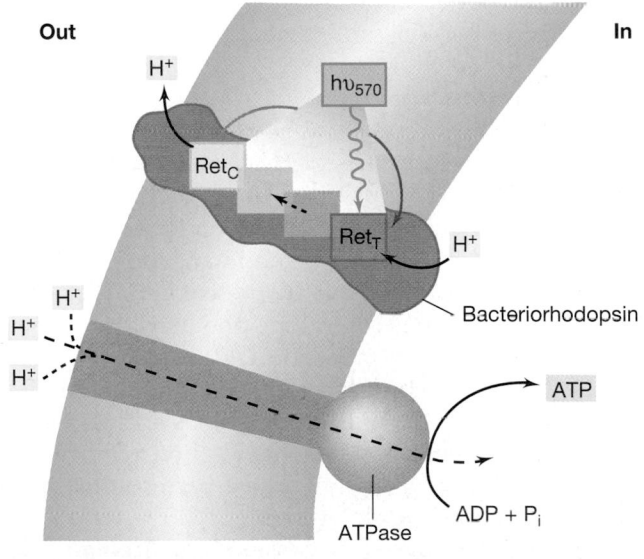

Figure 13.4 Model for the mechanism of bacteriorhodopsin activity. Light near 570 nm ($h\upsilon_{570}$) converts the protonated retinal of bacteriorhodopsin from the *trans* form (Ret_T) to the *cis* form (Ret_C), along with translocation of a proton to the outer surface of the membrane, thus establishing a proton motive force. ATPase activity (⚭ Section 5.12) is driven by the proton motive force.

TABLE 13.3	Concentration of ions in cells of *Halobacterium salinarum*	
Ion	**Concentration in medium (M)**	**Concentration in cells (M)**
Na^+	3.3	0.8
K^+	0.05	5.3
Mg^{2+}	0.13	0.12
Cl^-	3.3	3.3

reddish-purple as they synthesize bacteriorhodopsin and insert it into the cytoplasmic membrane.

Bacteriorhodopsin absorbs light strongly in the green region of the spectrum at about 570 nm. The retinal of bacteriorhodopsin, which normally exists in an *all-trans* configuration, becomes excited and converted to the *cis* form following the absorption of light (Figure 13.4). This transformation results in the translocation of protons to the *outside* surface of the membrane. The retinal molecule then returns to its more stable all-trans isomer in the dark along with the uptake of a proton from the cytoplasm, thus completing the cycle (Figure 13.4). As protons accumulate on the outer surface of the membrane, the proton motive force (Section 5.12) increases until the membrane is sufficiently "charged" to drive ATP synthesis through action of a proton translocating ATPase (Figure 13.4).

Light-mediated ATP production in *Halobacterium salinarum* has been shown to support slow growth of this organism anaerobically under nutritional conditions in which other energy-generating reactions do not occur, and light has been shown to maintain the viability of anoxic cultures of *Halobacterium* incubated in the absence of organic energy sources. The light-stimulated proton pump of *H. salinarum* also functions to pump Na^+ out of the cell by action of a Na^+/H^+ antiport system (Section 4.6) and to drive the uptake of a variety of nutrients, including the K^+ needed for osmotic balance. The uptake of amino acids by *H. salinarum* has been shown to be indirectly driven by light because the transport of amino acids occurs with Na^+ uptake by an amino acid–Na^+ symporter (Section 4.6). Continued uptake depends on the removal of Na^+ via the (light-driven) Na^+/H^+ antiporter.

Other Rhodopsins

At least three other rhodopsins besides bacteriorhodopsin are present in membranes of *Halobacterium salinarium*. **Halorhodopsin** is a light-driven pump that pumps chloride (Cl^-) into the cell as a counterion for K^+. The retinal of halorhodopsin binds Cl^- and transports it into the cell. Two **sensory rhodopsins** are present in *H. salinarium*. These are light sensors controlling phototaxis (the movement towards light, Section 4.12) in this organism. Through the interaction of a cascade of proteins similar to what drives chemotaxis (Sections 4.12 and 8.11), sensory rhodopsins affect flagellar rotation to move cells of *H. salinarium* toward light where bacteriorhodopsin can then function to make ATP (Figure 13.4).

We will learn in our discussion of marine microbiology (Section 19.6) that species of *Bacteria* exist in open ocean waters that contain bacteriorhodopsin-like proteins called *proteorhodopsins*. As far as is known, proteorhodopsin functions like bacteriorhodopsin, except that several different spectral forms exist, each tuned to the absorption of different wavelengths of light. Prote-

orhodopsin as a mechanism for energy conservation in marine prokaryotes makes good ecological sense, since concentrations of dissolved organic matter in the oceans are typically very low (Section 19.6) and thus strict chemoorganotrophy would be a difficult lifestyle.

✓ 13.3 Concept Check

Extremely halophilic *Archaea* are prokaryotes that require large amounts of NaCl for growth. These organisms pump large amounts of KCl into their cytoplasm as compatible solute. These salts affect cell wall stability and enzyme activity. The light-mediated proton pump bacteriorhodopsin helps extreme halophiles make ATP.

✓ What is the major physiological difference between the organisms *Halobacterium* and *Natronobacterium*?

✓ If cells of *Halobacterium* require high levels of Na^+, why is this not true of *cytoplasmic enzymes* from *Halobacterium* (see Table 13.3).

✓ What benefit does bacteriorhodopsin confer on a cell of *H. salinarum*?

John A. Breznak

Figure 13.5 The "Volta Experiment" as demonstrated in the Microbial Diversity Summer Course in Woods Hole, Massachusetts. A large inverted funnel was placed over freshwater sediments in which anaerobic decomposition was occurring in Cedar Swamp, Woods Hole. After water displaced the air in the funnel, the funnel was capped and the sediments were mixed with a stick, allowing trapped bubbles of methane to collect in the inverted funnel. Immediately after uncapping the funnel, a flame was held near the funnel port, thus igniting the methane. This experiment was performed over 200 years ago by the Italian physicist Alessandro Volta, which led him to describe methane as "combustible air."

13.4 Methane-Producing *Archaea*: Methanogens

Key Genera

Methanobacterium
Methanocaldococcus
Methanosarcina

A large number of Euryarchaeota produce methane (CH_4) as an integral part of their energy metabolism. Such organisms are called *methanogens* and the process of methane formation *methanogenesis*. Methane was first discovered as a type of "combustible air" by the Italian physicist Alessandro Volta, who collected gas from marsh sediments and showed that it was flammable (the "Volta experiment" can be easily reproduced if methane, trapped in freshwater sediments, is collected and *carefully* ignited) (Figure 13.5). In later chapters we will study the biochemically unique and amazingly complex process of methane formation (∞ Section 17.17) and then learn how methanogenesis is the terminal step in the biodegradation of organic matter in many anoxic habitats in nature, such as the swamp shown in Figure 13.5 (∞ Section 19.10). The major sources of biogenic methane are listed in Table 13.4.

Diversity and Physiology of Methanogens

A wide variety of morphological types of methanogens have been described (Figure 13.6 and Table 13.5). Their taxonomy is based on both phenotypic as well as phylogenetic (comparative 16S rRNA sequencing) analyses (Table 13.5), with several orders being recognized (in taxonomy, an order contains groups of related families, each of which contain one or more genera; ∞ Table 11.6). Structurally, methanogens are prokaryotic cells that show a diversity of cell wall chemistries. The latter include the pseudopep-

TABLE 13.4 Habitats of Methanogens

I. Anoxic sediments: marsh, swamp (see Figure 13.5), and lake sediments, paddy fields, moist landfills (∞ Section 19.10)
II. Animal digestive tracts:
   (a) Rumen of ruminant animals such as cattle, sheep, elk, deer, and camels (∞ Section 19.11)
   (b) Cecum of cecal animals such as horses and rabbits (∞ Section 19.11)
   (c) Large intestine of monogastric animals such as humans, swine, and dogs (∞ Section 21.4)
   (d) Hindgut of cellulolytic insects (for example, termites)
III. Geothermal sources of $H_2 + CO_2$: hydrothermal vents (∞ Section 19.8)
IV. Artificial biodegradation facilities: sewage sludge digestors (∞ Section 28.2)
V. Endosymbionts of various anaerobic protozoa (∞ Figure 19.26)

tidoglycan walls of *Methanobacterium* species and relatives (Figure 13.7a), the methanochondroitin (so named because of its structural resemblance to chondroitin, the connective tissue polymer of vertebrate animals) walls of *Methanosarcina* and relatives (Figure 13.7b), the protein or glycoprotein walls of *Methanocaldococcus* (Figure 13.8a), and *Methanoplanus* species, respectively, and the S-layer walls of *Methanospirillum* (Figure 13.6; ∞ Section 4.8).

Physiologically, methanogens are obligate anaerobes, and strict anoxic techniques are necessary to culture them. Depending on the species, cultures of methanogens can be established in a mineral salts medium under an atmosphere of H_2 plus CO_2 (in the ratio of 4:1, see the reaction shown later), or in complex media. Most known methanogens are mesophilic, although "extremophilic" species growing optimally at very high (see Figures 13.8 and 13.13) or very low temperature, or at very high salt concentration, have also been described.

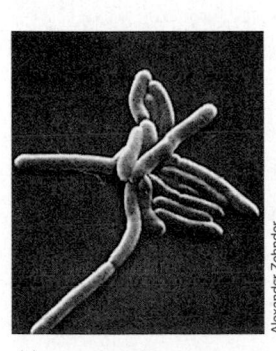
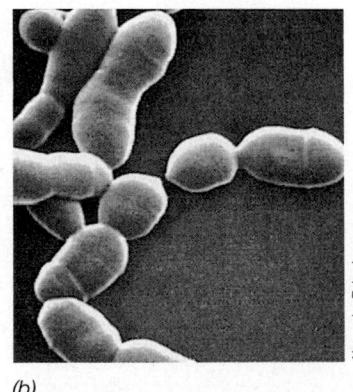
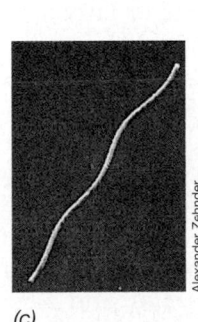
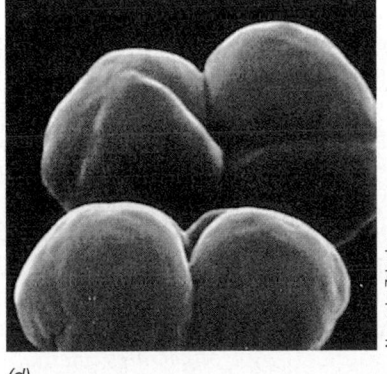

(a)   (b)   (c)   (d)

Figure 13.6 Scanning electron micrographs of cells of methanogenic *Archaea*, showing the considerable morphological diversity. (a) *Methanobrevibacter ruminantium*. A cell is about 0.7 µm in diameter. (b) *Methanobrevibacter arboriphilus*. A cell is about 1 µm in diameter. (c) *Methanospirillum hungatii*. A cell is about 0.4 µm in diameter. (d) *Methanosarcina barkeri*. A cell is about 1.7 µm wide.

TABLE 13.5 Characteristics of methanogenic *Archaea*[a]

Genus	Morphology	Number of species	Substrates for methanogenesis	DNA (mol % GC)
Methanobacteriales				
Methanobacterium	Long rods	8	$H_2 + CO_2$, formate	30–55
Methanobrevibacter	Short rods	7	$H_2 + CO_2$, formate	27–31
Methanosphaera	Cocci	2	Methanol + H_2 (both needed)	23–26
Methanothermus	Rods	2	$H_2 + CO_2$; can also reduce S^0; hyperthermophile	33
Methanothermobacter	Rods	6	$H_2 + CO_2$, formate, thermophiles	32–61
Methanococcales				
Methanococcus	Irregular cocci	3	$H_2 + CO_2$, pyruvate + CO_2, formate	29–35
Methanothermococcus	Cocci	1	$H_2 + CO_2$, formate	31–34
Methanocaldococcus	Cocci	4	$H_2 + CO_2$	31–33
Methanotorris	Cocci	1	$H_2 + CO_2$	31
Methanomicrobiales				
Methanomicrobium	Short rods	1	$H_2 + CO_2$, formate	49
Methanogenium	Irregular cocci	4	$H_2 + CO_2$, formate	47–52
Methanospirillum	Spirilla	1	$H_2 + CO_2$, formate	45–50
Methanoplanus	Plate-shaped cells—occurring as thin plates with sharp edges	3	$H_2 + CO_2$, formate	39–50
Methanocorpusculum	Irregular cocci	4	$H_2 + CO_2$, formate, alcohols	48–52
Methanoculleus	Irregular cocci	6	$H_2 + CO_2$, alcohols, formate	49–61
Methanofollis	Irregular cocci	2	$H_2 + CO_2$, formate	54–60
Methanolacinia	Irregular rods	1	$H_2 + CO_2$, alcohols	38–45
Methanosarcinales				
Methanosarcina	Large irregular cocci in packets	5	$H_2 + CO_2$, methanol, methylamines, acetate	36–43
Methanolobus	Irregular cocci in aggregates	5	Methanol, methylamines	39–46
Methanohalobium	Irregular cocci	1	Methanol, methylamines; halophilic	37
Methanococcoides	Irregular cocci	2	Methanol, methylamines	42
Methanohalophilus	Irregular cocci	3	Methanol, methylamines, methyl sulfides; halophile	39–41
Methanosaeta	Long rods to filaments	2	Acetate	52–61
Methanosalsum	Irregular cocci	1	Methanol, methylamines, dimethylsulfide	38–40
Methanopyrales				
Methanopyrus	Rods in chains	1	$H_2 + CO_2$; hyperthermophile, growth at 110°C	60

[a] Taxonomic orders are listed in bold.

Substrates for Methanogenesis

At least 11 substrates have been shown to be converted to methane by pure cultures of methanogens (Table 13.6). Interestingly, these substrates do *not* include such common compounds as glucose and organic or fatty acids (other than acetate and pyruvate). This is not to say that a compound like glucose can never be converted to methane. We will discuss in Chapter 19 how, in cooperative reactions involving methanogens and other anaerobic bacteria, virtually any organic compound can be converted to methane and CO_2 (∞ Section 19.10).

Three classes of compounds make up the list of 10 methanogenic substrates shown in Table 13.6; these include *CO₂-type substrates, methyl substrates*, and *acetotrophic substrates*. The first class includes the important substrate CO_2, substrate which is reduced to methane using H_2 as electron donor:

$$CO_2 + 4\ H_2 \rightarrow CH_4 + 2\ H_2O \qquad \Delta G^{0\prime} = -131\,kJ$$

TABLE 13.6 Substrates converted to methane by various methanogenic *Archaea*

I. **CO₂-type substrates**
   Carbon dioxide, CO_2 (with electrons derived from H_2, certain alcohols, or pyruvate)
   Formate, $HCOO^-$
   Carbon monoxide, CO

II. **Methyl substrates**
   Methanol, CH_3OH
   Methylamine, $CH_3NH_3^+$
   Dimethylamine, $(CH_3)_2NH_2^+$
   Trimethylamine, $(CH_3)_3NH^+$
   Methylmercaptan, CH_3SH
   Dimethylsulfide, $(CH_3)_2S$

III. **Acetotrophic substrates**
   Acetate, CH_3COO^-
   Pyruvate, CH_3COCOO^-

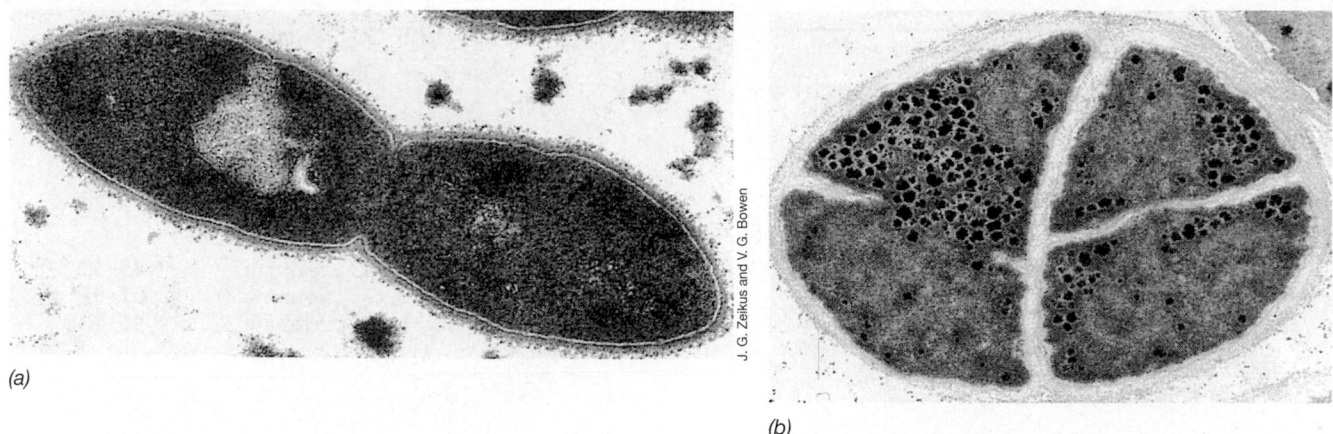

Figure 13.7 Transmission electron micrographs of thin sections of methanogenic *Archaea*. (a) *Methanobrevibacter ruminantium*. A cell is 0.7 μm in diameter. (b) *Methanosarcina barkeri*, showing the thick cell wall and the manner of cell segmentation and cross-wall formation. A cell is 1.7 μm in diameter. The cell wall of *M. ruminantium* contains pseudopeptidoglycan (⌁ Figure 4.34*a*), while the *M. barkeri* wall consists of protein and polysaccharides.

Other substrates here include formate (which is simply $CO_2 + H_2$ in combined form) and CO, carbon monoxide.

The second class of methanogenic substrates are methyl group substances (Table 13.6). Using CH_3OH as a model methyl substrate here, the formation of CH_4 can occur in either of two ways. First, CH_3OH can be reduced using an external electron donor such as H_2:

$$CH_3OH + H_2 \rightarrow CH_4 + H_2O \qquad \Delta G^{0\prime} = -113\,kJ$$

Alternatively, in the absence of H_2, some CH_3OH can be oxidized to CO_2 in order to generate the electrons needed to reduce other molecules of CH_3OH to CH_4:

$$4\,CH_3OH \rightarrow 3\,CH_4 + CO_2 + 2\,H_2O \qquad \Delta G^{0\prime} = -319\,kJ$$

The final methanogenic process is the cleavage of acetate to CO_2 plus CH_4, called the *acetotrophic* reaction:

$$CH_3COO^- + H_2O \rightarrow CH_4 + HCO_3^- \qquad \Delta G^{0\prime} = -31\,kJ$$

Only a very few methanogens are acetotrophic (Table 13.5), yet experimental measurements of methane formation in methanogenic habitats such as sewage sludge have shown that about two-thirds of the methane formed there originates from acetate and one-third from $H_2 + CO_2$; thus, although lacking in terms of known diversity, acetotrophic methanogens are very ecologically significant in nature.

As can be seen by inspection of each of the above reactions, they are all exergonic and can thus be used to synthesize ATP. We reserve until Chapter 17 the biochemical details of methanogenesis and only note here that the process is coupled to proton motive force formation and ATP synthesis by normal chemiosmotic (⌁ Section 5.12) mechanisms. Substrate-level phos-phorylation, typical of fermentative bacteria, apparently does not occur in methanogens.

Methanocaldococcus jannaschii as a Model Methanogen/Archaeon

As will be discussed in Section 15.3, the complete genome sequence of the hyperthermophilic methanogen *Methanocaldococcus jannaschii* (Figure 13.8*a*) has been determined. The 1.66-megabase circular genome of *M. jannaschii* contains about 1700 genes, and from sequence analyses, genes encoding enzymes of methanogenesis and several other key cell functions have been identified. Interestingly, the majority of *M. jannaschii* genes encoding functions like central metabolic pathways and cell division are similar to those in *Bacteria*, while most of the genes encoding core molecular processes like transcription and translation more closely resemble those of eukaryotes. These findings support the evolutionary tree of life that shows the domain *Archaea* positioned *between* the domains *Bacteria* and *Eukarya* (⌁ Section 11.7 and Figure 11.13). However, sequence analyses also showed that nearly 50% of *M. jannaschii* genes have no counterparts in known genes from *Bacteria or Eukarya*, suggesting that there may be many new cellular functions encoded in archaeal DNA that have yet to be discovered.

✓ 13.4 Concept Check

Methanogenic *Archaea* are strictly anaerobic prokaryotes whose metabolism is tied to the production of methane (CH_4).

✓ What did the Volta experiment demonstrate?

✓ What are the major substrates for methanogenesis?

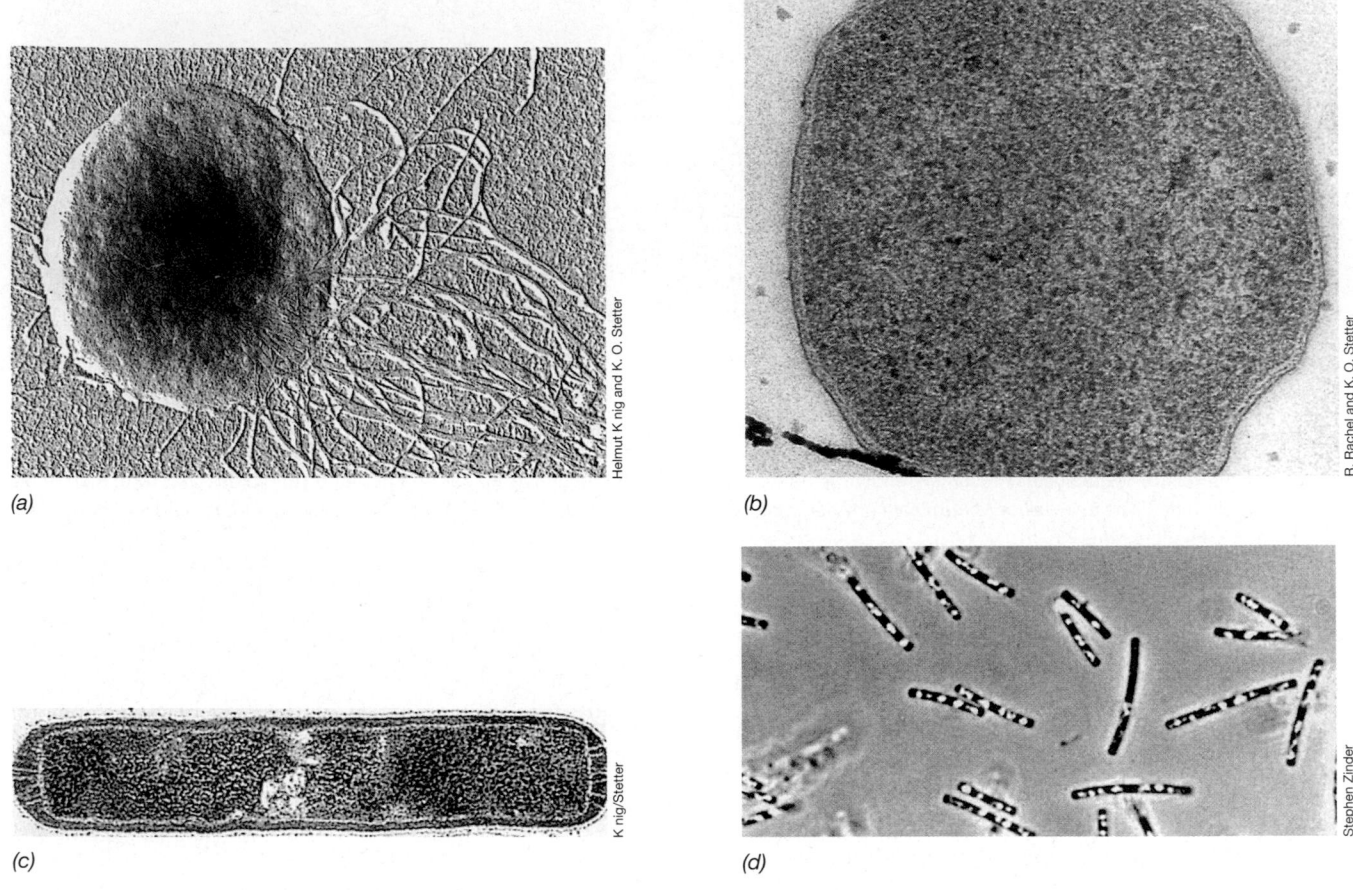

Figure 13.8 Hyperthermophilic and thermophilic methanogens. (a) *Methanocaldococcus jannaschii* (temperature optimum, 85°C), shadowed preparation electron micrograph. A cell is about 1 μm in diameter. (b) *Methanotorris igneus* (temperature optimum, 88°C), thin section. A cell is about 1 μm in diameter. (c) *Methanothermus fervidus* (temperature optimum, 88°C), thin-sectioned electron micrograph. A cell is about 0.4 μm in diameter. (d) *Methanosaeta thermophila* (temperature optimum, 60°C), phase contrast micrograph. A cell is about 1 μm in diameter. The refractile bodies inside the cells are gas vesicles.

13.5 Thermoplasmatales: *Thermoplasma, Ferroplasma,* and *Picrophilus*

Key Genera

Thermoplasma
Picrophilus

A phylogenetically distinct line of *Archaea* contains three thermophilic and extremely acidophilic prokaryotes, *Thermoplasma, Ferroplasma,* and *Picrophilus* (Figure 13.1). These organisms are among the most acidophilic of all known prokaryotes and, in the case of *Picrophilus,* even capable of growth below pH 0! They also form their own *order* of prokaryotes within the Euryarchaeota, the Thermoplasmatales. We begin with a description of the mycoplasma-like organism *Thermoplasma.*

Thermoplasma and *Ferroplasma*: Cell Wall-less Archaeons

Thermoplasma (Figure 13.9*a*) is a chemoorganotroph and grows optimally at 55°C and pH 2 in complex media; two species have been described, *T. acidophilum* and *T. volcanium*. Species of *Thermoplasma* are facultative aerobes, growing either aerobically or anaerobically by sulfur respiration. From a morphological point of view *Thermoplasma* resembles the mycoplasmas (Section 12.22) in that it lacks a cell wall. Most strains of *Thermoplasma* have been obtained from self-heating coal refuse piles (Figure 13.10). Coal refuse contains coal fragments, pyrite, and other organic materials extracted from coal, and when dumped into piles in coalmining operations, tends to self-heat by spontaneous combustion (Figure 13.10). This sets the stage for growth of *Thermoplasma,* which apparently metabolizes organic compounds leached from the hot coal refuse. A

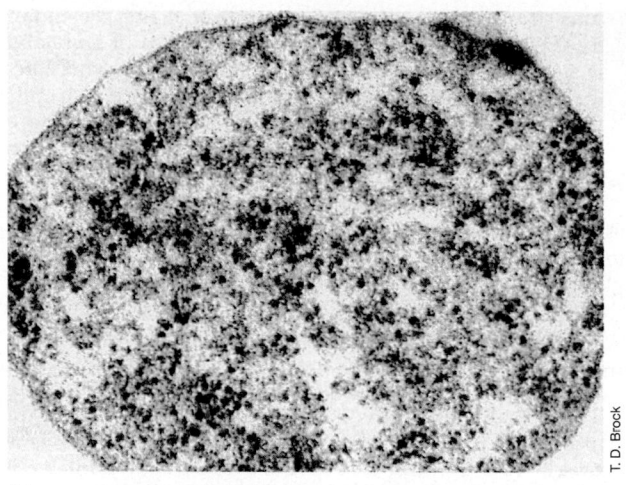

(a)

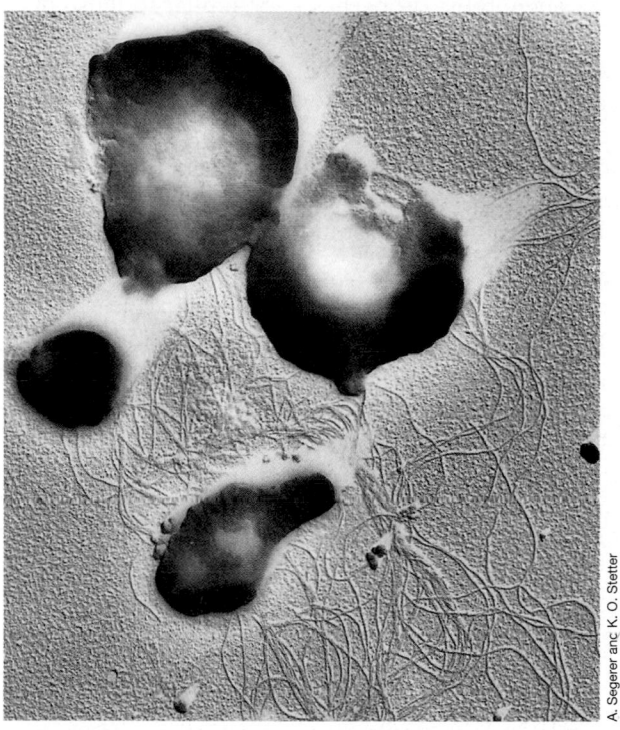

(b)

Figure 13.9 *Thermoplasma* species. (a) *Thermoplasma acidophilum,* an acidophilic, thermophilic mycoplasma-like archaeon. Electron micrograph of a thin section. The diameter of cells is highly variable from 0.2 to 5 μm. The cell shown is about 1 μm in diameter. (b) Shadowed preparation of cells of *Thermoplasma volcanium* isolated from hot springs. Cells are 1–2 μm in diameter. Notice abundant flagella.

second species of *Thermoplasma, T. volcanium,* has been isolated from solfatara fields throughout the world and is highly motile by multiple flagella (Figure 13.9*b*).

To survive the osmotic stresses of life without a cell wall and to withstand the dual environmental extremes of low pH and high temperature, *Thermoplasma* has evolved a cell membrane of chemically unique structure. The membrane contains a lipopolysaccharide-like material

(a)

(b)

Figure 13.10 A typical self-heating coal refuse pile, habitat of *Thermoplasma.* (a) Spontaneous heat production can ignite nearby vegetation. (b) Photo of a large hot refuse pile.

(referred to as *lipoglycan* in mycoplasmas) (∞ Section 12.22) consisting of a *tetraether* lipid with mannose and glucose units (Figure 13.11). This molecule constitutes a major fraction of the total lipid composition of *Thermoplasma.* The membrane also contains glycoproteins but not sterols. Together these and other molecules render the *Thermoplasma* membrane stable to hot acid conditions.

The genome of *Thermoplasma* is of interest. Like other mycoplasmas (∞ Sections 12.22 and 15.3), *Thermoplasma* contains a small genome (1.5 mB). In ad-

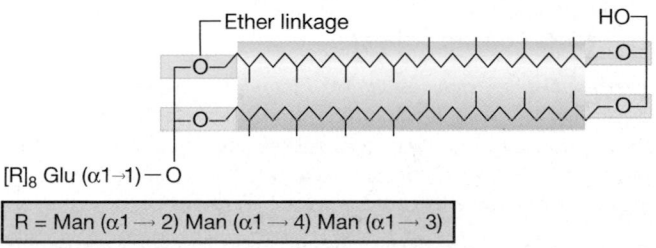

$[R]_8$ Glu $(\alpha 1 \rightarrow 1) - O$

R = Man $(\alpha 1 \rightarrow 2)$ Man $(\alpha 1 \rightarrow 4)$ Man $(\alpha 1 \rightarrow 3)$

Figure 13.11 Structure of the tetraether lipoglycan of *Thermoplasma acidophilum.* Glu, Glucose; Man, mannose. Note the ether linkages (shown in green) (∞ compare with Figure 4.20*b*).

dition, the DNA is complexed with a highly basic DNA binding protein that organizes the DNA into globular particles resembling the nucleosomes of eukaryotic cells (∞ Section 7.3 discusses the arrangement of DNA in eukaryotes). This protein is homologous to the basic histone proteins of eukaryotic cells. Similar DNA-binding histonelike proteins have been found in several hyperthermophilic Euryarchaeota (see Section 13.11).

Ferroplasma is a chemolithotrophic relative of *Thermoplasma*. Like *Themoplasma*, it lacks a cell wall and has an acidophilic lifestyle, but is not a thermophile, growing optimally at 35°C. *Ferroplasma* (∞ Figure 19.38) oxidizes Fe^{2+} to Fe^{3+} to obtain energy (this reaction generates acid; ∞ Section 19.14 and see legend to Figure 13.15) and uses CO_2 as its carbon source (autotrophy). *Ferroplasma* grows in mine tailings containing pyrite (FeS), its energy source. The extreme acidophily of *Ferroplasma* allows it to drive down the pH of its habitat to extremely low values. It is believed that after moderate acidity is generated from Fe^{2+} oxidation by organisms such as *Thiobacillus ferrooxidans* and *Leptospirillum ferrooxidans*, *Ferroplasma* becomes active and subsequently generates the very low pH values typical of acid mine drainage (∞ Figure 19.38 and Sections 19.14 and 19.15).

Picrophilus

A phylogenetic relative of *Thermoplasma* and *Ferroplasma* is *Picrophilus*. Although *Thermoplasma* and *Ferroplasma* are extreme acidophiles, *Picrophilus* is even more so, growing optimally at pH 0.7 and capable of growth to as low as pH −0.06! *Picrophilus* differs from *Thermoplasma* and *Ferroplasma* in other ways as well, including having a cell wall (an S-layer, ∞ Section 4.13) and has a much lower DNA GC base ratio. Two species of *Picrophilus* have been isolated from acidic Japanese solfataras and like *Thermoplasma*, both grow heterotrophically on complex media. The physiology of *Picrophilus* is clearly of interest as a model for acid tolerance. Studies of its cytoplasmic membrane suggest an unusual arrangement of lipids that form an extremely acid impermeable membrane at optimal pH values. By contrast, at only moderately acidic pH such as pH 4, the membranes of cells of *Picrophilus* quickly become leaky and literally disintegrate, clearly indicating that this organism has evolved to survive only in highly acidic habitats.

✓ 13.5 Concept Check

Thermoplasma, *Ferroplasma* and *Picrophilus* are extremely acidophilic thermophiles that form their own phylogenetic family of *Archaea* inhabiting coal refuse piles and highly acidic solfataras. Cells of *Thermoplasma* and *Ferroplasma* lack cell walls and thus resemble the mycoplasmas in this regard.

- ✓ In what ways are *Thermoplasma* and *Picrophilus* similar? In what ways do they differ?
- ✓ How does *Thermoplasma* strengthen its cell membrane to survive life in the absence of a cell wall?
- ✓ How does *Ferroplasma* obtain energy for growth?

13.6 Hyperthermophilic Euryarchaeota: Thermococcales and *Methanopyrus*

Key Genera

Thermococcus
Pyrococcus
Methanopyrus

A few euryarchaeotes thrive in thermal environments and some are extremely thermophilic; those with growth temperature optima over 80°C are called *hyperthermophiles*. We consider here three hyperthermophilic euryarchaeotes that branch on the archael tree (Figure 13.1) very near to the root. Two of these organisms, *Thermococcus* and *Pyrococcus*, show phenotypic properties very similar to those of hyperthermophilic crenarchaeotes to be discussed in Sections 13.8–13.10 and form a distinct order: the Thermococcales (see Table 13.9). The other, *Methanopyrus*, is a methanogen and closely resembles the methanogens (see Section 13.4 and Table 13.5) in its basic physiology but is unusual in its extreme requirement for heat and phylogenetic position, distinct from that of other methanogens (Figure 13.1).

Thermococcus and Pyrococcus

Thermococcus is a spherical hyperthermophilic euryarchaeote indigenous to anoxic thermal waters in various locations throughout the world. The spherical cells contain a tuft of polar flagella and are thus highly motile (Figure 13.12*a*). *Thermococcus* is an obligately anaerobic chemoorganotroph that grows on proteins and other complex organic mixtures (including some sugars) with S^0 as electron acceptor at temperatures from 70–95°C.

Pyrococcus is morphologically similar to *Thermococcus* (Figure 13.12*b*). *Pyrococcus* (the Latin derivation literally means "fireball") differs from *Thermococcus* primarily by

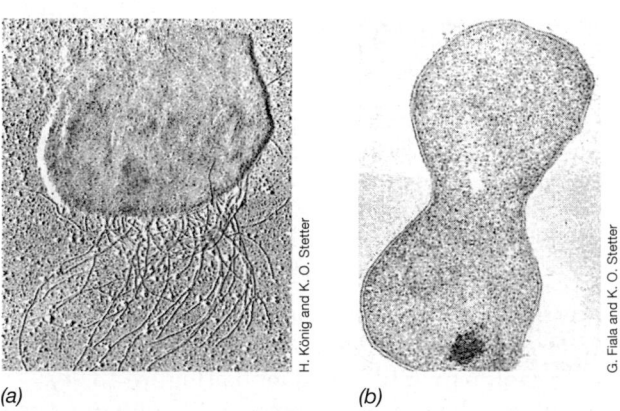

(a)   *(b)*

Figure 13.12 Spherical hyperthermophilic *Archaea* from submarine volcanic areas. (a) *Thermococcus celer*. Electron micrograph of shadowed cells (note tuft of flagella). (b) Dividing cell of *Pyrococcus furiosus*. Electron micrograph of thin section. Cells of both organisms are about 0.8 μm in diameter.

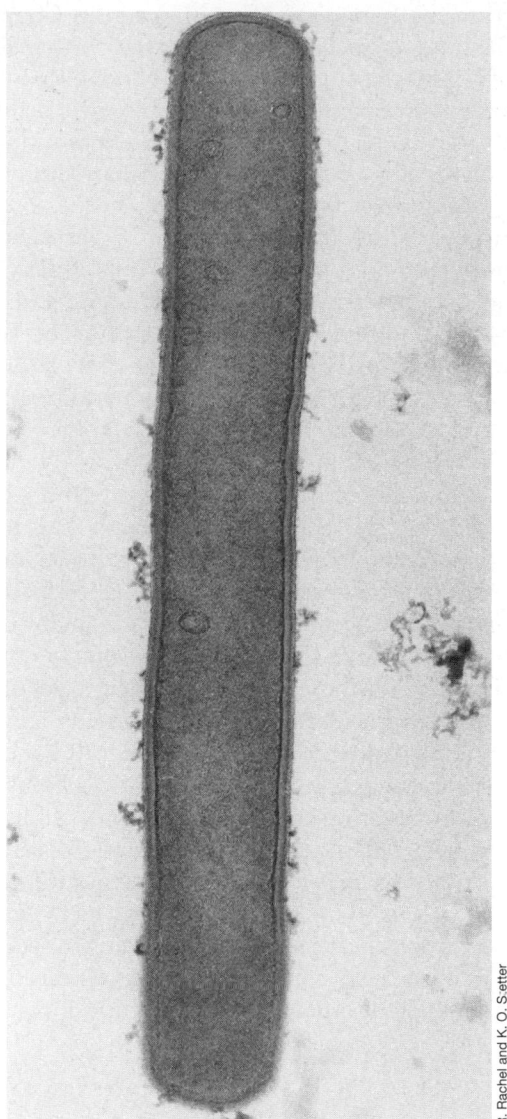

(a)

(b)

Figure 13.13 *Methanopyrus. Methanopyrus* grows optimally at 100°C and can make CH_4 only from $CO_2 + H_2$. (a) Electron micrograph of a cell of *Methanopyrus kandleri*, the most thermophilic of all known methanogens (upper temperature limit, 110°C). This cell measures 0.5×8 μm. (b) Structure of the novel lipid of *M. kandleri*. This is the normal ether-linked lipid of the *Archaea* (∞ Section 4.5) with the exception that the side chains are an *unsaturated* form of phytanyl called *geranylgeraniol*. It is thought that this unusual lipid predated the appearance of saturated phytanyl lipids, and that *Methanopyrus* is therefore a descendant of a very ancient lineage of *Archaea*; the data of Figure 13.1 support this hypothesis.

its higher temperature requirements; *Pyrococcus* grows between 70 and 106°C with an optimum of 100°C. *Thermococcus* and *Pyrococcus* are also metabolically quite similar: Proteins, starch, or maltose are oxidized as electron donors and S^0 is the terminal acceptor and is reduced to H_2S. More properties of *Thermococcus* and *Pyrococcus* are described in Tables 13.8 and 13.9.

Methanopyrus

Methanopyrus is a rod-shaped hyperthermophilic methanogen (Figure 13.13) that has been isolated from sediments near submarine hydrothermal vents and from the walls of "black smoker" hydrothermal vent chimneys (∞ Section 19.8 for a discussion of thermal vents). *Methanopyrus* occupies a unique phylogenetic position on the tree of *Archaea* (Figure 13.1); it is one of the most ancient (least derived) of all known hyperthermophilic *Archaea* and shares phenotypic properties with both the hyperthermophiles (growth temperature maximum, 110°C) and the methanogens (it carries out the reaction of $4 H_2 + CO_2 \rightarrow CH_4 + 2 H_2O$).

Methanopyrus produces methane from $H_2 + CO_2$ only and grows rapidly (generation time less than 1 h) at its temperature optimum of 100°C. Unlike mesophilic methanogens, however, cells of *Methanopyrus* have large amounts of the glycolytic derivative, cyclic 2,3-diphosphoglycerate, dissolved in the cytoplasm. This compound, present at more than 1 *M* concentration in *Methanopyrus*, is thought to function as a thermostabilizing agent to prevent denaturation of enzymes and DNA inside the cell (see Section 13.11). In addition, *Methanopyrus* contains a type of membrane lipid found in no other known organism; this ether lipid is an unsaturated form of the otherwise saturated dibiphytanyl tetraethers found in other hyperthermophiles (Figure 13.13*b*; ∞ Section 4.5 and Figure 4.20).

The discovery of *Methanopyrus* may explain the origin of hydrocarbon-like materials in hot oceanic sediments previously thought to be too hot to support biogenic methanogenesis. In addition, at the depth at which *Methanopyrus* was found, approximately 2000 m, water remains liquid at temperatures up to 350°C, suggesting that other hyperthermophilic methanogens may exist capable of growth at 110°C or at perhaps even higher temperatures.

13.7 Hyperthermophilic Euryarchaeota: The Archaeoglobales

Key Genera

Archaeoglobus
Ferroglobus

We will see later that a number of hyperthermophilic Crenarchaeota carry out anaerobic respirations in which S^0 is used as an electron acceptor, being reduced to H_2S (see Table 13.8). Curiously, none of these organisms are *sulfate-reducers*, that is, capable of reducing SO_4^{2-} to H_2S. However, one hyperthermophilic euryarchaeote, *Archaeoglobus*, is a true sulfate reducer and forms a phylogenetically distinct lineage within the Euryarchaeota (Figure 13.1).

Archaeoglobus and Its Genome

Archaeoglobus was isolated from hot marine sediments near hydrothermal vents and couples the oxidation of H_2, lactate, pyruvate, glucose, or complex organic compounds to the reduction of sulfate to sulfide. Cells of *Archaeoglobus* are irregular cocci (Figure 13.14*a*) and cultures grow optimally at 83°C. However, one of the most remarkable features of *Archaeoglobus* is its rela-

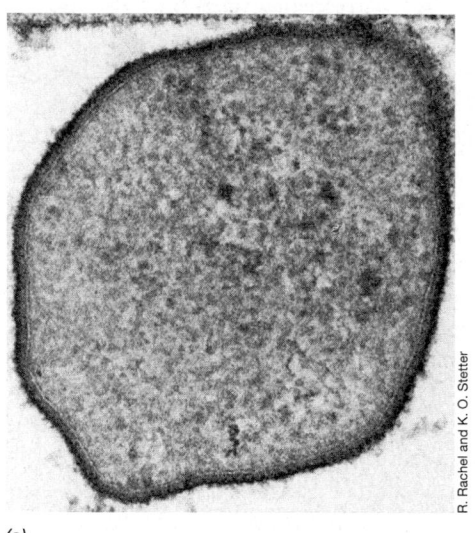

(a)

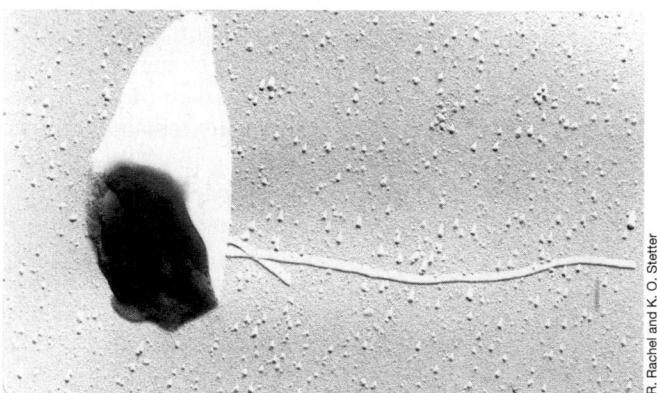

(b)

Figure 13.14 Archaeoglobales. (a) Transmission electron micrograph of the sulfate-reducing hyperthermophile *Archaeoglobus fulgidis*. The cell measures 0.7 μm in diameter. (b) Freeze-etched electron micrograph of *Ferroglobus placidus*, a ferrous iron-oxidizing, nitrate-reducing hyperthermophile. Cells measure about 0.8 μm in diameter.

tionship to methanogens. We will learn in Chapter 17 about the unique biochemistry associated with methanogenesis. Briefly, this involves the use of novel coenzymes like Factor-420, coenzyme M, and many others; thus far, these novel biomolecules have only been found in cells of methanogens (∞ Section 17.17). Surprisingly, however, *Archaeoglobus* contains traces of these coenzymes and cultures of this organism actually produce small amounts of methane during growth. Thus, *Archaeoglobus*, which also shows a rather close phylogenetic relationship to methanogens (Figure 13.1), may represent an evolutionary transitional type of organism among the *Archaea*, one that bridged the energy-generating processes of sulfur reduction and methanogenesis.

More information on methanogenesis by *Archaeoglobus* has emerged from the complete genome sequence of this organism. The sequence has revealed that although *Archaeoglobus* can make methane, it lacks genes for a key enzyme of methanogenesis found in methanogens, *methyl-CoM reductase* (∞ Section 17.17). Thus, the small amounts of methane that are made by *Archaeoglobus* do not come from the activity of this enzyme, and their exact origin is unclear. The genome of *Archaeoglobus* contains about 2400 genes and, interestingly, shares a number of genes with species of methanogens (see Section 13.4). However, the *Archaeoglobus* genome sequence also reveals a number of unique genes. This suggests that genetic diversity in *Archaea* is quite extensive and that analysis of the genome of *Archaea* will likely mimic the findings of the genome projects of various *Bacteria*: Each new organism sequenced has about one quarter of its genes as unique genes, never before seen in other organisms (∞ Section 15.3).

Ferroglobus

Ferroglobus (Figure 13.14*b*) is related to *Archaeoglobus* but is not a sulfate-reducing bacterium. Instead, *Ferroglobus* is an iron-oxidizing chemolithotrophic autotroph, conserving energy from the oxidation of Fe^{2+} to Fe^{3+} coupled to the reduction of NO_3^- to NO_2^- plus NO (see Table 13.8). *Ferroglobus* can also use H_2 or H_2S as electron donors in its energy metabolism. *Ferroglobus* was isolated from a shallow marine hydrothermal vent and grows optimally at 85°C. *Ferroglobus* is interesting for several reasons, but especially so because of its ability to produce Fe^{3+} from Fe^{2+} under anoxic conditions. The mechanism for the formation of Fe^{3+} in ancient rocks, previously assumed to have been from the oxidation of Fe^{2+} by O_2 produced by cyanobacteria (∞ Section 11.1), is now being questioned with the discovery that anoxic routes to the production of Fe^{3+} exist from the activities of organisms like *Ferroglobus* and certain anoxygenic phototrophic bacteria (∞ Section 17.11).

✓ **13.6–13.7 Concept Check**

Hyperthermophilic euryarchaeota include the Thermococcales, Archaeoglobales, and *Methanopyrus*. All organisms in this group have growth temperature optima above 80°C.

✓ How does the metabolism of *Thermococcus* differ from that of *Methanopyrus*?

✓ Which of the hyperthermophilic Euryarchaeota is a true sulfate-reducer?

✓ How does the metabolism of *Ferroglobus* differ from that of *Archaeoglobus*?

III PHYLUM CRENARCHAEOTA

Crenarchaeotes are phylogenetically distinct from Euryarchaeotes and contain representatives that live at both ends of nature's temperature extremes: boiling water and freezing water. Most cultured crenarchaeotes are hyperthermophiles (growth temperature optima above 80°C) and some actually have optima *above* the boiling point of water. We start with an overview of the habitats and energy metabolism of crenarchaeotes and then describe the properties of some key genera.

13.8 Habitats and Energy Metabolism of Crenarchaeotes

A summary of the habitats of Crenarchaeota is shown in Table 13.7. They include very hot and very cold environments. Most hyperthermophilic *Archaea* have been isolated from geothermally heated soils or waters containing elemental sulfur and sulfides (Figure 13.15) and most species metabolize sulfur in one way or another. In terrestrial environments, sulfur-rich springs, boiling mud, and soils may have temperatures up to 100°C and are mildly to extremely acidic owing to production of sulfuric acid (H_2SO_4) from the biological oxidation of H_2S and S^0 (∞ Sections 17.10 and 19.13). Such hot, sulfur-rich environments are called **solfataras**, and are found throughout the world (Figure 13.15); extensive solfataras are present in Italy, Iceland, New Zealand, and Yellowstone National Park in Wyoming (USA). Depending on the surrounding geology, solfataras can be slightly alkaline to mildly acidic, pH 5–8, or extremely acidic, with pH values below 1 not uncommon. Hyperthermophiles have been obtained from both types of environments, but the majority of these organisms inhabit neutral or mildly acidic habitats. In addition to these natural habitats, hyperthermophilic *Archaea* also thrive within artificial thermal habitats, in particular the boiling outflows of geothermal power plants.

Hyperthermophilic Crenarchaeota also thrive in undersea hot springs called *hydrothermal vents*. We discuss the geology and microbiology of these habitats in Section 19.8; here it is only necessary to note that submarine waters can be and often are much hotter than surface waters because the water is under pressure. Indeed, all hyperthermophiles with growth temperature optima above 100°C have come from submarine sources. The latter include both shallow (2–10 meters) vents such as those off the coast of Vulcano, Italy, to deep (2500–4000 meters) vents near ocean spreading centers (∞ Section 19.8). The latter are the hottest habitats so far known to yield living organisms.

Cold-Dwelling Crenarchaeotes

By contrast with the hyperthermophiles, cold-dwelling crenarchaeotes (Table 13.7) have been identified from community sampling of ribosomal RNA genes from many nonthermal environments. Using fluorescent phylogenetic probes (∞ Section 11.6 and 18.4), crenarchaeotes have been found in marine waters worldwide (Figure 13.16b). In stark contrast to the hyperthermophiles, marine crenarchaeotes

---

TABLE 13.7 Habitats of Crenarchaeota

| Characteristic | Thermal area[a] | | Nonthermal area[b] |
	Terrestrial	Marine	
Locations	Solfataras (hot springs, fumaroles, mudpots, steam-heated soils); geothermal power plants; deep in Earth's crust	Submarine hot springs, hot sediments and vents ("black smokers"), deep oil reservoirs	Planktonic in oceans worldwide; near-shore and deep Antarctic waters; sea ice; symbionts of marine sponges
Temperature	Surface to 100°C; subsurface, above 100°C	Up to 400°C (smokers)	-2 to +4°C
Salinity/pH	Usually less than 1% NaCl; pH 0.5–9	Moderate, about 3% NaCl; pH 5–9	3–8% NaCl; pH 7–9
Gases and other nutrients	CO_2, CO, CH_4, H_2, H_2S, S^0, $S_2O_3^{2-}$, SO_4^{2-}, NH_4^+, N_2	Same as for terrestrial	CO_2, N_2, O_2; chemolithotrophic substrates such as NH_4^+

[a] See Figures 13.10 and 13.15 (∞ Figures 19.19–19.21).
[b] See Figure 13.16.

(a)

(b)

(c)

(d)

Figure 13.15 Habitats of hyperthermophilic *Archaea*. (a) A typical solfatara in Yellowstone National Park. Steam rich in hydrogen sulfide rises to the surface of the earth. Because of the heat and acidity, only prokaryotes are present. (b) Sulfur-rich hot spring, a habitat containing dense populations of *Sulfolobus*. The acidity in solfataras and sulfur springs comes from the oxidation of H_2S and S^0 to H_2SO_4 (sulfuric acid) by *Sulfolobus* and related prokaryotes. (c) A typical boiling spring of neutral pH in Yellowstone Park; Imperial Geyser. Many different species of hyperthermophilic *Archaea* may reside in such a habitat. (d) An acidic iron-rich geothermal spring, another *Sulfolobus* habitat. Here the oxidation of Fe^{2+} to Fe^{3+} causes acidic conditions [$Fe^{3+} + 3\,H_2O \rightarrow Fe(OH)_3 + 3\,H^+$].

(a)

(b)

Figure 13.16 Cold-dwelling Crenarchaeota. (a) Photo of the Antarctic peninsula taken from shipboard. The frigid waters that lie under the surface ice shown here are typical habitats for cold-dwelling crenarchaeotes. (b) Fluorescence photomicrograph of seawater treated with a phylogenetic stain composed of a green fluorescing dye attached to an oligonucleotide complementary to a signature sequence in the 16S rRNA of species of Crenarchaeota (⚬⚬ Section 11.6 for a description of the methods used here). Blue cells are stained with DAPI, a stain that stains all cells (⚬⚬ Section 18.3 and Figure 18.6). Cells stained green are thus cold-dwelling Crenarchaeota. See Section 19.6 and Figure 19.13 for a description of *Bacteria* and *Archaea* in open ocean waters.

thrive even in frigid waters, such as those of the Antarctic (Figure 13.16a). These organisms are planktonic (suspended freely or attached to suspended particles in the water column) and occur in significant numbers ($\sim 10^4$/ml) for waters that are both nutrient poor and very cold. Although their physiology is still a mystery, lipid analyses of marine crenarchaeotes filtered from seawater have shown that they contain ether-linked lipids, the hallmark of the *Archaea* (∞ Section 4.5). The ecology of marine crenarchaeotes is discussed in Section 19.6.

Energy Metabolism

With a few exceptions, hyperthermophilic Crenarchaeota are obligate anaerobes. Their energy-yielding metabolism can be either chemoorganotrophic or chemolithotrophic (or both, for example in *Sulfolobus*) and involves a wide diversity of different electron donors and acceptors (Table 13.8). Fermentation is rare and most bioenergetic strategies are either aerobic or anaerobic respirations (Table 13.8). Energy conservation during these respiratory processes occurs by the same general mechanism widespread in *Bacteria*: electron transfer within the cytoplasmic membrane leading to the formation of a proton motive force from which ATP is made by way of proton-translocating ATPases (∞ Section 5.12).

Many hyperthermophilic crenarchaeotes can grow chemolithotrophically under anoxic conditions with H_2 as electron donor and S^0 or NO_3^- as electron acceptor; a few can also oxidize H_2 aerobically (Table 13.8). H_2 respiration with ferric iron (Fe^{3+}) as electron acceptor also occurs in several hyperthermophiles (Table 13.8). Other chemo-

lithotrophic lifestyles include the oxidation of S^0 and Fe^{2+} aerobically or Fe^{2+} anaerobically with NO_3^- as acceptor (Table 13.8). Only one sulfate-reducing hyperthermophile is known (the euryarchaeote *Archaeoglobus*). The only bioenergetic option apparently ruled out is photosynthesis. The most thermophilic phototrophic organism known can grow up to about 70°C (∞ Table 6.1)—too cold for most hyperthermophiles (see Table 13.9)!

Armed with a basic understanding of the habitats and energy metabolism of hyperthermophilic crenarchaeotes, let us now look at representative organisms.

13.9 Hyperthermophiles from Terrestrial Volcanic Habitats: Sulfolobales and Thermoproteales

Key Genera

Sulfolobus
Acidianus
Thermoproteus

Volcanic habitats can have temperatures as high as 100°C and are thus suitable for hyperthermophilic *Archaea*. Two phylogenetically related organisms isolated from these environments include *Sulfolobus* and *Acidianus*. These genera form the heart of an order called the Sulfolobales (Table 13.9).

TABLE 13.8 Energy-yielding reactions of hyperthermophilic *Archaea*

Nutritional class	Energy-yielding reaction	Metabolic[a] type	Example
Chemoorganotrophic	Organic compound $+ S^0 \rightarrow H_2S + CO_2$	AnR	*Thermoproteus, Thermococcus, Desulfurococcus, Thermofilum, Pyrococcus*
	Organic compound $+ SO_4^{2-} \rightarrow H_2S + CO_2$	AnR	*Archaeoglobus*
	Organic compound $+ O_2 \rightarrow H_2O + CO_2$	AeR	*Sulfolobus*
	Organic compound $\rightarrow CO_2 + H_2 +$ fatty acids	F	*Staphylothermus, Pyrodictium*
	Pyruvate $\rightarrow CO_2 + H_2 +$ acetate	F	*Pyrococcus*
Chemolithotrophic	$H_2 + S^0 \rightarrow H_2S$	AnR	*Acidianus, Pyrodictium, Thermoproteus, Stygiolobus, Ignicoccus*
	$H_2 + NO_3^- \rightarrow NO_2^- + H_2O$ (NO_2^- is reduced to N_2 by some species)	AnR	*Pyrobaculum*
	$4 H_2 + NO_3^- + H^+ \rightarrow NH_4^+ + 2 H_2O + OH^-$	AnR	*Pyrolobus*
	$H_2 + 2 Fe^{3+} \rightarrow 2 Fe^{2+} + 2 H^+$	AnR	*Pyrobaculum, Pyrodictium, Archaeoglobus*
	$2 H_2 + O_2 \rightarrow 2 H_2O$	AeR	*Acidianus, Sulfolobus, Pyrobaculum*
	$2 S^0 + 3 O_2 + 2 H_2O \rightarrow 2 H_2SO_4$	AeR	*Sulfolobus, Acidianus*
	$2 FeS_2 + 7 O_2 + 2 H_2O \rightarrow 2 FeSO_4 + 2 H_2SO_4$	AeR	*Sulfolobus, Acidianus, Metallosphaera*
	$2 FeCO_3 + NO_3^- + 6 H_2O \rightarrow 2 Fe(OH)_3 + NO_2^- + 2 HCO_3^- + 2 H^+ + H_2O$	AnR	*Ferroglobus*
	$4 H_2 + SO_4^{2-} + 2 H^+ \rightarrow 4 H_2O + H_2S$	AnR	*Archaeoglobus*
	$4 H_2 + CO_2 \rightarrow CH_4 + 2 H_2O$	AnR	*Methanopyrus, Methanocaldococcus, Methanothermus*

[a]AnR, anaerobic respiration; AeR, aerobic respiration; F, fermentation

TABLE 13.9   Properties of hyperthermophilic Crenarchaeota

Group/Genus[a]	Morphology	Number of species	Relationship to O_2[b]	DNA (mol% GC)	Temperature(°C) Minimum	Temperature(°C) Optimum	Temperature(°C) Maximum	Optimum pH
Sulfolobales								
Sulfolobus	Lobed coccus	6	Ae	37	55	75	87	2–3
Acidianus	Coccus	3	Fac	31	60	88	95	2
Metallosphaera	Coccus	2	Ae	45	50	75	80	2
Stygiolobus	Lobed coccus	1	An	38	57	80	89	3
Aeropyrum	Coccus	1	Ae	56	70	95	100	7
Stetteria	Coccus	1	An	65	68	95	102	6
Sulfophobococcus	Disc-shaped	1	An	54–56	70	85	95	7.5
Thermosphaera	Coccus	1	An	46	67	85	90	7
Thermoproteales								
Thermoproteus	Rod	2	An	56	60	88	96	6
Thermophilum	Rod	1	An	57	70	88	95	5.5
Pyrobaculum	Rod	3	Fac	46	74	100	102	6
Caldivirga	Rod	1	An	43	60	85	92	4
Thermocladium	Rods	1	An	52	60	75	80	4.2
Desulfurococcales								
Desulfurococcus	Coccus	2	An	51	70	85	95	6
Staphylothermus	Cocci in clusters	1	An	35	65	92	98	6–7
Pyrodictium	Disc-shaped with filaments	3	An	62	82	105	110	6
Pyrolobus	Lobed coccus	1	Fac	53	90	106	113	5.5
Thermodiscus	Disc-shaped	1	An	49	75	90	98	5.5
Ignicoccus	Irregular coccus	2	An	35	65	90	103	5
Hyperthermus	Irregular coccus	1	An	56	75	102	108	7
Sulfurisphaera	Coccus	1	Fac	33	63	84	92	2
Sulfurococcus	Coccus	2	Ae	43–46	40	75	85	2.5
Archaeoglobales[c]								
Archaeoglobus	Coccus	3	An	46	64	83	95	7
Ferroglobus	Irregular coccus	1	An	43	65	85	95	7
Thermococcales[c]								
Thermococcus	Coccus	14	An	38–57	70	88	98	6–7
Pyrococcus	Coccus	4	An	38	70	100	106	6–8

[a] The group names ending in "ales" are order names (⬭⬭ Section 11.10).

[b] Ae, aerobe; An, anaerobe; Fac, facultative.

[c] Phylogenetically, genera in this order of hyperthermophiles are members of the Euryarchaeota (see Sections 13.6 and 13.7).

Sulfolobales

Sulfolobus grows in sulfur-rich hot acid springs (Figure 13.15b) at temperatures up to 90°C and at pH values of 1–5.* *Sulfolobus* (Figure 13.17a) is an aerobic chemolithotroph that oxidizes H_2S or S^0 to H_2SO_4 and fixes CO_2 as sole carbon source. *Sulfolobus* can also grow chemoorganotrophically. Cells of *Sulfolobus* are more or less spherical but form distinct lobes (Fig-

ure 13.17a). Cells adhere tightly to sulfur crystals where they can be seen microscopically using fluorescent dyes (⬭⬭ Figure 17.26b). Besides the aerobic respiration of sulfur or organic compounds, *Sulfolobus* can also oxidize Fe^{2+} to Fe^{3+}, and this has been applied in the high-temperature leaching of iron and copper ores (⬭⬭ Section 19.16).

A facultative aerobe resembling *Sulfolobus* also lives in acidic solfataric springs. This organism, named *Acidianus* (Figure 13.17b), differs from *Sulfolobus* most clearly by its ability to grow anaerobically. Remarkably, *Acidianus* is able to use S^0 in both its aerobic and anaerobic metabolism. Under aerobic conditions the organism uses S^0 as an electron *donor*, oxidizing S^0 to H_2SO_4. Anaerobically, *Acidianus* uses S^0 as an electron *acceptor* (with H_2 as electron donor) forming H_2S as the reduced

Historical note: Sulfolobus was first discovered by Thomas Brock and colleagues in 1970 and formally described in 1972. The discovery of *Sulfolobus*, along with the previously isolated *Thermus aquaticus* (source of the extremely thermostable *Taq* DNA polymerase, ⬭⬭ Section 10.17), is generally credited with launching the field of hyperthermophilic microbiology. Thomas Brock was the senior author of the first seven editions of this book. In the 1980s to the present, Karl Stetter and colleagues in Germany have greatly expanded the field of hyperthermophilic microbiology with the discovery of many new genera and species.

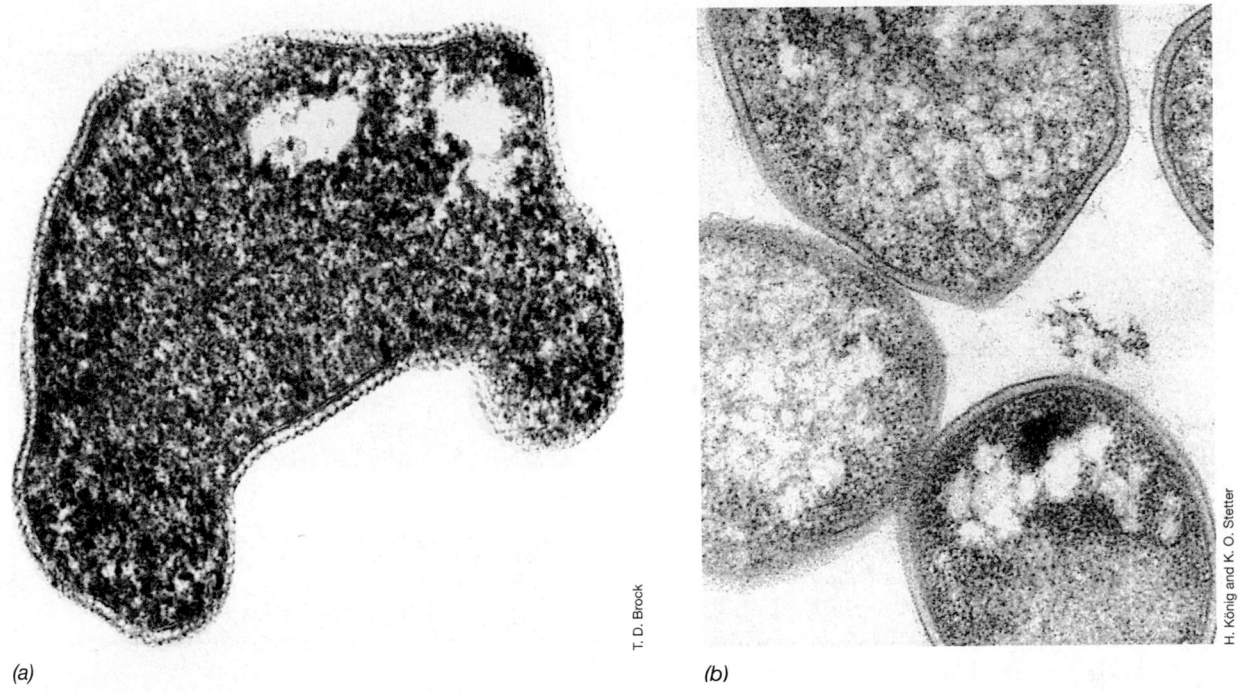

(a)

(b)

T. D. Brock

H. König and K. O. Stetter

Figure 13.17 Acidophilic hyperthermophilic *Archaea*, the Sulfolobales. (a) *Sulfolobus acidocaldarius*. Electron micrograph of a thin section. (b) *Acidianus infernus*. Electron micrograph of a thin section. Cells of both organisms vary from 0.8 to 2 μm in diameter.

product. Thus, the metabolic fate of S^0 in cultures of *Acidianus* depends on the presence or absence of O_2. Like *Sulfolobus*, *Acidianus* is roughly spherical in shape but is not as lobed (Figure 13.17b). It grows at temperatures from 65°C up to a maximum of 95°C, with an optimum of about 90°C.

Thermoproteales

The genera *Thermoproteus* and *Thermofilum* consist of *rod-shaped* cells that inhabit neutral or slightly acidic hot springs. Cells of *Thermoproteus* are stiff rods about 0.5 μm in diameter and are highly variable in length, ranging

from short cells of 1–2 μm up to filaments 70–80 μm long (Figure 13.18a). Filaments of *Thermofilum* are thinner, some 0.17–0.35 μm wide with filament lengths ranging up to 100 μm (Figure 13.18b). Both *Thermoproteus* and *Thermofilum* are strict anaerobes that carry out a S^0-based anaerobic respiration (Table 13.8). Unlike most hyperthermophiles, the oxygen sensitivity of *Thermoproteus* and *Thermofilum* is extreme, comparable to that of the methanogens (see Section 13.4); thus, strict precautions must be taken in their culture. Most *Thermoproteus* isolates can grow chemolithotrophically on H_2 or chemoorganotrophically on complex carbon substrates such as

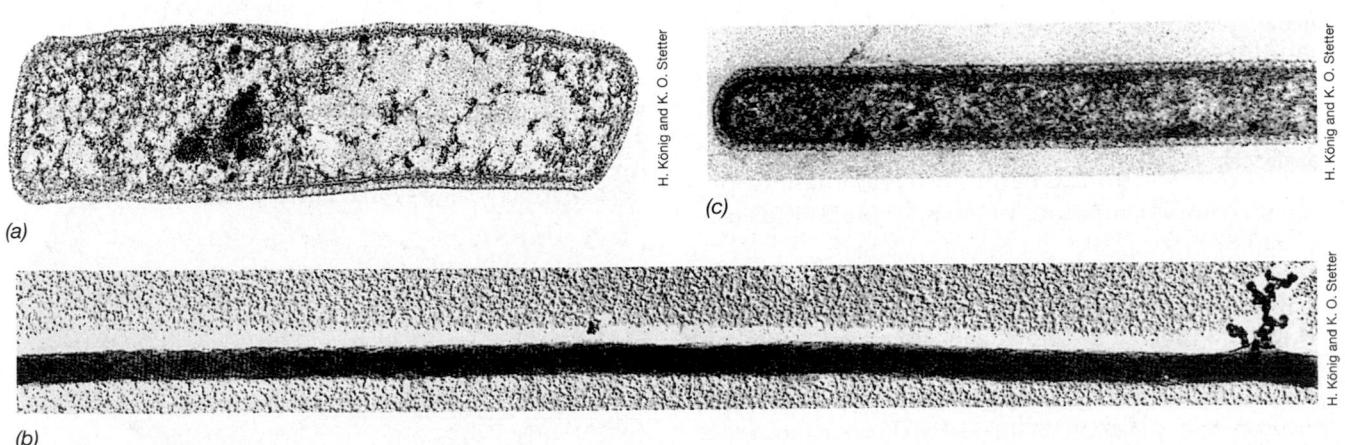

(a)

(b)

(c)

H. König and K. O. Stetter

Figure 13.18 Rod-shaped hyperthermophilic *Archaea*: Thermoproteales. (a) *Thermoproteus neutrophilus*. Electron micrograph of a thin section. A cell is about 0.5 μm in diameter. (b) *Thermofilum librum*. A cell is only about 0.25 μm in diameter. Electron micrograph of shadowed cells. (c) *Thermofilum librum*. Electron micrograph of a thin section.

yeast extract, small peptides, starch, glucose, ethanol, malate, fumarate, or formate (see Table 13.8). *Thermoproteus* and *Thermofilum* have similar GC base ratios (56–58% GC) but are phylogenetically distinct by nucleic acid hybridization analyses (∞ Section 11.9).

13.10 Hyperthermophiles from Submarine Volcanic Habitats: Desulfurococcales

Key Genera

Pyrodictium
Pyrolobus
Ignicoccus
Staphylothermus

We now turn to the microbiology of *submarine* volcanic habitats, homes to the most thermophilic of all known *Archaea*. These habitats include both shallow water thermal springs and deep-sea hydrothermal vents. We discuss the geology of these fascinating microbial habitats in Section 19.8. The organisms to be described here make up an order-level group of *Archaea* called the Desulfurococcales (Table 13.9).

Pyrodictium and *Pyrolobus*

Pyrodictium, *Pyrolobus*, and *Pyrobaculum* are interesting examples of prokaryotes whose growth temperature optimum lies at or even above 100°C; the optimum for *Pyrodictium* is 105°C and for *Pyrolobus* is 106°C. Cells of *Pyrodictium* are irregularly disc-shaped and grow in culture as a mycelium-like layer attached to crystals of elemental sulfur; the cell mass consists of a network of fibers to which individual cells are attached (Figure 13.19*a,b*). The fibers are hollow and consist of protein arranged in a fashion similar to that of bacterial flagella (∞ Section 4.10). The filaments do not function in motility but instead as organs of attachment (see Figure 13.22), and the cell walls of *Pyrodictium* are composed of glycoprotein. Physiologically, *Pyrodictium* is a strict anaerobe that grows chemolithotrophically on H_2 with S^0 as electron acceptor or chemoorganotrophically on complex mixtures of organic compounds (see Table 13.8).

Pyrolobus fumarii (Figure 13.19*c*) currently holds the record for the most thermophilic of all known

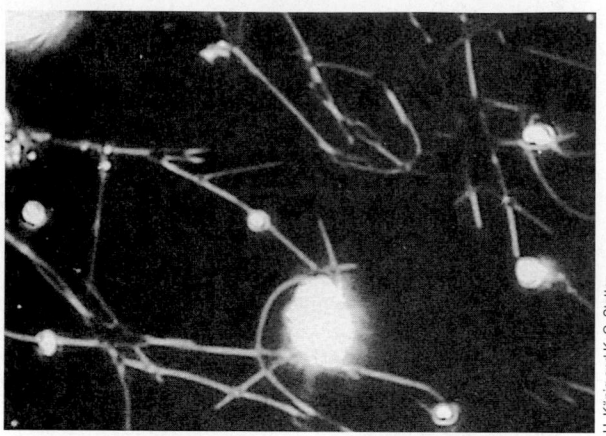

(a)

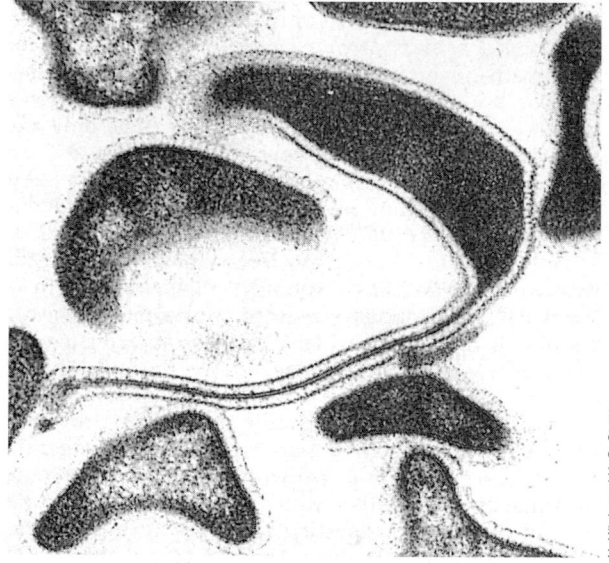

(b)

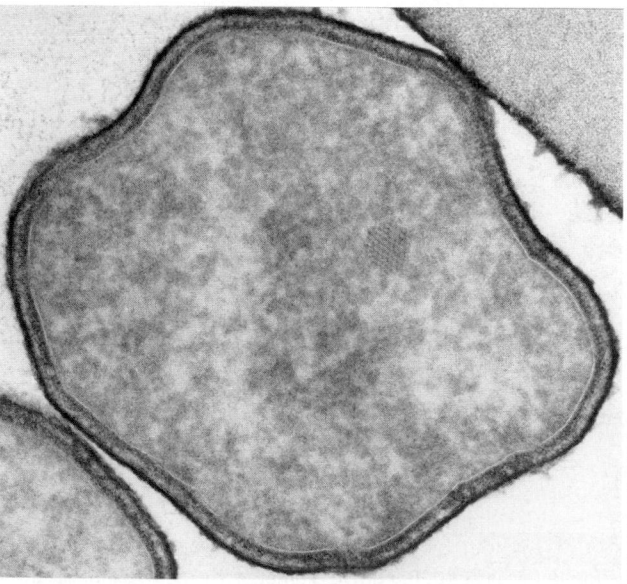

(c)

Figure 13.19 Desulfurococcales with growth temperature optima >100°C. (a) *Pyrodictium occultum* (growth temperature optimum, 105°C), dark-field micrograph. (b) Thin section electron micrograph of *P. occultum*. Cells are highly variable in diameter from 0.3 to 2.5 μm. (c) Thin section of a cell of *Pyrolobus fumarii*, the most thermophilic of all known prokaryotes (growth temperature optimum, 106°C); a single cell is about 1.4 μm in diameter.

organisms—its growth temperature maximum is 113°C (Table 13.9) (notably, *P. fumarii* is unable to grow below 90°C because it is too cold!). *Pyrolobus* lives in the walls of "black smoker" hydrothermal vent chimneys (∞ Section 19.8 and Figure 19.20) where, because of its autotrophic abilities, it is possibly a major source of primary productivity in this otherwise inorganic environment. *Pyrolobus* cells are coccoid-shaped (Figure 13.19*c*) and the cell wall is composed of protein. The organism is an obligate H_2 chemolithotroph, growing at the expense of H_2 oxidation coupled to the reduction of NO_3^- (to NH_4^+), $S_2O_3^{2-}$ (to H_2S), or very low concentrations of O_2 (to H_2O). Besides its extremely thermophilic nature, *Pyrolobus* is also resistant to temperatures substantially above its growth temperature maximum. For example, cultures of *P. fumarii* survive autoclaving (121°C) for 1 h, a condition that even bacterial endospores (∞ Section 4.15) cannot withstand. Considering the hydrothermal vent habitat of this organism, where periodic shifts in the temperature may be a regular occurrence, high heat resistance would be of great survival value.

Desulfurococcus, Ignicoccus, and *Pyrobaculum*

Other notable members of this group of hyperthermophiles include *Desulfurococcus,* the genus for which the order is named (Figure 13.20*b*). *Desulfurococcus* is a strictly anaerobic S^0-reducing bacterium like *Pyrodictium* but differs from this organism in that it is much less thermophilic, growing optimally at about 85°C. *Pyrobaculum* (Figure 13.20*a*) is a physiologically unique hyperthermophilic archaeon. Some species are capable of both aerobic respiration and anaerobic respiration with NO_3^-, Fe^{3+}, or S^0 as electron acceptors and H_2 as electron donor (that is, chemolithotrophic and autotrophic growth), while other species grow anaerobically on organic electron donors, reducing S^0 to H_2S. The growth temperature optimum of *Pyrobaculum* is right at the boiling point of water at sea level (100°C), and species of this organism have been isolated from both terrestrial hot springs as well as from hydrothermal vents.

Ignicoccus (Figure 13.20*c*) is a novel hyperthermophile that may contain a true "outer membrane", similar to the outer membrane of *Bacteria* (∞ Section 4.9). The outer membrane of *Ignicoccus* is unusual, however, in that it is formed at some distance from the cytoplasm of the cell; this arrangement yields an unusually large periplasm (Figure 13.20*c*). Indeed, the volume of the *periplasm* of *Ignicoccus* is some 2–3 times that of its *cytoplasm,* in contrast to that of gramnegative *Bacteria,* where periplasmic volume is about 25% that of the cytoplasm. The periplasm of *Ignicoccus* also contains membrane-bound vesicles (Figure 13.20*c*) that may function in exporting substances outside the

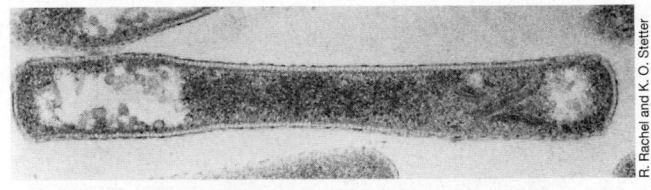

(a)

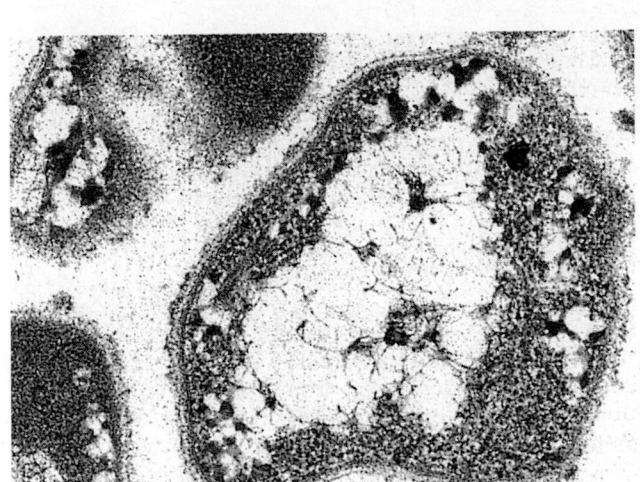

(b)

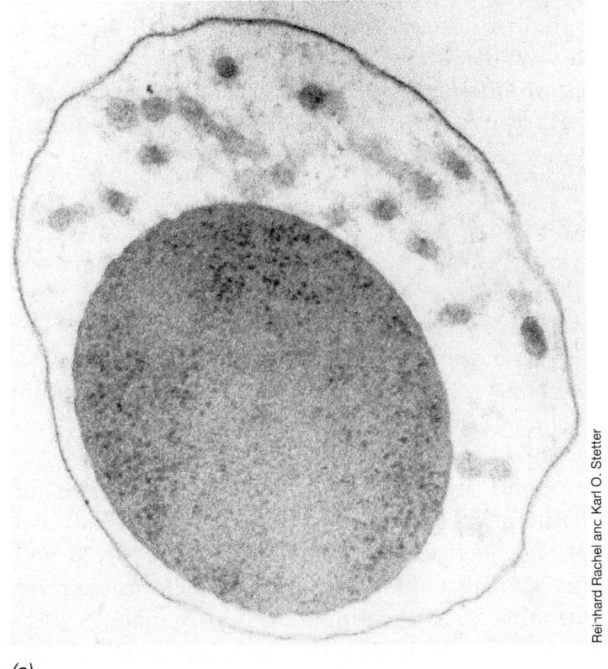

(c)

Figure 13.20 Examples of Desulfurococcales with growth temperature optima below the boiling point. (a) Transmission electron micrograph of a thin section of *Pyrobaculum aerophilum*; a cell measures $0.5 \times 3.5\,\mu m$. (b) Thin section of a cell of *Desulfurococcus saccharovorans*; a cell is 0.7 μm in diameter. (c) Thin section of a cell of *Ignicoccus islandicus.* The cell proper is surrounded by an extremely large periplasm (∞ Section 4.9). The cell itself measures about 1 μm in diameter but cell plus periplasm measures 1.4 μm.

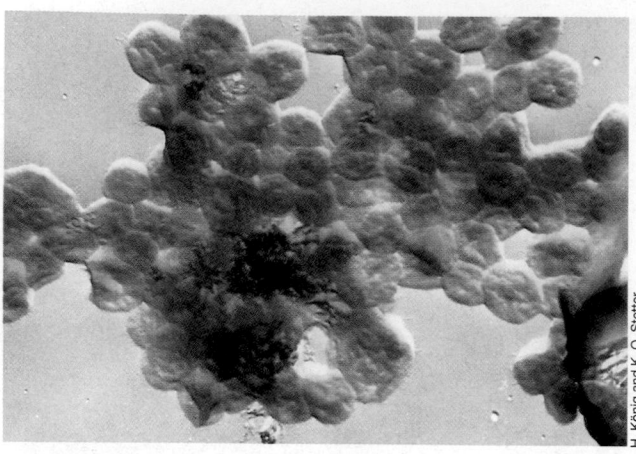

H. König and K. O. Stetter

Figure 13.21 The hyperthermophile *Staphylothermus marinus* (growth temperature optimum, 92°C). Electron micrograph of shadowed cells. A single cell is about 1 μm in diameter.

cell. *Ignicoccus* grows optimally at 90°C, and its metabolism is H_2/S^0 based, as is that of so many hyperthermophilic *Archaea* (Table 13.8). The function of the unusual periplasm of *Ignicoccus* is unclear, and chemical studies of its outer membrane are needed to determine whether it contains a lipopolysaccharide component similar to that of *Bacteria* (⚭ Section 4.9). However, regardless of the outcome of such studies, *Ignicoccus* stands out among hyperthermophilic *Archaea* (and indeed, all known prokaryotes) as being structurally unique.

Staphylothermus

A morphologically unusual member of the order Desulfurococcales is the genus *Staphylothermus*. Cells of *Staphylothermus* are spherical and about 1 μm in diameter, and form aggregates of up to 100 cells, resembling those of *Staphylococcus* (compare Figure 13.21 with Figure 4.4*b*). *Staphylothermus* is a chemoorganotroph that grows optimally at 92°C; energy is obtained from the fermentation of peptides, producing fatty acids like acetate and isovalerate as fermentation products (Table 13.8). Isolates of *Staphylothermus* have been obtained from both shallow marine vents, as well as black smoker hydrothermal vents, indicating that this organism is widely distributed in submarine thermal areas and thus may be an ecologically significant consumer of proteinaceous materials in these environments.

✓ 13.8–13.10 Concept Check

Hyperthermophilic Crenarchaeota inhabit the hottest habitats currently known to support life. Cold-dwelling phylogenetic relatives of these organisms are also known. A variety of different morphological types of Crenarchaeota are known and several different metabolic strategies are used to support growth.

✓ What are the major differences between the organisms *Sulfolobus* and *Pyrolobus*? *Staphylothermus* and *Ignicoccus*?

✓ What is unusual about the metabolic properties of *Acidianus* regarding elemental sulfur(S^0)?

✓ What energy class of organisms are those that use H_2 as electron donor? What Crenarchaeota use H_2, and what do they use as electron acceptors?

IV EVOLUTION AND LIFE AT HIGH TEMPERATURES

The hyperthermophilic *Archaea* grow at temperatures far higher than those supporting the growth of other prokaryotes. What is the nature of this extreme heat tolerance and what are the temperature limits beyond which life is impossible? Moreover, what can hyperthermophilic prokaryotes tell us about Earth's early history? We briefly examine these issues here.

13.11 Heat Stability of Biomolecules

Protein and DNA stability in hyperthermophiles is critical to surviving high temperature. Most proteins denature at high temperatures, so much research has been done to identify the properties of thermostable proteins. The amino acid composition of thermostable proteins from hyperthermophiles is not particularly unusual. In fact, enzymes from hyperthermophiles often contain the same major structural features as their heat labile counterparts from mesophilic bacteria. However, thermostable proteins do tend to have highly hydrophobic cores, which probably decreases their tendency to unfold, and in general have more ionic interactions on their surfaces. Ultimately, however, it is the *folding* of the protein that affects its heat resistance; thus, subtle changes in amino acid sequence are apparently sufficient to render heat stable an otherwise heat-labile protein.

Like all cells, hyperthermophiles produce special proteins called *chaperonins* (heat-shock proteins, ⚭ Section 7.16) that function to refold partially denatured proteins. In cells of *Pyrodictium* (Figure 13.22) the major chaperonin is called the *thermosome*. This protein functions to keep the cell's other proteins properly folded and functional and can help cells survive above their maximal growth temperature; for example, cells of *Pyrodictium* can survive 1 h in the autoclave (121°C).

DNA Stability

How does DNA stay intact at high temperatures and keep from melting apart into single strands? A variety of mechanisms may be involved here. The cytoplasm of hy-

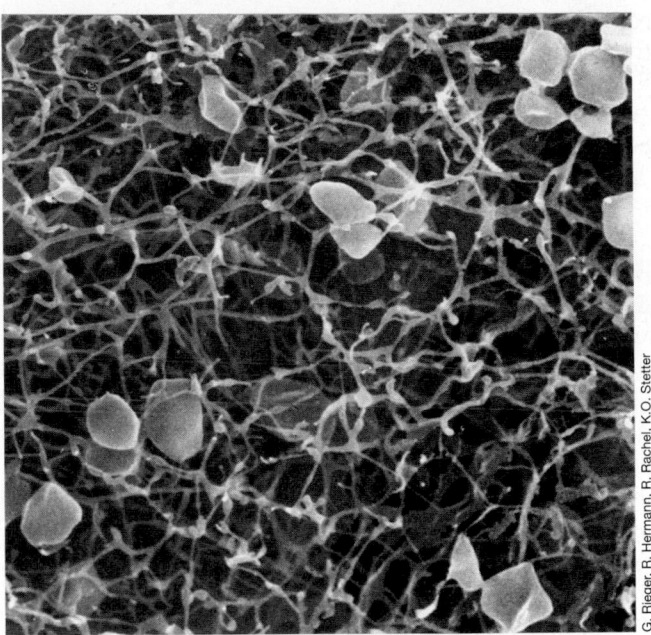

Figure 13.22 wrote *Pyrodictium abyssi*, scanning electron micrograph. *Pyrodictium* has been studied as a model of macromolecular stability at high temperatures. Cells are enmeshed in a sticky glycoprotein matrix that binds them together.

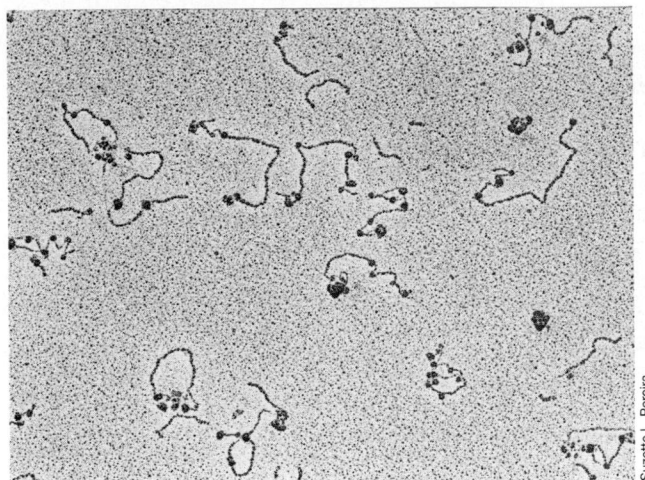

Figure 13.23 Archaeal histones and nucleosomes. Electron micrograph of linearized plasmid DNA wrapped around copies of archaeal histone Hmf (from the hyperthermophilic methanogen *Methanothermus fervidus*, see Figure 13.8c) to form the roughly spherical, darkly stained nucleosome structures. Compare this micrograph with an artist's depiction of the histones and nucleosomes of *Eukarya* shown in Figure 7.11.

perthermophilic methanogens contains large amounts of potassium cyclic 2,3-diphosphoglycerate. This solute prevents chemical damage, like depurination, that can occur to DNA at high temperatures. Not all hyperthermophiles produce this substance, however, so it is suspected that other DNA stabilizing mechanisms also exist. In this regard it has been found that all hyperthermophiles produce a unique form of DNA topoisomerase called *reverse DNA gyrase* (∞ Section 7.3). Reverse gyrase introduces *positive* supercoils into DNA (in contrast to the *negative* supercoils introduced by "normal" DNA gyrase, ∞ Section 7.3), and positive supercoiling has been shown to greatly stabilize DNA to heat denaturation.

In addition to salts and DNA gyrase, other proteins in hyperthermophiles function to maintain the integrity of duplex DNA. For example, a small heat-stable DNA binding protein called *Sac7d* has been found in cells of *Sulfolobus*, which binds to the minor groove of DNA in a nonspecific manner and increases its melting temperature by some 40°C. Protein Sac7d sharply kinks the DNA and thus may also be involved in gene regulation (∞ bent DNA, Section 8.4). Although Sac7d-like proteins are limited to species of crenarchaeotes, euryarchaeotes also contain DNA binding proteins. But unlike Sac7d, these are highly basic proteins that show strong amino acid sequence homology with the core histones of the *Eukarya* (∞ Section 7.3). These *archaeal histones* have been particularly well studied from the hyperthermophilic methanogen *Methanothermus fervidus*

where they wind and compact DNA into nucleosome-like structures (Figure 13.23) (∞ Section 7.3) that maintains the DNA in a double-stranded form at very high temperatures.

Lipid Stability

What about cellular lipids? How do hyperthermophiles prevent their membranes from peeling apart at high temperatures? Virtually all hyperthermophiles contain lipids constructed upon the dibiphytanyl tetraether model (∞ Section 4.5 for the structure of such lipids). Dibiphytanyl tetraether lipids are naturally heat resistant because the covalent bond between phytanyl units forms a lipid *monolayer* cytoplasmic membrane instead of the normal lipid *bilayer* (∞ Section 4.5 and Figure 4.20); this structure resists the tendency of heat to pull apart a lipid bilayer constructed of fatty acids.

Stability of Monomers

Besides the stability of *macromolecules*, the thermal lability of *monomers* is important in governing the upper temperature limits for growth of hyperthermophilic bacteria. No matter how stable the macromolecules, life is impossible if the basic building blocks themselves are unstable. The thermal stability of small molecules rather than macromolecules may therefore dictate the upper temperature for life, and even at temperatures as "low" as 120°C, some important small molecules are destroyed at significant rates. For example, two key molecules in energy metabolism, ATP and NAD^+, hydrolyze quite rapidly at these temperatures; their half-life is less than

30 minutes at 120°C and shortens dramatically at temperatures above this. With this in mind, what is the upper temperature for life?

The Limits to Microbial Existence

Because life as we know it depends on liquid water, microbial habitats hotter than 100°C are only found in environments under *pressure*, such as the sea floor. Indeed, all hyperthermophilic *Archaea* capable of growth over 100°C seem to be restricted to these superheated environments (see Table 13.9). But how hot can these environments be and still support life?

An answer to this question has been sought from the study of deep-sea hydrothermal vent black smokers as models for superheated natural environments. Smokers emit hydrothermal fluid at 250–350°C and form upright metallic structures called *chimneys* from metal sulfides in the fluid (∞ Figure 19.19). Although hyperthermophiles have been isolated from smoker chimney walls (which show a gradient in temperature from 250°C inside to 2°C outside), studies of the 250°C water itself show it to be sterile; this is consistent with laboratory measurements that have clearly shown that important biomolecules would be instantly destroyed at such a high temperature.

Laboratory experiments suggest that living processes could be maintained at temperatures as high as 140–150°C, but that above this temperature organisms would probably not be able to overcome the heat lability of the essential biomolecules of life. For example, if organisms exist that are capable of growth at temperatures even as "low" as 150°C, their energy economy would almost certainly have to be based on something other than ATP, the energy currency of all other forms of life, because ATP is unstable at this temperature. Thus, if "super" hyperthermophiles capable of growth at or above 150°C are ever discovered, they will likely demonstrate some fundamental new principles in biology and biochemistry.

13.12 Hyperthermophilic *Archaea* and Microbial Evolution

An interesting question concerning the *Archaea* is why so many of them inhabit extreme environments. Although molecular probing of nonextreme environments (like soil and water) indicate that *Archaea* are all around us, a common theme running through cultured representatives of the *Archaea* is *adaptation to environmental extremes*. As we have seen in this chapter, various species growing above the boiling point or at extremes of pH or salt are known. Is this a mere coincidence or a reflection of Earth's evolutionary history?

Extreme environments of various types existed on early Earth as they do today, and it is within such environments that life may first have flourished. At the time that cellular life first evolved it is almost certain that Earth was far hotter than it is today (∞ Section 11.1), and the hyperthermophiles in particular stand out as the best extant relatives of Earth's earliest life-forms. What does the phylogenetic tree of *Archaea* tell us about these hyperthermophiles?

Hyperthermophiles and Their Slow Evolutionary Clocks

Molecular sequencing of rRNA suggests that many *Archaea* evolved at slower rates than most *Bacteria* or the *Eukarya*. This is especially true of the hyperthermophilic *Archaea*, both euryarchaeotes and crenarchaeotes (see Figure 13.1). Much the same can be said about hyperthermophilic *Bacteria* such as *Thermotoga* and *Aquifex* (∞ Figure 12.1). This conclusion emerges from inspection of the evolutionary trees; lines to the hyperthermophiles are typically rather short and branch near the root (∞ Figure 12.1; see also Figure 13.1).

It is not known why heat-loving *Archaea* have such slow "evolutionary clocks," but one hypothesis is that this is a consequence of their inhabiting such extreme environments. Organisms living at very high temperatures may be under unusually strong evolutionary pressure to maintain those genes that specify phenotypic characteristics critical to life there. That is, beyond a certain point, survival of hyperthermophiles may not be furthered by additional genetic change. If this hypothesis is correct, then hyperthermophilic *Archaea* are indeed living relics of the earliest life forms, and their continued study should reveal important principles of early Earth biology.

"Korarchaeota"

The "slow clock" idea can be further explored in the phylum of the "Korarchaeota", which are shown in the phylogenetic tree in Figure 13.1 as the least derived of all *Archaea*. Unfortunately, we know very little about this group. Representatives of the "Korarchaeota" have thus far been identified only from a single hot spring, Obsidian Pool, in Yellowstone National Park, and the group is not recognized taxonomically (thus the use of quotes around the name). However, mixed laboratory cultures of "Korarchaeota" exist (Figure 13.24) that clearly indicate that they are hyperthermophiles, as would be expected of prokaryotes close to the root of the tree of *Archaea* (Figure 13.1). Further progress on this interesting group is anticipated and may yield secrets about the early Earth and early life forms.

Hydrogen as a Primitive Energy Source

Before we leave this chapter we wish to point out how often H_2 *metabolism* enters into the metabolic picture; many *Archaea* grow anaerobically with H_2 as an electron donor and one or more of the electron acceptors

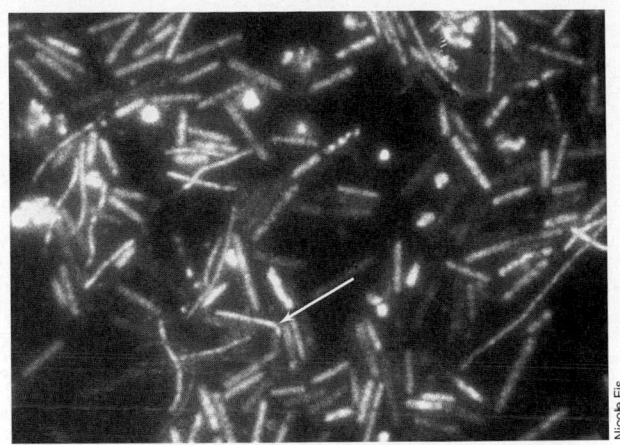

Nicole Eis

Figure 13.24 "Korarchaeota." Phase contrast photomicrograph of a stable enrichment culture containing cells of "Korarchaeota" from a Yellowstone hot spring and grown in the laboratory at 85°C. Various *Archaea* are present in this enrichment with the korarchaeotans being less than 1% of the total. Korarchaeotan cells are slightly curved rods 5–10 µm long (arrow). See also Figure 11.14.

S^0, NO_3^-, and Fe^{3+} (see Table 13.8). H_2 metabolism is likely a physiological relic of ancient metabolic schemes (∞ Figure 11.5 gives an example of one) that first evolved in primitive organisms because of the ready availability of these inorganic compounds in their primordial environment. The diversity of H_2-oxidizing hyperthermophilic *Archaea* (and *Bacteria*, ∞ Sections 12.5 and 12.37) today, and the fact that so many mesophilic *Bacteria* (distant rel-

atives of the hyperthermophilic H_2-oxidizers, ∞ Figure 12.1) also use H_2, attest to the evolution of H_2 oxidation as a metabolic success story. Indeed, the use of H_2 by hyperthermophiles and various physiological groups of prokaryotes living deep in the Earth (∞ Section 19.4) are prime examples of this metabolic strategy. Both hydrothermal vents, the habitat of many of the most extreme hyperthermophiles, and Earth's hot, deep subsurface, have been suggested as the types of environments where life could have first arisen. These environments contain an abundance of inorganic electron donors and acceptors and are free from exposure to ultraviolet radiation. It is thus possible that the hyperthermophiles discussed in this chapter are truly relics of the early Earth.

✓ *13.11–13.12 Concept Check*

Although hyperthermophiles live at very high temperatures, in some cases above the boiling point of water, there are temperature limits beyond which no living organism can survive. This limit is likely about 140–150°C.

- ✓ How do hyperthermophiles keep important macromolecules such as proteins and DNA from being destroyed by high heat?
- ✓ List at least two reasons why an upper temperature limit to life undoubtedly exists.
- ✓ What evidence suggests that extant hyperthermophiles most closely resemble ancient organisms?
- ✓ What form of energy metabolism was likely key to energy economies of ancient organisms?

Review Questions

1. What are some features that all *Archaea* have in common?
2. Which organism, *Pyrodictium*, *Thermoplasma*, or *Methanosarcina*, is the closest relative of the extreme halophile *Halobacterium*. (Hint: The answer lies in Figure 13.1).
3. How can organisms such as *Halobacterium* survive in a high-salt environment whereas an organism such as *Escherichia coli* cannot?
4. Contrast the roles of bacteriorhodopsin, halorhodopsin, and sensory rhodopsin in *Halobacterium salinarium*.
5. What is the electron *donor* for methanogenesis when CO_2 is reduced to CH_4?
6. What one major physiological feature unifies all members of the Thermoplasmatales? Why does this allow some of them to successfully colonize mine tailings?
7. What is physiologically unique about *Methanopyrus* compared with another methanogen such as *Methanobacterium*?
8. What is the major "claim to fame" of the organism *Pyrolobus fumarii*?
9. What is *reverse gyrase* and why is it important to hyperthermophiles?

Application Questions

1. Using the data of Figure 13.1 as a guide, discuss why bacteriorhodopsin was likely a late evolutionary invention.
2. Defend or refute the following statement: The upper temperature limit to life is unrelated to the stability of macromolecules such as proteins or nucleic acids.

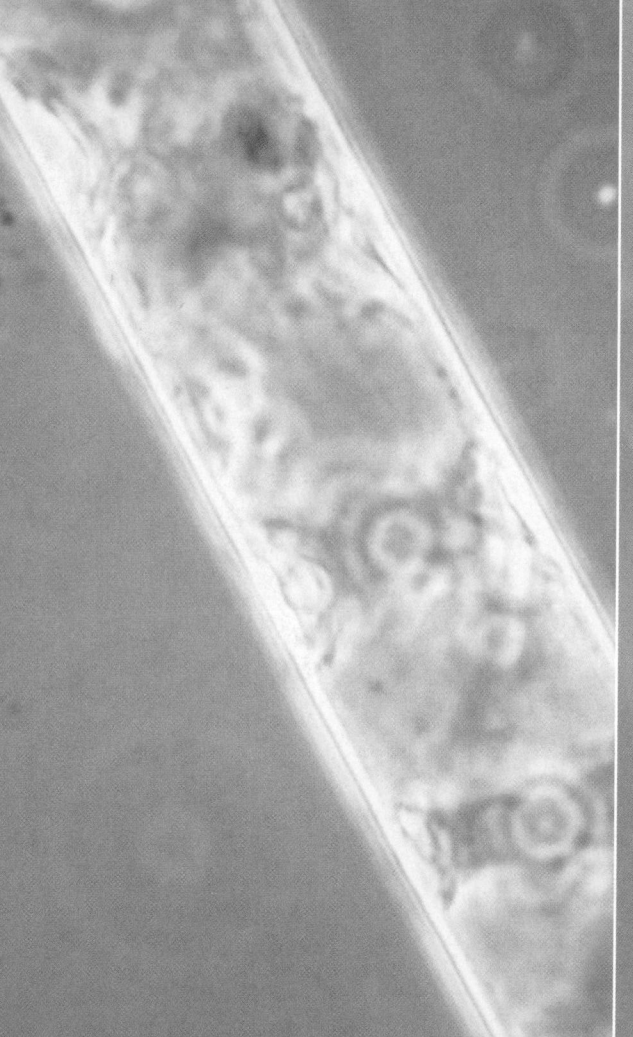

E ukaryotic cells include both microbial forms as well as plants and animals. The green alga *Spirogyra*, shown here, contains the typical organelles of a phototrophic eukaryotic cell, including a membrane-enclosed nucleus, mitochondria, and chloroplasts. Other microbial eukaryotes include the protozoa, fungi, and slime molds. All of these organisms are genetic chimeras because they contain genes from two sources. These include nuclear genes, of course, but also genes that reside in the chloroplast and the mitochondrion. The latter are remnants of the genomes of free-living *Bacteria* that established stable relationships with microbial eukaryotes eons ago in the process of *endosymbiosis*.

EUKARYOTIC CELL BIOLOGY AND EUKARYOTIC MICROORGANISMS

14

Working Glossary

Algae phototrophic eukaryotic microorganisms

Amoebold movement a type of motility in which cytoplasmic streaming moves the organism forward

Chitin a polymer of *N*-acetylglucosamine commonly found in the cell walls of algae and fungi

Chloroplast the photosynthetic organelle of eukaryotic phototrophs

Ciliates a group of protozoa characterized by rapid motility driven by numerous short appendages called cilia

Conidia asexual spores of fungi

Cytoskeleton cellular scaffolding typical of eukaryotic cells, in which an array of microfilaments define the cell's shape

Endosymbiosis the theory that the major organelles of eukaryotic cells arose from the uptake of free-living *Bacteria*

Eukarya all eukaryotic organisms

Flagellates a group of protozoa characterized by motility driven by the whip like action of one or more long thin appendages called flagella

Fungi nonphototrophic eukaryotic microorganisms that contain rigid cell walls

Hydrogenosome an organelle of endosymbiotic origin present in certain anaerobic eukaryotic microorganisms that functions to oxidize pyruvate to H_2, CO_2, and acetate along with the production of one ATP

Meiosis the process of nuclear division during gamete formation when the diploid number of chromosomes is halved to the haploid number, present in the gametes

Mitochondrion the respiratory organelle of eukaryotic organisms

Mitosis during cell division in eukaryotic cells, the process by which chromosomes are replicated and partitioned to each daughter cell

Molds filamentous fungi

Mushrooms filamentous fungi that produce large, often edible structures called fruiting bodies

Phagocytosis a mechanism for ingesting particulate food in which a portion of the cell membrane surrounds the particle and brings it into the cell

Protozoa nonphototrophic unicellular eukaryotic microorganisms that lack cell walls

Slime molds nonphototrophic eukaryotic microorganisms that lack cell walls and that aggregate to form fruiting structures (cellular slime molds) or masses of protoplasm (acellular slime molds)

Sporozoa nonmotile parasitic protozoa

Thylakoids layers of membranes containing the photosynthetic pigments in chloroplasts

Yeasts unicellular fungi

In this chapter we consider the structure, phylogeny, and diversity of eukaryotic microorganisms. We briefly encountered eukaryotic microorganisms in Chapter 2, and in Chapter 7 discussed some of the unique molecular properties of this group, such as the presence of introns in genes and complex transcriptional requirements. We will see here that while some eukaryotic microorganisms resemble prokaryotes in internal structure, others are structurally much more complex. The basic principles of eukaryotic cell biology will unfold in this chapter, and the material will be useful in an applied sense later on as we discuss the topics of host-parasite interactions, immunology, and disease (see Units IV and V), which intimately involve the activities of eukaryotic cells.

I EUKARYOTIC CELL STRUCTURE/FUNCTION AND GENETICS

The following six sections explore the structure of the eukaryotic cell, the ancestral link between organelles and *Bacteria*, and the basic genetics of eukaryotes using yeast as the model system.

14.1 Eukaryotic Cell Structure and the Nucleus

The typical eukaryotic cell is structurally more complex than a prokaryotic cell (Figure 14.1). The DNA in a prokaryote is not partitioned away within the cell (however, Section 12.29 provides an exception), while eukaryotes contain a *membrane-enclosed nucleus*. Most eukaryotes have other types of intracellular structures as well, the complement of structures depending on the organism. For example, mitochondria are nearly universal among eukaryotic cells, while chloroplasts are found only in phototrophic cells. Eukaryotic cells may have a cell wall (plant cells, algae, fungi) or may not (animal cells, protozoa). Several other internal structures, such as the Golgi apparatus, microsomes, endoplasmic reticula, and microtubles, are also present in a typical eukaryotic cell (Figure 14.1), and their functions will be covered shortly.

Nucleus

The nucleus contains the cell's genome (Figure 14.2). In eukaryotes, DNA within the nucleus is in the form of chromosomes (the intricate packaging of DNA necessary to form these structures was described in Chapter 7). In many eukaryotic cells the nucleus is many micrometers in diameter, easily visible with the light microscope even without staining (⌘ Figure 4.7a). In smaller eukaryotes, however, special staining procedures are often required to see the nucleus.

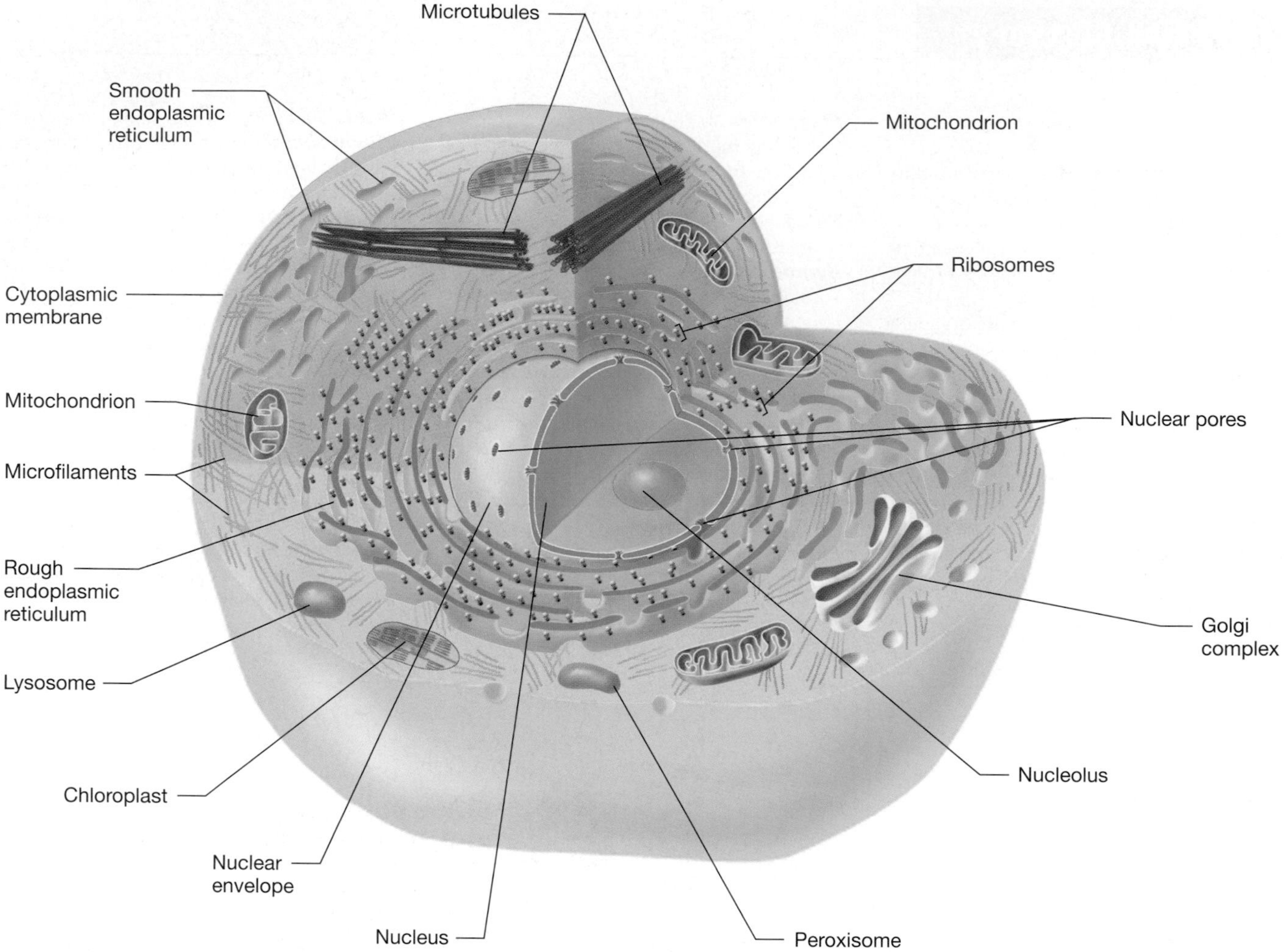

Microtubules

Smooth endoplasmic reticulum

Mitochondrion

Cytoplasmic membrane

Ribosomes

Mitochondrion

Microfilaments

Nuclear pores

Rough endoplasmic reticulum

Lysosome

Golgi complex

Chloroplast

Nucleolus

Nuclear envelope

Nucleus

Peroxisome

Figure 14.1   Schematic, cut-away view of a eukaryotic cell. Although all eukaryotic cells contain a nucleus, not all organelles or other structures shown are present in all eukaryotic cells.

The nuclear membrane consists of a *pair* of unit membranes separated by a space of variable thickness. The inner membrane is usually a simple sac, but the outermost membrane is in many places continuous with the endoplasmic reticulum. This dual-membrane arrangement facilitates functional specificity; the inner and outermost membranes specialize in interactions with the nucleoplasm and the cytoplasm, respectively.

The nuclear membrane contains pores (Figure 14.2) that are formed from holes in both unit membranes at places where the two membranes are joined. The pores consist of a complex of several proteins whose function is to import and export substances into and out of the nucleus. Like transport across the cytoplasmic membrane (⬅ Section 4.7), nuclear transport events require energy and this comes from the hydrolysis of guanosine triphosphate (GTP).

A structure often visible within the nucleus is the *nucleolus*, an area rich in RNA that is the site of ribosomal RNA synthesis. Ribosomal proteins synthesized in the cytoplasm are transported into the nucleolus and combined with ribosomal RNA to form the small and large subunits of the eukaryotic ribosome. These are then exported to the cytoplasm where they associate to form the intact ribosome and function in protein synthesis.

14.2 Respiratory Organelles: The Mitochondrion and the Hydrogenosome

The mitochondrion and the hydrogenosome are organelles that specialize in chemotrophic energy metabolism. We focus on these respiratory organelles here and consider chloroplasts in the next section.

Mitochondria

In aerobic eukaryotic cells the processes of respiration and oxidative phosphorylation (a mechanism of ATP formation) (⬅ Sections 5.11 and 5.12) are localized in membrane-enclosed organelles called **mitochondria** (sin-

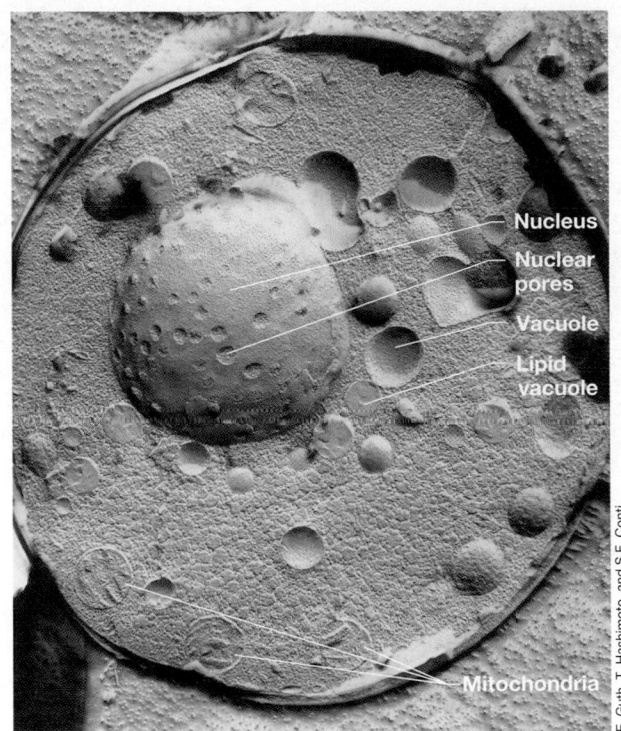

(a)

E. Guth, T. Hashimoto, and S.F. Conti

(b)

D. W. Fawcett

Figure 14.2 The nucleus and nuclear pores. (a) Electron micrograph of a yeast cell by the freeze-etch technique, showing a surface view of the nucleus. Freeze-etching is a form of transmission electron microscopy (🔎 Section 4.3), in which the specimen is frozen, fractured, and a thin replica made that is observed under the microscope. The cell is about 8 μm wide. (b) Thin section of mouse adipose tissue showing a portion of the nucleus and several mitochondria. The nucleus is about 2 μm wide. Note the pores in the nuclear membrane in both (a) and (b).

gular, **mitochondrion**). Mitochondria are of prokaryotic size and can be rod-shaped or nearly spherical (Figure 14.3). A typical animal cell can contain over 1000 mitochondria, but the number per cell depends somewhat on the cell type and size; a yeast cell may have many fewer mitochondria per cell (see Figure 14.2). The mitochondrial membrane, which lacks sterols, is much less rigid than the eukaryotic cell's cytoplasmic membrane, which does contain sterols. Mitochondria thus show a considerable plasticity, which makes their shape as seen in electron micrographs highly variable (Figure 14.3*b, c*).

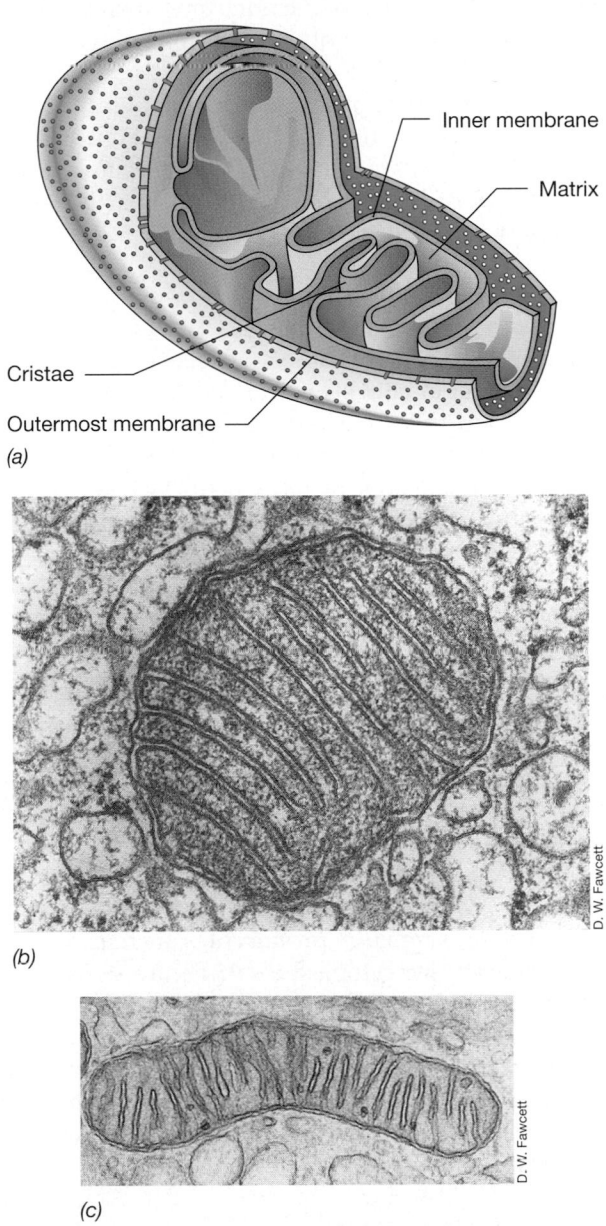

(a)

(b)

D. W. Fawcett

(c)

D. W. Fawcett

Figure 14.3 Structure of the mitochondrion. (a) Diagram showing the overall structure of the mitochondrion. Note inner and outermost membranes. (b,c) Transmission electron micrographs of mitochondria from rat tissue, showing the variability in morphology of typical mitochondria.

The mitochondrial membrane is constructed in a manner similar to other unit membranes: a bilayer of phospholipid with embedded proteins. However, unlike the cytoplasmic membrane (∞ Section 4.5), the mitochondrial membrane is rather permeable, having numerous minute channels that allow passage of ions and small organic molecules. Through these channels ATP, produced within the mitochondrion, moves to the cytoplasm, where it is used in energy-requiring reactions. Mitochondria also possess a system of folded inner membranes called *cristae*. These inner membranes, formed by invagination of the outermost membrane, are the sites of enzymes involved in respiration and ATP production. Cristae also contain specific transport proteins that regulate the passage of metabolites into and out of the *matrix* of the mitochondrion (Figure 14.3*a*). The matrix contains a number of enzymes involved in the oxidation of organic compounds, in particular, enzymes of the citric acid cycle (∞ Section 5.13).

The Hydrogenosome

Some eukaryotic microorganisms lack mitochondria and contain instead a membrane-enclosed respiratory organelle distinct in both structure and function from the mitochondrion, called the **hydrogenosome** (Figure 14.4*a*). Although about the size of mitochondria, the hydrogenosome lacks cristae and the citric acid cycle enzymes of the mitochondrion (Figure 14.3). A variety of microbial eukaryotes contain hydrogenosomes, all either obligate or aerotolerant anaerobes whose metabolism is strictly fermentative. These include a number of parasites, such as the flagellate *Trichomonas*, and ciliated protozoa that inhabit the rumen of ruminant animals (∞ Section 19.11) and anoxic muds and sediments.

The biochemical reactions in the hydrogenosome are focused on pyruvate oxidation to yield $H_2 + CO_2$ and acetate (Figure 14.4*b*). Pyruvate is oxidized to the high-energy intermediate acetyl-CoA (from which additional ATP is made during the formation of acetate), along with H_2 plus CO_2 (Figure 14.4*b*). In some anaerobic eukaryotes, H_2-consuming symbiotic prokaryotes such as methanogens reside in their cytoplasm (∞ Figure 19.26*b,c*) and consume the H_2 and CO_2 produced by the hydrogenosome, yielding methane. Because hydrogenosomes lack an electron transport chain and citric acid cycle, they cannot oxidize the acetate produced from pyruvate oxidation like mitochondria can, and acetate is therefore excreted into the cytoplasm of the host cell (Figure 14.4*b*).

✓ 14.2 Concept Check

The mitochondrion and the hydrogenosome are the respiratory organelles of eukaryotic cells. Mitochondria are involved in aerobic respiration, whereas the hydrogenosome, found only in certain anaerobic eukaryotes, ferments pyruvate to yield H_2 plus CO_2, acetate, and ATP.

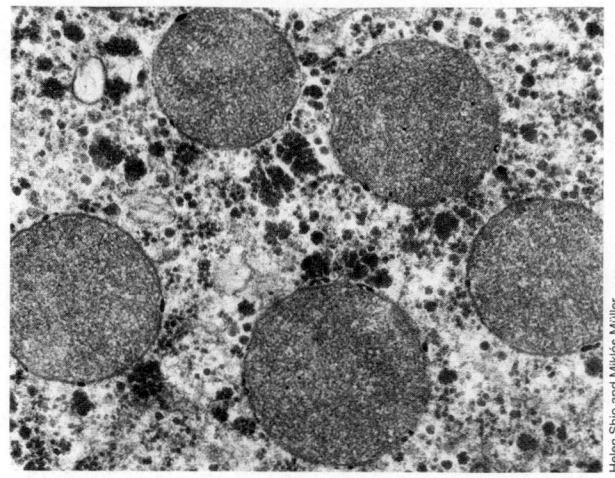

(a)

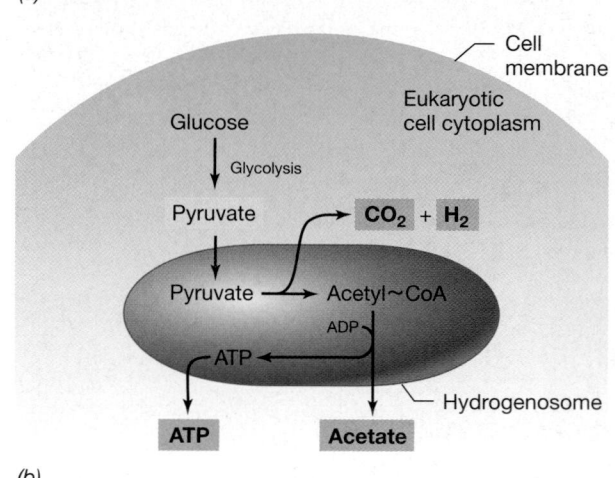

(b)

Figure 14.4 The hydrogenosome. (a) Electron micrograph of a thin section through a cell of the anaerobic flagellate, *Trichomonas vaginalis*, showing five hydrogenosomes. (b) Biochemistry of the hydrogenosome. Pyruvate is taken up by the hydrogenosome and H_2, CO_2, acetate, and ATP are produced. The key enzymes of the hydrogenosome are *pyruvate:ferredoxin oxidoreductase* and *hydrogenase*. Endosymbiotic methanogens (∞ Section 13.4) are often present in the cytoplasm of hydrogenosome-containing eukaryotes, producing methane from the $H_2 + CO_2$ (∞ Figure 19.26*c*).

✓ The mitochondrial membrane allows passage of what substances and why?

✓ Compare and contrast the metabolic fate of pyruvate in mitochondria and the hydrogenosome.

14.3 Photosynthetic Organelle: The Chloroplast

Chloroplasts are chlorophyll-containing organelles found in eukaryotes capable of photosynthesis (algae, plants). Chloroplasts of many algae are relatively large and readily visible with the light microscope (Figure 14.5). The

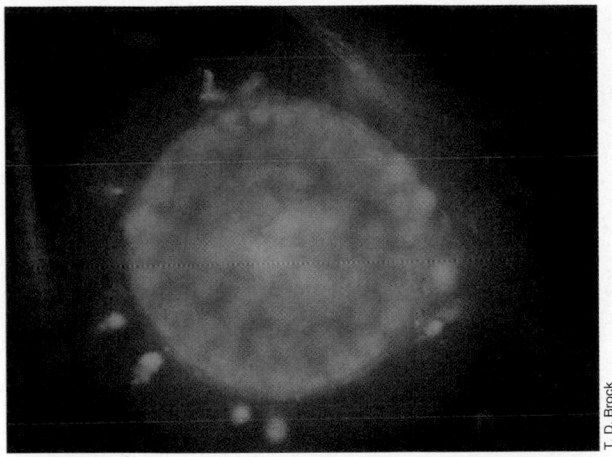

(a)

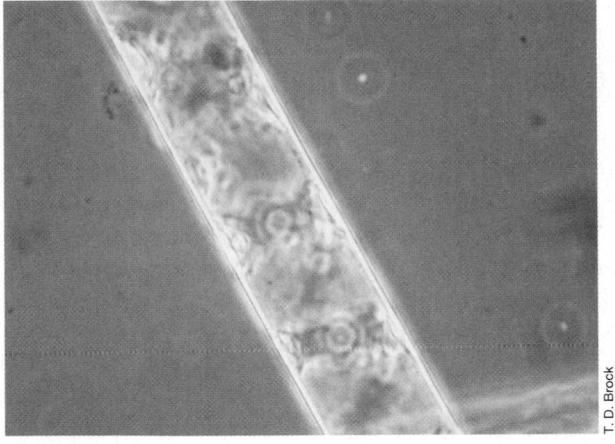

(b)

Figure 14.5 Photomicrographs of algal cells showing chloroplasts. (a) Fluorescence photomicrograph of the diatom *Stephanodiscus*. The chlorophyll in the chloroplasts absorbs light and fluoresces red. (b) Phase contrast photomicrograph of *Spirogyra* showing the characteristic spiral-shaped chloroplasts of this phototroph.

size, shape, and number of chloroplasts per cell vary markedly, and in contrast to mitochondria, chloroplasts are typically much larger than bacteria.

Like mitochondria, chloroplasts have a permeable outermost membrane, a much less permeable inner membrane, and an intermembrane space. The inner membrane surrounds the lumen of the chloroplast, called the *stroma*, but is not folded into cristae like the inner membrane of the mitochondrion. Instead, chlorophyll and all other components needed for photosynthesis are located in a series of flattened membrane discs called **thylakoids** (Figure 14.6). The thylakoid membrane is highly impermeable to ions and other metabolites because its function is to establish the proton motive force necessary for ATP synthesis (∞ Section 5.12). In green algae and green plants, thylakoids are usually associated in stacks of discrete structural units called *grana* (∞ Figure 17.5).

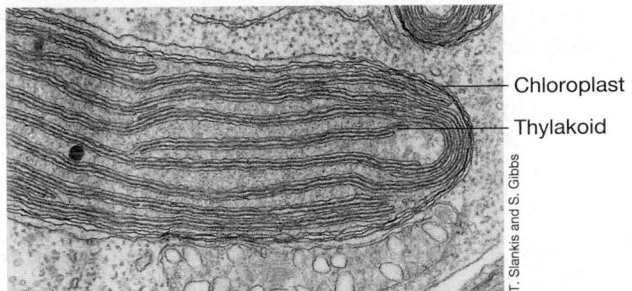

Figure 14.6 Transmission electron micrograph showing a chloroplast of the golden brown alga *Ochromonas danica*. Note the thylakoids.

The chloroplast stroma contains large amounts of the enzyme *ribulose bisphosphate carboxylase*, called *RubisCO* for short. This is the key enzyme of the *Calvin cycle*, the series of reactions by which most photosynthetic organisms convert CO_2 to organic form (∞ Section 17.6). RubisCO makes up over 50% of the total chloroplast protein and catalyzes the formation of phosphoglyceric acid, a key compound in the biosynthesis of glucose (∞ Sections 5.10 and 5.15). The permeability of the outermost chloroplast membrane allows glucose and ATP produced during photosynthesis to diffuse into the cytoplasm where they can be used to build new cell material.

✓ 14.3 Concept Check

Chloroplasts are the site of photosynthetic energy production and CO_2 fixation in eukaryotic phototrophs.

✓ Differentiate the stroma from thylakoids.
✓ What is the function of RubisCO and where is it found?

14.4 Relationships of Mitochondria and Chloroplasts to *Bacteria*

On the basis of their relative autonomy and morphological resemblance to bacteria, it was suggested long ago that mitochondria and chloroplasts were descendants of ancient prokaryotic organisms. This theory of *endosymbiosis* (endo means "within") states that eukaryotes arose from the engulfment of a prokaryotic cell by a larger cell ("the host") (∞ Sections 11.3). This theory predicts that the eukaryotic cell is a genetic chimera, containing DNA from two different sources, the endosymbiont and the host cell. Several lines of molecular evidence have confirmed this prediction and the endosymbiosis theory in general, and we summarize these here:

1. *Mitochondria and chloroplasts contain DNA.* Although most of their functions are encoded by nuclear DNA, a few organellar components are encoded by a genome present within the organelle

itself. These include ribosomal RNA, transfer RNAs, and certain proteins of the respiratory chain. Moreover, mitochondrial and chloroplast DNA exists in a covalently closed *circular* form, as it does in prokaryotes (∞ Sections 2.2, 7.2, and 7.3). Mitochondrial DNA can be seen in cells by using special staining methods (Figure 14.7).

2. **The eukaryotic nucleus contains bacterially derived genes.** Genomic sequencing (∞ Chapter 15) and other genetic studies have shown clearly that several nuclear genes exist that direct the activities of organelles. Because the sequence of these genes more closely resembles counterparts from species of *Bacteria* than species of *Archaea* or eukaryotes, many scientists believe these genes were transferred to the nucleus from bacterial symbionts during the evolution of the modern day organelle.

3. **Mitochondria and chloroplasts contain their own ribosomes.** Ribosomes, cell structures involved in protein synthesis (∞ Section 7.15), exist in either a large form [80 Svedberg (S) units], typical of the cytoplasm of eukaryotic cells, or in a smaller form (70S), unique to prokaryotes. Mitochondria and chloroplasts contain ribosomes, and they are 70S in size, the same as those of prokaryotes.

4. **Antibiotic specificity.** Several antibiotics (streptomycin is one example) kill or inhibit *Bacteria* by specifically interfering with 70S-ribosome function; these same antibiotics also inhibit protein synthesis in mitochondria and chloroplasts.

5. **Molecular phylogeny.** Phylogenetic studies using comparative ribosomal RNA sequencing methods (∞ Section 11.4–11.8) have shown convincingly that the chloroplast and mitochondrion are closely related to *Bacteria*. These studies clearly point to the modern eukaryotic cell as having arisen from an association of two organisms, presumably by endosymbiosis (∞ Section 11.3 and Figure 11.7).

Evidence that hydrogenosomes are endosymbionts has been found as well. For example, in the obligately anaerobic ciliated protozoan *Nyctotherus ovalis* that lives in the hindgut of termites (∞ Section 19.10), hydrogenosomes have been identified that contain DNA and ribosomes. Moreover, the nucleus of hydrogenosome-containing eukaryotes contains genes encoding specific proteins known to be of bacterial origin. Thus, hydrogenosomes probably originated by endosymbiosis of a pyruvate-fermenting bacterium (such as a *Clostridium* species) with an anaerobic microbial eukaryote, presumably as a way to extract a bit more energy out of a strictly fermentative lifestyle (Figure 14.4*b*). If this scenario is true, the mitochondrion and the hydrogenosome

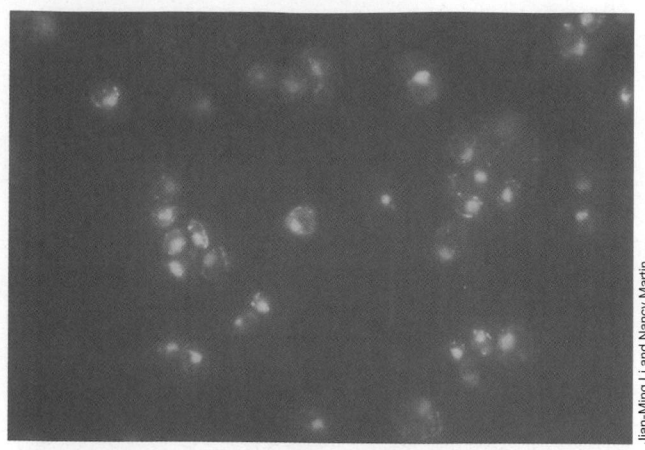

Figure 14.7 Cells of the yeast *Saccharomyces cerevisiae* stained with 4′6-diamidine-2′-phenylindole dihydrochloride (DAPI) (∞ Section 18.3) to show mitochondrial DNA. Each mitochondrion has two to four circular chromosomes that stain blue with the fluorescent dye used.

can be viewed as functionally related organelles that simply show different metabolic patterns.

In summary, many forms of evidence point to organelles as having arisen from the endosymbiotic uptake of free-living *Bacteria* by eukaryotic host cells. In this cozy arrangement, host cells obtained partners specializing in energy generation while the symbionts received a stable and supportive growth environment. That endosymbiosis was an evolutionary success is attested to by the presence, with rare exception, of energy-producing organelles such as mitochondria or chloroplasts within eukaryotic cells today.

✓ 14.4 Concept Check

Key metabolic organelles of eukaryotes are the chloroplast, involved in photosynthesis, and the mitochondrion or hydrogenosome, involved in respiration or fermentation. It is likely that these organelles were originally *Bacteria* that established permanent residence inside another cell (endosymbiosis).

✓ Summarize the molecular evidence that supports the relationship of organelles to *Bacteria*.

✓ Why might an organism such as *Nyctotherus* be better off with hydrogenosomes than with mitochondria?

14.5 Other Organelles and Eukaryotic Cell Structures

A variety of other cytoplasmic structures are typically present in eukaryotic cells, including the endoplasmic reticulum, Golgi apparatus, lysosomes, and peroxisomes. In contrast to mitochondria and chloroplasts, however, these structures lack DNA and ribosomes and are not of endosymbiotic origin.

Endoplasmic Reticulum and Golgi Apparatus

The **endoplasmic reticulum** is a network of membranes continuous with the nuclear membrane. Two types of endoplasmic reticulum (ER) are recognized: *rough*, which contains attached ribosomes, and *smooth*, which does not (Figure 14.1). Smooth ER participates in the synthesis of lipids and in some aspects of carbohydrate metabolism. Rough ER, through the activity of its ribosomes, is a major producer of glycoproteins, and also produces new membrane material that is transported throughout the cell to enlarge the various cell membrane systems (see Figure 14.1) before cell division.

The **Golgi apparatus** consists of a stack of membranes (Figure 14.1) that function in concert with the ER. In the Golgi complex products of the ER are chemically modified and sorted into those destined to be secreted, for example, hormones or digestive enzymes, and those that function in other membranous structures in the cell.

Lysosomes and Peroxisomes

Lysosomes (Figure 14.1) are membrane-enclosed sacs that contain various digestive enzymes that the cell uses to digest macromolecules such as proteins, fats, and polysaccharides. The internal pH of the lysosome is about 5, two units lower than that of the cytoplasm, and the hydrolytic enzymes within the lysosome function optimally at this pH. Because these hydrolytic enzymes are nonspecific in their action and could potentially destroy key cellular macromolecules if not contained, the lysosome is a structure that allows lytic activities to be partitioned away from the cytoplasm proper. Following hydrolysis of macromolecules in the lysosome, the resulting monomers pass from the lysosome into the cytoplasm as nutrients for the cell.

The **peroxisome** is a membrane-enclosed structure (Figure 14.1) whose function is to produce hydrogen peroxide (H_2O_2) from the reduction of O_2 by various hydrogen donors including alcohols and long chain fatty acids. The H_2O_2 produced in the peroxisome is degraded to H_2O and O_2 by the enzyme catalase (∞ Section 6.13). Peroxisomes play other roles as well, such as synthesizing bile salts that aid in the absorption and digestion of fats. Peroxisomes originate in the cell by incorporating their proteins and lipids from the cytoplasm, eventually becoming a membrane-enclosed entity that can enlarge and divide in synchrony with the cell.

Microfilaments and Microtubules

The large size of eukaryotic cells and their ability to move requires structural reinforcement. This support comes from proteins that arrange themselves into filamentous structures called *microfilaments* and *micro-*

tubules. Microfilaments are about 8 nm in diameter and are polymers composed of subunits of the protein *actin*. These fibers form scaffolds throughout the cell, defining and maintaining the shape of the cell (Figure 14.1). Microtubules are larger filaments, about 25 nm in diameter, and composed of the protein *tubulin*. Microtubules assist microfilaments in maintaining cell structure, but also play an important role in cell motility, both in the movement of internal cell structures (for example in the separation of chromosomes during cell division), and in movement of the organism itself (for example in movement of the flagellum in flagellated eukaryotic cells). Together, this system of microfilament and microtubule fibers is called the **cell cytoskeleton**. We saw in an earlier chapter that prokaryotes have homologs of actin and tubulin in the form of the proteins MreB and FtsZ, respectively (∞ Section 6.1). Thus, the eukaryotic cytoskeleton has deep evolutionary roots.

✓ 14.5 Concept Check

Besides the major organelles of eukaryotes, several other structures with defined functions are present in the cytoplasm. These include the endoplasmic reticulum, the site of ribosomes and cellular lipid synthesis; the Golgi apparatus involved in secretion; lysosomes, which play a role in macromolecular digestion; and the peroxisome, an organelle involved in H_2O_2 production. In addition, various proteinaceous tubes called microfilaments and microtubules are present, forming the cell's cytoskeleton.

- ✓ How does smooth ER differ from rough ER?
- ✓ Why are the activities that occur in the lysosome best partitioned away from the cytoplasm proper?
- ✓ Besides scaffolding, what other functions do microtubules have?

14.6 Overview of Eukaryotic Genetics

Here we briefly consider some of the idiosyncrasies of genetics in eukaryotes, using the baker's yeast, *Saccharomyces cerevisiae* as our model organism. Eukaryotic cells undergo sexual reproduction, with male and female each donating gametes to form the zygote. This is fundamentally different from the unidirectional mating processes in prokaryotes (∞ Chapter 10), and so we focus on these events in our discussion.

Mitosis and Meiosis

From a genetic standpoint, eukaryotic cells can exist in two forms: haploid or diploid. Diploid cells have two of each chromosome whereas haploid cells have but one. Cells of *S. cerevisiae* can exist indefinitely in the haploid (vegetative) stage (containing 16 chromosomes). But occasionally, two haploid yeast cells can fuse (mate) to yield a diploid cell (32 chromosomes) (Figure 14.8). This process

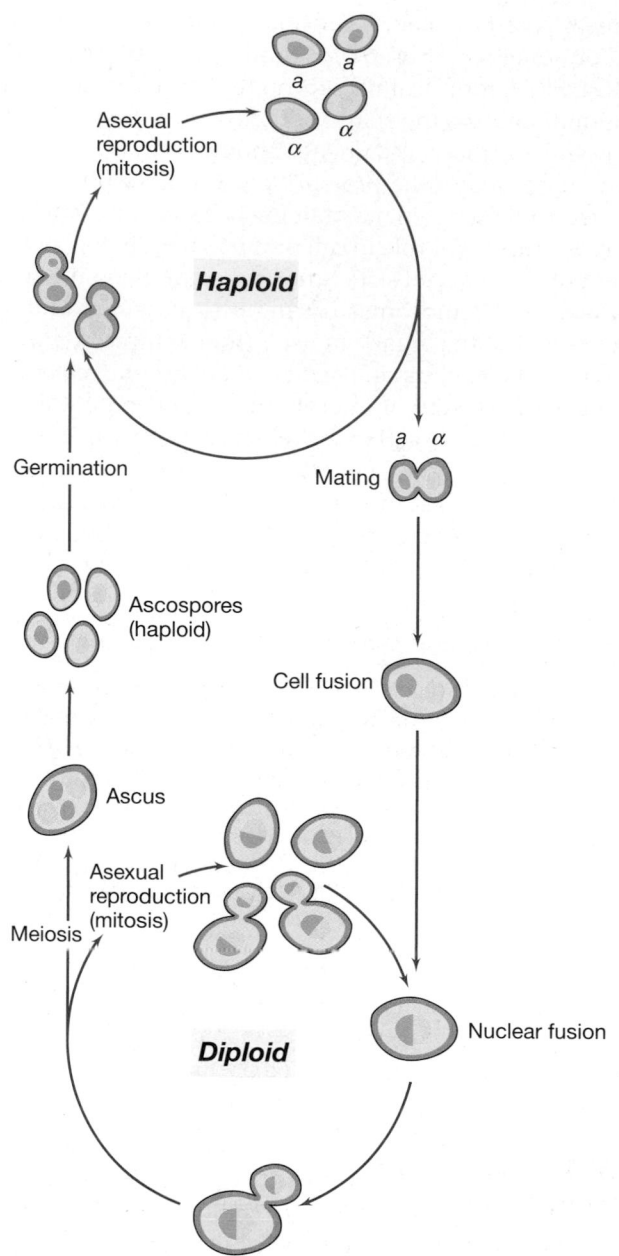

Figure 14.8 Life cycle of a typical yeast, *Saccharomyces cerevisiae*. A haploid cell of *S. cerevisiae* contains 16 chromosomes.

should not be confused with the state of affairs in cells of metazoans, such as the plants and animals. In these organisms the diploid phase is present in the organism proper, and the haploid phase occurs only in the gametes.

Recall that **mitosis** is the process following DNA replication in a cell in which chromosomes condense, divide, and are separated into two sets, one for each daughter cell (∞ Section 2.2). In haploid *S. cerevisiae* cells (Figure 14.8), mitosis occurs before each cell division to maintain the 16 chromosomes per cell. **Meiosis** is the process by which the conversion from the diploid to the haploid stage occurs. Meiosis involves two divisions. In the first meiotic divi-

sion, homologous chromosomes are segregated into separate cells, changing the genetic state from diploid to haploid. The second meiotic division is essentially the same as mitosis, as the two haploid cells divide to form a total of four haploid gametes (ascospores, Figure 14.8).

The Mating Process in Yeast: Mating Types

Yeasts have two different forms of haploid cells called *mating types*; these can be considered analogous to male and female gametes. On mating of opposite types, a diploid cell is formed. From a single diploid cell, a structure containing four gametes is formed, two of each mating type. The cell in which the gametes are formed is called an *ascus*, and the cells within the ascus are called *ascospores* (Figure 14.8).

The two mating types of *Saccharomyces cerevisiae* are designated α and a. Cells of type a mate only with cells of type α, and whether a cell is a or α is determined genetically. Some haploid strains of *S. cerevisiae* remain a or α, while other strains are able to *switch* their mating type from one to the other. This switch in mating type is caused by an inserted gene and the phenomenon is illustrated in Figure 14.9.

There is a single location on one of the *S. cerevisiae* chromosomes called the MAT (for *mating type*) locus, at which either gene a or gene α can be inserted. At this locus, the MAT promoter controls transcription of whichever gene is present; if gene a is at that locus, then the cell is mating type a, whereas if gene α is at that locus, the cell is mating type α. Elsewhere in the yeast genome are copies of genes a and α that are not expressed. These *silent copies* are the source of the inserted gene. When the switch occurs (Figure 14.9), the appropriate gene, a or α,

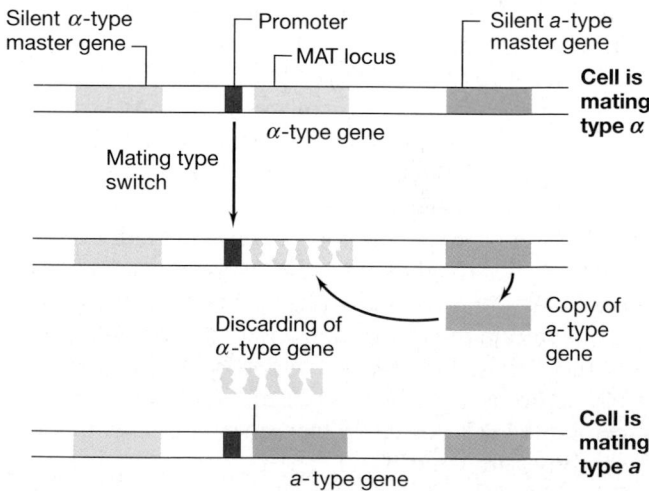

Figure 14.9 The cassette mechanism involved in the switch in yeast from mating type α to a and back again. Whichever "cassette" is inserted at the active locus (reading head) determines the mating type. The process shown is reversible, so type a can also revert to type α.

is copied from its silent site and inserted into the MAT location, *replacing* the gene already present. The old mating type gene is excised and discarded, and the new gene is inserted. This process has been called the *cassette mechanism* because the *a* and *α* genes can be considered analogous to tape cassettes; whichever gene is inserted in the MAT locus is the one that will be transcribed.

The *α* and *a* genes of *S. cerevisiae* are regulatory genes. Among other things, they regulate the production of the peptide hormones *α factor* or *a factor*, which are excreted by yeast cells undergoing mating. These hormones bind to cells of opposite mating type and bring about changes in their cell surfaces so that fusion can occur (Figure 14.10). Once cells of opposite mating type have associated, fusion of both the cells and their nuclei occurs, resulting in the formation of a diploid zygote (Figure 14.10). From the latter, meiosis regenerates the haploid vegetative form and the life cycle is complete (Figure 14.8).

✓ 14.6 Concept Check

Eukaryotic organisms can mate and exchange DNA during sexual reproduction. Mitosis ensures appropriate segregation of the chromosomes during asexual cell division. Haploid cells formed by meiosis can fuse to form a diploid zygote. There are two mating types in yeast, and yeast cells can convert from one type to the other.

✓ What microbial eukaryote has been the model organism for the study of eukaryotic genetics?

✓ If the diploid number of chromosomes in human cells is 46, how many chromosomes are present in a human sperm cell?

✓ Explain how a *single* haploid cell of *Saccharomyces* can eventually yield a diploid cell.

II EUKARYOTIC MICROBIAL DIVERSITY

We now explore the diverse groups of microbial eukaryotes. These include the protozoa, the fungi, the slime molds, and the algae. We preface this tour with a phylogenetic consideration of these organisms, where we will learn that some eukaryotes are phylogenetically highly derived while others have apparently retained key properties of ancient eukaryotes.

14.7 Phylogenetic Overview of *Eukarya*

We previewed the phylogeny of *Eukarya* within the context of the universal phylogenetic tree in Figure 11.13. From this we learned that *Eukarya* form their own phylogenetic domain and as a group are more closely related to *Archaea* than they are to *Bacteria*. This phylogeny of *Eukarya* is deduced from comparative sequencing of 18S ribosomal RNA, obtained from cytoplasmic ribosomes (∞ Section 11.5). But before we examine this, let us briefly consider the classical picture of phylogenetic diversity.

Five Kingdoms or Three Domains?

Eukarya have traditionally been grouped into four *kingdoms* of organisms: *plants, animals, fungi*, and the remaining microbial *Eukarya*, called the *protists*. Before the era of molecular phylogeny these four groups, along with all prokaryotes (that were simply lumped into one phylogenetic unit), formed the so-called *five kingdoms of living organisms*. However, molecular sequencing (∞ Chapter 11) has shown that the five-kingdom system greatly magnified the evolutionary importance of metazoans (multicellular plants and animals). Indeed, as we have already seen (∞ Chapters 11–13), the evolutionary picture that has emerged from ribosomal RNA sequencing, that of *three domains of cellular life*, differs significantly from the five-kingdom hypothesis,

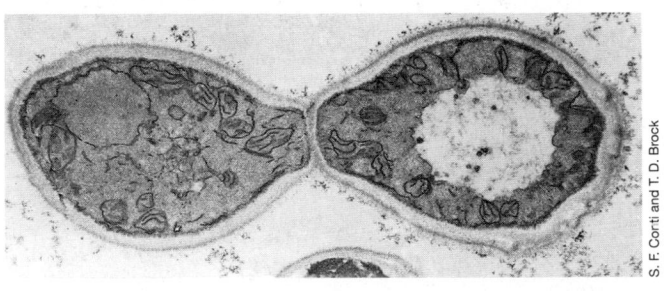

(a)

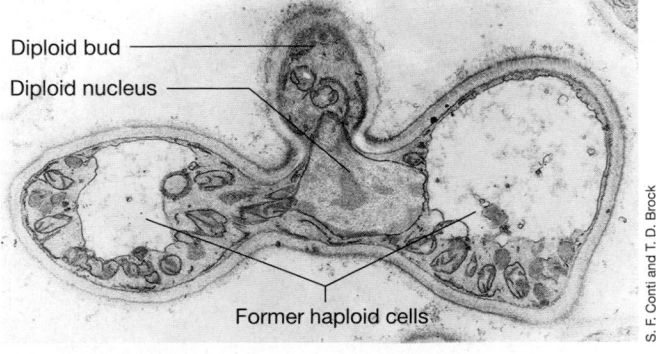

Diploid bud

Diploid nucleus

Former haploid cells

(b)

Figure 14.10 Electron micrographs of the mating process in a yeast, *Hansenula wingei*. (a) Two cells have fused at the point of contact and have sent out protuberances toward each other. (b) Late stage of mating. The nuclei of the two cells have fused, and the diploid bud has formed at a right angle to the conjugation tube. This bud eventually separates and becomes the fore-runner of a diploid cell line. A single cell of *Hansenula* is about 10 μm in diameter.

in particular by showing that the bulk of evolutionary diversity on Earth resides within the *microbial* world.

Figure 14.11 shows details of the *Eukarya* domain of the tree of life. We can see that several microbial eukaryotes occupy deep lineages on the tree, while metazoans (multicellular organisms) are highly derived organisms, an observation supported by the rather late appearance of metazoans in the fossil record (⚭ Section 11.1). Algae are scattered within the eukaryal tree (mainly in relatively recent lineages) while the fungi (except for the Oomycetes) form a very tight, and rather recent, phylogenetic unit (Figure 14.11).

Early Eukaryotes

Of great interest to microbiologists today is the identity of evolutionarily "early" eukaryotes—modern day organisms that are the least derived (evolved) from their ancient ancestors. The ribosomal RNA tree in Figure 14.11 clearly identifies diplomonads such as *Giardia*, microsporidia such as *Encephalitozoon*, and trichomonads such as *Trichomonas* as "early eukaryotes." In what ways do these organisms resemble primitive eukaryotes?

Giardia, Trichomonas, and *Encephalitozoon* are all *amitochondriate* eukaryotes: All have a membrane-enclosed nucleus but lack mitochondria. Some of these organisms, such as *Trichomonas*, contain hydrogenosomes (Figure 14.4), while others do not. Are these modern relics of ancient eukaryotes that never underwent endosymbiosis? This is one possibility, but genetic studies don't support this hypothesis. Using sensitive nucleic acid identification methods, it has been shown that amitochondriate eukaryotes contain genes in their nuclei that originated from *Bacteria*, just as mitochondrial genes remain today in the nucleus of aerobic eukaryotes (see Section 14.4). This suggests that organisms such as *Trichomonas* once harbored endosymbionts, but for some reason discarded them later, leaving only a trace of their former existence as bacterial-specific genes in the nucleus.

The microsporidia are an evolutionary enigma. These organisms lack *all* organelles, including the Golgi complex and hydrogenosomes, and contain very small genomes (the genome of *Encephalitozoon*, for example, is only 2.9 megabases, which is 1.5 megabases smaller than that of the bacterium *Escherichia coli*). Indeed, microsporidia have all the features one would predict to be present in an "early eukaryote." However, because this group has apparently undergone very rapid evolution (this can be seen in the long line in the tree in Figure 14.11), its phylogenetic position is hard to pin down using ribosomal RNA sequencing methods alone. Indeed, molecular sequencing of other genes and proteins from microsporidia tell a completely different story. Collective molecular sequencing studies strongly suggest that microsporidia are relatives of the fungi (rather highly derived eukaryotes, Figure 14.11), despite having many possibly "primitive" properties. Further work on the phylogenetic position of this interesting group of microbial eukaryotes is in progress.

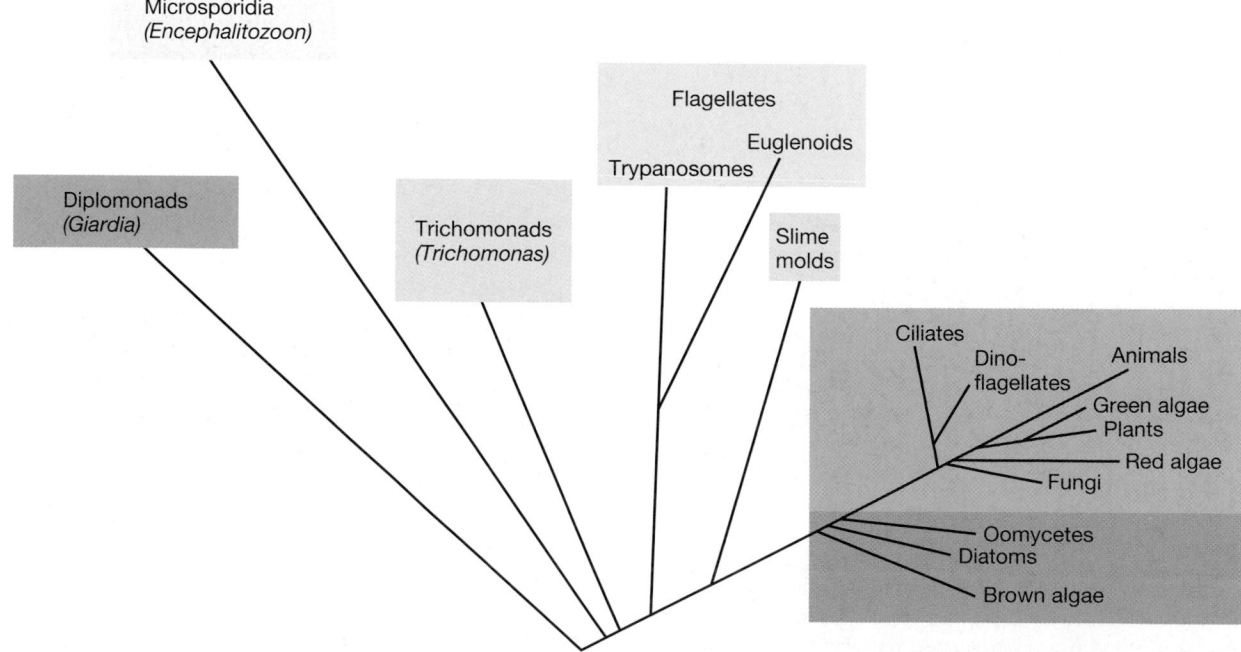

Figure 14.11 Phylogenetic tree of *Eukarya* based on comparative 18S ribosomal RNA sequences. Figure 11.13 places the eukaryal tree within the context of the universal tree. The phylogenetic position of Microsporidia such as *Encephalitozoon* is uncertain (see text).

The question remains: Do modern relatives of eukaryotic microorganisms exist today that never experienced endosymbiosis? If such organisms exist they are predicted to lack organelles and evidence for bacterially derived genes in their nuclei and to have branched earlier than the known "early eukaryotes" on the ribosomal RNA phylogenetic tree of eukaryotes (Figure 14.11). Whether such cells exist is unknown and a prime challenge for scientists studying the microbial diversity of eukaryotes today.

✓ 14.7 Concept Check

Eukaryotic cells form their own major line of evolutionary descent (the *Eukarya*). Some microbial eukaryotes, such as *Giardia* and *Trichomonas*, are early branching species, while the eukaryotic "crown" of the tree contains the multicellular plants and animals.

- ✓ From a *cellular* perspective, what is wrong with the five-kingdom hypothesis for grouping the eukaryotes?
- ✓ Summarize the evidence that the organism *Giardia* is a closer relative of primitive eukaryotes than is a human cell.

14.8 Protozoa

Key Genera

Amoeba
Paramecium
Trypanosoma

Protozoa are unicellular eukaryotic microorganisms that lack cell walls (Figure 14.12). They are generally colorless and motile. **Protozoa** are distinguished from prokaryotes by their eukaryotic nature and usually greater size, from algae by lacking chlorophyll, from yeasts and other fungi by their motility and absence of a cell wall, and from the slime molds by their inability to form fruiting bodies. Also, as previously mentioned, protozoa are phylogenetically diverse, appearing in several lineages on the *Eukarya* tree (Figure 14.11). Protozoa are found in a variety of freshwater and marine habitats; a large number are parasitic in other animals, including humans, and some are found growing in soil or in aerial habitats, such as on the surface of trees.

Most protozoa feed by ingesting particulate materials, usually other cells, by **phagocytosis**, a process of surrounding a food particle with a portion of their flexible cell membrane to engulf the particle and bring it into the cell. Some protozoa can literally swallow bacterial cells or small eukaryotic cells by operation of a special structure called a gullet (see Figure 14.16).

As is appropriate for organisms that "catch" their own food, most protozoa are motile. Indeed, their mechanisms of motility are key characteristics used to divide them into taxonomic groups (Table 14.1). Protozoa that move by ameboid motion are called Sarcodina; those using flagella, the Mastigophora; and those using cilia, the Ciliophora. The Apicomplexans, a fourth group, are generally nonmotile and are all parasitic for higher animals.

Mastigophora: The Flagellates

Members of this protozoal group are motile by the action of flagella (Figure 14.12c and Figure 14.13). Although many flagellated protozoa are free-living organisms, a number are parasitic in, or pathogenic for, animals, including humans. The most important pathogenic Mastigophora are the *trypanosomes*. These organisms cause a number of serious diseases in humans and vertebrate animals, including the feared disease *African sleeping sickness*. In *Trypanosoma*, the genus infecting humans, the protozoa are rather small, about 20 μm in

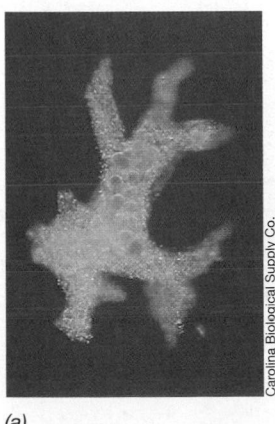
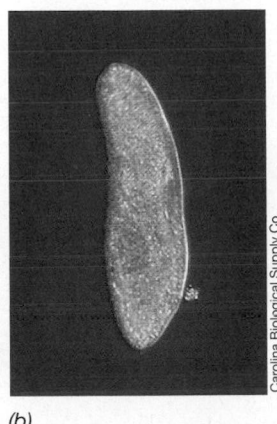
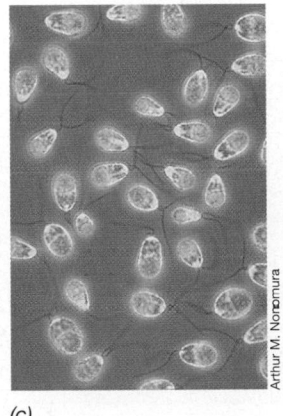
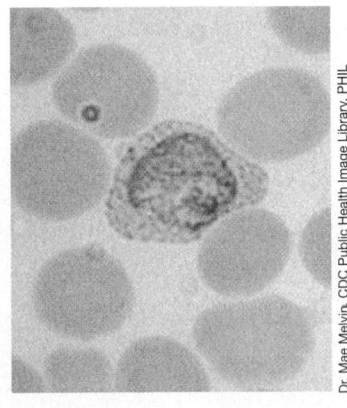

(a)   (b)   (c)   (d)

Figure 14.12 Typical protozoa. (a) *Amoeba*. (b) A typical ciliate, *Paramecium*. (c) A flagellate, *Dunaliella* (this flagellate contains chloroplasts and thus can also be considered an alga). (d) *Plasmodium vivax*, an apicomplexan sporozoan, growing in a human red blood cell.

TABLE 14.1 Characteristics of the major groups of protozoa

Group	Common name	Typical representatives	Habitats	Common diseases
Mastigophora	Flagellates	*Trypanosoma, Giardia, Leishmania, Trichomonas*	Freshwater; parasites of animals	African sleeping sickness, giardiasis, leishmaniasis
Euglenoids[a]	Phototrophic flagellates	*Euglena*	Freshwater; some marine	None known
Sarcodina	Amebas	*Amoeba, Entamoeba*	Freshwater and marine; animal parasites	Amebic dysentery (amebiasis)
Ciliophora	Ciliates	*Balantidium, Paramecium*	Freshwater and marine; animal parasites; rumen	Dysentery
Apicomplexa	Sporozoans	*Plasmodium, Toxoplasma*	Primarily animal parasites; insects (vectors for parasitic diseases)	Malaria, toxoplasmosis

[a] This group is also considered with the algae (see Section 14.11 and Table 14.3).

length, and are thin, crescent-shaped organisms. They have a single flagellum that originates in a basal body and folds back laterally across the cell where it is enclosed by a flap of surface membrane (Figure 14.13). Both the flagellum and the membrane participate in propelling the organism, making effective movement possible even in blood, which is rather viscous. *Trypanosoma gambiense* is the species that causes the chronic and usually fatal African sleeping sickness. In humans, the parasite lives and grows primarily in the bloodstream, but in the later stages of the disease, invasion of the central nervous system occurs, causing an inflammation of the brain and spinal cord that is responsible for the characteristic neurological symptoms of the disease. The parasite is transmitted from host to host by the tsetse fly, *Glossina* sp., a bloodsucking fly found only in certain parts of Africa. The parasite proliferates in the intestinal tract of the fly and invades the insect's salivary glands and mouthparts, from which it can be transferred to a new human host following a fly bite.

We describe *phototrophic* flagellates in Section 14.11. These are the *euglenoids*, flagellates that contain chloroplasts, which allow for photosynthetic growth. However, in darkness, cells of *Euglena* (see Figure 14.27a), a typical euglenoid, can survive and grow as chemoorganotrophs, and as such these organisms become phenotypically indistinguishable from protozoa. Many euglenoids are known and they are exclusively aquatic, inhabiting primarily fresh waters. Unlike other flagellated protozoa, the euglenoids are nonpathogenic.

Sarcodina: The Amebas

Among the sarcodines are organisms such as *Amoeba*, which are always naked in the vegetative phase (see Figure 14.12a), and the foraminifera, amebas that secrete a shell during vegetative growth (Figure 14.14). A wide variety of naked amebas are parasites of humans and other vertebrates, and their usual habitat is the oral cavity or the intestinal tract. They move in these habitats by *ameboid movement* (Figure 14.15), a mechanism also em-

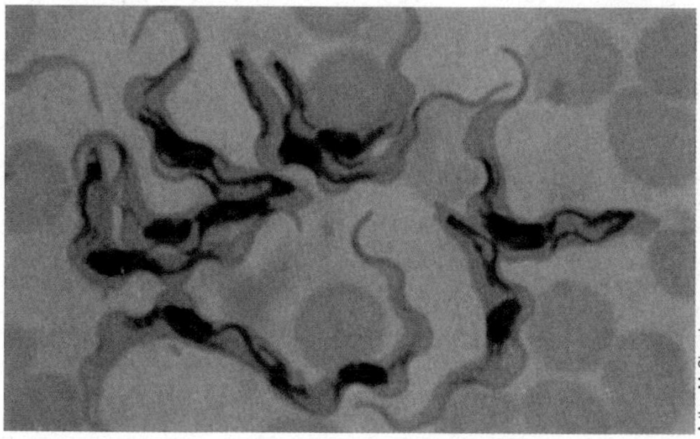

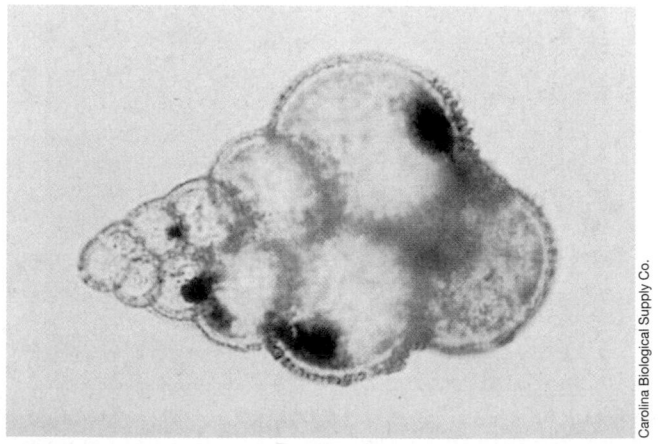

Figure 14.13 Photomicrograph of the flagellated protozoan *Trypanosoma gambiense*, the causative agent of African sleeping sickness, from a blood smear.

Figure 14.14 Shelled amebas: foraminifera. Note the ornate and multilobed test.

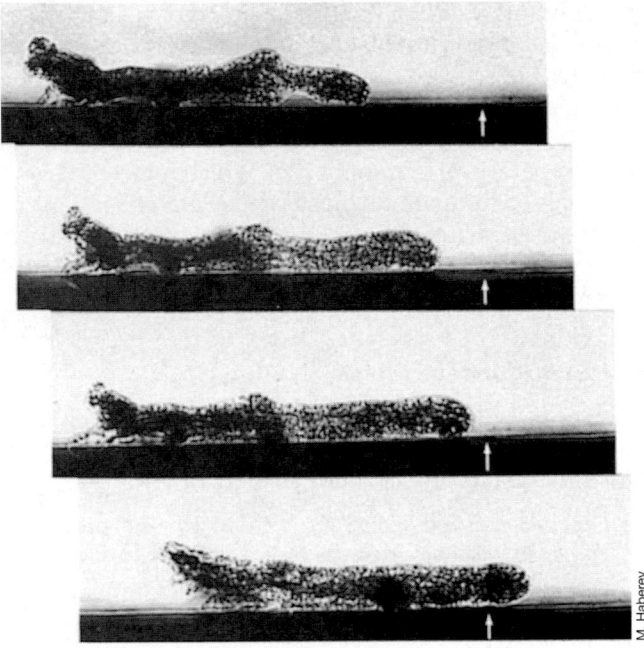

Figure 14.15 Side view of a moving ameba, *Amoeba proteus*, taken from a film, the time interval between frames being 2s. The arrows point to a fixed spot on the surface. A single cell is about 80 μm in diameter.

M. Haberey

ployed by the acellular slime molds (see Section 14.10). *Entamoeba histolytica* (∞ Figure 28.13) is a good example of a parasitic ameba. In many cases infection causes no obvious symptoms, but in some individuals it produces ulceration of the intestinal tract, which results in a diarrheal condition called *amebic dysentery* (amebiasis). The organism is transmitted from person to person in the cyst form by fecal contamination of water and food. We discuss the etiology and pathogenesis of amebic dysentery in Section 28.8.

Shelled sarcodines present a variety of interesting morphological forms. The best-known of the shelled forms are the *foraminifera*. Foraminifera are exclusively marine organisms, living primarily in coastal waters. The shells, called *tests*, of different species show distinctive characteristics and are often quite ornate (Figure 14.14). Tests are usually made of calcium carbonate. The cell is not firmly attached to the test, and the ameba cell may extend partway out of the shell during feeding. However, because of the weight of the test, the cell usually sinks to the bottom, and it is thought that the organisms feed on particulate deposits in the sediments, primarily bacteria and detritus. The shells of foraminifera are relatively resistant to decay and hence readily become fossilized (the White Cliffs of Dover, England, are composed to a great extent of foraminiferal shells). Because of the excellent fossil record these organisms leave, we have a better idea of their distribution through geological time than for virtually any other protozoa.

Ciliophora: The Ciliates

Ciliates are those protozoa that, in some stage of their life cycle, possess cilia (Figure 14.16). They are also unique among protozoa in having *two kinds* of nuclei: the *micronucleus*, which is concerned only with inheritance and sexual reproduction; and the *macronucleus*, which is involved only in the production of messenger ribonucleic acid (mRNA) for various aspects of cell growth and function (see Figure 14.17).

Probably the best-known and most widely distributed of the ciliates are those of the genus *Paramecium* (Figure 14.16), which will be used here as an example of the group. Most ciliates obtain their food by ingesting particulate materials through a distinct oral region or mouth connected to an underlying gullet (Figure 14.16b). Once inside, the food particles are carried down the gullet and into the cytoplasm where they are enclosed in a food vacuole, a structure into which digestive enzymes are secreted. In addition to cilia, which function in motility, many ciliates have *trichocysts*, which are long, thin filaments of a contractile nature, anchored beneath the surface of the outer cell layer. These enable the protozoa to attach to a surface and can aid in defense by signaling the cell that it is being attacked by a predator, thus stimulating cellular defenses. In the case of predatory ciliates, such as *Didinium*, trichocysts hold onto and paralyze the prey as a prelude to ingestion.

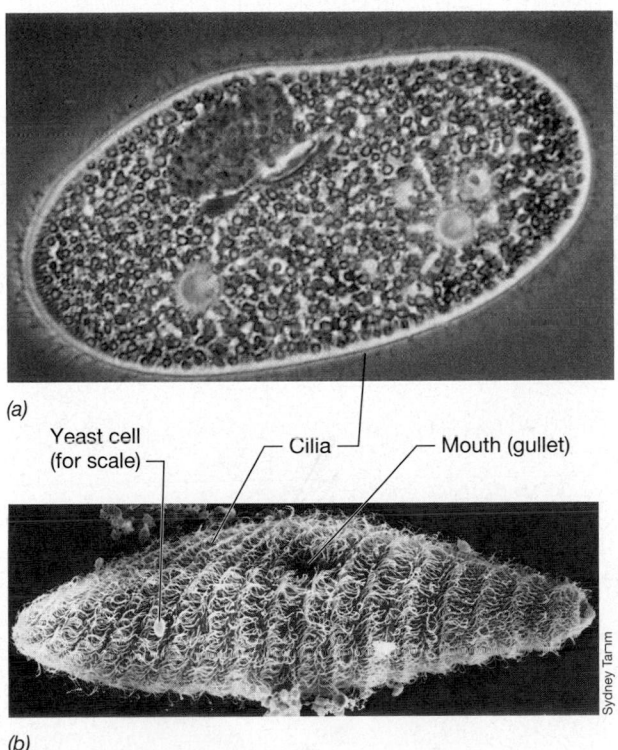

(a)

Yeast cell (for scale) — Cilia — Mouth (gullet)

(b)

Sydney Tamm

Figure 14.16 *Paramecium*, a ciliated protozoan. (a) Phase photomicrograph. (b) Scanning electron micrograph. Note the cilia in both micrographs. A single *Paramecium* cell is about 60 μm in diameter.

Many *Paramecium* species (as well as many other protozoa) are hosts for endosymbiotic bacteria living in the cytoplasm or the macronucleus. In some cases, evidence exists that these endosymbionts play a nutritional role for the host, synthesizing vitamins or other growth factors that would otherwise have to be obtained from the environment. In the case of protozoa existing in the termite gut, we previously discussed how endosymbiotic methanogens remove H_2 produced from pyruvate oxidation in the hydrogenosome (Figure 14.4), yielding CH_4, which is released to the atmosphere (∞ Section 19.10 and Figure 19.26*b,c*).

Although a few ciliates are parasitic for animals, this mode of existence is less extensively developed in the ciliates than it is in other groups of protozoa. The species *Balantidium coli* (Figure 14.17) is primarily a parasite of domestic animals, but occasionally it infects the intestinal tract of humans, producing intestinal dysentery symptoms similar to those caused by *Entamoeba histolytica*. In addition, there is usually a characteristic fauna of *obligately anaerobic* ciliates in the rumen, the forestomach of ruminant animals (∞ Section 19.11); these protozoa play a beneficial role in the digestive and fermentative processes that occur there.

Apicomplexa (Sporozoans)

The Apicomplexa, or *sporozoans* (Figure 14.12*d*), comprise a large group of protozoa, all of which are obligate parasites. They are characterized by a *lack* of motile adult stages and by a nutritional mode of life in which food is absorbed in soluble form through the outer wall, such as occurs in prokaryotes and in fungi. Although the name *sporozoa* implies the formation of spores, these organisms do not form true resting spores, like those of bacteria, algae, and fungi, but instead produce analogous structures called *sporozoites*, which are involved in transmission to a new host. Numerous kinds of vertebrates and

invertebrates are hosts for Sporozoa, and in some cases an alternation of hosts takes place, with some stages of the life cycle occurring in one host and some in another. The most important members of the sporozoa are the coccidia, usually parasites of birds, and the plasmodia (malaria parasites) (Figure 14.12*d*), which infect birds and mammals, including humans. Because malaria is a major disease of humans, especially in developing countries, we devote a considerable discussion to this disease and the properties of malarial parasites in Section 27.5.

✓ 14.8 Concept Check

Protozoa are unicellular microbial *Eukarya* that lack cell walls and are usually motile by various means. Many protozoa are pathogenic to humans and other animals.

- ✓ List at least two major ways in which the protozoan *Paramecium* differs from the protozoan *Trypanosoma*.
- ✓ How do the Sporozoa differ from all other protozoa?
- ✓ Why is the alga *Euglena* also considered a protozoan?

14.9 Fungi

Key Genera

Penicillium
Aspergillis
Saccharomyces
Candida

In contrast to the protozoa, fungi contain cell walls and produce spores, among many other differences, and most described species form a relatively tight phylogenetic cluster (Figure 14.11). Three major groups of fungi are recognized: the *molds*, the *yeasts*, and the *mushrooms*.

The habitats of fungi are quite diverse. Some are aquatic, living primarily in fresh water, and a few marine fungi are also known. Most fungi, however, have terrestrial habitats, in soil or on dead plant matter, and these types often play crucial roles in the mineralization of organic carbon in nature. A large number of fungi are parasites of terrestrial plants. Indeed, fungi cause the majority of economically significant diseases of crop plants (Table 14.2). A few fungi are parasitic on animals, including humans, although in general fungi are less significant as animal pathogens than are bacteria and viruses (∞ Section 27.7 for a discussion of pathogenic fungi).

Cell Walls and Metabolism

Fungal cell walls resemble plant cell walls architecturally, but not chemically. Although cellulose is present in the walls of certain fungi, most fungi have noncellulosic walls. **Chitin**, a polymer of the glucose derivative, *N*-acetylglucosamine (∞ Figure 3.5), is a common con-

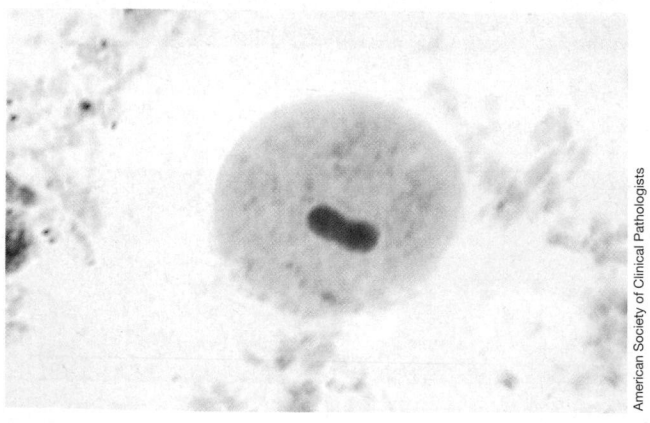

Figure 14.17 *Balantidium coli*, a ciliated protozoan that causes a dysentery-like disease in humans. The dark blue stained structure is the macronucleus.

American Society of Clinical Pathologists

TABLE 14.2 Classification and major properties of fungi[a]

Group	Common name	Hyphae	Typical representatives	Type of sexual spore	Habitats	Common diseases
Ascomycetes	Sac fungi	Septate	*Neurospora, Saccharomyces, Morchella* (morels)	Ascospore	Soil, decaying plant material	Dutch elm, chestnut blight, ergot, rots
Basidiomycetes	Club fungi, mushrooms	Septate	*Amanita* (poisonous mushroom), *Agaricus* (edible mushroom)	Basidiospore	Soil, decaying plant material	Black stem, wheat rust, corn smut
Zygomycetes	Bread molds	Coenocytic	*Mucor, Rhizopus* (common bread mold)	Zygospore	Soil, decaying plant material	Food spoilage; rarely involved in parasitic disease
Oomycetes	Water molds	Coenocytic	*Allomyces*	Oospore	Aquatic	Potato blight, certain fish diseases
Deuteromycetes	Fungi imperfecti	Septate	*Penicillium, Aspergillus, Candida*	None known	Soil, decaying plant material, surfaces of animal bodies	Plant wilt, infections of animals such as ringworm, athlete's foot, surface or systemic infections (*Candida*)

[a] With the exception of the Oomycetes, which are phylogenetically distinct, the other groups of fungi are closely related (see Figure 14.11).

stituent of fungal cell walls. It is laid down in microfibrillar bundles like cellulose; other glucans such as mannans, galactosans, and chitosans replace chitin in some fungal cell walls. Fungal cell walls are generally 80–90% polysaccharide, with proteins, lipids, polyphosphates, and inorganic ions making up the wall-cementing matrix. An understanding of fungal cell wall chemistry is important because of the extensive biotechnological uses of fungi (Chapter 30) and because the chemical nature of the fungal cell wall has been useful in classifying fungi for research and industrial purposes.

Fungi are chemoorganotrophs and generally have simple nutritional requirements. Many species can grow at environmental extremes of low pH or high temperature (up to 62°C) and this, coupled with the ubiquitous nature of fungal spores, makes these organisms common contaminants of food products, microbial culture media, and most surfaces. Molds and yeasts are not, however, classified on physiological grounds but instead by their diverse array of life cycle patterns including the formation of a variety of different sexual spores (see Table 14.2).

Filamentous Fungi: Molds

The molds are *filamentous fungi*. They are widespread in nature and are commonly seen on stale bread, cheese, or fruit. Each filament grows mainly at the tip, by extension of the terminal cell (Figure 14.18). A single filament is called a *hypha* (plural, *hyphae*). Hyphae usually grow together across a surface and form compact tufts, collectively called a *mycelium*, which can be seen easily without a microscope. The mycelium arises because the individual hyphae form branches as they grow, and these branches intertwine, resulting in a compact mat. In most cases, the vegetative cell of a fungal hypha contains more than one nucleus—often hundreds of nuclei are present. Thus, a typical hypha is a nucleated tube containing cytoplasm (referred to as *coenocytic*).

From the fungal mycelium, other hyphal branches may reach up into the air above the surface, and on these aerial branches spores called **conidia** are formed (Figure 14.18a). Conidia are *asexual* spores (their formation does *not* involve the fusion of gametes), often highly pigmented (Figure 14.19;   Figure 10.1a) and resistant to drying, and function in the dispersal of the fungus to new habitats. When conidia form, the white color of the mycelium changes, taking on the color of the conidia, which may be black, blue-green, red, yellow, or brown. The presence of these spores gives the mycelial mat a rather dusty appearance (Figure 14.19a).

Some molds also produce *sexual* spores, formed as a result of sexual reproduction (Table 14.2). The latter occur from the fusion either of unicellular gametes or of specialized hyphae called *gametangia*. Alternatively, sexual spores can originate from the fusion of two haploid

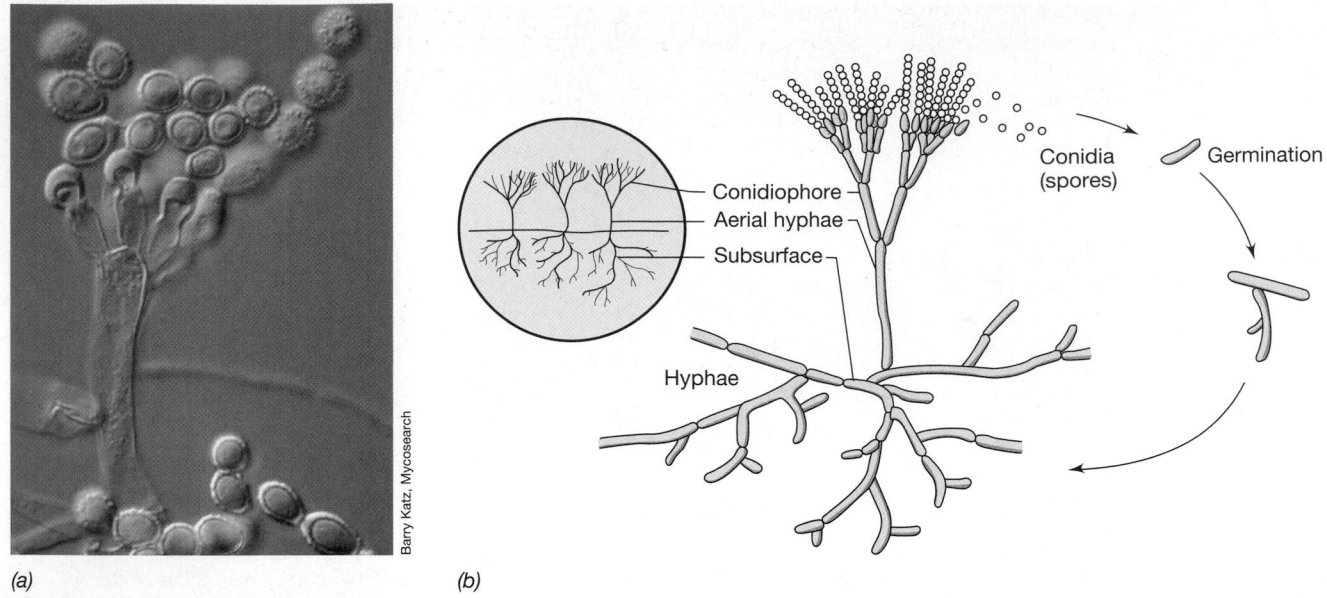

Barry Katz, Mycosearch

(a)

(b)

Figure 14.18 Mold structure and growth. (a) Photomicrograph of a typical mold. Conidia are seen as the spherical structures at the ends of aerial hyphae. (b) Diagram of a mold life cycle.

cells to yield a diploid cell, which then undergoes meiosis and mitosis to yield individual spores. Depending on the group to which a particular fungus belongs (see Table 14.2), different types of sexual spores are produced. Spores formed within an enclosed sac (*ascus*) are called *ascospores* (see Figure 14.8), and those produced on the ends of a club-shaped structure (*basidium*) are *basidiospores* (Table 14.2). *Zygospores*, produced by zygomycetous fungi like the common bread mold *Rhizopus*, are macroscopically visible structures and the result of the fusion of hyphae and genetic exchange. Eventually the zygospore matures and produces asexual spores that are dispersed by air and germinate to form new fungal mycelia.

Sexual spores of fungi are usually resistant to drying, heating, freezing, and some chemical agents. However, fungal sexual spores are not as resistant to heat as are bacterial endospores (∞ Section 4.15). Either an asexual or a sexual spore of a fungus can germinate and develop into a new hypha and mycelium.

A major ecological activity of many fungi, especially members of the Basidiomycetes (see Table 14.2), is the decomposition of wood, paper, cloth, and other products derived from natural sources. Basidiomycetes that attack these products are able to use cellulose or lignin from the product as carbon and energy sources. Lignin is a complex polymer in which the building blocks are phenolic compounds. It is an important constituent of woody plants, and in association with cellulose it confers rigidity on them. The decomposition of lignin in nature occurs almost exclusively through the action of certain Basidiomycetes called *wood-rotting fungi*. Two types of

wood rots are known: *brown rot*, in which the cellulose is attacked preferentially and the lignin left unchanged, and *white rot*, in which both cellulose and lignin are decomposed. The white rot fungi are of considerable ecological interest because they play such an important role in decomposing woody material in forests.

Macroscopic Fungi: Mushrooms

Mushrooms are filamentous basidiomycetes that form large *fruiting bodies*, the edible part of the mushroom (Figure 14.20a–c). We will discuss in Section 30.14 the commercial growth of mushrooms as a food source. During most of its existence, the mushroom fungus lives as a simple mycelium, growing in soil, leaf litter, or decaying logs. However, when environmental conditions are favorable, usually following periods of wet and cool weather, the fruiting body develops, beginning first as a small button-shaped structure underground and then expanding into the full-grown fruiting body that we see above ground (Figure 14.20c). Sexual spores, called **basidiospores** (Figure 14.20d) are formed, borne on the underside of the fruiting body on flat plates called **gills**, which are attached to the cap of the mushroom (Figure 14.20b,c). The mushroom basidiospores are dispersed by wind and eventually light on a favorable, usually moist and organic rich soil, and begin the cycle again (Figure 14.20b).

Unicellular Fungi: Yeasts

The yeasts are *unicellular fungi*, and most of them are classified with the Ascomycetes. Yeast cells are usually spherical, oval, or cylindrical, and cell division generally takes place by budding (Figure 14.21). In the budding

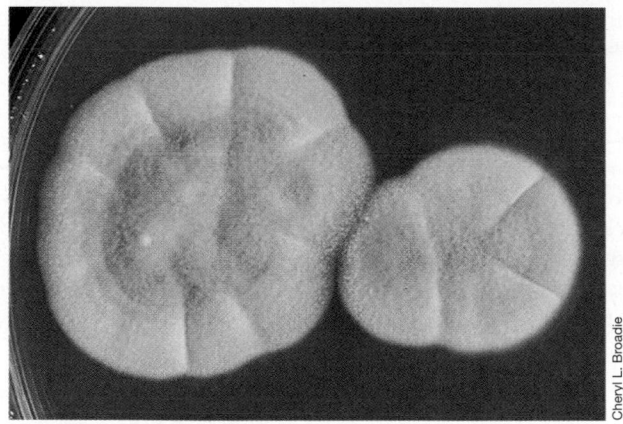

(a)

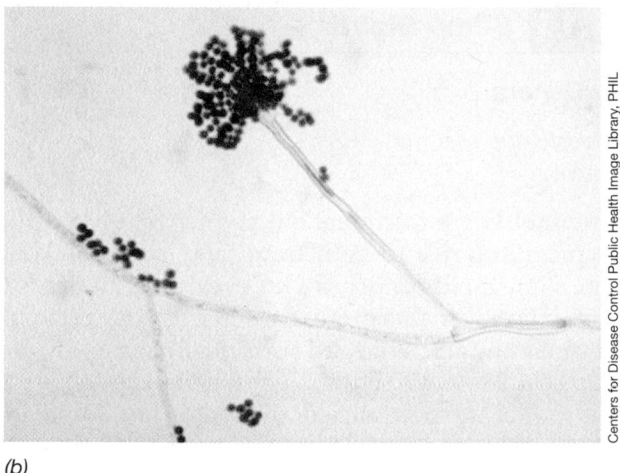

(b)

Centers for Disease Control Public Health Image Library, PHIL

Figure 14.19 Fungi. (a) Colonies of an *Aspergillis* species growing on an agar plate. Note the appearance of the masses of filamentous cells (the mycelium) and asexual spores (see Figure 14.18*b*) that give the colonies a dusty, matted appearance. (b) Conidiophore and conidia of *Aspergillis fumigatus*.

process, a new cell forms as a small outgrowth of the old cell; the bud gradually enlarges and then separates (Figures 14.21 and 14.22). Although most yeasts reproduce only as single cells, under some conditions some yeasts can form filaments. And in these species, certain characteristics may be expressed only by the filamentous form. For example, the filamentous phase is essential for pathogenicity in *Candida albicans*, a yeast that can cause vaginal, oral, or lung infections and, in acquired immunodeficiency (AIDS) patients, systemic tissue damage (∞ Section 26.14).

Figure 14.20 Mushrooms. (a) *Amanita*, a highly poisonous mushroom. (b) Life cycle of a typical mushroom. (c) Gills on the underside of the mushroom fruiting body contain the spore-bearing basidia. (d) Scanning electron micrograph of basidiospores released from mushroom basidia. The production of mushrooms as a food source will be covered in Section 30.14.

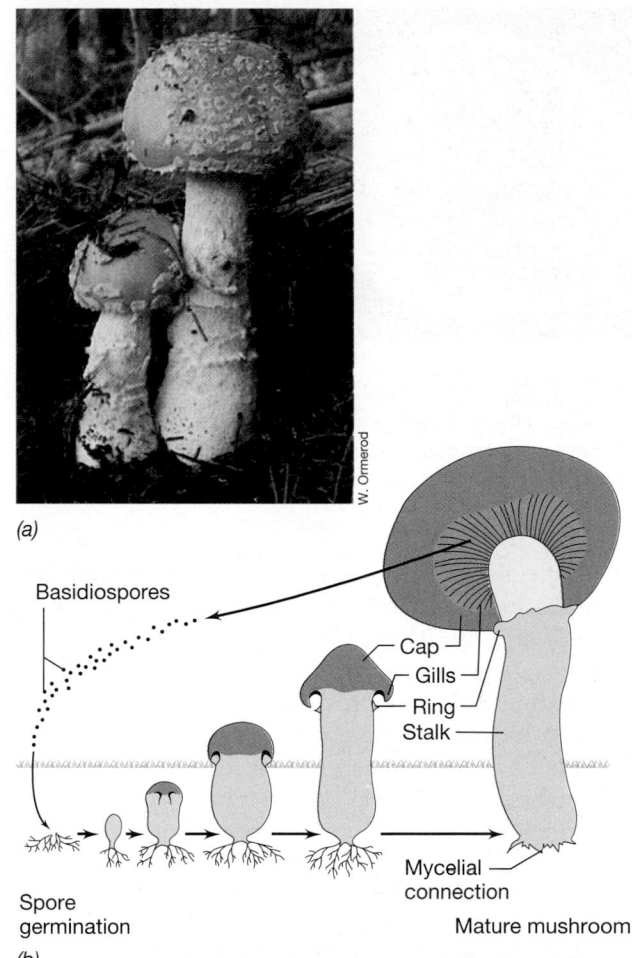

(a)

W. Ormerod

(b)

(c)

USDA

(d)

S. L. Fleger

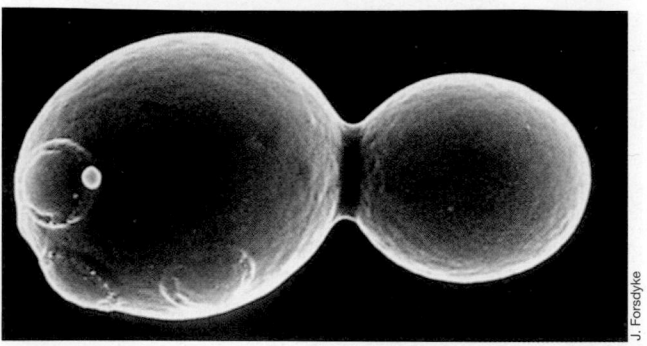

Figure 14.21 Scanning electron micrograph of the common baker's and brewer's yeast *Saccharomyces cerevisiae*. Note the budding division and previous bud scars. A single large cell is about 8 μm in diameter.

Yeast cells are much larger than bacterial cells and can be distinguished microscopically from bacteria by their size and by the obvious presence of internal cell structures, such as the nucleus (Figure 14.22). Some yeasts exhibit sexual reproduction by a process called *mating*, in which two yeast cells fuse. Within the fused cell, called a *zygote*, ascospores are eventually formed. We discussed the sexual cycle of a typical yeast, *Saccharomyces*, including the important property of *mating types*, in Section 14.6 (see Figures 14.8 and 14.9).

Yeasts flourish in habitats where sugars are present, such as fruits, flowers, and the bark of trees. A number of yeast species live symbiotically with animals, especially insects, and a few species are pathogenetic for animals and humans (∞ Section 27.7). The most important commercial yeasts are the baker's and brewer's yeasts, which are members of the genus *Saccharomyces* (∞ box, The Products of Yeast Fermentation, Chapter 5). The original habitats of these yeasts were undoubtedly fruits and fruit juices, but the commercial yeasts of today are probably quite different from wild strains because they have been greatly improved through the years by careful selection and genetic manipulation by industrial microbiologists. Indeed, *S. cerevisiae* has been studied as a model eukaryote for many years (see Section 14.6) and was the first eu-

karyote to have its genome completely sequenced (∞ Section 15.6).

✓ **14.9 Concept Check**

Fungi include the molds and yeasts and as a group differ from protozoa by virtue of their rigid cell wall, production of spores, lack of motility, and phylogenetic position. Mushrooms are large, often edible fungi that produce fruiting bodies containing basidiospores.

✓ How do *molds* differ from *yeasts*?

✓ What is *chitin* and what is its function in fungi?

✓ In molds, how do *conidia* differ from *ascospores*?

14.10 Slime Molds

Key Genera

Dictyostelium
Physarum

Slime molds are microbial eukaryotes that have phenotypic similarity to both fungi and protozoa. Like fungi, slime molds undergo a life cycle and can produce spores. However, like protozoa, slime molds are motile and can move across a solid surface rather rapidly (see Figures 14.23–14.25). From a phylogenetic perspective slime molds are more ancient than fungi and some protozoa (such as the ciliates), but more derived than flagellated protozoa and their evolutionary predecessors (Figure 14.11).

The slime molds can be divided into two groups, the *cellular slime molds*, whose vegetative forms are composed of single amebalike cells, and the *acellular slime molds*, whose vegetative forms are masses of protoplasm of indefinite size and shape called *plasmodia* (Figure 14.23).

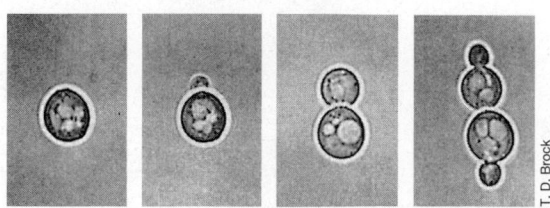

Figure 14.22 Growth by budding division in *Saccharomyces cerevisiae*. Note the pronounced nucleus. Phase-contrast micrograph. A single cell of *S. cerevisiae* is about 8 μm in diameter.

Figure 14.23 Slime molds. Plasmodia of the acellular slime mold *Physarum* growing on an agar surface.

Slime molds live primarily on decaying plant matter, such as leaf litter, logs, and soil. Their food consists mainly of other microorganisms, especially bacteria, which they ingest by phagocytosis. Slime molds can maintain themselves in a vegetative state for long periods, but eventually form differentiated spore-like structures that germinate and once again generate the active amoeboid state.

Acellular Slime Molds

In the vegetative phase, acellular slime molds such as *Physarum* exist as a mass of protoplasm of indefinite size (Figure 14.23). This structure is actively motile by *amoeboid motion*, the plasmodium flowing over the surface of the substratum, engulfing food particles as it moves. Ameboid motion is the result of cytoplasmic streaming. Cytoplasm flows forward because the tip of the plasmodium is less contracted and viscous, and thus cytoplasm takes the path of least resistance. In eukaryotic cells, cytoplasmic streaming is facilitated by microfilaments, which exist in a thin layer just beneath the cytoplasmic membrane of eukaryotic cells (see Section 14.4 and Figure 14.1). In acellular slime molds, cytoplasmic streaming occurs in definite strands, each surrounded by the thin cytoplasmic membrane (Figure 14.23). The

streaming itself is a mechanism for distributing cellular metabolites.

The acellular slime mold plasmodium (Figure 14.23) is genetically diploid. From this mass of protoplasm a sporangium and haploid spores can be produced. Under favorable conditions, spores germinate to yield haploid swarm cells. The fusion of two swarm cells then regenerates the diploid plasmodium (Figure 14.23).

Cellular Slime Molds: *Dictyostelium*

Dictyostelium discoideum, a cellular slime mold, undergoes a remarkable life cycle in which vegetative cells aggregate, migrate as a cell mass, and eventually produce fruiting bodies in which cells differentiate and form spores (Figures 14.24 and 14.25). As cells of *Dictyostelium* become starved, they aggregate and form a *pseudoplasmodium*, a structure in which the cells lose their individuality but do not fuse (Figures 14.24 and 14.25). This aggregation is triggered by the production of two compounds, *cyclic adenosine monophosphate* (cAMP) and a specific glycoprotein, both of which function as chemotactic agents (we discussed the involvement of cAMP in various regulatory systems in prokaryotes in Section 8.7). Those cells that are the first to produce these compounds

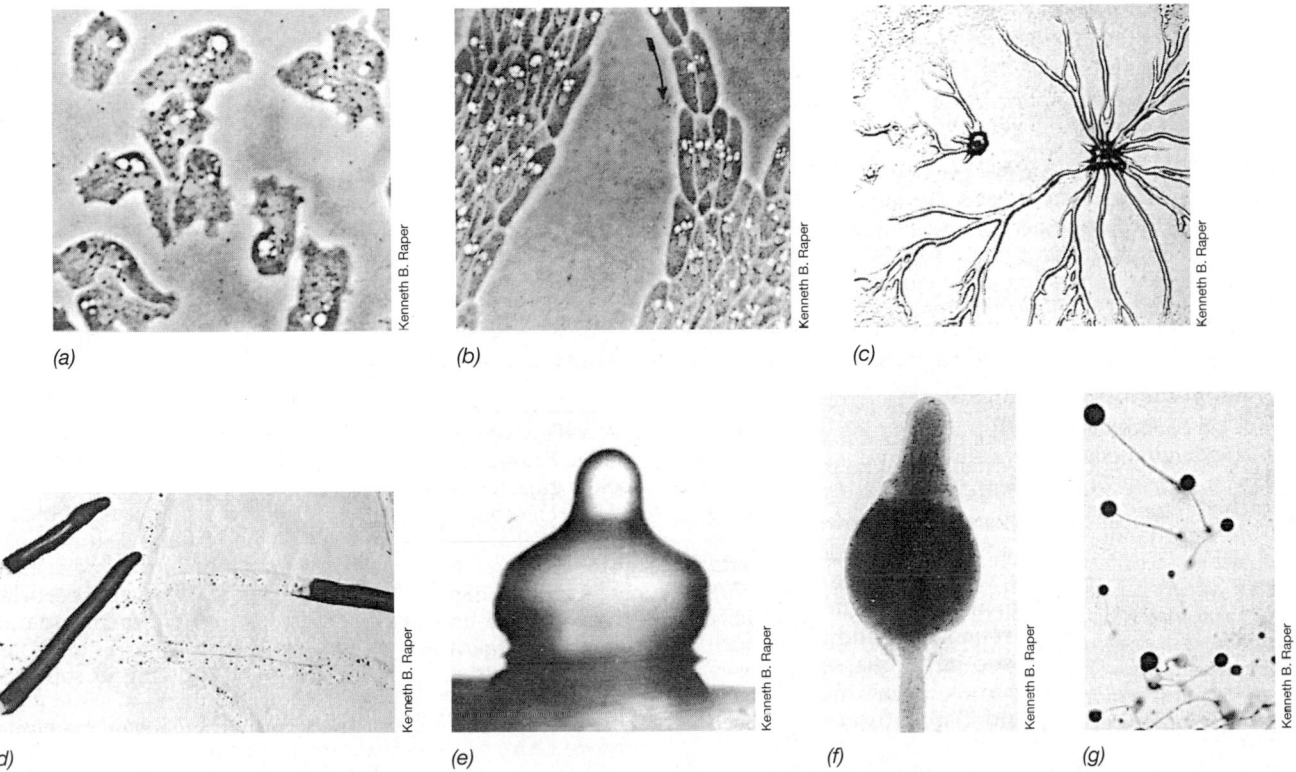

(a) (b) (c)

(d) (e) (f) (g)

Figure 14.24 Photomicrographs of various stages in the life cycle of the cellular slime mold *Dictyostelium discoideum*. (a) Amebas in preaggregation stage. Note irregular shape and lack of orientation. (b) Aggregating amebas. Notice the regular shape and orientation. The cells are moving in streams in one direction. (c) Low power view of aggregating amebas. (d) Migrating pseudoplasmodia (slugs) moving on an agar surface and leaving trails of slime in their wake. (e, f) Early stage of fruiting body. (g) Mature fruiting bodies. See Figure 14.25 for sizes of these structures.

Figure 14.25 Stages in fruiting-body formation in the cellular slime mold *Dictyostelium discoideum*. (A-C) Aggregation of amebas; (D-G) migration of the slug formed from aggregated ameba; (H-L) culmination and formation of the fruiting body; (M) mature fruiting body. The fruiting body contains spores that can regenerate vegetative cells and begin the life cycle anew.

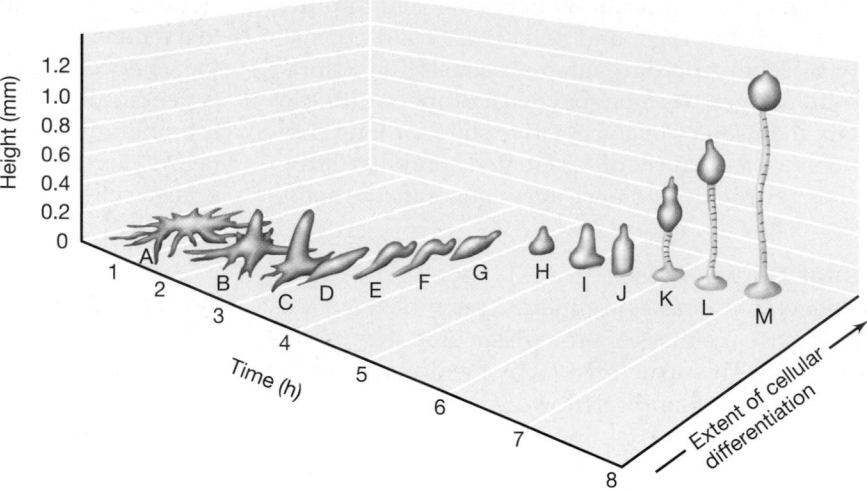

become centers for the attraction of other vegetative cells, leading to aggregating masses of cells that come together to form a slimy migrating mass referred to as a *slug* (Figure 14.24).

Fruiting-body formation begins when the slug ceases to migrate and becomes vertically oriented (Figures 14.24 and 14.25). The fruiting body then becomes differentiated into a stalk and a head; cells in the forward end of the slug become stalk cells, and those in the posterior end become spores. Cells that form stalk cells begin to secrete cellulose, which provides the rigidity of the stalk. Cells from the rear of the slug swarm up the stalk to the tip and form the head. Most of these posterior cells differentiate into spores. On maturation of the head the spores are released and dispersed. Each spore then germinates and becomes a vegetative ameba.

The cycle of fruiting-body and spore formation in *Dictyostelium* is an *asexual* process. However, *sexual* spores called *macrocysts* can also be produced. Macrocysts are formed from aggregates of amebas that become enclosed in a cellulose wall. Following the conjugation of two amebas, a single large ameba develops and proceeds to phagocytize the remaining amebas. At this point a thick cellulose wall forms around this giant ameba forming the mature macrocyst, which can remain dormant for long periods. Eventually, the diploid nucleus undergoes meiosis to form haploid nuclei that become integrated into new amebas that can once again initiate vegetative growth (Figure 14.24).

✓ **14.10 Concept Check**

Acellular slime molds are masses of motile protoplasm while cellular slime molds are masses of individual cells that aggregate to form fruiting bodies from which spores are released.

✓ In what ways are the slime molds similar to *fungi* and in what way are they similar to *protozoa*?

✓ What is a *macrocyst*?

14.11 Algae

Key Genera

Chlamydomonas
Euglena
Gonyaulax

Algae are a large and diverse group of eukaryotic organisms (Figures 14.26 and 14.27) that contain *chlorophyll* and carry out oxygenic photosynthesis (∞ Section 17.5). Algae should not be confused with *cyanobacteria*, which are also oxygenic phototrophs but which are *Bacteria* and thus evolutionarily quite distinct from algae (∞ Section 12.26). Although most algae are of microscopic size and hence are clearly microorganisms, a number of forms are macroscopic, some seaweeds growing to over 30 m in length.

Algal Diversity

Algae are either unicellular (Figure 14.26*a* and *c*) or colonial, the latter occurring as aggregates of cells (Figure 14.26*b*). When the cells are arranged end to end, the alga is said to be filamentous (Figure 14.26*d*). Among the filamentous forms, both unbranched filaments and more intricate branched filaments occur. Algae contain chlorophyll and are thus green in color. However, a few kinds of common algae are not green but appear brown or red because in addition to chlorophyll, other pigments such as xanthophylls are present that mask the green color. Algal cells contain one or more **chloroplasts**, membranous structures that house the photosynthetic pigments (see Section 14.3). Chloroplasts can often be recognized microscopically within algal cells by their distinct green color (Figure 14.26). We discussed the general structure and properties of chloroplasts in Section 14.3, and the phylogeny of the chloroplast in Sections 12.27 and 14.4.

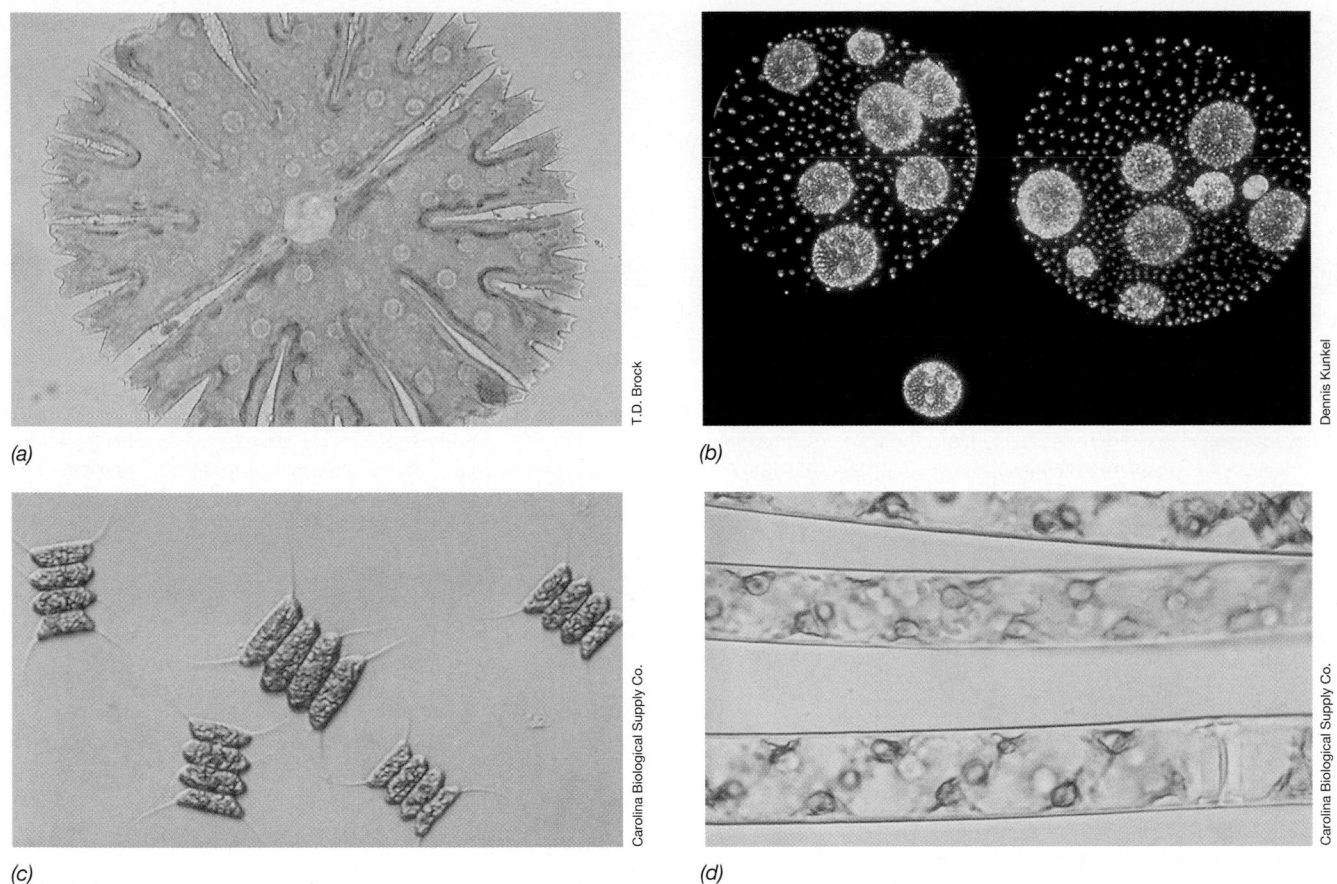

(a) T.D. Brock

(b) Dennis Kunkel

(c) Carolina Biological Supply Co.

(d) Carolina Biological Supply Co.

Figure 14.26 Light micrographs of representative green algae. (a) *Micrasterias*. A single cell. (b) *Volvox* colony, containing a large number of cells. (c) *Scenedesmus*. A packet of four cells. (d) *Spirogyra*. A filamentous alga. Note the green spiral-shaped chloroplasts.

Examination of Figure 14.11 shows the algae to constitute a phylogenetically heterogeneous group. Although green algae (Chlorophyta, Table 14.3) and to a lesser extent red algae (Rhodophyta, Table 14.3) are quite closely related to green plants, other algal groups such as the brown algae and diatoms constitute earlier lineages (Figure 14.11). Even less derived are the euglenoids, such as the alga *Euglena* (Figure 14.27a). Euglenoids show a phylogenetic relationship to flagellated protozoa (Figure 14.11), and in fact, cells of *Euglena* can spontaneously lose their chloroplasts and exist as completely heterotrophic organisms (see Section 14.8). Red algae (Figure 14.27b) are noteworthy in that their chloroplasts contain *phycobiliproteins*, the major light-harvesting pigments of the cyanobacteria (⊂⊃ Section 17.3). It is thus not surprising that the red algal chloroplast, an endosymbiont of course (see Section 14.4), shows close phylogenetic affinities to the cyanobacteria (⊂⊃ Section 12.25); their host cell, by contrast, forms a distinct line of Eukarya (see Figure 14.11).

Dinoflagellates are flagellated, primarily marine algae that are close phylogenetic relatives of ciliated protozoa (Figure 14.11). Some dinoflagellates are free-living (Figure 14.27f) while others live a symbiotic existence with animals that make up coral reefs, obtaining a sheltered and protected habitat in exchange for supplying photosynthetically fixed carbon as a food source for the reef.

Toxic Dinoflagellates

Dinoflagellates such as *Gonyaulax* occasionally grow in dense suspensions called *blooms* that are bright red in color from the xanthophylls present as accessory pigments in this organism. These blooms are the classic "red tides" (Figure 14.28a, page 496) that occasionally develop in warm, usually polluted, coastal waters and that are associated with fish kills and poisoning in humans following consumption of contaminated water or shellfish. Toxicity occurs because of a potent neurotoxin produced by *Gonyaulax* that is sufficiently toxic to kill fish at nanogram levels.

Pfiesteria is another genus of toxic dinoflagellates. Although capable of photosynthesis, this organism is more notable as a fish pathogen and occasional pathogen of humans. Toxic spores of *P. piscicida* (Figure 14.28b) infect fish and eventually kill them from neurotoxins that affect movement and destroy skin. Lesions form on areas of the

TABLE 14.3   Properties of major groups of algae

Algal group	Common name	Morphology	Pigments	Typical represen-tative	Carbon reserve materials	Cell wall	Major habitats
Chlorophyta	Green algae	Unicellular to leafy	Chlorophylls *a* and *b*	*Chlamydomonas*	Starch (α-1,4-glucan), sucrose	Cellulose	Freshwater, soils, a few marine
Euglenophyta[a]	Euglenoids	Unicellular flagellated	Chlorophylls *a* and *b*	*Euglena*	Paramylon (β-1,2-glucan)	No wall present	Freshwater, a few marine
Dinoflagellata	Dinoflagellates	Unicellular, flagellated	Chlorophylls *a* and *c*, xanthophylls	*Gonyaulax* *Pfiesteria*	Starch (α-1,4-glucan)	Cellulose	Mainly marine
Chrysophyta	Golden-brown algae, diatoms	Unicellular	Chlorophylls *a* and *c*	*Nitzschia*	Lipids	Two over-lapping components made of silica	Freshwater, marine, soil
Phaeophyta	Brown algae	Filamentous to leafy, occasionally massive and plantlike	Chlorophylls *a* and *c*, xanthophylls	*Laminaria*	Lammarin (β-1,3-glucan), mannitol	Cellulose	Marine
Rhodophyta	Red algae	Unicellular, filamentous to leafy	Chlorophylls *a* and *d*, phycocyanin, phycoerythrin	*Polysiphonia*	Floridean starch (α-1,4- and α-1,6-glucan)	Cellulose	Marine

[a] This group is also considered with the protozoa (see Section 14.8).

fish where the skin has been removed, and this allows the growth of opportunistic bacterial pathogens (Figure 14.28c). Fish killed by *P. piscicida* infection are then fed upon by the amoeboid form of *Pfiesteria*, which is itself a harmless saprophyte. All told, over 20 different "stages" are known in the life cycle of *Pfiesteria*, only one of which, the toxic spore, seems capable of causing disease. It appears that the presence of fish in some way triggers formation of the toxic stage in the *Pfiesteria* life cycle, and thus fish are the main target of infection. *P. piscicida* can cause massive fish kills (Figure 14.28c). For example, over 1 billion fish were killed by a *Pfiesteria* outbreak in the Nuese Estuary, North Carolina, in 1991, and other outbreaks along the eastern coast of the United States have been suspected.

Humans can also be harmed by *P. piscicida*, and those inflicted display a variety of seemingly unrelated symptoms. These include respiratory problems, headaches, aching joints and muscles, disorientation, and memory loss. As mentioned previously, in other toxic dinoflagellates, such as *Gonyaulax*, human disease occurs when people come in contact with contaminated water or fish, especially shellfish. But this is not true of *Pfiesteria*. Instead, the organism produces a volatile neurotoxin that can cause symptoms even when no direct contact with infected material has occurred. Contaminated water is also a vehicle for transmitting disease, although shellfish harvested from *Pfiesteria*-contaminated waters ap-

parently do not transmit disease. *Pfiesteria* is thus a very interesting and complex "alga", whose ecology may be tied much more to a lifestyle as a fish pathogen than as a phototroph.

Pigments, Energy Metabolism, and Reserve Polymers

Several characteristics are used to classify algae, including the nature of the chlorophyll(s) present, the carbon reserve polymers produced (for example, starch and several starch derivatives), the cell wall structure, and the type of motility. All algae contain chlorophyll *a*. Some, however, also contain other chlorophylls that differ in minor ways from chlorophyll *a*. The presence of these additional chlorophylls is characteristic of particular algal groups. The distribution of chlorophylls and other photosynthetic pigments in algae is summarized in Table 14.3.

All algae carry out oxygen-evolving (oxygenic) photosynthesis, using H_2O as an electron donor (∞ Section 17.5). In addition, some algae carry out photosynthesis *without* yielding oxygen, using H_2 as an electron donor. Many algae are obligate phototrophs and are thus unable to grow in darkness on organic carbon compounds. However, some species can grow chemoorganotrophically and catabolize simple sugars or organic acids in the dark. One of the organic compounds most widely used by algae is acetate, which can be used as a sole carbon and energy

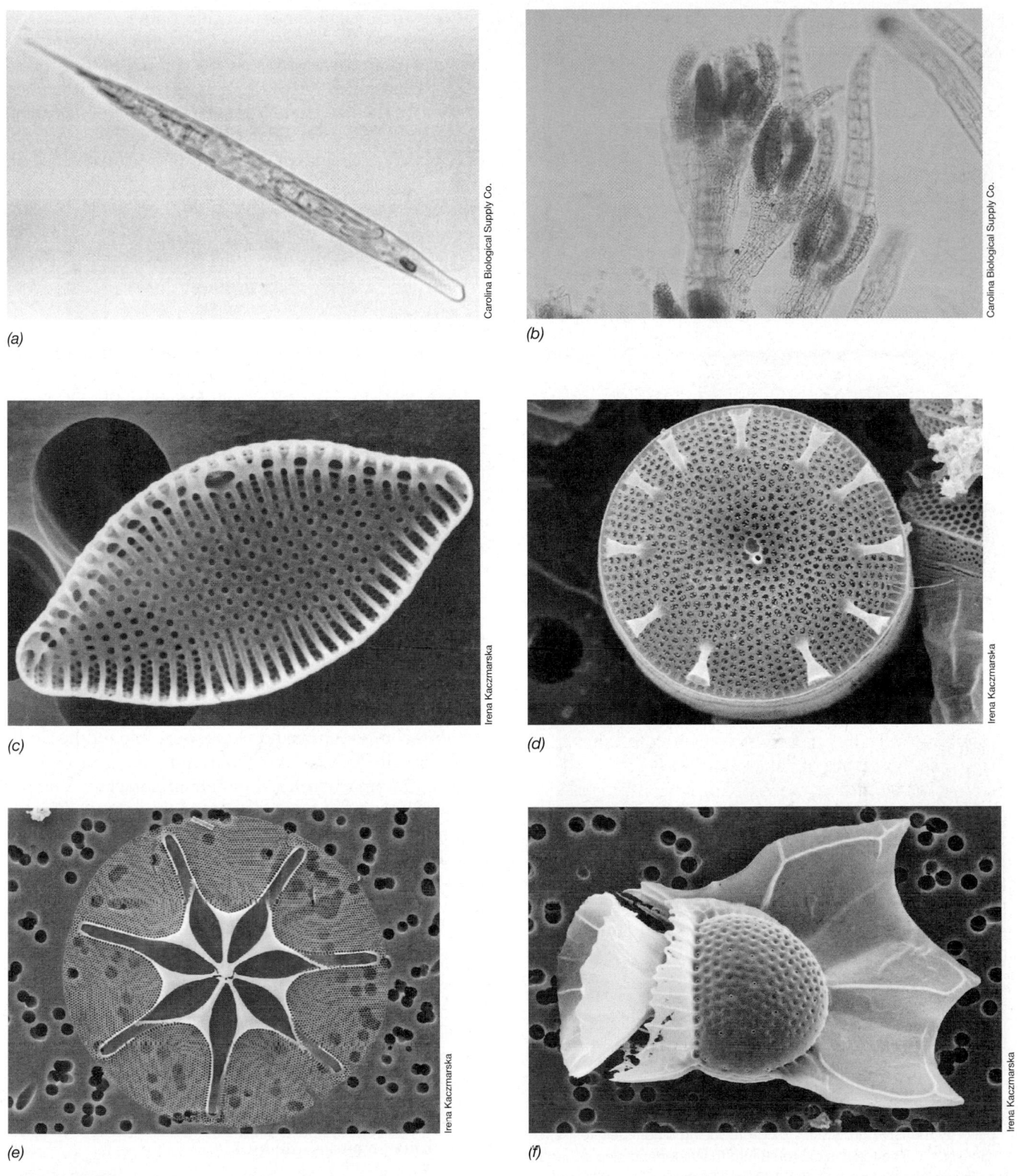

(a)

(b)

(c)

(d)

(e)

(f)

Figure 14.27 Algae other than green algae (Chlorophyta). (a–b) Light micrographs. (a) *Euglena*, a member of the Euglenophyta. This organism is phylogenetically much less derived than green algae (see Figure 14.11), and shares a number of other properties in common with flagellated protozoa (see text). (b) *Polysiphonia*, a marine red alga (Rhodophyta) that grows attached to the surfaces of marine plants. (c–f) Scanning electron micrographs. (c) Frustule of the marine diatom *Nitzschia*, showing pennate symmetry. (d) Frustule of the marine diatom *Thalassiosira*, showing radial symmetry. (e) Frustule of the marine diatom *Asteriolampra*, showing radial symmetry. (f) Cell of the marine dinoflagellate, *Ornithocercus magnificus*. The cell proper is the globular central structure; the ornate structures attached are called *lists*.

Rita R. Colwell

(a)

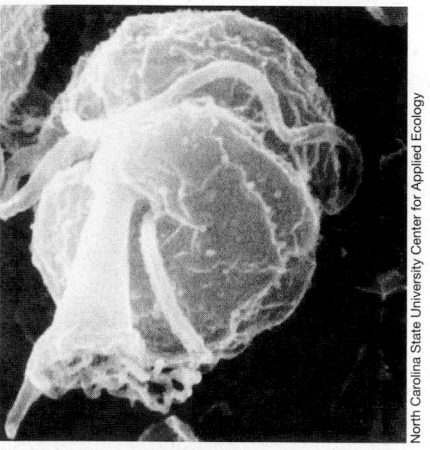

North Carolina State University Center for Applied Ecology

(b)

North Carolina State University Center for Applied Ecology

(c)

Figure 14.28 Toxic dinoflagellates. (a) Photograph of a "red tide" caused by massive growth of toxin-producing dinoflagellates, such as *Gonyaulax*. The toxin is excreted into the water and also accumulates in shellfish that feed on the dinoflagellates. The toxin is harmless to the shellfish but can poison fish and humans that eat infected shellfish. (b–c) Toxic *Pfiesteria*. (b) Scanning electron micrograph of a toxic spore of *P. piscicida*. (c) Fish killed by *P. piscicida*. Note the flesh-decaying lesions. Massive fish kills linked to *Pfiesteria* have been recorded along the eastern coast of the United States in recent years.

source by many flagellates and chlorophytes. In addition, some algae can assimilate simple organic compounds in the light (photoheterotrophy; ⚮ Figure 17.1 and Section 17.4) but cannot grow on them as sole energy sources.

One of the key characteristics used in the classification of algal groups is the nature of the **reserve polymer** synthesized as a result of photosynthesis. Algae of the division Chlorophyta produce starch (α-1,4-linked glucose) in a form very similar to that of higher plants. By contrast, algae of other groups produce a variety of reserve substances, some polymeric and some as free monomers, and the major ones are listed in Table 14.3.

Cell Walls of Algae

Algae show considerable diversity in the structure and chemistry of their cell walls. In many cases the cell wall is composed of a network of cellulose fibrils, but it is usually modified by the addition of other polysaccharides such as pectin, xylans, mannans, alginic acids, or fucinic acid. In some algae, the wall is additionally strengthened by the deposition of calcium carbonate; these forms are often called "calcareous" or "coralline" (coral-like) algae. Sometimes chitin (a polymer of N-acetylglucosamine, ⚮ Figure 3.5) is also present in the cell wall. In euglenoids a cell wall is absent.

In diatoms (Figure 14.27c–e) the cell wall is composed of *silica*, to which protein and polysaccharide are added. Even after the cell dies and the organic materials have disappeared, the external structure, called the *frustule*, (Figure 14.27c–e) often remains, showing that the siliceous component is indeed responsible for the rigidity of the cell. Because of the extreme resistance to decay of these diatom frustules, they remain intact for long periods of time and constitute some of the best algal fossils ever found. From this excellent fossil record, it is known that diatoms first appeared on Earth about 200 million years ago. Diatoms typically show some form of symmetry in the morphology of their frustules. Examples of symmetry include pennate (having similar parts arranged on opposite sides of an axis) (Figure 14.27c), and radial (Figure 14.27d, e).

Algal cell walls contain pores some 3–5 nm wide that are large enough to pass only low-molecular-weight substances such as water, inorganic ions, gases, and other small nutrient molecules needed for metabolism and growth. Phagocytic activities are thus not possible, and in this regard, algae differ from the actively phagocytic protozoa and slime molds.

Motility and Ecology of Algae

A number of algae are motile, usually because of flagella; cilia do not occur in algae. Simple flagellate forms, such as *Euglena* (Figure 14.27a), usually have a single polar flagellum, whereas flagellated representatives of the Chlorophyta have either two or four polar flagella.

Dinoflagellates (Figure 14.27*f*) have two flagella of different lengths and with different points of insertion into the cell. The transverse flagellum is attached laterally, whereas the longitudinal flagellum originates from the lateral groove of the cell and extends lengthwise. In many cases, algae are nonmotile in the vegetative state and form motile gametes only during sexual reproduction.

Algae abound in nature in aquatic habitats, both freshwater and marine. Algae are also found in artificial aquatic habitats like fish tanks and swimming pools, and in temporary pools of water formed from rainwater runoff. Algae are also common in most soils while a few species thrive in dry and even extremely dry soils where the water potential (∞ Section 6.12) can be very low. Algae are often the dominant or sole phototrophic microorganisms in acidic habitats as well; below pH 4–5 cyanobacteria are absent and several algal species flourish.

Endolithic Algal Communities

Some algal species have been found growing *inside* rocks. These endolithic (*endo* means "inside") organisms exist in porous rocks, for example, those containing quartz (Figure 14.29*a*), and are usually found in layers near the surface. Endolithic phototrophic communities are most common in dry environments such as deserts, (Figure 14.29), or cold dry environments like the Antarctic. For example, in the Antarctic Dry Valleys, where temperatures and humidity are very low, life within a rock has its advantages. Rocks are heated by the sun and water from snow melt can be absorbed and retained for relatively long periods, supplying needed moisture for growth. Moreover, water absorbed by a porous rock makes the rock more transparent, thus supplying more light to the algal layers.

A wide variety of phototrophs can form endolithic communities, including cyanobacteria (Figure 14.29*b*) and various green algae. In addition to free-living phototrophs, green algae and cyanobacteria coexist with fungi in endolithic lichen communities (∞ Section 19.20 for discussion of the lichen symbiosis). Metabolism and growth of these internal rock communities slowly weathers the rock, allowing gaps to develop where water can enter, freeze and thaw, and eventually crack the rock, producing new habitats for microbial colonization.

✓ 14.11 Concept Check

Algae are phototrophic *Eukarya* that contain photosynthetic pigments within a structure called the chloroplast.

✓ How do *algae* differ from *cyanobacteria*?

✓ What is unusual about red algae?

✓ What is the primary habitat of algae?

(a)

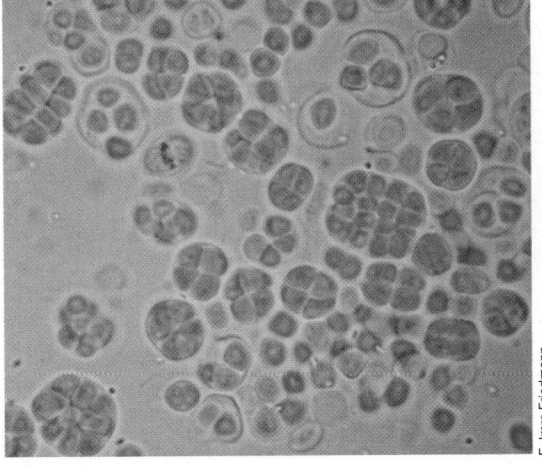

(b)

Figure 14.29 Endolithic cyanobacteria. (a) Photograph of a limestone rock from Makhtesh Gadol, Negev Desert, Israel, showing a layer of cells of the cyanobacterium *Chroococcidiopsis*. (b) Photomicrograph of cells of *Chroococcidiopsis* isolated from a sandstone rock in the Negev Desert.

Review Questions

1. List at least three features of eukaryotic cells that clearly differentiate them from prokaryotic cells.

2. How are the mitochondrion and the hydrogenosome similar structurally? How do they differ? How do they differ metabolically?

3. What does knowing that an antibiotic like streptomycin blocks protein synthesis in organelles tell us about their relationship to *Bacteria*?

4. What are the chemical differences between *microfilaments* and *microtubules*? Are such structures really absent from prokaryotic cells? (You will want to review Section 6.1 before answering this question.)

5. Compare and contrast the processes of *mitosis* and *meiosis*. Which process is absolutely necessary for growth of *Saccharomyces cerevisiae* and why?

6. How is the mating type of a yeast cell determined?

7. What are the common names of some major groups of protozoa?

8. Compare and contrast protozoa with fungi and algae, listing at least two ways you could differentiate a member of one group from the other. Which of these groups most closely resemble slime molds and why?

9. What major differences exist between the organisms *Physarum* and *Dictyostelium*?

10. What are some unusual features of dinoflagellates?

11. Examine Figure 14.29a. Why is the growth zone very near the surface of the rock? (*Hint:* Think of the *metabolism* of cyanobacteria. It is very similar to that of green algae.)

Application Questions

1. If organelles such as the mitochondrion and chloroplast had originally been free-living *eukaryotic* cells, how would the molecular properties of organelles listed in Section 4.5 differ?

2. In the next chapter you will learn that the genomes of eukaryotic cells are, in general, much larger than those of prokaryotic cells. List three reasons why this is not surprising.

K nowledge of the complete genome sequence of an organism reveals its complement of genes, but also tells us how the organism functions and its evolutionary history. Genomic sequences also allow for the facile study of *gene expression*. The traditional approach for studying gene expression was to focus on a single gene or group of related genes. The power of genomic analysis is such that expression of the entire complement of an organism's genes can be examined in a single experiment. DNA chips, shown here, facilitate analyses of gene expression by quickly revealing which genes are transcribed in cells grown under a particular condition. By comparing gene expression patterns, it is possible to determine which proteins in the cell are required for growth under a given set of conditions.

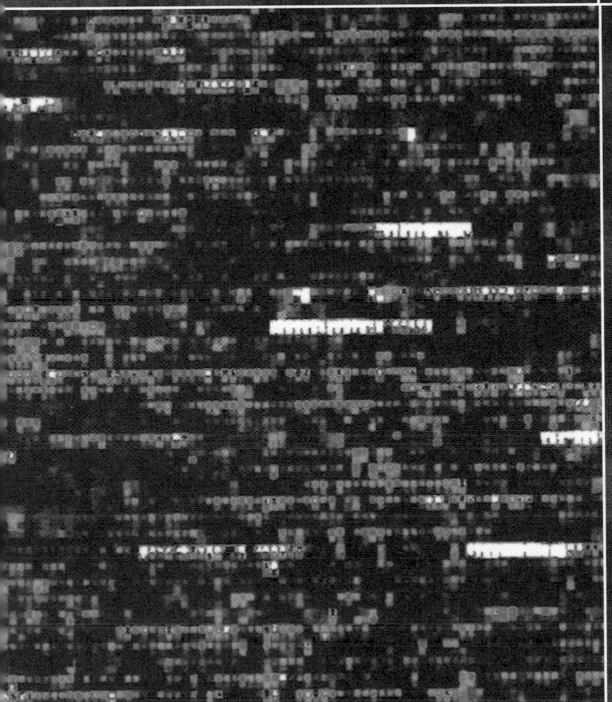

MICROBIAL GENOMICS 15

Working Glossary

Artificial chromosomes cloning vectors that can carry very large inserts of foreign DNA and exist in the cell very much like a cellular chromosome. The most widely used are bacterial artificial chromosomes (BACs) and yeast artificial chromosomes (YACs)

Bioinformatics the use of computer programs to analyze, store, and access DNA and protein sequences

Gene family genes that are related by sequence to other genes within the organism

Genome the total complement of genes of a cell or a virus.

Genomics the discipline involving mapping, sequencing, and analyzing genomes

Horizontal gene transfer the presence of a gene in an organism which originated in another organism

in silico the use of computers to perform sophisticated analysis

Open reading frame (ORF) a sequence of DNA that, if transcribed, could be translated to yield a protein of known length and composition. A *functional* ORF is one that actually encodes a protein in the cell

Orthologs genes found in one organism that are similar to those in another organism, but differ because of speciation (see also *Paralogs*)

Paralogs genes within an organism whose similarity is the result of gene duplication at some time during the evolution of the organism (see also *Orthologs*)

Proteome the total complement of proteins present in a cell, tissue, or organism at any one time

Proteomics the large-scale or genome-wide study of the structure, function, and regulation of the proteins of an organism

RNA editing modification of the RNA transcript of a protein-encoding gene, by a process other than splicing, to achieve the required coding properties

---

As discussed in Chapter 10, since the mid-1940s scientists have used a variety of genetic techniques to study the genes in microorganisms. These studies involved mapping genes as well as studying their function and regulation. Beginning in the 1970s the techniques of choice for genetic studies in many organisms have increasingly involved the *in vitro* manipulation of DNA (⚭ Sections 10.12–10.18). While individual scientists might pursue studies of small regions of a genome, or even single genes, the overall scientific goal in many cases has been to understand the total organism. The power of molecular cloning and DNA sequencing (⚭ Sections 10.13 and 10.14) was such that it became possible to sequence and analyze entire **genomes**, the total complement of an organism's genes (⚭ Section 7.4). In the mid-1980s the word **genomics** was coined to describe the discipline of mapping, sequencing, and analyzing genomes.

The first genome to be sequenced was the 3569-nucleotide RNA genome of the virus MS2 (⚭ Section 16.1) in 1976. The first DNA genome to be sequenced was the 5386-nucleotide sequence of the small, single-stranded DNA virus, ϕX174 (⚭ Section 16.2) in 1977. This feat was accomplished by a group led by Frederick Sanger (⚭ Section 10.13). The first cellular genome to be sequenced was the 1,830,137-bp chromosome of *Haemophilus influenzae* published in 1995 by Hamilton O. Smith, J. Craig Venter, and their colleagues. By spring of the year 2000 a "rough draft" had been established of the sequence of the haploid human genome, which is about 3 billion bp.

It would not have been possible to sequence and analyze such large and complex genomes without a corresponding improvement in technology. Part of this improvement has been the automation of DNA sequencing (⚭ Section 10.13). Another has been the development and use of computer programs to analyze, store, and access DNA and protein sequences, a field called **bioinformatics**. In this chapter we will discuss some microbial genomes and some of the techniques used to analyze these genomes.

I GENOMIC CLONING TECHNIQUES

Most of the techniques we will discuss in this section are modifications or expansions of the *in vitro* techniques that we have discussed in Chapter 10. The principles for these *in vitro* technologies remain the same, but now the emphasis is on studying the entire genome of an organism.

15.1 Vectors for Genomic Cloning and Sequencing

We have previously discussed some plasmid and viral vectors used for molecular cloning (⚭ Sections 10.15 and 10.16). Although the vectors we described, such as pBR322 and the Charon derivatives of bacteriophage lambda, have been extensively used for cloning and sequencing, including genomic analysis, other more specialized vectors have also been developed. These include bacteriophage M13 and bacterial and yeast artificial chromosomes.

Vectors Derived from Bacteriophage M13

M13 is a filamentous phage containing single-stranded DNA and replicates without killing its host (⚭ Section 16.3). Mature particles of M13 are released from host cells by a budding process, and it is possible to obtain infected cultures that can provide continuous sources of phage DNA. An important feature of M13 is its single-stranded DNA. In the Sanger procedure for DNA sequencing (∞ Section 10.13), single-stranded DNA is needed and DNA cloned into M13 thus provides a ready source of this single-stranded DNA (although single-stranded DNA can also be generated by denaturation). Also, single-stranded DNA is very useful as a probe for detecting other nucleic acid sequences in transfer procedures such as the Southern blot analysis, and M13 permits ready production of such single-stranded DNA probes (∞ Section 10.12).

However, in order to use M13 for cloning, a double-stranded form must be available because restriction enzymes work only on double-stranded DNA. Double-stranded M13 DNA can be obtained from infected cells because M13 replicates in the host as a double-stranded *replicative form* (∞ Section 16.3).

Most of the genome of wild-type M13 contains genetic information essential for virus replication. However, there is a small region called the intergenic sequence that can be used as a cloning site. Variable lengths of foreign DNA, up to about 5 kilobase pairs (kbp), can be cloned without affecting phage viability—as the genome gets larger, the virion gets longer. M13mp18 is a derivative of M13 in which the intergenic region has been modified to facilitate cloning. A map of this vector is shown in Figure 15.1*a*.

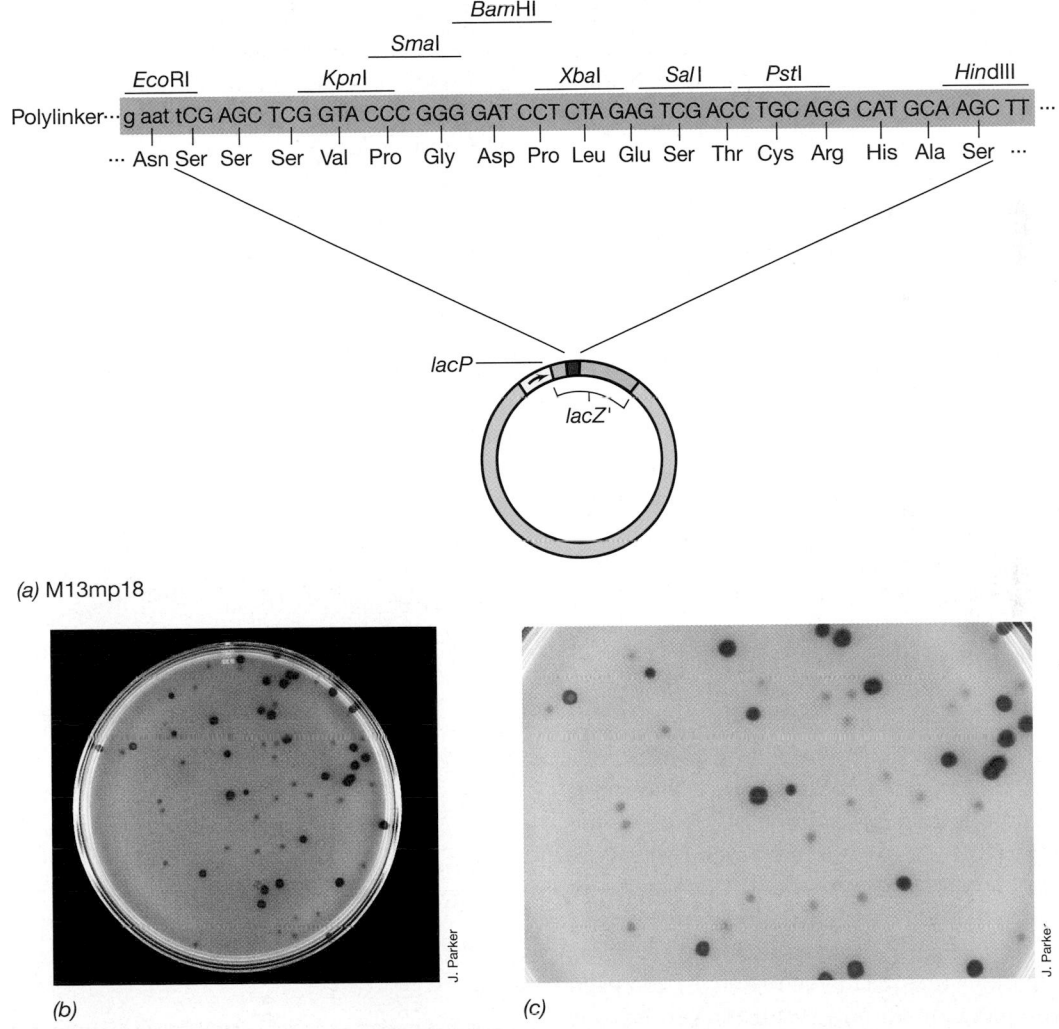

(a) M13mp18

(b)

(c)

Figure 15.1 (a) A partial map of M13mp18, a derivative of M13 constructed for use as a cloning vector. The vector contains the *lac* promoter and a gene, *lacZ'*, which encodes a functional part of β-galactosidase. At the beginning of this gene is a polylinker that contains several restriction sites but maintains the proper reading frame. The amino acids encoded by the polylinker are shown. Most DNA fragments cloned into the polylinker disrupt the *lacZ'* gene and abolish β-galactosidase activity. (b) A plate with plaques formed by M13mp18 and by clones made using this vector on a lawn of sensitive bacteria plated on a medium containing the chemical 5-bromo-4-chloro-3-indolyl-β-D-galactopyranoside, called X-gal. When β-galactosidase hydrolyzes X-gal, it releases a relatively insoluble blue dye. Many plaques on this plate are blue, indicating the presence of vector without cloned DNA. Many other plaques are colorless, indicating that foreign DNA has been inserted into the vector and the *lacZ'* gene has been disrupted (c) An enlargement of a portion of this plate.

One modification is the insertion of a functional fragment of *lacZ*, the *Escherichia coli* gene that encodes the enzyme β-galactosidase. Therefore, cells infected with M13mp18 can be easily detected by their color on indicator plates (Figure 15.1*b*). This *lacZ* gene has itself been modified to contain a 54-base-pair DNA fragment called a *polylinker*. The polylinker contains several restriction sites unique in M13 and can therefore be used for cloning. The polylinker is inserted into the beginning of the coding portion of the *lacZ* gene. This small insertion is in-frame, and the extra 18 amino acids do not affect the activity of the enzyme encoded by the gene. However, insertion of additional DNA into the polylinker during cloning inactivates the gene. Phages that contain additional DNA inserts give rise to colorless plaques, and it is therefore very simple to identify clones (Figure 15.1*b* and *c*). Similar constructs are used in lambda cloning vectors and plasmid cloning vectors to allow identification of cells containing cloned DNA.

Use of M13 in Molecular Cloning

To clone DNA in M13 vectors, the replicative double-stranded DNA is isolated from the infected host and treated with a restriction enzyme. The foreign DNA, also double-stranded, is treated with the same restriction enzyme. On ligation, double-stranded M13 molecules are obtained that contain the foreign DNA. When these molecules are introduced into the cell by transformation, they replicate and in time produce mature bacteriophage particles containing single-stranded DNA molecules. Only one strand of DNA is packaged into mature phage. Which of the two foreign strands the mature phage contains depends on the *orientation* in which the strand was inserted. Because foreign DNA can be inserted (in separate phage molecules) in either orientation, *both* strands of the foreign DNA can be cloned.

The single-stranded M13 DNA containing the foreign DNA can then be used in DNA sequencing. Since the base sequence where the foreign DNA is inserted is known (based on the specificity of the restriction enzyme used), it is possible to use a synthetic oligonucleotide complementary to this region as a primer and hence determine the sequence of the whole DNA downstream from this point. In this way, M13 derivatives have proved extremely useful in sequencing foreign DNA, even rather long molecules. M13 vectors have been used as a tool in the sequencing of many viral and prokaryotic genomes.

Artificial Chromosomes

Vectors like M13, or plasmid vectors that hold about 2 kb of cloned DNA, are adequate for making gene libraries (∞ Section 10.14) for genomic sequencing of prokaryotes. Sometimes, however, one uses lambda vectors, which hold 20 kb or more (∞ Section 10.16). Howev-

er, as the size of the genome to be sequenced increases, so will the number of clones needed to obtain a complete sequence. Therefore, for making libraries of DNA from eukaryotic microorganisms or from higher eukaryotes, such as from humans, it is useful to have available vectors that can carry very large segments of DNA. This allows the size of the initial library to be limited. Such vectors have been developed and are called **artificial chromosomes**.

Remember that many *Bacteria* contain large plasmids that are stably replicated (∞ Section 7.4). The F plasmid of *Escherichia coli* is such a plasmid (∞ Section 10.8). It is very stably replicated in *E. coli* and naturally occurring derivatives, called F' plasmids, are known that can carry large amounts of chromosomal DNA (∞ Section 10.9). The F plasmid has been used to construct cloning vectors called **bacterial artificial chromosomes** or **BACs**. The BAC shown in Figure 15.2 is 6.7 kb, not 99.2 kb like F itself. It contains only a few genes from F, including *oriS* and *repE*, which are necessary for replication, and *sopA* and *sopB*, which keep the copy number very low (∞ Section 10.8). Inserted into the plasmid is the *cat* gene, which confers chloramphenicol resistance and a cloning region that includes several restriction sites for cloning DNA. Foreign DNA of greater than 300 kb can be inserted and stably maintained in BAC vectors. The host for a BAC carrying cloned DNA is typically a mutant strain of *E. coli* that is missing the normal restriction and modification systems (∞ Section 9.6). Often the strain is also defective in some recombination pathways

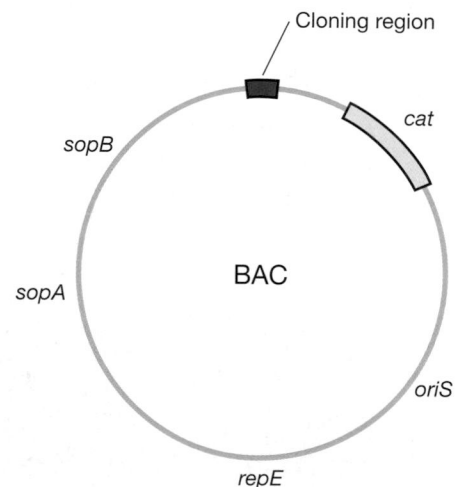

Figure 15.2 Genetic map of a bacterial artificial chromosome. BACs are derivatives of the F plasmid of *Escherichia coli*. The BAC diagrammed is 6.7 kb. At the top of the map is a cloning region that contains several unique restriction enzyme sites. The *cat* gene confers resistance to the antibiotic chloramphenicol. The other genes shown are involved in plasmid replication. In the 99.2-kb F plasmid itself (∞ Figure 10.17) all these genes are located in a relatively compact replication region. Therefore, BACs contain only a small fraction of the entire F plasmid.

Figure 15.3 Diagram of a yeast artificial chromosome (YAC) containing foreign DNA. The foreign DNA was cloned into the vector at a *Not*I restriction site. The telomeres are labeled TEL and the centromere CEN. The origin of replication is labeled ARS (for autonomous replication sequence). The gene used for selection is called URA3. The host into which the clone is transformed has a mutation in that gene so that it normally requires uracil for growth (Ura⁻). Host cells containing this YAC become Ura⁺. The diagram is not drawn to scale; the inserted DNA would normally be 200–800 kilobase pairs long and the vector about 10 kilobase pairs.

(⌘ Section 10.5). This prevents recombination and rearrangements of the cloned DNA.

The term artificial chromosome originated not with bacteria but with the development of **yeast artificial chromosomes** or **YACs**. These vectors have been designed to replicate in yeast like normal chromosomes, but they have sites where DNA can be inserted. To function like normal eukaryotic chromosomes, YACs must have an *origin of DNA replication, telomeres* at the ends of the chromosome (⌘ Section 7.7), and a *centromere* (the section of the chromosome required for segregation during mitosis). They must also contain a cloning site and a gene that can be used for selection after transformation into the host. Figure 15.3 shows a diagram of a YAC vector into which foreign DNA has been cloned. YAC vectors are themselves only about 10 kilobase pairs, but they can have 200–800 kilobase pairs of cloned DNA inserted. After identifying a particular region in the cloned DNA on a BAC or a YAC, this region can be *subcloned* into a plasmid or bacteriophage vector for more detailed analysis or sequencing. Although YACs can hold larger DNA inserts than BACs, there is a greater problem with recombination and rearrangement of the cloned DNA within yeast than within *E. coli*. For this reason BACs are now more widely used than YACs.

✓ 15.1 Concept Check

Specialized cloning vectors have been constructed that are useful in studying genomes. Some, like the M13 derivatives, are useful for DNA sequencing. Others, like artificial chromosomes, are useful for cloning very large fragments of DNA, fragments approaching a megabase.

✓ What property of M13 makes it useful for DNA sequencing?

✓ The yeast artificial chromosome behaves like a chromosome in a yeast cell. What makes this possible?

15.2 Cloning and Mapping Genomes

The analysis of a genome begins with the molecular cloning of fragments of its DNA (⌘ Section 10.14) and then sequencing this DNA (⌘ Section 10.13). However, the methodology used can be quite different when the goal is obtaining a complete genomic sequence rather than just the sequence of a particular gene or region (just as the methods can differ when the goal is obtaining a specific gene product; ⌘ Chapter 31).

Shotgun Cloning Bacterial DNA

One method for sequencing an entire genome, called *map-directed sequencing*, consists of developing an ordered library of overlapping or contiguous molecular clones, called **contigs**, physically mapping these clones using restriction enzymes (which also guide the cloning process), and then sequencing the cloned DNA. For the original set of clones, a vector such as lambda might be used. From these clones, smaller *subclones* are often obtained in vectors such as M13 so that DNA can be easily sequenced. (Typical sequencing reactions give on the order of 500 nucleotides whereas a lambda vector may contain many kilobases of cloned DNA.) Constructing restriction maps and ordering the clones in a gene library can be laborious and time consuming. However, once an ordered library is available (in plasmids, phage, or artificial chromosomes), sequencing becomes a straightforward process. When using map-directed sequencing it is relatively easy to determine if there are any gaps in the sequence and if one knows where the sequence of a particular clone should go on the total chromosome map.

Another method, call *shotgun sequencing*, made possible by automated sequencing, robotics, and powerful computer programs, involves shotgun cloning (⌘ Section 10.14) the entire genome and sequencing the clones randomly. That is, the clones are sequenced without knowing the order or orientation of the cloned DNA, or even if one has already sequenced this region. The sequences obtained will then be analyzed by computer, and it is the computer program that will seek overlapping sequences (and sequences from each strand) to construct a complete sequence of the entire genome. To ensure that all sequences are obtained (essentially a statistical process) it is necessary to sequence a very large number of clones, many of which will be identical or nearly identical. Typically there will be 7 to 10 sequences over any given part of the genome. This seven- to tenfold *coverage* also eliminates or at least greatly reduces errors in the sequence.

For shotgun sequencing to work effectively, it is essential that the cloning itself be efficient (one needs a large number of clones) and that, in so far as possible, the DNA cloned is randomly generated. Restriction sites are not random, but by cutting incompletely with an enzyme that recognizes a short sequence (which occurs commonly in the DNA), one can approach random digestion. To obtain more truly random fragments, DNA can be mechanically sheared.

One reason it took many years to discover that DNA was the genetic material in cells was because the DNA molecules isolated seemed too small. It was then shown that, quite surprisingly, the phosphodiester bonds of long DNA molecules in solution could be broken by shearing stresses during the process of purification, such as by pipetting or vigorously shaking the solutions. To prevent this, steps were taken to purify DNA using gentle procedures. This made it possible to isolate much longer molecules. Now, however, scientists take advantage of shear forces to generate fragments that are cleaved randomly (that is, with no sequence specificity) by forcing DNA through a *nebulizer*. This is a device with a small opening, which reduces the solution to a spray and shears the DNA. The DNA fragments can be purified by size using gel electrophoresis if necessary and then cloned into a vector and transformed into a host.

The procedure of picking colonies can also be automated, making it far easier to identify and pick cells containing the vector with inserted DNA. This can be done by *selection* if a vector is used that contains an antibiotic resistance gene and a cloning site in a gene that encodes a lethal product unless disrupted (Section 10.15). Note that the *lacZ* system, as in the M13 vectors (see Figure 15.1), makes it simple to *screen* for plasmids with inserts but not to *select* for them.

Modern automated sequencers can simultaneously sequence several fragments, and since different fluorescent dyes are used for each base (Figure 10.39), it is not necessary to run the reactions separately. Sometimes shotgun cloning and random sequencing will not give a complete sequence; that is, sometimes there will be gaps in the sequence. In such a situation, clones can be sought that are thought to cover the gap. One method of doing this is to perform PCR reactions (Section 10.17) on total genomic DNA to search for (and clone) fragments that have been missed.

Scientists are not primarily interested in the sequence itself but in the genes that are present and how they function and are regulated. The next step in mapping a genome is to identify possible genes and other functional regions, a process called *annotation*.

Locating Probable Genes

Recall that the great majority of genes of any organism encode proteins and that in most microbial genomes the great majority of the genome consists of coding sequences. Because microbial organisms have few introns (and prokaryotes almost none), microbial genomes usually consist of uninterrupted **open reading frames (ORFs)** (Section 7.13). A *functional ORF* is one that actually encodes a protein in the cell. Typically only a few of the proteins from a particular organism will themselves have been sequenced previously. Thus the simplest way to locate protein-encoding genes, or potential protein-encoding genes, is to have a computer search the sequence of the genome for ORFs.

In the cell, ribosomes establish a reading frame by initiating translation at a start codon, usually an AUG (Section 7.13). The ribosome then proceeds until it reaches an in-frame stop codon (Section 7.13). The identification of possible *functional* ORFs is, however, typically more complex than just searching for in-frame start and stop codons, since these will appear randomly with reasonable frequency.

One clue that an ORF is functional will be size. Most proteins contain 100 or more amino acids, so most functional ORFs are longer than 100 codons. However, simply programming the computer to ignore ORFs shorter than 100 codons will miss many functional genes. Because most organisms show preferences among synonymous codons (Section 7.13), *codon bias* can also give a clue as to whether an ORF is functional. Also remember that prokaryotic ribosomes start translation not at the first (most 5') possible start codon, but at one immediately downstream of a Shine–Dalgarno sequence on the mRNA (Section 7.15). Therefore, searching the DNA sequence of a prokaryotic genome for potential Shine–Dalgarno sequences can help establish both whether an ORF is functional and which potential start codon is actually used. Figure 15.4a shows a region of DNA containing the type of ORF being sought by the computer program. Although any given gene is always transcribed from a single strand, in all but the smallest plasmid or viral genomes, both strands are transcribed in some part of the genome. Figure 15.4b shows a diagram of a region of a genome in which one gene is transcribed from one strand and one from the other.

Of course, some genes encode transfer RNAs and ribosomal RNAs and these are not recognized by programs that search only for ORFs. Genes for tRNAs and rRNAs can usually be located on genome sequence because the sequences of these RNAs are very highly conserved (Sections 7.14 and 11.5–11.7).

An ORF is also likely to be functional if its sequence is similar to sequences of ORFs obtained from the genomes of other organisms (regardless of whether they encode known proteins), or if some part of the ORF has a sequence known to encode a protein functional domain. The computer can search for such se-

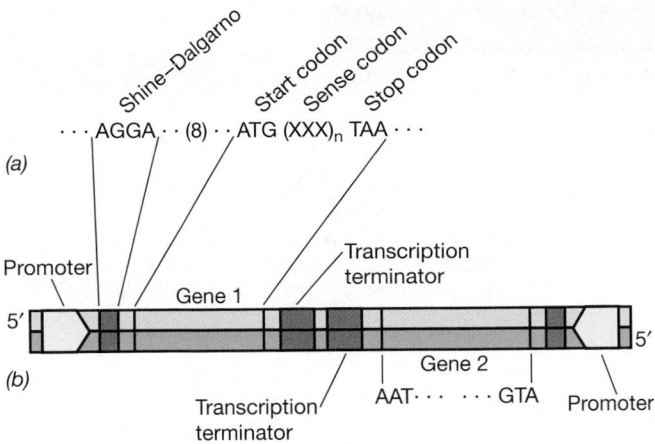

(a)

$\cdots$ AGGA $\cdots$ (8) $\cdots$ ATG (XXX)$_n$ TAA $\cdots$

Shine–Dalgarno Start codon Sense codon Stop codon

Promoter Gene 1 Transcription terminator

(b)

Transcription terminator AAT $\cdots$ $\cdots$ GTA Gene 2 Promoter

Figure 15.4 Locating possible functional open reading frames. (a) The sequence shown from one strand includes bases for a potential Shine–Dalgarno sequence (AGGA) separated from a start codon (ATG) by about 8 bases. The bases encoding the start codon are followed by about 100 sense codons and then a stop codon (UAA is the most commonly used stop codon). (b) The DNA molecule with the ORF described in (a) as Gene 1. Note that upstream of this ORF is a promoter and downstream a transcription terminator. Also shown on the DNA fragment is another gene that has the same components, but that would be transcribed in the opposite direction. Shine–Dalgarno sequences, start codons, sense codons, and stop codons function only at the level of RNA: Here we are showing bases (base pairs) that encode these functional sequences.

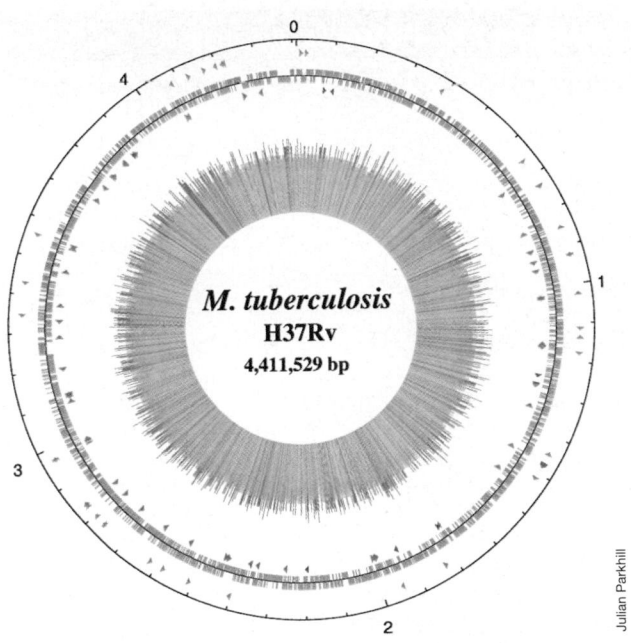

M. tuberculosis
H37Rv
4,411,529 bp

Julian Parkhill

Figure 15.5 Chromosome map of *Mycobacterium tuberculosis*. The outer circle shows the size of the chromosome in Mb with 0 placed at the point of origin of DNA replication. Inside this is a ring showing the locations of tRNA genes (blue) and the single rRNA operon (orange). The next ring shows ORFs transcribed clockwise (dark green) and counterclockwise (light green). Beneath this is a ring showing the location of some repetitive DNA elements, including insertion sequences (orange). The histogram in the center shows regional G + C content (Section 11.9) with yellow rays showing regions below 65% G + C and orange rays showing regions above 65% G + C. The figure was generated with software from DNASTAR and is reprinted by permission from *Nature 393*: 537–544 (1998). ©Macmillan Magazines Ltd. For additional discussion of the organism *M. tuberculosis*, see Section 12.23, and for the disease tuberculosis, Section 26.5.

quences also. It would be almost impossible to assemble the sequence of even a prokaryotic genome and to locate genes and other important functional DNA sequences without the use of sophisticated computer programs able to handle large databases. Sometimes the use of computers to do sophisticated analysis is referred to as working *in silico*, a term similar to the terms *in vivo* and *in vitro*, for "in living" and "in glass," respectively. Figure 15.5 shows a map constructed by a computer program using the complete sequence of *Mycobacterium tuberculosis*, the causative agent of tuberculosis (Section 26.5). In this case ORFs are indicated as well as genes encoding tRNAs and rRNAs and a few other sequences.

✓ *15.2 Concept Check*

Shotgun cloning followed by random sequencing of clones is one method used to obtain sequences of microbial genomes. After such sequences are assembled by computer programs, other programs search for ORFs and genes encoding tRNAs and rRNAs as part of an annotation process.

✓ For genomic analysis, one does not typically sequence directly from cloned DNA carried in a lambda vector. Why not?

✓ What is an *open reading frame (ORF)?*

II MICROBIAL GENOMES

Many microbial genomes have now been sequenced, too many to deal with individually. In this part of the chapter we will discuss a few of the genomes that have been sequenced and what the analysis of these genomes tells us.

15.3 Prokaryotic Genomes

Over 100 prokaryotic genomes have been completely sequenced and many more are currently being completed. Table 15.1 lists 20 of these from both *Archaea* and *Bacteria*. (A few prokaryotes with completely sequenced genomes are also given in Table 7.2.) This list is not random, nor are efforts to sequence genomes random. Although the sequencing of prokaryotic genomes

TABLE 15.1   Prokaryotic chromosomes[a]

Organism[b]	Size (base pairs)[c]	ORFs[d]	Comments
Mycoplasma genitalium (B)	580,070	470	Smallest known cellular genome (⬡ Section 12.21)
Mycoplasma pneumoniae (B)	816,394	677	Causes pneumonia (⬡ Section 12.21)
Borrelia burgdorferi (B)	910,725	853	Spirochete, has linear chromosome [e] causes Lyme disease (⬡ Sections 12.33 and 27.4)
Chlamydia trachomatis (B)	1,042,519	894	Obligate intracellular parasite, common human pathogen (⬡ Sections 12.27 and 26.13)
Rickettsia prowazekii (B)	1,111,523	834	Obligate intracellular parasite, causes epidemic typhus (⬡ Sections 12.13 and 27.3)
Treponema pallidum (B)	1,138,006	1041	Spirochete, causes syphilis (⬡ Sections 12.33 and 26.12)
Methanocaldococcus jannaschii (A)	1,664,976	1738	Methanogen (⬡ Section 13.4)
Helicobacter pylori (B)	1,667,867	1590	Causes peptic ulcers (⬡ Section 26.10)
Pyrococcus horikoshii (A)	1,738,505	2061	Hyperthermophile (⬡ Section 13.10)
Methanothermobacter thermoautotrophicus (A)	1,751,377	1855	Methanogen (⬡ Section 13.4)
Thermotoga maritima (B)	1,860,725	1877	Hyperthermophile (⬡ Section 12.36)
Archaeoglobus fulgidus (A)	2,178,400	2436	Hyperthermophile (⬡ Section 13.7)
Staphylococcus aureus (B)	2,814,816	2593	Major cause of nosocomial infections (⬡ Sections 12.19 and 25.7)
Synechocystis sp. (B)	3,573,470	3168	Cyanobacterium (⬡ Section 12.25)
Caulobacter crescentus (B)	4,016,942	3767	Complex life cycle (⬡ Section 12.16)
Bacillus subtilis (B)	4,214,810	4100	Gram-positive genetic model (⬡ Section 12.20)
Mycobacterium tuberculosis (B)	4,411,529	3924	Causes tuberculosis (⬡ Sections 12.23 and 26.5)
Escherichia coli (B)	4,639,221	4288	Gram-negative genetic model (⬡ Section 12.11)
Pseudomonas aeruginosa (B)	6,264,403	5570	Metabolically versatile opportunistic pathogen (⬡ Section 12.7)
Streptomyces coelicolor (B)	8,667,507	7846	Has linear chromosome,[e] produces antibiotics (⬡ Section 12.24)

[a] One sometimes finds the word "genome" used to refer to a prokaryotic chromosome. While in some cases this may actually be correct, remember that the genome actually includes *all* the genes found in an organism, even those of resident plasmids and viruses. In some cases these elements may not be found, and in some cases they may be very small or the genes carried by them may be of no importance to the organism's overall metabolism. However, this need not always be the case. For example, *Borrelia burgdorferi* contains 17 different plasmids consisting of 533 kilobase pairs (⬡ Table 7.2).

[b] In parentheses following the name of the species is an A if the organism is a member of the *Archaea* and a B if it is a member of the *Bacteria*.

[c] Information on these and other genomes can be found in the TIGR Microbial Database (www.tigr.org/tdb/mdb/mdbcomplete.html), a Web Site maintained by The Institute for Genomic Research (TIGR), Rockville, MD, a not-for-profit research institute.

[d] Open reading frames. The purpose of reporting ORFs is to predict the total number of proteins that an organism might encode. Therefore, constraints are placed on what is reported as an ORF. Of course, genes encoding known proteins are included, as are all ORF's that could encode a protein greater than 100 amino acid residues. Smaller ORFs are typically not included unless they show similarity to a gene from another organism or unless the codon usage is typical of the organism being studied.

[e] All chromosomes in this list are circular except those of *Borrelia burgdorferi* and *Streptomyces coelicolor*.

is becoming more rapid and routine, sequencing and analysis of a complete genome is still a considerable undertaking, and typically there are important scientific and/or societal reasons for undertaking such an effort.

Note that the list in Table 15.1 contains nine pathogens, causing widespread and severe illness. The three hyperthermophiles on the list may have important uses in biotechnology since the enzymes in these organisms will be heat stable (⬡ Sections 6.10, 10.17, and 13.11). Indeed, the needs of the biomedical and biotechnology industries are important considerations for generating the interest and the funding to pursue genomic sequencing. However, the list also includes organisms such as *Bacillus subtilis* and *Escherichia coli*, which remain widely studied genetic model systems, and *Caulobacter crescentus*, a model system for cell differentiation (⬡ Section 12.16).

Sizes of Prokaryotic Genomes

Analyzing genomic sequences allows us to answer some fundamental biological questions. For example, analysis of the small genomes of *Mycoplasma genitalium* and *Mycoplasma pneumoniae* shows that all 470 open reading frames (ORFs) found in *M. genitalium* are also present in *M. pneumoniae*. By seeking similar genes in other organisms, and by determining which genes are not essential using transposon mutagenesis (⬡ Section 10.11), scientists have evidence that between 265 and 350 protein-encoding genes may be sufficient to maintain a cellular existence. Perhaps someday we will find, or *construct*, an organism with such a minimal genome! (Construction of such an organism is not an insuperable challenge with the genetic tools discussed in Chapter 10.) Note that *Methanocaldococcus jannaschii* is an *autotroph* (Sections 13.4 and 17.17) and contains only 1738 ORFs. This is enough, however, to enable it to be not only free living, but also to synthesize all of its organic cellular components from CO_2 (Section 13.4).

It is no coincidence that Table 15.1 lists several relatively small genomes. Sequencing strategies are simpler for small genomes, so there was some bias for small genomes in early sequencing efforts. Note that the genome of *Mycoplasma genitalium* is smaller than that of the *Chlorella* viruses (Section 16.6) and Bacteriophage G (Section 9.2). However, *Pseudomonas aeruginosa* has almost as many ORFs as the yeast *Saccharomyces cerevisiae*, a eukaryote (see Section 15.6). The bacterium *Streptomyces coelicolor*, whose linear chromosome of over 8 Mb is typical of the *Streptomyces*, has 7846 ORFs, almost 1800 more than *S. cerevisiae*. Some even larger prokaryotic genomes, such as the 9.5-Mb chromosome of *Myxococcus xanthus*, have not yet been sequenced. In summary, prokaryotic genomes range in size from those of very large viruses to those of very small eukaryotes.

Genomic Analysis

Without bioinformatics it would not be possible to analyze and compare genomes. One of the first steps in genomic analysis is to compare the sequence obtained with that of previously sequenced genomes and with the sequences of genes of known function. This helps us to determine a gene's function and which genes are likely to be essential to a particular organism. One might imagine, for instance, that organisms such as *Treponema pallidum*, which are obligately parasitic (Section 26.12), might require relatively few genes for amino acid biosynthesis since all the amino acids they need can be supplied by their hosts. In fact this is the case, as *T. pallidum* has no genes involved in amino acid biosynthesis, although it encodes several proteases that can convert peptides taken up from the host into free amino acids. In contrast, *E. coli* has 131 genes involved in amino acid biosynthesis and metabolism, and *B. subtilis*, a soil bacterium, has over 200.

Comparative analyses such as these are useful in the search for genes that encode enzymes that presumably must exist because of the known properties of an organism. *Thermotoga maritima*, for example, is a hyperthermophile found in heated marine sediments and known to be able to metabolize a large number of sugars. Figure 15.6 diagrams some of the metabolic pathways and transport systems of *T. maritima* that have been derived from analysis of its genome. As might be expected, the genome is particularly rich in genes involved in transport, particularly for carbohydrates and amino acids. A large portion of its genome (7%) is also involved in metabolism of simple and complex sugars. All this suggests that *T. maritima* exists in an environment rich in organic material.

Some information on the division of genes and activities in prokaryotes is given in Table 15.2 and similar information is available for all the organisms whose genomes have been sequenced.

Although there are differences from organism to organism, in most cases the number of genes that have actually been identified is about 60% or less of the number of ORFs shown. (An ORF whose product does not resemble any known protein is sometimes referred to as a *URF*, an *unidentified reading frame*.) This does not mean that these unidentified ORFs are not actual protein-encoding genes. Rather it reflects the fact that there is still much we don't know about prokaryotic genomes. Searching a genomic database to discover new genes or new functions is called "mining" and can lead to unusual and exciting discoveries (see Section 15.5). The number of identified genes will continue to increase as more becomes known about a particular cell's biology. In this regard, it is interesting that in the very well-studied *E. coli*, functions have been assigned to about 2700 of its almost 4300 genes. Many of the genes involved in macromolecular syntheses and central metabolism essential for growth of *E. coli* have been identified. Therefore, as the functions of more URFs are identified, it is likely that most of them will be nonessential, and the percentage of *E. coli* genes involved in key functions in macromolecular synthesis or central metabolism will decrease (Table 15.2).

Genomic analysis can also provide new insights into the ecology of an organism. For instance, *Helicobacter pylori* encodes proteins that contain twice the amount of the basic amino acids arginine and lysine (Section 3.6) than do typical proteins from other prokaryotes. Presumably this helps cells of *H. pylori* survive in the acid environment of the stomach. Also, many of the genomes of pathogens, for example, *Borrelia burgdorferi*, *Mycobacterium tuberculosis*, and *Treponema pallidum*, that have been sequenced show the presence of (unrelated) gene families that encode proteins that might be involved in antigenic variation and hence protect the organism from the host's immune system (*e.g.*, syphilis, Section 26.12). In *B. burgdorferi* these genes are located on plasmids. Information derived from genomics can be critical in designing new strategies to protect humans and other animals from these pathogens. The ability to compare thousands of genes has resulted in considerable insight into the metabolism of the organisms listed in Table 15.1. However, at this stage there are many questions that remain unanswered. For example, there are no obvious "rules" for protein folding leading to thermostability that can be derived from

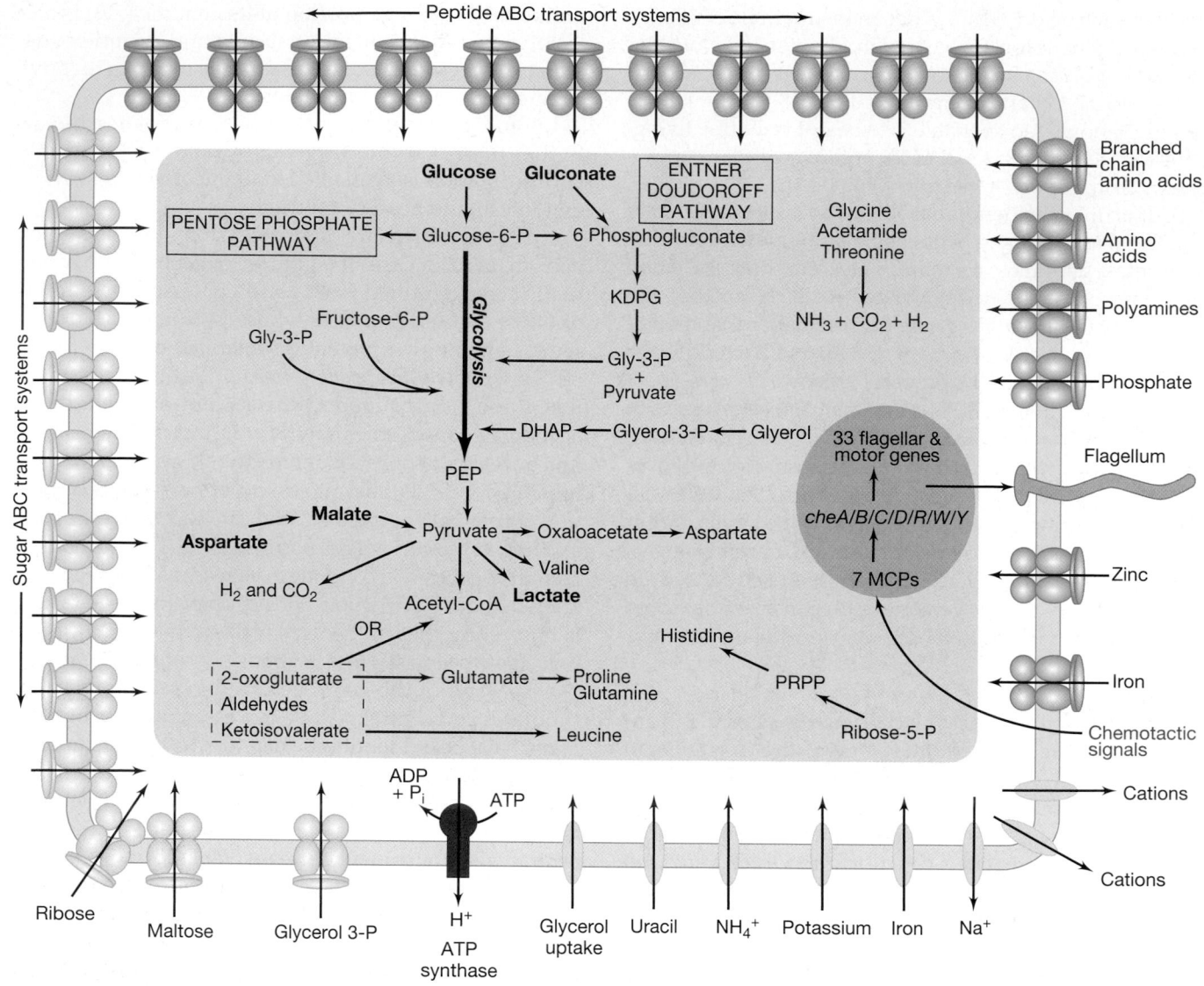

Figure 15.6 Overview of some metabolic pathways and transport systems of *Thermotoga maritma*, a member of the domain *Bacteria*. The figure shows a blueprint for the metabolic capabilities of this organism. These include some of the pathways for energy production and the metabolism of certain organic compounds, as well as proteins involved in transport, that were identified from analysis of the genomic sequence. Gene names have not been shown. The genome contains several ABC transport systems (⬤⬤ Section 4.7). There are 12 for carbohydrates, 14 for peptides and amino acids, and still others for ions. These are shown as multisubunit structures in the figure. Other types of transport proteins have also been identified and are shown as simple ovals. The flagellum is shown and this organism has seven transducers (MCPs) and several chemotaxis (*che*) genes (⬤⬤ Section 8.11) as well as several genes involved in flagellar assembly (⬤⬤ Section 4.10). A few aspects of sugar metabolism are also shown. This figure is adapted from work published by The Institute for Genomic Research (TIGR).

comparisons of the genomes of the hyperthermophiles that have been sequenced.

✓ 15.3 Concept Check

Sequenced prokaryotic genomes range in size from 0.58 Mb to 8.7 Mb. The smallest prokaryotic genomes are the size of the largest viruses, and the largest prokaryotic genomes have more genes than some eukaryotic microorganisms. Many genes can

be identified by their sequence similarity to genes found in other organisms. However, a large percentage of sequenced genes are of unknown function. Bioinformatics not only helps us to analyze the genes present but also to characterize the metabolism of an organism and its ecological environment.

✓ Compare an *ORF* with a *URF*.

✓ What type of information allows one to make predictions about an organism's environment?

TABLE 15.2	Gene function in bacterial genomes		
Functional categories	**Percentage of genes on chromosome[a] in that category**		
	Escherichia coli	*Haemophilus influenzae*	*Mycoplasma genitalium*
Metabolism	21.0	19.0	14.6
Structural	5.5	4.7	3.6
Transport	10.0	7.0	7.3
Regulation	8.5	6.6	6.0
Translation	4.5	8.0	21.6
Transcription	1.3	1.5	2.6
Replication	2.7	4.9	6.8
Other, known	8.5	5.2	5.8
Unknown	38.1	43.0	32.0

[a] For the size of the chromosome of each of these species and the number of open reading frames that each contains, see Table 15.1

15.4 Evolution and Gene Families

As we mentioned, the first priority of genomics is to determine the number and function of genes in an organism. However, there is more to genomics than counting genes and helping us understand how an individual organism copes with its environment. Comparative genomics also allows us to understand the evolutionary relationships between different organisms. Reconstructing evolutionary relationships helps to distinguish primitive from derived biological characteristics. This knowledge will help us better understand early life forms and, hopefully, help us eventually answer the question of how life first arose (Section 11.1).

Gene Duplication and Gene Families

When describing the chromosome of *E. coli* (Section 10.19), we discussed the fact that it contains **gene families**, genes that are related to other genes within the organism. This is true of most of the organisms listed in Table 15.1. Although large gene families are not the rule, comparative genomics has shown that many genes have arisen by duplication of other genes. For instance, 47% of the genes in *Bacillus subtilis* are related to one or more other genes on the chromosome. Such genes are called **paralogs**, genes whose similarity is the result of gene duplication at some time in the evolution of an organism. (Genes found in one organism that are similar to genes in another organism, but differ because of speciation, are called **orthologs**.) The study of genes having sequence similarity is one of the most important and complex in comparative genomics. Because chromosomes from many different kinds of prokaryotes have been sequenced, such comparisons can yield information on the earliest events in the evolution of cells. One interesting fact is that, as expected from ribosomal RNA sequencing studies (Chapter 11), the genes in the *Archaea* that are involved in DNA replication, transcription, and translation, are more similar to those of *Eukarya* than to those of *Bacteria*. Unexpectedly however, many of the rest of the genes in *Archaea* are more similar to those of *Bacteria*, than to those of *Eukarya*. (Of course, in the aggregate, genes from one archaeon are typically most similar to those in other *Archaea*.) These results lend further support to the phylogenetic picture of life deduced originally by comparative ribosomal RNA sequence analysis (Section 11.5) and suggest that many genes in all organisms have common evolutionary roots.

Horizontal Gene Transfer

Evolutionary arguments envision common ancestors and events that occurred in the very distant past. *Horizontal (lateral) gene transfer*, however, seems to be an ongoing phenomenon, and it can complicate evolutionary studies (Section 11.7). As mentioned (Section 10.19) **horizontal gene transfer** is an event in which a gene has apparently been picked up from a different organism. To be detectable, the difference between the organisms must be rather large. For example, several genes with eukaryotic origins have been found in *Chlamydia* (see Section 15.3). The bacterium *Thermotoga maritima* has over 400 genes (greater than 20% of its genome), which seem to have originated in the *Archaea*, and 81 of these are found in discrete clusters, not spread out around the genome. This argues that they have been obtained by horizontal gene transfer from thermophilic *Archaea* found in the environment also occupied by *Thermotoga*, not through evolution from a common ancestor.

Two organisms listed in Table 15.1 are obligate intracellular pathogens, *Chlamydia trachomatis* (Section 12.27) and *Rickettsia prowazekii* (Section 12.13). Their genomes are also missing many genes whose products would be involved in biosynthesizing substances that could be supplied by a host. As noted, several genes found in the *C. trachomatis* chromosome seem to be the result of horizontal gene transfer from a *eukaryotic* source. This is the opposite of the situation where genes in the eukaryotic nucleus seem to have been transferred there from the ancestor of the mitochondrion, an organelle found in eukaryotic cells derived evolutionarily by endosymbiosis of a bacterium (∞ Section 11.3; see also Section 15.7). *Rickettsia* are related to the *Bacteria* that are believed to be the ancestor of the mitochondrion. The genome of *R. prowazekii* seems to be undergoing a reduction similar to that which resulted in the very small mitochondrial genomes. Many of the functions retained by the rickettsial genome are involved in energy production, a function associated with mitochondria. Interestingly, 24% of the *R. prowazekii* genome is noncoding, the highest noncoding fraction found in any prokaryotic genome yet examined. It seems likely that these sequences are the remnants of functional genes that have not yet been eliminated from the genome.

Evolutionary considerations also have a practical component. The ability to identify genes that are clearly prokaryotic (since the genomes of eukaryotes are also being sequenced, see Section 15.6) or, in particular, genes characteristic of pathogenic bacteria, could lead to the design of very powerful and extremely specific therapeutic agents for use in clinical medicine.

✓ 15.4 Concept Check

Genomics can be used to study the evolutionary history of an organism. Organisms often contain gene families, genes with related sequences. If these arose because of gene duplication, the genes are said to be paralogs. Organisms may acquire genes from other organisms in their environment by a process called horizontal gene transfer.

- ✓ Contrast gene *paralogs* with gene *orthologs*.
- ✓ Describe a mechanism by which *horizontal gene transfer* might occur in a prokaryote.

15.5 Genomic Mining

As we have discussed, genomic analysis provides very valuable insight into the metabolism of an organism and the nature of its environment. Often, however, the analysis indicates one or more of the genes usually associated with a pathway seem to be missing. This could mean that the organism uses a novel pathway or that at least one novel reaction is involved. It would, therefore, be of interest to search for a gene encoding a protein that can carry out the missing function.

Searching through a database containing the complete sequence of an organism's genome to find a new gene is referred to as **genomic mining**. The search for the DNA polymerase of the cyanobacterium *Synechocystis* is a good example.

All *Bacteria* contain a DNA polymerase similar to DNA polymerase III of *Escherichia coli*, which is used as the primary polymerase in DNA replication (∞ Section 7.6). This is a complex enzyme containing many different subunits, and the subunit that catalyzes the actual polymerase reaction is DnaE, a product of the *dnaE* gene. Because DNA polymerases are highly conserved proteins and of prime importance to all cells, one would predict that inspection of the genome of *any* member of the *Bacteria* would yield a gene with sequence similarity (an *orthologous* gene) to the *dnaE* gene of *E. coli*. Interestingly, however, the original search of the genome of *Synechocystis* (see Table 15.1) failed to identify such a gene. Instead, what was found were two open reading frames (ORFs), which if combined, would form a gene with high similarity to *dnaE*. But these two ORFs were over 700,000 bp apart on the *Synechocystis* chromosome and located on opposite DNA strands, indicating that they were transcribed in opposite directions. Could these two ORFs be part of *dnaE*?

Careful sequence analysis of the two ORFs revealed that they each also encoded complementary halves of an **intein**, a self-splicing protein (∞ box, Protein Processing, Chapter 8). The two *Synechocystis* ORFs were then cloned, and it has now been demonstrated that they can be transcribed and translated and that this split intein can catalyze a splicing reaction between the two halves of DnaE to form a complete DnaE (Figure 15.7) and, presumably, a functional DNA polymerase. It is likely that this reaction also occurs in the *Synechocystis* cell.

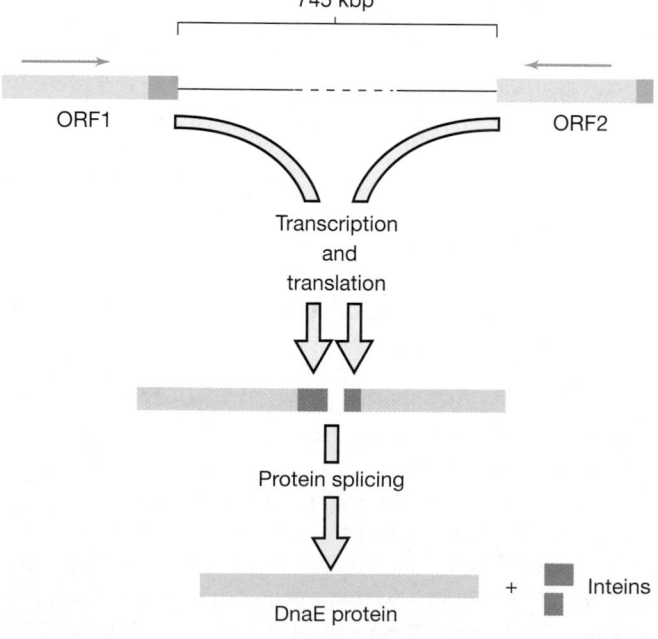

Figure 15.7 The split *dnaE* gene of *Synechocystis*. The two ORFs are transcribed in opposite directions, as shown by the orange arrows. The protein products of these ORFs each contain part of an intein (note the protein product of ORF2 has been drawn in the standard amino terminal to carboxyl terminal orientation).

Why does *Synechocystis* arrange its *dnaE* gene in this split fashion? This can only be guessed at this point, but one hypothesis involves regulation. Since there are only a few molecules of DNA polymerase III present in even a fast-growing cell like *E. coli*, it is likely that there are *very few* copies present in a much slower-growing cell like *Synechocystis*. Thus, perhaps the unusual encoding of the *Synechocystis dnaE* gene in some way helps this organism regulate the level of this crucial enzyme. Of course, it is also possible that this split gene/intein system for *dnaE*/DnaE is simply fortuitous and will be found in many other prokaryotes as well; only more complete genome sequences will confirm this possibility. But either way, it is difficult to imagine how this mechanism would have been discovered without being able to scan and compare entire genome sequences, as can be done routinely today. Mining genomes is thus likely to yield lots of surprises, and may also lead to discoveries that have practical applications as well.

✓ 15.5 Concept Check

Often it is necessary to search carefully through a genomic database to find a particular gene, a process called genomic mining. This can be done to find novel genes or to find genes that one predicts must be present.

✓ How does an intein differ from a self-splicing intron?

15.6 Eukaryotic Microbial Genomes

A large number of microorganisms are eukaryotic (∞ Chapter 14). Although many are well studied, all of the several eukaryotic genomes that have been se-

quenced and published (such as those of the fruit fly *Drosophila melanogaster* and the plant *Arabidopsis*) have been from multicellular organisms, except for that of the yeast *Saccharomyces cerevisiae*. *S. cerevisiae* is an extremely important organism both because of its widespread use in industry (∞ Chapter 30) and its use as a model organism for studying eukaryotic biochemistry and cell biology. It has long been studied from both a genetic and biochemical standpoint.

The Yeast Genome

The haploid yeast genome contains 16 chromosomes ranging in size from chromosome I at 220 kb to chromosome XII, which is about 2352 kb. The total yeast nuclear genome (excluding the mitochondria and some plasmid and viruslike genetic elements) is approximately 13,392 kb. You may wonder why the words "about" and "approximately" are used when this genome has been completely sequenced. Yeast, like many other eukaryotes, has a large amount of repetitive DNA (∞ Section 7.4). When the yeast genome was published in 1997, not all of the "identical" repeats had been sequenced. It is extremely difficult to sequence a very long run of identical or nearly identical sequences. Yeast chromosome XII contains a stretch of approximately 1260 kb containing 100–200 repeats of the yeast rRNA "operon" (∞ Section 7.11), also called rDNA. Another repeated sequence follows this long series of rDNA repeats. Because this entire region has not actually been completely sequenced, the precise size of the genome is not known.

Table 15.3 compares certain aspects of the yeast genome with that of other sequenced eukaryotes, in this case all from multicellular organisms. Although the yeast

TABLE 15.3 Eukaryotic nuclear genomes[a]

Organism	Comments	Genome size	Chromosome number	Protein-encoding genes[b]
Saccharomyces cerevisiae	This yeast is an industrially important organism (∞ Sections 30.12 and 30.13) that is also a model for biochemical and genetic studies	13 Mb	16	5,570
Caenorhabditis elegans	This roundworm is an important model for studying animal development	97 Mb	6	19,099
Drosophila melanogaster	The fruit fly is an intensively studied model organism	180 Mb	4	13,601
Arabidopsis thaliana	This plant serves as a model organism for genetic studies	125 Mb	5	25,498
Homo sapiens	The human genome is available only in draft form	3000 Mb	23	25,000–35,000[c]

[a] All data are for the haploid nuclear genomes of these organisms.

[b] The number of protein-encoding genes is in all cases an estimate based on the number of known genes and sequences that seem likely to encode functional proteins.

[c] There is still debate over the number of genes in the human genome, despite the release of "draft" sequences by two different groups. The number of genes could be as high as 60,000.

genome is considerably smaller than that of the other listed eukaryotes, it is far from the smallest known, as we will discuss shortly.

In addition to having 100–200 identical copies of the rRNA operons, the yeast nuclear genome has 275 genes for tRNAs (only a few are identical) and 80 genes for other types of noncoding RNAs (∞ Section 7.16). There were 6340 possible protein-encoding genes originally predicted, but more detailed analysis has reduced the hypothetical upper limit of different proteins encoded by yeast to 5570. Of these, almost 3400 encode known proteins, i.e., proteins whose functions are truly known or proteins/genes that are closely related to those found in other organisms. The wide array of genetic and biochemical techniques available for studying this organism have resulted in significant advances in the understanding of the function of these proteins.

Another mechanism for exploring protein function involves obtaining knockout mutations (∞ Section 10.18), that is, mutations in which a gene has been rendered nonfunctional. Ordinarily such mutations cannot be obtained in a gene essential for cell viability in a haploid organism. However, yeast can also be grown in a diploid state (∞ Section 14.6). By obtaining such mutations in the diploid and investigating whether they can also exist in the haploid, it is possible to determine whether a particular gene is essential for a cell's viability. Using such techniques, it has been shown that at least 877 of the yeast ORFs are essential, while 3121 clearly are not. Note that the number of essential genes is considerably greater than the approximately 300 that are considered the minimal number required for cellular existence in prokaryotes (see Section 15.3).

In the protein-encoding genes of yeast there are only a total of 225 introns. Most genes that have introns have a single small intron near the 5′ end of the gene. This situation is much different than in the other eukaryotic genomes shown in Table 15.3. In *Caenorhabditis elegans* the average gene has 5 introns and in *Drosophila* the average gene has 4 introns. Introns are also very common in *Arabidopsis*, which also has 5 exons per gene, and over 75% of the genes have introns. In humans almost all protein-encoding genes have introns, and it is not uncommon for a single gene to have 10 or more, nor for the introns to be much larger than the exons.

Other Eukaryotic Microorganisms

The genomes of some other eukaryotic microorganisms have been sequenced: the pathogens *Plasmodium falciparum*, *Encephalitozoon cuniculi*, and *Ustilago maydis*. *P. falciparum* causes malaria (∞ Section 27.5), and the sequencing of its genome was an international effort. The 27-Mb genome consists of 14 chromosomes ranging in size from 0.7 to 3.4 Mb. *E. cuniculi* is an intracellular pathogen of humans and other animals. It lacks mito-

chondria, and although its haploid genome contains 11 chromosomes, the genome size is only 2.9 Mb, making it smaller than many prokaryotic genomes (see Table 15.1). *U. maydis* is a phytopathogenic fungus with a genome of approximately 20 Mb. It causes smut disease in corn (maize), a disease with a large economic impact.

Considerable progress has been make in sequencing the genomes of other eukaryotic microorganisms, including those of the pathogens of *Leishmania major*, *Candida albicans* (∞ Section 27.7), *Entamoeba histolytica* (amebic dysentery; ∞ Section 14.9), *Giardia lamblia* (giardiasis; ∞ Section 28.6), and *Pneumocystis carinii* (AIDS-associated pneumonia; ∞ Section 26.14).

The sequences of some genomes are not immediately published. Knowledge of the genome of a pathogen may be useful in designing new and specific drugs. These applications would be of considerable potential commercial value. It is precisely this potential value that influences some commercial enterprises to withhold sequencing information from the public, at least for a period of time.

✓ 15.6 Concept Check

The complete genomic sequence of the yeast *Saccharomyces cerevisiae* has been determined. Yeast may encode up to 5777 proteins of which only 877 are essential for viability. Relatively few of the protein-encoding genes of yeast contain introns.

- ✓ Describe how one might show a gene is essential.
- ✓ Under what circumstances can you imagine a genomic sequence not being released to the public?

15.7 Genomes of Organelles

In earlier sections of this chapter we discussed some of the properties of the prokaryotic genomes and the nuclear genomes of eukaryotic microbes. In Chapters 9 and 16 the genomes of some viruses are discussed. Here we briefly discuss the genomes of the organelles found within eukaryotic cells: chloroplasts and mitochondria. These are sometimes referred to as *extranuclear genomes*.

Both of these organelles originated from prokaryotes by endosymbiosis (∞ Sections 11.3, 14.2 and 14.3). Therefore, it is not surprising that the proteins they encode are more closely related (by sequence) to those of prokaryotes than to those of the *Eukarya*.

Chloroplasts

The chloroplast is an organelle that carries out photosynthesis and is found in green plants (∞ Section 14.3). Our concern here is not with the chloroplast itself, but with its genome. Known chloroplast genomes are all circular DNA molecules and, although there are several copies in each chloroplast, each contains the total complement of genes. The typical chloroplast genome is

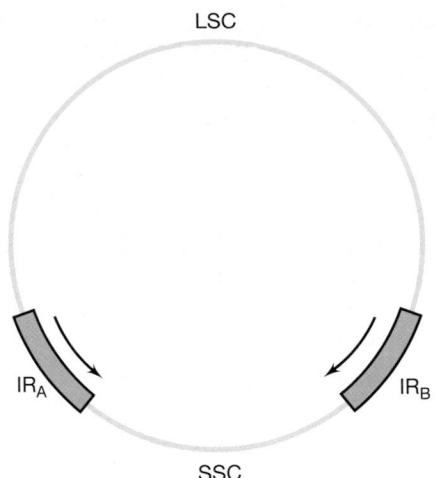

LSC

IR$_A$

IR$_B$

SSC

Figure 15.8 The map of a typical chloroplast genome. The genomes of chloroplasts are circular double-stranded DNA molecules. Most contain two inverted repeat regions (IR$_A$ and IR$_B$), which form the borders of a small single copy region (SSC) and a large single copy region (LSC).

about 120 to 160 kb and contains two inverted repeats of 6 to 76 kb (Figure 15.8). Several chloroplast genomes have been completely sequenced, and a few of these are shown in Table 15.4. The flagellated protozoan *Mesostigma viride* belongs to the earliest diverging green plant lineage. Its chloroplast contains more protein encoding genes and tRNA genes than any other so far known and has the typical genome structure illustrated in Figure 15.8.

Many of the genes in the chloroplast genome encode proteins involved in photosynthesis. However, the chloroplast genome also encodes the rRNAs used in chloroplast ribosomes, the tRNAs used by the translational apparatus, and a few of the proteins used in transcription and translation as well as some other proteins. There are also remnant chloroplast genomes known in some parasitic plants that have lost many of the genes required for photosynthesis.

Introns are much more common in chloroplast genes than in prokaryotic genes, but it is important to remember that these are not the type of introns found in eukaryotic protein-encoding genes (∞ Section 7.12). Although some chloroplasts have no introns, that of the protozoan *Euglena* (∞ Section 14.11) has 155 introns, making up 38% of its genome.

What has analysis of chloroplast genomes shown us, beyond giving evidence for the endosymbiotic hypothesis? For one thing, this analysis has helped us describe the evolutionary relationships of the plants themselves. Some chloroplast genomes have also been shown to contain genes that are very similar in sequence to *Escherichia coli* genes, whose protein products are involved in cell division, suggesting that the mechanism of chloroplast division may be similar to that of *Bacteria*. Similarly, the genes involved in protein transport through membranes are related to those of *Bacteria*.

Mitochondria

Mitochondria are involved in energy production and are found in most eukaryotic organisms. Recall from Section 15.3 that the intracellular parasite *Rickettsia prowazekii* is from the same phylogenetic lineage as the ancestor of the mitochondria.

Mitochondrial genomes encode proteins involved in oxidative phosphorylation and, as is the case of chloroplast genomes, also encode rRNAs and tRNAs used in the mitochondria and typically a few of the proteins involved in translation (in most cases they depend on the nucleus for all the proteins involved in transcription). Most mitochondrial genomes encode fewer proteins than do those of chloroplasts. Well over 200 mitochondrial genomes have been sequenced. The largest mitochondrial genome has 62 protein-encoding genes, but the others known encode between 3 and 37 proteins. The mitochondria of almost all animals encode only 13 proteins (plus 22 tRNAs and 2 rRNAs) and that of *Saccharomyces cerevisiae* only 8 proteins. Subsequently, these genomes

TABLE 15.4	Some chloroplast genomes[a]		Genes encoding			
Organism		Size (bp)	Proteins[c]	tRNA	rRNA[d]	Inverted repeats[b]
Chlorella vulgaris	Green alga	150,613	77	31	1	Absent
Euglena gracilis	Protozoan	143,170	67	27	3	Absent
Mesostigma viride	Protozoan	118,360	92	37	2	Present
Pinus thunbergii	Black pine	119,707	72	32	1	Present[e]
Oryza sativa	Rice	134,525	70	30	2	Present
Zea mays	Corn	140,387	70	30	2	Present

[a] All chloroplast genomes are circular, double-stranded DNA (∞ Section 14.4).

[b] See Figure 15.8

[c] These include genes encoding proteins of known function and ORFs that might be functional.

[d] Each unit is an rRNA operon, containing genes for each of the rRNAs (∞ Section 7.11).

[e] Although the inverted repeats are present, they are greatly truncated.

are usually smaller than those of chloroplasts. Large mitochondrial genomes have extremely large amounts of noncoding DNA, not more genes. The mitochondrion with the most protein-encoding genes, that of the protozoan *Reclinomonas americana*, is only 69,034 bp. In addition, whereas chloroplasts use the "universal" genetic code, remember that mitochondria use slightly different, simplified codes (∞ Section 7.13). These seem to have resulted from the selection pressures for smaller genomes. The standard set of 22 mitochondrial tRNAs is insufficient to read the genetic code even with the "standard" wobble pairing (∞ Section 7.13) used by all other translational systems. Therefore, base-pairing between anticodon and codon is even more flexible in mitochondria (∞ Figure 7.33).

It is not easy to discuss a "typical" mitochondrial genome. Unlike chloroplast genomes, which are all single, circular DNA molecules, the genomes of mitochondria are quite diverse. Although it seems likely that most mitochondrial genomes consist of a circular double-stranded DNA molecule, there are now a large number of organisms known that have linear mitochondrial genomes, including some algae, protozoans, and fungi. In some cases, such as in the yeast *Saccharomyces cerevisiae*, although genetic and restriction mapping have shown the mitochondrial genome to be circular, it seems that the major *in vivo* form is linear. (Recall that bacteriophage T4 has a genetically circular genome structure but is physically linear, ∞ Section 9.9). Some organisms even have more than one mitochondrial chromosome.

Animal mitochondrial genomes typically contain only 13 protein-encoding genes. Figure 15.9 shows a map of the 16,569-bp human mitochondrial genome, which was sequenced in 1981. The yeast mitochondrial genome is larger (85,779 bp) but has only 8 protein-encoding genes. Outside of the genes encoding for the RNAs and proteins, the genome of yeast mitochondria contains large stretches of extremely A + T−rich DNA that serves no apparent function.

Finally, it should be noted that in several organisms the mitochondria are known to contain small plasmids as well as their "chromosome," making mitochondrial genome analysis even more complicated. An additional complication in the case of some mitochondrial genomes is that it is sometimes difficult to find the gene for a particular protein even when the sequence of both the protein and the mitochondrial DNA are both known. This is because of RNA editing, a subject to which we now turn.

RNA Editing

In Chapter 7 we discussed the fact that certain genes had coding regions that were split by noncoding regions called introns. Remember that introns are removed typically after transcription during a processing step that forms mRNA called *splicing* (∞ Section 7.12).

This is true even for the type of introns that are found in mitochondria and chloroplasts, although they are not the same as those found in nuclear protein-encoding genes (∞ Section 7.12). (In a few cases it is the protein itself that is processed; ∞ box, Protein Processing, Chapter 8). There is a phenomenon found primarily in the genomes of organelles that is almost the opposite of splicing: *RNA editing*. **RNA editing** involves either the programmed insertion or deletion of nucleotides into the final mRNA that were not present in the DNA transcribed or the modification of a base in the mRNA that changes it to another.

In the mitochondria of trypanosomes and related protozoa (∞ Section 14.8) some mitochondrial transcripts have large numbers (hundreds in some cases) of uridylates added or, more rarely, deleted. An example

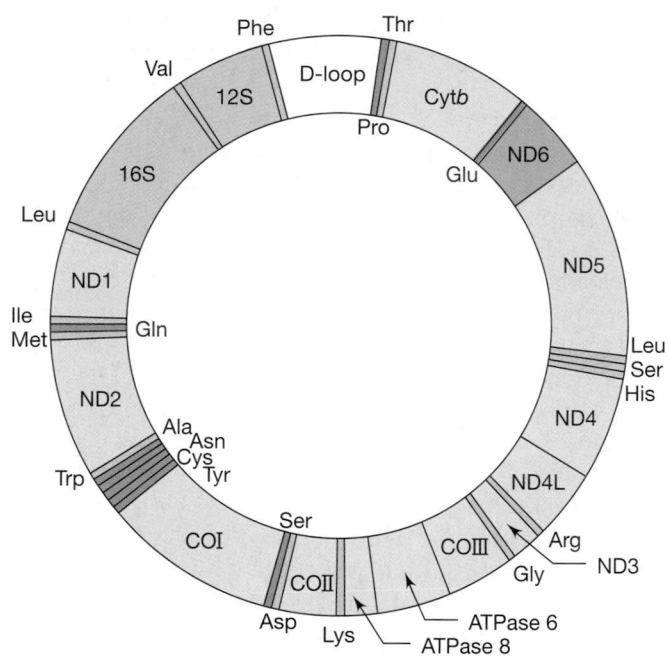

Figure 15.9 Map of the human mitochondrial genome. The circular genome of the human mitochondrion contains 16,569 bp. The genome encodes the 16S and 12S rRNA (corresponding to the prokaryotic 23S and 16S rRNA) and 22 tRNAs. These genes are indicated in two shades of orange, a darker orange for genes that are transcribed counterclockwise from the map as drawn and those in lighter orange that are transcribed clockwise. (The amino acid designations for the tRNA are also on the outside for counterclockwise-transcribed genes and on the inside of the map for clockwise-transcribed genes.) The 13 protein-encoding genes are shown in green (once again with darker green indicating those transcribed counterclockwise and lighter green for those transcribed clockwise). The genes encode: *Cytb*, cytochrome *b*; ND1-6, components of the NADH dehydrogenase complex; COI-III, subunits of the cytochrome oxidase complex; ATPase 6 and 8 polypeptides of the mitochondrial ATPase complex. The two promoters are in the region called the D-loop, a region also involved in DNA replication.

of this type of RNA editing is shown in Figure 15.10. This type of editing is precisely controlled by small RNA "guide" sequences that are present in the mitochondria and guide the enzymatic machinery involved. Obviously this process must be very precisely controlled. Inserting too many or too few bases would give a frameshift product that is unlikely to be functional.

The other type of editing, the changing of one base into another, is fairly common in the mitochondria and chloroplasts of higher plants. At specific sites in some transcripts a C will be converted to a U by oxidative deamination (the opposite modification is more rare). There are at least 25 sites of C to U conversion in the maize chloroplast. Although mostly found in organellar genomes, an example of the programmed conversion of a C to a U is also known for a mammalian nuclear gene.

RNA editing can make it very difficult to predict from a DNA sequence what proteins a genome encodes. That did not prove to be a great obstacle in analyzing the genomes of mitochondria and chloroplasts, however, because the number of proteins they encode is small and these proteins are highly conserved.

Organelles and the Nuclear Genome

Chloroplasts and mitochondria require more proteins than they encode. In both organelles, far more proteins are involved in translation than are encoded by the organelles. As mentioned, mitochondria encode none of the proteins required for transcription. Organelle genomes are also missing many genes that encode required structural proteins. In mitochondria even some of the proteins involved in oxidative phosphorylation are not encoded by the mitochondria.

It is estimated that the yeast mitochondrion includes over 400 different proteins, and as we have seen, only 8 of them are encoded by the yeast mito-

chondrion. This means that almost all proteins required by the yeast mitochondrion are encoded by genes in the nucleus. Although it might seem reasonable to think the same genes (and proteins) that serve these functions for the eukaryotic nucleus/cytoplasm would be put to use in the organelles, this is not so. Whereas the genes for the organelle proteins are found in the nucleus (and are transcribed there and translated in the cell's cytoplasm), these are genes whose products are used specifically by the organelles and must be transported there.

Among the first mitochondrial proteins to be studied were those involved in translation, and it was clear that these nuclear-encoded mitochondrial proteins were closely related to their counterparts in *Bacteria*, not to those in *Eukarya*. It seemed probable that most genes encoding mitochondrial proteins had been transferred from the mitochondrion to the nucleus.

Until the advent of genomics it was very difficult to gain further information on this situation. What is required is a sequence of the host nuclear genome, that of a species of *Bacteria* closely related to the mitochondrial genome, and genomic sequences of other *Bacteria* for comparative purposes. Knowledge of the mitochondrial genome itself was not required (since it encodes so few proteins), but a knowledge of the proteins making up the mitochondria was needed. All these requirements were met in the case of *Saccharomyces cerevisiue*, and the analysis has been performed. Surprisingly, of the 400 nuclear genes encoding mitochondrial proteins, only about 50 were closely related to the phylogentic lineage in *Bacteria* that led to mitochondria. Another 150 were clearly related to proteins of *Bacteria*. However, the remaining 200 were encoded by genes that seem to have no homologues in the *Bacteria*. The "bacterial" proteins were mostly involved in energy conversions, translation, and biosynthesis, whereas the "eukaryotic proteins" are mostly involved in membranes, regulation, and transport. Thus, genomics has shown that the evolution of mitochondria is more complicated than had been expected.

✓ *15.7 Concept Check*

Chloroplasts and mitochondria have small genomes independent of nuclear genomes. These genomes encode rRNAs and tRNAs and several proteins. The chloroplast genome encodes proteins involved in photosynthesis and the mitochondrial genome encodes proteins involved in oxidative phosphorylation. Although the genomes of the organelles are independent of the nuclear genome, the organelles themselves are not. Many genes in the nucleus encode proteins required for organellar function.

✓ Describe the structure of the organellar genomes.

✓ Describe how genes that encode mitochondrial function may have become part of the nuclear genome.

Protein	...Leu	Cys	Phe	Trp	Phe	Arg	Phe	Phe	Cys...
mRNA	...uuG	uGu	UUU	UGG	uuu	AGG	uuu	uuu	uGu...
DNA	...	G G	TTT	TCC		AGG		G	...
	...	C C	AAA	AGG		TCC		C	...

Figure 15.10 RNA editing. The upper part of the figure shows a portion of the amino acid sequence of subunit III of the enzyme cytochrome oxidase from the protozoan *Trypanosoma brucei* (⊂⊃ Section 14.8). This protein is encoded by mitochondria. Beneath the amino acid sequence, the sequence of the messenger RNA (mRNA) for this region is shown. The bases in upper case letters are those transcribed from the gene, which is shown below. The bases in the mRNA in lowercase have been inserted into the transcript by RNA editing. Although the DNA has many *informational gaps*, there are no actual gaps in the molecule itself. The spaces between the base pairs are simply for convenience in visualization.

III GENE FUNCTION AND REGULATION

15.8 Proteomics

The aim of genomic studies is not to simply assess an organism's genetic complement, but also to determine which genes are expressed (and under what conditions) and to determine the function of their protein products. The preceding section described the total complement of proteins found in the yeast mitochondrion. This complement could be called the *yeast mitochondrial proteome*. A **proteome** is all the proteins present in a cell, tissue, or organism at any one time. Note that the proteome is not simply "all the proteins encoded by an organism's genome." Although the genes of an organism are constantly present and relatively fixed, because gene expression is regulated (∞ Chapter 8) the number and types of proteins present are constantly changing in response to the organism's environment or its development. The large scale or genomewide study of the structure, function, and regulation of proteins of an organism is called **proteomics** or **functional genomics**.

One early approach to proteomics was the use of *two-dimensional polyacrylamide gel electrophoresis* to separate, identify, and measure all the proteins present in a sample of cells (from a culture, in the case of microorganisms). Like other types of electrophoresis, the separation is based on the charge of the molecule. A two-dimensional gel separating proteins from *Escherichia coli* is shown in Figure 15.11. In the first dimension (the horizontal dimension in the figure) the proteins are separated by their isoelectric points, that is they migrate in a pH gradient until the net charge on a particular protein reaches 0. In the second dimension, the proteins are denatured in a way that gives each amino acid residue a fixed charge. The proteins will then separate by size (in much the same way DNA molecules are separated by size; ∞ Figure 10.35). In the 1970s Frederick C. Neidhardt and his colleagues began an intensive study of the proteins of *E. coli* and their regulation. Many hundreds of proteins have been identified by biochemical or genetic means and their regulation studied in a variety of conditions. Biochemical methods of separation and analysis, such as chromatography and mass spectrometry, and genetic techniques, such as transposon and site-directed mutagenesis (∞ Sections 10.11 and 10.18) and the construction of libraries of gene fusions (∞ Section 31.4), are important tools. Proteomics still requires experimentation, in addition to extensive computer analysis of the results.

Although important clues can often be gained about a protein's function from *in silico* analysis, well over half of the genes in any sequenced organism have no known function, and many of these are specific to the organism (a situation that will likely change as more and more genomes are sequenced).

Functional and Structural Genomics

Although proteomics often requires intensive experimentation, one cannot dismiss *in silico* techniques. After obtaining the sequence of an organism's genome, initial analysis will compare its sequence to that of other organisms to locate and identify genes that are already known. The "sequence" here that is most important is the amino acid sequence of the protein. Because of degeneracy of the genetic code (∞ Section 7.13), differences in the DNA sequence may not necessarily lead to differences in the amino acid sequence (Figure 15.12).

There is strong evidence that proteins with greater than 50% sequence identity usually have quite similar functions. However, there are known exceptions to this rule. For instance, there are two proteins known in *Pseudomonas* that have 98% sequence identity but that catalyze quite different reactions. One important clue to the possible function of a protein is the identification of regions in the coding sequence that are present in other proteins and that encode domains of the protein having known functions, such as metal-binding domains or GTP-binding domains. These domains have specific sequences, but what makes them *functional* is their three-dimensional structure (∞ Section 3.8).

M_r (kDa)

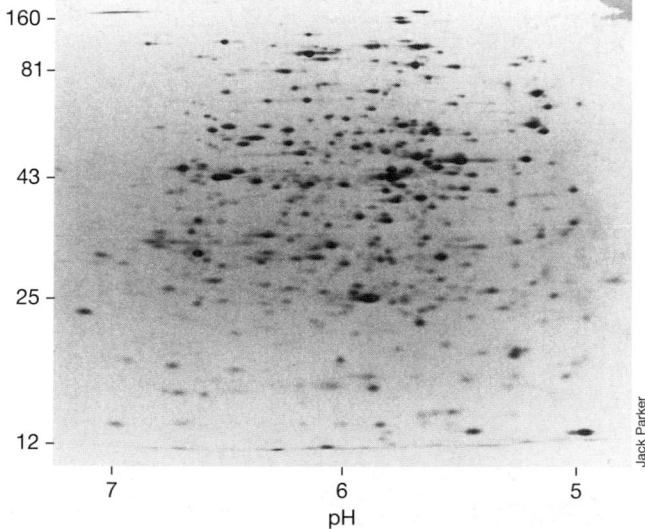

Figure 15.11 Two-dimensional polyacrylamide gel electrophoresis of proteins. The figure shows an autoradiogram of the proteins of a culture of *Escherichia coli*. Each spot on the gel is a particular protein. The proteins were labeled with radioactive methionine during growth to allow for visualization and quantitation. The proteins were first separated by isoelectric focusing under denaturing conditions, concentrating on the pH range 5 to 7 where most *E. coli* proteins are found (the basic ribosomal proteins are missing from this separation). The second dimension separates denatured proteins by their mass (M_r; given here in kilodaltons), with the largest proteins being toward the top of the gel.

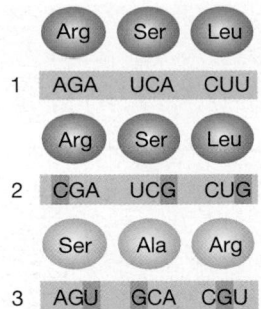

Figure 15.12 Comparison of nucleic acid and amino acid sequence similarities. Three different nucleotide sequences are shown (for convenience RNA is shown). Both sequence 2 and sequence 3 differ from sequence 1 in only three positions. However, the amino acid sequence encoded by 1 and 2 are identical whereas that encoded by sequence 3 is unrelated to the other two.

Structural genomics is the determination of the three-dimensional structures of proteins representative of the range of protein structure and function found in an organism. The ultimate aim is to build a body of structural information that will allow *in silico* techniques to predict the probable structure and potential function for almost any protein from knowledge of its coding sequence.

Coupling proteomics to genomics will yield important information allowing us to begin to clearly understand how a single organism integrates the expression of its entire genome to grow and replicate and how it responds to environmental stimuli. The prospects for basic science are enormous. However, much of the impetus for functional genomic studies in many organisms has been aimed toward applications. This can include, for example, the attempt to find potential therapeutic targets for the development of new antibiotics.

✓ **15.8 Concept Check**

A proteome is all the proteins present in an organism at any one time. The aim of proteomics is to study these proteins to learn their structure, function, and regulation.

✓ Can an organism have more than one proteome?

15.9 Microarrays

Although functional and structural genomics are still labor-intensive activities, they have been immensely aided by rapid changes in technology. One aim of proteomics is the study of the regulation of gene expression. Remember that gene expression is often regulated at the level of transcription (⚮ Chapter 8). A new and powerful technique has become available that allows the transcription of the entire genome to be analyzed: *microarrays*.

For a gene to be expressed, it must be transcribed. Knowing under what conditions a gene is transcribed may also give information about a gene's function. For example, the gene specific for maltose metabolism will be among those found when *Escherichia coli* is growing on maltose, but not when it is growing on glucose (⚮ Section 8.6).

In Chapter 10 we discussed how nucleic acid hybridization techniques help to locate genes on specific fragments of DNA (⚮ Section 10.12). Hybridization techniques have also long been used to measure the expression of genes by hybridizing mRNA to specific DNA fragments. This technique has been radically enhanced with the development of *microarrays* or *DNA chips*, in which genes or portions of genes representing the entire genome are spatially arrayed on a small solid-state support. These genes (or gene fragments) can then be hybridized using a variety of probes and then scanned and analyzed by a computer.

There are several methods of making such microarrays, and one of these is shown in Figure 15.13. The figure also illustrates how such an array can be used. Figure 15.14 shows a part of an actual chip used to assay expression of the *Saccharomyces cerevisiae* genome. This particular chip is made with silica-chip technology.

In the early 1990s it was discovered that the same process used to produce computer chips (photolithography) could be adapted to produce silica chips, each of which contained thousands of different DNA fragments. Currently, hundreds of thousands of different DNA strands of known sequence can be synthesized in an array on a 1-2 cm silica chip (see Figure 15.13). DNA chips have been manufactured so that multiple sequences representing each of the 5600 protein-encoding genes from the yeast *Saccharomyces cerevisiae* is represented in a known location. These DNA chips can then be hybridized with DNA or RNA probes that have been tagged with a fluorescent dye or other labels. The tagged DNA or RNA binds only to the probe that is complementary to its sequence. This chip is then "read" with a laser and analyzed by a computer. Distinct patterns of hybridization will be observed, depending upon which complementary DNA or RNA sequences are present in the sample.

We previously discussed hybridization using a single probe to find a corresponding gene (⚮ Section 10.12). With the DNA chip, one can use very complex probes. For instance, one can ask which of the over 5000 *Saccharomyces* genes are being expressed under a certain condition by using the entire population of mRNA as a probe. Of course, one can also compare expression of different genes under different conditions. The ability to relatively easily analyze the simultaneous expression of thousands of genes both qualitatively and quantitatively has tremendous potential for helping us to understand the

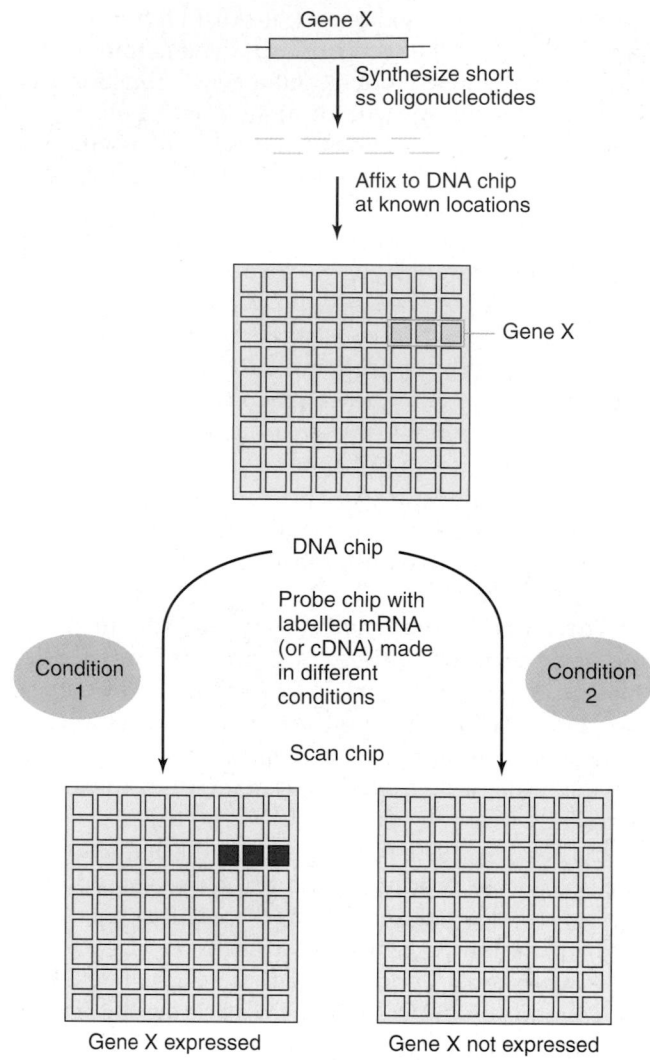

Figure 15.13 Making and using DNA chips. Short single-stranded (ss) oligonucleotides corresponding to all the genes of an organism are synthesized individually (⊙⊙ Section 10.13) and affixed at known and identifiable locations to make a DNA chip or microarray. This is exemplified here with a single gene. These DNA chips are assayed by using labeled probes made by the organism in question under a variety of conditions to determine patterns of gene expression.

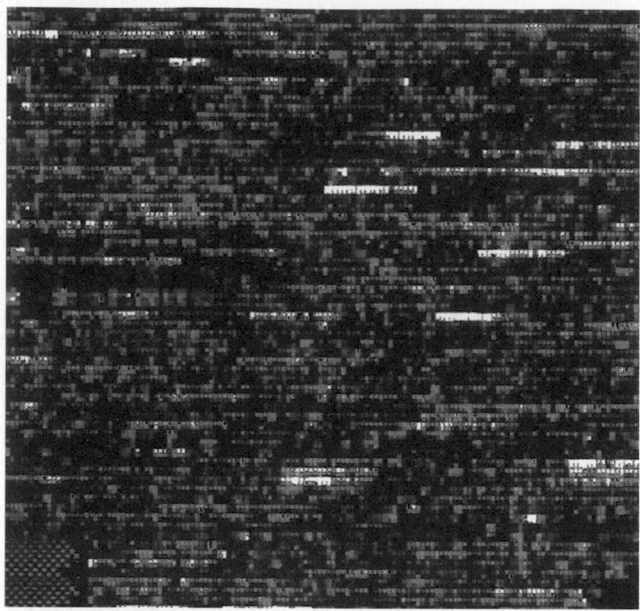

Figure 15.14 DNA chips and their use to assay gene expression. The photo shows fragments from one-fourth of the entire genome of the yeast, *Saccharomyces cerevisiae* affixed to a silica chip. Each gene is present in several copies and has been probed with fluorescently labeled mRNA obtained from yeast cells grown under a specific condition. The background of the chip is blue. Locations where the RNA has hybridized to the DNA is indicated by a gradation of colors up to maximum hybridization, which shows as white. Because the location of different genes on the chip is known, once the chip is scanned it will reveal which specific genes were expressed.

extreme complexity of metabolism and regulation in organisms as "simple" as *Escherichia coli* with its 4000 genes, or as complex as ourselves, with our 25,000–35,000 different genes.

DNA chips can also be used to *identify* organisms. One can use fragmented genomic DNA from a particular organism as a probe and differentiate between closely related strains by differences in their hybridization patterns. This allows for very rapid identification of pathogenic viruses or bacteria, and for the detection of specific strains or mutants of these organisms. This technology is also useful for identifying and quantifying specific organisms in microbial communities in the environment, a common goal of the microbial ecologist (⊙⊙ Chapter 19).

✓ 15.9 Concept Check

Microarrays are genes or gene fragments attached to a solid support in identifiable order. These arrays can be used to hybridize to probes and then analyzed to determine patterns of gene expression. The arrays are large enough and dense enough that the transcription pattern of the entire genome can be analyzed.

✓ How can microarrays be used to monitor the expression of a gene?

✓ Why might it be useful to know how gene expression of the entire genome responds to a particular condition?

Review Questions

1. Some genomic sequencing projects are "map direct-ed" but others rely on "shotgun" methods. Explain the difference between these two methodologies.

2. In *Bacteria* and *Archaea* the acronym ORF is almost a synonym for the word "gene." However, in the higher eukaryotes this is not, strictly speaking, true. Explain.

3. The organisms in Table 15.1 are listed in ascending order of chromosome size. However, chromosome size may not always be an accurate reflection of genome size. Explain.

4. The gene encoding the beta (β) subunit of RNA polymerase from *Escherichia coli* is said to be orthol-ogous to the *rpoB* gene of *Bacillus subtilis*. What does

that mean about the relationship between the two genes? What protein do you suppose the *rpoB* gene of *Bacillus subtilis* encodes? The genes for the differ-ent sigma factors (see Table 8.2) of *Escherichia coli* are paralogous. What does that say about the relationship between these genes?

5. Explain how horizontal gene transfer can compli-cate evolutionary studies.

6. Nuclear genes may encode proteins that function in the mitochondria or chloroplasts. How do these proteins get into the organelles (Section 7.16)?

7. Distinguish between genome, genomics, proteome, and proteomics.

Application Questions

1. When using restriction enzymes to attempt to clone "random" fragments, one tactic is to partially digest with a restriction enzyme that is predicted to have many recognition sites in the DNA of the organism in question. What properties of the restriction enzyme recognition site and the DNA of the organ-ism in question are important in predicting how fre-quently the enzyme's recognition sites might appear in the DNA (Section 10.12)?

2. One approach to obtaining random fragments of DNA is to do a partial digestion with a restriction enzyme. To obtain "random" fragments of about 10 kb, why would it be better to use an enzyme that recognizes a 4-bp sequence rather than one that rec-ognized a 6- or 8-bp sequence?

3. Assume you have cloned a DNA fragment obtained using the restriction enzyme *Bam*HI into the *Bam*HI site of M13mp18. For an individual clone, the strand of cloned DNA found in the mature single-stranded

phage depends on the orientation of the cloned fragment. Explain using diagrams.

4. When using shotgun sequencing, it is possible that there may be "gaps" in the sequence. Describe how it might be possible to try to close these gaps using a library of BAC clones.

5. The sequence of the yeast nuclear genome was pub-lished, but the entire sequence was never actually completely determined. The yeast mitochondrial genomes proved very difficult to sequence accurate-ly. Describe in both cases the practical difficulties that were encountered in the sequencing.

6. How might one experimentally study which pro-teins in *Escherichia coli* are repressed (Section 8.5) when a culture is shifted from a minimal medi-um, which contains only a single carbon source, to a rich medium, containing a large number of amino acids, bases, and vitamins?

V iruses infect all types of organisms, from bacteria to humans. Bacterial viruses (bacteriophages) are often studied as model systems for
molecular processes which occur in cellular organisms. Viruses that
infect plants are usually studied because of their impact on plant agriculture.
Viruses that infect humans are studied because of their impact on human health.
Some of the great scourges of mankind, such as smallpox and polio (a computer model of poliovirus is shown here), are caused by viruses. However, different viruses often display very different replication strategies. A major means
for classifying viruses has to do with the details of their replication schemes.

16

BACTERIAL, PLANT, AND ANIMAL VIRUSES

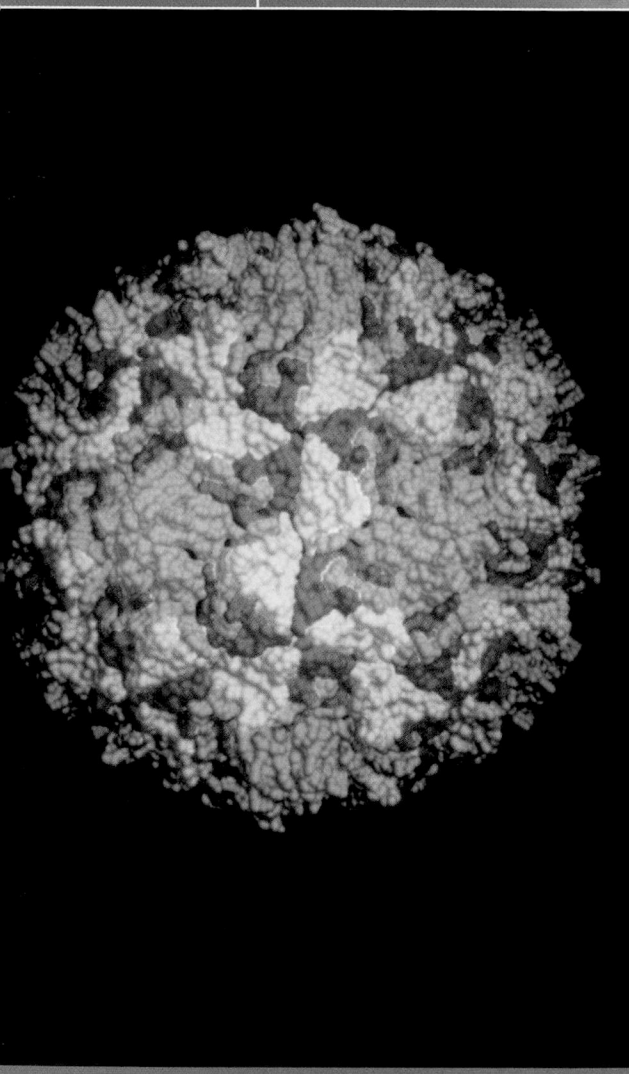

Working Glossary

Bacteriophage a virus that infects pro-karyotic cells

Minus (negative)-strand nucleic acid an RNA or DNA strand that has the opposite sense of (is complementary to) the mRNA of a virus

Plus (positive)-strand nucleic acid an RNA or DNA strand that has the same sense as the mRNA of a virus

Replicative form a double-stranded DNA molecule that is an intermediate in the replication of single-stranded DNA viruses

Retrovirus a virus whose RNA genome has a DNA intermediate as part of its replication cycle

Reverse transcription the process of copying information found in RNA into DNA

Virion the complete virus particle; the nucleic acid surrounded by a protein coat and, in some cases, other material

Virus a genetic element containing either RNA or DNA that replicates in cells but is characterized by having an extracellular state

This chapter on viruses is part of a unit on microbial diversity. It complements Chapter 9, in which we introduced the principles of virology. However, in that chapter we gave detailed information on only a very few of the different types of viruses known. In this chapter we will discuss several different types of viruses.

The previous chapters in this unit have explored the enormous diversity of the *Bacteria* and the *Archaea* and have touched on the diversity of the *Eukarya*. In the following unit we will see how this genetic richness manifests itself in a diversity of metabolic strategies and habitats. It is important to remember before going further in this chapter that viruses have no independent metabolism, and their reproductive habitat is a living cell. However, this is not to indicate that viruses are not diverse: far from it. Different viruses are known that infect most types of organisms (hosts, ∞ Section 9.3), and they have adapted to the lifestyle of the hosts. However, possibly the most interesting area of diversity in viruses is the diversity of their genomes. Whereas all cells have double-stranded DNA as their genetic material, viruses are known that have single-stranded RNA, double-stranded RNA, single-stranded DNA, or double-stranded DNA as their genetic material (∞ Section 9.1). Indeed, the Baltimore Classification of viruses uses the types of viral genomes and their replication strategies to divide viruses into various classes (∞ Section 9.7). In this chapter we will illustrate virus diversity by discussing particular types of viruses in the various classifications. However, we will further separate our discussion into viruses that infect prokaryotes and those that infect eukaryotes. The differences between these two types of cells are often reflected in the life cycles of the viruses that infect them (∞ Section 9.11).

I VIRUSES OF PROKARYOTES

Many viruses are known that infect *Bacteria* and increasing numbers are known which infect *Archaea*. However, we will confine our discussion to a few types that infect *Bacteria*. We mentioned in Chapter 9 that most bacteriophages known have double-stranded DNA genomes (∞ Section 9.8). Even so, there are many bacteriophages known with other types of genomes. The simplest are those with RNA genomes.

16.1 RNA Bacteriophages

Many bacteriophages are known that have RNA genomes of the plus configuration. Interestingly, the bacterial RNA viruses known in the enteric bacteria group infect only bacterial cells that contain a type of plasmid, called a *conjugative plasmid* (∞ Section 10.9), which allows the bacterial cell to function as a *donor* (sometimes referred to as a "male") in a certain type of genetic exchange (the interesting concept of male and female bacteria is discussed in Chapter 10). This restriction to male bacterial cells arises because these viruses infect bacteria by attaching to *pili* (Figure 16.1), which are encoded by the plasmid. Since such pili are absent from cells that do

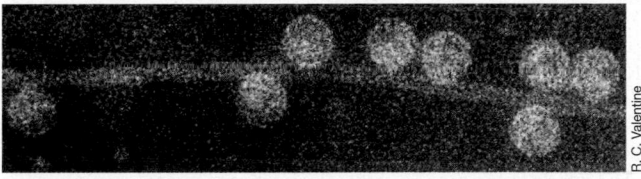

Figure 16.1 Electron micrograph of the pilus of a "male" bacterial cell of *Escherichia coli* showing virions of a small RNA phage attached to the pilus.

not carry the plasmid, these RNA viruses are unable to attach to such cells and hence do not initiate infection in these so-called female cells.

The bacterial RNA viruses are all quite small, about 26 nm in size, and they are all icosahedral, with 180 copies of coat protein per virus particle. The complete nucleotide sequences of several RNA phage genomes are known. The genome of the RNA phage MS2, which infects *Escherichia coli*, is 3569 nucleotides long. The RNA strand in the virion acts directly as mRNA on entry into the cell (∞ Figure 9.11).

The genetic map of MS2 is shown in Figure 16.2*a*, and the flow of events of MS2 multiplication is shown in Figure 16.2*b*. The small genome encodes only four pro-

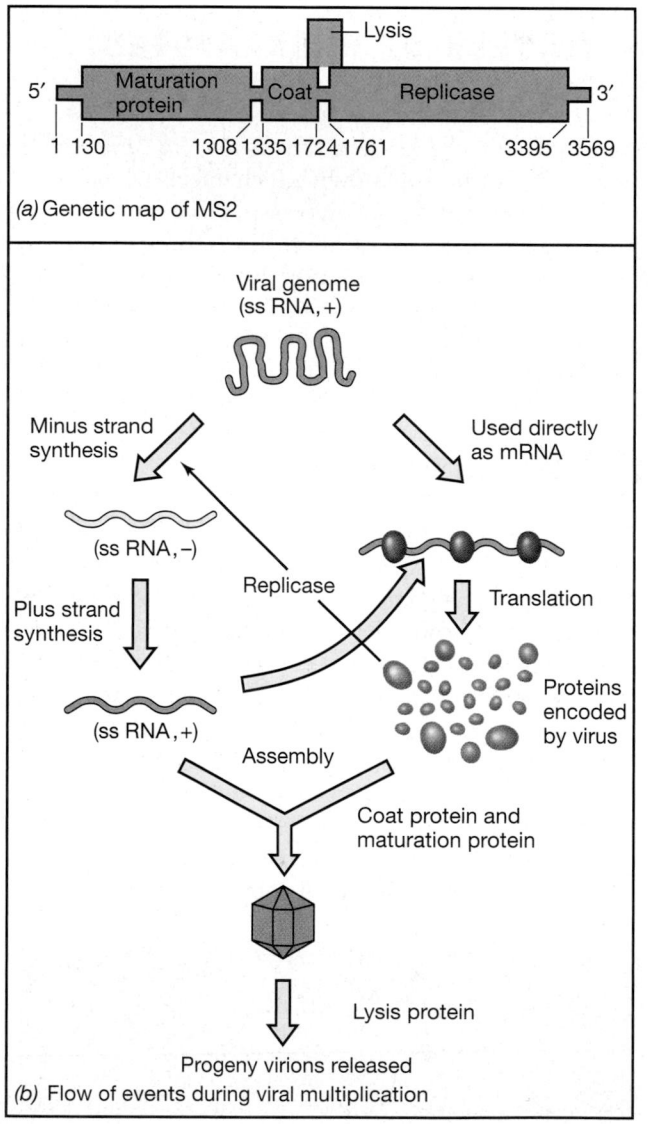

(a) Genetic map of MS2

(b) Flow of events during viral multiplication

Figure 16.2 (a) Genetic map of the RNA bacteriophage MS2. (b) Flow of events during multiplication. The numbers in (a) refer to the nucleotide positions on the RNA.

teins. These are the **maturation protein** (present in the mature virus particle as a single copy), **coat protein, lysis protein** (involved in the lysis process that results in release of mature virus particles), and a subunit of **RNA replicase**, the enzyme that brings about replication of the viral RNA. Interestingly, the RNA replicase is a composite protein, composed partly of the virus-encoded polypeptide and partly of host polypeptides. The host proteins involved in the formation of active viral replicase are part of the cell's normal translational machinery. Thus, the virus appears to employ host proteins that normally have entirely distinct functions and use them to make an active viral replicase.

As noted, the viral RNA is of the plus sense and can thus be translated immediately. After RNA replicase is synthesized, it in turn can synthesize RNA of minus sense using the infecting RNA as template (see Figure 9.11). After minus RNA has been synthesized, more plus RNA is made using this minus RNA as a template. The newly made plus RNA strands then serve as messengers for continued virus protein synthesis. The gene for the maturation protein is at the 5′-end of the RNA. Translation of the gene coding for the maturation protein (needed in only one copy per virus particle), occurs only from the nascent form of the plus-strand RNA as the replication process occurs. In this way, the amount of maturation protein needed is limited. The virus RNA is folded into a complex form with extensive secondary structure. Of the four AUG start sites, the most accessible to the translation process is that for the coat protein, and translation begins there very early. The replicase mRNA is also translated early. As coat protein molecules increase in number in the cell, they combine with the RNA around the AUG start site for the replicase protein, effectively turning off synthesis of replicase. The major virus protein synthesized is coat protein, which is needed in the highest amounts.

Another interesting feature of bacteriophage MS2 is that the fourth virus protein, the *lysis* protein, is encoded by a gene that *overlaps* with both the coat protein gene and the replicase gene (see genetic map in Figure 16.2*a*). The phenomenon of **overlapping genes** is quite common in very small genomes (see Section 16.2). It should be noted that although the use of overlapping genes makes possible more efficient use of genetic information, it seriously complicates the evolutionary process because a mutation in a region of gene overlap may affect two genes simultaneously. The start codon of this lysis gene is not easily accessible to ribosomes because of the secondary structure found in the RNA. When the ribosome terminates synthesis of the coat protein gene, the secondary structure in this region is disrupted, and sometimes this disruption allows a ribosome to begin reading the lysis gene. By restricting the efficiency of translation in this way, premature lysis of the cell is probably avoided. Only after sufficient copies of coat protein are available for the assembly of mature virus particles does lysis commence.

Ultimately, phage assembly takes place, and release of virions from the cell occurs as a result of cell lysis. The features of replication of these simple RNA viruses are themselves fairly simple. The viral RNA itself functions as an mRNA, and regulation occurs primarily by way of controlling access of ribosomes to the appropriate start sites on the viral RNA.

Although there are several different positive-strand RNA bacteriophages known, no bacteriophage has yet been identified with a negative-strand RNA genome. However, there are a few bacteriophages known that have segmented, double-stranded RNA as their genetic material. The best studied of these is $\phi 6$, whose host is *Pseudomonas syringae*. This virus, which is enveloped (∞ Figure 9.12), seems very closely related to the re-oviruses (see Section 16.9), which infect eukaryotes.

✓ 16.1 Concept Check

A variety of RNA viruses that infect bacteria are known. The small RNA genome of these bacterial viruses acts directly as mRNA and encodes only a few proteins.

✓ What is the difference between positive-strand RNA viruses and negative-strand RNA viruses? Explain.

✓ Describe what is meant by *overlapping genes*.

16.2 Single-Stranded DNA Bacteriophages: Icosahedral Virion

There are a number of bacteriophages known whose virions contain a single-stranded DNA genome. These are Class II viruses in the Baltimore Classification System. Before such a genome can be transcribed, a complementary strand of DNA must be synthesized (∞ Section 9.7 and Figure 9.11). All Class II bacteriophages package only the positive-strand DNA (that is, there are no negative-strand DNA bacteriophages known). In this section we will discuss one of these phages, ϕX174, which was important in studies of DNA replication. In the following section we will discuss the single-strand DNA bacteriophage M13, which is an important tool in genetic engineering and whose life cycle is much different from that of ϕX174.

Bacteriophage ϕX174 is one of a number of similar viruses with a circular single-stranded DNA genome and an icosahedral virion. These viruses are very small, about 25 nm in diameter, and the principal building block of the protein coat is a single protein present in 60 copies (the minimum number of protein subunits possible in an icosahedral virus), to which are attached at the vertices of the icosahedron several other proteins that make up spikelike structures (∞ Figure 9.12). These small DNA viruses possess only a limited amount of genetic information in their genomes, and the host cell DNA replication machinery is used in the replication of virus DNA.

The Genome of Phage ϕX174

Phage ϕX174 infects *Escherichia coli*. Its genome consists of a circular single-stranded molecule of 5386 nucleotide residues. The DNA of ϕX174 was the first DNA to be completely sequenced, a remarkable achievement when it was accomplished by Frederick Sanger and colleagues in 1977. Now, DNA sequencing is a routine procedure (∞ Section 10.13). ϕX174 is also of special interest because it was the first genetic element shown to have *overlapping genes* (see also Section 16.1). In very small viruses such as ϕX174 there is insufficient DNA to code for all virus-specific proteins unless parts of certain virus nucleotide sequences are read more than once in different reading frames (Figure 16.3).

As seen in the genetic map of ϕX174, the sequences of genes D and E overlap each other, gene E being contained completely *within* gene D. In addition, the termination codon of gene D overlaps the initiation codon of gene J. Several other instances of gene overlap occur in the ϕX174 genome (Figure 16.3). Additionally, a small protein, called A*-protein, is formed by *reinitiation of translation* (not transcription) within the mRNA of gene A, with A*-protein being read and terminated from the same mRNA reading frame as A-protein but starting at a different in-frame start codon.

DNA Replication by the Rolling Circle Mechanism

The replication process of such a circular single-stranded DNA molecule is of considerable general interest because cellular DNA always replicates in the double-stranded configuration (∞ Section 7.5). The DNA strand in single-stranded DNA phages is of the plus sense. On infection, this DNA strand becomes separated from the protein coat, and entrance into the cell is accompanied by the conversion of this single-stranded DNA to a double-stranded form called **replicative form** (RF) DNA (Figure 16.3b). Cell-coded proteins involved in the conversion of viral DNA to RF DNA include the enzymes *primase, DNA polymerase, ligase*, and *gyrase*. No virus-encoded proteins are involved in this process.

In cells, replication of the lagging strand involves the formation of short *RNA primers* by action of an enzyme called *primase* (∞ Section 7.5). These RNA primers are made at intervals on the lagging strand and are then removed and replaced with DNA by DNA polymerase (∞ Figure 7.19). The situation with ϕX174 DNA is similar but, of course the DNA is a single-stranded closed circle. To begin replication of this DNA,

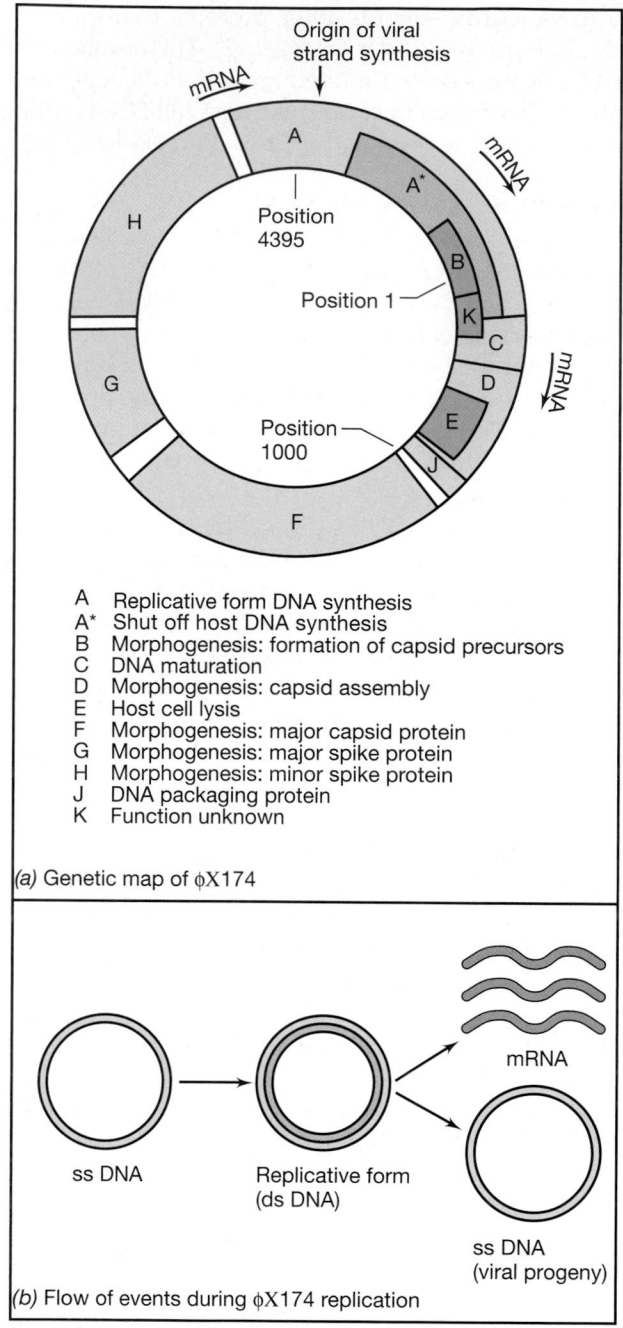

A   Replicative form DNA synthesis
A*   Shut off host DNA synthesis
B   Morphogenesis: formation of capsid precursors
C   DNA maturation
D   Morphogenesis: capsid assembly
E   Host cell lysis
F   Morphogenesis: major capsid protein
G   Morphogenesis: major spike protein
H   Morphogenesis: minor spike protein
J   DNA packaging protein
K   Function unknown

(a) Genetic map of φX174

(b) Flow of events during φX174 replication

Figure 16.3 Bacteriophage φX174, a single-stranded DNA phage. (a) Genetic map. Note the regions of gene overlap (A/B, K/B, K/C, K/A, A/C, and D/E). Intergenic regions are not colored. Protein A* is formed using only part of the coding sequence of gene A by reinitiation of translation (see text). (b) Flow of events in φX174 multiplication. The production of progeny ss DNA from replicative form ds DNA involves rolling circle replication and is shown in more detail in Figure 16.4.

primase brings about the synthesis of a short RNA primer at one or more specific initiation sites. DNA is then synthesized by DNA polymerase III, and the primer is removed and replaced by DNA using DNA polymerase I, exactly as in the case of a lagging strand.

This results in the formation of the complete, circular, double-stranded RF DNA.

Once the RF is formed, DNA replication occurs by conventional semiconservative replication, involving theta-form intermediates (⚭ Figure 7.16) and resulting in the formation of new RF DNA molecules. However, the formation of single-stranded viral genomes involves a different type of replication mechanism called **rolling circle replication** (Figure 16.4). The rolling circle arises because one strand is nicked and the 3'-end of this nick is used to prime synthesis of a new strand. Continued rotation of the circle leads to the synthesis of a linear, single-stranded structure. Note that synthesis is asymmetric because only one of the strands is serving as template. Contrast this with lambda, which replicates both strands late in infection via a rolling circle mechanism (⚭ Figure 9.20). In φX174, synthesis begins when the protein encoded by gene A, called *gene A protein*, cleaves the plus strand of the RF. When the growing viral strand reaches unit length (5386 residues for φX174), gene A protein cleaves and then ligates the two ends of the newly synthesized single strand to give a circular single-stranded DNA.

Transcription and Translation for φX174

Viral mRNA synthesis is directed by the RF DNA. Synthesis of mRNA begins at several major promoters and terminates at a number of sites (see map, Figure 16.3). The polycistronic mRNA molecules (⚭ Section 7.11) are then translated into the various phage proteins. As we have noted, several proteins are made from mRNA transcripts formed from different reading frames from the same DNA sequences (overlapping genes). One can truly be impressed by the efficiency with which such a small genome as that of φX174 can have multiple uses.

Ultimately, assembly of mature virus particles occurs. Release of virions from the cell takes place as a result of cell lysis, which involves the participation of gene E protein. Interestingly, protein E acts by inhibiting the activity of one of the enzymes involved in peptidoglycan synthesis in the cell wall. Because of the resulting weakness in newly synthesized cell wall material, the cell lyses.

✓ **16.2 Concept Check**

The single-stranded DNA genome of the virus φX174 is so small that only the presence of overlapping genes allows it to encode all its essential proteins. This virus provided the first example of overlapping genes. The production of progeny viral DNA involves a rolling circle mechanism.

✓ If the nucleic acid genome of φX174 is single-stranded and in the plus configuration, why can't it be used directly as mRNA?

✓ How does the replicative form of the viral nucleic acid differ from the form found in the virion?

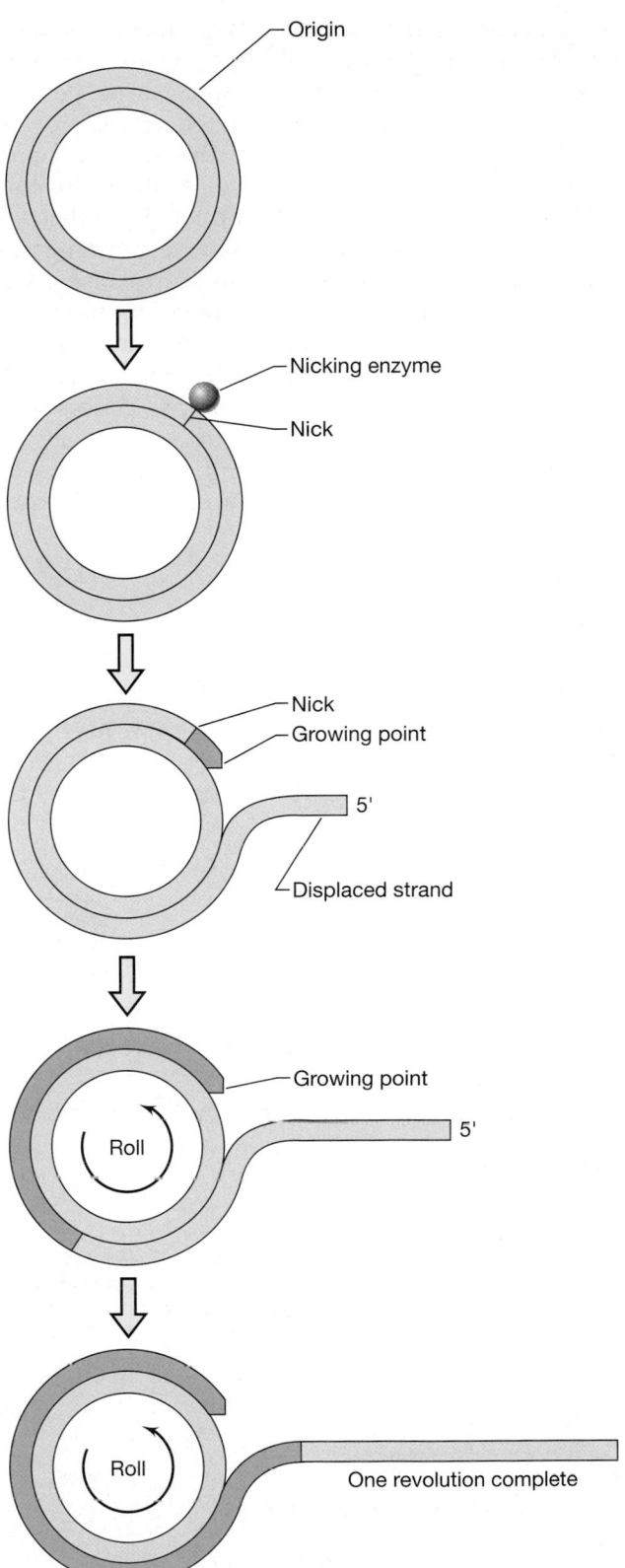

Figure 16.4 Rolling circle replication. Replication begins at the origin by nicking one strand of DNA (in φX174, the nicked strand is the *plus* strand, and the gene A protein makes the nick). After one new progeny strand has been synthesized (one revolution of the circle), the gene A protein cleaves the new strand and ligates its two ends.

16.3 Single-Stranded DNA Bacteriophages: Filamentous Virion

Quite distinct from φX174 are the filamentous DNA phages, which have helical rather than icosahedral symmetry. The most studied member of this group is phage M13, which infects *Escherichia coli*, but related phages include f1 and fd. As with the small RNA bacteriophages, these filamentous DNA phages infect only male cells, entering after attachment to the male-specific pilus (∞ Section 10.9). Even though these phages are linear (filamentous) in shape they possess *circular* single-stranded DNA. The DNA is not self-complementary, however, so the two adjacent halves of the molecule that run up and down the virus particle form loops at the ends but exhibit very little if any base pairing. Phage M13 has found extensive use as a cloning vector and DNA sequencing vehicle in genetic engineering (∞ Section 15.1). The virion of M13 is only 6 nm in diameter but is 860 nm long. These filamentous DNA phages have the additional interesting property of being released from the cell *without* killing the host cell. Thus, a cell infected with phage M13 can continue to grow, all the while releasing virus particles. Virus infection causes a slowing of cell growth, but otherwise a cell is able to coexist with its virus. Plaques are thus seen only as areas of reduced cell growth in the bacterial lawn.

Many aspects of DNA replication in filamentous phages are similar to that of φX174. The property of release without cell killing occurs by a budding process in which the virus particle is always released from the cell with the end containing the A-protein first (Figure 16.5). There is no accumulation of intra-cellular virus particles; the assembly of mature virions occurs on the inner surface of the cytoplasmic membrane, and virus assembly is coupled with the budding process.

Several features of these phages make them useful as cloning and DNA sequencing vehicles. First, they have single-stranded DNA, which means that sequencing can be carried out by the Sanger dideoxynucleotide method (∞ Section 10.13). Second, as long as infected cells are kept in the growing state, they can be maintained indefinitely with cloned DNA so a continuous source of the cloned DNA is available. Third, there is an intergenic space that does not code for protein and can be replaced by variable amounts of foreign DNA.

✓ 16.3 Concept Check

Some single-stranded DNA viruses, such as M13, have filamentous virions. These viruses have found great usefulness as a tool for DNA sequencing. They also have the very interesting property of being able to reproduce and be released without actually killing the host.

✓ How the M13 can be released without killing the infected host cell?

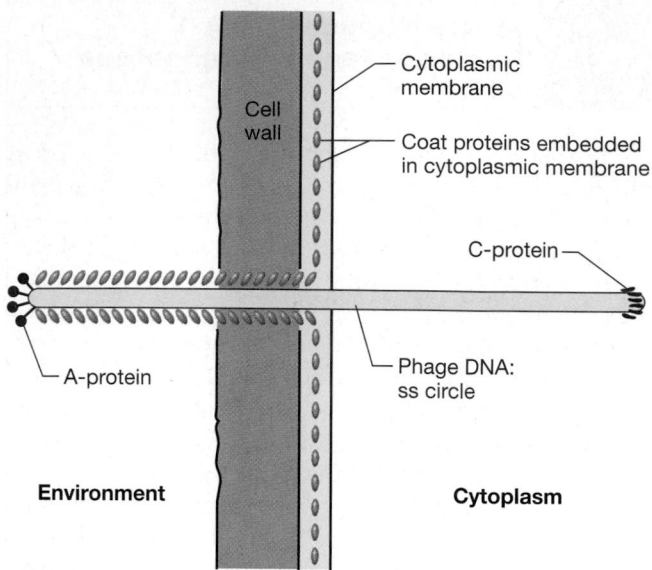

Figure 16.5 Illustration of the manner in which the virion of a filamentous single-stranded phage (such as M13 or fd) leaves an infected cell without lysis. The A-protein passes first through the membrane at a site on the membrane where coat protein molecules have first become embedded. The intracellular circular DNA is coated with dimers of another phage protein, which is displaced by coat protein as the DNA passes through the intact cytoplasmic membrane.

16.4 Double-Stranded DNA Bacteriophages: T7

The double-stranded DNA bacteriophages are among the best-studied organisms, and we have already discussed two of them, T4 and lambda (∞ Sections 9.9 and 9.10). Because of their importance as models for understanding molecular biology and gene regulation, we will discuss two more examples, T7 in this section and Mu in the next section.

Replication of Bacteriophage T7

Bacteriophage T7 and its close relative T3 are relatively small DNA viruses that infect *Escherichia coli*. (Some strains of *Shigella* and *Pasteurella* are also hosts for phage T7.) The virion has an icosahedral head and a very small tail (∞ Figure 9.12).

The T7 genome is a linear double-stranded DNA molecule of 39,936 base pairs. About 92% of the DNA of T7 encodes proteins. Gene overlap also occurs in the T7 genome, as do other translational strategies such as internal translational reinitiation and internal frame-shifts within certain genes, all apparently to maximize genetic economy.

The genetic map of T7 is shown in Figure 16.6. The order of the genes on the T7 chromosome influences the regulation of virus multiplication. When the virion at-

taches to the bacterial cell, the DNA is injected in a linear fashion, with the genes at the "left end" of the genetic map always entering the cell first. Several genes at the left end of the DNA are transcribed immediately by the cellular RNA polymerase, using three closely spaced promoters. One of these early proteins inhibits the host restriction system. Note that this protein is synthesized before the entire T7 genome enters the cell. Another one of these early proteins is a viral RNA polymerase, called T7 RNA polymerase. Two other early mRNA molecules code for proteins that stop the action of host RNA polymerase, thus turning off the transcription of the early genes as well as the transcription of host genes. Thus, the host RNA polymerase is used just to transcribe the first few genes and to make the mRNA that codes for the phage-specific RNA polymerase and a few other pro-

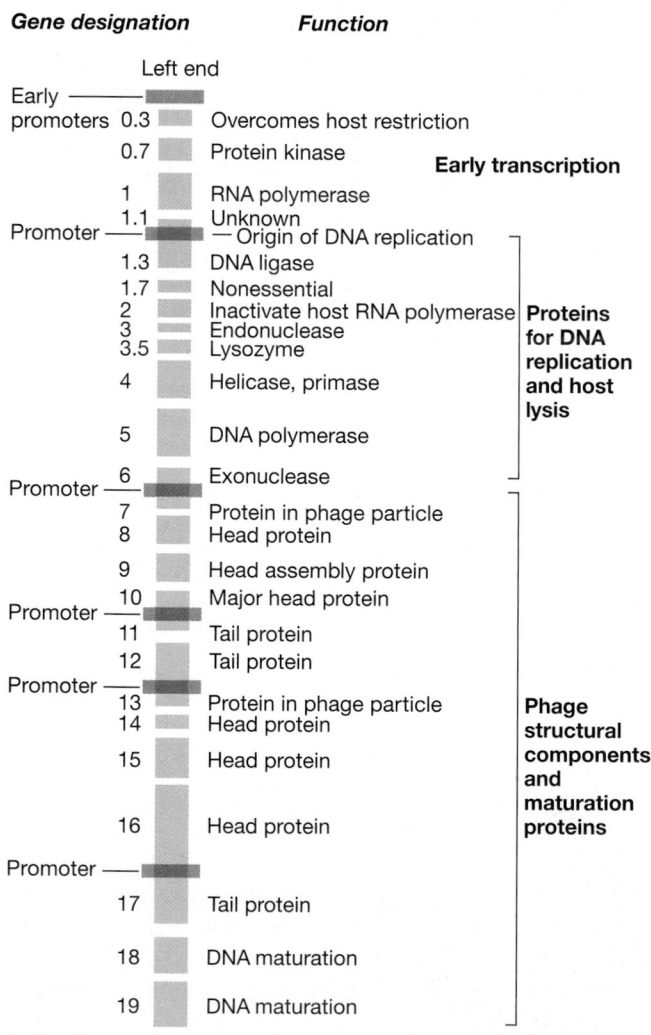

Figure 16.6 Genetic map of phage T7, showing gene numbers, approximate sizes, and functions of the gene products. Transcription from the early promoters involves host RNA polymerase. Transcription from all other promoters involves T7 RNA polymerase. The genes are designated by numbers.

teins. The phage-specific RNA polymerase is then involved in the major transcription processes of the phage. This T7 RNA polymerase uses only phage-specific promoters that are distributed along the left-center and center portions of the genome (see Figure 16.6). The T7 RNA polymerase is very specific for these promoters (whose sequence is unrelated to typical *Escherichia coli* promoters) and is also an extremely efficient enzyme. (Genetic engineers have taken advantage of this to fashion genes that can be highly expressed; ⚭ Section 31.4.) Note that this situation is unlike that in T4, which uses the host RNA polymerase and modified RNA polymerase throughout infection (⚭ Section 9.9).

DNA replication in T7 begins at a single origin of replication (shown in Figure 16.6) and proceeds *bidirectionally* from this origin (Figure 16.7). Replicating molecules of T7 DNA can be recognized under the electron microscope by their characteristic structures. Because the origin of replication is near the left end, Y-shaped molecules are frequently seen, and earlier in replication, bubble-shaped molecules appear (Figure 16.7). Several virus-encoded proteins are involved in T7 DNA replication, unlike the situation we described for φX174 (see Section 16.2).

A structural feature of the T7 DNA that is important in DNA replication is that there is a *direct terminal repeat* of 160 base pairs at the ends of the molecule. To replicate DNA near the 5′-terminus, RNA primer molecules have to be removed before replication is complete. There is thus an unreplicated portion of the T7 DNA at the 5′-terminus of each strand (see lower part of Figure 16.7*a*). As discussed in Section 7.7, genetic elements with linear DNA genomes have a variety of strategies for solving this problem in DNA replication. The strategy employed by T7 is similar to that used by T4 and involves the repeated sequence at its ends (⚭ Section 9.9). The opposite single 3′-strands on two separate DNA molecules, being complementary, can pair with these 5′-strands, forming a DNA molecule twice as long as the original T7 DNA (Figure 16.7*b*). The unreplicated portions of this end-to-end bimolecular structure are then completed through the action of DNA polymerase and DNA ligase, resulting in a *linear bimolecule* called a *concatamer*. Continued replication and recombination can

Figure 16.7 Replication of the linear, double-stranded DNA genome of bacteriophage T7. (a) Bidirectional replication of DNA giving rise to intermediate "eye" and "Y" forms. (b) Formation of concatamers by joining DNA molecules at the unreplicated terminal ends. The designation of the genes is arbitrary. (c) Production of mature viral DNA molecules from T7 concatamers by action of cutting enzyme, an endonuclease. Left: The enzyme makes single-stranded cuts of specific sequences (arrows); center: DNA polymerase completes the single-stranded ends; right: the mature T7 molecule with terminal repeats. Bacteriophage T4 also uses recombination to help replicate its linear genome; compare Figure 16.7*c* with Figure 9.13.

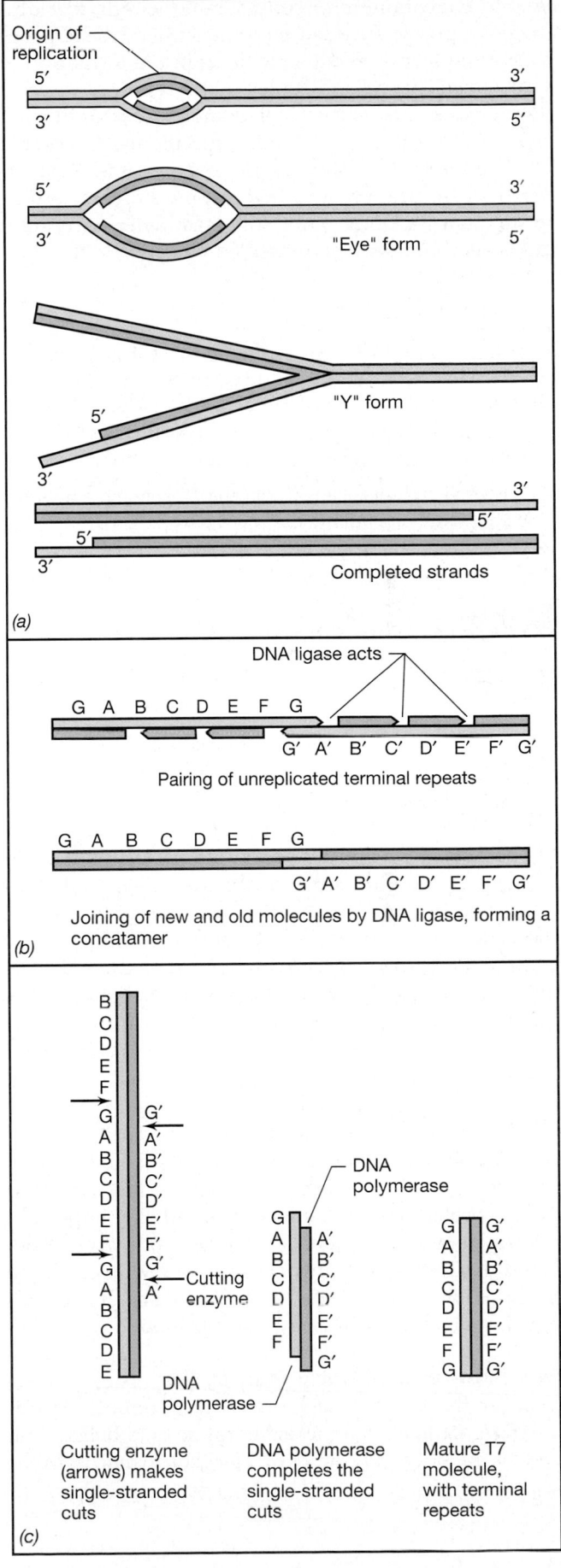

lead to concatamers of considerable length, but ultimately a phage-encoded endonuclease cuts each concatamer at a specific site, resulting in the formation of virus-sized linear molecules with terminal repeats (Figure 16.7c). Because T7, like lambda but unlike T4, cuts the DNA to be packaged at specific sequences, the DNA sequence it packages is always the same. Remember that T4 cuts using a headful mechanism, which means that its DNA is not only terminally redundant but is also circularly permuted (⚭ Section 9.9).

✓ 16.4 Concept Check

The bacteriophage T7 double-stranded DNA genome always enters the host cell in the same orientation. The late genes in T7 are transcribed by a virus-encoded RNA polymerase. The replication strategy of the linear T7 genome involves terminal repeats and recombination.

- ✓ Of what significance is it that the T7 genome enters the cell in only one orientation?
- ✓ What is meant by *terminal repeats*?

16.5 Double-Stranded DNA Bacteriophages: A Transposable Virus

One of the more interesting bacteriophages is the one called **Mu**. This virus is temperate, like lambda, but has the unusual property of replicating as a transposable element (⚭ Sections 7.4 and 10.11). This phage is called Mu because it is a *mutator* phage, inducing mutations in a host genome into which it becomes integrated. This mutagenic property of Mu arises because the genome of the virus can become inserted into the middle of host genes, causing these genes to become inactive (and hence, the host that has become infected with Mu behaves as a mutant). Mu is a useful phage because it can be used to generate a wide variety of bacterial mutants very easily. Also, as we will discuss in Chapter 31, Mu can be used in genetic engineering.

Transposable elements are sequences of DNA that have the ability to move from one location on their host genome to another as discrete elements. They are found in both prokaryotes and eukaryotes and play important roles in genetic variation (⚭ Section 10.11 gives a detailed discussion of transposition). Mu is a very large transposable element but a very efficient one. It is a virus which replicates its DNA by transposition.

Structure and Genetic Map of Mu
Structurally, bacteriophage Mu is a large double-stranded DNA virus with an icosahedral head, a helical tail, and six tail fibers (Figure 16.8). The genetic map of Mu is

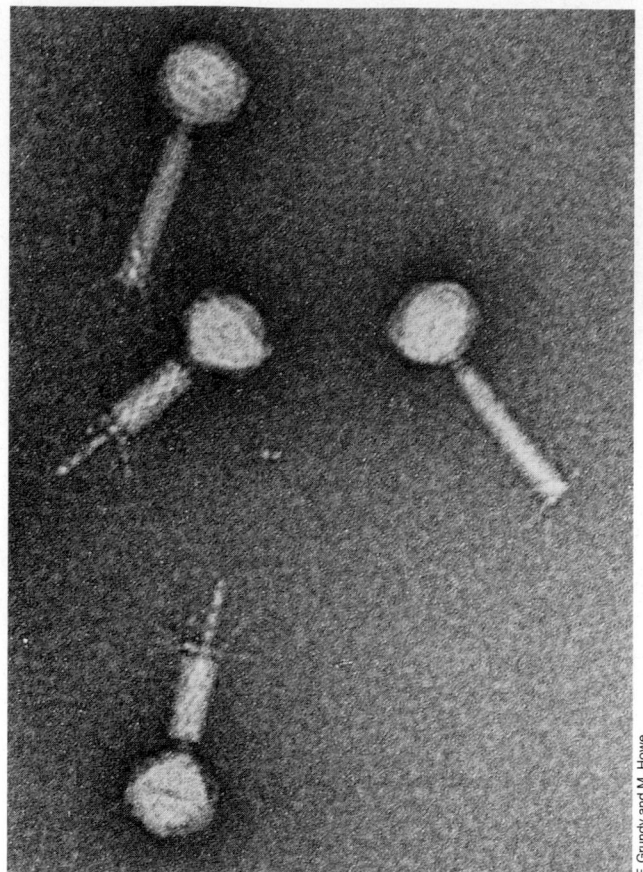

F. Grundy and M. Howe

Figure 16.8 Electron micrograph of virions of bacteriophage Mu, the mutator phage.

shown in Figure 16.9a. It can be seen that the bulk of the genetic information is involved in the synthesis of the head and tail proteins, but that important genes at each end are involved in replication and immunity. The DNA molecule found within the virion is approximately 39 kilobase pairs long, but only 37.2 kilobase pairs make up the actual Mu genome. This is because both ends of this DNA molecule contain host DNA. At the left end of the Mu DNA are 50–150 base pairs of host DNA, and at the right end are 1–2 kilobase pairs of host DNA. These host DNA sequences are not unique and represent DNA adjacent to the location where Mu was inserted into the genome of its previous host. When a Mu phage particle is formed, a length of DNA containing the Mu genome just large enough to fill the phage head is cut out of the host, beginning at the left end. The DNA is rolled in until the head is full, but the place at the right end where the DNA is cut varies from one phage particle to another. For that reason, as shown on the genetic map, there is a variable sequence of host DNA at the right-hand end of the phage (right of the *attR* site) that represents the *host* DNA

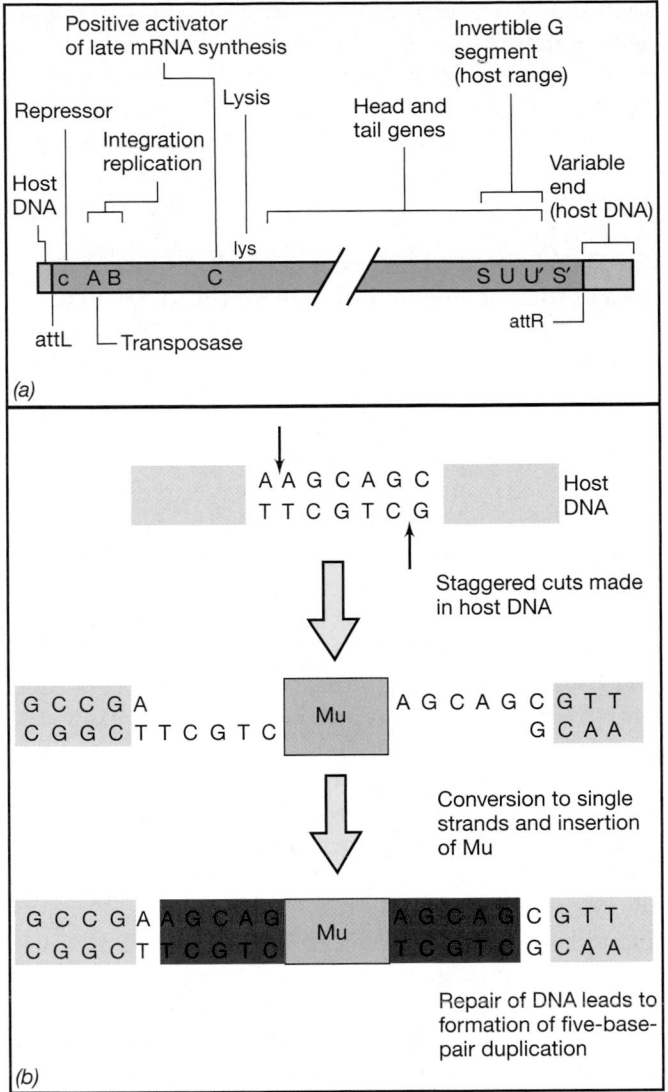

Figure 16.9 Bacteriophage Mu. (a) Genetic map of Mu. See text for details. Note that there is a lowercase *c* gene, which encodes a repressor, and an uppercase *C* gene, which encodes an activator protein. The region encoding the head and tail genes is not drawn to scale. (b) Integration of Mu into the host DNA, showing the generation of a five-base-pair duplication of host DNA.

that has become packaged into the phage head. Each virion arising from a single infected cell will have a different amount of host DNA, and the host DNA base sequence in each virion from the same cell will be different.

Note that the bacteriophages we have discussed that have double-stranded DNA genomes, T4 (⬭ Section 9.9), T7 (see Section 16.4), lambda (⬭ Section 9.10), and Mu, all have linear genomes but use three quite different strategies to replicate the ends (⬭ Section 7.7). Both T4 and T7 have terminal repeats and use recombination to form concatemers; lambda circularizes its

genome after infection; and Mu is always replicated as part of a larger DNA molecule.

As shown on the genetic map (Figure 16.9), a specific segment of the Mu genome called G (distinct from the G gene) is *invertible*, being present either in the orientation designated G^+ or in the inverted orientation G^-. The orientation of this segment determines the kind of tail fibers that are made for the phage. Since adsorption to the host cell is controlled by the specificity of the tail fibers, the host range of Mu is determined by which orientation of this invertible segment is present in the phage. If the G segment is in the orientation designated G^+, then the phage particle will infect *Escherichia coli* strain K12. If the G segment is in the G^- orientation, then the phage particle will infect *E. coli* strain C or several other species of enteric bacteria. The two tail fiber proteins are encoded on opposite strands within this small G segment. Left of the G segment is a promoter that directs transcription into the G segment. In the orientation G^+, the promoter directing transcription of S and U is active, whereas in the orientation G^-, a different promoter directs transcription of genes S' and U' on the opposite strand. Regulation involving rearrangement of DNA sequences is known in other viruses as well as both prokaryotic and eukaryotic cells.

Replication of Mu

On infection of a host cell by Mu, the DNA is injected and is protected from host restriction by a modification system in which about 15% of the adenine residues are acetoamidated. In contrast with lambda, integration of Mu DNA into the host genome is essential for both lytic and lysogenic growth. Integration requires the activity of the gene A product, which is a transposase enzyme. At the site where the Mu DNA becomes integrated, a five-base-pair duplication of the host DNA arises at the target site. As shown in Figure 16.9*b*, this host DNA duplication arises because staggered cuts are made in the host DNA at the point where Mu is inserted, and the resulting single-stranded segments are converted to the double-stranded form as part of the integration process. Duplication of short stretches of host DNA is typical of transposable element insertion (⬭ Section 10.11).

Lytic growth of Mu can occur either on initial infection, if the Mu repressor (the product of the *c* gene) is not formed, or by induction of a lysogen. In either case, replication of Mu DNA involves repeated transposition of Mu to multiple sites on the host genome. Initially, transcription of only the early genes of Mu occurs, but after gene C protein, a positive activator of late RNA synthesis, is expressed, the synthesis of the Mu head and tail proteins occurs. Eventually, expression of the lytic function occurs and mature phage particles are released.

✓ 16.5 Concept Check

The bacteriophage Mu is a temperate virus that is also a transposable element. In either the lytic or lysogenic pathway, its genome is integrated into the host chromosome. Even in the lytic pathway, its genome is replicated as part of a larger DNA molecule. The genome is packaged into the virion in such a way that there are short sequences of host DNA at either end.

✓ What is a transposable element?

✓ What mechanism does Mu use to ensure that the ends of its linear genome are completely replicated?

II VIRUSES OF EUKARYOTES

The differences between prokaryotic cells and eukaryotic cells place some constraints on the viruses that infect them. For example, in prokaryotes transcription and translation can be coupled. In eukaryotes, in contrast, transcription (and DNA replication) occur in the nucleus and translation occurs in the cytoplasm. We discussed some of these issues in Chapter 9 in an overview of animal viruses (∞ Section 9.11). In that chapter, however, we dealt very briefly with a few kinds of viruses and then only with those that infect animals. Our interest in animal viruses springs in part from their role in human disease. We shall discuss more animal viruses in the remainder of this chapter. However, not all eukaryotes are animals, and in the next section we will briefly discuss plant viruses.

16.6 Plant Viruses

Plant cell walls are extremely thick and strong, yet there are a number of plant viruses that can infect plants and, in multicellular plants, spread from the infected cell to neighboring cells. The great majority of plant viruses known are positive-strand RNA viruses, and it has been postulated that these small genomes facilitate transfer from cell to cell within the plant.

Tobacco Mosaic Virus

In 1892, the Russian scientist Dmitri Ivanovsky showed that the causative agent of tobacco mosaic disease could pass through filters that retain bacteria. In 1898, the Dutch microbiologist Martinus Beijerink (∞ Section 1.6) showed that this agent was not only filterable, but that it had many of the properties of a living organism. This agent, the first virus to be recognized, was tobacco mosaic virus (TMV). TMV has a virion with helical symmetry (∞ Section 9.2) and contains 2130 copies of a coat protein and a single copy of the positive-strand RNA genome (∞ Figure 9.2). It was with

TMV that it was first shown that RNA could be the genetic material, just as DNA was in other viruses and in cells. TMV remains a serious agricultural problem because it infects tomato plants as well as tobacco (where resistant plant lines are available). TMV infection of a plant requires damage to the wall of some plant cell(s) through which the virion can enter. Uncoating takes place in the cell.

The genome of TMV contains 6395 nucleotides; a map of the genome is shown in Figure 16.10. Like the bacteriophage MS2 (see Section 16.1), TMV encodes four proteins. The genome has a 5' cap (∞ Section 7.12) so that it can be translated, and the 3' end folds into a tRNA-like structure. However, eukaryotes cannot translate polycistronic mRNA (∞ Section 7.11), so the expression of all these genes is somewhat more complex than is the case with MS2. The first gene encodes a protein, MTH, with two functional domains: One a methyl transferase that functions in capping RNA and the other an RNA helicase. The next gene encodes the RNA-dependent RNA polymerase that the virus must synthesize to replicate a negative-strand copy from which it can then make more copies of the genomic RNA. This protein is only synthesized as part of a polyprotein including the MTH domains and only when a ribosome reads through the stop codon at the end of the MTH gene. Since this happens reasonably infrequently, this longer protein is made in low amounts. The remaining two genes encode the movement protein (MP) and the coat protein (CP). Because eukaryotic ribosomes cannot use polycistronic mRNA, these two genes are not translated directly from the genomic RNA. Instead, two subgenomic mRNAs are made from the negative-strand RNA, and each is translated to yield one of these proteins. Like coat proteins of other viruses, that of TMV is essential to the formation of the virion. The virion is essential for infecting new plants. However, it is the movement protein that enables TMV to infect neighboring cells in an already infected plant.

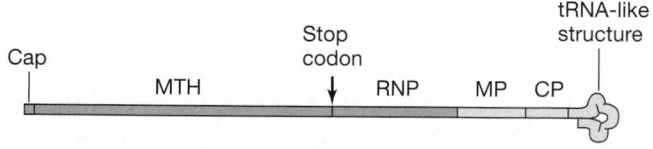

Figure 16.10 Genetic map of tobacco mosaic virus. The genome is a positive-strand RNA that is capped at its 5' end and has a tRNA-like structure (drawn larger than scale) at its 3' end. The MTH gene encodes a protein with both methyltransferase and RNA helicase activity. The RNP gene encodes an RNA-dependent RNA polymerase that is only translated as a polyprotein by ribosomes reading through the stop codon following the MTH gene. The MP gene encodes a movement protein, and the CP gene encodes the coat protein. These two proteins are translated from subgenomic RNAs that are synthesized using the negative-strand RNA after it has been made during infection.

Plant cells have structures called *plasmodesmata*, which are intercellular connections that span cell walls and connect the plant cells. Plasmodesmata have very narrow channels, so narrow in fact that neither the TMV virion nor free RNA can easily transverse these openings. The MP binds to the new genomic RNA and forms a complex that is extremely thin (2.5 nm) and it is these complexes that can move through the plasmodesmata.

As was mentioned before, there are many plant positive-strand RNA viruses known, and there are also positive-strand RNA viruses of *Bacteria* (see Section 16.1) and animals (see Section 16.7). Sequence analysis indicates that some of these viruses are relatively closely related, and all their RNA-dependent RNA polymerases are related. Nonetheless, they have slightly different replication strategies, often clearly related to differences in the host cells.

Chlorella Viruses

We often think of plant viruses as viruses that infect agricultural plants or other large multicellular plants. There are, however, a very large number of unicellular plants in our environment, none more widespread than the unicellular algae (⌘ Section 14.11). Among these are *Chlorella*, a very widely distributed genus of green alga. Most *Chlorella* are free-living but some *Chlorella*-like algae are endosymbionts in freshwater or marine animals, including protozoa such as *Paramecium* (⌘ Section 14.8). Many of these latter, however, can be grown in the laboratory independently of the organism in which they normally grow. There are a number of viruses that are known to infect these "exsymbiotic" *Chlorella*-like algae, but cannot infect the endosymbiotic form. These viruses are often referred to as *Chlorella* viruses. The best studied is *Paramecium bursaria* chlorella virus 1 (PBCV-1).

These viruses have large icosahedral virions (Figure 16.11) and double-stranded DNA genomes. The virions have a lipid component that is essential for infectivity but it is inside the capsid, therefore, the virion is not enveloped. The genomes of the *Chlorella* viruses are extremely large: All are over 300 kb (for a comparison with other viruses see Table 9.1). In many cases the DNA is also extensively methylated, up to 45% of the cytosine being present as 5-methylcytosine and over 20% of the adenine being present as 6-methyladenine. However, in PBCV-1 these modified bases are each less than 2% of the unmodified forms. The genome of PBCV-1 has been completely sequenced. This 330,742-bp genome encodes over 370 different proteins and 10 transfer RNAs. The ends of the genome are incompletely base-paired hairpin loops (⌘ Figure 7.8) very similar to those found in poxviruses (see Section 16.12).

These viruses enter cells somewhat like bacteriophage do. They bind specifically to the cell wall of their

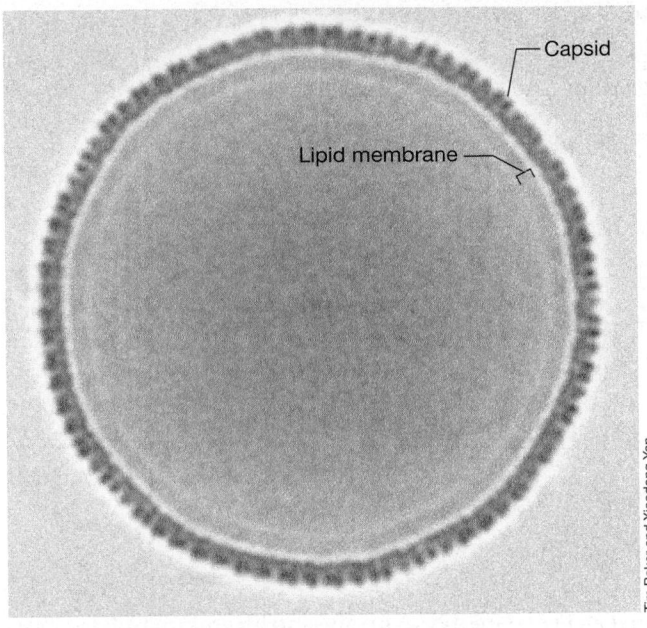

Figure 16.11 An equatorial cross-section of the reconstruction density maps of the virion of the *Chlorella* (green algae; ⌘ Section 14.11) virus PBCV-1. A lipid bilayer membrane is visible beneath the capsid shell. The virion has a diameter of approximately 170 nm.

host, then at least five different enzymes carried by the virion digest away the cell wall at the point of contact. Viral DNA is then released into the cell leaving the empty capsid behind.

As is the case for most double-stranded DNA viruses of eukaryotes, the DNA of PBCV is replicated in the nucleus, and RNA is synthesized there. PBCV-1 encodes several enzymes involved in DNA replication, including a DNA polymerase, but, although it encodes some transcription factors, it does not encode its own RNA polymerase or any of the core subunits (⌘ Section 7.8). A few of the genes of PBCV-1 contain introns that must be processed (⌘ Section 7.12). The virus mRNA is capped, and some of the enzymes involved are virus-encoded. Early mRNA, but not late mRNA, has poly-A tails. Like bacteriophage T4, PBCV-1 encodes several tRNAs. Unlike any other known type of virus (but presumably like the other *Chlorella* viruses), PBCV-1 encodes an elongation factor used in translation.

Although currently the known hosts for these viruses are only laboratory-isolated exsymbiotes, the viruses themselves are quite widespread in nature. Therefore, there must be abundant natural hosts. In addition, the *Chlorella* viruses are related to other viruses that infect algae. All of these latter have very large double-stranded DNA genomes, but some of the genomes are circular.

There is one other extremely interesting fact to note about the *Chlorella* viruses: They encode several restriction and modification enzyme systems (⌘ Sections 9.6 and 10.12). Indeed, they are the only source of restriction enzymes outside of the prokaryotes.

Most plant viruses have positive-strand RNA genomes. One example of this type of virus is tobacco mosaic virus (TMV), the first virus discovered. Although these viruses enter the cell through breaks in the plant cell wall, their genomes can move from the infected cell to neighboring cells through intercellular connections that span the cell walls. Other types of plant viruses are also known, including the *Chlorella* viruses that have very large double-stranded DNA genomes.

✓ Most plant RNA viruses encode a *movement protein*. What is its role in infection?

✓ Although the TMV genome is used as a messenger RNA, not all the proteins encoded by the virus can be translated from it. Explain.

16.7 Positive-Strand RNA Viruses of Animals

Just as there are many known positive-strand RNA bacteriophages and plant viruses, there are many different kinds of animal viruses with a positive-strand RNA genome. One important group of positive-strand RNA animal viruses is the *picornavirus family*, which contains such important viruses infecting humans as the *polioviruses*, the *rhinoviruses* that cause the common cold, and the *hepatitis A virus*. The first animal virus discovered, foot-and-mouth disease virus, is also a picornavirus. These viruses are called *picornaviruses* because they are very small viruses (30 nm in diameter) (*pico* means "small") and contain single-stranded RNA. The virus particle has a simple icosahedral structure with 60 morphological units per virion, each unit consisting of four distinct proteins (Figure 16.12*a*).

In poliovirus, the RNA is a linear molecule about 7500 bases in length (∞ Table 9.1). At the 5'-terminus of the viral RNA is a protein, called the *VPg protein*, that is attached covalently to the RNA. At the 3'-terminus of the RNA is a poly-A tail.

An overview of the manner of multiplication of poliovirus is illustrated in Figure 16.12*b*. The RNA of the virus acts directly as a messenger RNA. Interestingly, this is so even though the RNA is not capped (∞ Section 7.12). The 5'-end of poliovirus RNA has a

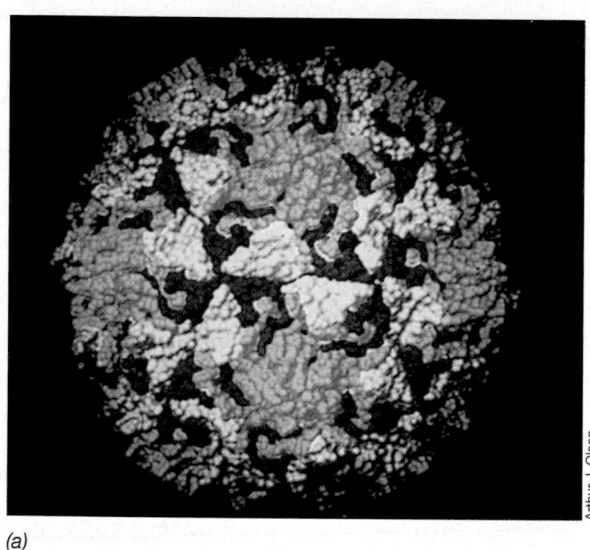

(a)

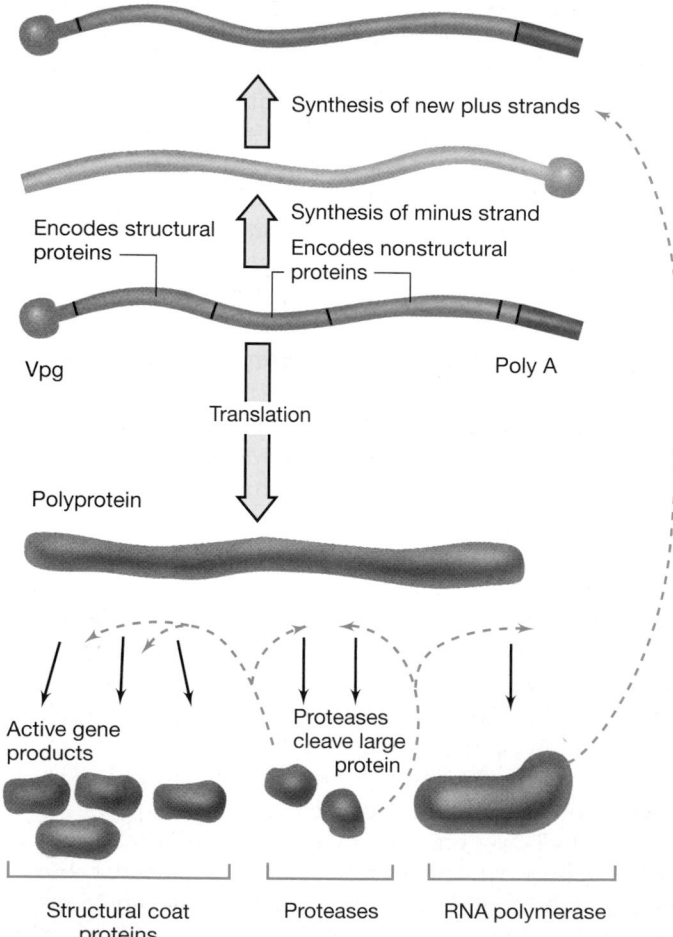

(b)

Figure 16.12 (a) The poliovirus particle. This is a computer model based on electron diffraction analysis of virus crystals. The various structural proteins are shown in distinct colors. (b) The reproduction of poliovirus. The single-stranded RNA of the virus is translated directly as a messenger RNA, with the production of one large protein molecule. This protein is cleaved, leading to production of the active viral proteins, including the structural coat proteins and the RNA polymerase that brings about replication of the poliovirus RNA. The assembly of intact poliovirus from coat protein molecules and RNA then follows. Translation of an mRNA to yield a polyprotein that is subsequently cleaved is a strategy also used by other viruses (see Section 16.14).

long sequence that can fold into several stem-loops (∞ Section 7.10). Somehow these permit binding of the eukaryotic ribosome. The virus RNA is monocistronic but codes for all the proteins of the virus in a single *polyprotein* that is later cleaved into the individual proteins. The whole replication process occurs in the cytoplasm.

At initial infection, the virus particle attaches to a specific receptor on the surface of a sensitive cell and enters the cell. Once inside the cell, the virus particle is uncoated, and the free RNA associates with ribosomes. The viral RNA is then translated from a single start codon into the large polyprotein mentioned earlier. This giant protein (about 2200 amino acid residues) then undergoes self-cleavage into about 20 smaller proteins (including cleavage intermediates), among which are the four *structural proteins* of the virus particle, the RNA-linked *VPg protein*, an *RNA polymerase* responsible for synthesis of minus-strand RNA, and at least one *virus-encoded protease*, which carries out the cleavage process. This cleavage process, called *posttranslational cleavage*, occurs in a wide variety of animal viruses as well as in normal cell metabolism in animal cells (∞ box, Protein Processing, Chapter 8).

Replication of Poliovirus

Replication of viral RNA begins within a short time after infection and is catalyzed by the RNA-dependent RNA polymerase (replicase) made in the process described in the previous paragraph. This replicase transcribes the viral RNA, of plus complementarity, into an RNA molecule of minus complementarity. This minus strand then serves as a template for repeated transcription of progeny plus strands. Some of the progeny plus strands may again be transcribed into minus strands, and as many as 1000 minus strands may subsequently be present in the cell. From these minus strands, as many as a million plus strands may ultimately be formed. Both the plus and the minus strands become covalently linked to the tiny VPg protein (only 22 amino acids long), which may serve as a primer for transcription.

Once virus multiplication begins, host RNA and protein syntheses are inhibited. Host protein synthesis is inhibited as a result of destruction of a host protein, the cap-binding protein required for translation of capped mRNAs (∞ Section 7.12).

At one time, polio was a major infectious disease of humans, but the development of an effective vaccine (∞ Section 22.11) has brought the disease almost completely under control. The World Health Organization has a vaccination program intended to eradicate the disease, and by late in the year 2000 there were reports of the virus in only a few countries in Africa and in the Indian subcontinent.

✔ **16.7 Concept Check**

In small RNA viruses such as poliovirus, the viral RNA acts directly as a single messenger RNA, causing the production of a long polyprotein that is broken down by enzymes into the numerous small proteins necessary for nucleic acid multiplication and virus assembly.

✔ What is a *cap* and what is its normal function?

✔ How can poliovirus RNA be synthesized in the cytoplasm whereas host RNA must be synthesized in the nucleus?

16.8 Negative-Strand RNA Viruses of Animals

In a number of RNA viruses of animals, the RNA does not serve directly as a messenger, but is transcribed into a complement that functions as the mRNA. As discussed in Section 9.7, it is conventional to express the configuration of the mRNA as *plus*, so if the viral genomic RNA is of opposite complementarity, it is called *minus*. This group of viruses is then called minus-strand or *negative-strand RNA viruses* and are in Class V of the Baltimore Classification (∞ Table 9.2). We discuss here two important negative-strand viruses: rhabdoviruses, including rabies virus, and orthomyxoviruses, including influenza virus. The Ebola virus, a human pathogen responsible for an emerging infectious disease (∞ Section 25.10) is also a negative-strand RNA virus.

Rhabdoviruses

One of the most important human pathogens that is a negative-stranded RNA virus is the rabies virus, which causes the important disease rabies in animals and humans (∞ Section 27.1). Rabies virus is called a *rhabdovirus*, from *rhabdo* meaning "rod," which refers to the shape of the virus particle. Another rhabdovirus that has been extensively studied is vesicular stomatitis virus (VSV) (Figure 16.13), a virus that causes the disease *vesicular stomatitis* in cattle, pigs, horses, and sometimes humans. Many rhabdoviruses, such as potato yellow dwarf virus, infect both insects and plants and can cause important agricultural problems. (There are no known negative-strand RNA bacteriophage.)

The rhabdoviruses are enveloped viruses, with an extensive and rather complex lipid envelope surrounding the nucleocapsid (Figure 16.13). In animal rhabdoviruses the virus particle is bullet-shaped, about 70 nm in diameter and 175 nm long. The nucleocapsid is helically symmetric and makes up only a small part of the virus particle weight (about 2–3% of the virion is RNA). The virion contains several enzymes that are essential for the infection process. One of these is an *RNA-dependent RNA polymerase*. As discussed in Section 9.7, the presence of RNA polymerase is essential because the

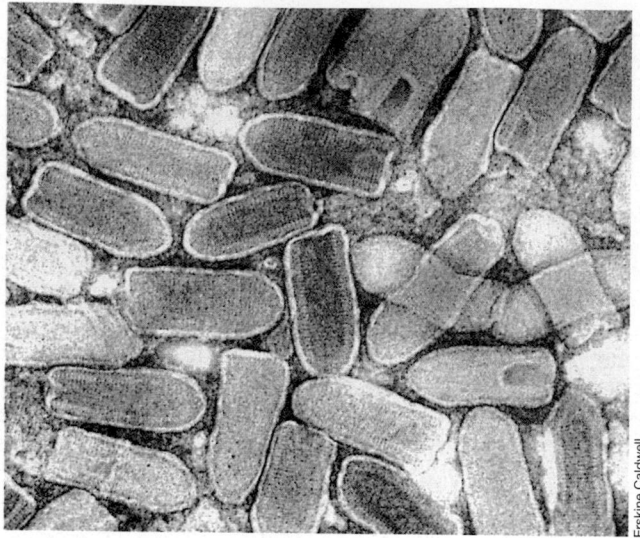

Figure 16.13 Electron micrograph of a rhabdovirus (vesicular stomatitis virus). A particle is about 65 nm in diameter.

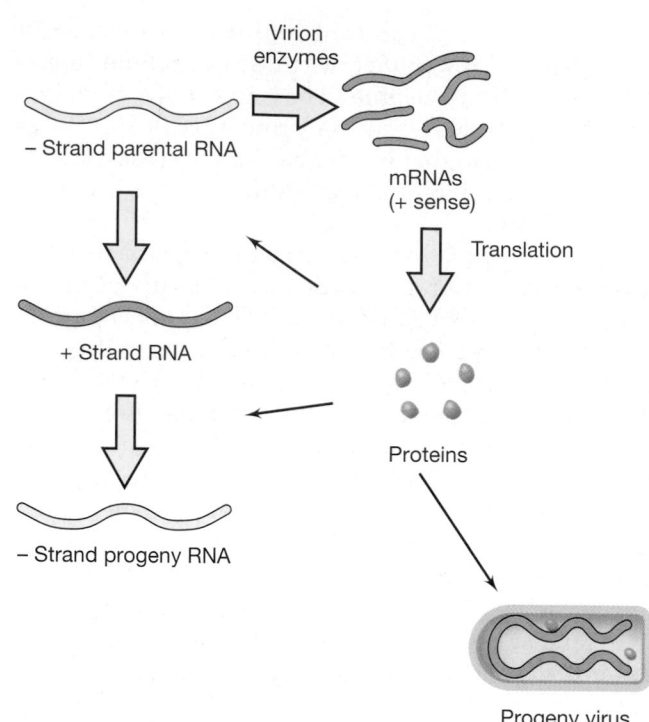

Figure 16.14 Flow of events during multiplication of a negative-strand RNA virus.

genome of these negative-strand viruses cannot act as messengers directly, but must first be transcribed into the plus complement, and host enzymes that transcribe RNA into RNA do not exist.

The RNA of the rhabdoviruses is transcribed inside the cytoplasm of the cell into two distinct kinds of RNA (Figure 16.14). The first type of RNA synthesis results in a series of messenger RNAs made from the various genes of the virus (VSV has five). The second is a plus-strand RNA that is a *copy* of the complete viral genome (the VSV genome is 11,162 nucleotides long). These long plus-strand RNAs then serve as templates for synthesis of the *negative-strand* RNA molecules, which will serve as genomes of progeny virions. Each mRNA is monocistronic, coding for a single protein. Once the mRNA for the virus RNA polymerase is made in this primary transcription process, synthesis of the virus RNA polymerase can begin, leading to the formation of many plus-strand RNA molecules, both messengers and full-length genomic (viral) RNA templates.

Translation of viral mRNAs leads to the synthesis of viral coat proteins. Assembly of an enveloped virus is considerably more complex than assembly of a simple virus particle. Two kinds of coat proteins are formed, *nucleocapsid proteins* and *envelope proteins*. The nucleocapsid is formed first by association of the nucleocapsid protein molecules around the viral RNA.

The envelope proteins that possess hydrophobic amino-acid leader sequences at their amino-terminal ends (⌾ Section 7.16) are synthesized on ribosome complexes that are themselves associated with membranes. As these proteins are synthesized, sugar residues are added, leading to the formation of *glycoproteins*. Such glycoproteins, characteristic of membrane-associated pro-

teins, are transported to the cytoplasmic membrane (and the leader sequences are removed), where they replace host membrane proteins. Nucleocapsids then migrate to the areas on the cytoplasmic membrane where these virus-specific glycoproteins exist, recognizing the virus glycoproteins with great specificity. The nucleocapsids then become aligned with the glycoproteins and bud through them, becoming coated by the glycoproteins in the process. The final result is an enveloped virion with a nucleocapsid center and a surrounding membrane whose lipid is derived from the host cell but whose membrane proteins are encoded by the virus. The budding process itself does not cause detectable damage to the cell, which may continue to release virions in this way for a considerable period of time. (Host damage does occur but is brought about by other unknown factors.)

Influenza and Other Orthomyxoviruses

Another group of negative-strand viruses of great importance is the group called the *orthomyxoviruses*, which contains the important human virus *influenza*. The term *myxo* refers to the fact that these viruses interact with the *mucus* or *slime* of cell surfaces. In the case of influenza virus, this mucus is at the mucous membrane of the respiratory tract, because these viruses are transmitted primarily by the respiratory route (⌾ Section 26.8). The term *ortho* has been added to the influenza virus group to distinguish this group from another group of negative-strand viruses, the *paramyxovirus group*. The paramyxoviruses, which include such important human viruses as those causing mumps and measles, are actually similar in molecular biology to rhabdoviruses. The orthomyxoviruses have been extensively studied over many years, beginning with early work during and after the 1918 pandemic of influenza that caused the deaths of millions of people worldwide (Section 26.8).

The orthomyxoviruses are enveloped viruses in which the viral RNA is present in the virion in a number of separate pieces. The genome of the orthomyxoviruses is thus said to be a **segmented genome**. In the case of influenza A virus, the genome is segmented into *eight* linear single-stranded molecules ranging in size from 890 to 2341 nucleotides. The influenza virus nucleocapsid is of helical symmetry, about 6–9 nm in diameter and about 60 nm long. This nucleocapsid is embedded in an envelope that has a number of virus-specific proteins as well as lipid derived from the host (Figure 16.15).

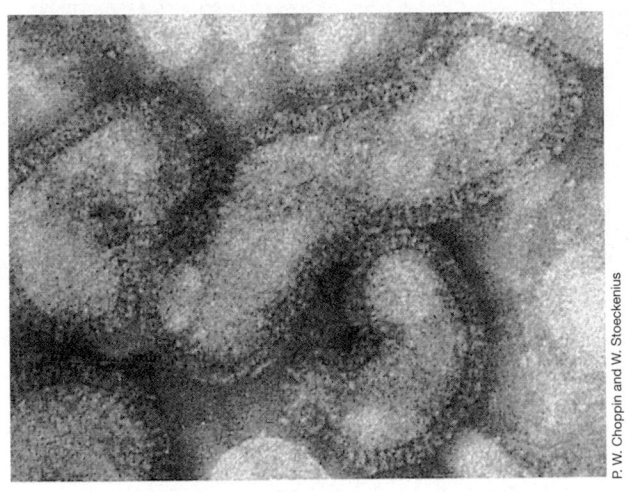

(a)

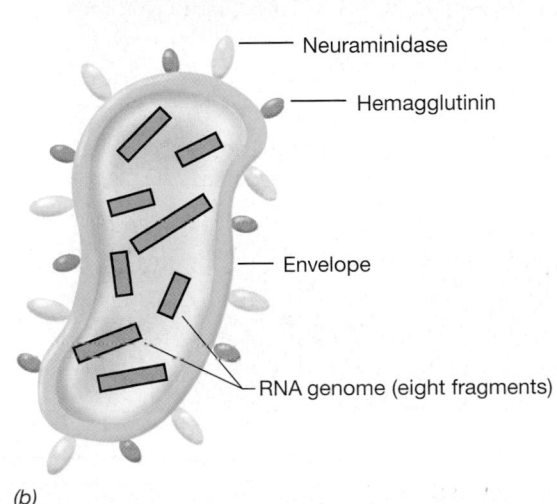

Neuraminidase

Hemagglutinin

Envelope

RNA genome (eight fragments)

(b)

Figure 16.15 Influenza virion structure. (a) Electron micrograph. (b) Diagram, showing some of the components.

Because of the way influenza virus buds as it leaves the cell, the virus has no defined shape and is said to be *polymorphic* (Figure 16.15a). There are spikes on the outside of the envelope that interact with the host cell surface. One spike is called a *hemagglutinin* because it causes agglutination of red blood cells. (Agglutination is a process by which cells are caused to clump when they are mixed with an antibody or other protein or polysaccharide molecule that combines specifically with a substance on the cell surface; Section 24.8.) If the cells undergoing agglutination are red blood cells, then the process is called *hemagglutination*. (*Hema* is the combining form referring to *blood*.) The red blood cell is not the type of host cell the virus normally infects but contains on its surface the same type of membrane component, chemically characterized as *sialic acid*, that the mucous membrane cells of the respiratory tract contain. Thus, the red blood cell is merely a convenient cell type for measurement of agglutination activity. An important feature of the influenza virus hemagglutinin is that antibody directed against this hemagglutinin *prevents* the virus from infecting a cell. Thus, antibody directed against the hemagglutinin *neutralizes* the virus, and this is the mechanism by which immunity to influenza is brought about during the immunization process (Sections 22.11 and 26.8).

A second type of spike on the virus surface is an enzyme called *neuraminidase* (see Figure 16.15). Neuraminidase breaks down the sialic acid component of the cytoplasmic membrane, which is a derivative of neuraminic acid. Neuraminidase appears to function primarily in the virus assembly process, destroying host membrane sialic acid that would otherwise block assembly or become incorporated into the mature virus particle.

In addition to the neuraminidase, the virus possesses two other enzymes, *RNA-dependent RNA polymerase*, which is involved in the conversion of a negative to a positive strand (as already discussed for the rhabdoviruses), and an *RNA endonuclease*, which cuts a primer from capped mRNA precursors.

The virus particle enters via the process of *endocytosis*. Once inside the cytoplasm, the nucleocapsid becomes separated from the envelope and migrates to the nucleus. Replication of the viral nucleic acid then occurs in the nucleus. Uncoating results in activation of the virus RNA polymerase. The mRNA molecules are then transcribed in the nucleus from the virus RNA, using oligonucleotide primers cut from the 5′-ends of newly synthesized capped cellular mRNAs. Thus, the viral mRNAs have 5′-caps. The poly-A tails of the viral mRNAs are added, and the virus mRNA molecules move to the cytoplasm.

Although influenza virus RNA replicates in the nucleus, influenza virus proteins are synthesized in the cytoplasm. *Ten* virus proteins are encoded by the *eight* segments of the virus genome. The mRNAs transcribed from six segments each encode a single protein, whereas the other two segments each encode two proteins. This is not done by using true polycistronic mRNA as in prokaryotes because eukaryotic ribosomes typically recognize only the AUG codon closest to the 5'-end of the mRNA as a start codon (Section 7.15). Therefore, they can make only one protein from a given RNA. The original full-length mRNAs transcribed from these two segments are each translated to give one protein. In each case, an additional protein is translated from these messages after they have been processed by the host's RNA splicing machinery. Like overlapping genes, this is another example of how RNA viruses make maximum use of their small genome size.

Some of these proteins are involved in virus RNA replication, and others are structural proteins of the virion. The overall strategy of virus RNA synthesis resembles that of the rhabdoviruses, with primary transcription resulting in the formation of *plus*-strand templates for the formation of progeny *minus*-strand molecules. Details of assembly are still uncertain. The formation of the complete enveloped virus particle occurs by a budding-out process, as was described for the rhabdoviruses.

The segmented genome of the influenza virus has some important practical consequences. Influenza virus and other viruses of this family exhibit a phenomenon called **antigenic shift** in which pieces of the RNA genome from two genetically distinct strains that have infected the same cell become reassorted. This results in a change in the surface antigens (membrane proteins) of the virus, making the virus resistant to antibody that has been formed as a result of an immunization process (Section 26.8). This antigenic shift makes it possible for the newly formed virus to infect hosts that the parent could not have infected. Antigenic shift is thought to bring about major epidemics of influenza.

✓ *16.8 Concept Check*

In a number of RNA viruses, called negative-strand viruses, the virus RNA does not act directly as messenger but is copied into mRNA by an RNA-dependent RNA polymerase present in the virion. An important negative-strand virus is influenza virus.

- ✓ Why is it essential that negative-strand viruses carry an enzyme in their virions?
- ✓ What is a *segmented genome*?

16.9 Double-Stranded RNA Viruses: Reoviruses

The reoviruses (or *Reoviridae*), an important family of animal viruses, have a genome consisting of *double-stranded RNA*. The name *reovirus* is an acronym, derived from the terms *r*espiratory, *e*nteric, and *o*rphan. The term *orphan* was

applied because the first viruses of this group to be isolated from humans were not associated with any specific disease syndrome. However, *rotavirus*, a member of the reovirus family, is probably the most common cause of diarrhea in infants from 6 to 24 months of age. Rotaviruses are also known to cause diarrhea in young animals.

As noted, the RNA of the reoviruses is double-stranded. This is the largest group of animal viruses with double-stranded RNA; most other RNA virus groups (including all those shown in Figure 9.22) have single-stranded RNA. There are also reoviruses known that infect plants, and we have mentioned that the bacteriophage φ6 seems to be closely related to the reoviruses (see Section 16.1).

The reovirus particle consists of a nonenveloped nucleocapsid 60–80 nm in diameter, with a *double* shell of icosahedral symmetry (Figure 16.16). Predictably, these double-stranded RNA viruses contain within the virion the virus-encoded enzymes necessary to synthesize RNA.

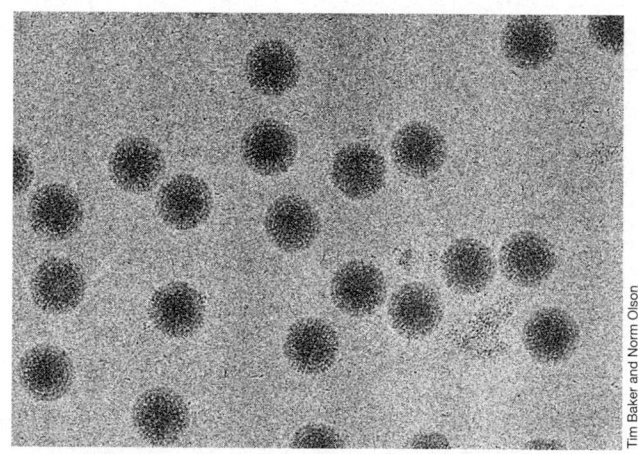

(a)

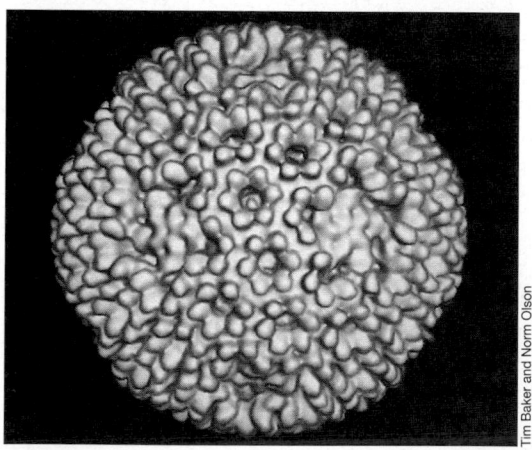

(b)

Figure 16.16 (a) An electron micrograph showing a number of individual reovirus virions (each with a diameter of about 70 nm). (b) Three-dimensional reconstruction of a reovirus virion calculated from images of frozen-hydrated virions.

The genome of reoviruses consists of 10–12 molecules of double-stranded RNA. Double-stranded RNA is more difficult to unwind than double-stranded DNA. Replication of the RNA occurs by an asymmetric method. First, one strand is used as the template, and then the single-stranded product serves as a template to form a double helix. The difficulty in unwinding double-stranded RNA and the susceptibility of single-stranded RNA to cleavage within the cell greatly limits the size of a double-stranded molecule. Segmenting the genome into several molecules seems to be an adaptation to circumvent these problems.

Replication of reovirus RNA occurs exclusively in the cytoplasm of the host. The double-stranded RNA is inactive as mRNA, and the first step in reovirus replication is transcription, using the minus strand as a template to make mRNA. Generally, each molecule of RNA in the genome codes for a single protein, although in a few cases the protein formed is cleaved to give the final product. However, one of the mRNAs produced actually encodes two proteins, and the RNA does *not* have to be modified to do so. Apparently a ribosome sometimes "misses" the start codon for the first gene in this message and travels on to the start codon of the second gene. Therefore, there are exceptions to the generalization that eukaryotic ribosomes initiate at the first AUG codon in mRNA.

Replication of the reovirus seems to occur within an intracellular equivalent of the viral core, called the *subviral particle*, which remains intact in the cell. Each of the 10 capped, single-stranded plus RNAs is assembled into this double-stranded RNA-synthesizing body. The capped single-stranded plus RNAs act as templates for the synthesis of the progeny minus genomic RNAs, yielding progeny double-stranded viral RNAs. The progeny double-stranded RNAs are further encapsidated, and when enough viral capsid proteins are present, mature virions are assembled.

In the initial infection process, the virion binds to a cellular protein. Once attachment has occurred, the virus enters the cell and is transported into lysosomes. Within the lysosome the outer shell of the virus particle is modified by removal of two proteins and cleavage of another by lysosomal enzymes. This uncoating process activates the viral RNA-dependent RNA polymerase and hence initiates the virus replication process.

✓ 16.9 Concept Check

Reoviruses are a large group of viruses whose genome consists of segmented double-stranded RNA. Like negative-strand RNA viruses, double-stranded RNA viruses contain an RNA-dependent RNA polymerase in the virion.

✓ Why might viruses with double-stranded RNA genomes also have segmented genomes?

16.10 Replication of Double-Stranded DNA Viruses of Animals

In the Baltimore Classification System (Section 9.7), double-stranded DNA viruses are in Class I and single-stranded DNA viruses are in Class II. While there are several different Class II bacteriophages known (see Section 16.2 and Section 16.3), and some plant viruses in this class, there are relatively few known that infect animals. The largest family of Class II animal viruses is the *Parvoviridae* (parvoviruses), some of which infect vertebrates and some infect insects. However, there are a very large number of Class I viruses infecting animals.

Among these double-stranded DNA viruses there are several families with members that can infect humans. These include the *Polyomaviridae* and *Papillomaviridae* (these two families were formerly both known as *Papoviridae*), *Herpesviridae* (herpesviruses), *Poxviridae* (pox viruses), and the *Adenoviridae* (adenoviruses). Of these, all replicate in the nucleus except for the pox viruses, which replicate in the cytoplasm. In this and the following sections, we discuss the replication of each of these families briefly.

Polyomaviruses: SV40

Some viruses of the polyomavirus family induce tumors in animals; indeed, *oma* means tumor. One of these DNA tumor viruses was first isolated from monkeys, and it was thus called simian virus 40 or SV40. (In the 1950s many millions of people were inadvertently inoculated with SV40, which was a contaminate in early preparations of poliovirus vaccine. Fortunately there seems to have been no medical complications arising from this.) It was one of the first genetic elements to be studied by genetic engineering techniques and has been extensively used as a *vector* for moving genes into eukaryotic cells (Section 31.4).

The SV40 virion is a simple, nonenveloped particle 45 nm in diameter with an icosahedral head containing 72 protein subunits. There are no enzymes in the virion. In addition to the capsid proteins, however, there are host-derived *histone proteins* found complexed with the viral DNA. We have mentioned histone proteins during our discussion of chromosome structure (Section 7.3) and have noted that histones play a role in neutralizing the negative charge originating from the phosphates of DNA and aid in packing of the DNA into more compact configurations.

The genome of SV40 consists of one molecule of double-stranded DNA of 5243 base pairs. The DNA is circular (Figure 16.17) and exists in a supercoiled configuration within the virion. The complete base sequence of SV40 has been determined, and a genetic map is shown in Figure 16.18.

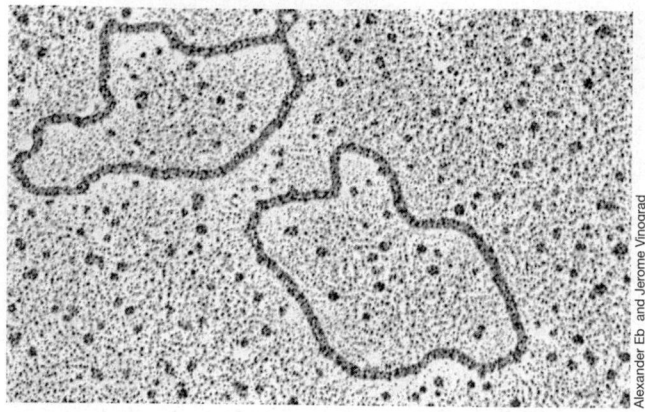

Figure 16.17 Electron micrograph of circular DNA from a tumor virus. The contour length of each circle is about 1.5 μm.

The nucleic acid is synthesized in the nucleus, but the proteins are synthesized in the cytoplasm. Final assembly of the virus particle occurs in the nucleus. The replication of these viruses can be divided into two distinct stages, *early* and *late*. During the early stage the *early region* of the viral DNA is transcribed (Figure 16.18). A single RNA molecule, the primary transcript, is made by cellular RNA polymerase, but it is processed into *two species of mRNA*, a large one and a small one. The DNA of SV40 has *introns* that are excised out of the primary RNA transcript. In the cytoplasm, viral mRNA is translated with the formation of two proteins. One of these proteins, the T-antigen, binds to the site on the parental DNA that is the *origin of replication*.

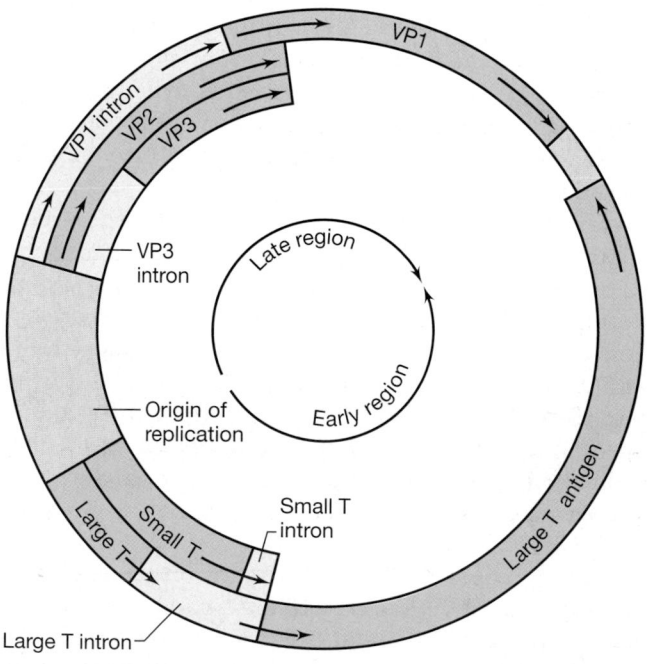

Figure 16.18 Genetic map of the papovavirus SV40. VP1, VP2, and VP3 are the genes coding for the three proteins that make up the coat of SV40. The arrows show the direction of transcription.

The viral DNA of SV40 is too small to code for its own DNA polymerases; host DNA polymerases are used. Replication occurs in a bidirectional fashion (so-called *theta* replication, ⚭ Figure 7.16) from a single replication origin. The process involves the same events that have already been described for host cell DNA replication (⚭ Sections 7.5 and 7.6).

Late SV40 mRNA molecules are synthesized using the strand complementary to that used for early mRNA synthesis (see Figure 16.18). Transcription begins at a promoter near the origin of replication. This late RNA is then processed by splicing and polyadenylation to give multiple forms of mRNA corresponding to the three coat proteins. These genes overlap; part of the nucleotide sequence contains information for all three proteins. These late mRNA molecules are transported to the cytoplasm and translated into the viral coat proteins. These proteins are then transported back into the nucleus where virion assembly takes place.

When a virus of the polyomavirus group infects a host cell, one of two modes of replication can occur, depending on the type of host cell. In some types of host cells, known as *permissive* cells, virus infection results in the formation of new virions and the lysis of the host cell. In other types of host cells, known as *nonpermissive*, efficient multiplication does not occur, but the virus DNA becomes integrated into some of the host cells, thereby creating new, genetically altered cells. Such cells frequently show loss of growth inhibition and are called *transformed* or tumor cells.

In nonpermissive hosts, transformation can take place if the early proteins can be expressed, but the viral DNA cannot be replicated independently. Instead, in the transformed cell, the viral DNA becomes stably integrated into the DNA of the host cell (Figure 16.19). Integration can occur at many sites in the cellular and viral genome. In this integrated form, two viral proteins are made that are essential for the maintenance of a stably integrated viral DNA, but no viral structural proteins are synthesized. Some transformed cells can be converted to cells capable of producing virus, a process that probably involves excision of the viral genome from the host genome.

A study of the manner of replication of SV40 virus has given some important insights into the manner by which viruses bring about the cancerous state in host cells. However, we note that DNA viruses of other groups can also cause the cancerous transformation, for example, human papillomavirus, a member of the closely related family *Papillomaviridae*. Also, the important family of viruses, the *retroviruses*, are also cancer viruses but have a completely different mode of replication (see Section 16.14).

✓ 16.10 Concept Check

Most double-stranded DNA animal viruses replicate in the nucleus. Some viruses causing cancer and other conditions in animal cells are double-stranded DNA viruses. In a permissive

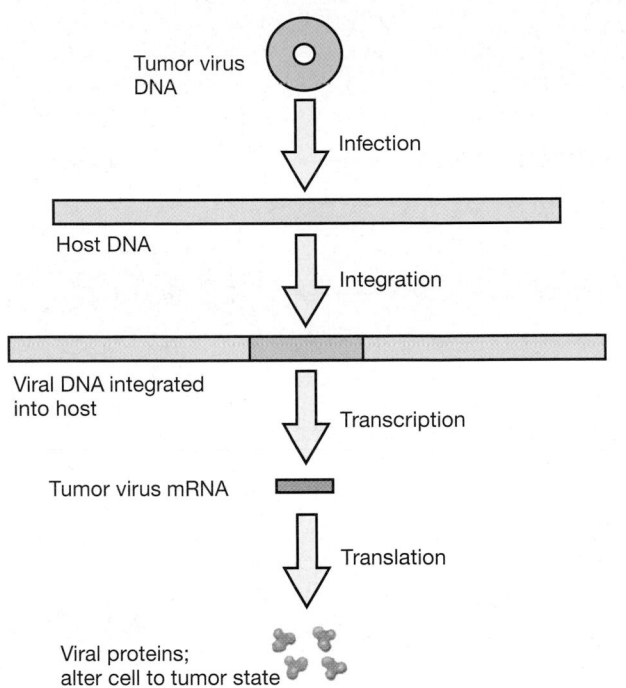

Tumor virus DNA

Infection

Host DNA

Integration

Viral DNA integrated into host

Transcription

Tumor virus mRNA

Translation

Viral proteins; alter cell to tumor state

Figure 16.19 General scheme of molecular events involved in cell transformation by a DNA tumor virus such as SV40. All or portions of viral DNA are incorporated into host cell DNA. The viral genes that encode transforming information are transcribed and processed to viral mRNA molecules, which are transported to the cytoplasm. Here they are translated to form transforming proteins or T-antigens that code for functions that can convert host cells into cancer cells.

host, the virus may cause death and lysis, but in a nonpermissive host, it may cause transformation of the infected host cell to a tumor cell.

✓ Why wouldn't a virus like SV40 be expected to carry enzymes in the virion?

✓ How can one transcript yield more than one mRNA?

16.11 Double-Stranded DNA Viruses: Herpesviruses

The herpesviruses (or *Herpesviridae*) are a large group of double-stranded DNA viruses that cause a wide variety of diseases in humans and animals, including fever blisters (cold sores), venereal herpes, chickenpox, shingles, and infectious mononucleosis; a number of these diseases are discussed in Chapter 26. One of the interesting features of some herpesviruses is their ability to remain *latent* in the body for long periods of time, becoming active only under conditions of stress. Both *herpes simplex*, the virus that causes *fever blisters*, and *varicella-zoster virus*, the cause of *chickenpox* and *shingles*, are able to remain latent in the neurons of the sensory ganglia, from which they are able to emerge to cause infections of the skin.

An important group of herpesviruses are tumorigenic, causing clinical forms of cancer. One herpesvirus that is tumorigenic is the *Epstein-Barr virus*, which is one of the underlying causes of *Burkitt's lymphoma*, a common tumor among children in Central Africa and New Guinea. Burkitt's lymphoma was among the first human cancers to be linked to virus infection (Table 9.3).

Molecular Biology of Herpesviruses

The *herpesvirus particle* is structurally complex, consisting of four distinct morphologic units. In herpes simplex type I, an enveloped virus about 150 nm in diameter (Figure 16.20a), the center of the virus is an electron-dense *core* consisting of double-stranded DNA. Surrounding this core is the nucleocapsid, of icosahedral symmetry, which consists of 162 capsomeres, each of which is composed of a number of distinct proteins. Outside the nucleocapsid is an amorphous layer that is called the *tegument*, a fibrous structure unique to the herpesviruses. Surrounding the tegument is an *envelope* whose outer surface contains many small *spikes*. A large number of separate proteins are present within the virion, but not all of them have been characterized.

The genome of herpes simplex type I virus consists of one large linear double-stranded DNA molecule of 152,260 base pairs (about 30 times larger than the SV40 genome) and encodes at least 84 separate proteins.

Infection occurs by attachment of virus particles to specific cell receptors, and, following fusion of the cytoplasmic membrane with the virus envelope, the nucleocapsid is released into the cell. The nucleocapsids are transported to the nucleus, where viral DNA is uncoated. Components of the virus particle inhibit macromolecular synthesis by the host.

There are *three classes of messenger RNAs*: immediate early (also called alpha, α), which codes for five regulatory proteins; delayed early (also called beta, β), which codes for DNA replication proteins, including thymidine kinase and DNA polymerase; and late (also called gamma, γ), which codes for the proteins of the virus particle (Figure 16.20b). During the *immediate early stage*, about one-third of the viral genome is transcribed by a host cell RNA polymerase. Early mRNA codes for certain positive-acting regulatory proteins that appear to stimulate the synthesis of the delayed early proteins. The second stage, *delayed early*, occurs only after the early proteins have been made. During this stage, about 40% of the viral genome is transcribed. Among the 10 proteins characterized from the delayed early stage are a *DNA polymerase*, enzymes involved in synthesis of deoxyribonucleotides, and a *DNA binding protein*. These enzymes are all involved in the process of viral DNA replication.

Herpes viral DNA synthesis itself takes place in the nucleus. After infection, the herpes genome apparently circularizes (remarkably like bacteriophage lambda) (Section 9.10) and replicates by a rolling circle

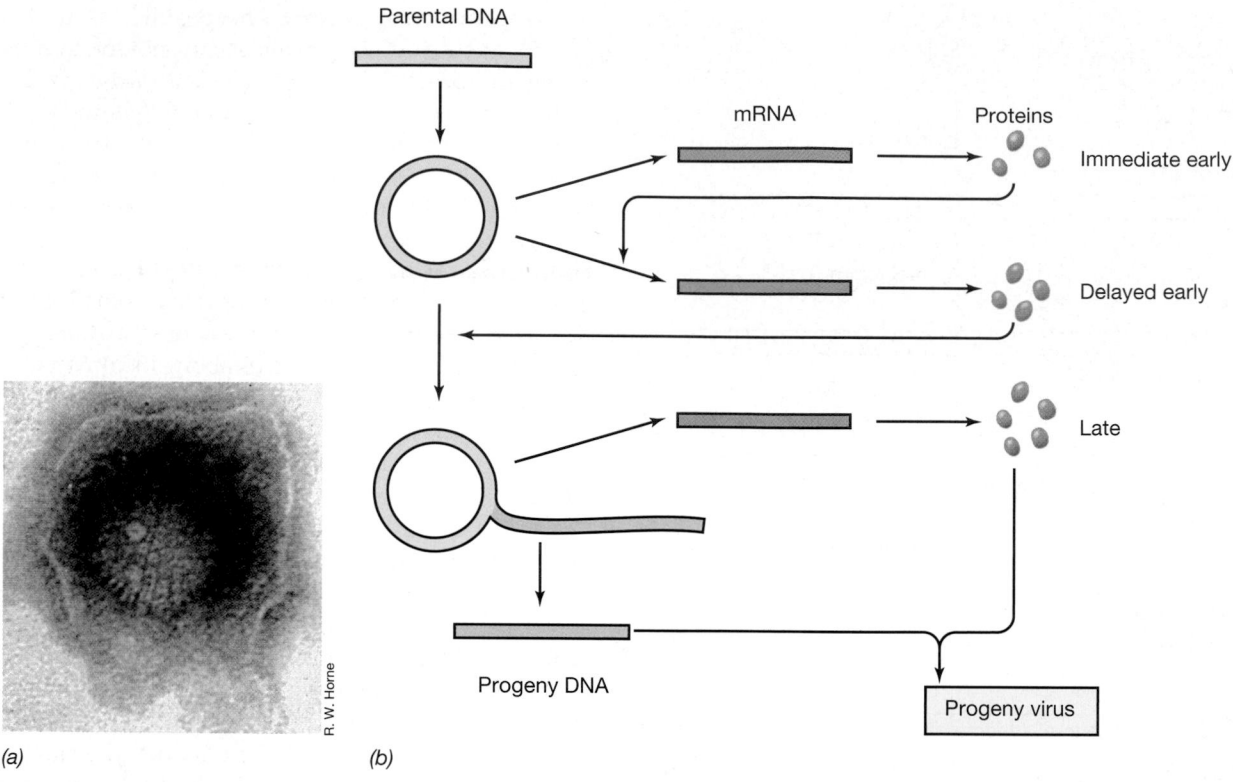

Parental DNA

mRNA

Proteins

Immediate early

Delayed early

Late

Progeny DNA

Progeny virus

R. W. Horne

(a)

(b)

Figure 16.20 Herpesvirus. (a) Electron micrograph of a herpesvirus particle. The diameter of the particle is about 150 nm. (b) Flow of events in multiplication of herpes simplex virus.

mechanism (∞ Figure 9.20). However, three origins seem to be involved. Long concatamers are formed that become processed to virus-length DNA during the assembly process itself (in a manner similar to that described for DNA bacteriophages) (∞ Sections 9.9 and 16.4).

Viral nucleocapsids are assembled in the nucleus, and acquisition of the virus envelope occurs via a budding process through the *inner membrane of the nucleus*. Mature virions are subsequently released through the endoplasmic reticulum to the outside of the cell. Thus, the assembly of this enveloped DNA virus differs markedly from that of the enveloped RNA viruses, which were assembled on the *cytoplasmic membrane* instead of the nuclear membrane.

✓ 16.11 Concept Check

A large number of different herpes viruses are known and several of them cause human disease. As in the case of the bacteriophage lambda, the large double-stranded DNA of these viruses circularizes and is apparently replicated by a rolling circle mechanism. Also like bacteriophage lambda, these bacteriophage can sometimes maintain themselves in a latent state in the host.

- ✓ Of what use is it to the herpes viruses to circularize their DNA prior to replication?
- ✓ Where is the nucleocapsid of a herpes virus assembled?

16.12 Double-Stranded DNA Viruses: Pox Viruses

Pox viruses are among the most complex and largest animal viruses known (Figure 16.21) and have some characteristics that approach those of primitive cells. However, the pox viruses, like all viruses, are not able to metabolize and thus depend on the host for the complete machinery of protein synthesis. These viruses are also almost unique in that they are DNA viruses that replicate in the *cytoplasm*. (Another group of double-stranded DNA viruses, the iridoviruses, some of which have genomes even larger than those of the pox viruses, also replicate in the cytoplasm.) Thus, a host cell infected with a pox virus exhibits *DNA synthesis outside the nucleus*, something that otherwise occurs only in intracellular organelles such as mitochondria.

General Properties of Pox Viruses

Pox viruses have been important medically as well as historically. *Smallpox* was the first virus to be studied in any detail and was the first virus for which a vaccine was developed (described by Edward Jenner in 1798). By diligent application of this vaccine on a worldwide basis, the disease smallpox has been *eradicated* in the wild, the first infectious disease to be elim-

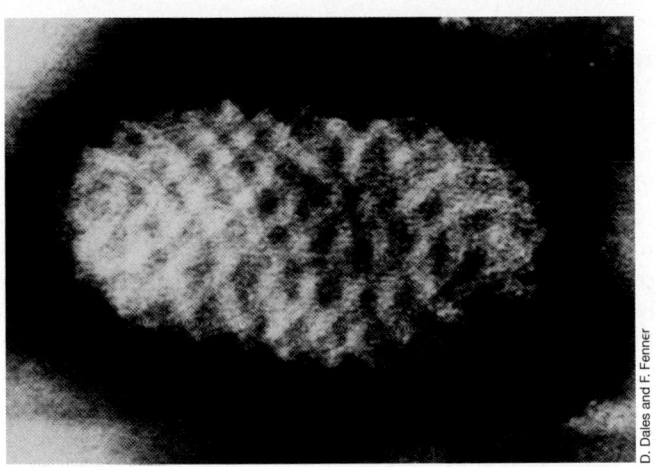

D. Dales and F. Fenner

Figure 16.21 Electron micrograph of a negatively stained vaccinia virus virion. The virion is approximately 400 nm (0.4 µm) long.

inated in this fashion. Other pox viruses of importance are *cowpox* and *rabbit myxomatosis virus*, an important infectious agent of rabbits and one that was intentionally used in an attempt to control the Australian rabbit population (Section 25.5). Some pox viruses also cause tumors.

The pox viruses are very large, so large that they can actually be seen under the light microscope. Most research has been done with *vaccinia virus*, the source of smallpox vaccine. The vaccinia virion is a brick-shaped structure about $400 \times 240 \times 200$ nm in size. The virion is covered on its outer surface with tubules or filaments arranged in a membranelike pattern, although the virus does not have a lipid membrane because the outer envelope consists of protein. Within the virion there are two lateral bodies of unknown composition and a core, the *nucleocapsid*, which contains DNA bounded by a layer of protein subunits.

The pox virus genome consists of double-stranded DNA. The vaccinia virus genome has about 185 kilobase pairs and contains between 150 and 200 genes. The pox DNA is interesting because the two strands of the double helix are cross-linked at the ends as a result of the formation of phosphodiester bonds between adjacent strands, as in the hairpin structure shown in Figure 7.8. The ends of pox virus DNA are therefore very similar to those of the *Chlorella* viruses (see Section 16.6).

Replication of Pox Viruses

Vaccinia virions are taken up into cells via a phagocytic process from which the cores are liberated into the cytoplasm. Interestingly, uncoating of the virus genome requires the action of a new protein that is synthesized after infection. This protein is encoded by viral DNA, and the gene specifying this protein is transcribed by an RNA polymerase present *within* the virus particle. In addition to this uncoating gene, a number of other viral genes are transcribed. The primary transcripts are turned into mRNAs by capping and polyadenylation while they are still inside the virus core.

Once the vaccinia DNA is fully uncoated, the formation of *inclusion bodies* within the cytoplasm begins. Within these inclusion bodies, transcription, replication, and encapsidation into progeny virus particles occur. Each infecting virion initiates its own inclusion body, so the number of inclusions depends on the multiplicity of infection. Progeny DNA molecules form a pool from which individual molecules are incorporated into virions. Mature virions accumulate in the cytoplasm. There seems to be no specific release mechanism, and most virions are released only when the infected cell disintegrates.

Pox Viruses and Recombinant Vaccines

Vaccinia virus has been used as a host for genetically altered proteins of other viruses, permitting the construction of genetically engineered vaccines. As we will see in Chapter 22, a vaccine is a substance capable of eliciting an immune response in an animal and serves to protect the animal from future infection with the same agent. Vaccinia virus causes no serious health effects in humans but is highly immunogenic. Molecular cloning methods have been used to express key viral proteins of influenza virus, rabies virus, herpes simplex type I virus, and hepatitis B virus in vaccinia virus virions, and then the latter used to develop a vaccine (Section 31.6).

A similar vaccine delivery system using adenovirus (see next section) as a vehicle has been developed because, like vaccinia virus, adenoviruses are of health consequence to humans.

✓ 16.12 Concept Check

The pox viruses, unlike the other DNA viruses we have discussed, replicate entirely in the cytoplasm. They are very large viruses whose virions carry several enzymes. These viruses also are responsible for several human diseases, but a vaccination campaign has eradicated the smallpox virus in the wild.

✓ Why is it notable that pox viruses replicate their DNA in the cytoplasm?

16.13 Double-Stranded DNA Viruses: Adenoviruses

The adenoviruses (or *Adenoviridae*) are a major family of icosahedral DNA-containing viruses. The term *adeno* is derived from the Latin for "gland" and refers to the fact that these viruses were first isolated from the tonsils and adenoid glands of humans. Adenoviruses cause generally mild respiratory infections in humans, and a number of such viruses are isolated from apparently healthy individuals.

The genomes of the adenoviruses consist of linear double-stranded DNA of about 36 kilobase pairs. Attached in covalent linkage to the 5'-terminus of the DNA is a protein component essential for infectivity of the DNA. The DNA has inverted terminal repeats of 100–1800 base pairs (this varies with the virus strain).

Replication of the viral DNA occurs in the nucleus (Figure 16.22). After the virus particle has been transported to the nucleus, the core is released and converted to a viral DNA-histone complex. Early transcription is carried out by an RNA polymerase of the host, and a number of primary transcripts are made. The transcripts are spliced, capped, and polyadenylated, giving several different mRNAs.

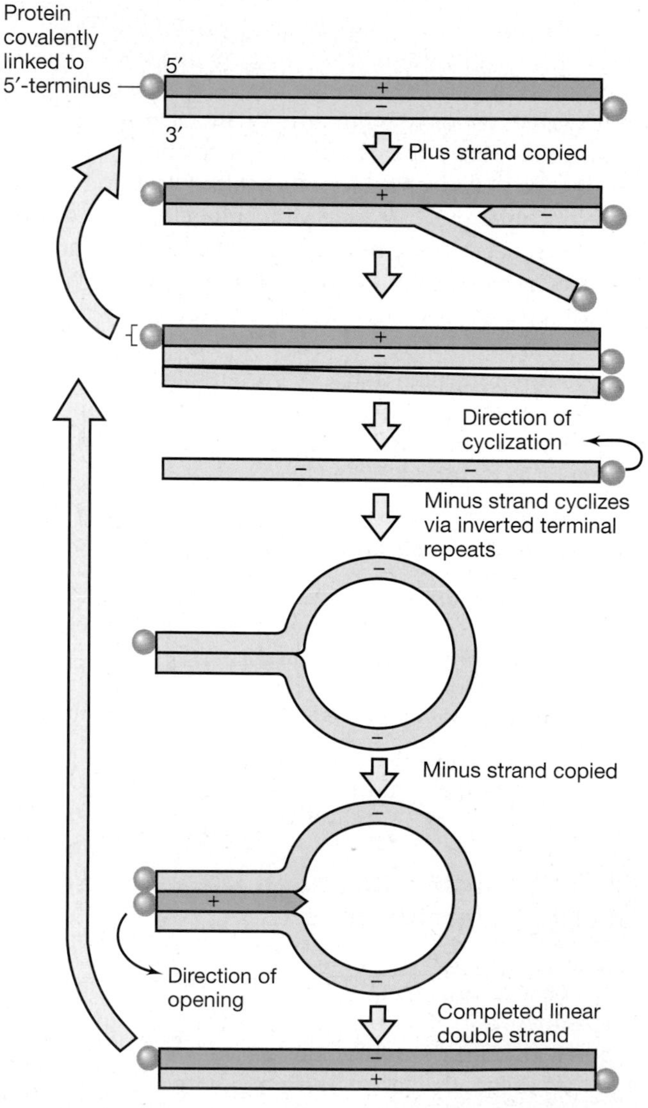

The early proteins are involved in regulation of DNA replication; the later proteins are the virus coat protein. Viral DNA replication uses a virus-encoded *protein* as a primer and another virus-encoded protein as DNA polymerase. For the replication of a linear double-stranded DNA molecule such as that of adenovirus, initiation of replication can begin at either end or at both ends simultaneously (⌔ Figure 7.24). In the case of the adenoviruses, replication begins at *either* end, the two strands being replicated asynchronously. The products of a round of replication are double- and single-stranded molecules. The latter then cyclizes by means of the inverted terminal repeats, and a new complementary strand is synthesized beginning from the 5'-end, the products being another double-stranded molecule (Figure 16.22). This mechanism of replication is interesting because it does not involve the formation of discontinuous fragments of DNA on the lagging strand, as occurs in conventional DNA replication (⌔ Section 7.6).

✓ 16.13 Concept Check

Different double-stranded DNA animal viruses have different DNA replication strategies. That of the adenoviruses involves protein primers and a mode of replication that avoids the synthesis of a lagging strand.

✓ Describe how adenovirus replicates its genome without the synthesis of a lagging strand.

16.14 Viruses Using Reverse Transcriptase

In Section 9.12 we discussed the retroviruses, a group of viruses that replicate using the enzyme reverse transcriptase. The retroviruses (family *Retroviridae*) are one of a number of genetic elements that use reverse transcriptase for replication. We mentioned that there are genetic elements found in some eukaryotic cells called retrotransposons, which are transposons that use reverse transcriptase (⌔ Section 9.12). In addition, some fungal "viruses" (which do not have an extracellular stage), such as the Ty1 and Ty3 elements of the yeast *Saccharomyces cerevisiae*, replicate very much like the retroviruses. There are even genetic elements in *Bacteria*, such as the *retron*, found in myxobacteria and *Escherichia coli*, which replicate using reverse transcriptase.

There are essentially two different types of viruses that use reverse transcriptase, and they are most easily differentiated by the type of nucleic acid genome in their virions. The retroviruses are considered Class VI viruses in the Baltimore scheme and have *RNA genomes* in their virions. However, there are also Class VII viruses that replicate using reverse transcriptase but have DNA

Figure 16.22 Replication of adenovirus DNA. See text for details.

genomes in their virions. These are exemplified by the Hepadnaviruses (*Hepadnaviridae*) in animals (◯ Figure 9.22) and the *Caulimoviridae* in plants (of which the most studied is cauliflower mosaic virus). In this section we will deal briefly with both RNA reverse-transcribing viruses and DNA reverse-transcribing viruses.

RNA Reverse-Transcribing Viruses: Retroviruses

This information is meant to complement that found in Section 9.12. Remember that retroviruses have enveloped virions that contain *two* separate copies of the RNA genome (◯ Figure 9.24). The virion also contains several enzymes, including reverse transcriptase, and also a specific tRNA. The reason that enzymes are required is that, although the genome of the retroviruses is of the plus sense and is capped and tailed, it is *not* used directly as messenger RNA. Instead, one of the copies of the genome is converted to DNA by reverse transcriptase and is integrated into the host genome. Although we covered retroviral replication in Section 9.12 (◯ Figure 9.25), we gave few details.

The DNA formed is a linear double-stranded molecule and is synthesized in the cytoplasm within an uncoated viral core particle. An outline of the reverse transcription of viral RNA into DNA is given in Figure 16.23.

The enzyme reverse transcriptase is essentially a DNA polymerase, but it actually shows *three* enzymatic activities: (1) synthesis of DNA with an RNA template (reverse transcription), (2) synthesis of DNA with a DNA template, and (3) ribonuclease H activity (an activity that degrades the RNA strand of an RNA:DNA hybrid). Like all DNA polymerases, reverse transcriptase needs a primer for DNA synthesis. The primer for retrovirus reverse transcription is a specific *cellular transfer RNA (tRNA)*. The type of tRNA used as primer depends on the virus and is brought into the viron from the previous host cell.

Using the tRNA primer, the 100 or so nucleotides at the 5′-terminus of the RNA are reverse-transcribed into DNA. Once transcription reaches the 5′-end of the RNA, the transcription process stops. To copy the remaining RNA, which is the bulk of the RNA of the virus, a different mechanism comes into play. First, terminally redundant RNA sequences at the 5′-end of the molecule are removed by the action of another enzymatic activity of reverse transcriptase, *ribonuclease H*. This leads to the formation of a small, single-stranded DNA that is complementary to the RNA segment at the *other end* of the viral RNA. The small, single-stranded piece of DNA then hybridizes with the other end of the viral RNA molecule, where copying of the viral RNA sequences continues. As summarized in Figure 16.23, continued action of reverse transcriptase and ribonuclease H leads to the formation of a double-stranded DNA molecule with long

terminal repeats (LTRs) at each end. These LTRs contain strong promoters of transcription and are involved in the integration process. The *integration* of the viral DNA into the host genome is analogous to the integration of Mu (see Section 16.5) or a bacterial transposon into a bacterial genome. Integration can occur anywhere in the cellular DNA, and once integrated, the element, now called a *provirus*, is a stable genetic element. As a provirus its genetic information may be expressed, or it may remain in a latent state and not be expressed.

If the promoters in the right LTR are activated, the integrated proviral DNA is transcribed by a cellular RNA polymerase into transcripts that are capped and polyadenylated. These RNA transcripts either may be encapsidated into virus particles or may be processed and translated into virus proteins.

Figure 9.24 shows simplified genetic maps of two different retroviruses, and we also briefly discussed gene expression in Section 9.12. All functional retroviruses have at least the three genes *gag*, *pol*, and *env*, in that order. Transcription from the integrated DNA copy gives rise to full-length genomic RNA, which is capped and tailed and can be used as mRNA or packaged into new virions.

The translation and processing of such an mRNA from a simple retrovirus is shown in Figure 16.24. Note the *gag* "gene" at the 5′ end of the mRNA actually encodes a number of small proteins. These are synthesized as a polyprotein that is subsequently processed by a protease (which itself is a part of the polyprotein). These proteins make up the capsid (and the protease becomes packaged in the virion). The *pol* gene is next, and it is translated as a polyprotein with the *gag* proteins. Interestingly, it is made in small amounts because, depending on the retrovirus, a ribosome that translates the *pol* gene must either read through the stop codon at the end of the *gag* gene or make a precise switch to a different reading frame in this region. (Almost all genetic elements that encode a reverse transcriptase make it in small amounts and as part of a polyprotein.) This protein must also be processed, first to remove it from the *gag* proteins, and then *pol* itself encodes both the reverse transcriptase and an integrase protein that is involved in DNA integration (and is also packaged in the virion). For the *env* gene to be translated, the full-length mRNA is processed as shown. Note then, from the point of view of the *env* gene, the *gag* and *pol* region is an intron! The *env* product is itself processed into two envelope proteins.

Although this pattern may seem complex, this is the pattern of a "simple" retrovirus. The human immunodeficiency virus 1 (HIV-1) genome is quite complex and includes several more small genes. Its expression not only involves extensive protein processing by proteases but also complex patterns of alternative splicing of

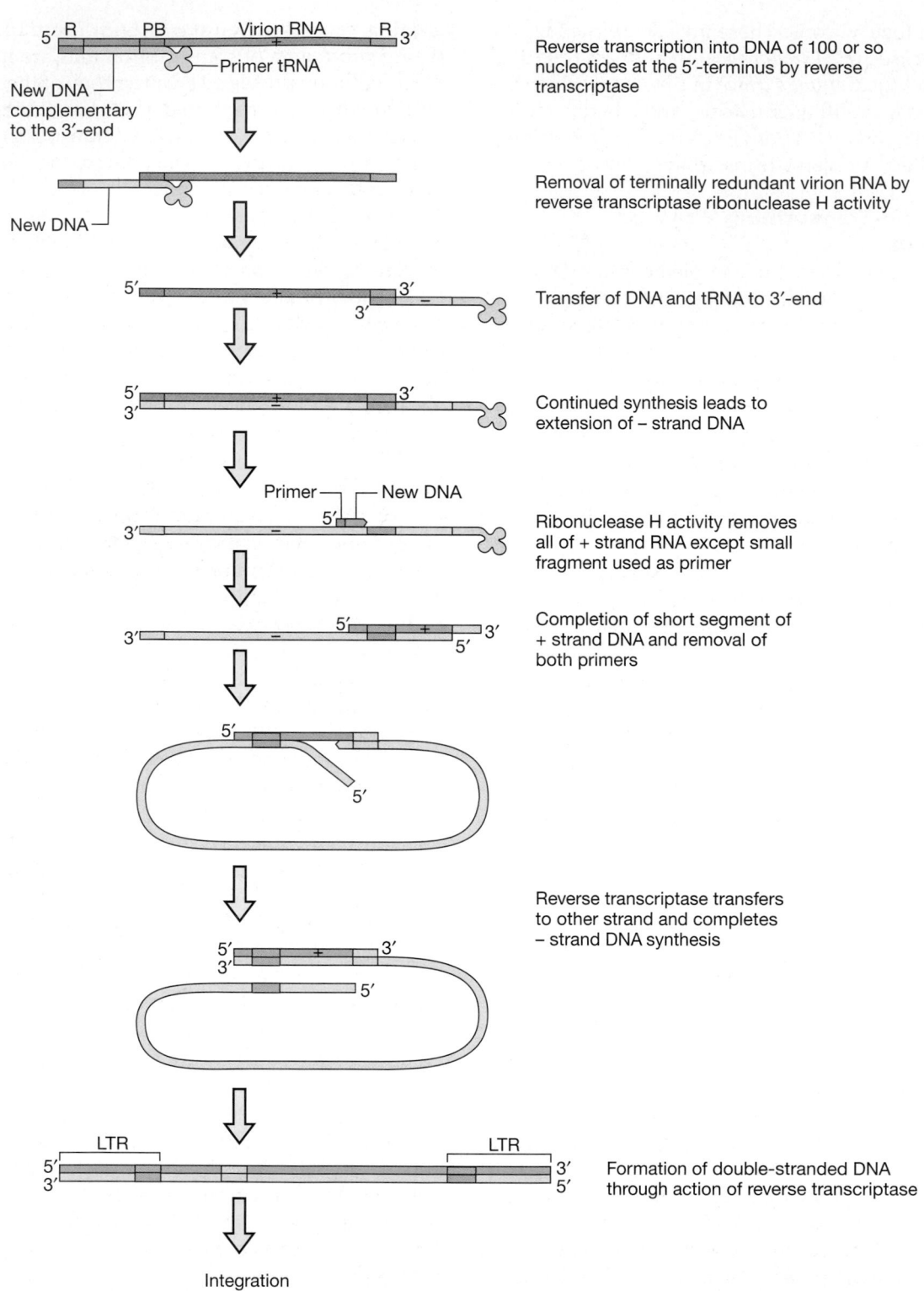

Figure 16.23 Overall steps in the formation of double-stranded DNA from retrovirus single-stranded RNA. The sequences labeled R on the RNA are direct repeats found at either end. The sequence labeled PB is where the primer (tRNA) binds. Note that the process of DNA synthesis has yielded longer direct repeats on the DNA than were originally on the RNA. These are called *long terminal repeats (LTRs)*.

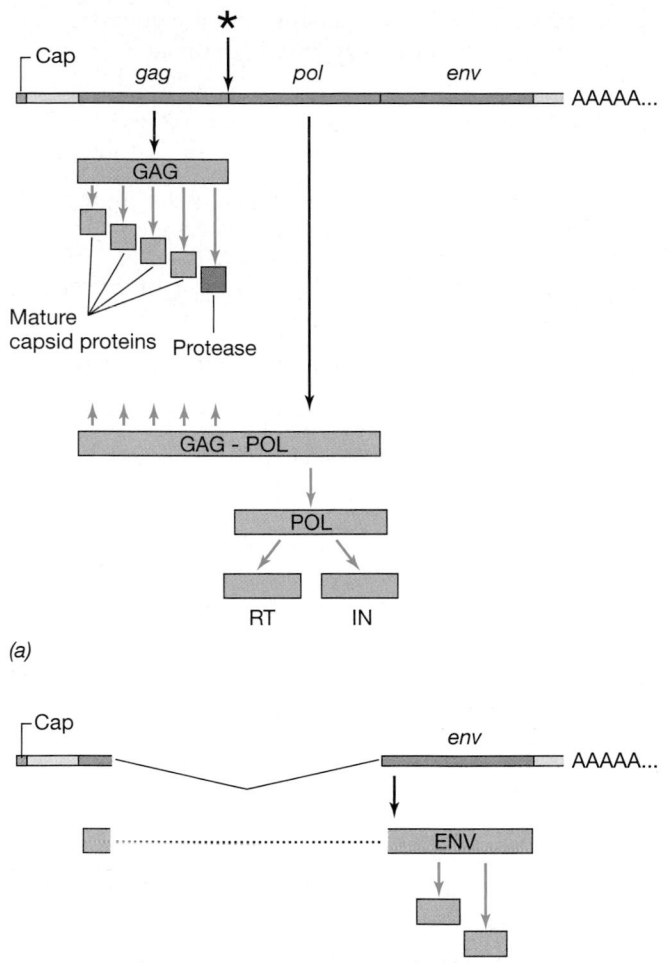

(a)

(b)

Figure 16.24 Translation of retrovirus mRNA and processing of the proteins. (a) The full-length mRNA with the three genes *gag, pol,* and *env* is shown at the top. The asterisk shows the site where a ribosome must read through a stop codon or do a precise shift of reading frame to synthesize the GAG-POL polyprotein. The black arrows indicate translation, while the blue arrows indicate protein processing events. One of the *gag* gene products is a protease. The POL product is processed to give reverse transcriptase (RT) and integrase (IN). (b) The mRNA has been processed to remove most of the *gag-pol* region. This shortened message is translated to give the ENV polyprotein, which is cleaved into two envelope proteins.

introns. However, in all these viruses many protease steps are involved in forming mature proteins. Therefore, specific protease inhibitors are now being used to treat retroviral infections, including HIV-1 infections.

DNA Reverse-Transcribing Viruses: Hepadnaviruses

The life cycles of viruses contain a variety of unexpected genome structures and information transfers. Many of these seem exemplified by the hepadnaviruses, such as human hepatitis B virus. The genomes of these fas-

cinating viruses are among the smallest known for a virus, but the virus life cycle is very complex. Like the retroviruses, these viruses use reverse transcriptase in their life cycle. In the case of hepadnaviruses the genome in the virion is DNA, but this DNA is replicated using an *RNA* intermediate, the opposite of what occurs in retroviruses.

The genomic DNA of hepadnaviruses is only *partially* double-stranded (Figure 16.25). One strand is incomplete, and both strands have breaks or gaps, but they are held in a circular form by hydrogen bonding. On entering the cytoplasm a viral polymerase carried in the virion completes the replication of this molecule. (This polymerase is quite a versatile protein; it contains DNA polymerase activity and reverse transcriptase activity and is the protein primer for synthesis of one of the DNA strands.) The figure showing the genetic map indicates the incredible compactness of the genome. Despite the small size (an average of 3200 base pairs), the genome encodes several proteins. Not only do genes overlap, but every base is part of a codon for at least one protein!

Replication of this genome involves transcription by host RNA polymerase (in the nucleus), yielding a transcript with terminal repeats. (The repeats occur because the polymerase proceeds slightly more than once around the circular molecule.) The viral polymerase then copies this into DNA, very much like the replication of retroviruses, but in this case the DNA becomes packaged into new virus particles.

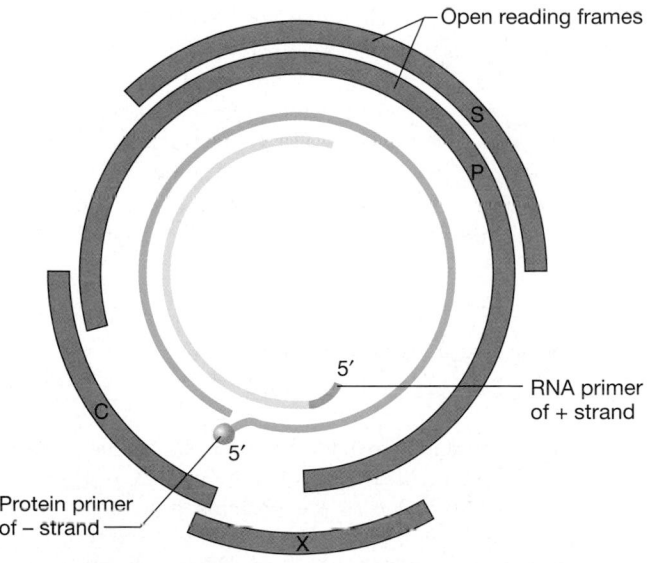

Figure 16.25 The partially double-stranded genome of human hepatitis B virus is shown in green. Note that the positive strand is not complete. The sizes of the open reading frames C, P, S, and X are also shown. Notice that all these genes overlap and that they cover every base in the genome.

Remarkably, the human hepatitis B virus is a "helper virus" for the *delta agent*. The delta agent is a "subvirus" and requires hepatitis B virus to supply the proteins necessary for its coat. Therefore, the delta agent is the parasite of a parasite! The delta agent has a circular, negative-strand RNA genome of 1679 bases. Like the viroids (∞ Section 9.13), it seems to be able to base-pair into a rodlike structure that can then be transcribed by the host RNA polymerase. However, unlike the situation with viroids, the transcript of the delta agent encodes at least one protein, an RNA binding protein carried by the delta agent virion. Incredibly, the RNA is also a ribozyme (∞ Section 7.12) that can cleave itself. This self-cleavage may be involved in mRNA formation.

Very small "viruses" can be very interesting indeed. The hepadnaviruses and the delta agent employ several different strategies for maximizing the genetic information carried in their very small genomes.

✓ 16.14 Concept Check

The retroviruses contain RNA genomes in the virion but use reverse transcriptase to make a DNA copy during their life cycle. The hepadnaviruses contain DNA genomes in the virion but use reverse transcriptase to make this DNA from an RNA copy. Both types of viruses have complex patterns of gene expression.

✓ Why might protease inhibitors be effective in limiting retrovirus multiplication?

✓ Describe the different molecules used by retroviruses and hepadnaviruses as primers for DNA synthesis.

Review Questions

1. Describe the types of genomes found in viruses. Give an example of at least one virus (or group of viruses) with each type of genome.

2. What are *overlapping* genes? Give examples of viruses that have overlapping genes.

3. Positive-strand RNA viruses are known for *Bacteria*, for plants, and for animals. What are the important distinctions between gene expression in these viruses, particularly between those that infect *Bacteria* and those that infect eukaryotes?

4. Figure 16.17 shows circular DNA that has a contour length of 1.5 μm. How many base pairs are there in this DNA molecule (∞ Section 7.2)?

5. Many bacteriophages have single-stranded DNA genomes. Describe how the genes carried by these viruses can be transcribed and translated.

6. One end of the T7 genome always enters the cell first during infection. This is necessary for the infection process to proceed. Explain.

7. The rhabdoviruses have an unusual genome structure. What is the nature of their genome and why *must* the virion contain an enzyme involved in its replication?

8. The *Chlorella* viruses have an unusual mechanism for infecting cells compared with other eukaryotic viruses. How do they infect cells and what other types of viruses use such a mechanism?

9. Although animal viruses are usually not referred to as temperate, as are some bacteriophages (∞ Section 9.10), several can set up latent infections. Describe any two such types of animal viruses.

10. Many viruses that have double-stranded DNA have linear genomes. For any three such viruses describe the mechanism they use to ensure that the 5' ends of their DNA strands can be replicated (∞ Section 7.7).

11. What is there about expression of retroviral genes that might make virus replication sensitive to protease inhibitors?

Application Questions

1. Not all proteins are made from the RNA genome of bacteriophage MS2 in the same amounts. Can you explain why? One of the proteins functions very much like a repressor (∞ Section 8.5), but it functions at the translational level. Which protein is it and how does it function?

2. Most known viruses that infect multicellular plants are RNA viruses. Why might this be so?

3. Give a mechanistic explanation of why the genomes of double-stranded RNA viruses might be segmented.

4. The mechanism of replication of both strands of DNA in some viruses, such as adenoviruses, is continuous. Show how this can be without violating the "rule" learned in Chapter 7 that all DNA synthesis occurs in the overall direction of 5' → 3'.

5. Most genetic elements that express reverse transcriptase make it in small amounts and/or as part of a polyprotein. Can you think of any reason(s) why this might be so?

T he enormous genetic diversity of prokaryotes is mirrored by an equally enormous metabolic diversity. Collectively, prokaryotes can carry out all of the metabolic reactions typical of eukaryotes. But in addition, the metabolic capacities of certain prokaryotes far exceeds anything possible by eukaryotes. These include conserving energy from chemical reactions under anoxic (oxygen-free) conditions, from reactions involving various inorganic compounds, as shown here with organisms that oxidize reduced iron to iron oxides, and from a virtually inexhaustible list of organic chemicals unusable by higher organisms. Indeed, when it comes to metabolic diversity, prokaryotes have exploited every conceivable means of making a living consistent with the laws of thermodynamics.

METABOLIC DIVERSITY

17

Working Glossary

Anaerobic respiration respiration in which some substance such as SO_4^{2-} or NO_3^- serves as terminal electron acceptor instead of O_2

Anammox anoxic ammonia oxidation

Anoxic oxygen-free

Anoxygenic photosynthesis photosynthesis in which O_2 is not produced

Antenna in reference to photocomplexes, light harvesting pigment molecules that funnel energy to the reaction center

Autotrophy the use of CO_2 as sole carbon source

Bacteriochlorophyll the chlorophyll pigment of anoxygenic phototrophs

Calvin cycle the biochemical route of CO_2 fixation in many autotrophic organisms

Carotenoids hydrophobic accessory pigments present along with chlorophyll in photosynthetic membranes

Chemolithotroph a microorganism capable of oxidizing inorganic compounds as energy sources

Chlorophyll the light-sensitive, Mg-containing porphyrin of photosynthetic organisms that initiates the process of photophosphorylation

Chlorosome cigar-shaped structures present in the periphery of cells of green sulfur and green nonsulfur bacteria and that contain the antenna bacteriochlorophylls (*c, d,* or *e*)

Cyclic photophosphorylation the reactions of light-driven ATP synthesis in phototrophic organisms in which electrons travel in a closed loop, establishing a proton motive force

Denitrification anaerobic respiration in which NO_3^- is reduced to gaseous nitrogen compounds, primarily N_2

Dioxygenase an enzyme that introduces both atoms of oxygen from O_2 into a substrate

Disproportionation splitting of a compound into two new compounds, one more oxidized and one more reduced than the original compound

Fermentation anaerobic catabolism of an organic compound in which the compound serves as both an electron donor and an electron acceptor and in which ATP is produced by substrate-level phosphorylation

Homoacetogenesis (acetogenesis) energy metabolism involving the production of acetate from either H_2 plus CO_2 or from organic compounds

Hydrogenase an enzyme, widely distributed in anaerobic microorganisms, capable of taking up or evolving H_2

Methanogenesis the biological production of methane (CH_4)

Methanotrophy methane oxidation

Methylotrophy energy metabolism in which methyl groups or methane are oxidized as electron donors

Mixotrophic a nutritional state in which an inorganic compound serves as energy source (electron donor) and organic compounds serve as carbon source

Monooxygenase an enzyme that catalyzes the incorporation of one atom of O_2 into a substrate while the other atom is reduced to H_2O

Nitrification the microbial conversion of NH_3 to NO_3^-

Nitrogenase an enzyme capable of reducing N_2 to NH_3 in the process of nitrogen fixation.

Nitrogen fixation the biological reduction of N_2 to NH_3 by nitrogenase

Oxygenic photosynthesis photosynthesis carried out by cyanobacteria and green plants in which O_2 is evolved

Photophosphorylation the production of ATP in photosynthesis

Photosynthesis the series of reactions in which ATP is synthesized by light-driven reactions and CO_2 is fixed into cell material

Phototroph an organism capable of using light as an energy source

Phycobiliprotein the accessory pigment complex in cyanobacteria that contains a molecule of phycocyanin or phycoerythrin coupled to proteins

Reaction center a photosynthetic complex containing chlorophyll (or bacteriochlorophyll) and several other components within which occurs the initial electron transfer reactions of photosynthetic electron flow

Recalcitrant resistant to microbial attack

Reductive dechlorination an aerobic respiration where a chlorinated organic compound is used as an electron acceptor, usually with release of Cl^-

Reverse electron transport the energy-dependent movement of electrons against the thermodynamic gradient to form a strong reductant from a weaker electron donor

Syntrophy a process whereby two or more microorganisms cooperate to degrade a substance neither can degrade alone

Thylakoids sheetlike photosynthetic membrane systems in cyanobacteria and chloroplasts

The preceding unit considered the great phylogenetic diversity of microbial life on Earth. In this chapter we focus on the *metabolic diversity* of these organisms, with special emphasis on the biochemical processes behind this diversity. We will then proceed to a study of microbial ecology—the interactions of microorganisms with each other and their environment. Metabolic diversity and microbial ecology go hand in hand; the metabolic diversity to be described in this chapter fuels the nutrient cycles and other critical microbial activities we will discuss in Chapters 18 and 19. Indeed, as we will see, microorganisms and their metabolic reactions play key roles in maintaining the web of life on Earth and are of crucial importance in agriculture and other aspects of human affairs (∞ Section 1.4).

I THE PHOTOTROPHIC WAY OF LIFE

Phototrophy, the use of light as an energy source, is widespread in the microbial world. In the first five sections we discuss the major forms of phototrophy, including that which forms the very oxygen we breathe, and then proceed to see how most phototrophs satisfy all of their carbon needs from CO_2.

17.1 Photosynthesis

One of the most important biological processes on Earth is **photosynthesis**, the conversion of light energy to chemical energy. Organisms that can carry out photosynthesis are called *phototrophs* (Figure 17.1). Most phototrophic organisms are also *autotrophs*, capable of growing with CO_2 as sole carbon source. Energy from light is thus used in the reduction of CO_2 to organic compounds.

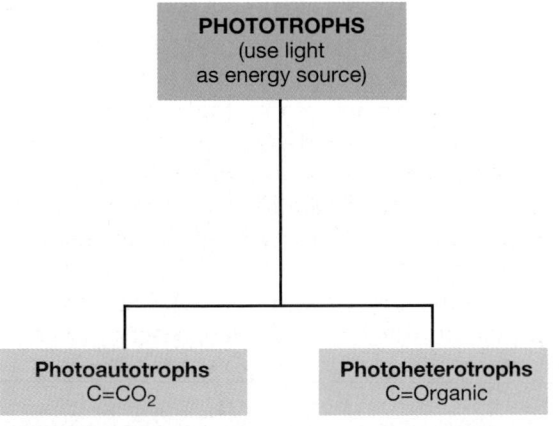

Figure 17.1 Classification of phototrophic organisms in terms of energy and carbon sources.

There are, as well, some phototrophs that use light as their energy source, but organic carbon as their carbon source; this lifestyle is called *photoheterotrophy* (Figure 17.1). The ability to photosynthesize depends on the presence of light-sensitive pigments, the *chlorophylls*, found in plants, algae, and some bacteria. Light reaches phototrophic organisms in distinct units of energy called *quanta*. Absorption of light quanta by chlorophylls begins the process of photosynthetic energy conversion.

Light and Dark Reactions

The growth of a photoautotroph can be characterized by two distinct sets of reactions: the **light reactions**, in which light energy is conserved as chemical energy, and the **dark reactions**, in which this chemical energy is used to reduce CO_2 to organic compounds. For autotrophic growth, energy is supplied in the form of adenosine triphosphate (ATP), while electrons for the reduction of CO_2 come from NADH or NADPH*. The latter is produced by the reduction of NAD^+ or $NADP^+$ by electrons originating from various electron donors to be discussed later.

The light reactions thus conserve some of the energy in light in a form of chemical energy, ATP, that can be used by cells. To drive autotrophic reactions, some phototrophic bacteria obtain reducing power from electron donors in their environment, typically reduced sulfur sources (H_2S, S^0, $S_2O_3^{2-}$) or H_2. By contrast, green plants, algae, and cyanobacteria use H_2O, an inherently poor electron donor (∞ Figure 5.9), as a source of reducing power to reduce $NADP^+$ to NADPH, producing molecular oxygen, O_2, as a by-product (Figure 17.2). Because O_2 is produced, photosynthesis in these organisms is called **oxygenic**. In other phototrophic bacteria no O_2 is produced, and thus the process is called **anoxygenic** photosynthesis. We will see later that whereas the production of NADH from substances like H_2S by anoxygenic phototrophs may or may not be directly driven by light, the oxidation of H_2O to O_2 by oxygenic phototrophs is always driven by light (Figure 17.2); thus these organisms require light for *both* reducing power and energy conservation.

✓ 17.1 Concept Check

Two classes of energy metabolism are recognized: *phototrophs* obtain energy from light, whereas *chemotrophs* derive their energy from chemical sources. Photosynthesis can be considered in terms of the light reactions, where ATP is generated, and the dark reactions, where ATP is consumed in the fixation of CO_2.

✓ How are *photoautotrophs* and *photoheterotrophs* similar and how do they differ?

✓ What is the fundamental difference between an *oxygenic* and an *anoxygenic* phototroph?

---

* In oxygenic phototrophs the reduced substance is NADPH, while in anoxygenic phototrophs it is NADH.

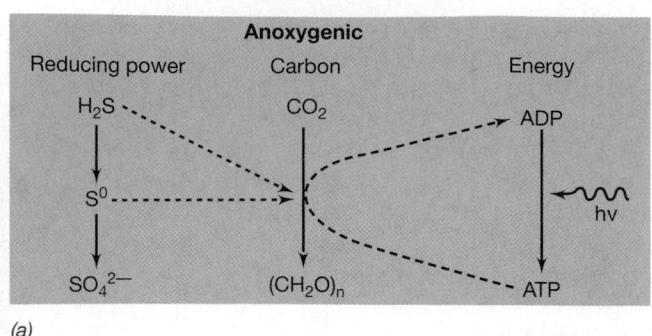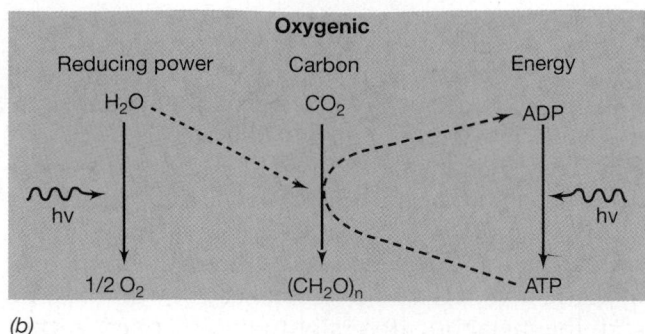

Figure 17.2 Energy and reducing power synthesis in (a) anoxygenic versus (b) oxygenic phototrophs. Although both types of phototrophs obtain their energy from light (hv), in oxygenic phototrophs light also drives the oxidation of water to oxygen.

17.2 The Role of Chlorophyll and Bacteriochlorophyll in Photosynthesis

Photosynthesis occurs only in organisms that possess some type of **chlorophyll**. Chlorophyll is a porphyrin, as are the cytochromes (∞ Section 5.11), but unlike the cytochromes, chlorophyll contains a *magnesium* atom instead of an iron atom at the center of the porphyrin ring. Chlorophyll also contains specific substituents bonded to the porphyrin ring, as well as a hydrophobic alcohol side chain that enables chlorophyll to associate with lipid and hydrophobic proteins of photosynthetic membranes.

The structure of chlorophyll *a*, the principal chlorophyll of higher plants, most algae, and the cyanobacteria, is shown in Figure 17.3. Chlorophyll *a* is green because it *absorbs* red and blue light preferentially and *transmits* green light. We mentioned the electromagnetic spectrum in Section 10.3. The spectral properties of any pigment can best be expressed by its *absorption spectrum*, which indicates the degree to which the pigment absorbs light of different wavelengths. The absorption spectrum of cells containing chlorophyll *a* shows strong absorption of red light (maximum absorption at a wavelength of 680 nm) and blue light (maximum at 430 nm) (Figure 17.3a).

There are a number of chemically different chlorophylls that are distinguished by their different absorption spectra. Chlorophyll *b*, for instance, absorbs maximally at 660 nm rather than at 680 nm. Many plants have more than one chlorophyll, but the most common are chlorophylls *a* and *b*. Among prokaryotes, cyanobacteria have chlorophyll *a* but anoxygenic phototrophs, such as the purple and green bacteria, can have any of a number of bacteriochlorophylls (Figures 17.3 and 17.4). Bacteriochlorophyll *a* (Figure 17.3b), present in most purple bacteria (∞ Section 12.2), absorbs maximally between 800–925 nm, depending on the species of phototroph. Different species have different pigment-

binding proteins and the absorption maxima of bacteriochlorophyll *a* depends to some degree on the nature of these proteins and how they are arranged in photocomplexes. Other bacteriochlorophylls, whose distribution runs along phylogenetic lines, absorb in other regions of the visible and infrared spectrum (Figure 17.4).

Why do organisms have several kinds of chlorophylls absorbing light at different wavelengths? One reason may be to make better use of the energy of the electromagnetic spectrum. Only light energy that is *absorbed* can be used to make energy; by having different pigments, two unrelated organisms can coexist in a habitat, each using wavelengths of light that the other is not using. Thus, pigment diversity has ecological significance.

Photosynthetic Membranes and Chloroplasts
The chlorophyll pigments and all the other components of the light-gathering apparatus are found inside the cell within special membrane systems, the **photosynthetic membranes**. The location of the photosynthetic membranes differs between prokaryotic and eukaryotic microorganisms. In eukaryotes, photosynthesis is associated with special intracellular organelles, the **chloroplasts** (Figure 17.5a, page 553; ∞ Section 14.3). The chlorophyll pigments are attached to sheetlike (lamellar) membrane structures of the chloroplast (Figure 17.5b). These photosynthetic membrane systems are called **thylakoids**; stacks of thylakoids are called *grana* (Figure 17.5b). The thylakoids are so arranged that the chloroplast is divided into two regions, the matrix space that surrounds the thylakoids and the inner space within the thylakoid array (Figure 17.5b). This arrangement makes possible the development of a light-driven proton motive force that can be used to synthesize ATP, as will be described in Section 17.5.

In prokaryotes, chloroplasts are absent and photosynthetic pigments are integrated into internal membrane systems that arise (1) from invagination of the

Figure 17.3 Structures of chlorophyll *a* and bacteriochlorophyll *a*, both of which are magnesium tetrapyrroles. The two molecules are identical except for those portions contrasted in yellow and green. The central Mg atom is shown in blue. The absorption spectra alongside each molecule are: (a) cells of the green alga *Chlamydomonas* (⌾ Section 4.11). The peaks at 680 and 430 nm are due to chlorophyll *a*; the peak at 480 nm is due to carotenoids; (b) cells of the phototrophic purple bacterium *Rhodopseudomonas palustris* (⌾ Section 12.2). Peaks at 870, 800, 590, and 360 nm are due to bacteriochlorophyll *a* while peaks at 525 and 475 nm are due to carotenoids. Notice how ring II (upper right) is *reduced* in bacteriochlorophyll *a* compared to chlorophyll *a*.

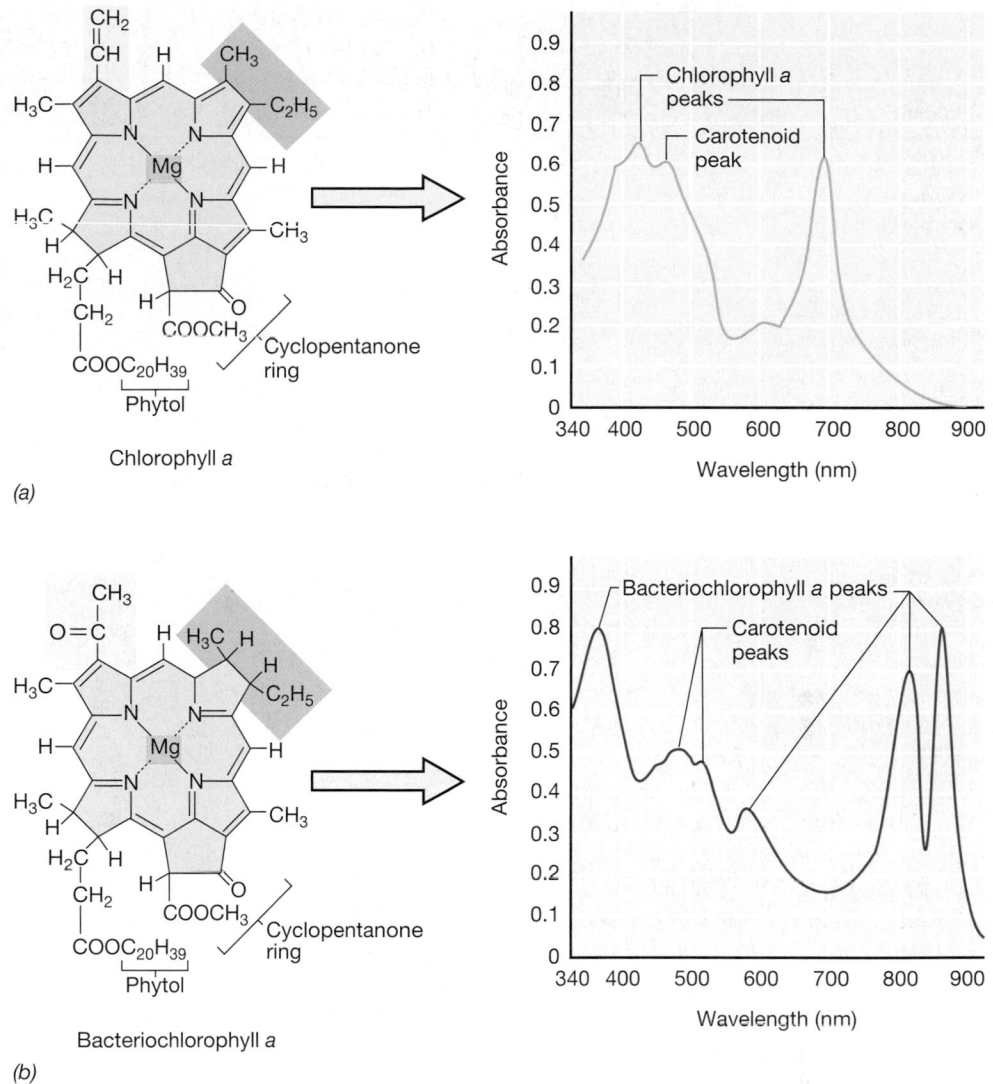

cytoplasmic membrane (purple bacteria) (for example, ⌾ Figures 12.2*a* and *b*; see also Figure 17.12), (2) from the cytoplasmic membrane itself (heliobacteria; ⌾ Section 12.20), (3) in both the cytoplasmic membrane and specialized non-unit membrane-enclosed structures called *chlorosomes* (green bacteria; see Figure 17.7), or (4) in thylakoid membranes in cyanobacteria.

Reaction Centers and Antenna Pigments

Within a photosynthetic membrane chlorophyll or bacteriochlorophyll molecules are associated with proteins to form complexes consisting of anywhere from 50 to 300 molecules (Figure 17.6). Only a very small number of these pigment molecules participate directly in the conversion of light energy to ATP—these are the **reaction center** chlorophylls or bacteriochlorophylls (Figure 17.6). These are surrounded by the more numerous **light-**

harvesting or **antenna** chlorophylls. The antenna pigments function to harvest light and funnel the energy of light to the reaction center. At the low light intensities that often prevail in nature, this arrangement of pigment molecules allows for the capture and utilization of photons that would otherwise be insufficient to drive reaction center photochemistry by themselves.

The ultimate in low-light efficiency is found in the **chlorosome** of green sulfur bacteria and *Chloroflexus* (Figure 17.7, page 554). This structure functions as a giant antenna system, but unlike the antenna of purple bacteria, the bacteriochlorophyll molecules in the chlorosome are not associated with proteins and instead function much like a solid state circuit. Chlorosomes absorb low light intensities and transfer the energy to bacteriochlorophyll *a* in the reaction center located in the cytoplasmic membrane (Figure 17.7). This arrangement is highly

Pigment	R$_1$	R$_2$	R$_3$	R$_4$	R$_5$	R$_6$	R$_7$	In vivo	Extract (methanol)
								Absorption maxima (nm)	
Bacterio-chlorophyll a (purple bacteria)	—C(=O)—CH$_3$	—CH$_3$[b]	—CH$_2$—CH$_3$	—CH$_3$	—C(=O)—O—CH$_3$	P/Gg[a]	—H	805 830–890	771
Bacterio-chlorophyll b (purple bacteria)	—C(=O)—CH$_3$	—CH$_3$[c]	=C(H)—CH$_3$	—CH$_3$	—C(=O)—O—CH$_3$	P	—H	835–850 1020–1040	794
Bacterio-chlorophyll c (green sulfur bacteria)	—C(H)(OH)—CH$_3$	—CH$_3$	—C$_2$H$_5$ —C$_3$H$_7$[d] —C$_4$H$_9$	—C$_2$H$_5$ —CH$_3$	—H	F	—CH$_3$	745–755	660–669
Bacterio-chlorophyll c$_s$ (green nonsulfur bacteria)	—C(H)(OH)—CH$_3$	—CH$_3$	—C$_2$H$_5$	—CH$_3$	—H	S	—CH$_3$	740	667
Bacterio-chlorophyll d (green sulfur bacteria)	—C(H)(OH)—CH$_3$	—CH$_3$	—C$_2$H$_5$ —C$_3$H$_7$ —C$_4$H$_9$	—C$_2$H$_5$ —CH$_3$	—H	F	—H	705–740	654
Bacterio-chlorophyll e (green sulfur bacteria)	—C(H)(OH)—CH$_3$	—C(=O)—H	—C$_2$H$_5$ —C$_3$H$_7$ —C$_4$H$_9$	—C$_2$H$_5$	—H	F	—CH$_3$	719–726	646
Bacterio-chlorophyll g (heliobacteria)	—C(H)=CH$_2$	—CH$_3$[b]	—C$_2$H$_5$	—CH$_3$	—C(=O)—O—CH$_3$	F	—H	670, 788	765

[a]P, Phytyl ester (C$_{20}$H$_{39}$O—); F, farnesyl ester (C$_{15}$H$_{25}$O—); Gg, geranylgeraniol ester (C$_{10}$H$_{17}$O—); S, stearyl alcohol (C$_{18}$H$_{37}$O—).

[b]No double bond between C$_3$ and C$_4$; additional H atoms are in positions C$_3$ and C$_4$.

[c]No double bond between C$_3$ and C$_4$; an additional H atom is in position C$_3$.

[d]Bacteriochlorophylls c, d, and e consist of isomeric mixtures with the different substituents on R$_3$ as shown.

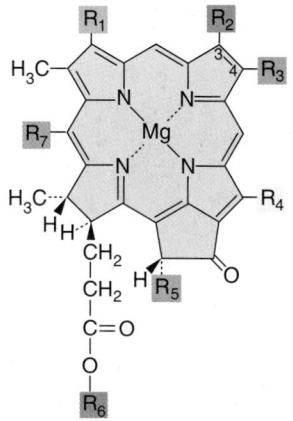

Figure 17.4 Structure of all known bacteriochlorophylls. The different substituents present in the positions R$_1$ to R$_7$ are given in the accompanying table. *In vivo* absorption properties can be determined by suspending intact cells in a viscous liquid such as 60% sucrose (this reduces light scattering and tends to smooth out spectra) and running absorption spectra as shown in Figure 17.3. *In vivo* absorption maxima are the *physiologically relevant* absorption peaks. Spectra of cell extracts are of pigments stripped from their proteins and dissolved in an organic solvent, such as methanol. Spectra of cell extracts reveal the absorption properties of the pure bacteriochlorophyll before it gets assembled into photocomplexes within membranes (see Figures 17.13 and 17.15).

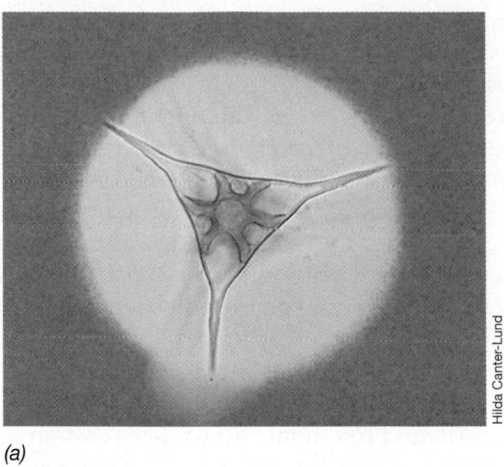

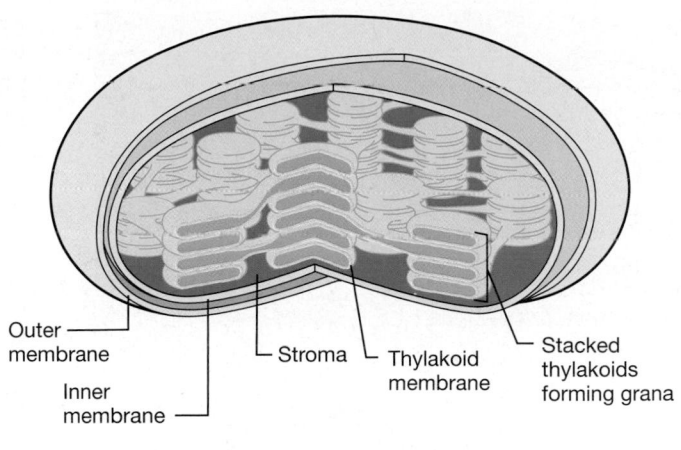

Outer membrane

Inner membrane

Stroma

Thylakoid membrane

Stacked thylakoids forming grana

(a)

(b)

Hilda Canter-Lund

Figure 17.5 The chloroplast. (a) Photomicrograph of an algal cell showing chloroplasts. (b) Details of chloroplast structure, showing how the convolutions of the thylakoid membranes define an inner space called the stroma and form membrane stacks called grana.

efficient for absorbing light at low intensities, and, indeed, it has been shown that green sulfur bacteria can grow at the lowest light intensities of any known phototrophs.

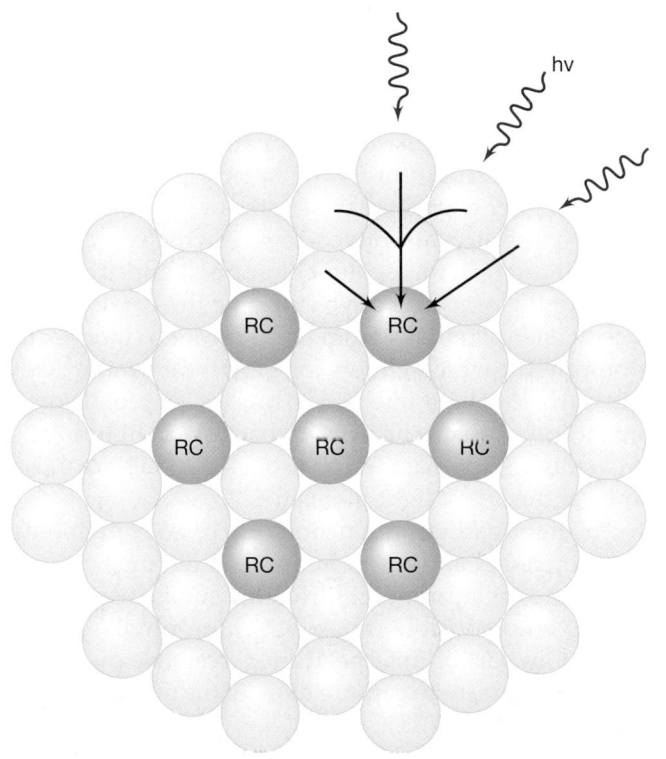

hv

Figure 17.6 Model for the arrangement of light-harvesting chlorophylls/bacteriochlorophylls versus reaction centers within a photosynthetic membrane. Light energy, absorbed by light-harvesting molecules (light green), is transferred to the reaction centers (dark green, RC) where photosynthetic electron transport reactions begin. All pigment molecules are held in place within the membrane by specific pigment-binding proteins. Compare this figure to Figures 17.13 and 17.15.

✓ **17.2 Concept Check**

The central pigment of photosynthesis is chlorophyll (or bacteriochlorophyll). Chlorophylls are located in photosynthetic membranes where the light reactions of photosynthesis are carried out. Antenna chlorophyll molecules function to harvest light energy and transfer it on to reaction center chlorophylls.

✓ Considering their function, why is it necessary for chlorophyll pigments to be located in membranes?

✓ What is the difference between the numbers of *antenna* and *reaction center* chlorophyll molecules in a photosynthetic membrane, and why?

✓ What pigments are found within the *chlorosome*?

17.3 Carotenoids and Phycobilins

Although chlorophyll or bacteriochlorophyll is obligatory for photosynthesis, phototrophic organisms have various accessory pigments involved in the capture and processing of light energy. These include the **carotenoids** and the **phycobilins**. These pigments play primarily a photoprotective role (carotenoids) or function in light-harvesting (phycobilins). We consider each of these groups of pigments now.

Carotenoids

The most widespread accessory pigments are the **carotenoids**, which are always found in phototrophic organisms. Carotenoids are hydrophobic pigments firmly embedded in the membrane; the structure of a typical carotenoid is shown in Figure 17.8. Carotenoids have long hydrocarbon chains with alternating C—C and C=C bonds, an arrangement called a *conjugated* double-bond system. As a rule, carotenoids are yellow, red, brown, or green (∞ Figure 12.2) and absorb light in

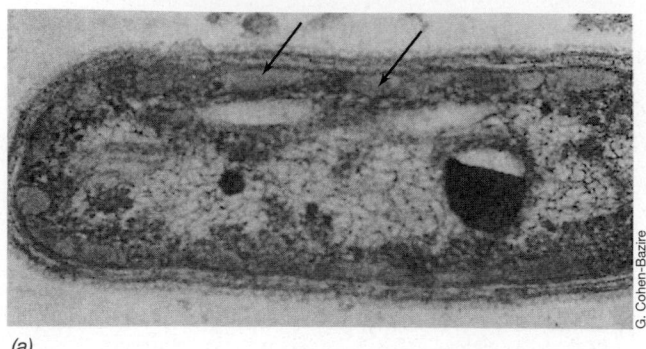

(a)

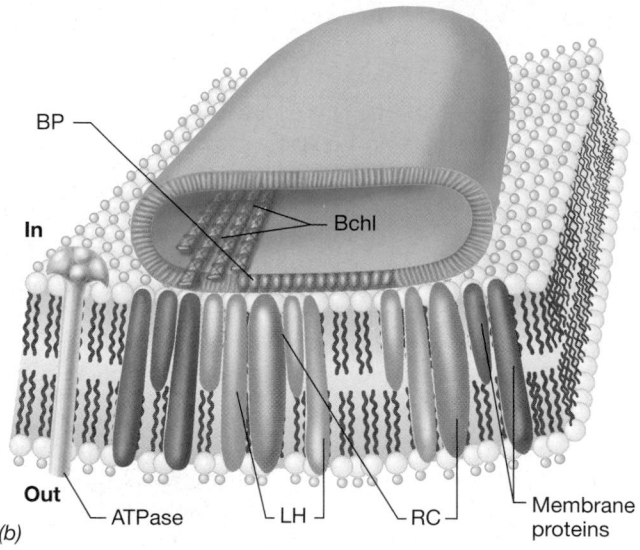

BP

In

Bchl

Out

ATPase — LH — RC — Membrane proteins

(b)

Figure 17.7 The chlorosome of green sulfur and green nonsulfur bacteria. (a) Electron micrograph of a cell of the green sulfur bacterium *Pelodictyon clathratiforme*. Note the chlorosomes (arrows). (b) Model of chlorosome structure. The chlorosome (green) lies appressed to the inside surface of the cytoplasmic membrane. Antenna bacteriochlorophyll (Bchl) molecules (Bchls *c*, *d*, or *e*) are arranged in tube-like arrays inside the chlorosome, and energy is transferred from these bacteriochlorophylls through light-harvesting Bchl *a* molecules (LH) to reaction center (RC) Bchl *a* in the cytoplasmic membrane (blue). Base plate (BP) proteins function as connectors between the chlorosome and the cytoplasmic membrane.

the blue region of the spectrum (see Figure 17.3). The types and structures of carotenoids of various phototrophs have been well studied, and the major carotenoids of anoxygenic phototrophs are shown in Figure 17.9. These pigments are responsible for the brilliant colors of red, purple, pink, green, yellow, or brown that are observed in different species of anoxygenic phototrophs (Figures 12.2 and 12.5).

Carotenoids are closely associated with chlorophyll or bacteriochlorophyll in the photosynthetic membrane but do not function directly in photophosphorylation reactions. They can, however, transfer energy to the reaction center, and this transferred energy may be used in pho-

tophosphorylation in the same way as light energy captured directly by chlorophyll. Carotenoids also function as photoprotective agents. Bright light can often be harmful to cells in that it causes photooxidation reactions that can lead to the production of toxic oxygen species such as singlet oxygen ($^1O_2$) (Section 6.13) and destruction of the photosynthetic apparatus itself. Carotenoids quench toxic oxygen species and absorb much of this harmful light. Because phototrophic organisms must by their very nature live in the light, the photoprotective role of carotenoids is thus an obvious advantage.

Phycobilins and Phycobilisomes

Cyanobacteria and red algal chloroplasts contain **phycobiliproteins**, which are the main light-harvesting pigments of these organisms. Phycobiliproteins are red or blue and consist of open-chain tetrapyrroles coupled to proteins (Figure 17.10a). The red pigment, called *phycoerythrin*, absorbs light most strongly at wavelengths around 550 nm, whereas the blue pigment, *phycocyanin* (Figure 17.10a), absorbs most strongly at 620 nm (Figure 17.11). A third pigment, called *allophycocyanin* absorbs at about 650 nm.

Phycobiliproteins occur as high-molecular-weight aggregates, called **phycobilisomes** attached to the photosynthetic membranes (Figure 17.10b). Phycobilisomes are constructed in such a way that the allophycocyanin molecules make physical contact with the photosynthetic membrane and are surrounded by molecules of phycocyanin and phycoerythrin. The latter pigments absorb shorter (higher energy) wavelengths of light and transfer the energy to allophycocyanin, which is closely linked to the reaction center chlorophyll and transfers energy to this site. The phycobilisome thus yields very efficient energy transfer from the biliprotein complex to chlorophyll *a*, which allows for growth of cyanobacteria at fairly low light intensities. Indeed, phycobilisome content *increases* in cells of cyanobacteria as light intensity *decreases*, such that phycobilisome-rich cells are those grown at the *lowest* light intensities.

The light-gathering function of accessory pigments like carotenoids and phycobilins is an obvious advantage for the organism. Light from the sun is distributed over the whole visible range, yet chlorophylls absorb well in only part of this spectrum. Accessory pigments allow the organism to capture more of the available light (Figures 17.3 and 17.11).

H₃C CH₃   CH₃   CH₃   H₃C

CH₃   CH₃   CH₃ H₃C CH₃

Figure 17.8 Structure of β-carotene, a typical carotenoid. The conjugated double-bond system is highlighted in orange.

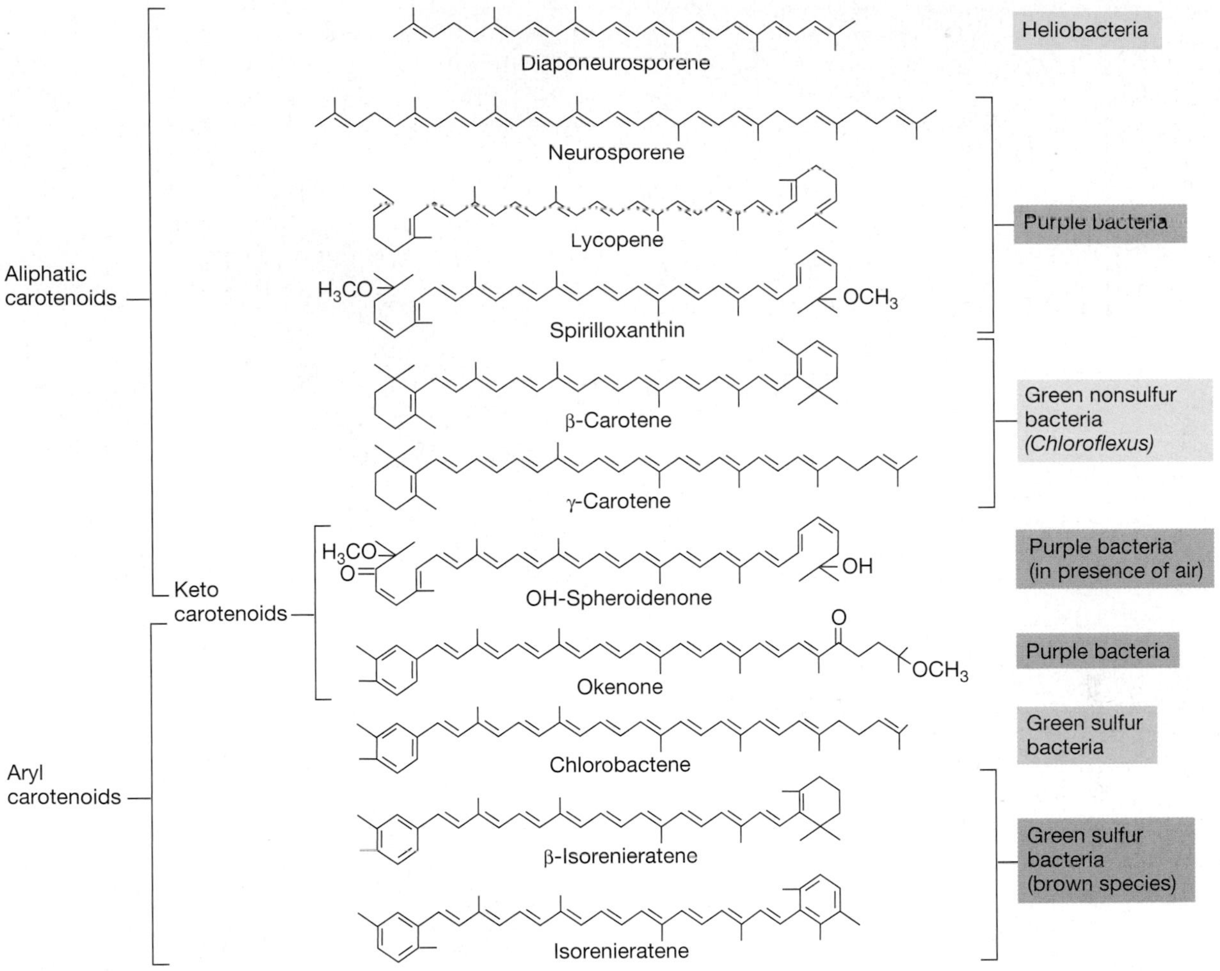

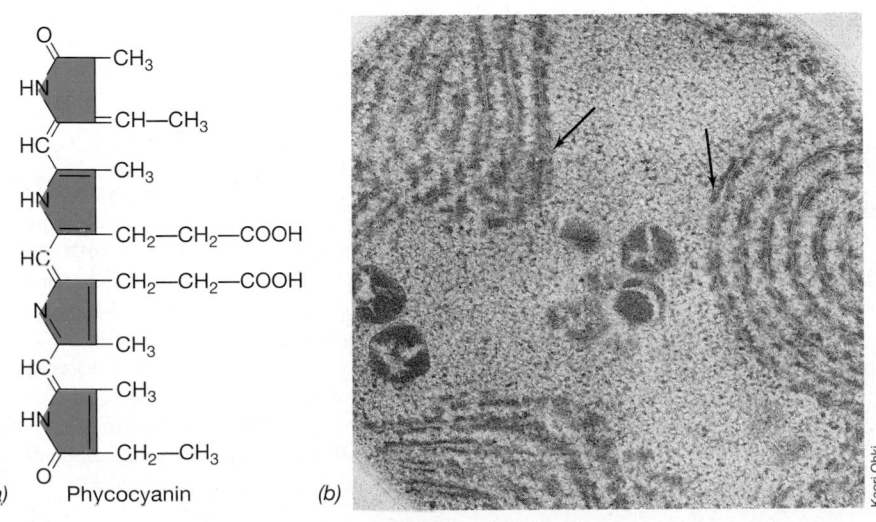

Figure 17.9 Structures of some common carotenoids found in anoxygenic phototrophs. Compare the structure of β-carotene shown in Figure 17.8 with how it is drawn here. For simplicity, in the structures shown here, methyl (CH₃) groups are designated by their bonds only. Aryl carotenoids are distinguished from alicyclic carotenoids in that aryl carotenoids contain an aromatic ring on one end. Purple bacteria are discussed in Section 12.2, heliobacteria in Section 12.20, green sulfur bacteria in Section 12.32, and green nonsulfur bacteria in Section 12.35.

(a) Phycocyanin

(b)

Kaori Ohki

Figure 17.10 Phycobilins and phycobilisomes. (a) A typical phycobilin. This compound is an open-chain tetrapyrrole derived biosynthetically from a closed porphyrin ring by loss of one carbon atom as carbon monoxide. The structure shown is the prosthetic group of phycocyanin, a proteinaceous pigment found in cyanobacteria (◯◯◯ Section 12.25) and red algae (◯◯◯ Section 14.11). (b) Electron micrograph of a thin section of the cyanobacterium *Synechocystis*. Note the darkly staining ball-like phycobilisomes (arrows) attached to the lamellar membranes.

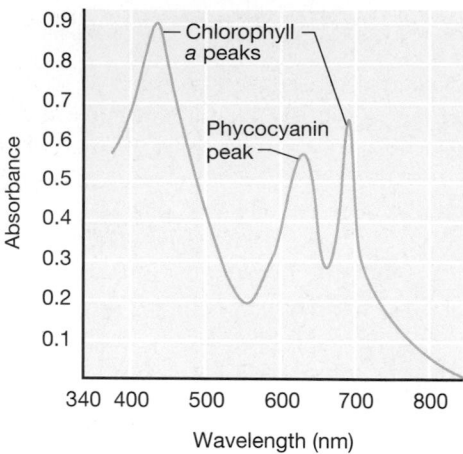

Figure 17.11 The absorption spectrum of a cyanobacterium that has a phycobiliprotein (phycocyanin) as an accessory pigment. Note how the presence of phycocyanin broadens the wavelengths of usable light energy (between 600 and 700 nm). Compare with Figure 17.3.

✓ 17.3 Concept Check

Accessory pigments such as carotenoids and phycobilins absorb light and transfer the energy to reaction center chlorophyll, thus broadening the wavelengths of light usable in photosynthesis. Carotenoids also play an important photoprotective role in preventing photooxidative damage to cells.

- ✓ In what organisms are phycobiliproteins found?
- ✓ How does the structure of a phycobilin compare with that of a chlorophyll? With a carotenoid?
- ✓ Phycocyanin is blue-green in color; what wavelengths of light is it absorbing (⟨⟨⟨∙ Figure 10.6)?

17.4 Anoxygenic Photosynthesis

The process of light-mediated ATP synthesis in all phototrophic organisms involves electron transport through a series of electron carriers. These electron carriers are arranged in the photosynthetic membrane in series from those with electronegative to those with more electropositive reduction potentials. We now consider the structure of the photosynthetic apparatus in anoxygenic phototrophs and the details of photosynthetic electron flow in purple bacteria, where much is known concerning the molecular events of photosynthesis.

Structure of the Purple Bacterial Photosynthetic Apparatus

The photosynthetic apparatus of purple phototrophic bacteria is contained in intracytoplasmic membrane systems of various morphologies. Membrane vesicles (chromatophores) or lamellae (Figure 17.12) are commonly observed membrane types. The photosynthetic apparatus consists of four membrane-integrated pigment-

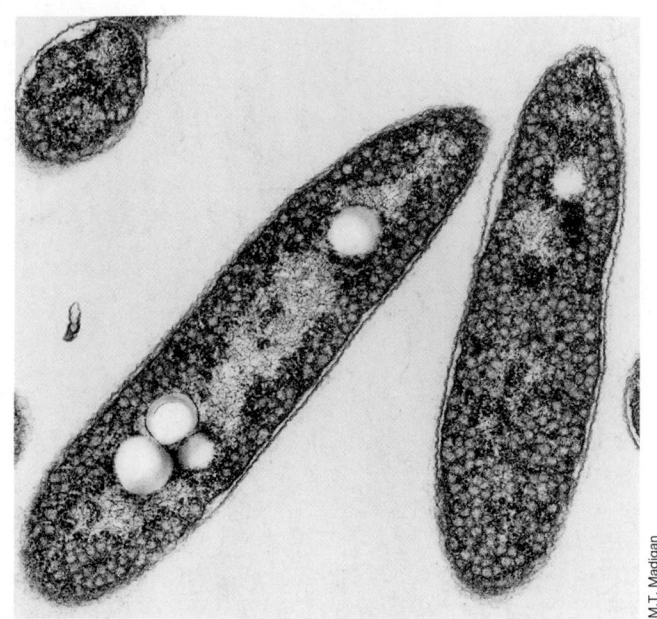

(a)

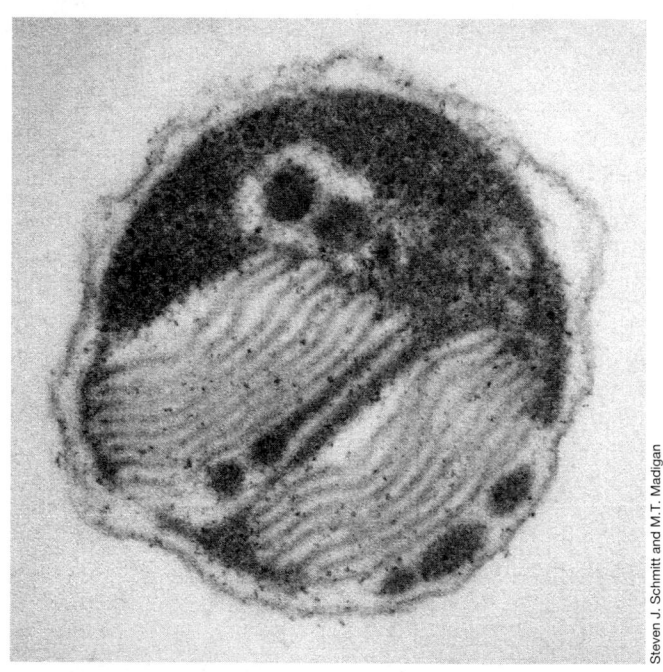

(b)

Figure 17.12 Membranes in anoxygenic phototrophs. (a) Chromatophores. Section through a cell of the phototrophic purple bacterium *Rhodobacter capsulatus* containing an abundance of vesicular photosynthetic membranes. The vesicles arise by invagination of the cytoplasmic membrane. The clear areas are regions in the cell in which the reserve polymer, poly-β-hydroxybutyrate (⟨⟨⟨∙ Section 4.13), was stored. A cell is about 1 μm wide. (b) Lamellar membranes in a halophilic purple bacterium. A cell is about 1.5 μm wide. These membranes also arise from invagination of the cytoplasmic membrane, but instead of forming vesicles, they become arranged as membrane stacks, similar to the thylakoids of cyanobacteria (Figure 17.5).

protein complexes plus an ATPase complex that drives ATP synthesis at the expense of a proton motive force (Section 5.12). Three of the four complexes specific to photosynthesis are the *reaction center, light-harvesting I*, and *light-harvesting II* components. The fourth complex of the photosynthetic apparatus, the *cytochrome bc_1 complex*, is common to both respiratory and photosynthetic electron flow. We discussed the structure and function of the cytochrome bc_1 complex in Section 5.11.

The purple bacterial photosynthetic reaction center has been crystallized and its structure determined to atomic resolution by X-ray diffraction (Figure 17.13*a*). Reaction centers of purple bacteria contain three polypeptides, designated the L, M, and H subunits. These proteins are firmly embedded in the photosynthetic membrane and traverse the membrane several times (Figure 17.13*b*). The L, M, and H polypeptides bind the reaction center photochemical complex, which consists of two molecules of bacteriochlorophyll *a* called the *special pair*, two additional bacteriochlorophyll *a* molecules whose function is unknown, two molecules of *bacteriopheophytin* (bacteriochlorophyll *a* minus its magnesium atom), two molecules of quinone, and two molecules of a carotenoid pigment. All components of the reaction center are integrated in such a way that they can interact in very fast electron transfer reactions that, as we will see, ultimately result in ATP production.

Photosynthetic Electron Flow in Purple Bacteria

Recall that the photosynthetic reaction center is surrounded by light-harvesting antenna bacteriochlorophyll *a* molecules that function to funnel light energy to the reaction center (see Figure 17.6). Light energy is transferred from the antenna to the reaction center in packets called *excitons*, mobile electronic states that migrate through the antenna pigments to the reaction center at high efficiency. Photosynthesis begins when exciton energy strikes the special pair of bacteriochlorophyll *a* molecules (Figure 17.13*a*). The absorption of energy excites the special pair, converting it to a strong electron donor with a very low reduction potential (E_0') (Section 5.4). Once this strong donor has been produced, the remaining steps in photosynthetic electron flow function to conserve the energy released when electrons are transported through a membrane from carriers of low E_0' to those of high E_0' (Figure 17.14).

Before excitation, the bacterial reaction center, which is referred to as *P870*, has an E_0' of about +0.5 V; after excitation it has a potential of about −1.0 V (Figure 17.14). The excited electron within P870 proceeds to reduce a molecule of bacteriopheophytin within the reaction center (Figures 17.13*b* and 17.14). This transition takes place incredibly fast, taking about three-trillionths of a second (3×10^{-12} s) to occur. Once reduced, bacteriopheophytin

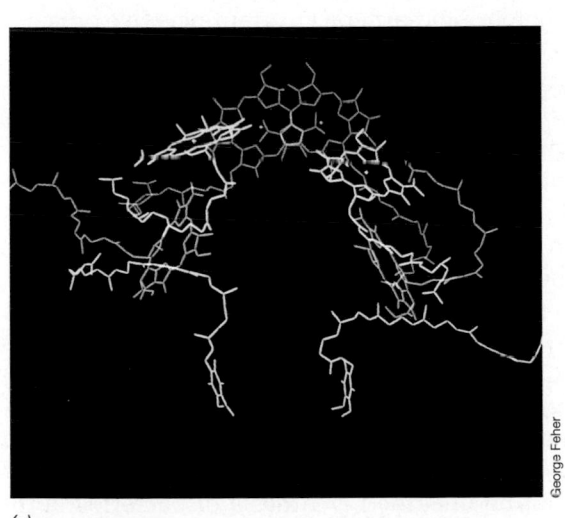

(a)

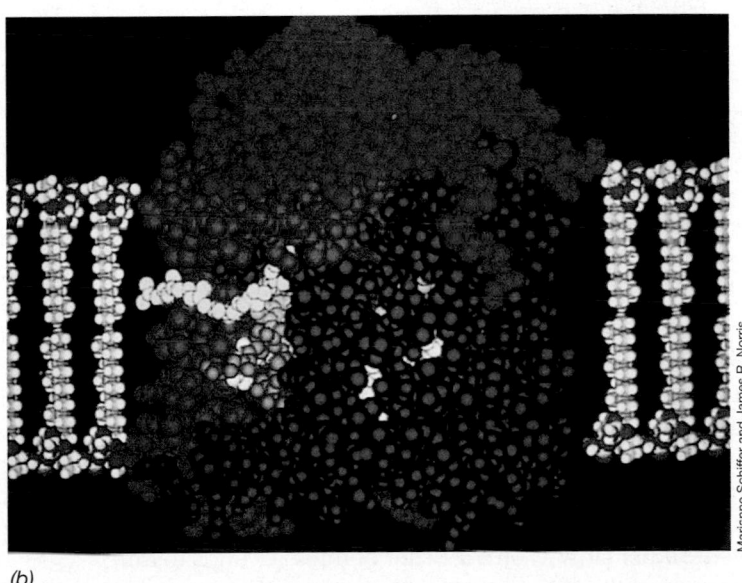

(b)

Figure 17.13 Structure of the reaction center of purple phototrophic bacteria. (a) Arrangement of components in the reaction center. The "special pair" of bacteriochlorophyll molecules are overlapping and shown in red, and molecules of quinone are in dark yellow and point downward in the figure. The accessory bacteriochlorophylls are in lighter yellow near the special pair, and the bacteriopheophytin molecules are shown in blue. (b) Molecular model of the protein structure of the reaction center. The pigments discussed in (a) are bound to membranes by three reaction center proteins called protein H (blue), protein M (red), and protein L (green). The reaction center pigment-protein complex is integrated into the lipid bilayer.

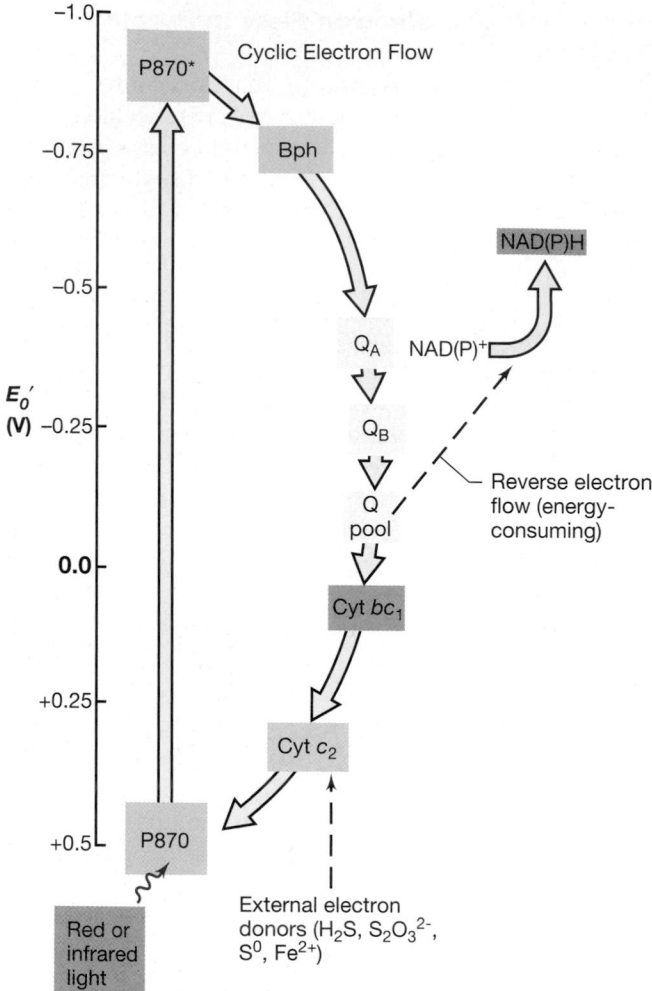

Figure 17.14 General scheme of electron flow in anoxygenic photosynthesis in a purple bacterium. Only a single light reaction occurs. Note how light energy converts a weak electron donor, P870, into a very strong electron donor, P870*, and that following this event, the remaining steps in photosynthetic electron flow are much the same as that of respiratory electron flow. RC, reaction center; Bchl, bacteriochlorophyll; Bph, bacteriopheophytin; Q_A, Q_B, intermediate quinones; Q pool, quinone pool in membrane; Cyt, cytochrome.

a reduces several intermediate quinone molecules, with the electron eventually reducing a quinone in the "quinone pool" within the membrane. This transition is also very fast, taking less than one-billionth of a second (Figures 17.14 and 17.15). Relative to what has happened in the reaction center, further electron transport reactions occur rather slowly, on the order of microseconds to milliseconds. From the quinone, electrons are transported in the membrane through a series of iron-sulfur proteins and cytochromes (Figures 17.14 and 17.15), eventually returning to the reaction center. Key electron transport proteins include cytochrome bc_1 and cytochrome c_2 (Figure 17.14). Cytochrome c_2 is a periplasmic cytochrome and serves as an electron shuttle between the membrane-

bound bc_1 complex and the reaction center (∞ Section 5.11 and Figures 5.20, 17.14, and 17.15).

Photophosphorylation

Synthesis of ATP during photosynthetic electron flow occurs as a result of the formation of a *proton motive force* generated by proton extrusion during electron transport and the activity of ATPases in coupling the dissipation of the proton motive force to ATP formation (∞ Section 5.12). The reaction series is completed when cytochrome c_2 donates an electron to the special pair bacteriochlorophylls (Figure 17.14), returning these molecules to their original ground state potential ($E_0' = +0.5$ V). The reaction center is then capable of absorbing new energy and repeating the process. This method of making ATP is called **cyclic photophosphorylation** because electrons are repeatedly moved around a closed circle. Cyclic photophosphorylation resembles respiration in that electron flow through the membrane establishes a proton motive force. However, unlike respiration, in cyclic photophosphorylation *there is no net input or consumption of electrons*; electrons simply travel a closed route.

The spatial relationships of the electron transport components in the bacterial photosynthetic membrane are illustrated in Figure 17.15. Note that as in respiratory electron flow (∞ Section 5.11), the cytochrome bc_1 complex interacts with the quinone pool during photosynthetic electron flow (∞ Figure 5.20) as a major means of establishing the proton motive force used to drive ATP synthesis (Figure 17.15).

Genetics of Bacterial Photosynthesis

Purple phototrophic bacteria are gram-negative prokaryotes (∞ Section 12.2), and certain species are readily amenable to genetic manipulation. Species of the genus *Rhodobacter*, especially *R. capsulatus* and *R. sphaeroides*, have been the main subjects of genetic research in bacterial photosynthesis. In *R. capsulatus*, most genes involved in photosynthesis are clustered in several operons that span a 50-kb region of the chromosome that has been called the **photosynthetic gene cluster** (Figure 17.16). Genes in the photosynthetic gene cluster encode proteins involved in (1) bacteriochlorophyll biosynthesis (*bch* genes), (2) carotenoid biosynthesis (*crt* genes), and (3) polypeptides that bind pigment molecules in the reaction center and light-harvesting complexes (*puf* and *puh* genes) (Figure 17.16).

As can be imagined, synthesis of bacteriochlorophyll, carotenoids, and pigment binding proteins in phototrophic bacteria must be a highly coordinated process. When new photosynthetic complexes are synthesized, the correct proportions of each of the components of the complex must be available within the cell for final assembly. Biochemical and genetic analyses of photosynthesis in *Rhodobacter capsulatus* have shown that coordinate ex-

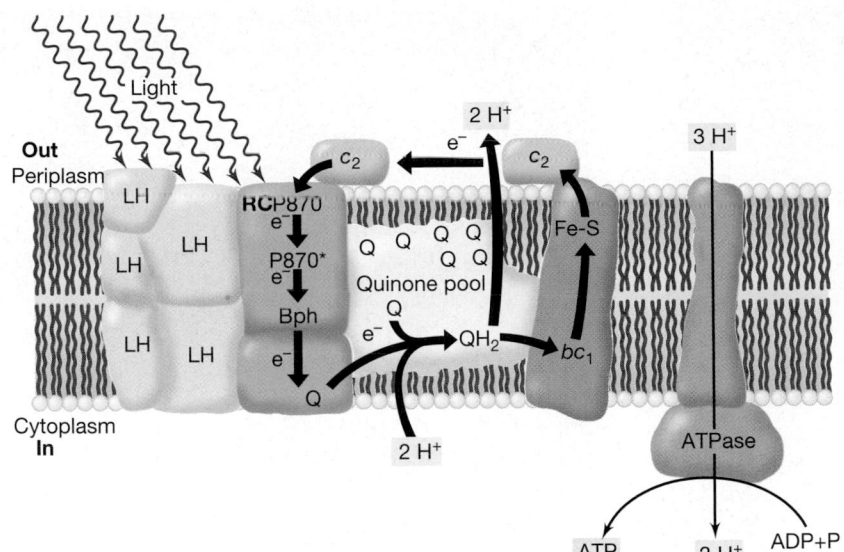

Figure 17.15 Arrangement of protein complexes in the photosynthetic membrane of a purple phototrophic bacterium. The light-generated proton gradient is used in the synthesis of ATP by the ATP synthase (ATPase). LH, Light-harvesting bacteriochlorophyll complexes; RC, reaction center; Bchl, bacteriochlorophyll; Bph, bacteriopheophytin; Q, quinone; FeS, iron-sulfur protein; bc_1, cytochrome bc_1 complex; c_2, cytochrome c_2. For description of the functioning of the cytochrome bc_1 complex and the Q cycle, the details of which are not shown here, see Figure 5.20.

pression of photosynthetic components does indeed occur because the operons are arranged to form **superoperons**. Instead of terminating at the end of one operon, transcripts of pigment biosynthesis operons (*bch* and *crt*) (Figure 17.16) extend through the promoters and structural genes encoding the polypeptides of the photosynthetic complexes, yielding large transcripts encoding many proteins. Photosynthesis superoperons thus allow for transcription of many functionally related genes whose products interact and form the photosynthetic complexes that eventually integrate into the membrane. The master regulatory signal governing transcription of the photosynthetic gene cluster in these organisms is O_2. Molecular oxygen represses pigment synthesis such that photosynthesis in anoxygenic phototrophs occurs only under *anoxic* conditions.

Genetic analysis of photosynthesis in purple bacteria has been greatly assisted by the bioenergetic diversity of *Rhodobacter* species; in addition to their capacity for photosynthesis, these organisms can grow in darkness by respiration in the presence or absence of oxygen. Thus, mutants unable to photosynthesize are easily obtained and have been used in genetic exchange experi-

ments (Chapter 10) to characterize the number, organization, and expression of photosynthesis genes.

Autotrophy in Purple Bacteria: Electron Donors and Reverse Electron Flow

For a purple bacterium to grow autotrophically, the formation of ATP is not enough. Reducing power (NADH or NADPH) must also be made so that CO_2 can be reduced to the level of cell material. As previously mentioned, for purple sulfur bacteria this is usually H_2S, although S^0, $S_2O_3^{2-}$ and even Fe^{2+} can be used by various species. When H_2S is the electron donor in purple sulfur bacteria, globules of S^0 are stored inside the cells (Figure 17.17a). How do these substances reduce NAD^+ to NADH?

Reduced subtances like **hydrogen sulfide** (H_2S) or **thiosulfate** ($S_2O_3^{2-}$) are oxidized by cytochromes of the c type and electrons from them eventually end up in the "quinone pool" of the photosynthetic membrane (see Figure 17.14). However, the E_0' (about 0 volts) of quinone is insufficiently negative to reduce NAD^+ (−0.32 volts) directly, such that electrons from the quinone pool must be forced *backward*, against the thermodynamic gradient, to reduce NAD^+ to NADH (Figure 17.14). This energy-requiring process is called *reversed electron flow* and is dri-

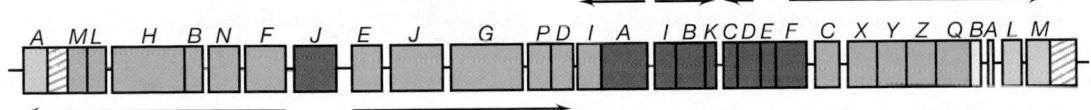

Figure 17.16 Map of the photosynthetic gene cluster of the purple phototrophic bacterium, *Rhodobacter capsulatus*. Genes are arranged in superoperons where transcripts of pigment biosynthesis operons extend through to include transcription of polypeptides of the photosynthetic complexes. The *bch* genes, which encode bacteriochlorophyll synthesis proteins, are shown in green, while *crt* genes, which encode proteins that synthesize carotenoids, are shown in red. Genes encoding reaction center polypeptides (*puh* and *puf* genes) are shown in blue, and genes encoding light-harvesting I polypeptides (B870 complex) (*puf* genes) are shown in yellow. Genes shown by diagonal lines are of unknown function. Not all genes have been given letter designations. Arrows indicate direction of transcription.

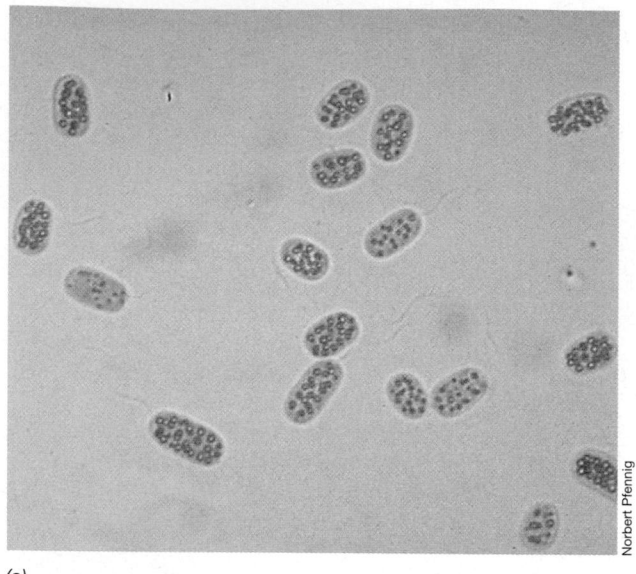

(a)

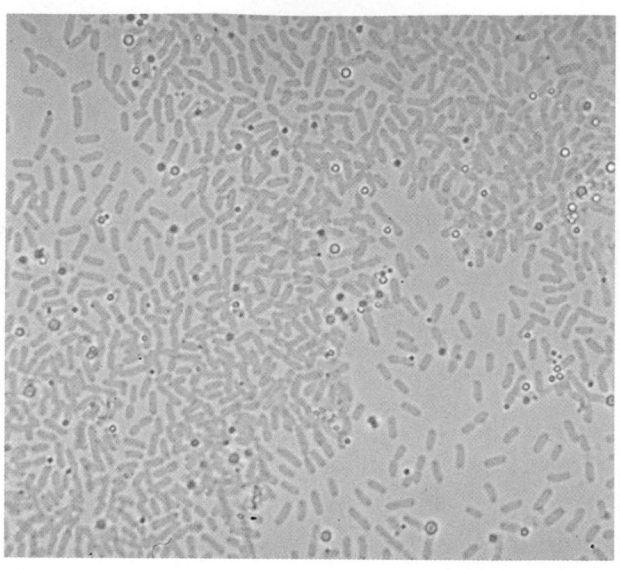

(b)

Figure 17.17 Photomicrographs of phototrophic bacteria taken by bright-field microscopy. (a) Purple bacterium: *Chromatium okenii*. Notice the sulfur granules deposited *inside* the cell. (b) Green bacterium: *Chlorobium limicola*. The refractile bodies are sulfur granules deposited *outside* the cell. In both cases the sulfur granules arise from the oxidation of H_2S to obtain reducing power.

ven by the energy inherent in the proton motive force. Reverse electron flow is also the mechanism by which chemolithotrophs make their reducing power, oftentimes from electron donors of extremely positive E_0' (see Sections 17.8–17.12).

Comparative Photosynthetic Electron Flow and Reducing Power Synthesis in Other Anoxygenic Phototrophs

Our discussion of photosynthetic electron flow has thus far focused on purple bacteria. Although similar membrane components drive photophosphorylation in other anoxygenic phototrophs, there are differences in certain photochemical reactions that impact on reducing power biosynthesis. Figure 17.18 compares the light reactions of purple and green bacteria and the heliobacteria. Note that in the latter two groups the excited state of the reaction center bacteriochlorophylls resides at a significantly more negative E_0', and that actual *chlorophyll a* (green bacteria) or a structurally modified form of chlorophyll *a* called *hydroxychlorophyll a* (heliobacteria) are present in the reaction center. This indicates that unlike purple bacteria, where the first stable acceptor molecule (quinone) has an E_0' of about 0 volts, the acceptors of green bacteria and heliobacteria (FeS proteins) have a much more electronegative E_0' than NADH. In green bacteria, ferredoxin (reduced by the FeS protein, Figure 17.18) serves directly as electron donor for CO_2 fixation reactions in the reverse citric acid cycle (see Figure 17.24*a*). Thus, like oxygenic phototrophs (to be discussed next), in green and heliobacteria both ATP *and* reducing power are direct

products of the light reactions. When sulfide is the electron donor for reducing power synthesis in green bacteria, globules of S^0 are produced as in purple bacteria, but the globules remain *outside*, rather than *inside* the cells (Figure 17.17*b*).

✓ 17.4 Concept Check

A series of electron transport reactions occur in the photosynthetic reaction center of anoxygenic phototrophs, resulting in the formation of a proton motive force and the synthesis of ATP. Reducing power for CO_2 fixation comes from reductants present in the environment and requires reverse electron transport in purple phototrophs.

✓ How does photophosphorylation compare with electron transport phosphorylation in respiration?

✓ What is the value of arranging photosynthesis genes into *superoperons*?

✓ What is *reverse electron flow* and why is it necessary?

17.5 Oxygenic Photosynthesis

In contrast to anoxygenic phototrophs, electron flow in oxygenic phototrophs involves two distinct, but interconnected, photochemical reactions. Oxygenic phototrophs use light to generate both ATP *and* NADPH, the electrons for the latter arising from the splitting of *water* into oxygen and electrons (see Figure 17.2). The two systems of light reactions are called *photosystem I* and *photosystem II*, each photosystem having a spectrally distinct

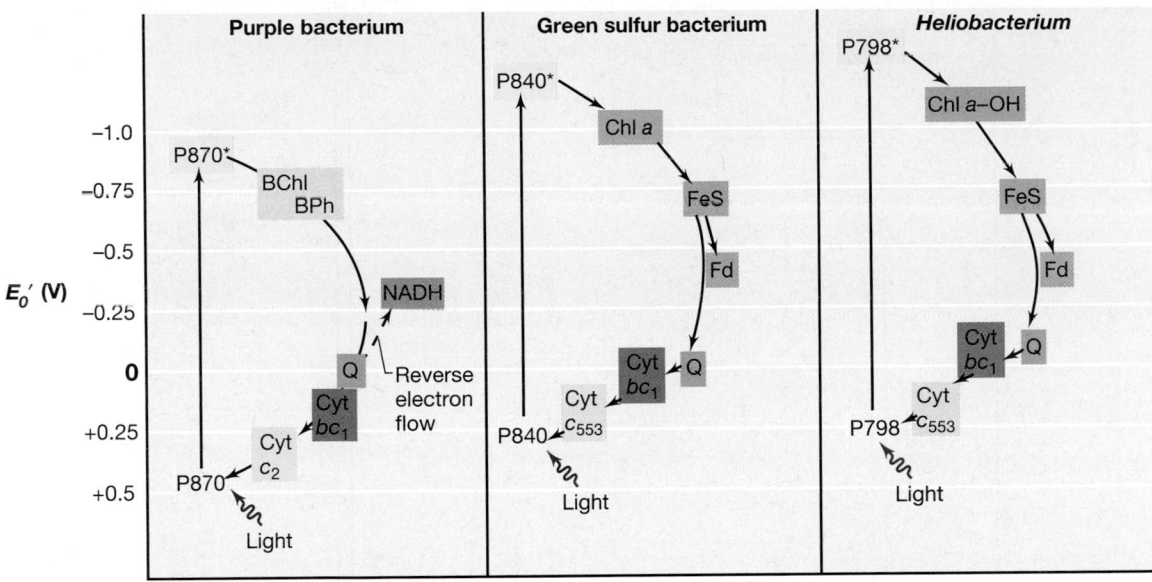

Figure 17.18 A comparison of electron flow in purple bacteria, green sulfur bacteria, and heliobacteria. Note how reverse electron flow in purple bacteria is necessitated by the fact that the primary acceptor (quinone, Q) is more positive in potential than the NAD^+/NADH couple. In green and heliobacteria ferredoxin (Fd), whose E_0' is actually more negative than that of NADH, is produced by light-driven reactions. In green bacteria ferredoxin drives the carboxylation reactions of autotrophy (see Figure 17.24a). Bchl, Bacteriochlorophyll; BPh, bacteriopheophytin. P870 and P840 are reaction centers of purple and green bacteria, respectively, and consist of Bchl a. The reaction center of heliobacteria (P798) contains Bchl g. The reaction center of Chloroflexus is similar to that of purple bacteria. Note the presence of forms of chlorophyll a in the reaction centers of green bacteria and heliobacteria.

form of reaction center chlorophyll *a*. Photosystem I chlorophyll, called P700, absorbs light at long wavelengths (far red light), whereas photosystem II chlorophyll, called P680, absorbs at shorter wavelengths (near red light). Like anoxygenic photosynthesis, oxygenic photochemical reactions occur in membranes. In eukaryotic cells, these membranes are found in the *chloroplast* (Figure 17.5), whereas in cyanobacteria, photosynthetic membranes are arranged in stacks within the cytoplasm (Figure 17.6). In both groups of phototrophs the membranes are arranged in a similar way and the two forms of chlorophyll *a* are attached to specific proteins in the membrane and interact as shown in Figure 17.19.

Electron Flow in Oxygenic Photosynthesis

The path of electron flow in oxygenic phototrophs roughly resembles the letter Z turned on its side, and scientists studying oxygenic photosynthesis have come to refer to the electron flow of oxygenic phototrophs as the "Z" scheme. We should first note that the reduction potential of the P680 chlorophyll *a* molecule in photosystem II is very electropositive, slightly more positive than that of the O_2/H_2O couple. This facilitates the first step in oxygenic electron flow, the *splitting* of water into oxygen and hydrogen atoms (Figure 17.19), a thermodynamically unfavorable reaction. An electron from water is donated to the oxidized P680 molecule following the absorption of a quantum of light near 680 nm.

Light energy converts P680 into a moderately strong reductant, capable of reducing an intermediary molecule of E_0' about -0.5 V. The nature of this molecule is uncertain, but it is likely a pheophytin *a* molecule (chlorophyll *a* without the magnesium atom). From here the electron travels through several membrane carriers including quinones, cytochromes, and a copper-containing protein called *plastocyanin*; the latter donates electrons to photosystem I. The electron is accepted by the reaction center chlorophyll of photosystem I, P700, which has previously absorbed light quanta and begun the steps that lead to the reduction of $NADP^+$ (Figure 17.18). These involve electron transfer through several carriers of increasing E_0', terminating with the reduction of $NADP^+$ (Figure 17.19).

ATP Synthesis in Oxygenic Photosynthesis

Besides the net synthesis of reducing power (that is, NADPH), other important events take place while electrons flow in the membrane from one photosystem to another. During transfer of an electron from the acceptor in photosystem II to the reaction center chlorophyll molecule in photosystem I, electron transport occurs in a thermodynamically favorable (negative-to-positive) direction. This generates a proton motive force from which ATP can be produced. This type of ATP generation has been called *noncyclic* photophosphorylation because electrons whose transport results in ATP formation do not cycle back to

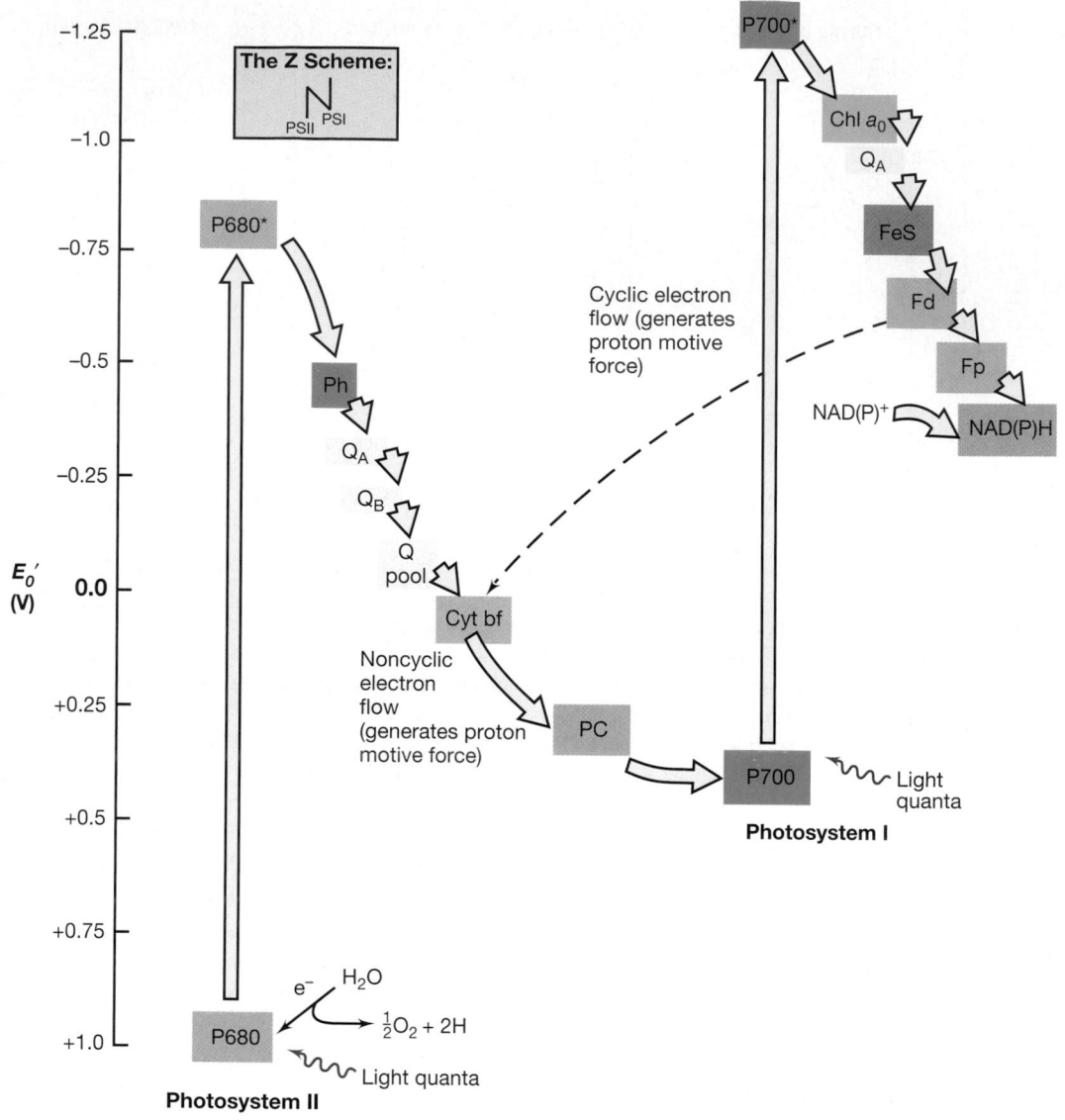

Figure 17.19 Electron flow in oxygenic (green plant) photosynthesis, the "Z" scheme. Two photosystems (PS) are involved, PS I and PS II. Ph, Pheophytin; Q, quinone; Chl, chlorophyll *a*; Cyt, cytochrome; PC, plastocyanin; FeS, nonheme iron-sulfur protein; Fd, ferredoxin; Fp, flavoprotein; P680 and P700 are the reaction center chlorophylls of PS II and PS I, respectively. Compare with Figure 17.14.

reduce the oxidized P680; they are ultimately used in the reduction of $NADP^+$. When sufficient reducing power is present, ATP can also be produced in oxygenic phototrophs by *cyclic* photophosphorylation involving only photosystem I (Figure 17.18). This occurs when electrons travel from ferredoxin to the cytochrome *bf* complex from which electron transport returns the electron to P700. This flow creates a membrane potential and synthesis of additional ATP (see dashed line in Figure 17.19).

Anoxygenic Photosynthesis in Oxygenic Phototrophs

Photosystems I and II normally function together in the oxygenic process. However, under certain conditions some algae and cyanobacteria can carry out cyclic pho-

tophosphorylation using *only* photosystem I, obtaining reducing power for CO_2 reduction from sources other than water. This is, in effect, photosynthesizing *anoxygenically*, as purple and green bacteria do.

A number of cyanobacteria can use H_2S as an electron donor for anoxygenic photosynthesis, while green algae use H_2. When H_2S is used, it is oxidized to elemental sulfur (S^0), and sulfur granules similar to those produced by green sulfur bacteria (see Figure 17.17*b*) are deposited outside the cells; an example of this is shown with the cyanobacterium *Oscillatoria limnetica* in Figure 17.20. This filamentous cyanobacterium lives in sulfide-rich saline ponds where it carries out anoxygenic photosynthesis along with photosynthetic green and purple bacteria and produces sulfur as an oxidation

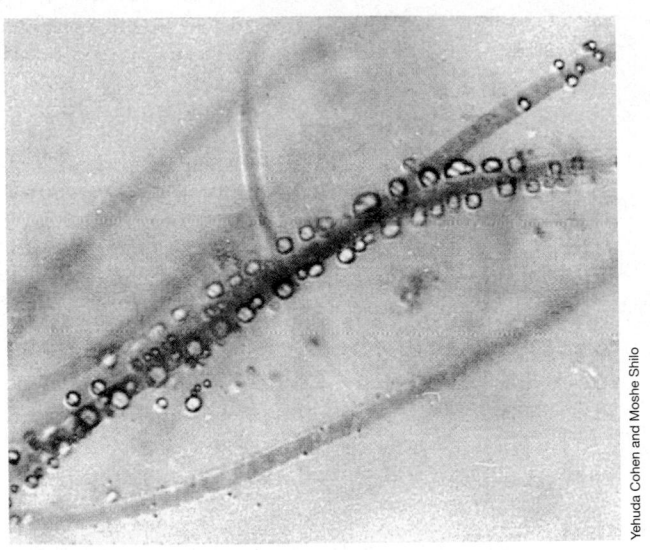

Figure 17.20 Cells of the filamentous cyanobacterium *Oscillatoria limnetica* grown anaerobically on sulfide as photosynthetic electron donor. Note the globules of elemental sulfur, the oxidation product of sulfide, formed *outside* the cells.

product of sulfide (Figure 17.20). In cultures of *O. limnetica*, electron flow from photosystem II is strongly inhibited by H_2S, thus necessitating anoxygenic photosynthesis if the organism is to survive.

From an evolutionary point of view, the existence of cyclic photophosphorylation in both oxygenic and anoxygenic phototrophs indicates their close relationship. Although organisms such as *Oscillatoria limnetica* that carry out oxygenic photosynthesis have acquired photosystem II, and hence the ability to split H_2O, they still retain the ability under certain conditions to use photosystem I alone.

✓ 17.5 Concept Check

In oxygenic photosynthesis, water donates electrons to drive autotrophy and oxygen is produced as a by-product. Two separate light reactions are involved, photosystems I and II. Photosystem I resembles the system in anoxygenic photosynthesis. Photosystem II is responsible for splitting H_2O to yield O_2.

✓ Why is the term *noncyclic* electron flow used in reference to oxygenic photosynthesis?

✓ What is a major difference between the two reaction center chlorophyll molecules in photosystems I and II?

17.6 Autotrophic CO₂ Fixation: The Calvin Cycle

Several biochemical mechanisms are known for the fixation of CO_2 into cell material. This section considers the most widespread of these systems, the **Calvin cycle**, named for its discoverer, Melvin Calvin. The Calvin cycle requires NAD(P)H and ATP and two key enzymes, *ribulose bisphosphate carboxylase* and *phosphoribulokinase*. The remainder of the cycle is driven by a series of enzymes that are present in many organisms, both autotrophs and heterotrophs.

RubisCO and the Formation of PGA

The first step in CO_2 reduction in the Calvin cycle is catalyzed by the enzyme ribulose bisphosphate carboxylase, or *RubisCO* for short. RubisCO is widely distributed, being present in purple bacteria, cyanobacteria, algae, and green plants, most chemolithotrophic *Bacteria*, and even in some *Archaea*, including structurally unique forms in hyperthermophilic *Archaea*. RubisCO catalyzes the formation of two molecules of *3-phosphoglyceric acid* (PGA) from ribulose bisphosphate and CO_2 as shown in Figure 17.21. The PGA is then phosphorylated and reduced to a key intermediate of glycolysis, *glyceraldehyde 3-phosphate* (∞ Section 5.10). From here, glucose can be formed by reversal of the early steps in glycolysis. But thus far only one molecule of CO_2 and one molecule of ribulose bisphosphate have been consumed. How does the cycle create a full glucose molecule, and how do we regenerate the acceptor, ribulose bisphosphate?

Stoichiometry of the Calvin Cycle

We now consider reactions of the Calvin cycle based on the incorporation of 6 molecules of CO_2. For RubisCO to incorporate 6 molecules of CO_2, 6 ribulose bisphosphate molecules are required as acceptor molecules (Figure 17.22). This yields 12 molecules of 3-phosphoglyceric acid (a total of 36 carbon atoms). These 12 molecules serve as carbon skeletons to form 6 *new* molecules of ribulose bisphosphate (a total of 30 carbon atoms), and 1 molecule of hexose for cell biosynthesis. A complex series of rearrangements involving C_3, C_4, C_5, C_6, and C_7 intermediates finally yields the 6 molecules of ribulose 5-phosphate from which the 6 ribulose bisphosphates are generated. The final step in this regeneration is the phosphorylation of ribulose 5-phosphate with ATP by the enzyme *phosphoribulokinase* (Figure 17.21c). Like RubisCO, this enzyme is also unique to the Calvin cycle.

Let us consider now the *overall* stoichiometry for conversion of 6 molecules of CO_2 into 1 molecule of fructose 6-phosphate (Figure 17.22). Twelve molecules each of ATP and NADPH are required for the reduction of 12 molecules of phosphoglyceric acid (PGA) to glyceraldehyde phosphate, and 6 ATP molecules are required for conversion of ribulose phosphate to ribulose bisphosphate. Thus, *12 NADPH and 18 ATP are required to synthesize 1 hexose molecule from 6 molecules of CO_2 by the Calvin cycle*. The hexose molecules can be converted to *storage polymers* such as glycogen, starch, or poly-β-hydroxyalkanoates (∞ Section 4.13) during periods when ATP and NADPH are abundant and then can be used later to build new cell material.

Carboxysomes

Several autotrophic prokaryotes that use the Calvin cycle for CO_2 fixation produce polyhedral cell inclusions

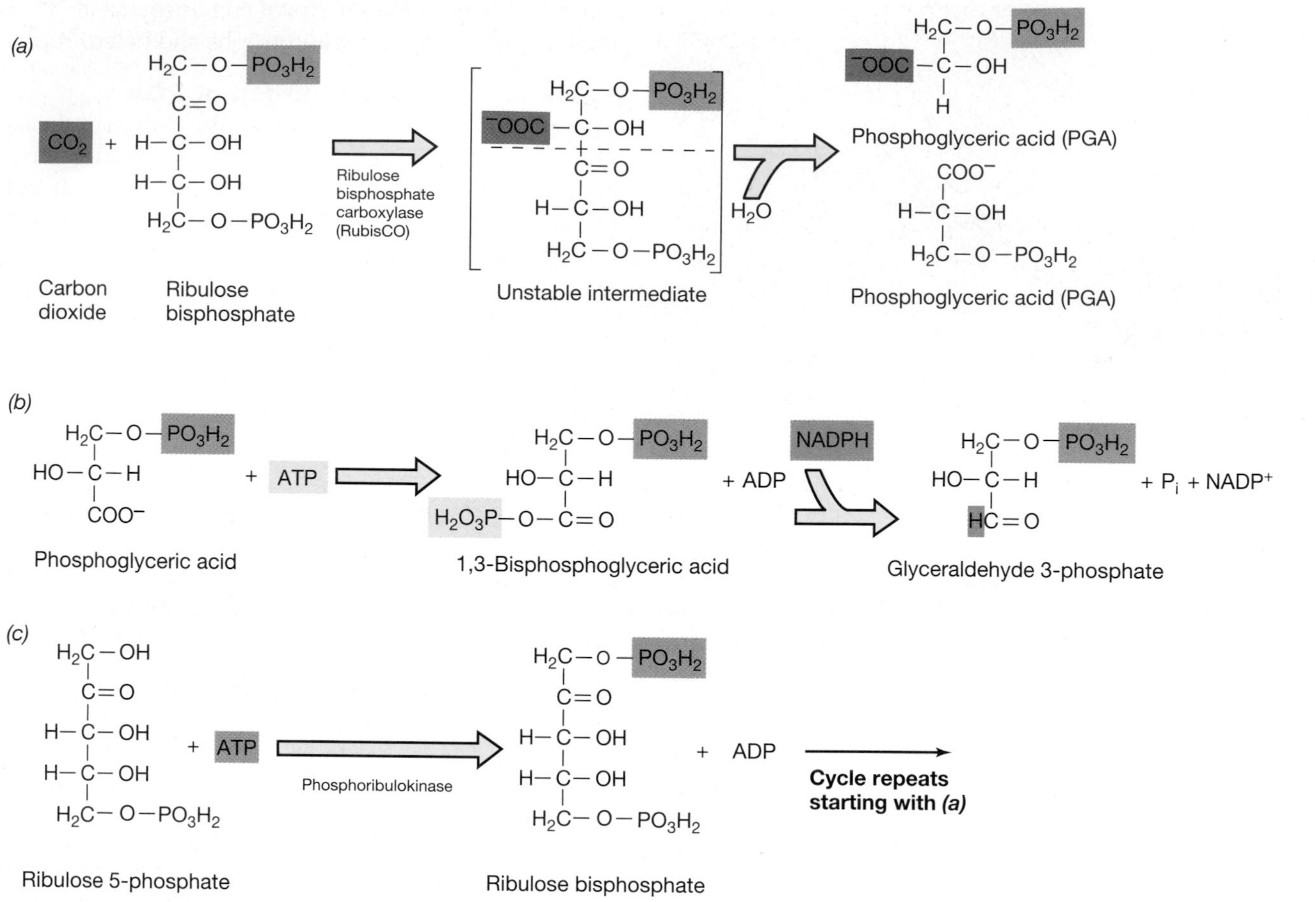

Figure 17.21 Key enzyme reactions of the Calvin cycle. (a) Reaction of the enzyme *ribulose bisphosphate carboxylase*. (b) Steps in the conversion of 3-phosphoglyceric acid (PGA) to glyceraldehyde 3-phosphate. Note that both ATP and NADPH are required. (c) Conversion of ribulose 5-phosphate to the CO_2 acceptor molecule ribulose bisphosphate by the enzyme *phosphoribulokinase*.

called *carboxysomes*. The inclusions, about 100 nm in diameter, are surrounded by a thin, nonunit membrane, and consist of a tightly packed crystalline array of molecules of RubisCO (Figure 17.23). It is thought that carboxysomes are a mechanism to increase the amount of RubisCO in the cell to allow for more rapid CO_2 fixation without affecting the osmolarity of the cytoplasm (osmotic pressure is not affected because the carboxysome is insoluble). Carboxysomes have been found in obligately chemolithotrophic sulfur-oxidizing bacteria, the nitrifying bacteria, and cyanobacteria and prochlorophytes (∞ Sections 12.3, 12.4, 12.26 and 12.27, respectively). They are not present in facultative autotrophs (organisms that can grow either as autotrophs or as heterotrophs) like purple anoxygenic phototrophs, despite the fact that when these organisms grow as photoautotrophs, they use the Calvin cycle to fix CO_2. Thus, the carboxysome may be an evolutionary adaptation to life under strictly autotrophic conditions.

✓ 17.6 Concept Check

The fixation of CO_2 by most phototrophic and other autotrophic organisms occurs via the Calvin cycle, in which the enzyme ribulose bisphosphate carboxylase (RubisCO) plays a key role. The Calvin cycle is an energy-demanding process in which CO_2 is converted into sugar.

✓ What reaction does the enzyme *ribulose bisphosphate carboxylase* carry out?

✓ Why is reducing power needed for autotrophic growth?

✓ What is a *carboxysome*?

17.7 Autotrophic CO₂ Fixation: Reverse Citric Acid Cycle and the Hydroxypropionate Cycle

Alternative mechanisms of autotrophic CO_2 fixation are present in green sulfur bacteria and green nonsulfur bacteria. In the green sulfur bacterium *Chlorobium* (see Figures 17.7a and 17.17b), CO_2 fixation occurs by a reversal of steps in the citric acid cycle (∞ Figure 5.22), a pathway referred to as the *reverse citric acid cycle* (Figure 17.24a). *Chlorobium* contains two ferredoxin-linked enzymes that catalyze the reductive fixation of CO_2 into intermediates of the citric acid cycle. The two ferredoxin-linked reactions involve the carboxylation of succinyl-CoA to

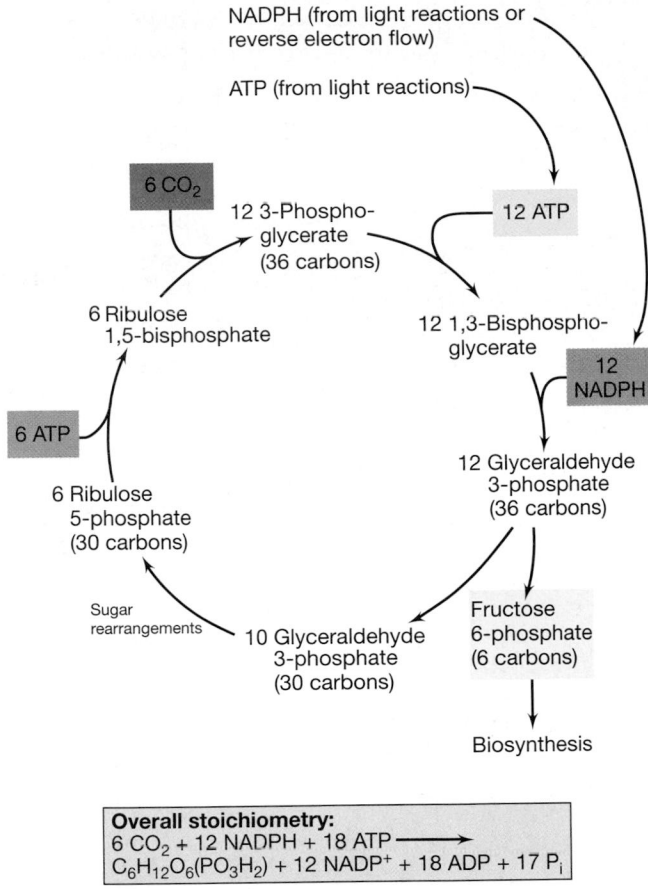

Overall stoichiometry:
$$6\ CO_2 + 12\ NADPH + 18\ ATP \longrightarrow$$
$$C_6H_{12}O_6(PO_3H_2) + 12\ NADP^+ + 18\ ADP + 17\ P_i$$

Figure 17.22 The Calvin cycle. For each *six* molecules of CO_2 incorporated, *one* fructose 6-phosphate is produced. Use the color coding here to follow the biochemical reactions occurring in Figure 17.21.

α-ketoglutarate and the carboxylation of acetyl-CoA to pyruvate (Figure 17.24*a*). Most of the other reactions of the reverse citric acid cycle are catalyzed by enzymes working in reverse of the normal oxidative direction of the cycle. One exception is *citrate lyase*, an ATP-dependent enzyme that cleaves citrate into acetyl-CoA and oxaloacetate in green sulfur bacteria. In the oxidative direction of the cycle, citrate is produced from these same precursors by the enzyme *citrate synthase* (∞ Figure 5.22).

The reverse citric acid cycle as a mechanism of autotrophy has also been found in certain nonphototrophic hyperthermophiles, including the *Archaea Sulfolobus* and *Thermoproteus* (∞ Section 13.9) and *Aquifex*, a very early branching autotroph on the phylogenetic tree of *Bacteria* (∞ Figure 12.1 and Section 12.37). These findings, of course, leave open the possibility that this pathway is more widespread among autotrophic prokaryotes than previously thought and suggest that it may have been an "early" form of autotrophy.

Autotrophy in *Chloroflexus*

The green nonsulfur phototroph *Chloroflexus* (∞ Section 12.35) grows autotrophically with either H_2 or H_2S as electron donors. However, neither the Calvin cycle nor the reverse citric acid cycle operates in this organism. Instead, two molecules of CO_2 are reduced to glyoxylate by a unique autotrophic pathway, the *hydroxypropionate pathway* (Figure 17.24*b*). This pathway leads to the synthesis of hydroxypropionate as a key intermediate. Thus far, the hydroxypropionate pathway has been confirmed only in *Chloroflexus*, and this is of evolutionary interest considering that this organism is the earliest branching anoxyphototroph on the tree of *Bacteria* (∞ Figure 12.1). This suggests that the hydroxypropionate pathway may have been the first attempt at autotrophy in anoxygenic phototrophs and, because it is widely believed that these organisms evolved long before the cyanobacteria, perhaps was the first autotrophic pathway in *any* phototrophic organism.

✓ *17.7* Concept Check

The reverse citric acid cycle and the hydroxypropionate cycle are pathways of CO_2 fixation found in green sulfur and green nonsulfur bacteria, respectively.

✓ Including the route of CO_2 fixation, discuss at least three ways that you could distinguish a purple sulfur bacterium from a green sulfur bacterium.

✓ Including the route of CO_2 fixation, what similarities and differences exist between green *sulfur* and green *nonsulfur* bacteria?

II CHEMOLITHOTROPHY: ENERGY FROM THE OXIDATION OF INORGANIC ELECTRON DONORS

Chapter 5 focused on microorganisms whose energy metabolism was based on chemoorganotrophy, the use of organic compounds as energy sources. In the next four sections we focus on the chemolithotrophs, highlighting the strategies, problems, and advantages of a

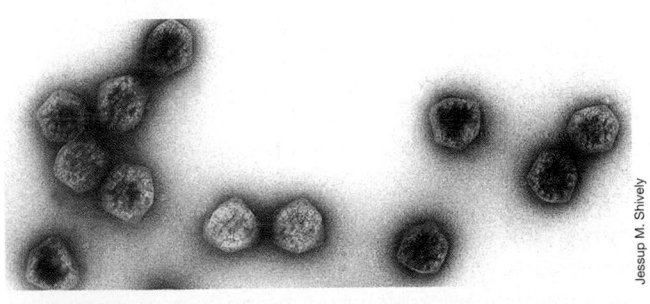

Figure 17.23 Carboxysomes purified from the chemolithotrophic sulfur oxidizer *Thiobacillus neapolitanus*. The structures are about 100 nm in diameter.

Jessup M. Shively

Figure 17.24 Unique autotrophic pathways in phototrophic green bacteria. (a) The reverse citric acid cycle is the mechanism of CO_2 fixation in the green sulfur bacterium *Chlorobium* (see also Figure 17.17b; ⚬⚬ Section 12.32). Ferredoxin$_{red}$ indicates carboxylation reactions requiring reduced ferredoxin (2 H each). Reduced ferredoxin is generated in *Chlorobium* by light-driven reactions (see Figure 17.18). Starting from oxalacetate, each turn of the cycle results in three molecules of CO_2 being incorporated and pyruvate as the product. The cleavage of citrate by the ATP-dependent enzyme citrate lyase regenerates the C_4 acceptor oxalacetate and produces acetyl-CoA for biosynthesis. The conversion of pyruvate to phosphoenolpyruvate consumes two high-energy phosphate bond equivalents. (b) The hydroxypropionate pathway is the means of autotrophy in the green nonsulfur bacterium *Chloroflexus* (⚬⚬ Section 12.35). Acetyl-CoA is carboxylated twice to yield methylmalonyl-CoA. This intermediate is rearranged to yield acetyl-CoA and glyoxylate. The latter is converted to cell material probably through a serine or glycine intermediate.

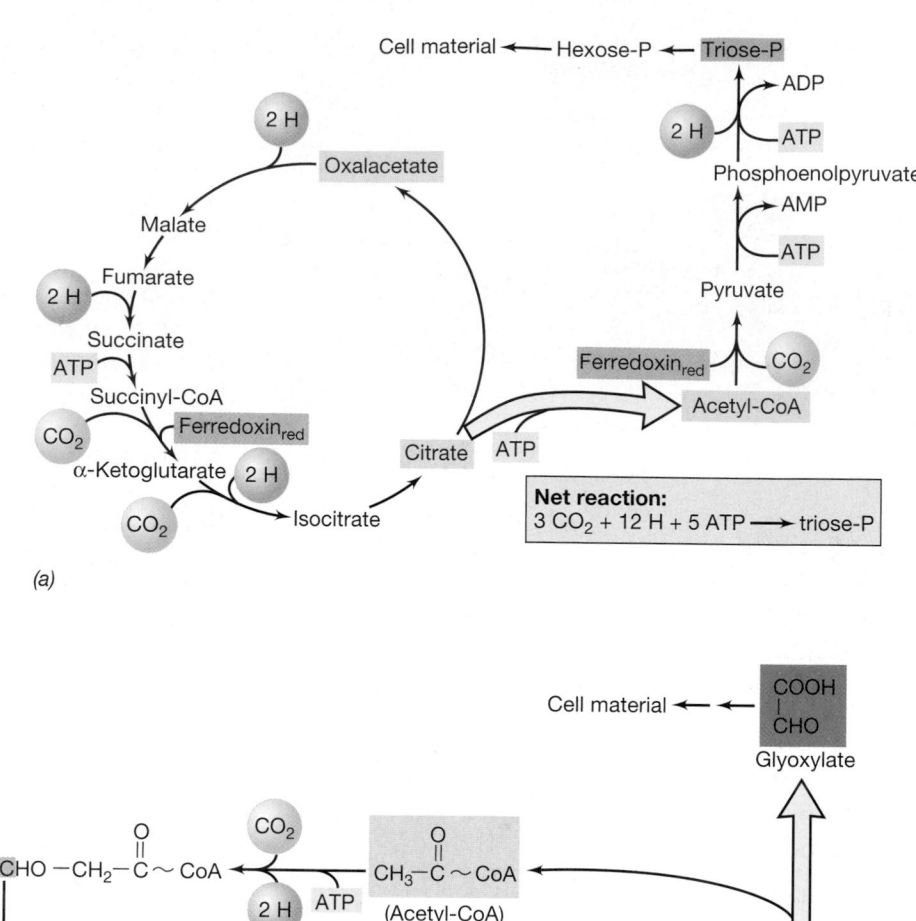

(a)

Net reaction:
$3 CO_2 + 12 H + 5 ATP \longrightarrow$ triose-P

(b)

Net reaction:
$2 CO_2 + 4 H + 3 ATP \longrightarrow$ glyoxylate

lifestyle geared toward the use of *inorganic* chemicals as energy sources.

17.8 Inorganic Electron Donors and Energetics

Organisms that obtain energy from the oxidation of *inorganic* compounds are called **chemolithotrophs**. Most chemolithotrophic bacteria are also able to obtain all their carbon from CO_2, so they are also autotrophs. As we have noted, for growth on CO_2 as sole carbon source, an organism needs (1) energy (ATP), and (2) reducing power. In chemolithotrophs, ATP generation is in principle similar to that in chemoorganotrophs, except that the electron donor is *inorganic* rather than organic. ATP synthesis is coupled to *oxidation* of the electron donor. Reducing power in chemolithotrophs is obtained either directly from the inorganic compound, if it has a sufficiently low reduction potential, or by *reverse electron transport reactions*, as discussed for phototrophic purple bacteria in Section 17.5.

Sources of Inorganic Electron Donors

There are many sources of inorganic electron donors for chemolithotrophs, which may be geological, biological, or anthropogenic in nature. Volcanic activity is a major source of reduced sulfur compounds, primarily H_2S.

Agricultural and mining operations add inorganic electron donors to the environment, especially nitrogen and iron compounds, as does the burning of fossil fuels and the input of industrial wastes. Biological sources are also quite extensive, especially the production of H_2S, H_2, and NH_3. The ecological success (∞ Chapter 19) and metabolic diversity (this chapter) of chemolithotrophs underscores the abundance of sources and supplies of inorganic electron donors in nature.

Energetics of Chemolithotrophy

A review of reduction potentials listed in Table A1.2 reveals that a number of inorganic compounds can provide sufficient energy for ATP synthesis when O_2 is used as electron acceptor. Recall from Chapter 5 that the further apart two half reactions are in terms of E_0', the greater the amount of energy released. For instance, the difference in reduction potential between the H^+/H_2 couple and the $\frac{1}{2}O_2/H_2O$ couple is -1.23 V, which is equivalent to a free-energy yield of -237 kJ/mol (see Appendix 1 for calculations). On the other hand, the potential difference between the H^+/H_2 couple and the NO_3^-/NO_2^- couple is less, -0.84 V, equivalent to a free-energy yield of -163 kJ/mol. This is still quite sufficient for the production of ATP (the high energy phosphate bond of ATP has a free energy of about -31.8 kJ/mol, see Table 17.6). However, a similar calculation will show that there is insufficient energy available from the oxidation of H_2S using CO_2 as electron acceptor.

Such energy calculations make it possible to predict the kinds of chemolithotrophs that might be found in nature. Since organisms must obey the laws of thermodynamics, only reactions that are thermodynamically favorable are potential energy-yielding reactions. Table 17.1 summarizes energy yields for some reactions known to be carried out by chemolithotrophic microorganisms. We discuss some of these processes in this chapter, and the organisms themselves were considered in Chapters 12 and 13. We also examine ecological aspects of chemolithotrophy in Chapter 19.

✓ 17.8 Concept Check

Chemolithotrophs are able to oxidize inorganic chemicals as their sole sources of energy and reducing power. Most chemolithotrophs are also able to grow autotrophically.

✓ For what two purposes are inorganic compounds used by chemolithotrophs?

✓ Why does the oxidation of H_2 yield more energy with O_2 as electron acceptor than with SO_4^{2-} as electron acceptor?

17.9 Hydrogen Oxidation

Hydrogen, H_2, is a common product of microbial metabolism, and a number of chemolithotrophs are able to use it as an electron donor in energy metabolism. A wide variety of anaerobic H_2-oxidizing *Bacteria* and *Archaea* are known, differing in the electron acceptor they use (for example, nitrate, sulfate, ferric iron, and others), and these organisms are discussed later in this chapter (see Sections 17.14–17.18). Here we consider only the *aerobic* H_2-oxidizing bacteria. Also, certain species of H_2 bacteria can grow by oxidizing carbon monoxide ($CO + \frac{1}{2}O_2 \rightarrow CO_2$), and these organisms were discussed in Section 12.5.

Energetics of H_2 Oxidation

Generation of ATP during H_2 oxidation results from the oxidation of H_2 by O_2 leading to formation of a proton motive force. The overall reaction:

$$H_2 + \frac{1}{2}O_2 \rightarrow H_2O \qquad \Delta G^{0'} = -237 \text{ kJ}$$

is highly exergonic and can support the synthesis of at least one ATP. The reaction is catalyzed by the enzyme **hydrogenase**, the electrons from H_2 initially being

---

TABLE 17.1 Energy yields from the oxidation of various inorganic electron donors[a]

Electron donor	Reaction	Type of chemolithotroph	E_0' of couple (V)	$\Delta G^{0'}$ (kJ/reaction)	Number of electrons	$\Delta G^{0'}$ (kJ/2 e$^-$)
Phosphite[b]	$4 HPO_3^{2-} + SO_4^{2-} + H^+ \rightarrow 4 HPO_4^{2-} + HS^-$	Phosphite bacteria	-0.69	-91	2	-91
Hydrogen	$H_2 + \frac{1}{2}O_2 \rightarrow H_2O$	Hydrogen bacteria	0.42	-237.2	2	-237.2
Sulfide	$HS^- + H^+ + \frac{1}{2}O_2 \rightarrow S^0 + H_2O$	Sulfur bacteria	-0.27	-209.4	2	-209.4
Sulfur	$S^0 + 1\frac{1}{2}O_2 + H_2O \rightarrow SO_4^{2-} + 2 H^+$	Sulfur bacteria	-0.20	-587.1	6	-195.7
Ammonium	$NH_4^+ + 1\frac{1}{2}O_2 \rightarrow NO_2^- + 2 H^+ + H_2O$	Nitrifying bacteria	$+0.34$	-274.7	6	-91.6
Nitrite	$NO_2^- + \frac{1}{2}O_2 \rightarrow NO_3^-$	Nitrifying bacteria	$+0.43$	-74.1	2	-74.1
Ferrous iron	$Fe^{2+} + H^+ + \frac{1}{4}O_2 \rightarrow Fe^{3+} + \frac{1}{2}H_2O$	Iron bacteria	$+0.77$	-32.9	1	-65.8

[a] Data calculated from values in Appendix 1; values for Fe^{2+} are for pH 2, and others are for pH 7. At pH 7 the Fe^{3+}/Fe^{2+} couple is about $+0.2$ V.

[b] Except for phosphite, all reactions are shown coupled to O_2 as electron acceptor. The only known phosphite oxidizer couples to SO_4^{2-} as electron acceptor.

transferred to a quinone acceptor. From here electrons pass through a series of cytochromes to eventually reduce O_2 to water (Figure 17.25). Some hydrogen bacteria have two hydrogenases, one soluble and one membrane-bound (∞ Table 12.6). In this case, the membrane-bound enzyme is involved in energetics while the soluble hydrogenase takes up H_2 and reduces NAD^+ to NADH directly; the reduction potential of H_2 (−0.42 V) is so low that reverse electron flow as a means of making reducing power is unnecessary (Figure 17.25). The organism *Ralstonia eutropha* has been a model for studying aerobic H_2 oxidation, and we discussed some of the properties of this organism in Section 12.5.

Autotrophy in H_2 Bacteria

Although most hydrogen bacteria can also grow as chemoorganotrophs, when growing chemolithotrophically these organisms fix CO_2 by the Calvin cycle (see Section 17.6). The stoichiometry observed here is:

$$6 H_2 + 2 O_2 + CO_2 \rightarrow (CH_2O) + 5 H_2O$$

where (CH_2O) represents cell material. However, when readily useable organic compounds such as glucose are present, synthesis of Calvin cycle and hydrogenase enzymes in typical aerobic H_2-oxidizing bacteria is repressed. Thus, in nature, where H_2 levels in oxic environments are transient and low at best, it is likely that aerobic hydrogen bacteria must closely regulate their catabolic enzymes and shift between chemoorganotrophy and chemolithotrophy often, depending on lev-

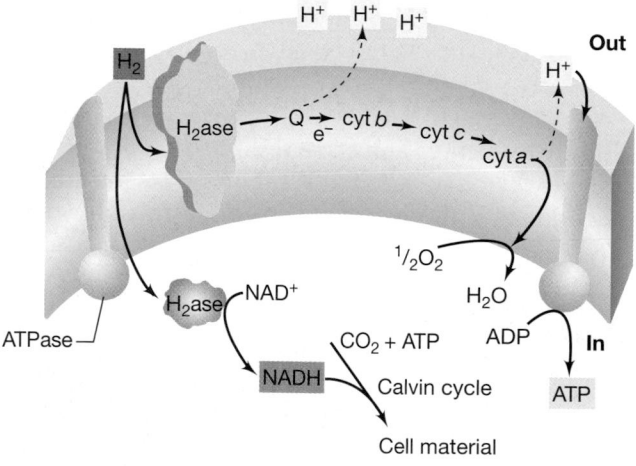

Figure 17.25 Bioenergetics and function of the two hydrogenases of aerobic H_2 bacteria. In *Ralstonia eutropha*, where two hydrogenases are present, the membrane-bound hydrogenase is involved in energetics while the cytoplasmic hydrogenase makes NADH for the Calvin cycle. Note how the membrane-bound hydrogenase begins the flow of electrons leading to formation of a proton motive force. Some H_2 bacteria have only the membrane-bound hydrogenase, and in these organisms reducing power synthesis occurs from reverse electron flow. H_2ase, hydrogenase; cyt, cytochrome; Q, quinone.

els of useable organic compounds and H_2 in their habitats. Moreover, because many aerobic H_2 bacteria grow best microaerobically, it is likely that these organisms are most successful as H_2 chemolithotrophs in oxic/anoxic interfaces where H_2 from fermentative metabolism would be in greater and more continuous supply than in highly oxic habitats.

17.10 Oxidation of Reduced Sulfur Compounds

Many reduced sulfur compounds can be used as electron donors by a variety of colorless sulfur bacteria [called "colorless" to distinguish them from the bacteriochlorophyll-containing (pigmented) green and purple sulfur bacteria discussed earlier in this chapter, see Figure 17.17]. Indeed, the concept of chemolithotrophy emerged from studies of the sulfur bacteria, as the great Russian microbiologist Winogradsky first proposed the idea of chemolithotrophy from studies of these organisms (see the box, Winogradsky and Chemolithoautotrophy, and ∞ Section 1.6).

Energetics of Sulfur Oxidation

The most common sulfur compounds used as electron donors are hydrogen sulfide (H_2S), elemental sulfur (S^0), and thiosulfate ($S_2O_3^{2-}$). The final product of sulfur oxidation in most cases is sulfate (SO_4^{2-}), and the total number of electrons involved between H_2S (oxidation state, −2) and sulfate (oxidation state, +6) is eight (see Table 17.3 for a summary of sulfur oxidation states). Less energy is available when one of the intermediate sulfur oxidation states is used:

$$H_2S + 2 O_2 \rightarrow SO_4^{2-} + 2 H^+$$
$$\Delta G^{0\prime} = -798.2 \text{ kJ/reaction}$$

$$HS^- + \tfrac{1}{2} O_2 + H^+ \rightarrow S^0 + H_2O$$
$$-209.4 \text{ kJ/reaction}$$

$$S^0 + H_2O + 1\tfrac{1}{2} O_2 \rightarrow SO_4^{2-} + 2 H^+$$
$$-587.1 \text{ kJ/reaction}$$

$$S_2O_3^{2-} + H_2O + 2 O_2 \rightarrow 2 SO_4^{2-} + 2 H^+$$
$$-818.3 \text{ kJ/reaction}$$
$$(-409.1 \text{ kJ/S atom oxidized})$$

The oxidation of the most reduced sulfur compound, H_2S, occurs in stages, and the first oxidation step results in the formation of elemental sulfur, S^0. Some H_2S-oxidizing bacteria deposit this elemental sulfur inside the cell (Figure 17.26a). The sulfur deposited as a result of the initial oxidation is an energy reserve, and when the supply of H_2S has been depleted, additional energy can be obtained from the oxidation of sulfur to sulfate.

Learning From the Past ... | Winogradsky and Chemolithoautotrophy

The concept of chemolithotrophic autotrophy was first developed by the great Russian microbiologist Sergei Winogradsky. Winogradsky studied sulfur bacteria because certain colorless sulfur bacteria (*Beggiatoa*, *Thiothrix*) are very large (see Fig. 1) and hence are easy to investigate even in the absence of pure cultures. Springs with waters rich in H_2S are fairly common around the world, and Winogradsky studied several such springs in the Bernese Oberland district of Switzerland. In the outflow channels of sulfur springs, vast populations of *Beggiatoa* and *Thiothrix* develop, and suitable material for microscopic and physiological studies could be obtained by merely lifting up the white filamentous masses (👓 Section 12.4 and Figures 12.11 and 12.13).

Winogradsky first showed that the colorless sulfur bacteria were present only in water containing H_2S. As the water flowed away from the source, the H_2S gradually dissipated, and sulfur bacteria were no longer present. This suggested to him that their development was dependent on the presence of H_2S. Winogradsky then showed that by starving *Beggiatoa* filaments for a while, they lost their sulfur granules; he found, however, that the granules were rapidly restored if a small amount of H_2S was added (Fig. 1). He thus concluded that H_2S was being oxidized to elemental sulfur. But what happened to the sulfur granules when the filaments were starved of H_2S? Winogradsky showed by some clever microchemical tests that when the sulfur granules disappeared, sulfate appeared in the medium. Thus, he formulated the idea that *Beggiatoa* (and by inference other colorless sulfur bacteria) oxidize H_2S to elemental sulfur and subsequently to sulfate ($H_2S \rightarrow S^0 \rightarrow SO_4^{2-}$). Because this organism seemed to require H_2S for development in the springs, he further postulated that this oxidation was the principal source of *energy* for these organisms.

Thus, studies on *Beggiatoa* provided the first evidence that an organism could oxidize an *inorganic* substance as an energy source, and this was the origin of the concept of chemolithotrophy. From these beginnings, Winogradsky turned to a study of the nitrifying bacteria, and it was with this group that he first showed that autotrophic fixation of CO_2 was coupled to the oxidation of an inorganic compound. The process of nitrification had been known before Winogradsky's work from studies on the fate of sewage when added to soil. For example, ammonia-rich sewage passed through a soil column showed conversion to nitrate. Winogradsky picked up from here and proceeded to isolate nitrifying bacteria using completely mineral media in which CO_2 was the sole carbon source and ammonia was the sole electron donor. Because ammonia is chemically stable, it was easy to show that the oxidation of ammonia to nitrite, and subsequently to nitrate, was a strictly bacterial process. In fact, Winogradsky further showed that nitrification was a *two-step* process, with one group of organisms converting NH_4^+ to NO_2^- and a second, NO_2^- to NO_3^- (see Section 17.12). Because no organic materials were present in the medium, it was also possible to show that organic matter (the bacterial cell material) was formed only from CO_2. When the ammonia or nitrite was left out of the medium, no growth occurred. Careful chemical analyses showed that the amount of organic matter formed by the bacteria was proportional to the amount of ammonia or nitrite they oxidized. Winogradsky concluded, "This [process] is contradictory to that fundamental doctrine of physiology which states that a complete synthesis of organic matter cannot take place in nature except through chlorophyll-containing plants by the action of light." At least in one way, however, autotrophy in most chemolithotrophs and phototrophs is similar in that in both processes the pathway of CO_2 fixation follows the same biochemical steps (the Calvin cycle) involving the enzyme ribulose bisphosphate carboxylase (see Section 17.7). ■

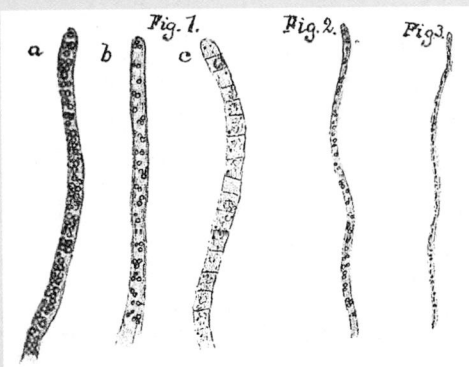

Figure 1 Drawings made by Winogradsky of *Beggiatoa* and translation (from the French) of the legend accompanying these figures. "Fig. 1. The tip of a filament of *Beggiatoa alba*: (a) in sulfurous [sulfide-containing] water, (b) after 24 h in water nearly depleted in H_2S, (c) after 48 h in water without H_2S [note depletion of sulfur globules with time]. Fig. 2. The tip of a filament of *Beggiatoa media*. Fig. 3. The tip of a filament of *Beggiatoa minima*." From Winogradsky, S. 1949. *Microbiologie du Sol*. Masson, Paris.

When elemental sulfur is provided externally as an electron donor, the organism must grow attached to the sulfur particle because of the extreme insolubility of elemental sulfur (Figure 17.26*b*). By adhering to the particle, the organism can efficiently obtain the atoms of sulfur needed. This is thought to occur through the action of membrane or periplasmic proteins that solubilize the sulfur, probably by reduction of S^0 to HS^-, from which it is transported into the cell and enters chemolithotrophic metabolism.

One of the products of the sulfur oxidation reactions is H^+. Production of protons lowers the pH; consequently one result of the oxidation of reduced sulfur compounds is the acidification of the medium. The acid

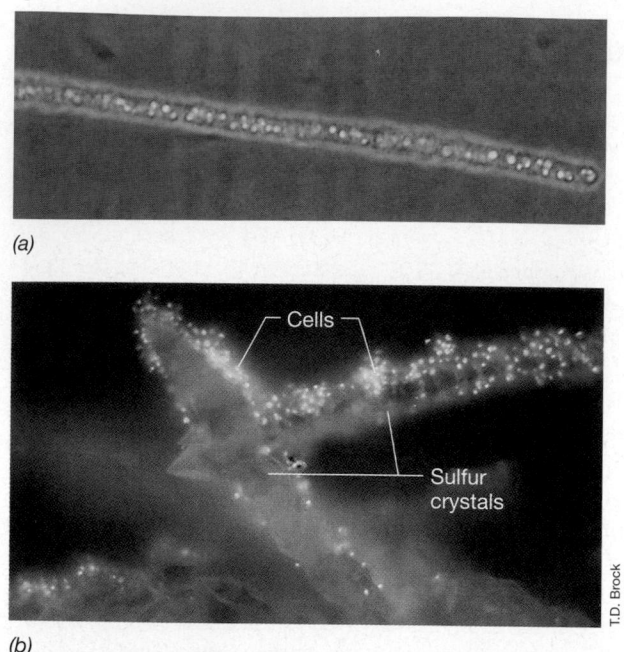

(a)

(b)

Figure 17.26 Sulfur bacteria. (a) Deposition of internal sulfur granules by *Beggiatoa*. (b) Attachment of the sulfur-oxidizing archaeon *Sulfolobus acidocaldarius* to a crystal of elemental sulfur. Cells are visualized by fluorescence microscopy after staining the cells with the dye acridine orange. The sulfur crystal does not fluoresce.

formed by the sulfur bacteria is *sulfuric acid*, H_2SO_4, and sulfur bacteria are often able to bring about a marked reduction in the pH of the medium.

Biochemistry and Energetics of Sulfur Oxidation

The biochemical steps in the oxidation of various sulfur compounds are summarized in Figure 17.27. Starting with sulfide, sulfite (SO_3^{2-}) is produced, a six electron oxidation. If S^0 is the starting substrate, sulfite is also produced, although the S^0 must first be reduced to sulfide (Figure 17.27a). There are two ways in which SO_3^{2-} can be oxidized to SO_4^{2-}. The most widespread system is that employing the enzyme *sulfite oxidase*. Sulfite oxidase transfers electrons from SO_3^{2-} directly to cytochrome *c*, and ATP is made from this during electron transport and proton motive force formation (Figure 17.27b). In addition to sulfite oxidase, a few sulfur chemolithotrophs oxidize SO_3^{2-} to SO_4^{2-} via a reversal of the activity of *adenosine phosphosulfate (APS) reductase*, an enzyme critical to the metabolism of sulfate-reducing bacteria (compare Figures 17.27a and 17.38). This reaction, run in the direction of SO_4^{2-} production by sulfur chemolithotrophs, produces one high-energy phosphate bond when AMP is converted to ADP (Figure 17.27a). When thiosulfate is the electron donor for sulfur chemolithotrophs, it is split into S^0 and SO_3^{2-}, both of which are eventually oxidized to SO_4^{2-}.

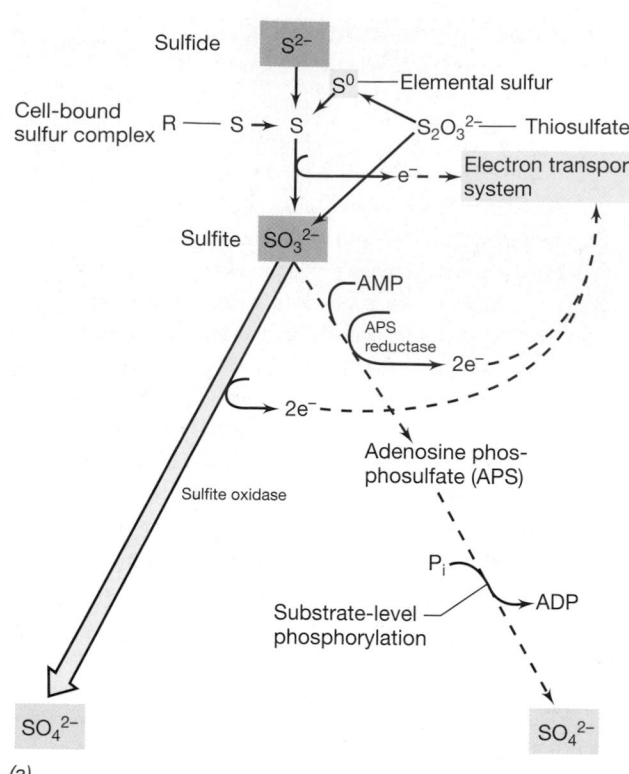

(a)

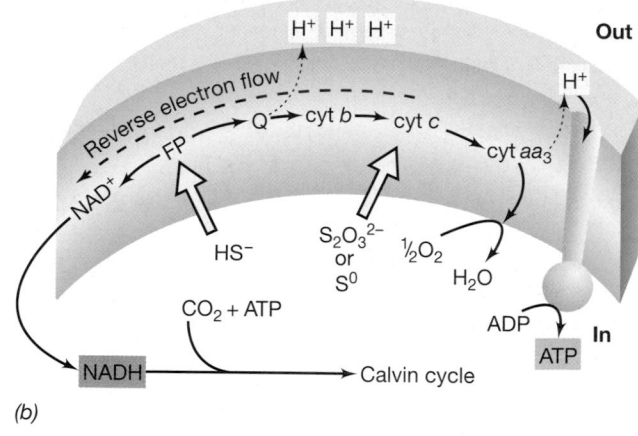

(b)

Figure 17.27 Oxidation of reduced sulfur compounds by sulfur chemolithotrophs. (a) Steps in the oxidation of different compounds. The sulfite oxidase pathway accounts for the majority of sulfite oxidized. (b) Electrons from sulfur compounds feed into the electron transport chain to drive a proton motive force; electrons from thiosulfate and elemental sulfur enter at the level of cytochrome *c*. NADH must be made by energy-consuming reactions of reverse electron flow since the electron donors have a more electropositive E_0' than does $NAD^+/NADH$. Cyt, cytochrome; FP, flavoprotein; Q, quinone.

All the electrons from reduced sulfur compounds eventually reach the electron transport system as shown in Figure 17.27b. Depending on the E_0' of the couple, electrons enter at either the flavoprotein ($E_0' = \sim -0.2$) or cytochrome *c* ($E_0' = +0.3$) level and are transported to O_2, generating a proton motive force that leads to ATP synthesis by ATPase. Electrons for autotrophic CO_2 fixation come from reverse electron flow (see Section 17.5), eventually yielding NADH, and CO_2 is actually fixed via the Calvin cycle (Figure 17.27*b*). Although the sulfur chemolithotrophs are primarily an aerobic group (∞ Section 12.4), some species can grow anaerobically using nitrate as an electron acceptor; *Thiobacillus denitrificans* is a classic example of this lifestyle.

✓ 17.9–17.10 Concept Checks

Hydrogen (H_2) and reduced sulfur compounds such as H_2S and S^0 are excellent electron donors for chemolithotrophs. These compounds can be oxidized by the hydrogen bacteria or the sulfur bacteria, respectively, thereby generating a proton motive force and ATP synthesis. These chemolithotrophs are also autotrophs and fix CO_2 by the Calvin cycle.

- ✓ What special enzyme is needed for growth on H_2?
- ✓ How many electrons are available from the oxidation of H_2S if S^0 is the final product? If SO_4^{2-} is the final product?

17.11 Iron Oxidation

The aerobic oxidation of iron from the ferrous (Fe^{2+}) to the ferric (Fe^{3+}) state is an energy-yielding reaction for a few bacteria. Only a small amount of energy is available from this oxidation (see Table 17.1), and for this reason the iron bacteria must oxidize large amounts of iron in order to grow. The ferric iron produced forms insoluble ferric hydroxide [$Fe(OH)_3$] precipitates in water (Figure 17.28*a*). This is in part because at neutral pH ferrous iron rapidly oxidizes nonbiologically to the ferric state and is thus stable for long periods only under

(a)

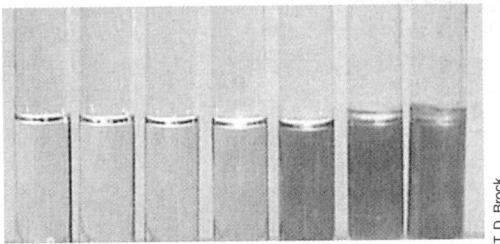

(b)

(c)

Figure 17.28 Iron-oxidizing bacteria. (a) Acid mine drainage, showing the confluence of a normal river and a creek draining a coal-mining area. The acidic creek is very high in ferrous iron. At low pH values, ferrous iron does not oxidize spontaneously in air, but *Thiobacillus ferrooxidans* carries out the oxidation. Insoluble ferric hydroxide and complex ferric salts precipitate, forming the precipitate called "yellow boy" by coal miners. (b) Cultures of *Thiobacillus ferrooxidans*. Shown is a dilution series, with no growth in the tube on the left and increasing amounts of growth from left to right. Growth is evident from the production of Fe^{3+}, which readily forms $Fe(OH)_3$, leading to the yellow-orange color. (c) Extensive development of insoluble ferric hydroxide in a small pool draining a bog in Iceland. Iron deposits such as this are widespread in cooler parts of the world and are modern counterparts of the extensive bog iron deposits of earlier geological eras. These ancient deposits are now the sources of much commercially mined iron ore. In the water-saturated bog soil, facultatively aerobic bacteria reduce ferric iron to the more soluble ferrous state. The ferrous iron leaches into the drainage area surrounding the bog, where oxidation occurs, either spontaneously or through the agency of iron-oxidizing bacteria (*Gallionella* and *Leptothrix*), and the insoluble ferric hydroxide deposit is formed.

anoxic conditions. At acid pH, however, ferrous iron is stable under oxic conditions. This explains why most iron-oxidizing bacteria are obligately acidophilic.

The best-known iron-oxidizing bacteria, *Thiobacillus ferrooxidans* and *Leptospirillum ferrooxidans*, are both able to grow autotrophically using ferrous iron (Figure 17.28*b*) as electron donor. These organisms are very common in acid-polluted environments such as coal-mining dumps (Figure 17.28*a*), and we will discuss their role in acid-mine pollution and mineral oxidation in Sections 19.14 and 19.16.

Despite the instability of Fe^{2+} at neutral pH, there are a number of iron-oxidizing bacteria that thrive in such environments, but these are situations where ferrous iron is moving from anoxic to oxic conditions. At interfaces between these zones iron bacteria can oxidize Fe^{2+} as it comes from an anoxic source before the Fe^{2+} oxidizes spontaneously. *Gallionella ferruginea* and *Sphaerotilus natans* are examples of organisms that live at these interfaces and are generally seen mixed in with the characteristic deposits they form (Figure 17.29; ∞ also Figure 12.43*a*). We discussed the taxonomy of these interesting organisms in Sections 12.15 and 12.16.

Energy from Ferrous Iron Oxidation

The bioenergetics of iron oxidation by *Thiobacillus ferrooxidans* is of interest because of the very electropositive reduction potential of the Fe^{3+}/Fe^{2+} couple (+0.77 V at pH 2). The respiratory chain of *T. ferrooxidans* contains cytochromes of the *c* and *a* types and a periplasmic copper-containing protein called *rusticyanin* (Figure 17.30). Because the reduction potential of the Fe^{3+}/Fe^{2+} couple is so high, the route of electron transport to oxygen ($\frac{1}{2}O_2/H_2O$, $E_0' = +0.82$ V) is obviously going to be very short. Ferrous iron oxidation begins in the periplasm where rusticyanin oxidizes Fe^{2+} to Fe^{3+}, a one electron transition. This protein then reduces cytochrome *c*, and this subsequently reduces cytochrome *a*. The latter in-

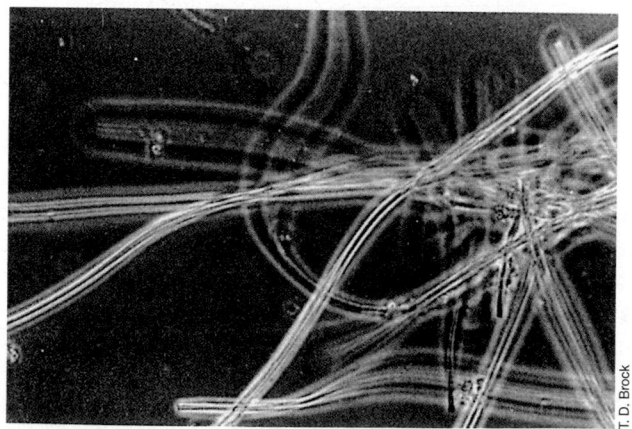

Figure 17.29 Phase contrast photomicrograph of empty iron-encrusted sheaths of *Sphaerotilus* collected from seepage at the edge of a small swamp.

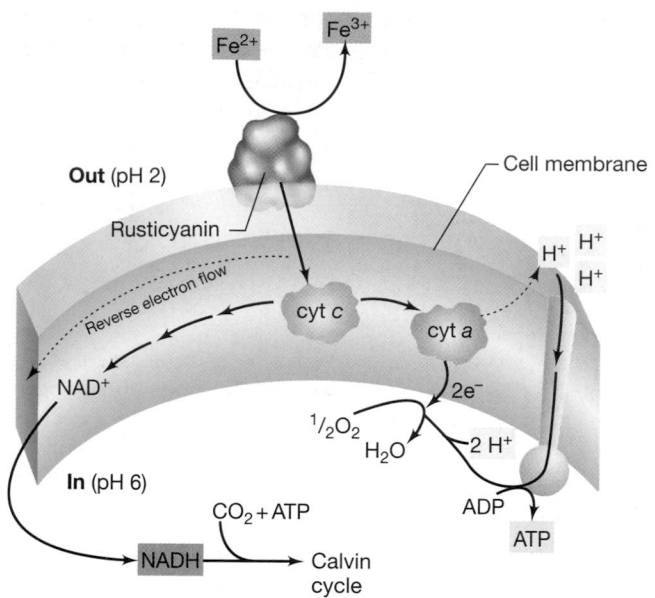

Figure 17.30 Electron flow during Fe^{2+} oxidation by the acidophile *Thiobacillus ferrooxidans*. The periplasmic copper-containing protein rusticyanin is the immediate acceptor of electrons from Fe^{2+}. From here, electrons travel a short electron transport chain resulting in the reduction of O_2 to H_2O. Reducing power to drive the Calvin cycle comes from reactions of reverse electron flow. Note the steep pH gradient (4–5 units) across the membrane.

teracts directly with O_2 to form H_2O (Figure 17.30). ATP is then synthesized from proton-translocating ATPases in the membrane, and ATP yields are relatively low because of the high potential of the electron donor.

Because of the large natural gradient of protons across the *T. ferrooxidans* membrane (the periplasm is pH 1–2 while the cytoplasm is pH 5.5–6), protons entering the cytoplasm via the ATPase must be consumed in order to maintain the internal pH within acceptable limits (Figure 17.30). The protons are consumed during the production of H_2O, but this reaction also requires electrons; these come from Fe^{2+} as follows:

$$2\ Fe^{2+} + \tfrac{1}{2}\ O_2 + 2\ H^+ \rightarrow 2\ Fe^{3+} + H_2O$$

Thus, as long as *T. ferrooxidans* has Fe^{2+} available, ATP synthesis can occur largely at the expense of the natural proton motive force that exists across the cytoplasmic membrane (Figure 17.30).

Autotrophy in *T. ferrooxidans* is driven by the Calvin cycle, and because of the high potential of the electron donor, Fe^{2+}, much energy is consumed in reverse electron flow reactions to obtain the reducing power necessary to drive CO_2 fixation. A relatively poor energetic picture coupled with large energetic demands means that *T. ferrooxidans* must oxidize large amounts of Fe^{2+} in order to produce even a very small amount of cell material. Thus, in environments where acidophilic Fe^{2+}-oxidizing bacteria are living, their presence is signaled not by the formation of much cell material but by the presence of large amounts of ferric iron (Figures 17.28; ⟳ Figure 19.37). We consider the important ecological processes connected with the iron-oxidizing bacteria in Sections 19.14 and 19.16.

Ferrous Iron Oxidation by Anoxygenic Phototrophs

Ferrous iron can be oxidized under *anoxic* conditions by certain anoxygenic phototrophic bacteria (Figure 17.31). The ferrous iron is used in this case not as an electron donor in energy metabolism, but as an electron donor for CO_2 reduction (autotrophy). At neutral pH where these organisms thrive, the Fe^{3+}/Fe^{2+} couple is much less electropositive than at pH 2, about +0.2 V, and thus electrons from Fe^{2+} can reduce cytochrome *c* in the photosystem of purple bacteria (see Section 17.5 for a discussion of anoxygenic photosynthesis). The organisms involved, which are species of photohophic purple bacteria (Figure 17.31*b*), can also use FeS; under these conditions both Fe^{2+} and S^{2-} are oxidized as electron donors. Certain phototrophic green sulfur bacteria (genus *Chlorobium*; ⟳ Section 12.33) can also use Fe^{2+} as a photosynthetic electron donor. Moreover, various chemotrophic denitrifying bacteria have been isolated that can couple the oxidation of Fe^{2+} to the reduction of NO_3^- to N_2 and grow under anoxic conditions. However, like the aerobic iron bacteria, in these organisms iron is an electron donor for *both* energy and reducing power needs.

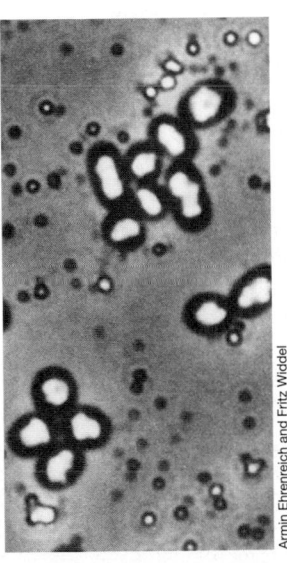

(a) *(b)*

Figure 17.31 Ferrous iron oxidation by anoxygenic phototrophic bacteria. (a) Fe^{2+} oxidation in anoxic tube cultures. Left to right: Sterile medium, inoculated medium, a growing culture. The brown-red color is mainly due to $Fe(OH)_3$ precipitate. (b) Phase contrast photomicrograph of an iron-oxidizing purple bacterium. The bright refractile areas within cells are gas vesicles (⟳ Section 4.14). The granules outside the cells are iron precipitates. This organism is phylogenetically related to the purple sulfur bacterium *Chromatium* (⟳ Section 12.2).

The discovery of Fe^{2+}-oxidizing phototrophs has important implications for both understanding the evolution of photosynthesis and explaining the large deposits of ferric iron found in ancient sediments. Such ferric iron was previously thought to have been formed from the oxidation of Fe^{2+} by O_2 produced by oxygenic phototrophs (⟳ Section 11.1). However, because of the age of these sediments, it is more likely that the ferric iron was formed by anoxygenic phototrophs oxidizing Fe^{2+} in anoxic environments.

✓ 17.11 Concept Check

The iron bacteria are chemolithotrophs able to use ferrous iron (Fe^{2+}) as sole energy source. Most iron bacteria grow only at acid pH and are often associated with acid pollution from mineral and coal mining. Some phototrophic purple bacteria can oxidize Fe^{2+} to Fe^{3+} anoxically.

- ✓ Why is only a very small amount of energy available from the oxidation of Fe^{2+} to Fe^{3+} at acidic pH?
- ✓ What is the function of *rusticyanin* and where is it found in the cell?
- ✓ How can Fe^{2+} be oxidized anoxically?

17.12 Nitrification and Anammox

The most common *inorganic nitrogen compounds* used as electron donors are ammonia (NH_3) and nitrite (NO_2^-), which are oxidized aerobically by the chemolithotrophic **nitrifying bacteria** (⟳ Section 12.3). The nitrifying bacteria are widely distributed in soil and water. One group of organisms, the *nitrosofyers* (*Nitrosomonas* is one genus), oxidizes ammonia to nitrite, and another group (*Nitrobacter*) oxidizes nitrite to nitrate; the complete oxidation of ammonia to nitrate, an eight-electron transfer, is thus carried out by members of these two groups of organisms acting in sequence (see the Winogradsky box).

Bioenergetics and Enzymology of Nitrification

The electrons from nitrogen compounds enter an electron transport chain, and electron flow establishes a membrane potential and proton motive force linked to ATP synthesis. However, because of the reduction potential of their electron donors, nitrifying bacteria are faced with bioenergetic problems similar to those of the sulfur chemolithotrophs. The E_0' of the NO_2^-/NH_3 couple, the first step in the oxidation of NH_3, is +0.34 V. The E_0' of the NO_3^-/NO_2^- couple is even higher, about +0.43 V. These relatively high reduction potentials mean that nitrifying bacteria must donate electrons to their electron transport chains at rather late steps in the overall process. This effectively limits the amount of ATP that can be produced from each pair of electrons introduced.

Several key enzymes are involved in oxidizing reduced nitrogen compounds. In ammonia-oxidizing bacteria, NH_3 is oxidized by **ammonia monooxygenase** (see Section 17.22 for a discussion of monooxygenase enzymes) that produces NH_2OH and H_2O (Figure 17.32). *Hydroxylamine oxidoreductase* then oxidizes NH_2OH to NO_2^-, removing *four* electrons in the process. Ammonia monooxygenase is an integral membrane protein, whereas hydroxylamine oxidoreductase is periplasmic (Figure 17.32). In the reaction carried out by ammonia monooxygenase,

$$NH_3 + O_2 + 2\,H^+ + 2\,e^- \rightarrow NH_2OH + H_2O$$

there is a need for two exogenously supplied electrons plus two protons to reduce one atom of dioxygen to water. These electrons originate from the oxidation of hydroxylamine and are supplied to ammonia monooxygenase from hydroxylamine oxidoreductase via cytochrome c and ubiquinone (Figure 17.32). Thus, for every four electrons generated from the oxidation of NH_3 to NO_2^-, only two actually reach the terminal oxidase (cytochrome aa_3, Figure 17.32).

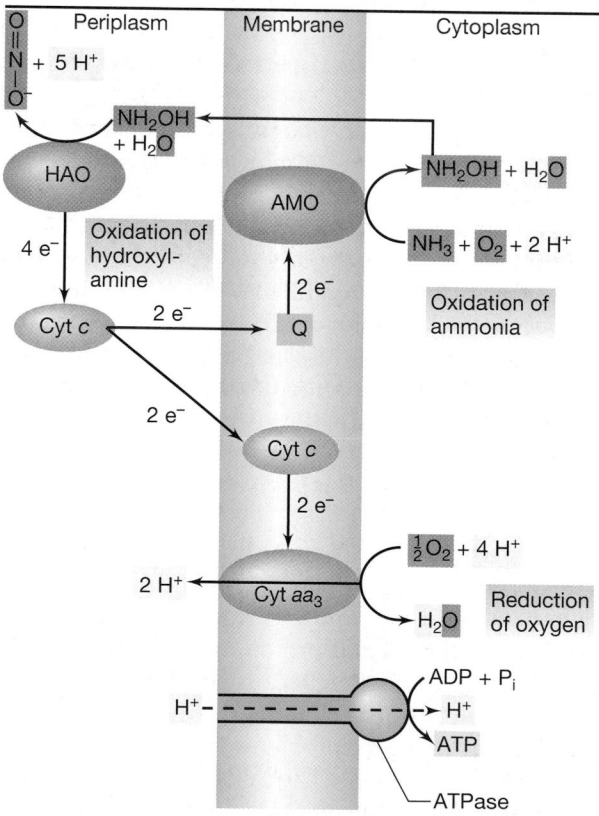

Figure 17.32 Oxidation of ammonia and electron flow in ammonia-oxidizing bacteria. The reactants and the products of this reaction series are highlighted. The cytochrome c (cyt c) in the periplasm is a different form of cyt c than that in the membrane. AMO, Ammonia monooxygenase; HAO, hydroxylamine oxidoreductase; Q, ubiquinone.

Nitrite-oxidizing bacteria employ the enzyme *nitrite oxidoreductase* to oxidize nitrite to nitrate, with electrons traveling a very short electron transport chain (because of the high potential of the NO_3^- / NO_2^- couple) to the terminal oxidase (Figure 17.33). Cytochromes of the a and c types are present in the electron transport chain of nitrite oxidizers, and generation of a proton motive force (which ultimately drives ATP synthesis) occurs through the action of cytochromes aa_3 (Figure 17.33). And like the situation with iron oxidation (see Section 17.11), only small amounts of energy are available. Thus, growth yields of nitrifying bacteria are relatively low.

Carbon Metabolism in Nitrifying Bacteria

Like sulfur- and iron-oxidizing chemolithotrophs, aerobic nitrifying bacteria employ the Calvin cycle for CO_2 fixation, and the ATP and reducing power requirements of this process place additional burdens on an already relatively low-yielding energy-generating system (NADH to drive the Calvin cycle is formed by reverse electron flow). The energetic constraints are particularly severe for nitrite oxidizers, and it is perhaps for this reason that most nitrite oxidizers can also grow chemoorganotrophically on glucose and certain other organic substrates (∞ Section 12.3).

Anoxic Ammonia Oxidation: Anammox

Although classical nitrifying bacteria are *strict aerobes*, at least when growing on their reduced nitrogen substrates, ammonia can also be oxidized under anoxic conditions. This process, known as **anammox** (for *anoxic ammo*nia

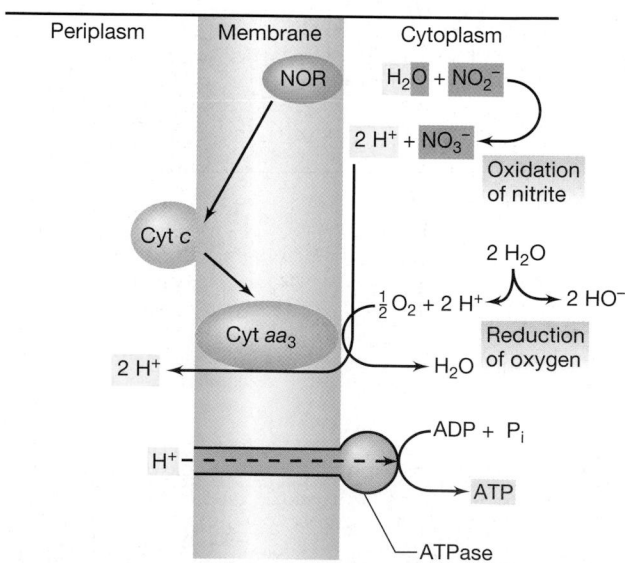

Figure 17.33 Oxidation of nitrite to nitrate by nitrifying bacteria. The reactants and products of this reaction series are highlighted. NOR, Nitrite oxidoreductase.

*oxi*dation), is highly exergonic and linked to energy metabolism of the organisms involved. Anammox involves the oxidation of ammonia with nitrite as the electron acceptor to yield gaseous nitrogen:

$$NH_4^+ + NO_2^- \rightarrow N_2 + 2\,H_2O \qquad \Delta G^{0\prime} = -357\,kJ$$

The organism that catalyzes anammox, *Brocadia anammoxidans*, is a phylogenetically distinct member of the Planctomycetes phylum of *Bacteria* (∞ Section 12.28) (Figure 17.34). Planctomycetes are unusual members of the *Bacteria*, lacking peptidoglycan and containing membrane-enclosed compartments inside the cell, including a structure analogous to the nucleus of eukaryotic cells (∞ Section 12.28). In *B. anammoxidans* a major compartment is present called the *anammoxosome;* this is the location of the anammox reaction (Figure 17.34).

The source of NO_2^- in the anammox reaction is the product of ammonia oxidation by aerobic nitrifying bacteria (Figure 17.32). The two groups of nitrifyers, aerobic (for example, *Nitrosomomas*) and anaerobic (*Brocadia*), live together in ammonia-rich habitats such as sewage and other wastewaters. In these environments suspended particles are present that contain both oxic and anoxic zones where the two groups of ammonia oxidizers can coexist. In mixed laboratory cultures, high levels of oxygen inhibit anammox and favors classic nitrification, and thus it is likely that in nature the extent of anammox-mediated ammonia oxidation is governed by the concentration of O_2 in the system.

Like classic nitrifying bacteria, *Brocadia anammoxidans* is also an autotroph. The anammox organism can grow with CO_2 as sole carbon source and uses nitrite as electron donor here to produce cell material:

$$CO_2 + 2\,NO_2^- + H_2O \rightarrow CH_2O + 2\,NO_3^-$$

Although an autotrophic nitrifyer, it appears that *B. anammoxidans* lacks Calvin cycle enzymes, and the mechanism of CO_2 fixation is unclear. However, the use of nitrite as electron donor for CO_2 fixation along with the production of nitrate, is exactly the same reaction carried out by aerobic nitrifyers like *Nitrobacter* (see Figure 17.33). Thus, although phylogenetically distinct, aerobic nitrifyers and the anammox organism share common substrates and ecology.

The discovery of anammox has greatly improved our understanding of the nitrogen cycle. Before anammox was understood it was thought that ammonia was stable in anoxic environments. Now, however, it is clear that ammonia can be oxidized in the absence of O_2. From an environmental standpoint, anammox holds great promise for the treatment of anoxic wastewaters to remove ammonia and amines, and research to stimulate the activities of anammox for these purposes is currently underway.

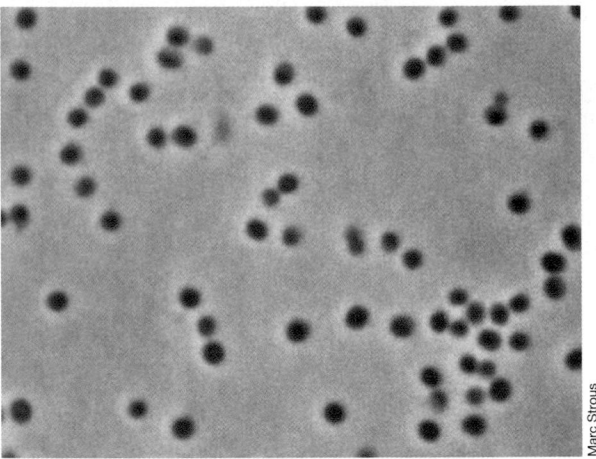

(a)

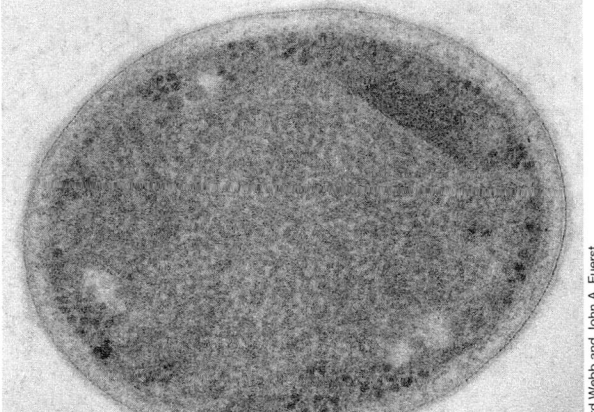

(b)

Marc Strous

Richard Webb and John A. Fuerst

Figure 17.34 The anammox organism, *Brocadia anammoxidans*. (a) Phase-contrast photomicrograph. A single cell is about 1 μm in diameter. (b) Transmission electron micrograph of a cell. Notice the membrane enclosed compartments including the large fibrillar anammoxosome in the center portion of the cell. *Brocadia* is phylogenetically related to the *Planctomyces*, organisms that contain several different types of membrane-enclosed compartments within their cytoplasm (∞ Section 12.28 and Figure 12.87).

✓ *17.12* Concept Check

Ammonia (NH_3) and nitrite (NO_2^-) can be used as electron donors by the nitrifying bacteria. The ammonia-oxidizing bacteria produce nitrite, which is then oxidized by the nitrite-oxidizing bacteria to nitrate (NO_3^-). Anoxic NH_3 oxidation is coupled to both N_2 and NO_3^- production.

✓ What is the inorganic electron donor for *Nitrosomonas*? For *Nitrobacter*?

✓ What are the substrates for the enzyme *ammonia monooxygenase*?

✓ What do nitrifying bacteria use as a carbon source?

✓ What is the *anammox* reaction and how does it differ from aerobic nitrification?

III THE ANAEROBIC WAY OF LIFE

Anaerobic metabolism is a hallmark of prokaryotes. In eukaryotes, anaerobic growth is rare. In prokaryotes, by contrast, anaerobic growth is common, and the mechanisms of anaerobic metabolism highly diverse. In the next several sections we survey anaerobic metabolism and see the many ways by which prokaryotes can make a living in the absence of air.

17.13 Anaerobic Respiration

We examined the process of *aerobic* respiration in some detail in Chapter 5. As we noted there, molecular oxygen (O_2) functions as a terminal electron acceptor, accepting electrons from electron carriers by way of an electron transport chain. However, we also noted that a variety of other electron acceptors can be used instead of O_2, in which case the process is called **anaerobic respiration**. We now consider some of these processes.

The bacteria carrying out anaerobic respiration generally possess electron transport systems containing cytochromes, quinones, iron-sulfur proteins and other typical electron transport proteins. Their respiratory systems are thus analogous to those of conventional aerobes. In some cases, such as with the denitrifying bacteria, the anaerobic respiration process competes in the same organism with an aerobic one. In such cases, if O_2 is present, aerobic respiration is usually favored, and when O_2 is depleted from the environment, the alternate electron acceptor is reduced. Other organisms carrying out anaerobic respiration are *obligate* anaerobes and are unable to use O_2.

Alternative Electron Acceptors and the Electron Tower

The energy released from the oxidation of an electron donor using O_2 as electron acceptor is higher than if the same compound is oxidized with an alternate electron acceptor (∞ Figure 5.9). These energy differences are apparent if the reduction potentials of each acceptor are examined (Figure 17.35). Because the O_2/H_2O couple is the most electropositive, more energy is available when O_2 is used than when another electron acceptor is used. Other electron acceptors that are near O_2 are Fe^{3+}, NO_3^-, and NO_2^-. More electronegative acceptors are S^0, CO_2, and SO_4^{2-}. A summary of the most common types of anaerobic respiration is given in Figure 17.35.

Assimilative and Dissimilative Metabolism

Inorganic compounds such as NO_3^-, SO_4^{2-}, and CO_2 are reduced by many organisms as sources of cellular nitrogen, sulfur, and carbon, respectively. The end prod-

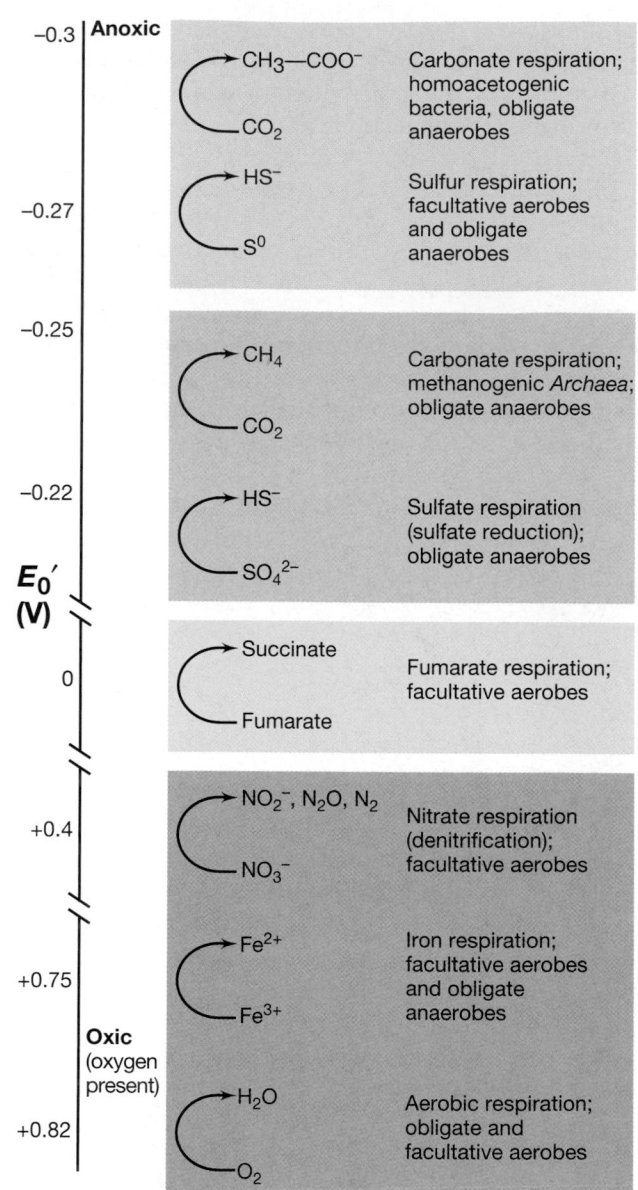

Figure 17.35 Examples of anaerobic respirations. The couples are arranged in order from most electronegative E_0' (top) to most electropositive E_0' (bottom).

ucts of such reductions are primarily amino groups (—NH_2), sulfhydryl groups (—SH), and organic carbon compounds, respectively. We examined the *nutrition* of microorganisms in Sections 5.1–5.3 and noted that all organisms need sources of N, S, and C for growth. When an inorganic compound such as NO_3^-, SO_4^{2-}, or CO_2 is reduced for use as a nutrient source, it is said to be *assimilated*, and the reduction process is called *assimilative* metabolism. We emphasize here that assimilative metabolism of NO_3^-, SO_4^{2-}, and CO_2 is conceptually quite different from the use of these compounds as electron acceptors for *energy* metabolism in anaerobic respiration. To distinguish these two kinds of reduction processes, the use of these compounds as electron acceptors in energy metabolism is called *dissimilative* metabolism.

Assimilative and dissimilative metabolism differ markedly. In assimilative metabolism, only enough of the compound (NO_3^-, SO_4^{2-}, or CO_2) is reduced to satisfy the needs for cell growth. The reduced atoms are eventually converted to cell material in the form of macromolecules. In dissimilative metabolism, a comparatively large amount of the electron acceptor is reduced, and the reduced product is *excreted* into the environment. Many organisms carry out assimilative metabolism of compounds such as NO_3^-, SO_4^{2-}, and CO_2 (for example, many *Bacteria, Archaea*, fungi, algae, and higher plants), whereas only a restricted variety of organisms, primarily prokaryotes, carry out dissimilative metabolism.

✓ 17.13 Concept Check

Although oxygen (O_2) is the most widely used electron acceptor in energy-yielding metabolism, a number of other compounds can be used as electron acceptors. This process of anaerobic respiration is less energy efficient but makes it possible for respiration to occur in environments where oxygen is absent.

✓ What is anaerobic respiration?

✓ With H_2 as electron donor, why is the reduction of NO_3^- a more favorable reaction than the reduction of S^0?

17.14 Nitrate Reduction and the Denitrification Process

Inorganic nitrogen compounds are some of the most common electron acceptors in anaerobic respiration. A summary of the various inorganic nitrogen species with their oxidation states is given in Table 17.2. The most widespread inorganic nitrogen species in nature are ammonia and nitrate, both of which are formed in the atmosphere by inorganic chemical processes, and nitrogen gas, N_2, also an atmospheric gas, which is the most stable form of nitrogen in nature. We discuss *nitrogen*

fixation, the use of N_2 as a biosynthetic nitrogen source, later in this chapter.

One of the most common alternative electron acceptors is nitrate, NO_3^-, which is reduced to N_2O, NO, and N_2. Because these products of nitrate reduction are all gaseous, they can easily be lost from the environment, and because of this the process is called **denitrification** (Figure 17.36). The process is the main means by which gaseous N_2 is formed biologically, and because N_2 is much less readily available to organisms than nitrate as a source of nitrogen, for agricultural purposes at least, denitrification is a detrimental process. For sewage treatment (∞ Section 28.3), however, denitrification is beneficial because it converts NO_3^- to N_2, effectively decreasing the amount of available nitrogen in the sewage treatment effluent that can stimulate algal growth.

Biochemistry of Dissimilative Nitrate Reduction

The enzyme involved in the first step of dissimilative nitrate reduction, *nitrate reductase*, is a molybdenum-containing membrane-integrated enzyme whose synthesis is repressed by molecular oxygen. All subsequent enzymes of the pathway (Figure 17.37) are coordinately regulated and thus also repressed by O_2, but in addition to anoxic conditions, nitrate must also be present before these enzymes are fully expressed.

The first product of nitrate reduction is nitrite (NO_2^-), and the enzyme *nitrite reductase* reduces it to nitric oxide (NO) (Figure 17.37c). Some organisms can reduce NO_2^- to ammonia (NH_3) in a dissimilative process, but it is the production of gaseous products, *denitrification*, that is of the greatest global significance. This is

TABLE 17.2 Oxidation states of key nitrogen compounds	
Compound	**Oxidation state**
Organic N (R— NH_2)	−3
Ammonia (NH_3)	−3
Nitrogen gas (N_2)	0
Nitrous oxide (N_2O)	+1 (average per N)
Nitrogen oxide (NO)	+2
Nitrite (NO_2^-)	+3
Nitrogen dioxide (NO_2)	+4
Nitrate (NO_3^-)	+5

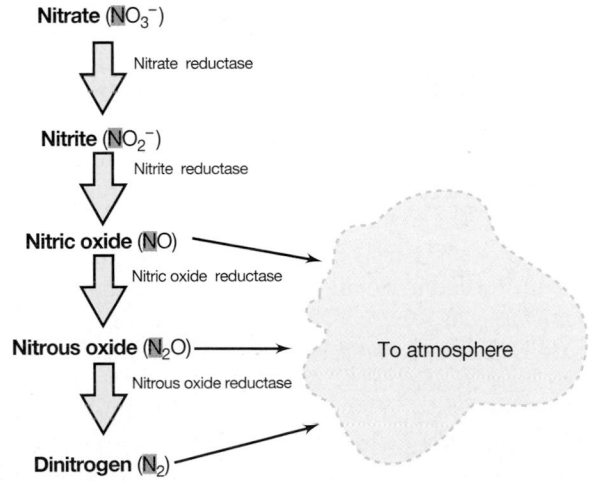

Figure 17.36 Steps in the dissimilative reduction of nitrate. Some organisms, for example *Escherichia coli*, can carry out only the first step. All enzymes involved are derepressed by anoxic conditions. Also, some prokaryotes are known that can reduce NO_3^- to NH_4^+ in dissimilative metabolism.

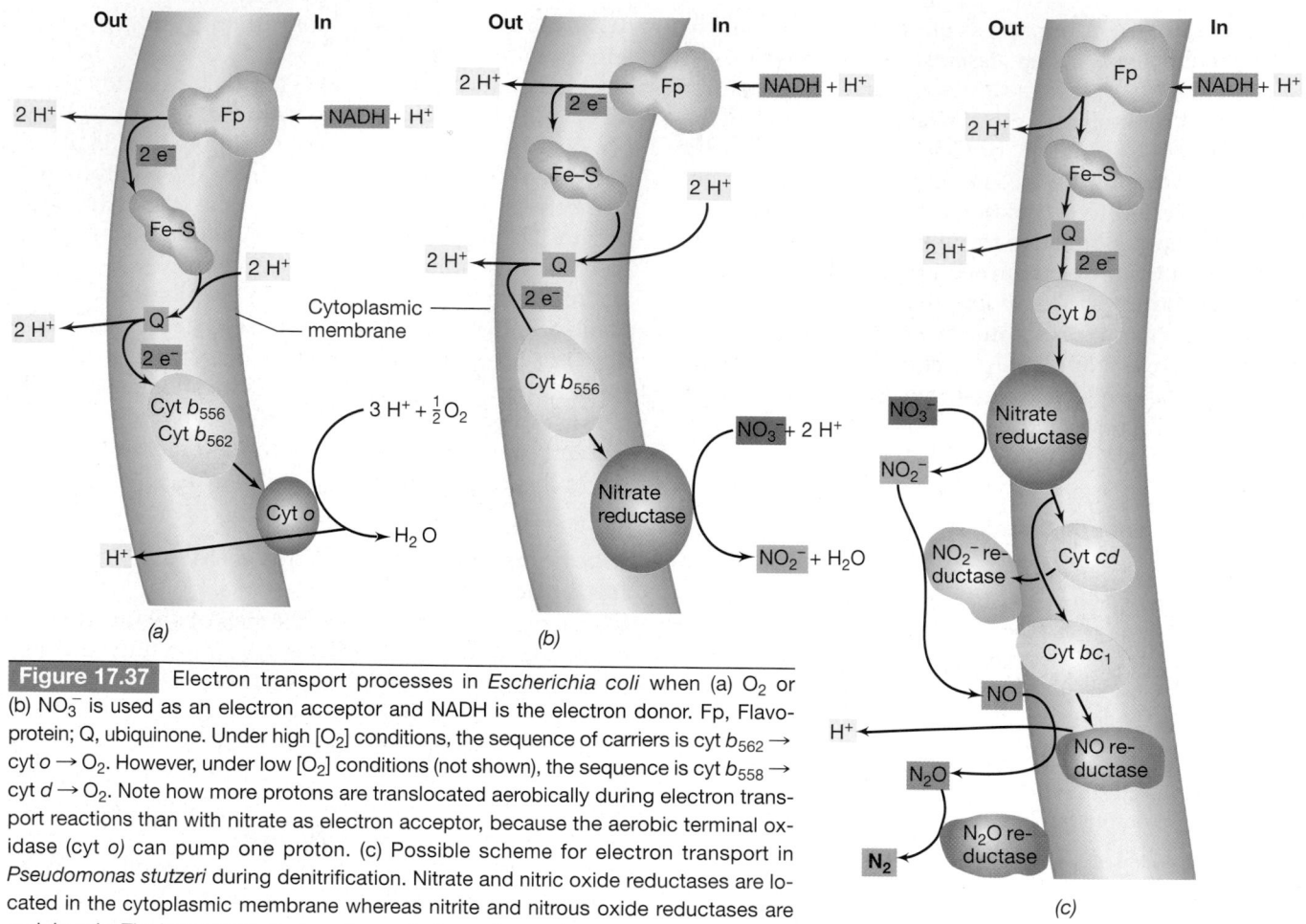

Figure 17.37 Electron transport processes in *Escherichia coli* when (a) O_2 or (b) NO_3^- is used as an electron acceptor and NADH is the electron donor. Fp, Flavoprotein; Q, ubiquinone. Under high $[O_2]$ conditions, the sequence of carriers is cyt $b_{562} \rightarrow$ cyt $o \rightarrow O_2$. However, under low $[O_2]$ conditions (not shown), the sequence is cyt $b_{558} \rightarrow$ cyt $d \rightarrow O_2$. Note how more protons are translocated aerobically during electron transport reactions than with nitrate as electron acceptor, because the aerobic terminal oxidase (cyt o) can pump one proton. (c) Possible scheme for electron transport in *Pseudomonas stutzeri* during denitrification. Nitrate and nitric oxide reductases are located in the cytoplasmic membrane whereas nitrite and nitrous oxide reductases are periplasmic. The immediate electron donors to the various reductases, with the exception of nitrate reductase, have not been definitively identified.

because it consumes a fixed form of nitrogen (NO_3^-) readily available to plants and produces gaseous nitrogen compounds, some of which are of environmental significance (N_2O can be converted to NO by sunlight, and NO reacts with ozone in the upper atmosphere to form nitrite, which returns to Earth as acid rain). The remaining biochemical steps in denitrification are shown in Figure 17.37c.

The biochemistry of dissimilative nitrate reduction has been studied in detail in several organisms including *Escherichia coli*, where NO_3^- is reduced only to NO_2^- (Figure 17.37b) and *Paracoccus denitrificans* and *Pseudomonas stutzeri*, where true denitrification occurs (Figure 17.37c). The *E. coli* nitrate reductase accepts electrons from a *b*-type cytochrome, and a comparison of the electron transport chains in aerobic versus anaerobic respiration in *E. coli* is shown in Figure 17.37a and *b*. As seen here, because of the reduction potential of the NO_3^- / NO_2^- couple (+0.43 V), only two proton translocating steps occur during nitrate reduction while three can occur in aerobic respiration ($\frac{1}{2}O_2/H_2O$, +0.82 V). In *P. denitrificans* and *P.*

stutzeri nitrogen oxides are formed from nitrite by a series of enzymes including *nitrite reductase, nitric oxide reductase*, and *nitrous oxide reductase*, as summarized in Figure 17.37c. During electron transport, a proton motive force is established and ATP is produced by ATPase in the usual fashion. Additional ATP is available when NO_3^- is reduced to N_2 because the NO reductase is linked to proton extrusion (Figure 17.37c).

Other Properties of Denitrifying Prokaryotes

Most denitrifying prokaryotes are phylogenetically members of the Proteobacteria (∞ Sections 12.2–12.19) and are facultative aerobes; aerobic respiration occurs when air is present, even if nitrate is also present in the medium. Many denitrifying bacteria will also reduce other electron acceptors anaerobically such as ferric iron (Fe^{3+}) and certain organic electron acceptors (see Section 17.18). In addition, many denitrifying bacteria can grow by fermentation. Thus, the denitrifying bacteria are quite metabolically diverse in terms of alternative energy-generating mechanisms.

✓ 17.14 Concept Check

Nitrate is a commonly used electron acceptor in anaerobic respiration. Its use requires the enzyme nitrate reductase that reduces nitrate to nitrite. Many bacteria that use nitrate in anaerobic respiration eventually produce nitrogen gas (N_2), a process called *denitrification*.

- ✓ For *E. coli*, why is more energy released in aerobic respiration than during NO_3^- reduction?
- ✓ Where is the dissimilative nitrate reductase found in the cell? What metal(s) does it contain?
- ✓ Why does an organism like *Pseudomonas stutzeri* derive more energy from NO_3^- respiration than does *Escherichia coli*?

17.15 Sulfate Reduction

Several inorganic sulfur compounds are important electron acceptors in anaerobic respiration. A summary of the oxidation states of the key sulfur compounds is given in Table 17.3. Sulfate, the most oxidized form of sulfur, is one of the major anions in seawater and is used by the *sulfate-reducing bacteria*, a group that is widely distributed in nature. The end product of sulfate reduction is H_2S, an important natural product that participates in many biogeochemical processes (∞ Section 19.13). Again, as with nitrogen, it is important to distinguish between assimilative and dissimilative sulfate reduction. Many organisms, including higher plants, algae, fungi, and most prokaryotes, use sulfate as a sulfur source for biosynthesis. The ability to use sulfate as an *electron acceptor* for energy-generating processes, however, involves a large-scale reduction of SO_4^{2-} and is restricted to the sulfate-reducing bacteria. In assimilative sulfate reduction, the H_2S formed is immediately converted into organic sulfur in the form of amino acids, and so on, but in dissimilative sulfate reduction, the H_2S is excreted.

As the reduction potential in Table A1.2 and Figure 17.35 shows, sulfate is a much less favorable electron acceptor than either O_2 or NO_3^-. However, sufficient energy to make ATP is available when an electron donor that yields NADH or FADH is used. Because of the less favorable energetics, growth yields are lower for an organism growing on SO_4^{2-} than for one growing on O_2 or NO_3^-. A list of some of the electron donors used by sulfate-reducing bacteria is given in Table 17.3. The first three compounds listed, H_2, lactate, and pyruvate, are used by a wide variety of sulfate-reducing bacteria; the others have more restricted use. However, a large variety of morphological and physiological types of sulfate-reducing bacteria are known (∞ Section 12.18).

Biochemistry and Energetics of Sulfate Reduction

The reduction of SO_4^{2-} to H_2S, an eight-electron reduction (Table 17.3), proceeds through a number of intermediate stages. The sulfate ion is stable and cannot be reduced without first being activated. Sulfate is activated by means of ATP. The enzyme *ATP sulfurylase* catalyzes the attachment of the sulfate ion to a phosphate of ATP, leading to the formation of **adenosine phosphosulfate (APS)** as shown in Figure 17.38. In dissimilative sulfate reduction, the sulfate moiety of APS is reduced directly to sulfite (SO_3^{2-}) by the enzyme *APS reductase* with the release of AMP. In assimilative reduction, another P is added to APS to form *phosphoadenosine phosphosulfate (PAPS)* (Figure 17.38b), and only then is the sulfate moiety reduced. In both cases, the first product of sulfate reduction is *sulfite*, SO_3^{2-}. Once SO_3^{2-} is formed, sulfide is formed by the enzyme *sulfite reductase* (Figure 17.38b).

In the process of dissimilative sulfate reduction, electron transport reactions occur leading to proton motive force formation, and this drives ATP synthesis by ATPase. A major electron carrier is cytochrome c_3, a periplasmic low potential cytochrome (Figure 17.39). Cytochrome c_3 accepts electrons from a periplasmically located hydrogenase (see below) and transfers these electrons to a membrane-associated protein complex called Hmc that carries them across the cytoplasmic membrane, thus making them available to APS reductase and sulfite reductase, which are cytoplasmic enzymes (Figure 17.39).

The enzyme hydrogenase appears to play a central role in sulfate reduction whether an organism like *Desulfovibrio desulfuricans* is growing on H_2, *per se*, or on an organic compound, like lactate. Evidence suggests that lactate is converted through pyruvate to acetate (the

TABLE 17.3	Sulfur compounds and electron donors for sulfate reduction	
Compound		**Oxidation state**
Oxidation states of key sulfur compounds		
Organic S (R—SH)		−2
Sulfide (H_2S)		−2
Elemental sulfur (S^0)		0
Thiosulfate ($S_2O_3^{2-}$)		+2 (average per S)
Sulfur dioxide (SO_2)		+4
Sulfite (SO_3^{2-})		+4
Sulfate (SO_4^{2-})		+6
Some electron donors used for sulfate reduction		
H_2	Acetate	
Lactate	Propionate	
Pyruvate	Butyrate	
Ethanol and other alcohols	Long-chain fatty acids	
Fumarate	Benzoate	
Malate	Indole	
Choline	Hexadecane	

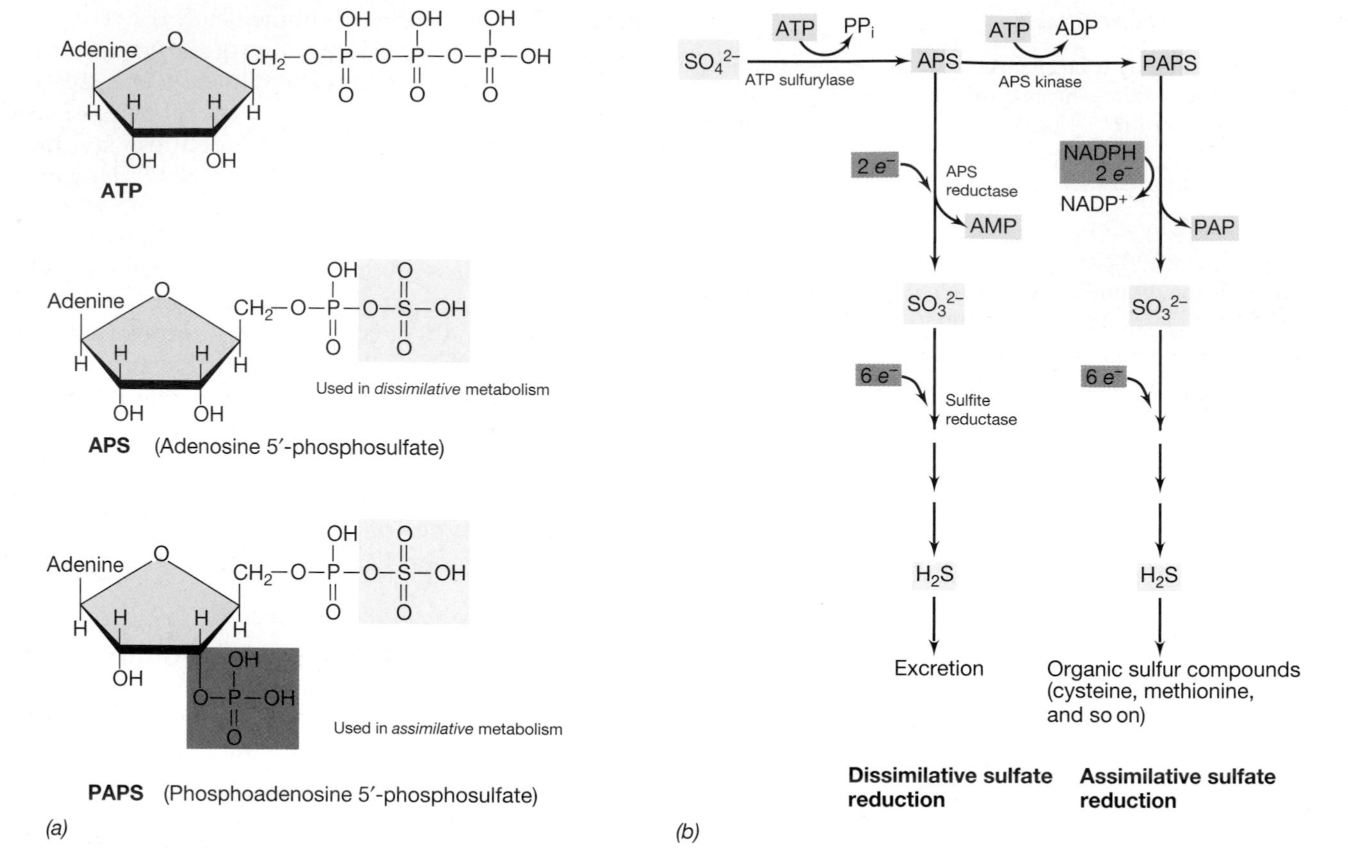

Figure 17.38 Biochemistry of sulfate reduction. (a) Two forms of *active sulfate* can be made, adenosine 5'-phosphosulfate (APS) and phosphoadenosine 5'-phosphosulfate (PAPS). (b) Schemes of assimilative and dissimilative sulfate reduction.

latter is mainly excreted because *D. desulfuricans* is a nonacetate-oxidizing sulfate reducer, ∞ Section 12.18), with the production of H_2. The H_2 produced crosses the cytoplasmic membrane and is oxidized by the periplasmic hydrogenase to initiate a proton motive force (Figure 17.39). Growth yields of sulfate-reducing bacteria suggest that one ATP is produced for each SO_4^{2-} reduced to HS^-. With H_2 the reaction is:

$$4\,H_2 + SO_4^{2-} + H^+ \rightarrow HS^- + 4\,H_2O \qquad \Delta G^{0\prime} = -152\ kJ$$

When lactate or pyruvate is the electron donor, not only is ATP produced from the proton motive force, but additional ATP is produced during the oxidation of pyruvate to acetate plus CO_2 via acetyl-CoA and acetyl-phosphate (see Section 17.19 for further discussion of this).

Acetate Use and Autotrophy

Many sulfate reducers can completely oxidize acetate to CO_2 as an electron donor for sulfate reduction (∞ Section 12.18):

$$CH_3COO^- + SO_4^{2-} + 3\,H^+ \rightarrow$$
$$2\,CO_2 + H_2S + 2\,H_2O \quad \Delta G^{0\prime} = -57.5\ kJ$$

Although the energetics of this process are not as well understood as H_2 or lactate metabolism by *D. desulfuricans*,

the mechanism for acetate oxidation is well understood. With few exceptions acetate is oxidized to CO_2 via the acetyl-CoA pathway, a series of reversible reactions used by a wide variety of anaerobes for acetate synthesis or acetate oxidation (see Section 17.16). This pathway employs the key enzyme *carbon monoxide dehydrogenase* and was first discovered in homoacetogenic bacteria that make acetate from $H_2 + CO_2$ as a mechanism of energy conservation (see Section 17.16). A few sulfate-reducing bacteria can also grow autotrophically in an anoxic mineral salts medium containing H_2 (as electron donor), SO_4^{2-} (as electron acceptor), and CO_2 (as carbon source). When growing under these conditions, autotrophic sulfate reducers use the acetyl-CoA pathway as a means of producing cell material. The acetate-oxidizing sulfate-reducing bacterium *Desulfobacter* lacks acetyl-CoA pathway enzymes and oxidizes acetate through the citric acid cycle, but this seems to be the exception rather than the rule.

Sulfur Disproportionation

Certain sulfate-reducing bacteria are capable of a unique form of energy metabolism called *disproportionation*, using sulfur compounds of intermediate oxidation state. The term disproportionation refers to the splitting of a compound into two new compounds, one of which is

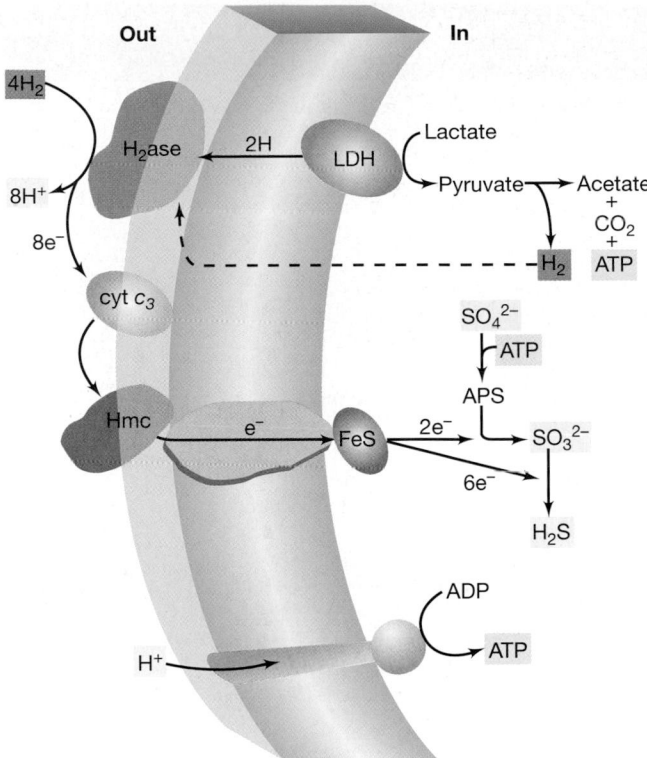

Figure 17.39 Electron transport and energy conservation in sulfate-reducing bacteria. In addition to external hydrogen (H_2), H_2 originating from the catabolism of organic compounds such as lactate and pyruvate can fuel hydrogenase. The enzymes hydrogenase, cytochrome c_3, and a cytochrome complex (Hmc) are periplasmic proteins. A separate protein functions to shuttle electrons across the cytoplasmic membrane from Hmc to a cytoplasmic iron-sulfur protein that supplies electrons to APS reductase (forming SO_3^{2-}) and sulfite reductase (forming H_2S).

more oxidized and one of which is *more reduced* than the original substrate. In the present discussion, we describe the disproportionation of thiosulfate ($S_2O_3^{2-}$), sulfite (SO_3^{2-}), and sulfur (S^0).

Desulfovibrio sulfodismutans can disproportionate sulfur compounds as follows:

$$S_2O_3^{2-} + H_2O \rightarrow SO_4^{2-} + H_2S$$

$$\Delta G^{0\prime} = -21.9 \text{ kJ/reaction}$$

Note that one sulfur atom of $S_2O_3^{2-}$ becomes more oxidized (forming SO_4^{2-}) and the other more reduced (forming H_2S). The oxidation of thiosulfate by *D. sulfodismutans* drives formation of a proton motive force that is used by the organism to make ATP. Other reduced sulfur compounds such as sulfite (SO_3^{2-}) and sulfur (S^0) can also be disproportionated by one or another sulfate reducer. These forms of metabolism may be ways by which sulfate-reducing bacteria can recover energy from sulfur intermediates produced from the oxidation of H_2S by sulfur chemolithotrophs that coexist with them in nature (Section 19.13).

Phosphite Oxidation

At least one sulfate-reducing bacterium has been isolated that is capable of coupling phosphite (HPO_3^-) oxidation to sulfate reduction. The reaction observed is a chemolithotrophic one, and the products are phosphate and sulfide:

$$4 \, HPO_3^- + SO_4^{2-} + H^+ \rightarrow 4 \, HPO_4^{2-} + HS^-$$

$$\Delta G^{0\prime} = -364 \text{ kJ}$$

The organism involved, *Desulfotignum phosphitoxidans*, requires only CO_2 for its carbon needs (it is thus an *autotroph*) and is a strict anaerobe. The latter is not surprising since phosphite spontaneously oxidizes in air and thus aerobic phosphite oxidizers likely do not exist. Although the source of phosphite in nature is not clear, the existence of an organism capable of using phosphite as sole energy source suggests that it is produced in anoxic environments, perhaps from the degradation of organophosphates. Along with sulfur disproportionation, also a chemolithotrophic process, and H_2 utilization, phosphite oxidation emphasizes the metabolic diversity of sulfate-reducing bacteria with respect to chemolithotrophy.

✓ 17.15 Concept Check

The sulfate-reducing bacteria reduce sulfate to hydrogen sulfide. The reduction of sulfate first requires activation by a reaction with ATP to form the compound adenosine phosphosulfate (APS). Electron donors for sulfate reduction include H_2, organic compounds, and even phosphite. Disproportionation of sulfur compounds is an additional energy-yielding strategy for certain members of this group.

- ✓ Define the following: S^0, SO_4^{2-}, SO_3^{2-}, $S_2O_3^{2-}$, H_2S.
- ✓ How is sulfate converted to sulfite?
- ✓ Why is H_2 of importance to sulfate-reducing bacteria?
- ✓ Give an example of disproportionation.

17.16 Acetogenesis

Carbon dioxide, CO_2, is common in nature and usually abundant in anoxic habitats because it is a major product of energy metabolism of chemoorganotrophs. Two major groups of strictly anaerobic prokaryotes can use CO_2 as an electron acceptor in energy metabolism, *homoacetogens* and *methanogens*. Hydrogen (H_2) is a major electron donor for both of these organisms, and an overview of the processes of methanogenesis and acetogenesis is shown in Figure 17.40. Both processes result in the generation of ion gradients either of H^+ or Na^+, which drives ATPases in the membrane, while acetogenesis also involves energy conservation via substrate-level phosphorylation. In this section we focus on acetogenesis and in the next section, methanogenesis.

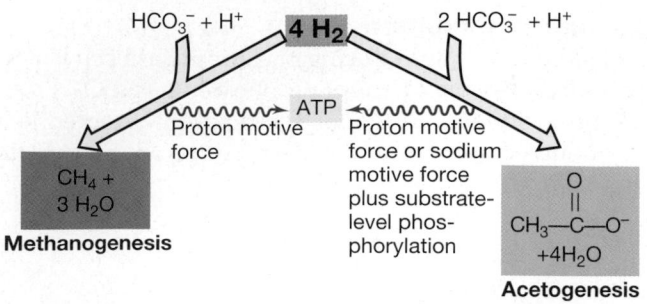

Figure 17.40 The contrasting processes of methanogenesis and acetogenesis. The free energy ($\Delta G^{0\prime}$) released in each reaction as drawn is methanogenesis, -136 kJ, and acetogenesis, -105 kJ.

Organisms and Pathway

Homoacetogens can carry out the reaction:

$$4 H_2 + H^+ + 2 HCO_3^- \rightarrow CH_3COO^- + 4 H_2O$$

In addition to H_2, electron donors for acetogenesis include a variety of C_1 compounds, sugars, organic and amino acids, alcohols, and certain nitrogen bases, depending on the organism. Many homoacetogens can also reduce NO_3^- and $S_2O_3^{2-}$; however, CO_2 reduction is probably the major reaction of ecological significance.

The major unifying thread among homoacetogens is the pathway of CO_2 reduction. Homoacetogens convert CO_2 to acetate by the *acetyl-CoA pathway* (see later), and in many homoacetogens autotrophic growth via this pathway also occurs. The acetyl-CoA pathway is also known as the *Ljungdahl-Wood pathway* in honor of its discoverers, Lars Ljungdahl and Harland Wood. A list of the major organisms that produce acetate or oxidize acetate via the acetyl-CoA pathway is given in Table 17.4. Organisms such as *Acetobacterium woodii* and *Clostridium aceticum* can grow either chemoorganotrophically by fermentation of sugars (reaction 1) or chemolithotrophically through the reduction of CO_2 to acetate with H_2 (reaction 2) as electron donor; in either case the major product is acetate:

(1) $C_6H_{12}O_6 \rightarrow 3 CH_3COO^- + 3 H^+$

(2) $2 HCO_3^- + 4 H_2 + H^+ \rightarrow CH_3COO^- + 4 H_2O$

Homoacetogens ferment glucose via the glycolytic pathway converting glucose to two molecules of pyruvate and two molecules of NADH (the equivalent of 4 H). From this point, two molecules of acetate are produced as follows:

(3) $2 \text{ pyruvate}^- \rightarrow 2 \text{ acetate}^- + 2 CO_2 + 4 H$

The third acetate of the homoacetate fermentation comes from the reduction of the two molecules of CO_2 generated in reaction (3), using the four electrons generated from glycolysis *plus* the four electrons produced from the oxidation of two pyruvates to two acetates [reaction

Table 17.4   Organisms employing the acetyl-CoA pathway of CO_2 fixation

I. **Acetate synthesis, the result of energy metabolism**
   Acetoanaerobium noterae
   Acetobacterium woodii
   Acetobacterium wieringae
   Acetogenium kivui
   Acetitomaculum ruminis
   Clostridium aceticum
   Clostridium thermoaceticum
   Clostridium formicoaceticum
   Desulfotomaculum orientis
   Sporomusa paucivorans
   Eubacterium limosum (also produces butyrate)
   Treponema sp. strains ZAS-1 and ZAS-2
       (termite gut spirochetes)
II. **Acetate synthesis in autotrophic metabolism**
   Autotrophic homoacetogenic bacteria
   Autotrophic methanogens
   Autotrophic sulfate-reducing bacteria
III. **Acetate oxidation in energy metabolism**
   Reaction: Acetate + $2 H_2O \rightarrow 2 CO_2 + 8 H$
       Group II sulfate reducers (other than *Desulfobacter*)
   Reaction: Acetate $\rightarrow CO_2 + CH_4$
       Acetotrophic methanogens (*Methanosarcina*,
       Methanosaeta)

(3)]. Starting from pyruvate, then, the overall production of acetate can be written as

$$2 \text{ pyruvate}^- + 4 H \rightarrow 3 \text{ acetate}^- + H^+$$

Most homoacetogenic bacteria that produce and excrete acetate in energy metabolism are gram-positive, and many are classified in the genus *Clostridium*. A few other gram-positive and many different gram-negative bacteria use the acetyl-CoA pathway for autotrophic purposes, reducing CO_2 to acetate for cell carbon. The acetyl-CoA pathway functions in autotrophic growth for certain sulfate-reducing bacteria (∞ Sections 12.18 and 19.13), and is also used by the methanogens, most of which can grow autotrophically on $H_2 + CO_2$ (∞ Sections 13.4, 17.17, and 19.10). By contrast, certain bacteria employ the reactions of the acetyl-CoA pathway primarily in the *reverse* direction as a means of oxidizing acetate to CO_2. These include acetotrophic methanogens (∞ Section 13.4) and sulfate-reducing bacteria (∞ Section 12.18).

Reactions of the Acetyl-CoA Pathway

Unlike other autotrophic pathways such as the Calvin cycle (see Section 17.6) or the reverse citric acid cycle (see Section 17.7), the acetyl-CoA pathway of CO_2 fixation is *not* a cycle. Instead it involves the reduction of CO_2 via two linear pathways—one molecule of CO_2 is reduced to the methyl group of acetate, and the other molecule of CO_2 is reduced to the carbonyl group—followed by their assembly at the end to form acetyl-CoA (Figure 17.41). A key

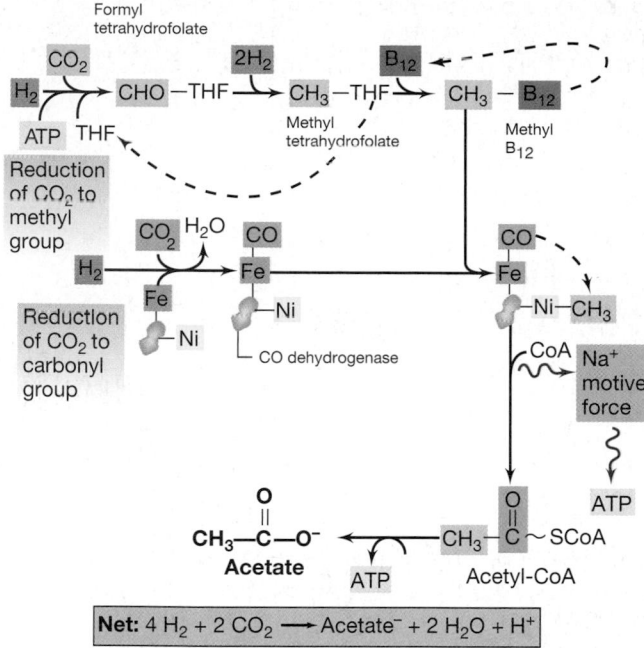

Figure 17.41 Reactions of the acetyl-CoA pathway. THF, Tetrahydrofolate; B_{12}, vitamin B_{12} in an enzyme-bound intermediate. CO is bound to an Fe atom in CO dehydrogenase, and the CH_3 group to a nickel atom in an organic nickel compound in CO dehydrogenase. Note how formation of acetate powers a Na^+ pump that is used to drive ATP synthesis and that ATP synthesis also occurs in the conversion of acetyl-CoA to acetate.

enzyme of the acetyl-CoA pathway is *carbon monoxide (CO) dehydrogenase*. CO dehydrogenase is a complex enzyme that contains the metals Ni, Zn, and Fe as cofactors. CO dehydrogenase catalyzes the following reaction:

$$CO_2 + H_2 \rightarrow CO + H_2O$$

and the CO that is produced ends up in the *carbonyl* ($—COO^-$) position of acetate (Figure 17.41). The methyl group of acetate originates from the reduction of CO_2 by a series of reactions involving the coenzyme *tetrahydrofolate* (Figure 17.41). The methyl group that is formed is then transferred from tetrahydrofolate to an enzyme containing vitamin B_{12} as cofactor (Figure 17.41). In the final step of the pathway, the CH_3 group is combined with CO in CO dehydrogenase to form acetate. Interestingly, the reaction mechanism here involves the CH_3 group, which is attached to an atom of nickel in the enzyme, combining with CO, which is bound to an atom of Fe in the enzyme, along with coenzyme A to form the final product, acetyl-CoA. The significance of this resides in the fact that this reaction mechanism was the first alkyl nickel reaction to be discovered in biochemistry.

Because homoacetogens can grow at the expense of reactions of the acetyl-CoA pathway, this reaction sequence must be overall an energy-conserving one

(Figure 17.41). One site of ATP synthesis is during the conversion of acetyl-CoA to acetate plus ATP (via acetyl-P) (see Section 17.19). However, additional energy-conserving steps occur because a Na^+ gradient (sodium motive force, analogous to a proton motive force but involving Na^+ instead of H^+) is established across the cytoplasmic membrane during acetogenesis. This energized state of the membrane allows for energy conservation via the action of a Na^+-powered ATPase. A similar situation occurs in the succinate fermenter *Propionigenium*, whose energy conservation from Na^+ gradients will be discussed in Section 17.20.

✓ 17.16 Concept Check

Homoacetogens are anaerobes that reduce CO_2 to acetate, usually with H_2 as electron donor. The mechanism of acetate formation is the acetyl-CoA pathway, a series of reactions widely distributed in obligate anaerobes as either a mechanism of autotrophy or for acetate catabolism.

✓ Draw the structure of acetate and identify the carbonyl group and the methyl group. What key enzyme of the acetyl-CoA pathway produces the *carbonyl* group of acetate?

✓ How do homoacetogens make ATP from the synthesis of acetate?

✓ If catabolism of fructose via glycolysis produces only *two* molecules of acetate, how can *Clostridium aceticum* ferment fructose by this pathway and produce *three* molecules of acetate?

17.17 Methanogenesis

The biological production of methane is carried out by a group of strictly anaerobic *Archaea* called the *methanogens*. We considered the basic properties and taxonomy of the methanogens in Section 13.4, here we focus on the biochemistry and bioenergetics of methanogenesis. Research on methanogenesis has revealed that the biological production of methane occurs through a series of reactions involving novel coenzymes and amazing complexity. We begin our discussion with a consideration of the coenzymes, as they are central to the reactions we will consider, and on the production of methane starting from $H_2 + CO_2$.

C_1 Carriers in Methanogenesis

The key coenzymes in methanogenesis can be divided into two classes, those involved in carrying the C_1 unit from the initial substrate, CO_2, to the final product, CH_4, and those that function in redox reactions to supply the electrons necessary for the reduction of CO_2 to CH_4 (Figure 17.42, and see Figure 17.44).

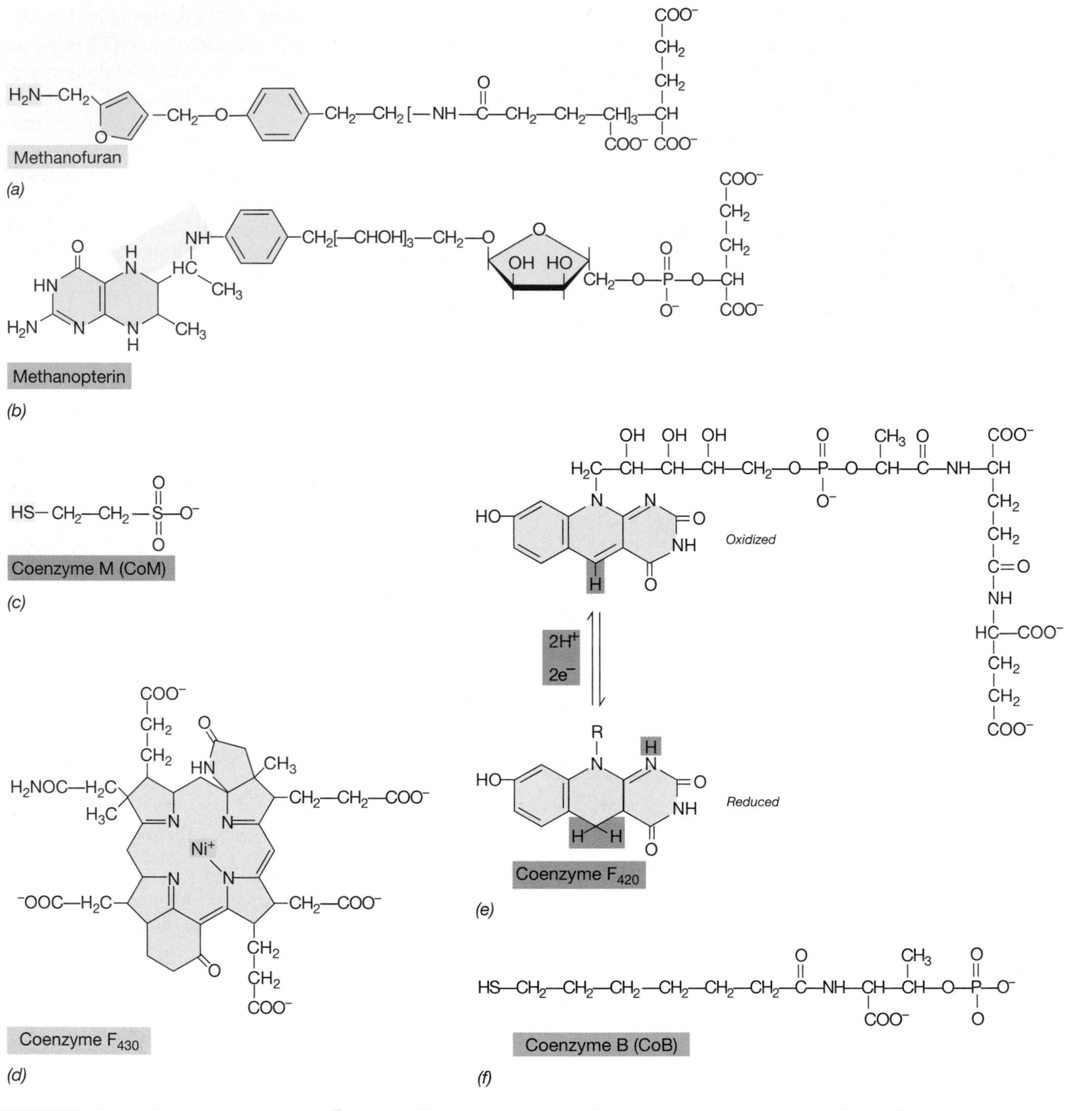

Figure 17.42 Coenzymes of methanogenic *Archaea*. The atoms shaded in brown or yellow are the sites of oxidation-reduction reactions (F_{420}—brown) or the position to which the C_1 moiety is attached during the reduction of CO_2 to CH_4 (methanofuran, methanopterin, and coenzyme M—yellow). The colors used to highlight a particular coenzyme itself (CoB is orange, for example) are used throughout in Figures 17.44–17.46 and can be used to follow the reactions in each figure.

The coenzyme **methanofuran** is involved in the first step of methanogenesis. Methanofuran contains the five-membered furan ring and an amino nitrogen atom that binds CO_2 (Figure 17.42*a*). **Methanopterin** (Figure 17.42*b*) is a methanogenic coenzyme that resembles the vitamin folic acid (∞ Figure 20.16*c*) and is the C_1 carrier in the intermediate steps of CO_2 reduction to CH_4. **Coenzyme M (CoM)** (Figure 17.42*c*) is a small molecule that is involved in the terminal step of methanogenesis, the conversion of a methyl group (CH_3) to CH_4. Although not

a C_1 carrier, the nickel-containing tetrapyrrole **coenzyme F_{430}** (Figure 17.42*d*) is also involved in the terminal step of methanogenesis as part of the methyl reductase enzyme complex (see later).

Redox Coenzymes

The coenzymes F_{420} and **7-mercaptoheptanoyl-threonine phosphate**, or **coenzyme B (CoB)**, are electron *donors* in methanogenesis. Coenzyme F_{420} (Figure 17.42*e*) is a flavin derivative, structurally resembling the common flavin coenzyme FMN (∞ Figure 5.15). F_{420} also plays a role in methanogenesis as the electron donor in several steps of CO_2 reduction (see Figure 17.44). The oxidized form of F_{420} absorbs light at 420 nm and fluoresces blue-green (Figure 17.43), and such fluorescence is a useful tool for microscopic identification of an organism as a methanogen. CoB is involved in the terminal step of methanogenesis catalyzed by the **methyl reductase enzyme complex**. As shown in Figure 17.42*f*, the structure of CoB is rather simple and the coenzyme resembles the vitamin pantothenic acid (part of acetyl-CoA) (∞ Figure 5.12). With this overview of the coenzymes of methanogenesis, let us now look at the reactions involved in the reduction of CO_2 to CH_4.

Biochemistry of CO_2 Reduction to CH_4

The reduction of CO_2 to CH_4 is generally H_2 dependent, but formate, carbon monoxide, and even certain organic compounds such as alcohols can supply the electrons for CO_2 reduction. For example, 2-propanol can be oxidized to acetone, yielding electrons for methanogenesis in some species. But in general, the production of CH_4 from CO_2 is driven by molecular hydrogen (H_2).

The steps in CO_2 reduction, shown in Figure 17.44, are summarized as follows:

1. CO_2 is activated by a methanofuran-containing enzyme and subsequently reduced to the formyl level.

2. The formyl group is transferred from methanofuran to an enzyme containing methanopterin (MP in Figure 17.44) and subsequently dehydrated and reduced in two separate steps to the methylene and methyl levels.

3. The methyl group is transferred from methanopterin to an enzyme containing CoM.

4. Methyl-CoM is reduced to methane by the methyl reductase system in which F_{430} and CoB are intimately

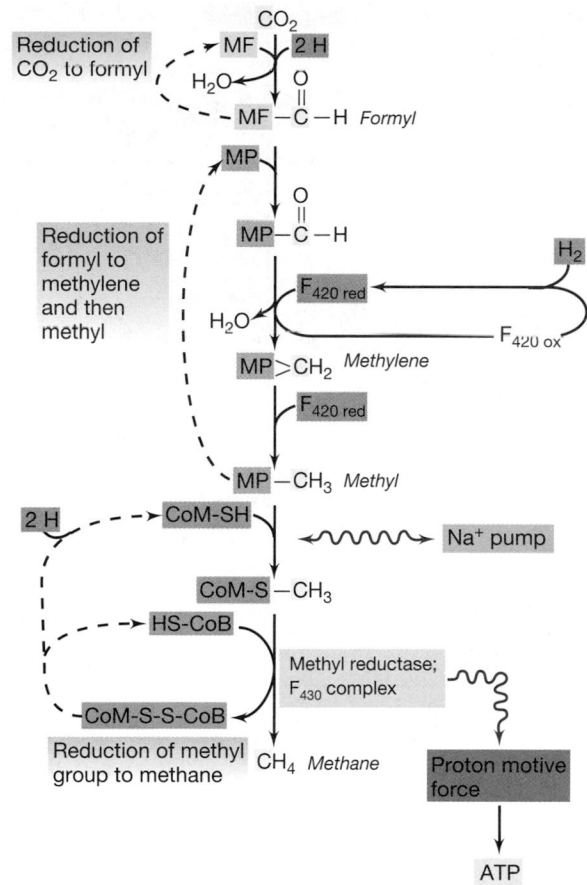

Figure 17.44 Pathway of methanogenesis from CO_2. MF, Methanofuran; MP, methanopterin; CoM, coenzyme M; F_{420red}, reduced coenzyme F_{420}; F_{430}, coenzyme F_{430}; CoB, coenzyme B. The carbon atom reduced is shown in yellow, and the source of electrons is highlighted in brown. See Figure 17.42 for the structures of the coenzymes and the text for discussion of the reversible Na^+ pump. The immediate electron donor in the first step of methanogenesis is unknown. Electrons for CO_2 reduction generally come from H_2 but in certain methanogens, a few organic compounds can be oxidized to yield electrons for CO_2 reduction.

mately involved. Coenzyme F_{430} removes the CH_3 group from CH_3-CoM, forming a Ni^{2+}-CH_3 complex. This is reduced by electrons from CoB, generating CH_4 and a disulfide complex of CoM and CoB (CoM-S-S—CoB). Free CoM and CoB are regenerated by reduction of this complex with H_2, and as we will see, it is this reaction that allows for energy conservation in methanogenesis.

Methanogenesis from Methyl Compounds and Acetate

We learned in Section 13.4 that in addition to $H_2 + CO_2$, methane can be formed from a variety of methylated compounds. Methyl compounds such as methanol are catabolized by donating methyl groups to a corrinoid protein to form CH_3-corrinoid (Figure 17.45). Corrinoids are the parent structures of such compounds as vitamin

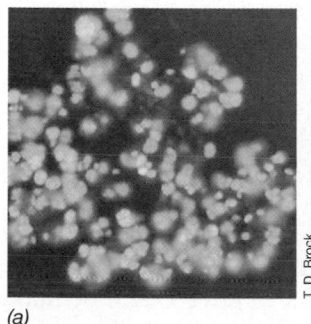
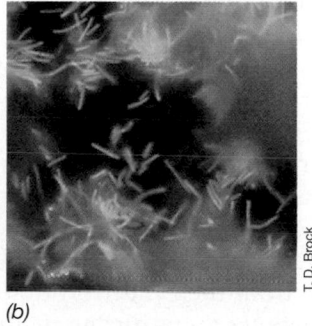

(a)   *(b)*

Figure 17.43 (a) Autofluorescence in cells of the methanogen *Methanosarcina barkeri* due to the presence of the unique electron carrier F_{420}. A single cell is about 1.7 μm in diameter. The organisms were made visible with blue light in a fluorescence microscope (∞ Section 4.1 and Figure 4.6). (b) F_{420} fluorescence in cells of the methanogen *Methanobacterium formicicum*. A single cell is about 0.6 μm in diameter.

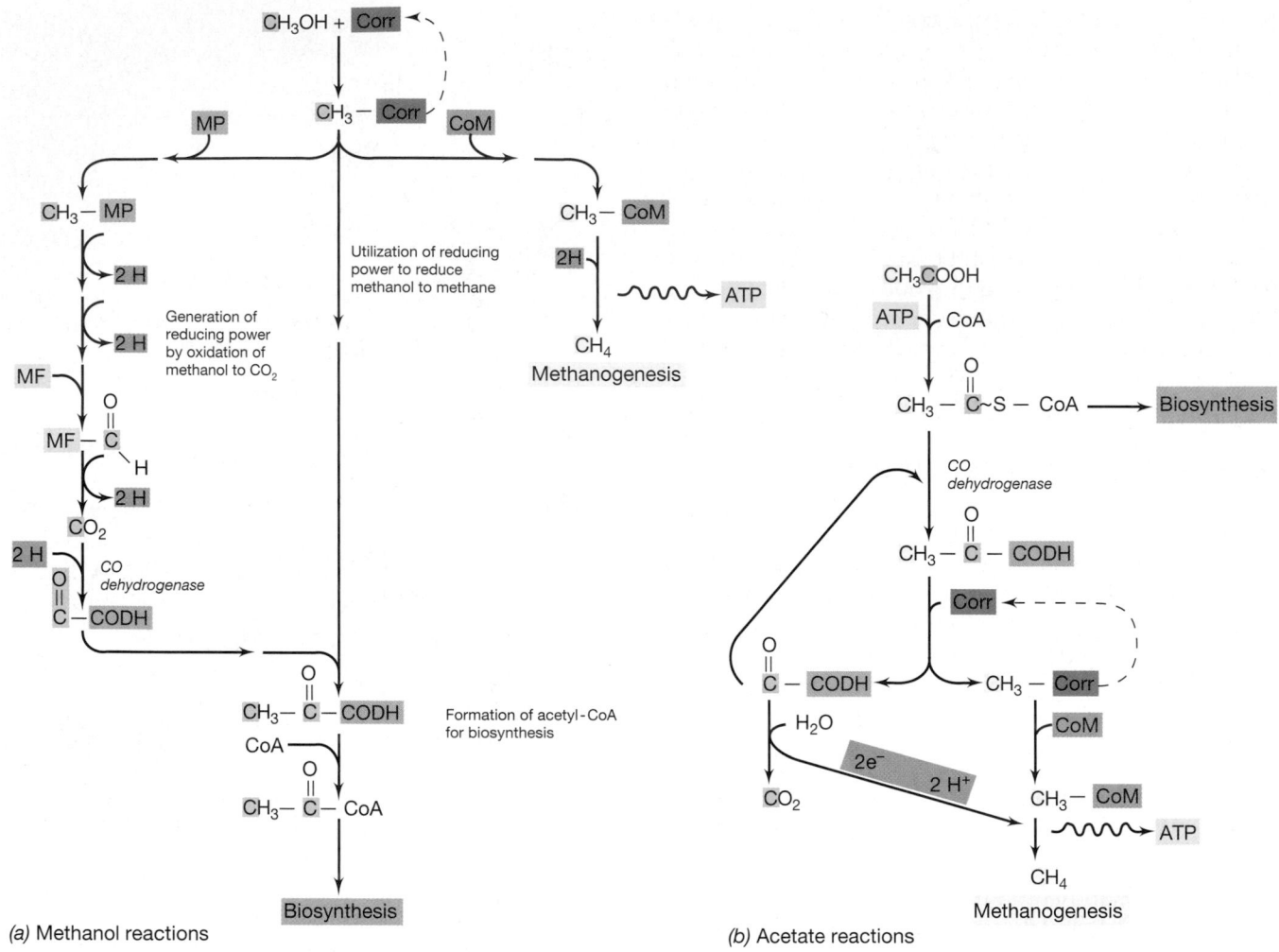

(a) Methanol reactions

(b) Acetate reactions

Figure 17.45 Utilization of reactions of the acetyl-CoA pathway during growth on methanol (a) or acetate (b) by methanogenic *Archaea*. For growth on methanol, most methanol carbon is converted to CH$_4$, while a smaller amount is converted to either CO$_2$ or, via formation of acetyl-CoA, is assimilated into cell material. Abbreviations and color coding are as in Figures 17.42 and 17.44; Corr, corrinoid-containing protein.

B$_{12}$ and contain a porphyrin-like corrin ring with a central cobalt atom (∞ Figure 30.12). The CH$_3$-corrinoid complex donates the methyl group to CoM, yielding CH$_3$-CoM from which methane is formed in the same way as in the terminal step of CO$_2$ reduction just described (compare Figures 17.44 and 17.45a). If reducing power (such as H$_2$) is not available to drive the terminal step, some of the methanol must be oxidized to CO$_2$ to yield electrons, and this occurs by reversal of steps in methanogenesis (Figure 17.45a).

When *acetate* is the substrate for methanogenesis it is first activated to acetyl-CoA, which can interact with carbon monoxide dehydrogenase of the acetyl-CoA pathway (see Section 17.16). Then, the methyl group of acetate is transferred to the corrinoid enzyme to yield CH$_3$-corrinoid, and from there it goes through the CoM-mediated terminal step of methanogenesis (Figure 17.45b).

Autotrophy

Autotrophy in methanogens occurs via the reactions of the acetyl-CoA pathway discussed in Section 17.16, and as we have seen, parts of this pathway are already integrated into the catabolism of methanol and acetate (Figure 17.45). However, methanogens lack the tetrahydrofolate-driven series of reactions of the acetyl-CoA pathway that lead to the production of a methyl group (see Figure 17.41). These, of course, are unnecessary, since methanogens either derive methyl groups directly from their electron donors (Figure 17.45) or make methyl groups during methanogenesis from H$_2$ + CO$_2$ (Figure 17.44); thus methyl groups are present in abundance in the cell to begin with. The carbonyl group of the acetate produced during autotrophic growth of methanogens is derived from CO dehydrogenase, and the terminal step in acetate synthesis occurs as described for homoacetogens (see Section 17.16 and Figure 17.41).

Energy Conservation in Methanogenesis

Under standard conditions the free energy change in the reduction of CO_2 to CH_4 with H_2 is $-131 \, kJ/mol$. This is sufficient for the synthesis of at least one ATP. We mentioned that energy conservation in methanogenesis is linked to the terminal step, the *methyl reductase* step (Figure 17.44). The interaction of CoB with CH_3—CoM in this terminal step forms CH_4 and a heterodisulfide, CoM-S—S-CoB. The latter complex is then reduced with electrons supplied from F_{420} to yield CoM-SH and CoB-SH (Figure 17.44). This reduction, carried out by the enzyme *heterodisulfide reductase*, is exergonic and is associated with the extrusion of protons across the membrane, creating a proton motive force (Figure 17.46). Dissipation of the proton gradient by a proton-translocating ATPase (⌖ Section 5.12 discusses ATPases) drives ATP synthesis during methanogenesis in the same way that this process occurs in other forms of respiration. Electron flow to the heterodisulfide reductase involves a unique membrane-integrated electron carrier, a phenazine compound called *methanophenazine* (Figure 17.46). In the electron transport process in the terminal step, methanophenazine is alternately reduced (by F_{420}) and then oxidized (by a *b*-type cytochrome), which is the ultimate donor of electrons to the heterodisulfide reductase (Figure 17.46).

Methanogenesis from methyl compounds is also linked to the heterodisulfide reductase proton pump, but an additional factor is involved. As previously mentioned, in the absence of H_2, methanogenesis from compounds like CH_3OH requires that some of the CH_3OH be oxidized to CO_2 to generate the electrons needed for methyl reduction to methane. This requires an energy input and occurs at the expense of a Na^+ motive force (a Na^+ energized membrane potential, see Section 17.20). The energy inherent in this potential is derived from the conversion of CH_3-MP to CH_3-CoM during methanogenesis; the reverse reaction consumes energy and is driven by the Na^+ pump. Further oxidative steps in the conversion of methyl groups of CO_2 proceed by reversal of the enzymatic steps leading to CH_4 formation from CO_2 (see Figure 17.44). Thus, in methanogens we see *two* types of ion pumps: a typical proton pump used to drive adenosine triphosphate (ATP) synthesis and a reversible Na^+ pump that can function to drive methyl group oxidation.

✓ 17.17 Concept Check

Methanogenesis is the biological production of CH_4 from either CO_2 reduction with H_2 or from methylated compounds. A variety of unique coenzymes are involved in methanogenesis and the process is strictly anaerobic. Energy conservation in methanogenesis involves both proton and sodium ion gradients.

- ✓ What coenzymes function as C_1 carriers in methanogenesis? As electron donors?
- ✓ Why are the steps in CH_3OH catabolism during methanogenesis different if H_2 is present than when it is not present?
- ✓ Concerning autotrophy, why are only some but not all of the enzymes of the acetyl-CoA pathway present in methanogens?
- ✓ How is a proton motive force produced in methanogenesis?

17.18 Ferric Iron, Manganese, Chlorate, and Organic Electron Acceptors

In addition to the electron acceptors for anaerobic respiration discussed thus far, ferric iron (Fe^{3+}), manganic ion (Mn^{4+}), chlorate (ClO_3^-), and various organic compounds are important electron acceptors for bacteria in nature (Figure 17.47). A wide diversity of bacteria are able to reduce these acceptors, especially Fe^{3+}, and many are able to reduce other acceptors as well, such as NO_3^- and S^0 (see Sections 17.14 and 17.15).

Ferric Iron Reduction

Ferric iron is an electron acceptor for energy metabolism in a wide variety of both chemoorganotrophic and chemolithotrophic bacteria, and because Fe^{3+} is abundant in

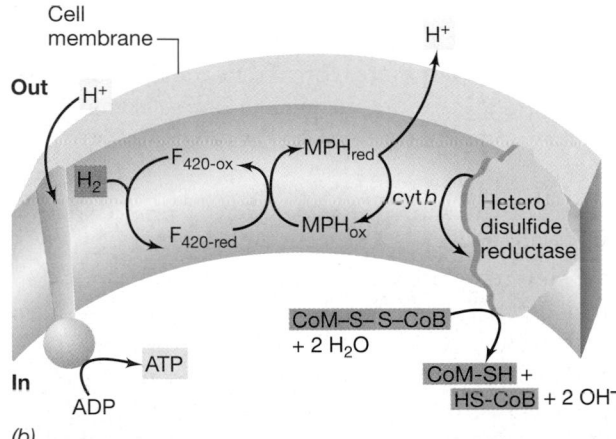

(a)

(b)

Figure 17.46 Energy conservation in methanogenesis. (a) Structure of methanophenazine (MPH in part b), an electron carrier in the electron transport chain leading to ATP synthesis. The central ring of the molecule can be alternately reduced and oxidized. (b) Steps in electron transport. Electrons originating from H_2 reduce F_{420} and then methanophenazine. The latter, through a cytochrome of the *b* type, reduces heterodisulfide reductase with the extrusion of protons to the outside of the membrane. In the final step, heterodisulfide reductase reduces Co-M-S—S-CoB to HS-CoM and HS-CoB. Refer to Figure 17.42 for the structures of CoM and CoB.

Acceptor	Reaction	E_0' of couple (V)	Product
Chlorate	$ClO_3^- \xrightarrow[\text{6 H}^+]{\text{6 e}^-} Cl^- + 3\ H_2O$	+1.03	Chloride
Manganic ion	$Mn^{4+} \xrightarrow{\text{2 e}^-} Mn^{2+}$	+0.798	Manganous ion
Selenate	$^-O-\overset{\overset{O}{\|\|}}{\underset{\underset{O}{\|\|}}{Se}}-O^- \xrightarrow[\text{2 H}^+]{\text{2 e}^-} \overset{O^-}{\underset{O^-}{Se}}=O + H_2O$	+0.475	Selenite
Ferric ion	$Fe^{3+} \xrightarrow{\text{e}^-} Fe^{2+}$	+0.2	Ferrous ion
Dimethyl sulfoxide (DMSO)	$H_3C-\overset{\overset{}{\underset{\downarrow}{S}}}{\underset{O}{}}-CH_3 \xrightarrow[\text{2 H}^+]{\text{2 e}^-} (CH_3)_2S + H_2O$	+0.16	Dimethyl sulfide (DMS)
Arsenate	$^-O-\overset{O^-}{\underset{O^-}{As}}=O \xrightarrow[\text{2 H}^+]{\text{2 e}^-} \overset{O^-}{\underset{O_\bullet}{As}}-O^- + H_2O$	+0.139	Arsenite
Trimethylamine-*N*-oxide (TMAO)	$H_3C-\overset{\overset{CH_3}{\|}}{\underset{\underset{O}{\downarrow}}{N}}-CH_3 \xrightarrow[\text{2 H}^+]{\text{2 e}^-} (CH_3)_3N + H_2O$	+0.13	Trimethylamine (TMA)
Fumarate	$\overset{O}{\underset{^-O}{C}}-\overset{H}{\underset{}{C}}=\overset{}{\underset{H}{C}}-\overset{O}{\underset{O^-}{C}} \xrightarrow[\text{2 H}^+]{\text{2 e}^-} \overset{O}{\underset{^-O}{C}}-CH_2-CH_2-\overset{O}{\underset{O^-}{C}}$	+0.03	Succinate

Figure 17.47 Some alternative electron acceptors for anaerobic respirations.

nature, its reduction is a major form of anaerobic respiration. The reduction potential of the Fe^{3+}/Fe^{2+} couple is somewhat electropositive ($E_0' = +0.2$ V at pH 7), and because of this, Fe^{3+} reduction can be coupled to the oxidation of several organic and inorganic electron donors. Various organic compounds, including aromatic compounds, can be oxidized anaerobically by ferric iron reducers with electrons traveling through electron transport chains that terminate in a ferric iron reductase system. Such electron flow establishes a proton motive force that can be used to generate ATP. Much research on the energetics of ferric iron reduction has been done in the gram-negative bacterium *Shewanella putrefaciens*, in which Fe^{3+}-dependent anaerobic growth occurs with various organic electron donors. Other important Fe^{3+} reducers include *Geobacter*, *Geospirillum*, and *Geovibrio*.

Geobacter metallireducens has been a model for study of the physiology of Fe^{3+} reduction. This organism can oxidize acetate with Fe^{3+} as an acceptor as follows:

$$Acetate^- + 8\ Fe^{3+} + 4\ H_2O \rightarrow$$
$$2\ HCO_3^- + 8\ Fe^{2+} + 9\ H^+ \qquad \Delta G^{0'} = -233\,kJ$$

Geobacter can also use H_2 or other organic electron donors including the aromatic hydrocarbon toluene (see Figure 17.56*d* for the structure of toluene). This may be of environmental significance because toluene from accidental spills or leakage from hydrocarbon storage tanks often contaminates ferric-rich aquifers, and it has been suggested that organisms like *Geobacter* may be natural cleanup agents in such environments.

Reduction of Manganese (Mn^{4+}) and Other Inorganic Substances

The metal manganese has a number of oxidation states, of which Mn^{4+} and Mn^{2+} are the most stable and biologically relevant. Anoxic reduction of Mn^{4+} to Mn^{2+} is carried out by a variety of microorganisms, mostly chemoorganotrophs. In *Shewanella putrefaciens* and a few other bacteria, anoxic growth on acetate and several other nonfermentable carbon sources occurs with Mn^{4+} as electron acceptor. The reduction potential of the Mn^{4+}/Mn^{2+} couple is extremely high (Figure 17.47); thus, several compounds should be able to donate electrons to Mn^{4+} reduction. This is also the case for chlorate, because its reduction potential is actually *more positive* than that of the O_2/H_2O couple (Figure 17.47). Several chlorate-reducing bacteria have been isolated, and most of them are facultative and thus also capable of aerobic growth.

Other inorganic substances can function as electron acceptors for anaerobic respiration. These include sele-

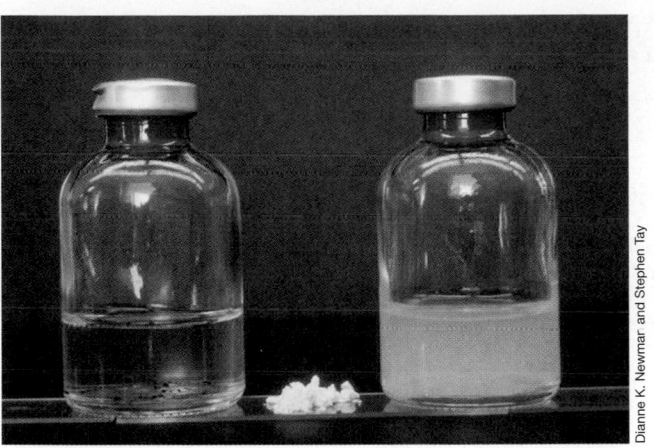

Figure 17.48 Production of the mineral arsenic trisulfide (As_2S_3) during arsenate reduction by the sulfate-reducing bacterium *Desulfotomaculum auripigmentum*. Left, appearance of culture bottle after inoculation. Right, following growth for two weeks. Center, synthetic sample of As_2S_3. Mineral production by microorganisms is called *biomineralization*.

nium and arsenic compounds (Figure 17.47). Although usually not present in large amounts in natural systems, arsenic and selenium compounds are occasional pollutants and can support anoxic growth of various bacteria. The reduction of SeO_4^{2-} to SeO_3^{2-} and eventually to Se^0 (metallic selenium) is an important method of selenium removal from water and has been used as a means of cleaning up (bioremediation) (∞ Section 19.19) of selenium-contaminated soils.

In the case of arsenic, the sulfate-reducing bacterium *Desulfotomaculum* can reduce arsenate (AsO_4^{3-}) to arsenite (AsO_3^{3-}), along with SO_4^{2-} (to HS^-), and in the process causes a mineral complex of arsenic and sulfide, As_2S_3, to precipitate spontaneously (Figure 17.48). The mineral is formed both intracellularly as well as extracellularly, and its formation is an example of *biomineralization*, the formation of a mineral by bacterial action. In this case As_2S_3 formation (Figure 17.48) also functions as a means of detoxifying what would otherwise be a toxic compound (arsenic in this case), and such activities may have practical applications for the microbial cleanup of toxic wastes.

Organic Electron Acceptors

Several organic compounds can participate as external electron acceptors in anaerobic respirations. Of those listed in Figure 17.47, the compound that has been most extensively studied is **fumarate**, which is reduced to **succinate**. An examination of the *citric acid cycle* (∞ Figure 5.22) will indicate that fumarate and succinate are important intermediates. The role of fumarate as an electron acceptor for anaerobic respiration derives from the fact that the fumarate-succinate couple has a reduction potential near 0 V (see Figure 17.35), which allows coupling of fumarate reduction to NADH or H_2 oxidation. The energy yield is sufficient for the synthesis of one ATP.

Bacteria able to use fumarate as an electron acceptor include *Wolinella succinogenes* (which can grow on H_2 as electron donor using fumarate as electron acceptor), *Desulfovibrio gigas* (a sulfate-reducing bacterium that can also grow under non-sulfate-reducing conditions), some clostridia, *Escherichia coli*, and many other bacteria.

The compound **trimethylamine oxide** shown in Figure 17.47 is an interesting organic electron acceptor. Trimethylamine oxide (TMAO) is an important osmotic solute in marine fish, where it functions in these animals as a means of excreting excess nitrogen. A variety of bacteria are able to reduce TMAO to trimethylamine (TMA), which has a strong odor and flavor. Some of the odor that occurs in spoiled marine fish is due to TMA produced by bacterial action. A variety of facultatively aerobic bacteria are able to use TMAO as an alternate electron acceptor. In addition, several phototrophic purple bacteria (∞ Section 12.2) are able to use TMAO as an electron acceptor for anaerobic metabolism in darkness. A compound analogous to TMAO is **dimethyl sulfoxide** (DMSO), which is reduced by a variety of bacteria to dimethyl sulfide (DMS). DMSO is a common natural product and is found in both marine and freshwater environments. DMS has a strong, pungent odor, and bacterial reduction of DMSO to DMS is signaled by the presence of the characteristic odor of DMS. A variety of bacteria, including *Campylobacter, Escherichia*, and many purple bacteria, are able to use DMSO as an electron acceptor in energy generation (∞ Section 19.13 discusses DMSO metabolism).

The reduction potentials of the TMAO/TMA and DMSO/DMS couples are similar, near +0.15 V, which means that any electron transport chain that ends with TMAO or DMSO reduction must be rather brief. As in fumarate reduction, in most instances of TMAO and DMSO reduction, cytochromes of the *b* type (with reduction potentials near 0 V) have been identified as terminal oxidases.

Halogenated Compounds as Electron Acceptors: Reductive Dechlorination

Several chlorinated compounds can function as electron acceptors for anaerobic respiration in the process known as *reductive dechlorination*. For example, the sulfate-reducing bacterium *Desulfomonile* grows anaerobically on H_2 or organic compounds as electron donors with chlorobenzoate as an electron acceptor:

$$C_7H_4O_2Cl^- + 2\ H \rightarrow C_7H_5O_2^- + HCl$$

3-Chlorobenzoate   Benzoate

Besides *Desulfomonile*, which is also a sulfate-reducing bacterium (Table 17.5), a wide variety of other bacteria can reductively dechlorinate; some of these are restricted to chlorinated compounds as electron acceptors (Table 17.5). Many of the chlorinated compounds reduced are toxic to fish and other animal life while many of the products of dechlorination are less toxic or completely

TABLE 17.5 Characteristics of major genera of bacteria capable of reductive dechlorination

Property	Genus			
	Dehalobacter	*Desulfomonile*	*Desulfitobacterium*	*Dehalococcoides*
Electron donors	H_2	H_2, formate, pyruvate, lactate, benzoate	H_2, formate, pyruvate, lactate	H_2, lactate
Electron acceptors	Trichloroethylene, tetrachloroethylene	Metachlorobenzoates, tetrachloroethylene, SO_4^{2-}, SO_3^{2-}, $S_2O_3^{2-}$	Ortho-, meta- or para-chlorophenols, NO_3^-, fumarate, SO_3^{2-}, $S_2O_3^{2-}$, S^0	Trichloroethylene, tetrachloroethylene
Product of reduction of tetrachloroethylene	Dichloroethylene	Dichloroethylene	Trichloroethylene	Ethene
Other properties[a]	Contains cytochrome *b*	Contains cytochrome c_3; requires organic carbon source; can grow by fermentation of pyruvate	Can also grow by fermentation	Lacks peptidoglycan
Phylogeny[b]	Related to low GC gram-positive *Bacteria*	Related to delta Proteobacteria	Related to low GC gram-positive *Bacteria*	Unique lineage of *Bacteria*

[a] All organisms are obligate anaerobes.
[b] See Chapter 12 and Figure 12.1 for discussion of prokaryotic phylogeny.

nontoxic. For example, the bacterium *Dehalococcoides* reduces tri- and tetrachloroethylene to the harmless gas ethene (Table 17.5). *Dehalobacterium* converts the toxic compound dichloromethane (CH_2Cl_2) into the fatty acid acetate plus formate (Table 17.5). Thus, reductive dechlorination is not only a form of energy metabolism, but an environmentally significant bioremedial process (∞ Section 19.19). Many reductive dechlorinators are also capable of reducing nitrate or various reduced sulfur compounds (Table 17.5), and thus the group consists of both specialist and opportunist species.

✓ 17.18 Concept Check

Besides inorganic nitrogen and sulfur compounds or CO_2, a variety of other substances, both organic and inorganic, can function as electron acceptors for anaerobic respiration. These include in particular Fe^{3+}, Mn^{4+}, fumarate, and certain chlorinated compounds.

- ✓ With H_2 as electron donor why is reduction of Fe^{3+} a more favorable reaction than reduction of fumarate?
- ✓ Give an example of *biomineralization*.
- ✓ Where in nature might TMAO-degrading bacteria be abundant? Why?
- ✓ What is reductive dechlorination?

17.19 Fermentations: Energetic and Redox Considerations

Because oxygen is not highly soluble (9.6 mg/l distilled water in equilibrium with air at 25°C), many environments easily become anoxic. In such environments,

decomposition of organic materials occurs anaerobically. If adequate supplies of electron acceptors like SO_4^{2-}, NO_3^-, Fe^{3+}, and the others considered in previous sections are not available in such anoxic environments, much of the carbon will be catabolized by fermentation. (CO_2 as an electron acceptor is an exception. It is rarely limiting in anoxic habitats but its conversion to methane requires H_2, and the latter is itself a product of fermentative reactions). We discussed the overall process of fermentation in Sections 5.9 and 5.10, demonstrating that it is an internally balanced oxidation-reduction process in which carbon from the same external organic compound was partially oxidized and partially reduced (Figure 17.49).

There are two problems an organism faces if it is to catabolize organic compounds in energy-yielding metabolism: (1) conserving some of the energy released as ATP, and (2) disposing of electrons removed from the electron donor. In fermentation, ATP synthesis general-

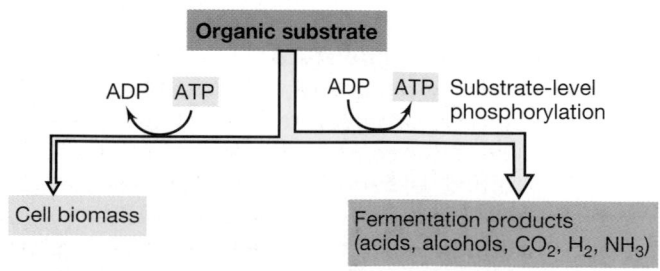

Figure 17.49 Overall process of fermentation. In a typical fermentation, most of the carbon is excreted as a partially reduced end product of energy metabolism and only a small amount is used in biosynthesis.

TABLE 17.6 Energy-rich compounds involved in substrate-level phosphorylation[a]

Compound	Free energy of hydrolysis, $\Delta G^{0\prime}$ (kJ/mol)[b]
Acetyl-CoA	−35.7
Propionyl-CoA	−35.6
Butyryl-CoA	−35.6
Succinyl-CoA	−35.1
Acetylphosphate	−44.8
Butyrylphosphate	−44.8
1,3-Bisphosphoglycerate	−51.9
Carbamyl phosphate	−39.3
Phosphoenolpyruvate	−51.6
Adenosine-phosphosulfate (APS)	−88
N^{10}-formyltetrahydrofolate	−23.4
Energy of hydrolysis of ATP ($ATP \rightarrow ADP + P_i$)	−31.8

[a] Data from Thauer, R. K., K. Jungermann, and K. Decker. 1977. Energy conservation in chemotrophic anaerobic bacteria. *Bacteriol. Rev.* 41: 100–180.

[b] The $\Delta G^{0\prime}$ values shown here are for "standard conditions," which are not necessarily those of cells. Including heat loss, the energy costs of making an ATP are more like 60 kJ than 32 kJ, and the energy of hydrolysis of the high energy compounds shown here are thus likely higher. But for simplicity and comparative purposes, the values in this table will be taken as the actual energy released per reaction.

ly occurs by way of *substrate-level phosphorylation*, a mechanism by which high energy phosphate bonds from organic intermediates of the fermentation are transferred to ADP (Section 5.9). The second problem, that of redox balance, is solved by production and excretion by the organism of *fermentation products* generated from the original substrate (Figures 17.49 and 17.50). We now consider these basic principles of fermentation in more detail and highlight the enormous diversity of microbial fermentations known.

High Energy Compounds and Substrate-Level Phosphorylation

Energy conservation from substrate-level phosphorylation can occur in many different ways. However, central to the mechanism of ATP synthesis is the production of one or another *high energy compound*. These are generally organic compounds containing a phosphate group or a coenzyme-A molecule, the hydrolysis of which is highly exergonic. A list of the major high energy intermediates is given in Table 17.6. This list is not complete, but it includes most recognized high energy intermediates known to be formed during biochemical processes. Because most of the compounds listed in Table 17.6 can couple directly to ATP synthesis (−31.8 kJ/mol), if an organism can form one or another of these compounds during fermentative metabolism, it can make ATP. Substrate-level phosphorylation is a more direct way of making ATP than via a proton motive force (Figure 5.13) but requires that the energy source couple directly to a high energy intermediate.

Pathways for the anaerobic breakdown of various fermentable substances to high energy intermediates are summarized in Figure 17.50. Note that this figure is organized by the high energy compounds listed in Table 17.6 and that either one of these compounds, or a related derivative, is generated in each case and leads to ATP synthesis. Thus, Figure 17.50 and Table 17.6 should be examined together.

Energy Yields of Fermentative Organisms

How much ATP can be produced by a fermentative organism? As we have seen, glucose fermenters produce 2–3 ATPs per glucose fermented in glycolysis (Figure 5.14). This is about the maximum amount of ATP produced by fermentation; many other substrates provide less energy. The potential energy released from a particular fermentation can be calculated from the balanced reaction and from the free energy values given in Appendix 1. For instance, the fermentation of glucose to ethanol and CO_2 has a theoretical energy yield of −235 kJ/mol, enough to produce about 7 ATPs. However, only 2 ATPs are actually produced, which implies that the organism operates at considerably less than 100% efficiency, some energy being lost as heat.

Oxidation-Reduction Balance

In any fermentation reaction, there must be a *balance* between oxidation and reduction. The total number of electrons in the products on the right side of the equation must balance the number in the substrates on the left side of the equation. When fermentations are studied experimentally in the laboratory, it is conventional to calculate a *fermentation balance* to make certain that no products are missed. The fermentation balance can also be calculated theoretically from the oxidation states of the substrates and products (see Appendix 1 for the procedure for calculating oxidation states).

In a number of fermentations, electron balance is maintained by the production of molecular hydrogen, H_2. In H_2 production, protons (H^+) derived from water serve as electron acceptors. Production of H_2 is generally associated with the presence in the organism of an iron-sulfur protein called *ferredoxin*, a very low potential electron carrier. The transfer of electrons from ferredoxin to H^+ is catalyzed by the enzyme **hydrogenase**, as illustrated in Figure 17.51. The energetics of hydrogen production are actually somewhat unfavorable, and so most fermentative organisms produce only a relatively small amount of hydrogen along with other fermentation products. Hydrogen production thus functions primarily to maintain redox balance.

Numerous anaerobic bacteria produce *acetate* as one of the products of fermentation. The production of acetate or certain other fatty acids (see Table 17.6) is energetically advantageous because it allows the organism to make ATP by substrate-level phosphorylation. The key intermediate

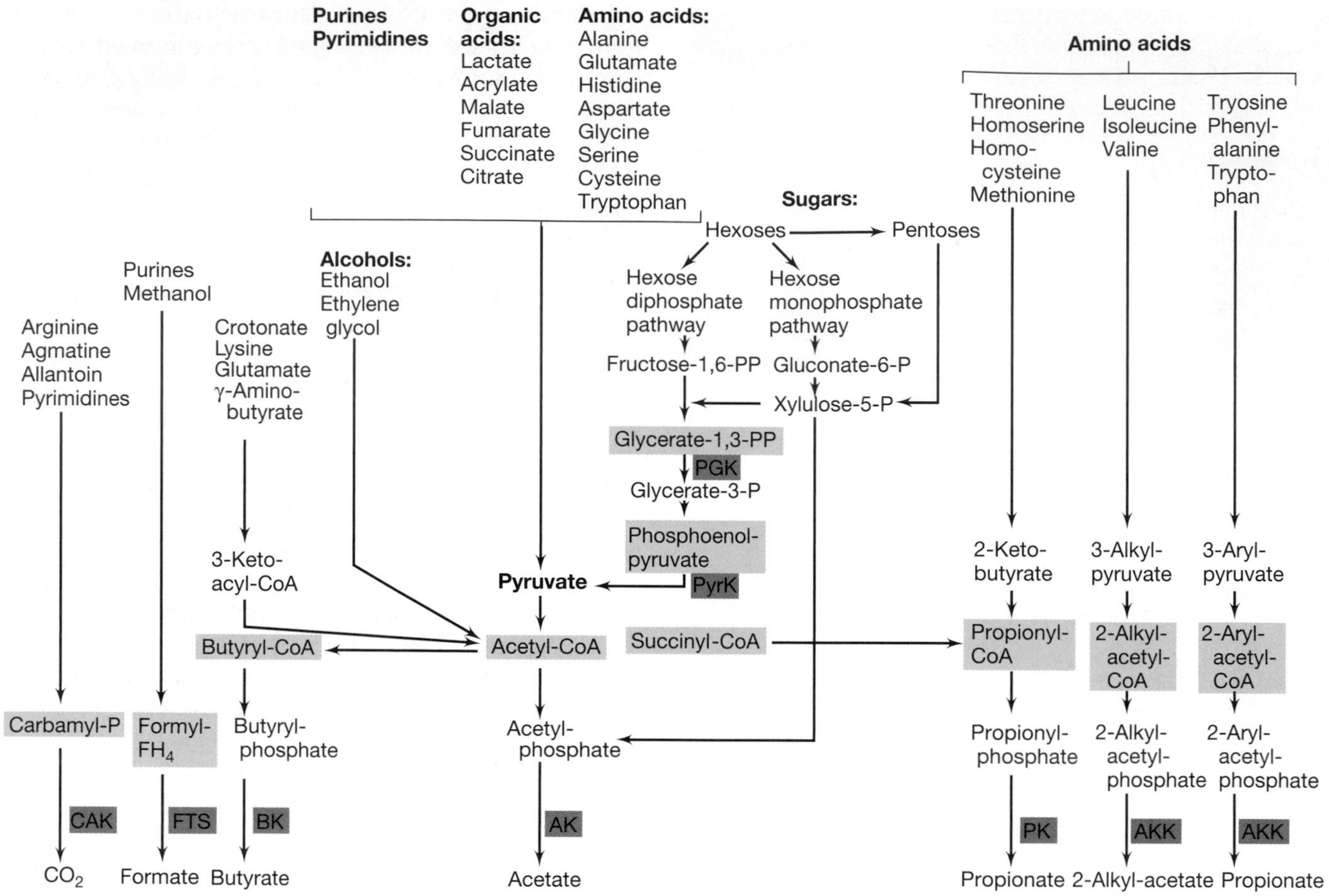

Figure 17.50 Major routes of the anaerobic breakdown of various fermentable substances. The sites of substrate-level phosphorylation are shown by the abbreviations of the enzymes involved. CAK, carbamyl phosphate kinase; FTS, formyltetrahydrofolate synthetase; AK, acetate kinase; PK, propionate kinase; BK, butyrate kinase; AKK, alkyl (aryl) acetate kinase; PGK, phosphoglycerate kinase; PyrK, pyruvate kinase. High energy CoA derivatives and other key high energy compounds are highlighted in blue while enzymes are in red. Refer back to Table 17.6.

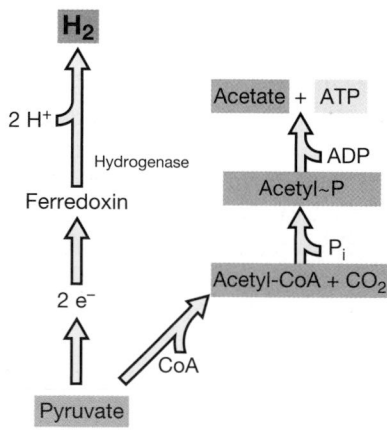

Figure 17.51 Production of molecular hydrogen and acetate from pyruvate. Note how production of acetate leads to ATP synthesis from hydrolysis of the high energy intermediate, acetyl phosphate (see Table 17.6).

generated in acetate production is acetyl-CoA (see Table 17.6 and Figure 17.50), a high-energy intermediate. Acetyl-CoA can be converted to acetyl phosphate (also listed in Table 17.6), and the high energy phosphate group of acetyl phosphate subsequently transferred to adenosine diphosphate (ADP) by acetate kinase, yielding ATP. One of the main substrates that is converted to acetyl-CoA is pyruvate, a major product of glycolysis. The conversion of pyruvate to acetyl-CoA is an oxidation reaction (Figure 17.51) and the excess electrons generated must be used either to make a more reduced end product or in the production of H_2 as discussed previously.

✓ 17.19 Concept Check

In the absence of an external electron acceptor, organic compounds can be catabolized only by fermentation. Only certain compounds are fermentable, and a requirement for most fermentations is that an energy-rich organic intermediate be formed that can yield ATP by substrate-level phosphorylation. Redox balance must also be achieved in fermentations, and H_2 production is one means of disposing of excess electrons.

✓ What is *substrate-level phosphorylation*?
✓ Why is acetate formation in fermentation energetically beneficial?

17.20 Fermentative Diversity

Fermentations are classified either in terms of the substrate fermented or in terms of the fermentation products formed. Many of the specific fermentation reactions of bacteria were discussed when the individual groups were examined in Chapters 12 and 13. Here we present an overview of common fermentations.

Diversity of Fermentations

Table 17.7 summarizes some of the main types of fermentations as classified on the basis of *products formed*. Note some of the broad categories, such as alcohol, lactic acid, propionic acid, mixed acid, butyric acid, and homoacetic acid. A number of fermentations are classified on the basis of the *substrate fermented* rather than the fermentation product. For instance, many of the spore-forming anaerobic bacteria (genus *Clostridium*) ferment *amino acids* with the production of acetate, ammonia, and H_2 (⚙ Figure 12.59). Other *Clostridium* species, such as *C. acidi-urici* and *C. purinolyticum*, ferment *purines* such as xanthine or adenine with the formation of acetate, formate, CO_2, and ammonia. Still other anaerobes ferment

aromatic compounds. For example, the bacterium *Pelobacter acidigallici* ferments the aromatic compound *phloroglucinol* (1,3,5-benzenetriol, $C_6H_6O_3$) via the following overall pathway:

Phloroglucinol ($C_6H_6O_3$) + 3 $H_2O \rightarrow$ 3 acetate$^-$ + 3 H^+
$$\Delta G^{0\prime} = -142.5 \text{ kJ/reaction}$$

Many unusual fermentations are carried out by only a very restricted group of anaerobes and, in some cases, by only a single known bacterium. Some examples are listed in Table 17.8. Many of these bacteria can be considered metabolic specialists, having evolved biochemical capabilities to catabolize a substrate or substrates not catabolized by other bacteria. However, as for the substances listed in Table 17.7, successful fermentation of these more unusual substrates requires that the organism be able to produce a high energy intermediate, usually a coenzyme-A derivative of the type listed in Table 17.6, during the fermentation in order to conserve some of the energy released as ATP.

Fermentations Without Substrate-Level Phosphorylation: Decarboxylations of Organic Acids

With certain substrates there is insufficient energy released to couple to the synthesis of ATP directly by substrate-level phosphorylation, yet these compounds still support fermentative growth of an organism. In these

TABLE 17.7 Examples of common bacterial fermentations and some of the organisms carrying them out

Type	Overall reaction[a]	Organisms
Alcoholic fermentation	Hexose $\rightarrow$ 2 Ethanol + 2 CO_2	Yeast *Zymomonas*
Homolactic fermentation	Hexose $\rightarrow$ 2 Lactate$^-$ + 2 H^+	*Streptococcus* Some *Lactobacillus*
Heterolactic fermentation	Hexose $\rightarrow$ Lactate + Ethanol + CO_2 + H^+	*Leuconostoc* Some *Lactobacillus*
Propionic acid	Lactate$^- \rightarrow$ Propionate$^-$ + Acetate$^-$ + CO_2	*Propionibacterium* *Clostridium propionicum*
Mixed acid	Hexose $\rightarrow$ Ethanol + 2, 3-Butanediol + Succinate^{2-} + Lactate$^-$ + Acetate$^-$ + Formate$^-$ + H_2 + CO_2	Enteric bacteria *Escherichia* *Salmonella* *Shigella* *Klebsiella* *Enterobacter*
Butyric acid	Hexose $\rightarrow$ Butyrate$^-$ + Acetate$^-$ + H_2 + CO_2	*Clostridium butyricum*
Butanol	Hexose $\rightarrow$ Butanol + Acetate$^-$ + Acetone + Ethanol + H_2 + CO_2	*Clostridium acetobutylicum*
Caproate	Ethanol + Acetate$^-$ + $CO_2 \rightarrow$ Caproate$^-$ + Butyrate$^-$ + H_2	*Clostridium kluyveri*
Homoacetogenic	Fructose $\rightarrow$ 3 Acetate$^-$ + 3 H^+ 4 H_2 + 2 CO_2 + $H^+ \rightarrow$ Acetate$^-$ + 2 H_2O	*Clostridium aceticum* *Acetobacterium*
Methanogenic	Acetate$^-$ + $H_2O \rightarrow CH_4$ + HCO_3^-	*Methanosaeta* *Methanosarcina*

[a] Reactions are intended as an overview of the process and are not necessarily balanced.

TABLE 17.8 Some unusual bacterial fermentations

Type	Overall balanced reaction	Organisms
Acetylene	$2\,C_2H_2 + 3\,H_2O \rightarrow$ ethanol + acetate$^-$ + H$^+$	*Pelobacter acetylenicus*
Glycerol	$4\,$Glycerol + $2\,HCO_3^- \rightarrow 7\,$acetate$^-$ + $5\,H^+$ + $4\,H_2O$	*Acetobacterium* spp.
Resorcinol (an aromatic compound)	$2\,C_6H_4(OH)_2 + 6\,H_2O \rightarrow 4\,$acetate$^-$ + butyrate$^-$ + $5\,H^+$	*Clostridium* spp.
Phloroglucinol (an aromatic compound)	$C_6H_6O_3 + 3\,H_2O \rightarrow 3\,$acetate$^-$ + $3\,H^+$	*Pelobacter massiliensis* *Pelobacter acidigallici*
Putrescine	$10\,C_4H_{12}N_2 + 26\,H_2O \rightarrow 6\,$acetate$^-$ + $7\,$butyrate$^-$ $+20\,NH_4^+ + 16\,H_2 + 13\,H^+$	Unclassified gram-positive nonsporing anaerobes
Citrate	Citrate^{3-} + $2\,H_2O \rightarrow$ formate$^-$ + $2\,$acetate$^-$ + HCO_3^- + H$^+$	*Bacteroides* sp.
Aconitate	Aconitate^{3-} + H$^+$ + $2\,H_2O \rightarrow 2\,CO_2$ + $2\,$acetate$^-$ + H_2	*Acidaminococcus fermentans*
Glyoxylate	$4\,$Glyoxylate$^-$ + $3\,H^+$ + $3\,H_2O \rightarrow 6\,CO_2$ + $5\,H_2$ + glycolate$^-$	Unclassified gram-negative bacterium
Succinate	Succinate^{2-} + $H_2O \rightarrow$ propionate$^-$ + HCO_3^-	*Propionigenium modestum*
Oxalate	Oxalate^{2-} + $H_2O \rightarrow$ formate$^-$ + HCO_3^-	*Oxalobacter formigenes*
Malonate	Malonate^{2-} + $H_2O \rightarrow$ acetate$^-$ + HCO_3^-	*Malonomonas rubra* *Sporomusa malonica*

cases, catabolism of the substrate is linked to ion pumps that establish a proton or sodium gradient across the membrane. Examples of this include the fermentations by *Propionigenium modestum* and *Oxalobacter formigenes*; both of these organisms couple the fermentation of dicarboxylic acids to membrane-bound energy-linked ion pumps. *Propionigenium modestum* carries out the following reaction:

$$\text{Succinate}^{2-} + H_2O \rightarrow \text{propionate}^- + HCO_3^-$$
$$\Delta G^{0\prime} = -20.5 \text{ kJ/reaction}$$

This overall reaction yields insufficient free energy to couple to ATP synthesis directly by substrate-level phosphorylation, but nevertheless it serves as the sole energy-yielding reaction for growth of the organism. This is possible because the decarboxylation of succinate (via methylmalonyl-CoA and its membrane-bound decarboxylase) by *Propionigenium modestum* is coupled to the export of Na$^+$ across the cytoplasmic membrane (Figure 17.52*a*). A Na$^+$-translocating ATPase in the membrane of *P. modestum* employs this Na$^+$ gradient to drive ATP synthesis (Figure 17.52*a*).

Oxalobacter formigenes carries out the fermentation of oxalate:

$$\text{Oxalate}^{2-} + H_2O \rightarrow \text{formate}^- + HCO_3^-$$
$$\Delta G^{0\prime} = -26.7 \text{ kJ/reaction}$$

At neutral pH, oxalate exists in the ionized form as oxalate^{2-}, and its decarboxylation to formate$^-$ consumes

one proton. The subsequent export of formate from the cell then builds a proton motive force that can be coupled to ATP synthesis by a proton-translocating ATPase in the membrane (Figure 17.52*b*).

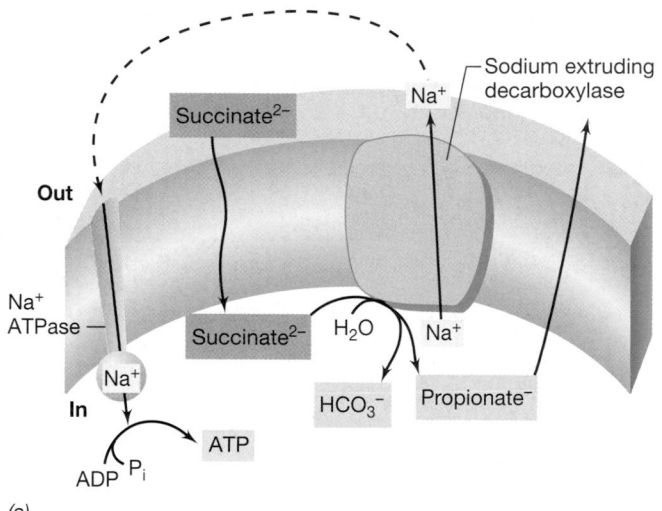

(a)

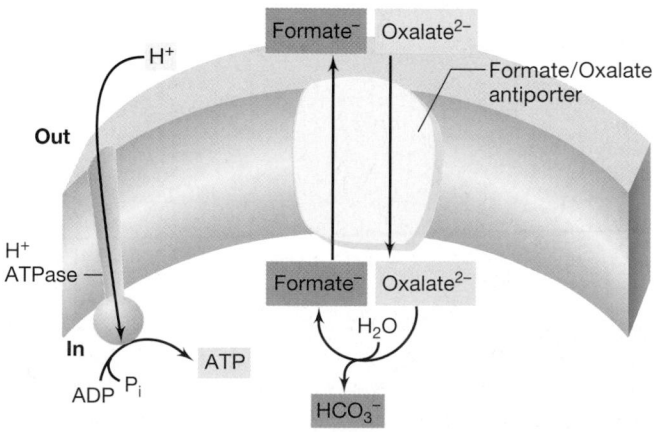

(b)

Figure 17.52 The unique fermentations of succinate and oxalate. (a) Succinate fermentation by *Propionigenium modestum*. A sodium-translocating ATPase produces ATP; sodium export is linked to the energy released by succinate decarboxylation. (b) Oxalate fermentation by *Oxalobacter formigenes*. Oxalate import and formate export by a formate-oxalate antiporter consume protons. ATP synthesis is linked to a proton-driven ATPase. All substrates and products of a given reaction are shown in contrasting colors.

The interesting and unique aspect of the metabolism of both *Propionigenium modestum* and *Oxalobacter formigenes* is that ATP synthesis occurs without substrate-level phosphorylation *or* electron transport occurring; however, chemiosmotic ATP formation still occurs as a result of a Na^+ or H^+ pump linked to decarboxylation of organic acids. Thus, any chemical reaction that yields less than the 31.8 kJ required to make one ATP (see Table 17.6) cannot be ruled out as a potential growth-supporting reaction for a bacterium. If the reaction can be coupled to an ion gradient, ATP production (and subsequently growth) remains a possibility. However, because the influx of approximately $3 H^+$ (or $3 Na^+$) is required to drive ATP formation by a membrane-associated ATPase, a reaction must yield at least the energy required to pump a single H^+ or Na^+ ion to the outside of the cell membrane to be theoretically capable of supporting growth.

✓ 17.20 Concept Check

A wide variety of fermentations are known, and in many cases the product of one organism's fermentation is fermented by a second organism. Some fermentations employ ion gradients (H^+ or Na^+) as the basis of their energetics.

✓ What is unusual about the fermentation of succinate and oxalate?

✓ A common product of the fermentations shown in Table 17.7 and 17.8 is a fatty acid such as acetate. Why is this important energetically?

17.21 Syntrophy

There are many examples in microbiology of *syntrophy*, a situation where two different organisms can together degrade some substance—and conserve energy doing it—that neither could degrade separately. We will see in Section 19.10 how syntrophy is extremely important in anoxic catabolism leading to the production of CH_4. Here we consider the microbiology and energetic aspects of syntrophy.

Hydrogen Consumption in Syntrophic Reactions

In most cases the nature of a syntrophic reaction involves hydrogen gas (H_2) being produced by one partner in the syntrophic relationship and being consumed by the other. Thus, syntrophy has also been called *interspecies H_2 transfer*. The H_2 consumer can be any of a number of organisms we have already considered, denitrifiers, sulfate-reducing bacteria, homoacetogens, and methanogens. Consider the case of syntrophy involving ethanol fermentation to acetate and eventual production of methane (Figure 17.53). As seen, the ethanol fermenter carries out a reaction that has an unfavorable (that is, positive) standard free energy change. However, the H_2 produced by the ethanol fermenter is a valuable electron

Ethanol fermentation

$$2 \, CH_3CH_2OH + 2 \, H_2O \longrightarrow \boxed{4 \, H_2} + 2 \, CH_3COO^- + 2 \, H^+$$
Ethanol   Acetate $\Delta G^{0\prime} = + 19.4 \text{ kJ/reaction}$

Methanogenesis

$$\boxed{4 \, H_2} + CO_2 \longrightarrow CH_4 + 2 \, H_2O$$
Methane $\Delta G^{0\prime} = - 130.7 \text{ kJ/reaction}$

Syntrophic, coupled reaction

$$2 \, CH_3CH_2OH + CO_2 \longrightarrow CH_4 + 2 \, CH_3COO^- + 2 \, H^+$$
$\Delta G^{0\prime} = - 111.3 \text{ kJ/reaction}$

Figure 17.53 Fermentation of ethanol to methane and acetate by syntrophic association of an ethanol-oxidizing bacterium and a H_2-consuming partner bacterium, in this case, a methanogen. Note how although the oxidation of ethanol to acetate plus H_2 is energetically unfavorable, the reaction becomes favorable when coupled to H_2 consumption by the methanogens. The two organisms thus share the energy released in the coupled reaction.

donor for methanogenesis by a methanogen (Figure 17.53). And when the two reactions are summed, the overall reaction is exergonic (Figure 17.53) and supports growth of both partners in the syntrophic mixture. Another good example of syntrophy is the oxidation of butyrate to acetate plus H_2 by the fatty acid-oxidizing syntroph *Syntrophomonas* (Figure 17.54):

$$Butyrate^- + 2 \, H_2O \rightarrow 2 \, acetate^- + H^+ + 2 \, H_2$$
$$\Delta G^{0\prime} = +48.2 \text{ kJ}$$

The free energy change of this reaction is highly unfavorable, and in pure culture *Syntrophomonas* will not grow on butyrate. But if the H_2 produced in the reaction is immediately consumed by a partner organism (like a methanogen), *Syntrophomonas* grows luxuriantly in coculture with the H_2 consumer. How can chemical reactions whose free energy changes are positive support growth of an organism?

Energetics of Syntrophy

If growth of syntrophic organisms occurs only when H_2 is removed by a partner organism, the removal itself must obviously affect the energetics of the reaction. How does this happen? A brief review of the principles of free energy given in Appendix 1 indicates that the *actual* concentration of reactants and products in a reaction can drastically change the energetics. This is because $\Delta G^{0\prime}$ is calculated on the basis of *standard* conditions—one molar concentration of products and reactants—whereas the related term ΔG is calculated on the basis of *actual* concentrations of products and reactants that are present. In anoxic habitats, the products of fatty acid oxidation, especially H_2 (because it is such a powerful electron donor for anaerobic respirations), are immediately consumed, and this keeps H_2 concentrations extremely low, usually below 10^{-4} atm (⚬⚬ Table 19.4). Thus, if the concentration of H_2 is very low, and ΔG is used to calculate

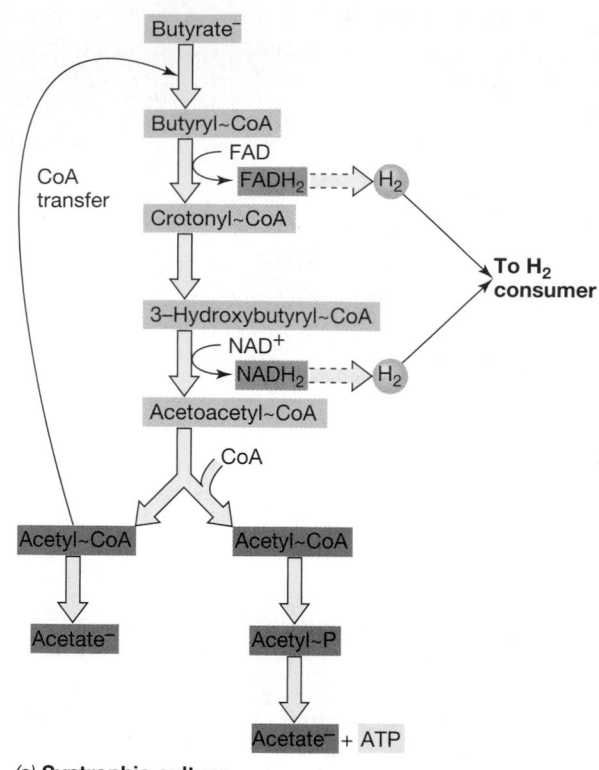

(a) **Syntrophic culture**

Figure 17.54 Energetics of growth of *Syntrophomonas wolfei* in syntrophic culture (a) and in pure culture (b). In syntrophic culture, growth is dependent on the presence of a H_2-consuming organism such as a methanogen. H_2 production probably involves proton motive force-driven reverse electron flow because the E_0' of FAD/FADH$_2$ or NAD$^+$/NADH is more electropositive than that of 2 H$^+$/H$_2$ (∞ Figure 5.9). In pure culture, energy production is linked to crotonate reduction to butyrate.

(b) **Pure culture**

the energetics of the reaction with this in mind (see Appendix 1 for how to calculate ΔG), the free energy change associated with the oxidation of ethanol or fatty acids to acetate plus H_2 becomes exergonic, indicating that energy is released. For example, if the concentration of H_2 is kept extremely low by the activities of a H_2-consuming partner organism, the oxidation of butyrate by *Syntrophomonas* yields about -18 kJ.

The mechanism of ATP production by syntrophs involves both substrate-level and oxidative phosphorylations, depending on how the organisms are grown. Substrate-level phosphorylation can occur during the conversion of acetyl-CoA (generated by β-oxidation of ethanol or the fatty acid) to acetate (Figure 17.54a) and may be the *only* way ATP is produced under syntrophic conditions. However, *Syntrophomonas* can also carry out an anaerobic respiration because it can be grown in pure culture on certain unsaturated fatty acids. Crotonate, for example, supports growth of *Syntrophomonas*, with some of the crotonate being oxidized to acetate and some being reduced to butyrate (Figure 17.54b). Crotonate reduction by *Syntrophomonas* is coupled to formation of a proton motive force and ATP synthesis as in other anaerobic respirations employing organic electron acceptors, such as fumarate reduction to succinate (see Section 17.18).

Regardless of how ATP is made in the syntrophic situation, there are additional energetic burdens for *Syntrophomonas* in this growth mode because the organism must somehow produce H_2 from more elec-

tropositive electron donors such as FADH$_2$ and NADH$_2$ (Figure 17.54a). This suggests that part of the ATP produced may be needed to drive reverse electron flow reactions (see Section 17.5) to yield H_2. Clearly, syntrophic alcohol and fatty-acid oxidizing bacteria are, from an energetic standpoint, living on the edge of existence.

Syntrophic bacteria have evolved effective systems for using the highly reduced fermentation products of primary fermenters and for cooperating with other anaerobes to supply them with a necessary substrate. By contrast, for *aerobic* organisms syntrophy is not an issue, because with O_2 as electron acceptor the energetics of the reaction is much more favorable than for fermentative catabolism of the same substrate. Thus, syntrophic relationships are characteristic of anoxic transformations in which the energy available is only very small and the organisms highly specialized for exploiting energetically marginal reactions.

✓ 17.21 Concept Check

Syntrophy involves two organisms working together to degrade some compound that neither can degrade alone. This process usually involves H_2 produced by one organism being consumed by the partner. H_2 consumption can affect the energetics of the reaction carried out by the H_2 producer, allowing it to make ATP where it otherwise could not.

✓ Give an example of syntrophy. Why can it be said that both organisms benefit in this example?

✓ Predict how ATP is made during the syntrophic degradation of ethanol shown in Figure 17.53.

✓ Can a fatty acid-oxidizing syntroph be grown in pure culture? Give an example of this and describe how the organism makes ATP under these conditions.

IV HYDROCARBON OXIDATION AND THE ROLE OF O₂ IN THE CATABOLISM OF ORGANIC COMPOUNDS

Hydrocarbons are both abundant on Earth and excellent substrates for various chemoorganotrophs. Their aerobic degradation, however, poses special biochemical problems that we discuss here. In addition, we briefly consider the aerobic degradation of polysaccharides, fats, and a few other substances as further examples of metabolic diversity.

17.22 Molecular Oxygen (O_2) as a Reactant in Biochemical Processes

We have discussed the role of O_2 as an *electron acceptor* in energy-generating reactions (∞ Section 5.11). Although this is by far the most important role of O_2 in cellular metabolism, O_2 plays an interesting and important role as a *direct reactant* in certain types of anabolic and catabolic processes.

Oxygenases

Oxygenases are enzymes that catalyze the incorporation of oxygen from O_2 into organic compounds. There are two classes of oxygenases: *dioxygenases*, which catalyze the incorporation of *both* atoms of O_2 into the molecule, and *monooxygenases*, which catalyze the transfer of *only one* of the two oxygen atoms in O_2 to an organic compound as a hydroxyl (OH) group, with the second atom of O_2 being reduced to water, H_2O. Because monooxygenases catalyze the formation of hydroxyl groups (OH) in organic compounds, they are sometimes called *hydroxylases*. In most monooxygenases, the electron donor is NADH or NADPH, although the direct coupling to O_2 is through a flavin that is reduced by the NADH or NADPH donor. In the case of ammonia monooxygenase discussed previously (see Section 17.12), the electron donor is cytochrome *c*.

Several types of reactions in living organisms require O_2 as a reactant. One of the best examples is the involvement of O_2 in sterol biosynthesis. The formation of the fused sterol ring system (∞ Figure 4.18) requires the participation of molecular oxygen. Such a reaction can obviously not take place under anoxic conditions so organisms that grow anaerobically must either dispense with this reaction or obtain the required substance (sterol) preformed from their environment. The requirement of O_2 as a reactant in biosynthesis is of considerable evolutionary significance, as molecular O_2 was originally absent from the atmosphere of Earth when life evolved and became available only after the evolution of cyanobacteria, the first phototrophic organisms to produce O_2 (∞ Chapter 11). The role of O_2 in hydrocarbon utilization is discussed next.

✓ 17.22 Concept Check

In addition to its role as an electron acceptor, oxygen (O_2) is also a chemical reactant in certain biochemical processes. Enzymes called oxygenases introduce O_2 into a biochemical compound.

✓ How do *monooxygenases* differ in function from *dioxygenases*?

17.23 Hydrocarbon Oxidation

Hydrocarbons are organic compounds containing only carbon and hydrogen and are highly insoluble in water. Low-molecular-weight hydrocarbons are gases, whereas those of higher molecular weight are liquids or solids at room temperature. Some hydrocarbons are *aliphatic compounds*, a class of carbon compounds in which the carbon atoms are joined in open chains. There is a tremendous variation among aliphatic hydrocarbons in chain length, degree of branching, and number of double bonds. Another important group of hydrocarbons contains the aromatic ring and can be viewed as derivatives of benzene. Although hydrocarbons can be degraded anaerobically by certain bacteria (∞ Section 19.18), most of our discussion here will be on the *aerobic* degradation of hydrocarbons.

Aliphatic Hydrocarbons

A number of bacteria and even molds and yeasts can use hydrocarbons for growth under aerobic conditions. The initial oxidation step of saturated aliphatic hydrocarbons involves molecular oxygen (O_2) as a reactant, and one of the atoms of the oxygen molecule is incorporated into the oxidized hydrocarbon. This reaction is carried out by a monooxygenase (see Section 17.22), and a typical reaction sequence is that shown in Figure 17.55. The end product of the reaction sequence is acetyl-CoA. However, the initial oxidation is not at the terminal carbon in all cases. Oxidation may sometimes occur at the second carbon, and then quite different subsequent reactions occur. *Unsaturated* aliphatic hydrocarbons containing a terminal double bond are not refractory to anoxic decomposition and can be oxidized by certain sulfate-reducing and other anaerobic bacteria.

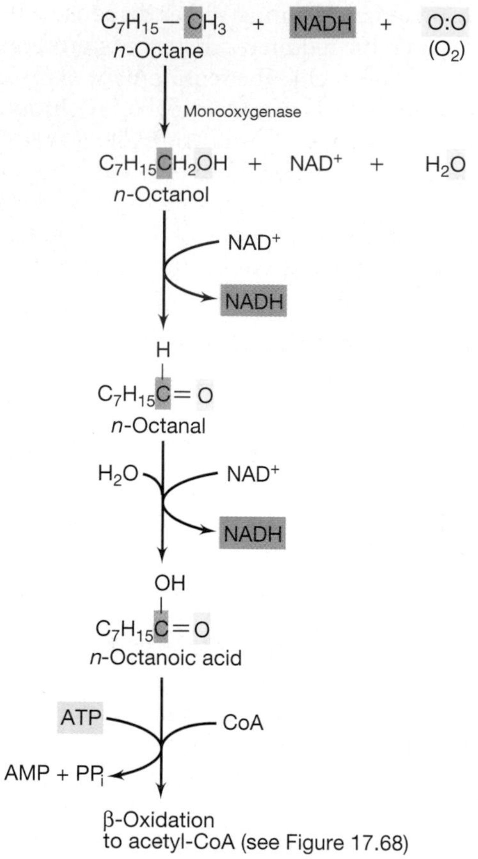

$$C_7H_{15}-CH_3 \quad + \quad \boxed{NADH} \quad + \quad O:O$$
$$\text{n-Octane} \qquad\qquad\qquad\qquad (O_2)$$

Monooxygenase

$$C_7H_{15}CH_2OH \quad + \quad NAD^+ \quad + \quad H_2O$$
$$\text{n-Octanol}$$

NAD⁺

NADH

$$\begin{array}{c} H \\ C_7H_{15}C=O \end{array}$$
n-Octanal

$$H_2O \qquad NAD^+$$

NADH

$$\begin{array}{c} OH \\ C_7H_{15}C=O \end{array}$$
n-Octanoic acid

ATP — CoA

AMP + PP$_i$

β-Oxidation
to acetyl-CoA (see Figure 17.68)

Figure 17.55 Steps in oxidation of an aliphatic hydrocarbon, the first of which is catalyzed by a monooxygenase.

Aromatic Hydrocarbons

Many aromatic hydrocarbons can be used as electron donors aerobically by microorganisms, of which bacteria of the genus *Pseudomonas* have been the best studied. It has been demonstrated that the metabolism of these compounds, some of which are quite large molecules, frequently has as its initial stage the formation of either *protocatechuate* or *catechol*, or a structurally related compound, as shown in Figure 17.56a.

These single-ring compounds are referred to as *starting substrates* because oxidative catabolism proceeds only after the large aromatic molecules have been converted to these more simple forms. Protocatechuate and catechol may then be further degraded to compounds that can enter the citric acid cycle: succinate, acetyl-CoA, and pyruvate (∞ Figure 5.22). Several steps in the catabolism of aromatic hydrocarbons usually require oxygenases. Figures 17.56b, c, and d show three different oxygenase-catalyzed reactions, one using a monooxygenase and two using a dioxygenase.

Anoxic Degradation of Aromatic Compounds

Aromatic compounds can also be degraded anaerobically by facultatively aerobic bacteria if the aromatic compound already contains an atom of oxygen. Such phenolic compounds are degraded by certain denitrifying, phototrophic, ferric iron-reducing, and sulfate-reducing bacteria. From a biochemical standpoint, the anoxic catabolism of aromatic compounds proceeds by *reductive* rather than oxidative ring cleavage (Figure 17.57). This involves *ring reduction* followed by *ring cleavage* to yield a straight-chain fatty acid or dicarboxylic acid. These intermediates can be converted to acetyl-CoA and used for both biosynthetic and energy-yielding purposes. Benzoate and benzoate derivatives are common natural products and are readily degraded anaerobically.

Evidence for the anoxic degradation of benzene and toluene, aromatic compounds lacking an oxygen atom (see Figure 17.56 for structures), has also been obtained. Catabolism of benzene occurs by certain denitriying bacteria and toluene by certain ferric iron-reducing and denitrifying bacteria. Biochemically speaking, toluene is eventually converted to the benzoate derivative benzoyl-CoA and then presumably further catabolized via ring reduction as shown in Figure 17.57.

✓ 17.23 Concept Check

Many microorganisms can degrade aliphatic and aromatic hydrocarbons. Catabolism aerobically involves the action of oxygenase enzymes. Anoxic aromatic degradation proceeds by reductive rather than oxidative pathways.

✓ Draw the chemical structure for benzene. Do the same for benzoate. Are these compounds aliphatic or aromatic?

✓ What fundamental difference exists in the *anaerobic* degradation of an aromatic compound compared with its *aerobic* metabolism? Give an example of this.

17.24 Methanotrophy and Methylotrophy

Section 12.6 considered the unique situation, relative to carbon metabolism, of methanotrophs and methylotrophs. These organisms, although not autotrophs, use C_1 compounds for energy metabolism and biosynthesis. We focus here on the details of these two major processes using *methanotrophy*, the utilization of methane (CH_4, the simplest of all hydrocarbons) as both electron donor and carbon source, as the prime example.

Biochemistry of Methane Oxidation

The individual steps in methane oxidation to CO_2 can be summarized as:

$$CH_4 \rightarrow CH_3OH \rightarrow \mathbf{CH_2O} \rightarrow HCOO^- \rightarrow CO_2$$

Interestingly, methanotrophs, those methylotrophs that can use CH_4, assimilate either all or one-half of their carbon (depending on the pathway used) at the

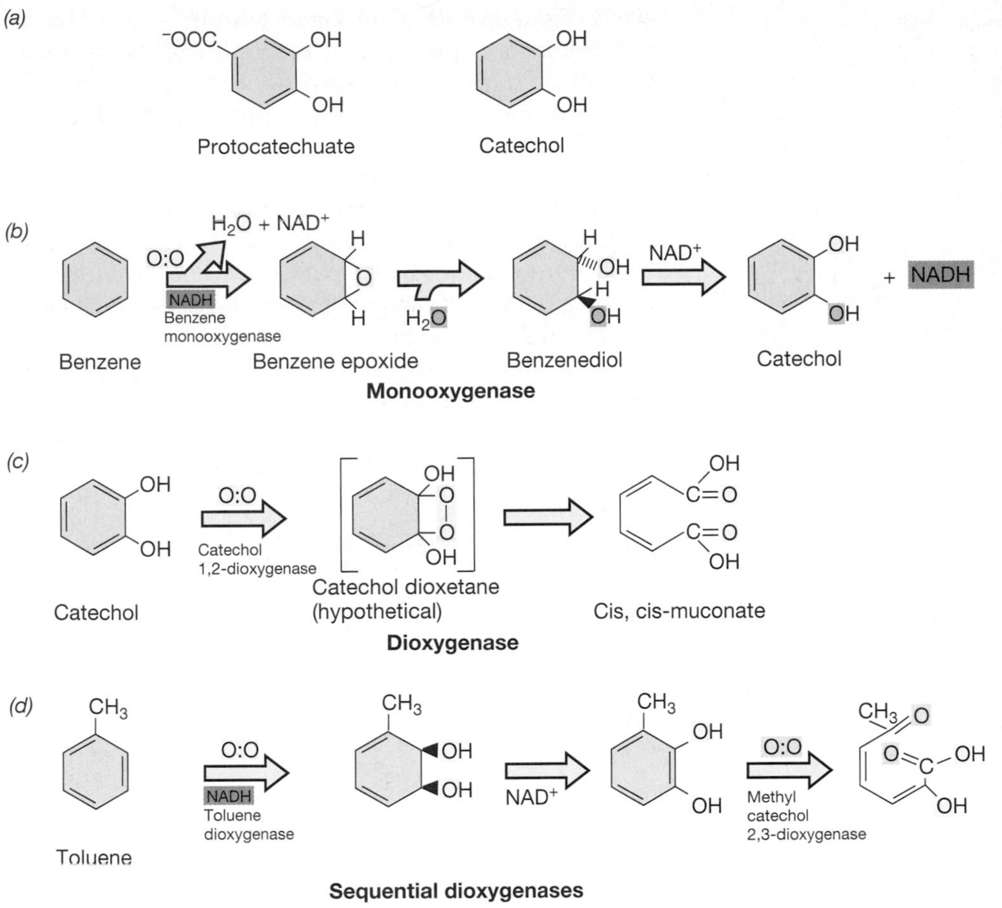

(a)

Protocatechuate   Catechol

(b)

Benzene   Benzene epoxide   Benzenediol   Catechol

Monooxygenase

(c)

Catechol   Catechol dioxetane (hypothetical)   Cis, cis-muconate

Dioxygenase

(d)

Toluene

Sequential dioxygenases

Figure 17.56 Roles of oxygenases in catabolism of aromatic compounds. (a) Protocatechuate and catechol, two common oxidation products of aromatic compounds. (b) Hydroxylation of benzene to catechol by a monooxygenase in which NADH is an electron donor. (c) Cleavage of catechol to *cis,cis*-muconate by a dioxygenase. Reactant oxygen atoms are shown in color in both reactions to demonstrate the different mechanisms. (d) The activity of toluene dioxygenase and methyl catechol 2,3-dioxygenase in the degradation of toluene. The oxygen atoms that each enzyme introduces are distinguished by different colors.

level of formaldehyde (bold in the methane oxidation reaction shown), and we will see that this affects a major energy savings compared with autotrophs, where carbon is assimilated from the more oxidized CO_2. But here our focus is on energy conservation.

The initial step in the oxidation of methane involves an enzyme called **methane monooxygenase**. As we discussed in Section 17.22, oxygenase enzymes catalyze the incorporation of oxygen from O_2 into carbon compounds (and some nitrogen compounds, see Section 17.12), and seem to be widely involved in the metabolism of hydrocarbons. Monooxygenases incorporate one atom of O_2 into the substrate while the second atom is reduced to H_2O. In the methanotroph

Methylosinus, where the methane oxidation process has been most thoroughly studied, the electrons needed for the oxidation of CH_4 to CH_3OH come from cytochrome *c* (Figure 17.58). The requirement for reducing power in the first step precludes synthesis of ATP during the oxidation of methane to methanol, a fact consistent with growth yields (grams of cells produced per mole of substrate consumed) of methanotrophs, which are the same whether methane or methanol is used as the growth substrate. The other oxidation steps from CH_3OH to CO_2 supply electrons to the electron transport chain, and, from these, ATP is made from the resulting electron transport and proton motive force generated (Figure 17.58).

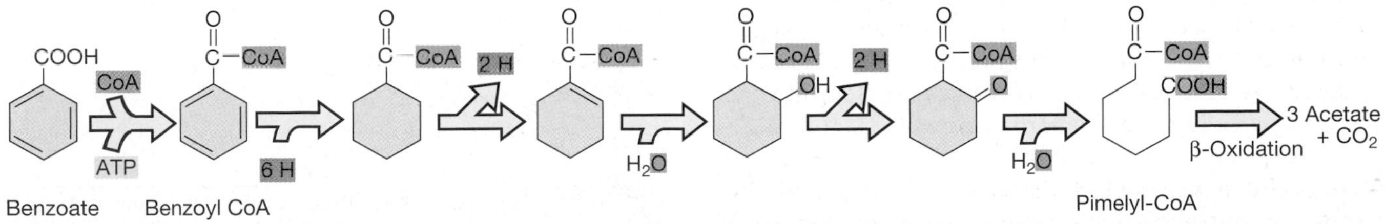

Benzoate   Benzoyl CoA   Pimelyl-CoA   3 Acetate + CO_2

Figure 17.57 Anoxic degradation of benzoate by reductive-ring cleavage. Note that all intermediates of the pathway are bound to coenzyme A. The acetate produced is further catabolized in the citric acid cycle (⌾ Section 5.13).

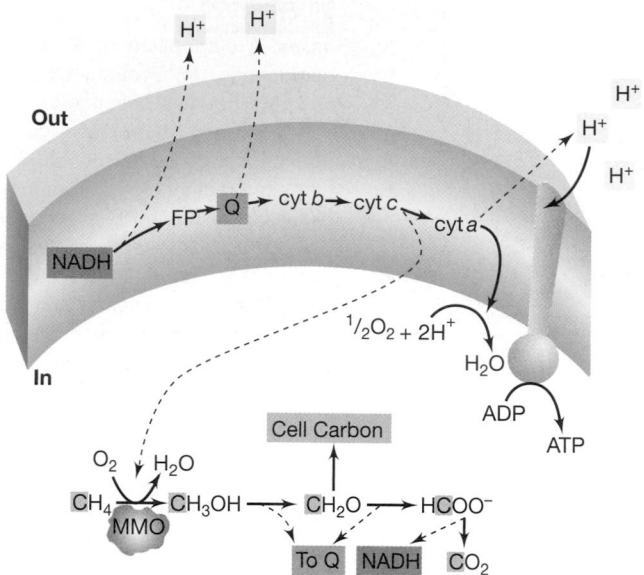

Figure 17.58 Oxidation of methane by methanotrophic bacteria. Methane (CH_4) is converted to methanol (CH_3OH) by the enzyme *methane monooxygenase*. The electrons needed to drive this first step come from cytochrome *c*, and no energy is conserved in this reaction. A proton motive force is established from electron flow in the membrane, and this fuels ATPase. Note how carbon for biosynthesis comes primarily from formaldehyde (CH_2O). MMO, methane monooxygenase; cyt, cytochrome; Q, quinone. Although not depicted as such, MMO is actually a membrane-associated enzyme.

C₁ Assimilation into Cell Material

As was noted in Section 12.6, two classes of methanotrophs are known, *Type I* and *Type II*. Members of each class share a number of properties in common, and each also shows a distinct mechanism for C_1 assimilation; these are the **ribulose monophosphate pathway** (Type I) and the **serine pathway** (Type II).

The serine pathway is outlined in Figure 17.59. In this pathway, a two-carbon unit, acetyl-CoA, is synthesized from one molecule of formaldehyde (produced from the oxidation of CH_4, see Figure 17.58) and one molecule of CO_2. The pathway requires the introduction of reducing power and energy in the form of two molecules each of NADH and ATP for each acetyl-CoA synthesized. The serine pathway employs a number of enzymes of the citric acid cycle and at least one enzyme, *serine transhydroxymethylase*, unique to the pathway (Figure 17.59).

The ribulose monophosphate pathway of Type I methanotrophs, is outlined in Figure 17.60. It is more efficient than the serine pathway because *all* of the carbon atoms for cell material are derived from formaldehyde. And, since formaldehyde is at the same oxidation level as cell material, no reducing power is needed. The ribulose monophosphate pathway requires the introduction of energy in the form of one molecule of ATP for each molecule of glyceraldehyde-3-phosphate synthe-

sized (Figure 17.60). Consistent with the lower energy requirements of the ribulose monophosphate pathway, the cell yield (grams of cells produced per mole of CH_4 oxidized) of Type I methanotrophs is higher than for Type IIs.

The enzymes *hexulosephosphate synthase*, that condenses one molecule of formaldehyde with one molecule of ribulose-5-phosphate, and *hexulose-6-P isomerase* (Figure 17.60) are unique to this pathway, while the remaining enzymes of the cycle are involved in sugar rearrangements in many different organisms. It should also be noted that the substrate for the initial reaction in this pathway, ribulose-5-P, is almost identical to the C_1 acceptor of the Calvin cycle (ribulose 1,5-bisphosphate), suggesting that these two cycles may share evolutionary roots.

✓ 17.24 Concept Check

Methanotrophy is the use of CH_4 as a carbon and energy source. The enzyme *methane monooxygenase* is a key enzyme in the catabolism of methane. C_1 units get assimilated into cell material

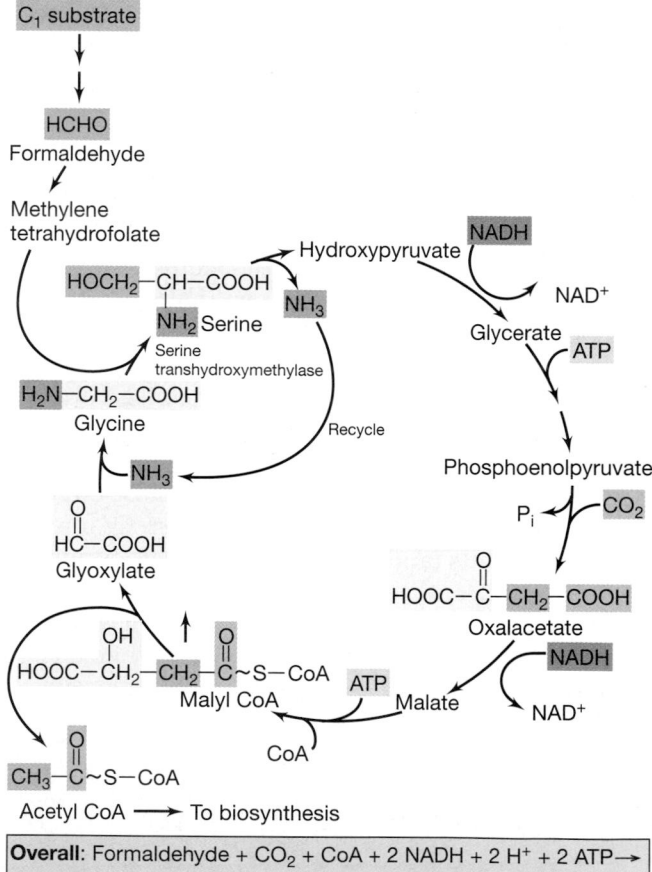

Overall: Formaldehyde + CO_2 + CoA + 2 NADH + 2 H^+ + 2 ATP → Acetyl~CoA + 2 NAD^+ + 2 ADP + 2 P_i + 2 H_2O

Figure 17.59 The serine pathway for the assimilation of C_1 units into cell material by Type II methylotrophic bacteria. The product of the pathway, acetyl-CoA, is used as the starting point for making new cell material. The key enzyme of the pathway is *serine transhydroxymethylase*.

at the level of formaldehyde in methanotrophs by either the ribulose monophosphate pathway or the serine pathway.

✓ What are the energy and reducing power requirements for the ribulose monophosphate pathway? For the serine pathway? Why do they differ?

✓ Why does the oxidation of CH_4 to CH_3OH require reducing power?

✓ Which pathway, the Calvin cycle or the ribulose monophosphate pathway, requires the greater energy input? Why?

17.25 Hexose, Pentose, and Polysaccharide Utilization

We complete our discussion of chemoorganotrophic metabolism with consideration of a few special aspects of the catabolism of organic compounds, especially the use of polymeric substances that must first be hydrolyzed to monomeric units before energy-generating mechanisms can be employed. We begin with the microbial degradation of polysaccharides.

Hexose and Polysaccharide Utilization

Sugars with six carbon atoms, called **hexoses**, are the most important electron donors for many chemoorganotrophs and are also important structural components of microbial cell walls, capsules, slimes, and storage products. The most common hexose sources in nature are listed in Table 17.9, from which it can be seen that most are polysaccharides, although a few are disaccharides. *Cellulose* and *starch* are two of the most important natural polysaccharides.

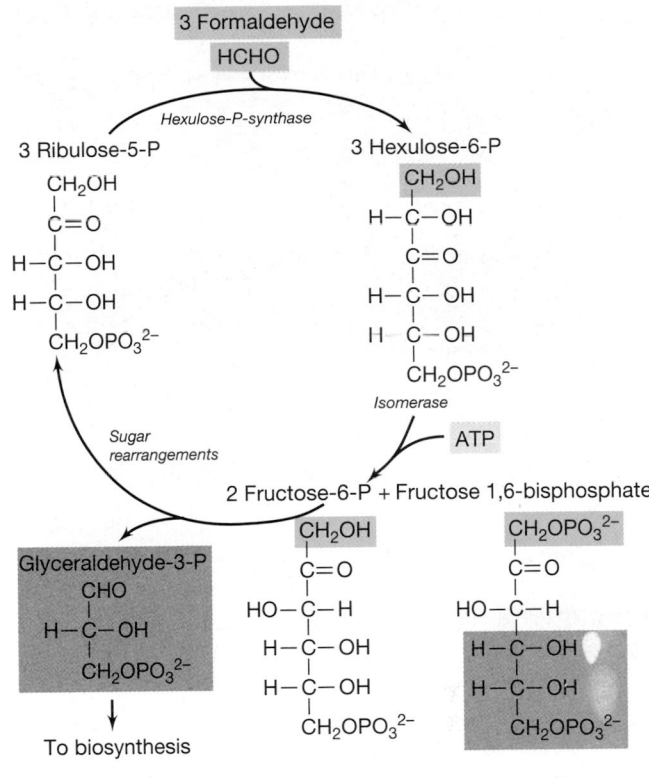

Overall: 3 Formaldehyde + ATP ⟶ glyceraldehyde-3-P + ADP

Figure 17.60 The ribulose monophosphate pathway for assimilation of one-carbon compounds, as found in Type I methylotrophic bacteria. The complete name of the hexulose sugar is D-erythro-L-glycero-3-hexulose 6-phosphate. Three formaldehydes are needed to carry the cycle to completion, with the net result being one molecule of glyceraldehyde-3-P. The key enzyme of this pathway is *hexulose-P-synthase*.

TABLE 17.9 Naturally occurring polysaccharides yielding hexose and pentose sugars[a]

Substance	Composition	Sources	Catabolic enzymes
Cellulose	Glucose polymer (β-1,4-)	Plants (leaves, stems)	Cellulases (β,1-4-glucanases)
Starch	Glucose polymer (α-1,4-)	Plants (leaves, seeds)	Amylase
Glycogen	Glucose polymer (α-1,4- and α-1,6-)	Animals (muscle)	Amylase, phosphorylase
Laminarin	Glucose polymer (β-1,3-)	Marine algae (Phaeophyta)	β-1,3-Glucanase (laminarinase)
Paramylon	Glucose polymer (β-1,3-)	Algae (Euglenophyta and Xanthophyta)	β-1,3-Glucanase
Agar	Galactose and galacturonic acid polymer	Marine algae (Rhodophyta)	Agarase
Chitin	N-Acetylglucosamine polymer (β-1,4-)	Fungi (cell walls) Insects (exoskeletons)	Chitinase
Pectin	Galacturonic acid polymer (from galactose)	Plants (leaves, seeds)	Pectinase (polygalacturonase)
Dextran	Glucose polymer	Capsules or slime layers of bacteria	Dextranase
Xylan	Heteropolymer of xylose and other sugars (β-1,4- and α-1,2 or α-1,3 side groups)	Plants	Xylanases
Sucrose	Glucose-fructose disaccharide	Plants (fruits, vegetables)	Invertase
Lactose	Glucose-galactose disaccharide	Milk	β-Galactosidase

[a] Each of these is subject to degradation by microorganisms.

Cellulose fiber — Bacteria

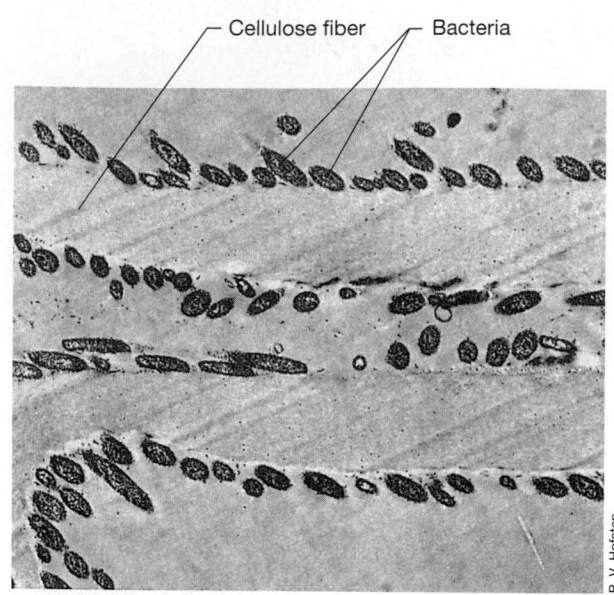

B. V. Hofsten

Figure 17.61 Transmission electron micrograph showing attachment of cellulose-digesting bacteria, *Sporocytophaga myxococcoides*, to cellulose fibers. Cells are about 0.5 μm in diameter.

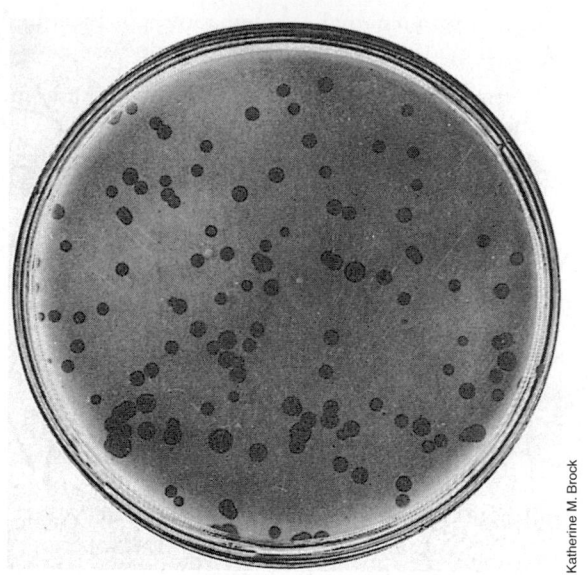

Katherine M. Brock

Figure 17.62 *Cytophaga hutchinsonii* colonies on a cellulose-agar plate. Clear areas are where cellulose has been digested.

Although both starch and cellulose are composed of glucose units, they are connected differently (Table 17.9, ∞ Figure 3.6), and this profoundly affects their properties. Cellulose is much more insoluble than starch and is usually less rapidly digested. Cellulose forms long fibrils, and organisms that digest cellulose are often found closely associated with them (Figure 17.61). Many fungi are able to digest cellulose, and these are mainly responsible for decomposition of plant materials on the forest floor. Among bacteria, however, cellulose digestion is restricted to relatively few groups, of which the gliding bacteria such as *Sporocytophaga* and *Cytophaga* (Figures 17.61 and 17.62; ∞ Figure 12.90), clostridia, and actinomycetes are among the most common. Anoxic digestion of cellulose is carried out by a few *Clostridium* species, which are common in lake sediments, animal intestinal tracts, and systems for anoxic sewage digestion. Cellulose digestion is also a major process in the rumen of ruminant animals where *Fibrobacter* and *Ruminococcus* species actively degrade cellulose (∞ Section 19.11).

Starch is digestible by many fungi and bacteria; this is illustrated for a laboratory culture in Figure 17.63. Starch-digesting enzymes, called *amylases*, are of considerable practical utility in many industrial situations where starch must be digested, such as the textile, laundry, paper, and food industries, and fungi and bacteria are the commercial sources of these enzymes (∞ Section 30.9).

All the polysaccharides occurring extracellularly and used as substrates are broken down to monomeric units by hydrolysis. In contrast, the polysaccharides formed within cells as storage products are broken down not by hydrolysis, but by **phosphorolysis**. This process, involving the addition of *inorganic* phosphate, results in the formation of hexose phosphate rather than the free hexose and may be summarized as follows for the degradation of starch, an α-1,4 polymer of glucose:

$$(C_6H_{12}O_6)_n + P_i \rightarrow (C_6H_{12}O_6)_{n-1} + \text{glucose 1-phosphate}$$

Because glucose 1-phosphate can be easily converted to glucose 6-phosphate, a key intermediate in glycolysis

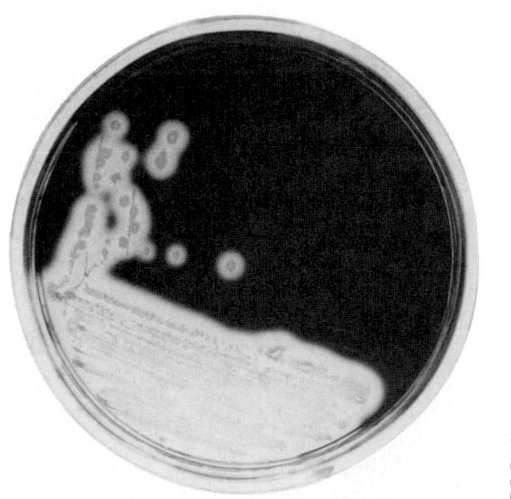

T. D. Brock

Figure 17.63 Demonstration of hydrolysis of starch by colonies of *Bacillus subtilis*. After incubation, the plate was flooded with Lugol's iodine solution. Where starch hydrolysis occurred, the characteristic purple color of the starch-iodine complex is absent. Hydrolysis of starch occurs at some distance from the bacterial colonies because of the production of extracellular amylase, which diffuses into the surrounding medium.

(∞ Figure 5.14), and no ATP is required to form it, phosphorolysis represents a net energy savings to the cell.

Disaccharides

Many microorganisms can use *disaccharides* for growth (Table 17.9). *Lactose* utilization by microorganisms is of considerable economic importance because milk-souring organisms produce lactic acid from lactose. *Sucrose*, the common disaccharide of higher plants, is usually first hydrolyzed to its component monosaccharides (glucose and fructose) by the enzyme *invertase*, and the monomers are then metabolized by normal pathways. *Cellobiose*, β-1,4-diglucose and a major product of cellulose digestion, is also readily degraded by a variety of bacteria that cannot degrade the cellulose polymer itself.

The microbial polysaccharide *dextran* is synthesized by some bacteria using the enzyme *dextransucrase* and sucrose as starting material:

$$n \text{ sucrose} \rightarrow \underset{\text{dextran}}{(\text{glucose})_n} + n \text{ fructose}$$

Dextran is formed in this way by the bacterium *Leuconostoc mesenteroides* and a few others, and the polymer formed accumulates around the cells as a massive slime or capsule (Figure 17.64). Because sucrose is required for dextran formation, no dextran is formed when the bacterium is cultured on a medium with glucose or fructose. In nature, when cells containing dextran or other polysaccharide capsules die, these materials once again become available for attack by fermentative or other chemoorganotrophic microorganisms.

✓ 17.25 Concept Check

Polysaccharides are abundant in nature and can be broken down, usually by phosphorolysis, into hexose or pentose monomers and used as energy sources. Starch and cellulose are common polysaccharides.

✓ What is *phosphorolysis*?
✓ What disaccharides are common in nature?

Figure 17.64 Slimy colony formed by the dextran-producing bacterium, *Leuconostoc mesenteroides*, growing on a sucrose-containing medium. When the same organism is grown on glucose, the colonies are small and not slimy because dextran synthesis requires sucrose (∞ Section 21.3 for further discussion).

17.26 Organic Acid Metabolism

A variety of organic acids can be used by microorganisms as carbon sources and electron donors. The acids of the citric acid cycle, such as *citrate, malate, fumarate,* and *succinate,* are common natural products formed by plants and are also fermentation products of microorganisms. Because the citric acid cycle has major *biosynthetic* (∞ Section 5.15) as well as *energetic* (∞ Section 5.13) functions, the complete cycle or major portions of it are nearly universal in microorganisms. Thus, it is not surprising that many microorganisms are able to use these acids as electron donors and carbon sources. Aerobic utilization of four-, five-, and six-carbon acids can be accomplished by means of enzymes of the citric acid cycle, with ATP formation by oxidative phosphorylation.

Anaerobic utilization of organic acids usually involves conversion to pyruvate followed by formation of acetate via acetyl phosphate with consequent ATP production by substrate-level phosphorylation (see Section 17.19).

Glyoxylate Cycle

Utilization of two- or three-carbon acids as carbon sources cannot occur by means of the citric acid cycle alone. This cycle can continue to operate only if the acceptor molecule, the four-carbon acid *oxalacetate,* is regenerated at each turn of the cycle; any removal of carbon compounds for biosynthetic reactions would prevent completion of the cycle. When acetate is used, the oxalacetate needed to continue the cycle is produced through the **glyoxylate cycle** (Figure 17.65), so called because glyoxylate is a key intermediate. This cycle is composed of most of the citric acid cycle reactions plus two additional enzymes: *isocitrate lyase,* which splits isocitrate to succinate and glyoxylate, and *malate synthase,* which converts glyoxylate and acetyl-CoA to malate (Figure 17.65).

Biosynthesis through the glyoxylate cycle occurs as follows. The splitting of isocitrate into succinate and glyoxylate allows the succinate molecule (or another citric acid cycle intermediate derived from it) to be drawn off for biosynthesis because glyoxylate combines with acetyl-CoA to yield malate. Malate can be converted to oxalacetate to maintain the cyclic nature of the citric acid cycle despite the fact that a C_4 intermediate (succinate) has been drawn off. The succinate molecule can be used directly in the production of porphyrins, be oxidized to oxalacetate and serve as a carbon skeleton for C_4 amino acids, or be converted (via oxalacetate and phosphoenolpyruvate) to glucose.

Pyruvate and C₃ Utilization

Three-carbon compounds such as pyruvate or compounds that can be converted to pyruvate (for example,

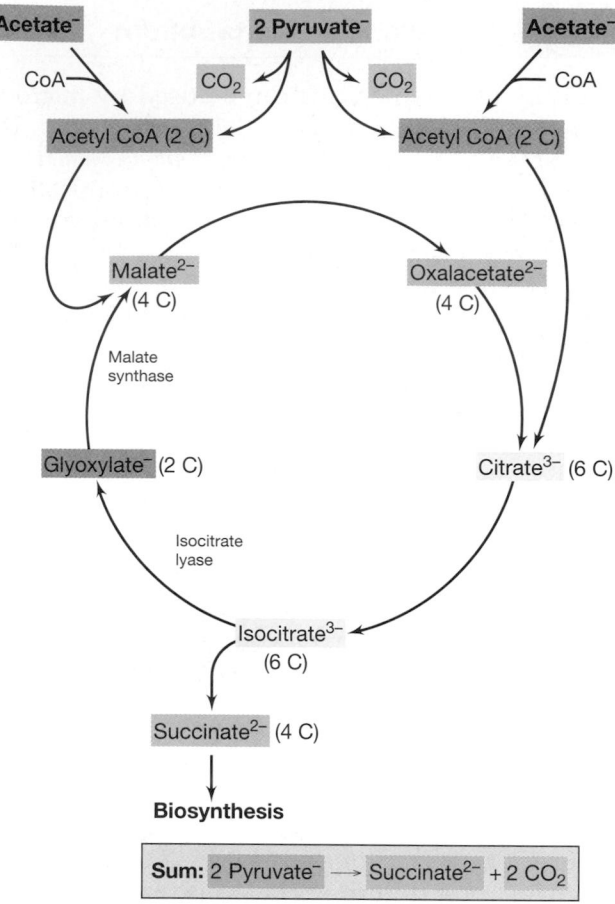

Figure 17.65 The glyoxylate cycle, leading to the synthesis of oxalacetate from acetate. Two unique enzymes, *isocitrate lyase* and *malate synthase*, operate with a majority of the citric acid cycle reactions. In addition to growth on pyruvate, the glyoxylate cycle can also operate during growth on acetate. All compounds containing the same number of carbons are shown by the same color.

lactate or carbohydrates) also cannot be used as energy sources through the citric acid cycle alone. Because some of the citric acid cycle intermediates are used for biosynthesis, the oxalacetate needed to keep the cycle going is synthesized from pyruvate or phosphoenolpyruvate by the addition of a carbon atom from CO_2. In some organisms this step is catalyzed by the enzyme *pyruvate carboxylase*:

$$\text{Pyruvate} + \text{ATP} + CO_2 \rightarrow \text{oxalacetate} + \text{ADP} + P_i$$

whereas in others it is catalyzed by *phosphoenolpyruvate carboxylase*:

$$\text{Phosphoenolpyruvate} + CO_2 \rightarrow \text{oxalacetate} + P_i$$

These reactions replace oxalacetate that is lost when intermediates of the citric acid cycle are removed for use in biosynthesis, and the cycle can continue to function.

✔ **17.26 Concept Check**

Organic acids are frequently metabolized through the citric acid cycle or through the glyoxylate cycle. Isocitrate lyase and malate synthase are the key enzymes of the glyoxylate cycle.

✔ Why is the glyoxylate cycle necessary for growth on acetate but not on succinate?

17.27 Lipids as Microbial Nutrients

Lipids are abundant in nature. The cytoplasmic membranes of all cells contain lipids, and many microorganisms as well as macroorganisms produce lipid storage materials. These substances are all biodegradable and can be excellent substrates for microbial energy-yielding metabolism.

Fat and Phospholipid Hydrolysis

Fats are esters of glycerol and fatty acids (∞ Section 3.4). Microorganisms use fats only after hydrolysis of the ester bond, and extracellular enzymes called **lipases** are re-

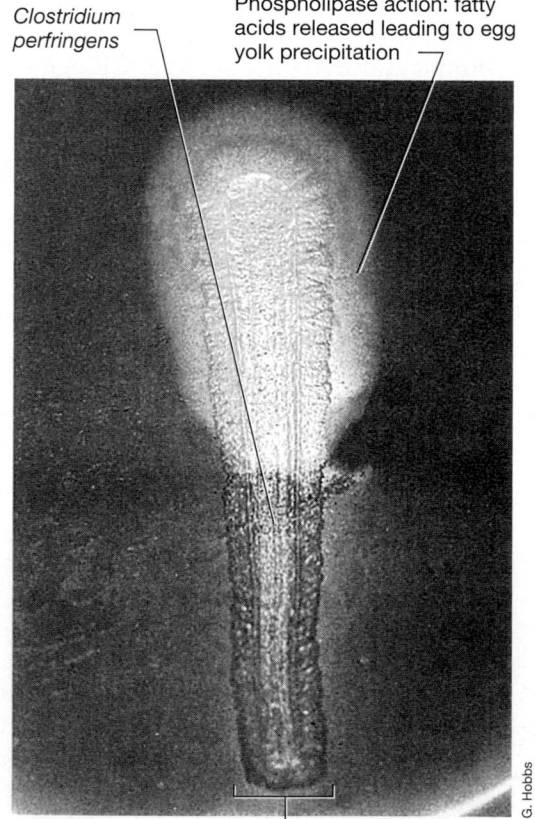

Figure 17.66 Action of phospholipase around a streak of *Clostridium perfringens* growing on an agar medium containing egg yolk. On half of the plate, an inhibitor of phospholipase was added, preventing action of the enzyme.

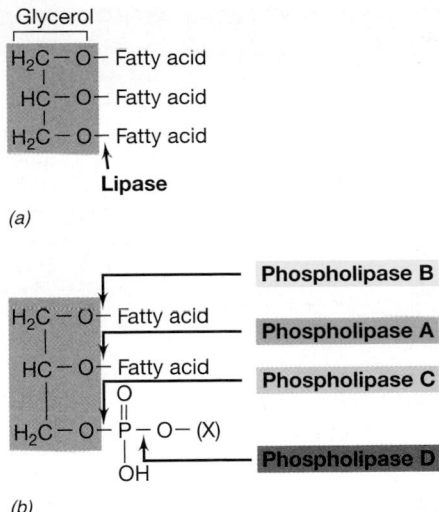

(a)

(b)

Figure 17.67 (a) Action of lipase on a fat. (b) Phospholipase action on phospholipid. The sites of action of the four distinct phospholipases A, B, C, and D are shown. X refers to a number of small organic molecules that may be at this position in different phospholipids. Compare this diagram to the more complete figure of a phospholipid in Figure 3.7.

sponsible for the reaction (Figure 17.66). The end result is formation of glycerol and free fatty acids (Figures 17.66 and 17.67). Lipases are not highly specific and attack fats containing fatty acids of various chain lengths. Phospholipids are hydrolyzed by specific enzymes called *phospholipases*, given different letter designations depending on which ester bond they cleave (Figure 17.67). Phospholipases A and B cleave fatty acid esters and thus resemble the lipases described earlier, but phospholipases C and D cleave phosphate ester linkages and hence are quite different types of enzymes. The result of lipase action is the release of free fatty acids and glycerol, and all these substances can be attacked both anaerobically as well as aerobically by various chemoorganotrophic microorganisms.

Fatty Acid Oxidation

Fatty acids are oxidized by a process called *beta oxidation*, in which two carbons of the fatty acid are split off at a time (Figure 17.68). In eukaryotes the enzymes are in the mitochondria, whereas in prokaryotes they are cytoplasmic. The fatty acid is first activated with coenzyme A; oxidation results in the release of *acetyl-CoA* and the formation of a fatty acid shorter by two carbons (Figure 17.68). The process of beta oxidation is then repeated, and another acetyl-CoA molecule is released. Two separate dehydrogenation reactions occur. In the first, electrons are transferred to flavin-adenine dinucleotide (FAD), whereas in the second they are transferred to NAD$^+$. Most fatty acids have an even number of carbon atoms, and complete oxidation yields only acetyl-CoA. The acetyl-CoA formed is then oxidized by way of the citric acid cycle or is converted to hexose and other cell

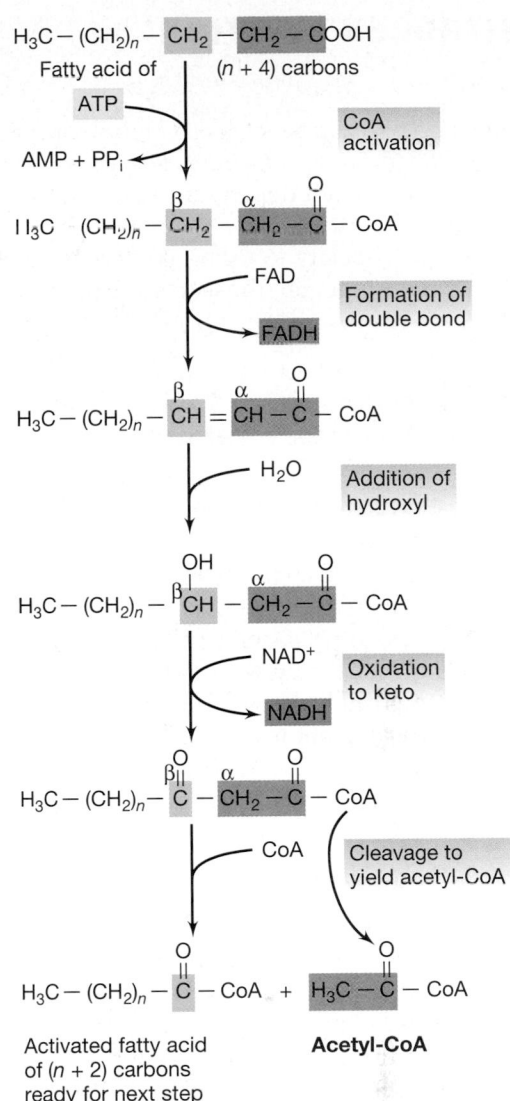

Figure 17.68 Mechanism of beta oxidation of a fatty acid, which leads to successive formation of two-carbon fragments of acetyl-CoA.

constituents via the glyoxylate cycle. Fatty acids are good electron donors. For example, the anaerobic oxidation of the 16-carbon fatty acid palmitic acid results in the net synthesis of 129 ATP molecules; these include electron transport phosphorylation from electrons generated during the formation of acetyl-CoA from beta oxidations and from oxidation of the acetyl-CoA units themselves through the citric acid cycle (⟳ Section 5.13).

✓ 17.27 Concept Check

Fats are metabolized by hydrolysis by lipases or phospholipases to free fatty acids. The latter are oxidized by beta oxidation to acetyl-CoA units, which are subsequently oxidized to CO$_2$ by the citric acid cycle.

✓ What are the functions of phospholipases?
✓ What is meant by the term *β-oxidation*?

V NITROGEN FIXATION

The utilization of dinitrogen gas (N_2) as a source of cell nitrogen is called *nitrogen fixation*. The ability to fix N_2 frees an organism from dependence on fixed forms of nitrogen. Since the latter are in high demand in microbial ecosystems, the capacity to fix N_2 confers a significant ecological advantage on those cells capable of the process.

17.28 Nitrogenase and the Process of Nitrogen Fixation

Nitrogen fixation is restricted to only certain prokaryotes. An abbreviated list of nitrogen-fixing organisms is given in Table 17.10, from which it can be seen that a variety of prokaryotes, both anaerobic and aerobic, fix nitrogen. In addition, there are some bacteria, called *symbiotic*, that fix nitrogen only in association with certain plants. As far as is currently known, no eukaryotic organisms fix nitrogen. Symbiotic nitrogen fixation, a process of enormous agricultural importance (for example, soybeans and alfalfa), will be discussed in Section 19.22.

Nitrogenase

In the fixation process, N_2 is *reduced* to ammonium and the ammonium converted to organic form. The reduction process is catalyzed by the enzyme complex **nitrogenase**, which consists of two separate proteins called *dinitrogenase* and *dinitrogenase reductase*. Both components contain iron, and dinitrogenase contains molybdenum as well. The iron and molybdenum in dinitrogenase are contained in a cofactor known as *FeMo-co* (Figure 17.69), and the actual reduction of N_2 occurs on this iron-molybdenum center. The composition of FeMo-co is $MoFe_7S_8$ homocitrate (Figure 17.69), and FeMo-co is present in two copies per molecule of nitrogenase.

Owing to the stability of the N≡N triple bond (which has a dissociation energy of 940 kJ compared with 493 kJ for the double bond in O_2), N_2 is extremely inert and its activation is a very energy-demanding process. Six electrons must be transferred to reduce N_2 to $2\,NH_3$, and several intermediate steps might be visualized; it is thought that the three successive reduction steps occur directly on nitrogenase with no free intermediates accumulating (Figure 17.70). Nitrogen fixation is highly reductive in nature, and the process is inhibited by oxygen because dinitrogenase reductase is rapidly and irreversibly inactivated by O_2 (even when the enzyme is isolated from *aerobic* nitrogen fixers). In aerobic nitrogen-fixing bacteria, N_2 fixation occurs in the presence of O_2 in whole cells but not in purified enzyme preparations, and

nitrogenase in such organisms is protected from O_2 inactivation by one of several different mechanisms, including the rapid removal of O_2 by respiration, the production of O_2-retarding slime layers (Figure 17.71, page 609), or, in certain cyanobacteria, by compartmentalization of nitrogenase in a special type of cell (the heterocyst) (⮑ Section 12.26). In addition, although N_2 fixation does not occur in oxic cell extracts, in aerobic nitrogen fixers like *Azotobacter*, nitrogenase is protected from oxygen inactivation by complexing with a specific protein; this has been referred to as *conformational protection*.

TABLE 17.10 Some nitrogen-fixing organisms

Free-living aerobes

Chemo-organotrophs	Phototrophs	Chemo-lithotrophs
Bacteria:	Cyanobacteria	*Alcaligenes*
Azotobacter spp.	(various, but not all)	*Thiobacillus*
Azomonas		(some species)
Klebsiella[a]		*Streptomyces*
Beijerinckia		*thermoau-*
Bacillus polymyxa		*totrophicus*
Mycobacterium flavum		
Azospirillum lipoferum		
Citrobacter freundii		
Acetobacter diazotrophicus		
Methylomonas		
Methylococcus		

Free-living anaerobes

Chemo-organotrophs	Phototrophs	Chemo-lithotrophs
Bacteria:	*Bacteria:*	*Archaea:*
Clostridium spp.	*Chromatium*	*Methanosarcina*
Desulfovibrio	*Thiocapsa*	*Methanococcus*
Desulfotomaculum	*Chlorobium*	*Methanobacterium*
	Rhodospirillum	*Methanospirillum*
	Rhodopseudomonas	*Methanolobus*
	Rhodomicrobium	
	Rhodopila	
	Rhodobacter	
	Heliobacterium	
	Heliobacillus	
	Heliophilum	

Symbiotic

Leguminous plants	Nonleguminous plants
Soybeans, peas, clover, locust, and so on, in association with a bacterium of the genus *Rhizobium, Bradyrhizobium, Sinorhizobium,* or *Azorhizobium*	*Alnus, Myrica, Ceanothus, Comptonia, Casuarina;* in association with actinomycetes of the genus *Frankia*

[a] N_2 fixation occurs only under anoxic conditions.

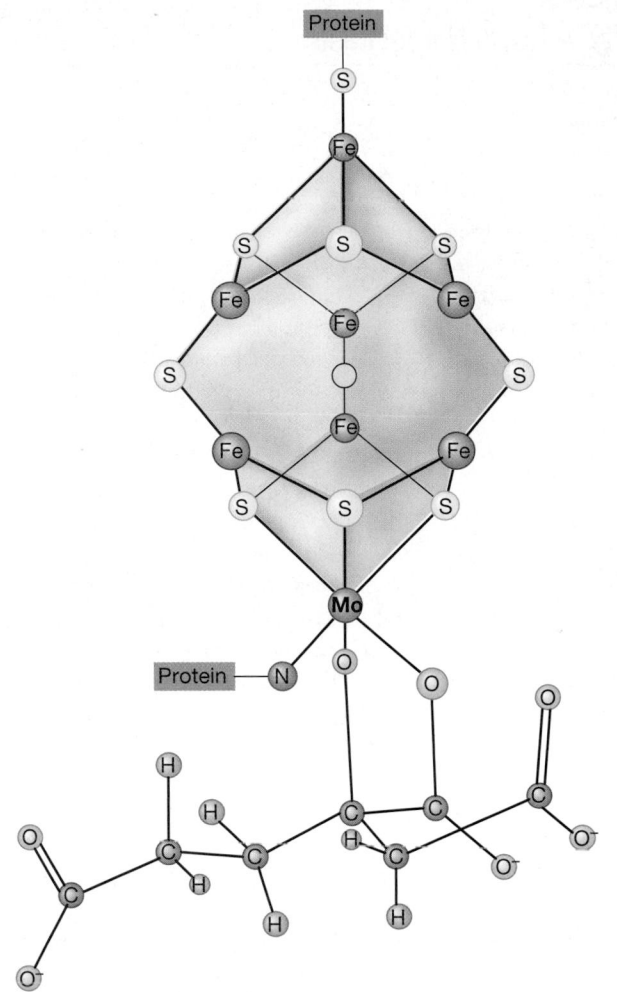

Figure 17.69 Structure of FeMo-co, the iron-molybdenum cofactor from nitrogenase. On the top is the Fe_7S_8 cube that binds to molybdenum along with oxygen atoms from homocitrate (bottom, all oxygen atoms shown in green) and N and S atoms from dinitrogenase. Two molecules of FeMo-co are present per molecule of nitrogenase.

Alternative Nitrogenases

Some nitrogen-fixing bacteria can synthesize non-molybdenum nitrogenases under certain growth conditions, and these so-called *alternative nitrogenases* do not contain molybdenum but instead contain either vanadium (and iron) or iron only. Cofactors similar to FeMo-co are present in both alternative nitrogenases as well: FeVa-co in the vanadium nitrogenase and an iron-sulfur cluster resembling FeMo-co and FeVa-co but lacking both Mo and Va, in the iron nitrogenase. Alternative nitrogenases are not synthesized when sufficient molybdenum is present, as the molybdenum nitrogenase is generally the main nitrogenase in the cell. Alternative nitrogenases presumably serve as a backup mechanism to ensure that N_2 fixation can still occur when molybdenum is limiting in the habitat (∞ Section 12.9). A structurally and func-

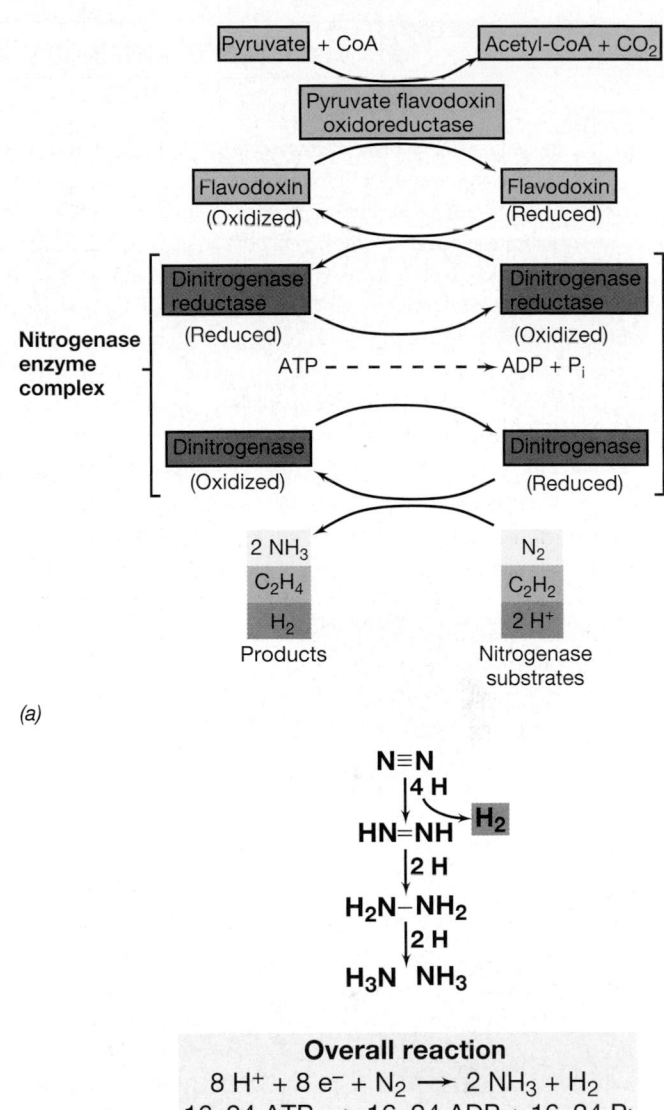

(a)

(b)

Figure 17.70 Nitrogenase function. (a) Shown are the steps in N_2 fixation starting from pyruvate. Electrons are supplied from dinitrogenase reductase to dinitrogenase one at a time, and each electron supplied is associated with the hydrolysis of 2–3 ATPs. (b) Hypothetical steps in N_2 reduction showing the H_2 evolution step, and a summary of nitrogenase activity.

tionally novel molybdenum nitrogenase has been discovered in the thermophilic streptomycete, *Streptomyces thermoautotrophicus*; its properties are compared with those of classic Mo nitrogenases in the box, The Power of Metabolic Diversity: A Novel Nitrogenase.

Electron Flow in Nitrogen Fixation

The sequence of electron transfer in nitrogenase is as follows: electron donor → dinitrogenase reductase → dinitrogenase → N_2 (Figure 17.70). The electrons for nitrogen

A Focus On ... The Power of Metabolic Diversity: A Novel Nitrogenase

Nitrogenases have been characterized from a wide variety of prokaryotes, including some *Archaea*, and all of them show significant sequence homology at both the gene and polypeptide level: All of them, that is, until the nitrogenase from the streptomycete *Streptomyces thermoautotrophicus* was characterized.[1]

S. thermoautotrophicus is a thermophilic (optimum temperature 65°C), gram-positive, filamentous prokaryote that occurs naturally in burning compost and charcoal piles (Figure 1). The organism is an aerobic chemolithotrophic H_2 bacterium that can also use carbon monoxide (CO) as an electron donor. Although *S. thermoautotrophicus* has been known to be a nitrogen fixer for some time, some unusual properties of its nitrogen fixation system (like the fact that ammonia did not repress nitrogenase synthesis and that the enzyme did not reduce acetylene) prompted a more detailed examination of its nitrogenase. What was found represents a totally new paradigm for N_2 fixation.

The *S. thermoautotrophicus* nitrogenase contains Mo, but unlike classic Mo nitrogenase, it is completely *insensitive* to O_2. The dinitrogenase component of the *S. thermoautotrophicus* nitrogenase, called *Str1*, contains three different polypeptides that show some structural similarity to dinitrogenase polypeptides from other nitrogen-fixers, but the dinitrogenase reductase component, called *Str2*, shows no similarity to other dinitrogenase reductases. However, Str2 shows *very high* sequence similarity to manganese-containing superoxide dismutases. In fact Str2 *is* a superoxide dismutase! Recall from Chapter 6 (◯◯◯ Section 6.13) that superoxide dismutases function in the cell to consume superoxide (O_2^-), forming O_2 in the process and thus preventing oxidative damage to cell components. But what does superoxide have to do with nitrogen fixation?

It has been shown that Str2 supplies electrons to Str1. The source of the electrons is O_2^-, and the O_2^- is formed from the reduction of O_2 by a CO dehydrogenase (Figure 2). Thus, in analogy to the pyruvate → flavodoxin → dinitrogenase reductase → dinitrogenase sequence in classical nitrogen-fixation (see Figure 17.70), in *S. thermoautotrophicus* the sequence is $CO → O_2^- → Str2 → Str1$. And, astonishingly, instead of O_2 *inhibiting* nitrogenase (as it does in every nitrogenase that has ever been examined), in *S. thermoautotrophicus* O_2 is actually *required* in the reaction mechanism of the enzyme!

Clearly, *S. thermoautotrophicus* nitrogenase is a structurally and functionally unique nitrogen-fixing system. How widespread such a system is and whether its primary function in the cell is actually to fix N_2, is as yet unknown. However, the *S. thermoautotrophicus* nitrogenase system is a good example of the power of metabolic diversity in prokaryotes: Even well-studied systems in which conformity prevails can occasionally yield big surprises and totally new concepts. Discovery of the *S. thermoautotrophicus* nitrogenase system has also renewed hope for the eventual genetic engineering of nitrogenase into agricultural crops such as corn. The fact that this nitrogenase is not oxygen labile and its energy requirements are much lower than those of classical nitrogenases (measurements show the *S. thermoautotrophicus* system to use only 25–50% of the ATP of classic Mo nitrogenases, see Figure 17.70) could make the dream of nitrogen-fixing row crops a reality someday. ■

Ortwin Meyer

Figure 1 Two burning charcoal piles in the Bavarian forest, Germany, containing cells of the N_2-fixing bacterium, *Streptomyces thermoautotrophicus*. The scientist shown is doing temperature measurements at various points in the piles. The piles emit the gases CO_2, CO, CH_4, and C_2H_2, and vary in temperature with depth. The surface to about 15 cm into the piles has temperatures of less than 100°C but deeper into the piles, temperatures can be over 300°C. *S. thermoautotrophicus* is active up to about 75°C, and oxidizes CO in both its energy metabolism and during N_2 fixation (see Figure 2).

[1]Ribbe, M., D. Gadkari, and O. Meyer, 1997. *J. Biol. Chem.* 272: 26627–26633.

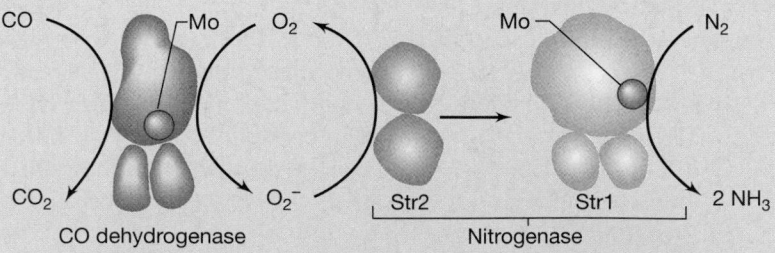

Figure 2 Reactions of nitrogen fixation in *S. thermoautotrophicus*. Although quite different proteins than dinitrogenase reductase and dinitrogenase, Str2 and Str1 are functionally equivalent, respectively, to these proteins.

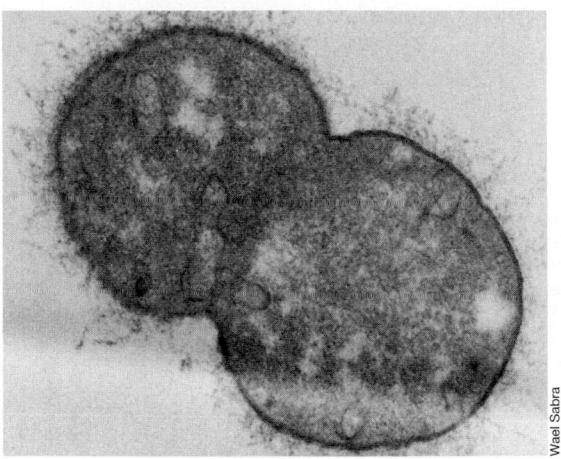

(a)

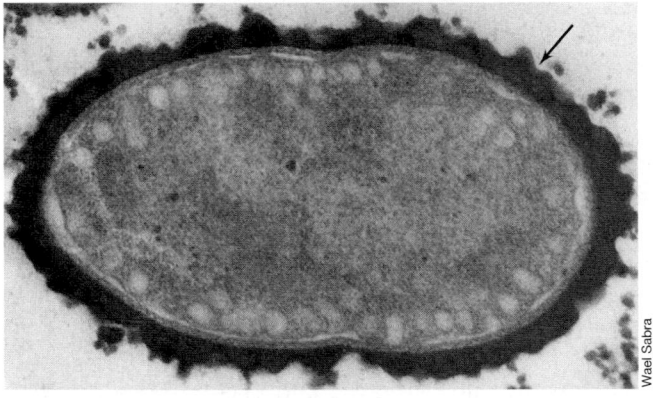

(b)

Figure 17.71 Induction of slime formation by O₂ in nitrogen-fixing cells of *Azotobacter vinelandii*. (a) Transmission electron micrograph of cells grown under microxic conditions on 2.5% O₂; very little slime is evident. (b) Cells grown in air (21% O₂). Note the extensive darkly staining slime layer (arrow). The slime retards diffusion of O₂ into the cell, thus preventing nitrogenase inactivation by oxygen.

reduction are transferred to dinitrogenase reductase from ferredoxin or flavodoxin, low potential iron-sulfur proteins. In *Clostridium pasteurianum*, ferredoxin is the electron donor and is reduced by phosphoroclastic splitting of pyruvate to acetyl-CoA + CO_2. In addition to reduced ferredoxin or flavodoxin, ATP is required for N_2 fixation. In each cycle of electron transfer, dinitrogenase reductase is reduced by ferredoxin/flavodoxin and binds two molecules of ATP. ATP binding alters the conformation of dinitrogenase reductase and lowers its reduction potential, allowing it to interact with dinitrogenase. Upon electron transfer to dinitrogenase, the ATP is hydrolyzed and dinitrogenase reductase dissociates from dinitrogenase and begins another cycle of reduction and ATP binding (Figure 17.70). When appropriately reduced, dinitrogenase then reduces N_2 to NH_3, with the actual reduction occurring at the FeMo-co center. Although only *six* electrons are nec-

essary to reduce N_2 to two NH_3, *eight* electrons are actually consumed in the process, *two* electrons being lost as hydrogen (H_2), for each mole of N_2 reduced (Figure 17.70). The reason for this apparent waste is not known, but evidence is strong that H_2 evolution is an intimate part of the reaction mechanism of nitrogenase.

Assaying Nitrogenase: Acetylene Reduction

Nitrogenase is not entirely specific for N_2, because it also reduces cyanide (CN^-), acetylene ($HC\equiv CH$), and several other triply bonded compounds (for an exception, see the box). The reduction of acetylene by nitrogenase is only a two-electron process, and *ethylene* ($H_2C=CH_2$) is produced. The reduction of acetylene probably serves no useful purpose to the cell, but it does provide the experimenter with a simple and rapid way of measuring the activity of nitrogen-fixing systems because it is fairly easy to measure the reduction of acetylene to ethylene by gas chromatography (Figure 17.72). This technique is now widely used to detect nitrogen fixation in unknown systems. Previously, it was not easy to prove that an organism fixed N_2; indeed, many claims for nitrogen fixation in microorganisms were shown to be erroneous. The growth of an organism in a medium to which no nitrogen compounds have been added does not mean that the organism is fixing nitrogen, because traces of fixed nitrogen compounds often occur as contaminants in the ingredients of culture media or enter the media in air or as dust particles.

Definitive proof of N_2 fixation is obtained using an isotope of nitrogen, ^{15}N, as a tracer. (^{15}N is not a radioisotope but a stable isotope. It is detected with a mass spectrometer.) The gas phase of a culture is enriched with ^{15}N, and after incubation, the cells and medium are digested, the ammonia produced being distilled off and assayed for its ^{15}N content. If there has been a significant production of ^{15}N-labeled NH_3, it is proof of nitrogen fixation. However, the acetylene reduction method is a more rapid and sensitive, albeit indirect, way of measuring nitrogen fixation. The sample, which may be soil, water, a culture, or a cell extract, is incubated with acetylene, and the gas phase of the reaction mixture is later analyzed by gas chromatography for production of ethylene (Figure 17.72). This method is far simpler and faster than other methods and can easily be adapted for field use in ecological studies of N_2-fixing bacteria directly in their habitats.

17.29 Genetics and Regulation of N₂ Fixation

Because the process of N_2 fixation is highly energy demanding, the synthesis and activity of nitrogenase is highly regulated. These regulatory events as well as the

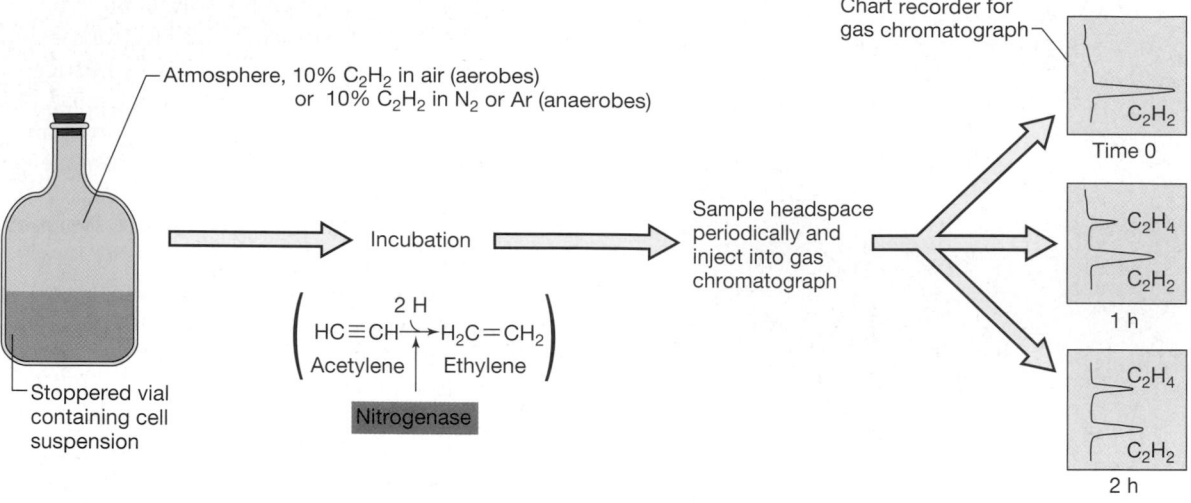

Figure 17.72 The acetylene reduction assay for nitrogenase activity. The results show no C_2H_4 when the experiment begins (time 0), but increasing production of C_2H_4 as the assay proceeds. Note how as C_2H_4 is produced, C_2H_2 is consumed. If the vial contained an enzyme extract, conditions would be anoxic, even if the nitrogenase were from an aerobic bacterium.

enzymes involved in N_2 fixation itself (Figure 17.70) require many genes to encode them, which we discuss here.

Genetics of Nitrogen Fixation

The genes for dinitrogenase and dinitrogenase reductase in *Klebsiella pneumoniae*, a well-studied N_2 fixer, are part of a complex regulon (a large network of operons) called the *nif regulon* (Figure 17.73); the *K. pneumoniae nif* regulon spans 24 kb of DNA and contains 20 genes arranged in several transcriptional units (Figure 17.73). In addition to nitrogenase structural genes, the genes for FeMo-co, genes controlling the electron transport proteins, and a number of regulatory genes are also present in the *nif* regulon. Dinitrogenase is a complex protein made up of two subunits, α (product of the *nifD* gene) and β (product of the *nifK* gene), each of which is present in two copies. Dinitrogenase reductase is a protein dimer consisting of two identical subunits, the product of *nifH*. FeMo-co is synthesized through the participation of several genes, including *nifN, V, Z, W, E*, and *B*, as well as *Q*, which encodes a product involved in molybdenum processing. The *nifA* gene encodes a positive regulatory protein that serves to activate transcription of other *nif* genes.

Nitrogenase is a highly conserved protein and the nitrogen structural genes *nifHDK* have been used as molecular probes to screen DNA from various prokaryotes for the presence of homologous genes. In all nitrogen fixers examined (except for the highly unusual *Streptomyces thermoautotrophicus*, see the box), *nifHDK*-like genes are present, suggesting that the genetic requirements for nitrogenase are rather specific. Alternative nitrogenases are encoded by their own genes, *vnfHDK* in the vanadium system, and *anfHDK* in the iron-only system, that show significant sequence homology to *nifHDK*.

Regulation of Nitrogen Fixation

Nitrogenase is subject to strict regulatory controls. Nitrogen fixation is blocked by O_2 and by fixed nitrogen, including NH_3, NO_3^-, and certain amino acids (but see the box). A major part of this regulation is at the level of transcription. The various transcriptional units of the *nif* regulon are shown in Figure 17.72. While transcription of the *nif* structural genes is *activated* by the NifA protein (positive regulation), NifL is a *negative* regulator of *nif* gene expression and contains a molecule of FAD (recall that FAD is a redox coenzyme for flavoproteins, ⌘ Section 5.11) that is critical for O_2-sensing by the protein. In the presence of sufficient O_2, NifL functions to shut down transcription of *nif* genes in order to prevent synthesis of the oxygen labile nitrogenase. Ammonia represses N_2 fixation through a second protein, called NtrC, whose activity is regulated by the nitrogen status of the cell. When ammonia is limiting, NtrC is active and promotes transcription of *nifA*. This produces NifA, the nitrogen fixation activator protein, and *nif* transcription begins.

The ammonia produced by nitrogenase does not repress enzyme synthesis because as soon as it is made, it is incorporated into organic form and used in biosynthesis. But when ammonia is in excess (as in environments high in ammonia), nitrogenase synthesis is quickly repressed. In this way, ATP is not wasted in making a product already present in ample amounts. In certain nitrogen-fixing bacteria, nitrogenase *activity* is also regulated by ammonia, a phenomenon called the ammonia "switch-off" effect. In this case, excess ammonia causes a

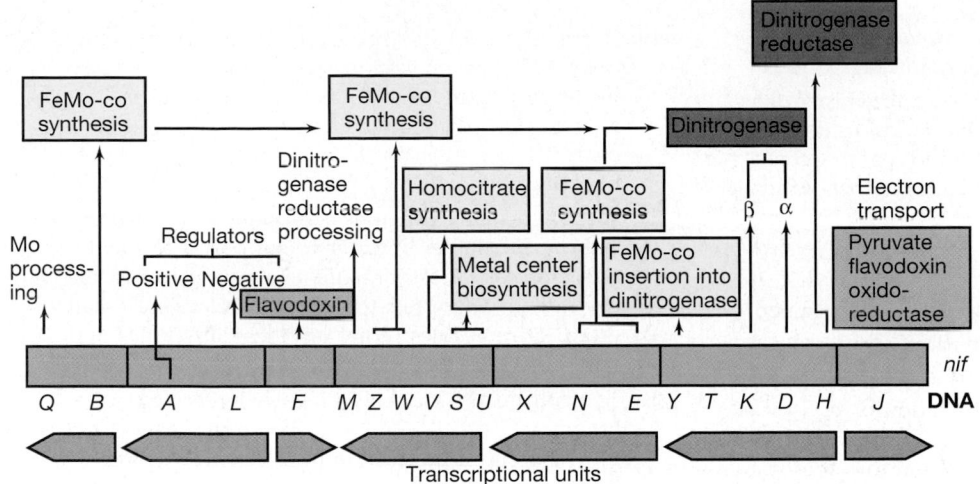

Figure 17.73 The *nif* regulon in *Klebsiella pneumoniae*, the best-studied nitrogen-fixing organism. The functions of some of the genes are uncertain. The mRNA transcripts (transcriptional units) are shown below the genes; arrows indicate the direction of transcription. Proteins involved in FeMo-co synthesis are shown in yellow. Other colors match those of Figure 17.70. Although capable of growth aerobically with ammonia as nitrogen source, nitrogen fixation in *Klebsiella* only occurs under anoxic conditions.

covalent modification of dinitrogenase reductase, which results in a loss of enzyme activity. When ammonia again becomes limiting, this modified protein is converted back to the active form and N_2 fixation resumes. Ammonia switch-off is thus a rapid and reversible method of controlling ATP consumption by nitrogenase. Not all nitrogen-fixing bacteria have the switch-off system, but switch-off is widespread in certain groups such as the phototrophic bacteria, where it was first discovered.

✓ 17.28–17.29 Concept Checks

Nitrogen fixation, the reduction of N_2 to NH_3, involves a complex enzyme system called nitrogenase, which consists of dinitrogenase and dinitrogenase reductase. Most nitrogenases

contain molybdenum or vanadium and iron as metal cofactors, and the process of nitrogen fixation is highly energy-demanding. Nitrogenase and most associated regulatory proteins are encoded by the *nif* regulon. Certain artificial substrates that are structurally similar to N_2, such as acetylene and cyanide, are also reduced by nitrogenase. Nitrogenase and nitrogen fixation are both highly regulated processes.

✓ Write a balanced equation for the reaction carried out by the enzyme *nitrogenase*.

✓ What is *FeMo-co*?

✓ What metals are necessary for nitrogen fixation?

✓ How is C_2H_2 useful for studies of nitrogen fixation?

✓ What chemical and physical factors affect the function of nitrogenase? How does the *Streptomyces thermoautotrophicus* nitrogenase system differ in this regard?

✓ How can ammonia affect the nitrogen fixation process?

Review Questions

1. In what nutritional class would you place an organism that uses *glucose* as sole carbon and energy source? *Elemental sulfur* as an energy source? How would we refer to the latter organism if it grew with CO_2 as sole carbon source? What is the energy source for *phototrophic* organisms?

2. What is the role of light in the photosynthetic process of green and purple bacteria? Of cyanobacteria? Compare and contrast the photosynthesis process in these two groups of prokaryotes.

3. What are the functions of light-harvesting and reaction center chlorophylls? Why would a mutant incapable of making light-harvesting chlorophylls (such mutants can be readily isolated in the laboratory) probably not be a successful competitor in nature?

4. Where are the photosynthetic pigments located in a purple bacterium? A cyanobacterium? A green alga? Considering the function of chlorophyll pigments, why can't they be located elsewhere in the cell, for example, in the cytoplasm or in the cell wall?

5. How does light result in ATP production in an *anoxygenic phototroph*? In what ways are photosynthetic and respiratory electron flow similar? In what ways do they differ?

6. How is reducing power made for autotrophic growth in a purple bacterium? In a cyanobacterium?

7. How does the reduction potential of chlorophyll *a* in photosystem I and photosystem II differ? Why must the reduction potential of photosystem II chlorophyll be so highly electropositive?

8. What is the major function of carotenoids and phycobilins in phototrophic microorganisms?

9. What two enzymes are unique to organisms that carry out the Calvin cycle? What reactions do these enzymes carry out? What would be the consequences if a mutant arose that lacked either of these enzymes?

10. For conversion of 6 molecules of CO_2 into 1 fructose molecule, 18 molecules of ATP are required. Where in the Calvin cycle reactions are these ATPs consumed?

11. What organisms employ the hydroxypropionate or reverse citric acid cycles as autotrophic pathways?

12. Compare and contrast the utilization of H_2S by a purple phototrophic bacterium and by a colorless sulfur bacterium like *Beggiatoa*. What role does H_2S play in the metabolism of each organism?

13. Discuss why the growth yield (grams of cells per mole of substrate) of *Thiobacillus ferrooxidans* is considerably greater when the organism is growing aerobically on elemental sulfur than on ferrous iron as electron donor (assume the organism is growing autotrophically in both cases).

14. In *Escherichia coli* synthesis of the enzyme *nitrate reductase* is repressed by oxygen. On the basis of bioenergetic arguments, why do you think this repression phenomenon might have evolved?

15. Discuss at least three major differences between *assimilative* and *dissimilative* nitrate reduction.

16. Why is hydrogenase a constitutive enzyme in *Desulfovibrio*?

17. Compare and contrast ferric iron reduction with reductive dechlorination in terms of (1) product of the reduction, and (2) environmental significance.

18. Compare and contrast *homoacetogens* with *methanogens* in terms of (1) substrates and products of their energy metabolism, (2) ability to use organic compounds as electron donors in energy metabolism, (3) mechanism of autotrophy, and (4) ability to grow by aerobic respiration. (You may want to review material in Section 13.4 before answering this question.)

19. Define the term *substrate-level phosphorylation*. How does it differ from *oxidative phosphorylation*? Assuming an organism is facultative, what basic nutritional conditions dictate whether the organism obtains energy from substrate-level rather than oxidative phosphorylation?

20. Although many different compounds are theoretically fermentable, in order to support a fermentative process, most organic compounds must be eventually converted to one of a relatively small group of molecules. What are these molecules and why must they be produced?

21. To a culture of *Escherichia coli* growing fermentatively you add 1 g/l of $NaNO_3$. Would you expect the growth *yield* of the culture to increase or decrease? Why?

22. How can fermentations occur in the absence of substrate-level phosphorylation?

23. Why have hydrocarbons accumulated in large reservoirs on Earth despite the fact that they are readily degradable microbiologically under certain conditions?

24. How does a *methanotroph* differ from a *methanogen*? How do Type I and Type II methanotrophs differ in their carbon assimilation patterns?

25. How do *monooxygenases* differ from *dioxygenases* in the reactions they catalyze?

26. Compare and contrast the conversion of cellulose and intracellular starch to glucose units. What enzymes are involved and which process is the more energy-efficient?

27. Write out the reaction catalyzed by the enzyme *nitrogenase*. How many electrons are required in this reaction? How many are actually used? Explain.

28. What metals are found in nitrogenase? Are all nitrogenases oxygen-sensitive?

29. How does the *Streptomyces thermoautotrophicus* nitrogenase differ from that of *Azotobacter*?

Application Questions

1. Compare and contrast the absorption spectrum of chlorophyll *a* and bacteriochlorophyll *a*. What wavelengths are preferentially absorbed by each pigment and how do the absorption properties of these molecules compare with the regions of the spectrum visible to our eye? Why are most plants green?

2. The growth rate of the phototrophic purple bacterium *Rhodobacter* is about twice as fast when the organism is grown phototrophically in a medium containing malate as carbon source as when it is grown with CO_2 as carbon source (with H_2 as electron donor). Discuss the reasons why this is true and list the nutritional class we would place *Rhodobacter* in when growing under each of the two different conditions.

3. Discuss the nature of the evidence obtained from studies on the photosynthetic process of certain cyanobacteria that supports the hypothesis that these organisms evolved from anoxygenic phototrophs.

4. Although physiologically distinct, chemolithotrophs and chemoorganotrophs share a number of features with respect to the production of ATP. Discuss these common features along with reasons why the growth yield (grams of cells per mole of substrate) of a chemoorganotroph respiring glucose is so much higher than for a chemolithotroph respiring sulfur.

5. Why is the following statement, if taken literally, incorrect? "Anaerobic respiration is simply a process where an alternative electron acceptor is substituted for O_2 in the respiratory process."

6. When methane is made from CO_2 (plus H_2) or from methanol (in the absence of H_2), various steps in the pathway shown in Figures 17.44 and 17.45 are used. Compare and contrast methanogenesis from these two substrates and discuss why they must be metabolized in *opposite* directions?

7. Although dextran is a glucose polymer, glucose cannot be used to make dextran. Explain. How is dextran synthesis important in oral hygiene (∞ Section 21.3)?

8. *Pseudomonas fluorescens* can grow on benzoate aerobically, whereas the phototrophic bacterium *Rhodopseudomonas palustris* can grow on benzoate anaerobically. Compare and contrast the metabolism of benzoate by these two species, focusing on the following considerations: requirement for oxygenases, initial reactions leading to the opening of the ring, and product(s) formed that can feed into central metabolic pathways.

Fluorescent stains attached to nucleic acid probes that target ribosomal RNA, are called phylogenetic stains. These stains can penetrate cells and bind to their ribosomes. The specificity of the probes can be made to vary from general to highly specific. This allows one to distinguish the microorganisms present in an ecosystem in either a general way, for example, *Bacteria* versus *Archaea*, or in a very detailed way, such as distinguishing between different genera within a particular phylogenetic lineage. Shown here are two types of nitrifying bacteria that form granules in sewage sludge. The organisms are closely related yet sufficiently distinct that only one of the two ribosomal RNA probes, one labeled with a red dye and one with a green dye, will bind to each population of cells. Microbial ecologists can exploit these genetic differences in microbial cells to identify and track them in natural microbial ecosystems.

18

METHODS IN MICROBIAL ECOLOGY

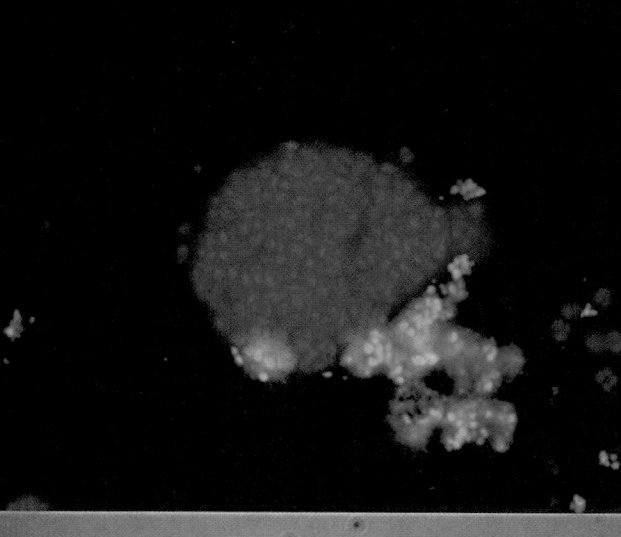

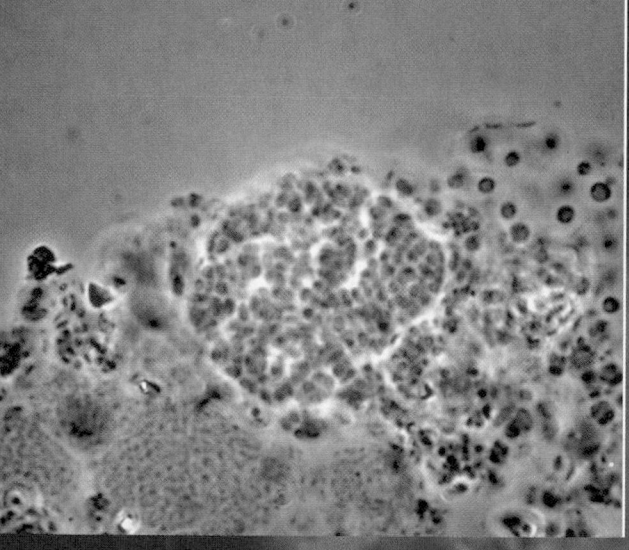

Working Glossary

Denaturing gradient gel electrophoresis (DGGE) an electrophoretic technique capable of separating out nucleic acid fragments of the same size but that differ in sequence

DAPI a nonspecific fluorescent dye used to stain bacteria in a natural sample

Enrichment bias a problem with liquid enrichment cultures in which the results are skewed toward the isolation of "weed" species to the detriment of the most abundant or ecologically significant organisms in a habitat

Enrichment culture a means of obtaining laboratory cultures of microorganisms from a natural sample by using highly selective culture methods

Fluorescent in situ hybridization (FISH) a method employing a fluorescent dye fused to a specific nucleic acid probe for identifying or tracking organisms in the environment

Isotope fractionation discrimination by enzymes against the *heavier* isotope of the various isotopes of C or S, leading to enrichment in the lighter isotopes of each during the metabolism of CO_2 or SO_4^{2-}

Laser tweezers a device used to obtain pure cultures in which a single cell is optically trapped with a laser beam and moved away from surrounding cells into sterile growth medium

Microbial ecology study of the interaction of microorganisms with each other and their environment

Microelectrode a small glass electrode for measuring pH or specific compounds such as O_2 or H_2S that can be immersed into a microbial habitat at microscale intervals

Most probable number (MPN) serial dilution of a natural sample to determine the highest dilution yielding growth

Nucleic acid probe an oligonucleotide, usually 10–20 bases in length, complementary in base sequence to a sequence in a target gene

Winogradsky column a column packed with mud and overlaid with water to mimic a lake, in which various bacteria develop over a period of months

---

I n this chapter we begin our study of **microbial ecology**. Like general ecology, microbial ecology deals with how organisms, in this case microorganisms, interact with each other and with their environment. Microbial ecologists primarily focus on two areas of study: (1) *biodiversity*, including the isolation, identification, and quantification of microorganisms in various habitats, and (2) *microbial activity*, that is, what microorganisms are *doing* in their habitats.

This chapter explores the methodology of microbial ecology—how microbial ecologists assess biodiversity and microbial activity in a given habitat. These methods include *enrichment* and *isolation*, both nonspecific and specific *cell-staining* methods, *gene isolation and characterization*, and measuring the *activities* of microorganisms *in situ* (in the natural environment). Chapter 19 will discuss the ecology of both large-scale microbial communities (for example, soil or water) and more specialized microbial habitats (for example, the rumen of ruminant animals or the root nodules of leguminous plants).

I CULTURE-DEPENDENT ANALYSIS OF MICROBIAL COMMUNITIES

In the next two sections we consider how microbiologists isolate microorganisms from nature. The process involves teasing the organisms of interest out from a microbial community, a procedure called *enrichment*, and then isolating the organism of interest in pure culture. Isolation is important because it yields cultures for de-

tailed and controlled laboratory studies and for applications in the fields of biotechnology and industrial and environmental microbiology.

18.1 Enrichment and Isolation

Rarely does a natural environment contain only a single type of microorganism. Instead *microbial communities* exist in nature, and a major activity of the microbial ecologist is to devise methods and procedures for the isolation and culture of organisms of interest. The most common approach to this goal is the **enrichment culture technique**. In an enrichment culture, a medium and a set of incubation conditions are chosen that are *selective* for the desired organism and *counterselective* for undesired organisms. The enrichment culture strategy is to duplicate as closely as possible the resources and conditions of a particular niche and then probe for organisms from a sample of that niche that are able to develop under the chosen enrichment conditions. Table 18.1 provides an overview of some successful enrichment culture procedures.

Successful enrichment cultures require an appropriate *inoculum* (sample containing the organism). Thus, any enrichment culture protocol begins by sampling the appropriate habitat (see Table 18.1). Enrichment cultures can be established by placing the inoculum directly into a highly selective medium; many common prokaryotes are isolated in this way. For example, the great Dutch microbiologist Martinus Beijerinck, founder of the enrichment culture technique (∞ Section 1.6), first isolated the aerobic N_2-fixing bacterium *Azotobacter* using what has become a classic

TABLE 18.1 Some enrichment culture methods for prokaryotes[a]

Light-phototrophic bacteria: main C source, CO_2

Incubation in air	Organisms enriched	Inoculum
N_2 as nitrogen source	Cyanobacteria	Pond or lake water; sulfide-rich muds; stagnant water; raw sewage; moist, decomposing leaf litter; moist soil exposed to light; pasteurized soil (heliobacteria)
NO_3^- as nitrogen source, 55°C	Thermophilic cyanobacteria	Hot spring microbial mat
Anoxic incubation		
H_2 or organic acids; N_2 as sole nitrogen source	Purple nonsulfur bacteria, heliobacteria	
H_2S as electron donor	Purple and green sulfur bacteria	

Dark-chemolithotrophic bacteria: main C source, CO_2 (medium must lack organic C)

Incubation in air: aerobic respiration			
Electron donor	Electron acceptor	Organisms enriched	Inoculum
NH_4^+	O_2	Nitrosifying bacteria (*Nitrosomonas*)	Soil, mud; sewage effluent
NO_2^-	O_2	Nitrifying bacteria (*Nitrobacter*)	
H_2	O_2	Hydrogen bacteria (various genera)	
H_2S, S^0, $S_2O_3^{2-}$	O_2	*Thiobacillus* spp.	
Fe^{2+}, low pH	O_2	*Thiobacillus ferrooxidans*	
Anoxic incubation			**Inoculum**
S^0, $S_2O_3^{2-}$	NO_3^-	*Thiobacillus denitrificans*	Mud, lake sediments, soil
H_2	NO_3^- + yeast extract	*Paracoccus denitrificans*	

Dark-chemoorganotrophic bacteria and methanogens: main C source, organic compounds

Incubation in air: aerobic respiration			
Electron donor and nitrogen source	Electron acceptor	Typical organisms enriched	Inoculum
Lactate + NH_4^+	O_2	*Pseudomonas fluorescens*	Soil, mud; lake sediments; decaying vegetation; pasteurize inoculum (80°C for 15 min) for all *Bacillus* enrichments
Benzoate + NH_4^+	O_2	*Pseudomonas fluorescens*	
Starch + NH_4^+	O_2	*Bacillus polymyxa*, other *Bacillus* spp.	
Ethanol (4%) + 1 % yeast extract, pH 6.0	O_2	*Acetobacter, Gluconobacter*	
Urea (5%) + 1% yeast extract	O_2	*Sporosarcina ureae*	
Hydrocarbons (e.g., mineral oil, gasoline, toluene) + NH_4^+	O_2	*Mycobacterium, Nocardia, Pseudomonas* (see Figure 19.45)	
Cellulose + NH_4^+	O_2	*Cytophaga, Sporocytophaga* (see Figure 12.90)	
Mannitol or benzoate, N_2 as N source	O_2	*Azotobacter*	

enrichment strategy (Figure 18.1). Because *Azotobacter* is a rapidly growing N_2-fixing bacterium capable of N_2 fixation in air, enrichment using media devoid of ammonia or other forms of fixed nitrogen selects strongly for this organism; non–nitrogen fixers or anaerobic nitrogen fixers are counterselected in this instance (Figure 18.1).

Literally hundreds of different enrichment strategies have been devised; some of the more dependable ones are listed in Table 18.1. We wish to emphasize, however, that both the culture medium *and* the incubation conditions are critical to a successful enrichment culture; that is, both *resources* and *conditions* must be optimized to have the best chance of enriching the organism of interest.

TABLE 18.1 Some enrichment culture methods for prokaryotes[a] (*continued*)

Anoxic incubation: anaerobic respiration

Main ingredients	Electron acceptor	Organisms enriched	Inoculum
Organic acids	KNO_3 (0.2%)	*Pseudomonas* (denitrifying species)	Soil, mud; lake sediments
Yeast extract	KNO_3 (1%)	*Bacillus* (denitrifying species)	
Organic acids	Na_2SO_4	*Desulfovibrio, Desulfotomaculum*	
Acetate, propionate, butyrate	Na_2SO_4	Fatty-acid oxidizing sulfate reducers	As above; or sewage digestor sludge; rumen contents
Acetate, ethanol	S^0	*Desulfuromonas*	
Acetate	Fe^{3+}	*Geobacter, Geospirillum*	
Acetate	ClO_3^-	Various chlorate-reducing bacteria	
H_2	Na_2CO_3	Methanogens (chemolithotrophic species only), homoacetogens	Mud, sediments, sewage sludge
CH_3OH	Na_2CO_3	*Methanosarcina barkeri*	
CH_3NH_2	KNO_3	*Hyphomicrobium*	

Anoxic incubation: fermentation

Electron donor and nitrogen source	Electron acceptor	Organisms enriched	Inoculum
Glutamate or histidine	No exogenous electron acceptors added	*Clostridium tetanomorphum*	Mud, lake sediments; rotting plant material; dairy products (lactic and propionic acid bacteria); rumen or intestinal contents (enteric bacteria); sewage sludge; soil
Starch + NH_4^+	None	*Clostridium* spp.	
Starch, N_2 as N source	None	*Clostridium pasteurianum*	
Lactate + yeast extract	None	*Veillonella* spp.	
Glucose or lactose + NH_4^+	None	*Escherichia, Enterobacter,* other fermentative organisms	
Glucose + yeast extract (pH 5)	None	Lactic acid bacteria (*Lactobacillus*)	
Lactate + yeast extract	None	Propionic acid bacteria	
Succinate + NaCl	None	*Propionigenium*	
Oxalate	None	*Oxalobacter*	

[a] All media must contain an assortment of mineral salts including N, P, S, Mg^{2+}, Mn^{2+}, Fe^{2+}, Ca^{2+}, and other trace elements (⬭ Sections 5.1–5.3). This table is meant as an overview of enrichment methods and does not speak to the effect incubation temperature might have in isolating thermophilic (high temperature), hyperthermophilic (very high temperature), and psychrophilic (low temperature) species, or the effect that extremes of pH or salinity might have, assuming an appropriate inoculum was available.

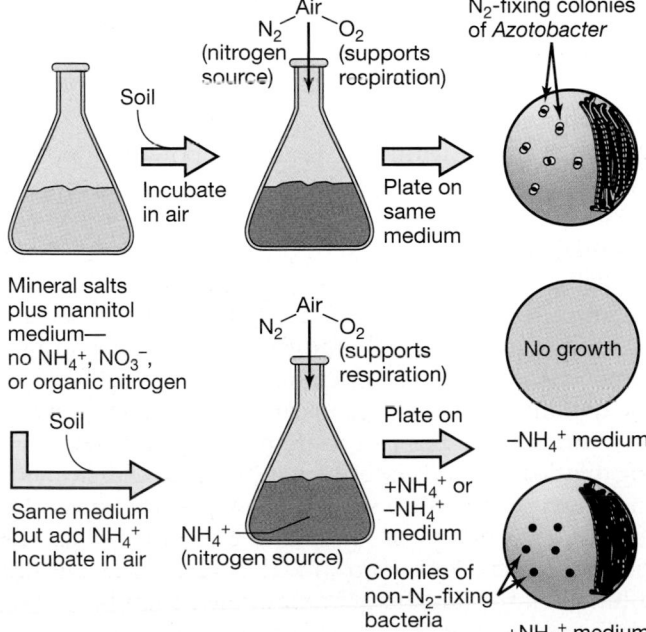

The Winogradsky Column

For isolation of purple and green phototrophic bacteria, sulfate-reducing bacteria, and many other anaerobes, the **Winogradsky column** has traditionally been used as an enrichment vehicle (Figure 18.2). Named for the famous Russian microbiologist Sergei Winogradsky (⬭ the box, Winogradsky's Legacy, Chapter 17, and Section 1.6), the column was first used by him in the late nineteenth century to study soil microorganisms. The Winogradsky column is a miniature anoxic ecosystem that can be used as a long-term source of bacteria for enrichment culture purposes.

A Winogradsky column is prepared by filling a large glass cylinder about one-half full with organic-rich, preferably sulfide-containing, mud (Figure 18.2). Carbon

Figure 18.1 The isolation of *Azotobacter*. Selection for aerobic N_2-fixing bacteria usually results in *Azotobacter* or its relatives. By contrast, enrichment with fixed forms of nitrogen such as NH_4^+ rarely results in N_2-fixing bacteria.

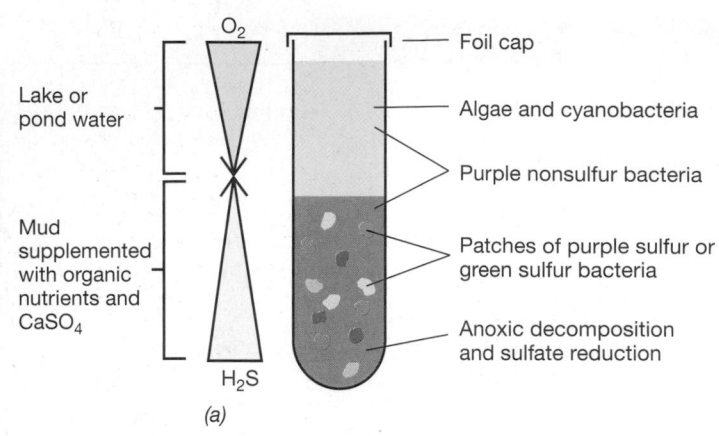

(b)

Figure 18.2 The Winogradsky column. (a) Schematic view of a typical column. The column is placed to receive subdued sunlight. Chemoorganotrophic bacteria grow throughout the column, aerobes and microaerophiles in the upper regions, anaerobes in the zones containing H_2S. Anoxic decomposition leading to sulfate reduction creates the gradient of H_2S. Green and purple sulfur bacteria stratify according to their tolerance for H_2S. (b) Photo of Winogradsky columns that have remained anoxic up to the top where blooms of three different phototrophic bacteria have occurred in the mud and up into the water column. Left to right: *Thiospirillum jenense*, *Chromatium okenii*, (both purple sulfur bacteria, Section 12.2) and *Chlorobium limicola* (green sulfur bacteria, Section 12.32).

substrates are first mixed into the mud. The substrates will determine which organisms are enriched, but fermentative substrates (that can lead to excessive gas formation), are usually avoided. The mud is supplemented with $CaCO_3$ and $CaSO_4$ as a buffer and as a source of sulfate, respectively. The mud is packed tightly in the container, taking care to avoid trapping air (Figure 18.2). The mud is then covered with lake, pond, or ditch water (or seawater if a marine column is established), and the top of the cylinder is covered to prevent evaporation. The cylinder is placed near a window receiving muted sunlight and left to develop for a period of weeks.

In a typical Winogradsky column a mixture of organisms develops. Algae and cyanobacteria appear quickly in the upper portions of the water column, and by producing O_2 these organisms help to keep this zone oxic. Decomposition processes in the mud quickly lead to the production of organic acids, alcohols, and H_2, suitable substrates for sulfate-reducing bacteria (Sections 12.18, 17.15, and 19.13). Sulfide from the sulfate reducers triggers development of purple and green sulfur bacteria (anoxygenic phototrophs, Sections 12.2 and 12.33). These organisms typically appear in patches on the sides of the mud in the column but may bloom in the water itself if oxygenic phototrophs are scarce (Figure 18.2). Sampling of the column for phototrophic bacteria is performed by inserting a long, thin pipet and removing some colored mud or water. Such samples can be used for microscopy and to inoculate enrichment media (Table 18.1) for isolation and characterization.

Winogradsky columns have been used to enrich a variety of prokaryotes, both aerobes and anaerobes. Besides producing a ready supply of inocula for enrichment cultures, columns also can be supplemented with a particular compound to test the hypothesis that an organism or organisms exist in the inoculum that can degrade it. Once a crude enrichment has been established, pure cultures can be pursued as will be discussed in Section 18.2.

Enrichment Bias

Although the enrichment culture is a powerful tool, in most enrichments there exists a bias, and sometimes a very severe bias, in the outcome. In typical liquid enrichment cultures the most fit organism for the particular set of conditions chosen becomes the dominant member in the enrichment. Unfortunately, however, the most fit organism for laboratory culture is often only a minor component of the microbial ecosystem rather than the dominant or most ecologically relevant component.

Enrichment bias can be demonstrated by comparing dilution methods (see Section 18.2) with classical liquid enrichment: Dilution of the inoculum followed by enrichment often yields a different organism than enrichment using undiluted inoculum. It is thought that dilution eliminates the quantitatively insignificant but rapidly growing "weed" species to allow other members of the microbial community a chance to develop. Thus, dilution of the inoculum is a common practice in enrichment culture microbiology today.

✓ 18.1 Concept Check

The enrichment culture technique is a means of obtaining microorganisms from natural samples. Specific reactions can be investigated by enrichment methods by choosing media and incubation conditions selective for particular organisms.

✓ Describe the enrichment rationale behind Beijerinck's isolation of *Azotobacter*.

✓ Why is sulfate (SO_4^{2-}) added to a Winogradsky column?

✓ What is *enrichment bias*?

18.2 Isolation in Pure Culture

Pure cultures can be obtained from enrichments in many ways; frequently employed means include the streak plate, the agar shake, and liquid dilution. For organisms that grow well on agar plates, the streak plate is quick, easy, and the method of choice (Figure 18.3*a*). By repeated picking and restreaking of a well-isolated colony, a pure culture can be obtained that can then be transferred to a liquid medium. With proper incubation facilities (for example, anoxic jars for anaerobes ⸺ Section 6.13), it is possible to purify both aerobes and anaerobes on agar plates by the streak plate method.

Agar Shakes and Most Probable Numbers (MPN)

The agar shake-tube method involves the dilution of a mixed culture in tubes of molten agar, resulting in colonies *embedded* in the agar rather than growing on the surface of an agar plate (Figure 18.3*b*). Shake tubes are useful for purifying several types of anaerobic microorganisms, such as phototrophic sulfur bacteria and sulfate-reducing bacteria from Winogradsky columns (see Figure 18.2). Purification is obtained by successive dilutions of cell suspensions in tubes of liquid medium (Figure 18.3*b*). By repeating this procedure using a colony from a tube at the highest dilution as inoculum for a new set of dilutions, pure cultures are eventually obtained.

A similar procedure but one not using agar is serial dilution of an inoculum in a liquid medium until the final tube or tubes in the series show no growth (Figure 18.4). Using 10-fold serial dilutions, for example, the last tube showing growth should have originated from 10 or fewer cells. By repeating this process several times, pure cultures can be obtained. This method, known as the *most probable number* (or *MPN*) technique (Figure 18.4), has been used for estimating the numbers of microor-

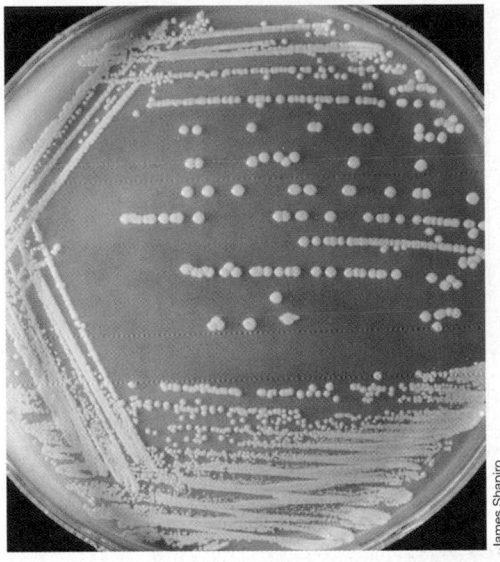

(a)

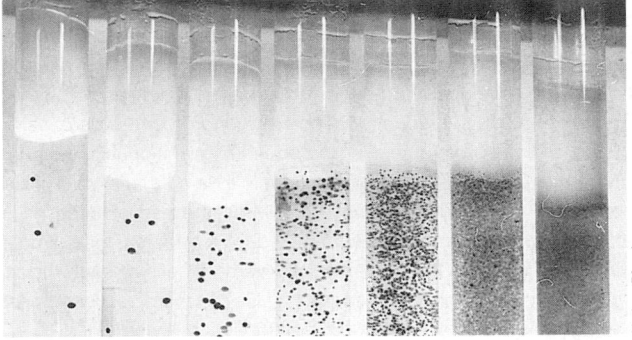

(b)

Figure 18.3 Pure culture methods. (a) Organisms that form distinct colonies on plates such as those shown here are usually very easy to purify. For oxygen-tolerant anaerobes, plate isolations also work well if the plates are put under anoxic conditions immediately after streaking. (b) Colonies of oxygen-sensitive anaerobes using the agar shake technique. A dilution series was established from right to left, eventually yielding well-isolated colonies. The tubes are sealed with a sterile mixture of paraffin and mineral oil to maintain anaerobiosis.

ganisms in foods, wastewater, and other samples, where cell numbers need to be assessed routinely. This can be done using highly selective media and incubation conditions to target one or a small group of organisms, such as a particular pathogen, or using complex media to get an estimate of total cell numbers.

Regardless of the method(s) used to purify a culture, once a putative pure culture has been obtained, it is essential to check its purity. This should be done through a combination of microscopy, the observation of colony characteristics on plates or in shake tubes, and by testing the culture for growth in media in which the culture grows poorly but contaminants grow vigorously. The microscopic observation of a single morphological type of

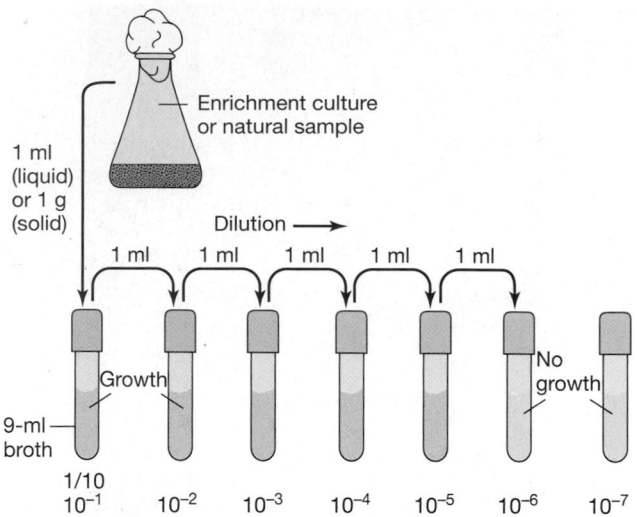

Figure 18.4 Procedure for an "extincting dilution" enrichment or a most-probable number (MPN) analysis. An existing enrichment culture or a natural sample of water, soil, etc., is added to a selective culture medium, and serial dilutions are made from them. The last tube showing growth (in the example shown it is the 10^{-5} dilution) should have developed from 10 or fewer cells. As a means of obtaining pure cultures, the highest dilution showing growth is used as inoculum, and the procedure is repeated. As a means of estimating the numbers of a particular type of bacteria in a sample, the MPN is defined as the highest dilution showing growth. If the medium used in this case were highly selective for sulfate-reducing bacteria, for example, one could conclude that 10^5–10^6 recoverable cells of such organisms were present per gram of sample. Use of several replicate tubes at each dilution improves accuracy of the final MPN obtained.

cell, coupled with uniform colony characteristics and lack of contamination in tests with various culture media, can be taken as evidence that a culture is *axenic* (pure).

High-Tech Pure Cultures—The Laser Tweezers

In addition to the more classical methods just described, technological advances in microscopy have yielded new tools for obtaining pure cultures, such as the **laser (optical) tweezers**.

Laser tweezers consist of an inverted light microscope equipped with a strongly focused infrared laser and a micromanipulation device (Figure 18.5). A single cell in a viewing field can be optically trapped by the laser beam and moved away from contaminating organisms. Trapping occurs because the laser creates a force that pushes down on small objects (like microbial cells) and holds them in place; when the laser beam is moved, the trapped bacterial cell moves along with it. If the sample is in a capillary tube, a single cell can be trapped and then isolated by breaking the tube at a point between the cell and the rest of the population; the trapped cell can then be flushed into a small tube of sterile medium to initiate a pure culture (Figure 18.5).

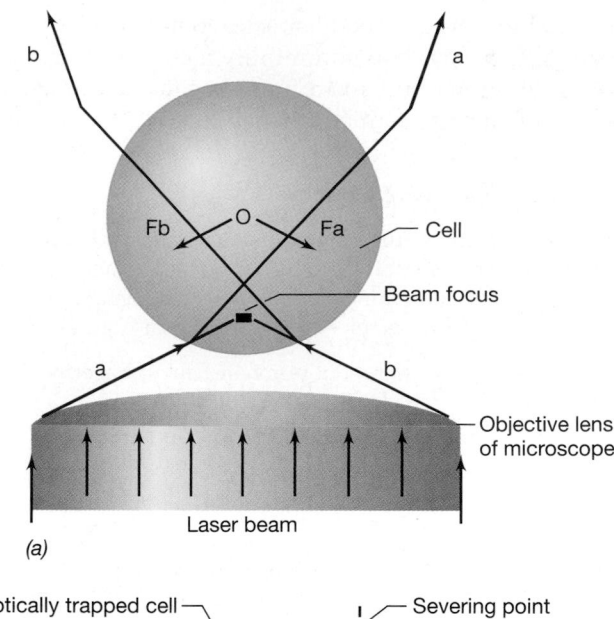

Figure 18.5 The laser tweezers for the isolation of bacteria. (a) Principle of the optical tweezers. A strongly focused laser beam on an object as small as a bacterial cell creates downward radiation forces (F_a, F_b) on the cell that allows the cell to be dragged in any direction as long as the beam force remains on it. (b) Isolation. The laser beam can lock onto a single cell present in a mixture in a capillary tube and drag the optically trapped cell (in the example here, the desired cell is dragged from right to left) away from the other cells. Once the desired cell is far enough away from the other cells, the capillary is severed, the laser turned off, and the cell flushed into a tube of sterile medium.

Laser tweezers technology is especially useful for isolating slow-growing bacteria that may be overgrown in typical enrichment cultures or for organisms present in such low numbers that they would be missed using dilution-based enrichment methods. And, coupled with the use of specific dyes that can identify particular organisms in a microscope field (see Section 18.4), the laser tweezers can be used to select these organisms from a mixture for purification and further laboratory study.

✓ 18.2 Concept Check

Once a successful enrichment culture has been established, a pure culture can be obtained by use of a variety of conventional microbiological procedures including streak plates, agar shakes, and dilution methods. The laser tweezers allow one to "pick" a cell from a microscope field and literally move it away from contaminants.

✓ How does the agar shake method differ from streaking to obtain isolated colonies?

✓ In an MPN (see Figure 18.4), why is the actual number of cells per sample likely to be between the reciprocal of the highest dilution showing growth and the next highest dilution?

✓ How might you isolate a morphologically unique bacterium present in an enrichment culture in relatively low numbers?

II MOLECULAR ANALYSIS OF MICROBIAL COMMUNITIES

Microbial ecologists often wish to quantitate cells in a microbial habitat or know whether particular organisms are present and in what numbers. Such information yields insight on the dynamics of an ecosystem and which organisms are ecologically most relevant. Cell stains help microbiologists obtain these numbers and we detail these methods here. Organisms in natural environments can also be detected by assaying for their *genes*. Genes encoding either ribosomal RNA (⚭ Sections 11.4–11.8) or a specific metabolic process are the usual targets in these studies, and we discuss these as well as methods for "gene prospecting."

18.3 Viability and Quantification Using Staining Techniques

A number of staining methods have been devised in microbiology, and some are suitable for quantifying microorganisms in natural samples. Although these methods say little if anything about the physiological or phylogenetic composition of the populations, they are widely used and are fairly reliable for obtaining quantitative information about the *total numbers* of cells in a particular habitat.

Fluorescent Staining Using DAPI
Fluorescent dyes have been widely used to stain microorganisms in opaque habitats. **DAPI** (4′,6-diamido-2-phenylindole) is a popular stain for this purpose; cells stained with DAPI fluoresce bright blue and are very easy to see and enumerate (Figure 18.6). DAPI staining is widely used for the enumeration of microorganisms in environmental, food, and clinical samples.

Depending on the sample, nonspecific background staining can be an occasional problem with fluorescent stains, but DAPI, which stains nucleic acids, for the most part is unreactive with inert matter. Thus, for many samples, soil as well as aquatic, DAPI staining can give a

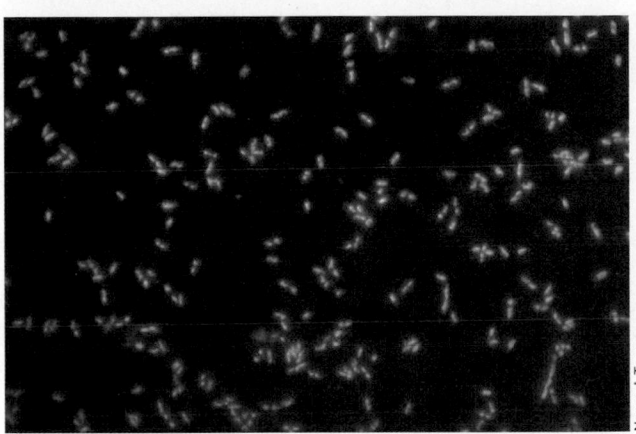

Figure 18.6 Cells of *Escherichia coli* made fluorescent by staining with DAPI. This staining technique is frequently used to make total microbial counts.

reasonable estimate of the cell numbers present. For aquatic samples cells can be stained on the surface of a filter after filtering a given volume of liquid. These simple staining techniques have on the one hand the advantage of being nonspecific—they stain *all* microorganisms in a sample—but on the other hand have the disadvantages of not differentiating between living and dead cells, nor allowing for tracking *specific* organisms in an environment.

Viability Staining
A fluorescent staining method has been developed for differentiating live from dead bacterial cells. This gives information on not only the number of organisms present in a sample but also on their viability. The basis of the differentiation concerns whether the cytoplasmic membrane of the cells is intact. Two dyes, colored green and red, are added to a sample; the green fluorescing dye penetrates all cells, viable or not, while the red dye, which contains propidium iodide, penetrates only those cells whose cell membrane is no longer intact and that are therefore dead. Thus, green cells are alive and red cells are dead, giving an instant assessment of viability (Figure 18.7). Although useful for research situations using laboratory cultures of bacteria, the live/dead staining method is not suitable for use in the microscopic examination of natural habitats because of problems with nonspecific staining of background materials.

Fluorescent Antibodies
Fluorescent-staining techniques can be made more specific by using *fluorescent antibodies*. We discuss the theory of fluorescent antibodies in Section 24.8. The great specificity of antibodies against surface constituents of a particular organism can be exploited as a means of identifying or tracking an organism in a complex habitat

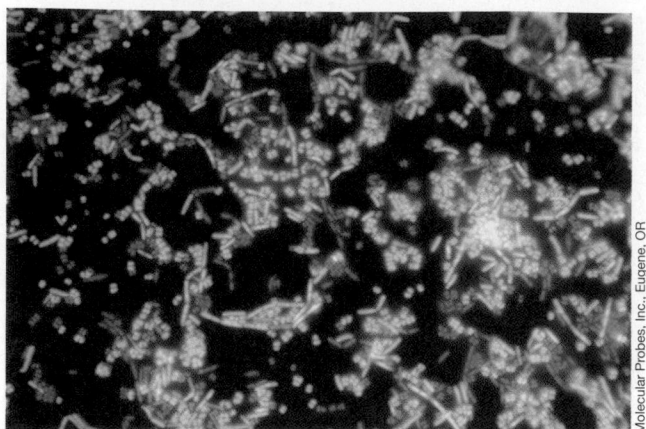

Figure 18.7 Viability staining. Live (green) and dead (red) cells of *Micrococcus luteus* (cocci) and *Bacillus cereus* (rods) stained by the LIVE/DEAD Bac Light™ Bacterial Viability Stain.

containing a mixture of many organisms, such as soil (Figure 18.8) or a clinical sample. The method requires the preparation of specific antibodies against the organism of interest, often a time-consuming and laborious procedure. For clinically relevant microorganisms, however, highly specific antibodies can be purchased commercially and are therefore widely used in clinical microbiology laboratories (⊂⊃ Section 24.8 and Figure 24.16).

Green Fluorescent Protein as a Cell Tag

Bacterial cells can be altered by genetic engineering to make them fluoresce. As will be discussed in Chapter 31, a gene encoding a protein called the *green fluorescent protein* (GFP) can be inserted into the genome of virtually any bacterium. When expressed, GFP-containing cells appear green when observed with an ultraviolet microscope (Figure 18.9).

Although not useful for the study of natural populations (because these cells lack the GFP gene), GFP-

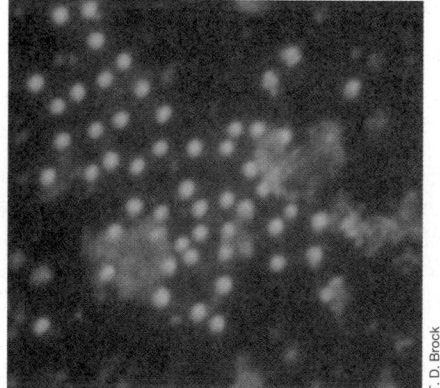

Figure 18.8 Fluorescent antibodies as a cell tag. Visualization of bacterial microcolonies (bacterial cells appear as greenish-yellow dots) of the hyperthermophilic archaeon, *Sulfolobus acidocaldarius* on the surface of solfatara soil particles by use of the fluorescent antibody technique. Cells are about 1 µm in diameter.

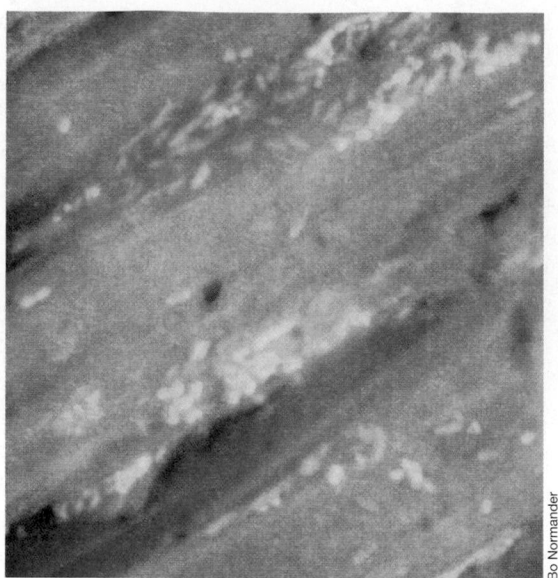

Figure 18.9 Cells of *Pseudomonas fluorescens* genetically engineered to express the green fluorescent protein (GFP) and observed by confocal laser microscopy (⊂⊃ Section 4.2). The cells are attached to barley roots and fluoresce green. The blue-staining cells are part of the normal flora of barley roots and have been stained with the general stain DAPI (see Section 18.3 and Figure 18.6). Using GFP one can track cells introduced into the environment.

tagged cells can be *introduced* into an environment, such as plant roots (Figure 18.9) and microscopy used to follow the tagged strain. Using this method, microbial ecologists can study microbial competition between the native microflora and the GFP-tagged introduced strain *in situ*, or assess the effect of perturbations in the environment on introduced bacteria. The GFP has also been used extensively in laboratory cultures of various bacteria as a "reporter" gene. When fused with an operon under the control of a specific repressor, transcriptional control of the genes can be studied using fluorescence as an assay of transcription. That is, when the genes containing the fused GFP gene are transcribed, the GFP gene is also transcribed. Following translation to produce the green fluorescent protein, cells fluoresce green (⊂⊃ Section 31.4).

Limitations of Microscopy

It should be clear by now that the microscope is a valuable tool for enumerating and identifying microorganisms in natural samples. But can the microscope suffice for the study of microorganisms in their natural habitats? The answer to this is "no," and for several reasons. Small cells can be a major problem. When observing natural samples, very small prokaryotes may be overlooked, especially if the sample contains much particulate matter or high numbers of larger cells. Also, it is often difficult to differentiate live cells from dead cells and cells from nonliving matter.

A major problem with the microscope and the stains we have discussed thus far, however, is that they are unable to assess the *genetic* diversity of microorganisms in a natural habitat. In this regard, it is easy to be fooled by a microscopic observation. For example, two morphologically quite similar cells may in fact be genetically distinct (Figure 18.10) and thus likely to play different roles in the ecosystem. Indeed, when using the microscope in ecological studies one should recall the old adage "all that glitters is not gold." In other words, when viewing a natural population of microorganisms under the microscope,

it should be with the realization that the sample almost certainly contains a genetically diverse population of cells, even if many of them "look" the same.

Thus, although the microscope is an important tool of the microbial ecologist, it must be supplemented with cultural or molecular tools, preferably both, if the ecology of a microbial ecosystem is to be understood. Molecular methods target specific genes that are linked to specific microorganisms; the presence of the gene infers the presence of the organism. We consider these molecular methods now, some of which also involve microscopy (see Figure 18.10).

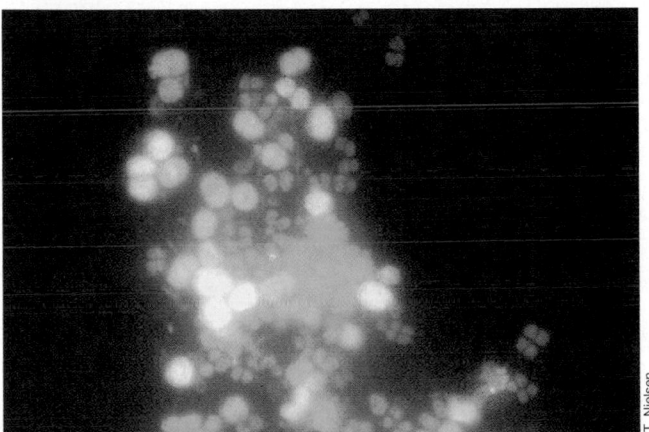

(a)

(b)

Figure 18.10 Morphology and genetic diversity. The tandem photos shown here, (a) phase contrast, and (b) phylogenetic staining (see Section 18.4), are of the same field of cells. Although the large oval-shaped cells are of a rather unusual morphology and size for prokaryotic cells and look similar by phase contrast microscopy, the phylogenetic stains reveal that these are two genetically distinct cell types. Simple light microscopic observations of natural samples must therefore be interpreted with caution, because it is easy to assume that cells of an unusual morphology in a single microscopic field are all of the same organism. The oval cells stained in blue or yellow are about 2.25 μm in diameter.

✓ 18.3 Concept Check

DAPI is a general stain for identifying microorganisms in natural samples. Some stains can differentiate live versus dead cells, and fluorescent antibodies that are specific for one or a small group of related cells can be prepared. The green fluorescent protein makes cells autofluorescent and is a means for tracking cells introduced into the environment. Unlike pure cultures, in natural samples, morphologically similar cells may actually be quite different genetically.

✓ Why is it incorrect to say that the GFP is a "staining" method?

✓ How do stains like DAPI differ from fluorescent antibodies in terms of specificity?

18.4 Genetic Stains

Nucleic acid probes offer powerful methods for identifying and quantitating microorganisms in nature. Recall that a **nucleic acid probe** is a DNA or RNA oligonucleotide complementary to a sequence in a target gene (∞ Section 10.12). As discussed in Section 11.6, oligonucleotide probes can be made fluorescent with certain dyes; these now-fluorescent probes can be used to identify organisms containing a nucleic acid sequence complementary to the probe. This technique is called **fluorescent *in situ* hybridization** (**FISH**, ∞ Section 11.6), and three different applications of this method are described here.

Phylogenetic Staining Using FISH

Phylogenetic stains are fluorescing oligonucleotides complementary in base sequence to "signature" sequences in 16S (prokaryotes) or 18S (eukaryotes) ribosomal RNA (∞ Section 11.6). Phylogenetic stains work by penetrating the cell and hybridizing with ribosomal RNA directly in the ribosomes (Figure 18.11). Because signature phylogenetic probes have been identified for individual microbial species as well as for entire domains of organisms, the degree of specificity of a phylogenetic stain can be controlled by the sequence of the probe fused to

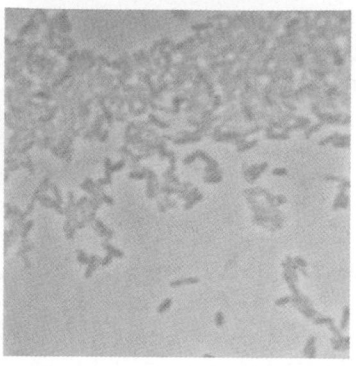

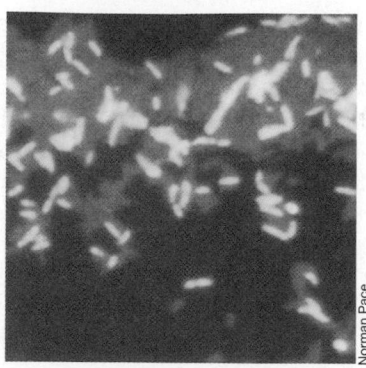

Figure 18.11 Nucleic acid probe methods for identifying microorganisms. Differentiation of closely related gram-negative *Bacteria*. Left, phase micrograph of mixture of *Proteus vulgaris* and a related bacterium isolated from an insect. Center, same field stained with a 16S rRNA probe specific for all members of the *Bacteria*. Right, same field stained with a probe specific for the bacterium from the insect. The cells in both cases are about 0.6 μm in diameter.

the dye. Using FISH, one can then identify or track a particular organism (or group of related organisms, depending on the specificity of the phylogenetic probe) in a natural sample.

Natural samples can also be subject to *multiple* phylogenetic probing by FISH. Using a suite of probes, each designed to react with a particular organism or group and containing its own fluorescent dye, FISH can be used to phylogenetically characterize an entire habitat (Figure 18.12). If FISH is combined with confocal microscopy (👓 Section 3.2), it is even possible to explore microbial populations in a sample *with depth*, as for example, in a biofilm (👓 Section 19.3 and Figures 18.12 and 19.4). The applications of FISH technology thus seem almost endless, and the technique has become popular beyond microbial ecology as a tool to identify specific pathogens in the food industry and in clinical diagnostics.

Chromosome Painting and In Situ Reverse Transcription

Two other FISH technologies are *chromosome painting* and *in situ reverse transcription*. The former technique is used to identify specific genes in the microflora of natural samples, whereas the latter targets specific mRNA (that is, *expressed* genes).

Chromosome painting begins with fluorescent labeling of an oligonucleotide probe or an entire gene, set of genes, or even an entire genome digested in such a

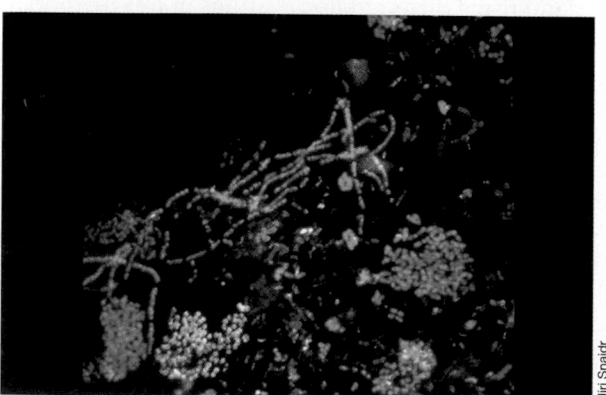

(a)

(b)

Figure 18.12 Use of multiple FISH probes on sewage samples to assess their microbial diversity. (a) Nitrifying bacteria in sewage sludge. Red, ammonia-oxidizing ($NH_3 \rightarrow NO_2^-$) bacteria; green, nitrite-oxidizing ($NO_2^- \rightarrow NO_3^-$) bacteria. Note the close proximity of the two metabolic types to allow for cross-feeding of NO_2^-. (b) Confocal laser scanning micrograph of a sewage sludge sample. The sample was treated with three phylogenetic probes, each containing a different fluorescent dye (green, red, or purple) and each targeting a different group of Proteobacteria (one of several major lineages in the domain *Bacteria*, 👓 Chapter 12). Green-, red-, or purple-stained cells reacted with only a single probe while blue- and yellow-stained cells reacted with two probes to give the different colors. The genetic diversity of this sample is obvious from the many colors, but keep in mind that the probes were only designed to react with a relatively restricted group of Proteobacteria. No *Archaea* (👓 Chapter 13) or other groups of *Bacteria* (👓 Chapter 12) would have shown up under these conditions. Thus, the actual genetic diversity in the sample is undoubtedly far greater than appears here.

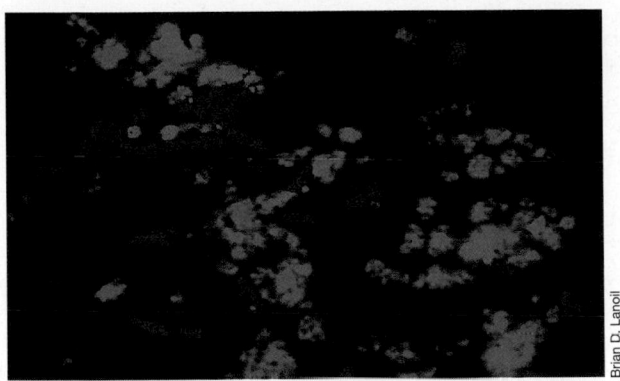

(a)

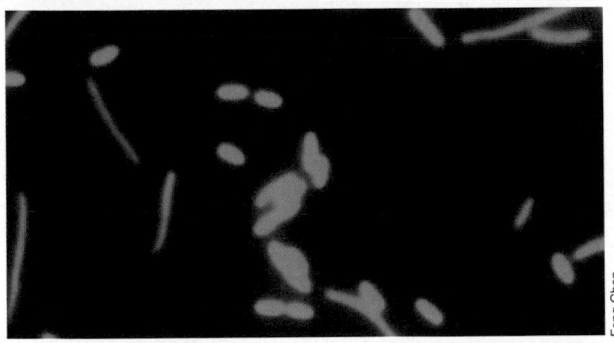

(b)

(c)

Figure 18.13 Fluorescent techniques employing bacterial chromosome painting and *in situ* reverse transcription (ISRT). (a) Chromosome painting. Mixture of cells of *Escherichia coli* (red) and *Oceanospirillum linum* (blue). Each fluorescent dye was linked to DNA oligonucleotides formed from genomic DNA from the respective organism. Because of the significant sequence differences in the genomes of each organism, the labeled genes only hybridize with genomic DNA from their respective organisms. (b, c) ISRT. The two photos show a mixed culture of the bacteria *Microbulbifer hydrolyticus* (short, fat rods) and *Sagittula stellata* (long, thin rods). Reverse transcription (Section 9.12) of a specific mRNA produced only by *M. hydrolyticus* led to production of a cDNA that was subsequently amplified by PCR and hybridized by a fluorescently labeled probe. (b) All cells stained with DAPI, a nonspecific stain (see Figure 18.6). (c) Cells stained with ISRT probe. ISRT data are useful for understanding which organism(s) in a microbial community are carring out a specific metabolic activity.

way as to yield small probes. The probes are then used to ask which organism(s) in a sample contains the gene or genes present in the probe(s) (Figure 18.13*a*). For example, chromosome painting can be used to identify bacteria that contain specific genes, such as those that encode the enzyme nitrogenase, responsible for nitrogen fixation, (Section 17.28), components of the photosynthetic reaction center (Section 17.4), or specific autotrophic pathways (Sections 17.6 and 17.7). The numbers of fluorescing cells in each case would yield an estimate of the numbers of nitrogen-fixing bacteria, phototrophic bacteria, or autotrophic bacteria, respectively, in a natural sample.

Unlike chromosome painting, *in situ* reverse transcription (ISRT) FISH asks whether an organism is *expressing* a particular gene or genes in nature at a given time (Figure 18.13*b,c*). The method involves the use of a probe that hybridizes with a specific mRNA (Section 7.8) previously transcribed by cells in a natural sample. Once the probe reacts, reverse transcription (Section 9.12) is performed to produce a complementary DNA, and the latter is amplified by the polymerase chain reaction (PCR, Section 10.17). The amplified DNA is then allowed to react with a fluorescent probe (Figure 18.13*b* and *c*). The main advantage of ISRT FISH over chromosome painting is that it allows the experimenter to explore the factors that control gene expression in natural populations and to observe how experimental perturbations in a sample affect the transcription of specific genes.

✓ **18.4 Concept Check**

A variety of fluorescent-staining methods employ the power of nucleic acid probes and thus are highly specific in their staining properties. These include phylogenetic staining, chromosome painting, and reverse transcription fluorescent *in situ* hybridization (FISH).

✓ What part of the cell is the target for fluorescent probes in phylogenetic FISH?

✓ Chromosome painting and ISRT FISH reveal different things about cells in nature. Explain.

18.5 PCR: Linking Specific Genes to Specific Organisms

For many studies in microbial ecology it is unnecessary to isolate organisms, or even to identify them microscopically using the stains described in Sections 18.3 and 18.4. The objective of a study is often to survey the biodiversity of a habitat *without* culturing or otherwise observing the cells. For these purposes, specific *genes* are isolated and characterized as a measure of biodiversity in the microbial community. The rationale here is that because specific genes are often linked to specific organisms, detection of the *gene* implies that the *organism* is present. The major techniques employed in this type of microbial community analysis are the polymerase chain reaction (PCR), denaturing gel electrophoresis (DGGE) or molecular cloning, and DNA sequencing and analysis.

Polymerase Chain Reaction and Microbial Community Analysis

We discussed the principle of PCR in Chapter 10 (∞ Section 10.17). Recall the major steps in PCR: (1) a specific nucleic acid probe hybridizes to a complementary sequence in a target gene, (2) DNA polymerase copies the target gene, and (3) multiple copies of the target gene are made by repeated melting of complementary strands, binding of primers, and new synthesis (∞ Figure 10.45). From a single copy of a gene, several million copies can be made and the amplified DNA visualized on a gel (see Figure 18.14).

What genes can be used as target genes? A commonly amplified gene is that encoding 16S ribosomal RNA. Recall that sequence analyses of genes encoding 16S rRNA can be used to derive phylogenetic relationships (∞ Section 11.7 and Figure 11.13). Likewise, sequence analyses of 16S genes amplified from a microbial community can paint a phylogenetic picture of that community (Figure 18.14). Alternatively, metabolic genes that encode proteins unique to the metabolism of a specific organism (or group of related organisms) can be the targets. Regardless of the gene (or genes) chosen for PCR amplification, strong specificity of the primers is essential. With natural samples as sources of DNA, primer design is critical for obtaining PCR results that are unambiguous and readily interpretable.

In a typical experiment, total community DNA is isolated from the habitat to be studied (Figure 18.14a). Commercially available kits are available for this purpose that yield PCR-grade DNA while removing substances from the sample that could interfere with PCR (for example, soil, metals, and other physical and chemical components of the habitat). The DNA obtained is a mixture of genomic DNA from all of the microorganisms that were originally present in the habitat (Figure 18.14a). It is the job of PCR to "fish out" the targeted gene(s) of interest from this mixture and make copies sufficient for further analysis.

The series of events in molecular microbial community analysis is shown in Figure 18.14. The end result is bands of the target gene on a gel (Figures 18.14 and 18.15). Ideally only a *single* band of a given size will appear (the size of the DNA fragment amplified is determined by the distance between the primers, ∞ Figure 10.45). If sequencing is the ultimate goal, the PCR products, which usually represent a mixture of the variants of the target gene that were present in the sample, need to be sorted out. A key method for separating PCR products is called *denaturing gradient gel electrophoresis*, to which we turn now.

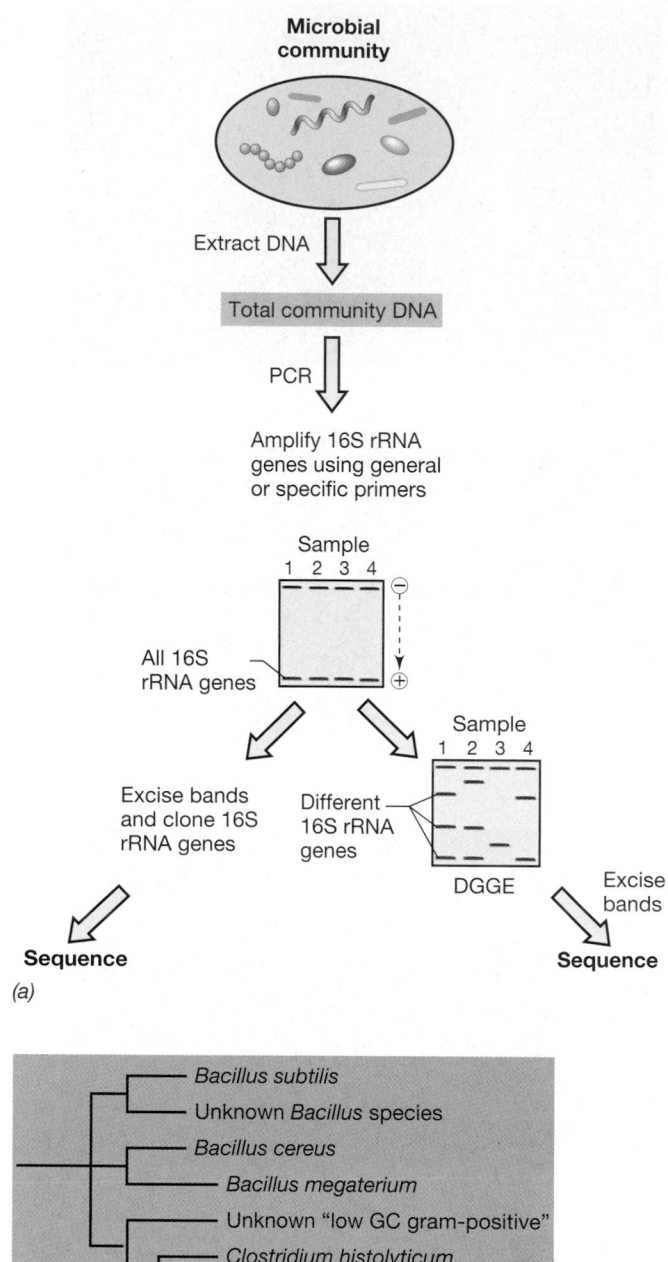

Figure 18.14 Steps in biodiversity analysis of a microbial community using phylogenetic probes. (a) Total community DNA is used with PCR to amplify 16S rRNA genes using universally conserved primers for *Bacteria* or primers that will target only a particular phylum of *Bacteria*. The PCR bands are excised and the different 16S genes separated by either cloning or by DGGE (see Figure 18.15). Note how in the DGGE gel, samples 1, 2, and 4 share a common band (gene), whereas samples 2 and 3 each contain one unique band. Following sequencing, a phylogenetic tree is generated. (b) Example of a phylogenetic tree that might be generated if primers used were specific for the "low GC gram-positive" lineage of the phylogenetic domain Bacteria, which includes species of *Bacillus* and *Clostridium* (∞ Section 12.21). Note that many of the sequences do not match the 16S rRNA sequence of any recognized species and therefore are "new" species that await isolation.

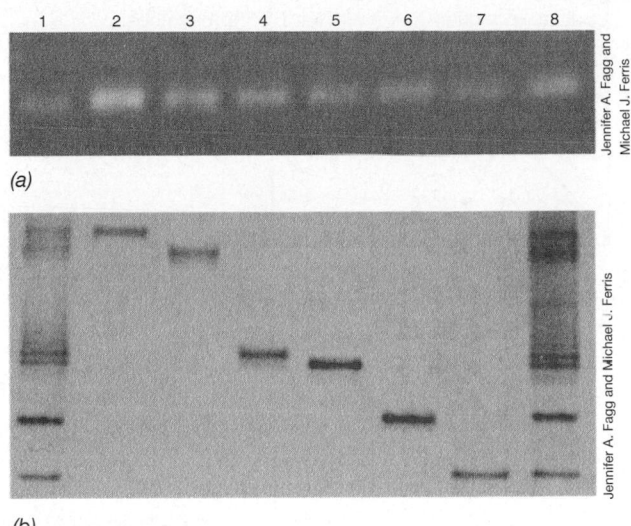

(a)

(b)

Jennifer A. Fagg and Michael J. Ferris

Figure 18.15 PCR and DGGE analyses. Bulk DNA was isolated from a microbial community and PCR amplified using primers for 16S rRNA genes of *Bacteria* (a; lanes 1 and 8). The PCR products yielded only a single band, but these products actually consisted of six distinct 16S rRNA gene sequences as determined by DGGE (b; lanes 1 and 8). Each of these six bands was then purified, reamplified by PCR (a; lanes 2–7), and then run on DGGE (b; lanes 2–7). Note how all bands run to the same location in the PCR gel (a), since although they differ in *sequence*, they are all of the same *size*. Each band could also be sequenced and a phylogenetic tree generated as shown in Figure 18.14.

Denaturing Gradient Gel Electrophoresis (DGGE): Separating Similar Genes

As just discussed, PCR amplification from natural microbial communities using a single set of primers usually generates a single gel band containing amplified DNA fragments of a single size (Figures 18.14 and 18.15). But despite the appearance of purity, the band can contain many highly related but not identical genes. This is because although priming sites in the target gene are highly conserved, the nucleotide sequence *between* the priming sites can vary as a result of evolutionary divergence of the target gene in the different species of organism that contain it. Thus, if the goal is to sequence amplified genes, as in phylogenetic analyses, an additional step is needed to resolve the different forms of the gene before sequencing can proceed. One way to do this is by molecular cloning (Figure 18.14, ∞ Section 10.14). Alternatively, one can use **denaturing gradient gel electrophoresis (DGGE)**.

DGGE is a gel electrophoresis method that separates genes of the same *size* that differ in *base sequence* (Figures 18.14 and 18.15). The technique employs a gradient of a DNA denaturant, such as a mixture of urea and formamide. When a double-stranded DNA fragment moving through the gel reaches a region containing sufficient denaturant, the strands begin to "melt," at which point migration stops (Figures 18.14 and 18.15).

Differences in melting properties are to a large degree controlled by differences in base sequence. Thus, the different bands observed in a DGGE gel are different forms of a given gene that vary, sometimes only very slightly, in their sequences.

Once DGGE has been performed, individual bands can be excised and sequenced (Figure 18.14). Using 16S rRNA genes, for example, DGGE analyses yield a detailed picture of the number of phylotypes (distinct 16S rRNA genes) present in a habitat (Figure 18.15). By sequencing these bands, the actual species present in the community can be determined by comparison of the sequences with those of known species available from appropriate data bases (Figure 18.14*b*). Using other genes—for example, metabolic genes—information can be obtained in the same way about the number of different types of organisms present in the community that contain the specific gene. But even without sequencing, DGGE analysis reveals the biocomplexity of a habitat with respect to a specific gene (Figures 18.14 and 18.15) and can therefore be used to guide further community analyses, such as enrichments and FISH (see Sections 18.1 and 18.4).

Results of PCR Phylogenetic Analyses

PCR phylogenetic analyses of microbial communities have yielded some surprising results. Using 16S rRNA as the target gene, most natural microbial communities have been shown to contain several phylogenetically distinct organisms whose ribosomal RNA sequences do not match any of those from laboratory cultures (Figures 18.14 and 18.15). In fact, in many cases where these methods have been refined to allow for both qualitative (that is, "who's out there?") as well as quantitative (that is, "how many of each type are out there?") analyses, it turns out that the most abundant members of the natural community are species that have so far defied laboratory culture! This problem suggests that our knowledge of microbial diversity from enrichment culture studies is still very incomplete and that enrichment bias (see Section 18.10) is likely a serious problem in biodiversity studies.

✓ 18.5 Concept Check

The polymerase chain reaction (PCR) can be used to amplify specific target genes, useful for ecological analyses, such as small subunit ribosomal RNA genes or key metabolic genes. Denaturing gradient gel electrophoresis (DGGE) can be used to resolve slightly or greatly different versions of these genes present in the different species inhabiting a natural sample.

✓ Why can it be said that primer specificity is the key to successful PCR?

✓ What could you conclude from PCR/DGGE analysis of a sample that yielded one band by PCR and one band by DGGE? One band by PCR and four bands by DGGE?

✓ What surprising finding has come out of many molecular studies of natural habitats using 16S ribosomal RNA as the target gene?

III MEASURING MICROBIAL ACTIVITIES IN NATURE

We began this chapter by pointing out that microbial ecologists are concerned with biodiversity and *in situ* microbial activities. Up to this point our discussion has centered on biodiversity. We turn now to measurements of microbial activities, particularly the use of (1) radioisotopes, (2) microelectrodes, and (3) stable isotope analyses.

We reiterate that any activity measurement in a natural sample is a *collective* estimate for the entire microbial community. Despite this limitation, well-designed activity measurements can reveal both the *types* and *rates* of major metabolic reactions in a habitat. Along with biodiversity estimates, these parameters define the microbial ecology of that habitat and also provide valuable information for the design of enrichment cultures.

18.6 Radioisotopes and Microelectrodes

In many situations, direct chemical measurements of microbial transformations are sufficient for assessing microbial transformations in an environment. For example, the fate of lactate oxidation by sulfate-reducing bacteria in a sediment sample can easily be followed; lactate is consumed and SO_4^{2-} is reduced to H_2S (Figure 18.16a). However, when very high sensitivity is required, or turnover rates need to be determined, or the fate of portions of a molecule need to be followed, radioisotopes are very useful. For instance, if photoautotrophy is to be measured, the light-dependent uptake of $^{14}CO_2$ into microbial cells can be measured (Figure 18.16b); if sulfate reduction is of interest, the rate of conversion of $^{35}SO_4^{2-}$ to $H_2^{35}S$ can be assessed (Figure 18.16c). Chemoorganotrophic activity can easily be assessed by tracking the release of $^{14}CO_2$ from ^{14}C-labeled organic compounds (Figure 18.16d).

Isotope methods are widely used in microbial ecology. These must, however, employ proper controls because there is always the possibility that some transformation of labeled compound might be due to a strictly chemical (rather than a microbial) process. The *killed cell control* is key here. It is absolutely essential to know that the transformation being measured is prevented by chemical agents or by heat treatments that either block microbial action or actually kill the organisms. Formalin at a final concentration of 4% is commonly used as a chemical sterilant in microbial ecology studies. This level of formalin kills all cells, and transformations of radiolabeled materials that occur in its presence can be ascribed to nonliving processes (Figure 18.16d).

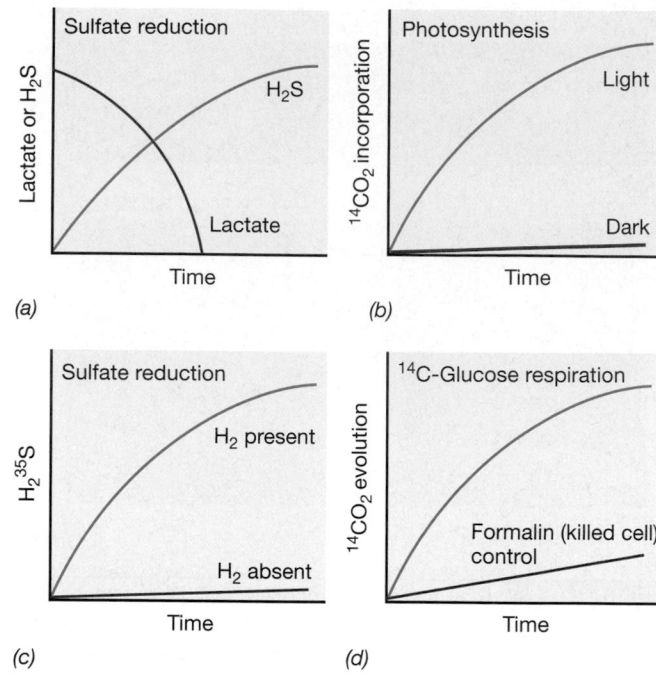

Figure 18.16 Microbial activity measurements. (a) Chemical assay for lactate and H_2S during sulfate reduction. (b–d) Use of radioisotopes. (b) Photosynthesis measured in natural seawater with $^{14}CO_2$. (c) Sulfate reduction in mud measured with $^{35}SO_4^{2-}$. (d) Production of $^{14}CO_2$ from ^{14}C-glucose.

Microelectrodes

Microbial ecologists have used small glass electrodes, referred to as **microelectrodes**, to study the activity of microorganisms in nature. The most useful microelectrodes have been those that measure pH, oxygen, N_2O, CO_2, H_2, or H_2S. As the name implies, microelec-

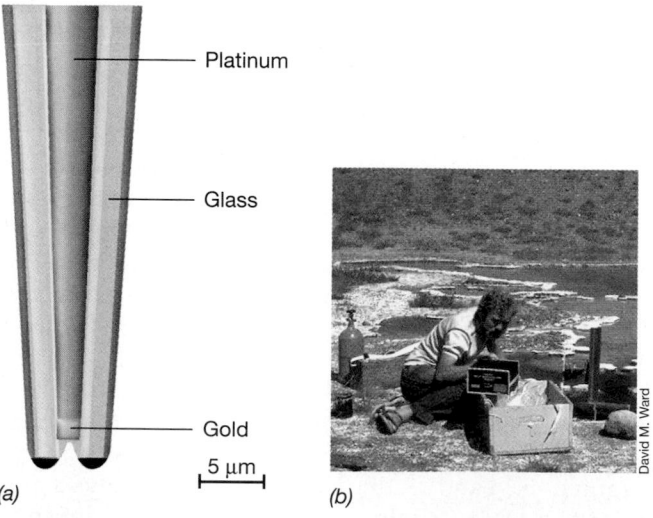

Figure 18.17 Microelectrodes. (a) Schematic drawing of an oxygen microelectrode. Note the scale of the electrode. (b) Photo of microelectrodes being used in a microbial mat (see Figure 18.18a).

trodes are very small, the tips of the electrode ranging in diameter from 2 to 100 μm (Figure 18.17a). The electrodes must be carefully inserted into the habitat using a micromanipulator, a device that allows for precise movement of the electrode through distances of a millimeter or less (Figure 18.17b).

Microelectrodes have been used extensively in the study of chemical transformations and photosynthesis in *microbial mats*. Microbial mats are layered microbial communities usually containing cyanobacteria in the uppermost layers, anoxygenic phototrophic bacteria in subsequent layers (until the mat becomes light-limited) and chemoorganotrophic, especially sulfate-reducing, bacteria in the lower layers (Figures 18.17b and 18.18). Microbial mats develop in a variety of environments, especially in hot springs (Figure 18.18) and marine intertidal zones.

With the use of O_2 microelectrodes, oxygen concentrations in microbial mats (Figure 18.18b) or soil particles (∞ Figure 19.2) can be sensitively measured over

(a)

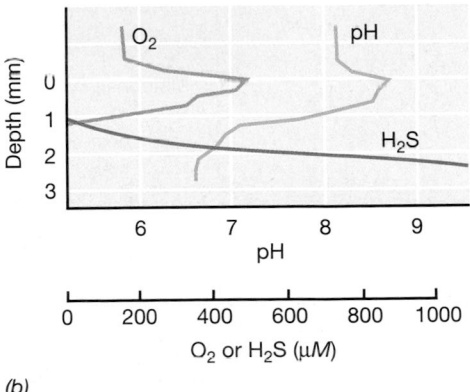

(b)

Figure 18.18 Microbial mats and the use of microelectrodes to study them. (a) Photograph of a core taken through the kind of hot spring microbial mat used in the experiment shown in part (b). Upper layer (dark green) contains cyanobacteria (∞ Section 12.25), beneath which are several layers of anoxygenic phototrophic bacteria (orange), primarily *Chloroflexus* (∞ Section 12.35). The whole thickness of the mat is about 2 cm. (b) Oxygen, sulfide, and pH microprofiles in a hot spring microbial mat. Note the millimeter scale on the ordinate.

extremely fine intervals. A *micromanipulator* is used to immerse electrodes gradually through a sample such that readings can be taken every 0.1 mm (100 μm) or less (Figures 18.17b and 18.18b). With a bank of microelectrodes, each sensitive to a different chemical, simultaneous measurements of several microbial transformations can be made at one time. Oxygen and sulfide measurements are often taken together because gradients of both form in many microbial environments as a result of photosynthesis and sulfate reduction, respectively (Figure 18.18b). Near the zone where O_2 and H_2S begin to mix, intense activity by phototrophic and chemolithotrophic sulfur bacteria (∞ Sections 12.2 and 12.4) lead to consumption of both to the limits of detection of the microelectrodes (Figure 18.18b).

✓ 18.6 Concept Check

The activity of microorganisms in natural samples can be assessed very sensitively using radioisotopes and/or microelectrodes. In most cases, measurements are of the net activity of a microbial community rather than of a population of a single species.

✓ Why are radioisotopes so useful in measuring microbial activities?

✓ Explain how a microelectrode works.

18.7 Stable Isotopes

For many elements different isotopes exist, differing in the number of neutrons present. Certain isotopes are unstable and break down as a result of radioactive decay. Others, called *stable isotopes*, are not radioactive but are acted upon differentially by microorganisms and can be used to study various microbial transformations in nature.

The two elements that have proven most useful for stable isotope studies in microbial ecology are *carbon* and *sulfur*. Carbon exists in nature primarily as ^{12}C, but a small amount of carbon (about 5%) exists as ^{13}C. Likewise, sulfur exists primarily as ^{32}S, although some sulfur is found as ^{34}S. The difference in natural abundance of these isotopes changes when C or S is metabolized by organisms because (for reasons involving the binding of substrates to enzymes) biochemical reactions tend to favor the *lighter* isotope. That is, the heavier isotope is discriminated against relative to the lighter isotope when the elements are acted on by enzymes (Figure 18.19). Thus, when CO_2 is fed to a phototrophic organism, for example, the cellular carbon becomes *enriched* in ^{12}C and *depleted* in ^{13}C, relative to a carbon standard. Likewise, sulfide produced from the bacterial reduction of sulfate is "lighter" than sulfide that has formed geochemically. This phenomenon is known as **isotopic fractionation** (Figure 18.19).

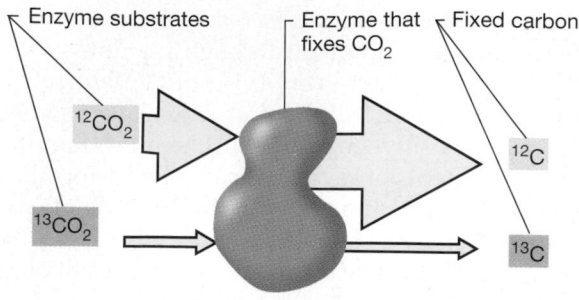

Figure 18.19 Mechanism of isotopic fractionation using carbon as an example. Although the ratio of natural abundance of $^{12}CO_2$ to $^{13}CO_2$ is about 19:1, enzymes that fix CO_2 preferentially fix the lighter isotope (^{12}C). This results in fixed carbon being enriched in ^{12}C and depleted in ^{13}C relative to the starting substrate. The degree of ^{13}C depletion is calculated as an isotopic fractionation (see legend to Figure 18.20 for calculation). The size of the arrows indicates the relative abundance of each isotope of carbon.

Use of Isotopic Fractionation in Microbial Ecology

The isotopic composition of a sample contains a record of its past biological activity (Figures 18.20 and 18.21). In the case of carbon, plant material and petroleum (which is derived from plant material) have similar isotopic compositions; carbon from both sources is isotopically lighter than a nonbiological standard because it was fixed by a path-

way that discriminated against $^{13}CO_2$ (Figure 18.20). Moreover, methane of microbiological origin is extremely light, while marine carbonates are clearly of nonbiological origin (Figure 18.19). Because of the differences in the proportion of ^{12}C and ^{13}C in carbon of biological and geological origin, the $^{13}C/^{12}C$ isotopic ratio of geological strata has been used to detect the onset of living processes in ancient rocks. Interestingly, organic carbon in rocks as old as 3.5 billion years shows some fractionation (Figure 18.20), supporting the idea that life existed at this time.

Isotope fractionation has several applications in microbial ecology. The activities of sulfate-reducing bacteria in nature are easy to recognize from the fractionation of $^{34}S/^{32}S$ (Figure 18.21). As compared with a sulfide standard, sedimentary sulfide is highly enriched in ^{32}S, while nonbiogenic sulfide (as, for instance, in igneous rocks from volcanic deposits) is significantly heavier (Figure 18.21). Sulfur isotope analyses have also been used as evidence for the lack of life on the Moon. For example, the data in Figure 18.21 show that the isotopic composition of sulfides in lunar rocks closely approximates that of the sulfide standard and not that of biogenic sulfide. Knowledge of the fractionations expected from sulfate reduction and the subsequent oxidation of sulfide by phototrophic or chemolithotrophic bacteria allows monitoring of these important links in the sulfur cycle. As previously mentioned, carbon isotopic analyses have been used to distinguish biogenic from abiogenic organic matter, and oxygen analyses (using $^{18}O/^{16}O$) of rocks of various ages have been used to

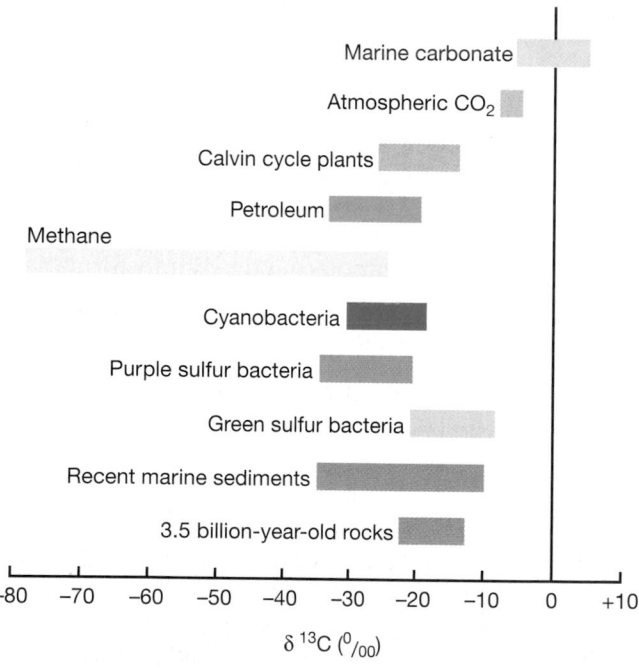

Figure 18.20 Carbon isotopic compositions of various substances. The values are given in parts per thousand (‰) and were calculated using the formula

$$\frac{(^{13}C/^{12}C \text{ sample}) - (^{13}C/^{12}C \text{ standard})}{(^{13}C/^{12}C \text{ standard})} \times 1000$$

The standard is a belemnite sample from the PeeDee rock formation. Note that carbon fixed by autotrophic organisms is enriched in ^{12}C.

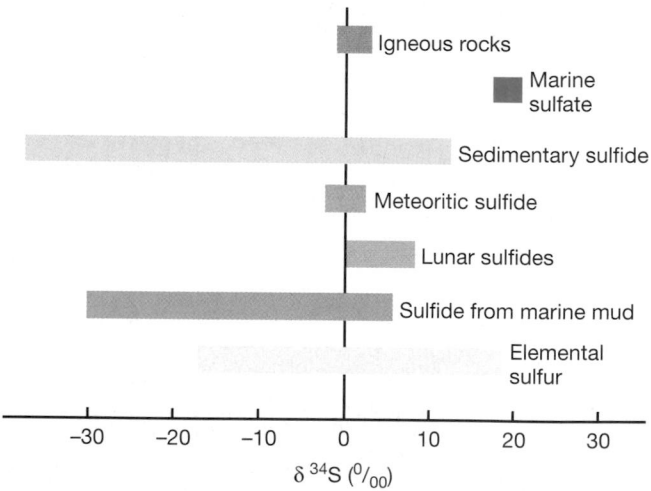

Figure 18.21 Summary of the isotope geochemistry of sulfur, indicating the range of values for ^{34}S and ^{32}S in various sulfur-containing substances. The values are given in parts per thousand (‰) and were calculated using the formula

$$\frac{(^{34}S/^{32}S \text{ sample}) - (^{34}S/^{32}S \text{ standard})}{(^{34}S/^{32}S \text{ standard})} \times 1000$$

The standard is an iron sulfide mineral from the Canyon Diablo meteorite. Note that sulfide and sulfur of biogenic origin tend to be depleted in ^{34}S (enriched in ^{32}S).

trace Earth's transition from an anoxic to an oxic environment (Earth's molecular oxygen originated from oxygenic photosynthesis by cyanobacteria, ⟳ Section 11.1).

With an understanding of some of the techniques in the microbial ecologist's toolbox, we now move on to the chapter on microbial ecology. In Chapter 19 we will consider the nature of microbial communities and the major activities of microorganisms in nature. In many instances we will use one of the methods described in this chapter to explore these relationships and reveal the contributions of microbial components to the big picture.

✓ **18.7 Concept Check**

Isotopic fractionation can reveal the biological origin of various substances. Fractionation is a result of the activity of enzymes that discriminate against the heavier form of an element when binding their substrates.

✓ How can the $^{12}C/^{13}C$ composition of a substance reveal its possible biological origin?

✓ What is the simplest explanation for why lunar sulfides are isotopically heavy?

Review Questions

1. What is the basis of the *enrichment culture technique*? Why is an enrichment medium usually suitable for the enrichment of only a certain group or groups of bacteria?

2. What is the principle of the Winogradsky column and what types of organisms does it serve to enrich? How might a Winogradsky column be used to enrich organisms present in an extreme environment, like a hot spring microbial mat?

3. Describe the principle of MPN for enumerating bacteria from a natural sample.

4. Why would the laser tweezers be a superior method to dilution and liquid enrichment for obtaining an organism present in a sample in *low* numbers?

5. Compare and contrast the use of fluorescent dyes and fluorescent antibodies for use in enumerating bacteria in natural environments. What advantages and limitations do each of these methods have?

6. Can nucleic acid probes in microbial ecology be as sensitive as culturing methods? What advantages do nucleic acid methods have over culture methods? What disadvantages?

7. How does *chromosomal painting* work? Why might it be a good method for enumerating microorganisms in a habitat capable of carrying out some specific metabolic process?

8. What is the *green fluorescent protein*? In what ways does a green fluorescing cell differ from a cell fluorescing from, for example, staining with a phylogenetic strain?

9. How can a phylogenetic picture of a microbial community be obtained without culturing the inhabitants?

10. After PCR amplification of total community DNA using a specific primer set, why is it usually necessary to either clone or run DGGE on the products before sequencing them?

11. What are the major advantages of *radioisotopic methods* in the study of microbial ecology? What type of controls (discuss at least two) would you include in a radioisotopic experiment to show $^{14}CO_2$ incorporation by phototrophic bacteria or to show $^{35}SO_4^{2-}$ reduction by sulfate-reducing bacteria?

12. List two reasons why a microelectrode study of a microbial mat is more relevant than a microelectrode profile through a similar depth of a water column.

13. Will autotrophic organisms contain *more* or *less* ^{12}C than the CO_2 that feeds them?

Application Questions

1. Why is the enrichment for nitrifying bacteria as described in Table 18.1? Why are each of the resources and conditions necessary? (You might want to review material in Section 17.14 before answering.)

2. How do stains such as DAPI differ from phylogenetic stains in assessing microbial diversity of a habitat?

3. Design an experiment for measuring the activity of sulfur-oxidizing bacteria in soil. How would you prove that your activity measurement was due to biological activity?

4. You wish to identify whether organisms exist in various soil samples capable of autotrophic growth using the Calvin cycle (⟳ Section 17.6). This pathway requires a unique enzyme, ribulose bisphosphate carboxylase (RubisCO). Describe two ways you could do this, one using a microscopic method, and another using a method that does not involve either culturing or microscopic methods.

Microorganisms interact in nature with other microorganisms and with plants and animals. A particularly beneficial association is seen between leguminous plants and bacteria of the genus *Rhizobium* and relatives. The bacteria enter the plant through the roots and stimulate the production of tumor-like structures called *root nodules*, as shown here formed on the roots of a soybean plant. Within the nodule the bacteria convert gaseous nitrogen into ammonia, a process called *nitrogen fixation*. Most of this ammonia ends up in plant protein; because of this, the plant is able to grow in soils deficient in fixed nitrogen. Besides quite intimate associations such as these, microorganisms play critical roles in nature as primary producers, decomposers, and recyclers of key elements needed by higher organisms.

19

MICROBIAL HABITATS, NUTRIENT CYCLES, AND INTERACTIONS WITH PLANTS AND ANIMALS

Working Glossary

Acid mine drainage acidic water containing H_2SO_4 derived from the microbial oxidation of iron sulfide minerals

Anoxic an oxygen-free environment, usually also highly reducing (low E_0')

Bacteroid morphologically misshapen *Rhizobium* cells inside a leguminous plant root nodule; can fix N_2

Barophilic an organism that grows best when placed under a pressure greater than 1 atm

Barotolerant an organism that can grow under elevated pressures but that grows best at atmospheric pressure

Biochemical oxygen demand (BOD) the microbial oxygen-consuming properties of a water sample

Biofilm colonies of microbial cells encased in slime and attached to a surface

Biogeochemistry study of biologically mediated chemical transformations

Black smoker an extremely hot (250–350°C) deep-sea hot spring emitting both hot water and various minerals

Cometabolism metabolism of a compound in the presence of a second organic compound, which is used as the primary energy source

Denitrification the microbial process by which NO_3^- is reduced to gaseous nitrogenous compounds

Ecosystem a community of organisms and their natural environment

Guild a population of metabolically related microorganisms

Hydrothermal vent a deep-sea warm or hot spring

Infection thread in the formation of root nodules, a cellulosic tube through which *Rhizobium* cells can travel to reach and infect root cells

Interspecies hydrogen transfer the production and subsequent consumption of H_2 by different groups of microorganisms that interact closely during anaerobic catabolism

Leaching solubilization and removal of metals from an ore by microbial attack

Leghemoglobin an O_2-binding protein found in root nodules

Lichen a fungus and an alga (or cyanobacterium) living in symbiotic association

Microbial leaching the removal of valuable metals such as copper from sulfide ores by microbial activities

Microbial plastics (bioplastics) biodegradable polymeric materials obtained from microorganisms that have properties similar to those of synthetic plastics

Microenvironment the immediate environmental surroundings of a microbial cell or group of cells

Mycorrhiza a symbiotic association between a fungus and the roots of a plant

Nitrification the process by which NH_3 is oxidized to NO_3^-

Nitrogen fixation the microbiological reduction of N_2 to NH_3

Oxic an oxygen-containing environment frequently possessing a high E_0'

Primary producer an organism that uses light to synthesize new organic material from CO_2

Proteorhodopsin a light-sensitive protein present in some open ocean *Bacteria* that catalyzes ATP formation

Pyrite a common iron-containing ore, FeS_2

Reductive dechlorination removal of Cl as Cl^- from an organic compound by reducing the carbon atom from C—Cl to C—H

Rhizosphere the region immediately adjacent to plant roots

Root nodule a tumorlike growth on plant roots that contains symbiotic nitrogen-fixing bacteria

Rumen the first vessel in the multichambered stomach of ruminant animals in which cellulose digestion occurs

Syntrophy a metabolic process in which two or more microorganisms cooperate to carry out a process that could not be carried out by a single organism

Ti plasmid a conjugative plasmid present in the bacterium *Agrobacterium tumefaciens* that can transfer genes into plants

Xenobiotic a totally synthetic product not naturally occurring in nature

---

Picking up from technique-oriented Chapter 18, we now examine what microorganisms actually *do* in their habitats. Each microorganism in an ecosystem interacts with both its surroundings and with other organisms. These interactions can result in significant chemical changes in the environment, and in some cases, changes that are detrimental to higher organisms, for example, in acid mine drainage. In other cases, the reactions that microorganisms carry out are absolutely essential to higher organisms, for example, the nutrient cycles that generate important forms of inorganic nutrients for plants. Either way—harmful or beneficial—microorganisms greatly control the workings of the biosphere, and we examine some key reactions here.

I MICROBIAL ECOSYSTEMS

We begin by examining general features of microbial ecosystems, paying special attention to the microenvironment and to surfaces as habitats for microbial growth. We consider the conditions best suited to growth and how this knowledge may be exploited in industry and medicine.

19.1 Populations, Guilds, and Communities

In a microbial ecosystem, individual cells grow to form *populations*. Metabolically related populations are called *guilds*, and sets of guilds interact to form *microbial communities* (Figure 19.1). Microbial communities interact with communities of macroorganisms and the environment to define the entire ecosystem.

Energy enters ecosystems in the form of sunlight, organic carbon, and reduced inorganic substances. Light is used by phototrophic organisms (Sections 17.1–17.5) to synthesize new organic matter (Figure 19.1) containing not only carbon but also nitrogen, sulfur, phosphorus, iron, and a host of other elements. This newly synthesized organic material, along with organic matter that enters the ecosystem from the outside (*allochthonous* organic matter) drives the metabolic activities of chemoorganotrophic organisms. Chemolithotrophs, by contrast, obtain their energy from inorganic electron donors, such as H_2, Fe^{2+}, S^o, and NH_3 (Chapter 17).

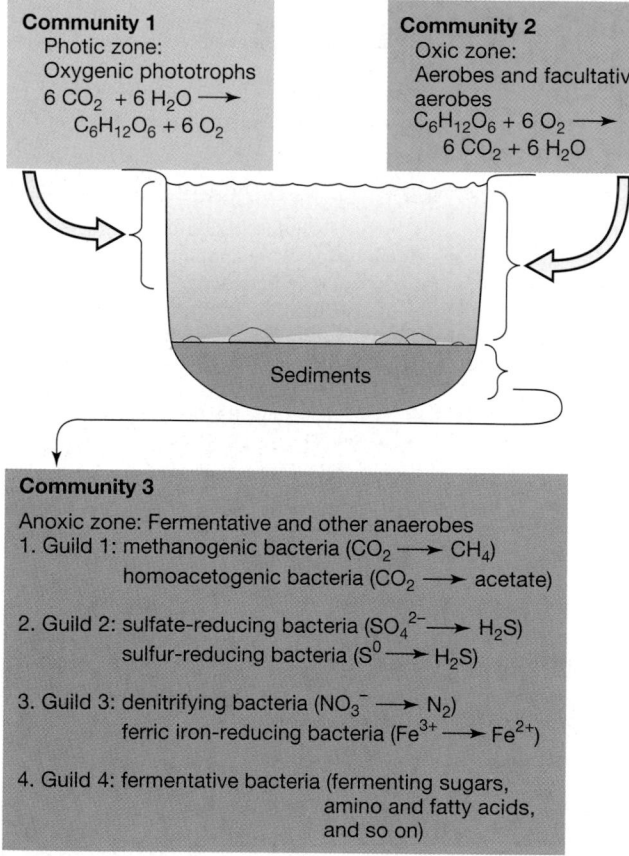

Figure 19.1 Populations, guilds, and communities—an example of microbial community structure in a lake ecosystem. Microbial communities consist of populations of cells of various species. In a lake ecosystem various communities would exist as shown here. For the sediment community, major guilds are indicated.

Biogeochemical Cycles

Microorganisms play an important role in the recycling of several elements, particularly carbon, sulfur, nitrogen, and iron. A **biogeochemical cycle** describes the transformations caused by both biological and chemical processes during the cycling of these key elements of living systems. Often these cycles involve oxidation–reduction reactions (Section 5.6) as the element moves through the ecosystem (Figure 19.1). Sulfur, for example, in the form of hydrogen sulfide (H_2S), can be oxidized by a variety of microorganisms, both phototrophic and chemotrophic, to sulfur (S^0) and sulfate (SO_4^{2-}). Sulfate is a key nutrient for plants. Sulfate, in turn, can be reduced to sulfide (and thereby closing the biogeochemical cycle), by activities of the sulfate-reducing bacteria. Microorganisms of various types are intimately involved in biogeochemical cycling and in many instances are the only biological agents capable of regenerating forms of the elements needed by other organisms, particularly plants. We will discuss many biogeochemical cycles in this chapter including those of carbon, nitrogen, sulfur, and iron.

✓ 19.1 Concept Check

Microbial communities consist of guilds of metabolically related organisms. Microorganisms play major roles in energy transformations and biogeochemical processes that result in the recycling of elements essential to living systems.

✓ How does a microbial *guild* differ from a *microbial community*?

✓ What is a *biogeochemical cycle*?

19.2 Environments and Microenvironments

The natural habitats of microorganisms are exceedingly diverse. Any habitat suitable for the growth of higher organisms can also support growth of microorganisms. But in addition, there are many habitats where, because of some physical or chemical extreme, higher organisms are absent yet microorganisms exist and occasionally even flourish. Microorganisms inhabit the surfaces of higher organisms and in some cases actually live *within* plants and animals. Microorganisms frequently reach large numbers in such habitats and may benefit the plant or animal in a nutritionally significant way. We focus here on the microbial habitat from the standpoint of the microorganism and emphasize the heterogeneous and rapidly changing nature of typical microbial habitats.

The Microorganism and the Microenvironment

As in laboratory culture, the growth of microorganisms in nature depends on the *resources* (nutrients) available and on the growth *conditions* (Table 19.1). Differences in

TABLE 19.1 Major resources and conditions that govern microbial growth in nature

Resources	Conditions
Carbon (organic, CO_2)	Temperature: cold → warm → hot
Nitrogen (organic, inorganic)	Water potential: dry → moist → wet
Other macronutrients (S, P, K, Mg)	pH: 0 → 7 → 14
Micronutrients (Fe, Mn, Co, Cu, Zn, Mn, Ni)	O_2: oxic → microoxic → anoxic
O_2 and other electron acceptors (NO_3^-, SO_4^{2-}, Fe^{3+}, etc.)	Light: bright light → dim light → dark
Inorganic electron donors (H_2, H_2S, Fe^{2+}, NH_4^+, NO_2^-, etc.)	Osmotic conditions: freshwater → marine → hypersaline

the type and quantity of different resources and the physiochemical conditions of a habitat define the *niche* for each particular microorganism. Ecological theory states that for every organism there exists at least one niche, the *prime* niche, in which that organism is most successful. The organism may also inhabit other niches, but in these it is less ecologically successful than in its prime niche. Countless microbial niches exist on Earth and are in part responsible for evolution of the great metabolic diversity (∞ Chapter 17) and biodiversity (∞ Chapters 12–14) of microorganisms we see today.

Because microorganisms are so small, their habitats are also small. A microbiologist must therefore learn to "think small" when considering the microorganism in its environment. For example, for a typical 3-μm rod-shaped bacterium, a distance of 3 mm in its habitat is equivalent to that which a human experiences over a distance of 2 km! And across that 3-mm distance chemical and physical gradients might exist that could greatly affect the organism. Thus, extreme precision is required in the characterization of a microorganism's habitat, and microbial ecologists use the term **microenvironment** to describe where a microorganism lives and metabolizes within its habitat.

In a 3-mm particle of soil, for example, several different microenvironments could exist, differing chemically and physically in many ways. Visualize, for example, the distribution of an important microbial nutrient like oxygen in a soil particle. Using microelectrodes (∞ Section 18.10) it is possible to measure oxygen concentrations throughout small soil particles. As shown in data from an actual experiment (Figure 19.2), soil particles are not homogeneous in terms of their oxygen content. The outer zones of a small soil particle may be fully oxic, whereas the center, only a very short distance away, can remain completely anoxic (Figure 19.2). This finding shows that different niches can exist across a very small spatial dimension and explains how various physiological types of microorganisms could coexist in such a soil particle. Anaerobic organisms could be active near the center of the particle shown in Figure 19.2, microaerophiles (aerobes that require very low oxygen levels) could be active further out, and obligately aerobic organisms could metabolize in the outer 2–3 mm of the particle; facultative-

ly aerobic bacteria could be distributed throughout the particle (∞ Section 6.13).

Physiochemical conditions in the microenvironment can change rapidly in terms of both time and space. Oxygen concentrations shown in Figure 19.2 represent only "instantaneous" measurements; oxygen measurements taken following a period of microbial respiration or after an increase in soil water content could show a drastically different gradient of oxygen across the microenvironment. It can thus be said that microenvironments are *heterogeneous* and that conditions in a given microenvironment can change rapidly. Microenvironments thus promote high microbial diversity in a relatively small physical area.

Nutrient Levels and Growth Rates

Nutrients (or in ecological terms, *resources*, see Table 19.1) often enter an ecosystem in intermittent fashion. A large pulse of nutrients—for example, an input of leaf litter or

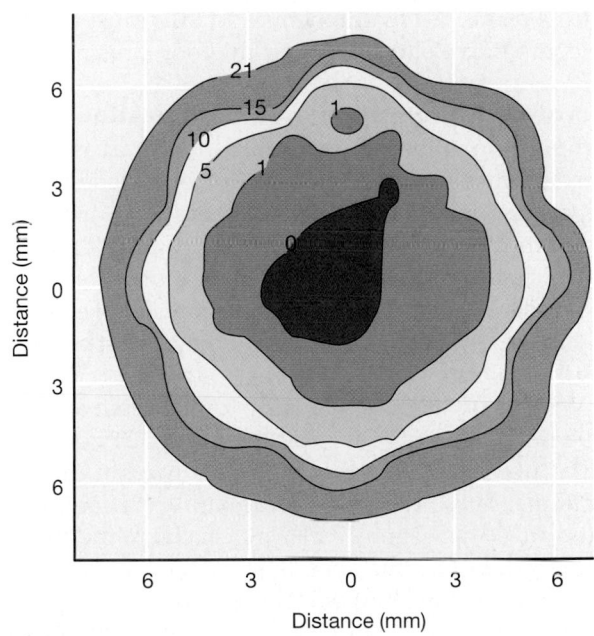

Figure 19.2 Contour map of O_2 concentrations in a soil particle. The axes show the dimensions of the particle. The numbers on the contours are O_2 concentrations (in percent; air is 21% O_2). In terms of oxygen relationships for microorganisms, each zone can be considered a different microenvironment.

the carcass of a dead fish or animal—may be followed by a period of severe nutrient deprivation. Microorganisms in nature are thus often faced with a "feast-or-famine" type of existence, and many microorganisms have evolved biochemical systems for production of storage polymers as reserve materials. Reserve polymers store excess nutrients present under favorable growth conditions for use during periods of nutrient deprivation. Examples of reserve materials are poly- β-hydroxybutyrate and other alkanoates, polysaccharides, polyphosphate, and so on (Section 4.13).

Extended periods of exponential growth of microorganisms in nature are rare. Growth more often occurs in spurts, linked closely to the availability of nutrients. Because physiochemical conditions in nature are rarely optimal all at the same time, growth rates of microorganisms in nature are generally well below the maximum growth rates recorded in the laboratory. For instance, the generation time of *Escherichia coli* in the intestinal tract is about 12 h (two doublings per day), whereas in pure culture it grows much faster, with a minimum generation time of 20 min under the best of conditions. Estimates of the growth rate of certain soil bacteria have shown that they grow in nature at less than 1% of the maximal growth rate measured in the laboratory. On average, these slow growth rates reflect the fact that (1) nutrients (resources, Table 19.1) are frequently in low supply, (2) the distribution of nutrients throughout the microbial habitat is not uniform, and (3) except for rare instances, microorganisms do not grow in pure culture in natural environments and thus must deal with the competitive effects of other microorganisms.

Microbial Competition and Cooperation

Competition among microorganisms for available resources may be intense, with the outcome dependent on several factors, including rates of nutrient uptake, inherent metabolic rates, and, ultimately, growth rates. A typical habitat will house a mixture of different microorganisms (Figure 19.1), with the density of each population dependent on how closely the habitat resembles its prime niche.

Instead of directly competing for the same resources, some microorganisms work together to carry out a particular transformation that neither organism can carry out alone. These types of microbial interactions, called *syntrophy* (Section 17.21), are crucial to the competitive success of certain anaerobic bacteria, as will be described in Section 19.10. Metabolic cooperation can also be seen in the activities of organisms that carry out *complementary* metabolisms. For example, in Chapter 17 we discussed metabolic transformations that involved two distinct groups of organisms, such as those of the *nitrosifying* and the *nitrifying* bacteria, which combine to oxidize NH_3 to NO_3^-, although neither group is capable

of doing this alone (Section 17.12). Because the product of the nitrosofying bacteria (NO_2^-) is the substrate for the nitrifying bacteria, such organisms typically live in tight association (Figure 18.12*a*).

✓ **19.2 Concept Check**

Microorganisms are very small, and their natural environments are likewise small. The microenvironment is the place in which the microorganism specifically lives. Microorganisms in nature often live a feast-or-famine existence such that only the best-adapted species survive in a given niche. Cooperation among microorganisms is also important in many microbial interrelationships.

✓ What aspects define the niche of a particular microorganism?

✓ Why can many different physiological groups of organisms live in a single habitat?

19.3 Microbial Growth on Surfaces and Biofilms

Surfaces are important microbial habitats because nutrients can adsorb to them; in the microenvironment of a surface, nutrient levels may be much higher than they are in the bulk solution. As a consequence, microbial numbers and activities are usually much greater on surfaces than in water.

Microscope slides can be used as experimental surfaces to which organisms can attach and grow. When a slide is immersed in a microbial habitat, left for a period of time, and then retrieved and examined by microscopy, the importance of the surface to microbial development is apparent (Figure 19.3*a*). Microcolonies readily develop on such surfaces, much as they do on natural surfaces in nature. In fact, periodic microscopic examination of immersed microscope slides is used to measure growth rates of attached organisms in nature.

A surface may itself also be a nutrient, such as a particle of organic matter, where attached microorganisms catabolize nutrients directly from the surface of the particle. Dead plant material, for example, is rapidly colonized by microorganisms in soil, and simple staining techniques can detect microbial populations attached to the solid surface (Figure 19.3*b*).

Biofilms: Structure

Microorganisms grow on surfaces enclosed in **biofilms**. These are microcolonies of bacterial cells attached to a surface and encased in adhesive polysaccharides excreted by the cells (Figure 19.4). Biofilms trap nutrients for growth of the microbial population and help prevent detachment of cells on surfaces present in flowing systems (Figure 19.5). Biofilms typically contain many layers, and microscopic examination of the microorganisms in each layer can be done using scanning laser confocal microscopy (∞ Section 3.2) (Figure 19.4).

Cell-to-cell communication is critical in the development and maintenance of a biofilm. Attachment of a cell to a surface is a signal for the expression of biofilm-specific genes. These genes encode proteins that synthesize cell-to-cell signaling molecules and that begin polysaccharide formation (Fig. 19.5). In *Pseudomonas aeruginosa*, a notorious biofilm former (Figure 19.4*a*), the major signaling molecules are compounds called *homoserine lactones*. As these molecules accumulate, they function as chemotactic agents to recruit nearby *P. aeruginosa* cells (a mechanism called *quorum sensing*, ∞ Section 8.10), and the biofilm develops (Figure 19.4*a*). *P. aeruginosa* has been implicated in the disease *cystic fibrosis*, where a tenacious biofilm forms in the lungs, leading to symptoms of pneumonia.

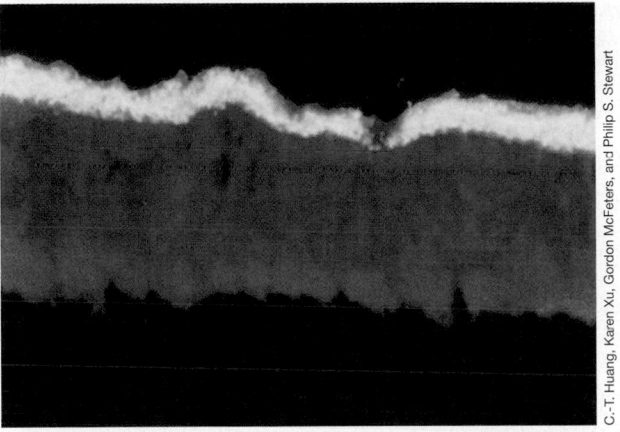

(a)

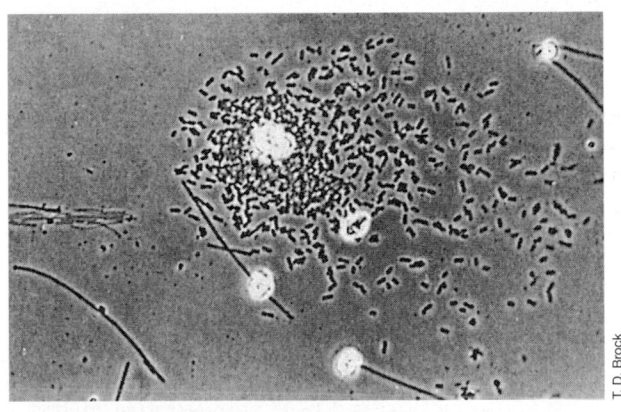

(a)

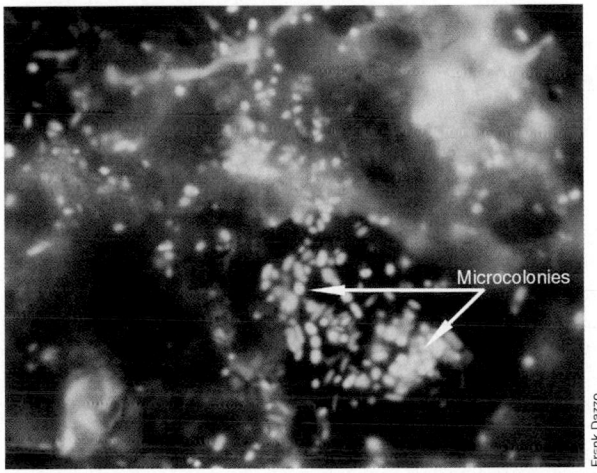

(b)

Figure 19.3 Microorganisms on surfaces. (a) Bacterial microcolonies developing on a microscope slide immersed in a small river. The bright particles are mineral matter. The short, rod-shaped cells are about 3 μm long. (b) Fluorescence photomicrograph of a natural microbial community colonizing plant roots in soil. Note microcolony development. The preparation has been stained with acridine orange.

Figure 19.4 Microbial biofilms. (a) A cross-sectional view of an experimental biofilm made up of cells of *Pseudomonas aeruginosa*. The yellow-green layer (about 15 μm in depth) contains cells and is stained by an enzyme activity stain for the enzyme alkaline phosphatase. (b) Confocal laser scanning microscopy (∞ Section 4.2) of a natural biofilm (top view) that developed on a leaf surface. The color of the cells indicates their depth in the biofilm: red, cells on the surface; green, 9 μm depth; blue, 18 μm depth.

Biofilms: Consequences and Control

Biofilms have significant implications in human medicine and commerce. In the body, bacterial cells within a biofilm are protected from attack by the immune system, and antibiotics and other antimicrobial agents often fail to pierce the biofilm. Biofilms have been implicated in a variety of medical and dental conditions besides cystic fibrosis, including periodontal disease, kidney stones, tuberculosis, Legionnaire's disease, and *Staphylococcus* infections. Medical implants are, unfortunately, excellent surfaces for biofilm development. These include both short-term devices, such as a urinary catheter, as well as

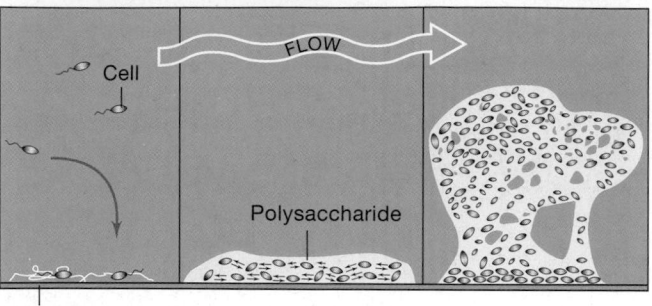

Attachment (adhesion of a few cells to a suitable solid surface)

Colonization (intercellular communication, growth and polysaccharide formation)

Development (more growth and polysaccharide)

Cell

FLOW

Polysaccharide

Surface

Figure 19.5 Formation of a bacterial biofilm. The biofilm begins with attachment of a few cells, after which growth and intercellular communication occurs. Polysaccharide formation occurs and becomes more extensive as the biofilm grows.

long-term implants, such as artificial joints. Estimates are that as many as 10 million people a year in the United States alone experience biofilm infections from implants or intrusive medical procedures. Biofilms are the reason why routine oral hygiene is so important. Dental plaque is a typical biofilm, and contains acid-producing bacteria responsible for dental caries (∞ Section 21.3).

In industrial situations biofilms can slow the flow of water, oil, or other liquids through pipelines and can accelerate corrosion of the pipes themselves. Biofilms also initiate degradation of submerged objects, such as structural components of offshore oil rigs, boats, and shoreline installations. Drinking water standards may be compromised by biofilms that develop in water distribution pipes, many of which in the United States are 50–100 years old. Although water pipe biofilms mostly contain harmless bacteria, if pathogens successfully colonize the biofilm, standard chlorination practices may be insufficient to kill them. Periodic releases of cells can then lead to outbreaks of disease. There is some concern that *Vibrio cholerae*, the causative agent of cholera (∞ Section 28.6), may be promulgated in this manner.

Biofilm control is a big business and thus far only a limited repertoire of tools are available for dealing with the problem. Collectively, industries spend billions of dollars each year treating pipes and other surfaces to keep them free of biofilms. Strategies for fighting biofilms in clinical medicine include the discoveries of new antibiotics that can penetrate biofilms and drugs that prevent biofilm formation by interfering with intercellular communication. In the latter regard, a class of chemicals called *furanones* have shown promise as biofilm preventatives in tests on abiotic surfaces. Because furanones are stable and nontoxic in humans, they may

also have applications as antibiofilm agents in human medicine.

✓ **19.3 Concept Check**

Biofilms are microcolonies of bacteria encased in slime that form on surfaces. Biofilms can lead to the destruction of inert as well as living surfaces from the excretory products of the bacterial cells. Biofilm formation is a complex process involving cell-to-cell communication.

✓ What is the chemical nature of the biofilm matrix?

✓ Why might a biofilm be a good habitat for bacterial cells living in a flowing system?

✓ Give an example of a medically relevant biofilm that most humans likely harbor.

II SOIL AND FRESHWATER MICROBIAL HABITATS

Among the major habitats of microorganisms are soils and freshwater, including lakes, ponds, and streams. Why is it that a soil or water sample taken from one habitat will harbor large numbers of microorganisms while a sample from another contains relatively few? Soils and waters vary in their physical structure, nutrient composition, temperature, and water potential. All of these factors combine to influence the types and numbers of microorganisms present, so our discussion here will focus on these factors and how they control microbial populations in soil and water.

19.4 Terrestrial Environments

The consideration of terrestrial environments inevitably turns to *soil* and *plants* because it is within the soil and on or near plants that many of the key processes occur that influence the functioning of the ecosystem. The process of soil development involves complex interactions among the parent material (rock, sand, glacial drift, and so on), topography, climate, and living organisms. Soils can be divided into two broad groups—**mineral soils** and **organic soils**—depending on whether they derive initially from the weathering of rock and other inorganic material or from sedimentation in bogs and marshes, respectively. Our discussion will concentrate on mineral soils, the predominant soil in most areas.

Soil Formation

Soils form as a result of combined physical, chemical, and biological processes. An examination of almost any exposed rock reveals the presence of algae, lichens (see Figure 19.54), or mosses. These organisms are dormant on dry rock and then grow when moisture is present. They are phototrophic and produce organic matter, which supports the growth of chemoorganotrophic bacteria and fungi. The numbers of chemoorganotrophs increase directly with the degree of plant cover of the rocks. Carbon dioxide produced during respiration by chemoorganotrophs is converted to carbonic acid ($CO_2 + H_2O \rightleftharpoons H_2CO_3$), which is an important agent in the dissolution of rocks, especially limestone. Many chemoorganotrophs also excrete organic acids, which further promote the dissolution of rock into smaller particles.

Freezing and thawing and other physical processes lead to the development of cracks in the rocks. In these crevices, a raw soil forms in which pioneering higher plants can develop. The plant roots penetrate farther into crevices and increase the fragmentation of the rock, and their excretions promote the development of a **rhizosphere** (the soil that surrounds plant roots) microflora. When the plants die, their remains are added to the soil and become nutrients for an even more extensive microbial development. Minerals are further rendered soluble, and as water percolates, it carries some of these chemical substances deeper. As weathering proceeds, the soil increases in depth, thus permitting the development of larger plants and trees. Soil animals become established and play an important role in keeping the upper layers of the soil mixed and aerated. Eventually the movement of materials downward results in the formation of layers, and a typical *soil profile* becomes evident (Figure 19.6). The rate of development of a typical soil profile depends on climatic and other factors, but it is usually very slow, taking hundreds of years.

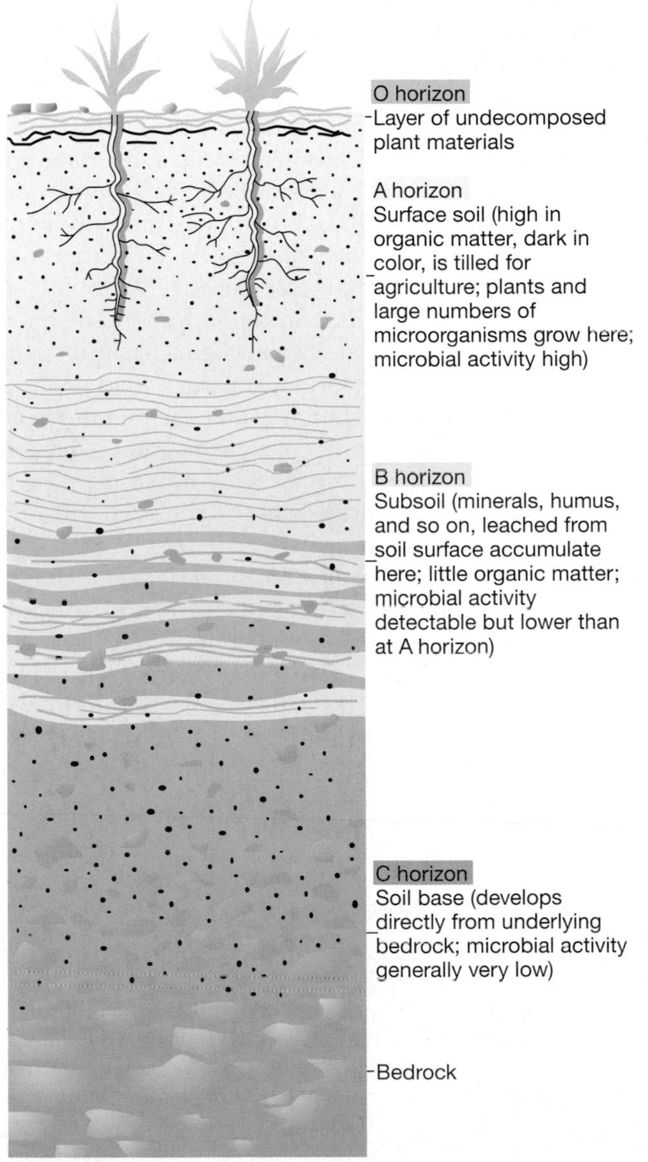

Soil as a Microbial Habitat

The most extensive microbial growth takes place on the *surfaces* of soil particles, usually within the rhizosphere (see Figures 19.2 and 19.3*b*). As pointed out in Section 19.2, even a small soil aggregate can have many differing microenvironments (compare Figures 19.2 and 19.7), and thus, several different types of microorganisms may be present. To examine soil particles directly for microorganisms, fluorescence microscopes are often used, the organisms in the soil being stained with a dye that fluoresces (see Figure 19.3*b*). To observe a *specific* microorganism in a soil particle, fluorescent-antibody staining (∽ Figure 18.8) or phylogenetic stains (∽ Section 18.4) can be used. Microorganisms can also be seen on such opaque surfaces as soil by means of the scanning electron microscope (Figure 19.8). The scanning electron microscope gives excellent information on the morphology of soil bacteria

O horizon
-Layer of undecomposed plant materials

A horizon
Surface soil (high in organic matter, dark in color, is tilled for agriculture; plants and large numbers of microorganisms grow here; microbial activity high)

B horizon
Subsoil (minerals, humus, and so on, leached from soil surface accumulate here; little organic matter; microbial activity detectable but lower than at A horizon)

C horizon
Soil base (develops directly from underlying bedrock; microbial activity generally very low)

-Bedrock

Figure 19.6 Profile of a mature soil. The soil horizons are soil zones as defined by soil scientists.

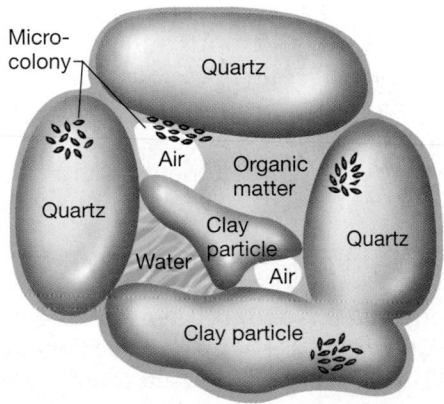

Figure 19.7 A soil aggregate composed of mineral and organic components, showing the localization of soil microorganisms. Very few microorganisms are found free in the soil solution; most of them occur as microcolonies attached to the soil particles.

Figure 19.8 Visualization of microorganisms on the surface of soil particles by use of the scanning electron microscope. (a) A microcolony of short, rod-shaped bacteria. (b) Actinomycete spores (∞ Section 12.25). The cells in (a) and the spores in (b) are about 1–2 μm wide. (c) Fungus hyphae. The fungal hyphae are about 4 μm wide and are coated with mineral matter.

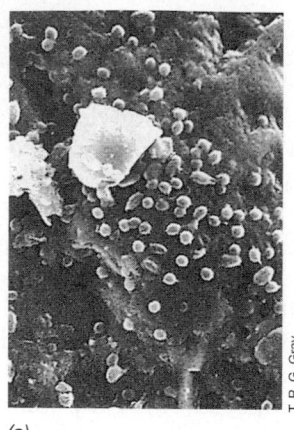

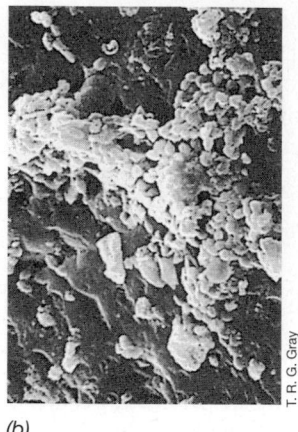

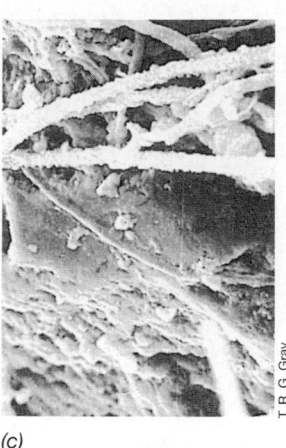

(a)

(b)

(c)

and can also be used to enumerate cells on soil particle surfaces.

One of the major factors affecting microbial activity in soil is the availability of *water*, and we have previously discussed the importance of water to microbial growth (∞ Section 6.12). Water is a highly variable component of soil, its presence depending on soil composition, rainfall, drainage, and plant cover. Water is held in the soil in two ways, by adsorption onto surfaces or as free water existing in thin sheets or films between soil particles. The water present in soils has a variety of materials dissolved in it, the whole mixture being referred to as the *soil solution*. In well-drained soils, air penetrates readily and oxygen concentrations can be high. In waterlogged soils, however, the only oxygen present is that dissolved in the water, and this is soon consumed by microorganisms. Such soils quickly become anoxic, showing profound changes in their biological properties. We discussed oxygen relationships in soil microenvironments in Section 19.2 and Figure 19.2.

The *nutrient status* (resources, Table 19.1) of a soil is the other major factor affecting microbial activity. The greatest microbial activity is in the organic-rich surface layers, especially in and around the rhizosphere. The numbers and activity of soil microorganisms depend to a great extent on the balance of nutrients present. In some soils carbon is not the limiting nutrient, but instead the availability of *inorganic* nutrients such as phosphorus and nitrogen limit microbial productivity.

Deep Subsurface Microbiology

Interest in the chemistry of groundwater and the potential leaching of pollutants and their transfer in groundwater aquifers has led to studies of the role microorganisms play in the *deep subsurface* terrestrial environment. The deep soil subsurface, which can extend for several hundred meters below the soil surface, is not a biological wasteland. Although cell numbers are much lower than in the A horizon, a variety of microorganisms, primarily bacteria, are present in most deep underground soils. For example, in

samples collected aseptically from bore holes drilled down to 300 m, a diverse array of bacteria have been found including anaerobes such as sulfate-reducing bacteria, methanogens, and homoacetogens (∞ Sections 17.15–17.17), and various aerobes and facultative aerobes. Microorganisms in the deep subsurface presumably have access to nutrients because groundwater flows through their habitats, but activity measurements suggest that metabolic rates of these bacteria are rather low in their natural habitats (see the box, Microbial Life Deep Underground). Compared to microorganisms in the upper layers of soil, the biogeochemical significance of deep subsurface microorganisms may thus be minimal. However, there is evidence that the metabolic activities of these buried microorganisms may over very long periods be responsible for slow mineralization of organic compounds and release of products into the groundwater.

The potential for *in situ* bioremediation (see Section 19.18) of toxic substances leached from soil into groundwater (for example, benzenes and agricultural chemicals) by deep subsurface microorganisms is of particular current interest. This could include either the addition of inorganic nutrients to stimulate biodegradation of these chemicals or the introduction of specific microorganisms into contaminated aquifers to promote biodegradative activities, or both.

✓ 19.4 Concept Check

The soil is a complex habitat with numerous microenvironments and niches. Microorganisms are present in the soil primarily attached to soil particles. The most important factor influencing microbial activity in surface soil is the availability of water, whereas in deep soil (the subsurface environment) nutrient availability plays a major role.

✓ What is the difference between *mineral* soils and *organic* soils?

✓ What factors govern the extent and type of microbial activity in soils?

| A Focus On ... | Microbial Life Deep Underground |

Microbiologists studying the deep terrestrial subsurface have found viable prokaryotes at depths of several thousand meters below the surface, existing in different physical and chemical environments. How are these microorganisms making a living? Initial findings indicated that these buried microorganisms were very slow growing chemoorganotrophic bacteria surviving by the slow catabolism of organic carbon deposited within the sediments. However, studies on the microbial ecology of deep basalt aquifers[1] have shown that sluggish chemoorganotrophs are not the only prokaryotes that live deep underground.

Basalts are iron-rich volcanic rocks that are essentially devoid of organic matter. In certain basalts up to 1500 m deep from the Columbia River Basin (Washington, USA), large numbers of anaerobic, *chemolithotrophic* bacteria have been found (Fig. 1a), including sulfate-reducing bacteria, methanogens, and homoacetogens.[1] The methanogens were shown to be metabolically active by carbon-stable isotope analyses of methane (CH_4) present in the rocks and surrounding groundwaters; measurements showed a strong enrichment in the lighter isotope of carbon (^{12}C), typical of biological methanogenesis (∞ Section 18.11). If the methanogens are active, it is likely that the

other groups are active too. Their products, H_2S from the sulfate-reducers and acetate from the homoacetogens (Fig. 1b), do not accumulate because they are actively consumed by other microorganisms.

A common metabolic link among these anaerobes is a thirst for H_2, an excellent electron donor for their various energy-yielding metabolisms (∞ Sections 17.9 and 17.14–17.18 and Fig. 1b below). Hydrogen is a common product of the fermentative decomposition of organic matter (∞ Sections 17.19–17.21). But if basalts contain very little organic material, where does the H_2 come from to support metabolism of the H_2 consumers? Interestingly, H_2 in the Columbia basalts apparently originates from the *strictly chemical* interaction of water with iron minerals in the rocks (see proposed reaction in Fig. 1b). Such reactions are known from inorganic chemistry, and in laboratory studies in which crushed Columbia basalt was mixed with sterile water under anoxic conditions, rapid H_2 evolution occurred.[1] H_2 was also detected *in situ* in groundwater percolating through the basalts.

From analyses of these experimental results it was hypothesized that H_2 formed in deep underground basalts is indeed the electron donor that supports growth of the substantial populations of anaerobic pro-

karyotes found there. If this is true, these organisms would be living a strictly *geochemical* existence because their electron acceptor (CO_2 in the case of methanogens and homoacetogens and SO_4^{2-} in the case of sulfate-reducing bacteria) and electron donor, H_2, are all derived from inorganic materials. These buried chemolithotrophs would also be novel because of their total independence of photosynthetic primary production, which generates the molecular oxygen and/or organic matter required by virtually all surface-dwelling organisms. However, as sometimes happens in research, other findings have questioned this interpretation. Although there is not much organic matter in these basalts, there is some, and thus fermentation could be the source of some (or even most) of the H_2 used by the various subsurface H_2 chemolithotrophs.[2]

How widespread and ecologically significant H_2-based microbial ecosystems deep underground are, regardless of the source of H_2, obviously awaits further research. But it is clear that metabolically active H_2-consuming microorganisms inhabit the deep subsurface, and this in turn suggests that they may be important biogeochemical agents in these environments. ■

[1]Stevens, T. O., and J. P. McKinley. 1995. Lithoautotrophic microbial ecosystems in deep basalt aquifers. *Science* 270:450–454.

[2]Anderson, R. T., F. H. Chapelle, and D. R. Lovely. 1998. Evidence against hydrogen-based microbial ecosystems in basalt aquifers. *Science* 281:976–977.

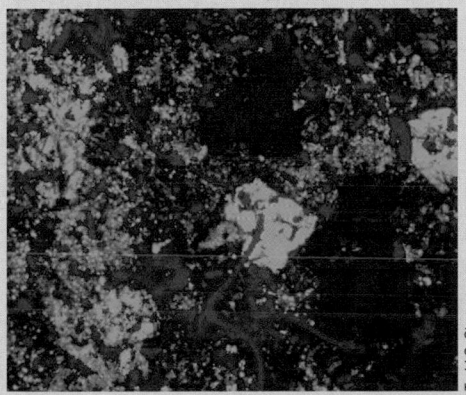

Todd O. Stevens

(a)

Figure 1 **(a) Laser confocal photomicrograph of a microbial biofilm attached to the surface of basalt chips. Green is reflected light from the basalt surface while the red color is from Nile-red stained bacterial cells. The cells in the biofilm were grown on H_2 from basalt as depicted in the bottom reaction of Fig. 1b. (b) Key metabolic reactions of anaerobic prokaryotes growing in anoxic deep basalt aquifers.**

Methanogenesis: $4 H_2 + CO_2 \rightarrow CH_4 + 2 H_2O$

Acetogenesis: $4 H_2 + 2 HCO_3^- + H^+ \rightarrow CH_3COO^- + 4 H_2O$

Sulfate reduction: $4 H_2 + SO_4^{2-} + H^+ \rightarrow HS^- + 4 H_2O$

Proposed inorganic
H_2 production: $FeO + H_2O \rightarrow H_2 + FeO_2$

(b)

19.5 Freshwater Environments

Typical freshwater environments are lakes, ponds, rivers, and springs. Freshwater environments differ considerably in chemical and physical properties, and it is not surprising that their microbial species compositions also differ. The predominant phototrophic organisms in most aquatic environments are microorganisms; in oxic areas cyanobacteria and algae prevail, and in anoxic areas anoxygenic phototrophic bacteria are preponderant. Algae floating or suspended freely in the water are called **phytoplankton**; those attached to the bottom or sides are called **benthic algae**. Because these phototrophic organisms use energy from light in the initial production of organic matter, they are called **primary producers**.

In the final analysis, the microbiological activity of an aquatic ecosystem is dependent on the rate of primary production by the phototrophic organisms. Oxygenic phototrophs produce both new organic material as well as oxygen. If photosynthetic activity is very high, the excessive organic matter leads to O_2 depletion and anoxic conditions. This in turn triggers alternative forms of metabolism, in particular, anaerobic respiration and fermentation (∞ Sections 17.13–17.21).

Oxygen Relationships in Lakes and Rivers

We discussed oxygen requirements and anaerobiosis in Section 6.13, the production of oxygen via photosynthesis in Section 17.5, and oxygen in microenvironments in Section 19.2. Although oxygen is one of the most plentiful gases in the atmosphere (~21% of air), it has limited solubility in water, and in a large water mass its exchange with the atmosphere is slow. Significant photosynthetic production of oxygen occurs only in the *surface layers* of a lake or ocean, where light is available (see Figure 19.1). Organic matter that is not consumed in these surface layers sinks to the depths and is decomposed by facultative microorganisms, using oxygen dissolved in the water. In lakes, once the oxygen is consumed, the deep layers become anoxic; here strictly aerobic organisms such as higher plants and animals cannot grow, and the bottom layers have a species composition restricted to anaerobic bacteria and a few kinds of microaerophilic animals. In addition, there is a conversion from respiratory to fermentative and methanogenic metabolisms, with important consequences for the carbon cycle and other nutrient cycles (see Sections 19.9–19.11 for a consideration of this).

Whether a body of water becomes depleted of oxygen depends on several factors. If organic matter is sparse, as it is in unproductive lakes or in the open ocean, there may be insufficient substrate available for chemoorganotrophs to consume all the oxygen. Also important is how rapidly the water from the depths exchanges

with surface water. Where strong currents or turbulence occurs, the water mass may be well mixed, and consequently oxygen may be transferred to the deeper layers. However, in many lakes in temperate climates, the water mass becomes *stratified* during the summer, with the warmer and less dense surface layers, called the *epilimnion*, separated from the colder and denser bottom layers (the *hypolimnion*) (Figure 19.9). After stratification sets in, usually in early summer, the bottom layers become anoxic (Figure 19.9). In the late fall and early winter, the surface waters become colder and thus more dense than the bottom layers, and the water "turns over," leading to a reaeration of the bottom. Most lakes in temperate climates show this annual cycle in which the bottom layers of water pass from oxic to anoxic and back to oxic. Microbial activity changes along with these changes in oxygen content, but other factors, especially temperature, strongly govern microbial activity as well.

Rivers

The oxygen relations in a river are of particular interest, especially in regions where rivers receive much organic matter in the form of sewage and industrial pollution. Even though a river may be well mixed because of rapid water flow and turbulence, large amounts of added organic matter can lead to a marked oxygen deficit from bacterial respiration. This is illustrated in Figure 19.10. As the water moves away from a sewage outfall, organic matter is gradually consumed and the oxygen content

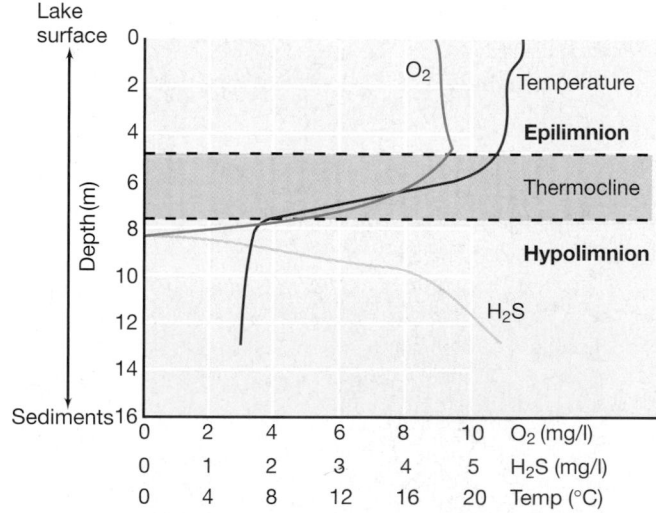

Figure 19.9 Development of anoxic conditions in the depths of a temperate climate lake as a result of summer stratification. The colder bottom waters are more dense and contain H_2S from bacterial sulfate reduction. The zone of rapid temperature change is referred to as the thermocline. Typically, as surface waters cool in the fall and early winter they reach the temperature and density of hypolimnetic waters and sink, displacing bottom waters and effecting "lake turnover."

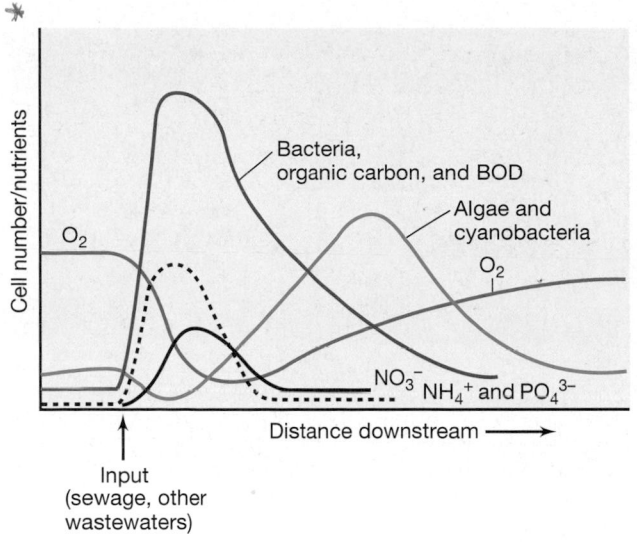

(a)

(b)

Figure 19.10 Effect of input of sewage or other organic-rich waste-waters into aquatic systems. (a) In a river, an increase in heterotrophic bacterial numbers and a decrease in O_2 levels occur immediately upon a spike of organic matter. If NH_4^+ is present in the input, for example, from sewage, NH_4^+ is oxidized to NO_3^- by nitrifying bacteria (⚬⚬ Sections 12.3, 17.12, and 19.12). Note how the rise in NH_4^+ is followed shortly by the rise in NO_3^-, as the two-stage process of nitrification proceeds. The rise in numbers of algae and cyanobacteria is primarily a response to inorganic nutrients, especially PO_4^{3-}. Oxygen levels return to their pre-input levels once most of the oxidizable organic and inorganic compounds are depleted. (b) A eutrophic (nutrient-rich) lake, Lake Mendota, Madison, Wisconsin, showing algae, cyanobacteria, and macrophytes (aquatic weeds) that develop in response to pollution by inorganic nutrients, much of which results from agricultural runoff into the lake watershed. Dead plant and algal material contains a very large BOD, such that although these phototrophs contribute oxygen by photosynthesis, their degradation results in large-scale O_2 depletion.

returns to normal. Oxygen depletion in a body of water is undesirable because many aquatic animals die under even very temporary anoxic conditions. Furthermore, conversion to anoxia results in the production by anaerobic bacteria of odoriferous compounds (for example, amines, H_2S, mercaptans, fatty acids), some of which are also toxic to higher organisms.

Biochemical Oxygen Demand

Sanitary engineers term the microbial oxygen-consuming property of a body of water its **biochemical oxygen demand (BOD)**. The BOD is determined by taking a sample of water, aerating it well, placing it in a sealed bottle, incubating for a standard period of time (usually 5 days at 20°C), and determining the residual oxygen in the water at the end of incubation. A BOD determination thus gives a measure of the amount of organic material in the water that could be oxidized by microorganisms. As a river recovers from contamination with an organic pollutant, the drop in BOD is accompanied by a corresponding increase in dissolved oxygen (Figure 19.10a).

We thus see that in a water body, the oxygen and carbon cycles are interrelated, with the concentrations of organic carbon and oxygen in general being inversely related. This is particularly evident in *anoxic* environments, which are frequently rich in organic carbon (see Sections 19.9–19.11).

✓ 19.5 Concept Check

In aquatic ecosystems phototrophic microorganisms are usually the main primary producers. Most of the organic matter produced is consumed by bacteria, which can lead to depletion of oxygen in the environment. BOD is a measure of the oxygen-consuming properties of a water sample.

✓ What is a *primary producer*?

✓ In a lake, where is the *epilimnion* and where is the *hypolimnion*?

✓ Will addition of organic matter to a water sample *increase* or *decrease* its BOD?

III MARINE MICROBIOLOGY

The oceans differ from freshwater environments in many ways, including salinity, average temperature, and nutrient status. Through use of the molecular tools of microbial ecology, especially genetic stains and gene isolation and sequencing (⚬⚬ Sections 18.4–18.6), much new information is emerging about marine microorganisms, and we focus here on three key topics: prokaryotes in the open oceans, deep-sea microbiology, and the extensive microbial ecosystems supporting animal life near deep sea hydrothermal vents.

19.6 Marine Habitats and Microbial Distribution

Compared with typical freshwater environments, nutrient levels in the open ocean are often limiting, especially those of the key inorganic nutrients of nitrogen, phosphorus, and iron. The activity of primary producers (mainly, photosynthetic CO_2 fixation) is limited by these nutrient deficiencies, and thus, cell numbers are typically lower in the ocean than in freshwater ecosystems. However, because the oceans are so large, the collective photosynthesis and oxygen production that occurs there are major factors in Earth's carbon balance and influence everything from the marine food chain to the Earth's climate.

Primary Productivity

Much of the primary productivity in the open oceans, even at significant depths, is due to the photosynthetic activities of *prochlorophytes*, tiny oxygenic prokaryotes that contain chlorophylls *a* and *b*; the organism *Prochlorococcus* (a relative of the cyanobacteria) is a particularly important primary producer (Figure 19.11*a* and ⌒⌒ Section 12.27). In tropical and subtropical oceans the planktonic filamentous marine cyanobacterium *Trichodesmium* (Figure 19.11*c*) is also widespread. Cells of *Trichodesmium* form tufts of filaments that constitute a significant fraction of the biomass suspended in these waters. *Trichodesmium* is a nitrogen-fixing cyanobacterium, and the production of fixed nitrogen by this organism is thought to be a major link in the nitrogen cycle (see Section 19.12) in marine environments. In coastal waters even very small phototrophic *eukaryotes* have been found, and some of these are among the smallest eukaryotic cells known. *Ostreococcus*, for example, is an extremely small alga, measuring only about 0.7 µm in diameter (Figure 19.11*b*), smaller than a cell of the prokaryote *Escherichia coli*.

Inshore ocean areas are typically more nutritionally fertile than open ocean waters and therefore support more dense populations of phototrophic microorganisms (Figure 19.12). This in turn supports higher densities of chemotrophic bacteria and aquatic animals, such as fish and shellfish. Marine bays and inlets receiving high levels of nutrients from sewage or industrial waste runoff can have very high phytoplankton and bacterial populations. If the pollution is severe enough, shallow marine waters can become anoxic from the consumption of O_2 by bacteria, and toxic to marine life from the production of H_2S by sulfate-reducing bacteria that quickly develop in anoxic seawater (see Section 19.13).

Open Ocean Microbiology

Despite low inorganic nutrient and organic carbon levels, significant numbers, between 10^5 and 10^6/ml, of prokaryotic cells have been found suspended in open

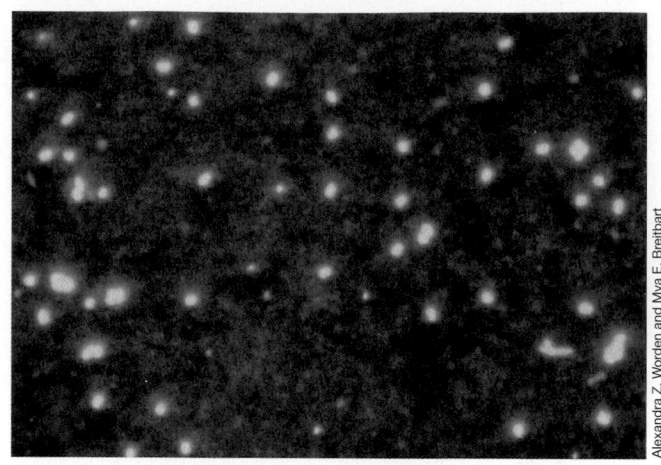
(a)

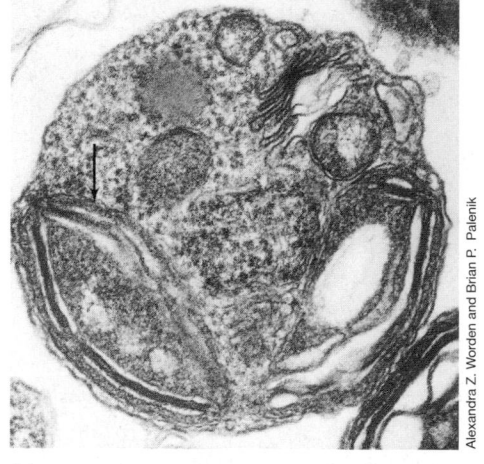
(b)

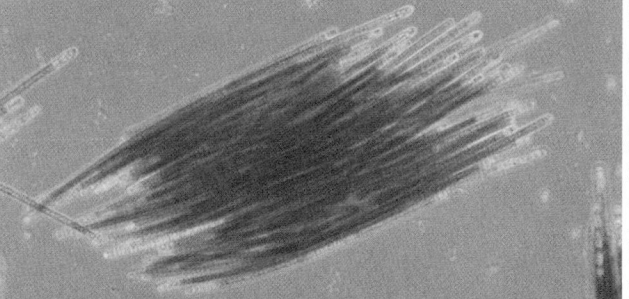
(c)

Figure 19.11 Marine oxygenic phototrophs. (a) FISH phylogenetic stain (⌒⌒ Section 18.4) of cells of the prochlorophyte *Prochlorococcus*, the dominant phototroph in subtropical open oceanic waters. (b) Transmission electron micrograph of a cell of *Ostreococcus*, a very small alga (eukaryote), found in substantial numbers in marine coastal waters. The arrow points to the chloroplast. Cells of both *Prochlorococcus* and *Ostreococcus* are about 0.7 µm in diameter. (c) The nitrogen-fixing cyanobacterium *Trichodesmium*. Cells form tufts that actively fix N_2 in tropical waters worldwide. A filament of *Trichodesmium* is about 6 µm in diameter. For more coverage of prochlorophytes, algae, and cyanobacteria, see Sections 12.26, 14.11, and 12.25, respectively.

Figure 19.12 Distribution of chlorophyll in the western North Atlantic Ocean as recorded by satellite. The east coast of the United States from mid-Florida to northern Maine is shown in dotted outline. Near the center of the photograph is Chesapeake Bay; the Great Lakes are at the upper left. Areas rich in phytoplankton are shown in red (>1 mg chlorophyll/m^3); blue and purple areas have lower chlorophyll concentrations (<0.01 mg/m^3). Note the high primary productivity of coastal areas and the Great Lakes.

ocean waters. In addition, upward of 10^4 very small eukaryotic cells are present per milliliter. How do these organisms make a living? Accumulating evidence points to light-driven energy metabolism as the key to supporting much of this open ocean microflora. Some of these organisms, such as *Prochlorococcus*, *Trichodesmium*, and the tiny phototrophic eukaryotes have already been discussed (see Figure 19.11). Other phototrophs are present that carry out *anoxygenic* photosynthesis (⌘ Sections 12.2 and 17.4). These organisms may be related to the "aerobic" phototrophic bacteria that have been isolated from inshore marine waters (⌘ Section 12.2).

But in addition to these phototrophs, a major surprise has been the discovery that many bacteria in the ocean's photic zone (top 300 meters) contain a form of the visual pigment rhodopsin and are able to use this pigment to convert light energy into ATP. In Section 13.3 we discussed the well-studied case of *bacteriorhodopsin*, present in the extremely halophilic archaeon *Halobacterium*, and how this molecule was involved in ATP synthesis. The form of rhodopsin found in open ocean prokaryotes is very similar to bacteriorhodopsin but is present in cells that are phylogenetically *Bacteria*, not

Archaea. Called **proteorhodopsin** because the organisms that contain it are Proteobacteria (⌘ Section 12.1), this molecule is presumably the basis of energy metabolism for these marine bacteria and releases them from a dependence on organic carbon for an energy source. Thus, light-mediated mechanisms of ATP synthesis, whether photosynthesis in the classic sense or otherwise, seem to be widely distributed in prokaryotes that inhabit surface waters in the world's oceans.

Archaea/Bacteria Distribution

Numbers of prokaryotes in the open oceans decrease with depth. In surface waters cell numbers average between 10^5–10^6 cells/ml. Below 1000 meters, however, total numbers fall to between 10^3–10^5/ml. Using phylogenetic stains to differentiate cells of each phylogenetic domain (⌘ Section 18.4), an interesting distribution of *Bacteria* and *Archaea* with depth has been found in the open oceans. In general, species of *Bacteria* predominate in upper waters (<1000 meters), while numbers are about equal or show a slight predominance of *Archaea* in lower waters (Figure 19.13). The *Archaea* present in

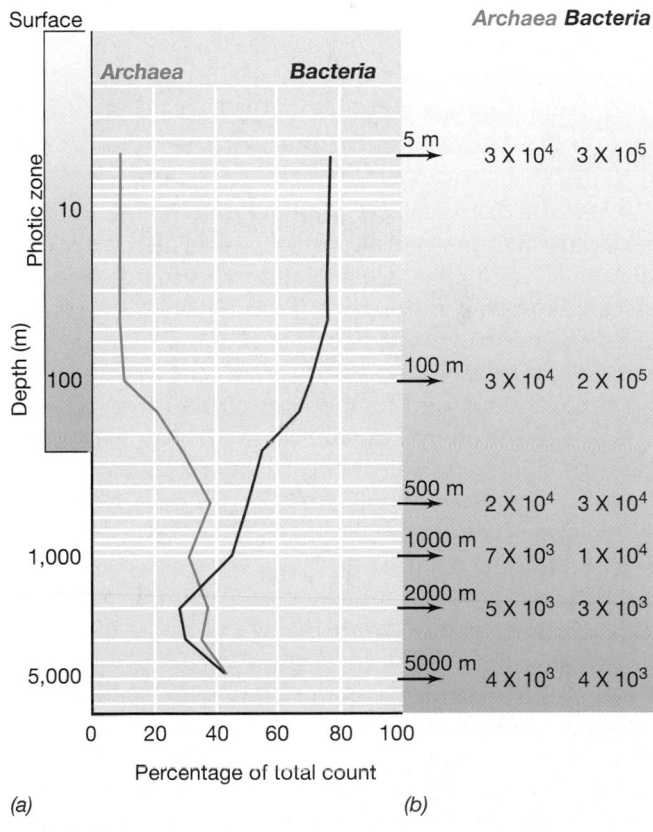

Figure 19.13 Percentage of total prokaryotes belonging to either the *Archaea* or the *Bacteria* in North Pacific ocean water. (a) Distribution of *Archaea* and *Bacteria* with depth. (b) Absolute numbers of *Archaea* and *Bacteria* with depth (per milliliter). Adapted from *Nature* 409:507–510 (2000).

lower waters are almost exclusively species of the Cren-archaeota, a phylum of *Archaea* that includes the hyper-thermophiles (∞ Sections 13.8–13.10). Extrapolating from the data of Figure 19.13, it has been estimated that 1.3×10^{28} and 3.1×10^{28} cells of *Archaea* and *Bacteria*, respectively, exist in the world's oceans. This indicates that the oceans contain the largest amount of microbial biomass on the surface of the Earth (∞ Section 1.3).

✓ 19.6 Concept Check

Marine waters are more nutrient deficient than many fresh-waters, yet substantial numbers of microorganisms exist there. Many of these depend on phototrophic metabolism of one sort or the other. In terms of prokaryotes, species of the domain *Bacteria* tend to predominate in oceanic surface waters where-as *Archaea* are more prevalent in deeper waters.

✓ How does the organism *Prochlorococcus* contribute to both the carbon and oxygen cycles in the oceans?

✓ What is proteorhodopsin and how was it named? How did the discovery of proteorhodopsin help solve the mystery of how large numbers of bacterial cells could exist in open ocean waters essentially devoid of dissolved organic matter?

✓ How are *Bacteria* and *Archaea* differentially distributed in ocean waters?

19.7 Deep-Sea Microbiology

What is the deep sea and what of microbiological interest occurs there? Visible light penetrates no further than about 300 m in open ocean waters; this upper region is referred to as the **photic zone** (Figure 19.13). Beneath the photic zone, down to a depth of about 1000 m, considerable biological activity still occurs as a result of the action of animals and chemoorganotrophic microorganisms. Water at depths greater than 1000 m is, by comparison, relatively biologically inactive and has come to be known as the "deep sea." Greater than 75% of all ocean water is in the deep sea, primarily at depths between 1000 and 6000 m.

Conditions in the Deep Sea

Organisms that inhabit the deep sea are faced with three major environmental extremes: low temperature, high pressure, and low nutrient levels. Below depths of about 100 m ocean water stays a constant 2–3°C. We discussed the responses of microorganisms to changes in temperature in Section 6.8. As would be expected, bacteria isolated from depths below 100 m are *psychrophilic* (cold-loving). Some are *extreme* psychrophiles, growing only in a narrow range near the *in situ* temperature. Deep-sea microorganisms must also be able to withstand the enormous hydrostatic pressures associated with

great depths. Pressure increases by *1 atm* for every *10 m* depth. Thus, an organism growing at a depth of 5000 m must be able to withstand pressures of 500 atm.

Barotolerant and Barophilic Bacteria

Do deep-sea bacteria simply *tolerate* high pressure (that is, are they *barotolerant*) or are they actually *dependent* on pressure (that is, are they *barophilic*)? Studies of cultures of deep-sea bacteria have shown that both types exist, and that the distribution of barotolerant and barophilic bacteria is basically a function of depth. Organisms isolated from depths down to about 3000 m and tested for growth or metabolic activity as a function of pressure are **barotolerant**; higher metabolic rates are observed at 1 atm than at 300 atm, although growth rates at the two pressures are about the same (Figure 19.14). However, barotolerant isolates do not grow at pressures above 500 atm. By contrast, cultures derived from samples taken at greater depths, 4000–6000 m, are often **barophilic**, growing optimally at pressures of about 400 atm (Figure 19.14). Note that although barophiles grow best under pressure, they retain the ability to grow at 1 atm (Figure 19.14).

Samples from even deeper water (10,000 m) have yielded **extreme (obligate) barophiles**. One strain studied in detail grew fastest at a pressure of 700–800 atm and grew nearly as well at 1035 atm, the pressure it was experiencing in its natural habitat (see Figure 19.15). The unique aspect of this extremely barophilic isolate was that it not only *tolerated* pressure but actually *required* pressure for growth; the isolate would not grow at pressures of less than about 400 atm (Figure 19.14). Interestingly, this extreme barophile was not killed by decompression be-

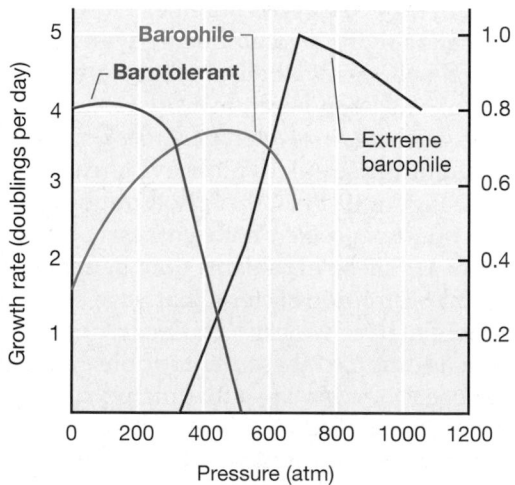

Figure 19.14 Growth of barotolerant, barophilic, and extremely barophilic bacteria. The extreme barophile was isolated from the Mariana Trench (see Figure 19.15). Note the much slower growth rate (at any pressure) of the extreme barophile (right ordinate) as compared to the barotolerant and barophilic bacteria (left ordinate). Note also the inability of the extreme barophile to grow at low pressures.

Figure 19.15 The sampling arm of the unmanned submersible *Kaiko* inserting a sampling tube into sediment on the sea floor of the Mariana Trench (off the Philippines, Pacific Ocean) at a depth of 10,897 meters and collecting a sample. The tubes of sediment are then retrieved and used for enrichment and isolation of barophilic bacteria.

cause it could tolerate moderate periods of decompression; however, viability was lost when the culture was left for several hours in a decompressed state.

Barotolerant and barophilic bacteria are also cold-loving, that is, psychrophilic. This property appears to be more prevalent among extremely barophilic isolates. The extreme barophile described in Figure 19.14 was found to be sensitive to temperature; the optimal growth temperature was determined to be the environmental temperature of 2°C, and temperatures above 10°C significantly reduced viability.

Molecular Effects of High Pressure

Pressure affects cellular physiology and biochemistry. It has been established that increased pressure decreases the binding capacity of enzymes for their substrates. Thus, the enzymes of extreme barophiles must be folded in such a way as to minimize these pressure-related effects. Other potential pressure-sensitive phenomena include protein synthesis and membrane activities such as transport. An organism grown under high pressure has a higher proportion of unsaturated fatty acids in its cytoplasmic membrane. This change is presumably of adaptive significance because it makes the membrane less likely to gel at high pressures. The rather slow growth rates of extreme barophiles (see Figure 19.14) are probably due to a combination of pressure effects on cellular biochemistry and their growth only at low temperatures, where reaction rates are decreased considerably to begin with.

The use of molecular genetic tools have given new insight into the physiology of barophilism. In gram-negative barophiles capable of growth up to 500–600 atm, it has been shown that growth at high pressure is accompanied by changes in the protein composition of the cell wall outer membrane. In one barophile studied in detail, a specific outer membrane protein called the OmpH protein (*outer membrane protein* H) is synthesized in cells grown at high pressures but not in cells grown at 1 atm pressure. The OmpH protein is a type of *porin*. Porins are structural proteins that form channels for the diffusion of organic molecules through the outer membrane and into the periplasm (⚭ Section 4.9 and Figure 4.36*b*). Presumably, the porin present in cells of the barophile grown at low pressures cannot function properly at high pressure, and thus a new type of porin molecule must be synthesized.

Studies of OmpH thus show that pressure can affect gene expression in barophilic bacteria. How this happens is unclear but could involve the activities of pressure-sensitive repressor proteins or pressure-dependent gene activators (⚭ Chapter 8). However, it appears that relatively few proteins are controlled by pressure in barophiles because many proteins are the same in cells grown at both high and low pressure; cell wall and related structural proteins and transport proteins seem to be the major variable components.

✓ **19.7 Concept Check**

The deep sea is a cold, dark habitat where high hydrostatic pressure and low nutrient availability prevails. Barophiles grow best under pressure, and extreme barophiles, obtained from the greatest depths, require high pressures for growth.

✓ How does pressure change with depth?
✓ What molecular adaptations occur in barophiles that allow them to grow optimally under pressure?

19.8 Hydrothermal Vents

The conception of the deep sea as a remote, low temperature, high pressure environment capable of supporting only the slow growth of barotolerant and barophilic bacteria is generally correct, but there are some amazing exceptions. A number of dense, thriving *animal* communities supported by the activities of microorganisms exist clustered about thermal springs in deep waters throughout the world. Geologically, these springs are associated with *sea floor spreading centers* (*rifts*), regions where hot basalt and magma near the sea floor cause the floor to slowly drift apart. Seawater seeping into these cracked regions mixes with hot minerals and is emitted from the springs (Figure 19.16); because of their unique properties these underwater hot springs have come to be known as **hydrothermal vents**.

Two major types of vents have been found. *Warm vents* emit hydrothermal fluid at temperatures of 6–23°C (into seawater at 2°C). *Hot vents*, referred to as "black smokers" because the mineral-rich hot water forms a dark cloud of precipitated material on mixing with

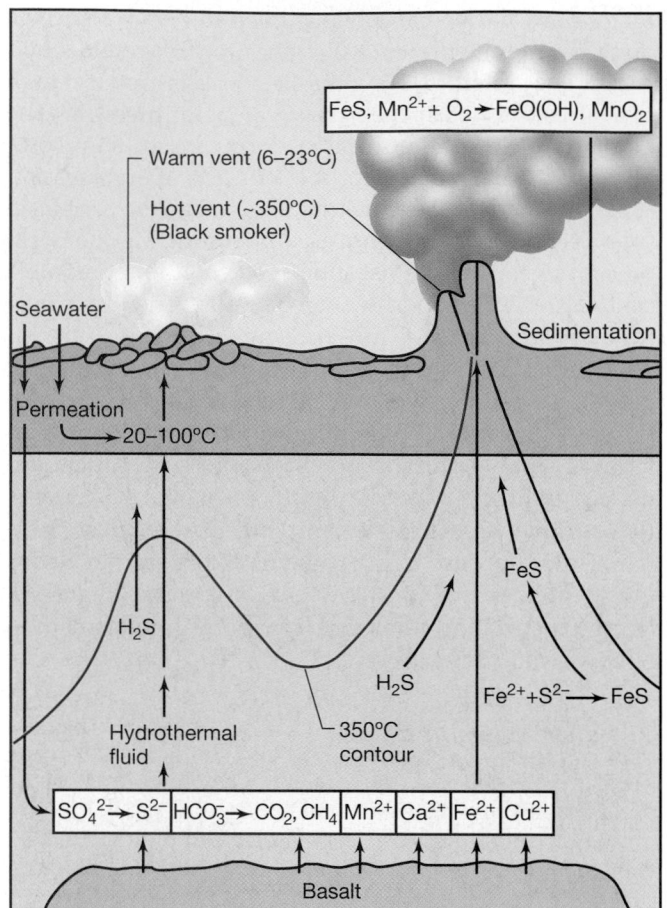

Figure 19.16 Schematic diagram showing the geological formations and major chemical species occurring at warm vents and black smokers. At warm vents, the hot hydrothermal fluid is cooled by cold (2–3°C) seawater permeating the sediments. In black smokers, hot hydrothermal fluid near 350°C reaches the sea floor directly. Warm vents and black smokers are typically found at about 2000 meters in depth. Such depths can be explored by humans in small submersibles such as *Alvin*, used by researchers at the Woods Hole Oceanographic Institution, Woods Hole, MA.

seawater, emit hydrothermal fluid at 270–380°C (Figure 19.16). Each of these two vent types has a characteristic flow rate; warm vents emit fluid at 0.5–2 cm/s, whereas hot vents have higher flow rates, about 1–2 m/s.

Animals Living at Hydrothermal Vents

Using small pressurized submarines, it is possible to visit hydrothermal vents and study the organisms associated with them. Thriving invertebrate communities are present, including *tube worms* over 2 m in length and large numbers of giant clams and mussels (Figure 19.17). Considering that other locations in the deep sea are so biologically unproductive, how do dense animal communities exist in the absence of phototrophic primary producers? What is the nature of the energy source(s)?

(a)

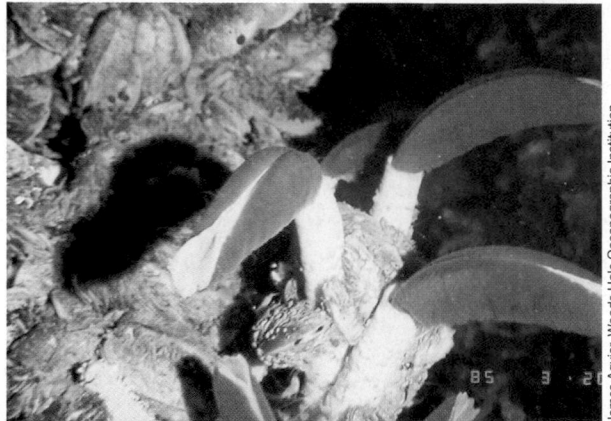

(b)

(c)

Figure 19.17 Invertebrates from habitats near deep-sea thermal vents. (a) Tube worms (family *Pogonophora*), showing the sheath (white) and plume (red) of the worm bodies. (b) Close-up photograph showing worm plume. The plume collects nutrients from hydrothermal vents, including in particular, H_2S. The latter is transported to symbiotic sulfide-oxidizing bacteria that live within the worm. See text for details. (c) Mussel bed in vicinity of a warm vent. Note yellow deposition of elemental sulfur (from the oxidation of H_2S by chemolithotrophic symbionts) in and around mussels.

Chemical analyses of hydrothermal fluid show large amounts of reduced inorganic materials, including H_2S, Mn^{2+}, H_2, and CO. Some vents contain little H_2S but have high levels of NH_4^+. Organic matter is not present in the fluid emitted from any of the hydrothermal vents thus far examined. From studies of vent chemistry and associated microbial processes, it is clear that the animals are dependent on the activities of chemolithotrophic bacteria (Sections 17.8–17.12), which grow at the expense of inorganic energy sources emitted from the vents. Carbon dioxide, abundant in seawater as CO_3^{2-} and HCO_3^-, is fixed into organic carbon by the chemolithotrophs, and the latter form the base of an extremely short food chain for hydrothermal vent animals.

Microorganisms in Hydrothermal Vents

Large numbers of sulfur-oxidizing chemolithotrophs such as *Thiobacillus*, *Thiomicrospira*, *Thiothrix*, and *Beggiatoa* (Sections 12.4 and 17.10) are present in and around the vents. Samples collected from near the vents have yielded cultures of these organisms, and *in situ* experiments have shown fixation of CO_2 and oxidation of H_2S and $S_2O_3^{2-}$ by natural populations of these bacteria. Some vents contain nitrifying bacteria, hydrogen-oxidizing bacteria, iron- and manganese-oxidizing bacteria, and methylotrophic bacteria, the latter presumably growing on the methane and carbon monoxide (CO) emitted from the vents (Chapters 12–13 discuss most of these physiological groups). Table 19.2 summarizes the electron donors and electron acceptors for chemolithotrophs suspected of playing a role in hydrothermal vent animal ecology. However, there is no evidence that the vent animals actually *eat* these chemolithotrophic bacteria. Instead, it is the autotrophic capacities of these organisms that provide nourishment for the animals.

Nutrition of Animals Living Near Hydrothermal Vents

Various chemolithotrophs have been found to live in symbiotic association with animals of the thermal vents. For example, the 2-m-long tube worms (see Figure 19.17)

lack a mouth, gut, or anus but contain a modified gastrointestinal tract consisting primarily of spongy tissue called the **trophosome**. Making up about 50% of the weight of the worm, trophosome tissue is loaded with sulfur granules, and microscopy of trophosome tissue shows large numbers of symbiotic prokaryotic cells (Figure 19.18), an average of 3.7×10^9 cells/g of trophosome tissue. The large spherical cells observed in the trophosome are structurally similar to the marine sulfur-oxidizing bacterium *Thiovulum*. Trophosome tissue also shows activity of the enzyme RubisCO and other enzymes of the *Calvin cycle*, the pathway by which most autotrophic organisms fix CO_2 into cellular material (Section 17.6).

The chemolithotrophic bacteria supply the worm with its nourishment, the animal living off the excretory products and dead cells of its chemolithotrophic symbionts. The bright red plume (Figure 19.17b) is rich in blood vessels and serves as a trap for O_2 and H_2S (see later) for transport to chemolithotrophs in the trophosome. Similar conclusions can be reached concerning the nutrition of the giant clams and mussels (see Figure 19.17c) present around the vents because sulfur-oxidizing bacterial communities are found in the gill tissues of these animals as well. Use of nucleic acid sequencing (Chapter 10) and phylogenetic analyses (Chapter 11) has shown that each vent animal harbors only one major species of bacterial symbiont and that the species of symbiont varies among the different animal types.

Further study of the tube worms has shown that these animals contain unusual hemoglobins that bind H_2S as well as O_2 and transport both substrates to the trophosome where they are released to the bacterial symbiont; trapping and transporting sulfide are necessary to prevent the H_2S from poisoning the animal. The CO_2 content of tube worm blood is also high, some 20–30 mM, and presumably this is released in the trophosome as a carbon source for the symbionts. In addition, stable isotope analyses (Section 18.11) of the elemental sulfur found within the bacterial symbionts have shown the

TABLE 19.2	Chemolithotrophic prokaryotes of potential significance to hydrothermal vent primary productivity[a]		
Chemolithotroph	Electron donor	Electron acceptor	Product from donor
Sulfur-oxidizing	HS^-, S^0, $S_2O_3^{2-}$	O_2, NO_3	S^0, SO_4
Nitrifying	NH_4^+, NO_2^-	O_2	NO_2^-, NO_3^-
Sulfate-reducing	H_2	S^0, SO_4^{2-}	H_2S
Methanogenic	H_2	CO_2	CH_4
Hydrogen-oxidizing	H_2	O_2, NO_3^-	H_2O
Iron and manganese-oxidizing	Fe^{2+}, Mn^{2+}	O_2	Fe^{3+}, Mn^{4+}
Methylotrophic	CH_4, CO	O_2	CO_2

[a] Sections 17.8–17.12 for a discussion of chemolithotrophic metabolism.

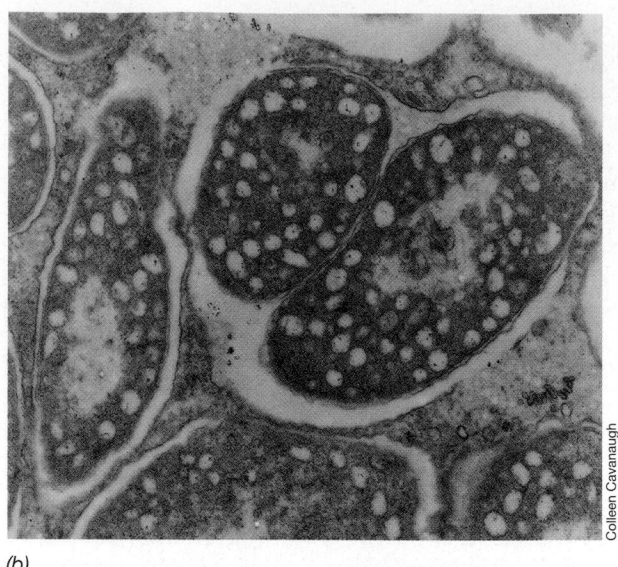

(a)   Colleen Cavanaugh   *(b)*   Colleen Cavanaugh

Figure 19.18 Chemolithotrophic sulfur-oxidizing bacteria associated with the trophosome tissue of tube worms from hydrothermal vents. (a) Scanning electron microscopy of trophosome tissue showing spherical chemolithotrophic sulfur-oxidizing bacteria. Cells are 3–5 µm in diameter. Reprinted with permission from *Science 213*:340–342 (1981), © AAAS. (b) Transmission electron micrograph of bacteria in sectioned trophosome tissue. The cells are frequently enclosed in pairs by an outer membrane of unknown origin.

$^{34}S/^{32}S$ isotope composition to be the same as the sulfide emitted from the vent. This ratio is distinctly different from that of seawater sulfate and serves as additional proof that geothermal sulfide is entering the worm.

A link between animal nutrition and other physiological groups of chemolithotrophs (for example, H_2 oxidizers and nitrifying bacteria) has also been suggested. Methanotrophic symbionts have been shown to play a nutritional role for animals living in symbiotic association with giant clams near natural gas seeps at relatively shallow depths in the Gulf of Mexico (∞ Figure 12.16). Although not truly autotrophs (CH_4 is an organic compound), these symbionts support growth of the animal, in this case by the oxidation of CH_4 as an energy source.

Superheated Water: Black Smokers and Sea Mounts

The great depths of the deep sea create huge hydrostatic pressures that affect the physical properties of water. At a depth of 2600 m, water does not boil until it reaches a temperature of about 450°C. At certain vent sites superheated (but not boiling) hydrothermal fluid is emitted at temperatures of 270–380°C (see Figure 19.16) and could theoretically be a habitat for hyperthermophilic bacteria (∞ Section 6.10 and Chapter 13). The hydrothermal fluid emitted from black smokers contains abundant metal sulfides, especially iron sulfides, and cools quickly as it emerges into cold seawater. The precipitated metal sulfides form a tower called a "chimney" about the source (Figure 19.19). Although prokaryotes

do not live in the superheated hydrothermal fluid itself, thermophilic and hyperthermophilic bacteria live in the seawater or hydrothermal fluid *gradient* that forms as the hot water blends with cold ocean water. For example, the walls of smoker chimneys are teeming with hyperthermophilic prokaryotes such as *Methanopyrus*, a methanogenic archaeon that oxidizes H_2 and is of great evolutionary interest (∞ Chapter 13). Using FISH technology (∞ Section 18.4), the presence of both *Bacteria* and *Archaea* in smoker chimney walls can be readily detected (Figure 19.20).

In addition to black smokers, *sea mounts* are habitats for prokaryotes. These are underwater volcanoes, located primarily in the Pacific Ocean on the Pacific tectonic plate, that spew out ferric minerals and very hot water. Although not as extensively studied as black smoker chimneys, good evidence exists that hyperthermophilic prokaryotes exist in and about these sea mounts and are released into ocean water during eruptions.

Metazoans and Vent Chimneys

Surprisingly, some hydrothermal vent chimneys are also colonized by metazoans, in particular by the small colonial tube worm *Alvinella* (Figure 19.21a). *Alvinella*, also known as the *Pompeii worm*, grows on the outer surfaces of chimneys and, interestingly, is able to tolerate extremely hot water emitted from the vent. Measurements taken in their natural habitat have shown that temperatures inside the worm can get as high as 80°C, making the Pompeii worm the most thermotolerant of all known animals. The surface

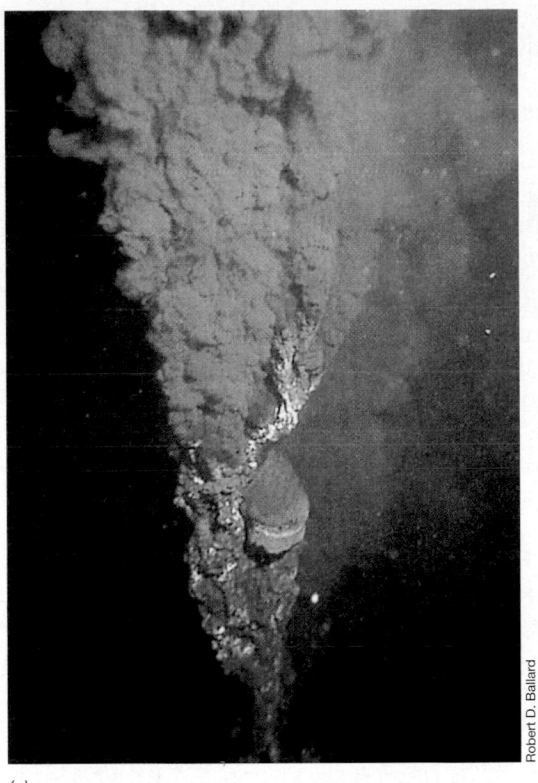

(b)

(a)

Robert D. Ballard

Dudley Foster, Woods Hole Oceanographic Institution

Figure 19.19 Black smokers emitting sulfide- and mineral-rich water at temperatures of 350°C. (a) This chimney is quite large, about 1 m in length. (b) This chimney is much smaller. Note the scientific equipment near the smoker in (b), indicating the relatively small size of the chimney. The walls of the black smoker chimneys display a steep temperature gradient and contain several types of prokaryotes (see Figure 19.20).

of the Pompeii worm is coated with symbiotic filamentous bacteria, possibly chemolithotrophs (Figure 19.21*b*). Although it is not known whether these symbionts nourish the worm like the symbionts that live in the large tube worms and mussels (Figure 19.17), the filamentous bacteria are clearly a food source for the Pompeii worm; scientists have observed the worms extending out of their tubes and feeding directly on the epibionts in the surrounding worm colony.

The mechanism of heat resistance in the Pompeii worm is unknown. But the very existence of such a heat tolerant animal proves that eukaryotic cell structures can withstand far higher temperatures than previously thought. This in turn suggests that more thermophilic eukaryotic microorganisms may exist, although the most thermophilic of all known eukaryotes, a fungus, grows only to 62°C (⊂⊃ Section 6.10).

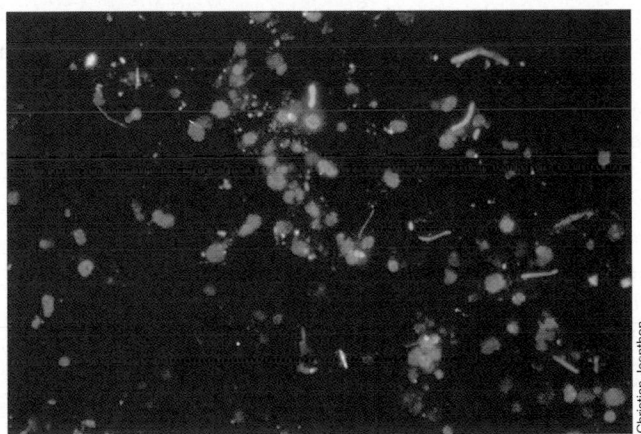

Christian Jeanthon

Figure 19.20 Phylogenetic staining of black smoker chimney material from a smoker at the Snake-Pit vent field of the Mid-Atlantic Ridge (3500 meters deep). A green fluorescing dye was conjugated to a probe that reacts with the 16S rRNA of all *Bacteria* and a red dye to a 16S rRNA probe for *Archaea* (⊂⊃ Sections 11.6 and 18.4). Cell numbers tend to be highest in the outer regions of the chimney wall (closest to 2°C seawater) and lowest in the inner region. The hydrothermal fluid going through the center of the chimney was 300°C.

✓ 19.8 Concept Check

Hydrothermal vents are deep-sea hot springs where volcanic activity generates fluids containing large amounts of inorganic energy sources that can be used by chemolithotrophic bacteria. The latter fix CO_2 into organic carbon, some of which is then used by the deep-sea animals. The deep-sea hydrothermal vents are thus habitats where the primary producers are chemolithotrophic rather than phototrophic.

✓ How does a *warm hydrothermal vent* differ from a *black smoker*, both chemically and physically?

✓ How do giant tube worms receive their nutrition?

✓ What evidence is there that microorganisms can grow at temperatures >100°C?

(a)

(b)

Figure 19.21 The black smoker chimney worm *Alvinella*. (a) A single specimen of *A. pompejana*; the worm is about 6 cm long. (b) Fluorescent photomicrograph of filamentous bacteria growing on the surface of *A. pompejana*. The cells were stained with two different phylogenetic probes, yellow-green, and red, each of which reacts with 16S rRNA from a different group of Proteobacteria (a major division of *Bacteria*, ⟁ Chapter 12). The diameter of the filaments is about 5 μm.

IV THE CARBON AND OXYGEN CYCLES

Global carbon cycling involves the activities of both microorganisms and macroorganisms and is intimately tied to the oxygen cycle. Due to the acceleration of CO_2 inputs by human activities, the carbon cycle recently has become a subject of renewed interest. Scientists seek to better understand the magnitude of carbon reservoirs, the major sinks for CO_2, and the rates of cycling within and between compartments, in order to prevent planetary catastrophes, such as severe global warming. Here we focus on the general principles of microbial carbon cycling with special reference to anoxic processes.

19.9 The Carbon Cycle

On a global basis, carbon is cycled through all Earth's major carbon reservoirs: the atmosphere, the land, the oceans and other aquatic environments, sediments and rocks, and biomass (Figure 19.22). As we will see, the carbon and oxygen cycles are intimately intertwined, since CO_2 fixation by oxygenic phototrophs releases O_2 and much organic matter is oxidized back to CO_2 by aerobic respiration (Figure 19.22).

Carbon Reservoirs
The largest carbon reservoir is present in the sediments and rocks of Earth's crust (Table 19.3), but the turnover time is so long that flux out of this compartment is relatively insignificant on a human scale. From the viewpoint of living organisms, a large amount of organic carbon is found in land plants (but see Section 1.3). This represents the carbon of forests and grasslands and constitutes the major site of photosynthetic CO_2 fixation. However, more carbon is present in dead organic material, called *humus*, than in living organisms. **Humus** is a complex mixture of organic materials. It is derived partly from the protoplasmic constituents of soil microorganisms that have resisted decomposition and partly from resistant plant material. Some humic substances are fairly stable, with a global turnover time of about 40 years, although certain other humic components decompose much more rapidly than this. For example, some humic substances can be electron acceptors for anaerobic respiration (⟁ Section 17.13).

The most rapid means of global transfer of carbon is via the CO_2 of the atmosphere. Carbon dioxide is removed from the atmosphere primarily by photosynthesis of land plants and is returned to the atmosphere by respiration of animals and chemoorganotrophic microorganisms. An analysis of the various processes suggests that the single most important contribution of CO_2 to the atmosphere is via microbial decomposition of dead organic material, including humus. In recent times, however, human activities have added tremendously to the atmospheric CO_2 reservoir. For example, in the past 40 years, CO_2 levels have risen 12%.

Importance of Photosynthesis in the Carbon Cycle
The only major ways in which new organic carbon is synthesized on Earth are via photosynthesis and chemosynthesis (CO_2 fixation by chemolithotrophs); most or-

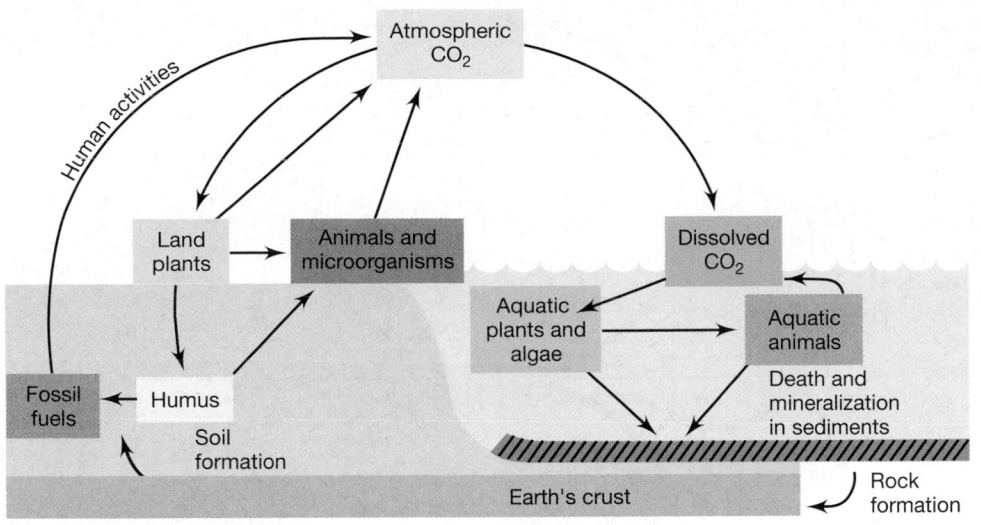

Figure 19.22 The carbon cycle. The carbon and oxygen cycles are closely connected, as oxygenic photosynthesis both removes CO_2 and produces O_2 while respiratory processes both produce CO_2 and remove O_2 (see also Section 19.5). Carbon cycling in the deep subsurface may be as significant as carbon cycling on the Earth's surface (see the box, Microbial Life Deep Underground), but no reliable estimates of this are available.

ganic carbon likely comes from photosynthesis. Phototrophic organisms are therefore the basis of the carbon cycle (Figure 19.22). Phototrophic organisms are found in nature almost exclusively in habitats where light is available. Thus, the deep sea and other permanently dark habitats are devoid of indigenous phototrophs. Oxygenic phototrophic organisms can be divided into two major groups: higher plants and microorganisms. Higher plants are the dominant phototrophic organisms of terrestrial environments, whereas phototrophic microorganisms are the most abundant photosynthesizers of aquatic environments.

The redox cycle for carbon is shown in Figure 19.23. We begin with photosynthesis. The overall equation for oxygenic photosynthesis is

$$CO_2 + H_2O \rightarrow (CH_2O) + O_2$$
light

where (CH_2O) represents organic matter at the oxidation state of cell material such as polysaccharides (the main form in which photosynthesized organic matter is stored in the cell). Phototrophic organisms also carry out respiration, both in the light and the dark. The overall equation for respiration is the reverse of the preceding equation:

$$(CH_2O) + O_2 \rightarrow CO_2 + H_2O$$
light or dark

where (CH_2O) again represents storage polysaccharides. If an organism is to grow (that is, increase in cell number or mass) phototrophically, then the rate of photosynthesis must exceed the rate of respiration. If this occurs, then some of the carbon fixed from CO_2 into polysaccharide can become the starting material for biosynthesis. The whole carbon cycle is built on a net positive balance of the rate of photosynthesis over the rate of respiration.

Decomposition

Photosynthetically fixed carbon is eventually degraded by microorganisms and two major oxidation states of carbon result: methane (CH_4) and carbon dioxide (CO_2) (Figure 19.23). These two gaseous products are formed from the activities of methanogens (CH_4) or from various chemoorganotrophs via fermentation, anaerobic respiration, or aerobic respiration (CO_2). In anoxic habitats CH_4 is produced from both the reduction of CO_2 with H_2 and from certain organic compounds like acetate. However, virtually *any* organic compound can eventually be converted to CH_4 from the combined activities of syntrophic bacteria and methanogens: H_2 generated from the fermentative degradation of organic compounds gets consumed by methanogens (Section 17.17 and see next section). Methane produced in anoxic habitats is highly insoluble and thus is easily transported to oxic environments where it is oxidized to CO_2 by methanotrophs (Figure 19.23). Hence, all organic carbon eventually returns to CO_2 from which autotrophic metabolism once again begins the carbon cycle.

TABLE 19.3	Major carbon reservoirs on Earth	
Reservoir	**Carbon (gigatons)[a]**	**Percent of total carbon on Earth**
Oceans	38×10^3 (>95% is inorganic C)	0.05
Rocks and sediments	75×10^6 (>80% is inorganic C)	>99.5
Terrestrial biosphere	2×10^3	0.003
Aquatic biosphere	1–2	0.000002
Fossil fuels	4.2×10^3	0.006
Methane hydrates	10^4	0.014

[a] One gigaton is 10^9 tons. Data adapted from *Science* 290:291–295 (2000).

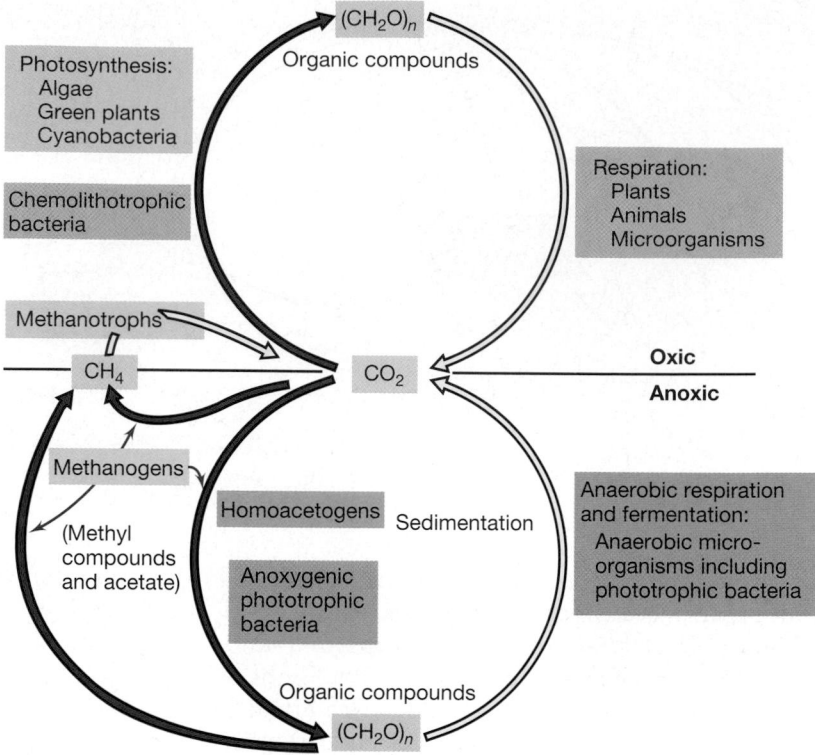

Figure 19.23 Redox cycle for carbon; note in particular the contrasts between autotrophic (CO$_2$ → organic compounds) and heterotrophic processes. Yellow arrows indicate oxidations; red arrows indicate reductions. Photosynthesis in oxic habitats is mainly oxygenic, whereas in anoxic environments it is mainly anoxygenic from the activities of purple and green bacteria. Under anoxic conditions, besides homoacetogens and methanogens, certain sulfate-reducing and nitrate-reducing bacteria are also autotrophic. Methanogens make methane, whereas methanotrophs consume methane.

The balance between the oxidative and reductive portions of the carbon cycle is critical; the products of metabolism of some organisms are the substrates for others. Thus, the cycle needs to keep in balance if it is to continue as it has for many billions of years. Any significant changes in levels of gaseous forms of carbon may have serious global consequences (as we are already experiencing in the form of global warming from the increasing CO$_2$ levels in the atmosphere caused by deforestation and the burning of fossil fuels). In terms of decomposition, CO$_2$ release by microbial activities far exceeds that of eukaryotes, and this is especially true of anoxic environments, which we consider next.

✓ 19.9 Concept Check

The oxygen and carbon cycles are highly interrelated through the complementary activities of autotrophic and heterotrophic organisms. Microbial decomposition is the single largest source of CO$_2$ released to the atmosphere.

- ✓ How is new organic matter made?
- ✓ In what ways are oxygenic photosynthesis and respiration related?

19.10 Syntrophy and Methanogenesis

Methane from biological methanogenesis is of great importance to carbon flow in many anoxic habitats. Methanogenesis is carried out by a group of *Archaea*,

the methanogens, which are strict anaerobes. We discussed the biochemistry of methanogenesis in Section 17.17 and methanogens themselves in Section 13.4. Most methanogens use CO$_2$ as a terminal electron acceptor in anaerobic respiration, reducing it to methane with H$_2$ (Figure 19.23). Only a very few other substrates, acetate being chief among them, can be directly converted to methane by methanogens. Thus, for the conversion of most organic compounds to CH$_4$, methanogens must team up with partner organisms that can supply them with their needed substrates. This is the job of the syntrophs.

In Section 17.21 we discussed the concept of **syntrophy**, a process in which two or more organisms cooperate in the degradation of some compound, and focused on the energetics behind the process. Here we consider the *ecological interactions* of syntrophic bacteria with other organisms and their significance for the whole anoxic carbon cycle. High-molecular-weight substances such as polysaccharides, proteins, and fats, are converted to CH$_4$ and CO$_2$ by the cooperative interaction of several physiological groups of prokaryotes. For the breakdown of a typical polysaccharide such as cellulose, for example (Figure 19.24 and Table 19.4), the process begins with *cellulolytic bacteria*, which cleave the high-molecular-weight cellulose molecule into cellobiose (glucose—glucose) and into free glucose. Glucose is then fermented by *primary fermenters* into a variety of fermentation products, with acetate, propionate, butyrate, succinate, alcohols, H$_2$, and CO$_2$ being the major prod-

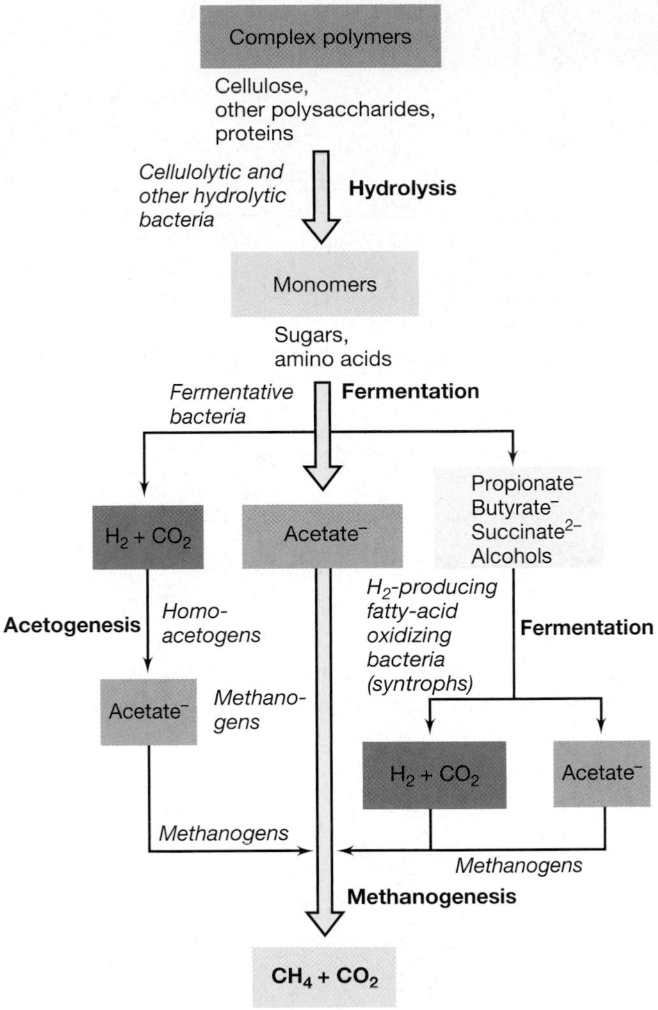

Figure 19.24 Overall process of anoxic decomposition, showing the manner in which various groups of fermentative anaerobes cooperate in the conversion of complex organic materials ultimately to methane (CH_4) and CO_2. Acetate and $H_2 + CO_2$ from primary fermentations can be directly converted to methane, although $H_2 + CO_2$ can also be consumed by homoacetogens. But note how the syntrophs play a key role in anoxic decomposition by consuming highly reduced fermentation products in a secondary fermentation. By activities of the syntrophs, fatty acids and alcohols are converted to the substrates for methanogenesis and acetogenesis. This picture holds for environments in which sulfate-reducing bacteria play only a minor role, for example, in freshwater lake sediments, sewage sludge bioreactors, or the rumen. If alternative electron acceptors are abundant, as for example, sulfate in marine sediments, anaerobic respiration prevails, as syntrophs cannot compete for fatty acids/alcohols with sulfate-reducing bacteria or bacteria carrying out other forms of anaerobic respiration.

ucts. Any H_2 produced in primary fermentation is immediately consumed by *H_2 consumers*, such as methanogens, homoacetogens, or sulfate-reducing bacteria (in environments containing sufficient levels of sulfate). In addition, acetate can be converted to methane by certain methanogens. But this leaves a large amount of carbon in the form of fatty acids and alcohols. The catabolism of these compounds occurs by way of the syntrophs.

Role of the Syntrophs

Key organisms in the conversion of complex organic materials to methane are *secondary fermenters*, especially the H_2-producing fatty acid-oxidizing syntrophic bacteria. For example, *Syntrophomonas wolfei* oxidizes C_4 to C_8 fatty acids yielding acetate, CO_2 (if the fatty acid contained an odd number of carbon atoms), and H_2 (Table 19.4). Other species of *Syntrophomonas* use fatty acids up to C_{18}, including some unsaturated fatty acids. *Syntrophobacter wolinii* specializes in propionate oxidation and generates acetate, CO_2, and H_2, while *Syntrophus gentianae* degrades the aromatic compound benzoate to acetate, H_2, plus CO_2 (Table 19.4). These reactions support luxuriant growth of the syntrophs in coculture with a H_2-consuming partner, but not in pure culture. Why should this be so?

As was explained in Section 17.21, H_2 consumption by the partner organism is crucial for the growth of fatty acid-oxidizing syntrophic bacteria. When the reactions in Table 19.4 are written with all reactants at standard conditions (solutes, 1 *M*; gases, 1 atm), the reactions yield free-energy changes that are positive in sign. That is, the $\Delta G^{0\prime}$ (Section 5.4) of these reactions do not release free energy (Table 19.4). But H_2 consumption by partner bacteria dramatically changes the energetic picture, allowing for sufficient energy conservation in the syntroph to support growth. This can be seen in Table 19.4 where the ΔG values (free-energy change measured under *actual conditions* in the habitat) are favorable for energy conservation if H_2 concentrations are kept very low by the activities of the partner organism.

Thus, through the combined action of primary fermenters, syntrophs, and their H_2-consuming partners,

TABLE 19.4 Major reactions occurring in the anoxic conversion of organic compounds to methane[a]

Reaction type	Reaction	Free-energy change (kJ/reaction)	
		$\Delta G^{0,b}$	ΔG^c
Fermentation of glucose to acetate, H_2, and CO_2	Glucose + 4 $H_2O \rightarrow$ 2 acetate$^-$ + 2 HCO_3^- + 4 H^+ + 4 H_2	−207	−319
Fermentation of glucose to butyrate, CO_2, and H_2	Glucose + 2 $H_2O \rightarrow$ butyrate$^-$ + 2 HCO_3^- + 2 H_2 + 3 H^+	−135	−284
Fermentation of butyrate to acetate and H_2	Butyrate$^-$ + 2 $H_2O \rightarrow$ 2 acetate$^-$ + H^+ + 2 H_2	+48.2	−17.6
Fermentation of propionate to acetate, CO_2, and H_2	Propionate$^-$ + 3 $H_2O \rightarrow$ acetate$^-$ + HCO_3^- + H^+ + H_2	+76.2	−5.5
Fermentation of ethanol to acetate and H_2	2 Ethanol + 2 $H_2O \rightarrow$ 2 acetate$^-$ + 4 H_2 + 2 H^+	+19.4	−37
Fermentation of benzoate to acetate, CO_2, and H_2	Benzoate$^-$ + 6 $H_2O \rightarrow$ 3 acetate$^-$ + 2 H^+ + CO_2 + 3 H_2	+47	−18
Methanogenesis from H_2 + CO_2	4 H_2 + HCO_3^- + $H^+ \rightarrow CH_4$ + 3 H_2O	−136	−3.2
Methanogenesis from acetate	Acetate$^-$ + $H_2O \rightarrow CH_4$ + HCO_3^-	−31	−24.7
Acetogenesis from H_2 + CO_2	4 H_2 + 2 HCO_3^- + $H^+ \rightarrow$ acetate$^-$ + 4 H_2O	−105	−7.1

[a] Data adapted from Zinder, S. 1984. Microbiology of anaerobic conversion of organic wastes to methane: Recent developments. *Am. Soc. Microbiol. News* 50:294–298.

[b] Standard conditions: solutes, 1 M; gases, 1 atm.

[c] Concentrations of reactants in typical anoxic freshwater ecosystem: fatty acids, 1 mM; HCO_3^-, 20 mM; glucose, 10 μM; CH_4, 0.6 atm; H_2, 10^{-4} atm. For calculating ΔG from $\Delta G^{0\prime}$, refer to Appendix 1.

virtually any organic compound can be degraded in anoxic habitats. The final products are CO_2 and CH_4. This includes even saturated hydrocarbons, as we see now.

Anaerobic Metabolism of Hydrocarbons

Although hydrocarbons cannot be fermented, they can be catabolized under anoxic conditions by a variety of anaerobic bacteria. These include denitrifying, iron-reducing, and sulfate-reducing bacteria (∞ Sections 17.13–17.18). In addition, mixtures of anaerobic prokaryotes containing methanogens can degrade hydrocarbons, yielding CH_4 as a final product. Using hexadecane as a model hydrocarbon, it has been shown that CH_4 formation from hydrocarbons begins with acetogenic bacteria (∞ Section 17.16) producing acetate plus H_2 from the hexadecane; methanogens then convert the acetate and H_2 (plus CO_2) to CH_4.

As we saw earlier, hydrocarbon oxidation under oxic conditions proceeds via oxygenase enzymes (∞ Section 17.22). Under anoxic conditions oxygenases obviously cannot function and thus the mechanism of O_2 introduction into the hydrocarbon is of interest to microbial biochemists. Although the mechanism for this is still unclear, it may involve a carboxylation reaction in which the hydrocarbon is converted into a fatty acid:

$$CH_3(CH_2)_{14}CH_3 + \overset{*}{C}O_2 \rightarrow CH_3(CH_2)_{14}CH_2\overset{*}{C}OOH$$

Acetate could be produced from such a molecule by beta oxidation as previously described for aerobic hydrocarbon oxidizers (∞ Section 17.23).

Methane, itself a hydrocarbon, can be oxidized under anoxic conditions in marine sediments by cell aggregates that contain sulfate-reducing bacteria and methanogens

(Figure 19.25*a*). How methane is oxidized by the aggregates is not yet clear, but two alternatives have been suggested. In one, the methanogens oxidize methane to CO_2 by reversing the steps of methanogenesis (∞ Section

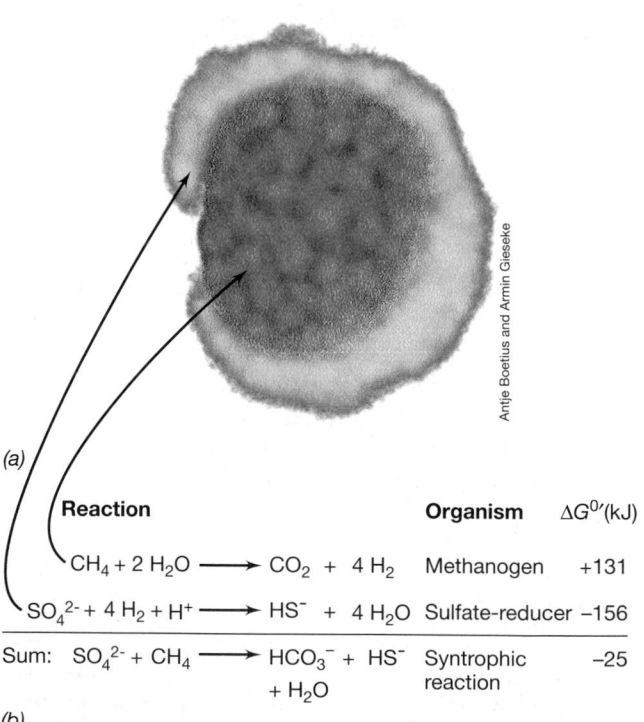

(a)

Reaction			Organism	$\Delta G^{0\prime}$(kJ)
$CH_4 + 2 H_2O$	$\longrightarrow$	$CO_2 + 4 H_2$	Methanogen	+131
$SO_4^{2-} + 4 H_2 + H^+$	$\longrightarrow$	$HS^- + 4 H_2O$	Sulfate-reducer	−156
Sum: $SO_4^{2-} + CH_4$	$\longrightarrow$	$HCO_3^- + HS^- + H_2O$	Syntrophic reaction	−25

(b)

Figure 19.25 Anoxic methane oxidation. (a) Methane-oxidizing cell aggregates from marine sediments. The aggregates contain methanogenic bacteria (red) surrounded by sulfate-reducing bacteria (green). Each cell type is stained by a different phylogenetic FISH stain (∞ Section 18.4). (b) Possible mechanism for syntrophic anoxic methane oxidation in the cell aggregates.

17.17; Figure 19.25*b*). However, this reaction is energetically unfavorable unless the H_2 produced is removed by a second organism, a syntrophic type of metabolism. This is presumably the job of the sulfate-reducer, consuming H_2 in the production of H_2S (Figure 19.25). Alternatively, the methanogenic component may convert CH_4 into acetate with the sulfate-reducing bacterium oxidizing the acetate to CO_2 during sulfate reduction. However this partnership works, anoxic methane oxidation is an interesting syntrophic process whereby two organisms cooperate to carry out a reaction that neither could do alone.

Methanogenic/Acetogenic Habitats

Despite the obligate anaerobiosis and specialized metabolism of methanogens, they are quite widespread on Earth. Although high levels of methanogenesis occur only in anoxic environments, such as swamps and marshes, or in the rumen (see Section 19.11), the process also occurs in habitats that normally might be considered oxic, such as forest and grassland soils. In such habitats methanogenesis occurs in anoxic microenvironments, for example, in the midst of soil crumbs (see Figure 19.2). An overview of the rates of methanogenesis in different kinds of habitats is given in Table 19.5. Note that biogenic production of methane by the methanogenic *Archaea* exceeds the production rate from gas

TABLE 19.5 Estimates of CH_4 released into the atmosphere[a]

Source	CH_4 emission (10^{12} g/year)	
Biogenic		
Ruminants	80–100	
Termites	25–150[b]	
Paddy fields	70–120	
Natural wetlands	120–200	
Landfills	5–70	
Oceans and lakes	1–20	
Tundra	1–5	
Abiogenic		
Coal mining	10–35	
Natural gas flaring and venting	10–30	
Industrial and pipeline losses	15–45	
Biomass burning	10–40	
Methane hydrates	2–4	
Volcanoes	0.5	
Automobiles	0.5	
Total	350–820	
Total biogenic	302–665	81–86% of total
Total abiogenic	48–155	13–19% of total

[a] Data adapted from estimates of Tyler in Tyler, S. C. 1991. The global methane budget, pp. 7–58, in E. J. Rogers and W. B. Whitman (eds.), *Microbial Production and Consumption of Greenhouse Gases: Methane, Nitrogen Oxides, and Halomethanes*, American Society for Microbiology, Washington, DC.

[b] More recent estimates indicate that the lower value is probably the more accurate.

wells and other abiogenic sources. Eructation by ruminants (see Section 19.11) and CH_4 released from termites, paddy fields, and natural wetlands are the largest sources of biogenic methane.

Methanogens have also been found living as endosymbionts of certain protozoa. Several types of protozoa, including free-living aquatic amebas and flagellates found in the insect gut, have been shown to harbor methanogens. In termites, for example, methanogens are present primarily within cells of trichomonal protozoa inhabiting the termite hindgut (Figure 19.26). Methanogenic symbionts of protozoa resemble rod-shaped species of the genus *Methanobacterium* or *Methanobrevibacter* (Section 13.4), but their exact relationship to other methanogens is unclear. In the termite hindgut, endosymbiotic methanogens are thought to benefit their protozoan hosts by consuming H_2 generated from glucose fermentation by cellulolytic protozoans. As shown in Table 16.4, termites can be a major source of biogenic CH_4.

A competing process to methanogenesis is *acetogenesis* (Section 17.16). In some habitats, for example the rumen (see Section 19.11), homoacetogens appear to be poor competitors with methanogens and thus methanogenesis is the dominant H_2-consuming process. But in other habitats, such as the termite hindgut, where some methanogenesis occurs (Table 19.5 and Figure 19.26), acetogenesis is a quantitatively more important process. What governs the dominant form of anaerobic metabolism is not clearly understood; based on simple energetic considerations, methanogenesis from H_2 is a more favorable process than is acetogenesis (−131kJ versus −105kJ, respectively) and thus methanogens should have a competitive advantage. However, in the hindgut of termites acetogenesis remains the dominant process possibly because (1) homoacetogens are in some way able to position themselves in the termite gut nearer to the source of H_2 than can methanogens and thus the homoacetogens can consume the majority of H_2 produced from cellulose fermentation; (2) unlike methanogens, homoacetogens can grow chemoorganotrophically on glucose (from cellulose) as well as by CO_2 reduction by H_2 (Section 17.16); and (3) the nature of the food eaten by termites, that is, wood, contains a high content of lignin mixed in with cellulose, and this type of material may be degraded in such a way that favors acetogenesis.

Methanogenesis versus Sulfidogenesis

Methanogenesis and acetogenesis are more extensive in anoxic freshwater and terrestrial environments than in marine sediments. This is because sulfate-reducing bacteria, which are abundant in marine sediments where sulfate levels are high, can outcompete methanogens and homoacetogens for H_2 produced either by primary

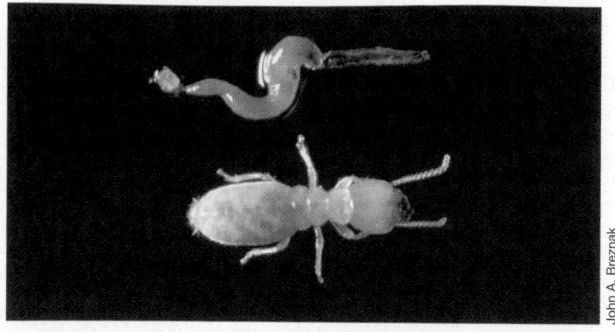

(a)

John A. Breznak

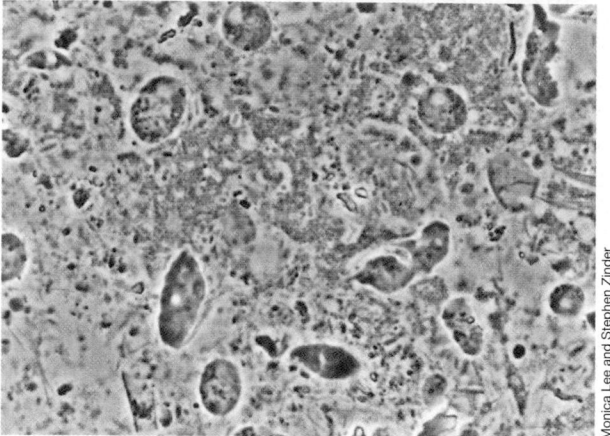

(b)

Monica Lee and Stephen Zinder

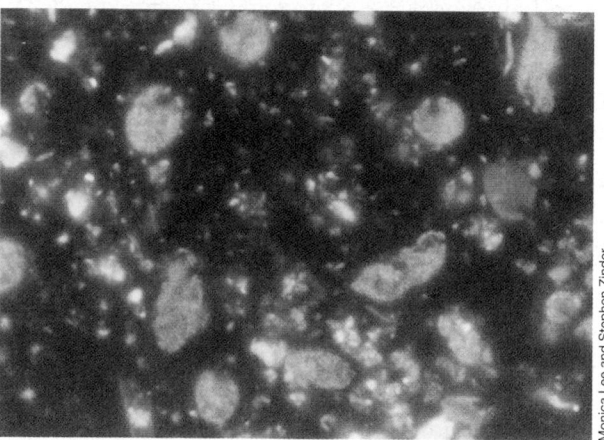

(c)

Monica Lee and Stephen Zinder

Figure 19.26 Termites and their carbon metabolism. (a) A common eastern (USA) subterranean termite worker larva shown beneath a hindgut extracted from a separate worker. The animal is about 0.5 cm long. Acetogenesis is the major form of anoxic carbon metabolism in these termites although methanogenesis also occurs. (b) Microorganisms from the hindgut of the termite *Zootermopsis angusticolis*. A single microscope field was photographed by two different methods. (b) Phase contrast. (c) Epifluorescence, showing color typical of methanogens due to the high content of the fluorescent coenzyme F_{420} (∞ Section 17.17 and Figure 17.42). The methanogens are inside cells of the protozoan *Tricercomitis* sp. Plant particles fluoresce yellow. The average diameter of a protozoan cell is 15–20 μm. See Section 4.1 for an explanation of florescence microscopy.

fermenters or by syntrophs. The biochemical basis for this is complex and has to do with the fact that the energetics of sulfate reduction with H_2 is significantly greater than the reduction of CO_2 with H_2 to either CH_4 or acetate (∞ Sections 17.15–17.17). In freshwater, however, where sulfidogenesis is typically very low because sulfate is limiting, methanogenesis and acetogenesis dominate.

✓ 19.10 Concept Check

Under anoxic conditions, organic matter is degraded principally to CH_4 and CO_2. Much CH_4 is formed from the reduction of CO_2 by H_2 supplied by H_2-producing syntrophic bacteria; these organisms depend on H_2 consumption to balance their energetics. On a global basis, biogenic CH_4 is a much larger source than abiogenic CH_4.

✓ What kinds of organisms can grow in coculture with *Syntrophomonas*?

✓ What is the final product of acetogenesis? What anoxic habitat shows greater acetogenesis than methanogenesis?

✓ Why is methanogenesis from H_2 not a major process in ocean sediments?

✓ How is CH_4 oxidized in the absence of O_2?

19.11 Carbon Cycling in Ruminant Animals

Ruminants are herbivorous mammals that possess a special digestive vessel, the **rumen**, within which the digestion of cellulose and other plant polysaccharides occurs through the activity of microbial populations. Some of the most important domestic animals, the cow, sheep, and goat, are ruminants. Because the human food economy depends to a great extent on these animals, rumen microbiology is of considerable economic significance.

Rumen Anatomy and Action

The bulk of the organic matter in terrestrial plants is present in insoluble polysaccharides, of which *cellulose* is the most important. Mammals, and indeed almost all animals, lack the enzymes necessary to digest cellulose, but all mammals that subsist primarily on grasses and leafy plants can metabolize cellulose by making use of microorganisms as digestive agents. Unique features of the rumen as a site of cellulose digestion are its relatively large size (100–150 liters in a cow, 6 liters in a sheep) and its position in the alimentary tract as the organ where ingested food goes *before* reaching the acidic stomach. The high constant temperature (39°C), constant pH (6.5), and anoxic nature of the rumen are also important factors in overall rumen function. The rumen operates in a more-or-less continuous fashion and in some ways can be considered analogous to a microbial chemostat (∞ Section 6.7).

The relationship of the rumen to other parts of the ruminant digestive system is shown in Figure 19.27. The digestive processes and microbiology of the rumen are well understood, in part because it is possible to create a sampling port, called a *fistula*, into the rumen of a cow (Figure 19.27b) or a sheep and remove samples for analysis. Food first enters the reticulum and is quickly pumped into the rumen where it is mixed with saliva containing bicarbonate and is churned in a rotary motion during which the microbial fermentation occurs. This peristaltic action grinds the cellulose into a fine suspension, which assists in microbial attachment. The food mass then migrates into the reticulum where it is formed into small clumps called *cuds*, which are regurgitated into the mouth and chewed again. The now finely divided solids, well mixed with saliva, are swallowed again, but this time the material passes mainly to the omasum, and from there to the abomasum, an organ more like a true (acidic) stomach. Here chemical digestive processes begin that continue in the small and large intestine.

Microbial Fermentation in the Rumen

Food remains in the rumen about 9–12 h. During this period cellulolytic bacteria and cellulolytic protozoa hydrolyze cellulose to the disaccharide cellobiose and to free glucose units. The released glucose then undergoes a bacterial fermentation with the production of **volatile fatty acids** (VFAs), primarily *acetic, propionic*, and *butyric*, and the gases *carbon dioxide* and *methane* (Figure 19.28). The fatty acids pass through the rumen wall into the bloodstream and are oxidized by the animal as its main source of energy. In addition to their digestive functions, rumen microorganisms synthesize amino acids and vitamins that are the main source of these essential nutrients for the animal. The rumen contains enormous numbers of prokaryotes (10^{10}–10^{11} bacteria/g rumen fluid). Most of the bacteria are adhered tightly to plant materials and feed particles. These materials proceed through the gastrointestinal tract of the animal where they undergo further digestive processes similar to those of nonruminants. Many microbial cells from the rumen are digested in the abomasum and thus are a major source of proteins and vitamins for the animal. Because microbial protein can be recovered, a ruminant is thus nutritionally superior to a nonruminant when subsisting on foods that are deficient in protein, such as grasses.

Rumen Bacteria

The biochemical reactions occurring in the rumen are complex and involve the combined activities of a variety of microorganisms. Because the rumen is anoxic, anaerobic bacteria naturally dominate. Furthermore, because the conversion of cellulose to CO_2 and CH_4 involves a multistep microbial food chain, a variety of anaerobes can be expected (Table 19.6).

Several different rumen bacteria hydrolyze polymers such as cellulose to sugars and ferment the sugars to volatile fatty acids. *Fibrobacter succinogenes* and *Ruminococcus albus* are the two most abundant cellulolytic rumen anaerobes. Although both organisms produce cellulases, *Fibrobacter*, a gram-negative bacterium, contains a periplasmic cellulase (Section 4.9) to break down cellulose; thus, the organism must remain attached to the cellulose fibril while digesting it. *Ruminococcus*, on the other hand, produces a cellulase that is excreted into the rumen contents (thus, it is an *exoenzyme*), where it degrades cellulose outside the bacterial cell proper. However, the end result is the same in both cases: Free glucose is made available for fermentative anaerobes. If a ruminant is gradually switched from cellulose to a diet high in starch (grain, for instance), then starch-digesting

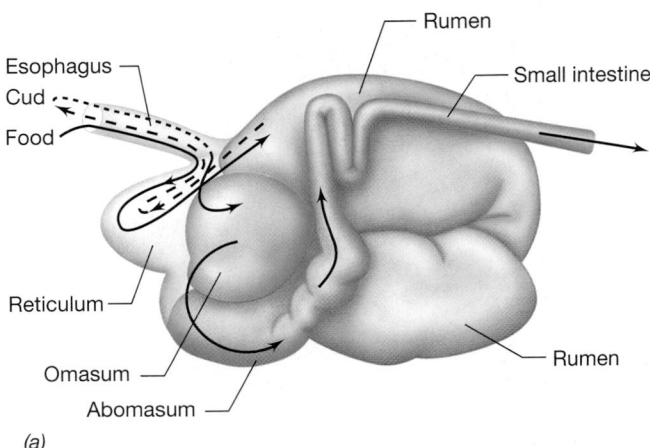

(a)

Sharise D. Beek, Dept. Animal Science. Southern Illinois Univ.

(b)

Figure 19.27 The rumen. (a) Schematic diagram of the rumen and gastrointestinal system of a cow. Food travels from the esophagus to the rumen and is then regurgitated and travels to the reticulum, omasum, abomasum, and intestines, in that order. The abomasum is an acidic vessel, analogous to the stomach of monogastric animals like pigs and humans. (b) Photo of a fistulated Holstein cow. The fistula, shown unplugged, is a sampling port that allows access to the rumen. Fistulated cows and sheep have been very useful for the study of both rumen microbiology and ruminant nutrition.

Figure 19.28 Biochemical reactions in the rumen. The major substrate, glucose, and end products are highlighted; dashed lines indicate minor pathways. Approximate steady state rumen levels of volatile fatty acids (VFAs) are acetate, 60 mM; propionate, 20 mM; butyrate, 10 mM. VFAs are consumed by the ruminant and converted into animal proteins. For this reason, syntrophic fatty acid-oxidizing bacteria (see Section 19.10) are not significant components of the rumen microbual ecosystem.

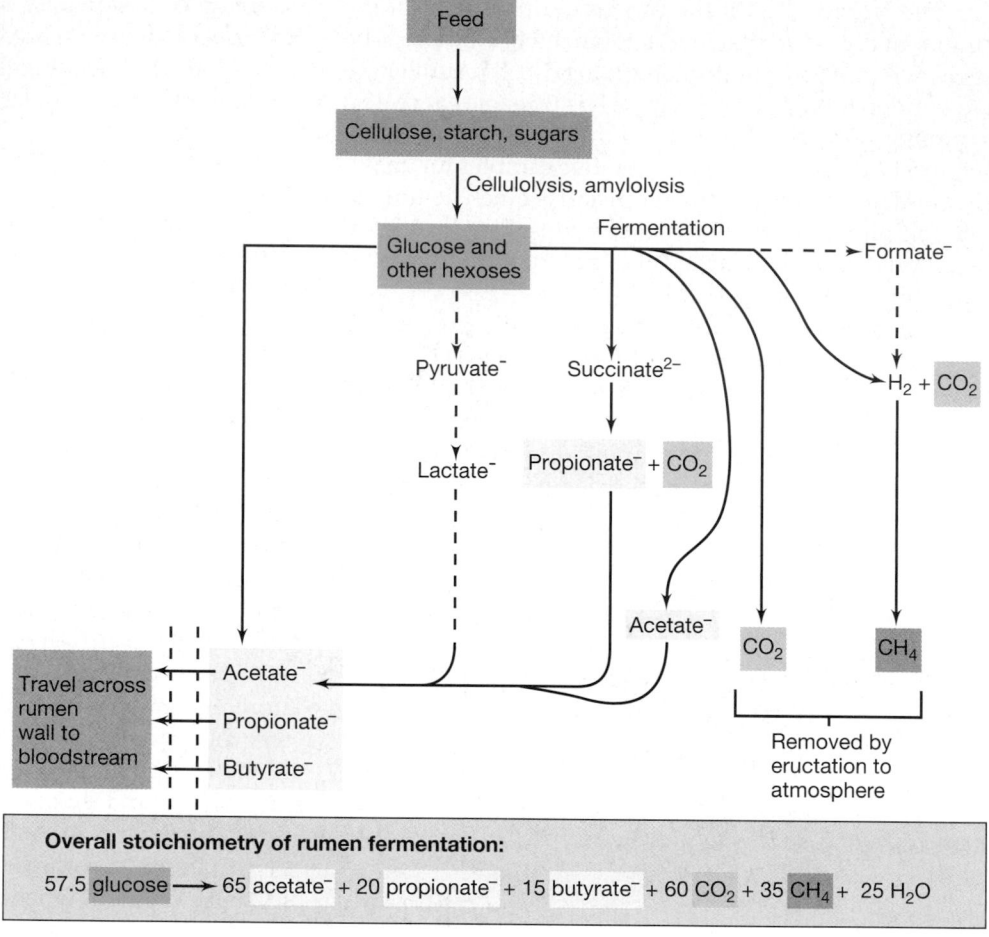

Overall stoichiometry of rumen fermentation:

57.5 glucose ⟶ 65 acetate⁻ + 20 propionate⁻ + 15 butyrate⁻ + 60 CO_2 + 35 CH_4 + 25 H_2O

bacteria such as *Ruminobacter amylophilus* and *Succinomonas amylolytica* develop; on a low starch diet these organisms are in a minority. If an animal is fed legume hay, which is high in pectin, then the pectin-digesting bacterium *Lachnospira multiparus* (Table 19.6) is a common member of the rumen flora.

Some of the fermentation products of the saccharolytic rumen microflora are used as energy sources by other rumen bacteria. Thus, *succinate* is converted to *propionate* and *CO₂* (Figure 19.28) by *Schwartzia*, and *lactate* is fermented to *acetic* and other acids by *Selenomonas* and *Megasphaera* (Table 19.6). Hydrogen produced in the rumen by fermentative processes never accumulates because it is quickly used to reduce CO_2 to CH_4 by methanogens. Another source of H_2 and CO_2 for methanogens is *formate* (Figure 19.28). Large amounts of CH_4 and CO_2 accumulate in the rumen, the average gas composition being about 65% CO_2 and 35% CH_4. These gases leave the ruminant during eructation (belching). Acetate is not converted to CH_4 in the rumen because the retention time is too short for development of acetotrophic methanogens, which typically grow slowly (⟨⟨∞⟩⟩ Section 13.4). In addition, syntrophic fatty acid-degrading bacteria (⟨⟨∞⟩⟩ Section 19.10) do not play a major role in the rumen be-

cause the ruminant itself is the major sink for fatty acids (Figure 19.28). That is, with propionate and butyrate being consumed by the animal (Figure 19.28), syntrophic conversion processes (Table 19.4 and Figure 19.24) are unnecessary in the rumen.

Rumen Protozoa and Fungi

In addition to prokaryotes, the rumen has a characteristic protozoal fauna (about 10^6/ml), composed almost exclusively of ciliates (⟨⟨∞⟩⟩ Section 14.8). Many of these protozoa are obligate anaerobes, a property that is rare among eukaryotes. Although protozoa are not essential for the rumen fermentation, they definitely contribute to the overall process. Some are able to hydrolyze cellulose and starch and ferment glucose with the production of the same organic acids formed by the bacteria (Table 19.6). Rumen protozoa also ingest rumen bacteria as food sources and are thought to play a role in controlling bacterial densities in the rumen.

Anaerobic fungi also inhabit the rumen and are known to play a role in ruminal digestive processes. Rumen fungi are generally species that alternate between a flagellated and a thallus form, and studies with pure cultures show that they can ferment cellulose to

Table 19.6   Characteristics of some rumen prokaryotes

Organism	Gram stain	Phylogenetic domain[a]	Morphology	Motility	Fermentation products	DNA (mol % GC)
Cellulose decomposers						
Fibrobacter succinogenes[b]	Negative	B	Rod	−	Succinate, acetate, formate	45–51
Butyrivibrio fibrisolvens[c]	Negative	B	Curved rod	+	Acetate, formate, lactate, butyrate, H_2, CO_2	41
Ruminococcus albus[b]	Positive	B	Coccus	−	Acetate, formate, H_2, CO_2	43–46
Clostridium lochheadii	Positive	B	Rod (spores)	+	Acetate, formate, butyrate, H_2, CO_2	—
Starch decomposers						
Prevotella ruminicola	Negative	B	Rod	−	Formate, acetate, succinate	40–42
Ruminobacter amylophilus	Negative	B	Rod	−	Formate, acetate, succinate	49
Selenomonas ruminantium	Negative	B	Curved rod	+	Acetate, propionate, lactate	49
Succinomonas amylolytica	Negative	B	Oval	+	Acetate, propionate, succinate	—
Streptococcus bovis	Positive	B	Coccus	−	Lactate	37–39
Lactate decomposers						
Selenomonas lactilytica	Negative	B	Curved rod	+	Acetate, succinate	50
Megasphaera elsdenii	Positive	B	Coccus	−	Acetate, propionate, butyrate, valerate, caproate, H_2, CO_2	54
Succinate decomposer						
Schwartzia succinovorans	Negative	B	Rod	+	Propionate, CO_2	46
Pectin decomposer						
Lachnospira multiparus	Positive	B	Curved rod	+	Acetate, formate, lactate, H_2, CO_2	—
Methanogens						
Methanobrevibacter ruminantium	Positive	A	Rod	−	CH_4 (from H_2 + CO_2 or formate)	31
Methanomicrobium mobile	Negative	A	Rod	+	CH_4 (from H_2 + CO_2 or formate)	49

[a] B, *Bacteria*; A, *Archaea*

[b] These species also degrade xylan, a major plant cell wall polysaccharide (⟳ Section 17.25).

[c] Also degrades starch

VFAs. *Neocalimastix*, for example, is an obligately anaerobic fungus that ferments glucose to formate, acetate, lactate, ethanol, CO_2 and H_2. Although a eukaryote, this fungus lacks mitochondria and cytochromes and thus lives an obligately fermentative existence. However, *Neocalimastix* cells do contain a redox organelle called the *hydrogenosome* that functions to evolve H_2 and has thus far only been found in evolutionarily primitive *Eukarya* (⟳ Sections 14.2, 14.4, and 14.8). Rumen fungi play an important role in the degradation of polysaccharides other than cellulose as well, including a partial degradation of lignin (the strengthening agent in the cell walls of woody plants), hemicellulose, and pectins.

Dynamics of the Rumen Ecosystem

A major feature of the rumen ecosystem is its *constancy*. Studies on various ruminant species in different parts of the world show that most animals contain the same major rumen bacterial species, with the proportions of each species varying somewhat with diet. In addition, the nature and proportions of the volatile fatty acids produced (Figure 19.28) and the levels of rumen CO_2 and CH_4 are relatively constant among different ruminant species.

Occasionally, changes in the microbial composition of the rumen cause illness or even death of the animal. For example, if a cow is changed abruptly from forage to a completely grain diet, an explosive growth of *Streptococcus bovis* is observed in the rumen; the normal level of *S. bovis*, about 10^7 cells/g (insignificant in terms of total rumen bacterial numbers), quickly expands to over 10^{10} cells/g. This occurs because *S. bovis* grows rapidly on starch and grain contains high levels of starch, whereas grasses contain mainly cellulose. Being a lactic acid bacterium (⟳ Section 12.19), *S. bovis* produces large amounts of lactate from the fermentation of starch, and this acidifies the rumen (a condition called *acidosis*), killing off the normal rumen flora. Severe acidosis can cause death of the animal. To avoid acidosis, animals are switched from forage rations to grain *gradually* over a period of a few days. A slow introduction of starch selects for volatile fatty acid–producing starch degraders (Table 19.6) instead of *S. bovis*, and thus normal rumen biochemical processes are not disrupted.

Other starch-fermenting bacteria secrete complex polysaccharides that tend to trap gas bubbles in the rumen, leading to a condition called "feedlot bloat." Severe bloat can compress the lungs and actually kill the animal if veterinary intervention is not immediate.

Ruminant Feeding and Foodborne Illness

In recent years, safety concerns have arisen over feeding practices with ruminant animals, especially cattle, and their impact on the animal's microflora. For economic reasons humans have drastically altered the diet of ruminant animals through the years from one of forage, the diet the rumen system evolved to handle, to more starch-based grain diets. Grain promotes rapid weight gain and tends to make meat from the carcass of the animal more tender. Thus, grain-fed animals get to market sooner, and their meat has more value. However, a high-starch diet not only affects the *rumen* microflora, as we have seen, but it also affects the microbial composition of the intestinal tract *downstream* of the rumen (Figure 19.27a). In particular, the pH of the intestinal tract is more acidic in grain-fed animals. This condition allows acid-tolerant enteropathogenic strains of *Escherichia coli*, like *E. coli* strain O157:H7 (⚬⚬ Section 29.7), to predominate over less acid-tolerant nonpathogenic strains. When grain-fed animals are slaughtered, these *E. coli* cells can adhere to the carcass and be introduced into meat products, particularly ground meat products like ground beef and beef sausage, thus increasing the risk of foodborne illness (⚬⚬ Section 29.7).

Other Herbivorous Animals: Cecal Animals

The familiar ruminants are cows and sheep. However, goats, camels, buffalo, deer, reindeer, caribou, and elk are also ruminants. Horses and rabbits are also herbivorous mammals, but they are not ruminants. Instead, these animals have only one stomach but use an organ called the **cecum**, a digestive organ located posterior to the large intestine (just before the anus), as their cellulolytic fermentation vessel. The cecum contains a cellulolytic microflora, and digestion of cellulose occurs there. The precise microflora of the cecum is not well understood, but the species involved are not thought to be the same as those in the rumen. Nutritionally, ruminants have an advantage over horses and rabbits in that the cellulolytic microflora of the ruminant eventually passes through a true (acidic) stomach and as such is killed and is a protein source for the animal. By contrast, in horses and rabbits the cellulolytic microflora is passed out of the animal in the feces.

✓ 19.11 Concept Check

Ruminants are animals that have a special digestive organ, the rumen, that is a unique ecosystem in which anaerobic microorganisms digest insoluble feed components such as cellulose and starch. Bacteria, protozoa, and fungi of the rumen produce volatile fatty acids that are used by the ruminant. In addition to their role in the digestive process, rumen microorganisms synthesize vitamins and amino acids used by the ruminant.

✓ What physical and chemical conditions prevail in the rumen?

✓ What are *VFAs* and of what value are they to the ruminant?

✓ Why is the metabolism of *Streptococcus bovis* of special concern to ruminant nutrition? How can foodborne illness in humans be due to ruminant feeding practices?

✓ How do cecal animals differ from ruminants in their digestive tract anatomy?

V OTHER KEY NUTRIENT CYCLES

Although on a quantitative basis carbon is the major cycled element in nature, many other elements important in the metabolism of plants and animals are cycled by microorganisms. These include nitrogen, sulfur, and iron, and we examine the cycling of these key nutrients now.

19.12 The Nitrogen Cycle

The element nitrogen, N, a key constituent of protoplasm, exists in a number of oxidation states (⚬⚬ Table 17.2). We discussed two major processes of microbial nitrogen transformation in Chapter 17: *nitrification* in Section 17.12 and *denitrification* in Section 17.14. These and several other nitrogen transformations are summarized in the redox cycle shown in Figure 19.29.

Nitrogen Fixation

Several of the key redox reactions of nitrogen are carried out in nature almost exclusively by microorganisms, and so microbial involvement in the nitrogen cycle is of great importance (Figure 19.29). Nitrogen gas, N_2, is the most stable form of nitrogen, and this explains why a major reservoir for nitrogen on Earth is the atmosphere. This is in contrast to carbon, for which the atmosphere is a minor reservoir (CO_2, CH_4). The high energy necessary to break the $N{\equiv}N$ bond of molecular nitrogen (⚬⚬ Section 17.28) means that the reduction of N_2 is an energy-demanding process. Only a relatively small number of organisms are able to use N_2 in the process called **nitrogen fixation** ($N_2 + 8 H^+ + 8 e^- \rightarrow 2 NH_3 + H_2$); thus, the recycling of nitrogen on Earth involves to a great extent the more easily available forms, ammonia and nitrate. However, because N_2 constitutes by far the greatest reservoir of nitrogen available to living organisms, the ability to use N_2 is of great ecological impor-

Key Processes and Prokaryotes in the Nitrogen Cycle

Processes	Example organisms
Nitrification ($NH_4^+ \rightarrow NO_3^-$)	
$NH_4^+ \rightarrow NO_2^-$	*Nitrosomonas*
$NO_2^- \rightarrow NO_3^-$	*Nitrobacter*
Denitrification ($NO_3^- \rightarrow N_2$)	*Bacillus, Paracoccus, Pseudomonas*
N_2 Fixation ($N_2 + 8H \rightarrow NH_3 + H_2$)	
Free-living	
Aerobic	*Azotobacter* Cyanobacteria
Anaerobic	*Clostridium*, purple and green bacteria
Symbiotic	*Rhizobium Bradyrhizobium Frankia*
Ammonification (organic-N $\rightarrow NH_4^+$)	
	Many organisms can do this

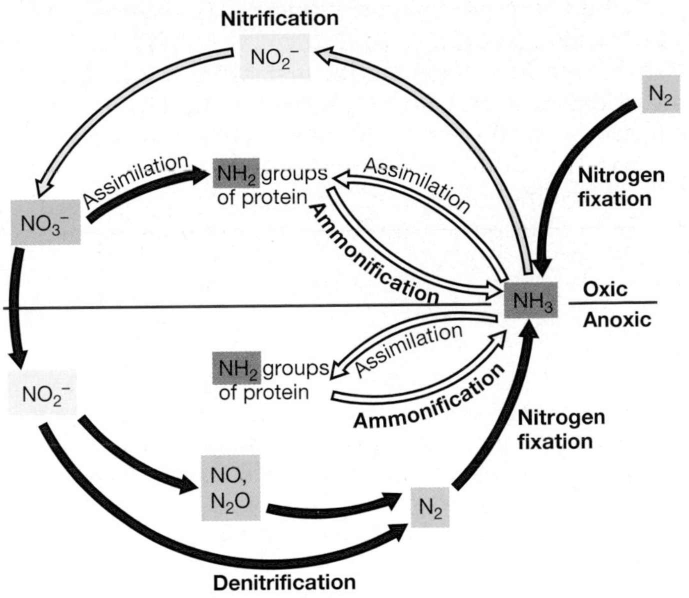

Figure 19.29 Redox cycle for nitrogen. Oxidation reactions are represented by yellow arrows and reductions in red.

tance. In many environments, productivity is limited by the short supply of combined nitrogen compounds, putting a premium on biological nitrogen fixation. We considered the basic process of N_2 fixation in Section 17.28 and revisit this important process in an agricultural context later in this chapter (Section 19.22) when we describe symbiotic N_2 fixation in leguminous plants.

Denitrification

We discussed the role of nitrate as an alternative electron acceptor in Section 17.14. Under most conditions, the end product of nitrate reduction is N_2 or N_2O, and the conversion of nitrate to gaseous nitrogen compounds is called **denitrification** (Figure 19.29). This process is the main means by which gaseous N_2 is formed biologically, and because N_2 is much less easily used by organisms than nitrate as a source of nitrogen, denitrification is a detrimental process because it removes fixed nitrogen from the environment. This can be a serious agricultural problem if fields fertilized with nitrate fertilizer become waterlogged following heavy spring rains. Anoxic conditions rapidly develop and denitrification can be extensive. By contrast, denitrification can be a very beneficial process in wastewater treatment (∞ Section 28.2), where nitrate can be removed from the water, thus minimizing algal growth when the water is discharged into lakes and streams.

Ammonia Fluxes and Nitrification

Ammonia is produced during the decomposition of organic nitrogen compounds (**ammonification**, Figure 19.29) such as amino acids and nucleotides, and exists at neutral pH as the ammonium ion (NH_4^+). Under anoxic conditions, ammonia is relatively stable (but see Section 17.12), and it is in this form that nitrogen predominates in most anoxic sediments. In soils, much of the ammonia released by aerobic decomposition is rapidly recycled and converted to amino acids in plants and microorganisms. Because ammonia is volatile, some loss can occur from soils (especially highly alkaline soils) by vaporization, and major losses of ammonia to the atmosphere occur in areas of dense animal populations (for example, cattle feedlots). However, on a global basis, ammonia constitutes only about 15% of the nitrogen released to the atmosphere, the rest being primarily in the form of N_2 or N_2O (from denitrification).

Nitrification, the oxidation of NH_3 to NO_3^-, is a major process in nature and occurs readily in well-drained soils at neutral pH by activities of the nitrifying bacteria (∞ Sections 12.3 and 17.12) (Figure 19.29). Note that while denitrification is the process of nitrate *consumption*, nitrification is the process of nitrate *production*. If materials high in protein, such as manure or sewage, are added to soils, the rate of nitrification is increased. Although nitrate is readily assimilated by plants, it is very water-soluble and is rapidly leached from soils receiving high rainfall. Consequently, nitrification is not beneficial in agricultural practice. Ammonia, on the other hand, is positively charged and consequently is strongly adsorbed to negatively charged clay minerals.

Anhydrous ammonia is used extensively as a nitrogen fertilizer, and chemicals are commonly added to the fertilizer to inhibit the nitrification process. One of the most common inhibitors of nitrification is a substituted pyridine compound called *nitrapyrin* (2-chloro-6-tri-chloromethylpyridine, also known as N-SERVE). Nitrapyrin specifically inhibits the first step in nitrification, the oxidation of NH_3 to NO_2^- (∞ Section 17.12), thus effectively inhibiting both steps in the nitrification process. The addition of nitrification inhibitors has greatly increased the efficiency of fertilization and has helped prevent pollution of waterways from nitrate leached from fertilized soils.

✓ 19.12 Concept Check

The principal form of nitrogen on Earth is nitrogen gas (N_2), which can be used as a nitrogen source only by the nitrogen-fixing bacteria. Ammonia produced by nitrogen fixation or by ammonification from organic nitrogen compounds can be assimilated into organic matter or it can be oxidized to nitrate by the nitrifying bacteria. Losses of nitrogen from the biosphere occur as a result of denitrification, in which nitrate is converted back to N_2.

✓ What is *nitrogen fixation* and why is it important?

✓ What is the process called that results in $NO_3^- \rightarrow N_2$?

✓ How does the compound *nitrapyrin* benefit both agriculture and the environment?

✓ How do the processes of *nitrification* and *denitrification* differ?

19.13 The Sulfur Cycle

Sulfur transformations are even more complex than those of nitrogen because of the variety of oxidation states of sulfur (Figure 19.30; ∞ Table 17.3) and the fact that some transformations occur at significant rates *chemically* as well as biologically. Sulfate reduction and chemolithotrophic sulfur oxidation were covered in Sections 17.15 and 17.10, respectively. The redox cycle for sulfur and the involvement of microorganisms in sulfur transformations are given in Figure 19.30. Although a number of oxidation states are possible, only three form significant amounts of sulfur in nature, −2 (sulfhydryl, R—SH, and sulfide, HS^-), 0 (elemental sulfur, S^0), and +6 (sulfate, SO_4^{2-}). The bulk of the sulfur of Earth is found in sediments and rocks in the form of sulfate minerals (primarily gypsum, $CaSO_4$) and sulfide minerals (primarily pyrite, FeS_2), although the oceans constitute the most significant reservoir of sulfur for the biosphere (in the form of inorganic sulfate). The global transport cycle for sulfur is given in Figure 19.31, and some of the components of this cycle are discussed later.

Hydrogen Sulfide and Sulfate Reduction

A major volatile sulfur gas is hydrogen sulfide (H_2S). This substance is produced from bacterial sulfate reduction ($SO_4^{2-} + 8 H \rightarrow H_2S + 2 H_2O + 2 OH^-$) (Figure 19.30) or is emitted from geochemical sources in sulfide springs and volcanoes. Although H_2S is volatile, the form of sulfide present in an environment is pH dependent; H_2S predominates below pH 7 while HS^- and S^{2-} are present above pH 7. Sulfate-reducing bacteria are widespread in nature, however in many anoxic habitats, such as freshwaters and many soils, their activities are limited by the low levels of sulfate present. And, because of the necessity for organic electron donors (or molecular hydrogen, which is a product of the fermen-

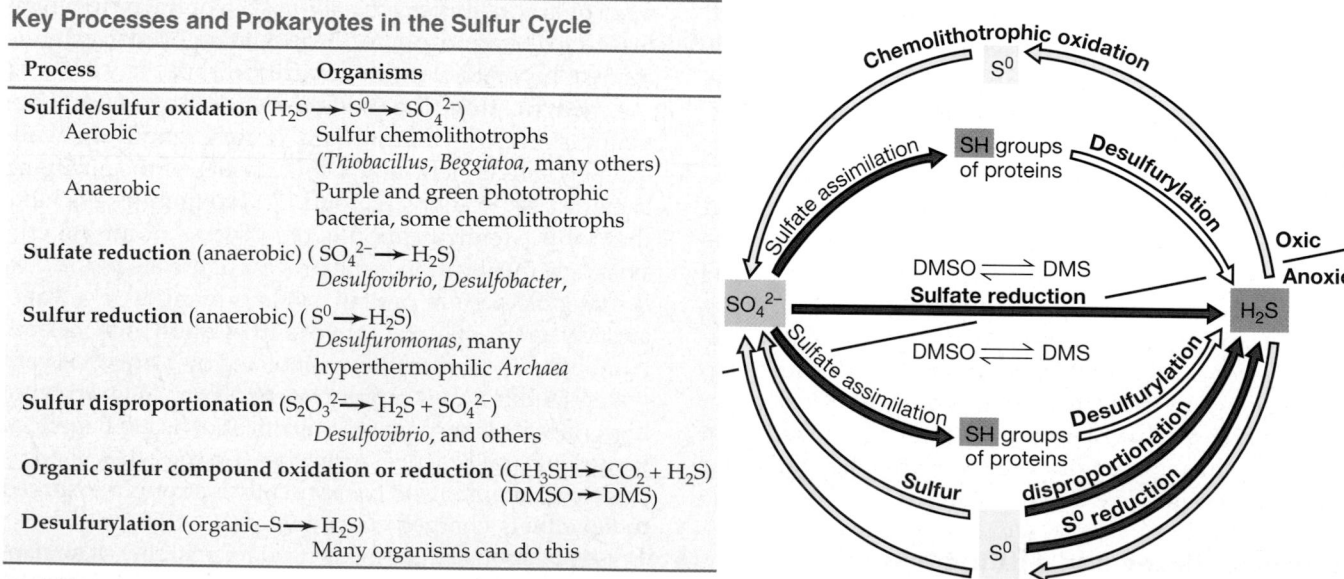

Key Processes and Prokaryotes in the Sulfur Cycle

Process	Organisms
Sulfide/sulfur oxidation ($H_2S \rightarrow S^0 \rightarrow SO_4^{2-}$)	
Aerobic	Sulfur chemolithotrophs (*Thiobacillus, Beggiatoa*, many others)
Anaerobic	Purple and green phototrophic bacteria, some chemolithotrophs
Sulfate reduction (anaerobic) ($SO_4^{2-} \rightarrow H_2S$)	*Desulfovibrio, Desulfobacter,*
Sulfur reduction (anaerobic) ($S^0 \rightarrow H_2S$)	*Desulfuromonas,* many hyperthermophilic *Archaea*
Sulfur disproportionation ($S_2O_3^{2-} \rightarrow H_2S + SO_4^{2-}$)	*Desulfovibrio,* and others
Organic sulfur compound oxidation or reduction ($CH_3SH \rightarrow CO_2 + H_2S$) ($DMSO \rightarrow DMS$)	
Desulfurylation (organic–$S \rightarrow H_2S$)	Many organisms can do this

Figure 19.30 Redox cycle for sulfur. Oxidations are shown in yellow arrows and reductions in red. DMSO, dimethylsulfoxide; DMS, dimethylsulfide.

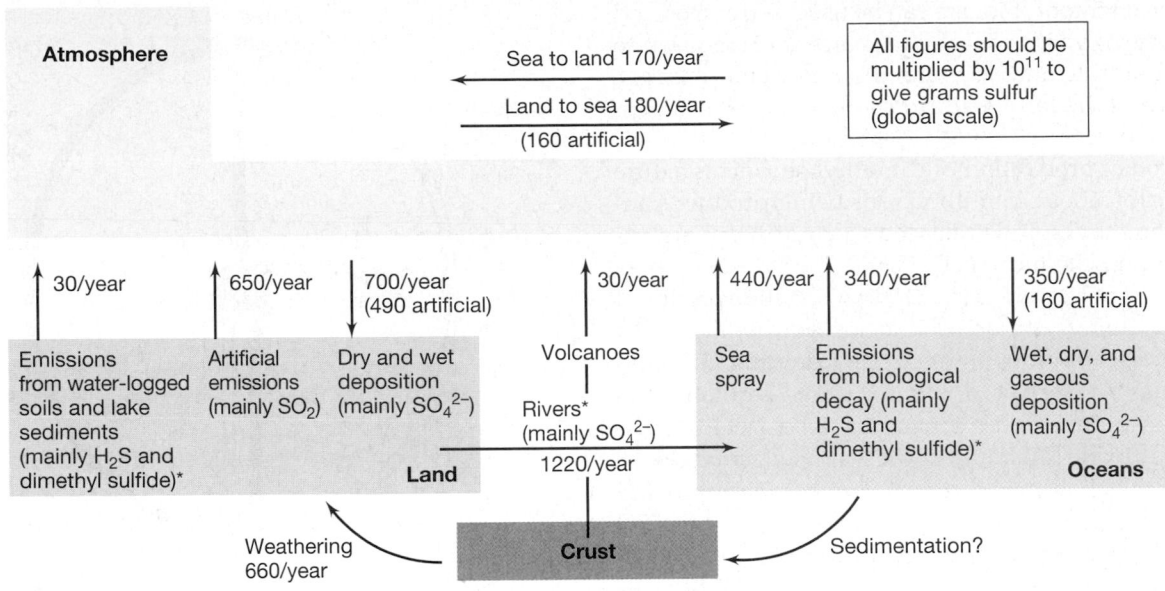

Figure 19.31 The global balance of sulfur. Artificial emissions are derived from human activities. An asterisk indicates a process that is partially or solely due to microbial action.

tation of organic compounds) to drive sulfate reduction, sulfide production only occurs where significant amounts of organic material are present. In many marine sediments, the rate of sulfate reduction is carbon-limited, and the rate can be greatly increased by the addition of organic matter. This is important because disposal of sewage, sewage sludge, and garbage in the sea can lead to marked increases in organic matter in the sediments, leading to marine pollution. Since sulfide (HS^-) is a toxic substance to many organisms, formation of HS^- by sulfate reduction is potentially detrimental. Sulfide is toxic because it combines with the iron of cytochromes and other essential iron-containing compounds in the cell. Sulfide is commonly detoxified in the environment by combination with iron, forming the insoluble FeS. The black color of many sediments where sulfate reduction is taking place is due to the accumulation of FeS.

Sulfide and Elemental Sulfur Oxidation/Reduction

Under oxic conditions, sulfide (HS^-) rapidly oxidizes spontaneously at neutral pH (∞ Section 17.10). Sulfur-oxidizing bacteria, most of which are aerobes, can catalyze the oxidation of sulfide, but because of the rapid spontaneous reaction, significant bacterial oxidation of sulfide occurs only in areas in which H_2S rising from anoxic areas meets O_2 descending from oxic areas. In addition, if light is available, anoxic oxidation of HS^- can also occur, catalyzed by the phototrophic sulfur bacteria (∞ Sections 12.2, 12.33, and 17.4).

Elemental sulfur, S^0, is chemically stable but is readily oxidized by sulfur-oxidizing bacteria, such as *Thio-*

bacillus (∞ Sections 12.4 and 17.10). Elemental sulfur is insoluble, and the bacteria that oxidize it attach firmly to the sulfur crystals (∞ Figure 17.26b). Oxidation of elemental sulfur results in the formation of sulfate (SO_4^{2-}) and hydrogen ions, and sulfur oxidation characteristically results in a *lowering* of the pH. Elemental sulfur is sometimes added to alkaline soils to effect a lowering of the pH, reliance being placed on the ubiquitous thiobacilli to carry out the acidification process.

S^0 can be reduced as well as oxidized. Sulfur reduction to sulfide (a form of anaerobic respiration, ∞ Section 17.15) is a major ecological process, especially among hyperthermophilic *Archaea* (∞ Chapter 13). Although sulfate-reducing bacteria can also carry out this reaction, the bulk of S^0 reduction in nature probably occurs by the phylogenetically distinct S^0 reducers that are unable to reduce SO_4^{2-} to H_2S. However, the habitats of the S^0 reducers are generally those of the sulfate reducers, so from an ecological standpoint, the two groups coexist.

Organic Sulfur Compounds

In addition to the *inorganic* forms of sulfur whose biogeochemistry was just discussed, a vast array of *organic* sulfur compounds are also synthesized by living organisms, and these enter into biogeochemical sulfur cycling as well (Figure 19.31). Many of these foul-smelling compounds are highly volatile and can thus enter the atmosphere. The most abundant organic sulfur compound in nature is dimethyl sulfide ($H_3C—S—CH_3$). It is produced primarily in marine environments as a degradation product of dimethylsulfoniopropionate, a major osmoregulatory solute in marine algae (∞ Section 6.12).

Dimethylsulfoniopropionate can be used as a carbon and energy source by microorganisms and is catabolized to dimethyl sulfide and acrylate; the latter compound, a derivative of the fatty acid propionate, is used to support growth.

Microbial production of dimethyl sulfide in nature is significant, about 45 million tons being produced annually. Dimethyl sulfide released to the atmosphere undergoes photochemical oxidation to methane sulfonate ($CH_3SO_3^-$), SO_2, and SO_4^{2-}. Dimethyl sulfide produced in anoxic habitats can be used as a substrate in at least three ways: (1) in methanogenesis (yielding CH_4 and H_2S), (2) as an electron donor for photosynthetic CO_2 fixation in phototrophic purple bacteria [yielding dimethyl sulfoxide (DMSO)], and (3) as an electron donor in energy metabolism in certain chemoorganotrophs and chemolithotrophs (also yielding DMSO). The DMSO produced can be an electron acceptor for anaerobic respiration (∞ Section 17.18), once again yielding dimethyl sulfide. Many other organic sulfur compounds impact the global sulfur cycle, including methanethiol (CH_3SH), dimethyl disulfide ($H_3C—S—S—CH_3$), and carbon disulfide (CS_2), but on a global basis, dimethyl sulfide is the most significant (Figure 19.31).

✓ 19.13 Concept Check

Bacteria play major roles in both the oxidative and reductive sides of the sulfur cycle. Sulfur- and sulfide-oxidizing bacteria *produce* sulfate, while sulfate-reducing bacteria *use* sulfate as electron acceptor in anaerobic respiration, producing hydrogen sulfide. Because sulfide is toxic and also reacts with various metals, sulfate reduction is an important biogeochemical process. Dimethyl sulfide is the major organic sulfur compound of ecological significance in nature.

- ✓ Is H_2S a *substrate* or a *product* of the sulfate-reducing bacteria?
- ✓ Why is acid generated from the bacterial oxidation of sulfur?
- ✓ What organic sulfur compound is most abundant in nature?

19.14 The Iron Cycle

Iron is one of the most abundant elements in Earth's crust. On the surface of the Earth, iron exists naturally in two oxidation states, ferrous (Fe^{2+}) and ferric (Fe^{3+}). Fe^0 is a major product of human activities in the smelting of ferrous or ferric iron ores to form cast iron. In nature then, iron cycles primarily between the ferrous and ferric forms, the reduction of Fe^{3+} occurring both chemically and as a form of anaerobic respiration, and the oxidation of Fe^{2+} occurring both chemically and as a form of chemolithotrophic metabolism (Figure 19.32).

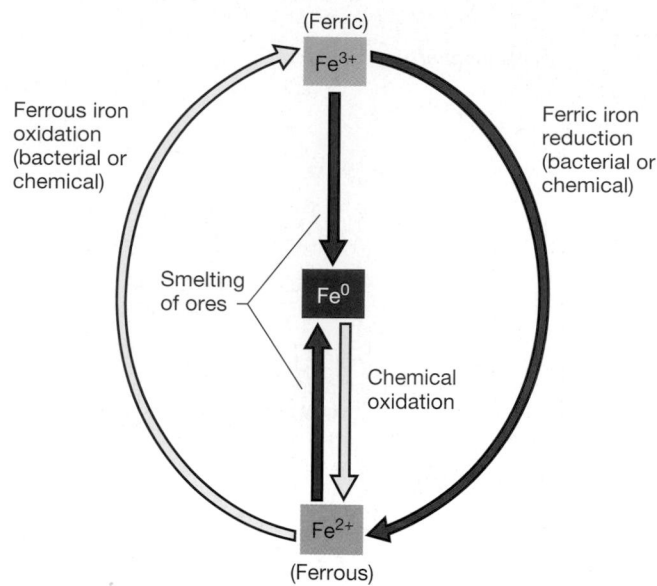

Figure 19.32 The redox cycle of iron. The major forms of iron in nature are Fe^{2+} and Fe^{3+}; Fe^0 is primarily a product of human activities in the smelting of iron ores. Ferrous iron oxidation occurs aerobically by the iron chemolithotrophs (or chemically at neutral pH) and anaerobically by certain anoxygenic phototrophic bacteria and denitrifying bacteria. Oxidations are shown with yellow arrows and reductions with red.

The only electron acceptor able to spontaneously oxidize Fe^{2+} is O_2; Section 17.11 discusses how this can occur at neutral pH yielding highly insoluble ferric iron precipitates. By contrast, at low pH, Fe^{2+} is stable, and this allows the prolific growth of acidophilic iron-oxidizing bacteria.

Bacterial Iron Reduction

A number of organisms can use ferric iron as an electron acceptor. Ferric iron reduction is common in waterlogged soils, bogs, and anoxic lake sediments. Movement of iron-rich groundwater from anoxic bogs or waterlogged soils can result in the transport of large amounts of ferrous iron. Once this iron-laden water reaches oxic regions, the ferrous iron is oxidized chemically or by iron bacteria (∞ Section 17.11) and ferric compounds precipitate, leading to the formation of brown iron deposits (∞ Figure 17.28c). The overall reaction of spontaneous ferrous iron oxidation is:

$$Fe^{2+} + \tfrac{1}{4} O_2 + 2\tfrac{1}{2} H_2O \rightarrow Fe(OH)_3 + 2 H^+$$

The ferric hydroxide precipitate can interact with other nonbiological substances, such as humics (see Section 19.9), to reduce Fe^{3+} back to Fe^{2+} (Figure 19.32). Ferric iron can also form complexes with various organic constituents, thus becoming solubilized and more available to ferric iron-reducing bacteria as an electron acceptor (∞ Section 17.18).

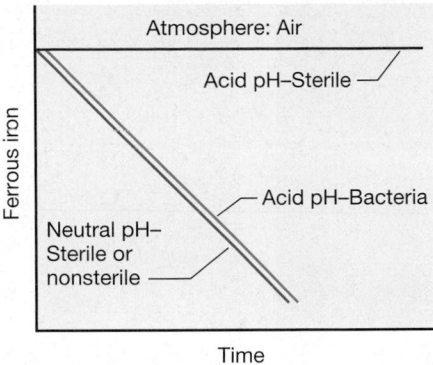

Figure 19.33 Oxidation of ferrous iron as a function of pH and the presence of the bacterium *Thiobacillus ferrooxidans*. Note how Fe^{2+} is stable under acidic conditions in the absence of the bacterial cells.

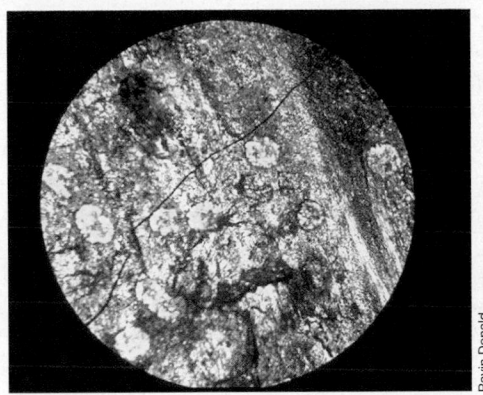

Figure 19.34 Pyrite in coal that can be oxidized by sulfur and iron-oxidizing bacteria. Section through a piece of coal from the Black Mesa formation in northern Arizona (USA). The gold-colored spherical discs (about 1 mm in diameter) are sectioned particles of pyrite, FeS_2.

Ferrous Iron and Pyrite Oxidation at Acid pH

In nonacidic habitats Fe^{2+} can be oxidized by iron bacteria such as *Gallionella* and *Leptothrix* (⟳ Sections 12.15 and 12.16). This occurs primarily at interfaces between ferrous-rich anoxic ground waters and air. However, it is at low pH, where Fe^{2+} is stable, that the most extensive bacterial iron oxidation occurs, and we consider this process now. The acidophilic chemolithotroph *Thiobacillus ferrooxidans* and related acidophilic iron oxidizers can oxidize Fe^{2+} to Fe^{3+} at extremely low pH (Figure 19.33). However, because very little energy is generated in the oxidation of ferrous to ferric iron (⟳ Section 17.11), these bacteria must oxidize large amounts of iron in order to grow, and consequently even a small number of cells can be responsible for precipitating a large amount of iron. This iron-oxidizing bacterium, which is a strict acidophile, is very common in acid mine drainages and in acid springs and is probably responsible for most of the ferric iron precipitated at acid pH values.

Thiobacillus ferrooxidans and *Leptospirillum ferrooxidans* live in environments in which sulfuric acid is the dominant acid and large amounts of sulfate are present. At 20–30°C and moderately acidic pH (2–4), *T. ferrooxidans* seems to dominate, while at 30–50°C and more acidic pH (1–2), *L. ferrooxidans* is the dominant organism. Under these conditions, ferric iron does not precipitate as the hydroxide but as a complex sulfate mineral called *jarosite* [$HFe_3(SO_4)_2(OH)_6$]. Jarosite is a yellowish or brownish precipitate and is responsible for one of the manifestations of acid mine drainage, an unsightly yellow stain called "yellow boy" by U.S. miners (⟳ Figure 17.28a and see also Figure 19.37).

One of the most common forms of iron in nature is *pyrite*, which has the overall formula FeS_2. Pyrite is formed from the reaction of sulfur with ferrous sulfide (FeS) to form a highly insoluble crystalline structure, and is very common in bituminous coals and in many ore bodies (Figure 19.34). The bacterial oxidation of pyrite is

of great significance in the development of acidic conditions in mining operations (Figure 19.35). Additionally, oxidation of pyrite by bacteria is of considerable importance in the process called *microbial leaching of ores* (see Section 19.15). The oxidation of pyrite is a combination of chemically and bacterially catalyzed reactions. Two electron acceptors are involved in this process: molecular oxygen (O_2) and ferric ions (Fe^{3+}).

When pyrite is first exposed, as in a mining operation, a slow chemical reaction with molecular oxygen occurs, as shown in Figure 19.36. This reaction, called the *initiator reaction*, leads to the oxidation of sulfide to sulfate and the development of acidic conditions under which the ferrous iron released is relatively stable in the presence of oxygen. *Thiobacillus ferrooxidans* and *L. ferrooxidans* then catalyze the oxidation of ferrous to ferric ions. The ferric ions formed under these acidic conditions, being soluble, can readily react spontaneously with more pyrite to oxidize the pyrite to ferrous ions plus sulfate ions:

$$FeS_2 + 14\,Fe^{3+} + 8\,H_2O \rightarrow 15\,Fe^{2+} + 2\,SO_4^{2-} + 16\,H^+$$

The ferrous ions formed are again oxidized to ferric ions by the bacteria, and these ferric ions again react with more pyrite. Thus, there is a progressive, rapidly increasing rate at which pyrite is oxidized, called the *propagation cycle*, as illustrated in Figure 19.36. Under natural conditions some of the ferrous iron generated by the bacteria leaches away, being carried by ground water into surrounding streams. However, because oxygen is present in the aerated drainage, bacterial oxidation of the ferrous iron takes place in these outflows and an insoluble ferric precipitate is formed.

Acid Mine Drainage

Bacterial oxidation of sulfide minerals is the major factor in the formation of **acid mine drainage**, a common environmental problem in coal-mining regions (Figure 19.37;

(a)   *(b)*   *(c)*

Figure 19.35 Pyrite-rich microbial habitats in bituminous coal- and copper-mining environments. (a) Bituminous coal-mining operation in a strip mine. The shovel is removing the soil (overburden) to reach the coal seam. (b) The coal seam. Removal of the bituminous coal exposes the environment to air. The pyrite associated with the coal (see Figure 19.34) is colonized with iron-oxidizing bacteria. (c) Copper ore deposit rich in pyrite. Mining of the copper ore results in exposure of the formation to air.

Figure 17.28a). Mixing of acidic mine waters with natural waters in rivers and lakes causes serious degradation in the quality of the natural water because both the acid and the dissolved metals are toxic to aquatic life (Figures 17.28a and see also Figure 19.37). In addition, such polluted waters are unsuitable for human consumption and industrial use. The breakdown of pyrite leads ultimately to the formation of sulfuric acid and ferrous iron, and pH values can be as low as pH 2. The acid formed attacks other minerals associated with the coal and pyrite, causing breakdown of the whole rock fabric. A major rock-forming element, aluminum, is soluble only at low pH, and often high levels of Al^{3+}, which can be toxic to aquatic organisms, are present per liter of acid mine waters.

The requirement for O_2 in the oxidation of ferrous to ferric iron helps to explain how acid mine drainage develops. As long as the coal is unmined, oxidation of pyrite cannot occur because neither air nor the bacteria can reach it. When the coal seam is exposed (Figure 19.35), it

quickly becomes contaminated with *Thiobacillus ferrooxidans*, and O_2 is introduced, making oxidation of pyrite possible. The acid formed can then leach into the surrounding streams (Figure 19.37).

Where acid mine drainage is extensive, a strongly acidophilic archaeon, *Ferroplasma acidophilum*, is active. This aerobic iron-oxidizing prokaryote is capable of growth at pH 0 and at temperatures to 50°C. *F. acido-*

$$FeS_2 \text{ (pyrite)} + 3\tfrac{1}{2} O_2 + H_2O \rightarrow Fe^{2+} + 2 SO_4^{2-} + 2 H^+$$

Initiator reaction

Slow spontaneous, bacteria catalyze

Fe^{2+}

O_2

Fast spontaneous (bacteria may also catalyze)

Spontaneous (bacteria may also catalyze)

FeS_2

Propagation cycle

Fe^{3+}

Figure 19.36 Role of iron-oxidizing bacteria in oxidation of the mineral pyrite.

Figure 19.37 Acid mine drainage from a bituminous coal region. Note the yellowish-red color due to precipitated iron oxides. (Figure 17.28a.)

philum forms into slimy steamers attached to pyrite surfaces in iron ore deposits (Figure 19.38). At Iron Mountain, California (Figure 19.38*a*), a particularly well-studied acid mine drainage site, the thick biofilm of *F. acidophilum* maintains the pH near 0. With iron concentrations of nearly 30 g/l at this site, the *F. acidophilum* mat is constantly bathed in fresh substrate (Fe^{2+}) from which acid is generated by the reactions described previously. *Ferroplasma* is a cell wall-less archaeon morphologically and phylogenetically related to *Thermoplasma* (Figure 19.38*b* and ⌘ Section 13.5).

✓ 19.14 Concept Check

Iron exists in nature primarily in two oxidation states, ferrous (Fe^{2+}) and ferric (Fe^{3+}), and bacterial and chemical transformation of these metals is of great geological and ecological importance. Bacterial ferric iron reduction occurs in anoxic environments and results in the mobilization of iron from swamps, bogs, and other iron-rich aquatic habitats. Bacterial oxidation of ferrous iron occurs on a large scale at low pH and is very common in coal-mining regions, where it results in a type of pollution called acid mine drainage.

✓ What oxidation state is iron in the mineral $Fe(OH)_3$? FeS? How is $Fe(OH)_3$ formed?

✓ Why does Fe^{2+} oxidation under *oxic* conditions occur mainly at *acidic* pH?

VI MICROBIAL BIOREMEDIATION

The collective biogeochemical potential of microorganisms is enormous. Indeed, microorganisms are the Earth's greatest chemists. And as such, they have been recruited to help extract valuable metals from low-grade ores (microbial leaching) and to clean up the environment (bioremediation). In the next four sections we examine several microbial processes that either yield valuable products or bioremediate the polluting activities of humans.

19.15 Microbial Leaching of Ores

We consider here how the acid production and metal solubility by acidophilic bacteria just discussed play a beneficial role in mining. Sulfide forms highly insoluble minerals with many metals, and many ores used as sources of these metals are sulfides. If the concentration of metal in the ore is low, it may not be economically feasible to concentrate the mineral by conventional chemical means. Under these conditions, **microbial leaching** is frequently practiced. Microbial leaching is especially useful for *copper* ores because copper sulfate, formed during oxidation of the copper sulfide ores, is very water-soluble. Indeed, approximately one-fourth of all copper mined worldwide is obtained from leaching processes.

We noted that sulfide itself, HS⁻, oxidizes spontaneously in air. Most metal sulfides also oxidize spontaneously, but much more slowly than free sulfide. Bacteria such as *Thiobacillus ferrooxidans* can catalyze a much faster rate of oxidation of the sulfide minerals, thus aiding in solubilization of the metal (Figure 19.36). The relative rates of oxidation of a copper mineral in the presence and absence of bacteria is illustrated in Figure 19.39. The susceptibility to oxidation also varies among minerals, and those minerals that are most readily oxidized are most amenable to microbial leaching. Thus, iron and copper sulfide ores such as pyrrhotite (FeS) and

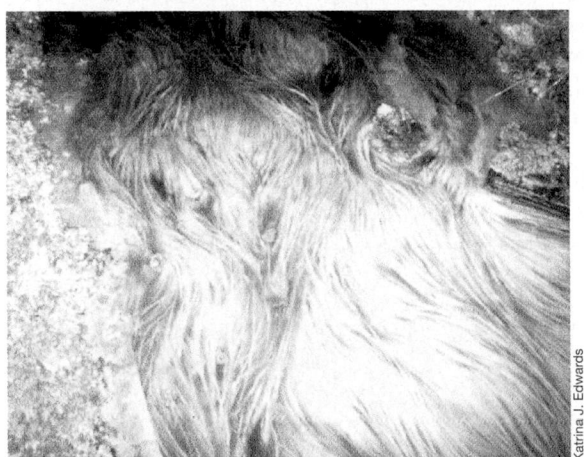

(a)

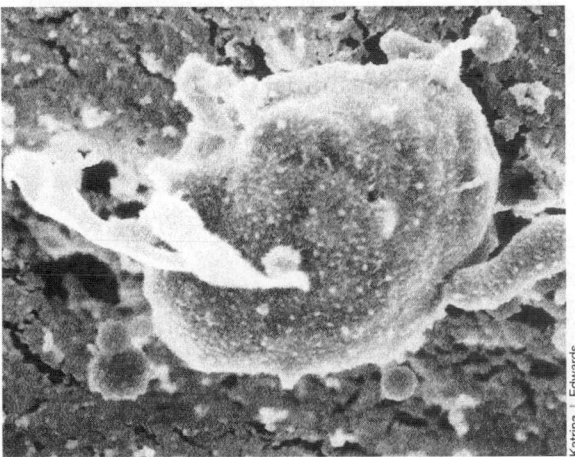

(b)

Figure 19.38 *Ferroplasma acidophilum*, an extremely acidophilic iron-oxidizing archaeon responsible for severe acid mine drainage. (a) Streamers of *Ferroplasma* cells growing in an acidic (pH near 0) mine drainage stream, Iron Mountain, California. (b) Scanning electron micrograph of a cell of *F. acidophilum* among mineral matter.

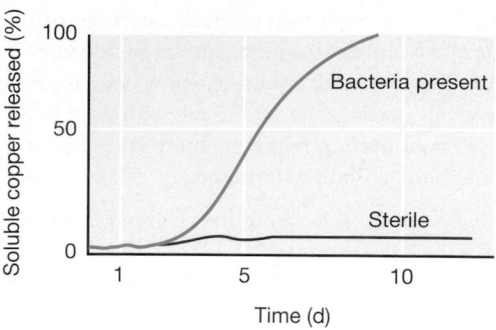

Figure 19.39 Effect of the bacterium *Thiobacillus ferrooxidans* on the leaching of copper from the mineral covellite. The leaching was done in a laboratory column, and the acid leach solution contained inorganic nutrients necessary for development of the bacterium. The leaching activity was monitored by assaying for soluble copper in the leach solution at the bottom of the column. The leach solution was continuously recirculated, maintaining an essentially closed system.

covellite (CuS) are readily leached, whereas lead and molybdenum ores are much less so.

The Leaching Process

In the microbial leaching process, low-grade ore is dumped in a large pile (the leach dump) and a dilute sulfuric acid solution (pH about 2) is percolated down through the pile (Figure 19.40a). The liquid coming out of the bottom of the pile (Figure 19.40b), rich in the mineral, is collected and transported to a precipitation plant (Figure 19.40c) where the metal is reprecipitated and purified. The liquid is then pumped back to the top of the pile and the cycle is repeated. As needed, more acid is added to maintain the low pH.

There are several mechanisms by which the bacteria can catalyze oxidation of the sulfide minerals. To illustrate, let's look at examples of the oxidation of two copper minerals: chalcocite, Cu_2S, in which copper has a valence of +1, and covellite, CuS, in which copper has a valence of +2. As illustrated in Figure 19.41, *Thiobacillus ferrooxidans* is able to oxidize Cu^+ in chalcocite (Cu_2S) to Cu^{2+}, thus removing some of the copper in the soluble form, Cu^{2+}, and forming the mineral covellite. Note that in this reaction there is no change in the valence of sulfide, the bacteria using the reaction Cu^+ to Cu^{2+} as a source of energy. This is analogous to the oxidation by the same bacterium of Fe^{2+} to Fe^{3+}. Covellite can then be oxidized, releasing sulfate and soluble Cu^{2+} as products (Figure 19.41).

A second mechanism, and probably the most important in most mining operations, involves *chemical* oxidation of the copper ore with *ferric* ions formed by the bacterial oxidation of ferrous ions (Figure 19.41). In al-

(a)

(b)

(c)

(d)

Figure 19.40 The leaching of low-grade copper ores using bacteria. (a) A typical leaching dump. The low-grade ore has been crushed and dumped in a large pile in such a way that the surface area exposed is as high as possible. Pipes distribute the acidic leach water over the surface of the pile. The acidic water slowly percolates through the pile and exits at the bottom. (b) Effluent from a copper leaching dump. The acidic water is very rich in dissolved copper. (c) Recovery of dissolved copper by passage of the copper-rich water over metallic iron in a long flume. (d) A small pile of recovered copper metal removed from the flume, ready for further purification.

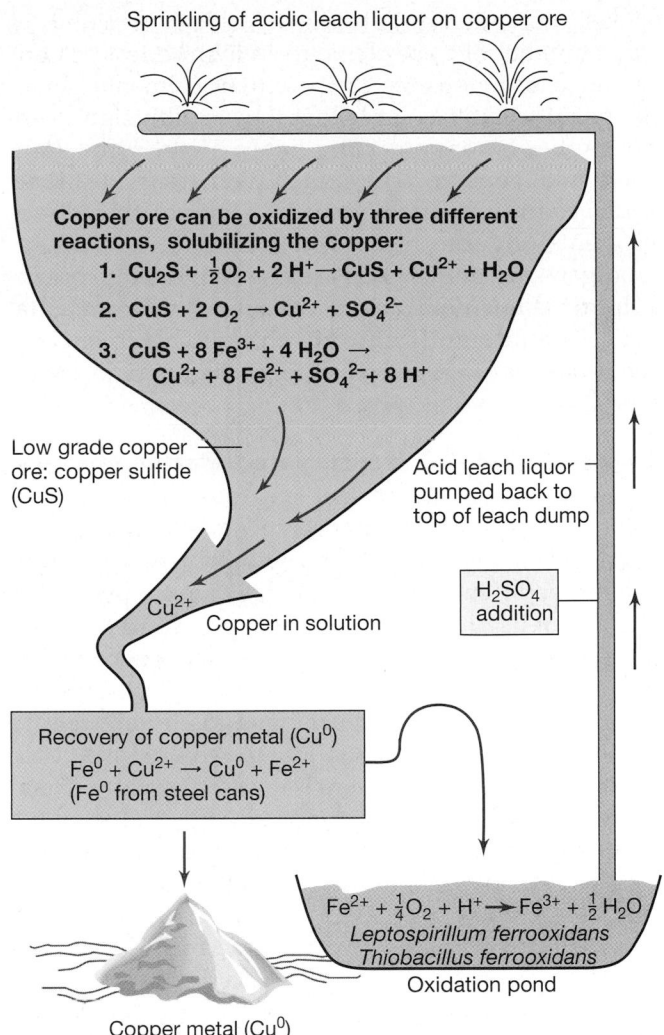

Sprinkling of acidic leach liquor on copper ore

Copper ore can be oxidized by three different reactions, solubilizing the copper:

1. $Cu_2S + \frac{1}{2}O_2 + 2 H^+ \rightarrow CuS + Cu^{2+} + H_2O$
2. $CuS + 2 O_2 \rightarrow Cu^{2+} + SO_4^{2-}$
3. $CuS + 8 Fe^{3+} + 4 H_2O \rightarrow$
   $Cu^{2+} + 8 Fe^{2+} + SO_4^{2-} + 8 H^+$

Low grade copper ore: copper sulfide (CuS)

Acid leach liquor pumped back to top of leach dump

Cu^{2+} Copper in solution

H_2SO_4 addition

Recovery of copper metal (Cu^0)
$Fe^0 + Cu^{2+} \rightarrow Cu^0 + Fe^{2+}$
(Fe^0 from steel cans)

$Fe^{2+} + \frac{1}{4}O_2 + H^+ \rightarrow Fe^{3+} + \frac{1}{2}H_2O$
Leptospirillum ferrooxidans
Thiobacillus ferrooxidans
Oxidation pond

Copper metal (Cu^0)

Figure 19.41 Arrangement of a leaching pile and reactions involved in the microbial leaching of copper sulfide minerals to yield Cu^0 (copper metal). Reaction 1 is primarily bacterial, while Reaction 2 occurs both biologically and chemically. Reaction 3 is strictly chemical, but is probably the most important reaction in copper-leaching processes. Note how it is essential for Reaction 3 to proceed that the Fe^{2+} produced (from the oxidation of sulfide in CuS to sulfate) be oxidized back to Fe^{3+} by *Thiobacillus ferrooxidans* and *Leptospirillum ferrooxidans* (*bottom of art*).

most any ore, pyrite is present, and its oxidation leads to the formation of ferric iron. Ferric iron is an excellent electron acceptor for sulfide minerals, and reaction of CuS with ferric iron results in solubilization of the copper and the formation of ferrous iron. In the presence of O_2, at the acid pH values involved, *Thiobacillus ferrooxidans* reoxidizes the ferrous iron back to the ferric form so it can oxidize more copper sulfide. Thus, the process is maintained by the oxidation of Fe^{2+} to Fe^{3+} by the bacterium.

Metal Recovery

Another source of iron in leaching operations is at the precipitation plant used in recovery of the soluble cop-

per from the leaching solution (Figure 19.40c and d). Scrap iron, Fe^0, is used to recover copper from the leach liquid by the reaction shown in the lower part of Figure 19.41, and this results in the formation of Fe^{2+}. In most leaching operations, this Fe^{2+}-rich liquid remaining after the copper is removed is transferred to an oxidation pond where *Thiobacillus ferrooxidans* proliferates and forms Fe^{3+}. Acid is added to the pond to keep the pH low, thus keeping the Fe^{3+} in solution, and this ferric-rich liquid is then pumped to the top of the pile and the Fe^{3+} is available to oxidize more sulfide mineral.

Because of the huge dimensions of copper leach dumps, penetration of oxygen from air is poor, and the interior of these piles usually becomes anoxic. Although many of the reactions written in Figure 19.41 require molecular O_2, because *Thiobacillus ferrooxidans* can use Fe^{3+} as an electron *acceptor* in the absence of O_2, the oxidation of copper minerals can also proceed anaerobically; the large amounts of Fe^{3+} added to the leach solution from scrap oxidized iron drive the process forward, even under anoxic conditions. High temperature can also be a problem with leaching operations. *T. ferrooxidans* is a mesophile, but temperatures inside a leach dump often rise spontaneously as a result of microbial activities. Thus, thermophilic iron-oxidizing chemolithotrophs such as thermophilic *Thiobacillus* species and *Leptospirillum ferrooxidans* and, at higher temperatures (60–80°C), the acidophilic archaeon *Sulfolobus* (∞ Section 13.9) may be important in the leaching of ores above 40°C.

Other Leaching Processes: Uranium and Gold

Microorganisms are also used in the leaching of uranium- and gold-containing ores. Although *T. ferrooxidans* can oxidize U^{4+} to U^{6+} with O_2 as an electron acceptor, it is likely that the uranium leaching process depends more on the chemical oxidation of uranium by Fe^{3+}, with *T. ferrooxidans* contributing mainly through the reoxidation of Fe^{2+} to Fe^{3+} as described for the leaching of copper ores (see Figure 19.41). The reaction observed is

$$UO_2 + Fe_2(SO_4)_3 \rightarrow UO_2SO_4 + 2 FeSO_4$$
$$(U^{4+}) \quad (Fe^{3+}) \quad (U^{6+}) \quad (Fe^{2+})$$

Unlike UO_2, the oxidized uranium mineral is soluble and can be recovered by other processes.

Gold is frequently found in nature associated with minerals containing arsenic and pyrite. In the microbial leaching of gold, *T. ferrooxidans* and its relatives are used to attack and solubilize the arsenopyrite minerals and, in the process, releasing the trapped gold (Au):

$$2 \, FeAsS[Au] + 7 \, O_2 + 2 \, H_2O + H_2SO_4 \rightarrow$$
$$Fe_2(SO_4)_3 + 2 \, H_3AsO_4 + [Au]$$

The gold is then complexed with cyanide by traditional gold-mining methods. Unlike copper, where leaching occurs in a huge leach dump (see Figure 19.40), gold leaching usually takes place in relatively small *bioreactor tanks* (Figure 19.42); bioleaching in this fashion has been shown to release greater than 95% of the trapped gold. And, although arsenic and cyanide are toxic residues from the mining process, both are removed in the bioreactor gold-leaching process, arsenic as a ferric precipitate and cyanide (CN^-) by its microbial oxidation to CO_2 and urea in later stages of the gold recovery process. Small scale microbial gold leaching as an alternative to large scale methods is becoming more popular and is beginning to replace the more costly and environmentally damaging conventional gold-mining techniques.

✓ 19.15 Concept Check

Oxidation of copper ores by bacteria can lead to the solubilization of copper, a process called *microbial leaching*. Leaching is important in the recovery of copper and uranium from low-grade ores. Bacterial oxidation of iron in the iron sulfide mineral pyrite is also an important part of the microbial leaching process because the ferric iron produced is itself an oxidant of ores.

- ✓ How is CuS oxidized under *anoxic* conditions?
- ✓ Why is it important to keep the leach liquor acidic in the copper ore leaching process?
- ✓ From the standpoint of metal oxidation, how does the copper leach process differ from the gold leach process?

19.16 Mercury and Heavy Metal Transformations

Trace elements are elements that are present in low concentrations in rocks, soils, waters, and the atmosphere. Some trace elements (for example, cobalt, copper, zinc,

Figure 19.42 Photo of the gold leaching tanks at the Ashanti Goldfields, Ghana, Africa. Within the tanks, a mixture of *Thiobacillus ferrooxidans*, *Thiobacillus thiooxidans*, and *Leptospirillum ferrooxidans* solubilizes the pyrite/arsenic mineral containing trapped gold and releases the gold. See text for the reactions involved.

nickel, molybdenum) are nutrients (∞ Section 5.1), but a number of trace elements in high concentrations are toxic to organisms. Of these toxic elements, several are sufficiently volatile that they exhibit significant atmospheric transport, and hence are of some environmental concern. These include mercury, lead, arsenic, cadmium, and selenium. Many of these trace elements undergo redox reactions catalyzed by microorganisms, and several are also converted to organic form via microbial action. Because of environmental concern and significant microbial involvement, we focus our discussion on the biogeochemistry of the element *mercury*.

Global Cycling of Mercury and Methylmercury

Although mercury is present in extremely low concentrations in most natural environments, averaging about 1 nanogram (ng)/liter, it is a widely used industrial product and is the active component of many pesticides. Because of its unusual propensity to concentrate in living tissues and its high toxicity, mercury is of considerable environmental importance. For example, the mining of mercury ores and the burning of fossil fuels release about 40,000 *tons* of mercury into the environment each year; an even greater amount is released by natural geochemical processes. Mercury is also a by-product of the electronics industry, especially battery and wiring production, the chemical industry, and the burning of municipal wastes.

The major form of mercury in the atmosphere is elemental mercury (Hg^0), which is volatile and is oxidized to mercuric ion (Hg^{2+}) photochemically; most of the mercury entering aquatic environments is thus Hg^{2+} (Figure 19.43). Mercuric ion readily adsorbs to particulate matter and can be metabolized from there by microorganisms. Microbial activity results in the *methylation* of mercury, yielding methylmercury, CH_3Hg^+ (Figure 19.43). Metabolically, methylation of mercury occurs by donation of methyl groups from methyl-B_{12}. Methylmercury is particularly toxic because it can be absorbed through the skin. But in addition, it is soluble and can be concentrated in the aquatic food chain, primarily in fish, or further methylated by microorganisms to yield the volatile compound dimethylmercury, $CH_3—Hg—CH_3$.

Both methylmercury and dimethylmercury tend to accumulate in animal tissues, especially muscle. Methylmercury is about 100 times more toxic than Hg^0 or Hg^{2+} and concentrates in fish, where it is a potent neurotoxin. Methylmercury is thus a major environmental toxin, and its accumulation seems to be a particular problem in freshwater lakes where enhanced levels of methylmercury have been observed in fish caught for human consumption. Mercury can also cause liver and kidney damage in humans and other animals.

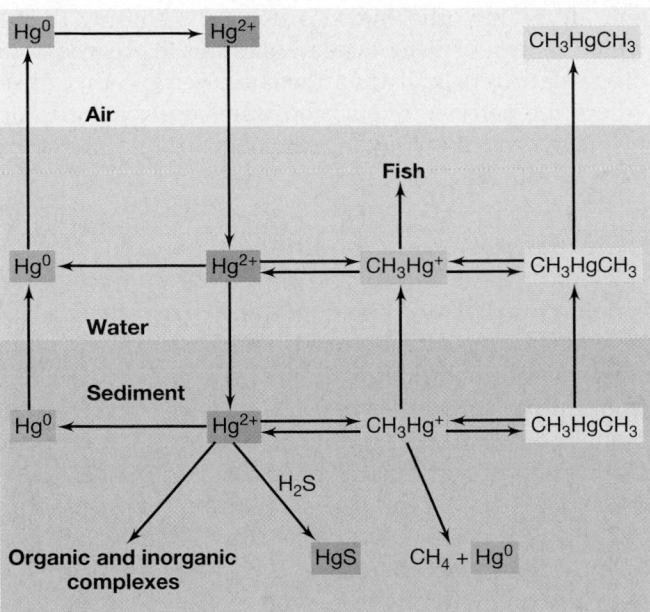

Figure 19.43 Biogeochemical cycling of mercury. The major reservoirs of mercury are in water and in sediments where it can be concentrated in animal tissues or precipitated out as HgS. The various forms of mercury commonly found in aquatic environments are each shown in a different color.

Several other mercury transformations occur on a global scale, including reactions involving sulfate-reducing bacteria ($H_2S + Hg^{2+} \rightarrow HgS$) and methanogens ($CH_3Hg^+ \rightarrow CH_4 + Hg^0$) (see Figure 19.43). The solubility of HgS is very low, so in anoxic sulfate-reducing sediments, most mercury is found as HgS. But on aeration, oxidation of HgS can occur, primarily by thiobacilli, leading to the formation of Hg^{2+} and eventually, methylmercury.

Mercuric Resistance

At sufficiently high concentrations, Hg^{2+} and CH_3Hg^+ can be toxic not only to higher organisms but also to microorganisms. Fortunately, several bacteria can carry out the biotransformation of toxic forms of mercury to nontoxic forms. In mercury-resistant gram-negative bacteria an NADPH-linked enzyme called *mercuric reductase* transfers two electrons to Hg^{2+}, reducing it to Hg^0. The Hg^0 produced in this reaction is volatile but is essentially nontoxic to humans and microorganisms, compared with Hg^{2+} or CH_3Hg^+. Bacterial conversion of Hg^{2+} to Hg^0 then allows more CH_3Hg^+ to be converted to Hg^{2+}.

Mercury resistance has been intensively studied in the gram-negative bacterium *Pseudomonas aeruginosa*, where genes for mercury resistance reside on a plasmid. These genes, called *mer* genes, are arranged in an operon and are under control of the regulatory protein MerR (the product of *merR*) (Figure 19.44a). Interestingly,

MerR functions as both a repressor and an activator (∞ Chapter 8). In the absence of Hg^{2+}, MerR binds to the operator region of the *mer* operon and prevents transcription of *mer TPCAD* genes (Figure 19.44a). However, when Hg^{2+} is present, it forms a complex with MerR, which then functions as an *activator* of transcription of the *mer* operon. The mercuric reductase, mentioned previously, is the product of the *merA* gene. MerD, the product of *merD*, also plays a regulatory role, whereas *merP* encodes a periplasmic Hg^{2+} binding protein (Figure 19.44b). This protein, MerP, binds Hg^{2+} and transfers it to a membrane protein MerT (the product of *merT*), which transports Hg^{2+} into the cell for reduction by mercuric reductase (Figure 19.44b). The final result is reduction of Hg^{2+} to Hg^0, which is volatile and is released from the cell (Figure 19.44b). In other organisms, additional *mer* genes have been found, but the ones described here seem to be conserved in all *mer* operons examined.

Resistance to Other Heavy Metals

A variety of plasmids (∞ Section 10.8) isolated from both gram-positive and gram-negative *Bacteria* have been found to encode resistance to the effects of heavy

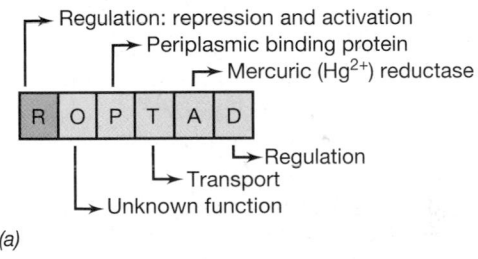

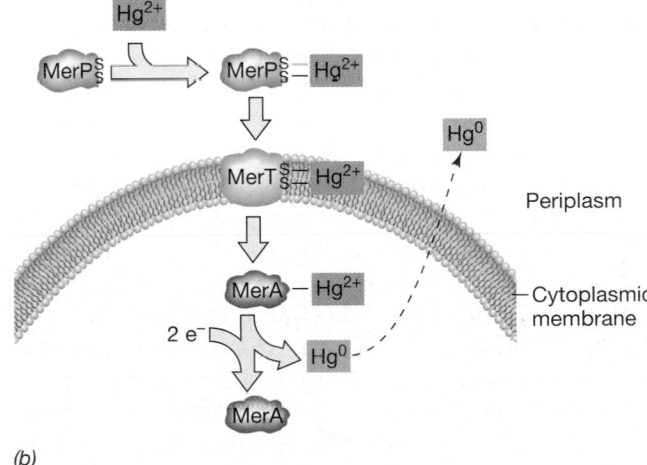

Figure 19.44 Mechanism of Hg^{2+} reduction to Hg^0 in *Pseudomonas aeruginosa*. (a) The *mer* operon. MerR can function as either a repressor (absence of Hg^{2+}) or transcriptional activator (presence of Hg^{2+}). (b) Transport and reduction of Hg^{2+}. Hg^{2+} is bound by cysteine residues in both proteins MerP and MerT.

metals. Certain antibiotic resistance plasmids also have genes for resistance to mercury and arsenic. Other plasmids encode only heavy-metal resistances. A large plasmid isolated from *Staphylococcus aureus* has been found to encode resistance to mercury, cadmium, arsenate, and arsenite. The mechanism of resistance to any specific metal varies. For example, arsenate and cadmium resistances are due to the action of enzymes that immediately pump out any arsenate or cadmium ions incorporated, thus preventing the metals from denaturing proteins.

Studies on nickel- and cobalt-resistant bacteria have shown that in most cases the resistance genes are plasmid-borne; resistance to both metals on a single plasmid is typical. Enrichment culture studies have shown that nickel-resistant bacteria are uncommon in soils and other environments in which this metal is absent in significant amounts. Bacteria highly resistant to nickel or other metals are most common in wastewaters of the metal processing industry or in mining operations where heavy metals are leached out along with iron or copper ores.

✓ *19.16* *Concept Check*

A major toxic form of mercury is methylmercury. The latter can yield Hg^{2+}, which is reduced by bacteria to Hg^0. The ability of bacteria to resist the toxicity of heavy metals is often due to the presence of specific plasmids that encode enzymes capable of detoxifying the metals.

✓ What forms of mercury are most toxic to organisms?

✓ How is mercury detoxified by bacteria?

19.17 **Petroleum Biodegradation**

Microbial decomposition of petroleum and petroleum products is of considerable economic and environmental importance. Because petroleum is a rich source of organic matter and the hydrocarbons within it are readily attacked aerobically by a variety of microorganisms, it is not surprising that when petroleum is brought into contact with air and moisture, it is subject to extensive microbial attack. Under some circumstances, such as in bulk storage tanks, microbial growth is not desirable. However, in other situations, such as in oil spills, microbial utilization of oil is desirable and even promoted by the addition of inorganic nutrients. The term *bioremediation* refers to the cleanup of oil or other pollutants by microorganisms, and in recent years the importance of bioremediation in oil spills has been amply demonstrated in several major crude oil spills in the marine environment (see Figure 19.46).

The biochemistry of petroleum biodegradation was covered in Section 17.23, where we emphasized the important role oxygenase enzymes play in introducing oxygen atoms into the hydrocarbon (∞ Figure 17.55), which allows further biochemical events to proceed. Our discussion here will thus focus on aerobic processes where the activity of such enzymes are relevant and where hydrocarbon oxidation in nature is quantitatively most significant.

Hydrocarbon Decomposition

A wide variety of bacteria, several molds and yeasts, and certain cyanobacteria and green algae have been shown to be able to oxidize hydrocarbons aerobically. Small-scale oil pollution of aquatic and terrestrial ecosystems from human as well as natural activities is very common, and there exists a diverse microbial community that uses hydrocarbons as an electron donor. Methane, the simplest hydrocarbon, is degraded by only a specialized group of bacteria, the *methanotrophic* bacteria (∞ Sections 12.6 and 17.24), and these organisms do not grow on higher hydrocarbons.

Hydrocarbon-oxidizing microorganisms develop rapidly on oil films and slicks, and oil-oxidizing activity is most extensive if environmental conditions such as temperature and inorganic nutrients (primarily N and P), are adequate (see Figure 19.46). Because oil is insoluble in water and is less dense, it floats to the surface and forms slicks. Hydrocarbon-oxidizing bacteria are able to attach to insoluble oil droplets and can often be seen there in large numbers (Figure 19.45). The action of these bacteria eventually leads to decomposition of the oil and dispersal of the slick.

Microorganisms participate in oil spill cleanups (Figure 19.46) by oxidizing the oil to CO_2. In large spills, volatile hydrocarbon fractions evaporate quickly, leaving longer-chain aliphatic and aromatic components for cleanup crews or microorganisms to tackle. In oil spills where careful bioremediation studies have been performed, it has been shown that hydrocarbon-oxidizing bacteria increase in number 10^3–10^6 times shortly after the oil spill.

Experiments using radioisotopic hydrocarbons as tracers or O_2 uptake as a measure of heterotrophic activity

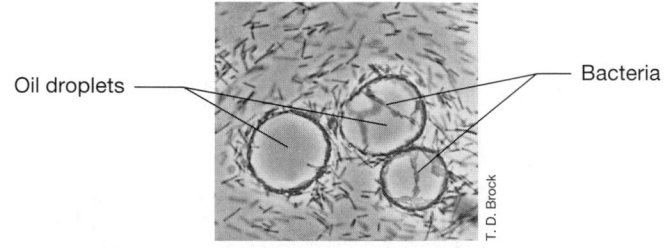

Figure 19.45 Hydrocarbon-oxidizing bacteria in association with oil droplets. The bacteria are concentrated in large numbers at the oil-water interface but are not within the droplet itself.

(a)

(b)

Figure 19.46 Environmental consequences of large oil spills in the marine environment and the effect of bioremediation. (a) A contaminated beach along the coast of Alaska containing oil from the *Exxon Valdez* spill of 1989. (b) The center rectangular plot (arrow) was treated with inorganic nutrients to stimulate bioremediation of spilled oil by microorganisms, whereas areas to the left and right were untreated.

have shown that under ideal conditions up to 80% of the nonvolatile components are oxidized by bacteria within a year of the spill. Certain fractions, such as branched-chain and polycyclic hydrocarbons, however, remain in the environment much longer. Spilled oil that travels to the sediments is only slowly degraded and can have a significant long-term impact on fisheries and related activities that depend on unpolluted waters for productive yields.

Interfaces where oil and water meet often occur on a large scale. For example, it is virtually impossible to keep moisture from accumulating in bulk-fuel storage tanks; water forms in a layer beneath the petroleum. Gasoline and crude oil storage tanks (Figure 19.47) are thus potential habitats for hydrocarbon-oxidizing microorganisms, which can accumulate and grow at the oil-water interface. If sufficient sulfate is present in the water, sulfate-reducing bacteria can grow, consuming

Figure 19.47 Bulk fuel storage tanks, where massive microbial growth may occur at oil-water interfaces.

hydrocarbons under totally anoxic conditions (∞ Section 17.23). The sulfide (H_2S) that is produced is highly corrosive and can cause pitting of the tanks.

Petroleum Production

Although microbial degradation of petroleum can be quite extensive, microbial *production* of hydrocarbons also occurs, particularly in certain green algae. For example, in the colonial alga *Botryococcus braunii*, growth of the alga is accompanied by the excretion of long-chain hydrocarbons (C_{30}–C_{36}) that have the consistency of oil (Figure 19.48). In *B. braunii* about 30% of the cell dry weight is petroleum, and there has been interest in using this and other oil-producing algae as renewable sources of petroleum. There is even evidence that oil in certain types of oil shale originated from green algae like *B. braunii* that grew in lake beds in ancient times.

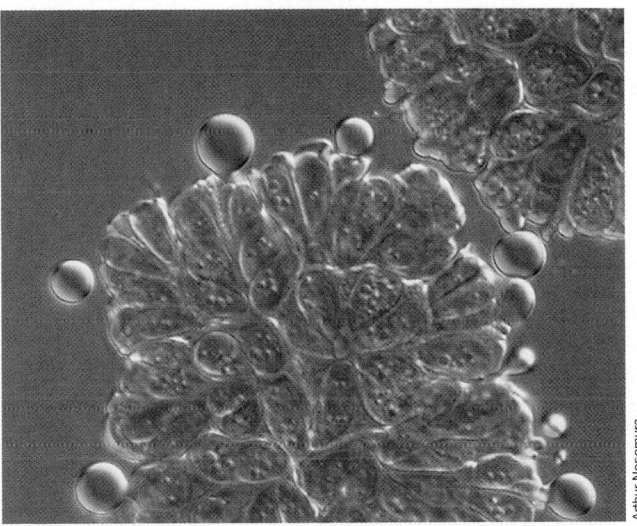

Figure 19.48 Photomicrograph by Nomarski interference contrast of cells of the green alga *Botryococcus braunii*. Note oil droplets, produced and excreted by this alga, at the margin of the cells.

19.17 Concept Check

Hydrocarbons are subject to microbial attack. Hydrocarbon-oxidizing microorganisms are used for bioremediation of spilled oil, and their activities assisted by addition of inorganic nutrients to balance carbon from the oil. Some algae can produce hydrocarbons.

✓ What is *bioremediation*?

✓ Why might the addition of inorganic nutrients stimulate oil degradation while the addition of glucose might not?

19.18 Biodegradation of Xenobiotics

Xenobiotics are chemically synthesized compounds that are not naturally occurring. Xenobiotics include a long list of compounds including pesticides, polychlorinated biphenyls (PCBs, used in the electric-generating and elated industries), munitions, dyes, and chlorinated solvents. Many xenobiotics are structurally related, sometimes closely structurally related, to natural compounds, and thus can be slowly degraded by enzymes that already exist to degrade these natural compounds. In other cases the compounds are structurally so different from anything organisms have previously experienced that their degradation rate in nature is extremely slow, if at all. Nevertheless, for many xenobiotic compounds microorganisms have been found that can degrade them, and we focus here on pesticides as an example of the potential of microbial degradation.

An important aspect of the xenobiotics field that goes beyond environmental cleanup, *per se*, is the quest for understanding of the *genetics* of xenobiotic-degrading microorganisms. Besides attacking applied problems, such as how humans can expedite the degradation of xenobiotics in nature, microbiologists are also interested in more basic problems, such as the nature and regulation of genes that encode the proteins that actually degrade these compounds. Organisms capable of bioremediation (see for example, Figure 19.50) obviously contain key catabolic genes, and it has not escaped the notice of biotechnologists that these genes could be introduced into other organisms, specifically plants. *Genetically modified organisms* (GMOs, ⟳ Section 31.7) armed with new or increased capacities for degrading various xenobiotics will likely have an important future in this field for at least two reasons. First, they will assist in cleaning up the environment, of course. But second, they will allow for the *continued use* of xenobiotics, particularly the herbicides (Table 19.7), in plant agriculture. Indeed, many of these compounds have proven very effective and have allowed current systems of intensive plant agriculture to be highly productive. Because of this, the continued use of xenobiotic pesticides is a reality that is unlikely to

TABLE 19.7 Persistence of herbicides and insecticides in soils	
Substance	**Time for 75–100% disappearance**
Chlorinated insecticides	
DDT [1,1,1-trichloro-2,2-bis-(*p*-chlorophenyl)ethane]*a*	4 years
Aldrin	3 years
Chlordane	5 years
Heptachlor	2 years
Lindane (hexachlorocyclohexane)	3 years
Organophosphate insecticides	
Diazinon	12 weeks
Malathion*a*	1 week
Parathion	1 week
Herbicides	
2,4-D (2,4-dichlorophenoxyacetic acid)	4 weeks
2,4,5-T (2,4,5-trichlorophenoxyacetic acid)	20 weeks
Dalapin	8 weeks
Atrazine*a*	40 weeks
Simazine	48 weeks
Propazine	1.5 years

*a*Structure shown in Figure 19.49.

change in the foreseeable future. Knowledge of the genetics of biodegradation is therefore an important aspect of the field of bioremediation and should assist in the development of new technologies for environmental cleanup and agricultural productivity.

Pesticides

Some of the most widely distributed xenobiotics are the **pesticides**, which are common components of toxic wastes. Over 1000 pesticides have been marketed for chemical pest control purposes. These include primarily *herbicides, insecticides,* and *fungicides*. Pesticides are of a wide variety of chemical types, including chlorinated compounds, aromatic rings, nitrogen- and phosphorus-containing compounds, and others (Figure 19.49). Some of these substances are suitable as carbon sources and electron donors for certain soil microorganisms, whereas others are not. If a substance can be attacked by microorganisms, it will eventually disappear from the soil. Such degradation in the soil is usually desirable because toxic accumulations of the compound are avoided. However, even closely related compounds may differ remarkably in their degradability, as is shown for the relative persistence rates of a number of herbicides in Table 19.7

The figures in Table 19.7 are only approximate because a variety of environmental factors, such as temperature, pH, aeration, and organic matter content of the soil, influence decomposition. Some of the chlorinated insecticides are so recalcitrant that they have persisted for over 10 years. Disappearance of a pesticide from an ecosystem does not necessarily mean that it was degraded by microorganisms because pesticide loss can

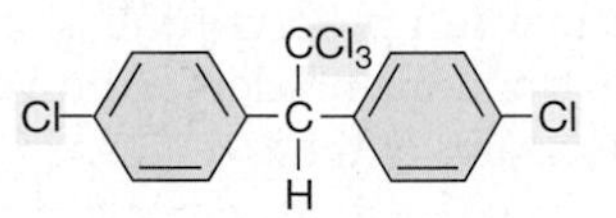

DDT; dichlorodiphenyltrichloroethane
(an organochlorine)

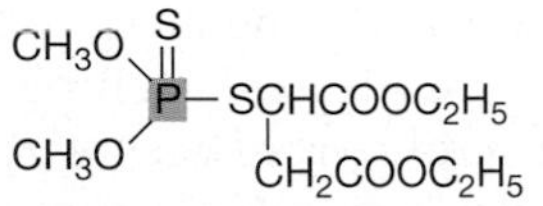

Malathion; mercaptosuccinic acid diethyl ester
(an organophosphate)

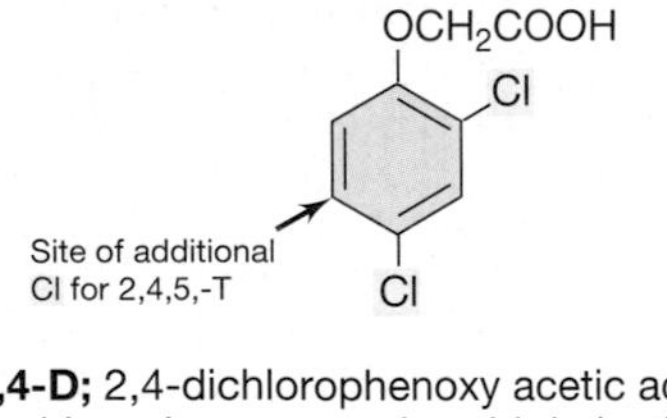

Site of additional
Cl for 2,4,5,-T

2,4-D; 2,4-dichlorophenoxy acetic acid
(a chlorophenoxy acetic acid derivative)

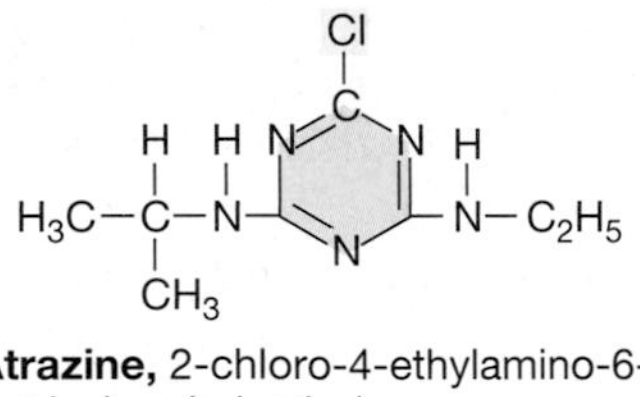

Atrazine, 2-chloro-4-ethylamino-6-isopropylaminotriazine
(a triazine derivative)

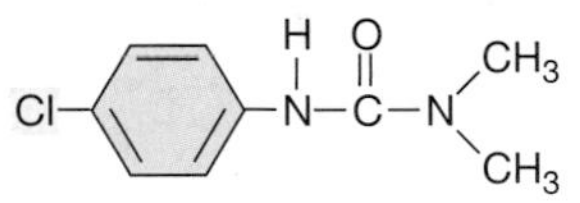

Monuron; 3-(4-chlorophenyl)-1,1-dimethylurea
(a substituted urea)

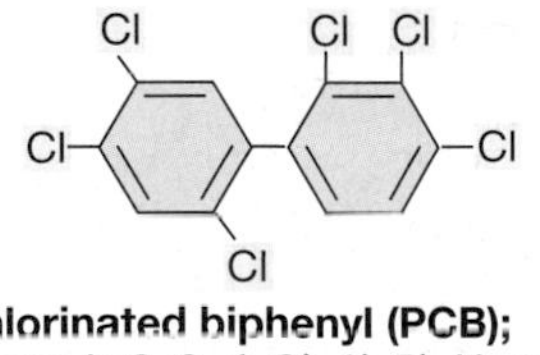

Chlorinated biphenyl (PCB);
shown is 2, 3, 4, 2′, 4′, 5′- Hexachlorobiphenyl

Trichloroethylene

Figure 19.49 Some xenobiotic compounds. Although none of these compounds exist naturally, various microorganisms exist that will break them down (see persistence data in Table 19.7).

also occur by volatilization, leaching, or spontaneous chemical breakdown. *Bioavailability* is also a factor governing microbial attack on xenobiotic compounds. Many xenobiotics are quite hydrophobic and thus not very soluble in water, and adsorption of these compounds to organic matter and clay in soils and sediments prevents access to the organism. Addition of surfactants or emulsifiers often increases bioavailability, and ultimately biodegradation, of the xenobiotic compound.

The organisms that are able to metabolize pesticides and herbicides are fairly diverse, including genera of both

bacteria and fungi. Some pesticides can be both carbon and energy sources and are oxidized completely to CO_2. However, other compounds are much more recalcitrant and are attacked only slightly or not at all, although they may often be degraded either partially or totally provided some other organic material is present as primary energy source, a phenomenon called **cometabolism** or **cooxidation**. However, when the breakdown is only partial, the microbial degradation product of a pesticide may sometimes be even more toxic than the original compound.

Reductive Dechlorination

In Chapter 17 we discussed a form of anaerobic respiration called *reductive dechlorination*, in which chlorinated organic compounds are used as terminal electron acceptors under anoxic conditions (∞ Section 17.18). A laboratory model for studying reductive dechlorination is the reduction of chlorobenzoate to benzoate by the bacterium *Desulfomonile*:

$$C_7H_4O_2Cl^- + 2\,H \rightarrow C_7H_5O_2^- + HCl$$

However, from the standpoint of bioremediation, other chlorinated compounds are ecologically more important than chlorobenzoate. For example, reductive dechlorination occurs with the compounds dichloro-, trichloro-, and tetrachloro- (perchloro-) ethylene, chloroform, dichloromethane, certain polychlorinated biphenyls (PCBs; Figure 19.49), and various brominated and fluorinated compounds. These toxic compounds, some of which (particularly trichloroethylene, Figure 19.49) are strongly suspected of being carcinogenic, are widely used as industrial solvents, degreasing agents, and insulators in electrical transformers. They enter anoxic environments through accidental spills or from slow leakage of storage containers or abandoned electrical transformers, and eventually migrate into groundwater, where they are the most frequently detected groundwater contaminants in the United States. Several genera of reductive dechlorinators are now known (∞ Section 17.18) that can transform a wide variety of chlorinated compounds into harmless metabolites. There is currently great interest in developing methods for stimulating the *in situ* activities of reductive dechlorinating prokaryotes as a strategy for the bioremediation of these highly toxic compounds in anoxic environments.

Aerobic Dechlorination

Aerobic dechlorination of chlorinated compounds is also possible (Figure 19.50) but here, of course, biochemical mechanisms involving O_2 can come into play. Under these conditions, the degradation of chlorinated aromatic compounds occurs by way of oxygenases (∞ Section 17.22). For example, in the aerobic degradation of the pesticide 2,4,5-T by pseudomonads, following dechlorination a dioxygenase enzyme breaks down the aromatic ring to generate compounds that can be metabolized by central metabolic pathways (Figure 19.50*b*). Although the aerobic breakdown of chlorinated xenobiotics is undoubtedly of ecological importance, reductive dechlorination is of particular environmental interest because of the rapidity with which anoxic conditions can develop in polluted microbial habitats in nature and the biochemical constraints (Figure 19.50*b*) this puts on aerobic organisms that could otherwise degrade the compound.

Biodegradation of Synthetic Polymers and Biodegradable Plastics

A major area of environmental concern besides the biodegradation of toxic wastes like pesticides and other chlorinated hydrocarbons, is the disposal of *solid wastes*,

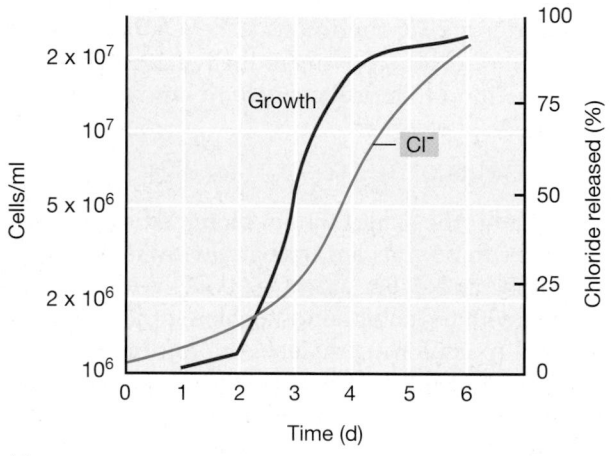

(a)

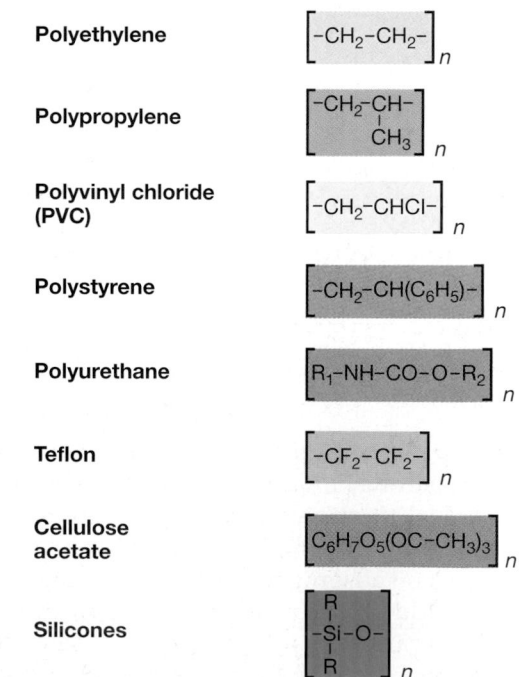

(b)

Figure 19.50 Biodegradation of the herbicide 2,4,5-T. (a) Growth of *Burkholderia* (formerly *Pseudomonas*) *cepacia* on 2,4,5-T as sole source of carbon and energy. The strain was enriched from nature using a chemostat to keep the concentration of herbicide low. Growth here is aerobic on 1.5 g/l of 2,4,5,-T. The release of chloride from the molecule is indicative of biodegradation. (b) Pathway of aerobic 2,4,5-T biodegradation. Note the steps in which Cl^- is released. The final products, succinate and acetate, are catabolized in the citric acid cycle (⟶ Figure 5.22). The mechanism of action of a dioxygenase is discussed in Chapter 17 (⟶ Figure 17.56c).

in particular, plastics. Typical landfills contain large amounts of paper, food, construction and demolition debris, and plastics. Degradation rates of these materials are frequently very low because the conditions, especially the lack of moisture and probably oxygen, as well, are not suitable to rapid microbial activity. But in addition, some of the products added to landfills are inherently recalcitrant in the first place. Plastics are a typical example.

The plastics industry currently produces nearly 40 *billion* kilograms of plastic per year, approximately 40% of which is discarded in landfills. Plastics are xenobiotic polymers of various types, polyethylene, polypropylene, and polystyrene are typical examples (Figure 19.51). Many of these synthetic polymers are highly recalcitrant and remain essentially unaltered for decades in landfills and refuse dumps. This problem has fueled the search for *biodegradable* alternatives (biopolymers) to the synthetic polymers now in use. Some success stories include photodegradable, starch-linked, and microbially synthesized plastics.

Photobiodegradable plastics are polymers whose structure is altered by exposure to ultraviolet radiation (from sunlight), generating modified polymers amenable to microbial attack. Starch-based plastics incorporate starch (⟶ Figure 3.6a) as a linker to connect short fragments of a second biodegradable polymer. This design accelerates biodegradation because starch-digesting bacteria in soil attack the starch, releasing polymer fragments that are then degraded by other microorganisms.

Polyethylene	$\left[-CH_2-CH_2-\right]_n$
Polypropylene	$\left[\begin{array}{l}-CH_2-CH-\\ \quad\quad CH_3\end{array}\right]_n$
Polyvinyl chloride (PVC)	$\left[-CH_2-CHCl-\right]_n$
Polystyrene	$\left[-CH_2-CH(C_6H_5)-\right]_n$
Polyurethane	$\left[R_1-NH-CO-O-R_2\right]_n$
Teflon	$\left[-CF_2-CF_2-\right]_n$
Cellulose acetate	$\left[C_6H_7O_5(OC-CH_3)_3\right]_n$
Silicones	$\left[\begin{array}{c}R\\ -Si-O-\\ R\end{array}\right]_n$

Figure 19.51 Chemistry of synthetic polymers. Monomeric structure of a number of common synthetic polymers.

Microbially synthesized plastics include the carbon storage polymer poly-β-hydroxyalkanoate (∞ Section 4.13) (Figure 19.52). Poly-β-hydroxyalkanoate (PHAs) have been considered as synthetic plastic substituents. PHAs have many of the general properties of synthetic plastics and can be synthesized by cells in various chemical forms. Depending on the length of the side chain in the monomers of the PHA polymer (a property that can be varied by modifying the growth medium or by genetically modifying the producing bacterium), PHAs of varying melting points, crystalinities, flexibilities, and tensile strengths are obtained, suitable for different plastics applications. A *copolymer* containing approximately equal amounts of PHB and PHV (Figure 19.52*a*), has had the greatest market success thus far.

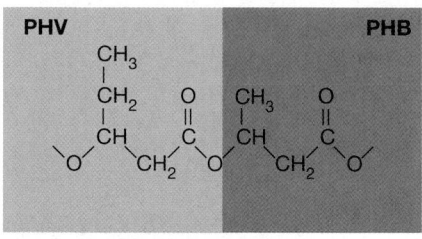

(a)

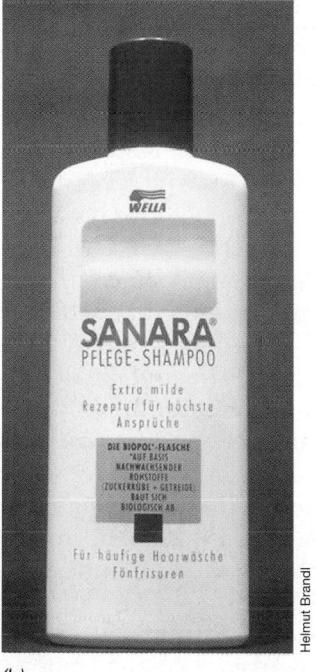

(b)

Figure 19.52 (a) PVH/PHB copolymer. (b) A brand of shampoo marketed in Europe and packaged in a bottle made of "bacterial plastic." The bottle consists of a copolymer of poly-β-hydroxybutyrate (PHB) and poly-β-hydroxyvalerate (PHV, see a). Because this material is a natural product, the bottle readily degrades both aerobically and anaerobically.

✓ **19.18 Concept Check**

Many chemically synthesized compounds such as insecticides, herbicides, and plastics (all called xenobiotics) are completely foreign to the environment but can often be degraded by one or another prokaryote. Both aerobic and anaerobic mechanisms are known.

✓ What chemical features are shared by the compounds shown in Figure 19.49?

✓ What is *reductive dechlorination* and how does it differ from the reactions shown in Figure 19.50?

✓ What advantages do *biopolymers* have over synthetic polymers?

VII MICROBIAL INTERACTIONS WITH PLANTS

We close this chapter with three examples of microbial interactions with plants. Many truly beneficial symbiotic associations exist, such as lichens, mycorrhizae, and the root nodules of leguminous plants. There are also destructive microorganisms that can cause plant diseases such as crown gall. Although many plant diseases are known and have been studied by plant pathologists, we consider here only crown gall because of the unique features of the disease and the microbial mode of transmission.

19.19 The Plant Environment

As microbial habitats, plants are clearly vastly different from animals. Compared with warm-blooded animals, plants vary greatly in temperature, both diurnally and throughout the year, and compared with the complex circulatory system of animals, the internal communication system of the plant is only poorly developed, and so transfer of microorganisms within the plant is relatively inefficient. The above-ground parts of the plant, especially the leaves and stems, are subjected to frequent drying, and for this reason many plants have developed waxy coatings that retain moisture and keep out microorganisms. The roots, on the other hand, exist in an environment in which moisture is less variable and nutrient concentrations are higher. For this reason, the roots of plants are a main area of microbial activity.

Rhizosphere and Phyllosphere

The **rhizosphere** is the region immediately outside the root; it is a zone where microbial activity is usually high. The **rhizoplane** is the actual root surface. The bacterial count is almost always higher in the rhizosphere/rhizoplane than it is in regions of the soil devoid of roots,

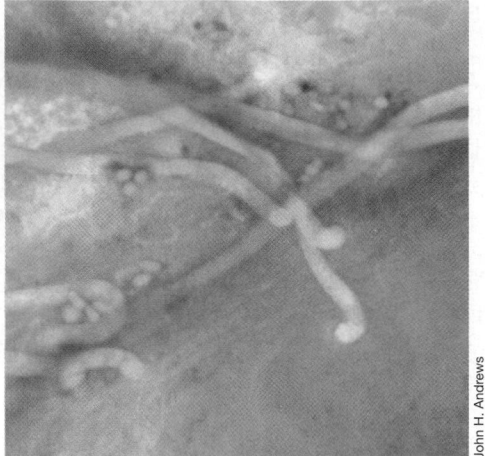

(a)

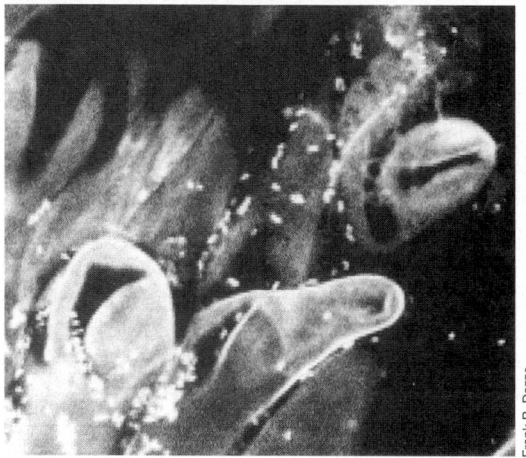

(b)

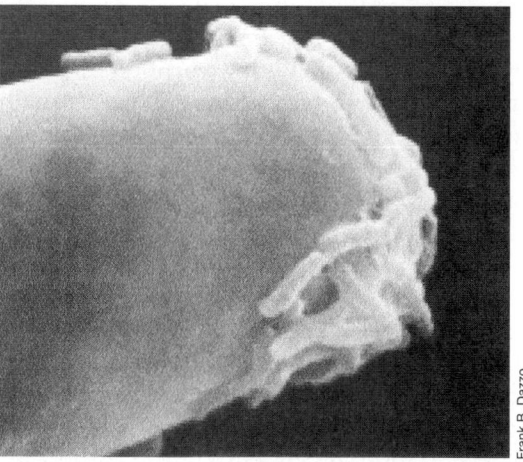

(c)

John H. Andrews

Frank B. Dazzo

Frank B. Dazzo

Figure 19.53 Examples of phyllosphere and rhizosphere microbial life. (a) Fluorescent micrograph of the fungus *Aureobasidium pullulans* on the phyloplane (leaf surface) of apple. Cells were stained by FISH (Section 18.4), and both fungal filaments and spores show the green stain. A fungal filament is about 7 μm in diameter. (b) Laser scanning confocal (Section 4.2) micrograph of bacterial cells on the rhizosphere/rhizoplane of clover. (c) Scanning electron micrograph of cells of *Rhizobium leguminosarum* biovar *trifolii* attached to the root hair of clover. The bacterial cells in (b) and (c) are about 1 μm in diameter.

often many times higher (Figure 19.53). This is because roots excrete significant amounts of sugars, amino acids, hormones, and vitamins, which promote such an extensive growth of bacteria and fungi that these organisms often form microcolonies on the root surface. The **phyllosphere** is the surface of the plant leaf, and under conditions of high humidity, as in wet forests in tropical and temperate zones, the microbial flora of leaves may be quite high, including fungi as well as bacteria (Figure 19.53*a*). Many of the bacteria on leaves fix nitrogen (Section 17.28, and see Section 19.22), and nitrogen fixation presumably aids these organisms in growing with the predominantly carbohydrate nutrients provided by leaves.

✓ 19.19 Concept Check

Key microbial habitats on plants include the rhizoplane/rhizosphere and the phyllosphere.

✓ Why might the roots of plants be a more attractive environment for bacteria than in free soil away from plants?

19.20 Lichens and Mycorrhizae

Lichens are leafy or encrusting growths that are widespread in nature and are often found growing on bare rocks, tree trunks, house roofs, and surfaces of bare soils (Figure 19.54). Lichens consist of a symbiosis of two organisms, a fungus and an alga. However, little specificity resides in the relationship, as a given fungus can establish the lichen symbiosis with several different algae, and vice versa. The alga is phototrophic and able to produce organic matter, which is then used for nutrition of the fungus. The fungus, unable to carry out photosynthesis, provides a firm anchor within which the alga can grow protected from erosion by rain or wind. In addition, the fungus facilitates the uptake of water and absorbs from the rock or other substrate on which the lichen is living the inorganic nutrients essential for the growth of the alga. Lichens are usually found on surfaces where other organisms do not grow, and their success in colonizing such environments is due to the mutual interrelationships between the alga and fungus partners.

Lichen Structure and Ecology

Lichens consist of a tight association of many fungal cells within which the algal cells are embedded (Figure 19.55). The shape of the lichen is determined primarily by the fungal partner, and a wide variety of fungi are able to form lichen associations. The diversity of algal types is much smaller, and many different kinds of lichens may have the same algal component. Some lichens contain cyanobacteria, frequently N_2-fixing species, instead of

(a)

(b)

Figure 19.54 Lichens. (a) A lichen growing on a branch of a dead tree. (b) Lichens coating the surface of a large rock.

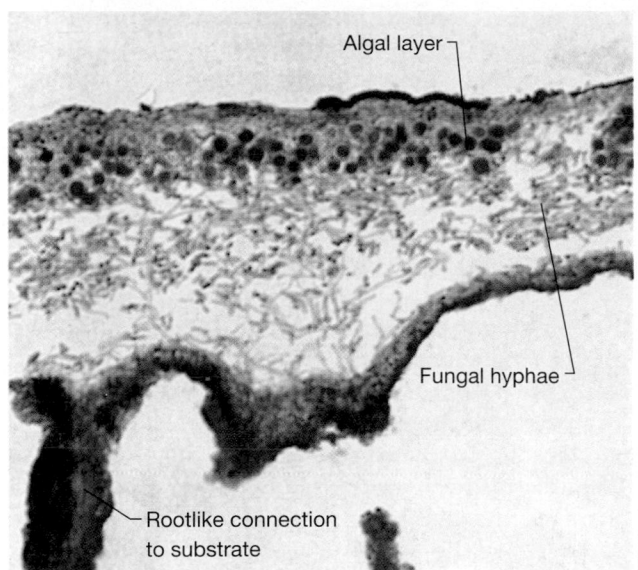

Figure 19.55 Photomicrograph of a cross section through a lichen.

algae as the phototrophic component. The algae or cyanobacteria are usually present in defined layers or clumps within the lichen structure (Figure 19.55).

The fungus clearly benefits from associating with the alga, but how does the alga benefit? *Lichen acids*, complex organic compounds excreted by the fungus, promote the dissolution and chelation of inorganic nutrients needed by the algae. Another role of the fungus is to protect the phototroph from drying; most of the habitats in which lichens live are dry (rock, bare soil, roof tops) (see Figure 19.54), and fungi are, in general, much better able to tolerate dry conditions than are algae.

Most lichens grow extremely slowly—a 2-cm lichen observed on the surface of a rock may actually be several years old. Measurements of lichen growth vary from 1 mm or less per year to over 3 cm/year, depending on the organisms composing the symbiosis, the amount of rainfall and sunlight received, and general weather conditions.

Mycorrhizae

Mycorrhiza literally means "root fungus" and refers to the symbiotic association between plant roots and fungi. Probably the roots of the majority of terrestrial plants

are mycorrhizal. There are two classes of mycorrhizae: **ectomycorrhizae**, in which fungal cells form an extensive sheath around the outside of the root with only little penetration into the root tissue itself, and **endomycorrhizae**, in which the fungal mycelium is embedded within the root tissue.

Ectomycorrhizae are found mainly in forest trees, especially conifers, beeches, and oaks, and are most highly developed in boreal and temperate forests. In such forests, almost every root of every tree is mycorrhizal. The root system of a mycorrhizal tree such as pine (genus *Pinus*) is composed of both long and short roots. The short roots, which are characteristically dichotomously branched in *Pinus* (Figure 19.56*a*), show typical fungal colonization, and long roots are also frequently colonized. Endomycorrhizae are even more common than ectomycorrhizae. *Arbuscular mycorrhizae*, a type of endomycorrhizae, are found in the roots of over 80% of all terrestrial plant species so far examined and are thus a nearly universal plant symbiosis.

Most mycorrhizal fungi do not attack cellulose and leaf litter but instead use simple carbohydrates for growth and usually have one or more vitamin requirements; they obtain carbon from root secretions but get inorganic minerals from the soil. Mycorrhizal fungi are rarely found in nature except in association with roots and hence most can be considered obligate symbionts. Mycorrhizal fungi produce plant growth substances that induce morphological alterations in the roots, stimulating formation of the mycorrhizal state. However, despite the close relationship between fungus and root, there is little species specificity involved; a single species of pine can form a mycorrhizal association with over 40 species of fungi.

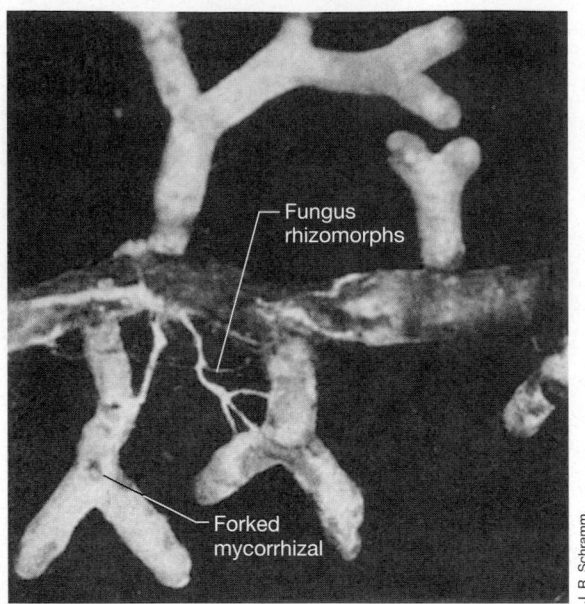

(a)

(b)

Figure 19.56 Mycorrhizae. (a) Typical ectomycorrhizal root of the pine, *Pinus rigida*, with rhizomorphs of the fungus *Thelophora terrestris*. (b) Seedling of *Pinus contorta* (lodgepole pine), showing extensive development of the absorptive mycelium of its fungal associate *Suillus bovinus*. This grows in fanlike formation from the ectomycorrhizal roots to capture nutrients from the soil.

The beneficial effect on the plant of the mycorrhizal fungus is best observed in poor soils, where trees that are mycorrhizal thrive but nonmycorrhizal ones do not. For example, when trees are planted in prairie soils, which ordinarily lack a suitable fungal inoculum, trees that are artificially inoculated at the time of planting grow much more rapidly than uninoculated trees (Figure 19.57). It is well established that the mycorrhizal plant is able to absorb nutrients from its environment more efficiently than a nonmycorrhizal one, and thus the mycorrhizal plant has a competitive advantage. This improved nutrient absorption is due to the greater surface area provided by the fungal mycelium; for example, in the pine seedling shown in Figure 19.56*b*, the ectomycorrhizal fungal mycelium constitutes the overwhelming part of the absorptive area of the plant root system.

But in addition to simply helping plants absorb nutrients, mycorrhizae also appear to play a significant role in controlling plant diversity. Indeed, field experiments have shown that there is a positive correlation between the abundance and diversity of mycorrhizae in a soil and the extent of the plant diversity that develops in it. Thus, mycorrhizae are a prime example of a plant-microorganism symbiosis that benefits both partners—the mycorrhizal plant is better able to function physiologically

Figure 19.57 Six-month-old seedlings of Monterey pine (*Pinus radiata*) growing in prairie soil: left, nonmycorrhizal; right, mycorrhizal.

and compete successfully in a species-rich plant community, while the fungus benefits from a steady supply of organic nutrients.

✓ 19.20 Concept Check

Lichens are symbiotic associations between a fungus and an alga or cyanobacterium. Mycorrhiza are fungi that associate with plant roots and improve their ability to absorb nutrients. Mycorrhiza have a great beneficial effect on plant health and competitiveness.

✓ How do *endomycorrhiza* differ from *ectomycorrhiza*?

✓ Why are mycorrhizal associations with plants considered a type of symbiosis?

19.21 Agrobacterium and Crown Gall Disease

Some microorganisms are plant pathogens; that is, they cause plant disease. The genus *Agrobacterium* comprises organisms that cause the formation of tumorous growths on a wide variety of plants. The two species most widely studied are *A. tumefaciens*, which cause *crown gall*, and *A. rhizogenes*, which causes *hairy root*.

Although plants often form benign accumulations of tissue, called a **callus**, when wounded, the growth induced by *A. tumefaciens* (Figure 19.58) is different in that the callus shows uncontrolled growth. It thus resembles tumor growth in animals, and considerable research on

crown gall has been carried out with the hope that it may provide a model for how malignant growths occur in humans. Interestingly, once induced, these tumors continue to grow in the absence of *Agrobacterium* cells. Thus, once *Agrobacterium* has brought about the induction of the tumorous condition, its presence is no longer necessary.

An overview of the events in crown gall formation is given in Figure 19.59. A large plasmid called the **Ti** (*tumor induction*) **plasmid** (Figure 19.60) must be present in the *Agrobacterium* cells if they are to induce tumor formation. In *Agrobacterium rhizogenes*, a similar plasmid called the *Ri plasmid* is necessary for induction of the disease called

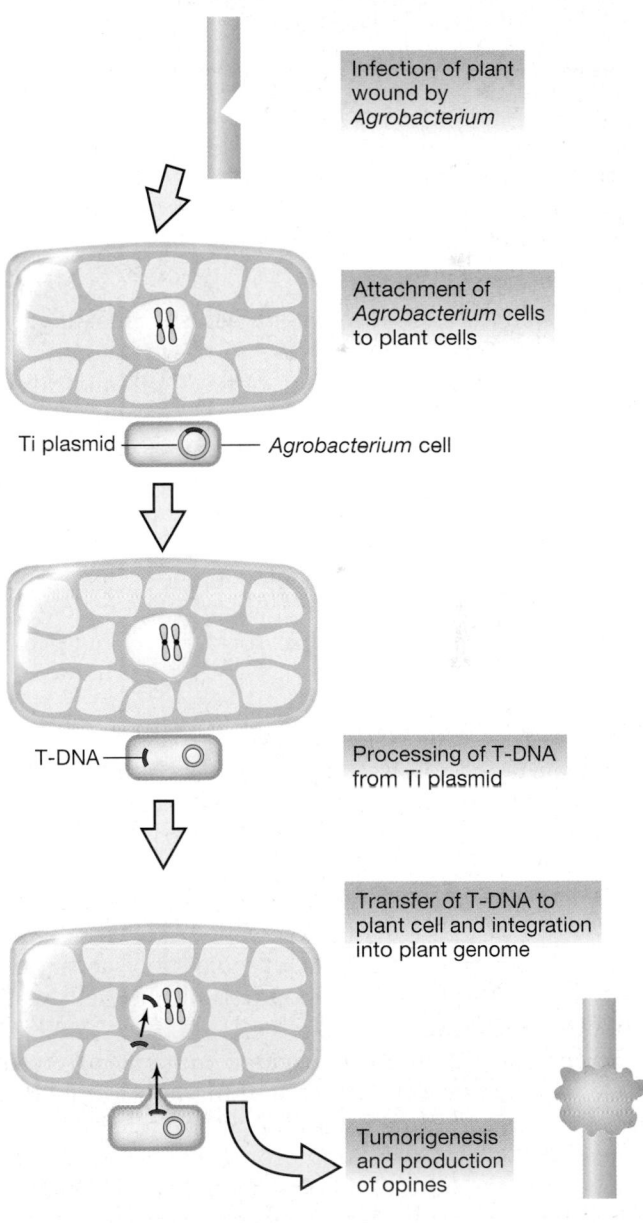

Figure 19.59 Overview of events of crown gall disease following infection of a susceptible plant by *Agrobacterium tumefaciens*. Note that it is only the T-DNA portion of the Ti plasmid that is transferred to the plant.

Figure 19.58 Photograph of a crown gall tumor on a tobacco plant caused by crown gall bacteria of the genus *Agrobacterium*.

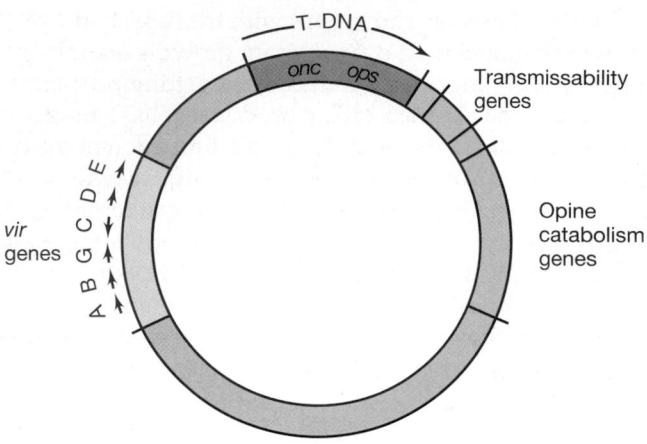

Figure 19.60 Structure of the Ti plasmid of *Agrobacterium tumefaciens*. T-DNA is the region actually transferred to the plant. *vir*, virulence genes; *onc*, oncogenes (tumorigenesis genes); *ops*, opine synthesis genes. Arrows indicate the direction of transcription of each gene. The entire Ti plasmid is about 200 kb of DNA, and the T-DNA about 20 kb (⌒⌒ Section 31.7 for further discussion of Ti).

hairy root. However, the best-studied system is that of *A. tumefaciens* and crown gall disease, so we focus on that here. Following infection, a part of the Ti plasmid called the *transfer DNA* (T-DNA), is integrated into the plant's genome. T-DNA carries the genes for tumor formation and also for the production of a number of modified amino acids called **opines**. *Octopine* [N^2-(1,3-dicarboxyethyl)-*L*-arginine] and *nopaline* [N^2-(1,3-dicarboxypropyl)-*L*-arginine] are the two most common opines. Opines are produced by plant cells transformed by T-DNA and are a source of carbon and nitrogen for *Agrobacterium* cells (Figure 19.59).

Recognition and DNA Transfer

To initiate the tumorous state, cells of *Agrobacterium* must first attach to a wound site on the plant. The recognition of *Agrobacterium* by plant tissue involves complementary receptor molecules on the surfaces of the bacterial and plant cells. It is thought that the plant receptor molecule is a type of *pectin* (a complex polysaccharide) and that the bacterial receptor is a type of polysaccharide containing β-glucans, embedded in the cell wall lipopolysaccharide.

Studies with nontumorigenic mutants of *Agrobacterium tumefaciens* have shown that most functions necessary for attachment of the bacterium to plant are borne on the bacterial chromosome. Following attachment, rapid synthesis of cellulose microfibrils by the bacterium anchors the cells to the wound site and literally entraps the bacterial cells, forming large bacterial aggregates on the plant cell surface. This sets the stage for plasmid transfer from bacterium to plant.

The structure of the Ti plasmid is given in Figure 19.60. Note that although a number of genes are

needed for infectivity, only a small portion of the Ti plasmid, a region called the *T-DNA* (Figures 19.59 and 19.60), is actually transferred to the plant. The T-DNA contains genes that induce tumorigenesis. The *vir* genes that reside on the Ti plasmid encode proteins that are essential for T-DNA transfer (Figure 19.60). *vir* gene expression is induced by plant signal molecules synthesized by wounded plant tissues. Some inducers that have been identified include the phenolic compounds acetosyringone, *p*-hydroxybenzoic acid, and vanillin.

The *vir* genes are the key to T-DNA transfer. The *virA* gene encodes a protein kinase that interacts with inducer molecules and then phosphorylates the product of the *virG* gene (Figure 19.61). The latter becomes activated by the phosphorylation event and functions to activate other *vir* genes. The product of the *virD* gene has endonuclease activity and nicks DNA in the Ti plasmid in a region adjacent to the T-DNA (Figures 19.60 and 19.61). The product of the *virE* gene is a single-stranded DNA binding protein that binds the *single strand* of T-DNA generated from endonuclease activity and transports this small fragment of DNA into the plant cell. The *virB* gene product is located in the bacterial membrane and mediates transfer of the single strand of DNA between bacterium and plant.

T-DNA transfer occurs in a process that resembles bacterial conjugation (Figures 19.59 and 19.61). The T-DNA becomes inserted into the nuclear genome of the plant (plant cell organellar genomes are not infected) where integration can occur at any number of sites where specific inverted or direct tandem repeats are present. The tumorigenesis (*onc*) genes of the Ti plasmid (Figure 19.60) encode enzymes involved in plant hormone production and at least one key enzyme of opine biosynthesis. Expression of these genes leads to tumor formation. The Ri plasmid involved in hairy root disease also contains *onc* genes. However, in this case the genes confer increased auxin responsiveness to the plant, which leads to overproduction of root tissue, resulting in the symptoms of the disease. The Ri plasmid also encodes several opine biosynthetic enzymes.

Genetic Engineering with the Ti Plasmid

From the standpoint of microbiology and plant pathology both crown gall and hairy root disease involve a unique type of plant-bacterial interaction in which bacterial DNA is physically transferred to plant cells. However, once the mechanism of DNA transmission was understood, it quickly became clear that the Ti system could be used by scientists as a vector for the introduction of genetically engineered DNA into plants—that is, that Ti was a natural plant transformation system. Thus, through the years the focus of the Ti/crown gall system has shifted away from the disease to new applications in the plant biotechnology industry.

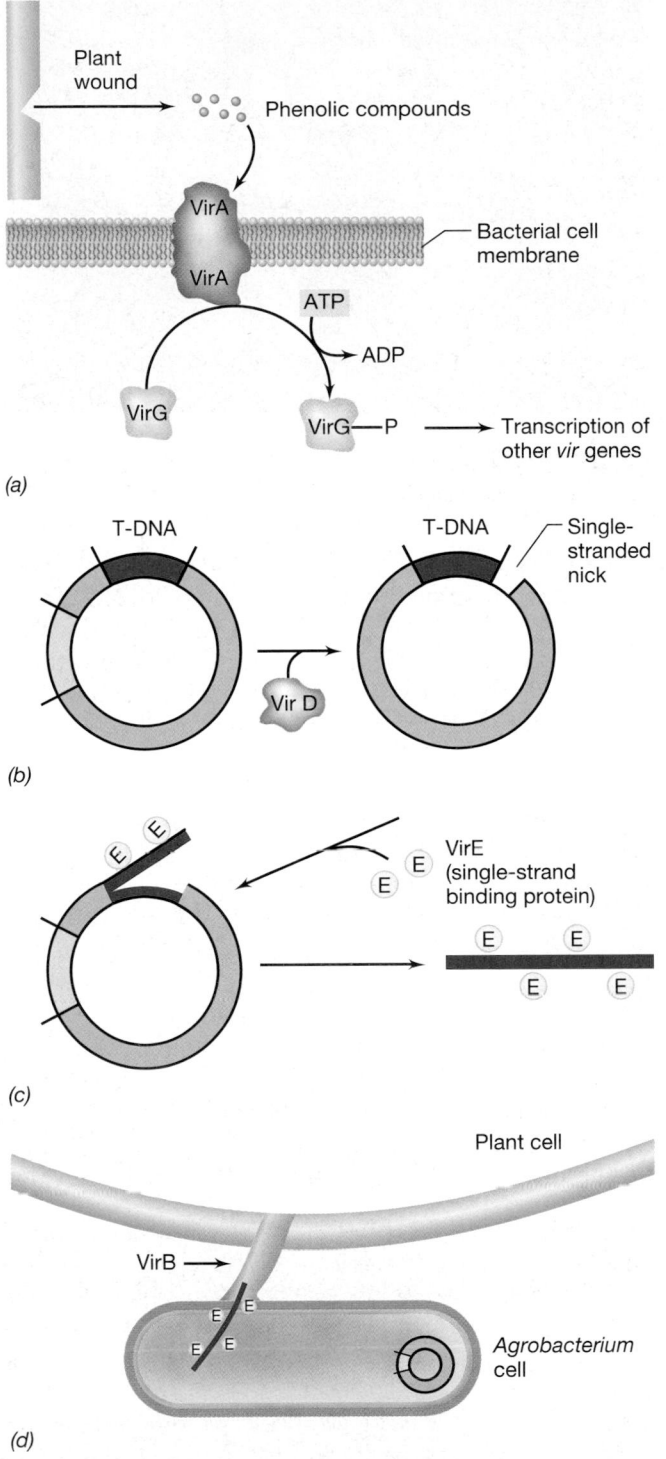

(a)

(b)

(c)

(d)

Figure 19.61 Mechanism of transfer of T-DNA to the plant cell by *Agrobacterium tumefaciens*. (a) Levels of VirA protein increase dramatically upon stimulation with plant phenolic inducer molecules. VirA activates VirG by phosphorylation, and VirG activates transcription of other *vir* genes. (b) VirD is an endonuclease. (c) VirE is a single-stranded DNA-binding protein. (d) VirB acts as a conjugation bridge between *Agrobacterium* and the plant cell. Plant DNA polymerase produces the complementary strand to the single strand of T-DNA that is transferred before the DNA gets integrated into the plant genome.

Experience has shown that plants are fairly difficult to transform. But through the power of genetic engineering a host of modified ("disarmed") Ti plasmids are now available for the facile production of transgenic plants without transfer of disease genes. Success stories are already recorded in the areas of herbicide and insect resistance, and many other areas remain to be explored. We discuss the use of the Ti plasmid as a vector in plant biotechnology in more detail in Section 31.7.

✓ **19.21 Concept Check**

The crown gall bacterium *Agrobacterium* enters into a unique relationship with higher plants. A plasmid in the bacterium (the Ti plasmid) is able to transfer part of itself to the genome of the plant, in this way bringing about the production of the crown gall disease. The crown gall plasmid has also found extensive use in the genetic engineering of crop plants.

✓ What are *opines* and why are they produced?

✓ How do the *vir* genes differ from *T-DNA* in the Ti plasmid?

✓ How has an understanding of crown gall disease benefited the area of plant molecular biology?

19.22 Root Nodule Bacteria and Symbiosis with Legumes

One of the most interesting and important plant bacterial interactions is that between leguminous plants and certain gram-negative nitrogen-fixing bacteria. Legumes are a large group that include such economically important plants as soybeans, clover, alfalfa, beans, and peas, and are defined as plants that bear seeds in pods. *Rhizobium, Bradyrhizobium, Sinorhizobium, Mesorhizobium,* and *Azorhizobium* are gram-negative motile rods. Infection of the roots of a leguminous plant with the appropriate species of one of these genera leads to the formation of **root nodules** (Figure 19.62) that are able to convert gaseous nitrogen to combined nitrogen, a process called *nitrogen fixation* (⌾ Section 17.28); *Azorhizobium* forms root nodules as well as stem nodules (see later). Nitrogen fixation by the legume-*Rhizobium*, symbiosis is of considerable agricultural importance, as it leads to very significant increases in combined nitrogen in the soil. Because nitrogen deficiencies often occur in unfertilized bare soils, nodulated legumes are at a selective advantage under such conditions and can grow well in areas where other plants cannot (Figure 19.63).

Leghemoglobin and Cross-Inoculation Groups
Under normal conditions, neither legume nor *Rhizobium* can fix nitrogen; yet the interaction between the two leads to the development of nitrogen-fixing ability. In

Figure 19.62 Soybean root nodules. The nodules develop by infection with *Bradyrhizobium japonicum*.

pure culture, the *Rhizobium* is able to fix N_2 alone when grown under strictly controlled *microaerophilic* conditions. Apparently *Rhizobium* needs some O_2 to generate energy for N_2 fixation, but its nitrogenase (like those of other nitrogen-fixing organisms, ⚭ Section 17.28) is inactivated by O_2. In the nodule, precise O_2 levels are controlled by the O_2 binding protein **leghemoglobin**. This is a red, iron-containing protein that is always found in healthy N_2-fixing nodules (Figure 19.64). Neither plant nor *Rhizobium* alone synthesizes leghemoglobin, but formation is thought to be induced through the interaction

Figure 19.63 A field of unnodulated (left) and nodulated (right) soybean plants growing in nitrogen-poor soil.

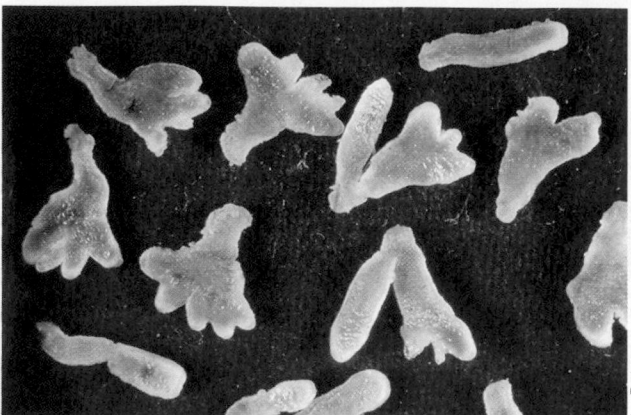

Figure 19.64 Sections of root nodules from the legume *Coronilla varia*, showing the reddish pigment leghemoglobin.

of these two organisms. Leghemoglobin functions as an "oxygen buffer" cycling between the oxidized (Fe^{3+}) and reduced (Fe^{2+}) forms to keep free O_2 levels within the nodule at a low but constant level. The ratio of leghemoglobin-bound O_2 to free O_2 in the root nodule is on the order of 10,000:1.

About 90% of all leguminous plant species are capable of becoming nodulated. However, there is a marked specificity between species of legume and strains of *Rhizobium*. A single *Rhizobium* strain is generally able to infect certain species of legumes and not others. A group of *Rhizobium* strains able to infect a group of related legumes is called a *cross-inoculation group*. The major rhizobial cross-inoculation groups are listed in Table 19.8. Even if a *Rhizobium* strain infects a certain legume, it is not always able to bring about the production of nitrogen-fixing nodules. If the strain is *ineffective*, the nodules formed will be small, greenish-white, and incapable of fixing nitrogen; if the strain is *effective*, on the other hand, the nodule will be large, reddish

TABLE 19.8	Major cross-inoculation groups of leguminous plants
Host plant	**Nodulated by**
Pea	*Rhizobium leguminosarum* biovar *viciae*[a]
Bean	*Rhizobium leguminosarum* biovar *phaseoli*[a]
Bean	*Rhizobium tropici*
Lotus	*Mesorhizobium loti*
Clover	*Rhizobium leguminosarum* biovar *trifolii*[a]
Alfalfa	*Sinorhizobium meliloti*
Soybean	*Bradyrhizobium japonicum*
Soybean	*Bradyrhizobium elkanii*
Soybean	*Rhizobium fredii*
Sesbania rostrata (a tropical legume)	*Azorhizobium caulinodans*

[a] Several varieties (biovars) of *Rhizobium leguminosarum* exist, each capable of nodulating a different legume.

(Figure 19.64), and nitrogen-fixing. Effectiveness is determined by genes in the bacterium (see the discussion of *nod* genes later in this section).

Stages in Root Nodule Formation

The stages in the infection and development of root nodules are now fairly well understood (Figure 19.65). They include

1. **recognition** of the correct partner on the part of both plant and bacterium and **attachment** of the bacterium to root hairs.

2. **excretion** of nod factors by the bacterium.

3. **invasion** of the root hair by the bacterial formation of an infection thread.

4. **travel** to the main root via the infection thread.

5. formation of deformed bacterial cells, **bacteroids**, within the plant cells and development of the nitrogen-fixing state.

6. continued plant and bacterial division and formation of the mature **root nodule**.

We now explore some of these stages in nodule formation in more detail.

Attachment and Infection

The roots of leguminous plants secrete a variety of organic compounds that stimulate the growth of a rhizosphere microflora. This stimulation is not restricted to the rhizobia but occurs with a variety of rhizosphere bacteria. If there are rhizobia in the soil, they grow in the rhizosphere and build up to high population densities. Attachment of bacterium to plant in the legume-*Rhizobium* symbiosis is the first step in the formation of nodules (Figure 19.65). A specific adhesion protein called *rhicadhesin* is present on the surfaces of all species of *Rhizobium* and *Bradyrhizobium*. Rhicadhesin is a calcium-binding protein and may function by binding calcium complexes on the root hair surface. Other substances, such as carbohydrate-containing proteins called *lectins*, also play some role in plant-bacterium attachment.

Initial penetration of *Rhizobium* cells into the root hair is via the root hair tip. Following binding, the root hair curls as a result of the action of substances excreted by the bacterium called *nod factors* (see later) and the bacteria enter the root hair and induce formation by the plant of a cellulosic tube, called the **infection thread**, which spreads down the root hair. Root cells adjacent to the root hairs subsequently become infected by rhizobia, and nod factors stimulate plant cell division, eventually leading to formation of the nodule (Figures 19.66 and 19.67, see also Figures 19.62 and 19.64).

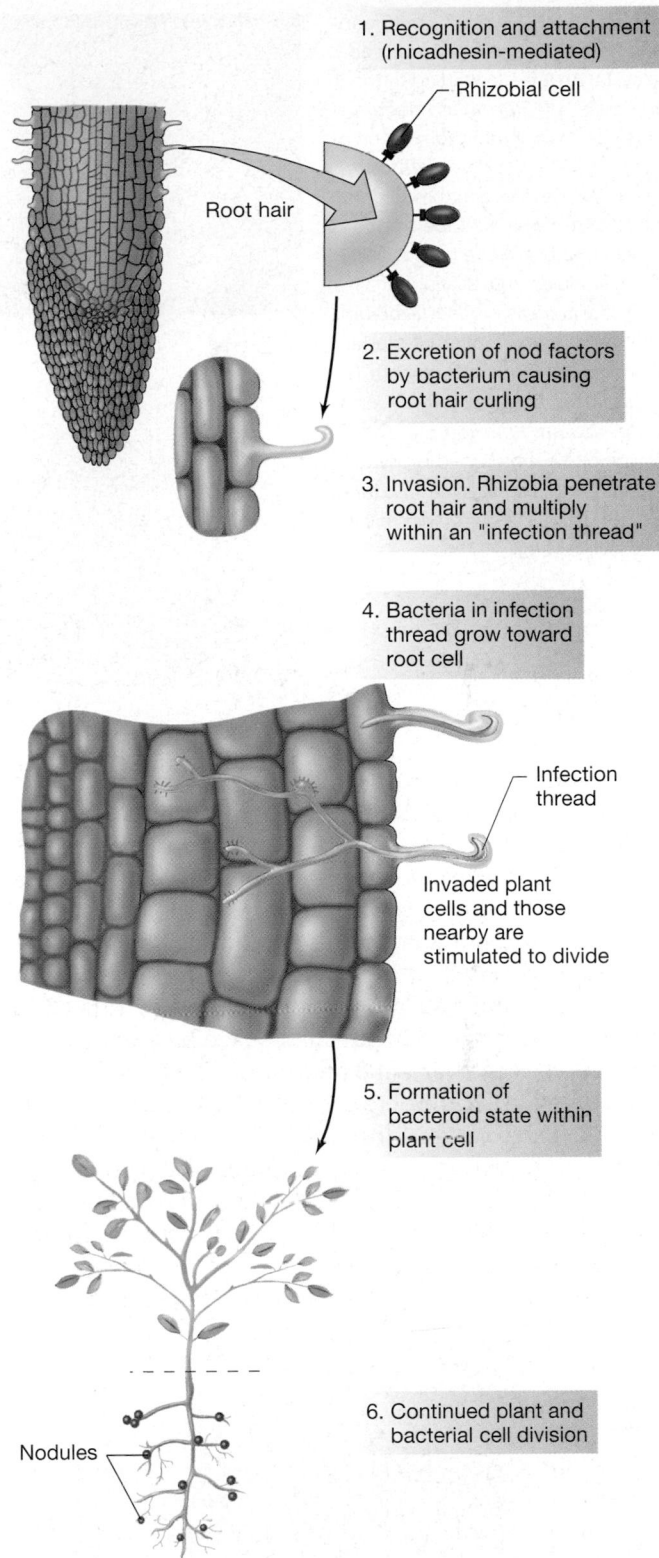

1. Recognition and attachment (rhicadhesin-mediated)
   — Rhizobial cell
   Root hair

2. Excretion of nod factors by bacterium causing root hair curling

3. Invasion. Rhizobia penetrate root hair and multiply within an "infection thread"

4. Bacteria in infection thread grow toward root cell

   Infection thread

   Invaded plant cells and those nearby are stimulated to divide

5. Formation of bacteroid state within plant cell

6. Continued plant and bacterial cell division

Nodules

Figure 19.65 Steps in the formation of a root nodule in a legume infected by *Rhizobium*. For a photomicrograph of an infection thread, see Figure 19.66a. For photos of root nodules, see Figures 19.62 and 19.64. Formation of the bacteroid state is a prerequisite for nitrogen fixation

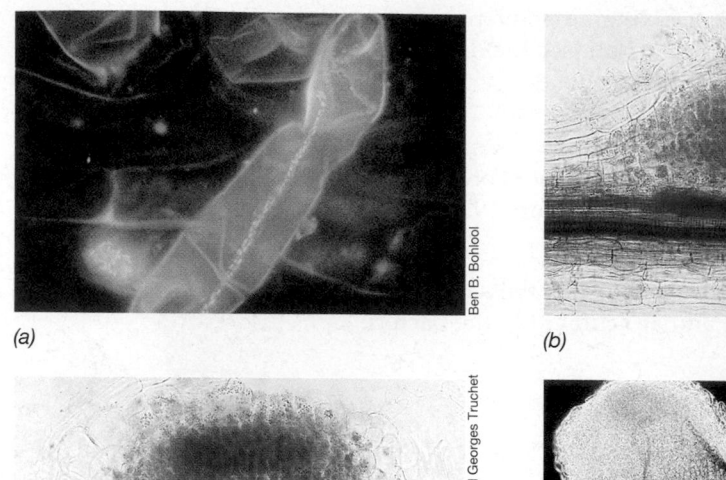

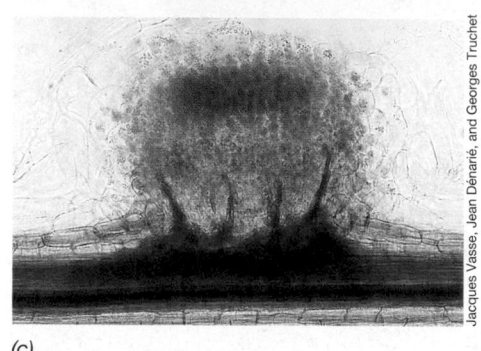

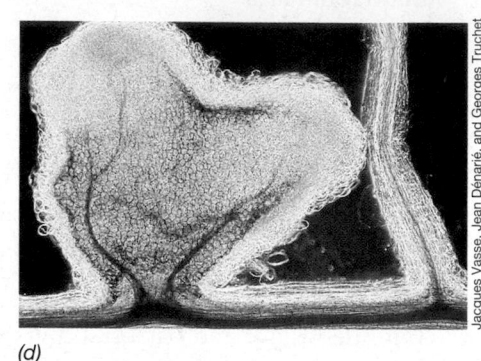

Figure 19.66 The infection thread and formation of root nodules. (a) An infection thread formed by cells of *Rhizobium leguminosarum* biovar *trifolii* formed on a root hair of white clover (*Trifolium repens*). The infection thread consists of a cellulosic tube through which bacteria move to root cells. (b–d) Nodules from alfalfa roots infected with cells of *Sinorhizobium meliloti* shown at different stages of development. Cells of both *R. leguminosarum* biovar *trifolii* and *Sinorhizobium meliloti* are about 2 μm long. Photos b–d reprinted with permission from *Nature* 351:670–673 (1991), © Macmillan Magazines Ltd.

Bacteroids

The *Rhizobium* bacteria multiply rapidly within the plant cells and are transformed into swollen, misshapen, and branched forms called **bacteroids**. Bacteroids become surrounded singly or in small groups by portions of the plant cell membrane to form structures called the **symbiosome** (see Figure 19.71). Only after the formation of the symbiosome does nitrogen fixation begin. (Effective nitrogen-fixing nodules can be detected by acetylene reduction, ∞ Section 17.28 and Figure 17.71). When the plant dies, the nodule deteriorates, releasing bacteria into the soil. The bacteroid forms are incapable of division, but there are always a small number of dormant rod-shaped cells present in the nodule. These now proliferate, using some of the products of the deteriorating nodule as nutrients, and the bacteria can initiate the infection in other roots or maintain a free-living existence in the soil.

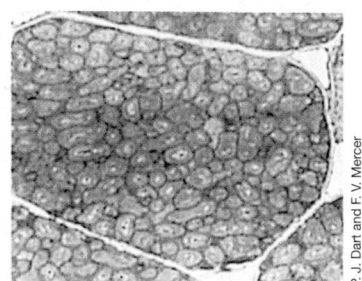

Figure 19.67 Cross section through a legume root nodule. Electron micrograph of a thin section through a single bacteria-filled cell of a subterranean clover nodule. Cells of *Rhizobium leguminosarum* biovar *trifolii* are about 2 μm long.

Nodule Formation: Nod Genes, Nod Proteins, and Nod Factors

Genes directing specific steps in nodulation of a legume by a strain of *Rhizobium* are called *nod genes* (Figure 19.68). Many *nod* genes from different *Rhizobium* species are highly conserved and are borne on large plasmids called *Sym plasmids*. In addition to *nod* genes, which direct specific nodulation events, Sym plasmids contain *specificity genes*, which restrict a strain of *Rhizobium* to a particular host plant. Indeed, cross-inoculation group specificity can be transferred across species of rhizobia by simply transferring the respective Sym plasmid.

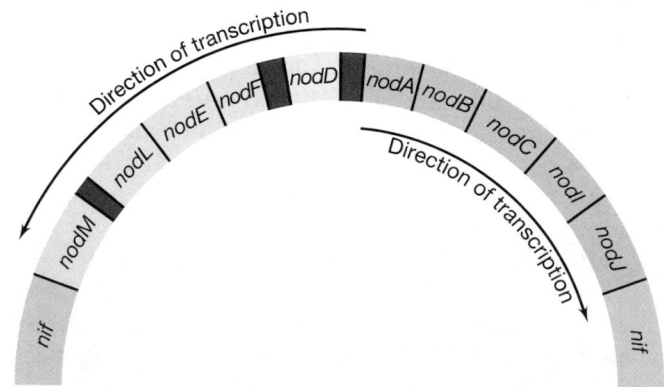

Figure 19.68 Organization of the *nod* gene cluster on the Sym plasmid of *Rhizobium leguminosarum* biovar *viciae*, the species that nodulates peas. The product of *nodD* controls transcription of other *nod* genes. The *nod* boxes are highlighted in red, and the arrows indicate the direction of transcription of *nod* genes. See text for further details.

In the Sym plasmid of *Rhizobium leguminosarum* biovar *viciae, nod* genes are located between two clusters of genes for nitrogen fixation, the *nif* genes (⊂⊃ Section 17.28; in this species and in certain other *Rhizobium* species, *nif* genes are plasmid-borne). The arrangement of *nod* genes in the *R. leguminosarum* Sym plasmid is shown in Figure 19.68. Ten *nod* genes have been identified in this species. The entire *nod* region has been sequenced, and the function of many Nod proteins is known. The *nodABC* genes are involved in the production of oligosaccharides, called *nod factors*, which induce root hair curling and trigger plant cell division, eventually leading to formation of the nodule (Figure 19.66). Chemically, nod factors consist of a backbone of *N*-acetylglucosamine to which various substituents are linked (Figure 19.69). Host specificity is determined by the precise structure of the nod factor produced by a given species of *Rhizobium*. Thus, besides the *nodABC* genes, whose products synthesize the nod backbone, each species of *Rhizobium* contains certain unique *nod* genes responsible for introducing chemical variations on the basic nod factor backbone.

In *Rhizobium leguminosarum* biovar *viciae,* the gene *nodD* encodes a regulatory protein, NodD, which controls transcription of other *nod* genes (Figure 19.68). NodF encodes a specific acyl carrier protein used by NodA to acylate the nod factor (Figure 19.69b), NodE and NodL are involved in host range (cross-inoculation groups), NodM is a glucosamine synthase involved in nod factor synthesis, and NodI and NodJ are membrane proteins that function to export nod factors from the bacterial cells.

NodD binds to DNA upstream of *nod* structural gene operons (regions called *nod boxes*) and, after interacting with inducer molecules, bends the DNA in this region, which promotes transcription (thus NodD is a type of positive regulator, ⊂⊃ Section 8.6). Several inducer molecules have been identified. Key inducers are plant *flavonoids*, complex organic molecules that are widespread plant products (Figure 19.70a and b). Flavonoids have many functions in plants, including growth regulation, attraction of pollinating insects, and pest control. However, leguminous plants are unusual because, unlike those of other plants, their *roots* secrete large amounts of flavonoids, presumably to trigger *nod* gene expression in nearby rhizobial cells in the soil. Interestingly, some flavonoids that are structurally very closely related to *nodD* inducers (Figure 19.70c) instead *inhibit* induction of *nod* genes in certain rhizobial species, suggesting that part of the specificity observed between plant and bacterium in the *Rhizobium*-legume symbiosis could lie in the chemical nature of the flavonoids excreted by a particular plant.

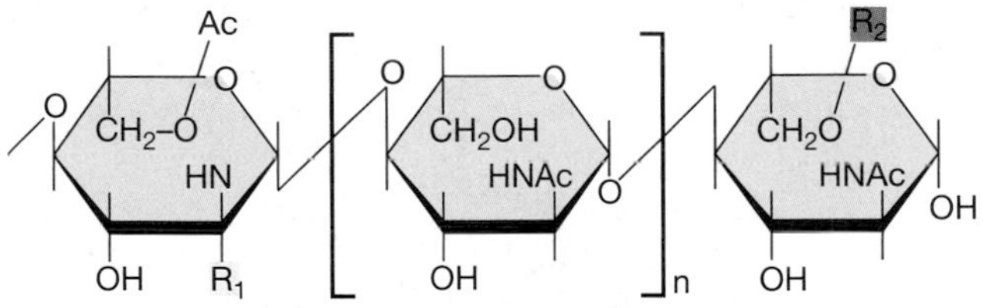

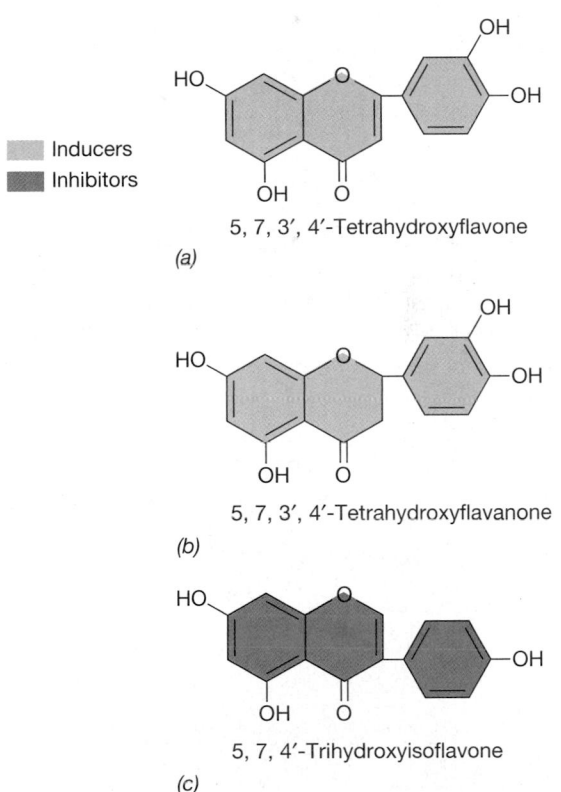

(a)

5, 7, 3′, 4′-Tetrahydroxyflavone

(b)

5, 7, 3′, 4′-Tetrahydroxyflavanone

(c)

5, 7, 4′-Trihydroxyisoflavone

Figure 19.70 Structure of flavonoid molecules that are (a, b) inducers of *nod* gene expression and (c) inhibitors of *nod* gene expression in *Rhizobium leguminosarum* biovar *viciae*. Note the similarities in the structures of all three molecules. The common name of the structure shown in (a) is *luteolin*, and it is a flavone derivative. The structure in (b) is called *eriodictyol* and is chemically a flavanone. The structure in (c) is called *genistein*, and it is an isoflavone derivative.

(a)

Species	R$_1$	R$_2$
Sinorhizobium meliloti	C16:2 or C16:3	SO$_4{}^{2-}$
Rhizobium leguminosarum biovar *viciae*	C18:1 or C18:4	H or Ac

(b)

Figure 19.69 Nod factors. (a) The general structure of the nod factor produced by *Sinorhizobium meliloti* and *Rhizobium leguminosarum* biovar *viciae* and (b) a table of the structural differences (R$_1$, R$_2$) that define the precise nod factor of each species. The central hexose unit can repeat up to three times. C16:2, palmitic acid with two double bonds; C16:3, palmitic acid with three double bonds; C18:1, oleic acid with one double bond; C18:4, oleic acid with four double bonds; Ac, acetyl.

Biochemistry of Nitrogen Fixation in Nodules

As discussed in Section 17.28, nitrogen fixation involves the activity of the enzyme **nitrogenase**, a large two-component protein containing iron and molybdenum. Nitrogenase in root nodules has characteristics similar to the enzyme from free-living N_2-fixing bacteria, including O_2 sensitivity and the ability to reduce acetylene as well as N_2 (∞ Figure 17.71). Nitrogenase is localized within the bacteroids themselves and is not released into the plant cytosol.

Bacteroids are totally dependent on the plant for supplying them with energy sources for N_2 fixation. The major organic compounds transported across the symbiosome membrane and into the bacteroid proper are citric acid cycle intermediates, in particular the C_4 acids *succinate, malate,* and *fumarate* (Figure 19.71). These are used as electron donors for ATP production and, fol-lowing conversion to pyruvate, as the ultimate source of electrons for the reduction of N_2 (∞ Figure 17.70).

The first stable product of N_2 fixation is *ammonia,* and several lines of evidence suggest that assimilation of ammonia into organic nitrogen compounds in the root nodule is carried out primarily by the plant. Although bacteroids can assimilate some ammonia into organic form, the levels of ammonia-assimilatory enzymes in bacteroids are quite low. By contrast, the ammonia-assimilating enzyme *glutamine synthetase* (∞ Figure 5.15b) is present in high levels in the plant cell cytoplasm. Hence, ammonia transported from the bacteroid to the plant cell can be assimilated by the plant as the amino acid *glutamine.* Besides glutamine, other nitrogenous compounds, in particular other amino acid amides, such as asparagine and 4-methylene glutamine, and the ureides *allantoin* and *allantoic acid,* are synthesized by the plant and subsequently transported to plant tissues (see Figure 19.71).

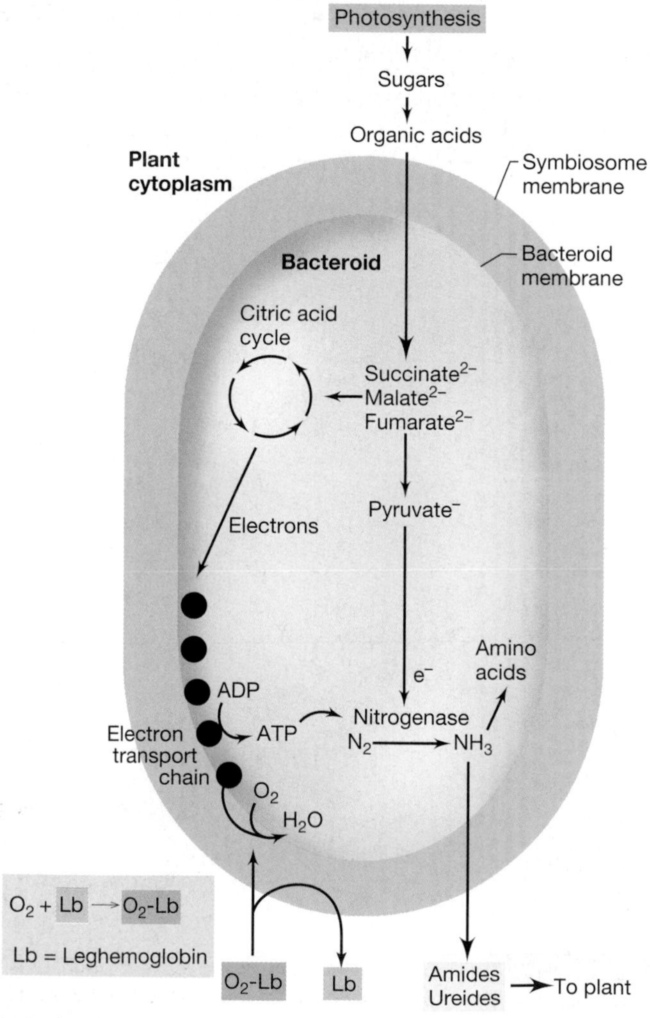

Stem-Nodulating Rhizobia and Other Rhizobial Associations

Although most leguminous plants form nitrogen-fixing nodules on their *roots,* a few legume species bear nodules on their *stems.* Stem-nodulated leguminous plants are quite widespread in tropical regions where soils are often nitrogen-deficient because of leaching and intense biological activity. The major experimental system here has been the tropical aquatic legume *Sesbania,* which is nodulated by the bacterium *Azorhizobium caulinodans* (Figure 19.72). Stem nodules usually form in the submerged portion of the stems or on portions of the stem just above the water level (Figure 19.72). From studies thus far, the general sequence of events in formation of stem nodules in

Figure 19.71 Schematic diagram of major metabolic reactions and nutrient exchanges occurring in the bacteroid. The symbiosome is a collection of bacteroids surrounded by a single membrane originating from the plant.

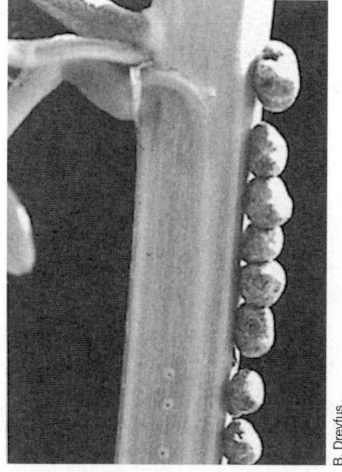

Figure 19.72 Stem nodules caused by stem-nodulating *Azorhizobium.* The photograph shows the stem of the tropical legume *Sesbania rostrata.* On the left side of the stem are uninoculated sites, on the right identical sites inoculated with stem-nodulating rhizobia.

Sesbania strongly resembles that of root nodules. An infection thread is formed, and bacteroid formation predates the N_2-fixing state.

One curious finding is that some stem-nodulating rhizobia produce bacteriochlorophyll *a* and thus may have the potential to carry out anoxygenic photosynthesis (∞ Section 17.4). Bacteriochlorophyll-containing rhizobia, grouped in the genus *Photorhizobium*, are widespread in nature, particularly in association with tropical legumes, and in these species, light could drive at least part of the energy-demanding process of N_2 fixation in the bacterium.

A third habitat, besides soil and the root nodules of clover plants, has been found for the rhizobial species *Rhizobium leguminosarum* biovar *trifolii*. In fields in which rice is grown in alternate years with clover, a specific association between rice roots and *R. leguminosarum* bv *trifolii* occurs (Figure 19.73). The association is one in which the rhizobial cells live an *inter*cellular existence within the rice roots and do not produce nodules. Nevertheless, rice plants containing the bacterial association grow faster and with higher yield than plants that lack *R. leguminosarum* biovar *trifolii* cells. Interestingly, the association does not result in nitrogen fixation in the rice roots. Instead, the benefit to the plant appears to come from improved nutrient uptake, water relations, and related factors.

The *Rhizobium*–Legume Symbiosis and Its Agricultural Importance

The *Rhizobium*–legume association is a true symbiosis, because each partner clearly contributes something to the other. Fixed nitrogen benefits the plant growing in nitrogen-deficient soils while the nodule is a physically protected and well-nourished environment for rhizobia. Nowhere in microbiology do we see an example of a beneficial plant/bacterial relationship so well developed or so well understood as in the legume/root nodule symbiosis. And the agricultural benefits of the more than 120 million tons of atmospheric nitrogen that is biologically fixed to ammonia each year, are enormous. The plants listed in Table 19.8 include soybean and alfalfa, key commodities for the soy-processing industry and the feeding of domesticated animals (alfalfa hay), respectively, as well as beans and peas, major vegetables for human nutrition. Several agricultural industries revolve around leguminous crops, and the ability of legumes to grow without nitrogen fertilizer saves farmers millions of dollars in fertilizer costs yearly.

Nonlegume Nitrogen-Fixing Symbioses

Nitrogen-fixing symbioses involving microorganisms other than rhizobia occur in a variety of nonleguminous plants. Nitrogen-fixing cyanobacteria form symbioses with a variety of plants. The water fern *Azolla* contains a species of heterocystous N_2-fixing cyanobacteria called *Anabaena azollae* within small pores of its fronds (Figure 19.74). *Azolla* has been used for centuries to enrich rice paddies with fixed nitrogen. Before planting rice, the farmer allows the surface of the rice paddy to become densely covered with the fern. As the rice plants grow, they eventually crowd out the *Azolla-Anabaena* mixture, leading to death of the fern and

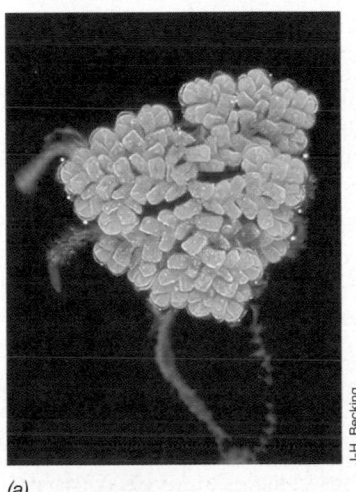

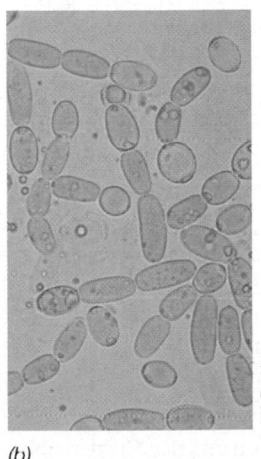

(a)   *(b)*

Figure 19.73 *Rhizobium leguminosarum* biovar *trifolii* and rice roots. Fluorescent photomicrograph of green fluorescent protein-containing (∞ Section 18.3) rhizobial cells (arrow) attached to and distributed within the roots of rice plants. Although they do not enter plant cells and fix nitrogen as in the root nodule symbiosis, rice plants with associated *R. leguminosarum* biovar *trifolii* are more vigorous and grow faster than those without.

Figure 19.74 *Azolla-Anabaena* symbiosis. (a) Intact association showing a single plant of *Azolla pinnata*. The diameter of the plant is approximately 1 cm. (b) Cyanobacterial symbiont *Anabaena azollae* as observed in crushed leaves of *A. pinnata*. Single cells of *Anabaena azollae* are about 5 μm wide. Note the oval-shaped *heterocysts* (lighter color), the site of nitrogen fixation in the cyanobacterium. For discussion of the heterocyst see Section 12.26 and Figure 12.80.

(a)

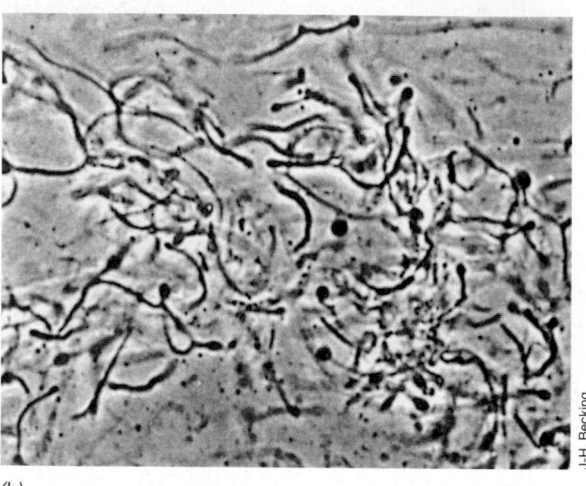

(b)

J-H. Becking

Figure 19.75 *Frankia* nodules and *Frankia* cells. (a) Root nodules of the common alder *Alnus glutinosa*. (b) *Frankia* culture purified from nodules of *Comptonia peregrina*. Note vesicles (spherical structures) on the tips of hyphal filaments.

release of fixed nitrogen, which is assimilated by the rice plants. By repeating this process each growing season, the farmer can obtain high yields of rice without the need for nitrogenous fertilizers.

The alder tree (genus *Alnus*) has nitrogen-fixing root nodules (Figure 19.75) that harbor a filamentous, streptomycete-like, nitrogen-fixing organism called *Frankia*. Although when assayed in cell extracts the nitrogenase of *Frankia* is sensitive to molecular oxygen, like intact cells of *Azotobacter* (Section 12.9), intact cells of *Frankia* fix N_2 at full oxygen tensions. This is because *Frankia* protects its nitrogenase by localizing it in terminal swellings on the cells called *vesicles* (Figure 19.75b). The vesicles contain thick walls that retard O_2 diffusion, thus maintaining the O_2 tension within vesicles at levels compati-

ble with nitrogenase activity. In this regard, *Frankia* vesicles resemble the heterocysts produced by some filamentous cyanobacteria as localized sites of N_2 fixation (see Figure 19.74b; Section 12.25 and Figure 12.80).

Alder is a characteristic pioneer tree able to colonize bare soils at nutrient-poor sites, probably because of its ability to enter into a symbiotic nitrogen-fixing relationship with *Frankia*. A number of other small woody plants are nodulated by *Frankia*. Unlike the *Rhizobium* legume relationship, a single strain of *Frankia* can form nodules on several different species of plants, suggesting that the *Frankia*/root nodule symbiosis is less specific in this regard.

Azospirillum lipoferum is a N_2-fixing bacterium that casually associates with roots of tropical grasses. It is found in the rhizosphere, where it grows on products excreted from the roots. It also has the ability to grow around the roots of cultivated grasses, such as corn (*Zea mays*), and inoculation of corn with *A. lipoferum* may lead to small increases (about 10%) in growth yield of the plant. In sugarcane, the organism *Acetobacter diazotrophicus* grows in plant vascular tissue and fixes substantial amounts of N_2, which benefits the plant. Cultures of *A. diazotrophicus* require high sucrose concentrations, suggesting that this plant-bacterium association may be more highly developed than the *Azospirillum* association. It is likely that many casual to tight associations exist between bacteria and plants, but only in the *Rhizobium* or *Frankia* nodule has direct evidence of a symbiotic relationship been found.

✓ **19.22 Concept Check**

One of the most widespread and important plant-microbial symbioses is that between legumes and certain nitrogen-fixing bacteria. The bacteria induce the formation of root nodules within which the nitrogen-fixing process occurs. The plant provides the organic energy source needed by the root nodule bacteria, and the bacteria provide fixed nitrogen for the growth of the plant. The root nodule bacteria play an important agricultural role because many important crop plants are legumes (alfalfa, clover, soybeans, peas, and so on).

✓ What effects do nod factors have on a plant?
✓ What is *leghemoglobin* and what is its function?
✓ What is a *bacteroid* and what occurs within it?
✓ What are the major similarities and differences between *Rhizobium* and *Frankia*?

Review Questions

1. Explain why both obligately *anaerobic* and obligately *aerobic* bacteria can be isolated from the same soil sample.

2. List some of the key *resources* that microorganisms need to thrive in their habitats. List some key *conditions* as well.

3. The surface of a rock in a flowing stream will often contain a biofilm. What advantage(s) could be conferred on bacteria growing in a biofilm compared with growth in a flowing stream?

4. How can biofilms complicate treatment of infectious diseases?

5. How and in what way does an input of organic matter, such as sewage, effect the O_2 content of a river or stream?

6. Assuming proteorhodopsin to be a functional analog of bacteriorhodopsin, describe how this molecule might benefit a cell living in a very organically deficient environment such as the open ocean. (You may want to refer back to Sections 12.2 and 17.1 before answering.)

7. What is the difference between *barotolerant* and *barophilic* bacteria? Between these two groups and *extreme barophiles*? What properties do barotolerant, barophilic, and extremely barophilic bacteria have in common?

8. What evidence from hydrothermal vents exists to support the idea that prokaryotes are growing at extremely high temperatures? After reviewing the data of Table 6.1, why is the marine worm *Alvinella* of interest to biologists?

9. How can organisms such as *Syntrophobacter* and *Syntrophomonas* grow when their metabolism is based on thermodynamically unfavorable reactions? How are these organisms grown in laboratory culture?

10. Why is sulfate reduction the main form of anaerobic respiration in marine environments, whereas methanogenesis dominates in fresh waters? Does any methanogenesis occur in the marine environment? If so, how?

11. What is the *rumen* and how do the digestive processes operate in the ruminant digestive tract? What are the major benefits and the disadvantages of a rumen system? How does a cecal animal compare with a ruminant?

12. Why can urea or ammonia be a nitrogen source for ruminants but not for humans?

13. Compare and contrast the processes of *nitrification* and *denitrification* in terms of the organisms involved, the environmental conditions that favor each process, and the changes in nutrient availability that accompany each process.

14. What organisms are involved in cycling sulfur compounds anoxically? If sulfur chemolithotrophs had never evolved, would there be a problem in the microbial cycling of sulfur compounds? What *organic* sulfur compounds are of interest in nature?

15. Why are most iron-oxidizing chemolithotrophs obligate aerobes and why are most iron oxidizers acidophilic?

16. Explain how spontaneous chemical reactions can acidify a coal seam and how both chemical reactions and *Thiobacillus ferrooxidans* continue the production of acid thereafter.

17. How is *Thiobacillus ferrooxidans* useful in the mining of copper ores? What crucial step in the indirect oxidation of copper ores is carried out by *T. ferrooxidans*? How is copper recovered from copper solutions produced by leaching?

18. How is mercury detoxified by the *mer* system?

19. What physical and chemical conditions are necessary for the rapid microbial degradation of oil in aquatic environments? Design an experiment that would allow you to test what conditions optimized the oil oxidation process.

20. What are *xenobiotic compounds* and why might microorganisms have difficulty catabolizing them?

21. How are reductive dechlorinators of benefit in solving environmental pollution problems?

22. Describe the chemical properties of "microbial plastic."

23. Compare and contrast the production of a plant tumor by *Agrobacterium tumefaciens* and a root nodule by a *Rhizobium* species. In what ways are these structures similar? In what ways are they different? Of what importance are plasmids to the development of both structures?

24. What genetic information resides on the Ti plasmid? Which gene(s) could be deleted from the Ti plasmid without affecting tumorigenesis?

25. What ecological advantages do leguminous plants have over nonlegumes?

26. What substances of plant origin are found in root nodules of legumes that are required for the rhizobial symbiont to fix nitrogen?

27. Describe the steps in the development of root nodules on a leguminous plant. What is the nature of the recognition between plant and bacterium? How does this compare with recognition in the *Agrobacterium*–plant system?

28. How does nitrogen fixed by *Rhizobium* become part of the plant's proteins?

29. Describe the *nod* operons of a typical *Rhizobium* species such as *R. leguminosarum*. Explain how *nod* genes are regulated. What are Nod factors and what do they do?

Application Questions

1. Compare and contrast a lake ecosystem with a hydrothermal vent ecosystem. How does energy enter each ecosystem? What is the basis of primary production in each ecosystem? What nutritional classes of organisms exist in each ecosystem, and how do they feed themselves?

2. Imagine a sewage plant that is releasing sewage containing high levels of ammonia and phosphate, but very little organic carbon. Growth of what type(s) of organisms might be triggered by this sewage, and how would the oxygen profile near and beyond the dump point differ from that shown in Figure 19.10*a*?

3. Look at the data of Figure 19.13. Keeping in mind that the open ocean waters are highly oxic, predict the possible metabolic lifestyles of open ocean *Archaea* and *Bacteria*. Why might proteorhodopsin be present only in one group of organisms but not the other?

4. Some have called the hydrothermal vent tube worms "autotrophic animals." Discuss the symbiosis occurring in the tube worms and why such a statement is in part true, but technically incorrect.

5. ^{14}C-Labeled cellulose is added to a vial containing a small amount of sewage sludge and sealed under anoxic conditions. A few hours later $^{14}CH_4$ appears in the vial. Discuss what has happened to yield such a result.

6. Suppose you have discovered a new animal that consumes only grass in its diet. You suspect it to be a ruminant and have available a specimen for anatomical inspection. If this animal is a ruminant, describe the position and basic components of the digestive tract you would expect to find and any key microorganisms and substances you might look for.

7. Acid mine drainage is in part a chemical process and in part a biological process. Discuss the chemistry and microbiology that leads up to acid mine drainage and point out the key reaction(s) that are biological.

8. Compare and contrast the processes by which bacteria interact with plants in crown gall disease and in root nodules. What parallels and differences exist in the two processes? In the latter, focus especially on what actually gets *into* the plant.

C ontrol of microorganisms is a major goal of experimental and clinical microbiologists, as well as an everyday battle in the home, in hospitals, restaurants, and anywhere sanitation is required. There are a number of physical and chemical means of microbial control, including chemical disinfectants and antiseptics, and, for *in vivo* use, very powerful antimicrobial drugs. However, the ultimate means of controlling microbial growth is sterilization, usually by physical methods such as pressurized steam and heat in the autoclave shown here.

MICROBIAL GROWTH CONTROL

Working Glossary

Aminoglycosides a group of antibiotics, including streptomycin, containing amino sugars linked by glycosidic bonds

Antibiotic chemical substance produced by a microorganism that kills or inhibits the growth of another microorganism

Antimicrobial drug resistance the acquired ability of a microorganism to grow in the presence of an antimicrobial drug to which the microorganism is usually sensitive

Antimicrobial agent a chemical compound that kills or inhibits the growth of microorganisms

Antiseptic antimicrobial agents that are sufficiently nontoxic to be applied on living tissues

Autoclave a sterilizer that destroys microorganisms with temperature and steam under pressure

β-Lactam antibiotics a group of antibiotics, including penicillin, that contains the four-membered heterocyclic β-lactam ring

Bacteriocidal agent an agent that kills bacteria

Bacteriostatic agent an agent that inhibits bacterial growth

Broad-spectrum antibiotic an antibiotic that acts on both gram-positive and gram-negative *Bacteria*

Chemotherapeutic agent an antimicrobial agent that can be used internally

Decontamination treatment that renders an object or inanimate surface safe to handle

Disinfectant an antimicrobial agent used only on inanimate objects

Disinfection the process of eliminating nearly all pathogens, but not all microorganisms, from inanimate objects or surfaces

Fungicidal agent an agent that kills fungi

Fungistatic agent an agent that inhibits fungal growth

Growth factor analog a chemical agent that is related to and blocks the uptake of a growth factor

Inhibition the reduction of microbial growth because of a decrease in the number of organisms present or alterations in the microbial environment

Interferons host-specific antiviral proteins, produced by virus-infected cells, that prevent viral infection of neighboring cells

Lysis loss of cellular integrity with release of cytoplasmic contents

MIC minimum inhibitory concentration—the minimum concentration of a substance necessary to prevent microbial growth

NNRTI non-nucleoside reverse transcriptase inhibitor

NRTI nucleoside reverse transcriptase inhibitor

Pasteurization destruction of all disease-producing microorganisms or a reduction in the number of spoilage microorganisms

PI protease inhibitor

Quinolones synthetic antibacterial compounds that interact with DNA gyrase and prevent supercoiling of bacterial DNA

Semisynthetic penicillin a natural penicillin that has been chemically altered

Sterilization the killing or removal of all living organisms and their viruses from a growth medium

Tetracycline an antibiotic characterized by a four-membered naphthacene ring

Viricidal agent an agent that stops viral replication and activity

Viristatic agent an agent that inhibits viral replication

This chapter begins a shift from our focus on microorganisms. Beginning with this section, we will be concerned with the relationships of microorganisms with humans. We start here by highlighting agents and methods used for control of microbial growth.

In general, control can be effected by limiting microbial growth, the process of **inhibition**, or by destroying the organisms with **sterilization**, the killing or removal of all viable organisms from a growth medium. Agents that destroy or kill bacteria are **bacteriocidal**. In practice, sterility is often not attainable, but in many cases we can inhibit the rapid growth of organisms through decontamination and disinfection methods. Agents that inhibit bacterial growth are said to be **bacteriostatic**.

Microbial control measures include **decontamination, disinfection**, and **sterilization**. We continually apply decontamination methods to inhibit microbial growth. For example, the casual act of wiping a table clean after a meal removes potential microbial nutrients and contaminating microbes, preventing microbial growth. More directed antimicrobial measures include disinfection with specialized chemical or physical agents, with the aim of inhibiting microbial growth or destroying microorganisms. For example, we routinely use disinfectant chemicals such as alcohol, to clean and disinfect wounds. Finally, when necessary, we use controlled sterilization methods to destroy all microorganisms. Sterilization, though difficult to achieve, completely prevents contamination and growth of microorganisms. Such measures are necessary, for instance, when making microbiological media or preparing surgical instruments. The goal for all of these procedures is to reduce the *microbial load*, or the number of viable microorganisms present.

Microbial control *in vivo* is, however, a much different matter: Clinically useful bacteriocidal or bacteriostatic agents must reduce or prevent microbial growth, while causing no harm to the host cell. This goal is achieved by a wide variety of natural and synthetic chemotherapeutic agents.

In this chapter we will first examine methods of microbial control that are used *in vitro*. We will then discuss antimicrobial drugs used *in vivo* in humans.

▌ PHYSICAL ANTIMICROBIAL CONTROL

Physical methods are often used to achieve microbial decontamination, disinfection, and sterilization. Heat, radiation, and filtration are standard methods used to destroy or remove undesirable microorganisms. Here we discuss the mechanisms of action and provide some practical examples that employ these methods to prevent microbial growth or to decontaminate areas or materials harboring microbes.

20.1 ▌ Heat Sterilization

Perhaps the most widespread method used for controlling microbial growth is the application of heat.

Kinetics of Heat Sterilization

For all microorganisms, there is a maximum temperature for growth, beyond which lethal effects occur. At very high temperatures, virtually all macromolecules lose their structure and their ability to function, a process known as *denaturation*. As shown in Figure 20.1, death from heating is an exponential (first-order) function and occurs more rapidly as the temperature rises. The first-order relationship shown in Figure 20.1 means that the rate of death is proportional at any time to the concentration of organisms at that time; the time taken for a definite fraction (for example, 90%) of the cells to be killed is independent of the initial concentration. These facts have important practical consequences. If we wish

to *sterilize* a microbial population, it will take longer at lower temperatures than at higher temperatures. Thus, it is necessary to adjust the time and temperature to achieve sterilization for each specific set of conditions. The nature of the heat is also important: Moist heat has better penetrating power than dry heat and produces a faster reduction in the number of living organisms at a given temperature.

The time required for a 10-fold reduction in the population density at a given temperature, called the *decimal reduction time* or *D*, is the most useful way to characterize heat sterilization. Over the range of temperatures usually used in food preparation (*i.e.*, cooking and canning), the relationship between *D* and temperature is essentially exponential. Thus, when the logarithm of *D* is plotted against temperature, a straight line is obtained (Figure 20.2). The slope of the line provides a measure of the sensitivity of the organism to heat under the conditions employed, and the graph can be used to calculate process times to achieve sterilization, such as in canning operations (⟨∞⟩ Section 29.2).

Determination of decimal reduction times is a fairly lengthy procedure requiring a number of viable count measurements (⟨∞⟩ Section 6.5). An easier way of characterizing the heat sensitivity of an organism is to determine the *thermal death time*, the time it takes to kill all cells at a given temperature. This is done simply by heating samples of a cell suspension for different times, mixing the heated suspensions with culture medium, and incubating. If all the cells have been killed, no growth is observed in the incubated samples. The thermal death time depends on the size of the population tested

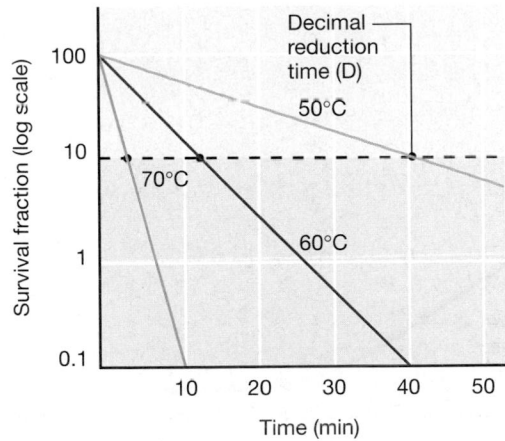

Figure 20.1 The effect of temperature on the viability of a mesophilic bacterium. The decimal reduction time, *D*, was obtained for the same mesophilic organism at three different temperatures. *D* is the time at which only 10% of the original population of organisms remain viable at a given temperature. For 70°C, *D* = 3 min; for 60°C, *D* = 12 min; for 50°C, *D* = 42 min.

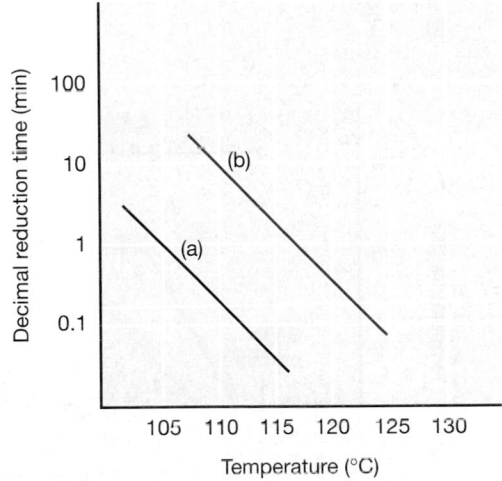

Figure 20.2 The relationship between the temperature and the rate of killing as indicated by the decimal reduction time for two different microorganisms. Data were obtained for decimal reduction times, *D*, at several different temperatures, as in Figure 20.1. For organism (a), a typical mesophile, exposure to 110°C for less than 20 s resulted in a decimal reduction, while for organism (b), a thermophile, 10 min were required to achieve a decimal reduction.

because a longer time is required to kill all cells in a large population than in a small one. When the number of cells is standardized, it is possible to compare the heat sensitivities of different organisms by comparing their thermal death times, as shown graphically in Figure 20.1.

Spores and Heat Sterilization

The heat resistance of vegetative cells and bacterial endospores from the same organism varies considerably. For instance, in the autoclave (see below) a temperature of 121°C is normally reached. Under these conditions, endospores may require 4–5 min for a decimal reduction, whereas vegetative cells may require only 0.1–0.5 min at 65°C. Thus, effective heat sterilization procedures are designed to destroy endospores.

Bacterial endospores (∞ Section 4.15) are the most heat-resistant structures known: They are able to survive heat that would rapidly kill vegetative cells of the same species. A major factor in heat resistance is the amount and state of *water* within the endospore. During endospore formation, the protoplasm is reduced to a minimum volume as a result of the accumulation of Ca^{2+}, *small acid-soluble spore proteins (SASPs)* and synthesis of dipicolinic acid, which lead to formation of a gel-like structure. At this stage, a thick cortex forms around the protoplast core. Contraction of the cortex results in a shrunken, dehydrated protoplast with a water content of only 10–30% of a vegetative cell. The water content of the protoplast coupled with the concentration of SASPs determines the heat resistance of the spore. If

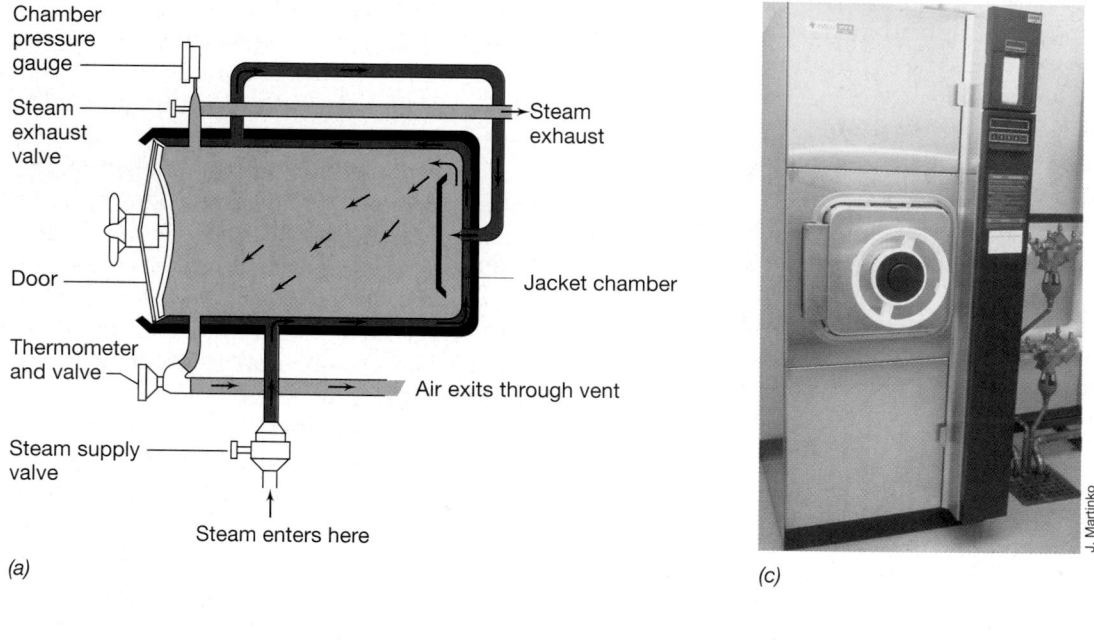

(a)

(c)

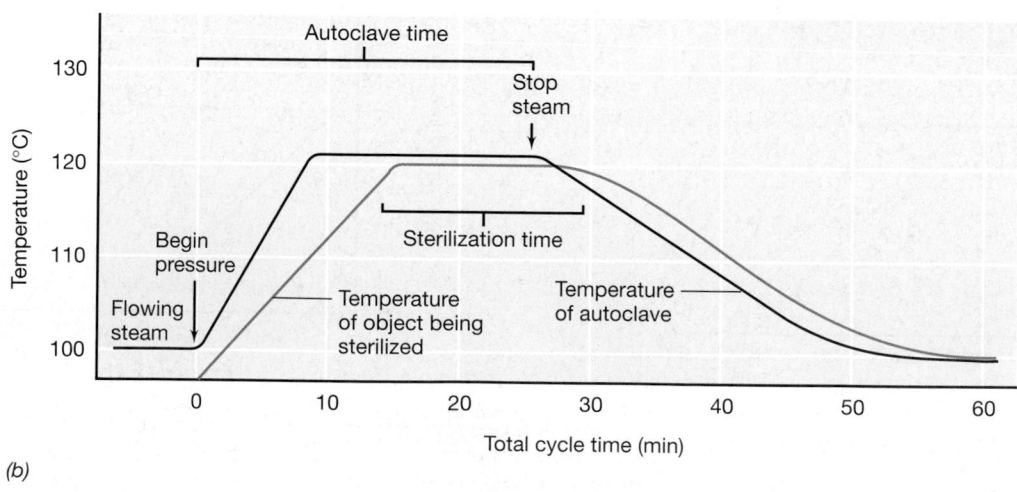

(b)

Figure 20.3 Use of the autoclave for sterilization. (a) The flow of steam through an autoclave. (b) A typical autoclave cycle. Shown is the sterilization of a fairly bulky object. The temperature of the object rises more slowly than the temperature of the autoclave. (c) A modern research autoclave. Note the pressure-lock door and the automatic cycle controls on the right panel. The steam inlet and exhaust fittings are on the right side of the autoclave.

Learning From the Past ... The Origin of Pasteurization

Pasteur's name is forever linked in the public mind with the process of pasteurization. The development of the pasteurization process has been nicely discussed by René Dubos in his book on Pasteur's life.[1] We quote from this book here:

The demonstration that microbes do not generate spontaneously encouraged the development of techniques to destroy them and to prevent or minimize subsequent contamination. Immediately these advances brought about profound technological changes in the preparation and preservation of food products and subsequently in other industrial processes as well. ...

It was soon discovered that the introduction of microorganisms in biological products can be minimized by an intelligent and rigorous control of the technological operations, but cannot be prevented entirely. The problem therefore was to inhibit the further development of these organisms after they had been introduced into the product. To this end, Pasteur first tried to add a variety of antiseptics, but the results were mediocre and, after much hesitation, he considered the possibility of using heat as a sterilizing agent.

Pasteur's first studies of heat as a preserving agent were carried out with wine. Pasteur had grown up in one of the best wine districts in France, and, as a connoisseur of the beverage, was much disturbed at the thought that heating might alter its flavor and bouquet. He therefore proceeded with very great caution and eventually convinced himself that heating at 55°C would not alter appreciably the bouquet of the wine. ... These considerations led to the process of partial sterilization, which soon became known the world over under the name of "pasteurization," and which was found applicable to wine, beer, cider, vinegar, milk, and countless other perishable beverages, foods, and organic products.

It was characteristic of Pasteur that he did not remain satisfied with formulating the theoretical basis of heat sterilization, but took an active interest in designing industrial equipment adapted to the heating of fluids in large volumes and at low cost. His treatises on vinegar, wine, and beer are illustrated with drawings and photographs of this type of equipment, and describe in detail the operations involved in the process. The word "pasteurization" is, indeed, a symbol of his scientific life; it recalls the part he played in establishing the theoretical basis of the germ theory, and the phenomenal effort that he devoted to making it useful to his fellow humans. It reminds us also of his well-known statement: "There are no such things as pure and applied science—there are only science, and the application of science." ■

[1]René Dubos, *Pasteur and Modern Science.* 1988. Science Tech. Madison, WI.

---

endospores have a low concentration of SASPs and high water content, they have low heat resistance; if they have a high concentration of SASPs and low water content, they have high heat resistance. Water moves freely in and out of spores, so it is not the impermeability of the spore coat that excludes water, but the gel-like material in the spore protoplast.

The nature of the medium in which heating takes place also influences the killing of both vegetative cells and spores. Microbial death is more rapid at acidic pH, and acid foods such as tomatoes, fruits, and pickles are much easier to sterilize than neutral pH foods such as corn and beans. High concentrations of sugars, proteins, and fats decrease heat penetration and usually increase the resistance of organisms to heat, whereas high salt concentrations may either increase or decrease heat resistance, depending on the organism. Dry cells (and spores) are more heat-resistant than moist ones; consequently, heat sterilization of dry objects always requires higher temperatures and longer times than sterilization of moist objects.

The Autoclave

The **autoclave** is a sealed device that allows the entrance of steam under pressure (Figure 20.3). Killing of the heat-resistant endospores requires heating at temperatures above boiling and the use of steam under pressure (Figure 20.3a). The usual procedure is to heat at 1.1 kilograms/square centimeter (kg/cm^2) [15 pounds/square inch (lb/in^2)] steam pressure, which yields a temperature of 121°C. At 121°C, the time to achieve sterilization is generally considered to be 10–15 min (Figure 20.3b). If bulky objects are being sterilized, heat transfer to the interior will be slow, and the heating time must be sufficiently long so that the *entire* object is at 121°C for 10–15 min. Extended times are also required when large volumes of liquids are being autoclaved because large volumes take longer to reach sterilization temperatures. Note that it is not the *pressure* of the autoclave that kills the microorganisms but the *high temperature* that can be achieved when steam is placed under pressure.

Pasteurization

Pasteurization is a process that reduces the microbial populations in milk and other heat-sensitive foods. It is named for Louis Pasteur, who first used heat for controlling the spoilage of wine (see the box, The Origin of Pasteurization). Pasteurization is not synonymous with sterilization because not all organisms are killed. Originally, pasteurization of milk was used to kill pathogenic bacteria, especially the organisms causing tuberculosis, brucellosis, Q fever, and typhoid fever, but the shelf life of milk was also improved following pasteurization. Although these pathogens are no longer common in food in developed countries, pasteurization prevents the spread of pathogens such as *Salmonella* species and *Escherichia coli* O157:H7 through common sources such as milk and juices. Pasteurization also prevents the growth of spoilage organisms, dramatically increasing the shelf life of perishable liquids (Sections 29.1 and 29.2).

Pasteurization of milk is usually achieved by passing the milk through a heat exchanger. The milk is fed through tubing that is in contact with a heat source. Careful control of the milk flow rate and the size and temperature of the heat source raises the temperature of the milk to 71°C for 15 sec. The milk is then rapidly cooled. The whole process is aptly called *flash pasteurization*. Milk can also be heated in large vats to 63–66°C for 30 min. However, this *bulk pasteurization* method is less satisfactory because the milk heats and cools slowly and must be held at high temperatures for longer times. Flash pasteurization alters the flavor less, kills heat-resistant organisms more effectively, and is normally done on a continuous-flow basis. The flash pasteurization method is more adaptable to large dairy operations, and modern dairies generally employ it, often with even shorter exposure times and higher temperatures.

✓ **20.1 Concept Check**

Sterilization is the killing of all organisms. The application of heat is the most widely used method for sterilization. The temperature for heat sterilization is selected to eliminate the most heat-resistant organisms in the material, usually bacterial endospores. For routine sterilization, an autoclave is used; this permits application of steam heat under pressure at temperatures above the boiling point of water. Pasteurization kills most pathogenic microorganisms, reduces microbial load, and reduces the growth of spoilage microorganisms, greatly extending shelf life.

✓ Why is heat an effective sterilizing agent?

✓ What steps are necessary to sterilize material that may have bacterial endospores?

20.2 Radiation Sterilization

An effective way to sterilize or reduce the microbial burden in almost any substance is through the use of electromagnetic radiation; heat is just one example. Microwaves, ultraviolet (UV) radiation, X-rays, gamma rays (γ-rays), and electrons are all forms of electromagnetic radiation (Section 10.3) that have the potential to control microbial growth. However, each type of radiation acts through a specific mechanism. For example,

the antimicrobial effects of microwaves are due, at least in part, to thermal effects. UV radiation, normally considered to be between 220 and 300 nm in wavelength, acts by a different mechanism. UV waves have sufficient energy to cause breaks in DNA, leading to the death of the exposed organism (Section 10.3). This "near-visible" light is useful for disinfecting surfaces, air, and other materials such as water that do not absorb the UV waves. For example, laboratory biological cabinets all come equipped with a "germicidal" UV light to decontaminate the surface after use (Figure 20.4). Ultraviolet radiation cannot penetrate solid, opaque, light-absorbing surfaces, and its usefulness is therefore limited to disinfection of exposed surfaces.

Ionizing Radiation

Ionizing radiation is electromagnetic radiation of sufficient energy to produce ions and other reactive molecular species from molecules with which the radiation particles collide. Ionizing radiation produces electrons, e^-, hydroxyl radicals, $OH\cdot$, and hydride radicals, $H\cdot$. Each of these reactive molecules is capable of degrading and altering biopolymers such as deoxyribonucleic acid (DNA) and protein. In addition, ionizing radiation can interact directly with DNA, causing breaks in the polymer. The ionization and subsequent degradation of biologically important molecules such as DNA and enzyme proteins leads to the death of irradiated cells. Several radiation sources are potentially useful for sterilization.

The unit of radiation is the *roentgen*, which is a measure of the radiation energy output from a source. The standard for biological applications such as sterilization is the *absorbed radiation dose*. The *absorbed dose* is the rad (100 erg/g), or the gray (1 Gy = 100 rad). Certain microorganisms are much more resistant to radiation than

Figure 20.4 A biological safety cabinet, shown with an ultraviolet (UV) radiation source (mercury vapor lamp), which is used for decontamination of the inside surfaces.

others. Table 20.1 shows the dose of radiation necessary to reduce the numbers of some representative microorganisms or biological functions 10-fold.

In general, microorganisms are much more resistant to ionizing radiation than multicellular organisms. For example, the amount of energy necessary to reduce the bacterial load by 10-fold is at least 200 Gy. By contrast, the *lethal* dose of radiation for humans is considered to be 10 Gy or less! Note that the doses shown in Table 20.1 are for a one log reduction in growth of a given organism. The *D10* or *decimal reduction value* gives information similar to the decimal reduction time for heat sterilization (see Section 20.1): The relationship of the survival fraction plotted on a semilogarithmic scale versus the radiation dosage in grays is essentially linear (Figure 20.5). A standard *killing* dose for radiation sterilization is 12 *D10* requirements. For destruction of the radioresistant endospores of *Clostridium botulinum*, for example, 39,600 Gy are required (Table 20.1).

Radiation Practice

Several sources of ionizing radiation are available, including X-ray machines, cathode ray tubes, and radioactive nuclides. These sources produce either X-rays or γ-rays, both of which have sufficient energy and penetrating power to efficiently inhibit microbial growth in solid and liquid media. The principal sources of commercially useful ionizing radiation are radioactive nuclides that emit γ rays. The two most commonly used radioisotopes are ^{60}Co and ^{137}Cs, both relatively inexpensive by-products of nuclear fission.

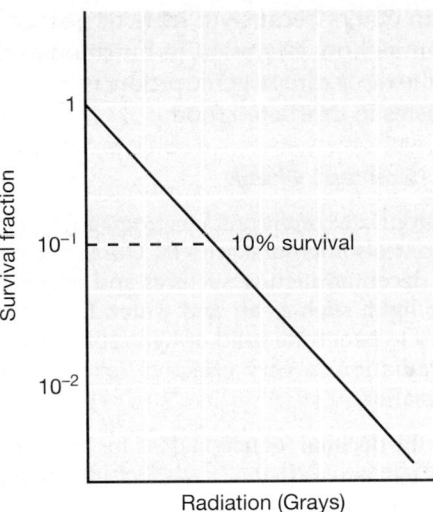

Figure 20.5 Relationship between the survival fraction and the radiation dose. The *D10*, or decimal reduction dose, can be interpolated from the data as shown. The dose in Grays differs significantly for each organism.

Radiation is currently used for sterilization and decontamination in the medical supplies and food industries. In the United States, the Food and Drug Administration has approved the use of radiation for sterilization of such diverse items as surgical supplies, disposable labware, drugs, and even tissue grafts (Table 20.2). However, because of the costs and hazards of radiation equipment, this type of sterilization is limited to large industrial applications or very specialized facilities. Food and food products are routinely irradiated (∞ Section 29.2). In addition to sterilization, pasteurization and insect deinfestation can be accomplished by adjusting the dose of radiation applied. Radiation is approved by the World Health Organization as a decontamination measure for use in foods particularly susceptible to microbial contamination, especially spices and fresh meat products such as hamburger and chicken, and is approved in the United States for decontamination of ground meat (∞ Section 29.2). The use of radiation for all these purposes is an established and accepted technology in many countries but has been slow to gain ac-

TABLE 20.1 Radiation sensitivity of microorganisms and biological functions

Species or function	Type of microorganism	D10ᵃ(Gy)
Clostridium botulinum	Gram-positive anaerobic sporulating *Bacteria*	3300
Clostridium tetani	Gram-positive anaerobic sporulating *Bacteria*	2400
Bacillus subtilis	Gram-positive aerobic sporulating *Bacteria*	600
Salmonella typhimurium	Gram-negative *Bacteria*	200
Lactobacillus brevis	Gram-positive *Bacteria*	1200
Deinococcus radiodurans	Gram-negative radiation-resistant *Bacteria*	2200
Aspergillus niger	Mold	500
Saccharomyces cerevisiae	Yeast	500
Foot-and-mouth	Virus	13,000
Coxsackie	Virus	4500
Enzyme inactivation	—	20,000–50,000
Insect deinfestation	—	1000–5000

ᵃD10 is the amount of radiation necessary to reduce the initial population or activity level 10-fold (one logarithm). Gy = grays. 1 gray = 100 rads.

TABLE 20.2 Medical and laboratory products sterilized by radiation

Tissue grafts	Drugs	Medical and laboratory supplies
Cartilage	Chloramphenicol	Disposable labware
Tendon	Ampicillin	Culture media
Skin	Tetracycline	Syringes
Heart valve	Atropine	Surgical equipment
	Vaccines	Sutures
	Ointments	

ceptance in others because of fears of possible radioactive contamination, alteration in nutritional value, production of toxic or carcinogenic products, and production of "off" tastes in irradiated food.

✓ 20.2 Concept Check

Under appropriate conditions, electromagnetic radiation effectively controls microbial growth. Ultraviolet radiation is useful for decontaminating surfaces and materials that do not absorb light, such as air and water. Ionizing radiation is necessary to penetrate solid or light-absorbing materials. Ionizing radiation is very effective for sterilization and decontamination.

✓ Define the decimal reduction dose for radiation.
✓ Why is ionizing radiation more effective than ultraviolet radiation for sterilization of food products?

20.3 Filter Sterilization

As we have just discussed in Section 20.1, heat is the most common and effective way of sterilizing liquids. However, *filtration* can be used for the sterilization of heat-sensitive liquids or gases. A filter is a device with pores too small for the passage of microorganisms but large enough to allow the passage of the liquid or gas. The size range of particles involved in sterilization is rather broad. Some of the largest microbial cells are greater than 10 μm in diameter, but at the lower end of the size scale certain bacteria are less than 0.3 μm in diameter. Historically, selective filtration methods were used to define and isolate infectious particles that were smaller than bacteria. These infectious particles, now known to be viruses, are very small and range from 28 nm to 200 nm in diameter (∞ Section 9.2).

Types of Filters

The three major types of filters are illustrated in Figure 20.6. One of the oldest types used is the *depth filter*. A depth filter is a fibrous sheet or mat made from a random array of overlapping paper, asbestos, or glass fibers (Figure 20.6*a*). The depth filter traps particles in the tortuous paths created throughout the depth of the structure. Because they are rather porous, depth filters are often used as *prefilters* to remove larger particles from a solution so that clogging does not occur in the final filter sterilization process. They are also used for the filter sterilization of air in industrial processes (∞ Section 30.4).

The most common type of filter for sterilization in the field of microbiology is the *membrane filter* (Figure 20.6*b*). Membrane filters are composed of polymers with high tensile strength such as cellulose acetate, cellulose nitrate, or polysulfone, manufactured to contain a large number of tiny holes. By adjusting the polymerization conditions during manufacture, the size of the holes in the membrane (and thus the size of the molecules that can pass through) can be precisely controlled. The membrane filter differs from the depth filter because it functions more like a sieve, trapping many of the particles on the filter surface. About 80–85% of the surface area of the membranes consists of the open pores. This openness provides for a relatively high fluid flow rate.

The third type of filter in common use is the **nucleation track (Nucleopore) filter**. These filters are created by treating very thin polycarbonate films (10 μm in thickness) with nuclear radiation and then etching the film with a chemical. The radiation causes localized damage in the film and the etching chemical enlarges these damaged locations into holes. The sizes of the holes can be precisely controlled by the strength of

Figure 20.6 The structure of (a) a depth filter, (b) a conventional membrane filter, and (c) a Nucleopore filter. Depth filters are used as prefilters and for the filtration of liquids with a high amount of suspended particles. Membrane filters are used in many applications in the laboratory and in industry because they are readily available in a wide variety of sizes and porosities, economical, and applicable to nearly all filtration needs. Nucleopore, or nucleation track, filters are useful for isolating specimens for microscopy because filtered material is captured in a single plane on the filter surface.

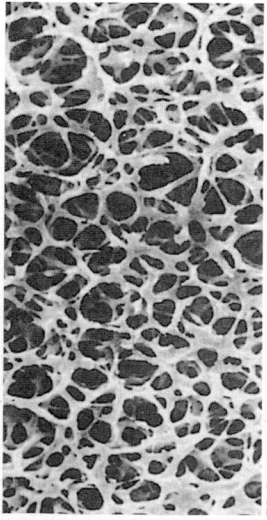
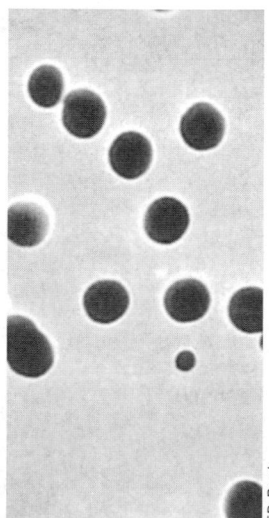

(a)   (b)   (c)

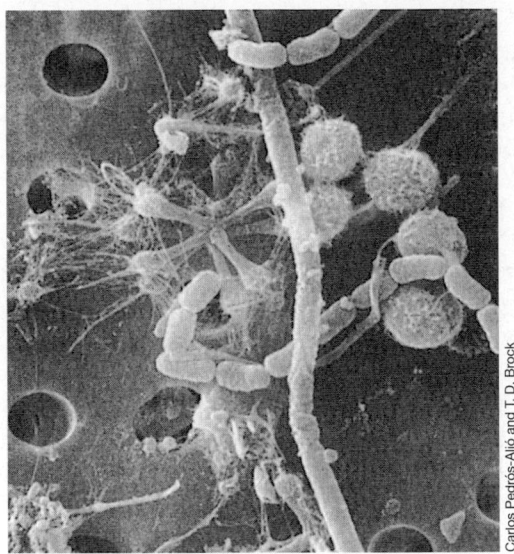

Figure 20.7 Scanning electron micrograph of aquatic bacteria and algae trapped on a Nucleopore filter. The pore size is 5 μm.

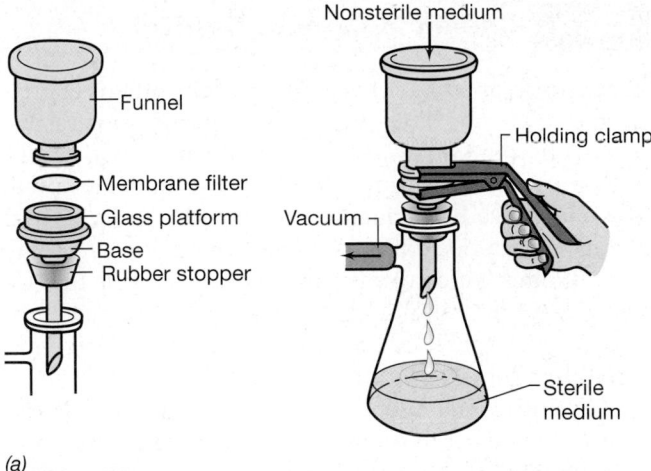

(a)

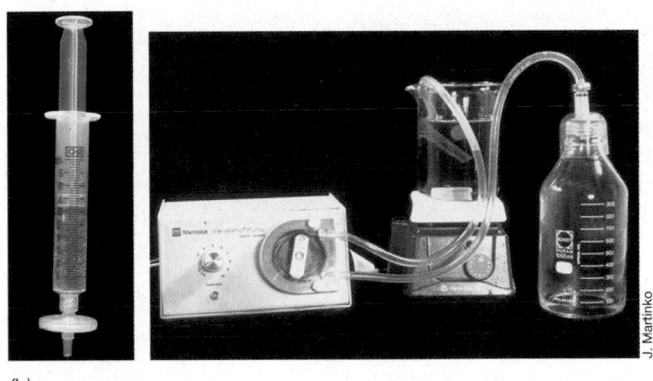

(b)

Figure 20.8 Membrane filters. (a) Assembly of a reusable membrane filter apparatus. (b) Disposable, presterilized, and assembled membrane filter units. Left: a filter system designed for small volumes. Right: a filter system designed for larger volumes.

the etching solution and the etching time. A typical Nucleopore filter has very uniform holes arranged almost vertically through the thin film (Figure 20.6c). Nucleopore filters are commonly used in scanning electron microscopy. An organism can be removed from liquid and concentrated in a single plane on the top of the filter, where it can be readily observed with the microscope (Figure 20.7).

Membrane filters for the sterilization of a liquid are illustrated in Figure 20.8. The filter apparatus is generally sterilized separately from the filter, and the apparatus assembled aseptically at the time of filtration. The arrangement shown in Figure 20.8a is suitable for small volumes of liquid. For large-volume sterile filtration, the membrane filter material is arranged in a cartridge and placed in a stainless steel housing. Large-volume filtration of heat-sensitive fluids is very commonly done in the pharmaceutical industry.

Presterilized membrane filter assemblies for sterilization of small to medium volumes are routinely used in most laboratories (Figure 20.8b). Filtration is accomplished by using a syringe, pump, or vacuum to force the liquid through the filtration apparatus into a sterile collection vessel.

✓ 20.3 Concept Check

Filter sterilization involves the *removal* of living microorganisms from liquids. Membrane filters are widely used for sterilization of heat-sensitive liquids in the laboratory.

✓ Why are depth filters not widely used for sterilization?

✓ What advantage does the membrane filter have over the nucleation filter in sterilization?

II CHEMICAL ANTIMICROBIAL CONTROL

We use a variety of chemicals in everyday situations to control microbial growth in the home and at work. The detergents and soaps we use to clean clothes and bathe are aimed, at least in part, at reducing the microbial load or killing microorganisms in clothes or on body surfaces. In the kitchen, we use a variety of chemicals to inhibit or destroy microorganisms on dishes, work surfaces, and utensils. In the microbiology laboratory and in industrial settings, chemical agents are used routinely to control unwanted microbial growth. Here we classify and discuss the mechanisms of action for a number of chemicals used to control microorganisms in our environment.

20.4 Chemical Growth Control

An **antimicrobial agent** is a natural or synthetic chemical that kills or inhibits the growth of microorganisms. Agents that kill organisms are often called *cidal agents*, with a prefix indicating the kind of organism killed. Thus, we have **bacteriocidal, fungicidal**, and **viricidal** agents. A bacteriocidal agent kills bacteria. It may or may not kill other kinds of microorganisms. Agents that do not kill but only inhibit growth are called *static agents*, and we can speak of **bacteriostatic, fungistatic**, and **viristatic** agents.

Antimicrobial agents vary with regard to *selective toxicity*. Some act in a rather nonselective manner and have similar effects on all types of cells. Others are far more selective and are more toxic to microorganisms than to animal tissues. Antimicrobial agents with selective toxicity are especially useful for treating infectious diseases because they can be used to kill disease-causing microorganisms *in vivo* without harming the host. They will be described later in this chapter.

Effect of Antimicrobial Agents on Growth

Three distinct effects can be observed when an antimicrobial agent is added to an exponentially growing bacterial culture: bacteriostatic, bacteriocidal, and bacteriolytic (Figure 20.9). A *bacteriostatic* effect is observed when growth is inhibited, but no killing occurs (Figure 20.9a). Bacteriostatic agents are frequently inhibitors of protein synthesis and act by binding to ribosomes. The binding, however, is not tight, and when the concentration of the agent is lowered, the agent becomes free from the ribosome and growth is resumed. Many antibiotics work by this mechanism and will be discussed in Sections 20.6–20.9. *Bacteriocidal* agents kill cells, but lysis or cell rupture does not occur (Figure 20.9b). Bacteriocidal agents are a class of chemical agents that generally bind tightly to their cellular targets and are not removed by dilution. *Bacteriolytic* agents induce killing by cell **lysis**, the breakdown of the cell, which is observed as a decrease in cell number or in turbidity after the agent is added (Figure 20.9c). Bacteriolytic agents include antibiotics that inhibit cell wall synthesis, such as penicillin (∞ Section 4.8 and see Section 20.8), as well as chemical agents that damage the cytoplasmic membrane.

Measuring Antimicrobial Activity

Antimicrobial activity is measured by determining the smallest amount of agent needed to inhibit the growth of a test organism, a value called the **minimum inhibitory concentration (MIC)**. To determine the MIC, a series of culture tubes is prepared, each containing medium with a different concentration of the agent, and all tubes of the series are inoculated. After incubation, the tubes are checked for visible growth (turbidity). The tube containing the *lowest concentration of agent that completely inhibits the growth of the test organism defines the MIC* (Figure 20.10). This simple and effective procedure is often called the *tube dilution technique*. The MIC is not a constant for a given agent, because it is affected by the nature of the test organism used, the inoculum size, the composition of the culture medium, the incubation time, and the conditions of incubation, such as temperature, pH, and aeration. When all conditions are rigorously standardized, it is possible to compare different antimicrobial agents and determine which is most effective against a given organism or to assess the activity of a single agent against a variety of organisms. This method

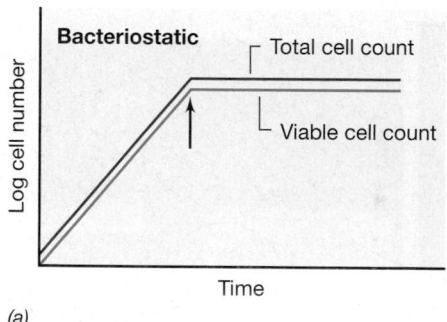

(a)

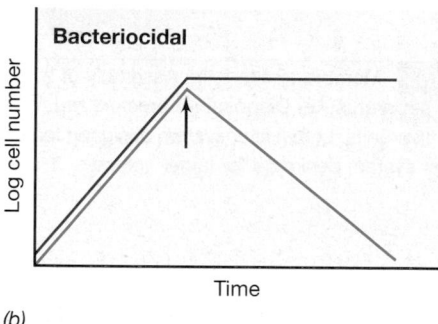

(b)

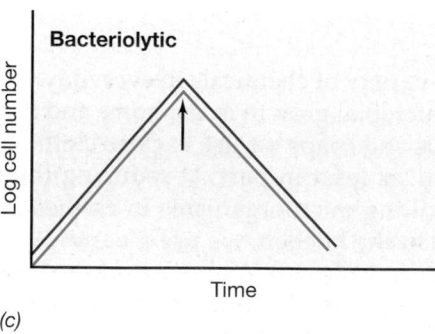

(c)

Figure 20.9 Three types of action of antimicrobial agents. At the time indicated by the arrow, a growth-inhibitory concentration was added to the exponentially growing culture. Note the relationships between viable and total cell counts.

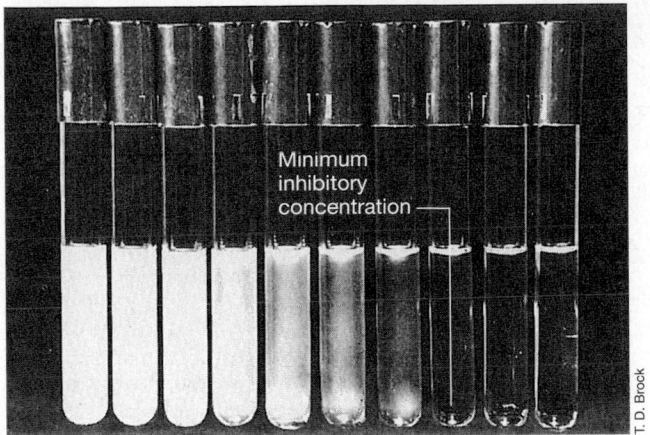

Figure 20.10 Antibiotic assay by tube dilution, permitting detection of the *minimum inhibitory concentration* (MIC). A series of increasing concentrations of antibiotic is prepared in the culture medium. Each tube is inoculated, and incubation is allowed to proceed. Growth (turbidity) occurs in those tubes with antibiotic concentrations below the MIC.

does not distinguish between a cidal and a static agent because the agent is present in the culture medium throughout the entire incubation period.

Another commonly used procedure for studying antimicrobial action is the *agar diffusion method* (Figure 20.11). A Petri plate containing an agar medium evenly inoculated with the test organism is prepared. Known amounts of the antimicrobial agent are added to filter paper discs, which are then placed on the surface of the agar. During incubation, the agent diffuses from the filter paper into the agar; the further it gets from the filter paper, the smaller the concentration of the agent. At some distance from the disc, the MIC is reached. Past this point growth occurs, but closer to the disc growth is absent. A *zone of inhibition* is thus created; the diameter of the zone is proportional to the amount of antimicrobial agent added to the disc, the solubility of the agent, the diffusion coefficient, and the overall effectiveness of the agent. This method is routinely used to test for antibiotic sensitivity in pathogens (⚙ Section 24.3).

✓ 20.4 Concept Check

Chemicals are often used to control microbial growth. Chemicals that kill organisms are called cidal agents; those that inhibit growth are called static agents. The effectiveness of an antibacterial chemical agent is assessed by determining the minimum concentration necessary to completely inhibit bacterial growth.

✓ With regard to antibacterial agents, what is meant by *selective toxicity*?

✓ Describe how the *minimum inhibitory concentration* of an antibacterial agent is determined.

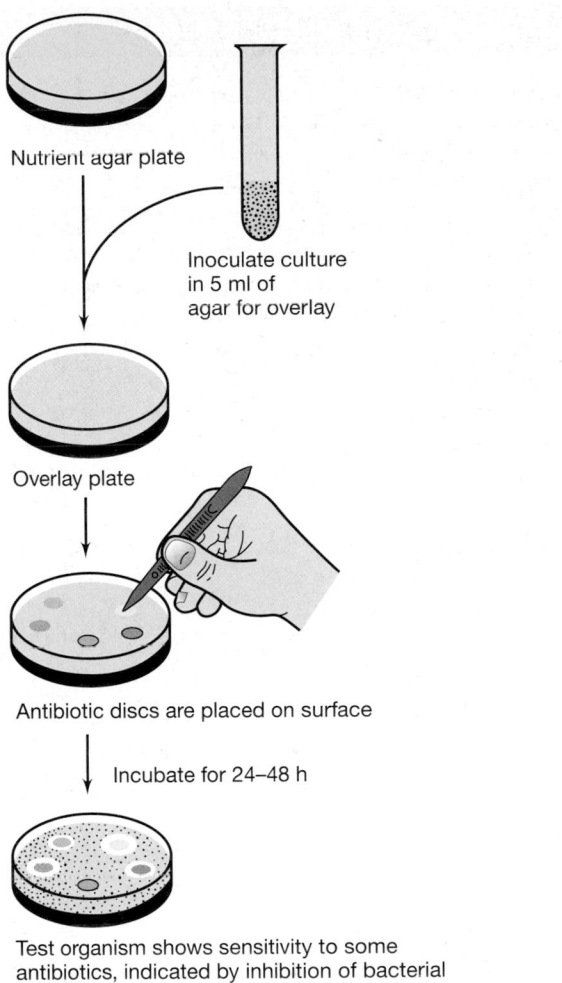

Nutrient agar plate

Inoculate culture in 5 ml of agar for overlay

Overlay plate

Antibiotic discs are placed on surface

Incubate for 24–48 h

Test organism shows sensitivity to some antibiotics, indicated by inhibition of bacterial growth around discs (zones of inhibition) after incubation

Figure 20.11 Agar diffusion method for assaying antibiotic activity.

20.5 Antiseptics, Disinfectants, and Sterilants

Antiseptics are chemical agents that kill or inhibit growth of microorganisms and are sufficiently nontoxic to be applied to living tissues. Most of the compounds that fall into this category are used for handwashing or for treating surface wounds (Table 20.3). Under some circumstances, some antiseptics are also effective disinfectants.

Disinfectants are chemicals that kill microorganisms and are used on inanimate objects. *Sterilants* are disinfectants that, under appropriate circumstances, can kill all microbial life and are used to sterilize inanimate objects and surfaces. Chemical disinfectant agents, which are frequently referred to as *germicides*, have wide use in situations where it is impractical to use heat (see Section 20.1) or radiation (see Section 20.2) for decontamination or sterilization. For example, hospitals and laboratories must be able to decontaminate floors, tables, bench tops, walls,

TABLE 20.3 Antiseptics, disinfectants, and sterilants

Agent	Use	Mode of action
Antiseptics		
Alcohol (60–85% ethanol or isopropanol in water) [a]	Skin	Lipid solvent and protein denaturant
Phenol-containing compounds (hexachlorophene, triclosan, chloroxylenol, chlorhexidine)	Soaps, lotions, cosmetics, body deodorants	Disrupts cell membrane
Cationic detergents, especially quaternary ammonium compounds (benzalkonium chloride)	Soaps, lotion	Interact with phospholipids of membrane
Hydrogen peroxide [a] (3% solution)	Skin	Oxidizing agent
Iodine-containing iodophor compounds in solution [a] (Betadine®)	Skin	Iodinates tyrosine residues of proteins; oxidizing agent
Silver nitrate	Eyes of newborn to prevent blindness due to infection by *Neisseria gonorrhoeae*	Protein precipitant
Disinfectants and sterilants		
Alcohol (60–85% ethanol or isopropanol in water) [a]	Disinfectant and sterilant for medical instruments, laboratory surfaces	Lipid solvent and protein denaturant
Cationic detergents (quaternary ammonium compounds)	Disinfectant for medical instruments, food and dairy equipment	Interact with phospholipids
Chlorine gas	Disinfectant for purification of water supplies	Oxidizing agent
Chlorine compounds (chloramines, sodium hypochlorite, chlorine dioxide)	Disinfectant for dairy and food industry equipment, and water supplies	Oxidizing agent
Copper sulfate	Algicide in swimming pools, water supplies (disinfectant)	Protein precipitant
Ethylene oxide (gas)	Sterilant for temperature-sensitive laboratory materials such as plastics	Alkylating agent
Formaldehyde	3%–8% solution used as surface disinfectant, 37% (formalin) or vapor used as sterilant	Alkylating agent
Glutaraldehyde	2% solution used as high-level disinfectant or sterilant	Alkylating agent
Hydrogen peroxide [a]	Vapor used as sterilant	Oxidizing agent
Iodine-containing iodophor compounds in solution [a] (Wescodyne®)	Disinfectant for medical instruments, laboratory surfaces	Iodinates tyrosine residues
Mercuric dichloride [b]	Disinfectant for laboratory surfaces	Combines with -SH groups
Ozone	Disinfectant for drinking water	Strong oxidizing agent
Peracetic acid	0.2% solution used as high-level disinfectant or sterilant	Strong oxidizing agent
Phenolic compounds [b]	Disinfectant for laboratory surfaces	Protein denaturant

[a] Alcohols, hydrogen peroxide, and iodine-containing iodophor compounds can act as antiseptics, disinfectants, or even sterilants depending on concentration, length of exposure, and form of delivery.

[b] Heavy metal (mercury) compound and phenolic compounds, produce environmentally hazardous waste products.

and so on. In the food industry, floors, walls, and surfaces of equipment must often be treated with germicides to reduce the microbial load and make the surfaces safe to handle. Hospitals and laboratories must also be able to sterilize heat-sensitive materials, such as thermometers, lensed instruments, polyethylene tubing, catheters, and reusable medical equipment such as respirometers. Usually, some form of *cold sterilization* is used for these purposes. Cold sterilization is performed in enclosed devices that resemble autoclaves, but employ a chemical agent such as ethylene oxide, formaldehyde, peracetic acid, or hydrogen peroxide. Drinking water is routinely disinfected with chlorine or chlorine compounds to eliminate potentially harmful organisms (Table 20.3).

Several factors affect the efficiency of various antiseptic and disinfectant procedures. For example, many germicides are neutralized by organic materials, inhibiting their ability to kill microorganisms by effectively reducing

TABLE 20.4 Industrial uses of disinfectants

Industry	Chemicals	Use
Paper	Organic mercurials, phenols	To prevent microbial growth during manufacture
Leather	Heavy metals, phenols	Antimicrobial agents are present in the final product
Plastic	Cationic detergents	To prevent growth of bacteria on aqueous dispersions of plastics
Textile	Heavy metals, phenols	To prevent microbial deterioration of fabrics exposed in the environment such as awnings, tents
Wood	Phenols	To prevent deterioration of wooden structures
Metal working	Cationic detergents	To prevent growth of bacteria in aqueous cutting emulsions
Petroleum	Mercurics, phenols, cationic detergents	To prevent growth of bacteria during recovery and storage of petroleum and petroleum products
Air conditioning	Chlorine, phenols	To prevent growth of bacteria (for example, *Legionella*) in cooling towers
Electrical power	Chlorine	To prevent growth of bacteria in condensors and cooling towers
Nuclear	Chlorine	To prevent growth of radiation-resistant bacteria in nuclear reactors

germicide concentrations. Further, pathogens are often encased in particles or grow in large numbers as *biofilms*, covering the surfaces of tissue with several layers of microbial cells (∞ Section 19.3). As a result, penetration of a chemical agent to the viable cells may be slowed or even completely prevented. In many cases, bacterial endospores are much more resistant to germicides than are vegetative cells because of their low water availability and reduced metabolism (see Section 20.1 and ∞ Section 4.15). Certain vegetative cells such as those of *Mycobacterium tuberculosis*, the causal agent of tuberculosis, are resistant to the action of germicides because of the complex nature of their cell wall (∞ Sections 12.24 and 26.5). As a result, total elimination of pathogens (sterilization) by germicidal treatment may not always occur. In practice, germicide effectiveness can be determined only under the actual conditions of use. A summary of the most widely used germicides and their modes of action is given in Table 20.3.

Antiseptic and disinfectant chemicals are used in many industrial applications, where they are used to prevent microbial deterioration of a number of organic materials. In some industries, the use of antimicrobial agents is very routine and quite extensive. This frequently leads to toxic waste problems when large amounts of antimicrobial agents, such as mercury and other heavy metal compounds (paper industry), or phenols (wood preservatives), are released into the environment. Table 20.4 summarizes some of the industrial applications for disinfectants used to control microbial growth.

✓ 20.5 Concept Check

Antiseptics can be used to decontaminate living tissues. Disinfectants are chemical compounds used to decontaminate or sterilize nonliving material. These compounds are used in many commercial, health care, and industrial applications.

✓ Distinguish between an antiseptic and a disinfectant.

✓ What disinfectants are used for sterilization of water?

III ANTIMICROBIAL AGENTS USED *IN VIVO*

A number of antimicrobial agents, both synthetic and naturally occurring, are used to treat microbial infections *in vivo*. These compounds revolutionized the treatment of infectious diseases.

20.6 Synthetic Antimicrobial Drugs

The preceding section dealt with chemical agents used to inhibit microbial growth *outside* the human body. Most of the chemicals mentioned are too toxic to be used in the body, although antiseptics can be used on the skin. For control of infectious disease, chemical compounds that can be used internally are essential. Such compounds are called **chemotherapeutic agents**, and they play major roles in clinical and veterinary medicine, as well as in agriculture.

Chemotherapeutic agents are categorized based on their structure (Figure 20.12) and mode of action (Figure 20.13). Each year, more than 500 metric tons of chemotherapeutic agents of various types are manufactured and used worldwide (Figure 20.14). The key requirement of a successful chemotherapeutic agent is *selective toxicity*, the ability to inhibit bacteria or other pathogenic agents without adversely affecting the host (see the box, Microbiology and "Magic Bullets," page 710). Each agent has a characteristic spectrum of antibacterial action (Figure 20.15, page 711). Chemotherapeutic agents fall into two general categories, the *synthetic agents* and the *antibiotics*. Here we will concentrate on the synthetic agents. We discuss antibiotics in the next three sections.

Antibiotic classification	Subclassification	Example	Representative structure
I. Carbohydrate-containing compounds	Pure sugars Aminoglycosides Orthosomycins *N*-Glycosides *C*-Glycosides Glycolipids	Nojirimycin Streptomycin Everninomicin Streptothricin Vancomycin Moenomycin	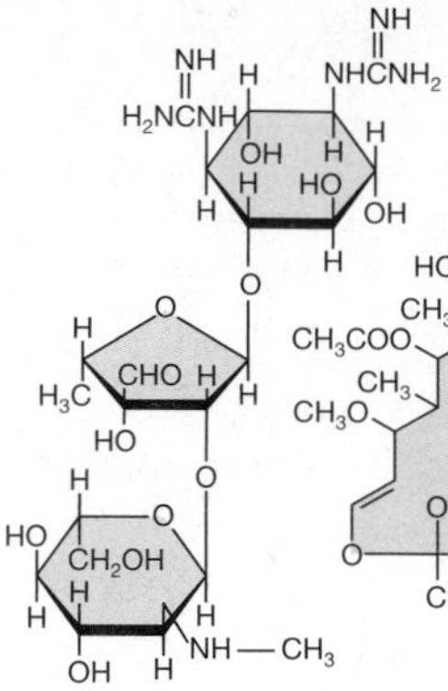 Streptomycin
II. Macrocyclic lactones	Macrolide antibiotics Polyene antibiotics Ansamycins Macrotetrolides	Erythromycin Candicidin Rifampin Tetranactin	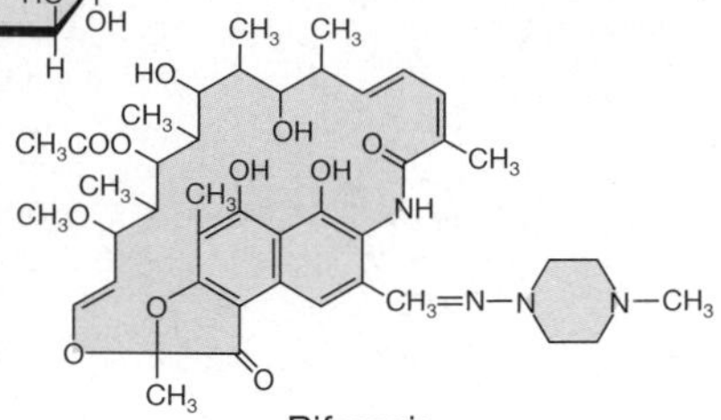 Rifampin
III. Quinones and related compounds	Tetracyclines Anthracyclines Naphthoquinones Benzoquinones	Tetracycline Adriamycin Actinorhodin Mitomycin	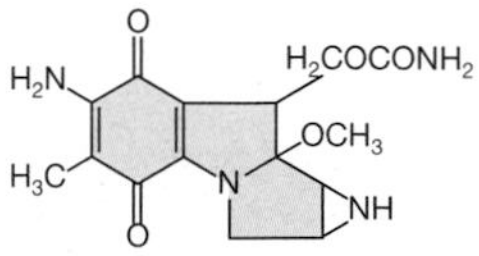 Mitomycin C
IV. Amino acid and peptide analogs	Amino acid derivatives β-Lactam antibiotics Peptide antibiotics Chromopeptides Depsipeptides Chelate-forming peptides	Cycloserine Penicillin, ceftriaxone Bacitracin Actinomycin Valinomycin Bleomycin	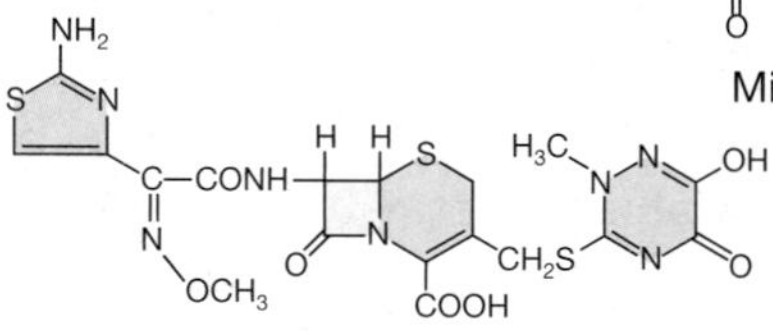 Ceftriaxone
V. Heterocyclic compounds containing nitrogen	Nucleoside antibiotics	Polyoxins	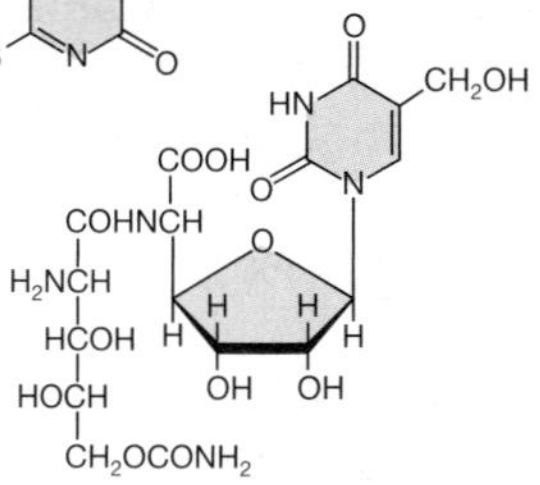 Polyoxin B
VI. Heterocyclic compounds containing oxygen	Polyether antibiotics	Monensin	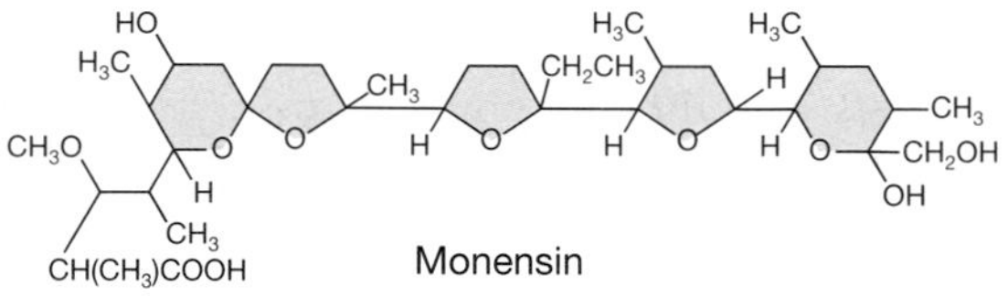 Monensin
VII. Alicyclic derivatives	Cycloalkane derivatives Steroid antibiotics	Cycloheximide Fusidic acid	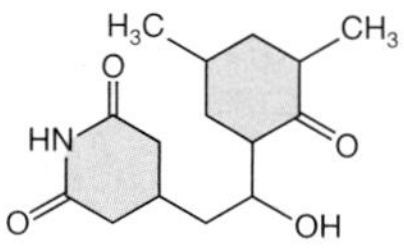 Cycloheximide
VIII. Aromatic compounds	Benzene derivatives Condensed aromatics Aromatic ether	Chloramphenicol Griseofulvin Novobiocin	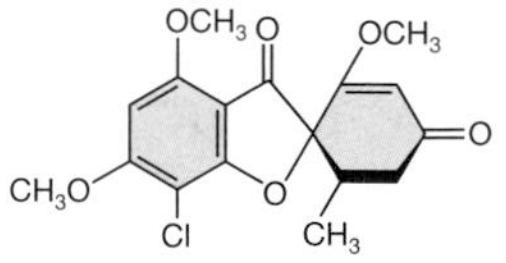 Griseofulvin
IX. Aliphatic compounds	Compounds containing phosphorus	Fosfomycin	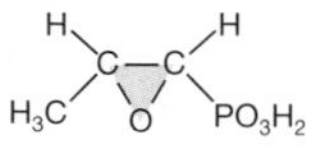 Fosfomycin
X. Quinolone compounds	4-Quinolone Fluoro-4-quinolones	Nalidixic acid Ciprofloxacin	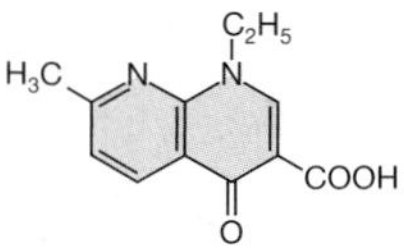 Nalidixic acid
XI. Oxazolidinone	Cyclic lactone	2-Oxazolidinone	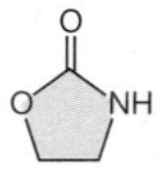 2-Oxazolidinone

Figure 20.12 Classification of antibacterial chemotherapeutic agents according to chemical structure. A representative example is shown for each group.

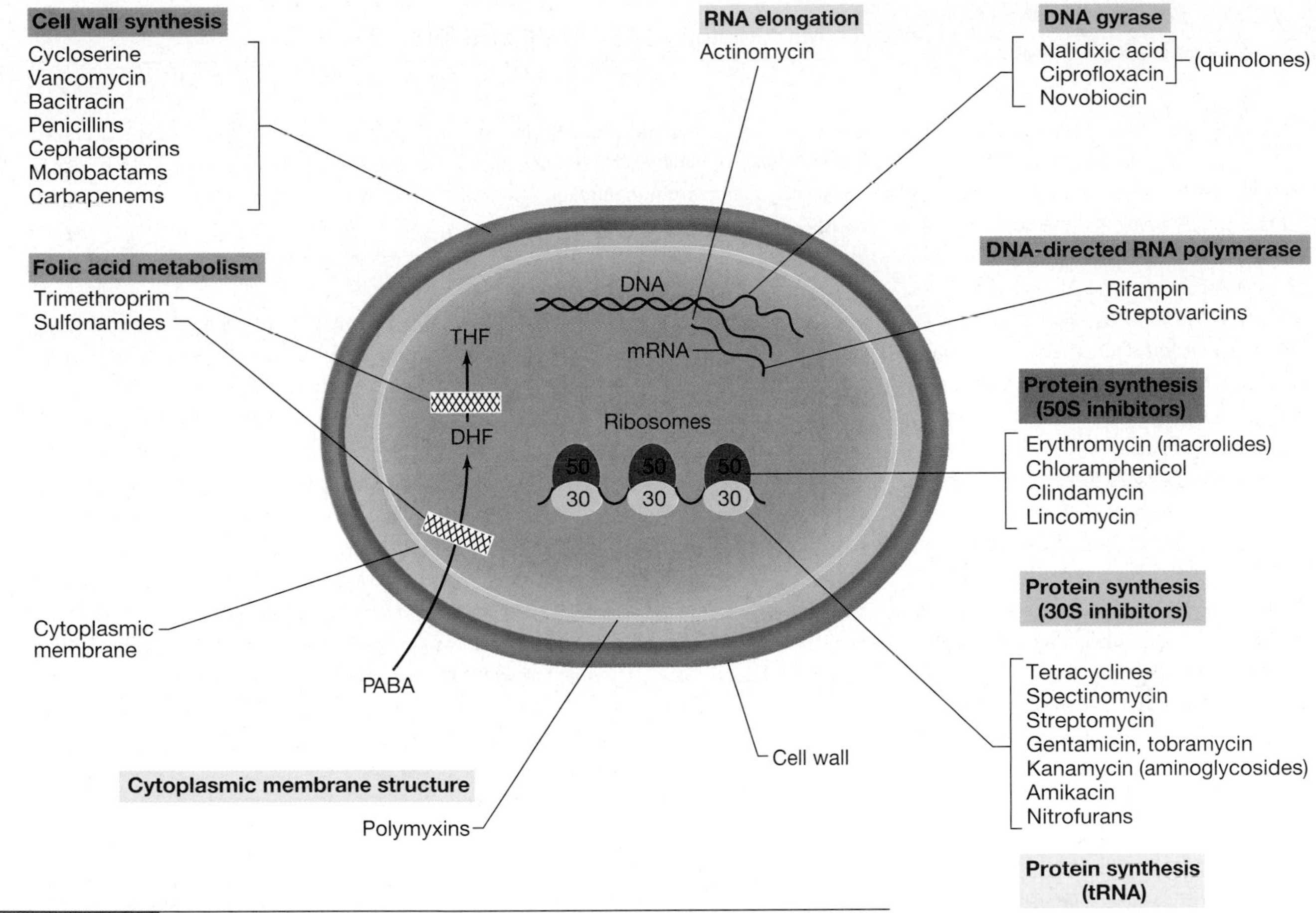

Cell wall synthesis

Cycloserine
Vancomycin
Bacitracin
Penicillins
Cephalosporins
Monobactams
Carbapenems

RNA elongation

Actinomycin

DNA gyrase

Nalidixic acid
Ciprofloxacin ┐ (quinolones)
Novobiocin

Folic acid metabolism

Trimethroprim
Sulfonamides

DNA

mRNA

THF

DHF

Ribosomes

50 50 50
30 30 30

DNA-directed RNA polymerase

Rifampin
Streptovaricins

Protein synthesis (50S inhibitors)

Erythromycin (macrolides)
Chloramphenicol
Clindamycin
Lincomycin

Protein synthesis (30S inhibitors)

Tetracyclines
Spectinomycin
Streptomycin
Gentamicin, tobramycin
Kanamycin (aminoglycosides)
Amikacin
Nitrofurans

Cytoplasmic
membrane

PABA

Cytoplasmic membrane structure

Polymyxins

Cell wall

Protein synthesis (tRNA)

Mupirocin
Puromycin

Figure 20.13 Mode of action of major antimicrobial chemotherapeutic agents. THF, Tetrahydrofolate; DHF, dihydrofolate; mRNA, messenger RNA; tRNA, transfer RNA.

Growth Factor Analogs

In Section 5.1 we defined growth factors as specific chemical substances *required* in the medium because the organisms cannot synthesize them. A substance that is related to a growth factor but blocks utilization of the growth factor is known as a **growth factor analog**.

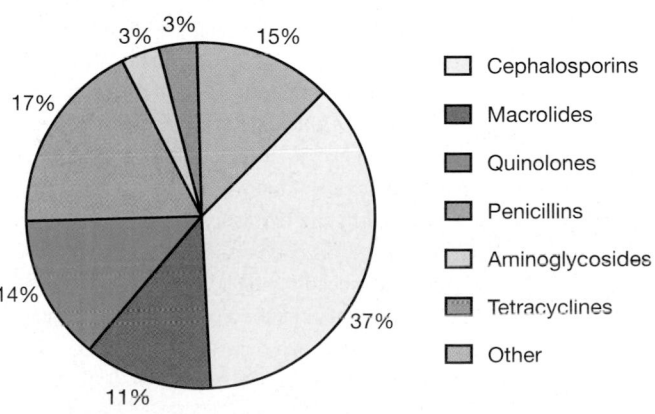

15%
3% 3%
17%
37%
14%
11%

☐ Cephalosporins
☐ Macrolides
☐ Quinolones
☐ Penicillins
☐ Aminoglycosides
☐ Tetracyclines
☐ Other

Figure 20.14 Annual worldwide production and use of antibiotics. Each year more than 500 metric tons of chemotherapeutic agents are manufactured.

Growth factor analogs are synthetic compounds that are structurally similar to the growth factors in question, but subtle structural differences in the analogs prevent them from duplicating the function of the natural growth factor in the cell. In addition to the bacterial growth factor analogs, which we will now discuss, there are a number of growth factor analogs that are active in the treatment of viral and fungal infections, and we will discuss these in Sections 20.10 and 20.11.

Sulfa Drugs

The *sulfa drugs* were the first widely used growth factor analogs to specifically inhibit the growth of bacteria (see the box). The simplest sulfa drug is *sulfanilamide* (Figure 20.16a). Sulfanilamide is an analog of *p*-aminobenzoic acid (Figure 20.16b), which is itself a part of the vitamin folic acid (Figure 20.16c). Sulfanilamide acts by blocking the synthesis of folic acid, a nucleic acid precursor. Sulfanilamide is active in *Bacteria* but not in higher animals because *Bacteria* synthesize their own folic acid, whereas higher animals obtain folic acid from their diet. Drug resistance to the clinically useful sulfanilamide derivatives, the sulfonamides, is quite common, usually because the resistant *Bacteria* have developed

Learning From the Past ...   Microbiology and "Magic Bullets"

Work on antimicrobial chemotherapeutic agents began with the German scientist Paul Ehrlich. In the early 1900s, Ehrlich developed the concept of selective toxicity. He began his work by studying the staining of microorganisms and observed that some dyes stained microorganisms but not animal tissue. He assumed that if a dye did not stain a tissue, the dye molecules were unable to combine with the cell constituents. He reasoned that if such a dye had toxic properties, it would not affect the animal cells because it could not combine with them, but it should attack the microbial cells. In an infected animal, chemicals of this sort should behave like "magic bullets," striking the pathogen but missing the host. Ehrlich tested large numbers of chemicals for selectivity and discovered the first chemotherapeutic agents, of which Salvarsan, an arsenic-containing drug for the cure of syphilis, was the most famous (see Fig. 1).

However, no chemical agents were discovered that affected the vast majority of infectious agents until the 1930s, when Gerhard Domagk discovered the sulfa drugs. The discovery of the sulfas came about through the large-scale screening of chemicals for activity in infectious diseases in experimental animals. Domagk, at the Bayer Chemical Company in Germany, tested a large variety of synthetic organic chemicals, mainly dyes, for their ability to cure streptococcal infections in mice. The first active compound was Prontosil, which was active in mice but had no activity against streptococci grown in the test tube. Domagk discovered that in the animal body, Prontosil broke down to sulfanilamide, which was the actual active agent. A program of synthesis based on the sulfanilamide structure yielded a large number of active drugs. D. D. Woods in England then showed that *p*-aminobenzoic acid specifically counteracted the inhibitory action of sulfanilamide, and he also showed that streptococci required *p*-aminobenzoic acid for growth. This led to the concept of the *growth factor analog*, which enabled chem-

ists to pursue the synthesis of a wide variety of chemical antimicrobial agents.

Although sulfa drugs controlled streptococcal infections, most infectious diseases were still not under control. However, the discovery of the first antibiotic, penicillin, by Alexander Fleming, a Scottish physician engaged in research at St. Mary's Hospital in London, led investigators to a new source of antimicrobial compounds. Antibiotics were novel compounds previuosly undetected in nature. It seem that, in order to successfully compete for limited nutrient and energy sources, many microorganisms produce and secrete one or more compounds toxic to microbial competitors in the environment (∞ Section 12.24). Fleming's first paper on penicillin, published in 1929, begins as follows:

While working with staphylococcus variants a number of culture plates were set aside on the laboratory bench and examined from time to time. In the examination these plates were necessarily exposed to the air and they became contaminated with various micro-organisms. It was noticed that around a large colony of contaminating mould the staphylococcus colonies became transparent and were obviously undergoing lysis. Subcultures of this mould were made and experiments conducted with a view of as-

certaining something of the properties of the bacteriolytic substance which had evidently been formed in the mould culture and which had diffused into the surrounding medium.

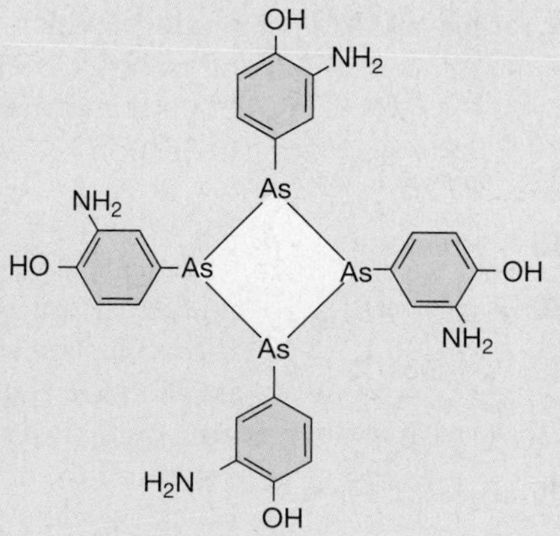

Fleming characterized the product, and since it was produced by a fungus of the genus *Penicillium*, gave it the name *penicillin*. His work, however, did not include a process for large-scale production nor did it show that penicillin was effective in the treatment of infectious disease. This was done by a group of British scientists at Oxford University, headed by Howard Florey in 1939, motivated in part by the impending World War II and the knowledge that infectious disease was the leading cause of death among soldiers on the battlefield. Florey and his colleagues developed methods for the analysis and testing of penicillin and for its production in large quantities. They then proceeded to test penicillin against bacterial infections in humans. Penicillin was dramatically effective in controlling staphylococcal and pneumococcal infections and was also more effective for streptococcal infections than the sulfa drugs. With the war in Europe becoming more intense, Florey brought cultures of the penicillin-producing fungus to the United States in 1941. He persuaded the U.S. government to create a large-scale research program, which led to a joint effort by the pharmaceutical industry, the U.S. Department of Agriculture at its laboratory in Peoria, Illinois, and several universities. By the end of World War II, penicillin was available for civilian as well as military use. As soon as the war was over, pharmaceutical companies entered into commercial production of penicillin on a competitive basis and began to look for other antibiotics. Success was quick and dramatic, and the impact on medicine has been close to phenomenal. Infant and child morbidity and mortality have been greatly reduced, and many diseases that formerly had high fatality rates are now no more than medical curiosities. ■

Figure 1 Salvarsan

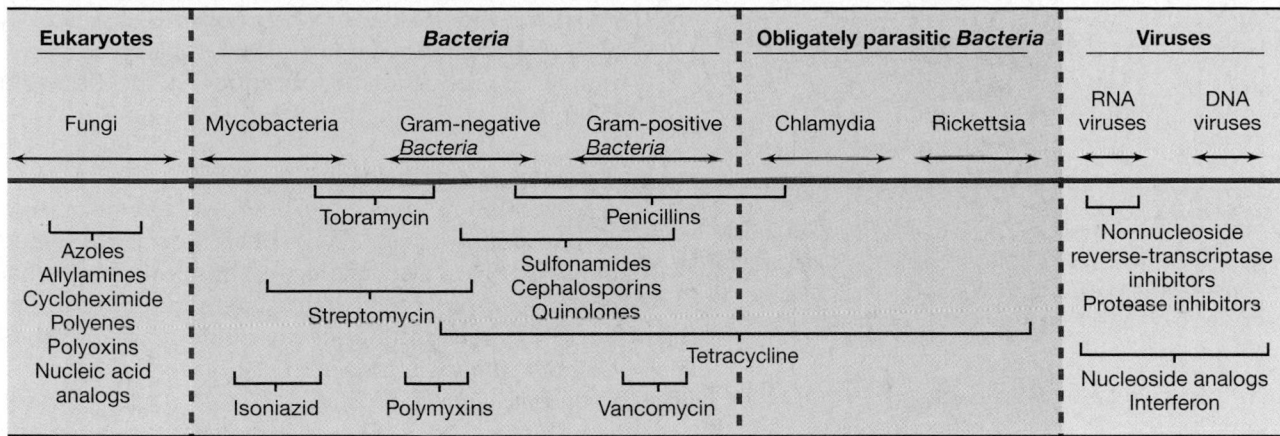

Eukaryotes	Bacteria			Obligately parasitic Bacteria		Viruses	
Fungi	Mycobacteria	Gram-negative Bacteria	Gram-positive Bacteria	Chlamydia	Rickettsia	RNA viruses	DNA viruses

Figure 20.15 Antimicrobial spectrum of action for selected chemotherapeutic agents.

the ability to use exogenous sources of preformed folic acid (Section 20.12).

Other Growth Factor Analogs

Analogs are now known for various vitamins, amino acids, purines, pyrimidines, and other compounds. These analogs resemble compounds found in a variety of organisms including many eukaryotes and viruses. An important example of a clinically useful growth factor analog is *isoniazid* (⌘ Figure 26.5). Isoniazid has a very narrow spectrum of activity (Figure 20.15) and is effective only against *Mycobacterium tuberculosis*, apparently interfering with the synthesis of the mycobacterial-specific mycolic acid cell wall material. This synthetic compound, a nicotinamide (vitamin) analog, has been the most effective single drug used for control and treatment of tuberculosis (Figure 20.15, ⌘ Section 26.5). In the examples shown in

Figure 20.17, analogs have been formed by addition of a fluorine or a bromine atom. Fluorine is a relatively small atom and does not alter the overall shape of the molecule, but it changes the chemical properties sufficiently so that the compound does not act normally in cell metabolism. Fluorouracil resembles the nucleic acid base uracil; bromouracil resembles another base, thymine. Growth factor analogs that resemble nucleic acids are used in the treatment of viral and

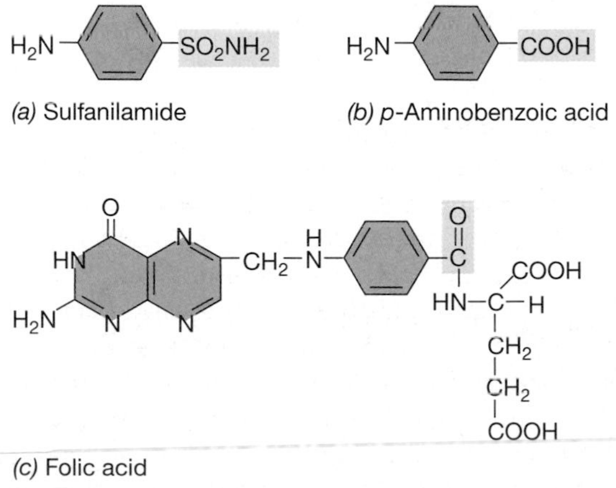

(a) Sulfanilamide

(b) p-Aminobenzoic acid

(c) Folic acid

Figure 20.16 (a) The simplest sulfa drug, sulfanilamide. (b) Sulfanilamide is an analog of p-aminobenzoic acid, which itself is part of (c) the growth factor folic acid (⌘ Section 5.1 discusses growth factors).

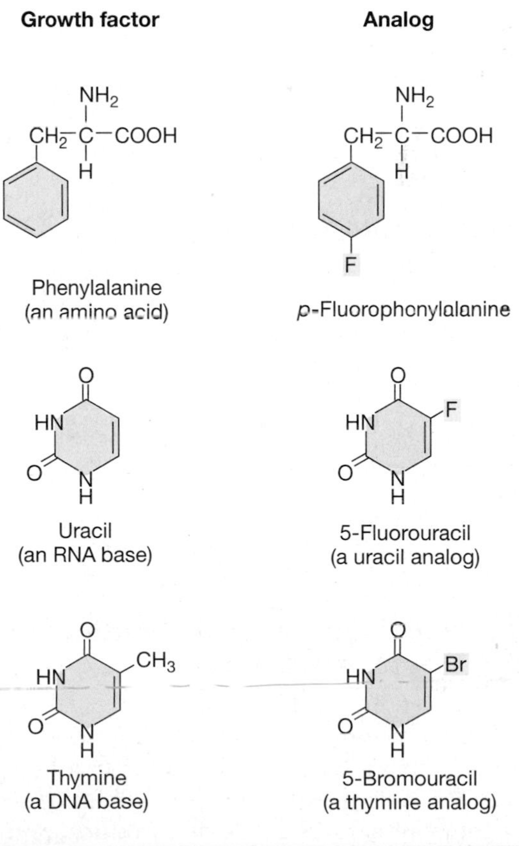

Growth factor

Analog

Phenylalanine (an amino acid)

p-Fluorophenylalanine

Uracil (an RNA base)

5-Fluorouracil (a uracil analog)

Thymine (a DNA base)

5-Bromouracil (a thymine analog)

Figure 20.17 Growth factors and structurally similar analogs.

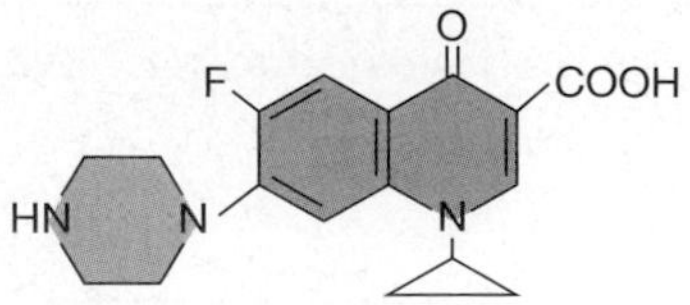

Figure 20.18 The structure of ciprofloxacin, a quinolone. Fluorinated derivatives of nalidixic acid (Figure 20.12) are more soluble than nalidixic acid and reach clinically therapeutic levels in blood and tissues. Ciprofloxacin is used to treat urinary tract infections and anthrax caused by penicillin-resistant *Bacillus anthracis*.

fungal infections and are also used as mutagens (see Sections 20.10 and 20.11, ⊙⊙ Section 10.3).

Quinolones

The quinolones are not growth factor analogs, but are a class of synthetic antibacterial compounds that interact with bacterial DNA gyrase and prevent the gyrase from supercoiling bacterial DNA, which is required for packaging of DNA in the bacterial cell (⊙⊙ Section 7.3). Nalidixic acid is the prototype quinolone (Figure 20.12 and Figure 20.13). Fluoroquinolone derivatives of nalidixic acid such as norfloxacin and ciprofloxacin (Figure 20.18) are routinely used to treat urinary tract infections in humans. Ciprofloxacin is also the drug of choice for treating infections from penicillin-resistant strains of *Bacillus anthracis*, the cause of anthrax (⊙⊙ Section 25.11). Because DNA gyrase is found in all *Bacteria*, the fluoroquinolones are effective for treating both gram-positive and gram-negative bacterial infections (Figure 20.15). The fluoroquinolones are also extensively used in the beef and poultry industries for prevention of respiratory diseases.

✓ 20.6 Concept Check

Synthetic chemotherapeutic agents exhibit selective toxicity for *Bacteria* and are safe to take internally. Growth factor analogs such as sulfanilamide are synthetic metabolic inhibitors. Quinolones inhibit the action of DNA gyrase in *Bacteria*.

✓ Identify at least two features that distinguish synthetic chemotherapeutic agents from antiseptics and disinfectants.

✓ What is *selective toxicity*?

20.7 Naturally Occurring Antimicrobial Drugs: Antibiotics

Antibiotics are chemical compounds produced by microorganisms that inhibit or kill other microorganisms. Antibiotics are distinguished from growth factor analogs because they are natural products (products of microbial activity) rather than synthetic chemicals. Antibiotics

constitute one of the most important classes of substances produced by large-scale microbial processes. The industrial production of antibiotics will be discussed in Chapter 30.

Targets of Antibiotics

A very large number of antibiotics have been discovered, but less than 1% have been of practical value in medicine. However, the useful antibiotics have had a dramatic impact on the treatment of infectious diseases. Further, many antibiotics are made more effective by chemical modifications in the laboratory; these are said to be *semisynthetic antibiotics*.

The sensitivity of microorganisms to antibiotics and other chemotherapeutic agents varies (Figure 20.15). Gram-positive *Bacteria* are usually more sensitive to antibiotics than are gram-negative *Bacteria*, although some antibiotics act only on gram-negative *Bacteria*. An antibiotic that acts on both gram-positive and gram-negative *Bacteria* is called a **broad-spectrum antibiotic**. In general, a broad-spectrum antibiotic finds wider medical use than a *narrow-spectrum antibiotic*, which acts on only a single group of organisms. An antibiotic with a limited spectrum of activity may, however, be quite valuable for the control of microorganisms that fail to respond to other antibiotics. An example is vancomycin, a glycopeptide that is a bacteriocidal agent that acts against gram-positive *Bacteria* of the genera *Staphylococcus, Bacillus*, and *Clostridium* (Figures 20.12, 20.13, and 20.15).

Antibiotics and other chemotherapeutic agents can be grouped based on chemical structure (Figure 20.12) or on mode of action (Figure 20.13). In *Bacteria*, the important targets of antibiotic action are the cell wall (for example, vancomycin), the cytoplasmic membrane (polymyxins), the biosynthetic processes of protein synthesis (the macrolides and tetracyclines), and nucleic acid synthesis (rifampin). Here we briefly discuss mechanisms for inhibiting protein synthesis and transcription.

Antibiotics Affecting Protein Synthesis

Many antibiotics inhibit protein synthesis by interacting with the ribosome (Figure 20.13). These interactions are quite specific and many involve rRNA. Several of these antibiotics are medically useful and several are also effective as research tools because they are specific for individual steps in protein synthesis (⊙⊙ Section 7.15). For instance, streptomycin inhibits protein chain *initiation*, whereas puromycin, chloramphenicol, cycloheximide, and tetracycline inhibit protein *elongation*. Even when two antibiotics inhibit the same step in protein synthesis, the mechanisms of inhibition can be quite different. For example, puromycin binds to the A site on the ribosome, and the growing polypeptide chain is transferred to the puromycin instead of to the amino acid–transfer RNA complex. The puromycin–peptide complex is then released from the ribosome, halting elongation prematurely. By contrast, chloramphenicol inhibits elongation by blocking formation of the peptide bond.

Many antibiotics specifically inhibit ribosomes of organisms from only one or two of the phylogenetic domains. For example, chlorampenicol and streptyomycin are specific for ribosomes of *Bacteria*, while cycloheximide affects only ribosomes from *Eukarya*. Since mitochondria and chloroplasts have ribosomes of the prokaryotic type, antibiotics that inhibit protein synthesis in *Bacteria* also inhibit protein synthesis in these organelles (Section 14.4). Several of these antibiotics, such as tetracycline, are medically useful because the eukaryotic mitochondria are not affected at the concentrations used for antimicrobial therapy.

Antibiotics Affecting RNA Polymerase

A number of antibiotics specifically inhibit RNA synthesis (Section 7.8 and Figure 20.13). For example, the rifamycins and streptovaricins inhibit RNA synthesis by attacking the β subunit of RNA polymerase. These antibiotics have specificity for *Bacteria*, chloroplasts, and mitochondria. Actinomycin inhibits RNA synthesis by combining with DNA and blocking RNA elongation. Actinomycin binds most strongly to DNA at guanine–cytosine base pairs, fitting into the major groove in the double strand where RNA is synthesized.

Some of the most useful antibiotics are directed against structural features such as cell walls that are unique to prokaryotes. We discuss a number of these antibiotics and their targets in the next section.

✓ *20.7 Concept Check*

Antibiotics are a chemically diverse group of bacteriocidal or bacteriostatic compounds that are produced by a variety of microorganisms. Most antibiotics function by inhibiting protein synthesis, transcription, or cell wall synthesis in target organisms. Although many antibiotics are useful for clinical applications, many others are not useful in humans or animals.

✓ Distinguish *antibiotics* from *growth factor analogs*.

✓ What is a *broad-spectrum antibiotic*?

20.8 β-Lactam Antibiotics: Penicillins and Cephalosporins

One of the most important groups of antibiotics, both historically and medically, is the β-lactam group. The **β-lactam antibiotics** include the penicillins, cephalosporins, and cephamycins, all medically useful antibiotics. These antibiotics all share the presence of a characteristic structural component, the β-lactam ring (Figure 20.19). Together, the penicillins and cephalosporins account for over one-half of all of the antibiotics pro-

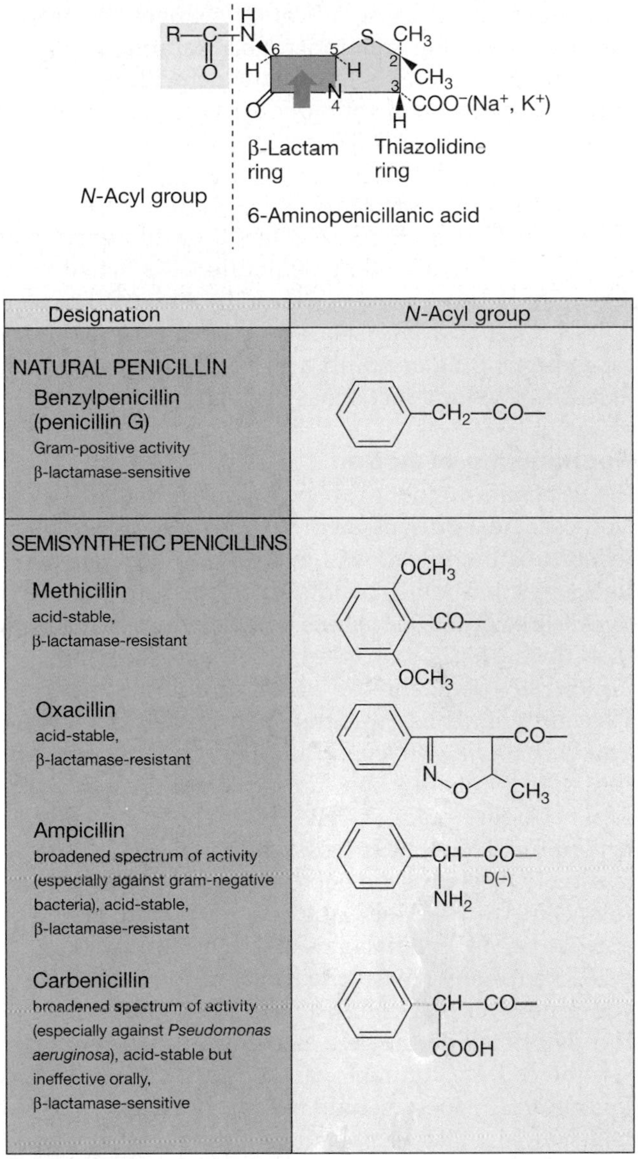

Figure 20.19 The structures of some important penicillins. The red arrow (top panel) is the site of action for most β-lactamases.

Designation	N-Acyl group
NATURAL PENICILLIN **Benzylpenicillin (penicillin G)** Gram-positive activity β-lactamase-sensitive	—CH₂—CO—
SEMISYNTHETIC PENICILLINS **Methicillin** acid-stable, β-lactamase-resistant	OCH₃ / —CO— / OCH₃
Oxacillin acid-stable, β-lactamase-resistant	—CO— CH₃
Ampicillin broadened spectrum of activity (especially against gram-negative bacteria), acid-stable, β-lactamase-resistant	—CH—CO— D(−) NH₂
Carbenicillin broadened spectrum of activity (especially against *Pseudomonas aeruginosa*), acid-stable but ineffective orally, β-lactamase-sensitive	—CH—CO— COOH

duced and used worldwide (Figure 20.14). Penicillin is produced by the fungus *Penicillium chrysogenum*, and cephalosporin is produced by the fungus *Cephalosporium* sp. (Section 30.6.)

Types of Penicillin

The first β-lactam antibiotic discovered, *penicillin G* (Figure 20.19), is active primarily against gram-positive *Bacteria*. Its action is restricted to gram-positive *Bacteria* primarily because gram-negative *Bacteria* are impermeable to the antibiotic. However, **semisynthetic penicillins** are constantly being developed and introduced, many of which are quite effective against gram-negative *Bacteria*.

Figure 20.19 shows the complex structures of some of these new penicillins. Modifications of the basic penicillin G structure significantly changes the properties of the resulting antibiotics. For example, ampicillin and carbenicillin are semisynthetic penicillins that have a broader spectrum of antibiotic activity, including some gram-negative *Bacteria*. The structural differences in the *N*-acyl groups of these semisynthetic penicillins allow them to be transported inside the gram-negative outer membrane (∞ Section 4.9) where they inhibit cell wall synthesis. Note also that penicillin G is sensitive to β-lactamase, an enzyme produced by a number of penicillin-resistant *Bacteria* (see Section 20.12). The semisynthetic penicillins oxacillin and methicillin are useful because they are β-lactamase resistant.

Mechanisms of Action

The β-lactam antibiotics are potent inhibitors of cell wall synthesis. As we discussed in Sections 4.8 and 6.2, an important feature of cell wall synthesis is the transpeptidation reaction, which results in the cross-linking of two glycan-linked peptide chains. The enzymes that accomplish this task, the transpeptidases, are also capable of binding to penicillin or other antibiotics with the β-lactam ring. Thus, these transpeptidases are known as *penicillin binding proteins* (PBPs). The PBPs bind very tightly to penicillin and can no longer catalyze the transpeptidase reaction. The cell wall continues to be formed but is no longer cross-linked and becomes progressively weaker as the peptidoglycan backbone is laid down. In addition, the antibiotic-PBP complex stimulates the release of autolysins that digest the existing cell wall. The result is a weakened, eventually degraded cell wall. Under normal circumstances, the osmotic pressure differences inside the cell, as compared with outside, lyse the cell. By contrast, vancomycin, a glycopeptide (Figure 20.12), does not bind to PBPs but acts directly on the terminal D-alanyl-D-alanine peptide on the peptidoglycan precursors (∞ Section 6.2), blocking the transpeptidase reaction. Because the cell wall and its synthesis mechanisms are unique to *Bacteria*, the β-lactam antibiotics have very high specificity and are not toxic to host cells. However, because of the complex structural configurations of these antibiotics, some individuals develop serious allergies to individual β-lactam compounds after repeated courses of antibiotic therapy. These allergic responses can be life-threatening (∞ Section 22.13).

Cephalosporins

The cephalosporins are another group of clinically important antibiotics that contain the β-lactam ring. They differ structurally from the penicillins because they have a six-member dihydrothiazine ring instead of the five-member thiazolidine ring. The cephalosporins have the same mode of action as the penicillins. That is, they bind irreversibly to the PBPs and prevent the cross-linking of peptidoglycan. The clinically important cephalosporins

are semisynthetic antibiotics that generally have a broader spectrum of antibiotic activity than the penicillins and are often more resistant to the action of enzymes that destroy β-lactam rings, the β-lactamases. For example, ceftriaxone (Figure 20.12), a widely used cephalosporin, is highly resistant to the β-lactamases and has now supplanted penicillin as the drug of choice for treatment of infections due to *Neisseria gonorrhoeae* (see Section 20.12, ∞ Section 26.12) because many *N. gonorrhoeae* strains have developed β-lactamases that cleave the β-lactam rings of penicillin.

✓ 20.8 Concept Check

The β-lactam compounds are the most important clinical antibiotics. This group includes the penicillins and the cephalosporins. These antibiotics are specific for the cell wall synthesis enzymes of *Bacteria* and, as a group, have very low host toxicity and a very broad spectrum of activity.

✓ Draw the structure of the β-lactam ring.

✓ How do the β-lactam antibiotics function?

20.9 Antibiotics from Prokaryotes

Many antibiotics active against prokaryotes are also produced by prokaryotes. These include the aminoglycosides, the macrolides, and the tetracyclines. Many of these antibiotics have major clinical applications, and thus we discuss their general properties here.

Aminoglycoside Antibiotics

Aminoglycoside antibiotics contain amino sugars bonded by glycosidic linkage (∞ Section 3.3) to other amino sugars. A number of clinically useful antibiotics are aminoglycosides, including *streptomycin* (Figure 20.12) and its relatives, *kanamycin* (Figure 20.20), *gentamicin*, and *neomycin*. The aminoglycosides act by inhibiting protein synthesis at the 30S subunit of the ribosome (Figure 20.13). The aminoglycoside antibiotics are used clinically against gram-negative *Bacteria*. Streptomycin has also been used extensively in the treatment of tuberculosis. Historically, the use of streptomycin for tuberculosis treatment was a major medical advance, as it was the first antibiotic capable of controlling this infectious disease. However, none of the aminoglycoside antibiotics are widely used today and together the aminoglycosides account for only about 3% of the total of all antibiotics produced and used (Figure 20.14). Streptomycin has been supplanted by several synthetic chemicals for tuberculosis treatment because streptomycin causes several serious side effects and bacterial resistance readily develops. The use of aminoglycosides for treatment of gram-negative infections has decreased since the development of the semisynthetic penicillins (see Section 20.8)

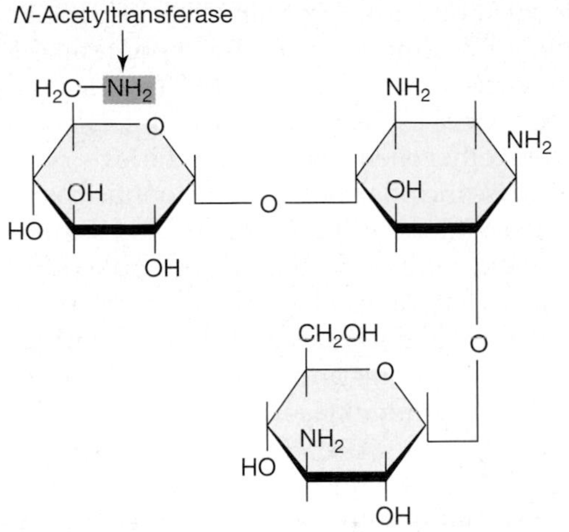

Figure 20.20 Structure of kanamycin, an aminoglycoside antibiotic. The amino sugars are in yellow. The site of modification by an *N*-acetyltransferase, encoded by a resistance plasmid, is indicated.

and the tetracyclines (see later in this section). The aminoglycoside antibiotics are now considered reserve antibiotics used primarily when other antibiotics fail.

Macrolide Antibiotics

Macrolide antibiotics contain large lactone rings connected to sugar moieties (Figure 20.21). Variations in both the macrolide ring and the sugar moieties are known, and so a large variety of macrolide antibiotics exist. The best-known macrolide antibiotic is *erythromycin*, but other macrolides include *oleandomycin, spiramycin*, and *tylosin*. Together, the macrolide antibiotics account for 11% of the total world production and use (Figure 20.14). Erythromycin acts as a protein synthesis inhibitor at the level of the 50S subunit of the ribosome (Figure 20.13). Erythromycin is commonly used clinically in place of penicillin in patients allergic to penicillin or other β-lactam antibiotics. Erythromycin has been particularly valuable in treating cases of legionellosis (➣ Section 28.7) because of the sensitivity of *Legionella pneumophila* to this antibiotic.

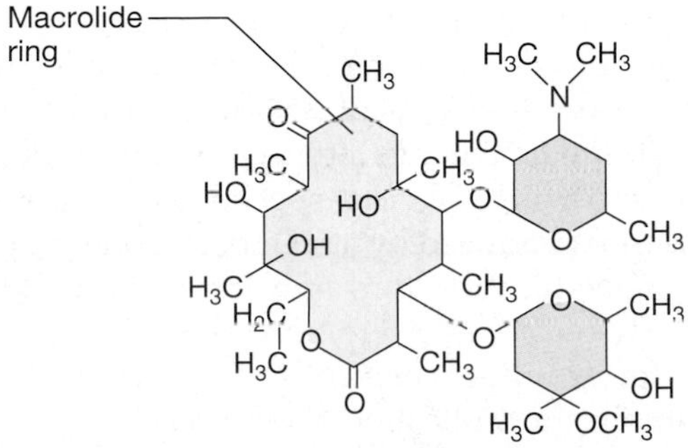

Figure 20.21 Structure of erythromycin, a macrolide antibiotic.

Tetracyclines

The **tetracyclines** are an important group of antibiotics that find widespread medical use in humans. They were some of the first broad-spectrum antibiotics, inhibiting almost all gram-positive and gram-negative *Bacteria*. The basic structure of the tetracyclines consists of a naphthacene ring system (Figure 20.22). The basic naphthacene ring structure can be substituted at several positions to form new tetracycline analogs. *Chlortetracycline*, for instance, has a chlorine atom, whereas *oxytetracycline* has an additional hydroxyl (OH) group and no chlorine (Figure 20.22). All three of these antibiotics are produced microbiologically, but there are also semisynthetic tetracyclines having synthetic substitutions in the naphthacene ring system. Like erythromycin and the aminoglycoside antibiotics, tetracycline is a protein synthesis inhibitor. It interferes with 30S ribosomal subunit function (see Figure 20.13).

The tetracyclines and the β-lactam antibiotics are the two most important groups of antibiotics in the medical field. The tetracyclines also find use in veterinary medicine, and in some countries are used as nutritional supplements for poultry and swine. Because extensive nonmedical uses of medically important antibiotics have resulted in wide-spread antibiotic resistance, this use is now discouraged (see Section 20.12).

✓ 20.9 Concept Check

The aminoglycosides, macrolides, and tetracycline antibiotics are structurally complex molecules produced by prokaryotes and are active against other prokaryotes. Erythromycin and the various tetracyclines are used widely in clinical medicine.

✓ What are the biological sources of the aminoglycoside, tetracycline, and macrolide antibiotics?

✓ What is the mechanism of action for each of these classes of antibiotics?

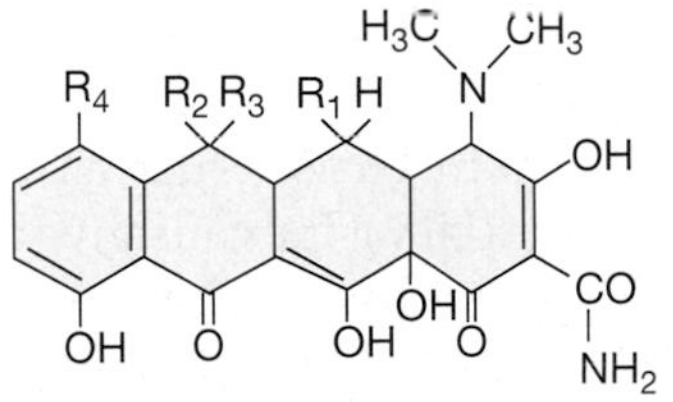

Tetracycline analog	R₁	R₂	R₃	R₄
Tetracycline	H	OH	CH₃	H
7-Chlortetracycline (aureomycin)	H	OH	CH₃	Cl
5-Oxytetracycline (terramycin)	OH	OH	CH₃	H

Figure 20.22 Structure of tetracycline and important semisynthetic analogs.

IV CONTROL OF VIRUSES AND EUKARYOTIC PATHOGENS

Drugs that control growth of viruses and eukaryotic pathogens such as fungi often affect the eukaryotic host cells. As a result, only compounds that preferentially affect pathogen-specific metabolic pathways or structural components are useful for treating these infections. There are a limited number of these drugs, and here we discuss the action of some of them.

20.10 Antiviral Drugs

Viruses use their eukaryotic hosts to reproduce and perform metabolic functions (∞ Chapter 9). Because the host and the virus use the host reproductive and metabolic pathways and structures, attempts to control viruses with drugs often result in toxicity for the host. However, several agents are more toxic for viruses than for the host, and a few agents produced by the host specifically target viruses.

Antiviral Chemotherapeutic Agents

Because viral structures and functions are so integrated into the functions of the host cell, the therapeutic successes achieved against bacteria using extremely selective antibacterial agents have not been matched by similar achievements with antiviral agents. However, largely because of efforts to find effective measures to control AIDS (∞ Section 26.14), some significant achievements have now been made in controlling viruses with chemical agents (Table 20.5).

The most successful and commonly used agents for antiviral chemotherapy are the nucleoside analogs (Table 20.5). The first compound to gain universal acceptance in this category was zidovudine, or azidothymidine (AZT). AZT inhibits retroviruses such as the human immunodeficiency virus (HIV), the causative agent of AIDS (∞ Figure 26.38). Azidothymidine is chemically related to thymidine but is a dideoxy derivative, lacking the 3'-hydroxyl group. AZT inhibits multiplication of retroviruses by blocking the synthesis of the DNA intermediate (reverse transcription) and successfully inhibits multiplication of HIV. A number of other nucleoside analogs having similar mechanisms have been developed for the treatment of HIV, as shown in Table 20.5. Nearly all nucleoside analogs, or *nucleoside reverse transcriptase inhibitors* (**NRTI**), work by the same mechanism, inhibiting elongation of the viral nucleic acid chain at the level of the host cell nucleic acid polymerase. Because the normal host cell function of nucleic acid replication is targeted, these drugs almost always exhibit

some level of host toxicity. Many also lose their antiviral potency with time due to the emergence of drug-resistant viruses (∞ Section 26.14). The *nucleotide analog* cidofovir works in the same way (Table 20.5).

Several other chemicals work at the level of viral polymerase. These include nevirapine, a nonnucleoside reverse transcriptase inhibitor that binds directly to reverse transcriptase and inhibits further action; phosphonoformic acid, which acts as an analog of inorganic pyrophosphate, inhibiting normal internucleotide linkage; and rifamycin, an antibiotic that binds and inhibits RNA polymerase.

A relatively novel class of antiviral drugs are the **protease inhibitors (PI)** (Table 20.5). These drugs are particularly effective for treatment of HIV. They prevent infection by binding the active site of HIV protease, inhibiting processing of viral polypeptides and virus maturation (∞ Section 26.14 and Section 20.13).

Interferon

Interferons are antiviral substances produced by many animal cells in response to infection by certain viruses. They are low-molecular-weight proteins (17,000 MW) that prevent viral multiplication in normal cells by stimulating the production of antiviral proteins. Interferons from virus-infected cells interact with receptors on noninfected cells, promoting the synthesis of antiviral proteins that function to prevent further virus infection. There are three molecular types: IFN-α, produced by leukocytes; IFN-β, produced by fibroblasts; and IFN-γ, produced by immune cells known as lymphocytes (∞ Section 23.10). All three types are effective viral inhibitors. They were first discovered in the course of studies on virus interference, a phenomenon whereby infection with one virus interferes with subsequent infection with another virus, hence the name *interferon*. Interferons are formed in response to live virus, viral nucleic acids, and also to virus inactivated by radiation. Interferon is produced in larger amounts by cells infected with viruses of low virulence, but little is produced against highly virulent viruses. Apparently highly virulent viruses inhibit cell protein synthesis before any interferon can be produced. Interferon is also induced by a variety of double-stranded RNA molecules, either natural or synthetic. Since double-stranded RNA does not exist in uninfected cells but exists as the replicative form in RNA virus-infected cells, double-stranded RNA may serve as a signal of virus infection in the animal cell and stimulate the interferon-producing system.

Interferons are not virus-specific but are *host*-specific. Interferon produced by a member of one species recognizes specific receptors only on cells of the same species. Therefore, interferon produced by one type of animal (for example, chicken) in response to influenza virus inhibits multiplication of other viruses in the same species but has no effect on the multiplication of any virus in other animal species.

TABLE 20.5 Antiviral chemotherapeutic compounds

Category/drug	Mechanism of action	Virus affected
Nucleoside analogs		
Acyclovir	Viral polymerase inhibitors	Herpes viruses, *Varicella zoster*
Ganciclovir		Cytomegalovirus
Trifluridine		Herpesvirus
Valacyclovir		Herpesvirus
Vidarabine		Herpesvirus, vaccinia, hepatitis B virus
Didanosine (dideoxyinosine or ddI)	Reverse transcriptase inhibitors	HIV[a]
Lamivudine (3TC)		HIV, hepatitis B virus
Stavudine (d4T)		HIV
Zalcitabine (ddC)		HIV
Zidovudine (AZT) (∞ Figure 26.28)		HIV
Ribavirin	Blocks capping of viral RNA	Respiratory syncytial virus, influenza A and B, Lassa fever
Synthetic amines		
Amantadine	Block uncoating of virus	Influenza A
Rimantadine		Influenza A
Nucleotide analog		
Cidofovir	Viral polymerase inhibitor	Cytomegalovirus, herpesviruses
Pyrophosphate Analog		
Phosphonoformic acid (Foscarnet)	Viral polymerase inhibitor	Herpesviruses, HIV, hepatitis B virus
Nonnucleoside reverse transcriptase inhibitor (NNRTI)		
Nevirapine	Reverse transcriptase inhibitor	HIV
RNA polymerase inhibitor		
Rifamycin	RNA polymerase inhibitor	Vaccinia, pox viruses
Protease inhibitors		
Indinavir (Figure 20.27)	Protease inhibitors	HIV
Ritonavir		HIV
Saquinavir (Figure 20.27)		HIV
Nelfinavir		HIV
Abacivir		HIV
Lopinavir		HIV
Interferons		
Interferon α	Induces proteins that inhibit viral replication	Broad spectrum (host specific)
Interferon β		
Interferon γ		

[a] Human immunodeficiency virus

Interferons have potential as possible antiviral and anticancer agents. There are now several approved recombinant interferon preparations available. However, the use of interferons as general chemotherapeutic agents is not widespread because interferon must be delivered locally in high concentrations to stimulate the production of antiviral proteins in uninfected host cells. Thus, the clinical utility of these antiviral agents depends on our ability to deliver interferon to local areas in the host through injections or aerosols. Alternatively, appropriate interferon-stimulating signals (*i.e.*, stimulation with viral nucleotides, nonvirulent viruses, or even synthetic nucleotides) given to host cells prior to viral infection might achieve the same goal.

✓ 20.10 Concept Check

Viruses use host cell metabolic machinery to replicate. However, several virus-specific enzymes and processes can be interrupted with chemotherapeutic agents to disrupt viral replication. Clinically effective antiviral agents include nucleoside analogs that work by inhibiting viral nucleotide polymerases. Several other agents, including the protease inhibitors, interfere with viral maturation steps. Host cells also produce antiviral proteins called *interferons* that stop viral replication.

✓ Why are there so few effective antiviral chemotherapeutic agents?

✓ What steps in the viral maturation process are inhibited by nucleoside analogs? By protease inhibitors? By interferons?

20.11 Antifungal Drugs

Like the viruses, fungi pose special problems for successful chemotherapy. Since fungi are *Eukarya*, much of their cellular machinery is the same as in animals and humans, and so chemotherapeutic agents that affect metabolic pathways in fungi often affect corresponding pathways in host cells, resulting in drug toxicity. Thus, many antifungal drugs can be used only for topical (surface) applications. However, some drugs are selectively toxic for fungi. Drugs for fungal treatment are becoming increasingly important as fungal infections in immunosuppressed individuals become more prevalent (∞ Sections 26.14 and 27.7). We will look in detail at the selective toxicity of several classes of chemicals that are effective against fungi.

Ergosterol Inhibitors

Two major groups of antifungal compounds work by interacting with ergosterol or inhibiting its synthesis (Table 20.6). In most fungi, ergosterol replaces the cholesterol component found in higher eukaryotic cell membranes (∞ Section 4.5). The first group includes the *polyenes*, a group of antibiotics produced by *Streptomyces* species. Polyenes bind to ergosterol, which disrupts membrane function, eventually causing membrane permeability and cell death (Figure 20.23). A second major group of antifungal compounds includes the *azoles* and the *allylamines*, agents that selectively inhibit ergosterol biosynthesis and therefore have broad antifungal activity. Treatment with azoles results in the inability to produce a normal membrane, leading to membrane damage and alteration of critical membrane activities such as nutrient transport. Allylamines also inhibit ergosterol biosynthesis but are useful only topically because they are not readily taken up by animal cells and tissues.

Other Antifungal Agents

A number of other antifungal drugs interfere with fungus-specific structures and functions (Table 20.6). For example, most fungal cell walls contain *chitin*, a polymer

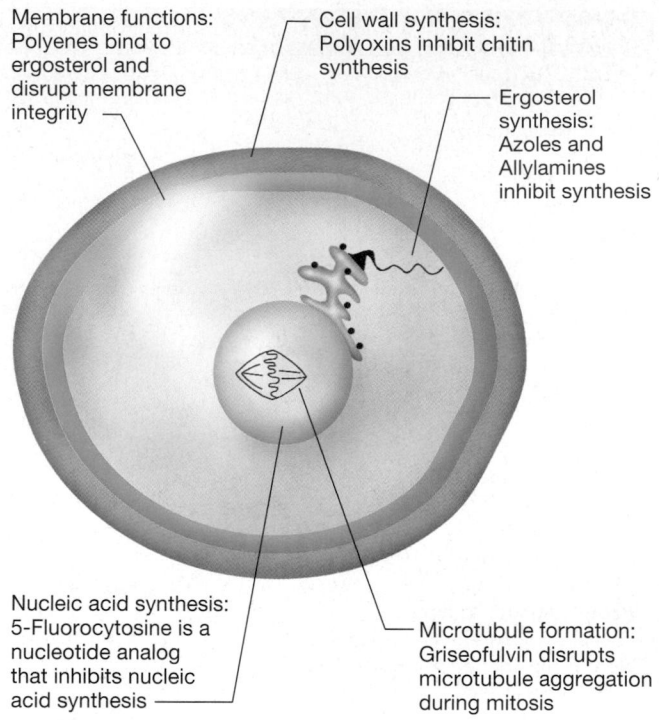

Figure 20.23 Sites of action of some antifungal chemotherapeutic agents. Because fungi are eukaryotic cells, antibacterial antibiotics are generally ineffective.

of *N*-acetylglucosamine found only in fungi and insects (∞ Section 17.25). Several *polyoxins* inhibit cell wall synthesis by interfering with chitin biosynthesis. Although the polyoxins are used as agricultural fungicides, no chitin-synthesis inhibitors are used clinically. Other drugs inhibit folate biosynthesis, interfere with DNA topology during replication, or, like *griseofulvin*, disrupt microtubule aggregation during mitosis (∞ Section 7.6). The nucleic acid analog *5-fluorocytosine* is an effective nucleic acid synthesis inhibitor in fungi. Some very effective antifungal drugs also have other biological applications. For example, *vincristine, vinblastin* and *toxol* are effective antifungal agents and have known anticancer properties.

TABLE 20.6	Antifungal drugs (fungicides)		
Category	**Target**	**Examples**	**Use**
Polyenes	Ergosterol synthesis	Amphotericin B	Oral
Nucleic acid analogs	DNA synthesis	5-Fluorocytosine	Oral
Polyoxins	Chitin synthesis	Polyoxin A	Agricultural
		Polyoxin B	Agricultural
Azoles	Ergosterol synthesis	Fluconazole	Oral
		Itraconazole	Oral
		Ketoconazole	Oral
		Clotrimazole	Topical
		Miconazole	Topical
		Voriconazole	Oral
Allylamines	Ergosterol synthesis	Terbenafine	Oral

Unfortunately, the use of antifungal drugs has predictably resulted in the emergence of populations of resistant fungi and the emergence of "new" fungal pathogens. For example, *Candida* species, which are normally not pathogenic, now produce disease in individuals who have been treated with antifungal drugs. These drug-resistant *Candida* pathogens are not treatable by employing any of the currently used antifungal agents.

As the use of chemotherapeutic agents, both antibacterial and antifungal, increases, the possibilities for opportunistic fungal infections and the corresponding need for specific fungal control agents will increase.

✓ 20.11 Concept Check

Antifungal agents fall into a wide variety of chemical categories. As with viruses, selective toxicity is hard to achieve, but there are some effective chemotherapeutic agents. Treatment of fungal infection is an emerging human health issue.

✓ Why are there very few clinically effective antifungal antibiotics?

✓ What factors are contributing to the apparent rise in fungal infections?

V ANTIMICROBIAL DRUG RESISTANCE AND DRUG DISCOVERY

Antimicrobial drug resistance is a major problem when dealing with many common pathogenic microorganisms. Here we explore some of the reasons for drug resistance and present several strategies for overcoming resistance.

20.12 Antimicrobial Drug Resistance

Antimicrobial drug resistance is the acquired ability of an organism to resist the effects of a chemotherapeutic agent to which it is normally susceptible. Most antimicrobial resistance involves *resistance genes* that are transferred by means of genetic exchange. To protect themselves, the antibiotic producers develop resistance mechanisms to neutralize or destroy their own antibiotics; genes encoding these resistance mechanisms can occasionally be transferred to other organisms.

Resistance Mechanisms

Not all antibiotics act against all microorganisms. Some microorganisms are *naturally resistant* to some antibiotics. There are several reasons why microorganisms may have an inherent resistance to an antibiotic. (1) The

organism may lack the structure an antibiotic inhibits. For instance, some bacteria, such as mycoplasmas, lack a typical bacterial cell wall and are resistant to penicillins. (2) The organism may be impermeable to the antibiotic. For example, most gram-negative *Bacteria* are impermeable to penicillin G. (3) The organism may be able to alter the antibiotic to an inactive form. Many staphylococci contain β-lactamases that cleave the β-lactam ring of most penicillins (Figure 20.24). (4) The organism may modify the *target* of the antibiotic. (5) By genetic change, alteration may occur in a metabolic pathway that the antimicrobial agent blocks. Thus, the organism develops a resistant biochemical pathway. For example, many

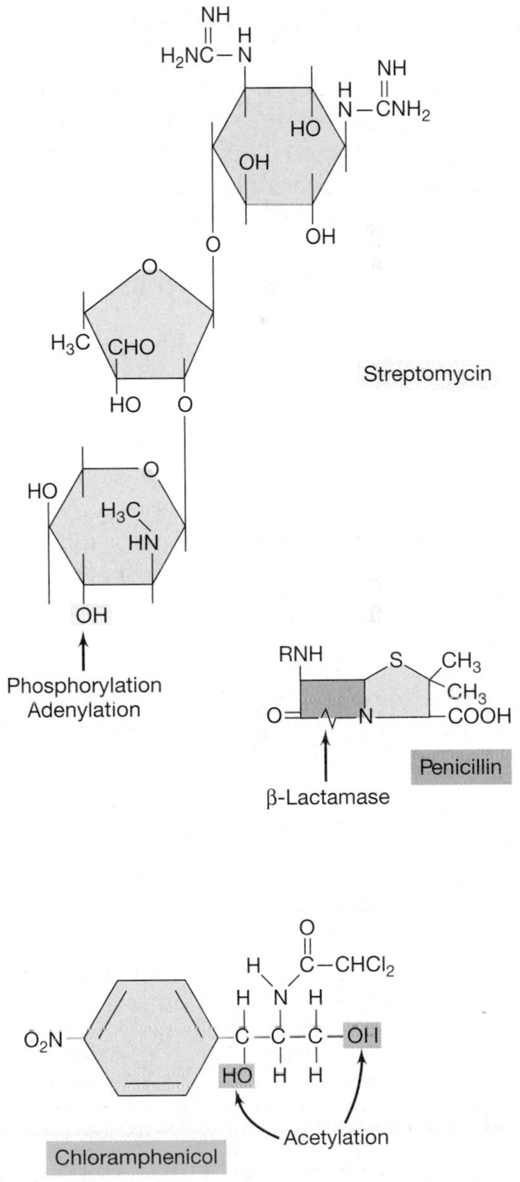

Figure 20.24 Sites at which antibiotics are attacked by enzymes encoded by R plasmid genes. In aminoglycoside antibiotics related to streptomycin, those with a free amino group may be inactivated by *N*-acetylation (see also Figure 20.20).

pathogens develop resistance to sulfonamide drugs (see Section 20.6 and Figure 20.16). Sulfonamides inhibit the production of folic acid in *Bacteria*, but resistant *Bacteria* modify their metabolism to take up preformed folic acid from the environment, avoiding the need for the pathway blocked by sulfonamides. (6) The organism may be able to pump out an antibiotic entering the cell (efflux). Some specific examples of bacterial resistance to antibiotics are shown in Table 20.7.

As discussed in Section 10.8, antibiotic resistance can be genetically encoded by the microorganism at either the chromosomal or the plasmid level on so-called *resistance plasmids* (*R factors*); specific types of resistance typically have a genetic basis in one location or the other (Table 20.7). Because of the development of antibiotic resistance, testing of bacteria isolated from clinical material for antibiotic sensitivity must be carried out using the minimum inhibitory concentration (MIC) method or the agar diffusion method (Section 20.4 and Figures 20.10 and 20.11). Details of the antibiotic sensitivity testing of clinical isolates are described in Section 23.3.

Mechanism of Resistance Mediated by R Plasmids

In the *laboratory*, antibiotic-resistant cells are often isolated from cultures that were predominantly antibiotic-sensitive. The resistance of these isolates is usually due to mutations in *chromosomal* genes. On the other hand, the majority of drug-resistant bacteria isolated from *patients* contain the drug-resistance genes on R plasmids. The mechanism of R plasmid resistance is different from that of chromosomal resistance. In most cases, antibiotic resistance mediated by chromosomal genes arises because of a modification of the *target* of antibiotic action (for example, a ribosome).

By contrast, R plasmid resistance is in most cases due to the presence in the R plasmid of genes encoding new enzymes that *inactivate* the drug (Figure 20.24) or genes that encode enzymes that either prevent uptake of the drug or actively pump it out. For instance, a number of known antibiotics have similar chemical structures containing aminoglycoside units. Among the aminoglycoside antibiotics are streptomycin, neomycin, kanamycin, and spectinomycin. Strains carrying resistance-conferring R plasmids contain enzymes that chemically modify the antibiotics either by phosphorylation, acetylation, or adenylylation. The modified drug then lacks antibiotic activity. In the case of the penicillins, R plasmid resistance is due to the formation of penicillinase (β-lactamase), which splits the β-lactam ring. Chloramphenicol resistance mediated by an R plasmid arises because of the presence of an enzyme that acetylates the antibiotic. Many R plasmids can confer multiple antibiotic resistance because a single R plasmid may contain several different genes, each encoding a different antibiotic-inactivating enzyme.

Origin of Resistance Plasmids

Although specific evidence for the origin of multiple drug resistance R plasmids is not available, a number of lines of circumstantial evidence suggest that R plasmids existed before the antibiotic era. The widespread use of antibiotics provided selective conditions for the spread of R plasmids with one or more antibiotic resistance genes (see the box, Nonmedical Uses of Antibiotics). For example, a strain of *Escherichia coli* that was freeze-dried in 1946 contained a plasmid with genes conferring resis-

TABLE 20.7	Mechanisms of bacterial resistance to antibiotics		
Resistance mechanism	**Antibiotic example**	**Genetic basis of resistance**	**Mechanism present in:**
Reduced permeability	Penicillins	Chromosomal	*Pseudomonas aeruginosa* Enteric *Bacteria*
Inactivation of antibiotic (for example, penicillinase; modifying enzymes	Penicillins	Plasmid and chromosomal	*Staphylococcus aureus* Enteric *Bacteria* *Neisseria gonorrhoeae*
methylases, acetylases, and phosphorylases;	Chloramphenicol	Plasmid and chromosomal	*Staphylococcus aureus* Enteric *Bacteria*
and others)	Aminoglycosides	Plasmid	*Staphylococcus aureus*
Alteration of target (for example, RNA polymerase, rifamycin; ribosome, erythromycin, and streptomycin; DNA gyrase, quinolones)	Erythromycin Rifamycin Streptomycin Norfloxacin	Chromosomal	*Staphylococcus aureus* Enteric *Bacteria* Enteric *Bacteria* Enteric *Bacteria* *Staphylococcus aureus*
Development of resistant biochemical pathway	Sulfonamides	Chromosomal	Enteric *Bacteria* *Staphylococcus aureus*
Efflux (pumping out of cell)	Tetracyclines Chloramphenicol	Plasmid Chromosomal	Enteric *Bacteria* *Staphylococcus aureus* *Bacillus subtilis*

A Focus On ... Nonmedical Uses of Antibiotics

Antibiotics are extensively used for nonmedical reasons in the United States and other developed countries as growth supplements in animal feed. The addition of low levels of antibiotics to animal feed stimulates animal growth, shortening the period required to get the animal to market. For example, addition of 25 milligrams (mg) of penicillin per pound of chicken feed saves 900 million kg (2 billion lb) of feed each year due to more rapid weight gains and feeding efficiency. The antibiotics probably act by inhibiting low-grade infections and reducing the resulting intestinal epithelial inflammation, allowing for more efficient uptake of nutrients. Studies with germ-free animals confirm this idea; the growth of germ-free animals is not accelerated by antibiotic-supplemented feed.

Unfortunately, low levels of antibiotics in animal feed select antibiotic-resistant microorganisms due to the constant exposure to antibiotics. Molecular studies in *Salmonella* isolated from poultry have shown that resistance is rapidly transferred between different bacterial species and even between different genera in the gut. In another study, over 80% of *Salmonella* strains recovered from meat in supermarkets were resistant to at least one antibiotic, and most resistant strains had acquired at least three resistance genes. Studies on microorganisms isolated from workers in animal husbandry have shown that many human strains had acquired resistance to the antimicrobial drugs used in animal feeds. The use of antibiotics in animal feed therefore expands the pool of antibiotic resistant microorganisms that can infect humans.

Another study indicates that some strains of *Enterococcus faecium*, already resistant to most antibiotics, acquired resistance to Synercid, a relatively new human drug that is used selectively for treatment of antibiotic-resistant infections, within two years of the drug's approval by the Federal Drug Administration. The investigators speculate that the resistant strains arose from livestock given a Synercid analog for the last twenty-five years!

Unfortunately, long-term studies on animals previously fed antibiotics and then put on antibiotic-free rations have shown that antibiotic-resistant bacteria are not quickly lost from the gut. The resistance genes are integrated into stable plasmids or chromosomes in the gut flora, and, in the absence of counterselective forces, will be maintained for some time, even if supplementation of feeds with antibiotics were to stop. Nonmedical use of antibiotics has therefore reinforced a simple lesson in microbial ecology: The environment selects the best-adapted species. However, in hope of reducing the spread of antibiotic resistance in Europe, some European countries have banned the use of clinically useful antibiotics in animal feeds. The World Health Organization has also drafted a policy discouraging the use of antibiotics in animal feed. In the United States large amounts of antibiotics continue to be used in the cattle, poultry, and swine industries. ■

tance to tetracycline and streptomycin, even though neither of these antibiotics were used clinically until several years later. Also, strains carrying R plasmid genes for resistance to semi-synthetic penicillins were shown to exist before the semi-synthetic penicillins had been synthesized. Of perhaps even more ecological significance, R plasmids conferring antibiotic resistance have been detected in some nonpathogenic gram-negative soil *Bacteria*. In the soil, resistance plasmids may confer selective advantages because major antibiotic-producing organisms (*Streptomyces*, *Penicillium*) are also normal soil organisms. Thus, it seems that R plasmids are not a recent phenomenon but existed in the natural microbial population before antibiotics were discovered and used in either medicine or agriculture. Later, widespread medical and agricultural use of antibiotics provided selective conditions for the rapid spread of these R plasmids. R plasmids are thus a predictable outcome of natural selection. They pose significant limits for the long-term use of any single antibiotic as an effective chemotherapeutic agent.

Spread of Antimicrobial Drug Resistance

Inappropriate, extensive use of antimicrobial drugs is leading to the rapid development of drug-specific resistance in disease-causing microorganisms. The discovery and clinical use of the many known antibiotics has been paralleled by the emergence of bacteria that resist their action. There are numerous examples of the overuse of antibiotics and the concomitant development of resistance. Figure 20.25*a* shows a correlation between the number of tons of antibiotics used and the percentage of bacteria resistant to each antibiotic.

Increasingly, the drug prescribed for treatment of a particular infection has changed because of increased resistance of the microorganism causing the disease. A classic example is the development of resistance to penicillin in *Neisseria gonorrhoeae*, the bacterium that causes gonorrhea (Figure 20.25*b*). Penicillin is no longer a useful antibiotic for treatment of gonorrhea because a large percentage of the clinical isolates produce β-lactamase and are resistant. Virtually all resistant strains have developed since 1980. The current drug of choice is ceftriaxone, but new treatment modalities are recommended nearly every year, simply to limit the effects of rapidly emerging resistance genes (∞ Section 26.12).

A number of surveys worldwide suggest that antibiotics are used in clinical practice far more often than is necessary. Data indicate that antibiotic treatment is warranted in 20% of individuals who are seen for clinical

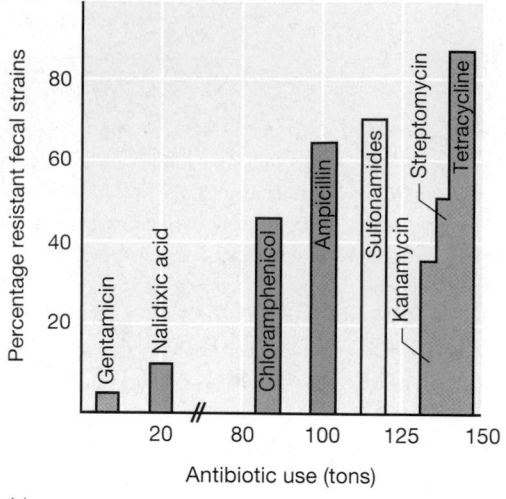

(a)

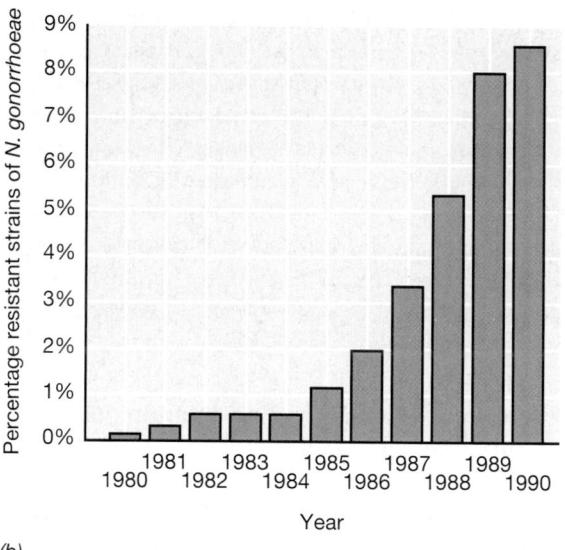

(b)

Figure 20.25 The emergence of antimicrobial drug-resistant bacteria. (a) Relationship between antibiotic use and the percentage of bacteria isolated from diarrheal patients resistant to the antibiotic. Those agents that have been used in the largest amounts, as indicated by the amount produced commercially, are those for which drug-resistant strains are most frequent. (b) Percentage of reported cases of gonorrhea caused by drug-resistant strains. The actual number of reported drug-resistant cases in 1985 was 9000. This number rose to 59,000 in 1990. Greater than 95% of the reported drug-resistant cases are due to penicillinase-producing strains of *Neisseria gonorrhoeae*. Since 1990, penicillin has not been recommended for treatment of gonorrhea because of emerging drug resistance. (Source: Centers for Disease Control, Atlanta, GA).

infectious disease. Yet, antibiotics are prescribed up to 80% of the time. Furthermore, in up to 50% of cases recommended doses or duration of treatments are not correct. This is compounded by patient noncompliance: Many patients stop taking medications, particularly an-

tibiotics, as soon as they "feel better." For example, the emergence of isoniazid-resistant tuberculosis correlates with a patient's failure to take the oral medication for six to nine months, as prescribed (⚬⚬ Section 26.5). Thus, virulent pathogens are often subjected to sublethal doses of antibiotics for short periods of time, selecting for the emergence of resistant organisms. Largely as a result of these failures to properly use and monitor antibiotic therapy, almost all pathogenic microorganisms have developed resistance to some chemotherapeutic agents since widespread use of antimicrobial chemotherapy began in the 1950s (Figure 20.26). Penicillin and sulfa drugs, the first widely used chemotherapeutic agents, are not as widely used today because many pathogens have acquired some resistance to them. Even the organisms that are still uniformly sensitive to penicillin, such as *Streptococcus pyogenes* (the bacterium that causes strep throat, scarlet fever, and rheumatic fever; ⚬⚬ Section 26.2), now need significantly more penicillin for successful treatment than a decade ago.

Other indiscriminant, nonessential uses of antibiotics contribute to the emergence of resistant strains. For example, antibiotics are used in agriculture both as growth-promoting substances in animal feeds and as prophylactics (to prevent the occurrence of disease rather than to treat an existing one). Several recent food infection outbreaks have been blamed on the use of an-

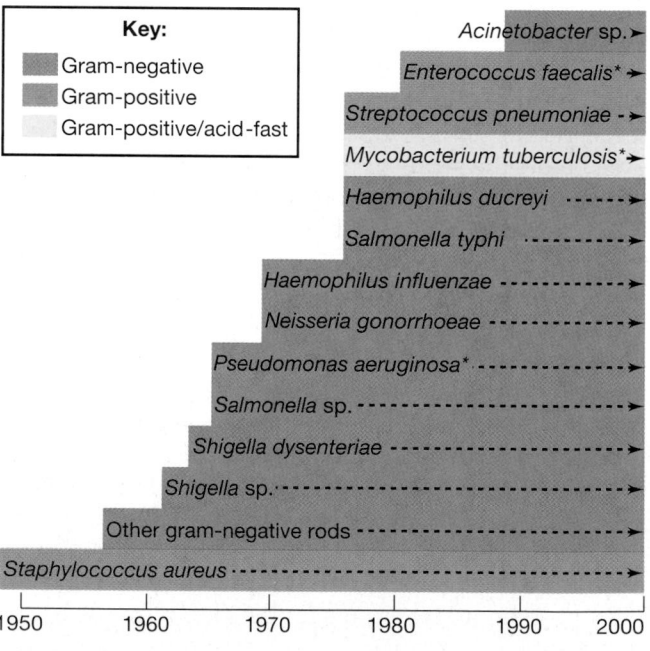

Figure 20.26 The appearance of antimicrobial drug resistance in some human pathogens. The *symbol indicates that some multi-drug resistant strains of these organisms are now untreatable with known antimicrobial drugs.

tibiotics in animal feeds. By overloading various environments with antibiotics, rapid development of drug resistance may result.

For example, fluoroquinolones have been extensively used for only about 10 years as growth-promoting and prophylactic agents in agriculture. However, fluoroquinolone-resistant *Campylobacter jejuni* have already emerged, presumably because of the practice of treating whole flocks of poultry with fluoroquinolones to prevent respiratory diseases. Voluntary guidelines are being used by both poultry and drug producers to monitor the use of second-generation fluoroquinolones in agriculture and prevent rapid emergence of resistance. Resistance can be minimized if drugs are used only for treatment of susceptible diseases and are given in sufficiently high doses and for sufficient lengths of time so that the microbial population is reduced before mutants have a chance to appear. This problem can be eliminated through physician and patient education. Resistance can also be minimized by combining two unrelated chemotherapeutic agents because it is likely that a mutant strain resistant to one antibiotic will still be sensitive to the other. However, with the increasing prevalence of R factors conferring multiple drug resistance in pathogenic bacteria, multiple antibiotic therapy is proving less attractive as a clinically useful strategem.

A few reports suggest that if the use of a particular antibiotic is stopped, the resistance to that antibiotic will be reversed over time. This information implies that resistance is reversible and that the efficacy of some antibiotics may be reestablished by withdrawing the antibiotic from use for a time, followed by a carefully monitored plan of prudent use. Finally, as we discuss below, new chemotherapeutic agents are constantly being produced using several methods for discovering and designing new drugs.

✓ **20.12 Concept Check**

An important side effect of the use of antimicrobial chemotherapeutic agents is the development of resistance by the targeted microorganisms. In many cases, resistance results from the selection of existing resistance genes through the improper and indiscriminate use of antimicrobial drugs. Many formerly useful antimicrobial agents are no longer useful because of drug resistance: A few organisms have developed resistance to all known antimicrobial drugs, prompting fears of a return to the preantibiotic era when infectious disease was untreatable.

✓ Why does antibiotic resistance occur?

✓ What practical steps should be taken to slow the development of antibiotic resistance?

20.13 The Search for New Antimicrobial Drugs

As we have just discussed, given sufficient drug exposure and time, resistance will develop to all known antimicrobial drugs. As a result, conservative, appropriate use of antibiotics is absolutely necessary to prolong the useful clinical life of these drugs. However, the long-term solution to microbial drug resistance is to develop new antimicrobial drugs. Several strategies are used to identify and produce useful analogs of existing compounds or to design or discover novel antimicrobial compounds.

New Analogs of Existing Antimicrobial Compounds

The production of new analogs of existing antimicrobial compounds is generally straightforward and often productive, largely because new compounds that are structural mimics of older ones have a predictable mechanism of action. In many cases, parameters such as solubility and affinity can be changed by introducing minor modifications to the chemical structure of a drug without altering structures critical to drug action. The new compound may actually be more potent than the parent compound, and, because resistance is based on structural recognition, the new compound may not be recognized by resistance factors. For example, Figure 20.22 shows the basic structure of tetracycline and two bioactive derivatives. Using tetracycline as a so-called *lead compound*, systematic chemical substitutions can be made at the four R group sites, generating an almost endless series of tetracycline analogs. Using this basic strategy, new β-lactam antibiotics (see Section 20.8), new tetracycline-related compounds (see Section 20.9), and new analogs of vancomycin (Figure 20.12), some up to 100 times as potent as the parent compound, are routinely synthesized and tested.

The application of automated robotic chemistry methods to drug discovery has dramatically increased our ability to rapidly generate potential new antimicrobial compounds. The automated robotic methods, referred to as *combinatorial chemistry*, apply systematic modifications of a known antimicrobial product to generate large numbers of new analogs. For instance, using automated combinatorial chemistry methods, and starting with the tetracycline lead compound (Figure 20.22), five different reagents might be used to introduce substitutions at the four different tetracycline R groups. The substituted sites would yield $5 \times 5 \times 5 \times 5$ (five derivatives at each of four sites), or 725 different tetracycline derivatives from only six different reagents, all in a few hours! These compounds are then assayed for *in vitro* biological activity on different microorganisms, also using automated techniques. The automated synthesis and screening processes dramatically shorten drug discovery time and increase the number of new candidate drugs by 10 times or more each year.

According to pharmaceutical industry estimates, about 7 million candidate compounds must be screened to yield a single useful clinical drug. As we have just discussed, this process starts with synthesis, isolation, and screening of candidate compounds. Screening involves *in vitro* tests used for antibiotic susceptibility (see Section 20.4). Drugs effective in the laboratory must then be tested for efficacy and toxicity in animals and finally in clinical trials in humans. Animal testing may take several years and multiple trials to ensure that the candidate drug is efficacious and safe for use. For the same reasons, clinical trials usually take years to complete for each drug. New drug discovery and development typically takes 10 years to 25 years before approval for clinical use. The cost of discovery and development is estimated to be about $500 million for each new approved drug.

Computerized Drug Design

Truly novel antimicrobial compounds are much more difficult to identify than analogs of existing drugs because new antimicrobial compounds must work at unique sites in metabolism and biosynthesis, or be structurally dissimilar to existing compounds to avoid existing resistance. To find these new compounds, candidate drugs had to be isolated from natural sources and systematically screened for antimicrobial activity. However, recent advances in computer and structural graphics technology now make it possible to design a drug to interact with specific known microbial structures. Drug discovery can now begin at the computer, where new drugs can be rapidly "created" and "tested" for binding and toxicity in the computer environment at relatively low cost (∞ Sections 30.4 and 26.8). One of the most dramatic recent successes in computer-directed drug design is the development of saquinavir, a protease inhibitor that is used to slow the growth of the human immunodeficiency virus (HIV) in infected individuals (Figure 20.27). HIV protease cleaves a virus-encoded precursor protein to produce the mature viral core and activate the reverse transcriptase enzyme necessary for replication (∞ Section 16.14). Saquinavir was designed by computer to fit the active site of HIV protease, based on the known three-dimensional structure of the protease-substrate complex; it is a peptide analog that displaces the HIV precursor protein, inhibiting virus maturation and slowing its growth in the human host. A number of other computer-designed protease inhibitors like saquinavir are in use as chemotherapeutic drugs for the treatment of AIDS (Table 20.5, Figure 20.27, and ∞ Figure 26.38). As this example demonstrates, com-

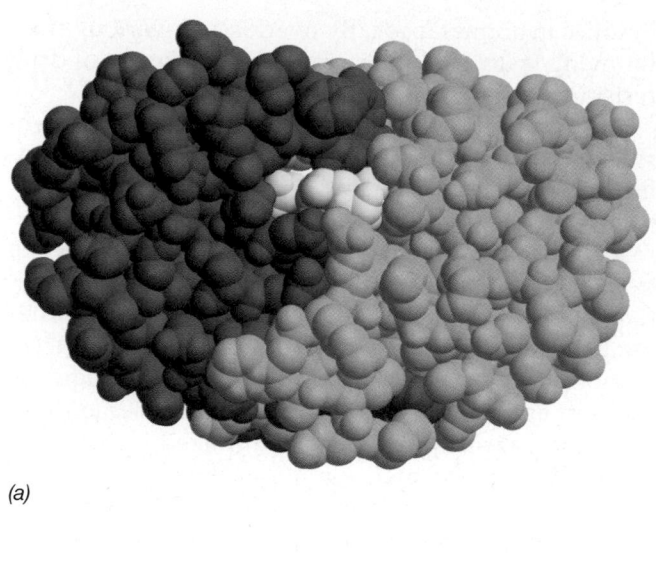

(a)

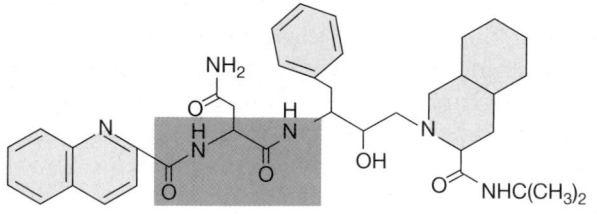

Saquinavir

Indinavir

(b)

Figure 20.27 Computer-generated antiviral drugs. (a) The HIV protease homodimer. Individual polypeptide chains are shown in green and blue. A peptide (yellow) is bound by the catalytic site. This protease cleaves an HIV precursor protein, a necessary step in virus maturation (∞ Section 16.14). Blocking of the protease site by the peptide shown inhibits precursor processing and HIV maturation. This structure is derived from information in the Protein Data Bank. (b) These anti-HIV drugs are peptide analogs that were designed by computer to block the active site of HIV protease. The areas highlighted in orange show the regions analogous to peptide bonds. Binding of these compounds by the HIV protease prevents HIV precursor processing and virus maturation. These compounds are representative of a class of therapeutic drugs known as non-nucleoside reverse transcriptase inhibitors (NNRTI). The concentration of these compounds in HIV-infected cells, coupled with their strong affinity for HIV reverse transcriptase, makes them very potent competitive inhibitors for the active sites of the transcriptase and prevents viral replication. These protease inhibitors are widely used for treatment of HIV infection (see Table 20.5 and ∞ Section 26.14).

puter design based on structural and biochemical modeling is a practical method for designing antimicrobial drugs, as well as being rapid and cost effective.

✓ **20.13 Concept Check**

New antimicrobial compounds are constantly being discovered and developed to deal with drug-resistant organisms and enhance our ability to treat infectious diseases. However, drug discovery and development are time consuming and expensive. Computer drug design is an important new tool for drug discovery.

✓ Analogs of existing drugs are often developed to be used as next generation antimicrobial compounds. Explain the advantages and disadvantages of this strategy for treatment of infectious diseases.

✓ How can computer drug design save time and money in the search for new drugs?

Review Questions

1. Why is the decimal reduction time (D) important in heat sterilization? How would the presence of bacterial endospores affect D?

2. Describe the effects of lethal ionizing irradiation at the molecular level.

3. What are the principal advantages of using membrane filters instead of depth filters?

4. Why are nucleopore filters particularly useful for isolating specimens for microscopy?

5. Describe the procedure for obtaining the minimum inhibitory concentration (MIC) for a chemical that is bacteriocidal for *Escherichia coli*.

6. Contrast the action of disinfectants and antiseptics. Why can't disinfectants normally be used on living tissue?

7. Growth factor analogs are generally distinguished from antibiotics by a single important criterion. Explain.

8. Most antibiotics are made by only certain groups of organisms. Is this statement true? What groups of organisms make antibiotics?

9. Describe the mechanisms of action that characterizes a β-lactam antibiotic.

10. Distinguish between the mode of action of at least three of the protein synthesis-inhibiting antibiotics.

11. Why do antiviral drugs generally exhibit host toxicity?

12. Define the fungi-specific targets that allow selective toxicity of chemotherapeutic agents in fungi.

13. Outline the major mechanisms responsible for antibiotic resistance.

14. What is the ultimate origin of bacterial antibiotic resistance genes?

15. Starting with a parent compound, describe *combinatorial chemistry* methods used for the production of new drug analogs.

Application Questions

1. Describe in a graph the experimental results you would expect for the decimal reduction time of a very heat-sensitive organism. How would this graph be affected if the vegetative cells were heat-sensitive but the organism formed heat-resistant endospores?

2. What are some potential drawbacks to the use of radiation in food preservation? Do you think these drawbacks could be manifested as health hazards? Why or why not? How would you distinguish between radiation-damaged and radiation-contaminated food?

3. Filtration is an acceptable means of pasteurization for some liquids. Design a filtration system for pasteurization of a heat-sensitive liquid. Why might a filtration system be desirable over a heat pasteurization system?

4. Design an experiment to distinguish between a cidal and a static agent. Can you use the minimum inhibitory concentration (MIC) test in your experiments? Explain.

5. What tests would you perform to decide whether a chemical agent could be used as an antiseptic? As a disinfectant? Some chemicals serve both purposes. Describe the properties of such a chemical and give an example.

6. Although growth factor analogs may inhibit microbial metabolism, only a few of these agents are practically useful. Many potential agents, and some that are in wide use, such as azidothymidine (see Table 20.5), exhibit significant host cell toxicity. Describe a growth factor analog that is effective and has low toxicity for host cells. Why is the toxicity low for the agent you chose? Also

describe a growth factor analog that is effective against an infectious disease but exhibits toxicity for host cells. Why might a toxic agent such as AZT still be used in certain situations to treat infectious diseases? What precautions would you take to limit the toxic effects of such a drug, while maximizing the therapeutic activity? Explain your answer.

7. Less than 1% of all known antibiotics have any practical value for either research or clinical use. Indicate why this might be so. Do you think it is important to expand and continue searches for new antibiotics? What alternatives to antibiotic treatments are, or could be, available for the treatment of microbial diseases in humans.

8. Design an experiment to examine microorganisms for the producrtion of novel antibiotics. Which group or groups of microorganisms would you choose? Where could you find and isolate these organisms in a natural environment? What advantage would the production of an antibiotic provide for these organisms in nature? What *in vitro* methods would you use to test the efficacy of your potential new antibiotics? (You may wish to review material in Sections 12.24 and 30.5 before answering.)

9. Although the β-lactam antibiotics demonstrate clear selective toxicity for *Bacteria*, many groups of *Bacteria* are innately resistant to their effects. Without invoking bacterial resistance genes, indicate why gram-negative *Bacteria* are resistant to the effects of most, but not all, β-lactam antibiotics. Further explain why some β-lactam antibiotics are useful against these organisms.

10. What potential advantages might the aminoglycosides, macrolides, and tetracyclines have over penicillin G for chemotherapy? Explain.

11. List the features of an ideal antiviral drug, especially with regard to selective toxicity. Do such drugs exist? What factors might limit use of such a drug?

12. Like viruses, fungi present special chemotherapeutic problems. Explain the problems inherent in chemotherapy of both groups and explain whether or not you agree with the preceding statement. Give specific examples and suggest at least one group of chemotherapeutic agents that might target both types of infectious agents.

13. Explain the genetic basis of acquired resistance to β-lactam antibiotics in *Staphylococcus aureus*. Design a set of experiments to reverse resistance to the β-lactam antibiotics. Do you think this can be done? Can your experiment be applied "in the field" to promote deselection of antibiotic-resistant organisms? Explain.

14. Design a drug for the inhibition of HIV protease activity that is based on the structure in Figure 20.27. Use a mechanism *different* than competitive inhibition at the enzyme active site.

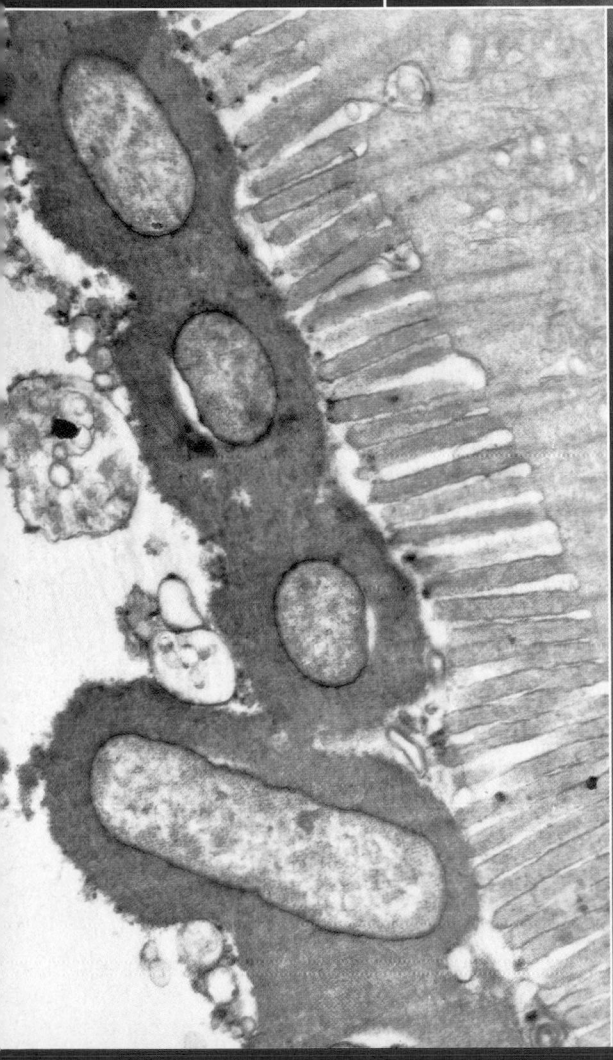

M icroorganisms are a constant and ubiquitous part of the human experience. We are continually exposed to microorganisms, and many have established residence in and on our bodies. Most do little or no harm, and a few are beneficial. A small number of microorganisms, however, cause significant disease, resulting in the potential destruction of body tissues and functions. In the image shown here, a common oral bacterium, *Streptococcus mutans*, has colonized tooth surfaces and is starting the decay process. The end result, dental caries, is a very common bacterial infectious disease.

HUMAN-MICROBE INTERACTIONS

21

Working Glossary

Attenuation decrease or loss of virulence

Bacteremia the presence of microorganisms in the blood

Capsule dense, well-defined polysaccharide or protein layer closely surrounding a cell

Colonization multiplication of a pathogen after it has gained access to host tissues

Dental caries tooth decay resulting from bacterial infection

Dental plaque bacterial cells encased in a matrix of extracellular polymers and salivary products, found on the teeth

Disease injury to the host that impairs host function

Endotoxin the lipopolysaccharide portion of the cell envelope of certain gram-negative *Bacteria*, which acts as a toxin when solubilized

Enterotoxin protein released extracellularly by a microorganism as it grows that produces immediate damage to the small intestine of the host

Exotoxin protein released extracellularly by a microorganism as it grows that produces immediate host cell damage

Fever an abnormal increase in body temperature

Glycocalyx a loose network of polymer fibers extending outward from the cell

Host an organism that harbors a parasite

Infection growth of organisms in the host

Inflammation host response to injury or infection, characterized by redness, swelling, heat, and pain

Invasiveness pathogenicity caused by the ability of a pathogen to enter the body and spread

Leukocytes nucleated cells found in the blood (white blood cells)

Lower respiratory tract trachea, bronchi, and lungs

Mucous membrane layers of epithelial cells that interact with the external environment

Mucus soluble glycoproteins secreted by epithelial cells that coat the mucous membranes

Normal flora microorganisms that are usually found associated with healthy body tissue

Parasite an organism that grows in or on a host and damages the host

Pathogen a microbial parasite that does harm to a host

Pathogenicity the ability of a parasite to inflict damage on the host

Slime layer a diffuse mat of polymer fibers surrounding cells that appear unattached to a single cell

Toxicity pathogenicity caused by toxins produced by a pathogen

Upper respiratory tract the nasopharynx, oral cavity, and throat

Virulence the degree of pathogenicity produced by a pathogen

In this chapter, we begin our discussions of the direct interactions of microorganisms with humans. Microorganisms normally grow on and in the human body in large numbers. The human gut has often been compared to a constant culture vessel, and the mucus membranes lining the mouth, gut, excretory, and reproductive systems have a normal population of microbes that are beneficial and sometimes required to maintain good health. However, a small but important group of microorganisms uses a variety of direct and indirect mechanisms to invade the human body and cause infections, disease, and damage to the host. For example, specialized, extremely potent biological toxins are used by some microorganisms to damage the host. However, humans have developed effective countermeasures to suppress or destroy most microbial invaders; a number of the nonspecific physical, anatomical, and biochemical processes make microbial infectious disease a relatively infrequent event in our lives.

❙ BENEFICIAL MICROBIAL INTERACTIONS WITH HUMANS

Microorganisms interact with humans in a variety of beneficial and occasionally harmful ways. Here we examine the microbes that inhabit the healthy human adult body under normal circumstances.

21.1 Overview of Human–Microbe Interactions

The human body is constantly exposed to microorganisms. Through normal everyday activities such as breathing, we are exposed to millions of microorganisms in the environment. Hundreds of species and billions of individual microorganisms, collectively referred to as the **normal flora**, grow on or in the human body. Most, but not all, microorganisms are benign.

Pathogens

Organisms that live on or in a host organism, causing damage to the host, are called **parasites**. Microbial parasites are called **pathogens**. The outcome of a host–parasite relationship depends on the *pathogenicity* of the parasite; that is, on the ability of the parasite to inflict damage on the host, and on the *resistance* or *susceptibility* of the host to the parasite.

Pathogenicity varies markedly for individual pathogens. The quantitative measure of pathogenicity is termed *virulence*. Virulence can be expressed as the cell number that will elicit a pathogenic response in a host within a given time period. Neither the virulence of the pathogen nor the relative resistance of the host are constant factors. The host–parasite interaction is a dynamic relationship between the two organisms; the virulence of the pathogen and the resistance of the host are constantly changing.

Infection and Disease

Infection refers to any situation in which a microorganism is established and growing in a host, whether or not the host is harmed. **Disease** is damage or injury to the host that impairs host function. *Infection is not synonymous with disease* because growth of a microorganism on a host does not always cause host damage. Thus, normal flora may produce microbial infections, but seldom cause disease. However, the normal flora sometimes cause disease if host resistance is compromised, as happens in diseases such as cancer and AIDS (∞ Section 26.14).

Host-Parasite Interactions

Animal bodies provide favorable environments for the growth of many microorganisms. They are rich in organic nutrients and growth factors required by chemoorganotrophs and provide relatively constant conditions of pH, osmotic pressure, and temperature. However, the animal body is not a uniform microbial environment. Each region or organ differs chemically and physically from others and thus provides a selective environment where the growth of certain microorganisms are favored. The skin, respiratory tract, gastrointestinal tract, and so on, provide a wide variety of chemical and physical environments in which different microorganisms can grow selectively. For example, the relatively dry environment of the skin favors the growth of gram-positive organisms such as *Staphylococcus aureus* (∞ Section 26.9); the highly oxygenated environment of the lungs favors the growth of the obligately aerobic *Mycobacterium tuberculosis* (∞ Section 26.5); and the anaerobic environment of the large intestine supports the growth of members of the obligately anaerobic *Clostridium* genus (∞ Section 12.21). Animals also possess a variety of defense mechanisms that collectively prevent or inhibit microbial invasion and growth. The microorganisms that ultimately colonize the host successfully are those that have developed ways of circumventing these defense mechanisms.

Infections frequently begin at sites in the animal's *mucous membranes*. Mucous membranes are found throughout the body including the mouth, pharynx, esophagus, and the urogenital, respiratory, and gastrointestinal tracts. Mucous membranes consist of single or multiple layers of *epithelial cells*, tightly packed cells that interface with the external environment. Mucous membranes are frequently coated with a protective layer of soluble glycoproteins known as *mucus*, which serves to protect epithelial cells. When bacteria contact host tissues at mucous membranes, they may associate either loosely or firmly. If they associate loosely with the mucosal surface, they are usually swept away by physical processes, but they may also adhere to the epithelial surface as a result of specific cell-cell recognition between pathogen and host. From there, actual tissue infection may follow. When this occurs, the mucosal barrier is breached, allowing the pathogen to invade deeper tissues (Figure 21.1).

Microorganisms are almost always found in those regions of the body exposed to the environment, such as the skin, oral cavity, respiratory tract, intestinal tract, and urogenital tract. They are not normally found in the organs, or in the blood, lymph, or nervous systems of the body; the growth of microorganisms in these usually sterile environments indicates serious infectious disease.

Table 21.1 shows some of the major types of microorganisms normally found in association with body surfaces. The most visible exposed body surface, the skin ($2\,\mathrm{m}^2$), has a number of normal microbial inhabitants. However, mucosal surfaces have an even larger variety

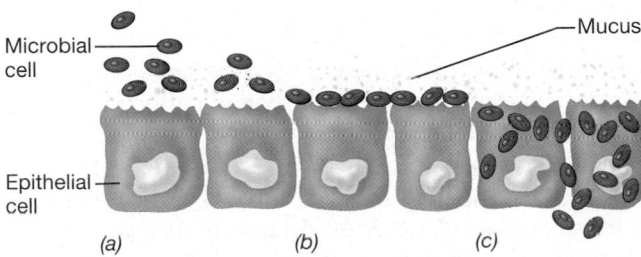

Figure 21.1 Bacterial interactions with mucous membranes. (a) Loose association. (b) Adhesion. (c) Invasion into submucosal epithelial cells.

TABLE 21.1	Representative genera of microorganisms in the normal flora of humans
Anatomical site	**Organism[a]**
Skin	*Staphylococcus, Corynebacterium, Acinetobacter, Pityrosporum* (yeast), *Propionibacterium, Micrococcus*
Mouth	*Streptococcus, Lactobacillus, Fusobacterium, Veillonella, Corynebacterium, Neisseria, Actinomyces*
Respiratory tract	*Streptococcus, Staphylococcus, Corynebacterium, Neisseria*
Gastrointestinal tract	*Lactobacillus, Streptococcus, Bacteroides, Bifidobacterium, Eubacterium, Peptococcus, Peptostreptococcus, Ruminococcus, Clostridium, Escherichia, Klebsiella, Proteus, Enterococcus, Staphylococcus*
Urogenital tract	*Escherichia, Klebsiella, Proteus, Neisseria, Lactobacillus* (vagina of mature females), *Corynebacterium, Staphylococcus, Candida, Provotella, Clostridium, Peptostreptococcus*

[a] This list is not meant to be exhaustive, and not all of these organisms are found in every individual. Most of these organisms can contribute to disease processes under certain conditions.

of associated microorganisms. This is due in part to the sheltered, moist environment of the various mucosal surfaces and also to the huge overall surface area of the mucosae ($400 m^2$). For example, the specialized function of a mucosal organ such as the small intestine requires a large specialized surface area for nutrient transport, and this surface is also a site for microbial growth. We shall now examine these normal microbial interactions in greater detail.

✓ 21.1 Concept Check

Animal bodies are favorable environments for the growth of many microorganisms, most of which do no harm. Microorganisms that grow in or on the body and cause harm are called *pathogens*. Initial pathogen growth is on the surface of a host, often on the mucous membranes, and may result in infection and disease. The ability to cause disease is influenced by complex host–parasite interactions. Pathogen actions are limited by host defense mechanisms.

- ✓ Distinguish between *infection* and *disease*.
- ✓ Why might one area of the body be more suitable for microbial growth than another?

21.2 Normal Flora of the Skin

An average human adult has about $2 m^2$ of skin surface that can vary greatly in chemical composition and moisture content. Figure 21.2 shows the anatomy of the skin and regions in which bacteria may live. The skin surface (epidermis) is not a favorable place for microbial growth, as it is subject to periodic drying.

Most skin microorganisms are associated directly or indirectly with the sweat glands. The *apocrine glands* are confined mainly to the underarm and genital regions, the

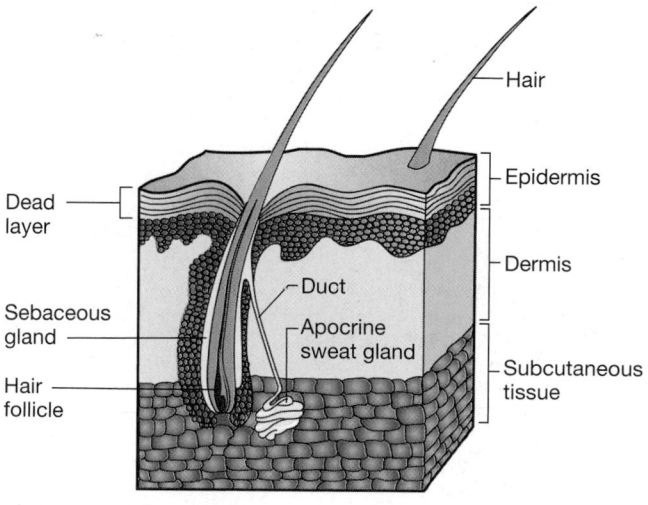

Figure 21.2 The human skin. Microorganisms are associated primarily with the sweat ducts and the hair follicles.

nipples, and the umbilicus. They are inactive in childhood and become fully functional only at puberty. Bacterial populations on the surface of the skin in these warm, humid places are relatively high, in contrast to the situation on the smooth, dry surface skin. *Underarm odor* develops as a result of bacterial activity in the apocrine secretions; aseptically collected apocrine secretion is odorless but develops odor on inoculation with bacteria. Each hair follicle is associated with a *sebaceous gland*, which secretes a lubricant fluid. Hair follicles provide an attractive habitat for microorganisms that inhabit these regions, mostly within the area just below the surface of the skin. The secretions of the skin glands are rich in microbial nutrients such as urea, amino acids, salts, lactic acid, and lipids. The pH of human secretions is almost always acidic, the usual range being between pH 4 and 6.

The normal flora of the skin consists of either *transient* or *resident* populations of microorganisms. The skin is continually being inoculated with transient microorganisms, virtually all of which are unable to multiply and usually die. Resident organisms are able to multiply, not merely survive, on the skin. The normal flora of the skin consists primarily of gram-positive *Bacteria* restricted to a few groups (Table 21.1). These include several species of *Staphylococcus* and a variety of both aerobic and anaerobic corynebacteria. Of the latter, *Propionibacterium acnes* can contribute to the condition known as *acne*. Gram-negative *Bacteria* are almost always minor constituents of the normal flora because such intestinal organisms as *Escherichia coli* are being continually inoculated onto the surface of the skin by fecal contamination. However, *Acinetobacter* is one of the few nonfecal gram-negative *Bacteria* commonly found on skin. Gram-negative *Bacteria* seldom grow on the skin, probably due to their inability to compete with gram-positive organisms that are better adapted to the dry conditions of the skin. Yeasts are uncommon in large numbers on the skin surface or mucous membranes. However, in the absence of host resistance (acquired immunodeficiency syndrome; AIDS) (⚬⚬ Section 26.14) or the absence of normal bacterial flora, fungi such as *Candida* spp. may grow and cause serious infections on the skin surface, but the lipophilic yeast *Pityrosporum ovalis* is occasionally found on the scalp.

Although the resident microflora remains more-or-less constant, various factors can affect the nature and extent of the normal flora: (1) The weather may cause an increase in skin temperature and moisture, which increases the density of the skin microflora. (2) Age has an effect, and young children have a more varied microflora and carry more of the potentially pathogenic gram-negative *Bacteria* than adults. (3) Personal hygiene influences the resident microflora, and unclean individuals usually have higher microbial population densities on their skin. Organisms that cannot survive on the skin generally succumb from either the skin's low moisture content or low pH (due to organic acid content).

21.2 Concept Check

The skin is a dry, acidic environment that is not conducive to the growth of most microorganisms. However, moist areas, especially around sweat glands, are colonized by gram-positive *Bacteria* and other members of the skin normal flora.

✓ How large is the surface area of the skin?

✓ Describe the properties of microorganisms that grow well on the skin.

21.3 Normal Flora of the Oral Cavity

The oral cavity contains one of the more complex and heterogeneous microbial habitats in the body. Although saliva is the most pervasive source of microbial nutrients in the oral cavity, it is not an especially good microbial culture medium because it contains very few nutrients and a number of antibacterial substances such as the enzymes *lysozyme* and *lactoperoxidase*. Lysozyme is an enzyme that cleaves glycosidic linkages in peptidoglycan in the bacterial cell wall, leading to weakening of the wall and cell lysis (∞ Section 4.8). Lactoperoxidase, an enzyme present in both milk and saliva, kills bacteria by a reaction in which singlet oxygen is generated (∞ Sections 6.13 and 22.2). Despite the activity of antibacterial substances, the presence of food particles and epithelial debris makes the oral cavity a very favorable microbial habitat.

The Teeth and Dental Plaque

The tooth consists of a mineral matrix of calcium phosphate crystals (enamel), within which the living tissue of the tooth (dentin and pulp) is present (Figure 21.3). Bacteria found in the mouth during the first year of life (when teeth are absent) are predominantly aerotolerant anaerobes such as streptococci and lactobacilli, but a variety of other bacteria, including some aerobes, occur in small numbers. When the teeth appear, there is a pronounced shift in the balance of the microflora toward anaerobes that are specifically adapted for growth on surfaces of the teeth and in the gingival crevices.

The bacterial colonization of tooth surfaces occurs as a result of attachment of single bacterial cells, followed by growth of microcolonies. Beginning with a freshly cleaned tooth surface, a thin organic film several micrometers thick forms as a result of the attachment of acidic glycoproteins from the saliva. This film provides a firmer attachment site for the colonization and growth of bacterial microcolonies (Figure 21.4). The colonization of this glycoprotein film is highly specific, and only a few species of *Streptococcus* (primarily *S. sanguis*, *S. so-*

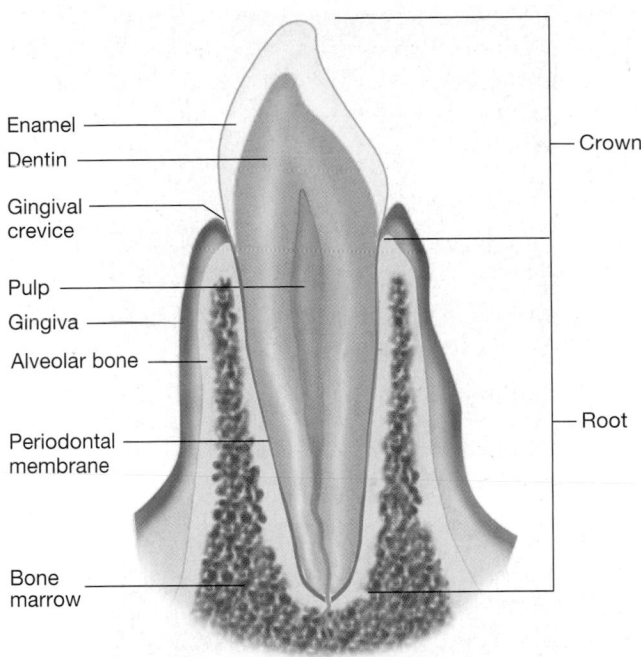

Figure 21.3 Section through a tooth showing the surrounding tissues that anchor the tooth in the gum.

brinus, *S. mutans*, and *S. mitis*) are involved. As a result of extensive growth of these organisms, a thick bacterial layer, called **dental plaque**, is formed (Figures 21.5 and 21.6). If plaque continues to form, filamentous *Fusobacterium* species begin to grow. The filamentous bacteria embed in the matrix formed by the streptococci and extend perpendicular to the tooth surface, making an ever-thicker bacterial layer. Associated with the filamentous bacteria, in addition to the streptococci, are spirochetes such as *Borrelia* species (∞ Section 12.33), gram-positive rods, and gram-negative cocci. In heavy plaque, filamentous organisms such as anaerobic *Actinomyces* species may predominate.

The anaerobic nature of the oral flora may seem surprising considering the availibility of oxygen in the mouth. However, anoxia probably develops due to the action of facultative bacteria growing aerobically on organic materials on the tooth. This plaque buildup produces a dense matrix that decreases oxygen diffusion onto the tooth surface, forming an anoxic microenvironment. Thus, the microbial populations of the dental plaque exist in a microenvironment of their own making and maintain themselves in the face of wide variations in the macroenvironment of the oral cavity.

Dental Caries

As dental plaque accumulates and acid products are formed, **dental caries** (tooth decay) results. Thus, tooth decay is an infectious disease caused by microorganisms. The smooth surfaces of the teeth that are exposed

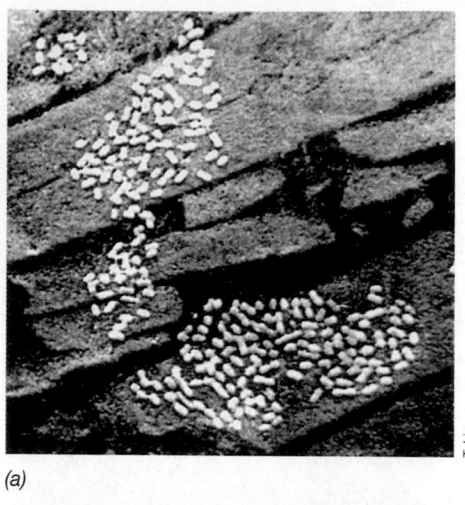

(a)

(b)

Figure 21.4 Microcolonies of bacteria growing on a model tooth surface inserted into the mouth for 6 h. (b) Higher magnification of the preparation in (a). Note the diverse morphology of the organisms present and the slime layer (arrows) holding the organisms together.

to frequent cleaning by the tongue, cheek, saliva, or toothbrush or to the abrasive action of food mastication are relatively resistant to dental caries. The tooth surfaces in crevices and near the gingival crevice, where food particles can be retained, are the sites where tooth decay usually occurs. Thus, dogs are highly resistant to tooth decay because the shape of their teeth does not favor retention of food. Diets high in sugars are especially *cariogenic* because lactic acid bacteria ferment the sugars to lactic acid, which causes decalcification of the enamel (see Figure 21.3) of the tooth. Once breakdown of the hard tissue (tooth enamel) has begun, proteolysis of the matrix of the tooth enamel occurs through the action of proteolytic enzymes released by bacteria. Microorganisms penetrate further into the decomposing matrix, but the later stages of the process may be exceedingly slow and are often highly complex. The structure of the calcified tis-

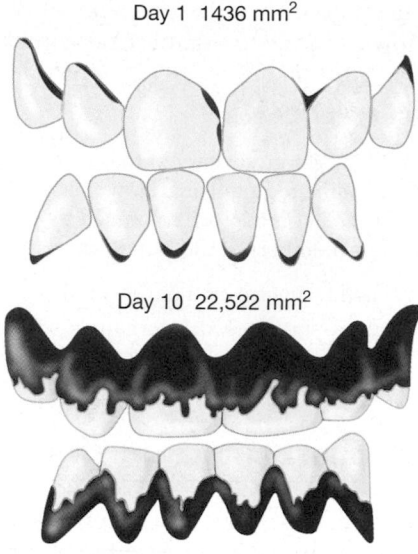

Day 1 1436 mm^2

Day 10 22,522 mm^2

Figure 21.5 Distribution of dental plaque, as revealed by use of a disclosing agent, on brushed (top) and unbrushed (bottom) teeth. The numbers give the total area of dental plaque. The red areas indicate plaque. Note that the plaque builds preferentially near the gum line, first occurring directly adjacent to the mucous membranes of the gingiva.

sue also plays an important role in the extent of dental caries. Incorporation of fluoride into the calcium phosphate crystal matrix makes the matrix more resistant to decalcification by acid. Thus, fluorides are used in drinking water and dentifrices to aid in controlling tooth decay.

Two organisms that have been implicated in dental caries are *Streptococcus sobrinus* and *Streptococcus mutans*, both lactic acid-producing bacteria. *S. sobrinus* is able to colonize smooth tooth surfaces because of its specific affinity for salivary glycoproteins (Figure 21.6) and is probably the primary organism involved in decay of smooth surfaces. *S. mutans* is found predominantly in crevices and small fissures, and its ability to attach to tooth surfaces is the result of its ability to produce a dextran polysaccharide that is strongly adhesive (Figure 21.7). *S. mutans* produces dextran only when sucrose is present, by means of the enzyme *dextransucrase*:

$$n\text{Sucrose} \xrightarrow{\text{Dextransucrase}} \text{Dextran } (n\text{Glucose}) + n \text{ Fructose}$$

Sucrose (common table sugar) is present in the diet of most individuals in developed countries and is highly cariogenic because it is a substrate for dextransucrase.

Susceptibility to tooth decay varies greatly and is affected by inherent traits in the individual as well as by diet and other extraneous factors. Studies of the distribution of oral streptococci have shown a direct correlation between the presence of *S. mutans*, and to a lesser degree *S. sobrinus*, and the extent of dental caries.

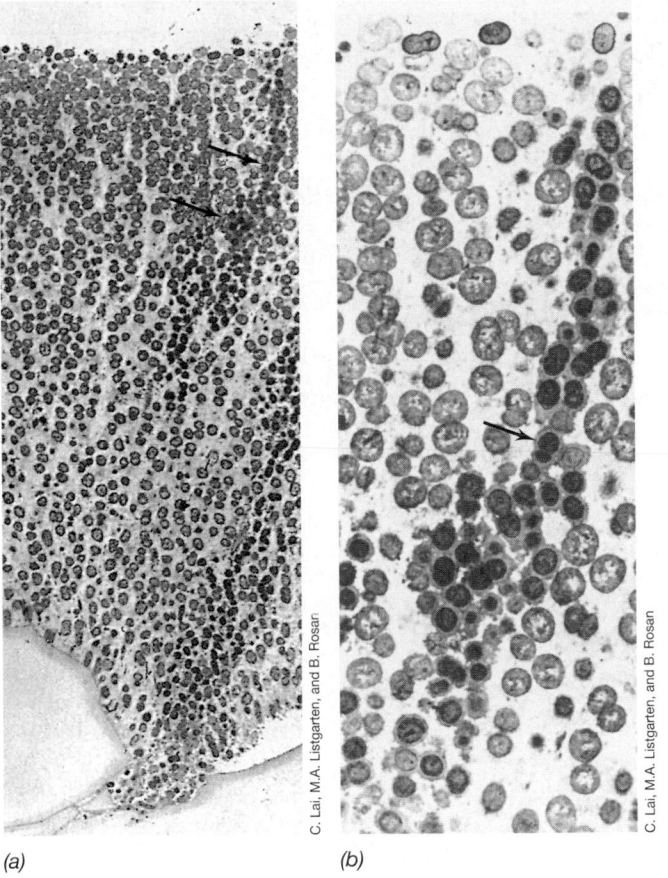

(a)   *(b)*

C. Lai, M.A. Listgarten, and B. Rosan

Figure 21.6 Electron micrographs of thin sections of dental plaque. Bottom is the base of the plaque; top is the portion exposed to the oral cavity. (a) Low power electron micrograph. Organisms are predominantly streptococci. The species *Streptococcus sobrinus* has been labeled by an antibody-microchemical technique, and these cells appear darker than the rest. They are seen as two distinct chains (arrows). The total thickness of the plaque layer shown is about 50 μm. (b) Higher power electron micrograph showing the region with *S. sobrinus* cells (dark, arrow). Note the extensive glycocalyx (see Section 21.6) surrounding the *S. sobrinus* cells.

In the United States and Western Europe, for example, 80–90% of all people have their teeth colonized by *S. mutans*, and dental caries is nearly universal. By contrast, dental caries do not occur in Tanzanian children, presumably because sucrose is almost completely absent in their diets, and *S. mutans* is absent from the plaque of these individuals.

Microorganisms in the mouth can also cause other infections. The areas along the periodontal membrane at or below the gingival crevice (*periodontal pockets*) (Figure 21.3) can become infected with a variety of microorganisms, causing inflammation (gingivitis) and more serious tissue and bone-destroying periodontal disease. Some of the genera involved are the fusiform anaerobic bacterium *Capnocytophaga* and the aerobic bacterium *Rothia*.

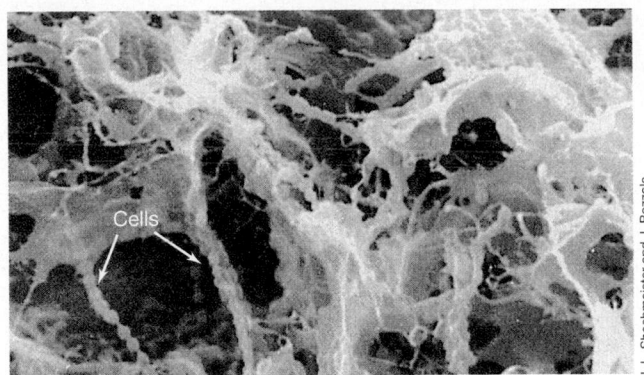

I. L. Shechmeister and J. Bozzcla

Figure 21.7 Scanning electron micrograph of the cariogenic bacterium *Streptococcus mutans*. The sticky dextran material holds the cells together as filaments. Individual cells are about 1 μm in diameter.

✓ 21.3 Concept Check

Bacteria can grow on tooth surfaces in thick layers called plaque. The microorganisms in plaque produce adherent substances that encourage further colonization. Acid produced by microorganisms in plaque damages tooth surfaces, and dental caries result. A variety of microorganisms contribute to caries and periodontal disease.

✓ How do anaerobic microorganisms become established in the mouth?

✓ Is dental caries an infectious disease? Give at least one reason for your answer.

21.4 Normal Flora of the Gastrointestinal Tract

The general anatomy of the gastrointestinal tract is shown in Figure 21.8. The human gastrointestinal tract, the site of food digestion, consists of the stomach, small intestine, and large intestine. The pH of stomach fluids is low, about pH 2. The stomach can thus be viewed as a chemical barrier against entry of foreign bacteria into the intestinal tract. The bacterial count of the stomach contents is generally very low, and so the human stomach is devoid of any significant normal flora. However, organisms such as *Helicobacter pylori* can colonize the stomach wall, resulting in ulcers (∞ Section 26.10).

The Intestinal Tract

The intestinal tract consists of the *small intestine* and the *large intestine*, each of which is further subdivided into different anatomical structures.

The *small intestine* is separated into two parts, the *duodenum* and the *ileum*, with the *jejunum* connecting them. The duodenum, adjacent to the stomach, is fairly acidic and resembles the stomach in its lack of a microbial flora. From the duodenum to the ileum, the pH

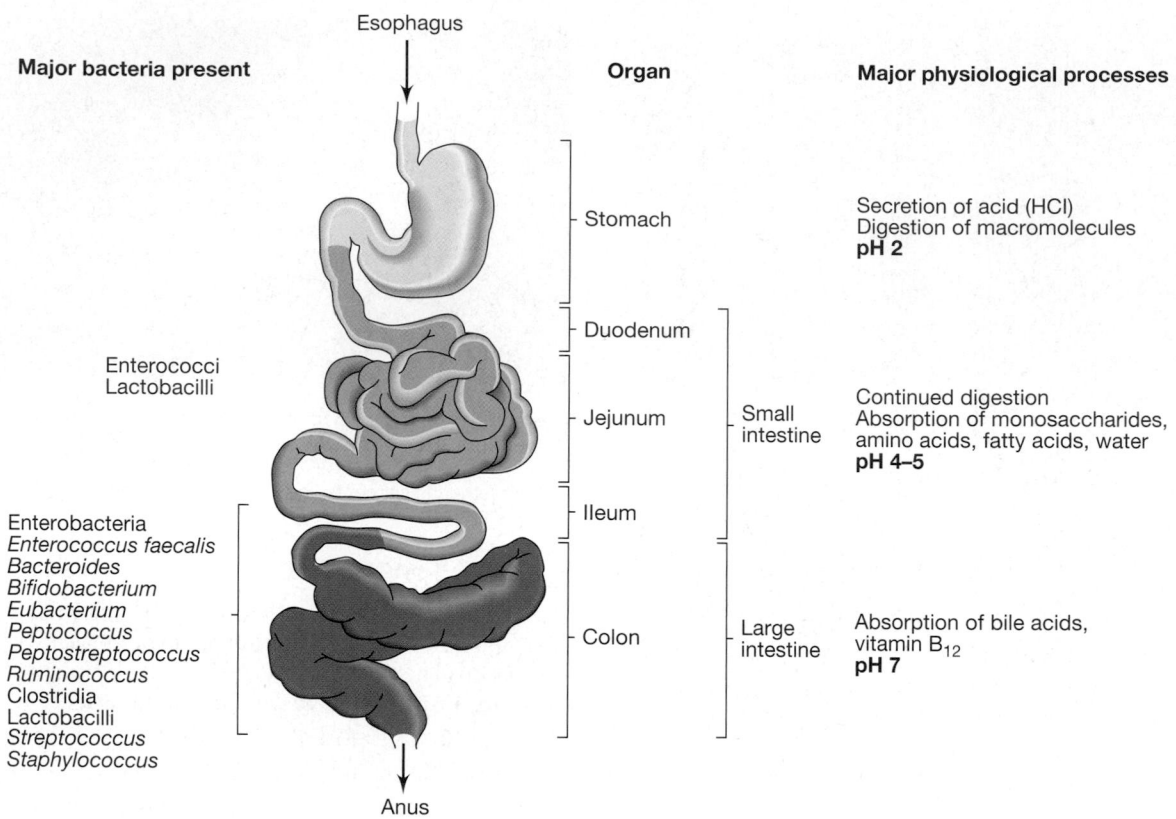

Major bacteria present

Esophagus

Organ

Major physiological processes

Enterococci
Lactobacilli

Stomach

Secretion of acid (HCl)
Digestion of macromolecules
pH 2

Duodenum

Jejunum — Small intestine

Continued digestion
Absorption of monosaccharides,
amino acids, fatty acids, water
pH 4–5

Ileum

Enterobacteria
Enterococcus faecalis
Bacteroides
Bifidobacterium
Eubacterium
Peptococcus
Peptostreptococcus
Ruminococcus
Clostridia
Lactobacilli
Streptococcus
Staphylococcus

Colon — Large intestine

Absorption of bile acids,
vitamin B_{12}
pH 7

Anus

Figure 21.8 The human gastrointestinal tract showing functions and the distribution of nonpathogenic microorganisms usually found in normal (healthy) individuals.

gradually becomes less acidic and bacterial numbers increase. In the lower ileum, bacteria are found in the intestinal cavity (the lumen), mixed with digestive material. Cell numbers of 10^5–10^7 per gram are common.

The ileum empties into the *cecum*, the first portion of the *large intestine*. The *colon* makes up the rest of the large intestine. In the colon, bacteria are present in enormous numbers, and this region can be viewed as a specialized fermentation vessel. Many bacteria live within the lumen itself, probably using as nutrients some products of the digestion of food (Table 21.1). Facultative aerobes such as *Escherichia coli* are present but in smaller numbers than other bacteria; total counts of facultative aerobes are generally less than 10^7 per gram of intestinal contents. The activities of facultative aerobes consume any oxygen present, making the environment of the large intestine strictly anaerobic and favorable for the profuse growth of obligate anaerobes. Many of these anaerobes are long, thin, gram-negative rods with tapering ends (called *fusiform*) and are attached end-on to small indentations in the intestinal wall (Figure 21.9). Other obligate anaerobes include species of *Clostridium* and *Bacteroides*. The total number of obligate anaerobes is enormous. Counts

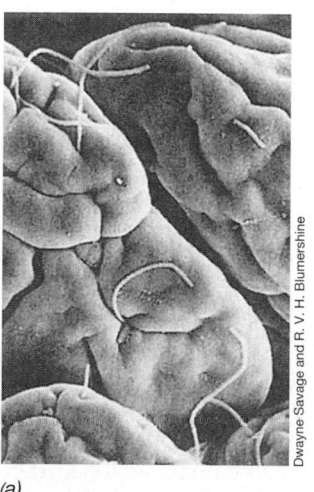
(a)

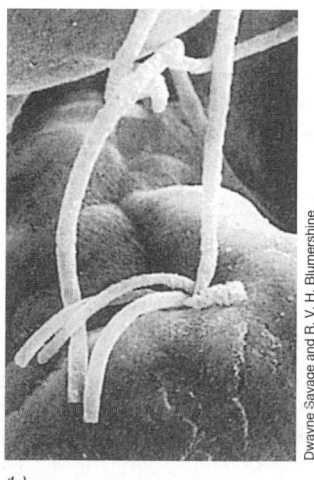
(b)

Figure 21.9 Scanning electron micrographs of the microbial community on the surface of the columnar epithelium in the mouse ileum. (a) An overview at low magnification. Note the long, filamentous *fusiform* bacteria lying on the surface. (b) Higher magnification, showing several filaments attached at a single depression. Note that the attachment is at the end of the filaments only. Individual cells are 10–15 μm long.

of 10^{10}–10^{11} cells/g of intestinal contents are not uncommon, with various species of *Bacteroides* accounting for the majority of intestinal obligate anaerobes. In addition, *Enterococcus faecalis* is almost always present in significant numbers.

The gut flora in humans can vary qualitatively depending on the diet. Persons who consume a considerable amount of meat show higher numbers of *Bacteroides* and lower numbers of coliforms and lactic acid bacteria than those with a vegetarian diet. An overview of microorganisms that inhabit the gastrointestinal tract is given in Figure 21.8.

The intestinal flora has a profound influence on host functions, carrying out a wide variety of metabolic reactions (Table 21.2). Among these are vitamin B_{12} and vitamin K production. These essential vitamins are not made by humans, but are absorbed from the gut, where they are made by the indigenous microbial flora. Steroids, produced in the liver and released into the intestine from the gall bladder as bile acids, are modified in the intestine by the microbial flora. The modified active steroid compounds are then absorbed from the gut. Other products generated by the action of fermentative and methanogenic microorganisms include gas (*flatus*) and the odor-producing substances listed in Table 21.2. For all of these metabolic products, the composition of the intestinal flora and the diet influence the type and amount of compounds produced.

During the passage of food through the gastrointestinal tract, water is absorbed from the digested material, which gradually becomes more concentrated and is converted to feces. *Bacteria* make up about one-third of the weight of fecal matter. Organisms living in the lumen of the large intestine are continuously displaced downward by the flow of material, and those bacteria that are lost must be replaced by new growth. Thus, the large intestine resembles a chemostat (Section 6.7). The time needed for passage of material through the complete gastrointestinal tract is about 24 h in humans;

the growth rate of bacteria in the lumen is one to two doublings per day.

When an antibiotic is given orally, it inhibits the growth of the normal flora as well as pathogens; continued movement of the intestinal contents then leads to loss of the preexisting bacteria and virtual sterilization of the intestinal tract. In the absence of the normal flora, opportunistic microorganisms such as antibiotic-resistant *Staphylococcus*, *Proteus*, or the yeast *Candida albicans* may become established. Occasionally, establishment of these opportunistic pathogens can lead to a harmful alteration in digestive function or even to disease. After antibiotic therapy stops, the normal flora eventually becomes reestablished, but often only after a considerable period.

Intestinal Gas

The gas produced within the intestines, called *flatus*, is the result of the action of fermentative and methanogenic microorganisms. Some foods can be metabolized by fermentative bacteria in the intestines, resulting in the production of hydrogen (H_2) and carbon dioxide (CO_2). Methanogens (Section 13.4) are found in the intestines of over one-third of normal adults. The methanogens convert H_2 and CO_2 produced by the other intestinal microorganisms to methane (CH_4). Normal adults expel several hundred milliliters of gas, of which about half is N_2 from swallowed air, from the intestines every day.

✓ 21.4 Concept Check

The stomach is very acidic and is a barrier to most microbial growth. The intestinal tract is slightly acid to neutral and supports a diverse population of microorganisms in a variety of nutritional and environmental conditions.

- ✓ Why might the small intestine be more suitable for growth of facultative aerobes than the large intestine?
- ✓ Identify several essential compounds made by indigenous intestinal microorganisms. What would happen if the microorganisms were eliminated from the body by the use of antibiotics?

21.5 Normal Flora of Other Body Regions

Each individual mucous membrane supports the growth of a specialized group of microorganisms. These organisms are part of the normal local environment and are characteristic of healthy tissue. In many cases, potentially pathogenic microorganisms cannot colonize mucous membranes because of the presence of the normal resident population of microorganisms. In this section, we discuss two such mucosal environments and their resident microorganisms.

TABLE 21.2 . Biochemical/metabolic contributions of intestinal microorganisms	
Vitamin synthesis	Product: thiamine, riboflavin, pyridoxine, B_{12}, K
Gas production	Product: CO_2, CH_4, H_2
Odor production	Product: H_2S, NH_3, amines, indole, skatole, butyric acid
Organic acid production	Product: acetic, propionic, butyric acids
Glycosidase reactions	Enzyme: β-glucuronidase, β-galactosidase, β-glucosidase, α-glucosidase, α-galactosidase
Steroid metabolism (bile acids)	Process: esterification, dehydroxylation, oxidation, reduction, inversion

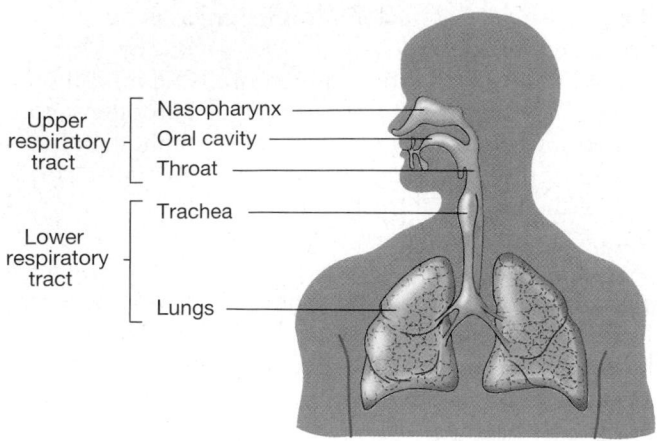

Figure 21.10 The respiratory tract. The upper respiratory tract is populated with a large variety and number of microorganisms, but the lower respiratory tract has relatively few microbial inhabitants, unless there is an ongoing active infection (∞ Figure 26.2).

Respiratory Tract

The anatomy of the respiratory tract is shown in Figure 21.10 (∞ Figure 26.2). In the **upper respiratory tract** (nasopharynx, oral cavity, and throat) microorganisms live primarily in areas bathed with the secretions of the mucous membranes. Bacteria enter the upper respiratory tract from the air during breathing, but most of them are trapped in the nasal passages and expelled again with the nasal secretions. The resident organisms most commonly found are staphylococci, streptococci, diphtheroid bacilli, and gram-negative cocci. Potentially harmful bacteria, such as *Staphylococcus aureus* and *Streptococcus pneumoniae*, are often part of the normal flora of the nasopharynx of healthy individuals (Table 21.1). These individuals are *carriers* of the pathogens but do not normally acquire disease, presumably because the other resident microorganisms compete successfully for resources and limit pathogen growth. The local immune system (∞ Section 22.8) is particularly active

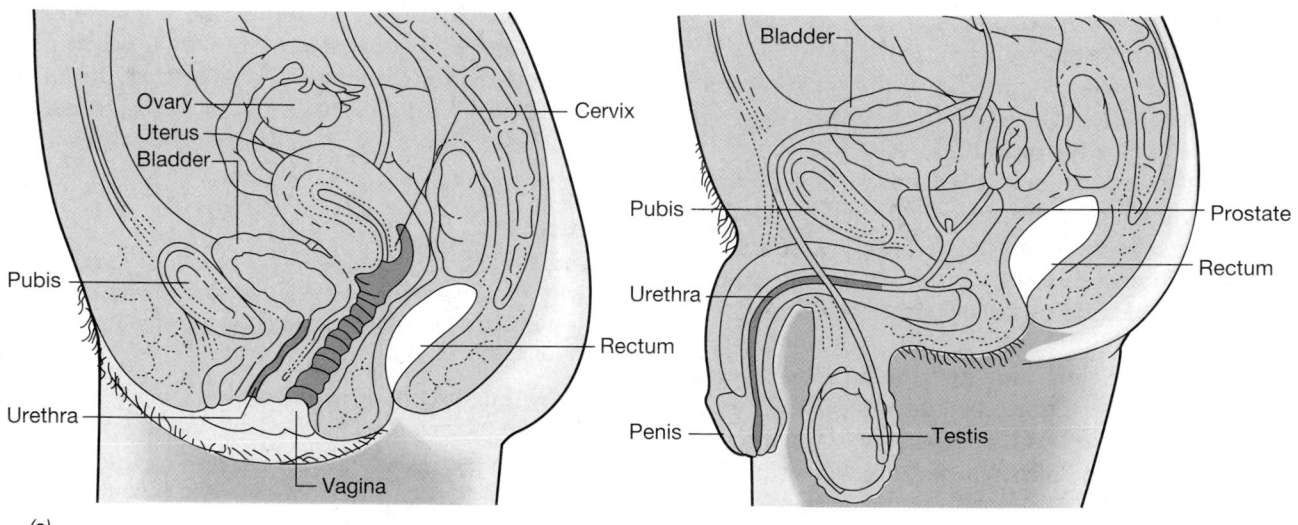

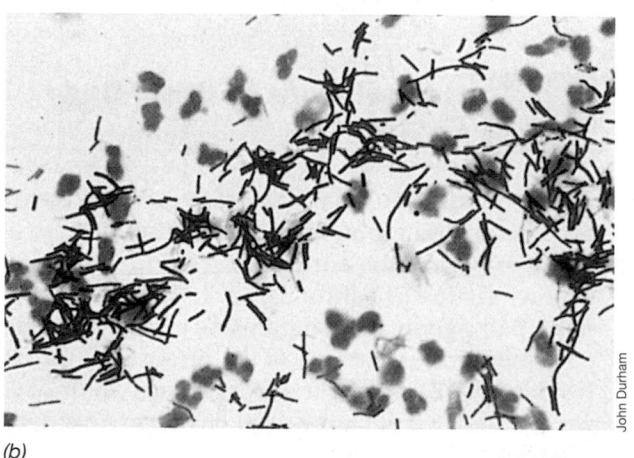

Figure 21.11 (a) The genitourinary tracts of the human female and male, showing regions (color) where microorganisms often grow. Note that the upper regions of the genitourinary tracts of both males and females is sterile in normal individuals. (b) Gram stain of *Lactobacillus acidophilus*, the predominant organism in the vagina of women between the onset of puberty and the end of menopause. Individual rods are 3–4 μm long. The genitourinary tract of older and younger women is less acidic and is populated by a much more heterogeneous group of microorganisms that grow at neutral to slightly alkaline pH.

at mucosal surfaces and may also inhibit the growth of pathogens.

The **lower respiratory tract** (trachea, bronchi, and lungs) is essentially sterile, despite the large numbers of organisms potentially able to reach this region during breathing. Dust particles, which are fairly large, settle out in the upper respiratory tract. As the air passes into the lower respiratory tract, its rate of flow decreases markedly, and organisms settle onto the walls of the passages. The walls of the entire respiratory tract are lined with ciliated epithelium, and the cilia, beating upward, push bacteria and other particulate matter toward the upper respiratory tract where they are then expelled in the saliva and nasal secretions. Only particles smaller than about 10 μm in diameter are able to reach the lungs.

Urogenital Tract

The main anatomical features of the male and female urogenital tracts are shown in Figure 21.11*a*. In both male and female, the bladder itself is usually sterile, but the epithelial cells lining the urethra are colonized by facultatively aerobic gram-negative rods and cocci (Table 21.1). These organisms, including *Escherichia coli*, *Proteus mirabilis*, and others, can occasionally become *opportunistic pathogens*. They are normally present in the body or in the local environment, but they are not pathogenic under normal circumstances. Changes in the body, such as local pH changes allow the organisms to multiply and become pathogenic. Such organisms frequently cause urinary tract infections, especially in women.

The vagina of the adult female generally is weakly acidic and contains significant amounts of the polysaccharide glycogen. *Lactobacillus acidophilus*, which is a resident organism in the vagina, ferments glycogen to produce lactic acid and lower the pH of the vagina (Figure 21.11) (⚭ Section 12.19). Other organisms—yeasts (*Torulopsis* and *Candida* species), streptococci, and *E. coli*—may also be present. Before puberty, the female vagina is alkaline and does not produce glycogen, *L. acidophilus* is absent, and the flora consists predominantly of staphylococci, streptococci, diphtheroids, and *E. coli*. After menopause, glycogen disappears, the pH rises, and the flora again resembles that found before puberty.

✓ 21.5 Concept Check

The presence of a population of normal nonpathogenic microorganisms in the respiratory and urogenital tracts is essential for normal organ function and often prevents the colonization of pathogens.

- ✓ Potential pathogens are sometimes found in the normal flora of the upper respiratory tract. Why do they not cause disease in some cases?
- ✓ Why is *Lactobacillus* found in the urogenital tract of normal adult women?

II HARMFUL MICROBIAL INTERACTIONS WITH HUMANS

Many microbial interactions are harmful to the host and cause disease. Here we will examine general mechanisms used by microorganisms to damage hosts. Pathogenesis, the ability of microorganisms to cause disease, starts with adherence of the microorganisms to host cells, followed by colonization and growth resulting in damage to the host. Disease-producing microorganisms elicit host changes using several different strategies to establish **virulence**, the relative ability of a pathogen to cause disease (Figure 21.12). We first consider the factors responsible for entry of a pathogen into a host.

21.6 Entry of the Pathogen into the Host

A pathogen must usually gain access to host tissues and multiply before damage can be done. In most cases, this requires that the organisms penetrate the skin, mucous

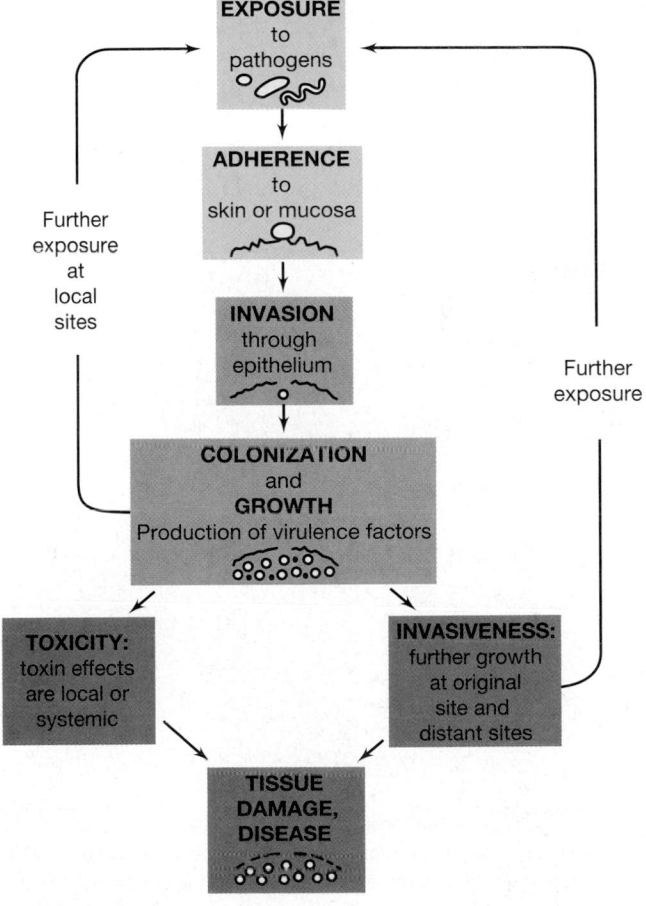

Figure 21.12 Microorganisms and pathogenesis. The presence and even growth of microorganisms on the host does not always lead to disease.

membranes, or intestinal epithelium, surfaces that normally act as microbial barriers.

Specific Adherence

Most microbial infections begin at breaks or wounds in the skin or on the mucous membranes of the respiratory, digestive, or genitourinary tract. Bacteria or viruses able to initiate infection often adhere *specifically* to epithelial cells (Figure 21.13) through protein-protein interactions on the surfaces of the pathogen and the host cell. An infecting microorganism does not adhere to all epithelial cells equally but selectively adheres to cells in the particular region of the body where it normally gains entrance. For example, *Neisseria gonorrhoeae*, the causative agent of the sexually transmitted disease gonorrhea (∞ Section 26.12), adheres much more strongly to urogenital epithelia than to other tissues using a surface protein called *Opa*. The host cells bind specifically to Opa with a protein called CD66, a protein found only on the surface of human epithelial cells. Thus, *N. gonorrhoeae* interacts exclusively with target cells through a cell surface receptor–ligand pair. This principle extends to host specificity. In many cases, a bacterial strain that normally infects humans adheres more strongly to the appropriate human cells than to similar cells in another animal (for example, the rat), whereas a strain that specifically colonizes the rat adheres more firmly to rat cells than to human cells.

Some bacterial adherence macromolecules are not covalently attached to the bacteria. These are usually polysaccharides synthesized and secreted by the bacteria (∞ Section 4.13). A polymer coat consisting of a dense, well-defined layer closely surrounding the cell is known as a **capsule** (∞ Figure 26.3). A loose network of polymer fibers extending outward from a cell is known as a **glycocalyx** (Figure 21.13*b*), while a diffuse mass of polymer fibers, seemingly unattached to any single cell, is called a **slime layer** (see Figure 21.4*b*). These structures may be important for adherence not only to host tissues, but also between other bacteria. In addition, these layers may protect the bacteria from host defense mechanisms such as phagocytosis by macrophages and other cells (∞ Section 22.2).

Fimbriae and *pili* (∞ Section 4.13) are bacterial cell surface protein structures (Figure 21.14) that may also function in the attachment process. For instance, the pili of *Neisseria gonorrhoeae* play a key role in the attachment of this organism to urogenital epithelium, and fimbriated strains of *Escherichia coli* (Figure 21.14) are much more frequent causes of urinary tract infections than strains lacking fimbriae. Among the best-characterized fimbriae are the so-called *type I fimbriae* of enteric bacteria (*Escherichia*, *Klebsiella*, *Salmonella*, *Shigella*). Type I fimbriae are uniformly distributed on the surface of cells. Pili are generally longer than fimbriae, with fewer pili found on the cell surface. Both pili and fimbriae function by binding host cell glycoproteins, initiating the attachment event.

Evidence for specific interactions between the mucosal epithelium and pathogens comes from studies of diarrhea caused by *Escherichia coli*. Most strains of *E. coli* are nonpathogenic and are part of the normal flora, inhabiting the cecum and the colon. Several strains are usually present in the body at the same time, and large numbers of these nonpathogens routinely pass through the body and are eliminated in fecal material. However,

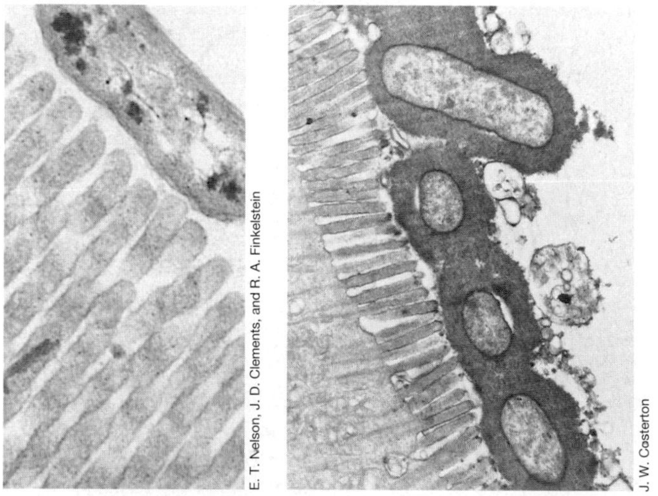

(a)   *(b)*

E. T. Nelson, J. D. Clements, and R. A. Finkelstein

J. W. Costerton

Figure 21.13 Adherence of pathogens to animal tissues. (a) Transmission electron micrograph of a thin section of *Vibrio cholerae* adhering to the brush border of rabbit villi. Note the absence of the glycocalyx. (b) Enteropathogenic *Escherichia coli* in a fatal model infection in the newborn calf. The bacterial cells are attached to the brush border of calf villi via an extensive glycocalyx. The rods are about 0.5 μm in diameter.

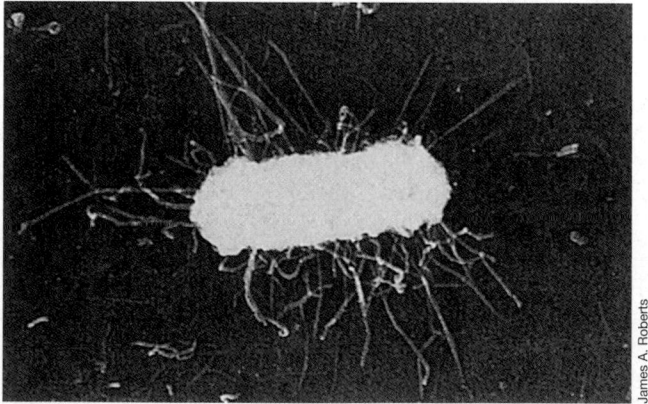

James A. Roberts

Figure 21.14 Shadow-cast electron micrograph of the bacterium *Escherichia coli*, showing type P fimbriae. Type P fimbriae resemble type I fimbriae but are somewhat longer. The cell shown is about 0.5 μm wide.

enteropathogenic strains of *E. coli* express fimbrial proteins called *CFA (colonization factor antigens)* that adhere specifically to cells in the *small* intestine, where they can colonize and produce enterotoxins (see Section 21.9) that cause diarrhea as well as other illnesses (Section 29.7). The nonpathogenic strains of *E. coli* seldom have the CFA proteins. Some major factors important in microbial adherence are shown in Table 21.3.

Invasion

A few microorganisms are pathogenic solely because of the toxins they produce. These organisms do not need to gain access to host tissues, and we will discuss them separately (see Sections 21.8 and 21.9). However, most pathogens penetrate the epithelium to initiate pathogenicity, a process called *invasion*. At the point of entry, usually at small breaks or lesions in the skin or in mucosal surfaces, growth is often established. Growth may also begin on intact mucosal surfaces, especially if the normal flora is altered or eliminated, for example, by antimicrobial chemotherapy. Pathogens may then more readily colonize the tissue and begin the invasion process. Pathogen growth may also be established at sites distant from the original point of entry. Access to distant, usually interior, sites is through the blood or lymphatic circulatory system (Section 22.1).

TABLE 21.3 Major adherence factors used to facilitate attachment of microbial pathogens to host tissues[a]

Factor	Example
Glycocalyx/capsule/ slime layer (Section 4.13 and Figure 21.13)	Pathogenic *Escherichia coli*—glycocalyx promotes adherence to the brush border of intestinal villi *Streptococcus mutans*—dextran glycocalyx promotes binding to tooth surfaces
Adherence proteins	*Streptococcus pyogenes*—M protein on the cell binds to receptors on respiratory mucosa *Neisseria gonorrhoeae*—Opa protein on the cell binds to receptors on urogenital epithelium
Lipoteichoic acid (Section 4.8 and Figure 4.32)	*Streptococcus pyogenes*—facilitates binding to respiratory mucosal receptor (along with M protein)
Fimbriae (pili) (Section 4.13 and Figure 21.14)	*Neisseria gonorrhoeae*—pili facilitate binding to urogenital epithelium *Salmonella* species—type I fimbriae facilitate binding to epithelium of small intestine Pathogenic *Escherichia coli*— colonization factor antigens (CFAs), which are fimbrial, facilitate binding to epithelium of small intestine

[a] Most receptor sites on host tissues are glycoproteins or complex lipids such as gangliosides or globosides.

✓ **21.6 Concept Check**

Pathogens usually gain access to host tissues by adherence to specific host molecules, often on mucosal surfaces. Invasion starts at the site of adherence and may spread throughout the host via the circulatory systems.

✓ How do CFA molecules on *Escherichia coli* influence adherence to mucosal tissues?

✓ How does adherence initiate invasion?

21.7 Colonization and Growth

If a pathogen gains access to tissues, it may multiply, a process called **colonization**. Because the initial inoculum of a pathogen is rarely sufficient to cause damage, a pathogen must find appropriate nutrients and environmental conditions in order to grow. Temperature, pH, and reduction potential are environmental factors that affect pathogen growth, but the availability of microbial nutrients in host tissues is most important. Although a vertebrate host might seem to be a nutritional paradise for microorganisms, not all nutrients are plentiful. Soluble nutrients such as sugars, amino acids, and organic acids are limited, and organisms able to use complex nutrient sources such as glycogen may be favored. Not all vitamins and growth factors are in adequate supply in all tissues at all times. *Brucella abortus*, for example, grows very slowly in most tissues of infected cattle but grows very rapidly in the placenta, where it causes abortion. This is due to the elevated concentration of erythritol found in the placenta, a nutrient that enhances growth of *B. abortus* (see Table 21.6).

Trace elements may also be in short supply and can influence establishment of the pathogen. For example, considerable evidence exists for the influence of *iron* on microbial growth. Specific proteins called *transferrin* and *lactoferrin*, present in animals, bind iron tightly and transfer it through the body. These proteins have such high affinity for iron that microbial iron deficiency may be common; a soluble iron salt fed or injected into an infected animal greatly increases the virulence of some pathogens. As we noted in Section 5.1, many bacteria produce iron-chelating compounds (*siderophores*) that help them obtain iron from the environment. Some iron chelators isolated from pathogenic bacteria are so efficient that they can remove iron from animal iron-binding proteins. For example, a siderophore called *aerobactin*, produced by certain strains of *Escherichia coli* and encoded by the Col V plasmid (Section 10.8), readily removes iron bound to transferrin.

Localization in the Body

After initial entry, the organism often remains localized and multiplies, producing a small focus of infection such as the boil, carbuncle, or pimple that commonly arises from *Staphylococcus* infections of the skin (Section

26.9). Alternatively, the organisms may pass through the lymphatic vessels and be deposited in lymph nodes. If an organism reaches the blood, it will be distributed to distant parts of the body, usually concentrating in the liver or spleen. Spread of the pathogen through the blood and lymph systems can result in a generalized (systemic) infection of the body, with the organism growing in a variety of tissues. If extensive bacterial growth in tissues occurs, some of the organisms are usually shed into the bloodstream in large numbers, a condition called **bacteremia**. Widespread infections of this type almost always start as a localized infection at a specific organ site.

✓ 21.7 Concept Check

A pathogen must gain access to nutrients and appropriate growth conditions before it can colonize and grow in substantial numbers in host tissue. Organisms may grow locally at the site of invasion, or may spread through the body.

✓ Why are colonization and growth necessary for the success of most pathogens?

✓ What host factors limit or accelerate colonization and growth of a microorganism at a local site?

21.8 Virulence

Virulence is the relative ability of a parasite to cause disease, and here we discuss some basic methods used to measure it. We will then provide specific examples of particularly virulent organisms and highlight the strategies and methods these organisms use to enhance virulence.

Measuring Virulence
The virulence of a pathogen can be estimated from experimental studies of the LD_{50}. The LD_{50}, or *lethal dose$_{50}$*, is the dose of an agent that kills 50% of the animals in a test group. Highly virulent pathogens frequently show little difference in the number of cells required to kill 100% of the population as compared with the number required to kill 50% of the population. This is illustrated in Figure 21.15 for experimental *Streptococcus* and *Salmonella* infections in mice. Only a few cells of *Streptococcus pneumoniae* are required to establish a fatal infection and kill all members of a test population once the virulence of a particular strain has been established. In fact, the LD_{50} for this organism is hard to determine because so few organisms are needed to produce a lethal infection. By contrast, the LD_{50} for *Salmonella typhimurium*, also a mouse pathogen but a much less virulent one, is much higher than for *S. pneumoniae*, and the number of cells required to kill 100% of the population is more than 100 times greater than the number of cells needed to achieve the LD_{50}.

When pathogens are kept in laboratory culture and not passed through animals, their virulence is often decreased or even completely lost. Such organisms are said to be **attenuated**. Attenuation probably occurs because

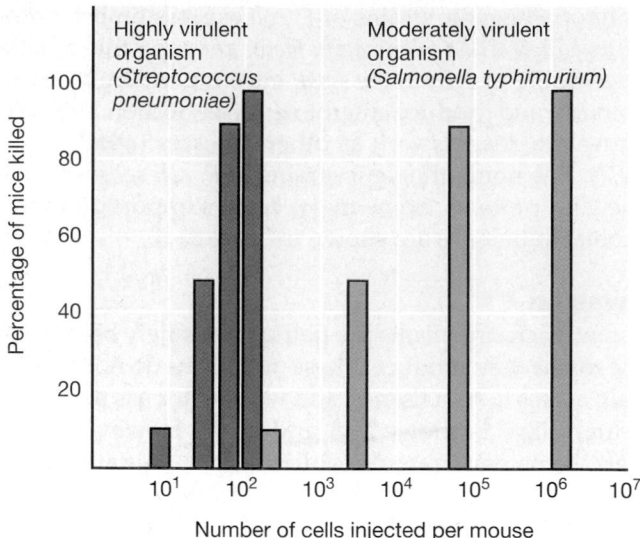

Figure 21.15 Comparison of differences in microbial virulence based on the number of cells of *Streptococcus pneumoniae* or *Salmonella typhimurium* required to kill mice.

nonvirulent mutants may grow faster and, through successive transfers to fresh media, such mutants are selectively favored. Attenuation often occurs more readily when culture conditions are not optimal for the species. If an attenuated culture is reinoculated into an animal, organisms sometimes regain virulence, but in many cases loss of virulence is permanent. Attenuated strains are often used in the production of vaccines, especially viral vaccines (∞ Section 22.11). Measles and mumps vaccines, for example, are composed of attenuated viruses.

Toxicity and Invasiveness
Virulence is due to the ability of a pathogen to cause host damage through toxicity and invasiveness. Each pathogen uses these properties to cause disease.

Toxicity is the ability of an organism to cause disease by means of a preformed toxin that inhibits host cell function or kills host cells. For example, the disease *tetanus* is produced by *Clostridium tetani* through a potent exotoxin (see Section 21.10). The cells of *C. tetani* rarely leave the wound where it was first deposited, growing relatively slowly at the wound site. Yet *C. tetani* is able to bring about disease because it produces tetanus toxin that moves to distant parts of the body and initiates irreversible muscle contraction and often death of the host.

Invasiveness is the ability of an organism to grow in host tissue in such large numbers that the pathogen inhibits host function. A microorganism may still be able to produce disease through invasiveness even if it produces no toxin. For example, the major virulence factor for *Streptococcus pneumoniae* is the polysaccharide capsule that prevents the phagocytosis of pathogenic strains (see Section 21.6, ∞ Section 22.2, and Figure 22.5), defeating a major defense mechanism used by the host to prevent invasion. Encapsulated strains of *S. pneumoniae* are able to cause extensive host damage because they are highly invasive; they grow in lung tissues in enormous numbers where they initiate host responses that lead to pneumonia (∞ Section 26.2). Nonencapsulated strains are quickly and efficiently taken up and destroyed by phagocytes.

Clostridium tetani and *Streptococcus pneumoniae* exemplify the extremes of toxicity and invasiveness; most successful pathogens fall between these extremes and use a combination of toxins and invasiveness to cause disease.

Virulence in *Salmonella*

Salmonella spp. employ a mixture of toxins, invasiveness, and other virulence factors to promote pathogenicity. First, several toxins contribute to the virulence of *Salmonella*, and at least three toxins are produced: enterotoxin (Table 21.4),

TABLE 21.4 Exotoxins and extracellular virulence factors produced by human pathogens

Organism	Disease	Toxin or factor[a]	Action
Bacillus anthracis	Anthrax	Lethal factor (LF) Edema factor (EF) Protective antigen (PA) (AB)	PA is the cell-binding B component, EF causes edema, LF causes cell death
Bacillus cereus	Food poisoning	Enterotoxin (?)	Induces fluid loss from intestinal cells
Bordetella pertussis	Whooping cough	Pertussis toxin (AB)	Blocks G protein signal transduction, kills cells
Clostridium botulinum	Botulism	Neurotoxin (AB)	Flaccid paralysis (see Figure 21.19)
Clostridium tetani	Tetanus	Neurotoxin (AB)	Spastic paralysis (see Figure 21.20)
Clostridium perfringens	Gas gangrene, food poisoning	α-Toxin (CT)	Hemolysis (lecithinase, see Figure 21.17*b*)
		β-Toxin (CT)	Hemolysis
		γ-Toxin (CT)	Hemolysis
		δ-Toxin (CT)	Hemolysis (cardiotoxin)
		κ-Toxin (E)	Collagenase
		λ-Toxin (E)	Protease
		Enterotoxin (CT)	Alters permeability of intestinal epithelium
Corynebacterium diphtheriae	Diphtheria	Diphtheria toxin (AB)	Inhibits protein synthesis in eukaryotes (Figure 21.18)
Escherichia coli (enteropathogenic strains only)	Gastroenteritis	Enterotoxin (AB)	Induces fluid loss from intestinal cells
Pseudomonas aeruginosa	*P. aeruginosa* infections	Exotoxin A (AB)	Inhibits protein synthesis
Salmonella spp.	Salmonellosis, typhoid fever, paratyphoid fever	Enterotoxin (AB)	Inhibits protein synthesis and lyses host cells
		Cytotoxin (CT)	Induces fluid loss from intestinal cells
Shigella dysenteriae	Bacterial dysentery	Shiga toxin (AB)	Inhibits protein synthesis
Staphylococcus aureus	Pyrogenic (pus-forming) infections (boils, and so on), respiratory infections, food poisoning, toxic shock syndrome, scalded skin syndrome	α-Toxin (CT)	Hemolysis
		Toxic shock syndrome toxin (SA)	Systemic shock
		Exfoliating toxin A and B (SA)	Peeling of skin, shock
		Leukocidin (CT)	Destroys leukocytes
		β-Toxin (CT)	Hemolysis
		γ-Toxin (CT)	Kills cells
		δ-Toxin (CT)	Hemolysis, leukolysis
		Enterotoxin A, B, C, D, and E (SA)	Induce vomiting, diarrhea, shock
		Coagulase (E)	Induces fibrin clotting
Streptococcus pyogenes	Pyrogenic infections, tonsillitis, scarlet fever	Streptolysin O (CT)	Hemolysin
		Streptolysin S (CT)	Hemolysin (see Figure 21.17)
		Erythrogenic toxin (SA)	Causes scarlet fever rash
		Streptokinase (E)	Dissolves fibrin clots
		Hyaluronidase (E)	Dissolves hyaluronic acid in connective tissue
Vibrio cholerae	Cholera	Enterotoxin (AB)	Induces fluid loss from intestinal cells (Figure 21.21)

[a] AB, A-B toxin; CT, cytolytic toxin; E, enzymatic virulence factor; SA, superantigen toxin;?, not classified.

endotoxin (∞ Section 4.9 and Figure 4.35), and *cytotoxin*. Cytotoxin acts by inhibiting host cell protein synthesis and allowing Ca^{2+} to escape from host cells. In addition, a number of virulence factors are involved in invasion. Adherence factors are the cell surface polysaccharide O antigen (∞ Figure 4.35) and the flagellar H antigen. Fimbriae may also enhance adherence. The capsular Vi polysaccharide inhibits complement binding and antibody-mediated killing (∞ Section 22.10). The *inv* genes of *Salmonella* encode at least 10 different proteins involved in invasion. For example, *invH* encodes a surface adhesin, while *invC, invG, invI,* and *invJ* encode proteins involved in assembly of the *surface appendages*, protein structures used for host-cell binding.

Salmonella readily establish infections through intracellular parasitism, growing inside the cells that line the intestine and even in macrophages, a group of white blood cells that normally ingest (phagocytize) and kill bacteria (∞ Section 22.2). The toxic oxygen products of macrophages are neutralized by proteins induced by the *Salmonella oxyR* gene. Macrophage-produced antibacterial molecules called *defensins* are neutralized by the products of the *phoP* and *phoQ* genes. Thus, the *oxy* and *pho* genes allow *Salmonella* to become intracellular pathogens, since their gene products neutralize host defense factors that limit intracellular bacterial growth. Several plasmid-borne virulence factors are also involved in persistence and spread in most *Salmonella* species. Finally, *Salmonella* also produce *siderophores*, bacterial iron-chelating proteins that sequester iron (∞ Section 5.1

and see Section 21.7), a valuable growth factor. Thus, *Salmonella*, and probably most other pathogens, use several virulence factors to initiate disease. Figure 21.16 summarizes the known virulence factors of *Salmonella*.

✓ 21.8 Concept Check

Virulence is determined by invasiveness, toxigenicity, and other factors produced by a pathogen. In most pathogens, a number of factors contribute to virulence. Attenuation is loss of virulence.

✓ Distinguish between toxicity and invasiveness.

✓ Explain how an organism may become attenuated. Discuss the role of attenuated organisms in vaccine production.

III VIRULENCE FACTORS AND TOXINS

Extracellular virulence factors and toxins produced by microorganisms promote pathogenesis. A wide variety of these proteins are produced by many different pathogens, but many share molecular characteristics and have similar mechanisms of action. Here we examine representative examples of several important categories of virulence factors and toxins at the mechanistic level.

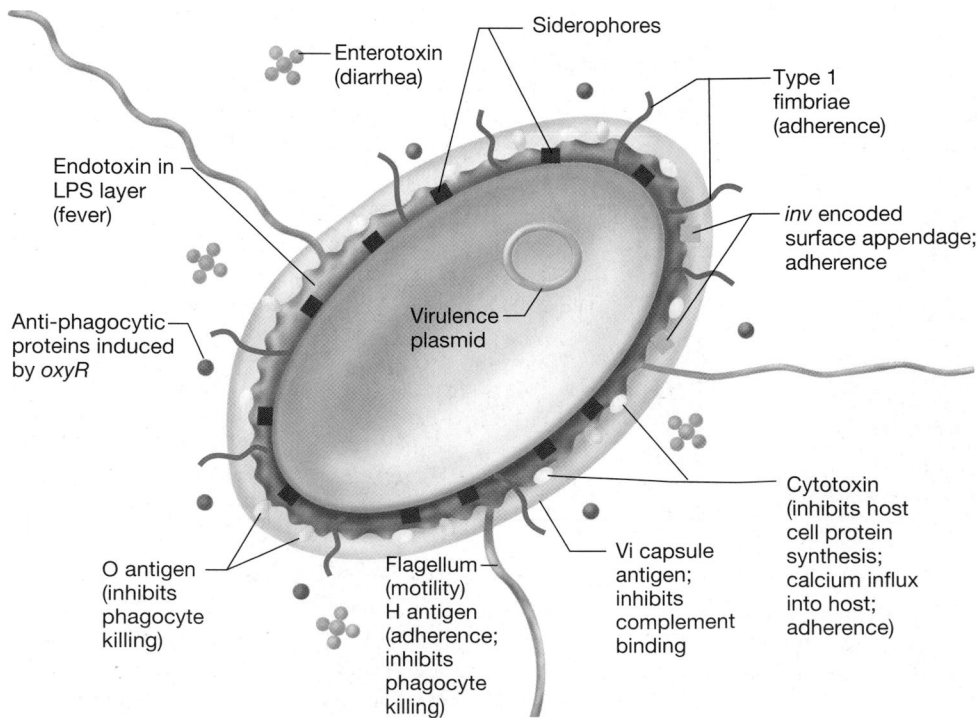

Figure 21.16 Virulence factors important in *Salmonella* pathogenesis. Structural elements known to be important in pathogenesis are shown. Protein products of the *pho* gene system have not been identified, but are known to neutralize the effects of the detergent-like, macrophage-produced defensins.

Siderophores

Enterotoxin (diarrhea)

Type 1 fimbriae (adherence)

Endotoxin in LPS layer (fever)

inv encoded surface appendage; adherence

Virulence plasmid

Anti-phagocytic proteins induced by *oxyR*

Cytotoxin (inhibits host cell protein synthesis; calcium influx into host; adherence)

O antigen (inhibits phagocyte killing)

Flagellum (motility) H antigen (adherence; inhibits phagocyte killing)

Vi capsule antigen; inhibits complement binding

21.9 Virulence Factors

A number of pathogen-produced extracellular proteins aid in the establishment and maintenance of disease. These proteins are called *virulence factors*. Most virulence factors are enzymes that help the pathogens colonize and grow. For example, streptococci, staphylococci, and certain clostridia produce hyaluronidase (Table 21.4), an enzyme that promotes spreading of organisms in tissues by breaking down hyaluronic acid, a polysaccharide that functions as intercellular cement. Production of hyaluronidase enables these organisms to spread from an initial site. Streptococci and staphylococci also produce a vast array of proteases, nucleases, and lipases that serve to depolymerize host proteins, nucleic acids, and lipids. Clostridia that cause gas gangrene produce collagenase, or *κ*-toxin (Table 21.4). *κ*-Toxin breaks down the collagen network supporting the tissues, enabling these organisms to spread through the body.

Fibrin, Clots, and Virulence

Fibrin clots are often formed at a site of microbial invasion by the host. The clotting mechanism, triggered by tissue injury, isolates the pathogens, limiting infection to a small region of the body. Some organisms are able to produce fibrinolytic enzymes to dissolve these clots and make further invasion possible. One such fibrinolytic substance produced by *Streptococcus pyogenes* is known as *streptokinase* (Table 21.4).

On the other hand, other organisms produce enzymes that *promote* the formation of fibrin clots, causing localization and protection of the organism rather than spread of the organism. The best-studied fibrin-clotting enzyme is *coagulase* (Table 21.4), produced by pathogenic *Staphylococcus aureus*. Coagulase causes fibrin to be deposited on the cocci and may protect coated bacteria from attack by host cells. The fibrin matrix produced as a result of coagulase activity probably accounts for the extremely localized nature of many staphylococcal infections, as in boils and pimples (∞ Figure 26.23).

✓ 21.9 Concept Check

Microorganisms produce a variety of different enzymes that enhance virulence by breaking down or altering host tissue to provide nutrients. Still other pathogen-produced virulence factors provide protection to the pathogen by interfering with normal host defense mechanisms. These factors enhance colonization and growth of the pathogen.

✓ What advantage does the pathogen gain by producing enzymes that digest structural components of host tissues?

✓ How can the action of coagulase work to help *or* hinder bacterial growth? Give specific examples.

21.10 Exotoxins

Exotoxins are proteins released extracellularly as the organism grows. These toxins may travel from a focus of infection to distant parts of the body and cause damage in regions far removed from the site of microbial growth. Table 21.4 provides a summary of the properties and actions of some of the best-known exotoxins as well as other extracellular virulence factors.

Most exotoxins fall into one of three categories; the *cytolytic toxins*, the *A-B toxins*, or the *superantigen toxins*. The cytolytic toxins work by enzymatically attacking cell constituents, causing lysis. The A-B toxins consist of two covalently bonded subunits, A and B. The B component generally binds to a cell surface receptor, allowing the transfer of the A subunit across the targeted cell membrane, where it functions to damage the cell. The superantigens work by stimulating large numbers of immune response cells, resulting in extensive inflammatory reactions (∞ Section 22.14).

Cytolytic Toxins

Various pathogens produce proteins that are able to act on the animal cytoplasmic membrane, causing cell lysis and hence cell death. The action of these toxins is most easily detected with red blood cells (erythrocytes), hence, they are often called *hemolysins* (Table 21.4). However, they also work on cells other than erythrocytes. The production of hemolysins is easily demonstrated in the laboratory by streaking the organism on a *blood agar plate*. During growth of the colonies, hemolysin is released and lyses the surrounding red blood cells, typically creating a zone of hemolysis (Figure 21.17a). Some hemolysins attack the phospholipid of the host cytoplasmic membrane. Because the phospholipid lecithin (phosphatidylcholine) is often used as a substrate, these enzymes are called *lecithinases* or *phospholipases* (Figure 21.17b). An example is the *α*-toxin of *Clostridium perfringens*, which is a lecithinase that dissolves membrane lipids, resulting in cell lysis (Table 21.4). Since the cytoplasmic membranes of all organisms contain phospholipids, phospholipases sometimes destroy bacterial as well as animal cytoplasmic membranes. Some hemolysins are not phospholipases, however. Streptolysin O, a hemolysin produced by streptococci, affects the sterols of the host cytoplasmic membrane. *Leukocidins* (Table 21.4) are lytic agents capable of lysing white blood cells and may decrease host resistance (∞ Section 22.1).

Diphtheria Toxin

Diphtheria toxin is produced by *Corynebacterium diphtheriae* and is an important factor in the pathogenesis of diphtheria, discussed in Section 26.3. Rats and mice are relatively resistant to diphtheria toxin, but humans,

Figure 21.17 (a) Zones of hemolysis around colonies of *Streptococcus pyogenes* growing on a blood agar plate. (b) Action of lecithinase, a phospholipase, around colonies of *Clostridium perfringens*, growing on an agar medium containing egg yolk, a source of lecithin. Lecithinase dissolves the membranes of red blood cells, leading to the clearing zones shown.

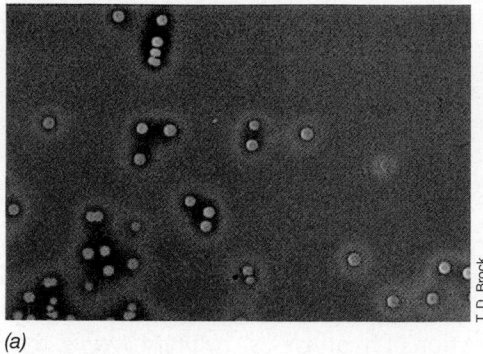
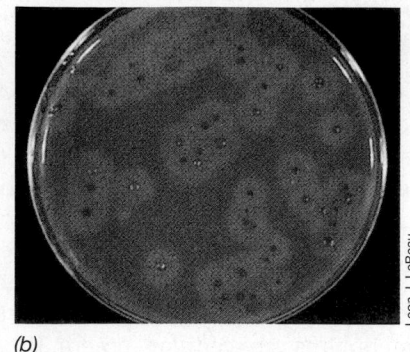

(a) (b)

rabbits, guinea pigs, and birds are susceptible, with only a single toxin molecule required to kill a cell in these species. Diphtheria toxin is an A-B toxin secreted by cells of *C. diphtheriae* as a polypeptide of 62,000 molecular weight. Fragment B promotes specific binding of the toxin to a host cell receptor (Figure 21.18). After binding, proteolytic cleavage between Fragment A and B allows entry of Fragment A (21,000 molecular weight) into the host cytoplasm. Fragment A then disrupts protein synthesis by blocking transfer of an amino acid from a transfer ribonucleic acid (tRNA) to the growing polypeptide chain. The toxin specifically inactivates elongation factor 2 (a protein involved in growth of the polypeptide chain) by catalyzing the attachment of adenosine diphosphate (ADP) ribose from NAD^+. Following ADP-ribosylation, the activity of the modified elongation factor 2 decreases dramatically and protein synthesis stops.

Diphtheria toxin is formed only by strains of *C. diphtheriae* that are lysogenized by a bacteriophage called phage β, which carries the toxin-encoding *tox* gene. Nontoxigenic, nonpathogenic strains of *C. diphtheriae* can be converted to pathogenic strains by infection with phage β (the process of *phage conversion*) (Section 10.7).

Another factor in diphtheria toxin production is the concentration of *iron* in the environment. In media containing sufficient iron for optimal growth, no toxin is produced. When the iron concentration is reduced to growth-limiting levels, toxin production occurs. The role of iron is to bind to a regulatory protein in *C. diphtheriae* (that is, act as a *negative control element*) (Section 8.5). The iron-binding protein then combines with a control region of the DNA of β phage and prevents expression of the diphtheria toxin. When iron is limiting, the regulatory protein does not act, and toxin synthesis occurs.

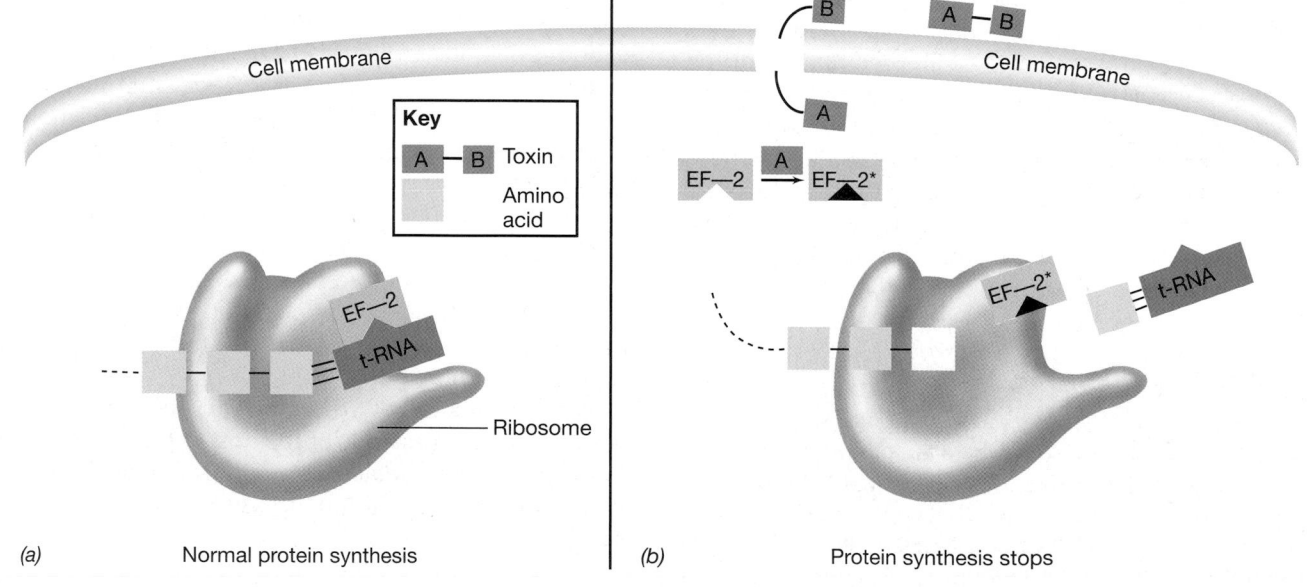

(a) Normal protein synthesis (b) Protein synthesis stops

Figure 21.18 The action of diphtheria toxin from *Corynebacterium diphtheriae*. (a) Elongation factor 2 (EF-2) normally binds to the ribosome and brings an amino acid-charged t-RNA to the ribosome, causing protein elongation. (b) Diphtheria toxin binds to the cell membrane, where it is cleaved and the A peptide is internalized. The A peptide catalyzes the ADP-ribosylation of elongation factor 2 (EF-2*). The modified elongation factor no longer aids transfer of amino acids to the growing polypeptide chain, resulting in shutdown of protein synthesis and death of the cell.

Exotoxin A of *Pseudomonas aeruginosa* (Table 21.4) functions similarly to diphtheria toxin, also modifying elongation factor 2 by ADP-ribosylation.

Tetanus and Botulinum Toxins

These toxins are produced by two species of obligately anaerobic bacteria, *Clostridium tetani* and *Clostridium botulinum*, which are normal soil organisms that occasionally cause disease in animals (∞ Section 27.8 and Section 29.5). *C. botulinum* rarely grows directly in the body, but it does grow and produce toxin in improperly preserved foods. Death from *botulism* is usually due to respiratory failure from flacid muscle paralysis. *C. tetani* grows in the body in deep wound punctures that become anoxic, and although *C. tetani* does not invade the body from the initial site of infection, the toxin can spread through the neural cells and cause spastic paralysis that is indicative of *tetanus* and can result in death.

Botulinum toxin is a series of seven related A-B toxins that are the most potent biological toxins known. One milligram of pure botulinum toxin is enough to kill more than 1 million guinea pigs. Of the seven distinct botulinum toxins known, at least two are encoded on lysogenic bacteriophages specific for *Clostridium botulinum*. The major toxin is a protein of about 150,000 molecular weight, which readily forms complexes with nontoxic botulinum proteins to give a bioactive toxin of almost 10^6 molecular weight. Toxicity occurs because the toxin binds to presynaptic membranes on the termini of the stimulatory motor neurons at the neuromuscular junction, blocking the release of acetylcholine. Because transmission of the nerve impulse to the muscle is through acetylcholine interaction with a muscle receptor, the botulinum-poisoned muscle cannot receive an excitatory signal and contraction is prevented, causing a flaccid paralysis (Figure 21.19).

Tetanus toxin is a protein of molecular weight 150,000, containing linked A-B polypeptides. On contact with the central nervous system this toxin is transported through the motor neurons back to the spinal cord, where it binds specifically to ganglioside lipids at the termini of the inhibitory interneurons. The inhibitory interneurons normally work by releasing an inhibitory neurotransmitter, usually glycine, that binds to receptors on the motor neurons. Normally, the glycine signal from the inhibitory interneurons stops the release of acetylcholine by the motor neurons and inhibits muscle contraction, allowing relaxation of the muscle fibers. However, if tetanus toxin blocks the release of glycine, the motor neurons cannot be inhibited, resulting in continual release of acetylcholine and uncontrolled contraction of the poisoned muscles (Figure 21.20). The outcome is a spastic, twitching paralysis, with affected muscles constantly contracted. If the muscles of the mouth are involved, the prolonged spasm restricts the mouth's movement, resulting in the condition known as

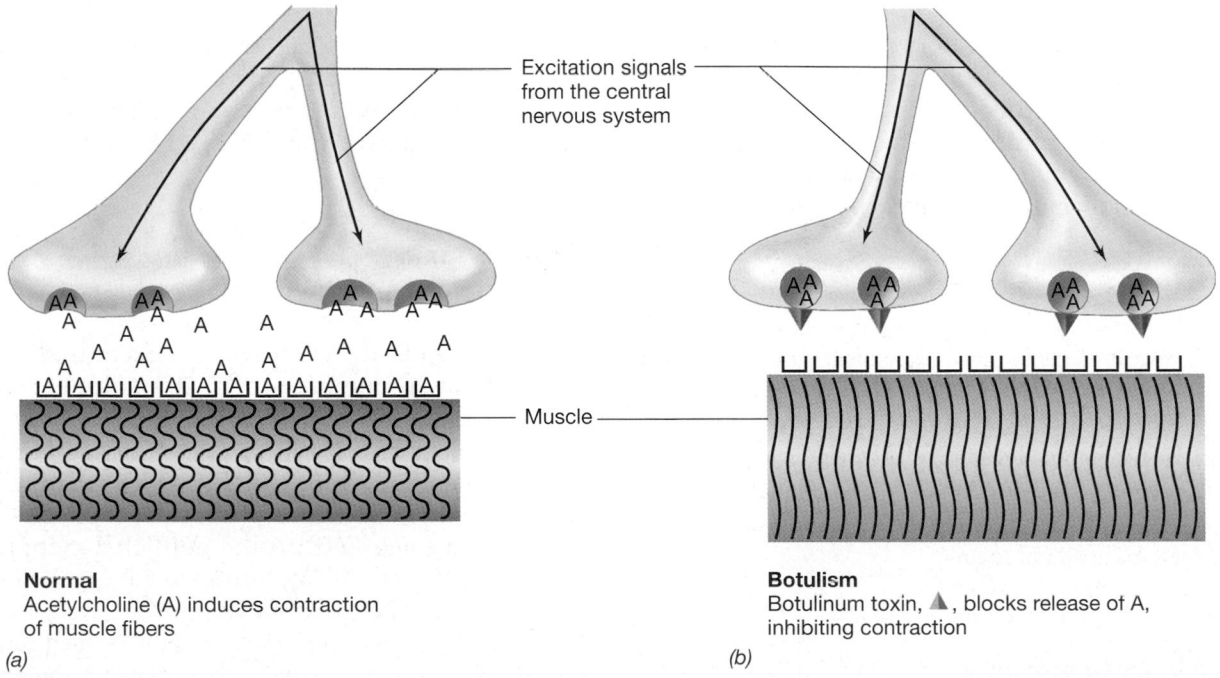

Normal
Acetylcholine (A) induces contraction of muscle fibers

(a)

Botulism
Botulinum toxin, ▲, blocks release of A, inhibiting contraction

(b)

Excitation signals from the central nervous system

Muscle

Figure 21.19 The action of botulinum toxin from *Clostridium botulinum*. (a) Upon central nervous stimulation, acetylcholine (A) is normally released from vesicles at the neural side of the motor end plate. Acetylcholine then binds to specific receptors on the muscle, inducing contraction. (b) Botulinum toxin acts at the motor end plate to prevent release of acetylcholine (A) from vesicles, resulting in a lack of stimulus to the muscle fibers, irreversible relaxation of the muscles, and flaccid paralysis. (∞ Section 29.5).

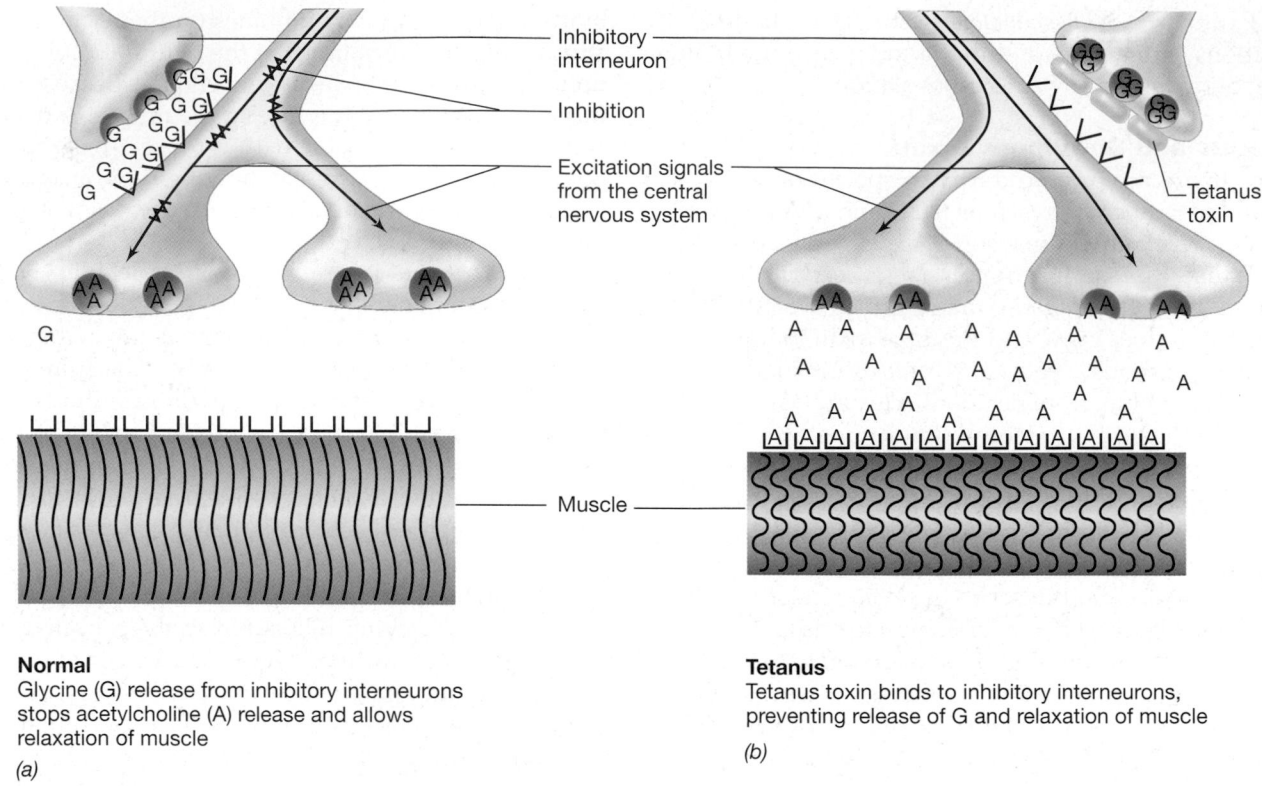

Normal
Glycine (G) release from inhibitory interneurons stops acetylcholine (A) release and allows relaxation of muscle

(a)

Tetanus
Tetanus toxin binds to inhibitory interneurons, preventing release of G and relaxation of muscle

(b)

Figure 21.20 The action of tetanus toxin from *Clostridium tetani*. (a) Muscle relaxation is normally induced by glycine (G) release from inhibitory interneurons. Glycine acts on the motor neurons to block excitation and release of acetylcholine (A) at the motor end plate. (b) Tetanus toxin binds to the interneuron to prevent release of glycine from vesicles, resulting in a lack of inhibitory signals to the motor neurons, constant release of acetylcholine to the muscle fibers, irreversible contraction of the muscles, and spastic paralysis. (Section 27.8 and Figure 27.19).

trismus or *lockjaw*. If respiratory muscles are involved, death may occur due to asphyxiation.

Tetanus toxin and botulinum toxin both block release of neurotransmitters involved in muscle control, but the outcome is quite different, depending on the particular neurotransmitters involved.

✓ 21.10 Concept Check

The most potent biological toxins known are the exotoxins produced by microorganisms. Each exotoxin acts on specific host cells or molecules, resulting in specific impairment of a major host cell function.

✓ What key features are shared by all exotoxins? By the A-B exotoxins?

✓ Are bacterial growth in the host and infection necessary for the production of toxins?

21.11 Enterotoxins

Enterotoxins are exotoxins that act on the *small* intestine, generally causing massive secretion of fluid into the intestinal lumen, leading to vomiting and diarrhea.

Enterotoxins are produced by a variety of bacteria, including the food-poisoning organisms *Staphylococcus aureus*, *Clostridium perfringens*, and *Bacillus cereus*, and the intestinal pathogens *Vibrio cholerae*, *Escherichia coli*, and *Salmonella enteritidis*.

Cholera Toxin

The enterotoxin produced by *Vibrio cholerae*, the causal agent of cholera, is the best understood enterotoxin. Cholera toxin is an A-B toxin (see Section 21.10) and consists of an A component of 27,200 molecular weight, with five B subunits, each of 11,600 molecular weight (82,200 molecular weight for the entire complex). The B subunit contains the binding site by which the cholera toxin combines specifically with the ganglioside GM1 (a complex glycolipid) in the epithelial cytoplasmic membrane (Figure 21.21), but the B subunit itself does not cause an alteration in membrane permeability. Rather, the toxic action is in the A chain, which activates the cellular enzyme *adenyl cyclase*, causing the conversion of adenosine triphosphate (ATP) to cyclic adenosine monophosphate (cAMP).

As we discussed in Section 8.7, cyclic AMP is a specific mediator of a variety of regulatory systems in cells.

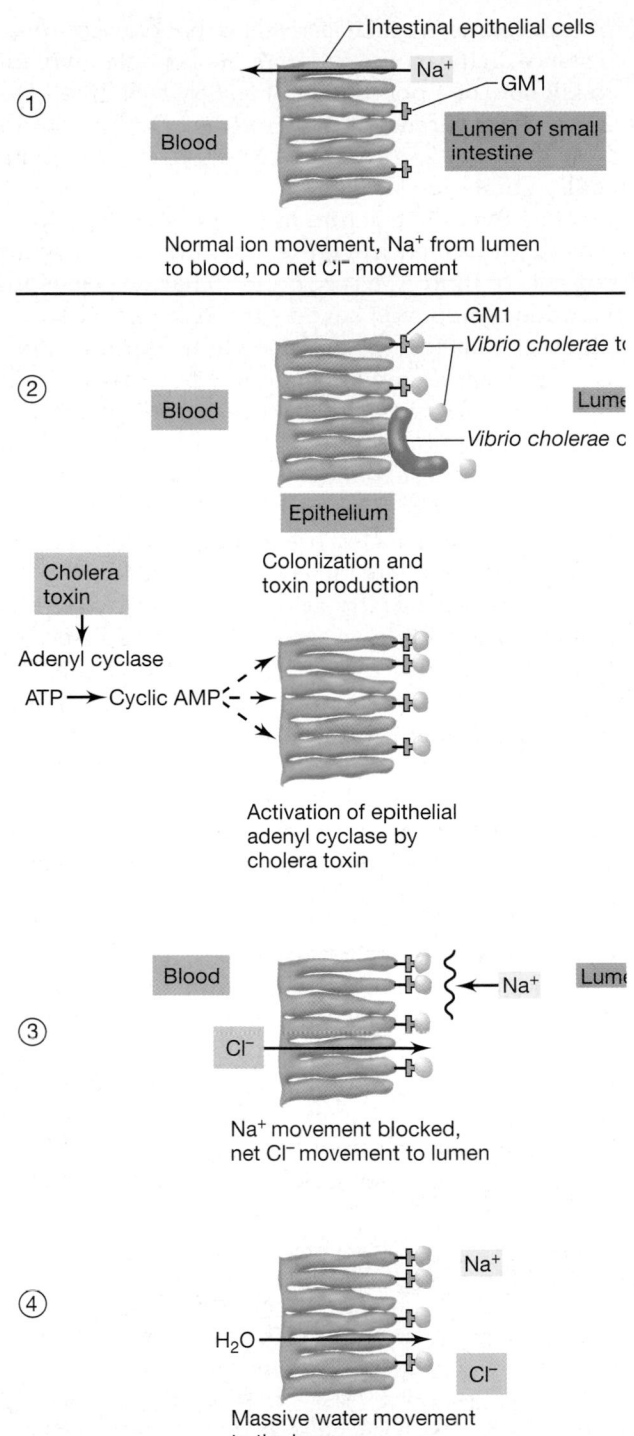

In mammals, cyclic AMP is involved in the action of a variety of hormones, as well as in synaptic transmission in the nervous system, and in inflammatory and immune reactions of tissues, including allergies. Although the A subunit of cholera toxin is responsible for activation of adenyl cyclase, A must first be activated by a cellular enzyme that requires NAD^+ and ATP. In the action of cholera enterotoxin, the increased cyclic AMP levels bring about the active secretion of chloride and bicarbonate ions from the mucosal cells into the intestinal lumen. This change in ion concentrations leads to the secretion of large amounts of water into the lumen (Figure 21.21). In the acute phase of cholera, the rate of water loss into the small intestine is greater than reabsorption of water by the large intestine, and so massive net fluid loss occurs. Cholera victims generally die from extreme dehydration, and the best treatment for the disease is the oral administration of electrolyte solutions containing solutes to replace lost fluid and ions.

Because cholera enterotoxin activates adenyl cyclase in a variety of cells and tissues, pathological manifestations of cholera toxin are related more to the specific site at which it binds, the epithelial cells of the small intestine, than to toxin activation of adenyl cyclase. Indeed, purified B subunits devoid of adenyl cyclase activity can actually *prevent* the action of cholera enterotoxin, if they are administered first, because they bind to the specific cholera receptors on the mucosal cells and block the binding of the complete toxin containing adenyl cyclase-activating activity.

Cholera enterotoxin is encoded by two genes, *ctxA* and *ctxB*. Expression of *ctxA* and *ctxB* is controlled by a positive regulatory element, a protein encoded by the *toxR* gene. The *toxR* gene product is a transmembrane protein that controls not only cholera toxin production but other important virulence factors, such as outer membrane proteins and pili required for successful colonization of *Vibrio cholerae* in the small intestine.

Other Enterotoxins

Several enterotoxins produced by enteropathogenic *Escherichia* and *Salmonella* have modes of action similar to that of cholera toxin, and antibody against cholera enterotoxin also inactivates these other enterotoxins, suggesting a similar structure. The sequence of the cholera toxin genes *ctxA* and *B* further supports this relationship, showing greater than 75% homology with the genes encoding the heat-labile enterotoxin produced by enteropathogenic *Escherichia coli*. However, enterotoxins produced by some of the food-poisoning bacteria (*Staphylococcus aureus*, *Clostridium perfringens*, *Bacillus cereus*) may be quite different in their modes of action. For example, *Clostridium perfringens* toxin is a cytotoxin and *Staphylococcus aureus* enterotoxin is a superantigen toxin (Table 21.4). Superantigens have a completely different

Figure 21.21 Action of cholera enterotoxin. (1) The normal process of ion movement in the intestine and colonization of the epithelium by *Vibrio cholerae* followed by binding of the enterotoxin by specific interaction with the GM1 ganglioside on host cells. (2) The A-B toxin acts by internalizing the toxic A component and activating adenyl cyclase, (3) leading to disruption of normal sodium (Na^+) in flux, (4) loss of H_2O into the lumen, and diarrhea. Treatment for cholera is by ion replacement and rehydration therapy. Antibiotic treatment may shorten the course of the disease by limiting *V. cholerae* growth, but has no effect on toxin that has already been produced.

mechanism of action, stimulating large numbers of immune lymphocytes and causing systemic as well as intestinal inflammatory responses (∞ Section 22.14).

✓ 21.11 Concept Check

Enterotoxins are exotoxins that act specifically on the small intestine, causing changes in intestinal permeability that lead to diarrhea. Enterotoxins include A-B toxins, cytotoxins, and superantigens.

- ✓ What key features are shared by all enterotoxins? By the A-B exotoxins?
- ✓ Describe the action of *Vibrio cholerae* toxin on the small intestine. Why does this lead to massive fluid loss?

21.12 Endotoxins

Gram-negative *Bacteria* produce lipopolysaccharides as part of the outer layer of their cell envelope (∞ Section 4.9), which under many conditions are toxic. These are called **endotoxins** because they are generally cell-bound and released in large amounts only when cells are lysed. Endotoxins have been studied primarily in *Escherichia*, *Shigella*, and especially *Salmonella*. The major differences between exotoxins and endotoxins are listed in Table 21.5.

Endotoxin Structure and Function

Endotoxins cause a variety of physiological effects. *Fever* is an almost universal symptom because endotoxin stimulates host cells to release proteins called *endogenous pyrogens*, which affect the temperature-controlling center of the brain. In addition, endotoxins can cause diarrhea, rapid decrease in lymphocyte, leukocyte, and platelet numbers, and generalized inflammation. Large doses of endotoxin can cause death, primarily through hemorrhagic shock and tissue necrosis. However, the toxicity of endotoxins is much *lower* than that of exotoxins. For instance, in the mouse the amount of endotoxin required to kill 50% of a population of test animals (the LD_{50}) is 200–400 μg per mouse, whereas the LD_{50} for botulinum toxin is about 25 picograms (pg) per mouse, about 10 million times less!

The overall structure of lipopolysaccharide (LPS) was diagrammed in Figure 4.35. Lipopolysaccharide consists of lipid A, a core polysaccharide consisting of ketodeoxyoctonate, seven-carbon sugars (heptoses), glucose, galactose, and *N*-acetylglucosamine, and the *O-polysaccharide*, a highly variable polymer that usually contains galactose, glucose, rhamnose, and mannose and one or more unusual dideoxy sugars such as abequose, colitose, paratose, or tyvelose. The sugars of the *O*-polysaccharide are connected in four- to five-sugar sequences (often branched), which then repeat to form the complete molecule (∞ Section 4.9). Lipid A is composed of fatty acids connected by ester linkages to *N*-acetylglucosamine. Fatty acids frequently found in the lipid include β-hydroxymyristic, lauric, myristic, and palmitic acids. Studies using fractions of the lipopolysaccharide indicate that the lipid fraction is responsible for toxicity, while the polysaccharide fraction makes the complex water soluble and immunogenic (∞ Section 22.3). Animal studies indicate that both the lipid and polysaccharide fractions are necessary for an *in vivo* toxic effect.

Limulus Assay for Endotoxin

Because endotoxins are pyrogens, pharmaceuticals such as antibiotics and intravenous solutions must be endotoxin-free. An endotoxin assay of very high sensitivity has been developed using lysates of amebocytes from the horseshoe crab, *Limulus polyphemus*. Endotoxin specifically causes lysis of amebocytes (Figure 21.22). In a laboratory test amebocyte extracts are mixed with the

TABLE 21.5 Properties of exotoxins and endotoxins		
Property	**Exotoxins**	**Endotoxins**
Chemical properties	Proteins, excreted by certain gram-positive or gram-negative *Bacteria*; generally heat-labile	Lipopolysaccharide–lipoprotein complexes (∞ Figure 4.35); released on cell lysis as part of the outer membrane of gram-negative *Bacteria*; extremely heat-stable
Mode of action; symptoms	Specific; usually binds to specific cell receptors or structures; either cytotoxin, enterotoxin, or neurotoxin with defined specific action on cells or tissues	General; fever, diarrhea, vomiting
Toxicity	Often highly toxic, sometimes fatal	Weakly toxic, rarely fatal
Immunogenicity	Highly immunogenic; stimulate the production of neutralizing antibody (antitoxin)	Relatively poor immunogen; immune response not sufficient to neutralize toxin
Toxoid potential	Treatment of toxin with formaldehyde will destroy toxicity, but treated toxin (toxoid) remains immunogenic	None
Fever potential	Do not produce fever in host	Pyrogenic, often induces fever in host

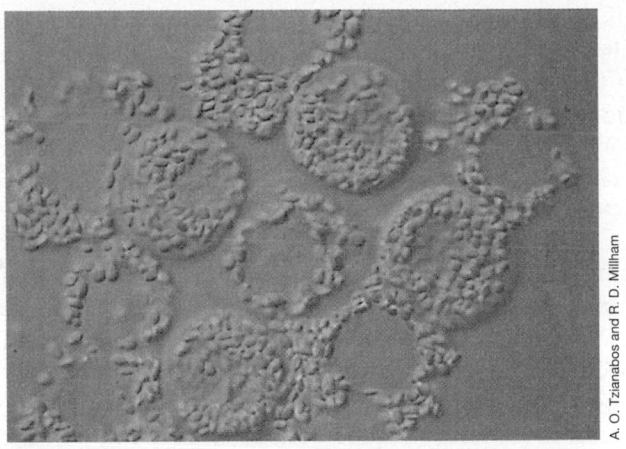

(a)

A. O. Tzianabos and R. D. Millham

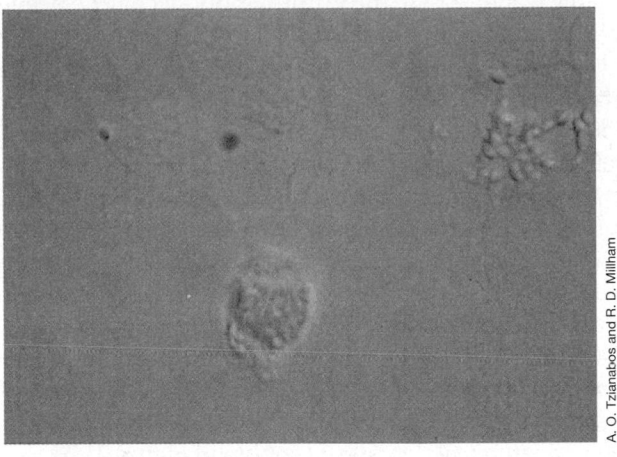

(b)

A. O. Tzianabos and R. D. Millham

Figure 21.22 Photomicrographs of *Limulus* amebocytes. (a) Normal amebocytes. (b) Amebocytes following exposure to bacterial lipopolysaccharide. Treatment with lipopolysaccharide causes degranulation of the cells, and this response can be used as an assay for lipopolysaccharide content.

solution to be tested. If endotoxin is present, the amebocyte extract gels and precipitates, causing a change in turbidity. This reaction can be measured quantitatively with a spectrophotometer. A measurable reaction can be obtained with as little as 10 pg/ml of lipopolysaccharide. The *Limulus* assay has been used to detect the presence of minute quantities of endotoxin in serum, cerebrospinal fluid, drinking water, and fluids used for injection and drug formulation.

The *Limulus* test is so sensitive that considerable care must be taken to avoid contamination of the equipment, solutions, and reagents with the gram-negative *Bacteria* in the laboratory and clinical environment, for example, as contaminants in distilled water. Detection of endotoxin by the *Limulus* assay in serum or cerebrospinal fluid is presumptive evidence of gram-negative infection of these body fluids.

Endotoxins are lipopolysaccharides derived from the outer membrane of gram-negative *Bacteria*. Released upon lysis of the *Bacteria*, endotoxins cause fever and other systemic toxic effects in the host. Endotoxins are generally less toxic than exotoxins.

✓ Why do gram-positive *Bacteria* not produce endotoxins?

✓ Why is it necessary to test drug preparations for endotoxin?

IV GENERAL HOST DEFENSE MECHANISMS

The innate "resistance factors" responsible for the suppression of pathogens can be divided into two categories: *nonspecific* host defenses directed against a broad variety of pathogens, and *specific* host factors directed against individual species or strains of pathogens. Here we consider some major nonspecific host defenses that are important for limiting infection and preserving the health of the host. In the following chapter, we consider the specific mechanisms that are important for host resistance to individual pathogens, the *immune response*.

21.13 Nonspecific Innate Resistance to Infection

The first line of host defense against pathogens involves a number of physical and chemical barriers innate to most animals that act nonspecifically to inhibit invasion by pathogens. These general mechanisms prevent almost all pathogens that we encounter from causing disease.

Natural Host Resistance

The ability of a particular pathogen to cause disease in an individual animal species is highly variable. In *rabies*, for instance, death usually occurs in all species of mammals once symptoms of the disease develop. Nevertheless, certain animal species are much more susceptible to rabies than others. Raccoons and skunks, for example, are extremely susceptible to rabies infection as compared with opossums, which rarely acquire the disease. *Anthrax* infects a variety of animals and causes disease symptoms varying from mild pustules in humans to a fatal blood poisoning in cattle. However, *pulmonary*, or airborne anthrax is fatal in humans (∞ Section 25.11). In contrast, birds are totally resistant to anthrax. In addition, diseases of warm-blooded animals are rarely transmitted to cold-blooded species, and vice versa.

Under certain circumstances, closely related species, or even members of the same species, may have different susceptibilities to a particular pathogen.

Age, Stress, and Diet

Age is an important factor for determining susceptibility to infectious disease. Infectious diseases are more common in the very young and in the very old. In the infant, for example, development of an intestinal microflora occurs quite quickly, but the normal flora of an infant is not the same as that of the adult. Before the development of an adult flora, and especially in the days immediately following birth, pathogens have a greater opportunity to become established and produce disease. Thus, diarrhea caused by pathogenic strains of *Escherichia coli* (Section 29.7) or *Pseudomonas aeruginosa* is frequently encountered in infants under the age of 1 year. Infant botulism is encountered only in very young infants because establishment of the intestinal normal flora in older children prevents intestinal infection with *Clostridium botulinum* (Section 29.5).

In individuals over 65 years of age, infectious diseases are much more common than in younger adults. For example, the elderly are much more susceptible to respiratory infections, particularly influenza (Section 26.8), probably because of a declining ability to make an effective immune response to respiratory pathogens. In addition, anatomical changes associated with age may also encourage infection. Enlargement of the prostate gland, a common condition in men over the age of 50, frequently leads to decreased urine flow, allowing pathogens to colonize the male urinary tract (Figure 21.11) more readily, leading to an increase in urinary tract infections in elderly men.

Stress can predispose a normally healthy individual to disease. In studies with rats and mice, fatigue, exertion, poor diet, dehydration, or drastic climatic changes, all sources of physiological stress, increase the incidence and severity of infectious diseases. For example, rats subjected to intense physical activity for long periods of time show a higher mortality rate from experimental *Salmonella* infections compared to well-rested control animals. Hormones that are produced under stress influence the immune system and may play a role in stress-mediated disease. The hormone *cortisone*, for example, is produced at much higher levels in times of stress than during normal periods, and this hormone is an effective anti-inflammatory agent. Suppression of inflammation removes one of the normal defenses against disease (see Section 21.14).

Diet plays a role in host resistance. The correlation between famine and infectious disease has been known for centuries. Diets low in protein may alter the composition of the normal flora, thus allowing opportunistic pathogens a better chance to multiply. For example,

the number of *Vibrio cholerae* necessary to produce cholera in an exposed individual is drastically reduced if the exposed individual is malnourished and is also drastically reduced when the *V. cholerae* is ingested in food, presumably because the food neutralizes the stomach acids that would normally destroy the pathogen (Section 28.5).

In some cases, lack of a particular dietary substance may *prevent* disease by depriving a pathogen of critical nutrients. The best example here is the effect sucrose has on the development of dental caries. As explained in Section 21.3, absence of sucrose from the diet (along with good oral hygiene) virtually eliminates tooth decay. Without a dietary source of sucrose, the highly cariogenic bacteria *Streptococcus mutans* and *S. sobrinus* are unable to synthesize the gummy polysaccharide needed to keep the bacterial cells attached to the teeth.

Physical and Chemical Defenses

The structural integrity of tissue surfaces poses a barrier to penetration by microorganisms. In the skin and mucosal tissues, potential pathogens must not only adhere to tissue surfaces but must also grow at these sites before traveling elsewhere in the body. Intact surfaces usually prevent colonization, but damaged surfaces (for example, abraded skin) are often readily colonized, promoting invasion. Resistance to colonization and invasion is due to the production of host defense substances and to various anatomical mechanisms that disrupt colonization. A summary of the major anatomical defenses is shown in Figure 21.23.

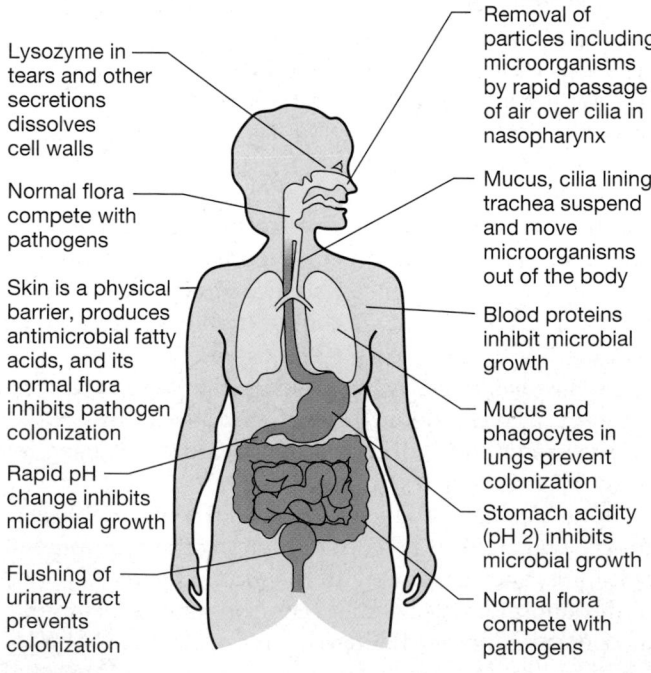

Lysozyme in tears and other secretions dissolves cell walls

Normal flora compete with pathogens

Skin is a physical barrier, produces antimicrobial fatty acids, and its normal flora inhibits pathogen colonization

Rapid pH change inhibits microbial growth

Flushing of urinary tract prevents colonization

Removal of particles including microorganisms by rapid passage of air over cilia in nasopharynx

Mucus, cilia lining trachea suspend and move microorganisms out of the body

Blood proteins inhibit microbial growth

Mucus and phagocytes in lungs prevent colonization

Stomach acidity (pH 2) inhibits microbial growth

Normal flora compete with pathogens

Figure 21.23 Physical, chemical, and anatomical barriers to infection.

The *skin* is an effective barrier to the penetration of microorganisms. *Sebaceous glands* in the skin (Figure 21.2) secrete fatty acids and lactic acid, which lower skin pH and inhibit colonization of pathogenic bacteria. Microorganisms inhaled through the *nose* or *mouth* are removed by the action of ciliated epithelial cells in the mucous surfaces of the nasopharynx and tracheal regions. Cilia push bacterial cells upward until they are caught in oral secretions and either are expectorated or are swallowed and killed in the stomach. Potential pathogens entering the stomach must survive the *acidity* (which is about pH 2) and then successfully compete with the increasingly abundant resident microflora present in the small intestine (which is about pH 5) and finally in the large intestine (pH 6–7). The large intestine contains bacterial numbers of about 10^{10} per gram of intestinal contents in a normal adult (see Section 21.4), making establishment of new microorganisms very difficult.

The kidney and the surface of the eye are constantly bathed with secretions containing *lysozyme*, an enzyme that markedly reduces microbial populations. Extracellular fluids such as blood plasma also contain bactericidal substances. For example, blood proteins called β-lysins act by binding and disrupting the bacterial cytoplasmic membrane, leading to leakage of cytoplasmic constituents and cell death.

Tissue Specificity

Most pathogens must first adhere and colonize at the site of exposure. If the site is not compatible with their nutritional and environmental needs, the organisms cannot multiply. Thus, if *Clostridium tetani* were ingested, it would not bring about tetanus because the pathogen is killed by the acidity of the stomach. If, on the other hand, *C. tetani* cells were introduced into a deep wound, the organism would grow and produce tetanus toxin in the anoxic zones created by localized tissue death (see Sections 21.10 and 27.8). By contrast, enteric bacteria such as *Salmonella* and *Shigella* do not cause wound infections but successfully colonize the intestinal tract.

Table 21.6 summarizes a number of examples of tissue specificity.

The Compromised Host

The term *compromised host* refers to hosts in which one or more resistance mechanisms are inactive and in which the probability of infection is therefore increased.

Many hospital patients with noninfectious diseases (for example, cancer and heart disease) acquire microbial infections because they are compromised hosts (∞ Section 25.7). Hospital procedures such as catheterization, hypodermic injection, spinal puncture, biopsy, and surgery may unintentionally introduce microorganisms into the patient. The stress of surgery may also lower patient resistance. Anti-inflammatory drugs given to reduce pain and swelling also reduce host resistance (see Section 21.13). Organ transplant patients are treated with immunosuppressive drugs. These drugs suppress rejection of the transplant but also lower the ability of the patient to resist infection.

Compromised hosts exist even outside the hospital. Smoking, excess consumption of alcohol, intravenous drug use, lack of sleep, poor nutrition, and current infection with another agent are conditions that compromise host resistance to infection. For example, infection with HIV (human immunodeficiency virus) predisposes a patient to a variety of infections from microorganisms that are not pathogens in uninfected individuals. HIV causes acquired immunodeficiency syndrome (AIDS) by destroying one type of immune cell, the CD4 T lymphocytes (∞ Section 22.7), involved in the immune response. Because of this preexisting host damage from HIV infection, AIDS patients are unable to mount effective resistance to infection; death is generally due to some infectious agent (∞ Sections 25.6 and 26.14).

Finally, certain genetic conditions compromise the host, such as genetic diseases that eliminate important parts of the immune system. Individuals with such conditions frequently die at an early age, not from the genetic condition itself but from microbial infection.

TABLE 21.6 Tissue specificity in infectious disease

Disease	Tissue infected	Organism
Acquired immunodeficiency syndrome (AIDS)	T helper lymphocytes	Human immunodeficiency virus (HIV)
Botulism	Motor end plate	*Clostridium botulinum*
Cholera	Small intestine epithelium	*Vibrio cholerae*
Dental caries	Oral epithelium	*Streptococcus mutans, S. sobrinus, S. sanguis, S. mitis*
Diphtheria	Throat epithelium	*Corynebacterium diphtheriae*
Gonorrhea	Urogenital epithelium	*Neisseria gonorrhoeae*
Malaria	Blood (erythrocytes)	*Plasmodium* sp.
Pyelonephritis	Kidney medulla	*Proteus* sp.
Spontaneous abortion (cattle)	Placenta	*Brucella abortus*
Tetanus	Inhibitory interneuron	*Clostridium tetani*

21.13 Concept Check

Nonspecific physical, anatomical, and chemical barriers prevent colonization of the host by most pathogens. Breakdown in these defenses results in a compromised host who is more susceptible to infection.

✓ How can diet and smoking influence host resistance to a pathogen?

✓ How might preexisting infection compromise an otherwise healthy host?

21.14 Inflammation and Fever

Inflammation is a nonspecific reaction to noxious stimuli such as toxins and pathogens. The characteristic inflammatory response results in redness, swelling, pain, and heat, localized at the site of infection. The mediators of inflammation include a group of proteins called *cytokines* (Section 23.10), which are produced by white blood cells or **leukocytes** (Section 22.1). Leukocytes are also involved in the pathogen-specific immune response, which we will discuss in Chapter 22. The most important outcome of the inflammatory response is the immediate localization of the pathogen, often by the production of a fibrin clot at the site of inflammation.

Inflammation is one of the most important and ubiquitous aspects of host defense against invading microorganisms. However, inflammation may also aid microbial pathogenesis because the inflammatory response elicited by an invading microorganism can result in considerable host damage, making nutrients available and providing access to host tissues.

Uncontrolled systemic inflammatory responses can also occur. Systemic inflammation is called *septic shock* and results in more generalized inflammatory symptoms such as severe edema (swelling) and uncontrollable fever. Septic shock is a life-threatening condition that occurs when infections and inflammatory responses are not contained at a local site and spread through the body via the circulatory or lymphatic systems (Section 22.1).

Fever

The healthy human body maintains a surprisingly constant temperature. Over an average 24-h period, body temperature varies over the narrow range of 1–1.5°C. Individuals vary in their "normal" temperatures, and although 37°C is considered normal, the actual normal temperature in some individuals may be as low as 36°C or as high as 38°C. Body temperature also varies considerably with physical activity and can be as much as 2°C below normal in sleep and as much as 4°C above normal during strenuous exercise.

Fever is an *abnormal* increase in body temperature. Although fever can be caused by noninfectious disease, most fevers are caused by infection. Fever occurs during many infections because certain products of pathogenic organisms are *pyrogenic* (fever-inducing). The best-studied pyrogenic agents are the endotoxins of gram-negative *Bacteria* (see Section 21.12). However, many organisms that do not produce endotoxins are also able to cause fever. These organisms cause the release proteins known as *endogenous pyrogens* from the leukocytes that destroy them (Section 22.2). Fevers may be beneficial to the host. Slight temperature increases benefit the host by accelerating phagocytic and antibody responses. However, strong fevers of 40°C (104°F) or greater usually benefit the pathogen because host tissues are further damaged.

Three fever patterns have been recognized in infectious disease. (1) *Continuous fever* is that condition in which the body temperature remains elevated over a whole 24-h period, and the total range of variation in temperature is less than 1°C. Continuous fever is seen in *typhoid fever* (Section 28.8) and *typhus fever* (Section 27.3). (2) A *remittent fever* is one in which the body temperature is abnormal over the whole of a 24-h period, and the daily range shows swings greater than 1°C. This occurs in some *pyogenic infections* (Sections 26.2 and 26.9) and in *tuberculosis* (Section 26.5). (3) An *intermittent fever* is one in which the temperature is normal for part of the day and then rises above normal. Most infectious diseases elicit some intermittent fever, and the condition is a diagnostic characteristic of malaria (Section 27.5), a protozoan infection. *Relapsing fever*, caused by various *Borrelia* species (Section 27.3) is an intermittent fever in which the temperature remains normal for a long period of time, followed by a new attack of fever. This is characteristic of an incomplete recovery from an infectious disease, the fever arising when the infection periodically reestablishes itself.

✓ **21.14 Concept Check**

Inflammation and fever are nonspecific responses to noxious stimuli such as pathogens. These host responses are designed to result in accelerated isolation and destruction of the pathogen but can lead to further damage to host tissue.

✓ Describe the chief symptoms of inflammation.

✓ Describe the three types of fever.

Review Questions

1. Distinguish between a *parasite* and a *pathogen*. Distinguish between infection and disease.

2. Which organs of the human body are normally colonized by microorganisms? What do these organs have in common? Which organs are normally devoid of microorganisms? What do these organs have in common?

3. Distinguish between *resident* and *transient* microorganisms at a body site. How could you distinguish between resident and transient microorganisms experimentally?

4. Why are members of the genus *Streptococcus* instrumental in forming dental caries? Why are they more capable of causing caries than other organisms?

5. How does pH affect the types of microorganisms that grow in each definable region of the gastrointestinal tract? How does oxygen concentration affect the types of microorganisms that grow in each definable region of the gastrointestinal tract?

6. Describe the relationship between *Lactobacillus acidophilus* and glycogen in the vaginal tract. Why do adult females have a different local flora than juvenile females in the vaginal tract?

7. Distinguish between glycocalyx, the capsule, and the slime layer. How do these unique structures contribute to microbial adherence?

8. Give an example of a microorganism that is pathogenic almost solely because of its toxin-producing ability. Define the toxin and its mode of action. Give an example of a microorganism that is pathogenic almost solely because of its invasive characteristics. What factor or factors confer invasive qualities on this microorganism?

9. Distinguish between A-B toxins, cytotoxins, and superantigens. Give an example of each category of toxin. How does each toxin category promote disease?

10. Define the mechanisms of action for tetanus toxin and botulinum toxin. With relation to their mechanism of action, why are these toxins so dangerous?

11. Review the mode of action of cholera enterotoxin. What is the appropriate therapy for this disease, and why is antibiotic treatment generally not effective? Review the mode of action of staphylococcal enterotoxin. What is the appropriate therapy for this disease, and why is antibiotic treatment generally not effective?

12. Describe the structure of a typical endotoxin. How does endotoxin induce fever? What microorganisms produce endotoxin?

13. How do temperature and pH work to limit bacterial infections? Where in or on the body might you find temperature and pH values that are different from standard body conditions? What organisms might benefit from differences in temperature or pH?

14. Distinguish between a continuous fever, a remittent fever, and an intermittent fever. Name at least one infectious agent that causes each type of fever. Which type most commonly occurs in infectious diseases?

Application Questions

1. How does mucus inhibit the growth of most microorganisms? Describe experiments to demonstrate the effects of mucus in protection against bacterial colonization.

2. Mucous membranes are effective barriers against colonization and growth of microorganisms. However, mucous membranes, for example in the throat, are colonized with a variety of different microorganisms that occasionally cause disease. Explain how normally nonpathogenic microorganisms can become pathogenic under certain circumstances. Be sure to describe at least one set of circumstances that might encourage pathogenicity.

3. What steps are involved in the formation of dental plaque? Describe and discuss experiments that demonstrate the buildup of plaque on toothlike surfaces and design experiments to illustrate biological methods for removal of plaque.

4. Would an antimicrobial dentifrice (toothpaste) be useful for preventing caries? What would be the advantages or disadvantages of such a caries-prevention strategy? How could you test the validity of this strategy in the laboratory?

5. Antibiotic therapy can significantly reduce the number of microorganisms residing in the gastrointestinal tract. What physiological symptoms might the reduction of normal flora produce in the host? Long-term antimicrobial therapy is often followed by infections due to opportunistic pathogens, many of which are caused by organisms causing opportunistic infections in individuals with AIDS (◁◯▷ Section 26.14). What pathogens might be involved? Why are individuals who have undergone antibiotic therapy susceptible to these pathogens?

6. Describe how enteropathogenic strains of *Escherichia coli* differ from nonpathogenic strains of *E. coli*. Include a discussion of structural and ecological variables.

7. Name at least three host factors that limit or accelerate colonization and growth of a microorganism at a local site. Incorporate each of these host factors into an *in vitro* experiment that tests the validity of your assumptions.

8. Identify potential selection pressures that might result in an increase in the virulence and pathogenicity of *Streptococcus pneumoniae*. Identify potential selection pressures that might result in an increase in the virulence and pathogenicity of *Clostridium tetani*. Would an increase in virulence by either of these organisms result in selective advantages for the organism? For each organism, be sure to consider the natural habitat.

9. Coagulase is a virulence factor for *Staphylococcus aureus* that acts by causing clot formation at the site of *S. aureus* growth. Streptokinase is a virulence factor for *Streptococcus pyogenes* that acts by dissolving clots at the site of *S. pyogenes* growth. Reconcile these opposing strategies for enhancing pathogenicity.

10. *Salmonella* species produce *at least* 10 different gene products that act as virulence factors and increase the virulence of the pathogen. *Streptococcus pneumo-niae*, arguably a more virulent pathogen (see Figure 21.15), relies on a single virulence factor. Review the critical virulence factors for each organism and explain why a single factor is so important and effective for *S. pneumoniae*. Why haven't all human pathogens evolved to use *S. pneumoniae*-like virulence strategies?

11. Vaccines to exotoxins such as diphtheria toxin and tetanus toxin are used to prevent disease (⌀ Section 22.11). Why are vaccines not developed for endotoxins?

12. Although mutants incapable of producing exotoxins are relatively easy to isolate, mutants incapable of producing endotoxins are much harder to isolate. From what you know of the structure and function of these types of toxins, explain the differences in mutant recovery.

13. Burns are some of the most difficult wounds to treat, partly because burns are easily infected, often with opportunistic pathogens. Explain why burns are easily infected based on your understanding of the physical, anatomical, and chemical processes that normally prevent infections.

14. Should fever always be treated? Give reasons for your answer based on the importance of the inflammatory host response in limiting infection.

The immune system is organized to prevent the infection and destruction of our bodies by individual microorganisms. Immunity starts with the recognition of dangerous microorganisms by a group of cells called *phagocytes*, shown here as blue-stained cells among the more numerous red blood cells. Phagocytes engulf and destroy pathogens, and then display on their own cell surfaces pieces of the pathogen to other immune-system cells called *lymphocytes*. The recruited lymphocytes trigger a long-lasting immune response that involves a number of different immune cells and recognition molecules, that together neutralize every succeeding challenge by that individual pathogen.

ESSENTIALS OF IMMUNOLOGY

22

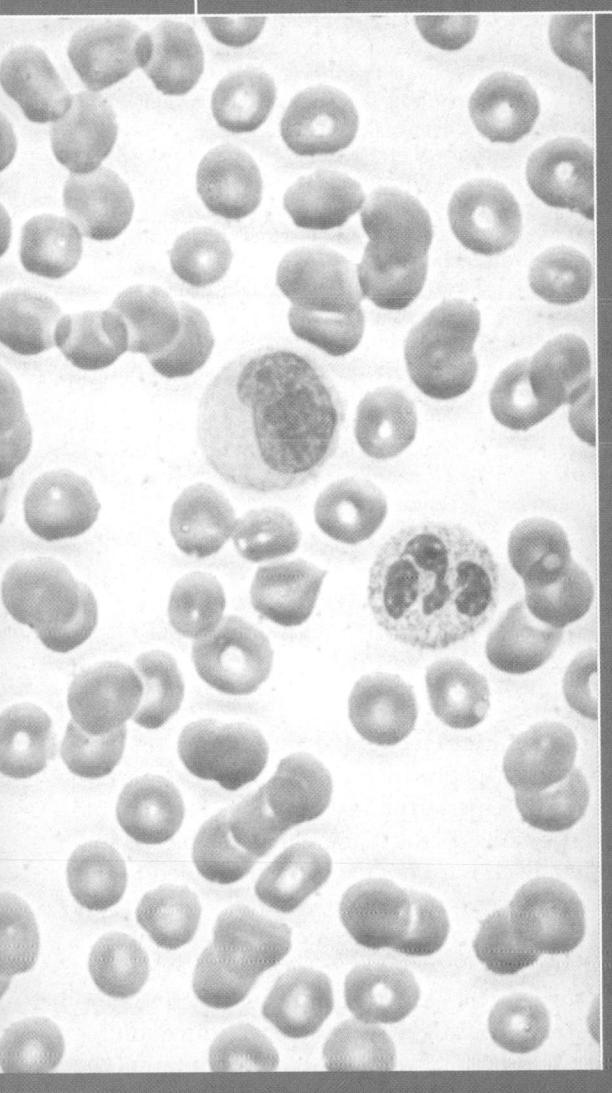

Working Glossary

Antibody a soluble protein, produced by B cells, that interacts with antigen; also called immunoglobulin

Antibody-mediated immunity (humoral immunity) immunity resulting from direct interaction with antibodies

Antigen a molecule capable of interacting with specific components of the immune system

Antigenic determinant that portion of an antigen that is reactive with a specific antibody or T-cell receptor; also called an epitope

Antigen-presenting cell (APC) any cell that functions primarily to present antigen to a T cell

Autoantibody an antibody that reacts to self antigens

B cell a lymphocyte that produces immunoglobulin

Cell-mediated immunity (CMI) immunity resulting from direct interaction with antigen-specific T cells

Class I MHC protein antigen-presenting molecule found on all nucleated vertebrate cells

Class II MHC protein antigen-presenting molecule found primarily on macrophages, B cells, and dendrocytes

Complement a series of proteins that react in a sequential manner with antibody-antigen complexes to amplify or potentiate their activity

Cytokine a soluble immune response modulator produced by leukocytes

Domain a region of a protein having a defined structure and function

Hapten a low-molecular-weight substance that combines with specific antibodies but that is incapable of eliciting an immune response by itself

Hypersensitivity an immune response leading to damage to host tissues sometimes referred to as allergies

Immunity the ability of an organism to resist infection

Immunization (vaccination) inoculation of a host with inactive or weakened pathogens or pathogen products to stimulate protective immunity

Immunogen a molecule capable of eliciting an immune response

Immunoglobulin (Ig) a soluble protein, produced by B cells, that interacts with antigens; also called antibody

Immunologic memory ability to rapidly produce large quantities of specific immune cells or antibodies after subsequent exposure to a previously encountered antigen

Interleukin (IL) soluble cytokine mediator secreted by leukocytes

Leukocytes nucleated cells found in the blood (white blood cells)

Lymph a fluid similar to blood but which lacks red blood cells and travels through a separate circulatory system (the lymphatic system) containing lymph nodes, which filter particulate materials such as bacterial cells

Lymphocytes a subset of nucleated cells found in the blood that are involved in the immune response

Macrophage a type of large leukocyte that has phagocytic properties

Major histocompatibility complex (MHC) a genetic complex responsible for encoding several cell surface proteins important in antigen presentation

Natural killer (NK) cell a specialized lymphocyte that recognizes and destroys foreign cells or infected host cells in a nonspecific manner

Phagocyte one of a group of cells that ingests and degrades pathogens and pathogen products

Plasma the liquid portion of the blood with cells removed and clotting proteins deactivated

Polymorphonuclear leukocyte (PMN) (neutrophil) a type of leukocyte exhibiting phagocytic properties, a granular cytoplasm, and a multilobed nucleus

Primary antibody response antibodies made on first exposure to antigen; mostly of the class IgM

Secondary antibody response antibody made on second (or any subsequent) exposure to antigen; mostly of the class IgG

Serology the study of antigen-antibody reactions *in vitro*

Serum the liquid portion of the blood with clotting proteins and cells removed

Specificity the ability of the immune response to interact with individual antigens

T cell a lymphocyte responsible for antigen-specific cellular interactions; T cells are divided into functional subsets including T_C cytotoxic T cells and T_H helper T cells. T_H cells are further subdivided into T_H1 inflammatory cells and T_H2 helper cells, which aid B cells in antibody formation

T-cell receptor (TCR) antigen-specific receptor protein on the surface of T cells

Tolerance inability to produce an immune response to specific antigens

---

Here we introduce **immunity**, the ability of higher organisms to resist infection. Immunity involves the interactions of a variety of cells and their products to defend against invasion and infection by pathogens. In Chapter 21, we discussed the physical and chemical processes that provide *nonspecific immunity*, the body's innate ability to resist infection. Nonspecific immunity is also a function of *phagocytes*, cells that engulf, digest, and destroy most pathogens. Unfortunately, phagocytes and other nonspecific defenses are not completely effective, and infections can sometimes still occur.

The phagocytes, however, activate another defense mechanism, that of *specific immunity*. Specific immunity is the acquired ability to recognize and destroy an individual pathogen or its products. Phagocytes use partially digested proteins from pathogens to activate *lymphocytes*, specialized cells that are responsible for specific immunity. Each lymphocyte is programmed to recognize a single protein, called an *antigen*, on the pathogen. When the lymphocyte recognizes its antigen, it grows and divides very rapidly, forming exact copies of itself, or *clones*. The clones of antigen-reactive lymphocytes destroy the antigen directly or produce soluble antigen-specific proteins called *antibodies* that target the antigens. Some of the antigen-reactive lymphocytes live for years. If we are again exposed to the same pathogen a second time, the lymphocytes expand rapidly and produce an immediate and vigorous immune response. The ability of the immune response to react more strongly to subsequent antigen exposures is called immune *memory*. While the immune system reacts very strongly with antigens from pathogens, it has a fail-safe mechanism that prevents destruction of proteins that are found on our own cells. The ability of the immune system to destroy nonself pathogen proteins while ignoring self proteins is called *tolerance*. Unfortunately, the tolerance mechanism occasionally fails, producing a variety of diseases.

In summary, the nonspecific phagocytes present antigens and activate lymphocytes, inducing a specific immune response. The activated immune lymphocytes grow to form clones that recognize individual antigens (specificity), respond vigorously when re-exposed to antigen (memory), and do not harm host cells (tolerance). Specific immunity has evolved to protect us from individual pathogens; without specific immunity, we cannot survive.

OVERVIEW OF THE IMMUNE SYSTEM

An immune response begins with recognition of a pathogen and culminates in destruction of the pathogen. Here we introduce some of the important features of nonspecific and specific host responses to pathogens. Nonspecific responses recognize all pathogens, while specific responses are directed at individual pathogens. We begin with the cells and organs of the immune system and then consider the cells and mechanisms involved in nonspecific immunity. We finish with an overview of specific, or adaptive, immunity, the focus of the rest of the chapter.

22.1 Cells and Organs of the Immune System

Specific and nonspecific immunity result from the actions of cells that circulate in the *blood* and *lymph*, body fluids that directly or indirectly interact with every major organ system. All of the cells involved in immunity share common precursors, or stem cells.

Blood and Lymph Components

Blood consists of cellular and noncellular components and contains many cells and molecules involved in the immune response. Because blood can be easily and safely obtained from patients, it is a valuable source of material for immune response assays. The most numerous cells in human blood are *erythrocytes* (red blood cells), which are nonnucleated cells that function to carry oxygen from the lungs to the tissues (Table 22.1). However, about 0.1% of the cells in blood are white blood cells, or

TABLE 22.1 Major cells found in normal human blood

Cell type	Cells per milliliter
Erythrocytes	$4.2–6.2 \times 10^9$
Leukocytes	$4.5–11 \times 10^6$
Lymphocytes	$1.0–4.8 \times 10^6$
Monocytes	Up to 8.0×10^5

Source: Henry, J. B. 1996. *Clinical Diagnosis and Management by Laboratory Methods*, 19th edition. W. B. Saunders, Philadelphia.

leukocytes. Leukocytes include a variety of phagocytic cells such as *monocytes*, as well as cells called *lymphocytes*, which are involved in antibody production and cell-mediated immunity. **Lymph** is a fluid similar to blood, but which lacks red blood cells.

As shown in Figure 22.1, common stem cells in the bone marrow are the progenitors of all the blood and lymph cells. Stem cells differentiate to produce mature cells under the influence of a group of soluble cell proteins known as **cytokines** (∞ Section 23.10).

When cells are removed from blood, the remaining fluid is called **plasma**. An important component of plasma is the protein fibrinogen, which undergoes a complex set of reactions during the formation of a blood clot. Clotting can be prevented by the addition of an anticoagulant such as potassium oxalate, potassium citrate, or heparin. Plasma is stable only when such an anticoagulant is added. If no anticoagulant is added, whole blood or plasma quickly forms a clot. The fluid left after clotting is called *serum*. Since serum contains a high concentration of antibody proteins, it is widely used in immunological investigations (∞ Section 24.7).

Blood and Lymph Circulation

Blood is pumped by the heart through a network of arteries and capillaries to various parts of the body and is returned through the veins (Figure 22.2*a*, *b*). Figure 22.2*b* and *c* shows the capillary beds, a site where leukocytes may pass to and from the blood into the *lymphatic system*, a separate circulatory system through which lymph circulates.

Lymph drains from extravascular tissues into lymphatic capillaries and then into **lymph nodes** (Figure 22.2*d*) found at various locations throughout the lymph system. Lymph nodes contain high concentrations of leukocytes, arranged in such a way that they filter out microorganisms and antigens. The spleen serves an analogous function in the blood circulatory system. Lymph nodes may become infected by organisms collected by the filtering mechanisms. The lymph nodes and the spleen are the sites of most immune responses. Lymph, carrying antibodies and immune cells, eventually flows back into the circulatory system via the thoracic lymph duct.

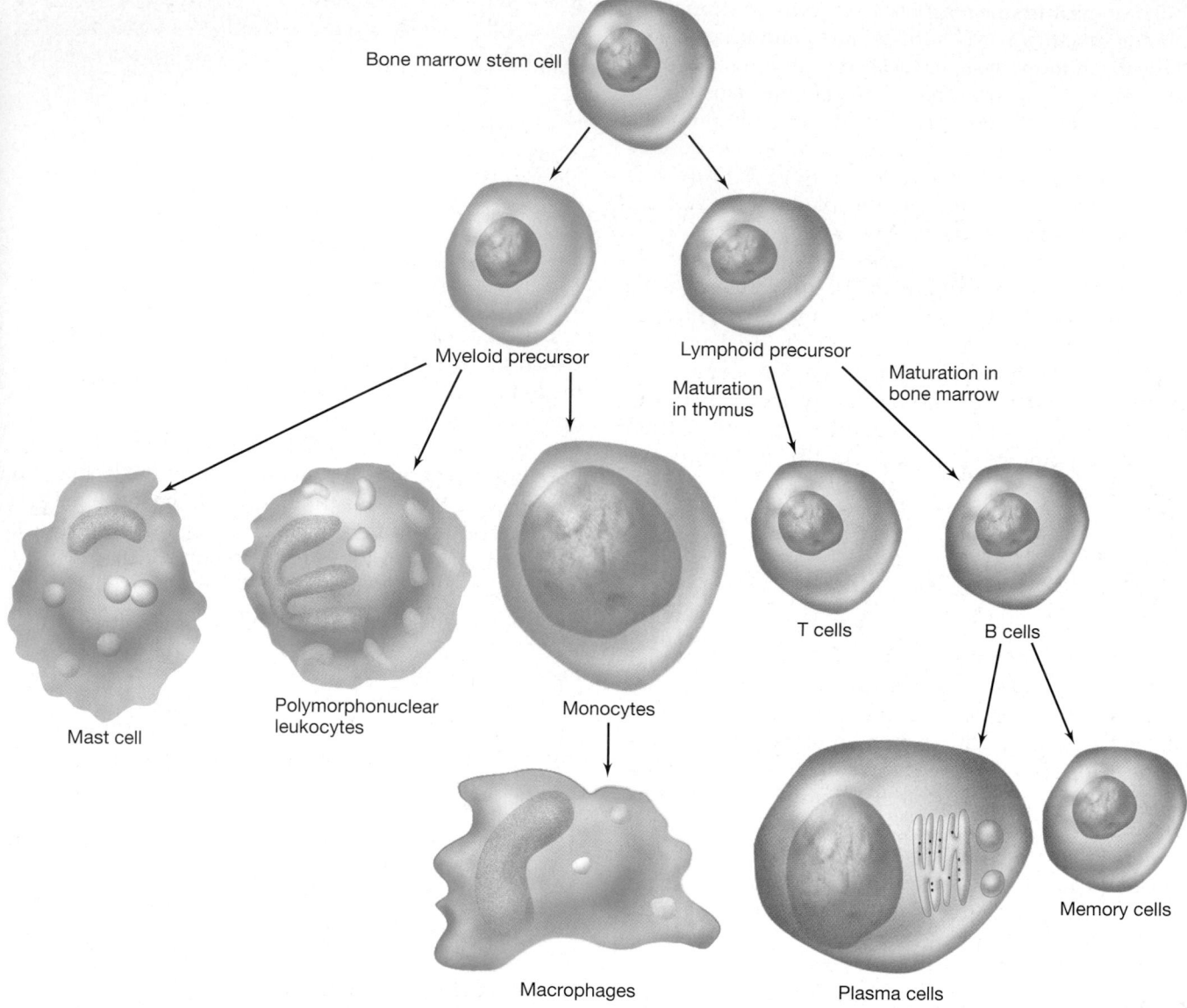

Bone marrow stem cell

Myeloid precursor

Lymphoid precursor

Maturation in thymus

Maturation in bone marrow

Mast cell

Polymorphonuclear leukocytes

Monocytes

T cells

B cells

Macrophages

Plasma cells

Memory cells

Figure 22.1 Origin of major cells involved in the immune response. The end cells produced by two major precursor lines, one generating phagocytic cells (myeloid precursors) and the other generating lymphocytic cells (lymphoid precursors), participate directly in immune responses.

Leukocytes

Leukocytes are nucleated white blood cells found in the blood and the lymph. There are several distinct kinds of leukocytes (Table 22.1), but all participate in nonspecific or specific immune functions. Specialized white blood cells called **macrophages** are found in abundance in tissues and lymph nodes and carry out the lymph filtering action, as will be described later (see Section 22.2). **Lymphocytes** are specialized leukocytes involved in the specific immune response. Mature lymphocytes are concentrated in the lymph nodes and spleen and interact there with antigens. The lymphocytes are further divided into **B cells** and **T cells** (Figure 22.1). *B cells* originate and mature in the bone marrow and are the precursors of the antibody-producing plasma cells. *T cells* originate in the bone mar-

row, but travel to the thymus to mature. Lymphocytes and other leukocytes can travel throughout the body and pass freely from blood to interstitial spaces to lymph and back, a process called *extravasation* (Figure 22.2).

✓ 22.1 Concept Check

All the cells involved in immunity originate from common stem cells in the bone marrow. The blood and lymph systems circulate cells and proteins that are important for a functional immune system. A variety of leukocytes participate in immune responses.

✓ Describe the circulation of a leukocyte from the blood to the lymph and back to the blood.

✓ Trace the development of B cells, T cells, and macrophages from the common stem cell.

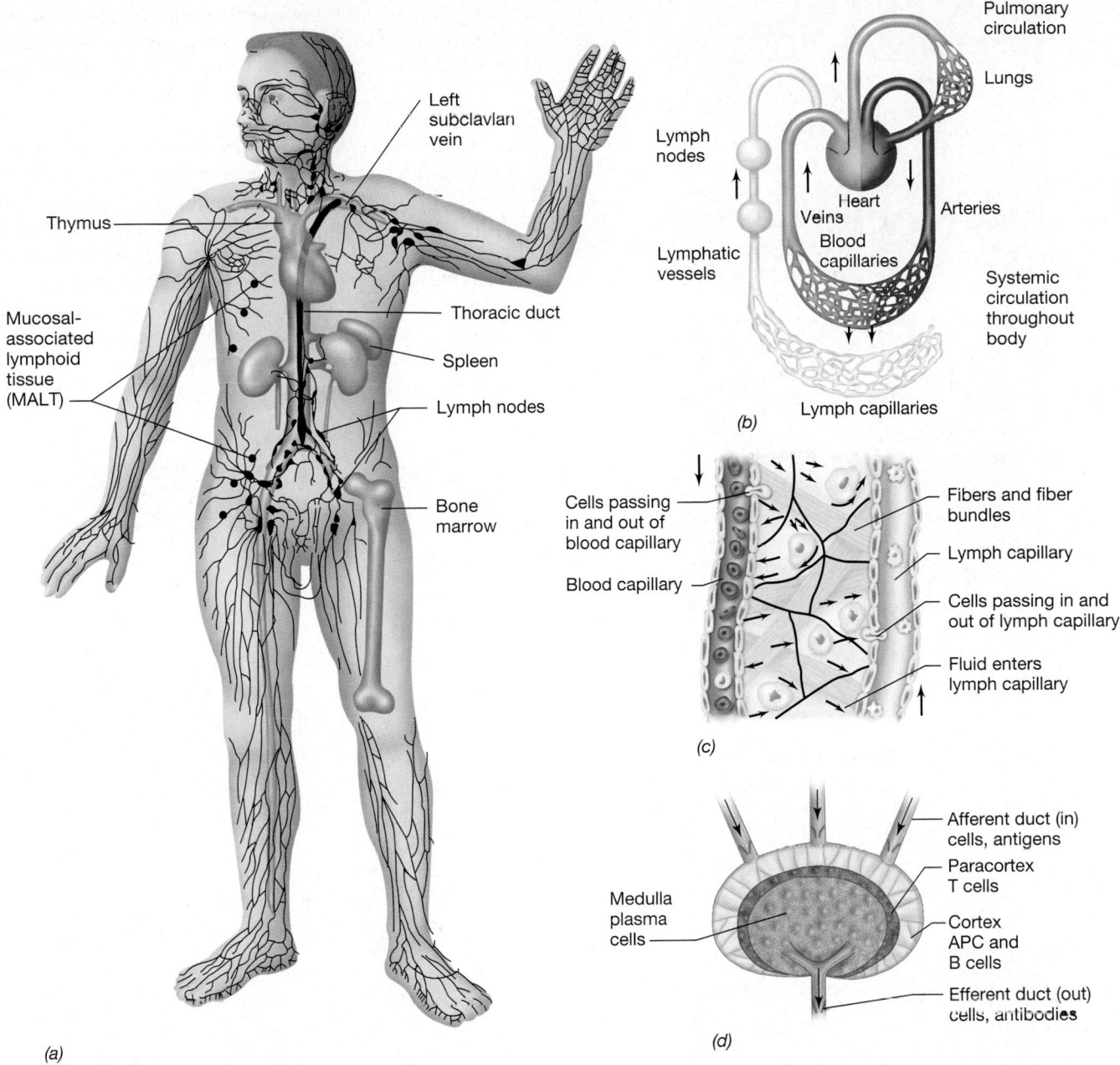

Figure 22.2 The blood and lymph systems. (a) Overall view of the lymph system, showing the locations of major organs. (b) Communication between the lymph and blood systems. Blood flows from the veins to the heart, then to the lungs where it becomes oxygenated, and then through the arteries to the tissues. (c) Connection between the blood and lymph systems is shown microscopically. Both blood and lymph capillaries are closed vessels, but cells and fluids can pass from one vessel to another by a process known as extravasation. (d) A lymph node. The diagram depicts major anatomical areas and the immune cells present.

22.2 Nonspecific Immunity

On rare occasions, pathogens break through the host physical and chemical defense mechanisms described in Chapter 21 (∞ Section 21.13). The pathogen can then infect the host (∞ Section 21.7). At this point, the immune system must be mobilized.

The starting point for immunity, whether the final effect is specific or nonspecific, cellular or humoral, is contact of a cell with the pathogen or a pathogen product such as a toxin. The cell type involved in this initial contact is a *phagocyte* (literally, a cell that eats). The primary function of a phagocyte is to engulf and destroy pathogens. In this process, some phagocytes act as *antigen-presenting cells (APCs)* and generate the peptide

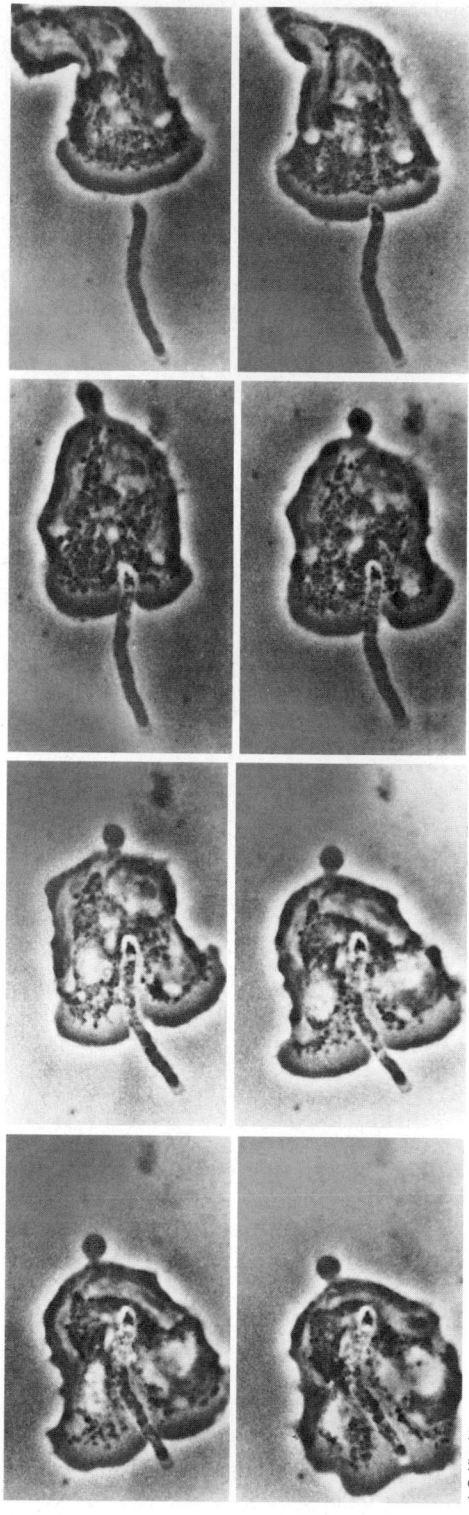

antigens that activate the specific immune response. In this section, we will examine some of the important phagocytic cells and their ability to neutralize pathogens.

Phagocytes

Some of the leukocytes found in blood are phagocytes, and phagocytes are also found in various tissues and fluids of the body. Phagocytes are usually motile and move by ameboid action. Most have granular inclusions called *lysosomes*, which contain bactericidal substances such as hydrogen peroxide, lysozyme, proteases, phosphatases, nucleases, and lipases. Phagocytes can trap a pathogen on a surface such as a blood vessel wall or a fibrin clot. The phagocyte's cytoplasmic membrane then engulfs the pathogen. The entire complex pinches off

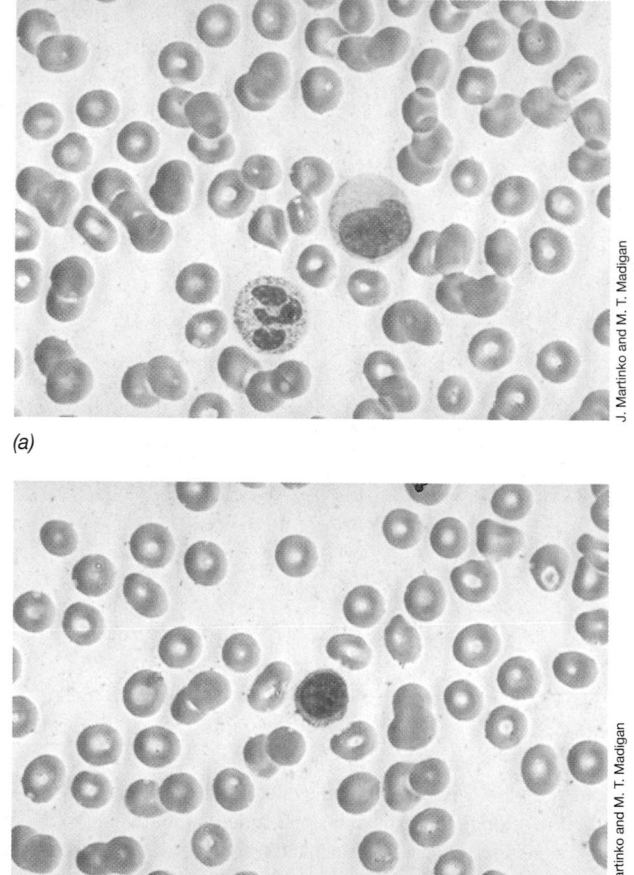

Figure 22.3 Phagocytosis: time-lapse photomicrographs of the engulfment and digestion of a chain of *Bacillus megaterium* cells by a human macrophage, observed by phase contrast microscopy. The bacterial chain is about 18–20 μm long. The macrophage is one of a group of cells that ingests and degrades pathogens and pathogen products.

Figure 22.4 Major immune cell types. (a) Phagocytic cells. The nucleated cell in the lower left center is a neutrophil (PMN), characterized by a segmented nucleus (violet stain) and granular cytoplasm. The nucleated cell to the right and slightly above the PMN is a monocyte. Phagocytes are 12–12 μm in diameter. The nonnucleated red blood cells are about 6 μm in diameter. (b) The nucleated cell is a circulating lymphocyte. The lymphocyte has almost no visible cytoplasm and is smaller than the phagocytes, about 10 μm in diameter.

and eventually fuses with the lysosomes, forming a new inclusion, a *phagolysosome*. The toxic substances and enzymes inside the phagolysosome usually kill and digest the engulfed microorganism (Figure 22.3).

One group of phagocytes, the **neutrophils**, or **polymorphonuclear leukocytes** (sometimes abbreviated PMNs), are actively motile cells containing large numbers of lysosomes (Figure 22.4a). PMNs are found predominantly in the bloodstream and bone marrow, but also can migrate to sites of active infection in tissues. They are attracted to bacteria and bacterial components. Large numbers of PMNs in the blood or at a site of inflammation usually indicate an active infection.

Macrophages and **monocytes** are the other major groups of phagocytic cells. Macrophages are large cells capable of ingesting and destroying most pathogens and antigens as well as cooperating with lymphocytes in the production of specific immunity. Monocytes are circulating cells that differentiate to become macrophages (Figures 22.4a and 22.3). Thus, the *macrophage* is a phagocyte that is fixed to tissue surfaces, and the term *monocyte* describes the circulating precursor. Macrophages are larger than monocytes and are abundant in lymphoid tissue and spleen, whereas monocytes circulate in the blood and lymph. Macrophages are important APCs that present peptide antigens to T cells, the first step in activating a specific immune response. This specialized feature of macrophages makes them a very important component of antigen-specific immunity, and we will examine their role as APCs in more detail in Section 22.5.

Ingestion of a pathogen stimulates the phagocytes to become more efficient, enhancing their ability to engulf and destroy pathogens.

Oxygen-Dependent Phagocytic Killing

As discussed in Section 6.13, various biochemical reactions can lead to the formation of toxic oxygen-containing compounds including hydrogen peroxide (H_2O_2), superoxide anions (O_2^-), hydroxyl radicals (OH •), singlet oxygen ($^1O_2$), hypochlorous acid (HOCl), and nitric oxide (NO). The acidic conditions found in the phagolysosome aid in the generation and reactivity of the toxic oxygen compounds. Phagocytic cells make use of toxic forms of oxygen to kill ingested bacterial cells. The combined action of these oxygen-dependent phagocyte enzymes forms sufficient levels of toxic oxygen compounds to kill ingested bacterial cells by oxidizing key cellular constituents. These reactions occur within the phagocytic cell itself, which is not damaged by the toxic oxygen products. The action of phagocytic cells in oxygen-mediated killing is summarized in Figure 22.5.

Inhibiting Phagocytes

In some cases, pathogens have developed mechanisms for neutralizing the effects of toxic phagocyte products, for killing the phagocyte, or for avoiding phagocytosis. For example, *Staphylococcus aureus* (oo Section 26.9). produces pigmented compounds called *carotenoids*, which quench singlet oxygen and prevent killing (oo Section 6.13). Intracellular pathogens such as *Mycobacterium tuberculosis* (tuberculosis bacillus) grow and persist within phagocytic cells (oo Section 26.5). They apparently use cell wall glycolipids (oo Section 12.23) to scavenge toxic oxygen compounds. These glycolipids remove hydroxyl radicals and superoxide anions, the most lethal toxic oxygen species produced by phagocytic cells.

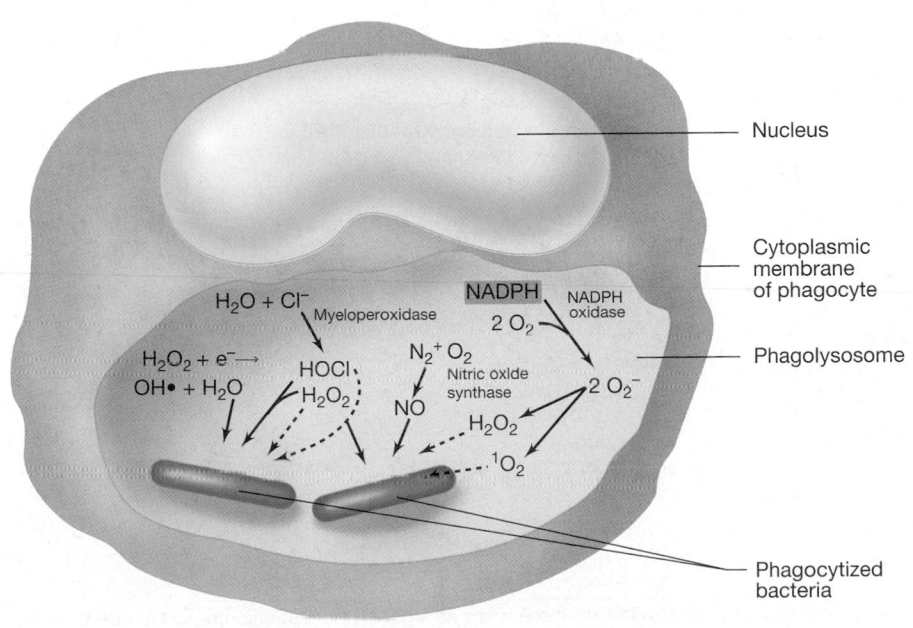

Figure 22.5 Action of phagocyte enzymes in generating toxic oxygen species. These include hydrogen peroxide (H_2O_2), the hydroxyl radical (OH·), hypochlorous acid (HOCl), the superoxide anion (O_2^-), singlet oxygen ($^1O_2$), and nitric oxide (NO). Formation of these toxic compounds requires a substantial increase in the uptake and utilization of molecular oxygen, O_2. This increase in oxygen uptake and consumption by activated phagocytes is known as the *respiratory burst*. For more discussion of toxic oxygen species, see Section 6.13.

Nucleus

Cytoplasmic membrane of phagocyte

Phagolysosome

Phagocytized bacteria

Some intracellular pathogens produce proteins called *leukocidins*, which destroy phagocytes. In such cases, the pathogen is ingested normally, but kills the phagocyte and is then released. *Streptococcus pyogenes* and *Staphylococcus aureus* are the major leukocidin producers. Dead phagocytes make up much of the material of *pus*; organisms that produce leukocidins are therefore usually *pyogenic* (pus-forming) and cause localized infections resulting in boils or abscesses (∞ Sections 21.7 and 26.9).

Another important microbial defense against phagocytosis is the bacterial capsule (∞ Section 4.13). Capsulated bacteria are often highly resistant to phagocytosis, apparently because the capsule prevents ad-

herence of the phagocyte to the bacterial cell. The clearest case of the importance of a capsule that prevents phagocytosis is that of *Streptococcus pneumoniae*. Fewer than 10 cells of a capsulated strain of *S. pneumoniae*, when injected, can kill a mouse in a few days. On the other hand, noncapsulated mutant strains are completely avirulent (∞ Figure 21.15). Surface components other than capsules can also inhibit phagocytosis. For instance, pathogenic *Streptococcus pyogenes* produces a specific substance, the M-protein (∞ Section 26.2). M-protein alters the surface of the bacterial cell and inhibits phagocytosis.

Antibodies to capsules or other cell surface molecules often reverse the protective effect of these bacteri-

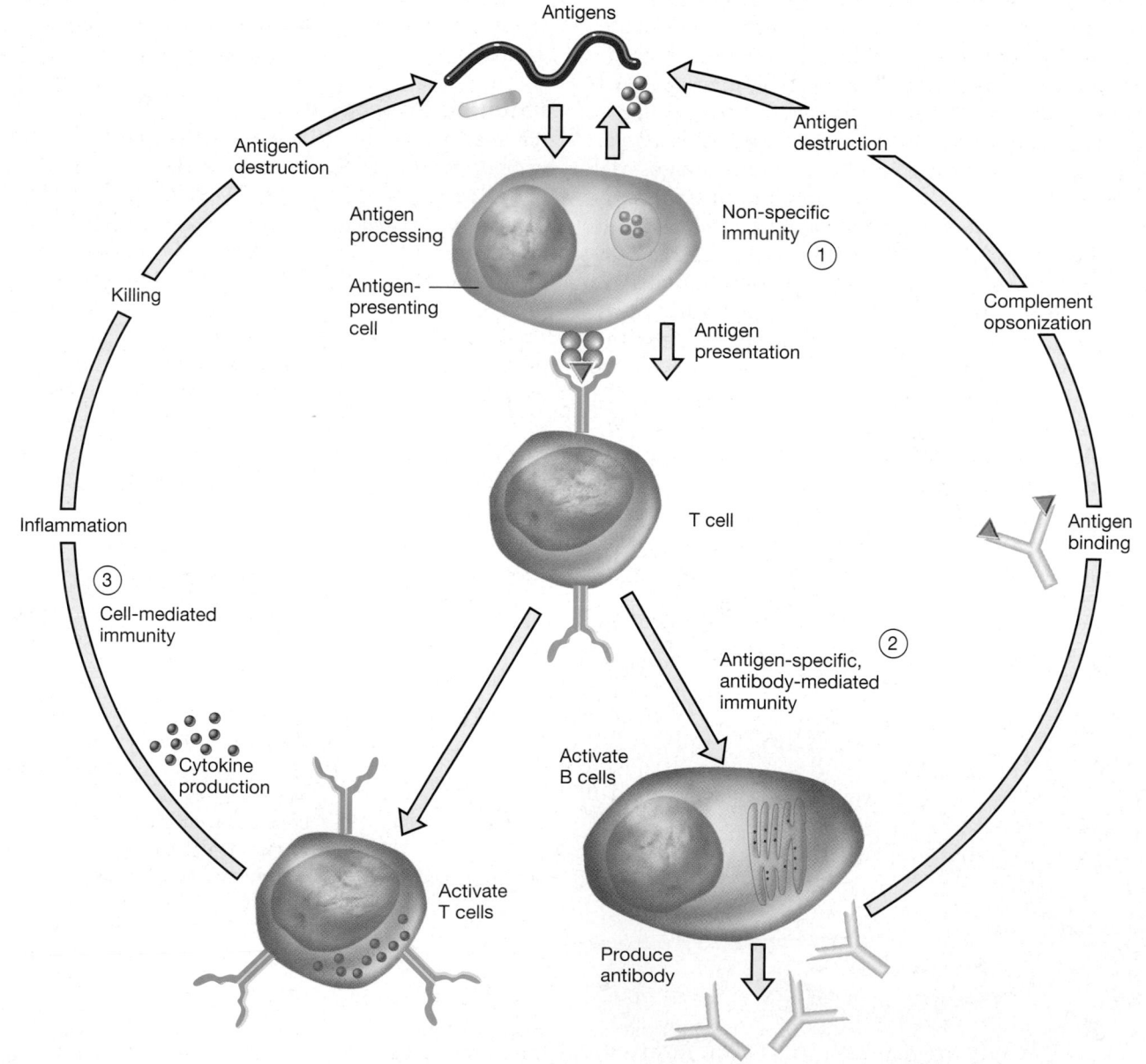

Figure 22.6 Overview of the immune response. Pathogens are destroyed by three mechanisms: (1) nonspecific immunity; (2) antibody-mediated immunity; and (3) cell-mediated immunity.

al defense mechanisms and enhance phagocytosis, a process known as *opsonization* (see Section 22.10).

22.3 The Specific Immune Response

The overall processes of the immune response are outlined in Figure 22.6. Phagocytes such as macrophages digest pathogens and present peptide antigen to lymphocytes called **T cells**. T cells recognize the peptide antigen through antigen-specific T-cell receptors (TCR) located on the surface of the T cells. The TCRs on each T cell interact specifically with a single peptide antigen. Some T cells, the T_C (T-cytotoxic) cells, *directly* attack and destroy antigen-bearing cells. Other antigen-activated T cells, the T_H or T-helper cells, act *indirectly* by secreting proteins called *cytokines* that activate other cells to destroy the antigen-bearing cells. Still other T_H cells interact with antigen-specific *B lymphocytes* or *B cells* and stimulate the B cells to make *antibodies* (also called *immunoglobulins*). Each B cell and its antibody product is specific for a single antigen. The antibodies are soluble proteins that interact specifically with antigen in the circulatory system or body fluids to neutralize or destroy the antigen.

Specific immune responses can be divided into two categories, **cell-mediated immunity**, and **antibody-mediated immunity (humoral immunity)**. Cell-mediated immunity leads to cell killing through recognition of antigens found on pathogen-infected cells such as virus-infected host cells. In contrast, antibody-mediated immunity is effective against pathogens such as viruses and bacteria in the blood or lymph and also against soluble pathogen products such as toxins.

Specific immunity is characterized by the properties of *specificity, memory,* and *tolerance.*

Specificity

The **specificity** of the antigen–antibody or antigen–T-cell interaction is unlike the other host resistance mechanisms we have discussed. The nonspecific host responses challenge virtually any invading microorganism, even those pathogens the host has never before encountered. In a specific immune response, no immunity can be detected for several days after the first contact with the pathogen. However, once the immune response occurs,

it is directed solely against the eliciting pathogen and its antigens (Figure 22.7a).

Memory

Once the immune system produces an antigen-specific antibody or T cell, further exposure to the same microorganism stimulates very rapid production of large quantities of antigen-reactive T cells or immunoglobulins that interact with the pathogen and destroy it. This capacity to respond more quickly and vigorously after further exposure to antigen is known as **immunologic memory** (Figure 22.7b). Memory allows the host to specifically resist reinfection by previously encountered pathogens. We take advantage of immunologic memory by immunizing (inoculating; vaccinating) susceptible individuals with dead or weakened pathogens or their products to artificially stimulate and enhance immunity for a number of dangerous pathogens (see Section 22.11).

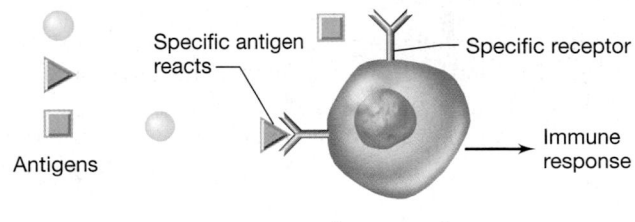

Specificity: Immune cells recognize and react with individual molecules (antigens) via direct molecular interactions.

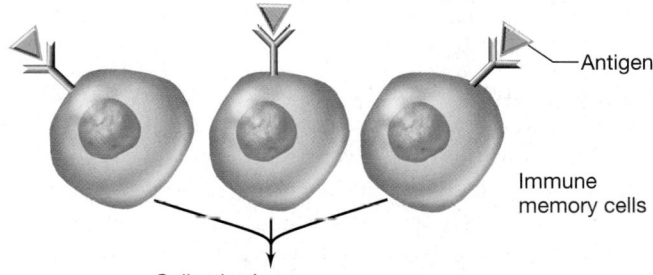

Memory: The immune response to a specific antigen is faster and stronger upon subsequent exposure because the initial antigen exposure induced growth and division of antigen-reactive cells, resulting in multiple copies of antigen-reactive cells.

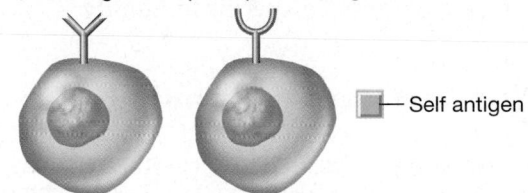

Tolerance: Immune cells are not able to react with self antigen. Self-reactive cells are destroyed during development of the immune response.

Figure 22.7 Key features of the specific immune response.

Tolerance

Tolerance is the acquired inability to make an immune response to certain antigens. Tolerance to certain molecules is necessary because all macromolecules in the host are potential antigens. As a result, the immune system must learn to *not* recognize host proteins because host cells would be damaged if they were recognized by antibodies or T cells (Figure 22.7c). The immune response has the ability to discriminate between foreign (nonself and dangerous) antigens and host (self and not dangerous) antigens and interact appropriately.

✓ 22.3 Concept Check

Nonspecific phagocytes present antigen to specific T cells, triggering the production of effector T cells and antibodies. Immune T cells and antibodies react directly or indirectly to neutralize or destroy the antigen. The immune response is characterized by *specificity* for the antigen, the ability to respond more vigorously when re-exposed to the same antigen (*memory*), and the ability to discriminate self antigens from nonself antigens (*tolerance*).

- ✓ Identify the antigen-specific cells involved in the cell-mediated and antibody-mediated immune response.
- ✓ What effects would a breakdown of specificity, memory, or tolerance have on our ability to respond to a pathogen?

II ANTIGENS, T CELLS, AND CELLULAR IMMUNITY

To understand immunity, we need to understand the molecules and structures recognized by the immune system. Therefore, we first discuss *antigens*, the molecules that are recognized by the specific immune response. Because the antigen-specific T cells are the first cells to specifically recognize antigen, we will next focus on the role of T cells in immunity. T cells are involved in various antigen-specific reactions such as cell-mediated killing, inflammatory responses, and providing "help" for antibody-producing B cells. In the absence of T cells, there is no effective antigen-specific immunity.

22.4 Immunogens and Antigens

Antigens are substances that react with either antibodies or the antigen-specific **T-cell receptors (TCRs)** found exclusively on T lymphocytes. Most antigens are **immunogens**, substances that induce an immune response. Here we examine the features of effective immunogens and then define the features of antigens that interact with antibodies and T-cell receptors.

Intrinsic Features of Immunogens

All immunogens share several common intrinsic properties that include *size, complexity,* and *form*.

Molecular size is an important component of immunogenicity. For example, low molecular weight antigens called **haptens** bind to antibodies, but cannot induce an immune response. Haptens include sugars, amino acids, and a variety of other small organic compounds. However, when coupled to a larger protein *carrier*, haptens become very effective immunogens. Most immunogens have a molecular weight of 10,000 or greater. Thus, sufficient *molecular size* is an indication of potential immunogenicity.

Complex, nonrepeating polymers such as proteins are usually effective immunogens. Complex carbohydrates can also be very effective immunogens. In contrast, nucleic acids and lipids, because they are composed of repeating monomers, tend to be very poor immunogens. Thus, the *molecular complexity* of a substance is a predictor of immunogenicity.

Complex macromolecules in insoluble or aggregated form (for example, proteins precipitated by heating) are usually excellent immunogens. The insoluble material is readily taken up by phagocytes, leading to an immune response. By contrast, the soluble form of the same molecule is often a very poor immunogen because the soluble molecule is not taken up by phagocytes. Thus, appropriate *physical form* is another good predictor of immunogenicity.

Extrinsic Features of Immunogens

Although many substances are intrinsically immunogenic, several *extrinsic* factors also influence immunogenicity. These include the *dose* of the immunogen, the *route* of administration, and finally the *foreign* nature of the immunogen with respect to the host.

The *dose* of an immunogen administered to a host is important for an effective immune response. There is a large range of potential immunogen doses that ordinarily provide satisfactory immunity. In general, doses of 10 μg to 1 g are appropriate and effective in almost any mammal. Doses of immunogen that are higher than 1 g or lower than 10 μg may not stimulate an immune response; extremely high or low doses may actually suppress a specific immune response and cause tolerance.

The *route* of administration of an immunogen is also important. Immunizations given by parenteral (outside of the gastrointestinal tract) routes, usually by injection, are normally more effective than those given topically or by mouth.

The final and most important extrinsic feature of an immunogen is the requirement that an effective immunogen must be *foreign* with respect to the host. As we have already seen, the immune system is designed to recognize and eliminate only foreign (nonself) antigens.

Many potential immunogens, however, are self proteins that are not recognized by the individual's immune response; we are *tolerant* to our own proteins, even though these same proteins have the features necessary to be immunogens in other hosts.

Antigen Binding by Antibodies and T-Cell Receptors

The antibody or TCR does not interact with the antigenic macromolecule as a whole but only against distinct portions of the molecule that are called **antigenic determinants** or **epitopes** (Figure 22.8). Antigenic determinants may include sugars, amino acids, and other organic molecules. Thus, haptens act as individual antigenic determinants. Most antibodies react with accessible surface determinants. A sequence of four to six amino acids is sufficient to define an antigenic determinant on a protein. Thus, the surface of a protein consists of a continuous array of overlapping antigenic determinants. In some cases, antibodies may even recognize epitopes that are composed of amino acids from two portions of the molecule that are distant in terms of their primary structure, but are brought together by the secondary, tertiary, or quaternary structure of the macromolecule (Figure 22.8, ⚬⚬ Sections 3.7 and 3.8). These **conformational determinants** add to the antigenic complexity of macromolecules. The surface of a bacterial cell or virus consists of a mosaic of proteins, polysaccharides, and other macromolecules, all with individual antigenic determinants. Thus, the antigenic makeup of a typical microorganism, or even of a single protein, is extremely complex.

Antibody specificity is sensitive enough to distinguish between closely related epitopes. For instance, antibodies can distinguish between the sugars glucose and galactose, which differ only in the orientation of a hydroxyl group. However, specificity is not absolute, and an antibody may react to some extent with other epitopes. The antigen that induced the antibody is called the **homologous antigen**, and other antigens that react with the antibody are called **heterologous antigens**. The interaction between an antibody and a heterologous antigen is called a *cross-reaction*.

While antibodies generally recognize epitopes expressed on macromolecular surfaces, TCRs recognize determinants only after the immunogens have been partially degraded. This degraded or "processed" antigen is then presented to T cells on the surface of specialized antigen-presenting cells (APCs) or target cells (see Section 22.5 and Figure 22.11). Since antigen processing and presentation normally destroy the conformational structure of an antigen, T cell epitopes consist of sequential linear determinants of proteins rather than the conformational determinants recognized by antibodies.

✓ 22.4 Concept Check

Immunogens are foreign macromolecules that induce an immune response. Size, complexity, and form are intrinsic properties of immunogens. When immunogens are introduced into a host in an appropriate dose and route, they initiate an immune response. Antigens are molecules recognized by antibodies or T-cell receptors. Antibodies recognize conformational antigens; T-cell receptors recognize linear peptide antigens.

✓ Distinguish between *immunogens* and *antigens*.

✓ Identify the intrinsic and extrinsic features of an immunogen.

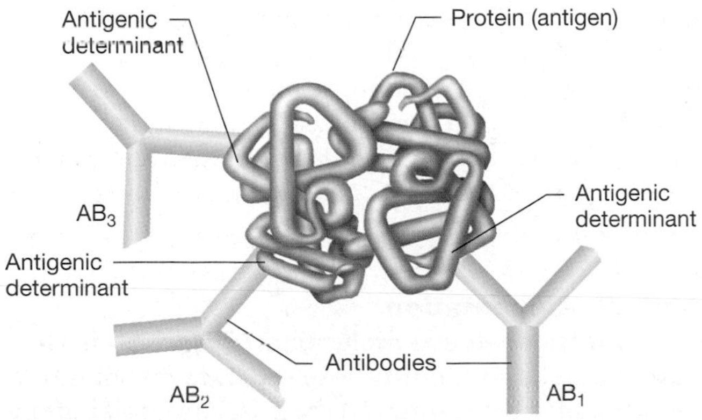

Figure 22.8 Antigens and antigenic determinants for antibodies. Antigens may contain several different antigenic determinants, each capable of reacting with a specific antibody. The antigenic determinant recognized by AB_1 is a *conformational determinant* consisting of two different parts of the same polypeptide chain. The polypeptide chain is folded to bring two distant parts of the protein together to make a single determinant.

22.5 Presentation of Antigen to T Lymphocytes

T lymphocytes, or **T cells**, are lymphocytes that interact specifically with antigen through the cell-surface *T-cell receptor (TCR)*. TCRs interact with antigens held in place on antigen-presenting cells through *major histocompatibility complex (MHC) proteins*.

The T-Cell Receptor

The T-cell receptor is a membrane-spanning protein that extends from the T cell surface into the extracellular environment. Each T cell has thousands of copies of the same TCR on its surface. A functional TCR consists of two proteins, an α chain and a β chain. Each of these chains has a variable (V) domain and a constant (C) domain (Figure 22.9). The $V\alpha$ and $V\beta$ domains cooperate to form a complete antigen-binding site. As we shall see (⚬⚬ Section 23.7), the immune system has the ability to

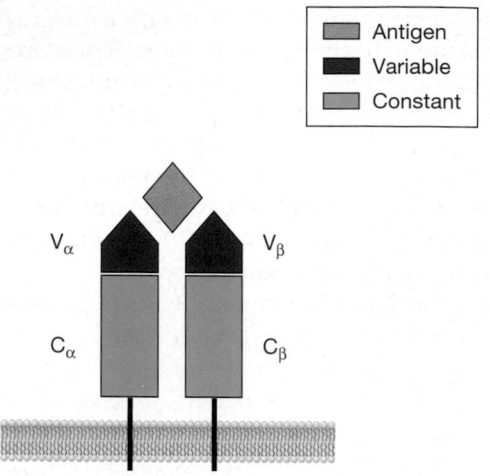

Figure 22.9 Structure of the T-cell receptor (TCR). The V domains of the α chain and β chain combine to form the peptide antigen-binding site.

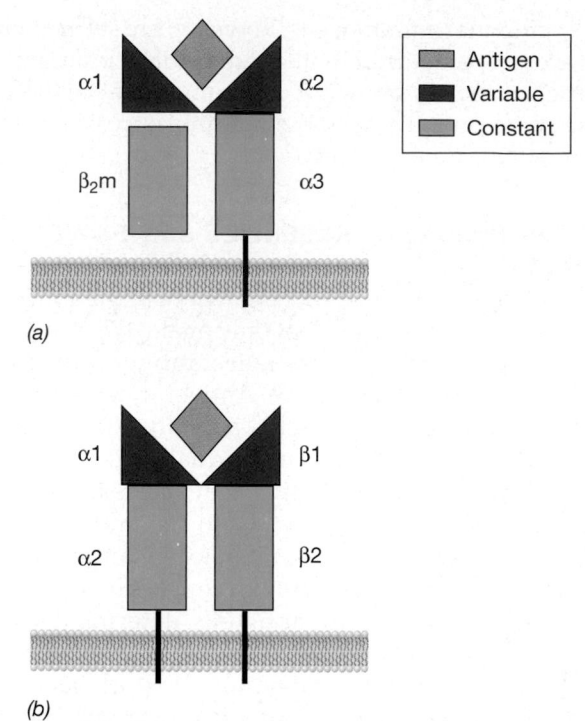

Figure 22.10 Structures of the MHC proteins. (a) Class I MHC protein. The $\alpha 1$ and $\alpha 2$ domains interact to form the peptide antigen-binding site. (b) Class II MHC protein. The $\alpha 1$ and $\beta 1$ domains combine to form the peptide antigen-binding site.

generate TCRs that will bind nearly every known peptide antigen. However, the TCR can only recognize and bind a peptide antigen if the peptide is first bound to *self* proteins known as the major histocompatibility complex (MHC) proteins.

Major Histocompatibility Complex (MHC) Proteins

MHC proteins are encoded by a genetic region, present in all vertebrates, called the **major histocompatibility complex (MHC)**. MHC proteins are produced by a number of genes in this complex and are collectively called *human leukocyte antigens* or HLAs. They were first discovered as the major target molecules for transplantation rejection; if tissues from one animal, a donor, are immunologically rejected when transplanted to another animal, a recipient, then their MHC proteins are different. We now know that MHC proteins function as antigen-presenting molecules and interact with both the antigen and the TCR. Thus, MHC proteins are antigen-binding molecules and play an integral role in the immune response. The MHC genes encode two distinct types of proteins, *class I* and *class II*. **Class I MHC proteins** are found on the surfaces of *all* nucleated cells. **Class II MHC proteins** are found only on the surface of B lymphocytes, macrophages, and other antigen-presenting cells (APCs). The reasons for this differential distribution will become apparent when we discuss the function of these molecules.

Class I MHC proteins consist of two polypeptides (Figure 22.10a), an alpha (α) chain encoded in the MHC gene region, and a smaller protein called β-2 *microglobulin* ($\beta_2 m$), encoded by a non-MHC gene. Class II MHC proteins consist of two noncovalently linked polypeptides, called α and β. Like class I molecules, these polypeptides are embedded in the cytoplasmic membrane and project outward from the cell surface (Figure 22.10b).

MHC proteins are not structurally identical *within* a given species. Different individuals often show subtle differences in the amino acid sequence of their MHC molecules. These limited sequence variations are called *polymorphisms*. There are several hundred different MHC genes in humans; the MHC proteins are the major reason why tissues transplanted from one individual to another seldom match, are recognized as nonself, and are rejected. The detailed molecular structure and genetic organization of the MHC genes and proteins are presented in Chapter 23.

Antigen Presentation

MHC proteins serve as molecular reference points that permit T cells to identify foreign antigens. In a normal animal, T cells, through their TCRs, constantly interact with proteins or other potential antigens; they must be able to discriminate self from nonself antigens. The T cell, through its TCR, binds to MHC molecules and can then recognize foreign antigens embedded in the MHC structure; a T cell cannot recognize a foreign antigen unless it is presented in the context of an MHC protein.

How does this happen? As we discussed previously, host phagocytes (here the antigen-presenting cells) take up and degrade (process) antigen. Processed peptide antigen then becomes embedded, or bound, to the MHC protein, and the complex is passed through the cytoplasmic membrane and presented on the surface of the cell. Two distinct antigen-processing schemes are known, one for class I antigen presentation and one for class II antigen presentation (Figure 22.11). In the class I scheme (Figure 22.11a), antigens that are manufactured by host degradation reactions in nonphagocytic cells are bound by class I proteins in the endoplasmic reticulum. The actual processed peptide is about 10 amino acids long. This method of antigen processing is very important in virus infections, where the host

cell manufactures and degrades viral proteins. Degraded viral peptides, which are nonself, then complex with class I proteins. The complex moves to the cell surface where it is presented to peptide-specific T cells through the TCRs.

Thus, the MHC molecules act as a *platform* on which the foreign viral antigen is bound. Next, the TCR on the surface of the T cell interacts with both antigen (nonself) and MHC protein (self) sites. This cell-cell interaction induces specialized T-cytotoxic cells to produce cytotoxic proteins called *perforins* (see Section 22.6) that kill the virus-infected *target cell*.

A second antigen presentation scheme involves the class II molecule (Figure 22.11b). In this case, class II proteins, complete with a self peptide called *Ii*, or

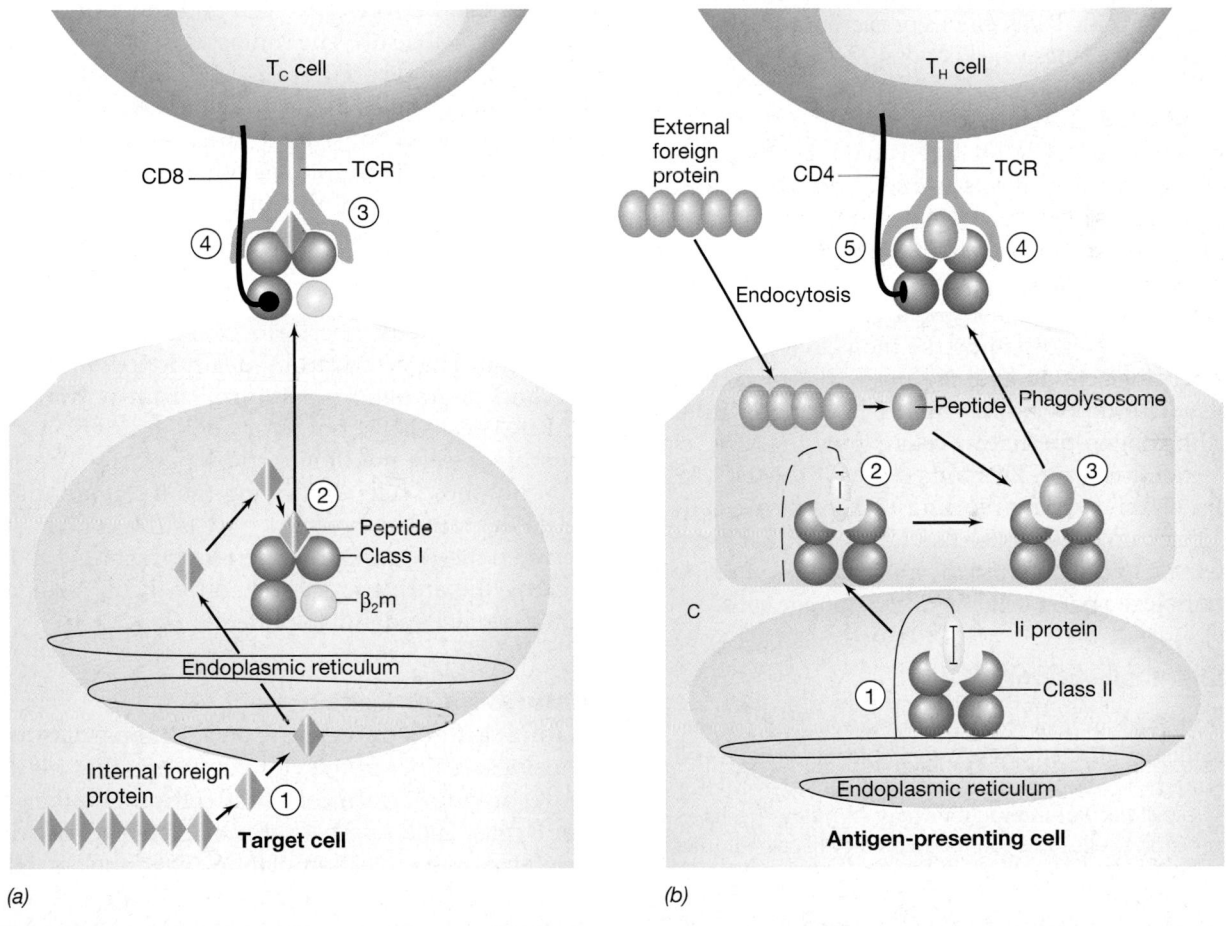

(a)   *(b)*

Figure 22.11 Antigen presentation by class I and class II major histocompatibility complex (MHC) proteins. (a) In the class I antigen presentation pathway, the class I proteins are made and assembled in the endoplasmic reticulum. ① Protein antigens manufactured within the cell, for instance from viruses or tumors, are degraded in the cytoplasm and transported across the endoplasmic reticulum membrane. ② The peptides then bind to class I, are transported to the cell surface, and ③ interact with T-cell receptors (TCRs) on the surface of T_C cells. ④ The CD8 coreceptor on the T_C cell also engages the class I MHC, resulting in a stronger complex. The T_C cells then release cytokines and cytotoxins, proteins that kill the target cell. Any nucleated cell can act as a target cell for T cells recognizing peptides embedded in class I proteins. (b) In the class II antigen presentation pathway, ① class II proteins are produced in the endoplasmic reticulum and are assembled with a blocking protein, Ii (invariant chain), which prevents class II from complexing with other peptides found in the endoplasmic reticulum. ② Class II then goes to the phagolysosome where the Ii and foreign proteins, imported from outside the cell (by endocytosis), are digested. ③ The class II protein then binds to the digested foreign peptides, and the complex is transported to the cell surface ④ where it interacts with TCRs and ⑤ the CD4 coreceptor on T_H cells. The T_H cells then release cytokines that act on other cells to activate an immune response. Only APCs can be targets for T cells recognizing peptides embedded in class II proteins. The APCs are monocytes, macrophages, dendritic cells, and B cells.

invariant chain, line the cell vacuoles (lysosomes) (see Section 22.2) that degrade antigens phagocytized by APCs. The phagosome containing the foreign antigen fuses with the lysosome forming a *phagolysosome*, and the antigens are digested by proteolytic enzymes along with the Ii. The foreign peptides, generally about 11 to 15 amino acids in length (slightly larger than class I-binding peptides), are then bound by the newly opened class II antigen-binding site, and the whole complex is eventually expressed on the external cytoplasmic membrane where it is presented to specialized T-helper, or T_H cells. The T_H cells, through the TCR, then recognize the class II MHC-foreign peptide complex on the surface of the APCs. The T_H cell is activated by contact with foreign antigen and secretes molecules that either stimulate antibody production by specific B cell clones or secrete a battery of inflammatory cytokines (see Sections 22.7 and 22.9).

CD4 and CD8 Coreceptors

In addition to the TCR, all T cells have another unique cell surface protein that acts as a *coreceptor*. For example, all T- helper cells express a CD4 protein coreceptor, while all T-cytotoxic cells express a CD8 protein coreceptor (Figure 22.11). When the TCR binds to the peptide–MHC complex, the coreceptor on the T cell also binds to the MHC protein, strengthening the molecular interactions and enhancing activation of the T cell. CD4 binds only to the class II protein, ensuring that T-helper cells interact only with antigen-presenting cells expressing MHC class II protein. Likewise, CD8 binds only to the MHC class I protein, enhancing the binding of T-cytotoxic cells to class I-bearing target cells. The CD4 and CD8 proteins also serve as very convenient markers used *in vitro* to differentiate T-helper cells from T-cytotoxic cells.

✓ 22.5 Concept Check

T cells recognize antigens presented by specialized phagocytes or by pathogen-infected cells. At the molecular level, T-cell receptors interact with peptide antigens presented by MHC proteins. These molecular interactions may stimulate T cells to kill antigen-bearing cells or to produce a battery of cell-stimulating proteins known as cytokines.

- ✓ Identify cells that have class I and class II MHC proteins as surface components.
- ✓ Define the process of antigen presentation for both intracellular and extracellular pathogens.

22.6 T-Cytotoxic Cells and Natural Killer Cells

In the previous section, we introduced two sets of T lymphocytes, the T-cytotoxic cells and the T-helper cells. Here we examine the killer function of the T cytotoxic cells in detail. We will also look at the *natural killer cell*, a cell that also kills infected target cells. We examine T-helper cell functions in Section 22.7.

T-Cytotoxic Cells

T-cytotoxic cells (T_C), also known as cytotoxic T lymphocytes (CTLs), are involved in the destruction of cells that display foreign antigens on their surfaces. As previously mentioned in connection with the major histocompatibility complex (MHC) (see Section 22.5), T_C cells recognize foreign antigens embedded in MHC class I molecules. Any cell carrying a foreign antigen, such as those introduced on cells infected with viruses, can be killed by T_C cells.

Contact between a T_C cell and the target cell is required for cell death. The contact is initiated by the TCR: Ag-MHC complex (Figure 22.11*a*). On contact with the target cell, granules in the T_C cell are drawn to the contact site and the contents of the granules are released (*degranulation*). The granules contain *perforin*, which enters the membrane of the target cell and forms a pore. In addition to the perforins, the T_C granules contain *granzymes*, proteins that cause *apoptosis*, or programmed cell death. When granzymes enter the target cell through the pores created by perforins, the target cell undergoes apoptosis, characterized by death followed by shrinking and degradation of the cell from within (Figure 22.12). The T_C cells, however, remain unaffected; their membranes are not damaged by perforin. The T_C cells kill only those cells displaying the foreign antigen because degranulation occurs only at the contact surface between the T_C and the antigen-bearing target cell. Cells lacking the antigen recognized by the T_C cells do not make contact and are not killed.

Natural Killer Cells

Natural killer (NK) **cells** are an additional class of lymphocytes distinct from cytotoxic T cells that play a role in destroying foreign cells. NK cells are neither T cells nor B cells. Their numbers are not enhanced, nor do they exhibit memory after stimulation. Nevertheless, NK cells resemble T_C cells in their ability to kill foreign cells. For example, NK cells also use perforin and granzymes to kill their targets. However, NK cells differ from T_C cells in that they kill targets in the *absence* of recognition of a specific antigen. NK cells are capable of destroying malignant and virus-infected cells *in vitro* without previous exposure or contact with the foreign antigen.

The molecular target for NK cells seems to be the *lack* of appropriate MHC class I proteins. NK cells recognize normal cells and their class I protein through a set of special class I receptors. Binding of these receptors to class I *deactivates* the killing mechanism, but in the absence of binding, the NK cell kills the unrecog-

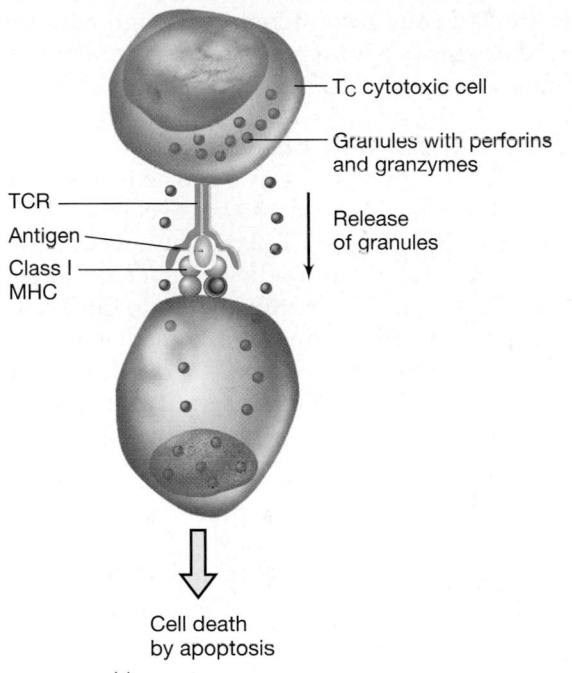

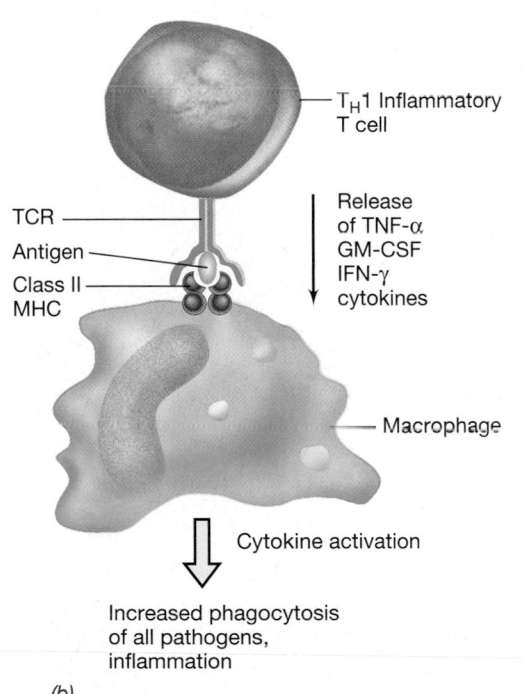

Figure 22.12 Effector T cells. (a) T-cytotoxic cells, or T_C cells, are activated by antigens presented on any cell in the context of MHC I protein. The T_C cells respond by releasing granules that contain perforin and granzymes, cytotoxins which perforate the target cell and cause apoptosis, respectively. (b) T-inflammatory cells, or T_H1 cells, are activated by antigens presented on macrophages in the context of MHC II protein. The T_H1 cells respond by producing cytokines that stimulate the macrophages to increase phagocyte activity and promote inflammation.

nized target. The main targets of NK cells, tumor cells and virus-infected cells, often have reduced or altered class I MHC protein expression.

✓ 22.6 Concept Check

T-cytotoxic (T_C) cells recognize antigens on virus-infected host cells and tumor cells through antigen-specific T-cell receptors. Antigen-specific recognition triggers killing via perforin and granzymes. Natural killer (NK) cells use the same effectors to kill virus-infected cells and tumors. However, NK cells do not require stimulation, nor do they exhibit memory. NK cells respond to the *absence* of MHC proteins.

- ✓ Identify and compare the target recognition mechanisms used by T_C and NK cells.
- ✓ Describe the common effector system used by cytotoxic cells.

22.7 T-Helper Cells: Activating the Immune Response

Here we will define the functions of the T-helper cells and their pivotal role in producing cytokines to foster an effective immune response. T-helper cells are divided into two subsets, the T_H1 cells and the T_H2 cells. T_H1 cells play a role in macrophage activation and T_H2 cells interact with B lymphocytes and stimulate antibody production.

T_H1 Cells and Macrophage Activation

Macrophages play a central role as **antigen-presenting cells (APCs)** in both antibody-mediated and cell-mediated immunity. As was illustrated in Figure 22.11*b*, macrophages bind, process, and present antigen to T_H cells. As phagocytic cells, however, macrophages also take up and kill certain foreign cells by themselves, an ability stimulated by T_H1 cells. A key property of activated macrophages is that they can kill intracellular bacteria that would normally multiply. As we have noted (see Section 22.2), some bacteria survive and multiply within macrophages, whereas most bacteria taken into macrophages are killed and digested. Bacteria multiplying within macrophages include *Mycobacterium tuberculosis, M. leprae* (causal agents of tuberculosis and leprosy, respectively), and *Listeria monocytogenes* (causal agent of listeriosis, ⬭ Section 29.9). Animals given a moderate dose of *M. tuberculosis* are able to overcome the infection and develop resisance because of the T cell-mediated immune response. The cells involved are the T-inflammatory cells, the T_H1 subset. They activate macrophages and other nonspecific phagocytes by secreting a number of cytokines, including IFN-γ (interferon gamma), GM-CSF (granulocyte-monocyte colony-stimulating factor), and TNF-α (tumor necrosis factor-alpha) (Figure 22.12). Surprisingly, such immunized animals also phagocytize and kill unrelated organisms such as *Listeria*, because macrophages in the immunized animal have been activated so that they more readily kill any secondary invader.

Macrophages not only kill foreign pathogens but are also involved in the destruction of foreign mammalian cells. This shows up in the development of transplantation immunity and is a major problem in the transplantation of organs and tissues from one person to another. Macrophages also target tumor cells in some cases. Tumor cells often have specific antigens not found on normal cells, and tumors function like self-inflicted transplants. There is considerable evidence that tumor cells are normally recognized as foreign and are destroyed primarily by macrophages that are activated by cytokines from T_H1 cells.

T_H2 Cells and B Cell Activation

T_H2 cells play a pivotal role in B cell activation and antibody production. As discussed in Section 22.3, B cells make antibodies. How are T_H2 cells involved? Mature B cells are coated with antibodies that act as antigen receptors. However, when antigen binds to the B cell antigen receptors, the B cell does not immediately produce soluble antibodies, but instead acts as an antigen-presenting cell (APC) and interacts with a T_H2 cell. More specifically, the bound antigen is first endocytosed and

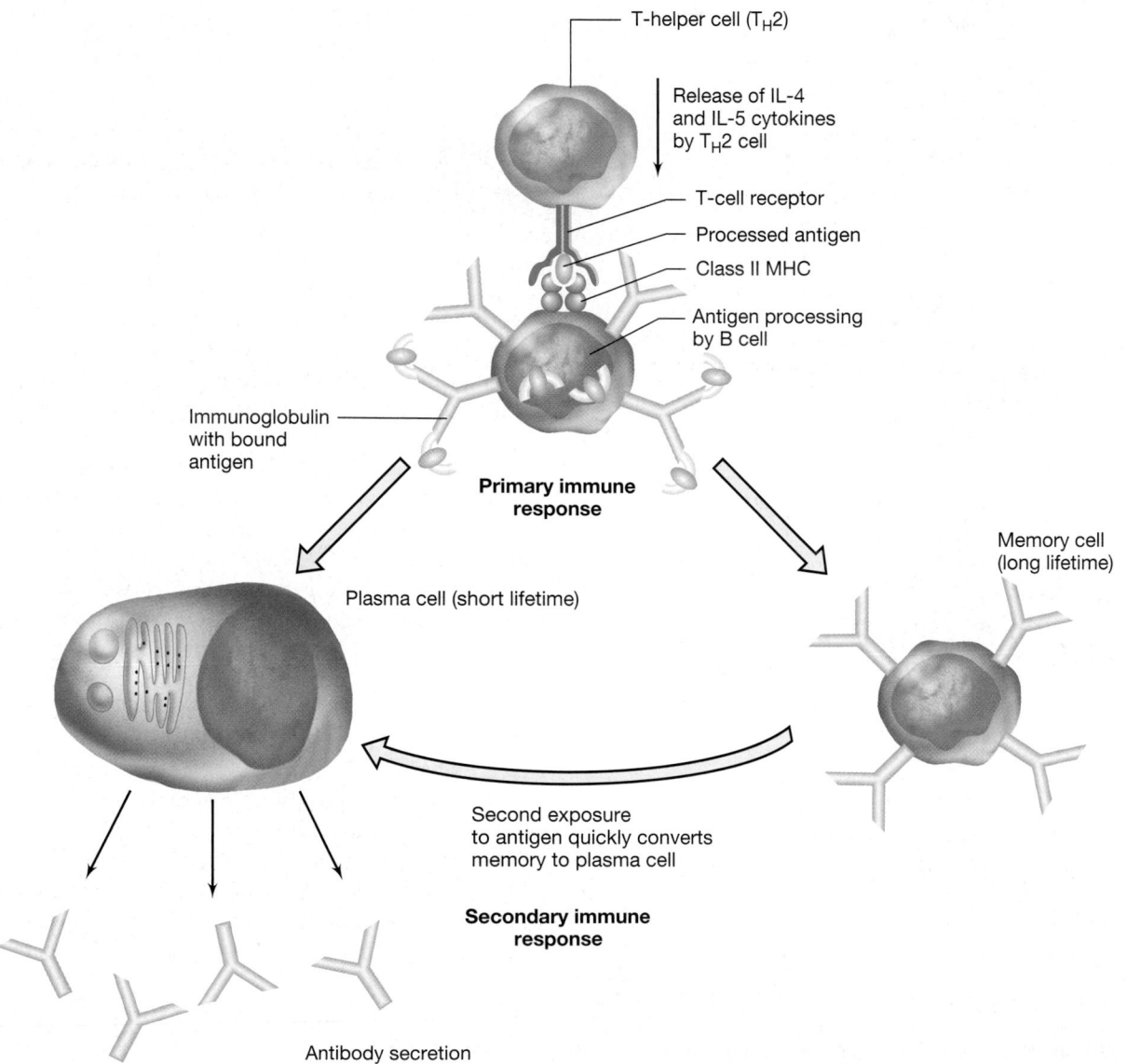

Figure 22.13 T cell–B cell interaction and antibody production. The B cell functions as an antigen-presenting cell (APC) and concentrates antigen using anigen-specific immunoglobulin receptors. After processing, antigen is presented to the T_H2 cell by a class II MHC molecule. The T_H2 cell then signals the same B cell to proliferate and form plasma cells (antibody producers) or memory cells. After subsequent antigen exposure, memory cells quickly convert to plasma cells.

degraded in the B cell. Peptides from the degraded antigen are then presented on a class II MHC protein to a T_H2 cell (Figures 22.11*b* and 22.13). The T_H2 cell then responds by producing IL-4 (*interleukin*-4) and IL-5, cytokines that act directly on the B cell. The B cell then makes and secretes soluble antibodies, as we will discuss in Section 22.9. Thus, the T_H2 cell is a "helper cell" that activates antibody production by B cells.

✓ 22.7 Concept Check

T_H1 and T_H2 cells play pivotal roles in cell-mediated and antibody-mediated immune responses. T_H1 and T_H2 cells each stimulate effector cells through the action of cytokines.

✓ Describe the role of T_H1 cells in activation of macrophages.
✓ Describe the role of T_H2 cells in the activation of B cells.

III ANTIBODIES AND IMMUNITY

Here we concentrate on the role of B cells and antibodies in immunity. Antibodies are important components of an effective immune response and provide antigen-specific immunity to extracellular pathogens and dangerous soluble proteins such as toxins. After considering the molecular nature of antibodies, we will look at the production of antibodies by B cells. We will conclude by discussing the ability of antibodies to neutralize or destroy antigens *in vivo*.

22.8 Antibodies (Immunoglobulins)

Antibodies, or **immunoglobulins** (Ig), are protein molecules that are able to combine with antigenic determinants. They are found in the serum and in other body fluids such as gastric secretions and milk. Serum containing antigen-specific antibody is often called **antiserum**. Immunoglobulins (Igs) can be separated into five major classes on the basis of their physical, chemical, and immunological properties: **IgG, IgA, IgM, IgD**, and **IgE** (Table 22.2). In most individuals about 80% of the serum immunoglobulins are IgG proteins, and these have been studied extensively.

Immunoglobulin G Structure

Immunoglobulin G (IgG) is the most common circulating antibody and is a very good structural model. IgG has a molecular weight of about 150,000 and is composed of four polypeptide chains (Figure 22.14). Interchain disulfide bridges (S–S bonds) connect the individual chains. In

TABLE 22.2   Properties of human Immunoglobulins

Class/ H chain[a]	Molecular weight/ formula[b]	Serum (mg/ml)	Antigen-binding sites	Properties	Distribution
IgG γ	150,000 2(H + L)	13.5	2	Major circulating antibody; four subclasses: IgG_1, IgG_2, IgG_3, IgG_4; IgG_1 and IgG_3 activate complement	Extracellular fluid; blood and lymph; crosses placenta
IgM μ	970,000 (pentamer) 5[2(H + L)] + J	1.5	10	First antibody to appear after immunization; strong complement activator	Blood and lymph; monomer is B cell-surface receptor
	175,000 (monomer) 2(H + L)	0	2		
IgA α	150,000 2(H + L)	3.5	2	Important circulating antibody	Secretions (saliva, colostrum, cellular and blood fluids); monomer in blood and dimer in secretions.
	385,000 (secreted dimer) 2[2(H + L)] + J + SC	0.05	4	Major secretory antibody	
IgD δ	180,000 2(H + L)	0.03	2	Minor circulating antibody	Blood and lymph; B lymphocyte surfaces
IgE ϵ	190,000 2(H + L)	0.00005	2	Involved in allergic reactions; C_H4 contains mast cell binding fragment	Blood and lymph; binds to mast cell surfaces

[a] All immunoglobulins may have either λ or κ light chain types, but not both.

[b] Based on the number and arrangement of heavy (H) and light (L) chains in each functional molecule. J is a joining protein present in serum IgM and secretory IgA. SC is the secretory component found in secreted IgA. See Figures 22.16, 22.17, and 22.18 for a diagram of each immunoglobulin.

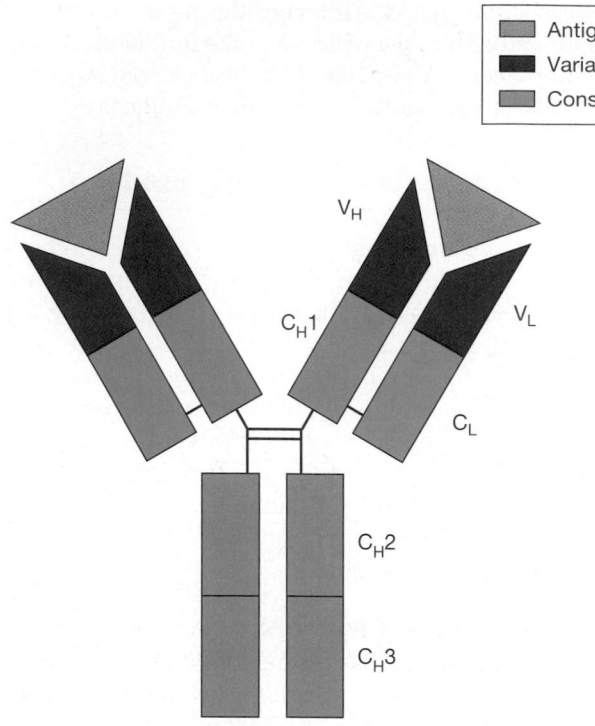

Figure 22.14 Immunoglobulin G structure. Immunoglobulin G (IgG) consists of two heavy chains (50,000 molecular weight) and two light chains (25,000 molecular weight), with a total molecular weight of 150,000. One heavy and one light chain interact to form an antigen-binding unit. The variable domains of the heavy and light chain (V_H and V_L) bind antigen and show sequence differences in each different immunoglobulin. The constant domains (C_H1, C_H2, C_H3) are identical in all IgG proteins. The chains are covalently joined by disulfide bonds.

an individual IgG protein, two identical light chains of 25,000 molecular weight are coupled to two identical heavy chains of 50,000 molecular weight. Each light chain has about 220 amino acids, and each heavy chain has about 440 amino acids. A functional IgG antibody consists of two antigen-binding units, each having one heavy chain and one light chain. Therefore, antibodies are *bivalent* and can bind two identical epitopes.

Light Chains

Each IgG light chain consists of two protein domains of equal size (Figure 22.14). The amino-terminal region is a *variable* **domain**, meaning that the amino acid sequence in this structural region differs in each different antibody. The variable domain is involved in antigen binding. By contrast, the carboxy-terminal domain is a *constant domain*; the amino acid sequence in this domain does not differ among light chains of the same type.

Heavy Chains

Each IgG heavy chain has an amino-terminal variable domain and three constant domains of about 110 amino acids each. As is the case for light chains, the three con-

stant domains of each heavy chain are identical in antibodies of the same class, while the variable region differs in each antibody.

The Antigen-Binding Site

The antigen-binding site of IgG and all other immunoglobulins is formed by cooperative interaction between the variable regions of both heavy and light chain (Figure 22.15). The variable regions of both chains in-

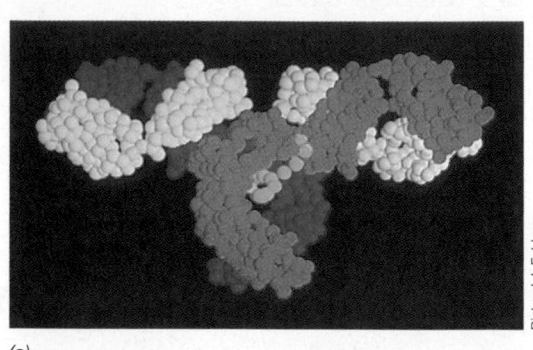

(a)

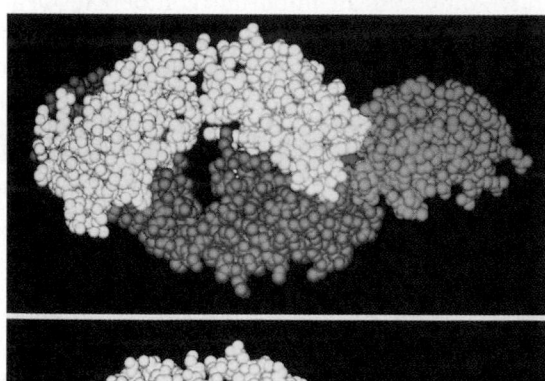

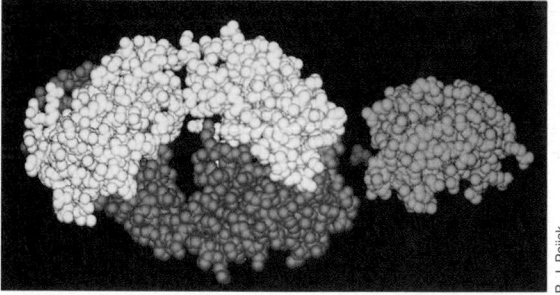

(b)

Figure 22.15 Immunoglobulin structure and the antigen-binding site. (a) Three-dimensional view of an immunoglobulin G (IgG) molecule. The heavy chains are shown in red and dark blue. The light chains are in green and light blue. (b) Structure of the combining site of an antigen and immunoglobulin. The antigen (lysozyme) is in green. The variable region of the immunoglobulin heavy chain is shown in blue, and the light chain in yellow. The amino acid shown in red is a glutamine residue of lysozyme. The glutamine residue fits into a pocket on the immunoglobulin molecule, but the overall antigen-antibody recognition involves contacts made between several other amino acids on both the immunoglobulin and the antigen. Reprinted with permission from *Science* 233:747 (1986) © AAAS.

teract to form a molecular site that binds strongly but noncovalently with the antigen. The strength of this binding is termed *antigen-binding affinity*. An antibody is said to bind with high affinity if the antigen interacts strongly and is held tightly by the antibody.

Each individual's immune system has the capacity to recognize, or bind, countless antigens, and each antigen is recognized by a unique antigen-binding site. To accommodate all possible antigens, each individual can produce more than 1 billion different antigen-binding sites in the form of antibodies. How is this diversity in the antigen-binding site generated? As we discuss in the next chapter, new antibodies are constantly created by randomly recombining and mutating the 300 or more genes that encode the light and heavy chain variable domains. B cells with the novel antibody (antigen-binding site) are then "selected" by interactions with antigen (⚬⚬ Section 23.5).

Other Classes of Immunoglobulins

How do immunoglobulins of the other classes differ from IgG? The heavy-chain *constant* domains of a given immunoglobulin molecule define its class and can have one of five amino acid sequences: gamma (γ), alpha (α), mu (μ), delta (δ), or epsilon (ϵ). These sequences constitute the carboxy-terminal three-fourths of the heavy chains of immunoglobulins of the class IgG, IgA, and IgD, re-

spectively and four-fifths of the heavy chains of IgM and IgE (Figure 22.16). Each antibody of the IgM class, for example, contains amino acids in its heavy-chain constant domains that constitute the *mu* sequence.

In a typical immune response, it is not uncommon to find two Igs of different classes, often IgM and IgG, that bind the same epitope. In this case, the *variable* (antigen-binding) domains of the heavy and light chains from both IgM and IgG are identical, but the class-determining *constant* domains of the heavy chains are different, with IgM expressing a mu heavy chain and IgG expressing a gamma heavy chain.

The structure of **immunoglobulin M** (IgM) is shown in Figure 22.17. IgM is usually found as an aggregate of five immunoglobulin molecules attached by at least one *J chain*. Each heavy chain of IgM contains a fourth constant domain (C_H4). IgM is the first class of immunoglobulin made in a typical immune response to a bacterial infection, but immunoglobulins of this class are generally of low affinity. Antigen-binding strength is enhanced to some degree, however, by the

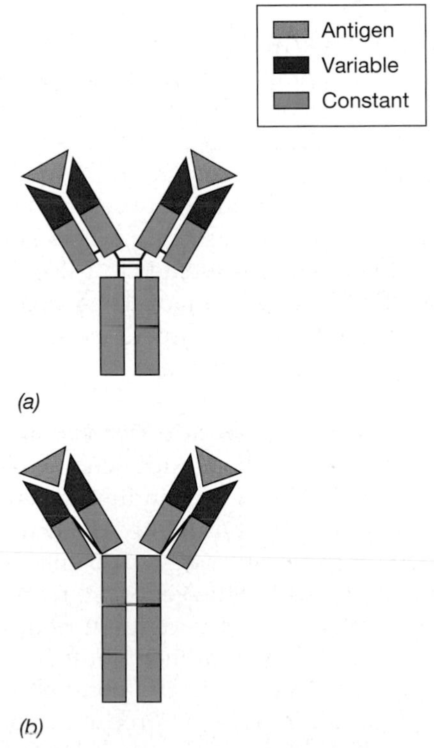

	Antigen
■	Variable
■	Constant

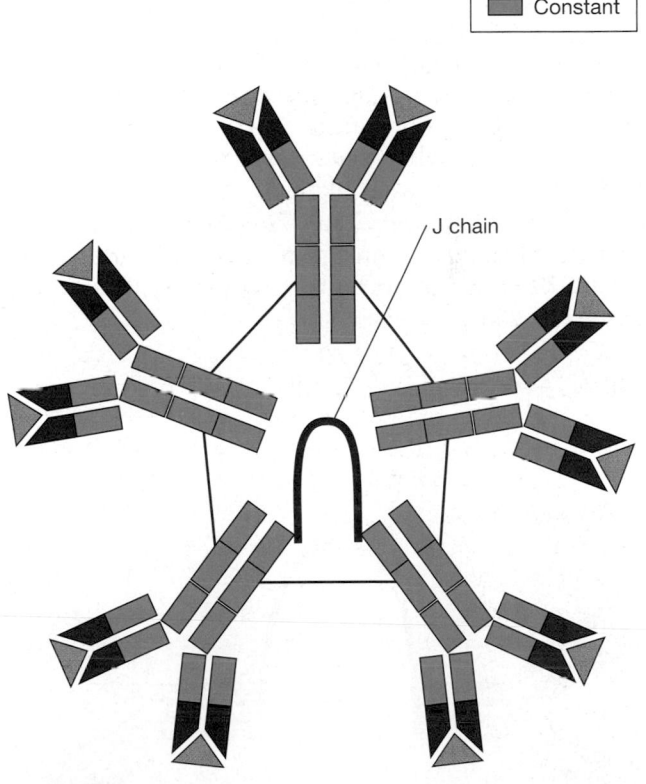

J chain

Figure 22.17 Immunoglobulin M. Immunoglobulin M (IgM) is found in serum as a pentameric protein consisting of five IgM proteins covalently linked to one another via disulfide bonds and a J chain protein. Because it is a pentamer, IgM can bind up to 10 antigens, as shown.

	Antigen
■	Variable
■	Constant

(a)

(b)

Figure 22.16 Immunoglobulin classes. All classes of immunoglobulins have V_H and V_L domains (red) that bind antigen (brown). (a) IgG, IgA, and IgD have three constant domains (blue). (b) IgM and IgE each have a fourth constant domain.

high *valency* of the pentameric IgM molecule; 10 binding sites are available for interaction with antigen (Table 22.2 and Figure 22.17). The term *avidity* is used to describe the *strength of binding* by multivalent antigen-binding molecules; thus, IgM is said to be of *low* affinity but *high* avidity. About 10% of serum antibodies are IgM. IgM monomers are also found on the surface of B cells, where they bind antigen.

Immunoglobulin A (IgA), in the dimeric form (Figure 22.18), is present in body secretions. IgA is present in saliva, tears, breast milk and colostrum, gastrointestinal secretions, and mucus secretions of the respiratory and genitourinary tracts. The mucosal surfaces of the human body total about 400 m². All these mucosal surfaces are associated with the MALT (Figure 22.2) that secrete IgA. As a result, the total amount of *secretory* IgA produced by the body is higher than the amount of *serum* IgG. IgA is also present in serum as a monomer (Table 22.2). The secretory form of IgA consists of a dimeric immunoglobulin attached to a protein *secretory piece*, and a J chain peptide (Figure 22.18). These proteins help hold the dimeric immunoglobulin molecule together and aid in the transport of IgA across membranes and into secretions.

Immunoglobulin E (IgE) is found in serum in extremely small amounts (about 1 of every 50,000 serum immunoglobulin molecules is IgE). Despite its low concentration, IgE is important because immediate-type hypersensitivities (allergies) are mediated by IgE (see Section 22.13). The molecular weight of an IgE molecule is significantly higher than most other immunoglobulins (Table 22.2) because, like IgM, IgE has a fourth constant domain (Figure 22.16). This additional constant region functions to bind IgE to mast cell surfaces (see Figure 22.23), an important prerequisite for certain allergic reactions.

Immunoglobulin D (IgD), present in serum in low concentrations, has no known function. However, IgD is abundant on the surfaces of B cells and plays a role along with monomeric IgM in binding antigen to B cells.

✔ 22.8 Concept Check

Immunoglobulin (antibody) proteins consist of four chains, two heavy and two light. The antigen-binding site is formed by the interaction of variable regions of heavy and light chains. Each class of immunoglobulin has different structural and functional characteristics.

✔ What immunoglobulin domains are involved in antigen binding?

✔ What functional and structural characteristics differentiate Ig classes?

22.9 B Lymphocytes and Antibody Production

Antibody production is a complex process. In this section, we will examine a typical antibody response, first reviewing the cell interactions necessary to produce effective immunity. We will then look at genetic control and examine the antibody response after exposure to antigen.

T Cell–B Cell Interactions

The production of immunoglobulins in response to antigen involves interactions between T cells and B cells through their respective antigen-specific cell surface molecules (Figure 22.13). Each step in the antibody production is *highly* specific: The antigen-presenting cell is the B cell with an antigen-specific Ig on its surface; the processed antigen is then presented to T_H2 cells with the antigen-specific TCR on its surface. These activated antigen-specific T_H2 cells then stimulate the same B cells to produce antigen-specific antibody.

Genetic Control of Antibody Production

The genetic control of antibody production is also a highly complex process. Although an individual is capable of making countless distinct antibody molecules, only a small number of genes are required to encode this immense antibody diversity; the genes are rearranged via somatic recombination and undergo mutation to produce the large number of immunoglobulins. During development of lymphocytes in the bone marrow, *gene rearrangements* occur in B cells to produce two complete mRNAs, one encoding the heavy chain and one encoding the light chain of the antibody molecule. The number of possible gene rearrangements is sufficient to account for the diversity of immunoglobulin molecules (∞ Section 23.5). Similar rearrangements occur during

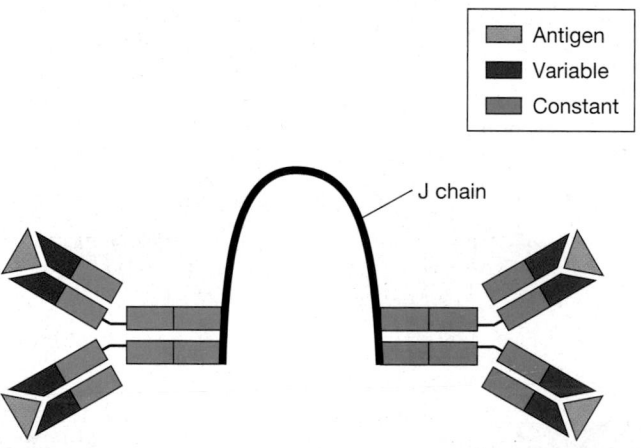

■ Antigen
■ Variable
■ Constant

J chain

Figure 22.18 Immunoglobulin A. Secretory immunoglobulin A (sIgA) is often found in body secretions as a dimer consisting of two IgA proteins covalently linked to one another via a J chain protein.

T-cell development and result in the generation of diversity in T-cell receptors (∞ Section 23.7).

Antibody Production

We now summarize the steps in antibody production, beginning with the introduction of an antigen and ending with the production of a specific antibody.

1. Antigens are spread via the lymphatic and blood circulatory systems to neighboring secondary lymphoid organs such as lymph nodes, spleen, or mucosal-associated lymphoid tissue (MALT) (Section 22.1 and Figure 22.2). Intravenously injected antigen travels via the blood to the spleen, where antibodies are formed. If antigen is introduced subcutaneously, intradermally, topically, or intraperitoneally, the lymphatic system carries antigen to the nearest lymph nodes. Likewise, antigen introduced to mucosal surfaces, for example by mouth, is delivered to the MALT lining the intestinal tract, resulting in antigen-specific IgA antibody production in the gut.

2. Following the initial antigen introduction, each antigen-stimulated B cell multiplies and differentiates to form both antibody-secreting *plasma cells* and *memory cells* (Figure 22.13). Plasma cells are relatively short-lived (less than 1 week), but produce and secrete large amounts of IgM antibody in this **primary antibody response** (Figure 22.19). There is a latent period before specific antibody appears in the blood, followed by a gradual increase in antibody titer

(quantity), and then a slow fall in the primary antibody response.

3. The memory cells generated by the initial exposure to antigen may live for several years. If re-exposure to the immunizing antigen occurs at a later time, memory cells need no T-cell activation; they quickly transform to plasma cells and begin producing IgG. The antibody titer rises rapidly to a level 10–100 times greater than the titer achieved following the first exposure. This rise in antibody titer is referred to as the **secondary antibody response** (Figure 22.19). The secondary response is the result of *immunologic memory* and produces a more rapid, more abundant antibody response when compared with the primary response. The secondary response is also characterized by a switch from IgM to IgG production, a phenomenon called *class switching* (Figure 22.19).

4. Over time the titer slowly decreases, but later exposures to the same antigen can cause another secondary response. The secondary response is the basis for the immunization procedure known as a "booster shot" (for example, the yearly rabies shot given to domestic animals). Periodic reimmunization maintains high levels of circulating antibody specific for a certain antigen and provides long-term active protection against infectious disease (see Section 22.11).

✓ 22.9 Concept Check

Antibody production is initiated by antigen contact through an APC, with an antigen-specific T_H2 cell. The activated T_H2 cell signals an antigen-specific B cell to produce antibody. Activated B cells live for years as memory cells and can quickly produce large quantities (high titers) of antibodies upon re-exposure to antigen.

✓ How do B cells act as APCs?

✓ How do T_H2 cells activate antigen-specific B cells?

22.10 Complement, Antibodies, and Pathogen Destruction

Complement plays an important role in the immune response. Complement is a series of proteins that are activated by interactions with antigen–antibody complexes. Several distinct complement proteins may react in sequence to cause *lysis* of bacterial cells. Complement proteins also enhance phagocyte recognition and destruction of antigens.

Complement Activation and Cell Damage

Complement is composed of a number of proteins, many with enzymatic activity. These proteins are activated in a

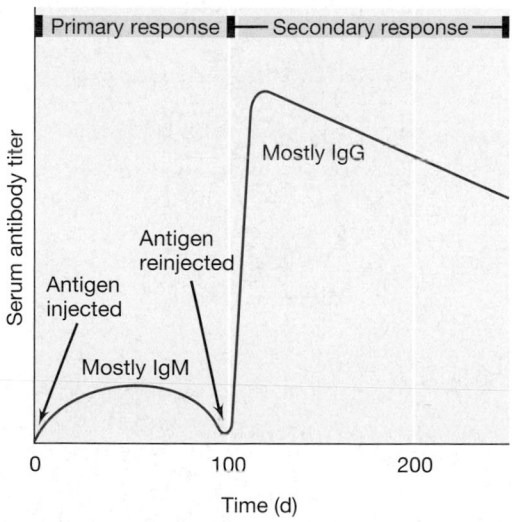

Figure 22.19 Primary and secondary antibody responses in serum. The antigen injected at day 0 and day 100 must be identical to induce the secondary response. The secondary response, also called a booster response, may be more than tenfold greater than the primary response. Note the class switch from IgM production in the primary response to IgG production in the secondary response.

sequential, ordered fashion by antigen–antibody complexes on bacterial cells, causing cell membrane damage and lysis or leakage of cellular contents. Complement proteins are found at comparable levels in the serum of all individuals. A major function of antibody is to recognize invading cells and activate the complement system for attack. Thus, many antibodies, each specific for a single antigen, can recruit the ubiquitous complement proteins; the immune system does not need individual enzymes to attack each invading agent.

The individual proteins of complement are designated C1, C2, C3, and so on. Activation of complement oc-

curs only with antibodies of the IgG and IgM classes (see Table 22.2). When IgG or IgM antibodies bind antigens, especially on cell surfaces, the antibodies fix (bind) the ever-present complement proteins (Figure 22.20). The complement proteins react in a sequential cascade, with activation of one complement component leading to activation of the next, and so on. The key steps start with (1) binding of antibody to antigen (initiation); (2) binding of C1 to the antibody–antigen complex, leading to C4–C2 binding at an adjacent membrane site, and activation and binding of C3; (3) membrane C3 catalyzes formation of a

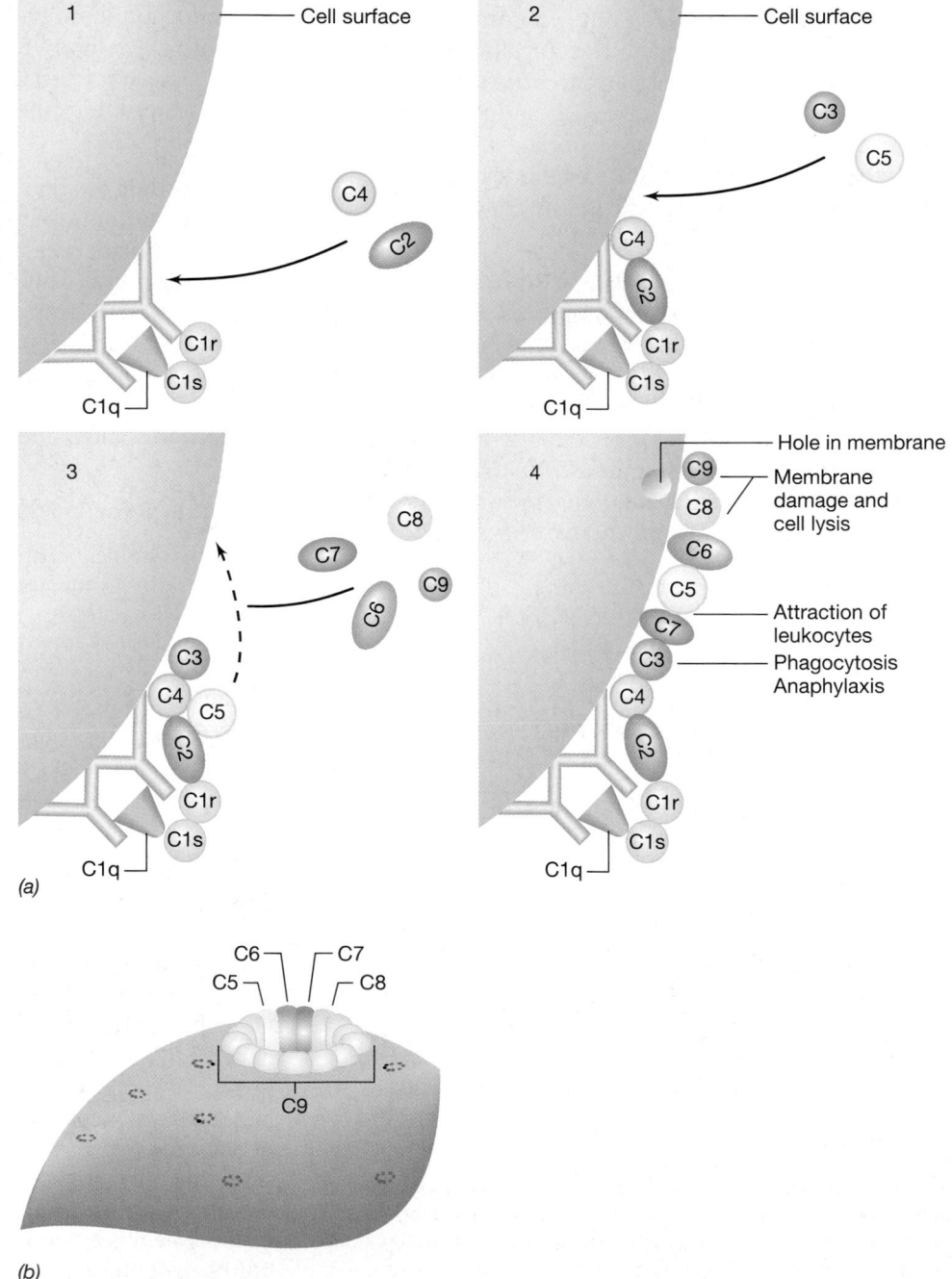

Figure 22.20 Complement system. (a) The sequence, orientation, and activity of the various components as they interact to lyse a cell. Panel 1: Binding of the antibody recognition unit C1q and other C1 proteins; panel 2: the C4-C2 complex; panel 3: the C4-C2-C3-C5 complex—after activation, the C5 unit travels to an adjacent membrane site; panel 4, binding of C6, C7, C8, and C9, resulting in membrane damage. (b) Three-dimensional view of the hole formed by complement components C5 through C9.

C5-C6-C7 complex at a second membrane site, and C8 and C9 are then deposited with the C5–C6–C7 complex, resulting in membrane damage and cell lysis (Figure 22.21).

Some of the by-products of complement activation are powerful chemoattractants called *anaphylatoxins* because they cause inflammatory reactions at the site of complement deposition (see Section 22.13). Reactions involving C3 result in chemotactic attraction and activation of phagocytes, resulting in increased phagocytosis. Reactions involving C5 lead to T-cell attraction and cytokine release.

Complement catalyzes the bactericidal and lytic actions of antibodies against many gram-negative bacteria. However, gram-positive bacteria are not killed by specific antibodies, even in the presence of complement. Gram-positive bacteria can, however, be destroyed through opsonization.

Opsonization

When antibody reacts with antigen on the surface of a cell, the cell is much more likely to be phagocytized. When complement fixes to this antibody–antigen complex on the surface of a cell, the cell is even more likely to be phagocytized. This is because most phagocytes, including macrophages and B cells, have antibody receptors as well as C3 receptors (C3R). These receptors bind the antibody constant domain and C3 complement protein, respectively. Normal phagocytic processes are enhanced by antibody binding and amplified even more by complement (C3) fixation (binding). The enhancement of phagocytosis by complement is called *opsonization*.

Gram-positive bacteria can be opsonized in this way, leading to enhanced phagocytosis and pathogen destruction. But, as was the case with gram-negative bacteria, antibodies must first be bound to surface antigens on the gram-positive cell to fix complement and promote opsonization.

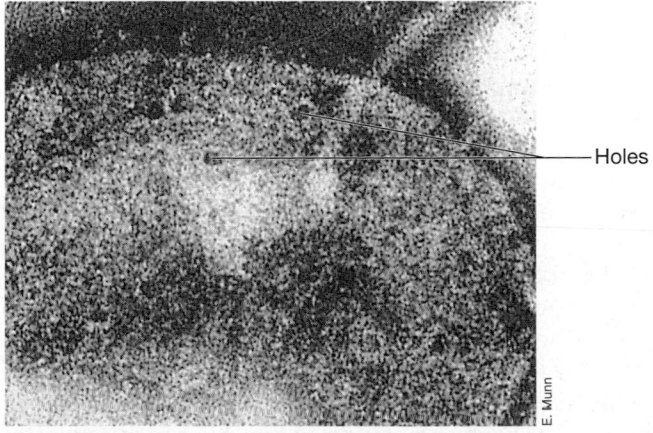

Holes

Figure 22.21 Electron micrograph of a negatively stained preparation of *Salmonella paratyphi*, showing holes created in the cell envelope as a result of a reaction involving cell envelope antigens, specific antibody, and complement.

The complement protein cascade is a nonspecific mechanism that enhances cell destruction and opsonization. Complement fixation is triggered by specific immune antibody interactions and is a critical component of host defense.

✓ Identify the immunoglobulin classes that fix complement.

✓ What complement components promote opsonization?

✓ What complement components are necessary to induce cell lysis?

IV USING THE IMMUNE RESPONSE TO PREVENT DISEASE

We now describe how the immune response prevents infectious disease through both naturally and artificially acquired immunity. We then discuss technical advances used for preparation of immunogens that may revolutionize our ability to safely induce specific immune responses to a variety of pathogens and their products.

22.11 Immunization to Prevent Disease

The major role of the immune response in the body is to protect the animal from the consequences of infection. The importance of antibodies in disease resistance is shown dramatically in individuals with the genetic disorder **agammaglobulinemia** in whom antibodies are not produced because their B cells are defective. Such individuals suffer from recurrent, life-threatening bacterial infections. The general lack of an antibody response is also observed in those suffering from acquired immunodeficiency syndrome (AIDS). In the case of AIDS, however, the problem is not due to a lack of B cells, but rather to a total lack of CD4 (T_H) cells (see Section 22.7). The crucial importance of T helper cells in immunity is clearly evident in AIDS patients: Lack of effective immunity results in death from infectious diseases (∞ Section 26.14).

The purposeful artificial induction of specific immunity to infectious diseases is a major contribution of microbiology to the treatment and prevention of infectious diseases. An animal or human may acquire immunity to a disease in several ways. (1) The individual may acquire infection and develop immunity. This is **natural active immunity** because the immunization was a natural outcome of infection in the infected individual producing the immune response. (2) The individual may be exposed to an antigen to induce formation of antibodies, a type of immunity known as **artificial active immunity** because

TABLE 22.3 Comparison of active and passive immunity

Active immunity	Passive immunity
Exposure to antigen; immunity achieved by injecting antigen	No exposure to antigen; immunity achieved by injecting antibodies or antigen-reactive T cells
Specific response made by individual achieving immunity	Specific immune response made in a secondary host
Immune system activated to antigen; immunological memory in effect	No immune system activation; no immunological memory
Immune response can be maintained via stimulation of memory cells (*i.e.*, booster immunization)	Immunity cannot be maintained and decays rapidly
Immune state develops over a period of weeks	Immunity develops immediately

the individual in question produced the antibodies. This process is commonly known as **vaccination**, properly termed **immunization**. (3) Alternatively, the individual may receive injections of an antiserum derived from another individual who has previously formed antibodies against the antigen in question. This is called **artificial passive immunity** because the individual receiving the antibodies played no active part in the antibody-producing process. (4) Finally, **natural passive immunity** also occurs. For several months after birth, newborns have maternal IgG antibodies in their blood. These antibodies, acquired across the placenta before birth, provide valuable disease protection while the immune system of the newborn is maturing. Active and passive immunity are contrasted in Table 22.3.

In *active* immunity, introduction of antigen produces fundamental changes in the host: The immune cells produce large quantities of antigen-specific immune effector molecules (Igs) and cells. A second ("booster") dose of the same antigen results in a faster, higher titer secondary response (see Figure 22.19). Active immunity often remains throughout life. A *passively* immunized individual never has more antibodies than received in the initial injection, and these antibodies gradually disappear from the body; moreover, a later exposure to the antigen does not elicit a secondary response. Active artificial immunity is usually used as a *prophylactic* measure, to protect a person against future attack by a pathogen. Passive artificial immunity is usually *therapeutic*, designed to cure a person who is presently suffering from the disease. For example, tetanus toxoid (see the following section) actively immunizes an individual against future encounters with *Clostridium tetani* exotoxin, whereas tetanus antiserum (antitoxin) (see later) is administered to passively immunize an individual suspected of acquiring a *C. tetani* infection.

Immunization

The material used in inducing artificial active immunity, the antigen or mixture of antigens, is known as a **vaccine** or an **immunogen**, and the process of generating such an immune response is **immunization**. Active introduction of immunity poses some risk when immu-

nizations against whole microorganisms are necessary. Inherently dangerous microorganisms must be altered to prevent infections. The microorganisms are usually killed by chemical agents such as phenol or formaldehyde or physical agents such as heat. The dead cells can then be used as an effective immunogen. Formaldehyde is also used to inactivate viruses for vaccines such as the *Salk polio vaccine*. Likewise, for toxin-caused diseases, the active form of the toxin is not used as an immunogen. Many exotoxins can be modified chemically so they retain their antigenicity but are no longer toxic. Such a modified exotoxin is called a **toxoid**. Toxoids are usually not such efficient antigens as the original exotoxin, but they can be given safely and in high doses.

Immunization with live cells or virus is usually more effective than with dead or inactivated material. Often it is possible to isolate a mutant strain of a pathogen that has lost its virulence but still retains the immunizing antigens; strains of this type are called **attenuated strains** (⌒ Section 21.8).

A summary of vaccines available for use in humans is given in Table 22.4. Most effective viral vaccines are in an attenuated form. Killed virus vaccines tend to provide only short-lived immune responses, without the desirable long-term memory response. Bacterial vaccines, on the other hand, are nearly always in killed form. The killed vaccines are adequate to induce long-term antibacterial protection in most cases. While the attenuated vaccines may be more effective (and in the case of many viruses are necessary), attenuated strains are difficult to select, standardize, and maintain. Live vaccines also have a limited shelf life and usually require refrigeration for adequate storage.

Immunization Practices

Infants possess antibodies derived from their mothers across the placenta or in breast milk and are relatively immune to infectious disease during the first 6 months of life. However, they are immunized for key infectious diseases as soon as possible so that their own *active* immunity can replace the maternal *passive* immunity. As discussed in Section 22.8, a single exposure to antigen does not lead to a high antibody titer; after the initial innoculation, a series of "booster" innoculations are given

TABLE 22.4 Available vaccines for infectious diseases in humans

Disease	Type of vaccine used
Bacterial diseases	
Anthrax	Toxoid
Diphtheria	Toxoid
Tetanus	Toxoid
Pertussis	Killed bacteria (*Bordetella pertussis*) or acellular proteins
Typhoid fever	Killed bacteria (*Salmonella typhi*)
Paratyphoid fever	Killed bacteria (*Salmonella paratyphi*)
Cholera	Killed cells or cell extract (*Vibrio cholerae*)
Plague	Killed cells or cell extract (*Yersinia pestis*)
Tuberculosis	Attenuated strain of *Mycobacterium tuberculosis* (BCG)
Meningitis	Purified polysaccharide from *Neisseria meningitidis*
Bacterial pneumonia	Purified polysaccharide from *Streptococcus pneumoniae*
Typhus fever	Killed bacteria (*Rickettsia prowazekii*)
Haemophilus influenzae meningitis	Conjugated vaccine (polysaccharide of *Haemophilus influenzae* conjugated to protein)
Lyme disease	Recombinant membrane protein of *Borrelia burgdorferi*
Viral diseases	
Yellow fever	Attenuated virus
Measles	Attenuated virus
Mumps	Attenuated virus
Rubella	Attenuated virus
Polio	Attenuated virus (Sabin) or inactivated virus (Salk)
Influenza	Inactivated virus
Rabies	Inactivated virus (human) *or* attenuated virus (dogs and other animals)
Smallpox	Cross-reacting virus (vaccinia) (∞ Section 16.12)
Hepatitis A	Recombinant DNA vaccine
Hepatitis B	Recombinant DNA vaccine *or* inactivated virus
Varicella (chickenpox)	Attenuated virus

to produce a high antibody titer. The recommended vaccination schedule for children in the United States from birth to adulthood is shown in Figure 22.22.

The importance of immunization in controlling infectious diseases is well established. Introduction of a specific immunization procedure into a population often dramatically reduces the incidence of the disease (∞ Figures 26.14 and 26.17). The degree of immunity obtained by vaccination varies greatly, depending on the individual and on the quality and quantity of the vaccine. However, lifelong immunity is rarely achieved by means of a single injection, or even a series of injec-

tions, and the immune cells induced by immunization gradually disappear from the body. On the other hand, antigenic stimulation often occurs even in the absence of immunization through asymptomatic or minor infections. A natural infection induces a secondary response, leading to an increase in production of antibody. In the complete absence of antigenic stimulation, the length of effective immunity varies considerably with different antigens. However, active immunity to certain immunogens such as tetanus toxoid may last many years.

Immunizations are not only beneficial to the individual but are effective public health measures because disease spreads poorly through a population with a large proportion of immune individuals (∞ Section 25.5).

Passive Immunity

The material used in inducing passive immunity—the serum containing antibodies—is known as a *serum*, an *antiserum*, or an *antitoxin* (antibodies directed against a toxin). Antisera are obtained either from immunized animals, such as horses, or from humans who have high antibody titers. These individuals are said to be **hyperimmune**. The antiserum or antitoxin is standardized to contain a known antibody titer; a sufficient number of units of antiserum must be injected to neutralize any antigen that might be present in the body. Sometimes the immunoglobulin fraction of pooled human serum is used as a source of antibodies. It contains a wide variety of antibodies that other individuals have formed by artificial or natural exposure to various antigens. Pooled sera are used when hyperimmune antisera are not available.

✓ **22.11 Concept Check**

Immunity to infectious disease can be either passive or active, natural or artificial. Immunization, a form of artificial active immunity, is widely employed to prevent infectious diseases. Most agents used for immunization are either attenuated or inactivated pathogens, or inactivated forms of natural microbial products.

✓ Provide an example of natural passive immunity. How does natural passive immunity benefit the immunized individual?

✓ Provide an example of artificial active immunity. How does artificial active immunity benefit the immunized individual?

22.12 New Immunization Strategies

Most immunization preparations are produced from whole organisms or toxoids, as described in the previous section. However, there are several other methods for producing antigens suitable for immunization.

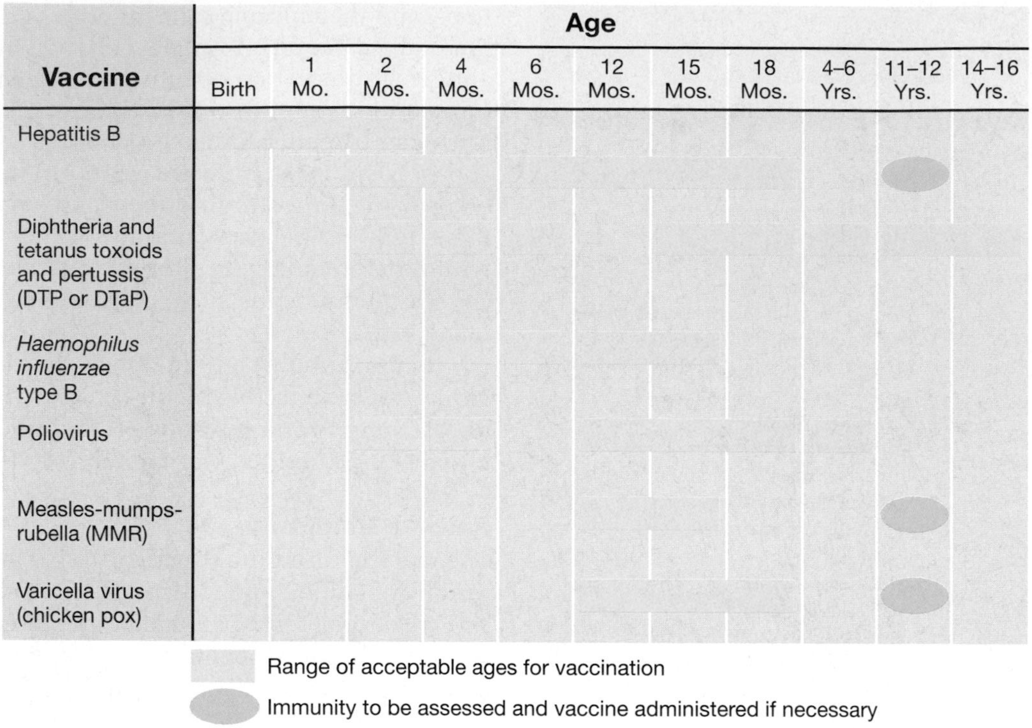

Vaccine	Age										
	Birth	1 Mo.	2 Mos.	4 Mos.	6 Mos.	12 Mos.	15 Mos.	18 Mos.	4–6 Yrs.	11–12 Yrs.	14–16 Yrs.
Hepatitis B										●	
Diphtheria and tetanus toxoids and pertussis (DTP or DTaP)											
Haemophilus influenzae type B											
Poliovirus											
Measles-mumps-rubella (MMR)										●	
Varicella virus (chicken pox)										●	

▭ Range of acceptable ages for vaccination

⬭ Immunity to be assessed and vaccine administered if necessary

Figure 22.22 Recommended childhood immunizations in the United States as specified by the Centers for Disease Control and Prevention, Atlanta, GA. For pertussis, the DTP vaccine contains a whole-cell *Bordetella pertussis* preparation while the DtaP vaccine is an acellular *Bordetella pertussis* preparation. Both vaccines are effective in preventing disease. For polio, killed poliovirus immunizations should be used at 2 and 4 months, while live attenuated preparations may be used for subsequent immunizations.

Synthetic and Genetically Engineered Immunizing Agents

The simplest alternate approach to vaccine development is the use of *synthetic peptides*. To make a vaccine, a peptide can be synthesized that corresponds to a known epitope on an infectious agent. For example, the structure of the protein antigen responsible for immunity to foot-and-mouth virus, an important animal pathogen, is known. A synthetic peptide of 20 amino acids constituting the antigenic determinant of the protein has been made and attached to suitable carrier molecules. This synthetic vaccine evokes an excellent neutralizing antibody response to foot-and-mouth virus. However, as a general method, this approach has one major problem: The entire antigenic profile of the protein must be known to make an effective vaccine. Although this condition has been met for foot-and-mouth virus, very few pathogens have such a well-defined antigenic profile.

Sophisticated molecular biology techniques can also be used to make vaccines. For example, genes that encode antigens from virtually any virus can be cloned into the vaccinia virus genome and expressed. Inoculation with the *genetically engineered* vaccinia virus can then be used to induce immunity to the product of the cloned gene. Such a preparation is a *recombinant-vector vaccine*.

This method depends on the availability of the cloned gene that encodes the antigen and also on the ability of the vaccinia virus to express the cloned gene as an antigenic protein. An effective recombinant vaccinia-rabies vaccine has been developed for use in animals. Recombinant DNA methods to develop vaccines will be discussed in Section 31.6.

Another immunization strategy involves the production of recombinant DNA proteins as immunogens. First, a pathogen gene must be cloned into a suitable host. The host, chosen for its ability to express protein encoded by the cloned gene, will then express the pathogen protein. The pathogen protein can then be harvested and used as a vaccine. This is known as a *recombinant antigen vaccine*. For example, the current hepatitis B virus vaccine is a major hepatitis surface protein antigen (HbsAg) expressed by yeast cells.

DNA Vaccines

A simple, novel method for immunization is based on expression of cloned genes in host cells. Bacterial plasmids containing cloned DNA are injected intramuscularly into a host animal. After several weeks, the host responds with T_C cells, T_H1 cells, and antibodies directed to the protein encoded by the cloned DNA. Apparently, the DNA is transcribed and translated to

TABLE 22.5 Classification of hypersensitivity diseases

Classification	Description	Immune mechanism	Time of latency	Examples
Type I	Immediate	IgE sensitization of mast cells	Minutes	Reaction to bee venom (sting) Hay fever
Type II	Cytotoxic[a]	IgG interaction with cell surface antigen	Hours	Drug allergies (Penicillin)
Type III	Immune complex[a]	IgG interaction with soluble or circulating antigen	Hours	Systemic lupus erythematosis (SLE)
Type IV	Delayed type	T_H1 inflammatory cells	Days	Poison ivy Tuberculin test

[a] Most autoimmune diseases are caused by Type II or Type III reactions.

produce an immunogenic protein, and this triggers an immune response. These plasmids are called DNA vaccines.

DNA vaccines provide considerable advantages over many conventional immunization protocols. For instance, since only a single foreign gene is injected, there is no chance of infection as there might be with an attenuated vaccine. Second, antigens and even single antigenic determinants can be used in a vaccine to target the immune response to a particular cell component such as a tumor-specific antigen. Finally, this is the only method known in which immunization with a single bioengineered antigen elicits a T_C response, in addition to the antibody response. This is presumably because the cloned gene causes expression of antigenic proteins inside the plasmid-infected host cells. The proteins are then processed and presented to T cells through the normal antigen presentation schemes, resulting in activation of T_C cells (see Sections 22.5 and 22.6).

✓ 22.12 Concept Check

Advances in biotechnology and immunology have resulted in alternate immunization strategies that eliminate exposure to microorganisms and, in some cases, even to antigen!

✓ Provide at least two examples of alternate immunization strategies. What is the advantage of each alternative immunization strategy over current immunization procedures?

V IMMUNE RESPONSE DISEASES

Immune reactions can cause host cell damage and disease. **Hypersensitivity** responses are inappropriate immune responses that result in host damage. **Superantigens** are proteins produced by bacteria and viruses that cause massive stimulation of immune cells, also resulting in host damage.

22.13 Allergy, Hypersensitivity, and Autoimmunity

Abnormal immune conditions that damage the host are classified as hypersensitivity diseases. Antibody-mediated *immediate hypersensitivity* is commonly called *allergy*. Cell-mediated reactions also cause disease in the form of *delayed-type hypersensitivity*. *Autoimmune diseases* are directed against self antigens.

Immediate Hypersensitivity (Type I Hypersensitivity)

Many hypersensitivity reactions occur within minutes after exposure to antigen. This form of hypersensitivity is referred to as **immediate hypersensitivity** and is mediated by *antibodies* (Table 22.5). Immediate hypersensitivities are commonly called **allergies**. Depending on the individual and the antigen, immediate hypersensitivity reactions may be very mild, or cause extremely severe, even life-threatening reactions, a process known as *anaphylaxis*. Antigens that cause these hypersensitivities are known as *allergens*.

Up to 20% of the population suffers from immediate hypersensitivities involving allergic (*anaphylactic*) reactions to specific allergens such as pollens, animal dander, and a variety of other agents (Table 22.6). In a typical anaphylactic reaction, an allergen elicits the production of immunoglobulins of the IgE class. Rather than circulating like IgG or IgM, IgE antibodies bind via a cell-binding constant domain (see Section 22.8) to specific receptors on mast cells (Figure 22.23). **Mast cells** are nonmotile connective tissue cells found adjacent to

TABLE 22.6 Common immediate-type hypersensitivity allergens

Pollen and fungal spores (hay fever)
Insect venoms (bee sting)
Certain foods
Animal dander
Mites in house dust

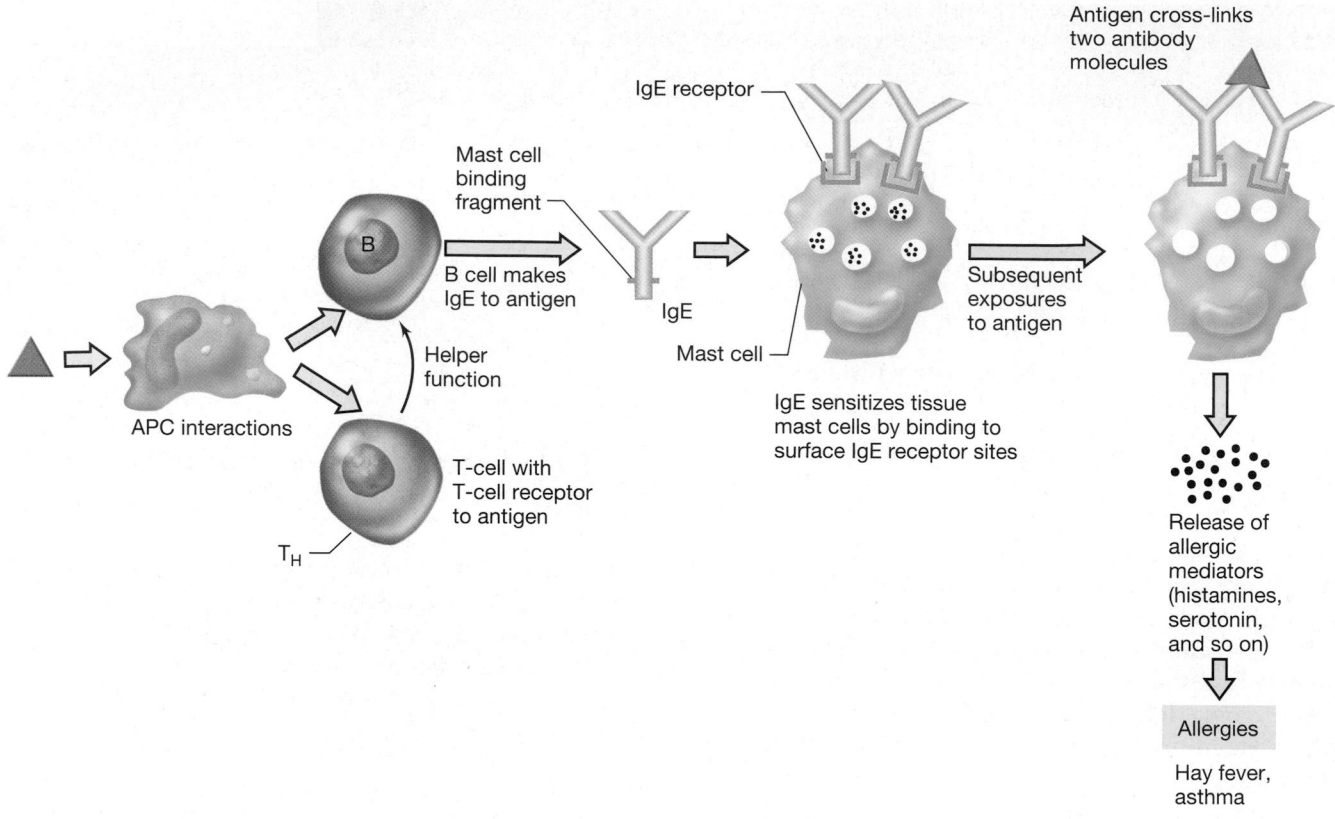

Figure 22.23 Immediate hypersensitivity. IgE binds to mast cells by means of a high-affinity surface receptor for the C_H4 domain. Binding arms the mast cell. Antigen contact and cross-linking of the IgE proteins initiates release of vasoactive mediators, resulting in systemic symptoms ranging from mild allergies to life-threatening anaphylaxis.

capillaries throughout the body. With the next exposure to antigen, the cell-bound IgE molecules bind the antigen. This binding triggers the release of several soluble allergic mediators from the mast cells.

The primary chemical mediators released from mast cells are **histamine** and **serotonin** (both are modified amino acids). The release of histamine and serotonin causes dilation of blood vessels and contraction of smooth muscle, initiating the typical symptoms of anaphylaxis. These symptoms include breathing difficulty, flushed skin, mucus production, sneezing, and itchy, watery eyes. In general, the symptoms are relatively short-lived, but once initially sensitized by an allergen, an individual can respond with each subsequent exposure to the antigen. The magnitude of the allergic reactions may vary from mild symptoms (or none), to severe symptoms such as **anaphylactic shock**. In humans, anaphylactic shock is characterized by severe respiratory distress, capillary dilation (causing a sharp drop in blood pressure), and flushing and itching of the skin. If severe cases of anaphylactic shock are not treated immediately with adrenalin to

counter smooth muscle contraction, increase blood pressure, and promote breathing, death can occur.

Delayed-Type Hypersensitivity (Type IV Hypersensitivity)

Delayed-type hypersensitivity (DTH) is cell-mediated and involves the activities of the T_H1 inflammatory cells (Table 22.5). Symptoms begin to appear several hours after secondary exposure to the eliciting antigen, with a maximal response usually occurring in 24–48 h. Typical antigens include certain microorganisms, a few self-antigens (Table 22.7), and a group of chemicals that covalently bind to the skin, creating new antigens. Hypersensitivity to these newly created antigens is known as **contact dermatitis** and is responsible for the allergic skin reactions to poison ivy, jewelry, cosmetics, and certain chemicals and drugs. Shortly after exposure to the agent, the skin feels itchy at the site of contact, and within several hours reddening and swelling appear, indicative of a general inflammatory response. Localized tissue destruction,

TABLE 22.7 Some autoimmune diseases of humans

Disease	Organ or area affected	Mechanism (hypersensitivity type)
Juvenile diabetes (insulin-dependent diabetes mellitus)	Pancreas	Cell-mediated immunity and autoantibodies against surface and cytoplasmic antigens of islets of Langerhans (II and IV)
Myasthenia gravis	Skeletal muscle	Autoantibodies against acetylcholine receptors on skeletal muscle (II)
Goodpasture's syndrome	Kidney	Autoantibodies against basement membrane of kidney glomeruli (II)
Rheumatoid arthritis	Cartilage	Autoantibodies against self IgG antibodies, which form complexes deposited in joint tissue, leading to inflammation and cartilage destruction (III)
Hashimoto's disease (hypothyroidism)	Thyroid	Autoantibodies to thyroid surface antigens (II)
Male infertility (some cases)	Sperm cells	Autoantibodies agglutinate host sperm cells (II)
Pernicious anemia	Intrinsic factor	Autoantibodies prevent absorption of vitamin B_{12} (III)
Systemic lupus erythematosis	DNA, cardiolipin, nucleoprotein, blood clotting proteins	Massive autoantibody response to various cellular constituents results in immune complex formation (III)
Addison's disease	Adrenal glands	Autoantibodies to adrenal cell antigens (II)
Allergic encephalomyelitis	Brain	Cell-mediated response against brain tissue (IV)
Multiple sclerosis	Brain	Cell-mediated and autoantibody response against central nervous system (II and IV)

often in the form of blistering, occurs as a result of the activities of immune cells.

A good example of a delayed-type hypersensitivity reaction is the development of immunity to the causal agent of tuberculosis, *Mycobacterium tuberculosis* (Figure 22.24). This cellular immune response was first discovered by Robert Koch during his classic work on tuberculosis (∞ Section 1.5) and has been widely studied. Antigens derived from the bacterium, when injected subcutaneously into an animal previously infected with *M. tuberculosis*, elicit a characteristic skin reaction that develops fully only after a period of 24–48 h. (In contrast, skin reactions to IgE-mediated responses as discussed previously, develop almost immediately after antigen injection.) In the region of the injected antigen, T_H1 cells become stimulated by the antigen and release cytokines that attract and activate large numbers of macrophages. The macrophages are responsible for the ingestion and destruction of the invading antigen. The characteristic skin reaction seen at the site of injection includes induration (hardening), edema (swelling), erythema (reddening), pain, and localized heating. These are common features of an inflammatory response resulting from the release of cytokines by activated, antigen-stimulated T_H1 cells. This skin response serves as the basis for the **tuberculin test** for determining prior exposure to *M. tuberculosis* (Figure 22.24).

A number of microbial infections elicit delayed-type hypersensitivity reactions. In addition to tuberculosis, these include leprosy, brucellosis, psittacosis (all caused by *Bacteria*), mumps (caused by a virus), and coccidioidomycosis, histoplasmosis, and blastomycosis (caused by fungi). In all these cellular immune reactions, specific skin reactions occur after injection of antigens derived from the pathogens, and these skin reactions are used to diagnose prior exposure to the pathogen.

Finally, T_H1 cells can also be involved in autoimmune responses directed against self-antigens. For example, T_H1 cells are involved in the pathogenesis of allergic encephalitis and Type I (juvenile) diabetes mellitus (Table 22.6). However, most autoimmune diseases are mediated by antibodies, as we shall now discuss.

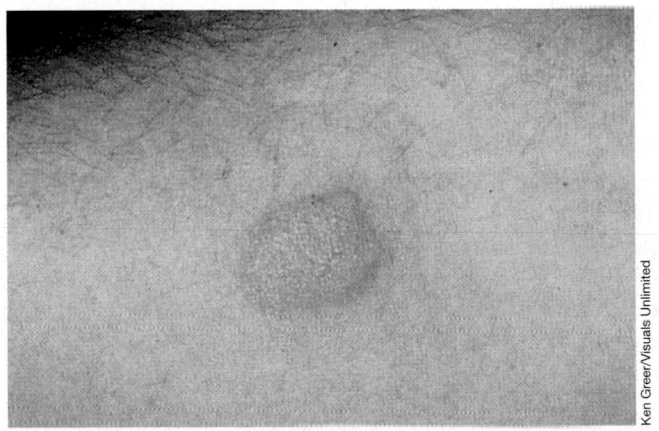

Ken Greer/Visuals Unlimited

Figure 22.24 Cell-mediated immunity. A positive tuberculin test, typical for delayed hypersensitivity, and the result of the action of T_H1 effector cells. The raised area of inflammation is 1.5 cm in diameter.

Autoimmune Diseases (Type II and Type III Hypersensitivities)

T and B cells destined to react with self antigens are normally eliminated during the process of lymphocyte maturation. However, in some individuals, T and B lymphocytes can be activated to produce immune reactions against self proteins, leading to *autoimmune disease* (Table 22.7).

Depending on the specific disorder, autoimmunity may involve autoantibodies or a cellular immune response to self constituents. Certain autoimmune diseases are highly organ-specific. For example, in *Hashimoto's disease*, **autoantibodies**, antibodies that interact with self antigens, are made against thyroglobulin and other thyroid proteins. The disease affects thyroid function and is classified as a Type II disorder because antibodies interact with antigens found on the surface of thyroid cells and cause destruction of the thyroid gland. In *juvenile diabetes* (insulin-dependent diabetes mellitus), autoantibodies against the insulin-producing cells in the islets of Langer-hans in the pancreas are observed, but tissue destruction occurs via T_H1 cells.

Systemic lupus erythematosis (SLE) involves a large-scale production of autoantibodies against many self constituents, including DNA. This disease and others like it are induced by circulating antigen–antibody complexes that deposit in several different body tissues, such as the kidney and the spleen. Complement fixation and the resulting lytic and inflammatory responses cause local but often severe cell damage at the site of the complex deposition. SLE is a clear example of a Type III immune complex disorder.

Organ-specific autoimmune diseases are sometimes more easily controlled clinically because the product of organ function, such as thyroxin in hypothyroidism or insulin in diabetes, can often be supplied in pure form from another source. More generalized diseases such as SLE can be controlled by immunosuppressive therapy such as steroid drugs, but this approach is risky because of the increased chance of opportunistic infections.

Heredity has an important influence on the incidence, type, and severity of autoimmune diseases. An inherited tendency to develop certain autoimmune diseases is known to exist; many autoimmune diseases correlate strongly with the presence of certain major histocompatibility complex (MHC) antigens (see Section 22.5; ∞ Section 23.2). Studies of model autoimmune diseases in mice support such a genetic link, but the precise conditions necessary for developing autoimmunity are also dependent on other factors, including hormone levels and exposure to certain bacterial or viral infections.

✓ 22.13 Concept Check

Hypersensitivity results when foreign antigens induce inappropriate cellular or humoral immune responses, leading to host tissue damage. Autoimmunity occurs when the immune response is directed against self antigens, resulting in damage to host tissue.

✓ Discriminate between *immediate hypersensitivity* and *delayed hypersensitivity* with respect to antigens and effectors.

✓ Identify the two main categories of autoimmune disease.

22.14 Superantigens

We discussed the mechanisms of action for several different categories of bacterial toxins in Chapter 21. Most exotoxins interact directly with cells to cause cell damage (∞ Sections 21.10 and 21.11). Endotoxins also interact directly with a large variety of cell types, causing release of endogenous pyrogens and other soluble mediators, and producing fever and general inflammation (∞ Secton 21.14). Here, we discuss another category of exotoxins, the **superantigens**. Superantigens do not act directly on host cells, but use a novel mechanism to cause extensive host tissue damage through the immune system.

Superantigen Activation of T Cells

Superantigens are proteins capable of eliciting a very strong response because they activate more T cells than a normal immune response. This happens because superantigens differ from conventional antigens in their mode of binding to the TCR. Most superantigens are produced by certain viruses and bacteria. Streptococci and stapylococci in particular produce a wide variety of different and very potent superantigens (∞ Table 21.4).

Superantigens work by activating large numbers of T cells, which in turn produce cytokines. The cytokines then stimulate other cells, in particular, macrophages and other phagocytes. Because of the large amount of cytokines produced, this cell-mediated response is characterized by *systemic* inflammatory reactions, often resulting in fever, diarrhea, vomiting, mucus production, and systemic shock. In extreme cases, exposure to superantigens can be fatal.

Conventional foreign antigens bind to the TCR at the variable (V) domain antigen-binding site. In a typical immune response, less than 0.01% of all available T cells interact with a conventional foreign antigen (see Section 22.5). However, superantigens bind to a site on the $V\beta$ domain of the TCR that is *outside* the normal TCR antigen-binding site. Many different TCRs share the same structure outside the antigen-binding site, and the superantigen binds to *all* T cells with the shared structure. A typical response to a superantigen will result in the binding and activation of 5% to 25% of all T cells. The superantigens also bind to class II MHC molecules on APCs, again at a specific site outside the normal peptide binding site (Figure 22.25). These cell surface interactions activate large numbers of T cells, resulting in the systemic inflammatory symptoms mentioned above.

Several diseases can be attributed to superantigens. *Staphylococcus aureus* food poisoning, characterized by fever, vomiting, and diarrhea, is caused by one of several superantigens that act as staphylococcal enterotoxins. *Staphylococcus aureus* also produces the superantigen responsible for toxic shock syndrome. *Streptococcus pyogenes* produces erythrogenic toxin, the superantigen responsible for scarlet fever. Several other superantigen-induced diseases were mentioned previously (∞ Section 21.10 and Table 21.4).

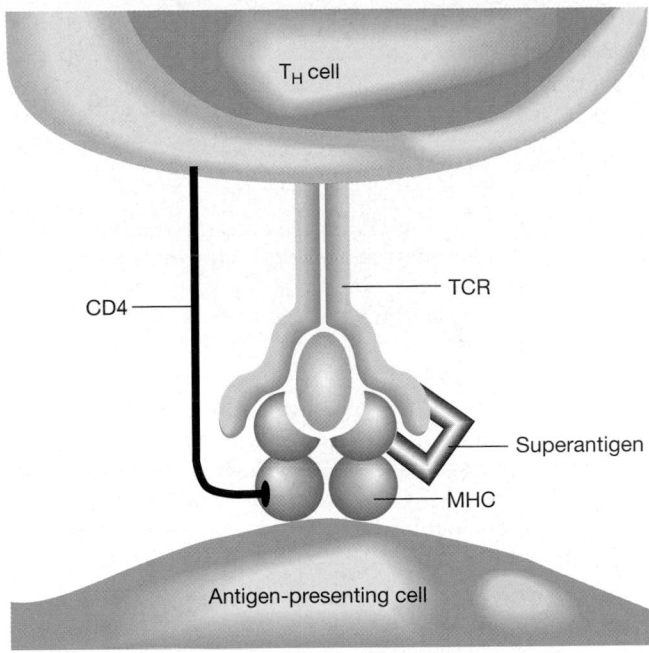

Figure 22.25 Bacterial superantigens act by binding to both the MHC protein and the TCR at positions outside the normal binding site. Because the superantigen-binding sites are found on conserved regions of MHC and TCR proteins, the superantigen can interact with large numbers of cells, stimulating massive T-cell activation, cytokine release, and even systemic inflammation.

✓ 22.14 Concept Check

Superantigens bind and activate large numbers of T cells in a novel fashion. Superantigen-activated T cells are capable of producing systemic diseases characterized by massive inflammatory reactions.

✓ Discriminate between normal and superantigen activation of T cells.

✓ Identify the binding site for superantigens on both T cells and APCs.

Review Questions

1. What is the origin of the phagocytic and antigen-specific cells involved in the immune response? Track the maturation of B cells and T cells.

2. Explain how phagocytes engulf and kill microorganisms, with particular attention to oxygen-dependent mechanisms.

3. Identify the most important features of the antigen-specific immune response.

4. What molecules induce immune responses? What properties are necessary for a molecule to induce an immune response?

5. Describe the basic structure of class I and class II major histocompatibility complex (MHC) proteins. In what functional ways do they differ?

6. Differentiate between T_C cells and NK cells. What is the activation signal for each cell type?

7. How do T_H cells differ from T_C cells? Differentiate between the functional roles of T_H1 and T_H2 cells.

8. Describe the structural and functional differences among the five major antibody classes.

9. Identify the cell interactions involved in production of antibodies by B cells.

10. Describe the complement cascade. Is the order of protein interaction important? Why or why not?

11. List the diseases for which you have been immunized. List the diseases for which you may have naturally acquired immunity.

12. Describe at least one of the biotechnology-based immunization strategies that we described. How does this strategy differ from a conventional immunization strategy?

13. Define the differences between immediate and delayed-type hypersensitivity in terms of immune effectors, target tissues, antigens, and clinical outcome.

14. Describe the general mechanism used by superantigens to activate T cells. How does superantigen activation differ from T-cell activation by conventional antigens?

Application Questions

1. Describe the potential problems that would arise if an individual had an acquired inability to phagocytize pathogens. Could the individual survive in a normal environment such as a college campus? Explain.

2. Specificity and tolerance are necessary qualities for an adaptive immune response. However, memory seems to be less critical, at least at first glance. Define the role of immune memory and explain how the production and maintenance of memory cells might benefit the host in the long term.

3. Trace the path of a pathogen that comes into the body via a wound in the left calf. What cells will recognize the pathogen first? Where is the pathogen most likely to cause an infection? What cells and organs are involved in destruction of the pathogen? What cells and organs are involved in production of specific immunity to the pathogen?

4. All immunogens are antigens. However, all antigens are not immunogens. Explain these statements using molecular examples to support your arguments.

5. Describe the potential problems that would arise if an individual had a hereditary deficiency that resulted in inability to present antigens to T_C cells. To T_H cells? To both T_C and T_H cells? What molecules might be deficient in each individual? Could any of these individuals survive in a normal environment? Explain.

6. Differentiate between the protective roles of T_C cells and NK cells. Is one cell type more important than the other for effective pathogen neutralization and destruction? Explain.

7. In certain situations, B cells are activated in the absence of T cells. Why might T cells not be able to "help" B cells with certain antigens? *Hint*: Think about the antigen-processing pathways. Are polysaccharide antigens processed in the same way as protein antigens?

8. Antibodies of the IgA class are probably more prevalent than those of the IgG class. Explain why this is true and what advantage this provides for the host.

9. How does TCR interaction with antigen differ from that of immunoglobulin (Ig) interaction? What fundamental difference does this make in the kinds of antigens recognized by the T cells and B cells?

10. Complement is regarded as a critical humoral defense mechanism. Do you agree with this statement? Explain your answer. What might happen to individuals who lack complement component C3? C5?

11. Would a passively administered antibody against an influenza viral surface protein be protective against the disease? Could this antibody protect individuals during an epidemic outbreak of the virus? Would you expect these antibodies to be effective over a long time period? Why or why not?

12. Many infectious diseases have no effective vaccines. Pick several of these diseases (for example, AIDS, malaria, the common cold) and explain why current vaccine strategies have not been effective. Prepare some alternate strategies for immunization against the diseases you have chosen.

13. Autoimmune diseases such as SLE and rheumatoid arthritis are very prevalent in women, and much less so in men. Moreover, individuals affected by these diseases generally show their first symptoms before the age of 40. Speculate concerning potential reasons for this sexual dimorphism with regard to certain autoimmune diseases.

14. For a pathogen, what is the evolutionary advantage for producing a superantigen? Give an example to support your answer. How might an infected host species avoid the problems of massive T cell activation associated with superantigens? Again, think in evolutionary terms.

The immune response works through a series of molecular interactions on the surface of cells. A critical first step in pathogen recognition is the presentation of pathogen-derived proteins in the context of a host protein, known as a *major histocompatibility complex protein*, shown here. The pathogen protein, now imbedded in the host protein, is recognized by pathogen-specific lymphocytes. Recognition triggers lymphocytes to target the imbedded pathogen protein and respond by destroying any cell displaying that protein; in so doing, the lymphocytes eliminate the pathogen from the host.

MOLECULAR IMMUNOLOGY

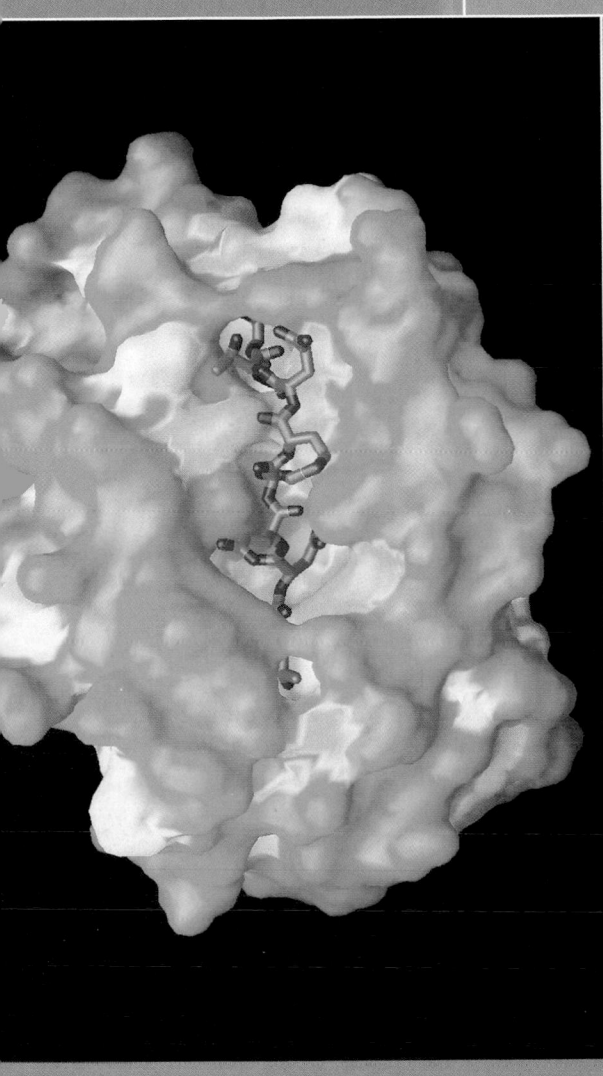

Working Glossary

Agretope the portion of a processed antigen that is recognized by MHC protein

Anergy the inability to produce an immune response to specific antigens due to neutralization of effector cells

Antibody a soluble protein, produced by B cells, that interacts with antigen; also called immunoglobulin

Apoptosis programmed cell death

Chemokine a small soluble protein produced by a variety of cells that modulates inflammatory reactions and immunity in target cells

Class I MHC protein antigen-presenting protein found on all nucleated vertebrate cells

Class II MHC protein antigen-presenting protein found on macrophages, B cells, and dendritic cells (antigen-presenting cells)

Clonal deletion for T-cell selection in the thymus, the killing of useless or self-reactive clones

Clonal selection each B or T cell produces copies of itself when stimulated with antigen

Complementarity-determining regions (CDRs) variations in amino acid sequence that occur within the variable domains of Igs or TCRs that provide most of the molecular contacts with antigen (also known as hypervariable regions)

Cytokine a small, soluble protein produced by a leukocyte that modulates inflammatory reactions and immunity in target cells

Domain a region of a protein having a defined structure and function

Epitope the portion of an antigen that is recognized by an immunoglobulin or a T-cell receptor

HLA human leukocyte antigen, the MHC of humans

Hypervariable regions variations in amino acid sequence that occur within the variable domains of Igs or TCRs that provide most of the molecular contacts with antigen (also known as complementarity-determining regions)

Immunoglobulin (Ig) a soluble protein, produced by B cells, that interacts with antigen; also called antibody

Immunoglobulin gene superfamily a family of genes that are evolutionarily, structurally, and functionally related to immunoglobulins

Major histocompatibility complex (MHC) a genetic region responsible for encoding several cell-surface proteins important for antigen processing and presentation

Motif a conserved amino acid sequence found in all peptide antigens that bind to a given MHC protein

Negative selection in T-cell selection, T cells that interact with self antigens in the thymus are deleted (see clonal deletion)

Polymorphism the occurrence of multiple alleles at a locus in frequencies that cannot be explained by the occurrence of recent random mutations

Positive selection in T-cell selection, T cells that interact with self MHC protein in the thymus are stimulated to grow and develop

T-cell receptor (TCR) antigen-specific receptor protein on the surface of T cells

Tolerance inability to produce an immune response to a specific antigen

---

The previous chapter introduced the essential features and functions of the immune response. In this chapter, we will build upon these principles and discuss the proteins and complex molecular interactions that underlie the immune response. We will first examine in detail the proteins that interact with antigens: the major histocompatibility complex (MHC) proteins, T-cell receptors (TCRs), and immunoglobulins (Igs). We will also examine the genetic basis for diversity and variation in all of these molecules. In the final section we will concentrate on the intercellular interactions that control and activate the immune response.

I THE IMMUNOGLOBULIN GENE SUPERFAMILY

The **immunoglobulin gene superfamily** is a group of genes and their protein products that share structural, evolutionary, and/or functional features with immunoglobulin genes and proteins. We discuss this superfamily here because the antigen-binding proteins involved in the immune response are members of this extended gene family.

23.1 Cell-Surface Receptors and Immunity

Three different cell-surface proteins interact specifically with antigens during an immune response, as we have discussed in Chapter 22. These are the **immunoglobulins** (Igs or antibodies), the **T-cell receptors** (TCRs), and products of the genes of the major histocompatibility complex (MHC), the **MHC proteins**. Although each of these proteins binds to antigen, each protein has a different location, structure, and function. MHC proteins, found on antigen-presenting cells and target cells, present antigens to TCRs, which are found exclusively on T cells (∞ Section 22.5). Immunoglobulins, which are found on B lymphocytes, interact directly with extracellular antigens (∞ Section 22.8).

Structure and Evolution of Antigen-Binding Proteins

The antigen-binding Ig, TCR, and MHC proteins share several structural features and are clearly evolutionarily related. Molecular and structural analysis indicate that these proteins belong to the Ig gene superfamily and have evolved by duplication and selection of primordial antigen-receptor genes. The basic features of selected Ig superfamily proteins are shown in Figure 23.1. Each protein has at least one region of highly conserved amino acid sequence, a so-called "constant" (C) **domain** of about 100 amino acids in length containing an intrachain disulfide bond spanning 50 to 70 amino acids. Beta-2 microglobulin (β_2m), part of the class I MHC protein (see Section 23.2), actually consists of a single C domain. The "variable" (V) domains of TCR, Ig, and MHC proteins are about the same size as the constant domains, but V domains are considerably different from one another and from the C domains. C domains provide structural integrity for the antigen-binding molecules, attach the V domains to the membrane, and give the proteins their characteristic shape. C domains also act as targets for other accessory molecules. For example, constant domains of IgG and IgM bind complement (◦◦ Section 22.10); MHC class I constant domains bind to the accessory CD8 protein on T-cytotoxic (T_C) cells, and homologous MHC class II constant domains bind CD4 on T_H cells (◦◦ Section 22.5). However, the V domains have evolved to interact with a wide variety of antigens.

TCR, Ig, and MHC proteins consist of two nonidentical polypeptides that associate to form a functional protein, and all are expressed on the surface of cells as antigen receptors. However, as we discussed in Chapter 22, the specific functions of these molecules are quite different. The Igs anchored on B-cell surfaces bind to proteins on bacteria and viruses and to bacterial products such as toxins; Igs are also produced in large quantities as soluble serum proteins. The TCRs, found exclusively on T lymphocytes, interact with antigenic peptides presented by

target cells or antigen-presenting cells (APCs). The *MHC proteins* bring antigenic peptides to the surface of the target cells or APCs. The peptide antigens embedded in MHC protein and presented on the surface of APCs are then bound by TCRs on T cells. Antigen-reactive T cells then kill the interacting cell, or produce proteins that activate the immune response (◦◦ Sections 22.6 and 22.7).

✓ 23.1 Concept Check

The immunoglobulin gene superfamily is composed of a family of genes that are evolutionarily, structurally, and functionally related to immunoglobulins. The antigen-binding immunoglobulins, T-cell receptors, and class I and class II MHC proteins are all members of this family, as are a number of other proteins involved in the immune response.

✓ Describe the structural features of the immunoglobulin C domain.

✓ Why do the V domains have structures that are dissimilar to the C domains?

II THE MAJOR HISTOCOMPATIBILITY COMPLEX (MHC)

The major histocompatibility complex (MHC) is a group of genes found in all vertebrates. The MHC spans about 4 million base pairs on human chromosome 6 and is

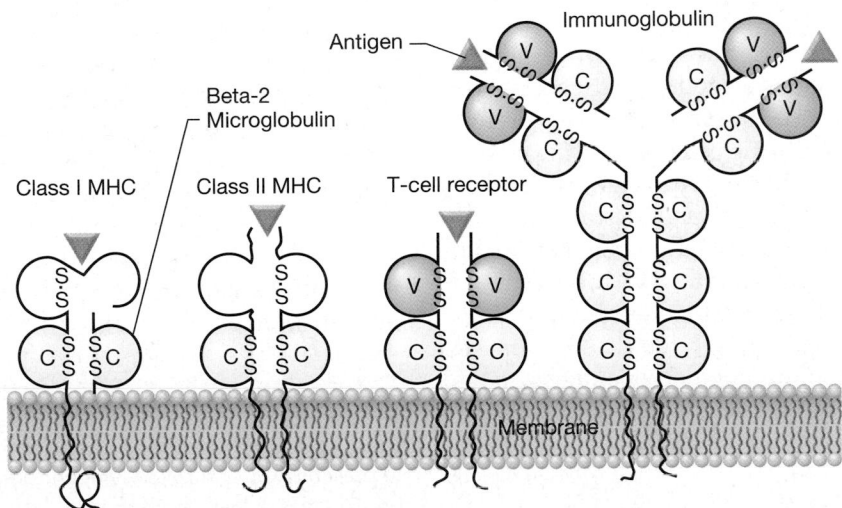

Figure 23.1 Proteins encoded by the immunoglobulin gene superfamily. Constant domains (C) are regions of homologous or nearly homologous amino acid sequence and higher-order structure. The presence of the Ig-like C domains identifies the proteins as members of the Ig gene superfamily and indicates evolutionary relationships. The variable domains (V) of Igs and TCRs and the antigen-binding domains of MHC class I and class II proteins cannot be identified as members of the Ig superfamily because of their highly individual, nonconserved structure. Presumably, these antigen-binding domains evolved from Ig-like domains, but because of evolutionary selection to bind certain forms of antigen, most recognizable structural homology to the Ig-like C domain has disappeared.

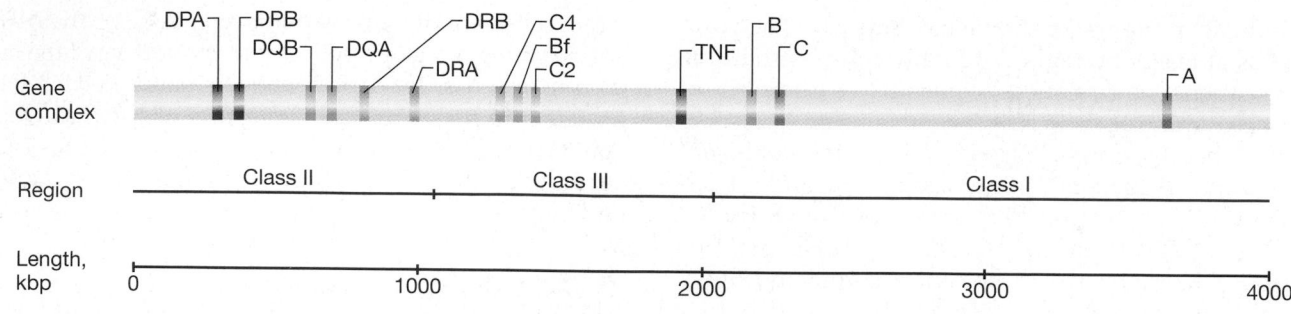

Figure 23.2 The human MHC gene map. The human MHC is called the HLA complex (human leukocyte antigen complex). The HLA complex is located on chromosome 6 and is more than 4 million bases in length. Class II genes DPA and DPB encode class II proteins DPα and DPβ; DQA and DQB encode DQα and DQβ; DRA and DRB encode DRα and DRβ; Class III genes encode several proteins associated with immune recognition functions, as well as other unrelated proteins. C4 and C2 encode complement proteins C4 and C2 (◯◯ Section 22.10). The TNF gene encodes a cytokine, tumor necrosis factor (see Section 23.10). The class I MHC proteins HLA-B, HLA-C, and HLA-A are encoded in the class I region by genes B, C, and A. Over 200 genes, many of which are involved in antigen recognition or processing, are located in the HLA complex.

known as the HLA (human leukocyte antigen) complex (Figure 23.2). The MHC products include many proteins that are involved in antigen processing and presentation. In this section, we will examine the structural and genetic features of the two MHC proteins that bind peptide antigen and the TCR.

23.2 MHC Protein Structure

MHC proteins were discovered because of their role as the major antigen barriers for tissue transplantation, hence their name. However, the real biological function of MHC proteins is to bind and present peptide antigens to T cells (◯◯ Section 22.5).

The MHC proteins are divided into two structural classes. Class I MHC proteins are found on the surfaces of *all* nucleated cells. As a rule, the class I proteins present peptide antigens to T-cytotoxic cells (T_C). If the class I-embedded peptides are recognized as foreign, the antigen-containing cell is targeted and directly destroyed by the T_C cell (◯◯ Section 22.6). Class II MHC proteins are found only on the surface of B lymphocytes, macrophages, and dendritic cells, the dedicated antigen-presenting cells (APCs) (◯◯ Section 22.5). Through the class II proteins, the APCs are involved in presentation of antigens to the T-helper cells, stimulating either inflammatory reactions or antibody responses (◯◯ Section 22.7).

Class I MHC proteins

Class I MHC proteins consist of two polypeptides (Figures 23.1 and 23.3). The membrane-integrated alpha (α) chain of 42,000 molecular weight is encoded in the MHC gene region on chromosome 6. The other class I polypeptide is the noncovalently associated 12,000-molecular-weight protein beta-2 microglobulin (β_2m).

The three-dimensional structure of class I MHC protein reveals a distinctive shape that suggests how this protein interacts with the antigen peptide and the TCR simultaneously (Figure 23.3). The class I α chain folds to form a large groove between the α1 and α2 domains, and it is within this groove that the MHC molecule binds to peptide antigens. The groove is closed on both ends, restricting the size of bound peptides to 8–10 amino acids. The surface of the MHC I protein consists of the two α-helices with the bound peptide and forms a contact area that interacts specifically with the TCR, as we will discuss later (see Section 23.6).

Class II MHC proteins

Class II MHC proteins consist of two noncovalently linked, membrane-integrated polypeptides, called α and β, of 33,000 and 28,000 molecular weight, respectively (Figures 23.1 and 23.3). Class II proteins are usually found in *pairs*, which enhances their ability to bind to TCRs. The α1 and β1 domains of the class II protein interact to form a peptide-binding site that is quite similar to the class I peptide-binding site. However, the ends of the groove are *open*, permitting the class II protein to bind and display peptides that are significantly longer than 10 amino acids (Figure 23.3).

✓ 23.2 Concept Check

Class I MHC proteins are expressed on all cells. They function to present cytosol-derived antigenic peptides to T-cell receptors on T_C cells. Class II MHC proteins are expressed only on antigen-presenting cells. They function to present externally derived peptide antigens to T-cell receptors on T_H cells.

- ✓ Compare the class I and class II MHC protein structures. How do they differ? How are they similar?
- ✓ Compare the peptide-binding sites of class I and class II MHC proteins. How do they differ? How are they similar?

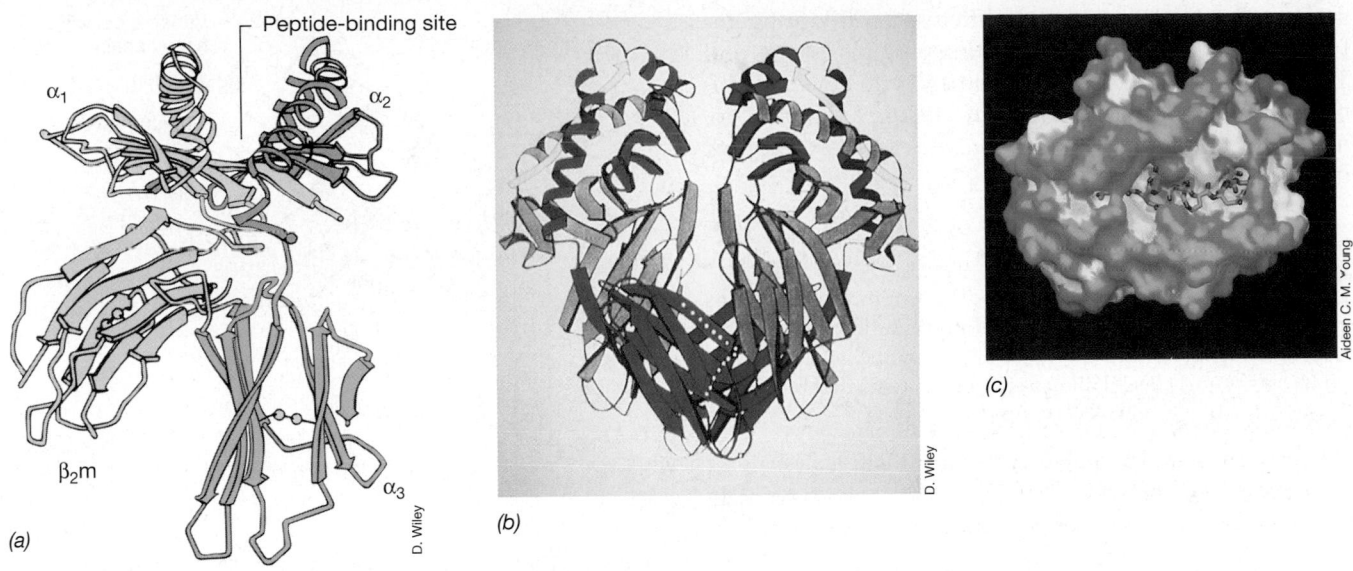

Peptide-binding site

α_1 α_2

β_2m α_3

(a)

(b)

(c)

D. Wiley

D. Wiley

Aideen C. M. Young

Figure 23.3 Three-dimensional structures of MHC proteins. (a) The class I protein. (b) The class II protein in its dimeric form. The yellow arrows indicate the peptide-binding positions. Reprinted with permission from *Nature* 364:33 (1993), © Macmillan Magazines. (c) A space-filling model of a class I protein with a bound peptide, as seen from above. A 9-amino acid peptide is shown as a stick structure, embedded in the human class I protein. As shown in the models, the class I proteins have closed ends, while the class II proteins are open at the ends. As a result, the class I proteins bind peptides of about 8 to 10 amino acids in length, and the class II proteins bind peptides of up to about 20 amino acids in length.

23.3 MHC Genes and Polymorphism

There are at least three gene loci for MHC class I genes, HLA-A, -B, and -C, and all show a high degree of *polymorphism*. **Polymorphism** is the occurrence of multiple alleles at a locus in frequencies that cannot be explained by the occurrence of recent random mutations. For example, there are 95, 207, and 50 different alleles at the HLA-A, HLA-B, and HLA-C loci, respectively. Thus, within the human species, there are multiple polymorphisms at each locus. However, each individual has only two of these alleles at each locus (one allele is of paternal origin, and one is of maternal origin). The two allelic variant proteins are expressed *codominantly* (equally). Likewise, highly polymorphic loci encode class II proteins HLA-DR, -DP, and -DQ. Again, the class II gene products are expressed codominantly. Thus, an individual usually displays *six* genetically and structurally distinct class I proteins and *six* distinct class II proteins. These polymorphic variations in MHC proteins are the major barriers to successful tissue transplants because the MHC proteins on the donor tissue (graft) are recognized as foreign antigens by the recipient's immune system. An immune response directed against the graft MHC proteins causes cell death and rejection of the graft.

The genetic organization and expression of MHC molecules is very simple: One gene encodes one protein. However, the polymorphic nature of the MHC genes en-

sures that, within the human population, there is a large and varied pool of MHC proteins available for antigen presentation. Genetic polymorphism implies that each MHC allele encodes a protein with a different amino acid sequence. The amino acid sequence variations in MHC proteins are concentrated in the peptide-binding groove (Figure 23.3), and each polymorphic variation of the MHC protein binds a different set of peptide antigens. The peptides bound by a single MHC protein share common structural patterns, or **motifs**, and each different MHC protein binds to a different motif. For example, all of the 10 amino acid peptides bound by a certain class I protein may have a tyrosine at position 2 and a leucine at position 7. Thus, all peptides with the sequence **X-tyrosine-X-X-X-X-leucine-X-X-X** (where X is *any* amino acid) would be bound and presented by that MHC protein. Another MHC class I protein encoded by a polymorphic allele might preferentially bind a different motif, with a valine at position 4 and a proline at position 8 (**X-X-X-valine-X-X-X-proline-X-X**). The invariant amino acids in each motif are known as *anchor residues*, because they bind directly and specifically within an individual MHC peptide-binding groove. Thus, an individual MHC protein can bind and present a large number of different peptides, as long as the peptides contain the appropriate anchor residues; it follows that each different MHC protein binds peptides with a different motif and, hence, different anchor residues. In this manner, a limited number of MHC proteins can bind and present a large number of different peptide antigens. However, the other antigen-binding proteins, the Igs and TCRs, must interact very specifically with a nearly infinite number of antigens and therefore use a different genetic mechanism to generate even greater receptor diversity (see Sections 23.5 and 23.7).

✓ **23.3 Concept Check**

The MHC is a group of genes encoding a number of proteins involved in antigen processing and presentation. Class I and class II MHC genes are the most polymorphic genes known. MHC class I and class II alleles encode proteins that bind peptides with specific conserved structural motifs.

✓ Define and explain polymorphism in MHC genes.

✓ How do polymorphic MHC proteins facilitate the presentation of a large number of peptides to T-cell receptors?

III   ANTIBODIES

Antibodies or **immunoglobulins (Ig)** are found as cell-surface antigen receptors on B cells, or in soluble form in high concentrations in serum and other body fluids, where Ig functions to neutralize and opsonize foreign antigens (Section 22.10). In this section, we will look at the structure, antigen-binding function, and genetic organization of the infinitely variable antibodies.

23.4   Antibody Proteins and Antigen Binding

As we discussed above (see Section 23.1) and in Chapter 22, functional immunoglobulins (Igs) consist of four polypeptides, two heavy chains and two light chains. The functional antigen-binding unit consists of a heavy-light chain heterodimer. The heavy and light chains are further divided into C (constant) and V (variable) domains, with the C domains responsible for common functions such as complement binding, while the V domains of H and L chains interact to form the antigen-binding site (Figure 23.4). Here we examine the structural features of the V domains and the antigen-binding site.

Variable Domains
Variations in amino acid sequence occur in the *variable domains* (V domains) of different Igs (Figure 23.4). This amino acid variability is especially apparent in several so-called **hypervariable regions**. These hypervariable regions, also termed **complementarity determining regions (CDR)**, provide most of the molecular contacts

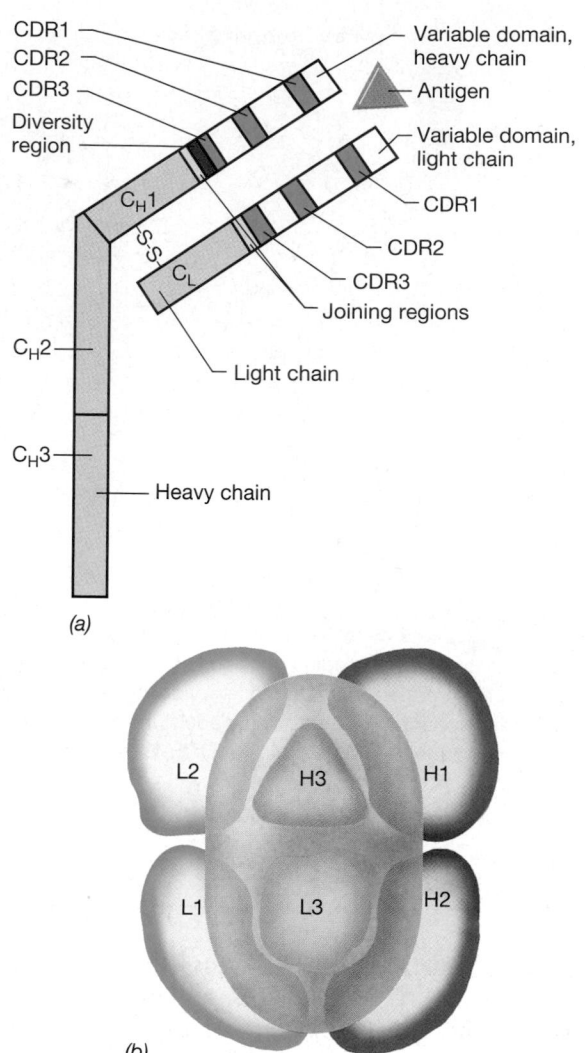

Figure 23.4   The variable regions of immunoglobulin light and heavy chains. (a) One-half of a typical immunoglobulin is shown schematically. C_H and C_L are constant domains of the heavy and light chain, respectively. (b) The complementarity-determining regions from both heavy and light chains in (a) are conformed to make a single antigen-binding site, or pocket, on the immunoglobulin. The site is shown from above. Red binding areas are from the heavy chain, and blue binding areas are from the light chain. Binding areas are numbered to indicate the CDR and chain. For example, L1 is the CDR1 from the light chain. The highly variable CDR3s from both heavy and light chains cooperate at the center of the site. An antigen is shown in gray, overlaying the site and contacting all CDRs. The actual shape of the site may be a shallow groove or a deep pocket, depending on the antibody-antigen pair chosen.

with antigen. Each V domain in the light and heavy chains has three CDRs. CDR1 and CDR2 domains differ somewhat between different immunoglobulins, but the CDR3s differ very dramatically from one another. The CDR3 of the heavy chain has a particularly complex structure. Amino acid–sequence studies indicate that the CDR3 consists of the carboxy-terminal portion of the V domain, followed by a short "diversity" (D) segment of about 3 amino acids, and a longer "joining" (J) region of about 13 to 15 amino acids in length. The light chain has a similar arrangement for its CDR3, but lacks the D region. All of the CDRs are involved in antigen binding.

Antigen Binding

An immunoglobulin three-dimensional structure was shown in Figure 22.15. The principle of all antibody reactions lies in the *specific* combination of determinants on the antigen with the *variable* region. The antigen-binding site of an antibody molecule is formed by the association of heavy and light chains, measures about 2 × 3 nm, and is capable of binding to an epitope that is approximately the size of a 10 to 15-amino acid peptide. The recognition of antigen is ultimately governed by the Ig folding pattern of the heavy and light polypeptide chains. The Ig folds of the V region bring all six CDRs (CDR1, 2, and 3 from both heavy and light chains) together at the end of the Ig protein. The result is a *unique* and *specific* antigen-binding site (Figure 22.15 and Figure 23.4b). In the following section, we will discuss the mechanisms used to produce the tremendous diversity found in the Ig proteins.

✓ 23.4 Concept Check

The antigen-binding site of an immunoglobulin is composed of the V (variable) domains of one heavy chain and one light chain. Each heavy and light chain contains three complementarity determining regions, or CDRs, that are folded together to form the final antigen-binding site.

- ✓ Describe the contributions of the V, D, and J regions of the heavy chain and the V and J regions of the light chain to the CDRs.
- ✓ Draw a complete immunoglobulin molecule and localize the antibody antigen-binding sites.

23.5 Antibody Genes and Diversity

We saw that genetic polymorphism in the MHC allowed the MHC proteins to bind a number of different antigens (see Section 23.3). For the MHC, one gene produces one antigen-binding protein. Following this line of reasoning, if one B lymphocyte produces one Ig, one gene should encode the light chain and one gene should encode the heavy chain. However, this is *not* the case. Because antibodies must specifically recognize and bind a nearly infinite variety of molecular structures using only a limited number of genes, another diversity-generating strategy is employed.

Genes in pieces

A single light or heavy chain is actually encoded by *several* genes that undergo a complex series of gene rearrangements (recombination followed by deletion of intervening

sequences) as B cells develop. This gene rearrangement strategy turns out to be more efficient for Igs (and TCRs) than the "one gene encodes one protein" strategy used for MHC and most other proteins. Gene rearrangement allows the formation of a virtually infinite number of antibodies from as few as 400 different genes. Molecular studies verified the "genes in pieces" hypothesis by demonstrating that genes for V, D, J, and C regions were separated from one another in the genome, but are brought together to form a mature Ig gene in the developing B cell (Figure 23.5). For the heavy chain, the V gene encodes CDR1 and CDR2, while CDR3 is encoded by a mosaic of the 3' end of the V gene, followed by the entire D and J genes. Finally, the class-defining constant domain of the Ig molecule is encoded by the C gene. Thus, four different genes, V, D, J, and C, recombine to form one functional heavy-chain gene. Similarly, light chains are encoded by recombination of light-chain V, J, and C genes.

The genes required for all Igs exist in each lymphocyte that develops from stem cells in the bone marrow. As shown in Figure 23.5, each B cell contains about 150 tandemly arranged light-chain V genes and 5 distinct J genes; about 200 tandem V genes, 50 D genes, and 4 J genes exist for the heavy chains. In addition, the heavy-chain constant domain (C_H) genes and the light-chain constant domain genes (C_L) are present. The V, D, J, and C genes are separated by noncoding sequences (introns) typical of gene arrangements in eukaryotes (Section 7.12). During maturation of B lymphocytes, genetic recombination occurs in each B cell. Randomly selected V, D, and J segments are fused by enzymes that delete all intervening DNA, resulting in an active heavy-chain gene and an active light-chain gene. The active gene (containing an intervening sequence between the VDJ gene segment and the C gene segment) is transcribed, and the resulting primary RNA transcript is spliced to yield the final messenger mRNA. The mRNA is then translated to make the heavy and light chains of the immunoglobulin molecule.

Somatic Diversity

Up to this point, all Ig diversity has been generated as a consequence of recombination of existing genes. However, *somatic* processes are also important for generating diversity. The final light-chain and heavy-chain genes expressed by a given B cell are largely a matter of *chance reassortment* of heavy chains and light chains. For example, based on the numbers of genes at the kappa (κ) light-chain loci, there are 150 V × 5 J possible rearrangements or 750 possible light chains. At the heavy-chain (H-chain) loci, there are approximately 200 V × 50 D × 4 J, or 4000 possible heavy chains. Assuming that each H-chain and light-chain has an equal chance to be expressed in each cell, there are 750 × 4000, or 3,000,000 possible antibodies that can be expressed (Figure 23.5).

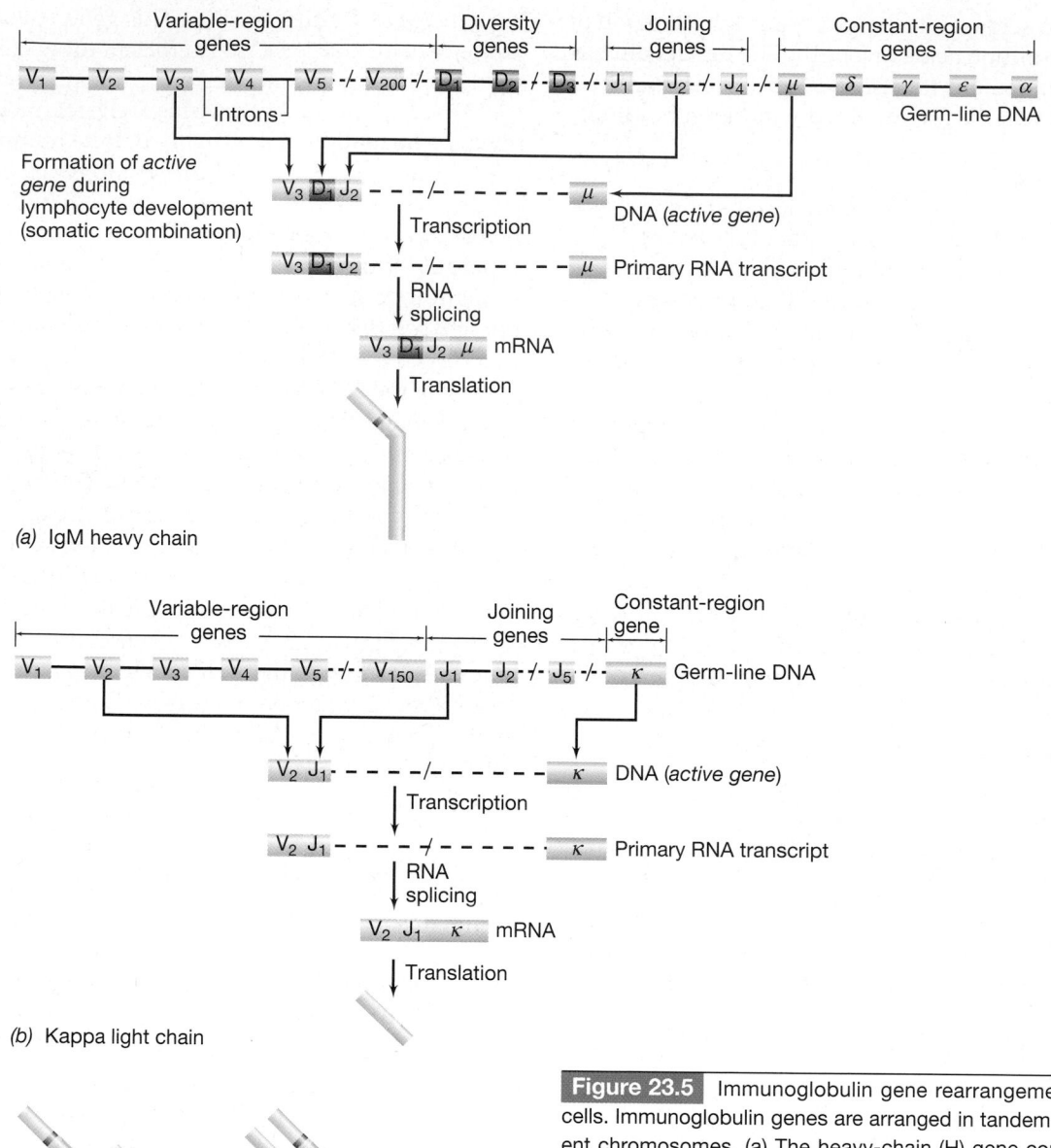

(a) IgM heavy chain

(b) Kappa light chain

(c) Sum of A and B

Figure 23.5 Immunoglobulin gene rearrangement in human B cells. Immunoglobulin genes are arranged in tandem on three different chromosomes. (a) The heavy-chain (H) gene complex on chromosome 14. The filled boxes represent immunoglobulin coding genes. The broken lines indicate intervening sequences and are not shown to scale. (b) The kappa (κ) light-chain complex on chromosome 2. The lambda (λ) light-chain genes are in a similar complex on chromosome 22. (c) Assembly of one-half of an antibody molecule.

Another somatic diversity-generating mechanism involves the DNA-joining mechanism. DNA joining of the V-D-J or V-J regions is imprecise and frequently varies the site of VDJ (or VJ) fusions by a few nucleotides. This genetic imprecision is sufficient to change an amino acid or two and leads to even greater antibody diversity. In each B cell, however, only a single protein-producing rearrangement occurs in the heavy- and light-chain genes. This principle of *allelic exclusion* ensures that *each B cell produces a single, unique Ig.*

Hypermutation

Additional antibody diversity is generated in B cells by *somatic hypermutation*, the mutation of Ig genes at higher rates than the mutation rates observed in other genes. Somatic hypermutation of Ig genes is usually evident after a second exposure to the immunizing antigen. As we saw, a second exposure to antigen results in a change in the predominant antibody class produced, with a switch from IgM to IgG production (∞ Section 22.9). In addition, class switching is often accompanied by an increase in antigen–binding strength (affinity). This *affinity maturation* is one of the factors responsible for the dramatically stronger secondary response in immunity (co Section 22.9 and Figure 22.19).

Compelling evidence for somatic mutations leading to affinity maturation comes from experimental studies on *abzymes*, antibodies that function as enzymes (Figure 23.6). Abzymes perform catalytic reactions on bound substrates, resulting in covalent modification of the substrate and formation of a product. Abzyme mechanisms seem to be identical to the mechanisms involved in enzyme-substrate reactions (co Section 5.5). Unfortunately, most abzymes have very low substrate affinity and do not efficiently convert substrate to product. However, when abzyme-producing B-cell clones are reexposed to antigen (secondary immunization), the substrate affinity of some of the abzymes increases dramatically (Figure 23.6). After repeated immunizations, some cloned abzymes bind substrate more than 10,000 times stronger than the original abzyme/antibody.

Studies of B-cell clones that evolved to produce a better abzyme after a series of immunizations revealed that mutations in the heavy- and light-chain V region genes correlated with increased substrate-binding affinity in the final high-affinity abzyme clone. Thus, antigen

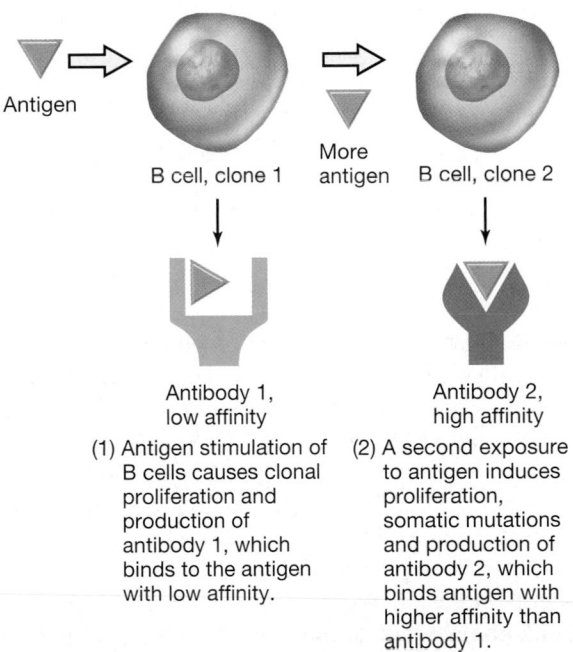

Figure 23.6 Abzymes are antibodies with enzymatic activity. Primary immunization with a substrate antigen induces antibodies with low affinity, such as those produced by clone 1. Further immunization with the same substrate antigen selects for mutations from clone 1 that have much higher affinity for the substrate, such as for clone 2. This seems to be a universal phenomenon for all antibodies: A hypermutation mechanism continuously works on antibody genes to generate new mutations, and antigen acts as a selecting force to choose those antibodies and clones that react most strongly with the antigen.

Antigen

B cell, clone 1

More antigen

B cell, clone 2

Antibody 1, low affinity

Antibody 2, high affinity

(1) Antigen stimulation of B cells causes clonal proliferation and production of antibody 1, which binds to the antigen with low affinity.

(2) A second exposure to antigen induces proliferation, somatic mutations and production of antibody 2, which binds antigen with higher affinity than antibody 1.

acted as a selection agent, and each successive dose of antigen induced somatic mutations leading to successively enhanced antibody responses. Coupled with the other diversity-generating mechanisms, the somatic mutation data suggest that the antibody diversity-generating capacity of an individual is essentially unlimited.

✓ 23.5 Concept Check

Recombination allows shuffling of various pieces of the final Ig genes. Random reassortment of the heavy- and light-chain genes maximizes genetically encoded diversity. Somatic mechanisms including hypermutation also contribute substantially to immunoglobulin diversity.

✓ Describe the recombination events that produce a mature heavy-chain gene.

✓ How many possible immunoglobulin binding sites can be produced?

IV T-CELL RECEPTORS

T-cell receptors or TCRs are found as cell-surface antigen receptors on T cells where they recognize peptide antigens embedded in MHC proteins (co Section 22.4). In this section, we will look at the structure, antigen-binding function, and genetic organization of the TCRs.

23.6 TCR Proteins and Antigen Binding

The **TCR** is a membrane-integrated protein found only on the surface of T cells. The TCR consists of two polypeptides, the alpha (α) chain and the beta (β) chain. The α chain has a molecular weight of about 27,000 and the β chain has a molecular weight of about 31,000. The α:β TCR specifically binds foreign peptides that are embedded in major histocompatibility complex (MHC) molecules on the surface of antigen-presenting cells (APCs) or target cells (co Section 22.5). Thus, *the TCR binds both a self-MHC protein and a foreign peptide*. The TCRs accomplish this dual binding function through a binding site composed of the V domains of the α chain and β chain. As with Igs, the α-chain and β-chain V domains of TCRs contain CDR1, CDR2, and CDR3 segments that bind to antigen.

The three-dimensional structure of the TCR-peptide-MHC protein complex is shown in Figure 23.7. Both TCR and MHC proteins bind directly to peptide antigen. The MHC protein binds one face of the peptide, the **agretope**, whereas the TCR interacts with the other peptide face, the **epitope**. The CDR regions are directly involved in the

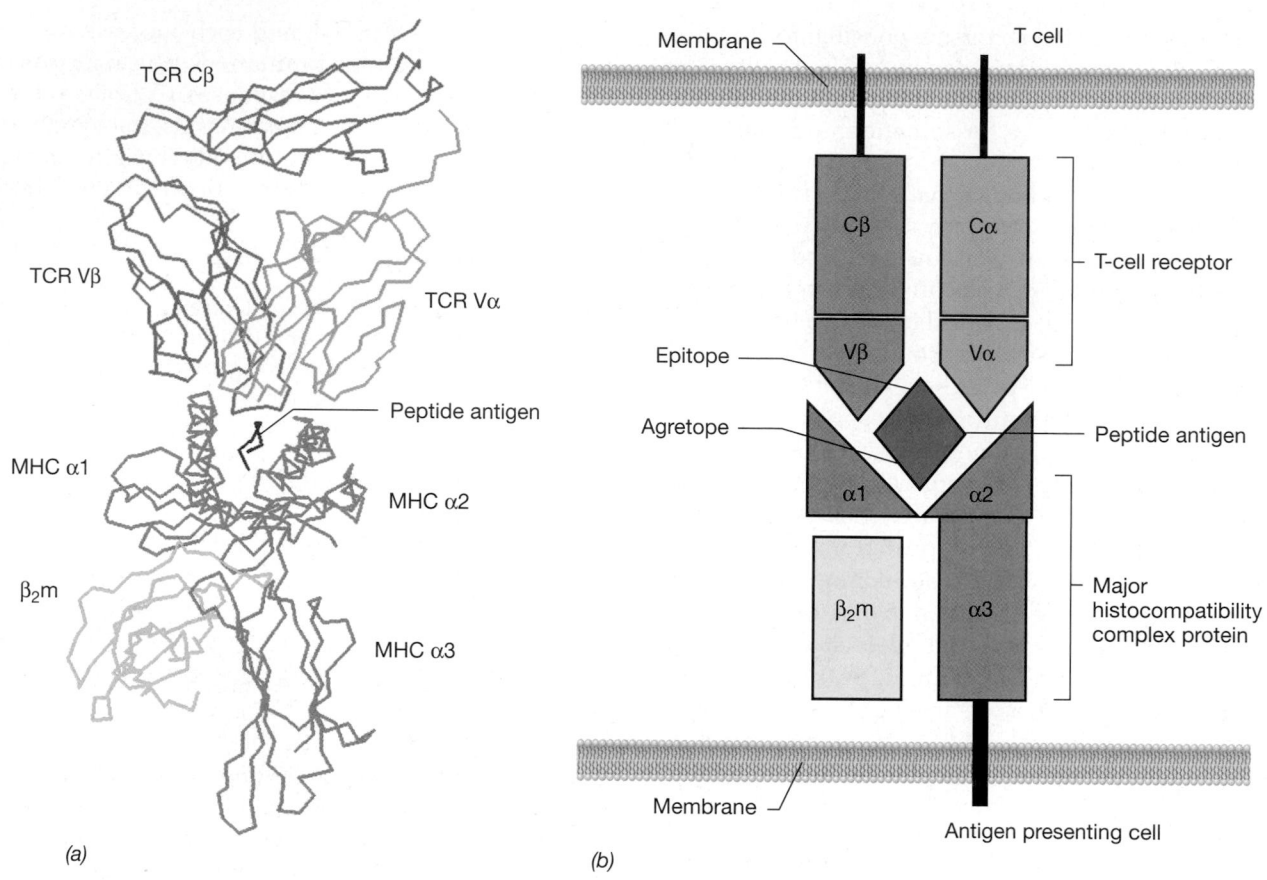

(a)   *(b)*

Figure 23.7 The TCR-peptide-MHC protein complex. (a) The three-dimensional structure, showing the orientation of TCR, peptide (red), and MHC. This structure was derived from data deposited in the Protein Data Bank. (b) A diagrammatic interpretation of the MHC-peptide-TCR structure.

binding of the MHC-peptide complex, and each CDR has a specific binding function. The CDR3 regions of the TCR α chain and β chain make contact with the peptide epitope. The CDR1 and CDR2 regions of the α and β chains contact the peptide-binding α-helices of the MHC proteins.

✓ **23.6 Concept Check**

The TCR is a protein that binds to peptide antigens presented by MHC proteins. The CDR3 regions of both the α chain and the β chain bind to the peptide epitope, whereas the CDR1 and CDR2 regions bind to the MHC protein.

✓ Distinguish among the functions of the TCR CDR1, CDR2, and CDR3 segments.

✓ Distinguish between an *agretope* and an *epitope*.

23.7 TCR Genes and Diversity

Like Ig genes, the genes encoding the TCR α and β chains include distinct genes for constant and variable domains. As for Igs, TCR V regions are encoded by a number of tandemly arranged genes. The α chain has about 80 vari-

able (V) genes and 61 joining (J) segments, whereas the β chain has 50 variable (V) genes, 2 diversity (D) genes, and 13 joining (J) segments (Figure 23.8). The β chain V, D, and J genes and the α chain V and J genes undergo recombination to form functional V-region genes. Additional diversity called *N-region diversity* is generated by random additions of one to six nucleotides between V and D gene segments and between D and J gene segments. Finally, the D region of the β chain can be transcribed in all three reading frames leading to production of three separate transcripts from each D-region gene and creating greater diversity than would be expected from the two D gene segments alone. As we discussed for reassortment of Ig H and L chains, individual α and β chains are produced by each T cell at random, and joined to form a complete α:β heterodimer. With all of these methods for creating diversity, the potential number of TCRs is enormous, probably on the order of 10^{15}.

✓ **23.7 Concept Check**

The V domain of the β chain of the TCR is encoded by V, D, and J gene segments. The V domain of the α chain of the TCR is encoded by V and J gene segments. Diversity generated by

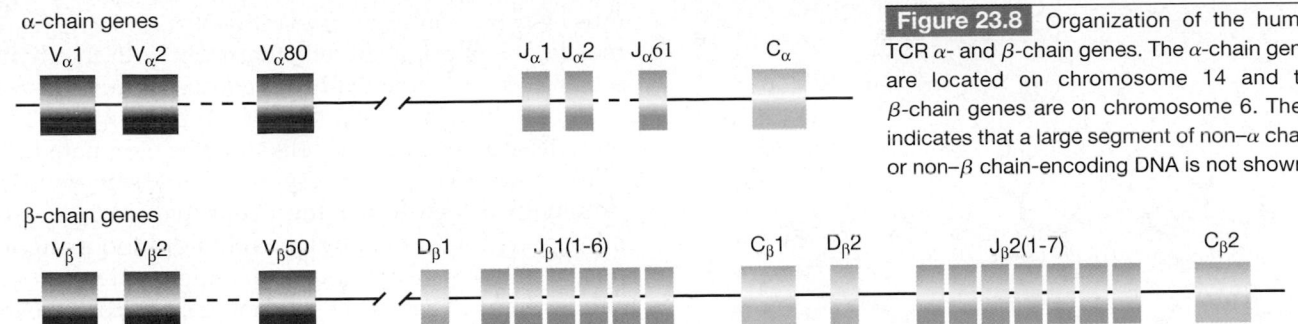

Figure 23.8 Organization of the human TCR α- and β-chain genes. The α-chain genes are located on chromosome 14 and the β-chain genes are on chromosome 6. The // indicates that a large segment of non–α chain- or non–β chain-encoding DNA is not shown.

recombination, reassortment of gene products, the possibility of transcribing D regions in three possible reading frames, and random N-nucleotide addition ensure that there are virtually unlimited possibilities for different antigen-binding TCRs.

✓ Explain the contributions to diversity for reading of the D region in all three reading frames.

✓ Explain the contributions of the N region additions to TCR diversity.

V MOLECULAR SIGNALS IN IMMUNITY

Antigen recognition through Igs and TCRs is the *first signal* and the most important molecular interaction for activating antigen-specific T and B lymphocytes, but other molecular signals need to be transmitted to lymphocytes before they can react appropriately to antigens. Here, we examine some of the molecular mechanisms that control activation of the immune response. First, we will explore the mechanisms that *select* immune cells to react with foreign antigens and to ignore self proteins. Next, we will examine the molecular *second signals* that are responsible for activating immune T cells. Finally, we will look at *cytokines* and *chemokines*, soluble proteins capable of activating other cells in the immune response.

23.8 Clonal Selection and Tolerance

In this section we examine the mechanisms used to select T cells for an effective immune response. T cells must learn to discriminate between the *dangerous non-self antigens* and the *nondangerous self antigens* that compose our body tissue. Thus, T cells must achieve **tolerance**, or specific unresponsiveness to self antigens. As we achieve tolerance, we select immune cells that interact only with dangerous non–self antigens.

Clonal Selection

Clonal selection is a hypothesis stating that each antigen-reactive B cell or T cell has a cell-surface receptor for a single antigen. When stimulated by interaction with that antigen, each cell can replicate, and antigen-stimulated B and T cells grow and differentiate. As a result, antigen-stimulated cells divide and produce a pool of cells that express the same antigen-specific receptors. These copies of the original antigen-reactive cell are known as *clones* (Figure 23.9). Cells that have not interacted with antigen do not grow.

To respond to the seemingly infinite variety of antigens, a large number of antigen-reactive cells are needed in the body, and each cell must be capable of expanding into an antigen-reactive clone. However, antigen-reactive cells must not provoke immune reactions against self antigens in the host. To meet these special requirements, the immune system is programmed to select clones that may be useful against non–self antigens and eliminate or suppress self-reactive clones.

T-Cell Selection and Tolerance

T cells undergo *selection for* antigen-reactive T cells and *selection against* clones that react with self antigens. Selection against self-reactive clones results in the development of *tolerance*, or specific immune unresponsiveness. Previously, we discussed the failure of tolerance and the subsequent development of autoimmunity (∞ Section 22.13).

T-cell selection occurs in the thymus, a primary lymphoid organ that plays a central role in the development of antigen-reactive T cells (∞ Figure 22.2).

Lymphocytes that will become T cells leave the bone marrow and enter the thymus from the lymphatic ducts (Figure 23.10). In the first T-cell maturation stage in the thymus, called **positive selection**, these immature T cells interact with the thymic epithelium, using their newly developed TCRs to bind to major histocompatibility complex (MHC) proteins on the thymus. The T cells that do not bind MHC proteins are programmed to die, a process called **apoptosis**; the T cells that bind thymic MHC proteins continue to grow. Thus, positive selection

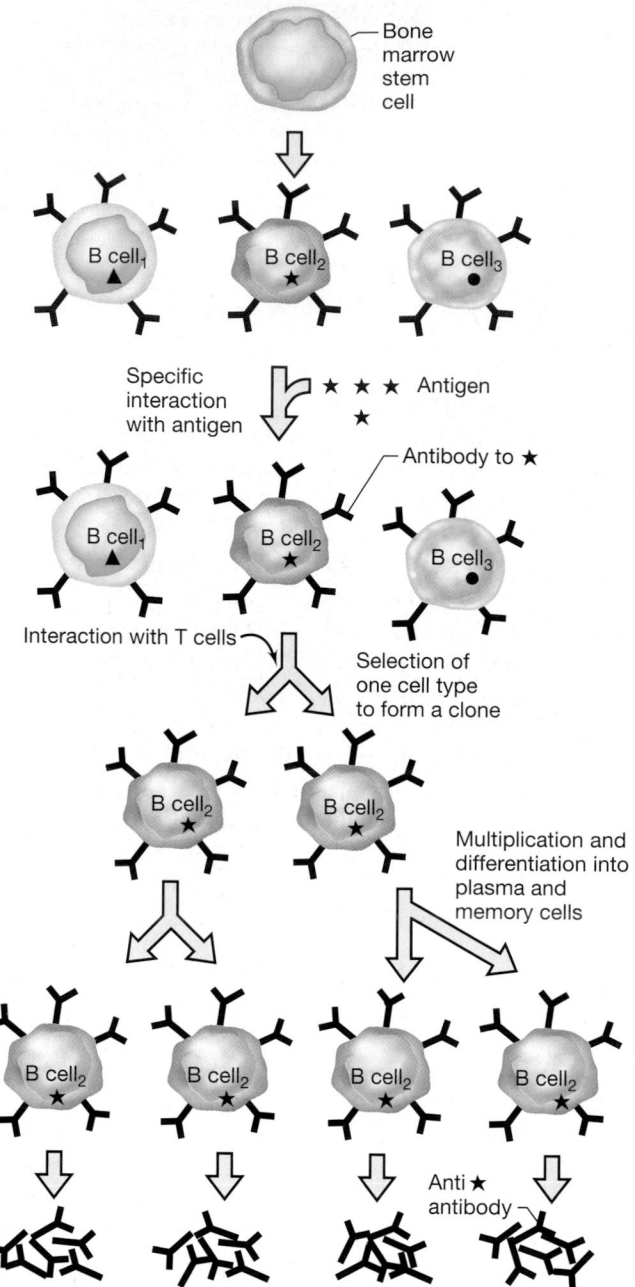

Figure 23.9 Clonal selection. Individual B cells, specific for a single antigen, proliferate and expand to form a clone after interaction with the specific antigen. The specific antigen acts as the selection agent, driving selection and then proliferation of the individual antigen-specific B cell. Clonal copies of the antigen-reactive cell all have the same antigen-specific surface receptor. Continued exposure to antigen results in continued expansion of the clone. An analogous situation exists for T cells.

retains T cells that recognize self-MHC proteins and deletes T cells that do not recognize self-MHC proteins.

The second stage of T-cell maturation is termed **negative selection**. After undergoing positive selection, the growing T-cell clones continue to react with MHC proteins, which are complexed with antigens in the thy-

mus (∞ Section 22.5). The antigens in the thymus are mostly of self origin. T cells that react with the thymic self antigens are potentially dangerous because they can lead to destruction of self tissue (autoimmunity). Therefore, these self-reactive T cells must be eliminated. The self-reactive T cells bind very tightly to the thymus and eventually die. However, the T cells that are destined to interact with non–self antigens do not bind as tightly, presumably because the non–self antigens are not found in the thymus. These T cells do not die, but leave the thymus and migrate to the spleen and lymph nodes where they can contact foreign antigens presented by B lymphocytes and other antigen-presenting cells.

This two-stage mechanism for selecting for antigen-reactive T cells while inducing tolerance is called **clonal deletion**; precursors of T-cell clones that are either useless or harmful die, and more than 99% of all T cells that enter the thymus and undergo selection do not survive.

T cells that survive positive and negative selection then leave the thymus and travel to the secondary lymphoid organs, where they can participate in the immune response to foreign antigens.

B-Cell Tolerance

The acquisition of immune tolerance in B cells is also necessary because antibodies produced by self-reactive B cells (autoantibodies) may damage host tissue (∞ Section 22.13). B cells also undergo clonal deletion; many self-reactive B cells are eliminated during development in the bone marrow, the primary lymphoid organ responsible for B-cell development in humans (∞ Section 22.1). In addition, *clonal anergy* (unresponsiveness) also plays a role; immature cells that are self reactive do not develop, even when exposed to high concentrations of self antigens in the bone marrow. These antigens may occupy the reactive B-cell Ig receptors, but the B cell is dependent on T cells for "help", as we shall see below (see Section 23.10). If no antigen-reactive T cells are available because they were eliminated during T-cell selection, the B cell remains in a constant state of anergy.

✓ **23.8 Concept Check**

The thymus is a primary lymphoid organ that provides an environment for the maturation of antigen-reactive T cells. Immature T cells that do not interact with MHC protein (positive selection) or react strongly with self antigens (negative selection) are eliminated by *clonal deletion* in the thymus. T cells that survive positive and negative selection leave the thymus and can participate in an effective immune response. B cells also undergo clonal deletion and selection.

✓ Provide an example of a self protein to which you are tolerant and another human protein to which you would *not* be tolerant.

✓ Distinguish between *positive* and *negative* T-cell selection.

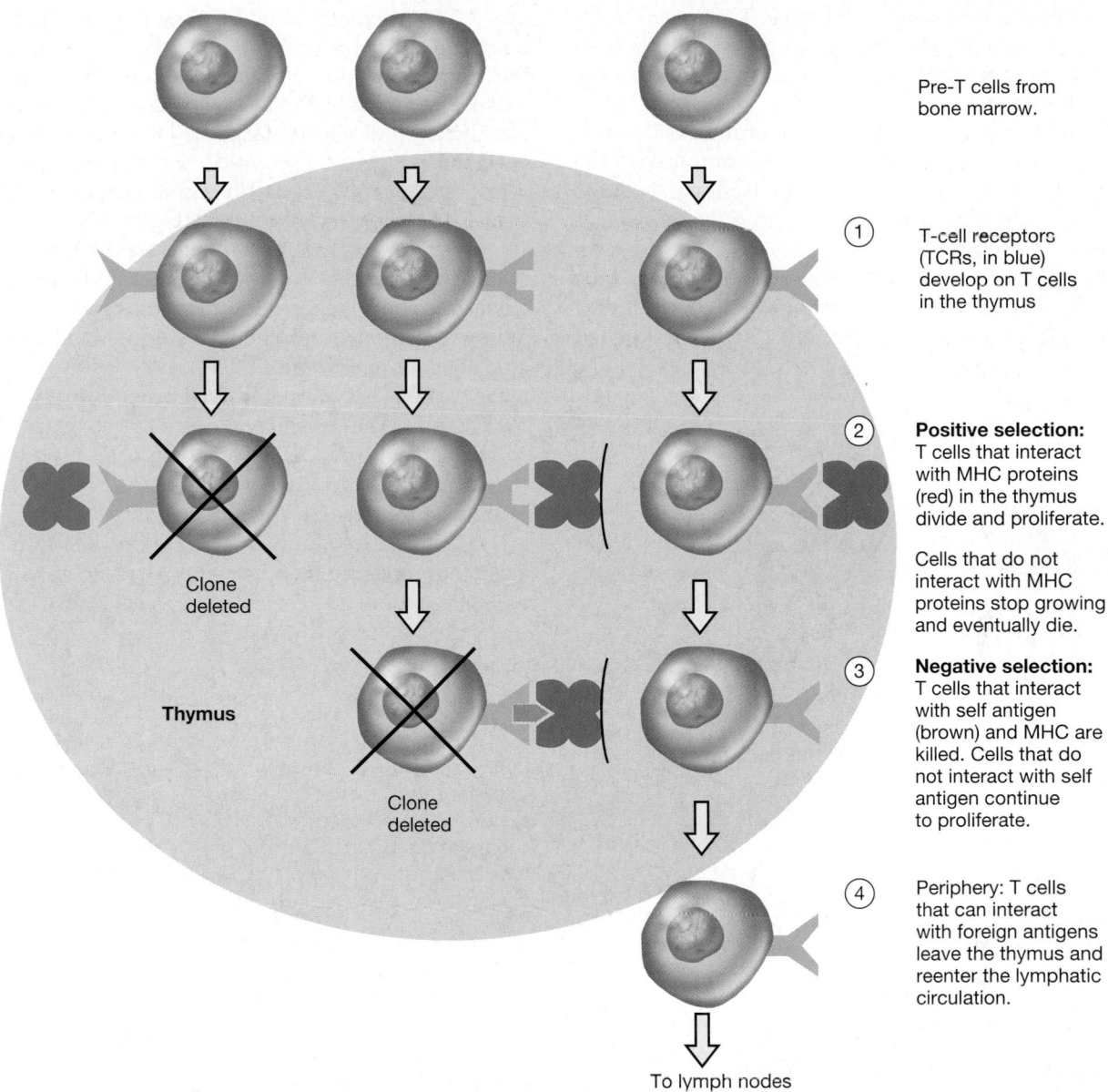

Pre-T cells from bone marrow.

① T-cell receptors (TCRs, in blue) develop on T cells in the thymus

② **Positive selection:** T cells that interact with MHC proteins (red) in the thymus divide and proliferate.

Cells that do not interact with MHC proteins stop growing and eventually die.

③ **Negative selection:** T cells that interact with self antigen (brown) and MHC are killed. Cells that do not interact with self antigen continue to proliferate.

④ Periphery: T cells that can interact with foreign antigens leave the thymus and reenter the lymphatic circulation.

Clone deleted

Thymus

Clone deleted

To lymph nodes

Figure 23.10 T-cell selection and clonal deletion. T cells undergo a stepwise selection process in the thymus. ① Pre-T cells enter the thymus and express T-cell receptors (TCRs). ② T cells are selected for the ability to bind to MHC proteins in the thymus. Cells that do not bind to MHC eventually die, a process known as *positive selection*. ③ Some of the positively selected T cells can then interact with self-antigens in the thymus. T cells that interact with self-antigens are deleted in the thymus, a process called *negative selection*. ④ T cells that survive positive and negative selection leave the thymus. The TCRs on the selected cells are capable of interacting with foreign antigens outside the thymus.

23.9 Second Signals

T cells selected for reactivity against dangerous non–self antigens in the thymus are also selected for unresponsiveness to self antigens; T cells that react with self antigens are deleted in the thymus. However, a very large number of self antigens are not expressed in the thymus. Self-reactive T-cell clones can avoid clonal deletion, but may become unresponsive to self antigen through interactions in the secondary lymphoid organs. The key

to inducing appropriate specific immune unresponsiveness, or **clonal anergy**, lies in the mechanism needed to activate the T cell.

T-Cell Activation and the Second Signal
When selected mature T cells leave the thymus, they travel to the secondary lymphoid organs (lymph nodes, spleen, and mucosal-associated lymphoid tissue, or MALT), where they take up residence. These T cells have not yet been exposed to antigen and are therefore

called naive or uncommitted T cells. Uncommitted T cells must be activated by antigen-presenting cells (APCs) (👓 Section 22.5) before they are fully competent as effector cells.

The first step in activation of uncommitted T cells is binding of the peptide:MHC protein complex on the APC by the TCR (Figure 23.11). This is signal 1 and is absolutely required for activation; without signal 1, a T cell cannot be activated.

The next step in activation of the T cell involves the interaction of two additional proteins, one found on the APC called *B7* and one found only on T cells, called

CD28. The interaction and binding of B7 to CD28 is signal 2 and activates the T cell; in the absence of signal 2, the T cell cannot be activated (Figure 23.11). The activated T cell then becomes an effector cell. A T_C cell that is activated will kill any target cell that displays antigen, even those target cells that do not display CD28. Activated effector cells need only signal 1 (peptide-MHC) to stimulate their effector activity (Figure 23.11).

This requirement for a second activation signal has major implications for establishing and maintaining clonal anergy. For example, an uncommitted T_C cell that interacts with a self antigen on a non-APC will receive only signal 1, since non-APCs do not display the B7 protein necessary to complete signal 2. In the absence of signal 2, this T_C is anergized and cannot be activated (Figure 23.11). Thus, the B7:CD28 second signal is absolutely required for activation; absence of signal 2 in the presence of signal 1 induces permanent anergy.

A different second signal is used to activate B cells and other immune response effectors, as we will see in the next section.

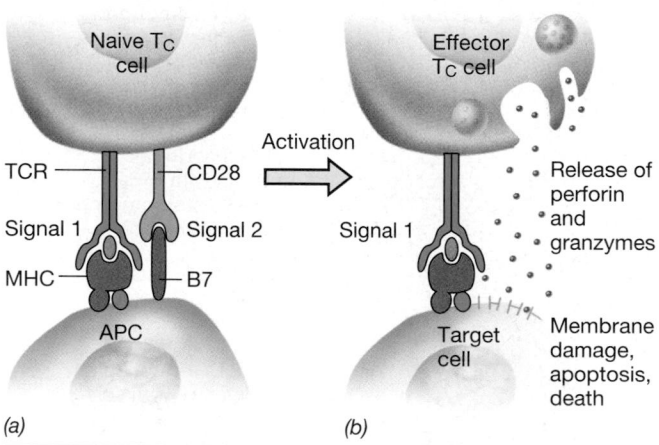

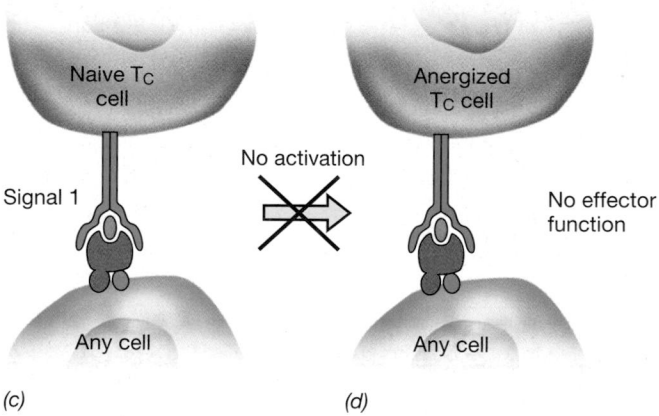

Figure 23.11 Signal 1 and signal 2 are required to activate naive T cells. (a) A naive T_C cell interacts via TCR with the peptide-MHC complex on an APC. This is signal 1. The T_C cell also has a CD28 protein that interacts with a B7 protein on the APC. This is signal 2. The simultaneous interactions of the T_C cell and APC via signal 1 and signal 2 activates the naive T cell. (b) This permanently activated T_C cell is then capable of killing any target cell as long as signal 1 interactions take place. (c) A naive T_C cell interacts via the TCR with the peptide-MHC complex on any cell. Although the conditions for signal 1 (interactions via TCR with the peptide-MHC complex) are met, signal 2 cannot be generated because only APCs display the B7 protein. (d) In the absence of signal 2, the T_C cell becomes permanently unresponsive, or anergized.

✓ **23.9 Concept Check**

Most self-reactive T cells are deleted during development and maturation in the thymus. Uncommitted T cells are activated in the secondary lymphoid organs by first binding peptide:MHC with their TCRs (signal 1), followed by binding of the B7 APC protein to the CD28 T-cell protein (signal 2). Uncommitted self-reactive T cells are anergized in the secondary lymphoid organs if they interact with signal 1 in the absence of signal 2.

✓ Define signal 1 and signal 2 for an *uncommitted* T cell.

✓ Identify the signal(s) necessary to induce effector function in an *activated* T cell.

23.10 Cytokines and Chemokines

To activate individual cells in the immune system, the cells must communicate, and a second and very general method for doing this is through soluble proteins known as **cytokines**. Cytokines are a group of soluble proteins produced by leukocytes that regulate cellular functions. Specialized cytokines produced by lymphocytes are known as *lymphokines*. In general, cytokines are secreted from one cell and bind to corresponding specific receptors on a target cell. Some cytokines bind to receptors on the cell that produced them. Thus, these cytokines have *autocrine* (self-stimulatory) abilities. The cytokine receptors are responsible for signal transduction (👓 Section 8.10). Cytokine-bound receptors relay information across the cell membrane to control activities such as protein synthesis and cell division. These signals can ultimately result in cell growth, differentiation, and clonal proliferation.

TABLE 23.1 Properties of some major cytokines and chemokines

Cytokine	Producers	Major targets	Effect
IL-1[a]	Monocytes	T_H	Activation
IL-2	Activated T cells	T cells	Growth, differentiation
IL-3	T_H1	Hematopoietic stem cells	Growth factor
IL-4	T_H2	B cells	IgG1 and IgE synthesis
IL-5	T_H2	B cells	IgA synthesis
IL-10	T_H2	T_H1	Inhibits T_H1
IL-12	Macrophages	T_H1	Differentiation, activation
IFN-α[b]	Leukocytes	Normal cells	Anti-viral
IFN-γ	T_H1	Macrophages	Activation
GM-CSF[c]	T_H1	Myeloid stem cells	Differentiation to granulo-cytes, monocytes
TGF-β[d]	T_H1 and T_H2	Macrophages	Inhibits activation
TNF-α[e]	T_H1	Macrophages	Activation
TNF-β	T_H1	Macrophages	Activation
Chemokines			
IL-8	Macrophages, fibroblasts, keratinocytes	Neutrophils, T cells	Attractant and activator
MCP-1[f]	Macrophages, fibroblasts, keratinocytes	Macrophages, T cells	Attractant and activator

[a] IL, interleukin; [b] IFN, interferon; [c] GM-CSF, granulocyte, monocyte-colony stimulating factor; [d] TGF, T-cell growth factor; [e] TNF, tumor necrosis factor, [f] MCP, macrophage chemoattractant protein.

Some cytokines are called *interleukins* (IL) because they are molecules that mediate interactions between leukocytes. Table 23.1 lists some important cytokines, their major producer cells, their most common target cells, and their biological effects. In all, there are nearly 40 cytokines, most of which are produced by either T_H cells or monocytes and macrophages. We now examine the action of three cytokines involved in the induction of an antigen-specific antibody-mediated immune response.

IL-1, IL-2, and IL-4

As we have seen, macrophages are responsible for antigen uptake, processing, and presentation (⚭ Section 22.5). In addition, they secrete a potent cytokine known as IL-1 that affects several different cell types (Table 23.1 and Figure 23.12). IL-1 is a key component of the immune response because T_H cells are activated by IL-1. Because of the proximity of T_H cells and macrophages in the lymph nodes, nearby T_H2 cells are very likely to be activated. IL-1 binding by the IL-1 receptors (IL-1R) on the T_H2 cell acts as an activation signal. The activated T_H2 cell, in turn, responds by producing IL-2, which is secreted and bound by the IL-2R on the surface of the T_H2 cells. Thus, IL-2 is autocrine; IL-2 can activate the same cell that secreted it. Under the influence of IL-2, the cell divides and makes clonal copies. In the process, the T_H2 cell also makes other cytokines, in this case IL-4, which then binds to the IL-4R on B cells. The IL-4:IL-4R complex stimulates the B cells to differentiate into plasma cells, which

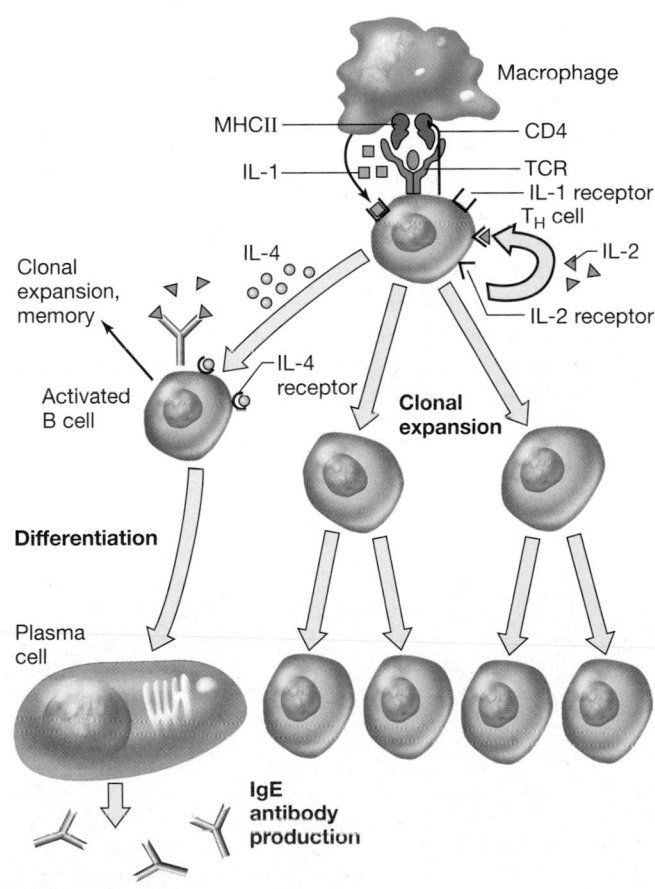

Figure 23.12 Some major effects and interactions of IL-1, IL-2, and IL-4 cytokines in the antibody response.

ultimately produce antibodies (⌀ Section 22.9). Thus, IL-1, IL-2, and IL-4 cytokines are soluble mediators and activators for macrophages, T lymphocytes, and B cells, the cells that interact to produce the antibody-mediated immune response.

Other Cytokines

Table 23.1 shows the action of several other cytokines. Many of these proteins affect cells involved in specific immunity. For example, IL-4 acts primarily on B cells. However, several cytokines do not affect T or B lymphocytes, but act on other cells; the cytokine-activated cells in turn serve as important modulators of nonspecific host responses. For example, interferons (IFN-α and IFN-γ) are produced by leukocytes and inhibit viral replication in virtually any cell in the body. Tumor necrosis factors (TNF-α and TNF-β) are capable of destroying a variety of tumors if the TNF-producing cells have access to the tumor. Interferons and TNFs appear to have no target cell specificity, but are produced by T cells and amplify the effects of immune cells.

Cytokines are important mediators of a variety of immune functions. Some cytokines have broader activities, interacting with a variety of normal cells. However, the unifying feature of these heterogeneous proteins is that they are all produced by leukocytes.

Chemokines

Chemokines are a group of small proteins that function as chemoattractants for phagocytic cells and T cells. They are produced by lymphocytes and a wide variety of other cells in response to bacterial products, viruses, and other agents that cause damage to host cells. Chemokines attract phagocytes and T cells to the site of injury, stimulating an inflammatory response as well as potentiating a specific immune response.

About 40 chemokines are known. Perhaps the best-studied chemokines are *interleukin-8* (IL-8) and *macro-phage chemoattractant protein-1* (MCP-1) (Table 23.1). IL-8 is produced by a wide variety of cell types including monocytes, macrophages, fibroblasts (connective tissue cells), and keratinocytes (skin cells) in response to tissue injury or contact with pathogens. IL-8 is secreted by the affected cells and binds to the surrounding tissue, where it is a chemoattractant for T cells and neutrophils (⌀ Section 22.1), resulting in a neutrophil-mediated inflammatory response followed by a specific immune response mediated by the attracted T cells. As is the case for the cytokine receptors, chemokine receptors on the target cells act through signal transduction pathways (⌀ Section 8.9) to induce activation of the target phagocytes or T cells.

MCP-1 is also produced by a variety of cells and attracts macrophages and T cells, again stimulating production of inflammatory mediators and potentially organizing an antigen-specific immune response. Thus, chemokines are potent initiators of nonspecific inflammatory reactions that lead to recruitment of T cells and antigen-specific immune reactions.

✓ 23.10 Concept Check

Cytokines are soluble mediators produced by leukocytes that regulate interactions between cells. Several cytokines such as IL-1, IL-2, and IL-4 affect leukocytes and are critical components in the generation of specific immune responses. Other cytokines such as IFN and TNF affect a wide variety of cell types. *Chemokines* are produced by a variety of cell types in response to injury and are potent attractants for nonspecific inflammatory cells and T cells.

✓ Compare the target cells for IL-1, IL-4, IL-12, IFN-γ, and TNF-β.

✓ How do cytokines and chemokines differ with respect to their cell sources? Their cell targets?

✓ What events stimulate cytokine production? What events stimulate chemokine production?

Review Questions

1. Define the criteria used to assign a gene and its encoded protein to the Ig gene superfamily.

2. Identify the major structural features of class I and class II MHC proteins.

3. Polymorphism implies that each different MHC protein binds a different peptide motif. For the HLA class I polymorphisms, how many different HLA proteins are expressed in an individual? By the entire human population?

4. Which Ig chains are used to construct a complete antigen-binding site? Which domains? Which CDRs?

5. Calculate the total number of V_H and V_L domains that can be constructed from the available Ig genes. How many complete Ig proteins can be produced from the reassortment of all possible heavy chains and light chains?

6. Describe the interaction of the TCR with peptide antigen and MHC protein. Be sure to identify the roles of the CDRs in the TCR.

7. In TCRs, diversity can be generated by recombination and reassortment events such as in Igs. However, additional diversity is accomplished by the use of N-region nucleotide additions and reading of the D (diversity) segment in all three reading frames. Explain how these diversity-generating mechanisms work.

8. Explain *positive* and *negative* selection of T cells.

9. What molecular interactions are necessary for activation of naive T cells? For activation of effector cells?

10. What are the chief effects of cytokines and chemokines? What are the chief differences between cytokines and chemokines?

Application Questions

1. Construct a table that lists the common features of proteins encoded by members of the Ig gene superfamily. For Igs, TCRs, and MHC proteins, identify the structural components that fit these common features.

2. What is the potential advantage of having a different binding site for class I and class II MHC proteins? *Hint*: Think in terms of the source of antigen peptides bound by each protein.

3. Polymorphism implies that each different MHC protein binds a different peptide motif. However, for the class I proteins, only 6 peptide motifs can be recognized in an individual, while over 350 motifs can be recognized by the entire human population. What advantage does this have for the population? For the individual?

4. What are the contributions of the individual CDRs to the Ig variable domains? What relation does this have to the binding sites?

5. While genetic recombination events are important for generating significant diversity in the antigen-binding site of Igs, post-recombination somatic events may be even more important in achieving overall Ig diversity. Do you agree or disagree with this statement? Explain.

6. Why must the TCR and MHC proteins recognize components of the same antigen? Can TCRs recognize antigens in the absence of MHC proteins?

7. TCR genes have the potential to produce an even more diverse antigen-binding repertoire than is produced by Ig genes. What advantage does this have for the immune system?

8. What would happen to the T-cell repertoire in the absence of positive selection? In the absence of negative selection?

9. What would be the result of activation of all peripheral T cells that contact antigen? How does the second signal scheme help to prevent this from happening?

10. Predict the outcome of an immune response due to signal 1 and signal 2 in the absence of IL-1, IL-2, or IL-4. Predict the outcome of the same immune response in the absence of IL-2 receptors on T cells or IL-4 receptors on B cells.

C linical microbiology involves a number of tests and procedures designed to isolate and identify pathogens. Shown here is a selective and differential microbial culture medium, *eosin-methylene blue (EMB) agar*, that allows growth of only a certain subset of organisms (gram-negative bacteria) but can also distinguish pathogens from nonpathogens isolated from the human intestinal tract, urinary tract, or other sites of bacterial infection. Microbial identification does not always require isolation and growth of the pathogen. Sophisticated methods employing the principles of immunology or molecular biology can be used to identify a pathogen both rapidly and reliably, and this oftentimes allows treatment to be more specific and to be initiated sooner.

CLINICAL MICROBIOLOGY AND IMMUNOLOGY

24

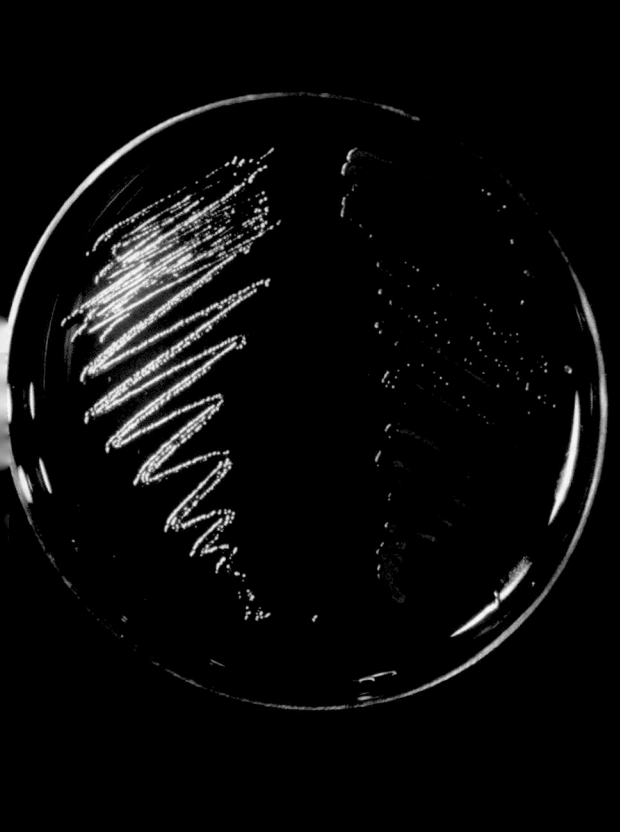

Working Glossary

Agglutination reaction between antibody and particle-bound antigen, resulting in visible clumping of the particles

Antibiogram a report indicating the sensitivity of clinically isolated microorganisms to the antibiotics in current use

Bacteremia the presence of bacteria in the blood

Differential media media that allow identification of microorganisms based on their appearance

ELISA enzyme-linked immunosorbent assay

Enriched media media that allow metabolically fastidious microorganisms to grow because of the addition of specific growth factors

Enrichment culture the use of selected culture media and incubation conditions to isolate microorganisms from natural samples

Fluorescent antibody covalent modification of an antibody molecule with a fluorescent dye; the dye makes the antibody visible under fluorescent light

General purpose media media that support the growth of most aerobic and facultatively anaerobic organisms

Immunoblot (Western blot) electrophoresis of proteins followed by transfer to a membrane and detection by addition of specific antibodies

Monoclonal antibody antibody made by a single B cell clone

Neutralization interaction of antibody with antigen that reduces or blocks the biological activity of the antigen

Nucleic acid probe in clinical microbiology, a short oligonucleotide of unique sequence used as a hybridization probe for identifying pathogens

Polyclonal antibodies antibodies made by many different B cell clones

Precipitation reaction between antibody and a soluble antigen resulting in a visible, insoluble complex

RIA radioimmunoassay

Selective media media that enhance the growth of certain organisms while retarding the growth of others due to an added media component

Sensitivity the lowest amount of antigen that can be detected

Septicemia blood infection

Serology the study of antigen–antibody reactions in vitro

Specificity the ability of an antibody to recognize a single antigen

Titer in an immunological context, the quantity of antibody present in a solution

Viral load the number of viral genome copies in the tissue of an infected patient

---

The most important activity of the microbiologist in medicine is to identify the agents that cause infectious disease. This major area of microbiology is called *clinical* or *diagnostic microbiology*. Clinical laboratories can isolate and identify most routinely encountered pathogenic bacteria within 48 h of sampling. However, recent advances in rapid diagnostic methods make it possible to identify some pathogens in minutes. Tests using immunology and molecular biology methods can identify many pathogens without culturing the organism. Rapid diagnostic methods are particularly important for the diagnosis of viral and protozoal infections, diseases that are typically difficult to identify because of the difficulty of culturing the causal agent.

▌ GROWTH-DEPENDENT CLINICAL DIAGNOSTIC METHODS

Here we discuss the principles involved in isolation and growth of pathogens from host tissue. Growth-dependent methods are an important and proven method of clinical diagnosis.

24.1 Isolation of Pathogens from Clinical Specimens

The physician, following clinical examination of a patient, may suspect that an infectious disease is present. Samples of infected tissues or fluids are then collected for microbiological, immunological, and molecular biological analyses (Figure 24.1). Depending on the kind of infection, materials collected may include blood, urine, feces, sputum, cerebrospinal fluid, or pus. A sterile swab may be used to sample a suspected infected area (Figure 24.2). The swab is then streaked over the surface of an agar plate or placed directly in a liquid culture medium. In some cases, small pieces of living tissue may be sampled for culture. Table 24.1 summarizes current recommendations for initial culture of organisms isolated from typical clinical specimens.

If clinically relevant organisms are to be isolated and identified, the specimen must be obtained properly. The clinician must ensure that the specimen is removed from the *actual site of the infection*. In addition, recovery of pathogens may not be possible if insufficient inoculum is taken. The sample must also be taken under aseptic conditions so that contamination is avoided. Care must also be taken to ensure that metabolic requirements for certain organisms, such as anoxic conditions, are maintained. Once obtained, the sample must be analyzed as soon as possible.

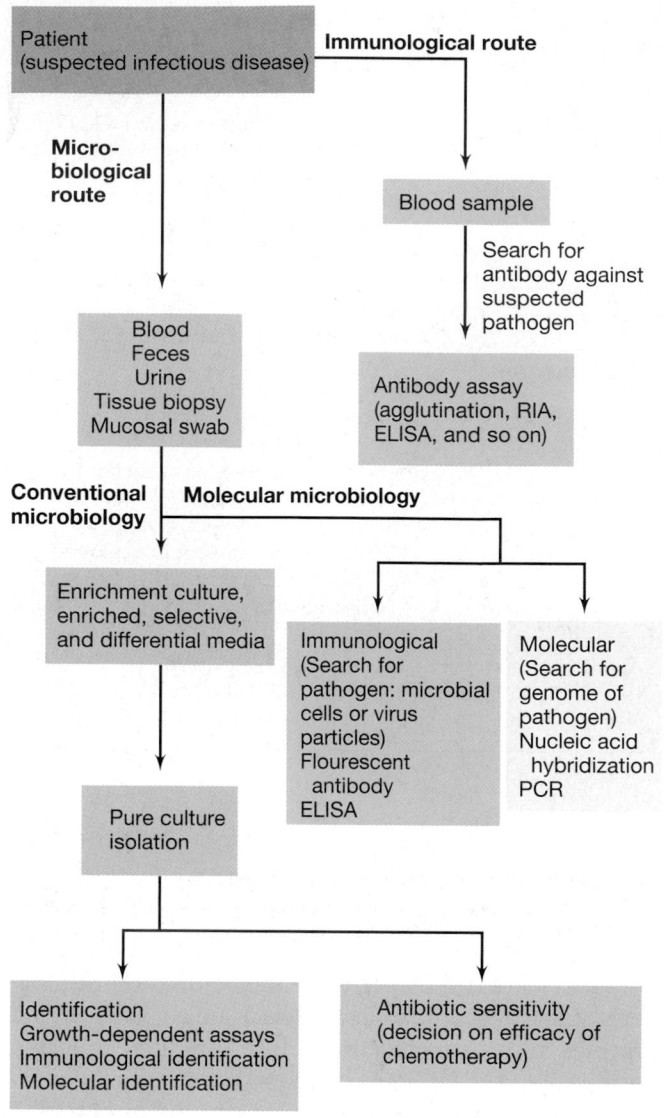

Figure 24.1 Clinical and diagnostic methods used for isolation and identification of infectious pathogens.

Growth Media and Culture

Enrichment culture, the use of selected culture media and incubation conditions to isolate microorganisms from natural samples, is an important part of clinical microbiology. Growth media used in the clinical laboratory are relatively specialized. Through the use of various growth media, most microorganisms of clinical importance can be grown, isolated, and identified. Most clinical samples are first grown on **general purpose media**, media such as blood agar (∞ Figure 21.17 and Table 24.1) that support the growth of most aerobic and facultatively anaerobic organisms. **Enriched media** that allow metabolically fastidious organisms to grow because of the addition of specific growth factors are often necessary to enhance the growth of certain pathogens, such as *Neisseria gonorrhoeae*, the organism that causes gonorrhea (Table 24.1). Certain media are **selective media**, media that enhance the growth of certain organisms while retarding the growth of others due to an added media component (Table 24.1). Finally, **differential media** are specialized media that allow identification of organisms based on their appearance on the media (see Figure 24.7).

Blood Cultures

Bacteremia is the presence of bacteria in the blood (∞ Section 21.7). Bacteria are normally cleared from the bloodstream rapidly. Therefore, bacteremia is uncommon in healthy individuals, and the presence of bacteria in the blood is generally indicative of systemic infection. The most common pathogens found in blood include *Pseudomonas aeruginosa*, enteric bacteria, especially *Escherichia coli* and *Klebsiella pneumoniae*, and the gram-positive cocci *Staphylococcus aureus* and *Streptococcus pyogenes*. **Septicemia** is a blood infection resulting from the growth of a virulent organism entering the blood from a focus of infection, multiplying, and traveling to various body tissues to initiate new infections. Septicemia is indicated by the presence of severe systemic symptoms, including fever and chills, followed by prostration. Severe cases of septicemia may result in *septic shock*, a life-threatening systemic condition char-

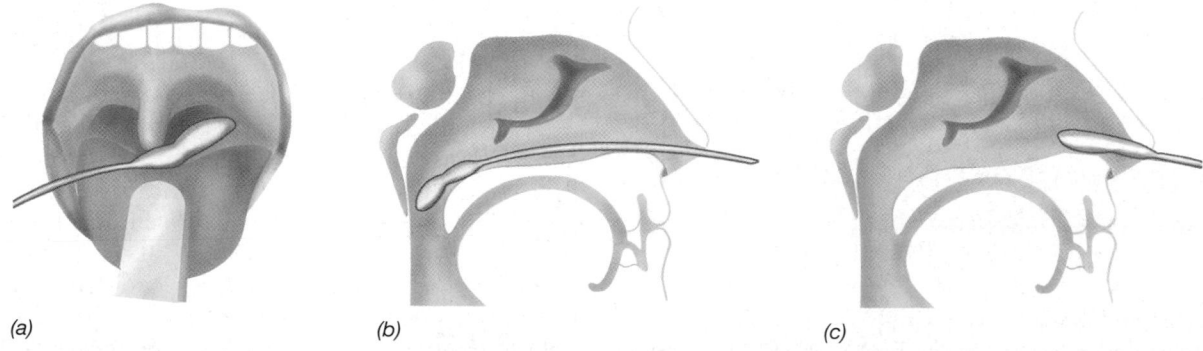

(a) *(b)* *(c)*

Figure 24.2 Methods for obtaining specimens from the upper respiratory tract. (a) Throat swab. (b) Nasopharyngeal swab passed through the nose. (c) Swabbing the inside of the nose.

TABLE 24.1 Recommended enriched and selective media for primary isolation of pathogens[a]					
	Media[b]				
Specimen	**Blood agar**	**Enteric agar**	**CA**	**MTM**	**ANA**
Fluids: chest, abdomen, pericardium	+	+	+	−	+
Feces: rectal swabs	+	+	+	−	−
Surgical tissue biopsies: lung, lymph nodes	+	+	−	−	+
Throat: swabs, sputum, tonsil, nasopharynx	+	+	+	−	−
Genitourinary swabs: urethra, vagina, cervix	+	+	+	+	−
Urine	+	+	−	−	−
Blood	+	−	−	−	+
Swabs: wounds, abscesses, exudates	+	+	+	−	+

[a] From Murray, P.R., E.J. Baron, M.A. Pfaller, F.C. Tenover, and R.H. Yolken. 1999. *Manual of Clinical Microbiology*, 7th edition. American Society for Microbiology, Washington, DC.

[b] Blood agar, 5% whole sheep blood added to trypticase soy agar; enteric agar, either eosin-methylene blue (EMB) agar or MacConkey agar; CA, chocolate (heated blood) agar; MTM, modified Thayer-Martin agar; ANA, anaerobic agar, thioglycolate-containing blood agar or supplemented thioglycolate agar incubated anaerobically.

acterized by severe reduction in blood pressure and multiple organ failures, including heart, kidneys, and lungs. In many disease situations, blood cultures provide the only immediate way of isolating and identifying the causal agent, and diagnosis therefore depends on careful and proper blood culture.

The standard blood culture procedure is to draw 10 ml of blood aseptically from a vein and inject it into two blood culture bottles containing an anticoagulant and an all-purpose culture medium. One bottle is incubated aerobically and one anaerobically. Blood culture bottles are incubated at 35°C and examined daily for up to 5 days in most automated systems. Most clinically significant bacteria are recovered within this period. Some blood isolation systems employ a chemical that lyses red and white blood cells, releasing potential intracellular pathogens that might otherwise be overlooked. Microorganisms in blood cultures are commonly detected by visual inspection (turbidity), microscopic examination, and subculture. Automated blood culture systems detect growth by continuously monitoring carbon dioxide production and turbidity. Readings can be taken every 10 minutes.

Because a certain amount of skin contamination is unavoidable when blood is drawn, a contamination rate of 2–3% can be expected. Contamination may be indicated if certain organisms commonly found on the skin are isolated, such as *Staphylococcus epidermidis*, coryneform bacteria, or propionibacteria, although even these organisms can occasionally cause infection of the heart (subacute bacterial endocarditis). Thus, considerable microbiological and clinical experience is necessary when interpreting blood culture results.

Urine Cultures

Urinary tract infections are very common, and because the causal agents are often identical to bacteria of the normal flora (for example, *Escherichia coli*), considerable care must be taken in the bacteriological analysis of urine. In most cases, urinary tract infection occurs as a result of an organism ascending the urethra from the outside, and the bladder may become infected. Urinary tract infections are the most common form of *nosocomial* (hospital-acquired) infection (∞ Section 25.7).

Significant urinary infection generally results in bacterial counts of 10^5 or more organisms per milliliter of a clean-voided midstream specimen. In the absence of infection, contamination of the urine from the external genitalia (almost unavoidable to some extent) results in less than 10^3 organisms per milliliter. The most common urinary tract pathogens are members of the enteric bacteria, with *E. coli* accounting for about 90% of the cases. Other urinary tract pathogens include *Klebsiella*, *Enterobacter*, *Proteus*, *Pseudomonas*, *Staphylococcus saprophyticus*, and *Enterococcus faecalis*. *Neisseria gonorrhoeae*, the causal agent of gonorrhea, does not grow in the urine itself, but in the urethral epithelium, and is diagnosed by different methods (see later).

Direct microscopic examination of urine may be used to indicate *bacturia*, the presence of abnormal numbers of bacteria in the urine. However, because nearly all urine contains some level of bacterial growth, significant bacturia is most commonly monitored by using a variety of commercially available dipstick tests. For example, one dipstick test monitors the reduction of nitrate by detecting the reduction product, nitrite. A positive test is indicated by a color change on the dipstick (Figure 24.3). Since significant nitrite production occurs only when large numbers ($>10^5$ per milliliter) of enteric organisms are present, the method is a virtually instantaneous check for urinary tract infections. Other dipstick tests for urinary tract infections, often used in conjunction with nitrate reduction, detect esterase (produced by leukocytes) (∞ Section 22.1) and peroxidase (produced by a variety of bacteria)

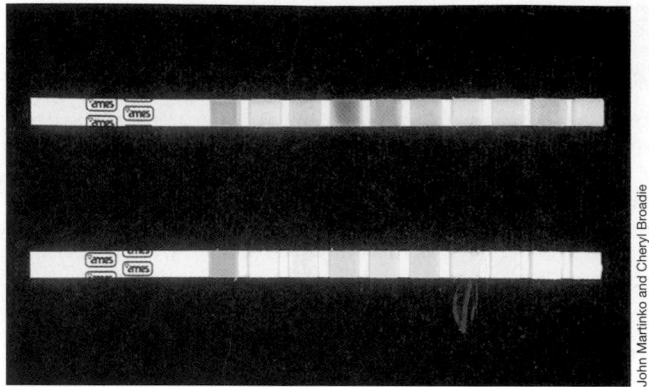

Figure 24.3 Urinalysis dipstick test. A control strip is shown underneath the test strip. From left to right, the strip measures abnormal levels of glucose, bilirubin, ketones, specific gravity, blood, pH, protein, urobilinogen, nitrite, and leukocytes (esterase) in a urine sample. Abnormal readings for esterase (trace positive, far right) and nitrite (strong positive, second from right) indicate bacturia. Subsequent culture of this sample indicated the presence of *Escherichia coli*.

(∞ Sections 6.13 and 17.22). Positive dipstick tests are followed by a urine culture.

To culture potential urinary tract pathogens, two media are used: (1) blood agar as a nonselective general medium, and (2) a medium selective for enteric bacteria, such as MacConkey or eosin-methylene blue agar (EMB) (see Section 24.2 and Figure 24.4). These specialized media permit the initial differentiation of lactose fermenters from nonfermenters, and the growth of gram-positive organisms such as *Staphylococcus* spp. (common skin contaminants) is inhibited. Experienced clinical microbiologists may make a tentative identification of an isolate by observing the color and morphology of colonies of the suspected pathogen growth on various media as described in Table 24.2. Such an identification must be followed with more detailed tests, but clinical microbiologists use this information in conjunction with further test results, discussed throughout the remainder of this chapter, to make a positive identification.

Finally, if no bacterial growth is obtained despite persistent urinary tract infection symptoms, a clinician may request direct cultures for a number of fastidious organisms, especially *Neisseria gonorrhoeae, Chlamydia trachomatis, Branhamella* spp., mycoplasma, or several anaerobic organisms.

Fecal Cultures

Proper collection and preservation of feces is important in the isolation of intestinal pathogens. During storage, fecal acidity increases so extended delay between sampling and sample processing must be avoided. This is especially critical for the isolation of *Shigella* and *Salmonella* species, both of which are rather sensitive to acid pH. Freshly collected fecal samples are

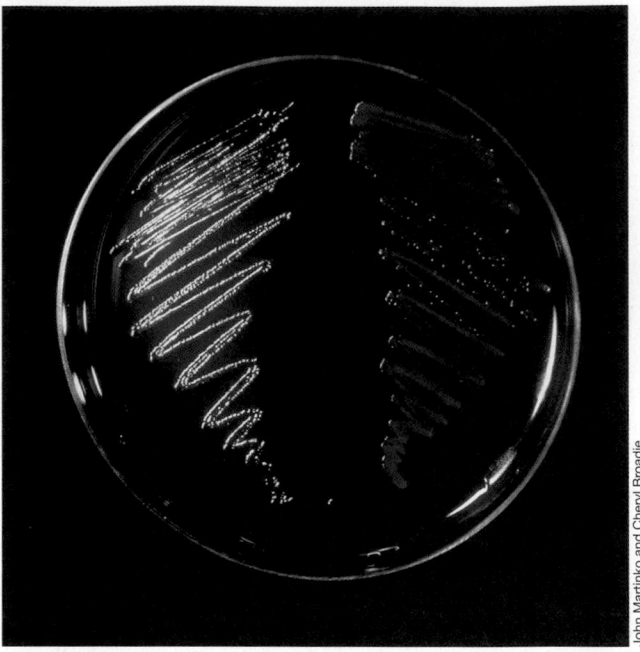

Figure 24.4 An eosin-methylene blue (EMB) agar plate showing a lactose fermenter, *Escherichia coli* (left), and a nonlactose fermenter, *Pseudomonas aeruginosa* (right). Note the green metallic sheen of the *E. coli* colonies.

placed in a vial containing phosphate buffer for transport to the lab. Bloody or pus-containing stools as well as stools from patients with suspected foodborne or waterborne infections should be inoculated into a variety of selective media (see Section 24.2) for isolation of individual bacteria. Intestinal parasites are identified microscopically in the stool sample rather than by culture methods (∞ Section 28.8).

Wounds and Abscesses

Infections associated with traumatic injuries such as animal or human bites, burns, cuts, or the penetration of foreign objects, must be carefully sampled in order to recover the relevant pathogen and results must be interpreted carefully. This is because wound infections and abscesses are frequently contaminated with normal flora. Swab samples of such lesions are frequently misleading. For abscesses and other purulent lesions, the best sampling method is to aspirate pus with a sterile syringe and needle following disinfection of the skin surface. Internal purulent lesions are usually sampled by biopsy or from tissues removed in surgery.

A variety of pathogens can be associated with wound infections, and because some of these are anaerobes, samples must be transported from the collection site under anaerobic conditions. A common pathogen associated with purulent discharges is *Staphylococcus aureus*, but enteric bacteria, *Pseudomonas aeruginosa*, and the anaerobes from the genera *Bacteroides* and *Clostridium*

TABLE 24.2 Colony characteristics of frequently isolated gram-negative rods cultured on various clinically useful media[a]

Organism	Agar media[b]			
	EMB	MC	SS	BS
Escherichia coli	Dark center with greenish metallic sheen (see Figure 24.4)	Red or pink	Red to pink	Mostly inhibited
Enterobacter	Similar to E. coli, but colonies are larger	Red or pink	White or beige	Mucoid colonies with silver sheen
Klebsiella	Large, mucoid, brownish	Pink	Red to pink	Mostly inhibited
Proteus	Translucent, colorless	Transparent, colorless	Black center, clear periphery	Green
Pseudomonas	Translucent, colorless to gold (see Figure 24.4)	Transparent, colorless	Mostly inhibited	No growth
Salmonella	Translucent, colorless to gold	Translucent, colorless	Opaque	Black to dark green
Shigella	Translucent, colorless to gold	Transparent, colorless	Opaque	Brown or inhibited

[a] Adapted from Murray, P. R., E. J. Baron, M. A. Pfaller, F. C. Tenover, and R. H. Yolken. 1999. *Manual of Clinical Microbiology*, 7th edition. American Society for Microbiology, Washington, DC.

[b] BS, Bismuth sulfite agar; EMB, eosin-methylene blue agar; MC, MacConkey agar; SS, *Salmonella-Shigella* agar.

are also commonly encountered. The major isolation media are blood agar, several selective media for enteric bacteria (Tables 24.1 and 24.2), and blood agar containing additional supplements and reducing agents for obligate anaerobes. Smears from such specimens are examined directly by microscopy.

Genital Specimens and the Laboratory Diagnosis of Gonorrhea

In males, a purulent urethral discharge is the classic symptom of the sexually transmitted disease gonorrhea (∞ Section 26.12). If no discharge is present, a sample can be obtained using a sterile narrow-diameter cotton swab that is inserted into the anterior urethra, left in place a few seconds to absorb any exudate, and then removed for culture of *Neisseria gonorrhoeae*, the causative agent of gonorrhea. Alternatively, a sample of the first early morning urine of an infected individual usually contains viable cells of *N. gonorrhoeae*. In females suspected of having gonorrhea or other genital infections, samples are usually obtained by swab from the cervix and the urethra.

Gonorrhea is a common sexually transmitted disease, and clinical microbiology procedures are central to its diagnosis. *N. gonorrhoeae* (referred to clinically as *gonococcus*) colonizes mucosal surfaces of the urethra, uterine cervix, anal canal, throat, and conjunctiva. The organism is quite sensitive to drying and therefore is transmitted almost exclusively by direct person-to-person contact, usually by sexual intercourse. The major goal of public health measures to control gonorrhea in-

volves identification of asymptomatic carriers, and this requires microbiological analysis.

Neisseria gonorrhoeae is usually found as gram-negative diplococci. No similar microorganisms are observed among the normal flora of the urogenital tract. Thus, direct microscopy of a gram-stained vaginal or cervical smear showing gram-negative diplococci is a probable indication of gonorrhea. In acute gonorrhea, microscopy usually reveals phagocytized gram-negative diplococci in the polymorphonuclear leukocytes (∞ Section 20.1), with virtually no other organisms present (Figure 24.5a).

Culture methods are more sensitive than microscopic analysis for identifying pathogens. Most nonselective enrichment media for the isolation of *N. gonorrhoeae* contain heat-lysed blood and are often called *chocolate agar* because of the deep brown appearance. The heated blood interacts with the media components, absorbing compounds that are normally toxic for *N. gonorrhoeae*. One of several selective media used for primary isolation is modified Thayer-Martin (MTM) agar (Figure 24.5). This medium incorporates vancomycin, nystatin, trimethoprim, and colistin to suppress the growth of normal flora, but these antibiotics do not suppress either *N. gonorrhoeae* or *N. meningitidis*, which causes bacterial meningitis.

After streaking, the plates must be incubated in a humid environment in an atmosphere containing 3–7% CO_2 (CO_2 is required for growth of gonococci). The plates are examined after 24 and 48 h, and should be immediately tested by the oxidase test because all *Neisseria* are

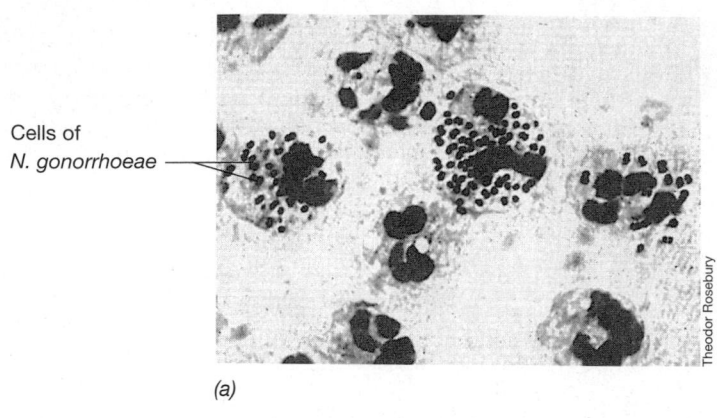

Cells of
N. gonorrhoeae

Theodor Rosebury

(a)

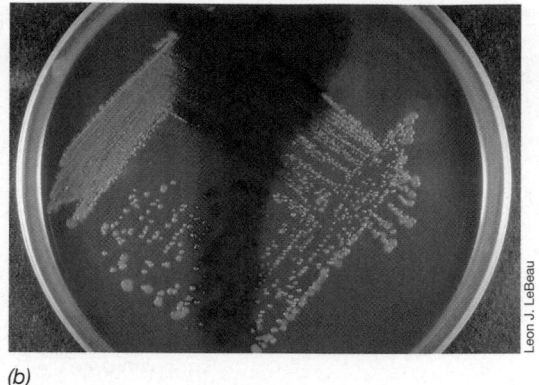

Leon J. LeBeau

(b)

Figure 24.5 (a) Photomicrograph of *Neisseria gonorrhoeae* within human polymorphonuclear leukocytes from a cervical smear. Note the paired diplococci (leaders). (b) *N. gonorrhoeae* growing on Thayer-Martin agar. The plate has been stained in the middle with a reagent that turns colonies blue if cells contain cytochrome *c* (the oxidase test). *N. gonorrhoeae* are oxidase-positive.

oxidase-positive (Figure 24.5*b*; see Section 24.2). Oxidase-positive gram-negative diplococci growing on chocolate agar or selective media are presumed to be gonococci if the inoculum was derived from genitourinary sources, but definitive identification requires determination of carbohydrate utilization patterns and immunological or nucleic acid probe tests (see Sections 24.5–24.14).

Culture of Anaerobes

Obligately anaerobic bacteria are common causes of infection and are completely overlooked in clinical diagnosis unless special isolation and culture methods are used. We have discussed anaerobes in general in Section 6.13, and we noted that many anaerobes are extremely susceptible to oxygen. Therefore, specimen collection, handling, and processing require special attention if obligate anaerobic organisms are to be recovered. There are several habitats in the body (for example, the oral cavity and the intestinal tract) (∞ Sections 21.3 and 21.4) that are generally anoxic and in which obligately anaerobic bacteria can be found as part of the normal flora. However, other parts of the body can become anoxic as a result of tissue injury or trauma, reducing blood supply to the injured site. These anaerobic sites can then be colonized by obligate anaerobes. In general, pathogenic anaerobic bacteria are part of the normal flora and are opportunistic pathogens. Two important exceptions are the pathogenic anaerobes *Clostridium tetani* (causal agent of tetanus) and *Clostridium perfringens* (causal agent of gas gangrene and one type of food poisoning), both endospore-forming *Bacteria* that are predominantly soil organisms.

With anaerobic culture, the microbiologist is presented not only with the usual problems of obtaining and maintaining an uncontaminated specimen but also with ensuring that the specimen not come in contact with air. Samples collected by suction or biopsy must be immediately placed in a tube containing oxygen-free gas, preferably containing a small amount of a dilute salts solution

with a reducing agent such as thioglycolate and the redox indicator dye *resazurin*. This dye is colorless when reduced and becomes pink when oxidized, indicating oxygen contamination of the specimen. If a proper anaerobic transport tube is not available, the syringe itself can be used to transport the specimen; the needle is inserted into a sterile rubber stopper so that no air enters the syringe.

For anaerobic incubation, agar plates are placed in a sealed jar, which is made anoxic by either replacing the atmosphere in the jar with an oxygen-free gas mixture (usually a mixture of N_2 and CO_2) or by adding some compound to the enclosed vessel that removes O_2 from the atmosphere. For example, as shown in Figure 24.6, H_2 is generated and, in the presence of a suitable catalyst, usually palladium, the H_2 is combined with free O_2

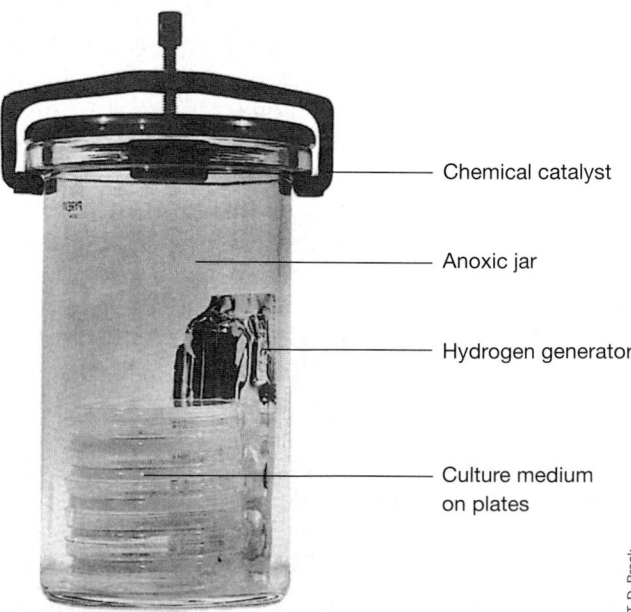

Chemical catalyst

Anoxic jar

Hydrogen generator

Culture medium
on plates

T. D. Brock

Figure 24.6 Sealed jar for incubating cultures under anoxic conditions. The catalyst and hydrogen generator packet produce and maintain a reducing (anoxic) environment.

to form H_2O, thus removing the contaminating oxygen. Alternative means for providing anaerobic conditions include the use of culture media containing reducing agents or the use of anoxic "glove boxes." Glove boxes are large gas-impermeable bags filled with an oxygen-free gas such as nitrogen or hydrogen and fitted with an airlock for inserting and removing cultures (Figure 6.26b). The advantage of an anoxic glove box is that manipulations can be done on a laboratory bench. However, because of their specialized nature, anoxic glove boxes are not employed extensively in clinical laboratories but are in widespread use in research laboratories that specialize in anaerobic microorganisms.

In general, media for anaerobes do not differ greatly from those used for aerobes, except that they are generally richer in organic constituents and contain reducing agents (usually cysteine or thioglycolate) and a redox indicator such as resazurin.

✓ 24.1 Concept Check

Proper sampling and culture of the suspected pathogen is the most reliable way to identify an organism that causes a disease. The selection of appropriate sampling and culture conditions requires knowledge of bacterial ecology, physiology, and nutrition.

✓ Why are urine cultures almost always positive for bacterial growth?

✓ Describe precautions required for successful isolation of anaerobic pathogens.

24.2 Growth-Dependent Identification Methods

If the inoculation of a primary medium results in bacterial growth, the clinical microbiologist must identify the organism or organisms present. Identification of a clinical isolate can frequently be made using a variety of growth-dependent assays. We discuss some of these methods here.

Growth on Selective and Differential Media

Based on growth characteristics on primary isolation media, an unknown pathogen is usually subcultured onto diagnostically useful media that are designed to measure one of many different biochemical reactions. The most important tests are summarized in Table 24.3. Many of these specialized media are available in miniaturized kits containing a number of media in separate wells, all of which can be inoculated at one time (Figure 24.7, page 814).

The battery of media employed are selective, differential, or both. A *selective medium* is one to which compounds have been added to selectively inhibit the growth of certain microorganisms but not others. A *differential medium* is one to which some sort of indicator, usually a dye, has been added, which allows the clinician to differentiate between various chemical reactions carried out during growth. Eosin-methylene blue (EMB) agar, for example, is a widely used selective *and* differential medium. EMB agar is used for the isolation of gram-negative enteric *Bacteria*. Methylene blue dye, present in small amounts, effectively inhibits the growth of most gram-positive *Bacteria*. Eosin is a dye that responds to changes in pH, going from colorless to black under acidic conditions. EMB agar medium contains lactose and sucrose, but not glucose, as energy sources. Lactose-fermenting (generally enteric) bacteria, such as *Escherichia coli*, *Klebsiella*, and *Enterobacter*, acidify the medium and the colonies appear black with a greenish sheen. Colonies of lactose nonfermenters, such as *Salmonella*, *Shigella*, and *Pseudomonas*, are translucent or pink (Figure 24.4). Thus, EMB preferentially *selects* for the growth of gram-negative *Bacteria*, and *differentiates* between several different genera of the selected gram-negative *Bacteria*.

These individual biochemical tests measure the presence or absence of *enzymes* involved in catabolism of the substrate or substrates in the differential medium. Fermentation of sugars is measured by incorporating pH indicator dyes that change color on acidification (Figure 24.7a). Production of hydrogen gas and/or carbon dioxide during sugar fermentation is assayed by observing gas production either in gas collection vials or in agar (Figure 24.7a and b). Hydrogen sulfide production is indicated following growth in a medium containing ferric iron. If sulfide is produced, ferric iron reacts with H_2S to form a black precipitate of iron sulfide (Figure 24.7b). Utilization of citric acid, a six-carbon acid containing three carboxylic acid groups, is accompanied by a pH increase, and a specific dye incorporated into the citric acid test medium changes color as conditions become alkaline (Figure 24.7c). Hundreds of differential tests have been developed for clinical use, but only about 20 are used routinely (Figure 24.7d).

The typical reaction patterns for various pathogens are stored in a computer data bank. The results of the differential tests on an unknown pathogen are entered, and the computer makes the best match by comparing the characteristics of the unknown with the known metabolic patterns of the species in the data bank. For many organisms, as few as three or four key tests are all that are required to make an unambiguous identification. In cases of a dubious match, however, more sophisticated identification procedures may be called for, especially if the chemotherapy regimens are different for several pathogens with similar growth characteristics.

Clinical Diagnosis

Many companies market their own versions of growth-dependent rapid identification systems (Figure 24.7). Such systems are frequently designed for use in identifying

TABLE 24.3 Important clinical diagnostic tests for bacteria

Test	Principle	Procedure	Most common use
Carbohydrate fermentation	Acid and/or gas produced during fermentative growth with sugars or sugar alcohols	Broth medium with carbohydrate and phenol red as pH indicator; inverted tube for gas	Enteric bacteria differentiation
Catalase	Enzyme decomposes hydrogen peroxide, H_2O_2	Add drop of H_2O_2 to dense culture and look for bubbles (O_2) (⟨⟨◯⟩⟩ Figure 6.29)	*Bacillus* (+) from *Clostridium* (−); *Streptococcus* (−) from *Micrococcus-Staphylococcus* (+)
Citrate utilization	Utilization of citrate as sole carbon source, results in alkalinization of medium	Citrate medium with bromthymol blue as pH indicator. Look for intense blue color (alkaline pH)	*Klebsiella-Enterobacter* (+) from *Escherichia*(−), *Edwardsiella* (−) from *Salmonella* (+) (Figure 24.7)
Coagulase	Enzyme causes clotting of blood plasma	Mix dense liquid suspension of bacteria with plasma, incubate, and look for fibrin clot	*Staphylococcus aureus* (+) *S. epidermidis* (−)
Decarboxylases (lysine, ornithine, arginine)	Decarboxylation of amino acid releases CO_2 and amine	Medium enriched with amino acids. Bromcresol purple pH indicator becomes purple (alkaline pH) if there is enzyme action	Aid in determining bacterial group among the enteric bacteria
β-Galactosidase (ONPG) test	Orthonitrophenyl-β-galactoside (ONPG) is an artificial substrate for the enzyme. When hydrolyzed, nitrophenol (yellow) is formed	Incubate heavy suspension of lysed culture with ONPG. Look for yellow color	*Citrobacter* and *Arizona* (+) *Salmonella* (−). Identifying some *Shigella* and *Pseudomonas* species
Gelatin liquefaction	Many proteases hydrolyze gelatin and destroy the gel	Incubate in broth with 12% gelatin. Cool to check for gel formation. If gelatin is hydrolyzed, tube remains liquid on cooling	To aid in identification of *Serratia*, *Pseudomonas, Flavobacterium*, *Clostridium*
Hydrogen sulfide (H_2S) production	H_2S produced by breakdown of sulfur amino acids or reduction of thiosulfate	H_2S detected in iron-rich medium from formation of black ferrous sulfide (many variants: Kliger's iron agar and triple sugar iron agar also detect carbohydrate fermentation)	In enteric bacteria, to aid in identifying *Salmonella, Arizona, Edwardsiella*, and *Proteus* (Figure 24.7)
Indole test	Tryptophan from proteins converted to indole	Detect indole in culture medium with dimethyl-aminobenzaldehyde (red color)	To distinguish *Escherichia* (+) from *Klebsiella* (−) and *Enterobacter* (−); *Edwardsiella* (+) from *Salmonella* (−)
Methyl red test	Mixed-acid fermenters produce sufficient acid to lower pH below 4.3	Glucose-broth medium. Add methyl red indicator to a sample after incubation	To differentiate *Escherichia* (+ , culture red) from *Enterobacter* and *Klebsiella* (usually −, culture yellow)

TABLE 24.3 Important clinical diagnostic tests for bacteria (continued)

Test	Principle	Procedure	Most common use
Nitrate reduction	Nitrate as alternate electron acceptor, reduced to NO_2^- or N_2	Broth with nitrate. After incubation, detect nitrite with α-naphthylamine-sulfanilic acid (red color). If negative, confirm that NO_3^- is still present by adding zinc dust to reduce NO_3^- to NO_2^-. If no color after zinc, then $NO_3^- \rightarrow N_2$	To aid in identification of enteric bacteria (usually +)
Oxidase test	Cytochrome c oxidizes artificial electron acceptor: tetramethyl (or dimethyl)-p-phenylenediamine	Broth or agar. Oxidase-positive colonies on agar can be detected by flooding plate with reagent and looking for blue or brown colonies	To separate *Neisseria* and *Moraxella* (+) from *Acinetobacter* (−). To separate enteric bacteria (all −) from pseudomonads (+). To aid in identification of *Aeromonas* (+)
Oxidation-fermentation (O/F) test	Some organisms produce acid only when growing aerobically	Acid production in top part of sugar-containing culture tube; soft agar used to restrict mixing during incubation	To differentiate *Micrococcus* (acid produced aerobically only) from *Staphylococcus* (acid produced anaerobically). To characterize *Pseudomonas* (aerobic acid production) from enteric bacteria (acid produced anaerobically)
Phenylalanine deaminase test	Deamination produces phenylpyruvic acid, which is detected in a colorimetric test	Medium enriched in phenylalanine. After growth, add ferric chloride reagent and look for green color	To characterize the genus *Proteus* and the *Providencia*
Starch hydrolysis	Iodine-iodide gives blue color with starch	Grow organism on plate containing starch. Flood plate with Gram's iodine and look for clear zones around colonies	To identify typical starch hydrolyzers such as *Bacillus* spp.
Urease test	Urea, $H_2N—CO—NH_2$ split to $2\ NH_3 + CO_2$	Medium with 2% urea and phenol red indicator. Ammonia release raises pH, intense pink-red color	To distinguish *Klebsiella* (+) from *Escherichia* (−). To distinguish *Proteus* (+) from *Providencia* (−)
Voges-Proskauer test	Acetoin produced from sugar fermentation	Chemical test for acetoin using α-naphthol	To separate *Klebsiella* and *Enterobacter* (+) from *Escherichia* (−). To characterize members of genus *Bacillus*

enteric bacteria because enterics are frequently implicated in routine urinary tract and intestinal infections (see Section 24.1).

Other growth-dependent rapid identification kits are available for other bacterial groups or even for single bacterial species. For example, commercial kits containing a battery of tests have been developed for *Staphylococcus aureus*, *Streptococcus pyogenes*, *Neisseria gonorrhoeae*, *Haemophilus influenzae*, and *Mycobacterium tuberculosis*. Other kits are available for identification of the pathogenic fungi *Candida albicans* and *Cryptococcus neoformans* (∽ Section 27.7).

The decision to use a specific diagnostic test is usually made by the clinical microbiologist. This individual takes into consideration the nature of the clinical specimen, basic characteristics (especially the Gram stain) of pure cultures obtained, and previous experience with similar cases.

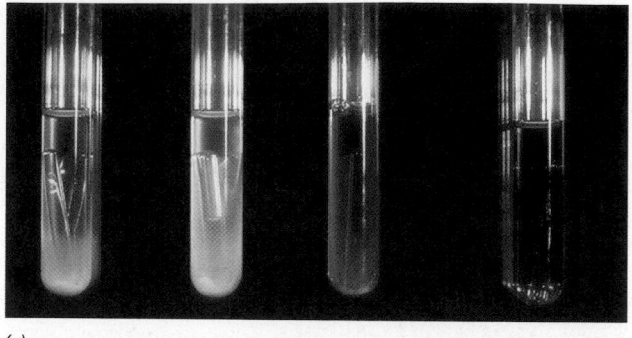

(a)

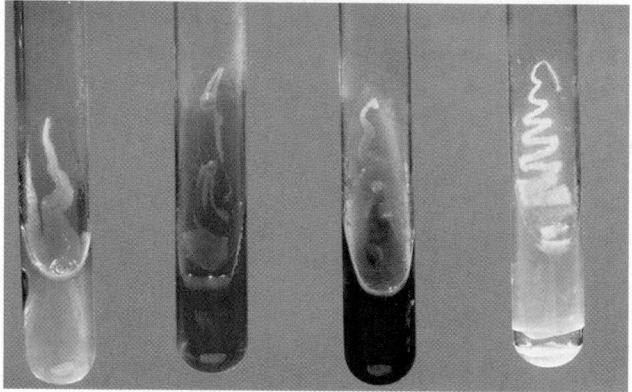

(b)

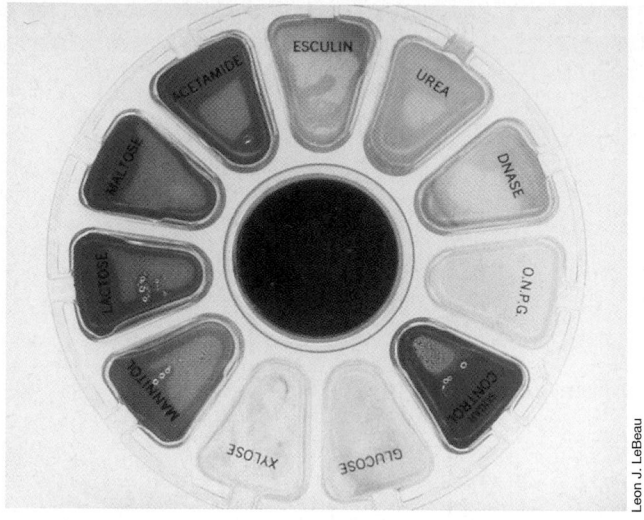

(e)

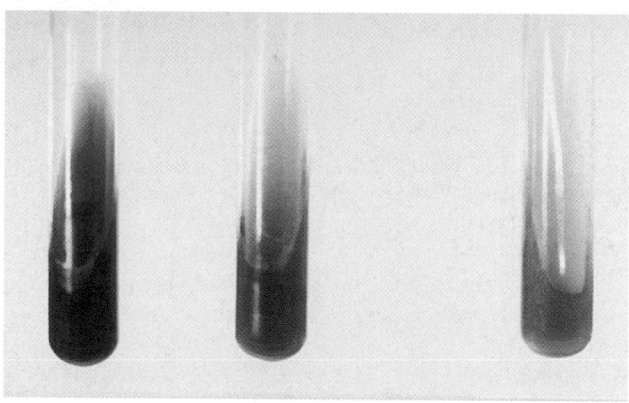

(c)

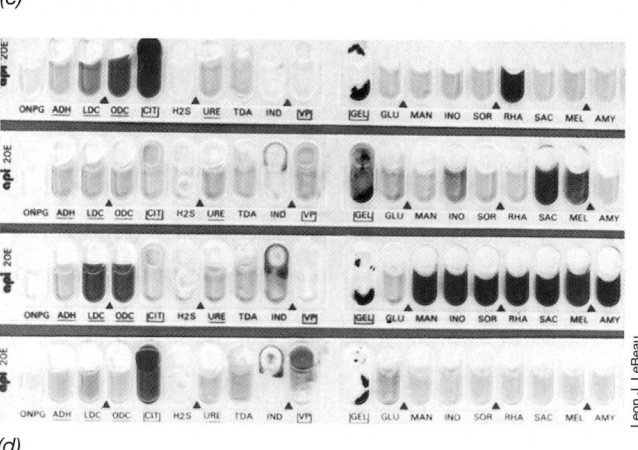

(d)

Leon J. LeBeau

Figure 24.7 Growth-dependent diagnostic methods used for the identification of clinical isolates by color changes in various diagnostic media. (a) Use of a differential medium to assess sugar fermentation. Acid production is indicated by color change of the pH-indicating dye added to the liquid medium. If gas production occurs, a bubble appears in the inverted vial in each tube. From left to right: acid, acid and gas, negative, uninoculated. (b) A conventional diagnostic test for enteric bacteria in a medium called *triple sugar iron (TSI) agar*. The medium is inoculated both on the surface of the slant and by stabbing into the solid agar butt. The medium contains a small amount of glucose and a large amount of lactose and sucrose. Organisms able to ferment only the glucose cause acid formation only in the butt, whereas lactose or sucrose-fermenting organisms also cause acid formation in the top. Gas formation is indicated by the breaking up of the agar in the butt. Hydrogen sulfide formation (either from protein degradation or from reduction of thiosulfate in the medium) is indicated by a blackening due to reaction of the H_2S with ferrous iron in the medium. From left to right: fermentation of glucose only; no reaction; hydrogen sulfide formation; fermentation of glucose and another sugar. (c) Measurement of citrate utilization by *Salmonella* on Simmons citrate agar. The change in pH causes a change in the color of the indicating dye. From left to right: positive, negative, uninoculated. (d) Media kits used for the rapid identification of clinical isolates. The principle is the same as in (a), but the whole arrangement has been miniaturized so that a number of tests can be run at the same time. Four separate strips, each with a separate culture, are shown. (e) Another arrangement of a miniaturized test kit. This one defines sugar utilization in nonfermentative organisms.

✓ **24.2 Concept Check**

Traditional methods for identifying pathogens depend on observing metabolic changes induced as a result of growth. These growth-dependent methods provide rapid and accurate methods for identifying pathogens.

✓ Distinguish between *selective* and *differential* media. Give an example of a medium used for each purpose.

✓ What parameters would a clinical microbiologist use to prescribe a specific diagnostic test kit for identification of an infectious agent?

24.3 Testing Cultures for Antimicrobial Drug Sensitivity

In medical practice, microbial cultures are isolated from diseased patients to confirm diagnoses and to aid in decisions on therapy. Determination of the sensitivity of microbial isolates to antimicrobial agents is one of the most important tasks of the clinical microbiologist.

We discussed the principles for the measurement of antimicrobial activity in Chapter 20. The sensitivity of a culture can be most easily determined by an agar diffusion method or by using a tube dilution technique to determine the *minimum inhibitory concentration* (MIC) of an agent that is necessary to inhibit growth

(∞ Section 20.4). Food and Drug Administration (FDA) regulations now control the procedures used for sensitivity testing in the United States, and similar regulations exist in other countries.

The MIC procedure for antibiotic sensitivity testing involves an *antibiotic dilution assay*, either in culture tubes (∞Figure 20.10) or in the wells of a microtiter plate (Figure 24.8e). A series of twofold dilutions of each antibiotic are made in the wells, and then all wells are inoculated with a standard amount of the same test organism. After incubation, growth in the presence of the various antibiotics is observed by measuring turbidity. Antibiotic sensitivity is usually expressed as the *highest dilution* (lowest concentration) of antibiotic that completely inhibits growth. The dilution assay, because it can be performed in microtiter plates, is readily automated.

A commonly used agar diffusion procedure that measures antimicrobial activity is called the *Kirby–Bauer method*, named after the workers who developed it (Figure 24.8). Culture media are inoculated by spreading a sample of culture evenly across the agar surface. Filter paper discs containing known concentrations of different antimicrobial agents are then placed on the plate. The concentration of each agent on the disc is specified, and after incubation, the presence and size of inhibition zones around the discs of the different agents are noted. Table 24.4 presents typical zone sizes for

TABLE 24.4 Zone sizes for some antimicrobial disc susceptibility tests

Antibiotic	Amount on disc	Inhibition zone diameter (mm)[a]		
		Resistant	Intermediate	Sensitive
Ampicillin[b]	10μg	11 or less	12–13	14 or more
Ampicillin[c]	10μg	28 or less	—	29 or more
Cephoxitin	30μg	14 or less	15–17	18 or more
Cephalothin	30μg	14 or less	15–17	18 or more
Chloramphenicol	30μg	12 or less	13–17	18 or more
Clindamycin	2μg	14 or less	15–16	17 or more
Erythromycin	15μg	13 or less	14–17	18 or more
Gentamicin	10μg	12 or less	13–14	15 or more
Kanamycin	30μg	13 or less	14–17	18 or more
Methicillin[c]	5μg	9 or less	10–13	14 or more
Neomycin	30μg	12 or less	13–16	17 or more
Nitrofurantoin	300μg	14 or less	15–16	17 or more
Penicillin G[d]	10 units	28 or less	—	29 or more
Penicillin G[e]	10 units	11 or less	12–21	22 or more
Polymyxin B	300 units	8 or less	9–11	12 or more
Streptomycin	10μg	11 or less	12–14	15 or more
Tetracycline	30μg	14 or less	15–18	19 or more
Trimethoprim-sulfamethoxazole	1.25/23.75μg	10 or less	11–15	16 or more
Tobramycin	10μg	12 or less	13–14	15 or more

[a] See Figure 24.8d for an illustration of a typical test.

[b] For gram-negative organisms and enterococci.

[c] For staphylococci and highly penicillin-sensitive organisms.

[d] For staphylococci.

[e] For organisms other than staphylococci. Includes some organisms, such as enterococci and some gram-negative rods, that may cause some systemic infections treatable with high doses of penicillin G.

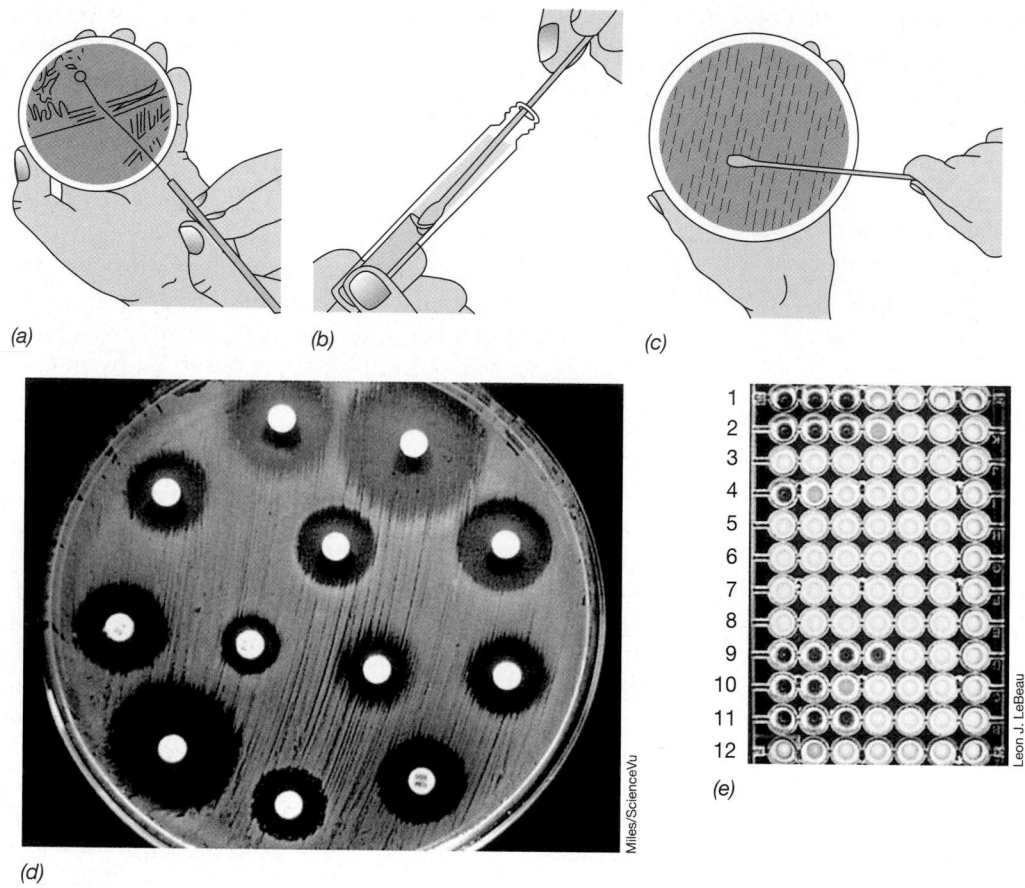

(a) (b) (c)

(d)

(e)

Miles/ScienceVu

Leon J. LeBeau

Figure 24.8 Antibiotic sensitivity testing. (a–d) The Kirby–Bauer procedure for determining the sensitivity of an organism to antibiotics. (a) A colony is picked from an agar plate. It is inoculated into a tube of liquid culture medium and allowed to grow to a specified density. (b) A swab is dipped in the liquid culture. (c) The swab is streaked evenly over a plate of sterile agar medium. (d) Discs containing known amounts of different antibiotics are placed on the plate. After incubation, inhibition zones are observed. The susceptibility of the organism is determined by reference to a chart of zone sizes (Table 24.4). (e) Antibiotic sensitivity determined by the dilution method. The organism is *Pseudomonas aeruginosa*. Each row has a different antibiotic. The use of the microtiter plate enables automation of these tests. The end point is read as the well with the lowest concentration of antibiotic that shows no evidence of bacterial growth. The highest concentration of antibiotic is in the well at the left; serial dilutions are made in the wells to the right. For example, in rows 1 and 2, the end point is the third well. In row 3, the antibiotic is ineffective at the concentrations tested, since there is bacterial growth in all the wells. In row 4, the end point is in the first well. The lowest concentration of antibiotic that completely inhibits bacterial growth defines the minimum inhibitory concentration (MIC) for that agent (∞ Section 20.4).

several antibiotics. Zones observed on the plate are measured and compared to standard data to determine if the isolate is truly sensitive to a given antibiotic.

Because of the widespread occurrence of antimicrobial drug resistance (∞ Section 20.12), an antibiotic sensitivity test is essential for pathogens isolated from each patient. Data such as those in Table 24.4 are useful for choosing the best antibiotic for a specific bacterial infection. Although many potentially serious pathogens are susceptible to a number of different antibiotics, some pathogens, for example, *Pseudomonas aeruginosa*, are sensitive to very few drugs. Other pathogens, such as some encountered in hospital environments, have developed antibiotic resistance and a few are completely resistant to all known antibiotics (∞ Section 20.12). Thus, antibiotic sensitivity testing for these organisms is ab-

solutely essential for effective chemotherapy. Using the drug sensitivity information gathered in this fashion, the clinical microbiologist generates periodic reports called **antibiograms** that indicate the sensitivity of clinically isolated organisms to the antibiotics in current use. Antibiograms are particularly valuable for tracking the emergence of antibiotic-resistant strains of pathogens in facilities such as hospitals and nursing homes.

✓ 24.3 Concept Check

Antimicrobial drugs are in wide use for the treatment of infectious diseases. Pathogens should be tested for sensitivity to individual antibiotics to ensure appropriate chemotherapy. This rigorous approach to antimicrobial drug treatment is usually applied only in hospital settings.

✓ Describe the Kirby–Bauer technique. What does it indicate?

✓ Why is antimicrobial drug sensitivity testing important for the clinical microbiologist, the physician, and the patient?

24.4 Safety in the Clinical Laboratory

Clinical laboratories are areas in which potentially dangerous biological specimens must be handled on a routine basis. Hence, a defined protocol for handling clinical samples must be established to avoid laboratory accidents. In the United States, every clinical and research institution that deals with human or primate tissue is required by law to have an occupational exposure control plan for the handling of all bloodborne pathogens. This law was specifically designed to protect workers from infection by hepatitis B virus (HBV) (∞ Section 26.11) and human immunodeficiency virus (HIV) (∞ Section 26.14), but effectively protects workers from infection by virtually all pathogens because of the stringent precautions.

Studies of laboratory-associated infections have indicated that most such infections do not result from known exposures or accidents but instead from routine handling of patient specimens. The two most common causes of laboratory accidents are ignorance and carelessness. Infectious aerosols, generated during processing of the specimen, are the most common cause of laboratory infections. In attempts to minimize the exposure of clinicians to infectious agents and to thereby reduce the number of nonaccident-associated laboratory infections, well-run clinical laboratories follow the safety rules outlined here. These rules, if applied stringently, ensure the prevention of pathogen spread and meet the requirements of United States law.

1. Laboratories handling hazardous materials must restrict access to laboratory and support personnel. These individuals must have knowledge of the biological risks involved in the laboratory and act accordingly.

2. Effective procedures for decontaminating infectious materials or wastes, including specimens, syringes and needles, inoculated media, bacterial cultures, tissue cultures, experimental animals, glassware, instruments, and surfaces must be in place and be practiced without compromise. A 5.25% (full strength) chlorine bleach solution or other approved disinfectant is recommended for decontaminating spilled infectious material. All potentially infectious waste must be burned in a certified incinerator or handled by a licensed waste handler.

3. Personnel working with hazardous infectious agents or vaccines (for example, rabies, polio, or diphtheria-pertussis-tetanus vaccines) must be properly vaccinated against the agent. Persons working with human or primate tissue must be vaccinated against HBV.

4. All clinical specimens should be considered infectious and handled appropriately. This is especially important for preventing laboratory-acquired hepatitis because of the relative frequency with which hepatitis viruses are present in clinical specimens (∞ Section 26.11).

5. All pipetting must be done with automatic pipetting devices (not by mouth).

6. Animals should be handled only by trained laboratory personnel, and anesthetics and/or tranquilizers should be used to avoid injury to both personnel and animals.

7. Laboratory personnel must wear laboratory coats or gowns, sealed shoes, rubber gloves, masks, eye protection, respiratory devices when needed, and other barrier protection as deemed appropriate by the level of exposure and the severity of the potential infection. These barrier devices must also be properly stored and decontaminated after use. Laboratory personnel must also practice good personal hygiene with respect to hand washing. Eating and drinking, applying cosmetics or lip balm, or wearing contact lenses is never permitted in the clinical laboratory.

8. Because of the special risks associated with AIDS, all clinical (human) specimens should be treated as if they contain HIV (which they might). Protective gloves should be worn whenever handling specimens of *any* kind. Masks and/or full-face shields must be worn any time there is a possibility of generating an aerosol during specimen preparation. Needles must not be resheathed, bent, or broken; they should be placed in a labeled container designated expressly for this purpose that can be sealed and decontaminated before disposal.

These safety rules should be the norm for all clinical laboratories. Specialized clinical laboratories may have additional rules to ensure a safe work environment. For example, if laboratory personnel handle extremely hazardous airborne pathogens (such as the causative agent of tuberculosis, *Mycobacterium tuberculosis*) on a routine basis, the laboratory should be fitted with special features, such as negatively pressurized rooms, biological safety cabinets (∞ Figure 20.4), and air filters, to prevent accidental release of the pathogen from the laboratory. In the final analysis, however, it is the attitude of the personnel that makes the laboratory a safe or an unsafe place to work. Any clinical laboratory is a potentially hazardous place for untrained personnel or those unwilling to take the necessary steps to prevent laboratory-acquired infection.

✓ 24.4 Concept Check

Safety in the clinical laboratory requires effective training, planning, and care to prevent the infection of laboratory workers with pathogens. Materials such as inoculated culture media, used hypodermic needles, and patient specimens require specific precautions for safe handling.

✓ What are the major precautions necessary to prevent spread of a bloodborne pathogen to laboratory personnel?

✓ What are the major causes of laboratory infections?

II IMMUNOLOGY AND CLINICAL DIAGNOSTIC METHODS

Immunoassays are widely used in the clinical laboratory for the detection of specific pathogens or pathogen products. They are used to confirm direct growth tests and also to identify infectious agents when growth tests fail. When growth tests for pathogens are not routinely available or are prohibitively difficult to perform, as is the case with most virus infections, including the human immunodeficiency virus (HIV), immunoassays may provide an effective and relatively simple means of identifying a specific pathogen.

24.5 Immunoassays for Infectious Disease

Immunoassays detect and measure specific immune responses to unique molecular portions of pathogens. In many cases, the immune response of a patient is used to indicate infection by a pathogen. In other cases, antibodies are used to design tests to identify pathogens *in vitro*.

Immunity to Infection

The immune response was discussed in Chapter 22. A summary of the major aspects of immunity is shown in Figure 22.6. The body responds to pathogens in a three-step process. For a pathogen that the body has never before encountered, the pathogen must first be recognized, usually by a group of cells called phagocytes (∞ Section 22.2). Fortunately, phagocytes ingest and destroy most pathogens (a process called *phagocytosis*). Phagocytosis is *nonspecific*, and the target may be any foreign substance, including the pathogens and their components.

In the second phase of immunity, the phagocytes present pathogen-derived *antigens* (proteins obtained from the destroyed pathogen) to antigen-specific lymphocytes known as T cells (∞ Section 22.1). Some T cells known as T helper (T_H) cells do not act directly on the pathogen but recruit and stimulate (help) other cells. One type of T_H cell, the antigen-specific T_H1 cell, attracts and activates phagocytes such as macrophages and neutrophils, causing an inflammatory reaction and limiting the infection (∞ Section 22.7). T_H2 cells, another T_H subset, activate other antigen-specific lymphocytes, the B cells. The B cells then respond by producing soluble, antigen-binding proteins known as *antibodies* (∞ Section 22.8 and 22.9). A *primary antibody response* generally occurs within 5 days, but antibodies do not reach peak quantities for several weeks. The antibody proteins are pathogen-specific and are critical components of the immune response.

The antibodies interact specifically with the antigen on target cells, but cannot kill the cells. A group of proteins, known collectively as complement (∞ Section 22.10), may attach to antibodies bound to the pathogen and lyse all cells with attached antibody. The complement–antibody interaction affects only those cells that have been targeted by antibodies. For example, antibodies specific for cell surface proteins of *Salmonella* spp. interact only with *Salmonella*: complement causes lysis of the antibody-sensitized *Salmonella* cell, but not of an *Escherichia coli* cell that is not antibody-sensitized. Thus, the immune response is *specific* for individual antigens, by virtue of specific antibodies, but the final effect may occur by means of nonspecific mechanisms such as complement.

In many cases, antibody-mediated immunity is not an effective mechanism for controlling the spread of infection. Some infectious agents parasitize the body from *within* cells. For example, animal viruses reproduce using host cell systems and, therefore, spend a large portion of their life cycle within the host cells (∞ Section 9.11). Likewise, bacteria such as *Mycobacterium tuberculosis*, the causative agent of tuberculosis, live in phagocytes (∞ Sections 22.2 and 26.5). Because antibodies are geared to recognize pathogens in the blood or at mucosal cell surfaces, the infected host cells must be identified and destroyed by other means, usually involving the cell-to-cell interactions of the *cell-mediated immunity*. Fortunately, intracellular pathogens produce antigens that are in turn presented on the surface of infected target cells. T cytotoxic cells (T_C) recognize the antigen and act directly on the infected target cell by secreting cytolytic proteins called *perforins*, which destroy the infected cell (∞ Section 22.6).

No specific immunity exists before exposure to antigen, but after the first antigen exposure, specific immune T and B cells are present and specific immunity may persist for years. More importantly, a second antigen stimulation of these cells generates a very rapid and very strong immune response that peaks within several days (◯◯ Section 22.9). This *secondary response* quickly targets and destroys the pathogen. Thus, the immune response has *memory*. Memory is characterized by a rapid rise in immune *titer*, or quantity. *Antibody titer* is routinely used to track infections.

Antibody Titers, Skin Tests, and the Diagnosis of Infectious Disease

In the diagnosis of an infectious disease, isolation of the pathogen is not always possible. One alternative is to measure antibody **titer** (quantity) for a suspected pathogen. As we discussed earlier, if an individual is infected with a suspected pathogen, the antibody titer to that pathogen should be elevated. Antibody titer can be measured by precipitation, agglutination, or any of the methods discussed in Sections 24.7–24.12. The general procedure is to set up dilutions of patient serum (usually twofold dilutions: 1:2, 1:4, 1:8, 1:16, 1:32, and so on) and to determine the *highest* dilution at which the antigen–antibody reaction occurs (Figure 24.9). These methods are all termed *serological tests* because they make use of patient *serum*.

A *single* measure of antibody titer does not indicate active infection. Many antibodies remain at high titer for long periods after infection; to establish that an acute illness is due to a particular pathogen, it is essential to show a *rise* in antibody titer in acute and convalescent serum samples from the same patient. Frequently, the antibody titer is low during the acute stage of the infection and rises during convalescence (Figure 24.9). Such a rise in antibody titer is the best indication that the illness is due to the suspected agent and is also useful in diagnosis of infectious diseases of a rather chronic nature, such as typhoid fever and brucellosis. In some cases, however, the mere presence of antibody may be sufficient to indicate infection. This is true for pathogens rarely found in a population: the presence of antibody is sufficient to indicate an infection. A relevant example here is acquired immunodeficiency syndrome (AIDS) (◯◯ Section 26.14). We will discuss methods for determining HIV antibody levels in Sections 24.11 and 24.12.

Unfortunately, not all infections result in formation of systemic antibody. If a pathogen is extremely localized, there may be little induction of an immunological response and no rise in antibody titer even if the pathogen is proliferating profusely at its site of infection. A good example is the disease gonorrhea. Infection with *Neisseria gonorrhoeae*, the causative agent of gonorrhea, does not elicit a systemic immune response, and thus re-

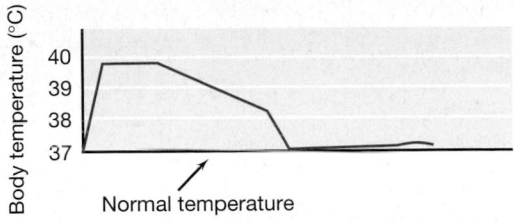

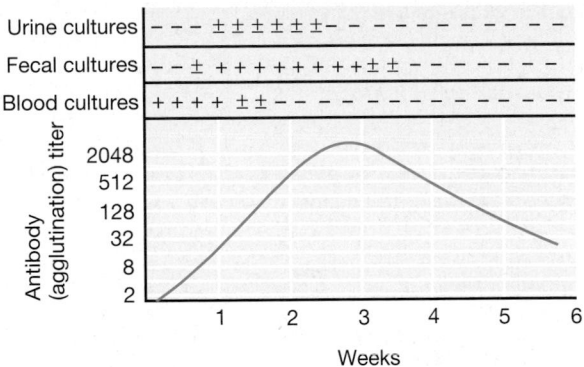

Figure 24.9 The course of infection in a typical untreated typhoid fever patient. Measurement of body temperature provides a measure of the course of clinical symptoms. The antibody titer was measured by determining the highest serum dilution (twofold series) causing agglutination of a test strain of *Salmonella typhi*. Titer is shown as the *reciprocal* of the highest dilution showing an agglutination reaction. Presence of viable bacteria in blood, feces, and urine was determined from periodic cultures. Note that the pathogen clears from the blood as the antibody titer rises, and clearance from feces and urine requires a longer time. Body temperature gradually drops to normal as the antibody titer rises. The data given do not represent a single patient but are a composite of the pattern seen in large numbers of patients.

infection of a cured individual is not uncommon (see Section 24.1; ◯◯ Section 26.12).

In other cases, the presence of antibody in the serum may be due to a recent immunization. In fact, measurement of the rise in antibody titer following immunization is one of the best ways of determining that the immunization is effective.

Skin testing is another method for determining exposure to a pathogen. The most commonly used skin test is the *tuberculin test*, which consists of an intradermal injection of a soluble extract of *Mycobacterium tuberculosis*. A positive inflammatory reaction at the site of injection within 48 hours indicates current infection or previous exposure to *M. tuberculosis*. This test identifies delayed-type hypersensitivity responses caused by pathogen-specific T_H1 cells (◯◯ Section 22.13). Skin tests are routinely used for diagnosing tuberculosis and leprosy (◯◯ Section 26.5), as well as a variety of fungal diseases (◯◯ Section 27.7).

Some of the most common immunodiagnostic tests for pathogens are shown in Table 24.5.

TABLE 24.5 Some clinical immunological procedures for identification of infectious agents

Pathogen/disease	Antigen	Procedure[a]
HIV (AIDS)	Human immunodeficiency virus (HIV)	ELISA
Borrelia burgdorferi (Lyme disease)	Flagellin	ELISA
	Surface proteins	Immunoblot
		Bactericidal test (Section 27.4)
Brucella	Cell wall antigen	Agglutination
Candida albicans (yeast infections)	Soluble extract of fungal proteins	Skin test
Corynebacterium diphtheriae (diphtheria)	Toxin	Skin test (Schick test)
Influenza virus (influenza)	Influenza virus suspensions	Complement-based assay
	Nasopharynx cells containing influenza virus	Immunofluorescence
Mycobacterium leprae (leprosy)	Lepromin (soluble extract of bacterial proteins)	Skin test
Mycobacterium tuberculosis (tuberculosis)	Tuberculin (purified protein derivative, PPD)	Skin test
Neisseria meningitidis (meningitis)	Capsular polysaccharide	Passive hemagglutination (*N. meningitidis* polysaccharide adsorbed to red cells)
Pneumocystis carinii (lung infection)	*P. carinii* cells	Immunofluorescence
Rickettsial diseases (Q fever, typhus, Rocky Mountain spotted fever)	Killed rickettsial cells	Complement-based assay or cell agglutination tests
		ELISA
Salmonella (gastroenteritis)	O and H antigen	Agglutination (Widal test)
		ELISA
Streptococcus (group A) (strep throat, scarlet fever)	Streptolysin O (exotoxin), DNase (extracellular protein)	Neutralization of hemolysis Neutralization of enzyme
Treponema pallidum (syphilis)	Cardiolipin-lecithin-cholesterol	Flocculation [Venereal Disease Research Laboratory (VDRL) test]
Vibrio cholerae (cholera)	O antigen	Agglutination
		Bactericidal test (in presence of complement)
		ELISA

[a] Immunofluorescence tests use preformed antibody to detect the presence of the indicated pathogen in a patient specimen. Skin tests for *C. albicans*, *M. tuberculosis*, and *M. leprae* indicate T_H1-mediated delayed-type hypersensitivity. The *C. diphtheriae* Schick test detects serum antibodies with a toxin-neutralization skin test. All other tests are used to measure serum antibody levels.

✓ 24.5 Concept Check

An immune response is a natural outcome of infection. Specific immune responses, particularly antibody titers and skin tests, can be monitored to provide valuable information concerning past infections as well as current infection and convalescence.

✓ Describe the development of a positive antibody titer to an infectious agent from the acute phase through the convalescent phase of the infection.

✓ Describe the method, time frame, and rationale for the TB skin test. What component of the immune response does this test detect?

24.6 Polyclonal and Monoclonal Antibodies

The immune response to an antigen (Section 22.4) results in the production of immunoglobulin molecules (Sections 22.8 and 22.9) directed at the numerous determinants present on the antigen. Only a few of the many immunoglobulins are directed toward each antigen determinant. The resulting antiserum, a mixture of different antibodies, is known as a **polyclonal** antiserum. Polyclonal antisera consist of antibody populations that provide adequate immune protection to the host, but are usually specific for a variety of determinants. These antisera are not precisely reproducible because they are the sum of the antibody response produced by an individual animal at a single time. Thus, polyclonal antisera, while often very potent, are extremely difficult to standardize for immunodiagnostic procedures.

However, each immunoglobulin is produced by a single B cell (Section 22.9), and B cells cloned *in vitro* can produce limitless supplies of a single monospecific immunoglobulin. Antibodies made by a single cloned B-cell are called **monoclonal** antibodies. The B-cell clones can be stored as frozen cell lines and later reconstituted, providing a reproducibile source of specific antibodies.

As a result, monoclonal antibody technology has supplanted standard polyclonal techniques for many immunodiagnostic applications. Table 24.6 compares the properties of polyclonal antibodies with the properties of monoclonal antibodies.

Monoclonal Antibodies and Hybridomas

Antibody-producing B lymphocytes normally die after several weeks in cell culture (*in vitro*). Therefore, antibody-producing B lymphocytes must be modified to enable them to live in cell culture. The antibody-producing B lymphocytes are fused with B-cell tumors called *myelomas*. Myeloma cells grow and divide indefinitely *in vitro*. The immortal cell lines that result from the B cell–myeloma fusion are hybrid cell lines called *hybridomas*. The hybridoma cell lines share the properties of both fusion partners: They grow indefinitely *in vitro* and produce antibodies (Figure 24.10).

To produce a monoclonal antibody, a mouse is immunized with the antigen of interest. During the next several weeks, antigen-specific B cells proliferate and begin producing antibody in the mouse (⚭ Section 22.9). Spleen tissue, rich in B-lymphocytes, is then removed from the mouse, and the B cells are fused with myeloma cells (Figure 24.10). However, viable hybridoma cell fusions are only a small fraction of the remaining cell population. The hybridomas are selected from other cells by addition of *h*ypoxanthine, *a*minopterin, and *t*hymidine to the *in vitro* cell culture medium (HAT medium). The HAT medium stops the growth of unfused myeloma cells because these cells, though able to grow indefinitely in cell culture, are unable to use the metabolites hypoxanthine and thymidine to bypass a metabolic block caused by aminopterin, a cell poison. By contrast, fused hybrid cells can use hypoxanthine and thymidine to bypass the aminopterin block and grow normally in HAT medium because they receive the genetic information for use of hypoxanthine and thymidine from the B cell fusion partner. Unfused B cells die in a few days because they cannot grow in culture for

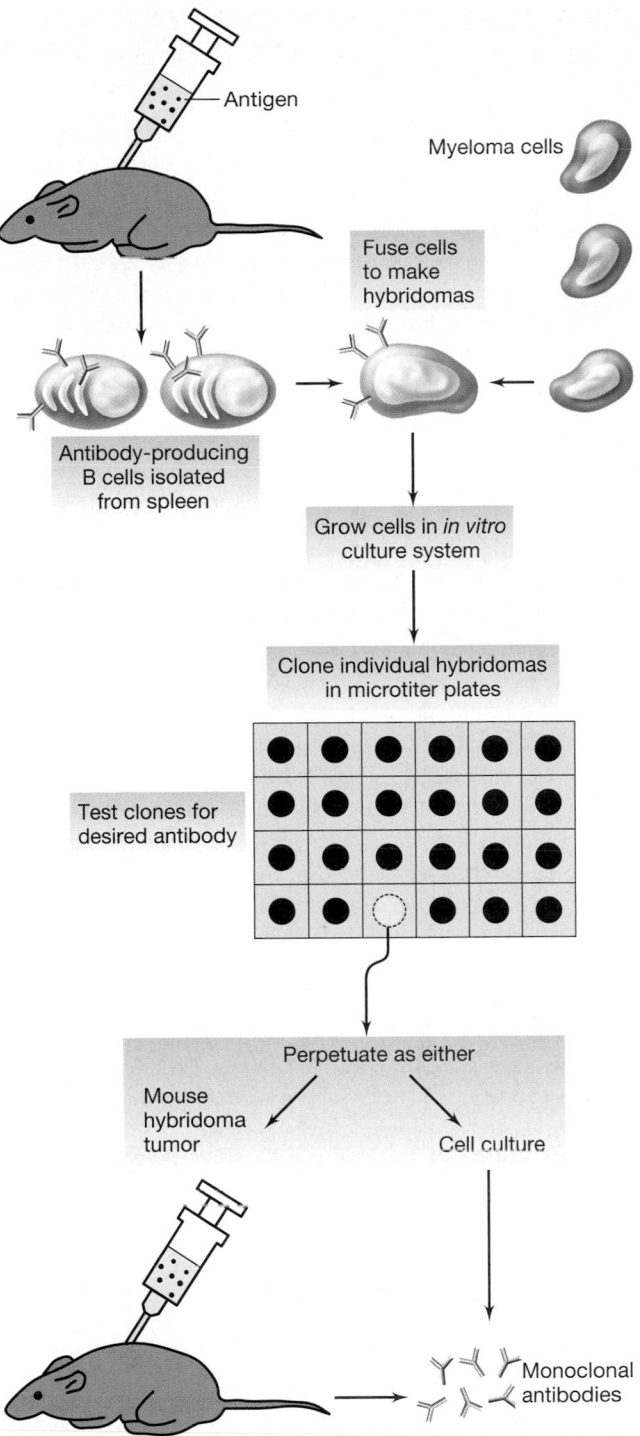

Figure 24.10 The hybridoma technique and production of monoclonal antibodies. The hybridoma can be indefinitely cultured or passed through animals as a tumor. The hybridoma cells can also be stored as frozen tumor cells and reconstituted when needed in tissue culture or in a suitable animal host.

more than a few cell divisions. Following fusion, the antibody-producing clones must be identified.

A sensitive ELISA (enzyme-linked immunoassay) test (see Section 24.11) can be used to identify hybridomas that produce monoclonal antibodies. From a typical fusion,

TABLE 24.6	Characteristics of monoclonal and polyclonal antibody production
Polyclonal	**Monoclonal**
Contains many antibodies recognizing many determinants on an antigen	Contains a single antibody recognizing only a single determinant
Various classes of antibodies are present (IgG, IgM, and so on)	Single class of antibody produced
Can make a specific antibody using only a highly purified antigen	Can make a specific antibody using an impure antigen
Reproducibility and standardization difficult	Highly reproducible

Techniques & Application ...   Urine Testing for Drug Abuse

Many government agencies and private employers carry out drug testing programs in an effort to control drug abuse in the workplace. If a person uses a drug, metabolites of the drug are excreted into the urine. Testing urine for the presence of such metabolites thus permits detection of drug use. Common drug tests include those for cannabinoids (found in marijuana), cocaine, phencyclidine, amphetamines, propoxyphene, benzodiazepine, opiates, steroids, and barbiturates.

Because of the minute quantities of drugs or drug metabolites that occur in the urine, extremely sensitive detection methods are needed, but these methods must also be very specific. Immunological procedures are among the most sensitive and specific methods known for testing urine. Two types of immunoassays are usually employed for drug testing, radioimmunoassay (RIA) and enzyme-linked immunosorbent assay (ELISA). For a drug immunoassay, an antibody must first be prepared against the drug or drug metabolite to be assayed. The antibody thus specifically recognizes an antigenic determinant of the drug. The assay is based on the principle of *competition* between a labeled antigen present in the drug testing system and an unlabeled antigen (the drug or drug metabolite in the urine) for binding sites on the specific antibody. In both the RIA and ELISA tests, the *greater* the amount of drug in the urine, the *less* of the labeled drug will be bound.

In RIA drug testing, known amounts of radiolabeled drug are added to a urine sample together with known amounts of an antibody specific for the drug being tested. The presence of the drug is measured by determining the radioactivity of the antibody bound to a solid substrate. As the concentration of drug in the urine goes up, the radioactivity bound goes down. Concentrations of drug as low as $1-5$ μg/ml can be detected in 1–5 h. The RIA test is very suitable for large-scale screening programs because automatic pipetting and counting procedures can be used.

In the ELISA drug testing procedure, the antigen (drug) is covalently linked to an enzyme, commonly glucose-6-phosphate dehydrogenase, whose activity can be detected by a simple and sensitive colorimetric assay. If the enzyme, via its linked drug ligand, becomes associated with the antibody, it loses its enzymatic activity, but free enzyme can react with the substrate. Urine is mixed with a reagent consisting of antibody to the drug, the glucose-6-phosphate dehydrogenase-drug derivative, and glucose 6-phosphate. If drug is present in the urine, it competes with the enzyme-drug derivative for specific antibody, and more enzyme is then free to react with its substrate. Thus, the more drug present in the subject's urine, the more intense the color. The ELISA procedure has the advantage that the analysis time is short, and the color change can be detected simply. However, the ELISA procedure is sometimes less sensitive than the RIA procedure. The ELISA procedure is especially valuable where a small number of samples are to be ana-

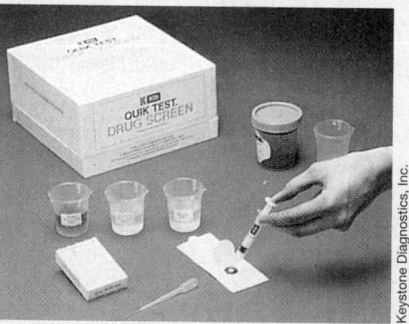

Keystone Diagnostics, Inc.

lyzed at the site where the urine is collected (for instance, away from a laboratory). In some cases, the reagents have been incorporated into paper strips, providing means for urine testing that can be used even by those lacking laboratory experience (see the photo). For large-scale testing, robotic devices have the capability of processing up to 18,000 urine samples per hour with each sample identified by a bar code identification sticker.

Either polyclonal or monoclonal antibodies can be used for urine testing. Monoclonal antibodies have the advantage of defined specificity but may not necessarily have as high an affinity for the drug as a selected polyclonal antibody. The selection of an antibody depends on the specificity and sensitivity needed, the cost of the product, and the ease of use.

One problem with the use of any of these immunological methods for urine testing is cross-reaction with other chemicals that might be present in the urine. For instance, the pain killer ibuprofen cross-reacts with marijuana metabolites in some tests, and the antihistamine drug diphenhydramine can cross-react in the ELISA test for methadone, a controlled narcotic. Because of such false-positive reactions, positive tests must be confirmed by independent tests, such as the use of gas chromatography or high performance liquid chromatography. Highly sensitive combined mass spectrometer-gas chromatographic methods are also available and are valuable for identifying exotic drugs or drugs that elicit a poor immune response.

Immunological tests for drugs in the urine have provided a new and potentially widespread technique for monitoring the use of illegal drugs in the workplace. Urine drug testing is another example of the power and utility of immunology in modern society. ■

---

several distinct clones are isolated, each making a monoclonal antibody to a different determinant on the antigen. Once the clones of interest are identified, they can be grown in cell culture and antibody can be harvested from the culture supernatant. Hybridomas can grow indefinitely or can be stored as frozen cells. Frozen hybridomas can then be thawed and grown in tissue culture to produce more of the specific monoclonal antibody.

Diagnostic and Therapeutic Uses

Monoclonal antibodies provide a limitless supply of highly specific biological reagents with unlimited applications (for an example, see the box, Urine Testing for Drug Abuse). Monoclonal antibodies are widely used for clinical diagnostic tests, immunological typing of bacteria, and for the identification of cells containing foreign surface antigens (for example, a

A Focus On ... Over-the-Counter Immunodiagnostic Kits

A number of manufacturers now market immunodiagnostic kits for use by the general public. For example, virtually every pharmacy now carries several brands of pregnancy tests and many also stock kits to determine the onset of ovulation in women.

These kits are all based on principles involving detection of a hormone excreted in the urine. All pregnancy tests detect the presence of human chorionic gonadotropin (HCG). When the ovum is fertilized, it produces HCG, which functions to maintain the corpus luteum and sustain pregnancy. As pregnancy continues, the new embryo produces and releases increasing amounts of HCG. The hormone, by this time being produced in massive amounts, is released into the bloodstream, removed by the kidneys, and excreted in the urine. Normal menstruation, which occurs about 10–14 days after ovulation if the ovum is not fertilized, does not take place, and at this time the level of HCG present in the urine of the pregnant woman is high enough to be detected easily by these tests. Thus, detectable HCG in the urine means that the individual is pregnant. The tests are advertised to be sensitive enough to detect pregnancy on the first day

of a missed menstrual period, that is, 10–14 days *after* fertilization, but *not* on the first day of a pregnancy.

The key to marketing these kits is their simplicity. Since most people using the tests are not trained in standard laboratory procedures, the test methods must be straightforward and the results must be easy to interpret. Typically, the patient must first provide a urine sample. In the easiest test, the urine sample is simply applied to a test strip. In others, the first step involves pouring the sample through a filter or immersing a "dipstick" in the urine, followed by exposing the filter or dipstick to a second solution. The results are interpreted as a simple color change: a positive test turns the strip, dipstick, or filter from white to pink (see figures) within 30 min or less. All tests share certain principles. At some point, the HCG in the urine is bound to an HCG-specific monoclonal antibody that is immobilized on a solid support—the dipstick, test strip, or filter. Next, a second HCG-specific monoclonal antibody, provided in a solution for the dipstick or filter test, and already soaked onto the test strip for the one-step test, reacts with the immobilized antibody–HCG complex. This second

antibody is chemically linked to colloidal gold particles. If this second antibody reacts with the HCG, it becomes immobilized and the whole complex appears colored.

These kinds of tests, all modifications of sol particle immunoassay (SPIA) tests, are also extensively used in doctor's offices to help diagnose diseases. For example, a number of available tests are specific for cell surface antigens on *Streptococcus pyogenes*, which causes strep throat. Formerly, the only way to positively identify a streptococcal infection was to culture and identify the organism, a process that involved skilled personnel and a laboratory. Because of the requirement for incubation of the culture, results were not available for at least 16 h. Today, results are available in minutes, using material from a throat swab as the antigen source. Several test kits are also available for detecting serum antibodies that react with self proteins. These kits are useful in diagnosing autoimmune diseases such as rheumatoid arthritis.

These tests have several advantages. They are simple to perform, require no technical expertise, are relatively inexpensive, and are reasonably accurate. They also have a long shelf life, and no special storage conditions are required. However, there are disadvantages. They are qualitative tests, which means that they can be used only when the outcome is absolute. For example, one is either pregnant or not pregnant. However, quantitative measures of antigen levels might be desirable in some cases, such as in the tests that identify autoimmune antibodies. In addition, many manufacturers do not provide adequate positive and negative controls to ensure that the test works properly. Thus, results should always be interpreted with caution, and negative or positive results may be in error because of improper test procedures or faulty test reagents. ■

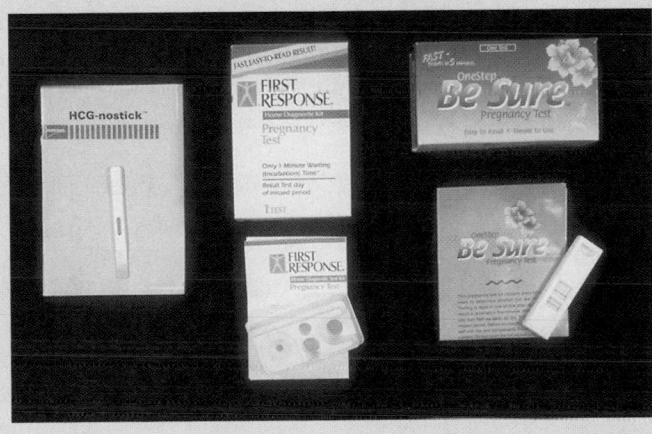

virus-infected cell). Monoclonal antibodies have also been used in genetic engineering for identifying and measuring levels of gene products not detectable by other methods and also show great promise for increasing the specificity of existing clinical tests including blood typing and even pregnancy determination.

For example, home pregnancy tests employ monoclonal antibodies specific for the hormones associated with pregnancy (see the box, Over-the-Counter Immunodiagnostic Kits).

Because of their specificity, monoclonal antibodies are now being used to detect and treat human cancer.

TABLE 24.7 Types of antigen–antibody reactions		
Location of antigen	**Accessory factors required**	**Reaction observed**
Soluble	None	Precipitation (Section 24.7)
On cell or inert particle	None	Agglutination (Section 24.8)
Flagellum	None	Immobilization or agglutination (Section 24.8)
On bacterial cell	Complement	Lysis (◯◯ Section 22.10)
On bacterial cell	Complement	Killing (◯◯ Section 22.10)
On erythrocyte	Complement	Hemolysis (◯◯ Section 22.10)
Toxin	None	Neutralization (Section 24.7)
Virus	None	Neutralization (Section 24.7)
On bacterial cell	Phagocyte, complement	Phagocytosis (opsonization; ◯◯ Section 22.10)

Malignant cells contain a variety of surface antigens not expressed by normal cells. These *tumor antigens* are unique, tumor-specific cell proteins. Monoclonal antibodies prepared against the tumor antigens specifically target the malignant cells and are used to deliver toxins directly and specifically to malignant cells. Tumor-specific monoclonal antibodies covalently linked to toxins are now undergoing clinical testing. The specificity of the monoclonal antibody treatment may greatly improve cancer chemotherapy by offering an alternative to chemical and radiation treatments that often damage normal host cells as well as cancer cells.

✓ 24.6 Concept Check

Polyclonal and monoclonal antibodies are used for research and clinical applications. Hybridoma technology provides reproducible, monospecific antibodies for a wide range of clinical, diagnostic, and research purposes.

- ✓ Can a single polyclonal antibody recognize a variety of determinants?
- ✓ What advantages do monoclonal antibodies have as compared with polyclonal antibodies?

24.7 *In Vitro* Antigen–Antibody Reactions: Serology

The study of antigen–antibody reactions *in vitro* is called **serology**. Serological reactions are the basis for all diagnostic immunology tests. The principles of antigen–antibody reactions lie in the specific interaction of determinants on the antigen with the *variable* region of the antibody molecule.

A variety of serological tests are used to identify antigens, depending on the properties of the antigen and on the conditions chosen for reaction (Table 24.7).

Specificity and Sensitivity

The usefulness of a serological test for diagnostic purposes is dependent on the test's specificity and sensitivity. **Specificity** is the ability of an antibody preparation to recognize a single antigen. An optimum level of specificity requires that the antibody is specific for a single antigen, will not cross-react with any other antigen, and therefore, will not provide *false positive* results. Specificity must be defined in terms of reactions with positive and negative control antigens. Specificity for each test must be determined experimentally and verified every time the test is used.

Sensitivity defines the lowest amount of an antigen that can be detected. The highest level of sensitivity requires that the antibody in a test be capable of identifying a single antigen molecule. High sensitivity prevents *false negative* reactions. The sensitivity of some common tests is shown in Table 24.8 in terms of the amount of antibody necessary to detect antigen. The amount of antigen detected by each test system is proportional to the amount of antibody used. For example, immune precipitation reactions require a large amount of the antibody and generally detect 0.1 to 1.0 mg quantities of antigen. Thus, precipitation tests are the least sensitive serological tests. At the other extreme, *enzyme-linked immunosorbent assays*, or *ELISA* tests (see Section 24.11) require 100,000 times less antibody and detect 1 million times less antigen (0.1 to 1.0 ng quantities) than precipitation antigens. Thus, ELISA tests are among the most sensitive serological tests.

TABLE 24.8 Sensitivity of immunodiagnostic assays	
Assay	**Sensitivity (μg antibody/ml)[a]**
Precipitin reaction	
In fluids	24–160
In gels (double immunodiffusion)	24–160
Agglutination reactions	
Direct	0.4
Passive	0.08
Radioimmunoassay (RIA)	0.0008–0.008
Enzyme-linked immunosorbent assay (ELISA)	0.0008–0.008
Immunofluorescence	8.0

[a] The smallest amount of antibody necessary to give a positive reaction in the presence of antigen.

Neutralization

Neutralization is the interaction of antibody with antigen to block or distort the antigen sufficiently to reduce or eliminate its biological activity. Neutralization reactions can occur *in vitro* or *in vivo*.

For example, neutralization of microbial toxins by specific antibody can occur when toxin molecules and antibody molecules directed against the toxin combine in such a way that the active portion of the toxin is blocked (Figure 24.11). Neutralization reactions of this type occur for many bacterial exotoxins, including most of those listed in Table 21.4. An antiserum containing an antibody that neutralizes a toxin is referred to as an *antitoxin*. Neutralization reactions may also occur when viruses are bound by specific antibodies. For example, antibodies directed against the hemagglutinin and neuraminidase proteins of influenza viruses prevent the adsorption of the viruses to specific receptors on host cells *in vitro* (∞ Section 26.8).

Neutralization reactions are seldom used for *in vitro* testing purposes in other than a few specialized laboratories because neutralization tests require biologically active systems.

Precipitation

Precipitation is the interaction of a soluble antibody with a soluble antigen to form an insoluble complex. Antibody molecules generally have two antigen-binding sites (that is, they are bivalent) (∞ Section 22.8). Therefore, it is possible for each site to combine with a separate antigen molecule. If the antigen also has more than one available determinant, a precipitate may develop consisting of aggregates of antibody and antigen molecules (Figure 24.12a). Because they are easily observed *in vitro*, precipitation reactions are very informative serological tests, especially for the quantitative measurement of antibody concentrations. Precipitation occurs maximally only when there are optimal proportions of the two reacting substances. The presence of either antigen or antibody in excess quantities results in the formation of only small, soluble immune complexes.

Precipitation reactions carried out in agar gels, referred to as *immunodiffusion* tests, are used to study the specificity of antigen–antibody reactions. Both antigen and antibody diffuse outward from separate wells cut in an agar gel, and precipitation bands form in the region where antibody and antigen interact in optimal proportions (Figure 24.12b). The shapes of the precipitation bands are characteristic for the reacting substances, and two antigens reacting with antibodies in an antiserum can be tested for relatedness by observing the bands formed when the two antigens are placed in adjacent wells near the antiserum well. For example, if two antigens in adjacent wells are identical, they will form a single, fused, precipitin band. This is referred to as a line of

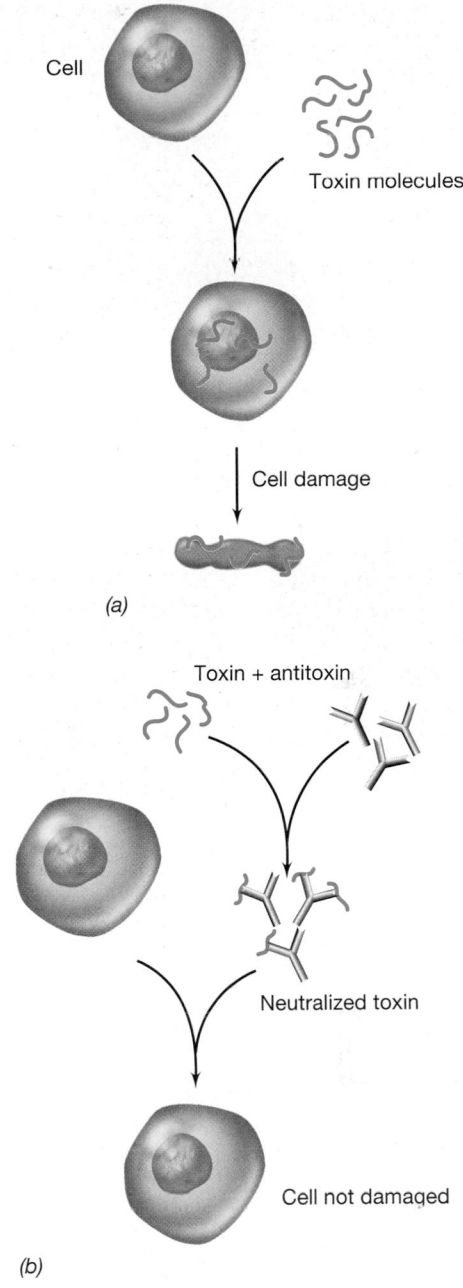

Cell

Toxin molecules

Cell damage

(a)

Toxin + antitoxin

Neutralized toxin

Cell not damaged

(b)

Figure 24.11 Neutralization of an exotoxin by antibody. Antitoxins are used to treat acute cases of diphtheria (∞ Section 26.3), tetanus (∞ Section 27.8), and botulism (∞ Section 29.5). (a) Untreated toxin results in cell destruction. (b) Antitoxin neutralizes toxin and prevents cell destruction.

identity. If, on the other hand, adjacent wells contain one antigen in common but one well contains a second antigen, a line of *partial identity* will form (Figure 24.12b). The small extended precipitin line (representing a reaction between the antiserum and the second antigen) is referred to as a *spur*. Immunodiffusion is used as a tool in biochemical research to assess the relatedness of proteins obtained from different sources.

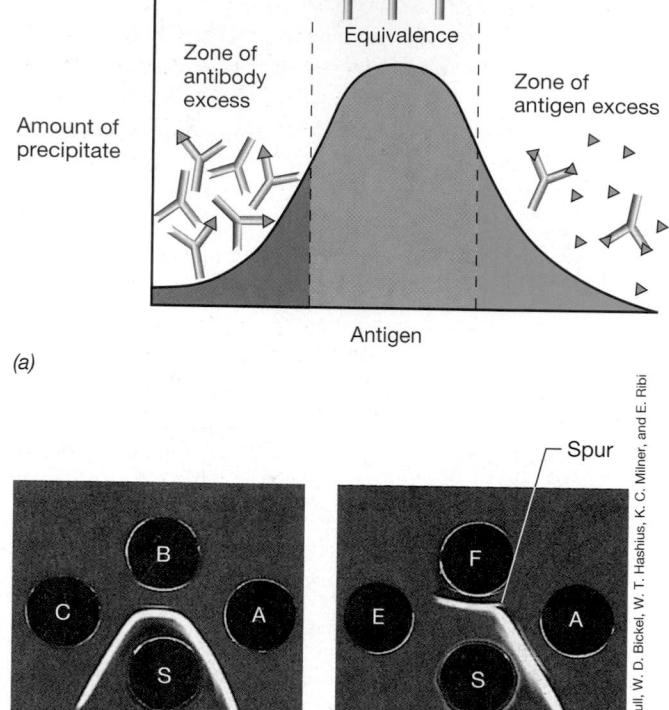

(a)

(c)

C. Weibull, W. D. Bickel, W. T. Hashius, K. C. Milner, and E. Ribi

Figure 24.12 Precipitation reaction between soluble antigen and antibody. The graph (a) shows the extent of precipitation as a function of antigen and antibody concentration. (b) Precipitation in agar gel, a process called immunodiffusion. Wells labeled S contain antibodies to cells of *Proteus mirabilis*. Wells labeled A, B, and C contain soluble extracts of *Proteus mirabilis*. A line of identity is observed in the wells on the left. On the right, antigen E does not react and antigen A shows partial identity with antigen F (see leader to spur).

Unfortunately, the highly visible precipitation reactions are not very sensitive. Microgram quantities of specific antibody are necessary to visualize a precipitate (Table 24.8) and most useful diagnostic tests require sensitivity at nanogram levels. Consequently, the precipitation reactions are normally used only in research and reference laboratories.

✓ 24.7 Concept Check

Antigen–antibody reactions all require that antibody bind to antigen. Specificity and sensitivity define the accuracy of individual serological tests. Neutralization and precipitation reactions are examples of antigen-binding tests that produce visible results involving antigen–antibody interactions.

✓ High specificity and high sensitivity in serological reactions prevent false positive and false negative reactions. Explain.

✓ What are the minimum antigen and antibody requirements for a precipitation reaction?

24.8 Agglutination

Agglutination is the visible clumping of a particulate antigen when mixed with antibodies specific for the particulate antigens. Although not as sensitive as some other serological tests, agglutination tests are about 100 times more sensitive than precipitation tests (Table 24.8). Agglutination tests are widely used in clinical and diagnostic laboratories because they are simple to perform, highly specific, inexpensive, and rapid. Standardized tests are available for the identification of blood group antigens and for identification of pathogens and pathogen products.

Direct Agglutination

Direct agglutination results when soluble antibody causes clumping due to interaction with an antigen that is an integral part of the surface of a cell or other insoluble particle. Perhaps the mostly widely used direct agglutination procedure is used for the classification of antigens found on the surface of red blood cells (erythrocytes). Agglutination of red blood cells is referred to as *hemagglutination*, and this diagnostic test is called *blood typing*.

Red blood cells contain a variety of antigens, and individuals vary considerably with respect to the antigens present on their red blood cells. The major human red cell antigens are called A, B, and D (also known as Rh).

A and B antigens and antibodies are the basis for the ABO blood typing assay. Antibodies specific for particular erythrocyte surface antigens cause red blood cells to visibly clump when a small drop of blood is mixed on a microscope slide with antiserum reactive against either the A or B antigens of human erythrocytes (Figure 24.13). The antisera are obtained from human donors who have been immunized to A or B antigens by natural or artificial means.

For the A, B, and O blood types, individuals express codominant A and B alleles as one of the following antigen phenotypes: A, B, AB (one allele expressing the A antigen and one expressing the B antigen), or O (the absence of either A or B antigens). In addition, individuals make antibodies to most nonself blood group antigens. Type A individuals make antibodies to group B antigens, while type B individuals make antibodies to group A antigens. Type AB individuals have neither A nor B antibodies, whereas type O individuals have antibodies to both A and B antigens (Figure 24.13). Antibodies against A and B antigens are called *natural antibodies* because they appear to be produced by most individuals in response to ubiquitous antigen sources such as enteric *Bacteria*.

Blood typing is always performed before blood transfusion to prevent red blood cell destruction. Red cell destruction would occur if antibodies in the recipi-

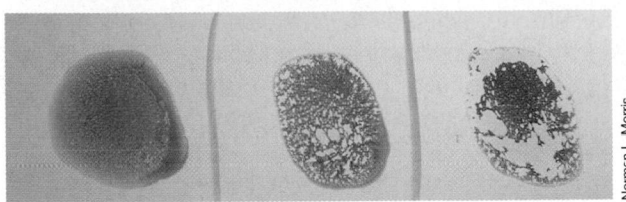

(a)

Blood type	Percentage of U.S. population	Serum	
		Anti A	Anti B
Type O	47	No aggl.	No aggl.
Type A	42	Aggl.	No aggl.
Type B	8	No aggl.	Aggl.
Type AB	3	Aggl.	Aggl.

(b)

Figure 24.13 Direct agglutination of human red blood cells for ABO blood typing. (a) The reaction on the left shows no agglutination. The reaction in the center shows the diffuse agglutination pattern that indicates a positive reaction for the B blood group. The reaction on the right shows the strong agglutination pattern with large, clumped agglutinates typical for the A blood group. (b) Table of expected blood grouping results for the U.S. population.

ent's blood reacted with the red blood cells in the transfused blood, or vice versa. If agglutination occurs, clumps of cells could lodge in blood vessels or arteries and block the flow of blood, causing serious illness or death. However, it is more likely that the antibodies could cause *hemolysis* (lysis of red blood cells) through the action of complement (Section 22.10), resulting in severe anemia.

Passive Agglutination

Passive agglutination is the agglutination of soluble antigens or antibodies that have been adsorbed or chemically coupled to cells or insoluble particles such as latex beads or charcoal particles. The insolubilized antigen or antibody can then be detected by agglutination reactions. The cell or particle serves as an inert carrier. Passive agglutination reactions are up to five times more sensitive than direct agglutination tests (Table 24.8) and greatly increase the ability to detect the presence of the soluble reactants.

The agglutination of antigen-coated or antibody-coated latex beads by complementary antibody or antigen from a patient is a typical method of rapid diagnosis. Small (0.8 μm) latex beads coated with a specific antigen or antibody are mixed with patient serum on a mi-

croscope slide and incubated for a short period. If patient antibody binds the antigen on the bead surface, the milky-white latex suspension will become visibly clumped, indicating a positive agglutination reaction. Latex agglutination is also used to detect bacterial surface antigens by mixing a small amount of a bacterial colony with antibody-coated latex beads. For example, a commercially available suspension of latex beads containing antibodies to protein A and clumping factor, two molecules found exclusively on the surface of *Staphylococcus aureus*, is virtually 100% accurate in identifying clinical isolates of *S. aureus*. Unlike traditional growth-dependent tests for *S. aureus*, identification of *S. aureus* by the latex bead assay takes only 30 sec (Figure 24.14). Other latex bead agglutination assays have been developed to identify *Streptococcus pyogenes*, *Neisseria gonorrhoeae*, *Hamemophilus influenza*, *Campylobacter* spp., and the fungi *Cryptococcus neoformans* and *Candida albicans*.

A very widely employed latex agglutination assay is that used for detecting specific serum antibodies for *rheumatoid factor*, an antibody directed against the body's own immunoglobulin and associated with the autoimmune disease *rheumatoid arthritis* (Section 22.13). Latex beads coated with human immunoglobulin are mixed with whole blood or serum, and agglutination is compared to agglutination with positive and negative control sera.

Passive agglutination assays are simple and specific. In addition, the inexpensive nature of the assays makes them suitable for large-scale screening purposes; the widespread use of the rheumatoid test is a good example. Because passive agglutination assays require no expensive equipment or particular expertise, they are widely used as diagnostic tools.

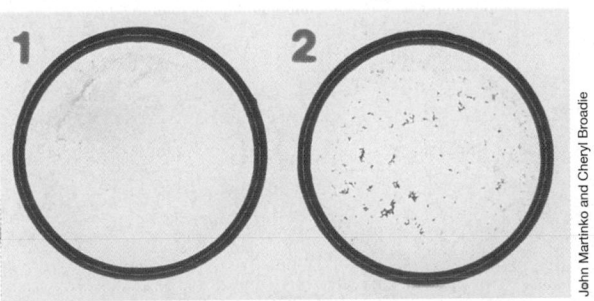

Figure 24.14 Latex bead agglutination test for *Staphylococcus aureus*. Panel 1 shows a negative control. Note the uniform pink color of the suspended latex beads coated with antibodies to protein A and clumping factor, two antigens found exclusively on the surface of *S. aureus* cells. Panel 2 shows the same suspension after a loopful of material from a bacterial colony was mixed into the suspension. The bright red clumps indicate a positive agglutination reaction took place and indicates that the colony is *S. aureus*.

✓ **24.8 Concept Check**

Direct agglutination tests are widely used for rapid determination of blood group antigens. A number of passive agglutination tests are available for identification of a variety of pathogens and pathogen-related products. Agglutination tests are rapid, relatively sensitive, highly specific, simple to perform, and inexpensive.

✓ Distinguish between *direct* and *passive* agglutination. Which tests are more sensitive?

✓ What advantages do agglutination tests have over other immunoassays? What disadvantages?

24.9 Immunoelectron Microscopy

Antibodies to which heavy metals have been chemically conjugated can be used to locate antigens in cells by electron microscopy. This is possible because heavy metals scatter the electron beam of the electron microscope.

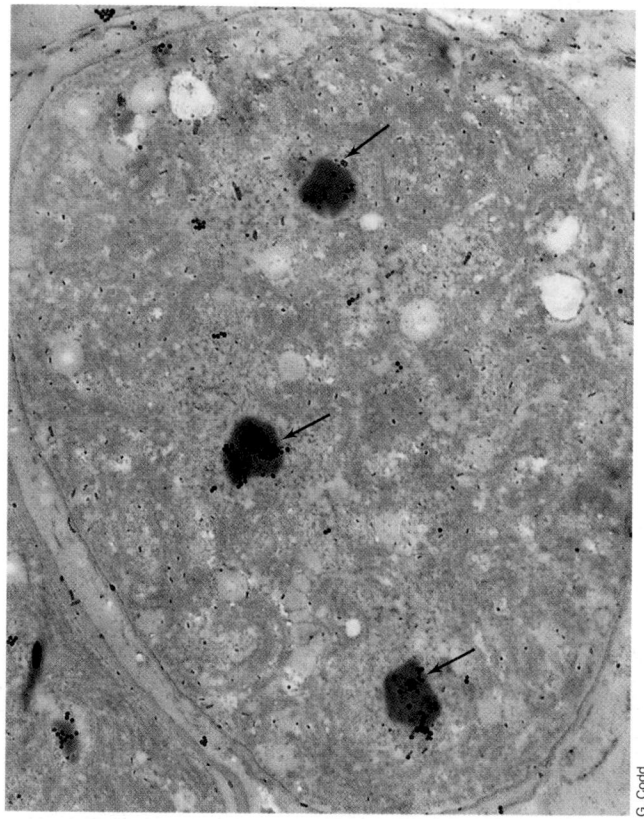

Figure 24.15 Immunoelectron microscopy. Antibodies made in rabbits to the enzyme ribulose-1,5-bisphosphate carboxylase from the cyanobacterium *Chlorogloeopsis fritschii* were added to thin sections of *C. fritschii* and the preparation treated with goat anti-rabbit IgG conjugated to 20-nm colloidal gold particles. The concentration of the particles around large inclusions called carboxysomes (arrows) indicate that these sites contain large amounts of the enzyme.

This technique, called *immunoelectron microscopy*, is used primarily in research where there is a need to determine where a specific antigen (usually a protein) is localized in a particular region of the cell (Figure 24.15). Cells, following chemical fixation and other preparations necessary for observation by the electron microscope, are treated with antibodies covalently conjugated to a heavy metal, usually gold or platinum. The electron-dense metals scatter electrons, and thus the presence of bound antibody can be detected by dense black spots in photographs of the preparation.

In immunoelectron microscopy, although the cell is dead and chemically fixed, most protein antigens retain sufficient native structure, and antibodies still react with little nonspecific cross-reaction. This technique has been used extensively to pinpoint the location of enzymes in cells, especially those suspected to be associated with the cytoplasmic membrane or some other internal structure (Figure 24.15).

Although immunoelectron microscopy can be used for identifying pathogens such as human immunodeficiency virus (HIV) in cells (Figure 26.35), the time, expense, expertise, and specialized equipment involved make it impractical for diagnostic procedures in all but the most specialized clinical research settings.

✓ **24.9 Concept Check**

Immunoelectron microscopy is a research tool used for localizing antigens in cells.

✓ Why are heavy metal-antibody conjugates used for immunoelectron microscopy?

24.10 Fluorescent Antibodies

Antibodies can be chemically modified with fluorescent dyes. These modified antibodies are used to detect antigens on intact cells. **Fluorescent antibodies** are widely used for diagnostic and research applications.

Fluorescent Methods

Antibodies can be covalently modified by fluorescent dyes such as rhodamine B, which fluoresces red, or fluorescein isothiocyanate, which fluoresces yellow-green. This does not alter the specificity of the antibody but makes it possible to detect the antibody bound to cell or tissue surface antigens by use of the fluorescence microscope (Figure 24.16). Cells with bound fluorescent antibodies emit a bright fluorescent color, usually red-orange or yellow-green, depending on the dye used. Fluorescent antibodies are used in diagnostic microbiology because they permit the identification of a microorganism directly in a patient specimen (*in situ*) and avoid the isolation and culturing of the organism (see below). The

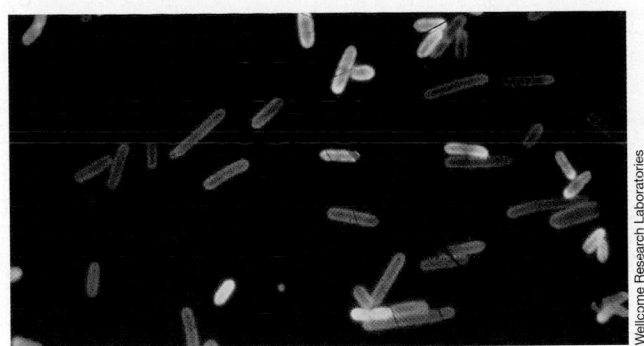

Figure 24.16 Fluorescent antibody reactions. Cells of *Clostridium septicum* were tested with antibody conjugated with fluorescein isothiocyanate, which fluoresces yellow-green. Cells of *Clostridium chauvei* were stained with antibody conjugated with rhodamine B, which fluoresces red-orange.

fluorescent antibody technique is also very useful in microbial ecology as a method for directly viewing and identifying microbial cells without the need to isolate and culture them.

Two distinct fluorescent antibody procedures, the *direct* and the *indirect* staining methods, are used. In the direct method, the antibody directed to the surface antigen is fluorescent. In the indirect method, the presence of a nonfluorescent antibody on the surface of the cell is detected by the use of a fluorescent antibody directed against the nonfluorescent antibody (Figure 24.17). This is done by immunizing one animal species, for example, a goat, with antibodies from a second species, for example, a rabbit, and then conjugating the fluorescent dye to the goat antibodies. The fluorescent goat anti-rabbit antibodies can then be used to detect the presence of any rabbit immunoglobulin that is bound to cells.

Clinical Applications

In a typical clinical test using fluorescent antibodies, a specimen containing a suspected pathogen is allowed to react with a specific fluorescent antibody and observed with a fluorescent microscope. If the pathogen contains surface antigens reactive with the antibody, the cells will fluoresce (Figure 24.18).

Fluorescent antibodies can be applied directly to infected host tissues, permitting diagnosis long before primary isolation techniques yield a suspected pathogen. For example, in diagnosing legionellosis (∞ Section 28.7), a positive identification can be made by staining biopsied lung tissue with fluorescent antibodies prepared against cell walls of *Legionella pneumophila* (Figure 24.18*a*). Likewise, a fluorescent antibody against the capsule of *Bacillus anthracis* can be used in the microscopic diagnosis for anthrax. Fluorescent antibody reactions can also be used in diagnosis of viral infection (Figure 24.18*b*) and in a variety of noninfectious diseases. For

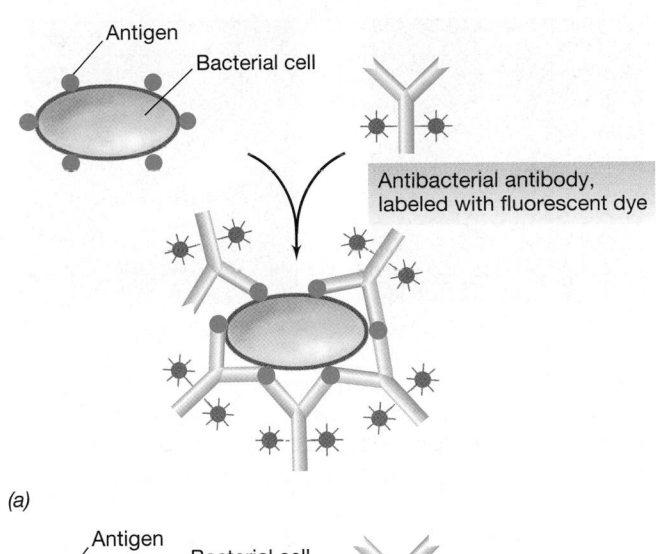

(a)

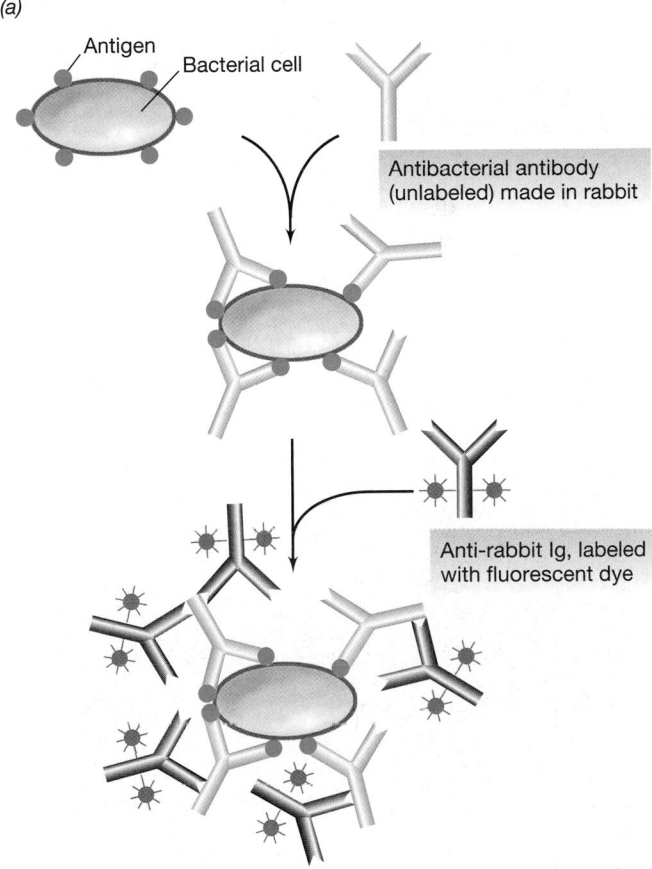

(b)

Figure 24.17 Fluorescent antibody methods for detection of microbial surface antigens. (a) Direct staining method. (b) Indirect staining method.

example, in identifying cell types expressing a particular antigen, such as on malignant cells, fluorescent antibodies may be very valuable in following the course of the disease (Figure 24.19).

Fluorescent antibodies can also be used to separate mixtures of cells into relatively pure populations or to define the numbers of certain cell types in complex

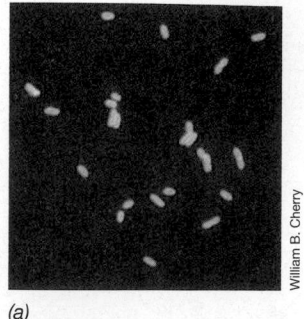
(a)
William B. Cherry

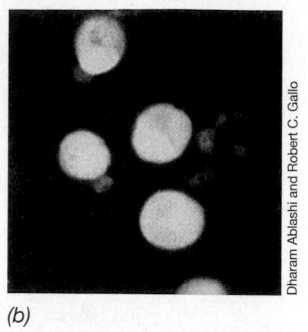
(b)
Dharam Ablashi and Robert C. Gallo

Figure 24.18 Examples of the use of fluorescent antibodies in clinical microbiology. (a) Immunofluorescent stained cells of *Legionella pneumophila*, the cause of legionellosis. The specimen was taken from biopsied lung tissue. The individual organisms are 2–5 µm in length. (b) Detection of virus-infected cells by immunofluorescence. Human B lymphotrophic virus (HBLV)-infected spleen cells were incubated with serum containing antibodies to HBLV from a patient with a lymphoproliferative disorder. Cells were then treated with fluorescein isothiocyanate-conjugated anti-human IgG antibodies. HBLV-infected cells fluoresce bright yellow. Cells in the background did not react with the patient's serum. Individual cells are about 10 µm in diameter.

mixtures such as blood. Fluorescent-labeled monoclonal antibodies directed against the CD4 and CD8 surface antigens of T lymphocytes (∞ Section 22.5) are routinely used to identify and enumerate these cells in the blood leukocyte population (Figure 24.20). For example, the definition of acquired immunodeficiency syndrome

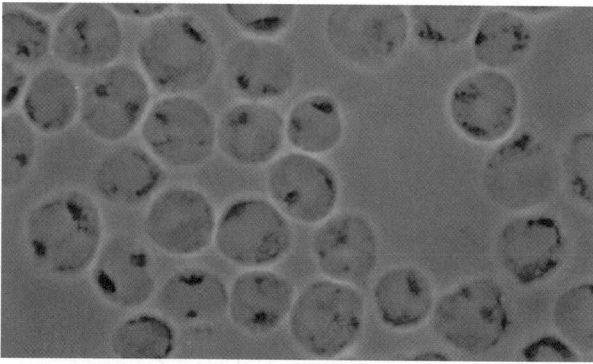
(a)

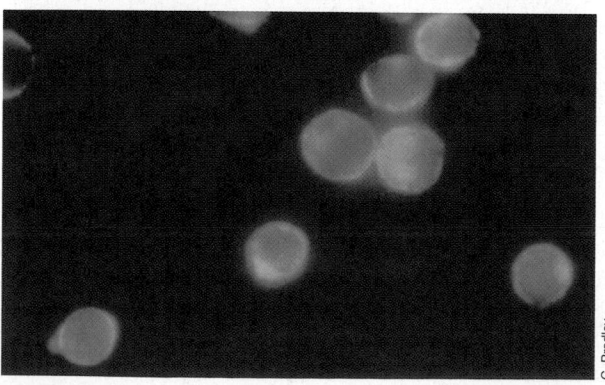
(b)
G. Bradley

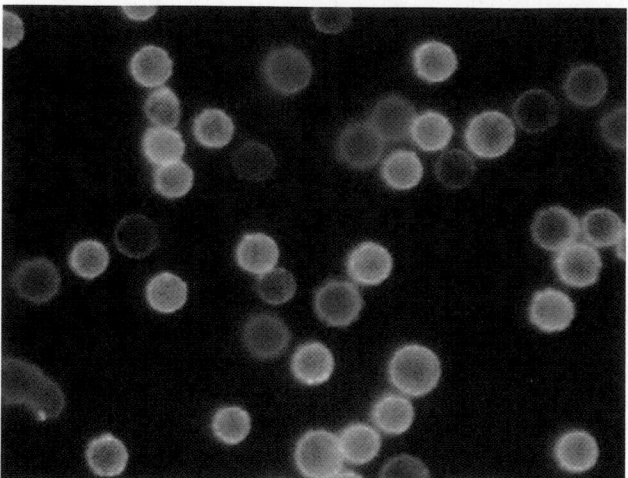
Richard Lewis

Figure 24.20 T lymphocytes stained with fluorescent-tagged monoclonal antibodies to specific surface markers. Yellow-green cells are cytotoxic (CD8) T cells; red-orange cells are T-helper (CD4) cells. The different-colored cells can be separated from one another by a fluorescence-activated cell sorter to yield enriched populations of different cell types. Individual cells are about 10–12 µm in diameter. Reprinted with permission from *Science* **239**: Cover (Feb. 12, 1988), © AAAS.

(AIDS) includes a reduction in CD4 cell numbers. In addition, the CD4 T cell number changes during the progression of AIDS and is diagnostic for the disease. Thus, by defining the CD4 numbers, the clinician can identify the reduction in CD4 cells compared with normal values and, with successive assays over time, can follow the progress of the disease (∞ Section 26.14).

Fluorescent antibody-labeled cells can be visualized, counted, and separated with an instrument called a *fluorescence spectrometer*, often referred to as a fluorescence-activated cell sorter (FACS). The FACS uses a laser beam to activate fluorescent antibody bound to cells, placing a charge on the labeled cells. An electric field is then applied to the cell mixture. The fluorescing and nonfluorescing cells are then deflected to opposite poles of the electric field where each cell population is counted and deposited in a tube. The use of several antibodies, each labeled with a different fluorescent dye, can result in the simultaneous identification of several cell markers. A typical application used for identifying CD3 and CD4 surface proteins on T cells in normal and AIDS patients is shown in Figure 24.21.

Figure 24.19 Use of fluorescent antibodies in noninfectious disease diagnostics. (a) Human leukemic cells, some of which are sensitive to a toxic anticancer drug and some of which are not, appear indistinguishable. (b) When the cells in (a) are treated with a fluorescent monoclonal antibody that binds specifically to a protein found only on the surface of drug-resistant cells, the latter fluoresce whereas drug-sensitive cells do not. Individual cells are about 10–12 µm in diameter.

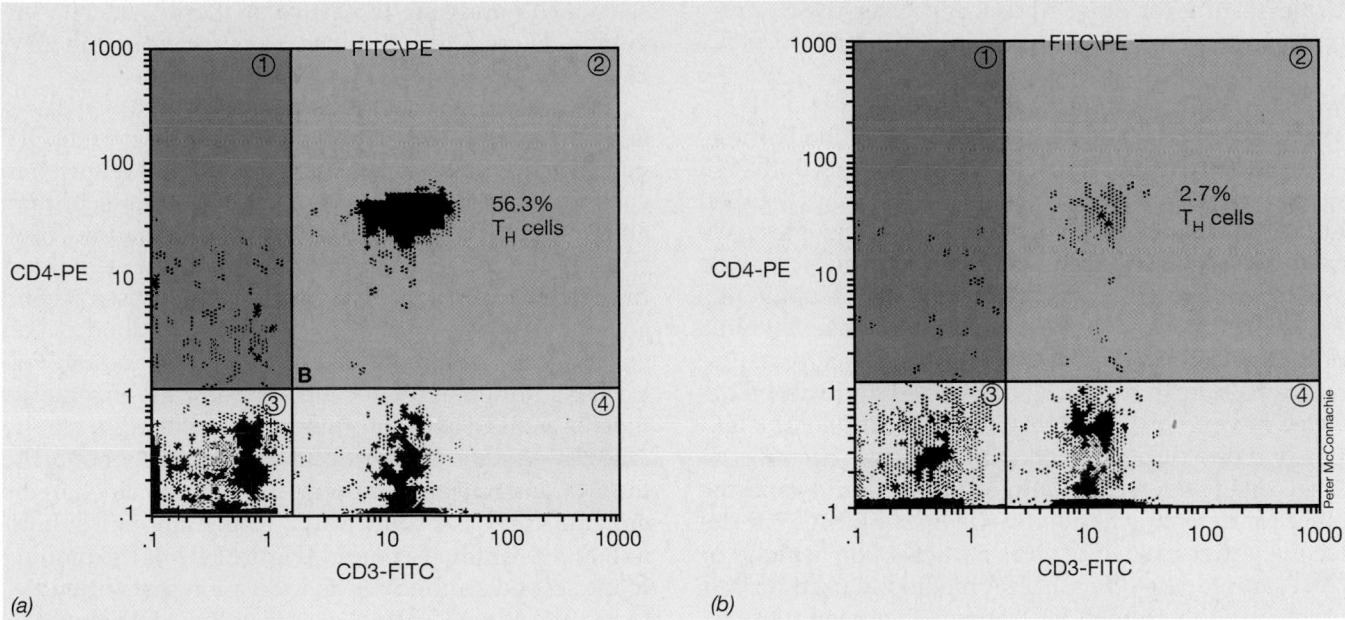

Figure 24.21 CD3 and CD4 cell enumeration from a healthy human (a) and from a human with acquired immunodeficiency syndrome (AIDS) (b) using a fluorescence-activated cell sorter (FACS). Each dot represents a single cell. Peripheral blood cells were simultaneously labeled with monoclonal antibody to CD4 conjugated to phycoerythrin (PE) and with monoclonal antibody to CD3 conjugated to fluorescein isothiocyanate (FITC). CD3 is found on all T cells. CD4 is found on T-helper (T_H) cells only. Quadrant 3 shows cells that were stained with neither antibody. Quadrant 1 shows cells stained with only anti-CD4. Quadrant 4 shows cells stained with only anti-CD3. Quadrant 2 shows cells stained with both anti-CD3 and anti-CD4. (a) Results from a healthy human. In this case, 56.3% of the T cells were T_H cells. Thus, quadrant 2 shows a dense staining pattern. (b) Results from a patient with clinical AIDS. In this case, only 2.7% of the total T cells are T_H cells. This is indicated by the very light staining pattern in quadrant 2. (Original data from Peter McConnachie, used with permission.)

FACS analysis is also useful for research applications. For example, immunologists routinely use FACS methods to separate complex mixtures of immune cells. They can then study the properties of the highly enriched cell populations.

Under appropriate conditions, fluorescent antibodies yield rapid, highly specific information about a variety of clinical conditions. However, fluorescent antibody techniques are not without their pitfalls. Nonspecific staining can be a problem because of surface antigens that may *cross react* between various bacterial species, some of which may be members of the normal flora. This is a major problem among enteric bacteria, where lipopolysaccharide antigens (∞ Section 4.9) are sufficiently similar among species to cause binding or partial binding of the fluorescent probe. The clinical microbiologist must therefore perform controls using nonspecific sera and confirm all positive immunofluorescent findings by other immunological or microbiological tests.

✓ 24.10 Concept Check

Fluorescent antibodies are used for quick, accurate identification of pathogens and other antigenic substances in tissue samples and other complex environments. Fluorescent antibodies can be used for quantitative enumeration and sorting of a variety of cell types.

- ✓ How do fluorescent antibodies recognize a single cell in a complex cell mixture such as a tissue sample?
- ✓ How are fluorescent antibodies used to identify specific cells in complex mixtures like blood?

24.11 Enzyme-Linked Immunosorbent Assay and Radioimmunoassay

The limiting factor in most of the immunological reactions discussed thus far is not *specificity* but *sensitivity* (see Section 24.5 and Table 24.6). Because of their high sensitivity, radioimmunoassay (RIA) and enzyme-linked immunosorbent assay (ELISA) are two widely used immunological techniques. RIA and ELISA employ radioisotopes and enzymes, respectively, to label antibody molecules used for antigen detection. The attachment of radioactive or enzyme ligands to antibody molecules decreases the amount of antigen–antibody complex required to detect a reaction. This increased sensitivity has been extremely helpful in clinical diagnostics and research and has allowed the development of a variety of new immunological tests (see boxes,

Urine Testing for Drug Abuse, and Over-the-Counter Immunodiagnostic Kits).

ELISA

The covalent attachment of enzymes to antibody molecules creates an immunological tool possessing both high specificity and high sensitivity. The technique, called **ELISA**, makes use of antibodies to which enzymes have been covalently bound such that the enzyme's catalytic properties and the antibody's specificity are unaltered. Typical bound enzymes include peroxidase, alkaline phosphatase, and β-galactosidase, all of which catalyze reactions whose products are colored and can be detected in very low amounts with a spectrophotometer.

Two basic ELISA methodologies have been developed, one for detecting antigen (*direct* ELISA) and the other for detecting antibodies (*indirect* ELISA). For detecting antigens such as virus particles from a blood or fecal sample, the direct ELISA method is used. In this procedure the antigen is "trapped" between two layers of antibodies (Figure 24.22). Thus, this method is sometimes called the *sandwich ELISA*. The specimen is added to the wells of a microtiter plate previously coated with antibodies specific for the antigen to be detected. If an antigen (virus particle) is present in a sample, it will be trapped by the antigen-binding sites on the antibodies. After washing unbound material away, a second antibody containing a conjugated enzyme is added. The second antibody is also specific for the antigen, and so it binds to any remaining exposed determinants. Following a wash, the enzyme activity of the bound material in each microtiter well is determined by adding the substrate for the enzyme. The color produced is proportional to the amount of antigen present (Figure 24.22).

To detect *antibodies* in human serum, an indirect ELISA is employed. An indirect ELISA test is widely used to detect antibodies to human immunodeficiency virus (HIV), and we will discuss this test in detail because the principles involved are applicable to all indirect ELISA tests.

The HIV-ELISA

The casual agent of AIDS, the *human immunodeficiency virus (HIV)* (⌘ Section 26.14), is transmitted by bodily fluids including blood. Sensitive, specific, rapid, and cost-effective screening tools are needed to test blood samples from individuals exposed to HIV and to ensure that HIV is not being inadvertently transmitted during blood transfusions or through the transfer of blood products. An ELISA test is used for the routine screening of blood for signs of exposure to HIV.

The HIV-ELISA test is an *indirect* ELISA designed to measure *antibodies* to HIV present in serum. Initial infection with HIV leads to the production of antibodies to several HIV antigens, in particular, those of the HIV envelope. These antibodies can be detected by the HIV-ELISA test (Figure 24.23).

To carry out an HIV-ELISA test, microtiter plates are first coated with a preparation of disrupted HIV particles; about 200 ng of disrupted HIV is required in each well. A diluted patient serum sample is then added, and the mixture is incubated to allow HIV-specific antibodies to bind to HIV antigens. To detect the presence of antigen–antibody complexes, a second antibody is then added. This second antibody is an enzyme-conjugated anti-human IgG preparation. Following addition of the second antibody, the enzyme activity is assayed (the anti-human IgG antibodies bind to any HIV-specific IgG antibodies previously bound to the HIV antigen preparation). The color obtained in the enzyme assay is proportional to the amount of anti-human IgG antibody bound (Figure 24.23). The binding of the second antibody is an indication that antibodies from the patient's serum recognized the HIV antigens and that the patient has antibodies to HIV, and therefore has been exposed to HIV. Control sera (known to be HIV-negative) are assayed in parallel with any samples to measure the extent of background absorbance in the assay.

The HIV-ELISA test is a rapid, highly sensitive, specific method for detecting exposure to HIV. Since ELISAs in general are highly adaptable to mass screening and automation, the HIV-ELISA test is used as a standard screening method for blood. However, this test method can give erroneous results under certain circumstances.

For example, the test occasionally gives false positive results. Because a number of factors can contribute to these results, none of which are related to exposure to HIV, all positive HIV-ELISA tests *must* be confirmed by another independent test, usually the Western blot (immunoblot) test (see Section 24.12). A positive HIV Western blot test after a positive HIV-ELISA test is considered proof of HIV infection.

The other drawback to the HIV-ELISA test is the possibility of obtaining false negative results. It takes the immune system some time to develop an effective antibody response and a detectable antibody titer (see Section 24.5). In the case of HIV infection, this lag time is estimated to be 6 weeks to a year. Therefore, individuals who have been recently infected with HIV may not yet be producing detectable amounts of antibody when they are tested. Another reason for a false negative result in the HIV-ELISA test is the total destruction of the immune system seen in advanced cases of AIDS; if no immune cells are left in the body, no antibodies can be made and the ELISA test is not useful. However, at this stage of disease, a diagnosis is possible based on clinical information (⌘ Section 26.14) and the ELISA test is useful only as a confirmatory indicator.

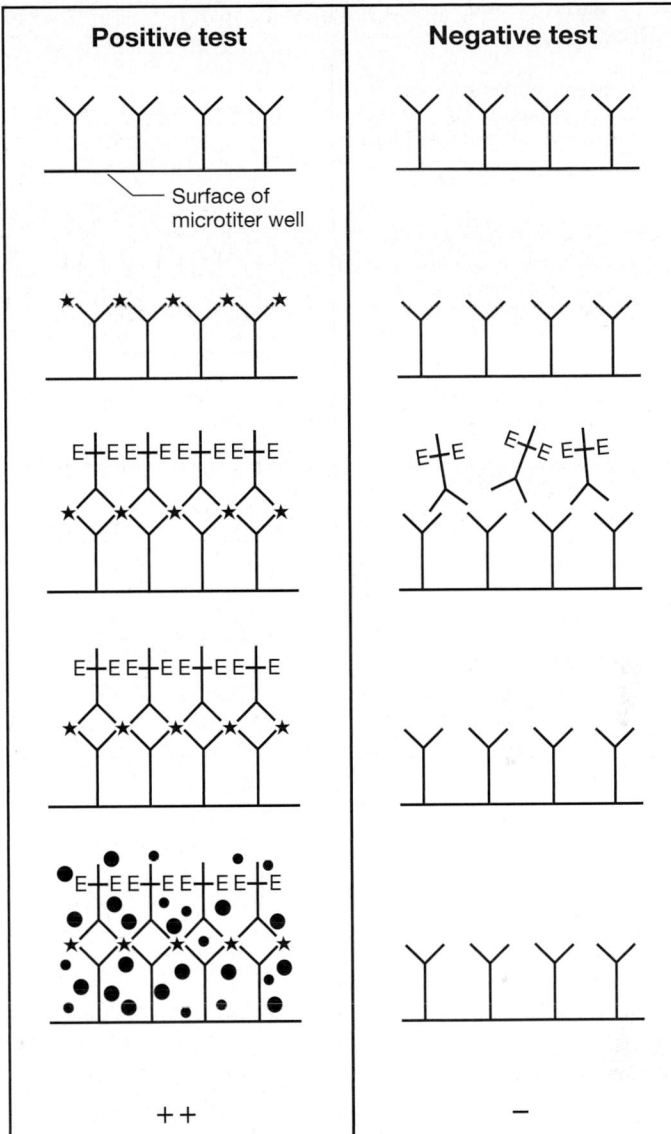

Procedure

1. Antibodies (Y) to virus (★) bound to wells of microtiter plate

2. Add patient sample (secretions, serum, and so on) suspected of containing virus particles or virus antigens and wash wells with buffer

3. Add antivirus antibody containing conjugated enzyme

   (E⊥E)

4. Wash with buffer

5. Add substrate for enzyme and measure amount of colored product (●).

Results

Colored product

Quantitation

Colored product produced is proportional to amount of antigen.

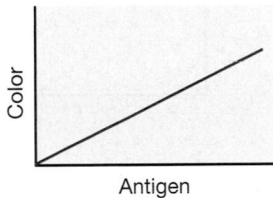

Figure 24.22 Detection of viruses by a direct ELISA test.

Other ELISA Tests of Clinical Importance

Besides the ELISA test for HIV, literally hundreds of clinically useful ELISAs have been developed. Some of these are direct ELISAs for detecting antigens, including bacterial toxins such as cholera toxin, enteropathogenic *Escherichia coli* toxin, and *Staphylococcus aureus* enterotoxin. Viruses currently detected using direct ELISA techniques include rotavirus, hepatitis viruses, rubella virus, bunyavirus, measles virus, mumps virus, and parainfluenza virus.

Indirect ELISAs have been developed for detecting antibodies to a variety of clinically important bacteria. Although not meant to be a complete list, ELISAs for detecting serum antibodies to *Salmonella* (gastrointestinal diseases), *Yersinia* (plague), *Brucella* (brucellosis), a variety of rickettsias (Rocky Mountain spotted

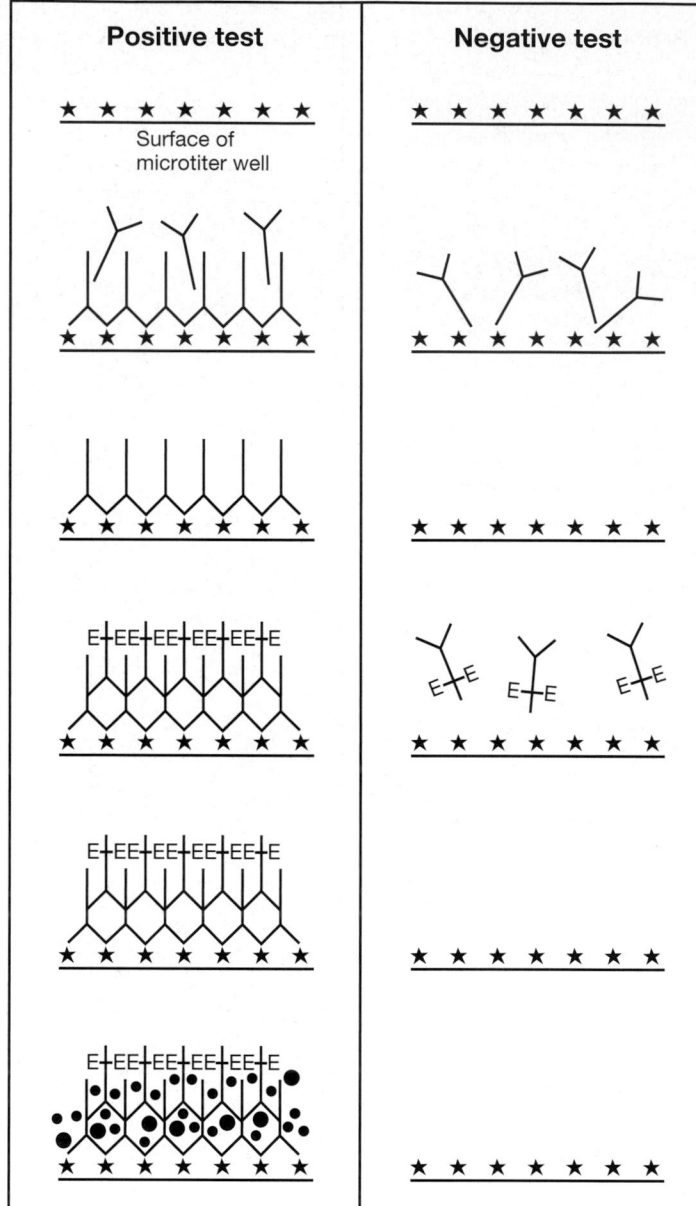

Procedure

1. Coat microtiter wells with antigen preparation from disrupted HIV particles (★)

2. Add patient serum sample. HIV-specific antibodies bind to HIV antigen. Other antibodies do not bind

3. Wash with buffer

4. Add human anti-IgG antibodies conjugated to enzyme (E⏉E)

5. Wash with buffer

6. Add substrate for enzyme and measure amount of colored product (●).

Results

Colored product

Quantitation

Colored product produced is proportional to the antibody concentration.

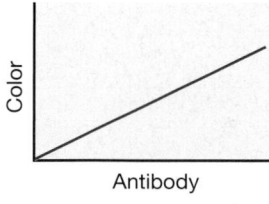

Figure 24.23 Indirect ELISA test for detecting antibodies to human immunodeficiency virus (HIV), the causal agent of acquired immunodeficiency syndrome (AIDS).

fever, typhus, Q fever), *Vibrio cholerae* (cholera), *Mycobacterium tuberculosis* (tuberculosis), *Mycobacterium leprae* (leprosy), *Legionella pneumophila* (legionellosis), *Borrelia burgdorferi* (Lyme disease), and *Treponema pallidum* (syphilis) have been developed. ELISAs have also been developed for detecting antibodies to *Candida* (yeast) and a variety of parasites, including those causing amebiasis, Chagas' disease, schistosomiasis, toxoplasmosis, and malaria.

The speed, low cost, lack of radioactive waste, and long shelf life make ELISA tests particularly attractive for many laboratories. But it is the high *sensitivity* of ELISAs that really make them important immunodiagnostic tools.

Radioimmunoassay

Radioimmunoassay (RIA) employs radioisotopes instead of enzymes as antibody conjugates. The isotope iodine-125 is the most commonly used detection system, because proteins can be readily iodinated without

disrupting their immune specificity. RIA is used clinically to measure rare serum proteins such as human growth hormone, glucagon, vasopressin, testosterone, and insulin present in humans in extremely small amounts (Figure 24.24) and also in some urine tests for drug abuse.

In most cases, a *direct* RIA is employed. The direct assay is a two-step procedure. First, radioactive antigen-specific antibodies are added to a series of microtiter wells containing known concentrations of pure antigen (such as a hormone), which is first bound to the wells. The radioactivity in each of these standard wells is then measured. These data establish a standard curve for antigen concentrations. Next, the antigen sample from a patient is allowed to bind to another well, and radioactive antibodies are added and measured, as before. The amount of radioactivity bound by the patient sample is then compared to a standard plot generated from the binding data obtained using the pure antigen, and the concentration of antigen in the

1. Bind insulin to wells of microtiter plate

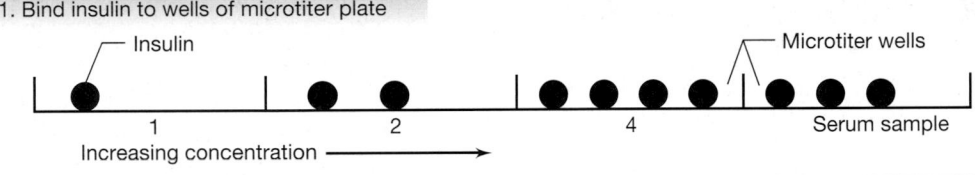

2. Add excess anti-insulin antibodies that are labeled with ^{125}I; wash to remove unbound antibody

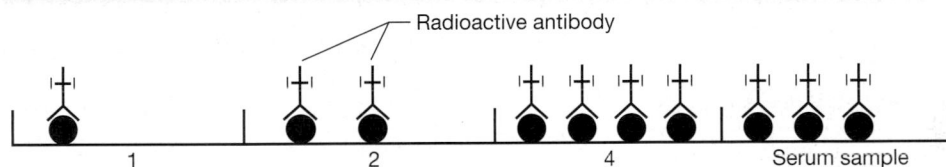

3. Count radioactivity in gamma radiation counter. Wells labeled 1, 2, and 4 establish a standard curve with known amounts of antigen (insulin). The radioactivity in the last well indicates, by comparison to the standard curve, how much insulin is present in a known amount of serum.

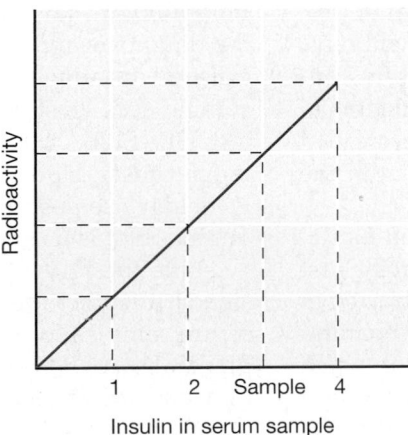

Figure 24.24 Radioimmunoassay (RIA). Using RIA to detect insulin levels in human serum. Following establishment of a standard curve, the insulin concentration in a serum sample can be estimated.

patient serum is interpolated from the standard plot (Figure 24.24).

RIA has the same sensitivity range as ELISA and can also be performed very rapidly. However, the instruments used to detect radioactivity are quite specialized and expensive. RIA generates a considerable amount of radioactive waste, and the radioactive decay time (half life) of the radioisotopes used for detection may limit the useful life of the test kit. As a result, RIA is often used only when ELISA is not sufficiently accurate or sensitive. For example, RIA is often more useful than ELISA for detecting serum protein levels (as described earlier) because some serum components may inhibit ELISA enzyme-substrate reactions or antibodies. Thus, for certain applications, each test system has clear advantages.

✓ 24.11 Concept Check

ELISA and RIA methods are the most sensitive immunoassay techniques. Both involve linking a detection system, either an enzyme or a radioactive molecule, to an antibody or antigen, enhancing sensitivity. ELISA and RIA are used for clinical and research work; tests have been designed to detect either antibody or antigen in a vast number of applications.

- ✓ Why are ELISA and RIA techniques more sensitive than standard immunoassays such as precipitation and agglutination?
- ✓ Compare ELISA and RIA with respect to their relative uses, advantages, and disadvantages.

24.12 Immunoblot Procedures

Antibodies can be also used in clinical diagnostics to identify individual specific *proteins* associated with specific pathogens. The procedure employs three techniques discussed previously: (1) the separation of proteins on polyacrylamide gels, (2) the transfer (blotting) of proteins from gels to a nitrocellulose or nylon membrane, and (3) identification of the proteins by specific antibodies. Protein blotting and the subsequent identification of the proteins by specific antibodies is also called the *Western blot* technique to distinguish it from the (DNA) *Southern blot* technique.

The **immunoblot** is a very sensitive method for detecting specific proteins in complex mixtures. In the first step of an immunoblot, a protein mixture is subjected to electrophoresis on a polyacrylamide gel. This separates the proteins into several distinct bands, each of which represents a single protein of specific molecular weight (Figure 24.25). The proteins are then transferred to the membrane by an electrophoretic transfer process that forces the proteins out of the gel and onto the membrane. At this point, antibodies raised against a protein or group

of proteins from a pathogen are added to the blot. Following a short incubation period to allow the antibodies to bind, a radioactive marker that binds antigen–antibody complexes is added. The most common radioactive marker used is *Staphylococcus* protein A iodinated with radioactive iodine, ^{125}I. Protein A has a strong affinity for antibody and binds firmly. Once the radioactive marker has bound, its vertical position on the blot can be detected by exposing the nitrocellulose blot to X-ray film; the gamma rays emitted by the ^{125}I expose the film only in the region where the radioactive antibody has bound to antigen–antibody complexes (Figure 24.25).

For many clinical applications, immunoblots employ enzyme-linked immunosorbent assay (ELISA) technology (see Section 24.11) for detection of bound antigen–antibody complexes. Following treatment of the blotted proteins with specific antibody, the membrane is washed and then treated with a second antibody, which binds to the first. For example, if antibodies from a human were used in the first step, then the second antibody could be a rabbit anti-human antibody. Covalently attached to this second antibody is an enzyme. The original antigen–antibody complexes are visualized when the enzyme is exposed to substrate: The product of the enzyme reaction leaves a colored product on the membrane at any spot where rabbit antibodies are bound to the human antibodies. By comparing the location of the color bands on the blot with the position of colored bands from control samples, a protein associated with a given pathogen can be positively identified.

The immunoblot procedure can be used to detect either antigen (*direct* evidence for pathogen presence) or antibody (*indirect* evidence for pathogen exposure). Thus, this very sensitive, extremely accurate method is analogous to the direct and indirect ELISA procedures detailed in Section 24.11.

The HIV Immunoblot

Immunoblots have had a significant clinical impact on the diagnosis and confirmation of HIV exposure. Because an immunoblot is more laborious, more time-consuming, less sensitive, and more costly than the ELISA test, HIV-ELISA tests have been widely used for screening purposes. However, the HIV-ELISA test occasionally yields false positive results. Thus, the highly specific immunoblot is used to confirm positive ELISA results.

Like the HIV-ELISA, the HIV immunoblot is designed to detect the presence of *antibodies* to HIV in a serum sample. To perform the immunoblot, a purified preparation of HIV is treated with the detergent sodium dodecyl sulfate (SDS), which solubilizes HIV proteins and also renders the virus inactive. HIV proteins are then resolved by polyacrylamide gel electrophoresis. The HIV proteins are then blotted from the gel onto membranes (Figure 24.25). At least seven major HIV proteins are re-

1. Denature proteins by boiling in detergent

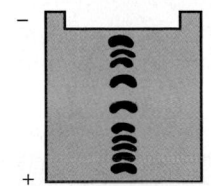

2. Subject to electrophoresis, proteins separate by molecular weight

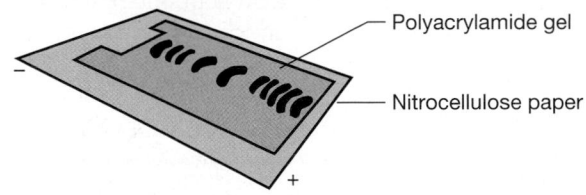

Polyacrylamide gel

Nitrocellulose paper

3. Blot the separated proteins from the gel to nitrocellulose paper

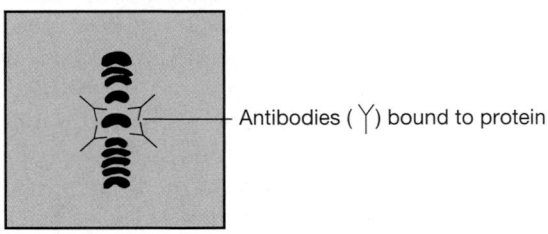

Antibodies (Y) bound to protein

4. Treat nitrocellulose paper containing blotted proteins with antibodies; each antibody recognizes and binds to a specific protein, labeling the protein for detection

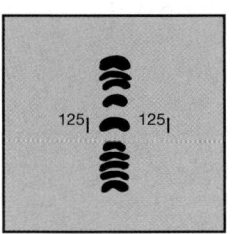 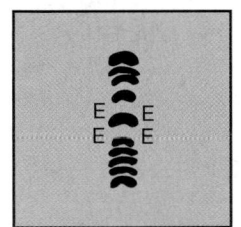

5. Add marker to bind to antigen–antibody complexes, either (left) radioactive *Staphylococcus* protein A–^{125}I, or (right) antibody containing conjugated enzyme

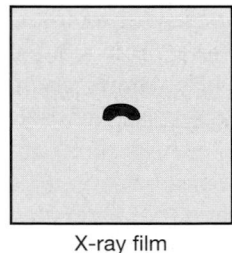 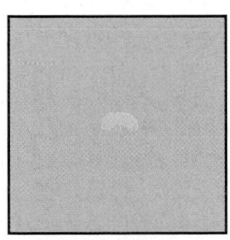

X-ray film

Nitrocellulose with enzyme-produced colored spot

(a)

Figure 24.25 The Western blot (immunoblot) and its use in the diagnosis of human immunodeficiency virus (HIV) infection. (a) Protocol for an immunoblot. (b) Developed HIV immunoblot. The proteins P24 and GP41–45 are coat proteins of the virus and are diagnostic for HIV. Lane 1, Positive control serum (from known AIDS patients); lane 2, negative control serum (from healthy volunteer); lane 3, strong positive from patient sample; lane 4, weak positive from patient sample; lane 5, reagent blank to check for background binding.

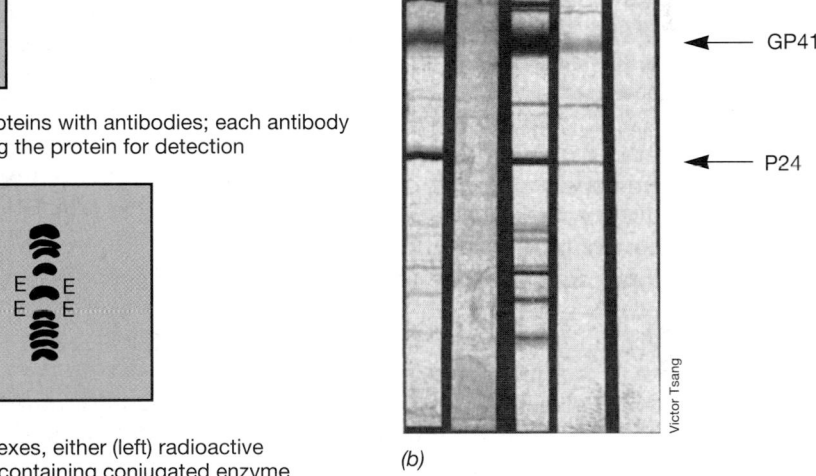

GP41-45

P24

Victor Tsang

(b)

solved by electrophoresis, and two of them, designated P24 and GP41-45, are used as specific diagnostic proteins in the AIDS immunoblot. Protein P24 is the HIV core protein, and proteins GP41-45 are HIV coat proteins (Section 16.16).

Following blotting of the proteins, the nitrocellulose strips are incubated with the test serum sample. If the sample is truly HIV-positive, antibodies against HIV proteins will be present and will bind to the HIV proteins separated on the membrane (Figure 24.25). To detect whether antibodies from the serum sample have bound to HIV antigens, a *detecting antibody*, anti-human IgG conjugated to the enzyme peroxidase, is added to the strips. If detecting antibody binds, the activity of the conjugated enzyme, after addition of substrate, will form a brown band on the strip at the site of antibody binding. The serum from the HIV-ELISA-positive patient is assayed in parallel with a positive control serum. The patient can be confirmed as HIV-positive if the position of the bands in the patient and the positive control sera are identical; negative control sera are also analyzed in parallel and must show no bands (Figure 24.25).

Although the intensity of the bands obtained in the HIV immunoblot varies somewhat from sample to sample (Figure 24.25*b*), the interpretation of an immunoblot is generally unequivocal, and thus the test is valuable for confirming HIV-ELISA positives and eliminating false positives. To make the HIV immunoblot clinically accessible, membrane strips containing inactivated HIV antigens (previously separated by electrophoresis) are available commercially. Separate strips can be incubated directly with the patient and control serum samples and subsequently treated with the detecting antibody.

The immunoblot technique is also used to confirm the specificity of antibody tests for Lyme disease (see Table 24.5). However, because of the expense, technical requirements, lower *sensitivity* (but higher *specificity*), and time involved, immunoblot tests are less useful than the rapid, low cost ELISA methods for general screening purposes.

✓ 24.12 Concept Check

Immunoblot procedures are used to detect antibodies to specific antigens or to detect the presence of the antigens themselves. The antigens are electrophoresed, transferred (blotted) to a membrane, and exposed to antibody. Immune complexes are visualized with enzyme-labeled or radioactive second antibodies. Immunoblots are extremely specific, but procedures are complex and time-consuming.

✓ What advantage does the immunoblot have over immunoassays such as ELISA and RIA?

✓ Why is the immunoblot not used for general screening for HIV exposure?

III MOLECULAR AND VISUAL METHODS IN DIAGNOSTICS

Extremely sensitive methods based on molecular biology principles or on electron microscopic techniques are available for detecting pathogens and pathogen genomes. These methods do not depend on pathogen isolation or growth, or on the detection of an immune response to the pathogen. While viruses are sometimes grown and identified in culture, growth-independent identification methods based on molecular or physical methods are widely used.

24.13 Nucleic Acid Methods

Molecular biology methods are widely used in diagnostic microbiology. These methods use *genotypic* rather than *phenotypic* characteristics to identify specific pathogens. The success of genetic or DNA-based diagnostic procedures is based on several facts: (1) Nucleic acids can be readily isolated from infected tissues; (2) nucleic acids can be readily visualized and measured; (3) the nucleic acid sequence of an individual pathogen genome is so unique that nucleic acid hybridization analysis can be used for unequivocal identification; and (4) nucleic acid sequences can be amplified to increase the amount of material available for analysis.

Nucleic Acid Probes

One of the most powerful analytical tools available to clinical microbiologists is *nucleic acid hybridization*. Instead of detecting a whole organism or its products (for example, antigens), hybridization detects the presence of *specific DNA sequences* associated with a specific organism. To identify a microorganism through DNA analysis, the clinical microbiologist must have available a *nucleic acid probe* to that microorganism, a *single strand* of DNA containing sequences unique to the organism. This **nucleic acid probe** may be up to several kilobases in length, but many synthetic oligonucleotides consist of 20 bases or less and are still highly specific. If a microorganism in a clinical specimen contains DNA sequences complementary to the probe, the two sequences can hybridize (following appropriate sample preparation to yield single-stranded DNA from the microorganism), forming a *double-stranded* molecule (Figure 24.26). To detect a reaction, the probe is labeled with a *reporter molecule*, a radioisotope, an enzyme, or a fluorescent compound that can be measured in small amounts following hybridization. Depending on the reporter used (radioisotopes are the most sensitive), as little as 0.25μg of DNA per sample can be detected.

Nucleic acid probes offer many advantages over clinical immunological assays. Nucleic acids are much more stable than proteins at high temperatures and at high pH, and are more resistant to organic solvents and other chemicals. This means that a clinical sample can be treated in a relatively harsh manner to destroy most interfering material, leaving behind the nucleic acid. Because of the relative chemical stability of the target nucleic acids, nucleic acid probe technology can even be used to positively identify organisms that are no longer alive. Additionally, some nucleic acid probes may be more specific than antibodies, since they can detect single base pair differences between DNA sequences. Finally, new probes can be synthesized whenever necessary in the laboratory, avoiding the complex biological systems necessary to produce antibodies.

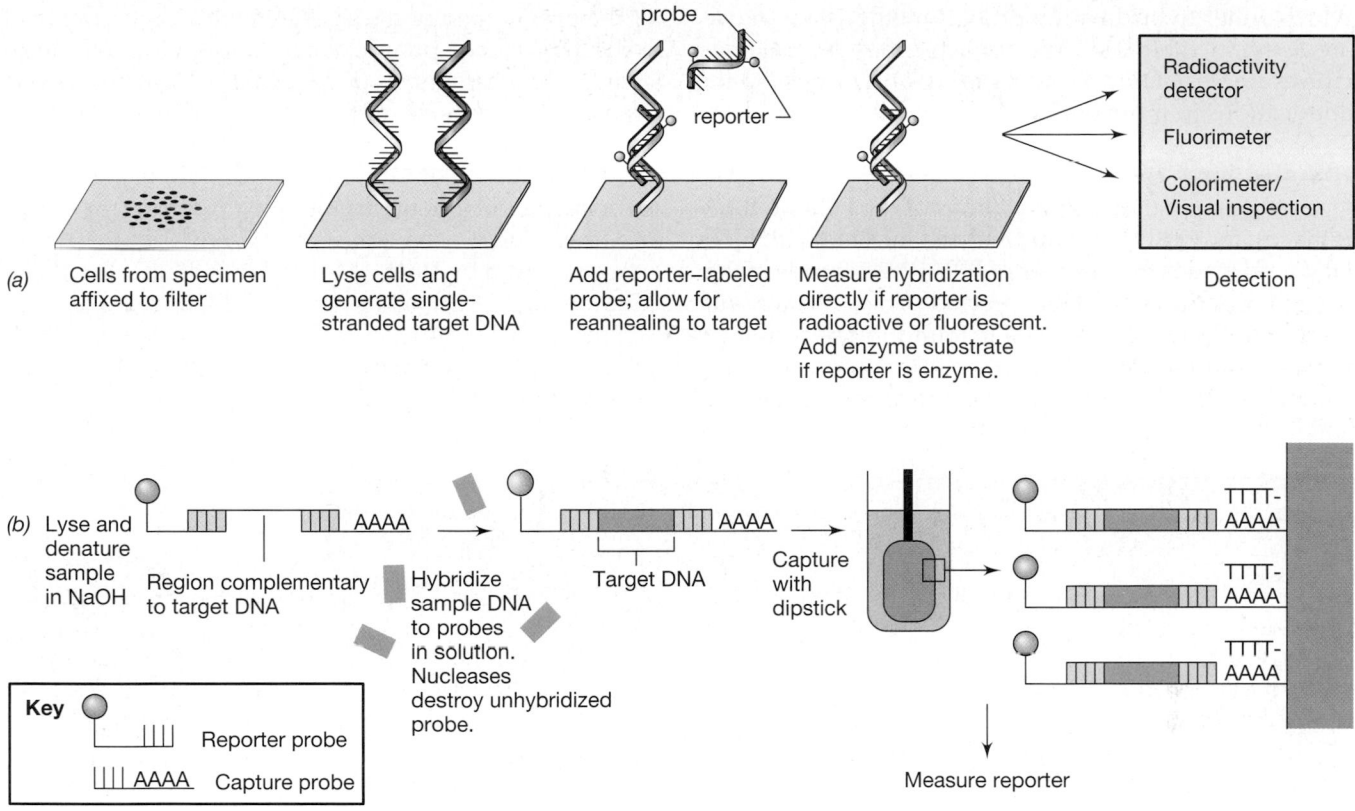

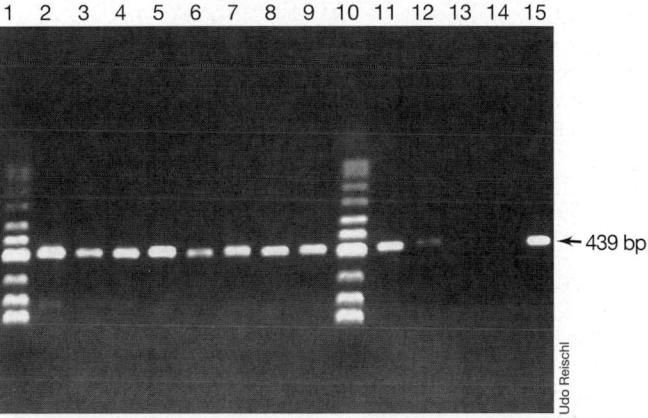

Figure 24.26 Nucleic acid probe methodology in clinical diagnostics. (a) Membrane filter assay. The detecting system (reporter) can be a radioisotope, a fluorescent dye, or an enzyme. (b) Dipstick assay. In the dipstick assay a dual reporter or capture probe is used. The capture probe contains a poly-dA tail that hybridizes to a poly-dT oligonucleotide affixed to the dipstick, binding the oligonucleotide–target–reporter complex. The complex can be detected as in (a) above.

Nucleic acid probes are also very sensitive. As mentioned above, it is possible to detect less than 1 μg of nucleic acid per sample. This translates into about 10^6 bacterial cells or virus particles. Although probes used in this fashion are not as sensitive as direct culture (where as few as 1–10 cells per sample can be detected), probe methods can be useful in situations where culture of the organism is difficult or even impossible.

PCR and Nucleic Acid Probes

Sequence-specific nucleic acid probes act as primers for the polymerase chain reaction (PCR) amplification of DNA or RNA from specific pathogens. PCR is a widely used clinical diagnostic tool (⚬ Section 10.17). DNA amplification of 1 million-fold or more allows theoretical detection of a single DNA molecule obtained from a single bacterial cell. For example, probes for a pathogen might be used to examine DNA derived from suspected infected tissue, even in the absence of an observable, culturable pathogen. These methods are particularly useful for identifying viral and intracellular infections. The presence of the appropriate amplified gene segment (Figure 24.27) confirms the presence of the pathogen. Several of the specific organisms for

Figure 24.27 Polymerase chain reaction (PCR) analysis of patient sputum for *Mycobacterium tuberculosis* in the diagnosis of tuberculosis. Sputum samples from patients were used as a source of DNA. Amplification was initiated with a primer pair, which produced the indicated 439-base pair product when a pure culture of *M. tuberculosis* was used as the DNA source (lane 15). Lanes 2–9, 11 and 12 are from sputums positive for *M. tuberculosis* (lane 12 is a weak positive). Lanes 13 and 14 are from *M. tuberculosis*-negative sputum samples. Lanes 1 and 10 are molecular weight reference markers. For a description of PCR technology, see Section 10.17.

which either hybridization or PCR methods are in use are listed in Table 24.9. We now consider several specific examples of nucleic acid probes and discuss some applications in more detail.

HIV and Viral Load

One of the most useful applications of the PCR method is the test for **viral load** in individuals infected with HIV. HIV viral load is the number of HIV RNA strands in the plasma or serum of an HIV-infected person. As we will discuss in Section 26.14, HIV is found in the serum of infected individuals at all times after infection. The amount of HIV present, or the viral load, is an indicator of the progression of the disease, with a high viral load indicating a poor prognosis and a low viral load indicating a good prognosis (Figure 24.28). Viral load is also an excellent measure of the efficacy of anti-HIV drug therapy (Figure 24.28; ∞ Sections 26.14 and 20.10).

Quantitation of viral load is possible using a commercially available quantitative PCR assay (for example, the widely used Amplicor HIV-1 Monitor Test®, manufactured by Roche Molecular Systems, Inc.). The method employed is known as RT-PCR (reverse-transcription PCR) and is used to make an amplifiable DNA template from RNA viral genomes (∞ Section 10.17). First, plasma samples are treated to lyse HIV

TABLE 24.9	Pathogens that can be identified with nucleic acid probes including PCR methods
Pathogen	**Diseases**
Bacteria	
Campylobacter spp.	Food infections
Chlamydia trachomatis	Venereal syndromes; trachoma
Enterococcus spp.	Nosocomial infections
Escherichia coli (enteropathogenic strains)	Gastrointestinal disease
Haemophilus influenzae	Infectious meningitis
Legionella pneumophila	Pneumonia
Listeria monocytogenes	Listeriosis
Mycobacterium avium	Tuberculosis
Mycobacterium tuberculosis	Tuberculosis
Mycoplasma hominis	Urinary tract infection; pelvic inflammatory disease
Mycoplasma pneumoniae	Pneumonia
Neisseria gonorrhoeae	Gonorrhea
Neisseria meningitidis	Meningitis
Rickettsia spp.	Typhus, hemmorhagic fever, etc.
Salmonella spp.	Gastrointestinal disease
Shigella spp.	Gastrointestinal disease
Staphylococcus aureus	Purulent discharges (boils, blisters, pus-forming skin infections)
Streptococcus pyogenes	Scarlet fever; rheumatic fever; strep throat
Streptococcus pneumoniae	Pneumonia
Treponema pallidum	Syphilis
Fungi	
Blastomyces dermatitidis	Blastomycosis
Candida spp.	Candidiasis, thrush
Coccidioides immitis	Coccidioidomycosis
Histoplasma capsulatum	Histoplasmosis
Viruses	
Cytomegalovirus	Congenital viral infections
Epstein-Barr virus	Burkitt's lymphoma; mononucleosis
Hepatitis viruses A, B, C, D, E	Hepatitis
Herpes virus (types I and II)	Cold sores; genital herpes
Human immunodeficiency virus (HIV)	Acquired immunodeficiency syndrome (AIDS)
Human papilloma virus	Genital warts; cervical cancer
Influenza	Respiratory disease
Polyoma virus	Neurological disease
Rotavirus	Gastrointestinal disease
Protozoa	
Leishmania donovani	Leishmaniasis
Plasmodium spp.	Malaria
Pneumocystis carinii	Pneumonia
Trichomonas vaginalis	Trichomoniasis
Trypanosoma spp.	Trypanosomiasis

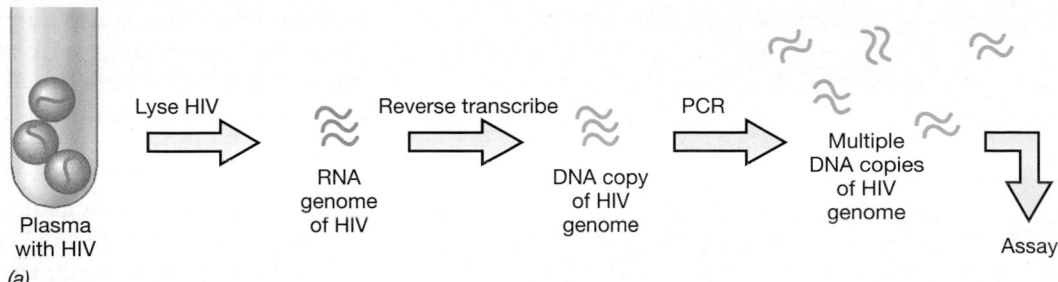

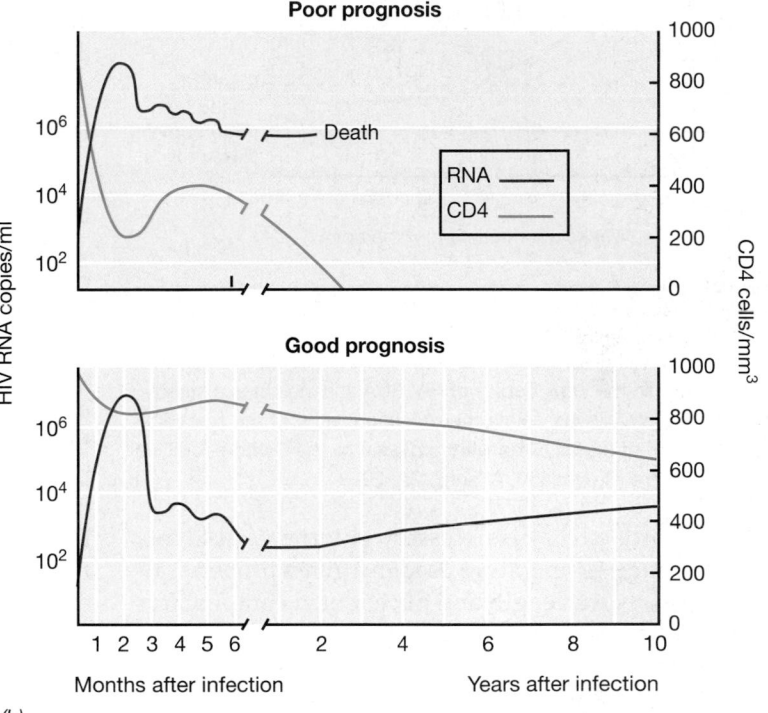

Figure 24.28 Monitoring of HIV load. (a) Detection of HIV via the RT-PCR test (reverse transcription-polymerase chain reaction). The HIV copies obtained are compared quantitatively with DNA copies from a control template that is amplified in the same PCR amplification. HIV load is expressed as the number of HIV copies per milliliter of patient plasma. For a discussion of reverse transcription and the PCR technique, see Sections 10.17 and 16.14, respectively. Time course for HIV infection as monitored by HIV RT-PCR. Progression of infection is estimated based on viral load at successive times after infection. CD4 T cell counts are measured in cells per cubic millimeter (see Figure 24.21). In the upper panel, a viral load of greater than 10^4 copies per milliliter correlates with below normal CD4 cell numbers (normal = 600–1500/mm^3), indicating a poor prognosis and early death of the patient. In the lower panel, a viral load of less than 10^4 copies per milliliter correlates with normal CD4 cell numbers, indicating a good prognosis and extended survival of the patient. This test is used to monitor the course of HIV infection and is particularly useful for tracking the efficacy of drug treatment protocols. Data are adapted from the Centers for Disease Control and Prevention, Atlanta, GA, USA.

and release the RNA viral genome. The RNA, after precipitation with isopropanol, is reverse transcribed with reverse transcriptase enzyme to make a DNA copy. The DNA copy then serves as the template in a PCR reaction with primers directed to a portion of the HIV *gag* gene (⟲ Section 16.14), yielding a 155-bp *amplicon* (amplified product) in HIV-positive samples. Using a standard DNA template of known quantity, the HIV signal present in the sample can be compared to the DNA standard, and the amount of HIV can be quantitated. By comparing viral load over time, a relatively accurate prognosis can be made for each patient (Figure 24.28).

The most important use of this technique is to monitor antiviral chemotherapy. HIV, after drug treatment, can be suppressed to virus levels below the limits of the test (less than 500 copies per milliliter of blood), as indicated in Figure 24.29. However, the few remaining HIV particles may acquire mutations that result in drug resistance and resurgence of the infec-

tion, resulting once again in a high viral load. The HIV monitoring system is useful for identifying the emergence of drug-resistant mutant HIV strains, necessitating a change in the chemotherapy regimen. Several other systems based on amplification techniques are also used to monitor HIV infections, and work is progressing to allow detection of HIV in the range of 0–500 copies per milliliter of blood, a necessary level for supporting any claims that HIV cure is possible using drug therapies.

Other Clinical Laboratory Probes
In most clinical probe assays, colonies from plates or pieces of infected tissue are treated with strong alkali, usually NaOH, to lyse the cells and partially denature the DNA, forming single-stranded molecules (Figure 24.26). This mixture is then affixed to a filter or left in solution (for dipstick assays, see later), and the labeled probe added. Hybridization is allowed to occur at a temperature at which sequence homology between target

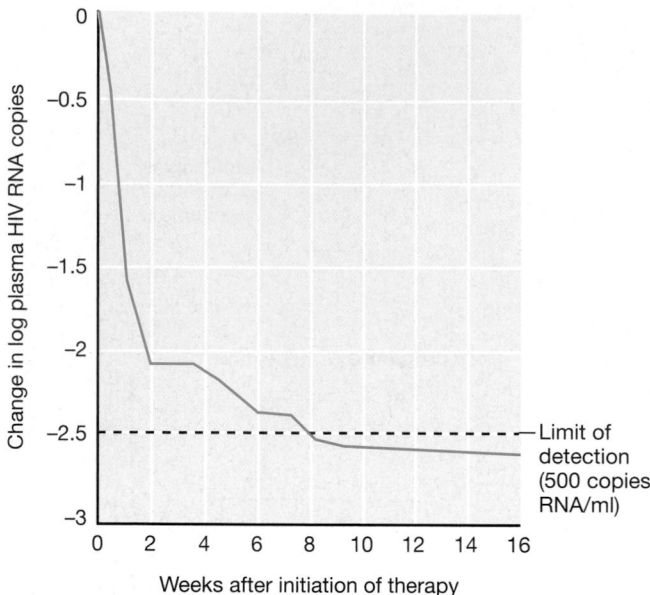

Figure 24.29 Rate of change in plasma HIV RNA copy numbers after initiation of antiretroviral drug therapy. The RT-PCR system was used to monitor viral load over a 16-week period. The data are shown as the change in the $\log_{10}$ copies of HIV RNA and indicates a steady, rapid decrease to levels below the detection limits of the test (500 copies/milliliter of blood). Data were provided by the Centers for Disease Control and Prevention, Atlanta, GA, USA.

DNA and probe DNA is necessary to form a stable duplex (the actual temperature used in a given probe assay is governed by the length and nucleic acid composition of the probe and target DNA). Following a wash to remove unhybridized probe DNA, the extent of hybridization is measured using the reporter molecule attached to the probe. Depending on how the probe was labeled, this involves measurement of radioactivity, enzyme activity, or fluorescence.

Nucleic acid probes have been marketed for the identification of several major microbial pathogens and are in widespread use for the detection of *Neisseria gonorrhoeae* and *Chlamydia trachomatis* (see Table 24.9 and ⌘ Sections 26.12 and 26.13). However, in addition to their clinical usefulness, probes are finding widespread application in food industries and in food regulatory agencies. Probe detection systems can be used to monitor foods for their content of important pathogens such as *Salmonella* and *Staphylococcus*. In probe assays of food, an enrichment period is usually employed to allow low numbers of cells in the food to multiply to a sufficient number to be detectable by the probe. However, the use of PCR gene amplification techniques eliminates the need for the enrichment period.

Probes designed for use in the food industry employ dipsticks coated with pathogen-specific DNA to hybridize with pathogen DNA from solution. Two component probes are used here, one serving as a *reporter* probe and the other as a *capture probe* (Figure 24.26). Following hybridization of the reporter or capture probe to target DNA, the dipstick, which contains a sequence complementary to the capture probe (usually poly dT to capture poly dA on the probe) (see Figure 24.26b) is inserted into the hybridization solution, and it traps hybridized DNA for removal and measurement.

Probes for detecting certain cancer viruses are also being developed. For example, a probe is now available to detect DNA sequences unique to human papilloma viruses. These viruses sometimes cause skin and cervical cancer in humans (⌘ Section 9.11), and a specific group of papilloma viruses causes genital warts. In women, an increased incidence of genital papilloma virus infection is associated with an increased risk of cervical cancer. The DNA probe developed for papilloma viruses can be used to search by hybridization for papilloma virus sequences in tissues removed during a cervical exam. Early detection and treatment of papilloma virus infections decreases the risk of cervical cancer.

Recent advances in our understanding of bacterial phylogenetics based on 16S ribosomal RNA (rRNA) sequences (⌘ Section 11.5) have allowed the construction of species-specific and strain-specific nucleic acid probes. Typing based on differences in rRNA sequences is called *ribotyping*. Ribotyping visualizes the unique DNA restriction patterns of the rRNA genes when DNA from a particular organism is digested by restriction endonucleases (⌘ Section 10.12). The digested DNA is separated on an agarose gel and a labeled rRNA probe is used to visualize the unique restriction patterns of the genes encoding rRNA (⌘ Section 11.10 and Figure 11.20). Within a *species*, and especially within a *strain*, the restriction pattern is highly conserved and is a molecular fingerprint for the organism. Because all organisms have ribosomal genes, ribotyping can be applied to a wide variety of clinical and phylogenetic studies.

✓ 24.13 Concept Check

Nucleic acid hybridization is a powerful laboratory tool used for identification of microorganisms. A nucleic acid sequence specific for the microorganism of interest must be available in order to design a probe. Perhaps the most widespread use of probe-based technology is in the application of the gene amplification (PCR) methods. Various probe-based methodologies are currently used in clinical, food, and environmental laboratories.

✓ What advantage does nucleic acid hybridization have over standard culture methods for identification of microorganisms? What disadvantages?

✓ Cite an example where information about a microorganism can be obtained with a nucleic acid probe in the absence of standard growth-dependent assays.

24.14 Diagnostic Virology

Identification and diagnosis of pathogenic animal viruses presents significantly different problems as compared with bacterial pathogens. Viruses are unique because they are all obligate intracellular parasites (∞ Chapter 8). Since no viruses live outside the cell, they cannot be directly cultured on artificial media, but must be grown in mammalian cells. However, viruses or their cytopathic effects can be observed directly and a variety of other *in vitro* test procedures are available for identification of viral pathogens by direct or indirect means.

Virus Growth *In Vitro*

Laboratory cultivation of animal viruses from clinical materials is more difficult, time-consuming, and specialized than the cultivation of most bacterial pathogens. This is because viruses grow only in living cells. We discussed the use of cell cultures for the growth of viruses in Section 9.3, and such cultures are commonly used in diagnostic virology. Several immortal human cell lines are available for propagation of viruses. These cell lines grow rapidly and indefinitely in cell culture medium and are easily infected by mammalian viruses.

In addition to these immortal cell lines, viruses can also be grown in Rhesus monkey kidney cell lines. Monkey kidney cell lines are called *primary* cell lines because they cannot be maintained indefinitely in the laboratory: They die after a limited number of cell divisions. Primary monkey kidney cells support growth of a number of pathogenic viruses and are valuable for the initial isolation of unknown viruses.

Electron Microscopy

Diagnostic virology can also be done by electron microscopy. Because many viruses have distinctive morphologies (∞ Chapter 9), their presence in clinical samples can often be detected by observing the sample with an electron microscope (Figure 24.30). In most specimens, the virus particles must be concentrated and separated from human tissues, and a variety of techniques, generally employing centrifugation and filtration methods, are used to obtain a sample enriched in virus particles. Although not as reliable as immunologic or nucleic acid probe methods, the observation of virus particles of a specific morphology in a particular type of human tissue is presumptive evidence for disease. Antibodies directed against particular viruses can be used to increase the sensitivity and specificity of this method: Agglutinating antibodies may cause the viral particles to aggregate, making them easier to distinguish from cellular debris under the electron microscope. Viruses can also be visualized by treatment of specimens with antiviral antibodies conjugated to

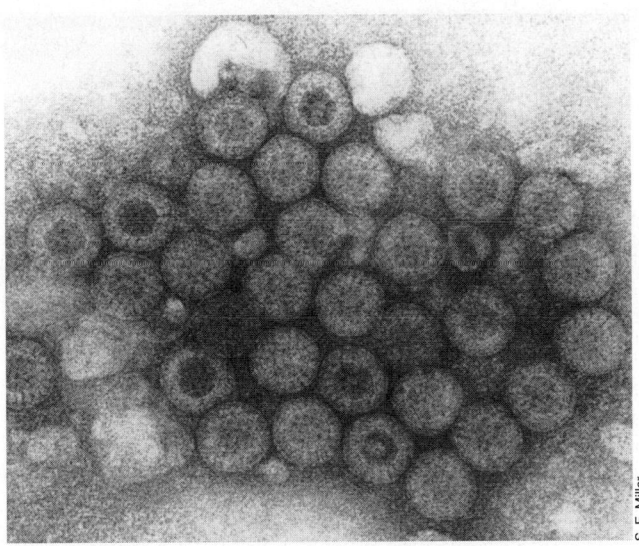

S. E. Miller

Figure 24.30 Electron microscopic observation of clinical specimens to detect viruses. Human rotavirus from a fecal sample. The distinct spherical nature of the virus when found in fecal matter is a highly diagnostic criterion. Each rotavirus particle is approximately 75 nm in diameter.

heavy metals (see Section 24.9). With the use of negative staining techniques (Figure 24.30), results from electron microscopic analysis can be available 20 min after collection of the specimen.

Other Virus Diagnostic Tests

A summary of some laboratory procedures used in diagnostic virology is given in Table 24.10. In many cases, immunodiagnostic methods (see Sections 24.5–24.12) and nucleic acid hybridization methods (see Section 24.13) are used for viral identification.

Many immunologic tests are either direct enzyme-linked immunosorbent assays (ELISAs) that detect viral particles (see Section 24.11 and Figure 24.22) or fluorescent antibody methods in which antibodies made against viral antigens are used to detect cells containing viruses (see Section 24.10 and Figure 24.18b). In indirect tests, the virus itself is not detected, but antibodies to the virus are regarded as presumptive evidence for viral infection. For example, routine testing for human immunodeficiency virus (HIV) infection involves the ELISA test and the immunoblot test (see Section 24.11 and Section 24.12). Both the ELISA and the immunoblot detect *antibodies* to HIV, not HIV itself. However, the *viral load* test discussed in Section 24.13 is used to detect the presence of HIV. Agglutination tests may also detect virus directly. Antiviral antibodies conjugated to latex beads or activated charcoal (see Section 24.8) can be used to test for agglutination of viral particles released from lysed tissue samples.

TABLE 24.10 Some laboratory procedures used in diagnostic virology[a]

Condition	Possible viral cause	Sample source	Inoculation procedure
Upper respiratory infection	Rhinovirus Coronavirus Adenovirus	Nasopharyngeal or tracheal fluid (aspirate)	Human fibroblast culture
Pneumonia	Influenza	Nasopharyngeal fluid or swab	Human fibroblast cultures or embryonated eggs
Measles	Measles virus	Nasopharyngeal fluid or swab	Monkey kidney cells
Vesicular rash	Herpes simplex	Vesicular fluid by aspiration	Human fibroblast culture
Diarrhea	Rotavirus (infants) Norwalk agent (adults)	Feces or rectal swab	Observe characteristic virus particles with the electron microscope (Figure 24.30)
Nonbacterial meningitis	Enterovirus Mumps Herpes simplex	Spinal fluid	Human fibroblast or monkey kidney cultures

[a] Immunological methods and nucleic acid probe methods are also widely used in the diagnosis of viral infections (see Sections 24.5–24.13).

Nucleic acid probes and PCR assays for detection and identification of clinically significant viruses are also a major diagnostic tool in clinical virology (see Section 24.13 and Table 24.9).

Many viral diagnostic procedures are used only under special circumstances. For routine virus infections, diagnoses are made by assessing clinical symptoms. Because of the technical expertise and expense involved in many viral diagnostic procedures (for example, laboratories that do primary culture must maintain or purchase Rhesus monkeys and have expertise in cell culture methods), most diagnostic virology laboratories are located at specialized government reference laboratories or major clinical research institutions.

✓ 24.14 Concept Check

Virus propagation *in vitro* can be accomplished only in tissue culture. Therefore, most diagnostic techniques for viral identification are not growth-dependent but routinely rely on immunoassays and nucleic acid hybridization techniques. Electron microscopy techniques are useful for direct observation of viruses in host samples.

✓ Why must viruses be grown in tissue culture and not on artificial, inert media?

✓ How can individual pathogenic viruses be identified in the laboratory?

Review Questions

1. Describe the standard procedure for obtaining and culturing a throat culture and a blood sample for bacteria. What special precautions must be taken while obtaining the blood culture?

2. Why are bacteria nearly always cultured from a urine specimen? Why is the *number* of bacterial cells in urine of significance? What organism is responsible for most urinary tract infections? Why?

3. Describe the procedures used for culturing anaerobic microorganisms. Why is it important to process all clinical specimens quickly? What special procedures and precautions are necessary for the isolation and culture of anaerobes?

4. Differentiate between *selective* and *differential* media. Is eosin-methylene blue (EMB) agar a selective medium or a differential medium? How and why is it used in a clinical laboratory?

5. Describe the Kirby–Bauer test for antibiotic sensitivity. Why should potential pathogens from patient isolates be tested by this method?

6. How are most laboratory-associated infections contracted? What action can be taken to prevent laboratory infections?

7. Why does the antibody titer rise after infection? Why is it necessary to obtain an acute and a convalescent blood sample to monitor infections? Is a high antibody titer indicative of an ongoing infection? Explain.

8. What advantages do *monoclonal* antibodies have over *polyclonal* antibody preparations, especially with regard to standardization of antibody preparations?

9. Describe a neutralization reaction with reference to microbial toxins and antisera.

10. Agglutination tests are more widely used for clinical diagnostic purposes than are precipitation tests. Why is this the case?

11. Explain how heavy metals are used to visualize viruses or subviral structures.

12. How are fluorescent antibodies used for the diagnosis of viral diseases? What advantages do fluorescent antibodies have over unlabeled antibodies?

13. Radioimmunoassay (RIA) and enzyme-linked immunosorbent assay (ELISA) tests are extremely sensitive, as compared with agglutination. Why is this the case?

14. Why is the immunoblot (Western blot) procedure used to confirm positive human immunodeficiency virus (HIV)-ELISA results?

15. What information is essential for the design of a pathogen-specific nucleotide probe? Where can one obtain such information? Is this information available for all pathogens?

16. What is a primary cell line? Why do animal viruses grow only in tissue culture?

Application Questions

1. From a blood culture, you obtain a culture positive for *Staphylococcus epidermidis*. Interpret and explain the results. Is it likely that the patient has *S. epidermidis* bacteremia? Why or why not?

2. Describe the microscopic and cultural evidence that would support a diagnosis of gonorrhea. Why is Thayer-Martin agar a more useful medium than "chocolate agar" for the isolation of *Neisseria gonorrhoeae*?

3. Compare changes in the color of pH-sensitive dyes in tests for carbohydrate fermentation and citrate utilization. Can the same dye be used in both tests? Why or why not?

4. With respect to both short-term and the long-term consequences, why is it a common medical practice to treat an infectious disease with antibiotics before isolating the suspected pathogen? After a pathogen has been isolated and identified, what further steps should be taken to confirm appropriate antibiotic sensitivity? Why are these measures rarely employed away from a hospital environment?

5. As a professional in a clinical laboratory, you are assigned the task of formalizing laboratory safety requirements to prevent infection of laboratory personnel from clinical specimens. Explain how you would monitor and enforce the biosafety recommendations outlined in Section 24.4.

6. Based on your knowledge of the primary and secondary antibody response (∞ Section 22.9), explain the rationale for collecting serum specimens during an acute infectious disease and about two weeks later during recovery. What information would you expect to obtain from the serum of the recovering patient? What classes of antibodies would you expect to see at each time, assuming that the patient had never before been exposed to the same disease?

7. How can administration of antigen influence the sensitivity or specificity of a polyclonal antibody preparation?

8. In the precipitation assay, soluble antigen–antibody complexes due to excess antibody, or conversely, due to excess antigen, occur because reagents are not present at optimum proportions. Will the phenomena of antigen excess and antibody excess be problematic for the agglutination assay? Explain.

9. What are the advantages of rapid identification systems such as agglutination tests as compared with growth-dependent clinical diagnostic procedures? What are the potential disadvantages of rapid, non–culture-based tests.

10. Design a fluorescent antibody assay for confirming an initial diagnosis of strep throat (*Streptococcus pyogenes* is the causative agent of strep throat). Discuss all aspects of the assay, including preparation of antisera, necessary controls, and clinical interpretation.

11. Design an ELISA test for detecting hepatitis A virus in fecal samples (∞ Section 26.11). Likewise, design an *indirect* test for detection of *exposure* to hepatitis A virus. List all reagents for each test and explain the fundamental differences between the direct and indirect tests. Can a single individual be positive for both tests? Why or why not?

12. What are the major advantages of using DNA probes in diagnostic microbiology? Where can you find information to design sequence-specific polymerase chain reaction (PCR) assay probes for the hepatitis A virus in Question 11?

13. Discuss the importance of *viral load* in assessing the treatment and progression of HIV infection.

14. Define the tests you would use to isolate a new viral pathogen. Be sure to include both growth-dependent assays and molecular assays. Where would you report your findings?

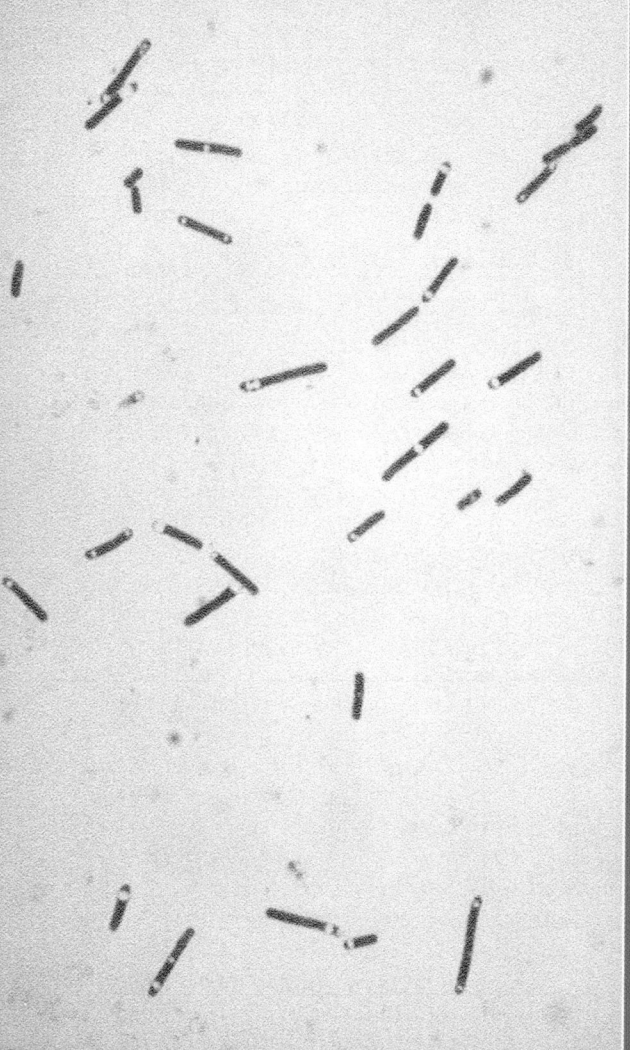

Epidemiology involves the tracking of current and emerging diseases. Epidemiologists must be continually alert to the occurrence of uncommon diseases, such as cutaneous or pulmonary anthrax caused by the gram-positive bacterium *Bacillus anthracis* shown here. Through the work of the epidemiologists, public health officials can assess disease threats, and formulate plans to deal with serious outbreaks. They can also define the sources of outbreaks and determine whether they occur naturally or, for example, through the activities of bioterrorists.

EPIDEMIOLOGY 25

Working Glossary

Acute short-term infection usually characterized by dramatic onset and rapid recovery

Biological warfare the use of biological agents to incapacitate or kill humans

Carrier subclinically infected individuals who may spread a disease

Chronic long-term infection

Common-source epidemic an epidemic resulting from infection of a large number of people from a single contaminated source

Emerging infection infectious disease whose incidence has increased in the past 20 years or whose incidence threatens to increase in the near future

Endemic disease constantly present, usually in low numbers

Epidemic the occurrence of a disease in unusually high numbers in a localized population

Epidemiology the study of the occurrence, distribution, and control of diseases

Fomites inanimate objects that, when contaminated with a viable pathogen, can transfer the pathogen to a host

Herd immunity resistance of a population to a pathogen as a result of the immunity of a large portion of the population

Host-to-host epidemic an epidemic resulting from person-to-person contact, characterized by a gradual rise and fall in numbers of cases

Incidence the number of cases of disease in a population

Morbidity incidence of illness in a population

Mortality incidence of death in a population

Nosocomial infection hospital-acquired infection

Outbreak the occurrence of a large number of cases of a disease in a short period of time

Pandemic a worldwide epidemic

Prevalence the proportion or percentage of individuals in the population having a disease

Public health the health of the population as a whole

Quarantine the practice of restricting the movement of individuals with highly contagious serious infections to prevent spread of the disease

Reemerging infections infectious diseases, thought to be under control, that produce new epidemics

Reservoir sites in which viable infectious agents remain and from which infection of individuals may occur

Surveillance observation, recognition, and reporting of diseases as they occur

Vector a living agent that transfers a pathogen (note alternative usage in Chapter 31)

Vehicle nonliving source of pathogens that infect large numbers of individuals; common vehicles are food and water

Zoonosis a disease that occurs primarily in animals but can be transmitted to humans

---

I PRINCIPLES OF EPIDEMIOLOGY

In this chapter, we consider how a pathogen spreads from an infected individual to others in a population. Thus, we are dealing here with *public health*. In the next four chapters we consider the diseases themselves. The principles put forward in this chapter are vital for understanding the spread and control of infectious disease.

One measure of our success in the control of infectious disease was shown by the data presented in Figure 1.18, which compares the most prevalent current causes of death in the United States with those at the beginning of the twentieth century. In most developed countries, microbial diseases are no longer perceived to be a significant threat to public health. However, a number of infectious diseases remain serious public health problems, even in developed countries, and other *new* infectious diseases are continuously emerging.

Worldwide, infectious diseases account for nearly 30 percent of the 56 million annual deaths. As we shall see, many of these diseases can be prevented. From a global perspective, infectious diseases continue to pose significant public health problems; even in the United States, deaths due to infectious diseases are increasing (Figure 25.1). Effective control of infectious diseases requires scientific, medical, economic, political, and educational solutions.

25.1 The Science of Epidemiology

Infectious diseases affect individuals, but infections are not acquired unless the individual is part of a *population*. Therefore, infectious diseases are studied in relation to

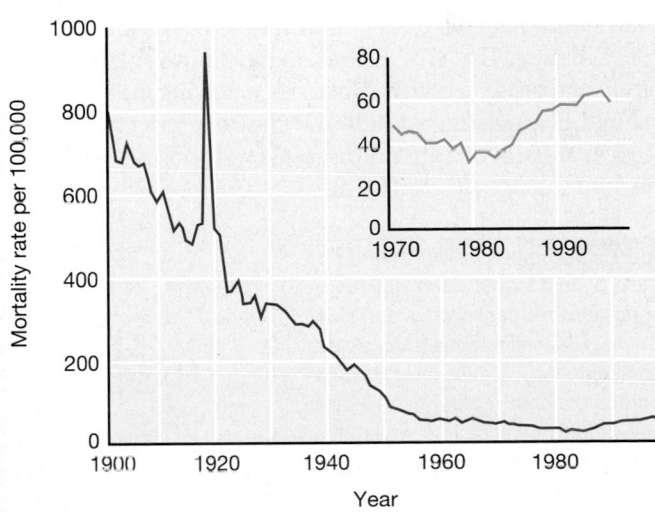

Figure 25.1 Deaths due to infectious disease in the United States. Although infectious disease death rates steadily declined throughout most of the twentieth century (except for the large numbers of deaths in 1918–1919 due to the influenza pandemic), the death rate has increased significantly since 1980. From Hughes, J.M. 2001. Emerging Infectious Diseases: A CDC Perspective. *Emerg. Infect. Dis.* 17: 494–496.

their effects on populations. **Epidemiology** is the study of the occurrence, distribution, and control of disease in populations.

A pathogen must be able to grow and reproduce. For this reason, an important aspect of the epidemiology of any disease is to track the natural history of the pathogen. In many cases the pathogen cannot grow outside the host, and if the host dies, the pathogen will also die. Pathogens that kill the host before they move to a new host should become extinct. How, then, do pathogens continue to exist if they kill the host? Actually, a well-adapted pathogen lives in synchrony with its host, taking what it needs for existence and causing only a minimum of harm. On the other hand, serious host damage often occurs when new natural pathogens arise for which the host has not developed resistance, or when the natural resistance of the host decreases because of factors such as poor diet, age, and other stressors (⌒⌒ Section 21.13). Thus, pathogens are selective forces in the evolution of the host, just as hosts act as selective forces in the evolution of pathogens. When equilibrium between host and pathogen exists, both host and pathogen survive. However, in cases where the pathogen is not dependent on the host for survival, the pathogen can cause devastating disease. Organisms in the genus *Clostridium*, for example, ubiquitous inhabitants of the soil (⌒⌒ Section 12.20), occasionally accidentally infect humans, causing such devastating diseases as tetanus and botulism (⌒⌒ Sections 21.10, 27.8, and 29.5).

The epidemiologist traces the spread of a disease to identify its origin and mode of transmission. Epidemiologic data are obtained from clinical studies, disease-reporting surveys, insurance questionnaires, and interviews with patients to define common factors that constitute a disease. This is in contrast to the clinical or laboratory study of disease, where the focus is on treating the *individual* patient. Knowledge of both the clinical aspects and ecological aspects of a given disease are important if public health measures to control diseases are to be effective.

✓   **25.1 Concept Check**

Epidemiology is the study of disease in populations. To understand infectious disease, the epidemiologist studies the interactions of the pathogen in the host population.

✓ How does an epidemiologist differ from a microbiologist?
✓ Why do epidemiologists acquire data for infectious diseases within a population?

25.2 The Vocabulary of Epidemiology

A number of common terms have specific meanings to the epidemiologist. The **prevalence** of a disease in a population is defined as the proportion of diseased individuals in a population in a given time period. The **incidence** of a disease is the *number* of cases of an individual disease in a population in a given time period. A disease is said to be **epidemic** when it occurs in an unusually high number of individuals in a population at the same time; a **pandemic** is a widely distributed epidemic (Figure 25.2). By contrast, an **endemic** disease is one that is constantly present, usually at low incidence, in a population. In an endemic disease, the pathogen may not be highly virulent, or the majority of the individuals may be immune, resulting in low disease incidence. However, as long as an endemic situation lasts, a few infected individuals remain who serve as *reservoirs* of infection.

Sporadic cases of a disease occur when individual cases are recorded in geographically separated areas, implying that the incidents are not related. A disease **outbreak**, on the other hand, occurs when a number of cases are observed, usually in a relatively short period of time, in an area previously experiencing only sporadic cases of the disease. Finally, diseased individuals who show no symptoms or only mild symptoms have *subclinical infections*. Subclinically infected individuals are frequently identified as **carriers** of a particular disease, because even though they themselves show few or per-

Figure 25.2 Classification of disease by incidence. Each dot represents several cases of a particular disease. Endemic diseases are always present in a population in a given geographical area. Epidemic diseases show high incidence in a wider area, usually developing from an endemic focus. Pandemic diseases are widely distributed and show no geographic focus.

(a) Endemic disease   *(b)* Epidemic disease   *(c)* Pandemic disease

haps no symptoms, they may still be actively carrying and shedding the pathogenic agent (see Section 25.3).

Mortality and Morbidity

The incidence and prevalence of disease is determined from statistical analysis of illnesses and deaths. From these data, a picture of the public health in a population can be obtained. The population might be the total global population of humans or the population of a localized region, such as a city, state, or country. Public health concerns vary with location and time; thus, assessment of public health at a given moment provides only a snapshot of a dynamic situation. By continuing to examine health statistics over many years, it is possible to assess the value of various public health policies that may influence the incidence of disease.

Mortality is the incidence of *death* in the population. Infectious diseases were the major causes of death in 1900 in developed countries (⌘ Figure 1.7), but they are now much less significant; noninfectious diseases such as heart disease and cancer are of greater importance. However, the current situation could rapidly change if a breakdown in public health measures were to occur. Worldwide, infectious diseases are still major killers (Table 25.1 and Section 25.9).

Morbidity refers to the incidence of *disease* in populations and includes both fatal and nonfatal diseases. Morbidity statistics define the health of the population more precisely than mortality statistics because many diseases that affect health have relatively low mortality.

The major causes of illness are quite different from the major causes of death. Major illnesses include acute respiratory diseases (the common cold, for instance) and acute digestive system conditions, which are often due to infectious agents and seldom cause death in developed countries.

Disease Progression

In terms of clinical symptoms, the course of a typical acute infectious disease can be divided into stages:

1. *Infection:* The organism begins to grow in the host.

2. *Incubation period:* The time between infection and the appearance of disease symptoms. Some diseases, like influenza (⌘ Section 26.8), have short incubation periods, measured in days; others, like AIDS, have longer ones, measured in years (⌘ Section 26.14). The incubation period for a given disease is determined by inoculum size, virulence and life cycle of the pathogen, resistance of the host, and distance of the site of entrance from the focus of infection (⌘ Section 21.8). At the end of incubation, the first symptoms, such as headache and a feeling of illness, appear.

3. *Acute period:* The disease is at its height, with overt symptoms such as fever and chills.

4. *Decline period:* Disease symptoms are subsiding, the temperature falls, usually following a period of intense sweating, and a feeling of well-being develops.

TABLE 25.1 Worldwide deaths due to infectious diseases, 1999

Disease	Deaths	Causative agents
Acute respiratory infections[*,a]	4,000,000	Bacteria, viruses, fungi
Acquired immunodeficiency syndrome (AIDS)	2,700,000	Virus
Diarrheal diseases	2,200,000	Bacteria, viruses
Tuberculosis[*]	1,700,000	Bacteria
Malaria	1,100,000	Protozoa
Measles[*]	875,000	Virus
Tetanus[*]	377,000	Bacterium
Pertussis (whooping cough)[*]	295,000	Bacterium
Meningitis, bacterial[*]	171,000	Bacterium
Syphilis	153,000	Bacterium
Hepatitis (all types)[*,b]	124,000	Viruses
Trypanosomiasis (sleeping sickness)	66,000	Protozoan
Leishmaniasis	57,000	Protozoan
Chlamydia	16,000	Bacterium
Intestinal nematode infections	16,000	Parasitic worms
Schistosomiasis	14,000	Parasitic worm
Dengue	13,000	Virus
Other communicable diseases	1,700,000	

Worldwide, there were about 56 million deaths from all causes. About 15.6 million deaths were from infectious diseases, nearly all in developing countries. Data are from the World Health Organization (WHO), Geneva, Switzerland.

[*] Diseases for which effective vaccines are available.

[a] For some acute respiratory agents such as influenza and *Streptococcus pneumoniae* there are effective vaccines; for others, such as colds, there are no vaccines.

[b] Vaccines are available for hepatitis A virus and hepatitis B virus. There are no vaccines for other hepatitis agents.

The decline may be rapid (within 1 day), in which case it is said to occur by *crisis*, or it may be slower, extending over several days, in which case it is said to be by *lysis*.

5. *Convalescent period:* The patient regains strength and returns to normal.

During the later stages of the infection cycle, the immune mechanisms of the host become increasingly important, and in most cases complete recovery from the disease requires and results in active immunity.

✓ 25.2 Concept Check

An endemic disease is constantly present at low incidence in a specific population. In epidemics, an unusually high incidence of disease occurs in a specific population. Infectious diseases cause morbidity (illness) and may cause mortality (death). An infectious disease follows a predictable clinical pattern in the host.

- ✓ Distinguish between an endemic disease, an epidemic disease, and a pandemic disease.
- ✓ Distinguish between morbidity and mortality. How might high host *morbidity* be advantageous for the pathogen?
- ✓ Is host *mortality* advantageous for the pathogen?

25.3 Disease Reservoirs and Epidemics

Reservoirs are sites in which viable infectious agents remain alive and from which infection of individuals may occur. Reservoirs may be either animate or inanimate. Table 25.2 lists some common human infectious diseases and their reservoirs. Some pathogens are primarily *saprophytic* (living on dead matter) and only incidentally infect humans and cause disease. For example, *Clostridium tetani* (the causal agent of tetanus) normally inhabits the soil. Infection of animals by this organism is an accidental event; infection of a host is not essential for its continued existence and in the absence of susceptible hosts, *C. tetani* would still survive in nature.

For many pathogens, other living organisms are their only reservoirs. In these cases, the reservoir is an essential component of the natural life cycle of the infectious agent. Some infections occur only in humans, and maintenance of the cycle involves person-to-person transmission (⊂⊃ Chapter 26). This type of pathogen cycle is common for viral and bacterial respiratory diseases, sexually transmitted diseases, staphylococcal and streptococcal infections, diphtheria, typhoid fever, and mumps. As we shall see, pathogens that live their entire life cycle dependent on a single host can be eradicated (see Section 25.8).

Zoonosis

A number of infectious diseases that occur in humans also occur in animals. A disease that occurs primarily in animals but is occasionally transmitted to humans is called a **zoonosis**. Because public health measures for animal populations are much less developed than for humans, the infection rate for many diseases is much higher in animals, and animal-to-animal transmission is the rule. Occasionally, transmission is from animal to human; in such cases, it is less likely for transmission to also occur from person to person. Thus, maintenance of the pathogen in nature depends on animal-to-animal transfer; humans are accidental hosts, such as in plague (⊂⊃ Section 27.6) However, control of a zoonosis in the human population in no way eliminates it as a public health problem. Control of the human disease can generally be achieved only through elimination of the disease in the animal reservoir. Considerable success has been achieved in the control of bovine tuberculosis, a disease that was often spread from infected cattle to humans. Control was achieved primarily by identifying and destroying infected animals. Pasteurization of milk was also of considerable importance because milk was the main vehicle of transmission for bovine tuberculosis.

Certain infectious diseases, particularly those caused by organisms such as protozoa, have more complex cycles, involving an obligate transfer from animal to human to animal (for example, malaria, ⊂⊃ Section 27.5). In such cases, the disease may potentially be controlled in either humans or the alternate animal host.

Carriers

A carrier is an infected individual with no obvious signs of clinical disease. Carriers are potential sources of infection for others and are critically important for the spread of disease. Carriers may be individuals in the incubation period of the disease, in which case the carrier state precedes the development of actual symptoms. These individuals are prime sources of respiratory infections because they are not yet aware of their infection and so are not taking any precautions against infecting others. Such persons are **acute carriers** because the carrier state lasts for only a short time. On the other hand, **chronic carriers** may remain infected and carry disease for extended periods of time. Chronic carriers usually appear perfectly healthy. They may be individuals who have recovered from a clinical disease, but still harbor viable pathogens, or they may be individuals with inapparent infections.

Carriers can be identified in populations by using a variety of diagnostic techniques, such as culture surveys (⊂⊃ Section 24.1) or serological (antibody) surveys (⊂⊃ Section 24.5). For example, tests using *Mycobacterium tuberculosis* antigens to test for delayed

TABLE 25.2 Epidemic diseases: Agents, sources, reservoirs, and control

Disease	Causative agent[a]	Infection sources	Reservoirs	Control measures
Common-source epidemics[b]				
Anthrax	*Bacillus anthracis* (B)	Milk or meat from infected animals	Cattle, swine, goats, sheep, horses	Destruction of infected animals
Bacillary dysentery	*Shigella dysenteriae* (B)	Fecal contamination of food and water	Humans	Detection and control of carriers; oversight of food handlers; decontamination of water supplies
Botulism	*Clostridium botulinum* (B)	Soil-contaminated food	Soil	Proper preservation of food
Brucellosis	*Brucella melitensis* (B)	Milk or meat from infected animals	Cattle, swine, goats, sheep, horses	Pasteurization of milk; control of infection in animals
Cholera	*Vibrio cholerae* (B)	Fecal contamination of food and water	Humans	Decontamination of public water sources; immunization
E. coli O157:H7 food infection	*Escherichia coli* O157:H7 (B)	Fecal contamination of food and water	Humans, cattle	Decontamination of public water sources; oversight of food handlers; pasteurization of beverages
Giardiasis	*Giardia* spp. (P)	Fecal contamination of water	Wild mammals	Decontamination of public water sources
Hepatitis	Hepatitis A, B, C, D, E (V)	Infected humans	Humans	Decontamination of contaminated fluids and fomites, immunization if available (A and B)
Legionnaire's disease	*Legionella pneumophila* (B)	Contaminated water	High-moisture environments	Decontamination of air conditioning cooling towers, etc.
Paratyphoid	*Salmonella paratyphi* (B)	Fecal contamination of food and water	Humans	Decontamination of public water sources; oversight of food handlers; immunization
Typhoid fever	*Salmonella typhi* (B)	Fecal contamination of food and water	Humans	Decontamination of public water sources; oversight of food handlers; pasteurization of milk; immunization
Host-to-host epidemics				
Respiratory diseases				
Diphtheria	*Corynebacterium diphtheriae* (B)	Human cases and carriers; infected food and fomites	Humans	Immunization; quarantine of infected individuals
Hantavirus pulmonary syndrome	Hantavirus (V)	Inhalation of contaminated fecal material; contact	Rodents	Control of rodent population and exposure
Hemorrhagic fever	Ebola virus (V)	Infected body fluids	Unknown	Quarantine of active cases
Meningicoccal meningitis	*Neisseria meningitidis* (B)	Human cases and carriers	Humans	Exposure treated with sulfadiazine for susceptible strains
Pneumococcal pneumonia	*Streptococcus pneumoniae* (B)	Human carriers	Humans	Antibiotic treatment; isolation of cases for period of communicability
Tuberculosis	*Mycobacterium tuberculosis* (B)	Sputum from human cases; contaminated milk	Humans, cattle	Treatment with isoniazid; pasteurization of milk
Whooping cough	*Bordetella pertussis* (B)	Human cases	Humans	Immunization; case isolation

(continues)

TABLE 25.2 Epidemic diseases: Agents, sources, reservoirs, and control (continued)

Disease	Causative agent[a]	Infection sources	Reservoirs	Control measures
German measles	Rubella virus (V)	Human cases	Humans	Immunization; avoid contact between infected individuals and pregnant women
Influenza	Influenza virus (V)	Human cases	Humans, animals	Immunization
Measles	Measles virus (V)	Human cases	Humans	Immunization
Sexually transmitted diseases[c]				
Acquired immunodeficiency syndrome (AIDS)	Human immunodeficiency virus (HIV)	Infected body fluids, especially blood and semen	Humans	Treatment with reverse transcriptase inhibitors, protease inhibitors (not curative) (∞ Section 26.14)
Chlamydia	*Chlamydia trachomatis* (B)	Urethral, vaginal, and anal secretions	Humans	Testing for organism during routine pelvic examinations; chemotherapy of carriers and potential contacts; case tracing and treatment
Gonorrhea	*Neisseria gonorrhoeae* (B)	Urethral and vaginal secretions	Humans	Chemotherapy of carriers and potential contacts; case tracing and treatment
Syphilis	*Treponema pallidum* (B)	Infected exudate or blood	Humans	Identification by serological tests; antibiotic treatment of seropositive individuals
Trichomoniasis	*Trichomonas vaginalis* (P)	Urethral, vaginal, prostate secretions	Humans	Chemotherapy of infected individuals and contacts
Vectorborne diseases				
Epidemic typhus	*Rickettsia prowazekii* (B)	Bite by infected louse	Humans, lice	Control louse population
Lyme disease	*Borrelia burgdorferi* (B)	Bite from infected tick	Rodents, deer, ticks	Avoid tick exposure; treat infected individuals with antibiotics
Malaria	*Plasmodium* spp. (P)	Bite from *Anopheles* mosquito	Humans, mosquito	Control mosquito population; treat infected humans with antimalarial drugs
Plague	*Yersinia pestis* (B)	Bite by flea	Wild rodents	Control rodent populations, immunization
Rocky Mountain spotted fever	*Rickettsia rickettsii* (B)	Bite by infected tick	Ticks, rabbits, mice	Avoid tick exposure; treat infected individuals with antibiotics
Direct-contact diseases				
Psittacosis	*Chlamydia psittaci* (B)	Contact with birds or bird excrement	Wild and domestic birds	Avoid contact with birds; treat infected individuals with antibiotics
Rabies	Rabies virus (V)	Bite by carnivores	Wild and domestic carnivores	Avoid animal bites; immunization of animal handlers and exposed individuals
Tularemia	*Francisella tularensis* (B)	Contact with rabbits	Rabbits	Avoid contact with rabbits; treat infected individuals with antibiotics

[a] B, Bacteria; V, virus; P, protozoan.

[b] Some common-source diseases can also be spread from host-to-host

[c] Sexually transmitted diseases can also be controlled by effective use of condoms and by sexual abstinence.

hypersensitivity (and thus exposure and previous or current infection) are also widely used to define the carrier state for tuberculosis (∞ Sections 22.13 and 26.5).

Diseases in which carriers are important for the spread of infection include hepatitis (∞ Section 26.11), tuberculosis (∞ Section 26.5), and typhoid fever (see the box, The Tragic Case of Typhoid Mary; ∞ Section 29.6). Surveys of food handlers and health care workers are sometimes used to identify individuals who are common sources of infection.

Learning From the Past ...   The Tragic Case of Typhoid Mary

The classic example of a chronic carrier was the woman known as "Typhoid Mary," a cook in New York City and Long Island in the early part of the twentieth century. Typhoid Mary (her real name was Mary Mallon) was employed in a number of households and institutions, and as a cook she was in a central position to infect large numbers of people. Extensive epidemiological investigation of a number of typhoid outbreaks by Dr. George Soper revealed that Mary was the likely source of contamination. When her feces were examined bacteriologically, she had very high numbers of the typhoid bacterium, *Salmonella typhi*. She remained a carrier throughout her life, probably because her gallbladder was infected, and organisms were continuously being excreted from there into her intestine. Public health authorities offered to remove her gallbladder, but she refused the operation, and to prevent her from continuing to serve as a source of infection, she was imprisoned. After almost 3 years in prison, she was released on the pledge that she would not cook or handle food for others and that she was to report to the health department every 3 months. She promptly disappeared, changed her name, and cooked in hotels, restaurants, and sanitariums, leaving behind a wake of typhoid fever. After 5 years she was identified as a result of the investigation of an epidemic at a New York hospital. She was again arrested and imprisoned and remained in custody on North Brother Island in the East River of New York City for 23 years. She died in 1938, 32 years after epidemiologists had first discovered she was a chronic typhoid carrier. ■

✓ 25.3 Concept Check

To understand how diseases spread, the pathogen reservoir must be known. Some pathogens exist in soil, water, or animals. Other pathogens exist only in humans and are maintained solely by person-to-person contact. An understanding of disease carriers and pathogen life cycles is critical for controlling disease.

✓ What is a *disease reservoir*?

✓ Distinguish between *acute* and *chronic* carriers.

25.4 Infectious Disease Transmission

Epidemiologists follow the transmission of a disease by correlating geographical, seasonal, and age group incidence of a disease with possible modes of transmission. A disease limited to a restricted geographical location may suggest a particular vector; malaria, for example, is transmitted by a mosquito found mainly in tropical regions. A marked seasonality to a disease is often indicative of certain modes of transmission, such as in the case of influenza, where the number of cases jumps sharply when children enter school and come in close contact.

Different pathogens have different modes of transmission, which are usually related to the habitats of the organisms in the body. For instance, respiratory pathogens are generally airborne, whereas intestinal pathogens are spread by food or water. If the pathogen is to survive, it must undergo transmission from one host to another. Even environmental factors may play a role in survival of the pathogen, and such variables as weather patterns may influence exposure to a pathogen. For example, California encephalitis, caused by RNA single-stranded Bunyaviruses (∞ Section 16.8), occurs primarily during the summer and fall months, and disappears every winter, in a predictable cyclical pattern (Figure 25.3). The virus is

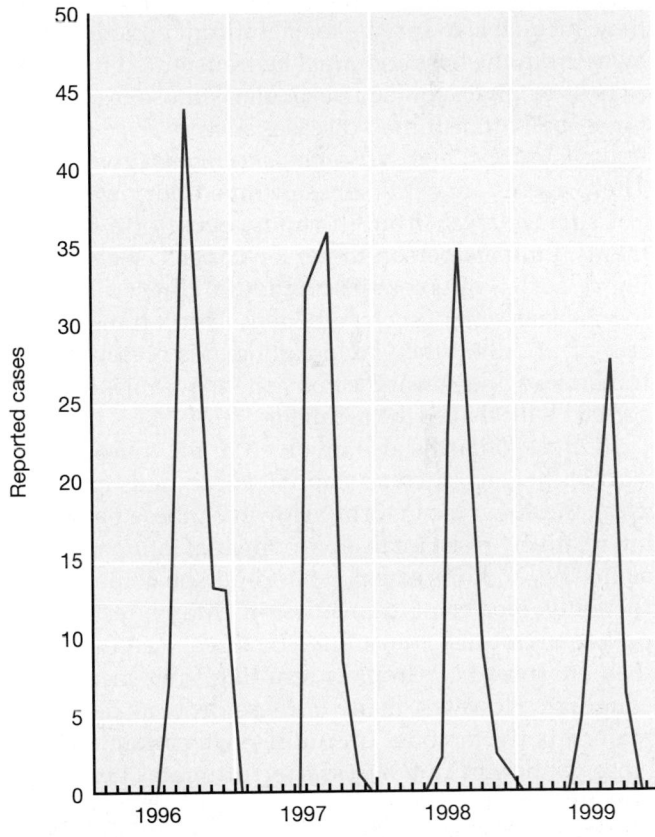

Figure 25.3 The incidence of California encephalitis, a viral disease transmissible through mosquitoes, in the United States, by month of onset. Note the sharp rise in cases in the summer, followed by a complete decline in winter. The disease cycle follows the yearly cycle of the mosquito vector prevalence. Data are from the Centers for Disease Control and Prevention, Atlanta, GA, USA. A number of mosquito-borne diseases follow similar patterns, including Western equine encephalitis, Eastern equine encephalitis, St. Louis encephalitis, and West Nile virus.

transmitted from mosquitoes that, of course, are prevalent only during warmer months. Disappearance of the insect vector causes the disease to disappear until the vector reappears and retransmits the virus in the summer months.

Pathogens can be classified by mode of transmission. Three stages of transmission are commonly evaluated: (1) escape from the host, (2) travel, and (3) entry into a new host. We give here a brief overview of transmission mechanisms.

Direct Host-to-Host Transmission

Host-to-host transmission occurs whenever an infected host transmits the disease to a susceptible host (◌◌ Chapter 26). Transmission by the respiratory route and by direct contact is very common. Transmission by infectious droplets is the most frequent means by which upper respiratory infections such as the common cold and influenza are propagated. However, some pathogens are so sensitive to environmental influences that they are unable to survive for significant periods of time away from the host and must be transmitted from host to host by direct contact, particularly those responsible for sexually transmitted diseases such as *Treponema pallidum* (syphilis) and *Neisseria gonorrhoeae* (gonorrhea). These agents are extremely sensitive to drying and do not survive away from the body, even for a few moments. Intimate person-to-person contact, such as kissing or sexual intercourse, provides a direct means for the transmission of such pathogens. Direct transmission can occur only if the viable pathogen is present on the transmitting person at the body site that comes in direct contact with that of the recipient.

Direct contact is also involved in the transmittal of skin pathogens, such as staphylococci (boils and pimples) and fungi (ringworm). However, these pathogens are relatively resistant to environmental influences such as drying, and intimate person-to-person contact is not the only means of transmission. Many respiratory pathogens are also transmitted by direct means because they are spread by droplets resulting from sneezing or coughing. However, many of these droplets do not remain airborne for long. Transmission, therefore, requires close, although not necessarily intimate, person-to-person contact.

Indirect Host-to-Host Transmission

Indirect transmission can be facilitated by either living or inanimate agents. Living agents transmitting pathogens are called **vectors;** they are generally arthropods (for example, insects, mites, ticks, or fleas) or vertebrates (for example, dogs, cats, or rodents) (◌◌ Chapter 27). Arthropod vectors may not be hosts for the disease but simply carry the agent from one host to another. Large numbers of arthropods obtain nourishment by biting,

and if the pathogen is present in the blood, the arthropod vector may ingest the pathogen and transmit it when biting another individual. In some cases, the pathogen actually replicates in the arthropod, which is then considered an alternate host. Such replication leads to an increase in pathogen numbers, increasing the probability that a subsequent bite will lead to infection.

Inanimate agents such as bedding, toys, books, and surgical instruments can also transmit disease. These inanimate objects are collectively referred to as **fomites.** Food and water are referred to as disease **vehicles.** Fomites can also be disease vehicles, but major epidemics originating from a single source are usually traced to food or water because these are actively consumed in large amounts by a number of individuals in a population (◌◌ Chapters 28 and 29).

Epidemics

Two major types of epidemics can be distinguished: common-source and host-to-host. These two types are contrasted in Figure 25.4. A **common-source epidemic** arises as the result of infection (or intoxication) of a large number of people from a contaminated common source, such

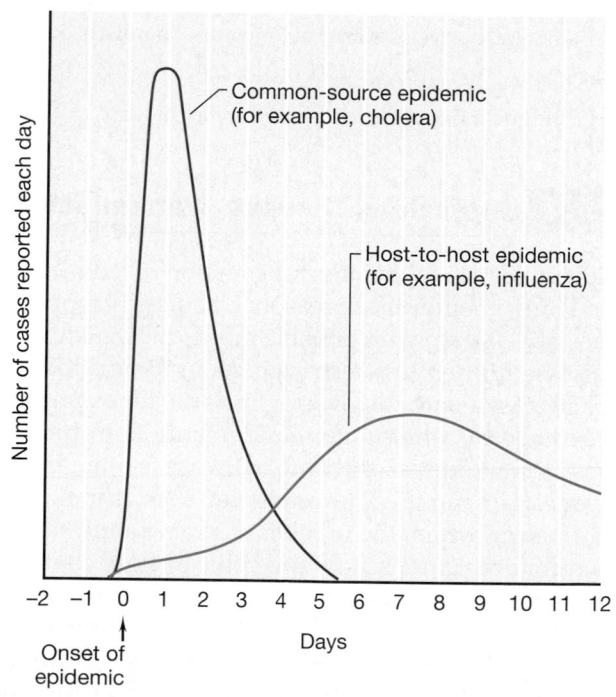

Figure 25.4 Origins of epidemics. The shape of the epidemic curve helps to distinguish the likely origin. In a common-source epidemic, such as from contaminated food or water, the curve is characterized by a sharp rise to a peak, with a rapid decline, which is less abrupt than the rise. Cases continue to be reported for a period approximately equal to the duration of one incubation period of the disease. In a host-to-host epidemic, the curve is characterized by a relatively slow, progressive rise, and cases continue to be reported over a period equivalent to several incubation periods of the disease.

Learning From the Past ...   Snow on Cholera

The importance of drinking water as a vehicle for the spread of cholera was first shown in 1855 by British physician John Snow, who at that time had no knowledge of the bacterial causation of the disease. Snow's study is one of the great classics of epidemiology and serves as a model for how a careful study can lead to clear and meaningful conclusions.

In London, the water supplies to different parts of the city were from different sources and were transmitted in different ways. In a large area south of the Thames River, across the river from Westminster Abbey and the Parliament Building, the water was supplied to houses by two competing private water companies, the Southwark and Vauxhall Company, and the Lambeth Company. It was the water of the former company that was the major vehicle for the transmission of cholera. When Snow began to suspect the water supply of the Southwark and Vauxhall Company, he made a careful survey of the residence of every cholera death in this district and determined which company supplied the water to that residence. In some parts of the area served by these two companies, each had a monopoly, but in a fairly large area the two companies competed directly, each having run independent water pipes along the various streets. Houses had the option of connecting with either supply, and the distribution of houses between the two

companies was random. The clear-cut results of Snow's survey were completely convincing, even to those skeptical about the importance of polluted water in the transmission of cholera. In the first seven weeks of the epidemic, there were 315 deaths per 10,000 houses supplied by the Southwark and Vauxhall Company, and only 37 per 10,000 houses supplied by the Lambeth Company. In the rest of London, there were 59 deaths per 10,000 houses, showing that those supplied by the Lambeth Company had fewer deaths than the general population. In the districts where each company had exclusive rights, it could of course be argued that it was not the water, but some other factor (soil, air, general layout of houses, and so on), that might have been responsible for the differences in disease incidence, but in the districts where the two companies competed, all of these other factors were the same, yet the incidence was high for those supplied with Southwark and Vauxhall water and low for those supplied with Lambeth water. Snow attempted to relate these differences in disease incidence to the sources of the waters used by the two companies. Since he suspected that the excrement and evacuations from cholera patients were highly infectious, he considered that sewage contamination of the water supply might exist. In those days, sewage treatment did not exist and raw sewage was dumped directly into the

Thames River. The Southwark and Vauxhall Company obtained its water supply from the Thames right in the heart of London, where sewage contamination could occur, while the Lambeth Company obtained its water from a point on the river considerably above the city, and hence was relatively free of pollution. It was this difference in source that accounted for the difference in disease incidence. In Snow's words:

As there is no difference whatever, either in the houses or the people receiving the supply of the two Water Companies, or in any of the physical conditions with which they are surrounded, it is obvious that no experiment could have been devised which would more thoroughly test the effect of water supply on the progress of cholera than this. . . . The experiment, too, was on the grandest scale. No fewer than three hundred thousand people of both sexes, of every age and occupation, and of every rank and station, from gentlefolk down to the very poor, were divided into two groups without their choice, and, in most cases, without their knowledge; one group being supplied with water containing the sewage of London, and, amongst it, whatever might have come from cholera patients, the other group having water quite free from such impurity. ■

as food or water. Usually such contamination occurs because of a malfunction in the sanitation of a central distribution system. Foodborne and waterborne diseases are primarily *intestinal* diseases; the pathogen leaves the body in fecal material, contaminates food or water via improper sanitary procedures, and then enters the intestinal tract of the recipient during ingestion. Because foodborne and waterborne diseases are some of those that are most amenable to control by public health measures, we shall discuss them in some detail in Chapters 28 and 29 (also see the boxes, Snow on Cholera, and The Tragic Case of Typhoid Mary in this chapter). The disease incidence for a common-source outbreak is characterized by a rapid rise to a peak because a large number of individuals ingest contaminated food or water and become ill within a relatively brief period of time (Figure 25.4). The common-source illness also declines rapidly, although the decline

is less rapid than the rise. Cases continue to be reported for a period of time approximately equal to the duration of one incubation period of the disease.

In a **host-to-host epidemic**, the disease incidence shows a relatively slow, progressive rise (Figure 25.4) and a gradual decline. Cases continue to be reported over a period of time equivalent to several incubation periods of the disease. A host-to-host epidemic can be initiated by the introduction of a single infected individual into a susceptible population, with this individual infecting one or more people in the population. The pathogen then replicates in susceptible individuals, reaches a communicable stage, and is transferred to other susceptible individuals, where it again replicates and becomes communicable. Table 25.2 summarizes some of the key epidemiological features of some major epidemic diseases observed today.

25.5 The Host Community

The colonization of a susceptible, unimmunized host by a parasite may first lead to explosive infections, transmission to uninfected hosts, and an epidemic. As the host population develops resistance, however, the spread of the parasite is checked, and eventually a *balance* is reached in which host and parasite are in equilibrium. In an extreme case, failure to reach equilibrium could result in death and eventual extinction of the host species. If the pathogen has no other host, then the extinction of the host could also result in extinction of the pathogen. Thus, the evolutionary success of a pathogen is best measured by its ability to establish a balanced equilibrium with the host, rather than by a pathogen's ability to destroy the host. In effect, the host and parasite are affecting each other's evolution; that is, the host and parasite are *coevolving*.

Coevolution of a Host and a Parasite

An excellent experimental example of coevolution of host and parasite occurred when a virus was intentionally introduced for purposes of population control in the wild rabbits of Australia.

Wild rabbits were introduced into Australia from Europe in 1859 and quickly spread until they were overrunning large parts of the continent. Myxoma virus was discovered in South American rabbits, which are a different species from the European rabbits in Australia. In South America the virus causes only a minor disease. However, this same virus is extremely virulent in the European rabbit and almost always causes a fatal infection. The virus is spread very rapidly from rabbit to rabbit by mosquitoes and other biting insects.

Myxoma virus was introduced into Australian rabbits in 1950 to control the rabbit population. Within several months, the virus spread over an area in Australia as large as all of Western Europe. The disease showed a marked seasonal pattern, rising to a peak in the summer when the mosquito vectors were present and declining in the winter. The epidemiology of myxoma virus was studied as a model of a virus-induced epidemic by Australian scientists. Virus was isolated from wild rabbits, and the isolated strains were characterized for virulence

with laboratory rabbits. At the same time, baby rabbits were removed from their dens before infection could occur and reared in the laboratory. Then these wild rabbits were infected with standard virulent strains of myxoma virus to determine their susceptibility. The results of this large-scale model study are shown in Figure 25.5.

During the first year of the epidemic, over 95% of the infected rabbits died. However, within 6 years both the virus and the rabbit population had changed. Over this interval, rabbit mortality dropped to about 84%, and the virus isolated was of decreased virulence. In addition, changes were noted in the resistance of the rabbit. In parts of Australia where the virus was first introduced, the resistance of the rabbits had increased dramatically (Figure 25.5). This resistance was due to acquired genetic changes in the rabbit population and not to immunological responses, for the rabbits tested had been removed from their mothers at birth and had never been in contact with the virus.

As a result of the introduction of myxoma virus, the Australian rabbit population was initially controlled, but the genetic changes in virus and host prevented complete eradication of the rabbit from Australia. In time, most of the surviving rabbits acquired the resistance factors, and by the 1980s the rabbit population in Australia was nearing the premyxomatosis levels, with widespread environmental destruction and pressure on native plants and animals. As a result, Australian authorities, starting in

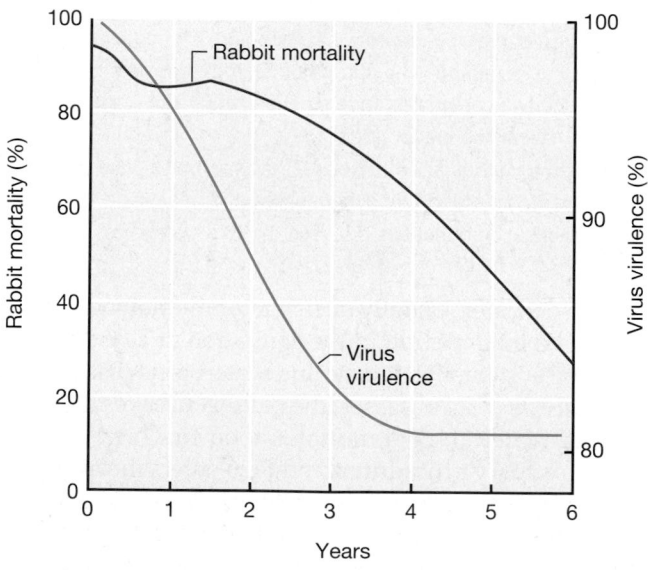

Figure 25.5 Changes in virulence of myxoma virus and in susceptibility of the Australian rabbit during the years after the virus was introduced into Australia in 1950. Virus virulence is given as the average mortality in standard laboratory rabbits for virus recovered from the field each year. Rabbit susceptibility was determined by removing young rabbits from their dens and infecting them with a virus strain of moderately high virulence, which killed 90–95% of normal laboratory rabbits.

1995, released another rabbit pathogen, the rabbit hemorrhagic disease virus (RHDV), a member of the Calciviridae, a single-stranded positive RNA virus (Section 16.9). This virus has been released at 80 specific sites via direct inoculation into rabbits, with eventual plans for up to 280 controlled releases. Because RHDV is spread largely by direct contact and kills animals usually within three days of initial infection, authorities believed the infections would be self-limiting, killing all the rabbits in a local population. Thus, they reasoned, the virus-host equilibrium could be monitored and controlled more reliably than the arthropod-borne myxomatosis virus; resistant rabbit populations should not arise and spread easily in the wild. However, although early reports indicate that RHDV was initially very effective at reducing local rabbit populations, instances of resistant rabbit populations have already been reported, suggesting that the host and pathogen are already coevolving. It seems that the RHDV release strategy is also working to rapidly reduce the rabbit population, but may also be destined to coevolve with the host, as was the case for myxomatosis, losing its lethality and ability to control the rabbit population.

While coevolution of host and pathogen may be the norm for diseases that rely on host-to-host transmission, for those pathogens that do not rely on host-to-host transmission, as we have already mentioned for *Clostridium* spp. (see Section 25.1), there is no selection for decreased virulence to support mutual coexistence. Vectorborne pathogens that are transmitted by the bite of arthropods or ticks are also under no evolutionary pressure to spare the human host. As long as the vector can obtain its blood meal before the host dies, the pathogen can maintain a high level of virulence, decimating the human host in the process of infection. For example, the malaria parasites *Plasmodium* spp. show antigenic variations in their coat proteins that aid in avoiding the immune response of the host. This genetic ability to avoid the host responses *increases* virulence within that specific host.

Other evidence for the phenomenon of continually increasing pathogen virulence comes from studies of super-virulent diarrheal diseases in newborns. In hospital situations, *Escherichia coli* can cause severe diarrheal illness and even death, and virulence seems to increase with each passage of the pathogen through a hospital patient. The *E. coli* organisms replicate in one host and are then transferred through carriers such as hospital attendants or fomites such as soiled bedding and furniture to another patient. Even if that host dies or cannot contact others to transfer the disease, the virulent *E. coli* strain infects others through means of transmission other than the person-to-person route. Extraordinary efforts such as completely washing the nursery and furniture with disinfectant and transferring staff may be necessary to interrupt the cycle of these super-virulent infections.

Herd Immunity

An analysis of the immune state of a group is of great importance in understanding the role of immunity in the development of epidemics. **Herd immunity** is the resistance of a group to infection due to immunity of a high proportion of the members of the group. If the proportion of immune individuals is sufficient, then the whole population will be protected. The fraction of resistant individuals necessary to prevent an epidemic is higher for a highly virulent agent or one with a long period of infectivity and lower for a mildly virulent agent or one with a brief period of infectivity.

The proportion of the population that must be immune to prevent infection of the rest of the population can be estimated from data on poliovirus immunization in the United States. From epidemiological studies of the incidence of polio in large populations, if a population is 70% immunized, polio will be essentially absent in the population. The immunized individuals protect the rest of the population because they cannot acquire and pass on the pathogen, thus breaking the cycle of infection (Figure 25.6). For a highly infectious disease such as influenza, the proportion of immune individuals necessary to confer herd immunity is higher, about 90–95%. A

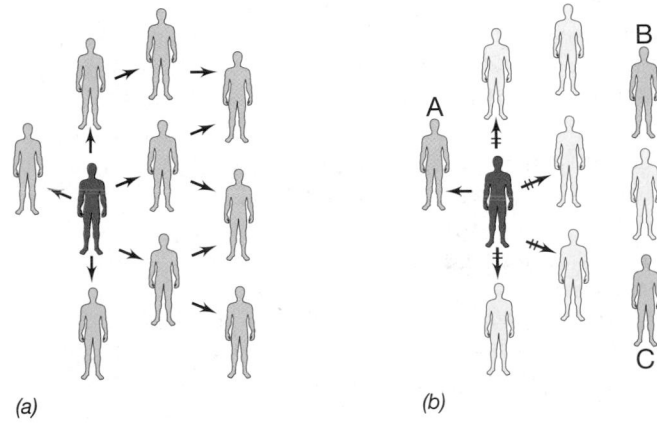

(a) (b)

Figure 25.6 Herd immunity and transmission of infection. Immunity in some individuals protects nonimmune individuals from infection. (a) In an unprotected population, an infected individual (red) can successfully infect (arrows) all of the susceptible individuals (blue). Newly infected individuals will in turn transfer the disease directly to other susceptible individuals. (b) For the case of a moderately transmissible pathogen such as *Corynebacterium diphtheriae* (diphtheria) in a population of moderate density, the infected individual (red) cannot transfer the disease to all susceptible individuals because resistant individuals (yellow), immune by virtue of previous exposure or immunization, break the cycle of pathogen transmission. Even if susceptible individual A (blue) acquires the disease, the other susceptible individuals, B and C, are protected. For a moderately transmissible pathogen such as *C. diphtheriae*, 70% immunity confers resistance to the entire population. For highly transmissible pathogens such as chickenpox (Section 26.7), higher levels of immunity, on the order of 90%, are needed to stop transmission.

value of about 70% has also been estimated for diphtheria, but further study of several small diphtheria outbreaks has shown that in densely populated areas a much higher proportion must be immunized to prevent development of an epidemic. Apparently, person-to-person transmission can occur even if the agent is not highly infectious if susceptible hosts come into constant contact with an infected individual. In the case of diphtheria, an additional complication arises because immunized persons can still harbor the pathogen (inapparent infection) and thus act as chronic carriers.

Cycles of Disease

The principles of epidemics and herd immunity explain why certain diseases occur in *cycles*. A good example of a cyclical disease is chickenpox, which occurs in a high proportion of school children. Because the chickenpox virus (∞ Section 16.12) is transmitted by the respiratory route (∞ Section 26.7), its infectivity is high in crowded situations such as schools. On entry into school at age 5, most children are susceptible, so that on the introduction of virus into the school, an explosive propagated epidemic results. Virtually every individual becomes infected and develops immunity, and as the immune population builds up, the epidemic dies down. Chickenpox shows an annual cycle probably because a new group of nonimmune children arrives each year; the epidemic starts again in the early fall, when children begin school and come into close contact with one another. In such a situation, a single infected child can initiate an epidemic causing disease in nearly all previously unexposed or unvaccinated children.

✓ 25.5 Concept Check

Hosts and pathogens coevolve with time and arrive at a steady state that favors the continued survival of both. With herd immunity, a large fraction of a population is immune to a given disease, and it is difficult for the disease to spread. Disease cycles occur when a large, recurring, nonimmune population such as children entering school is exposed to a pathogen.

✓ Explain coevolution of host and pathogen. Cite a specific example.

✓ How does herd immunity prevent a nonimmune individual from acquiring a disease?

II CURRENT EPIDEMIOLOGY

In this section we will examine data collected by several disease-tracking programs based on the epidemiological principles just described. These data document emerging disease patterns for AIDS and nosocomial infections and provide information to health care work-

ers so that prevention and treatment strategies may be implemented.

25.6 The AIDS Epidemic

Acquired immunodeficiency syndrome (AIDS) is a viral disease that attacks the immune system (∞ Section 26.14). The first reported cases were diagnosed in the United States in 1981. As of 2000, 765,559 cases have been reported in the United States, with a total of 442,882 deaths (Figure 25.7). Nearly 40,000 new AIDS cases continue to be diagnosed in the United States annually. Worldwide, from 1981 to 2000, 56 million individuals have been infected with human immunodeficiency virus (HIV), the virus that causes AIDS. Twenty million people have already died from AIDS, and 36 million are currently living with the disease. North America has almost 1 million infected individuals. Sub-Saharan Africa has 25.3 million infected people (Figure 25.8). In the African country of Botswana, 36% of the adult population is infected with HIV. AIDS causes nearly 3 million deaths annually (Table 25.1), the vast majority in developing countries.

Tracking an Epidemic

Initial case studies in the United States suggested an unusually high AIDS prevalence among homosexual men and intravenous drug abusers. This indicated a transmissible agent, presumably transferred during sexual

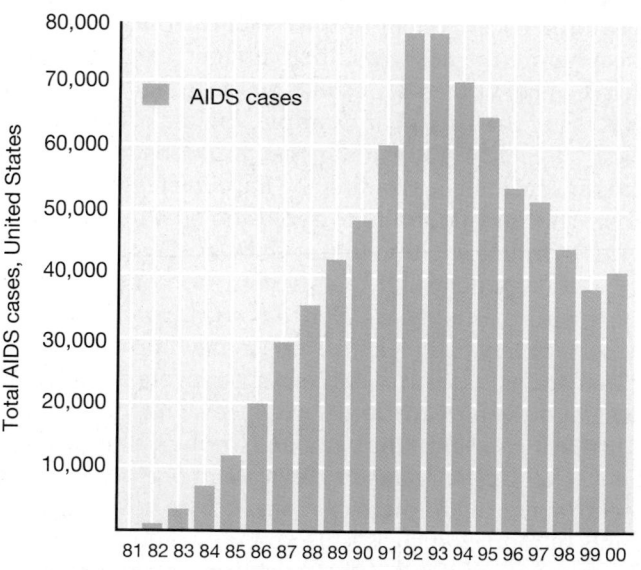

Figure 25.7 Annual diagnosed cases of acquired immunodeficiency syndrome (AIDS), since 1981 in the United States. Cumulatively there have been 765,559 cases of AIDS and 442,882 deaths due to AIDS through 2000. Data are from the Centers for Disease Control and Prevention, Division of HIV/AIDS Prevention.

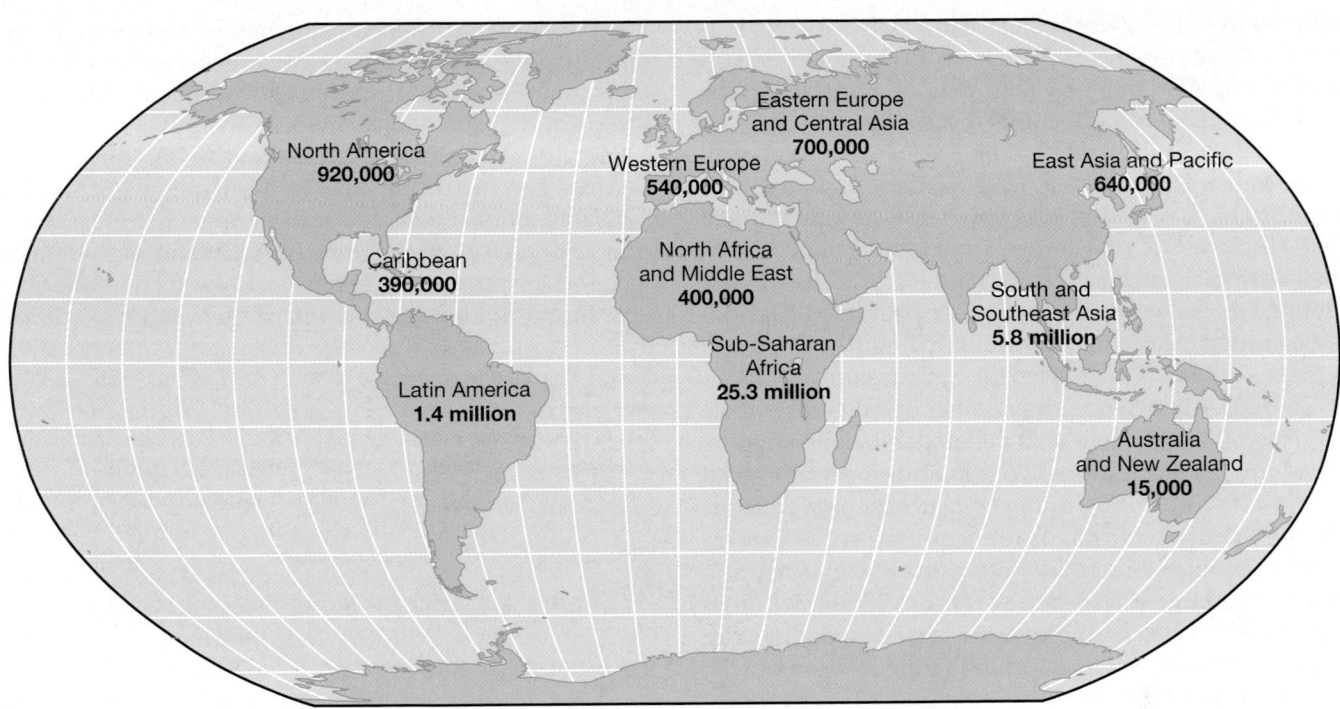

Figure 25.8 Number of HIV-AIDS infections worldwide, 2000. The total number is estimated to be 36.1 million. Data are from the Joint United Nations Program on AIDS.

activity or by contaminated needles. Individuals receiving blood or blood products were also at high risk: Hemophiliacs who required infusions of blood products and a small number of individuals who received blood transfusions or tissue transplants before 1982 acquired AIDS (today less than 1% of the total current AIDS cases are attributable to these modes of transmission) (Figure 25.9). The connection between transfer of AIDS with blood or tissue further reinforced the case for an infectious transmissible agent.

Soon after the discovery of HIV, laboratory tests were developed to detect antibodies to the virus in serum (⌀ Sections 24.11 and 24.12). This allowed extensive surveys of HIV incidence in different populations and also served as a screening method to ensure that new cases of AIDS were not transmitted by blood transfusions. Such tests revealed a fourth group of individuals at high risk for AIDS—the children of mothers who are themselves at high risk for AIDS. Since the beginning of the AIDS epidemic in the United States, there

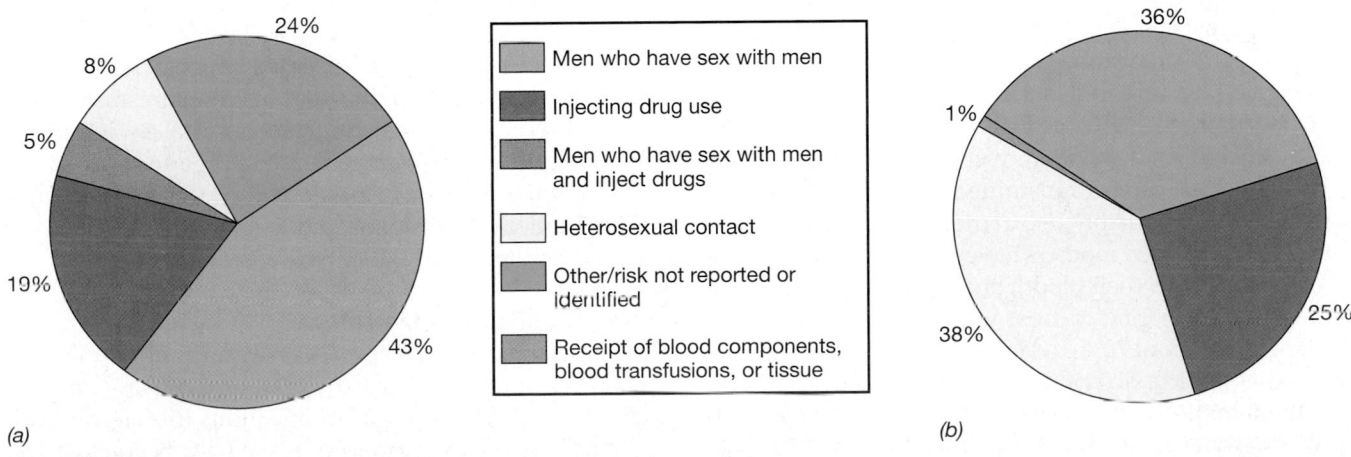

Figure 25.9 Distribution of AIDS cases by risk group and sex in adolescents and adults in the United States for the year 2000. The total number of cases reported was 41,960. (a) AIDS in men, N = 31,501. (b) AIDS in women, N = 10,459. Data are from the Centers for Disease Control and Prevention, Division of HIV/AIDS Prevention.

have been over 8000 cases of pediatric AIDS. In over 90% of these cases, the mothers' behavior (Figure 25.9b) was the only identifiable risk factor, again supporting the idea that AIDS is an infectious disease.

The pattern illustrated in Figure 25.9 is typical of an agent transmissible by sexual activity or by blood. The identification of well-defined high-risk groups implied that AIDS was *not* transmitted from person to person by casual contact, such as the respiratory route, or by contaminated food or water. Instead, epidemiological findings pointed clearly to body fluids, primarily blood and semen, as the major vehicles for transmission of HIV.

In the United States, the number of AIDS cases have been disproportionately high in homosexual men (Figure 25.9a) but the patterns in women and in certain racial and ethnic minorities indicate that homosexuality is not the predisposing factor for acquiring AIDS. For example, among women the heterosexual contact group is now the largest risk group (Figure 25.9b), while in African-American and Hispanic men, intravenous drug use is nearly as often implicated in transmission of HIV as homosexual activity. If we consider all risk groups, heterosexual activity is the fastest growing risk category for new AIDS cases among adults. This diverse array of individuals who are at high risk for acquiring AIDS prompts us to look for a set of factors common to all, and virtually all individuals who are at risk for acquiring HIV share two specific behavior patterns. First, they engage in sex or drug use that involves *transfer of body fluids*, usually semen or blood. Second, individuals who acquire HIV often exchange body fluids with *multiple partners*, either through sexual activity or through needle-sharing drug activity (or both). Thus, they increase their probability of exchanging body fluids with an HIV-infected individual and therefore their chance of acquiring HIV infection and AIDS.

The incidence of AIDS in hemophiliac and blood transfusion recipients has been nearly eliminated in recent years (Figure 25.9). This is due not only to screening of the blood supply but also because many blood clotting factors needed by hemophiliacs can withstand a heat treatment sufficient to inactivate HIV. Pediatric AIDS cases are still a concern. In 2000, there were 87 new cases among this group. HIV can be transmitted to the fetus by infected mothers and probably also in mother's milk. All infants born to HIV-infected mothers have maternally derived antibodies to HIV in their blood, but a positive diagnosis of HIV infection in infants must wait a year or more after birth because about 70% of infants showing maternal HIV antibodies at birth do not develop HIV infection.

Epidemiological studies of AIDS in Africa have clearly shown that transmission of AIDS is not linked to particular sexual practices, such as homosexuality, but instead to person-to-person transfer of HIV-infected fluids. In Africa heterosexual transmission of AIDS is the norm, with about equal numbers of men and women infected. The identification of high risk groups through epidemiological studies led to the development of health education campaigns to inform the public of how AIDS is transmitted and what activities constitute high risk behavior (☞ box, Sexual Activity and AIDS, Chapter 26). Because no cure or effective immunization for AIDS is available, public health education offers the most effective approach to the control of AIDS and is the major weapon for preventing the spread of infection. We discuss the pathology and therapy of AIDS in Section 26.14.

✓ **25.6 Concept Check**

AIDS is one of the most thoroughly studied disease pandemics. AIDS will continue to be a major public health problem, especially in developing countries. There is no effective cure or immunization to prevent AIDS, although we now know a great deal about its pathology and spread.

✓ Describe the major risk factors for acquiring AIDS. Tailor your answer to your country of origin.

✓ Estimate the total number of individuals in the United States who now have AIDS and predict how many will be living with AIDS in the year 2005.

25.7 Hospital-Acquired (Nosocomial) Infections

A hospital may not only be a place where sick people get well but may also be a place where sick people get sicker. Cross-infection from patient to patient or from hospital personnel to patients presents a constant hazard. Hospital infections are called *nosocomial infections* (*nosocomium* is the Latin word for "hospital") and occur in about 5% of all patients admitted. In certain clinical services, such as intensive care units, up to 10% of the patients acquire a nosocomial infection. In all, there are about 2 million nosocomial infections each year in the United States, leading directly or indirectly to 80,000 deaths. Hospital infections are partly due to the prevalence of diseased patients but are often due to the presence of pathogenic microorganisms that are selected and maintained within the hospital environment. For example, multiple-drug-resistant organisms are often spread from host to host as part of the normal flora in hospital situations: Nosocomial pathogens are often found as normal flora in either patients or hospital staff.

The Hospital Environment

Infectious diseases are spread easily and rapidly in hospital environments for several reasons. (1) Many patients have weakened resistance to infectious disease because of their illness (compromised hosts) (☞ Section 21.13). (2) Hospitals treat patients suffering from infectious disease, and these patients may be reservoirs of highly virulent pathogens. (3) The housing of multiple patients in rooms and wards increases the chance of cross-infection.

(4) Hospital personnel move from patient to patient, increasing the probability of transfer of pathogens. (5) Many hospital procedures, such as catheterization, hypodermic injection, spinal puncture, and removal of tissue samples (biopsy) or fluids (drawing blood), breach the skin barrier and carry with them the risk of introducing pathogens to the patient. (6) In maternity wards of hospitals, newborn infants are unusually susceptible to certain kinds of infection because they lack well-developed defense mech-

anisms. (7) Surgical procedures expose internal organs to sources of contamination and the stress of surgery often diminishes the resistance of the patient to infection. (8) Certain therapeutic drugs, such as steroid drugs used for controlling inflammation, increase susceptibility to infection. (9) Use of antibiotics to control infection selects for antibiotic-resistant organisms (∞ Section 20.12). Figure 25.10 summarizes information concerning the most prevalent hospital-acquired infections.

Hospital Pathogens

Hospital pathogens preferentially infect several sites, notably the urinary tract, blood, and the respiratory tract. A relatively small number of pathogens cause the majority of nosocomial infections at these sites (Figure 25.10).

One of the most important and widespread hospital pathogens is *Staphylococcus aureus*. It is the most common cause of pneumonia and the third most common cause of blood infections. *S. aureus* is also particularly problematic in nurseries. Many strains are unusually virulent and are also resistant to common antibiotics (∞ Section 20.12), making their treatment very difficult. In addition to *S. aureus*, other *Staphylococcus* species are now the largest collective cause of hospital-acquired blood infections and are also very prevalent as the causal agents of wound infections. Most members of the genus *Staphylococcus* are found in the upper respiratory tract or on the skin where they are a part of the normal flora in many individuals, including hospital patients and personnel.

Escherichia coli is the most common cause of urinary tract infections in hospitals, but *Enterococcus* species, *Pseudomonas aeruginosa*, *Candida albicans*, and *Klebsiella pneumoniae* infections are also very common. While *Enterococcus*, *E. coli*, and *K. pneumoniae* are normally found only in the human body, *Candida* and *Pseudomonas* are good examples of *opportunistic pathogens*: They are commonly found in the environment and cause disease only in individuals whose defenses are somehow debilitated (∞ Section 21.13). *P. aeruginosa* isolates from hospital infections are often resistant to multiple antibiotics, complicating treatment. *E. coli*, *Staphylococcus*, and *Enterococcus* also have the potential for multiple drug resistance (∞ Section 20.12).

✓ 25.7 Concept Check

Many common microorganisms have the potential to be pathogens in a hospital environment. Hospital patients are unusually susceptible to infectious disease and are exposed to a variety of infectious agents, including opportunistic pathogens, in the hospital environment. Treatment of these infections is complicated by antibiotic resistance.

- ✓ Why are hospital patients more susceptible than normal individuals to pathogens?
- ✓ Why is antibiotic resistance a major problem in hospital environments?
- ✓ What is the source of opportunistic pathogens?

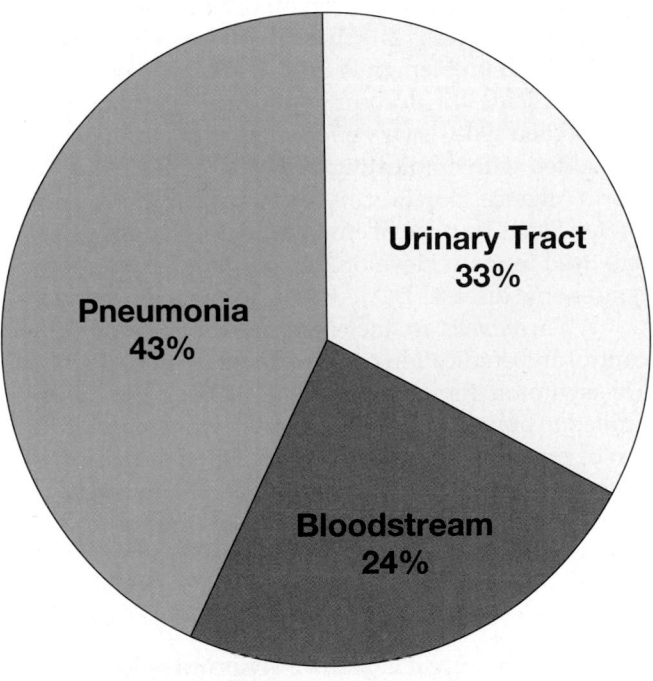

	Bloodstream	Pneumonia	Urinary Tract
Pathogen	No.	No.	No.
Enterobacter spp.	1083	4444	1560
Escherichia coli	514	1725	5393
Klebsiella pneumoniae	735	2865	1891
Haemophilus influenzae		1738	
Pseudomonas aeruginosa	841	6752	3365
Staphylococcus aureus	2758	7205	497
Staphylococcus spp.	8181		838
Enterococcus spp.	2967	682	4226
Candida albicans	1090	1862	4856
All other pathogens	3774	12,537	8075
Total no.	21,943	39,010	30,701

Figure 25.10 Nosocomial infections in intensive care units in the United States, by site and organism, 1992–1999. The total number of nosocomial infections during this time period was 92,454. Data are from the National Nosocomial Infections Surveillance System Report, Centers for Disease Control and Prevention, Atlanta, Georgia, USA.

III EPIDEMIOLOGY AND PUBLIC HEALTH

In this section, we will identify some of the methods used to identify, contain, and eradicate infectious diseases within populations. We will also identify some of the most important current and future public health threats from infectious diseases.

25.8 Public Health Measures for the Control of Disease

An understanding of the epidemiology of an infectious disease is necessary for control of the disease. **Public health** refers to the health of the general population and to the activities of public health authorities in the control of disease. The incidence of many infectious diseases has dropped dramatically over the past 100 years, especially in developed countries, not solely because of public health efforts but because of universal improvements in basic living conditions. Better nutrition, less crowded living quarters, and lighter work loads have probably done as much as public health measures to control diseases such as tuberculosis, primarily by reducing the risk factors related to disease (∞ Section 21.13). However, diseases such as typhoid fever, diphtheria, brucellosis, and poliomyelitis have been controlled largely by active and specific public health measures.

Controls Directed against the Reservoir

If the disease reservoir is primarily in *domestic animals,* then infection of humans can be prevented if the disease is eliminated from the infected animal population. Immunization or destruction of infected animals may eliminate the disease in animals and consequently, in humans. These procedures have nearly eliminated brucellosis and bovine tuberculosis in humans. Recently, these procedures have been used to eliminate bovine spongiform encephalitis (mad cow disease) in cattle in the United Kingdom. Not incidentally, the health of the domestic animal population is also enhanced, with likely long-term economic benefits to the farmer.

When the reservoir is a *wild animal,* eradication is much more difficult. *Rabies* is a disease that occurs in both wild and domestic animals but is transmitted to domestic animals primarily by wild animals. Thus *control* of rabies in domestic animals and in humans can be achieved by immunization of domestic animals. However, since the majority of rabies cases are in wild rather than domestic animals, at least in the United States (∞ Section 27.1), *eradication* of rabies would require the immunization or destruction of all wild ani-

mal reservoirs, including such diverse species as raccoons, bats, skunks, and foxes. Although oral rabies immunization is practical and recommended for rabies control in restricted wild animal populations, its efficacy is untested in large, diverse animal populations such as the wild animal reservoir in the United States.

If the reservoir is an *insect* (such as a mosquito in the case of malaria), effective control of the disease can be accomplished by eliminating the reservoir with chemical insecticides or other lethal agents. The use of toxic or carcinogenic chemicals, however, must be balanced with environmental concerns. In some cases the elimination of one public health problem only creates another. For example, the insecticide dichlorodiphenyltrichloroethane (DDT) (∞ Section 19.18) is very effective against mosquitoes and is credited with eradicating yellow fever and malaria in North America. However, its use is currently banned in the United States because of environmental concerns. DDT is still used in many developing countries to control mosquito-borne diseases but its use is declining worldwide.

When *humans* are the reservoir (for example, AIDS), control and eradication can be difficult, especially if there are asymptomatic carriers. On the other hand, diseases limited to humans that have no asymptomatic phase and can be prevented through immunization or treated with chemotherapy, can be eradicated if each case and all possible contacts are strictly quarantined, immunized, and treated. Such a strategy was successfully employed by the World Health Organization to eradicate smallpox and is currently being used to eradicate polio (see below).

Controls Directed against Transmission of the Pathogen

Pathogens transmitted via food or water can be eliminated by preventing contamination of these common-source vehicles by destroying the pathogen in the vehicle. Water purification methods (∞ Section 28.3) have dramatically reduced the incidence of typhoid fever, and the pasteurization of milk has helped control bovine tuberculosis in humans. Food protection laws have greatly decreased the probability of transmission of a number of enteric pathogens to humans (∞ Chapter 29). Transmission of respiratory pathogens is much more difficult to prevent. Attempts at chemical disinfection of air have been unsuccessful. Air filtration is a viable method, but is limited to small enclosed areas (∞ Section 20.3). In Japan, many individuals wear face masks when they have upper respiratory infections to prevent transmission to others, but such methods, although effective, are voluntary, and would be difficult to institute as public health measures.

Immunization

Smallpox, diphtheria, tetanus, pertussis (whooping cough), measles, mumps, rubella, and poliomyelitis have been controlled primarily by means of immunization.

As we discussed in Section 25.5, 100% immunization is not necessary in order to control the disease in a population, although the percentage needed to ensure disease control varies with the virulence of the pathogen and with the condition of the population (for example, crowding).

Measles offers an example of the importance of maintaining appropriate immunization levels for a given pathogen. Until 1963, the year an effective measles vaccine was licensed, nearly every child in the United States acquired measles through natural infections, resulting in over 400,000 annual cases. After introduction of the vaccine, the number of annual measles infections dropped precipitously (∞ Section 26.7). Case numbers reached a low of 1497 by 1983. However, by 1990, the percentage of children immunized against measles fell to 70%, and the number of new cases rose to 27,786. Within 3 years, a concerted effort to increase measles immunization levels to above 90% virtually eliminated indigenous measles transmission in the United States; a total of 312 measles cases were reported in 1993. Currently, about 100 cases of measles are reported each year in the United States, over half due to infections imported by visitors from other countries.

Although virtually all children are now adequately immunized, many adults lack effective immunity because immunity from childhood vaccinations gradually disappears with time. In the United States, up to 80% of adults, lack effective immunity to important infectious diseases. When these so-called childhood diseases occur in adults, they can have severe effects. If a woman contracts rubella (a viral disease) (∞ Section 26.7) during pregnancy, the unborn child can be affected by serious developmental and neurological disorders. Measles and polio are also much more serious diseases in adults than in children.

All adults are advised to review their immunization status, checking their medical records (if available) to ascertain dates of immunizations. *Tetanus* immunizations, for example, must be renewed at least every 10 years to provide effective immunity. Surveys of adult populations have shown that more than 10% of adults under the age of 40 and over 50% of those over 60 are not adequately immunized. *Measles* immunity in adults also needs to be reviewed. People born before 1957 probably had measles as children and are immune. Those born after 1956 may have been immunized, but the effectiveness of early vaccines was variable and effective immunity may not be present, especially if the immunization was given before 1 year of age. Reimmunization for polio is not recommended for adults unless they are traveling to countries in Africa and Asia where polio may still occur.

Recommendations for immunization were discussed in Section 22.11, and those for particular infections will be discussed in Chapters 26 through 29.

Quarantine

Quarantine involves restricting the movement of an individual with active infection to prevent spread of disease to other members of the population. The *time limit for quarantine is the longest period of communicability of the given disease*. Quarantine measures must prevent the infected individual from contacting unexposed individuals. Quarantine is not as severe a measure as strict isolation, which is used for unusually infectious diseases in hospital situations.

By international agreement, six diseases are considered quarantinable: smallpox, cholera, plague, yellow fever, typhoid fever, and relapsing fever. Although smallpox has been eliminated from the world, quarantine for the other five diseases is still mandated. Each is considered a highly serious, particularly communicable disease. Spread of certain other highly contagious diseases, such as Ebola hemorrhagic fever and meningitis, may also be controlled by quarantine (see Table 25.5 and Section 25.10).

Surveillance

Surveillance is the observation, recognition, and reporting of diseases as they occur. The diseases that are under surveillance in the United States are listed in Table 25.3. Note that several of the epidemic diseases listed in Table 25.2 and Table 25.5 are not on the surveillance list. Several diseases such as influenza, though not on the surveillance list, are surveyed through regional laboratories that identify *index cases*—those cases that exhibit new syndromes, characteristics, or pathogens indicating high potential for new epidemics.

Pathogen Eradication

Disease eradication can be accomplished in some specific cases and has been successful in the case of smallpox. As we mentioned earlier in this section, smallpox was a disease with a reservoir consisting solely of the individuals suffering from acute smallpox infections, and transmission of smallpox was exclusively person-to-person. Infected individuals transmitted the disease through direct contact with previously unexposed members of the population. Although smallpox, a viral disease, cannot be treated, immunization practices were very effective: Vaccination with a related viral strain conferred virtually complete immunity to lethal smallpox infection. In 1967, the World Health Organization (WHO) used this knowledge to develop a plan to eradicate smallpox. Because of successful vaccination programs throughout the developed countries, endemic smallpox was then largely confined to Africa, the Middle East, and the Indian subcontinent. After a preliminary program to vaccinate everyone in these few remaining endemic areas, every suspected smallpox outbreak was targeted by a team of WHO personnel who traveled to the outbreak site, quar-

TABLE 25.3 Reportable infectious diseases in the United States

Diseases caused by *Bacteria*	Diseases caused by *Bacteria*
Anthrax	Toxic shock syndrome
Botulism	Tuberculosis
Brucellosis	Tularemia
Chancroid	Typhoid fever
Chlamydia trachomatis, genital infections	**Diseases caused by fungi (molds, yeast)**
Cholera	Coccidiomycosis
Diphtheria	Cryptosporidiosis
Ehrlichiosis	**Diseases caused by viruses**
Enterohemorrhagic *Escherichia coli*	Acquired immunodeficiency syndrome
Escherichia coli O157:H7	(AIDS) and pediatric HIV infection
Gonorrhea	Encephalitis
Haemophilus influenzae, invasive disease	California serogroup
Hansen's disease (leprosy)	Eastern equine
Hemolytic uremic syndrome, post-diarrheal	St. Louis
Legionellosis	Western equine
Listeriosis	Hantavirus pulmonary syndrome
Lyme disease	Hepatitis A, B, C/non A, non B
Meningococcal disease	HIV infection
Pertussis	Adult
Plague	Pediatric (<13 yrs)
Psittacosis	Measles
Q fever	Mumps
Rocky Mountain spotted fever	Poliomyelitis, paralytic
Salmonellosis	Rabies, animal, human
Shigellosis	Rubella, acute and congenital syndrome
Streptococcal diseases, invasive, Group A	Varicella (deaths only)
Streptococcal toxic shock syndrome	Yellow fever
Streptococcus pneumoniae, drug-resistant invasive disease	**Diseases caused by a protozoan**
	Cyclosporiasis
Syphilis, acute and congenital	Malaria
Tetanus	**Diseases caused by a helminth**
	Trichinosis

antined individuals with active disease, and vaccinated all contacts and even the contacts of contacts. This aggressive policy resulted in the elimination of the active disease within a decade, and the WHO declared the eradication of smallpox in 1980.

Polio, another viral disease with a very effective immunization program, is also targeted for eradication (polio is already eradicated from the Western hemisphere). Using much the same strategy to target polio as was used for smallpox, the WHO immunized 420 million individuals in 1996 alone. By 2001, known polio outbreaks were restricted to sporadic cases in Mauritania, Egypt, Nigeria, and Somalia in Africa, as well as a few outbreaks in India, Pakistan, and Afghanistan.

Leprosy, another disease with a human reservoir, is also targeted for eradication. Active cases of leprosy can now be effectively treated with a multidrug therapy that cures the patient and also prevents spread of *Mycobacterium leprae*, the causal agent (∞ Section 26.5).

Other diseases that are being targeted for eradication are Chagas' disease (treat active cases and destroy the insect vector of this parasitic worm) and dracunculiasis (treat drinking water to prevent transmission of the Guinea worm parasite). Candidates for eradication also include syphilis (∞ Section 26.12) and rabies (∞ Section 27.1).

✓ 25.8 Concept Check

Food and water purity regulations, vector control, immunization, quarantine, disease surveillance, and pathogen eradication are public health measures that play a major role in reduction of disease incidence.

✓ Compare public measures for controlling infectious disease caused by insect reservoirs and by human carriers.

✓ What public health methods can be used to halt the spread of an epidemic disease once it has begun?

25.9 Global Health Considerations

The World Health Organization has divided the world into six geographic regions for the purpose of collecting and reporting health information, such as reports of morbidity and mortality. These geographic regions are Africa, the Americas (North America, the Caribbean,

Central America, and South America), the eastern Mediterranean, Europe, Southeast Asia, and the western Pacific. Here we will compare mortality data from a developed region, the Americas, to that from a developing region, Africa.

Infectious Disease in the Americas and Africa: A Comparison

About 800 million people live in the Americas. Each year there are about 5.7 million deaths, or about 7 deaths per 1000 inhabitants per year. In Africa, there are about 600 million people and about 10.4 million annual deaths, or about 17 deaths per 1000 inhabitants per year. Although these statistics alone are cause for concern, examination of the causes of mortality in these regions is even more disturbing. Figure 25.11 indicates that most African deaths are due to infectious diseases, while in the Americas, cancer and cardiovascular diseases are the leading causes of mortality. In Africa, death from infection is over 10 times more likely. Based on experiences in developed countries over the last century (∞ Figure 1.7), these differences in death rates from infection are due to differences in the availability and efficacy of public health services. Lack of resources in developing regions limits access to health care, safe food and water, and effective immunization programs.

Travel to Endemic Areas

The high incidence of disease in many parts of the world is a concern for people traveling to such areas. However, it is possible to be immunized against many of the diseases that are endemic in foreign countries. Some typical recommendations for immunization for those traveling abroad are shown in Table 25.4. Many foreign countries currently require immunization certificates for yellow fever, but most other immunizations are recommended only for people who are expected to be at high risk. There is also risk, in many parts of the world, of exposure to a variety of diseases (for example, Ebola hemorrhagic fever, dengue fever, amebiasis, encephalitis, malaria, and typhus) for which there are no effective immunizations. Travelers are advised to take reasonable precautions such as avoiding insect and animal bites, drinking only water that has been properly treated and eating food properly stored and prepared, and undergoing antibiotic and chemotherapeutic programs when exposure is suspected.

✓ 25.9 Concept Check

Infectious diseases account for nearly 30 percent of all worldwide mortality. Most infectious diseases occur in developing countries. Travelers to endemic disease areas should be immunized when possible and should take appropriate precautions to prevent infection.

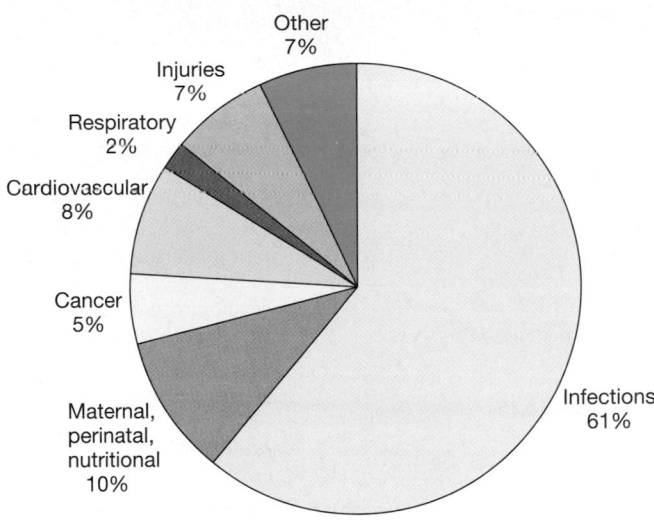

(a)

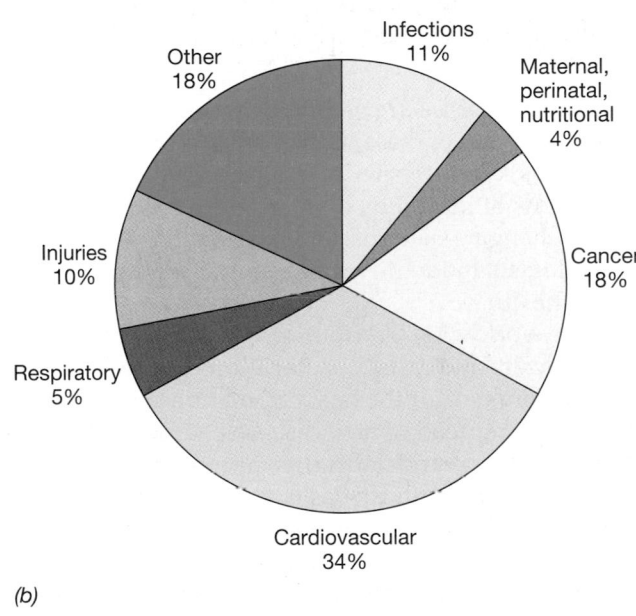

(b)

Figure 25.11 Causes of death in the Americas and Africa, 1999. (a) Africa; (b) the Americas. The charts show percentage of mortality estimates for 1999 by causes of death. There were 10.4 million deaths in Africa, 6.3 million due to infectious diseases. There were 5.7 million deaths in the Americas, 627,000 due to infectious diseases.

✓ Contrast mortality due to infectious diseases in Africa and the Americas.

✓ List a series of infectious diseases for which you have *not* been immunized and with which you could come into contact next year.

TABLE 25.4	Immunizations required or recommended for travel to developing countries[a]	
Disease	**Destination**	**Recommendation**
Cholera	Many central African nations, India, Pakistan, South Korea, Albania, Malta, endemic areas in South America	*Immunization recommended* if entering from or continuing to endemic areas
Yellow fever	Tropical and subtropical countries, worldwide	*Immunization often required* for entry; or if entering from or continuing to endemic areas
Plague	Mostly rural mountainous and upland areas of Africa, Asia, and South America	*Immunization recommended* if direct contact with rodents is anticipated
Infectious hepatitis (A)	Specific tropical areas and many developing countries	*Immunization recommended*
Serum hepatitis (B)	Africa, Indochina, eastern and southern Europe, countries in the former Soviet Union, Central and South America	*Immunization recommended*
Typhoid fever	Many African, Asian, Central and South American countries	*Immunization recommended*

[a] *Current Health Information for International Travelers*, U.S. Department of Health and Human Services.
Vaccinations are also recommended for diphtheria, pertussis, tetanus, polio, measles, mumps, and rubella. Most U.S. citizens are already immunized against these diseases through normal immunization practices.

25.10 Emerging and Reemerging Infectious Diseases

Infectious diseases are *global* health problems, and the scope and focus of these problems are constantly changing. In this section, we will examine some recent changes in patterns of infectious disease outbreaks, the reasons for the changing patterns, and the methods used by epidemiologists to identify and deal with new threats to public health.

The worldwide distribution of diseases can change dramatically and rapidly. Alterations in the pathogen, the environment, or the host population can contribute to the rapid spread of new diseases, with potential for high morbidity and mortality among infected individuals. We refer to diseases that suddenly become prevalent as **emerging** diseases. Emerging infections are not limited to "new" diseases but also include **reemergence** of diseases thought to be controlled, especially when antibiotics become less effective and public health systems fail. Some of the most recent, dramatic examples of emerging and reemerging disease are shown in Figure 25.12 on a global scale. Some recently occurring emerging and reemerging diseases are described in Table 25.5. In addition, the epidemic diseases listed in Table 25.2 all have the potential to cause widespread epidemics and even pandemics.

The phenomenon of suddenly emerging epidemic diseases is not new. Some of the diseases that suddenly emerged into prominence in the past were syphilis (caused by *Treponema pallidum*) (Section 26.12) and plague (caused by *Yersinia pestis*) (Section 27.6). In the Middle Ages, up to one-third of all living humans were killed by the plague epidemics that swept Europe, Asia, and Africa. Influenza caused a devastating worldwide epidemic in 1918–1919 (Figure 25.1 and Section 26.8). In the 1980s, legionellosis (caused by *Legionella pneumophila*) (Section 28.7), acquired immunodeficiency syndrome (AIDS) (Section 26.14), and Lyme disease (Section 27.4) emerged as major new diseases.

Emergence factors

Some factors responsible for emergence of new pathogens are (1) human demographics and behavior; (2) technology and industry; (3) economic development and land use; (4) international travel and commerce; (5) microbial adaptation and change; (6) breakdown of public health measures; and (7) abnormal natural occurrences that upset the usual host-pathogen balance.

The demographics of human populations have changed dramatically in the last two centuries. In 1800, less than 2% of the world's population lived in urban areas. By contrast, today nearly one-half of the world's population lives in cities. The numbers, sizes, and population densities of modern urban centers makes disease transmission much easier. For example, dengue fever (Table 25.5) is now recognized as a serious hemorrhagic disease in tropical cities, largely due to the spread of dengue virus in the mosquito *Aedes aegypti*. The disease now spreads as an epidemic in tropical urban areas. Prior to 1950, dengue fever was rare, presumably because the virus was not easily spread among a more dispersed, smaller population.

Human behavior, especially in large population centers, also contributes to disease spread. For example, sexually promiscuous practices in population centers have

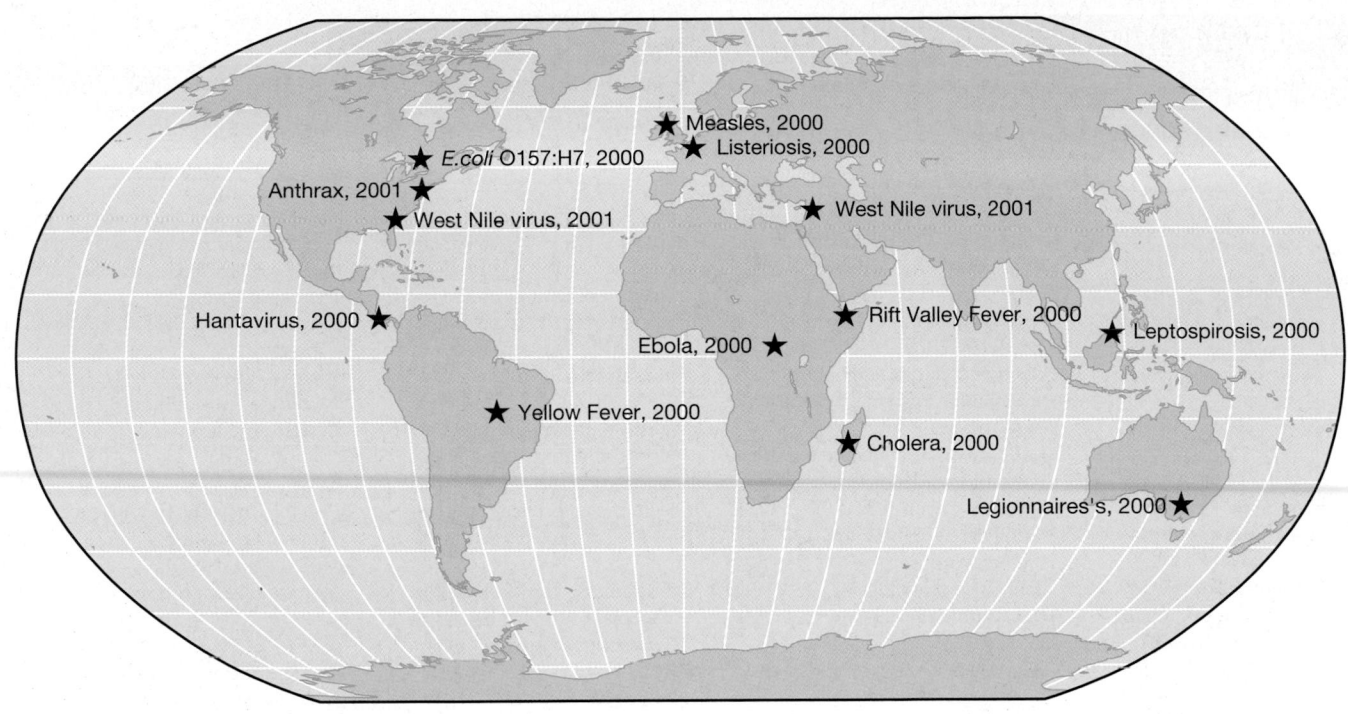

Figure 25.12 Recent outbreaks of emerging and reemerging infectious diseases.

been a major contributing factor to the spread of hepatitis and AIDS (Table 25.5; ⊙⊙ Sections 26.11 and 26.14).

Technological advances and industrial development have had a generally positive impact on living standards worldwide, but in some cases these advances have contributed to the spread of diseases. For example, while tremendous technological advances have been made in health care during the twentieth century, there has been a dramatic increase in nosocomial infections (Section 25.7). Antibiotic resistance in microorganisms is another negative outcome of modern health care practices; vancomycin-resistant enterococci and multiple drug-resistant *Streplococcus pneumoniae* have become important emerging diseases in developed countries.

Transportation, bulk processing, and central distribution methods have become increasingly important for quality assurance and economy in the food industry. However, these same factors can increase the potential for common-source epidemics when sanitation measures fail. For example, a single meat-processing plant spread *Escherichia coli* O157:H7 (Table 25.5) to at least 500 individuals in four states in the United States. The contaminated food source, ground beef, was recalled and the epidemic was eventually stopped, but not before several people died (⊙⊙ Section 29.7).

Economic development and changes in land use also can potentially promote disease spread. For example, Rift Valley fever, a mosquito-borne viral infection, has been on the increase since the completion of the Aswan

Dam in Egypt in 1970. The dam flooded 2 million acres, and the enlarged shore line increased breeding grounds for mosquitoes at the edge of the new reservoir. The first major epidemic of Rift Valley fever occurred in Egypt in 1977, when an estimated 200,000 people became ill and 598 died. Several epidemic outbreaks have occurred in the area since then, including one in 2000 (Figure 25.12), and the disease has become endemic near the reservoir.

Lyme disease, the most common vectorborne disease in the United States, is on the rise largely due to changes in land use patterns. Reforestation and the resulting increase in the numbers of deer and mice (the natural reservoirs for the disease-producing *Borrelia burgdorferi*) have resulted in greater numbers of infected ticks, the arthropod vector (⊙⊙ Section 27.4). In addition, larger numbers of homes and recreational areas in and near forests increase contact between the infected ticks and humans, consequently increasing disease incidence.

International travel and commerce also affects the spread of pathogens. For example, filoviruses (Filoviridae), a group of RNA viruses, causes fevers culminating in hemorrhagic disease in infected hosts. These untreatable viral diseases generally have a mortality rate of greater than 20%. Most outbreaks have been restricted to equatorial central Africa, where the still-unidentified natural hosts and vectors undoubtedly live. Travel of potential hosts to or from endemic areas is usually implicated in disease transmission. For example, one of these viruses was imported into Marburg, Germany,

TABLE 25.5 Some emerging and reemerging infectious diseases

Agent	Disease and symptoms	Mode of transmission	Cause(s) of emergence
Bacteria, Rickettsias, and Chlamydias			
Bacillus anthracis	Anthrax: respiratory distress, hemorrhage.	Inhalation or contact with spores	Bioterrorism
Borrelia burgdorferi	Lyme disease: rash, fever, neurological and cardiac abnormalities, arthritis	Bite of infective *Ixodes* tick	Increase in deer and human populations in wooded areas
Campylobacter jejuni	Campylobacter enteritis: abdominal pain, diarrhea, fever	Ingestion of contaminated food, water, or milk; fecal-oral spread from infected person or animal	Increased recognition; consumption of undercooked poultry
Chlamydia trachomatis	Trachoma, genital infections, conjunctivitis, infant pneumonia	Sexual intercourse	Increased sexual activity; changes in sanitation
Escherichia coli O157:H7	Hemorrhagic colitis; thrombocytopenia; hemolytic uremic syndrome	Ingestion of contaminated food, especially undercooked beef and raw milk	Development of a new pathogen
Haemophilus influenzae biogroup *aegyptus*	Brazilian purpuric fever; purulent conjunctivitis, fever, vomiting	Discharges of infected persons; flies are suspected vectors	Possible increase in virulence due to mutation
Helicobacter pylori	Gastritis, peptic ulcers, possibly stomach cancer	Contaminated food or water, especially unpasteurized milk; contact with infected pets	Increased recognition
Legionella pneumophila	Legionnaires' disease: malaise, myalgia, fever, headache, respiratory illness	Air-cooling systems, water supplies	Recognition in an epidemic situation
Mycobacterium tuberculosis	Tuberculosis: cough, weight loss, lung lesions; infection can spread to other organ systems	Sputum droplets (exhaled through a cough or sneeze) of a person with active disease	Immunosuppression, immunodeficiency
Neisseria meningitidis	Bacterial meningitis	Person-to-person contact	Urbanization, breakdown or lack of local public health surveillance
Staphylococcus aureus	Abscesses, pneumonia, endocarditis, toxic shock	Contact with the organism in a purulent lesion or on the hands	Recognition in an epidemic situation; possibly mutation
Streptococcus pyogenes	Scarlet fever, rheumatic fever, toxic shock	Direct contact with infected persons or carriers; ingestion of contaminated foods	Change in virulence of the bacteria; possibly mutation
Vibrio cholerae	Cholera: severe diarrhea, rapid dehydration	Water contaminated with the feces of infected persons; food exposed to contaminated water	Poor sanitation and hygiene; possibly introduced via bilge water from cargo ships
Viruses			
Dengue	Hemorrhagic fever	Bite of an infected mosquito (primarily *Aedes aegypti*)	Poor mosquito control; increased urbanization in tropics; increased air travel
Filoviruses (Marburg, Ebola)	Fulminant, high mortality, hemorrhagic fever	Direct contact with infected blood, organs, secretions, and semen	Unknown; in Europe and the United States, virus-infected monkeys shipped from developing countries via air
Hantaviruses	Abdominal pain, vomiting, hemorrhagic fever	Inhalation of aerosolized rodent urine and feces	Human intrusion into virus or rodent ecological niche
Hepatitis B	Nausea, vomiting, jaundice; chronic infection leads to hepatocellular carcinoma and cirrhosis	Contact with saliva, semen, blood, or vaginal fluids of an infected person; mode of transmission to children not known	Probably increased sexual activity and intravenous drug abuse; transfusion (before 1978)
Hepatitis C	Nausea, vomiting, jaundice; chronic infection leads to hepatocellular carcinoma and cirrhosis	Exposure (percutaneous) to contaminated blood or plasma; sexual transmission	Recognition through molecular virology applications; blood transfusion practices, especially in Japan
Hepatitis E	Fever, abdominal pain, jaundice	Contaminated water	Newly recognized

TABLE 25.5 Some emerging and reemerging infectious diseases (continued)

Agent	Disease and symptoms	Mode of transmission	Cause(s) of emergence
Viruses (cont.)			
Human immuno-deficiency viruses: HIV-1 and HIV-2	HIV disease, including AIDS: severe immune system dysfunction, opportunistic infections	Sexual contact with or exposure to blood or tissues of an infected person; vertical transmission	Urbanization; changes in lifestyle or mores; increased intravenous drug use; international travel; medical technology (transfusions and transplants)
Human papillomavirus	Skin and mucous membrane lesions (often, warts); strongly linked to cancer of the cervix and penis	Direct contact (sexual contact or contact with contaminated surfaces)	Newly recognized; perhaps changes in sexual lifestyle
Human T-cell lymphotrophic viruses (HTLV-I and HTLV-II)	Leukemias and lymphomas	Vertical transmission through blood or breast milk; exposure to contaminated blood products; sexual transmission	Increased intravenous drug abuse; medical technology (transfusion and transplantation)
Influenza pandemic	Fever, headache, cough, pneumonia	Airborne; especially in crowded, enclosed spaces	Animal-human virus reassortment; antigenic shift
Lassa	Fever, headache, sore throat, nausea	Contact with urine or feces of infected rodents	Urbanization and conditions favoring infestation by rodents
Measles	Fever, conjunctivitis, cough, red blotchy rash	Airborne; direct contact with respiratory secretions of infected persons	Deterioration of public health infrastructure supporting immunization
Monkey pox	Rash, lymphadenopathy, pulmonary distress	Direct contact with infected primates	Travel to endemic areas, consumption and handling of infected primates
Norwalk and Norwalk-like agents	Gastroenteritis; epidemic diarrhea	Most likely fecal-oral; vehicles may include drinking and swimming water, and uncooked foods	Increased recognition
Rabies	Acute viral encephalomyelitis	Bite of a rabid animal	Introduction of infected host reservoir to new areas
Rift Valley	Febrile illness	Bite of an infective mosquito	Importation of infected mosquitoes and/or animals; development (dams, irrigation)
Rotavirus	Enteritis: diarrhea, vomiting, dehydration, and low grade fever	Primarily fecal-oral; fecal-respiratory transmission can also occur	Increased recognition
Venezuelan equine encephalitis	Encephalitis	Bite of an infective mosquito	Movement of mosquitoes and hosts (horses)
West Nile virus	Meningitis, encephalitis	*Culex pipiens* mosquito and avian hosts	Agricultural development, increase in mosquito breeding areas
Yellow fever	Fever, headache, muscle pain, nausea, vomiting	Bite of an infective mosquito (*Aedes aegypti*)	Lack of effective mosquito control and widespread vaccination; urbanization in tropics; increased air travel
Protozoa and Fungi			
Candida	Candidiasis: fungal infections of the gastrointestinal tract, vagina, and oral cavity	Endogenous flora; contact with secretions or excretions from infected persons	Immunosuppression; medical management (catheters); antibiotic use
Cryptococcus	Meningitis; sometimes infections of the lungs, kidneys, prostate, liver	Inhalation	Immunosuppression
Cryptosporidium	Cryptosporidiosis: infection of epithelial cells in the gastrointestinal and respiratory tracts	Fecal-oral, person to person, waterborne	Development near watershed areas; immunosuppression

(continues)

TABLE 25.5 Some emerging and reemerging infectious diseases (continued)

Agent	Disease and symptoms	Mode of transmission	Cause(s) of emergence
Protozoa and Fungi (cont.)			
Giardia lamblia	Giardiasis; infection of the upper small intestine, diarrhea, bloating	Ingestion of fecally contaminated food or water	Inadequate control in some water supply systems; immunosuppression; international travel
Microsporidia	Gastrointestinal illness, diarrhea; wasting in immunosuppressed persons	Unknown; probably ingestion of fecally contaminated food or water	Immunosuppression; recognition
Plasmodium	Malaria	Bite of an infective *Anopheles* mosquito	Urbanization; changing parasite biology; environmental changes; drug resistance; air travel
Pneumocystis carinii	Acute pneumonia	Unknown; possibly reactivation of latent infection	Immunosuppression
Toxoplasma gondii	Toxoplasmosis; fever, lymphadenopathy, lymphocytosis	Exposure to feces of cats carrying the protozoan; sometimes foodborne	Immunosuppression; increase in cats as pets
Other Agents			
Bovine prions	Bovine spongiform encephalitis (animal and human)	Foodborne	Consumption of contaminated beef

with a shipment of African green monkeys used for laboratory work. The virus quickly spread from the primate vector to some of the human handlers. Twenty-five people were initially infected, and six more developed disease as a result of contact with the human cases. Seven people died in this outbreak of what became known as the Marburg virus. Another shipment of laboratory monkeys brought a different filovirus to Reston, Virginia, in the United States. Fortunately, the virus was not pathogenic for humans, but due to its respiratory transmission mode the Reston virus infected and killed most of the monkeys at the Reston facility within days. These two filoviruses are closely related to the Ebola virus (Table 25.5 and Figure 25.12). Sporadic Ebola outbreaks in central Africa, often characterized by mortality rates greater than 50%, highlight a group of pathogens for which there is no immunity or therapy. These pathogens could potentially be spread via air travel throughout the world in a matter of days. A highly contagious agent like the Reston virus that also possesses the high mortality potential of the Ebola virus could devastate population centers worldwide in a matter of weeks.

Microbial adaptation and change can contribute to pathogen emergence. For example, nearly all RNA viruses, including influenza and HIV, undergo rapid, unpredictable genetic mutations. RNA viruses lack correction mechanisms for replication steps, and so they incorporate mutations in their genome at an extremely high rate compared with most of the DNA viruses. The RNA viruses are considered to be major epidemiological problems because of their constantly changing genomes.

Bacterial genetic mechanisms are capable of enhancing virulence and promoting emergence of new epidemics. One group of virulence-enhancing mechanisms are the mobile genetic elements, bacteriophages, plasmids and transposons (∞ Sections 16.1–16.5, 10.8, and 10.11). Table 25.6 lists some virulence factors carried on these mobile genetic elements that contribute to pathogen emergence.

Antibiotic resistance is another factor in bacterial pathogen resurgence (∞ Section 20.12) and in virus emergence. Although several drugs are effective against certain viral diseases (∞ Section 20.10), resistance to these drugs is very common, especially among the RNA viruses. For example, many strains of HIV develop resistance to azidothymidine (AZT) unless it is used in combination with other drugs (∞ Section 26.14).

A breakdown of public health measures is sometimes responsible for the emergence or resurgence of diseases. For instance, cholera (caused by *Vibrio cholerae*, ∞ Section 28.5) can be adequately controlled, even in endemic areas, by providing proper sewage disposal and water treatment. However, in 1991 an outbreak of cholera due to contaminated municipal water supplies in Peru was one of the first indications that the current cholera pandemic had reached the Americas (∞ Section 28.5). In 1993, the municipal water supply of Milwaukee, Wisconsin, was contaminated with the chlorine-resistant protozoan *Cryptosporidium*, resulting in over 400,000 cases of intestinal disease, 4000 of which required hospitalization. Enhanced filtration systems were required to rid the water supply of the pathogen (∞ Section 28.6).

Inadequate public vaccination programs can lead to the resurgence of previously controlled diseases. For example, recent outbreaks of diphtheria (caused by *Coryne-*

TABLE 25.6	Virulence factors encoded by bacteriophages, plasmids, and transposons[a]	
Genetic element	**Organism**	**Virulence factors**
Bacteriophage	*Streptococcus pyogenes*	Erythrogenic toxin
	Escherichia coli	Shiga-like toxin
	Staphylococcus aureus	Enterotoxins A, D, E, staphylokinase, toxic shock syndrome toxin-1 (TSST-1)
	Clostridium botulinum	Neurotoxins C, D, E
	Corynebacterium diphtheriae	Diphtheria toxin
Plasmid	*Escherichia coli*	Enterotoxins, pili colonization factor, hemolysin, urease, serum resistance factor, adherence factors, cell invasion factors
	Bacillus anthracis	Edema factor, lethal factor, protective antigen, poly-D-glutamic acid capsule
	Yersinia pestis	Coagulase, fibrinolysin, murine toxin
Transposon	*Escherichia coli*	Heat-stable enterotoxins, aerobactin siderophores, hemolysin and pili operons
	Shigella dysenteriae	Shiga toxin
	Vibrio cholerae	Cholera toxin

[a]For discussion of bacteriophages, plasmids, and transposons, see Sections 9.8–9.10 and 16.1–16.5, 10.8, and 10.11, respectively.

bacterium diphtheriae) (∞ Section 26.3) in the former Soviet Union result from inadequate immunization of susceptible children due to the breakdown in public health infrastructures. Pertussis, another vaccine-preventable childhood respiratory disease (caused by *Bordetella pertussis*) (∞ Section 26.4), has increased recently in eastern Europe due to inadequate immunization.

Finally, abnormal natural occurrences sometimes upset the usual host-pathogen balance. For example, hantavirus is a well-known human pathogen that occurs in many rodent populations, even in laboratory animals (∞ Section 27.2). A number of lethal cases of hantavirus infection and disease were reported in 1993 in the American Southwest and were linked to exposure to wild animal droppings. The likelihood of exposure to mice and droppings was increased due to a larger than normal wild mouse population resulting from near-record rainfall, a long growing season, and a mild winter.

Addressing Emerging Diseases

Many of the emerging diseases we have discussed are absent from the official notifiable disease list for the United States (Table 25.3). How then do public health officials define emerging diseases and prevent major epidemics? The keys for addressing emerging diseases are *recognition* of the disease and *intervention* to prevent disease transmission.

The first step in disease recognition is *surveillance*. Epidemic diseases that exhibit particular *clinical syndromes* warrant intensive public health surveillance. These syndromes are (1) acute respiratory diseases, (2) encephalitis and aseptic meningitis, (3) hemorrhagic fever, (4) acute diarrhea, (5) clusterings of high fever cases, (6) unusual clusterings of any disease or deaths,

and (7) resistance to common drugs or treatment. Thus, new diseases are recognized because of their epidemic incidence, clusterings, and syndromes. As the prevalence and pathology of an emerging disease are recognized, it is added to the notifiable disease list. For example, AIDS was recognized as a disease in 1981 and was added to the notifiable disease list in 1984. Lyme disease was first recognized as a separate clinical disease in the 1980s and added to the notifiable disease list in 1991. Likewise, outbreaks of gastrointestinal disease due to enteropathogenic *Escherichia coli* O157:H7 have been increasing in recent years, and the strain was added to the notifiable disease list in 1995.

Intervention to prevent spread of emerging infections must be a public health response involving a variety of methods. Disease-specific intervention is the key to controlling individual outbreaks. Methods such as quarantine, immunization, and drug treatment must be applied to contain and isolate outbreaks of specific diseases. Finally, for vectorborne and zoonotic diseases, we must identify the nonhuman host or vector and intervene in the life cycle of the pathogen, ultimately interrupting transfer to humans.

✓ 25.10 Concept Check

Changes in host, vector, or pathogen conditions, whether natural or artificial, can result in conditions that encourage the explosive emergence or reemergence of certain infectious diseases. Global surveillance and intervention programs must be developed to prevent new epidemics and pandemics.

✓ What factors are important in the emergence or reemergence of potential pathogens?

✓ Indicate general and specific methods that would be useful for dealing with emerging infectious diseases.

25.11 Biological Warfare and Biological Weapons

Biological warfare is the use of biological agents to incapacitate or kill a military or civilian population in an act of war or terrorism. Biological warfare is receiving public attention because biological weapons and weapons-making facilities are known or suspected to be in the hands of several rogue governments and extremist groups.

Characteristics of Biological Weapons

Biological weapons must be biological agents such as organisms or toxins that are (1) easy to produce and deliver, (2) safe for use by the offensive soldiers, and (3) able to incapacitate or kill individuals under attack in a reproducible and consistent manner. Many organisms or biological toxins fit these rather general criteria, and we will discuss several of these below.

Although bioweapons are potentially useful in the hands of conventional military forces, the greatest likelihood of bioweapons use is probably by terrorist groups. This is in part because any well-trained microbiologist possesses the laboratory skills necessary for the propagation of many of the organisms useful for biological warfare; biological weapons are accessible to nearly every government and moderately well-financed private organization.

Candidate Biological Weapons

Virtually all pathogenic bacteria or viruses are potentially useful for biological warfare, and several of the most likely candidate organisms are relatively simple to grow and disseminate. The most commonly mentioned organism is *Bacillus anthracis*, the causal agent of anthrax. *B. anthracis* produces endospores (∞ Sections 4.15 and 12.20) that, when aerosolized, can be a very effective means of distributing the bacterium (Figure 25.13). Inhalation of the spores or the live bacteria results in pulmonary infections that have a mortality rate of nearly 100% when untreated, causing pulmonary and cerebral hemorrhage (Figure 25.14).

Smallpox virus (∞ Section 16.12) is another potential biological warfare agent. Although an extremely effective smallpox vaccine exists, it has not been in regular use for more than 20 years because wild smallpox was eradicated worldwide by 1980. As a result, over 90% of the current worldwide population is now inadequately vaccinated and susceptible to the disease. Could terrorist groups or even conventional military forces gain access to smallpox virus? There are several known remaining stocks of smallpox virus in the United States and in the former Soviet Union.

Other bacterial bioweapons candidates include *Yersinia pestis*, the organism responsible for plague (∞ Section 27.6), *Brucella abortus* (fever and bacteremia), *Francisella tularensis* ("rabbit fever"), and *Salmonella* (foodborne and waterborne illnesses) (∞ Section 29.6).

Viral pathogens with bioweapons potential include rabies virus (∞ Section 27.1) and Ebola virus (Section 25.10). These agents cause diseases associated with high death rates within days to weeks after exposure.

Bacterial toxins such as the botulinum toxin produced by *Clostridium botulinum* are also possible bioweapons (∞ Sections 21.10 and 29.5). Large amounts of the preformed toxin delivered to a population through a common vehicle such as drinking water could have devastating consequences. The lethal dose of botulinum toxin for a human is 2 μg or less.

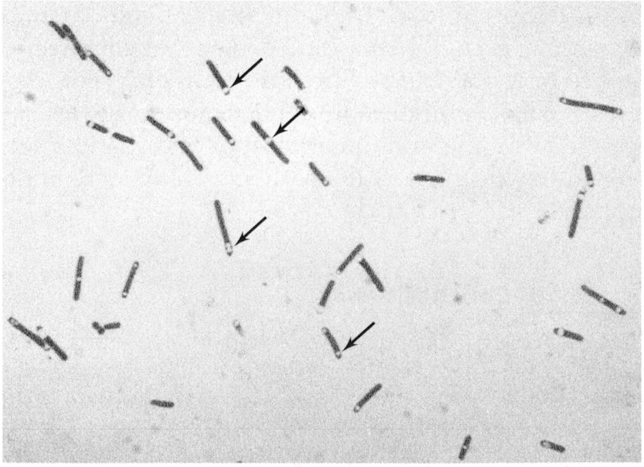

Figure 25.13 Gram stain of *Bacillus anthracis*. *B. anthracis* is a gram-positive spore-forming rod (∞ Section 12.20). Note the formation of endospores (arrows). Spore formation (∞ Section 4.15) enhances the ability to disseminate *B. anthracis* in aerosols.

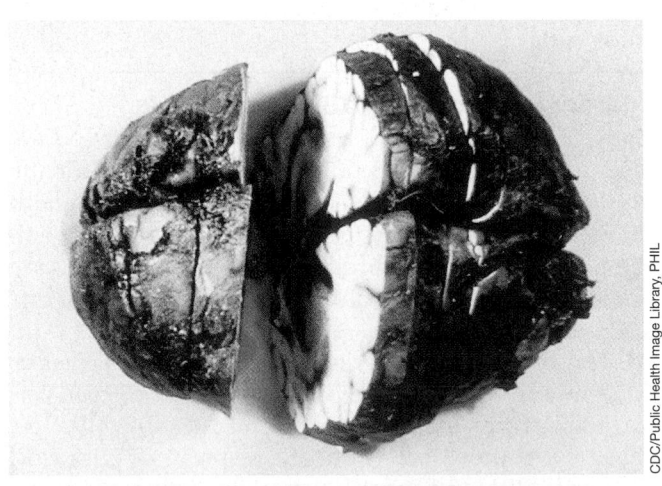

CDC/Public Health Image Library, PHIL

Figure 25.14 Fixed and sectioned brain showing hemorrhagic meningitis (dark coloration) due to a fatal case of inhalation anthrax.

Delivery of Biological Weapons

Most organisms suitable for bioweapons use can be spread in an aerosolized form, providing simple, rapid, widespread dissemination and infection. Examples of several exposures involving aerosols are instructive.

In 1962 one of the last outbreaks of smallpox in a developed country occurred in Germany. A German worker developed smallpox after returning from Pakistan, a country with endemic smallpox at the time. The individual was immediately hospitalized and quarantined, but the patient had a cough, and the aerosolized virus caused illness in 19 *vaccinated* individuals; at least one individual died from the resulting infection.

In another accident, *Bacillus anthracis* spores were inadvertently released into the atmosphere from a bioweapons facility in Sverdlovsk, Russia in 1979. Less than 1 gram of spores was released, and everyone in the area surrounding the weapons facility was immunized and given prophylactic antibiotic therapy as soon as the first anthrax case was diagnosed. However, 77 individuals outside the facility contracted pulmonary anthrax and 66 died.

Planned bioterrorist attacks have already occurred. In 1984 in The Dalles, Oregon (United States), cultists inoculated a salad bar with *Salmonella typhimurium* at 10 local restaurants, causing 751 cases of foodborne salmonellosis in a region that usually has less than 10 cases per year (ꚍ Section 29.6). In 1995, a radical political group released Saran nerve gas into a Tokyo subway, killing several people and injuring scores of others. Although this was a chemical weapon, this group also possessed anthrax cultures, bacteriological media, drone airplanes, and spray tanks.

Delivery of preformed bacterial toxins to large populations is impractical because the most potent exotoxins are proteins and would lose effectiveness as they are diluted or destroyed in common sources such as drinking water.

Prevention and Response to Biological Weapons

Proactive measures against bioweapons have already begun with efforts to update the international agreements of the 1972 Biological and Toxic Weapons Convention. At the practical level, governments are now supporting the large-scale production and distribution of vaccines and the development of strategic and tactical plans to prevent and contain the effects of bioweapons. Bioterrorism is a real threat in a world of rapid international travel and easily accessible technical information.

✓ 25.11 Concept Check

Infectious biological agents can be used as weapons by contemporary military forces or by terrorist groups. Aerosols are the most likely mode of inoculation. Prevention and containment measures rely on a well-prepared public health infrastructure.

✓ For *Bacillus anthracis*, what characteristics make this organism particularly useful as a bioweapon?

✓ Identify two other infectious agents that could be effective bioweapons. How could the agents be disseminated?

Review Questions

1. List the five most common causes of mortality due to infectious diseases throughout the world. Are any of these diseases preventable by immunization?

2. Distinguish between *mortality* and *morbidity*, *prevalence* and *incidence*, and *epidemic* and *pandemic*, as these terms relate to infectious disease.

3. Explain the difference between a *chronic* carrier and an *acute* carrier of an infectious disease.

4. Give examples of host-host transmission of disease via direct contact. Also give examples of indirect host-to-host transmission of disease via vector agents and fomites.

5. How can immunity to a pathogen by a large proportion of the population protect the nonimmune members of the population from acquiring a disease? Will this herd immunity work for diseases that have a common source, such as water? Why or why not?

6. Identify the major risk factors for acquiring human immunodeficiency virus (HIV) infection in the United States. Does this pattern hold for all geographic regions?

7. Hospital environments are conducive to the spread of infectious diseases. Review the reasons for the enhanced spread of infection in hospitals. What are the sources of most nosocomial infections?

8. Describe the major medical and public health measures developed in the twentieth century that were instrumental for controlling the spread of infectious diseases in developed countries.

9. Compare the role of infectious diseases on mortality in developed and developing countries.

10. Review the major reasons for the emergence of new infectious diseases. What methods are available for identifying and controlling the emergence of new infectious diseases?

11. Describe the general properties of an effective biological warfare agent. How does *Bacillus anthracis* meet these criteria?

Application Questions

1. How would an epidemiologist acquire data concerning a potential common-source epidemic? What resources are currently at the epidemiologist's disposal, and what resources must be enhanced to better define serious infectious disease outbreaks?

2. If an infectious disease causes high *mortality*, then *morbidity* may be quite low. On the other hand, diseases characterized by high morbidity often have very low mortality. Explain these statements and present examples to support (or refute) both the hypothesis and your explanation. Do common-source diseases such as tetanus (∞ Section 27.8) and amebiasis (∞ Section 28.8) fit these generalizations?

3. Smallpox, a disease that was limited to humans, was eradicated. Plague, a disease with a zoonotic reservoir in rodents (Table 25.2) can never be eradicated. Explain this statement and why you agree or disagree with the possibility of eradicating plague on a global scale. Devise a plan to eradicate plague in a limited environment such as a town or city. Be sure to use methods that involve the reservoir, the pathogen, and the host.

4. Transmission of many epidemic diseases is host to host, whereas other epidemics are spread via a common source. Some epidemic diseases can be transmitted by both routes. Explain how this might happen, using specific infectious agents (at least one bacterium and one virus) as examples.

5. What is the overall advantage of host-pathogen co-evolution in terms of species survival? Is it beneficial to the pathogen to cause high mortality in the host? Why or why not? What diseases in Table 25.2 have caused high mortality? Compare their reservoir or host to diseases with low mortality.

6. Acquired immunodeficiency syndrome (AIDS) transmission is spread by intimate person-to-person contact through exchange of body fluids. How did epidemiologists determine this fact? AIDS is a candidate for a disease that can be eliminated because it is propagated by person-to-person contact and there are no known animal reservoirs. Design a program for eliminating AIDS in a developed country and in a developing country. How would these programs differ from one another? What factors would work against the success of your program, both in terms of human behavior and in terms of the AIDS disease itself? Why are the numbers of HIV-infected and AIDS patients continuing to grow, especially in developing countries?

7. Why are diseases due to antibiotic-resistant pathogens more commonly found in hospital environments than in the general population? Why are diseases caused by so-called "normal flora" such as *Staphylococcus* species more common in hospital environments than in the general population? What special precautions must one take when diagnosing and treating infectious diseases in a hospital setting?

8. As a public health official, you are faced with a common-source epidemic, and you believe the source is the municipal water supply. How would you use your resources to stop the epidemic? (*Hint*: Do not focus on *treatment* of the disease unless you, as a public health officer, believe that treatment will stop disease spread.) List the steps you would take in priority order. Perform the same exercise for a host-to-host epidemic caused by a pathogen for which there are available vaccines and chemotherapeutic agents.

9. Travel to developing countries involves a certain amount of exposure to infectious diseases. What general precautions should you take before, during, and after visits to developing countries? Where can you obtain information on the infectious disease status in a specific foreign country? When you return from a foreign country, are you a disease risk to your family or your associates? Explain.

10. Although many factors may be involved in the emergence of an infectious disease, some diseases develop to the pandemic stage while others never get beyond localized epidemics. Examples of this are pandemic HIV and epidemic Ebola virus, which were both identified within the last 20 years. What factors do these viruses share that led to their emergence? What specific factors cause them to be quite different in terms of their spread, and how have these factors contributed to the worldwide spread of HIV, while limiting the spread of Ebola virus to very dramatic but sporadic and isolated outbreaks?

11. Identify a specific pathogen that you think would be a suitable agent for effective biological warfare. Describe the properties of the pathogen in the context of its use as a bioweapon. Describe conditions for growing large amounts of the pathogen. Identify a suitable delivery method. Since you will propagate and deliver the pathogen, describe the precautions you will take to protect yourself. Now reverse your role. As a public health official in a large city, describe how you would recognize and diagnose the disease caused by the agent. Indicate the measures you would take to treat the illnesses caused by the agent. How could you best limit the damage? Would quarantine and isolation methods be useful? What about immunization and antibiotics?

I nfectious microbial diseases are often rapidly spread from person to person among members of populations that are in close proximity. For example, influenza, colds, and even bacterial meningitis, a very serious neurological disease caused by the gram-negative bacterium *Neisseria meningitidis* shown here, often spread rapidly through schools, military bases, hospitals, and universities, because of the close association of infected individuals with a population of susceptible individuals.

PERSON-TO-PERSON MICROBIAL DISEASES

26

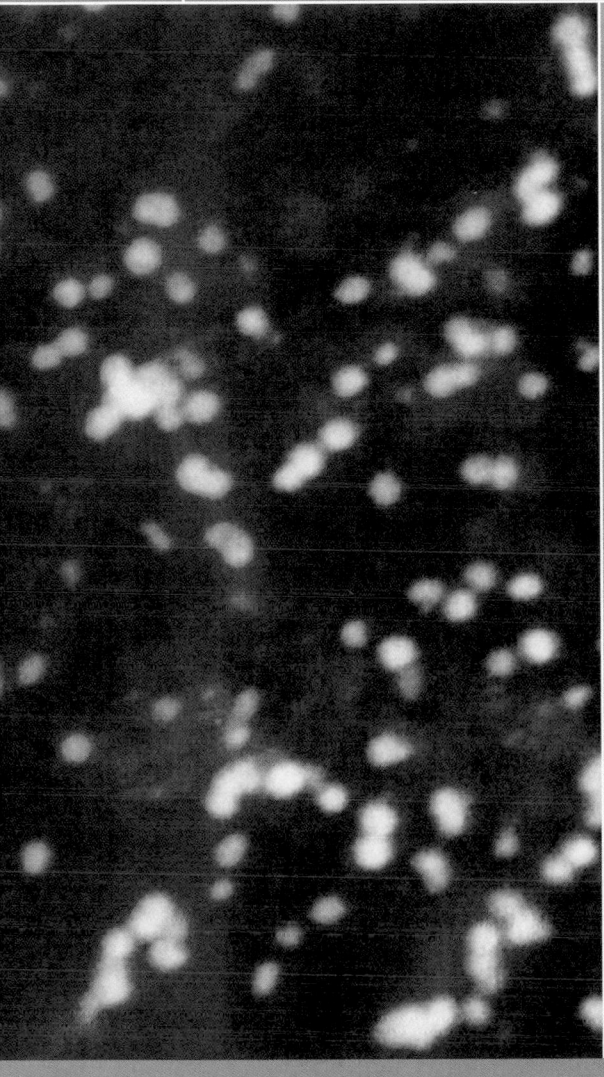

Working Glossary

Antigenic drift minor changes in antigens due to gene mutation in influenza virus

Antigenic shift major changes in antigens due to gene reassortment in influenza virus

Cirrhosis breakdown of the normal liver architecture resulting in fibrosis

Congenital syphilis syphilis contracted by an infant from its mother during birth

Hepatitis a liver inflammation commonly caused by an infectious agent

Jaundice production and release of excess bilirubin in the liver due to destruction of liver cells, resulting in yellowing of the skin and whites of the eye

Meningitis inflammation of the meninges (brain tissue), sometimes caused by *Neisseria meningitidis* and characterized by sudden onset of headache, vomiting, and stiff neck, often progressing to coma within hours

Meningococcemia fulminant disease caused by *Neisseria meningitidis* and characterized by septicemia, intravascular coagulation, and shock

Nonnucleoside reverse transcriptase inhibitor (NNRTI) a nonnucleoside compound that inhibits the action of viral reverse transcriptase by binding directly to the catalytic site

Nucleoside reverse transcriptase inhibitor (NRTI) a nucleoside analog compound that inhibits the action of viral reverse transcriptase by competing with nucleosides

Protease inhibitor a compound that inhibits the action of viral protease by binding directly to the catalytic site, preventing viral protein processing

Rheumatic fever an inflammatory autoimmune disease triggered by an immune response to infection by *Streptococcus pyogenes*

Scarlet fever characteristic reddish rash resulting from an exotoxin produced by *Streptococcus pyogenes*

Sexually transmitted disease (STD) a disease that is usually transmitted by sexual contact

Toxic shock syndrome (TSS) acute systemic shock resulting from a host response to an exotoxin produced by *Staphylococcus aureus*

Tuberculin test a skin test for previous infection with *Mycobacterium tuberculosis*

Viral load a quantitative assessment of the amount of virus in a host organism

---

At least 500,000 microbial species exist in nature, and probably many more (Section 11.7), but only a few hundred species are known human pathogens. Most microorganisms carry out essential life-supporting activities independently, and many others are closely associated with plants or animals in stable, beneficial relationships (Chapter 19). However, the pathogenic species have profoundly negative effects on host organisms and, as a result, have been intensively studied. In the next two chapters, we will examine representative examples of human pathogens and study the prevention, treatment, and pathology of the diseases they cause. We will first divide the pathogens based on their *mode of transmission*. We will then discuss representative infectious diseases in relation to the ecology of the pathogen. In this chapter we consider disease transmission via direct *person-to-person* interactions. In Chapters 27 through 29, we will consider diseases whose modes of transmission involve animal or arthropod vectors or common sources such as soil, water, or food.

By dividing pathogens according to their modes of transmission and the diseases they cause, we will connect seemingly unrelated organisms with one another. For example, influenza and streptococcal pharyngitis, which we will discuss together, are spread person-to-person via the respiratory route. They produce overlapping symptoms, although the causal agents, one viral and one bacterial, are markedly different. Using this approach, we hope to make connections among biologically diverse but ecologically and pathogenically related agents.

I AIRBORNE TRANSMISSION OF DISEASES

Aerosols, such as those generated by the sneeze shown in Figure 26.1, are an important means of person-to-person transmission for many infectious diseases. Most respiratory diseases are spread almost exclusively in this fashion. For example, *Mycobacterium tuberculosis*

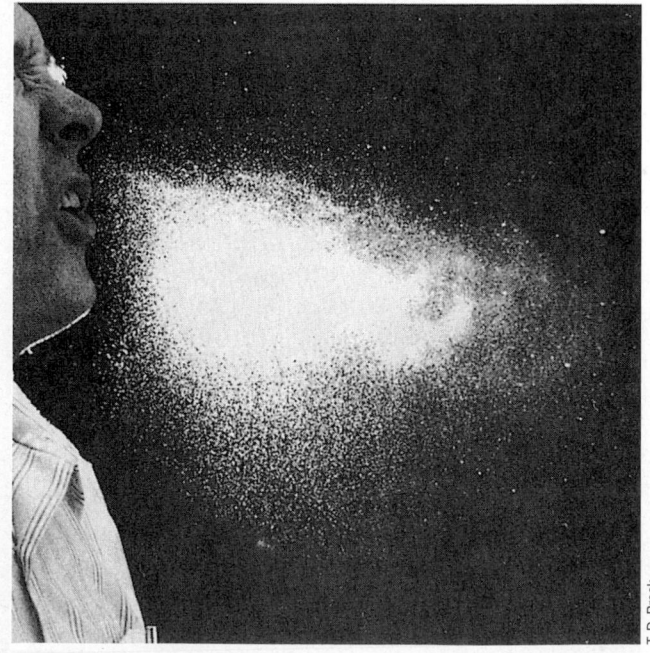

Figure 26.1 High speed photograph of an unstifled sneeze.

has successfully used this strategy to infect at least one-third of the world's population (∞ Section 25.1). Influenza and cold viruses are so successfully passed by respiratory routes that virtually everyone acquires more than one cold or case of influenza every year.

26.1 Airborne Pathogens

Air is not a suitable medium for the growth of microorganisms; organisms found in air are derived from soil, water, plants, animals, people, or other sources. In outdoor air, soil organisms predominate. Microbial numbers indoors are considerably higher than those outdoors, and the organisms are mostly those commonly found in the human respiratory tract.

Windblown dust carries with it significant microbial populations that can travel long distances. Most microorganisms survive poorly in air, and so effective transmittal to another human occurs only over short distances. However, certain human pathogens (*Staphylococcus, Streptococcus*) survive under dry conditions fairly well and remain alive in dust for long periods of time. Gram-positive *Bacteria* are in general more resistant to drying than gram-negative *Bacteria* because of their thicker, more rigid cell wall. Spore-forming *Bacteria* are extremely resistant to drying but are not generally passed from human to human in the spore form.

An enormous number of moisture droplets are expelled during sneezing (Figure 26.1), and a considerable number are expelled during coughing or talking. Each infectious droplet has a size of about 10 μm and contains one or two bacteria. The speed of the droplet movement is about 100 m/sec (more than 200 mi/h) in a sneeze and ranges from 16–48 m/sec during coughing or shouting. The number of bacteria in a single sneeze varies from 10,000 to 100,000. Because of the small size of the droplets, the moisture evaporates quickly in the air, leaving behind a nucleus of organic matter and mucus to which bacterial cells are attached.

Respiratory Infections

The average human breathes nearly 500 million liters of air in a lifetime, much of it containing microorganism-laden dust, which is a potential source of inoculum for respiratory infections. The speed at which air moves through the respiratory tract varies, and in the lower respiratory tract the rate is quite slow. As the air slows down, particles in it stop moving and settle. The larger particles settle first and the smaller ones later, and only particles smaller than 3 μm travel as far as the bronchioles in the lower respiratory tract (Figure 26.2). Different organisms reach different levels, thus accounting for the differences in the kinds of infections that occur in the upper and lower respiratory tracts.

Bacterial Respiratory Pathogens

A variety of bacterial pathogens affects the respiratory tract. Most inhabit only humans, and thus their mode of transmission must be person to person. A few respiratory pathogens such as *Legionella pneumophila* are found in

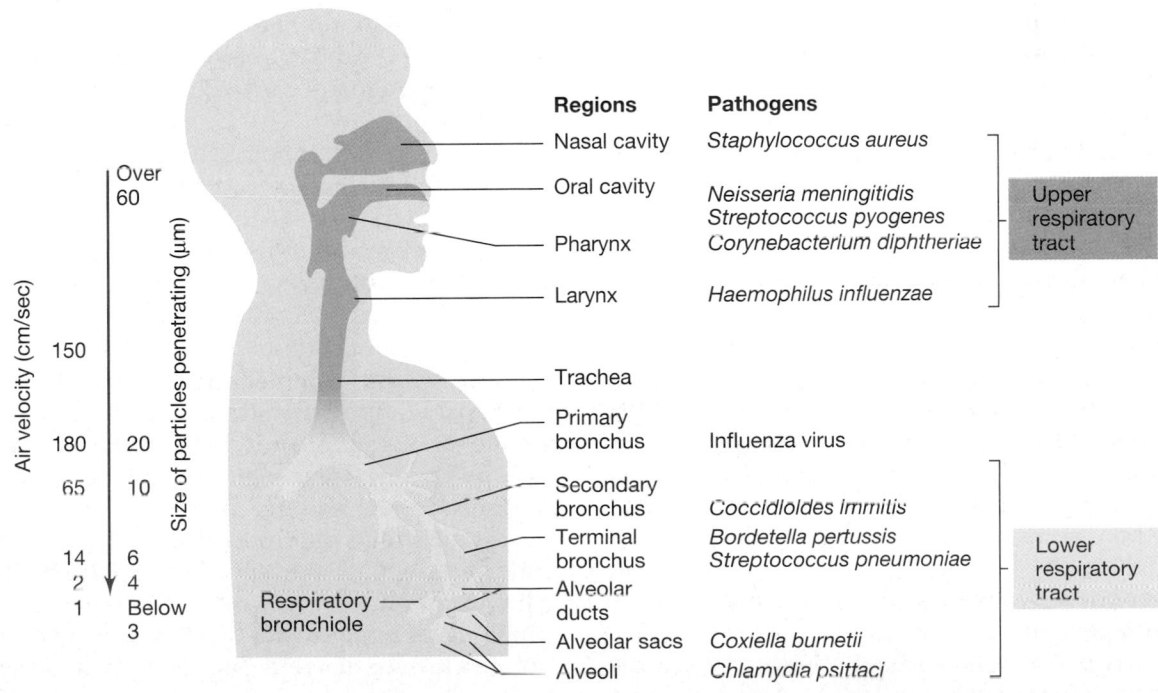

Figure 26.2 The respiratory system of humans and locations at which selected pathogenic microorganisms generally initiate infections.

water or soil; because they share common sources with other pathogens, they will be discussed later. As we explained here, many of the person-to-person respiratory pathogens are gram-positive *Bacteria* because the gram-positive *Bacteria* can survive in dry conditions outside the host for long periods of time by virtue of the thick gram-positive cell wall. Bacterial respiratory infections, while serious by themselves, often initiate secondary problems that can be life-threatening. Thus, it is critical to quickly and accurately diagnose and treat bacterial respiratory infections to limit host damage. Fortunately, most respiratory bacterial pathogens respond readily to antibiotic therapy and many can also be controlled by immunization. Nevertheless, bacterial respiratory infections are still rather common, and we begin here with a consideration of some common bacterial respiratory pathogens. We will then examine viral respiratory pathogens and the much less treatable and preventable diseases they cause.

✓ 26.1 Concept Check

Many respiratory pathogens are gram-positive *Bacteria*. Because gram-positive *Bacteria* are very resistant to drying, they are easily transmitted in air. Less hardy respiratory pathogens are transferred only from person to person via respiratory aerosols generated by coughing, sneezing, talking, or breathing.

✓ What physical features of gram-positive *Bacteria* allow them to survive for long periods in air?

✓ Why are certain pathogens more commonly found in the upper respiratory tract? Why are certain pathogens more commonly found in the lower respiratory tract?

26.2 Streptococcal Diseases

Streptococcus pyogenes and *Streptococcus pneumoniae* are potent human respiratory pathogens. *S. pyogenes* is transmitted by the respiratory route. *S. pneumoniae* is found in the respiratory flora of up to 40% of normal individuals, and endogenous strains can cause severe respiratory disease in weakened or otherwise compromised individuals.

Biology

Streptococci are nonsporulating, homofermentative, aerotolerant, anaerobic gram-positive cocci (◯◯ Section 12.19). *Streptococcus pyogenes* typically grow in elongated chains (◯◯ Figure 12.54). Pathogenic strains of *Streptococcus pneumoniae* typically grow in pairs or short chains and have an extensive polysaccharide capsule (Figure 26.3).

Streptococcus pyogenes: Epidemiology and Pathogenesis

Streptococcus pyogenes is frequently isolated from the upper respiratory tract of healthy adults. Although numbers of *S. pyogenes* are usually low here, if the host's de-

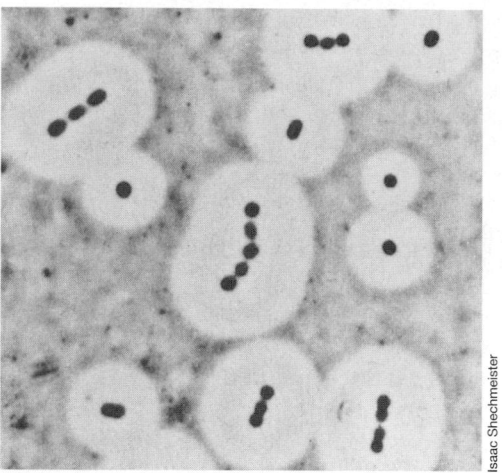

Figure 26.3 India ink negatively stained preparation of cells of *Streptococcus pneumoniae*. Note the extensive capsule surrounding the cells. The cells are about 0.5 μm in diameter.

fenses are weakened or a new, highly virulent strain is introduced, acute streptococcal bacterial infections are possible. *S. pyogenes* is the cause of streptococcal pharyngitis, so-called *strep throat* (the pharynx is the tube that connects the oral cavity to the larynx and the esophagus) (Figure 26.2). Most isolates from clinical cases of strep throat produce a toxin that lyses red blood cells, a condition called β-hemolysis (◯◯ Figure 21.17). Streptococcal pharyngitis is characterized by a severe sore throat, enlarged tonsils, tonsillar exudate, tender cervical lymph nodes, a mild fever, and a general feeling of malaise (tonsilitis). *S. pyogenes* can also cause related infections of the inner ear (otitis media), the mammary glands (mastitis), and infections of the superficial layers of the skin, a condition referred to as impetigo (many cases of impetigo are caused by *Staphylococcus aureus*) (Figure 26.4).

About half of the clinical cases of severe sore throat turn out to be due to *Streptococcus pyogenes*, the remainder being of viral origin. An accurate, prompt diagnosis is important because if the sore throat is due to a virus, treatment with antibacterial chemotherapeutic agents ("antibiotics") will be useless, whereas if the sore throat is due to *S. pyogenes*, immediate antibacterial therapy is indicated. Rapid, complete treatment of streptococcal sore throat is important because it can occasionally lead to more serious streptococcal syndromes such as scarlet fever, rheumatic fever, acute glomerulonephritis, and toxic shock syndrome (see later).

Certain strains of *Streptococcus pyogenes* carry a lysogenic bacteriophage that encodes production of erythrogenic toxin, an exotoxin responsible for most of the symptoms of **scarlet fever**. Erythrogenic toxin causes a pink-red rash to develop (Figure 26.5) and also acts to damage small blood vessels and initiate fever. The condition is acute and easily treated with antibacterial agents.

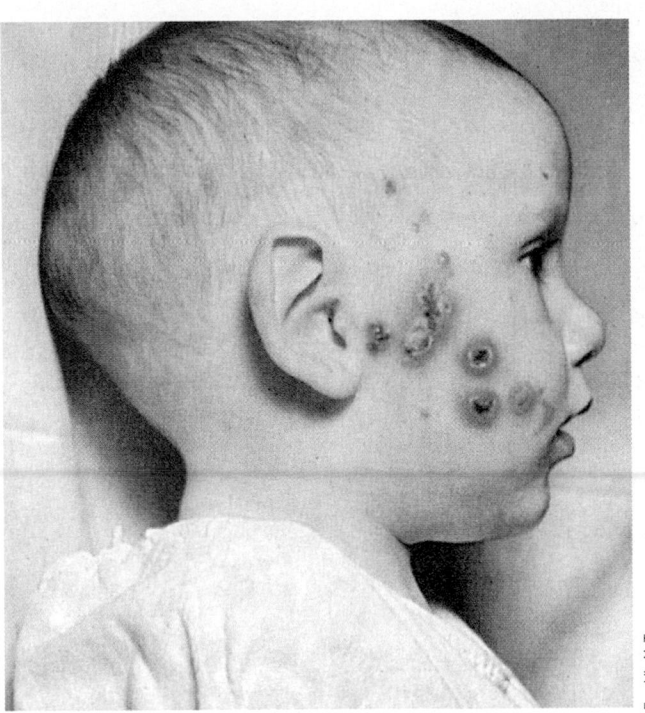

Figure 26.4 Typical lesions of impetigo, commonly caused by *Streptococcus pyogenes* or *Staphylococcus aureus*.

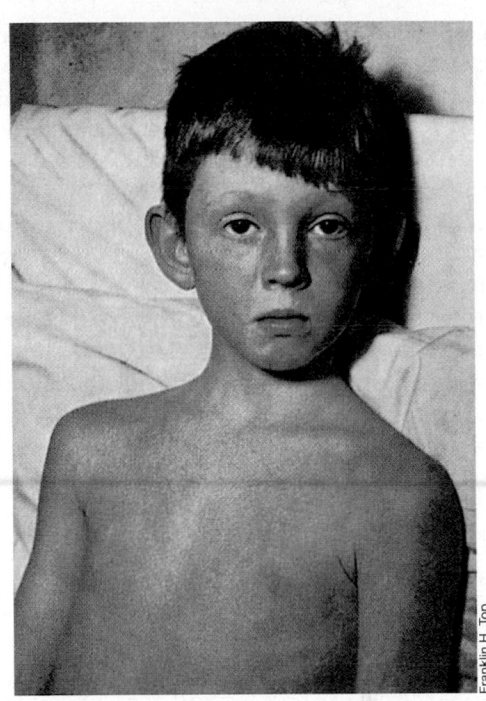

Figure 26.5 The typical rash of scarlet fever, resulting from the action of the erythrogenic toxin produced by *Streptococcus pyogenes*.

Occasionally, *Streptococcus pyogenes* may cause fulminant systemic infections, often marked by necrotizing fasciitis, a rapid and progressive infection resulting in extensive destruction of subcutaneous tissue. These infections are responsible for the dramatic, but fortunately rare, reports of "flesh-eating bacteria." In these cases, exotoxins A and B and the surface M-protein act as superantigens (⚬⚬ Section 22.14) that recruit massive numbers of T cells to the infected tissues. The T cells then secrete cytokines, which activate large numbers of effector cells, resulting in massive systemic inflammation, tissue destruction, and death in up to 30% of the cases.

Untreated or insufficiently treated cases of *Streptococcus pyogenes* infection may lead to severe *delayed sequelae*, or follow-up diseases. **Rheumatic fever**, one of these delayed sequelae, is caused by *rheumatogenic* strains of *S. pyogenes* containing cell-surface antigens that are similar to certain human cell-surface antigens. The immune response to the invading pathogen produces antibodies that cross-react with host tissues, in particular those of the heart, joints, and kidneys, resulting in tissue destruction. Rheumatic fever is a type of autoimmune disease, with antibodies reacting with self constituents (⚬⚬ Section 22.13). Damage may be permanent and is often compounded by later streptococcal infections and subsequent bouts of rheumatic fever.

Another potential delayed sequela of *Streptococcus pyogenes* infection is *acute glomerulonephritis*, a painful disease of the kidney. This is an immune complex disease (⚬⚬ Section 22.13) resulting from the formation of streptococcal antigen-antibody complexes in the bloodstream. The immune complexes lodge in the *glomeruli*, or filtration membranes of the kidney, causing inflammation of the kidney (*nephritis*) accompanied by severe kidney pain. Within several days, these complexes are usually dissolved and the patient quickly returns to normal. Unfortunately, even timely antibacterial treatment does not prevent glomerulonephritis. However, only a few strains of *S. pyogenes*—so-called *nephritogenic* strains—produce this painful disease.

Fortunately, reinfection by a particular *S. pyogenes* strain is rare, but there are over 60 different strains defined by the antigenically distinct, strain-specific, cell surface M-proteins. Thus, an individual can be infected over 60 times by different *S. pyogenes* strains. There are no available vaccines to prevent *S. pyogenes* infections.

Diagnosis: *Streptococcus pyogenes*

Because serious host damage following a streptococcal sore throat can occur, several rapid antigen detection (RAD) systems have been developed for identification of *S. pyogenes*. Surface antigens are first extracted by enzymatic or chemical means directly from a swab of the patient's throat. Immunological methods such as latex bead agglutination, enzyme-linked immunoassay (ELISA), or fluorescent antibody staining (⚬⚬ Sections 24.8, 24.11, and 24.10) using antibodies specific for surface proteins unique to *S. pyogenes* are employed.

Specimens are taken directly from a patient throat swab and are processed and analyzed in minutes. These rapid diagnostic procedures allow the physician to immediately initiate appropriate antibiotic therapy in order to avoid complications such as rheumatic fever. However, a more accurate confirmation of infection by pathogenic streptococci is a positive *S. pyogenes* culture from the throat. In general, the RAD tests are nearly as specific as throat cultures, but can be up to 40% less sensitive, leading to false negative reports (⟳ Section 24.5). Throat cultures may take up to two days to process, hence the popularity of the RAD tests. Finally, the most sensitive methods for identifying recent streptococcal infections are serology tests, where patients are examined for the presence or increase (rise in titer) of antibodies to various streptococcal antigens (⟳ Section 24.7). The presence of new antibodies or an increase in the quantity of an existing antibody confirms a very recent streptococcal infection.

Streptococcus pneumoniae

The other major pathogenic streptococcal species, *Streptococcus pneumoniae*, causes lung infections that often develop as secondary infections to other respiratory disorders. The capsule enables the cells to resist phagocytosis; capsulated strains of *S. pneumoniae* are very invasive. Cells invade alveolar tissues (lower respiratory tract) of the lung and elicit a strong host inflammatory response. Reduced lung function can result from accumulation of phagocytic cells and fluid, and the *S. pneumoniae* cells can spread from the focus of infection as a bacteremia, sometimes resulting in bone infections, inner ear infections, and endocarditis. Pneumococcal pneumonia is a serious infection and untreated cases have a mortality rate of about 30%. Even with aggressive antimicrobial treatment, individuals hospitalized with pneumococcal pneumonia have 5–10% mortality.

Laboratory diagnosis of *S. pneumoniae* involves the culture of the diagnostic gram-positive diplococci from either patient sputum or blood. There are 90 different serotypes or antigenic capsule variants, and infection leads to immunity to only the infecting strain of *S. pneumoniae*.

Prevention and Treatment

There are no effective available vaccines for prevention of *Streptococcus pyogenes* infections. However, an effective multivalent vaccine is available for prevention of infection by at least two-thirds of the 90 known strains of *Streptococcus pneumoniae*, including all common pathogenic strains. The vaccine consists of a mixture of the capsular polysaccharides from the most prevalent pathogenic strains. The vaccine is recommended for the elderly, health care providers, individuals with compromised immunity, and others at high risk for respiratory infections (⟳ Section 22.11).

Penicillin and its semisynthetc derivatives (⟳ Section 20.8) are the agents of choice for treating *S. pyogenes* infections. Erythromycin and other antibacterial drugs are used in individuals who have acquired penicillin allergies (⟳ Section 22.13).

Most strains of *S. pneumoniae* respond to penicillin therapy. However, there are penicillin-resistant strains, especially among strains causing hospital-acquired infections (⟳ Section 25.7), and individual isolates must be checked for penicillin sensitivity (⟳ Section 24.3). Erythromycin is the drug of choice for penicillin-resistant organisms, but cephalosporin, fluoroquinolone, ceftriaxone, cefotaxime, or vancomycin (⟳ Sections 20.6–20.9) may also be used. However, strains with resistance to each of these drugs, and strains with multiple drug resistance, have been found, underscoring the need to test each isolate individually.

✓ 26.2 Concept Check

Two respiratory diseases caused by streptococci are streptococcal sore throat and pneumococcal pneumonia. Under certain conditions, simple *Streptococcus pyogenes* infections can develop into more serious conditions such as scarlet fever and rheumatic fever. Pneumonia caused by *Streptococcus pneumoniae* is always a serious disease.

✓ How does *Streptococcus pyogenes* infection cause rheumatic fever?

✓ What is the primary virulence factor for *Streptococcus pneumoniae*?

26.3 Corynebacterium and Diphtheria

Corynebacterium diphtheriae is the organism that causes diphtheria, a severe respiratory disease that usually infects children. Diphtheria is preventable and treatable. *C. diphtheriae* is a gram-positive, nonmotile, aerobic bacterium that forms irregular rod-shaped or club-shaped cells during growth (⟳ Section 12.22).

Epidemiology and Pathology

Corynebacterium diphtheriae enters the body via the respiratory route with cells lodging in the throat and tonsils. Infection is usually spread from healthy carriers or infected individuals to susceptible individuals by airborne droplets. Previous infection or immunization (see below) provides complete resistance to infection. Although limited information is available concerning the mechanism of adherence of *C. diphtheriae* to these tissues, the organism produces a neuraminidase capable of splitting *N*-acetylneuraminic acid (a component of glycoproteins found on animal cell surfaces), and this may enhance the invasion process. The inflammatory response of throat tissues to *C. diphtheriae* infection results in formation of a characteristic lesion called a *pseudomembrane* (Figure 26.6), which consists of damaged host cells and cells of *C. diphtheriae*. As described in Section 21.9, certain strains of *C. diphtheriae* are lysogenized by bacteriophage β, and these strains produce a powerful exotoxin, the *diphtheria toxin*. Diphtheria toxin inhibits eukaryotic protein synthesis and thus kills cells (⌾ Section 21.10).

The pseudomembrane that forms in diphtheria may block the passage of air, and death from diphtheria is usually due to a combination of the effects of partial suffocation and tissue destruction by exotoxin. Although diphtheria was once a major childhood disease, it is now rarely encountered because an effective vaccine is available. Worldwide, there are still more than 50,000 cases of diphtheria per year, largely because of a lack of immu-

nization. Recent diphtheria outbreaks in Southeast Asia and in Eastern Europe have been attributed to a lack of vaccination programs or a breakdown in existing vaccination programs, respectively.

Diagnosis, Prevention, and Treatment

The death of tissue due to absorption of the toxin causes the appearance of the pseudomembrane in the patient's throat. *Corynebacterium diphtheriae* isolated from the throat is diagnostic for diphtheria. The toxin, if left untreated, can cause systemic damage to heart, kidneys, liver, and adrenal glands.

A patient diagnosed with diphtheria is treated simultaneously with antibiotics and diphtheria antitoxin (an antitoxin contains neutralizing antibodies formed in another animal) (⌾ Sections 24.7 and 22.11 discuss antitoxins). Penicillin, erythromycin, or gentamicin are generally effective for diphtheria therapy. Early administration of both antibiotics and antitoxin is necessary for effective treatment of the disease.

Prevention of diphtheria is accomplished through the use of a highly effective vaccine preparation. The vaccine is made by treating the diphtheria exotoxin with formalin to yield an immunogenic, yet nontoxic, toxoid. Diphtheria toxoid is part of the *DTP* (diphtheria, *tetanus*, pertussis) vaccine (⌾ Section 22.11).

✓ 26.3 Concept Check

Diphtheria is an acute respiratory disease caused by the gram-positive bacterium *Corynebacterium diphtheriae*. A standard early childhood vaccine (DTP) is very effective for preventing this very serious respiratory disease.

✓ Is the pathogenesis of diphtheria due to infection?

✓ How can the spread of diphtheria be prevented?

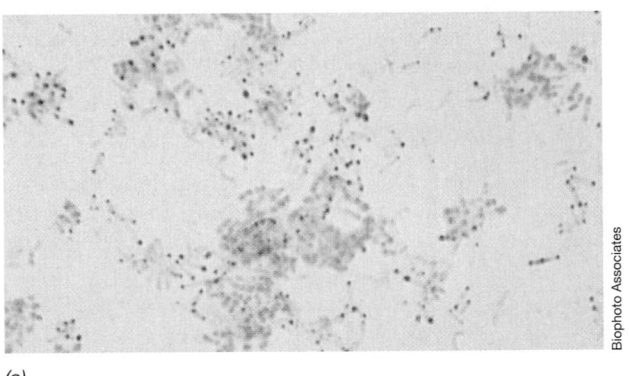

(a)

Biophoto Associates

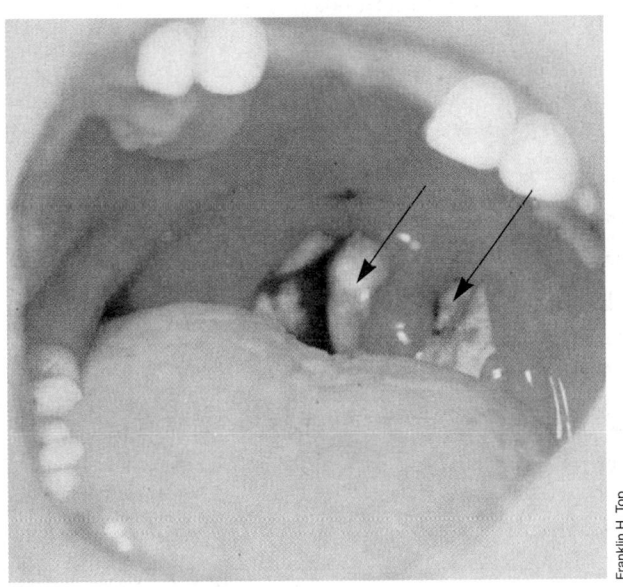

(b)

Franklin H. Top

Figure 26.6 Diphtheria. (a) Cells of *Corynebacterium diphtheriae* stained to show metachromatic (polyphosphate) granules. (b) Pseudomembrane (arrows) in an active case of diphtheria caused by the bacterium *C. diphtheriae*.

26.4 *Bordetella* and Whooping Cough

Whooping cough is a potentially serious childhood respiratory disease caused by infection with *Bordetella pertussis*. *B. pertussis* is a small, gram-negative, aerobic coccobacillus.

Epidemiology and Pathology

Whooping cough (pertussis) is an acute, highly infectious respiratory disease often observed in children under 5 years of age. *Bordetella pertussis* attaches to cells of the upper respiratory tract by producing a specific adherence factor called *filamentous hemagglutinin antigen*, which recognizes a complementary molecule on the surface of host cells. Once attached, *B. pertussis* grows and produces pertussis exotoxin that induces synthesis of cyclic adenosine monophosphate (cyclic AMP) (⌾ Section 8.7), which is at least partially responsible for the events that lead to host tissue damage. *Bordetella pertussis* also produces an endotoxin, which also may induce some of the symptoms of whooping cough. Clinically, whooping cough is characterized by a recurrent, violent cough that can last up to 6 weeks. The spasmodic coughing gives the disease its name, for a whooping sound results from the patient inhaling in deep breaths to obtain sufficient air.

Diagnosis

Diagnosis of whooping cough can be made by fluorescent antibody staining of throat smears or by culturing the organism. For best recovery of *Bordetella pertussis*, a nasopharyngeal aspirate is inoculated directly onto a blood–glycerol–potato extract agar plate (although not selective, this medium supports good recovery of *B. pertussis*). β-Hemolytic colonies containing small gram-negative coccobacilli are tested for *B. pertussis* by a latex bead agglutination test or are stained with an anti-*B. pertussis* fluorescent antibody for positive identification (∞ Sections 24.8 and 24.10

Prevention and Treatment

A vaccine consisting of killed whole cells or proteins derived from *Bordetella pertussis* is part of the routinely administered DTP vaccine. This vaccine, while normally very effective, must be given to susceptible individuals, usually children, at appropriate intervals beginning soon after birth (∞ Section 22.11). In the United States, up to 50% of children who acquire whooping cough have not been properly immunized and current immunization preparations are only 60–90% effective. The threat of this very communicable disease remains high, as illustrated by epidemic outbreaks in several cities in the United States and in the former Soviet Union, presumably as a result of breakdowns in public immunization programs. In recent years, there has been an alarming upward trend in the number of cases of pertussis in the United States (Figure 26.7) and up to 60%, or over 4000 cases per year, are in individuals over 5 years of age, including many adolescents and adults.

Because of undesirable side effects of pertussis vaccine, including local swelling and redness, fever, and occasional more serious problems such as encephalitis and convulsions, a "second generation" pertussis vaccine, containing purified cell fractions of *Bordetella pertussis* rather than whole cells, is now approved for use in the United States (∞ Section 22.11).

Cultures of *B. pertussis* are killed by ampicillin, tetracycline, and erythromycin, although antibiotics alone do not seem to be sufficient to kill the pathogen *in vivo*. Because a patient with whooping cough remains infectious for up to 2 weeks following commencement of antibiotic therapy, the immune response may be as important, if not more so, than antibiotics, in the elimination of *B. pertussis* from the body.

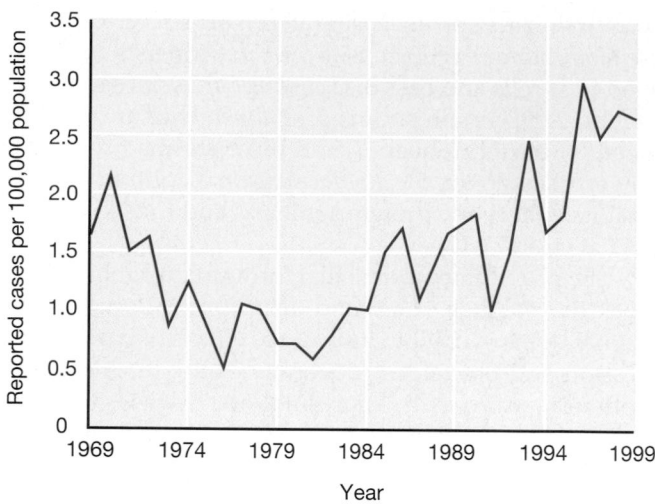

Figure 26.7 Pertussis (whooping cough) prevalence in the United States. Starting in the 1980s, there has been a consistent upward trend in the number of pertussis cases. In 1999, there were 7288 cases of pertussis in the United States. In 1976, the year of lowest prevalence and incidence, there were only 1010 cases of pertussis.

✓ 26.4 Concept Check

In the United States, there has been a disturbing increase in the number of annual cases of whooping cough in the last decade. From 1975 to 1982, there was an average of less than 2000 cases per year, but that number has now risen to more than 7000 cases per year. Up to 60% of preschool children in some cities are inadequately immunized, creating a major potential public health threat.

✓ What measures can be taken to decrease the current incidence of whooping cough in a population?

✓ Indicate potential problems with the use of whole-cell pertussis vaccines. Indicate the steps taken to alleviate those problems in new vaccine preparations.

26.5 *Mycobacterium* and Tuberculosis

Tuberculosis is caused by the gram-positive, acid-fast bacillus *Mycobacterium tuberculosis* (∞ Section 12.23). The famous German microbiologist Robert Koch isolated and described the causative agent of tuberculosis, *M. tuberculosis*, in 1882 (∞ Section 1.5 and see the box, Discoverers of the Main Bacterial Pathogens).

Epidemiology

The *Mycobacterium tuberculosis* tubercle bacilli are easily transmitted by the respiratory route, and even the simple act of talking can spread the organism from person to person. At one time, tuberculosis was the single most important infectious disease of humans and accounted for one-seventh of all deaths worldwide. At present in the United States, nearly 20,000 new cases of tuberculo-

Learning From the Past ... Discoverers of the Main Bacterial Pathogens

The history of the discovery of the microbial role in infectious disease was described in Chapter 1 (Section 1.5). Once the concept of specific microbial disease agents was clarified and the procedures for culture of microorganisms developed, it was a relatively simple procedure to isolate a large number of microbial pathogens. The decades surrounding the formulation of Koch's postulates (1884) were indeed fruitful for medical microbiology. The rapid development of this field is indicated by the accompanying table, which lists the main bacterial pathogens isolated during the "golden age of bacteriology." ■

Year	Disease	Organism	Discoverer
1873	Leprosy	*Mycobacterium leprae*	Hansen, G. A.
1877	Anthrax	*Bacillus anthracis*	Koch, R.
1878	Suppuration	*Staphylococcus*	Koch, R.
1879	Gonorrhea	*Neisseria gonorrhoeae*	Neisser, A. L. S.
1880	Typhoid fever	*Salmonella typhi*	Eberth, C. J.
1881	Suppuration	*Streptococcus*	Ogston, A.
1882	Tuberculosis	*Mycobacterium tuberculosis*	Koch, R.
1883	Cholera	*Vibrio cholerae*	Koch, R.
1883	Diphtheria	*Corynebacterium diphtheriae*	Klebs, T. A. E.
1884	Tetanus	*Clostridium tetani*	Nicolaier, A.
1885	Diarrhea	*Escherichia coli*	Escherich, T.
1886	Pneumonia	*Streptococcus pneumoniae*	Fraenkel, A.
1887	Meningitis	*Neisseria meningitidis*	Weichselbaum, A.
1888	Food poisoning	*Salmonella enteritidis*	Gaertner, A. A. H.
1892	Gas gangrene	*Clostridium perfringens*	Welch, W. H.
1894	Plague	*Yersinia pestis*	Kitasato, S., Yersin, A. J. E. (independently)
1896	Botulism	*Clostridium botulinum*	van Ermengem, E. M. P.
1898	Dysentery	*Shigella dysenteriae*	Shiga, K.
1900	Paratyphoid	*Salmonella paratyphi*	Schottmüller, H.
1903	Syphilis	*Treponema pallidum*	Schaudinn, F. R., and Hoffman, E.
1906	Whooping cough	*Bordetella pertussis*	Bordet, J., and Gengou, O.

sis are diagnosed each year. Worldwide, tuberculosis still accounts for almost *1.5 million deaths per year* and more than 10% of all deaths due to infectious disease. Up to one-third of the world's population have been infected with *M. tuberculosis* (Table 25.1). In recent years, many of the new tuberculosis cases in the United States result at least in part from the elevated incidence of tuberculosis in acquired immunodeficiency syndrome (AIDS) patients. Over 1000 people die every year from tuberculosis in the United States alone.

Pathology

The interaction of the human host and *Mycobacterium tuberculosis* is extremely complex, being determined in part by the virulence of the strain but also by the specific and nonspecific resistance of the host. Cell-mediated immunity plays an important role in the development of disease symptoms. It is convenient to distinguish between two kinds of human tuberculosis infections: *primary* and *postprimary* (or *reinfection*). Primary infection is the first infection that an individual acquires and usually results from inhalation of droplets containing viable bacteria from an individual with an active pulmonary infection. Dust particles that have become contaminated from sputum of tubercular individuals are another source of primary infection. The bacteria settle in the lungs and grow. A delayed-type hypersensitivity reaction (Section 22.13) results in the formation of aggregates of activated macrophages, called *tubercles*, characteristic of tuberculosis (Figure 1.13). However, the bacteria are often able to survive and grow to some extent within the macrophages. In individuals with low resistance, the bacteria are not effectively controlled, and an acute pulmonary infection occurs, which can lead to the extensive destruction of lung tissue, the spread of the bacteria to other parts of the body, and death.

In most cases of tuberculosis, however, acute infection does not occur, and the infection remains localized and is usually inapparent; later it subsides. But this initial infection hypersensitizes the individual to the bacteria or their products and consequently alters the response of the individual to subsequent *M. tuberculosis* exposures. A diagnostic test, called the **tuberculin test**, can be used to measure this hypersensitivity. When *tuberculin*, a protein fraction extracted from *Mycobacterium tuberculosis*, is injected intradermally into a hypersensitive individual, it elicits a localized immune reaction within 1–3 days at the site of injection. The reaction is characterized by *induration* (hardening) and *edema* (swelling) (∞ Figure 22.24). An individual exhibiting this reaction is said to be *tuberculin-positive*, and many healthy adults give positive reactions as a result of previous inapparent infections. A positive tuberculin test does not indicate active disease but only that the individual has been exposed to the organism in the past and has generated a cell-mediated immune response.

For most individuals, this immunity is protective and life-long. However, some tuberculin-positive patients develop postprimary tuberculosis through reinfection from outside sources or as a result of reactivation of bacteria that have remained alive but dormant in lung macrophages, often for years. Factors such as aging, malnutrition, overcrowding, stress, and hormonal imbalance all play a role in predisposing individuals to reinfection by reducing effective immunity (∞ Section 21.13) and allowing reactivation of dormant infections.

Secondary pulmonary infections often progress to chronic infections that result in destruction of lung tissue, followed by partial healing and calcification at the infection site. Thus, chronic postprimary tuberculosis often results in a gradual spread of tubercular lesions in the lungs. Areas of destroyed tissue are seen by X-ray examination (Figure 26.8), but bacteria are found in the sputum only in individuals with extensive tissue destruction.

Prevention and Treatment

Individuals who have active cases of tuberculosis may spread the disease simply by coughing on or speaking to uninfected individuals. Because tuberculosis is so highly contagious, the United States Occupational Safety and Health Administration has stringent requirements for the protection of health care workers who are responsible for tuberculosis patient care. For example,

patients with infectious tuberculosis must be hospitalized in negative-pressure rooms. In addition, health care workers who have patient contact must be provided with personally fitted face masks with high-efficiency particulate air (HEPA) filters. These special filters prevent the passage of *Mycobacterium tuberculosis* in sputum or on dust particles.

Chemotherapy of tuberculosis has been a major factor in control of the disease. The initial success in chemotherapy occurred with the introduction of streptomycin, but the real revolution in tuberculosis treatment came with the discovery of isonicotinic acid hydrazide (isoniazid or INH) (Figure 26.9), a nicotinamide derivative virtually specific for mycobacteria. This agent is effective, nontoxic, inexpensive, and is readily absorbed when given orally. Although the mode of action of isoniazid is not completely understood, it apparently affects the synthesis of mycolic acid by *Mycobacterium* (mycolic acid is a complex lipid that complexes with the peptidoglycan of the mycobacterial cell wall) (∞ Section 12.23). Isoniazid may act by mimicking the activity of a structurally related molecule, nicotinamide (Figure 26.9), becoming incorporated in place of nicotinamide and thus inactivating enzymes requiring this compound for activity. Treatment of mycobacteria with very small amounts of isoniazid (as little as 5 picomoles [pmol] per 10^9 cells) results in complete inhibition of mycolic acid synthesis, and continued incubation results in a complete loss of outer membrane areas of the cell, a loss of cellular integrity, and death. Following treatment with isoniazid, mycobacteria lose their acid-alcohol fastness, in keeping with the role of mycolic acid in this staining property (∞ Section 12.23). However, mycobacterial resistance to isoniazid and other drugs is increasing at an alarming rate, especially in AIDS patients (see Section 26.14).

Treatment typically involves daily doses of isoniazid and rifampin for 2 months, followed by biweekly doses for a total time of 9 months, to eradicate the tubercle bacilli and prevent emergence of antibiotic-resistant organisms. Failure to complete the entire prescribed treatment plan may allow the infection to be reactivated, and the reactivated organisms have often acquired resistance to the original treatment drugs. Inadequate treatment encourages antibiotic resistance because a high number of mutations

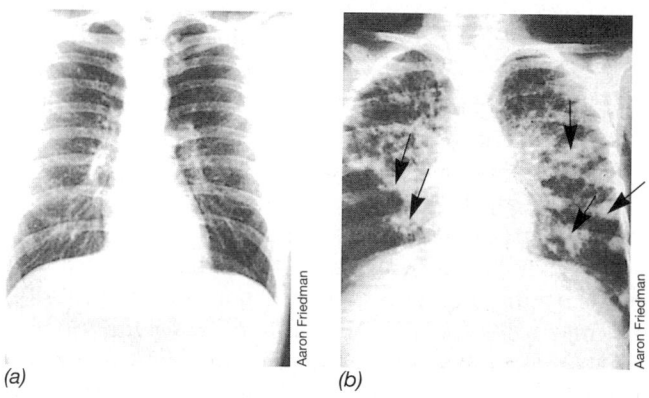

(a) *(b)*

Aaron Friedman

Figure 26.8 X-ray photographs. (a) Normal chest X-ray. The faint white lines are arteries and other blood vessels. The heart is visible as a white bulge in the lower right quadrant. (b) An advanced case of pulmonary tuberculosis; white patches (arrows) indicate areas of disease. These patches, or tubercles, may contain live *Mycobacterium tuberculosis*. Lung tissue and function is permanently destroyed by these lesions.

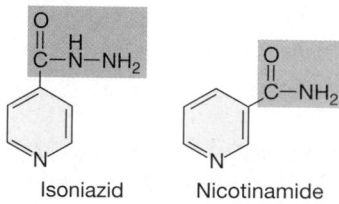

Isoniazid   Nicotinamide

Figure 26.9 Structure of isoniazid (isonicotinic acid hydrazide), an effective chemotherapeutic agent for tuberculosis. Note the structural similarity to nicotinamide.

spontaneously occur in *M. tuberculosis*, rapidly conferring resistance to single antibiotics. Adequate multiple drug therapy reduces the possibility that strains will emerge having resistance to both drugs. In populations such as hospitals and nursing homes, patients are routinely treated with up to four drugs for at least 6 months to discourage the emergence of drug-resistant tuberculosis.

Mycobacterium leprae and Hansen's Disease

Mycobacterium leprae is the causative agent of the ancient and dreaded *Hansen's disease*, or *leprosy*. *M. leprae* is the only *Mycobacterium* species that has not been grown on artificial media. The only experimental animal that has been successfully used to grow *M. leprae* and reproduce a similar disease is the armadillo. The most serious form of Hansen's disease is characterized by folded, bulblike lesions on the body, especially on the face and extremities (Figure 26.10) due to growth of *M. leprae* cells in the skin. The lesions contain up to 10^9 bacterial cells per gram of tissue. Like *M. tuberculosis*, *M. leprae* from the lesions stain deep red with carbol fuschin in the acid-fast staining procedure, providing a rapid, definitive demonstration of active infection (⊂⊃ Section 12.23). This *multibacillary* or *lepromatous* form of leprosy has a very poor prognosis. In severe cases the disfiguring lesions lead to destruction of peripheral nerves and loss of motor function. Many patients exhibit less pronounced lesions from which no bacterial cells can be recovered. These individuals have the *tubercular* or *paucibacillary* form of the disease. Tubercular leprosy is characterized by a vigorous delayed-type hypersensitivity response (⊂⊃ Section 22.13) and a good prognosis for spontaneous recovery. Hansen's disease of either form, and the continuum of intermediate forms between these extremes, is treated using a multiple drug therapy (MDT) protocol, which includes some combination of dapsone (4, 4'-sulfonylbisbenzeneamine), rifampin, and clofazimine. As in the case of tuberculosis,

drug-resistant organisms have appeared, especially after treatment with single drugs or inadequate treatment. Extended drug therapy of up to 1 year with a MDT protocol is required for eradication of the organism.

The pathogenicity of *M. leprae* is probably due to a combination of delayed hypersensitivity (⊂⊃ Section 22.13) and the invasiveness of the organism. Transmission probably involves both the direct contact and respiratory routes, and incubation time varies from several weeks to years or even decades. *M. leprae* grows within macrophages, causing an intracellular infection that can result in the enormous population of bacteria within the skin. In many areas of the world, the incidence of Hansen's disease is very low. For example, in the United States, fewer than 300 cases are diagnosed every year. Worldwide, areas of high disease incidence are concentrated in central and South America, Africa, and Southeast Asia. Leprosy has currently been diagnosed in 1.2 million people, with about 500,000 new cases occurring every year. However, leprosy may go unreported in as many as 12 million people.

Other Pathogenic *Mycobacterium* spp.

A common pathogen of dairy cattle, *Mycobacterium bovis*, is pathogenic for humans as well as other animals. *M. bovis* enters humans via the intestinal tract, typically from the ingestion of raw milk. After a localized intestinal infection, the organism eventually spreads to the respiratory tract and initiates the classic symptoms of tuberculosis. We do not know whether *M. bovis* is really a different organism from *M. tuberculosis* because the two organisms are nearly 100% homologous at the DNA level when examined and compared by DNA hybridization methods (⊂⊃ Section 11.9). Pasteurization of milk and elimination of diseased cattle have essentially eradicated bovine-to-human transmission of tuberculosis.

A number of other *Mycobacterium* species are also occasional human pathogens. For example, *M. kansasii*, *M. scrofulaceum*, *M. chelonae*, and other members of the genus (⊂⊃ Section 12.23) cause disease. Tuberculosis due to the *M. avium* complex (MAC) group of organisms is particularly prevalent in AIDS patients as compared with members of the normal population (see Section 26.14).

✓ 26.5 Concept Check

Tuberculosis is one of the most prevalent and dangerous single diseases in the world. Its incidence is on the increase in developed countries, in part because of the emergence of drug-resistant strains. The pathology of tuberculosis and leprosy is influenced by the cellular immune response.

✓ Why is *Mycobacterium tuberculosis* such a widespread respiratory pathogen?

✓ Describe factors that contribute to the incidence of drug resistance in mycobacterial infections.

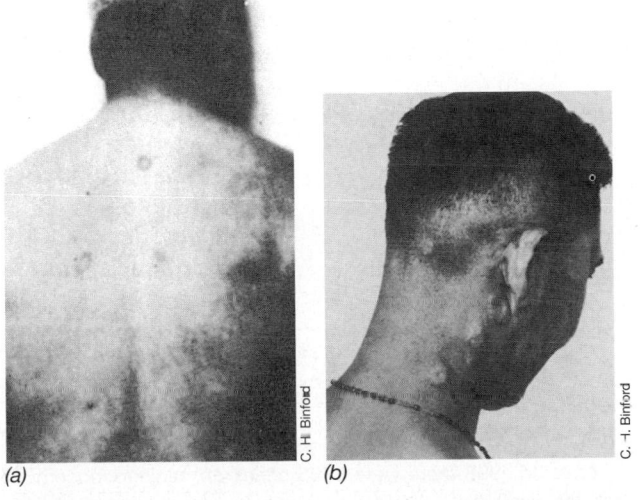

(a)   *(b)*

Figure 26.10 Leprosy lesions on the skin due to infection with *Mycobacterium leprae*.

26.6 Neisseria meningitidis, Meningitis, and Meningococcemia

Meningitis is an inflammation of the *meninges*, or the membranes that line the central nervous system, especially the spinal chord and brain. Meningitis can be caused by viral or bacterial infections, or may be *aseptic*, resulting from no infectious processes. Here we will deal with infectious bacterial meningitis caused by *Neisseria meningitidis* and a related infection, **meningococcemia**.

Biology

Neisseria meningitidis, often called meningococcus, is a gram-negative, nonsporulating, obligately aerobic, oxidase-positive, encapsulated diplococcus (Figure 26.11). At least 13 different pathogenic strains of *N. meningitidis* are recognized, based on antigenic differences in the capsular polysaccharides.

Epidemiology and Pathology

Menigococcal meningitis often occurs in epidemics, usually in closed populations such as military installations and college campuses. It normally affects older school-age children and young adults. Up to 30% of individuals normally carry *Neisseria meningitidis* in the nasopharynx with no apparent harmful effects. In epidemic situations, the prevalence of carriers may rise to 80%. The trigger for conversion from the asymptomatic carrier state to pathogenic acute infection is not known. In an acute meningococcus infection, the bacterium is transmitted to the host, usually via the airborne route, attaches to the cells of the nasopharynx, and then gains access to the bloodstream, causing bacteremia and simple upper respiratory tract symptoms. The bacteremia sometimes leads to fulminant menigococcemia, characterized by septicemia, intravascular coagulation, shock, and death in up to 17% of cases. Meningitis is another possible serious outcome of infection. Meningitis is characterized by sudden onset of headache, vomiting, and stiff neck, and can progress to coma and death in a matter of hours. Death occurs in up to 3% of acute meningococcal meningitis victims.

Figure 26.12 shows the prevalence of serious meningococcal infections in the United States over the last 30 years. There are over 2500 cases of serious meningococcal disease each year, with an overall mortality rate of about 10%.

Diagnosis

Specimens isolated from nasopharyngeal swabs, blood, or cerebrospinal fluid are inoculated onto Modified Thayer-Martin Medium (MTM) (∞ Section 24.1), a selective medium that suppresses the growth of most normal flora, but allows the growth of *Neisseria meningitidis* and *Neisseria gonnorheae*. Colonies showing a gram-negative diplococcus coupled with a positive oxidase test (∞ Table 24.3) are presumptively identified as *Neisseria*. However, due to the rapid onset of life-threatening symptoms, preliminary diagnosis is often based on clinical symptoms and treatment is started before culture tests confirm infection with *N. meningitidis*.

Prevention and Treatment

Penicillin G is the drug of choice for treatment of *Neisseria meningitidis* infections. However, resistant strains have been reported. Chlorampenicol is the accepted alternative agent for treatment of infections in penicillin-sensitive individuals. A number of broad-spectrum cephalosporins are also effective.

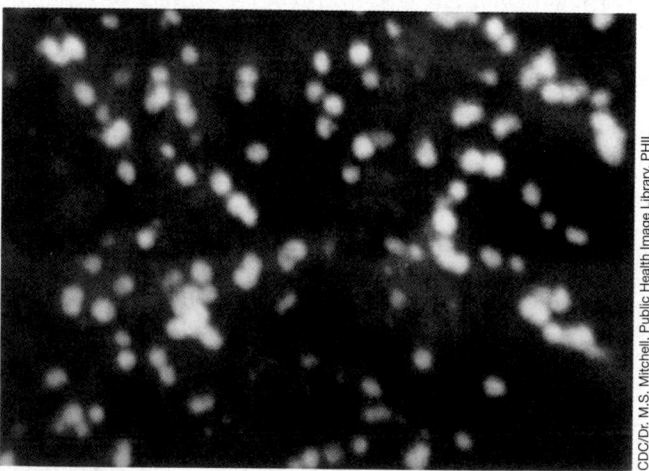

Figure 26.11 Fluorescent antibody stain of *Neisseria meningitidis*, the organism that causes adult meningitis and meningococcemia, from cerebrospinal fluid of an infected patient. The individual cocci are about 0.6–1.0 μm in diameter.

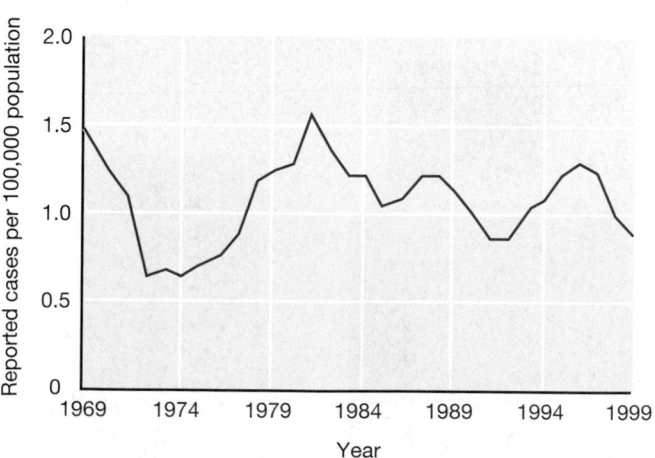

Figure 26.12 Prevalence of meningococcal infections in the United States. In 1998 there were 2725 cases of meningococcemia or meningitis, resulting in 234 deaths. Data are from the Centers for Disease Control and Prevention, Atlanta, GA, USA.

Naturally occurring strain-specific antibodies acquired by subclinical infections are effective for preventing infections in most adults. Unfortunately, there are no long-term vaccines available for the prevention of meningitis or meningococcemia. However, vaccines consisting of purified polysaccharides from some of the most prevalent pathogenic strains are available and are used to immunize close contacts when epidemics occur. The vaccine is also used to prevent infection in certain susceptible populations such as the military and is sometimes recommended for students living in dormitories. In addition, rifampin is often used as a chemoprophylactic antibiotic to prevent disease in close contacts and family members of infected individuals.

Other Causes of Meninigitis

A number of other organisms can also cause meningitis. Acute meningitis is usually caused by one of the pyogenic bacteria such as *Staphylococcus, Streptococcus,* or *Haemophilus influenzae. H. influenzae* primarily infects young children. An effective vaccine for preventing *H. influenzae* meningitis is available and is required in the United States for school-age children (∞ Section 22.11).

Several viruses also cause meningitis. Among these are herpes simplex virus (HSV), lymphocytic choriomeningitis virus (LCM), mumps virus, and the enteroviruses. In general, viral meningitis is less severe than bacterial menigitis.

✓ 26.6 Concept Check

Neisseria meningitidis is a common cause of meningococcemia and meningitis in young adults and occasionally occurs in epidemics in closed populations. Bacterial meningitis and meningococcemia are serious diseases with very high mortality rates. Treatment and prevention strategies are in place to deal with epidemic outbreaks, but an effective multivalent vaccine is not yet available.

✓ Describe the infection by *Neisseria meningitidis* and the resulting development of meningococcemia.

✓ Can meningococcemia be prevented? Explain.

26.7 Viruses and Respiratory Infections

As we discussed (∞ Section 20.10), viruses are less easily controlled by chemotherapeutic means than bacteria or other microorganisms because the growth of viruses is intimately tied to host cell functions. Most chemotherapeutic agents that specifically attack viruses cause at least some harm to host cells as well. Not surprisingly, therefore, the most prevalent infectious diseases, especially in developed countries, are of viral etiology. Most viral diseases are acute, self-limiting infections, but some

can be problematic in normal healthy adults. In addition, serious viral diseases such as smallpox and rabies have been effectively controlled by immunization. We begin here by describing measles, mumps, rubella, and chickenpox; these viral diseases are all transmitted in infectious droplets by an airborne route and all are controlled by immunization procedures.

Measles

Measles (rubeola) virus causes an acute highly infectious childhood disease characterized by nasal discharges, redness of the eyes, cough, and fever. The measles virus is a paramyxovirus (∞ Section 16.8) that enters the nose and throat by airborne transmission, quickly leading to systemic viremia. As the disease progresses, fever and cough appear and rapidly intensify, and a rash appears (Figure 26.13); in most cases measles lasts a total of 7–10 days. Circulating antibodies to measles virus are measurable about 5 days after initiation of infection, and both serum antibodies and cytotoxic T lymphocytes (∞ Sections 22.8 and 22.6) combine to eliminate the virus from the system. A variety of complications may occur due to measles infection, including inner ear infection, pneumonia, and, in rare cases, measles encephalomyelitis. Encephalomyelitis can cause neurological disorders and a form of epilepsy, and has a mortality rate of nearly 20%.

Although once a common childhood illness, measles generally occurs nowadays in rather isolated outbreaks because of widespread immunization programs begun in the mid-1960s (Figure 26.14*a*). Because of the highly infectious nature of the disease, all public school systems in the United States require proof of immunization before children can enroll. Active immunization is done with the

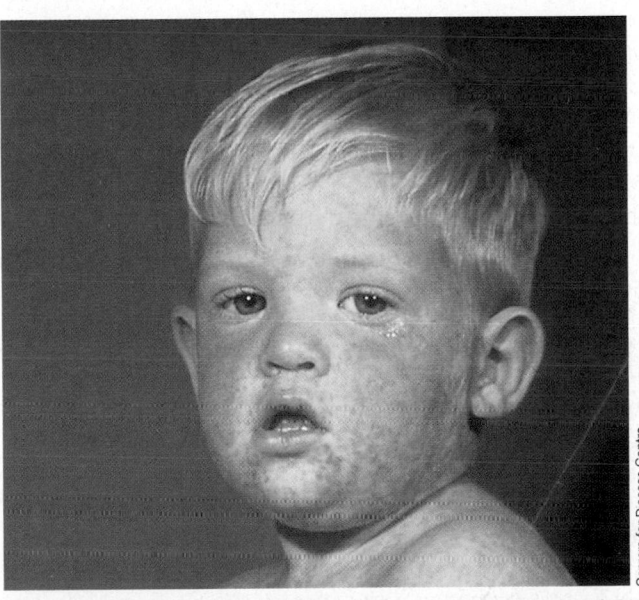

Figure 26.13 Typical rash associated with measles in children.

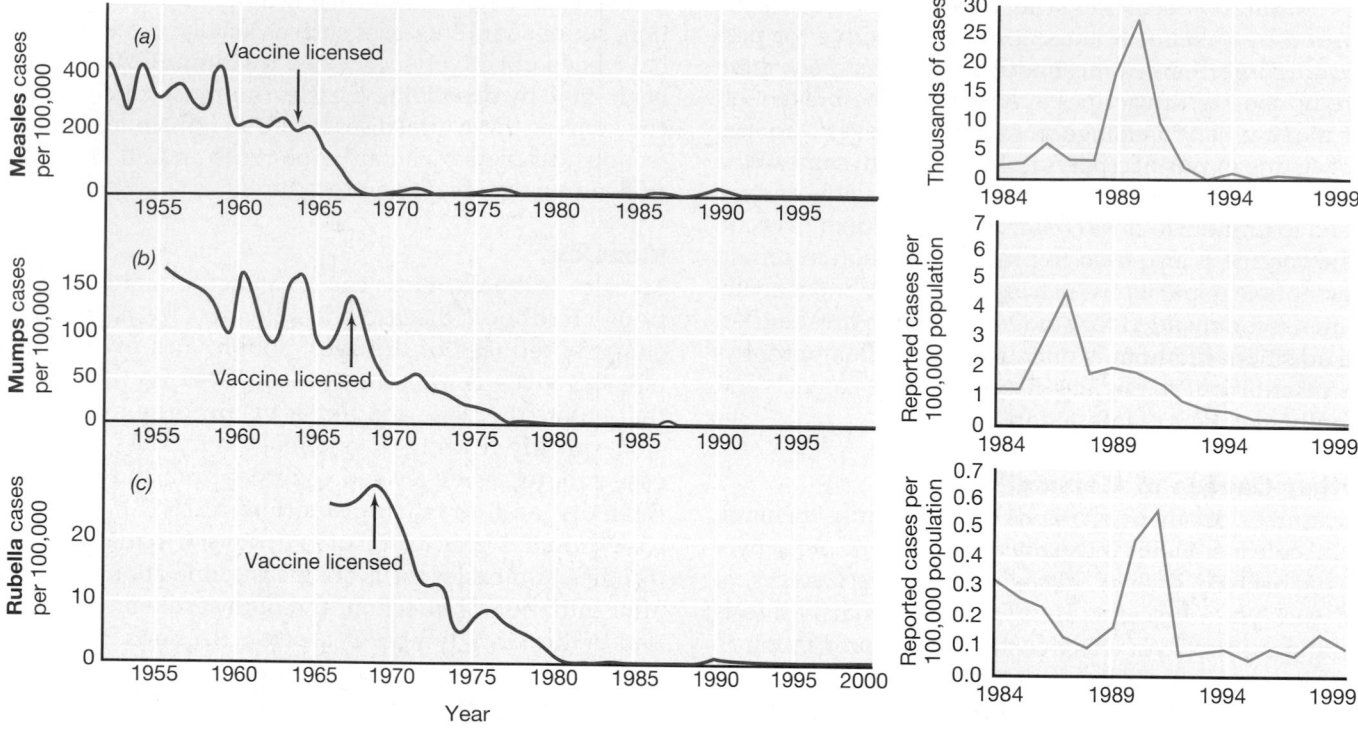

Figure 26.14 The effect of vaccines on the prevalence in the United States of the major childhood viral diseases now controlled by the MMR (measles, mumps, rubella) vaccine. (a) Measles. (b) Mumps. (c) Rubella. The insets show a more recent picture of these diseases. Data were obtained from the Centers for Disease Control and Prevention, Atlanta, GA, USA.

MMR (measles, mumps, and rubella) vaccine (☙ Section 22.11). A childhood case of measles generally confers life-long immunity to reinfection.

Mumps

Mumps is caused by a different paramyxovirus than that causing measles and is also highly infectious. Mumps is spread by airborne droplets, and the disease is characterized by inflammation of the salivary glands leading to swelling of the jaws and neck (Figure 26.15). The virus spreads through the bloodstream and may infect other organs including the brain, testes, and pancreas. Severe complications may include encephalitis and, very rarely, sterility. The host immune response produces antibodies to mumps virus surface proteins, and this generally leads to a quick recovery. An attenuated mumps vaccine is highly effective in preventing the disease (Figure 26.14b). Hence, like measles, the incidence of mumps in developed countries has been greatly reduced in the last three decades, with mumps epidemics usually restricted to those individuals who did not receive the MMR vaccine during childhood.

Rubella

Rubella (*German measles*) is caused by a single-stranded RNA virus of the togavirus group (☙ Section 16.7). The symptoms of the disease resemble those of measles but

are generally milder. Rubella is less contagious than true measles, and thus a good proportion of the population has never been infected. However, during the first 3 months of pregnancy, rubella virus can infect the fetus by placental transmission and cause serious fetal abnormalities. Rubella can cause stillbirth, deafness, heart and eye defects, and brain damage in live births. Thus, preg-

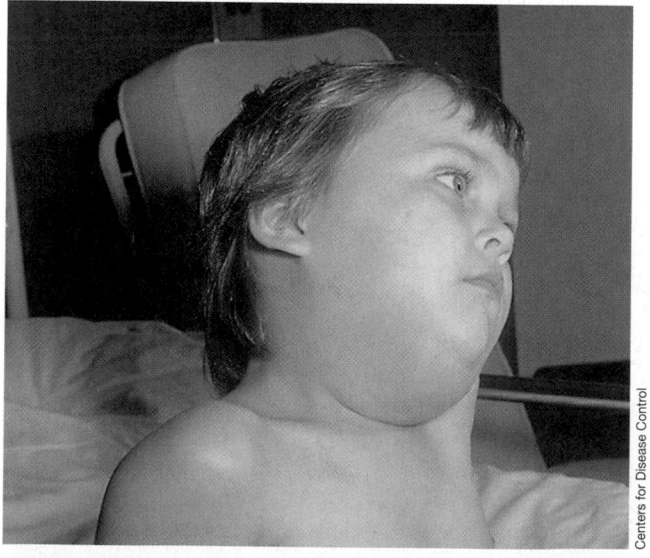

Figure 26.15 Typical glandular swelling associated with mumps.

nant women should not be immunized with the rubella vaccine, or contract rubella during this period. For this reason, routine childhood immunization against rubella should be practiced. An attenuated virus vaccine is administered with attenuated measles and mumps viruses in the MMR vaccine mentioned previously (see Figure 26.14c and ∞ Section 22.11).

Chickenpox and Shingles

Chickenpox (varicella) is a common childhood disease caused by a herpes virus (∞ Section 16.11). Chickenpox is highly contagious and is transmitted by infectious droplets, especially when susceptible individuals are in close contact. In school children, for example, close confinement during the winter months leads to the spread of chickenpox through airborne droplets from infected classmates and through contact with contaminated fomites. The virus enters the respiratory tract, multiplies, and is quickly disseminated via the bloodstream, resulting in a systemic papular rash that quickly heals, rarely leaving disfiguring marks (Figure 26.16). An attenuated virus vaccine is now recommended for use in the United States (∞ Section 22.11). The current reported annual incidence of chickenpox is now about one-third of the number of cases reported prior to 1994, the year immunization became widespread (Figure 26.17).

The chickenpox virus can remain dormant in nerve cells for years with no apparent symptoms. The virus occasionally migrates from this reservoir to the skin surface, causing a painful skin eruption referred to as *shingles* (zoster). Shingles most commonly strikes immunosuppressed individuals or the elderly. Studies with human volunteers suggest that T cells are important in destroying the virus. The prophylactic use of human hyperimmune globulin prepared against the virus is useful for preventing the onset of symptoms of shingles. Such therapy is advised only for patients where secondary infec-

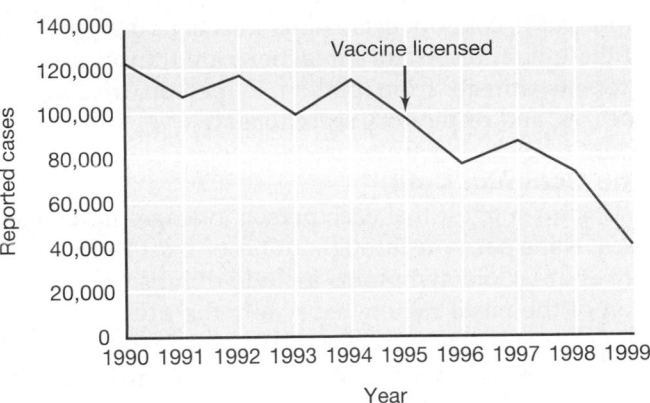

Figure 26.17 Reported incidence of chickenpox in the United States, 1990–1999. The chickenpox vaccine was introduced in 1995, and within 5 years the overall incidence decreased to about 30% of the total number of cases observed before the vaccine was available. Data was obtained from the Centers for Disease Control and Prevention, Atlanta, GA, USA.

tions occasionally associated with shingles, such as pneumonia or encephalitis, may be life-threatening.

✓ 26.7 Concept Check

Viral respiratory diseases are highly infectious and may cause serious health problems. However, the common childhood viral diseases measles, mumps, rubella, and chickenpox are all controllable with appropriate immunization procedures.

- ✓ Identify the date immunization was made available for measles, mumps, rubella, and chickenpox.
- ✓ Describe the potential serious outcomes of infection by these viruses.

26.8 Colds and Influenza

Colds and influenza are the most common infectious diseases. As shown in Figure 26.18, there are about two cases of influenza to every other infectious disease. There

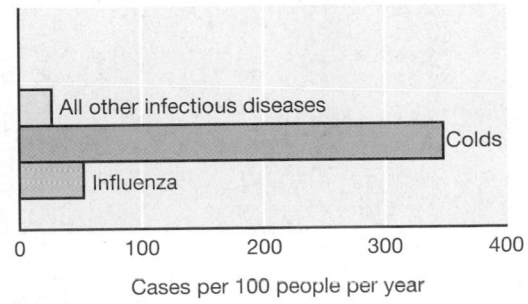

Figure 26.18 Cold and influenza viruses are the leading causes of acute infectious disease in the United States. These data are typical for recent years.

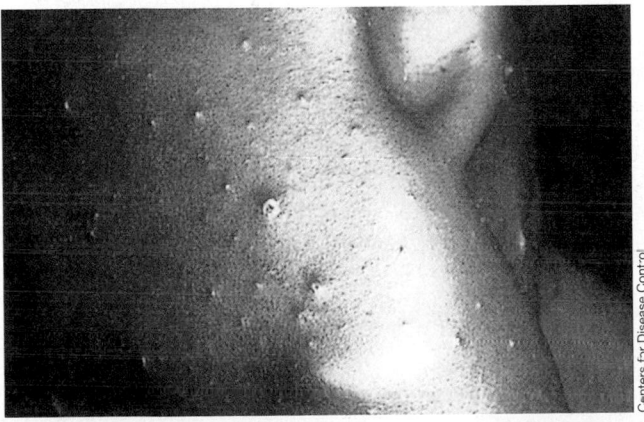

Centers for Disease Control

Figure 26.16 Mild papular rash associated with the disease chickenpox.

are about 15 colds for every other infectious disease. Both of these viral respiratory infections are transmitted via droplets spread from person to person in coughs, sneezes, and respiratory secretions.

The Common Cold

Estimates suggest that each person averages more than three colds per year throughout his or her lifetime (Figure 26.18). Cold symptoms include rhinitis (inflammation of the nasal region, especially the mucous membranes), nasal obstruction, watery nasal discharges, and a general feel of malaise, usually without an accompanying fever. Rhinoviruses, single-stranded ribonucleic acid (RNA) viruses of the picornavirus group (see Figure 26.19a and ∞ Section 16.7), are the most common

causes of colds. At least 115 different serotypes of rhinoviruses have been identified. Another group of single-stranded RNA viruses, the coronaviruses (Figure 26.19b), are responsible for about 15% of all colds in adults. A variety of other viruses including adenoviruses, coxsackie viruses, respiratory syncytial virus, and orthomyxoviruses, are responsible for about 10% of common colds. Colds generally induce a specific, local, neutralizing IgA response (∞ Section 22.8). However, the number of potential infectious agents makes immunity via immunization or previous exposure to a given virus very unlikely.

Aerosol transmission of the virus is probably the major means of spreading colds, although experiments with human volunteers suggest that direct contact and/or fomite contact is also an important method of transmission. Most antiviral drugs are ineffective, but a pyrazidine derivative (Figure 26.20a) has proven promising for preventing colds after virus exposure in volunteers. In addition, new experimental antiviral drugs are being designed based on information derived from three-dimensional structures. For example, the antirhinovirus drug WIN 52084 (Figure 26.20b) binds to the virus, changing its three-dimensional surface configuration and disrupting a cellular binding site, thus preventing infection. Interferon-α, a cytokine (∞ Section 23.10), is also effective in preventing the onset of colds. Thus, there are several experimental possibilities for cold prevention and treatment, although none are widely accepted as effective and safe. The accepted treatment for colds is to treat the symptoms, especially nasal discharges, with a variety of antihistamine and decongestant drugs.

Influenza

Influenza is caused by an RNA virus of the orthomyxovirus group (∞ Section 16.8). Influenza virus is a single-stranded, negative-sense, helical RNA genome surrounded by an envelope made up of protein, a lipid

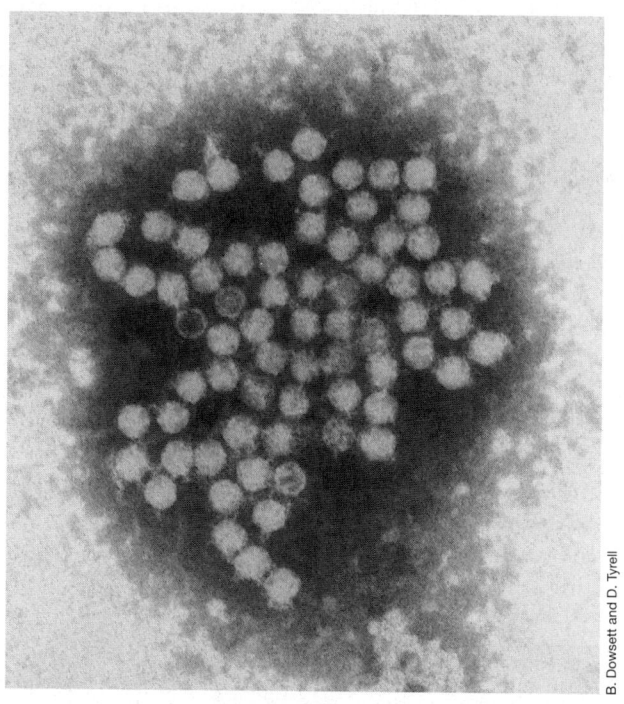

(a)

B. Dowsett and D. Tyrell

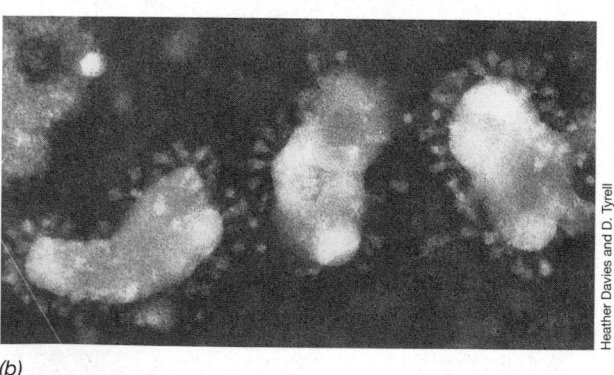

(b)

Heather Davies and D. Tyrell

Figure 26.19 Electron micrographs of some common cold viruses. (a) Human rhinovirus. (b) Human coronavirus. Each rhinovirus virion is about 30 nm in diameter. Each coronavirus virion is about 60 nm in diameter.

(a)

(b)

Figure 26.20 Experimental antirhinovirus drugs. (a) The structure of 3-methoxy-6-[4-(3-methylphenyl)]-1-piper-azinyl. (b) The structure of WIN 52084, a receptor-blocking drug.

A Focus On ... Is It a Cold or Is It the Flu?

The symptoms of a common cold and symptoms of "the flu" (influenza) often seem similar, but the two diseases are symptomatically distinct and caused by quite different viruses. A typical common cold caused by a rhinovirus is associated with nasal discharges, cough, chills, and perhaps a sore throat. Influenza, caused by an orthomyxovirus, is generally associated with a different set of symptoms. Although either condition may cause illness, colds are usually of shorter duration and the symptoms are milder. The following can serve as a guideline for determining whether you "caught a cold" or "have the flu." ■

Symptoms	Common cold	Influenza
Fever	Rare	Common (39–40°C); sudden onset
Headache	Rare	Common
General malaise	Slight	Common; often quite severe; can last several weeks
Nasal discharge	Common and abundant	Less common; usually not abundant
Sore throat	Common	Much less common
Vomiting and/or diarrhea	Rare	Common

bilayer, and external glycoproteins (Figure 26.21 and ⌀ Figure 16.15). Three different types of influenza viruses exist, influenza A, influenza B, and influenza C. Since it is the most important human pathogen, we limit our discussion here to influenza A.

The genetic material of influenza A virus, single-stranded RNA, is arranged in a highly unusual manner. As discussed in Section 16.8, the influenza virus genome is *segmented*, with genes found on each of eight distinct fragments of its single-stranded RNA (⌀ Figure 16.15 and Section 16.8). Such an arrangement allows the rapid and constant reassortment of genes between different strains of influenza virus because more than one strain of influenza virus can infect a cell at one time. Reassortment of genes results in the phenomenon called **antigen shift**. Antigenic shift refers to modifications in the protein coat of the virions, especially to two proteins im-

portant in the attachment and eventual release of virus from host cells, hemagglutinin and neuraminidase, respectively (⌀ Figures 16.15 and 26.21). The hemagglutinin and neuraminidase antigens also develop minor antigenic changes because of genetic mutations that result in the change of one or more amino acids. This phenomenon is known as **antigenic drift**.

Human influenza virus is transmitted from person to person through the air, primarily in droplets expelled during coughing and sneezing. The virus infects the mucous membranes of the upper respiratory tract and occasionally invades the lungs. Symptoms include a low-grade fever lasting for 3–7 days, chills, fatigue, headache, and general aching (see the box, Is It a Cold or Is It the Flu?). Recovery is usually spontaneous and rapid. Most of the serious consequences of influenza infection occur because bacterial invaders develop as secondary infections in persons whose resistance has been lowered by the influenza infection. Especially in infants and elderly people, influenza is often followed by bacterial pneumonia; death, if it occurs, is usually due to the bacterial infection. After infection, most individuals in an infected population become immune to the infecting virus, and it is impossible for a strain of similar antigenic type to cause an epidemic for about 2 to 3 years. Immunity is largely dependent on the production of secretory antibody (IgA) (⌀ Section 22.8), especially to antigenic determinants of the hemagglutinin and neuraminidase proteins.

Influenza outbreaks occur every year from late autumn through the winter months due to *endemic viral disease* (⌀ Section 25.3). Antigenic drift results in reduced immunity in the population and is responsible for the recurrence of *epidemics*, severe localized influenza outbreaks, occurring in a 2- to 3-year cycle. *Pandemics*, worldwide outbreaks, occur much less frequently, being about 10 to 40 years apart, and are the result of an antigenic shift.

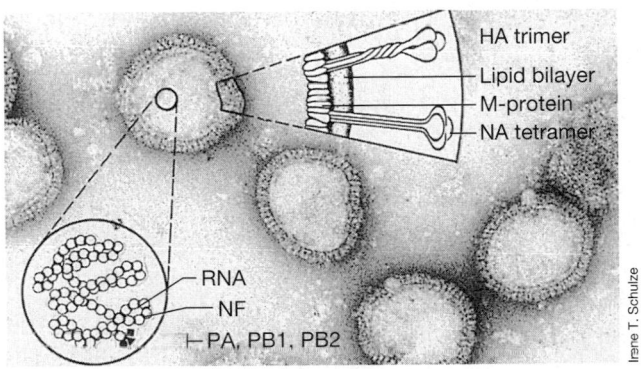

Irene T. Schulze

Figure 26.21 Electron micrograph of the influenza virus, showing the location of the major viral coat proteins and the nucleic acid. Each virion is about 100 nm in diameter. HA, Hemagglutinin (three copies make up the HA coat spike); NA, neuraminidase (four copies make up the NA coat spike); M, coat protein; NP, nucleoprotein; PA, PB1, PB2, other internal proteins, some of which may have enzymatic functions.

The 1957 worldwide outbreak of the so-called Asian flu provided an opportunity to study the development of a pandemic (Figure 26.22). The pandemic probably arose when a virulent mutant virus strain that differed antigenically from all previous strains appeared in the population. Since immunity to this strain was not present, the virus spread rapidly throughout the world. It first appeared in the interior of China in late February 1957 and by early April had been brought to Hong Kong by refugees. It spread from Hong Kong along air and naval routes and was apparently transferred to San Diego, California, by naval ships. In May, an outbreak occurred in Newport, Rhode Island, on a naval vessel. Other outbreaks occurred in various parts of the United States. Peak incidence occurred in the last 2 weeks of October, during which time 22 million new cases developed. In the pandemic of 1918, the influenza A strain may have originated from reassortment with a related virus that infects swine (swine flu). The 1957 strain may have also arisen from reassortment of influenza genes from an animal reservoir.

Influenza epidemics can be controlled by immunization. However, the choice of appropriate vaccines is complicated by the large number of existing strains and the continuing ability of these strains to undergo antigenic drift or antigenic shift. When new strains evolve, vaccines are not immediately available, but through careful worldwide surveillance (∞ Section 25.8), samples of the major emerging strains of influenza virus are usually obtained before epidemic outbreaks occur. In the United States, inactivated viral preparations from three candidate strains are mixed to prepare a polyvalent vaccine that is then used for immunization prior to the next influenza season, which usually starts in late autumn and continues through winter. Influenza immunization is recommended for those individuals most likely to succumb or be exposed to serious secondary illnesses, such as the elderly (more than 65 years of age), those suffering from chronic debilitating diseases (e.g., AIDS patients, see Section 26.14), and health care workers. The duration of effective artificial immunity from the inactivated influenza vaccine is usually only a few years, and of course, it is strain-specific. Therefore, immunization preparations are updated annually.

Influenza may also be controlled by use of the chemicals *amantadine* and *ramantadine* (∞ Section 20.10 and Table 20.5). These drugs inhibit viral replication and have been used as chemoprophylatic agents to prevent the spread of influenza to those at high risk. They are also used to treat ongoing influenza and shorten the course and severity of infection. The treatment of influenza with aspirin is not recommended, because there is evidence of a link between aspirin treatment of influenza and Reye's syndrome (a rare but occasionally fatal affliction involving the central nervous system) in children.

✓ **26.8 Concept Check**

Colds and influenza, or "flu," are the most common infectious diseases. While they are not usually life-threatening diseases by themselves, they can lead to serious secondary bacterial infections.

✓ Discuss the possibilities for effective immunization programs for colds and influenza.

✓ Define and compare the symptoms of influenza and of common colds.

Figure 26.22 Route of spread of a major influenza epidemic, the Asian flu pandemic of 1957.

II DIRECT CONTACT TRANSMISSION OF DISEASES

A number of diseases are spread primarily by direct contact with an infected person or by contact with blood or excreta from the infected person. Many of the respiratory diseases we have discussed can also be spread by direct contact. Here we discuss three diseases spread primarily person to person through direct contact with infected individuals.

26.9 *Staphylococcus*

The genus *Staphylococcus* contains common pathogens of humans and animals, occasionally causing life-threatening disease. Staphylococcal infections are common infections of the skin and wounds. Most staphylococcal infections result from the transfer of staphylococci in normal flora from an infected but asymptomatic individual to a susceptible individual.

Biology

Staphylococci are gram-positive cocci of about 0.8–1.0 μm in diameter that divide in several planes to form irregular clumps (∞ Section 12.19 and Figure 12.51). Staphylococci are nonsporulating but are resistant to drying and are readily dispersed in dust particles through the air and on surfaces. In humans, two species are important: *Staphylococcus epidermidis*, a nonpig-

mented form usually found on the skin or mucous membranes, and *Staphylococcus aureus*, a yellow-pigmented form. While both species are potential pathogens, *S. aureus* is much more commonly associated with human disease. Both species often occur as part of the normal microbial flora in the upper respiratory tract or on the skin (Figure 26.2).

Epidemiology and Pathogenesis

Staphylococci cause a variety of diseases including acne, boils (Figure 26.23), pimples, impetigo (Figure 26.4), pneumonia, osteomyelitis, carditis, meningitis, and arthritis. Many of these diseases cause the production of pus, so they are said to be *suppurative*, or pus-forming. The most common habitats of *Staphylococcus aureus* are the upper respiratory tract, especially the nose and throat, and the surface of the skin. Many healthy people are carriers, and in most cases resident staphylococci do not cause disease. However, infants often become infected during the first week of life from the mother or from another close human contact. Serious staphylococcal infections also occur when the resistance of the host is low because of hormonal changes, debilitating illness, wounds, or treatment with steroids or other drugs that compromise the immune system.

Those strains of *S. aureus* most frequently causing human disease produce a number of extracellular enzymes or toxins (∞ Section 21.9). At least four different *hemolysins* have been recognized, and a single strain often produces more than one. The production of hemolysins is responsible for the hemolysis seen around colonies on blood agar plates. *S. aureus* is also capable

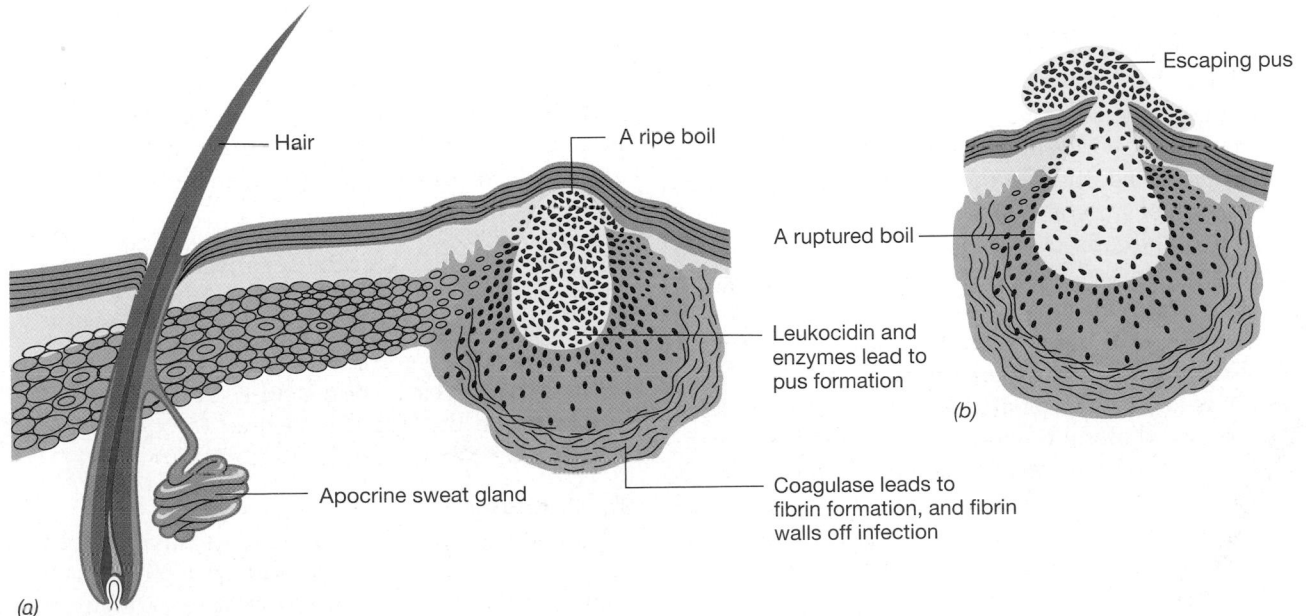

Figure 26.23 The structure of a boil. (a) Staphylococci initiate a localized infection of the skin and become walled off by coagulated blood and fibrin through the action of coagulase. (b) The rupture of the boil releases pus and bacteria.

of producing an *enterotoxin* associated with foodborne illness (⌖ Sections 21.9, 22.14, and 29.4).

Another substance produced by *S. aureus* is *coagulase*, an enzyme-like factor that causes fibrin to coagulate and form a clot (⌖ Section 21.9). The production of coagulase is generally associated with pathogenicity. Clotting induced by coagulase results in the accumulation of fibrin around the bacterial cells making it difficult for host defense agents to come into contact with the bacteria and making the staphylococci resistant to phagocytosis (Figure 26.23). Most *S. aureus* strains also produce leukocidin, which causes the destruction of leukocytes. Production of *leukocidin* in skin lesions such as boils and pimples results in considerable host cell destruction and is one of the factors responsible for pus formation (Figure 26.23). Some strains of *S. aureus* also produce other extracellular virulence factors, including proteolytic enzymes, hyaluronidase, fibrinolysin, lipase, ribonuclease, and deoxyribonuclease.

Certain strains of *S. aureus* have been implicated as the agents responsible for **toxic shock syndrome (TSS)**, a serious outcome of staphylococcal infection characterized by high fever, rash, vomiting, diarrhea, and occasionally death. Toxic shock was first recognized in menstruating women and was associated with use of tampons. In menstruating females, blood and mucus in the vagina can become colonized by hemolytic *S. aureus* from the skin, and the presence of a tampon concentrates this material, creating ideal microbial growth conditions. Largely through alterations in materials used in tampons and through education, toxic shock due to tampon use is now relatively rare. However, toxic shock syndrome is still seen in both men and women as a result of staphylococcal infections following surgery.

The symptoms of TSS result indirectly from an exotoxin called *toxic shock syndrome toxin (TSST)*. TSST is a superantigen (⌖ Section 22.14). Toxic shock toxin is released by the growing staphylococci, causing a massive T-cell reaction resulting in the inflammatory response characteristic of superantigen reactions. A TSS-like reaction may also be caused by different superantigens from other pathogenic *Bacteria*, including *Streptococcus pyogenes* (see Section 26.2).

Staphylococcal enterotoxin A causes a form of food poisoning and is also a superantigen. Presumably, after ingestion of toxin-contaminated food, the toxin stimulates T cells localized along the intestine, resulting in a massive T-cell response, release of mediators, and increased permeability of the intestine. The final outcome is the severe but short-lived diarrhea and vomiting associated with staphylococcal food poisoning (⌖ Section 29.4).

Treatment and Prevention

Extensive use of antibiotics has resulted in the natural selection of resistant strains of *Staphylococcus aureus* and *Staphylococcus epidermidis*. Hospital (nosocomial) infections with antibiotic-resistant staphylococci often occur in patients whose resistance is lowered due to other diseases, surgical procedures, or drug therapy (⌖ Section 25.7). Patients often acquire staphylococci from hospital personnel who are asymptomatic carriers of drug-resistant strains. Therefore, appropriate antimicrobial drug therapy for *S. aureus* infections is a problem. Although some community-acquired infections are treatable with penicillin, disease-producing isolates of *S. aureus* must be individually checked for antibiotic sensitivity (⌖ Section 24.3).

Prevention of staphylococcal infections is problematic because most individuals are asymptomatic carriers, and some diseases, such as acne, can be transmitted by simple contact with contaminated fingers. In hospital environments such as surgical wards and nurseries, carriers of known pathogenic strains must either be excluded or be treated with topical or systemic antimicrobial drugs to eradicate the carrier state.

✓ 26.9 Concept Check

Although staphylococci are usually harmless inhabitants of the upper respiratory tract and skin, several serious diseases can result from infection, including some caused by staphylococcal toxins that act as superantigens.

✓ What is the normal habitat of *Staphylococcus aureus*? How is *S. aureus* spread from person to person?

✓ List some of the diseases caused by infection with staphylococci.

26.10 *Helicobacter pylori* and Gastric Ulcers

Helicobacter pylori was first identified in human intestinal biopsies in 1983. This organism is a pathogen associated with gastritis, ulcers, and gastric cancers.

Biology

Helicobacter pylori is a gram-negative, highly motile, spiral-shaped bacterium that is related to *Campylobacter* (⌖ Section 29.8). It is 2.5–3.5 μm long and 0.5–1.0 μm in diameter and has one to six polar flagellae at one end. *H. pylori* colonizes the non–acid-secreting mucosa of the stomach and the upper intestinal tract, including the duodenum (⌖ Section 21.4).

Epidemiology

Up to 80% of gastric ulcer patients have concomitant *Helicobacter pylori* infections, and up to 50% of asymptomatic adults in developing countries are chronically infected. The mode of transmission has not been unequivocally established, but evidence suggests that host-to-host contact and ingestion of contaminated food or water transmit *H. pylori*. Although there is no known nonhuman reservoir of *H. pylori*, the organism has occasionally been recovered from cats kept as household pets, indicating that it can be spread to or from animals in close contact with humans. Infection occurs in high incidence in certain families, and the overall incidence in the population increases with age. These factors suggest a host-to-host type of transmission (Sections 25.3, 25.4, and 25.5). However, infections with *H. pylori* sometimes also occur in epidemic clusters, suggesting that a common source such as food or water may sometimes be involved.

Long-term epidemiology studies indicate that chronic gastritis due to untreated *H. pylori* infection may lead to the development of gastric cancers.

Pathology

Helicobacter pylori is slightly invasive, and colonizes the surfaces of the gastric mucosa. The organism is protected from the effects of stomach acids by the gastric mucus layer. After colonization of the mucosa, a combination of pathogen products and host responses causes inflammation, tissue destruction, and ulceration. Pathogen products such as *vacA* (a cytotoxin; Table 21.4), urease, and lipopolysaccharide may contribute to localized tissue destruction and ulceration. Antibodies to *H. pylori* are usually present in infected individuals, but are not protective and do not prevent colonization. Individuals who acquire *H. pylori* tend to have chronic, long-term infections unless they are treated with antibiotics.

Diagnosis

Definitive diagnosis involves the recovery and culture or observation of *Helicobacter pylori* from a biopsy of an ulcer. Serum antibodies indicate *H. pylori* infection, but since infections seem to be chronic and antibodies may persist for months after a given infection (Section 24.7), *H. pylori* antibodies are not reliable indicators of acute, active disease.

Treatment

Evidence for a causal association between *Helicobacter pylori* and many gastric ulcers comes from antibiotic treatments for the disease. Long-term treatment of ulcers with antacid preparations has seldom been suc-

cessful: Most patients relapse within 1 year. However, by treating ulcers as an infectious disease, permanent cures are often obtained.

Treatment of *H. pylori* infection usually consists of a combination of drugs including metranidazole, a second antibiotic such as tetracycline or amoxycillin, and a bismuth-containing antacid preparation. The combination treatment, administered for 14 days, abolishes the *H. pylori* infection and provides a long-term cure for the ulcers.

Although the causal relationship between *H. pylori* infection and ulcers has not been unequivocally established, current evidence indicates that *H. pylori* is a major preventable and treatable cause of many gastric ulcers.

✓ 26.10 Concept Check

Helicobacter pylori infection appears to be the most common cause of gastric ulcers. Treatment of gastric ulcers now involves antibiotics, which seem to promote a permanent cure.

- ✓ Describe the infection by *H. pylori* and the resulting development of an ulcer.
- ✓ How can ulcers due to *H. pylori* infection be permanently cured?
- ✓ Describe evidence indicating that *H. pylori* infections are spread from person to person.

26.11 Hepatitis Viruses

Hepatitis is a liver inflammation commonly caused by an infectious agent. Hepatitis sometimes results in acute illness followed by destruction of functional liver anatomy and cells, a condition known as **cirrhosis**. Hepatitis due to an infection can cause chronic or acute disease and some forms can lead to liver cancer. Although many viruses and a few bacteria can cause hepatitis, a restricted group of viruses is often associated with liver disease.

Biology and Epidemiology

Hepatitis viruses are a diverse group. Table 26.1 defines the five known hepatitis viruses. None of these viruses are related to one another, but all infect cells in the liver, causing various forms of hepatitis.

TABLE 26.1 Hepatitis viruses				
Disease	**Virus and genome**	**Vaccine**	**Disease**	**Route**
Hepatitis A	*Hepatovirus* (HAV) ss RNA	Yes	Acute	Enteric
Hepatitis B	*Orthohepadnavirus* (HBV) ds DNA	Yes	Acute, chronic, oncogenic	Parenteral, sexual
Hepatitis C	*Hepacivirus* (HCV) ss RNA	No	Chronic, oncogenic	Parenteral
Hepatitis D	*Deltavirus* (HDV) ss RNA	No	Fulminant, only with HBV	Parenteral
Hepatitis E	Calciviridae family (HEV) ss RNA	No	Fulminant disease in pregnant women	Enteric
Hepatitis G	Flaviviridae family (HGV) ss RNA	No	Asymptomatic	Parenteral

Hepatitis A virus (HAV) is transmitted from person-to-person or by ingestion of fecally contaminated food or water. The virus often causes mild, even subclinical infections, but rare cases of severe liver disease can occur. The most significant food vehicles for hepatitis A are shellfish, usually oysters and clams harvested from water polluted by human fecal material. While the general trend for numbers of HAV infections has moved downward (Figure 26.24), partly due to the availability of an effective vaccine, there were over 17,000 infections and 114 deaths attributed to HAV hepatitis in 1999 in the United States. HAV causes more cases of viral hepatitis than any other virus.

Infection due to hepatitis B virus (HBV) is often called serum hepatitis. The complete double-stranded DNA virus called the Dane particle is shown in Figure 26.25. HBV causes acute, often severe disease that can lead to liver failure and death. Chronic HBV infection can lead to cirrhosis and liver cancer. HBV is usually transmitted by a parenteral route, such as blood transfusion or through shared hypodermic needles contaminated with infected blood. HBV may also be transmitted through exchanges of body fluids, as in sexual intercourse. The numbers of new HBV infections is decreasing, again due to an effective vaccine, but there were still almost 8000 HBV cases and over 1000 deaths in 1999 in the United States.

Hepatitis D virus (HDV) is a *defective virus* that lacks genes for its own protein coat (∞ Section 16.14). HDV is also transmitted by parenteral routes, but since it is a defective virus, it cannot replicate and express a complete virus unless the cell is also infected with HBV. The HDV genome replicates independently but uses the protein coat of HBV for expression. Thus, HDV infections are always seen concomitantly with HBV infections.

Hepatitis C virus (HCV) is also transmitted parenterally. HCV generally produces a mild or even asymptomatic disease at first, but up to 85% of individuals develop chronic hepatitis, with up to 20% leading to chronic liver disease and cirrhosis. Chronic infection leads to hepatocarcinoma in 3–5% of infected individuals each year. The latency period for development of cancer can be several decades after the primary infection. In 1999, there were about 2500 *reported* new cases of HCV in the United States, but nearly that many deaths occurred due to chronic HCV infections that develop into liver cancer. HCV-induced liver disease is the most common liver disease currently seen in clinical settings in the United States and accounts for up to 40% of the 25,000 annual deaths due to chronic liver disease and cirrhosis.

Hepatitis E virus (HEV) transmits hepatitis via an enteric route. HEV causes an acute, self-limiting hepatitis that varies in severity from case to case, but is often the cause of rapid fulminant disease in pregnant women. HEV is endemic in Mexico as well as in tropical and subtropical regions of Africa and Asia.

Hepatitis G virus (HGV) is commonly found in the blood of patients with other forms of acute hepatitis, but HGV alone seems to cause very mild disease or is completely asymptomatic. Several different samples of volunteer blood donors tested as high as 8.1% positive for HGV, but since HGV is not associated with demonstrable clinical disease, the significance of these findings is not clear.

Pathology

Hepatitis is an acute disease of the liver. Symptoms include fever, **jaundice** (production and release of excess bilirubin by the liver due to destruction of liver cells, re-

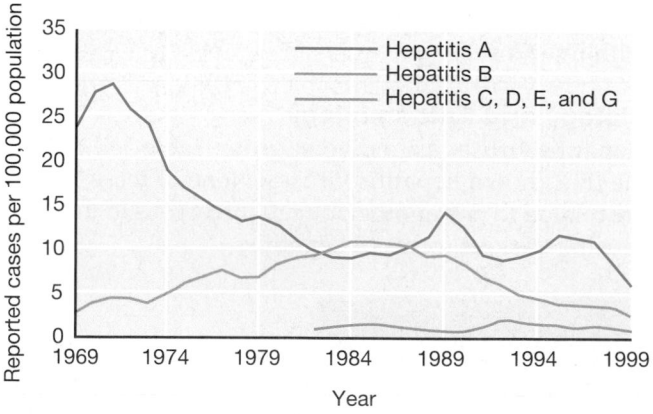

Figure 26.24 Hepatitis in the United States. The prevalence of hepatitis is shown by viral agent. In 1999 there were 17,047 cases of hepatitis A; 7694 cases of hepatitis B; and 3111 cases of other hepatitis, mostly caused by HCV. Data were obtained from the Centers for Disease Control and Prevention, Atlanta, GA, USA.

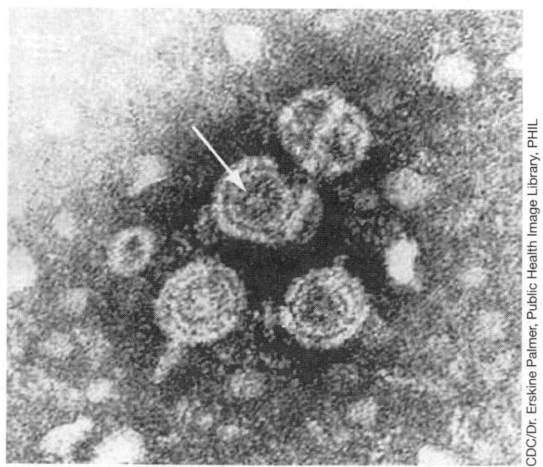

Figure 26.25 Hepatitis B virus (HBV). Arrows indicate the complete HBV particle, which is about 42 nm in diameter and is called the Dane particle.

sulting in yellowing of the skin and whites of the eye), *hepatomegaly* (liver enlargement), and *cirrhosis* (breakdown of the normal liver architecture with fibrosis). Mild hepatitis may be characterized by relatively minor elevation of liver enzymes such as alanine aminotransferase (ALT). Fulminant disease is characterized by rapid onset of severe symptoms such as jaundice and cirrhosis and is often a life-threatening condition. The various hepatitis viruses cause similar acute clinical disease and cannot be readily distinguished based on the clinical findings alone. Chronic hepatitis infections, usually caused by HBV or HCV, are often asymptomatic or produce very mild symptoms but can cause serious liver disease, even in the absence of *hepatocarcinoma* (liver cancer).

Diagnosis

Diagnosis of hepatitis is based primarily on clinical findings and laboratory tests that determine liver function problems. Cirrhosis is diagnosed by visual examination of biopsied liver tissue.

A number of virus-specific assays are also used to confirm diagnosis, identify the infectious agent, and determine a course of treatment. Direct culture of hepatitis viruses is usually not used for identification purposes, and HCV and HGV have not been successfully cultured.

Some of the most widely used methods for determining hepatitis identity are enzyme-linked immunoassay (EIA) tests (∞ Section 24.11). Most EIA tests are designed to identify viral proteins in blood specimens. However, tests are available that indicate the presence of IgM or IgG *antibodies* to HBV. IgM is associated with the primary immune response to HBV, and IgG is associated with the secondary response to HBV. Therefore, identification of the antibody class can determine whether or not the HBV infection is a new infection (IgM) or whether the antibody is due to a secondary response to a chronic or latent HBV infection (IgG) (∞ Section 22.9). Other immune-based tests used for the detection of hepatitis viruses include immunoblots (∞ Section 24.12), immunoelectron microscopy (IEM) (∞ Section 24.9), and immunofluorescence (∞ Section 24.10).

PCR-based tests and dot-blot DNA hybridization tests (∞ Section 24.13) are also used for the detection of the viral genome in blood or in liver tissue obtained by biopsy (∞ Section 24.13).

Prevention and Treatment

Infection with HAV or HBV can be prevented with effective vaccines. HBV vaccination is recommended and in most cases is required for school-age children in the United States (∞ Section 22.11). No effective vaccines are available for the other hepatitis viruses.

"Universal precautions" for prevention of disease spread by bloodborne pathogens were mandated by law and were designed primarily to protect against HBV infection and the human immunodeficiency virus (HIV) (∞ Section 24.4). These standards mandate precautions for personnel when handling infectious waste and body fluids and are designed to prevent infection by all parenterally transmitted hepatitis viruses (HBV, HCV, HDV, and HGV). The precautions prescribe a high level of vigilance and aseptic procedures to deal with patients, body fluids, and potentially infected waste materials.

Because hepatitis A can be spread through contamination of common sources such as food and water, major epidemic outbreaks of hepatitis A can be prevented by maintaining pathogen-free food and water supplies (∞ Chapter 28 and Chapter 29).

Postexposure treatment of hepatitis is sometimes successful. Pooled human immune gamma globulin can be used to prevent HAV infection if given soon after exposure. For postexposure prevention of HBV infection, specific hepatitis B immune globulin, coupled with administration of the HBV vaccine, has been effective (∞ Section 22.11).

However, most treatment of hepatitis is supportive, providing rest and time to allow liver damage to resolve and be repaired. In some cases, some antiviral drugs are effective for treatment (∞ Section 20.10). Interferon α is effective against HCV when combined with ribavirin in some patients. HBV can be treated with foscarnet, ribavirin, lamivudine, or ganciclovir.

✓ 26.11 Concept Check

Hepatitis caused by viruses can cause serious liver diseases. Vaccines are available for the hepatitis A virus and the hepatitis B virus. The overall prevalence of hepatitis has decreased significantly in the last 20 years in the United States, but hepatitis due to viruses is still a major infectious disease and public health problem because of the high infectivity of the viruses.

- ✓ Describe the mode of transmission for hepatitis A virus hepatitis B virus and hepatitis C virus.
- ✓ Describe potential prevention and treatment methods for hepatitis A virus and hepatitis B virus.

III SEXUALLY TRANSMITTED DISEASES

Several important human pathogens are transmitted almost exclusively by sexual contact. Therefore, the diseases these pathogens cause are known as **sexually transmitted diseases**, or STDs.

STDs or *venereal diseases* are caused by a wide variety of bacteria, viruses, protozoa, and even fungi (Table 26.2). Unlike respiratory pathogens that are shed constantly in large numbers by an infected individual, sexually

TABLE 26.2   Sexually transmitted diseases and treatment guidelines

Disease	Causative organisms[a]	Recommended treatment[b]
Gonorrhea	*Neisseria gonorrhoeae* (B)	Cefixime or ceftriaxone, *and* azithromycin or doxycycline
Syphilis	*Treponema pallidum* (B)	Benzathine penicillin G
Chlamydia trachomatis infections	*Chlamydia trachomatis* (B)	Doxycycline or azithromycin
Nongonococcal urethritis	*C. trachomatis* (B) or *Ureaplasma urealyticum* (B) or *Mycoplasma genitalium* (B) or *Trichomonas vaginalis* (P)	Azithromycin
Lymphogranuloma venereum	*C. trachomatis* (B)	Doxycycline
Chancroid	*Haemophilus ducreyi* (B)	Azithromycin
Genital herpes	Herpes simplex type 2 (V)	No known cure; symptoms can be controlled with topical application of acyclovir (Figure 26.32).
Genital warts	Papilloma virus (certain strains)	No known cure; symptomatic warts can be removed surgically, chemically, or by cryotherapy.
Trichomoniasis	*Trichomonas vaginalis* (P)	Metronidazole
Acquired immunodeficiency syndrome (AIDS)	Human immunodeficiency virus (HIV)	No known cure; nucleotide base analogs, protease inhibitors, and nonnucleoside reverse transcriptase inhibitors are clinically useful in some treatments (see Table 26.3).
Pelvic inflammatory disease	*N. gonorrhoeae* (B) or *C. trachomatis* (B)	Cefotefan
Vulvovaginal candidiasis	*Candida albicans* (F)	Butoconazole

[a] B, Bacterium; V, virus; P, protozoan; F, fungus.

[b] Recommendations of the U.S. Department of Health and Human Services, Public Health Service. For many drugs, there are a number of possible alternatives.

transmitted pathogens are generally only in body fluids exchanged during sexual activity. This is because sexually transmitted pathogens are very sensitive to drying and other environmental stresses such as heat and light. Their habitat, the human genitourinary tract, is a protected, moist environment. Thus, these organisms preferentially and sometimes exclusively colonize the genitourinary tract.

Effectively diagnosing and treating STDs is very difficult for a number of social and biological reasons. First, up to one-third of all STDs involve teenagers with multiple partners; it is hard to define and stop the sources and spread of infections. Second, many STDs have very minor symptoms and infected individuals do not seek treatment. Third, social stigmas still attached to STDs prevent many individuals from seeking prompt treatment. However, prompt effective treatment of STDs is desirable for a number of reasons. First, most STDs are curable and virtually all are controllable with appropriate medical intervention. Second, delay or lack of treatment can lead to long-term problems such as infertility, cancer, heart disease, degenerative nerve disease, birth defects, or stillbirth.

Because transmission of STDs is limited to intimate physical contact, generally during sexual intercourse, the spread of venereal diseases can be controlled by sexual abstinence (no exchange of body fluids) or by the use of barriers such as condoms that stop the exchange of body fluids during sexual activity.

STDs are very common and continue to pose social as well as medical problems. We deal here with a discussion of a number of prevalent STDs.

26.12   Gonorrhea and Syphilis

Gonorrhea and syphilis are both preventable and treatable bacterial STDs. Because of differences in their symptoms, the overall pattern of disease prevalence is quite different. Gonorrhea is very prevalent, and symptoms are often very mild and inapparent, especially in women. As a result, the disease is often unrecognized and goes untreated. Syphilis, on the other hand, now has a low prevalence. This is party because primary syphilis usually exhibits very obvious symptoms and infected individuals usually seek treatment (Figure 26.26).

Gonorrhea

Neisseria gonorrhoeae, often called *gonococcus*, causes gonorrhea. *N. gonorrhoeae* is a gram-negative, nonsporulating, obligately aerobic, oxidase-positive, encapsulated diplococcus (∞ Section 12.10) related biochemically and phylogenetically to *Neisseria meningitidis* (see Section 26.6). *N. gonorrhoeae* is very sensitive to drying and normally does not survive away from the mucus membranes of the genitourinary tract (Figure 26.27). Gonococci are killed rapidly by drying, sunlight, and ultraviolet light.

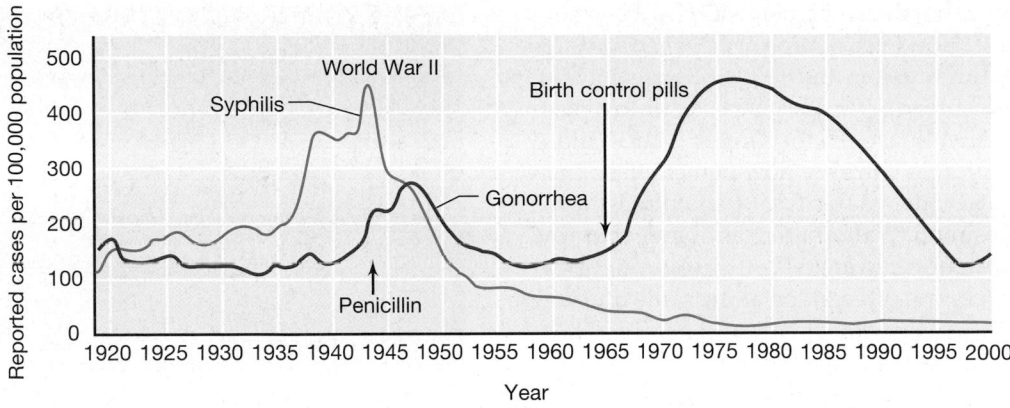

Figure 26.26 Reported cases of gonorrhea and syphilis (primary and secondary cases only) per 100,000 population in the United States. Note the downward trend in disease incidence after the introduction of antibiotics and the upward trend in the incidence of gonorrhea after the introduction of birth control pills. In 1999 there were over 360,000 cases of gonorrhea and only about 6600 cases of syphilis.

Because of its extreme sensitivity to environmental conditions, *N. gonorrhoeae* can only be transmitted by intimate person-to-person contact. The pathogen enters the body by way of the mucus membranes of the genitourinary tract.

The symptoms of gonorrhea are quite different in the male and female. In the female gonorrhea is characterized by a mild vaginitis that is difficult to distinguish from vaginal infections caused by other organisms, and thus, the infection may easily go unnoticed. In the male, however, the organism causes a painful infection of the urethral canal (☾☾ Figure 21.11). Complications from untreated gonorrhea include pelvic inflammatory disease and damage to heart valves and joint tissues.

In addition to gonorrhea, the organism also causes eye infections in newborns. Infants born of infected mothers may acquire eye infections during birth. Therefore, prophylactic treatment of the eyes of all newborns with an ointment containing erythromycin is generally mandatory to prevent gonococcal infection in infants. We discussed the clinical microbiology and diagnosis of gonorrhea in Section 24.1.

(a)

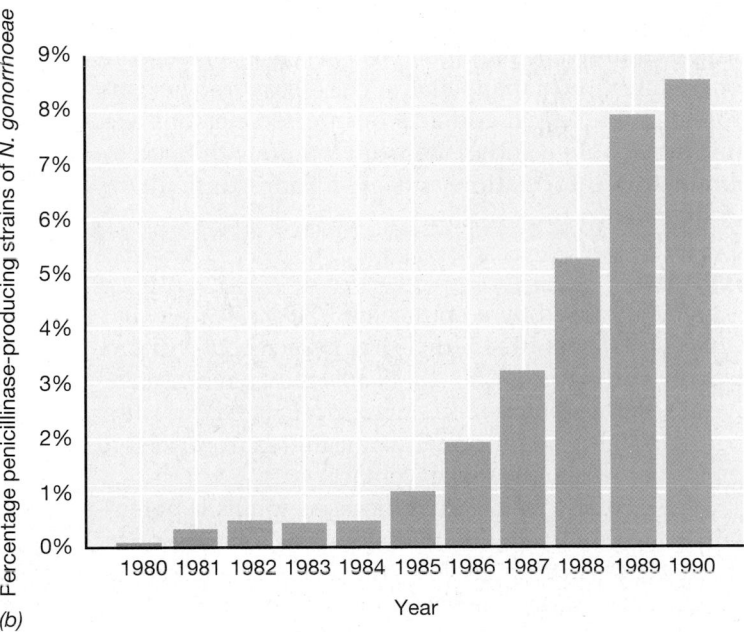

(b)

Figure 26.27 The causative agent of gonorrhea, *Neisseria gonorrhoeae*, and the prevalence of penicillinase production in this organism. (a) Scanning electron micrograph of the microvilli of human fallopian tube mucosa, showing how cells of *N. gonorrhoeae* attach to the surface of epithelial cells. Note the distinct diplococcus morphology. Cells of *N. gonorrhoeae* are about 0.8 μm in diameter. (b) Reported penicillinase-producing *N. gonorrhoeae* (PPNG) in the United States. Note the rapid rise in resistance. As a result, penicillin is no longer used for treatment of gonorrhea (see Table 26.2 for current treatment guidelines).

Treatment of gonorrhea with penicillin has been successful in the past. However, strains of *N. gonorrhoeae* resistant to penicillin arose in the 1970s (Figure 26.27*b*), and are now widespread. This resistance is due to a plasmid-encoded penicillinase. In the United States, more than 8% of all clinical isolates are now penicillinase-producing. Fortunately, the majority of penicillinase-producing strains respond to alternative antibiotic therapy, with a single dose of cefixime or ceftriaxone. Azithromycin or doxycycline is often given at the same time because they are antichlamydial agents and nearly 50% of gonorrhea patients are also infected with the harder-to-diagnose *Chlamydia trachomatis* organism (Table 26.2).

Despite the ease with which gonorrhea can be cured, the incidence of gonococcus infection remains relatively high. The reasons for this are threefold. (1) Acquired immunity does not exist; hence repeated reinfection is possible (whether this is due to lack of local immunity or to the fact that at least 16 distinct serotypes of *Neisseria gonorrhoeae* have been isolated is not understood). (2) The use of oral contraceptives alters the local mucosal environment in favor of the pathogen. Oral contraceptives induce the body to mimic pregnancy, which results, among other things, in a lack of glycogen production in the vagina and a raising of the vaginal pH. Lactic acid bacteria normally found in the adult vagina (∞ Section 21.5) fail to develop under such circumstances, and this allows *N. gonorrhoeae* transmitted from an infected partner to colonize more easily than in an acidic vagina. (3) Symptoms in the female are so mild that the disease may be unrecognized, and a promiscuous infected female can serve as a reservoir for the infection of many males. The disease can be controlled if the sexual contacts of infected persons are quickly identified and treated, but it is often difficult to obtain this information and even more difficult to arrange treatment.

Syphilis

Syphilis is caused by a spirochete, *Treponema pallidum*. *T. pallidum* is about 10–15 μm in length and about 0.15 μm in diameter (Figure 26.28). *T. pallidum* is extremely sensitive to environmental stress such as drying and exposure to heat or light and is normally transmitted from person to person by intimate sexual contact.

The sexually transmitted disease syphilis is potentially much more serious than gonorrhea, but because of differences in pathobiology, the incidence of syphilis in the United States has been much lower since the introduction of effective antibiotic therapy (Figure 26.26).

Syphilis exhibits variable symptoms. The organism does not pass through unbroken skin, and initial infection most probably takes place through tiny breaks in the epidermal layer. In the male, initial infection is usually on the penis; in the female it is most often in the

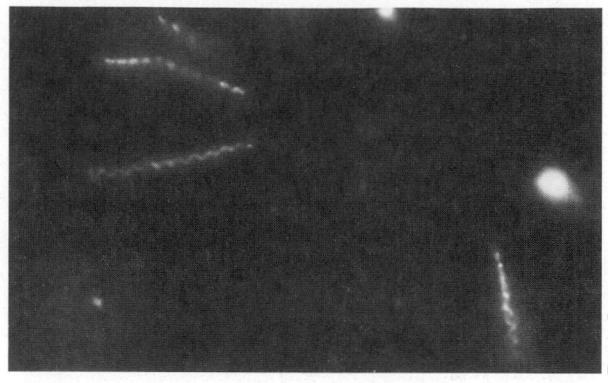
(a)

Theodor Rosebury

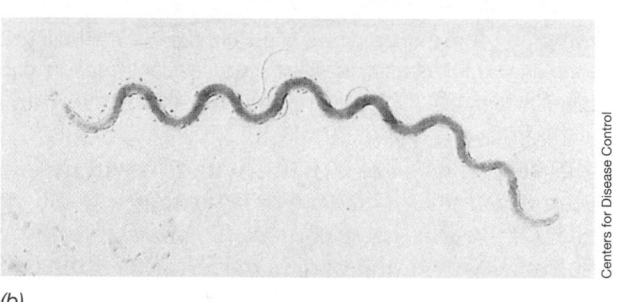
(b)

Centers for Disease Control

Figure 26.28 The spirochete of syphilis, *Treponema pallidum*. (a) Dark-field microscopy of an exudate. *Treponema pallidum* cells measure 0.15 μm wide and 10–15 μm long. (b) Shadow-cast electron micrograph of a cell of *T. pallidum*. Note the endoflagella, typical of spirochetes (∞ Section 12.33).

vagina, cervix, or perineal region. In about 10% of cases, infection is extragenital, usually in the oral region. During pregnancy, the organism can be transmitted from an infected woman to the fetus; the disease acquired in this way by an infant is called **congenital syphilis**. *T. pallidum* multiplies at the initial site of entry, and a characteristic *primary* lesion known as a *chancre* (Figure 26.29) is formed within 2 weeks to 2 months. Darkfield microscopy of the exudate from syphilitic chancres often reveals the actively motile spirochetes (Figure 26.28*a*). In most cases the chancre heals spontaneously and the organisms disappear from the site. Some cells, however, spread from the initial site to various parts of the body, such as the mucous membranes, the eyes, joints, bones, or central nervous system, and extensive multiplication occurs. A hypersensitivity reaction to the treponeme then often takes place, revealed by the development of a generalized skin rash; this rash is the key symptom of the *secondary* stage of the disease. The patient's condition may now be highly infectious, but eventually the organism disappears from secondary lesions and infectiousness ceases.

The subsequent course of the disease in the absence of treatment is highly variable. About one-fourth of infected individuals appear to undergo a cure as demon-

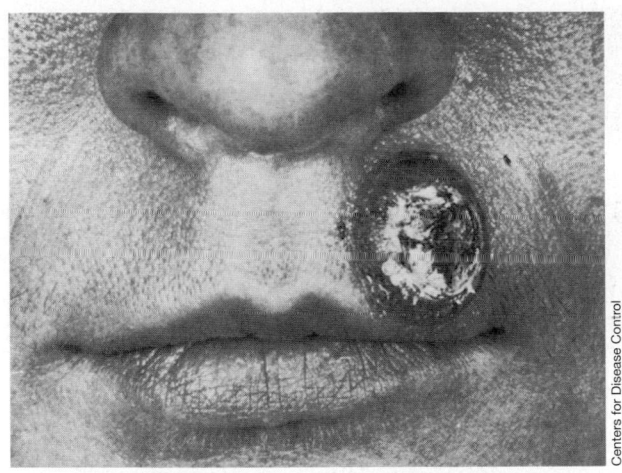

(a)

Centers for Disease Control

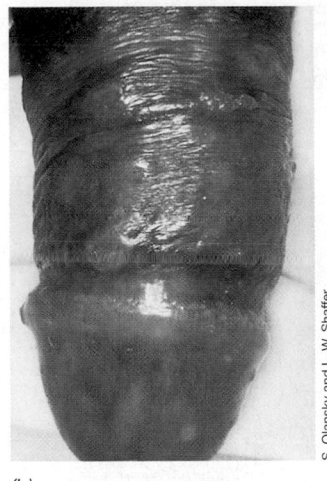

(b)

S. Olansky and L. W. Shaffer

Figure 26.29 Primary syphilis lesions. (a) Chancre on lip. (b) Several chancres on penis. The chancre is the characteristic lesion of primary syphilis at the site of infection by *Treponema pallidum*. Patients who acquire such lesions generally seek medical intervention, and the obvious chancre hastens diagnosis and treatment. Syphilis is treatable and curable with a single injection of penicillin G (⟳ Table 25.3).

strated by a decrease in antibody titer, and another one-fourth do not exhibit any further symptoms, although a demonstrable infection may persist, as demonstrated by a static or elevated antibody titer. In about half of the patients the disease enters the *tertiary* stage, with symptoms ranging from relatively mild infections of the skin and bone to serious or fatal infections of the cardiovascular system or central nervous system. Involvement of the nervous system is the most serious phase of the illness because generalized paralysis or other severe neurological damage may result. In the tertiary stage very few organisms are present, and most of the symptoms probably result from delayed hypersensitivity reactions (⟳ Sections 22.7 and 22.13) to the spirochetes.

We discussed the clinical immunology and microbiology as well as the laboratory diagnosis of syphilis in Section 24.1. The single most important physical sign of a primary syphilis infection, the chancre, is also diagnostic for the disease. Infected individuals generally seek treatment for syphilis because of the highly visible chancre.

Penicillin is highly effective in syphilis therapy, and the primary and secondary stages of the disease can usually be controlled by a single injection of benzathine penicillin G. In tertiary syphilis, penicillin treatment must extend for longer periods of time. The incidence of primary and secondary syphilis in the United States has decreased significantly over the last two decades and is now at the lowest level since record keeping began.

✓ 26.12 Concept Check

Gonorrhea and syphilis, caused by *Neisseria gonorrhoeae* and *Treponema pallidum*, respectively, are STDs with potential serious consequences if left untreated. Although the incidence of these diseases is at an all-time low, there are still over 300,000 cases of gonorrhea and 6000 cases of syphilis annually in the United States.

✓ Explain at least one potential reason for the high incidence of gonorrhea as compared with syphilis.

✓ Describe treatment methods for both gonorrhea and syphilis. Do these treatments effect a cure for each disease?

26.13 Chlamydia, Herpes, and Trichomoniasis

Chlamydia, herpes, and trichomoniasis are important STDs transmitted respectively by a bacterium, a virus, and a protozoan. These diseases are very prevalent in the population and are much more difficult to diagnose and treat than are syphilis and gonorrhea.

Chlamydia

A number of sexually transmitted diseases can be ascribed to infection by the obligate intracellular bacterium *Chlamydia trachomatis* (Figure 26.30 and ⟳ Section 12.27). The total incidence of sexually transmitted *C. trachomatis* infections (a reportable disease, ⟳ Table 25.3) probably greatly outnumbers the incidence of gonorrhea. There may be up to 3 million new *C. trachomatis* sexually transmitted infections every year, making this organism the most prevalent cause of venereal disease. *C. trachomatis* also causes a serious eye infection called trachoma (⟳ Section 12.27), but the strains of *C. trachomatis* responsible for venereal infections are distinct from those causing trachoma. Chlamydial infections may also be transmitted congenitally to the newborn from contamination in the birth canal, causing newborn conjunctivitis and pneumonia. Finally, chlamydial infections are now implicated in the development of arterial plaque and coronary artery disease.

Chlamydial nongonococcal urethritis (NGU) is one of the most frequently observed sexually transmitted diseases. *C. trachomatis* causes urethritis in males and urethritis, cervicitis, and pelvic inflammatory disease in

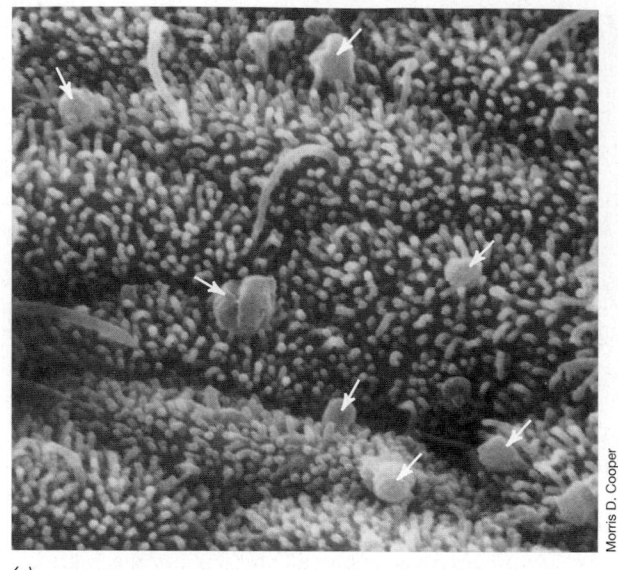

(a)

(b)

Figure 26.30 Cells of *Chlamydia trachomatis* (arrows) attached to human fallopian tube tissues. (a) Cells attached to the microvilli of a fallopian tube. (b) A damaged fallopian tube containing a cell of *C. trachomatis* (arrow) in the lesion.

females. In both the male and female, inapparent chlamydial infections are common. In a small percentage of cases, chlamydial NGU can lead to serious acute complications, including testicular swelling and prostate inflammation in men, and pelvic inflammatory diseases and fallopian tube damage in women; cells of *C. trachomatis* attach to microvilli of fallopian tube cells, enter, multiply, and eventually lyse the cells (Figure 26.30). Untreated disease can cause infertility.

Chlamydia NGU is relatively difficult to diagnose by traditional isolation and identification methods. To expedite diagnoses, a variety of immunological tests have been developed for identifying *C. trachomatis* from a vaginal or pelvic swab or from discharges. These clinical tests include fluorescent monoclonal antibodies and various enzyme-linked immunosorbent assay (ELISA) tests for detecting specific *C. trachomatis* antigens. If a chlamydial infection is suspected, treatment is initiated with azithromycin or doxycycline. Penicillin is ineffective against *C. trachomatis* because the organisms lack peptidoglycan, the target of penicillin (∞ Section 12.27 and Table 26.2).

Chlamydial NGU is frequently observed as a secondary infection following gonorrhea. If both *Neisseria gonorrhoeae* and *Chlamydia trachomatis* are transmitted to a new host in a single event, treatment of gonorrhea with cefixime or ceftriaxone is usually successful but does not eliminate the chlamydia. Although cured of gonorrhea, such patients are still infected with chlamydia and eventually experience an apparent recurrence of gonorrhea that is instead a case of chlamydial NGU. Thus, current recommendations are to *also* treat gonorrhea

patients with azithromycin or doxycycline to treat the potential coexisting, but usually undiagnosed, *C. trachomatis* infection.

Lymphogranuloma venereum is a sexually transmitted disease also caused by distinct *C. trachomatis*. The disease, which occurs most frequently in males, consists of a swelling of the lymph nodes in and about the groin. From the infected lymph nodes, chlamydial cells may travel to the rectum and cause a painful inflammation of rectal tissues called *proctitis*. Because of the potential for regional lymph node damage and the complications of proctitis, lymphogranuloma venereum is considered to be one of the most serious sexually transmitted chlamydial syndromes.

Herpes

Herpesviruses are a large group of complex double stranded DNA viruses (∞ Section 16.11), many of which are human pathogens. A subgroup of herpesviruses, the *herpes simplex viruses*, are responsible for cold sores and genital infections.

Herpes simplex virus type 1 (HSV-1) infects the epithelial cells particularly around the mouth and lips, causing cold sores and fever blisters (Figure 26.31). HSV-1 is generally spread via direct contact or through saliva. HSV-1 may, however, occasionally infect other body sites including the anogenital regions. The incubation period of HSV-1 infections is short (3–5 days), and the lesions heal without treatment in 2–3 weeks. Relapses of HSV-1 infections are relatively common, and it is thought that the virus is spread primarily via contact with infectious lesions. Latent herpes infections are ap-

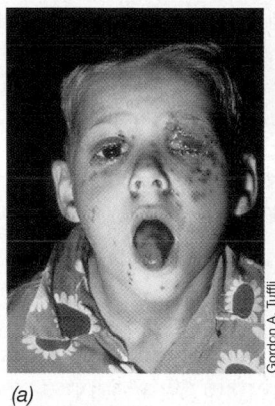
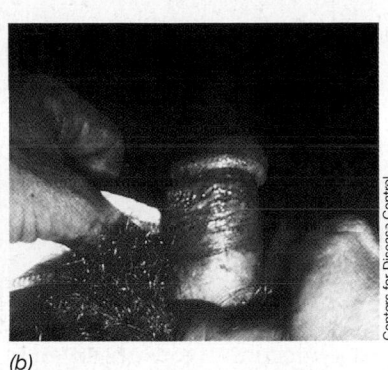
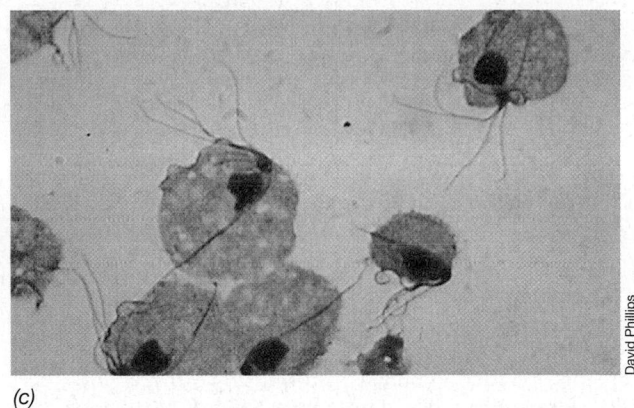

(a)   (b)   (c)

Figure 26.31 Nonbacterial sexually transmitted pathogens: herpesvirus and *Trichomonas*. (a) A severe case of herpes blisters on the face due to infection with herpes simplex virus type 1. (b) Genital herpes due to infection on the penis with herpes simplex virus type 2. (c) Cells of the flagellated protozoan *Trichomonas vaginalis*.

parently quite common, with the virus persisting in low numbers in nerve tissue. Recurrent acute herpes infections are due to a periodic triggering of virus activity by unknown causes.

Herpes simplex virus 2 (HSV-2) infections are associated primarily with the anogenital region, where the virus causes painful blisters on the penis of males or on the cervix, vulva, or vagina of females (Figure 26.31*b*). HSV-2 infections are transmitted by direct sexual contact, and the disease is most easily transmitted during the active blister stage rather than during periods of inapparent (presumably latent) infection. HSV-2 occasionally infects other body sites such as the mucous membranes of the mouth.

The long-term effects of genital herpes infections are not yet understood. Oral herpes is quite common and apparently has no harmful effects on the host beyond the oral blisters. However, epidemiological studies have shown a significant correlation between *genital* herpes infections and cervical cancer in females. In addition, HSV-2 can be transmitted to a newborn at birth by contact with herpetic lesions in the birth canal. The disease in the newborn varies from latent infections with no apparent damage to systemic disease that can result in brain damage or death. To avoid passing herpes infections to newborns, delivery by caesarean section is advised for pregnant women with genital herpes infections.

Genital herpes infections are presently incurable, although a limited number of drugs have been successful in controlling the infectious blister stages. The guanine analog **acyclovir** (Figure 26.32), given orally and also applied topically, is particularly effective in limiting the shed of active virus from blisters and promoting the healing of blistering lesions. Acyclovir specifically interferes with herpesvirus DNA polymerase, inhibiting viral DNA replication.

Trichomoniasis

Nongonococcal urethritis may also be caused by infections with the protozoan *Trichomonas vaginalis* (Figure 26.31*c*). Although many protozoa produce resting cells called *cysts*, *T. vaginalis* does not produce cysts. Thus transmission must be from person to person, generally by sexual intercourse. However, cells of *T. vaginalis* can survive a few hours outside the host, provided they do not dry out. Thus, transmission of *T. vaginalis* by contaminated toilet seats, sauna benches, and paper towels occasionally occurs. *T. vaginalis* infects the vagina in women, the prostate and seminal vesicles of men, and the urethra of both males and females.

Many cases of trichomoniasis are totally asymptomatic in males. In women trichomoniasis is characterized by a vaginal discharge, vaginitis, and painful urination. The infection is more common in females; surveys indicate that 25 to 50% of sexually active women are infected, while only about 5% of men are infected. The male partner of an infected female should be examined for *T. vaginalis* and treated if necessary because promiscuous asymptomatic males can serve as reservoirs, transmitting the infection to several females. Trichomoniasis is diagnosed by observation of the motile protozoan in a wet mount of fluid discharged from the patient (Figure 26.31*c*).

Guanine   H   Acyclovir   CH₂OCH₂CH₂OH

Figure 26.32 Structure of guanine and the guanine analog acyclovir. Acyclovir has been used therapeutically to control genital herpes (HSV-2) blisters.

The antiprotozoal drug *metronidazole* is particularly effective in treating trichomoniasis (Table 26.2).

✓ 26.13 Concept Check

Chlamydia is a STD caused by infection with the bacterium *Chlamydia trachomatis*. Chlamydia is the most prevalent STD. Herpes lesions can also be transmitted sexually and are caused by herpes simplex virus type 1 and herpes simplex virus type 2. HSV-2 is generally associated with sexual transmission and infection of the anogenital regions, although HSV-1 can also cause anogenital lesions. *Trichomonas vaginalis* is a protozoan responsible for trichomoniasis, another STD. In general, these STDs are widespread and are more difficult to diagnose and treat than gonorrhea or syphilis. There is no cure for herpes.

✓ Describe pertinent clinical features and treatment protocols for chlamydia, herpes, and trichomoniasis.

✓ Why are these diseases more difficult to diagnose than gonorrhea or syphilis?

26.14 Acquired Immunodeficiency Syndrome: AIDS and HIV

Acquired immunodeficiency syndrome (AIDS) was recognized as a distinct disease in 1981. More than 800,000 cases of AIDS have been reported since then in the United States alone, and more than 400,000 people have died (∞ Section 25.6). Over 900,000 people in the United States may now be infected with HIV. Worldwide, the outlook is even more serious, with more than 50 million people already infected with the human immunodeficiency virus (HIV), the causative agent of AIDS. Over 16 million people have already died from AIDS.

HIV is divided into two major types, HIV-1 and HIV-2. HIV-1 is genetically similar, but distinct from HIV-2. HIV-2, discovered in West Africa in 1985, has reduced virulence as compared with HIV-1, but also causes an AIDS-like disease. Currently, more than 99% of global AIDS cases are due to HIV-1, and, therefore, our discussions of AIDS will center on infections with HIV-1.

The numbers of HIV-infected individuals will continue to rise dramatically unless effective treatment or prevention methods are discovered. We have already discussed the epidemiology of AIDS (∞ Section 25.6), and the clinical diagnostic methods for identifying and tracking HIV infection (∞ Sections 24.11–24.12). In this section, we will concentrate on the pathogenesis of AIDS.

Human Immunodeficiency Virus

The disease AIDS is caused by human immunodeficiency virus. HIV-1 is a retrovirus (∞ Section 16.14) containing 9749 nucleotides in each of its two identical single-stranded RNA genomes. Using the enzyme *reverse transcriptase*, which is present in the intact virion, HIV forms a complementary single-stranded DNA molecule using RNA as a template and converts this complementary DNA (cDNA) into double-stranded DNA, which can enter the host cell genome. Here we consider the natural course of HIV infection and the effects of HIV on the immune system, leading to the development of AIDS.

A Definition of AIDS

AIDS was first suspected of being a disease affecting the immune system because a large number of opportunistic infections were observed in certain populations (∞ Section 25.6). *Opportunistic infections* are infections rarely observed in individuals with normal immune systems. This definition was adopted in 1993 by the Centers for Disease Control and Prevention (Atlanta, Georgia, USA) and is used to define AIDS cases in the United States.

The current case definition for acquired immunodeficiency syndrome (AIDS) includes individuals who test positive for human immunodeficiency virus (HIV) AND

1. have a CD4 T-cell number of less than 200/mm³ of whole blood (the normal count is 600–1000/mm³), or a CD4 T-cell/total lymphocytes percentage of less than 14%; OR

2. have a CD4 T-cell number of more than 200/mm³ and any of the following conditions: fungal diseases including candidiasis (Figure 26.33*a*), coccidiomycosis, cryptococcosis (Figure 26.33*b*), histoplasmosis (Figure 26.33*c*), isosporiasis, *Pneumocystis carinii* pneumonia (Figure 26.33*d*), cryptosporidiosis (Figure 26.33*e*), or toxoplasmosis (Figure 26.33*f*) of the brain; bacterial diseases including pulmonary tuberculosis or other *Mycobacterium* spp. infections (Figure 26.33*g*), or recurrent *Salmonella* septicemia; viral diseases including cytomegalovirus infection, HIV-related encephalopathy, HIV wasting syndrome, chronic ulcers, or bronchitis due to *herpes simplex* (∞ Section 16.11); or progressive multifocal leukoencephalopathy, malignant diseases such as invasive cervical cancer, Kaposi's sarcoma (Figure 26.34), Burkitt's lymphoma, primary lymphoma of the brain, or immunoblastic lymphoma; recurrent pneumonia due to any agent.

The most common of these opportunistic infections is *Pneumocystis carinii* pneumonia. Another very frequent disease in AIDS patients is Kaposi's sarcoma, an atypical cancer observed in high frequency in HIV-infected homosexual males. Kaposi's sarcoma is a cancer of the cells lining the blood vessels and is characterized by purple patches on the surface of the skin, especially in the extremities (Figure 26.34). Kaposi's sarcoma may be caused by coinfection of HIV and human herpesvirus 8

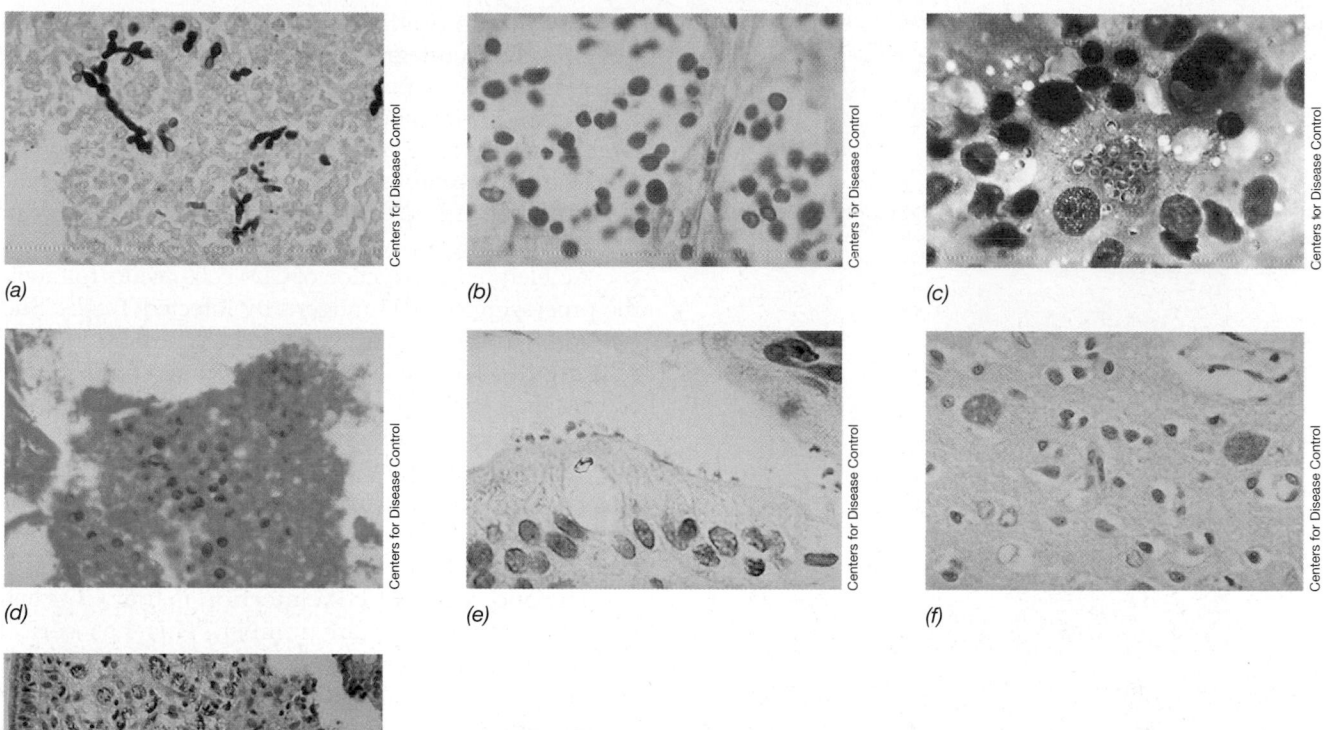

Figure 26.33 Opportunistic pathogens associated with cases of acquired immunodeficiency syndrome (AIDS). (a) *Candida albicans*, from heart tissue of patient with systemic *Candida* infection. (b) *Cryptococcus neoformans*, from liver tissue of a patient with cryptococcosis. (c) *Histoplasma capsulatum*, from liver tissue of patient with the histoplasmosis. (d) *Pneumocystis carinii*, from patient with pulmonary pneumocytosis. (e) *Cryptosporidium* sp. from small intestine of a patient with cryptosporidiosis. (f) *Toxoplasma gondii*, from brain tissue of patient with toxoplasmosis. (g) *Mycobacterium* spp. infection of small bowel, acid-fast stain.

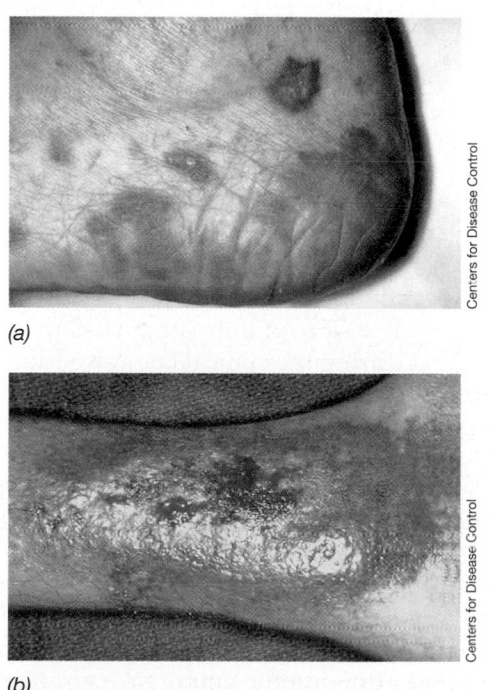

Figure 26.34 Kaposi's sarcoma lesions as they appear on (a) the heel and lateral foot, and (b) the distal leg and ankle.

(I II IV-8) and is 20,000-fold more prevalent in AIDS patients than in the general population.

In summary, an individual has AIDS if (1) he or she tests positive for HIV or HIV antibodies, *and* (2) has a drastically reduced T-helper lymphocyte count, *or* (3) has at least one of a number of opportunistic infections or atypical cancers.

HIV: Cell Interactions and Infection

HIV has the ability to infect cells displaying the CD4 cell-surface protein. Although a number of cells, including B cells and certain brain and intestinal cells, have very low levels of CD4 on their surfaces, the two cell types most commonly infected are macrophages and T-helper (T$_H$) cells, both of which are important components of the immune system (Sections 22.1 and 22.7). Infected macrophages and T cells produce and release large numbers of HIV particles, which in turn infect other cells that display CD4 (Figure 26.35).

HIV infection normally occurs first in macrophages, an antigen-presenting cell (APC) that has a very low level of CD4 on its surface (Section 22.1) (Figure 26.36). At the cell surface, the macrophage CD4 molecule binds to the gp120 protein of HIV. The viral

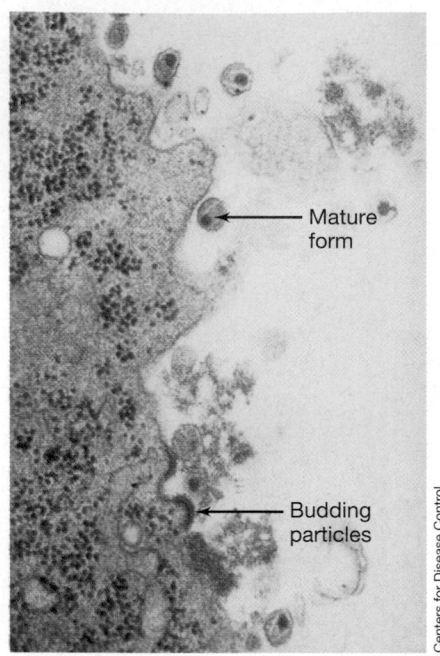

Centers for Disease Control

Figure 26.35 Transmission electron micrograph of a thin section of a lymphocyte releasing human immunodeficiency virus (HIV). Cells were from a hemophiliac patient who developed AIDS. HIV particles are 90–120 nm in diameter.

gp120 protein then interacts with another macrophage protein, the membrane-spanning chemokine receptor CCR5 (∞ Section 23.10). CCR5 acts as a coreceptor for HIV and, together with CD4, forms the docking site where the HIV envelope fuses with the host cell membrane, allowing insertion of the viral nucleocapsid.

The CCR5 coreceptor is required for HIV binding to macrophages. Individuals who express a variant CCR5 protein do not bind HIV and do not acquire HIV infection or AIDS.

After HIV has infected the macrophage APCs, a different form of gp120 is made, which in turn binds to a different coreceptor, the CXCR4 chemokine receptor on T cells. HIV then enters and destroys the CD4 T-helper lymphocytes, the T_H1 and T_H2 cells that are responsible for cell-mediated inflammatory responses and B-cell help, respectively (∞ Section 22.7). Thus, HIV starts as a macrophage infection and progresses to a T-cell infection. The net result of HIV infection is the systematic destruction of macrophages and T cells, leading to a catastrophic breakdown of immunity. Specific knowledge of the coreceptors involved in HIV infection in macrophages and T cells may be used to design HIV-specific chemokine-receptor blocking agents to prevent the attachment of HIV to either the macrophage or the T cell, thus preventing infection.

In cases of clinical AIDS, CD4 lymphocytes are greatly reduced in number. However, HIV does not immediately kill and lyse its host cell. Following reverse transcription to produce DNA from the RNA genome, the viral cDNA integrates into host chromosomal DNA where it exists as a provirus. The cell may show no outward sign of infection, and HIV DNA can remain in a latent state for long periods. Eventually, however, productive virus synthesis occurs and new HIV particles are produced and released from the cell. T cells producing HIV no longer divide and eventually die.

Accelerated destruction of CD4 cells occurs following the processing of HIV antigens by infected T cells. Such cells insert molecules of gp120 from HIV particles into their cell surfaces. The embedded gp120 protein on the infected cells then sticks to uninfected T cells by binding to the CD4 molecule. Eventually, numerous cells of each type fuse to produce multinucleate giant cells called *syncytia*. One HIV-infected T cell may eventually bind and fuse with up to 50 uninfected T cells. Shortly after syncytia formation occurs, the fused cells lose immune function and die.

The end result of HIV infection is that CD4 cells progressively decline in number. This has serious health consequences. In a normal human, CD4 cells constitute about 70% of the total T-cell pool; in AIDS patients, the number of CD4 cells steadily decreases, and by the time opportunistic infections become established, CD4 cells may be almost absent (∞ Figure 24.21 and Figure 26.37). As CD4 cells decline in number, there is a concomitant loss in the cytokines they produce, leading to the gradual reduction of uninfected T cells. Eventually all other lymphocyte production is shut down and the immune system in those suffering from clinical AIDS is effectively destroyed. This loss of both humoral and cellular immune function becomes apparent. Systemic infections by fungi and mycobacteria (Figure 26.33) point to a loss in T_H1 cellular immunity (∞ Section 22.7). Other opportunistic infections, such as the various viral and bacterial infections associated with AIDS, indicate the loss of humoral immunity; decline in antibody production is due to the loss of T_H2 cells necessary to stimulate antibody production by B cells (∞ Section 22.7).

The overall picture of untreated AIDS progression indicates that during the clinical latency period, a very active infectious process is proceeding. First, there is an intense immune response to HIV: About 1 billion virions are destroyed each day and HIV numbers drop. However, this means that HIV is replicating at a very high rate, and this replication results in the corresponding destruction of about 100 million CD4 T cells each day. Eventually, the immune response is simply overwhelmed, HIV levels increase, and the T cells are completely destroyed, crippling the immune response and allowing the emergence of opportunistic infections. The example in Figure 26.37 documents T-cell destruction and the increase in HIV over a typical time course for AIDS.

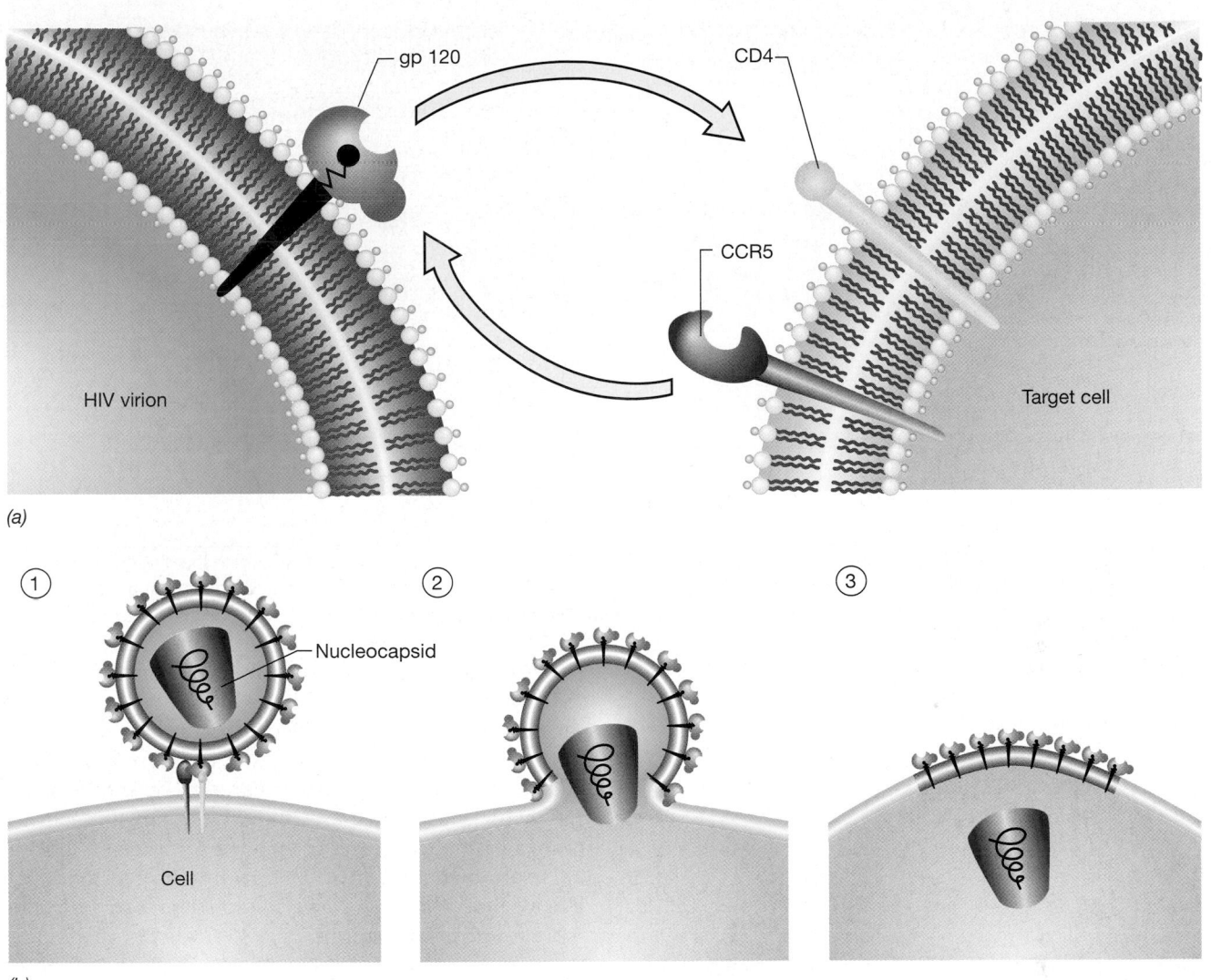

Figure 26.36 Infection of a CD4 target cell with the human immunodeficiency virus (HIV). (a) Interaction of HIV with a target cell through specific binding of HIV gp120 to the CD4 receptor and a coreceptor on target cells such as macrophages or T cells. The coreceptor shown is the membrane-spanning chemokine receptor CCR5 found on macrophages. A similar coreceptor, CXCR4, is found on CD4 T cells. (b) Fusion of the HIV envelope with the host cell membrane and insertion of the nucleocapsid. The actual site of virus entry is the site formed by the receptor and coreceptor. The coreceptor is necessary for viral insertion and infection because cells that do not express the coreceptor do not get infected with HIV.

Diagnosis of AIDS

Diagnosis of AIDS is based on clinical and laboratory findings, with the key elements being a positive test for HIV in the blood, a depressed level of CD4 T cells in the blood, and the presence or recent history of one or more opportunistic infections or atypical cancers. The presence of HIV antibodies in the blood is the usual indicator for HIV exposure and infection (⚭ Sections 24.11 and 24.12). In addition, several laboratory tests have been developed, based on the *reverse transcriptase-polymerase chain reaction (RT-PCR)* (⚭ Section 18.5) that identifies HIV RNA directly and quantitatively from blood samples (⚭ Section 24.13). The RT-PCR estimates the number of viruses present in the blood, or

the so-called **viral load**. The RT-PCR test indicates the magnitude of HIV replication and correlates with the rate of CD4 T cell destruction. The CD4 test indicates the extent of HIV-induced immune damage already suffered, a direct indicator of the magnitude of destruction of the immune system. The RT-PCR test is not routinely used to screen for HIV because it is costly and technically demanding. After initial discovery of infection, however, the RT-PCR test is used to monitor progression of AIDS and the effectiveness of chemotherapy (Figure 26.38).

The prognosis of an untreated HIV-infected individual is poor. Opportunistic pathogens or malignancies (Figures 26.33 and 26.34) eventually kill most AIDS

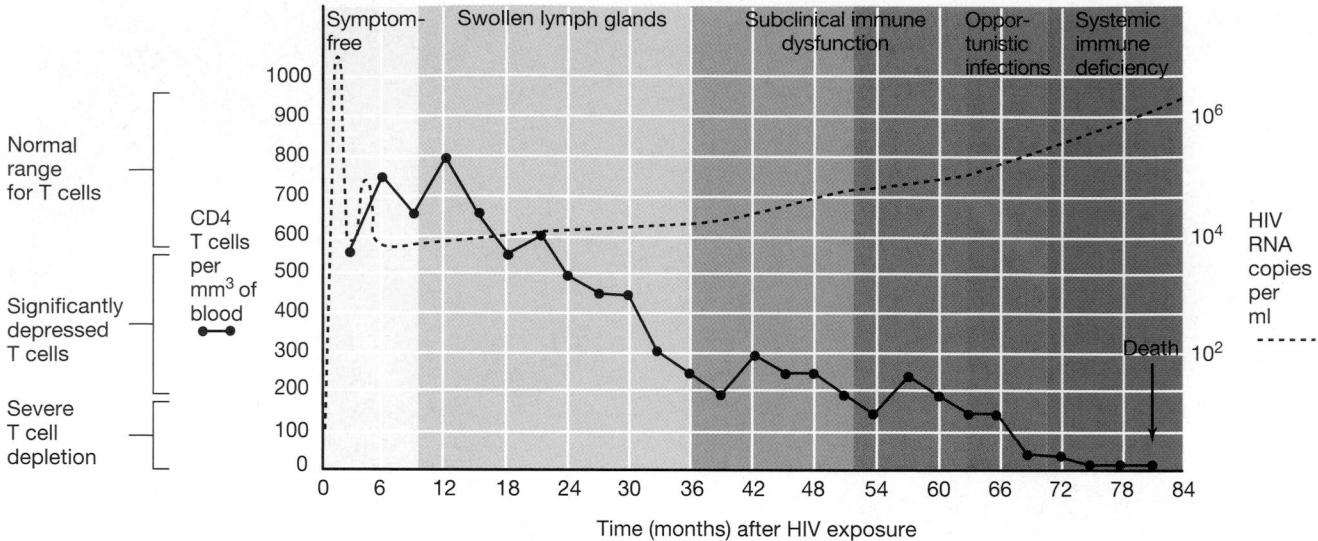

Figure 26.37 Decline of CD4 T lymphocytes and progress of HIV infection, based on viral numbers (viral load) in the blood of a typical un-treated AIDS patient (∞ Section 24.13). During the typical progression of AIDS, there is a gradual loss in the number and functional ability of the CD4 T cells, while the viral load, measured as HIV-specific RNA per milliliter of blood, gradually increases after an initial decline.

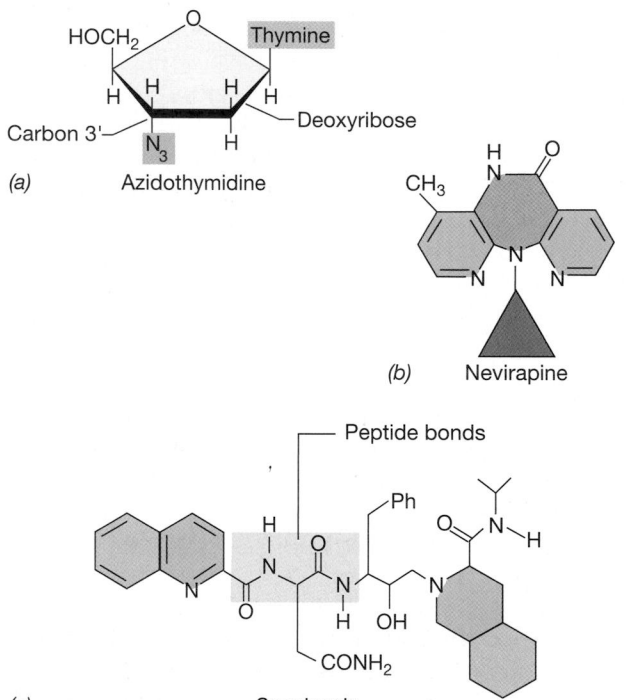

Figure 26.38 Examples of the three categories of HIV/AIDS chemotherapeutic drugs. (a) Azidothymidine (AZT), a nucleoside ana-log. The lack of an OH group on the 3′ carbon causes nucleotide chain elongation to cease when this analog is incorporated, inhibiting virus replication. (b) Nevirapine, a non-nucleoside reverse transcriptase in-hibitor, binds directly to the catalytic site, also inhibiting elongation of the nucleotide chain. (c) Saquinavir, a protease inhibitor, was designed by computer modeling to fit the active site of the HIV protease (∞ Section 20.13). Saquinavir is a peptide analog: The tan highlighted area shows the region analogous to peptide bonds. Blocking the ac-tivity of HIV protease prevents the processing of HIV proteins and maturation of the virus (∞ Figure 20.27).

patients. Long-term studies of AIDS patients indicate that the average person infected with HIV progresses through several stages of decreasing immune function, with CD4 cells dropping from a normal range of 600–1000/mm^3 of blood to near zero over a period of 5–7 years (Figure 26.37). Although the *rate* of decline in immune function varies from one HIV-infected individual to another, it is rare for an HIV-positive individual to live for more than 10 years without some form of chemotherapeutic intervention.

Treatment of AIDS

No *cure* for HIV infection is known, although research is very intensive in the areas of vaccine production and chemotherapy. Several drugs have been identified that delay symptoms of AIDS and in some cases prolong the life of those infected with HIV (Table 26.3).

Effective anti-HIV drugs fall into three major categories. The first of these are a group of *nucleoside analogs* that act as reverse transcriptase inhibitors. *Reverse transcriptase* is the enzyme that converts the single-stranded RNA genetic information into complementary DNA (∞ Section 16.14). The oldest effective anti-HIV drug *azidothymidine (AZT)*, is an effective inhibitor of HIV replication because it closely resembles the nucleoside thymidine but lacks the correct attachment site for the next base in an extending nucleotide chain, resulting in termination of the growing DNA chain. Thus, AZT and the other nucleoside analogs are **nucleoside reverse transcriptase inhibitors (NRTIs)**, effectively stopping HIV replication (Figure 26.38a, Table 26.3). When these drugs are given to HIV-infected individuals, the result is a rapid decrease in viral load. However, within several weeks, drug resistant strains of HIV arise in each patient

TABLE 26.3 Chemotherapeutic agents approved for HIV/AIDS treatment

Drug	Mechanism of action
Azidothymidine (AZT, ZDV, or Zidovudine)(Figure 26.38a)	*Nucleoside analog*; reverse transcriptase inhibitor; nucleotide chain synthesis terminator; increases survival time and reduces incidence of opportunistic infection in AIDS patients; toxic to bone marrow cells; may be used in combination with other drugs in multiple drug treatment protocols.
Dideoxycytidine (ddC or zalcitabine) Dideoxyinosine (ddI or didanosine) Stavudine (d4T) Lamivudine (3TC)	*Nucleoside analogs*; reverse transcriptase inhibitors; mechanism of action and effects are the same as AZT; may have less toxicity than AZT in some patients; may be used in combination with other drugs in multiple drug treatment protocols.
Efavirenz Nevirapine (Figure 26.38b) Delavirdine	*Nonnucleoside reverse transcriptase inhibitors* (NNRTIs); bind directly to reverse transcriptase and disrupts the catalytic site; do not compete with nucleosides; may be used in combination with other drugs in multiple drug treatment protocols.
Indinavir Nelfinavir Ritonavir Saquinavir (Figure 26.38c)	*Protease inhibitors*; computer-designed peptide analogs designed to bind to the active site of HIV protease, inhibiting processing of viral polypeptides and virus maturation; may be used in combination with other drugs in multiple drug treatment protocols.

as a result of mutation and selection. Although this process seems very rapid, HIV replicates very quickly (see above) and as little as four single-nucleotide mutations result in resistance to a given nucleoside analog.

The second category of anti-HIV drugs are the **nonnucleoside reverse transcriptase inhibitors (NNRTIs)** (Figure 26.38b; Table 26.3). These compounds directly inhibit the action of reverse transcriptase by interacting with the protein and altering the conformation of the catalytic site. Unfortunately, a single mutation in the reverse transcriptase gene is often sufficient to reduce the effectiveness of these drugs. The final category of anti-HIV drugs are the **protease inhibitors** (Figure 26.38c; Table 26.3). The protease inhibitors are computer-designed peptide analogs designed to bind to the active site of HIV protease, inhibiting processing of viral polypeptides (⌒⌒ Section 20.13); this inhibits virus maturation. However, as with the other enzyme-targeted chemotherapy strategy (reverse transcriptase; see above), a single mutation in the HIV protease gene is capable of rendering these drugs nonfunctional.

Because of the problems with drug resistance, a typical recommended protocol for treatment of an individual with established HIV infection includes at least one protease inhibitor *plus* a combination of two nucleoside analogs (Table 26.3). Multiple drug therapy (so-called "drug cocktails") reduces the possibility that a drug-resistant virus could emerge because the virus would need to develop resistance to three drugs simultaneously. This combination therapy is then monitored by the RT-PCR methods to track changes in viral load. An effective protocol reduces viral load to nondetectable levels (less than 500 copies of HIV per milliliter of blood) within several days. The therapy is continued and monitored for viral load indefinitely. If the viral load again reaches detectable limits, the drug cocktail is changed because an

increase in viral load indicates the emergence of a drug-resistant virus population (⌒⌒ Section 24.13).

In addition to drug resistance, some of the antiviral drugs are toxic to the host. In many cases, nucleoside analogs are not well tolerated by patients, presumably because they interfere with host functions such as cell division (Table 26.3). In general, the NNRTIs and the protease inhibitors are better tolerated because they interfere with only virus-specific functions. However, drug resistance coupled with host toxicity are major problems in HIV therapy, and new chemotherapeutic agents and drug protocols are constantly being developed.

AIDS Immunization

The genetic variability of HIV has thus far hampered the development of an AIDS vaccine. One strategy is to make antibodies to the envelope protein, gp120, and use these antibodies to block CD4-gp120 interactions (Figure 26.37) and thus block infection. However, this approach has not been successful thus far because the gene encoding gp120 mutates frequently, forming antigenic variants of the protein that are not recognized by antibodies made to a different form. The most impressive results from clinical immunization trials have emerged from *subunit vaccines* (⌒⌒ Section 31.6), where genes for several HIV envelope proteins have been engineered into vaccinia virus or adenovirus particles. Using these harmless viruses as expression vectors and vehicles for delivery of HIV antigens, several subunit vaccines elicit a potent humoral and cellular immune response to HIV. Clinical trials of subunit immunization procedures are under way.

Other potential immunization candidates include *killed intact HIV*. These inactivated vaccines are restricted to use in HIV-infected individuals because inactivation procedures may not kill 100% of the HIV; it would be un-

A Focus On ... Sexual Activity and AIDS

Sexual promiscuity has always been associated with sexually transmitted diseases, but the acquired immunodeficiency syndrome (AIDS) epidemic, discussed in this chapter and elsewhere in this book, has focused attention on the dangers of multiple sex partners and on the high risk associated with certain sex practices. AIDS, caused by the human immunodeficiency virus (HIV), is only one type of sexually transmitted disease. Others include gonorrhea, syphilis, herpes simplex, nonspecific urethritis (caused by *Chlamydia*), protozoal vaginitis (caused by *Trichomonas vaginalis*), fungal vaginitis (caused by *Candida albicans*), and venereal warts (caused by the human papilloma virus). Some of these sexually transmitted diseases have been associated with human society for all of recorded history. However, AIDS is unique. There are no drugs or immunizations to cure or prevent AIDS. Drugs used for treatment are costly and available only in developed countries, so 90% of the 50 million or more HIV-infected individuals do not have access to therapy. AIDS already kills more than 2 million people every year, and this number will grow as people who have harbored and spread HIV for up to 10 years develop full-blown AIDS.

Because AIDS is linked to certain sex practices, prevention means avoidance of these sex practices. The United States Surgeon General has issued a report that makes specific recommendations that individuals can follow if they wish to reduce the likelihood of AIDS infection. Among the recommendations are

1. Avoid mouth contact with penis, vagina, or rectum.
2. Avoid all sexual activities that could cause cuts or tears in the linings of the rectum, vagina, or penis.
3. Avoid sexual activities with individuals from high risk groups. These include prostitutes (both male and female), promiscuous homosexual men, bisexual individuals, and intravenous drug users.
4. If a person has had sex with a member of one of the high risk groups, a blood test should be done to determine if infection with HIV has occurred. If the test is positive, then it is essential that sexual partners of an HIV-positive individual be protected by use of a condom during sexual intercourse.

It is important to emphasize that AIDS is *not* just a disease of male homosexuals. In certain cultures, AIDS is as common in women as in men. The disease is linked to promiscuous sexual activities and other activities that involve exchange of body fluids, which include not only male homosexuality but also female prostitution and intravenous drug use.

Is it possible, then, to have sex without incurring the risk of AIDS? Certain sex practices are inherently much safer than others. Safe sex practices include dry kissing (mouths closed), mutual masturbation (in the absence of breaks in the skin), and intercourse protected by a condom. Dangerous sex practices include wet kissing (mouths open), masturbation where breaks in the skin occur, oral sex (either male or female), and unprotected sexual intercourse (either vaginal or anal). The U.S. Surgeon General has recommended that if the health status of the partner is unknown, a condom be used for all sex practices in which exchange of body fluids occurs.

The AIDS epidemic has focused new attention on the condom (see photo). Condoms have always played two roles in sexual activity: disease protection and prevention of pregnancy. Although the best way to avoid AIDS is to avoid dangerous sex practices, if sexual intercourse is to be carried out with an individual whose infection status is unknown, then a latex condom should be used. The U.S. Surgeon General strongly recommends the use of condoms for all extramarital sexual activity. In certain countries, advertising campaigns to promote the use of condoms are widespread.

Moralistic statements alone (prescriptions for monogamy, abstinence, avoidance of sexual activity outside of matrimony), will *not* control the AIDS epidemic. Epidemiological studies on all previously known sexually transmitted diseases have shown that fear of disease is not, by itself, sufficient to prevent sexual activities that put an individual at risk for a sexually transmitted disease. The sex drive in some individuals is so strong that it will suppress the fear of disease, even a disease like AIDS. Every individual must therefore take the responsibility for protecting himself or herself from this widespread and extremely dangerous infectious disease.

For more information about AIDS, the Public Health Service has a toll-free telephone resource, the PHS AIDS Hotline. The number is 800-342-2437. For the latest information on AIDS prevention, contact the CDC National Prevention Information Network at http://www.cdcnpin.org/. ■

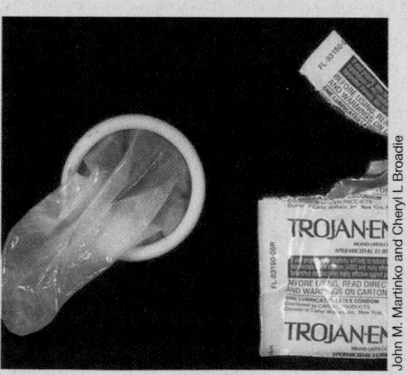

John M. Martinko and Cheryl L. Broadie

ethical to expose uninfected individuals to even a small risk of HIV infection. Next, some laboratories are exploring the possibilities of producing *live attenuated virus* for use as an immunizing agent. This strategy is bolstered by the finding that individuals infected with HIV-2, a related virus that causes a very mild form of AIDS with a very long latent period, prevents infection with HIV-1, the strain responsible for severe AIDS. However, there are serious potential risks. For example, integrated virus could cause cancer, mutations might reactivate virulence, and so on.

In all, there are about 20 different AIDS immunization protocols undergoing clinical trials, but no immunization has yet been found to be effective. Additionally,

AIDS immunization would most probably not be useful in *treating* most patients that already have the disease because these individuals lack significant immune function and could not respond to a vaccine. Thus, despite considerable advances in our molecular understanding of HIV and in our clinical understanding of the AIDS disease process, public education about AIDS and avoidance of high risk behavior are still the major tools to combat AIDS today (see the box, Sexual Activity and AIDS).

Detection of HIV Infection

Exposure to HIV can be diagnosed by immunological means. Both radioimmunoassay (RIA) and enzyme-linked immunosorbent assay (ELISA) tests (Section 24.11) have been developed for screening blood samples to detect anti-HIV antibodies. The ELISA test has proven particularly valuable for large-scale screening of donated blood to prevent transfusion-associated HIV. Statistics have shown that about 0.25% (2–3 per thousand) of all blood donated by volunteer donors in the United States tests HIV-positive in the ELISA assay. A positive HIV-ELISA test must be confirmed by a second procedure called immunoblotting (Western blotting), a technique that combines the analytical tools of protein purification and immunology (Section 24.12).

A number of other rapid tests are being developed and marketed to identify individuals who are infected with HIV. One test uses a single drop of patient blood and a single reagent. The reagent is a bioengineered antibody in which one binding site is directed to a red blood cell antigen, while the other site is directed to the gp41 HIV surface antigen. In a positive test, the bifunctional antibody cross-links the red blood cells to the HIV, resulting in a visible agglutination (Section 24.8). In another test, saliva is used as a source of secretory antibody to HIV. The saliva is expelled onto a cartridge containing immobilized HIV antigens. A second antibody, reactive with the bound antibody and conjugated to an enzyme, is then added. After addition of the enzyme substrate, a positive reaction shows a colored product, as in the ELISA methods (Section 24.11). The rapid tests are designed to provide maximum convenience, speed (*minutes* instead of the hours or days required for ELISA or immunoblot), extended shelf life, portability, and ease of application and interpretation. However, in general, the rapid tests are not as sensitive or accurate as the standard HIV-ELISA and HIV-immunoblot tests (Section 24.11 and 24.12).

All tests, no matter how sensitive or accurate, fail to detect HIV-positive individuals who have recently acquired the virus but have not yet made an antibody response, which may be 6 weeks to a year after exposure to HIV. In spite of this drawback, these tests ensure the general safety of the blood supply, and statistics indicate that the risk of contracting HIV through contaminated blood or blood products is now very low. Sexual promiscuity and group intravenous drug use are the major routes of HIV infection today (Figure 25.9).

✓ 26.14 Concept Check

AIDS is one of the most prevalent infectious diseases in the human population. HIV destroys the immune system, and opportunistic pathogens then kill the host. There is still no effective vaccine for HIV. However, several antiviral drugs are available that slow the progress of AIDS. The only prevention for the spread of HIV infection is through avoidance of behavior such as intravenous drug use (needle sharing) and unsafe sexual practices.

✓ Review the definition of AIDS. What diagnostic features are shared by all AIDS patients?

✓ What is the current treatment for HIV infection? Is it effective? What are the side effects?

Review Questions

1. Why do gram-positive bacteria cause respiratory diseases more frequently than gram-negative bacteria?

2. What are the typical symptoms of a streptococcal respiratory infection? Why should streptococcal infections be treated promptly?

3. Describe the causal agent and the symptoms of diphtheria. How is diphtheria prevented?

4. Describe the causal agent and the symptoms of whooping cough. Describe the changes in vaccine technology that have led to safer pertussis vaccines.

5. Describe the process of infection by *Mycobacterium tuberculosis*. Does infection always lead to active tuberculosis? Why or why not?

6. Describe the symptoms of meningococcemia and meningitis. How are these diseases treated? What is the prognosis for each?

7. Compare and contrast measles, mumps, and rubella. Include in your discussion a description of the pathogen, major symptoms encountered, and any potentially serious consequences of these infections. Why is it important that women be vaccinated against rubella *before* puberty?

8. Why are colds and influenza such common respiratory diseases? Give at least two reasons for the high incidence of each of these diseases.

9. Distinguish between pathogenic staphylococci and those that are part of the normal flora.

10. Describe the evidence linking *Helicobacter pylori* to gastric ulcers. How would you treat an ulcer patient?

11. Describe the major pathogenic hepatitis viruses. How are they related to one another?

12. Why did the incidence of gonorrhea rise dramatically in the mid-1960's, while the incidence of syphilis actually decreased at the same time?

13. For the sexually transmitted diseases chlamydia, herpes, and trichomoniasis, describe the methods of treating each infection. In each case, is the treatment an effective cure? Why or why not?

14. Describe how human immunodeficiency virus (HIV) effectively shuts down most aspects of both humoral immunity and cell-mediated immunity. Are there any parts of the immune system that remain functional in cases of acquired immunodeficiency syndrome (AIDS)?

Application Questions

1. Most sneezes generate an aerosol of mucous droplets containing microorganisms. Coughs also generate a considerable aerosol, as well as sputum. Why are infections transmitted by sneezes likely to be *upper* respiratory tract infections? Why are infections transmitted by coughs likely to be *lower* respiratory tract infections?

2. Explain the origin of *rheumatic fever* and the source of antigens involved in this *autoimmune* disease.

3. What is the reservoir for *Coynebacterium diphtheriae*? Can this disease be permanently eradicated in humans? How?

4. How can an epidemic of whooping cough be controlled? How can it be prevented? Since the incidence of this disease is rising, apply your prevention methods to the current "miniepidemic" (Figure 26.7). Would your methods be cost-effective?

5. Why does tuberculosis often lead to a permanent reduction in lung capacity, whereas most other respiratory diseases cause only temporary respiratory problems?

6. Your lab partner goes home for the weekend, becomes extremely ill, and is diagnosed with bacterial meningitis at a local hospital. Because he was away, university officials are not aware of his illness. What should you do to protect yourself against meningitis? Should you notify university health officials?

7. Measles, mumps, rubella, and chickenpox were once considered normal childhood diseases. However, they are now regarded by many as very serious infectious diseases. Explain this shift in attitudes in the context of disease prevalence, availability of vaccines, and the potential health consequences of each disease.

8. Discuss the molecular biology of *antigenic shift* in influenza viruses and comment on the immunologic consequences for the host. Why does antigenic shift prevent the production of a single universally effective vaccine for influenza control? Next, compare antigenic *shift* to antigenic *drift*. Which mechanism is more important for the evolution of the influenza virus? Which causes the greatest antigenic change? Which creates the biggest problems for vaccine developers? Why?

9. How can you eliminate antibiotic-resistant staphylococci from a hospital ward? Think in terms of physical changes to the ward as well as changes in personnel, personnel monitoring, and requirements for protective clothing, gloves, masks, and so on.

10. Using Koch's postulates (∞ Section 1.5) as a guide, design experiments to prove that *Helicobacter pylori* causes gastric ulcers.

11. Arrange the hepatitis viruses in order of disease severity, both in the short term and in the long term.

12. Despite the ease with which gonorrhea and syphilis can be diagnosed and cured, the incidence of gonorrhea remains very high while the incidence of syphilis is very low. Explain.

13. As the director of your dormitory's public health advisory group, you are charged to present information on chlamydia, herpes, and trichomoniasis, all STDs. Present information on prevention, symptoms, and treatment. Will your program for each disease overlap? For each of the diseases, discuss the social, legal, and public health issues that must be considered for reporting an occurrence of the disease to sexual partners of infected individuals.

14. In terms of chemotherapy, how is HIV infection currently treated? How is treatment monitored? Does treatment work? Why or why not? Describe the various vaccine possibilities for AIDS. How might each vaccine type induce immunity to HIV? Is passive immunity of potential use for treatment of AIDS? Why or why not? What risks may be involved?

Through the ages, arthropod-borne diseases such as plague and malaria have killed millions of humans and have actually influenced the course of human history and evolution. The arthropod-borne Lyme disease, caused by the bite of a tick of the genus *Ixodes* (shown here) that is infected with the Lyme pathogen *Borrelia burgdorferi*, is an important infectious disease that is now recognized throughout the world. Anyone in North America, Europe, or Asia has the potential to acquire this rapidly emerging disease if they come into contact with ticks.

27

ANIMAL-TRANSMITTED, ARTHROPOD-TRANSMITTED, AND SOILBORNE MICROBIAL DISEASES

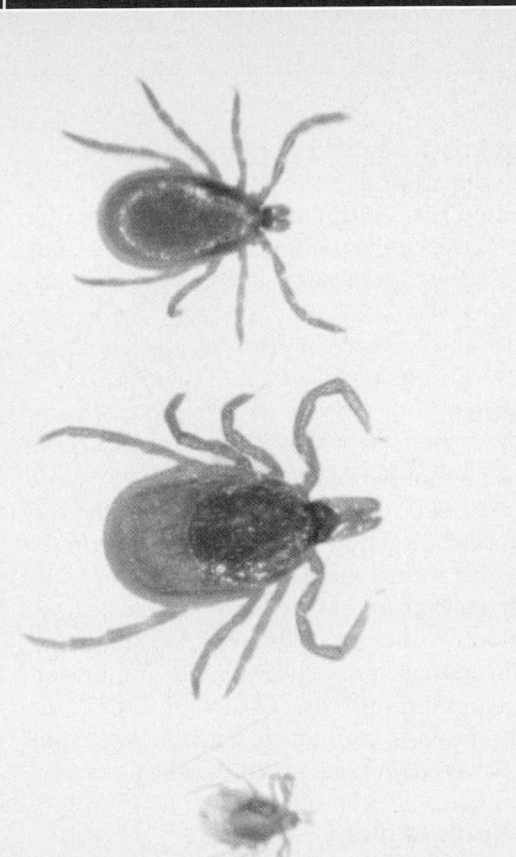

Working Glossary

Ehrlichiosis one of a group of emerging tick-transmitted diseases caused by rickettsias of the *Ehrlichia* genus

Hantavirus pulmonary syndrome (HPS) an emerging acute viral disease characterized by respiratory pneumonia, obtained by transmission of hantavirus from rodents

Lyme disease an emerging tick-transmitted disease caused by the spirochete *Borrelia burgdorferi*

Malaria an insect-transmitted disease characterized by recurrent episodes of fever and anemia caused by the protozoan *Plasmodium* spp., usually transmitted between mammals through the bite of the *Anopheles* mosquito

Mycoses infections cause by fungi

Plague an endemic disease in rodents caused by *Yersinia pestis* that is occasionally transferred to humans through the bite of a flea

Rabies a usually fatal neurological disease caused by the rabies virus that is usually transmitted by the bite or saliva of an infected carnivore

Rickettsias obligate intracellular parasites that cause a variety of diseases including typhus, Rocky Mountain spotted fever, and ehrlichiosis

Rocky Mountain spotted fever a tick-transmitted disease caused by *Rickettsia rickettsii*, causing fever, headache, rash, and gastrointestinal symptoms

Sickle cell anemia a genetic trait that confers resistance to malaria but causes a reduction in the efficiency of red blood cells by reducing the oxygen-binding affinity of hemoglobin

Thallasemia a genetic trait that confers resistance to malaria but causes a reduction in the efficiency of red blood cells by altering a red blood cell enzyme

Tetanus a disease involving rigid paralysis of the voluntary muscles, caused by an exotoxin produced by *Clostridium tetani*

Typhus a louse-transmitted disease caused by *Rickettsia prowazekii*, causing fever, headache, weakness, rash, and damage to the central nervous system and internal organs

Zoonoses animal diseases transmitted to humans

Insect-transmitted diseases such as plague and malaria have had a significant role in human history. Today, vectorborne diseases such as Lyme disease and hantavirus pulmonary syndrome are emerging as important infectious diseases, even in highly developed portions of the world. The diseases that we will discuss in this chapter have nonhuman reservoirs, such as the common wild deer mouse. This common forest mammal can harbor hantavirus as well as the Lyme disease bacteria. Our inabilty to control the populations of vector and reservoir animals makes control of certain diseases nearly impossible. Soilborne diseases present major problems because organisms in soil cannot be eradicated or effectively controlled.

I ANIMAL-TRANSMITTED DISEASES

Here we discuss several examples of animal-transmitted and vectorborne pathogens and the diseases they cause. The natural *host* for the animal-transmitted pathogens is a nonhuman vertebrate. Populations of wild animals act as reservoirs for these diseases, making pathogen irradication unlikely or impossible. As we shall see, when infected animal populations contact humans, the result is often human infection. The animal-transmitted diseases are generally spread to hosts such as humans by direct contact, aerosols, or bites. In the case of vectorborne

diseases, pathogens are spread to uninfected hosts via the bite of an arthropod *vector* that last fed on a pathogen-infected host. Humans are often accidental hosts in the life cycles of the pathogens, but infected humans may also act as a disease reservoir, as is the case with malaria.

27.1 Rabies

Animals contract a number of infectious diseases, some of which can be passed to humans. Fortunately, the use of effective immunization practices and good veterinary care prevents most infectious disease in domestic animals and limits the transfer of disease to humans. However, the situation is different in the wild (feral) animal population. Wild animals cannot be routinely immunized and do not receive veterinary care. Thus, animal diseases transmissible to humans (**zoonoses**) occur in wild animal populations, usually on a periodic, cyclic basis.

Biology and Epidemiology

Rabies is one of a handful of diseases that occurs primarily in animals but is spread to humans under certain conditions. The major reservoir of rabies in the United States is in wild animals, primarily raccoons, skunks, coyotes, foxes, and bats. However, a small but significant number of cases are also seen in domestic animals (Figure 27.1). Rabies is still a major disease in humans worldwide: Approximately 35,000 people die every year from this disease, primarily in developing countries where rabies is still endemic in domestic animals such as dogs, and about 1 million people worldwide receive rabies treat-

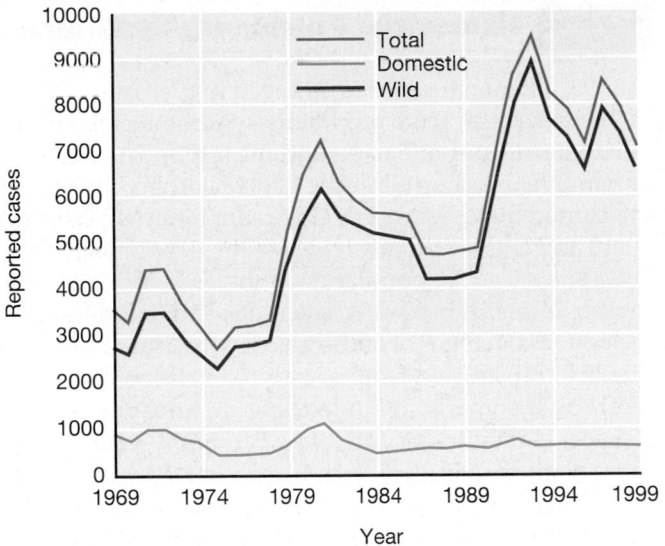

Figure 27.1 Rabies cases in wild and domestic animals in the United States. Since 1990, there have been 25 human rabies cases. Rabies is endemic in wild animal populations in the United States. The periodic rise in reported cases following several years of decline is the result of periodic reemergence of rabies in wild animals, especially in raccoons in the northeastern United States. Note that there are over 500 annual cases of rabies in domestic animals, presumably acquired due to contact with wild animals. Data are from the Centers for Disease Control and Prevention, Atlanta, GA, USA.

ment for animal bites. Rabies is caused by a member of the Rhabdovirus family (a negative-stranded RNA virus) (∞ Section 16.8) that infects cells in the central nervous system of most warm-blooded animals, almost invariably leading to death if not treated. The virus, present in the saliva of rabid animals, enters the body through a wound from a bite. Rabies virus multiplies at the site of inoculation and then travels to the central nervous system. The incubation period for the onset of symptoms is highly variable, depending on the size, location, and depth of the wound and the actual number of viral particles transmitted in the bite. In dogs, the incubation period averages 10–14 days. In humans, up to nine months may pass before rabies symptoms become apparent.

Pathogenesis

The virus proliferates in the brain (especially in the thalamus and hypothalamus), leading to fever, excitation, dilation of the pupils, excessive salivation, and anxiety. A fear of swallowing (hydrophobia) develops from uncontrollable spasms of the throat muscles; death eventually results from respiratory paralysis. In humans, an *untreated* rabies infection that progresses to the symptomatic stage is nearly always fatal.

Treatment, Diagnosis, and Prevention

Because of the lethal nature of rabies, any contact with potentially rabid animals must be taken seriously. A wild animal suspected of being rabid should be captured, sacrificed, and immediately examined for evidence of rabies. If a domestic animal, generally a dog, cat, or ferret, bites a human, especially if the bite is unprovoked, the animal is generally held 10 days for observation for development of clinical signs of rabies. If the animal is wild, exhibits rabies symptoms, or a determination cannot be made after 10 days, the patient will be passively immunized with rabies immune globulin (purified anti-rabies virus antibodies obtained from a hyperimmune individual) (∞ Section 22.11), injected at both the site of the bite and intramuscularly. The patient will also be immunized with an inactivated rabies virus preparation. A summary of guidelines for treating possible human exposure to rabies is shown in Table 27.1. Because of the very slow progression of rabies in humans, this combination therapy is nearly 100% effective, stopping the onset of the active disease. Rabies is diagnosed in the laboratory by examining tissue samples. Fluorescent antibody tests (∞ Section 24.10) that recognize rabies-infected brain or corneal tissue are used for confirming a clinical diagnosis of rabies, either in a potentially rabid animal or in postmortem examination of a human case. In addition, characteristic virus inclusion bodies in the

TABLE 27.1 Guidelines for treating possible human exposure to rabies virus	
Unprovoked bite by a domestic animal	
Animal suspected of rabies	*Animal not suspected of rabies*
1. Sacrifice animal and test for rabies.	1. Hold for 10 days. If no symptoms, do not treat human.
2. Begin treatment of human immediately.[a]	2. If symptoms develop, treat human immediately.[a]
Bite by wild carnivore (for example, skunk, bat, fox, raccoon, coyote)	
Regard animal as rabid	
1. Sacrifice animal and test for rabies.	
2. Begin treatment of human immediately.[a]	
Bite by wild rodent, squirrel, livestock, rabbit	
Consult local or state public health officials about possible recent cases of rabies transmitted by these animals (these animals rarely transmit rabies). If no reports, do not treat human.	

[a]All bites should be thoroughly cleansed with soap and water. Treatment is generally a combination of rabies immunoglobulin and human diploid cell rabies vaccine (five injections intramuscularly).

cytoplasm of nerve cells, called *Negri bodies*, also obtained by biopsy or postmortem sampling, are taken as confirmation of rabies.

Rabies is prevented largely through effective immunization practices. A number of effective inactivated rabies vaccines are used in the United States for both human and domestic animal immunizations, and a variety of effective inactivated and attenuated virus preparations are also used worldwide. As we discussed above, most cases of human rabies are preventable through prompt immunization: Because of the long incubation period, passive and active immunization of potentially exposed individuals is sufficient to prevent disease. Therefore, prophylactic rabies immunization is not generally recommended for humans, except for individuals at high risk such as veterinarians and animal control personnel.

The rabies prevention strategy just described has been extremely successful, and fewer than three cases of human rabies are reported in the United States each year, nearly always the result of a bite by a wild animal. Domestic animals are another matter. Since domestic animals often have exposure to wild animals, all dogs and cats should be vaccinated beginning at 3 months of age, and booster inoculations should be given yearly or triannually. Other domestic animals, including large farm animals, are also often immunized with rabies vaccines. However, the key to effective rabies prevention and possible eradication, at least in the United States, lies in control of the disease in the large wild animal reservoir of rabies virus (Figure 27.1). If all or even most members of the potential disease reservoir are immune, the disease can be stopped and possibly even eradicated. Currently, subunit vaccines consisting of rabies virus genes that encode rabies coat proteins expressed in vaccinia virus (∞ Section 31.6) are available. Because effective doses can be given orally, subunit vaccines can be included in food "baits" and used to immunize local populations of susceptible wild animals to reduce the incidence and spread of rabies. Such vaccines are a completely safe means of controlling rabies in the wild animal reservoir and, if successful in wild animal populations, could lead to eventual eradication of the disease.

✓ 27.1 Concept Check

Rabies occurs primarily in wild animals in the United States but can be transmitted to domestic animals or to humans. Vaccination of dogs and cats is of central importance for the control of rabies. In developing countries, rabies is often endemic in domestic animals and is still an important human disease.

✓ Why is rabies vaccine not given routinely to humans in the United States?

✓ What is the procedure for treating a patient bitten by an animal if the animal cannot be found?

✓ What advantages might an oral vaccine have over a parenteral vaccine for rabies control in wild animals?

27.2 Hantavirus Pulmonary Syndrome

In 1993, a hantavirus strain emerged in the United States as the cause of an acute respiratory syndrome now known as **hantavirus pulmonary syndrome** (HPS). The outbreak of hantavirus occurred in the "Four Corners" region of the United States (Arizona, Colorado, New Mexico, and Utah) and was eventually traced to the deer mouse (*Peromyscus maniculatis*) population in the area. The outbreak caused 32 deaths in 53 infected adults, and it underscored the potential danger of outbreaks due to diseases that are directly transmitted from a variety of different animal reservoirs, sometimes under new or unusual circumstances. Hantaviruses are related to many viruses that cause hemorrhagic fever, such Lassa fever virus and Ebola virus (∞ Section 25.10), all viruses that are occasionally transmitted to humans from animal reservoirs.

Biology and Epidemiology

The genus *Hantavirus* is a member of the Bunyaviridae, a family of enveloped single-stranded RNA viruses (∞ Section 16.8) (Figure 27.2). The family includes a number of viruses that cause either *hantavirus pulmonary syndrome (HPS)* or *hemorrhagic fever with renal syndrome (HFRS)*. The viruses are primarily found in rodents, including mice and rats of several species, lemmings, and voles, and are occasionally found in other animals. The HFRS strains, also found in rodents, are more common in the Eastern Hemisphere, and have been implicated in a number of HFRS outbreaks in recent years. Up to 200,000 cases per year are recognized, chiefly in China, Korea, and Russia. The HPS strains are more prevalent in the Western Hemisphere, and continued investigation of the ecology of these viruses is likely to uncover a number of other pathogenic strains.

Hantaviruses are most commonly transmitted by inhalation of virus-contaminated rodent excreta. Humans seem to be an accidental host and are infected only when they come into contact with rodents and their feces. For example, all of the individuals who acquired HPS in the Four Corners outbreak had been exposed to mice or their droppings, the result of a warm winter and an unusual increase in numbers of rodents in 1993 (∞ Section 25.10). The *aerosol route* of infection is most common, with most aerosols being in the form of dust generated from mouse droppings or dried urine. However, there are rare reports of person-to-person transmission, as well as a few incidents where HPS or HFRS was spread by a rodent bite.

Pathology, Diagnosis, Treatment and Prevention

Hantavirus pulmonary syndrome is characterized by a sudden onset of fever, myalgia, thrombocytopenia (reduction in the number of blood platelets), leukocytosis (an increase in the number of circulating lymphocytes),

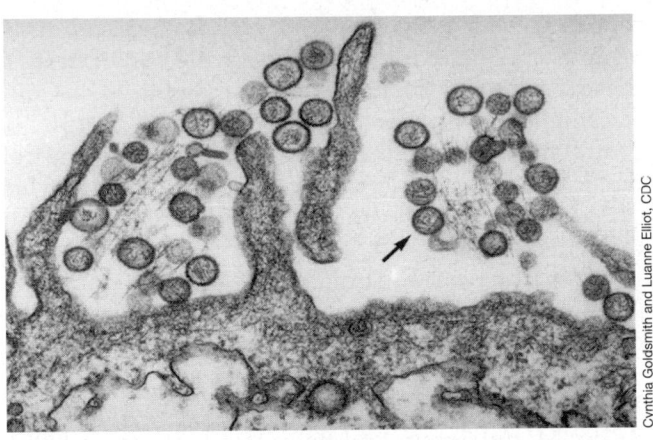

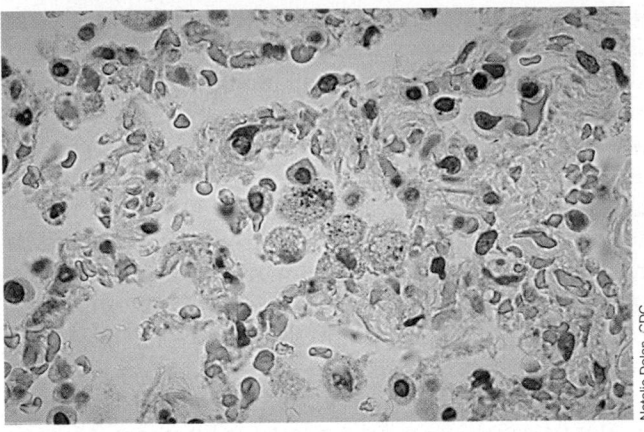

(a)

(b)

Figure 27.2 Hantavirus. (a) An electron micrograph of the Sin Nombre hantavirus. The arrow indicates one of several viruses. The virus is approximately 0.1 μm in diameter. (b) Immunostaining of Andes hantavirus antigens in alveolar macrophages. Each granular dark blue stained area indicates cellular infection of an individual macrophage (approximately 15 μm in diameter). Hantaviruses belong to the Bunyaviridae and, like many phylogenetically similar viruses (for example, Rift Valley virus and Ebola virus), cause hemorrhagic fevers with very high human mortality.

and pulmonary capillary leakage. Death occurs in a matter of several days in about 50% of cases, usually due to shock and cardiac complications precipitated by pulmonary edema (leakage of fluid into the lungs, causing suffocation). These symptoms are typical of the Sin Nombre virus, which caused the Four Corners outbreak, but a variety of other symptoms may be evident, depending on the strain of virus causing the disease. For example, the Bayou strain common in rodents in the southeastern United States also causes kidney failure.

If hantavirus from candidate infections can be grown in tissue culture (∞ Section 9.11), the strain can be identified by serological techniques including a virus plaque-reduction neutralization assay. More commonly, ELISAs (enzyme-linked immunoassays) are performed on patient blood to identify antibodies, indicating exposure and an immune response (∞ Section 24.11); or the presence of the viral genome, indicating infection, is detected based on a PCR (polymerase chain reaction) assay using patient tissue or blood specimens (∞ Section 24.1).

There is no virus-specific treatment or vaccine for any of the hantaviruses. However, the disease can be prevented by avoiding contact with rodents and rodent habitat, since its major mode of transmission is through exposure to rodent exreta. Reduction in exposure can be accomplished by destroying mouse habitat, restricting food supplies (for example, keeping food in sealed containers), and aggressive rodent extermination measures. The long-term prognosis for disease eradication is poor because a considerable percentage of the rodent population in a given geographical area is generally infected with the local hantavirus strain. For example, retrospective serological testing of the deer mice in the Four Corners area in 1993 indicated that 30% of the local mouse population carried the Sin Nombre hantavirus.

✓ **27.2 Concept Check**

Hantaviruses occur worldwide in rodent populations and cause serious diseases such as hantavirus pulmonary syndrome (HPS) when accidentally spread to humans. In the United States, hantavirus infections were not recognized until 1993.

✓ Why were hantaviruses not recognized as a major public health problem until 1993 in the United States?

✓ Describe the spread of hantaviruses to humans. What are some effective measures for preventing infection by hantaviruses?

II ARTHROPOD-TRANSMITTED DISEASES

Pathogens can be spread to hosts from the bite of a pathogen-infected arthropod vector. In many cases, such as in the rickettsial illnesses, humans are accidental hosts for the pathogen. However, infected humans can also play a critical role in the life cycle of the pathogen, as is the case for malaria.

27.3 Rickettsial Diseases

The **rickettsias** are small bacteria that have a strictly intracellular existence in vertebrates, usually in mammals, and are also associated at some point in their natural cycle with blood-sucking arthropods such as fleas, lice, or ticks. We discussed the biology of rickettsias in Section 12.13. Rickettsias cause a variety of diseases in humans and animals, of which the most important are typhus fever,

Rocky Mountain spotted fever, and ehrlichiosis. Rickettsias take their name from Howard Ricketts, a scientist at the University of Chicago who first provided evidence for their existence and who died from infection with the rickettsia that causes typhus fever, *Rickettsia prowazekii*. Rickettsias have not been cultured in artificial media but can be cultured in laboratory animals, lice, mammalian tissue culture cells, and the yolk sac of chick embryos. In animals, growth takes place primarily in phagocytic cells (∞ Section 22.2). Although the rickettsias have not been grown in pure culture, the 1.1 Mb genome of *Rickettsia prowazekii* has been sequenced (∞ Section 15.3). Based on homology with other genomic sequences, these intracellular parasites are closely related to human mitochondria. Like the mitochondria, the rickettsial genome has been reduced in size to contain a set of genes tailored for intracellular dependency. The rickettsias do not have many of the genes necessary for independent energy metabolism and structural biosynthesis. The rickettsial genome also contains a set of virulence genes closely related to the *virB* operon of the plant pathogen *Agrobacterium tumefaciens* (∞ Section 19.21). This operon encodes components of virulence factors involved in DNA transfer and protein export systems. Thus, the genomic sequence provides evidence for the intracellular dependence and the virulence of these pathogens.

Rickettsias are divided into three groups, based loosely on the clinical diseases they produce. The groups are (1) the *typhus group*, typified by *Rickettsia prowazekii*; (2) the *spotted fever group*, typified by *Rickettsia rickettsii*; and (3) the *ehrlichiosis group*, characterized by *Ehrlichia chaffeensis*. Here, we examine one example of a pathogen from each group.

The Typhus Group: *Rickettsia prowazekii*

Typhus is caused by *Rickettsia prowazekii*. Epidemic typhus is transmitted from human to human by the common body or head louse. The known mammalian reservoirs for typhus are humans. Typhus can be a serious disease. During World War I, an epidemic of typhus spread throughout eastern Europe and caused almost 3 million deaths. Typhus has frequently been a problem among military troops during wartime. Because of the unsanitary, cramped conditions characteristic of wartime military infantry operations, lice are spread easily among soldiers, and typhus is spread in epidemic proportions. Up until World War II, typhus caused more military deaths than combat.

Cells of *R. prowazekii* are introduced through the skin when the puncture caused by the louse bite becomes contaminated with louse feces, the major source of rickettsial cells. During an incubation period of 1–3 weeks, the organism multiplies inside cells lining the small blood vessels. Symptoms of typhus (fever, headache, and general body weakness) then begin to appear. Five to

nine days later a characteristic *rash* is observed in the armpits and generally spreads over the body *except* for the face, palms of the hands, and soles of the feet. Complications from untreated typhus involve damage to the central nervous system, lungs, kidneys, and heart. Epidemic typhus has a mortality rate of 6–30%. Tetracycline and chloramphenicol are most commonly used to control *R. prowazekii*.

Rickettsia typhi, the organism that causes murine typhus, is another important pathogen in the typhus group.

The Spotted Fever Group: *Rickettsia rickettsii*

Rocky Mountain spotted fever was first recognized in the western United States about 1900 but is more prevalent today in the southeastern United States. Rocky Mountain spotted fever is caused by *Rickettsia rickettsii* and is transmitted to humans by various species of ticks, most commonly the dog and wood ticks (Figure 27.3). Humans acquire the pathogen from tick fecal matter, which is injected into the body during a bite, or by rubbing infectious material into the skin by scratching. Cells of *R. rickettsii*, unlike other rickettsias, grow within the nucleus of the host cell as well as in host cell cytoplasm (Figures 27.3a and 27.3b). Following an incubation period of 3–12 days, an abrupt onset of symptoms occurs, including fever and a severe headache. Within 3 to 5 days, a rash occurs on the whole body (Figure 27.3c). Gastrointestinal problems such as diarrhea and vomiting are usually observed as well, and the clinical symptoms of Rocky Mountain spotted fever may exist for over 2 weeks if the disease is left untreated. Tetracycline or chloramphenicol generally promotes a prompt recovery from Rocky Mountain spotted fever if administered early in the course of the infection.

The Ehrlichiosis Group: *Ehrlichia*

The genus *Ehrlichia* (∞ Section 12.13) is responsible for two emerging tickborne diseases in the United States, **human monocytic ehrlichiosis (HME)** and **human granulocytic ehrlichiosis (HGE)**. The rickettsias that cause the diseases are *Ehrlichia chaffeensis* and an organism similar or identical to *Ehrlichia equi*, respectively. The onset of these clinically indistinguishable ehrlichioses is characterized by flulike symptoms including fever, headache, malaise, and frequently leukopenia (decreased number of leukocytes) and/or thrombocytopenia. Laboratory findings frequently document changes in liver function, characterized by an increase in the enzyme hepatic transaminase. In addition, peripheral blood leukocytes have visible inclusions of *Ehrlichia* cells (Figure 27.4). The symptoms, except for the inclusions, are similar to other rickettsioses, but the *Ehrlichia* genus is antigenically distinct from members of the other major rickettsial groups. Infections range from

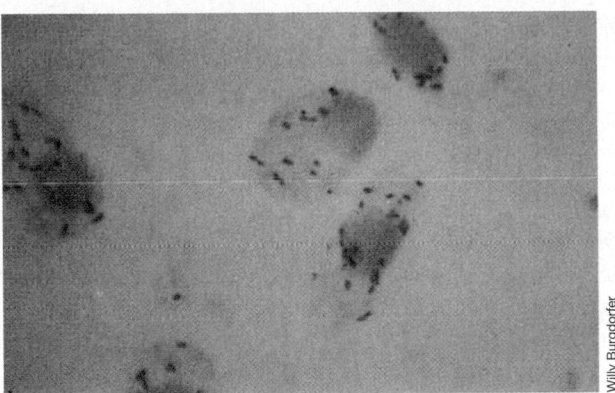

(a)

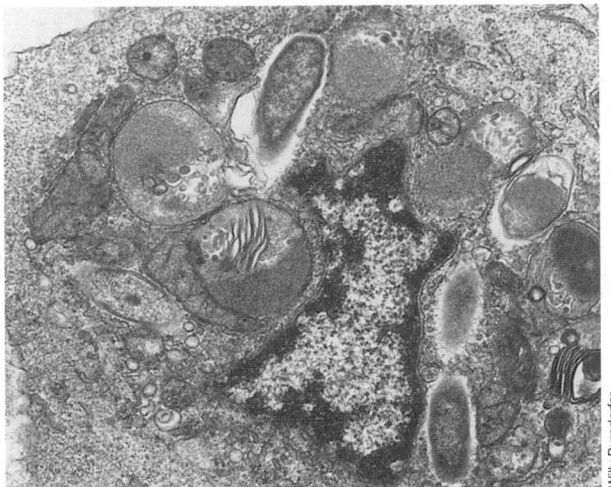

(b)

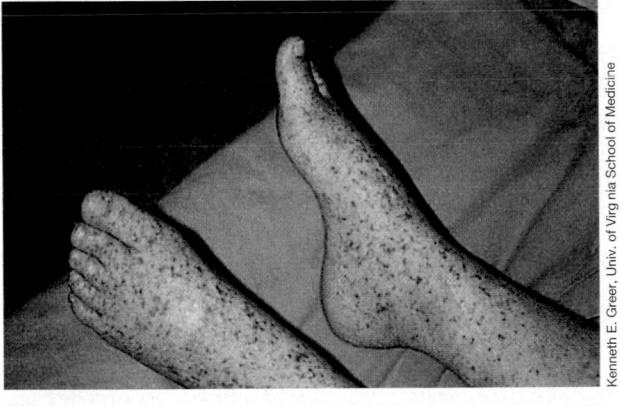

(c)

Figure 27.3 *Rickettsia rickettsii*, the causative agent of Rocky Mountain spotted fever. (a) Cells of *R. rickettsii*, growing in the cytoplasm and nucleus of tick hemocytes. Individual cells are about 0.4 μm in diameter. (b) Cells of *R. rickettsii* in a granular hemocyte of an infected wood tick, *Dermacentor andersoni*. Transmission electron micrograph. (c) Rash of the disease on the feet. The appearance of a rash covering the whole body is indicative of Rocky Mountain spotted fever and helps distinguish this disease clinically from typhus, in which the rash does not cover the whole body.

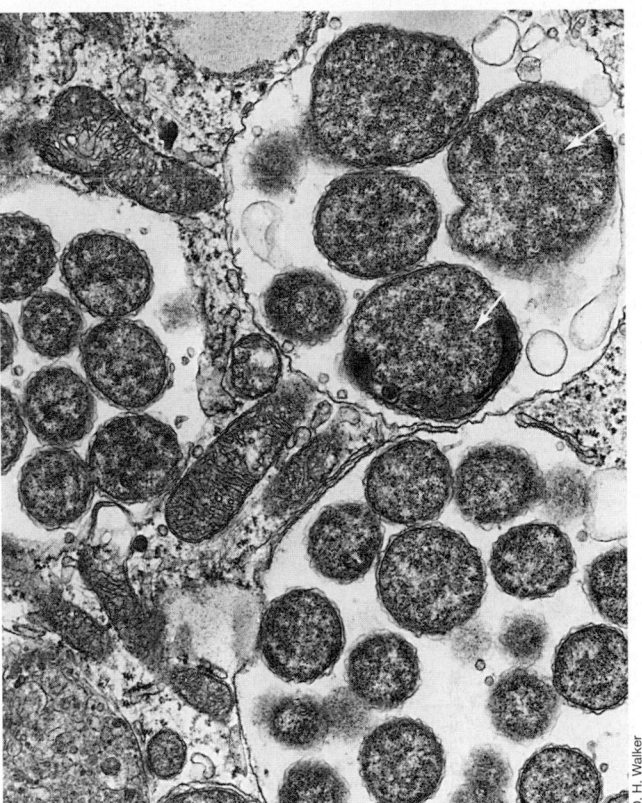

Figure 27.4 *Ehrlichia chaffeensis*, the causative agent of human monocytic ehrlichiosis (HME). The electron micrograph shows inclusions in a human monocyte which contains large numbers of *E. chaffeensis*. The arrows indicate two of the many bacteria in each inclusion. *E. chaffeensis* cells range from about 300–900 nm in diameter.

subclinical to fatal. Long-term complications for progressive untreated cases may include respiratory and renal insufficiency and serious neurological involvement. Diagnosis is based on an indirect fluorescence antibody assay (Section 24.10) of patient serum and also on polymerase chain reaction (PCR) tests of whole blood or serum to detect the presence of *Ehrlichia* DNA (Section 24.13).

The infection is spread by the bites of infected ticks; the mammalian reservoirs include deer and possibly rodents, in addition to the human hosts. Retrospective serological analyses in areas with relatively high incidence of tickborne disease indicate that HGE may be a more prevalent disease than Rocky Mountain spotted fever. Many infections are not properly identified because of the variable nature of the symptoms. However, since 1985 more than 800 cases in the United States have been verified by the Centers for Disease Control. Since 1998, ehrlichiosis has been a nationally reportable disease (Section 25.8), and in 1999, the first full year of reporting, more than 300 cases were diagnosed, with HME concentrated in the Northeast and mid-Atlantic states and HGE concentrated in the south Atlantic states and Arkansas. These patterns suggest that ehrlichiosis will be reported more frequently as physicians become more familiar with the disease. The numbers of cases are rising, and this represents, like Lyme disease (see Section 27.4), an emerging tickborne disease.

As with other tickborne illnesses, outdoor activities in tick-infested habitat are the major predisposing factors in acquiring ehrlichiosis: Golfers and hikers are particularly prone to infection. Prevention of ehrlichiosis at the individual level involves reducing exposure to ticks and tick bites by avoiding tick habitat, wearing tick-proof clothing, and applying appropriate insect repellents. At the community level, tick densities can be successfully reduced through areawide application of acaricides (chemicals specifically toxic for ticks and related arthropods) and removal of tick habitat such as leaves and brush. Doxycycline, a semisynthetic tetracycline derivative, is the antibiotic of choice for the treatment of ehrlichiosis.

Other Rickettsial Diseases

Q fever is a pneumonia-like infection caused by an obligate intracellular parasite, *Coxiella burnetii*, related to the rickettsias (∞ Section 12.13). Although not transmitted to *humans* directly by an insect bite, the agent of Q fever is transmitted to *animals* by insect bites, and various arthropod species serve as a reservoir of infection. Domestic animals generally have inapparent infections, but may shed large quantities of *C. burnetii* cells in their urine, feces, milk, and other body fluids. Contact with the animals or animal products serves as a source of infection for humans. The resulting influenza-like illness may progress to include prolonged fever, headache, chills, chest pains, pneumonia, and endocarditis. Confirmation of infection with *C. burnetti* can be made by a variety of immunologic tests designed to measure host antibodies to the pathogen. An immmunofluorescence antibody test (IFA) is the serological test of choice (∞ Section 24.10). *C. burnetii* infections respond to the antibiotic tetracycline, and therapy is usually begun quickly in any suspected human case of Q fever in order to prevent heart damage. Finally, Q fever is one of the infectious diseases that has been studied as a possible agent for biological warfare (∞ Section 25.11).

Scrub typhus, or *tsutsugamushi disease*, is restricted to Asia, the Indian subcontinent, and Australia, and is caused by *Orientia tsutsugamushi*. Although the disease is similar to typhus, it is transmitted by *mites* to its normal rodent hosts.

Diagnosis and Control of Rickettsial Diseases

In the past, rickettsial infections have been difficult to diagnose because the characteristic rash associated with many rickettsial diseases may be mistaken for measles, scarlet fever, or adverse drug reactions. Clinical confirmation of rickettsial diseases has now been greatly aided

by the introduction of specific immunological reagents. These include antibody-based tests that detect rickettsial surface antigens by latex bead agglutination assays, immunofluorescence assays, ELISA analyses (∞ Sections 24.8, 24.10 and 24.11), and PCR assays (∞ Section 24.13). Control of most rickettsial diseases requires control of the vectors: lice, fleas, and ticks. For humans traveling in wooded or grassy areas, the use of insect repellants usually prevents tick attachment. Firmly attached ticks should be removed gently with forceps, care being taken to remove all the mouth parts. A solvent such as ethanol applied to a tick with a saturated swab usually expedites removal. Although a vaccine is available for the prevention of typhus, the few cases reported do not warrant its general administration. No vaccines are currently available for the prevention of Rocky Mountain spotted fever or ehrlichiosis.

✓ **27.3 Concept Check**

Rickettsias are obligate intracellular parasitic *Bacteria* that are transmitted by arthropods. Most rickettsial infections are readily controlled by antibiotic therapy, but diagnosis and prompt recognition of these diseases is still difficult.

✓ What are the arthropod vectors for typhus, Rocky Mountain spotted fever, and ehrlichiosis?

✓ What are the normal mammalian hosts for these same diseases?

27.4 Lyme Disease

Lyme disease is a rapidly emerging tickborne disease that affects humans and other animals. Lyme disease was named for Old Lyme, Connecticut, where cases were first recognized, and has rapidly become the most prevalent tickborne disease in the United States.

Biology and Epidemiology

Lyme disease is caused by a spirochete, *Borrelia burgdorferi* (Figure 27.5; ∞ Section 12.33), which is spread primarily by the deer tick, *Ixodes scapularis* (Figure 27.6). The ticks that carry *B. burgdorferi* cells feed on the blood of birds, domesticated animals, various wild animals, and occasionally humans; deer and the white-footed field mouse are prime mammalian reservoirs of *B. burgdorferi* in the northeastern portions of the United States. However, in other parts of the country, different species of rodents and ticks are involved in the transmission of Lyme disease. In the western United States, *Ixodes pacificus* and the wood rat are potential vectors and hosts.

Lyme disease has also been identified in Europe and Asia. In Europe, the tick vector is *Ixodes ricinus*, which

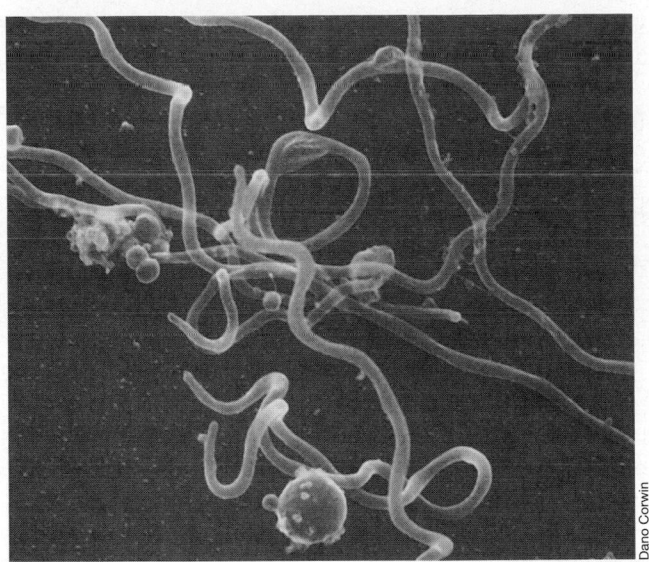

Figure 27.5 Electron micrograph of the Lyme spirochete, *Borrelia burgdorferi*. The diameter of a single cell is approximately 0.4 μm.

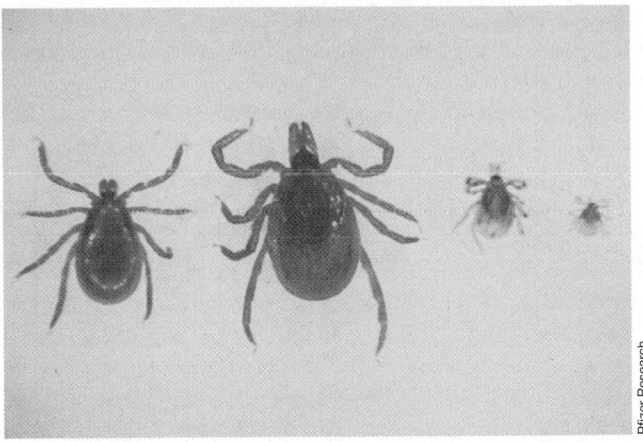

Figure 27.6 Deer ticks (*Ixodes scapularis*), the major vectors of Lyme disease. Left to right, male and female adult ticks, nymph, and larval forms. The length of an adult female is about 3 mm. All forms feed on humans and are capable of transmitting *Borrelia burgdorferi*.

may also harbor *Borellia garinii*, another organism that causes Lyme disease-like symptoms. In Asian countries, *Borellia afzelii* is transmitted by *Ixodes persulcatus* and causes Lyme disease. In all cases, different local rodent reservoirs have been identified. Thus, Lyme disease seems to have a broad geographic distribution and is transmitted to humans through a variety of closely related *Borrelia* species, rodent reservoirs, and tick vectors.

The deer tick and other members of the genus are much smaller than many other types of ticks and thus are easy to overlook (Figure 27.6). Unlike the case with other tickborne diseases, a very high percentage (up to 50% in certain regions of the Northeast) of deer ticks carry *B. burgdorferi* cells. Thus, extended contact with a vector gives a high probability of disease transmission.

In the United States most cases of Lyme disease have been reported from the Northeast and upper Midwest, but cases have been observed in nearly every state, and Lyme disease is spreading west and south. Figure 27.7 shows the recent spread of Lyme disease across the continental United States and the rapid rise in the total number of annual cases.

Pathogenesis

Cells of *Borrelia burgdorferi* are transmitted to humans while the tick is obtaining a blood meal (Figure 27.8). A systemic infection develops, leading to the main symptoms of Lyme disease, which include headache, backache, chills, and fatigue. In about 75% of all cases, a large rash, known as *erythema migrans (EM)*, is observed at the site of the tick bite (Figure 27.8). At this time, Lyme disease is treatable with tetracycline or penicillin. Untreated Lyme disease may progress to a chronic stage be-

ginning weeks to months after the initial tick bite. Chronic Lyme disease is characterized by arthritis in 40–60% of patients. Neurological involvement such as palsy, weakness in the limbs, and facial ticks occurs in 15–20% of patients. Cardiac damage occurs in about 8% of all cases. Treatment at the chronic stage generally requires intravenous antibiotics. The drug ceftriaxone, a highly active β-lactam antibiotic, is used to treat chronic Lyme disease because it is one of the few antibiotics that can cross the blood-brain barrier and attack spirochetes residing in the central nervous system. If no treatment is obtained, cells of *B. burgdorferi* infecting the central nervous system may lie dormant for long periods before eliciting a variety of additional chronic symptoms, including visual disturbances, facial paralysis, and seizures.

No toxins or other virulence factors have yet been identified in Lyme disease pathogenesis. In many respects the latent symptoms of Lyme disease resemble those of syphilis, caused by a different spirochete, *Treponema pallidum*. Indeed, some of the neurological symptoms of Lyme disease resemble those of chronic syphilis (∞ Section 26.12). However, unlike syphilis, Lyme disease is not spread by sexual intercourse or other types of human contact. Small numbers of *Borrelia burgdorferi* cells are shed in the urine of infected individuals, and there is some indication that Lyme disease can spread from domestic animal populations, particularly cattle, by infected urine.

Treatment and Prevention

Serological tests have been developed for detection of antibodies to *Borrelia burgdorferi*. Antibodies appear about 4–6 weeks after infection and can be detected by an indirect enzyme-linked immunosorbent assay (ELISA) or a fluorescent antibody assay (∞ Sections 24.11 and 24.10). However, the most definitive serological test for Lyme

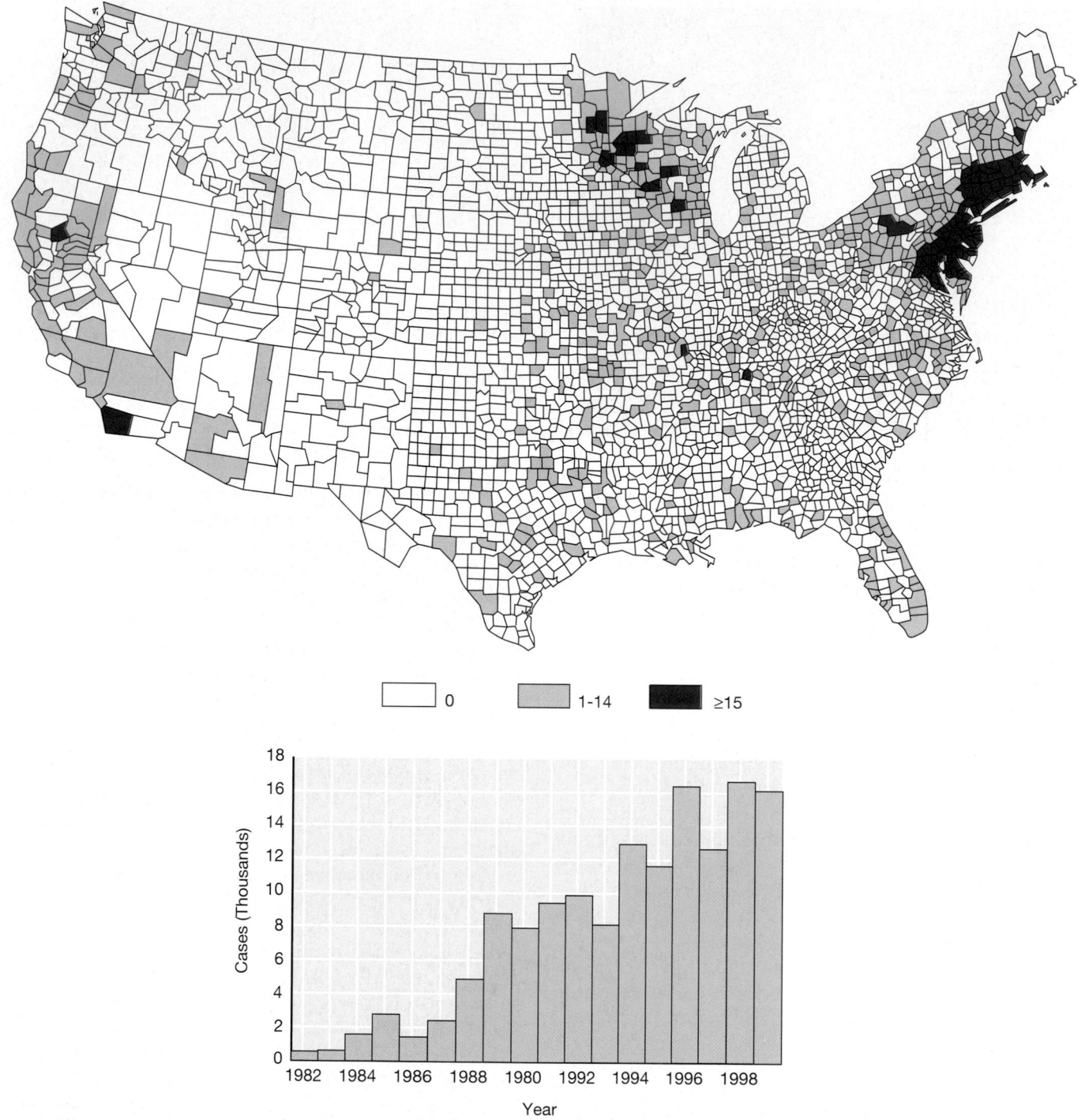

Figure 27.7 Lyme disease in the United States. (a) Incidence of Lyme disease in the United States in 1999. Each county that reported Lyme disease in 1999 is shaded, with counties reporting more than 15 cases shown in black. Lyme disease, while most prevalent in the Northeast and upper Midwest, is found throughout the United States. (b) Number of reported cases of Lyme disease by year in the United States. Lyme disease is reported through the National Notifiable Diseases Surveillance System of the Centers for Disease Control and Prevention.

disease is a Western blot (∞ Section 24.12). Because, antibodies to the Lyme spirochete antigens persist for years after infection, the presence of antibodies does not necessarily confer immunity to the disease or indicate recent infection.

A polymerase chain reaction (PCR) assay (∞ Section 24.13) has also been developed for the detection of *Borrelia burgdorferi* from many body fluids and tissues. While they are rapid and sensitive, the PCR methods cannot differentiate between live *B. burgdorferi* in active

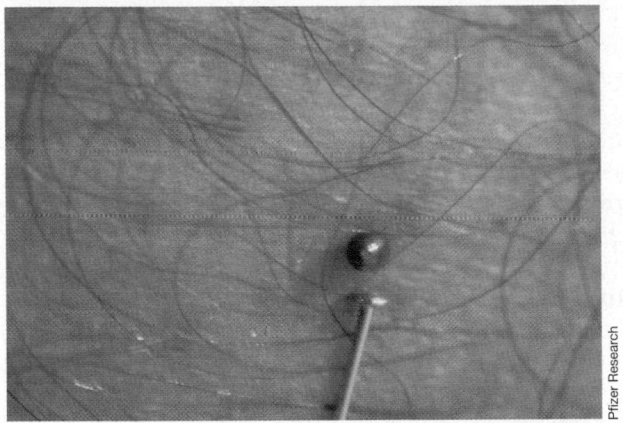

(a)

Pfizer Research

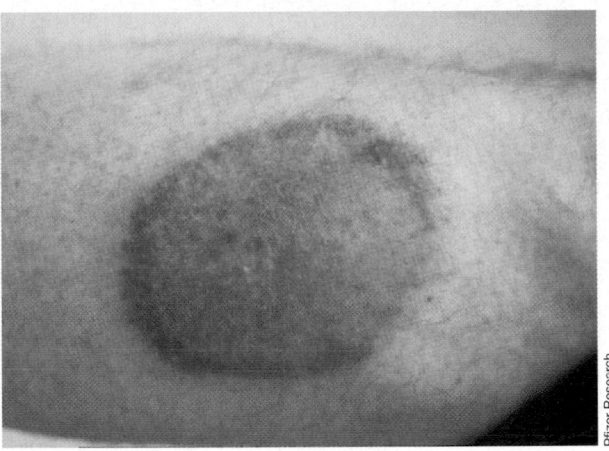

(b)

Pfizer Research

Figure 27.8 (a) Deer tick obtaining a blood meal from a human. (b) Characteristic circular rash associated with Lyme disease. The rash, known as *erythema migrans* (EM), typically starts at the site of the bite and grows in a circular fashion over a period of several days. This typical example is about 5 cm in diameter.

disease and dead *B. burgdorferi* found in treated or inactive disease.

B. burgdorferi can also be cultured from nearly 80% of the original erythema migrans lesions (Figure 27.8b), but is generally not done because of the long latent period before the organism grows on a highly specialized medium.

In the end, Lyme disease is usually diagnosed clinically: If a patient has Lyme disease symptoms and has had a recent tick exposure, especially if followed by erythema migrans, then a presumptive diagnosis of Lyme disease is made and antibiotic treatment is initiated.

Prevention of Lyme disease requires proper precautions to prevent tick attachment. In tick-infested areas such as woods, tall grass, and brush, it is advisable to wear protective clothing such as shoes, long pants, and a long-sleeved shirt with a snug collar and cuffs. Tucking the pants into tight-fitting socks worn with boots forms an effective barrier to tick attachment. After spending time in a tick-infested environment, individu-

als should check themselves carefully for ticks and gently remove any attached ticks (including the head). Insect repellants containing diethyl-*m*-toluamide (DEET) are very effective if applied to both skin and clothing.

An effective Lyme disease vaccine is available and is recommended for individuals who are occupationally or recreationally exposed to ticks or tick habitats. This highly protective vaccine (up to 90%) consists of a *B. burgdorferi* cell-surface protein antigen (OspA) that is expressed in *Escherichia coli* by a bio-engineered recombinant gene (∞ Sections 22.11 and 31.6).

✓ 27.4 Concept Check

Lyme disease is now the most prevalent arthropod-borne disease in the United States. It is transmitted from several mammalian host vectors to humans via ticks. Prevention and treatment of Lyme disease are straightforward, but proper diagnosis is a major problem.

- ✓ What are the primary symptoms of Lyme disease?
- ✓ What antibiotics can be used to treat Lyme disease?

27.5 Malaria

Malaria is a disease caused by a protozoan, a member of the Sporozoa group. We discussed sporozoa as a group in Section 14.6. The malaria parasite is one of the most important human pathogens and has played an extremely significant role in the development and spread of human culture. Indeed, as we will see, malaria has even affected human evolution. Malaria is still a significant human disease even though there are several effective treatments available: There are over *100 million* people worldwide who now have malaria, and each year over 1 million of these will die (∞ Section 25.1).

Epidemiology and Pathogenesis

The major mammalian reservoir for malaria is humans. Four species of sporozoa infect humans, of which the most widespread is *Plasmodium vivax* and the most serious is *Plasmodium falciparum*. This eukaryotic parasite carries out part of its life cycle in humans, and part in the mosquito vector, which spreads the parasite from person to person (∞ Section 14.9). Only female mosquitoes of the genus *Anopheles* transmit malaria. *Anopheles* mosquitoes inhabit primarily warmer parts of the world and, therefore, malaria occurs predominantly in the tropics and subtropics. Malaria did not exist in the northern regions of North America prior to settlement by Europeans but was a major problem in areas such as the southern United States, where appropriate breeding grounds for the mosquito existed. The disease is associated with swampy low-lying areas, and the name *malaria* is derived from the Italian words for "bad air."

The life cycle of the malaria parasite is complex (Figure 27.9). First, the human host is infected by plasmodial *sporozoites*, small, elongated cells produced in the mosquito, which localize in the salivary gland of the insect. The mosquito injects saliva (containing an anticoagulant) along with the sporozoites. The sporozoites travel through the bloodstream to the liver, where they may remain quiescent, or they may replicate and become enlarged in a stage known as a *schizont*. The schizonts then segment into a number of small cells called *merozoites*; these cells are liberated from the liver into the bloodstream. Some of the merozoites then infect red blood cells (erythrocytes). The cycle in erythrocytes proceeds as in the liver and usually repeats at regular intervals of 48 h in the case of *P. vivax*. It is during this period that the defining clinical symptoms of malaria occur, characterized by a fever of up to 40°C (104°F) followed by chills. The chills occur when a new generation of *P. vivax* cells is liberated from erythrocytes. Vomiting and severe headache may accompany the fever-chill cycles, and asymptomatic periods generally alternate with periods in which the characteristic symptoms are present. Because of the loss of red blood cells, malaria generally causes anemia and some enlargement of the spleen.

Not all protozoal cells liberated from red blood cells are able to infect other erythrocytes; those that cannot, called *gametocytes*, are infective only for the mosquito. These gametocytes are ingested when another *Anopheles* mosquito bites the infected person; they mature within the mosquito into *gametes*. Two gametes fuse, and a zygote is formed; the zygote migrates by ameboid motility to the outer wall of the insect's intestine where it enlarges and forms a number of sporozoites. These are released and some reach the salivary gland of the mosquito, from where they can be inoculated into another person; the cycle then begins again.

Diagnosis and Treatment

Conclusive diagnosis of malaria in humans requires the identification of *Plasmodium*-infected erythrocytes in blood smears (Figure 27.10).

Prophylaxis (when travelling to endemic areas) and treatment of malaria are usually accomplished with *chloroquine*. Choroquine is the drug of choice for treating parasites within red cells, but does not kill malarial parasites outside the cells. The closely related drug *primaquine* eliminates sporozoites, merozoites, and gametes outside the cells. Treatment with both chloroquine and primaquine produces a cure. However, even in individuals who have undergone drug treatment, malaria may recur years after the primary infection. Apparently, small numbers of sporozoites survive in the liver and reinitiate malaria months or years later by releasing merozoites.

In addition, in many parts of the world *Plasmodium* strains have developed resistance to chloroquine or primaquine or both, and some strains have developed resistance to other drugs as well. For use in areas with

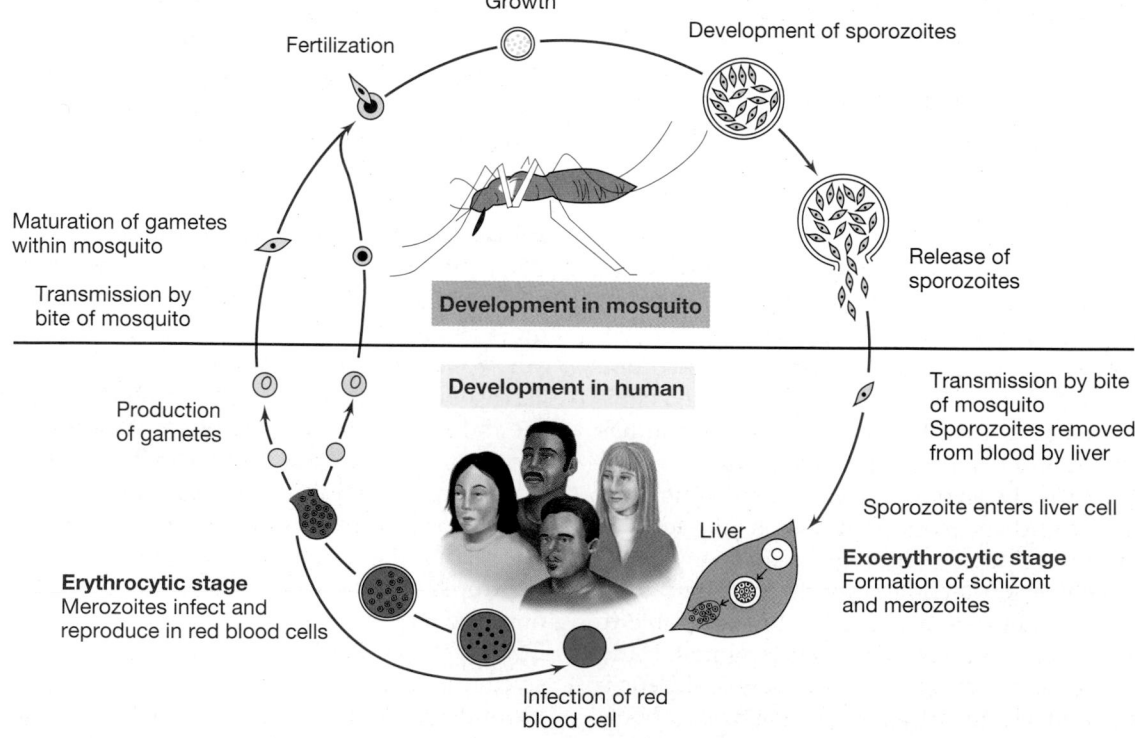

Figure 27.9 Life cycle of the malaria parasite, *Plasmodium vivax*.

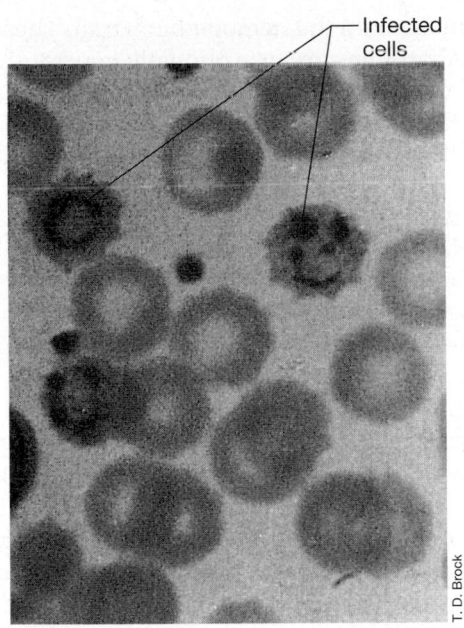

Figure 27.10 *Plasmodium vivax*, the causative agent of malaria, growing inside human red blood cells.

Infected cells

known drug-resistant strains, Melarone™, a combination of atovaquone and proguanil, is recommended for both treatment and prophylaxis.

Prevention and Eradication

Antimalarial drug treatment is an expensive and short-term solution to malaria prevention and control, and drug-resistant strains of the parasite complicate matters even further. The most effective control measure is to break the obligate life cycle of the parasite by eliminating one of the obligate hosts, the *Anopheles* mosquito.

Two approaches to mosquito control are possible: (1) elimination of habitat by drainage of swamps and similar breeding areas, (2) elimination of the mosquito by insecticides, and treating patients with primaquine, thereby breaking the *Plasmodium* life cycle. During the 1930s, about 33,000 miles of ditches were dug in 16 southern states in the United States, removing 544,000 acres of mosquito breeding area. Millions of gallons of oil were also spread on swamps to reduce the oxygen supply to mosquito larvae. With the discovery of the insecticide dichlorodiphenyltrichloroethane (DDT) (∞ Figure 19.49), chemical control of both larvae and adult mosquitoes was possible. During World War II, the Public Health Service organized an Office of Mosquito Control in War Areas, and because many U.S. military bases were in the southern states, this organization carried out an extensive eradication program in the United States as well as overseas. In 1946, a year in which there were 48,610 cases of malaria in the United States, Congress established a 5-year malaria eradication program. In endemic areas, the program involved

drug prophylaxis and treatment regimens for individuals, along with DDT treatment of mosquito infestations. By 1953 there were only 1310 malaria cases. In 1935 there were about 4000 deaths from malaria; in 1952 there were only 25 deaths. Thus, the overall public health threat from malaria in the United States is now minimal. However, *endemic* malaria has occurred in recent years, albeit in very low numbers, as far north as New York City. Increases in malaria incidence still occur as a result of influxes of imported malaria due to soldiers or immigrants coming from malaria-endemic areas (Figure 27.11), but there is, on average, less than one annual death due to malaria in the United States.

In other parts of the world, eradication has been much slower, but the same control measures are used and are still effective. Reduction of mosquito habitat, control of mosquitoes via insecticides (DDT is no longer used due to environmental concerns, but other effective insecticides are available), and treatment of infected individuals with drugs both for cure and prophylaxis, are still the major strategies for controlling malaria. Additionally, several experimental prophylactic vaccines are now in trial stages.

Malaria and Biochemical Evolution of Humans

Malaria has undoubtedly been endemic in Africa for thousands of years. In West Africans, resistance to malaria caused by *Plasmodium falciparum* is associated with an altered red blood cell protein, hemoglobin S, which differs from normal hemoglobin A at only a single amino acid. In the beta chain of hemoglobin S, the neutral amino acid *valine* is substituted for the *glutamic acid* of hemoglobin A. Hemoglobin S binds oxygen less efficiently than hemoglobin A. Under conditions of low oxygen concentration, hemoglobin S forms long, thin

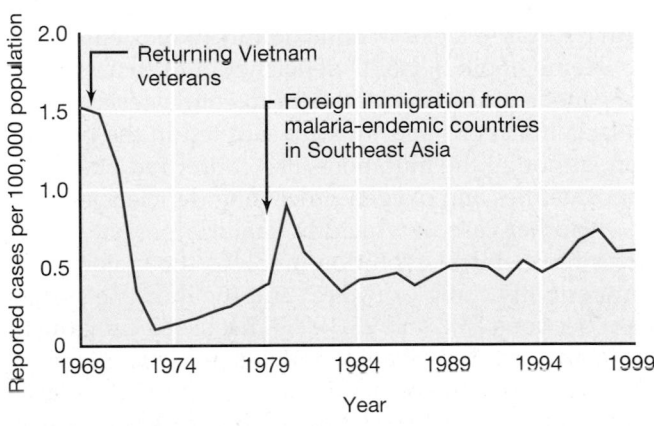

Figure 27.11 The prevalence of malaria in the United States. There are currently about 1000 cases of malaria per year. Nearly all cases are imported, with only a few cases from endemic sources. Prior to 1947, there were at least 48,000 cases per year, mostly from endemic sources. Data are from the Centers for Disease Control and Prevention, Atlanta, GA, USA.

aggregates that cause the red cell to change from a bi-concave round cell (∞ Figure 22.4) to an elongated C-shaped cell. Because of the shape of the cell, this condition is known as a *sickle cell*. Individuals who are *homozygous* for the sickle cell trait are particularly susceptible to changes in oxygen concentrations and suffer from **sickle cell anemia**.

Individuals who are heterozygous for hemoglobin S have the *sickle cell trait*, but have increased resistance to malaria. With the sickle cell trait, hemoglobin S can still produce sickled cells, but not as readily as in the case of the homozygotes. However, the growth of *P. falciparum* inside the red cell causes the heterozygous cells to sickle more easily than uninfected cells. The aggregated hemoglobin S in sickled cells apparently disrupts the membrane of red cells, allowing potassium to diffuse from the cell. *P. falciparum* cannot grow in the low-potassium environment of the disrupted cell. Thus, persons with the sickle cell trait are resistant to malaria.

In certain Mediterranean regions where malaria is endemic, resistance to *P. falciparum* is associated with a deficiency in the red blood cells of the enzyme glucose-6-phosphate dehydrogenase (G6PD), an enzyme that acts as an intracellular antioxidant (reducing) compound. The faulty G6PD leads to higher levels of intracellular oxidants such as H_2O_2 (actually produced inside the red cell by the growing *P. falciparum*). The increased levels of oxidants damage parasite membranes and limit parasite growth.

In many Mediterranean populations, a diverse group of genetic abnormalities affect hemoglobin production and efficiency. These are known collectively as the **thalassemias**. The thallasemias are also statistically and geographically associated with increased resistance to malaria, and, like the G6PD deficiency, are associated with decreased levels of antioxidants in red cells.

Hemoglobin S, G6PD deficiency, and thalassemias are genetic mutations that seem to confer resistance to malaria infections and thus are selected in the population, although the mutations also confer red blood cell abnormalities and oxygen-processing deficiencies.

Another case in which the malaria parasite influences biochemical evolution involves the major histocompatibility complex (MHC) and the immune system (∞ Sections 23.2 and 23.3). As discussed previously, the MHC class I and class II proteins present antigens to T cells for initiation of an immune response. In malaria-prone equatorial West Africa, individuals are very likely to have one particular MHC class I gene and one particular set of class II genes. These particular selected MHC genes are more common in the West African population and are virtually unknown in other human population groups. Individuals who express these genes have as much resistance to severe fatal malaria infec-

tions as those with the hemoglobin S trait. These particular MHC proteins are exceptionally good antigen-presenting molecules for certain malarial antigens and confer protective resistance to *Plasmodium* sp. infection as a result of the powerful immune response they help initiate.

Like the hemoglobin variants, the parasite acts as a strong selection agent for individual genes important for host survival: Individuals with particular MHC genes have a measurable survival advantage and are more likely to reproduce and pass the resistance-conferring genes to their progeny.

Thus, several lines of evidence indicate that malaria has been a selective agent in human evolution. Other pathogens, such as *Mycobacterium tuberculosis* (tuberculosis, ∞ Section 26.5) and *Yersinia pestis* (plague, see also Section 27.6), may also have also promoted selective changes in humans, but in no case is the evidence as clear as it is for malaria.

✓ 27.5 Concept Check

Malaria is a widespread, mosquitoborne infectious disease occurring mainly in tropical and subtropical portions of the world. It is a major cause of morbidity and mortality in developing countries and has been a selection factor for several resistance genes. The disease is preventable with a combination of public health and chemotherapy measures.

✓ How can malaria be prevented?
✓ Review genetic mechanisms responsible for malaria resistance. Why are antimalarial genes not found in all humans?
✓ What is the natural reservoir for *Plasmodium* spp.?

27.6 Plague

Pandemic occurrences of **plague** have been directly responsible for more human deaths than any other infectious diseases except malaria and tuberculosis. Plague killed between 25% and 33% of Europe's population in individual epidemics in the Middle Ages.

Biology and Epidemiology

Plague is caused by *Yersinia pestis*, a gram-negative facultatively aerobic rod that is a member of the enteric bacteria group (∞ Section 12.11). Plague is a natural disease of domestic and wild rodents; rats are the primary disease reservoir. Humans are only accidental hosts and are not critical for the maintenance of the disease. Fleas are intermediate hosts and act as vectors, spreading plague between the mammalian hosts (Figure 27.12). Most infected rats die soon after symptoms appear, but a low proportion develop a chronic infection and can serve as a source of virulent *Y. pestis*. The majority of

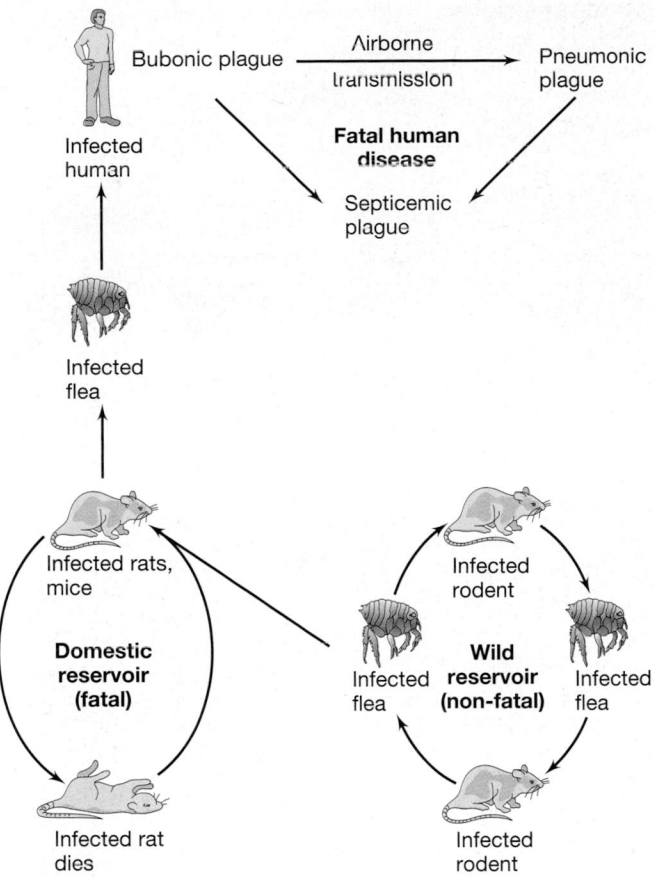

Figure 27.12 The epidemiology of plague due to *Yersinia pestis*. Plague in most wild rodents is generally a mild, self-limiting infection. Plague in rats and humans is frequently fatal. Infected fleas desert the dead host and look for other hosts, such as humans, an accidental host.

cases of human plague in the United States occur in the southwestern states, where the disease is endemic among wild rodents (sylvatic plague).

Plague is transmitted by the rat flea (*Xenopsylla cheopis*), which ingests *Y. pestis* cells by sucking blood from an infected animal. Cells multiply in the flea's intestine and can be transmitted to a healthy animal in the next bite. As the disease spreads, rat mortality becomes so great that infected fleas seek new hosts, including humans. Once in humans, cells of *Y. pestis* usually travel to the lymph nodes, where they cause the formation of swollen areas referred to as *buboes*. For this reason the disease is frequently referred to as *bubonic plague* (Figure 27.13*a*). The buboes become filled with *Y. pestis* and the capsule on cells of *Y. pestis* prevents phagocytosis by cells of the immune system. (Section 22.2). Secondary buboes form in peripheral lymph nodes, and cells eventually enter the bloodstream, causing a generalized septicemia. Multiple hemorrhages produce dark splotches on the skin giving plague its historical nickname, the "Black Death" (Figure 27.13*b*). If not treated prior to the septicemic stage, the symptoms of

plague, including extreme lymph node pain, prostration, shock, and delirium, progress and usually cause death within 3–5 days.

Pathology

The pathogenesis of plague is not clearly understood, but cells of *Yersinia pestis* produce a number of antigenically distinct molecules, including toxins, that undoubtedly contribute to the disease process. The V and W antigens of *Y. pestis* cell walls are protein-lipoprotein complexes that inhibit phagocytosis. Other envelope proteins are also present. An exotoxin called *murine toxin*, because of its extreme toxicity for mice, is produced by virulent strains of *Y. pestis*. Murine toxin is a respiratory inhibitor that blocks mitochondrial electron transport reactions at the point of coenzyme Q (Section 5.11). Although it is not clear that murine toxin is involved in the pathogenesis of human plague (murine toxin is highly toxic for certain animal species but not for others), it produces systemic shock, liver damage, and respiratory distress in mice. These symptoms are all seen in human plague. *Y. pestis* also produces a highly immunogenic *endotoxin* that may also play a role in the disease process.

Other Forms of Plague

Pneumonic plague occurs when cells of *Yersinia pestis* are either inhaled directly or reach the lungs during bubonic plague (Figures 27.12 and 27.13*c*). Symptoms are usually absent until the last day or two of the disease when large amounts of bloody sputum are produced. Untreated individuals rarely survive more than 2 days. Pneumonic plague is highly contagious and can spread rapidly via the person-to-person respiratory route if infected individuals are not immediately quarantined. *Septicemic plague* involves the rapid spread of *Y. pestis* throughout the body via the bloodstream without the formation of buboes and usually causes death before a diagnosis can be made.

Treatment and Control

Plague can be successfully treated if rapidly diagnosed. *Yersinia pestis* infection is preferentially treated with streptomycin, and alternatively with tetracycline. For individuals exposed to pneumonic plague, tetracycline is recommended for prophylaxis. If treatment is started promptly, mortality from bubonic plague can be reduced to 1–5% of those infected. Pneumonic and septicemic plague can also be treated, but these forms progress so rapidly that antibiotic therapy, even if begun when symptoms first appear, is usually too late. Although potentially a devastating disease, there have been only 89 cases of plague in the United States in the last decade (Figure 27.14). Unfortunately, 8 of these patients died (a fatality rate of 9%). Worldwide, there are fewer than 1500 confirmed cases and fewer than 300 deaths per year.

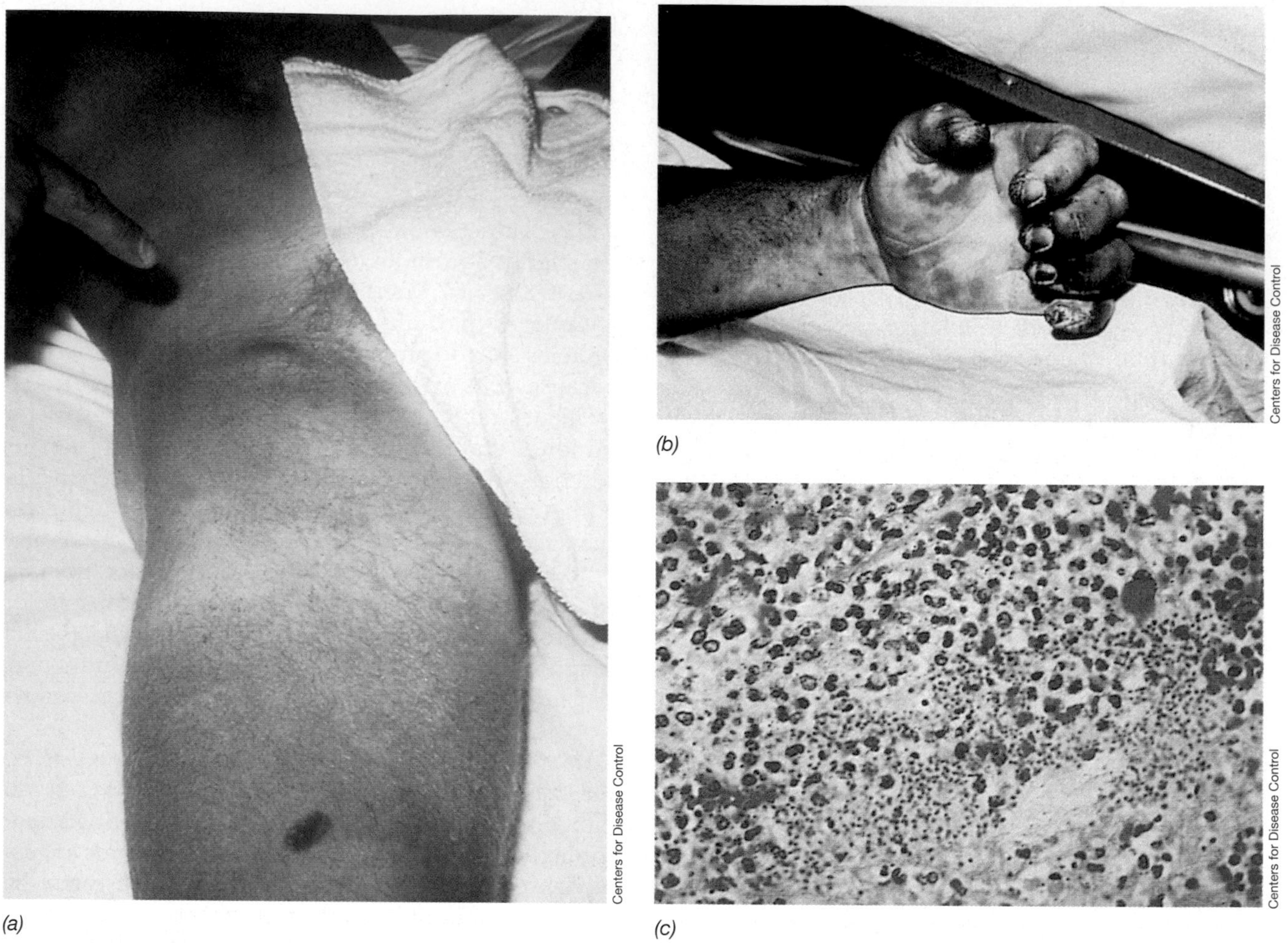

(a)

(b)

(c)

Centers for Disease Control

Figure 27.13 Plague in humans. (a) Bubo formed in the groin. (b) Gangrene and sloughing of skin in hand. (c) *Yersinia pestis*, the causative agent of plague. Cells are seen as very small blue cells from the lung tissue of a pneumonic plague victim.

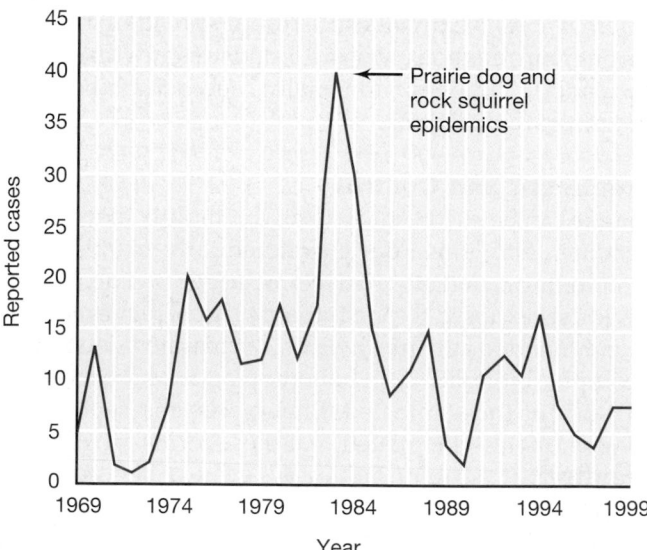

Figure 27.14 Plague in the United States. From 1990–1999, only 89 cases and eight deaths were reported. Data are from the Centers for Disease Control and Prevention, Atlanta, GA, USA.

Plague control is accomplished through surveillance of infected animals, vectors, and human contacts. Plague-infected animal populations must be destroyed when identified. Undoubtedly, improved public health practices and the overall control of rodent populations have limited human exposure to plague. Prevention can also be accomplished by immunization with a formalin-killed vaccine, but, because the incidence of disease is low, immunization is only recommended for individuals with high risk of exposure.

✓ 27.6 Concept Check

Plague is largely confined to individuals who come into contact with rodent populations that are endemic reservoirs for *Yersinia pestis*. A disseminated systemic infection often leads to rapid death, but localized infections are treatable with antibiotics.

✓ Distinguish among *bubonic, septicemic,* and *pneumonic* plague.

✓ What is the insect reservoir, natural host, and treatment for plague?

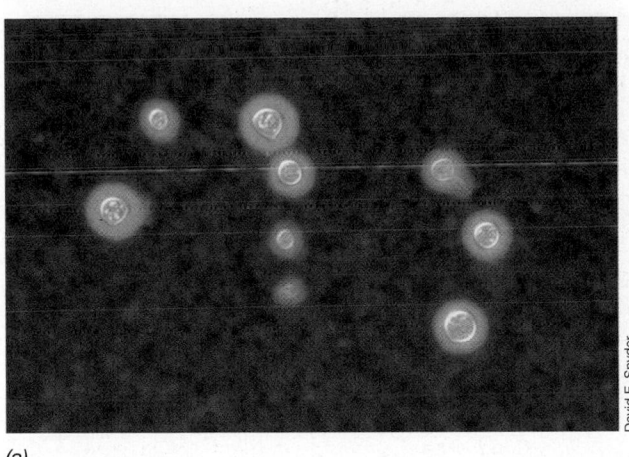

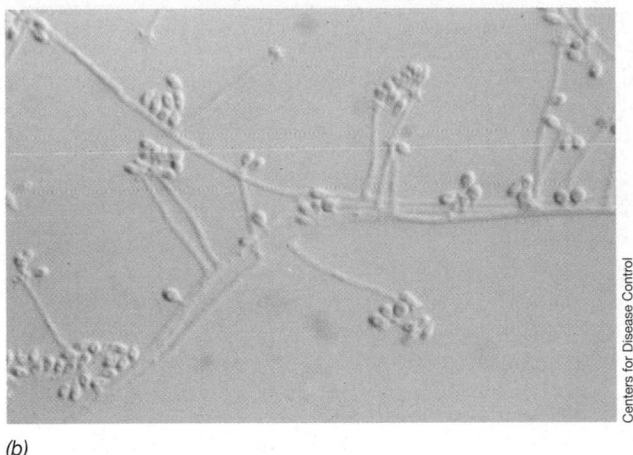

David E. Snyder

Centers for Disease Control

(a)   (b)

Figure 27.15 Typical forms of pathogenic fungi. (a) Yeast form of *Cryptococcus neoformans*, stained with india ink to show the capsule. The cells are from 4 to 20 μm. (b) *Sporothrix schenckii*, showing the branching, or hyphae, characteristic of the mold form of fungi. The round basidiospores are about 2 μm in diameter.

III SOILBORNE DISEASES

We now investigate several diseases that are normally spread through soil. Fungi are common and ubiquitous members of the community of microorganisms found worldwide in soil, and a few can act as pathogens. Some bacteria are also important soilborne pathogens. In contrast to many person-to-person or vectorborne pathogens, soilborne pathogens are accidental agents of infection, with no life-cycle dependency on the accidental host. Keep in mind that soil is an unlimited reservoir of pathogens and these pathogens cannot be eliminated because of this vast reservoir.

27.7 The Pathogenic Fungi

Many fungi occur in nature as free-living saprophytes. Among these are a few that occasionally act as accidental, often opportunistic, pathogens.

Biology and Epidemiology

Fungi grow in some form in nearly every ecological niche, but are particularly prevalent growing on the nonliving organic materials in soil. The fungi include the eukaryotic organisms commonly known as *yeasts*, which normally grow as single cells (Figure 27.15a), and *molds*, which grow in branching chains called *hyphae* (Figure 27.15b). The taxonomy and biological diversity of these organisms were discussed in Section 14.9. Fortunately, most fungi are harmless to humans. Only about 50 species cause human disease, and the overall incidence of serious fungal infections is rather low, although certain superficial fungal infections are quite common.

Fungi cause disease through three major mechanisms. First, some fungi cause immune responses that can result in allergic (hypersensitivity) reactions following exposure to specific fungal antigens (∞ Section 22.13). Exposure to fungi, whether growing on the host or in the environment, may cause allergic symptoms on reexposure. For example, *Aspergillus* spp. (∞ Figure 14.19), a common saprophyte often found in nature as a leaf mold, is a potent and common allergen, often causing asthma and other hypersensitivity reactions. As we shall see later, *Aspergillus* also has other mechanisms for producing disease.

A second fungal disease-producing mechanism involves the production and action of *mycotoxins*, a large, diverse group of fungal exotoxins. The best-known examples of mycotoxins are produced by *Aspergillus flavus*, an organism that commonly grows on improperly stored food such as grain. The toxins produced by *A. flavus* are known as *aflatoxins* (Figure 27.16). Aflatoxins are highly toxic and induce tumors in some animals, especially in birds that feed on contaminated grain. Their direct role in human disease is not well defined.

The third fungal disease-producing mechanism is through infection. The growth of a fungus on or in the body is called a **mycosis** (plural, **mycoses**). Mycoses can

Figure 27.16 Structure of aflatoxin B1. This toxin is one of a group of related compounds produced by *Aspergillus flavus*.

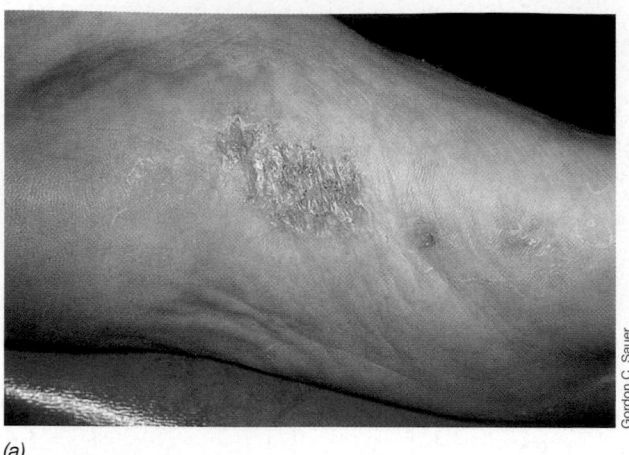

(a)

Gordon C. Sauer

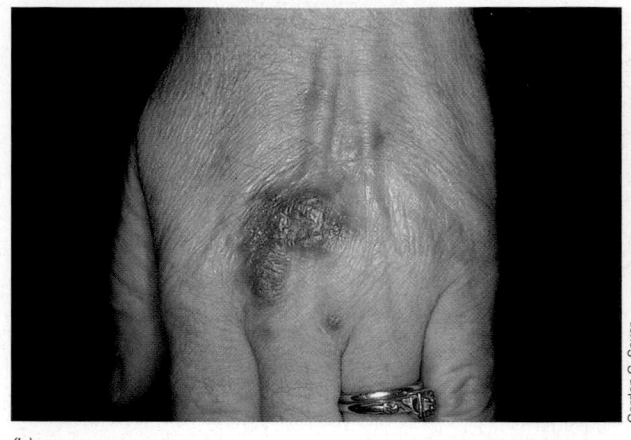

(b)

Gordon C. Sauer

Figure 27.17 Fungal infections. (a) Superficial mycosis of the foot (athlete's foot) due to infection with *Trichophyton rubrum*. (b) Sporotrichosis, a subcutaneous infection due to *Sporothrix schenckii*.

range in severity from relatively innocuous, superficial infections to serious, life-threatening diseases.

Mycoses

Mycoses are subdivided into three categories. The first of these is the *superficial mycoses*. These diseases involve colonization of the skin, hair, or nails and infect only the surface layers (Figure 27.17*a*). Table 27.2 lists some of the common superficial fungi. In general, these diseases are relatively benign and self-limiting. Some, such as *Trichophyton* infections of the feet (athlete's foot), are quite common. Spread is by personal contact with an infected person or by contact with contaminated surfaces such as bathtubs or shower stalls or other contaminated shared articles such as towels or bed linens. Treatment for severe cases is with topical application of miconazole nitrate or griseofulvin. Griseofulvin is also administered orally. After enter-

ing the bloodstream, it passes to the skin where it can inhibit fungal growth.

The *subcutaneous mycoses* involve deeper layers of skin (Figure 27.17*b*) and a different group of organisms (Table 27.2). One disease in this category is *sporotrichosis*, an occupational hazard of agricultural workers, miners, and other workers who come into contact with the soil. The causative organism is found as a ubiquitous saprophyte on wood and in soil. The lesions are usually initiated by fungal infection of a small wound or abrasion. *Sporothrix schenckii* can readily be isolated from the lesion and cultured *in vitro*. Treatment is with oral potassium iodide or oral ketoconazole.

The *systemic mycoses* involve fungal growth in internal organs of the body. These are subclassified as *primary* or *secondary* infections. A primary infection is one resulting directly from the fungal pathogen in otherwise normal, healthy individuals. A secondary infection involves

TABLE 27.2	Some pathogenic fungi and the diseases they cause	
Disease	**Causal organism**	**Main disease foci**
Superficial mycoses (dermatomycoses)		
Ringworm	*Microsporum*	Scalp of children
Favus	*Trichophyton*	Scalp
Athlete's foot	*Epidermophyton, Trichophyton*	Between toes, skin
Jock itch	*Trichophyton, Epidermophyton*	Genital region
Subcutaneous mycoses		
Sporotrichosis	*Sporothrix schenckii*	Arms, hands
Chromoblastomycosis	Several fungal genera	Legs, feet
Systemic mycoses		
Cryptococcosis[a]	*Cryptococcus neoformans*	Lungs, meninges
Coccidioidomycosis[a]	*Coccidioides immitis*	Lungs
Histoplasmosis[a]	*Histoplasma capsulatum*	Lungs
Blastomycosis	*Blastomyces dermatitidis*	Lungs, skin
Candidiasis[a]	*Candida albicans*	Oral cavity, intestinal tract

[a] Considered opportunistic pathogens and have been implicated in the pathogenesis of AIDS (∞ Section 26.14).

infection with the pathogen only in hosts with a predisposing condition such as antibiotic therapy or immunosuppression. In the United States, the most widespread primary fungal infections are *histoplasmosis*, caused by *Histoplasma capsulatum*, and *coccidioidomycosis* (San Joaquin Valley fever), caused by *Coccidioides immitis*. Both of these organisms normally live in soil. These are both respiratory diseases in which the host becomes infected by breathing in airborne spores, which germinate and grow in the lungs. Histoplasmosis is primarily a disease of rural areas in the midwestern United States, especially in the Ohio and Mississippi river valleys. Most cases are mild and are often mistaken for more common respiratory infections. San Joaquin Valley fever is generally restricted to the desert regions of the southwestern United States. The fungus lives in desert soils, and the spores are disseminated on dry, windblown particles that are inhaled. In some areas in the southwestern United States, as many as 80% of the inhabitants may be infected, although most individuals suffer no apparent ill effects.

A number of fungal infections, including histoplasmosis and coccidioidomycosis, are especially serious and common in individuals whose immune systems have been impaired, for example, by acquired immunodeficiency syndrome (AIDS) (Section 26.14) or by immunosuppressive drugs. These are secondary fungal diseases because normal individuals either do not get the disease or generally have a less severe form. Examples of other fungal organisms involved as secondary pathogens are given in Table 27.2. These fungi are known as *opportunistic* pathogens because of their particular ability to cause serious infections only in individuals with impaired defense mechanisms (in particular AIDS patients) (Section 26.14).

Treatment and Control

Effective chemotherapy against systemic fungal infections is very difficult (Section 20.11). Most antibiotics that inhibit fungi also harm other eukaryotic organisms, including the human host. One of the most effective antibiotics, amphotericin B, is widely used to treat systemic fungal infections of humans, but serious side effects such as kidney toxicity may occur.

Control of infections by elimination of fungal pathogens from the environment is impractical. As with many common-source pathogens, control of fungal growth is very difficult because there is a limitless reservoir: Exposure to fungi cannot be eliminated, except through efforts such as decontamination and air filtration in restricted local environments.

✓ 27.7 Concept Check

A variety of soilborne fungi produce disease in humans. Superficial, subcutaneous, and systemic mycoses are infections that are difficult to control because of a lack of specific anti-

fungal drugs. Fungal infections may cause serious systemic disease, often in individuals with impaired immunity, such as in AIDS patients.

✓ Describe superficial, subcutaneous, and systemic mycoses.

✓ Why are antifungal drugs difficult to develop?

27.8 Tetanus

Tetanus is a serious, often life-threatening disease. Although it is totally preventable, 473 individuals have acquired tetanus in the United States within the last decade (Figure 27.18) and 68 have died.

Biology and Epidemiology

Tetanus is caused by an exotoxin produced by *Clostridium tetani*, a motile, anoxic, spore-forming rod (Section 12.20). The natural reservoir of *C. tetani* is the soil, where it is a ubiquitous resident, although it is occasionally found in the gut of mammals, as are other *Clostridium* spp. *C. tetani* normally gains access to the body through a soil-contaminated wound, typically a deep puncture. In the wound, anoxic conditions allow germination of spores and growth of the organism. *C. tetani* produces a potent exotoxin, the *tetanus toxin*. The organism is noninvasive, and so its sole method of causing disease is through the direct action of the toxin on host cells. The incubation time is variable and may take from 4 days to several weeks, depending on the number of spores inoculated at the time of injury. Tetanus is not transmitted from person to person.

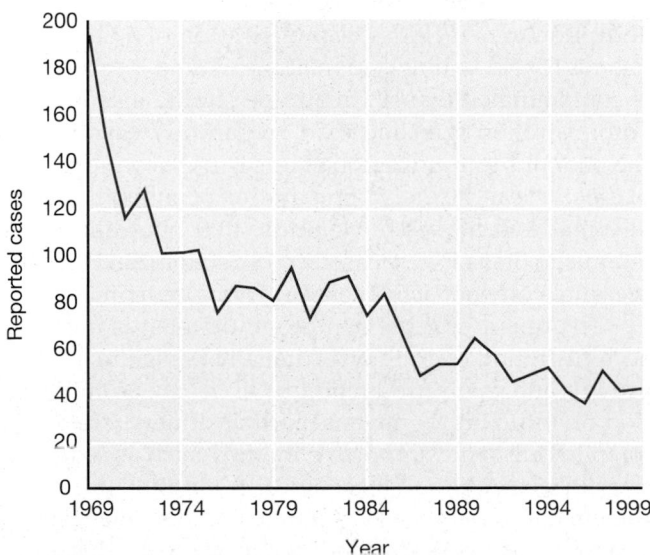

Figure 27.18 Annual incidence of tetanus in the United States by total number of annual cases. The overall numbers are decreasing, but the mortality rate is 15% over the last decade. Data are from the Centers for Disease Control and Prevention, Atlanta, GA, USA.

Pathogenesis

We have already examined the action of tetanus toxin at the cellular and molecular level (⚏ Section 21.10). The toxin directly effects the release of inhibitory signaling molecules in the nervous system. These inhibitory signals control the "relaxation" phase of muscle contraction. The overall result is rigid paralysis of the voluntary muscles, often called *lockjaw* because it is observed first in the muscles of the jaw and face (Figure 27.19). Death is usually due to respiratory failure, and the mortality rate is very high (15% over the last decade in the United States).

Diagnosis, Control, Prevention, and Treatment

Diagnosis of tetanus is based on exposure, clinical symptoms, and, rarely, identification of the toxin in the blood or tissues of the patient. The organism may also be cultured from the wound, but success is highly variable.

The natural reservoir of *Clostridium tetani* is the soil. Since *C. tetani* is an accidental pathogen in humans and is not dependent on humans or other animals for its propagation, there is no possibility for eradication. Therefore, control measures must focus on methods of prevention.

Tetanus is a preventable disease. The existing toxoid vaccine (⚏ Section 22.11) is completely effective for disease prevention. Virtually all tetanus cases occur in individuals who were inadequately immunized. The fastest growing age group for contracting tetanus is, surprisingly, individuals from 25–59 years of age, presumably because public health immunization programs target infants, school-age individuals, and seniors 60 years of age and older.

Appropriate treatment of serious cuts, lacerations, and punctures includes administration of a "booster" tetanus toxoid immunization. If the wound is severe and is contaminated by soil, treatment should also include administration of an antitoxin preparation, especially if the patient's immunization status is unknown or is out of date. The antitoxin is a preparation of antibodies, usually made in horses for tetanus, that neutralizes the tetanus toxin as it is released (⚏ Section 22.11). These measures prevent active tetanus from occurring.

Treatment of acute symptomatic tetanus involves administration of antibiotics, usually penicillin, to stop growth and toxin production by *C. tetani*; administration of antitoxin to prevent binding of newly released toxin to cells; and supportive therapy such as sedation, muscle relaxants, and mechanical respiration to control the effects of paralysis. Treatment at this level cannot provide a reversal of symptoms, since toxin that is al-

Figure 27.19 A soldier dying from tetanus. Note the rigid paralysis. Painting by Charles Bell in the Royal College of Surgeons, Edinburgh, Scotland.

ready bound to tissues cannot be neutralized. Even with antitoxin, antibiotics, and supportive therapy, tetanus patients have significant morbidity and mortality.

A number of members of the genus *Clostridium* are pathogens and virtually all exist as normal soil microbial community members. All cause disease because of their production of potent exotoxins. Elimination of the Clostridia, because of their common source in soil, is impossible. While *Clostridium tetani* is found almost exclusively in soil, *C. botulinum* and *C. difficile* are also occasionally found in the gut of humans and other animals as part of the normal flora (⚏ Section 21.4). *C. perfringens* and *C. botulinum*, as we shall see in Chapter 29, are important potential pathogens in common-source foodborne diseases. In each case, the source of the *Clostridium* is food contaminated with soil containing clostridial spores. The spores then germinate because of inadequate decontamination and food preservation methods, and produce potent exotoxins that cause disease symptoms.

✓ 27.8 Concept Check

Clostridium tetani is a ubiquitous soilborne microorganism that can cause *tetanus*, a disease characterized by toxin production and rigid paralysis. Tetanus is preventable with appropriate immunization. Treatment for acute tetanus is generally unsatisfactory, with significant morbidity and mortality.

✓ Describe infection by *C. tetani* and the elaboration of tetanus toxin.

✓ What is the overall effect of tetanus toxin on the host?

✓ Outline the steps taken to prevent tetanus in an individual who has sustained a puncture wound by stepping on a rusty nail.

Review Questions

1. What animals are likely to carry rabies in the United States? What immunization programs are in place for the treatment of rabies? For the prevention of rabies?

2. Why has hantavirus pulmonary syndrome (HPS) emerged as a human pathogen in the United States? How can HPS be prevented?

3. What are the three major categories of organisms that cause rickettsial diseases? For typhus, Rocky Mountain spotted fever, and ehrlichiosis, identify the most common reservoir and vector.

4. Identify the most common reservoir and vector for Lyme disease in the United States. How can the spread of Lyme disease be controlled? How can Lyme disease be treated?

5. Malaria involves severe long-term fever followed by chills. These symptoms are related to activities of the pathogen *Plasmodium vivax* or *Plasmodium falci-*

parum. Describe the growth stages of this pathogen in the human host and relate them to the fever–chill pattern. Why might a person of western European descent be more susceptible to malaria than a person of African or Mediterranean descent?

6. Distinguish between bubonic, septicemic, and pneumonic plague. Which is most serious? How is each acquired?

7. What is the reservoir for most fungal pathogens? How can fungal exposure be controlled? What particular problems, especially in terms of therapy, do fungi pose for the clinician?

8. Describe the invasiveness and toxicity of *Clostridium tetani*. Discuss the major mechanism of pathogenesis for tetanus and define measures for prevention and treatment of tetanus.

Application Questions

1. Describe the sequence of events you would take if a child received a bite (provoked or unprovoked) from a stray dog with no record of rabies immunization. Present one scenario where you were able to capture and detain the dog, and another for a dog that escaped. How would these procedures differ from a situation in which the child was bitten by a dog that had documented, up-to-date rabies immunizations?

2. Why are diseases like hantavirus pulmonary syndrome (HPS) emerging as important infectious diseases, even in developing countries? Include in your considerations a discussion of social, economic, environmental, and public health issues.

3. Discuss at least three common properties of the disease agents and review the disease process for Rocky Mountain spotted fever, typhus, and ehrlichiosis. Why is ehrlichiosis emerging as an important rickettsial disease? Compare its emergence to that of Lyme disease.

4. Discuss the causative agent, mode of transmission, symptoms, therapy, prevention, and diagnosis of Lyme disease. Predict the future history of this disease in the United States for the next decade. Keep in mind that diagnosis of Lyme disease is very difficult, but an effective vaccine is available.

5. Malaria eradication has been a goal of public health programs for at least 100 years. What factors preclude our ability to eradicate malaria? If an effective

vaccine was developed, could malaria be eradicated? Compare this possibility to the possibility of eradicating plague.

6. Bubonic plague may have killed up to 30% of the entire human population during several pandemics occurring as recently as the 19th century. Why are plague pandemics no longer occurring? Speculate on the possibility that humans have evolved to resist plague. Can you supply any evidence for this hypothesis? Identify and discuss similar issues for emerging diseases such as AIDS and Ebola hemorrhagic fever.

7. With regard to infections, why are fungi often secondary opportunistic pathogens in immunologically compromised individuals? Devise an environment that eliminates all exposure to fungi. Is such an environment a realistic possibility for isolation of patients with immune deficiency? Do you know of examples where this has been accomplished?

8. Formulate a plan to ensure that tetanus will be eradicated as an infectious disease in the United States by 2010. What methods would you use to eradicate the disease? On a national level, how might the goals be accomplished? Assuming that immunization is the key to your plan, identify procedures to ensure that all individuals receive immunizations.

9. Indicate the common human immunization practices in place for protection against rabies, plague, and Lyme disease. Why are immunizations for these diseases not required as they are for tetanus?

Effective water treatment has been the most important single factor influencing advances in public health in the last century. Purification of drinking water makes available for general consumption water free of infectious microorganisms and many undesirable chemicals. Decontamination of wastewater (shown here) is routinely practiced on an industrial scale in developing counties. Wastewater must be treated to reduce or eliminate pathogens and nutrients that stimulate algal growth before release into surface waters. The ultimate goal of wastewater treatment is to release water that is acceptable for intake directly to a drinking water treatment plant.

WASTEWATER TREATMENT, WATER PURIFICATION, AND WATERBORNE MICROBIAL DISEASES

28

Working Glossary

Biochemical oxygen demand (BOD) the amount of dissolved oxygen consumed by microorganisms for complete oxidation of organic and inorganic material in a water sample

Chloramine a chemical manufactured on site by combining chlorine and ammonia at precise ratios

Chlorine a chemical fed in its gaseous state to disinfect water. A residual level is maintained throughout the distribution system

Clarifier (coagulation basin) a reservoir in which the suspended solids of raw water are coagulated and removed

Coagulation the formation of large insoluble particles from much smaller, colloidal particles by the addition of aluminum sulfate and anionic polymers

Coliforms all aerobic and facultative aerobic, gram-negative, nonspore-forming lactose-fermenting bacteria

Cyst an infectious form of a protozoan parasite that is encased in a thick-walled chemically and physically resistant coating

Distribution system water pipes, storage reservoirs, tanks, and other means used to deliver drinking water to consumers or store it before delivery

Filtration the removal of suspended particles from water by passing it through one or more permeable membranes or media (e.g., sand, anthracite, or diatomaceous earth)

Finished water water delivered to the distribution system after treatment

Flocculation the water-treatment process after coagulation that uses gentle stirring to cause suspended particles to form larger, aggregated masses (floc)

Meningoencephalitis invasion, inflammation, and destruction of brain tissue, by the ameba *Naegleria fowlerii* or a variety of other pathogens

Polymer in water purification, a chemical in liquid form used as a coagulant in the clarification process to produce flocculation

Potable drinkable; safe for human consumption

Raw water surface water or groundwater that has not been treated in any way (also called untreated water)

Sewage liquid effluents contaminated with human or animal fecal material

Turbidity a measurement of fine, suspended particles in water

Untreated water surface water or groundwater that has not been treated in any way (also called raw water)

Wastewater liquid effluents derived from domestic sewage or industrial sources, which cannot be discarded in untreated form into lakes or streams

---

Clean, pure water is absolutely essential to public health, and we must have procedures to assess water quality and to assess the effectiveness of the microbial and chemical methods used to treat water. Unfortunately, water quality can break down, often with dramatic and even life-threatening spread of infectious disease.

I WASTEWATER MICROBIOLOGY AND WATER PURIFICATION

Water is the most important potential common source of infectious diseases, and water purification is the most important single measure available for ensuring public health. The methods used to assess microbial water quality are standardized microbiology procedures. The methods commonly used to treat water and make it fit for consumption or other human use include a variety of methods that use microorganisms to remove pollutants, in addition to physical and chemical purification procedures.

28.1 Public Health and Water Quality

How can we routinely ensure that drinking water is safe? Even water that looks clear and clean may be contaminated with pathogenic microorganisms and may pose a serious health hazard. Unfortunately, it is not generally practical to examine water to detect each of the pathogenic organisms that may be present. It is, however, practical to sample water supplies for the overall presence of microorganisms. We discuss here the general methods used to identify potentially harmful microorganisms in water.

Coliforms

While a few nonpathogenic microorganisms might be tolerable in a water supply, the presence of specific *indicator organisms* signals that a given water source might be contaminated with pathogens. These indicator organisms are usually associated with the intestinal tract; their presence indicates fecal contamination of the water source. The most widely used indicator is the *coliform* group of organisms. The **coliforms** are used as indicators of water contamination because they commonly inhabit the intestinal tract of humans and other animals in large numbers. Coliforms are defined in water bacteriology as aerobic and facultatively aerobic, gram-negative, nonspore-forming, rod-shaped *Bacteria* that ferment lactose with gas formation within 48 h at 35°C. This is an operational rather than a taxonomic definition, and the coliform group includes a variety of organisms, but most are members of the enteric bacterial group (∞ Section 12.11). For example, the coliform group includes the organism *Escherichia coli*, a common intestinal organism, and the organism *Klebsiella pneumoniae*, a less common pathogenic intestinal inhabitant. However, *Enterobacter aerogenes*, an organism not found in the enteric group or in the intestine, is also classified as a coliform because of its fermentative properties.

In general, we assume that the presence of coliform organisms in a water sample indicates fecal contamination and makes the water unsafe for human consumption. When excreted into water, the coliforms eventually die, but they do not die as quickly as some pathogens. The coliforms and the pathogens behave similarly during water purification.

The Coliform Test

Two procedures are commonly used to test for coliforms in water samples. These are the *most-probable-number (MPN)* procedure and the *membrane filter (MF)* procedure. The MPN procedure employs liquid culture medium in test tubes, in which samples of drinking water added to the media. Growth in the culture vessels indicates microbial contamination of the water supply. For the more common MF procedure, at least 100 ml of the water sample is passed through a sterile membrane filter, which removes the bacteria (∞ Section 20.3). The filter is placed on the surface of a plate of eosin-methylene blue (EMB) culture medium, which is highly selective for coliform organisms (Figure 28.1; ∞ Section 24.2). The coliform colonies are counted, and from this value the number of coliform *Bacteria* in the original water sample can be determined. In well-regulated water supply systems, coliform tests should be negative. If coliform tests are positive, a breakdown in the system has occurred in purification procedures or distribution systems.

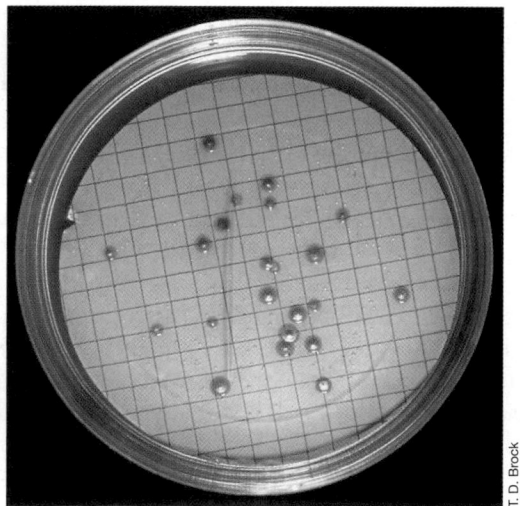

Figure 28.1 Coliform colonies growing on a membrane filter. A drinking water sample has been passed through the filter. The filter was placed on eosin-methylene blue (EMB) media that is both selective and differential for lactose-fermenting bacteria (coliforms) (∞ Section 24.2). The dark color of the colonies is characteristic of coliforms. Each colony represents one viable coliform isolated from the original sample.

Drinking water standards in the United States are specified under the Safe Drinking Water Act, which provides a framework for the development of drinking water standards. For the membrane filter (MF) technique, 100-ml samples are filtered. To be considered safe, the number of coliform bacteria in drinking water samples cannot exceed any of the following levels: (1) 1/100 ml as the arithmetic mean of all samples examined per month; (2) 4/100 ml in more than one sample when fewer than 20 are examined per month; or (3) 4/100 ml in more than 5% of the samples when 20 or more samples are examined per month. Water utilities report their results to the United States Environmental Protection Agency, and if they do not meet the prescribed standards, the utilities must notify the public and take steps to correct the problem. Many smaller communities and even large cities sometimes fail to meet these standards.

Public Health and Drinking Water Purification

Today the incidence of waterborne disease in developed countries is so low that it is difficult to directly measure the effectiveness of treatment practices and maintenance of drinking water standards. Most intestinal infections in developed countries are no longer due to transmission by water but via food (∞ Chapter 29). Effective water treatment practices were not in place until the twentieth century. Microbial culture methods for evaluating the health significance of polluted drinking water were not practiced until coliform counting procedures were developed and adapted in about 1905. Until then, water purification was limited to filtration to reduce turbidity (∞ Chapter 25, box, Snow on Cholera). Although filtration significantly decreases the microbial load of water, many microorganisms still passed through the filters. In about 1910, **chlorine** was discovered to be an extremely efficient disinfectant for large water supplies. Chlorine was so effective and so inexpensive that its use spread widely and was of major significance in reducing the incidence of waterborne disease. Figure 28.2 illustrates the dramatic drop in incidence of typhoid fever (infection with *Salmonella typhi*) in a major American city after filtration and chlorination purification procedures were introduced. Similar results were obtained in other major cities. The dramatic improvement in the health of the American people in the early decades of the twentieth century was largely due to the establishment of water purification procedures, and the effectiveness of chlorination could not have been assessed if standard methods for determining the coliform content of drinking water had not been developed. Thus, public works engineering, microbiology, and public health moved forward together.

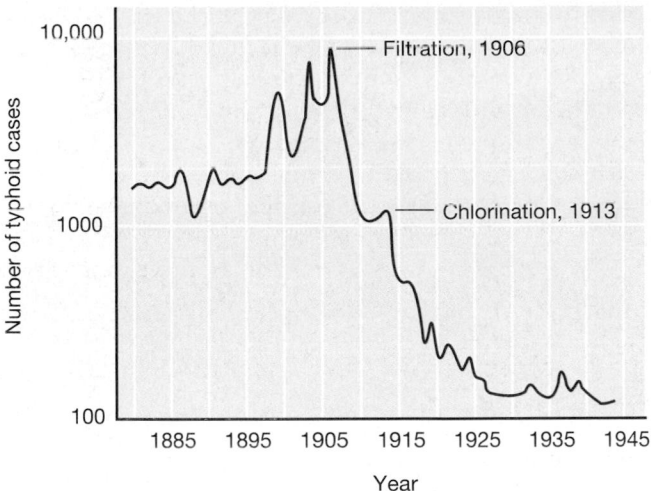

Figure 28.2 The effect of water purification on the incidence of waterborne disease. The graph shows the incidence of typhoid fever in Philadelphia (Pennsylvania, USA) before and after the introduction of filtration and chlorination processes for drinking water supplies. Note the dramatic reduction in the incidence of typhoid fever after the introduction of both filtration and chlorination. This result parallels the results obtained for reduction of other waterborne diseases.

✓ 28.1 Concept Check

Drinking water quality is determined by counting coliform bacteria. Strict adherence to uniform microbiologic standards make this method a reliable and reproducible indicator of fecal contamination in all public water supplies in the United States. Filtration and chlorination of water supplies significantly decreases microbial load. Application of water purification methods to drinking water is the most important public health measure ever devised.

✓ Why do the bacterial colonies recovered from drinking water and grown on EMB media indicate fecal contamination of the water supply?

✓ What procedures are used to reduce microbial load in water supplies?

28.2 Wastewater and Sewage Treatment

Wastewater and sewage treatment involves a large-scale use of microorganisms and can be considered a type of industrial-scale bioconversion. Wastewater enters a treatment plant and, following treatment, the effluent water is suitable for release into rivers and streams or to drinking water purification facilities.

Wastewater

Wastewater is liquid effluent derived from domestic sewage or industrial sources that cannot be discarded in untreated form into lakes or streams due to public health,

economic, and aesthetic considerations. **Sewage** is liquid effluent contaminated with human or animal fecal materials. Wastewater commonly contains potentially harmful inorganic and organic compounds as well as pathogenic microorganisms. As we shall see, complete wastewater treatment involves chemical and biological (microbiological) treatments to remove or neutralize contaminants.

About 15,000 wastewater treatment facilities exist in the United States. The vast majority of them are fairly small, treating 1 million gallons (3.8 million liters) or less of wastewater per day. Collectively, however, these plants treat nearly 40 *billion* gallons of wastewater every day. Wastewater plants are usually constructed to handle both domestic and industrial wastes. Domestic wastewater is made up of sewage, "gray water" (the water resulting from washing, bathing, and cooking), and wastewater from food processing. Industrial wastewater includes effluent from the petrochemical, pesticide, food and dairy, plastics, pharmaceutical, and metallurgical industries.

Industrial wastes may contain toxic substances that must be pretreated before they can be released for wastewater treatment. Pretreatment is generally a mechanical process in which debris that could clog equipment in the wastewater treatment plant is removed. However, certain wastewaters are pretreated biologically to remove highly poisonous substances such as cyanide and heavy metals. These substances can be converted to less toxic forms through the action of specific microorganisms capable of neutralizing, oxidizing, precipitating, or volatilizing toxic or infectious wastes. We now discuss the processes involved in a typical wastewater treatment facility.

Wastewater Treatment and Biochemical Oxygen Demand

The goal of a wastewater treatment facility is to reduce organic and inorganic materials in wastewater to a level that no longer supports microbial growth and to eliminate other potentially toxic materials. The efficiency of treatment is expressed in terms of a reduction in the **biochemical oxygen demand (BOD)**, the relative amount of dissolved oxygen consumed by microorganisms to completely oxidize all organic and inorganic matter in a water sample (⊂⊃ Section 19.5). Higher levels of oxidizable organic and inorganic materials in the wastewater result in a higher BOD. Typical values for domestic wastewater, including sewage, are approximately 200 BOD units. For industrial wastewater, for example from sources such as dairy plants, the values can be as high as 1500 BOD units. An efficient wastewater treatment facility reduces levels to less than 5 BOD units in the water released from the treatment plant.

A typical wastewater facility must treat both sewage and industrial wastes. Treatment is a multistep operation employing a number of independent physical and biological processes (Figure 28.3). Primary, secondary, and sometimes tertiary treatments are employed to reduce fecal and chemical contamination in the incoming water. Each level of treatment employs more complex and more expensive technologies.

Primary Wastewater Treatment

Primary treatment of wastewater consists only of physical separations. Wastewater entering the treatment plant is passed through a series of grates and screens that remove large objects. The effluent is left to settle for a number of hours to allow suspended solids to sediment (Figure 28.4).

Municipalities that provide only primary treatment suffer from extremely polluted water when the effluent is discharged into adjacent waterways because high levels of organic matter and other nutrients remain in water following primary treatment. Therefore, most treatment plants employ secondary treatment to reduce the organic content of the wastewater before release to natural waterways. Secondary treatment is intimately tied to microbiological processes.

Figure 28.4 Primary treatment of wastewater. Wastewater is pumped into the reservoir (left) where settling of solids occurs. As the water level rises, the water spills through the grates to successively lower levels. Water at the the lowest level, now free of solids, enters the spillway (arrow) and is pumped to a secondary treatment facility.

Anoxic Secondary Wastewater Treatment

Anoxic wastewater treatment involves a series of digestive and fermentative reactions carried out by a number of bacterial species and is usually employed to treat materials that have large amounts of insoluble organic matter (and, hence, very high BOD), such as fiber and cellulose waste from food- and dairy-processing plants. The anoxic degradation process itself is carried out in large enclosed tanks called *sludge digesters* or *bioreactors* (Figure 28.5a and b) and requires the collective activities of many different types of microorganisms; the reactions are summarized in Figure 28.5c. Through the action of the resident anoxic microorganisms, the macromolecular waste components are first digested by polysaccharases, proteases, and lipases into soluble components. These soluble components are then fermented to yield a mixture of fatty acids, H_2, and CO_2, and the fatty acids are further fermented to acetate, CO_2, and H_2. These products are then used as substrates by methanogenic bacteria (∞ Sections 12.6 and 17.17), which are capable of carrying out the reactions $CH_3COOH \rightarrow CH_4 + CO_2$ and $4 H_2 + CO_2 \rightarrow CH_4 + 2 H_2O$ (Figure 28.5c). Thus, major products of anoxic sewage treatment are CH_4 (methane) and CO_2. The methane can be collected and either burned off or used as fuel to heat and power the treatment plant.

Aerobic Secondary Treatment

In general, nonindustrial wastewater can be treated efficiently using only aerobic secondary treatment. Several kinds of aerobic decomposition processes are used for wastewater treatment, but the trickling filter and activated sludge methods are the most common. A trickling filter (Figure 28.6a) is a bed of crushed rocks, about

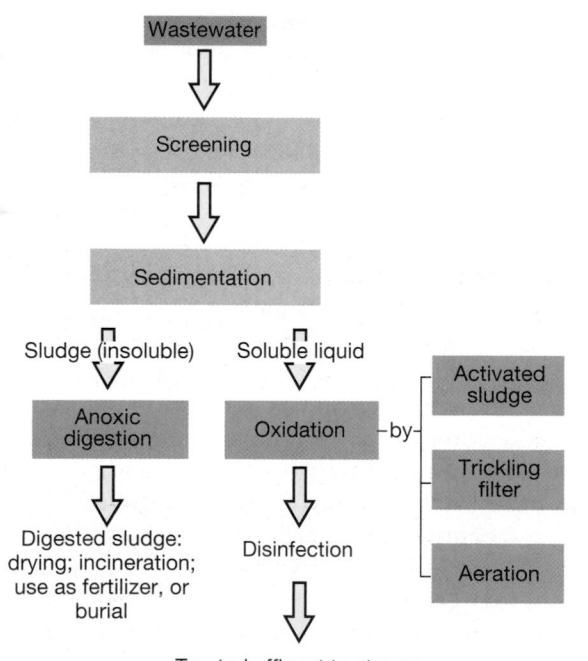

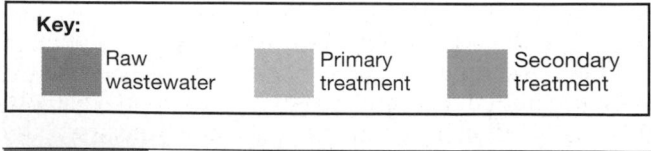

Figure 28.3 Wastewater treatment processes.

(a)

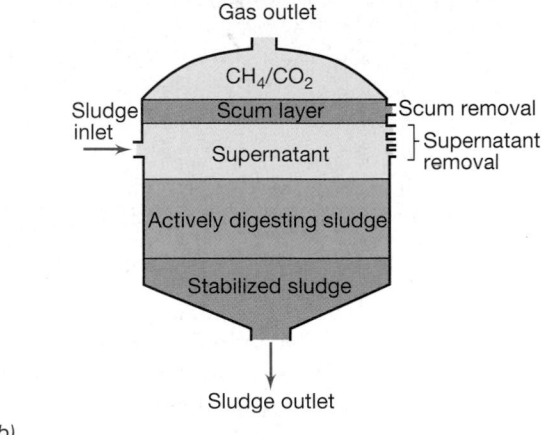

(b)

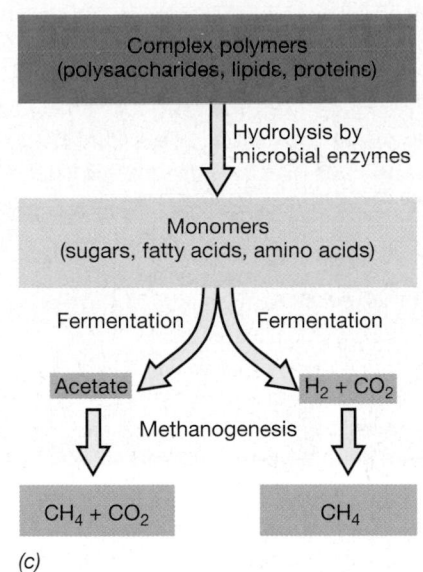

(c)

T.D. Brock

Figure 28.5 Anoxic secondary wastewater treatment. (a) Anoxic sludge digester. Only the top of the tank is shown; the remainder is underground. (b) Inner workings of a sludge digestor. (c) Major microbial processes occurring during anoxic sludge digestion. Methane (CH_4) and carbon dioxide (CO_2) are the major products of anaerobic biodegradation (⬭ Section 17.17).

2 m thick, on top of which the wastewater is sprayed. The liquid slowly passes through the bed, the organic matter adsorbs to the rocks, and microbial growth occurs on the rocks. The complete mineralization of organic matter to carbon dioxide, ammonia, nitrate, sulfate, and phosphate takes place in the microbial biofilm on the rocks. The most common aerobic treatment system is the activated sludge process. Here, the wastewater to be treated is mixed and aerated in a large tank (Figure 28.6b). Slime-forming bacteria, including *Zoogloea ramigera*, among others, grow and form *flocs* (larger, aggregated masses) (Figure 28.7), and these flocs form the substratum to which protozoa and small animals attach. Occasionally, filamentous bacteria and fungi are also present. The basic process of oxidation is similar to a trickling filter. The effluent containing the flocs is pumped into a holding tank or *clarifier* where the flocs settle. Some of the floc material (called *activated sludge*) is then returned to the aerator to serve as inoculum, and the rest is sent to the anoxic sludge digestor (see above) (Figure 28.5) or is removed, dried, and burned or used for fertilizer.

Wastewater normally stays in an activated sludge tank for 5 to 10 h, a time too short for complete oxidation of all organic matter. However, during this time much of the soluble organic matter is adsorbed to the floc and is incorporated into microbial cells. The BOD of the liquid effluent is considerably reduced (by up to 95%) by this process, with most of the BOD now contained in the settled flocs, and the goal of BOD reduction in the water is achieved. Nearly complete BOD reduction can occur if the flocs are then transferred to the anoxic sludge digestor.

Most treatment plants now chlorinate the effluent (to further reduce the possibility of biological contamination) and discharge the treated water to streams or lakes. A few plants, however, process wastewater through a tertiary stage.

Tertiary Treatment

Tertiary treatment is the most complete method of treating sewage but has not been widely adopted because it is very expensive. Tertiary treatment is a physicochemical process employing precipitation, filtration, and chlorination procedures similar to those employed for drinking water purification (see Section 28.3) to sharply reduce the levels of inorganic nutrients, especially phosphate and nitrate, from the final effluent.

(a)

(b)

(c)

Figure 28.6 Aerobic secondary wastewater treatment processes. (a) Trickling filter. The booms rotate, distributing wastewater slowly and evenly on the rock bed. The rocks are 10–15 cm in diameter, and the bed is 2 m deep. (b–c) The activated sludge process. (b) Aeration tank of an activated sludge installation in a metropolitan wastewater treatment plant. The tank is 30 m long, 10 m wide, and 5 m deep. (c) Wastewater flow through an activated sludge installation. Recirculation of activated sludge to the aeration tank introduces microorganisms responsible for oxic digestion of the organic components of the wastewater. The anoxic sludge digestor is detailed in Figure 28.5.

Wastewater receiving proper tertiary treatment is so free of nutrients that it is unable to support extensive microbial growth.

✓ 28.2 Concept Check

Wastewater treatment is primarily concerned with removing sewage and industrial wastes, thereby reducing the BOD (biochemical oxygen demand) to acceptable levels. Primary, secondary, and tertiary treatment of water involves physical, biological, and physicochemical methods, respectively. After tertiary treatment, water may be suitable for release directly to a water purification plant.

- ✓ What is biochemical oxygen demand (BOD)? Why is BOD reduction necessary in wastewater treatment?
- ✓ Identify the main processes in primary, secondary, and tertiary sewage treatment.
- ✓ Define the main by-products of wastewater treatment. How are these by-products handled?

28.3 Drinking Water Purification

Wastewaters treated by secondary methods are generally of such quality that they can be directly discharged into rivers and streams. However, such water is not **potable**; that is, safe for human consumption. The production of potable water requires further treatment to remove potentially pathogenic microorganisms, decrease **turbidity** (a measure of suspended particles), eliminate taste and odor, and reduce nuisance chemicals such as iron and manganese.

Physical and Chemical Purification

A typical drinking water treatment installation for a small city is shown in Figure 28.8a. Figure 28.8b traces the flow of **raw water (untreated water)** through a typical treatment scheme. Raw water is first pumped from the source, in this case a lake, to a sedimentation basin where anionic **polymers**, alum (aluminum sulfate), and chlorine are

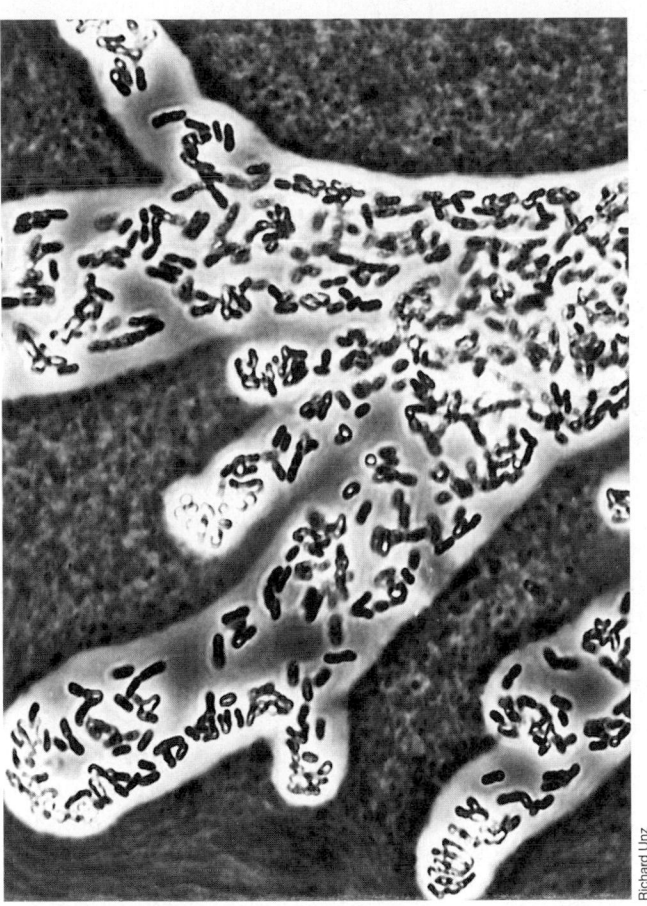

Figure 28.7 A floc formed by the bacterium *Zoogloea ramigera*, a characteristic organism found in the activated sludge process. The floc consists of a large number of small, rod-shaped cells of *Z. ramigera* surrounded by a polysaccharide slime layer, arranged in characteristic fingerlike projections. Negative stain using india ink.

added. Sand, gravel, and other large particles settle out. This pretreated water is then pumped to a **clarifier** or **coagulation basin**, a large holding tank where **coagulation** takes place: The alum and anionic polymers form larger suspended particles from the much smaller suspended colloidal particles. After mixing, the particles continue to interact, forming large, aggregated masses, a process known as **flocculation**. The large, aggregated particles, called floc, settle out by gravity, trapping any remaining microorganisms and absorbing organic matter and sediment. After coagulation and flocculation, the clarified water undergoes **filtration**. The water is passed through a series of filters designed to remove the remaining suspended particles and microorganisms. The filters usually consist of thick layers of sand and ionic filtration media. When combined with previous purification steps, the filtered water is free of all particulate matter, most organic and inorganic chemicals, and all microorganisms.

Disinfection

Clarified, filtered water must then be disinfected before it is released to the supply system as pure, potable **finished water**. Chlorination is the most common method of disinfection. In sufficient doses, chlorine kills microorganisms within 30 minutes (certain pathogenic protozoa such as *Cryptosporidium* are not easily killed by chlorine treatment and thus can be important waterborne pathogens; see Section 28.6). In addition to killing microorganisms, chlorine reacts with organic compounds, oxidizing and effectively neutralizing them. Therefore, since most taste- and odor-producing compounds are organic in nature, chlorine treatment also improves water taste and smell. Chlorine is added to water either from a concentrated

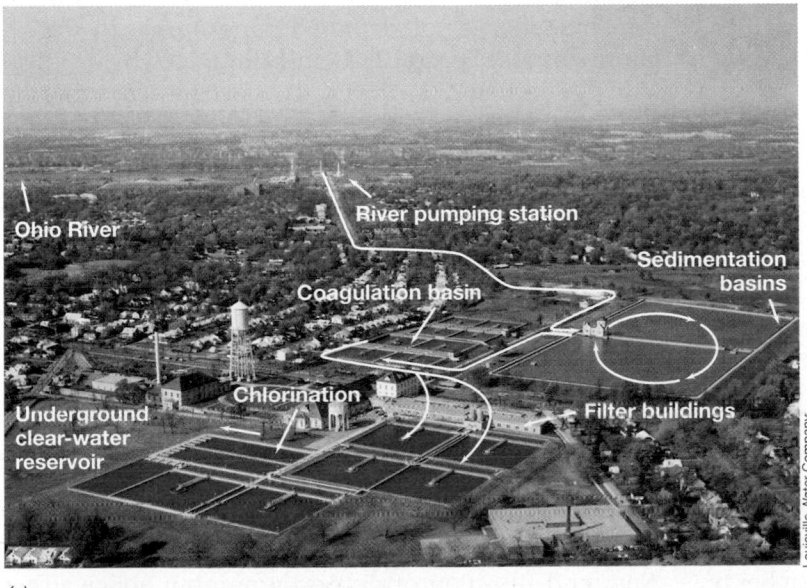

(a)

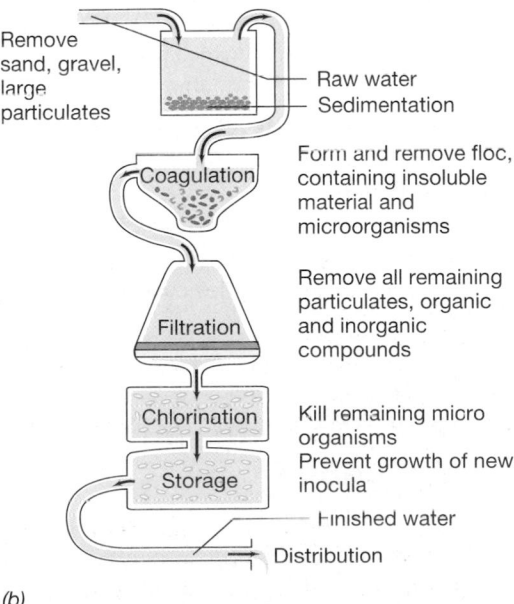

Remove sand, gravel, large particulates

Raw water
Sedimentation

Coagulation

Form and remove floc, containing insoluble material and microorganisms

Filtration

Remove all remaining particulates, organic and inorganic compounds

Chlorination

Kill remaining micro organisms
Prevent growth of new inocula

Storage

Finished water

Distribution

(b)

Figure 28.8 Water purification plant. (a) Aerial view of a water treatment plant in Louisville, Kentucky, USA. The arrows indicate direction of flow of water through the plant. (b) Schematic overview of a typical community water purification system.

solution of sodium or calcium hypochlorite or as a gas from pressurized tanks. The latter method is used most commonly in large water treatment plants because it is most amenable to automatic control.

Chlorine is consumed when it reacts with organic materials. Therefore, sufficient quantities of chlorine must be added to water containing organic materials so that a residual amount remains to react with the microorganisms after all reactions with organic materials have occurred. The water plant operator performs chlorine analyses on the treated water to determine the residual level of chlorine. A residual chlorine level of 0.2–0.6 μg/ml is suitable for most water supplies. After chlorine treatment, the now-potable water is pumped to storage tanks from which it flows by gravity or pumps through the **distribution system** of storage tanks and supply lines to the consumer. Residual chlorine levels ensure that the finished water will reach the consumer without becoming contaminated (assuming that there is no catastrophic failure, such as a broken pipe, in the distribution system). Chlorine gas, even when dissolved in water, is extremely volatile and can dissipate within hours from treated water. To further ensure that residual chlorine levels are maintained throughout the distribution system, most municipal water treatment plants also introduce ammonia gas with the chlorine to form the stable, nonvolatile chlorine-containing compound **chloramine** ($HOCl + NH_3 \rightarrow NH_2Cl + H_2O$).

✓ 28.3 Concept Check

Water treatment plants employ industrial-scale physical and chemical systems that remove or neutralize biological, inorganic, and organic contaminants from a variety of community and industrial wastewater sources. Water purification plants employ clarification, filtration, and disinfection processes to produce potable water. Finished potable water is free of chemical and biological contamination.

✓ Trace the treatment of water through a drinking water treatment plant, from the inlet to the final distribution point (faucet).

✓ What specific purposes do *sedimentation*, *coagulation*, and *filtration* accomplish in the drinking water treatment process?

✓ Identify the specific roles chlorine plays in water purification.

II WATERBORNE MICROBIAL DISEASES

Common-source infectious diseases are caused by microbial contamination of materials shared by a large number of individuals. Such common sources of disease are contaminated food and water. We discuss foodborne diseases in

Chapter 29. Common-source waterborne diseases are a very significant source of morbidity and mortality, especially in developing countries. A variety of bacteria, viruses, and protozoa cause waterborne infectious diseases.

Waterborne diseases begin as infections. Water may cause infection even if only a small number of microorganisms are present; the exact numbers of pathogens necessary to cause disease are functions of the virulence of the pathogen and the general ability of the host to resist infection (∞ Sections 21.8 and 21.13).

28.4 Sources of Waterborne Infection

Human pathogens can be transmitted through improperly treated water used for drinking and cooking. Another very common source of disease transmission is through pathogen-contaminated water used for swimming and bathing.

Drinking Water

Because everyone consumes water, drinking water is a common source of pathogen dissemination and has a very high potential for the catastrophic spread of epidemic disease. As we discussed above, water supplies in developed countries usually meet rigid quality standards, effectively limiting the spread of waterborne diseases, but there are still widespread waterborne disease outbreaks in developing countries, and developed countries occasionally have severe outbreaks of waterborne disease due to lapses in water purity.

Microorganisms transmitted in water generally grow in the intestines and leave the body in feces. Fecal pollution of water supplies may then occur, and if the contamination is not identified (for example, by the coliform test; see Section 28.1) and eliminated by disinfection (see Section 28.3), then a new host may consume the water and the pathogen may colonize the intestine and cause disease. In the United States, a number of different bacterial and protozoan pathogens are occasionally transmitted in drinking water (Table 28.1). We will discuss giardiasis and cryptosporidiosis in Section 28.7, but we will save our discussions of shigellosis and *Escherichia coli* O157:H7 infections for Chapter 29, where we introduce these pathogens as common agents of foodborne infections.

Recreational Water

Recreational water includes freshwater recreational areas such as ponds, streams, and lakes as well as public swimming and wading pools. The operation, disinfection, and filtration of public swimming and wading pools are regulated by state and local health departments. In the United States, the Environmental Protec-

TABLE 28.1 Waterborne infectious disease outbreaks associated with drinking water in the United States[a]

Disease	Agent	Outbreaks	Cases
Shigellosis	*Shigella sonnei*	1	183
Giardiasis	*Giardia lamblia*	4	159
Cryptoporidiosis	*Cryptosporidium parvum*	2	1432
Gastroenteritis	*Escherichia coli* O157:H7	3	164
Acute gastro-intestinal illness	Unknown	5	163

[a]Compiled from data provided by the Centers for Disease Control and Prevention for 1997–1998. There were a total of 15 outbreaks and 2001 cases of infectious disease due to drinking water contamination.

tion Agency has established guidelines for fresh recreational water (monthly geometric mean of ≤33/100 ml for enterococci or ≤126/100 ml for *E. coli*), although local and state authorities can set standards above or below the guidelines. Private and therefore unregulated swimming pools, spas, and hot tubs are also occasional sources of outbreaks of waterborne diseases.

Over the last decade, about 15 waterborne disease outbreaks have occurred annually from regulated recreational waters in the United States. Table 28.2 categorizes these outbreaks according to the diseases produced.

Waterborne Infections in Developing Countries

Worldwide, waterborne infections are a much larger problem than in the United States and other developed countries. Developing countries often have poorly developed water and sewage treatment facilities, and access to safe potable water is limited. As a result, diseases such as cholera (Section 28.5), typhoid fever, and amebiasis (Section 28.8) are important public health problems worldwide.

TABLE 28.2 Recreational waterborne disease outbreaks in the United States, 1989–1998

Disease	Number of outbreaks	Percent
Gastroenteritis[a]	74	49.0
Dermatitis	50	33.1
Meningoencephalitis	18	11.9
Other	9	6.0
Total	151	100.0

[a]Most case of gastroenteritis were due to *Cryptosporidium parvum* (Section 28.6), *Escherichia coli* O157:H7 (Section 29.7), or a Norwalk-like virus (Section 28.8). Most cases of dermatitis were caused by *Pseudomonas aeruginosa*. Meningoencephalitis was caused by the ameba *Naegleria fowleri* (Section 28.8). Other diseases caused by microorganisms include leptospirosis and Pontiac fever due to infection by *Legionella* (Section 28.7).

✓ **28.4 Concept Check**

Drinking water and recreational water may both be sources of waterborne pathogens. In the United States, the number of disease outbreaks due to either of these sources is relatively small in relation to the large number of exposures to water. Worldwide, lack of adequate water treatment facilities and access to clean water contributes significantly to the spread of infectious diseases.

✓ Identify the microorganism most commonly responsible for disease outbreaks due to drinking water contamination.

✓ Identify the disease most commonly caused by exposure to contaminated recreational water.

28.5 Cholera

Cholera is a severe diarrheal disease that is now largely restricted to the developing parts of the world. Cholera is an example of a major waterborne disease that can be controlled by application of appropriate water treatment measures.

Biology and Epidemiology

The disease cholera is caused by *Vibrio cholerae*, a gram-negative, curved rod (∞ Section 12.12) transmitted almost exclusively via contaminated water. However, as with many waterborne diseases, cholera is also associated with food consumption. In the Americas, consumption of raw shellfish and raw vegetables has also been associated with cholera. Presumably, the vegetables were washed in contaminated water and the shellfish beds were contaminated by untreated sewage.

Since 1817, cholera has swept the world in seven major pandemics. The seventh began in 1961 and continues to the present. Two biotypes of *Vibrio cholerae* have been recognized, the classic and the *El Tor* types. The *classic* strain of *V. cholerae* was first isolated by Robert Koch in 1883 and was the prevalent biotype causing cholera pandemics before 1961. The El Tor biotype is responsible for the seventh pandemic that started in 1961. At least 5 million cases of cholera have been reported since 1961, with over 250,000 deaths. In 1999, there were over 400,000 cases and 9000 deaths worldwide.

Cholera is endemic in Africa, Southeast Asia, the Indian subcontinent, and Central and South America, especially in areas where sewage treatment is either inadequate or absent. Travelers to endemic areas should consider being immunized for cholera. Even in developed countries the disease is a threat. Sporadic outbreaks of cholera (fewer than 250 total cases) have been reported in the United States, mostly along the Gulf Coast, in the last 10 years. Raw shellfish may be the vehicle; *Vibrio cholerae* organisms appear to be endemic and free-living in coastal waters, adhering to normal flora (Figure 28.9).

Control of cholera depends primarily on adequate sanitation measures; *Vibrio cholerae* is completely eliminated from wastewater during proper sewage treatment and water purification.

Pathogenesis

Following ingestion of a substantial inoculum, the *Vibrio cholerae* cells take up residence in the *small* intestine. Studies in human volunteers have shown that stomach acidity is responsible for the large inoculum needed to initiate cholera. The ingestion of $10^8–10^9$ cholera vibrios is generally required to cause disease, but human volunteers given bicarbonate to neutralize gastric acidity developed cholera when given as few as 10^4 cells. Even lower cell numbers can initiate infection if *V. cholerae* is ingested with food, presumably because the food protects the vibrios from stomach acidity.

When the cholera vibrios attach to epithelial cells in the small intestine, they grow and release enterotoxin (∞ Section 21.11). Cholera enterotoxin causes diarrhea that can result in dehydration and death unless the patient is given fluid and electrolyte therapy. The enterotoxin causes fluid losses of up to 20 l (20 kg or 44 lb) per day. The mortality rate from untreated cholera can reach 60%.

Diagnosis and Treatment

Cholera is diagnosed by the presence of the gram-negative comma-shaped *Vibrio cholerae* bacilli in the "rice water" stools of patients with severe diarrhea. Intravenous or oral liquid and electrolyte replacement therapy (20 g glucose, 4.2 g NaCl, 4.0 g $NaHCO_3$, 1.8 g KCl, dissolved in 1.0 l H_2O) is the most effective means of cholera treatment. Oral treatment is preferred since no special equipment or sterile precautions are necessary.

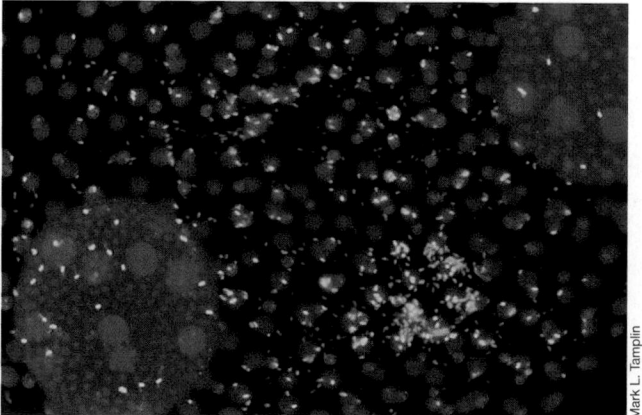

Figure 28.9 Cells of *Vibrio cholerae* attached to the surface of *Volvox*, a freshwater alga. The isolate was from a cholera-endemic area in Bangladesh. The *V. cholerae* cells are stained green by a monoclonal antibody to bacterial cell surface proteins. The red color is due to the fluorescence of chlorophyll *a* in the algae.

Effective treatment reduces the mortality rate to about 1%. Streptomycin or tetracycline may shorten the course of cholera, but antibiotics are of little benefit without simultaneous fluid and electrolyte replacement.

✓ 28.5 Concept Check

Vibrio cholerae is a pathogen that causes cholera, an acute diarrheal disease resulting in severe dehydration. Cholera occurs in pandemics. The current pandemic has endemic foci in the Americas, the Indian subcontinent, Asia, and Africa.

✓ Identify measures for preventing cholera in endemic areas.

✓ Identify specific measures for treating cholera.

28.6 Giardiasis and Cryptosporidiosis

Giardiasis and cryptosporidiosis are diseases caused by the protozoans *Giardia lamblia* and *Cryptosporidium parvum*, respectively. These organisms continue to be problematic even in well-regulated water supplies because they are found in nearly all surface waters and are highly resistant to chlorine.

Giardiasis

Giardia lamblia is a flagellated protozoan (∞ Section 14.8) that is usually transmitted to humans in fecally contaminated water, although foodborne and sexual transmission of giardiasis has been documented. *Giardiasis* is an acute gastroenteritis cause by the protozoan parasite (Figure 28.10). The protozoal cells, called trophozoites (Figure 28.10*a*) produce a resting stage called a **cyst** (Figure 28.10*b*). The cyst has a thick protective wall that allows the pathogen to resist drying and chemical disinfection. After ingestion in contaminated water, the cysts germinate, attach to the intestinal wall, and cause the symptoms of giardiasis: an explosive, foul smelling, watery diarrhea, intestinal cramps, flatulence, nausea, weight loss, and malaise. Symptoms may be acute or chronic. The foul-smelling diarrhea and the absence of blood or mucus in the stool are diagnostically helpful in distinguishing giardiasis from diarrhea of bacterial or viral origin. Many individuals who are infected exhibit no symptoms, simply acting as infected carriers.

Giardia was implicated in 4 of the 15 recent drinking water infectious disease outbreaks in the United States (Table 28.1). Giardiasis can also occur after accidental ingestion of water from infected swimming pools or lakes. *Giardia* cysts have been found in 97% of surface water sources (lakes, ponds, and streams) in the United States. The thick-walled cysts are resistant to chlorine, and many outbreaks have been associated with water systems that use only chlorination as a means of water purification. Water subjected to proper clarification and

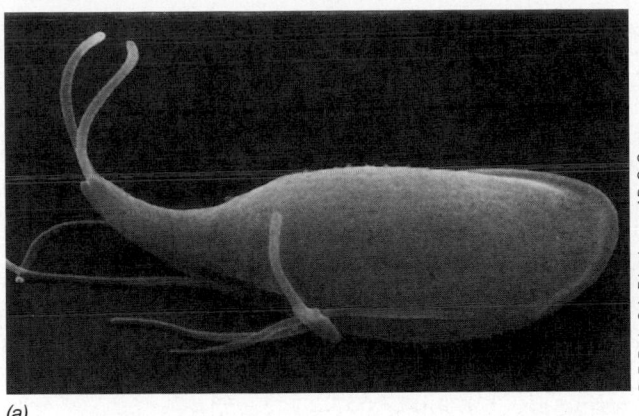
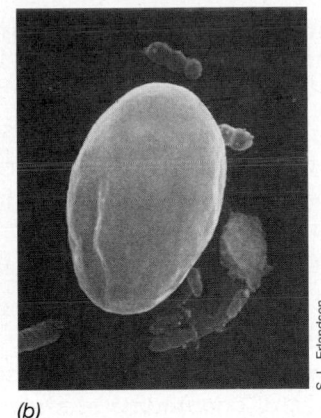

(a)

(b)

Figure 28.10 Scanning electron micrographs of the parasite *Giardia*. (a) Motile trophozoite. The trophozoite is about 15 μm in length. (b) Cyst. The cyst is about 11 μm in length.

filtration followed by chlorination (see Section 28.3) is generally free of *Giardia* cysts.

Isolated cases of giardiasis have also been associated with untreated drinking water in wilderness areas. Beavers and muskrats are frequent carriers of *Giardia* and may transmit cells or cysts to water supplies, causing human infection when ingested. As a safety precaution, all water consumed from rivers and streams, for example, during a camping or hiking trip, should be filtered *and* treated with iodine or chlorine, or filtered *and* boiled. Boiling is the preferred method of rendering water microbially safe.

Laboratory diagnostic methods include the demonstration of *Giardia* cysts in the stool or the demonstration of *Giardia* antigens in the stool using a direct ELISA (enzyme-linked immunosorbent assay; ⟳ Section 24.11). The drugs quinacrine, furazolidone, and metronidazole are useful in treating the acute disease.

Cryptosporidiosis

The protozoan *Cryptosporidium parvum* occurs in nature as a parasite in a variety of warm-blooded animals. The parasites are small (2–5 μm), round *coccidia* that invade and grow intracellularly in mucosal epithelial cells of the stomach and intestine. The protozoan produces thick-walled, chlorine-resistant, infective oocysts, which are shed into water in high numbers in the feces. The infection is passed on when other animals consume the fecally contaminated water. *Cryptosporidium* cysts are highly resistant to chlorine (up to 14 times as resistant as the chlorine-resistant *Giardia*), and therefore sedimentation and filtration methods must be used to remove *Cryptosporidium* from water supplies.

Cryptosporidium parvum was responsible for the largest single common-source outbreak of a waterborne disease ever recorded. In Milwaukee, Wisconsin, in the spring of 1993, about 400,000 people developed a diarrheal illness that was traced to the municipal water supply. Spring rains and runoff from surrounding farmland

had drained into Lake Michigan and overburdened the water supply system, leading to contamination by the protozoan. *C. parvum* is a significant intestinal parasite in dairy cattle, the most likely source of this outbreak.

Cryptosporidiosis is characterized by mild diarrhea. In normal individuals the diarrhea is self-limiting and most people recover without incident within 2 weeks. However, individuals with impaired immunity such as acquired immunodeficiency syndrome (AIDS) patients or the very young or old can develop serious complications. In the Milwaukee outbreak, about 4400 people required hospital care, and several died of complications from the disease, including severe dehydration.

The Milwaukee outbreak highlights the vulnerability of water purification systems, the need for constant water monitoring and surveillance, and the catastrophic consequences of the failure of a large water supply system.

Laboratory diagnostic methods for cryptosporidiosis include the demonstration of *Cryptosporidium* oocysts in the stool. Treatment is unnecessary for those with normal immunity. For individuals undergoing immunosuppressive therapy (e.g., prednisone), discontinuation of immunosuppressive drugs is indicated. Immunocompromised individuals should be given supportive therapy (for example, intravenous fluids and electrolytes).

✓ 28.6 Concept Check

Giardiasis and cryptosporidiosis are spread by the chlorine-resistant cysts of *Giardia lamblia* and *Cryptosporidium parvum*, respectively, in water contaminated by the feces of infected humans or animals. These diseases are occasionally propagated in drinking water and recreational water sources. Infection with either parasite causes diarrhea.

✓ Explain the importance of cysts in the survival and infectivity of both *Giardia lamblia* and *Cryptosporidium parvum*.

✓ Why are protozoans often associated with waterborne diseases, even in developed countries? Outline steps to reduce their impact.

28.7 Legionellosis (Legionnaires' Disease)

Legionella pneumophila, the bacterium that causes legionellosis, is a waterborne pathogen normally transmitted in aerosols rather than through drinking water or recreational water and has become an important emerging waterborne pathogen, especially in developed countries (∞ Section 25.10).

Biology and Epidemiology

Legionella pneumophila was first recognized as a pathogen due to an outbreak of pneumonia occurring during an American Legion convention in the summer of 1976. *Legionella* is a thin, gram-negative rod (Figure 28.11) with complex nutritional requirements, including an unusually high iron requirement, and is not related to any other pathogen associated with respiratory infections. *Legionella* can be detected by immunofluorescence techniques (∞ Section 24.10) and can be isolated from terrestrial and aquatic habitats as well as from patients suffering from legionellosis.

Legionella is present in small numbers in lakes, streams, and soil. It is relatively resistant to heating and chlorination, so it can spread through water distribution systems. It is commonly found in large numbers in cooling towers and evaporative condensers of large air conditioning systems. The pathogen grows in the water and is disseminated in humidified aerosols. Human infection is via airborne droplets, but the infection is not spread person to person. Multiple outbreaks of legionellosis tend to peak in mid to late summer months when air conditioners are extensively used.

Legionella has also been found in hot water tanks and whirlpool spas where it grows to high numbers in warm (35–45°C), stagnant water. Epidemiological studies now indicate that *Legionella* infections occur at all times of the year, primarily as a result of aerosols generated by heating/cooling systems and common practices such as showering or bathing. Overall, the incidence of *reported* cases of legionellosis has been steady (Figure 28.12), but up to 90% of cases are probably not diagnosed or properly reported. Prevention of legionellosis can be accomplished by improving the maintenance and design of water-dependent cooling and heating systems and water delivery systems *Legionella pneumophila* can be eliminated from water supplies by hyperchlorination or by heating to >63°C.

Pathogenesis

Legionella is an intracellular parasite that invades and grows in alveolar macrophages and monocytes (∞ Section 22.2). Infections are often asymptomatic or produce a mild cough, sore throat, mild headache, and fever. These mild, self-limiting cases, called *Pontiac fever*, need no treatment, and resolve in 2–5 days. However, more serious infections resulting in cases of pneumonia often occur in elderly individuals whose resistance has been previously compromised. Certain serotypes of *Legionella* (more than 10 are known) are strongly associated with the pneumonic form of the infection. Prior to the onset of pneumonia, intestinal disorders are common, followed by high fever, chills, and muscle aches. These symptoms precede the dry cough and chest and abdominal pains typical of legionellosis. Death occurs in up to 10% of cases and is usually due to respiratory failure.

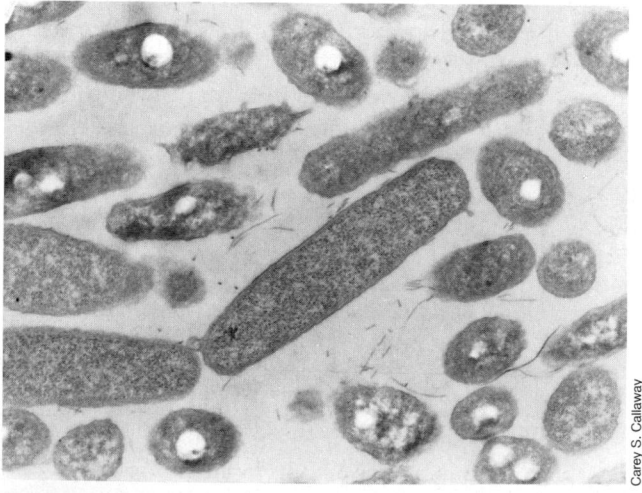

Figure 28.11 Transmission electron micrograph of *Legionella pneumophila*. Cells are approximately 0.6 μm in diameter.

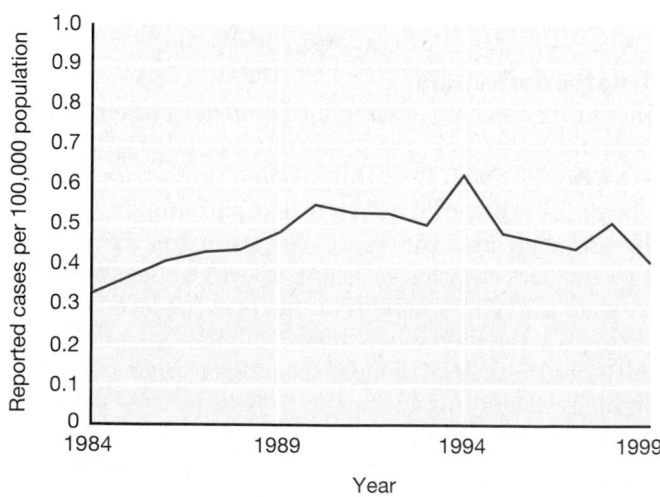

Figure 28.12 The annual prevalence of legionellosis in the United States. Note the steady trend with no significant decrease in the total incidence in recent years. In 1999 the total number of reported cases of legionellosis was 1108. Data are from the Centers for Disease Control and Prevention, Atlanta, GA, USA.

Diagnosis and Treatment

Clinical detection of *Legionella pneumophila* is usually done by culture from bronchial washings, pleural fluid, or other body fluids. Serological (antibody) tests are used as retrospective evidence for *Legionella* infection (∞ Section 24.7). As an aid in diagnosis, *Legionella* antigens can sometimes be detected in patient urine. *Legionella pneumophila* is sensitive to the antibiotics rifampin and erythromycin. Intravenous administration of erythromycin is the treatment of choice in most cases.

✓ 28.7 Concept Check

Legionella pneumophila is a respiratory pathogen that causes legionellosis and Pontiac fever. *L. pneumophila* grows to high numbers in warm water and is spread via aerosols. The prevalence of legionellosis is not decreasing.

- ✓ Indicate the source of *Legionella pneumophila*.
- ✓ Identify specific measures for control of *Legionella pneumophila*.

28.8 Typhoid Fever and Other Waterborne Diseases

A variety of bacteria, viruses, and protozoa can transmit common-source waterborne diseases. These diseases are a significant source of morbidity, especially in developing countries. Here we briefly discuss several other important waterborne diseases.

Typhoid Fever

On a global scale, probably the most important pathogenic *Bacteria* transmitted by the water route are *Salmonella typhi*, the organism causing typhoid fever, and *Vibrio cholerae*, the organism causing cholera, as discussed above. Although *Salmonella typhi* may also be transmitted by contaminated food (∞ Section 29.6) and by direct contact from infected individuals (∞ Chapter 25, box, The Tragic Case of Typhoid Mary), the most common and serious means of transmission is the water route. Fortunately, typhoid fever has been virtually eliminated in developed countries, primarily due to effective water treatment procedures (see Figure 28.2). In the United States, there are fewer than 400 cases in most years, but, as we discussed in Section 28.1, typhoid fever was a major public health threat before drinking water was routinely filtered and chlorinated. However, breakdown of water treatment methods, contamination of water during floods, earthquakes, and other disasters, or cross-contamination of water supply pipes from leaking sewer lines can still lead to epidemics of typhoid fever.

Viruses

Viruses can also be transmitted in water and cause human disease. Quite commonly, enteroviruses such as poliovirus (∞ Section 16.7), Norwalk-like virus (∞ Table 25.5), and hepatitis A virus (∞ Section 26.11) are shed into the water in fecal material. The most serious of these was poliovirus, but wild poliovirus has been eliminated from the Western Hemisphere (∞ Section 25.8). Although viruses can survive in water for relatively long periods, they are easily neutralized by ordinary water purification schemes such as chlorination. The maintenance of 0.6 parts per million (ppm) of chlorine in water (Section 28.3) ensures the neutralization of viruses in the water supply.

Amebiasis

A number of amebas inhabit the tissues of humans and other vertebrates, usually in the oral cavity or intestinal tract, and some of these are pathogenic. We discussed the general properties of ameboid protozoa in Section 14.8.

Worldwide, *Entamoeba hystolytica* is a common pathogenic protozoan transmitted to humans primarily by contaminated water and occasionally by the foodborne route (Figure 28.13). *E. histolytica* is an *anaerobic* ameba; the trophozoites lack mitochondria. Like *Giardia*, the trophozoites of *E. histolytica* produce cysts. The cysts cause infestation, and cyst germination occurs in the intestine, where amebic cells grow both on and in intestinal mucosal cells. Many infections are asymptomatic, but continued growth may lead to invasion and ulceration of the intestinal mucosa, causing diarrhea and severe intestinal cramps. Diarrhea can progress to invasion of the intestinal wall, a condition known as *dysentery*, characterized by intestinal inflammation, fever, and the passage of intestinal exudates, including blood and mucus. If not treated, invasive trophozoites of *E. histolytica* can invade the liver, and sometimes the lung and brain. Growth in these tissues can cause severe abscesses and death. Worldwide, up to 100,000 individuals die

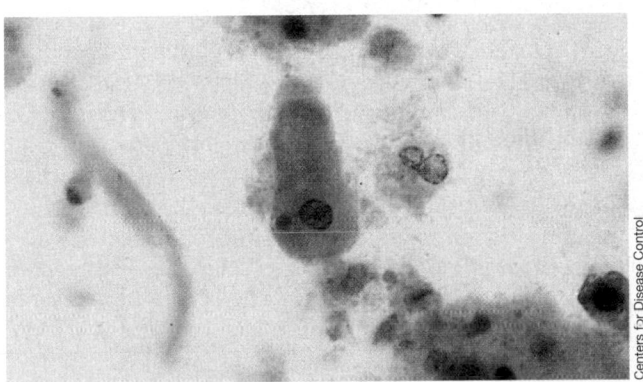

Figure 28.13 The trophozoite of *Entamoeba histolytica*, the ameba that causes amebiasis. The small red structures are red blood cells.

each year from invasive amebic dysentery. The disease is extremely common in tropical and subtropical countries worldwide, with at least 50 million people developing symptomatic diarrhea annually and up to tenfold more having asymptomatic disease. In the United States, there are several hundred cases per year, mostly occurring near international borders in the southwest.

E. histolytica amebiasis can be treated with the drug dehydroemetine for invasive disease and diloxanide furoate for certain asymptomatic cases, such as in immune-compromised individuals, but amebicidal drugs are not universally effective. Spontaneous cures do occur, suggesting that the host immune system plays some role in ending the infection. However, protective immunity is not an outcome of primary infection and reinfection is common. The disease occurs at very low incidence in regions that practice adequate sewage treatment. Ineffective sewage treatment and use of untreated surface waters for drinking purposes are the usual scenarios for cases of amebiasis.

Laboratory diagnosis involves the demonstration of *E. histolytica* cysts in the stool, trophozoites in tissue, or the demonstration of antibodies to *E. histolytica* in the blood using an ELISA (enzyme-linked immunosorbent assay; ∞ Section 24.11).

Naegleria fowleri can also cause amebiasis, but in a very different form. *N. fowleri* is a free-living ameba found in soil and in water runoff. *N. fowleri* infections usually result from swimming or bathing in warm, soil-contaminated water sources such as hot springs, or lakes and streams in the summer. The free-living ameba enters the body through the nose and burrows directly into the brain. Here, the ameba propagates, causing extensive hemorrhage and brain damage. This condition is called **meningoencephalitis**. Death usually results within a week. In the past decade, there have been 18 outbreaks of meningoencephalitis due to exposure to contaminated recreational waters in the United States (Table 28.2). In 1998, there were 4 outbreaks, each involving a single child (ages 3 to 14) who swam or waded in lakes or streams in summer. All four victims died.

Diagnosis of *N. fowleri* infection requires observation of the ameba in the cerebrospinal fluid. If a definitive diagnosis can be done quickly, amphotericin B can be successfully used to treat infections.

✓ 28.8 Concept Check

Typhoid fever, viral infections, and amebiasis are important waterborne diseases. Waterborne typhoid fever and viral illnesses, while still common diseases in developing countries, have been controlled by effective water treatment in developed countries. Amebic dysentery caused by *Entamoeba histolytica* is a worldwide problem that affects millions of people. Meningoencephalitis is a rare but serious condition caused by *Naegleria* amebiasis.

✓ Explain the impact of effective water hygiene on the spread of human diseases such as typhoid fever and polio spread by fecal contamination of water supplies.

✓ Describe public health measures that could be used to eliminate or reduce the number of cases of meningoencephalitis.

Review Questions

1. Define coliforms and explain the coliform test. Why is the coliform test used to assess the purity of drinking water?

2. Trace the purification of wastewater in a typical treatment plant. What is the overall reduction in the BOD?

3. Identify (stepwise) the methods used to process drinking water. What important contaminants are targeted by each step in the process?

4. Identify the major sources of waterborne infection. Why are common sources of infection so dangerous to public health?

5. Why are antibiotics ineffective for the treatment of cholera? What methods are commonly used to treat cholera victims?

6. Giardiasis and cyptospordiosis remain significant public health problems even in areas with stringent water quality standards. Explain.

7. Describe the main features of legionellosis. What distinguishes this disease from other waterborne diseases?

8. A number of waterborne diseases have been controlled by effective water hygiene plans. Identify at least two of these diseases and indicate how they have been controlled.

Application Questions

1. For the coliform test, explain why the test procedure does not identify pathogens. Would it be feasible to perform tests on water for all known waterborne pathogen? Why or why not?

2. Why is reduction in the BOD of wastewater the primary goal of waterwater treatment? What occurs if the BOD of wastewater is not significantly reduced before it is distributed to local water sources such as lakes or streams?

3. Identify and explain the procedures used to eliminate chemical and biological contaminants from drinking water. What problems would occur if either chemical or biological purification methods failed? Remember to take into account the purity of the raw water.

4. The federal government has defined a strict set of drinking water standards. These standards are enforced by law. However, recreational water standards are not regulated as tightly. Explain why recreational water standards are more flexible and devise recreational water standards for the area in which you live.

5. Discuss whether or not *Vibrio cholerae* could cause a cholera epidemic in the United States. Justify your answer. Remember that *V. cholerae* is found in the waters of the Gulf of Mexico.

6. Why are surface waters contaminated with the cysts of various protozoans? What steps might public health officials take to remedy this problem?

7. Identify the potential problems in the design of air-conditioning units or showerheads that lead to contamination by and growth of *Legionella pneumophila*.

8. Discuss the wastewater and drinking water treatment schemes that must be in place to control such diseases as typhoid fever. Is it possible to eliminate *Salmonella typhi* as has been effectively done for the polio virus? Would it be possible or desirable to eliminate *Naegleria fowleri*? Explain.

Foodborne disease is an important emerging area of concern to microbiologists. The widespread consumption of prepared, prepackaged, and ready-to-eat food items places a tremendous burden on food production and distribution facilities to maintain high standards and prevent contamination by such common organisms as *Clostridium perfringens*, shown here. Food infections or food poisonings originating from a common production source can affect hundreds of individuals.

FOOD PRESERVATION AND FOODBORNE MICROBIAL DISEASES

29

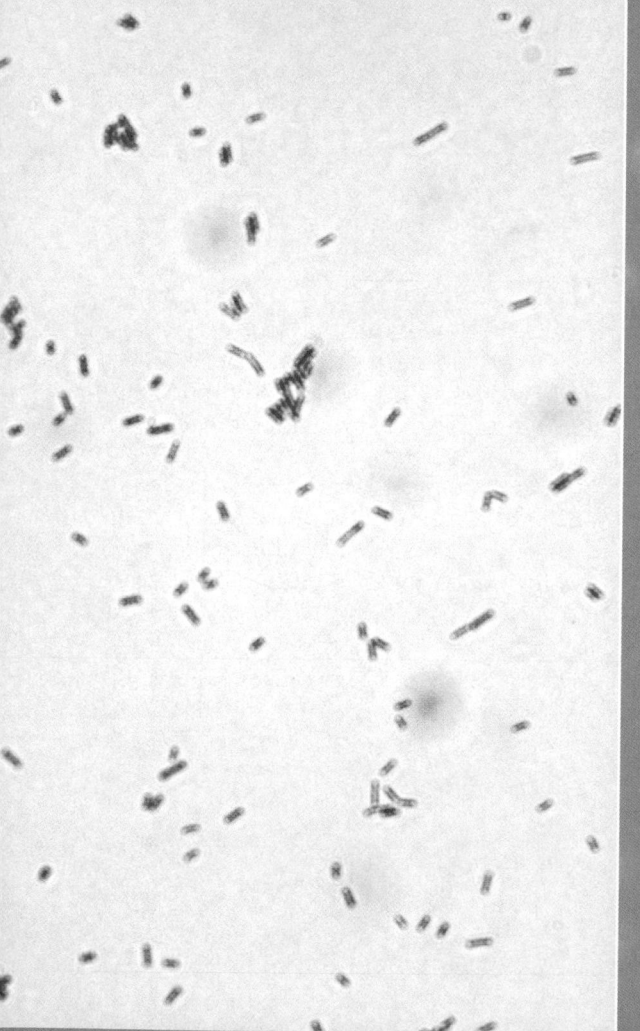

Working Glossary

Botulism food poisoning due to ingestion of food containing botulinum toxin produced by *Clostridium botulinum*

Canning the process of sealing food in a closed container and heating to destroy living organisms

Food infection disease caused by active infection resulting from ingestion of pathogen-contaminated food

Food poisoning (food intoxication) disease caused by the ingestion of food that contains preformed microbial toxins

Food spoilage a change in the appearance, smell, or taste of a food that makes it unacceptable to the consumer

Irradiation the exposure of food to ionizing radiation for the purpose of inhibiting growth of microorganisms and insect pests, or to retard growth or ripening

Listeriosis gastrointestinal food infection caused by *Listeria monocytogenes* that may lead to bacteremia and meningitis

Lyophilization (freeze-drying) the process of removing all water from frozen food under vacuum

Nonperishable (stable) foods foods of low water activity that have an extended shelf life and are resistant to spoilage by microorganisms

Perishable foods fresh foods generally of high water activity, that have a very short shelf life due to potential for spoilage by growth of microorganisms

Pickling the process of acidifying food to prevent microbial growth and spoilage

Salmonellosis enterocolitis caused by any of 1400 serotypes of *Salmonella* spp.

Semiperishable foods foods of intermediate water activity that have a limited shelf life due to potential for spoilage by growth of microorganisms

Water activity (a_w) the availablity of water for use in metabolic processes

---

Microorganisms are important factors in our food supply. A variety of foods that we consume are produced or enhanced by microbial action. For example, dairy products such as cheese, buttermilk, sour cream, and yogurt are all produced by microbial fermentation. Sauerkraut is a fermented vegetable food. Meat products including certain sausages, pates, and liver spreads are produced using microbial fermentation techniques. Cider vinegar is produced by lactic acid bacteria (∞ Section 30.10), and alcoholic beverages are produced by fermentation processes using yeast (∞ Section 30.13). We discuss the use of microorganisms to produce edible foods on an industrial scale in Chapter 30.

In this chapter, we will concentrate on the negative aspects of microbial growth in food. Uncontrolled and unwanted microbial growth destroys vast quantities of food, causing significant economic loss as well as a tremendous loss of nutrients. Consumption of food contaminated with particular microorganisms or microbial products can also cause serious illnesses such as food infections or food poisoning.

■ FOOD PRESERVATION AND MICROBIAL GROWTH

Microorganisms are ubiquitous in our environment and can be found in water, air, and especially in food. Fresh food, most prepared foods, and sometimes even preserved foods are contaminated with microorganisms. First we discuss the microorganisms that are important

spoilage agents in foods and we then present a variety of methods used to control their growth and preserve the food supply.

29.1 Microbial Growth and Food Spoilage

A wide variety of microorganisms colonize and grow on common foods. Many foods provide a suitable medium for the growth of microorganisms, and microbial growth usually reduces food quality and availability.

Food Spoilage

Food spoilage is any change in the appearance, smell, or taste of a food product that makes it unacceptable to the consumer. Spoiled food is not necessarily food unsafe to eat, but in some cases pathogenic organisms may cause spoilage. Spoiled food is generally regarded as unpalatable and will not be purchased or consumed. Food spoilage causes economic loss to producers, distributors, and even consumers in the form of higher prices and constricted supply.

Since foods are organic material, they provide nutrients for the growth of a wide variety of chemoorganotrophic bacteria. The physical and chemical characteristics of the food determine its degree of susceptibility to microbial activity. With respect to spoilage, foods can be classified into three major categories: (1) **perishable foods**, including many fresh foods; (2) **semiperishable foods**, such as potatoes and nuts; and (3) **stable** or **nonperishable foods** such as flour and sugar (Table 29.1). These food categories differ largely with regard to *moisture content*, which is related, as we saw in Section 6.12, to water activity, a_w. Nonperishable foods have low water activity and can

TABLE 29.1 Food classification by storage potential

Food classification	Examples
Highly perishable	Meats, fish, poultry, eggs, milk, most fruits and vegetables
Semiperishable	Potatoes, some apples, and nuts
Nonperishable	Sugar, flour, rice, and dry beans

generally be stored for considerable lengths of time without deterioration. Perishable and semiperishable foods are those with higher water activity. These foods must be stored under conditions that slow or stop microbial growth.

Fresh foods are spoiled by a variety of bacteria and fungi, and each type of fresh food is typically colonized by particular microorganisms (Table 29.2). Because the chemical properties of foods vary widely, different foods are colonized by the indigenous spoilage organisms that are best able to use the available nutrients.

For example, enteric bacteria such as *Salmonella*, *Shigella*, and *Escherichia*, all potential pathogens that live in the gut of animals, are rarely implicated in fruit or vegetable spoilage, but often contaminate and spoil meat. At slaughter, intestinal contents, including the living bacteria, can leak and contaminate the meat. Likewise, lactic acid bacteria, the most common microorganisms in dairy products, are the major spoilers of milk and milk products. *Pseudomonas* species are found in both soil and animals and are thus widely involved in the spoilage of fresh foods.

Microbial growth in foods follows the standard pattern for bacterial growth (➡ Sections 6.3 and 6.4). The lag phase may be of variable duration in a food, depending on the contaminating organism and its previous growth history. The rate of growth during the exponential phase depends on the *temperature*, the *nutrient value of the food*, and other conditions of growth. The time required for

the population density to reach a significant level in a given food product depends on both the size of the initial inoculum and the rate of growth during the exponential phase. Only when the microbial population density reaches a substantial level are spoilage effects usually observed. Throughout much of the exponential growth phase, population densities may be so low that no effect can be observed. Only the last doubling or two leads to observable spoilage (➡ Section 6.4). Thus, for much of the period of microbial growth in a food, there is no visible or easily detectable change in food quality.

✓ 29.1 Concept Check

Foods often spoil due to contamination by microorganisms. Foods vary considerably in their sensitivity to microbial growth, depending on their nutrient content and water content. Individual categories of food have specific spoilage patterns and spoilage organisms. Many food spoilage microorganisms are also potential pathogens.

- ✓ List the major categories of food and define them with respect to water content.
- ✓ Identify at least three bacterial genera that cause both food spoilage and human disease.

29.2 Food Preservation

We now examine a number of food storage and preservation processes that inhibit or stop microbial growth in food, stopping the growth of spoilage microorganisms and human pathogens.

Temperature

Besides moisture, one of the most crucial factors affecting microbial growth in food is temperature (➡ Section 6.9). In general, a *lower* storage temperature results in a retarded spoilage rate. A number of *psychrotolerant* (cold

TABLE 29.2 Microbial spoilage of fresh food[a]

Food product	Type of microorganism	Common spoilage organisms, by genus
Fruits and vegetables	Bacteria	*Erwinia*, **Pseudomonas**, **Corynebacterium** (mainly vegetable pathogens; rarely spoil fruit)
	Fungi	*Aspergillus*, *Botrytis*, *Geotrichium*, *Rhizopus*, *Penicillium*, *Cladosporium*, *Alternaria*, *Phytophora*, various yeasts
Fresh meat, poultry, and seafood	Bacteria	*Acinetobacter*, *Aeromonas*, **Pseudomonas**, *Micrococcus*, *Achromobacter*, *Flavobacterium*, **Proteus**, **Salmonella**, **Escherichia**, **Campylobacter**, **Listeria**
	Fungi	*Cladosporium*, *Mucor*, *Rhizopus*, *Penicillium*, *Geotrichium*, **Sporotrichium**, **Candida**, *Torula*, *Rhodotorula*
Milk	Bacteria	**Streptococcus**, *Leuconostoc*, *Lactococcus*, *Lactobacillus*, **Pseudomonas**, **Proteus**
High-sugar foods	Bacteria	**Clostridium**, **Bacillus**, *Flavobacterium*
	Fungi	*Saccharomyces*, *Torula*, *Penicillium*

[a] The organisms listed are the most commonly observed spoilage agents of fresh, perishable foods. Genera in bold face include possible human pathogens.

tolerant) microorganisms, however, can survive and grow at refrigerator temperatures. Therefore, storage of perishable food products for long periods of time (greater than several days) is possible only at temperatures below freezing. Freezing and thawing alter the physical structure of many foods. Therefore, freezing is not an acceptable preservation method for many fresh foods, but is widely used for the preservation of meats and many fruits and vegetables. Freezers providing a temperature of −20°C are most commonly used. At −20°C, storage for weeks or months is possible, but microbial growth may still occur in pockets of liquid water trapped within the frozen mass. For long-term storage, temperatures such as −80°C (dry ice temperature) are necessary. Maintenance of such low temperatures is expensive and consequently is not used for routine food storage.

Acidity

Another major factor affecting microbial growth in food is pH or acidity. Foods vary somewhat in pH, but most are neutral or acidic. Microorganisms differ in their ability to grow under acidic conditions, but conditions of pH 5 or less inhibit the growth of most spoilage organisms (∞ Section 6.11). Therefore, acid is often used in food preservation, a process called **pickling**. Vinegar, which is dilute acetic acid, is usually added in the pickling process (vinegar is a fermentation product of the acetic acid bacteria; its industrial production will be discussed in Section 30.10). In addition to vinegar, pickling methods usually include the addition of large amounts of salt or sugar to decrease water availability (see below) and further inhibit microbial growth. Common pickled foods include cucumbers (sweet, sour, and dill pickles), peppers, meats, fish, and fruits. In some cases, acid can develop in the food as a result of microbial action, and the product is called a *fermented food*, such as in sauerkraut (cabbage), yogurt, cheese, and sour cream. The microorganisms involved in food fermentations include the lactic acid bacteria, the acetic acid bacteria, and the propionic acid bacteria. These bacteria do not grow below about pH 4, so the food fermentation is a self-limiting process.

Water activity

Water activity (a_w) is the availability of water for use by microorganisms in metabolic processes. Because microorganisms do not grow under condition of low water activity (low water availability), microbial growth can be controlled by lowering the available water content of the food by drying or by adding high concentrations of a solute such as salt or sugar (∞ Section 6.12). Natural or artificial heat is often used for drying, but the least damaging physical method used to dry foods is the process of **lyophilization (freeze-drying)**, where foods are frozen and water is removed under vacuum. Milk,

meat, fish, vegetables, fruit, eggs, and other economically important foods are commonly preserved by some form of drying.

A number of foods are preserved by addition of salt or sugar to reduce water activity. Foods preserved by addition of sugar are mainly fruits (jams, jellies, and preserves). Salted products are primarily meats and fish. Sausage and ham are preserved by salt, although individual products vary in water activity depending on how much salt is added and how much the meat has been dried.

Canning

Canning is a process in which a food is sealed in a container such as a can or glass jar and heated to kill all living organisms, or at least to ensure that there will be no growth of residual organisms. When the can is properly sealed and heated, the food should remain stable and unspoiled indefinitely, even when stored in the absence of refrigeration.

The temperature–time relationships for canning depend on the type of food, its pH, the size of the container, and the consistency or density of the food. Because heat must penetrate completely to the center of the food within the can, heating times must be longer for large cans or very dense foods. Acid foods can often be canned effectively by heating just to boiling, 100°C, whereas nonacid foods must be heated to autoclave temperatures (121°C). Unfortunately, heating times long enough to guarantee absolute sterility of every can (∞ Section 20.1) would change the food so greatly that it would likely be unpalatable and lose nutritional value. Therefore, properly canned foods may not be sterile.

The environment inside a can is anoxic, and microbial growth in a canned food is frequently the result of the growth of organisms that produce extensive amounts of gas. This can cause pressure build-up inside the can, resulting in bulges or, in severe cases, even an explosion of the can (Figure 29.1). Because some of the anoxic bacteria that grow in canned foods are toxin producers of the genus *Clostridium* (∞ Sections 21.10 and 29.5), food from a can that is visibly altered should never be eaten. However, the lack of obvious gas production is not an absolute guarantee that the food is safe to consume.

Chemical Food Preservation

A number of chemicals are used commercially to control microbial growth in food. These are classified by the U.S. Food and Drug Administration as "generally recognized as safe" and find wide application in the food industry (Table 29.3). Many of these chemicals, like sodium propionate, have been used for many years with no evidence of human toxicity. Others, like nitrites (carcinogen precursor), ethylene or propylene oxides (mutagens;

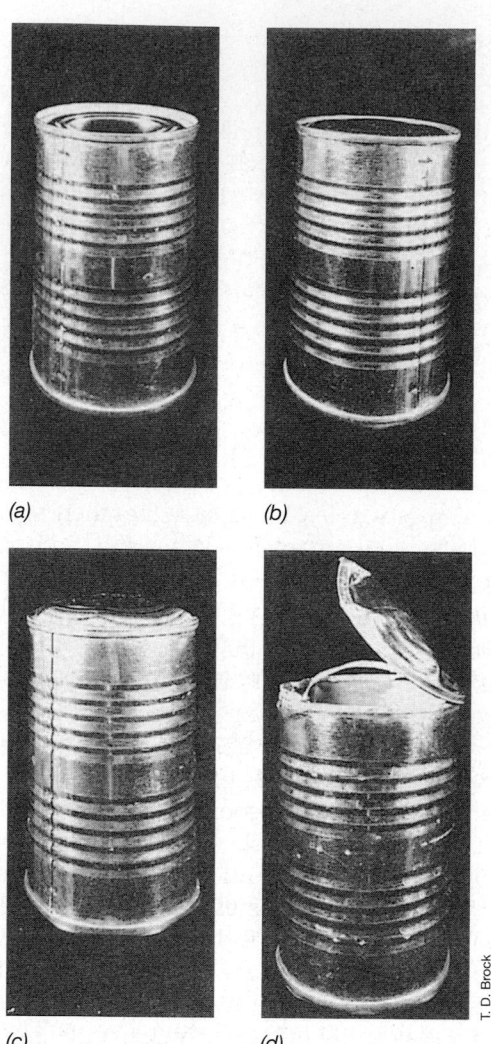

(a)

(b)

(c)

(d)

T. D. Brock

Figure 29.1 Changes in cans as a result of microbial spoilage. (a) Normal can; the top of the can is slightly indented due to negative pressure (vacuum) inside. (b) Swelling resulting from minimal gas-production. Note that the top is slightly distended. (c) Severe swelling due to extensive gas production. (d) The can shown in (c) was dropped and the gas pressure resulted in a violent explosion, tearing the lid apart.

TABLE 29.3 Chemical food preservatives

Chemical	Foods
Sodium or calcium propionate	Bread
Sodium benzoate	Carbonated beverages, fruit, fruit juices, pickles, margarine, preserves
Sorbic acid	Citrus products, cheese, pickles, salads
Sulfur dioxide, sulfites, bisulfites	Dried fruits and vegetables; wine
Formaldehyde (from food-smoking process)	Meat, fish
Ethylene and propylene oxides	Spices, dried fruits, nuts
Sodium nitrite	Smoked ham, bacon

 Section 10.4), or antibiotics (development of antibiotic-resistant pathogens; Section 20.12), are more controversial food additives because of evidence that these compounds may be detrimental to human health.

Because of lengthy and costly testing programs for any new chemical proposed as a food preservative today, it is unlikely that new compounds will be added to the list of chemical food preservatives in Table 29.3 in the near future.

Irradiation

Irradiation of food using ionizing irradiation is now a standard method for reducing contamination by bacteria, fungi, and even insects (Section 20.2). Table 29.4 lists some foods for which radiation treatment has been approved. Foods such as spices are routinely irradiated. In the United States, fresh meat products such as hamburger and poultry can now be irradiated to limit contamination by *Escherichia coli* O157:H7 and other enteric pathogens (hamburger) and *Campylobacter jejuni* (poultry). For food irradiation, gamma rays are generated from ^{60}Co or ^{137}Cs sources. The food products receive a controlled radiation dose. This dose varies considerably by each food category and purpose. For example, a dose of 44 kilo Grays (kGys) is used to sterilize meat products used on United States NASA space flights and is nearly 10 times higher than the 4.5-kGy dose used for control of pathogens in hamburger (Table 29.4). In the United States, a consumer product information label must be affixed to foods that are irradiated.

✓ 29.2 Concept Check

Food microbiology deals with methods for limiting spoilage and the growth of disease-causing microorganisms in food during processing and storage. Foods vary considerably in their sensitivity to microbial growth, depending on their nutrient content, water availability, and pH. The growth of microorganisms in perishable foods can be controlled by refrigeration, freezing, canning, pickling, dehydration, chemical preservation, or irradiation.

✓ Outline at least four methods of food preservation. How does each method limit growth of microorganisms?

✓ Are food spoilage microorganisms also pathogens? Give examples to support your answer.

II FOODBORNE DISEASES

Failure to adequately decontaminate and preserve food may allow the growth of pathogens, resulting in diseases with significant morbidity and mortality. Like waterborne diseases, foodborne diseases are common-source diseases; a single contaminated food source, for instance

TABLE 29.4 Irradiated foods by category and purpose[a]

Food category	Purpose for irradiation
Fresh pork	Control of *Trichinella spiralis* parasite
Fresh fruits and vegetables	Inhibition of growth and maturation (ripening)
Dried spices, herbs, and flavoring mixtures	Microbial disinfection
Refrigerated or frozen uncooked meat products, including ground meat	Control of foodborne pathogens
	Extension of shelf life
	Sterilization
Packaged frozen meats used in the National Aeronautics and Space Administration (NASA) flight program	
Dry or dehydrated enzyme preparations (e.g., meat tenderizer)	Microbial disinfection
Frozen, uncooked poultry and poultry products	Control of foodborne pathogens

[a] Consumer labeling laws in the United States require that all irradiated foods must be conspicuously labeled "Treated with radiation" or "Treated by irradiation" in addition to information required by other regulations.

at a food-processing plant or a restaurant, may affect a number of individuals.

29.3 Foodborne Diseases and Microbial Sampling

A summary of the most prevalent foodborne diseases and the microorganisms that cause them is shown in Table 29.5. These common diseases can be separated into two categories, *food poisoning* and *food infection*. Specialized microbial sampling techniques are necessary to isolate the pathogens responsible for foodborne diseases.

Food Poisoning

Food poisoning or **food intoxication** is disease that results from ingestion of foods containing preformed microbial toxins. The microorganisms that produced the toxins do not have to grow in the host and are often not alive at the time the contaminated food is consumed. The illness is due to ingestion and subsequent action of preformed bioactive toxin. We previously discussed some of these toxins, notably the exotoxin of *Clostridium botulinum* (∞ Section 21.10) and the superantigen toxins of *Staphylococcus aureus* (∞ Section 22.14).

Food Infection

Food infections are active infections resulting from ingestion of pathogen-contaminated food. In addition to the passive transfer of microbial toxins, food may contain sufficient numbers of viable pathogens to cause infection and disease in the host. Food infection is a very common type of foodborne illness (Table 29.5), and *Salmonella* food infection is a typical example (see Section 29.6). Many of these infectious agents also cause waterborne diseases (∞ Chapter 28).

Microbial Sampling of Foods

As discussed in Section 29.1, microorganisms are always present in fresh foods. Because pathogens may be present

along with many harmless organisms, methods have been developed to detect important pathogens such as *Escherichia coli* O157:H7, *Salmonella*, *Staphylococcus*, and *Clostridium botulinum*. We discussed in Section 24.13 the use of nucleic acid probes for the detection of specific foodborne pathogens. For growth studies of nonliquid foods, preliminary treatment is usually required to suspend microorganisms embedded or entrapped within the food. The most suitable method is high-speed blending. The food should be examined as soon after sampling as possible; if examination cannot begin within 1 h of sampling, the food should be refrigerated. A frozen food should be thawed in its original container in a refrigerator and examined or cultured as soon as thawing is complete. Samples can be inoculated onto enriched media, followed by transfer to differential or selective media for isolation and identification, as we described for human pathogens (∞ Section 24.1), or probed directly for pathogen presence using nucleic acid-based methods such as the polymerase chain reaction (PCR) (∞ Section 24.13).

✓ 29.3 Concept Check

Foodborne diseases include food poisoning resulting from the action of microbial toxins and food infections due to the growth and invasion of microorganisms in the body. Specialized techniques are used to sample microorganisms in food.

- ✓ Distinguish between food infection and food poisoning.
- ✓ Describe the sampling of a solid food such as meat for the presence of microorganisms.

29.4 Staphylococcal Food Poisoning

A very common food *poisoning* is caused by the gram-positive coccus, *Staphylococcus aureus*.

Biology and Epidemiology

Staphylococcus aureus and other members of the genus are small, gram-positive cocci (∞ Section 12.19). As we

TABLE 29.5	Annual foodborne disease estimates for the United States		
Organism	Disease[a]	Number per year	Foods
Bacteria			
Bacillus cereus	FP	27,000	Rice and starchy foods, high-sugar foods
Campylobacter jejuni	FI	1,963,000	Poultry, dairy
Clostridium perfringens	FP	248,000	Cooked and reheated meats and meat products
Escherichia coli O157:H7	FI	63,000	Meat, especially ground meat
Other enteropathogenic *Escherichia coli*	FI	110,000	Meat, especially ground meat
Listeria monocytogenes	FI	2,500	Meat and dairy
Salmonella spp.	FI	1,340,000	Poultry, meat, dairy, eggs
Staphylococcus aureus	FP	185,000	Meat, desserts
Streptococcus	FI	50,000	Dairy, meat
Yersinia enterocolitica	FI	87,000	Pork, milk
All other bacteria	FP and FI	102,000	
Total bacterial		**4,177,500**	
Parasites			
Cryptosporidium parvum	FI	30,000	Raw and undercooked meat
Cyclospora cayetanensis	FI	16,000	Fresh produce
Giardia lamblia	FI	200,000	Contaminated or infected meat
Toxoplasma gondii	FI	113,000	Raw and undercooked meat
Total parasites		**359,000**	
Viruses			
Norwalk-like viruses	FI	9,200,000	Shellfish, many other foods
All other viruses	FI	82,000	
Total viruses		**9,282,000**	
Total Annual Foodborne Diseases		**13,818,500**	

[a]FP, food poisoning: FI, food infection. Estimates are based on data provided by the Centers for Disease Control and Prevention, Atlanta, GA, USA.

discussed in Section 26.9, staphylococci are found on the skin and in the respiratory tract of nearly all humans, and are often opportunistic pathogens. *Staphylococcus aureus* is frequently associated with food poisoning because, as it grows, this organism produces several heat-stable protein enterotoxins (⌒⌒ Section 21.11). The enterotoxins are released into the surrounding medium or food; if food that contains toxin is ingested, gastroenteritis characterized by nausea, vomiting, and diarrhea, occurs within 1–6 h.

Each year, an estimated 185,000 cases of staphylococcal food poisoning occur in the United States (Table 29.5). The foods most commonly involved are custard- and cream-filled baked goods, poultry, meat and meat products, gravies, egg and meat salads, puddings, and creamy salad dressings. If such foods are kept refrigerated after preparation, they usually remain safe, because *S. aureus* growth is significantly reduced at low temperatures. Foods of this type, however, are often kept in kitchens at room temperature or outdoors at picnics. The food, if inoculated with *S. aureus* from an infected food handler, supports rapid bacterial growth and enterotoxin production. Even if the toxin-containing foods are reheated again before eating, the toxin is relatively heat-stable and may remain active.

Pathogenesis

Staphylococcus aureus produces seven different enterotoxins: A, B, C1, C2, C3, D, and E. Enterotoxin A, a superantigen, is most frequently associated with staphylococcal food poisoning (⌒⌒ Section 22.14). Superantigens work by stimulating large numbers of T cells, which in turn release intercellular mediators called cytokines, activating a general inflammatory response in the intestine that results in gastroenteritis, including massive loss of fluids from the intestine.

S. aureus enterotoxin A is a small single peptide of 30,000 molecular weight that is encoded by a chromosomal gene. Cloning and sequencing of this gene, the *entA* gene, and of several other *S. aureus* enterotoxin genes, show that the toxins are genetically related. Although the *entA* gene is on the bacterial chromosome, type B and C *S. aureus* enterotoxins may be encoded on plasmids, transposons, or lysogenic bacteriophages. We discussed the importance of accessory genetic elements such as plasmids and bacteriophages as vectors for toxin production in Sections 21.10 and 21.11.

Diagnosis, Treatment, and Prevention

Several assays based on the detection of either enterotoxin (ELISA detection of enterotoxin; ⌒⌒ Section 24.11) or *S. aureus* exonuclease (an enzyme that degrades DNA) are available to detect dangerous levels of *S. aureus* in food. However, these rapid tests are qualitative, confirming only the presence or absence of *S. aureus* above the detection limits of the assay. To obtain quantitative data and determine the extent of bacterial contamination, bacterial plate counts are required. For staphylococcal counts, a high-salt medium (either sodium chloride or lithium chloride at a final concentration of 7.5%) is used. Of the organisms present in foods, staphylococci are the only common ones tolerant of such levels of salt.

S. aureus food poisoning can be quite severe, but is self-limiting, usually resolving within 48 h after onset. Severe cases may require treatment for dehydration. Treatment with antibiotics is not useful because the disease is caused by a preformed toxin, not an active infection.

Staphylococcal food poisoning can be prevented by careful sanitation and hygiene measures both in production and food preparation steps and by storage of foods at low temperatures to inhibit bacterial growth. Foods susceptible to colonization by *S. aureus* and kept for several hours above 4°C (refrigerator temperature) should be discarded rather than eaten.

✓ 29.4 Concept Check

Staphylococcal food poisoning results from the ingestion of preformed enterotoxin A, a superantigen produced by *Staphylococcus aureus* when growing in foods. In many cases, *S. aureus* cannot be cultured from the contaminated food.

- ✓ Explain the symptoms of staphylococcal food poisoning and the action of enterotoxin A.
- ✓ Will antibiotics effect the outcome or the severity of staphylococcal food poisoning? Explain.

29.5 Clostridial Food Poisoning

Both *Clostridium perfringens* and *Clostridium botulinum* cause serious food poisoning. Members of the *Clostridium* genus are anaerobic spore formers. Canning and cooking procedures kill living organisms but do not kill spores. Under appropriate anaerobic conditions, the spores germinate and toxin is produced.

Clostridium perfringens Food Poisoning

Clostridium perfringens is an anaerobic, gram-positive spore-forming rod commonly found in soil (Figure 29.2). It also lives in small numbers in the intestinal tract of many animals and is therefore found in sewage (∞ Section 12.21). *C. perfringens* is the most prevalent reported cause of *food poisoning* in the United States, with an estimated 248,000 annual cases (Table 29.5).

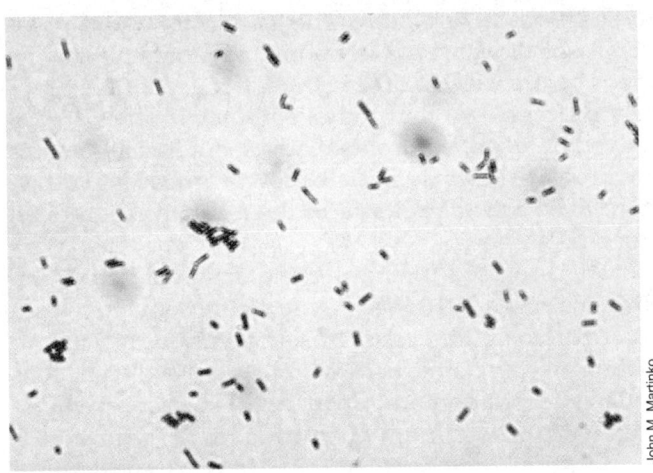

Figure 29.2 Gram stain of *Clostridium perfringens*. The individual gram-positive bacilli are about 1 µm in diameter.

The disease results from the ingestion of a large dose of *Clostridium perfringens* ($>10^8$ cells) in contaminated cooked and uncooked foods, especially meat, poultry, and fish. Large numbers of *C. perfringens* can grow in meat dishes cooked in bulk (heat penetration in these situations is often slow and insufficient) and then left at 20–40°C for short time periods. Spores of *C. perfringens* germinate under anoxic conditions, such as in a sealed container, and grow quickly in the meat. However, the toxin is not yet present.

After consumption of the contaminated food, the living *C. perfringens* begins to sporulate in the intestine, triggering production of the perfringens enterotoxin (∞ Table 21.4). The enterotoxin alters the permeability of the intestinal epithelium, leading to diarrhea and intestinal cramps, usually with no fever or vomiting. The onset of perfringens food poisoning begins about 7–15 h after consumption of the contaminated food, but usually resolves within 24 h, and fatalities are rare.

Diagnosis, Treatment, and Prevention

Diagnosis of perfringens food poisoning is made by isolation of *C. perfringens* from the gut or, more reliably, by a direct enzyme-linked immunosorbent assay (ELISA) to detect *C. perfringens* enterotoxin in feces (∞ Section 24.11). Because *C. perfringens* food poisoning is self-limiting, treatment is usually not necessary, although antitoxins are available (∞ Section 22.11). Prevention of perfringens food poisoning requires measures to prevent contamination of raw and cooked foods and control of cooking and canning procedures to ensure proper heat treatment of all foods.

Botulism

Botulism is a severe food poisoning; it is often fatal and occurs following the consumption of food containing the exotoxin produced by the anaerobic, gram-positive rod *Clostridium botulinum*. This bacterium normally inhabits soil or water, but its spores may contaminate raw foods before harvest or slaughter. If the foods are properly processed so that the *C. botulinum* spores are removed or killed, no problem arises; but if viable spores are present, they may initiate growth and toxin production. Even a small amount of the resultant neurotoxin can be poisonous.

We discussed the nature and action of botulinum toxin in Section 21.10 (∞ Figure 21.19). Botulinum toxin is a neurotoxin that causes flacid paralysis, usually affecting the autonomic nerves that control body functions such as respiration and heart beat. At least seven distinct types of botulinum toxin are known. The toxins are destroyed by high heat (80°C for 10 min), and so thoroughly cooked food, even if contaminated with toxin, *may* be harmless.

Most cases of botulism occur as a result of eating foods that are not cooked after processing (Figure 29.3*a*). For example, nonacid, home-canned vegetables (e.g., home-canned corn and beans) are often used without cooking when making cold salads. Similarly, smoked and fresh fish, vacuum packed in plastic, are often eaten without cooking. Under such conditions, *C. botulinum* spores germinate, and the resulting cells produce toxin. If these foods are consumed, then ingestion of even a small amount will result in this severe and highly dangerous type of food poisoning.

Infant botulism occurs when spores of *Clostridium botulinum* are ingested, often from raw honey (Figure 29.3*b*). If the infant's normal flora is not well developed or if the infant is undergoing antibiotic therapy, the spores may germinate and *C. botulinum* cells may grow and release toxin. Most cases of infant botulism occur between the first week of life and 2 months of age; infant botulism is rare in children older than 6 months when the normal intestinal flora is more developed (∞ Section 21.4).

All forms of botulism are quite rare and usually less than 150 total cases occur each year in the United States, but up to 25% are fatal. Death occurs from respiratory paralysis or cardiac arrest due to the paralyzing action of the botulinum neurotoxin (∞ Section 21.10).

Diagnosis, Treatment, and Prevention

Diagnosis is by demonstrating toxin in patient serum, or by finding toxin or *Clostridium botulinum* in suspected food products. Laboratory findings are coupled with clinical observations including neurological signs of localized paralysis (impaired vision and speech) beginning 18–24 h after ingestion of contaminated food. Treatment involves administration of antitoxin (∞ Section 22.11) and mechanical ventilation. Prevention requires maintaining careful controls over canning and preservation methods and heating susceptible foods to destroy spores, or boiling for 20 minutes to destroy the toxins.

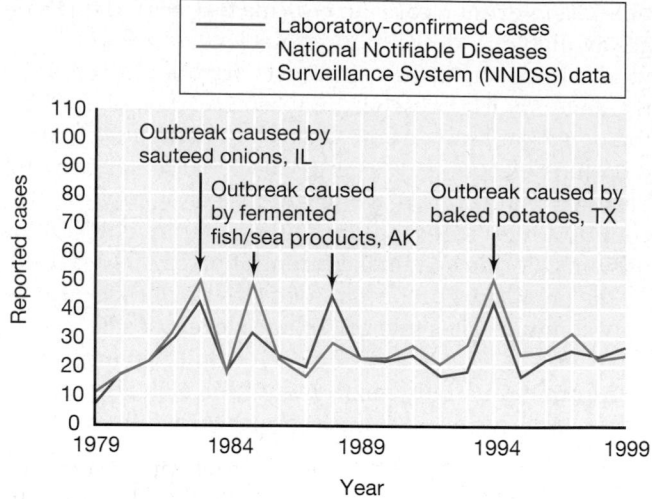

(a) Foodborne botulism

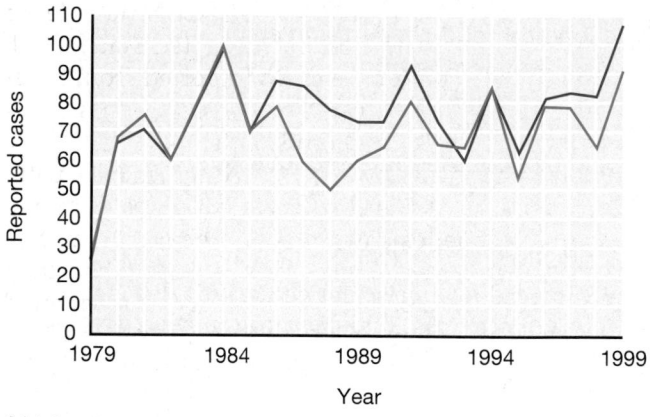

(b) Infant botulism

Figure 29.3 The incidence of botulism in the United States. (a) Foodborne botulism. In years with high numbers of cases, major outbreaks that account for the increase are indicated. (b) Infant botulism. More than half of the cases of infant botulism in the United States occur in California. Data are from the Centers for Disease Control and Prevention, Atlanta, GA, USA.

In infant botulism, *C. botulism* and toxin are often found in bowel contents. Infant botulism is usually self-limiting and most infants recover with only supportive therapy, such as assisted ventilation. Occasional deaths may occur due to respiratory failure. Honey may be a source of *C. botulinum* spores. Therefore, feeding honey to children under 2 years of age is generally discouraged.

✓ 29.5 Concept Check

Clostridium food poisoning results from ingestion of toxins produced by microbial growth in foods or due to microbial growth and toxin production in the body. Perfringens food poisoning is quite common, and is usually a self-limiting gastrointestinal disease. Botulism is a rare but very serious disease, with significant mortality.

✓ Describe the events that lead to perfringens food poisoning. What is the likely outcome of the poisoning?

✓ Describe the events that lead to botulism. What is the likely outcome of botulism?

29.6 Salmonellosis

Although sometimes called food poisoning, **salmonellosis** is a gastrointestinal disease due to foodborne *Salmonella* infection. Symptoms begin only after the pathogen colonizes the intestinal epithelium.

Biology and Epidemiology

Salmonella spp. are gram-negative facultative aerobic rods related to *Escherichia coli*, *Shigella* spp., and other enteric *Bacteria* (∞ Section 12.11). *Salmonella* normally inhabit the gut of animals and are thus found in sewage. Virtually all *Salmonella* are pathogenic for humans: One, *S. typhi*, causes the serious human disease typhoid fever, but is fortunately very rare in the United States, with most of the 500 foodborne cases imported from other countries. However, a number of *Salmonella* species cause foodborne gastroenteritis. In all, over 1400 serotypes of various *Salmonella* species are known to be pathogenic for humans. *S. typhimurium* is the most common cause of salmonellosis in humans. The incidence and prevalence of *reported* salmonellosis has been very steady over the last decade, with about 45,000 documented cases each year (Figure 29.4). However, less than 4% of the total cases of salmonellosis are probably reported,

and estimates place the actual number at over 1,300,000 cases of salmonellosis every year (Table 29.5).

The ultimate sources of the foodborne salmonellas are the intestinal tracts of humans and warm-blooded animals. The organism may reach food by fecal contamination from food handlers. Food production animals such as chickens and cattle may also harbor *Salmonella* strains that are pathogenic to humans and may pass the bacteria to finished fresh foods such as eggs, meat, and dairy products. *Salmonella* food infections are often traced to products such as custards, cream cakes, meringues, pies, and eggnog made with uncooked eggs. Other foods commonly implicated in salmonellosis outbreaks are meats and meat products such as meat pies, cured but uncooked sausages and meats, poultry, milk, and milk products.

Pathogenesis

The most common salmonellosis is a *Salmonella*-induced *enterocolitis*. Ingestion of food containing $10^5–10^8$ viable *Salmonella* results in colonization of the small and large intestine. Onset of the disease occurs 8–48 h after ingestion. Symptoms include the sudden onset of headache, chills, vomiting, and diarrhea, followed by a fever that lasts a few days. The disease normally resolves without intervention in 2 to 3 days. However, even after recovery, patients shed *Salmonella* in feces for several weeks. Some patients recover and remain asymptomatic, but shed organisms for months or even years, resulting in a chronic carrier condition (∞ the box, the Tragic Case of Typhoid Mary, Chapter 25).

Salmonellosis may also cause septicemia, a blood infection, enteric fever, or *typhoid fever*, a disease characterized by systemic infection and high fever lasting several weeks. Mortality can approach 15% in untreated typhoid fever.

Diagnosis, Treatment, and Prevention

Diagnosis is made by observation of clinical symptoms, history of recent food consumption, and by culture of the organism from feces. Several selective media are available (∞ Section 24.2), and tests for the presence of *Salmonella* are commonly done on animal food products, such as raw meat, poultry, eggs, and powdered milk, because *Salmonella* from production animals is the usual source of food contamination.

For enterocolitis, treatment is usually unnecessary, and antibiotic treatment does not shorten the course of the disease or eliminate the carrier state. Antibiotic treatment, however, significantly reduces the length and severity of septicemia and typhoid fever. Mortality due to typhoid fever can be reduced to less than 1% with appropriate antibiotic therapy.

Cooked foods heated to 70°C for at least 10 minutes are considered safe if consumed immediately, or if held

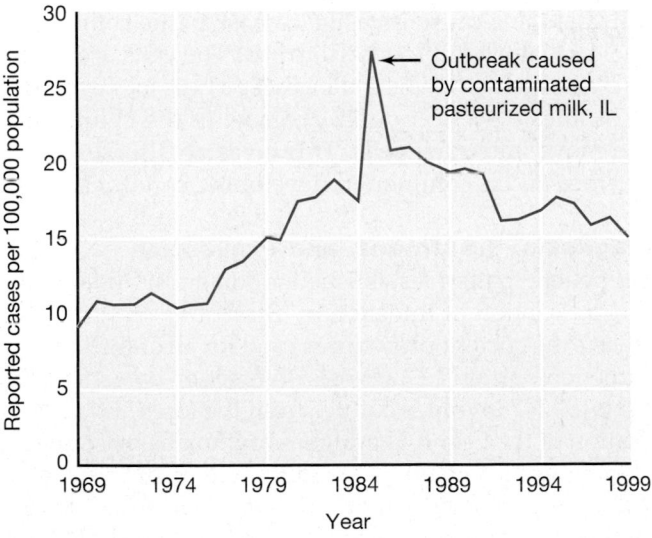

Figure 29.4 The prevalence of reported cases of salmonellosis in the United States. The total number of *reported* cases is between 40,000 and 45,000 per year. Epidemiologic investigations suggest that only about 3% of all cases of salmonellosis are properly identified and reported. Data are from the Centers for Disease Control and Prevention, Atlanta, GA, USA.

at 50°C or stored at 10°C or less. Cooked or canned foods that become contaminated by an infected food handler can support the growth of *Salmonella* if the foods are held for long periods of time without heating or refrigeration. *Salmonella* infections are more common in summer than in winter, probably because warm environmental conditions are more favorable for growth of microorganisms in foods (Figure 29.4).

Although local laws and enforcement vary, because of the lengthy carrier state, infected individuals are often banned from work as food handlers until their feces are negative for *Salmonella* in three successive cultures.

✓ 29.6 Concept Check

Salmonellosis, an extremely common foodborne infection, results from infection with ingested *Salmonella* spp. *Salmonella* can enter the food chain via production animals or food handlers.

✓ Describe the three kinds of salmonellosis food infection. Which is most common?

✓ How might *Salmonella* contamination of production animals be contained?

29.7 Pathogenic *Escherichia coli*

Several strains of *Escherichia coli* are potential foodborne pathogens. All pathogenic strains act first on the intestine and several are characterized by their ability to produce potent enterotoxins (co Section 21.11).

Biology, Epidemiology, and Pathogenesis

Escherichia coli is a common inhabitant of the animal gut. The short, gram-negative rods are classified as enteric *Bacteria* (co Section 12.11). There are about 200 known pathogenic *E. coli* that can cause life-threatening diarrheal disease and urinary tract infections. The pathogenic strains are divided into several categories, based primarily on the toxins they produce and the diseases they cause.

Enterohemorrhagic *E. coli* (EHEC) produce *verotoxin*, an enterotoxin similar to one produced by *Shigella dysenteriae*, the Shiga toxin (co Table 21.4). After ingestion of food or water containing one particular EHEC strain, *E. coli* O157:H7, the organism grows in the small intestine and produces the verotoxin. Verotoxin causes both hemorrhagic (bloody) diarrhea and kidney failure. *E. coli* O157:H7 causes at least 60,000 infections and 50 deaths each year from foodborne disease (Table 29.5) and is a leading cause of kidney failure in children. The most common cause of this infection is the consumption of contaminated uncooked or undercooked meat, particularly mass-processed ground meat. In several major outbreaks involving *E. coli* O157:H7–infected ground beef,

regional distribution centers were the source of the contaminated meat and the infected product caused disease in several states. Another outbreak involved processed, cured, but uncooked beef in ready-to-eat sausages. The major source of contamination was the beef, and the *E. coli* O157:H7 probably originated from the slaughtered beef carcasses. Since *E. coli* O.157:H7 grows in the intestines and is found in fecal material, it is also a potential source of waterborne disease. There have been several cases of serious *E. coli* O157:H7 infections from fecally contaminated public swimming areas.

Diarrheal disease often occurs in children in developing countries. It also occurs as "traveler's diarrhea," an extremely common enteric infection causing watery diarrhea in travelers to developing countries. The primary causal agents are the enterotoxigenic *Escherichia coli* (ETEC). The ETEC strains make two heat-labile diarrhea-producing enterotoxins. In studies done with U.S. citizens traveling in Mexico, the infection rate with ETEC is often greater than 50%. The prime vehicles are foods such as fresh vegetables (for example, lettuce in salads) and water. The very high infection rate in travelers is due to contamination of local public water supplies. The local population is usually resistant to the infecting strains, undoubtedly because they have lived with the agent for a long period of time. Secretory antibodies (co Section 22.8) present in the bowel may prevent successful colonization of the pathogen in local residents, but the organism readily colonizes the intestine of a nonimmune person, causing disease.

Enteropathogenic *E. coli* (EPEC) cause diarrheal diseases in infants and small children, but does not cause invasive disease or produce toxins. Enteroinvasive *E. coli* (EIEC) strains cause invasive disease in the colon, producing watery to bloody diarrhea. The cells are taken up by phagocytes, where they escape lysis in the phagolysosomes (co Section 22.2), grow in the cytoplasm, and move into other cells. This invasive disease causes diarrhea and is common in developing countries.

Diagnosis, Treatment, and Prevention

The general pattern established for diagnosis, treatment, and prevention of infection by *Escherichia coli* O157:H7 reflects the current procedures used for all of the various pathogenic strains. Diagnosis of infection by *Escherichia coli* O157:H7 involves culture from the feces and identification of the O and H antigens and toxins by serology (co Sections 4.9 and 24.7). Subtyping of strains is also done using molecular methods such as restriction fragment length polymorphism (RFLP) and pulse field gel electrophoresis (co the box, DNA Fingerprinting, Chapter 31). *Escherichia coli* O157:H7 illness is a nationally reportable infectious disease (co Table 25.3).

Treatment of all pathogenic *E. coli* infections involves supportive therapy and, in severe cases, antimicrobial drugs to shorten and eliminate infection.

The most effective way to prevent infection with foodborne enteropathogenic *E. coli* O157:H7 is to make sure that meat is cooked thoroughly, which means that it should appear gray or brown and juices should be clear. As we discussed above (Section 29.2), the United States has approved the irradiation of ground meat as an acceptable means of eliminating or reducing food infection bacteria, largely because *E. coli* O157:H7 has been implicated in several foodborne epidemics. Penetrating radiation is considered the only effective means to ensure decontamination after the grinding process because grinding may distribute the pathogens throughout the meat, not simply on the surface. In general, proper food handling, water purification, and hygiene habits will prevent the spread of pathogenic *E. coli*. Traveler's diarrhea can be prevented by avoiding local water sources and fresh foods.

✓ 29.7 Concept Check

Enteropathogenic *Escherichia coli* can cause serious food infections. Specific measures, such as radiation of ground beef, have been implemented to curb spread of these pathogens.

- ✓ Describe the pathology of enteropathogenic *Escherichia coli* food infection. What is the likely outcome?
- ✓ How might *Escherichia coli* contamination of production animals be prevented?

29.8 *Campylobacter*

Campylobacter spp. cause the most prevalent *bacterial foodborne infections* in the United States.

Biology and Epidemiology

Campylobacter species are gram-negative, motile, curved rods that grow at reduced oxygen tension, that is, as microaerophiles (Section 12.14). Several pathogenic species, *Campylobacter jejuni*, *C. coli*, and *C. fetus*, are recognized. *C. jejuni* and *C. coli* account for nearly 2 million annual cases of bacterial diarrhea (Table 29.5). *Campylobacter fetus* is economically important because it is a major cause of sterility and spontaneous abortion in cattle and sheep.

Campylobacter is transmitted to humans via contaminated food, most frequently in poultry, pork, raw clams, and other shellfish, or in surface waters not subjected to chlorination. *C. jejuni* is a normal resident in the intestinal tract of poultry, and virtually all chicken and turkey carcasses contain this organism. Beef, on the other hand, is rarely a vehicle. *Campylobacter* species also infect domestic animals such as dogs, causing a milder form of diarrhea than that observed in humans. Infant cases of *Campylobacter* infection are frequently traced to infected domestic animals, especially dogs.

Pathogenesis

After ingesting food contaminated with at least 10^4 *Campylobacter*, the organism multiplies in the small intestine, invades the epithelium, and causes inflammation, resulting in disease. The symptoms of *Campylobacter* infection include a high fever (usually greater than 104°F or 40°C), headache, malaise, nausea, abdominal cramps, and profuse diarrhea with watery, frequently bloody, stools. The disease subsides in about 1 week, and recovery is complete and spontaneous.

Diagnosis, Treatment, and Prevention

Diagnosis requires isolation of the organism from stool samples and identification by growth-dependent tests or immunological assays. Because of the frequency with which *C. jejuni* infections are observed in infants, a variety of selective media and highly specific immunological methods have been developed for positive identification of this organism. Treatment of infections with erythromycin do not shorten the acute diarrhea, but may shorten the time during which patients shed *Campylobacter* in their feces. Personal hygiene, proper washing of uncooked poultry (and any utensils coming in contact with uncooked poultry), and thorough cooking of the meat eliminate the possibility of *Campylobacter* infection.

✓ 29.8 Concept Check

Campylobacter infection is by far the most prevalent foodborne bacterial infection. Though usually self-limiting, this disease affects nearly 2 million people per year.

- ✓ Describe the pathology of *Campylobacter* food infection. What is the likely outcome?
- ✓ How might *Campylobacter* contamination of production animals be controlled?

29.9 Listeriosis

Listeria monocytogenes is emerging as an important foodborne pathogen. *L. monocytogenes* causes *listeriosis*, a gastrointestinal food infection that may lead to bacteremia and meningitis.

Biology and Epidemiology

Listeria monocytogenes is an acid-tolerant, *psychrotolerant* (cold-tolerant), facultatively anaerobic, and salt-tolerant bacterium. It is a short, gram-positive, nonspore-forming rod (Section 12.19) (Figure 29.5). It is found widely in soil and water, and virtually no fresh food source is safe from possible *L. monocytogenes* contamination. Fresh food can become contaminated at any stage during food growth or processing. Methods such as refrigeration, which ordinarily slow microbial growth, are ineffective

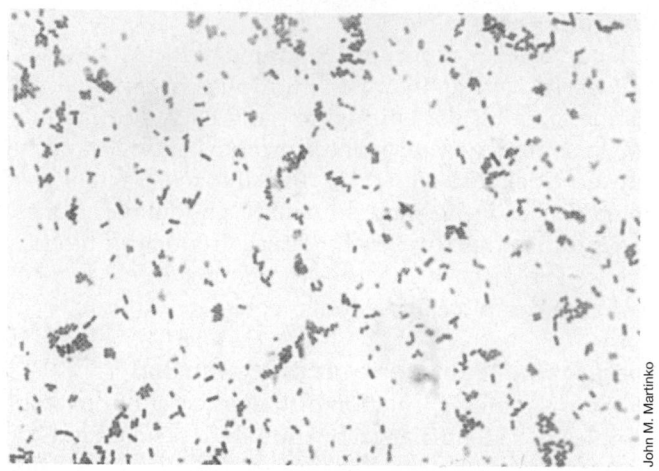

Figure 29.5 Gram stain of *Listeria monocytogenes*. The short, gram-positive bacilli are about 0.5 μm in diameter.

in limiting growth of this psychrotolerant organism. Thus, meat, dairy products, and fresh produce can be contaminated with this pathogen. Common sources of listeriosis outbreaks are ready-to-eat processed foods such as meat products and unpasteurized dairy products that are stored for long periods at refrigerator temperature (4°C).

Pathogenesis

Listeria monocytogenes is an intracellular pathogen. It enters the body through the gastrointestinal tract after ingestion of contaminated food. Uptake of the pathogen by phagocytes results in growth and proliferation of the bacterium, lysis of the phagocyte, and spread to surrounding cells. Immunity to *L. monocytogenes* is mainly cell-mediated via T_H1 cells (∞ Section 22.7). Individuals having weakened cellular immunity, including the elderly, neonates, patients undergoing immunosuppressive drug treatment (e.g., steroid treatment), or those who have immunosuppressive diseases such as AIDS, have increased susceptibility to listeriosis (∞ Section 26.14).

Although exposure to *L. monocytogenes* is undoubtedly very common, acute *listeriosis* is quite rare. The acute disease is usually characterized by bacteremia and meningitis and has a mortality rate of about 20%. Although there are only about 2500 cases of acute listeriosis each year, about 500 cases end in death. Nearly all require hospitalization.

Diagnosis, Treatment, and Prevention

The diagnosis of listeriosis is accomplished by culturing *Listeria monocytogenes* from the blood or spinal fluid. *L. monocytogenes* can be identified in food by direct culture or by a variety of molecular methods such as ribotyping (∞ Section 11.9) and the polymerase chain reaction (PCR) (∞ Section 10.17, Section 24.13, and

Table 24.9). Antibiotic treatment with trimethoprim-sulfamethoxazole is effective.

Prevention measures include recalling contaminated food and taking steps to limit *L. monocytogenes* contamination at the food-processing site. Since *L. monocytogenes* is susceptible to heat and radiation, raw food and food handling equipment can be readily decontaminated. However, without sterilizing the finished food product, the risk of food contamination cannot be completely eliminated because of the widespread distribution of the pathogen. Individuals who are immunocompromised should avoid nonpasteurized dairy products and pay careful attention to expiration dates when consuming ready-to-eat processed foods.

✓ **29.9 Concept Check**

Listeria monocytogenes is ubiquitous in the environment. In normal individuals, *Listeria* causes no infection. However, in immunocompromised individuals, *Listeria* can cause serious disease and even death.

✓ Describe the pathology of *Listeria* food infection. What is the likely outcome in normal individuals?

✓ Why are immunocompromised individuals extremely susceptible to life-threatening *Listeria* infections?

29.10 Other Foodborne Infectious Diseases

A number of other microorganisms and infectious agents contribute to foodborne diseases.

Bacteria

Table 29.5 lists several other bacteria that cause human foodborne disease. *Yersinia enterocolitica* is commonly found in the intestines of domestic animals and causes foodborne infections due to contaminated meat and dairy products. *Y. enterocolitica* causes enteric fever, a severe life-threatening infection. *Bacillus cereus* produces an enterotoxin that causes diarrhea and vomiting. *B. cereus* grows in high-carbohydrate foods such as rice. Spores of this gram-positive rod germinate and, as the organism grows in food that is left at room temperature, pathogenic amounts of toxin are produced. Reheating may kill the *B. cereus*, but the toxin remains active. *Shigella* spp. cause nearly 100,000 cases of severe foodborne invasive gastroenteritis called *shigellosis* each year. Several members of the *Vibrio* genus cause food poisoning after consumption of contaminated shellfish.

Viruses

The largest number of annual foodborne infections are thought to be caused by viruses. In general, viral foodborne illness consists of gastroenteritis characterized by diarrhea, often accompanied by nausea and vomiting. Recovery is spontaneous and rapid, usually within 24–48 h ("24-hour flu"). Norwalk-like viruses (∞ Table 25.5 and Section 28.8) are responsible for most of the foodborne infectious disease in the United States (Table 29.5), accounting for over 9 million of the estimated 13 million cases of food infection per year. Rotavirus, astrovirus, and hepatitis A (∞ Section 26.11) collectively cause 100,000 cases of foodborne disease each year. These viruses inhabit the gut and are often transmitted to food or water with fecal matter. As with many foodborne infections, proper food handling, handwashing, and a source of clean water to prepare fresh foods are essential to prevent infection.

Parasites

Important foodborne parasitic diseases are listed in Table 29.5. Parasites including *Giardia lamblia*, *Cryptosporidium parvum*, and *Cyclospora cayetanensis* can be spread via food, presumably contaminated by fecal matter in untreated water used to wash, irrigate, or spray crops. Fresh foods such as fruits are often implicated as the source of these parasites. We discussed giardiasis and cryptosporidiosis in the previous chapter (∞ Section 28.6). Cyclosporiasis is an acute gastroenteritis and is an important emerging disease. Most cases appear to be the result of eating imported fresh produce.

Toxoplasma gondii is a parasite spread through cat feces, but also found in raw or undercooked meat. In most individuals, the *toxoplasmosis* infection causes a mild, self-limiting gastroenteritis. However, prenatal infection can lead to a variety of complications, including blindness and stillbirth. Immunocompromised patients also exhibit signs of acute toxoplasmosis.

Prions

Prions are proteins, presumably of host origin, that adopt novel conformations, inhibiting normal protein function and causing disruption in neural tissue. The foodborne variety of prion disease in humans is known as "variant Creutzfeldt-Jakob Disease" (vCJD). VCJD is a slow-acting degenerative nervous system disorder and has affected several hundred people in the United Kingdom and other European countries (∞ Section 9.13). The disease appears to be spread by eating meat products from cattle afflicted with bovine spongiform encephalitis (BSE), a prion disease commonly called mad cow disease. Although the mechanism of replication is not entirely clear, BSE prions consumed in the meat products from affected cattle somehow trigger structurally and functionally related human proteins to assume an altered conformation, resulting in protein dysfunction and disease. BSE has not yet been discovered in cattle nor has vCJD been observed in humans in the United States. In Europe, all cattle known or suspected to have BSE have been destroyed. Bans on feeding cattle with animal meal appear to have stopped the development of new cases of BSE in Europe.

✓ 29.10 Concept Check

Over 200 different infectious agents cause foodborne disease. Viruses cause the vast majority of foodborne illnesses. A number of other bacteria, parasites, and prions also cause foodborne illnesses.

- ✓ Identify the viruses most likely to be involved in foodborne illnesses.
- ✓ How might *prion* contamination of production animals be prevented in the United States?

Review Questions

1. Identify and define the three major categories of food with respect to their perishability.
2. Identify the major methods used to preserve food. Provide an example of a food preserved by each method.
3. Distinguish between foodborne *infection* and foodborne *poisoning*.
4. Outline the pathogenesis of staphylococcal food poisoning. Suggest methods for prevention of this disease.
5. Identify the two major types of clostridial food poisoning. Which is most prevalent? Which is most dangerous?
6. What are the possible sources of *Salmonella* spp. that cause food infections?
7. What measures are used to control the growth of *Escherichia coli* O157:H7 in ground meat?
8. *Campylobacter* causes more foodborne infections than any other bacterium. Identify at least one reason why this is true.
9. Identify the food sources of *Listeria monocytogenes* infections. Identify the individuals who are at high risk for listeriosis.
10. Why are viral agents so commonly associated with foodborne disease?

Application Questions

1. Identify optimum storage conditions for perishable, semiperishable and nonperishable food products. Consider economic factors such as the cost of storage and the value of the food item.

2. For a food of your choice, devise a way to preserve the food without freezing or drying.

3. Perfringens food poisoning involves ingestion of *Clostridium perfringens* followed by growth and sporulation in the intestine of the host. Sporulation triggers toxin production. Is this disease truly a food poisoning, or might it be classified as a food infection? Give reasons for your answer.

4. Mayonnaise, an egg-based salad dressing, is commonly used to prepare potato salad. Improperly handled potato salads are often the source of staphylococcal food poisoning, or salmonellosis. Explain several methods by which a potato salad could become inoculated with either *Staphylococcus aureus* or *Salmonella* spp.

5. *Clostridium botulinum* requires an anoxic environment for production of botulinum toxin. Identify methods of food preservation that create the anoxic environment necessary for growth of *Clostridium botulinum*. Conversely, identify methods of food preservation that create an oxic environment and discourage the growth of *Clostridium botulinum*.

6. Why aren't antibiotics generally used to treat salmonellosis? Explain your answer based on the habitat of the organism and access to antibiotics. Also consider the issue of potential antibiotic resistance.

7. Indicate the precautions you need to prevent infection with enteropathogenic *Escherichia coli*. Concentrate on safe food handling, cooking, and consumption.

8. Devise a plan to eliminate *Campylobacter* contamination from a poultry flock. Explain the benefits of *Campylobacter*-free poultry and explain the problems that your plan might create.

9. Listeriosis normally occurs only when there is a breakdown in T_H1 cell-mediated immunity. Indicate why this is so. Devise a vaccine to protect against listeriosis. Would your vaccine be of use in the listeriosis-prone population?

10. Indicate the reasons for the high incidence of viral foodborne disease. Devise a plan to eliminate Norwalk-like virus from the food supply.

I n industrial microbiology, economical production is only possible if it is done on a very large scale. In order to make a competitively priced product by microbial fermentation, the growth medium used must be inexpensive and the growth vessels extremely large. Shown here are large outdoor fermentors for the growth of yeast to produce alcohol in Japan. Unlike many biocatalytic products, such as antibiotics, ethanol is a commodity chemical that sells in large quantities at a relatively cheap price. Therefore, the microbial process must be highly refined and run at peak efficiency on a large scale if it is to generate product in the quantity necessary to be successful in the marketplace.

30

INDUSTRIAL MICROBIOLOGY/BIOCATALYSIS

Working Glossary

Aminoglycosides a group of antibiotics including streptomycin, containing amino sugars linked by glycosidic bonds

β-Lactam antibiotics a group of antibiotics including penicillin, containing the four-membered heterocyclic β-lactam ring

Biocatalysis the use of microorganisms to carry out a specific chemical transformation

Bioconversion the use of microorganisms to carry out a chemical reaction that is more costly or not feasible nonbiologically

Biosynthetic penicillin production of a particular form of penicillin by supplying the organism with specific side chain precursors

Brewing the manufacture of alcoholic beverages such as beer from the fermentation of malted grains

Commodity chemicals chemicals such as ethanol that have low monetary value and are thus sold primarily in bulk

Distilled beverage a beverage containing alcohol concentrated by distillation

Extremozyme an enzyme able to function in the presence of one or more chemical or physical extremes, for example, high temperature or low pH

Fermentation in an industrial context, any large-scale microbial process whether carried out aerobically or anaerobically

Fermentor the tank in which an industrial fermentation is carried out

Immobilized enzyme an enzyme attached to a solid support over which substrate is passed and converted to product

Primary metabolite a metabolite excreted during the growth phase

Protease an enzyme that can degrade proteins by hydrolysis

Scale-up conversion of an industrial process from a small laboratory setup to a large commercial fermentation

Secondary metabolite a metabolite excreted at the end of the primary growth phase and into the stationary phase

Secondary treatment in sewage treatment, either the aerobic or anoxic decomposition of sewage following the removal of nondegradable objects by primary treatment

Semisynthetic penicillin a penicillin produced using components derived from both microbial fermentation and chemical syntheses

Tetracyclines a class of antibiotics containing the four-membered naphthacene ring

I INDUSTRIAL MICROORGANISMS, PRODUCTS, AND PRODUCT FORMATION

Industrial microbiology is the discipline that uses microorganisms, usually grown on a large scale, to produce valuable commercial products or carry out important chemical transformations. Industrial microbiology originated with alcoholic fermentation processes, such as those for making beer and wine. Subsequently, microbial processes were developed for the production of pharmaceuticals (such as antibiotics), food additives (such as amino acids), enzymes, and chemicals such as butanol and citric acid. All these industrial microbiological processes are enhancements of metabolic reactions that microorganisms were already capable of carrying out, with the goal in most cases of simply overproducing the product of interest. But now, in addition to traditional industrial microbiology, a new era has unfolded—that of *microbial biotechnology*. In biotechnology, methods for gene manipulation have given rise to new microbial products, most of which are not naturally produced by microorganisms (∞ Chapter 31).

The term *biocatalysis* has been used to describe the reactions carried out by microorganisms in industrial microbiology. In this chapter we discuss several industrial biocatalyses along with some of the problems large-scale microbial culture entails and the solutions to these problems worked out through the years by industrial microbiologists. We begin with an overview of industrial organisms and products.

30.1 Industrial Microorganisms and Products

The major organisms used in biocatalytic processes are fungi (yeasts and molds) and certain prokaryotes, in particular members of the genus *Streptomyces*. To a great extent, industrial microorganisms are metabolic specialists capable of being manipulated in large-scale culture to produce one or more products in high yield. In order to achieve this high metabolic specialization, strains of industrial microorganisms are often genetically altered by mutation or recombination, with the focus usually remaining on the *yield* of the particular product that a given strain can produce.

The ultimate source of all strains of microorganisms used in biocatalytic processes is nature. However, actual industrial strains are usually far removed from the "wild-type" condition that existed when the strain was first isolated. Once valuable industrial microorganisms have been developed, they are conserved by both microbiology laboratories in industry as well as in large national microbial culture collections, such as the American Type Culture Collection (ATCC) in the United States or the Deutsche Sammlung von Mikroorganismen und Zellkulturen (DSMZ) in Germany. In these collections, cultures of microorganisms are supplied for educational, research, and industrial purposes. When a new biocatalytic process is patented, a strain capable of carrying out the process must be deposited into one of these collections. However, for several reasons, proprietary rights chief among them, the strains deposited are *not* the

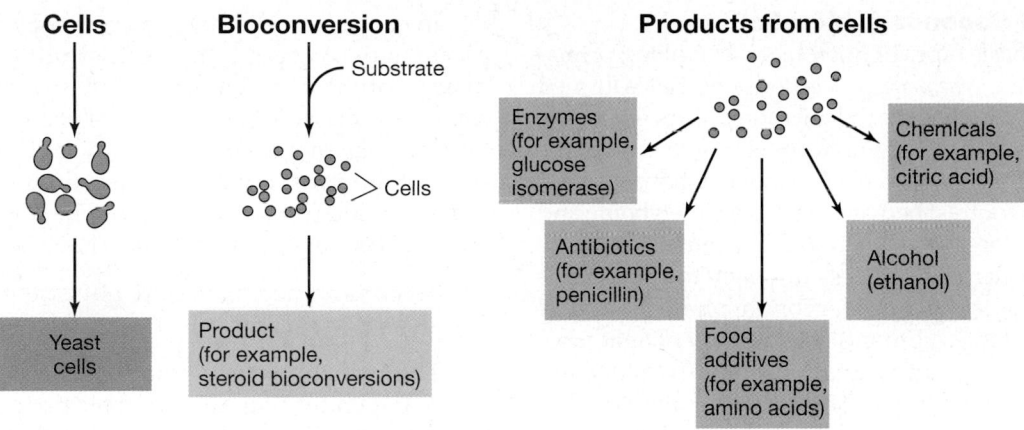

Figure 30.1 Products of industrial microbiology/biocatalysis. The products may be the cells themselves or products made from cells. In the case of bioconversion, cells are used to chemically convert a specific substance from one form to another.

actual high-yielding production strain(s) but instead a strain or strains that carry out the process at lower yield.

Properties of a Useful Industrial Microorganism

A microorganism suitable for industrial use must produce the substance of interest, of course, but there is much more to it than that. The organism must be capable of growth and product formation in large-scale culture. Moreover, the organism should preferably produce spores or some other reproductive cell form so that it can be easily inoculated into large fermentors.

Another important characteristic of an industrial organism is that it grow rapidly and produce the desired product in a relatively short period of time. The organism must also be able to grow in a relatively inexpensive liquid culture medium obtainable in bulk quantities. Many industrial microbiological processes use waste carbon from other industries as major or supplemental ingredients for large-scale culture media. These include *corn steep liquor* (a product of the corn wet milling industry that is rich in nitrogen and growth factors), *whey* (a waste liquid of the dairy industry containing lactose and minerals), and other industrial waste materials having high organic carbon contents. In addition, an industrial microorganism should not be pathogenic, especially to humans or economically important animals or plants. Because of the large population size in the industrial fermentor and the virtual impossibility of avoiding contamination of the environment outside the fermentor, a pathogen would present potentially disastrous problems. Finally, an industrial microorganism should be amenable to genetic manipulation. In industrial microbiology, increased yields have often been obtained genetically by means of mutation and selection. A genetically stable and easily manipulable producing organism is thus a clear advantage.

Examples of Industrial Products

Microbial products of industrial interest are of several major types (Figure 30.1). These include the microbial cells themselves, for example, yeast cultivated for food, baking, or brewing (see Figure 30.20), and substances produced by cells. Examples of the latter include enzymes such as glucose isomerase, pharmacologically active agents such as antibiotics, steroids, and alkaloids, specialty chemicals and food additives such as the currently popular aspartame food and drink sweetener, and commodity chemicals, such as ethanol. A summary of some important industrial products, many of which will be discussed in more detail later, is given in Figure 30.1.

✓ 30.1 Concept Check

An industrial microorganism must produce the product of interest in high yield, grow rapidly on inexpensive culture media available in bulk quantities, be amenable to genetic manipulation, and, if possible, be nonpathogenic. Industrial products are many and include both cells and substances made by cells.

✓ Why should industrial microorganisms be genetically manipulable?

✓ List three important products of industrial biocatalysis.

30.2 Growth and Product Formation in Biocatalyses

In Section 6.1 we discussed the microbial growth process and described the various stages: *lag, exponential,* and *stationary.* Here we describe microbial growth and product formation in an industrial context, posing the question: "When in the growth cycle is the industrially useful metabolite produced?"

Primary and Secondary Metabolites

There are two basic types of microbial metabolites: primary and secondary. A *primary metabolite* is one that is formed during the growth phase of the microorganism, whereas a *secondary metabolite* is one that is formed near the end of the growth phase, frequently at, near, or in the stationary phase of growth. The contrast between a primary metabolite and a secondary metabolite is illustrated in Figure 30.2.

A typical microbial process in which the product is formed during the primary growth phase is *alcohol (ethanol) fermentation:** Ethanol is a product of anaerobic metabolism of yeast and certain bacteria (⇌ Section 5.10) and is formed as part of energy metabolism. Because growth can occur only if energy production can occur, ethanol formation takes place in parallel with growth (Figure 30.2*a*).

---

*In industrial microbiology, the term *fermentation* refers to *any* large-scale microbial process, whether or not it is biochemically a fermentation. In fact, most industrial fermentations are aerobic. The *tank* in which the industrial fermentation is carried out is called a fermentor, the microorganism involved is the fermenter.

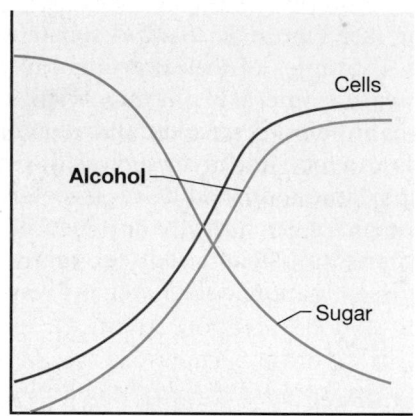

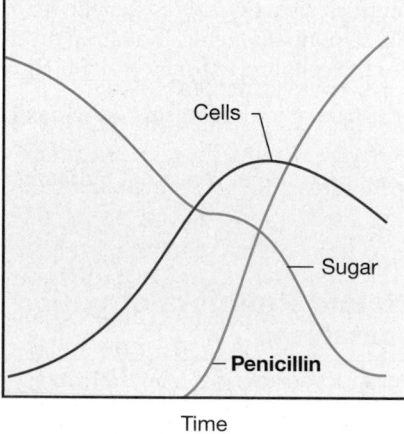

Figure 30.2 Contrast between production of primary and secondary metabolites. (a) Alcohol formation by yeast—an example of a *primary* metabolite. (b) Penicillin production by the mold *Penicillium chrysogenum*—an example of a *secondary* metabolite. Note how penicillin is not made until after mid-log phase (⇌ Figure 6.8).

In contrast to ethanol production by yeast, in some biocatalytic processes the desired product is not produced during the active growth phase but instead in the *stationary* phase. Metabolites produced during the stationary phase are called **secondary metabolites** and are some of the most common and important metabolites of industrial interest (Figure 30.2*b*). The following characteristics of secondary metabolites have been recognized:

1. Secondary metabolites are not essential for growth and reproduction.

2. The formation of secondary metabolites is extremely dependent on growth conditions, especially on the composition of the medium. Repression of secondary metabolite formation frequently occurs.

3. Secondary metabolites are often produced as a group of closely related structures. For instance, a single strain of a species of *Streptomyces* has been found to produce over 30 related but different anthracycline antibiotics.

4. It is often possible to get dramatic *overproduction* of secondary metabolites, whereas primary metabolites, linked as they are to primary metabolism, usually cannot be overproduced in such a dramatic manner (Figure 30.2).

Relationship between Primary and Secondary Metabolism

Most secondary metabolites are complex organic molecules that require a large number of specific enzymatic reactions for synthesis. For instance, it is known that at least 72 separate enzymatic steps are involved in synthesis of the antibiotic *tetracycline* (see Section 30.6) and over 25 steps in the synthesis of *erythromycin*, none of which are reactions occurring during primary metabolism. However, the metabolic pathways of these secondary metabolites do arise out of primary metabolism, because the starting materials for secondary metabolism come from the major biosynthetic pathways. This is summarized in Figure 30.3, which shows the interrelationship of the main primary metabolic pathway for aromatic amino acid synthesis with the secondary metabolic pathways for a variety of antibiotics. As can be seen, many structurally complex secondary metabolites originate from structurally quite similar precursor molecules (Figure 30.3).

✓ 30.2 Concept Check

Primary and secondary metabolites are produced during active cell growth or near the onset of stationary phase, respectively. Many economically valuable microbial products are secondary metabolites.

✓ Is penicillin a *primary* or a *secondary* metabolite? Why?

✓ What type of metabolite, primary or secondary, can be more easily overproduced? Why?

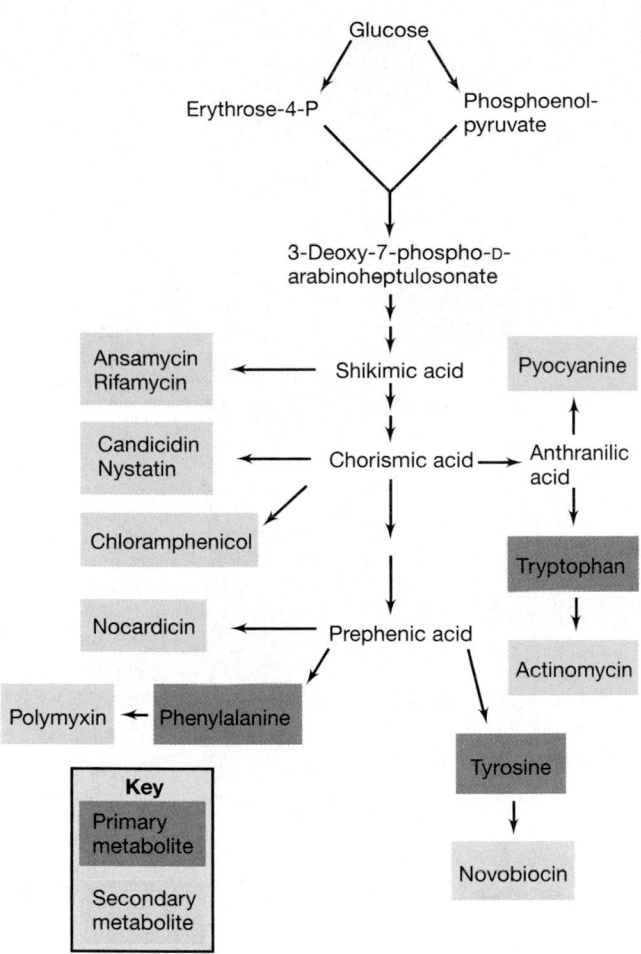

Figure 30.3 Relationship of the primary metabolic pathway for the synthesis of aromatic amino acids (⊂⊃ Section 5.15) and formation of a variety of secondary metabolite antibiotics containing aromatic rings. This is a composite scheme of processes occurring in a variety of microorganisms: No one organism produces all these secondary metabolites, and many individual steps exist between amino acid and antibiotic in all cases.

30.3 Characteristics of Large-Scale Fermentations

The vessel in which the industrial process is carried out is called a *fermentor*. Fermentors can vary in size from the small 5- to 10-liter laboratory scale (Figure 30.4*a*) to the enormous 500,000-liter industrial scale. The size of the fermentor used depends on the process and how it is operated. A summary of fermentor sizes for some common microbial fermentations is given in Table 30.1.

Industrial fermentors can be divided into two major classes, those for *anaerobic* processes and those for *aerobic* processes. Anaerobic fermentors require little special equipment except for removal of heat generated during the fermentation, whereas aerobic fermentors require much more elaborate equipment to ensure that mixing

and adequate aeration are achieved. Because most industrial fermentations are aerobic, the present discussion will be confined to aerobic fermentors.

Construction of an Aerobic Fermentor

Large-scale industrial fermentors are almost always constructed of stainless steel. Such a fermentor is essentially a large cylinder, closed at the top and bottom, into which various pipes and valves have been fitted (Figure 30.4*b*). Because sterilization of the culture medium and removal of heat are vital for successful operation, the fermentor generally has an external *cooling jacket* through which steam or cooling water can be run. For very large fermentors, insufficient heat transfer occurs through the jacket, and so *internal coils* must be provided through which either steam or cooling water can be piped.

A critical part of the fermentor is the *aeration system*. With large-scale equipment, transfer of oxygen from the gas to the liquid is critical and elaborate precautions must be taken to ensure proper aeration. Oxygen is poorly soluble in water, and in a fermentor with a high microbial population density, there is a tremendous oxygen demand by the culture. Two separate installations are used to ensure adequate aeration: an aeration device, called a *sparger*, and a stirring device, called an *impeller* (Figure 30.4*b*). The sparger is a device, often a series of holes in a metal ring or a nozzle, through which filter-sterilized air can be passed into the fermentor under high pressure. The air enters the fermentor as a series of tiny bubbles from which the oxygen passes by diffusion into the liquid.

In small fermentors use of a sparger alone may be sufficient to ensure adequate aeration, but in industrial-size fermentors, *stirring* of the fermentor with an impeller is essential (Figure 30.4*c*). Stirring accomplishes two things: It mixes the gas bubbles through the liquid and it mixes the organism through the liquid, thus ensuring uniform access of microbial cells to the nutrients.

A typical large scale fermentor is shown in Figure 30.5.

Fermentation Control and Monitoring

Any microbial fermentation must be monitored to ensure that it is proceeding properly, but it is especially important that industrial fermentors be monitored carefully because there is such a major expense involved. In most cases, it is necessary not only to measure growth and product formation but also to *control* the process by altering environmental parameters as the process proceeds. Environmental factors that are frequently controlled include temperature, oxygen concentration, pH, cell mass, and product concentration.

During the growth and product formation process in a large-scale fermentation it is essential, if the fermentation is to be operated properly, that data be obtained on the process in real time. For instance, it may be desirable to change one of the environmental parameters as the fermentation progresses, or to feed a nutrient at a rate

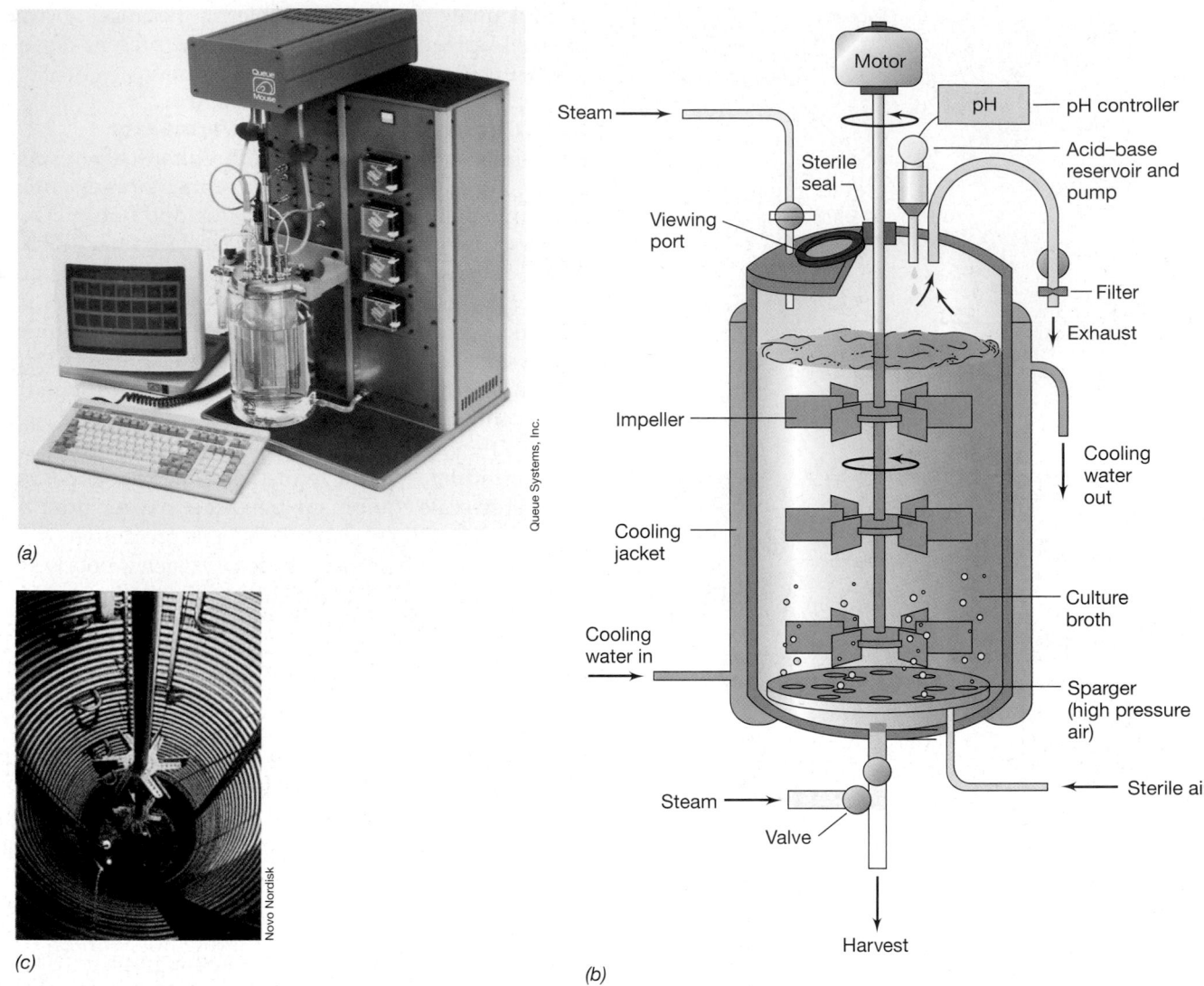

Figure 30.4 Fermentors. (a) A small research fermentor. The volume is 5 liters. (b) Diagram of a fermentor, illustrating construction and facilities for aeration and process control. (c) Inside of an industrial fermentor, showing the impeller and internal heating and cooling coils. In a typical industrial fermentation, aeration and cooling are the key components for online monitoring and adjustment. Nutrient levels and pH are also closely monitored and adjustments made automatically when needed.

TABLE 30.1	Fermentor sizes for various industrial processes
Size of fermentor (liters)	**Product**
1–20,000	Diagnostic enzymes, substances for molecular biology
40–80,000	Some enzymes, antibiotics
100–150,000	Penicillin, aminoglycoside antibiotics, proteases, amylases, steroid transformations, amino acids, wine, beer
200,000–500,000	Amino acids (glutamic acid), wine, beer

that exactly balances growth. Computers process these types of data on-line and then respond accordingly by adding nutrients to avoid diversion of nutrient from the desired product into unwanted products.

Computers are also used to *model* fermentation processes. Mathematical models test the effect of various parameters on growth and product yield quickly and interactively, and then modify the parameters to see how each affects the process. In this way, many variations in the fermentation can be studied inexpensively on screen, rather than expensively at the pilot plant stage or in the industrial plant (see next section).

(a)

(b)

Figure 30.5 (a) A large industrial fermentation plant. Only the tops of the fermentors, which can be several stories high, are visible. (b) Computer control room for a large fermentation plant.

✓ 30.3 Concept Check

Large-scale industrial fermentations present several engineering problems. Aerobic processes require mechanisms for stirring and aeration. The microbial process must be continuously monitored to ensure satisfactory yields of the desired product.

- ✓ What types of devices are used to ensure proper aeration in a large-scale fermentation?
- ✓ What parameters in an industrial fermentation need to be monitored and what adjustments need to be made by computerized control?

30.4 Fermentation Scale-Up

One of the most important aspects of industrial microbiology is the transfer of a process from small-scale laboratory equipment to large-scale commercial equipment, a procedure called **scale-up**. An understanding of the problems in scale-up is extremely important because rarely does a biocatalytic process behave the same way

in large-scale fermentors as in small-scale laboratory equipment (Figure 30.6).

Why does a microbial process differ between large-scale and small-scale equipment? For one thing, mixing and aeration are much easier to accomplish in the small laboratory flask than in the large industrial fermentor. Oxygen transfer especially is much more difficult to obtain in a large fermentor, and because most industrial fermentations are aerobic, effective oxygen transfer is essential. With the rich culture media used in industrial processes, a high biomass is obtained, leading to a high oxygen demand. If aeration is reduced, even for a short period, the culture may experience temporary anoxic conditions, with serious consequences in terms of product yield. Scale-up of an industrial process is the task of the *biochemical engineer*, one who is familiar with gas transfer, fluid dynamics, mixing, and thermodynamics.

The Scale-Up Process

In transferring an industrial process from the laboratory to the commercial fermentor, several stages occur. (1) Experiments in the *laboratory flask*, which are generally the first indication that a process of commercial interest is possible. (2) The *laboratory fermentor*, a small-scale fermentor, generally of glass and of 1- to 10-liters in size, in which the first efforts at scale-up are made (Figures 30.4a and 30.6a). In the laboratory fermentor, it is possible to test variations in medium, temperature, pH, and so on, inexpensively because little cost is involved for either equipment or culture medium. (3) When tests in the laboratory fermentor are successful, the process moves into the *pilot plant stage*, usually carried out in equipment 300–3000 liters in size. Here, the conditions more closely approach the commercial scale; however, cost is not yet a major factor. (4) Finally, the process is moved to the *commercial fermentor* itself, generally 10,000–500,000 liters (Figures 30.5a and 30.6b). In all stages, aeration is very closely monitored. As scale-up proceeds from flask to production fermentor, O_2 dynamics are carefully measured at each step to determine how volume increases affect O_2 demand in the fermentation.

In summary, scale-up of a biocatalytic process can be enormously complex and requires knowledge not only of the biology of the producer organism, but also of the physics of fermentor design and operation.

✓ 30.4 Concept Check

Scale-up is the process of gradually converting a useful industrial fermentation from laboratory scale to production scale. Aeration is a particularly critical aspect to monitor during scale-up studies.

- ✓ What are the differences in size among a typical laboratory fermentor, a pilot plant fermentor, and a commercial fermentor?

(a)

(b)

Figure 30.6 (a) A bank of small research fermentors, used in process development. The fermentors are the glass vessels with the stainless steel tops. The small plastic bottles are to collect overflow. (b) A large bank of outdoor industrial-scale fermentors (240 m^3) used in commercial production of alcohol in Japan. Because of the great difference in their size, the same microbial fermentation would probably operate quite differently in the two different sizes of fermentors.

II MAJOR PRODUCTS OF INDUSTRIAL MICROBIOLOGY

We now consider the industrial production of microbial products, beginning with the antibiotics. Antibiotic production is a huge industry worldwide and one where many important principles of large-scale microbial cultures were first developed.

30.5 Antibiotics: Isolation and Characterization

Of the microbial products manufactured commercially, probably the most important are the antibiotics. As discussed in Chapter 20, antibiotics are chemical substances produced by microorganisms that kill or inhibit the growth of other microorganisms. The development of antibiotics as agents for treatment of infectious disease has probably had more impact on the practice of medicine than any other single development. Antibiotics are typical secondary metabolites (see Section 30.2). Commercially useful antibiotics are produced primarily by filamentous fungi and by *Bacteria* of the actinomycete group (∞ Section 12.24). A listing of the most important antibiotics produced by large-scale industrial fermentation is given in Table 30.2.

Search for New Antibiotics

Although pharmaceutical companies currently do much of their drug discovery using computer modeling (∞ Section 20.13), the traditional way in which new antibiotics are discovered is by *screening*. In the screening approach, a large number of isolates of possible antibiotic-producing microorganisms are obtained from nature in pure culture (Figure 30.7a), and these isolates

TABLE 30.2 Some antibiotics produced commercially

Antibiotic	Producing microorganism[a]
Bacitracin	*Bacillus licheniformis* (EFB)
Cephalosporin	*Cephalosporium* sp (F)
Chloramphenicol	Chemical synthesis (formerly produced microbially by *Streptomyces venezuelae*) (A)
Cycloheximide	*Streptomyces griseus* (A)
Cycloserine	*Streptomyces orchidaceus* (A)
Erythromycin	*Streptomyces erythreus* (A)
Griseofulvin	*Penicillium griseofulvin* (F)
Kanamycin	*Streptomyces kanamyceticus* (A)
Lincomycin	*Streptomyces lincolnensis* (A)
Neomycin	*Streptomyces fradiae* (A)
Nystatin	*Streptomyces noursei* (A)
Penicillin	*Penicillium chrysogenum* (F)
Polymyxin B	*Bacillus polymyxa* (EFB)
Streptomycin	*Streptomyces griseus* (A)
Tetracycline	*Streptomyces rimosus* (A)

[a]EFB, endospore-forming bacterium; F, fungus; A, actinomycete.

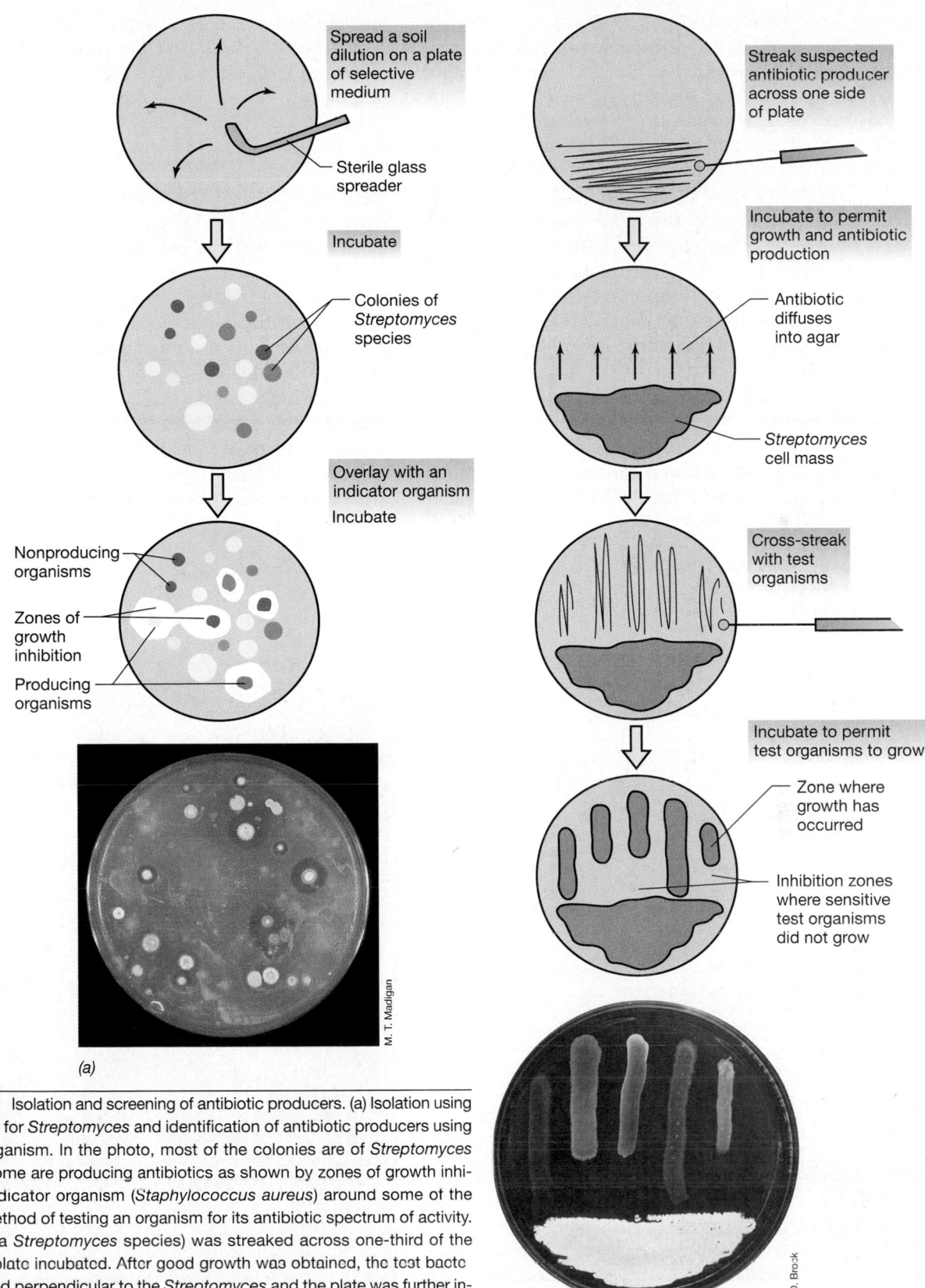

Figure 30.7 Isolation and screening of antibiotic producers. (a) Isolation using media selective for *Streptomyces* and identification of antibiotic producers using an indicator organism. In the photo, most of the colonies are of *Streptomyces* species, and some are producing antibiotics as shown by zones of growth inhibition of the indicator organism (*Staphylococcus aureus*) around some of the colonies. (b) Method of testing an organism for its antibiotic spectrum of activity. The producer (a *Streptomyces* species) was streaked across one-third of the plate, and the plate incubated. After good growth was obtained, the test bacteria were streaked perpendicular to the *Streptomyces* and the plate was further incubated. The failure of several organisms to grow near the mass growth of *Streptomyces* indicates that the *Streptomyces* produced an antibiotic active against these bacteria. Test organisms (left to right): *Escherichia coli*, *Bacillus subtilis*, *Staphylococcus aureus*, *Klebsiella pneumoniae*, *Mycobacterium smegmatis*.

are then tested for antibiotic production by seeing whether they produce any diffusible materials that are inhibitory to the growth of test bacteria. The test bacteria used are selected from a variety of bacterial types but are chosen to be representative of, or related to, bacterial pathogens. The classical procedure for testing new microbial isolates for antibiotic production is the cross-streak method, first used by Fleming in his pioneering studies on penicillin (∞ the box, Microbiology and "Magic Bullets" in Chapter 20 and Figure 30.7b). Those isolates that show evidence of antibiotic production are then studied further to determine if the antibiotics they produce are new. Most of the isolates obtained produce *known* antibiotics, so the industrial microbiologist must quickly identify such organisms so that time and resources are not wasted in studying them. Once an organism producing a *new* antibiotic is discovered, the antibiotic is produced in sufficient amounts for structural analyses and then tested for toxicity and therapeutic activity in infected animals. Unfortunately, most

new antibiotics *fail* these animal tests, but a few prove to be medically useful and are produced commercially. However, with estimates of the number of different antibiotics produced by species of the genus *Streptomyces* alone at over 100,000, research to discover new antibiotics occurs on a continuous bases.

Purification and Increased Yield

An antibiotic that is to be produced commercially must first be produced successfully in large-scale industrial fermentors. We discussed the general problem of scale-up in Section 30.4. The next challenge is the development of efficient purification methods. Because of the relatively small amounts of antibiotic present in the fermentation liquid, elaborate methods for extraction and purification of the antibiotic are necessary (Figure 30.8). If the antibiotic is soluble in an organic solvent, it may be relatively simple to purify it by extraction into a small volume of the solvent. If the antibiotic is not solvent-soluble, then it must be removed from the fermentation liquid by ad-

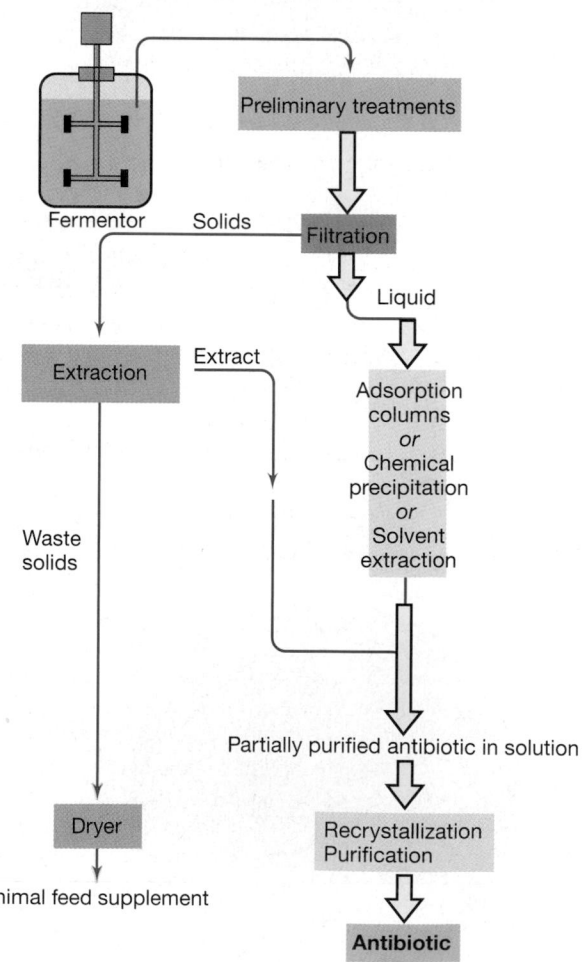

(a)

(b)

Figure 30.8 Purification of an antibiotic. (a) Overall process of extraction and purification. (b) Installation for the solvent extraction of an antibiotic from fermentation broth. Effective engineering is as important as microbiological factors in the successful production of an antibiotic.

sorption, ion exchange, or chemical precipitation (Figure 30.8). In all cases, the goal is to eventually obtain a crystalline product of high purity.

Rarely do antibiotic-producing strains just isolated from nature produce the desired antibiotic at sufficiently high concentration that commercial production can begin immediately. One of the major tasks of the industrial microbiologist is thus to isolate *high-yielding strains*. Strain selection involves mutagenesis of the initial culture, plating of mutant types, and testing of these mutants for antibiotic production. However, genetic engineering has greatly improved this process. For example, the technique of *gene amplification* makes it possible to insert additional copies of genes of interest into a cell by means of a vector such as a plasmid (◌◦ Chapter 31). Alterations in regulatory processes also may permit increased yields. However, one difficulty with using genetic engineering procedures for increasing antibiotic yield is that the biosynthetic pathways for the synthesis of most antibiotics involve large numbers of steps with many genes, and in many cases it is not clear which genes should be altered or increased in number to increase yields. Thus, it is critical that the rate-limiting step in a given biochemical pathway first be identified by basic research.

Final yield is a critical issue with virtually all pharmaceuticals. Even after commercial production of an antibiotic or other product has begun, research often continues to identify or produce higher-yielding strains or to modify the process in some way so as to increase yield. Although pharmaceuticals are produced on a small scale compared with bulk chemicals or agricultural products, there are nevertheless good economic reasons to reach the highest yields possible in the shortest period of time.

✓ 30.5 Concept Check

The industrial production of antibiotics begins with screening for antibiotic producers. Once new producers are identified, purification and chemical analyses of the antimicrobial agent are made. If the new antibiotic is biologically active *in vivo*, the industrial microbiologist may genetically modify the producing strain to increase yields to levels acceptable for commercial development.

- ✓ What is the natural habitat of most antibiotic-producing microorganisms?
- ✓ What is meant by the word *screening* in the context of finding new antibiotics?

30.6 Industrial Production of Penicillins and Tetracyclines

Once an antibiotic has been structurally characterized, has been proven medically effective in tests on experimental animals and sufficiently nontoxic, and, finally, has passed clinical trials (this sequence of events can take several years in actual practice), it is ready to be produced commercially and marketed. For antibiotics like penicillin and tetracycline, these hurdles were passed long ago; today, literally tons of these antibiotics are produced for medical and veterinary use. We thus focus here on the industrial production of these two antibiotics as examples of antibiotic production in general.

β-Lactam Antibiotics: Penicillin and Its Relatives

As discussed in Section 20.8, the penicillins are a class of antibiotics characterized by the *β-lactam ring* and are produced by a variety of molds (eukaryotes) of the genera *Penicillium* and *Aspergillis*, and by certain prokaryotes (Figure 30.9). Among the clinically useful penicillins several different forms are known, and these derivatives may be the result of both biocatalytic reactions and later chemical modification by the organic chemist to produce a penicillin with specific clinical properties.

The basic structure of all penicillins is *6-aminopenicilanic acid* (6-APA), which consists of a thiazolidine

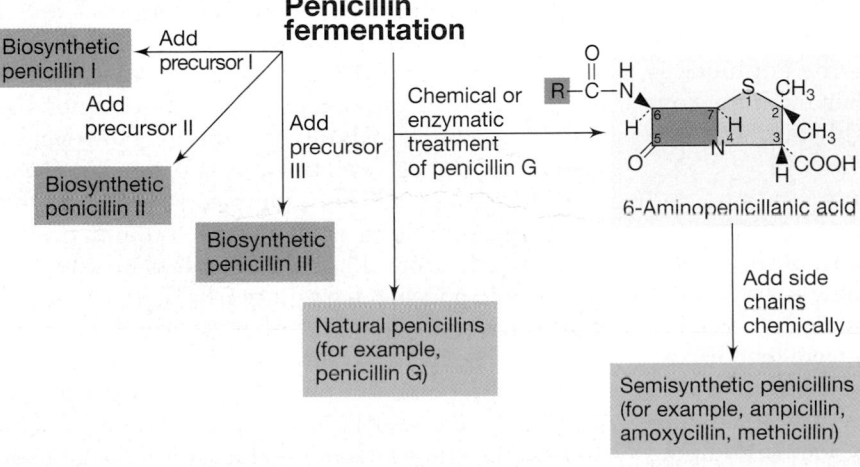

Figure 30.9 Industrial production of penicillins. The β-lactam ring is shown in dark brown. The normal fermentation leads to the natural penicillins. If specific precursors are added during the fermentation, various biosynthetic penicillins are formed. Semisynthetic penicillins are produced by chemically adding a specific side chain to the 6-aminopenicillanic acid nucleus on the "R" group shown in purple. Semisynthetic penicillins have the greatest clinical usefulness because they are typically active against gram-negative bacteria and can be administered orally.

ring with a condensed β-lactam ring (Figure 30.9). The 6-APA carries a variable side chain in position 6. If the penicillin fermentation is carried out without addition of side-chain precursors, the **natural penicillins** are produced (Figure 30.9). The fermentation can be more directed by adding to the broth a *side-chain precursor* so only one desired penicillin is produced. The product formed under these conditions is referred to as a **biosynthetic penicillin** (Figure 30.9). However, in order to produce the most useful penicillins, those with activity against *gram-negative Bacteria*, a combined fermentation and chemical approach is used that leads to the production of **semisynthetic penicillins** (Figure 30.9). In this case, a microbially produced natural penicillin is split either chemically or enzymatically to yield 6-APA and the latter chemically modified by the addition of a side chain (Figure 30.9). Semisynthetic penicillins have many significant clinical advantages in terms of their spectrum of activity and the fact that many of them, for example, ampicillin, can be taken orally and thus do not require injection. For these reasons, semisynthetic penicillins make up the bulk of the penicillin market today.

Production Methods for β-Lactam Antibiotics

Penicillin G is produced in 40,000- to 200,000-liter fermentors. Penicillin production is a highly aerobic process, and efficient aeration is thus necessary. Penicillin is a typical secondary metabolite. During the growth phase, very little penicillin is produced, but once the carbon source has been nearly exhausted, the penicillin production phase begins (Figure 30.10). By feeding with various culture medium components, the production phase can be extended for several days (Figure 30.10).

A major ingredient of most penicillin production media is **corn steep liquor**. This substance contains the nitrogen source as well as other growth factors. The carbon source is generally *lactose* (Figure 30.10). Penicillin is excreted into the medium, and after the cells are removed by filtration, the pH of the medium is lowered and the antibiotic extracted with an organic solvent. After concentration into the solvent, the antibiotic is back-extracted into an alkaline aqueous medium, concentrated further, and crystallized. Highly purified penicillin can be readily obtained in this way.

Other β-Lactam Antibiotics

A variety of other β-lactam antibiotics are known. *Cephalosporins* are β-lactam antibiotics containing a dihydrothiazine instead of a thiazolidine ring system (⬡ Section 20.8 and Figure 20.12). Cephalosporins were first discovered as products of the fungus *Cephalosporium acremonium*, but a number of other fungi as well as some prokaryotes also produce antibiotics with this ring system. In addition,

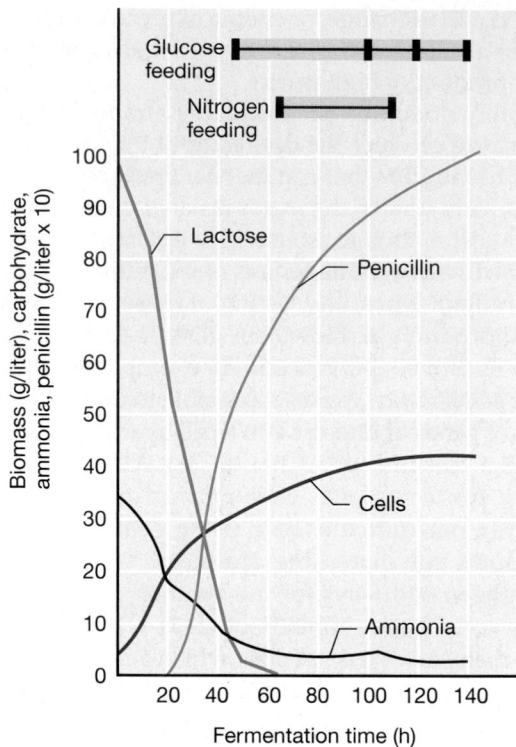

Figure 30.10 Kinetics of the penicillin fermentation with *Penicillium chrysogenum*. Note how the production of penicillin occurs as cells are entering stationary phase and most of the carbon and nitrogen is exhausted. Nutrient "feedings" keep penicillin production high.

a number of semisynthetic cephalosporins are produced. Cephalosporins are valued clinically not only because of their low toxicity but also because they are *broad-spectrum* antibiotics (⬡ Section 20.8), useful against a wide variety of bacterial pathogens.

Production of Tetracyclines

The biosynthesis of a tetracycline involves a large number of enzymatic steps. In the case of chlortetracycline (Figure 30.11), as many as 72 intermediate products may be involved, most of which are known in only a general way. Studies on the genetics of *Streptomyces aureofaciens*, the producer of chlortetracycline, have shown that a total of more than 300 genes are involved! With such a large number of genes, regulation of biosynthesis of this antibiotic is obviously quite complex. However, a few regulatory signals are known and production schemes are well worked out. For example, repression of chlortetracycline synthesis by both glucose and phosphate is known to occur. Phosphate repression is especially significant, and so the medium used in commercial production contains low phosphate concentrations. A production scheme for chlortetracycline is shown in Figure 30.11. As in penicillin production, corn steep liquor is used in the large-scale production of chlortetracycline but sucrose (not lactose) is the typical carbon source.

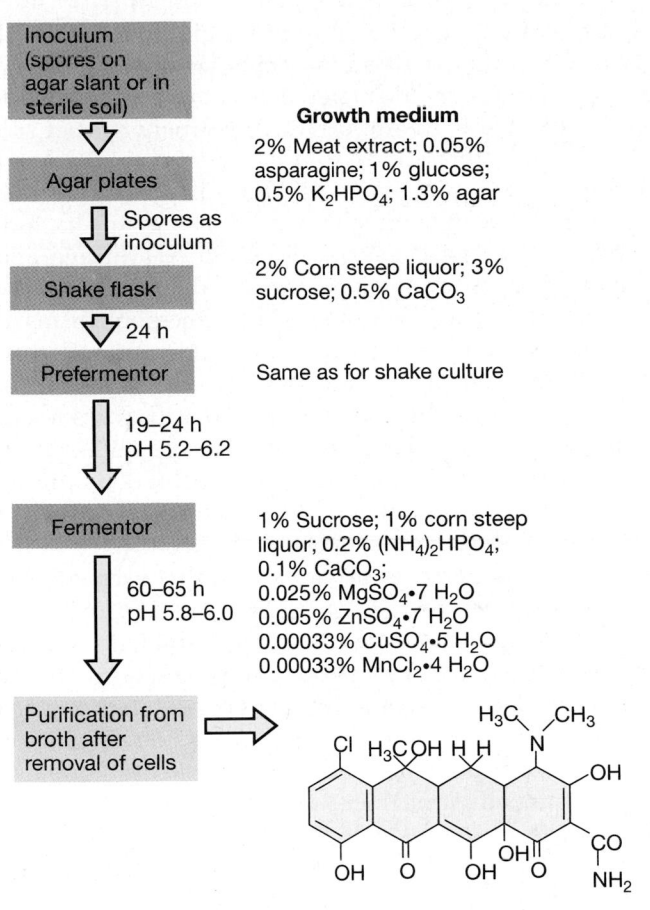

Growth medium

2% Meat extract; 0.05% asparagine; 1% glucose; 0.5% K_2HPO_4; 1.3% agar

2% Corn steep liquor; 3% sucrose; 0.5% $CaCO_3$

Same as for shake culture

1% Sucrose; 1% corn steep liquor; 0.2% $(NH_4)_2HPO_4$; 0.1% $CaCO_3$; 0.025% $MgSO_4$•7 H_2O 0.005% $ZnSO_4$•7 H_2O 0.00033% $CuSO_4$•5 H_2O 0.00033% $MnCl_2$•4 H_2O

Figure 30.11 Production scheme for chlortetracycline with *Streptomyces aureofaciens*. The structure of chlortetracycline is shown on the bottom right. Growth temperature, 28°C throughout.

Glucose is avoided because it causes catabolite repression (⟳ Section 8.7) of antibiotic production.

✓ 30.6 Concept Check

Major antibiotics of clinical significance include the β-lactam antibiotics penicillin and cephalosporin, and the tetracyclines. All of these antibiotics are typical secondary metabolites, and their industrial production is well worked out despite the fact that the biochemistry and genetics of their biosynthesis are only partially understood.

✓ What chemical structure is common to both penicillin and cephalosporin?

✓ In terms of penicillin production, what is meant by the term *semisynthetic? Biosynthetic?*

30.7 Vitamins and Amino Acids

Vitamins and amino acids are growth factors that are often used pharmaceutically or are added to foods. Several important vitamins and amino acids are produced commercially by biocatalytic processes.

Vitamins

Vitamins are used as supplements for human food and animal feeds, and production of vitamins is second only to that of antibiotics in terms of total sales of pharmaceuticals. Most vitamins are made commercially by chemical synthesis. However, a few are too complicated to be synthesized inexpensively but can be made by biocatalysis. Vitamin B_{12} and riboflavin are the most important of this class of vitamins.

Vitamin B_{12} (Figure 30.12*a*) is synthesized in nature exclusively by microorganisms. As a coenzyme, vitamin B_{12} plays an important role in animal biochemistry in various intramolecular rearrangements in which a hydrogen atom on one carbon atom and a substituent on

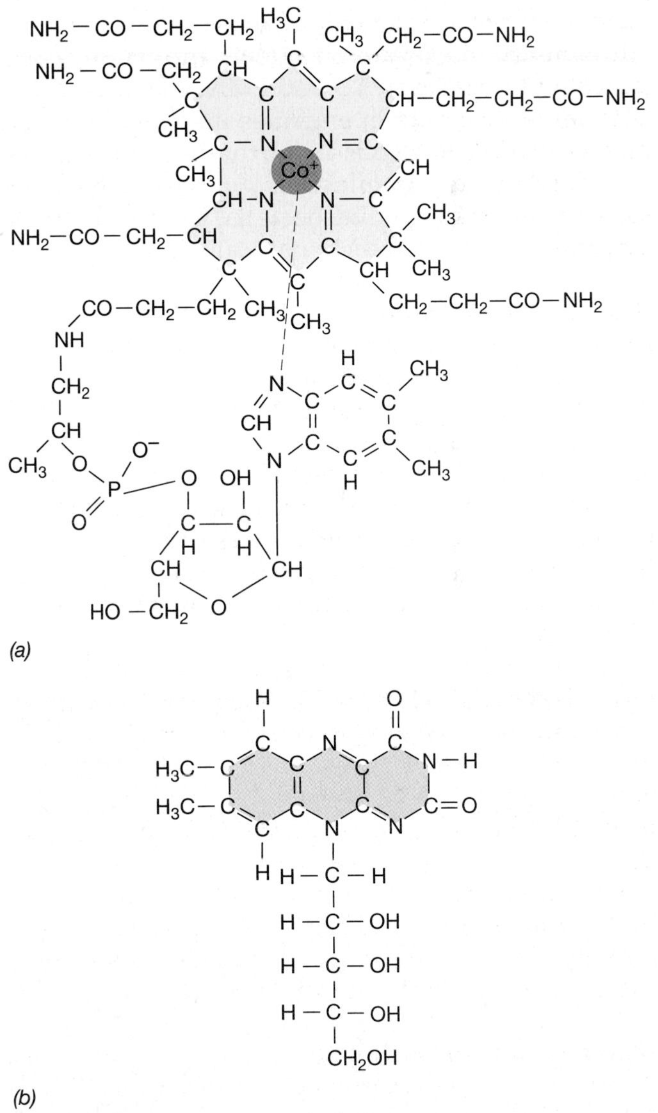

(a)

(b)

Figure 30.12 Vitamins produced by microorganisms on an industrial scale. (a) Vitamin B_{12}. Shown is the structure of cobalamin; note the central cobalt atom. The coenzyme form of vitamin B_{12} contains a deoxyadenosyl group attached to Co above the plane of the ring. (b) Riboflavin (vitamin B_2; ⟳ Section 5.11 and Figure 5.15).

the adjacent carbon atom exchange places. In humans, a major deficiency of vitamin B_{12} leads to a severe condition called *pernicious anemia*, characterized by low production of red blood cells and nervous system disorders. The requirements of animals for vitamin B_{12} are satisfied by food intake or by absorption of the vitamin produced in the gut of the animal by intestinal microorganisms. Plants do not produce or use vitamin B_{12}.

For industrial production of vitamin B_{12}, microbial strains are employed that have been specifically selected for their high yields of the vitamin. Members of the bacterial genera *Propionibacterium* and *Pseudomonas* are the main commercial producers. Cobalt is an important component of the structure of vitamin B_{12} (Figure 30.12*a*), and yields of the vitamin are greatly increased by addition of cobalt to the culture medium.

Riboflavin (Figure 30.12*b*) is the parent compound of the flavins, FAD and FMN, coenzymes that play important roles in enzymes involved in oxidation-reduction reactions in virtually all organisms (☜ Section 5.11). Riboflavin is synthesized by many microorganisms, including bacteria, yeasts, and fungi. The fungus *Ashbya gossypii* naturally produces huge amounts of this vitamin (up to 7 g/l) and is therefore used for most of the microbial production processes. In spite of this good yield, there is great economic competition between this microbiological process and strictly chemical synthesis.

Amino Acids

Amino acids have extensive uses in the food industry, as feed additives, in medicine, and as starting materials in the chemical industry (Table 30.3). The most important commercial amino acid is **glutamic acid**, which is used as a flavor enhancer [monosodium glutamate (MSG)]. Two other important amino acids, **aspartic acid** and **phenylalanine**, are the ingredients of the artificial sweetener **aspartame**, an important constituent of diet soft drinks and other foods sold as sugar-free products. **Lysine**, an essential amino acid for humans and certain farm animals, is commercially produced by the bacterium *Brevibacterium flavum* for use as a food additive, and we focus on its production here.

Because amino acids are used by microorganisms as building blocks of proteins, strict cellular regulation of their production generally occurs (☜ Chapter 8). However, for the industrial production of an amino acid, ways to circumvent these regulatory mechanisms are necessary in order to obtain an *overproducing strain* capable of producing the amino acid economically. The production of lysine in *Brevibacterium flavum*, is biochemically controlled at the level of the enzyme aspartokinase, in that excess lysine feedback inhibits activity of this enzyme (Figure 30.13*a*) (the general phenomenon of feedback inhibition was described in Section 8.2). However, overproduction of lysine can be obtained by isolating mutants of *B. flavum* in which aspartokinase is no longer subject to feedback inhibition. This is done by isolating mutants resistant to the lysine analog *S*-aminoethylcysteine (AEC), which binds to the allosteric site of aspartokinase and shuts down activity of the enzyme (Figure 30.13*b*). AEC-resistant mutants, which are easily obtained by positive selection, produce a modified form of aspartokinase with an allosteric site that no longer recognizes AEC *or* lysine, and thus feedback inhibition by lysine is greatly reduced. Such mutants of *B. flavum* can produce over 60 g of lysine per liter in industrial fermentors, a concentration sufficiently high to make the process commercially viable.

Amino acid[b]	Annual production worldwide (metric tons)	Uses	Purpose
L-Glutamate (monosodium glutamate, MSG)	370,000	Various foods	Flavor enhancer; meat tenderizer
L-Aspartate and alanine	5,000	Fruit juices	"Round off" taste
Glycine	6,000	Sweetened foods	Improve flavor; starting point for organic syntheses
L-Cysteine	700	Bread	Improves quality
		Fruit juices	Antioxidant
L-Tryptophan + L-Histidine	400	Various foods, dried milk	Antioxidant, prevents rancidity; nutritive additive
Aspartame (made from L-phenylalanine + L-aspartic acid)	7,000	Soft drinks	Low-calorie sweetener
L-Lysine	70,000	Bread (Japan), feed additives	Nutritive additive
DL-Methionine	70,000	Soy products, feed additives	Nutritive additive

TABLE 30.3 Amino acids used in the food industry[a]

[a]Data from Glazer, A. N., and H. Mikaido. 1995. *Microbial Biotechnology*, W. H. Freeman, New York.
[b]The structures of these amino acids are shown in Figure 3.12.

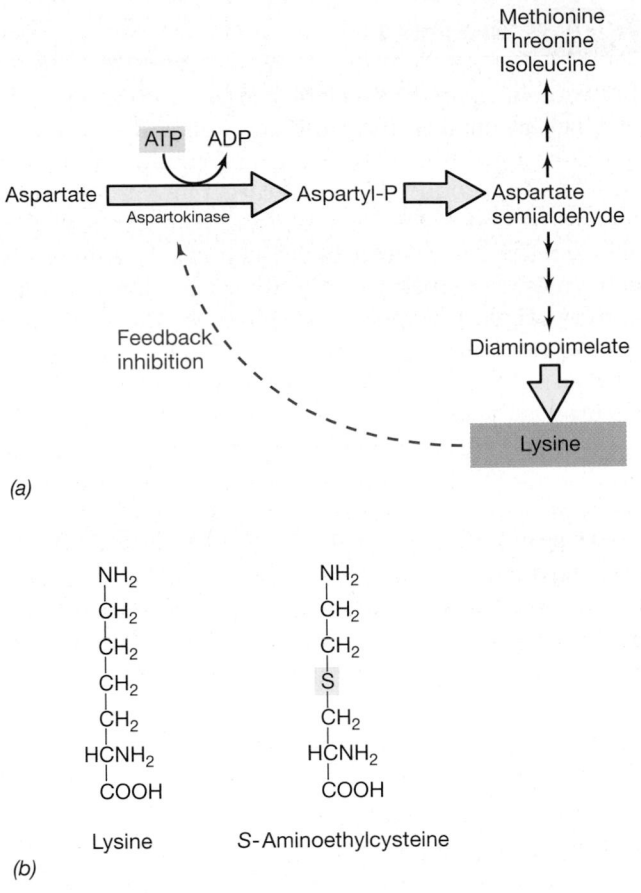

(a)

(b)

Figure 30.13 Industrial production of lysine using *Brevibacterium flavum*. (a) Biochemical pathway leading from aspartate to lysine; note that lysine can feedback-inhibit (Section 8.2) activity of the enzyme aspartokinase, leading to cessation of lysine production. (b) Structure of lysine and the lysine analog *S*-aminoethylcysteine (AEC). AEC normally inhibits growth, but AEC-resistant mutants of *B. flavum* have an altered allosteric site on their aspartokinase and grow and overproduce lysine because feedback inhibition no longer occurs.

✓ **30.7 Concept Check**

Vitamins produced microbially include vitamin B_{12} and riboflavin, whereas the most important amino acids produced commercially are glutamic acid, aspartic acid, phenylalanine, and lysine. High yields of amino acids are obtained by modifying regulatory signals that control synthesis of the particular amino acid such that overproduction occurs.

✓ What amino acid is commercially produced in the greatest amounts?

✓ How can overcoming feedback inhibition improve the yield of an amino acid?

30.8 Microbial Bioconversion

Microorganisms can be used to biocatalyze specific chemical reactions beyond the capabilities of organic chemistry. The use of microorganisms for this purpose is called **bioconversion** or **biotransformation** and involves growth of the organism in large fermentors, followed by the addition at the appropriate time of the chemical to be converted. Following a further incubation period during which the chemical is acted on by the organism, the fermentation broth is extracted and the desired product purified. Although in principle bioconversion may be used for a wide variety of processes, its major industrial use has been in the production of certain steroid hormones (Figure 30.14).

We discussed the role of sterols in eukaryotic membranes in Section 4.5. Steroids, which are derivatives of sterols, are important hormones in animals that regulate various metabolic processes. Some steroids are also used as drugs in human medicine. Members of one group, the *adrenal cortical steroids*, reduce inflammation and hence, are effective in controlling the symptoms of arthritis and allergy. Members of another group, the estrogens and androgenic steroids, are involved in human fertility, and some of them can be used in the control of fertility. Steroids can be obtained by complete chemical synthesis, but this is a complicated and expensive process.

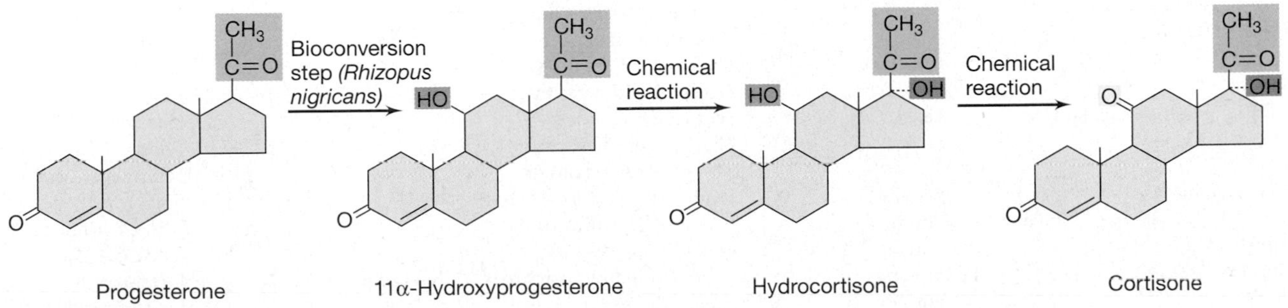

Figure 30.14 Cortisone production using a microorganism. The first reaction is a typical microbial bioconversion, the formation of 11α-hydroxyprogesterone from progesterone. This highly specific oxidation, carried out by the fungus *Rhizopus nigricans*, bypasses a difficult chemical synthesis. All the other steps from progesterone to the steroid hormone cortisone are performed chemically.

Certain key steps in chemical synthesis can be carried out more efficiently by microorganisms, and commercial production of steroids usually has at least one microbial step.

Cortisone and Hydrocortisone

In the production of hydrocortisone and cortisone, steroids used to reduce swelling and itching from minor skin irritations, the fungus *Rhizopus nigricans* carries out a key bioconversion, the stereospecific hydroxylation of a cortisone precursor (Figure 30.14). Most steroid bioconversions involve hydroxylations of this type, and a variety of different fungi are used industrially to carry out one or another specific hydroxylation. Steroid production is currently a big business, as world wide sales of the four major steroids, hydrocortisone, cortisone, prednisone and prednisolone, amount to over 800 tons/year.

✓ 30.8 Concept Check

Microbial bioconversion employs microorganisms to biocatalyze a specific step or steps in an otherwise strictly chemical synthesis.

- ✓ Give an example of a microbial bioconversion. Why is this bioconversion necessary?
- ✓ Describe two ways in which a bioconversion differs from a typical fermentation, such as the production of antibiotics.

30.9 Enzymes

Each organism produces a large variety of enzymes, most of which are made in only small amounts and are involved in cellular processes (⚮ Section 5.5). However, certain enzymes are produced in much larger amounts by some organisms, and instead of being held within the cell, they are excreted into the medium. *Extracellular enzymes* (exoenzymes) are capable of digesting insoluble polymers such as cellulose, protein, and starch, the products of digestion then being transported into the cell where they are used as nutrients for growth. Some of these exoenzymes are used in the food, dairy, pharmaceutical, and textile industries and are produced in large amounts by microbial synthesis (Table 30.4). Enzymes are especially useful biocatalysts because they often act on single chemical functional groups, they can easily distinguish between similar functional groups on a single molecule, and in many cases, they catalyze reactions in a stereospecific manner producing only one of two possible enantiomers (for example, a D-sugar or an L-amino acid; ⚮ Section 3.6).

Proteases, Amylases, and High Fructose Syrup

Enzymes are produced commercially from both fungi and bacteria. The microbial enzymes produced in the largest amounts on an industrial basis are the bacterial proteases, used as additives in laundry detergents. Most

TABLE 30.4 Microbial enzymes and their applications

Enzyme	Source	Application	Industry
Amylase (starch-digesting)	Fungi	Bread	Baking
	Bacteria	Starch coatings	Paper
	Fungi	Syrup and glucose manufacture	Food
	Bacteria	Cold-swelling laundry starch	Starch
	Fungi	Digestive aid	Pharmaceutical
	Bacteria	Removal of coatings (desizing)	Textile
	Bacteria	Removal of stains; detergents	Laundry
Protease (protein-digesting)	Fungi	Bread	Baking
	Bacteria	Spot removal	Dry cleaning
	Bacteria	Meat tenderizing	Meat
	Bacteria	Wound cleansing	Medicine
	Bacteria	Desizing	Textile
	Bacteria	Household detergent	Laundry
Invertase (sucrose-digesting)	Yeast	Soft-center candies	Candy
Glucose oxidase	Fungi	Glucose removal, oxygen removal	Food
		Test paper for diabetes	Pharmaceutical
Glucose isomerase	*Bacteria*	High-fructose corn syrup	Soft drink
Pectinase	Fungi	Pressing, clarification	Wine, fruit juice
Rennin	Fungi	Coagulation of milk	Cheese
Cellulase	*Bacteria*	Fabric softening, brightening; detergent	Laundry
Lipase	Fungi	Breaks down fat	Dairy, laundry
Lactase	Fungi	Breaks down lactose to glucose and galactose	Dairy, health foods
DNA polymerase	*Bacteria*	DNA replication in polymerase chain	Biological research;
	Archaea	reaction (PCR) technique (⚮ Section 10.17)	forensics

laundry detergents today contain enzymes, chiefly proteases, but also amylases, lipases, reductases, and others. Many of these enzymes are isolated from alkaliphilic bacteria (∞ Section 6.11), mainly species of *Bacillus* such as *Bacillus licheniformis* (Table 30.4). These enzymes, which have pH optima between 9 and 10, remain active at the alkaline pH of laundry detergent solutions.

Other important enzymes manufactured commercially are amylases and glucoamylases, which are used in the production of glucose from starch. The glucose so produced can then be converted by the enzyme glucose isomerase to produce fructose (which is sweeter than either glucose or sucrose), resulting in the final production of a high fructose sweetener from corn, wheat, or potato starch. The use of this process in the food industry is big business, primarily for use in the production of soft drinks.

Extremozymes: Enzymes from Prokaryotes That Inhabit Extreme Environments

In Chapters 2 and 6 we considered aspects of microbial growth at high temperature and discovered that some prokaryotes, called *hyperthermophiles*, grow optimally at very high temperatures including, in some cases, above the boiling point of water. Hyperthermophiles are able to grow at such high temperatures because they produce heat-stable macromolecules (∞ Section 6.10) including enzymes, like some of those listed in Table 30.4 but which function at very high temperatures (Figure 30.15*b*). The term **extremozyme** has been coined to refer to enzymes that function at extremely high temperature (or enzymes that function optimally under *any* environmental extreme; for example, in the cold, in very high salt, or at very acid [Figure 30.15*a*] or alkaline pH). The organisms that produce extremozymes are called **extremophiles** (∞ Table 2.1) to indicate that they are organisms that grow best under conditions unsuitable for most microorganisms.

Because many industrial processes operate best at high temperatures, extremozymes from hyperthermophiles are becoming increasingly attractive as biocatalysts for the industrial applications shown in Table 30.4 and also for many research applications that require enzymes. Besides the *Taq* and *Pfu* DNA polymerases for use in the polymerase chain reaction (PCR) described in Section 10.17, extremely thermostable proteases, amylases, cellulases, pullulanases (Figure 30.15*b*), and xylanases have been isolated and characterized from various hyperthermophiles. Such temperature-resistant biocatalysts as well as extremozymes that are cold-active (from psychrophiles), active in the presence of high salt (from halophiles), or active at high or low pH (from alkaliphiles and acidophiles, respectively) will undoubtedly find more industrial applications in the coming years in situations that call for biocatalysis under extreme conditions. Indeed, the great specificity of enzymes and their ability to distinguish between chiral isomers make

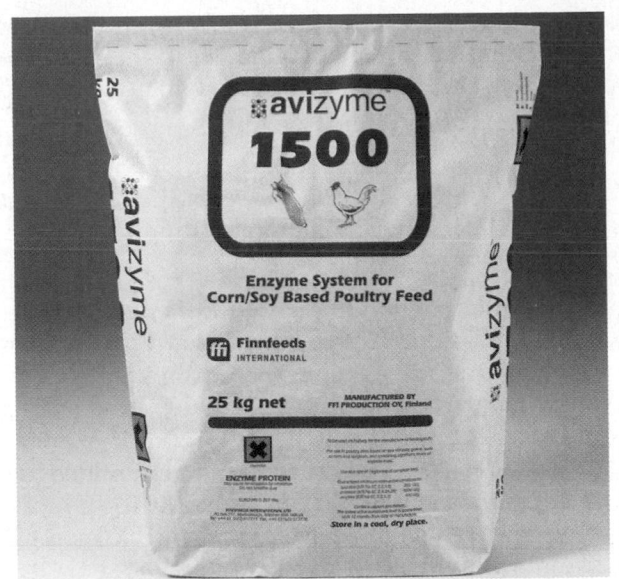

(a)

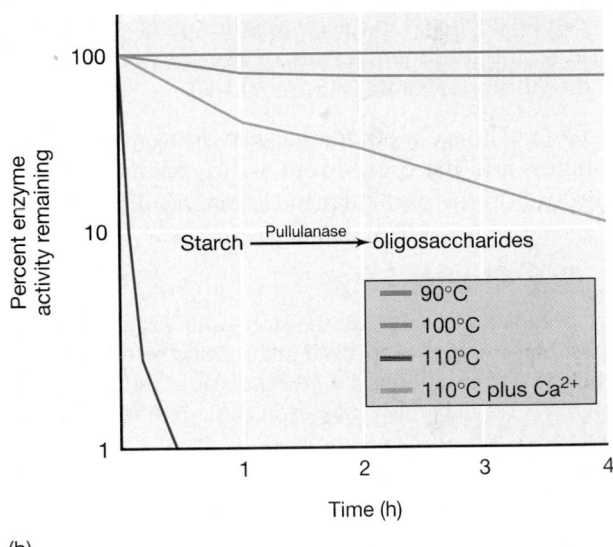

Starch $\xrightarrow{\text{Pullulanase}}$ oligosaccharides

- 90°C
- 100°C
- 110°C
- 110°C plus Ca^{2+}

(b)

Figure 30.15 Examples of extremozymes, enzymes that function under environmentally extreme conditions. (a) Acid-tolerant enzymes. An enzyme mixture used as a feed supplement for poultry. The enzymes function in the stomach of the bird to digest fibrous materials in the feed, thereby improving the nutritional value of the feed and promoting more rapid growth of the bird. (b) Thermostable enzymes. Thermostability of the enzyme pullulanase from *Pyrococcus woesei*, a hyperthermophile whose growth temperature optimum is 100°C (∞ Section 13.9). Calcium improves the heat stability of this enzyme.

those that also function at environmental extremes particularly important to the chemical industry.

Immobilized Enzymes

For some biocatalytic processes it is desirable to convert soluble enzymes into an immobilized state. Immobilization not only makes it easier to carry out the enzymatic reaction under large-scale continuous conditions, but also helps stabilize the enzyme to denaturation. There are three basic approaches to enzyme immobilization (Figure 30.16):

1. **Bonding** of the enzyme to a carrier. The bonding can be through adsorption, ionic bonding, or covalent bonding. Carriers used include modified celluloses, activated carbon, clay minerals, aluminum oxide, and glass beads (Figure 30.16).

2. **Cross-linkage (polymerization)** of enzyme molecules. Linkage of enzyme molecules with each other is usually done by chemical reaction with a cross-linking agent such as glutaraldehyde. Cross-linking of enzymes involves the chemical reaction of amino groups of the enzyme protein with glutaraldehyde. If the reaction is carried out properly, the enzyme molecules can be linked in such a way that most enzymatic activity is maintained.

3. **Enzyme inclusion**, which involves incorporation of the enzyme into a *semipermeable membrane*. Enzymes can be enclosed in microcapsules, gels, semipermeable polymer membranes, or fibrous polymers such as cellulose acetate (Figure 30.16).

Each of these methods has advantages and disadvantages, and the procedure used depends on the enzyme and on the particular industrial application.

✓ 30.9 Concept Check

Microorganisms are ideal for the large-scale production of enzymes. Many enzymes are used in the laundry industry to remove stains from clothing, and thermostable and alkali-stable enzymes have many advantages in these markets. Enzymes from extremophiles are desirable for biocatalyses under extreme conditions. When an enzyme is used in a large-scale process, it may be desirable to immobilize it by bonding it to an inert substrate.

✓ How are enzymes of use in the laundry industry?
✓ Summarize the enzymatic steps that lead to the high-fructose syrup found in soft drinks.
✓ What is an *extremozyme*?

30.10 Vinegar

Vinegar is the product resulting from the conversion of ethyl alcohol to acetic acid by **acetic acid bacteria**, members of the genera *Acetobacter* and *Gluconobacter*. Vinegar can be produced from any substance that contains ethanol, although the usual starting material is wine, beer, or alcoholic apple juice (cider). Vinegar can also be produced from a mixture of pure alcohol in water, in which case it is called *distilled vinegar*, the term *distilled* referring to the alcohol from which the product is made rather than the vinegar itself. Vinegar is used as a flavoring ingredient in salads and other foods, and because of its acidity, it is also used in pickling. Meats and vegetables properly pickled in vinegar can be stored unrefrigerated for years.

The acetic acid bacteria are an interesting group of prokaryotes (∞ Section 12.8). These are strictly aerobic bacteria that differ from most other aerobes in that some of them, such as species of *Gluconobacter*, do not oxidize their organic electron donors completely to CO_2 and water (Figure 30.17). Thus, when provided with ethyl alcohol as electron donor, they oxidize it via quinones only to acetic acid, which accumulates in the medium. Acetic acid bacteria are quite acid-tolerant and are not killed by the acidity that they produce. There is a high oxygen demand during growth, and the main problem in the production of vinegar is to ensure sufficient aeration of the medium.

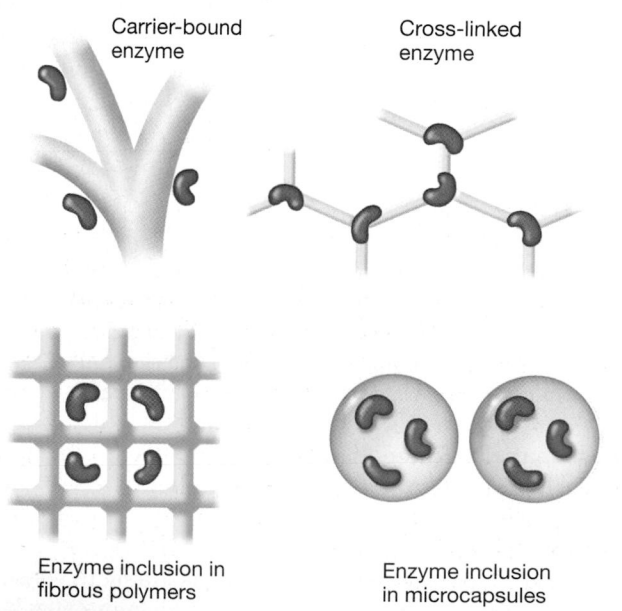

Carrier-bound enzyme

Cross-linked enzyme

Enzyme inclusion in fibrous polymers

Enzyme inclusion in microcapsules

Figure 30.16 Procedures for the immobilization of enzymes. In all cases, enzyme molecules are shown in red.

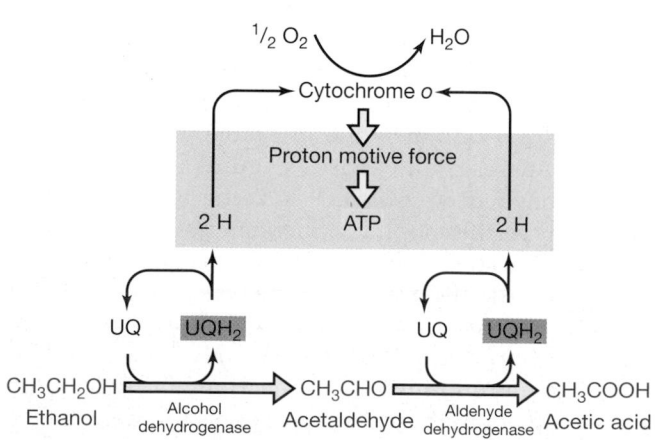

Figure 30.17 Oxidation of ethanol to acetic acid, the key process in the production of vinegar. UQ, ubiquinone.

Vinegar Production

There are three different processes for the production of vinegar. The **open-vat** or **Orleans method** was the original process and is still used in France where it was developed. Wine is placed in shallow vats with considerable exposure to the air, and the acetic acid bacteria develop as a slimy layer on the top of the liquid. This process is not very efficient because the only place that the bacteria come in contact with both the air and the substrate is at the surface. The second process is the **trickle (Quick vinegar) method**, in which the contact between the bacteria, air, and substrate is increased by trickling the alcoholic liquid over beechwood twigs or wood shavings packed loosely in a vat or column while a stream of air enters at the bottom and passes upward. The bacteria grow on the surface of the wood shavings and thus are maximally exposed both to air and liquid. The vat is called a *vinegar generator* (Figure 30.18), and the whole process is operated in a continuous fashion. The life of the wood shavings in a vinegar generator is long, from 5 to 30 years, depending on the kind of alcoholic liquid used in the process.

The third vinegar process is the **bubble method**. This is basically a submerged fermentation process such as was already described for antibiotic production. With proper aeration, the efficiency of the bubble method is high, and 90–98% of the alcohol is converted to acid.

Although acetic acid can be easily made chemically from alcohol, the microbial product, *vinegar*, is a distinctive material, the flavor being due in part to other substances present in the starting material and produced in the fermentation. For this reason, the microbial process, especially using the vinegar generator, has not been supplanted by a chemical process.

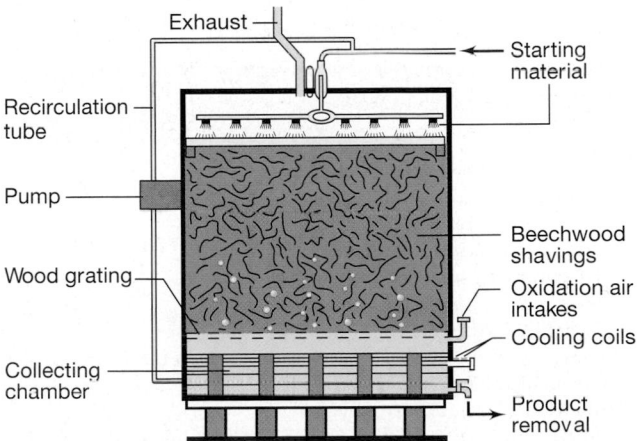

Figure 30.18 Diagram of a vinegar generator. The alcoholic juice is allowed to trickle through the wood shavings, and air is passed up through the shavings from the bottom. Acetic acid bacteria develop on the wood shavings and convert alcohol to acetic acid. The acetic acid solution accumulates in the collecting chamber and is recycled through the generator until the acetic acid content reaches at least 4%, the minimum for a product to be labeled as "vinegar."

The active ingredient in vinegar is acetic acid, which is produced by an acetic acid bacterium oxidizing an alcohol-containing fruit juice. Adequate aeration is the most important consideration in ensuring a successful vinegar process.

✓ Why is O_2 necessary in vinegar production?

✓ Why does vinegar produced by the trickle method have a more distinctive taste than vinegar produced by the bubble method?

30.11 Citric Acid and Other Organic Compounds

Many organic chemicals are produced by microorganisms in sufficient yields that they can be manufactured commercially by fermentation. For example, *citric acid*, used widely in foods and beverages, *itaconic acid*, used in the manufacture of acrylic resins, and *gluconic acid*, used in the form of calcium gluconate to treat calcium deficiencies in humans and industrially as a washing and water-softening agent, are produced by fungi. In addition, *sorbose*, which is produced when *Acetobacter* oxidizes sorbitol, is used in the manufacture of *ascorbic acid* (vitamin C), and *lactic acid*, used in the food industry to acidify foods and beverages, is produced by lactic acid bacteria (⟳ Section 12.19). We focus here on citric acid formation.

Citric Acid

Citric acid is produced microbiologically by a fermentation using the mold *Aspergillus niger*. Although citric acid is normally considered in connection with the citric acid cycle (⟳ Section 5.13), in certain organisms such as *A. niger*, excretion of large amounts of citric acid can be obtained. The fermentation is carried out aerobically in large fermentors, and a key requirement for high citric acid yield is that the medium be *iron-deficient* because citric acid is overproduced by the fungus as a chelator to scavenge iron (Figure 30.19a). Therefore, the medium used for citric acid production is treated to remove most of the iron, and the fermentors themselves are made of stainless steel to prevent leaching of iron from the fermentor walls at the low pH values generated by citric acid accumulation. The media used for citric acid production contain any of a variety of starting materials including starch from potatoes, starch hydrolysates, glucose syrup from saccharified starch, sucrose (Figure 30.19b), sugarcane syrup, sugarcane molasses, and sugar beet molasses. If starch is used, amylase (Table 30.4), formed by the producing fungus or added to the fermentation broth, hydrolyzes the starch to sugars. The sugars are catabolized through the glycolytic pathway (⟳ Section 5.10) and enter the citric acid cycle where citrate production occurs.

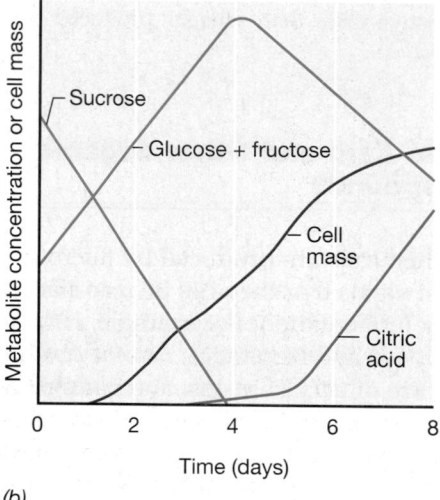

(a)

(b)

Figure 30.19 Citric acid fermentation. (a) Structure of citric acid. Note how the ionized form, citrate, contains three carboxylic acid groups, which can chelate ferric iron (Fe^{3+}). (b) Kinetics of citric acid fermentation. Sucrose is degraded by the enzyme *sucrase* to yield glucose plus fructose. See text for further details.

Most citric acid today is produced by submerged processes in large fermentors. Because *A. niger* is a strict aerobe, it is crucial to this fermentation to make sure that the culture stays properly aerated. Citric acid is produced in this way as a typical secondary metabolite. During the growth phase, sucrose is broken down into glucose and fructose, and by the time stationary phase is reached, large amounts of these hexoses remain and are converted to citric acid to counter iron starvation (Figure 30.19).

Historically, the development of a submerged process for citric acid was of great importance because it was the first *aerobic* industrial fermentation. The technology for manufacturing aerobic fermentors was perfected with the citric acid process. This technology was then applied to penicillin and the other important antibiotic fermentations. Thus, we owe some of our current success with large-scale production of antibiotics to the pioneering work done on the citric acid fermentation.

✓ 30.11 Concept Check

A number of organic chemicals are produced commercially by use of microorganisms, of which the most important economically is citric acid, produced by certain fungi.

✓ Why is citric acid produced by *Aspergillus niger* considered a secondary metabolite (see Figure 30.19b)?

✓ What is the relationship between iron and citric acid production by *A. niger*?

30.12 Yeast as an Agent of Fermentation and as a Food Supplement

Yeasts are among the most extensively used microorganisms in industry. They are cultured for the cells themselves, for cell components, and for the end products they produce during the alcoholic fermentation (Figure 30.1 and Table 30.5). Yeast cells are also used in the manufacture of bread and also as sources of food, vitamins, and other growth factors. Large-scale fermentation by yeast is responsible for the production of alcohol for industrial purposes, but yeast is better known for its role in the manufacture of alcoholic beverages: beer, wine, and liquors (◌◌ box, The Products of Yeast Fermentation, Chapter 5).

Production of yeast cells and production of alcohol by yeast are two quite different processes industrially, in that the first process requires the presence of oxygen for maximum production of cell material whereas the alcoholic fermentation is anaerobic. However, the same or similar species of yeasts, derivatives of *Saccharomyces cerevisiae*, are used in virtually all industrial processes.

Yeast Cell Production

Bakers use yeast as a leavening agent in the rising of the dough prior to baking. A secondary contribution of yeast to bread is its flavor. In the leavening process, the yeast is mixed with the moist dough in the presence of a small amount of sugar. The yeast converts the sugar to alcohol and CO_2, and the gaseous CO_2 expands, causing the

TABLE 30.5	Industrial uses of yeast and yeast products[a]

Production of yeast cells
 Baker's yeast, for bread making
 Dried food yeast, for food supplements
 Dried feed yeast, for animal feeds
Yeast products
 Yeast extract, for microbial culture media
 B vitamins, vitamin D
 Enzymes for food industry: invertase (sucrase), galactosidase
 Biochemicals for research: ATP, NAD^+, RNA
Fermentation products from yeast
 Ethanol, for industrial alcohol and as a gasoline extender
 Glycerol
Beverage alcohol
 Beer, Wine
Distilled beverages
 Whiskey, Brandy, Vodka, Rum

[a] ◌◌ box, The Products of Yeast Fermentation, Chapter 5.

dough to rise. When the bread is baked, the heat drives off the CO_2 and the alcohol and holes are left within the bread, giving it its characteristic light texture.

Yeast for baking or nutritional purposes is cultured in large aerated fermentors in a medium containing molasses as a major ingredient. Molasses contains large amounts of sugar that serves as the source of carbon and energy, and also contains minerals, vitamins, and amino acids used by the yeast. To make a complete medium for yeast growth, phosphoric acid (a phosphorus source) and ammonium sulfate (a source of nitrogen and sulfur) are added.

Fermentation vessels for yeast production range from 40,000 to 200,000 liters. Beginning with the pure stock culture, several intermediate stages are needed to scale up the inoculum to a size sufficient to inoculate the final stage (Figure 30.20a). It is undesirable to add all the molasses to the fermentor at once because this results in a sugar excess and the yeast ferments some of this surplus sugar to alcohol plus CO_2 rather than turning it into yeast cells. Therefore, only a small amount of the molasses is added initially, and then as the yeast culture grows and consumes this sugar, more is added.

At the end of the growth period, the yeast cells are recovered from the broth by centrifugation. The cells are then washed by dilution with water and recentrifuged until they are light in color. Baker's yeast is marketed in two ways, either as compressed cakes or as a dry powder. *Compressed yeast* cakes (Figure 30.20b) are made by mixing the centrifuged yeast with emulsifying agents, starch, and other additives that give it a suitable consistency and reasonable shelf life, and the product is then formed into cubes or blocks of various sizes for domestic or commercial use. A compressed yeast cake contains about 70% moisture and thus must be stored in the refrigerator so its activity is maintained. Yeast marketed in the dry state for baking is usually called *active dry yeast* (Figure 30.20b). The washed yeast is mixed with additives and dried under vacuum until its moisture is reduced to about 8%. It is then packed in airtight containers such as fiber drums, cartons, or multiwall bags, sometimes under a nitrogen atmosphere to promote long shelf life. Active dry yeast does not exhibit as great a leavening action as compressed fresh yeast but has a much longer shelf life. *Nutritional yeast*, marketed as a food supplement (Figure 30.20b), is heat-killed and usually dried. Yeast cells are rich in B vitamins and in protein, except for sulfur-containing amino acids. Yeast is added to wheat or corn flour to increase the nutritional

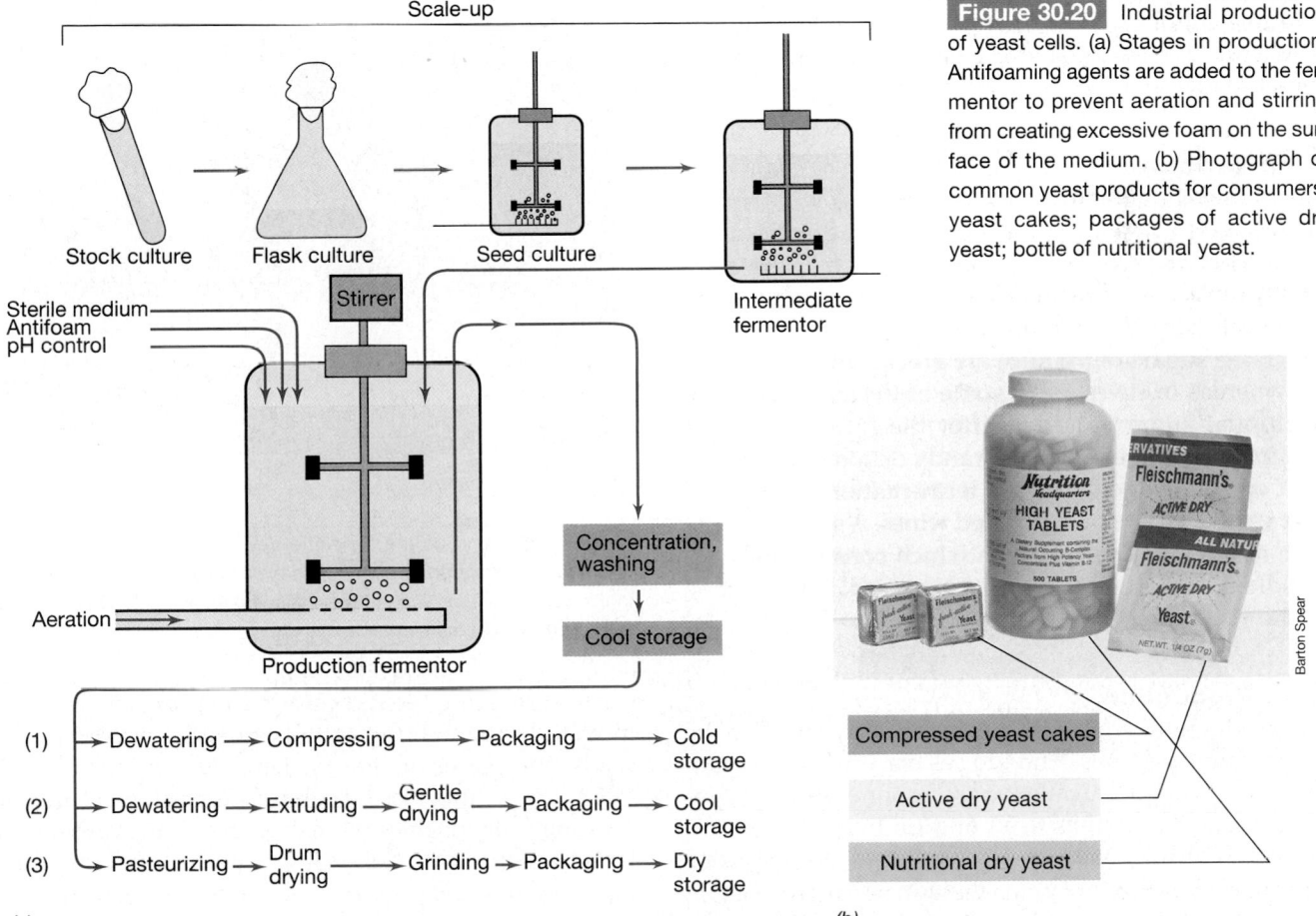

Figure 30.20 Industrial production of yeast cells. (a) Stages in production. Antifoaming agents are added to the fermentor to prevent aeration and stirring from creating excessive foam on the surface of the medium. (b) Photograph of common yeast products for consumers: yeast cakes; packages of active dry yeast; bottle of nutritional yeast.

(a)

(b)

value of these foods and is also sold in pelleted form as a health food (Figure 30.20*b*).

Yeast cells are grown for use in the baking and food industries. Commercial yeast is produced in large-scale aerated fermentors using molasses as the main carbon and energy source.

✓ Write a balanced chemical reaction that accounts for the action of yeast in bread making.

✓ Why is it important when growing yeast *for cells* to maintain oxic conditions in the fermentor?

30.13 Alcohol and Alcoholic Beverages

The use of yeast in the production of alcoholic beverages is an ancient process. Most fruit juices undergo a natural fermentation caused by wild yeasts that are present on the fruit. From these natural fermentations, yeasts have been selected for more controlled production, and today alcoholic beverage production is a large industry world-wide. The most important alcoholic beverages are *wine*, produced by the fermentation of fruit juice; *beer*, or *ale*, produced by the fermentation of malted grains; and *distilled beverages*, produced by concentrating alcohol from a fermentation by distillation. The biochemistry of alcohol fermentation by yeast was discussed in Section 5.10.

Wine Varieties

Most wine is made from grapes, and thus wine manufacture occurs in parts of the world where grapes can be most economically grown (Figure 30.21). There are a great number of different wines, and their quality and character vary considerably. *Dry wines* are wines in which the sugars of the juice are practically all fermented, whereas in *sweet wines*, some of the sugar is left or additional sugar is added after the fermentation. A *fortified wine* is one to which brandy or some other alcoholic spirit is added after the fermentation; sherry and port are the best-known fortified wines. A *sparkling wine*, such as champagne, is one in which considerable carbon dioxide is present, arising from a final fermentation by the yeast directly in the sealed bottle.

Wine Production

The production of wine begins in the early fall with the harvesting of grapes. The grapes are crushed by machine, and the juice, called *must*, is squeezed out. Depending on the grapes used and on how the must is prepared, either white or red wine may be produced (Figure 30.22). A white wine is made either from white grapes or from the juice of red grapes from which the

(a)

(b)

(c)

Figure 30.21 Commercial wine making in California (USA). (a) Equipment for transporting grapes to the winery for crushing. (b) Large tanks where the main wine fermentation takes place. (c) Barrels where the aging process takes place.

skins, containing the red coloring matter, have been removed. In the making of red wine, the *pomace* (skins, seeds, and pieces of stem) is left in during the fermentation. In addition to the color difference, red wine has a stronger flavor than white because of the presence of larger amounts of chemicals called *tannins*, which are extracted into the juice from the grape skins during the fermentation.

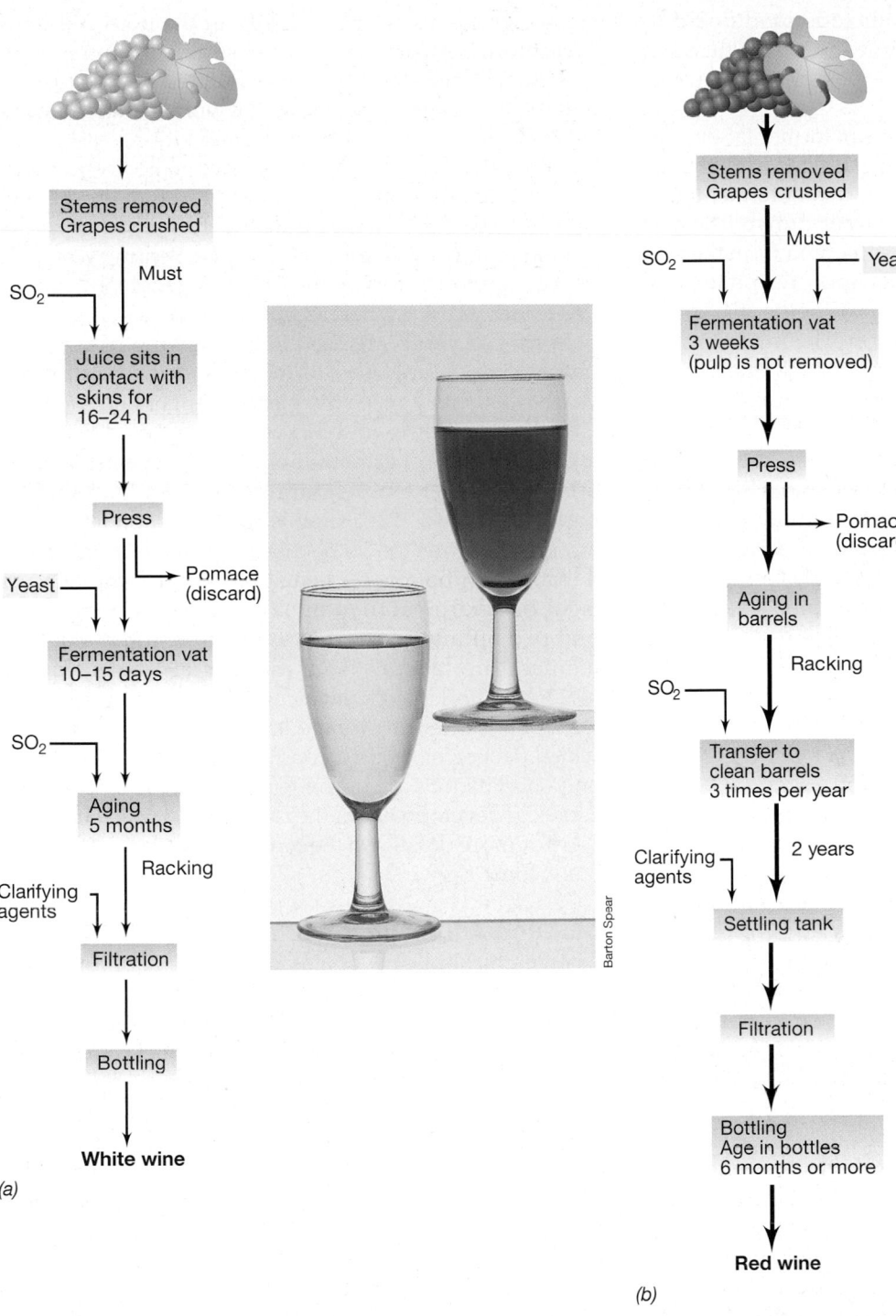

Figure 30.22 Wine production. (a) White wine. White wines can vary from nearly colorless to straw-colored depending on the grapes used. (b) Red wine. Red wines can vary in color from a faint red to a deep, rich burgundy. The photograph shows a glass of Chenin Blanc, a typical white wine (left), and a glass of a light red wine (rosé, right). Other typical varieties of white wine include chablis, Rhine wine, sauterne, and chardonnay, while other typical red wines include burgundy, chianti, claret, zinfandel, cabernet, and merlot.

The yeasts involved in wine fermentation are of two types: the so-called wild yeasts, which are present on the grapes as they are taken from the field and are transferred to the juice, and the cultivated wine yeast, *Saccharomyces ellipsoideus*, which is added to the juice to begin the fermentation. Wild yeasts are less alcohol-tolerant than commercial wine yeasts and can also produce undesirable compounds affecting quality of the final product. Thus, it is the practice in most wineries to kill the wild yeasts present in the must by adding sulfur dioxide (listed on the bottle as "sulfites") at a level of about 100 parts per million (ppm). *Saccharomyces ellipsoideus* is resistant to this concentration of sulfur dioxide and is added as a starter culture from a pure culture grown on sterilized grape juice. The fermentation is carried out in vats of various sizes, from 200-l casks to 200,000-l tanks made of oak, cement, stone, or glass-lined metal (see Figure 30.21b). The fermentor must be constructed so that the large amount of carbon dioxide produced during the fermentation can escape but air cannot enter, and this is accomplished by fitting the vessel with a special one-way valve.

With a red wine, after 3–5 days of fermentation, sufficient tannin and color have been extracted from the pomace and the wine is drawn off for further fermentation in a new tank, usually for another week or two. The next step is called *racking;* the wine is separated from the sediment, which contains yeast cells and precipitate, and then stored at lower temperature for aging, flavor development, and further clarification. The final clarification may be hastened by the addition of materials called *fining agents,* such as casein, tannin, or bentonite clay, or the wine may be filtered through diatomaceous earth, asbestos, or membrane filters. The wine is then bottled and either stored for further aging or sold. Red wine is usually aged for several years or more (Figure 30.21*c*), but white wine is usually sold without much aging. During the aging process, complex chemical changes occur, including reduction of bitter components, resulting in improvement in flavor and odor, or *bouquet.* The final alcohol content of wine varies from 6–14%, depending on the sugar content of the grapes, length of the fermentation, and strain of wine yeast used.

Brewing

The manufacture of alcoholic beverages made from malted grains is called *brewing.* Typical malt beverages include beer, ale, porter, and stout. *Malt* is prepared from germinated barley seeds, and it contains natural enzymes that digest the starch of grains and convert it to sugar. Since brewing yeasts are unable to digest starch (⌒ Figure 3.6*b*), the malting process is essential for the preparation of a fermentable material from cereal grains.

The fermentable liquid from which beer and ale are made is prepared by a process called *mashing.* The grain of the mash may consist only of malt, or other grains such as corn, rice, or wheat may be added. The mixture of ingredients in the mash is cooked and allowed to steep in a large mash tub at warm temperatures. During the heating period, enzymes from the malt cause digestion of the starches and liberate sugars, which will be fermented by the yeast. Proteins and amino acids are also liberated into the liquid, as are other nutrient ingredients necessary for the growth of yeast.

After cooking, the aqueous extract, called *wort,* is separated by filtration from the husks and other grain residues of the mash. *Hops,* an herb derived from the female flowers of the hops plant, are added to the wort at this stage. Hops is a flavoring ingredient, but it also has antimicrobial properties, which probably help to prevent contamination in the subsequent fermentation. The wort is then boiled for several hours, usually in large copper kettles (Figure 30.23*a,b*), during which time de-

sired ingredients are extracted from the hops, proteins present in the wort that are undesirable from the point of view of beer stability are coagulated and removed, and the wort is sterilized. Then the wort is filtered again, cooled, and transferred to the fermentation vessel.

Brewery yeast strains are of two major types: top-fermenting and bottom-fermenting. The main distinction between the two is that **top-fermenting yeasts** remain uniformly distributed in the fermenting wort and are carried to the top by the CO_2 gas generated during the fermentation, whereas **bottom yeasts** settle to the bottom. Top yeasts are used in the brewing of ales, and bottom yeasts are used to make lager beers. Bottom yeasts are usually given the species designation *Saccharomyces carlsbergensis,* and top yeasts are called *Saccharomyces cerevisiae.* Fermentation by top yeasts usually occurs at higher temperatures (14–23°C) than that by bottom yeasts (6–12°C) and is accomplished in a shorter period of time (5–7 days for top fermentation versus 8–14 days for bottom fermentation). After completion of lager beer fermentation by bottom yeasts, the beer is pumped off into large tanks where it is stored at a cold temperature (about −1°C) for several weeks (Figure 30.23*c*). Following the lagering process, the beer is filtered and placed in storage tanks (Figure 30.23*d*) from which packaging occurs. Top-fermented ale is stored for only short periods at a higher temperature (4–8°C), which assists in development of the characteristic ale flavor.

For more details on the brewing process, refer to the box, Home Brew.

Distilled Alcoholic Beverages

Distilled alcoholic beverages are made by heating a fermented liquid at a high temperature that volatilizes most of the alcohol. The alcohol is then condensed and collected, a process called *distilling.* A product much higher in alcohol content can be obtained by this process than is possible by direct fermentation. Virtually any alcoholic liquid can be distilled, and each yields a characteristic distilled beverage. The distillation of malt brews yields *whiskey,* distilled wine yields *brandy,* distillation of fermented molasses yields *rum,* distillation of fermented grain or potatoes yields *vodka,* and distillation of grain and juniper berries yields *gin* (Figure 30.24).

The distillate contains not only alcohol but also other volatile products arising either from the yeast fermentation or from the ingredients themselves. Some of these other products are desirable flavor ingredients, whereas others are undesirable. To eliminate the latter, the distilled product is almost always aged, usually in wood barrels. During the aging process, undesirable products are removed and desirable new flavor ingredients develop. The fresh distillate is usually colorless, whereas the aged product is often brown or yellow (Figure 30.24). The character of the final product is

(a)

(b)

(c)

(d)

Figure 30.23 Brewing beer in a commercial brewery. (a, b) The copper brew kettle is where the wort is mixed with hops and then boiled. From the brew kettle the liquid is passed to large fermentation tanks where yeast ferments glucose to ethanol plus CO_2. (c) If a lager, the beer is then stored for several weeks at low temperature in tanks where settling of particulate matter including yeast cells occurs. (d) The beer is then filtered and placed in storage tanks from which it is packaged into kegs, bottles, or cans.

partly determined by the manner and length of aging (aging times of 10 years or more are not uncommon for some distilled spirits), and the whole process of manufacturing distilled alcoholic beverages is highly complex. To a great extent, the process is carried out by traditional methods that have been found to yield a particular product, rather than by scientifically proven methods.

Commodity Ethanol

Production of ethanol as a commodity chemical is a major biocatalytic process, and today over one billion gallons (3.8 billion liters) of alcohol are produced yearly in the United States, primarily from the fermentation of corn-starch. This ethanol is used as an industrial solvent and also for the production of *gasohol*, a lead-free fuel containing 10% ethanol in gasoline. The combustion of gasohol produces lower amounts of carbon monoxide and nitrogen oxides than pure gasoline; thus gasohol is marketed as a cleaner burning fuel, and its use is encouraged in major cities where automobile pollution is extensive. Various yeasts have been used in commodity ethanol production, including species of *Saccharomyces*, *Kluyveromyces*, and *Candida*, but most ethanol in the United States is produced by *Saccharomyces*.

✓ **30.13 Concept Check**

Alcoholic beverages are produced by yeast from the fermentation of sugar to ethyl alcohol and CO_2. Wine is produced from grape juice, beer from malted grain, and distilled spirits from the distillation of fermented solutions. Commodity alcohol is used as a gasoline additive and industrial solvent.

✓ How do wines differ from beer in terms of the amount of alcohol they contain?

✓ What are the major differences between a *beer* and an *ale*?

Figure 30.24 Color of distilled spirits. Gin or vodka (not shown) are colorless. However, aging in wood casks yields a distinctive amber or yellow color to certain distilled spirits. Left to right, dark rum, brandy, whiskey.

Techniques & Application ...   Home Brew[1]

The amateur brewer can make many kinds of beer, from English bitters and India pale ale to German bock and Russian Imperial stout. The necessary equipment and supplies, including yeast, can be purchased from a local beer and winemakers shop (the Home Wine and Beer Trade Association, 604 N. Miller Road, Valrico, FL 33594, can supply the address of a nearby shop).

The brewing process can be divided into three basic stages: making the wort, carrying out the fermentation, and bottling and aging. The character of a brew depends on many factors: the proportion of malt, sugar, hops, and grain; the kind of yeast; the temperature and duration of the fermentation; and how the aging process is carried out. The fermentor itself consists of a 20-l (5-gal) glass jar or carboy that can be fitted with a tightly fitting closure. In order to have a good quality beer, it is essential that *everything* be sterilized that comes into contact with the wort. This includes the fermentor, tubing, stirring spoon, and bottles. The best procedure is to use a sterilizing rinse consisting of 50–60 ml of liquid bleach in 20 l of water. Soak the items for 15 min and then rinse lightly with hot water or air-dry.

[1]Primary reference source: Burch, B. 1992. *Brewing Quality Beers—The Home Brewer's Essential Guidebook*, 2nd edition. Joby Books, Fulton, CA.

1. **Making the wort** (see Figure 1). In commercial brewing, the wort is made by extracting fermentable sugars and yeast nutrients from malt, sugar, and hops. The process is complex and relatively difficult to carry out satisfactorily. Many home brewers make their own wort from malt, but a reasonably satisfactory beer can be made with hop-flavored malt extract purchased ready-made. Malt extracts come in a variety of flavors and colors, and the kind of beer depends on the type of malt extract used. A simple recipe for making the wort uses 5–6 lb (2.25–2.75 kg) of hop-flavored malt extract and 20 l of water. The malt extract and 5–6 l of water are brought to a boil for 15 min in an enamel or stainless steel container (aluminum heating kettles must be avoid-

Bryon Burch

Figure 1

ed because of the inhibiting action of metals leached from aluminum containers) (see Figure 1). The hot wort is then poured into 14–15 l of clean, cold water that has already been added to the fermentor. After the temperature has dropped below 30°C the yeast is added to initiate the fermentation.

2. **Carrying out the fermentation** (see Figure 2). Add two packs of fresh beer yeast to the cooled wort and cover the fermentor with a rubber stopper into which a plastic hose has been inserted. The hose is directed into a bucket containing water. During the initial 2–3 days of the fermentation, large amounts of CO_2 will be given off, which will exit through the hose. The water trap is to prevent wild yeasts or bacteria from the

Bryon Burch

Figure 2

30.14 Mushrooms as a Food Source

Several kinds of *fungi* are sources of human food, of which the most important are the mushrooms. Mushrooms are a group of filamentous fungi that form large, edible structures called **fruiting bodies** (Figure 30.25). The fruiting body is commonly called the *mushroom* and is formed through the association of a large number of individual hyphae to form a mycelium. We discussed the basic biology of mushrooms in Section 14.10.

Commercial Growth of Mushrooms

The mushroom commercially grown in most parts of the world is *Agaricus bisporus*, and it is generally cultivated in mushroom farms. The organism is grown in special beds, usually in buildings where temperature and humidity are carefully controlled (Figure 30.25a). Beds are prepared by mixing soil with a material very rich in organic matter, such as horse manure, and the beds are then inoculated with mushroom *spawn*. The spawn is actually a pure culture of the mushroom fungus that has been grown in large bottles on an organic-rich medium. In the bed, the mycelium grows and spreads through the substrate, and after several weeks it is ready for the next step, the induction of mushroom formation. This is accomplished by adding to the surface of the bed a layer of soil called *casing soil*. The appearance of mushrooms on the surface of the bed is called a *flush* (Figure 30.25a), and when flushing occurs, the mushrooms must be collected immediately while

air from getting back into the fermentor. After about 3 days, the activity will diminish as the fermentable sugars are used up. At this time, the rubber stopper and hose are replaced with an inexpensive fermentation lock. The fermentation lock, which can be purchased at the home brew store, prevents contamination while permitting the small amount of gas still being produced to escape. Allow the beer to ferment for 7–10 days at 10–15°C or higher.

3. **Bottling and aging** (see Figure 3). The fermentation should be allowed to proceed for the full 7–10 days, even if the vigorous fermentation action ceases earlier. By this time, most of the yeast should have settled to the bottom of the fermentor. Carefully siphon the beer off the yeast layer, allowing it to run into sanitized glass beer bottles. Take care that the yeast at the bottom of the fermentor

Figure 3

is not stirred up and leave the yeast-rich liquid at the bottom. The bottles used should accept standard crown caps, and new, clean caps should be used. Before capping, add ¾ teaspoon (but no more) of corn syrup to each 12–16 ounce (350–465-ml) bottle. Once the bottles are capped, turn each one upside down once to mix the sugar syrup and then allow the beer to age upright at room temperature for at least 7–10 days. After this aging period, the beer may be stored at a cooler temperature.

All homemade beer has a natural yeast sediment in the bottom of the bottles. The beer will improve if it is allowed to age for several weeks. Aging tends to make beer smoother. Using the same basic production equipment, several different types of beer can be made, each with its own distinctive taste and character (the reference listed in the footnote contains a number of beer recipes). Dark beers, which generally contain more alcohol than lighter beers, require more malt for their production and are usually brewed from a combination of different malts such as ones obtained from darker varieties of grain or ones that have been roasted to caramelize the sugars and yield a darker color. A typical American style light lager (see Figure 4, left) contains about 3.5% alcohol

(by volume) whereas a Munich style dark (Figure 4, right) contains 4.25% alcohol, and bock beers contain about 5% alcohol.

The trend toward "individuality" in beer can be attested to not only by the growing number of home brewers but also by the fact that major brewers in the United States are feeling more and more competition from new, usually very small, breweries called *microbreweries*. Although total production by a microbrewery may pale by comparison to that of a major brewer, the products themselves often have their own distinctive character and local appeal. Part of these differences probably has to do with the smaller scale on which the brewing takes place but also undoubtedly has to do with the use of different sources of ingredients, water, yeast strains, and brewing times. ■

Figure 4

(a)

(b)

Figure 30.25 Commercial mushroom production. (a) Close-up of a mushroom flush. (b) Close-up of the shiitake mushroom.

still fresh. After collection they are packaged and kept cool until brought to market.

Another widely cultured mushroom is **shiitake**, *Lentinus edulus*. The most widely cultivated mushroom in the Far East, shiitake is now finding expanding demand in North America. Shiitake is a cellulose-digesting fungus that grows well on hardwood trees and is cultivated on small logs (Figure 30.25b). The logs are soaked in water to hydrate them and then inoculated by inserting plugs of spawn into small holes drilled in them. The fungus grows through the log, and after about a year forms a flush of fruiting bodies (Figure 30.25b). The shi-

itake mushroom is generally considered to have a superior taste to *Agaricus bisporus* and because of this, commands a substantially greater price.

✓ 30.14 Concept Check

The most important food produced from a microorganism is the mushroom, which is produced not for its protein but for its flavor.

- ✓ Why are mushrooms considered microorganisms?
- ✓ What is a *mushroom flush?*

Review Questions

1. In what ways do industrial microorganisms differ from conventional microorganisms? In what ways are they similar?

2. Describe some of the techniques that can be used to improve strains of industrial microorganisms.

3. List three major types of industrial products that can be obtained with microorganisms and give two examples of each.

4. Give an example of a *commodity chemical* produced by a microorganism and describe briefly the process by which this chemical is manufactured.

5. Compare and contrast *primary* and *secondary* metabolites and give an example of each. List at least two molecular explanations for why some metabolites are secondary rather than primary.

6. How does an industrial fermentor differ from a laboratory culture vessel? How does a fermentor differ from a fermenter?

7. Discuss the problems of scale-up from the viewpoints of *aeration*, *sterilization*, and *process control*. Why is sterility so important in an industrial fermentor?

8. List three examples of *antibiotics* that are important industrially. For each of these antibiotics, list the producing organisms, the general chemical structure, and the mode of action.

9. Compare and contrast the production of *natural*, *biosynthetic*, and *semisynthetic* β-lactam antibiotics.

10. Addition of what metal to the fermentation medium can markedly improve production of vitamin B$_{12}$?

11. What unusual characteristics must an organism have if it is to overproduce and excrete an amino acid such as *lysine*?

12. Define *microbial bioconversion* and give an example. Explain why the chemical reactions involved in microbial bioconversions are preferably carried out microbially rather than chemically.

13. List three different kinds of enzymes that are produced commercially. For each enzyme, list the organism used in commercial production, the action of the enzyme, and how the enzyme is used in commerce.

14. What is *high-fructose syrup*, how is it produced, and what is it used for in the food industry?

15. What are extremozymes? What industrial uses do they have?

16. Give two reasons why stainless steel fermentors are used in the industrial production of citric acid.

17. Why are yeasts of such great industrial importance?

18. In what way is the manufacture of *beer* similar to the manufacture of *wine*? In what ways do these two processes differ? How does the production of *distilled alcoholic beverages* differ from that of beer and wine?

19. What part of the mushroom is actually consumed as food? What is contained within this structure?

Application Questions

1. As a researcher in a pharmaceutical company you are assigned the task of finding and developing an antibiotic effective against a new bacterial pathogen. Outline a plan for this process, starting from isolation of the low-yield producing organism to high-yield industrial production of the new antibiotic.

2. A partially consumed bottle of an "organic" (containing no preservatives) red wine is recapped and stored under refrigeration for 2 months. On tasting the wine again, you notice a distinct bitter taste, making the wine undrinkable. Using information presented in this chapter and in Section 12.8, describe (a) what microbe-mediated process occurred in the wine, and (b) a very easy way in which this process could have been prevented.

3. You wish to produce high yields of the amino acid phenylalanine for use in production of the sweetener *aspartame*. The overproducing organism you wish to use is not subject to feedback inhibition by phenylalanine but is subject to typical repression of phenylalanine biosynthesis enzymes by excess phenylalanine. Applying the principles of enzyme regulation studied in Chapter 8 and microbial genetics in Chapter 10, describe two classes of mutants you could isolate that would overcome this problem and detail the genetic lesions each would have.

From a scientific standpoint, the ability to manipulate DNA in a test tube and then place the altered DNA back into an organism and have it expressed, is an extremely useful and powerful technique. Practical applications of such technology include harnessing microorganisms to produce mammalian proteins and new vaccines. Other applications, however, can be used to answer questions related to basic science. Here the gene that encodes a fluorescent protein called the *green fluorescent protein* has been fused to another gene in the yeast *Saccharomyces cerevisiae* so that studies on protein localization can be performed. When the protein of interest is made, the green fluorescent protein is also made and travels with it to its target in the cell, revealing where in the cell the protein of interest is active.

GENETIC ENGINEERING AND BIOTECHNOLOGY

Working Glossary

Biotechnology use of living organisms to carry out defined chemical processes for industrial application

DNA fingerprinting use of the techniques of genetic engineering to determine the origin of DNA in a sample of tissue

Expression vector a cloning vector that contains the necessary regulatory sequences to allow transcription and translation of cloned genes

Gene therapy treatment of a disease caused by a dysfunctional gene by introduction of a normally functioning copy of the gene

Genetically modified organism (GMO) an organism whose genome has been altered using genetic engineering. The abbreviation *GM* is also used in constructions such as *GM crops* and *GM foods*

Genetic engineering the use of *in vitro* techniques in the isolation, manipulation, recombination, and expression of DNA, and in the development of *genetically modified organisms*

Integrating vector a cloning vector that becomes integrated into a host chromosome

Molecular cloning isolation and incorporation of a fragment of DNA into a vector where it can be replicated

Nucleic acid probe a strand of nucleic acid that can be labeled and used to hybridize to a complementary molecule from a mixture of other nucleic acids

Reporter gene a gene incorporated into a vector because the product it encodes is easy to detect

Reverse translation the mental process of using a codon table and the amino acid sequence of a protein to obtain a possible sequence of the mRNA or the gene that encoded the protein

Shuttle vector a cloning vector that can replicate in two or more dissimilar hosts

T-DNA the segment of the *Agrobacterium* Ti plasmid that is transferred to plant cells

Ti plasmid a plasmid in *Agrobacterium* species capable of transferring genes from bacteria to plants

Transgenic organisms plants or animals that stably pass on cloned DNA that has been inserted into them

Biotechnology is the use of living organisms to carry out defined chemical processes for industrial or commercial application. In that sense, biotechnology refers to all industrial microbiology, such as that discussed in Chapter 30, and much more. However, the usage of the word biotechnology today often implies that the organism used to carry out the process has been manipulated using *in vitro* genetic techniques; techniques such as were discussed in Chapter 10.

These techniques make it possible to isolate, manipulate, and sequence DNA as well as to control its expression. In Chapter 15 we saw how these techniques can be used to examine entire genomes. The use of these *in vitro* techniques, particularly molecular cloning (⧉ Section 10.14), that result in *genetically modified organisms* is often referred to as **genetic engineering**. In Chapters 10 and 15 our discussion was largely confined to the uses of these techniques in basic research. However, genetic engineering also has found very important *practical* applications. These include the development of microbial cultures capable of producing valuable products such as human insulin, human growth hormone, interferon, vaccines, and industrial enzymes. In this chapter we will discuss a few of the many possible applications of genetic engineering.

▌ THE TECHNIQUES OF GENETIC ENGINEERING

The basic techniques of genetic engineering are also the basic techniques of modern microbial genetics, and we introduced most of them in Chapter 10 (⧉ Sections 10.12–10.18). Genomics and bioinformatics also are beginning to play increasingly important roles in biotechnology (⧉ Section 15.6). We cannot review all these techniques here but we can summarize some of the principles underlying them. In addition, we will expand our discussion of certain techniques because they are more directly applied to biotechnology.

31.1 Review of Principles Underlying Genetic Engineering

Much of genetic engineering is based on *molecular cloning* (⧉ Section 10.14). In molecular cloning, a DNA fragment from essentially any type of genetic element composed of double-stranded DNA is recombined with a vector and introduced into a suitable host. Commonly employed cloning vectors include plasmids and bacteriophages (⧉ Sections 10.15 and 10.16). Molecular cloning developed because of our increased understanding of certain aspects of molecular genetics. In addition, in many cases the specific biotechnological application depends not just on the ability to identify and clone a gene, but also on manipulating the expression of the gene to produce, identify, and purify its protein product.

The following developments were essential for the development of genetic engineering, their interrelationships are diagrammed in Figure 31.1.

1. **DNA chemistry**: development of procedures for isolation, sequencing, and synthesis of DNA (∞ Section 10.13).

2. **DNA enzymology**: discovery of restriction endonucleases, DNA ligases, and DNA polymerases (∞ Sections 7.5 and 10.12).

3. **DNA replication**: understanding how DNA replication occurs, the importance of DNA cloning vectors capable of independent replication and the development of the **Polymerase Chain Reaction (PCR)** (∞ Sections 7.5–7.7 and 10.15–10.17).

4. **Plasmids and conjugation**: discovery of plasmids, determination of the mechanisms by which plasmids replicate, and how some can transfer from cell to cell by conjugation (∞ Sections 10.8 and 10.9).

5. **Temperate bacteriophage**: understanding how replication and/or integration is controlled in temperate bacteriophages and how specialized transducing phages are formed (∞ Sections 9.10 and 10.7).

6. **Transformation**: discovery of methods for getting free DNA into cells (∞ Section 10.6).

7. **RNA chemistry and enzymology**: understanding how to work with messenger RNA, how eukaryotic mRNA is constructed, and the importance of RNA processing in the formation of mature eukaryotic mRNA (∞ Section 7.12).

8. **Reverse transcription**: discovery of the enzyme *reverse transcriptase* in retroviruses and its development as a means for transcribing information from mRNA back into DNA (∞ Sections 9.12 and 16.14).

Figure 31.1 Summary of the fundamental processes underlying genetic engineering.

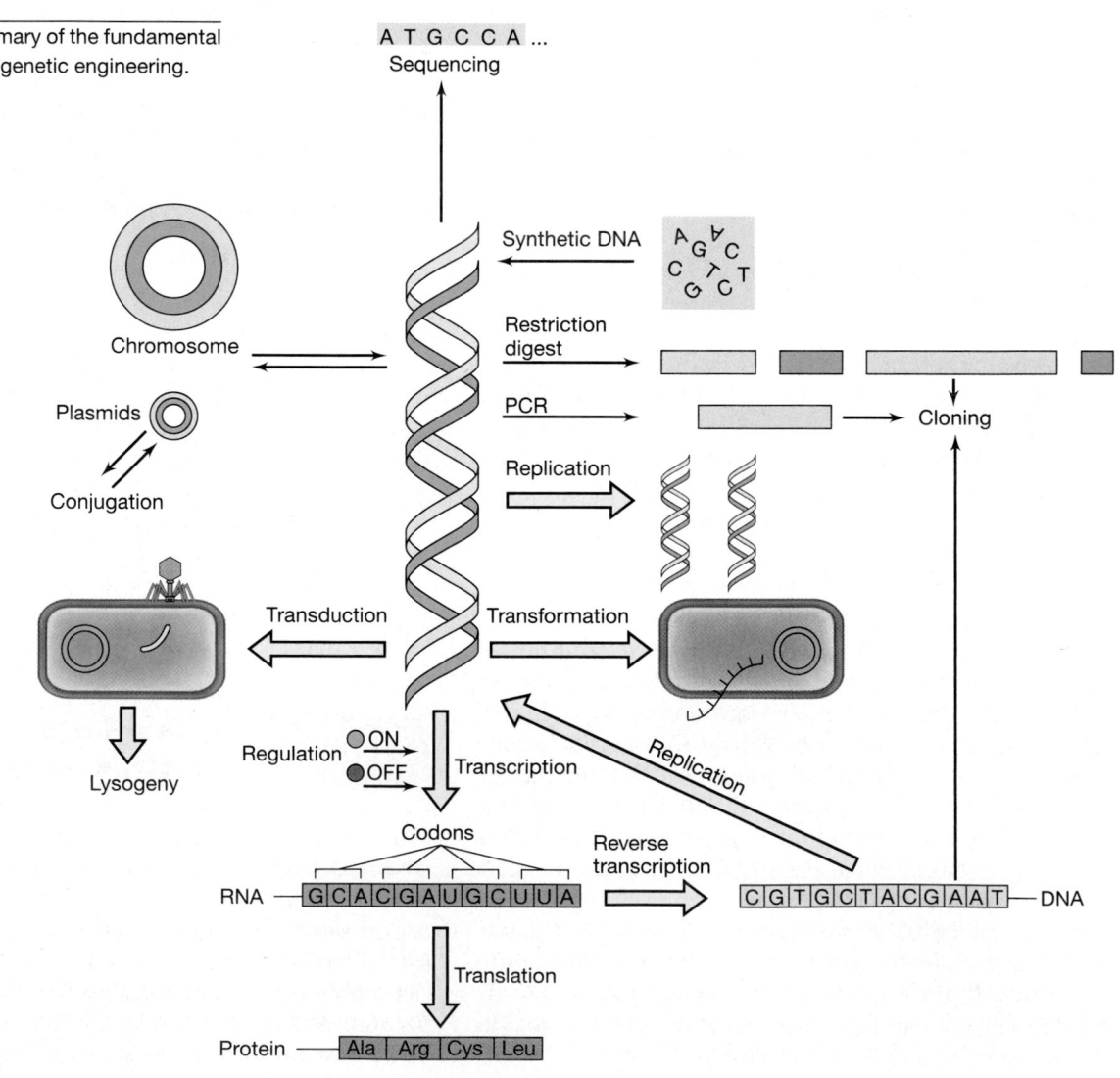

9. **Regulation**: understanding the factors involved in the regulation of transcription, including the discovery of promoter sites and operon control (∞ Sections 7.8–7.11 and Chapter 8).

10. **Translation**: understanding the steps involved in translation, the importance of ribosome binding sites on mRNA, the role of the initiation codon, and the importance of a proper reading frame (∞ Section 7.15).

11. **Protein chemistry**: development of methods for isolation, purification, assay, and sequencing of proteins.

12. **Protein excretion and posttranslational modification**: understanding how proteins are built with signal sequences that are removed during or after excretion (∞ Section 7.16). Discovery of other kinds of posttranslational modification of proteins.

13. **The genetic code**: elucidation of the genetic code and the determination that it was the same in almost all organisms but that certain codons were less frequently used in some organisms than in others (∞ Section 7.13).

✓ 31.1 Concept Check

The techniques of genetic engineering are based on fundamental discoveries in molecular genetics and biochemistry. Successful genetic engineering depends not only on being able to carry out molecular cloning, but also knowledge of replication, transcription, and translation.

✓ What is a *cloning vector*?

✓ Which enzyme(s) is (are) involved in the technique known as *PCR*?

31.2 Hosts for Cloning Vectors

When *in vitro* genetic techniques were introduced in Chapter 10 it was in the context of the genetic analysis of *Bacteria*. There the goal was to understand the organism under study. By contrast, in biotechnology we wish to use genetic engineering for a specific commercial end. It might be to produce large amounts of a specific protein efficiently, to produce a new vaccine, or to introduce a specific gene into a plant or an animal. In these cases it is essential to consider the *host* that we will use, which will also guide our choice of a vector.

For simply obtaining large amounts of cloned DNA, the ideal characteristics of a host are rapid growth, capability of growth in an inexpensive culture medium, nonpathogenicity, ability to take up DNA, and stability in culture. The host must have the appropriate enzymes to allow replication of the vector. The most useful hosts for cloning are typically microorganisms that grow well and for which we have both sufficient genetic information and the tools for genetic manipulation. These include the *Bacteria Escherichia coli* and *Bacillus subtilis* and the yeast *Saccharomyces cerevisiae*. However, many basic scientific questions can be answered only if the cloned DNA can be returned to the species of organism from which it originated. This is particularly true in studies involving gene regulation. Finally, if recombinant DNA itself is to be used therapeutically to treat human disease, the host must be a human being.

Prokaryotic Hosts

Although most molecular cloning has been done in *Escherichia coli*, there are some potential disadvantages in using this host. *E. coli* presents dangers for large-scale production of products derived from cloned DNA because it is found in the human intestinal tract and wild-type strains are potentially pathogenic. Also, even nonpathogenic strains produce endotoxins that can contaminate products, an especially bad situation with pharmaceutical injectables. Finally, *E. coli* retains extracellular proteins in the periplasmic space, making isolation and purification potentially difficult. However, modified *E. coli* strains have been developed for which most of these problems have been eliminated. Because of the extensive knowledge of its genetics and biochemistry, *E. coli* remains the organism of choice for most cloning studies.

The gram-positive organism *Bacillus subtilis* can also be used as a host. *B. subtilis* is not potentially pathogenic, does not produce endotoxin, and secretes proteins into the medium. Although the technology for cloning in *B. subtilis* is not nearly as well developed as that for *E. coli*, plasmids and phages suitable for cloning have been developed and transformation is a well-developed procedure in *B. subtilis*. Disadvantages of using *B. subtilis* as a cloning host exist, however. Plasmid instability remains a problem: It can be difficult to maintain plasmid replication over many culture transfers. Also, foreign DNA is not well maintained in *B. subtilis* cells and so the cloned DNA is often unexpectedly lost. Adapting a bacterium for use as a host for cloning experiments is not always simple.

Often organisms used as hosts for cloning must have specific genotypes to be effective. For instance, if the vector carries the gene for β-galactosidase (∞ Sections 10.16 and 15.1), then the host must have a mutation disabling this gene. Because M13 infects only bacteria with F pili (∞ Sections 10.8 and 16.3), hosts used with M13-derived vectors contain the F plasmid. These types of considerations, and others such as the ability to select for transformants, must be taken into account whether the host is prokaryotic or eukaryotic.

Eukaryotic Hosts

Cloning in *eukaryotic microorganisms* has some important uses, especially in understanding the details of gene regulation in eukaryotic systems. The yeast *Saccharomyces cerevisiae* is the best known genetically and is being extensively used as a cloning host. Plasmid vectors, as well as YACs (∞ Section 15.1), have been developed for yeast.

For many applications, gene cloning in *mammalian cells* is desirable. Mammalian cell culture systems can be handled in some ways like microbial cultures and find wide use in research on human genetics, cancer, infectious disease, and physiology.

One important advantage of eukaryotic cells as hosts for cloning vectors is that they already possess the complex RNA and posttranslational processing systems involved in the production of gene products in higher organisms, and so these systems do not have to be engineered into the vector as they need to be when production of the desired product is to be carried out in a prokaryote (posttranslational processing, in particular, can create some molecular cloning problems; see Section 31.5).

A disadvantage of mammalian cells as hosts is that they are expensive and difficult to produce under large-scale conditions and expression levels of cloned genes are often low. Insect cell lines are simpler to grow, and vectors have been developed from an insect DNA virus, the baculovirus (see Section 31.4).

Of course, for many applications it is important that the eukaryotic host be a plant cell or a plant. Indeed, there are a very large number of applications in plant agriculture for genetic engineering (see Section 31.7). It is necessary that the host be able to take up DNA. We discussed the methodology used for doing this with prokaryotic hosts earlier (∞ Section 10.6), and here we turn our attention to eukaryotic hosts.

Transformation (Transfection) of Eukaryotic Cells

Eukaryotic microorganisms and animal and plant cells can take up DNA in a process that resembles bacterial transformation. Because the word *transformation* in mammalian cells is used to describe the conversion of cells to the malignant (tumorous, cancerous) state (∞ Section 9.11), the introduction of DNA into mammalian cells has been called *transfection* (a term with another meaning in bacterial systems; see earlier discussion).

Transfection of cultured animal cells was originally accomplished by precipitating DNA in such a way that the cells would take it up by phagocytosis (∞ Section 22.2) because they do not have cell walls. In yeast, which is an important organism for genetic engineering, transfection at low efficiencies can be mediated by various methods of inducing artificial competence. However, as in the case of prokaryotes, *electroporation* (∞ Section 10.6)

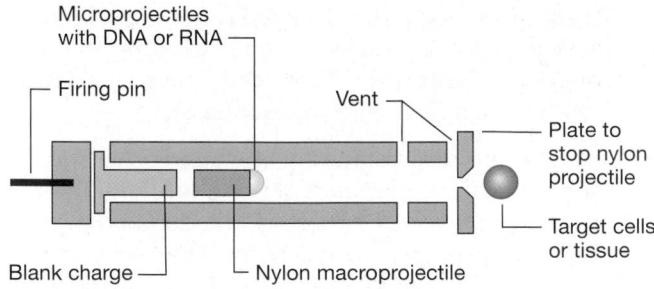

Figure 31.2 Nucleic acid gun for transfection of certain eukaryotic cells. The inner workings of the gun show how nucleic acids attached to metal pellets are projected at target cells.

is becoming widely applied to all types of eukaryotic cells and can be used whether or not the cell wall is removed.

In addition to electroporation, a high velocity microprojectile "gun" has been developed for incorporating DNA into cells. The original **particle gun** operates somewhat like a conventional shotgun. A small steel cylinder containing a gunpowder charge is used to fire nucleic acid-coated particles at the target cells (Figure 31.2). The particles bombard the cell, piercing cell walls and membranes without actually killing the cells. The nucleic acid entering the cells can then recombine with host DNA. The particle gun has been used successfully to transfect yeast, algae, a variety of plant cells, and even mitochondria and chloroplasts. The particle gun is very useful because, unlike electroporation, it can be used on intact tissue such as plant seeds. Finally, in animal cells DNA can be injected into the nucleus using micropipets, a technique called *microinjection*.

✓ 31.2 Concept Check

Molecular cloning requires a cell or organism to serve as the host for the cloned DNA. In biotechnology, the choice of a host depends on the final application. In many cases the host can be a prokaryote, but in others it is essential that the host be eukaryotic. To serve as a host it is necessary for the cell or organism to take up DNA, and there are a variety of techniques by which this can be accomplished.

✓ Why does molecular cloning require a host?
✓ Describe three mechanisms by which cells can take up DNA.

31.3 Finding the Right Clone

The practice of genetic engineering usually begins with the isolation of a clone containing a gene of interest. Later in this chapter we shall discuss a variety of manipulations that can be performed on this gene, depending on the particular application that is to be developed using it. Here we shall begin by discussing some of the ways that can be used to identify the original clone.

There are a variety of methodologies used to clone DNA; for example, one can make gene libraries from total genomic DNA (⚬⚬ Section 10.14) or one might simply clone a DNA fragment made by PCR (⚬⚬ Section 10.17). One usually clones from PCR products when the gene of interest has already been identified and the goal is simply to obtain a single clone. However, in a gene library there are thousands or tens of thousands of clones, and typically, only one or a few may be the genes of interest. We have discussed how one can select for hosts containing a plasmid vector by selecting for a vector marker, such as antibiotic resistance, so only these cells form colonies (⚬⚬ Section 10.15). For host cells containing a viral vector, one simply looks for plaques (⚬⚬ Section 10.16). We also discussed how these colonies or plaques can be screened for vectors that contain foreign DNA inserts by looking for the inactivation of a vector gene (⚬⚬ Figures 10.42 and 15.1). If one was cloning a single DNA fragment generated by PCR or purified by some other means, these simple selections or screenings should be sufficient. However, if a very heterogenous collection of DNA fragments were being used, as in the formation of a gene library, identifying cells carrying cloned DNA is only the first step. One is then left with the biggest challenge: *finding the clone* that has the gene of interest. Procedures must be available for examining colonies of bacteria or plaques of infected cells growing on agar plates and detecting those few that contain the gene of interest. It is the purpose of the present section to discuss possible approaches to finding the right clone. We consider first the situation in which the gene is *expressed* (that is, the protein is synthesized) in the cloning host. Then we discuss the situation, rather common, in which the gene is not expressed and we must look for the DNA itself.

If the Foreign Gene is Expressed in the Cloning Host

If the foreign gene is expressed (that is, the protein product is synthesized) in the cloning host, then procedures can be used that look for the presence of this protein in recombinant colonies. The cloning host itself must *not* produce the protein being studied. If we are looking for clones that express the gene, then we are looking for the rare colonies in which this protein is present. If the protein is one that the cloning host normally produces, then the host used must be defective, that is, mutant, for the gene of interest. Then, when the foreign gene is incorporated, the expression of this foreign gene can be detected by complementation (⚬⚬ Section 10.10). If the function is required, a complementing clone can be *selected*, greatly facilitating the process. Clearly, if the host already expressed a protein with the same activity, there will be a large background of this activity against which the protein produced via the foreign gene cannot be detected. If the protein is not normally produced in host

bacteria, then the host may be naturally defective. In some cases these novel activities are expressed and can be detected. A striking example is the cloning of luciferase genes into *Escherichia coli*, causing colonies containing such clones to glow in the dark (⚬⚬ Figure 8.22).

Antibody as a Method of Detecting the Protein

If the protein does not have a readily detectable function, then a different approach is needed. It involves the use of an antibody as a reagent that is specific for the protein of interest. We discussed antibodies and immunology in Chapter 22. Remember that an antibody is a serum protein produced by a mammalian system that combines in a highly specific way with another protein, the *antigen* (⚬⚬ Section 22.4). In the present case, the protein product of the cloned gene is the antigen, and this protein is used to produce an antibody in an experimental animal. Since the antibody combines specifically with the antigen, when the antigen is present in one or more colonies on the plate, then the locations of these colonies can be determined by observing the binding of the antibody. Because only a small amount of the protein (antigen) is present in the colonies, only a small amount of antibody is bound, and so a highly sensitive procedure for detecting bound antibody must be available. In practice, this is done using a system involving a radioactive agent, an agent that emits light, or an agent with a specific enzyme attached to it. The radioactivity or light can be detected by autoradiography using X-ray film. The enzymes used typically convert a colorless substrate into a colored one whose absorbance can be measured very sensitively. These and other extremely sensitive techniques for detecting antigens were previously discussed (⚬⚬ Sections 24.7–24.12).

Note that this method of detection involves *screening*, not selection, and so thousands of clones must be examined. These can be colonies containing plasmids or plaques containing viruses that produce the cloned product. The whole procedure, utilizing plasmids and radioactive detection, is outlined in Figure 31.3a. As seen, the replica plating procedure (⚬⚬ Figure 10.2) is used to make a duplicate of the master plate, but the duplication is done onto a membrane filter and all the manipulations are done with this filter. After the duplicate colonies have grown up, they are lysed to release the protein (antigen) of interest. (Screening for expression in phage vectors eliminates this step because the bacteria are already lysed.) The antibody is then added, and the antibody–antigen reaction allowed to proceed. Unbound antibody is then washed off, and a radioactive agent is then added that is specific for the antibody. A piece of X-ray film is placed over the filter and exposed. If a radioactive colony is present, a spot on the X-ray film will be observed after it is developed. The location of

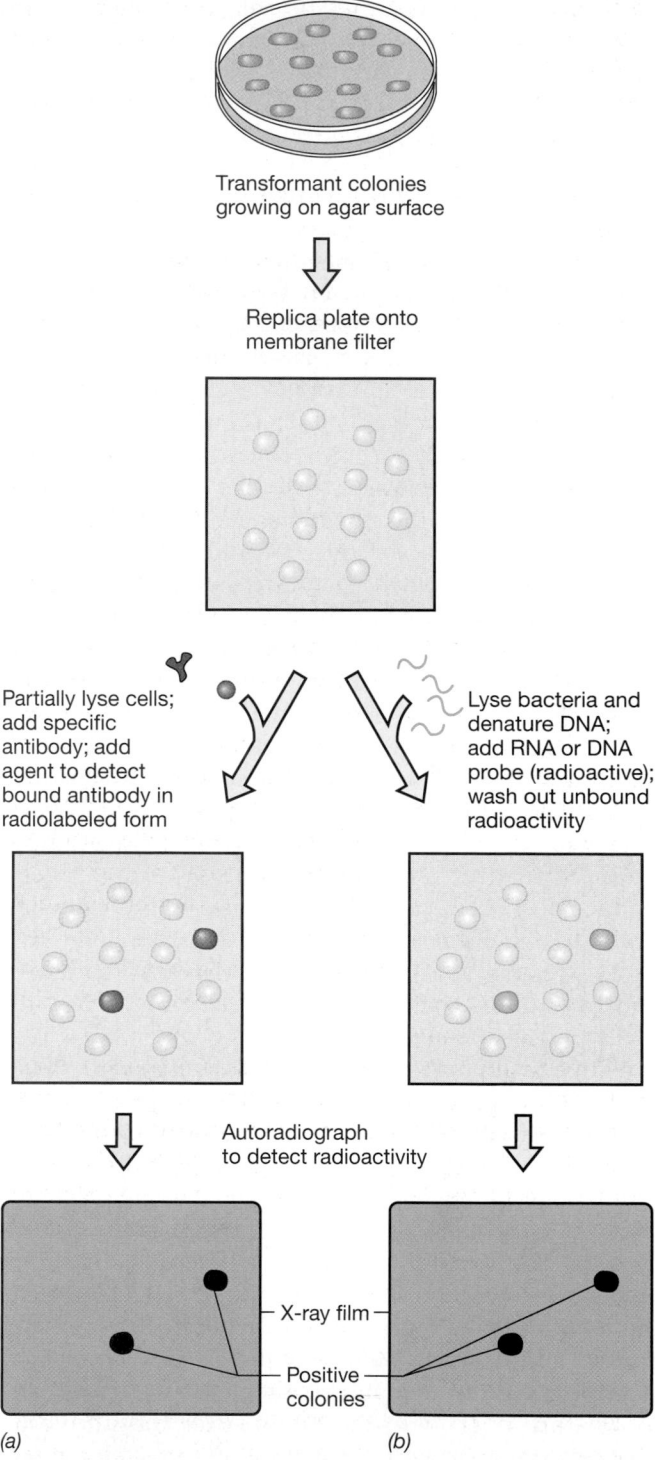

Transformant colonies
growing on agar surface

Replica plate onto
membrane filter

Partially lyse cells;
add specific
antibody; add
agent to detect
bound antibody in
radiolabeled form

Lyse bacteria and
denature DNA;
add RNA or DNA
probe (radioactive);
wash out unbound
radioactivity

Autoradiograph
to detect radioactivity

X-ray film

Positive
colonies

(a)   (b)

Figure 31.3 Finding the right clone. (a) Method for detecting production of protein by use of specific antibody. (b) Method for detection of recombinant clones by colony hybridization with a radioactive nucleic acid probe. Although both parts of the figure show detection involving radioactivity, many other types of nonradioactive detection systems are now being employed.

this spot on the film corresponds to a location on the master plate where a colony is present that produces the protein. This colony can then be picked from the master plate and cultured.

One limitation of this procedure is that an antibody must be available that is *specific* for the protein in question. As we saw in Chapter 22, antibody can be readily produced by injecting the protein (antigen) into an animal, but the protein injected must be pure; otherwise more than one antibody will be formed. Thus, one must have previously purified the protein.

Nucleic Acid Probes: Searching for the Gene Itself

Suppose that the gene is not expressed in the cloning host or that no assay or antibody is available for the gene product. How does one detect its presence in colonies? The most general way is to use a **nucleic acid probe** containing a key part of the base sequence of the gene of interest. As we have discussed (∞ Section 10.12), nucleic acid hybridization can be used as a specific means of detecting polynucleotides with specific sequences. Either DNA or RNA can be employed as a probe. The general procedure is to label the nucleic acid probe, often with radioactive phosphate, but nonisotopic techniques are increasingly being used and allow a single-stranded probe to hybridize with single-stranded nucleic acid derived from the cloned DNA. Because of specific complementary base pairing, two single-stranded polynucleotides will hybridize only if they are fairly complementary. By using appropriate hybridization conditions, it is possible to obtain binding of the radioactive probe only to the nucleic acid of interest.

The way in which a nucleic acid probe can be used to detect the presence of recombinant DNA in colonies is shown in Figure 31.3*b*. The procedure, **colony hybridization**, again makes use of replica plating to produce a duplicate of the master plate on a membrane filter. (The same procedure can be carried out with virus vectors by blotting the plaques onto a membrane.) The cells on the filter are lysed in place to release their nucleic acid and to convert the DNA into a single-stranded form and fix it to the filter. This filter is then treated with a radioactive nucleic acid probe (either RNA or DNA) to allow hybridization, and after removal of unbound radioactive nucleic acid, the filter is subjected to autoradiography. After development, the X-ray film is examined for spots. These correspond to locations on the membrane where the radioactive probe hybridized the DNA from a particular colony. Colonies corresponding to these spots are then picked and studied further. A modification of this procedure, avoiding the use of a radioactive probe, has been developed for clinical microbiology (∞ Section 24.13).

Special procedures are needed for detecting the foreign gene in the cloning host. If the gene is expressed, the presence of the foreign protein itself, as detected either by its activity or by reaction with specific antibodies, is evidence that the gene is present. However, if the gene is not expressed, then its presence can be detected by use of a nucleic acid probe.

✓ Does use of nucleic acid probes depend on gene expression? Explain.

✓ Why is it necessary to lyse cells containing plasmids in order to detect the product of the cloned gene?

31.4 Specialized Vectors

If the purpose of cloning a gene is to achieve a high level of its expression in a suitable host, then it is insufficient to simply locate a cloned copy of the gene. Just as specialized vectors have been developed for working with large fragments of DNA, such as bacterial artificial chromosomes and yeast artificial chromosomes (BACs and YACs, ∞ Section 15.1), so specialized vectors have been developed for use as specific tools in biotechnology.

For example, there are *shuttle vectors* that allow cloned DNA to be moved between unrelated organisms. A **shuttle vector** is a cloning vector that can stably replicate in two different organisms. Shuttle vectors have been developed that replicate in both *Escherichia coli* and *Bacillus subtilis*, *E. coli* and yeast, and *E. coli* and mammalian cells, as well as in many other pairs of organisms.

Even more important than shuttle vectors for many purposes are cloning vectors that facilitate the *expression* of cloned DNA. Organisms have complex regulatory systems (∞ Chapter 8), and one would expect that many cloned genes will not be expressed, or efficiently expressed, in a foreign host. This obstacle can be overcome by the use of *expression vectors*. An **expression vector** is a vector that can be used not only to clone the desired gene, but also contains the necessary regulatory sequences so that expression of the gene can be subjected to experimental manipulation. Many factors are very important in regulating gene expression, but one of the most fundamental is regulation of transcription.

Regulation of Transcription from Expression Vectors

One of the most important elements in the expression vector is a system that allows transcription of the cloned gene. Typically, it is also important that transcription be very tightly controlled. For very high levels of expression, it is essential to produce high levels of mRNA. The promoter region is the site at which binding of RNA polymerase first occurs (∞ Section 7.9). For *Bacteria*, the DNA region around 10 and 35 nucleotides before the start of transcription (called the −10 and −35 regions)

(∞ Figure 7.27) is especially important in the promoter. A cloned gene's native promoter may work very poorly or not at all in the new host. Promoters from eukaryotes and some other prokaryotes function poorly or not at all in *Escherichia coli*. Even some *E. coli* promoters function at low levels in *E. coli* because their sequences are not close to the consensus sequence (∞ Section 7.9). For this reason, the expression vector must contain a promoter that will function efficiently in the host and one that is correctly positioned so that it can permit the transcription of the cloned gene. Promoters from *E. coli* that have been used in the construction of expression vectors include *lac* (the *lac* operon promoter), *trp* (the *trp* operon promoter), *tac* and *trc* (synthetic hybrids of the *trp* and *lac* promoters), and lambda P_L (the leftward lambda promoter; ∞ Section 9.10). Note that each of these promoters can be regulated (∞ Sections 8.5, 8.8, and 9.10).

In almost all cases it is important to be able to regulate the expression of the cloned gene. That is, although one typically wants to produce very high levels of mRNA (and have it translated), it is usually undesirable to design a vector that permits the gene to be transcribed to high levels at all times. Indeed, some proteins that are of commercial value are toxic to the host, and in the early stages of growth of the culture it may be important that the gene not be transcribed at all. The ideal situation is to be able to grow the culture containing the expression vector until a large population of cells is obtained, each containing a large copy number of the vector, and then turn on expression in all copies simultaneously by manipulation of a regulatory switch. Therefore, transcription needs to be tightly regulated.

We discussed regulatory controls of gene expression in Chapter 8. Recall the major importance of the repressor-operator system in regulating gene transcription (∞ Section 8.5). A strong repressor can completely block the synthesis of the proteins under its control by binding to the operator region. Repressor function can be turned off at the chosen time by adding an inducer, allowing transcription of the genes controlled by the operator.

For the repressor-operator system to work as a regulatory switch for the production of a foreign protein, the expression vector must contain the operator controlled by the repressor to which the cloned gene is fused. This permits proper arrangement of the sequence of genetic elements: promoter-operator-ribosome binding site-structural gene, so efficient transcription and translation can occur. In most cases, the operator and promoter correspond to each other (for instance, the *lac* operator is used with the *lac* promoter), but this is not always the case. A vector could easily be constructed to contain a *trp* promoter under the control of a *lac* operator.

For vectors using the *lac* operator, the promoter is switched on by inducers such as lactose or related β-galactosides (◯ Section 8.5). Phasing of cell growth and protein synthesis can thus be achieved by allowing growth to proceed in the absence of inducer until a suitable cell density is achieved, and then adding inducer to bring about synthesis of the desired proteins. If the vector is present in the cell at a high copy number (◯ Section 10.8), the normal level of repressor encoded by the chromosomal gene may be inadequate to keep the promoter from functioning. Figure 31.4 shows an expression vector that uses the *trc* promoter under the control of the *lac* operator (*lacO*). This plasmid also contains a copy of the *lacI* gene that encodes the *lac* repressor. The level of repressor in a cell containing this plasmid is sufficient to prevent transcription from the *trc* promoter until inducer is added.

Transcription Controls from Bacteriophage

Vectors using bacteriophage lambda promoter P_L (and the corresponding operator, O_L) are controlled by having the lambda repressor protein in the cell (◯ Section 9.10). Typically the lambda repressor is encoded by a mutant

gene (carried by the vector or by a prophage in the host) and is temperature-sensitive. By raising the temperature of the culture to the proper value (usually 8–10°C higher than the growth temperature), the lambda repressor is inactivated and transcription from P_L begins.

In some cases the control system used may not be a normal part of the host at all. An excellent example of this is the use of the bacteriophage T7 promoter and RNA polymerase as a regulatory system in an expression vector. When T7 infects *Escherichia coli*, it codes for its own RNA polymerase, which recognizes only T7 promoters, thus effectively shutting down host transcription (◯ Section 16.4). In expression vectors it is possible to place expression of cloned genes under control of a T7 promoter. However, when this is done, it is necessary to engineer into the plasmid the gene for T7 RNA polymerase as well. The latter is placed under control of an easily regulated promoter such as that of lambda or *lac*. Expression of the cloned gene(s) occurs shortly after T7 RNA polymerase transcription has been switched on. Because it recognizes only T7 promoters, T7 RNA polymerase transcribes only the cloned genes; all other host genes remain untranscribed. Because this system is so powerful and specific, induction of the T7 promoter/RNA polymerase system will cause the host to stop growing.

In addition to a strong and regulable promoter, most expression vectors contain an effective transcription terminator. This prevents transcription of the entire vector, which could interfere with vector stability. The expression vector shown in Figure 31.4 has strong transcription terminators to terminate transcription downstream from the cloned gene. In addition, the secondary structure of some transcription terminators seems to increase message stability (◯ Section 7.10). It is important to not only make large quantities of mRNA from the cloned gene, but also to attempt to ensure that this mRNA be as stable as possible.

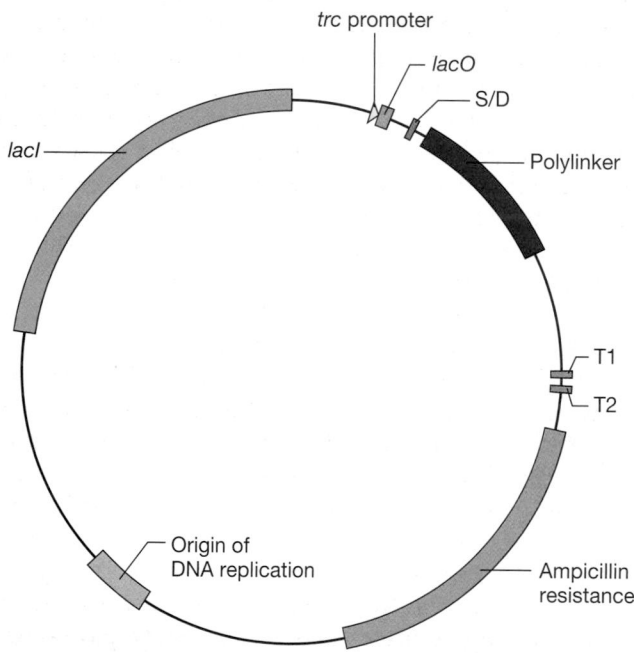

Figure 31.4 A partial genetic map of the expression vector pSE420. This vector is sold by Invitrogen Corp., a genetic engineering company. The polylinker is a site containing many different restriction enzyme recognition sequences to facilitate cloning (◯ Section 10.15). This region (and the cloned gene) would be transcribed by the *trc* promoter, which is immediately upstream of the *lac* operator (*lacO*). Immediately upstream of the polylinker is a sequence that encodes a Shine–Dalgarno site on the resulting mRNA (◯ Section 7.15). Downstream of the polylinker are two transcription terminators (T1 and T2). The plasmid also contains the *lacI* gene, which encodes the *lac* repressor, and a gene conferring resistance to the antibiotic ampicillin. These two genes are under the control of their own promoters, which are not shown.

Translation of the Cloned Gene

Expression vectors must also be designed to ensure that the mRNA produced can be efficiently translated. To synthesize protein from an mRNA, it is essential that the ribosomes bind at the correct site and begin reading in the correct frame. In prokaryotes this is accomplished by having a ribosome binding site (Shine–Dalgarno sequence) (◯ Section 7.15) and a nearby start codon on the mRNA. Bacterial ribosome binding sites are not found in eukaryotic genes, and it is thus essential that the bacterial region be present in the vector if high levels of gene expression are to be obtained. Once again, the vector shown in Figure 31.4 has such a site. Part of the requirement for proper ribosome binding is the necessity for a proper distance between the ribosome binding site and the translation initiation codon. If these sites are too close or too far apart, the gene will be translated at low efficiency. In some cases, even the initiation codon for the gene to be cloned is part of the expression vector. Interestingly, translation is sometimes more efficient if the cloned gene is the second gene in a bicistronic mRNA, a phenomenon called *translational coupling*. The upstream gene often encodes a small peptide, which is rapidly degraded.

Often, adjustments have to be made to ensure high efficiency translation *after* the gene has been cloned. For instance, it is important that the Shine–Dalgarno site mentioned above not be involved in a region of secondary structure, and this cannot always be predicted until the sequence of the cloned gene is known. There are also sometimes difficulties having to do with *codon usage*.

There is more than one codon for most of the 20 amino acids (◌◌ Table 7.3), and some codons are used more frequently than others: This preference may differ considerably between different organisms. Codon usage is partly a function of the concentration of the appropriate tRNA in the cell. Therefore, a gene whose codon usage pattern is considerably different from that of its new host may be translated inefficiently. Insertion of the appropriate codons would be difficult because it would have to be changed at all locations in the gene. However, this can be done if necessary by using synthetic DNA and site-directed mutagenesis (◌◌ Sections 10.13 and 10.18) to create a gene more amenable to the codon usage patterns of the host.

Finally, if the cloned gene contains introns (◌◌ Section 7.12), the correct protein product will not be made if the host is a prokaryote. This problem can also be corrected by site-specific mutagenesis or synthetic DNA, but there are other methods for creating an intron-free gene (see Section 31.5).

Eukaryotic Vectors

All of the previous discussion deals with vectors that replicate in *Bacteria*. However, it is often very desirable to clone and express genes in eukaryotes. A similar wide range of cloning vectors are available.

We briefly discussed the use of YAC cloning vectors in the yeast *Saccharomyces cerevisiae* to clone very large fragments of DNA (◌◌ Section 15.1). Many other vectors are also available for cloning into yeast. Yeast is one of the few eukaryotes that has a plasmid, called the *two-micron circle* (2-μ circle), because of its size, and many yeast vectors are based on this plasmid. Although yeast is an extremely useful organism both for genetic studies and commercial applications, it is often important to use other eukaryotes as hosts for cloned DNA. Cloning vectors have been developed for many different eukaryotes, including plants (see Section 31.7).

Most vectors used in the higher eukaryotes are virus vectors. The DNA virus SV40 (◌◌ Section 16.10), a virus causing tumors in primates, has been developed as a cloning vector for human tissue culture lines. Derivatives of SV40 that do not induce tumors have been developed for cloning mammalian genes and also for the expression of these genes. SV40 and other mammalian cloning vectors are proving very useful for understanding the mechanisms involved in gene expression in these complex organisms.

There are mammalian vectors that utilize *adenovirus* (◌◌ Section 16.13) and *vaccinia virus* (◌◌ Section 16.12). Vaccinia virus vectors have been used in the development of new vaccines (see Section 31.6). A variety of eukaryotic expression vectors have also been developed and are essentially of two kinds. One type is designed to produce a particular protein for commercial purposes. Vectors derived from *baculovirus*, a DNA virus that replicates in insect cells, can be used to make large quantities of the products of cloned genes.

Other expression vectors have been developed so a cloned gene can be stably maintained and expressed in an organism or tissue. These **integrating vectors** are obviously maintained in very low copy number (typically one copy per genome) and have been developed in eukaryotes ranging from yeast to mammalian cells, as well as in prokaryotes. Integrating vectors have uses in basic science and in applications such as gene therapy (see Section 31.8). The *retroviruses* (◌◌ Sections 9.12 and 16.14) can be used to introduce genes into mammalian cells because these viruses replicate through a DNA form that becomes integrated into the host chromosome.

Reporter Genes

Just as there are specialized vectors for certain applications, there are also specialized genes or portions of genes that are useful in genetic engineering. We have already seen how certain promoters and operators can be used in the construction of expression vectors. We also discussed the usefulness of having a cloning site within a gene whose activity was very easy to assay. The system we described in most detail was one in which the gene encoding β-galactosidase was used, *lacZ* (◌◌ Figure 15.1). The activity of the enzyme β-galactosidase can be detected easily on indicator plates or by simple assays. Therefore, *lacZ* can be used as a *reporter gene*.

Reporter genes are genes incorporated into vectors because they encode proteins that are simple to detect. They can be used to signal the presence or absence of a particular genetic element or its location. They can also be fused to other genes or to the promoter of other genes so that expression can be studied. There are many choices for reporter genes besides *lacZ*. The luciferase enzyme, which we discussed earlier, makes cells expressing it luminescent (◌◌ Figure 8.22). One can then detect colonies containing this reporter system on agar plates by their luminescence against a large background of other colonies. Production and detection of luciferase usually

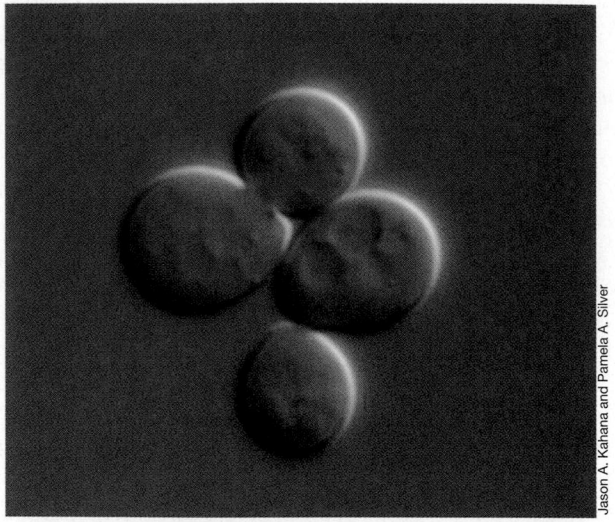

(a)

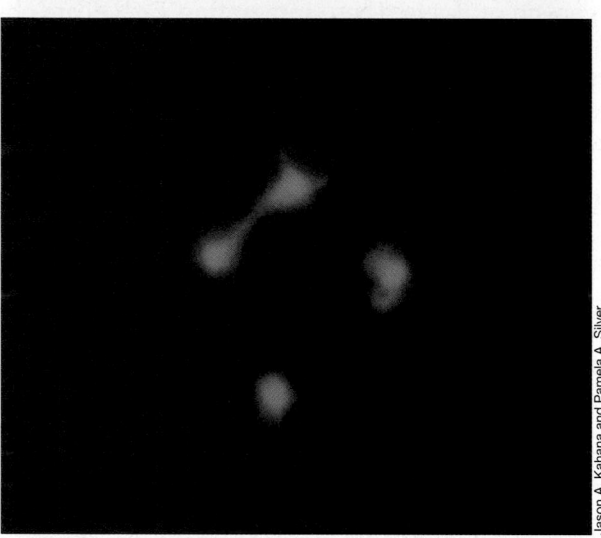

(b)

(c)

involves more than one gene and accessory factors. Recently a gene encoding a fluorescent protein, called the *green fluorescent protein* (GFP), has been isolated from the jellyfish *Aequorea victoria* and cloned. GFP needs no accessory factors and has been used as a reporter or tag in a wide variety of organisms (Figure 31.5).

✓ 31.4 Concept Check

Many cloned genes will not normally be expressed efficiently in a new host. Expression vectors have been developed that contain a variety of genes or portions of genes that will increase the level of transcription of the cloned gene and make its transcription subject to specific regulation. Signals to improve the efficiency of translation may also be present in the expression vector. Expression vectors and other specialized vectors have been developed for both prokaryotic and eukaryotic hosts.

✓ Describe some of the elements on an expression vector that will improve transcription of the cloned gene.

✓ Reporter genes are often included as a part of a cloning vectors. What is a reporter gene?

31.5 Expression of Mammalian Genes in Bacteria

In the past, one of the greatest challenges facing the genetic engineer who wished to clone and express a particular mammalian gene was simply to find the correct clone. We have discussed some of the methods that can be used to do this (see Section 31.3). However, the sequencing and annotation of the human genome, as part of the Human Genome Project, is well advanced. There will still be problems of determining the functions of genes, but it is now possible to use sequence information to find clones of specific interest in gene libraries. Other mammalian genomes, such as that of the mouse, are also being sequenced. Even if the genome of the organism of interest has not been sequenced, one can use sequence information on related genes to design DNA primers and clone the gene using PCR (⊂⊃ Section 10.17).

However, there are still obstacles to be faced, even if one had a cloned mammalian gene in an expression vector. One of the foremost obstacles is the presence of in-

Figure 31.5 The green fluorescent protein (GFP) can be used as a tag for protein localization *in vivo*. The gene encoding Pho2, a DNA binding protein from the yeast *Saccharomyces cerevisiae*, was fused to the gene encoding GFP. The recombinant gene was transformed into budding yeast cells, which could then express the fluorescent fusion protein that has localized to the nucleus. (a) Cells expressing Pho2-GFP were observed by differential interference contrast microscopy and (b) by epifluorescence microscopy (⊂⊃ Sections 4.1 and 4.2). (c) This panel is an overlay of the images in a and b.

trons (∞ Section 7.12). An intron in a protein-encoding mammalian gene simply cannot be removed by a prokaryotic host. Since almost all mammalian genes contain introns they will be nonfunctional in a prokaryotic host. Many mammalian genes have 50 or more introns, and these may contain tens or even hundreds of thousands of base pairs, so powerful and efficient methods are needed to remove them. We discuss two such methods here. In both cases the introns are not removed from the cloned genes, but instead are removed before cloning.

Reaching the Gene via Messenger RNA

One approach to isolating a functional gene is to get to it through its mRNA. A major advantage of using mRNA is that the noncoding information present in the DNA (introns) has been removed (∞ Section 7.12). The isolated mRNA is used to make complementary DNA (cDNA) by means of reverse transcription (∞ Sections 9.12 and 16.14). It is likely that a tissue expressing the gene contains large amounts of the desired mRNA, although except in rare cases this certainly is not the only mRNA, produced. In a fortunate situation, where a single mRNA dominates a tissue type, extraction of mRNA from that tissue provides a useful starting point for gene cloning.

In a typical mammalian cell, about 80–85% of the RNA is ribosomal, 10–15% is transfer RNA and other low-molecular-weight RNAs, and 1–5% is messenger RNA. Although low in abundance, the mRNA in a eukaryote is identifiable because of the poly-A tails found at the 3'-end (∞ Section 7.12). By passing a poly-A-rich RNA extract over a chromatographic column containing poly-T fragments (linked to a cellulose support), most of the mRNA of the cell can be separated from the other cellular RNA by the specific pairing of A and T bases. Elution of the RNA from the column then gives a preparation greatly enriched in mRNA.

Once the RNA message has been isolated, it is necessary to convert the information to DNA. This is accomplished by use of the enzyme *reverse transcriptase*, (∞ Sections 9.12 and 16.14). This remarkable enzyme, an essential component of retrovirus replication, copies information from RNA into DNA (Figure 31.6). As we noted, this enzyme requires a primer in order for it to begin working (in retrovirus infection the primer is a tRNA). In the present procedure, an oligo-dT primer is used that is complementary to the poly-A tail of the isolated mRNA. The oligo-dT primer is hybridized with the mRNA, and then reverse transcriptase is allowed to act (Figure 31.6). As seen, the newly synthesized DNA copy has a hairpin loop at its end, which is synthesized because after the enzyme completes copying the mRNA it starts to copy the newly synthesized DNA. This hairpin loop, which is probably an artifact of the test tube reaction, provides a convenient primer for synthesis of

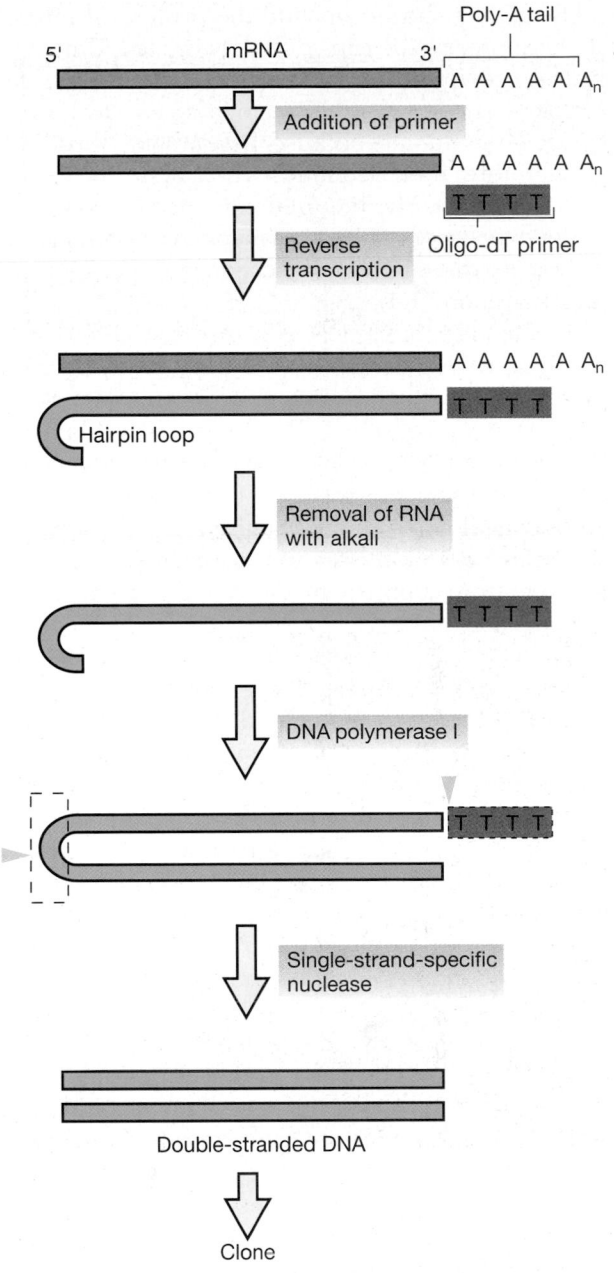

Figure 31.6 Steps in the synthesis of complementary DNA (cDNA) from an isolated mRNA using the retroviral enzyme reverse transcriptase.

the second DNA strand. The resultant double-stranded DNA, with the hairpin loop intact, is then cleaved by a single-strand-specific nuclease to produce the desired double-stranded DNA, one strand of which is complementary to the mRNA. This double-stranded DNA (the gene of interest) can then be inserted into a plasmid or other vector for cloning. The detection of specific clones makes use of the procedures discussed in Section 31.3. As mentioned in Section 10.17, one can also use RT-PCR (reverse transcriptase-PCR) to synthesize large amounts of cDNA without having to clone it.

The cDNA should encode the protein of interest and, therefore, can be considered its "gene." Unlike the "natural" gene on the mammalian chromosome, this one contains no introns. Although there is a start codon, there are no promoters because these are not transcribed and, therefore, their sequence won't be in the mRNA (⚭ Section 7.8). The requirements for achieving high levels of expression with genes made in this manner are simply those discussed in the section on expression vectors (see Section 31.4).

One can also make *cDNA libraries* from different tissues when seeking genes whose expression is specific to those tissues. This can be extremely useful because the actual gene (chromosomal DNA) is found in almost all cells, but the mRNA is found only in cells actively producing the protein. For instance, the gene encoding the hormone insulin is found throughout the body, but insulin mRNA is found only in certain cells in the pancreas. Therefore, a library made from pancreatic cells is enriched for cDNA corresponding to the insulin gene.

Reaching the Gene via the Protein

Recall at the beginning of this section that we mentioned that knowledge of the sequence of a gene can facilitate detecting it in a gene library or in cloning it using PCR. This is made possible by the fact that sequence information can be used to make a synthetic DNA molecule to use as a probe (⚭ Section 10.13). This method can also be used to design a probe to find the mRNA or cDNA made from it. However, this information might be used differently to actually *construct* a gene.

In Chapter 15 we discussed how analysis of DNA sequences, specifically the search for open reading frames (ORFs), can be used to detect possible genes (⚭ Section 15.2). As we discussed, however, the fact that an ORF exists does not mean that it is functional, i.e., that it actually encodes a protein. One means of proof that an ORF is a functional gene is finding a protein whose amino acid sequence is predicted by the nucleotide sequence of the ORF.

All cells process genetic information according to the central dogma (DNA → RNA → protein; ⚭ Section 7.1). However, it is possible for humans to reverse this information flow. That is, a genetic engineer could use the amino acid sequence of a protein to design and synthesize a nucleotide sequence that could encode it. This process is called *reverse translation*. The procedure of reverse translation is illustrated in Figure 31.7. From the genetic code, the nucleotide sequence of a section of the DNA is deduced, and this piece of DNA is synthesized. Unfortunately, degeneracy of the genetic code (⚭ Section 7.13) somewhat complicates the problem. Most amino acids are encoded by more than one codon, and codon usage varies from organism to organism. The best section of DNA to synthesize is one that corresponds to a part of the protein rich in

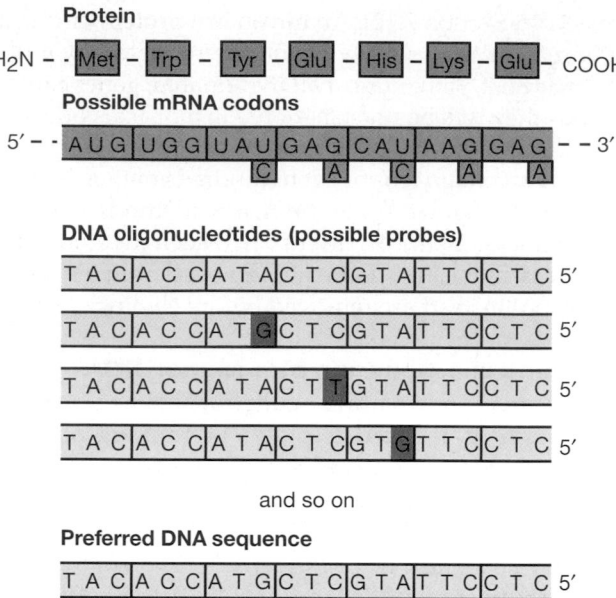

Figure 31.7 Reverse translation: deducing the best sequence of an oligonucleotide probe from the amino acid sequence of the protein. Because of degeneracy, many probes are possible. If codon usage by the same organism is known, then a preferred sequence can be selected. It is not essential that complete accuracy be achieved because a small amount of mismatch can be tolerated.

amino acids specified by only a single codon (methionine, AUG; tryptophan, UGG) or by two codons (for example, phenylalanine, UUU, UUC; tyrosine, UAU, UAC; histidine, CAU, CAC) because this increases the chances that the synthesized DNA will be complementary or nearly complementary to the mRNA of interest. If the complete amino acid sequence of the protein is not known, then the sequence used is generally one at the amino terminus of the protein because it is at the amino terminus that sequencing of the protein begins.

For making a probe to identify the actual gene, its mRNA, or cDNA made from it, it is important that the synthetic DNA be as exactly complementary as possible. However, it is unnecessary to have the complete sequence of the protein encoded by the gene. By contrast, this information is required if you wish to construct an entire *synthetic* gene. In constructing a gene it is also important to not use just any of the degenerate codons for a particular amino acid, but rather to choose codons preferred by the organism that will *express* the cloned gene. Why would one choose to synthesize an entire gene?

Many mammalian proteins (including peptide hormones) are the products of posttranslational processing (⚭ box, Protein Processing, Chapter 8) and might, therefore, be quite small. If one was interested in producing only a peptide hormone, it would be more efficient to construct a gene that encoded just the final hormone and not the entire protein from which it was derived. It could also be that even for a larger protein, changing codon usage for efficient translation in the specified host might be economically justified. In addition, chemical synthesis not only permits the acquisition of genes that cannot be obtained otherwise but also allows synthesis of *modified genes* that may make new proteins of utility. Techniques for the synthesis of DNA molecules are now well developed, and it is possible to synthesize genes coding for proteins 100–200 amino acid residues in length (300–600 nucleotides). The synthetic approach was used for production of the human hormone insulin in bacteria, as will be discussed in Section 31.6.

With the use of all these techniques, a large number of different human proteins have been expressed at high yield under the control of bacterial regulatory systems, including human growth hormone, insulin, virus antigens, interferon, and somatostatin (see Section 31.6).

Protein Folding and Stability

Unfortunately, sometimes just being able to synthesize a protein in a new host is insufficient. Some proteins are susceptible to degradation by intracellular proteases and may be destroyed before they can be isolated. Some eukaryotic proteins are toxic to the prokaryotic host, and the host for the cloning vector may be killed before a sufficient amount of the product is synthesized. Further engineering of either the host or the vector may be necessary to eliminate these problems.

Sometimes when foreign proteins are massively overproduced they form inclusion bodies inside the host. Although inclusion bodies are relatively easy to purify because of their size, the protein found in these bodies can be very difficult to solubilize. It seems that in many cases these bodies form because the protein is incorrect-

ly folded. One potential solution to this problem is to use a host that overproduces molecular chaperones that aid in folding (∞ Section 7.16). Interestingly, this problem, and some others, can sometimes be solved if the protein from the cloned gene be made as a fusion product with a protein encoded by the vector. This not only stabilizes the protein but might simplify purification if the portion encoded by the vector is a protein for which rapid, simple, inexpensive purification techniques are known. Several special fusion vectors are now available. The "cloned protein" is released from the fusion protein after purification by special proteases. Figure 31.8 shows an example of a *fusion vector* that is also an expression vector.

In some cases the desired protein can also be removed from the fusion protein by chemical means. Fusion systems can also be used for purposes other than achieving increased protein stability. One advantage of making a fusion protein is that the bacterial portion can contain the bacterial sequence coding for the *signal peptide* that enables transport of the protein across the cytoplasmic membrane (∞ Section 7.16), making possible the development of a bacterial system that not only synthesizes the mammalian protein but also actually excretes it.

Even with the best-designed vector, some genes are poorly expressed in a particular cell. In some cases these problems can be rectified by using a mutant host. For instance, some "foreign mRNAs" are degraded very rapidly in wild-type *Escherichia coli* but not in particular mutant strains. Using such expression systems, one can typically produce very high levels of mammalian proteins in *E. coli*. In many cases the desired protein exceeds 200,000 molecules per cell and can make up as much as 40% of the protein molecules in a cell.

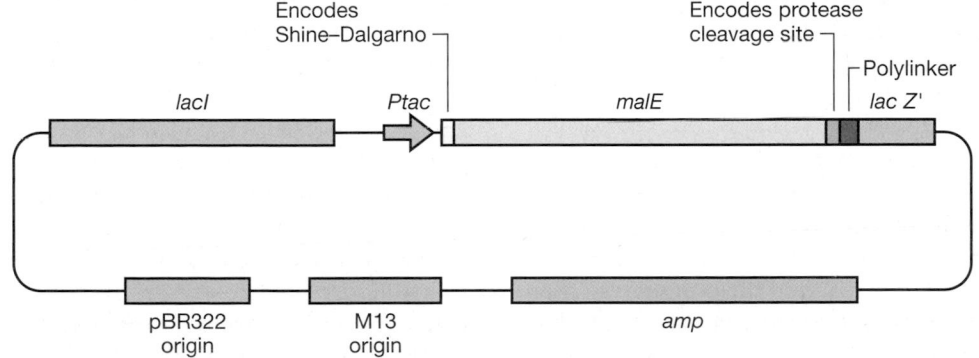

Figure 31.8 An expression vector for fusions. This vector was developed by the New England Biolabs Company. The gene to be cloned is inserted at the polylinker site (∞ Figure 15.1) so it is in frame with the *malE* gene, which encodes the maltose binding protein. This insertion inactivates the *lacZ'* gene (∞ Figure 15.1). The fused gene is under control of the hybrid *tac* promoter (*Ptac*). The plasmid also contains the *lacI* gene, which encodes the *lac* repressor. Therefore, an inducer must be added to the cells in order to turn on the *tac* promoter. The fusion protein is easily purified by methods involving the affinity of the protein for maltose. Once purified, the two portions of the fusion protein can be separated by a very specific protease (factor Xa). The plasmid contains a gene conferring ampicillin resistance on its host. In addition to the plasmid origin of replication, there is a bacteriophage M13 origin. Therefore, this is a *phagemid* and can be propagated either as a plasmid or as a phage.

✓ **31.5** *Concept Check*

It is possible to achieve very high levels of expression of mammalian genes in *Bacteria*. However, these genes are almost invariably different from the native gene, as it is essential that it contain no introns. This can be accomplished by using reverse transcriptase to synthesize complementary DNA (cDNA) from messenger RNA (mRNA). It can also be accomplished by synthesizing an entirely synthetic gene using the knowledge of the amino acid sequence of the protein one wishes it to encode.

✓ Why doesn't mRNA contain introns?

✓ Of what use can it be if the protein of interest is made as part of a fusion protein?

II PRACTICAL APPLICATIONS OF GENETIC ENGINEERING

In this unit we will discuss only a few of the many exciting applications of genetic engineering to biotechnology. Many other areas will not be covered here at all. These include the use of genetic engineering to enhance microbial fermentations, particularly in the production of antibiotics (∞ Section 30.6). We will also not discuss the use of genetically engineered organisms in environmental biotechnology. It should be remembered that the metabolic diversity of the prokaryotes is enormous (∞ Chapter 17), and many of these metabolic pathways can be combined in new and useful ways. Thus, in the next three sections we shall glimpse not only many exciting possibilities, but actual existing applications that are of great importance in agriculture and medicine.

31.6 Production of Mammalian Products and Vaccines by Genetically Engineered Microorganisms

One of the first practical applications of genetic engineering was the use of easily grown bacteria to produce useful proteins whose genes were from organisms that are more difficult or expensive to grow. Although the special DNA polymerases used in the polymerase chain reaction were originally isolated from thermophilic bacteria (∞ Section 10.17), they are now produced in *Escherichia coli* from cloned genes. Most restriction enzymes are also produced in *E. coli* from cloned genes. Note that *E. coli* is not necessarily easier to grow than some of the bacteria that normally synthesize a particular restriction enzyme. However, cloning allows manipulation of expression levels (see Section 31.4) and always using the same host is more efficient, since a company can employ similar or identical culture conditions when making different products. Similarly, many proteins used industrially are now produced from cloned genes, and in some cases, the protein itself has been altered by using site-specific mutagenesis (∞ Section 10.18) to change the cloned gene.

Many proteins and peptides from mammalian cells have high pharmaceutical value. However, these proteins are usually present in very small amounts in normal tissue, and it is therefore extremely costly to purify them. In addition, even if the protein can be produced in cell cultures (∞ Section 9.3), this is a much more expensive technique than growing microbial cultures. Therefore, another early effort of the biotechnology industry was to use genetic engineering to produce these proteins in microorganisms.

For many early applications, such as the production of insulin, it was known that the product would have great commercial value because of its established therapeutic value in treating a reasonably well-understood disease (in this case, diabetes). However, such success is not always guaranteed, even when the protein can be produced and purified. Often this is because the disease process is complex and not well understood or the product has unexpected side effects. Nonetheless, by the late 1990s, biotechnology companies had hundreds of products in clinical trials. Several different classes of therapeutic protein products have been produced. These include hormones, interferons, growth factors, and vaccines. A few examples of these and other products are shown in Table 31.1.

Production of Insulin

One of the earliest and most dramatic commercial successes was the production of the hormone **insulin**. Many hormones are peptides or small proteins. These molecules are extremely important in controlling mammalian metabolism and have important therapeutic uses. Insulin is a protein produced in the pancreas that is vital for the regulation of carbohydrate metabolism in the body. Diabetes, a disease characterized by insulin deficiency, afflicts millions of people. The standard treatment for diabetes is periodic injections or oral administration of insulin. Because insulins of most mammals are similar in structure, it is possible to treat human diabetes by use of insulin isolated commercially from beef or pork pancreas. However, nonhuman insulin is not as effective as *human insulin*, and the isolation process is expensive and complex. Cloning of a human insulin "gene" in bacteria has hence been carried out.

Producing hormones such as insulin in genetically engineered microorganisms is not simply a matter of cloning a gene (or even a cDNA) (see Section 31.5) in an expression vector. This is because many of these hormones are only small fragments of the polypeptides encoded by the gene. Insulin in its active form consists of

TABLE 31.1	A few therapeutic products made by genetic engineering
Product	**Function**
Blood proteins	
Erythropoietin	Treats certain types of anemia
Factors VII, VIII, IX	Promote clotting
Tissue plasminogen activator	Dissolves clots
Urokinase	Blood clotting
Human hormones	
Epidermal growth factor	Wound healing
Follicle stimulating hormone	Treatment of reproductive disorders
Insulin	Treatment of diabetes
Nerve growth factor	Possible treatment of degenerative neurological disorders and stroke
Relaxin	Facilitates childbirth
Somatotropin (growth hormone)	Treatment of some types of growth failure and short stature
Immune modulators	
α-Interferon	Antiviral, antitumor, agent
β-Interferon	Treatment of multiple sclerosis
Colony stimulating factor	Treatment of infections and cancer
Interleukin-2	Treatment of certain cancers
Lysozyme	Anti-inflammatory
Tumor necrosis factor	Antitumor agent, potential treatment of arthritis
Replacement Enzymes	
β-glucocerebrosidase	Treatment of Gaucher disease, an inherited neurological disease
Vaccines	
Hepatitis B	Prevention of serum hepatitis
Lyme disease	Prevention of infection
Measles	Prevention of measles
Rabies	Prevention of rabies

two polypeptides (A and B) connected by disulfide bridges (Figure 31.9a). These two polypeptides are coded by separate parts of a single insulin gene. The insulin gene codes for *preproinsulin*, a longer polypeptide containing a signal sequence (involved in excretion of the protein) (∞ Section 7.16), the A and B polypeptides of the active insulin molecule, and a connecting polypeptide that is absent from mature insulin. *Proinsulin* is formed from preproinsulin, and the conversion of proinsulin to insulin involves enzymatic cleavage of the connecting polypeptide from the A and B chains (∞ box, Protein Processing in Chapter 8).

Two approaches have been used to obtain production of human insulin in bacteria: (1) production of proinsulin and conversion to insulin by chemical cleavage, and (2) production of the A and B chains in two separate bacterial cultures, and joining of the two chains chemically to produce insulin. Because the insulin protein is fairly small, it was more convenient with either

approach to synthesize the proper DNA sequence chemically (∞ Section 10.13) rather than attempt to isolate the insulin gene from human tissue. There are 63 bases encoding the A chain and 90 bases encoding the B chain. In proinsulin there are an additional 105 bases for the peptide connecting the A and B chains (Figure 31.9b). When the polynucleotides were synthesized, suitable restriction enzyme sites were placed at each end so the polynucleotides could be ligated into a plasmid vector. To obtain effective expression, the synthesized genes were inserted downstream from a suitable *Escherichia coli* promoter but in a manner such that the insulin fragment was synthesized as part of a *fusion protein* (see Section 31.5). An important advantage of making the fusion protein is that the fusion product is much more stable in *E. coli* than insulin itself. Finally, a nucleotide triplet coding for methionine was placed at the junction joining the insulin gene to the upstream part of the fusion gene. The reason for this is that the chemical reagent *cyanogen bromide* specifically cleaves polypeptide chains at methionine residues, permitting recovery of the insulin product once the fused protein has been isolated from the bacteria. Insulin itself does not contain methionine and hence, is unaffected by cyanogen bromide treatment.

When the proinsulin route is used, the proinsulin isolated from the bacteria via cyanogen bromide treatment is converted to insulin by disulfide bond formation, followed by enzymatic removal of the connecting peptide of proinsulin. Proinsulin naturally folds so the cysteine residues are opposite each other (Figure 31.9a), and chemical treatment then causes the formation of disulfide cross-links. Once this has been accomplished, the connecting peptide can be removed by treatment with the proteases trypsin and carboxypeptidase B, which have no effect on insulin itself.

When insulin is produced by way of the separate A and B peptides, each of the fusion proteins is isolated from a separate bacterial culture and the chains released by cyanogen bromide cleavage. The cleaved chains are then connected by use of chemical treatment that results in disulfide bond formation.

The final product, biosynthetic human insulin, is identical in all respects to insulin purified from the human pancreas. Microbially produced human insulin is less expensive to make and just as effective as porcine or bovine insulin, the major source of insulin for diabetics before the advent of biotechnology.

Recombinant Vaccines

Vaccines are suspensions of killed or modified pathogenic microorganisms or specific fractions isolated from the microorganisms that when injected into an animal produce immunity to a particular disease. Often the substance that elicits the immune response is a surface protein, for instance, a virus coat protein. Genetic engineering

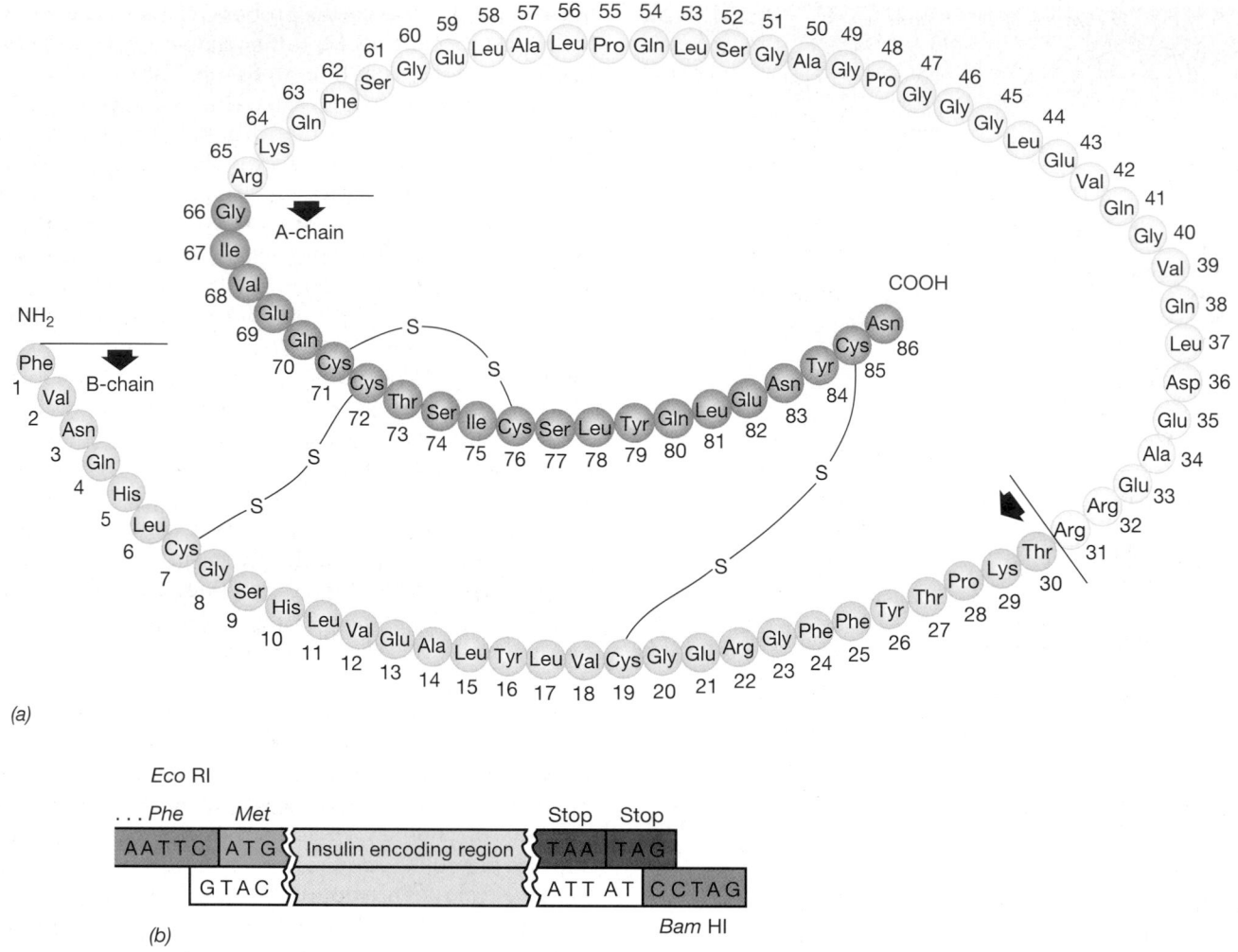

(a)

(b)

Figure 31.9 Genetic engineering for the production of human insulin in bacteria. (a) Structure of human proinsulin. The peptide shown in yellow must be removed from between the A and B chains in order to make insulin. (b) Chemical synthesis of the insulin gene and suitable linkers, permitting cloning and expression. The synthesized fragments were linked via restriction sites *Eco*RI and *Bam*HI in a plasmid vector in such a way that the insulin chains are formed as a fusion protein (see Section 31.5) with a portion of a gene found on the vector (note that the *Eco*RI site is part of this coding region). The methionine coding sequence was inserted to permit chemical cleavage of the A and B chains from the fused protein made in the bacteria because the reagent cyanogen bromide specifically cleaves at methionine residues and insulin does not contain methionine. Two stop codons were incorporated at the downstream end of the coding sequence.

can be applied in many different ways to the production of vaccines.

Recombinant DNA techniques can make it much easier to modify a pathogen simply because *in vitro* genetic techniques are often more precise and more powerful than traditional *in vivo* techniques (⚬ Chapter 10). For instance, in some cases one can simply delete genes involved in virulence but leave those whose products elicit an immune response, giving one a *recombinant live attenuated vaccine*. Of course, using recombinant technology one can add genes to a virus that will specifically confer immunity to a viral disease. In this latter category is a live recombinant virus vaccine that offers protection in poultry against both fowlpox (a disease that reduces weight gain and egg production) and Newcastle

disease (a viral disease that is often lethal). The fowlpox virus, a typical pox virus (⚬ Section 16.12), was first modified to delete genes that cause disease (but not those that elicit immunity). Then immunity-inducing genes from the Newcastle virus were added. This resulted in a *polyvalent vaccine*, in this case a single virus that can confer immunity to two different important diseases.

One vector used to prepare live recombinant vaccines is *vaccinia virus* (⚬ Section 16.12). Cloning in vaccinia virus is done using an *Escherichia coli* plasmid containing a fragment of the vaccinia virus thymidine kinase gene (Figure 31.10*a*). An appropriate foreign DNA is inserted into this plasmid, and the recombinant plasmid is transformed into a host cell whose own thymidine kinase gene is inactive, but that has previously been

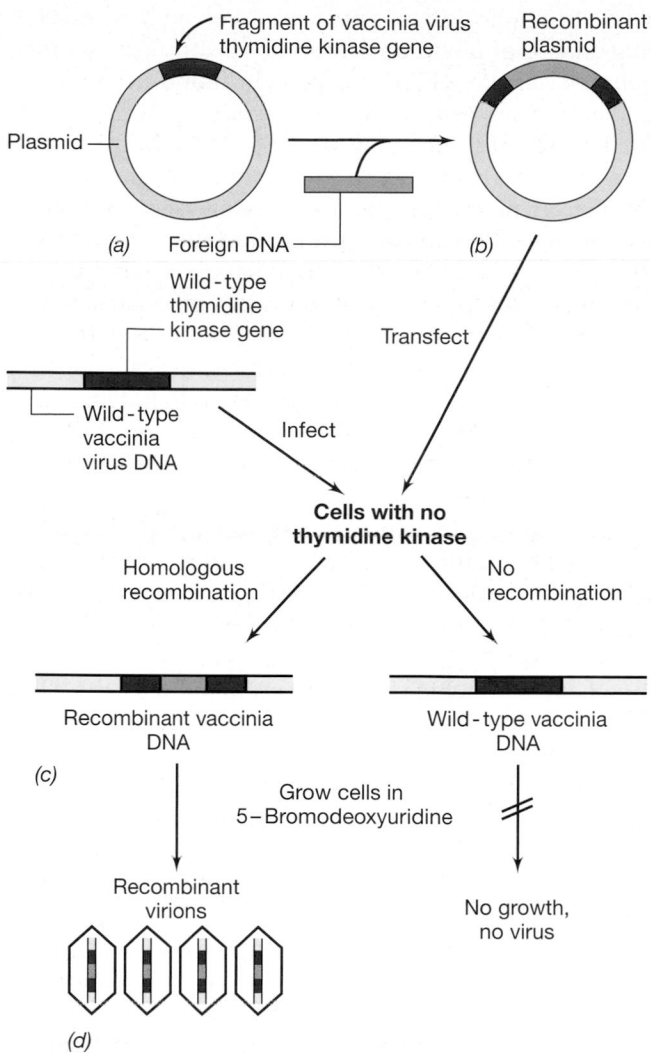

Figure 31.10 Production of recombinant vaccinia virus. (a) Foreign DNA is cloned into a plasmid containing a small piece of the vaccinia virus thymidine kinase gene. (b) Recombinant plasmid formed. (c) The recombinant plasmid is then used to transfect host cells already infected with wild-type vaccinia. If recombination occurs, recombinant vaccinia DNA can be produced. (d) The cells are then placed in the presence of 5-bromodeoxyuridine, a compound that is toxic to cells having an active thymidine kinase. Only recombinant virions develop under these conditions. If the recombinant vaccinia virions contain genes for other viral coat proteins, these may be expressed.

infected with wild-type vaccinia virus (Figure 31.10*b*). If homologous recombination occurs between plasmid DNA and vaccinia genomic DNA (Figure 31.10*c*), recombinant virions can be obtained containing an *inactivated* thymidine kinase gene. An *active* thymidine kinase leads to growth inhibition by the compound 5-bromodeoxyuridine. Therefore, recombinant vaccinia virions can be selected by allowing viral replication to occur in the presence of this inhibitor (Figure 31.10*d*). Although such recombinant viruses no longer express thymidine kinase, they can still infect human cells, and

they do express the foreign genes that have been cloned into them. Indeed, some recombinant vaccinia viruses can carry genes from four different viruses!

Vaccinia virus itself is generally not pathogenic for humans (vaccinia virus was originally used as a vaccine against the related virus smallpox). However, vaccinia virus is not completely benign (it causes severe complications in some people), and therefore, more research must be done before such vaccines can be used in humans. This research includes using other viral vectors.

Vaccines do not have to include the entire pathogenic organism. *Subunit vaccines* contain only a specific subunit from a pathogenic organism. For viruses this is often the coat protein. The highly immunogenic coat proteins are purified and used in high dosage to elicit a rapid and high level of immunity. Genetic engineering has proved invaluable in the development of subunit vaccines, because recombinant DNA techniques can be used to produce large amounts of these subunits and there is no possibility that the purified products will contain the entire pathogenic organism in even minute amounts.

Subunit Vaccines

The steps for viral gene cloning are those outlined in the previous sections: fragmentation of viral DNA by restriction enzymes; cloning viral coat protein genes into a suitable vector; providing for proper promoters, reading frame, and ribosome binding sites; and reinsertion and expression of the viral genes in a microorganism. Sometimes only certain *domains* of the protein are expressed, rather than the entire protein.

Unfortunately, when *Escherichia coli* is used as the cloning host, the vaccines are often poorly immunogenic and fail to protect animals from subsequent infection with the virus. The problem involves the fact that many key viral coat proteins are posttranslationally modified, generally by the addition of sugar residues (glycosylation), when the virus replicates in its normal host. However, the recombinant proteins produced by *E. coli* or other *Bacteria* are unglycosylated, and apparently glycosylation is necessary for the proteins to be immunologically active. Therefore, a eukaryotic host is used.

The first recombinant subunit vaccine approved for use in humans was made using yeast. The gene encoding a surface protein from hepatitis B virus was cloned and expressed in yeast. The protein was produced and formed aggregates very similar to those found in patients infected with the virus. These aggregates were purified and used to vaccinate people against hepatitis B virus. Subunit vaccines against a large variety of viruses and pathogenic organisms are being developed employing genetic engineering. Insect cells and cultured mammalian cells are now being used as hosts to prepare recombinant vaccines. To obtain the correct pattern of glycosylation or other modifications of the protein, it is often important to use a host that is closely related to humans. However, vaccines can also be produced in plants (see Section 31.7).

Genetically engineered vaccines are likely to become increasingly common because (1) they are safer than normal attenuated or killed vaccines, (2) they are more reproducible because their genetic makeup can be carefully monitored, and (3) they can be administered in high doses without fear of side effects.

Vaccines produced using genetic engineering can also usually be made much faster than those produced by more traditional methods. Recombinant vaccines made with cloned influenza virus hemagglutinin genes (⚭ Section 16.7) can be made in just 2 or 3 months rather than 6–9 months. This could be of real advantage during an epidemic caused by a new strain of influenza virus (⚭ Section 26.8). Recombinant vaccines are also sometimes much less expensive than those produced in other ways, and this in itself can allow for new uses. Bait carrying a recombinant rabies vaccine has been distributed over large tracts of land in Europe and has triggered a dramatic decline in the incidence of rabies among wild foxes. Such a method of vaccination would previously have been too expensive.

DNA Vaccines

Although vaccines of various types have been extremely successful against a wide variety of infectious diseases, there are other important diseases for which successful vaccines have been difficult to produce. These include such important diseases as malaria (⚭ Section 27.5) and AIDS (⚭ Section 26.14). Although "standard" genetic engineering techniques might yet yield effective vaccines against these diseases, an exciting new direction has been the development of *DNA vaccines*, also known as *genetic vaccines*. **DNA vaccines** use the genetic material of the pathogen itself to immunize. This genetic material may be in the form of defined fragments of the genome of the pathogen, or genes from the pathogen cloned into a plasmid or viral vector. Even in the later cases, though, it is the DNA (or RNA) that is used as the vaccine.

In some cases it appears that if a particular gene is delivered to an animal in such a way that it is taken up by the cells, the protein will be produced and the animal will develop immunity. A number of such DNA vaccines are undergoing clinical trials and many more are being developed. DNA vaccines would be both safe and inexpensive. Unlike viral vaccines, they would also escape surveillance by the host immune system.

Other Proteins and Other Products

Table 31.1 lists a few of the mammalian proteins and products that are being produced by recombinant DNA techniques. These include a number of hormones in ad-

dition to insulin and also proteins involved in blood clotting and other blood processes. For example, *tissue plasminogen activator* (TPA) is a protein found in the blood that acts to scavenge and dissolve old blood clots in the final stages of the healing process. The clinical usefulness of TPA is primarily in heart patients or anyone suffering from poor circulation because of excessive clotting tendencies. TPA can be administered following cardiac bypass, transplant, or other open heart surgeries to prevent the development of pulmonary embolisms, which are often life-threatening. Heart disease is a leading cause of death in many developed countries, so microbially produced TPA promises to be in high demand.

In contrast to TPA, blood clotting factors VII, VIII, and IX are critically important for the *formation* of blood clots. Hemophiliacs suffer from a deficiency of one or more clotting factors and can be readily treated with the microbially produced product. Recombinant clotting factors take on added significance when one considers that hemophiliacs have in the past been treated with concentrated clotting factor extracts from pooled human blood, some of which was contaminated with the AIDS virus: This had put hemophiliacs at high risk for contracting AIDS.

A number of proteins have roles as anticancer agents or immune modulators. *Interferons* are a series of proteins made by animal cells in response to viral infection (⚭ Section 23.10) or immune activation in the case of one type of interferon.

Some mammalian proteins made via recombinant DNA technology do not fit in the categories listed in Table 31.1. For instance, human DNase I is being produced and used to treat the build-up of DNA-containing mucus in patients with cystic fibrosis. Even monoclonal antibodies are now produced in microorganisms by genetic engineering (we discussed monoclonal antibodies in Section 24.6).

Certainly not all the proteins produced using recombinant DNA techniques have therapeutic uses. Many commercial enzymes (⚭ Section 30.9) are produced in this way as well. In addition, even hormones can have other uses. *Bovine somatotropin* produced using these techniques is used to increase milk production in cattle in the United States. Often the "benefits" of genetic engineering can be quite unexpected. *Rennin*, which is used to make cheese, is an animal product. In Great Britain a "vegetarian cheese" containing a recombinant protein produced in a microorganism is being marketed and is finding wide acceptance. The first products made by genetic engineering were primarily the protein products of cloned genes, and there is still a great deal to be done with this approach. Added applications come from being able to use site-directed mutagenesis on the cloned gene so that new products with new attributes can be generated. It must also be remembered that molecules such as antibiotics are synthesized in cells in biochemical pathways using a series of enzymes (proteins). These enzymes can be modified so new antibiotics can be developed.

✓ 31.6 Concept Check

The first human protein made commercially using engineered bacteria was human insulin, but numerous other hormones and other human proteins are now being produced. Many proteins found in humans that were formerly extremely expensive to produce because they were found in human tissues in only small amounts can now be made in very large amounts from the cloned gene in a suitable expression system. In addition to useful pharmaceuticals such as anticancer agents and immune modulators, even vaccines can be produced using genetic engineering.

- ✓ Why is it sometimes important to produce proteins used for therapeutic purposes in a host closely related to humans?
- ✓ Explain why recombinant vaccines might be safer than some vaccines produced by traditional methods.

31.7 Genetic Engineering in Plant Agriculture

Genetic improvement of plants has traditionally been a slow and difficult task, but recombinant DNA technology has lead to revolutionary changes. Now it is possible to use *in vitro* genetic techniques to modify plant DNA and then to transform plant cells with free DNA by either electroporation or particle gun methods (∞ Section 10.6 and see Section 31.2) or by using vectors from the bacterium *Agrobacterium tumefaciens*, which can transfer DNA directly to certain plants (∞ Section 19.21). It is possible to use plant tissue culture techniques to select clones of plant cells that have been genetically altered using *in vitro* techniques and then, with proper treatments, induce these cell cultures to make whole plants that can be propagated vegetatively or by seeds.

The plants that are the result of these *in vitro* genetic manipulations are often referred to as **genetically modified organisms** (GMOs), or *GM plants*. Interestingly, organisms whose modifications have been accomplished by more traditional *in vivo* methods are not usually designated in this way, and most organisms used in industry, agriculture, or medicine have been genetically manipulated. The difference is that those isolated after using *in vitro* techniques often contain genes from other organisms, that is they are **transgenic organisms**. Although the techniques to generate *transgenic plants*, or *transgenic animals*, are essentially identical to those used to generate microorganisms expressing foreign genes, the use of the term transgenic organism is confined to multicellular organisms. In this section we will see how the foreign gene (sometimes called a *transgene*) is inserted into a plant genome and how transgenic plants can be used.

Vectors for Cloning in Plants

The gram-negative plant pathogen *Agrobacterium tumefaciens* contains a large plasmid called the **Ti plasmid** which is responsible for its virulence. The plasmid contains genes that mobilize DNA for a transfer to the plant (for details of the disease process and genetic events, ∞ Section 19.21). The segment of the Ti plasmid DNA that is actually transferred to the plant is called **T-DNA**. The sequences at the ends of the T-DNA are essential for transfer, and the DNA to be transferred must be between these ends. One type of vector system that has been constructed and is commonly used for the actual transfer of genes to plants is called a *binary vector system*. The word *binary* means "consisting of two parts," and a binary vector system involves two plasmids. One is the actual vector into which the foreign DNA is cloned. This vector contains the two ends of the T-DNA on either side of the site used for cloning and an antibiotic resistance marker that can be used in plants. It also contains an origin of replication so that it can replicate in both *A. tumefaciens* and *Escherichia coli* (the latter serves as a host for cloning work), and another antibiotic resistance marker that is expressed in bacteria (Figure 31.11). The DNA to be cloned is inserted into the vector, which is then transformed into *E. coli*. It is then transferred to *A. tumefaciens* (usually by conjugation; ∞ Section 10.9).

This cloning vector does not contain the genes necessary for transfer of T-DNA to a plant, so the *Agrobacterium* into which it is transferred must contain the other member of the binary vector system. This other plasmid contains the virulence (*vir*) region of a Ti plasmid but it has been "disarmed." Although it can direct the transfer of DNA into a plant, it no longer has genes that cause disease. This disarmed plasmid, D-Ti, will supply all the genes needed to transfer the T-DNA from the cloning vector. The cloned DNA and the kanamycin resistance marker of the vector can be mobilized by the D-Ti plasmid and transferred into a plant cell (Figure 31.11). Following recombination with a host chromosome, the foreign DNA can be expressed to confer new properties on the plant. Many genes are not expressed efficiently in plants unless they are cloned into an expression vector containing a plant promoter. Some of the promoters that have been used in constructing plant expression vectors include some normally found in T-DNA and a promoter from cauliflower mosaic virus, a plant DNA virus (∞ Table 9.1).

With the use of *Agrobacterium tumefaciens*, a number of transgenic plants have been produced. Most successes have come with broadleaf plants (dicots) such as

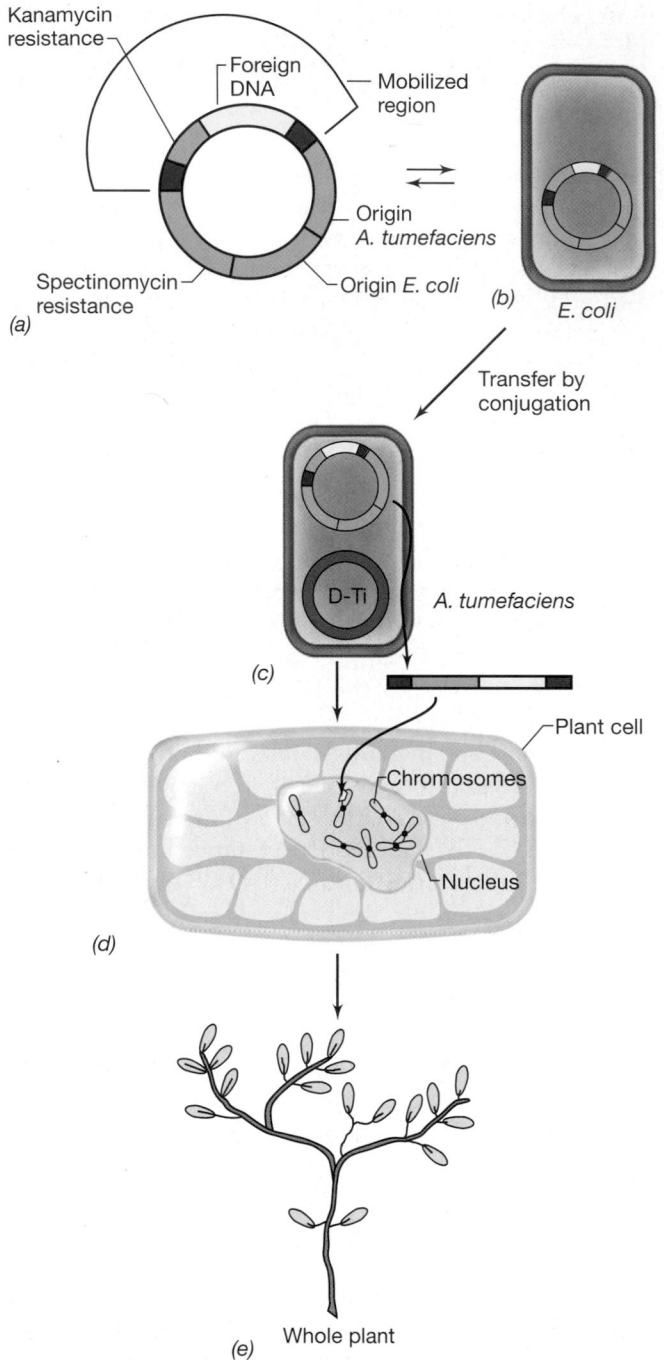

Figure 31.11 Production of transgenic plants using a binary vector system in *Agrobacterium tumefaciens*. (a) Generalized plant cloning vector containing ends of T-DNA (in red), foreign DNA (in yellow), origin of replication elements for both *E. coli* and *A. tumefaciens*, and spectinomycin and kanamycin resistance markers. The kanamycin resistance marker can be selected for in plants. (b) The vector can be put into cells of *E. coli* for cloning purposes and then transferred to *A. tumefaciens* by conjugation. (c) The resident Ti plasmid used for transferring the vector to the plant (D-Ti) is itself genetically engineered to remove key pathogenesis genes. (d) However, D-Ti can mobilize the T-DNA region of the vector for transfer to plant cells grown in tissue culture. From the recombinant cell, whole plants can be regenerated.

tomato, potato, tobacco, soybean, alfalfa, and cotton, but *A. tumefaciens* has also been used to produce transgenic woody dicots such as walnut and apple. Transgenic crop plants from the grass family (monocots) have been more difficult to generate using *A. tumefaciens*, but other methods of introducing DNA, such as a particle gun (see Section 31.2) have been used successfully.

Applications in Plant Biotechnology

Major areas targeted for genetic improvement in plants include herbicide, insect, and microbial disease resistance, as well as improved product quality. More than 1000 different field trials on more than 30 different plant species have been carried out in the last decade. The first genetically modified crop (GM crop) to be grown commercially was tobacco grown in China in 1992. By the year 2000, over 100 million acres (44 million hectares) of GM crops were planted worldwide. Of these, 58% were in soybeans, 23% were in corn, 12% were in cotton, and 6% were in canola. Almost all the GM soybeans and canola planted were herbicide resistant, whereas corn and cotton were herbicide resistant or insect resistant or both.

Herbicide Resistance

Herbicide resistance can be obtained by genetically engineering the crop plant to no longer respond to the toxic chemical. Many herbicides act by inhibiting a key plant enzyme or protein necessary for growth. For example, the herbicide *glyphosate* kills plants by inhibiting the activity of an enzyme necessary for making aromatic amino acids. Such an herbicide kills both weeds and crop plants and thus must be used as a "preemergence herbicide," that is, before the crop plants emerge from the ground. However, some bacteria contain an enzyme that is naturally resistant to glyphosate. A gene encoding a resistant enzyme from *Agrobacterium* has been cloned, modified for expression in plants, and transferred into crop plants. When sprayed with glyphosate, plants containing the bacterial gene grow as well as unsprayed control plants (Figure 31.12). Soybeans expressing glyphosate resistance have been developed by Monsanto Company.

Insect and Virus Resistance

Novel means of insect resistance have also been genetically introduced into plants. One already in wide use has been the toxic protein genes of *Bacillus thuringiensis*. This organism produces a crystalline protein (∞ Section 12.20), called *Bt-toxin*, that is toxic to moth and butterfly larvae, and certain strains of *B. thuringiensis* produce additional proteins toxic to beetle and fly larvae and mosquitoes. Biotechnologists are using several different approaches to enhance the use of Bt-toxin for pest control in plants.

One approach is to develop a single Bt-toxin that is effective against many different insects. This can be done because the protein has separate domains for its speci-

Stephen R. Padgette, Monsanto Company

Figure 31.12 The photograph shows a portion of a field of soybeans that has been treated with *Roundup*, a glyphosate-based herbicide manufactured by Monsanto. The plants on the right are normal soybeans; those on the left have been genetically engineered to express glyphosate resistance.

ficity and its toxic function. The toxic domain is highly conserved in all the various Bt-toxins. Genetic engineers can make a gene that encodes a Bt-toxin carrying one toxic domain and several different specificity domains. In Section 31.5, we dealt with how one might have to modify a gene from a higher eukaryote before it could be efficiently expressed in a bacteria. Sometimes one also has to modify bacterial genes before being expressed in a eukaryote. For instance, bacterial genes may contain sequences that lead to accidental splicing of the mRNA produced or the codon usage may have the wrong bias. Because of all the modification necessary, Bt-toxin genes now in use are typically completely synthetic.

An effective approach to achieve gene expression and stability is to transfer the gene directly into the plant genome. For example, a natural Bt-toxin gene has been cloned into a plasmid vector under control of a chloroplast rRNA promoter and transferred into tobacco plant chloroplasts by particle bombardment. With this methodology plants were obtained that expressed this protein at levels that were extremely toxic to insect larvae from a number of species (Figure 31.13).

However, there have been reports of insects that have acquired resistance to Bt-toxin. Resistance to insecticides and herbicides is a common problem in agriculture, and the fact that a product has been produced by genetic engineering does not give it any magical properties. This emphasizes that many approaches must be used for pest control, and Bt-toxin is only one of many being developed by biotechnologists.

Genetic engineering has also been used to protect plants from virus infection. For example, it has been discovered that transgenic plants that express the coat protein gene of a virus become resistant to infection by that virus. Although the mechanism of resistance is unknown, the presence of viral coat protein in plant cells apparently interferes with the uncoating of viral particles containing that coat protein, and this interrupts the virus replication cycle.

Other Uses of Plant Biotechnology

Not all genetic engineering is directed toward making plants disease-resistant. Genetic engineering can be used in a variety of ways for developing mutant strains of plants with desired characteristics. For instance, the first GM food grown for sale in the U.S. market was a tomato with delayed spoilage. In addition, transgenic plants can be genetically engineered to produce commercial or pharmaceutical products, as has been done with microorganisms (see Section 31.6) and animals (see Section 31.8). Crop plants such as tobacco and tomatoes have been engineered to produce a number of different products, such as the human protein interferon. Transgenic crop plants can be used to produce human antibodies efficiently and inexpensively (such plant-made antibodies are sometimes called "plantibodies"). These antibodies have potential as anticancer or antivirus drugs, and some are undergoing clinical trials.

Plant hosts can be useful in producing these types of products because plants typically modify proteins correctly and because crop plants can be efficiently grown and harvested.

Crop plants are also being developed for the production of vaccines. For instance, a recombinant tobacco mosaic virus has been engineered whose coat contains antigens of *Plasmodium vivax*, the organism that causes malaria (∞ Section 27.5). This recombinant virus could be used to develop a malaria vaccine that can be produced in very large amounts at very low cost by harvesting tobacco grown in fields. Another very interesting approach is to produce a vaccine in an edible plant product. Such *edible vaccines* are now under development that could immunize against diseases caused by enteric bacteria, including cholera and diarrhea (∞ Section 28.5).

Although public acceptance of GM crops remains high in the United States, there have been recent concerns, some stemming from the identification in human food of GM corn approved only for animal food.

In Europe there has been considerable public concern and several countries have considered bans on growing GM crops and legislation to require all GM foods to be labeled. Therefore, even good science and potential commercial viability does not always ensure public acceptance. Because of increased public concern, the amounts of acreage planted in many GM crops has leveled off or even decreased.

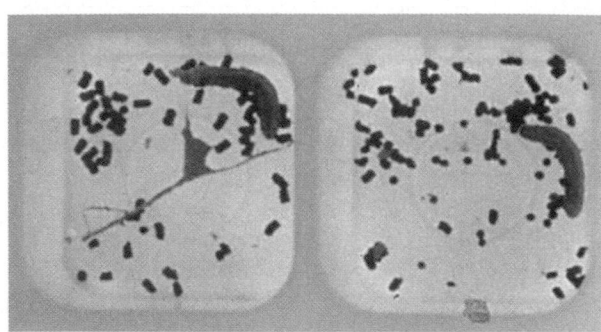

(a)

(b)

Kevin McBride, Calgene, Inc.

Figure 31.13 Panel (a) shows the results of two different assays to determine the effect of beet armyworm larvae on tobacco leaves from normal plants. Panel (b) shows the results of similar assays but using tobacco leaves taken from transgenic plants that express Bt-toxin in their chloroplasts.

✓ 31.7 Concept Check

Genetic engineering is being employed to make plants resistant to disease, to improve product quality, and to use crop plants as a source of recombinant proteins and even vaccines. One commonly used cloning vector for plants is the Ti plasmid of the bacterium *Agrobacterium tumefaciens*. This plasmid can transfer DNA into plant cells. Plants whose genomes have been modified using *in vitro* genetic techniques are called *genetically modified organisms* or *GMOs*.

✓ Transfer of DNA by the Ti plasmid most resembles what form of bacterial gene transfer (👓 Chapter 10)?

✓ What is a *transgenic plant*?

31.8 Genetic Engineering in Animal and Human Genetics

This section covers only a few highlights of some of the huge number of uses of genetic engineering in animal and human genetics. Some of these applications have to do with producing products, but more have to do with understanding gene function in mammals and in curing or treating genetic disease.

Transgenic Animals

With the use of recombinant DNA technology and microinjection techniques to deliver cloned genes to fertilized eggs, many foreign genes have been expressed in both laboratory research animals and in species important in commercial animal industries. Such animals are called *transgenic animals* and have become increasingly important in basic biomedical research for studying gene regulation and developmental biology. However, many applied aspects of transgenic animals are also of interest.

One approach is to improve the productivity or disease resistance of the animal, as in the case of transgenic plants used in agriculture. However, transgenic animals are also being used to produce proteins of pharmaceutical value—a process some scientists have called "pharming." Transgenic animals may be useful for producing human proteins that require specific posttranslational modifications for activity, such as certain blood clotting enzymes; many proteins of this type are not produced in an active form by microorganisms or by plants. Also, some proteins have been engineered to be secreted into the animals' milk, which can be readily collected and processed. These proteins include α-1-antitrypsin (used to treat lung disease) produced in sheep and tissue plasminogen activator (used to dissolve blood clots) produced in goats. The production of transgenic animals for research and commercial purposes seems likely to continue as an important area in biotechnology.

Human Genetics

Conventional genetics, involving genetic crosses or mutagenesis, cannot be done with humans. Therefore, our understanding of human genetics had lagged considerably behind our understanding of the genetics of many other organisms until the recent revolution brought about by genetic engineering. A detailed discussion of human genetics is beyond the scope of this book, but a few general remarks on the utility of recombinant DNA technology can be made.

We mentioned in Chapter 15 that a rough draft of the sequence of the human genome was released in 2000. A considerable amount of work will be necessary to complete the sequence. This draft greatly facilitates studies in understanding the structure and function of human genes. However, as in other areas of genomics, the human genome sequence will also have great practical applications.

Already the techniques of *in vitro* engineering are widely used to identify individuals in a process called **DNA fingerprinting** (see the box, DNA Fingerprinting). One of the primary goals of sequencing the human genome was to better understand human genetic diseases. Therefore, some applications of genetic engineering of the human genome are directed at treating human disease.

| # DNA fingerprinting

he techniques used in molecular genetics and genetic engineering can be used to differentiate between very closely related organisms. In fact, such techniques are now commonly used to distinguish between individuals, a process often called **DNA fingerprinting**. DNA fingerprinting is made possible both by the technology that allows precise detection and amplification of very small amounts of DNA and by the fact that higher organisms contain repetitive DNA sequences that can exist in different numbers and patterns in the genome.

As discussed (⟳ Section 7.4), the genomes of higher eukaryotes contain a very large amount of repetitive DNA. Some of these repeats exist in families of related sequences scattered around the genome, and members of these families have been cloned and sequenced. To be useful for identification purposes, a DNA sequence must have a reasonable chance of differing among different groups in a population of organisms. One type of these repetitive sequences was found to vary not simply as to sequence but also as to how many repeats of an individual sequence occurred at a single site on a particular chromosome. These sequences are called *variable number of tandem repeats* (VNTR), and several have been identified.

The use of VNTRs in DNA fingerprinting is illustrated in Figure 1. It shows two different alleles (alternative states of the same gene) of a eukaryotic chromosome that differ only in how many copies of the repeated sequence are present. Since the VNTR DNA has been sequenced, it is known which restriction enzyme sites are *not* found in a particular VNTR. Digestion with such an enzyme then releases the complete VNTR intact. When the DNA from two chromosomes with a different number of repeats at this particular locus is digested, the restriction fragments containing this DNA differ in size. (Such a difference is called a *restriction fragment length polymorphism*, or RFLP.) This DNA can be separated by gel electrophoresis, and the VNTR-containing fragments detected by Southern blotting (⟳ Section 10.12) using a probe made from the cloned VNTR sequence. Even if multiple alleles are present in a population (rather than just two as in the example), differences at a single site are insufficient to identify an individual. Many individuals in a population would be expected to have the same pattern at this site. However, it is possible to probe the digested DNA simultaneously for several different VNTR markers. With the use of these methods, the probability of identifying a particular individual by comparing two different DNA samples is extremely high.

Notice that the polymerase chain reaction (PCR; ⟳ Section 10.17) does not need to be used in DNA fingerprinting. However, the use of PCR is essential when the amount of DNA in the sample is very small—such as that found in the cells on the root of a single hair. The use of PCR in DNA fingerprinting is also shown in the figure. To use PCR, it is necessary to know the sequences surrounding a particular site that contains a VNTR so primers can be synthesized. However, since PCR amplifies only the DNA between the primers, one does not have to cut the DNA with restriction enzymes before running it on a gel. With enough cycles of amplification, it is also sometimes possible to detect the PCR-generated bands by simply staining the gels rather than using a hybridization probe.

VNTRs are not restricted to higher organisms. The genome of *Bacillus anthracis*, the species of *Bacteria* that causes the disease anthrax (⟳ Section 25.11) contains several different VNTRs. This allows strains used as biological weapons to be identified and tracked. ■

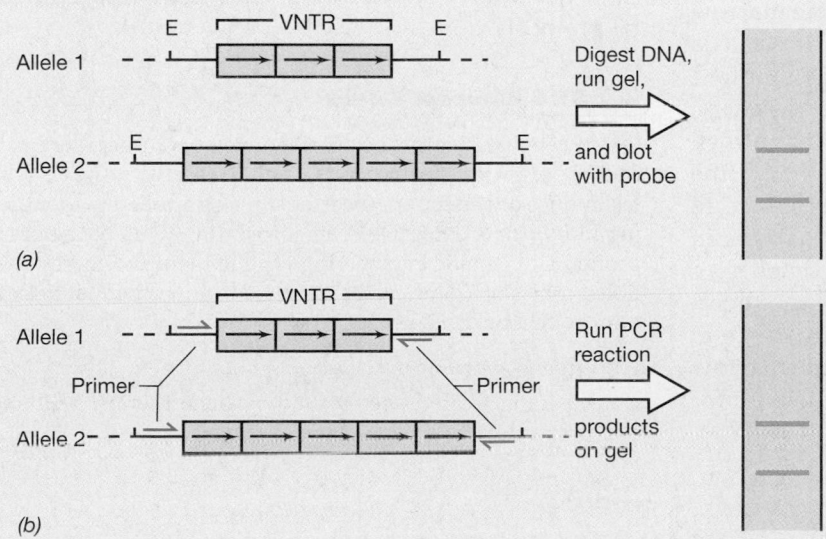

(a)

(b)

Figure 1 **DNA fingerprinting. (a) Two different alleles of a region of a single chromosome. The alleles differ only in the number of repeats in the VNTR. DNA from cells containing these chromosomes can be cut with the restriction enzyme *Eco*RI (which does not cut within the VNTR) and the fragments separated on an agarose gel. The fragments containing the VNTR are then identified after Southern blotting by hybridization with a probe specific to the VNTR. The figure shows only the result from individuals whose two chromosomes have different alleles at this site. If an individual had the same allele on both chromosomes, only a single band would be observed. (b) The same alleles, but with primers that could be used to amplify the VNTR segments by PCR. The products of the PCR reaction can be loaded directly onto the gel without restriction digestion.**

A vast number of genetic diseases are known, but except in rare cases, little was known until recently about their molecular bases. By use of recombinant DNA technology, coupled with conventional genetic studies (following family inheritance, and so on), it is possible to localize particular defects to particular chromosomes and to particular locations on chromosomes. With the use of recombinant DNA technology, it is possible to clone the region containing the genetic defect and then to make comparisons between the base sequence in the normal gene and the defective gene. From such studies, even in the absence of knowledge of the enzyme defect, it has been possible to obtain information about the genetic change. Many genes, including those for Huntington's disease, cystic fibrosis, and Duchenne's muscular dystrophy, have been localized with these techniques and the mutation(s) in the defective genes identified.

Gene Therapy

Genetic engineering is employed to attempt to provide treatment for some of these diseases using **gene therapy**. In gene therapy, a nonfunctional or dysfunctional gene is augmented or replaced by a functional gene. Not all gene therapy is designed toward treating genetic disease; a considerable effort has been directed toward protocols for treating cancer. Major obstacles to this approach exist in trying to target the correct cells for gene therapy and in successfully transfecting cell lines that will perpetuate the genetic alteration.

The first genetic disease for which an approved gene therapy technique was used was one form of severe combined immune deficiency (SCID) caused by the absence of adenosine deaminase (ADA), an enzyme involved in purine metabolism, in bone marrow cells. The procedure involved using a retrovirus as a vector to insert a good copy of the ADA gene into T lymphocytes (cells that are part of the immune system; ∞ Section 22.1) removed from the patient and then placing these "corrected" cells back in the body (the retrovirus also carries a marker gene, resistance to neomycin, so cells carrying the inserted retrovirus can be selected and identified). Since T lymphocytes have a limited life span, it is necessary to repeat the therapy every month or two. Attempts are being made to insert the gene into the stem cells of the bone marrow (which continue to divide) and effect a true cure for the disease.

Several other gene therapy treatments, some using other virus vectors, are currently being tested with various levels of success. Since the first gene therapy experiment with ADA in 1990, there have been no striking practical breakthroughs until 2000. Another form of SCID, involving a different gene, was successfully treated in three different patients. It seems likely that this very rare form of the disease can now be successfully treated using gene therapy. Though gene therapy has tremendous practical potential, most applications still remain a distant prospect.

Some of the current difficulties are related to the vectors being used. Although transduction using retroviral vectors gives stable integration of the gene, the site of insertion is unpredictable, the amount of cloned DNA is limited, and expression of the cloned gene is often transient. The vectors also have limited infectivity and are rapidly inactivated in the host. Many nonretroviral vectors, such as the adenovirus vector (∞ Section 16.13), have similar problems, and adverse reactions to the vector can also be a severe problem. Promising vectors include human artificial chromosomes and highly modified versions of the HIV virus (∞ Sections 9.12 and 16.14).

It is important to note that in the protocols being tested, the defective gene is not replaced. Rather, the retrovirus (and the good copy of the gene) simply integrates somewhere in the human genome of these cells. Actual gene replacements in germ line cells (cells that give rise to gametes) can be accomplished with some mammals, although the techniques of isolating individual animals with these changes cannot readily be applied to humans. However, attempts to change the germ cells of an individual would raise many ethical and societal questions beyond simple questions of experimental protocols.

✓ 31.8 Concept Check

Genetic engineering can be used to develop transgenic animals capable of producing proteins of pharmaceutical value. The techniques of genetic engineering are also applied to identifying individuals using DNA fingerprinting. One of the great promises of genetic engineering is gene therapy, where functional copies of a gene can be supplied to an individual to treat human genetic disease.

✓ What is *pharming*?

✓ A person treated successfully by gene therapy will still have a defective copy of the gene. Explain.

Review Questions

1. Genetic engineering depends on many different *in vitro* genetic techniques. Describe the techniques used in molecular cloning. Include in your description the terms cloning vector, restriction endonuclease, and transformation.

2. The technique known as PCR has many applications in genetic engineering. What do the letters of the acronym mean? Describe how PCR is carried out.

3. Describe the similarities and differences between expression vectors, shuttle vectors, and integrating vectors.

4. What is a *reporter gene*?

5. Explain why the use of a regulatory switch is desirable for the large-scale production of a protein.

6. How has bacteriophage T7 been used in expressing foreign genes in *Escherichia coli* and what desirable features does this regulatory system possess?

7. What is a subunit vaccine and why are subunit vaccines considered a safer way of conferring immunity to viral pathogens than attenuated virus vaccines?

8. What is the Ti plasmid and how has it been of use in genetic engineering?

9. How do transgenic plants and animals differ from plants and animals modified by conventional breeding techniques?

10. What advantages might there be in using a transgenic plant rather than a transgenic animal to produce a protein?

Application Questions

1. Suppose you are given the task of constructing a plasmid expression vector suitable for molecular cloning in an organism of industrial interest. List the characteristics such a plasmid should have. List the steps you would use to develop such a plasmid.

2. Suppose you have just determined the DNA base sequence for an especially strong promoter in *Escherichia coli* and you are interested in incorporating this sequence into an expression vector. Describe the steps you would use. What precautions would be necessary to be sure that this promoter actually works as expected in its new location?

3. Making cDNA libraries from cells undergoing a particular regulatory response can be a powerful way of detecting the genes involved. However, until recently it has been much easier to isolate mRNA that is not contaminated with other types of RNA (for example, rRNA and tRNA) from eukaryotes than from prokaryotes. The technique used in eukaryotes is "positive," in that mRNA is isolated directly. In prokaryotes the techniques involve removing the other RNAs. Describe how one would isolate mRNA from eukaryotes. Why isn't this method used with prokaryotic mRNA?

4. You have just discovered a protein in mice that may be an effective cure for cancer but it is present only in extremely small amounts. Describe the steps you would use to obtain production of this protein in *Escherichia coli*. If the protein produced in *E. coli* proved to be only marginally effective in clinical trials, what might you do to find a protein that is more effective?

5. Gene therapy is used to treat people who have a genetic disease and, if successful, will cure them. However, such people will still be able to pass on the genetic disease to their offspring. Explain. Why do you believe this might be an area of research that is not attracting as much attention as treatment of the individual?

The information in Appendix 1 is intended to help students calculate changes in free energy accompanying chemical reactions carried out by microorganisms. It begins with definitions of the terms required to make such calculations and proceeds to show how knowledge of redox state, atomic and charge balance, and other factors are necessary to calculate free-energy problems successfully.

Definitions

1. ΔG^0 = standard free-energy change of the reaction at 1 atm pressure and 1 M concentrations; ΔG = free-energy change under the conditions specified; $\Delta G^{0\prime}$ = free-energy change under standard conditions at pH 7.
2. Calculation of ΔG^0 for a chemical reaction from the free energy of formation, G_f^0, of products and reactants:

$$\Delta G^0 = \sum \Delta G_f^0 \text{(products)} - \sum \Delta G_f^0 \text{ (reactants)}$$

That is, sum the ΔG_f^0 of products, sum the ΔG_f^0 of reactants, and subtract the latter from the former.

3. For energy-yielding reactions involving H^+, converting from standard conditions (pH 0) to biochemical conditions (pH 7):

$$\Delta G^{0\prime} = \Delta G^0 + m\Delta G_f^\prime (H^+)$$

where m is the net number of protons in the reaction (m is negative when more protons are consumed than formed) and $\Delta G_f^\prime (H^+)$ is the free energy of formation of a proton at pH 7 = -39.83 kJ at 25°C.

4. Effect of concentrations on ΔG: With soluble substrates, the concentration ratios of products formed to exogenous substrates used are generally equal to or greater than 10^{-2} at the beginning of growth and equal to or less than 10^{-2} at the end of growth. From the relation between ΔG and the equilibrium constant (see item 8), it can be calculated that ΔG for the free-energy yield in practical situations differs from the free-energy yield under standard conditions by at most 11.7 kJ, a rather small amount, and so for a first approximation, standard free-energy yields can be used in most situations. However, with H_2 as a product, H_2-consuming bacteria present may keep the concentration of H_2 so low that the free-energy yield is significantly affected. Thus, in the fermentation of ethanol to acetate and $H_2(C_2H_5OH + H_2O \rightarrow C_2H_3O_2^- + 2H_2 + H^+)$, the $\Delta G^{0\prime}$ at 1 atm H_2 is $+9.68$ kJ, but at 10^{-4} atm H_2 it is -36.03 kJ. With H_2-consuming bacteria present, therefore, the ethanol fermentation becomes useful. (See also item 9.)

5. Reduction potentials: by convention, electrode equations are written in the direction, oxidant $+ ne^- \rightarrow$ reductant (that is, as reductions), where n is the number of electrons transferred. The standard potential (E_0) of the hydrogen electrode, $2H^+ + 2e^- \rightarrow H_2$ is set by definition at 0.0 V at 1.0 atm pressure of H_2 gas and 1.0 M H^+, at 25°C. $E_0^\prime$ is the standard reduction potential at pH 7. See also Table A1.2.

6. Relation of free energy to reduction potential:

$$\Delta G^{0\prime} = -nF\Delta E_0^\prime$$

where n is the number of electrons transferred, F is the Faraday constant (96.48 kJ/V), and $\Delta E_0^\prime$ is the $E_0^\prime$ of the electron-accepting couple minus the $E_0^\prime$ of the electron-donating couple.

7. Equilibrium constant, K. For the generalized reaction $aA + bB \rightleftharpoons cC + dD$,

$$K = \frac{[C]^c[D]^d}{[A]^a[B]^b}$$

where A, B, C, and D represent reactants and products; a, b c, and d represent number of molecules of each; and brackets indicate concentrations. This is true only when the chemical system is in equilibrium.

8. Relation of equilibrium constant, K, to free-energy change. At constant temperature, pressure, and pH,

$$\Delta G = \Delta G^{0\prime} + RT\ln K$$

where R is a constant (8.29 J/mol/°K) and T is the absolute temperature (in °K).

9. Two substances can react in a redox reaction even if the standard potentials are unfavorable, provided that the concentrations are appropriate.

Assume that normally the reduced form of A would donate electrons to the oxidized form of B. However, if the concentration of the reduced form of A were low and the concentration of the reduced form of B were high, it would be possible for the reduced form of B to donate electrons to the oxidized form of A. Thus, the reaction would proceed in the direction opposite that predicted from standard potentials. A practical example of this is the utilization of H^+ as an electron acceptor to produce H_2. Normally, H_2 production in fermentative bacteria is not extensive because H^+ is a poor electron acceptor; the $E_0^\prime$ of the $2H^+/H_2$ pair is -0.41 V. However, if the concentration of H_2 is kept low by continually removing it (a process done by methanogenic prokaryotes, which use $H_2 + CO_2$ to produce methane, CH_4, or by many other anaerobes capable of consuming H_2 anaerobically), the potential will be more positive and then H^+ will serve as a suitable electron acceptor.

TABLE A1.1 Free energies of formation, G_f^0, for some substances (kJ/mol)[a]

Carbon compound	Metal	Nonmetal	Nitrogen compound
CO, −137.34	Cu$^+$, +50.28	H$_2$, 0	N$_2$, 0
CO$_2$, −394.4	Cu^{2+}, +64.94	H$^+$, 0 at pH 0;	NO, +86.57
CH$_4$, −50.75	CuS, −49.02	−39.83 at pH 7	NO$_2$, +51.95
H$_2$CO$_3$, −623.16	Fe^{2+}, −78.87	(−5.69 per pH unit)	NO$_2^-$, −37.2
HCO$_3^-$, −586.85	Fe^{3+}, −4.6	O$_2$, 0	NO$_3^-$, −111.34
CO$_3^{2-}$, −527.90	FeCO$_3$, −673.23	OH$^-$, −157.3 at pH 14;	NH$_3$, −26.57
Acetate, −369.41	FeS$_2$, −150.84	−198.76 at pH 7;	NH$_4^+$, −79.37
Alanine, −371.54	FeSO$_4$, −829.62	−237.57 at pH 0	N$_2$O, +104.18
Aspartate, −700.4	PbS, −92.59	H$_2$O, −237.17	
Benzoic acid, −245.6	Mn^{2+}, −227.93	H$_2$O$_2$, −134.1	
Butyrate, −352.63	Mn^{3+}, −82.12	PO$_4^{3-}$, −1026.55	
Caproate, −335.96	MnO$_4^{2-}$, −506.57	Se^0, 0	
Citrate, −1168.34	MnO$_2$, −456.71	H$_2$Se, −77.09	
Crotonate, −277.4	MnSO$_4$, −955.32	SeO$_4^{2-}$, −439.95	
Cysteine, −339.8	Hgs, −49.02	S^0, 0	
Ethanol, −181.75	MoS$_2$, −225.42	SO$_3^{2-}$, −486.6	
Formaldehyde, −130.54	ZnS, −198.60	SO$_4^{2-}$, −744.6	
Formate, −351.04		S$_2$O$_3^{2-}$, −513.4	
Fructose, −915.38		H$_2$S, −27.87	
Fumarate, −604.21		HS$^-$, +12.05	
Gluconate, −1128.3		S^{2-}, +85.8	
Glucose, −917.22			
Glutamate, −699.6			
Glycerate, −658.1			
Glycerol, −488.52			
Glycine, −314.96			
Glycolate, −530.95			
Guanine, +46.99			
Lactate, −517.81			
Lactose, −1515.24			
Malate, −845.08			
Mannitol, −942.61			
Methanol, −175.39			
Oxalate, −674.04			
Phenol, −47.6			
n-Propanol, −175.81			
Propionate, −361.08			
Pyruvate, −474.63			
Ribose, −757.3			
Succinate, −690.23			
Sucrose, −370.90			
Urea, −203.76			
Valerate, −344.34			

[a] Values for free energy of formation of various compounds can be found in Dean, J. A. 1973. *Lange's Handbook of Chemistry*, 11th edition. McGraw-Hill, New York; Garrels, R. M., and C. L. Christ. 1965. *Solutions, Minerals, and Equilibria*. Harper and Row; New York; Burton, K. 1957. In Krebs, H. A., and H. L. Kornberg. Energy transformations in living matter, *Ergebnisse der Physiologie* (appendix): Springer-Verlag, Berlin; and Thauer, R. K., K. Jungermann, and H. Decker. 1977. Energy conservation in anaerobic chemotrophic bacteria. *Bacteriol Rev* 41:100–180.

Oxidation State or Number

1. The oxidation state of an element in an elementary substance (for example, H$_2$, O$_2$) is zero.

2. The oxidation state of the ion of an element is equal to its charge (for example, Na$^+$ = +1, Fe^{3+} = +3, O^{2-} = −2).

3. The sum of oxidation numbers of all atoms in a neutral molecule is zero. Thus, H$_2$O is neutral because it has two H at +1 each and one O at −2.

4. In an ion, the sum of oxidation numbers of all atoms is equal to the charge on that ion. Thus, in the OH$^-$ ion, O(−2) + H(+1) = −1.

5. In compounds, the oxidation state of O is virtually always −2, and that of H is +1.

6. In simple carbon compounds, the oxidation state of C can be calculated by adding up the H and O atoms present and using the oxidation states of these elements as given in item 5, because in a neutral compound the sum of all oxi-

dation numbers must be zero. Thus, the oxidation state of carbon in methane, CH_4, is -4 (4 H at $+1$ each $= +4$); in carbon dioxide, CO_2, the oxidation state of carbon is $+4$ (2 O at -2 each $= -4$).

7. In organic compounds with more than one C atom, it may not be possible to assign a specific oxidation number to each C atom, but it is still useful to calculate the oxidation state of the compound as a whole. The same conventions are used. Thus, the oxidation state of carbon in glucose, $C_6H_{12}O_6$, is zero (12 H at $+1 = 12$; 6 O at $-2 = -12$) and the oxidation state of carbon in ethanol, C_2H_6O, is -2 each (6 H at $+1 = +6$; one O at -2).

8. In all oxidation-reduction reactions there is a balance between the oxidized and reduced products. To calculate an oxidation-reduction balance, the number of molecules of each product is multiplied by its oxidation state. For instance, in calculating the oxidation-reduction balance for the alcoholic fermentation, there are two molecules of ethanol at $-4 = -8$ and two molecules of CO_2 at $+4 = +8$ so the net balance is zero. When constructing model reactions, it is useful to calculate redox balances to be certain that the reaction is possible.

Calculating Free-Energy Yields for Hypothetical Reactions

Energy yields can be calculated either from free energies of formation of the reactants and products or from differences in reduction potentials of electron-donating and electron-accepting partial reactions.

Calculations from Free Energy

Free energies of formation are given in Table A1.1. The procedure to use for calculating energy yields of reactions follows.

1. *Balancing reactions.* In all cases, it is essential to ascertain that the coupled oxidation-reduction reaction is balanced. Balancing involves three things; (*a*) the total number of each kind of atom must be identical on both sides of the equation; (*b*) there must be an ionic balance so that when positive and negative ions are added up on the right side of the equation, the total ionic charge (whether positive, negative, or neutral) exactly balances the ionic charge on the left side of the equation; and (*c*) there must be an oxidation-reduction balance so that all the electrons removed from one substance must be transferred to another substance. In general, when constructing balanced reactions, one proceeds in the reverse of the three steps just listed. Usually, if steps (*c*) and (*b*) have been properly handled, step (*a*) becomes correct automatically.

2. *Examples:* (*a*) What is the balanced reaction for the oxidation of H_2S to SO_4^{2-} with O_2? First, decide how many electrons are involved in the oxidation of H_2S to SO_4^{2-}. This can be most easily calculated from the oxidation states of the compounds, using the rules given previously. Because H has an oxidation state of $+1$, the oxidation state of S in H_2S is -2. Because O has an oxidation state of -2, the oxidation state of S in SO_4^{2-} is $+6$ (because it is an ion, using the rules given in items 4 and 5 of the previous section). Thus, the oxidation of H_2S to SO_4^{2-} involves an eight-elec-

tron transfer (from -2 to $+6$). Because each O atom can accept two electrons (the oxidation state of O in O_2 is zero, but in H_2O is -2), this means that two molecules of molecular oxygen, O_2, are required to provide sufficient electron-accepting capacity. Thus, at this point, we know that the reaction requires 1 H_2S and 2 O_2 on the left side of the equation, and 1 SO_4^{2-} on the right side. To achieve an ionic balance, we must have two positive charges on the right side of the equation to balance the two negative charges of SO_4^{2-}. Thus, 2 H^+ must be added to the right side of the equation, making the overall reaction

$$H_2S + 2 O_2 \rightarrow SO_4^{2-} + 2 H^+$$

By inspection, it can be seen that this equation is also balanced in terms of the total number of atoms of each kind on each side of the equation.

(*b*) What is the balanced reaction for the oxidation of H_2S to SO_4^{2-} with Fe^{3+} as electron acceptor? We have just ascertained that the oxidation of H_2S to SO_4^{2-} is an eight-electron transfer. Because the reduction of Fe^{3+} to Fe^{2+} is only a one-electron transfer, 8 Fe^{3+} will be required. At this point, the reaction looks like

$$H_2S + 8 Fe^{3+} \rightarrow 8 Fe^{2+} + SO_4^{2-} \quad \text{(not balanced)}$$

We note that the ionic balance is incorrect. We have 24 positive charges on the left and 14 positive charges on the right (16+ from Fe, 2− from sulfate). To equalize the charges, we add 10 H^+ on the right. Now our equation looks like

$$H_2S + 8 Fe^{3+} \rightarrow 8 Fe^{2+} + 10 H^+ + SO_4^{2-}$$
$$\text{(not balanced)}$$

To provide the necessary hydrogen for the H^+ and oxygen for the sulfate, we add 4 H_2O to the left and find that the equation is now balanced:

$$H_2S + 4 H_2O + 8 Fe^{3+} \rightarrow 8 Fe^{2+} + 10 H^+ + SO_4^{2-}$$
$$\text{(balanced)}$$

In general, in microbiological reactions, ionic balance can be achieved by adding H^+ or OH^- to the left or right side of the equation, and because all reactions take place in an aqueous medium, H_2O molecules can be added where needed. Whether H^+ or OH^- is added generally depends on whether the reaction is taking place under acid or alkaline conditions.

3. *Calculation of energy yield for balanced equations from free energies of formation.* Once an equation has been balanced, the free-energy yield can be calculated by inserting the values for the free energy of formation of each reactant and product from Table A1.1 and using the formula in item 2 of the first section of this appendix.

For instance, for the equation

$$H_2S + 2 O_2 \rightarrow SO_4^{2-} + 2 H^+$$

G'_f values $\rightarrow (-27.87) + (0) (-744.6) + 2(-39.83)$
$$\text{(assuming pH 7)}$$

$$\Delta G^{0'} = -796.39 \text{ kJ/rx}$$

The G'_f values for the products (right side of equation) are summed and subtracted from the G'_f values for the reactants (left side of equation), taking care to ensure that the arithmetic signs are correct. From the data in Table A1.1, a wide variety of free-energy yields for reactions of microbiological interest can be calculated.

Calculation of Free-Energy Yield from Reduction Potential

Reduction potentials of some important redox pairs are given in Table A1.2. The amount of energy that can be released from two half reactions can be calculated from the differences in reduction potentials of the two reactions and from the number of electrons transferred. The further apart the two half reactions are, and the greater the number of electrons, the more energy released. The conversion of potential difference to free energy is given by the formula $\Delta G^{0\prime} = -nF\Delta E_0'$ where n is the number of electrons, F is the Faraday constant (96.48 kJ/V), and $\Delta E_0'$ is the difference in potentials. Thus, the $2\,H^+/H_2$ couple has a potential of -0.41 V and the $\frac{1}{2}O_2/H_2O$ pair has a potential of $+0.82$ V, and so the potential difference is 1.23, which (because two electrons are involved) is equivalent to a free-energy yield (ΔG^0) of -237.34 kJ. On the other hand, the potential difference between the $2\,H^+/H_2$ and the NO_3^-/NO_2^- reactions is less, 0.84 V, which is equivalent to a free-energy yield of -162.08 kJ.

Because many biochemical reactions are two-electron transfers, it is often useful to give energy yields for two-electron reactions, even if more electrons are involved. Thus, the SO_4^{2-}/H_2 redox pair involves eight electrons, and complete reduction of SO_4^{2-} with H_2 requires $4\,H_2$ (equivalent to eight electrons). From the reduction potential difference between $2\,H^+/H_2$ and SO_4^{2-}/H_2S (0.19 V), a free-energy yield of -146.64 kJ is calculated, or -36.66 kJ per two electrons. By convention, reduction potentials are given for conditions in which equal concentrations of oxidized and reduced forms are present. In actual practice, the concentrations of these two forms may be quite different. As discussed earlier in this appendix (item 9, first section), it is possible to couple half reactions even if the potential difference is unfavorable, providing the concentrations of the reacting species are appropriate.

TABLE A1.2 Microbiologically important reduction potentials[a]

Redox pair	E_0' (V)
SO_4^{2-}/HSO_3^-	−0.52
CO_2/formate	−0.43
$2\,H^+/H_2$	−0.41
$S_2O_3^{2-}/HS^- + HSO_3^-$	−0.40
Ferredoxin ox/red	−0.39
Flavodoxin ox/red[b]	−0.37
$NAD^+/NADH$	−0.32
Cytochrome c_3 ox/red	−0.29
CO_2/acetate$^-$	−0.29
S^0/HS^-	−0.27
CO_2/CH_4	−0.24
FAD/FADH	−0.22
SO_4^{2-}/HS^-	−0.217
Acetaldehyde/ethanol	−0.197
Pyruvate$^-$/lactate$^-$	−0.19
FMN/FMNH	−0.19
Dihydroxyacetone phosphate/glycerolphosphate	−0.19
$HSO_3^-/S_3O_6^{2-}$	−0.17
Flavodoxin ox/red[b]	−0.12
HSO_3^-/HS^-	−0.116
Menaquinone ox/red	−0.075
APS/AMP $+HSO_3^-$	−0.060
Rubredoxin ox/red	−0.057
Acrylyl-CoA/propionyl-CoA	−0.015
Glycine/acetate$^-$ + NH_4^+	−0.010
$S_4O_6^{2-}/S_2O_3^{2-}$	+0.024
Fumarate^{2-}/succinate^{2-}	+0.033
Cytochrome b ox/red	+0.035
Ubiquinone ox/red	+0.113
AsO_4^{3-}/AsO_3^{3-}	+0.139
Dimethyl sulfoxide (DMSO)/dimethylsulfide (DMS)	+0.16
$Fe(OH)_3 + HCO_3^-/FeCO_3$	+0.20
$S_3O_6^{2-}/S_2O_3^{2-} + HSO_3^-$	+0.225
Cytochrome c_1 ox/red	+0.23
NO_2^-/NO	+0.36
Cytochrome a_3 ox/red	+0.385
Chlorobenzoate$^-$/benzoate$^-$ + HCl	+0.297
NO_3^-/NO_2^-	+0.43
SeO_4^{2-}/SeO_3^{2-}	+0.475
Fe^{3+}/Fe^{2+}	+0.77
Mn^{4+}/Mn^{2+}	+0.798
O_2/H_2O	+0.82
$ClO_3^-Cl^-$	+1.03
NO/N_2O	+1.18
N_2O/N_2	+1.36

[a]Data from Thauer, R. K., K. Jungermann, and K. Decker, 1977. Energy conservation in anaerobic chemotrophic bacteria. *Bacteriol. Rev.* 41:100–180.

[b]Separate potentials are given for each electron transfer in this potentially two-electron transfer.

VOLUME 1
The *Archaea*, Cyanobacteria, Phototrophs and Deeply
Branching Genera of *Bacteria*

THE ARCHAEA

Phylum I: *Crenarchaeota*
SECTION I Thermoprotei, sulfolobi and barophiles
  Class I: *Thermoprotei*
    Order I: *Thermoproteales*
      Family I: *Thermoproteaceae*

Pyrobaculum	3 sp.
Sulfophobococcus	1 sp.
Thermocladium	1 sp.
Thermoproteus	2 sp.

      Family II: *Thermofilaceae*

Thermofilum	1 sp.

    Order II: *Desulfurococcales*
      Family I: *Desulfurococcaceae*

Aeropyrum	1 sp.
Desulfurococcus	2 sp.
Igniococcus	1 sp.
Staphylothermus	1 sp.
Stetteria	1 sp.
Sulfophobococcus	1 sp.
Thermodiscus	1 sp.
Thermosphaera	1 sp.

      Family II: *Pyrodictiaceae*

Hyperthermus	1 sp.
Pyrodictium	3 sp.
Pyrolobus	1 sp.

    Order II: *Sulfolobales*
      Family I: *Sulfolobaceae*

Acidianus	3 sp.
Metallosphaera	2 sp.
Stygiolobus	1 sp.
Sulfolobus	6 sp.
Sulfurisphaera	1 sp.
Sulfurococcus	2 sp.

SECTION II The Methanogens
Phylum II *Euryarchaeota*
  Class I: *Methanobacteria*
    Order I: *Methanobacteriales*
      Family I: *Methanobacteriaceae*

Methanobacterium	19 sp.
Methanobrevibacter	7 sp.
Methanosphaera	2 sp.
Methanothermus	2 sp.

  Class II: *Methanococci*
    Order I: *Methanococcales*
      Family I: *Methanococcaceae*

Methanococcus	11 sp.
Methanothermococcus	1 sp.

      Family II: *Methanocaldococcaceae*

Methanocaldococcus	4 sp.
Methanotorris	1 sp.

    Order II: *Methanomicrobiales*
      Family I: *Methanomicrobiaceae*

Methanoculleus	6 sp.
Methanogenium	11 sp.
Methanolacinia	1 sp.
Methanomicrobium	2 sp.
Methanoplanus	3 sp.
Methanofollis	2 sp.

      Family II: *Methanocorpusculaceae*

Methanocorpusculum	4 sp.

    Order III: *Methanosarcinales*
      Family I: *Methanosarcinaceae*

Methanococcoides	2 sp.
Methanohalobium	1 sp.
Methanohalophilus	5 sp.
Methanolobus	5 sp.
Methanosarcina	8 sp.
Methanosalsum	1 sp.

      Family II: Methanosaetaceae

Methanosaeta	2 sp.

SECTION III The Halobacteria
  Class I: *Halobacteria*
    Order I: *Halobacteriales*
      Family I: *Halobacteriaceae*

Haloarcula	10 sp.
Halobacterium	1 sp.
Halobaculum	1 sp.
Halococcus	3 sp.
Haloferax	4 sp.
Halogeometricum	1 sp.
Halorubrum	7 sp.
Haloterrigena	1 sp.
Natrialba	2 sp.
Natrinema	2 sp.
Natronobacterium	1 sp.
Natronococcus	2 sp.
Natronomonas	1 sp.
Natronorubrum	2 sp.

SECTION IV The Thermoplasmas
  Class I: 'Thermoplasma'
    Order I: *Thermoplasmatales*
      Family I: *Thermoplasmataceae*

Thermoplasma	2 sp.

      Family II: *Picrophilaceae*

Picrophilus	2 sp.

SECTION V The Thermococci
  Class I: *Thermococci*
    Order I: *Archaeoglobales*
      Family I: *Archaeoglobaceae*

Archaeoglobus	3 sp.
Ferroglobus	1 sp.

    Order II: *Thermococcales*
      Family I: *Thermococcaceae*

Pyrococcus	2 sp.
Thermococcus	12 sp.

  Class II: *Methanopyri*
    Order I: *Methanopyrales*
      Family I: *Methanopyraceae*

Methanopyrus	1 sp.

THE DEEPLY BRANCHING GENERA OF *BACTERIA*

SECTION VI *Aquifex* and relatives
  Class I: *Aquificae*
    Order I: *Aquificales*
      Family I: 'Aquificaceae'

Aquifex	2 sp.
Calderobacterium	1 sp.
Hydrogenobacter	2 sp.
Thermocrinus	1 sp.

*The list of genera and higher order taxa shown here are the organisms currently recognized, with the numbers besides each genus name giving the number of species recognized as of 2001. Genera or higher order taxa in quotation marks or set in roman instead of italics are recognized taxa whose names have not yet been validated. Because bacterial taxonomy is a work in progress, updates to the list shown here occur as new genera and species are described and as new data support new taxonomic arrangements.

SECTION VII Thermotogas and Geotogas
    Class I: *Thermotogae*
        Order I: *Thermotogales*
            Family I: *Thermotogaceae*
                Fervidobacterium   4 sp.
                Geotoga   2 sp.
                Petrotoga   2 sp.
                Thermosipho   2 sp.
                Thermotoga   5 sp.
    Class II: *Thermodesulfobacteria*
        Order I: *Thermodesulfobacteriales*
            Family I: *Thermodesulfobacteriaceae*
                Thermodesulfobacterium   2 sp.
SECTION VIII The Deinococci
    Class I: *Deinococci*
        Order I: *Deinococcales*
            Family I: *Deinococcaceae*
                Deinococcus   8 sp.
SECTION IX *Thermus* and relatives
    Class I: *Thermi*
        Order I: *Thermales*
            Family I: *Thermaceae*
                Meiothermus   4 sp.
                Thermus   6 sp.
SECTION X Chrysiogenes
    Class I: *Chrysiogenetes*
        Order I: *Chrysiogenales*
            Family I: *Chrysiogenaceae*
                Chrysiogenes   1 sp.
SECTION XI The Chloroflexi and Herpetosiphons
    Class I: 'Chloroflexi'
        Order I: 'Chloroflexales'
            Family I: 'Chloroflexaceae'
                Chloroflexus   2 sp.
                Chloronema   1 sp.
                Heliothrix   1 sp.
                Oscillochloris   2 sp.
        Order II: 'Herpetosiphonales'
            Family I: 'Herpetosiphonaceae'
                Herpetosiphon   5 sp.
SECTION XII Thermomicrobia
    Class I: 'Thermomicrobia'
        Order I: 'Thermomicrobiales'
            Family I: 'Thermomicrobiaceae'
                Thermomicrobium   2 sp.
SECTION XIII *Nitrospira* and relatives
    Class I: 'Nitrospira'
        Order I: *Nitrospirales*
            Family I: *Nitrospiralaceae*
                Nitrospira   1 sp.
                Thermodesulfovibrio   1 sp.
                'Leptospirillum'   1 sp.
                'Magnetobacterium'   1 sp.
SECTION XIV *Deferribacter* and relatives
    Class I: *Deferribacteraceae*
        Order I: *Defferibacterales*
            Family I: *Defferibacteraceae*
                Defferibacter   1 sp.
                Flexistipes   1 sp.
                Geovibrio   1 sp.
                Synergistes   1 sp.
SECTION XV The Cyanobacteria
    Class I: 'Prochlorophyta'
        Order I: 'Chroococcales'
            Family I:
                'Chamaesiphon'   2 sp.
                'Chroococcus'   1 sp.
                'Cyanobacterium'   1 sp.
                'Cyanothece'   1 sp.
                'Dactylococcopsis (Myxobaktron)'   1 sp.
                'Gloeobacter'   1 sp.
                'Gloeocapsa'   1 sp.
                'Gloeothece'   1 sp.
                'Microcystis'   0 sp.

'Prochlorococcus'   1 sp.
Prochloron   1 sp.
'Synechococcus'   3 sp.
'Synechocystis'   1 sp.
        Order II: 'Pleurocapsales'
            Family I:
                'Chroococcidiopsis'   2 sp.
                'Cyanocystis'   1 sp.
                'Dermocarpa'   1 sp.
                'Dermocarpella'   1 sp.
                'Hyella'   1 sp.
                'Myxosarcina'   1 sp.
                'Pleurocapsa'   1 sp.
                'Stanieria'   1 sp.
                'Xenococcus'   1 sp.
        Order III: 'Oscillatoriales'
            Family I:
                'Arthrospira'   2 sp.
                'Borzia'   1 sp.
                'Crinalium'   1 sp.
                'Geitlerinema'   1 sp.
                'Hormoscilla'   1 sp.
                'Isocystis'   1 sp.
                'Leptolyngbia'   1 sp.
                'Limnothrix'   1 sp.
                'Lyngbya'   1 sp.
                'Microcoleus'   2 sp.
                'Oscillatoria'   6 sp.
                'Planktothrix'   1 sp.
                Prochlorothrix   1 sp.
                'Pseudoanabaena'   1 sp.
                'Spirulina'   1 sp.
                'Starria'   1 sp.
                'Symploca'   1 sp.
                'Trichodesmium'   1 sp.
                'Tychonema'   1 sp.
        Order IV: 'Nostocales'
            Family I: 'Nostocaceae'
                'Anabaena'   1 sp.
                'Anabaenopsis'   1 sp.
                'Aphanizomenon'   1 sp.
                'Cyanospira'   1 sp.
                'Cylindrospermum'   2 sp.
                'Microchaete'   1 sp.
                'Nodularia'   1 sp.
                'Nostoc'   4 sp.
            Family II: 'Scytonemataceae'
                'Scytonema'   1 sp.
                'Tolypothrix'   2 sp.
            Family III: 'Rivulariaceae'
                'Calothrix'   1 sp.
                'Rivularia'   1 sp.
        Order V: 'Stigonematales'
            Family IV:
                'Chlorogloeopsis'   1 sp.
                'Fischerella'   1 sp.
                'Geitleria'   1 sp.
                'Hapalosiphon'   1 sp.
                'Loriella'   1 sp.
                'Mastigocladus'   1 sp.
                'Mastigocoleus'   1 sp.
                'Matteia'   1 sp.
                'Nostochopsis'   1 sp.
                'Stigonema'   1 sp.
SECTION XVI
    Chlorobia and other green-colored anoxygenic phototrophs
    Class I: 'Chlorobia'
        Order I: *Chlorobiales*
            Family I: *Chlorobiaceae*
                Ancalochloris   1 sp.
                Chlorobium   7 sp.
                Chloroherpeton   1 sp.
                Pelodictyon   4 sp.
                Prosthecochloris   1 sp.

Class II: *'Clostridia'*
  Order I: *Clostridiales*
    Family: *Heliobacteriaceae*
      Heliobacterium — 3 sp.
      Heliobacillus — 1 sp.
      Heliophilum — 1 sp.
      Heliorestis — 1 sp.

VOLUME 2
The Proteobacteria

THE BACTERIA

Phylum VII: *'Proteobacteria'*
SECTION XVII The α-Proteobacteria
  Class I: *'Rhodospirilli'*
    Order I: *Rhodospirillales*
      Family I: *'Rhodospirillaceae'*
        Azospirillum — 7 sp.
        Magnetospirillum — 2 sp.
        Phaeospirillum — 2 sp.
        Rhodocista — 1 sp.
        Rhodospira — 1 sp.
        Rhodospirillum — 9 sp.
        Rhodothalassium — 1 sp.
        Rhodovibrio — 2 sp.
        Roseospira — 1 sp.
      Family II: *Acetobacteraceae*
        Acetobacter — 20 sp.
        Acidiphilium — 8 sp.
        Acidocella — 2 sp.
        Acidomonas — 1 sp.
        Craurococcus — 1 sp.
        Frateuria — 1 sp.
        Gluconacetobacter — 6 sp.
        Gluconobacter — 8 sp.
        Paracraurococcus — 1 sp.
        Rhodopila — 1 sp.
        Roseococcus — 1 sp.
        Stella — 2 sp.
        Zavarzinia — 1 sp.
    Order II: Rickettsiales
      Family I: *Rickettsiaceae*
        Orientia — 1 sp.
        Rickettsia — 22 sp.
      Family II: *Ehrlichiaceae*
        Aegyptianella — 1 sp.
        Anaplasma — 4 sp.
        Cowdria — 1 sp.
        Ehrlichia — 8 sp.
        Eperythrozoon — 5 sp.
        Haemobartonella — 3 sp.
        Neorickettsia — 1 sp.
      Family III: *'Holosporaceae'*
        Caedibacter — 5 sp.
        Holospora — 4 sp.
        Lyticum — 2 sp.
        Polynucleobacter — 1 sp.
        Pseudocaedibacter — 3 sp.
        Symbiotes — 1 sp.
        Tectibacter — 1 sp.
    Order III: *'Rhodobacterales'*
      Family I: *'Rhodobacteraceae'*
        Amaricoccus — 4 sp.
        Antarctobacter — 1 sp.
        Gemmobacter — 1 sp.
        Hirschia — 1 sp.
        Hyphomonas — 5 sp.
        Octadecabacter — 2 sp.
        Paracoccus — 13 sp.
        Rhodobacter — 8 sp.
        Rhodovulum — 4 sp.
        Roseobacter — 4 sp.
        Sagittula — 1 sp.

        Sulfitobacter — 1 sp.
        Ahrensia — 1 sp.
        Roseovarius — 1 sp.
        Rubrimonas — 1 sp.
        Ruegeria — 3 sp.
        Stappia — 2 sp.
    Order IV: *'Sphingomonadales'*
      Family I: *'Sphingomonadaceae'*
        Blastomonas — 1 sp.
        Erythrobacter — 2 sp.
        Erythromicrobium — 1 sp.
        Erythromonas — 1 sp.
        Porphyrobacter — 2 sp.
        Rhizomonas — 1 sp.
        Sandaracinobacter — 1 sp.
        Sphingomonas — 19 sp.
        Zymomonas — 2 sp.
        Rhodanobacter — 1 sp.
    Order V: *Caulobacterales*
      Family I: *Caulobacteraceae*
        Asticcacaulis — 2 sp.
        Brevundimonas — 2 sp.
        Caulobacter — 11 sp.
    Order VI: *'Rhizobiales'*
      Family I: *Rhizobiaceae*
        Agrobacterium — 10 sp.
        Carbophilus — 1 sp.
        Chelatobacter — 1 sp.
        Ensifer — 1 sp.
        Rhizobium — 20 sp.
        Sinorhizobium — 6 sp.
      Family II: *'Phyllobacteriaceae'*
        Mesorhizobium — 7 sp.
        Phyllobacterium — 2 sp.
      Family III: *Bartonellaceae*
        Bartonella — 14 sp.
      Family IV: *Brucellaceae*
        Brucella — 6 sp.
        Mycoplana — 4 sp.
        Ochrobactrum — 2 sp.
      Family VI: *'Rhodobiaceae'*
        Rhodobium — 2 sp.
      Family V: *Methylobacteriacea*
        Methylorhabdus — 1 sp.
        Protomonas — 1 sp.
        Roseomonas — 3 sp.
      Family VI: *'Methylocystaceae'*
        Methylocystis — 2 sp.
        Methylosinus — 2 sp.
      Family VII: *'Beijerinckiaceae'*
        Beijerinckia — 6 sp.
        Chelatococcus — 1 sp.
        Derxia — 1 sp.
      Family VIII: *Hyphomicrobiaceae*
        Ancalomicrobium — 1 sp.
        Ancylobacter — 1 sp.
        Angulomicrobium — 1 sp.
        Aquabacter — 1 sp.
        Azorhizobium — 1 sp.
        Blastochloris — 2 sp.
        Devosia — 1 sp.
        Dichotomicrobium — 1 sp.
        Filomicrobium — 1 sp.
        Gemmiger — 1 sp.
        Hyphomicrobium — 12 sp.
        Labrys — 1 sp.
        Nevskia — 1 sp.
        Pedomicrobium — 4 sp.
        Prosthecomicrobium — 4 sp.
        Rhodomicrobium — 1 sp.
        Rhodoplanes — 2 sp.
        Seliberia — 1 sp.
        Xanthobacter — 4 sp.

Family IX: 'Bradyrhizobiaceae'
   Afipia — 3 sp.
   Blastobacter — 5 sp.
   Bosea — 1 sp.
   Bradyrhizobium — 3 sp.
   Nitrobacter — 1 sp.
   Nitrococcus — 1 sp.
   Nitrospina — 1 sp.
   Oligotropha — 1 sp.
   Rhodopseudomonas — 15 sp.
SECTION XVIII The β-Proteobacteria
  Class I: 'Neisseriae'
   Order I: 'Neisseriales'
    Family I: Neisseriaceae
     Alysiella — 1 sp.
     Aquaspirillum — 21 sp.
     Catenococcus — 1 sp.
     Chromobacterium — 2 sp.
     Eikenella — 1 sp.
     Iodobacter — 1 sp.
     Kingella — 4 sp.
     Microvirgula — 1 sp.
     Neisseria — 24 sp.
     Prolinoborus — 1 sp.
     Simonsiella — 3 sp.
     Vogesella — 1 sp.
   Order II: 'Burkholderiales'
    Family I: 'Burkholderiaceae'
     Burkholderia — 20 sp.
     Cupriavidus — 1 sp.
     Ralstonia — 3 sp.
     Thermothrix — 2 sp.
    Family II: 'Oxalobacteraceae'
     Duganella — 1 sp.
     Janthinobacterium — 1 sp.
     Lautropia — 1 sp.
     Oxalobacter — 2 sp.
     Telluria — 2 sp.
    Family III: Alcaligenaceae
     Achromobacter — 4 sp.
     Alcaligenes — 16 sp.
     Bordetella — 7 sp.
     Pelistega — 1 sp.
     Sutterella — 1 sp.
     Taylorella — 1 sp.
    Family VI: Comamonadaceae
     Acidovorax — 7 sp.
     Brachymonas — 1 sp.
     Comamonas — 3 sp.
     Herbaspirillum — 2 sp.
     Hydrogenophaga — 4 sp.
     Ideonella — 1 sp.
     Leptothrix — 5 sp.
     Polaromonas — 1 sp.
     Rhodoferax — 1 sp.
     Rubrivivax — 1 sp.
     Sphaerotilus — 1 sp.
     Thiomonas — 4 sp.
     Variovorax — 1 sp.
   Order III: 'Rhodocyclales'
    Family I: 'Rhodocyclaceae'
     Azoarcus — 5 sp.
     Rhodocyclus — 3 sp.
     Thauera — 4 sp.
     Zoogloea — 1 sp.
   Order IV: 'Nitrosomonadales'
    Family I: 'Nitrosomonadaceae'
     Nitrosomonas — 1 sp.
     Nitrosospira — 3 sp.
    Family II: Spirillaceae
     Spirillum — 1 sp.
     Thiobacillus — 21 sp.
    Family III: Gallionellacea
     Gallionella — 1 sp.

Order V: 'Methylophilales'
  Family I: 'Methylophilaceae'
   Methylobacillus — 2 sp.
   Methylophaga — 3 sp.
   Methylophilus — 1 sp.
   Methylovorus — 1 sp.
SECTION XIX The γ-Proteobacteria
  Class I: 'Zymobacteria'
   Order I: 'Chromatiales'
    Family I: Chromatiaceae
     Allochromatium — 3 sp.
     Amoebobacter — 4 sp.
     Chromatium — 13 sp.
     Halochromatium — 2 sp.
     Isochromatium — 1 sp.
     Lamprobacter — 1 sp.
     Lamprocystis — 1 sp.
     Marichromatium — 2 sp.
     Nitrosococcus — 2 sp.
     Rhabdochromatium — 1 sp.
     Thermochromatium — 1 sp.
     Thiocapsa — 5 sp.
     Thiococcus — 1 sp.
     Thiocystis — 4 sp.
     Thiodictyon — 2 sp.
     Thiohalocapsa — 1 sp.
     Thiolamprovum — 1 sp.
     Thiopedia — 1 sp.
     Thiorhodococcus — 1 sp.
     Thiorhodovibrio — 1 sp.
     Thiospirillum — 1 sp.
    Family II: Ectothiorhodospiraceae
     Arhodomonas — 1 sp.
     Ectothiorhodospira — 9 sp.
     Halorhodospira — 3 sp.
   Order II: 'Xanthomonadales'
    Family I: 'Xanthomonadaceae'
     Stenotrophomonas — 2 sp.
     Xanthomonas — 24 sp.
     Xylella — 1 sp.
   Order III: 'Cardiobacteriales'
    Family I: Cardiobacteriaceae
     Cardiobacterium — 1 sp.
     Dichelobacter — 1 sp.
     Suttonella — 1 sp.
   Order IV: 'Thiotrichales'
    Family I: 'Thiotrichaceae'
     Achromatium — 1 sp.
     Beggiatoa — 1 sp.
     Leucothrix — 1 sp.
     Macromonas — 2 sp.
     Thiobacterium — 1 sp.
     Thioploca — 4 sp.
     Thiospira — 1 sp.
     Thiothrix — 1 sp.
     Vitreoscilla — 3 sp.
    Family II: Piscirickettsiaceae
     Cycloclasticus — 1 sp.
     Hydrogenovibrio — 1 sp.
     Piscirickettsia — 1 sp.
     Thiomicrospira — 4 sp.
    Family III: 'Francisellaceae'
     Francisella — 5 sp.
   Order V: 'Legionellales'
    Family I: Legionellaceae
     Legionella — 44 sp.
    Family II: 'Coxiellaceae'
     Coxiella — 1 sp.
     Rickettsiella — 4 sp.
   Order VI: 'Methylococcales'
    Family I: Methylococcaceae
     Methylobacter — 6 sp.
     Methylobacterium — 9 sp.
     Methylocaldum — 3 sp.

Methylococcus	8 sp.
Methylomicrobium	3 sp.
Methylomonas	4 sp.
Methylosphaera	1 sp.
Order VII: 'Oceanospirillales'	
Family I: 'Oceanospirillaceae'	
Balneatrix	1 sp.
Marinomonas	2 sp.
Marinospirillum	2 sp.
Oceanospirillum	16 sp.
Family II: Halomonadaceae	
Alcanivorax	1 sp.
Carnimonas	1 sp.
Chromohalobacter	1 sp.
Deleya	8 sp.
Halomonas	19 sp.
Zymobacter	1 sp.
Order VIII: Pseudomonadales	
Family I: Pseudomonadaceae	
Agromonas	1 sp.
Aminobacter	3 sp.
Azomonas	3 sp.
Azotobacter	19 sp.
Cellvibrio	2 sp.
Chryseomonas	2 sp.
Flavimonas	1 sp.
Lampropedia	1 sp.
Lysobacter	5 sp.
Mesophilobacter	1 sp.
Morococcus	1 sp.
Oligella	2 sp.
Phenylobacterium	1 sp.
Pseudomonas	117 sp.
Rhizobacter	1 sp.
Rugamonas	1 sp.
Serpens	1 sp.
Thermoleophilum	2 sp.
Xylophilus	1 sp.
Family II: Moraxellaceae	
Acinetobacter	7 sp.
Moraxella	8 sp.
Moraxella (Branhamella)	4 sp.
Moraxella (Moraxella)	6 sp.
Psychrobacter	5 sp.
Order IX: 'Alteromonadales'	
Family I: 'Alteromonadaceae'	
Alteromonas	21 sp.
Colwellia	7 sp.
Ferrimonas	1 sp.
Marinobacter	1 sp.
Marinobacterium	1 sp.
Microbulbifer	1 sp.
Pseudoalteromonas	17 sp.
Shewanella	10 sp.
Order X: 'Vibrionales'	
Family I: Vibrionaceae	
Allomonas	1 sp.
Enhydrobacter	1 sp.
Listonella	3 sp.
Photobacterium	10 sp.
Salinivibrio	1 sp.
Vibrio	46 sp.
Order XI: 'Aeromonadales'	
Family I: Aeromonadaceae	
Aeromonas	23 sp.
Ruminobacter	1 sp.
Tolumonas	1 sp.
Order XII: 'Enterobacteriales'	
Family I: Enterobacteriaceae	
Arsenophonus	1 sp.
Buchnera	1 sp.
Budvicia	1 sp.
Buttiauxella	7 sp.
Calymmatobacterium	1 sp.

Cedecea	3 sp.
Citrobacter	10 sp.
Edwardsiella	4 sp.
Enterobacter	15 sp.
Erwinia	30 sp.
Escherichia	6 sp.
Ewingella	1 sp.
Hafnia	1 sp.
Klebsiella	11 sp.
Kluyvera	4 sp.
Leclercia	1 sp.
Leminorella	2 sp.
Moellerella	1 sp.
Morganella	2 sp.
Obesumbacterium	1 sp.
Pantoea	8 sp.
Photorhabdus	1 sp.
Plesiomonas	1 sp.
Pragia	1 sp.
Proteus	7 sp.
Providencia	6 sp.
Rahnella	1 sp.
Saccharobacter	1 sp.
Salmonella	12 sp.
Serratia	12 sp.
Shigella	4 sp.
Tatumella	1 sp.
Trabulsiella	1 sp.
Wigglesworthia	1 sp.
Xenorhabdus	9 sp.
Yersinia	12 sp.
Yokenella	1 sp.
Brenneria	6 sp.
Pectobacterium	11 sp.
Sodalis	1 sp.
Order XIII: 'Pasteurellales'	
Family I: Pasteurellaceae	
Actinobacillus	17 sp.
Haemophilus	20 sp.
Lonepinella	1 sp.
Pasteurella	23 sp.
Mannheimia	5 sp.
SECTION XX The δ-Proteobacteria	
Class I: 'Predibacteria'	
Order I: 'Desulfovibrionales'	
Family I: 'Desulfovibrionaceae'	
Bilophila	1 sp.
Desulfomicrobium	4 sp.
Desulfomonas	1 sp.
Desulfonatronovibrio	1 sp.
Desulfovibrio	29 sp.
Lawsonia	1 sp.
Family II: 'Desulfurellaceae'	
Desulfurella	4 sp.
Family III: 'Desulfobacteraceae'	
Desulfobacter	6 sp.
Desulfobacterium	7 sp.
Desulfococcus	2 sp.
Desulfosarcina	1 sp.
Desulfospira	1 sp.
Desulfocella	1 sp.
Family IV: 'Desulfobulbaceae'	
Desulfacinum	1 sp.
Desulfobulbus	3 sp.
Desulfocapsa	1 sp.
Desulfofustis	1 sp.
Desulfomonile	1 sp.
Desulforhabdus	1 sp.
Syntrophobacter	3 sp.
Thermodesulforhabdus	1 sp.
Family V: 'Geobacteraceae'	
Desulfohalobium	1 sp.
Desulfonema	2 sp.
Desulfuromonas	4 sp.

Desulfuromusa	3 sp.
Geobacter	2 sp.
Pelobacter	6 sp.
Family VI: 'Bdellovibrionaceae'	
Bdellovibrio	3 sp.
Micavibrio	1 sp.
Vampirovibrio	1 sp.
Order II: *Myxococcales*	
Family I: *Myxococcaceae*	
Angiococcus	1 sp.
Myxococcus	8 sp.
Family II: *Archangiaceae*	
Archangium	1 sp.
Family III: *Cystobacteraceae*	
Cystobacter	3 sp.
Melittangium	3 sp.
Stigmatella	2 sp.
Family IV: *Polyangiaceae*	
Chondromyces	5 sp.
Nannocystis	1 sp.
Polyangium	10 sp.

SECTION XXI The ε-Proteobacteria
Class I: 'Campylobacteres'
Order I: 'Campylobacterales'
Family I: *Campylobacteraceae*

Arcobacter	4 sp.
Campylobacter	28 sp.
Sulfurospirillum	2 sp.
Thiovulum	1 sp.
Family II: 'Helicobacteraceae'	
Helicobacter	18 sp.
Wolinella	3 sp.

VOLUME 3
The Low G + C Gram-Positive *Bacteria*

THE BACTERIA

Phylum VIII: 'Firmicutes'
SECTION XXII The *Clostridia* and relatives
Class I: 'Clostridia'
Order I: *Clostridiales*
Family I: *Clostridiaceae*

Anaerobacter	1 sp.
Anaerofilum	2 sp.
Caloramator	3 sp.
Clostridium	146 sp.
Ilyobacter	3 sp.
Oxobacter	1 sp.
Sarcina	2 sp.
Sporobacter	1 sp.
Thermobrachium	1 sp.
Family II: 'Peptostreptococcace'	
Filifactor	1 sp.
Peptostreptococcus	18 sp.
Tissierella	3 sp.
Family III: 'Eubacteriaceae'	
Eubacterium	52 sp.
Helcococcus	1 sp.
Pseudoramibacter	1 sp.
Family IV: 'Lachnospiraceae'	
Acetitomaculum	1 sp.
Butyrivibrio	2 sp.
Catonella	1 sp.
Coprococcus	3 sp.
Johnsonella	1 sp.
Lachnospira	2 sp.
Pseudobutyrivibrio	1 sp.
Roseburia	1 sp.
Ruminococcus	14 sp.
Family V: *Peptococcaceae*	
Acetonema	1 sp.

Acidaminococcus	1 sp.
Ammonifex	1 sp.
Desulfitobacterium	3 sp.
Desulfosporosinus	1 sp.
Desulfotomaculum	18 sp.
Dialister	1 sp.
Heliobacillus	1 sp.
Heliobacterium	3 sp.
Heliophilum	1 sp.
Megasphaera	2 sp.
Mitsuokella	2 sp.
Pectinatus	2 sp.
Peptococcus	8 sp.
Phascolarctobacterium	1 sp.
Propionispira	1 sp.
Quinella	1 sp.
Schwartzia	1 sp.
Selenomonas	11 sp.
Sporomusa	7 sp.
Succiniclasticum	1 sp.
Succinivibrio	1 sp.
Syntrophobotulus	1 sp.
Thermoterrabacterium	1 sp.
Veillonella	14 sp.
Zymophilus	2 sp.
Family VI: *Syntrophomonadaceae*	
Acetogenium	1 sp.
Anaerobaculum	1 sp.
Anaerobranca	1 sp.
Caldicellulosiruptor	3 sp.
Dethiosulfovibrio	1 sp.
Moorella	3 sp.
Sporotomaculum	1 sp.
Syntrophomonas	3 sp.
Syntrophospora	1 sp.
Thermoanaerobacter	13 sp.
Thermoanaerobacterium	5 sp.
Thermoanaerobium	2 sp.
Thermohydrogenium	1 sp.
Thermosyntropha	1 sp.
Order II: *Haloanaerobiales*	
Family I: *Haloanaerobiaceae*	
Haloanaerobium	8 sp.
Halocella	1 sp.
Halothermothrix	1 sp.
Natroniella	1 sp.
Family II: *Halobacteroidaceae*	
Acetohalobium	1 sp.
Haloanaerobacter	3 sp.
Halobacteroides	4 sp.
Orenia	1 sp.
Sporohalobacter	2 sp.

SECTION XXIII The Mollicutes
Class I: *Mollicutes*
Order I: *Mycoplasmatales*
Family I: *Mycoplasmataceae*

Mycoplasma	110 sp.
Ureaplasma	6 sp.
Family II: 'Spiroplasma group'	
Entomoplasma	6 sp.
Mesoplasma	12 sp.
Spiroplasma	33 sp.
Family III: *Acholeplasmataceae*	
Acholeplasma	16 sp.
Anaeroplasma	4 sp.
Family IV: 'Asteroleplasmataceae'	
Asteroleplasma	1 sp.
Family V: 'Erysipelothrichaeceae'	
Erysipelothrix	2 sp.
Holdemania	1 sp.

SECTION XXIV The Bacilli and Lactobacilli
Class I: 'Bacilli'
Order I: *Bacillales*

Family I: *Bacillaceae*
    Ammoniphilus   2 sp.
    Aneurinibacillus   3 sp.
    Bacillus   114 sp.
    Brevibacillus   10 sp.
    Exiguobacterium   2 sp.
    Halobacillus   3 sp.
    Oxalophagus   1 sp.
    Saccharococcus   1 sp.
    Virgibacillus   1 sp.
Family II: *Planococcaceae*
    Filibacter   1 sp.
    Kurthia   3 sp.
    Planococcus   5 sp.
    Sporosarcina   2 sp.
Family III: *Caryophanaceae*
    Caryophanon   2 sp.
Family IV: '*Sporolactobacillaceae*'
    Marinococcus   3 sp.
    Sporolactobacillus   6 sp.
Family V: '*Paenibacillaceae*'
    Paenibacillus   27 sp.
Family VI: '*Alicyclobacillaceae*'
    Alicyclobacillus   3 sp.
    Sulfobacillus   3 sp.
Family VII: '*Thermoactinomycetaceae*'
    Thermoactinomyces   8 sp.
Order II: '*Lactobacillales*'
  Family I: *Lactobacillaceae*
    Lactobacillus   100 sp.
  Family II: '*Leuconostocaceae*'
    Leuconostoc   15 sp.
    Oenococcus   1 sp.
    Weissella   7 sp.
  Family III: *Streptococcaceae*
    Lactococcus   7 sp.
    Pediococcus   8 sp.
    Streptococcus   68 sp.
  Family IV: '*Enterococcaceae*'
    Enterococcus   20 sp.
    Melissococcus   1 sp.
    Vagococcus   2 sp.
  Family V: '*Carnobacteriaceae*'
    Agitococcus   1 sp.
    Alloiococcus   1 sp.
    Carnobacterium   6 sp.
    Dolosigranulum   1 sp.
    Lactosphaera   1 sp.
    Desemzia   1 sp.
  Family VI: '*Aerococcaceae*'
    Abiotrophia   3 sp.
    Aerococcus   2 sp.
    Facklamia   2 sp.
    Globicatella   1 sp.
    Tetragenococcus   2 sp.
    Ignavigranum   1 sp.
  Family VII: '*Listeriaceae*'
    Brochothrix   2 sp.
    Listeria   9 sp.
  Family VIII: '*Staphylococcaceae*'
    Gemella   4 sp.
    Macrococcus   4 sp.
    Salinicoccus   2 sp.
    Staphylococcus   47 sp.
  Family IX: '*Genera incertae sedis*
    Acetoanaerobium   1 sp.
    Acetobacterium   7 sp.
    Amphibacillus   1 sp.
    Oscillospira   1 sp.
    Pasteuria   4 sp.
    Syntrophococcus   1 sp.
    Trichococcus   1 sp.

VOLUME 4
The High G + C Gram-Positive *Bacteria*

THE BACTERIA

Phylum VIII: '*Firmicutes*'
SECTION XXV   The Actinobacteria
  Class I: *Actinobacteria*
    Subclass I: *Acidimicrobidae*
      Order I: *Acidimicrobiales*
        Suborder I: *Acidimicrobidae*
          Family I: *Acidimicrobiaceae*
            Acidimicrobium   1 sp.
    Subclass II: *Rubrobacteridae*
      Order II: *Rubrobacterales*
        Suborder I: *Rubrobacteridae*
          Family I: *Rubrobacteraceae*
            Rubrobacter   2 sp.
    Subclass III: *Coriobacteridae*
      Order III: *Coriobacteriales*
        Suborder I: *Coriobacteridae*
          Family I: *Coriobacteriaceae*
            Atopobium   3 sp.
            Coriobacterium   1 sp.
    Subclass IV: *Sphaerobacteridae*
      Order IV: '*Sphaerobacterales*'
        Suborder I: *Sphaerobacteridae*
          Family I: *Sphaerobacteraceae*
            Sphaerobacter   1 sp.
    Subclass V: *Actinobacteridae*
      Order V: *Actinomycetales*
        Suborder I: *Actinomycineae*
          Family I: *Actinomycetaceae*
            Actinobaculum   2 sp.
            Actinomyces   23 sp.
            Arcanobacterium   4 sp.
            Mobiluncus   3 sp.
        Suborder II: *Micrococcineae*
          Family II: *Micrococcaceae*
            Arthrobacter   32 sp.
            Bogoriella   1 sp.
            Demetria   1 sp.
            Kocuria   6 sp.
            Leucobacter   1 sp.
            Micrococcus   9 sp.
            Nesterenkonia   1 sp.
            Renibacterium   1 sp.
            Rothia   1 sp.
            Stomatococcus   1 sp.
            Terracoccus   1 sp.
          Family III: '*Brevibacteraceae*'
            Brevibacterium   26 sp.
          Family IV: *Cellulomonadaceae*
            Cellulomonas   11 sp.
            Oerskovia   2 sp.
            Rarobacter   2 sp.
          Family V: *Dermabacteraceae*
            Brachybacterium   7 sp.
            Dermabacter   1 sp.
          Family VI: *Dermatophilaceae*
            Dermacoccus   1 sp.
            Dermatophilus   2 sp.
            Kytococcus   1 sp.
          Family VII: *Intrasporangiaceae*
            Intrasporangium   1 sp.
            Janibacter   1 sp.
            Sanguibacter   3 sp.
            Terrabacter   1 sp.
          Family VIII: *Jonesiaceae*
            Jonesia   1 sp.
          Family IX: *Microbacteriaceae*
            Agrococcus   1 sp.
            Agromyces   6 sp.

Aureobacterium	14 sp.
Clavibacter	11 sp.
Cryobacterium	1 sp.
Curtobacterium	8 sp.
Microbacterium	27 sp.
Rathayibacter	4 sp.
Family X: *Promicromonosporaceae*	
Promicromonospora	3 sp.
Suborder III: *Corynebacterineae*	
Family I: *Corynebacteriaceae*	
Corynebacterium	67 sp.
Family II: *Dietziaceae*	
Dietzia	2 sp.
Family III: *Gordoniaceae*	
Gordonia	9 sp.
Skermania	1 sp.
Family IV: *Mycobacteriaceae*	
Mycobacterium	85 sp.
Family V: *Nocardiaceae*	
Micropolyspora	5 sp.
Nocardia	29 sp.
Rhodococcus	25 sp.
Family VI: *Tsukamurellaceae*	
Tsukamurella	5 sp.
Suborder IV: *Micromonosporaceae*	
Family I: *Micromonosporaceae*	
Actinoplanes	23 sp.
Catellatospora	5 sp.
Catenuloplanes	6 sp.
Couchioplanes	2 sp.
Dactylosporangium	6 sp.
Micromonospora	19 sp.
Pilimelia	4 sp.
Spirilliplanes	1 sp.
Verrucosispora	1 sp.
Suborder VI: *Propionibacterineae*	
Family I: *Propionibacteriaceae*	
Luteococcus	1 sp.
Microlunatus	1 sp.
Propionibacterium	12 sp.
Propioniferax	1 sp.
Family II: *Nocardioidaceae*	
Aeromicrobium	2 sp.
Friedmanniella	1 sp.
Nocardioides	7 sp.
Suborder VII: *Pseudonocardineae*	
Family I: *Pseudonocardiaceae*	
Actinopolyspora	3 sp.
Actinosynnema	3 sp.
Amycolatopsis	11 sp.
Kibdelosporangium	4 sp.
Kutzneria	3 sp.
Lentzea	1 sp.
Micropolyspora	5 sp.
Pseudonocardia	12 sp.
Saccharomonospora	5 sp.
Saccharopolyspora	10 sp.
Saccharothrix	15 sp.
Streptoalloteichus	1 sp.
Thermobispora	1 sp.
Thermocrispum	2 sp.
Suborder VIII: *Streptomycineae*	
Family I: *Streptomycetaceae*	
Streptomyces	509 sp.
Suborder IX: *Streptosporangineae*	
Family I: *Streptosporangiaceae*	
Herbidospora	1 sp.
Microbispora	15 sp.
Micropolyspora	5 sp.
Microtetraspora	20 sp.
Nonomuria	15 sp.
Planobispora	2 sp.
Planomonospora	5 sp.

Planopolyspora	1 sp.
Planotetraspora	1 sp.
Streptosporangium	19 sp.
Family II: *Nocardiopsaceae*	
Nocardiopsis	17 sp.
Thermobifida	2 sp.
Family III: *Thermomonosporaceae*	
Actinomadura	51 sp.
Spirillospora	2 sp.
Thermomonospora	7 sp.
Suborder X: *Frankineae*	
Family I: *Frankiaceae*	
Frankia	1 sp.
Family II: *Geodermatophilaceae*	
Blastococcus	1 sp.
Geodermatophilus	1 sp.
Family III: *Microsphaeraceae*	
Microsphaera	1 sp.
Family IV: *Sporichthyaceae*	
Sporichthya	1 sp.
Family V: *Acidothermaceae*	
Acidothermus	1 sp.
Family VI: 'Incertae sedis'	
Cryptosporangium	2 sp.
Kineococcus	1 sp.
Kineosporia	5 sp.
Suborder XI: *Glycomycineae*	
Family I: *Glycomycetaceae*	
Glycomyces	3 sp.
Order VI: *Bifidobacteriales*	
Family I: *Bifidobacteriaceae*	
Bifidobacterium	33 sp.
Falcivibrio	2 sp.
Gardnerella	1 sp.
Family II: 'Unknown Affiliation'	
Actinobispora	1 sp.
Actinocorallia	1 sp.
Actinokineospora	5 sp.
Excellospora	1 sp.
Pelczaria	1 sp.
Turicella	1 sp.

VOLUME 5
The Planctomycetes, Spirochaetes, Fibrobacteres, Bacteroides and Fusobacteria

THE BACTERIA

Phylum IX: 'Wall-less forms'	
SECTION XXVI The Planctomycetes and Chlamydia	
Class I: 'Planctomycetacia'	
Order I: *Planctomycetales*	
Family I: *Planctomycetaceae*	
Gemmata	1 sp.
Isosphaera	1 sp.
Pirellula	2 sp.
Planctomyces	6 sp.
Order II: *Chlamydiales*	
Family I: *Chlamydiaceae*	
Chlamydia	4 sp.
SECTION XXVII The Spirochetes	
Phylum X: 'Spirochetes'	
Class I: 'Spirochaetes'	
Order I: 'Spirochaetales'	
Family I: *Spirochaetaceae*	
Borrelia	30 sp.
Brevinema	1 sp.
Clevelandina	1 sp.
Cristispira	1 sp.
Diplocalyx	1 sp.
Hollandina	1 sp.
Pillotina	1 sp.

Spirochaeta	14 sp.		Riemerella	2 sp.
Treponema	18 sp.		Weeksella	2 sp.
Family II: Serpulinaceae			Psychroflexus	2 sp.
Brachyspira	5 sp.		Family II: 'Myroideaceae'	
Serpulina	6 sp.		Myroides	2 sp.
Family III: Leptospiraceae			Psychromonas	1 sp.
Leptonema	1 sp.		Family III: 'Blattabacteriaceae'	
Leptospira	12 sp.		Blattabacterium	1 sp.

SECTION XXVIII The Fibrobacters
Phylum XI: 'Fibrobacter'
  Class I: 'Fibrobacteres'
    Order I: 'Fibrobacterales'
      Family I: 'Fibrobacteraceae'

Fibrobacter	3 sp.			

Family II: 'Acidobacteriaceae'

Acidobacterium	1 sp.	
Holophaga	1 sp.	

SECTION XXIX The Bacteroides
Phylum XII: 'Bacteroids'
  Class I: 'Bacteroides'
    Order I: 'Bacteroidales'
      Family I: Bacteroidaceae

Acetivibrio	4 sp.
Acetofilamentum	1 sp.
Acetomicrobium	2 sp.
Acetothermus	1 sp.
Acidaminobacter	1 sp.
Anaerobiospirillum	2 sp.
Anaerorhabdus	1 sp.
Anaerovibrio	3 sp.
Bacteroides	65 sp.
Centipeda	1 sp.
Formivibrio	1 sp.
Malonomonas	1 sp.
Megamonas	1 sp.
Propionivibrio	1 sp.
Succinimonas	1 sp.
Syntrophus	2 sp.

Family II: 'Rikenellaceae'

Marinilabilia	2 sp.
Rikenella	1 sp.

Family III: 'Porphyromonadaceae'

Porphyromonas	13 sp.

Family IV: Prevotellaceae

Prevotella	26 sp.

SECTION XXX The Flavobacteria
Phylum XIII: 'Flavobacteria'
  Class I: 'Flavobacteria'
    Order I: 'Flavobacteriales'
      Family I: Flavobacteriaceae

Bergeyella	1 sp.
Capnocytophaga	7 sp.
Chryseobacterium	6 sp.
Empedobacter	1 sp.
Flavobacterium	40 sp.
Gelidibacter	1 sp.
Ornithobacterium	1 sp.
Polaribacter	4 sp.
Psychroserpens	1 sp.

SECTION XXXI The Sphingobacteria
Phylum XIV: 'Sphingobacteria'
  Class I: 'Sphingobacteria'
    Order I: 'Sphingobacteriales'
      Family I: Sphingobacteriaceae

Pedobacter	4 sp.
Sphingobacterium	8 sp.

Family II: 'Saprospiraceae'

Haliscomenobacter	1 sp.
Lewinella	3 sp.
Saprospira	1 sp.

Family III: 'Flexibacteraceae'

Cyclobacterium	1 sp.
Cytophaga	23 sp.
Flectobacillus	3 sp.
Flexibacter	17 sp.
Meniscus	1 sp.
Microscilla	1 sp.
Runella	1 sp.
Spirosoma	1 sp.
Sporocytophaga	1 sp.

Family IV: 'Flammeovirgaceae'

Flammeovirga	1 sp.
Persicobacter	1 sp.
Thermonema	2 sp.

Family V: 'Crenothrichaeceae'

Chitinophaga	1 sp.
Crenothrix	1 sp.
Flexithrix	1 sp.
Rhodothermus	2 sp.
Toxothrix	1 sp.

SECTION XXXII The Fusiforms
Phylum XV: 'Fusobacteria'
  Class I: 'Fusobacteria'
    Order I: 'Fusobacteriales'
      Family I: 'Fusobacteriaceae'

Fusobacterium	23 sp.
Leptotrichia	1 sp.
Propionigenium	2 sp.
Sebaldella	1 sp.
Streptobacillus	1 sp.

Family II: 'Genera incertae sedis'

Cetobacterium	1 sp.

SECTION XXXIII Verrucomicrobium and relatives
Phylum XVI: 'Verrucomicrobia'
  Class I: Verrucomicrobiae
    Order I: Verrucomicrobiales
      Family I: Verrucomicrobiaceae

Prosthecobacter	4 sp.
Verrucomicrobium	1 sp.

Only the major terms and concepts are included. If a term is not here, consult the index.

ABC transporter   A membrane transport system consisting of three proteins, one of which hydrolyzes ATP, one of which binds the substrate, and one of which functions as the transport channel through the membrane.

Abscess   A localized infection characterized by production of pus.

Acetotrophic   Splitting of acetate into CH_4 plus CO_2 by certain methanogens.

Acetyl-CoA (Ljungdahl-Wood) pathway   A pathway of autotrophic CO_2 fixation widespread in obligate anaerobes including methanogens, homoacetogens, and sulfate-reducing bacteria.

Acetylene reduction assay   Method of measuring activity of nitrogenase by substituting acetylene for the natural substrate of the enzyme, N_2. Acetylene is reduced to ethylene or ethane.

Acid fastness   A staining property of *Mycobacterium* species where cells stained with hot carbol fuschin do not decolorize with acid-alcohol.

Acid mine drainage   Acidic water containing H_2SO_4 derived from the microbial oxidation of iron sulfide minerals.

Acidophile   An organism that grows best at acidic pH values.

Activation energy   Energy needed to make substrate molecules more reactive; enzymes function by lowering activation energy.

Activator protein   A regulatory protein that binds to specific sites on DNA and stimulates transcription; involved in positive control.

Active immunity   An immune state achieved by self-production of antibodies. Compare with *Passive immunity*.

Active site   The portion of an enzyme that is directly involved in binding substrate(s).

Active transport   The energy-dependent process of transporting substances into or out of the cell in which the transported substances are chemically unchanged.

Acute   In reference to infections, short-term, usually characterized by dramatic onset and rapid recovery.

Adherence   A property of bacteria that allows them to stick to host surfaces.

Aerobe   An organism that grows in the presence of O_2; may be facultative, obligate, or microaerophilic.

Aerosol   Suspension of particles in airborne water droplets.

Aerotolerant   Of an anaerobe, not being inhibited by O_2.

Agglutination   Reaction between antibody and particle-bound antigen resulting in clumping of the particles.

Agretope   The portion of a processed antigen that is recognized by MHC protein.

Algae   Phototrophic eukaryotic microorganisms.

Alkaliphile   An organism that grows best at high pH.

Allergy   A harmful immune reaction, usually caused by a foreign antigen in food, pollen, or chemicals; immediate-type or delayed-type hypersensitivity.

Allosteric enzyme   An enzyme that contains two combining sites, the active site (where the substrate binds) and the allosteric site (where an effector molecule binds).

Ameboid movement   A type of motility in which cytoplasmic streaming moves the organism forward.

Aminoacyl-tRNA synthetase   An enzyme that catalyzes the attachment of the correct amino acid to the correct tRNA.

Aminoglycoside   An antibiotic such as streptomycin that consists of amino sugars linked by glycosidic bonds.

Anabolism   The biochemical processes involved in the synthesis of cell constituents from simpler molecules, usually requiring energy.

Anaerobe   An organism that grows in the absence of O_2; some may even be killed by O_2.

Anaerobic respiration   Use of an electron acceptor other than O_2 in an electron transport-based oxidation and leading to a proton motive force.

Anammox   The term that describes anoxic ammonia oxidation.

Anaphylatoxins   The C3a and C5a fractions of complement that act to mimic some of the reactions of anaphylaxis.

Anaphylaxis (anaphylactic shock)   A violent allergic reaction caused by an antigen–antibody reaction.

Anergy   The inability to produce an immune response to specific antigens due to neutralization of effector cells.

Anoxic   Absence of oxygen. Usually used in reference to a microbial habitat.

Anoxygenic photosynthesis   Use of light energy to synthesize ATP by cyclic photophosphorylation without O_2 production.

Antibiogram   A report indicating the sensitivity of clinically isolated microorganisms to the antibiotics in current use.

Antibiotic   A chemical agent produced by one organism that is harmful to other organisms.

Antibiotic resistance   The acquired ability of a microorganism to grow in the presence of an antibiotic to which the microorganism is usually sensitive.

Antibody   A protein present in serum or other body fluid that combines specifically with antigen. An immunoglobulin.

Antibody-mediated immunity   Immunity resulting from direct interaction with antibodies; also called humoral immunity.

Anticodon   A sequence of three bases in transfer RNA that base-pairs with a codon in messenger RNA during protein synthesis.

Antigen   A substance that interacts with a T cell receptor or an immunoglobulin.

Antigen-presenting cell (APC)   A cell that processes and presents antigen to T lymphocytes.

Antigenic determinant   The portion of an antigen that interacts with an immunoglobulin or T cell receptor. Also called an *epitope*.

Antigenic drift   In influenza virus, minor changes in viral proteins (antigens) due to gene mutation.

Antigenic shift   In influenza virus, major changes in viral proteins (antigens) due to gene reassortment.

Antimicrobial   Harmful to microorganisms by either killing or inhibiting growth.

Antimicrobial agent A chemical that kills or inhibits the growth of microorganisms.

Antimicrobial drug resistance The acquired ability of a microorganism to grow in the presence of an antimicrobial drug to which the microorganism is usually sensitive.

Antiparallel In reference to double-stranded DNA, one strand runs $5' \rightarrow 3'$, the other $3' \rightarrow 5'$.

Antiseptic An agent that kills or inhibits microbial growth but is not harmful to human tissue.

Antiserum A serum containing antibodies.

Antitoxin An antibody that specifically interacts with and neutralizes a toxin.

Apoptosis Programmed cell death.

Archaea A phylogenetic domain of prokaryotes consisting of the methanogens, most extreme halophiles and hyperthermophiles, and *Thermoplasma*.

Artificial chromosomes Cloning vectors which can carry very large inserts of foreign DNA and exist in the cell very much like a cellular chromosome. The most widely used are Bacterial Artificial Chromosomes (BACs) and Yeast Artificial Chromosomes (YACs).

Aseptic technique Manipulation of sterile instruments or culture media in such a way as to maintain sterility.

ATP Adenosine triphosphate, the principal energy carrier of the cell.

Attenuation Selection of nonvirulent strains of a pathogen still capable of immunizing. Also, a mechanism for controlling gene expression. Typically transcription is terminated after initiation but before a full-length mRNA is produced.

Autoantibody An antibody that reacts to self antigens.

Autoclave A sterilizer that destroys microorganisms by high temperature using steam under pressure.

Autoimmunity Immune reactions of a host against its own self antigens.

Autolysis The lysis of a cell brought about by the activity of the cell itself.

Autoradiography Detection of radioactivity in a sample, for example, a cell or gel, by placing it in contact with a photographic film.

Autotroph An organism able to utilize CO_2 as a sole source of carbon.

Auxotroph An organism that has developed a nutritional requirement through mutation. Contrast with a *Prototroph*.

B lymphocyte A cell of the immune system that differentiates into an immunoglobulin-producing cell.

Bacteremia The transient appearance of bacteria in the blood.

Bacteria All prokaryotes that are not members of the domain *Archaea*.

Bacteriocidal Capable of killing bacteria.

Bacteriocins Agents produced by certain bacteria that inhibit or kill closely related species.

Bacteriophage A virus that infects prokaryotic cells.

Bacteriorhodopsin A protein containing retinal that is found in the membranes of certain extremely halophilic *Archaea* and that is involved in light-mediated ATP synthesis.

Bacteriostatic Capable of inhibiting bacterial growth without killing.

Bacteroid A swollen, deformed *Rhizobium* cell found in the root nodule; capable of nitrogen fixation.

Barophile An organism that lives optimally at high hydrostatic pressure.

Barotolerant An organism able to tolerate high hydrostatic pressure, although growing better at 1 atm.

Base composition In reference to nucleic acids, the proportion of the total bases consisting of guanine plus cytosine or thymine plus adenine base pairs. Usually expressed as a guanine + cytosine $(G + C)$ value for example, 60% G + C.

Batch culture A closed-system microbial culture of fixed volume.

Beta-lactam An antibiotic such as penicillin that contains the four-membered heterocyclic beta-lactam ring.

Binomial system The system for naming organisms in which an organism is given a genus name and a species epithet.

Biocatalysis The use of microorganisms to synthesize a product or carry out a specific chemical transformation.

Biochemical oxygen demand (BOD) The amount of dissolved oxygen consumed by microorganisms for complete oxidation of organic and inorganic material in a water sample.

Bioconversion In industrial microbiology, use of microorganisms to convert a substance to a chemically modified form.

Biofilm Microbial colonies encased in an adhesive, usually polysaccharide material, and attached to a surface.

Biogeochemistry Study of microbially mediated chemical transformations of geochemical interest, for example, nitrogen or sulfur cycling.

Bioinformatics The use of computer programs to analyze, store, and access DNA and protein sequences.

Biological warfare The use of biological agents to kill or incapacitate a population.

Bioremediation Use of microorganisms to remove or detoxify toxic or unwanted chemicals in an environment.

Biosynthesis The production of needed cellular constituents from other (usually simpler) molecules.

Biosynthetic penicillin Production of a particular form of penicillin by supplying the producing organism with specific side chain precursors.

Biotechnology The use of living organisms to carry out defined chemical processes for industrial application.

Black smoker A deep-sea hydrothermal vent emitting very hot (250–400°C) water and minerals.

Botulism Food poisoning due to ingestion of food containing botulism toxin produced by *Clostridium botulinum*.

Brewing The manufacture of alcoholic beverages such as beer from the fermentation of malted grains.

Broad-spectrum antibiotic An antibiotic that acts on both gram-positive and gram-negative *Bacteria*.

Calvin cycle The biochemical route of CO_2 fixation in many autotrophic organisms.

Canning The process of sealing food in a closed container and heating to destroy living organisms.

Capsid The protein coat of a virus.

Capsomere An individual protein subunit of the virus capsid.

Capsule Dense, well-defined polysaccharide or protein layer closely surrounding a cell.

Carboxysomes Polyhedral cellular inclusions of crystalline ribulose bisphosphate carboxylase (RubisCO), the key enzyme of the Calvin cycle.

Carcinogen A substance that causes the initiation of tumor formation. Frequently a mutagen.

Carrier An individual that harbors infectious organisms but does not show symptoms of disease.

Catabolism The biochemical processes involved in the breakdown of organic or inorganic compounds, usually leading to the production of energy.

Catabolite repression Repression of a variety of unrelated enzymes when cells are grown in a medium containing glucose.

Catalysis Increase in rate of a chemical reaction.

Catalyst A substance that promotes a chemical reaction without itself being changed in the end.

CD4 cells T helper cells. They are targets for HIV infection.

Cell The fundamental unit of life.

Cell-mediated immunity An immune response generated by the activities of nonantibody-producing cells such as T cells. Compare with *Humoral immunity*.

Chemiosmosis The use of ion gradients, especially proton gradients, across membranes to generate ATP. See *Proton motive force*.

Chemokine A low-molecular-weight soluble immune response modulator protein produced by a variety of cells.

Chemolithotroph An organism obtaining its energy from the oxidation of inorganic compounds.

Chemoorganotroph An organism obtaining its energy from the oxidation of organic compounds.

Chemostat A continuous culture device controlled by the concentration of limiting nutrient and dilution rate.

Chemotaxis Movement toward or away from a chemical.

Chemotherapeutic agent An antimicrobial agent that can be used internally.

Chemotherapy Treatment of infectious disease with chemicals or antibiotics.

Chloramine A water purification chemical made by combining chlorine and ammonia at precise ratios.

Chlorination A highly effective disinfectant procedure for drinking water using chlorine gas or other chlorine-containing compounds as disinfectant.

Chlorine A chemical fed in its gaseous state to disinfect water. A residual level is maintained throughout the distribution system.

Chlorophyll and bacteriochlorophyll Pigments of phototrophic organisms consisting of light-sensitive magnesium tetrapyrroles.

Chloroplast The chlorophyll-containing organelle of phototrophic eukaryotes.

Chlorosomes Cigar-shaped structures enclosed by a nonunit membrane and containing the light-harvesting bacteriochlorophyll (c, c_s, d, or e) in green sulfur bacteria and in *Chloroflexus*.

Chromogenic Producing color; a chromogenic colony is a pigmented colony.

Chromosome A genetic element carrying genes essential to cellular function. Prokaryotes typically have a single chromosome consisting of a circular DNA molecule. Eukaryotes typically have several chromosomes, each containing a linear DNA molecule.

Chronic Long-term.

Cidal Lethal or killing.

Cilium Short, filamentous structure that beats with many others to make a cell move.

Cirrhosis Breakdown of the normal liver architecture resulting in fibrosis.

Citric acid cycle A cyclical series of reactions resulting in the conversion of acetate to CO_2 and NADH. Also called the *Tricarboxylic acid cycle* or the *Kreb's cycle*.

Clarifier (Coagulation basin) A reservoir in which the suspended solids of raw water are coagulated and removed.

Class I MHC proteins Antigen-presenting molecules found on all nucleated vertebrate cells.

Class II MHC proteins Antigen-presenting molecules found primarily on macrophages and B lymphocytes in vertebrates.

Clonal deletion For T-cell selection in the thymus, the killing of useless or self-reactive clones.

Clonal selection A theory that each B or T lymphocyte, when stimulated by antigen, divides to form a clone of itself.

Clone A population of cells all descended from a single cell. Also, a number of copies of a DNA fragment obtained by allowing an inserted DNA fragment to be replicated by a phage or plasmid.

Cloning vectors Genetic elements into which genes can be recombined and replicated.

Coagulation The formation of large insoluble particles from much smaller, colloidal particles by the addition of aluminum sulfate and anionic polymers.

Coccoid Sphere-shaped.

Coccus A spherical bacterium.

Codon A sequence of three bases in messenger RNA that encodes a specific amino acid.

Coenzyme A low-molecular-weight molecule that participates in an enzymatic reaction by accepting and donating electrons or functional groups. Examples: NAD^+, FAD.

Coliforms Gram-negative, nonsporing, facultative rods that ferment lactose with gas formation within 48 h at 35°C.

Colonization Multiplication of a microorganism after it has attached to host tissues or other surfaces.

Colony A macroscopically visible population of cells growing on solid medium, arising from a single cell.

Cometabolism The metabolic transformation of a substance while a second substance serves as primary energy or carbon source.

Commodity chemicals Chemicals such as ethanol that have low monetary value and thus are sold primarily in bulk.

Common-source epidemic An epidemic resulting from infection of a large number of people from a single contaminated source.

Compatible solutes Organic compounds that serve as cytoplasmic solutes to balance water relations for cells growing in environments of high salt or sugar.

Competence Ability to take up DNA and become genetically transformed.

Complement A complex of proteins in the blood serum that interacts sequentially with specific antigen–antibody complexes.

Complement fixation The consumption of complement by an antibody–antigen reaction.

Complementary Nucleic acid sequences that can base-pair with each other.

Complementarity-determining regions (CDRs) Variations in amino acid sequence that occur within the variable domains of Igs or TCRs that provide most of the molecular contacts with antigen (also known as *hypervariable regions*).

Complex media Culture media whose precise chemical composition is unknown. Also called *undefined media*.

Concatamer A DNA molecule consisting of two or more separate molecules linked end to end to form a long, linear structure.

Congenital syphilis Syphilis contracted by an infant from its mother during birth.

Conjugation Transfer of genes from one prokaryotic cell to another by a mechanism involving cell-to-cell contact.

Consensus sequence A nucleic acid sequence in which the base present in a given position is that base most commonly found when many experimentally determined sequences are compared.

Consortium A two- (or more) membered bacterial culture (or natural assemblage) in which each organism benefits from the others.

Contagious Transmissible.

Cortex The region inside the spore coat of an endospore, around the core.

Covalent bond A nonionic chemical bond formed by a sharing of electrons between two atoms.

Crista Inner membrane in a mitochondrion, site of respiration.

Culture A particular strain or kind of organism growing in a laboratory medium.

Culture medium An aqueous solution of various nutrients suitable for the growth of microorganisms.

Cutaneous Relating to the skin.

Cyanobacteria Prokaryotic oxygenic phototrophs containing chlorophyll *a* and phycobilins.

Cyst A resting stage formed by some bacteria and protozoa in which the whole cell is surrounded by a protective layer; not the same as a spore.

Cytochrome Iron-containing porphyrin complexed with proteins, which functions as an electron carrier in the electron transport system.

Cytokine A soluble immune response modulator produced by cells other than lymphocytes, usually phagocytic cells.

Cytoplasm Cellular contents inside the cytoplasmic membrane, excluding the nucleus.

Cytoplasmic membrane The permeability barrier of the cell, separating the cytoplasm from the environment.

Cytoskeleton Cellular scaffolding in which microfilaments define the shape of eukaryotic cells.

Decontamination Treatment that renders an object or inanimate surface safe to handle.

Defined media Culture media whose exact chemical composition is known. Compare with *Complex media*.

Degeneracy In relation to the genetic code, the fact that more than one codon can code for the same amino acid.

Deletion Removal of a portion of a gene.

Denaturation Irreversible destruction of a macromolecule, as for example, the destruction of a protein by heat.

Denaturing gradient gel electrophoresis (DGGE) An electrophoretic technique that allows resolving nucleic acid fragments of the same size but that differ in sequence.

Denitrification Conversion of nitrate into nitrogen gases under anoxic conditions.

Dental caries Tooth decay resulting from bacterial infection.

Dental plaque Bacterial cells encased in a matrix of extracellular polymers, found on the teeth.

Deoxyribonucleic acid (DNA) A polymer of nucleotides connected via a phosphate–deoxyribose sugar backbone; the genetic material of cells and some viruses.

Desiccation Drying.

Dideoxynucleotide A nucleotide lacking the 3'-hydroxyl group on the deoxyribose sugar. Used in the Sanger method of DNA sequencing.

Differential media Media that allow identification of organisms based on their appearance.

Differentiation The modification of a cell in terms of structure and/or function occurring during the course of development.

Diploid In eukaryotes, an organism or cell with two chromosome complements, one derived from each haploid gamete.

Disease Injury to the host that impairs host function.

Disinfectant An agent that kills microorganisms but may also be harmful to human tissue.

Disinfection The process of eliminating nearly all pathogens, but not all microorganisms, from inanimate objects or surfaces.

Disproportionation The splitting of a chemical compound into two new compounds, one more oxidized and one more reduced than the original compound.

Distribution system Water pipes, storage reservoirs, tanks, and other means used to deliver drinking water to consumers or store it before delivery.

DNA fingerprinting Use of genetic engineering to determine the origin of DNA in a sample of tissue.

DNA library A collection of cloned DNA fragments that in total contain genes from the entire genome of an organism; also called a *gene library*.

DNA polymerase An enzyme that synthesizes a new strand of DNA in the 5' → 3' direction using an antiparallel DNA strand as a template.

DNA gyrase An enzyme found in most prokaryotes that introduces negative supercoils in DNA.

Domain The highest level of biological classification. The three domains of biological organisms are the *Bacteria*, the *Archaea*, and the *Eukarya*. Also used to describe a region of a protein having a distinct function.

Doubling time The time needed for a population to double. See also *Generation time*.

Downstream position Refers to nucleic acid sequences on the 3' side of a given site on the DNA or RNA molecule. Compare with *Upstream position*.

Early protein Proteins synthesized soon after virus infection.

Ecology Study of the interrelationships between organisms and their environments.

Ecosystem A community of organisms and their natural environment.

Ecotype A population of genetically identical cells sharing a particular resource within an ecological niche.

Ehrlichiosis An emerging tick-transmitted disease caused by rickettsia of the *Ehrlichia* genus.

Electron acceptor A substance that accepts electrons during an oxidation–reduction reaction.

Electron donor A compound that donates electrons in an oxidation–reduction reaction.

Electron transport phosphorylation Synthesis of ATP involving a membrane-associated electron transport chain and the creation of a proton motive force. Also called *Oxidative phosphorylation*. See also *Chemiosmosis*.

Electrophoresis Separation of charged molecules in an electric field.

Electroporation The use of an electric pulse to enable cells to take up DNA.

ELISA Enzyme-linked immunosorbent assay. An immunoassay that uses specific antibodies to detect antigens or antibodies in body fluids. The antibody-containing complexes are visualized through enzyme coupled to the antibody. Addition of substrate to the enzyme–antibody–antigen complex results in a colored product.

Emerging infection An infectious disease that has increased in incidence in the last 20 years or threatens to increase in incidence in the future.

Enantiomer One form of a molecule that is the mirror image of another form of the same molecule.

Endemic A disease that is constantly present in low numbers in a population. Compare with *Epidemic*.

Endergonic reaction A chemical reaction requiring an input of energy to proceed.

Endocytosis A process in which a particle such as a virus is taken intact into an animal cell. Phagocytosis and pinocytosis are two kinds of endocytosis.

Endoplasmic reticulum An extensive array of internal membranes in eukaryotes.

Endospore A differentiated cell formed within the cells of certain gram-positive bacteria that is extremely resistant to heat as well as to other harmful agents.

Endosymbiosis The hypothesis that mitochondria, chloroplasts, and hydrogenosomes are the descendants of ancient prokaryotic organisms from the domain *Bacteria*.

Endotoxin A toxin not released from the cell; bound to the cell surface or intracellular. Compare with *Exotoxin*.

Enriched media Media that allow metabolically fastidious organisms to grow because of the addition of specific growth factors.

Enrichment bias The skewing of enrichment culture results towards "weed" species.

Enrichment culture Use of selective culture media and incubation conditions to isolate microorganisms from natural samples.

Enteric Intestinal.

Enterotoxin A toxin affecting the intestine.

Entropy A measure of the degree of disorder in a system; entropy always increases in a closed system.

Enzyme A catalyst, usually composed of protein, that promotes specific reactions or groups of reactions.

Epidemic A disease occurring in an unusually high number of individuals in a population at the same time. Compare with *Endemic*.

Epidemiology The study of the incidence and prevalence of disease in populations.

Epitope Antigenic determinant.

***Escherichia coli* O157:H7** An emerging enterotoxigenic strain of *E. coli* spread by fecal contamination of animal or human origin to food and water.

Eukarya The phylogenetic domain containing all eukaryotic organisms.

Eukaryote A cell or organism having a unit membrane-enclosed (true) nucleus and usually other organelles.

Evolution Change in a line of descent over time leading to the production of new species or varieties within a species.

Evolutionary distance In phylogenetic trees, the sum of the physical distance on a tree separating organisms; this distance is inversely proportional to evolutionary relatedness.

Exergonic reaction A chemical reaction that proceeds with the liberation of energy.

Exons The coding sequences in a split gene. Contrast with *Introns*, the intervening noncoding regions.

Exotoxin A toxin released extracellularly. Compare with *Endotoxin*.

Exponential growth Growth of a microorganism where the cell number doubles within a fixed time period.

Exponential phase A period during the growth cycle of a population in which growth increases at an exponential rate.

Expression The ability of a gene to function within a cell in such a way that the gene product is formed.

Expression vector A cloning vector that contains the necessary regulatory sequences allowing transcription and translation of a cloned gene or genes.

Extreme halophile An organism whose growth is dependent on large amounts (generally > 10%) of NaCl.

Extremophile An organism that grows optimally under one or more chemical or physical extremes, such as high or low temperature or pH.

Facultative A qualifying adjective indicating that an organism is able to grow in either the presence or absence of an environmental factor (for example, "facultative aerobe").

FAME Fatty acid methyl ester.

Feedback inhibition A decrease in the activity of the first enzyme of a biochemical pathway caused by buildup of the final product of the pathway.

Fermentation Catabolic reactions producing ATP in which organic compounds serve as both primary electron donor and ultimate electron acceptor and ATP is produced by substrate-level phosphorylation.

Fermentation (industrial) A large-scale microbial process.

Fermenter An organism that carries out the process of fermentation.

Fermentor A growth vessel, usually quite large, used to culture microorganisms for the production of some commercially valuable product.

Ferredoxin An electron carrier of low reduction potential; small protein containing iron-sulfur clusters.

Fever A rise of body temperature above normal.

Filamentous In the form of very long rods, many times longer than wide.

Filtration The removal of suspended particles from water by passing it through one or more permeable membranes or media (e.g., sand, anthracite, or diatomaceous earth).

Fimbria (plural fimbriae) Short, filamentous structures on a bacterial cell; although flagella-like in structure, generally present in many copies and not involved in motility. Plays a role in adherence to surfaces and in the formation of pellicles. See also *Pilus*.

Finished water   Water delivered to the distribution system after treatment.

FISH   Fluorescent *in-situ* hybridization; a process in which a cell is made fluorescent by labeling it with a specific nucleic acid probe that contains an attached fluorescent dye.

Flagellum (plural flagella)   A thin, filamentous organ of motility in prokaryotes that functions by rotating.

Flavoprotein   A protein containing a derivative of riboflavin, which functions as electron carrier in the electron transport system.

Flocculation   The water-treatment process after coagulation that uses gentle stirring to cause suspended particles to form larger, aggregated masses (floc).

Fluorescent   Having the ability to emit light of a certain wavelength when activated by light of another wavelength.

Fluorescent antibody   Immunoglobulin molecule that has been coupled with a fluorescent molecule so that it exhibits fluorescence.

Fomites   Inanimate objects that, when contaminated with a viable pathogen, can transfer the pathogen to a host.

Food infection   Microbial infection resulting from ingestion of contaminated food.

Food poisoning   Disease resulting from ingestion of food contaminated with a toxin produced by a microorganism.

Food spoilage   Any change in a food product that makes it unacceptable to the consumer.

Frame shift   A type of mutation. Because the genetic code is read three bases at a time, if reading begins at either the second or third base of a codon, a faulty product usually results.

Free energy   Energy available to do useful work.

Fruiting body   A macroscopic reproductive structure produced by some fungi (for example, mushrooms) and some *Bacteria* (for example, myxobacteria), each distinct in size, shape, and coloration.

Fungi   Nonphototrophic eukaryotic microorganisms that contain rigid cell walls.

Fungicidal agent   An agent that kills fungi.

Fungistatic agent   An agent that inhibits fungal growth.

Fusion protein   The result of translation of two or more genes joined such that they retain their correct reading frames but make a single protein.

G + C base ratio   In DNA (or RNA) from any organism, the percentage of the total nucleic acid that consists of guanine plus cytosine bases (expressed as mol % GC).

Gametes   In eukaryotes, the haploid germ cells that result from meiosis.

Gas vesicle   A gas-filled structure made of protein that confers ability on a cell to float.

Gel   An inert polymer, usually made of agarose or poly-acrylamide, used for separating macromolecules such as nucleic acids and proteins by electrophoresis.

Gene   A unit of heredity; a segment of DNA specifying a particular protein or polypeptide chain, a tRNA or an rRNA.

Gene cloning   See *Molecular cloning*.

Gene disruption   Use of genetic techniques to inactivate a gene by inserting within it a DNA fragment containing an easily selectable marker. The inserted fragment is called a *cassette*, and the process of insertion, *cassette mutagenesis*.

Gene family   Genes that are related by sequence to other genes within the organism.

Gene library   A collection of cloned DNA fragments that contains all the genetic information for a particular organism.

Gene therapy   Treatment of a disease caused by a dysfunctional gene by introduction of a normally functioning copy of the gene.

General purpose media   Media that support the growth of most aerobic and facultatively aerobic organisms.

Generation time   Time needed for a population to double. See also *Doubling time*.

Genetically modified organism (GMO)   An organism whose genome has been altered using genetic engineering. The abbreviation is also used in constructions such as *GM crops* and *GM foods.*

Genetic engineering   The use of *in vitro* techniques in the isolation, manipulation, recombination, and expression of DNA, and in the development of *genetically modified organisms* (GMO).

Genetic map   The arrangement of genes on a chromosome.

Genetics   Heredity and variation of organisms.

Genome   The complete set of genes present in an organism.

Genomics   The discipline involving mapping, sequencing, and analyzing genomes.

Genotype   The precise genetic constitution of an organism. Compare with *Phenotype*.

Genus   A taxonomic group of related species.

Germicide   A substance that inhibits or kills microorganisms.

Glycocalyx   A term that describes polysaccharide components outside of the bacterial cell wall; usually a loose network of polymer fibers extending outward from the cell.

Glycolysis   Reactions of the Embden–Meyerhof pathway in which glucose is converted to pyruvate.

Glycosidic bond   A type of covalent bond that links sugar units together in a polysaccharide.

Gonococcus   *Neisseria gonorrhoeae*, the gram-negative diplococcus that causes the disease gonorrhea.

Gram-negative cell   A prokaryotic cell whose cell wall contains relatively little peptidoglycan but has an outer membrane composed of lipopolysaccharide, lipoprotein, and other complex macromolecules.

Gram-positive cell   A prokaryotic cell whose cell wall consists chiefly of peptidoglycan and lacks the outer membrane of gram-negative cells.

Growth   In microbiology, an increase in cell number.

Growth-factor analog   A chemical agent that is related to and blocks the uptake or utilization of a growth factor.

Growth rate   The rate at which growth occurs, usually expressed as the generation time.

Guild   A group of metabolically related organisms.

Habitat   The location in nature where an organism resides.

Halophile   An organism requiring salt (NaCl) for growth.

Halotolerant   Capable of growing in the presence of NaCl, but not requiring it.

Hantavirus pulmonary syndrome (HPS)   An emerging acute viral disease characterized by respiratory pneumonia, obtained by transmission of hantavirus from rodents.

Haploid An organism or cell containing only one set of chromosomes.

Hapten A low-molecular-weight substance not inducing antibody formation itself but still able to combine with a specific antibody.

Helix A spiral structure in a macromolecule that contains a repeating pattern.

Hemagglutination Agglutination of red blood cells.

Hemolysins Bacterial toxins capable of lysing red blood cells.

Hemolysis Lysis of red blood cells.

Hepatitis Liver inflammation commonly caused by an infectious agent.

Herd immunity Resistance of a group to a pathogen as a result of the immunity of a large proportion of the group to that pathogen.

Heterocyst A differentiated cyanobacterial cell that carries out nitrogen fixation.

Heteroduplex A double-stranded DNA in which one strand is from one source and the other strand is from another, usually related, source.

Heterofermentation Fermentation of glucose or another sugar to a mixture of reduced products.

Heterotroph Chemoorganotroph.

HLA Human leukocyte antigen, the MHC of humans.

Homoacetogens *Bacteria* that produce acetate as the sole product of sugar fermentation or from $H_2 + CO_2$.

Homofermentation Fermentation of glucose or other sugar leading to a single product, lactic acid.

Homologous antigen An antigen that reacts with the antibody it has induced.

Horizontal (lateral) gene transfer A gene in an organism that originated by transfer from another organism.

Host An organism capable of supporting the growth of a virus or other parasite.

Host-to-host epidemic An epidemic resulting from host-to-host contact, characterized by a gradual rise and fall in disease incidence.

Humoral immunity An immune response involving antibodies.

Hybridization The natural formation or artificial construction of a duplex nucleic acid molecule by complementary base pairing between two nucleic acid strands derived from different sources.

Hybridoma The fusion of an immortal (tumor) cell with a lymphocyte to produce an immortal lymphocyte.

Hydrogen bond A weak chemical bond between a hydrogen atom and a second, more electronegative element, usually an oxygen or nitrogen atom.

Hydrogenosome An organelle of endosymbiotic origin in the cytoplasm of certain anaerobic eukaryotes that functions to oxidize pyruvate to $H_2 + CO_2 +$ acetate.

Hydrolysis Breakdown of a polymer into smaller units, usually monomers, by addition of water; digestion.

Hydrophobic interactions Attractive forces between molecules due to the close positioning of nonhydrophilic portions of the two molecules.

Hydrothermal vents Warm or hot water-emitting springs associated with crustal spreading centers on the sea floor.

Hypersensitivity An immune reaction, usually harmful to the animal, caused either by antigen-antibody reactions or cellular immune processes. See *Allergy*.

Hyperthermophile A prokaryote having a growth temperature optimum of 80°C or higher.

Hypervariable regions Variations in amino acid sequence that occur within the variable domains of Igs or TCRs that provide most of the molecular contacts with antigen (also known as *complementarity-determining regions*).

Isosahedron A geometrical shape occurring in many virus particles, with 20 triangular faces and 12 corners.

Immobilized enzyme An enzyme attached to a solid support over which substrate is passed and converted to product.

Immune Able to resist infectious disease.

Immunity The ability of an organism to resist infection.

Immunization Induction of specific immunity by injecting antigens, antibodies, or immune cells into an animal.

Immunoblot (Western blot) Detection of proteins immobilized on a filter by complementary reaction with specific antibody. Compare with *Southern blot* and *Northern blot*.

Immunodeficiency Having dysfunctional or completely nonfunctional immune system.

Immunogen An antigen that can induce the production of an immune response.

Immunoglobulin Antibody.

Immunoglobulin gene superfamily A family of genes that are evolutionarily, structurally, and functionally related to immunoglobulins.

Immunological memory The ability to rapidly produce large quantities of specific immune cells following reexposure to a previously encountered antigen.

In silico The use of computers to perform sophisticated analyses.

In vitro In glass, away from the living organism.

In vivo In the body, in a living organism.

Incidence In reference to disease transmission, the number of cases of the disease in a specific subset of the population.

Induced enzyme An enzyme subject to induction.

Induction The process by which an enzyme is synthesized in response to the presence of an external substance, the inducer.

Infection Growth of an organism within the body.

Infection thread In the formation of root nodules, a cellulosic tube through which *Rhizobium* cells travel to reach and infect root cells.

Inflammation Characteristic reaction to foreign particles and noxious stimuli, resulting in redness, swelling, heat, and pain.

Inhibition In reference to growth, the reduction of microbial growth because of a decrease in the number of organisms present or alterations in the microbial environment.

Inoculum Material used to initiate a microbial culture.

Insertion A genetic phenomenon in which a piece of DNA is inserted into the middle of a gene.

Insertion sequence (IS elements) The simplest type of transposable element. Has only genes involved in transposition.

Integrating vector A cloning vector that becomes integrated into a host chromosome.

Integration The process by which a DNA molecule becomes incorporated into another genome.

Interferons Host-specific antiviral proteins produced by virus-infected cells, which prevents viral infection of neighboring cells.

Interleukin (IL) Soluble cytokine mediator secreted by leukocytes.

Interspecies hydrogen transfer The process by which organic matter is degraded by the interaction of several groups of microorganisms in which H_2 production and H_2 consumption are closely coupled.

Introns The intervening noncoding sequences in a split gene. Contrasted with *Exons*, the coding sequences.

Invasiveness The degree to which an organism is able to spread through the body from a focus of infection.

Ionophore A compound that can cause the leakage of ions across membranes.

Irradiation In food microbiology, the exposure of food to ionizing radiation to inhibit microorganisms and insect pests, or to retard ripening.

Isotopes Different forms of the same element containing the same number of protons and electrons but differing in the number of neutrons.

Jaundice Production and release of excess bilirubin in the liver due to destruction of liver cells, resulting in yellowing of the skin and whites of the eye.

Joule (J) A unit of energy equal to 10^7 ergs; 1000 Joules equal 1 kilojoule (kJ).

Kilobase (kb) A 1000-base fragment of nucleic acid. A *kilobase pair* is a fragment containing 1000 base pairs.

Kinase An enzyme that adds a phosphoryl group, usually from ATP, to a compound.

Lag phase The period after inoculation of a culture before growth begins.

Laser tweezers A device used to obtain pure cultures in which a single cell is optically trapped with a laser and moved away from contaminating organisms into sterile growth medium.

Late protein Proteins synthesized toward the end of virus infection.

Latent virus A virus present in a cell, yet not causing any detectable effect.

Leaching Removal of valuable metals from ores by microbial action.

Leukocidin A substance able to destroy phagocytes.

Leukocyte A white blood cell.

Lichen A fungus and an alga (or a cyanobacterium) living in symbiotic association.

Lipid Water-insoluble organic molecules important in structure of the cytoplasmic membrane and (in some organisms) the cell wall. See also *Phospholipid*.

Lipopolysaccharide (LPS) Complex lipid structure containing unusual sugars and fatty acids found in most gramnegative *Bacteria* and constituting the chemical structure of the outer membrane.

Listeriosis Gastrointestinal food infection caused by *Listeria monocytogenes* that may lead to bacteremia and meningitis.

Lophotrichous Having a tuft of polar flagella.

Lower respiratory tract Trachea, bronchi, and lungs.

Luminescence Production of light.

Lyme disease An emerging tick-transmitted disease caused by the spirochete *Borrelia burgdorferi*.

Lymph A clear, yellowish fluid found in the lymphatic vessels that carries various white (but not red) blood cells.

Lymphocyte A white blood cell involved in antibody formation or cellular immune responses.

Lysin An antibody that induces lysis.

Lysis Rupture of a cell, resulting in a loss of cell contents.

Lysogen A prokaryote containing a prophage. See also *Temperate virus*.

Lysogenic pathway A series of steps in which after virus infection, leads to a genetic state (lysogeny) where the viral genome is replicated as a prophage along with that of the host.

Lysosome A cell organelle containing digestive enzymes.

Lytic pathway A series of steps after virus infection which lead to virus replication and the destruction (lysis) of the host cell.

Lyophilization (freeze-drying) The process of removing all water from frozen food under vacuum.

Macromolecule A large molecule (polymer) formed by the connection of a number of small molecules (monomers).

Macrophage Large, noncirculating phagocytic cells involved in both phagocytosis and the antibody production process.

Magnetosomes Small particles of Fe_3O_4 present in cells that exhibit magnetotaxis (magnetic bacteria).

Magnetotaxis Directed movement of bacterial cells by a magnetic field.

Major histocompatability complex (MHC) A cluster of genes encoding cell surface proteins important for antigen presentation to T cells in the immune response.

Malaria An insect-transmitted disease characterized by recurrent episodes of fever and anemia caused by the protozoan *Plasmodium* spp., usually transmitted between mammals through the bite of the *Anopheles* mosquito.

Malignant In reference to a tumor, an infiltrating metastasizing growth no longer under normal growth control.

Mast cells Tissue cells adjoining blood vessels throughout the body that contain granules with inflammatory mediators.

Medium (plural media) In microbiology, the nutrient solution(s) used to grow microorganisms.

Megabase (Mb) One million nucleotide bases.

Meiosis In eukaryotes, reduction division, the process by which the change from diploid to haploid occurs.

Membrane Any thin sheet or layer. See especially *Cytoplasmic membrane*.

Memory cell A differentiated B lymphocyte capable of rapid conversion to an antibody-producing plasma cell on subsequent stimulation with antigen.

Meningitis Inflammation of the meninges (brain tissue), sometimes caused by *Neisseria meningitidis* and characterized by sudden onset of headache, vomiting, and stiff neck, often progressing to coma within hours.

Meningococcemia Fulminant disease caused by *Neisseria meningitidis* and characterized by septicemia, intravascular coagulation, and shock.

Meningoencephalitis Invasion, inflammation, and destruction of brain tissue, by the ameba *Naegleria fowlerii* or a variety of other pathogens.

Mesophile Organism living in the temperature range near that of warm-blooded animals, and usually showing a growth temperature optimum between 25 and 40°C.

Messenger RNA (mRNA)   An RNA molecule transcribed from DNA that contains the genetic information necessary to encode a particular protein.

Metabolism   All biochemical reactions in a cell, both anabolic and catabolic.

Metazoa   Multicellular organisms.

Methanogen   A methane-producing prokaryote; member of the *Archaea*.

Methanogenesis   The biological production of methane (CH_4).

Methanotroph   An organism capable of oxidizing methane.

Methylotroph   An organism capable of oxidizing organic compounds that do not contain carbon–carbon bonds; if able to oxidize CH_4, also a methanotroph.

MIC   Minimum inhibitory concentration—the minimum concentration of a substance necessary to prevent microbial growth.

Microaerophilic   Requiring O_2 but at a level lower than atmospheric.

Microenvironment   The immediate physical and chemical surroundings of a microorganism.

Micrometer   One-millionth of a meter, or 10^{-6}m (abbreviated μm), the unit used for measuring microorganisms.

Microorganism   A microscopic organism consisting of a single cell or cell cluster, also including the viruses.

Microtubules   Tubes that are the structural entity for eukaryotic flagella, have a role in maintaining cell shape, and function as mitotic spindle fibers.

Minus (negative)-strand nucleic acid   An RNA or DNA strand that has the opposite sense of (would be complementary to) the mRNA of a virus.

Mitochondrion   Eukaryotic organelle responsible for the processes of respiration and electron transport phosphorylation.

Mitosis   A highly ordered process by which the nucleus divides in eukaryotes.

Mixotroph   An organism able to assimilate organic compounds as carbon sources while using inorganic compounds as electron donors for energy metabolism.

Molds   Filamentous fungi.

Molecular chaperone   A protein that helps other proteins fold or refold properly.

Molecular cloning   Isolation and incorporation of a fragment of DNA into a vector where it can be replicated.

Molecule   Two or more atoms chemically bonded to one another.

Monoclonal antibody   An antibody produced from a single clone of cells. This antibody has uniform structure and specificity.

Monocytes   Circulating white blood cells that contain many lysosomes and can differentiate into macrophages.

Monomer   A building block of a polymer.

Monotrichous   Having a single polar flagellum.

Morbidity   Incidence of disease in a population, including both fatal and nonfatal cases.

Mortality   Incidence of death in a population.

Most probable number (MPN)   Serial dilution of a natural sample to determine the highest dilution yielding growth.

Motif   A conserved amino acid sequence found in all peptide antigens that bind to a given MHC protein.

Motility   The property of movement of a cell under its own power.

Mucous membrane   Layers of epithelial cells that interact with the external environment.

Mucus   Soluble glycoproteins secreted by epithelial cells that coat the mucous membrane.

Mushrooms   Filamentous fungi that produce large, often edible structures called fruiting bodies.

Mutagen   An agent that induces mutation, such as radiation or certain chemicals.

Mutant   A strain differing from its parent because of mutation.

Mutation   An inheritable change in the base sequence of the genome of an organism.

Mycorrhiza   A symbiotic association between a fungus and the roots of a plant.

Mycoses   Infections caused by fungi.

Myeloma   A malignant tumor of a plasma cell (antibody-producing cell).

Natural killer (NK) cell   A specialized lymphocyte that recognizes and destroys foreign cells or infected host cells in a nonspecific manner.

Negative selection   In T-cell selection, T cells that interact with self antigens in the thymus are deleted. See *Clonal deletion*.

Neutralization   Interaction of antibody with antigen that reduces or blocks the biological activity of the antigen.

Neutrophil   A polymorphonuclear leukocyte.

Neutrophile   An organism that grows best around pH 7.

Nitrification   The microbiological conversion of ammonia to nitrate.

Nitrogen fixation   Reduction of nitrogen gas to ammonia by the enzyme nitrogenase.

Nodule   A tumorlike structure produced by the roots of symbiotic nitrogen-fixing plants. Contains the nitrogen-fixing microbial component of the symbiosis.

Nonnucleoside reverse transcriptase inhibitor (NNRTI)   A nonnucleoside compound that inhibits the action of retroviral reverse transcriptase by binding directly to the catalytic site.

Nonperishable (stable) foods   Foods of low water activity that have an extended shelf life and are resistant to spoilage by microorganisms.

Nonpolar   Possessing hydrophobic (water-repelling) characteristics and not easily dissolved in water.

Nonsense mutation   A mutation that changes a sense codon into one that does not code for an amino acid.

Normal flora   Microorganisms that are usually found associated with healthy body tissue.

Northern blot   Hybridization of a single strand of nucleic acid (DNA or RNA) to RNA fragments immobilized on a filter. Compare with *Southern blot* and *Western blot*.

Nosocomial infection   Hospital-acquired infection.

Nucleic acid   A polymer of nucleotides. See *Deoxyribonucleic acid* and *Ribonucleic acid*.

Nucleic acid probe   A strand of nucleic acid that can be labeled and used to hybridize to a complementary molecule from a mixture of other nucleic acids. In clinical microbiology or microbial ecology, short oligonucleotides of unique sequences used as hybridization probes for identifying pathogens or other organisms of interest.

Nucleoid   The aggregated mass of DNA that makes up the chromosome of prokaryotic cells.

Nucleoside   A nucleotide minus phosphate.

Nucleoside reverse transcriptase inhibitor (NRTI)   A nucleoside analog compound that inhibits the action of viral reverse transcriptase by competing with nucleosides.

Nucleotide   A monomeric unit of nucleic acid, consisting of a sugar, a phosphate, and a nitrogenous base.

Nucleus   A membrane-enclosed structure in eukaryotes containing the genetic material (DNA) organized in chromosomes.

Nutrient   A substance taken by a cell from its environment and used in catabolic or anabolic reactions.

Obligate   A qualifying adjective referring to an environmental 5 always required for growth (for example, "obligate anaerobe").

Oligonucleotide   A short nucleic acid molecule, either obtained from an organism or synthesized chemically.

Oligotrophic   Describing a habitat in which nutrients are in low supply.

Oncogene   A gene whose expression causes formation of a tumor.

Open reading frame (ORF)   A sequence of DNA which, if transcribed, could be translated to yield a protein of known length and composition. A *functional* ORF is one that actually encodes a protein in the cell.

Operator   A specific region of the DNA at the initial end of a gene, where the repressor protein binds and blocks mRNA synthesis.

Operon   A cluster of genes whose expression is controlled by a single operator. Typical of prokaryotic cells.

Opsonization   Promotion of phagocytosis by a specific antibody in combination with complement.

Organelle   A membrane-enclosed structure found in eukaryotic cells.

Orthologs   Genes found in one organism that are similar to those in another organism, but differ because of speciation. See also *Paralogs*.

Osmosis   Diffusion of water through a membrane from a region of low solute concentration to one of higher concentration.

Outbreak   The occurrence of a large number of cases of a disease in a short period of time.

Oxic   Containing oxygen; aerobic. Usually used in reference to a microbial habitat.

Oxidation   A process by which a compound gives up electrons (or H atoms) and becomes oxidized.

Oxidation-reduction (redox) reaction   A pair of reactions in which one compound becomes oxidized while another becomes reduced and takes up the electrons released in the oxidation reaction.

Oxidative (electron transport) phosphorylation   The nonphototrophic production of ATP at the expense of a proton motive force formed by electron transport.

Oxygenic photosynthesis   Use of light energy to synthesize ATP and NADPH by noncyclic photophosphorylation with the production of O_2 from water.

Palindrome   A nucleotide sequence on a DNA molecule in which the same sequence is found on each strand but in the opposite direction.

Pandemic   A worldwide epidemic.

Paralogs   Genes within an organism whose similarity is the result of gene duplication at some time during the evolution of the organism. See also *Orthologs*.

Parasite   An organism able to live on and cause damage to another organism.

Passive immunity   Immunity resulting from transfer of antibodies or immune cells from an immune to a nonimmune individual.

Pasteurization   Destruction, usually by heat treatment, of all disease-producing microorganisms along with a reduction in the number of spoilage microorganisms.

Pathogen   An organism able to inflict damage on a host it infects.

Pathogenicity   The ability of a parasite to inflict damage on the host.

Peptide bond   A type of covalent bond joining amino acids in a polypeptide.

Peptidoglycan   The rigid layer of the cell walls of *Bacteria*, a thin sheet composed of *N*-acetylglucosamine, *N*-acetylmuramic acid, and a few amino acids. Also called *murein*.

Periplasmic space   The area between the cytoplasmic membrane and the outer membrane in gram-negative *Bacteria*.

Perishable foods   Fresh foods generally of high water activity, having a very short shelf life due to potential for spoilage by growth of microorganisms.

Peritrichous flagellation   Condition of having flagella attached to many places on the cell surface.

Phage   See *Bacteriophage*.

Phagemid   A cloning vector that can replicate either as a plasmid or as a bacteriophage.

Phagocyte   A body cell able to ingest and digest foreign particles.

Phagocytosis   Ingestion of particulate material such as bacteria by protozoa and phagocytic cells of higher organisms.

Phenotype   The observable characteristics of an organism. Compare with *Genotype*.

Phosphodiester bond   A type of covalent bond linking nucleotides together in a polynucleotide.

Phospholipid   Lipids containing a substituted phosphate group and two fatty acid chains on a glycerol backbone.

Photoautotroph   An organism able to use light as its sole source of energy and CO_2 as its sole carbon source.

Photoheterotroph   An organism using light as a source of energy and organic materials as carbon source.

Photophosphorylation   Synthesis of high energy phosphate bonds as ATP, using light energy.

Photosynthesis   The use of light energy to drive the incorporation of CO_2 into cell material. See also *Anoxygenic photosynthesis* and *Oxygenic photosynthesis*.

Phototaxis   Movement toward light.

Phototroph   An organism that obtains energy from light.

Phylogeny   The ordering of species into higher taxa and the construction of evolutionary trees based on evolutionary (natural) relationships.

Phytanyl   A branched-chain hydrocarbon containing 20 carbon atoms, commonly found in the lipids of *Archaea*.

Pickling   The process of acidifying food, typically with acetic acid, to prevent microbial growth and spoilage.

Pilus   A fimbria-like structure that is present on fertile cells, both Hfr and F^+, and is involved in DNA transfer during conjugation. Sometimes called a *sex pilus*. See also *Fimbria*.

Pinocytosis   In eukaryotes, phagocytosis of soluble molecules.

Plague An endemic disease in rodents caused by *Yersinia pestis* that is occasionally transferred to humans through the bite of a flea.

Plaque A zone of lysis or cell inhibition caused by virus infection on a lawn of cells.

Plasma The noncellular portion of blood.

Plasma cell A large, differentiated, short-lived B lymphocyte specializing in abundant (but short-term) antibody production.

Plasmid An extrachromosomal genetic element that is not essential for growth and has no extracellular form.

Platelet A noncellular disc-shaped structure containing protoplasm found in large numbers in blood and functioning in the blood clotting process.

Plus-strand nucleic acid An RNA or DNA strand that has the same sense as the mRNA of a virus.

Point mutation A mutation that involves one or only a very few base pairs.

Polar Possessing hydrophilic characteristics and generally water-soluble.

Polar flagellation Condition of having flagella attached at one end or both ends of the cell.

Poly-β-hydroxybutyrate (PHB) A common storage material of prokaryotic cells consisting of a polymer of β-hydroxybutyrate (PHB) or other β-alkanoic acids (PHA).

Polyclonal antibodies A mixture of antibodies made by many different B cell clones.

Polyclonal antiserum A mixture of antibodies to a variety of antigens or to a variety of determinants on a single antigen.

Polymer A large molecule formed by polymerization of monomeric units.

Polymerase chain reaction (PCR) A method used to amplify a specific DNA sequence *in vitro* by repeated cycles of synthesis using specific primers and DNA polymerase.

Polymorphism The occurrence of multiple alleles at a locus in frequencies that cannot be explained by the occurrence of recent random mutations.

Polymorphonuclear leukocyte (PMN) Motile white blood cells containing many lysosomes and specializing in phagocytosis. Characterized by a distinct segmented nucleus. Also, a neutrophil.

Polynucleotide A polymer of nucleotides bonded to one another by phosphodiester bonds.

Polypeptide Several amino acids linked together by peptide bonds.

Polysaccharide A long chain of monosaccharides (sugars) linked by glycosidic bonds.

Porins Protein channels in the outer membrane of gram-negative *Bacteria* through which small to medium-sized molecules can flow.

Porters Membrane proteins that function to transport substances into and out of the cell.

Positive selection In T-cell selection, T cells that interact with self MHC protein in the thymus are stimulated to grow and develop.

Precipitation A reaction between antibody and soluble antigen resulting in visible antibody–antigen complexes.

Prevalence The *proportion* of individuals in a population having a disease.

Pribnow box The consensus sequence TATAAT located approximately 10 base pairs upstream from the transcriptional start site. A binding site for RNA polymerase.

Primary antibody response Antibodies made on first exposure to antigen; mostly of the class IgM.

Primary metabolite A metabolite excreted during the growth phase.

Primary producer An organism that uses light or an inorganic compound to synthesize new organic material from CO_2.

Primary structure In an informational macromolecule, such as a polypeptide or a nucleic acid, the *precise sequence* of monomeric units.

Primary transcript An unprocessed RNA molecule that is the direct product of transcription.

Primer A molecule (usually a polynucleotide) to which DNA polymerase can attach the first deoxyribonucleotide during DNA replication.

Prion An infectious agent whose extracellular form may contain no nucleic acid.

Probe See *Nucleic acid probe*.

Prochlorophyte A prokaryotic oxygenic phototroph that contains chlorophylls *a* and *b* but lacks phycobilins.

Prokaryote A cell or organism lacking a nucleus and other membrane-enclosed organelles, usually having its DNA in a single circular molecule.

Promoter The site on DNA where the RNA polymerase binds and begins transcription.

Prophage The state of the genome of a temperate virus when it is replicating in synchrony with that of the host, typically integrated into the host genome.

Prophylactic Treatment, usually immunological or chemotherapeutic, designed to protect an individual from a future attack by a pathogen.

Prostheca A cytoplasmic extrusion from a cell such as a bud, hypha, or stalk.

Prosthetic group The tightly bound, nonprotein portion of an enzyme; not the same as a *Coenzyme*.

Protease inhibitor A compound that inhibits the action of viral protease by binding directly to the catalytic site, preventing viral protein processing.

Protein A polymeric molecule consisting of one or more polypeptides.

Proteome The total complement of proteins present in a cell, tissue, or organism at any one time.

Proteomics The large scale or genome-wide study of the structure, function, and regulation of the proteins of an organism.

Proteorhodopsin A light-sensitive retinal-containing protein found in some marine *Bacteria* that catalyzes ATP formation.

Proton motive force An energized state of a membrane created by a protein gradient and usually formed through action of an electron transport chain. See also *Chemiosmosis*.

Protoplasm The complete cellular contents, cytoplasmic membrane, cytoplasm, and nucleus/nucleoid.

Protoplast A cell from which the wall has been removed.

Prototroph The parent from which an auxotrophic mutant has been derived. Contrast with *Auxotroph*.

Protozoa Unicellular eukaryotic microorganisms that lack cell walls.

Provirus See *Prophage*.

Psychrophile An organism able to grow at low temperatures and showing a growth temperature optimum of $<15°C$.

Psychrotolerant   Able to grow at low temperature but having a growth temperature optimum of <15°C.

Public health   The health of the population as a whole.

Pure culture   A culture containing a single kind of microorganism.

Pyogenic   Pus-forming: causing abscesses.

Pyrite   A common iron ore, FeS_2.

Pyrogenic   Fever-inducing.

Quarantine   The limitation on the freedom of movement of an infected individual to prevent spread of a disease to other members of a population.

Quaternary structure   In proteins, the number and arrangement of individual polypeptides in the final protein molecule.

Quinolones   Synthetic antibacterial compounds that interact with DNA gyrase and prevent supercoiling of bacterial DNA.

Quorum sensing   Regulatory pathways in prokaryotes that respond to population density.

Rabies   A usually fatal neurological disease caused by the rabies virus that is usually transmitted by the bite or saliva of an infected carnivore.

Radioimmunoassay   An immunological assay employing radioactive antibody or antigen for the detection of antigen or antibody binding.

Radioisotope   An isotope of an element that undergoes spontaneous decay with the release of radioactive particles.

Reaction center   A photosynthetic complex containing chlorophyll (or bacteriochlorophyll) and other components, within which occurs the initial electron transfer reactions of photophosphorylation.

Reading-frame shift   See *Frame shift*.

Recalcitrant   Resistant to microbial attack.

Recombinant DNA   A DNA molecule containing DNA originating from two or more sources.

Recombination   Process by which genetic elements in two separate genomes are brought together in one unit.

Redox   See *Oxidation–reduction reaction*.

Reduction   A process by which a compound accepts electrons to become reduced.

Reduction potential (E_0')   The inherent tendency, measured in volts, of the oxidized compound of a redox pair to become reduced.

Reductive dechlorination   Removal of Cl as Cl^- from an organic compound by reducing the carbon atom from C—Cl to C—H.

Reemergent infections   Infectious diseases, thought to be under control, that produce new epidemics.

Regulation   Processes that control the rates of synthesis of proteins. Induction and repression are examples of regulation.

Regulon   A set of operons that are all controlled by the same regulatory protein (repressor or activator).

Replacement vector   A cloning vector, such as a bacteriophage, in which some of the DNA of the vector can be replaced with foreign DNA.

Replication   Synthesis of DNA using DNA as a template.

Replicative form   A double-stranded DNA molecule that is an intermediate in the replication of single-stranded DNA viruses.

Reporter gene   A gene incorporated into a vector because the product it encodes is easy to detect.

Repression   The process by which the synthesis of an enzyme is inhibited by the presence of an external substance, the repressor.

Repressor protein   A regulatory protein that binds to specific sites on DNA and blocks transcription; involved in negative control.

Reservoir   In epidemiology, the organism or environment that normally harbors a pathogen.

Respiration   Catabolic reactions producing ATP in which either organic or inorganic compounds are primary electron donors and organic or inorganic compounds are ultimate electron acceptors.

Response regulator protein   One of the members of a two-component system; a regulatory protein that is phosphorylated by a sensor protein (see *Sensor protein*).

Restriction endonucleases (restriction enzymes)   Enzymes that recognize and cleave specific DNA sequences, generating either blunt or single-stranded (sticky) ends.

Retrovirus   A virus containing single-stranded RNA as its genetic material, which produces a complementary DNA by activity of the enzyme reverse transcriptase.

Reverse electron transport   The energy-dependent movement of electrons against the thermodynamic gradient to form a strong electron donor from a weaker electron donor.

Reverse transcription   The process of copying information found in RNA into DNA.

Reverse translation   The mental process of using a codon table and the amino acid sequence of a protein to obtain a possible sequence of the mRNA or the gene that encoded the protein.

Rheumatic fever   An inflammatory autoimmune disease triggered by an immune response to infection by *Streptococcus pyogenes*.

Rhizosphere   The region immediately adjacent to plant roots.

Ribonucleic acid (RNA)   A polymer of nucleotides connected via a phosphate–ribose backbone; involved in protein synthesis or as genetic material of some viruses.

Ribosomal RNA (rRNA)   Type of RNA found in the ribosome; some rRNAs participate actively in the process of protein synthesis.

Ribosome   A cytoplasmic particle composed of ribosomal RNA and protein, which is part of the protein-synthesizing machinery of the cell.

Ribozyme   An RNA molecule that can catalyze a chemical reaction.

Rickettsias   Obligate intracellular parasites that cause a variety of diseases, including typhus, Rocky Mountain spotted fever, and ehrlichiosis.

RNA editing   Modification of the RNA transcript of a protein-encoding gene, by a process other than splicing, yielding a molecule with the required coding properties.

RNA life   A hypothetical ancient life form lacking DNA and protein, in which RNA had both a genetic coding and a catalytic function.

RNA polymerase   An enzyme that synthesizes RNA in the $5' \rightarrow 3'$ direction using an antiparallel $3' \rightarrow 5'$ DNA strand as a template.

RNA processing   The conversion of a precursor RNA to its mature form.

Rocky Mountain spotted fever   A tick-transmitted disease caused by *Rickettsia ricketsii*, causing fever, headache, rash, and gastrointestinal symptoms.

Root nodule A tumorlike growth on certain plant roots that contains symbiotic nitrogen-fixing bacteria.

Rumen The forestomach of ruminant animals in which cellulose digestion occurs.

S-layer A paracrystalline outer wall layer composed of protein or glycoprotein and found in many prokaryotes.

Salmonellosis Enterocolitis caused by any of 1400 serotypes of *Salmonella* species.

Scale-up Conversion of an industrial process from a small laboratory setup to a large commercial fermentation.

Scarlet fever Disease characterized by high fever and a reddish skin rash resulting from an exotoxin produced by cells of *Streptococcus pyogenes*.

Screening Any of a number of procedures that permits the sorting of organisms by phenotype or genotype by allowing growth of some types but not others.

Secondary antibody response Antibody made on second (subsequent) exposure to antigen; mostly of the class IgG.

Secondary metabolite A product excreted by a microorganism near the end of the growth phase or during the stationary phase.

Secondary structure The initial pattern of folding of a polypeptide or a polynucleotide, usually the result of hydrogen bonding.

Secondary treatment In sewage treatment, either the aerobic or anaerobic decomposition of sewage following the removal of nondegradable objects by primary treatment.

Secretion vector A DNA vector in which the protein product is both expressed in and secreted (excreted) from the cell.

Selection Placing organisms under conditions where the growth of those with a particular genotype will be favored.

Selective media Media that enhance the growth of certain organisms while inhibiting the growth of others due to an added media component.

Semiconservative replication DNA synthesis yielding new double helices, each consisting of one parental and one progeny strand.

Semiperishable foods Foods of intermediate water activity that have a limited shelf life due to potential for spoilage by growth of microorganisms.

Semisynthetic penicillin A natural penicillin that has been chemically altered.

Sensitivity In immunodiagnostics, the lowest amount of antigen that can be detected in an immunological assay.

Sensor protein One of the members of a two-component system; a kinase found in the cell membrane that phosphorylates itself in response to an external signal and then passes the phosphoryl group to a response regulator protein (see *Response regulator protein*).

Septicemia Infection of the bloodstream by microorganisms.

Serology The study of antigen–antibody reactions *in vitro*.

Serum Fluid portion of blood remaining after the blood cells and materials responsible for clotting are removed.

Sexually transmitted disease (STD) A disease whose usual means of transmission is by sexual contact.

Shine–Dalgarno sequence A short stretch of nucleotides on a prokaryotic mRNA molecule upstream of the transla-

tional start site that binds to ribosomal RNA and thereby brings the ribosome to the initiation codon on the mRNA.

Shuttle vector A cloning vector that can replicate in two different organisms; used for moving DNA between unrelated organisms.

Sickle-cell anemia A genetic trait that confers resistance to malaria but causes a reduction in the number and efficiency of red blood cells.

Siderophore An iron chelator that can bind iron present at very low concentrations.

Signal sequence A short stretch of amino acids found at the beginning of proteins destined to be excreted from the cell. The signal sequence is usually rich in hydrophobic amino acids, which helps transport the entire polypeptide through the membrane.

Signature sequence Short oligonucleotides of unique sequence found in 16S ribosomal RNA of a particular group of prokaryotes.

Single-cell protein Protein derived from microbial cells for use as food or a food supplement.

Site-directed mutagenesis A technique whereby a gene with a specific mutation can be constructed *in vitro*.

16S rRNA A large polynucleotide (~1500 bases) that functions as a part of the small subunit of the ribosome of prokaryotes (*Bacteria* and *Archaea*) and from whose sequence evolutionary relationships can be obtained; eukaryotic counterpart, 18S rRNA.

Slime layer A diffuse mat of polymer fibers surrounding cells that appear unattached to a single cell.

Slime molds Nonphototrophic eukaryotic microorganisms lacking cell walls, which aggregate to form fruiting structures (cellular slime molds) or simply masses of protoplasm (acellular slime molds).

Solfatara A hot, sulfur-rich, generally acidic environment commonly inhabited by hyperthermophilic *Archaea*.

Southern blot Hybridization of a single strand of nucleic acid (DNA or RNA) to DNA fragments immobilized on a filter. Compare with *Northern blot* and *Western blot*.

Species Of prokaryotes, a collection of closely related (>97% 16S rRNA sequence homology and >70% genomic hybridization) strains sufficiently different from all other strains to be recognized as a distinct unit.

Specificity The ability of the immune response to interact with individual antigens.

Spheroplast A spherical, osmotically sensitive cell derived from a bacterium by loss of some but not all of the rigid wall layer. If all the rigid wall layer has been completely lost, the structure is called a *Protoplast*.

Splicing The RNA processing step by which introns are removed and exons joined.

Spontaneous generation The hypothesis that living organisms can originate from nonliving matter.

Spore A general term for resistant resting structures formed by many prokaryotes and fungi.

Sporozoa Nonmotile parasitic protozoa.

Stalk An elongate structure, either cellular or excreted, that anchors a cell to a surface.

Static Inhibitory.

Stationary phase The period during the growth cycle of a microbial population in which growth ceases.

Stereoisomers   Mirror image forms of two molecules having the same molecular and structural formulas.

Sterile   Free of living organisms and viruses.

Sterilization   The killing or removal of all living organisms and their viruses from a growth medium.

Sterols   Hydrophobic multiringed structures that strengthen the cytoplasmic membrane of eukaryotic cells and a few prokaryotes.

Strain   A population of cells of a single species all descended from a single cell; a clone.

Stromatolites   Laminated microbial mats, typically built from layers of filamentous and other microorganisms which can become fossilized.

Substrate   The molecule that undergoes a specific reaction with an enzyme.

Substrate-level phosphorylation   Synthesis of high-energy phosphate bonds through reaction of inorganic phosphate with an activated organic substrate.

Supercoil   Highly twisted form of circular DNA.

Superoxide anion (O_2^-)   A derivative of O_2 capable of oxidative destruction of cell components.

Suppressor   A mutation that restores a wild-type phenotype without altering the original mutation, usually arising by mutation in another gene.

Surveillance   Observation, recognition, and reporting of diseases as they occur.

Symbiosis   A relationship between two organisms.

Synthetic DNA   A DNA molecule that has been made by a chemical process in a laboratory.

Syntrophy   A nutritional situation in which two or more organisms combine their metabolic capabilities to catabolize a substance not capable of being catabolized by either one alone.

Systemic   Not localized in the body; an infection disseminated widely through the body.

T cell   A lymphocyte responsible for antigen-specific cellular interactions. T cells are divided into functional subsets including T_C (cytotoxic) T cells and T_H (helper) T cells. T_H cells are further subdivided into T_H1 (inflammatory) and T_H2 (helper cells) that aid B cells in antibody formation.

T-cell receptor   The antigen-specific receptor on the surface of T lymphocytes.

T-DNA   The segment of the *Agrobacterium* Ti plasmid that is transferred to plant cells.

Taxis   Movement toward or away from a stimulus.

Taxonomy   The study of scientific classification and nomenclature.

Temperate virus   A virus that on infection of a host does not necessarily cause lysis but whose genome may replicate in synchrony with that of the host. See *Lysogen*.

Tertiary structure   The final folded structure of a polypeptide that has previously attained secondary structure.

Tetanus   A disease involving rigid paralysis of the voluntary muscles, caused by an exotoxin produced by *Clostridium tetani*.

Tetracycline   A member of a class of antibiotics containing the four-membered naphthacene ring.

Thallasemia   A genetic trait that confers resistance to malaria, but causes a reduction in the efficiency of red blood cells by altering a red blood cell enzyme.

Thermocline   Zone of water in a stratified lake in which temperature and oxygen concentration drop precipitously with depth.

Thermophile   An organism with a growth temperature optimum between 45 and 80°C.

Thylakoids   Layers of membranes containing the photosynthetic pigments in chloroplasts and in cyanobacteria.

Ti plasmid   A conjugative plasmid present in the bacterium *Agrobacterium tumefaciens* that can transfer genes into plants.

Titer   Measure of antibody quantity.

Tolerance   Inability to produce an immune response to a specific antigen.

Toxic shock syndrome   Acute systemic shock resulting from host response to an exotoxin produced by *Staphylococcus aureus*.

Toxicity   The degree of pathogenicity caused by toxins produced by a pathogen.

Toxigenicity   The degree to which an organism is able to elicit toxic symptoms.

Toxin   A microbial substance able to induce host damage.

Toxoid   A toxin modified so that it is no longer toxic but is still able to induce antibody formation.

Transcription   Synthesis of an RNA molecule complementary to one of the two strands of a double-stranded DNA molecule.

Transduction   Transfer of host genes from one cell to another by a virus.

Transfection   The transformation of a prokaryotic cell by DNA or RNA from a virus. Used also to describe the process of genetic transformation in eukaryotic cells.

Transfer RNA (tRNA)   A type of RNA that carries amino acids to the ribosome during translation; contains the anticodon.

Transformation   Transfer of genetic information via free DNA. Also, a process, sometimes initiated by infection with certain viruses, whereby a normal animal cell becomes a cancer cell.

Transgenic organisms   Plants or animals that stably pass on cloned DNA that has been inserted into them.

Translation   The synthesis of protein using the genetic information in a messenger RNA as a template.

Transpeptidation   The formation of peptide bonds between the short peptides present in the cell wall polymer, peptidoglycan.

Transposable element   A genetic element that has the ability to move (transpose) from one site on a chromosome to another.

Transposon   A type of transposable element that, in addition to genes involved in transposition, carries other genes; often genes conferring selectable phenotypes such as antibiotic resistance.

Transposon mutagenesis   Insertion of a transposon into a gene; this inactivates the host gene, leading to a mutant phenotype, and also confers the phenotype associated with the transposon gene.

Tuberculin test   A test for previous infection with *Mycobacterium tuberculosis* characterized by an inflammatory cell-mediated immune response.

Two-component system   A regulatory system containing a sensor protein and a response regulator protein (see *Sensor protein* and *Response regulator protein*).

Typhus A louse-transmitted disease caused by *Rickettsia prowazekii*, causing fever, headache, weakness, rash, and damage to the central nervous system and internal organs.

Upper respiratory tract The nasopharynx, oral cavity, and throat.

Upstream position Refers to nucleic acid sequences on the 5′ side of a given site on a DNA or RNA molecule. Compare with *Downstream position*.

Vaccination Inoculation of a host with inactive, killed, or weakened pathogens or pathogen products to stimulate protective immunity.

Vaccine Material used to induce specific protective immunity to a pathogen.

Vacuole A small space in a cell that contains fluid and is surrounded by a membrane. In contrast to a vesicle, a vacuole is not rigid.

Vector An agent, usually an insect or other animal, able to carry pathogens from one host to another. Also, a genetic element able to incorporate DNA and cause it to be replicated in another cell.

Vehicle Nonliving source of pathogens that infect large numbers of individuals; common vehicles are food and water.

Viable Alive; able to reproduce.

Viable count Measurement of the concentration of live cells in a microbial population.

Viral load The number of viral genome copies in the tissue of an infected host, providing a quantitative assessment of the amount of virus in the host.

Virion A virus particle; the virus nucleic acid surrounded by a protein coat and in some cases other material.

Viricidal agent An agent that stops viral replication and activity.

Viristatic agent An agent that inhibits viral replication.

Viroid A small RNA molecule with viruslike properties.

Virulence Degree of pathogenicity of a parasite.

Virus A genetic element containing either DNA or RNA that replicates in cells but is characterized by having an extracellular state.

Water activity (a_w) An expression of the relative availability of water in a substance. Pure water has an a_w of 1.000.

Western blot See *Immunoblot*.

Wild type A strain of microorganism isolated from nature. The usual or native form of a gene or organism.

Wobble In reference to reading the genetic code, the concept that nonstandard base pairing is allowed between the anticodon and the third position of the codon.

Xenobiotic A completely synthetic chemical compound not naturally occurring on Earth.

Xerophile An organism adapted to growth at very low water potentials.

Yeasts Unicellular fungi.

Zoonoses Diseases primarily of animals that are occasionally transmitted to humans.

Zygote In eukaryotes, the single diploid cell resulting from the union of two haploid gametes.

PHYLOGENY OF THE LIVING WORLD—*ARCHAEA*

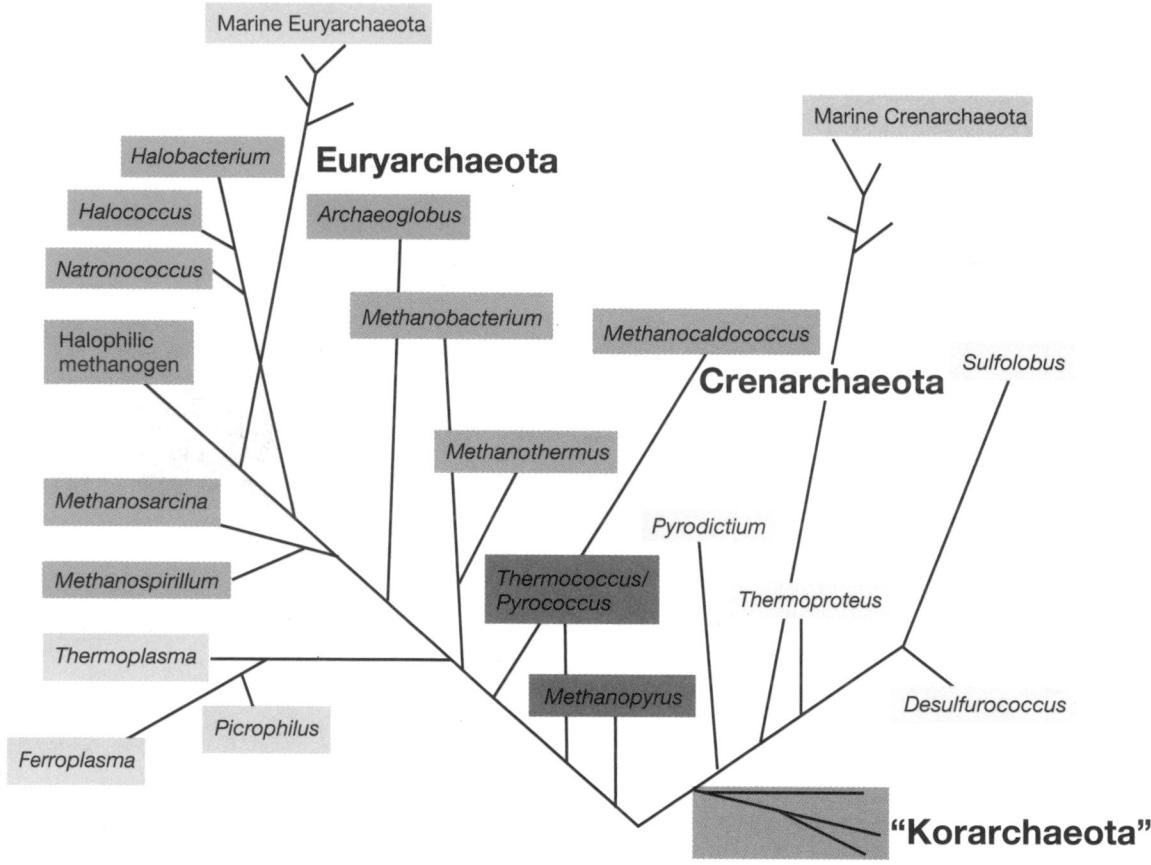

PHYLOGENETIC TREE OF *ARCHAEA*. This tree is derived from 16S ribosomal RNA sequences. Three major phyla of *Archaea* can be defined: the Crenarchaeota, which consists of both hyperthermophiles and cold-dwelling species; the Euryarchaeota, which contains methanogenic and extremely halophilic prokaryotes; and the "Korarchaeota," which are, as far as is known, hyperthermophiles. See Sections 11.4–11.8 for further information on ribosomal RNA-based phylogenies. *Data for the tree obtained from the Ribosomal Database project* http://rdp.cme.msu.edu

PHYLOGENY OF THE LIVING WORLD–*EUKARYA*

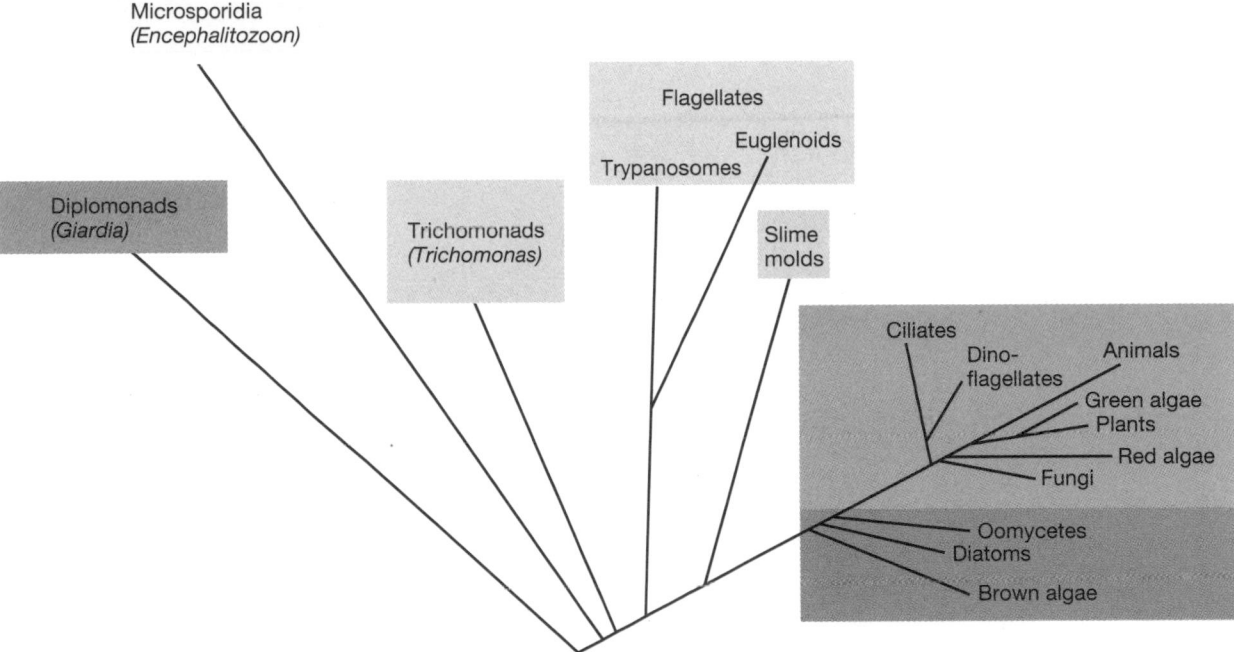

PHYLOGENETIC TREE OF *EUKARYA*, THE EUKARYOTES. This tree is based on 18S rRNA sequences obtained from the small subunit of cytoplasmic ribosomes. Note the close phylogenetic relationship between animals and plants but the distant relationship between these groups and organisms such as *Giardia*. See Sections 11.4-11.8 for further information on ribosomal RNA-based phylogenies. *Data for the tree obtained from the Ribosomal Database project* http://rdp.cme.msu.edu